Materials Properties Handbook: Titanium Alloys

Rodney Boyer
Boeing Commercial Airplane Company

Gerhard Welsch
Case Western Reserve University

E.W. Collings
Battelle Memorial Institute (Columbus)

Dr. William W. Scott, Jr., Director of Technical Publications
Scott D. Henry, Manager of Handbook Development
Steve Lampman, Handbook Editor
Veronica Flint, Acquisitions and Review

Production Assistance
Nancy M. Sobie
Ann-Marie O'Loughlin
Randall L. Boring
Patricia Eland
William J. O'Brien
Jeff Fenstermaker

Editorial Assistance
Nikki D. Wheaton
Judith Woodruff
Terri Weintraub

The Materials Information Society

Copyright © 1994
by
ASM International®
All rights reserved

No part of this book may be reproduced, stored in a retrieval system, or transmitted, in any form or by any means, electronic, mechanical, photocopying, recording, or otherwise, without the written permission of the copyright owner.

First printing, June 1994
Second printing, January 1998

This book is a collective effort involving hundreds of technical specialists. It brings together a wealth of information from worldwide sources to help scientists, engineers, and technicians solve current and longrange problems.

Great care is taken in the compilation and production of this Volume, but it should be made clear that NO WARRANTIES, EXPRESS OR IMPLIED, INCLUDING, WITHOUT LIMITATION, WARRANTIES OF MERCHANTABILITY OR FITNESS FOR A PARTICULAR PURPOSE, ARE GIVEN IN CONNECTION WITH THIS PUBLICATION. Although this information is believed to be accurate by ASM, ASM cannot guarantee that favorable results will be obtained from the use of this publication alone. This publication is intended for use by persons having technical skill, at their sole discretion and risk. Since the conditions of product or material use are outside of ASM's control, ASM assumes no liability or obligation in connection with any use of this information. No claim of any kind, whether as to products or information in this publication, and whether or not based on negligence, shall be greater in amount than the purchase price of this product or publication in respect of which damages are claimed. THE REMEDY HEREBY PROVIDED SHALL BE THE EXCLUSIVE AND SOLE REMEDY OF BUYER, AND IN NO EVENT SHALL EITHER PARTY BE LIABLE FOR SPECIAL, INDIRECT OR CONSEQUENTIAL DAMAGES WHETHER OR NOT CAUSED BY OR RESULTING FROM THE NEGLIGENCE OF SUCH PARTY. As with any material, evaluation of the material under enduse conditions prior to specification is essential. Therefore, specific testing under actual conditions is recommended.

Nothing contained in this book shall be construed as a grant of any right of manufacture, sale, use, or reproduction, in connection with any method, process, apparatus, product, composition, or system, whether or not covered by letters patent, copyright, or trademark, and nothing contained in this book shall be construed as a defense against any alleged infringement of letters patent, copyright, or trademark, or as a defense against liability for such infringement.

Comments, criticisms, and suggestions are invited, and should be forwarded to ASM International.

Library of Congress Cataloging-in-Publication Data

Materials properties handbook: titanium alloys /
editors, Rodney Boyer, Gerhard Welsch, E.W. Collings

p. cm.
ISBN 0-87170-481-1

1. Titanium alloys.
I. Welsch, Gerhard.
II. Boyer, Rodney
III. Collings, E.W.

TA480.T54M37   1994
620.1'89322—dc20   94-15791
                                      CIP
SAN No. 204-7586

ASM International®
Materials Park, OH 44073-0002

Printed in the United States of America

Preface

Titanium Alloys is the result of an ambitious effort to provide comprehensive property data in electronic form for not only databases but also print products such as the *Materials Properties Handbooks* series. In this endeavor, *Titanium Alloys* represents a "book-first" approach devoted to comprehensive, alloy-specific compilations of properties and processing information on engineering materials. This work has produced a substantial amount of titanium property data in electronic form, and follow-up efforts will determine which of the information is suitable for more structured and searchable electronic formats such as *MatDB*.

Titanium was chosen as the first topic in this "book-then-database" effort because the small number of major titanium alloys was a factor considered by the initial project managers. However, the scope was expanded and a substantial effort was expended in collecting a wide variety of information on different alloys and properties (with particular emphasis on the work horse alloy, Ti-6Al-4V). The amount of information was monumental, and the task of selecting and editing the data for subsequent production was pursued with the goal of providing comprehensive coverage on an alloy-specific basis. Whether this approach was prudent may be questionable in hindsight. However, this handbook provides a compilation of properties and fabrication procedures for virtually all of the alloys which have been developed over the 45-year time span of the titanium industry up to early 1993. The data is quite comprehensive for the more important alloys and not as complete for some of the lesser alloys, particularly those that never went into production. It is not intended to provide all the data in the literature, but to provide a quick, up-to-date assessment of the key information that is available. However, for those alloys and/or properties where more detail is required, references are cited to enable the reader to obtain further information.

This handbook will be a valuable addition to the library of anyone with more than a superficial involvement or interest in titanium in that in this single volume, the physical, thermal, mechanical, corrosion, fatigue, and fracture properties of almost all titanium alloys (except for alloys of the former Soviet Union), along with chapters on the basic metallurgy of titanium are compiled. This greatly facilitates comparison of alloy properties; thermomechanical and heat treatment effects on the properties of these alloys are also provided. This book will furnish a quick, state-of-the-art overview, which will provide the starting point from which a more detailed search of the literature can be initiated, leading to an intelligent assessment of the proper alloy for a specific application. It is truly unique to have a database this comprehensive for basically all alloys in a given alloy system contained in one volume. This one book will either provide the data you need, or provide references on where to find it, for any titanium alloy.

In addition, this volume also contains processing information such as forging, forming, casting, powder metallurgy, and welding. Recommended procedures/limits in these areas are provided, and where appropriate, the affects of some of these processing variables on the final properties are discussed.

This has been an international effort, with contributors from North America, Europe, and Asia. These contributors are leaders in the field, and represent all sectors of the industry including titanium producers, titanium fabricators, end users, governments, and academia. An effort of this magnitude represents a substantial commitment by ASM International and the efforts of hundreds of individuals in collection of the data, compilation into coherent chapters and sections, review of the assembled sections, and the painstaking efforts of producing and proofing graphics and page layouts. We would like to give them our heartfelt thanks, for without them this book would not have been possible.

R. Boyer and S. Lampman

Contributors and Reviewers

Stanley Abkowitz
  Dynamet Technology, Inc.
Susan Abkowitz
  Dynamet Technology, Inc.
Philip Adler
  Grumman Corporation
John E. Allison
  Ford Motor Company
Om Arora
  David Taylor Research Center
William Baeslack
  Ohio State University
C.C. Bampton
  Rockwell International Science Center
Paul Bania
  TIMET
Michael L. Bauccio
  Boeing Commercial Airplane Company
Joanne P. Beckman
  Crucible Compaction Metals
Gerald Beranger
  University of Compiegne
Howard Bomberger
Elihu Bradley
  Metallurgical Consulting Services
J. Breme
  Universität des Saarlandes
Charles Butler
  Charles E. Butler & Associates
Ivan Caplan
  David Taylor Research Center
James Costa
  Sandia National Laboratory
Bruce Craig
  Metallurgical Consultants, Inc.
Frank Crossley
Gopal Das
  Wright-Patterson AFB
David Davidson
  Southwest Research Institute
Charles Deak
  Analytical Associates, Inc.
Matthew Donachie, Jr.
  Pratt & Whitney
Paul Ducheyne
  University of Pennsylvania
T.W. Duerig
  Nitinol Development Corporation
M.L. Emiliani
  Pratt & Whitney
Gerald Friedman
  Precision CastParts Corporation

Roger Gilbert
  IMI Titanium Ltd.
Richard E. Goosey
  IMI Titanium Ltd.
Frank W. Gorsler
  Product/Process Integration
  General Electric Company
Domenic Grandemange
  University of Dayton
George Gray
  Los Alamos National Laboratory
David Gupta
  Thomas J. Watson Research Center
Edward Harper
  MetLab, Inc.
Dennis Hasson
  U.S. Naval Academy
James Howe
  Carnegie Mellon University
S.C. Huang
  Physical Metallurgy Laboratory
  General Electric Company
M.A. Imam
  Naval Research Laboratory
Allen Jackson
  Universal Energy Systems
Wayne Jones
  University of Michigan
Ali I. Kahveci
  ALCOA Technical Center
Toshiro Kobayashi
  Toyohashi University
Donald Koss
  Pennsylvania State University
G. William Kuhlman
  ALCOA, Forging Division
Frank Kustis
  Martin-Marietta Corporation
James Larson
  Wright-Patterson AFB
Yong-Tai Lee
  Korea Institute of Machinery and Materials
Zhendong Liu
  AECL Res. Co.
Michael Maguire
  Teledyne Wah Chang Albany
Murray Mahoney
  Rockwell International
Brian Marquardt
  General Electric Aircraft Engines
Dale L. McLellan
  Boeing Commercial Airplane Company

Edward Mild
  TIMET
Mohan Misra
  Martin-Marietta Corporation
Melvin A. Mettnick
  Textron Specialty Materials
Howard Nelson
  NASA-Ames Research Center
J.R. Newman
  IMI Titanium, Inc.
  TiTech Pomona Operation
Takashi Nishimura
  Special Metals Laboratory
  Kobe Steel Ltd.
Thomas E. O'Connell
  TIMET
Neil Paton
  Rockwell International
Alan R. Pelton
  Nitinol Development Corporation
Manfred Peters
  DLR-Institut für Werkstoff-Forschung
B.B. Rath
  Naval Research Laboratory
Cecil Rhodes
  Rockwell International
A. Grant Rowe
  Alloy Properties Laboratory
  General Electric Company
Patrick A. Russo
  RMI Titanium Company
C.R. Shannon
  Rolled Products Metallurgy
  Teledyne Allvac
David Snow
  United Technologies Research Center
John Thorne
  IMI Titanium, Inc.
  TiTech Pomona Operation
R. Terence Webster
  Teledyne Wah Chang
Isaac Weiss
  Wright State University
Yukio Yamada
  Tohpoka University

Summary Table of Titanium Alloys

ALLOY, UNS NUMBER, AND COMMON NAMES	GENERAL DESCRIPTION	APPLICATIONS
UNALLOYED TITANIUM		
High-purity Ti UNS: None Iodide or electrolytic Ti	High-purity Ti has one-half the oxygen content as commercially pure (ASTM Grade 1) titanium. High-purity Ti is produced from a special grade of sponge (<0.1 wt% oxygen).	Applications requiring minimum interstitials such as oxygen, nitrogen, carbon, and hydrogen
ASTM Grade 1 UNS: R50250 Grade 1	ASTM Grade 1 has the highest purity, lowest strength, and best room temperature ductility and formability of the four ASTM unalloyed Ti grades.	ASTM Grade 1 should be used where maximum formability is required, or where low iron and interstitials might enhance corrosion resistance. Used in continous service up to 425 °C (800 °F).
ASTM Grade 2 UNS: R50400 Grade 2	Grade 2 Ti is the workhorse for industrial applications requiring good ductility and corrosion resistance. The guaranteed minimum yield strength of 275 MPa (40 ksi) for Grade 2 is comparable to those of annealed austenitic stainless steels.	Grade 2 can be used in continous service up to 425 °C (800 °F) and in intermittent service up to 540 °C (1000 °F). Used in the fabrication of components for aircraft, reaction vessels, heat exchangers, and electrochemical processing equipment.
ASTM Grade 3 UNS: R50550 Grade 3	Like the other grades of Ti metals and alloys, Grade 3 bridges the design gap between aluminum and steel. Grade 3 has lower iron limits than Grade 4 Ti.	Same application areas as Grade 2. Higher strength than Grade 2.
ASTM Grade 4 UNS: R50700 Grade 4	Grade 4 has the highest strength of the four ASTM grades of unalloyed Ti. Moderate formability. Outstanding corrosion fatigue resistance in salt water.	Often used interchangeably with Grade 3. The strength-to-weight ratio of Grade 4 is higher than 301 stainless up to 315 °C (600 °F).
MODIFIED TITANIUM		
Ti-0.2Pd UNS: R52400 and R52250 ASTM Grade 7 (R52400) and Grade 11 (R52250)	The two Ti-0.2Pd ASTM Grades (7 and 11) have better resistance to crevice corrosion at low pH and high temperature than corresponding unalloyed Grades 1 and 2.	Used for special corrosion applications. Grade 7 is comparable to Grade 2 in strength, while Grade 11 is comparable to Grade 1 in strength.
Ti-0.3Mo-0.8Ni UNS:R53400 ASTM Grade 12	Grade 12 is stronger and more resistant to crevice corrosion than unalloyed Ti grades. Grade 12 is less expensive than Ti-0.2Pd grades, but it does not offer the same crevice corrosion resistance at low pH (<3 pH).	Grade 12 is used in applications requiring moderate strength and enhanced corrosion resistance, such as heat exchangers, pressure vessels, chlorine cells, and salt evaporators. In near-neutral brines, crevice corrosion resistance is similar to Ti-0.2Pd.
COMMON ALPHA AND NEAR-ALPHA ALLOYS		
Ti-3Al-2.5V UNS: R56320 Tubing alloy, ASTM Grade 9, and Half 6-4	Ti-3Al-2.5V has excellent cold formability and 20 to 50% higher strength than unalloyed Ti.	Available as foil, seamless tubing, pipe, forgings, and rolled products. Used mostly in tubular form in various aerospace and nonaerospace applications.

(continued)

ALLOY, UNS NUMBER, AND COMMON NAMES	GENERAL DESCRIPTION	APPLICATIONS (continued)
Ti-5Al-2.5Sn, **standard grade and extra-low interstitial grade(ELI) grade** UNS: R54520 and R54521(ELI) Ti-5-2.5 and Ti-5-2 1/2	Ti-5Al-2.5Sn is a medium-strength alloy with excellent weldability. It has good oxidation resistance and intermediate strength at service temperatures up to 480 °C (900 °F).	The primary use is for cyrogenic applications, in which the ELI grade has good strength and toughness for low temperatures. It has been replaced by Ti-6Al-4V in other applications.
Ti-6Al-2Nb-1Ta-0.8Mo UNS: R56210 Ti-6211 and Ti-621/0.8	Ti-6211 was developed in the 1950s for U.S. Navy submersibles. It has excellent fracture toughness in marine environments.	Ti-6211 is used for hulls of marine vehicles and for pressure vessel applications requiring high toughness.
Ti-6Al-2Sn-4Zr-2Mo-0.1Si UNS: R54620 Ti-6242S and Ti-6242Si	Ti-6242S is one of the most creep-resistant Ti alloys and is recommended for use up to 565 °C (1050 °F).	Ti-6242S is primarily used for turbine components but is also used in sheet form for afterburner structures and various "hot" airframe applications.
Ti-8Al-1Mo-1V UNS: R54810 Ti-811	Ti-811 is available from most Ti-alloy producers. It has the highest tensile modulus of all commercial Ti alloys and has good creep resistance up to 455 °C (850 °F).	Ti-811 is used for airframe and turbine engine applications. It has room-temperature strength comparable to Ti-6Al-4V.

SINGLE-SOURCE ALPHA AND NEAR-ALPHA TITANIUM ALLOYS

Ti-5Al-6Sn-2Zr-1Mo-0.25Si UNS: None Ti-5621S	Ti-5621S is a semicommercial alloy developed by RMI Titanium Company in the 1960s to extend the use of titanium applications to 540 °C (1000 °F).	Used for jet engine components
Ti-6Al-2Sn-1.5Zr-1Mo-0.35Bi-0.1Si UNS: None Ti-11	Developed by Timet for improved creep resistance	Ti-11 was not commercially marketed because qualification costs exceeded the technical benefit.
Ti-6Al-2.75Sn-4Zr-0.4Mo-0.45Si UNS: None TiMetal® 1100 and Ti-1100	Ti-1100 was developed for high temperature use up to 600 °C (1100 °F).	High pressure compressor dics, low-pressure turbine blades, automotive valves
Ti-2.5Cu UNS: None IMI 230	IMI 230 combines the the formability and weldability of unalloyed Ti with improved strength from precipitation hardening.	IMI 230 is used at temperatures up to 350 °C (660 °F) and is used in the annealed condition for fabrication of components such as bypass ducts of turbines.
Ti-5.8Al-4.0Sn-3.5Zr-0.7Nb-0.5Mo-0.35Si-0.06C UNS: None IMI 417	IMI 417 is the general engineering version of IMI 834. Product manufacture may differ for IMI 417 applications versus IMI 834 manufacture.	Used as cast or wrought parts for turbine or internal combustion engines. Ideal for high-temperature (600 °C, max) fatigue-sensitive applications.
Ti-11Sn-5Zr-2.25Al-1Mo-0.25Si UNS: None IMI 679	IMI 679 was introduced in the 1960s as a high-temperature alloy.	IMI 679 has been superceded by other high-temperature Ti-alloys such as Ti-6242S.
Ti-6Al-5Zr-0.5Mo-0.25Si UNS: None IMI 685	IMI 685 was introduced in 1969 and developed to meet aerospace engine requirements.	Turbine engine alloy with a maximum useful temperature of about 520 °C (970 °F)
Ti-5Al-3.5Sn-3.0Zr-1Nb-0.3Si UNS: None IMI 829 and Ti-5331S	IMI 829 is a medium-strength alloy with high-temperature capability up to ~ 540 °C (1000 °F).	Used for discs and blades in aerospace turbine compressors
Ti-5.8Al-4Sn-3.5Zr-0.7Nb-0.5Mo-0.35Si UNS: None IMI 834	IMI 834 is a medium-strength alloy having high-temperature capability up to 600 °C (1100 °F) combined with good fatigue resistance.	Major uses include compressor discs and blades in the aerospace industry

COMMON ALPHA-BETA ALLOYS

Ti-5Al-2Sn-2Zr-4Mo-4Cr UNS: R58650 Ti-17	Ti-17 has strength properties superior to those of Ti-6Al-4V and higher creep resistance at intermediate temperatures.	Ti-17 is a high-strength, deep hardenable alloy used for heavy section forgings up to 150 mm (6 in.) for gas turbine components.

(continued)

ALLOY, UNS NUMBER, AND COMMON NAMES	GENERAL DESCRIPTION	APPLICATIONS (continued)
Ti-6Al-2Sn-4Zr-6Mo UNS: R56260 Ti-6246	Designed to combine short-term strength of alpha-beta alloys with long-term creep strength. Used at lower temperatures than Ti-6242S alpha alloy, but considered for long-term load carrying up to 400 °C (750 °F) and short-term loading up to 540 °C (1000 °F)	Ti-6246 is used for forgings in turbine engines and also for seals and airframe components. Ti-6246 has also been evaluated for deep, sour-well applications.
Ti-6Al-4V and Ti-6Al-4V ELI UNS: R56400 and R56401(ELI) Ti-6-4 ASTM Grade 5	Ti-6Al-4V is the most widely used Ti alloy, accounting for 50% of all the titanium tonnage wordwide. The alloy is most commonly used in the annealed. Ti-6Al-4V is also hardenable in sections up to 25 mm (1 in.) with yield strengths as high as 1140 MPa (165 ksi).	The aerospace industry accounts for more than 80% of Ti-6Al-4V usage. The biggest user of Ti-6Al-4V outside of the aerospace is for medical protheses, which accounts for about 3% of the market. The alloy also has a variety of weight reducing applications in high performance automotive and marine equipment.
Ti-6Al-6V-2Sn UNS: R56620 Ti-662	Ti-662 offers ultimate tensile strengths of 1200 Mpa (175 ksi) in the heat treated condition in sizes up to 25 mm (1 in.). Special melting practices are required to minimize alloy segregation.	Ti-662 is used for airframe structures where higher strength than Ti-6Al-4V is required. Usage is generally limited to secondary structures.
Ti-7Al-4Mo UNS: R56740	Ti-7Al-4Mo offers improved creep resistance over Ti-6Al-4V up to about 480 °C (900 °F).	This alloy has limited use today. It is primarily used for horns on ultasonic equipment.

OTHER ALPHA-BETA ALLOYS

Ti-4Al-4Mo-2Sn-0.5Si UNS: None IMI 550 (previously, Hylite 50)	IMI 550 is a medium strength alloy with typical ultimate tensile strength of 1100 MPa (159 ksi) and temperature capability up to about 400 °C (750 °F)	IMI 550 is used for both airframe and engine components. IMI is age hardenable in sections up to 150 mm (6 in.).
Ti-4Al-4Mo-4Sn-0.5Si UNS: None IMI 551	IMI 551 is similar to IMI 550, but its higher alloy content increases strength. IMI 551 is one of the strongest Ti-alloys with room-temperature ultimate tensile strength of 1250 to 1400 MPa (181 to 203 ksi).	IMI 151 is primarily intended for airframe components such as mounting brackets, pump casings, and undercarriage parts. Also suitable for connecting rods and reciprocating components.
Ti-4.5Al-3V-2Mo-2Fe UNS: None SP-700	SP-700 is designed to offer superior superplastic formability properties. The alloy is age hardenable.	SP-700 is superplastically formed into components such as aerospace parts, golf club heads, metal ballons, working tools, wrist watch casings, automotive parts, and mountain climbing equipment.
Ti-6Al-1.7Fe-0.1Si UNS: None TiMetal® 62S and 62S	With iron as a beta stabilizer, alloy 62S has lower formulation costs with processing and property characteristics similar to Ti-6Al-4V.	Practical substitute to Ti-6Al-4V.
Ti-6Al-7Nb UNS: None IMI 367	IMI 367 was developed specifically for femoral components for hip protheses.	IMI 367 has excellent biocompatibilty for surgical implants.
Ti-6Al-2Sn-2Zr-2Mo-2Cr-0.25Si UNS: None Ti-6-22-22-S	Development work on this alloy has been revived, and a strong effort is underway to develop processing procedures to optimize strength and fracture resistance of Ti-6-22-22-S in sheet, plate, and forged form.	The first production applications of Ti-6-22-22-S will be for the F-22 fighter. The alloy provides improved damage tolerance with respect to strength in relation to Ti-6Al-4V.

MISCELLANEOUS ALPHA-BETA ALLOYS

Ti-4Al-3Mo-1V UNS: None Ti-431	Developed in the 1950s as a sheet alloy with a good combination of strength and temperature stability up to 480 °C (900 °F)	Was used in the 1960s for aircraft skins, stiffeners, and internal aircraft structures.
T-4.5Al-5Mo-1.5Cr UNS: None Corona 5	Corona 5 was produced for experimental Navy programs, but has not found any commercial applications.	As an experimental alloy intended for fracture-critical structural applications, Corona 5 has been superceded by Ti-10V-2Fe-3Al.
Ti-5Al-1.5Fe-1.4Cr-1.2Mo UNS: None Ti-155A	Dropped from production because it was made using ferromaster alloys	Obsolete

(continued)

ALLOY, UNS NUMBER, AND COMMON NAMES	GENERAL DESCRIPTION	APPLICATIONS (continued)
Ti-5Al-2.5Fe DIN 3.7110 Tikrutan LT 35	Ti-5Al-2.5Fe provides corrosion resistance, biocompatibility, amenability to osseointegration, and biofunctionality (high ratio of fatigue strength to Young's modulus) as a material for permanent implants.	Potential applications for this alloy include all types of joint prostheses, spinal implants, as well as bone nails, screws, and plates.
Ti-5Al-5Sn-2Zr-2Mo-0.25Si UNS: R54560 Ti-5522-S	Ti-5522-S is a beta-lean $\alpha+\beta$ alloy with good tensile, stress rupture, and creep resistant properties.	Ti-5522-S is a semicommercial alloy for high-temperature applications in the range of 425 to 540 °C (800 to 1000 °F).
Ti-6.4Al-1.2Fe UNS: None RMI Low-Cost Alloy	Low cost alternative to Ti-6Al-4V	No applications reported to date
Ti-2Fe-2Cr-2Mo UNS: None	Not produced since 1960, Ti-2Fe-2Cr-2Mo was developed as a sheet alloy.	Replaced by Ti-6Al-4V
Ti-8Mn UNS: R56080 8Mn	Ti-8Mn was originally developed for its excellent formability and intermediate strength. It has good strength and temperature stability up to 315 °C (600 °F).	Although no longer in production, Ti-8Mn was used in the F8 and F9 fighters. The alloy has been replaced by Ti-6Al-4V.

COMMON BETA ALLOYS

Ti-11.5Mo-6Zr-4.5Sn UNS: R58030 Beta III	Beta III was developed in the 1960s as a supplement to Ti-13V-11Cr-3Al. It has excellent formability and age-hardened mechanical properties, but is difficult to melt.	Beta III was used for aircraft fasteners, rivets, and sheet metal parts, where its cold formability and strength potential are used to advantage. Also used for springs and orthodontic appliances.
Ti-3Al-8V-6Cr-4Mo-4Zr UNS: R58640 Beta C™ and 38-6-44	Beta C™ has similar characteristics to Ti-13V-11Cr-3Al, but it is easier to melt. It is cold rollable and drawable, and is used mainly as bar and wire for springs.	Beta C™ is used in fasteners, springs, torsion bars, and as foil for sandwich structures. It is also used for tubulars and casings in oil, gas, and geothermal wells.
Ti-10V-2Fe-3Al UNS: None Ti-10-2-3	This near-beta alloy was developed primarily for high-strength and toughness applications for temperatures up 315 °C (600 °F) and tensile strengths of 1240 MPa (180 ksi) The alloy is capable of attaining a wide variety of strength levels through heat treatment.	Ti-10V-2Fe-3Al is used as high-strength forgings in airframe applications requiring uniformity of tensile properties at surface and center locations. It provides medium-to-high strength and high toughness in bar, plate, or forged sections up to 125 mm (5 in.) thick.
Ti-13V-11Cr-3Al UNS: R58010 Ti-13-11-3 and B120VCA	Ti-13V-11Cr-3Al was the first beta alloy developed in the mid-1950s. For several years, Ti-13-11-3 was the only beta alloy of commercial significance, until the advent of Ti-10-2-3, Ti-15-3, and Beta C™	Although used in limited quantities today, Ti-13-11-3 is still used for some aircraft sheet metal applications and springs. It is used where extremely high strengths are required for short periods of time.
Ti-15V-3Al-3Cr-3Sn UNS: None Ti-15-3	Although originally developed in the 1970s as a sheet alloy, Ti-15-3 is used in other forms such as fasteners, foil, tubing, castings, and forgings.	Ti-15-3 is used in a variety of airframe applications. It has also been evaluated for aerospace tankage applications, high-strength hydraulic tubing, and fasteners.

OTHER BETA ALLOYS

Ti-1.5Al-5.5Fe-6.8Mo[a] UNS: None Timetal® LCB [a]Datasheet not included in the *Materials Properties Handbook: Titanium Alloys*	Timetal® LCB is a low-cost beta (LCB) beta alloy that can be produced at less than half the cost of typical beta alloys due to lower formulation costs from using ferro-molybdenum master alloys.	Coil springs, leaf springs, and clock springs are potential uses. High-strength fasteners and armor applications are also being investigated.
Ti-5Al-2Sn-4Zr-4Mo-2Cr-1Fe UNS: None Beta CEZ®	Beta CEZ® is a high-strength, high-toughness near-beta alloy with processing flexibility for a variety of applications.	Applications of Beta CEZ® include springs, fasteners, and heavy section forgings for intermediate-temperature turbine disks.
Ti-8Mo-8V-2Fe-3Al UNS: None Ti-8823	Ti-8823 has good formability and age hardening characterisics like other beta alloys. Producers have found the metallurgical behavior of Ti-8823 more predictable than Ti-13V-11Cr-3Al.	Ti-8823 was developed primarily as a sheet alloy, but it is also hardenable in sections up to 150 mm (6 in.).

(continued)

ALLOY, UNS NUMBER, AND COMMON NAMES	GENERAL DESCRIPTION	(continued) APPLICATIONS
Ti-15Mo-3Al-2.7Nb-0.25Si UNS: R58210 TiMetal® 21S and Beta 21S	Beta 21S is a recently developed high-strength beta alloy designed specifically for improved oxidation resistance, creep resistance, and thermal stability.	The primary applications include foil for honeycomb, and sheet, plate and bar for high-temperature engine and nacelle applications. Being used for short times up to 650 °C (1200 °F). Considered for composite foil, protheses, and tubing for hot hydraulic fluids
Ti-15Mo-5Zr UNS: None	Ti-15Mo-5Zr is a cold formable, age hardening alloy which is available as bar, cold drawn wire, and cold rolled sheet up to 0.1 mm.	High-strength alloy with high corrosion resistance in reducing media. Corrosion resistance is superior to that of Ti-0.2Pd
Ti-15Mo-5Zr-3Al UNS: None	Compared with Ti-15Mo-5Zr, embrittlement caused by ω phase does not occur as predominately in this alloy due to the 3-Al addition.	Candidate material for sour gas well applications. It is currently used as an erosion shield for turbine blades.

MISCELLANOUS BETA ALLOYS

Ti-11.5V-2Al-2Sn-11Zr UNS: None Transage 129, T129	Transage 129 is a noncommercial, age-hardenable alloy especially recommended for cold formable sheet applications.	As an experimental alloy, Transage 129 was intended to improve structural efficiency in airframe and chemical applications.
Ti-12V-2.5Al-2Sn-6Zr UNS: None Transage 134	Noncommercial, age-hardenable alloy	Intended for high-strength applications
Ti-13V-2.7Al-7Sn-2Zr UNS: None Transage 175	Noncommercial, age-hardenable alloy	Intended for high-strength applications
Ti-8V-5Fe-1Al UNS: None	Generally supplied as bar and billet	Was used for fasteners or where shear and ultimate strength are critical
Ti-16V-2.5Al UNS: None	Available only by special order	Developed by RMI Titanium Co. for use in high-strength sheet applications.
Ti-35V-15Cr[a] UNS: None Tiadyne 3515 [a]Datasheet not included in the *Materials Properties Handbook: Titanium Alloys*	Developed by Pratt and Whitney and liscensed to Teledyne Wah Chang Albany. This alloy has markedly improved resistance to sustained combustion when ignited.	Available in wrought and powder form

ADVANCED MATERIALS

Alpha-2 (Ti₃Al) aluminide alloys UNS: None Ti₃Al alloys or α-2 alloys, α₂ alloys	Ti₃Al alloys developed to date include two-phase (α-2 +β/B2) alloys and alloys with an orthorhombic (O) phase. The O-phase alloys appear to improve fracture toughness and creep resistance.	Semicommercial and experimental alloys are under development for possible high-temperature applications.
Gamma (Ti-Al) aluminide alloys UNS: None Ti-Al intermetallics can be classified generally as either single phase (γ) or two-phase (γ + α-2) alloys.	Although not commercially developed, currently available γ alloys contain approximately 46 to 52 at.% Al and 1 to 10 at.% of lest one of the following: V, Cr, Mn, Mo, Nb, W, Ta.	Like other intermetallics, the brittleness of γ alloys is a limiting factor for the use of these materials in high-temperature applications.
Ti-Ni shape memory alloys UNS: None Nickel-titanium, titanium-nickel, Tee-Nee, Memorite™, Nitinol, Tinel™, and Flexon™.	Shape memory properties of the Ti-Ni alloys greatly depends on exact composition, processing history, and small ternary additions.	Shape memory alloys are used to recover strain, stresses, and work in a variety of devices such as couplings, actuators, and other components.
Particulate-reinforced Ti alloys UNS: None CERMETi®	CERMETi® composites are reinforced with a variety of ceramic or intermetallic particulates such as TiC, TiB₂, or TiAl.	CERMETi® improves the tensile strength and elastic modulus of titanium alloys, at both room and elevated temperatures. Hardness and wear resistance are also improved.

Alloy Data Sheet Contents

TITANIUM DATA SHEETS

High-Purity Ti 125
    Summary of Physical Properties 125
    Phases and Structures 126
    Phase Diagrams 129
    Damping Properties 131
    Elastic Properties 131
    Electrical Properties 136
    Magnetic Characteristics 138
    Chemical/Corrosion Properties 139
    Thermal Properties 143
    Transport Properties 145
    Tensile Properties 147
    Hardness 149
    Creep Properties 150
    Fatigue Properties 152
    Impact Strength 154
    Fracture Mechanisms 155
    Fracture Stress 156
    Plastic Deformation 158
    Flow Stress 160
    Cold-Impact Extrusion 162
    Heat Treatment 163

Commercially Pure and Modified Titanium 165
    Unalloyed Ti Grade 1, R50250 165
    Unalloyed Ti Grade 2, R50400 167
    Unalloyed Ti Grade 3, R50550 170
    Unalloyed Ti Grade 4, R50700 171
    Ti-0.2Pd, R52400 (Grade 7), R52250 (Grade 11) .. 173
    Ti-0.3Mo-0.8Ni, R53400 175
    Phases and Structures (CP and Modified Ti, all
        grades) 176
    Elastic Properties (all ASTM grades) 178
    Electrical Resistivity 180
    Chemical Reactivity 181
    Electrochemical Potentials 185
    General Corrosion 188
    Unalloyed Ti: General Corrosion by Media 200
    Modified Ti: General Corrosion by Media 209
    Crevice Corrosion 211
    Hydrogen Damage 214
    Stress-Corrosion Cracking 217
    Thermal Properties 219
    Mechanical Properties 224
    High-Temperature Strength 227
    Creep Properties 233
    Fatigue Properties 236
    Fracture Properties 237
    Stress Strain Curves 239
    Plastic Deformation 242
    Bulk Working 245
    Forming Properties 247
    Forming Limit Diagrams 252
    Forming Methods 255
    Heat Treatment 259
    Machining 260

ALPHA AND NEAR-ALPHA ALLOYS

Ti-3Al-2.5V 263
    Phases and Structures 265
    Physical Properties 266
    Chemical/Corrosion Properties 266
    Thermal Properties 269
    Mechanical Properties 270
    Hardness 270
    Microindentation Hardness 271
    Typical Room-Temperature Strength 272
    Strength vs. Temperature 273
    High-Temperature Strength 275
    Fatigue Properties 277
    Fracture Properties 279
    Fracture Toughness 279
    Seamless Tubing 280
    Forming 283
    Heat Treatment 285

Ti-5Al-2.5Sn 287
    Physical Properties 289
    Chemical/Corrosion Properties 291
    Stress-Corrosion Cracking 292
    Thermal Properties 295
    Mechanical Properties 297
    Low-Temperature Tensile Properties 301
    High-Temperature Strength 303
    Fatigue Life 307
    Fatigue Crack Growth 310
    Fracture Properties 311
    ELI Fracture Toughness 314
    Forging 315

Forming 316	**IMI 417** **419**
Fabrication 317	**IMI 679**
Heat Treatment 318	Ti-11Sn-5Zr-2.25Al-1Mo-0.25Si 421
Ti-6Al-2Nb-1Ta-0.8Mo **321**	Phases and Structures 421
Physical Properties 322	Physical Properties 422
Mechanical Properties 324	Tensile Properties 423
High-Temperature Strength 328	Creep Properties 426
Fatigue Properties 329	Fatigue Properties 427
Fracture Properties 330	Plastic Deformation 428
Forging 334	Processing 429
Other Fabrication 335	**IMI 685**
Ti-6Al-2Sn-4Zr-2Mo-0.08Si **337**	Ti-6Al-5Zr-0.5Mo-0.25Si 431
Phases and Structures 338	Physical Properties 431
Physical Properties 340	Mechanical Properties 432
Stress-Corrosion Cracking 342	Fatigue Properties 433
Thermal Properties 346	Processing and Heat Treatment 433
Mechanical Properties 349	**IMI 829**
High-Temperature Strength 353	Ti-5Al-3.5Sn-3.0Zr-1Nb-0.3Si 435
Creep Properties 357	Physical Properties 435
Fatigue Properties 361	Mechanical Properties 436
Fracture Properties 363	Processing 437
Typical Stress-Strain Curves 365	**IMI 834**
Flow Softening 366	Ti-5.8Al-4Sn-3.5Zr-0.7Nb-0.5Mo-0.35Si ... 439
Flow Stress 367	Physical Properties 439
Forging 370	Mechanical Properties 440
Thermomechanical Effects 371	High-Temperature Strength 440
Forming 373	Fatigue Properties 441
Heat Treatment 374	Processing 443
Ti-8Al-1Mo-1V **377**	**Ti-5Al-6Sn-2Zr-1Mo-0.25Si** **445**
Physical Properties 378	Physical Properties 445
Stress-Corrosion Cracking 380	Mechanical Properties 446
Thermal Properties 388	Processing 449
Mechanical Properties 389	**ALPHA-BETA ALLOYS**
High-Temperature Strength 392	**Ti-5Al-2Sn-2Zr-4Mo-4Cr** **453**
Fatigue Properties 395	Phases and Structures 454
Fatigue-Crack Growth 398	Physical Properties 455
Fracture Properties 400	Mechanical Properties 455
Stress-Strain Curves 402	Creep Properties 457
Forging 403	Fatigue Properties 459
Forming 403	Fracture Properties 460
Heat Treatment 406	Forging 462
Machining and Welding 407	Heat Treatment 463
Ti-11 .... **409**	**Ti-6Al-2Sn-4Zr-6Mo** **465**
TIMETAL® 1100 .... **411**	Phases and Structures 466
Phases and Structures 412	Physical Properties 466
Mechanical Properties 413	Chemical/Corrosion Properties 469
Processing 414	Thermal Properties 470
IMI 230 .... **415**	Mechanical Properties 472
Physical Properties 415	High-Temperature Strength 474
Mechanical Properties 416	
Processing 417	

Fracture Properties	477
Forging	479
Other Fabrication Methods	480
Heat Treatment	480

Ti-6Al-4V ... 483

Introduction	483
Phases and Structures	488
Damping Characteristics	491
Elastic Properties	493
Physical Properties	497
Corrosion/Chemical Properties	500
Hydrogen Damage	504
Stress-Corrosion Cracking	507
Potentials	512
Thermal Properties	513
Design Mechanical Properties	517
Typical Strengths	522
Typical Room-Temperature Tensile Properties	524
High-Temperature Strength	528
Creep Properties	530
General Fatigue Behavior	533
Low-Cycle Fatigue	534
Fatigue Limits and Endurance Ratios	537
Surface and Texture Effects on Fatigue	539
Influence of Mean Stress	542
Effect of Processing	542
Effect of Heat Treatment on Fatigue	544
Constant-Life Fatigue Diagrams	548
Unnotched Fatigue Strength	553
Strain Life	558
Notched Fatigue Strength	559
Cast and P/M Fatigue	565
Corrosion Fatigue	569
Fatigue Crack Growth in Air	572
Crack Growth and Corrosion	575
Impact Toughness	579
Fracture Toughness	581
Plastic Deformation	592
Flow Stress	593
Forging	594
Forming	599
Forming Limits	601
Superplastic Forming	602
Machining	603
Forming Examples:	605
Heat Treatment	606
Beta Heat Treated Microstructures	611
Alpha + Beta Annealed Microstructures	614
Effects of Working	617
Quenched and Aged Microstructures	617
Welding	620
Welding Processes	620
Weldment Microstructures	627
Weldment Fatigue	629
Casting	633

Ti-6Al-6V-2Sn ... 637

Phases and Structures	639
Physical Properties	641
Chemical/Corrosion Properties	641
Thermal Properties	643
Mechanical Properties	644
High-Temperature Strength	649
Low-Cycle Fatigue	653
High-Cycle Fatigue	653
Constant Lifetime Diagrams	656
Fatigue Crack Propagation	658
Fracture Properties	660
Plastic Deformation	662
Forging	663
Fabrication	664
Heat Treatment	665

Ti-7Al-4Mo ... 667

Physical Properties	668
Corrosion	669
Thermal Properties	669
Mechanical Properties	670
High-Temperature Strength	672
Fatigue and Fracture	674
Fabrication	676

TIMETAL® 62S ... 679

Physical Properties	679
Mechanical Properties	681
Processing	684

Ti-4.5Al-3V-2Mo-2Fe ... 685

Physical Properties	685
Mechanical Properties	686
Fatigue and Fracture Properties	688
Plastic Deformation	689
Heat Treatment	690

IMI 367
Ti-6Al-7Nb ... 693

IMI 550 ... 695

Physical Properties	695
Mechanical Properties	696
High-Temperature Strength	697
Fatigue Properties	698
Processing	699

IMI 551
Ti-4Al-4Mo-4Sn-0.5Si ... 701

Physical Properties	701
Mechanical Properties	702
Processing	704

Corona 5
Ti-4.5Al-5Mo-1.5Cr ... 705

Physical Properties	705

Mechanical Properties	706
Fatigue Life	708
Fatigue Crack Growth Rates	710
Fracture Toughness	711

Ti-6-22-22S
Ti-6Al-2Sn-2Zr-2Mo-2Cr-0.25Si **713**

Physical Properties	714
Tensile Properties	715
High-Temperature Strength	718
High- and Low-Cycle Fatigue	720
Fatigue-Crack Propagation	722
Fracture Properties	725
Plastic Deformation	726
Forging	728
Other Fabrication Methods	729
Heat Treatment	730

Ti-4Al-3Mo-1V **733**

Ti-5Al-1.5Fe-1.4Cr-1.2Mo
Ti-155A **735**

Ti-5Al-2.5Fe **737**

Phases and Structures	737
Physical Properties	738
Mechanical Properties	739
Fatigue and Fracture Properties	741
Plastic Deformation	743
Net Shaping	745
Treatments	745

Ti-5Al-5Sn-2Zr-2Mo-0.25Si **747**

Physical Properties	747
Mechanical Properties	748
Processing	749

Ti-6.4Al-1.2Fe
RMI Low-Cost Alloy **751**

Ti-2Fe-2Cr-2Mo **753**

Ti-8Mn **755**

Physical Properties	755
Corrosion Properties	757
Thermal Properties	758
Mechanical Properties	759
Fatigue	761
Plastic Deformation	763

BETA AND NEAR-BETA ALLOYS

Ti-11.5Mo-6Zr-4.5Sn **767**

Phases and Structures	768
Physical Properties	770
Chemical/Corrosion Properties	772
Thermal Properties	774
Mechanical Properties	775
High-Temperature Strength	780

Fatigue Properties	784
Fracture Properties	785
Stress-Strain Curves	787
Forging	788
Forming	789
Sheet	790
Fasteners/Springs	792
Heat Treatment	793

Ti-3Al-8V-6Cr-4Mo-4Zr (Beta C) **797**

Phases and Structures	798
Physical Properties	799
Chemical/Corrosion Properties	801
Stress-Corrosion Cracking	802
Thermal Properties	805
Mechanical Properties	806
Typical Tensile Properties	811
High-Temperature Strength	814
Fatigue Properties	819
Fracture Properties	821
Stress-Strain Curves	822
Forging	822
Forming	824
Heat Treatment	826

Ti-10V-2Fe-3Al **829**

Phases and Structures	830
Physical Properties	833
Mechanical Properties	834
Room-Temperature Tensile Properties	837
High-Temperature Strength	841
Fatigue (Smooth)	845
High-Cycle Notched Fatigue	849
High-Cycle Fatigue: P/M and Cast	850
Fatigue Crack Growth	852
Fracture Toughness	854
Stress-Strain Curves	858
Flow Stress	860
Forging	862
Thermomechanical Processing Effects	863
Heat Treatment	864

Ti-13V-11Cr-3Al **867**

Phases and Structures	868
Physical Properties	869
Chemical/Corrosion Properties	870
Stress-Corrosion Cracking	871
Thermal Properties	873
Mechanical Properties	875
Typical Tensile Properties	877
High-Temperature Strength	880
Creep Properties	883
Fatigue Properties	885
Fracture Properties	888
Deformation	889
Forging	890

Formability 891	Heat Treatment 956
Bending and Stretching Limits 892	
Spinning, Beading, and Dimpling 894	
Heat Treatment 896	

Ti-15V-3Cr-3Al-3Sn 899

- Phases and Structures 900
- Physical Properties 901
- Corrosion Properties 901
- Thermal Properties 902
- Mechanical Properties 904
- Sheet Mechanical Properties 905
- Cast Tensile Properties 907
- High-Temperature Strength 908
- Fatigue Properties 910
- Fracture Properties 911
- Phase Deformation 912
- Flow Stress 913
- Forging 914
- Forming 915
- Heat Treatment 918

TIMETAL® 21S 921

- Physical Properties 922
- Corrosion Properties 922
- Thermal Properties 923
- Tensile Properties 924
- High-Temperature Strength 926
- Crack Resistance 928
- Processing 928

Ti-5Al-2Sn-4Zr-4Mo-2Cr-1Fe Beta-CEZ® 931

- Physical Properties 931
- Mechanical Properties 932
- Fabrication 933

Ti-8Mo-8V-2Fe-3Al 935

- Physical Properties 935
- Mechanical Properties 936
- High-Temperature Strength 939
- Fatigue and Fracture 940
- Processing 940

Ti-15Mo-5Zr 943

- Physical Properties 943
- Tensile Properties 944
- Fatigue and Fracture 947
- Fabrication 947

Ti-15Mo-5Zr-3Al 949

- Physical Properties 949
- Mechanical Properties 950
- Fatigue Properties 953
- Fracture Properties 953
- Flow Stress 954
- Forming Properties 955

Ti-11.5V-2Al-2Sn-11Zr 957

- Phases and Structures 958
- Physical Properties 959
- Tensile Properties 959
- Creep Properties 961
- Fatigue Properties 962
- Fracture Properties 963
- Working 964
- Heat Treatment 967
- Welding 969

Ti-12V-2.5Al-2Sn-2Zr 971

- Physical Properties 971
- Typical Tensile Properties 973
- Fatigue Properties 975
- Fracture Toughness 976
- Mechanical Properties of Bar 977
- Heat Treatment 979

Ti-13V-2.7Al-7Sn-2Zr 979

- Mechanical Properties 980
- High-Temperature Strength 981
- Fatigue 983
- Forming 985
- Heat Treatment 989
- Weldments 990

Ti-8V-5Fe-1Al 993

- Physical Properties 993
- Mechanical Properties 994
- High-Temperature Strength 996
- Fabrication 997
- Heat Treatment 997

Ti-16V-2.5Al 999

- Elastic Properties 999
- Room-Temperature Strength 1001
- High-Temperature Strength 1002
- Fatigue Strength 1004
- Plastic Deformation 1006

ADVANCED MATERIALS

Titanium Aluminides 1011

- Ti_3Al (α_2 or α-2) 1013
- Ti-Al 1014
- Ti_2Al-Nb (O Phase) 1016

Ti_3Al Alloys 1019

- Phases and Structures 1019
- Oxidation 1021
- Mechanical Properties 1021
- High-Temperature Strength 1023
- Fracture Toughness 1026
- References 1027

Gamma (Ti-Al) Alloys **1029**
    Mechanical Properties 1029
    High-Temperature Strength 1031
    Fatigue Strength 1032
    Environment Resistance 1033
    References 1033

Ti-Ni Shape Memory Alloys **1035**
    Chemistry and Density 1036
    Phases and Structures 1037
    Physical Properties 1038
    Corrosion 1039
    Thermal Properties 1040
    Transition Temperatures 1040
    Tensile Properties 1042
    Superelasticity (Between M_s and M_d) 1043
    High-Temperature Behavior 1044
    Fatigue Properties 1044
    Fracture 1046
    Processing 1046
    References 1047

Technical Note Contents

TECHNICAL NOTES

Technical Note 1: Metallography and Microstructure 1051

    Alpha Structures 1051
    Beta Structures 1055
    Interface Phase 1057
    Metallography 1057

Technical Note 1 Appendix: Example of ω_{ISO} Formation 1061

    Experimental Procedure 1061
    Results and Discussion 1061

Technical Note 2: Corrosion 1065

    General Corrosion 1066
    General Corrosion in Specific Media 1068
    General Corrosion Testing 1070
    Crevice Corrosion 1070
    Pitting 1072
    Hydrogen Damage 1072
    Stress-Corrosion Cracking 1073
    Other Forms of Corrosion 1075

Technical Note 3: Casting 1079

    Molding Methods 1079
    Alloys 1080
    Casting Design 1081

Technical Note 4: Forging 1083

    Metal Temperatures 1084
    Forging Equipment 1085
    Ancillary Procedures 1086
    Tolerances 1088
    Precision Forgings 1089

Technical Note 5: Forming 1093

    Preparation of Sheet for Forming 1094
    Tool Materials and Lubricants 1095
    Blank Preparation 1095
    Forming Temperatures 1096
    Forming Methods 1097

Technical Note 5A: Superplastic Forming of Titanium Alloys 1101

    Strain-Rate Sensitivity 1102
    Grain Size 1103
    Phase Ratio 1104
    Alloy Composition 1105
    Temperature 1105
    Forming Processes 1107
    Superplastic Forming and Diffusion Bonding 1108

Technical Note 6: Heat Treating 1111

    Alloy Types and Response to Heat Treatment 1111
    Stress Relieving 1112
    Annealing 1113
    Solution Treating and Aging 1113
    Contamination During Heat Treatment 1115
    Growth During Heat Treatment 1116

Technical Note 7: Machining 1119

    Guidelines 1119
    Turning and Boring 1121
    Milling 1123
    Drilling 1129
    Tapping 1130
    Broaching 1132
    Grinding 1133
    Sawing 1134
    Nontraditional Machining Methods 1135

Technical Note 8: Powder Metallurgy 1137

    Elemental Powder 1137
    Prealloyed Powder 1138
    Consolidation 1139
    Applications 1140
    P/M Property Data CermeTi® Property Data 1141
    Ti-8V-5Fe-1Al P/M Data 1142
    Typical P/M Property Data 1143

Technical Note 9: Descaling and Special Surface Treatments 1145

    Removal of Surface Oxides 1145
    Cleaning 1149
    Plating of Titanium 1150
    Diffusion Treatments 1153
    Other Coatings 1155
    Thermomechanical Surface Treatment 1156

Technical Note 10: Welding and Brazing 1159

    Weldability 1159

Cleaning 1159
  Arc Welding 1160
  Other Welding Methods 1163
  Postweld Evaluation 1164
  Brazing 1165

Rolling 1167

  Alpha and Alpha-Beta Alloys 1167
  Ti-6Al-4V Rolling 1168
  Beta Alloys 1168

Friction and Wear of Titanium Alloys 1169

  Surface Modification Treatments 1169
  Physical Vapor Deposition 1170
  Thermochemical Conversion Surface Treatments 1172
  Solid Lubrication 1173

ABBREVIATIONS and SYMBOLS

A	anneal		GTAW	gas tungsten arc welding
AFML	Air Force Materials Laboratory		HE	hot extruded
Aged	aged		Heat	heat
AH	age hardened		HCF	high cycle fatigue
Ann	annealed		Hc	critical field strength
Apps	applications		HIP	hot isostatic pressing
AS Weld	as welded		HR	hot rolled
at.%	atomic percent		HT	heat treated
AQ	as quenched		HW	hot worked
BE	blended elemental (powder)		Imp	implant
Bil	billet		Inv	investment
BYS	bearing yield strength		IPTS	international practical temperature scale
Cp	specific heat at constant pressure		Jc	critical current density
Cv	specific heat at constant volume		K	bulk modulus, strain hardening coefficient, or Hall Petch grain size constant
CD	cold drawn			
CMG	high interstitial composition		Kc	plane stress fracture toughness
Conds	condenser		KIc	plane strain fracture toughness
Cor	corrosion		KIi	stress intensity for crack initiation
CR	cold rolled		ks	kiloseconds
CW	cold worked		L	longitudinal
CYS	compressive yield strength		LBW	laser beam welding
da/dN	change in crack length per fatigue cycle		LCF	low cycle fatigue
DA	duplex annealed		LT	long transverse
DCB	double cantilever beam		m	strain rate sensitivity factor
DCEN	direct current electrode negative		MA	mill annealed
Diam	diameter		Mach	machined
E	Young's modulus, tensile modulus of elasticity		Met	metal
Ec	compressive modulus of elasticity		MPIF	Metal Powder Industries Federation
EBW	electron beam welding		Mult	Multiple
El	electrode, elongation		n	strain hardening exponent
ELI	extra low interstitial		NA	not available
Exch	exchanger		NHT	not heat treated
Ext	extrusion		nom	nominal
$Fbru$	ultimate bearing strength		Norm	normalized
$Fbry$	bearing yield strength		OE	others each
Fcy	compressive yield strength		O_{eq}	oxygen equivalent
Ftu	ultimate tensile strength		OT	others total (for impurity content)
Fty	tensile yield strength		PA	prealloyed (powder)
FC	furnace cooled		PAW	plasma arc welding
FAC	forced air cooled		PH	precipitation hardened
FCGR	fatigue crack growth rate		PHT	precipitation heat treated
FCP	fatigue crack propagation		Pip	pipe
Fill	filler		Plt	plate
Frg	forging(s)		Powd	powder
G	shear modulus		ppm	parts per million, or 10^6
GMAW	gas metal arc welding		PREP	plasma rotating electrode process

Press	pressure	Strp	strip
PWHT	post weld heat treatment	Surg	surgical
QT	quenched and tempered	T	temperature
Quen	quenched	TA	triplex annealed
r	strain ratio, anisotropy ratio of n values	T_C	Curie temperature
RA	recrystallization anneal	T_c	critical temperature
Res	resistant	TD	transverse direction
Rng	ring	T_m	melting temperature in degrees kelvin
RS	rapid solidification	TEM	transmission electron microscopy
RT	room temperature	Temp	tempered
Sand	sand cast	TMP	thermomechanical processing
SCE	saturated calomel electrode	Tub	tube
SG	side grooved	TYS	tensile yield strength
Sh	sheet	UBS	ultimate bearing strength
Shp	shape	UTS	ultimate tensile strength
SHT	solution heat treated	Ves	vessel
Sint	sintered	Weld	welded or welding in context
Smls	seamless	Wir	wire
SR	stress relieved	yr	year
STA	solution treated and aged	YS	yield strength
Stab	stabilized	R^m	general symbol for martensite κ compressibility
SI	standard interstitial		
STAN	solution treat and anneal	μ	Poisson's ratio
STOA	solution treated and overaged		

Table of Contents

Preface iii
Contributors and Reviewers v
Summary Table of Titanium
  Alloys vii
Alloy Data Sheet Contents xiii
Technical Note Contents xix
Abbreviations and Symbols xxi

Section I: Physical Metallurgy of Titanium Alloys

Introduction 3
Classification of Titanium Alloys 5
Physical Properties 12
Equilibrium Phases 23
Nonequilibrium Phases 34
Deformation 49
Aging 56
Titanium Alloys for Low-Temperature Service 68
Evolution of Conventional (Ingot Metallurgy)
  High-Temperature Titanium Alloys 76
Powder Metallurgy and Rapid-Solidification
  Processing 81
Rapid-Solidification Processing of Precipitate and
  Dispersion-Strengthened Titanium Alloys 87
Mechanical Properties 94
References 112

Section II: Titanium Data Sheets

High-Purity Ti 125
Commercially Pure and Modified Ti 165

Section III: Alpha and Near-Alpha Alloys

Ti-3Al-2.5V 263
Ti-5Al-2.5Sn 287
Ti-6Al-2Nb-1Ta-0.8 Mo (Ti-6211) 321
Ti-6Al-2Sn-4Zr-2Mo-0.1Si (Ti-6242) Si 337
Ti-8Al-1Mo-1V 377
Ti-11 409
TIMETAL® 1100 411
IMI 230 415
IMI 417 419
IMI 679 421
IMI 685 431
IMI 829 435
IMI 834 439
Ti-5Al-6Sn-2Zr-1Mo-0.1Si 445

Section IV: Alpha-Beta Alloys

Ti-5Al-2Sn-2Zr-4Mo-4Cr (Ti-17) 453
Ti-6Al-2Sn-4Zr-6Mo (Ti-6246) 465
Ti-6Al-4V 483
Ti-6Al-6V-2Sn 637
Ti-7Al-4Mo 667
TIMETAL® 62S 679
Ti-4.5Al-3V-2Mo-2Fe (SP-700) 685
IMI 367 693
IMI 550 695
IMI 551 701
Corona 5 705
Ti-6-22-22-S 713
Ti-4Al-3Mo-1V 733
Ti-5Al-1.5Fe-1.4Cr-1.2Mo 735
Ti-5Al-2.5Fe 737
Ti-5Al-5Sn-2Zr-2Mo-0.25Si 747
Ti-6.4Al-1.2Fe (RMI Low-Cost Alloy) 751
Ti-2Fe-2Cr-2Mo 753
Ti-8Mn 755

Section V: Beta and Near-Beta Alloys

Ti-11.5Mo-6Zr-4.5Sn (Beta III) 767
Ti-8V-3Al-6Cr-4Mo-4Zr (Beta C) 797
Ti-10V-2Fe-3Al (Ti-10-2-3) 829
Ti-13V-11Cr-3Al 867
Ti-15V-3Al-3Cr-3Sn (Ti-15 3) 899
TIMETAL 21S 921
Beta CEZ® 931
Ti-8Mo-8V-2Fe-3Al 935
Ti-15Mo-5Zr 943
Ti-15Mo 5Zr-3Al 949
Ti-11.5V-2Al-2Sn-11Zr (Transage 129) 957
Ti-12V-2.5Al-2Sn-6Zr (Transage 134) 971
Ti-13V-2.7Al-7Sn-2Zr (Transage 175) 979
Ti-8V-5Fe-1Al 993
Ti-16V-2.5Al 999

Section VI: Advanced Materials

Titanium Aluminides 1009
Ti₃Al Alloys 1019
Gamma (Ti-Al) Alloys 1029

Ti-Ni Shape Memory Alloys..................... 1035

Section VII: Technical Notes

Technical Note 1: Metallography and
  Microstructure............................ 1051
Technical Note 1 Appendix: Example of ω_{iso}
  formation................................. 1065
Technical Note 2: Corrosion 1065
Technical Note 3: Casting 1079
Technical Note 4: Forging...................... 1083

Technical Note 5: Forming...................... 1093
Technical Note 5a: Superplastic Forming 1101
Technical note 6: Heat Treating 1111
Technical Note 7: Machining..................... 1119
Technical Note 8: Powder Metallurgy 1137
Technical Note 9: Surface Treatments 1145
Technical Note 10: Welding and Brazing........... 1159
Rolling... 1167
Friction and Wear of Titanium Alloys 1169

Physical Metallurgy of Titanium Alloys*

E.W. Collings, Battelle Memorial Institute, Columbus, Ohio, U.S.A.

*Revised from *The Physical Metallurgy of Titanium Alloys* (ASM International, 1984) and "Introduction to Titanium Alloy Design" in *Alloying* (ASM International, 1988)

1. Introduction

1.1 Origin and Uses of Titanium

Titanium is widely distributed throughout the universe. It has been discovered in the stars, in interstellar dust, in meteorites, and on the surface of the earth. Its concentration within the earth's crust of about 0.6% makes it the fourth most abundant of the structural metals (after aluminum, iron, and magnesium). It is 20 times more prevalent than chromium, 30 times more than nickel, 60 times more than copper, 100 times more than tungsten, and 600 times more than molybdenum. This abundance is to some extent illusory, however, in that titanium is not so frequently found in economically extractable concentrations. Concentrated sources of the metal are the minerals ilmenite, titanomagnetite, rutile, anatase, and brookite.

Ilmenite is haematite (Fe_2O_3) in which half of the iron has been replaced by titanium; titanomagnetite is magnetite (Fe_3O_4) in which one-third of the iron has been replaced by titanium. Rutile is TiO_2 (as are anatase and brookite). Naturally occurring (and titanium-deficient) ilmenite consists of haematite particles in a matrix of ilmenite; naturally occurring (and, again, titanium-deficient) titanomagnetite is magnetite containing laths of ilmenite. In short, the most important titanium minerals are ilmenite and rutile.

Titanium was first discovered in minerals now known as rutile by W. Gregor (England) and M.H. Klaproth (Germany) in about 1790. The first commercial mill products were produced by the Titanium Metals Company of America (TMCA) around 1950. From that time to the present, production of the metal has grown at an average annual rate of about 8%. Superimposed upon part of this temporal growth curve is a large fluctuating component, a reminder of the capriciousness of the materials demands of the aerospace industry, titanium's principal market during the early years. Fortunately for the titanium-production industry, the 13% annual growth rate exhibited by the civilian sector of the total market since the early 1960s has served to somewhat offset the decline in military demand during the same period, thereby yielding not only a net growth but a relatively steady one.

Titanium (meaning titanium and its alloys) has two principal virtues: (1) a high strength/weight ratio and (2) good corrosion resistance. At one time or another practically all aerospace structures —airframes, skin, and engine components—have benefited from the introduction of titanium. Nonaerospace applications include steam-turbine blades, hydrogen-storage media, high-current/high-field superconductors, condenser tubing for nuclear and fossil-fuel power generation, and other corrosion-resistant applications such as components for ocean thermal-energy conversion, offshore oil drilling, marine-submersible vessels, desalination plants, waste-treatment plants, the pulp-and-paper industry, and the chemical and petrochemical industries.

Interest in the properties of titanium and its alloys began to accelerate in the late 1940s [CRA49] and early 1950s as their potential as high-temperature, high-strength/weight materials with aeronautical applications became more and more widely recognized. The history of titanium and its development in alloyed form has been described in detail in the introduction to the first International Conference on the subject [JAF70] and in the introduction to ZWICKER's comprehensive metallurgical treatise *Titan und Titanlegierungen* [ZWI74]. As evidenced by the papers presented at the subsequent International Conferences, titanium and its alloys have by now found widespread use in the aerospace industry (for both frame and engine components) and in the chemical and related industries, where advantage can be taken of their corrosion resistance. According to WOOD [WOO72], by 1972 about 30 commercial alloys were already on the market in mill-product form. Of these, the eight most favored compositions, accounting for some 90% of the sales, were three grades of unalloyed titanium and the alloys Ti-5Al-2.5Sn, Ti-6Al-4V, Ti-8Al-1Mo-1V, Ti-6Al-6V-2Sn, and Ti-13V-11Cr-3Al. At that time also, interest in each of the alloys Ti-6Al-2Sn-4Zr-2Mo (i.e., "Ti-6242"), Ti-6Al-2Sn-4Zr-6Mo (i.e., "Ti-6246"), and Ti-11.5Mo-6Zr-4.5Sn (i.e., "β-III") was on the increase. Today the alloy Ti-6242 to which about 0.1% Si has been added is being used in titanium alloy forgings and has received extensive study and use in its role as a gas-turbine compressor-disc material. Finally it should be noted that Ti-10V-2Fe-3Al has been the beneficiary of the renewed interest being shown in so-called "near-β" titanium alloys [DUE80[a]] [TER80] [TOR80], while it is at last becoming recognized that Ti-50Nb, one of the most important of today's technical superconductors, is in fact a β-Ti alloy [COL81].

1.2 Extraction of Titanium

In order to cope with unexpected increases in the demand for a metal, it is helpful to be able to rely on a copious and stable supply of the basic ore. The titanium industry is fortunate in this regard. Titanium dioxide is produced in large quantities for many applications, so much so that in 1977, for example, only a few percent of the world's production of titanium ore was tapped for metallic sponge refinement (most of the mined ore being used to make paint pigment). Thus, since the overall demand for raw material is not subject to the same fluctuations as the demand for the metal, should the latter undergo a significant increase at any time, there is at least a strong raw-material base from which to draw.

Industry's growing awareness of the need for energy conservation has served to emphasize an unfortunate characteristic of the current methods of titanium metal refinement: their energy intensiveness. The energy required to produce a ton of sponge-titanium from its ore is 16 times that needed to produce a ton of steel, 3.7 times that needed for ferrochrome, 1.7 times that needed for aluminum production, and a little more than that needed for a 1-ton ingot of magnesium. Since, however, the heats of formation of rutile (~ –228 kcal/mol), haematite (~ –200 kcal/mol), and magnetite (~ –268 kcal/mol) are in the ratio of 1:0.88:1.18,

Table 1.1 Total Impurity Contents of Iodide- and Kroll-Process Titaniums (in wt%) [RAS72]

Element	Iodide Ti	Kroll Ti
Mg	0.01	0.13
Si	0.01	0.05
Al	0.02	...
Fe	0.01	0.20
Ni	0.01	...
Co	...	0.02
Cr	0.01	...
Mn	0.005	0.02
C	0.01	0.08
N	0.02	0.04
O	0.02	0.11

Table 1.2 Typical Interstitial Impurity Contents of Several Grades of Titanium

Grade of titanium	Interstitial content, ppm			Data source
	C	N	O	
MRC (MARZ-grade)	78	6	63	1
MRC (VP-grade)	150	40	350	2
TMC electrorefined sponge (grade ELXX)	...	40	370	3
Kroll-process (Toho sponge)	...	110	860	4
Kroll-process	800	400	1100	5
Iodide-process	100	200	200	5

(1) Materials Research Corp.: Zone-refined; supplied typical analysis. (2) Materials Research Corp.: Vacuum melted; supplied typical analysis. (3) Titanium Metals Corp.: See also [COL70]. (4) See [COL70]. (5) See Table 1.1.

there seems to be some scope for increasing the energy efficiency of the titanium-refinement process.

The most well-known method of titanium production is the Kroll process, which involves the reduction of $TiCl_4$ by magnesium. The first step in the process is the preparation of the tetrachloride itself, which is carried out by the chlorination of a mixture of carbon with rutile or ilmenite. The Kroll magnesium-reduction reaction takes place in a closed heated reactor vessel under an inert atmosphere. Liquid $TiCl_4$ is introduced to the liquid magnesium already present in the vessel, thereby initiating the reaction $2Mg + TiCl_4 \rightarrow 2MgCl_2 + Ti$. The reaction products are commercially pure sponge-titanium (in the form of a porous, gray, coke-like mass) and $MgCl_2$, most of which can be drained out of the reaction chamber as a liquid. The $MgCl_2$ is electrolytically recycled. The titanium sponge is consolidated by arc melting in a water-cooled copper crucible: this process involves several iterations of a procedure in which an arc is maintained between a consumable compacted-sponge-titanium electrode and a pool of molten sponge.

The highest purity titanium is prepared for research purposes by the iodide process. Crude titanium is first reacted with iodine in an inert atmosphere to form titanium iodide. This can then be decomposed at the surface of a heated titanium wire, which acts as nucleus for the growth of a long cylindrical bar of high-purity titanium crystals. Typical impurity contents of several grades of titanium are listed in Tables 1.1 and 1.2.

These and other standard commercial methods of titanium production, such as the sodium-reduction (or Hunter) process, the direct-oxide-reduction process, and the electrolytic process, have been described in detail by MCQUILLAN [MCQ56, Chap. 2], HOCH [HOC73b], and ZWICKER [ZWI74, pp. 21-27], while some new approaches developed in the Soviet Union have been outlined by REZNICHENKO and coworkers [REZ82, REZ82a].

2. Classification of Titanium Alloys

2.1 Systematics of Phase Stability

Pure titanium undergoes an allotropic transformation from hcp (α) to bcc (β) as its temperature is raised through 882.5 °C [MOL65][ZWI74]. Elements that when dissolved in titanium produce little change in the transformation temperature (e.g., tin) or cause it to increase (e.g., aluminum, oxygen) are known as "α stabilizers"; they are simple metals (SM) or interstitial elements [MOL65, p. 154]—generally nontransition elements. Alloying additions that decrease the phase-transformation temperature are referred to as "β stabilizers"; they are generally the transition metals (TM) (e.g., Mo and V) and noble metals—i.e., metals that, like titanium, have unfilled or just-filled d-electron bands. In the alloys, of course, the single-phase-α and single-phase-β regions are not in contact as they are in pure titanium; they are instead separated by a two-phase $\alpha + \beta$ region whose width increases with increasing solute concentration. Based on these considerations, technical alloys of titanium are classified as "α," "β," and "$\alpha + \beta$."

The question of lattice stability plays an important role in any discussion of the physics of pure metal or alloy systems. This is particularly true of titanium alloys, whose lattice stability (i.e., structural phase stability) has technical as well as fundamental significance. The crystal structures of the three long periods of transition elements change more or less systematically from hcp through fcc as the group number increases from IV to VIII. Whether or not there is an underlying physical significance to this, in the case of transition metals a useful correlation certainly exists between crystal structure and group number (in the case of elements) or crystal structure and average group number or electron/atom ratio (in the case of alloys). The existence of such correlations suggests that electronic structure plays an important role in the control of phase stability.

Numerous workers have attempted to define the factors that govern the existence of the α and β phases of titanium alloys. Solute atoms which lower the temperature of the allotropic $\alpha + \beta$ transformation, with respect to that of pure titanium, are referred to as β stabilizers. Conversely, α stabilizers raise that temperature. As pointed out by MCQUILLAN [MCQ63], the relatively more open bcc structure has a higher vibrational entropy than do the close-packed structures hcp and fcc. Consequently, during heating, the free energy of an imaginary bcc lattice will decrease more rapidly than those of the competing alternatives such that eventually a temperature will be reached whereat the lattice (if it does not melt) will transform from the low-temperature-stable close-packed structure (generally hcp, α) to bcc. Underlying this thermodynamic picture is an atomistic model involving electronic cohesive forces (directional or otherwise) and atomic-size effects. JAFFEE, in an early analysis of the situation [JAF58], suggested that atomic-size effect was the dominant factor; subsequently, he was able to conclude that, although size effect needed to be taken into consideration, the dominant phase-stabilizing mechanism was electronic in nature. MCQUILLAN also took this latter view [MCQ63], but pointed out that exceptions did of course exist—for example, the β-stabilizing tendencies of the solutes bismuth and lead were thought to be due to their relatively large atomic sizes [MCQ63].

Factors controlling the stabilization of the α and β phases in titanium alloys have also been discussed in several publications by COLLINGS and GEGEL [COL73a, COL73b, COL75a], with particular reference to the Ti-Al and Ti-Mo systems. Stability was qualitatively discussed from both electronic [COL73, COL82a] and thermodynamic (phenomenological) [COL75a] standpoints.

2.1.1 Electronic Considerations in Phase Stability

As a result of low-temperature specific heat measurements, it was noted that the more stable of a pair of allotropes was associated with the lower electronic density-of-states at the Fermi level, $n(E_F)$. This rule was exemplified using data for the following pairs of competing phases: α_2* and α; α and β; ω and β [COL73].

With transition metals, the electron/atom ratio, e/a, is the same as the average "group number"—referring to the numbers assigned to the groups of the periodic table. Thus, e/a takes on the values 4 through 10 when applied to the members of the seven columns of the TM block of the periodic table headed by the elements Ti through Ni. The e/a is a parameter in terms of which numerous physical and mechanical properties of binary TM alloys, particularly Ti-TM, can be conveniently displayed. Several important physical (including superconductive) properties may also be indexed in terms of quantities related to the above-mentioned conventional e/a, viz.: the atomic-volume-corrected "electron concentration" of JENSEN et al. [JEN65] or the "effective electron/atom ratio," N_{eff}, of DESORBO [DES65]. Another quantity advocated by LUKE et al. [LUK64] as being appropriate for the indexing of the compositional threshold for martensitic transformation in Ti-TM alloys is an average Pauling valence which, although equal to conventional e/a for the groups IV through VI transition elements, never exceeds the value 6 for elements of later groups. The crystal structures, particularly those of simple metals, have been justified from several fundamental standpoints. BREWER [BRE67] has related structure to the spectroscopic states of the individual participating atoms. PAULING [PAU67], in considering the metallic bond, has also utilized this as a basis for discussion. The OPW type of approach also utilized atomic spectroscopic states, but in a more satisfactory manner by starting with an array of bare ions and then replacing the electrons in such a way that their wavefunctions represent tightly bound electrons near the cores, and nearly free electrons in the spaces between. Although attempts to deal electronically with phase stability in transition metals have been made by INGLESFIELD [ING69] and PETTIFOR

---

*A hexagonal DO_{19} structure found in the Ti-Al system.

[PET72], the situation with regard to alloys is much more difficult.

Very successful calculations of the *electronic structures* of alloys, and in particular the manner in which the band density of states, $n(E)$, varies with energy, E, have been made using the coherent potential approximation (CPA) first applied by EHRENREICH and colleagues [KIR70] to the Cu-Ni system. The particular method used, since it took a tight-binding (TB) approach to the d-electrons and a nearly-free-electron (NFE) one to the other electrons in the band, has been referred to as the NFE-TB-CPA. Although it was especially applicable to Cu-Ni, it was the forerunner of more sophisticated methods, developed by others, of dealing with the energy-band structures of disordered alloys [FAU82]. In overcoming the limitations of the NFE-TB-CPA, a CPA method was developed which had some features in common with the old Korringa-Kohn-Rostocker (KKR) method. The first publication of a full KKR-CPA calculation, again as it applied to Cu-Ni alloys, was by STOCKS *et al.* [STO78]. The number of alloy systems to which such calculations have been applied, and for which the results have been compared with experiment (angular resolved photoemission is a favored method), has been quite limited.

However, it is still a large step from calculations of this kind to calculations of lattice-phase (crystal-structure) stability. PETTIFOR [PET79] has made considerable progress toward the calculations of the heats of formation of binary alloys by using a simple formalism, based on a Friedel expression for the binding energy per atom, in which the CPA played a fundamental role. As indicated above, it is a remarkable experimental fact that the crystal structures of $3d$, $4d$, and $5d$ transition metals, and their "adjacent" binary alloys, vary in a regular manner from hcp through bcc to fcc as a function of the e/a or average group number. MOTT and JONES' interpretation of one of the Hume-Rothery rules was an unsuccessful attempt to provide a crystal-structure/electron-concentration relationship for nontransition metals; other approaches have been more successful [BLA67]. So far the empirical crystal structure ("phase stability") versus e/a relationships as they apply to *transition metals* seem to exist without a general theoretical interpretation [FAU82, p. 186].

The closest approach to an exact calculation of phase stability in a transition-metal alloy system, in particular Zr-Nb, has been made by MYRON *et al.* [MYR75], who dealt not with equilibrium phases but with an electronic mechanism leading to the appearance of the metastable ω phase. Adequately discussed in their paper (see also SINHA and HARMON [SIN76]), the technique employed coupled a KKR band-structure and Fermi-surface calculation for bcc zirconium with the effects of "rigid-band" modifications of it brought about by the addition of niobium, in order to demonstrate that electronically instigated enhancement of the natural dip in the bcc-lattice phonon spectrum at $2/3\langle 111\rangle$ could lead, in a manner to be discussed below, to the ω-phase transformation.

2.1.2 Thermodynamic Considerations in Phase Stability

Purely electronic descriptions of equilibrium-phase stability have been strongly criticized from two standpoints by KAUFMAN and NESOR [KAU73]. They noted that: (1) in many treatments, competition between phases was completely ignored; and (2) when electronic property data acquired at low temperatures were used to justify high-temperature phase transformations, no account was taken of the entropy differences. KAUFMAN and NESOR recommended the use of a thermodynamic procedure, in which the energetic competition between candidate phases was fully taken into account, when attempting to define the lattice stabilities of metallic elements as well as alloy systems. Full discussion of a quantitative thermodynamic approach, leading to the computer-assisted calculation of binary and multicomponent phase diagrams, is to be found in the work of KAUFMAN [KAU70].

Pair-interaction-potential calculations based on the relative-vapor-pressure measurements of HOCH *et al.* [ROL71, ROL72], have divided the field of titanium-base alloys into two regimes: (1) β-stabilized Ti-TM alloys whose regular-solution thermodynamic interaction parameter, Ω_{ij}, is positive (indicative of clustering systems), and (2) α-stabilized Ti-SM alloys for which Ω_{ij} is negative (short-range-ordering systems) [COL75a].

2.2 Alpha Alloys

Unalloyed titanium and alloys of titanium with α stabilizers such as aluminum, gallium, and tin, either singly or in combination, as in the commercial alloy Ti-5Al-2.5Sn or the experimental Ti-Al-Ga alloys [HOC73][GEG73], are hcp at ordinary temperatures and as such are classified as α alloys. These alloys, according to WOOD [WOO72], are characterized by satisfactory strength, toughness, creep resistance, and weldability. Furthermore, the absence of a ductile-brittle transformation, a property of the bcc structure, renders α alloys (typified by Ti-5Al-2.5Sn) suitable for cryogenic applications [SAL79].

Alpha-stabilizing solutes are those which, as a function of concentration, elevate the temperature of the $(\alpha + \beta)/\alpha$ transus. Such solutes are generally nontransition metals (i.e., "simple metals", SM). An explanation of α stability based on electron-screening arguments proceeds as follows: When simple metals (e.g., aluminum) are dissolved in titanium, very few electrons appear at the Fermi level, most of them going to states within the lower part of the band. The titanium d-electrons tend to avoid the aluminum atoms, which thereby have the effect of diluting the titanium sublattice. The consequence of this is to emphasize any preexisting Ti-Ti bond directionality and thus to preserve the hcp structure characteristic of the titanium crystal. In general, when simple metals are added to titanium, the fields of titanium-like α stability are eventually terminated by intermetallic compounds, of compositions such as Ti_3SM, which are also hexagonal in structure. The bond argument is consistent with the observation that α stabilizers are quite rapid solution strengtheners either in hcp solid solution or when added to bcc alloys [GEG73a]. The classification of α-phase alloys into systems whose phase diagrams exhibit (1) peritectic transformations or (2) peritectoid transformations, according to Molchanova's simplified scheme, is considered in Section 2.5.

2.3 Beta Alloys

Transition-metal (TM) solutes are stabilizers of the bcc phase. Thus all-β alloys generally contain large amounts of one or more of the so-called "β-isomorphous"-forming additions—vanadium, niobium, tantalum (group-V TM's), and molybdenum (a group-VI TM). The systematics of β stabilization in binary and multicomponent titanium-base alloys has been discussed in detail by AGEEV and PETROVA [AGE70]. The archetypal binary β-stabilized titanium-base alloy, about which a great deal of physical and metallurgical information has been garnered over the years, is Ti-Mo. For a useful overview of the mechanical properties and aging characteristics of a pair of typical β alloys, Ti-15Mo-5Zr and Ti-15Mo-5Zr-3Al, the work of NISHIMURA *et al.* [NIS82] is recommended. There are several important commercial β alloys; three that have been attracting considerable attention recently are Ti-10V-2Fe-3Al, Ti-15V-3Cr-3Al-3Sn, and Ti-3Al-8V-6Cr-4Mo-4Zr [FRO73][PET73][VIG82][WIL82a]. Beta alloys, according to WOOD [WOO72], are extremely formable. They are, however, prone to ductile-brittle transformation [GOR73] and, along with other bcc-phase alloys, are unsuitable for low-temperature applications [SAL79].

The TM block of the periodic table may be regarded as commencing with group III: scandium, yttrium, and lanthanum (or, perhaps more precisely, lutetium). In this scheme, the alkaline-earth metals—calcium, strontium, and barium—may be regarded as "pretransition metals," and the noble metals—copper, silver, and gold—as "post-transition metals." As indicated in most periodic charts of the elements, the structures of the transition metals all change from hcp to bcc as e/a increases from 4 through 6. It is possible that stabilization of the bcc structure can be justified within the framework of a screening model in terms of which a high conduction-electron concentration, which enhances the screening of ion cores, may favor a symmetrical, hence cubic, structure. Thus an increase in electron density (as in groups V and VI elements), which tends to symmetrize the screening, increases the stability of the bcc structure. Symmetrization may also be accomplished through lattice vibrations; thus, all six of the groups III and IV elements transform to the bcc structure at high temperatures (as compared with their Debye temperatures). With regard to alloys, the addition of transition elements to titanium increases the electron density and consequently stabilizes the bcc or β structure. Thus, as a general rule, the transition elements are β stabilizers. The systematics of β stabilization by transition elements has been discussed in detail by AGEEV and PETROVA [AGE70], according to whom: (1) the β-stabilizing action of TM solutes is greater the "farther" they are from titanium in the periodic table; and (2) for the retention of the metastable-β solid solution during quenching, the β stabilizer has to provide for an electron/atom ratio of at least 4.2.

According to ZENER [ZEN48], and subsequently FISHER [FIS70, FIS75], who has considered the problem of bcc stability in considerable detail, the magnitude of the elastic shear modulus $C' = (C_{11} - C_{12})/2$ is a useful parameter for ranking the stabilities of bcc transition metals and alloys. The variation of C' with the conventional e/a ratio is plotted in Fig. 2.1, which shows that the alloying of group IV elements with other elements to the "right" of them in the periodic table increases the bcc stability, which rises to a maximum near $e/a = 6$ for the elements chromium, molybdenum, and tungsten. On the other hand, with decreasing e/a, the vanishing of C' for $e/a = 4.1$ corresponds to the compositional threshold for martensitic transformation—to be discussed below. As a result of the alloying of titanium with transition elements of higher group number, the continuous increase of bcc stability manifests itself as a lowering of the β/(α + β) transus temperature.

As indicated below, within the context of β stabilization two subclasses of phase diagrams exist—the "β isomorphous" and the "β eutectoid," depending on whether or not a solid-solution/compound eutectoid exists at a sufficiently elevated temperature. It is instructive in the present context to consider a group of simplified, compositionally truncated, binary Ti-TM equilibrium phase diagrams, arranged according to the positions that the solute elements occupy in the TM block of the periodic table. Some representative diagrams selected from such a postulated arrangement are presented in Fig. 2.2. In order to focus attention on the alloys of most interest, the limiting composition (in at.%) in each group (except group IV itself) has been selected such that $e/a \leq 5.0$. In so doing it has been assumed that the numbers of $s + d$ valence electrons belonging to the elements in the columns headed by Fe, Co, Ni, and Cu, are 8, 9, 10, and 11, respectively. An alternative way of deriving a reduced composition scale for intercomparison purposes, and one that would focus attention on alloy composition rather than electron density, might have been to normalize composition (i.e., stretch the composition scale) to that of the first β-eutectoidal intermetallic compound.

Interesting systematics to be noted in Fig. 2.2 are that: (1) as the solute element moves to the "right," the phase diagram changes from the β-isomorphous to the β-eutectoidal type; and (2) along the row Mn-Fe-Co-Ni-Cu, the eutectoid temperature increases monotonically. Extrapolating this trend to the "left" suggests that Ti-V can also be thought of as eutectoidal, but with an inaccessibly low eutectoid temperature.

2.4 Alpha + Beta Alloys

The α + β alloys are such that at equilibrium, usually at room temperature, they support a mixture of α and β phases. Although many binary β-stabilized alloys in thermodynamic equilibrium are two-phase, in practice the α + β alloys usually contain mixtures of both α and β stabilizers. The simplest of such alloys, and one upon which most attention has undoubtedly been lavished, is Ti-6Al-4V. Although this particular alloy is difficult to form, even in the annealed condition [SAL79], α + β alloys generally exhibit good fabricability as well as high room-temperature strength and moderate elevated-temperature strength. They may contain between 10 and 50% β phase at room temperature; if they contain more than 20%, they are not weldable. The properties of α + β alloys can be controlled by heat treatment, which is used to adjust the microstructural and precipitational states of the β component.

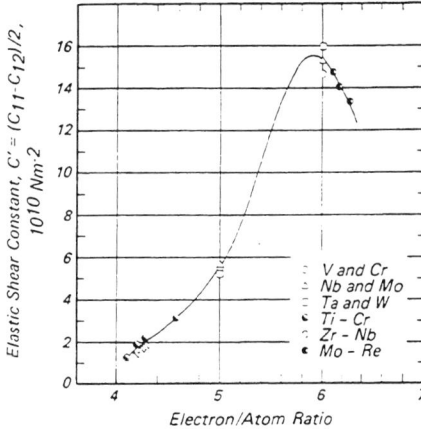

Fig. 2.1 Elastic shear modulus for bcc transition metals and some of their binary alloys as a function of e/a ratio [COL73ª].

2.5 Classification Schemes for Binary Alloys

All authors agree that the alloys of titanium can be assigned to one of two major categories*—α-stabilized or β-stabilized systems. MARGOLIN [MAR60] has recommended subdividing the former into two more groups according to the degree of α stabilization: (1) those of "limited α stability," in which decomposition of α takes place by peritectoid reaction into β plus a compound (e.g., Ti-B, Ti-C, and Ti-Al); and (2) those of "complete α stability," in which the α phase can coexist with the liquid (e.g., Ti-O and Ti-N). MARGOLIN has also recommended subdividing β-stabilized alloys into four categories in the following way: (1) β-isomorphous systems, such as Ti-Mo and Ti-Ta, which show restricted α- and extensive β-solubility ranges; (2) β-and-α-isomorphous systems, such as Ti-Zr, showing complete mutual solubilities in both the α and β phases; and (3) β-eutectoid systems, in which the β phase has a limited solubility range and is able to decompose into α and a compound (e.g., Ti-Cr and Ti-Cu)—this class being further subdivisible into two more depending on whether the β decomposition is rapid (e.g., Ti-Cu, Ti-Ni, and Ti-Sn) or sluggish (e.g., Ti-Cr, Ti-Mn, and Ti-Fe). KORNILOV [KOR82] has discussed a subdivision into what he refers to as four basic alloy types. Although the classes were untitled, their descriptions con-

---

*As indicated above, technical alloys are, of course, classified as α, β, and α + β according to their microstructural states when placed in service.

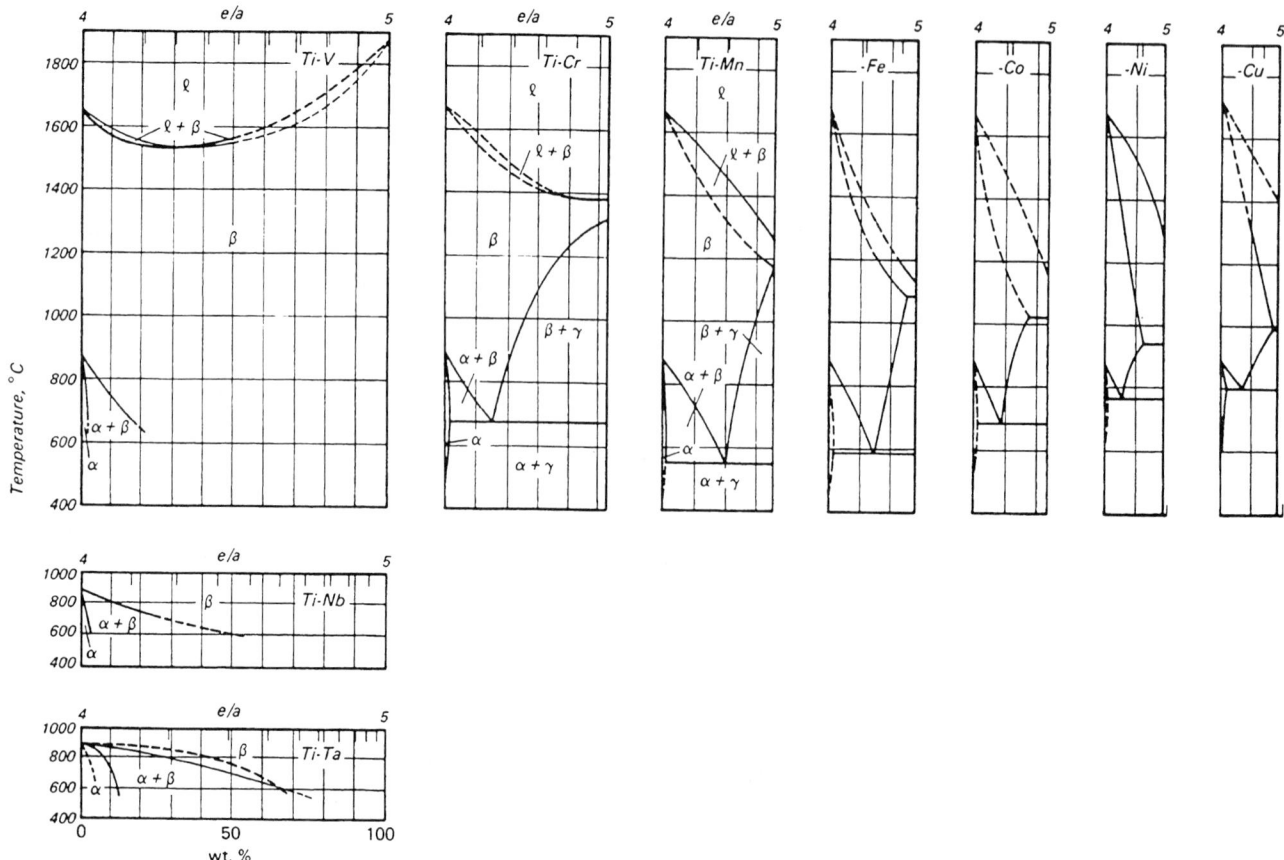

Fig. 2.2 Equilibrium phase diagrams for a representative group of binary Ti-TM alloys truncated at an e/a ratio of 5.0. Composition scales: 10 at.% intervals along the tops of the figures; 10 wt% along the grid lines.

formed to the categories: α phase, β-and-α isomorphous, β isomorphous, and β eutectoid, referred to above. In a survey of titanium alloy phases, MOLCHANOVA [MOL65, p. xiv] has offered a detailed subdivision of the equilibrium phase diagrams into two groups of three subcategories, and one group of four. The schematic phase diagrams which typify these 10 subcategories, and the solutes which give rise to each of them, are depicted in Fig. 2.3. The descriptions of the groupings are as follows:

Group I: Systems with continuous β solid solubility
   I(a) Complete miscibility in the α phase
   I(b) Partial miscibility in the α phase
   I(c) Partial miscibility in the α phase and eutectoid decomposition of the β phase

Group II: Eutectic systems
   II(a) Partial miscibility in the α and β phases; eutectoid decomposition of the β phase
   II(b) Partial miscibility in the α and β phases; peritectic decomposition of the β phase
   II(c) No detectable solid solubility

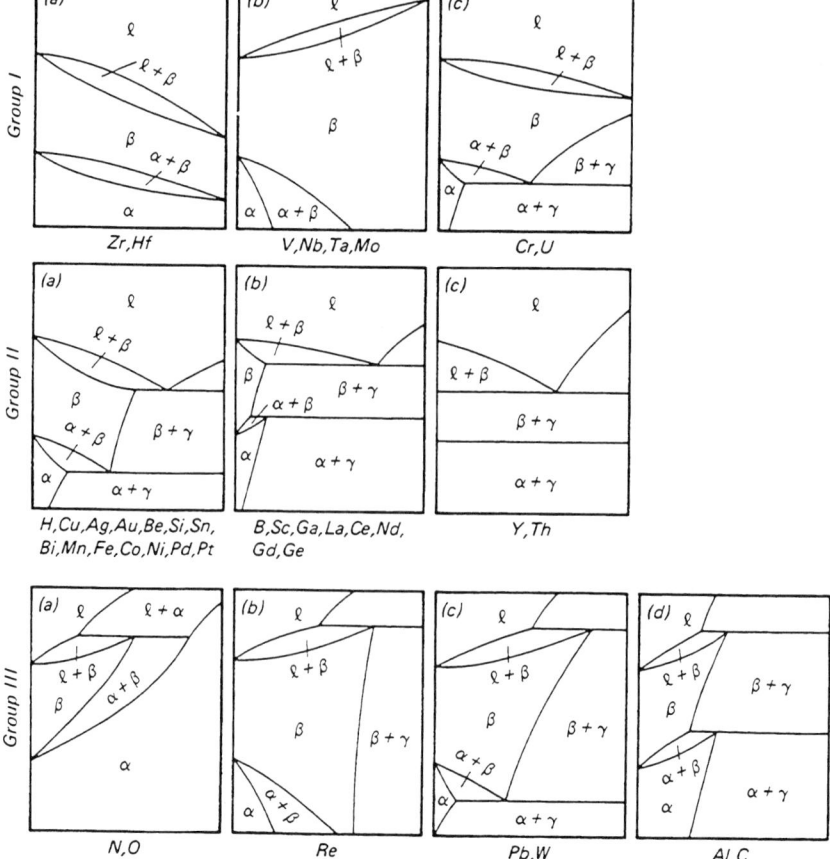

Fig. 2.3 Classification scheme for binary titanium alloy phase diagrams [MOL65, p. xiv].

Group III: Peritectic systems
  III(a) Simple peritectic
  III(b) Partial miscibility in the α and β phases
  III(c) Partial miscibility in the α and β phases; eutectoid decomposition of the β phase
  III(d) Partial miscibility in the α and β phases; peritectic dissociations of the β phase

In the same book, MOLCHENOVA [MOL65, p. 154] has also offered a simpler subdivision into the four categories depicted in Fig. 2.4.

2.6 Classification of Technical Multicomponent Alloys

2.6.1 Classification Schemes

Technical multicomponent alloys are generally composed of mixtures of α and β stabilizers (see Section 2.4), depending on the ratio of which they may be classified broadly as "α," "β," or "α + β." Within the last category are the subclasses "near-α" and "near-β," referring to alloys whose compositions place them near the α/(α + β) or (α + β)/β phase boundaries, respectively. A list of U.S. alloys subdivided into these categories is presented in Table 2.1. The compositions of some other U.S. alloys and of some technical British and Soviet alloys are listed in Tables 2.2, 2.3, and 2.4, respectively. These alloys may also be sorted into microstructural classifications with the aid of a scheme to be discussed in the following subsection.

In an alternative attempt at alloy classification, NISHIMURA et al. [NIS84] have mapped (Fig. 2.5) the locations of a series of U.S. technical alloys along the abscissa of a "β-isomorphous" (see Fig. 2.4) binary alloy phase diagram. Presented in this way, it is clear that position along the abscissa is controlled by the relative abundance of the alloy's α- and β-stabilizing components. This ratio can be most conveniently expressed in terms of α-stabilizing and β-stabilizing "equivalences."

2.6.2 Alpha-Stabilizing and Beta-Stabilizing Equivalences

The prototypical α-stabilizing and β-stabilizing additions to titanium are aluminum and molybdenum, respectively.

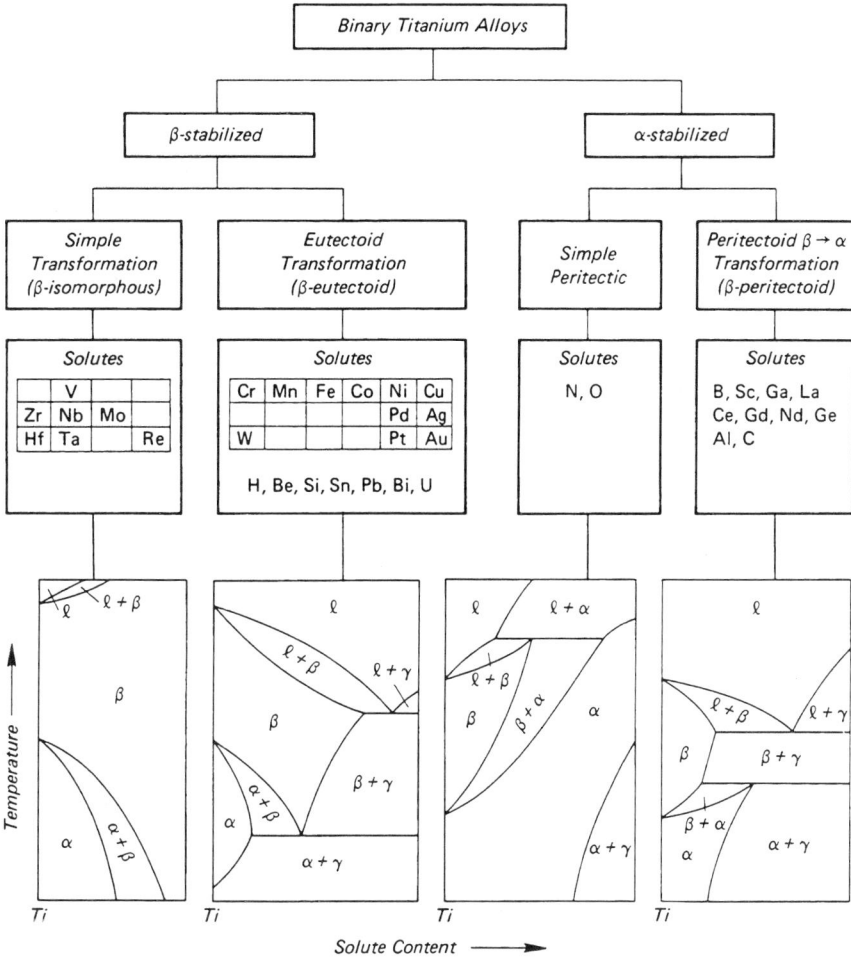

Fig. 2.4 Classification scheme for binary titanium alloy phase diagrams—an alternative to the scheme in Fig. 2.3. "α" and "β" are hcp and bcc solid-solution alloys, respectively, and "γ" represents an intermetallic compound [MOL65, p. 154].

Table 2.1 Classification of U.S. Technical Multicomponent Alloys [WOO72][STR82]

Composition, wt%	Classification
Ti-5Al-2.5Sn	α
Ti-8Al-1Mo-1V	Near-α
Ti-6Al-2Sn-4Zr-2Mo	
Ti-6Al-4V	
Ti-6Al-6V-2Sn	α + β
Ti-3Al-2.5V	
Ti-6Al-2Sn-4Zr-6Mo	
Ti-5Al-2Sn-2Zr-4Cr-4Mo	Near-β
Ti-10V-2Fe-3Al	
Ti-13V-11Cr-3Al	
Ti-15V-3Cr-3Al-3Sn	
Ti-3Al-8V-6Cr-4Mo-4Zr	β
Ti-8Mo-8V-2Fe-3Al	
Ti-11.5Mo-6Zr-4.5Sn	

Table 2.2 Commercial and Semicommercial U.S Titanium Alloys [FRO85] (Supplement to Table 2.1)

Composition, wt%	Classification
Ti-0.8Ni-0.3Mo	
Ti-6Al-2Nb-1Ta-0.8Mo	α and near-α
Ti-2.25Al-11Sn-5Zr-1Mo	
Ti-5Al-5Sn-2Zr-2Mo	
Ti-7Al-4Mo	
Ti-4.5Al-5Mo-1.5Cr	α + β
Ti-6Al-2Sn-2Zr-2Mo-2Cr	
Ti-8Mn	Metastable β

Table 2.3 British Technical Commercial Alloys [FRO85]

Designation	Composition, wt%
IMI 318	Ti-6Al-4V
IMI 550	Ti-4Al-4Mo-2Sn-0.5Si
IMI 679	Ti-11Sn-1Mo-5Zr-2.25Al-0.25Si
IMI 680	Ti-11Sn-4Mo-2.25Al-0.25Si
IMI 685	Ti-6Al-5Zr-0.5Mo-0.3Si
IMI 829	Ti-5.5Al-3.5Sn-3Zr-1Nb-0.3Mo-0.3Si
IMI 834	Ti-5.5Al-4Sn-4Zr-1Nb-0.3Mo-0.5Si

Table 2.4 Soviet Titanium Alloys(a) [FRO85]

Code	Composition, wt%
TG-00	99.7 Ti
TG-2	99.2 Ti
VT1-1	
VT1D-1	Commercial unalloyed grades
VT1-2	
VT-2	Ti-1.6Al-2.5Cr
VT-3	Ti-4.6Al-2.5Cr
VT-3-1	Ti-4.6Al-2Cr-1.7Mo-0.5Fe
VT-4	Ti-4.6Al-1.5Mn
VT-5	Ti-4.5Al
VT-5-1	Ti-5Al-2.5Sn
VT-6	Ti-6Al-4V
VT-8	Ti-6Al-3Mo
VT-14	Ti-4Al-3Mo-1V
VT-15	Ti-3Al-6.5Mo-11Cr
VT-16	Ti-2Al-7Mo
OT-4	Ti-3Al-1.5Mn
OT-4-1	Ti-1.7Al-1.4Mn
48-OT3	Ti-4Al-0.1Si-0.1Fe-0.005B
IRM-1	Ti-4Al-4Nb
IRM-2	Ti-4Al-4Nb-0.1Re
IRM-3	Ti-4Al-3.5Mo
IRM-4	Ti-3.5Al-3.5Mo-0.1Re
AT-2-1	Ti-Zr-(Mo or Nb or V)
AT-2-2	Ti-Zr-(Mo or Nb or V)
AT-2-4	Ti-Zr-(Mo or Nb or V)
AT-3	Ti-3Al-0.7Cr-0.4Fe-0.3Si-0.01B
AT-4	Ti-4Al-0.6Cr-0.23Fe-0.4Si-0.01B
AT-6	Ti-6Al-0.6Cr-0.4Fe-0.3Si-0.01B
AT-8	Ti-7Al-0.6Cr-0.2Fe-0.3Si-0.01B

(a) See also Table 9.2

Table 2.5 Concentration of Transition Elements Needed to Retain the β Phase at Room Temperature. After MOLCHANOVA [MOL65]

Group No.	Element	Critical concentration, wt%
V	V	15
	Nb	36
	Ta	50
VI	Cr	8
	Mo	10
	W	25
VII	Mn	6
VIII(a)	Fe	4
VIII(b)	Co	6
VIII(c)	Ni	8

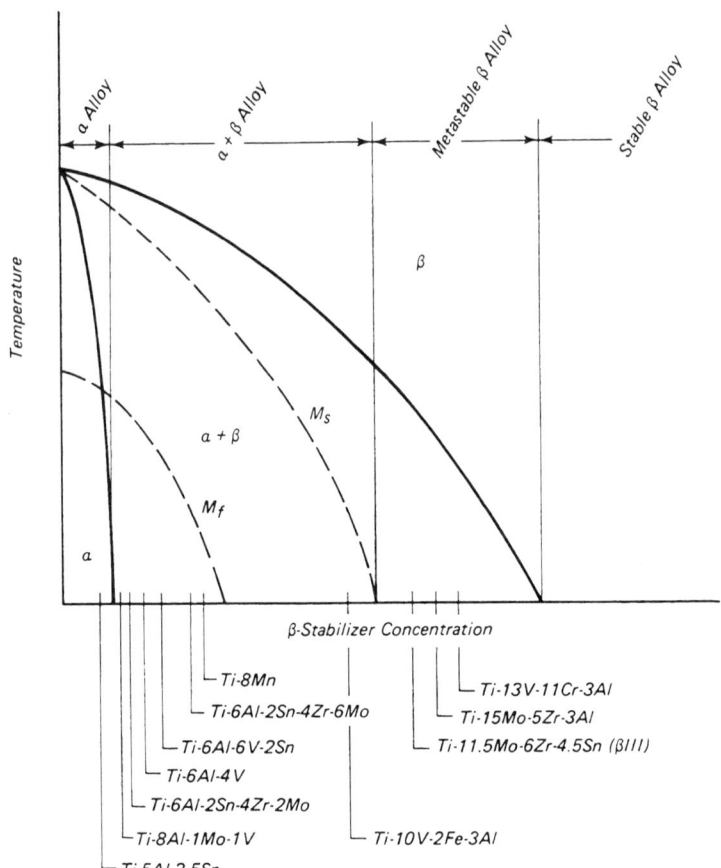

Fig. 2.5 Compositions of U.S. technical alloys mapped onto a pseudobinary β-isomorphous phase diagram [NIS84].

Accordingly, it is useful to be able to classify a multicomponent titanium-base alloy in terms of its equivalent aluminum and molybdenum contents.

Equivalent Aluminum Content. In that elements such as aluminum and oxygen elevate the $(\alpha + \beta)/\alpha$ transus when alloyed into titanium, they are regarded as strong stabilizers of the α phase. Tin is also an α stabilizer, although not a strong one. Since zirconium has the effect of lowering the temperature of the $(\alpha + \beta)/\alpha$ transus at a very low rate, it may from that standpoint be regarded as a neutral addition. On the other hand, zirconium occupies the same column of the periodic table as titanium (viz., group IV). As a consequence of this chemical similarity, zirconium may substitute for titanium in a multicomponent alloy and thereby add weight to its α-stabilizing component. For this reason it may be regarded as an α stabilizer.

Aluminum is the canonical α-stabilizing addition against which other such additives may be compared. According to ROSENBERG [ROS70], the equivalent aluminum content of an alloy containing aluminum, zirconium, tin, and oxygen is:

$$[Al]_{eq} = [Al] + \frac{[Zr]}{6} + \frac{[Sn]}{3} + 10[O] \quad (Eq\ 2.1)$$

where [x] indicates the concentration of element "x" in weight percent.

Equivalent Molybdenum Content. The β-stabilizing strength of transition-element additions to titanium can be gauged by the rates at which they lower the martensite transus and hence the degree to which they permit the retention of the β phase at room temperature. MOLCHANOVA has displayed this information in the form of a partial phase diagram [MOL65, p. 158] whose essential data are summarized in Table 2.5. An intercomparison of these data enables the Mo-equivalence of an alloy to be expressed in the form:

$$[Mo]_{eq} = [Mo] + \frac{[Ta]}{5} + \frac{[Nb]}{3.6} + \frac{[W]}{2.5}$$
$$+ \frac{[V]}{1.5} + 1.25[Cr] + 1.25[Ni] + 1.7[Mn]$$
$$+ 1.7[Co] + 2.5[Fe] \quad (Eq\ 2.2)$$

Transformation of a number of multicomponent titanium-base alloys into their Al- and Mo-equivalent formats provides a rationalization for their placement into one or another of the previously discussed phase-stability classifications (Table 2.6).

Table 2.6 Aluminum and Molybdenum Equivalences of a Series of U.S. Titanium Alloys

Alloy classification and composition, wt.%	Aluminum equivalency, wt.%				Molybdenum equivalency, wt.%									
	[Al]	$\frac{[Zr]}{6}$	$\frac{[Sn]}{3}$	$[Al]_{eq}$	[Mo]	$\frac{[Ta]}{5}$	$\frac{[Nb]}{3.6}$	$\frac{[V]}{1.5}$	1.25[Cr]	1.25[Ni]	1.7[Mn]	1.7[Co]	2.5[Fe]	$[Mo]_{eq}$
Alpha and near-alpha alloys														
Ti-0.8Ni-0.3Mo	...	...	...	...	0.3	...	...	...	...	1.0	...	...	...	1.3
Ti-5Al-2.5Sn	5.0	...	0.8	**5.8**	...	...	...	...	...	...	...	...	...	...
Ti-8Al-1Mo-1V	8.0	...	...	**8.0**	1.0	...	...	0.7	...	...	...	...	...	1.7
Ti-6Al-2Sn-4Zr-2Mo-0.1Si	6.0	0.7	0.7	**7.4**	2.0	...	...	...	...	...	...	...	...	2.0
Ti-6Al-2Nb-1Ta-0.8Mo	6.0	...	...	**6.0**	0.8	0.2	0.6	...	...	...	...	...	...	1.6
Ti-2.25Al-11Sn-5Zr-1Mo	2.3	0.8	3.7	**6.8**	1.0	...	...	...	...	...	...	...	...	1.0
Ti-5Al-5Sn-2Zr-2Mo	5.0	0.3	1.7	**7.0**	2.0	...	...	...	...	...	...	...	...	2.0
Alpha-beta alloys														
Ti-6Al-4V	6.0	...	...	**6.0**	...	...	...	2.7	...	...	...	...	...	2.7
Ti-6Al-6V-2Sn	6.0	...	0.7	**6.7**	...	...	...	4.0	...	...	...	...	...	4.0
Ti-7Al-4Mo	7.0	...	...	**7.0**	4.0	...	...	...	...	...	...	...	...	4.0
Ti-4.5Al-5Mo-1.5Cr	4.5	...	...	**4.5**	5.0	...	...	...	1.9	...	...	...	...	6.9
Ti-6Al-2Sn-4Zr-6Mo	6.0	0.7	0.7	**7.4**	6.0	...	...	...	...	...	...	...	...	6.0
Ti-5Al-2Sn-2Zr-4Mo-4Cr	5.0	0.3	0.7	**6.0**	4.0	...	...	...	5.0	...	...	...	...	9.0
Ti-6Al-2Sn-2Zr-2Mo-2Cr	6.0	0.3	0.7	**7.0**	2.0	...	...	...	2.5	...	...	...	...	4.5
Ti-3Al-2.5V	3.0	...	...	**3.0**	...	...	...	1.7	...	...	...	...	...	1.7
Ti-10V-2Fe-3Al	3.0	...	...	**3.0**	...	...	...	6.7	...	...	...	...	5.0	11.7
Beta alloys (metastable)														
Ti-8Mn	...	...	...	...	...	...	...	...	...	...	13.6	...	...	13.6
Ti-11.5Mo-6Zr-4.5Sn	...	1.0	0.4	**1.4**	11.5	...	...	...	...	...	...	...	...	11.5
Ti-10V-2Fe-3Al	3.0	...	...	**3.0**	...	...	...	6.7	...	...	...	...	5.0	11.7
Ti-15V-3Cr-3Al-3Sn	3.0	...	1.0	**4.0**	...	...	...	10.0	3.8	...	...	...	...	13.8
Ti-13V-11Cr-3Al	3.0	...	...	**3.0**	...	...	...	8.7	13.8	...	...	...	...	22.5
Ti-8Mo-8V-2Fe-3Al	3.0	...	...	**3.0**	8.0	...	...	5.3	...	...	...	...	5.0	18.3
Ti-3Al-8V-6Cr-4Mo-4Zr	3.0	0.7	...	**3.7**	4.0	...	...	5.3	7.5	...	...	...	...	16.8

3. Physical Properties

3.1 Introduction

3.1.1 Properties and Scope

Accompanying the development of titanium-base alloys and the associated mechanical-property measurements during the 1970s was a series of studies by COLLINGS and coworkers of one or more of the physical properties: electrical resistivity, Hall coefficient, magnetic susceptibility, and low-temperature specific heat. Representative α-phase alloys that have been subjected to these measurements are: Ti-Al, Ti-Ga, Ti-Sn, Ti-Al_x-Ga_x, Ti-Al-Ga (Ti_3X-type), and Ti-Al-Sn (Ti_3X-type). Representative β-phase alloys were: Ti-Mo, Ti-Mo-Al, Ti-Mo-Si, Ti-Mo-Fe, and Ti-Mo-Al-Fe. The numerical results themselves were reported in [COL71a], while a critical interpretation of these and other results acquired during the 1970s has been presented in [COL80]. FISHER and coworkers have conducted an extensive series of studies of the elastic properties of monocrystalline titanium-base and other transition-metal alloy systems [FIS70, FIS70a, FIS75][KAT79, KAT79a]. The elastic moduli of polycrystalline alloys, in relation to their compositions and microstructures, have been studied by FEDOTOV and coworkers and the results published in a series of papers spanning the period 1963 to 1973 [FED63, FED64, FED66, FED73]. Elastic modulus may of course be regarded as either a physical or a mechanical property. Again, with considerable emphasis on microstructural states and phase transformations, various physical properties such as electrical resistivity and superconducting transition temperature have been studied by POLONIS, coworkers, and students on research alloys such as Ti-Nb, Ti-Cr, Ti-Mo, Ti-Mo-Al, Ti-Nb-Al, and others [CHA73, CHA74][LUH68, LUH69, LUH70, LUH70a, LUH71, LUH72][POL69, POL70, POL71]. With regard to the physical properties of *technical* titanium-base alloys, a compendium of properties such as electrical resistivity, specific heat, thermal conductivity, and thermal expansion of unalloyed titanium, Ti-5Al-2.5Sn, Ti-6Al-4V, Ti-8Al-1Mo-1V, and Ti-13V-11Cr-3Al covering the temperature range from very low–(2 ~ 20 K) to room temperature has been assembled by SALMON [SAL79].

Within the space of a single chapter it is obviously not possible to do full justice to the literature referred to above. Instead, a brief review based on a representative collection of papers is presented in which the emphasis is placed on the manner in which physical-property measurements may be used as indicators of: (1) microscopic and macroscopic metallurgical states of titanium alloys, and (2) phase transformations and precipitation effects which take place in response to heat treatment. In other words, this chapter will take the form of a survey of physical-property measurements as they apply to metallurgical-property diagnosis.

Dynamic elastic modulus, regarded herein as a mechanical property, will be treated along with static elastic modulus and the plastic properties in Chapter 12. Properties included for discussion in this chapter are listed in the following five subsections.

3.1.2 Measurement of Electrical Resistivity

Measurements of electrical resistivity as functions of composition and temperature have provided useful metallurgical insights into certain strength and stability properties of α-phase and β-phase alloys. In α-Ti alloys (i.e., Ti-SM alloys where the solute is a so-called "simple metal"), a large specific solute resistivity (resistivity per at.% solute) is indicative of rapid solid-solution strengthening and is often accompanied by a rapid hardening coefficient. In β-Ti-TM alloys, an anomalous resistivity composition dependence is associated with the composition range over which isothermal- and athermal-ω phases are expected, with an anomalous resistivity temperature dependence *within* this composition range indicating the occurrence of reversible precipitation or associated structural fluctuations.

3.1.3 Measurement of Magnetic Susceptibility

Magnetic susceptibility is the sum of numerous terms, one of which, χP, the Pauli paramagnetism, is proportional to the density-of-states at the Fermi level, $n(E_F)$, an important fundamental electronic property. But, ignoring its underlying significance, magnetic susceptibility has been used to (1) delineate phase boundaries in quenched Ti-SM alloys, (2) investigate the $\alpha_2 \to \alpha$ (order-disorder) transformation in Ti-SM alloys, and transformations to the β phase in both Ti-SM and Ti-TM alloys, (3) augment electrical resistivity in studying reversible ω-phase precipitation in the temperature range 150 to 300 K in quenched Ti-TM alloys, and (4) monitor the course of ω-phase precipitation during the protracted moderate-temperature aging of initially quenched Ti-TM alloys.

3.1.4 Measurement of Low-Temperature Specific Heat

The specific heat at low temperatures, C, is generally the sum of two components: γT and βT^3, where T is the absolute temperature, γ is the electronic specific-heat coefficient (proportional to the density-of-states at the Fermi level, referred to above), and β, the lattice specific-heat coefficient, contains the Debye temperature, θ_D. In the case of Ti-TM alloys, a decrease of θ_D to low values, when plotted versus composition or electron/atom ratio, signifies lattice softening interpretable as a precursor to ω-phase precipitation. If the sample is a superconductor, another electronic property obtainable from low-temperature specific-heat measurements is T_c, the superconducting transition temperature. Both γ and T_c, together with the total magnetic susceptibility, χ, have been used to monitor isothermal ω-phase precipitation during aging. The electronic component of the low-temperature specific heat of a pure unstrained single-phase superconductor undergoes a sharp discontinuous jump at the superconducting transition. If, as a result of inadequate quenching, deliberate aging, or mechanical deformation, the superconductive specific-heat jump is severely rounded, the fitting of a distribution of sharp jumps to it can provide information relating to the microstructural constitution of a polyphase sample.

3.1.5 Measurement of AC Impedance

Low-temperature calorimetry provides a contactless means of studying the superconducting transition temperature of a bulk sample. AC impedance measurement is another such technique. In this method, as applied by LUHMAN [LUH70] to the study of metallurgical effects in Ti-TM alloys, particularly Ti-Cr, the sample is surrounded by a coil connected to an oscillator adjusted to some convenient frequency, say 1 kHz. An electronic voltmeter placed across the coil gives an indication proportional to the impedance of the coil + sample. Since this is sensitive to the permeability (hence AC susceptibility, dM/dH) of the sample, the voltmeter reading responds to the transition from the superconducting to the normal state as the temperature of the sample is increased. LUHMAN and others (e.g., [LUH70]) have exploited this technique in a study of the microstructural responses of several Ti-TM alloys to variations in composition and thermal treatment. Their work on Ti-Cr, a β-eutectoid alloy, might be regarded as an indirect companion to the comparable series of calorimetric studies performed by COLLINGS and HO on Ti-Mo, a related β-isomorphous alloy system.

3.1.6 Magnetization Measurements of Superconducting β-Ti-TM Alloys

When a magnetic field, H_a, is applied to a type-II superconducting material, it is excluded from its interior by circulating surface supercurrents until it reaches a value H_{c1}, the lower critical field. Penetration of the field to form what is known as the "mixed state"* then commences. As H_a increases, the normal fraction of the mixed state increases until the entire sample goes normal at H_{c2}, the upper critical field. If metallurgical defects of the kind which inhibit the ingress and egress of magnetic flux are absent, the magnetization is *reversible*, as in Fig. 3.1, curve (a); on the other hand, if flux-pinning sites such as precipitates or other metallurgical irregularities are present, some of the applied flux will remain trapped when the applied field is removed—i.e., the magnetization is *irreversible*, as in Fig. 3.1, curve (b). POLONIS and coworkers, particularly LUHMAN (e.g., [LUH70]), have employed this principle to study the precipitation of ω phase in Ti-Cr alloys as well as the (ω + β → β' + β) ω-reversion effect.

3.2 Electrical Resistivity

Studies of the composition- and temperature-dependences of electrical resistivity provide insights into strengthening mechanisms, phase stability, and the electronic structure of alloys. Employed as a diagnostic tool, electrical resistivity may be used to detect phase transformation during rapid quenching, and the measurement of relative resistivity during isothermal aging facilitates the construction of time-temperature-transformation (TTT) diagrams [SOE69][HOR73]. In studies of binary alloys of titanium, the favoring by solute atoms of either α-phase or β-phase stability subdivides titanium-base binary alloys into two classes: (1) alloys of titanium with simple metals or interstitial elements, and (2) alloys of titanium with transition metals. Figure 3.2, an example of this, shows the resistivity composition dependences of titanium-base alloys falling essentially onto two branches: an upper branch consisting of the Ti-SM alloys and a lower branch corresponding to the Ti-TM alloys. As demonstrated by COLLINGS et al. [GEG73ª][COL75ª] and

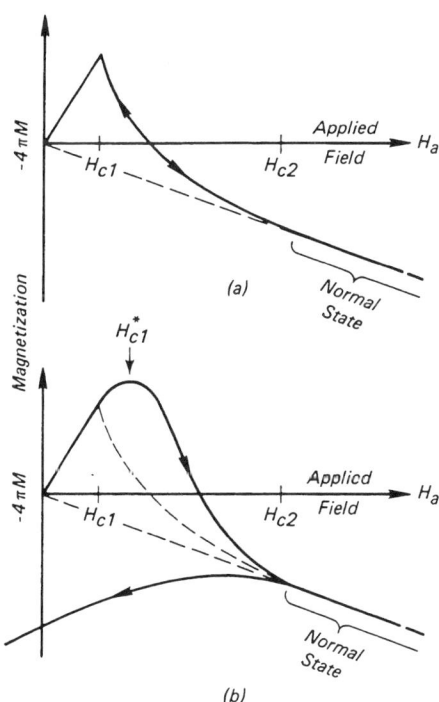

Fig. 3.1 Magnetization of a superconducting paramagnetic Ti-TM alloy— schematic diagrams of magnetization, $4\pi M$, versus the applied magnetic field, H_a: (a) reversible magnetization curve for an "ideal" or unpinned annealed sample; (b) irreversible (hysteretic) curve for a sample with a high density of flux-pinning sites.

pointed out by STERN [STE75], the rapid strengthening exhibited by simple metals in titanium is consistent with their being strong scatterers of the conduction electrons.

The resistivity of an alloy can be usefully separated into two terms, thus:

$$\rho_{total} = \rho_i + \rho_s \qquad (Eq\ 3.1)$$

where ρ_i, the "ideal" resistivity of the host, which may at high temperatures be expressed in the form [MEA65, p. 98]

$$\rho_i \to \frac{k_B T}{\theta_D^2} \qquad (Eq\ 3.2)$$

is a function of both the electronic structure of the alloy and thermal scattering, and the other term, ρ_s, represents the temperature-independent impurity scattering from the solute atoms.

In numerous low- or intermediate-concentration alloys, it has been discovered that ρ_i and ρ_s are independent. Evidence in support of this property, known as Matthiessen's rule, is the parallelism frequently noted among the $\rho(c)$ curves for members of an alloy series. Naturally, Matthiessen's rule breaks down when the presence of solute begins to influence ρ_i through its effect on $n(E_F)$ and θ_D, or for other reasons such as:

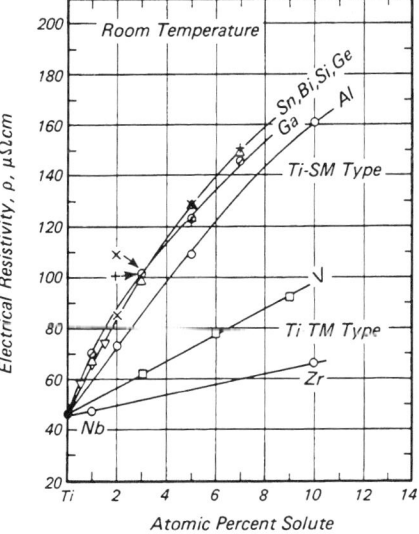

Fig. 3.2 Intercomparison between the composition dependences of electrical resistivity of two classes of titanium-base binary alloys: the Ti-SM type and the Ti-TM type. *Conditions*: as-cast: Ti-Sn (△), Ti-Ga (⌀), Ti-Al (O); 1 h/1000 °C/WQ: Ti-Ge (×), Ti-Bi (+), Ti-Si (▽), Ti-V (☐) [COL73ª, COL75ª].

---

* The mixed state is a microscopic ordered arrangement of normal and superconducting zones.

- When ρ_{total} becomes sufficiently large, as a result of either impurity scattering (ρ_s at high solute concentrations) or thermal scattering (ρ_i at high temperatures), further increments of solute or temperature, respectively, become relatively less effective. For example: (1) the specific resistivities of Al, Ga, Ge, or Sn in Ti-Mo (25 at.%) are on the average four times smaller than when the same elements are dissolved in pure titanium [GEG73a][COL75a]; (2) at high temperatures, the resistivity temperature dependences of some pure metals and alloys develop negative curvatures (see Fig. 3.3).

- The concentration dependence of resistivity of concentrated simple binary alloys not only decreases with increasing concentration, but passes through a maximum according to Nordheim's rule [MEA65, p. 113], which states that:

$$\rho_s \propto c(1-c) \qquad \text{(Eq 3.3)}$$

- Numerous Ti-SM and Ti-TM alloys exhibit negative temperature coefficients of resistivity. Such gross departures from Matthiessen's rule require detailed knowledge of the electronic structures, and/or the phonon spectra of the alloys concerned, for their explanations.

3.2.1 Anomalous Resistivity Temperature Dependence, $d\rho/dT$

Negative resistivity temperature dependence has attracted considerable attention over a prolonged period of time. Depending on temperature range and alloy type, the phenomenon has been attributed to: (1) the Kondo effect (dilute alloys at low temperatures [RIZ74]), (2) the increase with decreasing temperature of spin-disorder scattering from local moment clusters (e.g., concentrated Cu-Ni alloys [HOU70]), (3) an increase with decreasing temperature of the density of ω-phase precipitation itself [HO72][COL74, COL78] (as in Ti-V and Ti-Mo alloys and related alloy systems—see also references in [CHA74]), and (4) a smearing-out with increasing temperature of the density-of-states structure near E_F in certain classes of strong-scattering concentrated binary alloys [CHE72]. Mechanism-3, which in the spirit of the above three equations refers to the scattering contribution, ρ_s, and mechanism-4, which relies on an alloy density-of-states effect, are of particular significance in this context, the former being applicable to Ti-TM alloys and the latter to Ti-SM alloys.

3.2.2 Anomalous $d\rho/dT$ in Ti-SM Alloys

Ti-SM systems are strong-scattering alloys whose density-of-states functions possess considerable structure. CHEN et al. [CHE72], using the coherent potential approximation (CPA) [FAU82], have performed a model calculation on a concentrated binary alloy system, and have watched the changes in density-of-states, $n(E)$, which occur in response to: (1) change of solute concentration, (2) change of solute scattering strength, or (3) change of temperature. In order to do so, they have calculated the relative electrical conductivity as a function of band filling, and have been able to predict in a semiquantitative way the manner in which resistivity may change with temperature in two classes of concentrated binary alloys: (1) *virtual-crystal or weak-scattering alloys*, characterized by featureless parabolic $n(E)$ curves, whose resistivities increase with temperature in the "usual way," and (2) *strong-scattering alloys*, whose $n(E)$ curves possess deep minima or "pseudogaps," such that alloys whose compositions fall within the gap—which broadens and fills in with increasing temperature—have electrical conductivities that increase with temperature (i.e., negative values of $d\rho/dT$). The salient features of the model are illustrated in Figs. 3.4 and 3.5.

3.2.3 Anomalous Isothermal Resistivity Composition Dependence in Ti-TM Alloys

The resistivities of Ti-TM alloys exhibit isothermal resistivity-composition-dependence anomalies within which anomalous (i.e., negative) resistivity temperature dependences are located. The resistivity composition dependences of Ti-V, for example, at the temperatures 300, 200, and 77 K [COL74] are shown in Fig. 3.6. The corresponding quenched microstructures are also indicated in that figure. According to McCABE and SASS [MCC71], who have made a detailed TEM study of the system, ω phase is seen as a submicroscopic precipitate in the concentration range 13 through 25 at.% V, just that which includes the resistivity maximum. But although the sequence of sharp, then diffuse, electron-diffraction spots is confined to the above concentration range, diffuse haloes persist in gradually decreasing intensity all the way across to pure vanadium, a manifestation of a corresponding gradually decreasing lattice instability. The obvious conclusion is that the anomalous excess isothermal resistivity is closely associated in some way with the presence of both the athermal *and* the diffuse ω phases.

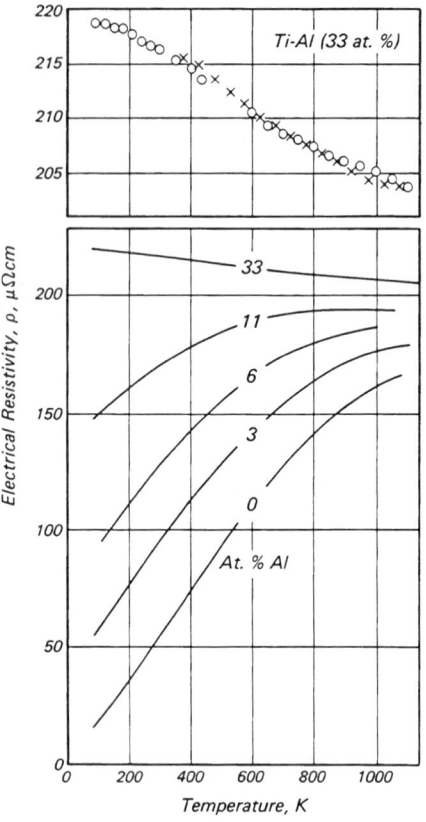

Fig. 3.3 Temperature dependences of the electrical resistivities, ρ, of unalloyed titanium and four Ti-Al alloys showing the tendency for $d\rho/dT$ to shift from strongly positive to weakly negative with increasing aluminum content [Moo73].

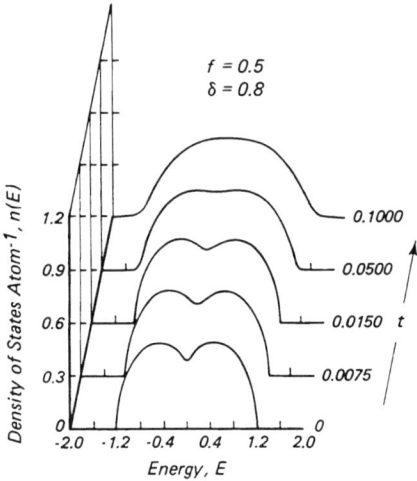

Fig. 3.4 Density-of-states, $n(E)$, versus energy, E, curves calculated using the CPA method for a model equiatomic (f = 0.5) strong-scattering (parameterized by δ, with δ = 0.8 on a scale of 0 to 1) binary alloy. Results for five values (0 to 0.1) of a reduced temperature, t, are indicated [CHE72].

3.2.4 Anomalous $d\rho/dT$ in Ti-TM Alloys

Figure 3.6 shows, in addition to the effect considered above, that the resistivity isothermals intersect in such a way as to establish a negative $d\rho/dT$ within the composition interval 20 to about 33 at.% V. Presented in this way, it appears that negative $d\rho/dT$ in alloys such as Ti-V is a minor perturbation of a much larger effect—the anomalous composition dependence—and, as such, is also related to the ω instability. Other Ti-TM systems in which negative $d\rho/dT$ has been studied are Ti-Nb [AME54][PRE74], Ti-Cr [LUH68][CHA73, CHA74], Ti-Mo [YOS56][HAK61][HO72][CHA73, CHA74], and Ti-Fe [HAK61][PRE76]. The question inevitably arose as to whether the negative $d\rho/dT$ was a consequence of reversible (athermal) ω-phase precipitation (as suggested in [HO72]) or a manifestation of the soft-phonon instability that gives rise to it [COL74]. Circumstantial evidence which could be taken in support of the former hypothesis can be presented in the form of Fig. 3.7, in which the anomalous reversible resistivity component, $\Delta\rho|_{77K}^{300}$, is juxtaposed against $\Delta f\omega|_{150K}^{300}$, a magnetically deduced reversible change of crystalline athermal ω-phase abundance. But since the athermal ω is expected to be associated with a fluctuation (or diffuse) component, the result was still inconclusive. The picture has been clarified by POLONIS et al. [CHA74] in an elegant series of experiments commencing with measurements on quenched Ti-Cr(20 at.%). Since both the as-quenched $\omega + \beta$-phase alloy and the 435 °C-reverted $\beta' + \beta$-phase alloy shared the same negative value of $d\rho/dT|_{77K}^{273}$, it became evident that the negative resistivity temperature dependence exhibited by Ti-Cr alloys was associated with the *insta-*

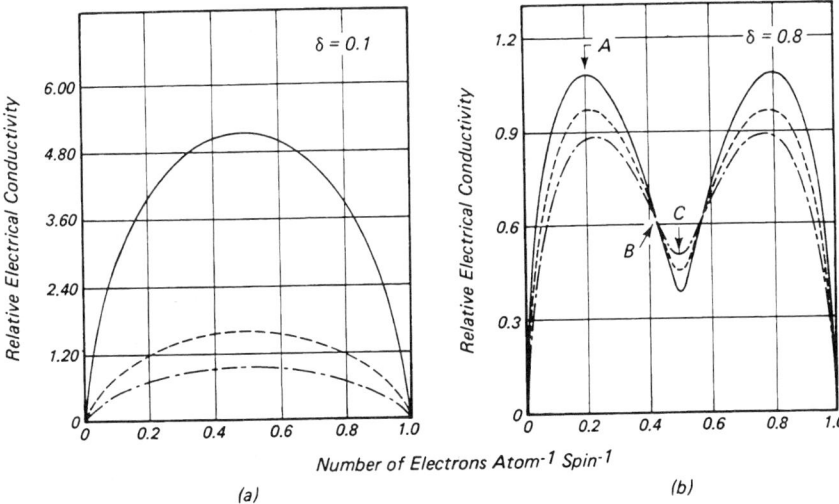

Fig. 3.5 Relative electrical conductivity as a function of band filling at three reduced temperatures for the model equiatomic alloy of Fig. 3.4: (a) weak-scattering case ($\delta = 0.1$); $t = 0.000$ (——), 0.006 (- - -), 0.012 (— · — · —); (b) strong-scattering case ($\delta = 0.8$); $t = 0.000$ (——), 0.0075 (- - -), 0.015 (— · — · —). In the strong-scattering case, which applies to Fig. 3.3, three $d\rho/dT$ signatures are possible depending on the level of band filling: $d\rho/dT$ is positive at A, zero at B, and negative at C [CHE72].

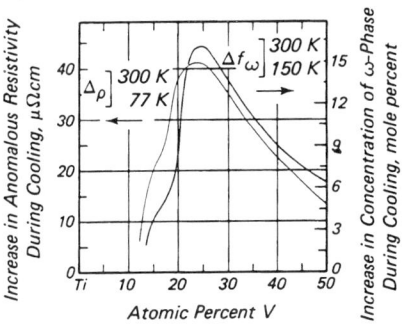

Fig. 3.7 Increase in anomalous resistivity, $\Delta\rho$, incurred on lowering the temperature of quenched Ti-V alloys from 300 to 77 K, compared with a magnetically derived estimate of the increase in ω-phase abundance that takes place as the temperature is lowered from 300 to 150 K [COL74, COL78].

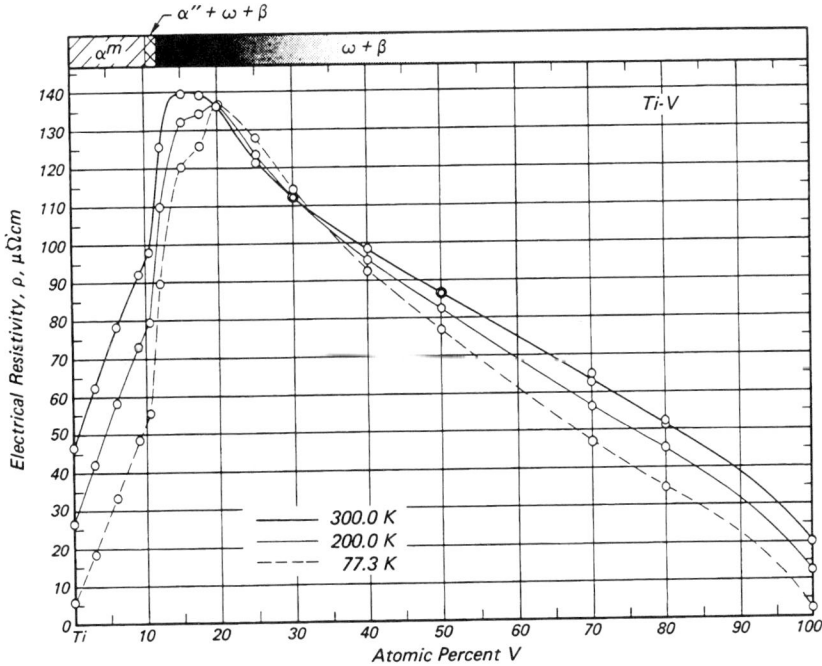

Fig. 3.6 Electrical resistivities of Ti-V alloys at three temperatures. Resistivities were measured at 77.3, 200 ± 1, and 298 ± 1 K. In the latter cases they were corrected to 200.0 and 300.0 K, respectively, using measured $d\rho/dT$ data. Negative $d\rho/dT$ is found within the composition range 20 ~ 30 at.% V between the points of intersection of the isothermals [COL74].

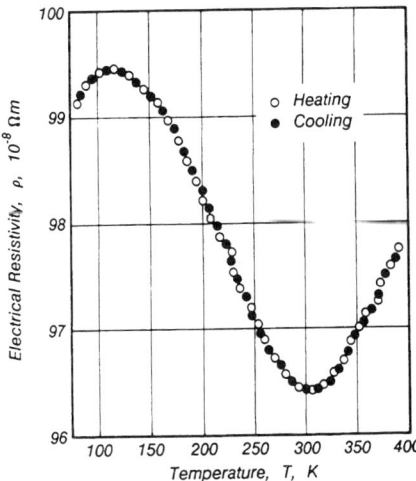

Fig. 3.8 Temperature dependence of resistivity of Ti-39.8Nb. See the original paper for the method of determining $\rho(T)$ without hysteresis. After IKEDE et al. [IKE88c].

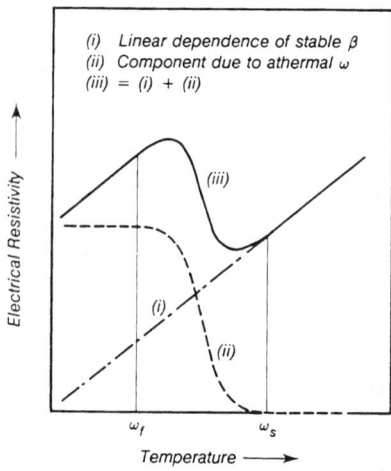

Fig. 3.9 Schematic representation of a ρ(T) curve formed from the sum of a linear component due to stable β and a sigmoidal component due to scattering from athermal ω. After IKEDA et al. [IKE88[b], IKE88[c]].

bility of the β phase itself, rather than its by-product, the ω-phase precipitate. The recent series of studies by IKEDA et al. represents a renewed interest in the anomalous resistivity temperature dependence of Ti-TM alloys and its interpretation in metallurgical terms. Alloy systems investigated were Ti-Mo(2-25 wt%) [IKE88[a], IKE88[b]], Ti-Nb(20-50 wt%) [IKE88[c], IKE89], and Ti-V(5-50 wt%) [IKE90]. In this work the characteristic ρ(T) exhibited by a typical "β-phase" alloy consisted (with decreasing temperature) of a minimum followed by a maximum as in Fig. 3.8. The authors have pointed out that such a curve could be regarded as having been generated by the superposition of a sigmoidal ρ(T) curve due to reversible athermal ω-phase formation onto a uniformly positive $d\rho/dT$ background coming from the unperturbed β phase (Fig. 3.9). The temperatures at which ρ(T) first departed from linearity with decreasing temperature, and that at which it resumed its linear descent, were designated ω_s and ω_f, respectively, for the starting and finishing temperatures of athermal-ω transformation.

3.3 Magnetic Susceptibility

As with many other physical properties, magnetic susceptibility not only provides useful information relating to the electronic structures of metals and alloys but, when calibrated against suitable metallographic bench marks, can aid in phase-diagram investigation and in the interpretation of aging experiments. As an ingredient of the theory of superconductivity, the magnitude of the Pauli spin susceptibility component of the total normal-state susceptibility may profoundly influence the value of the mixed-state upper critical field.

3.3.1 Total Magnetic Susceptibility

Components of the Total Susceptibility. It is convenient to treat the total magnetic susceptibility of a transition metal or alloy as the linear superposition of components representative of contributions to it from (1) electrons at the Fermi level, (2) states within the band, and (3) the individual ion cores thus:

$$c = \underbrace{\chi_P + \chi_L}_{(1)} + \underbrace{\chi_{so} + \chi_{orb}}_{(2)} + \underbrace{\chi_i}_{(3)} \quad \text{(Eq 3.4)}$$

where the terms are entitled, respectively, Pauli spin paramagnetism, Landau diamagnetism, spin-orbit susceptibility, orbital paramagnetism, and ion-core diamagnetism. The properties of these individual components have been adequately discussed elsewhere [COL70, COL71, COL80]. For the present purpose, it is sufficient simply to regard the total susceptibility as a macroscopic physical measurable and to consider its changes in response to changes in metallurgical variables.

Magnetic Diagnostic Methods. The total magnetic susceptibility of a system of two components A and B of susceptibilities, $\chi_{A,B}$, and relative abundances, $f_{A,B}$ (with $f_A = 1 - f_B$), is given by the usual continuity equation:

$$c = f_A \chi_A + f_B \chi_B \quad \text{(Eq 3.5)}$$

If χ_A and χ_B are known as a result of some preliminary investigation, Eq 3.5 can be manipulated so as to yield quantitative information relating to various metallurgical effects, processes, and properties such as (1) athermal ω-phase precipitation [COL78], (2) the precipitation of ω phase during isothermal aging [COL75[b]], and (3) the development of equilibrium phase diagrams [COL79]. Finally, and in a somewhat different vein, advantage may be taken of the magnetic anisotropy characteristic of α-phase titanium-base alloys in order to quantify the crystallographic textures which they acquire as a result of anisotropic cold deformation. Alloys which have been examined in this way are Ti-Al(0, 3, 5, and 10 at.%) [COL82].

3.3.2 Magnetic Studies of Athermal ω-Phase Precipitation

An expression for the temperature dependence of the total magnetic susceptibility can be obtained by differentiating Eq 3.5. Performing this differentiation, and writing ω for A and β for B, we find:

$$\frac{d\chi}{dT} = f_\omega \left(\frac{\partial \chi}{\partial T}\right)_\omega + f_\beta \left(\frac{\partial \chi}{\partial T}\right)_\beta \quad \text{(Eq 3.6)(a)}$$

$$+ (\chi_\omega - \chi_\beta) \left(\frac{\partial f}{\partial T}\right)_\omega \quad \text{(Eq 3.6)(b)}$$

The first pair of terms, (a), on the right-hand side of the equation is equivalent to $\partial \chi / \partial T$, the intrinsic temperature dependence of the total mean susceptibility. The second pair, (b), represents the change in susceptibility that takes place during reversible ω ↔ β allotropic transformation. The fraction of athermal ω phase, f_ω, is a reversible function of temperature whose value at any temperature, say T_i, is

$$f_\omega^{T_i} = \frac{(\chi_\beta - \chi)}{(\chi_\beta - \chi_\omega)} \quad \text{(Eq 3.7)}$$

according to Eq 3.5. An application of this analysis to the results of a susceptibility temperature dependence investigation of a series of Ti-V alloys has enabled $\Delta f_\omega \big|_{150K}^{300K} \equiv f_\omega^{150K} - f_\omega^{300K}$ to be calculated and plotted versus vanadium concentration as in Fig. 3.7. The quantity $\Delta f_\omega \big|_{150K}^{300K}$ is the mole-fraction of ω phase that appears and disappears reversibly as the temperature is cycled between 300 and 150 K [COL78].

3.3.3 Magnetic Studies of Isothermal θ-Phase Precipitation

During the isothermal aging of a Ti-TM alloy, within the metastable ω + β-phase regime, the magnetic response to the approach to ω + β meta-equilibrium can be described by means of the following equation, derived from Eq 3.5 and similar to Eq 3.6:

$$\Delta_\chi = -\frac{\Delta N}{N}\underbrace{\left[\left(\frac{\partial \chi}{\partial c}\right)_\omega - \left(\frac{\partial \chi}{\partial c}\right)_\beta\right]}_{(a)} + \underbrace{(\chi_\omega - \chi_\beta) \Delta f_\omega}_{(b)}$$
(Eq 3.8)

where ΔN (>1) represents the number of moles of solute that are transferred from ω to β during the aging of N moles of alloy. As before, the first term, (a), represents an intrinsic effect—this time, the difference between the susceptibility composition dependences of the ω and β phases. The second term, (b), represents the susceptibility change in response to an allotropic change in the alloy's structure between ω and β. Recognizing that χ_ω is always less than χ_β, Eq 3.8 shows that, if the composition dependences $\chi_\omega(c)$ and $\chi_\beta(c)$ are exactly parallel, the susceptibility change with aging (invariably a decrease, [HO73] [COL75[b]]) will be a direct result of the allo-

3. Physical Properties / 17

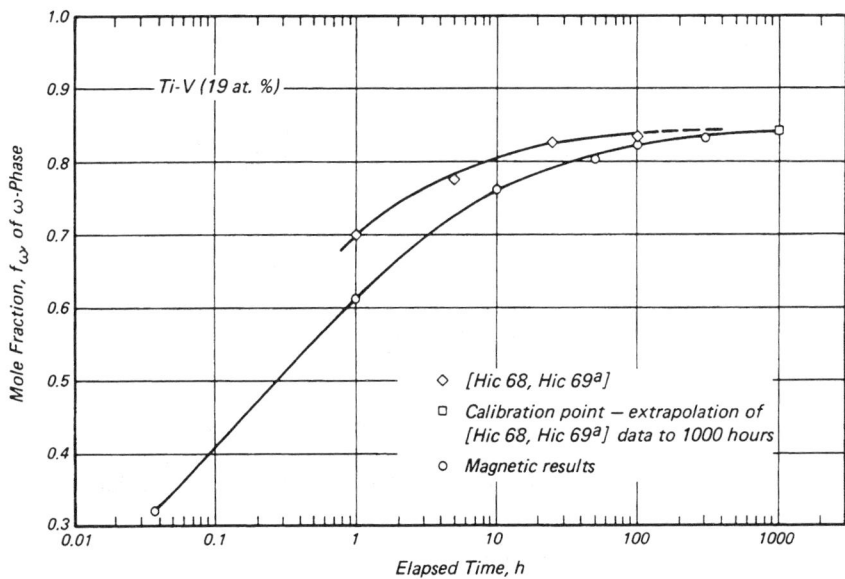

Fig. 3.10 Magnetic study of 300 °C-aging-induced ω-phase precipitation in a Ti-V alloy. The results are in good accord with those of HICKMAN, from whose work the calibration point, $f_\omega(1000\text{ h}/300\text{ °C}) = 0.84$, was taken [COL75^b].

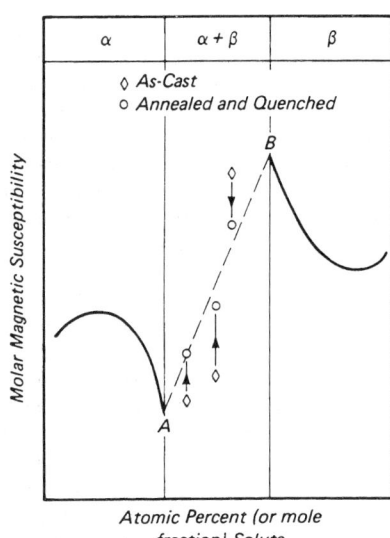

Fig. 3.11 Tie-line concept in the determination of equilibrium phase boundaries. The method requires well-defined "single-phase curves" (insensitive to annealing temperature), in the construction of which some extrapolation may be necessary near the phase boundaries. The concentration dependence of χ in the two-phase region is linear and is constructed either through datum points or on their "far sides" (with respect to some reference condition—e.g., as-cast) [COL79].

tropic β→ω transformation component, (b); otherwise, this decrease will be aggravated if $(\partial\chi/\partial c)_\omega$ is more positive than $(\partial\chi/\partial c)_\beta$ (as it turned out to be in Ti-V) or partially offset if the converse is true. By exploiting these principles it has, for example, been possible to obtain magnetic estimates of the responses of quenched Ti-V(15 at.%) and Ti-V(19 at.%) to aging at 300 °C [COL75^b]. A typical result is given in Fig. 3.10, in which a comparison has been made with the results of the more direct measurements of HICKMAN [HIC68, HIC69a].

3.3.4 Magnetic Studies of Phase Equilibria

In Ti-SM alloys, and particularly the Ti-Al system quenched from various temperatures, magnetic susceptibility procedures have assisted in the investigation of phase equilibria [YAO61] [COL70^a, COL79, COL82]. Particular attention has been devoted to Ti-Al in which single-phase-disordered (α) and long-range-ordered (α_2 and γ) regions alternate with two-phase fields whose boundaries can then be determined by the "tie-line" method [YAO61].

Although susceptibility-composition characteristics of single-phase alloys are generally curvilinear, any line crossing a two-phase field must be uncompromisingly straight, as in Fig. 3.11. That this is so can be demonstrated by combining Eq 3.5 and its compositional counterpart:

$$c = f_\omega c_\omega + (1-f_\omega)c_\beta \quad (\text{Eq 3.9})$$

in such a way that

$$c = \frac{c_B\chi_A - c_A\chi_B}{c_B - c_A} + \frac{\chi_B - \chi_A}{c_B - c_A} \cdot c \quad (\text{Eq 3.10})$$

which implies that a plot of χ versus c for a series of equilibrated two-phase alloys is indeed linear with intercept $(c_B\chi_A - c_A\chi_B)/(c_B - c_A)$ and slope $(\chi_B - \chi_A)/(c_B - c_A)$. This is the "tie-line," proper identification of whose endpoints can result in the accurate determination of a pair of phase boundaries. In practice, several series of alloys are prepared, equilibrated at a set of temperatures, and quenched. The quenched structure is assumed to reflect that at equilibrium, due regard being given to the possibility of athermal transformation (such as β → α' or α → α_2 for Ti-Al alloys) which, however, does not influence the position of the tie-line endpoints. From the family of magnetic "isothermals" so generated, loci of endpoints can be constructed to form the equilibrium phase boundaries. The results of applying this technique to a determination of the portion of the equilibrium phase diagram for Ti-Al within the composition range 30 to 57 at.% Al and the temperature range 900 to 1315 °C are presented in Chapter 4.

3.3.5 Magnetic Studies of Texture

In hexagonal close-packed crystals, magnetic susceptibility, as with other second-rank tensor properties, may be assigned two principal components, $\chi_\parallel$ and $\chi_\perp$. It follows that an average susceptibility, $\chi_{av} = \frac{1}{3}\chi_\parallel + \frac{2}{3}\chi_\perp$, may be obtained as the result of a single measurement of an ideal polycrystalline sample. But the large number of randomly oriented grains required may not be present in a small as-cast specimen. Grain size may, of course, be reduced to microscopic dimensions by cold work followed by recrystallization, but then randomness of orientation cannot be guaranteed. Deformation generally induces texturization, which may survive, or even be enhanced by, subsequent heat treatments. Provided proper precautions are followed, however, it is still possible to obtain a χ_{av} from measurements on a textured specimen. In addition, if $\chi_\parallel$ and $\chi_\perp$ values are available from measurements on a single crystal, it is possible to take advantage of the above effect by employing magnetic susceptibility to make quantitative estimates of *bulk* (as distinct from surface) texturization.

Thus with α-phase Ti-SM alloys, which are magnetically anisotropic, an

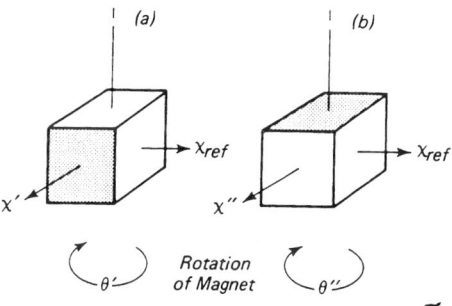

Fig. 3.12 "Double-rotation method" for the determination of three mutually orthogonal magnetic-susceptibility components. The average susceptibility is, of course, $\chi_{av} = (\chi_{ref} + \chi' + \chi'')/3$ [COL82].

opportunity exists for using magnetic-susceptibility techniques in the study of basal-pole texture. The technique recommended, now referred to as the "double-rotation method," was developed by COLLINGS and SMITH in 1968 [COL68] for the determination of the monocrystalline principal susceptibility components $\chi_\parallel$ (parallel to the c-axis) and $\chi_\perp$ (within the basal plane) of hcp crystals.

Determination of $\chi_\parallel$ and $\chi_\perp$. The measurement of $\chi_\parallel$ and $\chi_\perp$ is described with reference to Fig. 3.12 and 3.13. An arbitrary reference plane is ground on a single-crystal specimen of unrecorded orientation; it is then suspended, in turn, along each of the two directions that are orthogonal to the reference direction (along which the susceptibility is χ_{ref}) and rotated through angles θ' and θ''. The resulting susceptibility oscillations possess a common turning point, χ_{common}, which by geometry is always $\chi_\perp$. If when θ' and θ'' are equal to 90° the corresponding susceptibilities are χ' and χ'', respectively, then since

$$3\chi_{av} = \chi_{ref} + \chi' + \chi'' \quad (\text{Eq 3.11})$$

and

$$3\chi_{av} = \chi_\parallel + 2\chi_\perp = \chi_\parallel + 2\chi_{common} \quad (\text{Eq 3.12})$$

enough information is available with which to compute the remaining unknown, $\chi_\parallel$.

The double-rotation technique may also be applied to a textured sample regarded as a "pseudocrystal" characterized by new pseudoprincipal susceptibility components whose magnitudes are functions of the monocrystalline $\chi_\parallel$ and $\chi_\perp$ and the degree of texturization.

Determination of Texture. The simplest texture models are those in which the basal poles are distributed symmetrically about some preferred direction. If the susceptibility in that direction is designated by $\chi_\parallel'$, and that in the plane normal to it by $\chi_\perp'$, then it can be shown that

$$\chi_\parallel' = \chi_\parallel - \Delta\chi \quad (\text{Eq 3.13a})$$

and

$$\chi_\perp' = \chi_\perp + \Delta\chi/2 \quad (\text{Eq 3.13b})$$

where $\Delta\chi$ is a measure of the magnetic anisotropy introduced by the texture [COL82]. Within this context, two model distribution functions have been considered.

(1) *A rectangular (or step) distribution function* in which all basal pole directions lying within a cone of semivertical angle φ_c are equally probable. In this case:

$$\chi_\parallel' = \chi_\parallel - 2A(1-Q)/3 \quad (\text{Eq 3.14a})$$

and

$$\chi_\perp' = \chi_\perp + A(1-Q)/3 \quad (\text{Eq 3.14b})$$

where

$$A \equiv \chi_\parallel - \chi_\perp \quad (\text{Eq 3.15a})$$

and

$$2Q \equiv \cos\varphi_c(1 + \cos\varphi_c) \quad (\text{Eq 3.15b})$$

and where, of course, $2A(1-Q)/3$ plays the role of the $\Delta\chi$ in Eq 3.13a and 3.13b.

(2) *A cosine distribution function* between $\varphi = 0$ and $\pi/2$, φ being an angle that some direction makes with the preferred direction, in which case:

$$\chi_\parallel' = \chi_\parallel - A/2 \quad (\text{Eq 3.16a})$$

and

$$\chi_\perp' = \chi_\perp + A/4 \quad (\text{Eq 3.16b})$$

which are much simpler functions of the anisotropy, A, than those described in Eq 3.14a and 3.14b.

Development of a Texture Parameter. Double-rotation experiments similar to that depicted by Figs. 3.12 and 3.13 serve to determine first $\chi_\perp'$ (the common minimum) and then $\chi_\parallel'$. A single-rotation experiment could of course yield $\chi_\perp'$ immediately, and after insertion in Eq 3.14b yield a value for the texture parameter, Q (or φ_c), provided single-crystal data were available. This would, however, involve a comparison of $\chi_\perp'$ with $\chi_\perp$ (for the

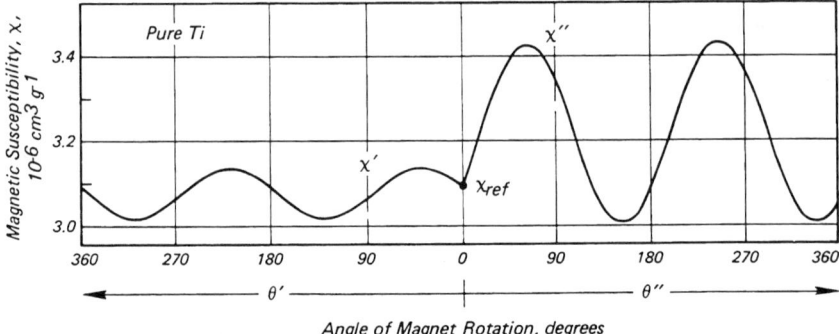

Fig. 3.13 Magnetic susceptibility of a sample suspended from a vertical fiber versus the angle (θ) of an applied magnetic field rotating about the sample-suspension as axis. The common minimum $\chi_\perp$; $\chi_\parallel$ is then given by $3\chi_{av} - 2\chi_\perp$, where χ_{av} is given (as before) by $(\chi_{ref} + \chi' + \chi'')/3$ [COL80].

Table 3.1 Texturization Parameters ("Isotropic Model") for Cold-Rolled Ti-Al Alloys [COL82]

Aluminum concentration, at.%	Reduction in thickness by cold rolling, %	Magnetic susceptibility components, 10^{-6} cm^3/g			Texturization parameters	
		χ_{av} $= (\chi_\parallel' + 2\chi_\perp')/3$	A $= (\chi_\parallel - \chi_\perp)$(a)	A' $= 3(\chi_{av} - \chi_\perp')$(b)	Q $= A'/A$	φ_c degrees (from Eq 3.15b)
0.0	25	3.16$_6$	0.51$_5$	0.14$_4$	0.28$_0$	66
	50	3.17$_3$		0.29$_7$	0.57$_7$	48
3.2	25	3.11$_7$	0.41$_5$	0.08$_4$	0.20$_2$	72
	50	3.11$_6$		0.29$_4$	0.70$_8$	38
5.5	25	3.11$_6$	0.35$_3$	0.21$_0$	0.59$_5$	46
	50	3.11$_3$		0.25$_5$	0.72$_2$	37
10.6	24	3.09$_8$	0.23$_3$	0.11$_7$	0.50$_2$	52

(a) From monocrystalline results. (b) From textured polycrystalline results; see [COL82] for further details.

single crystal) determined in a separate experiment, and would expose the result to uncertainties arising from positioning and other errors inherent in absolute susceptibility determination. These difficulties can be completely avoided by working in terms of magnetic *anisotropies*.

Full double-rotation measurements yield

$$A = \chi_\parallel - \chi_\perp \quad \text{(monocrystal)} \quad \text{(Eq 3.17a)}$$

and

$$A' = \chi'_\parallel - \chi'_\perp \quad \text{(textured sample)} \quad \text{(Eq 3.17b)}$$

These in turn yield the texture parameter, Q, which according to Eq 3.14a and 3.14b is none other than A'/A. The alternative texture index, φ_c, if needed, can then be obtained by solving Eq 3.15b.

Both methods are fully described in [COL82]. A set of results for a series of cold-rolled Ti-Al alloys is given in Table 3.1.

3.4 Low-Temperature Specific Heat

As indicated in Section 3.1.4, the specific heat, C, of a normal metal at low temperatures (below 6 ~ 10 K) can be expressed as the sum of an electronic component, $C_e = \gamma T$, and a lattice component, βT^3. Clearly C/T, when plotted versus T^2, is linear with intercept γ and slope β. In case the sample is a superconductor, however, the electronic specific heat acquires an additional component, C_{es}, at the transition temperature, T_c, such that according to BCS theory [BAR57]:

$$C_{es}\big|_{T_c} = 2.43\,\gamma T\big|_{T_c} \quad \text{(Eq 3.18a)}$$

or

$$\frac{\Delta C}{\gamma T}\bigg|_{T_c} = 1.43 \quad \text{(Eq 3.18b)}$$

Thus, as the sample temperature decreases, a sharp jump in specific heat takes place as soon as the transition temperature is encountered (see Fig. 3.14). The position of the jump gives, of course, the transition temperature, T_c, while its relative height, $\Delta C/\gamma T_c$, when compared with 1.43, yields a measure of the degree of "completeness" of the transition.*

3.4.1 Debye Temperature, θ_D

The lattice specific-heat coefficient, β, if expressed in units of $J \cdot \text{mole}^{-1} \cdot K^{-4}$, yields a low-temperature value of the Debye temperature via the formula:

$$\theta_D = \left(\frac{1.944 \times 10^3}{\beta}\right)^{1/3} \text{(K)} \quad \text{(Eq 3.19)}$$

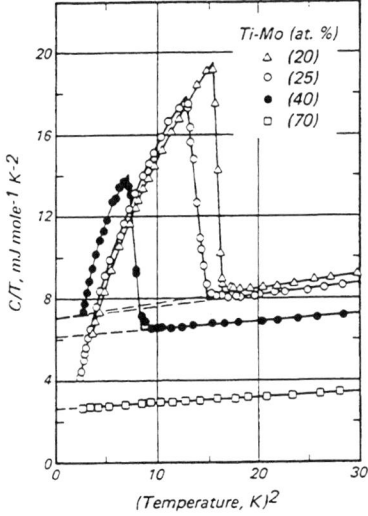

Fig. 3.14 Low-temperature specific-heat results for quenched Ti-Mo(20-70 at.%) alloys plotted in the usual format C/T versus T^2. The sharp jumps in the specific heat take place at the superconducting transition temperatures [COL70a, COL71d, COL72a] [HO73a].

θ_D may be regarded as a kind of bulk stiffness modulus. It is well known that the directional interatomic bonding favored by the majority of intermetallic compounds [COL71b] is associated with elastic stiffness, hardness maxima, and brittleness. Thus, it is not surprising to find in Ti-Al, a typical Ti-SM system, local maxima in θ_D corresponding to the positions of the brittle intermetallic compounds Ti_3Al and TiAl (Fig. 3.15).

Turning now to Ti-TM alloys, a comparable set of studies has also been undertaken on the prototype β-isomorphous system Ti-Mo. Figure 3.16, which displays the calorimetrically measured θ_D as a function of electron/atom ratio for a series of quenched alloys, shows: (1) a continuous softening of the bcc lattice with decreasing molybdenum concentration; (2) then, with further decrease of molybdenum concentration, a pronounced stiffening of the lattice due to the appearance of ω-phase precipitation, the occurrence of which is clearly related to the lattice-softening effect just referred to [COL72, COL74].

3.4.2 Superconducting Transition Temperature: Response to Aging

Low-temperature specific heat makes a useful tool for the monitoring of aging in Ti-TM superconductors. Studies have been conducted on Ti-Fe(7.5 at.%) aged for 1170 h at 175 °C followed by an additional 88 h at 300 °C [HO73]; Ti-Mo(10 at.%)

---

* That is, a measure of how much of the sample is actually participating in transition, particularly in the case of two-phase material.

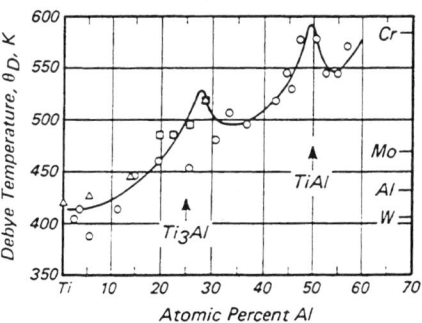

Fig. 3.15 Calorimetrically measured Debye temperature, θ_D, for Ti-Al alloys. *Condition*: as-cast (O); ordered (▫); various other heat treatments (△). The Debye temperatures of several pure metals are inserted for comparison [COL80, COL82a].

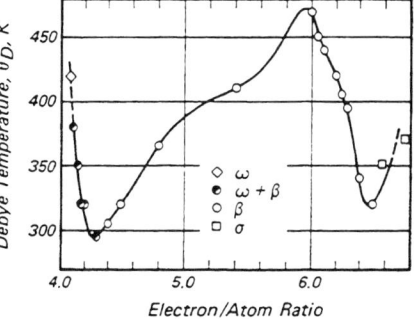

Fig. 3.16 Debye temperature as a function of e/a ratio for binary alloys of the Ti-Mo-Re sequence. Particularly noteworthy are that: (1) an e/a of 6 guarantees a maximum in the stiffness of the bcc alloys; (2) at sufficiently low e/a values, the occurrence of ω phase begins to stiffen the bcc lattice; and (3) at sufficiently high e/a values, the lattice is stiffened by a transformation to σ phase [COL73].

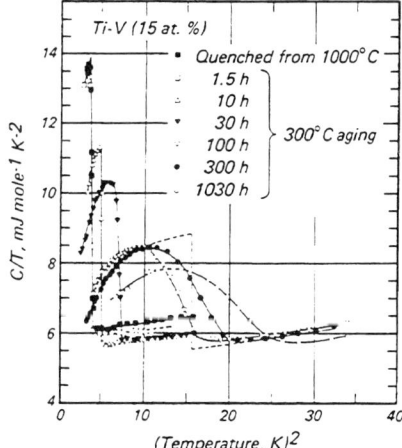

Fig. 3.17 Effects of prolonged aging at 300 °C on the low-temperature specific heat of Ti-V(15 at.%). The relative height of the specific-heat jump for specimens with broad transitions may be graphically estimated by extrapolating data above and below the superconductive transition to a vertical line positioned at the transition midpoint (see, for example, the dashed line for the 300-h aging result). As aging proceeds, the transition broadens, the jump height decreases, but T_c increases [COL75c].

aged for 880 h at 350 °C [COL72a][HO73a]; and Ti-V(15 and 19 at.%) aged for 1030 and 2200 h, respectively, at 300 °C [COL75c]. In Ti-Fe, as with Ti-Mo, T_c decreases with aging time and the transition remains fairly sharp, observations that are consistent with the maintenance of a complete proximity effect (precipitate radius < coherence length*) during the development of ω-phase precipitation. Just the opposite is true for the Ti-V alloys; according to Fig. 3.17, for example, the maximum T_c increases while the transition broadens and the volume fraction of superconducting phase (as gauged by the usual $\Delta C / \gamma T_c$ criterion) decreases. These facts can be explained in terms of a growth of precipitate size (radii becoming >> coherence length) accompanied by a solute enrichment of the β phase (hence an increase in its T_c) for the alloy compositions concerned [COL75c].

3.4.3 Superconducting Transition Temperature: Response to Deformation

This subject has not been investigated extensively. Among the few studies that have been made of the influence of deformation on the superconducting transition were the resistive measurements of tin by SWANSON and QUENNEVILLE [SWA73] and the calorimetric investigations of niobium by ZUBECK et al. [ZUB79]. Of particular interest in this context, however, are the results of the measurements of HO and COLLINGS of several plastically deformed Ti-TM alloys.

In alloys of titanium with 4.5 at.% Mo [COL70], 5 at.% Mo [HO71][COL71c], and 7 at.% Mo [HO71], it has been noted that T_c is raised as a result of deformation-induced-martensitic or twinning transformations; likewise, the addition of 1 or 3 at.% Al to Ti-Mo(5 at.%), which again influences martensitic transformation, results in an increase in T_c [COL76]. Following an earlier suggestion by STRONGIN et al. [STR68], the observed T_c enhancement was initially attributed to a mechanism that required localized soft-phonon modes to be associated with displaced atoms in the deformed structure [COL70b]. More recently, however, as a result of the computer fitting of an "asymmetrical-Gaussian-distributed" BCS-specific-heat function to the experimental calorimetric data in the vicinity of the transition (Fig. 3.18), it has been possible to advance a somewhat more plausible argument couched in metallurgical terms [COL78a]. For example, the specific-heat results for Ti-Mo(5 at.%), in which the deformation raises T_c

---

* The characteristic size of the superconducting quasiparticle (electron-pair).

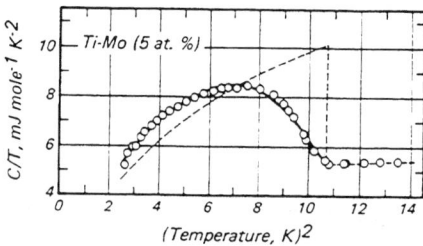

Fig. 3.18 Low-temperature specific-heat results for quenched-plus-deformed Ti-Mo(5 at.%) plotted in the usual format C/T versus T^2 and fitted with a Gaussian-rounded BCS-specific-heat function. Best fit to the data (solid line) was achieved with an "extreme negative" skew distribution (i.e., left-half Gaussian, or $f = -1.0$). The unrounded function is shown as a broken line [COL78a]; see also [WHI76]. (Note: The symmetrical Gaussian is parameterized by $f = 0.0$; see [WHI76].)

from 1 to about 3 K (Fig. 3.18), can be interpreted in terms of 68% α″ martensite with $T_c = 3.27$ K, 32% (ω + β) with $T_c = 1.0$ K, plus a proximity effect between the two. In Ti-Mo(7 at.%), in which *twinning* is believed to be the primary deformation product, the results of the fitting exercise can be interpreted in terms of 63% original ω + β phase, 27% low-T_c twin-boundary and highly defected material, and 10% high-T_c ω-deficient twin-boundary phase.

3.4.4 Superconducting Transition Temperature: Low-Concentration Quenched-Martensitic Ti-TM Alloys

Superconducting transitions associated with the quenched-martensitic ($\alpha^m \equiv \alpha'$ or α'') structure have been investigated calorimetrically in several systems, notably Ti-V and Ti-Nb [HEI64], Ti-Mn, and Ti-Co [HAK64], Ti-Fe [BAT64], and Ti-Mo [COL69]. Comparative studies of the superconducting transition in an extensive series of dilute Ti-TM alloys have been undertaken by BUCHER et al. [BUC65] (Ti-V,-Cr,-Mn,-Fe,-Nb,-Mo), who applied a Gaussian rounding technique to the analysis of the specific-heat jump in much the same manner as that referred to above, and subsequently by AGARWAL [AGA74] (Ti-Sc,-V,-Cr,-Mn,-Fe,-Co,-Ni,-Hf). Taken together, the results of both workers lead to the following conclusions: (1) the low-temperature specific heat of α^m-Ti-Mn has a temperature dependence characteristic of the localized-magnetic-moment behavior referred to earlier; (2) the specific-heat jumps in alloys such as Ti-V, Ti-Nb, and Ti-Mo are not unduly rounded; and (3) those of the alloys Ti-Cr, Ti-Fe, and Ti-Co are exceptionally broad. A subdivision of the alloys into two groups which include (1) Ti-V, Ti-Nb, and Ti-Mo on one hand and (2) Ti-Fe on the other, with Ti-Cr occupying an intermediate position, is apparent in Fig. 3.19.

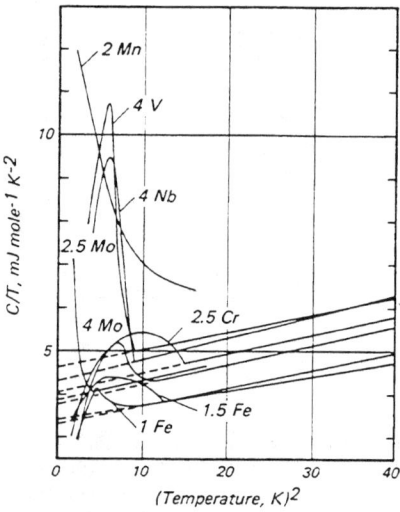

Fig. 3.19 Low-temperature specific heats in the vicinity of their superconducting transitions for low-concentration martensitic Ti-V, Ti-Nb, Ti-Mo, Ti-Cr, Ti-Mn, and Ti-Fe alloys, indicating a decrease in the abruptness of the specific-heat jump on proceeding from the "β-isomorphous" to the "β-eutectoid" class of alloys [BUC65].

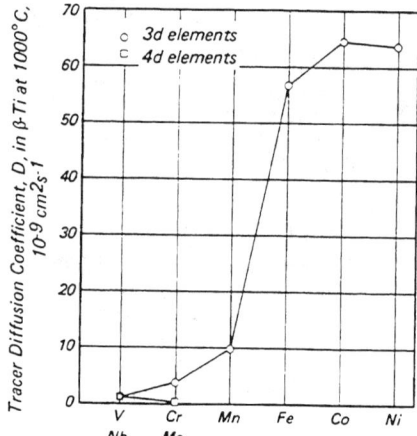

Fig. 3.20 Tracer diffusion coefficients for the 3d solutes V, Cr, Mn, Fe, Co, and Ni, and the 4d solutes Nb and Mo, in β-Ti at 1000 °C—computed from frequency-factor and activation-energy data of ZWICKER [ZWI74, p. 174].

For a rationalization of the above-mentioned behavior we turn again to a metallurgical explanation, this time in terms of the relative rates of diffusion of transition-element solutes in β-Ti. Figure 3.20, which intercompares tracer diffusivities [ZWI74, p. 108], indicates that of the alloys for which data are available, only Ti-V, Ti-Nb, and Ti-Mo have the opportunity to transform athermally to α^m during quenching from the β phase, while in alloys such as Ti-Fe, Ti-Co, and Ti-Ni, whose diffusivities are almost two orders of magnitude higher, significant levels of solute redistribution and Widmanstätten growth can be expected during the quenching of the moderately massive samples needed for the usual kind of low-temperature specific-heat measurement.

3.5 AC Impedance

LUHMAN and colleagues, using an AC inductive technique, have measured the position, width, and fine structure associated with the superconducting/normal transition in Ti-Cr alloys as part of an extensive study of precipitational effects associated with quenching, aging, and "up-quenching" [LUH69, LUH70, LUH70^a, LUH71]. As indicated in Section 3.1.5, impedance (inductance) changes in a sample/coil system were detected by measuring the voltage drop across a coil (surrounding the alloy sample) supplied with 1-kHz current from an oscillator. The experimental results were displayed as plots of "impedance" versus sample temperature [LUH70^a] or the first derivative of inductance, with respect to temperature, versus temperature [LUH69]. The manner in which the impedance results are interpretable can be described with the aid of Fig. 3.21 for as-quenched and quenched-plus-aged (51 min/300 °C) Ti-Cr(15 at.%). In part (a) of the figure, the asymmetric shape of the transition curve was ascribed to composition gradients in the all-bcc alloy. During aging at 300 °C, athermal ω-phase precipitation was supposed to take place. The existence of ω phase after 51 min/300 °C was responsible for the double superconducting transition barely detectable in part (b) of the figure. But double transitions are more easily detected and resolved in the first derivatives of the impedance-temperature curves. Some results for a Ti-Cr(10.3 at.%) alloy in the as-quenched and quenched-plus-aged (28 min/196 °C) conditions are given in Figs. 3.22(a) and (b), respectively. Peak A was interpreted as being due to ω phase. The double peak B,C was identifiable with the β phase, since it generally occurred independently of the presence of ω phase. The doublet nature of the peak was taken as an indication of the presence of solute gradients.

A particularly interesting phenomenon characteristic of ω + β-phase Ti-TM alloys is ω-phase reversion. The impedance-measurement technique has been used to study this effect in Ti-Cr(9.5 at.%), Ti-Cr(15 at.%), and Ti-V(24.4 at.%) [LUH70, LUH71]. Following an aging heat treatment at 300 °C to produce isothermal ω phase, both alloys were "up-quenched" to 450 °C, where they were held for 3 min prior to water quenching. The up-quenching of aged Ti-Cr(15 at.%) raised its T_c from 3.05 to 4.292 K, the latter being less than the 4.456 K of the as-quenched alloy. The up-quenching of Ti-V(24.4 at.%) yielded a T_c of 5.089 K, higher than the 4.382 K of the as-quenched sample. The observed differences in T_c between the as-quenched, ω + β-aged, and β' + β-reverted samples were interpreted in terms of composition fluctuations, screening of precipitation zones

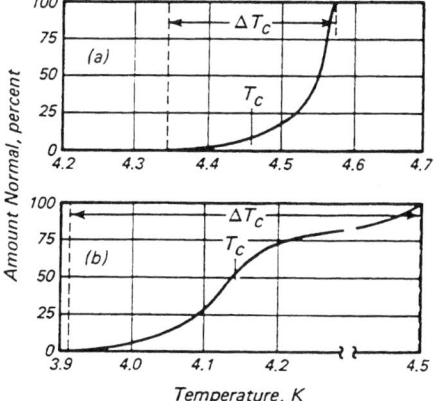

Fig. 3.21 Variation of impedance with temperature of a sample of Ti-Cr(15 at.%) in two metallurgical conditions: (a) solution treated 1 h/850 °C/WQ; (b) solution treated plus aged 51 min/300 °C [LUH70, p. 104; LUH70^a].

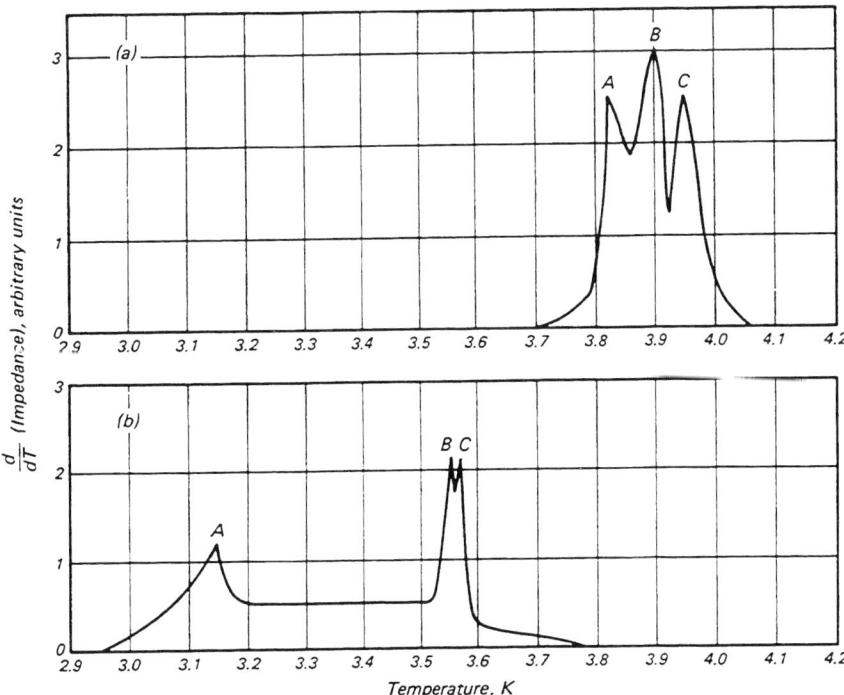

Fig. 3.22 First derivative of impedance with respect to temperature versus temperature for a sample of Ti-Cr(10.3 at.%) in two metallurgical conditions: (a) solution treated 1.5 h/980 °C/WQ; (b) solution treated plus aged 28 min/196 °C [LUH69].

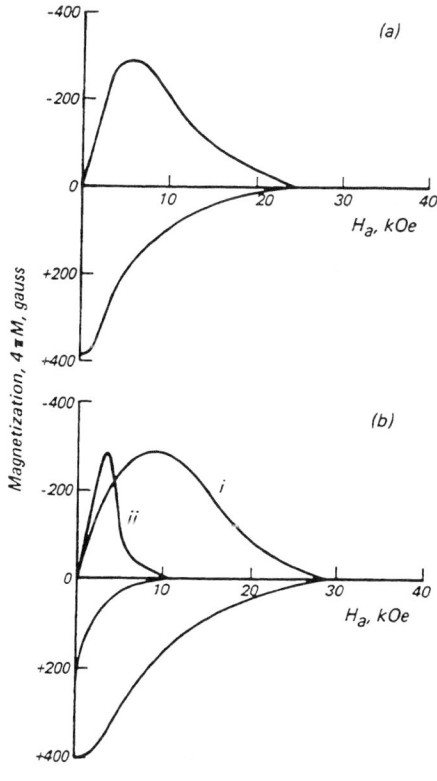

Fig. 3.23 Magnetization versus applied field curves (see Fig. 3.1) for a sample of Ti-V(28 at.%) in three metallurgical conditions: (a) solution treated 1 h/1000 °C/WQ plus aged 10 h/350 °C; (b) (i) solution treated plus aged 24 h/350 °C, (ii) same treatment followed by 3 min/550 °C [LUH70, Sect. V(K); LUH72].

by the surrounding matrix, the effects of coherency strain fields, and the T_c-composition dependences of the individual components.

3.6 Magnetization

Magnetization measurements, which yield values of the upper critical field, H_{c2}, and generally exhibit irreversibility, provide additional indirect information on precipitation and solute redistribution. LUHMAN and colleagues [LUH70, LUH72] have also employed this technique in studies of ω-phase precipitation and aging, ω-phase reversion, and α-phase precipitation in some representative Ti-V and Ti-Nb alloys. Some typical results are depicted in Fig. 3.23. In Fig. 3.23(a), representing Ti-V(28 at.%) aged 10 h/350 °C, the magnetic irreversibility is due to ω-phase precipitation. This is enhanced by extending the heat treatment to 24 h at the same temperature (Fig. 3.23b, curve i) and is diminished following a heat treatment of 3 min/550 °C (followed of course by quenching) which reverts the ω phase to β (curve ii). The hysteresis that remains is a consequence of flux pinning (field trapping) by the modulated β' + β structure.

4. Equilibrium Phases

The equilibrium phase diagrams of numerous titanium-base binary alloys have been presented and discussed by McQuillan and McQuillan [MCQ56], Imgram et al. [IMG61], and Zwicker [Zwi74], and the properties of several important systems have been reviewed by Jaffee [JAF58] and Margolin and Nielson [MAR60]. The most comprehensive compendium of binary phase diagrams has of course been provided by Molchanova [MOL65].

In what follows, two important alloy systems will be briefly reviewed: Ti-Al, the basis of technical α-Ti alloys; and Ti-Mo, a β-isomorphous alloy and the basis of several technical β-Ti alloys.

4.1 The Typical Alpha Alloy, Ti-Al

Equilibrium phase diagrams for Ti-Al, a representative α-stabilized titanium-base binary alloy, are to be found in McQuillan and McQuillan [MCQ56, p. 174], Molchanova [MOL65, p. 137], (see Fig. 4.1), and Zwicker [Zwi74, p. 147]. Research articles in which descriptions of portions of the equilibrium diagram have been discussed are listed in Table 4.1. Two of the most interesting and important regions of the diagram surround the ordered intermetallic compounds Ti_3Al and TiAl. Short-range and long-range order in Ti-Al alloys and the occurrence of the long-range ordered $α_2$ phase in the vicinity of the compound Ti_3Al were discussed by Blackburn [BLA67a], who offered an equilibrium partial phase diagram for the composition range 0 to 30 at.% (Fig. 4.2). Confirmatory evidence for the existences and ranges of the disordered and ordered phases in Ti-Al(20~30 at.%) is to be found in the results of the magnetic-susceptibility measurements of Collings et al. [COL75a]. Gehlen [GEH70] has investigated the crystallography of Ti_3Al, which was found to possess the DO_{19} structure—a unit cell apparently composed of four regular hcp cells apparently supported by covalent-like directional bonds connecting the aluminum and titanium atoms. Solution strengthening in Ti-Al alloys in terms of localized Ti-Al bonds, leading at sufficiently high solute concentrations to the above-mentioned long-range-ordered structure and electronic effects related to it, have been discussed by Collings and coworkers [COL70a, COL75a, COL82a]. Some physical properties of Ti-Al alloys attributable to the occurrence of brittle intermetallic compounds at compositions near 25 at.% (Ti_3Al), 50 at.% (TiAl), and ~70 at.% (~$TiAl_3$) have been discussed, with particular reference to electronic bonding and its relationship to intermetallic compound formation, by Collings [COL82a].

The wide discrepancies that exist among the numerous Ti-Al phase diagrams presently in existence are evident in Zwicker's collection of six equilibrium diagrams [Zwi74, p.147]. To shed further light on the position of a particularly important feature, the ($α_2$ + γ)/γ phase boundary, Collings [COL79] employed magnetic susceptibility techniques (augmented by optical metallography) in order to develop an equilibrium partial diagram for Ti-Al(30-57 at.%) within the tempera-

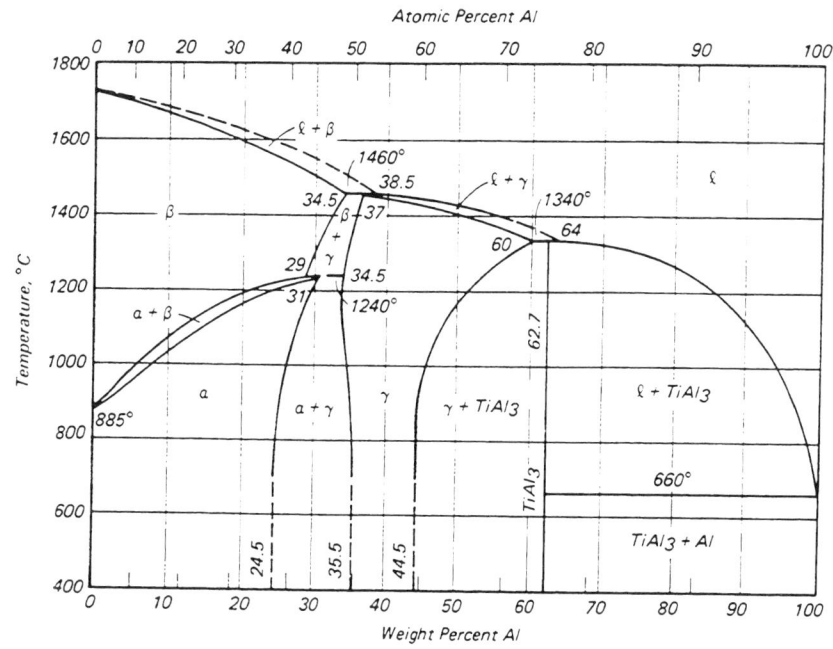

Fig. 4.1 Ti-Al phase diagram (but see Table 4.1) [MOL65, p. 137].

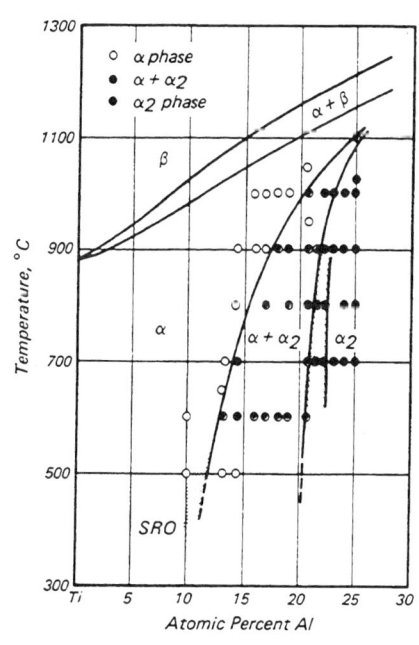

Fig. 4.2 Partial Ti-Al equilibrium phase diagram for the range 0 to 25 at.% Al [BLA67a].

Table 4.1 List of Investigations Directed Toward a Determination of the Ti-Al Equilibrium Phase Diagram

Aluminum concentration range, at.%	Temperature range for equilibrated solid alloys		Principal and auxiliary techniques described	Ref
	°C	°F		
0-64	750-1100	1382-2012	Optical metallography; with x-ray diffraction and thermal analysis	[Ogd51]
0-75	700-1400	1292-2552	Optical metallography; with x-ray diffraction and Vickers hardness	[Bum52]
0-75	700-1200	1292-2192	Optical metallography; with x-ray diffraction, Vickers hardness, thermal analysis, and centrifugal bend tests	[Kor56]
5-49	550-1050	1022-1922	Electrical resistivity, magnetic susceptibility; with optical metallography and x-ray diffraction	[Sag56]
0-63	450-1350	842-2462	Electrical resistivity; with optical metallography and x-ray diffraction	[Sat59]
0-48	800-1450	1472-2642	Optical metallography and x-ray diffraction	[Enc61]
5-38	400-1100	752-2012	Magnetic susceptibility	[Yao61]
0-38	550-1200	1022-2192	Optical metallography, electrical resistivity, and x-ray diffraction	[Cla63]
5-43	550-1200	1022-2192	Electrical resistivity and Vickers hardness; with thermal analysis, dilatometry, and x-ray diffraction	[Kor65]
7-35	550-1100	1022-2012	Optical metallography; with electron microscopy, x-ray diffraction, differential thermal analysis, electrical resistivity, and dilatometry	[Cro66]
5-25	500-1100	932-2012	Electron microscopy	[Bla67^a]
27-45	1025-1225	1877-2237	Electron microscopy	[Bla70]
7-19	200-900	392-1652	Electrical resistivity; with electron microscopy	[Nam73]
0-33	500-1100	932-2012	Differential thermal analysis, x-ray diffractometry, electrical resistivity, and hardness; with optical metallography	[Kor76]
30-57	900-1365	1652-2489	Magnetic susceptibility; with optical metallography	[Col79]
0-30	625-1100	1157-2012	Magnetic susceptibility; with electron microscopy	[Swa81]
7-44	450-1150	842-2102	Differential thermal analysis; with electron microscopy	[Shu85]

ture range 900 ~ 1300 °C (Fig. 4.3). SWARTZENDRUBER et al. [SWA81] also used magnetic susceptibility as a technique for studying a portion of the Ti-Al phase diagram. The most recent investigation of the Ti-Al system was by SHULL et al. [SHU85], who employed differential thermal analysis, assisted by transmission electron microscopy (TEM), to develop a phase diagram for the composition range 7 to 44 at.% Al (Fig. 4.4).

4.2 The Typical Beta-Isomorphous Alloy, Ti-Mo

Equilibrium phase diagrams for Ti-Mo have been developed by CRAIGHEAD and coworkers (1950),* HANSEN and coworkers (1951),* DUWEZ (1951),* MOLCHANOVA [MOL65, pp. 27-32], TERAUCHI et al. [TER82], and, most recently, HAYMAN [HAY85]. The diagram according to HANSEN et al. [HAN58, p. 977] is reproduced in Fig. 4.5. The alloys used in developing this diagram had been homogenized for 20 to 40 h at 1250 °C prior to being annealed at eight temperatures between 855 and 600 °C for times ranging from 90 to 650 h. Diagrams such as Fig. 4.5 are not usually continued below 600 °C owing to the difficulties that are always encountered in attempts to attain thermodynamic equilibrium in reactive alloys when the diffusion rates are low. According to MOLCHANOVA [MOL65, pp. 27-32], the β phase is stable at all temperatures in alloys containing more than 16 at.% Mo (28 wt%). The β/(α + β) phase boundary is almost linear and intersects the 650 °C line at 14 at.% Mo (24 wt%). The maximum solubility of molybdenum at 600 °C in α-Ti-Mo alloys was stated to be only about 0.4 at.% (0.8 wt%), with the cautionary note that there was some uncertainty associated with that number.

In numerous studies of quenched nonequilibrium β-phase titanium alloys and the effects of aging on them as they proceed toward thermodynamic equilibrium, it has been noted that within the equilibrium α + β field, and bordering the metaequilibrium ω + β zone, the aging of quenched β-stabilized alloys can result in a separation of the β phase into a solute-rich β matrix and a solute-lean β′ precipitate. The β′/β interfaces [LUH70], or the interiors of the β′ precipitates themselves

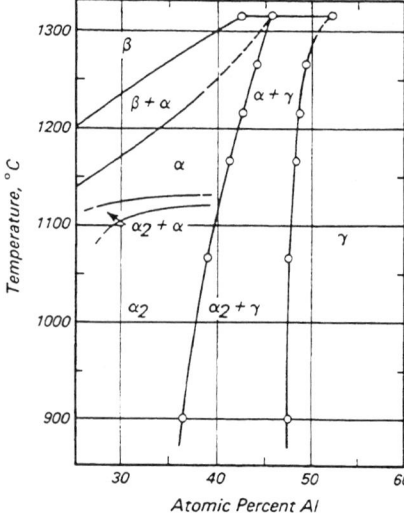

Fig. 4.3 Partial Ti-Al phase diagram for the range 25 to 57 at.% Al. The data points (○) were determined magnetically [Col79]; the boundaries of the β + α and α$_2$ + α fields were earlier established by BLACKBURN [Bla70].

* See references in MOLCHANOVA [MOL65, p. 32].

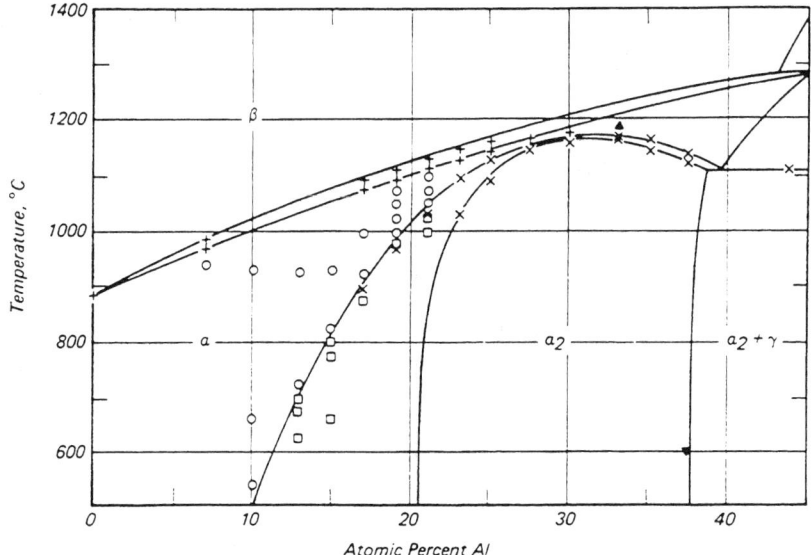

Fig. 4.4 Partial Ti-Al phase diagram for the range 0 to 45 at.% Al. The squares represent single-phase α; the circles indicate two-phase $\alpha + \alpha_2$ [SHU85].

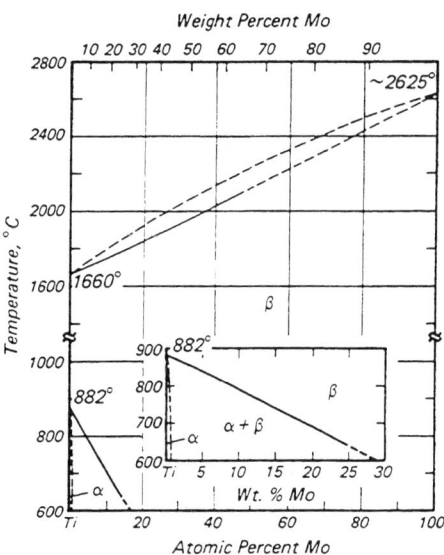

Fig. 4.5 Ti-Mo equilibrium phase diagram [HAN58, p. 977].

[WIL73], are the sites of α-phase precipitation during further aging. Clearly the phase-separated $\beta' + \beta$ is a nonequilibrium condition, and as such is discussed in Chapter 5. A double-bcc phase can, however, exist as an *equilibrium* two-phase state in some alloy systems. Referred to as "β-phase immiscibility," it occurs, for example, in Zr-Nb [LOV66] and related systems such as Ti-Zr-Nb [KIT70^a]. The purpose of this digression into the existences of phase-separated and phase-immiscible double-bcc phases is to provide a suitable context for introducing the results of some studies of the Ti-Mo system by TERAUCHI and colleagues [TER82]. From optical observations, electrical-resistivity measurements, x-ray diffractometry, lattice-parameter measurements, and TEM, those authors have deduced the existence of a pair of bcc phases—referred to as $\beta_1 + \beta_2$—occupying an area of the equilibrium phase diagram lying outside the $\alpha + \beta$ field (Fig. 4.6). Strong confirmation of the validity of TERAUCHI's diagram has recently been presented by HAYMAN [HAY85].

4.3 The Beta-Isomorphous Alloy, Ti-Nb

Since Ti-50Nb (~34 at.% Nb), as well as ternary and quaternary alloys based on it, have found use in the form of copper-matrix multifilamentary composites in the windings of large superconducting magnets, an accurate knowledge of the equilibrium and meta-equilibrium phase diagrams of the Ti-Nb system is particularly important. A composite equilibrium phase diagram for Ti-Nb, clearly also a β-isomorphous system, is reproduced in Fig. 4.7 based on the work of HANSEN et al. [HAN51] and IMGRAM et al. [IMG61] (see also RONAMI et al. [RON70] and JEPSON et al. [JEP70]). Since thermodynamic $\alpha + \beta$ phase equilibrium is difficult to achieve at temperatures below about 600 °C, the experimentally deduced line diagram is not continued below about that temperature. At temperatures below the $\beta/(\alpha + \beta)$ transus, or its projection, the approach to equilibrium is made via the decomposition of the metastable $\omega + \beta$ or β phases, the rate of which is accelerated in the presence of oxygen or the products of heavy deformation. The recent renewal of interest in the occurrence of α-Ti-Nb precipitation in rather concentrated Ti-Nb alloys of up to about 53.5 wt% (37 at.%) Nb, and its role in technical superconductivity, has not only rekindled a corresponding level of interest in the equilibrium phase diagram for temperatures in the vicinity of 400 °C, but has also led to the ready availability of heavily cold-worked and heat-treated samples upon which the necessary electron microscopy can be carried out. In this context, HILLMANN et al. [PFE68] and WILLBRAND and SCHLUMP [WIL75] have examined the occurrence and morphology of α-phase precipitation in alloys of 50 wt% Nb after moderate-time aging at temperatures near 380 °C, and more recently WEST and LARBALESTIER [WES80, WES82] have observed the presence of α-Ti-Nb precipitates in Ti-53.5Nb after two stages of heat treatment (separated by cold drawing) of 80 h/375 °C and 40 h/375 °C. The latter authors conducted a series of high-resolution TEM studies of a number of heavily cold-worked and heat-

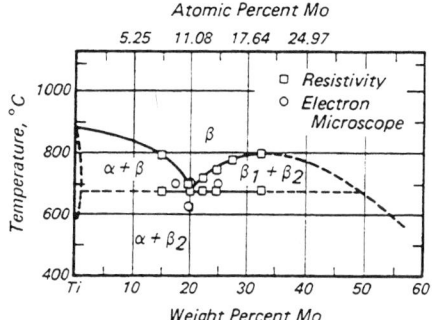

Fig. 4.6 Suggested partial Ti-Mo equilibrium phase diagram [TER82].

treated Ti-Nb alloy superconductors, detecting α-phase precipitation in alloys again with niobium concentrations as high as 53.5 wt% [WES83]. Subsequently, LEE et al. noted that Ti-52Nb yielded some 15 to 20% of α-phase precipitation in response to repeated heat treatments at about 400 °C interspersed with cold work [LEE87]. As the niobium concentration increased, precipitation became more and more sluggish, and less and less α-phase precipitated out—for example, Ti-56Nb was found to yield about 10% α-phase, while none at all could be detected in Ti-63Nb [LEE90].

On the other hand, in a well-known study of Ti-58Nb (analyzed composition), heavily deformed but aged usually for relatively short periods of time at temperatures of 350 to 500 °C, NEAL et al. [NEA71] were able to detect only *traces* of α-phase precipitation. The failure to ob-

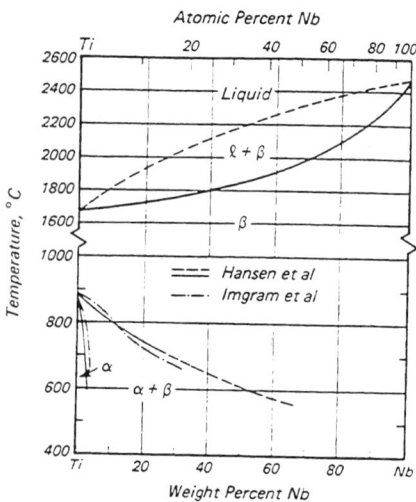

Fig. 4.7 Ti-Nb equilibrium phase diagram due to HANSEN et al. [HAN51] and IMGRAM et al. [IMG61], modified by the observation that no appreciable α-phase precipitation takes place during aging near 400 °C of Ti-Nb alloys with more than about 63 wt% Nb [LEE90].

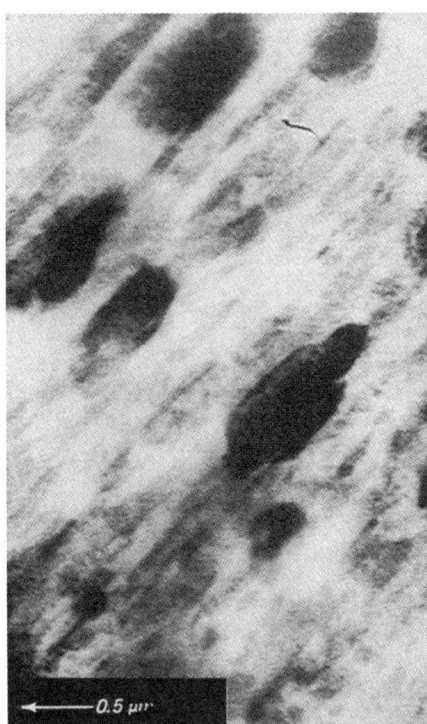
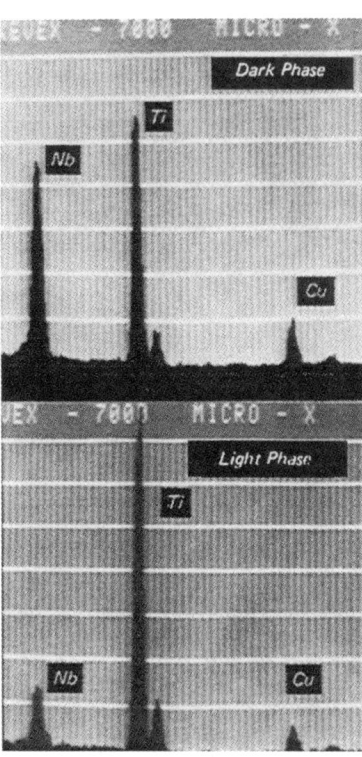

Fig. 4.8 Left side: STEM micrograph of a Ti-53.5 Nb alloy cold worked to a diameter of 3.66 mm and aged 80 h/375 °C, then cold worked to a diameter of 1.5 mm and aged 40 h/375 °C [WES82]. Right side: EDAX pictures of the "dark" (Nb-rich, matrix) and "light" (Nb-lean, α-phase precipitate) regions. Photographs courtesy of A.W. WEST (University of Wisconsin).

serve appreciable precipitation in this case may have been due partly to deficiencies in the detection techniques used and partly to the short aging times. With regard to the latter, the long-time aging of titanium-base alloys is fraught with the danger of oxygen contamination even under the most carefully regulated conditions. Since oxygen is an α stabilizer, its presence can always be interpreted as being partly responsible for the appearances of α precipitates in regions where they might otherwise not be expected. Although the heavy cold work experienced by wires that have been drawn down to small diameters facilitates the attainment of thermodynamic equilibrium in alloys aged at only moderate temperatures, it also makes the detection and identification of any resulting precipitates all the more difficult.

Precipitation in cold-worked superconductive Ti-Nb(36 at.%) alloys has been carefully studied by OSAMURA et al. [OSA80]. Specimens were in the form of: (1) *foils*—solution treated 1 h/800 °C/slow cooled, cold rolled to thickness reductions of up to 90% and aged at 380 °C; and (2) *fine wires*—obtained from copper-clad composites that had experienced reductions of 90 to 99.994%, aged at 380 °C. Small-angle x-ray scattering (SAXS) experiments were performed on the foil samples to determine the average diameter of the precipitated particles (twice R_G, the Guinier radius) and their interparticle spacing (derived from R_G assuming a close-packed arrangement of particles). TEM observations were then performed to obtain visual images and further information about the particles and their distributions. With regard to the wire samples, although SAXS measurements could be performed using bundles of them, direct observation of the precipitates by TEM was very difficult. Instead, wide-angle x-ray diffraction served to confirm that the dominant precipitate formed during aging was in fact α phase.

In discussing precipitate detection techniques for use in heavily cold-worked samples, WEST [WES82] has pointed out that since α-phase precipitates are not easy to identify in such structures, particularly since dark-field imaging is complicated by the close positioning of matrix and precipitate reflections in selected-area diffraction (SAD) patterns, the best analytical results are obtained through the use of scanning transmission electron microscopy (STEM) and associated energy-dispersive x-ray analysis (EDAX). The results of such observational methods applied to a sample of Ti-53.5Nb are given in Fig. 4.8.

With regard to the compositional range of α-phase precipitation, based on the combined results of the experiments published to date, it is concluded that 63 ± 1 wt% Nb can be reasonably taken as a practical boundary between the α + β and β phases at about 400 °C [LEE90]. This result has been inserted in Fig. 4.7. In so doing it was recognized that although α-phase precipitation from alloys of

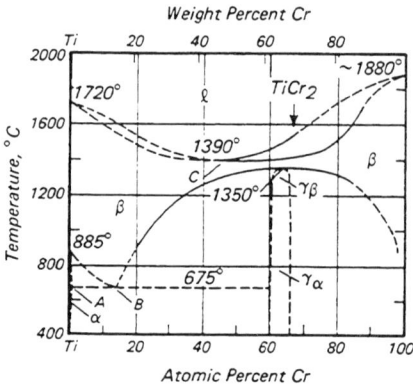

Fig. 4.9 Ti-Cr equilibrium phase diagram. The points indicated by A, B, and C are at concentrations of 0.5, 14, and ~45 at.% Cr, respectively [HAN58, p. 566].

concentrations greater than the above limit may be thermodynamically permissible at temperatures below about 400 °C, it may not be practically realizable.

4.4 A Representative Beta-Eutectoid Alloy System, Ti-Cr

Since chromium is an ingredient of technical alloys such as Ti-13V-11Cr-3Al and Ti-3Al-8V-6Cr-4Mo-4Zr, both metastable β-alloys, an understanding of its binary phase diagram with titanium is particularly important. The complete equilibrium phase diagram for Ti-Cr, a β-eutectoid system, is reproduced in Fig. 4.9. The source of that figure is HANSEN [HAN58, p. 566]; other standard reference sources such as MCQUILLAN and MCQUILLAN [MCQ56, p. 193] and MOLCHANOVA [MOL65, p. 33] have offered qualitatively similar diagrams, differing from Fig. 4.9 only in minor details. Of particular interest in systems of this type is the tendency for the titanium-rich bcc phase to decompose eutectoidally into a weak α-phase solid solution plus a compound. In the Ti-Cr system depicted here, Ti-Cr(14 at.%) decomposes very sluggishly at temperatures below 550 to 685 °C (675 °C is the value preferred by HANSEN [HAN58, p. 566] and SHUNK [SHU69, p. 282]). The presence of the interstitial elements H, N, and O increases the rate of eutectoid decomposition [MOL65, p. 34]. At higher and lower chromium levels, hyper- or hypoeutectoidal decomposition, respectively, can also take place. Once formed, the products of such decomposition are readily redissolved during heating in the β field. The intermetallic compound component of the eutectoidal decomposition—represented by the symbol γ in the appropriate diagram of Fig. 2.4—is of nominal composition TiCr$_2$ with a "homogeneity range" of some 2 percentage points. Its composition, as a function of temperature, is reviewed in [SHU69, p. 283]. Ti-Cr$_2$ is polymorphic, existing as the hcp (MgZn$_2$-structure) "β-TiCr$_2$" phase (labelled γ$_β$ in Fig. 4.9) at high temperatures and the fcc (MgCu$_2$ structure) "α-TiCr$_2$" phase (γ$_α$) at lower temperatures. The transformation temperature of TiCr$_2$ seems to be uncertain [SHU69, p. 283]: according to MOLCHANOVA [MOL65, p. 36], the hexagonal modification exists above 1300 °C and the cubic below 1000 °C, both phases coexisting in the intervening temperature range. As regards the high-temperature bcc solid solutions, a titanium-rich phase, β', and a chromium-rich phase, β'', coexist in a temperature-composition zone bounded by 1350 to 1400 °C and 50 to 70% (wt% or at.%) Cr [MOL65, p 34]. The coexistence in thermodynamic equilibrium of β' and β'' is comparable to the β' + β'' immiscibility exhibited by the Zr-Nb system, but should not be confused with β → β' + β phase separation, a nonequilibrium state of previously quenched alloys during moderate-temperature aging.

4.5 Multicomponent Titanium-Base Alloys

Once a binary alloy with properties more or less suitable for the application in mind has been identified, whether it be structural or superconductive, it can generally be improved by the carefully engineered addition of further alloying components. Thus, for example, commencing with Ti-Al, the addition of tin has led to the technical α alloy Ti-5Al-2.5Sn and the addition of vanadium to the popular α + β alloy Ti-6Al-4V. Substitutions of tantalum for niobium and/or zirconium for titanium have improved the superconductive properties of Ti-50Nb and resulted in technically important ternary and quaternary superconducting alloys. Substitutions of zirconium and tin into the basic β-stabilized Ti-12Mo have yielded the well-known technical alloy, β-III. In structural alloys, the additions are chosen to achieve improvements in mechanical properties such as strength and toughness, structural phase stability, and chemical stability.

4.5.1 Alpha Alloys

The Technical Alpha Alloy Ti-5Al-2Sn. The total α-stabilizing content, on an at.% basis, in Ti-5Al-2.5Sn is 9.7 at.%. Reference to the binary Ti-Al equilibrium phase diagram (Fig. 4.2) suggests that this ternary alloy possesses the highest level of solution strengthening possible while avoiding precipitation of the embrittling α$_2$ phase. The commercial alloy may, however, contain traces of β phase resulting from contamination by iron originating in the sponge-titanium used in its preparation [WOO72]. The following microstructures may be developed in Ti-5Al-2.5Sn by appropriate thermomechanical processing: (1) equiaxed α, obtained by annealing a mechanically worked alloy in the α field (below ~1025 °C); (2) sharp acicular α, obtained by water quenching from the bcc field (above ~1050 °C); and (3) structures immediate between these extremes, obtained by furnace cooling from the bcc field and by adjusting the prior β grain size through appropriate control of the annealing time in that field.

Advanced Alpha-Stabilized Alloys. HOCH et al. [HOC73], drawing an analogy with the nickel-base superalloys and their γ' (Ni$_3$Al) precipitates, recommended the use of highly alloyed α-phase alloys containing α$_2$-phase precipitates for high-temperature applications where creep resistance is important. The α$_2$ phase referred to was understood to be an ordered compound of variable stoichiometry, based on the DO_{19} compound Ti$_3$SM, where SM may be Al, Ga, In, or Sn.

The binary stoichiometric α$_2$-Ti$_3$Al phase is extremely brittle in tension (less so in compression, of course). Accordingly, it has been found to severely embrittle the two-phase Ti-Al(>12 at.%) alloys in which it occurs. Some degree of ductility can be acquired if the α$_2$ particles can be coarsened sufficiently to enable a dislocation bypass (looping) mechanism to operate, but the desired coarsening is difficult to achieve in practice. The goals of high-concentration α-phase alloy development have been to take the greatest possible advantage of solution- and precipitate-strengthening but at the same time to avoid the previously inevitable α$_2$-Ti$_3$Al particle embrittlement. With these goals in mind, considerable effort has been directed toward exploring the microstructural, physical, and mechanical properties of Ti-Al-Ga alloys. The situation has been discussed by GODDEN et al. [GOD73] and HOCH et al. [HOC73] with reference to some earlier relevant studies by BLACKBURN and WILLIAMS [BLA69] [WIL69] and LÜTJERING and WEISSMANN [LUT70, LUT70^a]. An equilibrium phase diagram depicting a corner of the Ti-Al-Ga system is given in Fig. 4.10.

With regard to the solution-strengthening aspects, COLLINGS and GEGEL [COL75^a] have shown that, as functions of total α-stabilizer content, the tensile strengths of Ti-Al$_x$-Ga$_x$ alloys were always greater than those of either Ti-Al$_{2x}$ or Ti-Ga$_{2x}$. The extra strengthening was attributed to secondary solid-solution strengthening arising from Al-Ga interaction. Two alternative approaches to the solving of the α$_2$-phase embrittlement problem have been discussed: (1) one involved the properties of the matrix and its ability to accommodate the presence of the precipitate particle; (2) the other in-

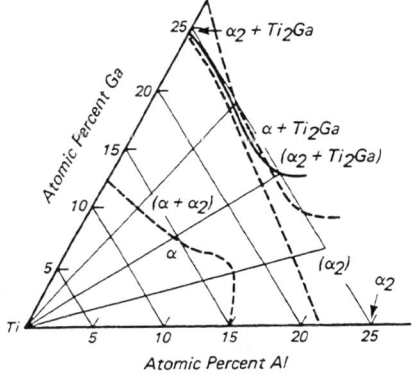

Fig. 4.10 Titanium-rich corner of a Ti-Al-Ga equilibrium phase diagram according to SAKAI [SAK69]. The dashed lines and phases in parentheses indicate the range of α$_2$ proposed by WILLIAMS and BLACKBURN [WIL69] but not observed by SAKAI, see also [HOC73, HOC73^a].

volved the properties of the precipitate particle itself. HOCH et al. [HOC73] tended to focus attention on alternative (1) above. They compared the systems Ti-Al-Sn and Ti-Al-Ga in terms of the parameter of misfit between the α_2 particle and the matrix, and on this basis judged that, whereas dislocation-cutting-type* deformation mechanisms could be expected in the former, a dislocation-bypass deformation mechanism operative in Ti-Al-Ga should have led to at least some measure of ductility. GODDEN and ROBERTS [GOD73] discarded the possible role played by α/α_2 mismatch and suggested that the increased ductility observed in Ti-Al-Ga might have been due to an alloying-induced improvement in the ductility of the α_2 itself. Basing this particular argument on the results of some order-disorder transition temperature (T_{OD}) measurements by COLLINGS et al. [COL71a], they suggested that with a T_{OD} of ~500 °C the α_2 component of an alloy such as Ti$_{75}$-Al$_{12.5}$-Ga$_{12.5}$, quenched from 1300 °C, may be present in the disordered state, and consequently intrinsically more ductile.

4.5.2 Alpha + Beta Alloys

In alloys such as Ti-Al, the two-phase $\alpha + \beta$ region is both *narrow* and *high in temperature*. With Ti-6Al(10 at.% Al), for example, the $\beta/(\alpha + \beta)$ and $(\alpha + \beta)/\alpha$ transformations take place at ~1010 and 970 °C, respectively. The introduction of vanadium at constant aluminum concentration, although it has a comparatively small influence on the position of the $\beta/(\alpha + \beta)$ transus, produced a rapid decrease in the $(\alpha + \beta)/\alpha$ transus. These effects are shown in Fig. 4.11, which plots transformation temperature versus vanadium concentration levels of aluminum, and Fig. 4.12, which performs a complementary function in terms of a continuous variation of the aluminum concentration for two fixed at four fixed levels of vanadium. The corresponding equilibrium ternary phase diagrams for Ti-Al-V are given in Fig. 4.13. Evidently, even in the presence of, say, 10 wt% Al, that of 8 wt% V is sufficient to permit the retention of a small β-phase component to temperatures as low as 600 °C. The popular alloy Ti-6Al-4V is a member of this system.

The $\alpha + \beta$ Alloy Ti-6Al-4V. The alloy Ti-6Al-4V could be regarded as being derived from unalloyed titanium by (1) the

---
*That is, fracture of the precipitate particle.

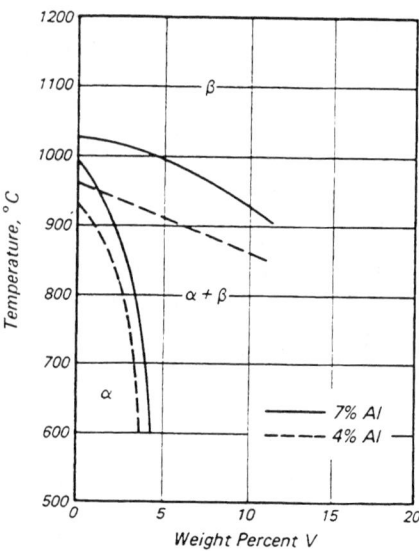

Fig. 4.11 "Vertical" sections of the Ti-Al-V versus T equilibrium-phase solid (right triangular prism) at 4 and 7 wt% Al [RAU56].

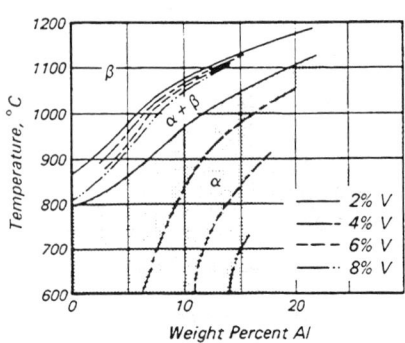

Fig. 4.12 "Vertical" sections of the Ti-Al-V versus T equilibrium-phase solid (right triangular prism) at 2, 4, 6, and 8 wt% V [RAU56].

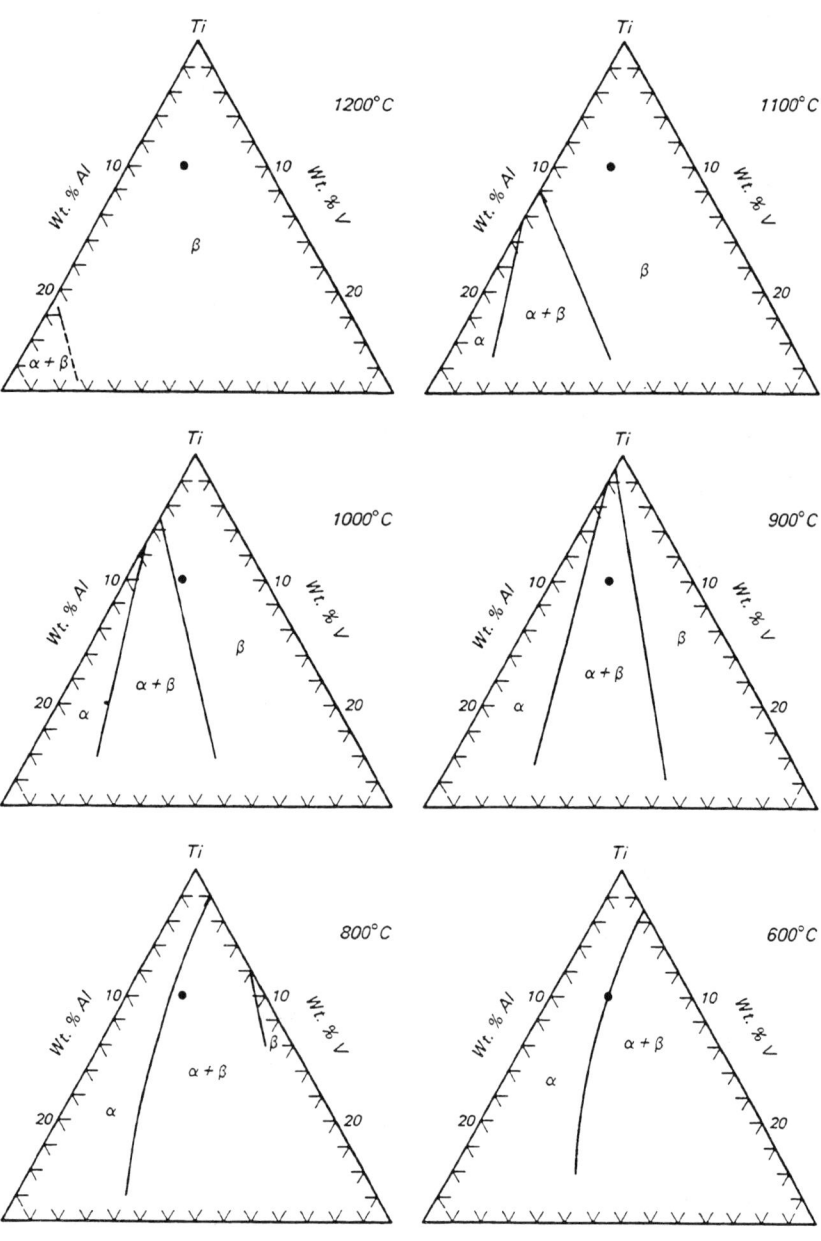

Fig. 4.13 Isothermal (i.e., "horizontal") sections of the Ti-Al-V versus T equilibrium-phase prism at the temperatures indicated [RAU56]. The solid circle (•) represents the alloy Ti-6Al-4V.

addition of aluminum to produce solution strengthening and raise what becomes the β/(α + β) transus, and (2) the addition of vanadium to lower the (α + β)/α transus. The equilibrium states of this alloy at various temperatures are indicated by special points in Fig. 4.13.

A wide range of processing techniques are applied to Ti-6Al-4V to produce numerous kinds of mill products exhibiting a wide range of microstructures. An example of the manner in which thermal processing can control the microstructure of Ti-6Al-4V is depicted schematically in Fig. 4.14. A set of typical microstructures is presented in Fig. 4.15.

The existence of a third phase in α + β Ti-6Al-4V was first reported by RHODES and WILLIAMS [RHO75]. Designated an "interface phase," it occurred either as a monocrystalline layer with an fcc structure, or as a striated layer about 200 μm thick consisting of platelets. Interface layers have also been reported in Ti-11.6Mo and Ti-14Mo-6Al by MARGOLIN et al. [MAR77], and, more recently, an fcc or fct phase has been observed separating the α and β phases of a β-cooled (<~400 °C/min) multicomponent alloy based on Ti-6Al-5Zr by HALLAM and HAMMOND [HAL80].

The Near-Alpha Alloy Ti-8Al-1Mo-1V. In the alloy Ti-8Al-1Mo-1V, with its relatively large amount of the α stabilizer, aluminum, the low concentrations of the β stabilizers, molybdenum and vanadium, which are present permit only small amounts of β phase to become stabilized [WOO72, p. 1-3:72-1]. The β/(α + β) transus is at approximately 1040 °C. On being quenched from within the α + β field, from just below the transus down to about 900 °C, the β component transforms to α'. While being annealed at lower temperatures, the β phase becomes sufficiently enriched in molybdenum and vanadium (see Fig. 4.11) that it resists transformation during quenching. At 870 and 815 °C, the alloy contains 16 and 14%, respectively, of the β phase. Although this alloy must be designated "α + β, near-α", it does exhibit several α-alloy characteristics, such as good weldability and elevated-temperature creep strength. Indeed, it was developed in the first place as a "super-α" alloy* for forged engine components [WOO72].

The Near-Alpha Alloy Ti-6Al-2Sn-4Zr-2Mo. The alloy Ti-6Al-2Sn-4Zr-2Mo (Ti-6242) was developed in order to extend the previously existing upper-temperature limit of titanium-alloy service. For this reason, like Ti-8Al-1Mo-1V, it can also be described as a "super-α" alloy. The development philosophy seems to have been to replace the "4V" of Ti-6Al-4V with

---

* The term "super" implies suitability for high-temperature service.

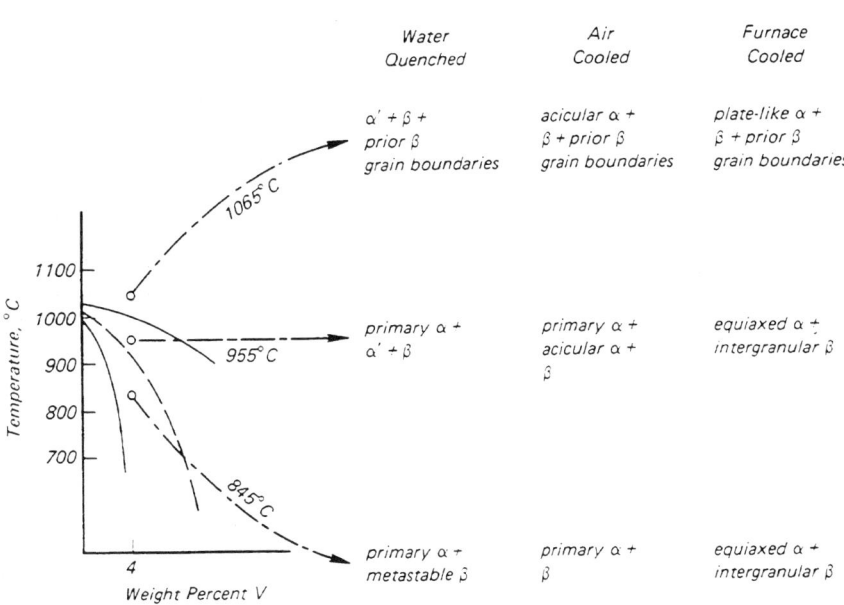

Fig. 4.14 Microstructural control of Ti-6Al-4V through thermal processing. After DONACHIE [DON82, p. 43].

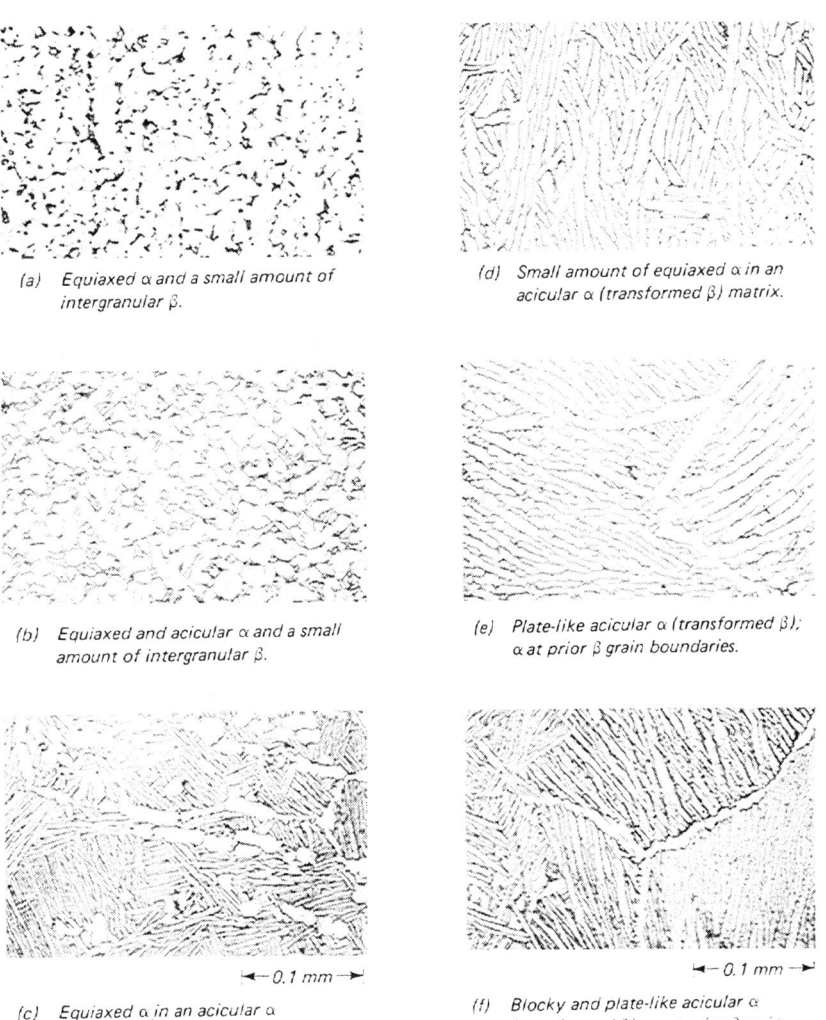

(a) Equiaxed α and a small amount of intergranular β.

(b) Equiaxed and acicular α and a small amount of intergranular β.

(c) Equiaxed α in an acicular α (transformed β) matrix.

(d) Small amount of equiaxed α in an acicular α (transformed β) matrix.

(e) Plate-like acicular α (transformed β); α at prior β grain boundaries.

(f) Blocky and plate-like acicular α (transformed β); α at prior β grain boundaries.

Fig. 4.15 Optical microstructures of Ti-6Al-4V in six representative metallurgical conditions [WOO72, pp. 1-4:72-4, 72-5].

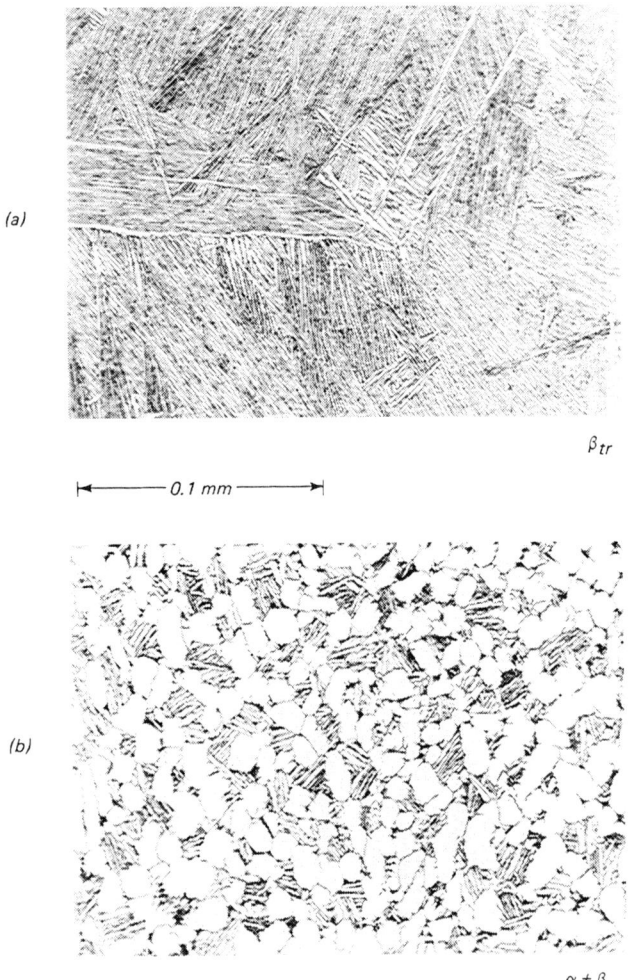

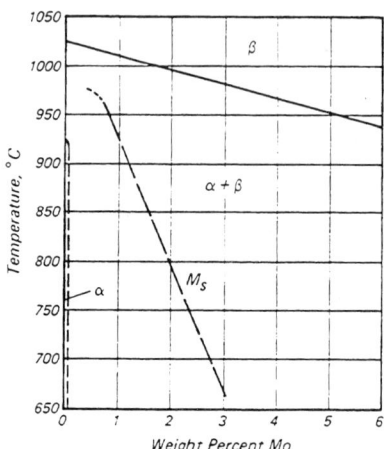

Fig. 4.16 Optical micrographs of Ti-6Al-2Sn-4Zr-2Mo-0.1Si (Ti-6242-Si) in two characteristic metallurgical conditions: (a) transformed β [$β_{tr}$, or Widmanstätten α + β, $(α + β)_W$], as a result of the heat treatment 2 h/1024 °C/air cool; (b) α + β, as a result of 2 h/968 °C/air cool. Micrographs courtesy of S.L. SEMIATIN.

Fig. 4.17 Pseudobinary equilibrium phase diagram for (Ti-6Al-2Sn-4Zr)-XMo for values of X (the wt% of Mo) between 0 and 6. Source: S.L. SEMIATIN et al., unpublished research.

"2Mo" (the latter being a stronger β stabilizer than vanadium) and to insert tin and zirconium, which are neutral in this regard. Tin is well known as a solution strengthener; zirconium contributed little strengthening but may have been included to improve the stability of the β phase (see the next subsection) and/or to assist with homogenization [FEE70] (see Section 3.5.4). As with the alloy described above, Ti-6242 functions as a near-α, α + β alloy.

Further improvement in the elevated-temperature creep strength of Ti-6242 has been obtained by addition of 0.2 wt% Si, which in the form of silicide precipitates provides dispersion hardening. The alloy so fortified, designated Ti-6242-Si, has been used extensively by the turbine engine industry for elevated-temperature applications.

Although the microstructures assumed by this alloy are characteristic of its class, and are generally quite similar to those exhibited by its ternary prototype, Ti-6Al-4V (see Fig. 4.15), a wide range of variants can be generated by adjusting the thermochemical-process parameters. In this manner a set of eight microstructural types has been generated by CHEN and COYNE as part of an investigation of the influence of microstructural variation on mechanical properties [CHE80]. Since mechanical properties, particularly at elevated temperatures, are very sensitive to microstructure, a great deal of attention has been given recently to the study of their interrelationships [CHE80]. The two basic microstructures, as depicted in Fig. 4.16, are interpretable with the aid of Fig. 4.17, a pseudobinary phase diagram with molybdenum as the variable. The β/(α + β) transus is at about 990 °C [WOO72, pp. 1-6:72-2]. On cooling from above this temperature, the transformed-β structure (Fig. 4.16a) varies from martensitic to Widmanstätten α + β as the cooling rate decreases from water-quench speeds to those characteristic of air cooling. The structure preserved after an anneal below the β/(α + β) transus (Fig. 4.16b) consists of primary α (the original α phase*) plus a transformed version (usually Widmanstätten) of the original β. Figure 4.18 shows a set of three typical microstructures selected to represent the effects of air cooling from anneals at temperatures of 900 and 980 °C. The highest annealing temperature yields the largest volume fraction of β (see Fig. 4.17), which, being the most unstable at ordinary temperatures, yields the finest transformed structure.

A knowledge of the compositions of the α needles and the β matrix of the Widmanstätten β-transformed structure (Fig. 4.16a), as well as those of any incoherent precipitates that may be present within them, is necessary for a proper interpretation of the mechanical properties. X-ray fluorescence analyses of the latter have been undertaken, and the results discussed by CHEN and COYNE [CHE80], while the compositions of the α and β components of the Widmanstätten structure itself have been measured by GEGEL and colleagues, whose results are presented in Fig. 4.19 and Table 4.2. Of particular interest are: (1) the compositional uniformity across the sample exhibited by the "neutral" elements zirconium and tin; (2) the anticipated preferences of aluminum for the α needles and molybdenum for the β matrix; and (3) the existence of an "interface phase" characterized by steep compositional gradients of the aluminum and the molybdenum.

The Near-Beta Alloy Ti-6Al-2Sn-4Zr-6Mo. The alloy Ti-6Al-2Sn-4Zr-6Mo (i.e., Ti-6246) is formed from the earlier Ti-6242 by increasing the molybdenum con-

---
* Also referred to as "globular-α."

tent in the interests of improved hardenability. The additional molybdenum increases the volume fraction of the β component at a given temperature, and the presence in it of tin and zirconium which according to Fig. 4.19 do not partition between the α and β phases, retards its transformation during cooling, thereby contributing to the hardenability. Detailed descriptions of metallurgy and properties are to be found in [WOO72, pp. 1-7:72-1 *et seq.*].

4.5.3 The Superconducting Ternary System Ti-Zr-Nb

The Ti-Zr-Nb system is of considerable interest from both practical and pedagogical standpoints in that within it are to be found examples of ω-phase and α-phase precipitation, β′ + β phase separations (near the Ti-Nb edge), and β′ + β″ immiscibility (in the Zr-Nb sector of the diagram). A set of equilibrium phase diagrams for this system, based on the work of ALEKSEEVSKII *et al.* [ALE67] and DOI *et al.* [DOI66], who had used the standard techniques of optical metallography and x-ray diffraction, is presented in Fig. 4.20. Rather good agreement between the results of these two research groups was obtained, as evidenced by the overall similarity of their diagrams* and, in particular, by the fact that the diagram for 570 °C due to DOI *et al.* fits logically between those for 550 and 600 °C obtained by ALEKSEEVSKII *et al.* At 1050 °C the ternary systems show unlimited mutual β solid solubility. Below 975 °C a region of two-β-phase immiscibility, β′ + β″, begins to develop from the Zr-Nb edge, and expands as the temperature falls. Between 700 and 600 °C an α + β region develops along the Zr-Nb edge separated from the β′ + β″ lobe first by a line and, at lower temperatures, by a region α + β′ + β″. By this time an α + β region is developing from the titanium corner, and another from the zirconium corner, severely restricting the remaining region of bcc stability. These changes in phase relationships take place very rapidly with temperature between 570 and 500 °C, at which point an α + β phase occupies almost the entire diagram.

* In analyzing the x-ray data, DOI *et al.* [DOI66] chose not to distinguish between the two β phases when they coexisted with the α phase, and labelled as α + β those regions which ALEKSEEVSKII *et al.* [ALE67] designated (using present terminology) α + β′ + β″. In Fig. 4.20, some reasonable liberties have been taken with the data of ALE for 570 °C, after which the diagram falls nicely into position with respect to the surrounding Soviet results.

4.5.4 Beta Alloys

The β alloys are characterized by sufficient β stabilizers (transition elements) to ensure the retention of the bcc phase on rapid cooling to room temperature. Strengthening by α-phase particles is then achieved during aging heat treatments. Advantages of β alloys are: (1) cold workability in contrast to the limited room-temperature ductility of the α– and α + β alloys; and (2) deep hardenability—*i.e.*, the ability of thick parts formed from them to be hardened [FRO73][PET73]. The first commercial heat-treatable β-Ti alloy was Ti-13V-11Cr-3Al. Subsequently, to avoid the difficulties that arose because of its chromium content, a molybdenum-containing replacement (β-III) was developed. Since then both of these alloys have been replaced by alloys such as Ti-10V-2Fe-3Al, Ti-3Al-8V-6Cr-4Mo-4Zr and Ti-15V-3Cr-3Al-3Sn.

Table 4.2 Compositions of the Component Phases in Widmanstätten α + β-Phase Ti-6242 [GEG80[a]]

Component	Composition in wt% (at.%)				
	Ti	Al	Sn	Zr	Mo
Average(a)	86.0 (85)	6.0 (11)	2.0 (1)	4.0 (2)	2.0 (1)
β platelet(b)	78.5 (87)	0.5 (1)	2.0 (1)	4.0 (2)	15.0 (8)
α platelet(b)	88.5 (88)	5.0 (8)	2.0 (1)	4.0 (2)	0.5 (<1)

(a) Nominal composition. (b) STEM/EDAX analysis

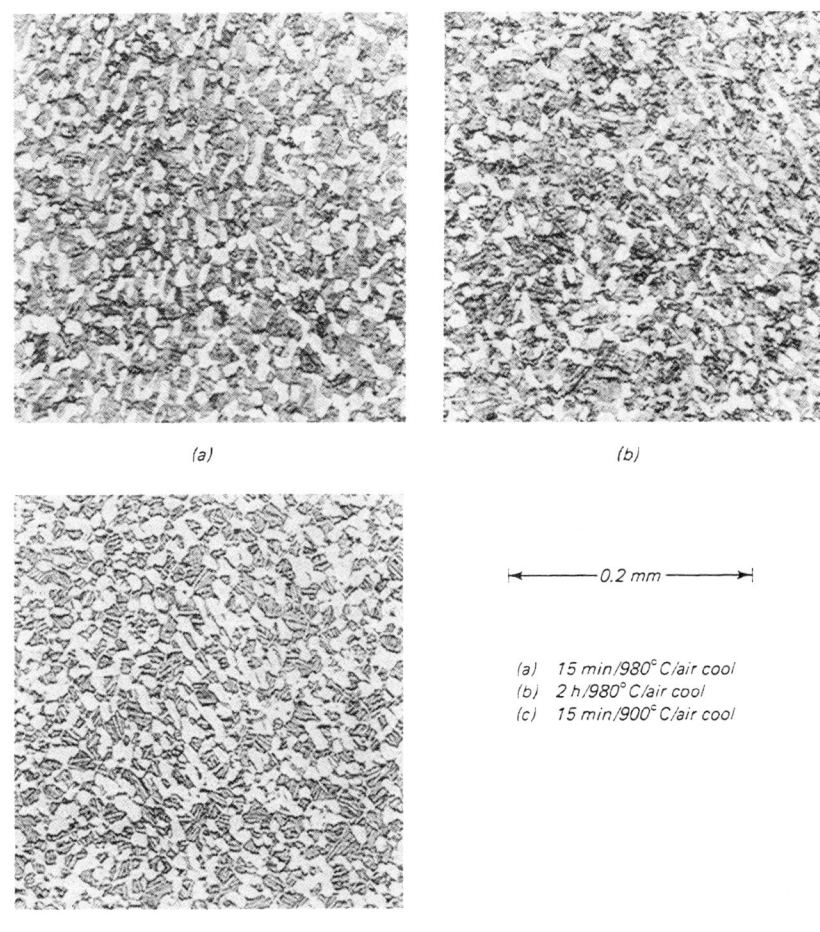

(a) 15 min/980° C/air cool
(b) 2 h/980° C/air cool
(c) 15 min/900° C/air cool

Fig. 4.18 Microstructures in α + β-phase Ti-6Al-2Sn-4Zr-2Mo-0.1Si (see Fig. 4.16b) after further exposure to the heat treatments specified. Micrographs courtesy of S.L. SEMIATIN.

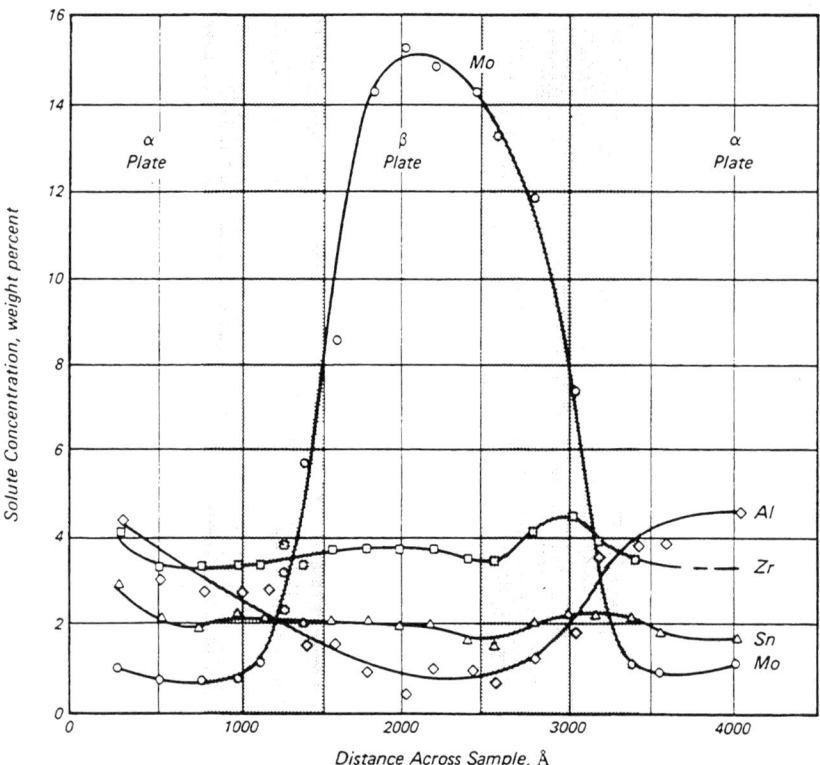

Fig. 4.19 X-ray analysis of the compositional distribution across the $(\alpha + \beta)_W$ structure in Ti-6Al-2Sn-4Zr-2Mo-0.1Si (see Fig. 4.16a) [MAH82].

The Beta Alloy Ti-13V-11Cr-3Al. The alloy Ti-13V-11Cr-3Al was for many years the only commercially important β alloy. As a consequence of the very high β-stabilizing content, the β phase is completely retained during even slow cooling (~0.5 h, for β-annealed material) [Woo72, p. 1-9:72-1]. A pseudobinary equilibrium phase diagram (variable chromium) is given in Fig. 4.21. Hardening to high strength levels is achieved by an aging heat treatment designed to precipitate α phase plus $TiCr_2$. The $TiCr_2$ precipitate derives of course from eutectoidal decomposition associated with the Ti-Cr alloy component (see Fig. 4.9); its occurrence in this alloy is responsible for a relatively low aged toughness [PET73] and embrittlement during service at temperatures above 300 °C [FEE70]. These drawbacks were to lead to the partial abandonment of "13-11-3" in favor of its chromium-free counterpart, to be described below.

The Beta Alloy Ti-11.5Mo-6Zr-4.5Sn: "β-III." The alloy Ti-11.5Mo-6Zr-4.5Sn (β-III) contains none of the β-eutectoid formers chromium, nickel, iron or copper and consequently does not suffer from the intermetallic-compound embrittlement experienced by its predecessor. β-III was developed in the late 1950s by the then Crucible Steel Company, which also devised melting practices to overcome homogeneity difficulties previously associated with the high melting point and density of molybdenum [PET71]. β-III is essentially a strengthened Ti-Mo, with which it shares many fundamental metallurgical properties. Solution strengthening of both the α and β phases, when they are deliberately produced in the alloy by heat treatment, is conferred by the additions of tin and zirconium (see Fig. 4.22). Zirconium also, apparently, helps to eliminate the inhomogeneity otherwise encountered during melting together of elements that differ in density as much as molybdenum and titanium [FEE70]. A useful review of Ti-Mo-base β alloys has been presented by OHTANI et al. of Kobe Steel Ltd., who have also patented a melting technique to overcome the homogeneity difficulties previously encountered in the melting of Ti-Mo alloys [OHT73]. The metallurgical behavior exhibited by β-III is similar to that of binary Ti-12Mo (~6 wt%) [BOY74]. Thus, decomposition of the bcc phase can take place by both displacive transformation (athermal ω phase) and diffusional transformation (isothermal ω phase and α phase) [VIG82]. Quenching from above the β/(α + β) transus (755 °C), yields a bcc matrix with a fine dispersion of hexagonal ω-phase precipitates no larger than 25 Å [FEE70][RAC70]. Binary Ti-Mo alloys with more than 11 wt% Mo (~6 at.%) exhibit transformation-aided ductility and are capable of more than 30% elongation at room temperature. For similar reasons β-III is also ductile at ordinary temperatures, its cold rollability, for example, being at least as good as that of commercial unalloyed Ti [Woo72, pp. 1-10: 72-2]. The deformation of β-III within the temperature range 77 K (liquid nitrogen) to 150 °C is assisted by twinning and to a larger extent by stress-induced martensitic transformation (to orthorhombic α'') [FEE70]. In β-III, which is also a deep-hardenable alloy, a fine α-phase dispersion is sought. The aging of β-III in order to avoid ω-phase embrittlement in service and to achieve α-dispersion hardening will be discussed in Section 7.14.2.

4. Equilibrium Phases / 33

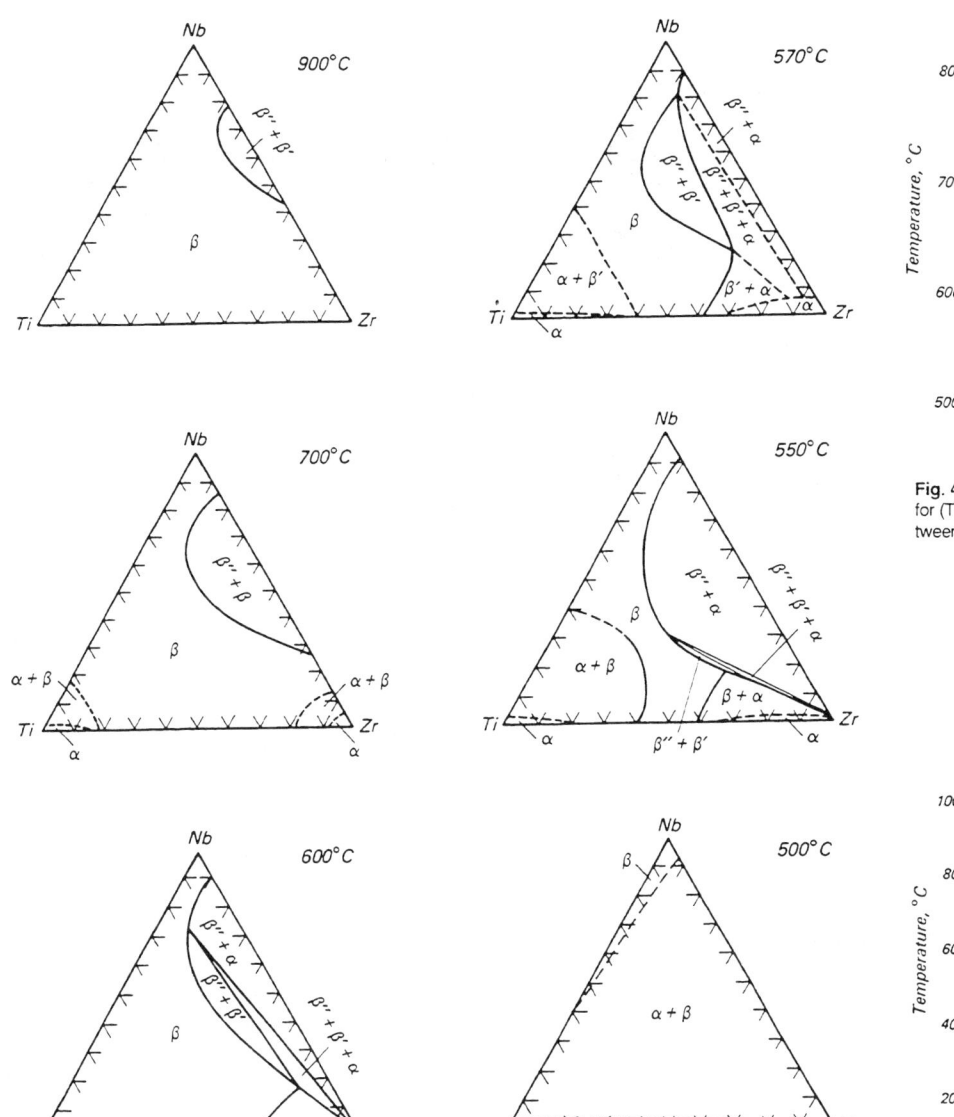

Fig. 4.20 Ti-Zr-Nb equilibrium phase diagram (at.% linear), for temperatures between 900 and 500 °C (1650 and 930 °F), based on the work of Doi et al. [Doi66] (900, 700, and 570 °C, or 1650, 1290, and 1060 °F) and ALEKSEEVSKII et al. [ALE67] (600, 550, and 500 °C, or 1110, 1020, and 930 °F), the latter two diagrams having been modified to take into consideration the absence of appreciable α-phase precipitation from moderate-temperature-aged Ti-Nb alloys with niobium concentration greater than ~40 at.% (see Fig. 4.7).

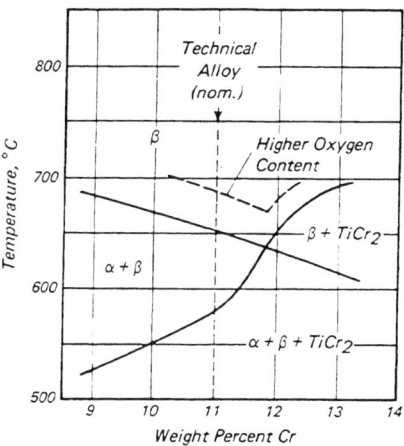

Fig. 4.21 Pseudobinary equilibrium phase diagram for (Ti-13V-3Al)-XCr for values of X (the wt% Cr) between about 9 and 14 [Woo72; p 1-9:72-2].

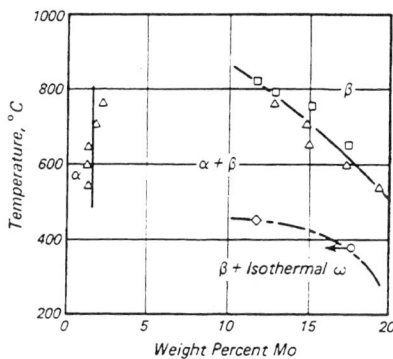

Fig. 4.22 Pseudobinary phase diagram for (Ti-6Zr-4.2Sn)-XMo for values of X (the wt% Mo) of up to about 19. Data sources: (△) partitioning data; (○) behavior of "17.3Mo–β–III" upon 370 °C (700 °F) aging; (◊) upper limit of isothermal ω in regular β-III; (□) β transus for 0.280 β-III type alloys [Fro73].

5. Nonequilibrium Phases

5.1 General Description

5.1.1 Introduction: Formation of Nonequilibrium Phases

Equilibrium phase diagrams of the type discussed in the previous chapter are usually developed by deducing the *initial states* of alloys that have been quenched to room temperature. The nonequilibrium phases to be considered herein represent the *final* states of such quenching processes. The preceding chapter could scarcely have been written in its present form without anticipating some of the results to be discussed below. In considering the near-α α + β alloys (Section 4.5.2), for example, it was necessary to point out that quenching from the β field rapidly through the equilibrium α + β region resulted in a martensitic structure, while less rapid cooling from the same initial temperature provided an opportunity for α-phase nucleation and growth to take place, giving rise to the characteristic Widmanstätten structure (Fig. 4.16a). The occurrence of these structures has been described in detail by WILLIAMS [WIL73, pp. 1435 *et seq.* and pp. 1460 *et seq.*].

The structure of α-stabilized alloys quenched from the β field is martensitic. When quenched from below the (α + β)/α transus, the structures found are of course simply the frozen-in untransformed results of equilibrium at the pre-quenched temperature.

The structures assumed by rapidly β-quenched binary Ti-TM alloys are mapped in Fig. 5.1. Below a start temperature, M_s, the bcc structure begins a spontaneous allotropic transformation by means of a complicated shearing process to a structure known as martensite and designated α′ or α″ depending upon whether the transformation product is hcp or orthorhombic. When the distinction between α′ and α″ is unimportant, the martensites are to be herein represented collectively by the notation α^m. Being of second order, the martensitic transformation is anticipated by a regime of structural fluctuations called diffuse ω phase. As represented in Fig. 5.1, the ω phase, as a result of very rapid quenching, exists as a crystalline precipitate plus a fluctuating component within a narrow composition range overlapping the boundary of the martensite phase. In practice, however, the range over which it occurs during the brine quenching of macroscopic samples is quite broad and is depicted in Fig. 5.1 as a region of gradually diminishing precipitate abundance. The free energy of α^m is lower than that of ω; consequently, during the partial martensitic transformation of an alloy in which ω phase is also able to form, the martensite needles generally consume any ω-phase precipitates that lie in their paths.

5.1.2 Quenching Process

In studies of quenched microstructures, an important but not always attainable goal is the control and quantification of the quench rate. If the quench is too slow, diffusional processes intervene to obscure the result. When the primary aim is to study microstructure (rather than the production of material for physical- or mechanical-property testing) and the highest possible quench rates are mandatory, thin foils are generally heated in a controlled environment and subjected to *in situ** gas or liquid quenching. In HICKMAN's method, for example, rolled strips self-heated under high vacuum to 1250 °C by the passage of direct current were quenched by admitting He gas to a pressure of 0.1 atm, restoring sample temperature, and then switching off the current [HIC68, HIC69a]. Such techniques are generally capable of quench rates of 50 to 2×10^4 °C/s, and in HICKMAN's case about 10^3 °C/s was claimed. Helium is about three times as effective a quench medium as argon under the same conditions.

BROWN *et al.* [BRO65] applied indirect heating under vacuum to specimens varying in thickness from 0.05 to 5 mm (0.002 to 0.20 in.); they then applied a 150- to 700-torr head pressure of argon to suppress the boiling of the iced water or refrigerated calcium chloride solution subsequently admitted to quench the sample. In this way, quench rates of 2.5×10^4 to 2×10^5 °C/s were achieved [JEP70]. BALCERZAK and SASS [BAL72], whose results are discussed below, attached rolled specimens (0.05 to 0.076 mm thick) to an Inconel specimen holder by means of which they could be transferred from the hot zone of a vacuum resistance furnace to a waiting pool of water-cooled silicone oil. Although quenching was known to be rapid, the quench rates achieved by this method were not specified. The quench rates achievable by all of these methods are of course much higher than those obtained during the ice-brine quenching of the massive samples (up to 40 g) needed for study of mechanical and physical properties (especially low-temperature specific heat). Accordingly, some discrepancies must be expected between the microstructural results obtained from thin foils and those derived from quenching of bulk specimens.

The measured M_s temperature for a given alloy composition is itself a function of quench rate. In Ti-Nb (5 at.%), for ex-

---

* In which the quenching medium is introduced into the furnace space containing a fixed sample.

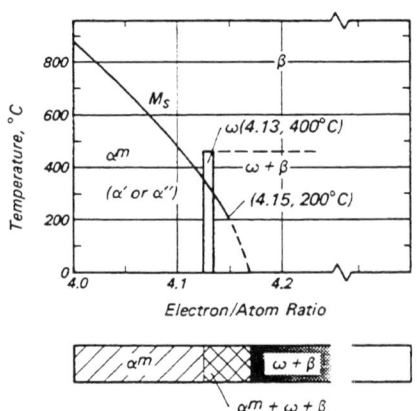

Fig. 5.1 Schematic representation of the occurrences of the martensitic phases α′ and α″ (i.e., α^m collectively) and the ω phase in Ti-TM alloys. Both "quenched data" and "aged data" are included, see also Table 5.1 (α^m phases) and Table 5.5(a) (ω phase).

ample, JEPSON et al. [JEP70] noted that the M_s temperature decreased from 760 to 710 °C as the cooling rate increased from 10^{-3} to 10 °C/s, but that once a critical cooling rate of 32 °C/s was exceeded, M_s was independent of the cooling rate. The critical threshold itself was a function of alloy composition and decreased from 200 to ~0.4 °C/s as the niobium content increased from 0 to 15 at.%.

In Ti-TM alloys, as with other systems, the quench rates necessary to achieve structural transformation while preserving compositional homogeneity are strongly constitution-dependent. Thus, whereas bulk dilute alloys of titanium with early transition elements can be water quenched without evidencing serious decomposition, the same is not true of alloys such as Ti-Fe, Ti-Ni, and Ti-Co, whose anomalous superconducting properties could be partially interpreted in terms of compositional, hence structural, segregation. The pronounced differences between the properties of the quenched dilute Ti-V, Ti-Nb, etc. alloys and those of Ti-Fe, Ti-Co, and Ti-Ni can be simply explained in terms of differences among the solute tracer diffusion coefficients in β-Ti at 1000 °C. As shown in Fig. 3.20, the diffusion coefficients of vanadium, niobium, and molybdenum are less than 1.3×10^{-9} cm²/s, while those of the Fe-group elements are 60×10^{-9} cm²/s. The extreme examples are cobalt on one hand and molybdenum on the other; their diffusion coefficients (β-Ti, 1000 °C) are in the ratio 200:1. This must be taken into consideration when comparing the properties of the two classes of titanium-base alloys, and in selecting a quenching technique.

5.1.3 Stability Limit of the β Phase in Ti-TM Alloys

The optical microstructures and composition ranges of the martensites characteristic of ice-brine-quenched massive samples of Ti-V, Ti-Nb, Ti-Mo, and Ti-Fe are exemplified by Fig. 5.2. Compositionally, Ti-Fe and Ti-Nb are extreme examples, their M_s curves bounding those for all other measured Ti-TM alloys and intersecting a 200 °C isothermal, for example, at 3.3 and 20.5 at.%, respectively [ZWI74, p. 174].

Table 5.1 lists the $M_{s,200\,°C}$ compositions of eight Ti-TM alloys bounded constitutionally by Ti-Fe and Ti-Nb, together with the corresponding conventional electron/atom ratios. Quite remarkable is the fact that all the e/a ratios, except for those of Ti-Co and Ti-Ni (which may be exceptional cases), lie within ±0.06 of a common value, 4.15, suggesting that the martensitic transformation in Ti-TM alloys is of common origin and related to electronic factors. Electronic and lattice properties

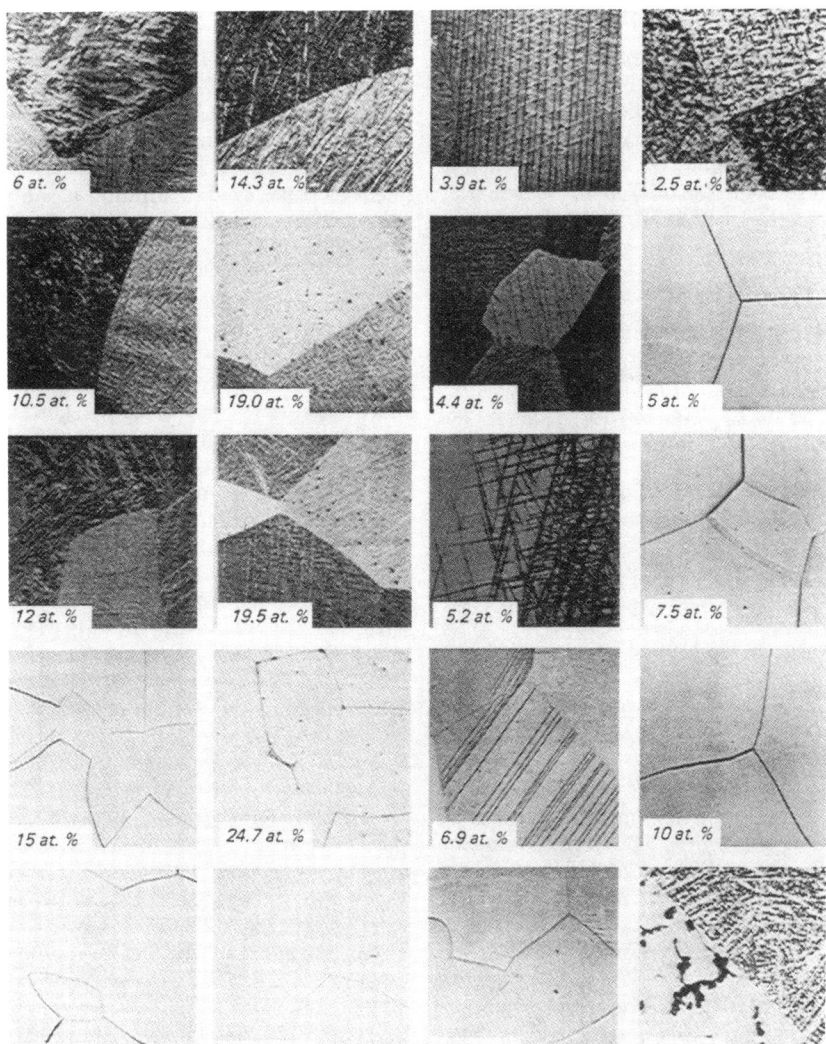

Fig. 5.2 Optical micrographs from massive tokens of Ti-TM alloys, quenched from the β phase into iced brine, showing the transitions from the α^m to the ω + β regimes with increase in solute concentration (or e/a ratio). Compositions of Ti-V and Ti-Fe are nominal, those of Ti-Nb and Ti-Mo are analyzed. Magnifications of the original 9 × 9 cm² micrographs were 50×; a 300× micrograph of Ti-Fe(20 at.%) is also shown [COL84].

Table 5.1 Compositions of the $M_{s,200\,°C}$ Intercepts Expressed in Terms of Conventional Electron/Atom Ratio

Solute group number, GN	Solute element	Concentration corresponding to M_s at 200 °C, at.%, c(a)	Conventional e/a based on group number(b)
V	V	13.3	4.13
	Nb	20.5	4.21
	Ta	19.1	4.19
VI	Cr	6.0	4.12
	Mo	6.7	4.13
	W	8.2	4.16
VII	Mn	5.0	4.15
	Fe	3.3	4.13
VIII	Co	6.0	4.24 – 4.30(c)
	Ni	7.6	4.30 – 4.46(c)
			Mean(d) 4.15 ± 0.03 (s.d.)

(a) After ZWICKER [ZWI74, p. 174]. (b) Calculated according to: $e/a = 4 + \Delta GN(c/100)$, where $\Delta GN = GN_{solute} - GN_{Ti} = GN_{solute} - 4$. (c) Based on number of valence (s + d) electrons. (d) Excluding cobalt and nickel

can, at least in principle, be coupled through calculations of the elastic constants. A connection between the end of bcc stability and the vanishing of the elastic shear modulus, $C' = (C_{11} - C_{12})/2$, has already been considered in Section 2.3, Fig. 2.1 [COL72, COL72a, COL73].

5.2 Occurrence and Structures of the Martensitic Phases

The word "martensite," named for Professor A. Martens, was originally adopted by metallurgists to define the acicular structure in quenched carbon steel that was responsible for its outstanding hardness [COH51]. The occurrence and structures of martensites in numerous other alloy systems have been described by COHEN [COH51] and by BILBY and CHRISTIAN [BIL56]. Detailed discussions of martensitic transformations, but with particular reference to titanium alloys, have been offered by the McQUILLANS [MCQ56, Chap. 9], MARGOLIN and NIELSEN [MAR60], HAMMOND and KELLY [HAM70], and OTTE [OTT70]. Unalloyed titanium transforms martensitically from bcc to hcp during cooling through its β→α allotropic transformation temperature, 882.5 °C. In alloys of titanium, the equilibrium α and β fields are separated by a two-phase, α + β region and the β→$α^m$ transformation temperature,* M_s, is composition-dependent. In α-stabilized alloys, typified by Ti-Al, M_s may lie a little below the (α + β)/α transus [JEP70]; in the β-stabilized alloys it always lies within the α + β field. For some recent information on martensitic transformations in the β-isomorphous systems Ti-V, Ti-Nb, and Ti-Mo, and the manner in which the transformed structures revert to β or decompose on aging, the papers of DAVIS, FLOWER, and WEST [DAV79, DAV79a] [FLO82] should be consulted.

5.2.1 Martensitic Transformation in Titanium Alloys

In contrast to nucleation-and-growth types of phase change which rely on thermally activated atomic diffusion, martensitic transformations involve a cooperative movement of atoms resulting in a microscopically homogeneous transformation of one crystal lattice into another. The ideal martensitic process itself is not thermally activated and takes place at high temperature-independent speeds; but in practice a clear-cut separation of transformation processes into "nucleation and growth" and "martensitic" is generally not possible. Although in pure metals, such as unalloyed titanium, a simple athermal martensitic transformation is conceivable; in an alloy the situation is more complicated. The *long-range effect* of alloying, described for example in terms of changes of the elastic parameters, may simply be to change the conditions under which athermal transformation takes place. The *local effect* of alloying, on the other hand, is to inhibit the movement of atomic planes and thereby: (1) to reduce the distances over which atomic regions can cooperate, which perturbs the microstructure of the transformation product; and (2) to reduce the speed of the transformation, thus bringing it into competition with the nucleation-and-growth mechanism. The influence of solute concentration on the kinetics of the transformation can be seen in the results of the experiments of JEPSON et al. [JEP70] on a series of Ti-Nb(0-17.5 at.%) alloys. In studies of the influence of cooling rate, say $r \equiv dT/dt$, on the transformation temperature, M_s, they were able to show that the critical cooling rate, r_c (at rates higher than which M_s became independent of r) decreased with increasing niobium concentration. Thus, for example, whereas in unalloyed titanium, $r_c = 10^2$ °C/s, the addition of 15 at.% Nb reduced it to 0.3 °C/s. In considering what might be referred to in Ti-TM alloys as a "solute-controlled transformation at solute-inhibited transformation speed," in which athermal shear competes with diffusion, a given quench rate cannot be expected to yield similar transformed structures for all concentrations and all types of Ti-TM alloys. The properties of the quenched product will depend on several variables, one of the most important of which is the diffusion coefficient of the particular solute species in β-Ti. Figure 3.20, which compares the 1000 °C-diffusion coefficients of eight transition elements in β-Ti shows that a quench rate that will completely suppress diffusion in Ti-Nb ($D_{Nb} = 1.3 \times 10^{-9}$ cm²/s) and Ti-Mo ($D_{Mo} = 0.3 \times 10^{-9}$ cm²/s) may be quite inadequate to do so in Ti-Fe, Ti-Ni, and Ti-Co (for which $\langle D \rangle_{Fe,Ni,Co} = 61.6 \times 10^{-9}$ cm²/s). Against this background it is possible to believe in the existence of a series of quenched structures extending from "bulk martensite" through "acicular martensite" to the Widmanstätten structure defined below.

5.2.2 Morphology of the Martensites

When conditions are particularly favorable, transformation from β to $α^m$ takes place completely, on a large scale, and with considerable structural coherence. The result is the so-called "massive martensite" (otherwise known as packet, or lath, martensite), which consists of large irregular zones on the scale of 50 to 100 μm, subdivided into parallel arrays of fine platelets less than 1 μm across (Fig. 5.3). In massive martensite, the lack of re-

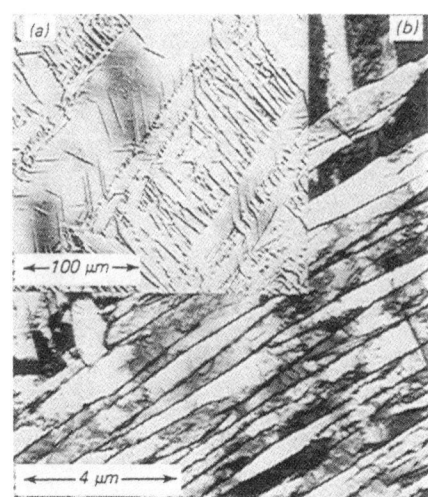

Fig. 5.3 An example of "massive martensite." Specimen: Ti-1.78Cu quenched from 900 °C (1650 °F). (a) Optical micrograph showing large colonies. (b) Electron micrograph showing individual plates within the colonies [WIL73]. Micrographs courtesy of J.C. Williams.

Table 5.2 Compositions and Electron/Atom Ratios of the α'/α" Boundary in Some Ti-TM Binary Alloys [BAG59]

	α'/α" Boundary			Conventional
Element	wt%	at.%	ΔGN(a)	e/a ratio(a)
V	9.4	8.9	1	4.09
Nb	10.5	5.7	1	4.06
Ta	26.5	8.7	1	4.09
Mo	4	2.0	2	4.04
W	8	2.2	2	4.04
Re	<10	<2.8	3	<4.08

(a) See Table 5.1 for calculation of e/a ratio in terms of Δ GN.

---

* The symbol $α^m$ is used herein as a shorthand notation for the product of martensitic transformation, whether it be α' or α". M_s refers to the start of the transformation during cooling; M_f usually designates its finish.

tained β phase prevents direct determination of the habit plane. With increasing solute concentration, the coherence between the platelets, which would otherwise make up a massive colony, is lost. The result of this is a partially disordered array of individual platelets referred to as "acicular martensite" (Fig. 5.4). Further increase in solute concentration prevents a complete transformation from taking place, and β phase trapped between the platelets of the acicular martensite enables direct habit-plane determination to be accomplished. Not far removed from the β-plus-acicular-martensite quenched structure is the Widmanstätten arrangement, consisting of groups of α-phase needles lying with their long axes parallel to the {110} planes of the parent retained β (see Fig. 5.5). Widmanstätten α + β (or Widmanstätten α, if the focus is primarily upon α-phase precipitation), which is characteristic of dilute Ti-TM alloys or near-α α + β alloys (such as Ti-6Al-4V) appropriately cooled, is usually treated as a product of α-phase nucleation and growth. It is introduced into this chapter on nonequilibrium phases since, when solute diffusion coefficients are sufficiently large, either intrinsically so (large frequency factor, D_0) or if the temperature is high, diffusional processes compete with diffusionless martensitic transformation during the quenching of β-Ti alloys [CHR65][JEP70][FLO82].

The optical microstructures of a series of β-quenched Ti-TM alloys are shown in Fig. 5.6. The five alloys Ti-Mo, Ti-V, Ti-Nb, Ti-Mn, and Ti-Fe had comparable e/a ratios and similar masses, and were ice-brine quenched under similar conditions. Noticeable in the photomicrographs of these alloys, which are arranged in ascending order of solute diffusion coefficient in β-Ti (Fig. 3.20), is what appears to be a gradual transition from a diffusionless to a diffusion-influenced as-quenched structure on proceeding from Ti-Mo to Ti-Fe. These results suggest that atomic diffusion coefficient must be introduced as a scaling factor when estimating the effects of quench rate on martensitically transformable Ti-TM alloys.

Metastable-bcc Ti-TM alloys will also transform under the application of mechanical stress. Although some confusion has arisen in the past over the structure of the deformation product, the situation has been adequately clarified by WILLIAMS [WIL73]. The β-quenched solute-lean but untransformed Ti-TM alloys are characterized by low shear moduli. In such an alloy, which may be represented in the phase diagram by a point close to M_s, the application of stress will trigger a transformation to "deformation-induced"

or "stress-induced" martensite. Farther away from the M_s transus, the bcc lattice responds directly to the influence of an applied stress by twinning. This, as pointed out by SUZUKI and WUTTIG [SUZ75], can in the present context be regarded as a special case of martensitic transformation to a crystallographically equivalent structure. Indistinguishable optically from deformation martensite, mechanical twins are easily identified by diffractometry—their structure, of course, being identical to that of the parent lattice. Stress-induced martensitic transformation and twinning in Ti-Mo alloys have been recently discussed by OKA and TANIGUCHI [OKA80]. Deformation-induced orthorhombic martensite produced by compressively deforming a β-quenched Ti-Mo (5 at.%) alloy is depicted in Fig. 5.7.

Fig. 5.4 An example of "acicular martensite." Specimen: Ti-12V quenched from 900 °C. (a) Optical micrograph. (b) Electron micrograph showing lenticular-shaped plates, some of which are internally twinned [WIL73]. Micrographs courtesy of J.C. Williams.

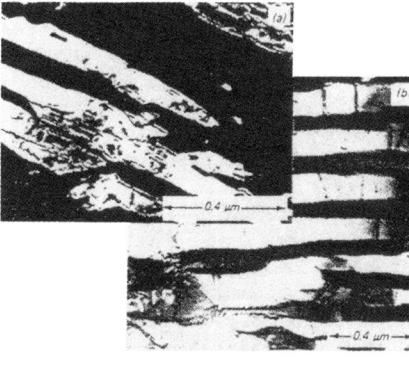

Fig. 5.5 An example of Widmanstätten α phase. This type of α phase is found in near-α α + β binary alloys and in alloys with substantial amounts of aluminum, e.g., Ti-6Al-4V [WIL73]. (a) Dark-field micrograph showing absence of internal structure in the platelets. (b) Bright-field micrograph showing the α platelets separated by the β matrix. Micrographs courtesy of J.C. Williams.

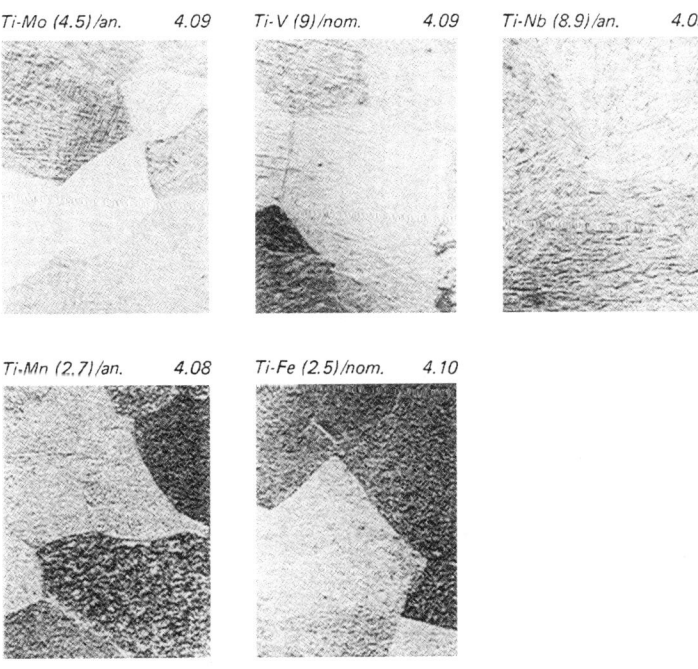

Fig. 5.6 Optical micrographs of five low-concentration Ti-TM alloys after quenching into iced brine from the β phase, arranged in ascending order of solute atomic diffusion coefficient in β-Ti (see Fig. 3.20). Indicated are the analyzed solute concentrations (nominal in the cases of Ti-V and Ti-Fe) and the conventional e/a ratio ($\cong 4.09$ = const.). Magnifications of the original 11.5×9 cm² micrographs, $50\times$.

5.2.3 Structure of the Martensites

The transformed structure assumed by the pure elements titanium, zirconium, and hafnium, and by the dilute Ti-TM alloys in general, is hcp and has been assigned the symbol α' [BAG59] [WIL73]. Otherwise, quenched Ti-TM alloys with compositions exceeding certain limits, which differ from system to system transform to an orthorhombic martensite designated by α''. The α'/α'' compositional boundaries for the alloy systems Ti-V, Ti-Nb, Ti-Ta, Ti-Mo, Ti-W, and Ti-Re have been determined by BAGARIATSKII et al. [BAG59], whose results are summarized in Table 5.2. The results of more recent studies of martensitic transformation in representative members of the Ti-V, Ti-Nb, and Ti-Mo systems by FLOWER et al. [FLO82] are in good agreement with this, α'' having been detected in Ti-10.6V(10.0 at.%),* Ti-14Nb(8 at.%), Ti-20Nb(11 at.%), and Ti-4Mo through Ti-8Mo (2 through 4 at.%). Not all authors agree that the higher-concentration Ti-V martensites are orthorhombic. WILLIAMS had not succeeded in detecting the α'' variant in Ti-11.6V(11.0 at.%) and attributed its presence, apparent or otherwise, in earlier experiments to difficulties encountered in the handling and study of thin foils in general [WIL73]. In Ti-V and the like, such difficulties are of course further exacerbated by the instabilities of the lattices under study.

Since stress-induced martensite can occur only in alloys not already transformed by quenching, it is obviously confined to the higher concentration ranges; in light of the foregoing observations on the structures of quenched martensites, it is, therefore, not at all surprising to find that deformation martensite is invariably of the orthorhombic variety [WIL73].

5.2.4 Crystallography

Details of the crystallographic $\beta \to \alpha^m$ transformation in titanium alloys have been reviewed and discussed by several authorities, notably OTTE [OTT70], HAMMOND and KELLY [HAM70], and WILLIAMS [WIL73]. The crystallographic approach focuses attention on the influence of various possible shear systems on the habit plane of the transformed plates with the hope of determining which of them are consistent with the results of habit-plane orientation measurements and, consequently, which dislocations dominate the transformation. By a process of elimination, the justification for which forms the basis for discussion and is the essence of the exercise, a multitude of possible dislocation paths are reduced to just a few. According to OTTE [OTT70], the entire $\beta \to \alpha'$ transformation process was reducible to an activation of the shear systems:

$$[111]_\beta(11\bar{2})_\beta \equiv [2\bar{1}\bar{1}3]_{\alpha'}(\bar{2}112)_{\alpha'}$$
and (Eq. 5.1)
$$[111]_\beta(\bar{1}01)_\beta \equiv [2\bar{1}\bar{1}3]_{\alpha'}(\bar{1}011)_{\alpha'}$$

It has been well established that the transformation from β to α' is characterized by a habit plane near $\{334\}_\beta$ [BLA70][WIL73][SHI77] and that the Burgers orientation relation is closely obeyed [WIL73] [SHI77] [DAV79] [FLO82]; stated in crystallographic terms, this implies: $(0001)_{\alpha'} \parallel (011)_\beta; \langle 11\bar{2}0\rangle_{\alpha'} \sim \parallel \langle 11\bar{1}\rangle_\beta$. The results of the study by DAVIS et al. [DAV79] of titanium and the alloys Ti-Mo (2, 4, 6, and 8 wt%) were entirely consistent with that general statement. Furthermore, they were able to show, for titanium and all four of the alloys, that the transformation of β to α' using the Burgers relationship can be achieved via:

- a 10% contraction along $[100]_\beta$ which corresponds to $[2\bar{1}\bar{1}0]_{\alpha'}$

- a 10% expansion along $[01\bar{1}]_\beta$ which corresponds to $[01\bar{1}0]_{\alpha'}$

- a 1% expansion along $[011]_\beta$ which corresponds to $[0001]_{\alpha'}$

The crystallography of the orthorhombic α'', as formed in quenched Ti-Nb(20 at.%), has been discussed by HATT and RIVLIN [HAT68], who concluded that the following orientation relationships existed (accuracy ± 0.5°):

- $[100]_{\alpha''}$ 2° from $\langle 001\rangle_\beta$

- $[010]_{\alpha''}$ 2° from $\langle 110\rangle_\beta$

- $[001]_{\alpha''} \parallel \langle 1\bar{1}0\rangle_\beta$

5.2.5 Transformations from and to the β Phase: Case Studies of Some Representative Titanium-Base Alloy Systems

In α-stabilized alloys the M_s temperature increases with solute content; in β-phase alloys it decreases. Thus, if the transformation has a diffusional component (temperature dependent), it will respond differently to increases in solute concentration in these two classes of systems. The martensitic transformation is characterized phenomenologically by the assignment of several temperatures: M_s, the start of the martensitic transformation during quenching; M_f, its finish; β_s, the start of the $\alpha^m \to \beta$ reversion on upquenching; T_0, the temperature at which α^m and β are in thermodynamic equilibrium (i.e., have the same free energies).

The Ti-Al System. The martensitic transformation in Ti-Al and the effect of cooling rate on it have been studied by JEPSON et al. [JEP70]. The various transformation temperatures encountered are listed in Table 5.3. For this system it was found impossible to suppress the $\beta \to \alpha'$

Fig. 5.7 An example of deformation martensite and/or twinning. Specimen: Ti-Mo(5 at.%) quenched from the β phase and deformed 23% by compression. Magnification of the original 9 × 9 cm² micrograph, 50×.

Table 5.3 Phase-Transformation and Equilibrium-Phase-Boundary Temperatures for Ti-Al Alloys [JEP70]

Al content, at.%	Transformation temperatures(a), °C (°F)			Phase-boundary temperatures, °C (°F)	
	T_0	M_s	β_s	$\alpha/(\alpha+\beta)$	$(\alpha+\beta)/\beta$
5	943 (1729)	918 (1684)	940 (1724)	925 (1697)	960 (1760)
10	1008 (1846)	960 (1760)	1010 (1850)	975 (1787)	1040 (1904)
15	1066 (1951)	1015 (1859)	1080 (1976)	1035 (1895)	1100 (2012)
20	1113 (2035)	1060 (1940)	1110 (2030)	1080 (1976)	1100 (2012)

(a) T_0 = Temperature corresponding to zero free-energy difference between the α' and β phases (calculated). M_s = Martensite start ($\beta \to \alpha'$) temperature. β_s = Martensite reversion ($\alpha' \to \beta$) temperature

---

* The spinodal decomposition of aged, previously martensitic, Ti-10.6V is good indirect evidence that its structure was originally α'' and not α' [FLO82]—see Section 5.25.

transformation at the cooling rates available, viz., $<2 \times 10^5$ °C/s. If the martensitic transformations were truly athermal, there would be no change in M_s with changing cooling rate. The fact that M_s did begin to drop at cooling rates in the vicinity of 5×10^4 °C/s indicated that the transformation was thermally activated. The closeness of M_s to β_s (Table 5.3) indicated that the driving force for the transformation was small.

The Ti-Mo System. As indicated earlier within the context of structural alloys, Ti-Mo is an important prototype system, establishing as it does the basis of several commercially interesting multicomponent formulations [OHT73], particularly β-III. Ti-Mo alloys have been the subjects of several recent investigations. DAVIS, FLOWER, and WEST studied the martensitic transformation itself [DAV79] as well as the decomposition of the Ti-Mo martensites during aging [DAV79a][FLO82], while crystallographic and microstructural studies of the stress-induced transformations have been conducted by OKA and TANIGUCHI [OKA80] and HIDA et al. [HID80], respectively.

The molybdenum concentrations studied by FLOWER et al. [FLO82] were 2, 4, 6, and 8 wt% Mo (1, 2, 3, and 4 at.%) and thereby spanned the composition, ~ 4 wt% Mo (~ 2 at.% Mo), at which the quenched structure changed from α' to α'' [BAG59]. With increasing molybdenum content there is a transition in the martensite morphology from massive to acicular (Section 5.2.2.). Ti-2Mo was already acicular with only a small proportion of the massive; the higher-concentration alloys were entirely acicular. In agreement with BAGARIATSKII et al. [BAG59], DAVIS et al. [DAV79] showed that, whereas the martensitic form of Ti-4Mo was hexagonal α', that of Ti-6Mo was orthorhombic α''. They also mentioned, with reference to earlier work, that whereas α'' was twinned, the α' was generally dislocated; but they modified the strength of this comparison by pointing out that twinning was not completely confined to the α'' variant [FLO82].

Layers of retained β were detected in Ti-2Mo and Ti-4Mo, but not in Ti-6Mo and Ti-8Mo. This observation supports the general conclusion that in dilute β-stabilized Ti-TM alloys, with their relatively high M_s temperatures, water quenching is insufficiently rapid to completely inhibit diffusional reactions. Some segregation of the molybdenum takes place, leading to the production of some untransformable β phase. It was suggested by DAVIS et al. [DAV79] that the existence of the diffusional component may have been a factor in the formation of massive martensite. As the solute concentration

increases, M_s decreases, the diffusional contribution becomes suppressed, and a full transformation to α^m is able to take place. This continues until M_f drops below the temperature of the quench bath, again enabling the retention of some β phase.

The results of the aging studies of FLOWER et al. [FLO82] were particularly interesting. It is not appropriate to describe them in detail here; however, it is useful simply to mention that, whereas α' reverts to β via a nucleation-and-growth process, the initial stage of α'' decomposition may be spinodal, to $\alpha'_{lean} + \alpha''_{rich}$, thereby yielding a characteristically modulated microstructure. Thus, through observation of the microstructure of the aged product, the nature of the original as-quenched structure can in some cases be deduced.

As indicated in Section 5.2.2, the application of stress to a metastable β-Ti-TM alloy will result in either martensitic transformation or twinning. The composition ranges of these two transformation products have been investigated by OKA and TANIGUCHI [OKA80] using a series of Ti-Mo alloys of compositions: 9.3, 10.5, 11, 13, 14, and 15.5 wt% Mo. Using x-ray diffraction and TEM they were able to demonstrate that the deformation of Ti-(9-11)Mo led to stress-induced α'', while that of Ti-(11-15.5)Mo resulted in {332} twinning in agreement with numerous earlier reported results.

The Ti-Nb System. (1) *Ranges of Occurrence of the Martensitic α' and α'' Phases.* For Ti-Nb, an upper-concentration limit for α' martensite of 6 at.% (10.5 wt%) has been reported by BAGARIATSKII et al. [BAG59] (Table 5.2). The limit of 3 at.% Nb referred to by HATT and RIVLIN [HAT68] and perpetuated by some subsequent authors may have originated in what seems to be a plotting error in Fig. 1(d) of [BAG59], although it had subsequently been corrected, by implication, in that same paper. At concentrations higher than ~11 at.% Nb, the quenched martensite structure is orthorhombic α'', which seems to persist, more or less, out to concentrations as high as 25 at.% Nb depending on the quenching conditions [HIC69a]. Indeed, the M_s transus of Fig. 5.8 intersects the 300-K axis at 26.5 at.% Nb. However, HICKMAN's rapidly quenched Ti-Nb(27 at.%) was single-phase β [HIC69a]. In other studies related to the high-concentration limit of α'', BALCERZAK and SASS [BAL72] showed that the structure of Ti-Nb(18.4 at.%), oil quenched from 900 °C, was α'', as was that of Ti-Nb(20 at.%), water quenched from 900 °C, according to BAKER and SUTTON [BAK69]. With Ti-Nb(20.7 at.%), HATT and RIVLIN [HAT68] showed that water quenching from 900 °C yielded α'' plus a trace of β. Ti-Nb(22 at.%), gas quenched at about 10^3 °C/s from 1200 ~ 1300 °C, yielded $\alpha'' + \beta$ [HIC69a], as did Ti-Nb(22.6 at.%) oil quenched from 800 °C [BAL72] and gas-quenched Ti-Nb(25 at.%) [HIC69a]. Ti-Nb(25.6 at.%) oil quenched from 900 °C was essentially $\omega + \beta$ [BAL72].

(2) *Morphology of the Ti-Nb Martensite.* In water-quenched Ti-0.5Nb(0.26 at.%), the α' martensite was predominantly massive, while that in Ti-1.0Nb(0.52 at.%) was predominantly

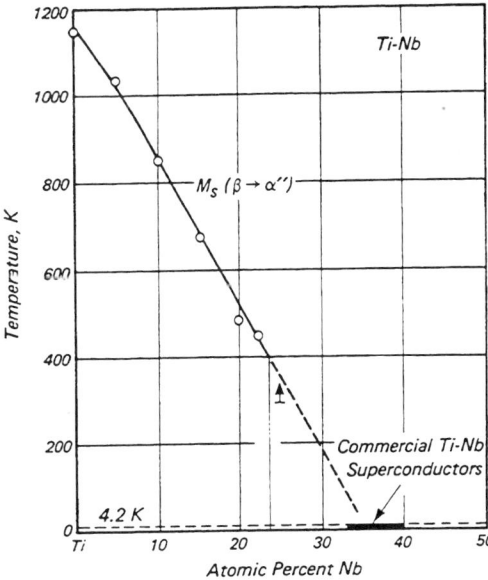

Fig. 5.8 The martensitic $\beta \rightarrow \alpha''$ transformation curve for Ti-Nb based on the data of DUWEZ [DUW53], BROWN et al. [BRO64], and BAKER [BAK71]. The extrapolation to liquid-He temperatures was suggested and discussed by KOCH and EASTON [KOC77].

acicular [FLO82]. In both cases, layers of retained β were present, an expected consequence of water quenching through the high-temperature segment of M_s. Both the occurrence and the microstructure of the quenched martensite are sensitive to the rate of cooling. The most detailed study of the effect of cooling rate was undertaken by JEPSON et al. [JEP70]. In experiments in which the cooling/quench rate was varied from 0.2 °C/s to 2×10^5 °C/s, considerable control over the microstructure of the product was possible. For example, the structure of Ti-Nb(17.5 at.%) quenched at 1.0×10^5 °C/s was equiaxed single-phase β. Naturally in less concentrated alloys, martensitic transformation was able to take place at quench rates fast enough to retain the β phase in the more concentrated ones. JEPSON et al. made the interesting discovery that the effect of increasing the niobium content was not only to decrease M_s but also to make the $β \rightarrow α^m$ transformation more sluggish. As with the Ti-Al alloys referred to above, the transformation temperatures in Ti-Nb(0-17.5 at.%) all decreased with increasing quench rates, especially at high rates, indicating that the transformations were actually thermally activated [JEP70].

(3) *Reversion and Aging of the Ti-Nb Martensites.* Numerous authors have investigated the thermal stabilities and modes of decomposition of tempered (aged) martensites. Using a thermal-arrest technique, JEPSON et al. [JEP70] have determined both the M_s temperature and the temperature, $β_s$, at which the reverse transformation commences. The results are given in Table 5.4. If M_s and $β_s$ are fairly close, the temperature T_0, corresponding to equality of the β and $α^m$ free energies, can be taken as their mean. Other workers, notably HATT and RIVLIN [HAT68], BAKER and SUTTON [BAK69], HICKMAN [HIC69a], and FLOWER et al. [DAV79a][FLO82], have considered the manner in which the $α^m$ decomposes and the natures of the products, which may be β, ω + β, or α + β, depending on the conditions. The situation can be loosely, but instructively, summarized with the aid of Fig. 5.9, a heuristic nonequilibrium/equilibrium phase diagram illustrating the relative positions of the $M_s^{α'}$ and $M_s^{α''}$ transi (after [FLO82]) and the isothermal ω + β phase region. Several classes of decomposition take place, depending on the composition of the Ti-Nb martensite and the temperature. Upon aging of the α' martensites (<11 wt%, 6 at.% Nb), the equilibrium β phase nucleates heterogeneously. As indicated in Fig. 5.9, the α'' martensites within a certain composition range are able to spinodally decompose into $α''_{lean} + α''_{rich}$ [FLO82] [DAV79a]. This

Table 5.4 Phase-Transformation and Equilibrium-Phase Boundary Temperatures for Ti-Nb Alloys

Nb content, at.%	Transformation temperatures(a) °C (°F)			β/(α + β) boundary, °C(°F)	
	M_s				
	[JEP70]	[DUW53]	$β_s$ [JEP70]	[JEP70]	[HAN51]
0	855 (1571)	855 (1571)	...	885 (1625)	885 (1625)
2.5	753 (1387)	...	...	...	...
5	720 (1328)	760 (1400)	...	760 (1400)	810 (1490)
7.5	619 (1146)	...	646 (1195)	...	...
9	567 (1053)	...	592 (1098)	...	...
10	560 (1040)	600 (1112)	540 (1004)	650 (1202)	765 (1409)
11	517 (963)	...	530 (986)	...	...
12.5	455 (851)	500 (932)	455 (851)	620 (1148)	740 (1364)
15	385 (725)	400 (752)	387 (729)	585 (1085)	725 (1337)
17.5	300 (572)	...	317 (603)	545 (1013)	705 (1301)

(a) See Table 5.3 for definitions.

process, which has been observed to take place in Ti-Nb(14-20 wt%, 8-11 at.% Nb) [FLO82], begins during the quench, continues on further aging, and eventually proceeds to α + β. Alloys sufficiently high in niobium content and aged at moderate temperatures revert to metastable β, which then decomposes into ω + β. This process has been noted in water-quenched α''-Ti-Nb(20 at.%) aged at 330 °C [BAK69], water-quenched α''-Ti-Nb(20.7 at.%) aged at 335 °C [HAT68], Ti-Nb(22 at.%) aged above 200 °C [HIC69a], and Ti-Nb(25 at.%) aged above 150 °C [HIC69a], both of the latter alloys having been previously gas quenched.

Quenched Multicomponent α + β Alloys. (1) *Ti-6Al-4V.* The various types of structures encountered in Ti-6Al-4V can be readily appreciated with the aid of Fig. 4.11. If the alloy is cooled rapidly from the bcc field, it will undergo a transformation to acicular α (referred to then as transformed β, $β_{tr}$) en route. The prior-β grain boundaries, decorated by α phase, will still be visible in the $β_{tr}$ structure where they demarcate the junctions between the rafts of α-phase needles.* If the alloy is equilibrated high in the α + β region, it will contain a large volume fraction of β phase in a concentration sufficiently low that it will again transform on rapid cooling. The lower the temperature of α + β equilibration, the smaller the volume fraction of β phase but the higher its vanadium concentration and, consequently, its β stability. Following such a heat treatment it is possible to obtain structures consisting of large equiaxed α grains with small amounts of retained β preserved intergranularly. All these features were depicted in Fig. 4.15.

(2) *Ti-6Al-2Sn-4Zr-2Mo.* The transformation kinetics of Ti-6242 have been studied by conventional quench techniques to produce TTT diagrams such as

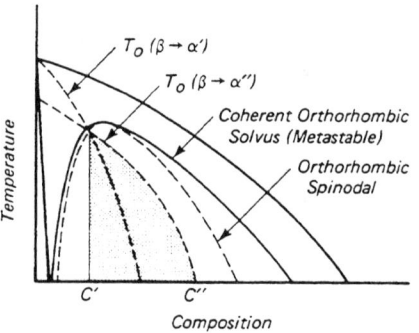

Fig. 5.9 Schematic representation of phase relationships in β-isomorphous Ti-TM alloys (i.e., Ti-V, Ti-Nb, and Ti-Mo) for the purpose of illustrating the modes of martensitic transformation and the decomposition of martensites by spinodal decomposition [FLO82]; see also [DAV79a].

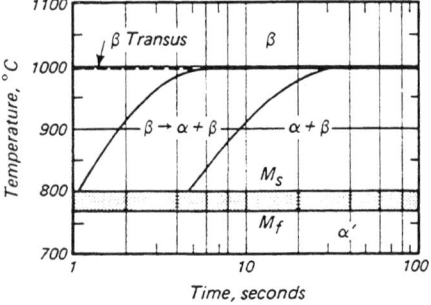

Fig. 5.10 Isothermal TTT diagram for Ti-6Al-2Sn-4Zr-2Mo [WOO72; p. 1-6:72-3].

* In alloys that contain a substantial concentration of aluminum, such as Ti-6Al-4V, α phase may nucleate directly from the metastable β in the form of plates or needles (acicular particles), aligned parallel to {110}β, yielding a "basketweave" type of structure. The Widmanstätten-α plates, as they are called, are separated by a β-phase matrix [WIL73].

Table 5.5(a) Compositions of the $\omega/(\omega + \beta)$ Data Points Expressed in Terms of Conventional Electron/Atom Ratio

Solute group number	Solute element	Saturation composition of ω phase (at.%) after aging at ~400 °C [HIC69[a]]	Solute concentration (at.%) in Ti alloys for which ω phase is formed on quenching [BAG59]	Conventional e/a based on group number (see Table 5.1)	
				"Aged data"	"Quenched data"
V	V	13.8 ± 0.3	13	4.14	4.13
	Nb	~9 ± 2	18	4.09	4.18
VI	Cr	6.5 ± 0.2	7	4.13 4.13	4.14
	Mo	4.3 ± 0.4	4.5	4.09	4.09
	W	...	7.5	4.15	4.15
VII	Mn	5.1 ± 0.2	5.5		4.16
	Re	...	4.5	4.17	4.14
VIII	Fe	4.3 ± 0.2	3	...	4.12
				Mean 4.13 ± 0.03	Mean 4.14 ± 0.03

that depicted in Fig. 5.10. Although full martensitic transformation is, in principle, possible during quenching from above the β transus (990 °C) transformation of the β phase to Widmanstätten α + β [occasionally designated (α + β)$_W$)] by nucleation and growth of α platelets takes place very rapidly. For example, α-phase precipitation will commence after the alloy has been held for a little more than 1 s at 800 °C. The properties of (α + β)$_W$, formed in a test sample of Ti-6242-0.1Si (condition: 2 h/1024 °C/air cool) in response to deformation and thermal cycling, have been examined in detail by GEGEL and colleagues [GEG80[a]]. They were able to show the transformed-β structure [designated β$_{tr}$ or (α + β)$_W$] possessed several interesting and important features: (1) a well-defined "interface phase" separating the α and β components of the Widmanstätten structure; (2) a characteristic distribution of solute species across α/interface/β (see Fig. 4.19); and (3) a variation of the composition-dependent e/a ratio across these three phases and the possible occurrence within one of them of a vanishing e/a-controlled elastic shear modulus C' (see Sections 2.3, 5.1.3, and 12.7).

The STEM-analyzed compositions of the α, β, and interface phases, as measured by GEGEL and colleagues, are given in Fig. 4.19 and Table 4.2. In order to translate composition into electron/atom ratio, use is made of the general formula:

$$(e/a) = 4.00 + \sum_{i}^{n} f_i \Delta(e/a)_i$$

where f_i is the mole (atomic) fraction of the ith solute and $\Delta(e/a)_i$ is the difference between its e/a ratio and that of Ti (= 4.00).

Table 5.5(b) Compositional Limits of the $\omega + \beta$ Phase in Quenched Alloys of Titanium with Other Transition Elements [GUS82]

Alloy		Solute concentration range, at.%		PAULING e/a-ratio range	
Solvent	Solute	Min	Max	Min	Max
Ti	V	14	21	4.14	4.21
Ti	Nb	12	20	4.12	4.20
Ti	Cr	...	11	...	4.20
Ti	Mo	6	10	4.11	4.18
Ti	Fe	4	10	4.07	4.18
Ti	Ru	5	10	4.09	4.18
Ti	Ni	7	...	4.12	...
Ti-V(2.5 at.%)	Ru	5	...	4.13	...
Ti-V(5 at.%)	Ru	3.5	7.5	4.13	4.20
Ti-V(7.5 at%)	Ru	2.5	6	4.13	4.20
Ti-V(10 at.%)	Ru	1.5	4	4.13	4.18
Ti-V(15 at.%)	Ru	...	2.5	...	4.20

For the phases and components listed in Table 4.2, then:

$$(e/a)_\alpha = 4.00 - (0.08 \times 1) + (0.005 \times 2) = 3.93$$

and

$$(e/a)_\beta = 4.00 - (0.01 \times 1) + (0.08 \times 2) = 4.15$$

Viewing this value for $(e/a)_\beta$ in light of Table 5.1, and the discussions of Sections 2.3 and 5.1.3, it is clear that the β phase separating the Widmanstätten platelets in the structure of quenched Ti-6242 (or Ti-6242-0.1Si) is just on the threshold of β stability. Furthermore, since the molybdenum concentration decreases toward the edges of the β layer, its shear modulus, C' (see Fig. 2.1), is likely to vanish close to the interface, thus creating a condition for mechanical instability there. The stability of the interface layer itself is not an issue since its structure, if analogous to that formed between the α and β phases in transformed Ti-6Al-4V [WIL82[b]], is probably α rather than β.

5.3 Occurrence and Structure of the Athermal (Quenched) ω Phase

5.3.1 Occurrence of the ω Phase

In discussing the occurrence of ω phase in titanium alloys, Fig. 5.1 makes a useful starting point. As represented therein, a narrow region exists in which ω phase appears *athermally* during very rapid quenching from the β phase. Over a broader composition range, ω phase will occur as a precipitation product of β decomposition during moderate-temperature (≤400 °C) *isothermal* aging; alloys *rapidly quenched* into this broader region are host to a "diffuse ω phase," so-called because of the existence of straight or curvilinear lines of diffuse intensity in selected-area electron diffractograms [LUH70][BAL71][SAS72]. The occurrence, composition, and structure of ω phase, and its relationship to the competing α and β phases, have been reviewed by HICKMAN [HIC69] and SASS [SAS72], who made reference to important earlier work, including the often-quoted studies of SILCOCK [SIL58] and BAGARIATSKII et al. [BAG59].

During moderate-temperature (≤400 °C) aging for several days, a metastable equilibrium ω + β state is attained, analysis of which yields the compositions of the ω (solute-lean) and β (solute-rich) phase boundaries. The ω-phase regime does not extend to zero solute concentration, but is tightly confined by the nearby lower-energy martensitic-phase field. As a result of the limited composition range of the athermal ω phase, good agreement is obtained between the "saturation composition of ω phase after aging at 400 °C" [HIC69a] and the "solute concentrations which yield ω phase on quenching" [BAG59], as shown in Tables 5.5(a) and (b). The latter is of course a difficult quantity to determine accurately, since away from the narrow zone of athermal ω phase, rapidly quenched alloys support the diffuse ω phase and more slowly quenched larger samples will contain isothermal ω.

Tables 5.5(a) and (b) give the conventionally calculated e/a ratios for ω phase, separated into two listings according to the source and the method of data acquisition. It is remarkable to note that in each case the critical e/a ratios for the eight alloys listed are constant [within 20% in terms of $\Delta(e/a)$], and that the two independent mean values (which differ by <0.01) agree within experimental scatter. Secondly, a comparison of Table 5.1 with Tables 5.5(a) and (b), which yield, respectively, $\langle e/a \rangle_{\alpha^m} = 4.15 \pm 0.03$ and (from an overall mean) $\langle e/a \rangle_\omega = 4.13 \pm 0.03$, emphasizes that athermal ω phase occurs at the threshold of martensitic transformation and suggests that the α and α^m transformations are interrelated through a common electronic mechanism. The results also reconfirm the validity of Fig. 5.1.

The results of a literature study of collected transformation data for nine Ti-TM alloys, conducted by LUKE et al. [LUK64], independently of that of BAGARIATSKII et al. [BAG59] and prior to that of HICKMAN [HIC60a], and indexed in terms of an e/a ratio based on PAULING valence [PAU56] rather than group number, led to a similar conclusion. In conducting this survey of the limits of bcc stability, LUKE et al. [LUK64] chose not to distinguish between martensite and ω phase as the transformation product. But after reviewing the original references cited therein and separating the results into "α^m-limited" and "ω-phase-limited" β phase as in Table 5.6, one is led again to the conclusion—supported by the data of Tables 5.1 and 5.5(a) and (b) considered jointly—that: (1) in terms of e/a, the threshold for the α^m transformation (presumably $M_s \equiv$ room temperature) is constant to within about ± 0.03; (2) the composition of athermal ω phase expressed as an electron/atom ratio is also constant to within about ± 0.03; and (3) the martensitic and ω-phase "thresholds" agree to well within that scatter.

The significance of e/a as an index of the limit of bcc stability in Ti-TM alloys was emphasized by LUKE et al. [LUK64], who were probably the first to point explicitly to the importance of electronic factors in the stability of bcc Ti-TM alloys.

The presence of athermal ω-phase particles, too small (~20 to 40 Å) to be detected optically, can be unequivocally diagnosed using TEM and selected-area electron diffraction (SAD). By way of example, their occurrence in ice-brine-quenched Ti-Mo (5 at.%), which is just on the edge of the quenched martensitic regime (Fig. 5.2), has been confirmed using these techniques, the results of which are shown in Fig. 5.11.

The occurrence of athermal ω phase in such close proximity to the martensitic phase boundary is a phenomenon of general validity and of fundamental importance in the theory of the ω transformation (to be considered later in more detail). Although from the standpoint of the crystallographer the absence of a habit-plane description precludes the defining of the ω-phase transformation as martensitic [WIL73], both transformations, according to SUZUKI and WUTTIG [SUZ75] and CLAPP [CLA73], possess a common ingredient in

Table 5.6 Solute Concentrations and Electron/Atom Ratios for 100% β Stabilization in Ti-TM Alloys—After LUKE et al. [LUK64]

Solute group number	Solute element	Critical solute concentration, at.%	PAULING valence [PAU56]	Conventional e/a based on group number (see Table 5.1)	PAULING e/a	Phase with respect to which the β phase was considered stable(a)	Ref
V	V	15	5	4.15	4.148	ω	[BRO55][SIL58]
	Ta	15-21(b)	5	4.15	4.148	m	[MAY53]
VI	Cr	7.5	6	4.15	4.147	ω	[SPA58]
	Mo	7.2	6	4.14	4.143	m	[DEL52]
		7.4	6	4.15	4.148	ω	[SIL58]
	W	7	6	4.14	4.14	m	[OTS61]
VII	Mn	5.5	6	4.17	4.118	ω	[JAF58]
		7.5	6	4.23	4.148	m or ω	[FRO54]
		8	6	4.24	4.158	ω	[LER60]
VIII	Fe	6-8(c)	6	4.28	4.139	ω	[POL55]
	Co	6	6	4.24	4.12	m	[SWA58]
		7.5	6	4.29	4.148	ω	[YAK61]
		5-7(c)	6	4.24	4.12	m	[ORR55]
	Ni	7	6	4.28	4.14	m	[MAR53]
		7-8(c)	6	4.29	4.15	ω	[BAR60]

Mean conventional $e/a \rightarrow 4.20 \pm 0.06$ or 4.22 ± 0.06
Mean PAULING $e/a \rightarrow 4.14 \pm 0.01$ or 4.14 ± 0.01

(a) That is, whether an M_s composition (m) or absence of ω phase (ω) was considered. (b) The lower value is taken. (c) Mean taken

the form of a soft-phonon instability. Arguments supporting this view are developed in subsequent sections.

The development of our current understanding of ω-phase precipitation in titanium-base and zirconium-base alloys can be traced with the aid of Fig. 5.12, a literature survey in diagrammatic form. During the 1950s, ω-phase precipitation was shown to occur athermally over a very narrow composition range during rapid quenching from the elevated-temperature β field of numerous Ti-TM alloys. Athermal ω phase is a diffusionless transformation product and as such cannot be suppressed no matter how rapid the quench rate (at least up to 1.1×10^4 °C/s) [DUE80]. Early x-ray studies showed its structure to be either hexagonal (SILCOCK [SIL58]) or trigonal (BAGARIATSKII et al. [BAG59]), depending (as a result of more recent work) on solute concentration. The compositions of athermal ω phase in Ti-V, -Nb, -Cr, -Mo, -W, -Mn, and -Re, as determined by BAGARIATSKII et al. [BAG59] on quenched alloys using x-ray techniques, are presented in Table 5.5(a). They can be seen to be in excellent agreement with results obtained a decade later by HICKMAN [HIC69a] in his studies of the constitutions of aged alloys. GUSEVA and DOLINSKAYA [GUS82], applying x-ray techniques to quenched alloys, extended the range of materials first investigated by BAGARIATSKII et al. [BAG59]. The results of this series of measurements, which included an investigation of some ternary alloy systems, are presented in Table 5.5(b).

In a series of investigations using single-crystal x-ray techniques, HICKMAN [HIC68, HIC69, HIC69a] had studied the occurrence of ω phase in the systems Ti-V, -Cr, -Fe, -Co, and -Ni, paying particular attention to the composition range over which it appeared during quenching, and to the effects of isothermal aging which led eventually to the establishment of an "equilibrium" ω + β state. During this period of renewed interest in ω phase, between about 1968 and 1972, the use of TEM and electron diffraction led rapidly through a period of phenomenological exploration, with which the names of BRAMMER and RHODES (Ti-Nb [BRA67]), BLACKBURN and WILLIAMS (Ti-V, Ti-Mo [BLA68]), and KRAMER and RHODES (Ti-Nb [KRA67]) are associated, to an elegant series of quantitative investigations by SASS and his students of the systems Ti-Zr [SAS69], Zr-Nb [DAW70], Ti-V [MCC71], and Ti-Nb [BAL72], which together with an important investigation of Ti-Mo by DEFONTAINE et al. [DEF71] laid the groundwork for our current understanding of athermal and "diffuse" quenched ω phase, and indeed the entire ω-phase precipitation phenomenon.

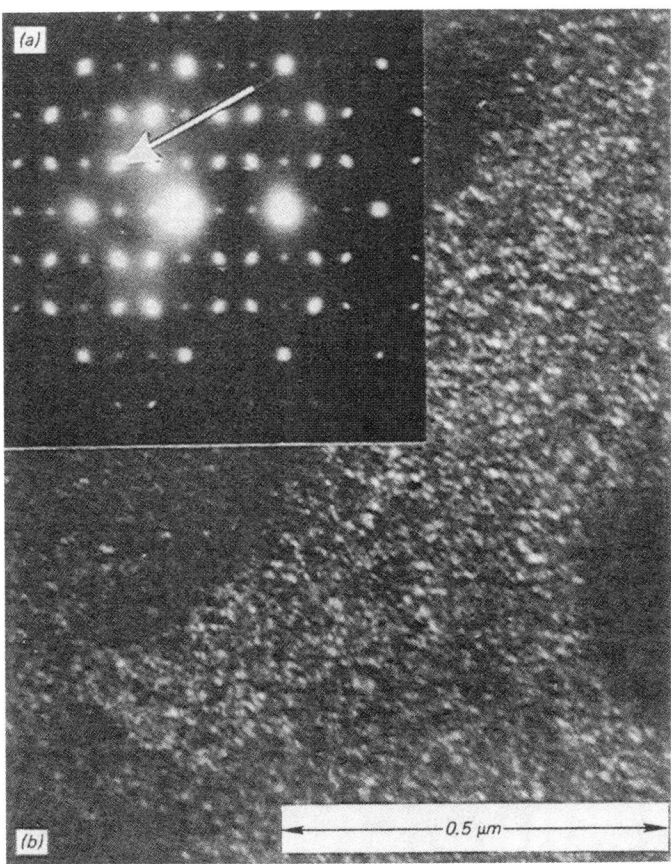

Fig. 5.11 Transmission electron micrographs of Ti-Mo(5 at.%) quenched into ice brine from the β field. (a) Electron diffractograph showing a superposition of two principal spot patterns: a rectangular arrangement of round spots originating from the β matrix, and groups of elongated spots originating from the ω phase. (b) Dark-field electron micrograph of the ω-phase precipitate originating from the diffraction spot indicated by the arrow in the diffractograph. Using the dark-field technique, the bright patches of the photograph are specific to the ω phase; however, only one-quarter of the precipitate is visualized at one time [COL71d].

The morphology of the ω phase has been investigated by BLACKBURN and WILLIAMS [BLA68, BLA70], who have shown that it forms as cubes when in the Ti-(3d)TM alloys Ti-V, Ti-Cr, Ti-Mn, and Ti-Fe, and as ellipsoids in the Ti-(4d)TM alloys Ti-Nb and Ti-Mo. The precipitate morphology is related to the Ti/TM atomic-volume ratios, hence to lattice misfit. Thus, it turns out that when the misfit is low (< 0.5%), the precipitate is ellipsoidal [BLA68][HIC69a]; otherwise, it is cubic.

5.3.2 Linear-Fault Mechanism for Athermal ω Phase

It has been well established that the ω phase is related to the bcc parent crystal according to $(0001)_\omega \parallel (111)_\beta$; $[21\overline{1}0]_\omega \parallel [110]_\beta$ implying the existence of four variants of it with respect to the bcc lattice [SIL58][BAG58][BAG72]. Omega phase may be regarded as being developed within that lattice by applying to pairs of adjacent $(110)_\beta$ planes equal and opposite shears, in the $\langle 111 \rangle_\beta$ direction, through distances about equal to one-sixth of the separation of the $(111)_\beta$ planes. The arrows in Fig. 5.13 indicate the planes, or rows, of atoms involved and the direction of the shears required. The coherent β/ω interface at the boundary $(110)_\beta$ plane is an important feature. If z is the separation of the $(111)_\beta$ planes $(= a\sqrt{3}/2$, where a is the bcc lattice parameter) then: (1) displacements of the A and B atoms by the amounts $\pm(1/6)z$ lead to the hexagonal structure of Fig. 5.13 proposed originally by SILCOCK [SIL58] as a result of measurements on a low solute-content ω phase; and (2) displacements of $\pm 0.15z$ yield the trigonal structure originally proposed by BAGARIATSKII (see [BAG59]) and now understood to be characteristic of higher-solute-content ω phases.

Assuming touching hard spheres of constant diameter (φ) in the simple geometrical model of Fig. 5.13, the separation of the $(110)_\beta$ planes, originally $2\sqrt{2}\,\phi$, shrinks to $(1 + \sqrt{3})\,\phi$ upon transformation to ω phase, a 3% contraction, which justifies the tendency for ω to form in titanium

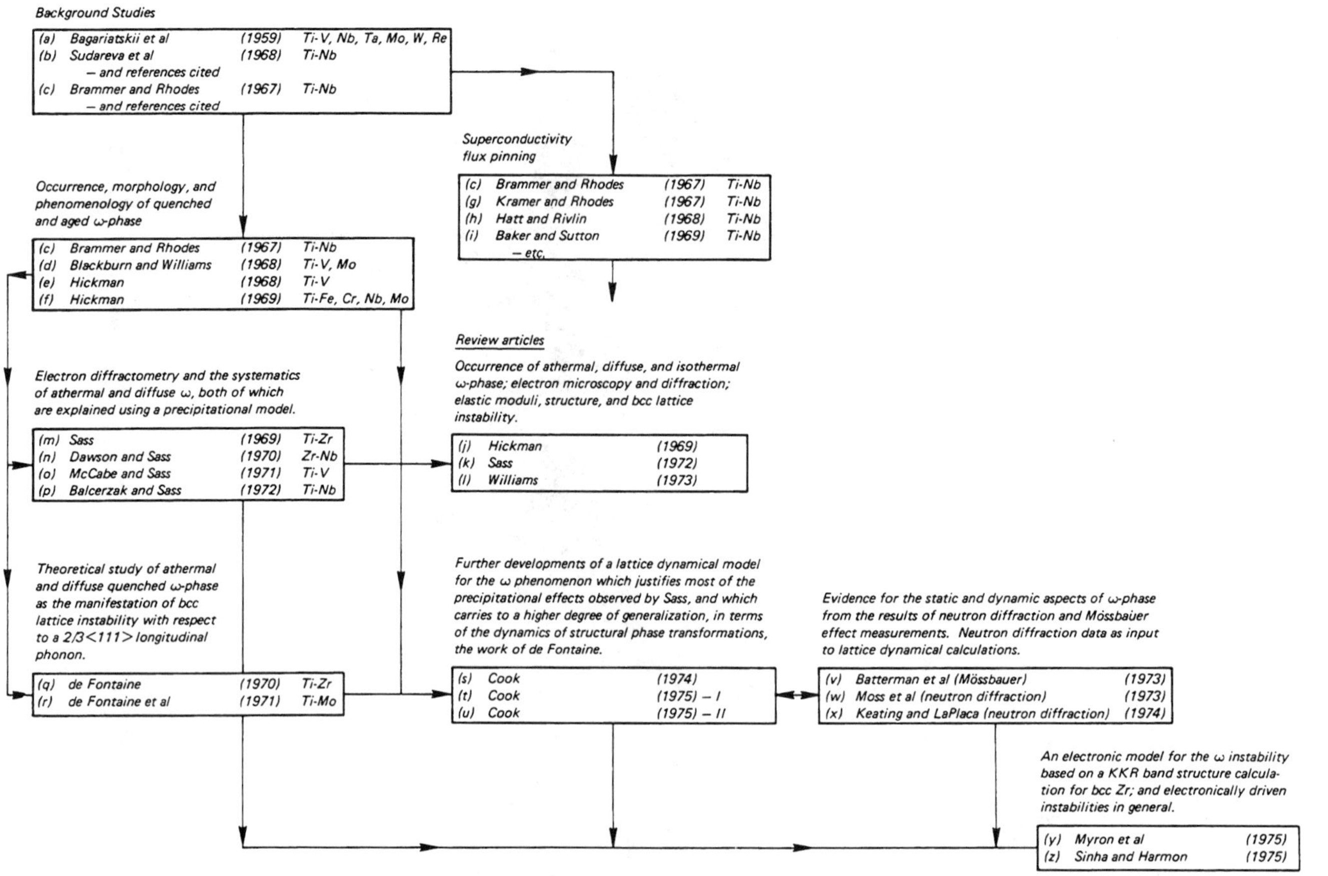

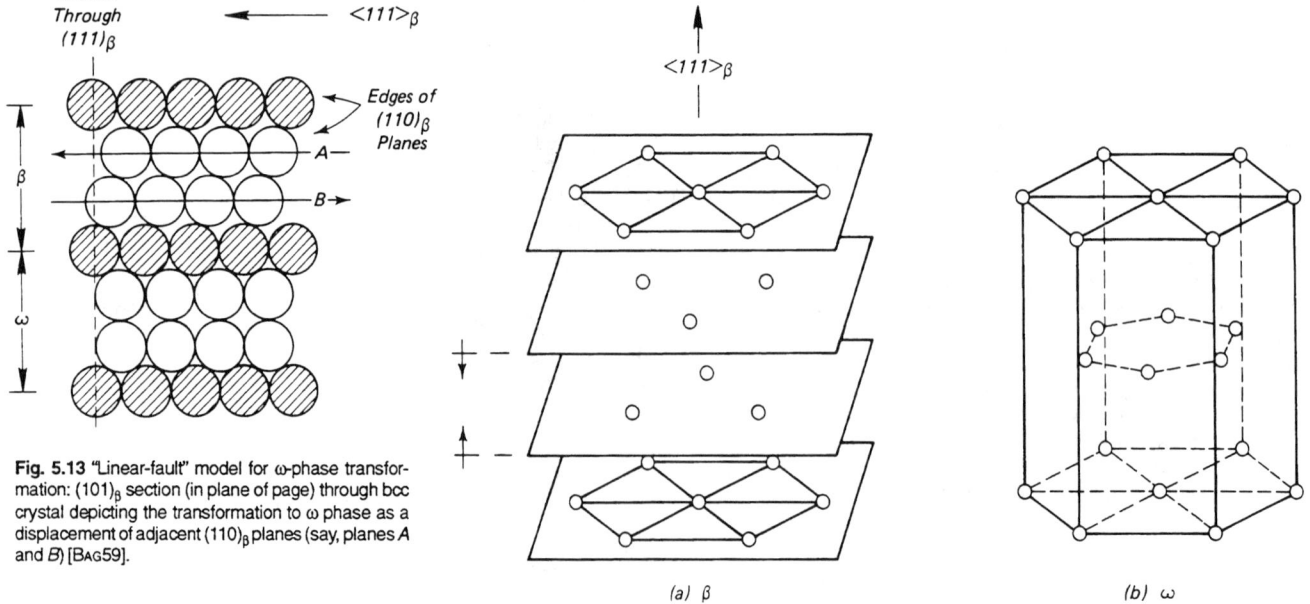

Fig. 5.12 Interrelationships among studies of ω-phase precipitation in Ti-TM alloys. The literature cited is: (a) [Bag59], (b) [Sud68], (c) [Bra67], (d) [Bla68], (e) [Hic68], (f) [Hic69ᵃ], (g) [Kra67], (h) [Hat68], (i) [Bak69], (j) [Hic69], (k) [Sas72], (l) [Wil73], (m) [Sas69], (n) [Daw70], (o) [Mcc71], (p) [Bal72], (q) [Def70], (r) [Def71], (s) [Coo74], (t) [Coo75], (u) [Coo75ᵃ], (v) [Bat73], (w) [Mos73], (x) [Kea74], (y) [Myr75], (z) [Sin75].

Fig. 5.13 "Linear-fault" model for ω-phase transformation: $(101)_\beta$ section (in plane of page) through bcc crystal depicting the transformation to ω phase as a displacement of adjacent $(110)_\beta$ planes (say, planes A and B) [Bag59].

Fig. 5.14 "Linear-fault" model for ω-phase transformation. The same displacement process as that depicted in Fig. 5.13 is represented here, but viewed from a different direction.

and zirconium under pressure [Jay63]. Adoption of this conventional crystallographic approach enabled most of the gross features of quenched ω phase, as determined by TEM and electron diffraction [Sas69][Daw70][Mcc71][Bal72], to be interpreted. Figure 5.13 suggests a ⟨111⟩ texture to the ω precipitation. Indeed, the model proposed by Sass et al. [Daw70][Mcc71][Bal72] was based upon

⟨111⟩ rows of particles, 10 to 15 Å in diameter and 15 to 25 Å apart. The way in which the displacements depicted in Fig. 5.13 lead to the collapsing together of selected $(111)_\beta$ planes, and hence to the ω transformation, is depicted in Fig. 5.14.

According to the static-particle model, athermal ω phase, whose electron diffractograms (Fig. 5.15) are characterized by sharp spots and straight "lines of intensity," was made up of clusters of such rows, while the broad reflections and either the straight or curved lines of intensity ("diffuse streaking") of "diffuse ω" were supposed to originate from either individual rows of particles or isolated particles, respectively.

However, the reversible nature of ω precipitation under temperature cycling between room temperature and 100 K calls for more than a static crystallographic interpretation. Figure 5.15 compares the composition dependence of the electron diffractograms of Ti-Nb with the temperature dependence of those of Ti-Mo. It seems that the effect of lowering temperature is similar to that of lowering solute concentration, in that in both cases curvilinear lines of diffuse intensity become straight and well defined. This figure serves to demonstrate, moreover, the reason for the uncertainties and arguments that have been associated with the assignment of compositional limits for athermal-ω-phase formation. As pointed out by WILLIAMS [WIL73]: (1) since the diffuse streaking tends to coincide with the positions of the ω-particle reflections when they are present, there is no sharp line of demarcation separating the regions of athermal and diffuse ω; and (2) the reversibility of the ω makes the specification of temperature particularly important, especially when relating structure to low-temperature physical properties such as the superconducting transition temperature. The soft-phonon mechanistic model of the ω-phase effect, originating with the work of DEFONTAINE [DEF71] and developed more fully by DE-FONTAINE et al. [DEF71], provided a satisfactory rationalization, in lattice-dynamical terms, for both the temperature- and composition-dependences of the athermal- and diffuse ω phases.

5.3.3 Phonon Mechanism for Athermal β Phase

The dynamical equivalent of the linear-fault crystallographic model centers about the proposed existence of a longitudinal phonon propagating in the ⟨111⟩ direction. The wavelength necessary to achieve (with the aid of anharmonicity) the necessary displacements of the A and B atoms (Fig. 5.13) is illustrated in Fig. 5.16, which represents a unit cell of the earlier figure. Clearly, if the shaded atoms are to remain unmoved, while A and B are to be shifted in opposite directions as shown, a longitudinal wave of wavelength equal to the separation of the $(111)_\beta$ planes is needed, viz., a longitudinal pho-

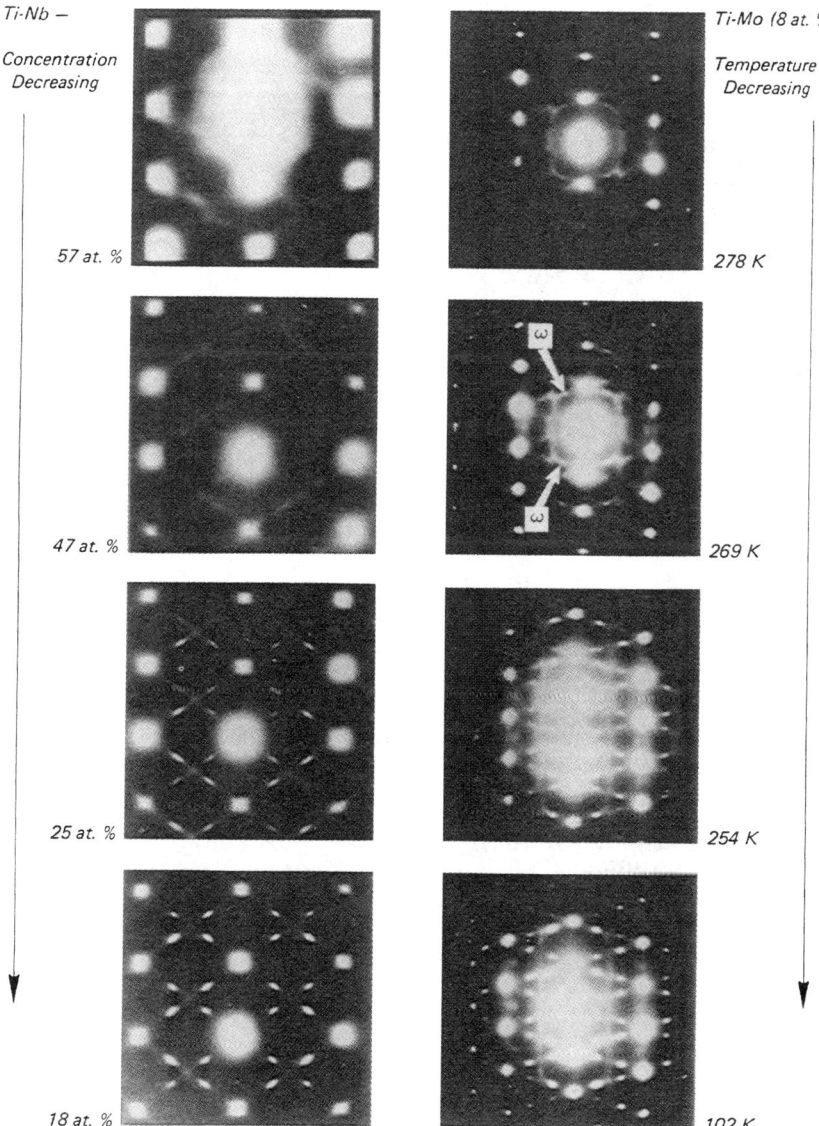

Fig 5.15 Changes from diffuse to sharp ω reflections from quenched Ti-TM alloys in response to either a decrease in solute content or a decrease in temperature. Left side: As-quenched Ti-Nb alloys in the (110) reciprocal-lattice section [BAL71, BAL72,] [SAS72]; photographs courtesy of S.L. Sass (Cornell University). Right side: As-quenched Ti-Mo(8 at.%) in the (131) reciprocal-lattice section [DEF71]; photographs courtesy of J.C.Williams.

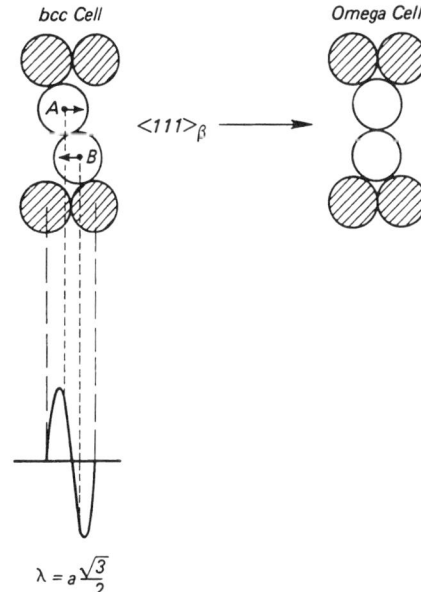

Fig. 5.16 Schematic representation of bcc → ω transformation induced by application of a 2/3 <111> longitudinal displacement wave to the bcc lattice [SAS72].

non with wave-vector $\frac{2}{3}\langle 111\rangle$. It is the instability of the bcc lattice to this disturbance that is responsible for the thermal transition. An application of the longitudinal-phonon model to the situation depicted in Fig. 5.14 leads to the possibility of a continuum of incomplete-ω states intermediate between the original β and the completely transformed ideal-ω phases. This is illustrated in Fig. 5.17.

If the crystal is removed from a region of instability by variations of either its composition or its temperature, the $\frac{2}{3}\langle 111\rangle$ phonon is responsible for the so-called reciprocal-lattice "streaking" effect, also known as "diffuse" ω (Fig. 5.15). With particular reference to the Ti-V system, Table 5.7 summarizes and compares the lattice-dynamical and linear-fault approaches as they would apply to quenched ω phases in Ti-TM alloys in general. Both philosophies are in agreement in the athermal-precipitate regime in that they represent, respectively, "cause" and "effect." They do not, however, agree in the diffuse regime: since diffuse-ω domains cannot be imaged in the electron microscope, the linear-fault model is incapable of explaining the curvilinear diffraction lines of intensity. The phonon model does so by invoking the coupling of displacement modes [DEF71]—the combined process being represented by a hyperbola asymptotic to the straight lines of the independent processes.

Further developments and generalization of the lattice-dynamical model for phase stability, with particular references to the quenched ω-phase phenomenon, have been made by COOK [COO75, COO75a] with help from the observations of SASS and DEFONTAINE, and the results of the neutron diffraction studies of Zr-Nb alloys by MOSS, KEATING and AXE [MOS73][KEA74]. Lattice fluctuation effects in the region of diffuse ω phase have been probed using both neutron diffraction [MOS73][KEA74] and Mössbauer-effect measurements [BAT73]. There seems little doubt that the $\frac{2}{3}\langle 111\rangle$ soft mode, already present in the pure "solute" elements, interacting with a lattice of temperature- and composition-dependent relative stability, is responsible not only for the athermal precipitate but also for the diffuse ω, which represents to varying degrees dynamical fluctuations between the crystalline bcc and ω phases.

It is important to remember that even the lattice-dynamical model is phenomenological, in the sense that a virtual-crystal approximation (identical atoms) is assumed and the electronic properties are disguised as force constants. A more fundamental understanding of the transformation requires an examination of what is generally referred to as the "electronic structure" of the alloy. It is neither possible nor appropriate to deal with the problem in this space. Suffice it to mention that SINHA and HARMON [SIN75] have constructed a dielectric screening model for

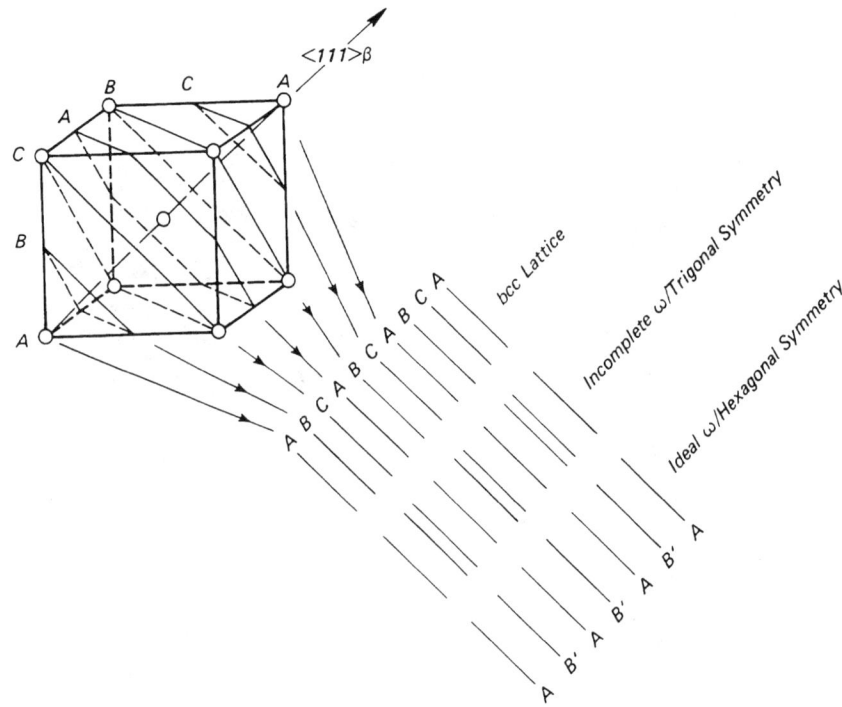

Fig. 5.17 Representation of the bcc → ω transformation in terms of the displacement-wave mechanism. The application of this mechanism (see also Fig. 5.16) raises the possibility of an incomplete ω phase.

Table 5.7 Electron Diffraction and Lattice-Dynamical Approaches to ω-Phase Formation in Group-IV-Base Binary TM Alloys
The Ti-V system has been selected as an example. The headings of the last two columns of the table indicate consecutively: technique, author, class of alloy, approach.

Typical solute concentration ranges in Ti-V	Diffraction effects as seen in electron microscopic studies of quenched alloys [SAS72]	•Electron microscopy •SASS [SAS72] •Real alloy •Static(a)	•Lattice-dynamical calculations •DEFONTAINE [DEF70, DEF71] •Virtual crystal •Dynamical(a)
13 at.% V	Sharp ω reflections and straight lines of intensity	Clusters of rows of particles plus individual rows	Propagating (long-range) $\frac{2}{3}\langle 111\rangle$ longitudinal wavelength of $(\sqrt{3/2}) \times$ (bcc lattice parameter)
13, 15, and 19 at.% V	Broad ω reflections, and straight lines of intensity	Rows of 20-Å particles; spaced 22 Å, typically 5 to a row	$\frac{2}{3}\langle 111\rangle$ vibrational modes tending to become more localized and less coherent
25 at.% V	Broad ω reflections, lines of intensity curved	Isolated 15-Å particles	
25-55 at.% V	Weak, broad lines of intensity, becoming diffuse and circular as solute concentration increases	Not intrepretable in terms of precipitation	Curvilinear diffuse reflections due to pronounced coupling between local vibrational modes

(a) These two columns are coupled by the following: (1) the dynamical effect yields a static (observable) atomic displacement through anharmonicity; (2) the ω phase becomes "frozen in" as a result of solute-solvent interdiffusion in a real alloy.

the treatment of general so-called "electronically driven" instabilities, while MYRON et al. [MYR75] have, with the aid of KKR calculations of the band structure and Fermi surface of bcc zirconium, and some assumptions regarding the perturbation of this structure during the addition of niobium, directly addressed the subject of the ω instability.

5.3.4 Relationship Between the Martensitic and ω-Phase Transformations

Underlying the phenomenological crystallographic theory of the martensitic transformation are several levels of semimechanical, mechanistic, and, finally, thermodynamic interpretations. Earlier in this section it was shown that similar e/a ratios characterized the limit of bcc stability with respect to either the ω-phase or martensitic transformation. The underlying common ingredient is a localized soft-phonon mode whose relationship to those two types of transformation can be traced through a series of papers by DEFONTAINE et al. [DEF70, DEF71], COOK [COO73, COO74, COO75a], CLAPP [CLA73], and SUZUKI and WITTIG [SUZ75]. In the last two papers a formalism was developed for comparing martensitic transformation to spinodal decomposition, which in turn has been coupled by others to ω-phase precipitation. A formal identification is achieved by comparing, in macroscopic terms, composition fluctuations with displacement fluctuations (i.e., deformation). Thus, just as spinodal decomposition, with which the names of HILLERT [HIL61] and CAHN [CAH61] are associated, can be treated as a free energy, g, instability with respect to a fluctuation in composition, c, the threshold of which is defined by $(\partial^2 g/\partial c^2) = 0$, the martensitic transformation, according to CLAPP [CLA73] and SUZUKI et al. [SUZ75], can be triggered by strain, ε, fluctuations in the vicinity of a strain spinodal, $(\partial^2 g/\partial \varepsilon^2) < 0$. The approach taken by CLAPP to describe the initial nucleation of the martensitic transformation commenced with an expression for the free-energy density as a function of strain to third order in the elastic constants. For a cubic lattice, the vanishing of one or more elements of a 6 × 6 matrix representing the second derivatives of g with respect to strain, viz.:

$$\partial^2 g/\partial \varepsilon_I \, \partial \varepsilon_J \equiv g_{IJ}(\varepsilon) \quad (I, J = 1, ..., 6)$$
(Eq 5.2)

defines a strain spinodal for the corresponding directions in strain space. In lattices which are prone to transform, g_{IJ}, already small, may be forced to zero by the application of a few percent strain (the deformation-martensite effect). The condition for static instability may also be approached via the introduction of lattice imperfections such as point and line defects, and even the surface itself, which may place the crystal close to one or another of its strain spinodals; an appropriately applied dynamic strain, in the form of one-half cycle of a lattice vibration, will then be able to nucleate a burst of transformation.

The discussion of the previous section led to the conclusion that the ω-phase and martensitic transformations are closely related. A connection between them in terms of soft-mode lattice dynamics is now apparent. As we shall see, the DEFONTAINE theory [DEF70, DEF71] of the ω phase required the bcc lattice to be unstable with respect to longitudinal phonons centered about $\frac{2}{3}\langle 111 \rangle$.

Finally, on general thermodynamic grounds, it can be shown that after quenching through an equilibrium phase boundary, the boundary of a metastable phase is encountered, just outside of which a region of statistical fluctuations is expected. Near a *structural* transformation, *strain* fluctuations would be appropriate. It is known from the neutron diffraction studies of MOSS et al. [MOS73] and the Mössbauer-effect measurements of BATTERMAN et al. [BAT73] that ω phase possesses a fluctuating component. It is, therefore, tempting to suggest that the dynamic (or "diffuse") ω is the critical opalescence of the athermal precipitate and/or martensite.

5.3.5 ω-Phase Precipitation in Binary and Multicomponent Alloys: Case Studies of Some Representative Titanium-Base Alloy Systems

(1) The Ti-Nb System. The occurrence and structure of the athermal ω phase in Ti-Nb has been investigated most thoroughly by BALCERZAK and SASS [BAL71, BAL72]. Some of their results have been presented in Fig. 5.15, a series of diffraction patterns that depict a gradual change in the ω reflections from sharp to diffuse with increasing niobium concentration. According to those authors, alloys of composition between 18 and 25 at.% Nb give rise to sharp ω reflections. This result agrees with that of BAGARIATSKII et al. [BAG59], who noted that the threshold of ω formation in quenched alloys was 18 at.% Nb. HICKMAN [HIC69a], as the result of a series of aging experiments, had shown that at 450 °C the isothermal ω phase eventually attained compositions of 6 to 11 at.% Nb, which should correspond to the isothermal threshold. In another investigation, GUSEVA and DOLINSKAYA [GUS82] noted, as a result of x-ray and other measurements on quenched alloys, that the minimum and maximum critical concentrations for the appearance of athermal ω phase were, respectively, 12 and 20 at.% Nb.

It is especially difficult to make categorical statements about the occurrence of ω phase in higher-concentration alloys, to some extent because the quenched product depends not only on the quench rate (as is generally true) but also on the temperature at which the sample is held immediately prior to the quench [BAL72]. Thus, whereas the structure of Ti-Nb(18.4 at.%) was ω + β after oil quenching from 1000 °C, it transformed to α″ when quenched from 900 °C. Likewise, Ti-Nb(22.5 at.%) became ω + β during a quench from either 900 or 1000 °C and α″ + β if the prequench temperature was reduced to 800 °C. From a heat-content and heat-transfer standpoint, it is reasonable to expect that the actual quench rate experienced by the initially hotter sample should be the lower one. Thus, in terms of their effects on the quenched microstructure there may not be very much of a distinction between "quench rate" and "prior annealing temperature." HICKMAN [HIC69a] achieved relatively high cooling rates (10^3 °C/s) during the He-gas quenching of resistively heated strips of alloy. In so doing, he was able to suppress ω phase in all the alloys examined, viz., Ti-Nb (≥22 at.%). In particular, Ti-Nb(22 at.%) yielded α″ + β, in agreement with the results of the 800 °C quench of BALCERZAK and SASS [BAL72]; gas-quenched Ti-Nb (25 at.%) was also α″ + β, whereas the structure of Ti-Nb (25.6 at.%) oil quenched from 900 °C (by BALCERZAK and SASS [BAL72]) was essentially ω + β. HICKMAN's Ti-Nb(27 at.%) was found to be "all β" [HIC69a], while for Ti-Nb (34 at.%) the quenched structure obtained by BALCERZAK and SASS [BAL72], although designated "β," yielded a SAD pattern exhibiting so-called "lines of intensity." These lines, which are also present in association with ω reflections in the lower-concentration alloys, persisted in decreasing intensity through Ti-Nb(57 at.%) and have actually been noted in pure niobium [SAS72]. It is permissible to regard the lines of intensity as arising from "diffuse ω" since they are believed to originate from the influence on the bcc lattice of that same vibrational state [SAS72] which gives rise, in the lower-concentration alloys, to the athermal ω phase itself.

(2) The Multicomponent β Alloys. The multicomponent β alloys have been discussed in Section 4.5.4 with particular reference to their equilibrium properties, which are of course achieved only after aging. The β alloys result from the inclusion in their formulations of sufficient β stabilizer to drop their M_s temperatures to well below room temperature. The properties of quenched multicomponent β alloys are comparable to those of their binary counterparts, e.g., Ti-12Mo in the case of β-III [BOY74]. It has long been known that the presence of solutes such as aluminum have a retarding influence on the β → ω reaction [JAF58, p. 141]. Thus, although the presence of iron and aluminum in Ti-10V-2Fe-3Al and of zirconium and tin in Ti-11.5Mo-6Zr-4.5Sn (β-III) does not completely prevent the formation of ω phase during quenching from above the β/(α + β) transus [DUE80^a][RAC70], the fact that the ω reflections in the latter alloy are diffuse rather than sharp, as they are in Ti-11.6Mo itself [BLA68], indicates that the simple metals do inhibit the transformation. Since the diffusionless athermal ω transformation is a response to bcc lattice instability, this inhibiting effect by SM additions in general [WIL71] is presumably a result of their increasing the stiffness of the bcc lattice.

(1) Ti-10V-2Fe-3Al. The commercial alloy Ti-10V-2Fe-3Al (i.e., "Ti-10-2-3"), whether or not it is "near-β" [TOR80] or "β" [DUE80^a], is certainly a modified Ti-V alloy just on the threshold of β stability. Its e/a ratio of 4.11_4* identifies it with Ti-V(11 at.%) whose β phase is not fully retained on quenching [MOL65, p. 14][COL75^b]. According to TORAN and BIEDERMAN [TOR80], the M_s temperature of Ti-10-2-3

---

* $(e/a) = 4.00 + (0.0954 \times 1) + (0.0186 \times 4) - (0.0559 \times 1) = 4.11_4$.

is 555 °C, and quenching from above the β transus (788 °C), yields both martensite and ellipsoidal ω-phase particles. On the other hand, DUERIG et al. [DUE80^a] claimed that quenching into water, presumably at room temperature, yielded β plus athermal ω phase. If, following DUERIG et al., room temperature lies between M_d (the transus for deformation martensite) and M_s, it is possible that mechanical stress, if it takes place during quenching, could be responsible for a partial martensitic transformation.

In any case, whether quenching stress or deliberately applied postquench mechanical stress is responsible for a martensitic transformation in this alloy, it is interesting to note that the transformed structure and the ω phase are coexistent. Thus, as DUERIG et al. [DUE80^a] have noted, the transformation to deformation α″ leaves the ω particles intact; of course, the ω phase is equally durable during mechanical twinning. In other systems it has been noted that martensitic transformation obliterates the ω phase. The energetics of its stability in this case have been discussed by DUERIG et al. [DUE80^a].

(2) β-III. The commercial alloy β-III (Ti-11.5Mo-6Zr-4.5Sn) is generally used in the α + β aged condition since athermal ω in the quenched alloy, and isothermal ω in an alloy aged in service, are precipitation hardeners which, although they increase the flow stress, reduce the ductility (i.e., they embrittle the alloy). Consequently, there has been little inducement to investigate the fundamental properties of as-quenched β-III. One such study has, however, been undertaken by RACK et al., the results of whose x-ray, TEM, and other observations of quenched and deformation-transformed material have been reported in considerable detail [RAC70]. Quenching of β-III from above the β transus (755 °C) resulted in a bcc lattice and a finely dispersed ω-phase precipitate with a mean particle size of <30 Å. No trace of martensitic transformation was observed either as a result of quenching into water, or after subsequent quenching into liquid nitrogen, indicating that M_s was below 77 K. Foils being thinned for TEM observation underwent the usual spontaneous transformation which, however, in this case did not obliterate the ω-phase particles. The athermal ω particles could not be distinctly imaged in the electron microscopy, making it impossible to specify their shape [RAC70]. Likewise, selected-area electron diffraction performed on the as-quenched alloys revealed the streaked reflections characteristic of diffuse ω. The occurrence of sharp and diffuse, respectively, athermal ω reflections and their implication with respect to ω-phase precipitation has been considered by SASS and colleagues, some of whose results have been summarized in the foregoing discussions. The morphology of β-III's isothermal ω, which has been photographed many times in aged samples (e.g., [BOY74] [WIL82^a]), is distinctly ellipsoidal and similar to that present in binary Ti-Mo.

6. Deformation

Probably the most common laboratory-scale deformation process is that experienced by samples in the final stages of uniaxial compressive or tensile testing. The most interesting properties to be encountered under such conditions are the so-called "anomalous tensile properties" (e.g., serrated yielding, pseudoelasticity, etc.), which are the subjects of the latter part of Chapter 12.

The manner in which a metal deforms after its yield strength has been exceeded by the applied stress depends on many factors. Parameters controlling the deformation process include the alloy's *composition*, its *class* (i.e., whether α, α + β, or β), its *condition* (i.e., whether quenched— e.g., $β_{tr}$, α + β-annealed, low-temperature aged, etc.), and the *rate* and *temperature* at which the deformation is carried out. Some *observables* or results of the deformation process include the *anomalous stress-strain behavior* alluded to above and discussed in Chapter 12, *phase transformation* under stress (i.e., transformation-assisted deformation), and *texturization* (i.e., the development of preferential crystal orientation or the formation of deformation cells or subbands in response to heavy cold work).

Some examples of these processes and effects are offered below under sections that deal with (1) ductility at low temperatures, (2) deformation at elevated temperatures (forming), (3) transformation-assisted plasticity, (4) deformation textures, and (5) deformation microstructures.

6.1 Low-Temperature Ductilities of Some Representative Technical Titanium-Base Alloys

Unalloyed titanium, the α alloy Ti-5Al-2.5Sn, and the near-α and α + β alloys Ti-8Al-1Mo-1V and Ti-6Al-4V, respectively, have properties that are suitable for a wide range of cryogenic applications, while β alloys such as Ti-13V-11Cr-3Al have a strong tendency to embrittle on cooling to cryogenic temperatures. The *low-temperature* mechanical and physical properties of these alloys have been specified in SALMON's *Low Temperature Data Handbook* [SAL79].

6.1.1 Unalloyed Titanium

All commercial grades of unalloyed titanium exhibit moderately good ductility at temperatures down to ~20 K. Their elongations to fracture ($δ_B$, %) actually increase as the temperature is decreased from 300 K, and pass through broad maxima ($δ_B$ = 40 ~ 50%) at about 77 K before descending rapidly as the temperature approaches 4.2 K. In some samples, $δ_B$ becomes negligibly small at liquid-He temperatures. Cold rolling increases the yield and ultimate strengths, but at the expense of ductility, as usual. The effects of interstitial elements on the strength of titanium have been considered in great detail by CONRAD and coworkers [CON67, CON70, CON75, CON81] [SAR72][OKA73] [TYS75]. In a long series of papers it has been pointed out that the solutes carbon, nitrogen, and oxygen, which bond in a covalent-like manner to the surrounding titanium atoms, have pronounced influences on the strength of otherwise unalloyed titanium at temperatures below about $0.5\,T_m$.

6.1.2 Ti-5Al-2.5Sn

The ductility of extra-low-interstitial (ELI) grade Ti-5Al-2.5Sn (with O, 1200; C, 800; N, 500 max ppm by wt) is fairly independent of temperature between room temperature and 20 K, $δ_B$ remaining at about 16 ± 1% throughout that range. The ductility of the normal-interstitial grade (O, 2000; C, 1500; N, 700 max ppm by wt) is considerably lower; in fact, $δ_B$ decreases monotonically between room temperature, dropping to 12% at 77 K and to only 5% at 20 K.

6.1.3 Ti-6Al-4V

The ductility of annealed Ti-6Al-4V is fairly temperature-independent between room temperature and 77 K. Below that, it decreases rapidly as the temperature continues to lower toward 20 K (Fig. 6.1). The ductility of the annealed alloy is twice as great as that of the solution-treated-and-aged (STA) material; e.g., $δ_{B,77K}$ = 11.4% as compared with 4.9% at normal-interstitial levels. Reducing the interstitial content influences the tensile properties only marginally, but improves the fracture toughness by 130% at room temperature and 40% at 20 K. For this reason, the ELI grade of this alloy is recommended for cryogenic service.

6.1.4 Ti-8Al-1Mo-1V

The near-α, α + β alloy Ti-8Al-1Mo-1V, although originally developed for high-temperature applications, can be used reliably down to moderate subambient temperatures in either the single-annealed (SA) or duplex-annealed (DA) condition.* The room-temperature ductilities of SA and DA alloys are similar ($δ_B$ ≅ 15%), but upon cooling, that of the SA alloy decreases, while that of the DA increases before passing through a maximum ($δ_B$ ≅ 22%) at about 77 K and dropping to low values at 20 K ($δ_B$ ≅ 1%).

Single-annealed: "mill annealed" (8 h/790 °C) and furnace cooled. *Duplex-annealed*: mill annealed plus 15 min/790 °C plus air cooled.

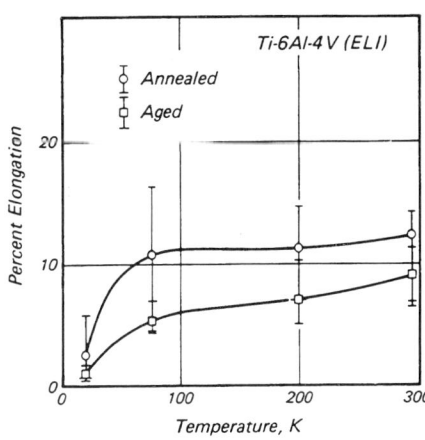

Fig. 6.1 Ductility of the ELI grade of Ti-6Al-4V as a function of temperature in the cryogenic to room temperature range [SAL79].

6.1.5 Ti-13V-11Cr-3Al

As a β-Ti alloy, Ti-13V-11Cr-3Al would be expected to possess poor low-temperature ductility. Indeed it does, the elongation-at-fracture of the STA material becoming insignificantly small below about 100 K. At 77 K, $\delta_B = 0.2\%$. Some improvement results if the aging stage of the STA heat treatment (20 ~ 100 h/430 ~ 500 °C) is omitted, in which case $\delta_{B,77K}$ becomes about 2%.

6.2 Deformation at Elevated Temperatures

As an introduction to the topic of elevated-temperature deformation, the forming requirements of the representative technical alloys of the previous section will be briefly reviewed.

Unalloyed Titanium (α/β Transition Temperature, 883 °C). Commercial grades of titanium are usually hot formed at 870 °C, just below the allotropic transformation temperature. If cold deformed, the material will exhibit "*springback*" and will also require some annealing.

Ti-5Al-2.5Sn (β Transus Temperature, ~ 1025 °C). The formability of Ti-5Al-2.5Sn is inferior to that of commercial unalloyed titanium. Forming operations are usually carried out at 200 ~ 650 °C, care being taken to minimize the time of exposure to temperatures above 540 °C.

Ti-6Al-4V (β Transus Temperature, ~ 995 °C). Ti-6Al-4V is difficult to form even after annealing—usually for ½ to 4 h at 700 to 820 °C. Primary fabrication operations, (ingot breakdown), are often initiated above the β transus and finished below it, in several steps. These steps could involve β and α/β forging, and will include some β recrystallization steps. Mill products are usually finished in the temperature range 870 ~ 980 °C—i.e., high in the α + β field. Other forging practices, and the properties that result from them, are discussed by WOOD [WOO72].

Ti-8Al-1Mo-1V (β Transus Temperature, 1040 °C). In the deformation metalworking of Ti-8Al-1Mo-1V, temperatures in the β field are generally avoided. It has been found preferable to perform metalworking operations high in the α + β field. Sheet metalworking (secondary fabrication) is generally conducted within the temperature range 650 ~ 800 °C, although sheet forming is not impossible at temperatures as low as room temperature.

Ti-13V-11Cr-3Al (Metastable β, β-Transus Temperature, 650~700 °C). Ti-13V-11Cr-3Al is normally fabricated to flat-rolled products at temperatures in the β-phase field [WOO72, p. 1-9:72-3]. As for secondary fabrication, because of the alloy's excellent bend ductility, any sheet-forming operation that is applicable to unalloyed titanium is suitable for annealed Ti-13V-11Cr-3Al.

6.2.1 Beta Forging of Titanium-Base Alloys

α-Forging Processes [HAM78]. Beta forging may be carried out isothermally with the billet and dies initially at the same temperature above the β transus. Otherwise, if only the billet has been heated, die chilling as the forging takes place may cool the workpiece to temperatures within the α + β field. Provided the first 25% of the reduction takes place within the β field, it is still permissible to refer to the operation as "β forging." If the initial temperature is sufficiently high in the β field, the forging can be accomplished as a single step. A second advantage of β forging is that at the high temperature at which it takes place, the silicon [in silicon-bearing alloys such as Ti-6242(Si)] can be retained in solid solution during the operation, thereby leading to a product with improved creep strength. On the other hand, the high temperatures associated with the β forging of α and many α + β alloys may result in: (1) a large β grain size, particularly if the alloy is allowed to anneal in the β regime before cooling below the transus, and (2) a coarse Widmanstätten structure on cooling to room temperature. Such an "aligned-α," or locally textured, structure has poor low-cycle fatigue properties. This disadvantage can be reduced if the aligned α is replaced by the "basket-weave" structure by increasing the cooling rate from the β field, or by refining the prior-β grain size through final forging high in the α + β field.

Superplasticity in β Forging. Isothermal β forging of β alloys can take advantage of the superplastic properties that have been exhibited by some of them. GRIFFITHS and HAMMOND [GRI73] showed that the alloys Ti-8Mn, Ti-15Mo, and Ti-13Cr-11V-3Al exhibited superplasticity when deformed at low strain rates at temperatures of about 0.6 T_m. Elongations of from 150 to 450% were observed, depending on the numbers of grains in the specimen cross sections. With grain sizes of several hundreds of μm, the sample could not have been exhibiting the normal kind of fine-grain (<10 μm) superplasticity. The *subgrains* (or deformation cells) were, however, fine, and of the above-mentioned size. The flow was interpreted as taking place by way of a "diffusion-creep" mechanism. In this, the vacancy source/sink property of the subgrain boundaries was supposed to combine with the anomalously high diffusion rates inherent in β-Ti alloys to permit the extensive low-strain-rate moderate-temperature deformation that characterizes superplasticity [GRI73][HAM78].

6.2.2 Alpha + Beta Forging of Titanium-Base Alloys

α + β Forging Process [HAM78]. Although during nonisothermal forging, as defined above, the billet may spend a significant fraction of its deformation time in the α + β field, in order to achieve a *uniform* α + β structure, the final working operations must be carried out in that field. The processing temperature is then, of course, upper limited by the β transus and lower bounded by press and materials constraints. Such a tight restriction on processing temperature range may require preheating the tooling to the initial billet temperature (i.e., the use of isothermal forging).

Superplasticity in α + β Forging. As indicated above, isothermal forging at low strain rates is conducive to superplastic deformation. In α + β alloys, superplasticity generally takes place via the fine-grain-size mechanisms. This is achieved in α + β alloys through heavy hot working in the α + β phase field. Continuous working and recrystallization produces a fine grain size, which is stabilized by the presence of a coarse dispersed phase, primarily α. Obviously, and as HAMMOND and NUTTING have pointed out [HAM78], there may not be a very sharp distinction between this kind of superplasticity and isothermal forging. The isothermal closed-die approach has been successfully applied to the superplastic forging of Ti-6Al-6V-2Sn and Ti-6Al-4V in the temperature range 900 to 950 °C [FIX73].

6.2.3 Thermomechanical Processing

The properties of α + β titanium-base alloys are strongly dependent on microstructure. CHEN and colleagues, for example, have devoted considerable attention to the relationship between the microstructures and the mechanical properties of Ti-6Al-2Sn-4Zr-2Mo~0.1Si forgings [CHE80]. Sections 6.2.1 and 6.2.2 have briefly indicated the characteristics of β and α + β forging, both isothermal and non-isothermal. Factors that must be taken into consideration in the design of forging operations in general are: (1) the starting microstructure, (2) the start and finish temperatures, (3) the extent of the deformation, and (4) the rate at which the deformation takes place [GEG80]. In achieving the desired final microstructure, hence mechanical properties, two more thermal variables are available for control and adjustment: (1) the cooling rate from the final working operation and (2) the final heat treatment. The entire sequence of operations is known as "thermomechanical processing" [WIL82b].

6.3 Transformation-Assisted Plasticity

The deformation of alloys can be facilitated if thermally induced or stress-induced transformations of one kind or another take place during the time that stress is being applied to the sample. Either high-temperature or low-temperature transformations are eligible for consideration in this section, which draws implicitly on the contents of Sections 5.2 and 5.3.4 (stress-induced martensitic transformation and twinning) and Sections 8.3 and 12.14 (low-temperature and high-temperature serrated yielding) as well as other sources.

6.3.1 Beta Alloys

Stress-assisted transformation permits some β alloys to achieve a degree of low-temperature ductility otherwise unexpected in a bcc structure. For example, Ti ~ 50Nb, whose deformation at low temperature is discussed in Sections 8.1.2 and 12.14.1, when tested to failure at 4.2 K, reveals the finely dimpled fracture surfaces characteristic of microscopic ductile fracture (Fig. 6.2).

Numerous binary β-Ti alloys have been shown to deform via martensitic transformation [ZWI74, p. 175] or twinning, depending on the solute concentration. In a typical basic study, that by OKA et al. [OKA80] of Ti-Mo (see Section 5.2.5), it was shown that at room temperature the stress-induced transformation product was martensitic provided the molybdenum concentration was between about 9 and 11 wt%, but took the form of $\{332\}_\beta$ twins when it was within the range 11 ~15.5 wt%. In a related study, HIDA et al. [HID80] investigated the relationship between the thermal instability of metastable quenched β-Ti-Mo and its plastic properties, observing again that the remarkable ductility exhibited by Ti-14Mo correlated with the formation of $\{332\}_\beta$ twins. Secondly, they noted that the measured linear work-hardening rate could be correlated with a continuous formation of new twins, these being the source of an increasing density of ω phase, a known hardener. As already mentioned in Section 5.3.5, the β alloy Ti-10V-2Fe-3Al will undergo stress-assisted transformation at *room temperature*, which lies part-way between M_d and M_s [DUE80a]. The β-solution-treated alloy contains a fine dispersion of athermal ω precipitation after quenching. Upon stressing at room temperature, the matrix distorts to α" martensite, leaving the ω precipitates intact—a phenomenon which imparts some ductility to an alloy that might otherwise be embrittled by ω phase.

6.3.2 Alpha- and Alpha + Beta Alloys at Elevated Temperatures

Since, as has been generally noted, a metal's resistance to plastic deformation decreases during a phase change, phase transformation can be brought in to assist in elevated-temperature deformation. GEGEL et al., in a study of the dynamics of flow and fracture during the isothermal forging of Ti-6Al-2Sn-4Zr-2Mo-0.1Si have noted that high-strain-rate deformation at temperatures some 50 to 150 °C below the β transus would induce the formation of microcracks [GEG80]. This crack formation could, however, be suppressed by reducing the strain rate below some critical value in order, it was thought, to permit the operation of a phase-transformation type of stress-relaxation mechanism. Another example is to be found in the work of KOT et al. with unalloyed titanium and Ti-6Al-4V [KOT70]. In a typical experiment, the sample was subjected to fixed loading while its temperature was cycled from a lower temperature through the β transus (650 to 925 °C for titanium, and 760 to 980 °C for Ti-6Al-4V). Each time the alloy underwent an α → β transformation, quasiviscous flow took place, accompanied by a large increment of plastic deformation. Upon repeating such cycling, stepwise elongations of titanium and Ti-6Al-4V in excess of 300% are possible and have been observed. In an interesting application of the principle, KOT et al. [KOT70] employed a traveling induction coil as a kind of "thermal die." Looped around a rod of titanium or Ti-6Al-4V maintained under tension it was capable of administering area reductions in excess of 50% during each of its passes along the rod.

6.4 Deformation Textures and Microstructures in Alpha- and Beta-Titanium Alloys

The two metalworking processes that administer the greatest amounts of deformation from ingot to finished product are *sheet rolling* and *wire drawing*. As a result of the heavy unidirectional deformation which these processes impart, the resulting sheets or wires acquire certain directional properties either at the atomic level (as gauged by the results of x-ray diffractometry) or at the microstructural level (as visualized by optical and electron microscopy).

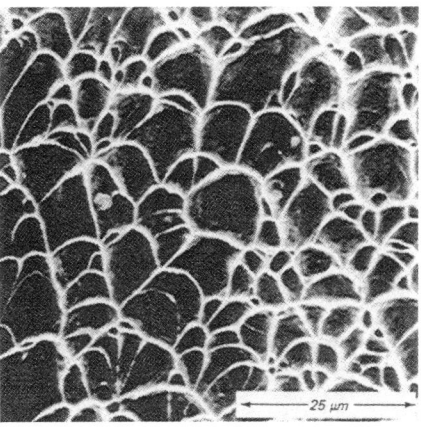

Fig. 6.2 Honeycombed or dimpled shear-fracture surface in Ti-50Nb (90% cold deformed) [ALB76]. Copyright © 1976, *Zeitschrift für Metallkunde*, reprinted with permission.

6.4.1 Texture

"Texture" refers to the tendency for the principal crystallographic directions in adjacent grains of a polycrystalline material to become aligned, or to assume a "preferred orientation" (Fig. 6.3). Texture is represented by a two-dimensional map, or pole figure, which (as the projection of a hemispherical surface onto a flat one) quantifies the angular clustering of selected crystallographic directions or poles. Since the existence of texture is amplified if the individual crystallites are themselves highly anisotropic, titanium-alloy texture has usually been studied in unalloyed titanium, dilute α-phase alloys (particularly Ti-Al), or the α + β alloy Ti-6Al-4V. The β alloys have not, of course, been neglected; in fact, binary alloys of titanium with vanadium, manganese, niobium, and molybdenum [MAR60] and alloys such as Ti-16V-2.5Al [LAR74] have been extensively studied. The subject of texture has been reviewed numerous times—the early work by authors such as JAFFEE [JAF58, p. 100-101] and MARCOLIN [MAR60, p. 269-271], and the more recent results by LARSON and ZARKADES [LAR74] and ZWICKER [ZWI74].

6.4.2 Deformation Microstructures

Whether or not a material becomes crystallographically textured, heavy deformation will produce aligned arrays of dislocation-cell walls (Fig. 6.4). Our present understanding of the effects of heavy deformation on the microstructures of metals can be traced back to an important series of papers by HIRSCH and coworkers

(Refs. 53 to 55) of [NAR66], the essential results of which, as they apply to bcc metals, having been elegantly summarized by NARLIKAR and DEW-HUGHES [NAR66]. With the aid of a "microbeam" Laue back-reflection technique, in which the x-ray beam could be collimated through a capillary <20 μm in diameter, HIRSCH and coworkers were able to recognize the presence of crystallites with diameters as small as 1 μm. The importance of this work lay in its being able to show quantitatively that: (1) heavy deformation did not lead to a uniformly disordered structure as had earlier been thought but rather, in their words, to a "foam"-like structure consisting of particles of low dislocation density embedded in a continuous three-dimensional net of highly dislocated material; and (2) the resulting structure, instead of being the product of grain disintegration (and consequently true grain refinement), was in fact generated by dislocation motion within the grains such that the resulting dislocation network formed low-angle subgrain boundaries. In a related study, EMBURY et al. [EMB66] investigated and compared the microstructures of several steels and a sample of 99.99% Cu in response to wire-drawing at room temperature to area reductions of up to 99%. Again, it was noted that all the samples developed fibrous or cellular structures, the cell walls acting as dislocation barriers in a manner analogous to the function of grain boundaries in this regard. A much greater degree of cell refinement was noted in the iron alloys than in the copper, due, it was claimed, to a difference between the rates of dynamic recovery in the two cases. It was thought that the presence of interstitial impurities in the former was responsible for the greater stability of the substructural boundaries once they were produced by the wire-drawing process. As a consequence of the wire-drawing and the resulting fine elongated cellular structure,

the steels became significantly strain hardened. For each steel the flow stress, σ_f, increased linearly with $1/\sqrt{d}$, where d is the cell diameter. To a first approximation, σ_f (kg mm^{-2}) = $40/\sqrt{d}$, if d is expressed in μm. In the English scientific literature these subgrains have also been referred to as "cells" or "subcells" and the dislocation network defining them as "cell/subcell walls/boundaries"; the German* and more recent English literatures seem to prefer the use of "subbands" and "subband boundaries" or "walls" to describe the same features. The presence of individual dislocation-free cells was disclosed by the appearance of spots on the Laue microdiffraction rings (Refs. 53 and 54) of [NAR66]). Later investigators, studying heavily cold-drawn [ARN74] or cold-swaged [LOH71] wire using both microbeam and conventional diffraction methods, have interpreted the spot patterns as being indicative of ⟨110⟩ texture in the direction of the wire axis.

6.5 Deformation-Induced Textures in Titanium-Base Alloys

During deformation metalworking, or as a result of recrystallization, grain growth, or phase transformation, the grains (crystallites) of polycrystalline metals may develop preferred orientations or texture. This topic has been subjected to a comprehensive general review by DILLAMORE and ROBERTS [DIL65] (407 references). With regard to titanium and its alloys, reviews have been offered by JAFFEE [JAF58], MARGOLIN and NIELSEN [MAR60], LARSON and ZARKADES [LAR74], and ZWICKER [ZWI74], authors who have focused attention primarily on the results of cold rolling, which, of course, is capable of developing strong, reproducible, preferred orientations.** The development of texture during the *forging* of titanium al-

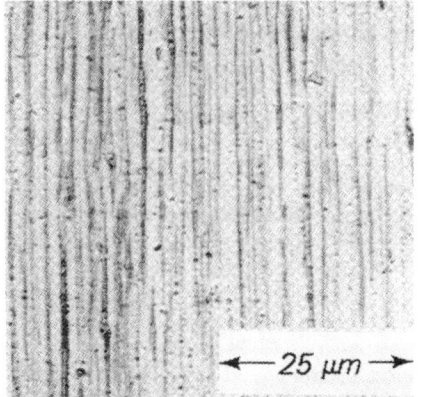

Fig. 6.4 Longitudinal section of a cold-swaged (97.2%) Ti-50Nb wire [LOH71]. Micrograph courtesy of U. Zwicker (Universität Erlangen-Nürnberg).

loy parts has been considered by ROMERO [ROM 71].

Texture is determined by x-ray diffraction and usually defined by means of a single-quadrant stereographic representation of the clustering of specified crystallographic orientations (the "pole figure"). Texture in hcp polycrystals is often described in terms of the distribution of c-axis or [0001] directions. The [0001] direction in the hcp crystal is referred to as the "basal pole"; it is also referred to as the "(0002) pole," the implication here being that the basal planes of an hcp crystal are responsible for (0002) Bragg reflections. In short, the terminologies "[0001]" and "(0002)" that occasionally appear together, and even interchangeably, in the literature of hcp texture refer to *crystalline direction* and *experimental method*, respectively. A pair of schematic basal-pole quadrant-diagrams is presented in Fig. 6.5.

6.5.1 Unalloyed Titanium

The cold-rolled texture of unalloyed titanium has been determined many times either as a study in itself (e.g., [KEE56]) or as part of an investigation into the influence of alloying on textural change (e.g., [THO73]). As with most low-c/a (<1.63) metals, the (0002) or basal poles are concentrated in regions ±30° to ±40° in the transverse direction away from the sheet normal (Fig. 6.6a). All authors agree that such a texture results from a competition between $\{0001\}\langle11\bar{2}0\rangle$ slip, which rotates the basal poles toward the sheet normal, and $\{11\bar{2}2\}$ twinning, which tends to rotate them into the transverse direction.

The annealing of high-purity titanium at various temperatures below the transformation temperature (883 °C) results in a sharpening and slight "rotation" of the texture; i.e., it leaves the (0002) pole figure essentially unchanged, but rotates the $(10\bar{1}0)$ pole figure about the sheet normal. Heating through the α/β transition and back again does not erase the texture in either phase, presumably because the Burgers relationship, $\{110\}_\beta \| (0001)_\alpha$, holds during both the positive and negative temperature excursions [JAF58, p. 101].

---

*According to HILLMANN [HIL73], "deformation bands" are 10^{-4} and 10^{-5} cm in width, while "subbands" formed as a result of still stronger deformation are 10^{-5} to 2×10^{-6} cm wide.

**Sheet textures can be defined by specifying the Miller indices of a plane parallel to the rolling plane (hkl) and a direction parallel to the rolling direction $[uvw]$ (see [LAR74]).

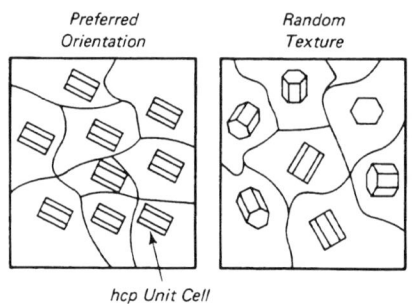

Fig. 6.3 Schematic representation of preferred and random textures in a polycrystalline hcp metal.

6.5.2 Alpha-Phase Binary Alloys

Textures in Ti-Zr (0.03-9.04 wt%), Ti-Sn (0.01-4.00 wt%), and TiAl (0.01-3.92 wt%) alloys have been measured by LARSON et al. [LAR71]. Ti-Al has, of course, been extensively studied—first by McHARGUE et al. [MCH53][SPA57] and subsequently by LARSON et al. (just mentioned) and by THORNBURG [THO73]. The early work showed that in the presence of sufficient aluminum, certainly with 3.8 wt% of it [MCH53], cold rolling produced an almost "ideal" basal texture (i.e., one in which the basal poles were clustered about the sheet normal), but left open the question as to whether the transition to this, from the split texture of unalloyed titanium, took place suddenly at some critical aluminum concentration or was a smooth, continuous function of it.* To answer this question, THORNBURG [THO73] prepared for measurement alloys of titanium with 0.25, 0.5, 1.0, 1.5, 2.0, 3.0, and 4.0 wt% Al and measured their textures after 20, 40, 60, 80, 90, and 95% area reduction by cold rolling. They discovered that not only did the texture maintain a fairly constant degree of splitting up to an aluminum concentration of 2 wt%, but that it transformed suddenly to basal at that concentration and remained that way as the aluminum concentration continued to increase. They also noted that,

---

* Between them, [MCH53] and [SPA57] treated Ti-Al (0.07, 0.47, 1.05, 1.43, and 3.8 wt%).

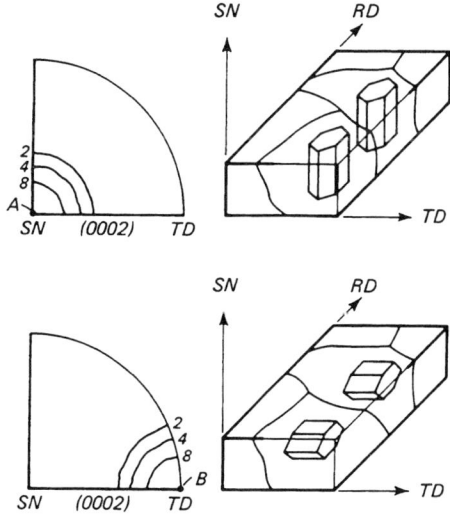

Fig. 6.5 Schematic representation of basal pole figures for textured hcp metals. Upper diagram: crystallites aligned normal to the rolling direction; point A would represent perfect alignment of all the crystallites. Lower diagram: crystallites aligned transverse to the rolling direction; point B represents perfect alignment in this case [LAR74].

with the exception of the threshold-concentration alloy (2 wt% Al), the texture had reached a stable saturated condition by 60% reduction in area. A series of (0002) pole figures taken at 95% reduction by cold rolling is given in Fig. 6.6(b) through (d).

6.5.3 Alpha + Beta-Phase Binary Ti-TM Alloys

The textures of binary Ti-TM alloys, hot worked in the α + β field and subsequently cold rolled 84% with intermediate annealing, were investigated by LARSON et al. [LAR71]. The entire Ti-(3d)TM series (TM = V to Ni) was measured, as were representatives of the two Ti-(4d)TM systems: Ti-Nb and Ti-Mo. Solute concentration ranges were chosen so as to provide maximum β-phase fractions of some 15 ~ 40%. Among all of the systems some striking similarities were to be seen in the manner in which the (0002) texture varied with solute concentration. A typical series of results, that for Ti-Mn (0.46, 1.42, 5.89, and 7.09 wt%), is given in Fig. 6.6(e) through (h). At low solute concentrations, the split basal-pole texture remained practically unchanged with increasing concentration except that, in some cases, it was slightly perturbed by the appearance of a small fraction of basal poles in the transverse direction.* Then, after what seemed to be a critical solute concentration (corresponding to the presence of 16 ~ 20% β phase in the α + β alloys), the split basal texture shifted to the rolling direction. The texture of cold-rolled Ti-Cu was not a member of the above class: ac-

---

*For further details regarding this peak, a $(11\bar{2}0)[10\bar{1}0]$ texture, and the $(10\bar{1}0)$ pole figures, the original literature [LAR71] should be consulted.

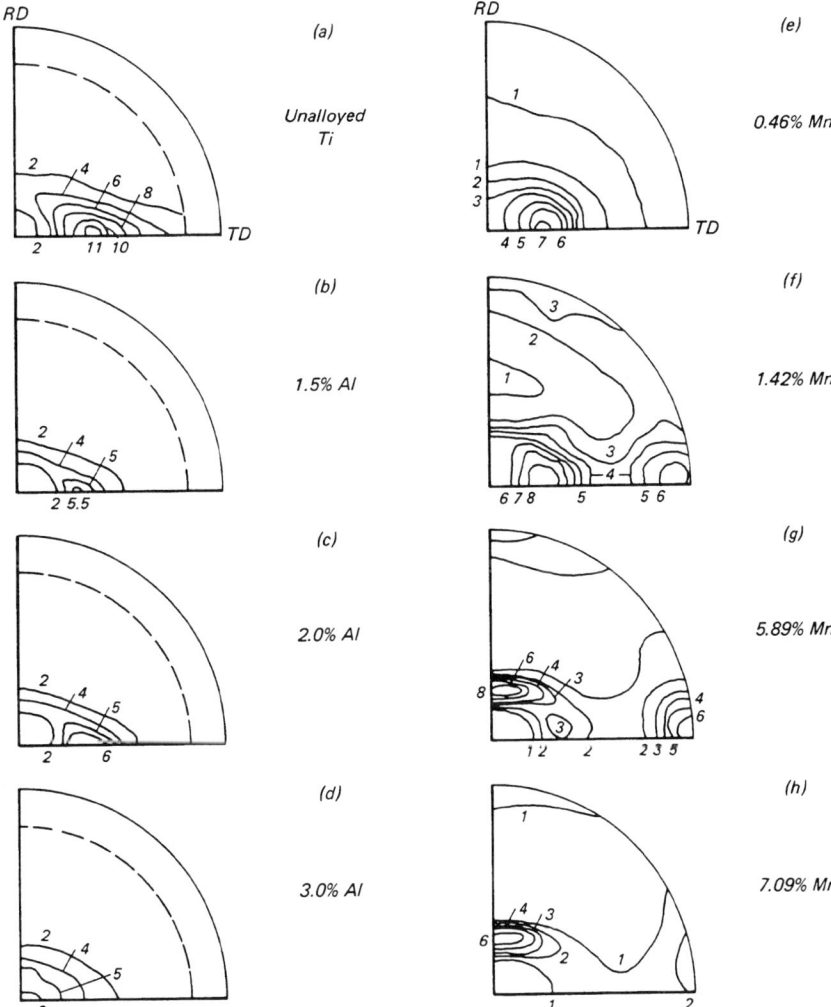

Fig. 6.6 Textures of cold-rolled Ti-Al and Ti-Mn alloys. *Sample conditions:* Ti-Al series: hot rolled 50 °C (90 °F) above β/(α + β), then hot rolled 50 °C (90 °F) below α/(α + β) and annealed there for 1 h; cold rolled 95% [THO73]. Ti-Mn series: forged at 840 °C (1545 °F), hot rolled at 815 °C (1500 °F), then cold rolled 84% with intermediate anneals at 730 °C (1345 °F) [LAR71].

cording to LARSON et al. [LAR71, LAR74], the addition of 0.55 wt% Cu to titanium resulted in an almost "ideal" $(0001)[10\bar{1}0]$-type texture.

6.6 Deformation Microstructures in β-Ti Alloys

6.6.1 Rolling-Induced Microstructures

Relatively little attention has been given to the study of rolling-induced deformation microstructures in β-Ti alloys. Such information as is available has been reviewed by ZWICKER [ZWI74, p. 134]. In an early study of type-II superconductivity, HAKE et al. [HAK63] noted that the cold rolling of Ti-Nb(60 at.%) to a thickness reduction ratio of 24:1 resulted in the formation of a pronounced laminar cell structure when viewed either transversely or longitudinally within the rolling plane. The cells, which appeared under optical magnifications to be continuous across the width of the sample, were about 2 μm thick. TEM observations of the in-plane structure of a sample of Ti-Nb(75 at.%), cold rolled to more than 90%, revealed a well-defined network of cells about 0.4 μm in size [NAR66]. In order to completely evaluate the cell structure of highly anisotropic cold-rolled Ti-Nb, BAKER and TAYLOR [BAK67] carried out a TEM study of 97% cold-rolled Ti-45Nb, taking observations both normal to the rolling plane and parallel to it along the rolling direction. The diameters of the in-plane cell cross sections, which were fairly equiaxed, were 0.25 μm or larger, in good agreement with the earlier result [NAR66]. But, as expected, the end view was one of severely flattened cells, about 0.1 μm thick and 2.5 μm or more in width.

6.6.2 Wire Flattening

Wire flattening consists of the light rolling of wire already deformed by drawing. In a cold-flattening study by BAKER and TAYLOR [BAK67], a wire, cold drawn 99.87% to 0.033 in. diam, was lightly cold rolled (4.7:1) to a thickness of 0.007 in. In consequence, an initial structure of parallel fibers ~0.1 μm in diameter underwent a coarsening when viewed normal to the rolling plane, and dislocation cells appeared within some of the fibers. It was expected that further cold reduction of this type would transform the structure to that observed in rolled materials, which has been shown to consist of equiaxed cells in the rolling plane and a fibrous structure at right angles to it. Indeed, in a similar experiment, in which heavily cold-drawn (99.994%) Ti-50Nb wire was reduced from a diameter of 0.25 mm to a thickness of 0.1 mm by cold rolling, PFEIFFER and HILLMANN [PFE68] noted that whereas the in-plane structure was smeared out, especially along the edges, in the section at right angles to the rolling plane the initial fibrous structure of the wire had scarcely been disturbed.

6.6.3 Wire Drawing

As a uniaxial deformation process, wire drawing produces uniform deformation across a plane at right angles to the wire axis. Numerous photomicrographs are available in the literature showing an equiaxed cell structure in cross section and the usual elongated bands in the axial direction. One such pair of micrographs—Figs. 6.7 and 6.8—show Ti-58Nb cold drawn to an area reduction ratio of 5×10^4:1 (99.998%) [NEA71]. These figures depict the microstructure of heavily cold-drawn wire as bundles of "pencil-shaped subcells" [NEA71] or "fibers" [BAK67] running parallel to the wire axis.

In studies of cold-drawn texture in β-Ti-Nb, since sample smallness precluded the use of normal x-ray pole-figure determination methods, ARNDT and EBELING [ARN74] [WIL75a] employed the electron beam (20 μm$^\Phi$) microdiffraction technique referred to above to view thinned cross sections of the wire. Although heat-treated wire, in which some cell growth had taken place, yielded single-crystal diffraction patterns, results interpretable in terms of texture could be obtained from wires with narrow and well-defined straight subbands. In such cases, the diffraction pattern of numerous sharp spots representative of a superposition of many crystal orientations suggested a strong ⟨110⟩ texture in the axial direction.* The results of other measurements of Ti-Nb [HIL81] as well as of Ti-50Nb-1Cu and Ti-Mo [ZWI74, p. 136] were in agreement with this. Copper, which is fcc, was reported to develop a ⟨100⟩ texture after plastic deformation [REE77].

Although the subbands in as-deformed wire are far from regular, it has been possible to quantify the decrease in fiber diameter that accompanies an increase in the degree of cold work as gauged by the area-reduction ratio, ARR. For example, PFEIFFER and HILLMANN [PFE68] showed that subband density (i.e., number of cells per cm^2 in the transverse section) of 1.3×10^{10} cm^{-2} in 99.78% cold-drawn (32 → 1.5 mm$^\Phi$) Ti-50Nb increased to 4.9×10^{11} cm^{-2} after 99.993% (32 → 1.5 mm$^\Phi$) reduction. The cell diameters decreased from 0.09 to 0.01 μm. Earlier results by BAKER and TAYLOR [BAK67]

---

*A fiber texture can be defined in terms of the crystallographic direction that points along the axis of the cylindrical product; the other crystallographic directions are disposed randomly about this axis [LAR74].

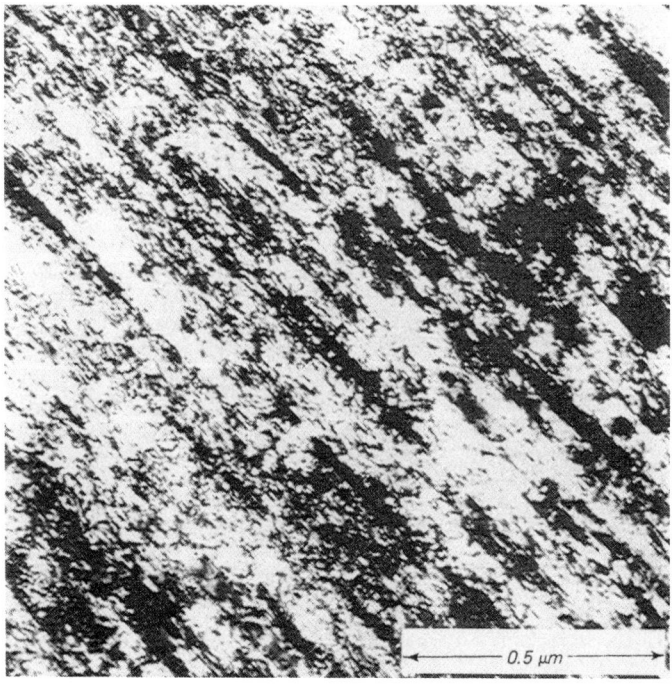

Fig. 6.7 Longitudinal section of a cold-drawn (ARR, 5×10^4:1) Ti-58Nb wire [NEA71]. Micrograph courtesy of D.F. Neal (Imperial Metal Industries); copyright © 1971, Pergamon Press, reprinted with permission.

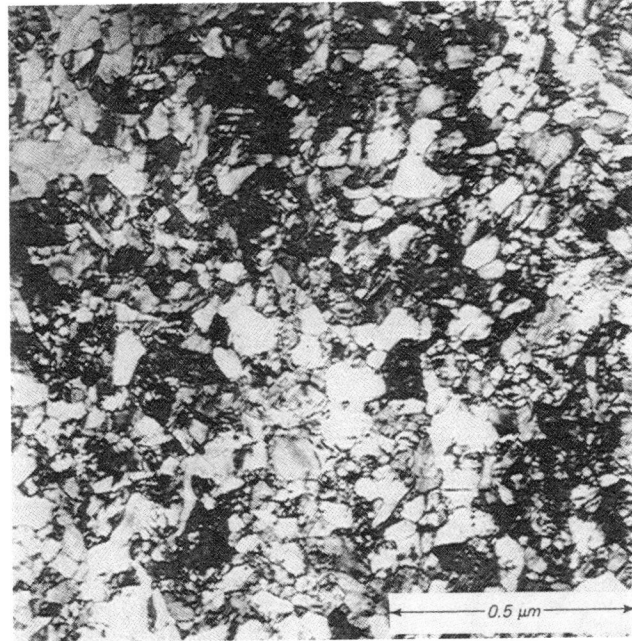

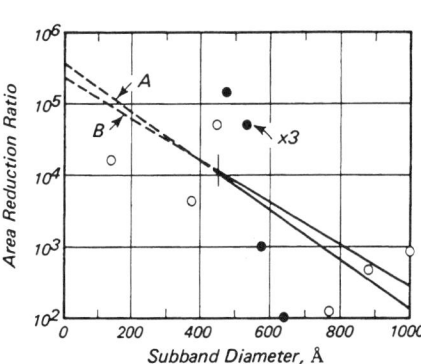

Fig. 6.9 Subband diameter as a function of area-reduction ratio (ARR) by cold drawing. The lines are least-squares fits to the data represented by the open (as-drawn) and closed (cold-drawn-plus-aged 385 °C, or 725 °F) circles. Lines A and B correspond to the omission and inclusion, respectively, of the point (140 Å, 1.6×10^4).

Fig. 6.8 Transverse section of the wire of Fig. 6.7 [NEA71]. Micrograph courtesy of D.F. Neal (Imperial Metal Industries); copyright © 1971, Pergamon Press, reprinted with permission.

for 99.87% cold-drawn Ti-55Nb with a cell diameter of about 0.1 µm were in good accord with this, as were later data by NEAL et al. for Ti-58Nb [NEA71] and by ARNDT and EBELING [ARN74] [WIL75[a]] again for Ti-50Nb. A survey of these results shows that, in spite of the variation in Ti-Nb alloy composition, a remarkably good picture of the inverse correlation between fiber diameter and degree of cold work can be assembled. Data from these sources, and others [WES80], when plotted semilogarithmically as in Fig. 6.9, indicate that subband diameter in the range 100 to 1000 Å is negatively correlated with $\log_{10}$ ARR. A least-squares fit to the data yields a semiquantitative relationship (correlation, –65%) between subband diameter, d (which is never much less than about 450 Å), and ARR of the form:

$$d = \frac{(5.49 \pm 0.10) - \log_{10}\text{ARR}}{(320 \pm 26) \times 10^{-5}} \ (\text{Å}) \quad (\text{Eq 6.1})$$

in which the "+" signs are to be taken together when the 140-Å datum point is *excluded* from the slope analysis and the "–" signs when it is *included*. Notice that the deformations imposed in order to develop these fine subband structures are orders of magnitude greater than those needed simply to texturize the sample (see Fig. 6.6).

7. Aging

Part 1: Microstructural Phenomenology

The terms "aging" and "tempering" refer to moderate-temperature heat treatments during which diffusion-controlled metallurgical processes take place within macroscopic periods of time, measured in minutes and hours. The aging of metastable alloys is accompanied by precipitation as it proceeds, generally by means of a nucleation-and-growth mechanism, toward thermodynamic equilibrium. An excellent treatment of the thermodynamics of such processes has been produced by HARDY and HEAL [HAR56]. Aging can also assist in the recovery of a deformed structure, again toward a lower-energy state — in this case regularly polycrystalline (e.g., equiaxed) and defect-free.

The equilibrium phases considered in Chapter 4 are of course achieved by the prolonged heating of previously metastable structures of the type considered in Chapter 5. Although heat treatment or annealing at any temperature will permit a metastable alloy to approach more closely its state of thermodynamic equilibrium, the term "aging" is generally understood to imply a heat treatment in the low- to moderate-temperature range. Of particular interest in this context is the aging of the martensitic and $\omega + \beta$ phases, and of course the aging of $\alpha + \beta$ and β alloys. It is inevitable that many of these topics have already been touched on in the previous chapters. The present chapter, therefore, restricts itself to summarizing and unifying some of the earlier discussions. Moreover, the existence of three important recent papers by WILLIAMS on precipitation and phase transformations in titanium alloys [WIL73, WIL76, WIL82b*] preempts for a time further detailed discussion of this subject.

The aging of α-phase alloys, and in particular the long-range ordering of the DO_{19} α_2-phase Ti$_3$SM-base structure,** is not treated here. What will be considered, though, with reference to binary Ti-TM alloys, are: (1) the transformation under aging conditions of the α' and α'' martensitic phases; (2) aging in the $\omega + \beta$-phase field and precipitation of the isothermal ω phase; (3) precipitation out of the β-phase field adjacent to the $\omega + \beta$ region, of a solute-lean bcc phase designated β'—the so-called "phase-separation" reaction; (4) the decomposition of β into $\alpha + \beta$ and practical methods of distinguishing between α and aged-ω precipitates; (5) the effects of various ternary additions on the kinetics and products of the aging reaction; and finally, (6) a different type of β decomposition, this time into a pair of *equilibrium* bcc phases designated β' (solute lean, again) and β'' (solute rich), characteristic of the phase diagram of, for example, Zr-Nb. The purpose of considering this reaction is primarily to draw attention to the distinction between the $\beta \to \beta' + \beta$ nonequilibrium reaction and the reaction $\beta \to \beta' + \beta''$, where $\beta/(\beta' + \beta'')$ represents an equilibrium solid-state miscibility gap.

7.1 Tempering of the Quenched Martensites

The term "tempering", often used to describe aging of quenched martensites, draws by implication an analogy with the heat treatment of quenched steels to whose microstructures the term "martensite" was originally applied. In discussing the aging of titanium martensites, WILLIAMS [WIL76] has catalogued the several processes that have been identified and has drawn attention to the disagreement that has arisen over the manner in which the early stages of α'' decomposition take place. In an interesting case study of the decomposition of hexagonal martensite, TRENOGINA and LERINMAN [TRE82] have investigated the tempering, at temperatures of 500 to 600 °C of the $\alpha + \beta$ alloy Ti-6.5Al-3.5Mo-2Zr-0.2Si after β quenching from 1050 and 1100 °C. Attempts to provide a unified mechanistic model of the decomposition processes in both α' and α'' Ti-TM alloys have been made by FLOWER, DAVIS, and WEST [FLO82] [DAV79a]. Although yet to be confirmed by others, and to be applied to Ti-TM alloys in general, their approach provides a very convincing description of the manner in which the α' and α'' variants of Ti-V, Ti-Nb, and Ti-Mo decompose, depending on the concentration ranges concerned. As pointed out in Section 5.2.5, in which Ti-Mo and Ti-Nb alloys are dealt with, and illustrated by Fig. 5.9, the α' variant of the β-isomorphous martensites transforms directly to $\alpha + \beta$ by the nucleation and growth of β-phase precipitates. The reaction is fast because the α' contains a high density of heterogeneous nucleation sites and since the aging temperature, $>\sim M_s^{\alpha'}$, is necessarily high. Within the intermediate-concentration range for α'' transformation, a spinodal decomposition of the α'' to $\alpha''_{lean} + \alpha''_{rich}$ is supposed to take place during the quench. This process, which via the accompanying compositional modulation gives rise to the reaction $\alpha''_{lean} + \alpha''_{rich} \to \alpha + \beta$, forms the basis for the $\alpha + \beta$ cellular reaction which has been observed to take place during the aging of, for example, Ti-Mo alloys within a specified concentration range. At higher concentrations, spinodal decomposition of the α'' does not take place, and it goes directly to β during the initial stage of aging. Since the aging temperature, if near $M_s^{\alpha'}$, is not necessarily low, the product of continued aging will frequently be $\alpha + \beta$.

7.2 The Isothermal Omega Phase: Aging of Omega + Beta-Phase Ti-TM Alloys

7.2.1 Kinetics and Morphology

Athermal ω phase has been shown to occur as a crystalline precipitate within a narrow composition range in quenched Ti-TM alloys. The sensitivity of this process to composition and experimental conditions is exemplified by the numerous studies which have been carried out on the Ti-Nb system.

While BAGARIATSKII et al. [BAG59], on the basis of x-ray and hardness data, claimed a composition of 18 at.% for ather-

---

* This paper, although published in 1982, was actually written in about 1975/76.

** SM implies a simple metal such as aluminum, tin, etc., or mixtures of them.

mal ω, a value which agrees with the results of an electron diffraction study by BALCERZAK and SASS [BAL71, BAL72], HATT and RIVLIN [HAT68] found α″ as the water-quenched product in Ti-Nb(20.7 at.%), as did HICKMAN [HIC69a] in gas-quenched ribbons of Ti-Nb(22 and 25 at.%). On the other hand, in a pair of papers discussing the structure and superconducting properties of Ti-Nb(22 at.%), BRAMMER and RHODES [BRA67] and KRAMER and RHODES [KRA67], respectively, showed that the compositional limit for athermal ω phase was already exceeded, and that the diffractographically defined "diffuse ω" with its unresolvable real-space counterpart was the quenched product. During the moderate-temperature aging (<~450 °C) of an alloy exhibiting the diffuse ω reflections, decomposition into the isothermal-ω (solute-lean) and β phases takes place. Numerous descriptions of this process accompanied by photomicrographs of the isothermal-ω precipitate can be found in the literature.

An early model [COU69] of the physics and kinetics of isothermal ω-phase development pictured an initial *structural* transformation of the lattice into ω and β (as for athermal ω phase in *its* composition regime), followed by an exchange of solute and solvent across the ω/β interface. DEFONTAINE et al. [DEF71], on the other hand, suggested that the first step was a *compositional* fluctuation, followed by a structural β → ω transformation within a solute-lean zone triggered by $\frac{2}{3}\langle 111\rangle$ longitudinal lattice vibrations, already an athermal property of the low-concentration bcc lattice (see Section 5.3). If such lattice vibrations are indeed responsible for the isothermal process, it is not surprising that during the aging of Ti-V [MCC71] and Ti-Nb [BAL71, BAL72] the cubic or ellipsoidal, respectively, precipitates that form appear to result from the growth of clusters of $\langle 111\rangle$ rows of particles.

Provided that the temperature is below about 400 °C, after prolonged aging, a metastable ω + β state is attained, characterized at a given temperature by a fixed volume fraction and composition of the ω and β endpoints [HIC69a]. A suggested metastable equilibrium phase diagram for Ti-Mo from the work of DEFONTAINE et al. [DEF71] is presented for further discussion in Fig. 7.1. Using x-ray techniques of the kind described by HICKMAN [HIC69, HIC69a], the volume fraction of the ω phase may be obtained from the corrected relative intensities of the ω and β reflections; if the rate of change of lattice parameter with composition is not too small, compositions can be calculated to ± 0.1 at.% from calibrated lattice-parameter measurements.

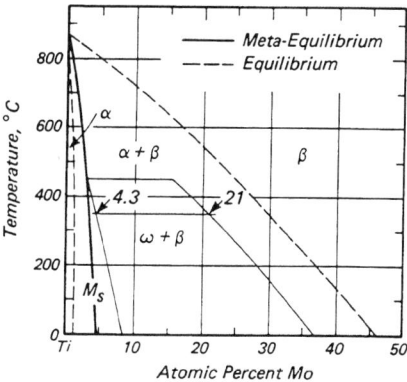

Fig. 7.1 Meta-equilibrium phase diagram for Ti-Mo indicating the ω/(ω + β) and (α + β)/β phase boundaries (fine full lines) and an M_s transus (heavy full line). The 350 °C (660 °F) isotherm is shown intersecting the transi at 4.3 and 21 at.% Mo. Also indicated (dashed lines) is a standard equilibrium phase diagram [DEF71].

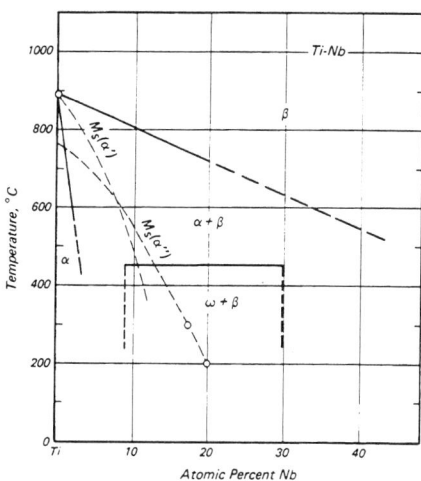

Fig. 7.2 Locations of the α- and β-equilibrium transi, the $M_s(\alpha')$ and $M_s(\alpha'')$ transi, and the regimes of occurrence of athermal and isothermal ω phases in Ti-Nb. Data sources: α and β transi, [MOL65, p. 20] (see also Fig. 4.10); M_s transi, [JEP70] and [FLO82]; ω + β phase data, [HIC69a].

Table 7.1 Time Needed for the Appearance of α-Phase Precipitation During Aging of Quenched Ti-TM Alloys

Alloy	Aging time (h) at temperature			Ref
	400 °C (750 °F)	450 °C (840 °F)	500 °C (930 °F)	
Ti-V(15, 19 at.%)	20-30	< 20		[HIC68]
Ti-V(19 at.%)			<4	[VET68]
Ti-V(25 at.%)	20-30	(No ω)		[HIC68]
Ti-Cr(9.3 at.%)	50			[HIC69a]
Ti-Mn(6.7 at.%)	68			[HIC69a]
Ti-Fe(6.0 at.%)	150	12	(No ω)	[HIC69a]
Ti-Nb(22 at.%)	72			[HAT68]
Ti-Mo(8 at.%)		320	50	[HIC69a]
Ti-Mo(10 at.%)		150	(No ω)	[HIC69a]

It has been determined that the *aged* product bears the same crystallographic relationship to the parent lattice as does the *athermal* ω phase [WIL76]. After sufficiently long aging times at 450 and 500 °C, α-phase precipitation can be expected, as indicated in Table 7.1. The relative sluggishness of the decomposition process in Ti-Mo compared with that in Ti-Fe is a reflection of the great difference in the diffusion coefficients of these two solute atoms (see Fig. 3.20).

The compositions of the meta-equilibrium isothermal ω phases have already been given in Table 5.5 within the context of a discussion of the limits of β-phase stability. With the aid of corresponding data for the β phase [HIC68, HIC69a], such as that presented in Table 7.2, a semiquantitative metastable equilibrium phase diagram for ω + β may be assembled. Figure 7.2 is such a diagram for Ti-Nb. In it we note that: (1) martensitic transformation supersedes athermal ω transformation in quenched alloys (cf., for example, the results of BAGARIATSKII et al. [BAG59] and HATT and RIVLIN [HAT68] mentioned earlier), and (2) at 450 °C the maximal niobium concentration for isothermal ω-phase precipitation is 30 at.%. This is probably also true for lower aging temperatures, if the results of the isothermal aging studies of Ti-V(15 at.%) [HIC68] and Ti-Cr(9.3 at.%) [HIC69a], at temperatures between 300 and 400 °C [which yielded practically vertical (ω + β)/β transii], can be accepted as having general significance. The isothermal precipitate particles assume one of two types of morphology—cubic or ellipsoidal—depending on

Table 7.2 Niobium Contents of the ω and β Phases in Aged (at 450 °C [840 °F]) Metastable-Equilibrium Ti-Nb Alloys [HIC69ª]

Average Nb concentration, at.%	Aging time, h	Volume fraction of ω phase	Nb concentration, at.% ω phase	β phase
22	10	0.36	6-11	29 ± 1
	30	0.34	5-10	30 ± 1
	50	0.33	6-11	31 ± 1
25	10	0.25	7.5-12	30 ± 1
	24	0.26	7.5-12	30 ± 1

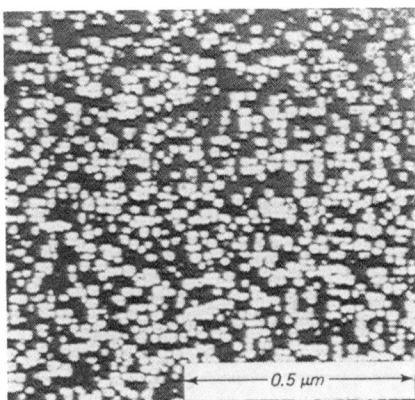

Fig. 7.3 Example of cubic ω phase. Specimen: Ti-10Fe [WIL73(corrected)]. Micrograph courtesy of J.C. Williams; copyright © 1973, Plenum Publishing Corporation, reprinted with permission.

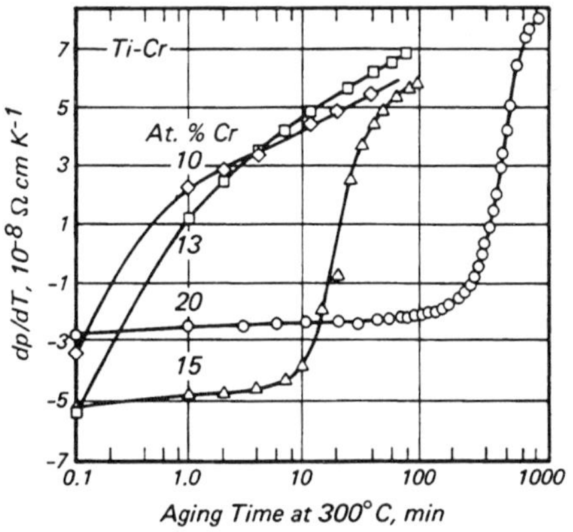

Fig. 7.4 Example of ellipsoidal ω phase. Specimen: Ti-11.5Mo-6Zr-4.5Sn [WIL73 (corrected), WIL76]. Micrograph courtesy of J.C. Williams; copyright © 1973, Plenum Publishing Corporation, reprinted with permission.

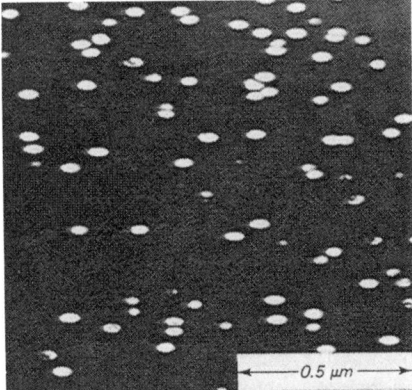

Fig. 7.5 Influence of aging time at 300 °C (570 °F) on the resistivity-temperature dependence of initially quenched Ti-Cr alloys [CHA74].

the linear lattice misfit $(V_\omega - V_\beta)/3V_\beta$, where V represents the unit-cell volume divided by the number of atoms per unit cell [HIC69ª]. If this is large (1 to 3%), as in Ti-V, Ti-Cr, Ti-Mn, and Ti-Fe [HIC69], minimization of elastic strains in the cubic matrix dictates a cubic morphology (Fig. 7.3). If the misfit is small (<0.5%), as in Ti-Mo and Ti-Nb [HIC69, HIC69ª], the morphology is dominated by surface-energy considerations, leading to the ellipsoidal particle shape depicted in Fig. 7.4. The influence of misfit on ω-particle morphology has been graphically demonstrated by WILLIAMS et al. [WIL71] in experiments in which the addition of 5.5 at.% Zr to Ti-V(20 at.%) resulted in a decrease in the misfit from 1.5-2.0% to ~0.25% and caused the precipitate shape to change from cubic to ellipsoidal.

7.2.2 Influence of Aging on Alloy Properties

Phenomenological studies of the effects of aging on the properties of ω + β Ti-TM alloys were conducted almost 40 years ago by FROST et al. [FRO54] and BROTZEN et al. [BRO55]. Their results, along with numerous others, have already been thoroughly reviewed by MCQUILLAN [MCQ63, pp. 51-57] and MARGOLIN and NIELSEN [MAR60, pp. 257-262].

It was the hardening and embrittling properties of ω phase that originally led to its discovery [PAR53]; subsequently it was found that the hardness, already characteristic of quenched ω + β-phase samples, increased with aging time [FRO54] [BRO55]. During aging, the tensile- and yield strengths increased and the ductility (elongation at fracture) decreased. Upon overaging, during which the ω phase dissolves and is replaced somehow by α phase, the ductility is restored. Most of the common transition elements (niobium is an exception) decrease the lattice parameter of the bcc phase; thus, during the isothermal aging of ω + β, the lattice generally shrinks as the β component becomes enriched in solute. If the aging temperature is increased and the alloy overaged to α + β, the lattice may expand to accompany a readjustment of the volume fraction of β phase to its equilibrium value [BRO55] [MAR60]. More recently, using Ti-Cr as a candidate system, POLONIS and coworkers have conducted an exhaustive investigation of isothermal aging and its effects on microstructure [CHA73ª, CHA78], electrical conductivity [CHA74], and superconducting properties [LUH69, LUH70ª]. During the same period of time, COLLINGS and HO applied physical-property measurement techniques to the study of isothermal ω + β aging in several previously β-quenched binary Ti-TM alloys. Techniques employed were: magnetic susceptibility (Ti-Fe [HO73], Ti-V [COL75ᵇ]); electrical resistivity (Ti-Mo [COL72], Ti-V [COL74]); and low-temperature specific heat (Ti-Fe [HO73], Ti-V [COL75ᶜ, COL82ᵇ], Ti-Mo [COL72ª, COL82ᵇ][HO73ª]).

Of particular interest are the results of all the electrical resistivity investigations. The temperature dependency of electrical resistivity, $d\rho/dT$, of as-quenched and quenched-plus-aged binary Ti-TM alloys have been studied by both of the above-mentioned research groups, and by a dozen or so previous workers whose results were briefly reviewed by CHANDRASEKARAN et al. [CHA74]. For alloys both within and outside the composition range for the occurrence of athermal ω phase, $d\rho/dT$ is negative. Electron scattering from thermally reversible athermal ω phase would obviously provide a mechanism for this effect. However, the observation of negative $d\rho/dT$ in quenched Ti-Cr(20 at.%), which on cooling in the electron microscope to −180 °C revealed no ω-phase reflections, suggested that the anomalous resistivity temperature dependence was associated with the instability of the β lattice itself. As shown in Fig. 7.5, for a series of quenched-and-aged Ti-Cr (10, 13, 15, 20 at.%) alloys during aging at 300 °C for more than 16 h, $d\rho/dT$ sigmoidally approached a common positive (normal) value as the precipitation of isothermal ω phase took place.

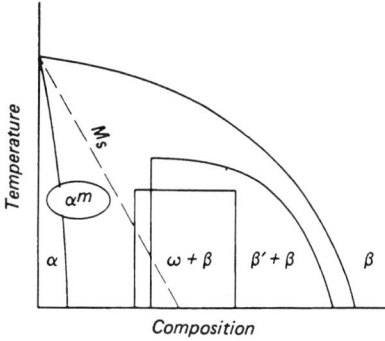

Fig. 7.6 Schematic representation of the locations of the two metastable phases ω + β and β' + β within the equilibrium α + β-phase field in a typical Ti-TM alloy [WIL71].

7.3 Beta-Phase Separation

7.3.1 Occurrence of the β' Precipitate

When the temperature is too high [LUH70][WIL71] or the alloys too concentrated [WIL76] to support the ω-phase precipitation, a solute-lean bcc precipitate, designated β', separates out. The relationship in temperature/composition space between the metastable ω + β- and β' + β-phase fields is indicated schematically in Fig. 7.6, which itself is based on Fig. 5.1 and 7.2 and some suggestions by WILLIAMS et al. [WIL71]. As with ω + β, the β' + β mixed phase is metastable; but unlike ω, which can be generated by a displacement wave in a virtual crystal if need be, the β' precipitate stems from the chemical differences between the solute and solvent atoms. Thus, the β → β' + β "phase-separation" reaction, as it is called, is a clustering reaction characteristic of alloy systems which show positive heats of mixing [CHA72a] or equivalent manifestations of a tendency for the alloying constituents to unmix. Partly as a consequence of the time-temperature conditions for β' formation and ω-phase precipitation, the former (or at least the precursor stages of it) was for a time confused with ω phase. This is no longer the case, the status of β' as a metastable bcc precipitate now being well established [WIL73, WIL76].

The β → β' + β reaction has been studied in considerable detail using Ti-Cr [LUH70][NAR71] and Ti-Mo [KOU70, KOU70a][NAR70][CHA72a] as model systems, although its appearance in more complicated alloys (e.g., Ti_{72}-Zr_8-V_{20}) is well documented [WIL73, WIL76]. Typical examples of β' precipitate morphology are given in Fig. 7.7. It is interesting to note that a structure very much similar to that illustrated has been discovered in the Soviet alloy 65 BT (composition about Ti_{40}-Zr_7-Nb_{53}) after water quenching from 1250 °C and aging 2 h/700 °C [BUY70]. Evidence for phase separation in Ti-35Nb(22 at.%) has been offered by MENDIRATTA et al. [MEN71]. Ti-15Cr, when quenched, appeared to be single-phase bcc [NAR71] and after 5½ min/300 °C exhibited the characteristics of diffuse ω phase. But after holding the temperature at 450 °C for 30 min, β' precipitation in the form of thin discs (~ 0.1 μm across) lying parallel to the {100} planes of the matrix [LUH70] were clearly discernible when photographed at a magnification of 52,000 ×. In this case, apparently, for aging times longer than 90 min at 450 °C, α precipitation commenced at the β'/β interfaces [LUH70].

Using Ti-V and Ti-Mo alloys as host systems, WILLIAMS and colleagues [WIL71] have considered the influences of temperature, composition, and third-element addition on the relative stabilities of the ω and β' phases and the precipitation processes that are associated with them. An interesting conclusion that can be drawn from their results is that α stabilizers such as aluminum, tin, and oxygen, as well as zirconium, when added to those bases in sufficient quantity, increased the stability of the bcc lattice, thereby suppressing ω-phase formation in favor of β-

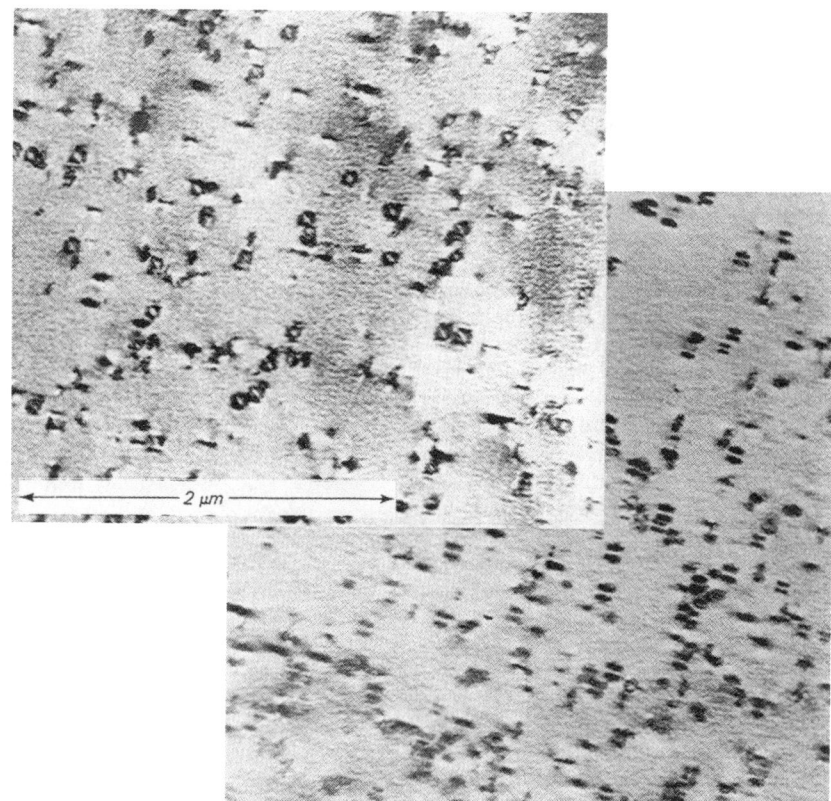

Fig. 7.7 Example of β' precipitation in a β matrix; two orientations of the same sample are represented. Specimen: Ti_{72}-V_{20}-Zr_8 [WIL73]. Micrographs courtesy of J.C. Williams.

phase separation. It was suggested on those grounds that the presence of aluminum in the β-stabilized commercial alloy Ti-8Mo-8V-2Fe-3Al and aluminum plus zirconium in Ti-3Al-8V-6Cr-4Mo-4Zr (β-C) would stimulate the phase-separation reaction as an intermediate step toward a very uniform distribution of α phase under suitable heat treatment (see below). Although their results were unable to be substantiated by later measurements [MOR80], RHODES and PATON [RHO77] amply confirmed the prediction of WILLIAMS et al. [WIL71] by detecting β' precipitation in β-C in response to aging at temperatures of 350 °C (precipitation after 24 h) to 500 °C (precipitation after ½ h).

7.3.2 Thermodynamics of the Phase-Separation Reaction

Although early studies indicated that Ti-Mo was a short-range-ordering system, the converse is now known to be true [CHA72a]. HOCH and VISWANATHAN [HOC71], using a relative-vapor-pressure technique, have determined the activity of molybdenum in titanium and shown it to be positive; and using the more direct technique of low-angle x-ray scattering, GEHLEN has shown that Ti-Mo is a clustering system [HO69]. Consequently, in a temperature-composition zone lying just outside the ω + β-phase field, the tendency for compositional modulation (which underlies the establishment of the equilibrium α + β field itself) stimulates in Ti-20Mo, for example, at 450 °C [CHA72a] a β → β' + β phase-separation reaction similar to that just noted for Ti-15Cr. These examples serve to illustrate the general clustering tendencies of solid solutions of titanium with transition metals—tendencies which may be gauged, for example, by the algebraic sign of the thermodynamic interaction parameter Ω_{ij}. A set of interaction-parameter values for some representative transition metals and simple metals against titanium has been assembled by COLLINGS and GEGEL [COL75a]. As pointed out by those authors, the simple metals, which yield negative interaction parameters, give rise to α-stabilized, short-range-ordered systems when alloyed with titanium. Conversely, the transition elements, which have positive interaction parameters, are clustering-type β stabilizers. Although KOUL and BREEDIS [KOU70, KOU70a], NARAYANAN and ARCHBOLD [NAR70], and GULLBERG et al. [GUL71] used a bimodal free-energy/composition curve to treat phase separation as an equilibrium event, it must be remembered that β' + β is a *metastable* mixture with regard to which only an inflected free-energy/composition relationship (implying spinodal decomposition, wherever $\partial^2 g/\partial c^2 < 0$) is a sufficient description (Fig. 7.8). A bimodal free-energy/composition curve, to be considered below, implies the existence of a miscibility gap in the β field. Whereas this is not a feature of any of the Ti-TM phase diagrams, the Zr-Nb system does possess an *elevated-temperature* field representing, in equilibrium, two immiscible β phases designated β' (solute-lean, as before) and β" (solute-rich). The equilibrium β' + β" phase is to be discussed below in connection with the Ti-Zr-Nb system.

7.4 Alpha-Phase Precipitation From Beta Alloys

7.4.1 Direct Precipitation

Outside the composition- and temperature ranges of the ω or β' phases, represented schematically in Fig. 7.6, but still within the equilibrium α + β field, α-phase precipitation must take place "directly" from the β phase. Mechanisms of direct α-phase precipitation have been discussed in considerable detail by WILLIAMS [WIL73, WIL76]. A well-known morphology is the Widmanstätten α phase which, since it forms as platelets parallel to $\{110\}_\beta$ (Fig. 5.5), assumes under optical magnifications a characteristic basketweave-like structure. The interface between the α plates and the matrix is then the source of further, apparently athermal [WIL76], α phase. Other forms of direct α-phase precipitation which have been discussed by WILLIAMS take on the form of large clumps >1 μm across [WIL73], or dispersions of fine particles ~400 Å in diameter [WIL76]. Also of great interest when considering the morphology and kinetics of α-phase precipitation are the ways in which α phase nucleates heterogeneously from "defects" such as grain boundaries, dislocations, and β'- and ω-phase precipitates.

7.4.2 Precipitation From the β' + β Phase

In Ti-15Cr, according to LUHMAN [LUH70], prolonged aging in the β' + β field (e.g., >90 min/450 °C) resulted in a nucleation of α phase at the β'/β interfaces. On the other hand, WILLIAMS et al. [WIL71] were careful to point out that for the Ti-Mo-Al alloy which they studied, nucleation of the α phase took place *within* the β' region, indicating that in this case the composition difference between precipitate and matrix was the controlling factor. With only two examples, insufficient information is available to establish any systematics concerning the relative roles of lattice mismatch and composition gradient in the β' → α nucleation reaction. An ability to nucleate α seems to be a general property of β' precipitates, but the subsequent growth of the α phase may be either acicular (a fine dispersion of starlike objects) or globular [WIL76].

7.4.3 Precipitation From the ω + β Phase

As indicated in Table 7.1, the overaging of ω + β generally results in α-phase precipitation. Several precipitation modes have been identified, none of which involves a direct conversion of ω phase to α. As a matter of fact, in only one alloy system, Ti-V, is the ω-phase precipitate itself involved in the conversion process; in this system, BLACKBURN and WILLIAMS [BLA68] demonstrated the growth of α plates at the ω/β interfaces. The exceptional role played by Ti-V in this regard is not surprising in view of the fact that the ω phase in Ti-V possesses one of the largest lattice misfits [HIC69]. In Ti-Mo and Ti-Nb, which support the lowest-misfit ω phases, α precipitation takes place elsewhere than near the ω-precipitate site. In Ti-Mo [BLA68], large-scale lamellae of α and β phases have been observed growing from a grain boundary, and α plates have been noted to nucleate at dislocations. Similarly, in Ti-Nb [BRA67] α phase nucleating at β grain boundaries grew to consume the β and ω phases.

With alloys such as Ti-Nb, lattice defects not only act as nucleation sites for α precipitation, but also provide strain energy to accelerate the transformation process. The influence of prior deformation on the aging process is considered below.

7.5 Down-Quenching and Up-Quenching: Omega Reversion

By taking an alloy previously aged isothermally in the ω + β field and, by raising the temperature, placing it in the β' + β field, it is possible to exert some control over the β-phase separation reaction [LUH70]. The isothermal ω phase "reverts" at the higher temperature to a β phase leaner in solute content than the matrix. The reaction is not reversible, the β-phase precipitates remaining in place as the alloy is returned to room temperature. The new phase could be thought of as β', since it is stabilized by the inherent tendency of phase separation (clustering) to occur in the parent β alloy [LUH70]. In studies of Ti-15Cr, LUHMAN [LUH70] observed that a larger number-density of finer precipitates resulted from up-quenching to 450 °C for 2½ min from isothermal ω + β, aged at 300 °C, than were found in the same alloy down-quenched to 450 °C from the β field. Since, as before, β' provides a site for the nucleation of α pre-

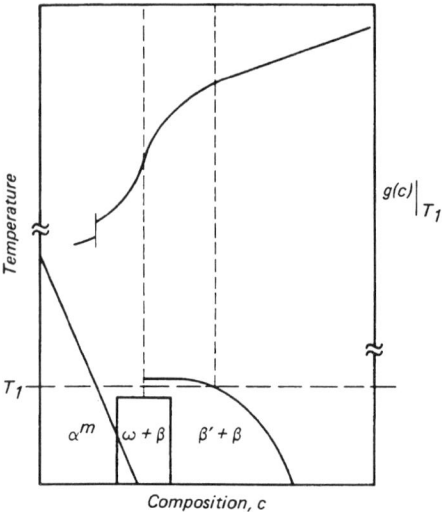

Fig. 7.8 Schematic metastable equilibrium phase diagram for quenched-and-aged Ti-TM alloys accompanied by a postulated free energy, g, versus composition, c, curve for some temperature, T_1, showing a segment with negative curvature corresponding to the phase-separation region of the phase diagram.

cipitates, these too can be produced as a fine dispersion as a result of this duplex aging treatment which, therefore, affords a means of obtaining the strength that is characteristic of fine uniform precipitation, but that is unencumbered by the embrittling effect of ω phase [LUH70]. It is useful to note in conclusion that, even prior to α-phase precipitation, the modulated β' + β structure itself is associated with enhanced strength compared with that of the quenched homogeneous β alloy, and greater ductility than that of ω + β [POL71].

7.6 Effect of Third-Element Additions on Precipitation in Quenched-and-Aged Ti-TM Alloys

Current fundamental understanding of the manner in which third-element additions influence the ω + β and β' + β decomposition reactions stems from an investigation by WILLIAMS et al. into the effects of zirconium, aluminum, tin, and oxygen additions on precipitation in Ti-V(20 at.%) and Ti-Mo(6 at.%). Their results can be readily appreciated in light of the preceding sections. As before, precipitation effects in the ω + β and β' + β regimes, respectively (see Fig. 7.6), will be treated separately.

7.6.1 The Ternary ω + β Regime

Ternary additions of aluminum and oxygen reduced the time of stability of the ω phase during aging by promoting earlier precipitation of α phase. In Ti-V(20 at.%), for example, addition of 3 at.% Al reduced the time required at 400 °C to produce a given volume fraction of α precipitation from 100 to 24 h; and whereas in the binary alloy, grain-boundary nucleation was preferred, aluminum in solid solution promoted the formation of a uniform dispersion of precipitates. $Ti_{74.5}$-V_{20}-$Zr_{5.5}$, although within the ω + β field, is close to its boundary with β' + β (to be discussed below). Thus, as compared with the results of aging in the binary system, that alloy yields a low volume fraction of isothermal ω phase. Furthermore, as a consequence of the now relatively small lattice misfit, the precipitate is ellipsoidal rather than cubic.

7.6.2 The Ternary β' + β Regime

As was noted above in connection with binary alloy decomposition, aging in the β' + β regime, which lies outside the ω + β field (Fig. 7.6), eventually results in β'-nucleated α precipitation. Additions to binary Ti-TM alloys of solutes such as zirconium and tin tend to stabilize the bcc structure when dissolved in it.* This is not to say they are "β stabilizers" in the conventional sense, but rather stabilizers of the bcc lattice against the ω instability. Zirconium and tin thus impose constraints on the ω + β regime to such an extent that alloys such as Ti_{72}-V_{20}-Zr_8 and Ti_{84}-Mo_{10}-Sn_6 are left outside it and consequently lie in the β' + β field. Aging then brings about the phase-separation reaction which results in the precipitation of β', the site for eventual precipitation of finely dispersed α phase.

Finally, it is important to recall that irrespective of whether the alloys were in the ω + β or β' + β field, the electron diffraction patterns were characterized by the reciprocal-lattice "streaking" effect discussed earlier in connection with the $\langle 111 \rangle_\beta$ longitudinal displacement-driven ω-phase reaction. These results confirm that whether or not an ω or a β' transformation is eventually to take place, bcc transition metal alloys (along with other classes of bcc metals; see for example SMITH et al. [SMI76]), support to some degree the $\frac{2}{3} \langle 111 \rangle$ soft phonon.

7.7 Beta-Phase Immiscibility

In contrast to the β → β' + β phase-separation reaction discussed above is a second type of β-decomposition process with which, in one way or another, it has been frequently confused. Below a transus delineating the upper boundary of a region referred to as a "miscibility gap," a previously homogeneous single-phase-β solid solution decomposes into a *thermodynamically stable* pair of bcc phases — a solute-lean β' and a solute-rich β".

An ideal example of β → β' + β" decomposition is to be found in the system Zr-Nb, whose equilibrium phase diagram is comparable to that of the general Ti-TM eutectoidal alloy system (Fig. 2.4) with the titanium replaced by zirconium and the intermetallic compound ($TiTM_x$, or simply γ) replaced by β". Indeed, common to precipitation from both the aged β' + β" (and in fact α + β") and the aged α + γ-phase fields is the lamellar, cellular, or pearlitic morphology characteristic of a classical discontinuous growth (see [CHR65, p. 472]). Precipitation from within the β' + β" field does, however, possess an additional feature (to be discussed in detail below) not shared by the α + γ precipitation process.

Other examples of β-phase immiscibility are to be found in ternary alloys based on Zr-Nb, such as the Ti-Zr-Nb system of Figs. 4.20 and 7.9. The development of the equilibrium β' + β" region from the Zr-Nb edge of the composition triangle is exemplified in Fig. 7.9, where it is juxtaposed against the appropriate isothermal section of the Zr-Nb binary phase diagram. Associated with the binary diagram is a

---

* The solution strengthening of both α-stabilized and β-stabilized Ti-TM alloys has been discussed by COLLINGS and GEGEL [COL75a] [GEG73a].

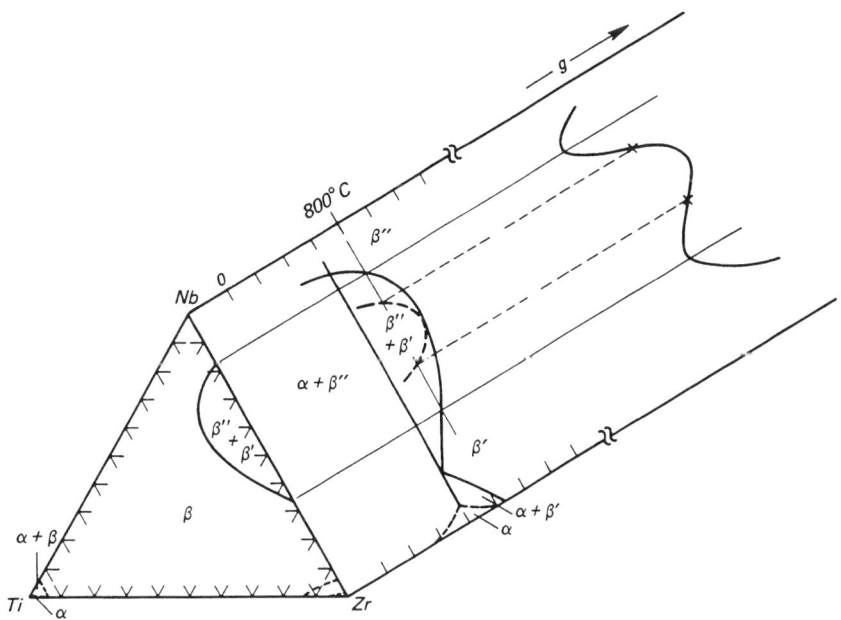

Fig. 7.9 Binary equilibrium phase diagram for Zr-Nb [Lov66] projected from a side of the ternary (at.% linear) Ti-Zr-Nb 800 °C-equilibrium phase diagram [Doi66]. The two minima in the bimodal *g(c)* curve define a region of equilibrium β-phase immiscibility, β' + β", within which is a region of spinodal decomposition delimited by the points of inflection on *g(c)*.

bimodal free-energy curve representing the existence *in equilibrium* of two β solid solutions. Such a curve has been incorrectly applied to the β′ + β *metastable* situation [KOU70,KOU70a] [NAR70] [GUL71]. Thermodynamic equilibrium requires a free-energy versus composition minimum, as distinct from the inflected type of free-energy versus composition relationship that characterizes the metastable phase-separated condition of Fig. 7.8. But in contrast to α + β phase equilibrium which is described by two independent free-energy parabolas, equilibrium phase separation with its absence of structural change must be described by a *continuous* curve with two minima. An important corollary to this is the necessary presence of an intervening maximum with its associated description in free-energy terms: $\partial^2 g/\partial c^2 < 0$. In other words, away from the edges of the two-phase field, the β → β′ + β″ equilibrium process is preceded by spinodal decomposition. In Fig. 7.9, two points on the spinodal (indicated by the dashed curve on the phase diagram) are determined by the two points marked "×" on the free-energy (g) versus composition (c) curve for which $\partial^2 g/\partial c^2 = 0$. Precipitation phenomena interpretable in this manner have been seen during metallographic studies of quenched-and-aged Ti_{10}-Zr_{40}-Nb_{50}. During the early stages of its aging, according to KITADA and DOI [KIT70], a very fine precipitate developed, presumably spinodally, throughout the grain interior such that after 10 h/700 °C the hardness had increased from 280 to 338 kg/mm². In the meantime, a lamellar or pearlitic precipitate of alternate β′ and β″ regions, of much lower hardness (~200 kg/mm²), had begun to develop outward from the grain boundaries.

Part 2: The Kinetics of Precipitation

The equilibrium and metastable phases which appear during aging of quenched β-Ti-TM alloys were considered in Chapters 4 and 5 and in Part 1 of this chapter. These *same phases* appear during aging of deformed alloys but at *different rates* determined by the degree of strain and the level of retained stress. The following sections, therefore, deal with some of the factors that control the *kinetics* of the aging process.

7.8 The TTT Diagram

The traditional descriptor of transformation kinetics is the time-temperature-transformation (TTT) diagram in which phase boundaries are plotted semilogarithmically against temperature

and time. Sets of the typically C-shaped TTT diagrams for the systems Ti-V, Ti-Cr, Ti-Mn, Ti-Fe, Ti-Ni, and Ti-Cu are to be found in [MAY61] and the further references contained therein. Such diagrams are usually obtained by detecting isothermal transformations in samples which had initially been quenched from single-phase elevated-temperature fields to the temperatures of observation. It must be borne in mind, however, that such curves will not be coincident with those obtained by performing isothermal aging on specimens previously quenched to room temperature [MAY61], although they will be of the same general form. The characteristic C-shape of the TTT diagram (Fig. 7.10) demands an explanation. Microstructural studies of samples quenched from various regions of the temperature-time diagram indicate that, at high temperatures, transformation begins at lattice defects such as grain boundaries and grows into the grains (discontinuous precipitation), while at the low-temperature extreme the precipitation tends to take place uniformly throughout the grains (continuous precipitation). These effects result from the fact that the driving force for nucleation (the Δg between the pairs of phases involved) scales with some measure of the degree to which the energy of the quenched alloy differs from its equilibrium value. Once nucleated, the precipitate growth rate increases with temperature; consequently, the transformation rate, which is proportional to the product of nucleation rate and growth rate, is maximum at some intermediate temperature [CHR65, p. 453]. The separation of the C-curve into a low-temperature nucleation-controlled region and a high-temperature growth-controlled region leads to an understanding of the manner in which stress, strain, and interstitial atoms influence its shape. Thus, if the density of nuclei for transformation is increased, by deformation or some other means, the effect will be to increase the transformation speed at high temperatures, where recovery tends to eliminate nucleation sites, much more than at low temperatures, where the driving force for nucleation is high and the alloy already well supplied with nucleation sites. The result of combining a more nearly uniform distribution (with temperature) of nucleation rates with a growth rate that increases with temperature is to shift the C-curve toward the temperature axis by an amount which increases uniformly with temperature; in doing so, this also has the effect of raising the temperature of the "nose" in the manner depicted in Fig. 7.10. Several specific examples of this effect are given below.

The (time, temperature) data pairs representing some selected fraction of

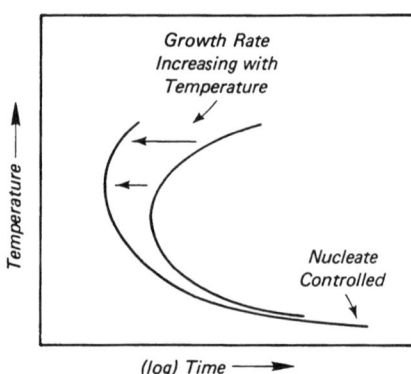

Fig. 7.10 Schematic representation of factors that control the form of the TTT curve. For a given nucleation rate, the indicated increase in reaction rate (= product of nucleation and growth rates) has the effect of raising the "nose" of the TTT curve and giving it a "forward tilt."

material transformed (e.g., 5% for the start of the transformation and 95% for its end) may be obtained either at-temperature, using a diagnostic technique such as dilatometry or electrical resistometry [HOR73], or from interpretation of quenched structures using hardness or metallography, which in any case should always be employed as a semidirect observation and control. Two mechanical variables—stress level and degree of plastic deformation, both of which have strong influences on the *decomposition kinetics* of β-Ti-TM alloys—are considered below. Also to be discussed are the influences of third-element additions. These have, of course, just been considered in connection with aging transformations in *quenched* β alloys, but inasmuch as substitutional elements have a pronounced tendency to accumulate at dislocations and grain boundaries during aging, they also play an important role in the phase decomposition of plastically deformed alloys.

7.9 Influence of Stress on the Transformation

Studies by GOLDENSTEIN *et al.* [GOL59] on a series of Ti-Cr alloys, which underwent the eutectoid transformation α + β → α + $TiCr_2$, provide a useful example of stress-assisted* decomposition. Application during isothermal aging of a relatively low level of stress (capable of producing 1% plastic strain in 1000 h) reduced the time needed to complete the eutectoid reaction in Ti-12Cr at 550 °C from 200 to 32 h. In a like manner, the so-called "internal stress fields" (see TYSON

---

* The term "stress" as used here implies elastic strain as distinct from the (heavy) plastic strain to be discussed later.

[TYS75]) of lattice defects may also speed the transformation process.

7.10 Heavy Plastic Deformation and Its Properties Under Heat Treatment

It is common practice in alloy-preparation laboratories to heavily deform, if possible, the as-cast billet to accelerate the equilibration reaction during a subsequent homogenization heat treatment. This section considers the effect of aging on microstructure and precipitation in heavily cold-worked β-Ti-TM alloys, particular emphasis being placed on the influence of deformation by wire drawing.

7.10.1 Influence of Aging on the Fibrous Cell Structure

During the early stages of moderate-temperature aging, the intracell dislocations migrate to the cell walls, then at higher aging temperatures extremely rapid cell growth sets in. In what follows, the mechanism and kinetics of this process will be discussed.

Dislocation Motion. A strongly cold-worked alloy with subband diameters in the range 100 to 700 Å may contain a dislocation density, within the cells, of about 10^{12} cm^{-2} [PFE68].* During moderate-temperature heat treatment (<~300 °C, [NEA71]), migration of the dislocations to the cell walls begins to take place by some sort of thermally activated slip process. Evidence for the occurrence of this perfectly general phenomenon, which is discussed in standard works on dislocations (e.g., [FRI64]), is to be found in the sharpening of the x-ray line pattern [CHA70]. Dislocation density is still higher after 1 h/385 °C in the large cells which result from low levels of cold deformation but much lower, according to NEAL et al. [NEA71], in the narrow cells of the heavily reduced materials with their higher levels of strain energy and shorter path lengths. Viewed in cross section, the cells possess curved boundaries and are irregular in size; then during the early stages of heat treatment the cell walls become straight [NEA71] as in the early stages of classical polygonization [FRI64] [CHA70]. Referring to the 1-h annealing of heavily cold-worked $(5 \times 10^4:1)$ Ti-58Nb, the intracell dislocation density is lowest at 400 °C [NEA71], but the cell structure itself is on the threshold of growth; at about 380 °C the deformation-cell structure seems to be moderately stable [NEA71].

---

* A cell of diameter 500 Å (area = 1.96×10^{-11} cm^2) would enclose 20 dislocations.

Cell Growth. Cell growth during heat treatment of deformed Ti-Nb alloys has been considered briefly by several authors [REU66] [NAR66] [PFE68] [BAK70] and investigated quantitatively by NEAL et al. [NEA71] and ARNDT and EBELING [ARN74] [WIL75], the results of whose work on this subject are combined in Figs. 7.11(a) and (b). During annealing at 385 °C, the cell diameter increased from 443 to 536 Å in the first hour, and then remained stable at 530 ± 6 Å at least for another 24 h [NEA71]. At 500 °C the cells grew to diameters of more than 2000 Å after just a few hours [ARN74] [Fig. 7.11(a)]. Although at temperatures below 385 °C the cells grow (<20%) to stable size within about 1 h, at 400 °C and above the cell size is strongly dependent on both aging time and temperature [Fig. 7.11(b)].

7.10.2 Influence of Heavy Deformation on the Kinetics of Precipitation

The literature abounds with descriptions of the microstructures and precipitates which occur during the aging of cold-formed alloys. For the purpose of this discussion, the alloy system Ti-Zr-Nb has been selected because of its ability to support several decomposition modes.

As indicated in Fig. 4.20, the Zr-Nb-rich members of the system may exist in any one of four equilibrium-phase fields: (1) α + β, (2) α + β' + β", (3) the miscibility-gap region, β' + β", or (4) the single-phase β region at elevated temperature. Figure 7.12, constructed from electrical resistivity data, describes the kinetics of transformation of previously deformed Ti$_{4.4}$-Zr$_{40.7}$-Nb$_{54.9}$ into two of these regions, viz: (a) into discontinuous β' + β" from predeformation equilibrium in the β regime at 1200 °C, and (b) into α + β from a predeformation equilibration in the miscibility-gap region at 700 °C, as functions of percent area reduction by cold drawing. Obviously the deformation introduced by the cold work has a major influence on the transformation speeds. The effect of deformation is greatest toward the upper temperature limits of each range of stability; thus, each C curve is shifted, and tilted toward the temperature axis, and the "nose" temperature is raised.

7.11 Influence of Interstitial-Element Additions on the Kinetics of Precipitation

Considered in this section are the ways in which the interstitial elements boron, carbon, nitrogen, and oxygen influence phase decomposition during aging of cold worked β-Ti-TM alloys. Precipitational effects associated with the presence of oxygen in Ti-Nb(35 at.%) and Ti-Nb(46 at.%) have been discussed by WITCOMB and DEW-HUGHES [WIT73] and REUTER et al. [REU66], and the influence of nitrogen on Ti-56Nb has been considered by BAKER [BAK70]. The kinetics of oxygen diffusion has been discussed by CHARLESWORTH and MADSEN [CHA70], and the influence of oxygen on the kinetics of β-phase decomposition by DELAZARO and ROSTOKER [DEL53] and WILLIAMS et al. [WIL71].

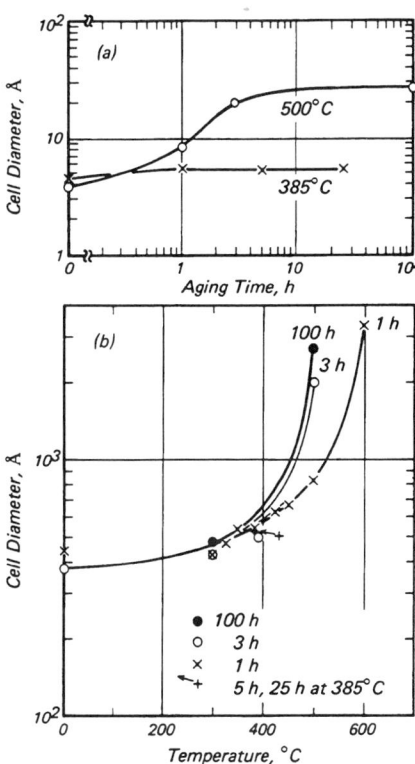

Fig. 7.11 Deformation-cell growth in β-Ti-Nb as a function of (a) time and (b) temperature. Data sources: ARNDT et al. [ARN74] [prior deformation 99.98% (o)] and NEAL et al. [NEA71] [prior deformation 99.998% (x)].

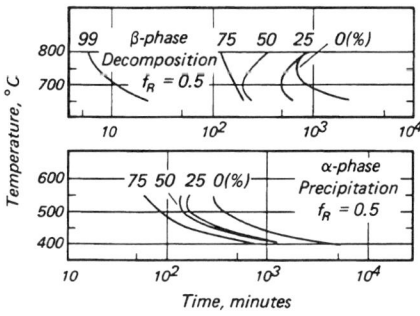

Fig. 7.12 TTT diagrams representing the β-phase decomposition and α-phase precipitation kinetics in Ti$_{4.4}$-Zr$_{40.7}$-Nb$_{54.9}$ cold drawn to the percentages indicated. The degree of completeness of the transformation, in resistometric terms, is $f_R = 0.5$ (where f_R is the fractional change in electrical resistance) [SOE69].

Interstitial elements are extremely potent strengtheners of titanium alloys as a consequence of their ability to pin dislocations (see Section 12.8.1). Although it is unlikely, in view of the high mobility of interstitials, that dislocation pinning takes place *at temperature*, several studies of oxygen-doped Ti-39Nb [COM67] have indicated that coarser deformation substructures can be expected to result from the presence of strengthening centers (see also Section 6.4.2).

The diffusion of interstitial and substitutional elements in titanium has been thoroughly reviewed by ZWICKER [ZWI74, Chap. 5], some of whose collected data have already been used in the construction of Fig. 3.20, which describes the relative rates of diffusion of the $3d$ and some $4d$ transition elements in β-Ti. Characteristic data (i.e., frequency factor and molar activation energies) describing the diffusion of C, N, O, and H in α- and β-Ti are available in the literature [ZWI74, p. 104]; those for oxygen in β-Ti are reproduced in Table 7.3. Their reliability is obviously open to question; for example, they lead to diffusivities for oxygen in β-Ti at 1000 °C of from 3×10^{-7} to 8×10^{-9} cm²/s. CHARLESWORTH and MADSEN [CHA70] have also assembled a set of diffusivity data (including an *atomic* activation energy expressed in eV) for C, N, O, and H in β-Ti as well as in unalloyed niobium at moderate temperatures (Table 7.4). In the absence of published diffusivity data for interstitials in Ti-Nb alloys, those authors assumed that an activation energy for diffusion in Ti-Nb(45 at.%), cited in this context as a typical β-isomorphous Ti-TM alloy, could be adequately extracted by linear interpolation between the published values for niobium and β-Ti. The values so obtained are included in Table 7.4. Bearing in mind the inaccuracy of diffusivity data for interstitial elements in transition metals, it was suggested by CHARLESWORTH and MADSEN that an atomic activation energy of 1.55 eV for C, N, and O in Ti-Nb(45 at.%) would reasonably well approximate the experimental situation. Finally, a value of 2.0×10^{-2} cm²/s (see the oxygen entry, Table 7.4) was suggested for the frequency factor of oxygen, the most mobile of all three interstitials in niobium. At a temperature of 380 °C, for example, these data then yielded a diffusivity for oxygen of 2.2×10^{-14} cm²/s. In order to establish a benchmark for comparison, Fig. 7.13, which presents a set of diffusivity data for ^{95}Nb in a series of Ti-Nb alloys (with appropriate extrapolations where necessary), may be consulted. If this is done, it will be found that oxygen diffuses in Ti-Nb(45 at.%) at 380 °C as rapidly as niobium does at about 800 °C. This leads immediately to the interesting conclusion that as cell growth in Ti-Nb begins to accelerate with increase in temperature above 400 °C (Fig. 7.11), the interstitial atoms will long since have diffused to, and become trapped by, the cell walls. Alternatively, using linear random walk theory, according to which an atom moves a root-mean-square distance from its starting point of $\sqrt{2Dt}$ cm in t seconds, it can be seen that an interstitial atom will migrate a distance of 1300 Å within the Ti-Nb alloy during a heat treatment of 1 h/380 °C; by the same token, the time taken for it to traverse one-half of a typical subband diameter at that temperature is about 1.5 min.

Interstitial-atom diffusion has an important influence on cell-wall precipitation. The elements B, C, N, and O are all stabilizers of the α phase in titanium and as such are expected to encourage α-phase precipitation as they accumulate at the cell boundaries or residual dislocations. Furthermore, as a solid solution strengthener, dissolved oxygen tends to stabilize the bcc lattice. Taken together, these comments agree with the observation that oxygen retards ω precipitation but accelerates β → α + β decomposition. The accelerating effect of oxygen on the ki-

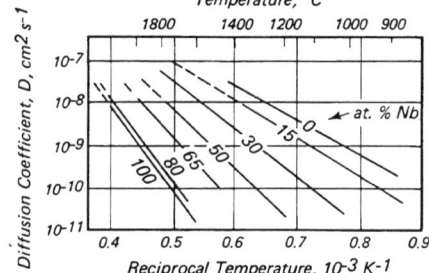

Fig. 7.13 Temperature dependence of the diffusion coefficients of ^{95}Nb in Ti-Nb alloys [ZWI74, p. 112].

Table 7.3 Diffusion Parameters for Oxygen in β-Ti [ZWI74, p. 104]

Temperature range °C	°F	Frequency factor, D_0, cm²/s	Molar activation energy, kcal/mol	Diffusivity at 1000 °C (1830 °F), D, cm²/s
930-1150	1706-2070	31,400	68	5.0×10^{-8}
932-1142	1678-2056	330	58	3.6×10^{-8}
950-1415	1742-2579	1.6	48	8.4×10^{-9}
907-1030	1665-1886	0.45	36	3.0×10^{-7}
1135-1355	2075-2471	8.3×10^{-2}	31	3.6×10^{-7}

Table 7.4 Diffusion Parameters for the Interstitial Elements C, N, and O in Nb, β-Ti (at elevated temperatures), and Ti-Nb(45 at.%) [CHA70]

Interstitial element	Solvent	Temperature range °C	°F	Frequency factor, D_0, cm²/s	Atomic activation energy, eV	Diffusivity at 380 °C (750 °F), cm²/s
Carbon	Nb	130-280	266-536	4×10^{-3}	1.43	3.7×10^{-14}
		150	302	1.5×10^{-2}	1.15	2.0×10^{-11}
	β-Ti	950-1150	1742-2102	108	2.1	...
	Ti-Nb(45 at.%)	...	...	...	1.68-1.80	...
Nitrogen	Nb	150-300	302-572	8.6×10^{-3}	1.51	1.9×10^{-14}
		300	572	9.8×10^{-2}	1.65	1.8×10^{-14}
	β-Ti	900-1570	1652-2858	3.5×10^{-2}	1.46	...
	Ti-Nb(45 at.%)	...	...	...	1.49-1.57	...
Oxygen	Nb	40-150	104-302	2.0×10^{-2}	1.17	1.9×10^{-11}
		150	302	1.5×10^{-2}	1.2	8.2×10^{-12}
	β-Ti	950-1414	1742-2577	1.6	2.09	...
	Ti-Nb(45 at.%)	...	...	...	1.68-1.69	...

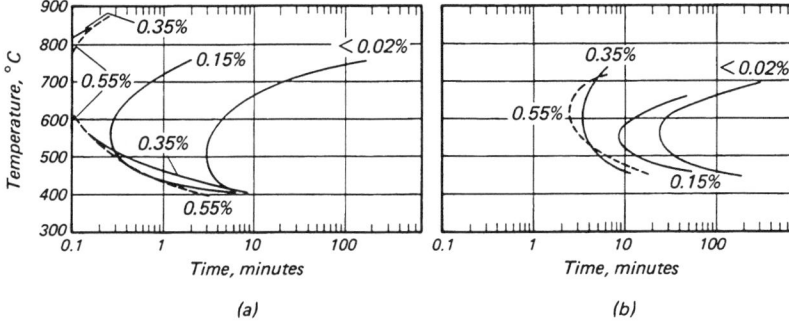

Fig. 7.14 Influence of oxygen (present at the percentages indicated) on the β → α + β transformation kinetics of the β-isomorphous alloy Ti-11Mo; (a) transformation start, (b) transformation finish [DEL53].

netics of α-phase precipitation in Ti-Mo has been graphically demonstrated by DE-LAZARO and ROSTOCKER [DEL53] (Fig. 7.14).

Part 3: Aging of Some Technical Titanium-Base Alloys

The aging effects encountered in commercial multicomponent alloys, although often complicated by the mixture of phases present in the starting materials depending on their compositions or thermal histories, can be interpreted in terms of the aging properties of binary alloys. As indicated above, further control over the precipitational processes and microstructures can be exerted by associating the heat treatments with mechanical deformation. The precipitational effects likely to be encountered during aging of technical titanium-base alloys are illustrated here by three case studies.

7.12 Aging of Ti-6Al-4V

The effect of heat treatment on the properties of Ti-6Al-4V, as catalogued by MCQUILLAN [MCQ63, p. 72] and WOOD [WOO72, pp. 1-4:72-11], can be appreciated to some extent with the aid of the equilibrium phase diagrams of Figs. 4.11 to 4.13. The aging behavior will of course depend on the initial state of the alloy. For example, depending on the temperature at which it was held prior to being quenched, the alloy could consist of either: (1) all martensite, (2) primary α plus martensite, or (3) primary α plus retained β [MCQ63]. In case the initial condition is martensitic: (1) the α' recovers toward an α possessing the appropriate equilibrium composition; and (2) the boundaries and internal structures (Section 5.2.2) of the plates provide heterogeneous nucleation sites for the precipitation of β [WIL73]. The vanadium content of the β and the aluminum content of the α naturally play important roles in the precipitation process. With regard to the aluminum component, the precipitation of α_2 can contribute to the strengthening of the alloy.

7.13 Aging of Ti-6Al-2Sn-4Zr-2Mo

Alloys such as Ti-6Al-2Sn-4Zr-2Mo provide enormous opportunities for studying the effects of thermomechanical processing on microstructure. Again, the aged properties will depend on the combination of initial heat treatment and mechanical deformation to which the material has been subjected. In actual practice, for this alloy the thermomechanical processing will be an integral part of a forging operation. Properties sought are a combination of high strength and fracture toughness and good creep resistance. The creep properties of Ti-6242(Si) alloys can be maximized in the following way [CHE80]: (1) for α + β forgings, by a β solution treatment* followed by aging for 8 h/593 °C/air cool; (2) for β forgings, the solution heat treatment should be conducted in the α + β field. The microstructural goal of these processing operations is to achieve a fine Widmanstätten structure and to avoid the occurrence of both martensitic α' and globular α, both of which seem to degrade creep resistance and fracture toughness [CHE80].

7.14 Aging of Two Technical Beta Alloys

7.14.1 Introduction

A useful insight into the aging/mechanical-property characteristics of β-Ti alloys in general can be acquired from a study of the properties of the alloy sequence Ti-15Mo, Ti-15Mo-5Zr, and Ti-15Mo-5Zr-3Al [PEN80] [NIS82]. The aging of Ti-15Mo and Ti-15Mo-5Zr at temperatures below about 400 °C resulted in pronounced ω-phase-induced hardening. In Ti-15Mo-5Zr, the isochronal curves of tensile strength versus temperature and elongation versus temperature were practically mirror images of each other about a suitably selected horizontal axis; at 400 °C, the tensile strengths rose to maxima while the elongations dropped to zero [NIS82]. In order to develop α-phase precipitation in Ti-15Mo-5Zr, the aging temperature had to be raised above 400 °C. The exchange of ω phase for α phase resulted in a recovery of ductility. NISHIMURA et al. [NIS82] found that a satisfactory combination of strength and ductility could be acquired by developing a suitable mixture of ω-phase and α-phase precipitates; they did not investigate the ω-reversion phenomenon (see Section 7.5). The presence of 3 wt% Al in the above alloy was sufficient to severely suppress the formation of ω phase. Although such precipitation could be induced to form at temperatures below 400 °C, some two hours of aging were needed to elicit its first appearance. The principal strengthener of Ti-15Mo-5Zr-3Al is α precipitation, which forms at temperatures above about 425 °C, and which is relatively rich in aluminum, a rapid hardener of titanium alloys.

In alloys such as Ti-15Mo, Ti-15Mo-5Zr, and Ti-15Mo-5Zr-3Al, the heat-treatment-induced microstructural mechanisms that control strength and ductility are not amenable to simple explanation. It is not sufficient simply to specify the relative abundances of the ω-phase and α-phase precipitates. Among the other factors that must be taken into consideration are: (1) the grain size of the β matrix [NIS82], and (2) the size and morphology of the α precipitates—for example, whether they: (a) grow out from the grain boundaries as platelets, (b) exist as fine distributions of ellipsoidal particles, or (c) assume a coarse lenticular Widmanstätten morphology [PEN80].

7.14.2 Aging of Ti-11.5Mo-6Zr-4.5Sn (β-III)

As pointed out in Section 4.5.4, the metallurgical behavior exhibited by β-III is similar to that found in Ti-12Mo [BOY74], which may, therefore, be regarded as its binary prototype. Aging of Ti-Mo (<~21 at.%) for several hours at temperatures below 400 °C results in nucleation and growth of isothermal ω-phase precipitates.** In particular, during aging of Ti-Mo(8 at.%) for 150 h at that temperature, HICKMAN [HIC69a] found that 68 vol% of the alloy had become ω

---

* The β transus is at 990 °C and M_s is at 800 °C (see Fig. 4.17).
** As pointed out by HO and COLLINGS [HO73a], the ω/(ω + β) and (ω + β)/β boundaries of the meta-equilibrium ω + β field at 350 °C are at 4.3 and 21 at.% Mo, respectively.

phase with a molybdenum concentration of 4.7 at.%, while the composition of the β matrix was 14.5 at.% Mo(25.4 wt%). The presence of ω phase severely embrittles any titanium-base alloy in which it occurs; for this reason, an early proclaimed disadvantage of β-III was its unsuitability for service at temperatures between about 200 and 425 °C, a very important temperature range within which any high-strength titanium alloy would be expected to perform reliably. It was, however, a simple matter to devise a stabilization heat treatment that would avoid this difficulty. In doing so, FROES et al. [FRO73] developed the pseudobinary equilibrium and meta-equilibrium diagram for β-III (Mo variation) depicted in Fig. 4.22. The effect of heat treatment on β-III, as discussed by FROES et al. [FRO73], BOYER et al. [BOY74], and WILLIAMS et al. [WIL82^a], can be appreciated with the aid of such a diagram.

The stabilization treatment suggested by FROES et al. [FRO73] was an aging for 8 h in the temperature range 550~600 °C to produce about 40% α phase and a β matrix too rich in molybdenum to be susceptible to ω-phase embrittlement during subsequent service (>~8000 h) at 200~425 °C. The molybdenum content of the β-III matrix was 17.2 wt% (9.4 at.%) after the stabilization heat treatment. POLONIS and colleagues (e.g., [LUH71]; see also [CHA78]) have conducted extensive investigations of the ω-reversion effect with particular reference to the Ti-Cr system in which the kinetics of the $\omega \xrightarrow{500\,°C} \beta$ reaction are quite rapid (see Section 7.5). While extending these studies to β-III, the strengths and ductilities of the variously thermomechanically processed alloy were correlated with the observed microstructures [BOY74]. It was hoped that the success previously obtained with Ti-Cr [CHA72] would be repeated for β-III. In the former case, the $\omega + \beta \xrightarrow{500\,°C} \beta' + \beta$ reversion was complete after 5 min; the strength of the original ω-strengthened aged alloy was preserved and the ductility was increased. But with β-III, after a similar treatment, the ω phase was only partially reverted. Some of the strengthening arising from the presence of β phase was retained, and some ductility, as a result of the increased volume fraction of β phase, was acquired. But it was estimated that if complete reversion had been achieved, the strength would have been no greater than that of the solution-treated-and-quenched alloy. Further interesting results were obtained by repeating the aging and reversion heat treatments on alloys that had been deformed after solution heat treatment. In these cases, the α phase which formed at dislocation sites during ω + β aging played a strengthening role at all stages of the heat treatment.

The experiments of FROES et al. [FRO73] and POLONIS et al. [BOY74] were subsequently extended in detail by WILLIAMS et al. [WIL82^a], who studied the microstructural and mechanical responses of β-III to: (1) variation of solution-treatment temperature, (2) variation of α + β aging temperature, (3) duplex aging (ω reversion), and (4) cold work administered prior to various heat treatments. Some of the results can be summarized, with reference again to Fig. 4.22 and to Table 7.5, as follows: (1) In alloys α + β aged 8 h/590 °C after solution treatment (ST) and quenching, a change of ST temperature from 790 °C [i.e., above β/(α + β)], in which case the α precipitation was uniformly distributed, to 720 °C [i.e., below β/(α + β)], which was responsible for a very irregular α-phase precipitate, had no significant effect on the mechanical properties [Table 7.5, cf. rows (a) and (b)]. (2) Aging at a lower temperature than 590 °C, in particular 8 h/480 °C, resulted in a closely spaced array of finer α-phase particles and an increase in strength [Table 7.5, cf. rows (b) and (c)]. (3) Rows (d) and (e) of Table 7.5 compare the effects of 8 h/510 °C aging on samples that had been quenched, and quenched plus ω + β aged, respectively. In the latter case partial ω reversion,* followed in time by α-phase precipitation at the remaining ω-phase sites, seemed to be taking place. As compared with the single aging treatment, the duplex aging, which included a preliminary ω + β aging, engendered a deterioration in the ductility with no improvement in strength.

7.14.3 Aging of Ti-3Al-8V-5Cr-4Mo-4Zr (β-C)

β-C is a commercial metastable β-Ti alloy characterized by high strength and toughness and hardenability by heat treatment. It resulted from the need for an alloy more β-stabilized than β-III, which itself is less β-stabilized than the Ti-13V-11Cr-3Al alloy which it has replaced (see Section 4.5.4).

As a result, presumably, of its containing the right amounts of aluminum and zirconium, this alloy does not precipitate isothermal ω during aging [WIL71] (see also Section 7.4). It does, however, undergo phase separation, the β' precipitates eventually giving way to α phase. The morphology of the latter, which is particularly interesting in this system, has been studied in detail by RHODES and PATON [RHO77]. Although those authors have unequivocally identified β' as the precipitate that formed during the early stages of moderate-temperature aging, MORGAN and HAMMOND [MOR80] in a similar study found only disc-shaped "zones." Otherwise, all authors agreed on the nature of the subsequently formed α-phase precipitates.

Following the work of RHODES and PATON [RHO77], it is convenient to treat the aging of the β-quenched (from >~790 °C) alloys in two temperature regimes, 350 and 500 °C: (1) At 350 °C, β → β' + β begins after 24 h and matures after about 7 days. Fine α-phase precipitates begin to develop about this time and agglomerate into "rafts" of particles after about 14 days. Subsequent aging coarsens the rafts. The precipitate is so-called Type-2 α. It is not Burgers related to the β matrix. (2) At 500 °C, a very short-lived β' precipitate is replaced after about 2 h by α-phase plates. These are Type-1 α plates which, being Burgers related to the β matrix and

---

* At a second-stage aging temperature of 540 °C, complete reversion of the ω phase was achieved and the hardness dropped to that of the as-ST material, in agreement with the earlier observation of POLONIS et al. [BOY74].

Table 7.5 Tensile Properties of Solution-Treated-And-Aged β-III [WIL82^a]

Row	Condition (a)	Tensile strength		Yield strength		Elongation,	Reduction in
		ksi	10^8 N/m^2	ksi	10^8 N/m^2	%	area, %
(a)	ST$_1$ + 8 h/590 °C	142	9.8	133	9.2	12	62
(b)	ST$_2$ + 8 h/590 °C	140	9.7	133	9.2	11.5	58
(c)	ST$_2$ + 8 h/480 °C	188	13.0	179	12.3	5	25
(d)	ST$_2$ + 8 h/510 °C	205	14.1	200	13.8	6.5	39
(e)	ST$_2$ + 50 h/370 °C + 8 h/510 °C	204	14.1	197	13.6	4	13

(a) ST$_1$ = 5 min/790 °C + water quench. ST$_2$ = 5 min/720 °C + water quench

sharing slip systems with it, are in principle able to be sheared during deformation. They are not stable, however, and decompose internally during the next few hours of aging into Type-2 α [RHO77] [MOR80]. As for the mechanical properties, RHODES and PATON [RHO77] found that the best combination of strength and ductility in β-C was achieved in response to the formation at temperatures above 500 °C of large noncoherent Type-2 α precipitates.

8. Titanium Alloys for Low-Temperature Service

Cryogenic applications of titanium alloys include rotors for superconducting generators, and components with aerospace applications such as reusable upper-stage spacecraft and high-pressure rocket engines (for craft such as the Space Shuttle, the Aerospace Plane, and "space tugs"). The reusable rocket engine is a particularly challenging application. Fuel pumps and engines operate at high pressures and must be small and lightweight; high turbine and fuel-pump-impeller tip speeds place high stresses on the components. The search for suitable alloys for cryogenic service is, as usual, the search for materials which under service conditions (in this case, at temperatures near 4.2 K) have adequate strength, ductility, and fracture toughness. The advantages of titanium alloys are high strength, high specific strength (strength/weight), and low thermal conductivity. The major disadvantage is lower fracture toughness than that of fcc ferrous materials under the same conditions.

8.1 Alloy Phase Selection for Low-Temperature Service

A summary of the mechanical properties of several representative α–phase, α + β phase, and β-phase titanium alloys is given in Table 8.1.

The immediate choice for an alloy intended for use at cryogenic temperatures would be an all-α alloy, since the bcc structure at low temperatures generally undergoes a ductile-to-brittle transition (bcc steel is a classical example). To improve the low-temperature ductility of α-phase alloys, the interstitial level is reduced, typically, to below about 0.1 wt% per element, giving rise to the extra-low interstitial (ELI) grade of alloy. But the reduction in interstitial level is accompanied by a reduction in strength.

With the quest for strength as their driving force, designers went from unalloyed titanium (hcp) to α-Ti alloys such as Ti-5Al-2.5Sn(ELI). Likewise, a trend has been to introduce α and α + β alloys having even higher strengths, such as Ti-8Al-1Mo-1V and Ti-6Al-4V, respectively, [HUB73] [SAL79], and in particular, the latter alloy suitably heat treated for maximum cryogenic toughness [NAG84, NAG85]. With this in mind, Ti-6Al-4V(ELI) has been the subject of recent investigations. Recognizing that Ti-6Al-4V is an α + β alloy, it is natural to inquire into the role played by the β-phase component. This will be dealt with subsequently in this chapter.

Due to their poor low-temperature ductility, bcc-phase alloys have never been seriously considered as structural materials for use at cryogenic temperatures. However, one such alloy, Ti-50Nb, necessarily finds widespread use in superconducting machinery. The low-temperature deformation of this alloy is considered below.

Seemingly to demonstrate a nonapplicability of a β-Ti alloy to cryogenic service, SALMON [SAL79] has presented the properties of Ti-13V-11Cr-3Al within the context of those of more cryogenically useful materials (see Table 8.1). The alloy Ti-13V-11Cr-3Al tends to embrittle below 170 K. The solution-treated (ST)-plus annealed alloy has an elongation of 5% at 20 K; the elongation of the ST-and-aged material (a two-phase mixture of α and $TiCr_2$) is 0.2% at 77 K, and evidently is nonexistent at liquid helium temperatures (Table 8.1 [SAL79]).

8.1.1 Unalloyed Titanium and Dilute Alloys of Titanium

Addition of Transition-Metal Solutes. The transition metals—i.e., the β stabilizers—are not rapid strengtheners of titanium. Accordingly, the strength-versus-temperature curves for various dilute α-phase Ti-TM alloys are not strong functions of solute concentration (Fig. 8.1).

Small levels of some β stabilizers are known to be detrimental to the extreme low-temperature properties [RYD85]. For this reason, iron and manganese, common contaminants of commercial titanium, must be removed from material intended

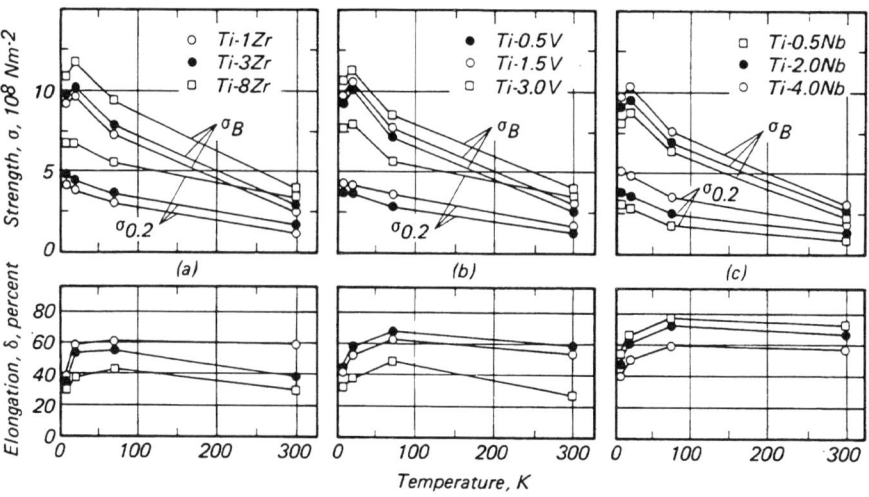

Fig. 8.1 Yield strength (0.2%) and tensile strength, with associated elongations at fracture, of low-concentration Ti-Zr, Ti-V, and Ti-Nb alloys as functions of test temperature [Mos70, Mos80].

for cryogenic service. The levels of these elements are adequately low in the ELI grades [RYD85]—e.g., Fe < 0.25 wt% in Ti-5Al-2.5Sn(ELI).

Addition of Yttrium. Yttrium in small amounts has been added to titanium alloys to reduce grain size and improve forgeability. The mechanical properties of titanium alloys may be deleteriously affected through segregation of yttrium. The possibility of segregation and precipitation or agglomeration of yttria must be considered a potential problem [RYD85]. As a result, yttrium additions are no longer permitted.

Addition of Interstitial-Element Solutes. The properties of unalloyed titanium at low temperatures have been considered by numerous workers [SAL79] [MOS80] [CON75, CON80, CON84] [NIS84] [YIN83]. Their studies had to do with the influence of the interstitial elements carbon, nitrogen, and oxygen on the plastic mechanical properties. Interstitial elements are potent strengtheners of titanium (Table 8.2). Some commercial grades, although exhibiting modest ductility down to 20 K, lose most of it upon cooling into the liquid-helium range [SAL79]. A reduction in the interstitial level is needed to restore low-temperature ductility. A penalty one pays for this is a reduction in strength [MOS80] [CON84] (Fig. 8.2). To compensate for this, it then becomes necessary to turn to solution-strengthened alloys of titanium, bearing in mind that steps must always be taken to preserve adequate fracture toughness [NAG84, NAG85].

Table 8.1 Tensile Strengths at Low Temperatures for Several Commercial Titanium-Base Alloys [SAL79]

Alloy	Condition	Test temperature K	Ultimate strength 10^8 N/m^2	Yield strength 10^8 N/m^2	Elongation, %
Ti-5Al-2.5Sn (5-2.5)	Annealed(a), normal interstitial	295	8.8	8.6	16
		200	11.0	10.6	14
		77	14.0	13.7	12
		20	16.9	16.8	5.1
	Annealed, extra-low interstitial (ELI)	295	7.6	7.1	17
		200	9.2	8.6	16
		77	12.6	11.9	17
		20	15.4	14.4	15
Ti-8Al-1Mo-1V (8-1-1)	Annealed(b)	295	10.3	9.7	16
		200	11.9	11.3	14
		77	15.6	14.4	13
		20	17.5	16.2	2.4
	Duplex annealed(c)	295	10.2	9.5	15
		200	11.2	10.3	15
		77	14.9	13.4	22
		20	16.9	16.1	1.2
Ti-6Al-4V (6-4)	Annealed(d), normal interstitial	295	9.9	8.9	12
		200	11.6	10.7	11
		77	15.3	14.3	11
		20	17.9	17.3	2.4
	Solution treated(e) and aged(f), normal interstitial	295	12.2	11.3	8
		200	13.2	12.8	6
		77	17.6	17.0	5
		20	20.4	19.9	0.7
	Annealed, ELI	295	9.9	9.3	12
		200	11.5	10.9	12
		77	15.1	14.6	10
		20	18.2	17.9	2.9
	Solution treated and aged, ELI	295	11.2	10.6	9
		200	13.2	13.2	7
		77	17.2	16.7	5
		20	19.6	19.6	1.0
Ti-13V-11Cr-3Al (13-11-3)	Annealed or solution treated(g)	295	9.7	9.4	19
		200	12.5	12.2	12
		77	19.5	18.9	2.1
		20	22.6	...	0.5
	Solution treated and aged(h)	295	13.6	12.4	7
		200	15.7	14.7	2.1
		77	16.5	...	0.2
		20	...	...	...

(a) 15 min to 4 h at 707-867 °C (1305-1593 °F), air cool. (b) 8 h at 787 °C (1449 °F), furnace cool. (c) 8 h at 787 °C (1449 °F), furnace cool, plus 15 min at 787 °C (1449 °F), air cool. (d) 30 min to 4 h at 707-817 °C (1305-1503 °F), air or furnace cool. (e) 847-957 °C (1557-1755 °F). (f) 1-10 h at 482-597 °C (900-1107 °F). (g)10-30 min at 757-787 °C (1395-1449 °F). (h) 20-100 h at 427-507 °C (801-945 °F)

Table 8.2 Solution Hardening of Titanium by Interstitial Elements and α Stabilizers [COL84]

Alloying addition	Concentration range, at.%	Condition	Law(a)	Slope, b, kg·mm^{-2}·at.%$^{-1/2}$ or kg·mm^{-2}·at.%$^{-1}$	Intercept, a, kg/mm^2	Correlation coefficient, %	Hardening rate, $\partial H_V/\partial c$, kg·mm^{-2}·at.%$^{-1}$ At 0.1 at.%	At 1.0 at.%
Simple-metal additions								
Aluminum	0-10	100 h/850 °C/IBQ	c	15	102	99.6	...	15
Gallium	0-5	As-cast	c	24	108	99	...	24
Tin	0-7	As-cast	c	24	112	99	...	24
Interstitial-element additions								
Boron	0-0.2	120 h/800 °C/IBQ	$c^{1/2}$	218	110	92	344	...
Carbon	0-0.5	120 h/800 °C/IBQ	$c^{1/2}$	170	104	99.9	269	...
Nitrogen	0-5	120 h/800 °C/IBQ	$c^{1/2}$	239	98	99.8	378	...
Oxygen	0-3	120 h/800 °C/IBQ	$c^{1/2}$	194	100	99.9	307	...

(a) Data fitted to either $H_V = a + bc$ or $H_V = a + bc^{1/2}$.

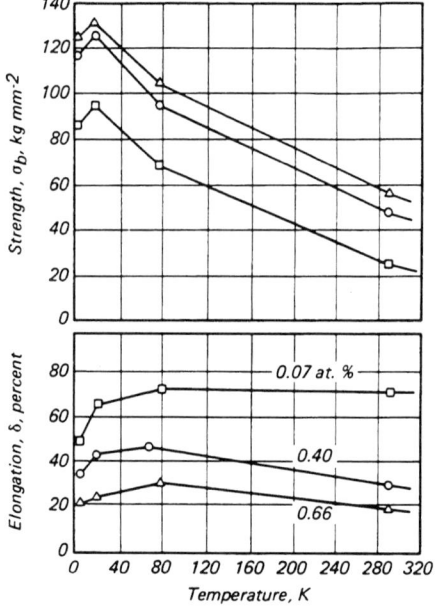

Fig. 8.2 Ultimate tensile strength and elongation as functions of test temperature for unalloyed titanium containing various levels of interstitial impurity expressed in terms of equivalent-oxygen content, O_{eq}. In this case, O_{eq} was given by: [O] + [H] + 2/3[C] + 2[N], all in at.% [Mos80] [Con84].

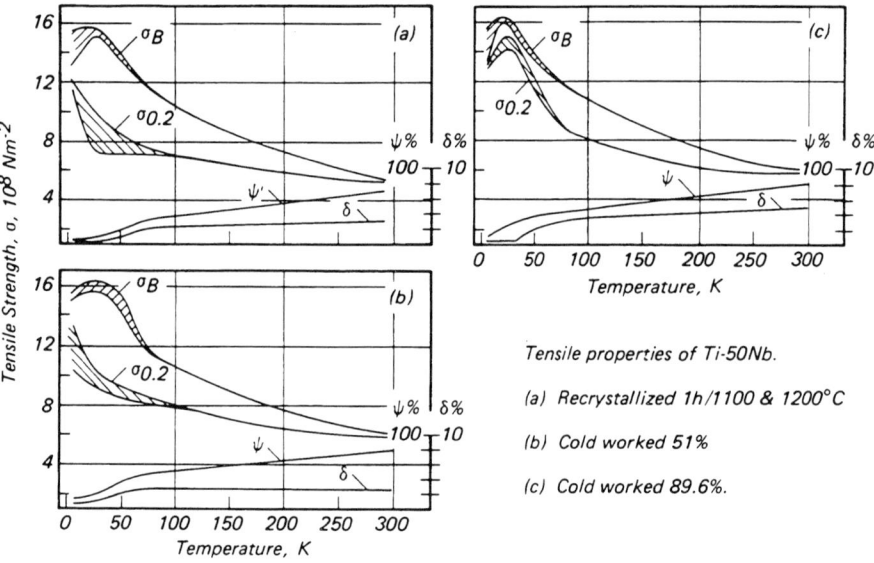

Fig. 8.3 Tensile properties of Ti-50Nb in various starting conditions as functions of test temperature (room temperature, 77K, 30 K, and 4.2 K) [Alb76] [Hil76].

Tensile properties of Ti-50Nb.

(a) Recrystallized 1h/1100 & 1200°C

(b) Cold worked 51%

(c) Cold worked 89.6%.

8.1.2 Beta-Phase Titanium Alloys

The bcc structure is not generally considered suitable for use at low temperatures since the most frequently used structural alloys undergo ductile-to-brittle transitions at sufficiently low temperatures. The effect is particularly pronounced in alloys of iron and chromium, but the fact that a ductile-to-brittle transition is not a property of sodium (a simple metal) or niobium (a transition metal) demonstrates that the phenomenon is not *necessarily* characteristic of the bcc lattice. β-Ti alloys are metastable (unless extraordinarily concentrated) and hence would not be expected to fail by conventional brittle cleavage, as is the case with bcc ferrous alloys. Nevertheless, as a comparison of Fig. 8.1 and 8.3 indicates, Ti-50Nb, an all-β alloy, has a very much lower ductility than α-phase titanium. By way of further example, the elongation of Ti-13V-11Cr-3Al [β-phase solution-treated (10 to 30 min at 760 to 790 °C)] at 20 K is 0.5% [Sal79], which may be compared with 14.7% for α-phase annealed (15 to 240 min at 700 to 870 °C) Ti-5Al-2.5Sn(ELI).

8.1.3 Alpha + Beta-Phase Titanium Alloys

Available in the literature is detailed information on the low-temperature properties of three representative α + β alloys: the "near-α" alloys Ti-8Al-1Mo-1V [Sal79] and Ti-6Al-3Nb-2Zr [Lav82] and the α + β alloy Ti-6Al-4V [Sal79][Nag84, Nag85]. The low-temperature ductilities of these three materials are listed in Table 8.3. Evidently it is possible to exert considerable control over the low-temperature ductility through appropriate variation (by heat treatment) of the two-phase microstructure. Detailed electron-microscope studies of low-temperature disloca-

Table 8.3 Low-Temperature Ductility of Some α + β Titanium Alloys

Composition	Condition(a)	Elongation(b)	Ref
Ti-8Al-1Mo-1V	8 h at 790 °C + FC + 15 min at 790 °C + AC	1.2% at 20 K	[SAL79]
Ti-6Al-3Nb-2Zr	1 h at 800 °C + AC	4 to 5% at 4.2 K	[LAV82]
Ti-6Al-4V	1 h at 1050 °C(c) + AC	4% at 4 K (ELI) 1.5% at 4 K (normal)	[NAG85]
Ti-6Al-4V	1/2 to 4 h at 710 to 820 °C	2.9% at 20 K (ELI) 2.4% at 20 K (normal)	[SAL79]

(a) FC = furnace cool; AC = air cool. (b) ELI = extra-low interstitial content. (c) β anneal

tion pile-ups at acicular β-phase precipitates in Ti-6Al-3Nb-2Zr were undertaken by LAVRENTEV et al. [LAV82]. NAGAI et al. [NAG84, NAG85] studied the influence of the optical microstructure on the plastic tensile properties and fracture toughness of variously heat-treated samples of Ti-6Al-4V.

8.1.4 Alpha-Phase Titanium Alloys

The α-Ti alloy generally selected for low-temperature service is Ti-5Al-2.5Sn (ELI). It possesses a considerably lower yield strength than, say, Ti-6Al-4V (ELI) (14.4×10^8 N/m² ksi compared with 17.9×10^8 N/m² ksi for annealed alloys; see Table 8.1), but the absence of β phase renders it considerably more ductile (15% elongation at 20 K compared with 2.9%, again for annealed alloys; see Table 8.1). The fracture toughness of Ti-5Al-2.5Sn is generally thought of as being greater than that of Ti-6Al-4V [SAL79]. However, recent studies have shown that the toughness of Ti-6Al-4V can be doubled in response to β annealing followed by suitably slow cooling [NAG84, NAG 85]; in such a metallurgical condition, and for some applications, it would then be preferable to an all-α alloy.

8.1.5 Summary

Unalloyed Titanium. All commercial grades of unalloyed titanium exhibit moderately good ductility at temperatures down to about 20 K. Their elongations at fracture actually increase as the temperature is decreased from 300 K, and pass through broad maxima (of about 40 to 50%) at about 77 K before descending rapidly as the temperature approaches 4.2 K. In some samples, the elongation, δ_B, becomes negligibly small at liquid-helium temperatures. Cold rolling increases the yield- and ultimate strengths but at the expense of ductility. The effects of interstitial elements on plastic properties have been considered in great detail by CONRAD and coworkers (e.g., [CON84]). It has been pointed out that the solutes carbon, nitrogen, and oxygen, which bond in a covalent-like manner to the surrounding titanium atoms, have pronounced influences on the strength of otherwise unalloyed titanium at temperatures below about one-half the melting point.

The α Alloy Ti-5Al-2.5Sn. The ductility of Ti-5Al-2.5Sn (ELI) is fairly independent of temperature between room temperature and 20 K, δ_B remaining at about 16% throughout that range. The ductility of the normal-interstitial grade is considerably lower; in fact, δ_B decreases monotonically below room temperature, dropping to 12% at 77 K and to only 5% at 20 K.

The α + β Alloy Ti-6Al-4V. The ductility of annealed Ti-6Al-4V is fairly independent of temperature between room temperature and 77 K. Below 77 K, it decreases rapidly as the temperature continues to fall toward 20 K. The ductility of the annealed alloy is twice as great as that of the solution-treated-and-aged material—e.g., $\delta_{B,77K}$ = 11.4% as compared with 4.9% (at normal interstitial levels). Reducing the interstitial content influences the tensile properties only marginally but improves the fracture toughness by 130% at room temperature, and by 40% at 20 K.

The Near-α Alloy α + β Ti-8Al-1Mo-1V. The near-α + α + β alloy Ti-8Al-1Mo-1V, although originally developed for high-temperature applications, can be used reliably down to moderate subambient temperatures in either the single-annealed (SA, "mill-annealed," 8 h/790 °C/FC) or duplex-annealed (DA, mill annealed + 15 min/790 °C/AC) condition. The room-temperature ductilities of SA and DA alloys are similar ($\delta_B \cong$ 15%), but upon cooling, that of the SA alloy decreases, while that of the DA alloy increases before passing through a maximum ($\delta_B \cong$ 22%) at about 77 K and dropping to low values at 20 K ($\delta_B \cong$ 1%).

The β Alloy Ti-13V-11Cr-3Al. As a β-Ti alloy, Ti-13V-11Cr-3Al would be expected to possess poor low-temperature ductility. Indeed it does, the elongation at fracture of the solution-treated-and-aged (STA) material becoming insignificantly small below about 100 K; at 77 K, δ_B = 0.2%. Some improvement results if the aging stage (i.e., 20-100 h/430-500 °C) of the STA heat treatment is omitted, in which case $\delta_{B,77K}$ becomes about 2%.

8.2 Physical Metallurgy and Low-Temperature Strength

8.2.1 Influence of Interstitial Content

Strengthening of titanium alloys by interstitial elements has been referred to above and also in Section 8.1.1. The interstitial solutes carbon, nitrogen, and oxygen are relatively immobile below about 300 °C, and provide stable solution strengthening below that temperature. The solution-strengthening potencies of boron, carbon, nitrogen, and oxygen are compared with those of the α stabilizers aluminum, gallium, and tin in Table 8.2. There it can be seen that the hardening rates produced by the interstitials are more than an order of magnitude greater than those produced by the α stabilizers, which are themselves potent strengtheners. Table 8.2 also shows that the hardening potency of the interstitials increases in the sequence C < O < B < N. These results agree qualitatively with those of CONRAD et al. [CON75, CON84], who showed that the measured Gibbs free energies of activation associated with thermally activated plastic flow of dilute Ti-interstitial alloys had the values 1.50, 1.64, and 1.73 eV, respectively, for Ti-C, Ti-O, and Ti-N alloys.

8.2.2 Influence of Interstitial Content and Grain Size

The influence of interstitial content and grain size on the low-temperature tensile properties of titanium alloys has been considered by CONRAD [CON84]. A set of typical results is given in Fig. 8.4. There it can be seen that at low temperatures (from 400 down to 77 K) the true fracture stress, ε_F, is generally less dependent on temperature and grain size than at higher temperatures. At temperatures above 400 K, the effect of interstitial content depends on grain size. CONRAD [CON84] also confirmed that an increase in the interstitial content leads to a decrease in ductility, with the rate of decrease in elongation with concentration increasing in the sequence C < O < N.

8.2.3 Influence of Heat Treatment (Alloy Phase Morphology)

The strength and toughness of an α + β titanium alloy can be adjusted by controlling the volume fractions and morphologies of the α- and β-phase components through suitable heat treatments. NAGAI et al. [NAG84, NAG 85] have shown how the structure of Ti-6Al-4V varies in response to β annealing (1 h/1050 °C in this case) followed by variable-time cooling from 1050 to 550 °C (in 48 s to 2.8 h) followed by water quenching to room temperature (Fig. 8.5).

The results of mechanical-property testing are given in Fig. 8.6, in which it can be seen that slow cooling (≥162 s) following β annealing brings about excellent fracture toughness and good fracture strain, with only a slight loss of strength at 4 K. The slow-cooled Widmanstätten or colony structure is characterized by a "basket-weave" arrangement of "packets" of α-phase plates. It turns out that a crack, although it will propagate in a straight line within a packet of similarly aligned α plates, becomes deflected or arrested at the boundary between packets. The inhibition of crack propagation seems to be responsible for the increased fracture toughness exhibited by the slow-cooled material.

8.3 Low-Temperature Deformation Modes (See Also Chapter 12, Part 3)

At low temperatures, titanium alloys generally exhibit serrated yielding in their stress-strain curves. This is true of both α-phase- and near-α-phase alloys (see Fig. 8.7 for commercial-purity titanium [CON84] and Ti-6Al-3Nb-2Zr [LAV82]) as well as of metastable β-phase alloys (see Fig. 8.8). The mechanisms of serrated yielding are different in the two cases: the α-phase alloys seem to undergo

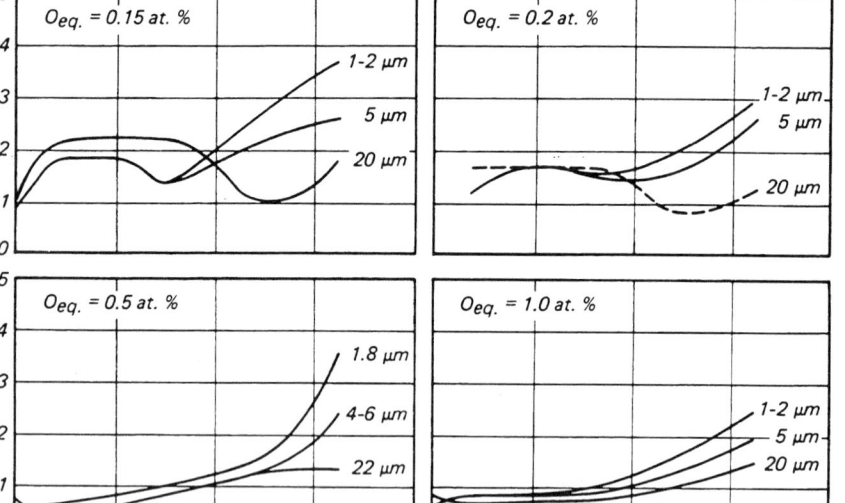

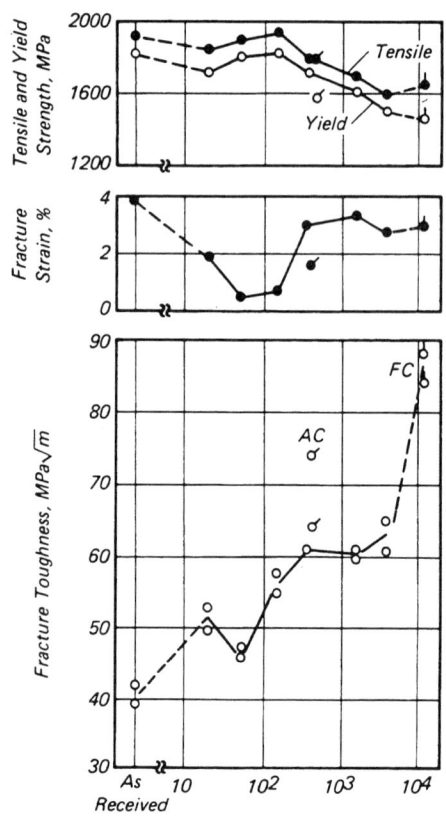

Fig. 8.4 Effects of temperature, interstitial level (expressed in terms of an equivalent-oxygen concentration, O_{eq}), and grain size on the true fracture strain of unalloyed titanium wires [CON84].

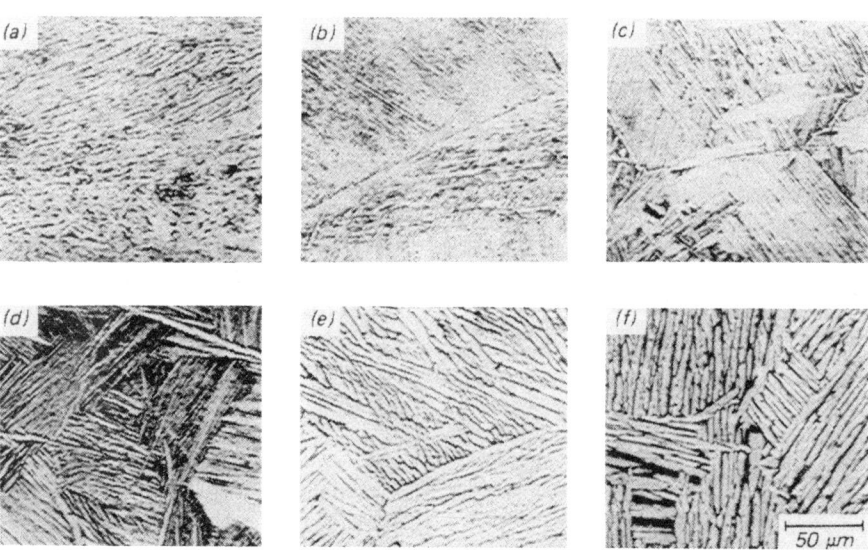

Fig. 8.5 Optical microstructure of normal-interstitial-level (as distinct from ELI-grade) Ti-6Al-4V as a function of heat treatment. Conditions: (a) as-received (a mill-annealed fine lamellar α + β structure) (cf. Fig. 10.4); (b) β-annealed 3.6 ks/1050 °C plus cooled in 48 s to 550 °C and water quenched; (c) cooled in 162 s; (d) cooled in 360 s; (e) cooled in 3600 s; (f) furnace cooled in 10.2 ks [NAG84, NAG85]. Micrographs courtesy of K. Nagai, National Research Institute for Metals, Tsukuba.

Fig. 8.6 Mechanical properties of normal-interstitial-level (as distinct from ELI-grade) Ti-6Al-4V in the as-received (mill-annealed) condition [see Fig. 8.5(a)] and as a function of cooling time from 1050 to 550 °C (1920 to 1020 °F) following a β anneal for 3.6 ks/1050 °C [NAG84, NAG85].

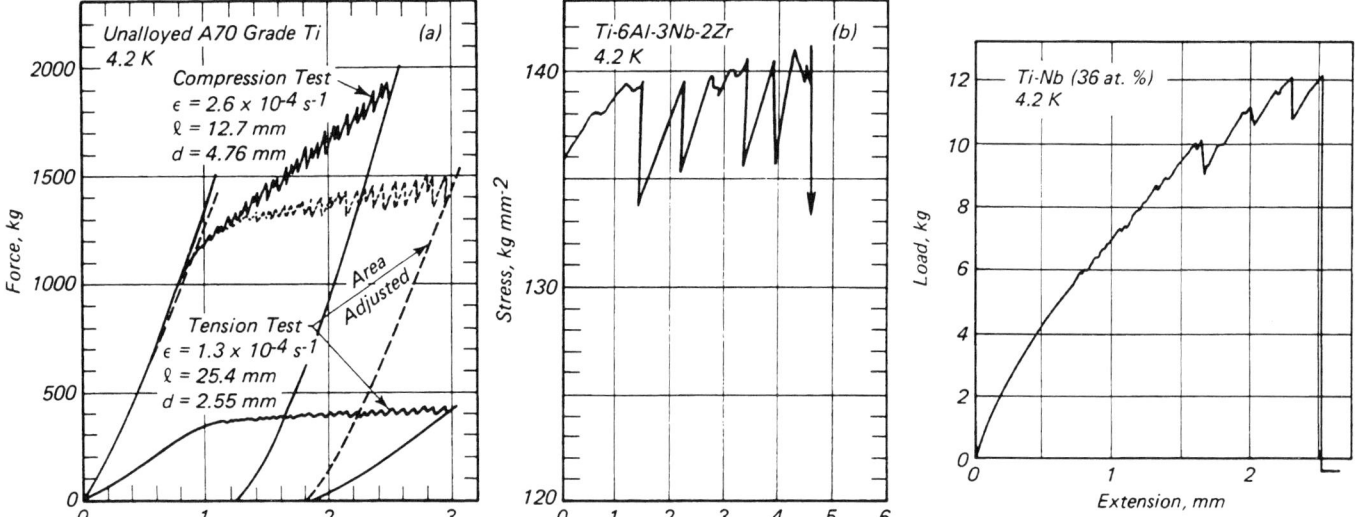

Fig. 8.7 Occurrence of serrated yielding during tensile testing at 4.2 K of (a) α-Ti ([Mad74]; see also [Con84]) and (b) the near-α alloy Ti-6Al-3Nb-2Zr [Lav82].

Fig. 8.8 Occurrence of serrated yielding during tensile testing at 4.2 K of the β alloy Ti-Nb(36 at.%) [Koc77].

conventional adiabatic-heating oscillations under tensile loading at low temperatures; in the β-phase alloys the serrated yielding is clearly a manifestation of lattice instability.

8.3.1 Serrated Yielding in α-Ti Alloys

Serrated yielding may be regarded as a quite general low-temperature mechanical effect which, when it occurs, arises from the following combination of factors: (1) adiabatic heating, (2) a strong negative temperature coefficient of the yield stress, and (3) the small specific heat of a metal at low temperatures. MOSKALENKO et al. [Mos70, Mos80], who studied unalloyed titanium (iodide and commercial), Ti-2.4Zr-1.2Mo, Ti-Al(1.5, 3, 5.5 wt%), Ti-Zr(1, 3, 8 wt%), Ti-V(0.5, 1.5, 3 wt%), and Ti-Nb(0.5, 2, 4 wt%), and CONRAD [Con84], who studied commercial titanium in considerable detail, agreed that the adiabatic heating was due to dislocation avalanching (which follows in response to the sequence: localized plastic flow/dislocation pile-up/stress increase). As mentioned above, serrated yielding is a frequently observed phenomenon in low-temperature testing. Since the discovery of the effect in aluminum and its alloys by BASINSKY [Bas57], it has been known to be *not necessarily* associated with either twinning or martensitic transformation. But in the commercial titanium studied by CONRAD, and in the dilute titanium alloys studied by MOSKALENKO and PUPSOVA [Mos70] (see Fig. 8.9), the serrated yielding seemed to be invariably accompanied by twinning, which, according to CONRAD, appears to assist in the nucleation of the dislocation avalanche.

Fig. 8.9 Microstructure of Ti-0.5Nb in response to tensile deformation at various temperatures: (a) 12% deformation at 293 K; (b) 25% at 293 K; (c) 12% at 77 K; (d) 13% at 4.2 K [Mos70]. Micrographs courtesy of the late B.I. Verkin, Physicotechnical Institute of Low Temperatures, Kharkov.

8.3.2 Serrated Yielding in β-Ti Alloys

Serrated yielding (Fig. 8.8) and twinning (Fig. 8.10) have both been observed in Ti-50Nb, a representative metastable β-Ti alloy. However, the mechanism in such systems has been identified as an "anomalous plastic property" [Col86, p. 100-108] related to thermoelasticity and pseudoelasticity (Fig. 8.11) and

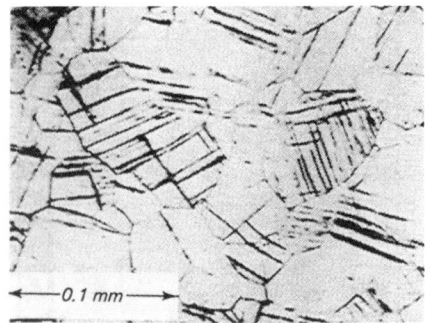

Fig. 8.10 Subsurface micrograph (original magnification, 600 ×) of Ti-45Nb (recrystallized at 800 °C and water quenched) after straining to fracture at 4 K, revealing a twin-like deformation structure [REA78]. Micrograph courtesy of D.T. Read, National Bureau of Standards, Boulder, CO, copyright 1978, Butterworth and Co. (Publishers) Ltd. (reproduced with permission).

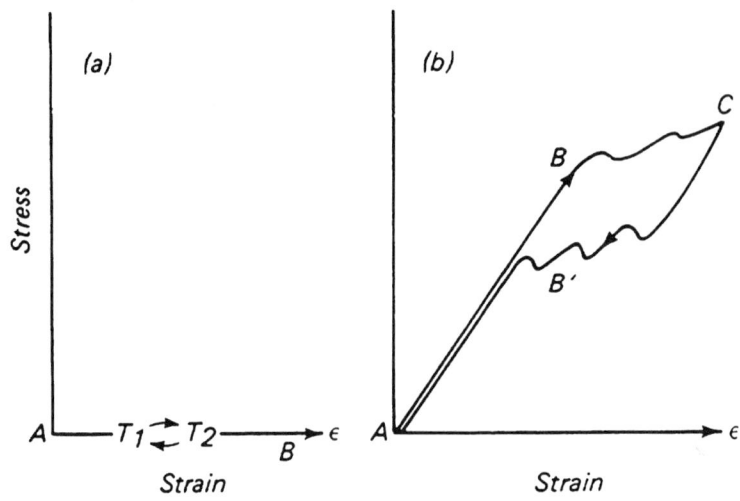

Fig. 8.11 Schematic representations of (a) thermoelasticity and (b) pseudoelasticity.

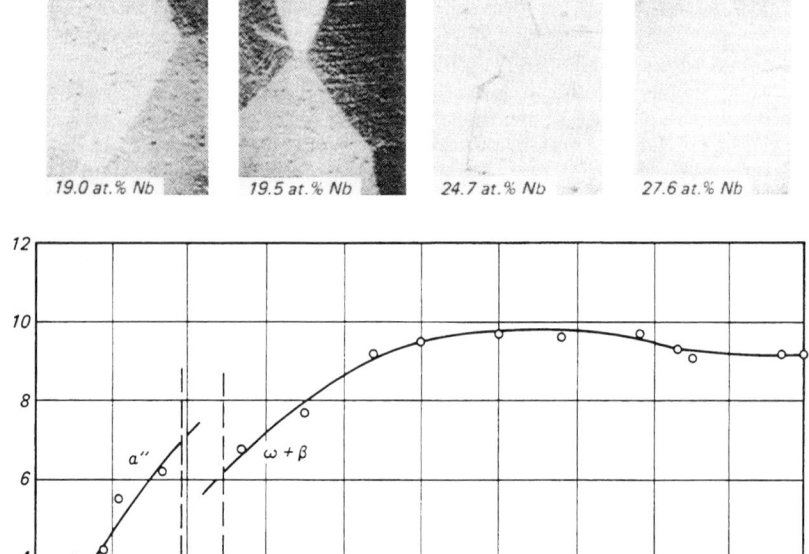

Fig. 8.12 Optical microstructures of quenched Ti-Nb alloys referred to the composition dependence of the superconducting transition temperature. The martensitic regime terminates at the same composition in each case; evidently no further transformation of the quenched alloys takes place on cooling them into the liquid-helium range.

traceable to the instability of the β-Ti lattice [HOC84, HOC85]. The starting point for a discussion of these properties is Fig. 5.8, which shows the measured M_s for Ti-Nb extrapolated into the liquid-helium temperature range. This figure suggests that β ↔ α″ transformation will take place whenever the extrapolated line is crossed —as a function of either composition or temperature. It turns out, however, that the transformation does not take place spontaneously. For example, a physical-property measurement sensitive to martensite has shown that no fresh martensite is produced when a water-quenched alloy is cooled to 4 K for measurement [COL80] (Fig. 8.12). Some extra energy, in the form of lattice strain energy, for example, is needed to initiate transformation. Thus the transformation will take place under either of the following two circumstances: (1) the crystal is prestrained and lowered in temperature through the extrapolated M_s [HOC84, HOC85], or (2) the crystal is lowered in temperature below M_s and then strained, as in a low-temperature tensile test. The effects in both cases are reversible (unless, as is usually the case in option 2, the sample is strained to fracture), thereby providing examples, respectively, of: (1) thermoelasticity and (2) pseudoelasticity (see Fig. 8.11).

Evidence for reversible martensitic transformation at zero applied stress (thermoelasticity) in room-temperature-strained Ti-Nb alloys has been provided by HOCHSTUHL, who demonstrated the existence of *negative*, reversible, thermal-expansion components in Ti-Nb(21 at.%) and other alloys [HOC84, HOC85]. With *decreasing* temperature, the sample's length *increased* monotonically as its temperature was lowered from room temperature into the liquid-helium range (the formation of athermal-ω would have resulted in enhanced lattice contraction; see Section 5.3.2). The effect, which was reversible, did not occur in the as-quenched samples, presumably because they lacked the strain energy needed to trigger the transformation.

Evidence for pseudoelasticity under tensile-test conditions has been provided by KOCH and EASTON (see [COL86, p. 107]). In most cases the tensile test is continued through the serrated-yielding region to sample failure (Fig. 8.8), which, when it takes place, is of the shear-rupture type (Fig. 8.13). If tensile unloading is commenced before the sample ruptures, evidence for reverse transformation, in the form of acoustic emission, is obtained as the stress is released [KOC77]. Just as was the case for α-Ti alloys, the β alloy's tensile strength drops as it enters the serrated-yielding regime (cf. Figs. 8.1 and 8.3).

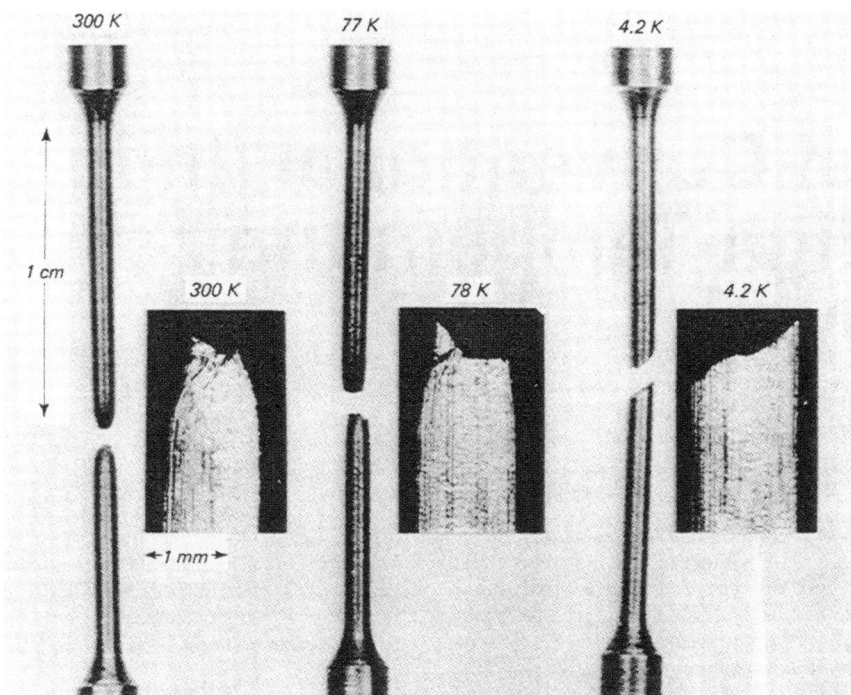

Fig. 8.13 Tensile specimens and fracture ends of samples after testing of fracture at the temperatures indicated. An intercomparison of the results of Koch and Easton [Koc77] on Ti-Nb (38.5 at.%) (tensile bars depicted) and Albert and Pfeiffer [Alb76] on Ti-Nb(34 at.%) (fracture ends depicted) demonstrates the reproducibility of the fracture behavior at each of the test temperatures. Photographs courtesy of D.S. Easton, Oak Ridge National Laboratory, and H. Hillmann, Vacuumschmeltze GmbH.

9. Evolution of Conventional (Ingot Metallurgy) High-Temperature Titanium Alloys

Whereas at low temperatures, elements in substitutional solid solution are supposed to contribute to the athermal component of flow stress, and interstitial elements to the thermal barriers, as the temperature rises *all* alloying species become more or less mobile and associate themselves with "atmosphere effects" to extents that depend on solute-atom diffusivities [ROS73]. Chemical effects such as oxidation and hot-salt stress corrosion may limit the service temperature of a titanium alloy in some applications; in others, mechanical degradation such as high-temperature creep will limit the service-temperature range.

As pointed out in Chapter 8, β alloys are unsuited for low-temperature applications. Whereas α alloys perform satisfactorily at low temperatures, where solid-solution strengthening is important, their mechanical properties decrease rapidly with increasing temperature. But α alloys may still be suitable for elevated-temperature service if they combine (1) thermal stability in the temperature range of interest with (2) reinforcement by precipitates, particles, or second phases.

As the temperature increases above room temperature, since the strengths of all the alpha alloys such as Ti-5Al-2.5Sn continue to decrease rapidly (Fig. 9.1), they must be abandoned in the intermediate-temperature range in favor of alloys such as those depicted in Fig. 9.2, in which the strengthening mechanisms are "microstructural" in nature. The chronological development of high-temperature (heat-resistant) titanium alloys prior to 1973 has been reviewed by ROSENBERG [ROS73]. According to that article, the design of α + β alloys for elevated-temperature service (less than 550 °C) is based on the following general principles: athermal strengthening of the β phase is achieved by the addition of molybdenum [ZEY71], and that of the α phase by aluminum [SAR70]. The effect of tin (and also gallium) is comparable to that of aluminum [COL75]; the effect of zirconium was also at the time supposed to be comparable to that of aluminum (although the results of more recent work might take issue with that statement), with which it was supposed to interact to some extent. By 1973, the alloys most widely used in the United States at temperature above about 430 °C were the so-called "super-α" alloys: Ti-8Al-1Mo-1V and Ti-6Al-2Sn-4Zr-2Mo (Ti-6242). Of these, the latter was the more heat resistant, being capable of service at temperatures as high as 480 ~ 510 °C. Although several other compositions had been developed since 1966, when Ti-6242 was introduced, they failed to achieve commercial acceptance either because of only marginal advantages over Ti-6242 or because, although exhibiting improved creep resistance, their stabilities were unsatisfactory [PAR73]. But one alloy, introduced commercially in 1973, did exhibit creep resistance markedly superior to that of its predecessor. Formed from Ti-6242 by the addition of silicon to improve "surface stability" and bismuth to improve creep strength, this alloy, referred to as Ti-11, was intended for use in aircraft gas-turbine engines.

9.1 Analytical Design of Conventional High-Temperature Titanium Alloys

The properties of technical titanium alloys can be generally understood in terms of the prototypical α- and β-stabilized binary alloys, Ti-Al and Ti-Mo, respectively [COL84]. The high-temperature metallurgical stability of titanium alloys (ignoring environmental or chemical effects) is limited by β ↔ α + β transformation, which in unalloyed titanium occurs at 883 °C. Thus, although titanium melts at 1670 °C, its service temperature, based on metallurgical stability, is governed by the so-called "homologous temperature"—i.e., the temperature adjusted to β ↔ α + β transformation rather than to melting. It would therefore seem that the addition of α stabilizers, of which alu-

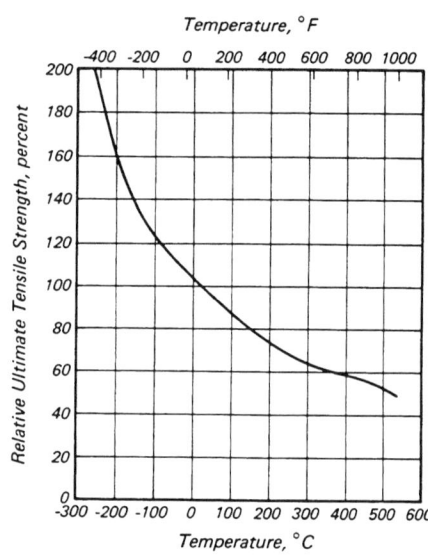

Fig. 9.1 Temperature dependence of the relative ultimate tensile strength of annealed Ti-5Al-2.5Sn (in the form of sheet) [WOO72, p. 5-2:72-4].

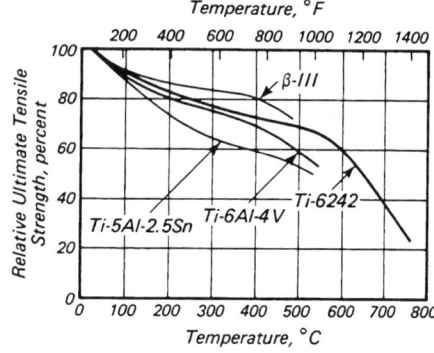

Fig. 9.2 Intercomparison of the temperature dependences of the relative ultimate tensile strengths of four commercial titanium alloys [WOO72].

minum is an excellent example, would not only provide potent solution strengthening, but would also increase the alloy's "heat resistance" by raising its transformation temperature (see COLLINGS [COL84, p. 56]). This prospect encouraged the early development of technical α alloys. Working on the assumption that if a modest amount of α stabilizer is good, even more would be better, early developers produced alloys such as Ti-5Al-5Sn-5Zr only to find that they became embrittled by α_2-phase precipitation during elevated-temperature service. After exploring the boundary of the single-phase-α solid solution, ROSENBERG [ROS70] recommended that the total "equivalent-Al" content of an alloy should be kept below 9 wt%, such that for an alloy containing aluminum, zirconium, tin, and oxygen, for example:

$$[Al]_{eq} = [Al] + \frac{[Zr]}{6} + \frac{[Sn]}{3} + 10[O] < 9$$

(Eq 9.1)

where [x] is the concentration of element x in weight percent. Based on more recent work we may now include gallium in terms of [Ga]/2. Alloys with $[Al]_{eq} > 9$ are the now-obsolete "super-alpha" alloys.

An alternative approach to the development of titanium-base alloys for high-temperature service is to "skip over" the β transus and to operate in the β-phase field. This approach led to the development of the β-Ti alloys typified by the U. S. alloy "β-III" of which the binary prototype is Ti-Mo (see Section 4.2). The guiding philosophy behind this approach can be appreciated by visualizing the typical β-isomorphous phase diagram whose transformation temperatures drop rapidly with solute concentration (see Sections 2.3 and 2.6). The β alloys tend to be metastable (see Section 7.14.2), but if suitably processed they can be operated successfully at moderately high temperatures.

An important advantage that β-Ti alloys, such as β-III, have over the α-Ti alloys is that their strengths tend to decrease much less rapidly with increasing temperature. In attempts to combine the better temperature-dependence characteristic of the β-Ti alloys with the metallurgical stability of the α-Ti alloys, and for other reasons associated with the details of thermomechanical processing, the two-phase α + β alloys were developed. Fine adjustments in the compositions of the α + β alloys over the years, in both the U.S. and the U.K., have resulted in alloys suitable for service at temperatures up to about 590 °C (IMI 834 and TIMETAL-1100). An important ingredient of both alloys is the 0.35 wt% Si. In the U.S., experiments with the alloy Ti-6Al-2Sn-4Zr-2Mo with varying silicon contents yielded a maximum in creep resistance at 0.1 wt% Si (Fig. 9.3), resulting in a limiting silicon-level specification in the U.S. of 0.1 to 0.2 wt%. But comparable experiments on other alloys have demonstrated a continuous increase in creep resistance with increasing silicon content up to the limit of solubility [SEA75]. Accordingly, alloys in the U.K. have tended to contain more silicon than their U.S. counterparts. Recently, however, two new alloys (Beta 21S and TIMETAL-1100) have been developed in the U.S. with silicon at or above 0.2 wt%.

The development chronology of conventional ingot metallurgy (I/M) heat-resistant alloys is depicted in Fig. 9.4, in which the service temperature is plotted as a function of year of introduction. Their relative heat resistances, as gauged by the Larson-Miller parameter, are plotted in Fig. 9.5. The Larson-Miller parameter (a measure of "time at temperature" for a given strain) is given by $T(C + \log t)$, where T is temperature in degrees Rankine (°F + 460), t is time in hours, and C is a constant (about 20 for a large number of alloys).

The compositions of the most heat resistant of the alloys depicted in the figures are listed in Table 9.1. The philosophy guiding the design of these alloys can be appreciated with reference to that table: (1) First of all, we notice that the alloys have almost the same $[Al]_{eq}$ (except IMI 550) which, moreover, has a mean value of 7.3, only 1.7 units less than the limiting value of 9 referred to in Eq. 9.1. The inclusion of 1700 wt ppm of O in an $[Al]_{eq} = 7.3$ alloy would raise it to the limit, and thus for all practical purposes the alloys could be said to be operating at the limit of α stability. (2) The level of β stabilizer is very low—sufficient to confer some microstructural strengthening, but not enough to engender metallurgical instability. It is notable that the level of heat resistance goes up as the β-stabilizer content decreases; in proceeding from Ti-6242 to IMI 834 and TIMETAL-1100, the molybdenum content steadily decreases. As pointed out by HOCH et al. [HOC73], the β-stabilizing strengths of the alloying additions niobium, vanadium, and molybde-

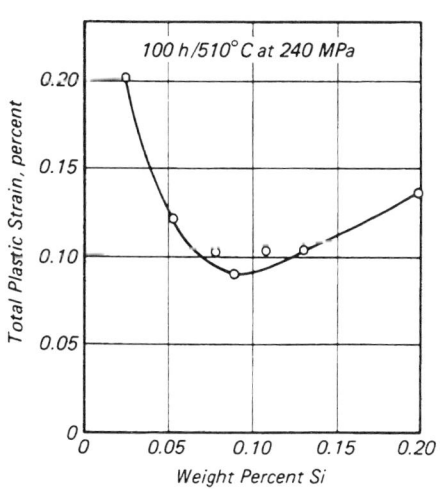

Fig. 9.3 Influence of silicon on the creep performance of Ti-6242. After SEAGLE et al. (see BLENKINSOP [BLE85]).

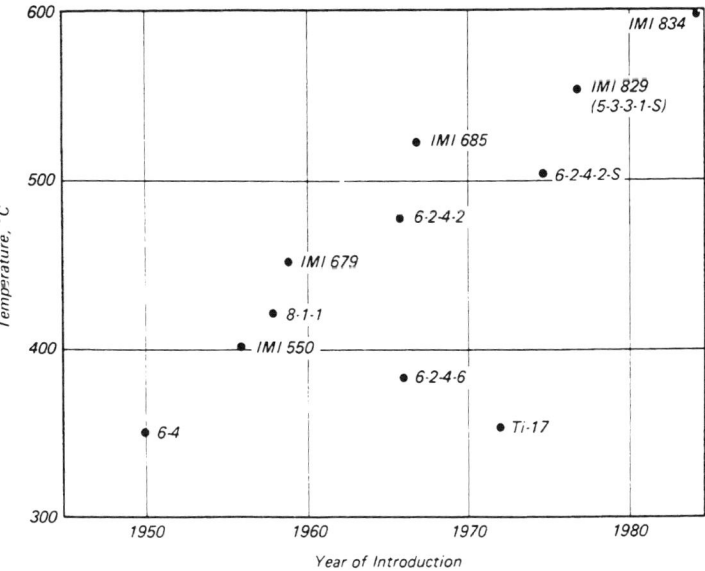

Fig. 9.4 High-temperature titanium-base alloys for aircraft-engine applications. Year of introduction and operating-temperature capability under optimum stress conditions. The suffix "S" indicates that a small amount of silicon (usually 0.25 wt%) has been added to the basic formulation [BLE85].

Table 9.1 Evolutionary Design(a) of Multicomponent α + β Titanium Alloys(b)

Alloy	α			[Al]$_{eq}$	β			Si
	Al	Sn	Zr		Nb	V	Mo	
IMI 834	5.5	4	4	7.5	1	...	0.3	0.5
Ti-1100	6	2.75	4	8.2	...	...	0.4	0.45
IMI 829	5.5	3.5	3	7.2	1	...	0.3	0.3
IMI 685	6	...	5	6.8	...	...	0.5	0.25
Ti-11	6	2	1.5	7.2	...	...	1	0.1
6242-Si	6	2	4	7.4	...	...	2	0.2
6242	6	2	4	7.4	...	...	2	...
IMI 679	2 ¼	11	5	6.8	...	...	1	0.2
811	8	...	...	8	...	1	1	...
IMI 550	4	2	...	4.7	...	...	4	0.5

(a) Analysis: (1) Maintain [Al]$_{eq}$; (2) reduce β content; (3) increase silicon content. (b) See also Table 2.6.

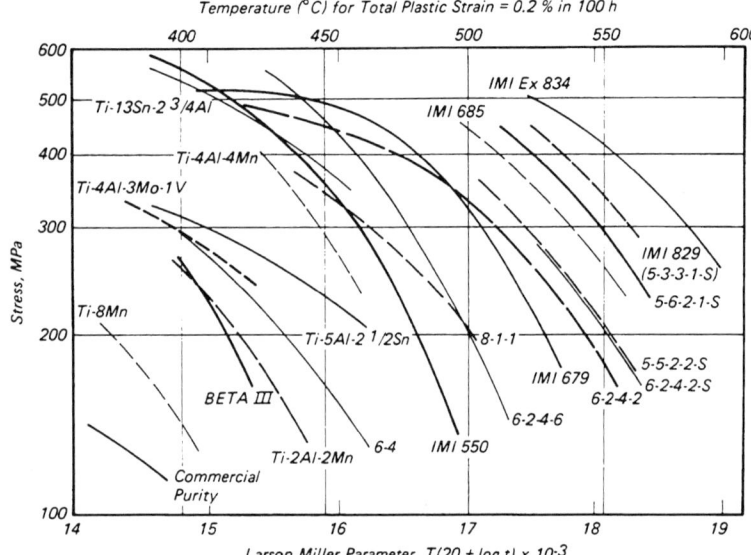

Fig. 9.5 Elevated-temperature creep performances of the alloys referred to in the previous figure (except Ti-17, vis. Ti-5Al-2Sn-2Zr-4Mo-4Cr, which was developed for applications up to 350 °C, or 660 °F) and some others [BLE85].

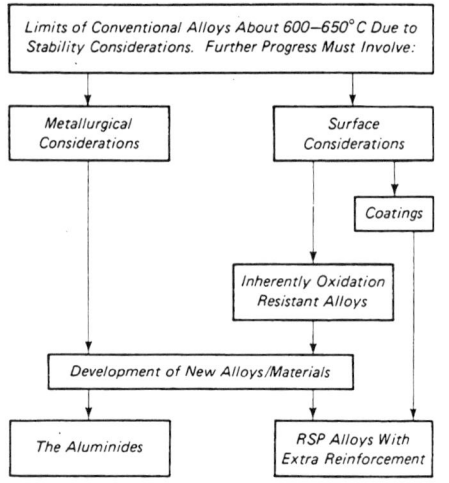

Fig. 9.6 Directions for further development of titanium-base alloys for very-high-temperature applications.

increases with increasing silicon content up to the saturation limit of 0.5 wt%.

Table 2.6 presents an analysis of the α-stabilizer and β-stabilizer contents of numerous U.S. alloys.

9.2 Directions for Advancement

9.2.1 Overall Picture

It seems that the temperature range for conventional unprotected (no surface coatings) solid-solution titanium alloys is upper limited to about 600 °C [BLE85] and that future development should include combinations of: (1) inherently oxidation-resistant alloys, (2) coatings, (3) further development of the α$_2$-phase and γ-phase aluminides and their variants, and (4) the use of rapidly solidified alloys. Figure 9.6 outlines the manner in which such developments proceed naturally out of the earlier work.

9.2.2 Prognosis for Alpha + Beta Alloys

The upper temperature limit of the α + β alloys is controlled as much by metallurgical stability as by surface or environmental stability (e.g, oxidation). The *service* temperature range designated for the next generation of titanium alloys is actually the temperature range currently being used to *process* the present α + β alloys. At temperatures between 850 and 950 °C, the alloy Ti-6242 (Table 9.1) undergoes volumetric hysteresis during temperature cycling in response to the hysteretic microstructural cycling of β and (α + β)$_W$ (where W indicates Widmanstätten precipitation [LAH80]). The use of such an α + β alloy in this temperature range would be accompanied by dimensional instability stemming from the volumetric differences between the α and β phases. Accordingly, attempts to develop

num decrease in the sequence Mo > V > Nb; thus, to provide β-phase microstructural strengthening with the least possible lowering of the β transus, it is best to substitute niobium for some of the molybdenum. This has been done in IMI 829 and IMI 834. As maximally α-stabilized alloys with minimal β-stabilizer contents, this family of α + β alloys is referred to as "near-α" α + β alloys. The picture is completed by noting that the heat resistance

the α + β alloys for very-high-temperature service is not recommended.

9.2.3 Prognosis for Beta Alloys

The metallurgical properties of the β alloys as they approach equilibrium have been discussed by COLLINGS [COL84, pp. 72-74, 206-210] (see also Section 7.14). In alloys such as Beta-III, so-called solution heat treatment at temperatures of 720 and 790 °C followed by aging at temperatures in the range 370 to 590 °C can be used to control the microstructure and mechanical properties. Likewise, the alloy Beta-C* can be made to undergo phase separation (at 350 °C) and α-phase precipitation (at 500 °C) by appropriate heat treatment at temperatures through which new-generation high-temperature titanium alloys would be expected to cycle during service. It follows that conventional β-Ti alloys would also be metallurgically unstable in use.

Such difficulties may possibly be averted by choosing very concentrated β alloys. But this tends to erode the density advantages that titanium alloys have over their competitors. It is then natural to ask whether this problem can be solved by selecting vanadium (whose density is only 35% greater than that of titanium) as the β stabilizer**; the answer seems to be no, the β/(α + β) transus in Ti-V tending to become rather flat at concentrations greater than about 20 wt% V. Furthermore, increasing vanadium decreases oxidation resistance. Indeed, both Ti-34V and Ti-39V begin to precipitate α phase after about 6 h at temperatures above 450 °C. It has been demonstrated that the alloy Ti-35V-15Cr-0.1C has both metallurgical stability and very good creep strength up to 500 °C, but the alloy has the expected oxidation resistance problems at the high end of its potential use range.

The requirement for a strip rollable alloy suitable for high-temperature use in metal matrix composites (MMC's) led to the development of TIMETAL-21S (Beta 21S). The alloy has creep resistance equivalent to Ti-6Al-4V combined with the advantages of a strip rollable, cold formable β alloy. It is finding many uses where its corrosion and oxidation resistance at elevated temperature are more important than the ability to withstand high stresses at elevated temperature. Applications up to 600 °C are anticipated. The alloy's oxidation resistance and moderate creep strength make it a candidate for MMC applications to 815 °C and low stress, monolithic applications up to 700 °C.

Again, metallurgical instability considerations militate against the use of β alloys for very-high-temperature service.

9.2.4 Prognosis for Alpha Alloys

It is well known that the effect of α-stabilizing elements is to preserve (in the case of tin) or substantially increase (in the cases of aluminum and particularly oxygen) the temperature of the (α + β)/α transus. With aluminum, at the single-phase solubility limit (at about 800 °C) of 9 wt% (i.e., 15 at.%), the temperature of the (α + β)/α transus has increased from 883 °C (α/β for pure titanium) to 1050 °C. The α-stabilizing potency of oxygen is even greater; however, hardness considerations have prevented full advantage from being taken of it, at least for intermediate-temperature operation.

The large negative yield strength temperature dependences of conventional α alloys have prevented them from entering high-temperature service. On the other hand, their intrinsic metallurgical stability at all temperatures up to the (α + β)/α transus makes them suitable as bases upon which to build (with the aid of dispersion or fiber strengthening) the next generation of high-temperature alloys.

A logical extension of the process of solid-solution strengthening of titanium by α-stabilizing elements such as interstitial elements or simple metals (aluminum in particular) leads to the formation of the intermetallic compounds. The two most well-known compounds in the Ti-Al system are the so-called "aluminides": α_2 (based on the compound Ti_3Al) and γ (based on the compound TiAl). These compounds are already showing great potential for high-temperature service: Ti_3Al undergoes an order-disorder reaction at 1100 °C and begins to decompose to β phase at about 1150 °C; TiAl melts at about 1450 °C.

Based on these considerations it was concluded that the best bases for the next generation of titanium alloys for elevated-temperature service are α-Ti solid solutions and the α_2-phase and γ-phase aluminides. But further developments are underway.

9.3 Ingot Metallurgy of Advanced Alpha-Phase Solid-Solution Alloys

Some of the most extensive studies of α-Ti alloys for high-temperature service were carried out in the 1950s at the Baikov Institute of Metallurgy in Moscow [KOR66]. Alloys studied for strength and high-temperature creep resistance included the six-component alloys AT-3, -4, -6, and -8 (whose compositions are listed in Table 9.2); some developments of them (so-called AT-10 and AT-12) containing seven to eight alloying elements; and, for reference purposes, an alloy based on Ti_3Al. As a result of both tensile and centrifugal (bend-type) creep tests, it was concluded that: (1) the AT-3 and AT-4 alloys were useful up to 400 to 500 °C, the AT-6 up to 550 °C, and the AT-8 up to 500 to 600 °C; and (2) to sustain temperatures in the range 600 to 700 °C it was necessary to increase the number of alloying elements to seven or eight (thereby forming the AT-10 and AT-12 alloy types). Although showing a great improvement over those of the six-component alloys, the properties of the seven- and eight-component α-phase alloys were still inferior to those of an alloy based on Ti_3Al.

The results of work such as this (1) are in accordance with the results of studies conducted in the U.S., which indicate that the aluminides hold great promise as high-temperature structural materials, and (2) indicate that, within the realm of solid-solution alloying, RSP multicomponent materials based on α-phase alloys also have potential for high-temperature applications.

Unfortunately, recent studies show that exposure to anticipated operating conditions between 600 and 700 °C severely reduce the tensile and fatigue properties of α_2-aluminides. The reduction in

---

*Beta-C is Ti-4Mo-8V-6Cr-4Zr-3Al.
** The alloy Ti-15V-6Al and its relatives have been considered as β-Ti candidate alloys and warrant further study as potential RSP (rapid solidification processed) β alloys.

Table 9.2 Compositions of Some Experimental Soviet Heat-Resistant Titanium Alloys [KOR66]

Alloy code(a)	Composition, wt%				
	Al	Cr	Fe	Si	B
AT-3	3.2	0.84	0.34	0.40	0.01
AT-4	4.8	0.86	0.36	0.29	0.01
AT-6	5.6	0.52	0.30	0.33	0.01
AT-8	6.8	0.98	0.40	0.59	0.01

(a) The six-component alloys specified had considerably lower rates of creep than other investigated industrial alloys (the so-called VT-1, OT-4, VT-5-1, and VT-14) and the experimental alloy OT-4-2 (viz., Ti-4.95Al-1.54Mn).

properties is attributed to the embrittling effect of increased oxygen in the surface layer and, in the case of super-α_2, bulk instability. To date, no solution to this problem has been identified within the normal aluminide composition range, but the orthorhombic alloys (based on Ti_2AlNb) appear to hold promise.

10. Powder Metallurgy and Rapid-Solidification Processing

The utility of powder metallurgy stems from its ability to yield relatively low-cost shapes conforming rather closely to those desired in the finished product (so-called "near-net-shape" processing). With the advent of advanced computer-aided design and manufacturing (CAD-CAM) technology, near-net shapes can also be produced in modern forging operations; nevertheless, powder metallurgy still offers certain advantages over ingot metallurgy (I/M). These advantages have to do with what might be termed enhanced "microstructural and macrostructural design flexibility." That is to say: (1) through the use of blended powders it is possible to adjust and optimize the metallurgical properties of various regions of the part in order to satisfy local mechanical-property requirements; and (2) modern methods of producing powder enable hitherto unobtainable alloy compositions and microstructures to be engineered into the finished product. Thus the starting stock for powder metallurgy may be in one of two forms: (1) mixed powder of different compositions—the so-called "blended-elemental," or BE, approach; or (2) powder of uniform composition, identical to that of the intended product—the "prealloyed powder," or PA, approach.

There are numerous methods of producing metallic powders, many of which are in no way related to rapid solidification. However, the generic techniques of powder metallurgy can be applied with advantage to powders produced by rapid-solidification processing (RSP). Powder metallurgy (P/M) as it applies to titanium alloys has been reviewed by FROES and EYLON [FRO85, p. 49]. Rapid-solidification processing of metals and alloys in general has been reviewed by SAVAGE and FROES [FRO85, p. 60], who have presented a comprehensive listing of RSP techniques along with their estimated cooling rates (ECR), [FRO85, pp. 68, 69]. Typical examples of free-flight methods are the centrifugal atomization technique with an ECR of 10^5 K/s, (Fig. 10.1a), free-flight melt spinning with an ECR of 10^2 to 10^4 K/s, the rotating-electrode method with an ECR of 10^2 K/s (Fig. 10.1b), and the gas atomization method with an ECR of 10^2 to 10^3 K/s. Typical examples of chill-block processes are chill-block melt spinning (ECR, 10^5 to 10^7 K/s) and the crucible and pendant-drop melt-extraction processes (ECR, 10^5 to 10^6 K/s) (Fig. 10.2a and b).

In practice, cooling rate is of course a function of, among several factors, the size of the quenched product—diameter in the case of a gas-cooled particle, and thickness in the case of a chill-block-quenched flake or ribbon. BRODERICK et al. [BRO85] have made a detailed study of the effect of quench rate on the microstructure of Ti-6Al-4V after rapid solidification by the techniques of hammer-and-anvil splat quenching, electron-beam splat quenching, pendant-drop melt extraction, and rotating-electrode processing. Coming out of this work was a relationship between the size of the prior-β grains (L, μm) and the calculated cooling rate (T, K/s) of the form:

$$L = 3.1 \times 10^6/T^{0.93 \pm 0.12} \quad (\mu m) \quad (Eq\ 10.1)$$

For example, measurements of a PDME Ti-6Al-4V fiber (see below) yielded a grain size of 35 μm and hence a cooling rate of 2×10^5 K/s, in agreement with the general estimates cited above.

After some introductory comments on P/M approaches, this chapter goes on to describe a case study of the microstructural and mechanical properties of pendant-drop melt-extracted Ti-6Al-4V powder and fiber. The purpose of that research was two-fold: (1) to develop a new approach to the production of high-purity titanium-alloy powder for conventional P/M processing; and (2) to explore the potential of RSP for the production of high strength/weight titanium-alloy fibers through grain refinement and solution and precipitation strengthening. This chapter serves as an introduction to Chapter 11, which goes on to deal with the rapid-solidification processing and design of alloys intended for high-temperature applications.

10.1 Blended-Elemental Powder Metallurgy

In titanium-alloy powder metallurgy, the blended-elemental approach implies the mixing of fine granular unalloyed titanium with a powdered master alloy. According to FROES et al. [FRO85], the method has been applied to alloys Ti-6Al-4V (predominantly), Ti-6Al-6V-2Sn, Ti-

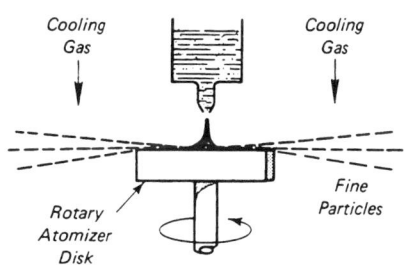

(a) Centrifugal Atomization Technique

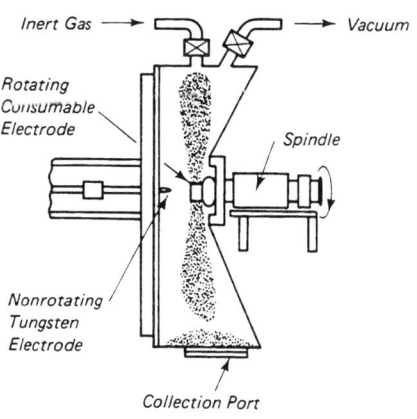

Fig. 10.1 Schematic representations of two rapid-solidification techniques [SAV84].

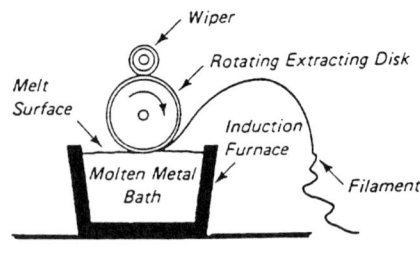

(a) Crucible Melt Extraction

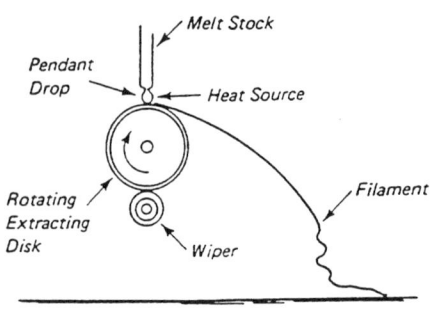

(b) Pendant Drop Melt Extraction

Fig. 10.2 Schematic representation of two melt-extraction techniques [SAV84].

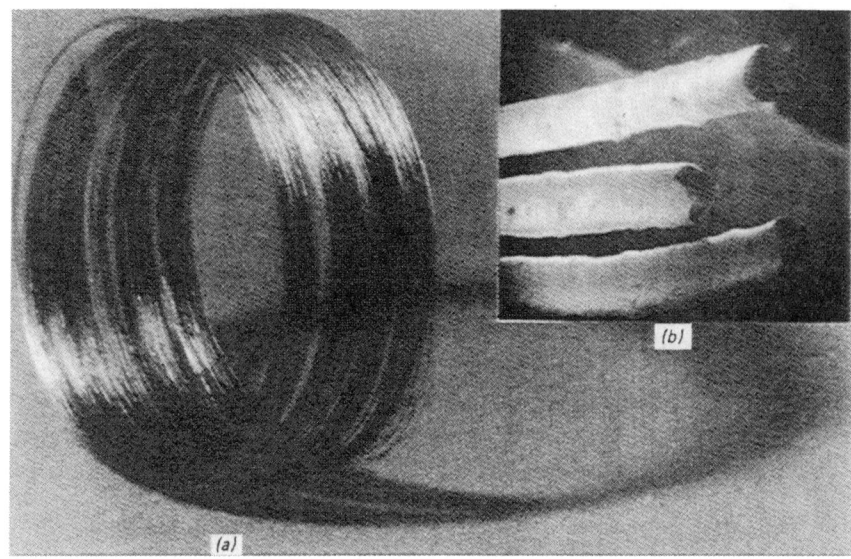

Fig. 10.3 Pendant-drop melt-extracted Ti-6Al-4V fiber. (a) One of several coils (about 14 cm in diameter) produced during the preliminary phase of a so-called "L/D-powder" program. (b) Scanning electron micrograph of the same fiber (width about 30 μm) showing the "wheel surface" and the "free surface" [COL78b].

6Al-2Sn-4Zr-2Mo, Ti-6Al-2Sn-4Zr-6Mo, and Ti-10V-2Fe-3Al. The BE approach could also imply the use of mixed oxides or salts as prereduced starting materials [FRO85, p. 52].

The generalized BE approach also finds application in the P/M processing of composite materials for high-temperature service—i.e., the term could be applied to the introduction of oxide particles or "unconstitutional precipitates" (e.g., refractory intermetallic compounds) as dispersion strengtheners, or of suitably coated SiC or refractory metallic whiskers for fiber reinforcement.

The most common BE compaction method involves cold isostatic pressing (CIP) followed by sintering. The CIP process molds a "green" shape using an elastomeric mold at room temperature. The sintering is accomplished at a high temperature to bond the powder particles and homogenize the chemistry of the blend. This will provide a compact with about 95% of theoretical density. PA compacts are most often accomplished by hot isostatic pressing (HIP) and vacuum hot pressing (VHP). BE shapes can be further densified by HIP or VHP. Further densification may also be accomplished by forging, i.e., using the pressed and sintered BE compact as a forging preform. The BE approach also lends itself to direct production of foil or sheet, in which case the constituent powders are poured into the "pinch" space between a pair of rolls.

10.2 Prealloyed Powder Metallurgy

The PA process begins with the preparation of powder that is uniform in particle size, shape (preferably spherical), and composition. Various methods are available for the production of PA-process starting material whether it be in the form of spherical particles, flake, or granules. Many of the available methods fall within the purview of RSP—see Chapter 11, and also FROES et al. [FRO85]. Briefly, PA powder-preparation methods could be categorized under the headings: (1) Hydride-dehydride processing, a method which relies on the severe embrittling effect of dissolved hydrogen to produce small granules of the alloy; (2) liquid-"atomization" methods of various kinds, such as the centrifugal atomization process (Fig. 10.1a) and the rotating-electrode process (REP) (Fig. 10.1b); and (3) melt-extraction methods (Fig. 10.2a and b), which are capable of yielding products in a variety of useful shapes, such as flake, elongated particles (so called "L/D powder"), and continuous fiber [COL78b]. Continuous fiber is suitable for fiber reinforcement applications—but even if not planned as a final product, it is a convenient form for measurement of representative tensile properties. According to FROES [FRO85], PA powder processing has been applied to the titanium-base alloys Ti-6Al-4V, Ti-6Al-2Sn-4Zr-6Mo, Ti-4.5Al-5Mo-1.5Cr ("CORONA 5"), Ti-11.5Mo-6Zr-4.5Sn ("β-III"), Ti-10V-2Fe-3Al, and the British alloys IMI 685 (Ti-6Al-5Zr-0.5Mo-0.25Si) and IMI 829 (Ti-5.5Al-3.5Sn-3Zr-1Nb-0.3Mo-0.3Si).

10.3 Pendant-Drop Melt Extraction of Continuous Ti-6Al-4V Fiber [MAR76] [COL78b]

10.3.1 Fiber Production and Microstructures

Long, continuous lengths of Ti-6Al-4V fiber have been produced by pendant-drop melt extraction using a polished-chisel-edged extraction disk. Views of the fiber are given in Fig. 10.3. Optical micrographs of sections of the fiber, in the as-spun condition and after two heat treatments, are given in Fig. 10.4(a). For heat treatment the fibers had been wrapped in tantalum foil and enclosed in quartz tubes along with getter-packages of titanium chips; the tubes were sealed off under argon at a pressure adjusted to be atmospheric at the heat treatment temperature.

The as-spun (hence rapidly quenched) structure is martensitic, whereas that of the mill-annealed sample is, in this case, Widmanstätten $(\alpha + \beta)_W$. The recrystallized structure is equilibrium $\alpha + \beta$, with highly elongated α-phase regions. The three classes of microstructures can be more easily distinguished in the replica electron micrographs in Fig. 10.4(b).

10.3.2 Mechanical Properties of the Ti-6Al-4V Fiber

Elastic Modulus. The elastic modulus was determined by combining the results of acoustic-wave velocity and density (aqueous buoyancy) measurements.

Tensile Strength. Tensile tests were carried out with the sample held between grips specially designed for accommodating wire specimens. The calculation of modulus and strength from the results of tensile measurements requires a knowledge of the specimen cross-sectional area. This was determined in two different ways: (1) from the measured mass-per-unit length and the density; and (2) by comparing the tensile Young's modulus with the acoustically measured value. The results of the mechanical-property measurements are summarized in Table 10.1.

10.4 Pendant-Drop Melt Extraction and Consolidation of Ti-6Al-4V Powder [MAR75, MAR76, MAR78[a]]
[COL78[b]]

10.4.1 Powder Production

By introducing notches into the rim of the extraction disk it was possible to interrupt the casting process in such a way as to produce a particulate product (Fig. 10.5a). By using first widely spaced and then more closely spaced notches, two classes of powder were produced for experimentation. These powders, which were referred to as L/D-1 and L/D-2, are depicted in Fig. 10.5(b) and (c), respectively. Their dimensions are defined in the caption of the figure.

10.4.2 Powder Consolidation by Vacuum Hot Pressing

Processing. Hot pressing of the (elongated) L/D-1 powder was carried out in vacuum under an axial load of 0.56 kg/mm² (800 psi) for 1 h/955 °C. The resulting flat disk (after light machining to 5.1 cm diam by 1.1 cm thick), together with a sample of the L/D-1 powder from which it was produced, are depicted in Fig. 10.6(a).

Microstructures. Replica electron micrographs of the VHP powder are presented in Fig. 10.7(a). By comparing this figure with Fig. 10.4(b) it can be seen that the VHP material is in the "recrystallized" condition. This is to be expected, since the first stage of recrystallization heat treatment is 4 h/930 °C.

Mechanical Properties. Samples of the VHP material were prepared for compression testing and tensile testing. The average tensile strength of the VHP material in the as-VHP condition was 10.4×10^8 N/m² (151.0 ksi), which may be compared with the 10.7×10^8 N/m² (155.4 ksi) of the wrought starting material after mill annealing, and the 9.9×10^8 N/m² (144.3

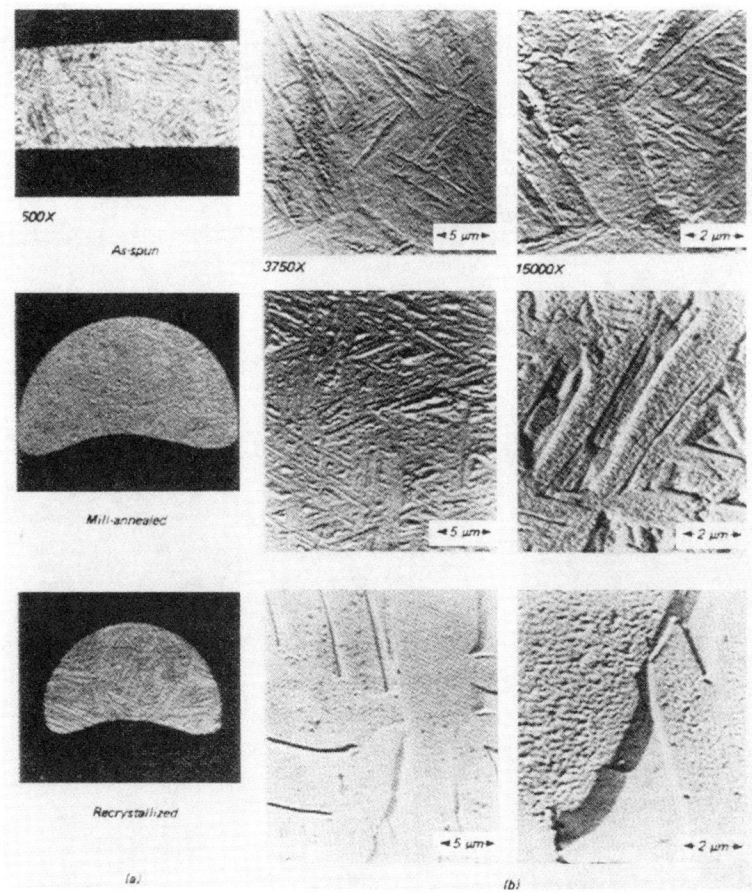

Fig. 10.4 Optical micrographs (a) and replica electron micrographs (b) of samples of pendant-drop melt-extracted continuous fibers of Ti-6Al-4V in three conditions. (Top) As-spun (martensitic α′). (Center) Mill-annealed 2 h/720 °C/AC (Widmanstätten α + β). (Bottom) Recrystallized 4 h/930 °C plus furnace cooled at 55 °C/h to 760 °C plus furnace cooled at 360 °C/h to 480 °C plus air cooled (equilibrium α + β) [COL78[B]].

Table 10.1 Summary of Tensile Properties of Ti-6Al-4V Fibers

Condition	Microstructure	Modulus(a), E, 10^3 ksi(b)	Average tensile strength, ksi(b), based on: A_1(c)	A_2(d)
As-spun	Martensitic α′	13.44	89.0	125.0
Mill-annealed(e)	Reverted α′	15.73	96.9	144.3
Recrystallized(f)	α + β	15.59	87.8	129.9

(a) From sound-velocity measurement. (b) To convert ksi to MPa, multiply by 6.8948. (c) $A_1 = W/\rho L$. (d) A_2 back calculated from acoustic elastic modulus. (e) Mill anneal (MA): 2 h/1350 °F/AC. (f) Recrystallization anneal: 4 h/1700 °F, furnace cool to 1400 °F (100 °F per hour — no faster), furnace cool to 900 °F (650 °F per hour—no slower), air cool

ksi) of the mill-annealed melt-extracted fiber.

10.4.3 Powder Consolidation by Hot Isostatic Pressing

Processing. Hot isostatic pressing of cold pressed pellets of L/D-1 and L/D-2 powders was carried out under conditions of 1 h at 950 °C and 6.9×10^7 N/m² (1 h at 1750 °F and 10 ksi). The HIP-compacted samples were machined in preparation for compression testing (see, for example, Fig. 10.6b).

Microstructures. Optical micrographs of as-HIP samples prepared from L/D-1 and L/D-2 powders showed the structure to be predominantly α phase—partly equiaxed and partly elongated. The replica electron micrographs in Fig. 10.7 (a) and (b) show that the prior-β interphase regions (β-transus temperature,

996 °C) must have undergone transformation to Widmanstätten $(\alpha + \beta)_W$ during cooldown in the HIP device.

Mechanical Properties. Samples of HIP L/D-1 and L/D-2 powders were prepared for compression testing at room temperature. The average compression yield stress (at 0.2% offset) of the as-HIPed L/D-1 material was 10.2×10^8 N/m² (148.2 ksi), and that of the as-HIPed L/D-2 material was 10.0×10^8 N/m² (145.7 ksi). These values may be compared with the 9.9×10^8 N/m² (144.0 ksi) of the wrought starting material after mill annealing.

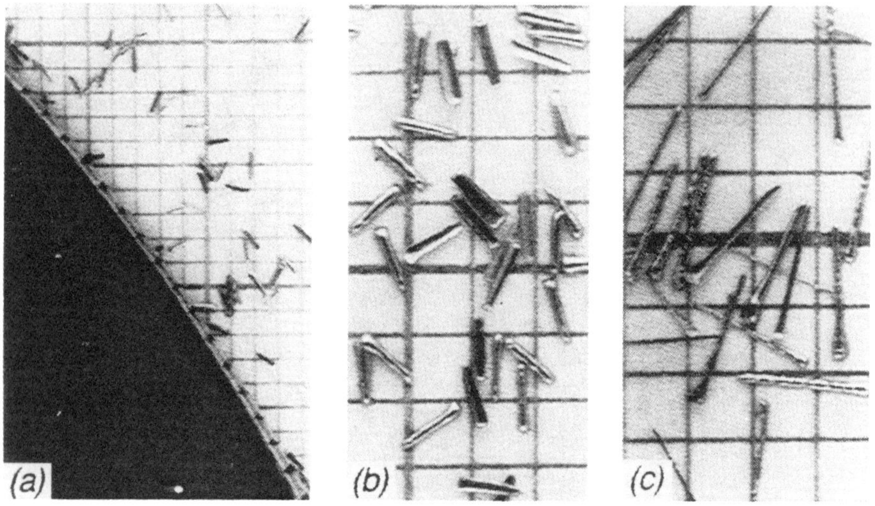

10.5 Pendant-Drop Melt Extraction of Precipitation-Strengthened Titanium-Alloy Fiber [COL78b, COL78c] [MAR79]

10.5.1 Introduction

Rapid quenching from the melt may lead to extreme grain-size refinement (if not complete amorphousness in suitable alloy formulations). In a pure substance the fine grain size is unstable against growth. But in two-phase material, some degree of elevated-temperature stability is obtainable, especially if the equilibrium phases differ substantially in composition. Reversion to equilibrium is in such cases inhibited by the necessity for a massive flux of atoms over a considerable distance. Although most readily obtainable in a two-phase eutectic or eutectoidal system (in which spinodal decomposition is very effective), a fine (1 to 5 µm), stable, equiaxed grain structure can sometimes be stabilized by grain boundary-pinning precipitate particles. A stable, refined grain structure is conducive to superplasticity.

An increase in the modulus of elasticity has been observed in some rapidly quenched materials as a result of the presence of a very stable dispersed phase. Increases in strength have also been noted as consequences of: (1) fine grain size, (2) a fine dispersion of second-phase particles,

Fig. 10.5 Melt extraction of Ti-6Al-4V in the form of so-called "L/D powder." (a) The beveled-and-notched extraction disk used for producing L/D-2 powder. (b) L/D-2 powder, about 0.08 by 0.2 by 1.3 mm. (c) L/D-1 powder, about 0.05 mm diam by 25 mm. All objects photographed against 1-mm-square grid paper [COL78b].

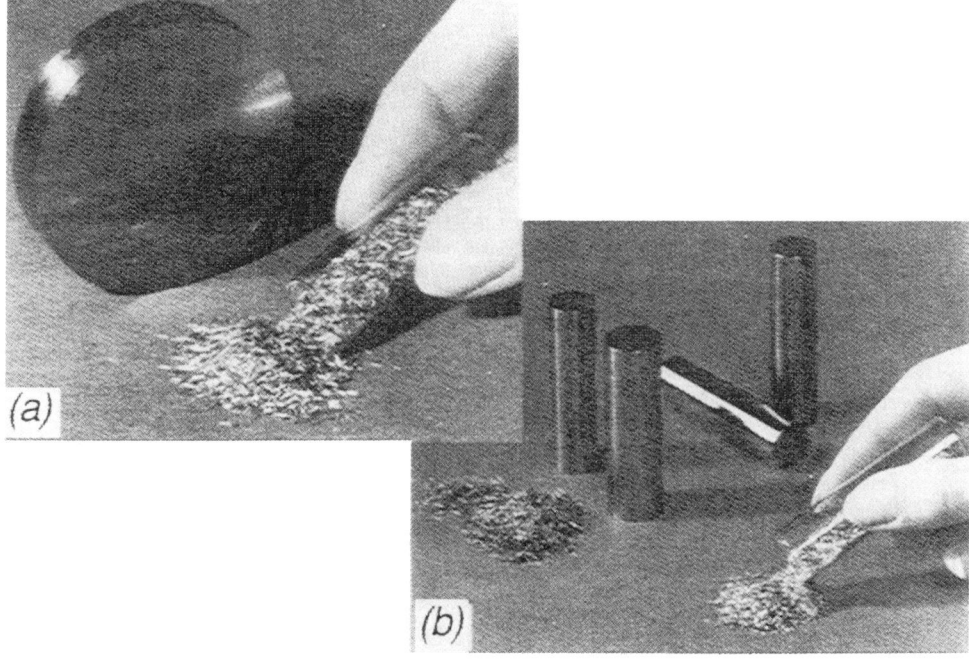

Fig. 10.6 Samples of consolidated Ti-6Al-4V L/D powder. (a) A vacuum hot pressed disk (1.1 by 5.1 cm) and an example of the L/D-1 powder from which it was processed. (b) Two pairs of hot HIPed and machined cylinders and examples of the L/D-1 powder (left) and L/D-2 powder (right) from which they were processed.

and (3) the generally high concentration of solution strengthener present as a consequence of the extended solubility range offered by the rapid-quench process.

10.5.2 *In Situ* Melted Pendant-Drop Melt-Extracted Alloys [COL78[b]].

For swiftly and economically obtaining large numbers of samples for microstructural and mechanical-property screening, the pendant-drop melt-extraction (PDME) method, operated in the *"in situ alloy-melting"* mode (reminiscent of the BE method of powder preparation), is a useful preparation technique. The *in situ* PDME method has been used for the preparation of multicomponent alloys based on Ti-6Al-4V. In one series of investigations, alloying materials in the form of wires or rods of various metals were fastened longitudinally to a rod of Ti-6Al-4V. Once melting of the rod had begun, the attached wires would usually melt and alloy smoothly into the pendant drop.

Alloys were also prepared by adding materials in powdered form to the host alloy. Longitudinal slots (about 1 mm wide and 1 mm deep) were cut into a Ti-6Al-4V alloy rod. Intermetallic-compound powders (in particular TiB_2, $TiSi_2$, or TiC) were mixed with alcohol and the resultant thick slurry smeared into the slots and allowed to dry. The dried cake had enough cohesion and strength to bond to the slots during the mounting and melt-extraction operations and alloyed smoothly into the basic stock. After some preliminary testing, several materials were selected for more detailed evaluation consisting of microstructural study, density measurements, acoustic and tensile measurement of elastic modulus, and tensile-strength testing. Some results for *in situ* PDME Ti-6Al-4V/TiB_2, which of all the compositions explored yielded the highest values of modulus and strength, are presented below.

Metallography. Two samples were chosen for metallographic evaluation: a very fine wire about 0.1 mm wide, and a ribbon-like fiber about 0.7 mm wide [Fig. 10.8(a) and (b), respectively]. The micron-size precipitate-defined grain structure is presumably responsible for the high strength values obtained.

Table 10.2 Mechanical Properties of Ti-6Al-4V/TiB_2

Average sound velocity, km/s	5.04
Density, g/cm^3	4.50
Acoustic elastic modulus, 10^3 ksi(a)	16.22
Ultimate tensile strength, ksi(a)	255

(a) To convert ksi to MPa, multiply by 6.8948.

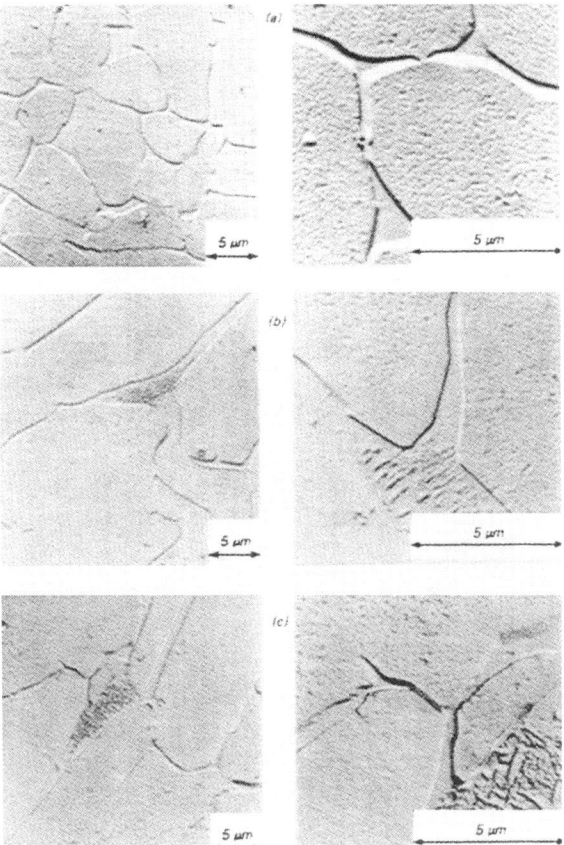

Fig. 10.7 Replica electron micrographs of consolidated Ti-6Al-4V L/D powder. (a) Vacuum hot pressed L/D-1 powder. (b) Hot isostatically pressed L/D-1 powder. (c) Hot isostatically pressed L/D-2 powder [COL78[b]].

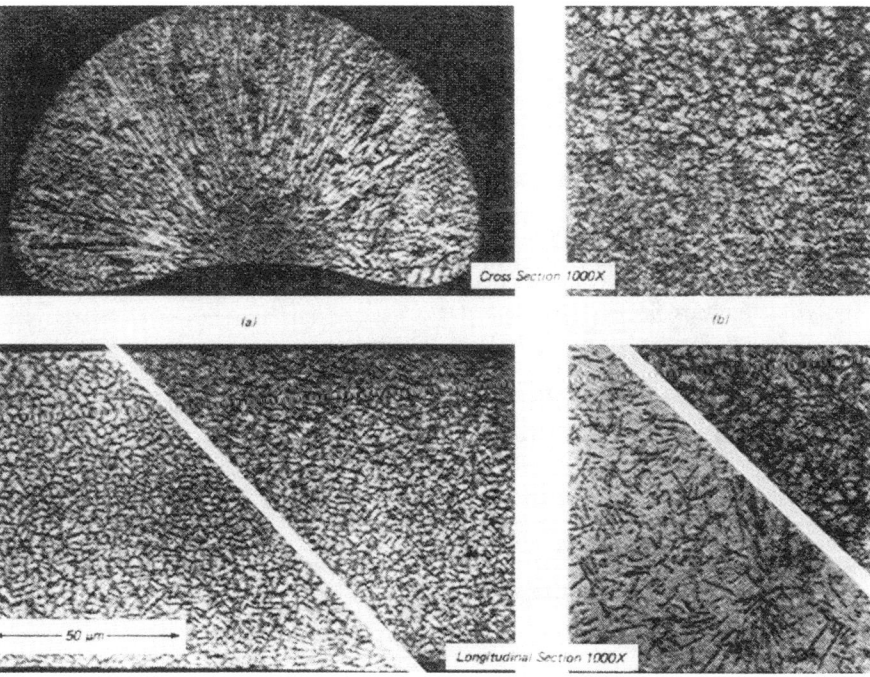

Fig. 10.8 Optical micrographs of transverse and longitudinal sections of two samples of pendant-drop melt-extracted fibers prepared by *in situ* alloying of TiB_2 into Ti-6Al-4V [COL78[b]].

Elastic Modulus. Values of elastic modulus were determined by combining the results of acoustic-wave velocity and density (aqueous buoyancy) measurements (Table 10.2).

Tensile Strength. Tensile tests were carried out with the sample held between grips specially designed for accommodating wire specimens. The calculation of modulus and strength values from the results of tensile measurements requires a knowledge of the area of the specimen cross section, which was very irregular in some cases. Instead of attempting to use directly measured filament cross sections, it was decided to assign an *effective cross-sectional area*—one which normalized the tensile-test modulus value to that obtained acoustically. The results of the tensile tests are given in Table 10.2. A comparison of the results presented there with those of Table 10.1 for the PDME control alloy emphasizes the considerable enhancement in strength that can be obtained by combining the addition of boron to the melt with the use of rapid solidification to achieve solution strengthening, precipitation strengthening, and a fine grain structure.

10.5.3 Advanced Multicomponent Prealloyed Pendant-Drop Melt-Extracted Alloys [COL78^c][MAR79]

Preparation and Basic Properties. Following the screening study outlined above, several five-component alloys, some based on Ti-6Al-4V, were arc melted and cast into rod form suitable for pendant-drop melt extraction. Three of the resulting materials were selected for detailed evaluation. These alloys, and their process conditions and properties, are listed in Table 10.3.

Elastic Modulus. Values of elastic modulus (Table 10.3) were determined as described previously.

Tensile Strength. In preliminary tests in which untreated as-spun fibers were mounted in conventional wire-holding grips, fracture occurred at the grips. This difficulty was greatly reduced after the design and construction of a holder specially suited for small wire and ribbon samples (Fig. 10.9). Filamentary wire samples were epoxied into the shanks of "eye-ended" solder lugs, and ribbon samples into the flattened shanks of such lugs. To facilitate handling right up to the time the tensile test actually began, the lugs themselves were attached to a C-shape cardboard yoke. The results of the tensile measurements are summarized in Table 10.3. Finally, by way of conclusion, Table 10.4 shows that the specific strengths of the three alloys under discussion compare favorably with data of TANNER and RAY [TAN77] for some commercial metallic glass fibers.

Fig. 10.9 Small sample of melt-spun ribbon mounted in preparation for tensile testing and photographed against 1-mm-square grid paper [COL78^b]

Table 10.3 Properties of Polycrystalline Titanium-Base Alloys Prepared by Pendant-Drop Melt Extraction

Alloy code(a)	ME-1	ME-2	ME-3
Extraction disk rim speed, m/s	3 to 4	4 to 13	4 to 9
Vickers hardness number, kg/mm^2	588 ± 28	542 ± 30	673 ± 43
Density, g/cm^3	4.694	4.686	4.326
Sound velocity, km/s	4.067	3.768	5.358
Acoustic modulus, 10^3 ksi(b)	11.04	9.46	17.66
Tensile strength, ksi(b)	340	346	242
Tensile strength, kg/mm^2	239	243	170

(a) Alloy compositions: ME-1, Ti-5.4Al-3.6V-8Fe-3Cu; ME-2, Ti-5.4Al-3.6V-6Fe-5Cu; ME-3, Ti-1.5Al-1V-3Be-2B.
(b) To convert ksi to MPa, multiply by 6.8948.

Table 10.4 Specific Strength (Strength/Density) of Polycrystalline Fibers Compared with Those of Some Commercial Metallic Glass Fibers

Material	Composition	Hardness, kg/mm^2	Strength, kg/mm^2	Density, g/cm^3	Specific strength, 10^5 cm
ME-1	Ti-5.4Al-3.6V-8Fe-3Cu	588	239	4.694	50.9
ME-2	Ti-5.4Al-3.6V-6Fe-5Cu	542	243	4.686	51.9
ME-3	Ti-1.5Al-1V-3Be-2B	673	170	4.326	39.3
Metglas 2615	$Fe_{80}P_{16}C_3B_1$	835	249	7.30	34.1
Metglas 2826A	$Ni_{36}Fe_{32}Cr_{14}P_{12}B_6$	880	278	7.46	36.9
Metglas 2605	$Fe_{80}B_{20}$	1100	370	7.40	50.0
Metglas 2204	$Ti_{50}Be_{40}Zr_{10}$	740	231(a)	4.13	56.0

(a) Calculated from strength = hardness/3.2

11. Rapid-Solidification Processing of Precipitate- and Dispersion-Strengthened Titanium Alloys

The applicability of conventional titanium alloys in aerospace engineering is limited by their relatively low operating temperature range, which to date is no higher than some 600 °C [EYL84]. Solid-solution strengthening, only partially effective even in the intermediate temperature range, cannot be looked to as a mechanism for high-temperature strengthening. At moderate-to-high temperatures, dislocations are thermally activated around point defects; also as a consequence of thermal activation, the alloy's microstructure or substructure, which is chiefly responsible for strengthening in the intermediate temperature range, is unstable at high temperatures. Dispersion strengthening offers a substitute for solution strengthening at high temperatures, and at the same time tends to stabilize the substructure. Through the use of dispersion strengthening, the operating temperature range of titanium alloys can be substantially increased. Moreover, although improvement of high-temperature properties is usually the primary goal of dispersion strengthening, the presence of dispersoids also enhances the flow stress throughout the entire temperature range; additional benefits claimed for dispersion strengthening are increases in creep resistance and stress-rupture life [SAS84, SAS84a]. In selecting a dispersoid it should be recognized that although its composition, as such, is not important, it should be chemically and physically stable in the matrix at elevated temperatures—i.e, insoluble, nonreactive, and resistant to coarsening. Dispersoids should be incoherent with the matrix crystal and resistant to deformation. Dispersoids should be closely spaced (e.g., number density on a per-unit-area basis, 2.6×10^6 mm^{-2} [ROW85]) and small in size (typical diameters, 0.05 to 0.5 μm [SAS84] [CHI85]); as such they are effective barriers to dislocation motion at both ambient- and high temperatures. Dispersoids also tend to "pin" grain- and subgrain boundaries, thereby stabilizing the alloy's substructure against change and inhibiting recrystallization during high-temperature exposure [SAS85]. The dispersion strengthening discussed in all except the last subsection of this presentation is what might be referred to as *"in situ,"* in that the strengthening ingredients are placed in the starting material prior to its first melting.

11.1 Dispersion Strengthening of Titanium Alloys

11.1.1 Dispersion Strengthening in Ingot Metallurgy

The application of dispersion strengthening to conventionally processed (ingot metallurgy, I/M) titanium alloys has been successful up to a point. Boron added to titanium alloys segregates to the interdendritic regions during ingot solidification and contributes substantially to grain refinement and stabilization [ROW85]. Small amounts of yttrium and erbium added to I/M titanium alloys have been shown to yield a uniform distribution of fine incoherent precipitates of the corresponding sesquioxides, Y_2O_3 and Er_2O_3. But attempts to obtain a suitably high density of fine dispersoids by increasing the levels of yttrium and erbium in the starting ingot have always been unsuccessful. The invariable result is a small number density of large-diameter (>1 μm) particles [SAS84, SAS85a] [CHI85]. The coarsening that takes place during the relatively slow cooling of the arc-melted ingots renders the precipitate particles ineffective for dispersion strengthening. They are in fact detrimental, degrading both the fracture toughness and the fatigue properties of the alloy. Eutectoid-forming elements such as silicon and the transition elements iron, nickel, and cop-

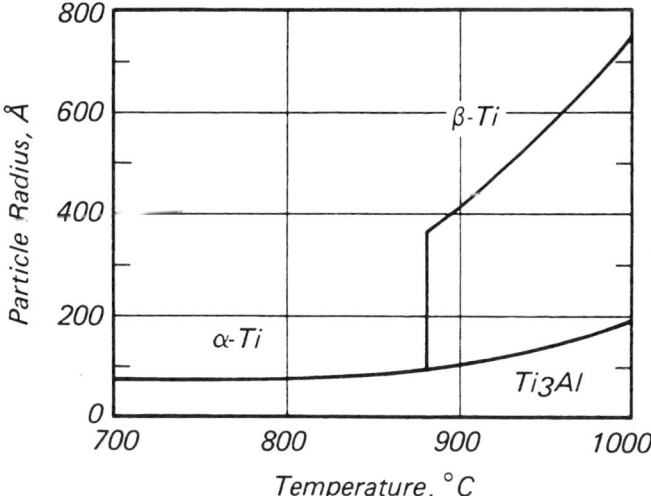

Fig. 11.1 Transmission electron micrograph of a sample of Ti-Fe (5 at.%) in the as-quenched condition after quenching from the melt by the hammer-and-anvil technique [WHA87].

Table 11.1 Comparison of Mechanical Properties of RSP and I/M Ti-Al-Er Alloys [Sas85]

Alloy	Heat treatment(a)	Yield stress MPa		Ultimate tensile stress, MPa		Total elongation, %	
		RSP	I/M	RSP	I/M	RSP	I/M
Ti-5Al-2Er	ST	670	469	735	536	27.0	...
Ti-7.5Al-2Er	ST	850	680	920	756	...	7.0
Ti-9Al-2Er	ST	880	750	928	790	11.0	0.1
T-5Al-2Er	STA (625 °C)	700	510	763	564	13.8	10.0
Ti-7.5Al-2Er	STA (625 °C)	952	815	973	843	7.7	6.0
Ti-9Al-2Er	STA (625 °C)	931	802	952	824	1.6	0.2
Ti-5Al-2Er	STA (550 °C)	714	515	780	590	54.0	18.0
Ti-7.5Al-2Er	STA (550 °C)	973	830	990	865	12.0	9.0
Ti-9Al-2Er	STA (550 °C)	...	810	...	835	...	0.3

(a) ST = solution treating: 3 h/860 °C/WQ. STA = ST plus aging: 25 h/625 °C or 500 h/550 °C

Table 11.2 Microstructural and Property Improvements of RSP Titanium Alloys [Sas83]

Alloy type	Alloy system	Problems with ingot metallurgy	Rapid-solidification microstructural modifications
Dispersion-strengthened alloys	Ti-RE(a)	Coarse particles	Extended solid solutions; fine incoherent dispersoids
Compound formers	Ti-B Ti-C	Limited solid solubility, coarse dispersoids	Grain refinement; titanium boride and carbide dispersoids
Eutectoid formers	Ti-Ni Ti-Si Ti-Fe	Segregation, coarse grains, and precipitates	Controlled eutectoid-decomposition products
Combined precipitates and dispersoids	Ti-Al-RE Ti-Al-Ni Ti-Al-B, C	Coarse dispersoids	Coherent, ordered precipitates and incoherent dispersoids
Conventional titanium alloys	Ti-6Al-4V Ti-8Al-1Mo-1V Ti-6Al-2Sn-4Zr-2Mo	Coarse, elongated grains	Fine martensite structure in as-rapidly-solidified alloys; fine equiaxed $\alpha + \beta$ grains upon annealing in a $\alpha + \beta$ field
Amorphous alloys	Ti-M-B(b) or Ti-M-Si	Cannot be made	Amorphous and microcrystalline structures
Intermetallic compounds	Ti_3Al TiAl	Coarse grains	Grain refinement, incoherent fine dispersions, possible decrease in long-range order

(a) RE = Er, Y Gd, Nd, Sc, La, Dy. (b) M = Mn, Nb, V, Cr

until the advent of rapid solidification processing (RSP). It has been discovered that the coarsened precipitates (in the case of the rare-earth oxides, see later) and grain-boundary precipitates (in the case of eutectoid formers) of I/M products are completely absent from the products of RSP.

But if the coarsening and grain-boundary segregation of I/M are a result of the slow cooling of the ingot ("aging during processing"), how can the products of RSP remain usefully stable at high temperatures, even if produced as fine dispersions in the first place? The answer has to do with the high diffusivities of many elements in β-Ti and the need for rapidly solidifying the alloy into the α phase and making sure that its subsequent processing and service conditions never take it above its α transus.

11.1.2 Dispersion Strengthening in Rapid-Solidification Processing

The problems encountered in attempts to provide *in situ* dispersion strengthening or precipitation strengthening in conventional I/M titanium alloys can be avoided by RSP [Sas85a]. The addition of dispersoid-forming elements to the starting material prior to RSP can lead to microstructural refinement and large number densities of fine dispersoids (e.g., rare earth oxides) or precipitates. As a result, significant improvements in both room-temperature- and elevated-temperature strengths, and in creep properties and stress-rupture lifetimes, have been noted. Table 11.1 is a comparison of some room-temperature mechanical properties of RSP and I/M-processed alloys. Elements that have been added to titanium alloys prior to RSP include: the interstitial elements boron and carbon (which under RSP conditions have contributed significant increases in modulus and yield strength [Sas85a]); the eutectoid formers silicon, chromium, manganese, iron, cobalt, nickel, and tungsten; and the rare earth elements (including yttrium) selected from the "La group" of the periodic table. With regard to the latter, although many rare earths (RE) have low room-temperature solubilities in titanium [Sas84] [Bom86], and have the potential for precipitating in metallic form during solidification, in practice they scavenge dissolved oxygen from the alloy and oxidize to RE_2O_3. The rare earths thus play a useful secondary role in RSP titanium powder metallurgy. Although some dissolved oxygen is desirable for solution strengthening at ordinary temperatures, too much oxygen (easily acquired during P/M processing) causes excessive hardening. The inclusion of RE elements in the alloy formulation can con-

per might be expected to be candidates for dispersion strengthening through fine intermetallic-compound formation. But these elements tend to segregate to grain boundaries during conventional I/M processing [Sas83, Sas85a]. The production of dispersoids by solid-state precipitation (precipitation hardening) has resulted in coarse, thermally unstable particles subject to overaging during processing. Thus, because of the difficulties that have been encountered in producing a suitable distribution of sufficiently small stable precipitates, *in situ* dispersion strengthening did not play a significant role in titanium alloy metallurgy, particularly in the U.S.,

trol to some extent, the final level of oxygen in solid solution. Rapid-solidification processing of Ti-RE alloys has resulted in precipitation of fine oxide particles (< 0.05 μm in diameter [SAS84]), which are very suitable for dispersion strengthening. In RSP alloys containing boron, TiB is the dispersoid species; the TiB coarsens rapidly at the grain boundaries, depleting adjacent regions of boron and averting grain-boundary embrittlement. Other types of precipitates that have been investigated as potential strengtheners include, for example, Ti_5Si_3, TiC, Al_4La, Al_3La, Ti_2Ni, CeS, and $Ce_2(SO_2)$ [SAS83] [WHA84] [ROW85].

Although conventional α-Ti alloys such as Ti-5Al-2.5Sn can be improved by RSP-induced dispersion strengthening, it has been claimed that the greatest advantage can be taken by RSP when it is applied to specially formulated alloys [SAS85a]. This approach has given rise to the many new alloy compositions discussed in the following sections. Rapid-solidification processing is also accompanied by a secondary benefit: When the product is in the form of powder or particles of various kinds, the subsequent consolidation and fabrication steps are accompanied by all of the advantages generally associated with powder metallurgy, with the proviso that any hot consolidation operations should take place at low temperatures and high pressures, rather than conversely, as in conventional P/M.

11.2 Systems for Dispersion Strengthening by Rapid-Solidification Processing

11.2.1 Review of Recent Advances

Conventional alloys such as Ti-8Al-1Mo-1V, Ti-6Al-4V, and Ti-6Al-2Sn-4Zr-2Mo, after rotating-electrode processing (REP) and plasma rotating-electrode processing (PREP) (both sometimes referred to as "conventional" powder-production methods, but see Chapter 10), have yielded elongated microstructures after consolidation by hot isostatic pressing (HIP) and vacuum hot pressing (VHP). On the other hand, RSP of the same alloys (into particulate form) has yielded particles of high dislocation density which recrystallize readily into material with a fine equiaxed grain structure, a property conducive to improved room-temperature mechanical properties and to high-temperature superplasticity [SAS83]. But as SASTRY has pointed out, advantage should be taken of RSP to produce completely new materials especially designed for *in situ* dispersion or precipitation strengthening. With this in mind, alloy systems yielding oxide-particle dispersions and intermetallic-compound precipitates, and systems based on the titanium aluminides, have been prepared by RSP and metallurgically examined [SAS83, SAS85a].

The classes of systems which have been examined are listed in Table 11.2. Alloying elements selected include (1) the interstitial elements boron and carbon (which yield intermetallic-compound precipitates); (2) the group IIIB elements scandium and yttrium, and the lanthanides lanthanum, cerium, neodymium, gadolinium, dysprosium, and erbium (which scavenge oxygen from the matrix to form sesquioxide dispersoids); and (3) β-eutectoid-forming elements such as silicon, iron, nickel, and copper (which yield intermetallic-compound precipitates or fine lamellar microstructures [FRO86]).

Bases for dispersion strengthening with interstitial elements and RE oxides have been: previously unalloyed titanium [SAS84a, SAS85a], Ti-Al alloys [CHI85] [SAS85], and commercial titanium alloys such as Ti-624 [ROW85, ROW85a] and Ti-6242 and Ti-633 [ROW85a].* Interstitial-element strengthening, particularly with boron, has been applied to unalloyed titanium [SAS85a] [CHI86] as well as to more complex systems such as Ti-8Al-1.5Er [SAS83] and Ti-6Zr-6Al-1Er [ROW85a]. Eutectoid-element strengthening studies have so far been confined principally (but not exclusively) to binary systems such as Ti-Fe [KRI84], Ti-Co [KRI85], Ti-Ni [SAS83] [BAE85], and Ti-Cu [KRI86].

In situ dispersion- and precipitation-strengthening of titanium alloys shows great promise for high-temperature applications. Recent work has demonstrated

---

* The compositions of these alloys are: Ti-624, Ti-6Al-2Sn-4Zr; Ti-6242, Ti-6Al-2Sn-4Zr-2Mo; and Ti-633, Ti-6Al-3Sn-3Zr.

Table 11.3 RSP Dispersion- and Precipitation-Strengthened Previously Unalloyed Titanium(a)

System	Ref	System	Ref
Interstitial-element additions		**"Eutectoid-forming" additions**	
Ti-B	[WHA84,WHA86,CHI86]	Ti-0.6Si	[KRI84]
Ti-0.5B	[SAS83,SAS85a,PEN85]	Ti-0.9Si	[KRI84]
Ti-1.0B	[SAS85]	Ti-2Si	[KRI84]
		$Ti-Si_6$	[WHA85]
Ti-1.0C	[SAS85a]		
(Ti-2Zr)-1C	[SAS85a]	(Ti-17.9Zr)-3.3Si	[CHI84]
		(Ti-18Zr)-4.4Si	[WHA84]
Rare-earth and related-element additions		$(Ti-Zr_{10})-Si_{0.8}$	[WHA85]
		Ti-3Cr	[KRI84]
Ti-Y	[SAS85a][CHI86][WHA86]	Ti-15Cr	[KRI84]
Ti-1.0Y	[SAS84]	Ti-30Cr	[KRI84]
Ti-1.5Y	[SAS84]		
		$Ti-Mn_{2.5}$	[WHA87]
Ti-La	[WHA86]	$Ti-Mn_5$	[WHA87]
Ti-2.0La	[SAS84]	$Ti-Mn_{10}$	[WHA87]
Ti-3La	[CHI86]		
		$Ti-Fe_{2.5}$	[WHA87]
Ti-Ce	[CHI86]	$Ti-Fe_{2.8}$	[WHA87]
Ti-1.0Ce	[SAS84]	$TiFe_{3.0}$	[WHA87]
		$Ti-Fe_5$	[WHA87]
Ti-1.5Nd	[SAS84,SAS84a,PEN85]	Ti-3Fe	[WHA87]
Ti-2.0Nd	[SAS85]	Ti-16Fe	[WHA87]
Ti-3.0Nd	[SAS83,SAS84,SAS84a]	Ti-22Fe	[WHA87]
Ti-1.5Gd	[SAS84]	Ti-9Co	[KRI85]
Ti-2.0Dy	[SAS84]	Ti-3Ni	[SAS83][ONE83]
		Ti-5.5Ni	[KRI85]
Ti-Er	[SAS85a]	Ti-7Ni	[SAS83][ONE83]
$Ti-Er_{0.4}$	[KON85]		
Ti-0.5Er	[SAS84][KON85]	Ti-3W	[KRI84]
$Ti-Er_{0.7}$	[KON83]	Ti-28W	[KRI84]
Ti-1.0Er	[SAS84]	Ti-36W	[KRI85a]
Ti-2.0Er	[SAS83,SAS84,SAS84a] [PEN85]	Ti-40W	[KRI84][BAE86]
		Ti-7Cu	[KRI86]

(a) In this and subsequent tables, numerical prefixes indicate composition in weight percent; numerical subscripts indicate atomic percent.

that RSP of RE-containing alloys is capable of yielding ultrafine-grain materials containing dispersoids as fine as 0.02 to 0.05 μm in diameter at number densities (on a per-unit-area basis) as high as 2.6×10^6 mm^{-2}. Some dispersoid species have been found to be stable at temperatures as high as 1000 °C [Row85].

11.2.2 Rapid-Solidification Processing of Titanium-Base Alloys

As indicated in Table 11.3, unalloyed titanium with additions of: (1) interstitial elements, (2) RE elements, and (3) eutectoid formers have been used as bases for rapid-solidification processing.

Although conventional processing yields coarse boride or carbide precipitates, RSP of Ti-1.1B and Ti-1.0C yielded a large number density of fine dispersoids [Sas83]. Of all the Ti-RE systems investigated [Sas84], Ti-Er and Ti-Nd showed particularly promising results [Sas84a]: Ti-Er yielded closely spaced, thermally stable, incoherent dispersoids less than 0.01 μm in diameter; and Ti-Nd yielded two classes of dispersoid (a "bimodal distribution")—very fine particles less than 0.01 μm in diameter, and coarse particles within the size range 0.1 to 1.0 μm—together with neodymium in solid solution. Sastry regarded Ti-Nd as being strengthened by a combination of dispersion- and solution strengthening, and Ti-Er as a purely dispersion-strengthened system. Alloys of titanium with the eutectoid formers are notable for the variety of their microstructures. Depending on solute concentration—i.e, whether hypoeutectoid (solute lean) or eutectoid—it is possible to obtain either a fine-grain material with uniform precipitates (Fig. 11.1) or a lamellar microstructure.

11.2.3 Rapid-Solidification Processing of Ti-Al-Base Alloys

Binary Ti-Al Alloy Bases. Table 11.4 lists many of the binary Ti-Al-base alloys that have been subjected to dispersion- or precipitation-strengthening by rapid-solidification processing. With Ti-Al-B alloys, RSP yielded high-aspect-ratio filamentary dispersions that coarsened during annealing to needle-shape precipitates [Sas85]. This high-aspect-ratio second-phase precipitate, in association with the fine grain size, resulted in significant improvements in modulus and strength [Sas85a]. TiB needle formation has also been noted in the heat treatment of Ti-6Zr-6Al-1Er-0.08B [Row85a]. As was the case with the binary alloys, fine incoherent dispersoids associated with fine grain sizes (1 to 5 μm) have been obtained in RSP of RE-containing ternaries [Sas85]. In Ti-Al alloys with eutectoid-forming additions such as silicon and nickel, rapid solidification followed by carefully controlled heat treatment can lead to fine-scale homogeneous microstructures [Sas85a]; consolidation temperatures must be kept as low as possible to prevent coarsening. The range of ternary materials in this category which have been studied also includes the so-called "super-α" alloys—ones in which the aluminum content is sufficiently high for some α_2-phase precipitation to take place [Sas85]. Although the presence of α_2 precipitation severely embrittles the I/M binary alloy whenever it occurs, the addition of RE solutes in association with RSP is responsible for refining the grain structure and enhancing the post-creep ductility [Bom86] [Fro86]. The presence of a finely dispersed α_2-phase precipitate in the RSP material was claimed to improve the high-temperature strengths.

Multicomponent and Commercial Alloy Bases. A representative selection of the numerous multicomponent and commercial alloys that, with the addition of strengthening elements, have undergone RSP is presented in Table 11.5. Strengthening elements represented in the list are: interstitial elements (boron), eutectoid formers (silicon), RE elements (lanthanum, cerium, and erbium), the metalloid germanium, and sulfur. The

Table 11.4 RSP Dispersion- and Precipitation-Strengthened Alloys Based on Ti-Al

System	Ref
Ti-8Al-2Y	[Sas85]
Ti-8Al-4Y	[Kon83a]
Ti-8.5Al-0.5Y	[Sas84a]
Ti-5Al-3La	[Chi86]
Ti-5Al-4.5La	[Lu85]
Ti-5Sn-3La	[Chi86]
Ti-9.5Sn-3La	[Chi86]
Ti-9.5Sn-5.3La	[Chi86]
Ti-8Al-2Nd	[Sas85]
Ti-5Al-2Er	[Sas85]
Ti-5Al-5.4Er	[Chi85]
Ti-7.5Al-2Er	[Sas85]
Ti-8Al-2Er	[Sas85]
Ti-9Al-2Er	[Sas85]
Ti-Al$_{10}$-Er$_{0.4}$	[Kon85]
Ti-Al$_{15}$-Er$_{0.4}$	[Kon85]
Ti-Al$_{24}$-Er$_{0.4}$	[Kon85]
Ti-8Al-1B	[Sas85]
Ti-8Al-1.5Er-0.25B	[Sas83]
Ti-6Zr-6Al-1Er-0.08B	[Row85a]
Ti-7.9Zr-3.5Al-1.4B	[Chi84]
Ti-5Al-2Si	[Wha86]
Ti-8Al-2.0Si	[Sas85]
Ti-8.5Al-0.2Si	[Sas85]
Ti-8.5Al-0.5Si	[Sas85]
Ti-8.5Al-1.0Si	[Sas85]
Ti-7.7Zr-3.4Al-3.6Si	[Wha84]

Table 11.5 RSP Dispersion- and Precipitation-Strengthened Alloys Based on Multicomponent and Commercial Titanium Alloys

System	Ref	System	Ref
Interstitial-element additions		**Rare earth and interstitial-element additions (cont.)**	
(Ti-5Al-2.5Sn)-0.2B	[Chi84]		
(Ti-5Al-2.5Sn)-1B	[Chi84]		
(Ti-6Al-4V)-1B	[Chi84]	(Ti-4Zr-6Al-2Sn-2Mo)-0.08Si	[Vog86]
Ti-7.5Zr-4Mo-1.3B	[Chi84]	(Ti-4Zr-6Al-2Sn-2Mo)-1Er	[Row85]
Ti-8.2Mo-2.3Al-1.4B	[Wha84]	(Ti-4Zr-6Al-2Sn-2Mo)-0.08Si-2Er	[Vog86]
		(Ti-4Zr-6Al-2Sn-2Mo)-0.08Si-3W	[Vog86]
(Ti-5Al-2.5Sn)-1C	[Chi84]	(Ti-4Zr-6Al-2Sn-2Mo)-0.4Si	[Vog86]
		(Ti-4Zr-6Al-2Sn-2Mo)-0.4Si-2Er	[Vog86]
Rare earth and interstitial-element additions			
		(Ti-4Zr-6Al-2Sn-6Mo)-1Er	[Sno84]
(Ti-5Al-2.5Sn)-2Y	[Chi84]	(Ti-4Zr-6Al-2Sn-6Mo)-2Er	[Vog86]
(Ti-5Al-2.5Sn)-3La	[Chi84]	(Ti-6Zr-6Al)-0.08B-1Er	[Row85]
(Ti-5Al-2.5Sn)-3Ce	[Wha84]	**"Eutectoid-forming" additions**	
(Ti-6Al-4V)-1Er	[Row85a]	(Ti-5Al-2.5Sn)-0.5Ge	[Jac85]
Ti-6Al-15V-2Er	[Fro86]	(Ti-5Al-2.5Sn)-7.5Ge	[Jac85]
Ti-25V-4Ce-0.6S	[Fro86]	(Ti-5Al-2.5Sn)-0.5Si	[Jac85]
		(Ti-5Al-2.5Sn)-5Si	[Jac85]
(Ti-4Zr-5Al-2.5Sn)-3La	[Wha84]	(Ti-6Al-4V)-2.2Si	[Wha84]
(Ti-4Zr-6Al-2Sn)-1Er	[Row85a]		
(Ti-4Zr-6Al-2Sn)-1Ce-0.15S	[Row85]	(Ti-7.4Zr)-3.9Mo-3.4Si	[Chi84]

Table 11.6 RSP Titanium Aluminides

System	Ref
$Ti_3Al + Er_{0.4}$	[KON85a]
$Ti_3Al + Er_{0.6}$	[ROW86,ROW86a]
	[SUI86]
$Ti_3Al + Nb$	[EYL86]
$Ti_3Al + Nb_5 + Ce_{0.6} + S_{0.2}$	[ROW86,ROW86a]
$Ti_3Al + Nb_{7.5} + Ce_{0.7} + S_{0.2}$	[ROW86,ROW86a][SUI86]
$Ti_3Al + Nb_5 + Er_{0.6}$	[ROW86,ROW86a][SUI86]
$Ti_3Al + Nb_{7.5} + Er_{0.6}$	[ROW86,ROW86a][SUI86]
$Ti_3Al + Nb_{10} + Er_{0.5}$	[SUI86]
TiAl	[MAR83]
TiAl + W	[MAR83]

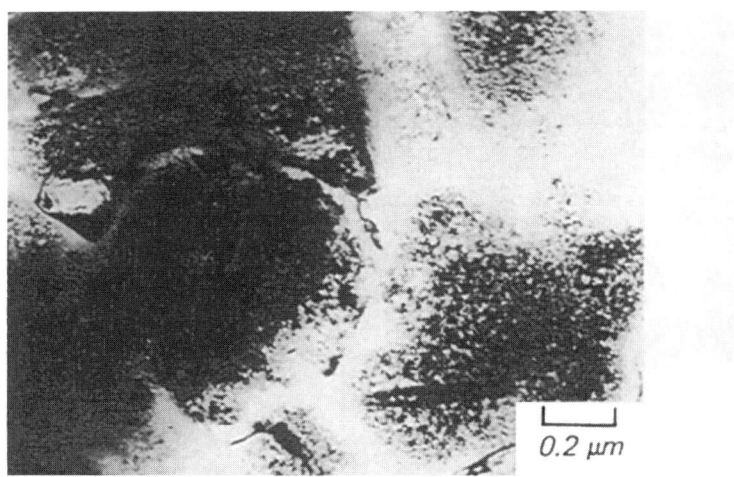

Fig. 11.2 Simulation of the change in radius of Er_2O_3 particles annealed for 10 h within the temperature range shown. Below about 880 °C (the β transus for titanium), the curves for α-Ti and Ti_3Al are continuous (since the oxygen diffusivity is assumed to be the same in each phase). Above 880 °C, the particles show marked coarsening in the β phase, whereas only a modest change of size in Ti_3Al [KON85].

presence of oxygen is assumed. In an important recent study, cerium and sulfur were added to Ti-6Al-2Sn-4Zr under RSP conditions. Sulfur is often regarded as an undesirable "tramp impurity" that by segregation to grain boundaries, tends to embrittle some conventional I/M-processed alloys. The rationale underlying its introduction, together with cerium, into RSP Ti-624 is that in a titanium environment the heats of formation of CeS and $Ce_2(SO_2)$ are greater than that of Er_2O_3, a favored dispersion hardener [ROW85].

11.2.4 Rapid-Solidification Processing of Titanium Aluminides

Attractive properties of the $α_2$-phase (based on Ti_3Al) and γ-phase (based on TiAl) aluminides are their high transus temperatures: 1100 °C for the $α_2 \to α + α2$ transus of Ti_3Al, and 1400 ± 60 °C for the melting point of the γ phase (although the useful temperature range of the latter tends to be limited by a brittle-to-ductile transition at 700 °C [LIP75]). Thus, at least from a phase-stability standpoint, the aluminides make suitable bases on which to design potentially useful high-temperature alloys. Table 11.7 refers to some recent studies of RSP-and-strengthened titanium aluminides. A second important advantage of the $α_2$- and γ-phase aluminides is their better oxidation resistance compared with conventional titanium alloys, a property which in the case of Ti_3Al has been improved even further by the addition of 5 to 10 wt% Nb [MEN80]. But, as ordered intermetallic compounds, both Ti_3Al and TiAl lack tensile ductility at ordinary temperatures, a property which has severely limited their applicability. Powder metallurgy of Ti_3Al [LIP80] and TiAl [MAR83] has yielded some promising results, and with the advent of RSP metallurgy in association with the introduction of RE and other third-element additions, the aluminides at last seem to be on the threshold of practical application.

Rapid-solidification processed Ti_3Al + Nb has been used successfully in experimental studies of Borsic-reinforced metal-matrix composites [EYL86]. The addition of 0.4 at.% Er to Ti_3Al led to a fine Er_2O_3 dispersion that seemed to be quite stable at 900 °C [KON85a]. Erbium added to Ti_3Al + Nb resulted in a refinement of the grain size, and hence to an improvement in ductility. Under extrusion, however, a rapid coarsening of the Er_2O_3 dispersoids was noted [ROW86]. In compounds containing CeS or $Ce_2(SO_2)$, it seems that the coarsening is less pronounced—a reflection of the high stability of these compounds, already considered above in connection with the dispersion strengthening of Ti-624—and hence that Ti_3Al with additions of cerium and sulfur should be considered for inclusion in any list of prospective alloys for high-temperature application.

11.3 Microstructural Stability of RSP Titanium Alloys

11.3.1 Precipitate Coarsening

Heat treatment (carefully controlled limited aging) of RSP titanium alloys with interstitial elements (e.g., boron and carbon) generally results in improved properties. In boron-containing alloys, aging results in high-aspect-ratio needle-shape precipitates of TiB. Unless prolonged exposure to elevated temperature allows them to coarsen excessively, these precipitates contribute a large increment of strength [SAS85a]. Heat treatment of titanium alloys containing carbon results in an increase in ductility (as carbon becomes removed from solid solution) and an increase in strength as the accompanying reduction in solution strengthening is more than compensated for by an increase in TiC precipitation strengthening.

Precipitates arising from the interstitial elements boron and carbon and the eutectoid-forming elements (including silicon) are all prone to excessive coarsening during prolonged exposure to high temperatures [WHA84][LU85][SAS85a]. WHANG has contrasted the behavior of boron, carbon, and silicon with that of the RE element lanthanum under high-temperature aging (at 800 °C). The relative stability of the La_2Sn dispersoids in Ti-5Sn-4.5La as compared with TiSi precipitates in Ti-5Al-2Si was attributed to the lower diffusivity of lanthanum as compared with silicon in α-Ti ($\sim 4 \times 10^{-14}$ and $\sim 1.2 \times 10^{-11}$ cm^2/s, respectively [LU85]). Comparisons among the RE dispersoids themselves have been made by SASTRY and colleagues [SAS84]. These authors noted that the rare earths could be subdivided into classes: (1) cerium, neodymium, and gadolinium, which have appreciable solubilities in titanium and which on isothermal aging yield precipitates that become relatively coarse (~0.2 to 2 μm); and (2) yttrium, lanthanum, dysprosium, and erbium, which have negligible solubilities in titanium and which under aging yield relatively fine (0.04 to 0.12 μm) dispersoids. The coarse dispersoids of the first group were RE sesquioxides, and the fine dispersoids of the second group were compounds of titanium, rare earths, oxygen, and carbon. Of all the RE elements in titanium, cerium yielded the coarsest dispersoids and erbium yielded the finest.

SASTRY's work was followed (or accompanied) by numerous other studies of erbium-containing RSP titanium alloys.

ROWE et al. [ROW85a] selected the system Ti-6Al-6Zr-1Er-0.08B for study. After aging, the usual TiB needles were noted. But of particular interest was the fact that only near the grain boundaries did the Er_2O_3 particles undergo coarsening, presumably as a result of grain-boundary diffusion. KONITZER et al. [KON85] undertook a comprehensive study of Er_2O_3 dispersoids in Ti and Ti-Al(10, 15, 24, at.%) alloys. During the 10-h aging at 900 °C of Ti-Al_{24}-$Er_{0.4}$ (after 10 h/700 °C to develop the dispersion) it was found that the Er_2O_3 particles were fairly resistant to coarsening. This tended to be true for all the α-Ti alloys below the α-transus temperature. But as indicated in Fig. 11.2, rapid coarsening could be expected for all Er_2O_3 particles lying within the β regions of a two-phase alloy, due to the much higher diffusivity of oxygen in the β phase. KONITZER and FRASER [KON85a] showed that in RSP Ti_3Al + 0.4 at.% Er a lack of significant oxide-precipitate—oarsening was exhibited after a 10-h exposure to temperatures as high as 800 to 900 °C. This performance emphasized the importance of a high transus temperature in dispersion coarsening oxide precipitates in binary Ti-Er alloys exhibited significant coarsening in response to heat treatment for 10 h at 900 °C. The fineness and stability of Er_2O_3 precipitates in the hexagonal phases of titanium alloys indicate that erbium should always be considered among the possible dispersion-strengthening additives to α-phase and $α_2$-phase titanium alloys.

Recent work by ROWE and KOCH [ROW85] has indicated that other additives besides erbium have important roles to play when, in addition to strengthening, grain refinement must be taken into consideration. As indicated in Section 11.2.3, these authors had estimated that sulfides and oxysulfides of cerium were more stable in a titanium environment that was Er_2O_3, the hitherto premier dispersion strengthener. Accordingly, they decided to introduce both cerium and sulfur into a titanium-alloy base; Ti-6Al-2Sn-4Zr was chosen as the test alloy. SASTRY's studies had indicated that cerium in titanium yielded the coarsest precipitates of all the RE elements [SAS84]. But CeS and $Ce_2(SO_2)$, according to ROWE et al., resisted coarsening at temperatures as high as 1000 °C except near the grain boundaries, where grain-boundary diffusion was likely to assist in the process. Particles in the grain interiors were about 0.03 to 0.04 μm in size, and those near the grain boundaries, about 0.15 to 0.20 μm.

11.3.2 Grain Growth

Alloys containing erbium were found unable to resist some grain growth during aging, a disadvantage which may outweigh their ability to yield ultrafine dispersions. On the other hand, alloys containing boron or silicon retain their fine as-RSP grain structures to high temperatures. In particular, ROWE and KOCH [ROW85] found that their consolidated sulfur-bearing alloy possessed a submicron grain structure which resisted growth at 1000 °C.

11.4 Mechanical Properties of RSP Titanium Alloys

11.4.1 Hardness and Tensile Strength

Contributions to strengthening in RSP titanium alloys are: (1) solid-solution strengthening arising from the extended solubilities that accompany the process, (2) fine-grain strengthening, and (3) Orowan strengthening from high-number-density arrays of fine incoherent precipitates. Provided the dispersoids resist coarsening, Orowan strengthening with its weak temperature dependence is the mechanism to be relied on in alloys for high-temperature service. Rapid-solidification processed alloys are generally subject to age hardening [CHI85, CHI86] as precipitates form from supersaturated solid solutions. Overaging refers to the excessive coarsening of the dispersoids, which takes place much more readily in β-phase and (α + β)-phase alloys than in α alloys due to the higher diffusivities of many solutes (particularly oxygen in this content) in β-Ti. Obviously an alloy's resistance to aging is closely related to its heat resistance—i.e., its ability to withstand high-temperature service conditions.

Many tensile-property studies of RSP alloys have been confined to the room-temperature testing of as-formed and/or age-hardened (moderate-temperature annealed) material. In this regard, SASTRY et al. [SAS85a] have investigated the properties of Ti-C and Ti-B. In the latter study it was found that the strengthening effect of boron also persisted to high temperatures [SAS85b] (in spite of the extensive coarsening that has been noted for TiB precipitates during exposure to temperatures in the range 800 to 900 °C [WHA84]). Boron added to Ti-8Al resulted in an alloy with a good combination of low density, high modulus, high room-temperature and elevated-temperature tensile strengths, and a potential for high-temperature applications.

The temperature dependence of the tensile properties of Ti-Nd and Ti-Er have been measured by SASTRY et al. [SAS84a]. As pointed out above, neodymium and erbium belong to the "coarse" and "fine," respectively, classes of dispersoid-forming elements, yet at 700 °C the yield strengths of Ti-1.5Nd and Ti-0.1Er (previously aged 2 h/700 °C) were the same. Very much greater strengths are exhibited by the ternary alloys based on Ti-Al. The room-temperature properties of various Ti-Al-Nd and Ti-Al-Er alloys have also been measured by SASTRY et al. [SAS85]. It is interesting to note that after Ti_3Al precipitation was caused to form in some of these alloys, the strengthening due to the incoherent dispersoids plus the Ti_3Al was less than that due to the incoherent dispersoids plus aluminum in solid solution. The high-temperature strengths of Ti-Al-Er alloys were anticipated to be greater than those of all conventional titanium alloys.

The relative qualities of erbium and lanthanum dispersoid-forming additions to RSP Ti-Al alloys were investigated by WHANG [CHI85]. Both of these RE elements are members of the "fine" class of dispersoid-forming additions. A distinction must be drawn between the room-temperature properties of the age-hardened alloys and their relative performances at elevated temperatures. In the former category, Ti-5Al-4.5La is superior to Ti-5Al-5.4Er after aging for 2 h at all temperatures up to 900 °C. However, in hot hardness tests, due to the rapid softening of Ti-5Al-4.5La at temperatures above about 600 °C, at 900 °C both alloys were equally hard.

For reasons outlined in the previous section on aging, cerium and sulfur in association hold considerable promise as high-temperature strengtheners of titanium alloys. Although tensile testing has not been carried out above 538 °C, metallographic studies of grain- and dispersoid growth have been conducted on alloys exposed to temperatures as high as 1000 °C, during which considerable microstructural stability was noted. The tensile work indicated that the sulfide and oxysulfide precipitates provided strengthening over the entire temperature range, yet at the same time permitted adequate room-temperature ductility [ROW85].

11.4.2 Creep

Relatively little has been written about the creep properties of RSP titanium alloys. They can, however, be qualitatively predicted from those of dispersion-strengthened alloys in general. The usual mechanisms of creep are associated with diffusion, grain-boundary sliding, and dislocation movement. The initial fine grain structure of RSP alloys tends to enhance creep; thus from a creep standpoint a certain amount of deliberately induced grain growth is advantageous. Creep resistance in RSP alloys at high temperatures relies primarily on the ability of the dispersoids to pin dislocations. But severe matrix softening is always to be expected in α-Ti solid-solution alloys at temperatures above about 900 °C [CHI85]. To combat this, the introduction of some form of fibrous reinforcement is recommended.

11.5 Summary

The operating temperature range of conventional I/M multicomponent alloys such as Ti-6242 (Ti-6Al-2Sn-4Zr-2Mo) and IMI 834 (Ti-5.5Al-4Sn-4Zr-1Nb-0.3Mo-0.5Si) is limited to 500 to 600 °C, above which microstructural instability becomes a problem. Furthermore, I/M alloys are not amenable to *in situ* dispersion or precipitation strengthening as a consequence of the coarsening which occurs during the alloy's long dwell time in the β-phase field during cooldown. To find a way out of the instability difficulty it is necessary to turn to materials which do not undergo phase transformation within the service-temperature range: stable β-phase alloys (not a practical solution), all-α alloys (especially with high α-transus temperatures), and the aluminides of titanium. The coarsening-during-processing difficulty is eliminated through the use of rapid solidification techniques; coarsening in service is eliminated by turning, again, to the nontransforming class of alloys.

Within the realm of all-α alloys, the requirements of solution strengthening, low density, and high α-transus temperature are simultaneously served if aluminum is selected as a solute. If the aluminum concentration exceeds about 9 wt%, a finely dispersed α_2-phase precipitate will be present (Eq 9.1). Although this severely embrittles I/M alloys whenever it occurs, its presence under RSP conditions has been claimed to improve the high-temperature strength. On the other hand, it has been determined that when incoherent dispersoids are present in it, the single-phase solid solutions are stronger materials than those containing α_2-phase precipitates.

High-temperature creep strength is enhanced through the introduction of a submicroscopic dispersed phase. It has been noted that dispersoids should be insoluble in the alloy matrix, incoherent, nonreactive, fine and closely spaced, resistant to coarsening, and resistant to deformation. Elements that have been considered as ingredients in RSP titanium alloys for dispersion- or precipitation strengthening are: (1) the interstitial elements boron and carbon (which yield intermetallic-compound precipitates); (2) the group IIIB elements scandium and yttrium, and the lanthanides lanthanum, cerium, neodymium, gadolinium, and erbium (which scavenge oxygen from the host alloy to form sesquioxide dispersoids); and (3) β-eutectoid-forming elements such as silicon, iron, nickel, and copper (which yield intermetallic-compound precipitates or fine lamellar microstructures). Insufficient information is known about the high-temperature mechanical properties of the alloys with β-eutectoid formers. The chemical reactivity of the rare earths can be turned to advantage—they scavenge excess oxygen from the alloy (which is particularly advantageous in P/M) and, in addition, after being converted to RE_2O_3, act as dispersion strengtheners.

Many published studies have focused attention on the interstitial element boron, and several of the RE elements. (1) *Boron additions:* It has been noted that RSP Ti-Al-B alloys contain high-aspect-ratio filamentary dispersoids which coarsen during annealing to needle-shape precipitates ideally suited to matrix reinforcement (see Fig. 10.8) unless prolonged exposure to very high temperatures causes them to coarsen excessively. The strengthening effect of boron at high temperatures, in spite of coarsening, has been noted: boron-doped Ti-8Al has been identified as an alloy with potential for high-temperature applications. (2) *Rare earth and other additions:* Comprehensive studies of the stability and effect of RE additions to titanium have indicated that the most promising ones are erbium and neodymium. Both yield very fine dispersions of RE_2O_3 with particle diameters of less than 0.01 μm; but neodymium also yields a crop of larger dispersoids within the size range of 0.1 to 1.0 μm. Cerium, on the other hand, yields the coarsest dispersoids of all the rare earths. As for stability, studies have shown that Er_2O_3 is fairly resistant to coarsening during high-temperature exposure (especially in the grain interiors, as distinct from the grain boundaries within which coarsening seems to be promoted by boundary diffusion). But if Er_2O_3 is fairly stable, the sulfide and oxysulfide of cerium, CeS and $Ce_2(SO_2)$, are even more so. Thus, in spite of the fact that cerium alone in titanium yields the coarsest precipitates of all the RE elements, its inclusion accompanied by sulfur yields a dispersoid system with considerable stability. Recent tensile work has indicated that CeS and $Ce_2(SO_2)$ precipitates are capable of providing strengthening at temperatures approaching 1000 °C.

Both Ti_3Al and TiAl, to which 5 to 10 wt% Nb has been added to improve ambient-temperature ductility and oxide-scale adherence at high temperatures (in the case of Ti_3Al), have assumed considerable importance as potential high-temperature alloys. To these, the addition of dispersoid formers should also be considered—in this case not for dispersion strengthening (there is little need for this), but rather to inhibit grain growth during RSP and in service; the establishment and maintenance of microcrystallinity in this way tends to contribute to ambient-temperature ductility.

12. Mechanical Properties

12.1 Elastic and Plastic Properties of Titanium Alloys at Low and High Temperatures

Technical titanium-base alloys fall into three categories: α, α + β, and β. Unalloyed titanium, α alloys such as Ti-5Al-2.5Sn, and near-α α + β alloys such as Ti-6Al-4V and Ti-8Al-1Mo-1V are preferred for service at low temperature where the β phase could otherwise cause embrittlement. Other α + β alloys find use in the medium-temperature and "high-temperature" (<500 °C) ranges. The α phase is stabilized by simple metals such as aluminum and tin and interstitial elements such as carbon, nitrogen, and oxygen—in other words, by nontransition elements. The α stabilizers are also rapid-solution strengtheners of the alloy in which they are dissolved. The β phase is stabilized by transition elements, which also provide weak solution strengthening. Although the transition elements are much less potent (rapid) strengtheners than the α stabilizers (on a per-atom basis), this deficiency is more than compensated for by their greater solid-solubility range. It therefore turns out that, whereas the strength of the α phase is limited to 80 ~ 100 ksi (550 ~ 690 MPa), that of the β phase may be as high as 100 ~ 120 ksi (690 ~ 825 MPa) [JAF73a]. To a first approximation, the strengths of α + β alloys are a mixture-rule average of those of their constituents [MAR60, p. 291]. The influence of microstructure on strength is a subject of perennial interest [JAF58, p. 149 *et seq.*] [MAR60, p. 291 *et seq.*][JAF73a] [CHE80]. This is particularly true of α + β alloys, for which thermomechanical process variation offers a wide range of microstructural states. On the other hand, apart from the special properties associated with precipitated α_2, the mechanical properties of the α alloys are not so sensitive to microstructure [JAF73a].

As pointed out above, the β alloys are unsuited for low-temperature applications; but whereas the α alloys perform satisfactorily at low temperatures their mechanical properties decrease rapidly with increasing temperature (Fig. 9.1). Some α + β alloys, such as Ti-6Al-4V and Ti-8Al-1Mo-1V, can also be regarded as having properties suitable for a wide range of cryogenic applications [SAL79].

As the temperature increases above room temperature, since the strengths of the all-α alloys such as Ti-5Al-2.5Sn continue to decrease rapidly, these alloys must be abandoned in the intermediate-temperature range in favor of the α + β or β alloys which are capable of maintaining theirs (see Fig. 9.2).

With regard to high-temperature service, although all-β technical alloys such as β-III and the more complex β-C are available, general discussions of heat-resistant alloys (e.g., [POS81]) do not emphasize their use. In fact, the tendency is not only to use an α + β alloy at elevated temperatures, but also to select one whose low β content places it in the near-α category. According to POSTANS and JEAL [POS81], for example, the alloys best suited to gas-turbine engine use (compressor disks and blades) are Ti-8Al-1V-1Mo (limit 400 °C), Ti-4Al-2Sn-4Mo-0.5Si (IMI 550, limit 450 °C), Ti-6Al-2Sn-4Zr-2Mo (limit 450 °C), Ti-6Al-2Sn-4Zr-2Mo-0.1Si (limit 510 °C), and Ti-6Al-5Zr-0.5Mo-0.2Si (IMI 685, limit 520 °C).

This chapter discusses the elastic properties (the moduli) and the plastic properties (the strengths) of titanium-base alloys in that order. In so doing, it deals with mechanical properties as measured using: (1) the static techniques of hardness measurement and tensile testing, and (2) the vibrational or acoustic techniques of dynamic elastic modulus measurement. The static modulus is an engineering number. It is introduced into this chapter in tabular form with no discussion. The results of dynamic modulus measurement lend themselves to discussion in terms of fundamental alloy theory and in this vein are considered in detail in the third section of this chapter. Two important strengthening mechanisms—solution strengthening and precipitation strengthening—are reviewed in Section 12.8 (see also Section 8.1). Hardness is considered in Section 12.9; it is a measurement, simple to perform, whose results are related to both elastic modulus and yield strength. In this sense it couples the results of the preceding two sections. It is logical to describe next some normal tensile properties of titanium alloys. Accordingly, Section 12.13, under the heading "Tensile Strengths of Some Commercial Titanium Alloys," confines itself to some yield-strength temperature-dependence data for a few representative technical alloys. The chapter concludes with a brief survey of "anomalous" tensile properties, viz., those which exhibit pronounced departures from stress-strain linearity and/or reversibility.

Part 1: Elastic Properties

12.2 Static Elastic Moduli

The Young's modulus, E, being the slope of the linear portion of the stress-strain, $\sigma(\varepsilon)$, curve may be obtained from the results of a static or quasistatic (i.e., very low frequency) tensile test. Alternatively, the slope about the origin of $\sigma(\varepsilon)$ may be obtained from a measurement of the velocity of sound in the sample via an appropriate form of the general relationship:

$$\text{sound velocity} = \sqrt{\frac{\text{modulus}}{\text{density}}} \quad \text{(Eq 12.1)}$$

"Dynamic moduli" obtained using such approaches will be discussed in the following sections.

Returning to the static moduli, the subject of this section, Fig. 12.1 serves as a reminder of the definitions of the engineering quantities: bulk modulus (K), shear modulus (G), Young's modulus (E), and Poisson's ratio (ν), which in an isotopically elastic solid (fine, randomly textured grains) are simply related to each other according to:

$$K = \frac{E}{3(1-2\nu)} \quad \text{(Eq 12.2)}$$

and

$$G = \frac{E}{2(1+\nu)} \quad \text{(Eq 12.3)}$$

Table 12.1 Elastic Moduli of Several Commercial Titanium-Base Alloys: Typical Room-Temperature Values [STR82]

Alloy name	Nominal composition	Condition	Young's modulus, E				Shear modulus, G	
			Tensile		Compressive			
			10^6 psi	10^{10} N/m^2	10^6 psi	10^{10} N/m^2	10^6 psi	10^{10} N/m^2
5-2.5	Ti-5Al-2.5Sn	Annealed (0.25-4 h/1300-1600 °F)	15.8	10.9	...	...	7.0	4.8
3-2.5	Ti-3Al-2.5V	Annealed (1-3 h)/1200-1400 °F)	14.5	10.0	15.0	10.3	...	...
6-2-1-1	Ti-6Al-2Nb-1Ta-1Mo	Annealed (0.25-2 h/1300-1700 °F)	16.8	11.6	18.0	12.4	3.4	2.4
8-1-1	Ti-8Al-1Mo-1V	Annealed (8 h/1450 °F)	17.5	12.1	18.0	12.4	6.7	4.6
Corona 5	Ti-4.5Al-5Mo-1.5Cr	α-β annealed after β processing	15.5-17.0	10.7-11.7	...	...	...	...
Ti-17	Ti-5Al-2Sn-2Zr-4Mo-4Cr	α-β or β processed plus aged	16.3	11.2	...	...	...	...
6-4	Ti-6Al-4V	Annealed (2 h/1300-1600 °F)	16.0	11.0	16.1	11.1	6.1	4.2
		Aged	16.5	11.4	16.6	11.4	6.1	4.2
6-6-2	Ti-6Al-6V-2Sn	Annealed (3 h/1300-1500 °F)	16.0	11.0	...	...	6.5	4.5
		Aged	17.0	11.7	17.5	12.1	6.5	4.5
6-2-4-2	Ti-6Al-2Sn-4Zr-2Mo	Annealed (4 h/1300-1550 °F)	16.5	11.4	18.0	12.4	...	...
6-2-4-6	Ti-6Al-2Sn-4Zr-6Mo	Annealed (2 h/1500-1600 °F)	16.5	11.4	18.0	12.4	...	...
6-22-22	Ti-6Al-2Sn-2Zr-2Mo-2Cr-0.25Si	α-β processed plus aged	15.7	10.8	16.1	11.1	6.7	4.6
10-2-3	Ti-10V-2Fe-3Al	Aged	15.9	11.0	16.3	11.2	...	...
15-3-3-3	Ti-15V-3Cr-3Sn-3Al	Aged	14.3	9.9	15.9	11.0	...	...
13-11-3	Ti-13V-11Cr-3Al	Annealed (0.5 h/1400-1500 °F)	14.3	9.9	15.2	10.5	6.2	4.3
		Aged	16.0	11.0	15.8	10.9	...	...
38-6-44	Ti-3Al-8V-6Cr-4Mo-4Zr	Annealed (0.5 h/1500-1700 °F)	12.5	8.6	...	...	...	...
		Aged	16.7	11.5	15.0	10.3	5.8	4.0
β-III	Ti-4.5Sn-6Zr-11.5Mo	Annealed (0.5 h/1300-1600 °F)	12.0	8.3	11.0	7.6	3.9	2.7
		Aged	15.0	10.3	16.0	11.0	5.9	4.1

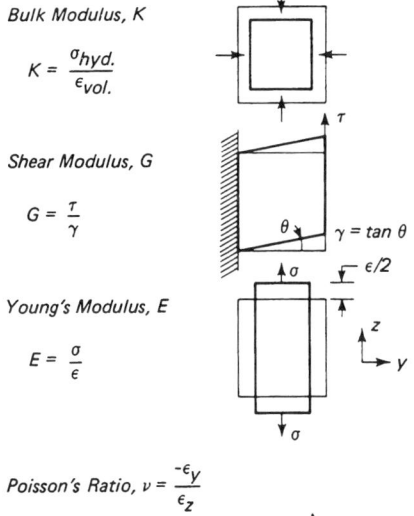

Fig. 12.1 Standard definitions of the elastic moduli of polycrystalline solids (see, for example, HAYDEN et al. [HAY65, p. 23]).

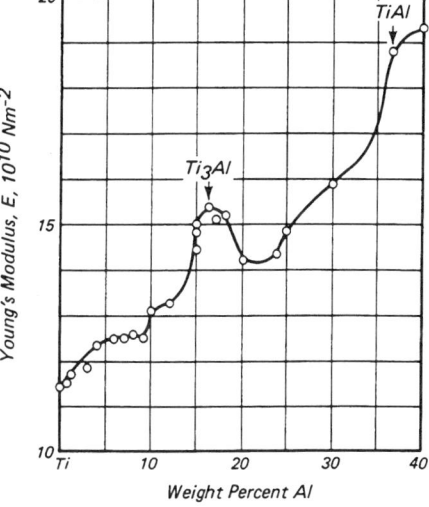

Fig. 12.2 Young's modulus E, of Ti-Al as a function of aluminum concentration [FED66, p. 208].

In the literature, mechanical properties are expressed in one or more of the following units: kg/mm^2, N/m^2, dyne/cm^2, Mbar, and GPa. The numerical relationships among them are:

$$1 \text{ N} = 10^5 \text{ dyne}$$
$$1 \text{ N/m}^2 = 1 \text{ Pa}$$
$$= 1.4504 \times 10^{-7} \text{ ksi}$$
$$10^7 \text{ N/m}^2 = 1.0197 \text{ kg/mm}^2$$
$$10^{11} \text{ N/m}^2 = 10^{12} \text{ dyne/cm}^2$$
$$= 100 \text{ GPa}$$
$$= 1 \text{ Mbar}$$

Table 12.1 is a listing of the Young's moduli of several technical titanium-base alloys.

12.3 Dynamic Elastic Moduli: Long-Wavelength Methods

The wavelengths of kHz-frequency vibrations propagated along a metallic bar or wire are commensurate with its length. In either its longitudinal (L) or transverse (T) vibrational modes, since unconstrained stretching and relaxation are taking place in either case, the wave velocity, v, along the sample is controlled simply by the "static" Young's modulus, E, according to $v_{L,T}$ (long wavelength) = $\sqrt{E/\rho_d}$, where ρ_d is the density. Torsional vibrations are required for the determination of the shear modulus according to $v_{torsion} = \sqrt{G/\rho_d}$. A commercially available device often used for this type of measurement is the "Elastomat" designed by F. Forster (see [FED63]). Alternatively, the Marx-oscillator technique may be employed. In recent measurements of Ti-Nb samples using the latter method [LED81], cylindrical-rod specimens, about 5 mm in diameter, were cemented to a matched pair of Y-cut, X-plated, rectangular-rod 50-kHz quartz crystals. Using suitable electronics, this three-component composite oscillator (the Marx oscillator) was swept in frequency until the half-wave resonance took place. This was detected by an oscilloscope and measured by a frequency meter. The Young's modulus could then be determined from the relationship:

$$E = 4\rho_d f^2 l^2 \quad \text{(Eq 12.4)}$$

where f is the resonant frequency of the rod of length l.

12.4 Systematic Variation of Elastic Moduli with Composition and Microstructure in Titanium-Base Alloys

Using the Elastomat method, FEDOTOV and colleagues have measured the

Young's and shear moduli of numerous series of α-stabilized Ti-SM and β-stabilized Ti-TM binary alloys as functions of composition, hence of composition-related microstructure. A representative group of results is presented and briefly discussed in the following two sections.

12.4.1 Elastic Moduli of Ti-Al Alloys: Long-Wavelength Results

The Young's modulus, E, and the shear modulus, G, have been measured by FEDOTOV [FED66] on a series of Ti-Al alloys whose composition range 0 to 40 wt% Al (i.e., 0 to 54 at.%), includes both the α_2-phase (Ti_3Al) and γ-phase (TiAl) intermetallic compounds. Since $E \cong 2.6\, G$ (according to Eq 12., assuming $\nu \cong 0.3$ = const.), it is to be expected that the curves of E and G versus some common parameter would be "parallel." This indeed turns out to be the case, and several examples of it are presented below. Only one of the moduli E is shown in Fig. 12.2, which emphasizes two important properties of the Ti-Al system: (1) the addition of aluminum is responsible for a rapid increase in modulus; and (2) anomalies appear at compositions corresponding to Ti_3Al (a local maximum) and TiAl (a point of inflexion). Similar anomalies have also appeared in the plot of θ_D versus aluminum concentration (Fig. 3.15). This modulus-θ_D parallelism is to be expected in view of the fact that values of θ_D can be calculated from the results of sound-velocity measurements in the manner to be discussed in Section 12.6. As suggested in Section 3.4.1, the actual or incipient maxima in E and θ are indicative of the lattice stiffening that occurs when the natural tendency for bond directionality characteristic of Ti-SM alloys "sharpens up" in the vicinity of the stoichiometric compositions [COL82a].

12.4.2 Elastic Moduli of Ti-TM Alloys: Long-Wavelength Results

The β-Isomorphous Alloys: Ti-V, Ti-Nb, and Ti-Mo. Using the long-wavelength transverse, longitudinal, and torsional resonances of alloy rods, FEDOTOV and colleagues [FED63, FED64, FED66, FED73] have measured as functions of composition the Young's moduli, E, and the shear moduli, G, of the β-isomorphous alloys—Ti-V, Ti-Nb, and Ti-Mo—in both the quenched (from 24 h/900 °C) and quenched-plus-annealed (200 h/700 °C plus 500 h/600 °C) conditions. The results for the quenched alloys are depicted in Fig. 12.3, where they can be compared with the accompanying equilibrium/nonequilibrium phase diagrams. Immediately obvious is the expected parallelism of the composition dependences of E and G (see Section 12.6). Further discussions

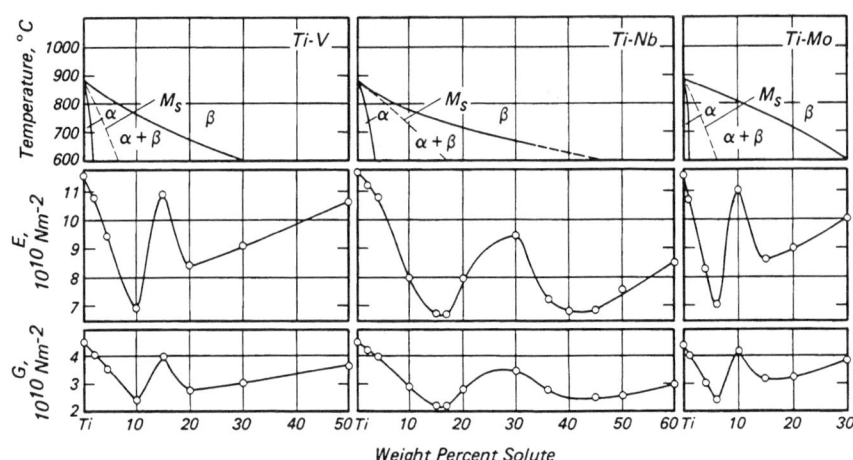

Fig. 12.3 Young's modulus, E, and shear modulus, G, as functions of composition-related microstructure in Ti-V, Ti-Nb, and Ti-Mo alloys [FED73].

of the figures must be in terms of the nonequilibrium phases α^m and athermal ω which form on quenching.

Attention is first of all drawn to the titanium-rich alloys whose structures are first α' and then α'' with increasing solute concentration.* In the martensitic regime, each alloy exhibits rapid softening as the solute concentration increases. Assuming the correctness of the data of Table 5.2* (for the composition of the α'/α'' phase boundary), it seems that whereas the change from the α' to the α'' structure does not interrupt the E_{α^m} versus composition curve for Ti-Mo and Ti-Nb, the same is not true for the Ti-V system. Of course, it would be useful to know to what extent the possible existence of mixed phases—and in particular the presence of ω phase—influences the upturn shown by each of the curves.

Transferring attention to the β-phase alloys—on the right-hand side of each diagram—as solute content is *decreased* the β phase becomes continuously softer until the product of that instability, ω-phase precipitation, eventually makes its presence felt by stiffening the lattice. According to BAGARIATSKII et al. [BAG59](Table 5.5a), the solute concentrations corresponding to which (athermal) ω phase is formed on quenching are: Ti-V, 13 at.% (14 wt%); Ti-Nb, 18 at.% (30 wt%); Ti-Mo, 4.5 at.% (8.6 wt%). These values are in excellent agreement with the positions of the E- and G-modulus peaks.

Numerous other physical properties respond in a like manner to the composition-related microstructures. Since the Debye temperature can be synthesized from the macroscopic elastic moduli (see below), a parallelism between θ_D and E or G is expected and, according to Fig. 12.4, is indeed observed. The behavior of the Vickers-hardness curve also exemplifies the connection (to be considered below)

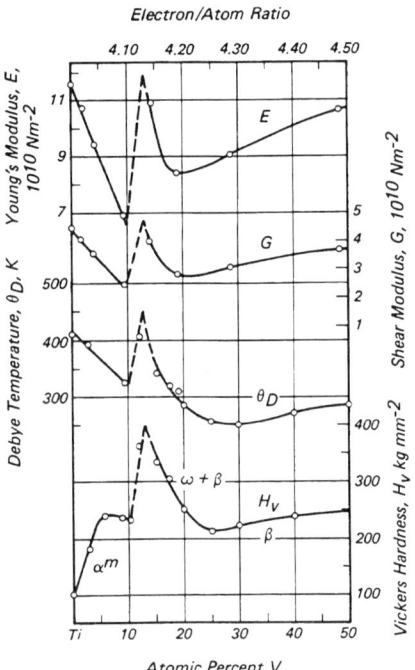

Fig. 12.4 Young's modulus, E, shear modulus, G, Debye temperature, θ_D, and Vickers hardness, H_V, as functions of composition in quenched Ti-V alloys. References: [FED73] (E and G), [COL84, p. 119] (5-kg diamond-pyramid hardness, H_V).

among hardness, strength, and modulus. As solute content decreases in the bcc field, the composition dependences of the four parameters plotted in Fig. 12.4 respond to the stiffening and hardening influences of ω-phase precipitation. On the α^m side, only the hardness data, particu-

---

*The compositions of the α'/α'' boundaries are given in Table 5.2. There is some disagreement about the existence of a quenched-α'' variant in Ti-V [WIL73][FLO82] (see Section 5.2.3).

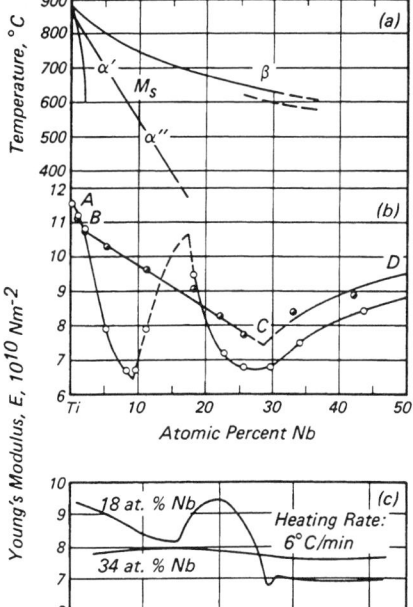

Fig. 12.5 Young's moduli, E, of quenched-and-aged Ti-Nb alloys as a function of metallurgical condition. (a) Equilibrium transi from standard sources (see Fig. 4.7) and an M_s line from [JEP70]. (b) Young's moduli of quenched (from 24 h/900 °C and quenched-plus-aged (100 h/800 °C) + 200 h/700 °C + 500 h/600 °C) Ti-Nb alloys [FED73] (see also [FED64]). (c) Change of modulus in response to heating at the rate of about 6 °C/min [FED64].

larly at low solute concentrations, exhibit departures from parallelism in a manifestation of some kind of competition between solution strengthening and lattice softening.

The influence of heat treatment on modulus is considered in Fig. 12.5 with reference to the equilibrium and metastable-equilibrium phase diagrams of Ti-Nb, constructed from data sources referred to in the caption. The form of the E versus composition curve for the 24 h/900 °C/WQ alloys has already been discussed. After the alloy series has been annealed according to the prescription 200 h/700 °C + 500 h/600 °C, the modulus which represents equilibrium-α phase, naturally follows the "as-quenched" data; the segment C-D, which is in the 600 °C equilibrium-β field, also follows the old data; while B-C, for the equilibrium-α + β field, is simply a "tie-line." The temperature-time results are also interesting: with a metastable alloy, depending on the decomposition kinetics, a temperature dependence experiment may also be a short-time aging experiment. For example, quenched Ti-Nb(18 at.%) possesses a high volume-fraction of ω phase (Table 5.5); upon heating through 300 °C, solute diffusion becomes active and additional isothermal precipitation takes place, resulting in a modulus peak centered about 400 °C, an optimal temperature for isothermal ω-phase precipitation. At higher temperatures some α precipitation commences, enriching the matrix with niobium and lowering E, although within the 1 h which elapses as the temperature is raised from 400 to 800 °C, thermodynamic equilibrium is not achieved. The experiment illustrates the relatively rapid kinetics of the ω-phase reaction. The equilibrium state of the other alloy represented in the figure, Ti-Nb(34 at.%), is α + β at room temperature. But since the reaction kinetics are so sluggish,* the β phase is retained on quenching and the experiment on initially quenched Ti-Nb(34 at.%) measures the actual E-modulus temperature dependence of the quenched β phase. This turns out to be relatively small.

The responses of the mechanical properties of three representative β-isomorphous Ti-TM alloys (viz., Ti-12V, Ti-17Nb, and Ti-8Mo) to quenching from various temperatures within the interval 600 to 900 °C have been investigated by JAMES and MOON [JAM70]. Hardness measurements and tensile tests were conducted. Young's modulus was determined using an Elastomat-type of instrument similar to that employed by FEDOTOV et al. (see above), while the internal friction (i.e., the imaginary component of the complex elastic modulus) was calculated from the decay time constant of the resonant vibrations of rods. The quantities listed generally turned out to be strongly dependent on the prequench temperature: In Ti-12V, for example, as that temperature was dropped from 895 to 659 °C, although the hardness remained fairly constant at about 200 kg/mm² down to 709 °C (prequench), it rose steeply to 405 kg/mm² at 700 °C, then went on to decrease monotonically with further reduction in the prequench temperature. The results were interpreted in the following way: (1) In response to quenching from temperatures above the (α + β)/β transus, the quenched product is α′ martensite plus retained interplatelet β; the resulting low strength and modulus are attributable to stress-induced martensitic transformation of the metastable-β phase. (2) When the prequench temperature is dropped below (α + β)/β, ω-phase precipitation in the β component is responsible for the observed pronounced increases in hardness, yield strength, and modulus. (3) The decrease in these quantities with further drop in prequench temperature corresponds to the establishment of an equilibrium-β component too rich to support any ω-phase precipitation.

The β-Eutectoid Alloys: Ti-Cr, Ti-Mn, Ti-Fe, Ti-Co, and Ti-Ni. The elastic properties of the β-eutectoid alloys of titanium with first row transition elements have been studied by FEDOTOV and colleagues [FED73]. The modulus-composition dependences of alloys quenched from 1000 °C are intercompared in Fig. 12.6. As before, with decreasing solute content, the alloys all show an elastic softening. This is followed by a rapid increase in stiffness in the composition range where athermal ω phase is expected. In the martensitic regime, the Young's modulus composition dependences of the β-eutectoid alloys are generally much smaller than those exhibited by the β-isomorphous series. Secondly, a trend manifests itself within the eutectoid series. Particularly for the Ti-Fe, Ti-Co, and Ti-Ni trio, the relatively flat composition dependences are believed to be due to departures from the single-phase martensitic structure as a consequence of partial decomposition of either (1) the β phase above M_s or (2) the martensitic phase below it during the quenching process, as a result of the very high diffusivities in titanium of iron, cobalt and nickel. The special positions occupied by these three elements, with regard to diffusivity, have already been discussed in Section 3.4.4 (see Fig. 3.20).

12.5 Dynamic Elastic Modulus: Ultrasonic Methods

12.5.1 Basic Theory

Using ultrasonic techniques in the 20 to 100-MHz frequency range it is possible to generate sound waves of wavelength $\sim 10^{-4}$ m, very much smaller than the physical dimensions of the usual specimen. By applying Eq 12.1 in the form $v_{ij} = \sqrt{C_{ij}/\rho_d}$ to a series of appropriately cut monocrystalline samples it is, in principle, possible to separately evaluate the individual components of the 6 × 6 stiffness matrix** $[C_{ij}]$ defined by:

$$[\sigma] = [C_{ij}][\varepsilon] \qquad \text{(Eq 12.5)}$$

where σ and ε have their usual meanings of stress and strain, respectively. Although the ultrasonic methods may be replaced by low-frequency techniques in the measurement of the elastic moduli of polycrystalline materials, their use is almost mandatory if the elastic moduli of single crystals are required. In general

---

*Hence the need for the 500-h equilibration time at 600 °C referred to above.

**Or a compliance matrix, $[S_{ij}]$, may be derived from Eq 12.5 by matrix inversion: $[\varepsilon] = [S_{ij}][\sigma]$.

linear elastic theory, the $[C_{ij}]$ or $[S_{ij}]$ may be represented in Voigt's contracted notation by the 6×6 matrix (Eq 12.6):

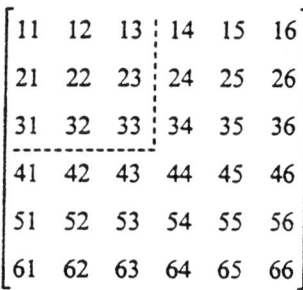

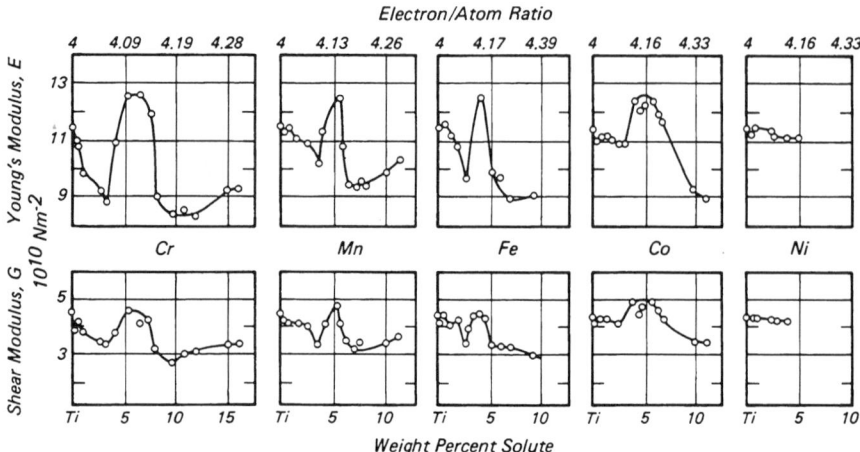

Fig. 12.6 Young's moduli, E, and shear moduli, G, of β-quenched alloys of titanium with the "β-eutectoid-forming" solutes chromium, manganese, iron, cobalt, and nickel.

For a crystal of cubic symmetry, this reduces to (Eq 12.7):

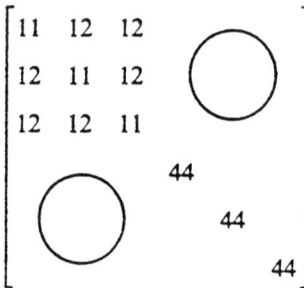

12.5.2 Stiffness Moduli of Cubic Monocrystals

As shown in matrix 12.7, a cubic monocrystal is characterized by three fundamental stiffness moduli: C_{11}, C_{44}, and C_{12}. In practice, the first two are measured directly, while C_{12} is obtained in association with the others. An important shear stiffness modulus is $C' = (C_{11} - C_{12})/2$ which, although it is made up of two of the fundamental moduli, is actually obtainable directly by experiment. The propagation descriptions of the ultrasonic waves needed for measurement of C_{11}, C_{44}, and C' are:

- C_{11}: a longitudinal wave in a ⟨100⟩ direction

- C_{44}: a transverse wave in a ⟨100⟩ direction polarized along ⟨010⟩, or a transverse wave in a ⟨110⟩ direction polarized along ⟨001⟩

- C': The other transverse wave in a ⟨110⟩ direction, polarized along ⟨1̄10⟩

12.5.3 The Significance of C' for Transition-Metal Alloys

Since C_{44} is governed by the transverse ⟨110⟩ wave, ⟨001⟩-polarized, and C' by the same wave, ⟨1̄10⟩-polarized, in an isotropic solid $C_{44} = C'$ (see also [FIS75, p. 202]). Departure from isotropy is gauged by the value of the Zener "anisotropy ratio," $A \equiv C_{44}/C'$. In bcc simple metals, A can be quite large (for example $A_{Na} = 7.5$), whereas for the bcc transition elements of the sixth group—chromium, molybdenum, and tungsten—the anisotropy ratio takes on the values 0.71, 0.72, and 1.01, respectively, [FIS75], indicating the influence of the d-band electrons in stabilizing the bcc lattice. FISHER has discussed in detail the physical significances of A and C' in relationship to electronic factors which govern bcc stability [FIS70, FIS75] [KAT79]. He has pointed out that the C_{44} shears are resisted primarily by nearest-neighbor forces. This explains its low values in simple metals (e.g., 0.0065×10^{11} N/m² for sodium) [FIS75]. On the other hand, additional cohesive contributions from d-electrons are thought to be responsible in transition elements for the observed large values of C'. The C' modulus appears to be interpretable as a bcc stability parameter; thus, for the highly stable bcc transition elements chromium, molybdenum, and tungsten, $C' \cong 1.5 \times 10^{11}$ N/m², but with decreasing electron/atom (e/a) ratio C' decreases rapidly, tending to zero at room temperature for alloys which possess ω-phase instabilities or transform martensitically to hcp at ordinary temperatures [COL73a]. This continuously decreasing bcc lattice stability with decreasing e/a is illustrated in Fig. 2.1 with data for the pure elements chromium, molybdenum, and tungsten (group VI); vanadium, niobium, and tantalum (group V); and members of the alloy systems Zr-Nb, Ti-Cr, and Mo-Re. It is interesting to note that C' seems to be a universal function of e/a [FIS70].

12.6 Ultrasonic Measurements of the Macroscopic Elastic Moduli

12.6.1 Isotropic Solid

All propagation directions being equivalent in the isotropic solid, the only variables are those embodied in the vibrational modes themselves. These, which may be either transverse or longitudinal, call for only *two* elastic moduli, best represented by the Lamé parameters λ and μ [JAE62, p. 54]. The modulus for longitudinal-wave propagation is $\lambda + 2\mu$ and that for transverse-wave propagation is μ. The macroscopic moduli are expressible in terms of the Lamé parameters in the following way:

$$K = \lambda + \frac{2}{3}\mu \qquad \text{(Eq 12.8)}$$

$$G = \mu \qquad \text{(Eq 12.9)}$$

$$E = \frac{9KG}{3K+G} = \frac{\mu(3\lambda + 2\mu)}{\lambda + \mu} \qquad \text{(Eq 12.10)}$$

$$\nu = \frac{E}{2G} - 1 = \frac{\lambda}{2(\lambda + \mu)} \qquad \text{(Eq 12.11)}$$

12.6.2 An Aggregation of Cubic Monocrystals: The VRH Approximation

VOIGT has shown how, starting with the monocrystalline elastic stiffness moduli, C_{ij}, it is possible to derive expressions for the macroscopic shear modulus, G_V, and the bulk modulus, K_V. For a macroscopically isotropic aggregate of cubic monocrystals, the VOIGT approach yields:

$$5G_V = C_{11} - C_{12} + 3C_{44} \qquad \text{(Eq 12.12)}$$

and

$$3K_V = C_{11} + 2C_{12} \quad \text{(Eq 12.13)}$$

In a parallel analysis, REUSS has expressed the macroscopic moduli, now G_R and K_R, respectively, in terms of the monocrystalline compliance moduli, S_{ij}. Thus, under conditions similar to the above:

$$\frac{5}{G_R} = 4S_{11} - 4S_{12} + 3S_{44} \quad \text{(Eq 12.14)}$$

and

$$\frac{1}{3K_R} = S_{11} + 2S_{12} \quad \text{(Eq 12.15)}$$

Next, through the application of some identities connecting S_{ij} and C_{ij} in cubic crystals, these relationships reduce still further to:

$$5G_V = 2C' + 3C_{44} \quad \text{(Eq 12.16)}$$

and

$$\frac{5}{G_R} = \frac{2}{C'} + \frac{3}{C_{44}} \quad \text{(Eq 12.17)}$$

and

$$K_V = K_R = K = \frac{1}{3}(C_{11} + 2C_{12}) \quad \text{(Eq 12.18)}$$

HILL showed that the VOIGT and REUSS approximations were, respectively, greater than and less than true polycrystalline moduli, which were then better represented, in what later became known as the VRH approximation, by the arithmetic means of these extremes.* Thus, as summarized by ANDERSON [AND63]:

$$\langle G \rangle = G_H = (G_V + G_R)/2 \quad \text{(Eq 12.19)}$$

and

$$\langle K \rangle = K_H = (K_V + K_R)/2 \quad \text{(Eq 12.20)}$$

in which the HILL arithmetic-mean approach has been adopted. After this, with the aid of Eq 12.16 to 12.18, E_H and v_H can be calculated by means of Eqs. 12.10 and 12.11, and a Debye temperature can be obtained in the manner to be outlined below.

12.6.3 An Aggregation of Cubic Monocrystals: The VRHG Approximation for θ_D

The Debye temperature, θ_D, obtainable experimentally from the lattice contribution to the low-temperature specific heat, may be regarded as the ultimate embodiment of all the elastic constants. ANDERSON [AND63] has described how a reliable value of θ_D for an elastically isotropic solid can be calculated from the monocrystalline moduli, using computed values of the macroscopic moduli as an intermediate step.

The expression for θ_D which he recommended is:

$$\theta_D = \frac{h}{k_B}\left(\frac{3N}{4\pi}\right)^{1/3}\left(\frac{\rho_d}{M}\right)^{1/3} v_m \quad \text{(K)} \quad \text{(Eq 12.21)}$$

in which the product of the first two factors (N is Avogadro's number) is 2.514×10^{-3} cgs units, ρ_d is the density (g/cm^3), and M is the mean atomic weight of the alloy. v_m (cm/s) is an average sound velocity; it may be evaluated using:**

$$v_m = \left\{\frac{1}{3}\left(\frac{2}{v_T^3} + \frac{1}{v_L^3}\right)\right\}^{-1/3} \quad \text{(Eq 12.22)}$$

after the "longitudinal" and "transverse" velocities v_L and v_T, respectively, have been deduced from:

$$v_L = \sqrt{[K + (4/3)G]/\rho_d} \quad \text{(Eq 12.23)}$$

and

$$v_T = \sqrt{G/\rho_d} \quad \text{(Eq 12.24)}$$

with K and G having been calculated from the monocrystalline elastic moduli using the procedures of the previous subsection. The title VRHG was applied by ANDERSON [AND63] to this VRH-based method of calculating θ_D, in recognition of the contribution that GILVARRY had made to the subject.

12.7 Monocrystalline Elastic Moduli of β-Ti-TM Alloys

Using techniques such as the pulse-superposition method and the pulse-echo-overlap method [KAT79], FISHER and colleagues have measured the ultrasonic elastic moduli of Ti-V [KAT79, KAT79a], Ti-Cr [FIS70, FIS70a], and Ti-Nb [REI73]. The results of these measurements, together with those of subsequent VRH calculations of E, are presented in Table 12.2.

Finally, as a test of the applicability of the VRHG method, the Debye tempera-

---

*References to bounds more sophisticated than VRH are given by LEDBETTER [LED80].
**LEDBETTER [LED80] has recommended that attempts be made to replace Eq 12.22 with a more elaborate formula in the interests of obtaining more accurate values of θ_D, especially in cases of large elastic anisotropy.

Table 12.2 Monocrystalline Elastic Stiffness Moduli (Elastic Constants) of Ti-V, Ti-Cr, and Ti-Nb

Crystal	Electron/atom ratio	Condition	Elastic constants(a) 10^{10} N m^{-2} (10^{11} dyne/cm^{-2})			VRH-calculated Young's modulus(b), E, 10^{10} N m^{-2}	VRHG-calculated Debye temperature(c)		
			C_{11}	C_{44}	C'		Molar weight	Density, g/cm^3	θ_D, K
Ti-V(29.4 at.%)	4.29		14.002	3.966	2.026	8.34	48.79	4.929	328
Ti-V(38.5 at.%)	4.39		14.896	4.095	2.421	9.09	49.07	5.044	340
Ti-V(53 at.%)	4.53		16.760	4.129	3.125	10.09	49.51	5.328	355
Ti-V(73 at.%)	4.73		19.227	4.148	4.060	11.22	50.12	5.040	369
Ti-Cr(6.98 at.%)	4.14	Extrapolated from > 900 °C to 20 °C	12.50	4.10	1.24	7.08	48.19	4.677	305
Ti-Cr(6.98 at.%)	4.14	Brine quench	15.59	5.54	3.67	12.29	48.19	4.677	409
Ti-Cr(9.36 at.%)	4.19	Brine quench	13.31	4.27	1.90	8.45	48.28	4.725	334
Ti-Cr(13.81 at.%)	4.28	Brine quench	13.99	4.42	2.18	9.08	48.47	4.834	345
Ti-Cr(28.37 at.%)	4.57	Brine quench	15.91	4.77	3.25	10.98	49.06	5.027	377
Ti-Nb(40.4 at.%)	4.40	Annealed 65 h/1700 °C	15.65	3.963	2.247	8.74	66.08	6.189	291
Nb	5.00		24.74	2.80	5.69	10.45	92.91	8.578	268

(a) Ti-V data, KATAHARA et al. [KAT79]; Ti-Cr data, FISHER and DEVER [FIS70, FIS70a]; Ti-Nb data, REID et al. [REI73]; Nb data, FISHER et al.[FIS75a]. (b) Calculated from Eq 12.10 with G and given by Eq 12.19 and 12.20. (c) Calculated from Eq 12.21 with the assistance in turn of Eq 12.22, 12.23, 12.24, 12.19, and 12.20. In this c.g.s. representation, the C_{ij} are in dyne/cm^2. Note: In both (b) and (c), $K_V = K_R$ are given by Eq 12.18 and G_V and G_R are given by Eq 12.16 and 12.17.

tures of a series of Ti-Cr alloys were calculated from the monocrystalline elastic moduli. The experimentally obtained moduli themselves are plotted as functions of chromium composition in Fig. 12.7(a); the VRHG-calculated θ_D's are listed in Table 12.2 and plotted in Fig. 12.7(b). Figure 12.7(c) displays the calculated θ_D's alongside the experimentally obtained and empirically extrapolated θ_D's of a series of Ti-Mo alloys [COL72a]. Taken together, the figures intercompare for each alloy system the measured properties of the actual quenched alloys with those predicted, using certain extrapolation procedures, for what has been referred to as "virtual-β" alloys (i.e., bcc alloys outside their compositional ranges of stability). The extrapolation procedures have been described elsewhere: that for Ti-Cr in [FIS70a] and that for Ti-Mo in [COL72a, COL73]. Figure 12.7(c) demonstrates that C' vanishes and θ_D drops to about 200 K (very low for a transition element) at a composition very close to that at which martensitic transformation takes place at room temperature in these systems. The ω-phase precipitate (a product of the precursor instability), which serves to stabilize the lattice, is responsible for the stiffening noted at low values of e/a in C_{11}, C_{44}, and θ_D.

Part 2: Normal Plastic Properties

12.8 Strengthening of Titanium Alloys

Although the principles of solution strengthening must be taken into full consideration in the design of titanium—especially α alloys—the strengths of the heat-treatable $\alpha + \beta$ and β alloys tend to be dominated by "microstructural effects." The different classes of microstructure, and the associated mechanical properties obtainable as a result of heat treating a single $\alpha + \beta$ alloy, are so numerous that strengthening in such systems, for this reason alone, eludes fundamental analysis. Furthermore, since little information is available on the extent to which alloying elements are distributed between the two phases in such alloys, it is generally not possible to apply solution-strengthening principles to them [JAF73a, p. 1680]. As a consequence of these two difficulties, strengthening in heat-treatable alloys can be systematized only from a phenomenological standpoint.

12.8.1 Solid-Solution Strengthening

The two important classes of solution strengtheners are the interstitial elements boron, carbon, nitrogen, and oxygen, and simple metals such as aluminum, gallium, and tin. Interstitial-element strengthening of titanium and other metals has been considered extensively by CONRAD and colleagues (see below). The substitutional α stabilizer/strengtheners, tin, gallium, and especially aluminum either singly or in combination, were the solutes considered by COLLINGS and GEGEL in their papers dealing with the electronic and thermodynamic aspects of solution strengthening [GEG73a] [COL73a, COL73b, COL75a, COL82a]. Transition elements cannot be regarded as true solution strengtheners of titanium; naturally they must contribute some measure of strengthening, but that is not their primary role.

Solution Strengthening by α Stabilizers. Solution strengthening by simple metals when dissolved in titanium can be qualitatively understood in terms of the formation of strong, local, and directional electronic bonds between the solute and the surrounding titanium atoms. A moving dislocation experiences a strong pinning force when a segment of its core becomes identified with the local environment of the SM solute atom. Guided by the responses to alloying of physical (electronic) properties such as electrical resistivity, Hall coefficient, and magnetic susceptibility, COLLINGS and coworkers have attempted a qualitative explanation of the nature of the local solute-solvent interaction. As a result, the following description has emerged: in the vicinity of isolated TM atoms dissolved in simple metals, the amplitudes of the d-wavefunctions are large; these are referred to as virtual bound states. This picture carries over to the converse situation, viz., isolated SM atoms dissolved in transition metals (in particular, titanium)—with the result that we again find maximum d-wavefunction amplitudes on the transition-metal sites and a tendency for the d-electrons to be excluded from the SM-atom positions. This avoidance of the SM atoms by the d-wavefunctions, a phenomenon which has been demonstrated experimentally in the case of V-Al by VAN OSTENBURG et al. [VAN64], has several consequences: (1) It leads to a representation of the alloy as a "diluted" or "expanded" transition metal, in this case titanium. This in turn, it can be argued, justifies the α-stabilizing property of such solute atoms. (2) It suggests that the solute atoms should be strong scatters of conduction electrons., i.e., should be associated with high specific (per atom) solute resistivities. This has indeed been ob-

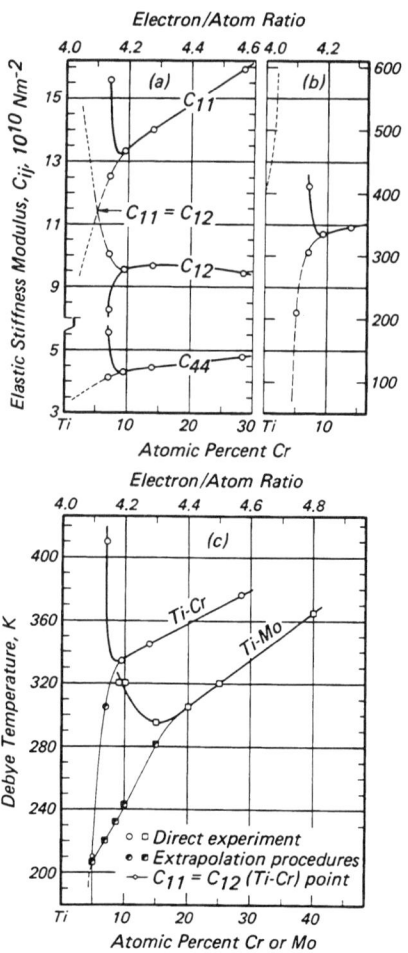

Fig. 12.7 (a) Elastic constants of Ti-Cr [FIS70] extrapolated smoothly to the 5 at.% Cr point (dashed lines) and thereafter linearly (chain lines). (b) VRHG-calculated θ_D's based on the actual and extrapolated (dashed and chain lines) elastic constants. (c) VRHG-calculated θ_D for Ti-Cr compared with the calorimetrically measured and extrapolated θ_D's for Ti-Mo; see [COL72] for details.

served. The results of STERN's tight-binding theory of disordered alloys [STE68], augmented by VAN OSTENBURG's experimental results for a single member of that class, suggest that in Ti-SM alloys, the s, p valence electrons of the SM atoms contribute to a low-lying band. The $n(E)$ versus E picture which thus emerges is one in which a conduction band (centered about E_F) and a valence band are separated by a minimum in $n(E)$ (see Fig. 3.5b). The low-lying band represents covalent bonding which, being highly directional, is responsible for local lattice stiffening in the dilute alloy and bulk stiffening—even brittleness—in the intermetallic compound. Although somewhat more refined, the conclusions arrived at by LYE regarding the electronic structures of intermetallic compounds of titanium with the elements carbon, nitrogen, and oxygen based on the results of band-structure calculations for

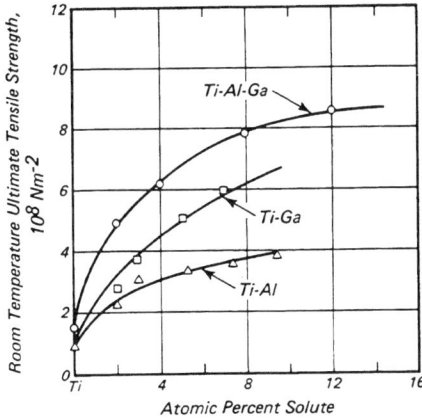

Fig. 12.8 Solid-solution strengthening of titanium with binary and ternary additions of aluminum and gallium [COL75^a].

Table 12.3 Solid-Solution Strengthening of Titanium by Transition Elements [HAM78]

Solid-solution strengthening rate, 10^6 N/m^2	Solute element							
	V	Cr	Mn	Fe	Co	Ni	Cu	Mo
Per wt%	19	21	34	46	48	35	14	27
Per at.%	20	23	39	54	59	43	18	54

TiC, are also in general agreement with this picture [LYE66]. Solution strengthening in ternary solid solutions can be thought of as two-stage application of this model: strengthening is due not only to the Ti/SM$_1$ and Ti/SM$_2$ interactions—in which case a mixture rule (weighted average) would be obeyed—but also to SM$_1$/SM$_2$ interactions which (for a given total solute concentration) allow the strength of a Ti-SM$_1$-SM$_2$ alloy to be greater than that of the stronger of Ti-SM$_1$ and Ti-SM$_2$. Figure 12.8 is an example of this synergistic effect.

Interstitial-Atom Strengthening. CONRAD and coworkers have conducted an extensive investigation of the influences of the so-called "interstitial elements"—carbon, nitrogen, and oxygen—on the plastic properties of titanium [CON67, CON70, CON75, CON81] [SAR72] [OKA73] [TYS75]. A detailed study of the influence of nitrogen, in particular, was undertaken by OKA et al. [OKA73], whose results were analyzed using the conventional activation-energy approach. An intercomparison of the effects of carbon, nitrogen, and oxygen is presented in [CON75], while the entire subject has been reviewed in [CON81]. The influence of silicon has been investigated by FLOWER et al. [FLO73], who also adopted an activation-energy approach when analyzing their temperature-dependent strengthening data.

Although CONRAD et al. have frequently justified their results in terms of conventional lattice-defect theory (including the size-misfit and modulus-defect formalisms), they have gone on to consider the effects of "chemical interaction" between the solute and solvent atoms. In doing so, the interaction mechanism was deduced, with the aid of atomic-orbital theory, to take the form of covalent bonding between the interstitial atom and the matrix. The approach adopted by CONRAD et al. [SAR72][CON75] represents a satisfactory unification of conventional and electronic theories, in that size-misfit and modulus-defect were regarded by them as manifestations of the underlying electronic interaction. The word "chemical" invokes "alloy chemistry" with its thermodynamic (e.g., heats of formation of the appropriate intermetallic compounds) as well as electronic implications (e.g., electronegativity difference, covalent bonding, etc.). No doubt the calculations by LYE referred to in the previous subsection could provide a useful theoretical basis for this model of local-covalent-bond-solid solution strengthening.

The picture of interstitial-atom hardening that emerged, especially from the paper by SARGENT and CONRAD [SAR72], was formally indistinguishable from the substitutional strengthening model of COLLINGS and GEGEL [COL75^a]. If differences do exist between interstitial and substitutional solid-solution strengthening, they are quantitative rather than conceptual and are, therefore, to be found in: (1) the energies of the respective covalent bonds, and (2) the diffusivities of the dissolved elements.

Solution Strengthening by Transition Metals. When transition elements, especially "nearby" ones (referring to the periodic table), are dissolved in titanium, the perturbation of electron states characteristic of the presence of simple metals and interstitials (so-called "s,p elements") does not take place. To a first approximation, the Ti-TM alloy may be regarded as a *new transition metal* with properties—particularly lattice stability—appropriate to the average group number or electron/atom ratio. Although some degree of solution strengthening is inevitably contributed by the presence of transition metals in β-phase solid solution, the dominant strengthening mechanisms in such alloys are precipitational effects (to be considered below). Of course, in small amounts TM solutes actually *lower* the modulus of titanium (see Fig. 12.3). According to Table 12.3, the solid-solution strengthening capacity of transition metals is on the average about 30 MN/m^2 per wt%. That of the "β-isomorphous" early transition elements is about 20 MN/m^2 per at.%; this is equivalent to a hardening rate of about 6 kg/mm^2 per at.%*—somewhat smaller than those of the substitutional α stabilizers and very much smaller than those of the interstitial elements (Table 8.2). For this reason, transition elements are regarded as "β stabilizers" and as precipitation hardeners rather than as solid-solution strengtheners.

Conclusion. Transition elements are stabilizers of the β phase in titanium rather than strengtheners of it. As for the α stabilizers it may be concluded that there is no conceptual difference between the ways in which the "substitutional" elements and the "interstitial" elements solution strengthen titanium. In both cases the mechanism consists of the establishment of local covalent-like bonds whose strengths may be gauged by any one of several measures of atomic-interaction strength, such as electronegativity difference, heats of solution, heats of formation of intermetallic compounds, and so on. The difference between interstitial- and substitutional element strengthening is principally one of degree. Taking oxygen and aluminum as examples, the rates of hardening at the 1 at.% level are in the ratio of 6.5:1. The heats of formation of the equiatomic compounds, TiO and TiAl, are in exactly the same ratio [KUB56, pp. 160 and 41, respectively], while the ratio of the GEGEL-calculated electronegativity differences between oxygen and titanium and aluminum and titanium, respectively, is 6.6:1 [COL75^a]. Further comparisons of this kind are listed in Table 8.2.

Solution strengthening by interstitial atoms is chiefly of importance at low temperatures where their diffusivities are low. The same is evidently true, but to a lesser extent, for substitutional (SM) solution strengtheners since, as is well known, the strengths of the α-solid-solution alloys decrease rapidly with increasing temperature (Fig. 9.1). If strength must be maintained to high temperatures, it is necessary to introduce precipitation strengthening in the form of α$_2$-phase particles or the numerous microstructural effects available with α + β alloying.

*Assuming that $H_V = 3Y$ (see Section 12.10).

Microstructural Strengthening

The temperature range within which solid-solution strengthening is fully effective does not extend far above room temperature. Thus, as indicated in Fig. 9.1, the ultimate tensile strength of Ti-5Al-2.5Sn drops 50% upon warming from 20 K to room temperature, and then to 60% of its room-temperature value upon further heating to 370 °C. As a consequence, alloy designers have turned to microstructural effects, including precipitation, to extend the temperature range of titanium-base alloys. Since very high-temperature service is equivalent to long-time aging, the most stable precipitation-strengthened, heat-resistant alloy is one that is in thermodynamic equilibrium at the service temperature. For this reason, considerable hope was attached to the possibility that the α_2 precipitate in Ti-Al would turn out to be a successful dispersion hardener, comparable in its properties to the γ' phase in nickel-base superalloys. Unfortunately, this did not turn out to be the case, although the possibility of reducing the embrittlement problems associated with α_2-Ti$_3$Al has been explored [HOC73][HAM78]. At the other extreme, highly alloyed all-β alloys are not only far removed from titanium-rich alloys in strength/weight ratio and other important "titanium-alloy" characteristics, but they are also single-phase (i.e., precipitate-free) when in thermodynamic equilibrium at high temperatures. Consequently, the tendency has been to turn to $\alpha + \beta$ alloys, and near-α ones at that, for high-temperature service. As indicated in the introduction to this section, when strengthening is due primarily to the presence of a complicated microstructure involving networks of interfaces between finely divided α and β phases, and precipitates within them, the categorization of strengthening sources is generally treated phenomenologically (i.e., in the form of a description of the features present) rather than mechanistically. It is only fair to state, however, that whenever possible serious attempts are made to analyze the strengthening processes in terms of well-established mechanisms such as: (1) direct dislocation-particle interaction, (2) coherency strains between α and β phases, and (3) coherency strains between the precipitate and the matrix. The work of RHODES and PATON [RHO77] is an excellent example of this.

It is clear, in light of the foregoing remarks, that many of the sections of Chapter 7 ("Aging") could be recast in the form of a discussion of "Precipitate Strengthening." There is no need to do this here; however, a summary will be presented, based on the contents of earlier chapters, of the precipitates which have been induced to form in $\alpha + \beta$ and β alloys during attempts to increase their strengths and service-temperature ranges while retaining their ductilities.

When quenched, solute-lean β phases yield a martensitic structure designated α', $\alpha''_{lean} + \alpha''_{rich}$, or α'', depending on the starting composition (see, for example, Section 5.2.5). During aging, all forms result in β-phase precipitation. Isothermal ω forms during the aging of quenched β within an appropriate time-temperature frame (Section 7.2). This is responsible for an increase in strength at the expense of ductility. By administering a short higher-temperature heat treatment, the ω phase can be induced to revert to a lean-β (β') precipitate, stable upon recooling. This confers on the alloy a strength greater than that of the original quenched β, but without the brittleness associated with the original ω phase (Section 7.5). The β' can also be the source of a fine dispersion of α-phase precipitation, which also strengthens the β alloy without embrittling it (Section 7.5). In some technical alloys, the $\omega + \beta \rightarrow \beta' + \beta$ reaction may not proceed to completion within a reasonable time; the remaining ω phase then contributed strength while the increased volume fraction of β phase ensures ductility (Section 7.14.1). Various heat treatments, administered to β alloys with and without cold work, are possible. Their results have also been considered in Section 7.14.1. The heat treatment of $\omega + \beta$ alloys, at temperatures above the ω-reversion limit, will cause α-phase precipitation to take place in the vicinity of the ω sites (Section 7.4.3). This fine dispersion of α phase which results is also an effective strengthener. Heat treatment of metastable β just outside the $\omega + \beta$ field will stimulate the phase-separation reaction and result in a modulated β structure, $\beta' + \beta$. Under heat treatment, the β' can also be the site of fine α-phase precipitation, with beneficial results (Section 7.4.2). In technical alloys such as β-C, the $\beta' \rightarrow \alpha$ reaction is complicated and in fact will yield two types of α phase, depending on the heat treatment time/temperature conditions. By adjusting the aging prescription it is possible to "fine-tune" the α-phase morphology; thus as RHODES and PATON [RHO] discovered, in β-C the best combination of strength and ductility resulted from the presence of large noncoherent so-called "Type-2" α-phase particles (Section 7.14.3).

12.9 Hardness

12.9.1 Vickers Hardness Test

The measurement of hardness is a simple but useful technique for characterizing mechanical properties and investigating phases in quenched-and-aged alloys. Conventional techniques currently in use for measuring hardness, as well as the history of that test, are fully discussed by HANKE [HAN54]. In the Vickers method, a weighted square pyramid, usually of diamond, is allowed to rest for a specified length of time on a polished surface of the specimen. Since the area of the impression (mean diagonal, d, mm) is proportional to the load (L, kg), a load-independent hardness number can be formed from the quotient L/d^2. According to the Vickers prescription, $H_V = 1.8544\, L/d^2$. Thus, for example, a 5-kg load resting on a surface of Vickers hardness $H_V = 150$ kg/mm^2 will produce a 0.25-mm (diagonal) impression (as in the studies of Ti-V and Ti-Nb referred to in Fig 12.9).

Of course if it is desired to investigate the individual grains of a fine-grain polycrystalline sample, miniaturized versions of the tests are needed. Using for example the Leitz Miniloader, loads in the range of 25 to 100 g produce measurable impressions less than 10 μm across. Since samples mounted and polished for optical metallography are ideally prepared for hardness measurement, the two investigations are frequently associated with each other in studies of precipitation and aging. Although very useful in tracing the course of an aging reaction, for example,

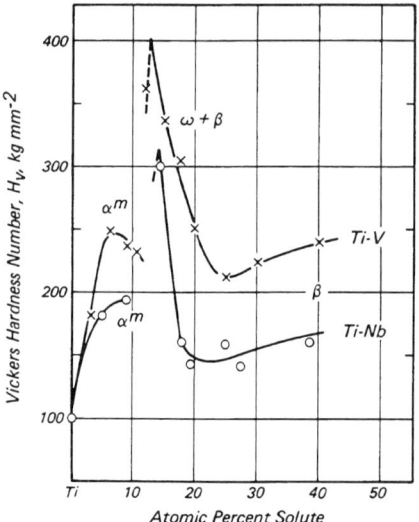

Fig. 12.9 Vickers hardness (5-kg load) of β-quenched Ti-V and Ti-Nb alloys as a function of solute concentration and (by implication) concentration-controlled microstructure [COL84, p. 138].

the hardness measurement is obviously not capable of identifying the nature of any precipitating phase. This must always be determined by a separate x-ray or TEM investigation, in the absence of which several alternative possibilities, particularly with regard to submicroscopic continuous precipitation processes (such as for example $\beta + \beta' + \beta \rightarrow \alpha + \beta$, $\beta + \beta' + \beta''$, and $\beta \rightarrow \omega + \beta$, all of which produce hardening), could easily be confused.

With the aid of formulas such as those due to HILL [HIL67, p. 213] and MARSH [MAR64, MAR64a], to be discussed below, Vickers hardness can be related to yield strength. This approach enables tests to be carried out on very small and irregular samples whose yield strengths would otherwise elude measurement.

12.9.2 Hardness of Quenched Ti-TM Alloys

Vickers hardness measurements, aided by x-ray diffraction studies, were employed by BAGARIATSKII et al. [BAG59] in their well-known investigation of the limits of the quenched α'-, α''-, and ω-phase regimes in the systems Ti-V, Ti-Nb, Ti-Ta, Ti-Mo, Ti-W, and Ti-Re. The athermal ω-phase regime was in each system characterized by a pronounced local maximum in the hardness ($\Delta H_V \cong 80$ kg/mm^2). In another investigation of alloy phases, this time in the Ti-Nb system, GUZEI et al. [GUZ66] measured hardness in a series of such alloys quenched from equilibrium at various temperatures. RASSMANN and ILLGEN [RAS72] also investigated hardness as function of composition in Ti-Nb(10-80 at.%) alloy samples, either as-cast or prepared in cold-rolled form. For both series of samples, the hardness increased rapidly with reduction of the niobium content below 30 at.%, as expected. The hardness of the cold-deformed alloy was much higher throughout the concentration range, presumably because of work hardening in general and deformation-martensitic transformation (to α'') at the titanium-rich end. The microstructural possibilities in response to 1-h annealing, which leads to a pronounced hardness increase at 300 °C, are even more complicated. The frequent attempts which have been made to explore such deformation, precipitation, and aging effects using only hardness as a tool exceeded the capabilities of the technique.

The results of some recent studies of hardness in Ti-V and Ti-Nb alloys are presented in Fig. 12.9. The alloys of both systems are seen to become extremely hard as the solute concentration decreases below about 13 at.% ($e/a = 4.13$). In contrast to early practice [BAG59][GUZ66], no attempt is made here to generate a smoothly rounded maximum from the low concentration data (see Fig. 12.3 and 12.4). The discontinuity shown, which is in keeping with the behavior of physical-property data such as magnetic susceptibility, electronic specific heat coefficient, and Debye temperature, expresses the strong hardening associated with a high volume fraction of athermal ω phase; to the left of the discontinuity the alloys have transformed martensitically.

12.9.3 Influence of α Stabilizers on the Hardness of Titanium and a Typical β-Isomorphous Ti-TM Alloy

Section 12.8 has discussed the principles whereby titanium becomes solution strengthened by: (1) the interstitial elements, (2) the substitutional α stabilizers, and (3) other transition metals. An extension of the same principles to the strengthening of Nb-Ti-TM alloys has been offered in [GEG73a] and [COL75a]. The known relationship between strength and hardness (see Section 12.10) has enabled intercomparisons to be made between the rates of strengthening produced by these three classes of solute when dissolved in titanium and β-Ti-TM.

Some experimental results are listed in Tables 8.2, 12.3, and 12.4, and depicted in Fig. 12.10. Solute species selected for the hardness test were: (1) the interstitial elements boron, carbon, nitrogen, and oxygen, (2) the valence-3 simple metals aluminum and gallium, and (3) the valence-4 metals and semimetals silicon, germanium, and tin. In presenting for comparison the hardening of a representative β-Ti-TM alloy by some of these same elements, the availability of data dictated the selection of aluminum and germanium as solutes and Ti-50Nb as the β-Ti-TM solvent (Table 12.4). Finally, as indicated in that table, the hardening produced by zirconium when dissolved in a β-Ti-TM base was also investigated.

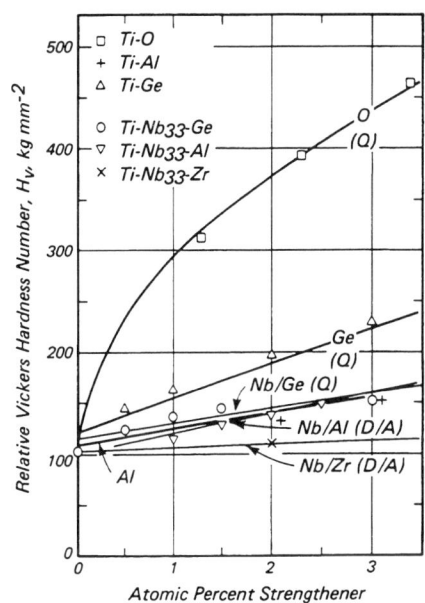

Fig. 12.10 Influence of various interstitial, substitutional-SM, and substitutional-TM additions on the Vickers hardness of titanium and Ti-50Nb in the conditions quenched (Q) and deformed-plus-aged (D/A). The ternary data have been adjusted to $H_V = 103$ kg/mm^2 at zero solute concentration. *Data sources:* binary alloys [COL84, p. 141], Ti-Nb-Ge [HEL71], Ti-Nb-Al/Ge [ZWI70].

Table 12.4 Solid-Solution Hardening of α'-Ti-50Nb by Simple Metals

Alloying addition	Concentration range, at.%	Condition	Law(a)	Slope, b, kg·mm^{-2}·at.%$^{-1}$	Intercept, a, kg/mm^2	Correlation coefficient, %	Hardening rate, $\delta H_V/\delta c$, kg·mm^{-2}·at.%$^{-1}$	Ref
Cu	0-5	3 h/1000 °C/WQ	c	14	193	99.5	14	[LOH71]
Ge	0-4	3 h/1000 or 1080 °C/WQ	c	14	165	93	14	[HEL71]
Al	0-2.5	Cold deformed plus 34 h/400 °C	c	19	195	99	19	[ZWI70]
Ge	0-2	Cold deformed plus 34 h/400 °C	c	60	198	99	60	[ZWI70]
Zr	0-6	Cold deformed plus 34 h/400 °C	c	3.7	198	99.9	3.7	[ZWI70]

(a) Data fitted to $H_V = a + bc$.

In least-squares fitting of the H_V versus concentration (c) data, it was noted that hardening by the interstitial elements closely followed a $c^{1/2}$ law (correlation coefficient in most cases > 99%) (Table 8.2). With the simple-metal solutes listed in that table, the hardening rates were sufficiently low that a linear relationship adequately described the data. Table 8.2 also shows that at the 0.1 at.% level the instantaneous hardening rate of previously unalloyed titanium by interstitials (viz., 269 to 378 kg/mm² per at.%) was up to 20 times greater than that by, say, aluminum, or up to nine times higher than that produced by silicon. The data of Tables 8.2 and 12.4 together indicate that aluminum hardens titanium at about the same rate that it does Ti-50Nb (15 and 19 kg/mm² per at.%, respectively). Finally, the above-mentioned tables and Fig. 12.10 demonstrate the relatively insignificant influence that zirconium exerts on the hardness of Ti-50Nb.

Numerical and graphical comparisons, such as those of Tables 8.2, 12.3, and 12.4 and Fig. 12.10, serve to illustrate that the solid-solution hardening of an α-Ti or β-Ti-TM base is subdivisible into three regimes according to whether the rate of hardening is "low" (neighboring transition elements such as zirconium), "moderate" (substitutional simple metals or semimetals such as aluminum, tin, silicon, and germanium), or "rapid" (the interstitial elements boron, carbon, nitrogen, and oxygen). Strengthening mechanisms associated with these three classes of solute have been described in Section 12.8.

Again anticipating the results of a discussion to be presented below concerning the proportionality between yield strength and hardness, it is expected that the addition of interstitial elements to titanium (and to Ti-TM alloys) will increase yield strength (at the 0.1 at.% addition level) at the rate of about 30% per 0.1 at.%. On a wt% basis, the strengthening rate due to oxygen is, of course, even greater than this.

12.10 Theoretical Relationships Between Hardness and Strength

A pair of complementary relationships between the hardness and strengths of metals have been developed by HILL [HIL67] and MARSH [MAR64, MAR64a]. An excellent review of these theories has been presented by DAVIS [DAV75] in connection with a study of the hardness/yield-strength ratio in metallic glasses.

12.10.1 Hill's Theory

The model assumed by HILL[HIL67] pictured an elastically rigid, yet plastic, body with the material displaced by the indenter squeezed up into a rim around the edges of the imprint. It led to the commonly observed relationship $H_V \cong 3Y$ between the hardness, H_V, and the yield strength, Y, when expressed in the same units (usually kg/mm²).

12.10.2 Marsh's Theory

The theory of MARSH, on the other hand, pictures a cavity being introduced into an elastic body. It results in the semiempirical formula:

$$H_V/Y = 0.28 + 0.6\, B \ln z \qquad \text{(Eq 12.25)}$$

where
$$B \equiv 3/(3 - l)$$
$$Z \equiv 3/(l + 3m - lm)$$
with
$$l \equiv (1 - 2\nu)(Y/E)$$
$$m \equiv (1 + \nu)(Y/E)$$

where ν is Poisson's ratio and E is Young's modulus.

12.10.3 Relationship Between the Models

The HILL and MARSH models represent extreme limits of the relative magnitudes of E and Y. It turns out that if 133 (Y/E) is "small" (<1) we have the plastic/elastically rigid situation discussed by HILL, whereas if 133 (Y/E) is "large" (>1) the situation is identifiable as the unyielding (nonplastic)/elastic condition required by MARSH. The crossover value of Y/E is 1/133.

With the aid of the following approximations, the MARSH equation becomes much simpler to apply in practice: For many polycrystalline metals, ν may be replaced by 0.3 and Y/E is generally found to be of the order 10^{-2}.* Using this information:

$$l = (1 - 2\nu)(Y/E) \cong 0.4 \times 10^{-2} \qquad \text{(Eq 12.26)}$$

hence
$$B \cong 1$$
Similarly,
$$m = (1 + \nu)(Y/E) \cong 1.3 \times 10^{-2}$$
Now by definition,
$$Z = 3/(l + 3m - lm)$$
But if
$$l \cong 0.4 \times 10^{-2}$$
and
$$3m \cong 4 \times 10^{-2}$$
then
$$lm \sim 10^{-4}$$
and is negligible compared to $l + 3m$. Thus,
$$Z \cong 3/(l + 3m)$$
which from the definitions of l and m becomes

$$Z \cong E/1.43Y \qquad \text{(Eq 12.27)}$$

Inserting Eq 12.26 and 12.27 into Eq 12.25 yields finally

$$H_V/Y \cong 0.065 + 0.6 \ln(E/Y) \qquad \text{(Eq 12.28)}$$

which is applicable, as mentioned above, for $E/Y < 133$, which according to numerical solution of the equation corresponds to $E/H_V < 44.3$.

12.11 Application of the Marsh Formulas to the Determination of the Yield Strength of a Wire

The microhardness, H_V, is determined on a metallographically mounted cross section of the wire. The Young's modulus, E, is determined using either a sonic or an ultrasonic sound-velocity measurement. If $E/H_V > 44.3$, the HILL relationship $Y = H_V/3$ is immediately accepted. Otherwise, the MARSH approach is adopted and Eq 12.28 solved transcendentally for E/Y, hence Y. The Newton-Raphson method, which when applied to Eq 12.28 converges very rapidly, is well suited for this purpose.

12.11.1 Newton-Raphson Solution of the Simplified Marsh Equation

Direct Method. If X_i approximately satisfied $F(X_i) = 0$, then a better root is X_{i+1}, where

$$X_{i+1} = X_i - \frac{F(X_i)}{F'(X_i)} \qquad \text{(Eq 12.29)}$$

By iterating Eq 12.29 a few times, excellent solutions can be obtained.

In the present case, values for H_V and E having been obtained, the root Y of the function $F(Y) \equiv 0.065 - H_V/Y + 0.6 \ln(E/Y)$ is required. This is quickly obtained by substituting $y = Y^{-1}$ and iterating

$$y_{i+1} = y_i - \frac{0.065 H_V y_i + 0.6 \ln(E y_i)}{-H_V + 0.6/y_1}$$

Alternative (Trial-and-Error) Method. A trial-and-error method commences with:

$$H_V/Y \cong 0.065 + 0.6 \ln(E/Y) \qquad \text{(Eq 12.28)}$$

If we let

$$E/H_V \equiv R \qquad \text{(Eq 12.30)}$$

then

$$H_V/Y = 0.065 + 0.6 \ln R + 0.6 \ln(H_V/Y)$$

so that the equation to be solved by trial-and-error for H_V/Y is:

$$H_V/Y - 0.6 \ln(H_V/Y) = R' \qquad \text{(Eq 12.31a)}$$

*In Ti-Nb(40 at.%), for example, $Y = 8.8 \times 10^8$ N/m² [KOC77] and $E = 8.0 \times 10^{10}$ N/m² [LED81].

Table 12.5 Estimation of Yield Strength from Measured Hardness Data Using the Models of HILL and MARSH Where Applicable: A Case Study of Three β-Ti-Nb Alloys

Alloy	Experimental input				Model regime based on the magnitude of the E/H_V ratio	Calculation of yield strength				Measured yield strength $Y_{meas.}$ 10^8 N/m²	Deviation, $100 \Delta Y/Y_{meas.}$ %
	Hardness, H_V, kg/mm²	Modulus, E, 10^{10} N/m²	10^6 g/mm²	E/H_V		Hardness/ strength ratio from application of model, H_V/Y	Yield strength calculated from hardness, $Y_{calc.}$				
							kg/mm²	10^8 N/m²			

Condition: "Annealed"

800 °C and water quenched

Ti-Nb(30 at.%)	145	6.6	6.7₃	46	HILL	3.0	48	4.7	4.5	4
Ti-Nb(39 at.%)	150	7.8	7.9₅	53	HILL	3.0	50	4.9	4.6	7
Ti-Nb(50 at.%)	147	8.5	8.6₇	59	HILL	3.0	49	4.8	4.1	17

Condition: "Processed"

Ti-Nb(30, 39 at.%): (800 °C/WQ/extruded 86%/48 h/375 °C)/drawn 56%
Ti-Nb(50 at.%): As received cold-worked condition

Ti-Nb(30 at.%)	293	7.1₅*	7.2₉*	25	MARSH	2.56	114	11.2	10.3	9
Ti-Nb(39 at.%)	174	7.8₈*	8.0₄*	46	HILL	3.0	58	5.7	7.0	19
Ti-Nb(50 at.%)	152	8.7₇*	8.9₄*	59	HILL	3.0	51	5.0	5.2	4

All input data and measured yield strengths are due to READ [REA78] except the asterisked modulus data, which have been derived from the dynamic-modulus results of ALBERT and PFEIFFER [ALB72] for Ti-Nb(34, 44 at.%) in the condition 95% cold worked plus 40 h/400 °C.

where

$$R' \equiv 0.065 + 0.6 \ln R \quad \text{(Eq 12.31b)}$$

12.11.2 Graphical Solution of the Marsh Equation

Figure 12.11 has been constructed from Eq 12.28, 12.30, and 12.31. Below the HILL lower bound ($E/H_V = 44.2$), values of H_V/Y or E/Y corresponding to measured values of E/H_V can be simply read from the graph.

whether E/H_V (as listed in the fifth column of Table 12.5) is greater than or less than 44.3. In all cases but one, $E/H_V > 44.3$, enabling Y to be obtained from the HILL relationship $Y = H_V/3$. The last three columns of Table 12.5 compare the calculated and measured values of Y. In most cases the agreement is remarkably good, the sole exception being Ti-Nb(39 at.%) ("processed"), whose calculated yield strength is 19% too low, a deficiency possibly traceable to a low measured hardness number.

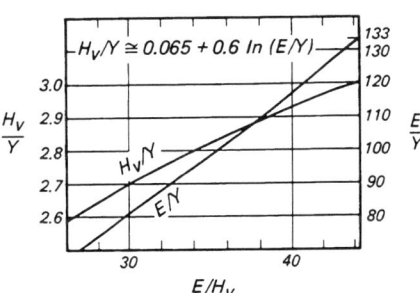

Fig.12.11 Parametric representation of the relationship $H_V/Y \cong 0.065 + 0.6 \ln(E/Y)$ (Eq 12.28).

12.12 Interrelationships Among Hardness, Young's Modulus, and Yield Strength In a β-Isomorphous Alloy Sequence: A Case Study

Ti-Nb, a typical β-isomorphous system, has been selected for this study. READ [REA78] has measured the hardness and yield strengths of the alloys Ti-Nb(30, 39, and 50 at.%) in the "annealed" and "processed" conditions as defined in Table 12.5, in which all the mechanical-property data are summarized. The moduli of the "annealed" alloys were also reported, but not those of the "processed" alloys. To replace the missing data, values of E were deduced by extrapolation and interpolation of the dynamic-modulus results of ALBERT and PFEIFFER [ALB72] for Ti-Nb(34 at.%) and Ti-Nb(44 at.%).

The first step in deriving Y from the measured H_V and E values is to determine

12.13 Tensile Strengths of Some Commercial Titanium Alloys

In selecting a titanium alloy for a given application, strength (and indeed the other mechanical properties) is not of course the only, or even the most important, consideration. Only strength properties are presented and discussed in this section. Other important properties which govern the selection of a titanium alloy for a given application are *weldability* and *heat resistance* (the latter having to do with the rate at which the strength decreases with temperature in the elevated-temperature regime, e.g., 315 to 540 °C). Of course, the properties of even a single alloy type can be altered by heat treatment (in the case of the α + β metastable β classes) and by slight variation or "fine-tuning" of its composition (e.g., at least six varieties of "Ti-6-4" can be produced in this way).

Ti-6Al-4V is the most commonly used titanium-base material (including unalloyed commercial titanium). As indicated above, it is produced in a number of varieties by slight alteration of the bare composition; it is also produced in numerous heat-treatment conditions in order to optimize its performance for particular applications within a wide range of service conditions. Ti-5Al-2.5Sn is a non-heat-treatable intermediate-strength alloy. The extra-low interstitial (ELI) grade is especially noted for its excellent properties at cryogenic temperatures; this alloy, which has also been found to possess useful strengths at intermediate-to-high temperatures (up to 370 to 425 °C), is suitable for aircraft engine and frame applications. Ti-3Al-2.5V is lower in strength than both Ti-5Al-2.5Sn and Ti-6Al-4V; its primary use is for aircraft hydraulic tubing. Ti-8Al-1Mo-1V, a near-α α + β alloy, has the highest aluminum content of any of the currently available commercial tita-

Table 12.6 Tensile Strengths of Several Commercial Titanium-Base Alloys: Typical Room-Temperature Values [STR82]

Alloy name	Nominal composition	Condition	Ultimate strength ksi	Ultimate strength 10^8 N/m^2	Yield strength ksi	Yield strength 10^8 N/m^2	Elongation %
5-2.5	Ti-5Al-2.5Sn	Annealed (0.25-4 h/1300-1600 °F)	120-130	8.3-9.0	115-120	7.9-8.3	13-18
3-2.5	Ti-3Al-2.5V	Annealed (1-3 h/1200-1400 °F)	95	6.5	90	6.2	22
6-2-1-1	Ti-6Al-2Nb-1Ta-1Mo	Annealed (0.25-2 h/1300-1700 °F)	125	8.6	110	7.6	14
8-1-1	Ti-8Al-1Mo-1V	Annealed (8 h/1450 °F)	145	10.0	135	9.3	12
Corona 5	Ti-4.5Al-5Mo-1.5Cr	α-β annealed after β processing	140-160	9.7-11.0	135-150	9.3-10.3	12-15
Ti-17	Ti-5Al-2Sn-2Zr-4Mo-4Cr	α-β or β processed plus aged	165	11.4	155	10.7	8
6-4	Ti-6Al-4V	Annealed (2 h/1300-1600 °F)	140	9.6	130	9.0	17
		Aged	170	11.7	160	11.0	12
6-6-2	Ti-6Al-6V-2Sn	Annealed (3 h/1300-1500 °F)	155	10.7	145	10.0	14
		Aged	185	12.8	175	12.1	10
6-2-4-2	Ti-6Al-2Sn-4Zr-2Mo	Annealed (4 h/1300-1550 °F)	145	10.0	135	9.3	15
6-2-4-6	Ti-6Al-2Sn-4Zr-6Mo	Annealed (2 h/1500-1600 °F)	150	10.3	140	9.7	11
		Aged	175	12.1	165	11.4	8
6-22-22	Ti-6Al-2Sn-2Zr-2Mo-2Cr-0.25Si	α-β processed plus aged	162	11.2	147	10.1	14
10-2-3	Ti-10V-2Fe-3Al	Annealed (1 h/1400 °F)	140	9.7	130	9.0	9
		Aged	180-195	12.4-13.4	165-180	11.4-12.4	7
15-3-3-3	Ti-15V-3Cr-3Sn-3Al	Annealed (0.25 h/1450 °F)	115	7.9	112	7.7	20-25
		Aged	165	11.4	155	10.7	8
13-11-3	Ti-13V-11Cr-3Al	Annealed (0.5 h/1400-1500 °F)	135-140	9.3-9.7	125	8.6	18
		Aged	175	12.1	165	11.4	7
38-6-44	Ti-3Al-8V-6Cr-4Mo-4Zr	Annealed (0.5 h/1500-1700 °F)	120-130	8.3-9.0	113-120	7.8-8.3	10-15
		Aged	180	12.4	170	11.7	7
β-III	Ti-4.5Sn-6Zr-11.5Mo	Annealed (0.5 h/1300-1600 °F)	100-110	6.9-7.6	95	6.5	23
		Aged	180	12.4	170	11.7	7

nium alloys. A moderately high-strength material, it is used for gas-turbine-engine cases, discs, and compressor blades, as well as other ambient-temperature aircraft applications. Ti-5Al-2Sn-2Zr-4Mo-4Cr (Ti-17) was developed specifically to meet the requirements of the gas-turbine-engine discs. Ti-6242 was developed in the mid-1960s as a gas-turbine material capable of service at 540 °C. Ti-6246 was developed in the late 1960s as an improved-performance material for use in gas-turbine-engine discs compressor components.

The room-temperature tensile and compressive strengths of a number of commercial titanium-base alloys are presented in Table 12.6.

The low-temperature tensile properties of a selected few are presented in Table 8.1. For useful extended summaries of the properties and applications of the alloys listed in these tables, a recently published Appendix to the *Structural Alloys Handbook* [STR82] is recommended.

Part 3: Anomalous Plastic Properties

12.14 Normal and Anomalous Tensile Properties

When a metal is stressed uniaxially, say in tension, it extends linearly and elastically for a time, then after passing a yield point begins to elongate plastically, still smoothly but at a much faster rate than before. For most purposes the information describing the results of such a tensile test are: the Young's modulus (E), the 0.2% offset yield strength (Y or $\sigma_{0.2}$), the ultimate strength (Y_{ULT}), the fracture strength (σ_B), and whether or not the fracture was brittle or ductile. These might be referred to as the "normal" tensile properties. All materials possess them; however, close attention to the details of the tensile and/or compression test and its results often reveals what might be termed "anomalous tensile properties." Properties to be defined and discussed below with particular reference to titanium alloys are: serrated yielding, acoustic emission, thermoelasticity, pseudoelasticity, shape-memory effect, and Bauschinger effect. They are illustrated in Fig. 12.12; see also Fig. 8.7 and 8.8. Depending on their purity (interstitial content), phase stability (proximity to a phase boundary), grain size, temperature, strain rate, and so on, some materials/samples exhibit anomalous properties more strongly than others.

12.14.1 Serrated Yielding

The jagged oscillation in stress at constant strain rate that characterized the $\sigma(\varepsilon)$ curves of many metals is referred to as "serrated yielding". At low temperatures, if the material being tested is susceptible to stress-induced martensitic transformation or twinning, the output trace of a screw-driven or other such extension-controlled tensile machine will exhibit a sharp load drop each time a spontaneous elongation of the sample takes place. At high temperatures, serrated yielding is the manifestation of a stress-strain relaxation oscillation. Microscopically, this consists of the cyclical repetition of a sequence of events in which short bursts of plastic flow are abruptly terminated by rapid-diffusion-induced local solution-strengthening, a phenomenon to which the term "dynamic strain aging" has been aptly applied. Depending on their composition, the temperature, and their condition, titanium alloys exhibit both classes of serrated yielding.

Serrated Yielding at Low Temperatures. At low temperatures where diffusion rates are low, serrated yielding, when it occurs, is generally a property of the crystal lattice. The low-temperature phenomenon has been well discussed by BASINSKI [BAS57], who, by using aluminum and aluminum alloys as samples, disassociated the general effect from both twinning and martensitic transformation. Necessary conditions for serrated yielding in BASINSKI's experiment were (1) a strong negative temperature dependence of the yield stress and (2) the usual small low-temperature specific heat. Since under such conditions local slip-nucleation is sufficient to trigger the thermomechanical instability, it was regarded as a quite general low-temperature mechanical property [BAS57]. The local mean temperature increase associated with a serration was calculated by BASINSKI to be ~ 60 K. He also obtained direct experimental evidence for the temperature increase in an aluminum test piece by not-

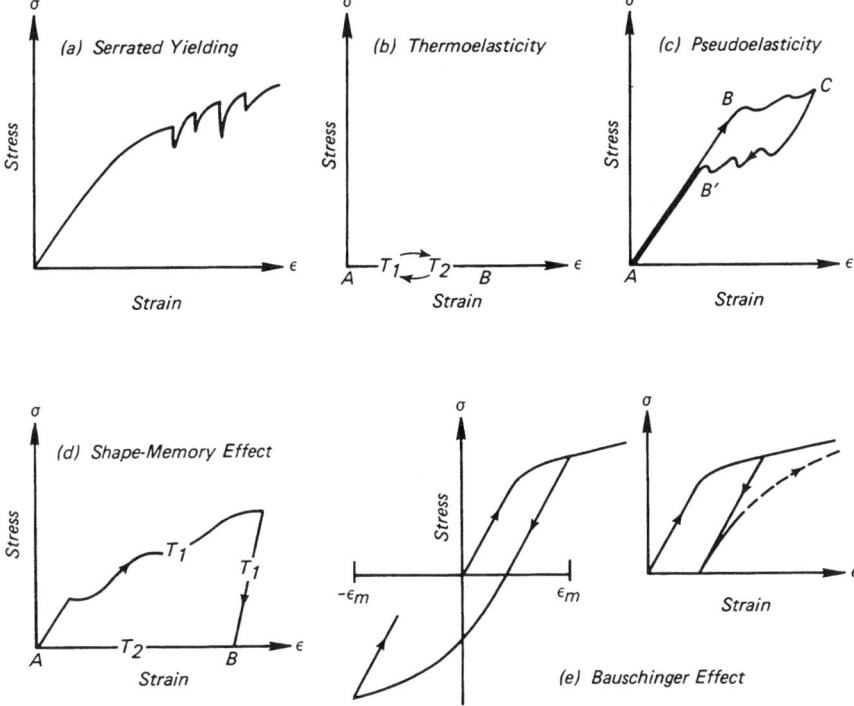

Fig. 12.12 Stress-strain signatures of five anomalous plastic properties.

for Ti-Nb (see Fig. 5.8), it has been noted [HIL76] that the transition from the "normal" parabolic σ(ε) region to the "anomalous" region characterized by intermittent load drops is gradual and not associated with a definite transition temperature. Detailed studies of the temperature dependences of the tensile properties by HILLMANN et al. [HIL76], one of the few groups to include 20 K in the measurement program, have shown that the entry into the serrated-yield region is accompanied by several important departures from the behavior expected simply by extrapolating from higher-temperature results. The σ_B and $\sigma_{0.2}$ versus T curves are characterized by: (1) considerable scatter in the values of $\sigma_B(T)$, $\sigma_{0.2}(T)$, and $\sigma_{0.01}(T)$ obtained from repeated experiments; (2) a $\sigma_B(T)$ which passes through a broad maximum in the vicinity of 20 K; (3) yield strengths, $\sigma_{0.2}(T)$ and $\sigma_{0.01}(T)$, which increase rapidly below about 30 K, but which for cold deformation >90% also maximize at about 20 K; and (4) elongations, δ_B, which with decreasing temperature drop rapidly from 5 ~ 6% to below 1%. All these effects, which are clearly related to each other and to the serrated-yielding effect in Ti-Nb, warrant further study.

READ [REA78] has detected serrated σ(ε) behavior in Ti-45Nb at 4 K (Fig. 8.10). The appearance of the fracture surface indicated that at least some of the deformation took place by irreversible twinning. EASTON and KOCH have noted serrated yielding in Ti-Nb(36 at.%) wire (0.0105 in.$^\phi$) at 4.2 K [EAS75][REA78] and in a sample of filament (~280 μm$^\phi$) of comparable composition etched from a commercial multifilamentary composite conductor. In agreement with SCHMIDT [SCH76], they recorded that serrations appeared not only with increasing but also with decreasing load, a reversibility which suggested the operation of a pseudoelastic effect such as the reversible stress-induced martensitic transformation or strain-reversible twinning.

In any investigation of serrated yielding, the possibility that the results are being perturbed by the presence of sample artifacts and the characteristics of the testing machine must be considered. SCHMIDT [SCH77] has considered in some detail the influence of the dynamics of the sample and load train on elongation and load drop during serrated yielding. With regard to the former, he pointed out that the rate of instantaneous *elongation* is determined in part by the inertia of the loading system and is, therefore, apparatus-dependent. As for the *load drop*, SCHMIDT noted that during the instant of localized yield the rest of the sample relaxes elastically within a few microseconds, a time negligible compared with the response time of a relatively massive load train

ing the transition to the normal state of a niobium wire embedded axially in it.

MOSKALENKO et al. [MOS80] have suggested a dislocation mechanism to explain the observed low-temperature (<4.2K) serrated yielding in unalloyed titanium (iodide and commercial) and Ti-2.4Zr-1.2Mo. The effect was not attributed to any kind of lattice instability; rather, a relaxation-oscillation mechanism was suggested based on the cycle: *dislocation pile-up / stress increase / dislocation avalanche / local adiabatic heating / load drop*—clearly a low-temperature version of that referred to above.

Serrations have been noted whenever β-Ti-Nb alloys (about 50 wt%, 34 at.%) were strained at liquid helium temperatures beyond the linear elastic region [EVA73][HIL73^a][EAS75, EAS77] [ALB76] [KOC77] [PAS78]. In a series of tensile tests of Ti-50Nb conducted at various temperatures from 300 down to 4 K, the onset temperature for serrated yielding was found to be about 28 K [ALB76]. EVANS [EVA73], in a search for a mechanism underlying the "training" phenomenon in superconducting magnets, seems to have been the first to apply to Ti-Nb, the general concepts of unstable low-temperature plastic flow previously observed in aluminum alloys. Those experiments, and those of MOSKALENKO et al. [MOS80] on unalloyed αTi and α-Ti alloys reported above, have demonstrated that no exotic deformation mechanisms were needed, provided that a sufficiently steep negative yield-stress temperature dependence was present. Thus, for an alloy such as Ti-50Nb with inherent lattice instability, manifesting itself as thermoelasticity, as pseudoelasticity, or as a shape-memory effect, low-temperature serrated yielding is ensured.

In studies of the deformation and fracture of Ti-50Nb at 4.2 K, HILLMANN attributed the serrations to spontaneous slip taking place on the {110} principal bcc glide planes [HIL73^a]. ALBERT and PFEIFFER [ALB76], in a test of cold-worked (>90%) Ti-50Nb wires possessing the usual ⟨110⟩ texture parallel to the wire axis, noted that the fracture surface was at 45° to the wire axis, and concluded that slip in these samples took place on the {100} planes. A conventional slip-plasticity model was adequate to explain the effects of prior deformation or load alternation. Within such a model, the addition of work-induced defects (work hardening) had the effect of inhibiting slip, thereby moving the serrations up to higher and higher levels of strain, in agreement with observations. Returning to the temperature dependence of the effect, it was mentioned above that with decreasing temperature the instabilities in Ti-Nb appeared for the first time at 28 K. Although this value is in good accord with the extrapolated M_s transformation curve

which, in effect, remains "rigid" during the process (see Ref 35 of [KOC77]). Accordingly, it was concluded that under such conditions the magnitude of the load drop should be independent of the apparatus.

Serrated Yielding at High Temperatures. Although the occurrence of dynamic strain aging via solute-atom diffusion is sufficient to ensure serrated yielding, it is not necessarily the only mechanism. Even remote from the low-temperature regime referred to in the previous subsection, serrated yielding in some alloys has been attributed to lattice instability. For example, β-phase instability was regarded as being responsible for the serrated yielding observed in Ti-3Al-16Mo-10Cr during tensile testing at 200 to 400 °C [BOG73], since the same alloy after α + β annealing exhibited no such effect. Similarly, β-III [OHT73] [RAC75] and some other β-Ti alloys (Ti-15Mo-5Zr, Ti-12Mo-6Sn, and Ti-15Mo-0.2Pd [OHT73]) during testing at 300 ~ 400 °C exhibited serrated yielding, due it was supposed to an isothermal β → ω transformation.

Dynamic strain aging is the most frequently cited mechanism of elevated-temperature serrated yielding. It is generally interpreted as the manifestation of a hardness oscillation brought about by the periodic movement of solute atoms sufficiently fast to slow down the moving dislocations with which they become associated [BRI70]. Since interstitials, which occur abundantly in titanium alloys, are the most mobile of the solute atoms, serrated yielding in titanium alloys is usually attributed to them. For example, GARDE et al. [GAR72] showed that dynamic strain aging in iodide titanium (C, N, O = 78, 6, 63 ppm, respectively) was very much less pronounced than in commercial Ti (C, N, O = 200, 100, 1360 ppm, respectively). In studies of dilute alloys of titanium with silicon [WIN73], zirconium [SAS75, SAS82], and the solutes hafnium, silver, vanadium, indium, tin, aluminum, niobium, and tantalum [SAS82], serrated yielding was again attributed primarily to the presence of interstitial atoms, especially since the amplitudes of the serrations were higher in those alloys which exhibited the highest internal-friction relaxation peaks [SAS82]. The presence of substitutional impurities was not, however, ignored; it was thought to assist in the dynamic strain-aging process by distorting the interstitial-atom site [WIN73][SAS75]. Furthermore, substitutional atoms can in their own right play a role in dynamic strain aging. Although their diffusivities would normally be too small to satisfy the conditions necessary to establish "atmospheres" about moving dislocations, this is no longer true in the presence of a high density of vacancies such as will become established during plastic flow [BRI70]. Thus, as pointed out by DÖNER and CONRAD [DON75] in a very important paper on elevated-temperature (0.3 to 0.6 T_m) deformation in Ti-5Al-2.5Sn (0.5 at.% O_{eq}), the dynamic strain aging which was observed in the temperature range 320 ~ 560 °C could very well have been due to the diffusion of the aluminum and/or the tin atoms under the conditions of the experiment.

12.14.2 Acoustic Emission

Introduction. In a comprehensive review of proposed and confirmed sources of emission JAFFREY [JAF79] has found it convenient to deal with acoustic emission in terms of: (1) phase transformation, (2) dislocation motion, and (3) fracture mechanics. The first of the above topical areas included elastic twinning, deformation twinning (e.g., the "cry" of tin), and martensitic transformation. BASINSKI, for example, during the low-temperature studies referred to earlier, discovered that the martensitic transformation of austenitic stainless steel at temperatures below 45 K was accompanied by tensile load drops and loud clicks [BAS57]. With regard to dislocation motion as a possible source, it has been pointed out that whereas acoustic energy is expected to be emitted from individual moving screw and edge dislocations, at 10^{-23} J per event it was below the detectability limit of contemporary equipment (~10^{-16} J, 1979). Similarly, although sound was also expected to be emitted during the annihilation of a single dislocation at a free surface, or of a pair of colliding oppositely signed dislocations, the per-event energy was again calculated to be below the detectability threshold [JAF79]. On the other hand, should several hundreds or thousands of dislocations move cooperatively, as in an avalanche, their combined acoustic output would rise above the instrumentational threshold. Thus, acoustic emission can occur in association with microplastic flow (see Ref 9 of [SCH77a]). It has also been detected in association with the normal plastic deformation of copper at strain levels above 0.3% during which noise in the 100 to 300-kHz frequency range was emitted continuously during straining [PAS78]. In the area of fracture mechanics, acoustic emission has been postulated and/or detected in connection with: (1) microcracks forming ahead of subcritical crack tips, (2) dislocation breakaway at the tips of cracks, and (3) void nucleation and growth.

Acoustic Emission in Titanium Alloys. Acoustic emissions from titanium alloys, as they have been studied and reported, fall within the general categories (1) and (2) referred to above.

The mechanical stability of Ti-50Nb, because of its use in the multifilamentary composite conductors of large superconducting magnets, has received considerable attention. Serrated yielding in that alloy has already been referred to. Audible clicking associated with load drops were occasionally noted by REED et al. [REE77] during the tensile testing of a Ti-Nb superconductor at 4.0 K. EASTON and KOCH [EAS75], in a more extensive study of audio-frequency emission during the alternate loading and unloading of a 2-mm$^\phi$ tensile-test sample of Ti-55Nb noted that: (1) while emission occurred almost immediately upon application of the stress at 4.2 K, none was detected at 77 or 300 K; (2) emission took place upon both loading and unloading and *bursts of noise (clicks) preceded any load drop that occurred;* (3) emission took place *almost immediately* upon application of stress at 4.2 K; and (4) during repeated loading, the emission *increased in amplitude and frequency* when the immediately preceding stress level was exceeded. It is clear from a close examination of these results and an intercomparison of the italicized statements that pseudoelastic serrated yielding is a sufficient but not necessary condition for acoustic emission in Ti-Nb at low temperatures.

These results were amplified and clarified in the reports of a subsequent series of experiments by PASZTOR and SCHMIDT [SCH77a][PAS78]. In studies of ultrasonic emission (100 to 300 kHz) during straining of Ti-50Nb wire at 4.2 K, the above authors were able to resolve the effect into reversible and irreversible components depending on whether or not emission would take place during the unloading half-cycle. The threshold for the *reversible* component was a strain level of about 0.4%, which also marked the onset of hysteretic stress-strain behavior of the type referred to as pseudoelastic by EASTON and KOCH [EAS75]. The *irreversible* component was present throughout the entire stress range, and was detectable even at room temperature using 6 db of additional amplified gain. Noise output took place in bursts of some 100 μs time duration, commencing at extremely low levels of strain (~0.005%). In repeated loading to successively higher levels of strain, the irreversible noise signals (and the only noise generated at all for strains <0.3 to 0.4 %) began only when the previous stress level was exceeded. The research of KOCH and EASTON [KOC77] [EAS75] and its extension by PASZTOR and SCHMIDT [SCH77a][PAS78] has demonstrated that in Ti-50Nb at 4.2 K both irreversible serrated yielding (accompanied by its acoustic emission), which began at about 1% strain [SCH77a] *and* pseudoelasticity (accompanied by stress-strain hysteresis and *reversible* acoustic emission [EAS75]), beginning at about 0.4% strain [SCH77a][EAS75], could take place at stress levels as low as ~50 ksi. Furthermore, the occurrence of additional *irreversible* acoustic emission, manifesting itself as steady streams of bursts beginning at extremely small levels of strain (<0.01%), was thought to be symptomatic of movement at the atomic level.

VODOLAZSKY et al. [VOD80] have used acoustic emission to study deformation processes in titanium alloys. The material selected for testing in tension as well as in compression at 850 °C was Ti-6Al-2Zr-1.5V-1.5Mo. As indicated above, plastic deformation of metals is known to be accompanied by the continuous generation of microcracks, which appear even during the initial stages of plastic deformation. By monitoring the acoustic emission associated with microcrack formation, an early warning of impending fracture, even during high-temperature testing, could be obtained.

12.14.3 Thermoelasticity

The thermoelastic martensitic transformation (Fig. 12.12b) is one in which a continuous transformation takes place as the temperature is lowered [DEL74]. In contrast to the more usual mode of the martensitic transformation, the thermoelastic kind is not accompanied by the sudden appearance, or bursts, of groups of platelets. The effect *is* reversible, the volume fraction of martensite decreasing continuously to zero as the temperature is restored to its original value. Implicit in the extrapolated martensitic transformation diagram of Fig. 5.8 is the suggestion that Ti-Nb alloys of compositions close to those used in commercial superconductors should be thermoelastic in the helium-temperature regime. Although an intriguing possibility, there is at present no direct microscopic evidence for such a low-temperature reversible transformation. Indeed, indirect evidence refutes such conjecture, and favors instead the reversible athermal ω-phase transformation. Comparative studies of the room-temperature and helium-temperature physical properties of Ti-TM alloys indicate that the composition of the martensitic-phase boundary remains fixed as the temperature decreases below room temperature. Thus, whereas a reasonable extrapolation of the M_s curve for Ti-V [ZWI74, p. 174] passes through for example, the point (−200 °C, 19 at.% V), a comparison of the room-temperature magnetic-susceptibility composition dependence with that of the superconducting transition temperature shows that the compositional range of the martensitic regime is no greater at helium

temperatures than it is at 298 K—certainly less than 12 at.% V in both cases [COL75b, COL75c, COL80, COL82b]; see also Section 8.3.2 for a comparable discussion of the Ti-Nb system). On the other hand, there is abundant irrefutable direct and indirect evidence for the reversibility of athermal ω phase in alloys of this type. Having rejected, with respect to Ti-TM alloys, reversible low-temperature martensitic transformation in favor of ω phase, one can only speculate at this stage as to *its* ability to lead to a measurable thermoelastic effect. A simple geometrical calculation leads to the prediction that a net 3% contraction takes place perpendicular to the $(\overline{1}12)_\beta$ plane during the β → ω transformation [LUH70] (Section 5.3.2). Since this is able to be accommodated by coherency strains [HIC69, HIC69a], a thermoelastic effect via this mechanism is also unlikely.

12.14.4 Pseudoelasticity

The pseudoelastic effect is illustrated schematically in Fig. 12.12(c). In it the sample, after a period of elastic extension to *B*, is shown deforming in an apparently plastic manner to *C*, and upon unloading returning spontaneously to *B'* with net dissipation of energy, then elastically back (or almost so) to the initial state, *A* [DEL74]. In this mechanical analog of the thermoelastic effect, martensitic transformation proceeds continuously with increasing applied stress and is reverted continuously when that stress is removed. At sufficiently low stress levels, the sample exhibits normal elastic properties. The apparent plastic deformation is thus the result of a (reversible) stress-induced martensitic transformation. The behavior in the plastic regime suggests that the effect would be more appropriately entitled "pseudoplasticity"; the term "pseudoelasticity" arises from the return of the sample to its original state after the removal of the stress, as in normal elastic behavior.

There is, of course, abundant evidence for stress-induced martensitic transformation in Ti-TM alloys. Section 5.3.4 has shown how the transformation can be triggered by the application of additional strain in the vicinity of a strain spinodal defined by $\partial^2 g/\partial \varepsilon^2 = 0$. That is to say, in a lattice which is prone to transform—a situation signalled for example by a low value of the elastic shear modulus $C' = (C_{11} - C_{12})/2$—the application of a few percent strain is sufficient to initiate the martensitic reaction. The special characteristic of pseudoelasticity is, of course, the reversible nature of the transformation. As explained by DELAEY et al. [DEL74], pseudoelasticity can also occur in the absence of martensitic phase change by some form of mutual accommo-

dation between pairs of twinned structures.

Pseudoelasticity in Titanium Alloys at Low Temperatures: Ti-Nb Alloys (see also Section 8.3). The extrapolated M_s curve of Fig. 5.8 [EAS77][KOC77], although perhaps not to be taken literally, does, however, suggest the possibility of a β → α″ transformation. This and other indicators of lattice instability, such as the relatively low shear modulus for alloys of this class (Section 12.7) and the proven existence of diffuse ω phase (Section 5.3.3), are evidence in support of a possible stress-induced martensitic transformability for alloys whose temperature/composition states lie just above the dashed line; this includes the commercial superconducting alloy, Ti-50Nb, at low temperatures. If the transformation is reversible (i.e., pseudoelastic), no trace will remain upon the removal of stress and the return of the sample to room temperature [EAS65][PAS78]. Evidence of its occurrence may be *indirect*—such as the characteristic signature of the σ(ε) curve. It may also be *direct*—as in the *in situ* cold-stage TEM experiments of OBST et al. [OBS80], who noted what appeared to be a reversible transformation in reversibly strained Ti-45Nb.

Stress-strain behavior, thought to be characteristic of pseudoelasticity, has been detected in Ti-55Nb at 4.2 K by EASTON and KOCH [EAS75][KOC77]. During the first loading cycle of a low-temperature tensile test, slip took place partially reversibly at a strain level of about 2.2% but the sample subsequently executed closed hysteresis loops with no further accumulation of residual strain up to 100 cycles of deformation. The hysteretic deformation, which set in above the 0.4% strain level [EAS75][PAS78], and which was accompanied by acoustic emission during both loading and unloading [KOC77] [EAS75,EAS77], was interpreted as a manifestation of the pseudoelastic martensitic transformation. At higher strain levels, serrated yielding was encountered.

Pseudoelasticity in Titanium Alloys at Room Temperature: Ti-V and Ti-6Al-4V. Since strain cycling is needed to elicit the characteristic response, pseudoelasticity, if present, will manifest itself during the initial stages of cyclic fatigue testing in the form of: (1) an irregularity in an otherwise smooth stress-strain hysteresis loop, or (2) an "anomalous Bauschinger effect." The Bauschinger effect itself (to be discussed below) is the premature yielding in compression (tension) of a material initially deformed in tension (compression). Clearly, a fully pseudoelastic material in which a removal of the tensile stress is sufficient to close the stress-strain hysteresis loop exhibits the greatest possible departure from normal Bauschinger behavior. Otherwise, partial pseudoelasticity will produce extreme premature reverse-stress yielding or cause undulations to develop in the reverse-stress-strain curve, either of which could be described as anomalous Bauschinger behavior.

CHAKRABORTTY et al. [CHA78a] studied the cyclic stress-strain responses of a series of Ti-V(24, 28, 32, and 36 wt%) alloys in the quenched + aged conditions. In only Ti-24V was an anomalous Bauschinger effect observed.* The mechanism in this case was deformation twinning under tension, followed by untwinning when the stress was removed. A prerequisite was a low level of forward strain, during which small undamaged twins were produced and the twin-energy was conserved. The anomaly tended to go away at large strain amplitudes as stress-induced twins became damaged by mutual interaction and by interaction with dislocations, and energy became dissipated.

Pseudoelasticity gives rise to an exaggerated Bauschinger effect. On this evidence, acquired during both tensile [STE76] and torsional [GIL76] cyclic fatigue tests, some pseudoelastic character has been attributed to Ti-6Al-4V. The practical evidence is that Ti-6Al-4V exhibits "pseudoelastic fatigue failure." This is brought about by pseudoelastic slip, which in turn initiates crack formation at stress levels well below the macroyield range. In the experiments of GILMORE et al. [GIL76], the slip appeared to be taking place at grain boundaries rather than within the grains.

12.14.5 The Shape-Memory Effect

The shape-memory effect (Fig. 12.12d) may be regarded as a combination of the previous two effects (Fig. 12.12b and c) in that after the initial stress-induced permanent transformation, the reverse transformation is achieved by raising the temperature of the sample [DEL74]. The effect is usually associated with alloys exhibiting martensitic thermoelasticity (socalled "marmen" alloys), whose behavior and properties have been considered fully in reviews assembled by PERKINS et al. [PER75, PER75a]. Thermoelastic martensitic transformation and the shape-memory effect have also been reviewed by TONG and WAYMAN [TON75], with particular reference to the alloys Au-Cd, Ag-Cd, and Ti-Ni, and their use as the "working substances" in heat engines. Although the most famous shape-memory material is,

---
*Quenched Ti-24V was β plus athermal ω. The other alloys were all-β and did not exhibit anomalous Bauschinger behavior.

of course, TiNi (so-called "NITINOL"), numerous alloys exhibit the effect, which, as the following brief discussion indicates, is not necessarily confined to martensitic structures.

NISHIHARA and IGUCHI [NIS76] detected what they referred to as shape-memory behavior in Ti-6Al-4V, which after superplastic deformation underwent partial length recovery as a result of $\alpha + \beta \rightarrow \beta$ transformation during stress-free heating. In experiments on Ti-Nb(22 at.%), which fully transforms to α'' on a quenching through ~175 °C, BAKER [BAK71] showed that 1% "permanent" strain could be almost fully (~90%) erased by heating to 184 °C. Similar effects were noted for the same alloy in the $\omega + \beta$ state after aging for 2 and 4 min at 400 °C, except that almost 100% recovery from 1.5% strain was achieved. Experiments have indicated that a shape-memory effect can occur either in an alloy undergoing a martensitic $\beta \rightarrow \alpha''$ transformation, or in a specimen whose initial structure is already fully martensitic, in which case the effect proceeds by way of a reorientation of the martensite plates [DEL74]. Since the $\beta \rightarrow \alpha''$ transformation is accompanied by considerable dilatation, its occurrence during a tensile test either isothermally (pseudoelastic effect) or adiabatically (shape-memory effect) will give rise to load drops and corresponding $\sigma(\varepsilon)$ anomalies (serrated yielding), the forms of which will be characteristic of both the sample properties and the testing-machine response times (see Section 12.14.1).

12.14.6 The Bauschinger Effect

It could be argued that since the Bauschinger effect is generally observed during the stress-strain cycling of metals, it is not anomalous and should, therefore, have been considered in Part 2 of this chapter. On the other hand, it *is* anomalous in the sense that it involves stress-strain asymmetry explicable only in microstructural terms. In contrast to the repeated unidirectional application-and-the release of stress, which leads to work-hardened elastic stress-strain reversal, cyclical loading/unloading/load-reversals lead to hysteretic stress-strain loops of the kind illustrated in Fig. 12.12(e). Furthermore, it is generally found that the third quarter-cycle is accompanied by yielding at prematurely low stress levels as gauged by the parameters of the first. This is the Bauschinger effect, which may, therefore, be defined as reversible stress softening which permits premature yielding of a material upon stress reversal after an initial half-cycle involving plastic deformation (in either the compressive or tensile mode). It has been discussed by GITTUS [GIT75, p. 76] (in terms of disloca-

tion effects) and TIMOSCHENKO [TIM56, p. 413] (in terms of an intergranular/intragranular mechanism). The effect is generally smaller the greater the strain during the initial half-cycle, and disappears completely with large amounts of strain hardening (Section 12.14.4) [TWE64, p. 74].

With regard to the *dislocation mechanism*, a "normal" Bauschinger effect would result when dislocation energy, stored during forward straining, adds to the applied stress during reverse loading. THEODORSKI and KOSS [THE82] have discussed the Bauschinger effect in Ti_{61}-V_{38}-Si_1 and the technical alloys Ti-8Mo-8V-2Fe-3Al (Ti-8823) and Ti-3Al-8V-6Cr-4Mo-4Zr (β-C, Ti-38-6-44) from this standpoint.

TIMOSHENKO, whose approach is specific to polycrystalline metals, developed a mechanical model for the Bauschinger effect based on *intragranular slip and its anisotropy*. Expanding on this and related ideas, MARGOLIN et al. [MAR78] have suggested that the Bauschinger effect could arise as a result of residual stresses which accumulate in the presence of grains with different yield stresses. Assuming a tension/compression sequence, grains with lower yield stresses become placed in compression by the other grains and thereby experience, in addition to the applied stress, an extra compressive stress. The theory was examined using Ti-15Mn as a test material [MAR78]. Presumably, the same mechanism is also operative but at a more macroscopic level, in certain bimetallic composites which by their very natures generally consist of pairs of metals with widely differing mechanical properties. An excellent example is the well-known Ti-50Nb/Cu superconductive multifilamentary composite, in which hundreds of fine filaments of Ti-50Nb, with a relatively high modulus, are embedded in annealed OFHC copper. The copper yields at a much lower stress than the Ti-50Nb; consequently, it is possible for the Ti-Nb still to be stretching elastically long after the copper has passed its yield point. As a result, on the return stroke of the testing machine, the copper is placed in compression by a combination of the applied stress and the elasticity of the Ti-Nb [HEI74]. A somewhat similar effect should be encountered in two-phase alloys which could in this context be regarded as *in situ* metal-matrix composites. The Bauschinger effect in two-phase Ti-Mn alloys has been considered by SALEH and MARGOLIN [SAL79a].

The above models of the Bauschinger effect have all invoked an additional hidden component of stress to augment the applied stress and bring about an "apparent" premature third-quadrant yielding. In an unstable lattice, stress-induced transformation or twinning is responsible for an "actual" premature softening and will result in an anomalous Bauschinger effect. As pointed out in Section 12.14.4, pseudoelastic materials provide examples of this. Anomalous Bauschinger effects are also exhibited by alloys such as Ti-6Al-4V, which after being strained in tension (from the as-received condition) to a 1% permanent set, suffer a 40% loss of compressive yield stress[Woo72, p. 1-4:72-9]. As mentioned above, the anomalous Bauschinger effect in this alloy is thought to be caused by yielding at the grain boundaries [GIL76].

References

[AGA74] AGARWAL, K.L. "Low Temperature Calorimetry of Transition Metal Alloys," Ph.D. Thesis, University of New Orleans, 1974

[AGE70] AGEEV, N.V. and PETROVA, L.A., The Theoretical Basis of the Development of the High-Strength Metastable β-Alloys of Titanium, in [JAF70], p. 809-814

[ALB72] ALBERT, H. and PFEIFFER, I., Uber die Temperaturabhängigkeit des Elastizitätsmoduls von Niob-Titan-Legierungen, *Z. Metallkd*, Vol 63, 1972, p. 126-131

[ALB76] ALBERT, H. and PFEIFFER, I., Temperaturabhängigkeit der Festigkeitseigenschaften des Hochfeldsupraleiters NbTi50, *Z. Metallkd*, Vol 67, 1976, p. 356-360

[ALE67] ALEKSEEVSKII, N.Ye., IVANOV, O.S., RAYEVSKIY, I.I., and STEPANOV, M.V., Constitution Diagram of the System Niobium-Titanium-Zirconium and Superconducting Properties of the Alloys, *Phys. Met. Metallogr. (USSR)*, Vol 23 (No. 1), 1967, p. 28-35 [transl. of *Fiz. Met. Metalloved.*, Vol 23, 1967, p. 28-36]

[AME54] AMES, S.L. and MCQUILLAN, A.D., The Resistivity-Temperature-Concentration Relationships in the System Niobium-Titanium, *Acta Metall.*, Vol 2, 1954, p. 831-836

[AND63] ANDERSON, O.L., A Simplified Method for Calculating the Debye Temperature from Elastic Constants, *J. Phys. Chem. Solids*, Vol 24, 1963, p. 909-917

[ARN74] ARNDT, R. and EBELING, R., Einfluss von Gefügeparametern auf die Stromtragfähigkeit von Niob-Titan-Supraleitern, *Z. Metallkd*, Vol 65, 1974, p. 364-373

[BAE85] BAESLACK, W.A. III, KRISHNAMURTHY, S., and FROES, F.H., Rapid Solidification and Aging of a Near-Eutectoid Titanium Nickel Alloy, in *Strength of Metals and Alloys*, H.J. McQueen, J.-P. Bailon, J.I. Dickson, J.J. Jonas, and M.G. Akben, Ed., Vol 2, Pergamon Press, 1985

[BAE86] BAESLACK, W.A. III, KRISHNAMURTHY, S., and FROES, F.H., A Study of Rapidly Quenched Microstructures in a Hypereutectoid Ti-40 wt.% W Alloy, *J. Mater. Sci. Lett.*, Vol 5, 1986, p. 315-318

[BAG58] BAGARIATSKII, Yu.A. and NOSOVA, G.I., A More Accurate Determination of Atomic Coordinates of the Metastable ω-Phase in Ti-Cr Alloys, *Soviet Phys. Crystallogr.*, Vol 3, 1958, [transl. of *Kristallografiya*, Vol 3, 1958, p. 17-28]

[BAG59] BAGARIATSKII, Yu.A., NOSOVA, G.I., and TAGUNOVA, T.V., Factors in the Formation of Metastable Phases in Titanium-Base Alloys, *Sov. Phys. Dokl.*, Vol 3, 1959, p. 1014-1018 [transl. of *Dokl. Akad. Nauk SSSR*, Vol 122, 1958, p. 593-596]

[BAK67] BAKER, C. and TAYLOR, M.T., Some Problems of Relating the Critical Current Density to Dislocation Distribution in Worked Superconducting Alloys, *Philos. Mag.*, Vol 16, 1967, 1129-1132

[BAK69] BAKER, C. and SUTTON, J., Correlation of Superconducting and Metallurgical Properties of a Ti-20at.% Nb Alloy, *Philos. Mag.*, Vol 19, 1969, p. 1223-1255

[BAK70] BAKER, C., The Effect of Heat-Treatment and Nitrogen Addition on the Critical Current Density of a Worked Niobium 44 wt% Titanium Superconducting Alloy, *J. Mater. Sci.*, Vol 5, 1970, p. 40-52

[BAK71] BAKER, C., The Shape-Memory Effect in a Titanium-35 wt.% Niobium Alloy, *Met. Sci.*, Vol 5, 1971, p. 92-100

[BAL71] BALCERZAK, A.T., "An Electron Microscope Study of the Ti-Nb System," M.S. thesis, Cornell University, Ithaca, NY, 1971

[BAL72] BALCERZAK, A.T. and SASS, S.L., The Formation of the ω Phase in Ti-Nb Alloys, *Metall. Trans.*, Vol 3, 1972, 1601-1605

[BAR57] BARDEEN, J., COOPER, L.N., and SCHRIEFFER, J.R., Theory of Superconductivity, *Phys. Rev.*, Vol 108, 1957, p. 1175-1201

[BAR60] BARTON, J.W., PURDY, G.R., TAGGART, R., and PARR, J.G., Structure and Properties of Titanium-Rich Titanium-Nickel Alloys, *Trans. TMS-AIME*, Vol 218, 1960, p. 844-849

[BAS57] BASINSKI, Z.S., The Instability of Plastic Flow of Metals at Very Low Temperatures, *Proc. R. Soc.*, Vol A240, 1957, p. 229-242

[BAT64] BATT, R.H., "Apparatus for Heat Capacity Measurements," Ph.D. thesis, University of California, Berkeley, 1964

[BAT73] BATTERMAN, B.W., MARACCI, G., MERLINI, A., and PACE, S., Diffuse Mössbauer Scattering Applied to Dynamics of Phase Transformations, *Phys. Rev. Lett.*, Vol 31, 1973, p. 227-230

[BIL56] BILBY, B.A. and CHRISTIAN, J.W., Martensitic Transformations, in *The Mechanisms of Phase Transformations in Metals*, The Institute of Metals, 1956, p. 121-172

[BLA67] BLANDIN, A.P., Theory of the Hume-Rothery Rules, in *Phase Stability in Metals and Alloys*, P.S. Rudman, J. Stringer, and R.I. Jaffee, Ed., McGraw-Hill, 1967, p. 115-124

[BLA67^a] BLACKBURN, M.J., The Ordering Transformation in Titanium:Aluminum Alloys Containing up to 25 at. pct. Aluminum, *Trans. TMS-AIME*, Vol 239, 1967, p. 1200-1208

[BLA68] BLACKBURN, M.J. and WILLIAMS, J.C., Phase Transformations in Ti-Mo and Ti-V Alloys, *Trans. TMS-AIME*, Vol 242, 1968, p. 2461-2469

[BLA69] BLACKBURN, M.J. and WILLIAMS, J.C., Strength, Deformation Modes, and Fracture in Titanium-Aluminum Alloys, *Trans. ASM*, Vol 62, 1969, p. 398-409

[BLA70] BLACKBURN, M.J., Some Aspects of Phase Transformations in Titanium Alloys, in [JAF70], p. 633-643

[BLE85] BLENKINSOP, P.A., Developments in High Temperature Alloys, in *Titanium Science and Technology* (Proc. 5th Int. Conf. on Titanium, Oberursel, West Germany, 1984), G. Lütjering, U. Zwicker, and W. Bunk, Ed., D.G. für Metallkd., 1985, p. 2323-2338

[BOG73] BOGATCHEV, I.N. and DYAKOVA, M.A., Titanium Alloys' Behaviour under Various Loading Conditions, in [JAF73], p. 1119-1130

[BOM86] BOMBERGER, H.B. and FROES, F.H., Prospects for Developing Novel Titanium Alloys Using Rapid Solidification, in *Titanium Rapid Solidification Technology*, F.H. Froes and D. Eylon, Ed., The Metallurgical Society, 1986, p. 21-43

[BOY74] BOYER, R.R., TAGGART, R., and POLONIS, D.H., Effect of Thermal and Mechanical Processes on the β-III Titanium Alloy, *Metallography*, Vol 7, 1974, p. 241-251

[BRA67] BRAMMER, W.G. Jr. and RHODES, C.G., Determination of Omega Phase Morphology in Ti-35%Nb by Transmission Electron Microscopy, *Philos. Mag.*, Vol 16, 1967, p. 477-486

[BRE67] BREWER, L., Viewpoints of Stability of Metallic Structures, in *Phase Stabil-*

ity in Metals and Alloys, P.S. Rudman, J. Stringer, and R.I. Jaffee, Ed., McGraw-Hill, 1967, p. 39-61

[BRI70] BRINDLEY, B.J. and WORTHINGTON, P.J., Yield-Point Phenomena in Substitutional Alloys, *Metall. Rev.*, Vol 15, 1970, p. 101-114

[BRO55] BROTZEN, F.R., HARMON, E.L. Jr., and TROIANO, A.R., Decomposition of Beta Titanium, *Trans. TMS-AIME*, Vol 203, 1955, p. 413-419

[BRO64] BROWN, A.R.G., CLARK, D., EASTABROOK, J., and JEPSON, K.S., The Titanium-Niobium System, *Nature*, Vol 201, 1964, p. 914-915

[BRO65] BROWN, A.R.G., JEPSON, K.S., and HEAVENS, J., High-Speed Quenching in Vacuum, *J. Inst. Met.*, Vol 93, 1965, p. 542-544

[BRO85] BRODERICK, T.F., JACKSON, A.G., JONES, H., and FROES, F.H., The Effect of Cooling Conditions on the Microstructure of Rapidly Solidified Ti-6Al-4V, *Metall. Trans.*, Vol 16A, 1985, p. 1951-1959

[BUC65] BUCHER, E., HEINIGER, F., and MÜLLER, J., Anomalies in the Superconducting Transition of HCP Titanium Alloys, in *Low Temperature Physics—LT9* (Proc. 9th Int. Conf., Columbus, OH, Sept 1964), J.G. Daunt et al., Ed., Plenum Press, 1965, p. 482-486

[BUM52] BUMPS, E.S., KESSLER, H.D., and HANSEN, M., Titanium-Aluminum System, *Trans. TMS-AIME*, Vol 194, 1952, p. 609-614

[BUY70] BUYNOV, N.N., VOZILKIN, V.A., and RAKIN, V.G., Structural Study of the Superconductive Alloy 65 BT, *Phys, Met. Metallogr. (USSR)*, Vol 29 (No. 5), 1970, p. 115-119 [transl. of *Fiz. Met. Metalloved.*, Vol 29, 1970, p. 1005-1009]

[CAH61] CAHN, J.W., On Spinoidal Decomposition, *Acta Metall.*, Vol 9, 1961, p. 795-801

[CHA70] CHARLESWORTH, J.P. and MADSEN, P.E., "Effect of Heat Treatment on the Superconducting Critical Current of Cold-Worked Titanium-45at.% Niobium: Part I," Report No. AERE-R-6534, Atomic Energy Research Establishment, Harwell, UK, Oct 1970

[CHA72] CHANDRASEKARAN, V., TAGGART, R., and POLONIS, D.H., Fracture Modes in a Binary Titanium Alloy, *Metallography*, Vol 5, 1972, p. 235-250

[CHA72[a]] CHANDRASEKARAN, V., TAGGART, R., and POLONIS, D.H., Phase Separation Processes in the Beta Phase of Ti-Mo Binary Alloys, *Metallography*, Vol 5, 1972, p. 393-398

[CHA73] CHANDRASEKARAN, V., COTTON, W., NARASIMHAN, S., TAGGART, R., and POLONIS, D.H., "Study of Phase Transformation and Superconductivity," Annual Progr. Report No. RLO-2225-T-13-19, University of Washington, Seattle, Oct 1973

[CHA73[a]] CHANDRASEKARAN, V., TAGGART, R., and POLONIS, D.H., Decomposition Processes Prior to Detection of the Omega Phase in Aged Ti-Cr Alloys, *Metallography*, Vol 6, 1973, p. 313-322

[CHA74] CHANDRASEKARAN, V., TAGGART, R., and POLONIS, D.H., The Influence of Constitution and Microstructure on the Temperature Coefficient of Resistivity in Ti-Base Alloys, *J. Mater. Sci.*, Vol 9, 1974, p. 961-968

[CHA78] CHANDRASEKARAN, V., TAGGART, R., and POLONIS, D.H., An Electron Microscopic Study of the Aged Omega Phase in Ti-Cr Alloys, *Metallography*, Vol 11, 1978, p. 183-198

[CHA78[a]] CHAKRABORTTY, T.K., MUKHOPADHYAY, T.K., and STARKE, E.A. Jr., The Cyclic Stress-Strain Response of Titanium-Vanadium Alloys, *Acta Metall.*, Vol 26, 1978, p. 909-920

[CHE72] CHEN, A.-B., WEISZ, G., and SHER, A., Temperature Dependence of the Electron States and dc Electrical Resistivity of Disordered Binary Alloys, *Phys. Rev.*, Vol 5, 1972, p. 2897-2924

[CHE80] CHEN, C.C. and COYNE, J.E., Relationships Between Microstructure and Mechanical Properties in Ti-6Al-2Sn-4Zr-2Mo~0.1Si Alloy Forgings, in [KIM80], p. 1197-1207

[CHI84] CHI, C.S. and WHANG, S.H., Rapidly Solidified Ti Alloys Containing Metalloids and Rare Earth Metals—Their Microstructure and Mechanical Properties, in *Rapidly Solidified Metastable Materials* (MRS 1983 Ann. Meet. Symp. Proc.), B.H. Kear and B.C. Giessen, Ed., Vol 28, North-Holland, 1984, p. 353-366

[CHI85] CHI, C.S. and WHANG, S.H., Microstructures and Mechanical Properties of Rapidly Solidified Ti-5Al-4.5La and Ti-5Al-5.4Er Alloys, in *Mechanical Behavior of Rapidly Solidified Materials*, S.M.L. Sastry and B.A. MacDonald, Ed., The Metallurgical Society of AIME, 1985, p. 231-245

[CHI86] CHI, C.S. and WHANG, S.H., Microstructural Characteristics of Rapidly Quenched Alpha-Ti Alloys Containing La (personal communication)

[CHR65] CHRISTIAN, J.W., Phase Transformations, in *Physical Metallurgy*, R.W. Cahn, Ed., North-Holland, 1965

[CLA63] CLARK, D., JEPSON, K.S., and LEWIS, G.I., A Study of the Titanium-Aluminum System up to 40 At.-% Aluminum, *J. Inst. Met.*, Vol 91, 1963, p. 197-203

[CLA73] CLAPP, P.C., A Localized Soft Mode Theory for Martensitic Transformations, *Phys. Status Solidi (b)*, Vol 57, 1973, p. 561-569

[COH51] COHEN, M., The Martensitic Transformation, in *Phase Transformations in Solids,* R. Smoluchowski, J.E. Mayer, and W.A. Weyl, Ed., John Wiley & Sons, 1951, p. 588-660

[COL68] COLLINGS, E.W. and SMITH, R.D., Measurement of the Principal Magnetic Susceptibility Components of Hexagonal-Close-Packed Crystals of Arbitrary Orientation, *J. Appl. Phys.*, Vol 39, 1968, p. 4462-4464

[COL69] COLLINGS, E.W. and HO, J.C., Enhancement of Superconducting Transition Temperatures in Martensitic Ti-Mo Alloys, *Phys. Lett.*, Vol 29A, 1969, p. 306-307

[COL70] COLLINGS, E.W. and HO, J.C., Magnetic Susceptibility and Low Temperature Specific Heat of High-Purity Titanium, *Phys. Rev.*, Vol 2, 1970, p. 235-244

[COL70[a]] COLLINGS, E.W. and HO, J.C., Physical Properties of Titanium Alloys, in [JAF70], p. 331-347

[COL70[b]] COLLINGS, E.W. and HO, J.C., Influence of Microstructure on the Superconductivity of a Dilute Ti-Mo Alloy, *Phys. Rev.*, Vol 1, 1970, p. 4289-4294

[COL71] COLLINGS, E.W. and HO, J.C., Magnetic Susceptibility and Low Temperature Specific Heat of Ti, Zr, and Hf, *Phys. Rev.*, Vol 4, 1971, p. 349-356

[COL71[a]] COLLINGS, E.W., HO, J.C., and JAFFEE, R.I., "Theory of Titanium Alloys for High-Temperature Strength," Technical Report AFML-TR-71-228, 1971

[COL71[b]] COLLINGS, E.W., ENDERBY, J.E., and HO, J.C., Influence of Generalized Order-Disorder on the Electron States in Five Classes of Binary Alloy System, in *Electronic Density of States* (Proc. 3rd Materials Research Symp.), Nat. Bur. Stand. (U.S.) Spec. Publ. 323, Dec 1971, p. 483-492

[COL71[c]] COLLINGS, E.W., BOYD, J.D., and HO, J.C., Enhancement of the Superconducting Transition Temperature in Deformation-Induced Structures, in *Low Temperature Physics—LT12* (Proc. 12th Int. Conf., Kyoto, Japan, Sept 1970), E. Kanda, Ed., Academic Press of Japan, 1971, p. 316

[COL71[d]] COLLINGS, E.W. and HO, J.C., Density of States of Transition Metal Binary Alloys in the Electron-to-Atom Ratio Range 4.0 to 6.0, in *Electronic Density of States* (Proc. 3rd Materials Research Symp.), Nat. Bur. Stand. (U.S.) Spec. Publ. 323, Dec 1971, p. 587-596

[COL72] COLLINGS, E.W., HO, J.C., and JAFFEE, R.I., Superconducting Transition Temperature, Lattice Instability, and Electron-to-Atom Ratio of Transition-Metal Binary Solid Solutions, *Phys. Rev.*, Vol 5, 1972, p. 4435-4449

[COL72[a]] COLLINGS, E.W., Experimental Studies of Density-of-States-Related Properties in Ti-Mo Alloys, *Proceedings, Michigan State University Summer School on Alloys,* 1972, p. 236-269

[COL73] COLLINGS, E.W., HO, J.C., and JAFFEE, R.I., Physics of Titanium Alloys—II: Fermi Density-of-States Properties and Phase Stability of Ti-Al and Ti-Mo, in [JAF73], p. 831-842

[COL73[a]] COLLINGS, E.W. and GEGEL, H.L., A Physical Basis for Solid Solution Strengthening and Phase Stability in Alloys of Titanium, *Scr. Metall.*, Vol 7, 1973, p. 437-443

[COL73[b]] COLLINGS, E.W., ENDERBY, J.E., GEGEL, H.L., and HO, J.C., Some Relationships Between the Electronic and Mechanical Properties of Ti Alloys Discussed from the Standpoint of

Fundamental Alloy Theory, in [JAF73], p. 801-814

[COL74] COLLINGS, E.W., Anomalous Electrical Resistivity, bcc Phase Stability, and Superconductivity in Titanium-Vanadium Alloys, *Phys. Rev.*, Vol 9, 1974, p. 3989-3999

[COL75] COLLINGS, E.W. and GEGEL, H.L., Ed., *Physics of Solid Solution Strengthening*, Plenum Press, 1975

[COL75^a] COLLINGS, E.W. and GEGEL, H.L., Physical Principles of Solid Solution Strengthening in Alloys, in [COL75], p. 147-182

[COL75^b] COLLINGS, E.W., Magnetic Studies of Omega-Phase Precipitation and Aging in Titanium-Vanadium Alloys, *J. Less-Common Met.*, Vol 39, 1975, p. 63-90

[COL75^c] COLLINGS, E.W., HO, J.C., and UPTON, P.E., Low-Temperature Calorimetric Studies of Superconductivity and Microstructure in Titanium-Vanadium Alloys, *J. Less-Common Met.*, Vol 42, 1975, p. 285-301

[COL76] COLLINGS, E.W. and HO, J.C., Solute-Induced Lattice Stability as it Relates to Superconductivity in Titanium-Molybdenum Alloys, *Solid State Commun.*, Vol 18, 1976, p. 1493-1495

[COL78] COLLINGS, E.W., Anomalous Electrical Resistivity and Magnetic Susceptibility Temperature Dependences in Ti-V Alloys Exhibiting Reversible Soft-Phonon-Induced Structural Inhomogeneities, in *Electrical Transport and Optical Properties of Inhomogeneous Media*. J.C. Garland and D.B. Tanner, Ed., AIP Conf. Proc. No. 40, American Institute of Physics, 1978, p. 410-415

[COL78^a] COLLINGS, E.W. and WHITE, J.J., Deformation- and Solute-Induced Microstructural Effects in the Superconductivity of Transition-Metal Alloys, in *Transition Metals 1977*, M.J.G. Lee, J.M. Perz, and E. Fawcett, Ed., The Institute of Physics, Conf. Ser. No. 39, 1978, p. 645-649

[COL78^b] COLLINGS, E.W., MARINGER, R.E., and MOBLEY, C.E., "Amorphous Glassy Metals and Microcrystalline Alloys for Aerospace Applications," Tech. Report AFML-TR-78-70 (for period Jan 1975 to Aug 1977), 1978

[COL78^c] COLLINGS, E.W., MOBLEY, C.E., MARINGER, R.E., and GEGEL, H.L., Selected Properties of Melt-Extracted Titanium-Base Polycrystalline Alloys, in *Rapidly Quenched Alloys III*, B. Cantor, Ed., Vol 1, The Metals Society (UK), 1978, p. 188-191

[COL79] COLLINGS, E.W., Magnetic Studies of Phase Equilibria in Ti-Al(30-57 at.%) Alloys, *Metall. Trans.*, Vol 10A, 1979, p. 463-474

[COL80] COLLINGS, E.W., The Metal Physics of Titanium Alloys: A Review, in [KIM80], p. 77-132

[COL81] COLLINGS, E.W., Titanium Alloy Superconductors, in *Titanium for Energy and Industrial Applications*, D. Eylon, Ed., The Metallurgical Society of AIME, 1981, p. 143-160

[COL82] COLLINGS, E.W., Magnetocrystalline Anisotropy in Monocrystalline and Textured Polycrystalline Ti-Al Alloys, in [WIL82], p. 1461-1473

[COL82^a] COLLINGS, E.W., Magnetic Investigations of Electronic Bonding and α through γ Phase Equilibria in the Titanium Aluminum System, in [WIL82], p. 1391-1402

[COL82^b] COLLINGS, E.W., Response of the Superconducting Transition Temperature and Other Physical Properties to Phase Transformation, Precipitation, and Aging in Titanium-Base Transition-Metal-Binary Alloys, in [WIL82], p. 1407-1419

[COL84] COLLINGS, E.W., *The Physical Metallurgy of Titanium Alloys*, American Society for Metals, 1984

[COL86] COLLINGS, E.W., *Applied Superconductivity, Metallurgy, and Physics of Titanium Alloys*, Vol 1, Plenum Press, 1986

[COM67] COMEY, K.R., "The Superconductivity of a Titanium-Niobium Alloy," M.S. thesis, Massachusetts Institute of Technology, Cambridge, MA, 1967

[CON67] CONRAD, H., Thermally Activated Deformation of α Titanium Below $0.4 T_M$, *Can. J. Phys.*, Vol 45, 1967, p. 581-590

[CON70] CONRAD, H. and JONES, R., Effects of Interstitial Content and Grain Size on the Mechanical Behavior of Alpha Titanium Below $0.4 T_m$, in [JAF70], p. 489-501

[CON75] CONRAD, H., DE MEESTER, B., DONER, M., and OKASAKI, K., Strengthening of Alpha Titanium by the Interstitial Solutes in C, N, and O, in [COL75], p. 1-45

[CON80] CONRAD, H. and WANG, K.K., Solution Strengthening of Titanium by Aluminum at Low Temperatures, in *Strength of Metals and Alloys* (Proc. 5th Int. Conf., Aachen, West Germany, Aug 1979), Vol 2, Pergamon Press, 1980, p. 1067-1072

[CON81] CONRAD, H., Effect of Interstitial Solutes on the Strength and Ductility of Titanium, *Prog. Mater. Sci.*, Vol 26, 1981, p. 123-403

[CON84] CONRAD, H., Plastic Flow and Fracture of Titanium at Low Temperatures, *Cryogenics*, Vol 24, 1984, p. 293-304

[COO73] COOK, H.E., On the Nature of the Omega Transformation, *Acta Metall.*, Vol 21, 1973, p. 1445-1449

[COO74] COOK, H.E., A Theory of the Omega Transformation, *Acta Metall.*, Vol 22, 1974, p. 239-247

[COO75] COOK, H.E., On First-Order Structural Phase Transitions—I. General Considerations of Pre-Transition and Nucleation Phenomena, *Acta Metall.*, Vol 23, 1975, p. 1027-1039

[COO75^a] COOK, H.E., On First-Order Structural Phase Transitions—II. The Omega Transformation, *Acta Metall.*, Vol 23, 1975, p. 1041-1054

[COU69] COURTNEY, T.H. and WULFF, J., Omega Phase Formation in Superconducting Ti Alloys, *Mater. Sci. Eng.*, Vol 4, 1969, p. 93-97

[CRA49] CRAIGHEAD, C.M., SIMMONS, O.W., MADDEX, P.J., GREENIDGE, C.T., and EASTWOOD, L.W., "Preparation and Evaluation of Titanium Alloys," Summary Report—Part III, July 30, 1949 to Air Material Command, Wright-Patterson Air Force Base, OH, on Contract No. W 33-038 ac-21229

[CRO66] CROSSLEY, F.A., Titanium-Rich End of the Titanium-Aluminum Equilibrium Diagram, *Trans. TMS-AIME*, Vol 236, 1966, p. 1174-1185

[DAV75] DAVIS, L.A., Hardness/Strength Ratio of Metallic Glasses, *Scr. Metall.*, Vol 9, 1975, p. 431-435

[DAV79] DAVIS, R., FLOWER, H.M., and WEST, D.R.F., Martensitic Transformations in Ti-Mo Alloys, *J. Mater. Sci.*, Vol 14, 1979, p. 712-722

[DAV79^a] DAVIS, R., FLOWER, H.M., and WEST, D.R.F., The Decomposition of Ti-Mo Alloy Martensites by Nucleation and Growth and Spinoidal Mechanisms, *Acta Metall.*, Vol 27, 1979, p. 1041-1052

[DAW70] DAWSON, C.W. and SASS, S.L., The As-Quenched Form of the Omega Phase in Zr-Nb Alloys, *Metall. Trans.*, Vol 1, 1970, p. 2225-2233

[DEF70] DE FONTAINE, D., Mechanical Instabilities in the b.c.c. Lattice and the Delta to Omega Phase Transformation, *Acta Metall.*, Vol 18, 1970, p. 275-279

[DEF71] DE FONTAINE, D., PATON, N.E., and WILLIAMS, J.C., The Omega Phase Transformation in Titanium Alloys as an Example of Displacement Controlled Reactions, *Acta Metall.*, Vol 19, 1971, p. 1153-1162

[DEL52] DELAZARO, D.J., HANSEN, M., RILEY, R.E., and ROSTOKER, W., Time, Temperature, Transformation Characteristics of Titanium-Molybdenum Alloys, *Trans. TMS-AIME*, Vol 194, 1952, p. 265-269

[DEL53] DELAZARO, D.J. and ROSTOKER, W., The Influence of Oxygen Contents on Transformations in a Titanium Alloy Containing 11 Per Cent Molybdenum, *Acta Metall.*, Vol 1, 1953, p. 676-677

[DEL74] DELAEY, L., KRISHNAN, R.V., TAS, H., and WARLIMONT, H., Thermoelasticity, Pseudoelasticity and the Memory Effects Associated with Martensitic Transformations, *J. Mater. Sci.*, Vol 9, 1974, p. 1521-1535

[DES65] DESORBO, W., Solute Size and Valence Effect in Some Superconducting Alloys of Transition Elements, *Phys. Rev.*, Vol 140, 1965, p. A914-A919

[DIL65] DILLAMORE, I.L. and ROBERTS, W.T., Preferred Orientation in Wrought and Annealed Metals, *Metall. Rev.*, Vol 10, 1965, p. 271-380

[DOI66] DOI, T., ISHIDA, H., and UMEZAWA, T., Study of Nb-Zr-Ti Phase Diagram (Studies of Hard Superconductor, II), *Nippon Kinzoku Gakkaishi*, Vol 30, 1966, p. 139-145

[DON75] DÖNER, M. and CONRAD, H., Deformation Mechanisms in Commercial Ti-5Al-2.5Sn(0.5 At. Pct O$_{eq}$) Alloy at Intermediate and High Temperatures (0.3-0.6 T$_m$), *Metall. Trans.*, Vol 6A, 1975, p. 853-861

[DON82] DONACHIE, M.T. Jr., Ed., *Titanium and Titanium Alloys Source Book*, American Society for Metals, 1982

[DUE80] DUERIG, T.W., TERLINDE, G.T., and WILLIAMS, J.C., The ω-Phase Reaction in Titanium Alloys, in [KIM80], p. 1299-1305

[DUE80^a] DUERIG, T.W., MIDDLETON, R.W., TERLINDE, G.T., and WILLIAMS, J.C., Stress Assisted Transformations in Ti-10V-2Fe-3Al, in [KIM80], p. 1503-1512

[DUW53] DUWEZ, P., The Martensitic Transformation Temperature in Titanium Binary Alloys, *Trans. ASM*, Vol 45, 1953, p. 934-940

[EAS75] EASTON, D.S. and KOCH, C.C., Tensile Properties of Superconducting Composite Conductors and Nb-Ti Alloys at 4.2 K, in [PER75], p. 431-444

[EAS77] EASTON, D.S. and KOCH, C.C., Mechanical Properties of Superconducting Nb-Ti Composites, *Adv. Cryo. Eng. (Materials)*, Vol 22, 1977, p. 453-462

[EMB66] EMBURY, J.D., KEH, A.S., and FISHER, R.M., Substructural Strengthening in Materials Subject to Large Plastic Strains, *Trans. TMS-AIME*, Vol 236, 1966, p. 1252-1260

[ENC61] ENCE, E. and MARGOLIN, H., Phase Relations in the Titanium-Aluminum System, *Trans. TMS-AIME*, Vol 221, 1961, p. 151-157

[EVA73] EVANS, D., "An Hypothesis Concerning the Training Phenomenon Observed in Superconducting Magnets," Report No. RL-73-092, Rutherford High Energy Lab., Chilton, England, Aug 1973

[EYL84] EYLON, D., FUJISHIRO, S., POSTANS, P.J., and FROES, F.H., High-Temperature Titanium Alloys—a Review, *J. Met.*, Vol 36, Nov 1984, p. 55-62

[EYL86] EYLON, D., COOKE, C.M., and FROES, F.H., Production of Metal Matrix Composites from Rapidly Solidified Titanium Alloy Foils, in *Titanium Rapid Solidification Technology*, F.H. Froes and D. Eylon, Ed., The Metallurgical Society, 1986, p. 311-322

[FAU82] FAULKNER, J.S., The Modern Theory of Alloys, *Prog. Mater. Sci.*, Vol 27, 1982, p. 1-187

[FED63] FEDOTOV, S.G. and BELOUSOV, O.K., The Elastic Properties of Alloys of Titanium with Molybdenum, Vanadium, and Niobium, *Sov. Phys. Dokl.*, Vol 8, 1963, p. 496-498 [transl. of *Dokl. Akad. Nauk SSSR*, Vol 150, 1963, p. 77-80]

[FED64] FEDOTOV, S.G. and BELOUSOV, O.K., Elastic Constants of the System Titanium-Niobium, *Phys. Met. Metallogr. (USSR)*, Vol 17 (No. 5), 1964, p. 83-86 [transl. of *Fiz. Met. Metalloved.*, Vol 17 (No. 5), 1964, p. 732-736]

[FED66] FEDOTOV, S.G., Dependence of the Elastic Properties of Titanium Alloys on Their Composition and Structure, in *Titanium and its Alloys*, I.I. Kornilov, Ed., Akademiya Nauk SSSR, (1963) [transl. Israel Program for Scientific Translations Ltd., IPST. Cat. No. 1454, 1966, p. 199-215]

[FED73] FEDOTOV, S.G., Peculiarities of Changes in Elastic Properties of Titanium Martensite, in [JAF73], p. 871-881

[FEE70] FEENEY, J.A. and BLACKBURN, M.J., Effect of Microstructure on the Strength, Toughness, and Stress-Corrosion Cracking Susceptibility of a Metastable Beta Titanium Alloy (Ti-11.5Mo-6Zr-4.5Sn), *Metall. Trans.*, Vol 1, 1970, p. 3309-3323

[FIS70] FISHER, E.S. and DEVER, D., Relation of the C' Elastic Modulus to Stability of b.c.c. Transition Metals, *Acta Metall.*, Vol 18, 1970, p. 265-269

[FIS70^a] FISHER, E.S. and DEVER, D., The Single Crystal Elastic Moduli of Beta-Titanium and Titanium-Chromium Alloys, in [JAF70], p. 373-381

[FIS75] FISHER, E.S., A Review of Solute Effects on the Elastic Moduli of bcc Transition Metals, in [COL75], p. 199-225

[FIS75^a] FISHER, E.S., WESTLAKE, D.G., and OCKERS, S.T., Effects of Hydrogen and Oxygen on the Elastic Moduli of Vanadium, Niobium, and Tantalum Single Crystals, *Phys. Status Solidi (a)*, Vol 28, 1975, p. 591-602

[FIX73] FIX, D.K., Titanium Precision Forgings, in [JAF73], p. 441-451

[FLO73] FLOWER, H.M., SWANN, P.R., and WEST, D.R.F., Thermally Activated Deformation in Ti 1% Si and Ti 5% Zr 1%Si, in [JAF73], p. 1143-1153

[FLO82] FLOWER, H.M., DAVIS, R., and WEST, D.R.F., Martensite Formation and Decomposition in Alloys of Titanium Containing β-Stabilizing Elements, in [WIL82], p. 1703-1715

[FRI64] FRIEDEL, J., *Dislocations*, Pergamon Press, 1964 [transl. of *Les Dislocations*, Gauthier-Villars, 1956]

[FRO54] FROST, P.D., PARRIS, W.M., HIRSCH, L.L., DOIG, J.R., and SCHWARTZ, C.M., Isothermal Transformation of Titanium-Manganese Alloys, *Trans. ASM*, Vol 46, 1954, p. 1056-1074

[FRO73] FROES, F.H., CAPENOS, J.M., and WELLS, M.G.H., Alloy Partitioning in Beta III and Effects on Aging Characteristics, in [JAF73], p. 1621-1633

[FRO85] FROES, F.H., EYLON, D., and BOMBERGER, H.B., Ed., *Titanium Technology: Present Status and Future Trends*, Titanium Development Association, 1985

[FRO86] FROES, F.H. and ROWE, R.G., Rapidly Solidified Titanium, in *Rapidly Solidified Alloys and Their Mechanical and Magnetic Properties*, B.C. Giessen, D.E. Polk, and A.I. Taub, Ed., Vol 58, Materials Research Society, 1986, p. 309-343

[GAR72] GARDE, A.M., SANTHANAM, A.T., and REED-HILL, R.E., The Significance of Dynamic Strain Aging in Titanium, *Acta Metall.*, Vol 20, 1972, p. 215-220

[GEG73] GEGEL, H.L. and HOCH, M., Thermodynamics of α-Stabilized Ti-X-Y Systems, in [JAF73], p. 923-931

[GEG73^a] GEGEL, H.L., HO, J.C., and COLLINGS, E.W., An Electronic Approach to Solid Solution Strengthening in Titanium Alloys, *Inst. Met. (London) Monogr. Rep. Ser.*, Vol 1 (PAP 116), 1973, p. 544-548

[GEG80] GEGEL, H., NADIV, S., and RAJ, R., Dynamic Effects on Flow and Fracture During Isothermal Forging of a Titanium Alloy, *Scr. Metall.*, Vol 14, 1980, p. 241-247

[GEG80^a] GEGEL, H.L., unpublished research

[GEH70] GEHLEN, P.C., The Crystallographic Structure of Ti-Al, in [JAF70], p. 349-357

[GIL76] GILMORE, C.M., FREEDMAN, M., and IMAM, M.A., The Relationship of Axial Strain Induced by Cyclic Torsion to Metal Stability and Fatigue Failure in Ti-6Al-4V, *Eng. Fract. Mech.*, Vol 8, 1976, p. 9-15

[GIT75] GITTUS, J., *Creep, Viscoelasticity, and Creep Fracture in Solids*, John Wiley & Sons, 1975

[GOD73] GODDEN, M.J. and ROBERTS, W.M., Ductility of Ti-Al-Ga Alloys, in [JAF73], p. 2207-2218

[GOL59] GOLDENSTEIN, A.W., METCALFE, A.G., and ROSTOKER, W., The Effect of Stress on the Eutectoid Decomposition of Titanium-Chromium Alloys, *Trans. ASM*, Vol 51, 1959, p. 1036-1052

[GOR73] GORYNIN, I.V., CHECHULIN, B.B., USHKOV, S.S., and BELOVA, O.S., A Study of the Nature of the Ductile-Brittle Transition in Beta Titanium Alloys, in [JAF73], p. 1109-1118

[GRI73] GRIFFITHS, P. and HAMMOND, C., Superplasticity in Large Grained Beta Titanium Alloys, in [JAF73], p. 1155-1167

[GUL71] GULLBERG, R.B., TAGGART, R., and POLONIS, D.H., On the Decomposition of the Beta Phase in Titanium Alloys, *J. Mater. Sci.*, Vol 6, 1971, p. 384-389

[GUS82] GUSEVA, L.N. and DOLINSKAYA, L.K., Metastable Phases in Quenched Titanium Alloys with Transition Elements, in [WIL82], p. 1559-1565

[GUZ66] GUZEI, L.S., SOKOLOVSKAYA, E.M., and GRIGOR'EV, A.T., Phase Diagram of the Niobium-Titanium System, *Vestn. Mosk. Univ., Khim.*, Vol 21 (No. 5), 1966, p. 79-82

[HAK61] HAKE, R.R., LESLIE, D.H., and BERLINCOURT, T.G., Electrical Resistivity, Hall Effect and Superconductivity of Some b.c.c. Titanium-Molybdenum Alloys, *J. Phys. Chem. Solids*, Vol 20, 1961, p. 177-186

[HAK63] HAKE, R.R., LESLIE, D.H., and RHODES, C.G., Giant Anisotropy in the High-Field Critical Currents of Cold-Rolled Transition Metal Alloy Superconductors, in *Low Temperature Physics—LT8* (Proc. 8th Int. Conf., London, 1962), R.O. Davies, Ed., Butterworths, 1963, p. 342-344

[HAK64] HAKE, R.R. and CAPE, J.A., Calorimetric Investigations of Localized Magnetic Moments and Superconductivity in Some Alloys of Titanium with Manganese and Cobalt, *Phys. Rev.*, Vol 135, 1964, p. A1151-1160

[HAL80] HALLAM, P. and HAMMOND, C., The Interface Phase in a Near-α Titanium Alloy, in [KIM80], p. 1435-1441

[HAM70] HAMMOND, C. and KELLY, P.M., Martensitic Transformations in Titanium Alloys, in [JAF70], p. 659-676

[HAM78] HAMMOND, C. and NUTTING, J., The Physical Metallurgy of Superalloys and Titanium Alloys, in *Forging and Properties of Aerospace Materials*, The Metals Society, 1978, p. 75-102

[HAN51] HANSEN, M., KAMEN, E.L., KESSLER, H.D., and McPHERSON, D.J., Systems Titanium-Molybdenum and Titanium-Columbium, *Trans. TMS-AIME*, Vol 191, 1951, p. 881-888

[HAN54] HANKE, H., *Prüfung Metallischer Werkstoffe*, VEB Verlag Technik, 1954

[HAN58] HANSEN, M., *Constitution of Binary Alloys*, 2nd ed. (with K. Anderko), McGraw-Hill, 1958

[HAR56] HARDY, H.K. and HEAL, T.J., Nucleation and Growth Processes in Metals and Alloys, in *The Mechanism of Phase Transformations in Metals*, Inst. of Metals Monograph and Report Ser. No. 18, 1956, p. 1-46

[HAT68] HATT, B.A. and RIVLIN, V.G., Phase Transformations in Superconducting Ti-Nb Alloys, *J. Phys. D, Appl. Phys.*, Vol 1, 1968, p. 1145-1149

[HAY65] HAYDEN, H.W., MOFFATT, W.G., and WULFF, J., *The Structure and Properties of Materials: Volume III, Mechanical Behavior*, John Wiley & Sons, 1965

[HAY85] HAYMAN, C., "Resolution of the Incommensurate Omega Phase Problem by Scattering from Multilayer Films," Thesis abstract, University of Minnesota, June 1985; see D. de Fontaine, Simple Models for the Omega Phase Transformation, *Met. Trans. A*

[HEI64] HEINIGER, F. and MÜLLER, J., Bulk Superconductivity in Dilute Hexagonal Titanium Alloys, *Phys. Rev.*, Vol 134, 1964, p. 1407-1409

[HEI74] HEIM, J.R., "Superconducting Coil Training and Instabilities Due to the Bauschinger Effect," Tech. Memo TM-334, Fermi National Accelerator Lab., 1974

[HEL71] HELLER, W., "Über den Einfluss von Germaniumzusätzen dritter Elemente auf das kritische Magnetfeld von Titan-Niob-Legierungen," D. Eng. dissertation, Universität Erlangen-Nürnberg, West Germany, 1971

[HIC68] HICKMAN, B.S., Precipitation of the Omega Phase in Titanium-Vanadium Alloys, *J. Inst. Met.*, Vol 96, 1968, p. 330-337

[HIC69] HICKMAN, B.S., The Formation of Omega Phase in Titanium and Zirconium Alloys: A Review, *J. Mater. Sci.*, Vol 4, 1969, p. 554-563

[HIC69a] HICKMAN, B.S., Omega Phase Precipitation in Alloys of Titanium with Transition Metals, *Trans. TMS-AIME*, Vol 245, 1969, p. 1329-1335

[HID80] HIDA, M., SUKEDAI, E., YOKOHARI, Y., and NAGAKAWA, A., Thermal Instability and Mechanical Properties of Beta Titanium-Molybdenum Alloys, in [KIM80], p. 1327-1334

[HIL61] HILLERT, M., A Solid-Solution Model for Inhomogeneous Systems, *Acta Metall.*, Vol 9, 1961, p. 525-535

[HIL67] HILL, R., *The Mathematical Theory of Plasticity*, Oxford University Press, 1967

[HIL73] HILLMANN, H., Entwicklung harter Supraleiter, vorzugsweise am Beispiel Nb-Ti Teil I, *Metall* (Berlin), Vol 27, 1973, p. 797-808

[HIL73a] HILLMANN, H., "Werkstoffe für supraleitende Wechselfeldmagnete mit Ummagnetisierungszeiten der Grössenordnung Sekunde," Forschungsbericht T 73-03, Vacuumschmeltze GmbH, Hanau, April 1973

[HIL76] HILLMANN, H., "Werkstoffe für dynamische beanspruchbare supraleitende Magnete," Forschungsbericht BMFT-FB T 76-13, Vacuumschmeltze, GmbH, Hanau, June 1976

[HIL81] HILLMANN, H., Personal communication

[HO69] HO, J.C., GEHLEN, P.C., COLLINGS, E.W., and JAFFEE, R.I., "Phase Stability and Solution Strengthening of Solid Solution Phase Titanium Alloys," Tech. Rep. AFML-TR-70-1, Battelle Memorial Institute, Sept 1969

[HO71] HO, J.C. and COLLINGS, E.W., Enhancement of the Superconducting Transition Temperatures of Ti-Mo(5, 7 at.%) Alloys by Mechanical Deformation, *J. Appl. Phys.*, Vol 42, 1971, p. 5144-5150

[HO72] HO, J.C. and COLLINGS, E.W., Anomalous Electrical Resistivity in Titanium-Molybdenum Alloys, *Phys. Rev. B*, Vol 6, 1972, p. 3727-3738

[HO73] HO, J.C. and COLLINGS, E.W., Calorimetric Studies of Superconductive Proximity Effects in a Two-Phase Ti-Fe(7.5 at.%) Alloy, in *Low Temperature Physics—LT13* (Proc. 13th Int. Conf., Boulder, CO, Aug 1972), K.D. Timmerhaus et al., Ed., Plenum Press, 1974, p. 403-407

[HO73a] HO, J.C. and COLLINGS, E.W., Physics of Titanium Alloys—I: Alloying and Microstructural Effects in the Superconductivity of Ti-Mo, in [JAF73], p. 815-830

[HOC71] HOCH, M. and VISWANATHAN, R., Thermodynamics of Titanium Alloys: III, The Ti-Mo System, *Metall. Trans.*, Vol 2, 1971, p. 2765-2767

[HOC73] HOCH, M., BIRLA, N.C., COLE, S.A., and GEGEL, H.L., "The Development of Heat-Resistant Titanium Alloys," Tech. Report AFML-TR-73-297, Air Force Materials Lab., Dec 1973

[HOC73a] HOCH, M., SAKAI, T., KRUPOWICZ, J.J., and DELAHANTY, M., The Titanium-Aluminum-Gallium System, in [JAF73], p. 935-949

[HOC73b] HOCH, M., Winning and Refining, a Critical Review, in [JAF73], p. 205-231

[HOC84] HOCHSTUHL, P. and OBST, B., Beta-Phase Instability in NbTi-Superconductors, in *Seventeenth International Conference on Low Temperature Physics, LT-17, Proceedings* (Karlsruhe, West Germany, Aug 1984), North-Holland, 1984, p. 1369-1370

[HOC85] HOCHSTUHL, P., "Lattice Instability and Metastable Phases in Niobium-Titanium Superconductors," *KfK Report No. 3931* (in German), Kernforschungszentrum Karlsruhe, West Germany, July 1985

[HOR73] HORIUCHI, T., MONJU, Y., TATARA, I., and NAGAI, N., Phase Transformation of Ti-30Nb-30Zr-7Ta Superconducting Alloy, *Nippon Kinzoku Gakkaishi*, Vol 37, 1973, p. 1057-1064

[HOU70] HOUGHTON, R.W., SARACHIK, M.P., and KOUVEL, J.S., Anomalous Electrical Resistivity and the Existence of Giant Magnetic Moments in Ni-Cu Alloys, *Phys. Rev. Lett.*, Vol 25, 1970, p. 238-239

[HUB73] HUBBARD, R.T.J., Low to Intermediate Temperature (Titanium) Alloys: A Critical Review, in [JAF73], p. 1887-1891

[IKE88a] IKEDA, M., KOMATSU, S.-Y., SUGIMOTO, T., and KAMEI, K., Abnormal Temperature Dependence of Resistivity in Ti-Mo Binary Alloys, *J. Jpn. Inst. Met.*, Vol 52, 1988, p. 144-149

[IKE88b] IKEDA, M., KOMATSU, S.-Y., SUGIMOTO, T., and KAMEI, K., Negative Temperature Dependence of Electrical Resistivity in Ti-Mo Binary Alloys, *Proc. Sixth World Conf. on Titanium, I* (Cannes, France, 6-9 June 1988), Les Editions de Physique, 1988, p. 313-318

[IKE88c] IKEDA, M., KOMATSU, S.-Y., SUGIMOTO, T., and KAMEI, K., Temperature Range of Formation of Athermal ω Phase in Quenched β Ti-Nb Alloys, *J. Jpn. Inst. Met.*, Vol 52, 1988, p. 1206-1211

[IKE89] IKEDA, M., KOMATSU, S.-Y., SUGIMOTO, T., and KAMEI, K., Reverse Transformation of α″ and Initial β Decomposition in Quenched Ti-Nb Binary Alloys, *J. Jpn. Inst. Met.*, Vol 53, 1989, p. 664-671

[IKE90] IKEDA, M., KOMATSU, S.-Y., SUGIMOTO, T., and KAMEI, K., Resistometric Estimation of Temperature Range of Athermal ω Phase Formation in Quenched Ti-V Alloys, *J. Jpn. Inst. Met.*, Vol 54, 1990, p. 743-751

[IMG61] IMGRAM, A.G., WILLIAMS, D.N., WOOD, R.A., OGDEN, H.R., and JAFFEE, R.I., "Metallurgical and Mechanical Characteristics of High-Purity Titanium-Base Alloys," WADC Technical Report 59-595, Part II, March 1961

[ING69] INGLESFIELD, J.E., Perturbation Theory and Alloying Behaviour, I. Formalism, *J. Phys. C, Solid State Phys.*, Vol 2, 1969, p. 1285-1292

[JAC85] JACKSON, A.G., BRODERICK, T.F., and FROES, F.H., Microstructures of Rapidly Solidified Ti-5Al-2.5Sn with Si or Ge Additions, in *Titanium Science and Technology* (Proc. 5th Int. Conf. on Titanium, Oberursel, West Germany, 1984, G. Lütjering, U. Zwicker, and W. Bunk, Ed., D.G. für Metallkd., 1985, p. 381-387

[JAE62] JAEGER, J.C., *Elastic Fracture and Flow*, 2nd ed., Methuen and Co., 1962

[JAF58] JAFFEE, R.I., The Physical Metallurgy of Titanium Alloys, *Prog. Met. Phys.*, Vol 7, 1958, p. 65-163

[JAF70] JAFFEE, R.I. and PROMISEL, N.E., Ed., *The Science, Technology and Application of Titanium* (Proc. First Int. Conf. on Titanium, London), Pergamon Press, 1970

[JAF73] JAFFEE, R.I. and BURTE, H.M., Ed., *Titanium Science and Technology* (Proc. Second Int. Conf. on Titanium, Boston), Plenum Press, 1973

[JAF73[a]] JAFFEE, R.I., Metallurgical Synthesis; A Critical Review, in [JAF73], p. 1665-1693

[JAF79] JAFFREY, D., Sources of Acoustic Emission (AE) in Metals—A Review, *Australas. Corros. Eng.*, 1979; Part 1, June, p. 9-15; Part 2, July/Aug, p. 9-15; Part 3, Sept/Oct, p. 25-32

[JAM70] JAMES, D.W. and MOON, D.M., The Martensitic Transformation in Titanium Binary Alloys and its Effect on Mechanical Properties, in [JAF70], p. 767-778

[JAY63] JAYARAMAN, A., KLEMENT, W., and KENNEDY, G.C., Solid-Solution Transitions in Titanium and Zirconium at High Pressures, *Phys. Rev.*, Vol 131, 1963, p. 644-649

[JEN65] JENSEN, M.A., MATTHIAS, B.T., and ANDRES, K., Electron Density and Electronic Properties in Noble-Metal Transition Elements, *Science*, Vol 150, 1965, p. 1448-1450

[JEP70] JEPSON, K.S., BROWN, A.R.G., and GRAY, J.A., The Effect of Cooling Rate on the Beta Transformation in Titanium-Niobium and Titanium-Aluminum Alloys, in [JAF70], p. 677-690

[KAT79] KATAHARA, K.W., MANGHNANI, M.H., and FISHER, E.S., Pressure Derivatives of the Elastic Moduli of BCC Ti-V-Cr, Nb-Mo, and Ta-W Alloys, *J. Phys. F, Met. Phys.*, Vol 9, 1979, p. 773-790

[KAT79[a]] KATAHARA, K.W., NIMALENDRAN, M., MANGHNANI, M.H., and FISHER, E.S., Elastic Moduli of Paramagnetic Chromium and Ti-V-Cr Alloys, *J. Phys. F, Met. Phys.*, Vol 9, 1979, p. 2167-2176

[KAU70] KAUFMAN, L. and BERNSTEIN, H., *Computer Calculations of Phase Diagrams*, Academic Press, 1970

[KAU73] KAUFMAN, L. and NESOR, H., Phase Stability and Equilibria as Affected by the Physical Properties and Electronic Structure of Titanium Alloys, in [JAF73], p. 773-800

[KEA74] KEATING, D.T. and LAPLACA, S.J., Neutron Diffraction Determination of the Number and Displacement of the Atoms in the Diffuse ω-Phase of $Zr_{0.8}Nb_{0.2}$, *J. Phys. Chem. Solids*, Vol 35, 1974, p. 879-891

[KEE56] KEELER, J.H. and GEISLER, A.H., Preferred Orientations in Rolled and Annealed Titanium, *Trans. TMS-AIME*, Vol 206, 1956, p. 80-90

[KIM80] KIMURA, H. and IZUMI, O., Ed., *Titanium '80: Science and Technology* (Proc. Fourth Int. Conf. on Titanium, Kyoto, Japan), The Metallurgical Society of AIME, 1980

[KIR70] KIRKPATRICK, S., VELICKY, B., and EHRENREICH, H., Paramagnetic Ni-Cu Alloys: Electronic Density of States in the Coherent-Potential Approximation, *Phys. Rev. B*, Vol 1, 1970, p. 3250-3263

[KIT56] KITTEL, C., *Introduction to Solid State Physics*, 2nd ed., John Wiley & Sons, 1956

[KIT70] KITADA, M. and DOI, T., Discontinuous Precipitation of Solution Treated Nb-40Zr-10Ti Superconducting Alloy, *Nippon Kinzoku Gakkaishi*, Vol 34, 1970, p. 361-365

[KIT70[a]] KITADA, M. and DOI, T., Precipitation and Superconducting Properties of Nb-40Zr-10Ti Alloy, *Nippon Kinzoku Gakkaishi*, Vol 34, 1970, p. 369-374

[KOC77] KOCH, C.C. and EASTON, D.S., A Review of Mechanical Behavior and Stress Effects in Hard Superconductors, *Cryogenics*, Vol 17, 1977, p. 391-413

[KON83] KONITZER, D.G., MUDDLE, B.C., and FRASER, H.L., Formation and Thermal Stability of an Oxide Dispersion in a Rapidly Solidified Ti-Er Alloy, *Scr. Metall.*, Vol 17, 1983, p. 963-966

[KON83[a]] KONITZER, D.G., MUDDLE, B.C., and FRASER, H.L., A Comparison of the Microstructure of As-Cast and Laser Surface Melted Ti-4Y, *Metall. Trans.*, Vol 14A, 1983, p. 1979-1988

[KON85] KONITZER, D.G., MUDDLE, B.C., FRASER, H.L., and KIRCHHEIM, R., Refined Dispersions of Rare Earth Oxides in Ti-Alloys Produced by Rapid Solidification Processing, in *Titanium Science and Technology* (Proc. 5th Int. Conf. on Titanium, Oberursel, West Germany, 1984), G. Lütjering, U. Zwicker, and W. Bunk, Ed., D.G. für Metallkd., 1985, p. 405-410

[KON85[a]] KONITZER, D.G. and FRASER, H.L., The Production and Thermal Stability of a Refined Dispersion of Er_2O_3 in Ti_3Al Using Rapid Solidification Processing, in *High Temperature Ordered Intermetallic Alloys* (MRS Symp. Proc.), C.C. Koch, C.T. Liu, and N.S. Stoloff, Ed., Vol 39, Materials Research Society, 1985, p. 437-442

[KOR56] KORNILOV, I.I., PYLAEVA, E.N., and VOLKOVA, M.A., Constitution Diagram of the Binary Titanium-Aluminum System, *Izv. Akad. Nauk SSSR, Otd. Khim. Nauk*, Vol 7, 1956, p. 771-778

[KOR65] KORNILOV, I.I., PYLAEVA, E.N., VOLKOVA, M.A., KRIPYAKEVICH, P.I., and MARKIV, V. YA., Phase Structure of Alloys in the Binary Ti-Al System Containing from 0-30% Al., *Dokl. Akad. Nauk SSSR*, Vol 161, 1965, p. 843-846

[KOR66] KORNILOV, I.I., Ed., *Titanium and Its Alloys*, Israel Program for Scientific Translation, Jerusalem, 1966, p. 250-261

[KOR76] KORNILOV, I.I., NARTOVA, T.T., and CHERNYSHOVA, S.P., The Titanium-Aluminum Phase Diagram in the Titanium-Rich Part, *Russ. Met. (Metally)*, Vol 6, 1976, p. 192-198

[KOR82] KORNILOV, I.I., Equilibrium Diagrams, Electronic and Crystalline Structures and Physical Properties of Titanium Alloys, in [WIL82], p. 1281-1305

[KOT70] KOT, R., KRAUSE, G., and WEISS, V., Transformation Plasticity of Titanium, in [JAF70], p. 597-605

[KOU70] KOUL, M.K. and BREEDIS, J.F., Phase Transformations in Beta Isomorphous Titanium Alloys, *Acta Metall.*, Vol 18, 1970, p. 579-588

[KOU70[a]] KOUL, M.K. and BREEDIS, J.F., Reply to Comments on "Phase Transformations in Beta Isomorphous Titanium Alloys," *Scr. Metall.*, Vol 4, 1970, p. 877-880

[KRA67] KRAMER, D. and RHODES, C.G., Omega Phase Precipitation and Superconducting Critical Transport Currents in Titanium-22 at.% Niobium (Columbium), *Trans. TMS-AIME*, Vol 239, 1967, p. 1612-1615

[KRI84] KRISHNAMURTHY, S., VOGT, R.G., EYLON, D., and FROES, F.H., Microstructures of Rapidly Solidified Titanium-Eutectoid Former Alloys, in *Rapidly Solidified Metastable Materials*, B.H. Kear and B.C. Giessen, Ed., Materials Research Society, Vol 28, Elsevier Science, 1984, p. 361-366

[KRI85] KRISHNAMURTHY, S., WEISS, I., EYLON, D., and FROES, F.H., Aging Response of a Rapidly Solidified Beta-Eutectoid Ti-9 wt.% Co Alloy, in *Strength of Metals and Alloys*, H.J. McQueen, J.-P. Bailon, J.I. Dickson, J.J. Jonas, and M.G. Akben, Ed., Vol 2, Pergamon Press, 1985, p. 1627-1632

[KRI85[a]] KRISHNAMURTHY, S., JACKSON, A.G., WEISS, I., and FROES, F.H., Aging Response of Rapidly Solidified Titanium-Tungsten Alloys with Nickel and Silicon Additions, in *Rapidly Solidified Materials*, P.W. Lee and R.S. Carbonara, Ed., American Society for Metals, 1985, p. 121-127

[KRI86] KRISHNAMURTHY, S. and FROES, F.H., Secondary Cooling Effects in Rapidly Solidified Titanium Alloys, in *Titanium Rapid Solidification Technology*, F.H. Froes and D. Eylon, Ed., The Metallurgical Society, 1986, p. 111-120

[KUB56] KUBASCHEWSKI, O. and CATTERALL, J.A., *Thermochemical Data of Alloys*, Pergamon Press, 1956

[LAH80] LAHOTI, G.D. and ALTAN, T., "Research to Develop Process Models for Producing a Dual Property Titanium Alloy Compressor Disc," Tech. Report AFWAL-TR-80-4162, Air Force Wright Aeronautical Laboratory, Aug 1979 to July 1980, p. 327

[LAR71] LARSON, F.R., ZARKADES, A., and AVERY, D.H., "Twinning and Texture Transitions in Titanium Solid-Solution Alloys," Technical Report 71-11, Accession Number DA 0A4716, Army Materials and Mechanics Research Center, June 1971

[LAR74] LARSON, F. and ZARKADES, A., "Properties of Textured Titanium Alloys," Report No. MCIC-74-20, Metals and Ceramics Information Center, Battelle, June 1974

[LAV82] LAVRENTEV, F.F., POKHIL, Yu.A., and DUDKO, P.P., The Evolution of a Defect Structure and its Relation to the Deformation Parameters in Ti-Al-Nb-Zr Alloy at 4.2 K, *Mater. Sci. Eng.*, Vol 56, 1982, p. 117-124

[LED80] LEDBETTER, H.M., Sound Velocities and Elastic-Constant Averaging for Polycrystalline Copper, *J. Phys. D, Appl. Phys.*, Vol 13, 1980, p. 1879-1884

[LED81] LEDBETTER, H.M., personal communication

[LEE87] LEE, P.J. and LARBALESTIER, D.D., Developments of Nanometer Scale Structures in Composites of Nb-Ti and Their Effect on the Superconducting Critical Current Density, *Acta Metall.*, Vol 35, 1987, p. 2523-2536

[LEE90] LEE, P.J., personal communication, March 1990

[LER60] LERINMAN, R.M., SHCHEGOLEVA, T.V., KUSHAKEVICH, S.A., and SELITSKAYA, S.I., Electron Microscope Investigation of the Structural Transformation in Titanium-Manganese and Titanium-Chromium Alloys, *Phys. Met. Metallogr. (USSR)*, Vol 9 (No. 3), 1960, p. 99 [transl. of *Fiz. Met. Metalloved.*, Vol 9, 1960, p. 437-440]

[LIP75] LIPSITT, H.A., SHECHTMAN, D., and SCHAFRICK, R.E., The Deformation and Fracture of TiAl at Elevated Temperatures, *Metall. Trans.*, Vol 6A, 1975, p. 1991-1996

[LIP80] LIPSITT, H.A., SHECHTMAN, D., and SCHAFRICK, R.E., The Deformation and Fracture of Ti_3Al at Elevated Temperatures, *Metall. Trans.*, Vol 11A, 1980, p. 1369-1375

[LOH71] LOHBERG, R., "Uber den Einfluss von Kupferzusätzen auf die supraleitenden Eigenschaften un Phasenumwandlungen von technischen Titan-Niob-Legierungen," D. Eng. dissertation, Universität Erlangen-Nürnberg, West Germany, 1971

[LOV66] LOVE, G.R. and PICKLESHEIMER, M.L., The Kinetics of Beta-Phase Decomposition in Niobium (Columbium)-Zirconium, *Trans. TMS-AIME*, Vol 236, 1966, p. 430-435

[LU85] LU, Y.Z., CHI, C.S., and WHANG, S.H., Second Phase Coarsening in Rapidly Solidified Ti-5Sn-4.5La System, in *Rapidly Quenched Metals* (Proc. 5th Int. Conf. on Rapidly Quenched Metals, Wurzburg, West Germany), S. Steeb and H. Warlimont, Ed., Vol 1, Elsevier Science, 1985, p. 949-952

[LUH68] LUHMAN, T.S., TAGGART, R., and POLONIS, D.H., A Resistance Anomaly in Beta Stabilized Ti-Cr Alloys, *Scr. Metall.*, Vol 2, 1968, p. 169-172

[LUH69] LUHMAN, T.S., TAGGART, R., and POLONIS, D.H., Correlation of Superconducting Properties with the Beta to Omega Phase Transformation in Ti-Cr Alloys, *Scr. Metall.*, Vol 3, 1969, p. 777-783

[LUH70] LUHMAN, T.S., "Superconductivity and Constitution of Titanium Base Transition Metal Alloys," Ph.D. thesis, University of Washington, Seattle, 1970

[LUH70a] LUHMAN, T.S., TAGGART, R., and POLONIS, D.H., The Effects of Step Quenching and Aging on the Superconducting Transition in Beta Stabilized Ti-Cr Alloys, *Scr. Metall.*, Vol 4, 1970, p. 611-615

[LUH71] LUHMAN, T.S., TAGGART, R., and POLONIS, D.H., The Effect of Omega Phase Reversion on the Superconducting Transition in Titanium-Base Alloys, *Scr. Metall.*, Vol 5, 1971, p. 81-86

[LUH72] LUHMAN, T.S., TAGGART, R., and POLONIS, D.H., Magnetic Hysteresis Studies of Superconducting Beta Stabilized Titanium Alloys, *Scr. Metall.*, Vol 6, 1972, p. 1055-1060

[LUK64] LUKE, C.A., TAGGART, R., and POLONIS, D.H., The Metastable Constitution of Quenched Titanium and Zirconium-Base Binary Alloys, *Trans. ASM*, Vol 57, 1964, p. 142-149

[LUT70] LÜTJERING, G. and WEISSMANN, S., Mechanical Properties of Age-Hardened Titanium-Aluminum Alloys, *Acta Metall.*, Vol 18, 1970, p. 785-795

[LUT70a] LÜTJERING, G. and WEISSMANN, S., Mechanical Properties and Structures of Age-Hardened Ti-Cu Alloys, *Metall. Trans.*, Vol 1, 1970, p. 1641-1649

[LYE66] LYE, R.G., "Band Structure and Bonding in Titanium Carbide," RIAS Technical Report to NASA; 2nd Tech. Rept. on Contr. NASw-1290, Nov 1966

[MAD74] MADHAVA, N.M. and ARMSTRONG, R.W., Discontinuous Twinning of Titanium at 4.2 K, *Metall. Trans.*, Vol 5, 1974, p. 1517-1519

[MAH82] MAHAJAN, Y., NADIV, S., and FUJISHIRO, S., Interface Phase in Ti-6242 Alloy, *Scr. Metall.*, Vol 16, 1982, p. 375-379

[MAR53] MARGOLIN, H., ENCE, E., and NIELSEN, J.P., Titanium-Nickel Phase Diagram, *Trans. TMS-AIME*, Vol 197, 1953, p. 243-247

[MAR60] MARGOLIN, H. and NIELSEN, J.P., Titanium Metallurgy, in *Modern Materials, Advances in Development and Application*, H.H. Hausner, Ed., Vol 2, Academic Press, 1960, p. 225-325

[MAR64] MARSH, D.M., Plastic Flow in Glass, *Proc. R. Soc.*, Vol A279, 1964, p. 420-435

[MAR64a] MARSH, D.M., Plastic Flow and Fracture of Glass, *Proc. R. Soc.*, Vol A282, 1964, p. 33-44

[MAR75] MARINGER, R.E., MOBLEY, C.E., and COLLINGS, E.W., Preparation and Properties of Compacts of Cast (Melt Extracted) Staple Fibers of Ti-6Al-4V, *Vacuum Metallurgy Conference Proceedings*, 1975, p. 336

[MAR76] MARINGER, R.E., MOBLEY, C.E., and COLLINGS, E.W., An Experimental Method for the Casting of Rapidly Quenched Filament and Fibers, in *Rapidly Quenched Metals*, N.J. Grant and B.C. Giessen, Ed., Section 1, MIT Press, 1976, p. 29-36

[MAR77] MARGOLIN, H., LEVINE, E., and YOUNG, M., The Interface Phase in Alpha-Beta Titanium Alloys, *Metall. Trans.*, Vol 8A, 1977, p. 373-377

[MAR78] MARGOLIN, H., HAZAVEH, F., and YAGUCHI, H., The Grain Boundary Contribution to the Bauschinger Effect, *Scr. Metall.*, Vol 12, 1978, p. 1141-1145

[MAR78a] MARINGER, R.E., MOBLEY, C.E., and COLLINGS, E.W., Preparation and Properties of Compacts of Cast Staple Fibers, in *A.I. Ch. E. Symposium: Spinning Wire from Molten Metal, Proceedings*, Vol 74, 1978, p. 111-116

[MAR79] MARINGER, R.E., COLLINGS, E.W., MOBLEY, C.E., and GEGEL, H.L., "High Specific Strength Polycrystalline Titanium-Based Alloys," U.S. Patent No. 4,149,884, April 17, 1979

[MAR83] MARTIN, P.L., MENDIRATTA, M.G., and LIPSITT, H.A., Creep Deformation of TiAl and TiAl+W Alloys, *Metall. Trans.*, Vol 14A, 1983, p. 2170-2174

[MAY53] MAYKUTH, D.J., OGDEN, H.R., and JAFFEE, R.I., Titanium-Tungsten and Titanium-Tantalum Systems, *Trans. TMS-AIME*, Vol 197, 1953, p. 231-237

[MAY61] MAYKUTH, D.J., HOLDEN, F.C., WILLIAMS, D.N., OGDEN, H.R., and JAFFEE, R.I., "The Effects of Alloying Elements in Titanium: Volume B. Physical and Chemical Properties, Deformation and Transformation Characteristics," DMIC Report 136B, Battelle Memorial Institute, May 29, 1961

[MCC71] MCCABE, K.K. and SASS, S.L., The Initial Stages of the Omega Phase Transformation in Ti-V Alloys, *Philos. Mag.*, Vol 23, 1971, p. 957-970

[MCH53] MCHARGUE, C.J., ADAIR, S.E., and HAMMOND, J.P., Effects of Solid Solution Alloying on the Cold-Rolled Texture of Titanium, *Trans. TMS-AIME*, Vol 197, 1953, p. 1199-1203

[MCQ56] MCQUILLAN, A.D. and MCQUILLAN, M.K., *Titanium*, Academic Press, 1956

[MCQ63] MCQUILLAN, M.K., Phase Transformations in Titanium and its Alloys, *Metall. Rev.*, Vol 8, 1963, p. 41-104

[MEA65] MEADEN, G.T., *Electrical Resistance of Metals*, Plenum Press, 1965

[MEN71] MENDIRATTA, M.G., LÜTJERING, G., and WEISSMANN, S., Strength Increase in Ti 35Wt Pct Nb Through Step-Aging, *Metall. Trans.*, Vol 2, 1971, p. 2599-2605

[MEN80] MENDIRATTA, M.G. and LIPSITT, H.A., Steady-State Creep Behaviour of Ti_3Al-Base Intermetallics, *J. Mater. Sci.*, Vol 15, 1980, p. 2985-2990

[MOL65] MOLCHANOVA, E.K., *Phase Diagrams of Titanium Alloys* [transl. of *Atlas*

Diagram Sostoyaniya Titanovyk Splavov], Israel Program for Scientific Translations, Jerusalem, 1965

[Moo73] MOOIJ, J.H., Electrical Conduction in Concentrated Disordered Transition Metal Alloys, *Phys. Status Solidi (a)*, Vol 17, 1973, p. 521-530

[MOR80] MORGAN, C.C. and HAMMOND, C., The Ageing Characteristics of Ti-3%Al-8%V-6%Cr-4%Mo-4%Zr (Ti-38644), in [KIM80], p. 1443-1451

[MOS70] MOSKALENKO, V.A. and PUPSOVA, B.H., "Influence of Alloying on the Plastic Deformation of Alpha-Titanium at Low Temperatures" (in Russian), report from the Physico-Technical Institute for Low Temperatures, Ukr. Acad. Sci., 1970

[MOS73] MOSS, S.C., KEATING, D.T., and AXE, J.D., Neutron Study of the Beta-to-Omega Instability in $Zr_{0.80}Nb_{0.20}$; in *Phase Transitions, 1973*, L.E. Cross, Ed., Pergamon Press, 1973, p. 179-188

[MOS80] MOSKALENKO, V.A., STARTSEV, V.I., and KOVALEVA, V.N., Low Temperature Peculiarities of Plastic Deformation in Titanium and its Alloys, in [KIM80], p. 821-830; see also *Cryogenics*, Vol 20, 1980, p. 503-508

[MYR75] MYRON, H.W., FREEMAN, A.J., and MOSS, S.C., Electronically Induced Lattice Instabilities in bcc Zr, *Solid State Commun.*, Vol 17, 1975, p. 1467-1470

[NAG84] NAGAI, K., HIRAGA, K., OGATA, T., and ISHIKAWA, K., Heat Treatments and Low Temperature Fracture Toughness of a Ti-6Al-4V Alloy, *Adv. Cryo. Eng. (Materials)*, Vol 30, 1984, p. 375-382

[NAG85] NAGAI, K., HIRAGA, K., OGATA, T., and ISHIKAWA, K., Cryogenic Temperature Mechanical Properties of β-Annealed Ti-6Al-4V, *Trans. Jpn. Inst. Met.*, Vol 26, 1985, p. 405-413

[NAM73] NAMBOOHIRI, T.K.G., MCMAHON, C.J., and HERMAN, H., Decompositions of the α-Phase in Titanium-Rich Ti-Al Alloys, *Metall. Trans.*, Vol 4, 1973, p. 1323-1331

[NAR66] NARLIKAR, A.V. and DEW-HUGHES, D., Superconductivity in Deformed Niobium Alloys, *J. Mater. Sci.*, Vol 1, 1966, p. 317-335

[NAR70] NARAYANAN, G.H., and ARCHBOLD, T.F., Comments on "Phase Transformations in Beta Isomorphous Titanium Alloys," *Scr. Metall.*, Vol 4, 1970, p. 873-876

[NAR71] NARAYANAN, G.H., LUHMAN, T.S., ARCHBOLD, T.F., TAGGART, F., and POLONIS, D.H., A Phase Separation Reaction in a Binary Titanium-Chromium Alloy, *Metallography*, Vol 4, 1971, p. 343-358

[NEA71] NEAL, D.F., BARBER, A.C., WOOLCOCK, A., and GIDLEY, J.A.F., Structure and Superconducting Properties of Nb 44 Percent Ti Wire, *Acta Metall.*, Vol 19, 1971, p. 143-149

[NIS76] NISHIHARA, T. and IGUCHI, N., The Shape-Memory Effect under Transformation Superplasticity of Ti-6Al-4V, *Nippon Kinzoku Gakkaishi*, Vol 40, 1976, p. 51-56

[NIS82] NISHIMURA, T., NISHIGAKI, M., and KUSAMICHI, H., Aging Characteristics of Beta Titanium Alloys, in [WIL82], p. 1675-1689

[NIS84] NISHIMURA, T., MIZOGUCHI, T., and ITOH, Y., Titanium Materials for Cryogenic Service, *Kobe Steel Engineering Reports*, Vol 34, 1984, p. 63-66

[OBS80] OBST, B., PATTANAYAK, D., and HOCHSTUHL, P., Structural Effects in the Superconductor NbTi65, *J. Low Temp. Phys.*, Vol 41, 1980, p. 595-609

[OGD51] OGDEN, H.R., MAYKUTH, D.J., FINLAY, W.L., and JAFFEE, R.I., Constitution of Titanium-Aluminum Alloys, *Trans. TMS-AIME*, Vol 191, 1951, p. 1150-1155

[OHT73] OHTANI, S., NISHIGAKI, M., and NISHIMURA, T., The Characteristics of Ti-Mo Beta Titanium Alloy, in [JAF73], p. 1945-1956

[OKA73] OKAZAKI, K., MOMOCHI, M., and CONRAD, H., Thermally Activated Deformation of Ti-N Alloys, in [JAF73], p. 1131-1142

[OKA80] OKA, M. and TANIGUCHI, Y., Crystallography of Stress-Induced Products in Metastable Beta Ti-Mo Alloys, in [KIM80], p. 709-715

[ONE83] O'NEAL, J.E., SASTRY, S.M.L., PENG, T.C., and TESSON, J.F., Microstructures of Rapidly Solidified Titanium Alloys, *Microstruc. Sci.*, Vol 11, 1983, p. 143-151

[ORR55] ORRELL, F.R. and FONTANA, M.G., The Titanium-Cobalt System, *Trans, ASM*, Vol 47, 1955, p. 554-564

[OSA80] OSAMURA, K., MATSUBARA, E., MIYATANI, T., MURUKAMI, Y., HORIUCHI, T., and MONJU, Y., Effect of Cold Working on Precipitation Behaviour in Superconducting Ti-Nb Alloys, *Philos. Mag. A*, Vol 42, 1980, p. 575-589

[OTS61] OTS PB 171424, DMIC Report 136 B, Battelle Memorial Institute, May 29, 1961, p. 106

[OTT70] OTTE, H.M., Mechanism of the Martensitic Transformation in Titanium and its Alloys, in [JAF70], p. 645-657

[PAR53] PARRIS, W.M., HIRSCH, L.L., and FROST, P.D., Low Temperature Aging in Titanium Alloys, *Trans. TMS-AIME*, Vol 197, 1953, p. 178-179

[PAR73] PARRIS, W.M. and RUSSELL, H.A., A New Titanium Alloy for Elevated Temperature Application, in [JAF73], p. 2219-2225

[PAS78] PASZTOR, G. and SCHMIDT, C., Dynamic Stress Effects in Technical Superconductors and the "Training" Problem of Superconducting Magnets, *J. Appl. Phys.*, Vol 49, 1978, p. 886-899

[PAU56] PAULING, L., The Electronic Structures of Metals and Alloys, in *Theory of Alloy Phases*, American Society for Metals, 1956, p. 220-242

[PAU67] PAULING, L., *The Chemical Bond*, Cornell University Press, 1967, Chap 11

[PEN80] PENNOCK, G.M., FLOWER, H.M., and WEST, D.R.F., The Control of α Precipitation by Two Step Ageing in β Ti-15%Mo, in [KIM80], p. 1343-1351

[PEN85] PENG, T.C., SASTRY, S.M.L., and O'NEAL, J.E., Rapid Solidification Processing of Titanium Alloys, in *Titanium Science and Technology* (Proc. 5th Int. Conf. on Titanium, Oberursel, West Germany, 1984), G. Lütjering, U. Zwicker, and W. Bunk, Ed., D.G. für Metallkd., 1985, p. 389-396

[PER75] PERKINS, J., Ed., *Shape-Memory Effect in Alloys*, Plenum Press, 1975

[PER75a] PERKINS, J., EDWARDS, G.R., SUCH, C.R., JOHNSON, J.M., and ALLEN, R.R., Thermomechanical Characteristics of Alloys Exhibiting Martensitic Thermoelasticity, in [PER75], p. 273-299

[PET71] PETERSON, V.C. and BUEHL, R.C., "Methods for Melting Titanium-Base Alloy," U.S. Patent No. 3,552,947, Jan 5, 1971

[PET72] PETTIFOR, D.G., Theory of the Crystal Structures of Transition Metals at Absolute Zero, in *Metallurgical Chemistry*, O. Kubaschewski, Ed., National Physical Laboratory, HMSO, 1972, p. 191-199

[PET73] PETERSON, V.C., FROES, F.H., and MALONE, R.F., Metallurgical Characteristics and Mechanical Properties of Beta III, A Heat-Treatable Titanium Alloy, in [JAF73], p. 1969-1980

[PET79] PETTIFOR, D.G., Theory of the Heats of Formation of Transition-Metal Alloys, *Phys. Rev. Lett.*, Vol 42, 1979, p. 846-850

[PFE68] PFEIFFER, I. and HILLMANN, H., Der Einfluss der Struktur auf die Supraleitungseigenschaften von NbTi50 und NbTi65, *Acta Metall.*, Vol 16, 1968, p. 1429-1439

[POL55] POLONIS, D.H. and PARR, J.G., Martensite Formation in Powders and Lump Specimens of Ti-Fe Alloys, *Trans. TMS-AIME*, Vol 203, 1955, p. 64

[POL69] POLONIS, D.H., "A Study of Phase Transformations and Superconductivity," Progress Report No. 10 and Report No. RLO-1375-18, University of Washington, Seattle, Oct 1969

[POL70] POLONIS, D.H., "A Study of Phase Transformations and Superconductivity," Annual Progress Report No. RLO-2225-T-13-6, University of Washington, Seattle, Oct 1970

[POL71] POLONIS, D.H., "A Study of Phase Transformations and Superconductivity," Annual Progress Report No. RLO-2225-T-13-9, N72-24786, University of Washington, Seattle, Oct 1971

[POS81] POSTANS, P.J. and JEAL, R.H., Titanium for Fuel Efficient Gas Turbines, in *Titanium for Energy and Industrial Applications*, D. Eylon, Ed., The Metallurgical Society of AIME, 1981, p. 183-197

[PRE74] PREKUL, A.E., RASSOKHIN, V.A., and VOLKENSHTEIN, N.V., Effect of Spin

Fluctuations on the Superconducting and Normal Properties of Ti Containing V, Nb, or Ta, *Sov. Phys. JETP*, Vol 40, 1974, p. 1134-1136 [transl. of *Zh. Eksp. Teor. Fiz.*, Vol 67, 1974, p. 2286-2292]

[PRE76] PREKUL, A.F., SHCHERBAKOV, A.S., and VOLKENSHTEIN, N.V., Resistivity and Anomalous Superconducting Transition in $Ti_{1-x}Fe_x$ Alloys ($0 < x \leq 0.2$), *Sov. J. Low Temp. Phys.*, Vol 2, 1976, p. 684-686 [transl. of *Fiz. Nizk. Temp.*, Vol 2, 1976, p. 1399-1404]

[RAC70] RACK, H.J., KALISH, D., and FIKE, K.D., Stability of As-Quenched Beta-III Titanium Alloy, *Mater. Sci. Eng.*, Vol 6, 1970, p. 181-198

[RAC75] RACK, H.J., Dynamic Strain Aging of Metastable Beta Titanium Alloys, *Scr. Metall.*, Vol 9, 1975, p. 829-831

[RAS72] RASSMANN, G. and ILLGEN, L., Zum Zusammenhang zwischen Gefüge und kritischer Stromdichte bei supraleitenden binären Titan-Niob-Legierungen, *Neue Hütte*, Vol 17, 1972, p. 321-328

[RAU56] RAUSCH, J.J., CROSSLEY, F.A., and KESSLER, H.D., Titanium-Rich Corner of the Ti-Al-V System, *Trans. AIME, J. Met.*, Vol 8, 1956, p. 211-214

[REA78] READ, D.T., Metallurgical Effects in Niobium-Titanium Alloys, *Cryogenics*, Vol 18, 1978, p. 579-584

[REE77] REED, R.P., MIKESELL, R.P., and CLARK, A.F., Low Temperature Tensile Behavior of Copper-Stabilized Niobium-Titanium Superconducting Wire, *Adv. Cryo. Eng. (Materials)*, Vol 22, 1977, p. 463-471

[REI73] REID, C.N., ROUTBORT, J.L., and MAYNARD, R.A., Elastic Constants of Ti-40 at.% Nb at 298 K, *J. Appl. Phys.*, Vol 44, 1973, p. 1398-1399

[REU66] REUTER, F.E., RALLS, K.M., and WULFE, J., Microstructure and Superconductivity of a 44.7 At.Pct Niobium (Columbium)-54.3 At.Pct Titanium Alloy Containing Oxygen, *Trans. TMS-AIME*, Vol 236, 1966, p. 1143-1151

[REZ82] REZNICHENKO, V.A., Physiochemical Principles and Research Trends in New Methods of Titanium Production, in [WIL82], p. 63-77

[REZ82a] REZNICHENKO, V.A., KARYASIN, I.A, ROGATIN, A.A., KIPRICJ, N.A., KASHKAROV, A.Z., ZHACHKIN, V.N., MENYAILOVA, G.A., and DENISOV, S.I., Preparation of High-Titanium Slags, in [WIL82], p. 79-100

[RHO75] RHODES, C.G. and WILLIAMS, J.C., Observations of an Interface Phase in the α/β Boundaries in Titanium Alloys, *Metall. Trans.*, Vol 6A, 1975, p. 1670-1671

[RHO77] RHODES, C.G. and PATON, N.E., The Influence of Microstructure on Mechanical Properties in Ti-3Al-8V-6Cr-4Mo-4Zr (Beta-C), *Metall. Trans.*, Vol 8A, 1977, p. 1749-1761

[RIZ74] RIZZUTO, C., Formation of Localized Moments in Metals; Experimental Bulk Properties, *Rep. Prog. Phys.*, Vol 37, 1974, p. 147-229

[ROL71] ROLINSKI, E.J., HOCH, M., and OBLINGER, C.J., Determination of Thermodynamic Interaction Parameters in Solid V-Ti Alloys Using the Mass Spectrometer, *Metall. Trans.*, Vol 2, 1971, p. 2613-2618

[ROL72] ROLINSKI, E.J., HOCH, M., and OBLINGER, C.J., Determination of Thermodynamic Interaction Parameters in Solid V-Ti-Cr Alloys Using the Mass Spectrometer, *Metall. Trans.*, Vol 3, 1972, p. 1413-1418

[ROM71] ROMERO, C.J., "The Effect of Microstructure and Preferred Orientation on the Mechanical Behavior of Titanium Alloy Forgings," LR 24347, Lockheed-California Co., June 1971

[RON70] RONAMI, G.N., KUZNETSOVA, S.M., FEDOTOV, S.G., and KONSTANTINOV, K.H., Determination of Phase Boundaries of Ti with V, Nb, Mo by Using the Method of Diffusing Layers, *Vestnik Mosk. Univ., Fiz.-Astron. (U.S.S.R.)*, No. 2, 1970, p. 186-189 [in Russian]

[ROS70] ROSENBERG, H.W., Titanium Alloying in Theory and Practice, in *The Science, Technology and Application of Titanium* (Proc. 1st Int. Conf. on Titanium, London), R.I. Jaffee and N.E. Promisel, Ed., Pergamon Press, 1970, p. 851-859

[ROS73] ROSENBERG, H.W., High Temperature Alloys: A Critical Review, in [JAF73], p. 2127-2140

[ROW85] ROWE, R.G. and KOCH, E.F., Rapidly Solidified Titanium Alloys Containing Cerium Sulfide and Oxysulfide Dispersions, in *Rapidly Solidified Materials*, P.W. Lee and R.S. Carbonara, Ed., American Society for Metals, 1985, p. 115-120

[ROW85a] ROWE, R.G., KOCH, E.F., BRODERICK, T.F., and FROES, F.H., Microstructural Study of Rapid Solidified Titanium Alloys Containing Erbium and Boron, in *Rapidly Solidified Materials*, P.W. Lee and R.S. Carbonara, Ed., American Society for Metals, 1985, p. 107-114

[ROW86] ROWE, R.G., SUTLIFF, J.A., and KOCH, E.F., Dispersion Modification of Ti_3Al-Nb Alloys, in *Rapidly Solidified Alloys and Their Mechanical and Magnetic Properties*, G.C. Giessen, D.E. Polk, and A.I. Taub, Ed., Vol 58, Materials Research Society, 1986, p. 359-364; see also J.A. Sutliff and R.G. Rowe, Rare Earth Oxide Dispersoid Stability and Microstructural Effects in Rapidly Solidified Ti_3Al and Ti_3Al-Nb, *ibid.*, p. 371-376

[ROW86a] ROWE, R.G., SUTLIFF, J.A., and KOCH, E.F., Comparison of Melt Spun and Consolidated Ti_3Al-Nb Alloys with and Without a Dispersoid, in *Titanium Rapid Solidification Technology*, F.H. Froes and D. Eylon, Ed., The Metallurgical Society, 1986, p. 239-248

[RYD85] RYDER, J.T. and WITZELL, W.E., Effect of Low Temperature on Fatigue and Fracture Properties of Ti-5Al-2.5Sn (ELI) for Use in Engine Components, in *Fatigue at Low Temperatures*, Proceedings (Louisville, KY, May 1983), American Society for Testing and Materials, 1985, p. 210-237

[SAG56] SAGEL, K., SCHULZ, E., and ZWICKER, U., Investigation of the Titanium-Aluminum System, *Z. Metallkd.*, Vol 46, 1956, p. 529-534

[SAK69] SAKAI, T., "Study of the Titanium-Rich Region of the Titanium-Aluminum-Gallium Ternary System," M.S. thesis, University of Cincinnati, Aug 1969

[SAL79] SALMON, D.R., *Low Temperature Data Handbook, Titanium and Titanium Alloys*, National Physical Laboratory, NPL Report QU53 (N 80 23448), May 1979

[SAL79a] SALEH, Y. and MARGOLIN, H., Bauschinger Effect During Cyclic Straining of Two Ductile Phase Alloys, *Acta Metall.*, Vol 27, 1979, p. 535-544

[SAR70] SARGENT, G.A. and CONRAD, H., Stress Relaxation and Thermally Activated Deformation in a Titanium-4 wt% Aluminum Alloy, *Scr. Metall.*, Vol 4, 1970, p. 129-133

[SAR72] SARGENT, G.A. and CONRAD, H., On the Strengthening of Titanium by Oxygen, *Scr. Metall.*, Vol 6, 1972, p. 1099-1101

[SAS69] SASS, S.L., The ω Phase in a Zr-25 at.%Ti Alloy, *Acta Metall.*, Vol 17, 1969, p. 813-820

[SAS72] SASS, S.L., The Structure and Decomposition of Zr and Ti b.c.c. Solid Solutions, *J. Less-Common Met.*, Vol 28, 1972, p. 157-173

[SAS75] SASANO, H. and KIMURA, H., Serrated Yielding in Ti-2 at.% Zr Alloy, *Nippon Kinzoku Gakkaishi*, Vol 39, 1975, p. 142-147

[SAS82] SASANO, H. and KIMURA, H., Serrated Flow in α-Titanium Alloys, in [WIL82], p. 539-551

[SAS83] SASTRY, S.M.L., PENG, T.C., MESCHTER, P.J., and O'NEAL, J.E., Rapid Solidification Processing of Titanium Alloys, *J. Met.*, Vol 35, Sept 1983, p. 21-28

[SAS84] SASTRY, S.M.L., MESCHTER, P.J., and O'NEAL, J.E., Structure and Properties of Rapidly Solidified Dispersion-Strengthened Titanium Alloys: Part I, Characterization of Dispersoid Distribution, Structure, and Chemistry, *Metall. Trans.*, Vol 15A, 1984, p. 1451-1463

[SAS84a] SASTRY, S.M.L., PENG, T.C., and BECKERMAN, L.P., Structure and Properties of Rapidly Solidified Dispersion-Strengthened Titanium Alloys: Part II, Tensile and Creep Properties, *Metall. Trans.*, Vol 15A, 1984, p. 1465-1474

[SAS85] SASTRY, S.M.L., BOWDEN, D.M., and LEDERICH, R.J., Dispersion Strengthening of Ti-Al Alloys by Rapid Solidification Technology, in *Titanium Science and Technology* (Proc. 5th Int. Conf. on Titanium, Oberursel, West Germany, 1984), G. Lütjering, U.

Zwicker, and W. Bunk, Ed., D.G. für Metallkd., 1985, p. 435-441

[SAS85ª] SASTRY, S.M.L., PENG, T.C., and O'NEAL, J.E., Design and Development of Advanced Titanium Alloys by Rapid Solidification Technology, in *Titanium Science and Technology* (Proc. 5th Int. Conf. on Titanium, Oberursel, West Germany, 1984), G. Lütjering, U. Zwicker, and W. Bunk, Ed., D.G. für Metallkd., 1985, p. 397-404

[SAS85ᵇ] SASTRY, S.M., PENG, T.C., and O'NEAL, J.E., Rapid Solidification and Powder Metallurgical Processing of Titanium Alloys, in *Modern Developments in Powder Metallurgy*, Vol 16, *Ferrous and Nonferrous Materials* (Toronto, Canada, June 1984), E.N. Aqua and C.I. Whitman, Ed., Metal Powder Industries Federation, 1985, p. 577-606

[SAT59] SATO, T., HUANG, Y., and KONDO, Y., Equilibrium Diagram of the System Ti-Al, *Trans. Jpn. Inst. Met.*, Vol 1, 1959, p. 456-460

[SAV84] SAVAGE, S.J. and FROES, F.H., Production of Rapidly Solidified Metals and Alloys, *J. Met.*, Vol 36, 1984, p. 20-33

[SCH76] SCHMIDT, C., Investigation of the Training Problem of Superconducting Magnets, *Appl. Phys. Lett.*, Vol 28, 1976, p. 463-465

[SCH77] SCHMIDT, C., Effect of Dynamic Stress on Commercial Superconductors: A Test Facility, *Rev. Sci. Instrum.*, Vol 48, 1977, p. 597-601

[SCH77ª] SCHMIDT, C., Superconductors under Dynamic Mechanical Stress, *IEEE Trans. Magn.*, MAG-13, 1977, p. 116-119

[SEA75] SEAGLE, S.R., HALL, G.S., and BOMBERGER, H.B., High Temperature Properties of Ti-6Al-2Sn-4Zr-2Mo-0.09Si, *Met. Eng. Quart.*, Feb 1975, p. 48-54

[SHI77] SHIBATA, M. and ONO, K., On the Minimization of Strain Energy in the Martensitic Transformation of Titanium, *Acta Metall.*, Vol 25, 1977, p. 35-42

[SHU69] SHUNK, F.A., *Constitution of Binary Alloys, Second Supplement*, McGraw-Hill, 1969

[SHU85] SHULL, R.D., MCALISTER, A.J., and RENO, R.C., Phase Equilibria in the Titanium-Aluminum System, in *Titanium Science and Technology* (Proc. 5th Int. Conf. on Titanium, Oberursel, West Germany, 1984), G. Lütjering, U. Zwicker, and W. Bunk, Ed., D.G. für Metallkd., 1985, p. 1495

[SIL58] SILCOCK, J.M., An X-Ray Examination of the ω Phase in TiV, TiMo, and TiCr Alloys, *Acta Metall.*, Vol 6, 1958, p. 481-492

[SIN75] SINHA, S.K. and HARMON, B.N., Electronically Driven Lattice Instabilities, *Phys. Rev. Lett.*, Vol 35, 1975, p. 1515-1518

[SIN76] SINHA, S.K. and HARMON, B.N., Phonon Anomalies in d-Band Metals and their Relationship to Superconductivity, in *Superconductivity in d- and f-Band Metals*, D.H. Douglass, Ed., Plenum Press, 1976, p. 269-296

[SMI76] SMITH, H.G., WAKABAYASHI, N., and MOSTOLLER, M., Phonon Anomalies in Transition Metals, Alloys and Compounds, in *Superconductivity in d- and f-Band Metals*, D.H. Douglass, Ed., Plenum Press, 1976, p. 223-249

[SNO84] SNOW, D.B., Structure and Mechanical Properties of Laser-Consolidated Ti-6Al-4V and Ti-6Al-2Sn-4Zr-6Mo with Rare Earth Element Additions, in *Laser Processing of Materials*, K. Mukherjee and J. Mazumder, Ed., The Metallurgical Society, 1984, p. 83-89

[SOE69] SOENO, K. and KURODA, T., Kinetics of Beta-Phase Decomposition and the Precipitation of Alpha-Zirconium in Nb-40at.%Zr-10at.%Ti Superconducting Alloy, *Nippon Kinzoku Gakkaishi*, Vol 33, 1969, p. 791-795

[SPA57] SPARKS, C.J. Jr., MCHARGUE, C.J., and HAMMOND, J.P., Effects of Aluminum on the Cold-Rolled Textures of Titanium, *Trans. TMS-AIME*, Vol 209, 1957, p. 49-50

[SPA58] SPACHNER, S.A., Comparison of Structure of Omega Transition Phase in Three Titanium Alloys, *Trans. TMS-AIME*, Vol 212, 1958, p. 57-59

[STE68] STERN, E.A., Requirements for a Theory of Disordered Alloys, in *Energy Bands in Metals and Alloys*, L.H. Bennett and J.T. Waber, Ed., Metallurgical Society Conferences Vol 45, Gordon and Breach, 1968, p. 151-173 (see also *Phys. Rev.*, Vol 144, 1966, p. 545; *Physics*, Vol 1, 1965, p. 255)

[STE75] STERN, E.A., Application of Alloy Physics to Solution Strengthening, in [COL75], p. 183-197

[STE76] STEELE, R.K. and MCEVILY, A.J., The High-Cycle Fatigue Behavior of Ti-6Al-4V Alloy, *Eng. Fract. Mech.*, Vol 8, 1976, p. 31-37

[STO78] STOCKS, G.M., TEMMERMAN, W.M., and GYORFFY, B.L., Complete Solution of the Korringa-Kohn-Rostoker Coherent-Potential Approximation Equations: Cu-Ni Alloys, *Phys. Rev. Lett.*, Vol 41, 1978, p. 339-341

[STR68] STRONGIN, M., KAMMERER, O.F., CROW, J.E., PARKS, R.D., DOUGLASS, D.H. Jr., and JENSEN, M.A., Enhanced Superconductivity in Layered Metallic Films, *Phys. Rev. Lett.*, Vol 21, 1968, p. 1320-1321

[STR82] *Structural Alloys Handbook: 1982 Supplement*, produced and published by Battelle's Columbus Laboratories, Columbus, OH

[SUD68] SUDEREVA, S.V., BUYNOV, N.N., and RAKIN, V.G., Electron Microscopic and X-ray Diffraction Analysis of the Quenched Alloy Ti-25 at.% Nb, *Phys. Met. Metallogr. (USSR)*, Vol 26 (No. 5), 1968, p. 14-20 [transl. of *Fiz. Met. Metalloved.*, Vol 26, 1968, p. 781-788]

[SUT86] SUTLIFF, J.A. and ROWE, R.G., Rare Earth Oxide Dispersoid Stability and Microstructural Effects in Rapidly Solidified Ti₃Al and Ti₃Al-Nb, in *Rapidly Solidified Alloys and Their Mechanical and Magnetic Properties*, B.C. Giessen, D.E. Polk, and A.I. Taub, Ed., Vol 58, Materials Research Society, 1986, p. 371-376

[SUZ75] SUZUKI, T. and WUTTIG, M., Analogy Between Spinodal Decomposition and Martensitic Transformation, *Acta Metall.*, Vol 23, 1975, p. 1069-1076

[SWA58] SWANN, P.R. and PARR, J.G., Phase Transformations in Titanium-Rich Alloys of Titanium and Cobalt, *Trans. TMS-AIME*, Vol 212, 1958, p. 276-279

[SWA73] SWANSON, M.L. and QUENNEVILLE, A.F., The Effect of Compressional Plastic Deformation on the Superconducting Transition Temperature of Tin Alloys, *Scr. Metall.*, Vol 7, 1973, p. 1011-1017

[SWA81] SWARTZENDRUBER, L.T., BENNETT, L.H., IVES, L.K., and SHULL, R.D., The Ti-Al Phase Diagram: The α-α_2 Phase Boundary, *Mater. Sci. Eng.*, Vol 51, 1981, p. 1-9

[TAN77] TANNER, L.E. and RAY, R., Physical Properties of Ti$_{50}$-Be$_{40}$-Zr$_{10}$ Glass, *Scr. Metall.*, Vol 11, 1977, p. 783-789

[TER80] TERLINDE, G.T., DUERIG, T.W., and WILLIAMS, J.C., The Effect of Heat Treatment on Microstructure and Tensile Properties of Ti-10V-2Fe-3Al, in [KIM80], p. 1571-1581

[TER82] TERAUCHI, S., MATSUMOTO, H., SUGIMOTO, T., and KAMEI, K., Investigation of the Titanium-Molybdenum Binary Phase Diagram, in [WIL82], p. 1335-1349

[THE82] THEODORSKI, G. and KOSS, D.A., The Cyclic Stress-Strain Response of Age-Hardenable Beta Titanium Alloys, in [WIL82], p. 553-567

[THO73] THORNBURG, D.R. and PIEHLER, H.R., Cold-Rolling Texture Development in Titanium and Titanium-Aluminum Alloys, in [JAF73], p. 1187-1197

[TIM56] TIMOSHENKO, S., *Strength of Materials, Part II, Advanced Theory and Problems*, 3rd ed., Robert E. Kreiger Publishing Co., 1956

[TON75] TONG, H.C. and WAYMAN, C.M., Thermodynamic Considerations of "Solid State Engines" Based on Thermoelastic Martensitic Transformations and the Shape Memory Effect, *Metall. Trans.*, Vol 6A, 1975, p. 29-32

[TOR80] TORAN, J.R. and BIEDERMAN, R.R., Phase Transformation Study of Ti-10V-2Fe-3Al, in [KIM80], p. 1491-1501

[TRE82] TRENOGINA, T.L. and LERINMAN, R.M., Decomposition of the Martensite in Two-Phase Titanium Alloys, in [WIL82], p. 1623-1632

[TWE64] TWEEDALE, J.G., *The Mechanical Properties of Metals, Assessment and Significance*, American Elsevier Publishing Co., 1964

[TYS75] TYSON, W.R., Solution Hardening by Interstitials in Close-Packed Metals, in [COL75], p. 47-77

[VAN64] VANOSTENBURG, D.O., LAM, D.J., TRAPP, H.D., PRACHT, D.W., and ROWLAND, T.J., Nuclear Magnetic Resonance and Magnetic Susceptibilities of Alloys of V with Al, *Phys. Rev.*, Vol 135-A, 1964, p. 455-459

[VET68] VETRANO, J.B., GUTHRIE, G.L., KISSINGER, H.E., BRIMHALL, J.L., and MASTEL, B., Superconductivity Critical Current Densities in Ti-V Alloys, *J. Appl. Phys.*, Vol 39, 1968, p. 2524-2528

[VIG82] VIGIER, G., MERLIN, J., and GOBIN, P.F., Decomposition of the Solid Solution in the All-Beta βIII, in [WIL82], p. 1691-1701

[VOD80] VODOLAZSKY, V.P., KATAJA, V.K., ALEKSANDROV, V.K., KAGANOVICH, A.Z., VOLKOV, V.A., and NEFED'EV, E.I., Study of Titanium Alloys' Deformation Process by Acoustic Emission, in [KIM80], p. 841-848

[VOG86] VOGT, R.G., EYLON, D., and FROES, F.H., Effect of Er, Si, and W Additions on Powder Metallurgy High Temperature Titanium Alloys, in *Titanium Rapid Solidification Technology*, F.H. Froes and D. Eylon, Ed., The Metallurgical Society, 1986, p. 177-194

[WES80] WEST, A.W. and LARBALESTIER, D.C., Transmission Electron Microscopy of Commercial Filamentary Nb-Ti Superconducting Composites, *Adv. Cryo. Eng. (Materials)*, Vol 26, 1980, p. 471-478

[WES82] WEST, A.W. and LARBALESTIER, D.C., α-Ti Precipitation in Niobium-Titanium Alloys, *Adv. Cryo. Eng. (Materials)*, Vol 28, 1982, p. 337-344

[WES83] WEST, A.W. and LARBALESTIER, D.C., α-Ti Precipitates in High Current Density Multifilamentary Niobium Titanium Composites, *IEEE Trans. Magn.*, Mag-19, 1983, p. 548-551

[WHA84] WHANG, S.H., Rapidly Solidified Ti Alloys Containing Novel Additives, *J. Met.*, Vol 36, April 1984, p. 34-40

[WHA85] WHANG, S.H., LU, Y.Z., and KIM, Y.W., Microstructure and Age Hardening of Rapidly Quenched Ti-Zr-Si Alloys, *J. Mater. Sci. Lett.*, Vol 4, 1985, p. 883-887

[WHA86] WHANG, S.H., Rapidly Solidified Titanium Alloys for High-Temperature Applications, *J. Mater. Sci.*, Vol 21, 1986, p. 2224-2238

[WHA87] WHANG, S.H., Occurrence of Metastable Phases in Binary Ti-Rich Alloys Quenched From the Melt, in *Undercooled Alloy Phases*, E.W. Collings and C.C. Koch, Ed., The Metallurgical Society, 1987, p. 163-183

[WHI76] WHITE, J.J. and COLLINGS, E.W., Analysis of Calorimetrically Observed Superconducting Transition Temperature Enhancement in Ti-Mo(5 at.%)-Based Alloys, *Magnetism and Magnetic Materials—1976* (Joint MMM-Intermag Conference, Pittsburgh), AIP Conference Proceedings No. 34, p. 75-77

[WIL69] WILLIAMS, J.C. and BLACKBURN, M.J., The Structure, Mechanical Properties and Deformation Behavior of Ti-Al and Ti-Al-X Alloys, *Proceedings of the Third Bolton Landing Conference*, Aug 27, 1969

[WIL71] WILLIAMS, J.C., HICKMAN, B.S., and LESLIE, D.H., The Effect of Ternary Additions on the Decomposition of Metastable β-Phase Titanium Alloys, *Metall. Trans.*, Vol 2, 1971, p. 477-484

[WIL73] WILLIAMS, J.C., Kinetics and Phase Transformations: A Critical Review, in [JAF73], p. 1433-1494

[WIL75] WILLBRAND, J. and SCHLUMP, W., Einfluss von Ausscheidungsdichte und Teilchengrösse auf die Stromtragfähigkeit von NbTi-Supraleitern, *Z. Metallkd.*, Vol 66, 1975, p. 714-719

[WIL75a] WILLBRAND, J., ARNDT, R., EBELING, R., and MOHS, R., "Optimierung supraleitender NbTi Legierungen," Forschungsbericht T 75-35, Fried. Krupp GmbH, Krupp Forschungsinstitut, Essen, Nov 1975

[WIL76] WILLIAMS, J.C., Precipitation in Titanium-Base Alloys, in *Precipitation Processes in Solids*, K.C. Russell and H.I. Aaronson, Ed., The Metallurgical Society of AIME, 1978, p. 191-224

[WIL82] WILLIAMS, J.C. and BELOV, A.F., Ed., *Titanium and Titanium Alloys, Scientific and Technological Aspects* (Proc. Third Int. Conf. on Titanium, Moscow), Plenum Press, 1982

[WIL82a] WILLIAMS, J.C., FROES, F.H. and FUJISHIRO, S., Microstructure and Properties of the Alloy Ti-11.5Mo-6Zr-4.5Sn (Beta III), in [WIL82], p. 1421-1436

[WIL82b] WILLIAMS, J.C., Phase Transformations in Ti Alloys—A Review of Recent Developments, in [WIL82], p. 1477-1498

[WIN73] WINSTONE, M.R., RAWLINGS, R.D., and WEST, D.R.F., Dynamic Strain Aging in Some Titanium-Silicon Alloys, *J. Less-Common Met.*, Vol 31, 1973, p. 143-150

[WIT73] WITCOMB, M.J. and DEW-HUGHES, D., Superconductivity of Heat-Treated Nb-65at.%Ti Alloy, *J. Mater. Sci.*, Vol 8, 1973, p. 1383-1400

[WOO72] WOOD, R.A., *Titanium Alloys Handbook*, Metals and Ceramics Information Center, Battelle, Publication No. MCIC-HB-02, Dec 1972

[YAK61] YAKYMYSHYN, F.W., PURDY, G.R., TAGGART, R., and PARR, J.G., The Relationship Between the Constitution and Mechanical Properties of Titanium-Rich Alloys of Titanium and Cobalt, *Trans. ASM*, Vol 53, 1961, p. 283-294

[YAO61] YAO, Y.L., Magnetic Susceptibilities of Titanium-Rich Titanium-Aluminum Alloys, *Trans. ASM*, Vol 54, 1961, p. 241-246

[YIN83] YIN, C.-A., DÖNER, M., and CONRAD, H., Deformation Kinetics of Commercial Ti-50A (0.5 At. Pct. O_{eq}) at Low Temperatures (T < $0.3T_m$), *Metall. Trans.*, Vol 14A, 1983, p. 2545-2555

[YOS56] YOSHIDA, S. and TSUYA, Y., The Temperature Dependence of the Electrical Resistivity of the β-Phase Titanium-Molybdenum Alloys, *J. Phys. Soc. Jpn.*, Vol 11, 1956, p. 1206-1207

[ZEN48] ZENER, C., *Elasticity and Anelasticity of Metals*, University of Chicago Press, 1948

[ZEY71] ZEYFANG, R. and CONRAD, H., Deformation Dynamics of a B.C.C. Titanium Alloy (15.2 at.%Mo) Below 650 K (~$0.4 T_m$), *Acta Metall.*, Vol 19, 1971, p. 985-990

[ZUB79] ZUBECK, R.B., BARBEE, T.W. Jr., GEBALLE, T.H., and CHILTON, F., Effects of Plastic Deformation on the Superconducting Specific-Heat Transition of Niobium, *J. Appl. Phys.*, Vol 50, 1979, p. 6423-6436

[ZWI70] ZWICKER, U., LÖHBERG, R., and HELLER, W., Metallkundliche Probleme und Supraleitung bei Legierungen auf Basis Titan-Niob, die als Werkstoffe für die Herstellung von supraleitenden Magneten dienen können, *Z. Metallkd.*, Vol 61, 1970, p. 836-847

[ZWI74] ZWICKER, U., *Titan und Titanlegierungen*, Springer-Verlag, 1974

Titanium Data Sheets

High-Purity Ti

Common Name: Iodide or electrolytic Ti
UNS Number: Unassigned

High-purity titanium has one-half the oxygen content as commercially pure (ASTM grade 1) titanium. High-purity titanium is produced from a special grade of titanium sponge (<0.1 wt% oxygen).

Chemistry. High-purity titanium generally has oxygen contents of about 500 ppm or less. Electrolytic methods are being used to produce a very hig- purity titanium sponge on a pilot-plant scale.

Product Forms. Pure titanium is supplied as single crystals, crystal bars, and polycrystalline wrought forms.

Density. α phase: 4.51 g/cm^3 (0.163 lb/in.^{3}) at 20 °C. β phase: 4.35 g/cm^3 at 885 °C (1625 °F) from indirect measurements (0.157 lb/in.^{3})

Product Conditions. Cold worked, stress relieved, annealed

Applications. Typical uses include experimentation and research, commercial applications requiring minimum interstitial alloying elements (oxygen, nitrogen, carbon, and hydrogen).

Specifications and Compositions. (see table)

High-purity Ti compositions

Specification designation	Form(s)	C	H	N	O	Cu	Fe	Mn	Sn	Si	Zr	Cl	Mg	Ti bal
Typical %, electrolytic Ti	...	0.008	...	0.004	0.037	0.007	0.009	<0.001	<0.020	0.002	<.001	0.073	<0.001	99.837
Typical %, Iodide Ti	Crystal bar	0.001	...	0.002	0.03-0.06	<0.001	0.002	0.003	0.001	0.005	0.050	0.002	0.003	99.87
IMI 110, max wt%	Sheet	0.02	0.004	0.005	0.05	0.02	0.025	0.05	0.02	0.02	...	...	...	...

Summary of Physical Properties

Atomic Properties

Atomic Number: 22
Atomic Weight: 47.88 on carbon scale (47.88 g/mole)
Electron Structure: $1s^2, 2s^2, 2p^6, 3s^2, 3p^6, 3d^2, 4s^2$

Ionization potentials

Ion	Outer electrons	Ionization potential, V
Ti$^+$	3d^{2}4s	6.83
Ti^{++}	3d^2	13.63
Ti^{+++}	3d	28.14

Electrical/Magnetic Properties

Electrical/Electronic Properties

Electrical resistivity(a)	
High purity	42 μΩ · cm
Commercial purity	55 μΩ · cm
Superconductivity, critical temperature	0.37 to 0.56 K
Hall coefficient	+1.82 ± 0.2 × 10^{-13} volt · cm/amp/oersted
Work function	4.17 eV
Band width (from soft X-ray spectra)	6.0 ± 0.5 eV
Electronic specific heat	8.00 × 10^{-4} cal/°C · mole

(a) Room-temperature propertites iodide titanium or relatively pure commerical titanium with a minimum yield strength of 275 to 380 mPa (40 to 55 ksi). Source: Lepkowski and Holladay, TML ReportNo. 73, Battelle, 1957. See subsequent sections of this datasheet for additional data on electrical resistivity and other physical properties.

Magnetic Properties

Magneto-resistance coefficient	6.6 × 10^{-13}/oersted^2
Magnetic susceptibility	3.17 × 10^{-6}(±0.03 × 10^{-6})emμgram
Relative magnetic permeability	1.00005

Source: Lepkowski and Holladay, TML Report No. 73, Battelle, 1957

Mass Properties

Compressibility. 0.80 × 10^{-6} cm^2/kgf at room temperature

Density. α phase, 4.507 g/cm^3 at 20 °C
β phase, 4.35 g/cm^3 at 885 °C

Lattice parameters of α: c = 0.468 nm; a = 0.295 nm

Lattice parameters of β: a = 0.332 nm at 900 °C

Sonic velocities in pure titanium

Sample description	Velocity of sound, km/s (ft/s), for:			Reference
	Longitudinal (Compressive) waves	Transverse (Shear) waves	Surface waves	
Unalloyed Ti (density unspecified)	~5(a) (16 390)	...	...	Fusfeld and Gilbert, *Phy. Rev.*, Series 2, Vol 77, 1950, p 302-303
Iodide titanium (crystal bar, 4.5 g/cm^3)	6.05 (19 850)	3.1 (10 170)	...	Fisher and Manghnani, *J. Phys. Chem. Solids*, Vol 32, 1971, p 657-667
Commercially pure titanium (4.5 g/cm^3)	6.10 (20 000)	3.12 (10 235)	2.79 (9150)	9th Edition *Metals Handbook*, Vol 17, p 235

(a) As reported by Lepkowski and Holladay (TML Report No. 73, Battelle, 1957) without specification of waveform type

Nuclear Properties

Thermal Neutron Cross Section. The absorption cross section for thermal neutrons is about 5.6 ± 0.4 barns (0.070 cm^2/g) for the normally occurring isotopic mixture of metallic titanium. Scattering cross sections for natural isotopes are:

Stable isotopes of titanium

Isotope	Natural abundance, %	Cross section, b(a)
^{46}Ti	7.95	0.6
^{47}Ti	7.75	1.6
^{48}Ti	73.43	8.0
^{49}Ti	5.51	1.8
^{50}Ti	5.34	0.2

(a) b, barns. Source: *The Reactor Handbook*, Vol I, Atomic Energy Commission, AECD 3645, 1955, p 352

Surface Properties

Surface Tension. 1.2 N/m (1200 dynes/cm)

Work Function. The work function is the energy necessary to remove an electron from the surface and is high if the cohesive energy is high. The work function is also affected by oxidation or other surface layers. Reported values range from 3.95 ± 0.02 eV to 4.45 ± 0.05 eV (W. Lepkowski and J. Holladay, TML Report No. 73, Battelle, 1957).

Total Hemispherical Emittance. 0.30 at 710 °C

Optical constants and reflectance of titanium

Wavelength, μm	Refractive index (n)	Absorption coefficient (k)	Reflectance (R) calculated, % (a)	Reflectance (R) measured, % (a)
436	2.04	2.85	53.0	51.3
546	2.53	3.33	57.0	55.8
578	2.64	3.42	57.0	56.5
650	3.03	3.65	59.0	58.5

(a) R = reflectance for normal incidence. Source: W.J. Lepkowski and J.W. Holladay, TML Report No. 73, Battelle, 1957, p 67

Thermal Properties

More detailed assessments of the following thermal properties are given in subsequent sections of this datasheet.

Melting Temperature. 1670 ± 5 °C

Boiling Point. 3260 °C (estimated)

Vapor Pressure. From 1587 to 1698 K:

$$\log P = 7.7960 - \frac{24\,644}{T} - 0.000227\,T$$

where P is in Pa and T is in K

Phase Transformation Temperature. α to β, 882 ± 2 °C (1620 °F)

Coefficient of Thermal Expansion. At 20 °C, 8.41×10^{-6}/°C

Specific Heat. 523 J/kg · °C at room temperature

Latent Heat of Fusion. ~292 kJ/kg

Latent Heat of Transformation. ~85 kJ/kg (estimated)

Latent Heat of Vaporization. 9.83 MJ/kg (estimated)

Thermal Conductivity. ~17 W/m · °C at room temperature (see the section "Transport Properties" for a more detailed assessment).

Phases and Structures

Transition Temperatures

Beta Transus. At atmospheric pressure, titanium has the α(cph) structure at low temperature and transforms at 882 ± 2 °C (1620 ± 3.5 °F) to the high-temperature β(bcc) structure. The assessed transformation temperature of 882 ± 2 °C is based on phase diagram data for binary Ti systems (see figure for Ti-O system) as well as on direct measurements (see table). When alloys of high purity

Pure titanium: Effect of oxygen on α – β phase boundaries

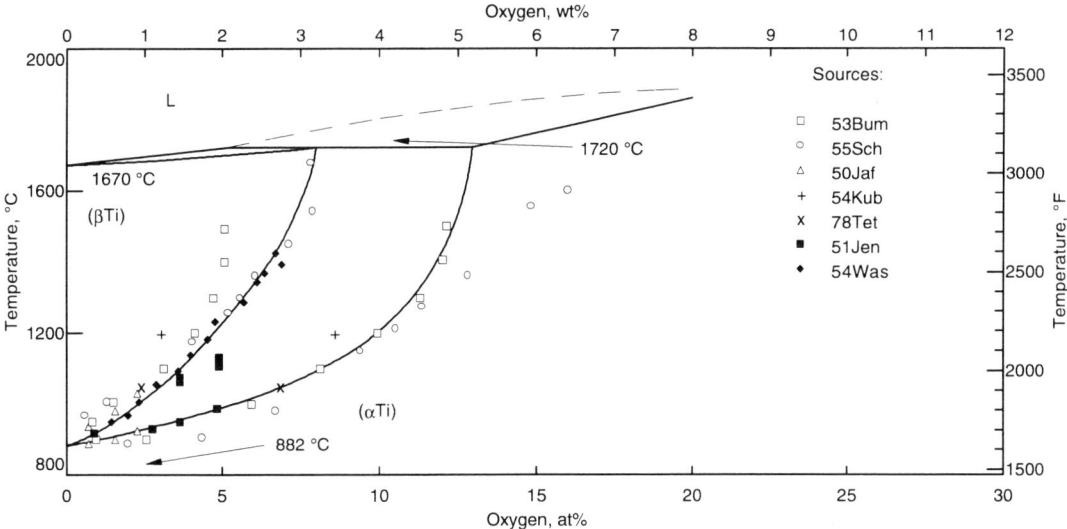

See reference for full citations of data sources.
Source: J. Murray, Ed., *Phase Diagrams of Binary Titanium Alloys*, ASM International, 1987, p 214

Effect of iron on M_s and reversion ($β_σ$) temperatures

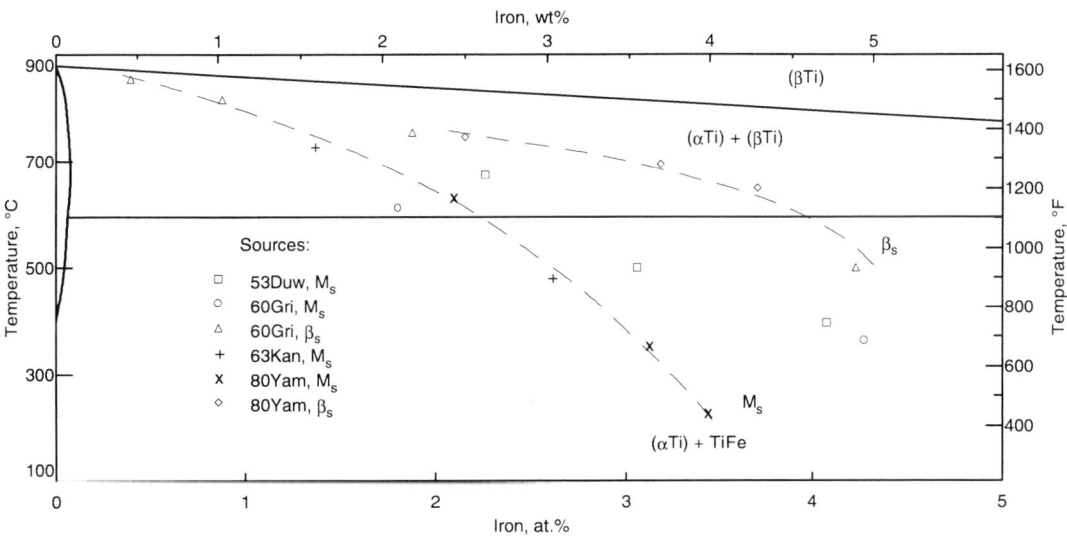

Full citations of data sources are given in the figure reference.
Source: J. Murray, Ed., *Phase Diagrams of Binary Titanium Alloys*, ASM International, 1987, p 104

are used, (αTi)/(βTi) phase boundary data are consistently found to extrapolate to 882 °C at pure Ti. The effect of O or N contamination is to raise the temperature of the transformation. Values as high as 900 °C have been reported in studies of binary systems, and the existence of an appreciable two phase αTi + βTi region in apparently pure Ti was due to contamination by O and/or N.

Martensite-Start Temperature. Above a critical cooling rate of 3000 °C/s (5400 °F/s), the (βTi) ↔ (αTi) transition is diffusionless (i.e., either martensitic or massive in mechanism). For unalloyed titanium, the martensite-start (M_s) temperature is influenced by oxygen and iron contents (see table and figure).

Martensite-start temperatures of Ti-O alloys

Composition, at.%	M_s temperature	
	°C	°F
0	802 ± 10	1475 ± 18
0.6	836	
0.9	861	
1.0	901(a)	
1.2	849	
1.5	832	

(a) The maximum in M_s at about 1 at.% O was not explained.
Source: M. Cormier and F. Claisse, "Beta-Alpha Transformation in Ti and Ti-O Alloys," *J. Less-Common Met.*, Vol 34, 1974, p 181-189

Lattice Parameters

Lattice Parameters of Pure α Titanium. Pure titanium has a c:a ratio of 1.587, which is significantly lower than that for other hexagonal metals such as magnesium, zinc, and cadmium. Lattice parameters are influenced by interstitial impurities (such as oxygen, principally) and temperature (see figures). Room-temperature lattice parameters (see table) include measurements on exceptionally pure titanium (90 to 4000 at.ppm O) by M. Dechamps et al., Scr. Metall., Vol 11, 1977, p 941-945 (77 Dec). Results were extrapolated to zero content and are accepted.

Lattice Parameter of Quenched ω Titanium. $a = 0.2813$ nm and $c = 0.4625$ nm.

Experimental determinations of the βTi ↔ αTi Transformation Temperature

Reference(a)	Transformation temperature (IPTS-68),°C	Experimental technique
[51Mcq]	883	Hydrogen pressure
[51Duw]	882 ± 4	Cooling curve
[52Kot]	881	Drop calorimetry
[57Sco]	883 ± 2	Adiabatic calorimetry
[67Rud]	882	DTA
[74Cor]	~882	Electrical resistivity
[76Etc]	882 ± 2	Dilatometry
[78Cez](b)	893 ± 6	Pulse heating and resistivity
[84Mca]	883 ± 2	DTA (cooling, heating)
Assessed	882 ± 2	

(a) Full citations given in *Phase Diagrams of Binary Titanium Alloys*, ASM International, 1987, p 1-4. (b) The transformation temperature value reported by [78Cez] is considered incorrect due to the use of pulse heating (2500 to 2700 °C/s).

Room-Temperature Lattice Parameters of αTi

Reference(a)	Lattice parameters, nm a	c	Temperature, °C
[49Cla]	0.29504	0.46833	25 ± 2
[49Gre]	0.29450	0.46845	25
[50Fin]	0.29504	0.46834	RT
[55Sza]	0.29506	0.46788	25
[59Spr]	0.29506	0.46797	22
[62Woo]	0.29511	0.46843	25 ± 2
[68Paw]	0.29508	0.46855	28
[68Sch]	0.29503	0.46810	20
[77Dec]	0.29512	0.46826	21 ± 1

(a) Citations listed in *Phase Diagrams of Binary Titanium Alloys*, J.L. Murray, Ed., ASM International, 1987, p 1-5

Pure β titanium: Effect of temperature on a

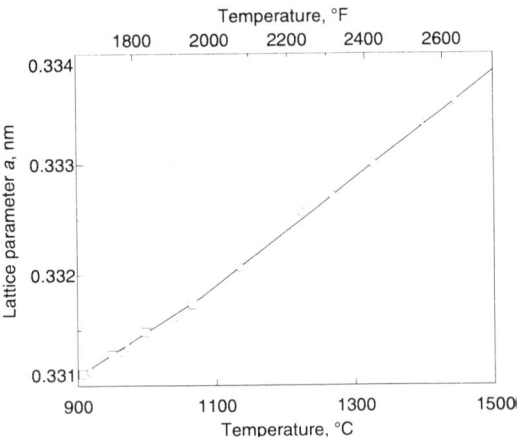

Source: J.L. Murray and H.A. Wriedt, Titanium, Phase Diagrams of Binary Titanium Alloys, J.L. Murray, Ed., ASM International, p 1-5

Pure α titanium: Effect of oxygen on lattice parameters

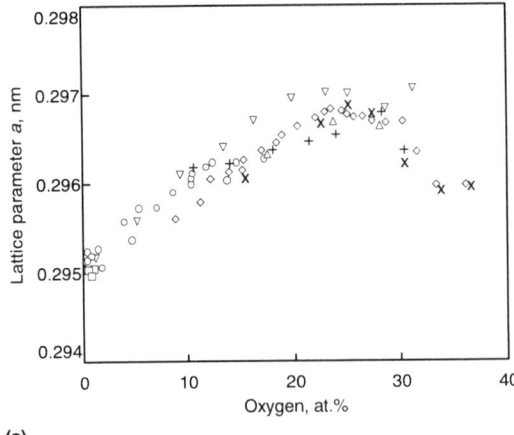

(a) Lattice parameter a. (b) Lattice parameter c
Source: Citations for the data sources are listed in *Phase Diagrams of Binary Titanium Alloys*, ASM International, 1987, p 211-229

Effect of temperature on lattice parameters

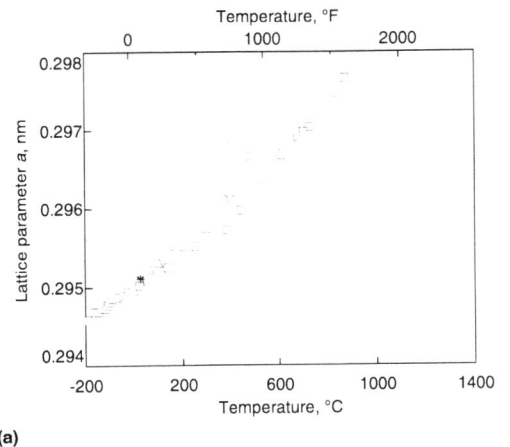
(a)

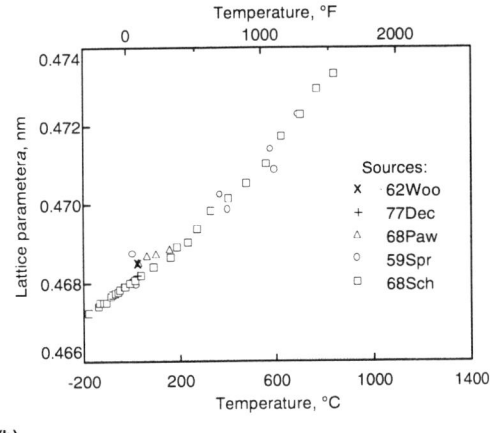
(b)

Source: *Phase Diagrams of Binary Titanium Alloys*, ASM International, 1987

Phase Diagrams

Pressure-Temperature Phase Diagram

Equilibrium crystal structure of pure titanium

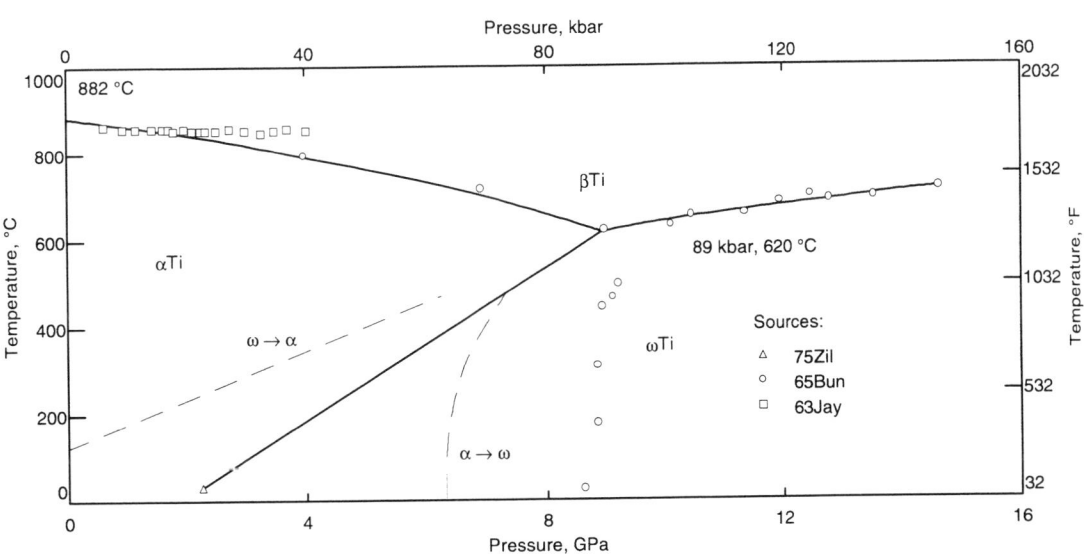

Full citations of data sources are given in the figure reference.
Source: *Phase Diagrams of Binary Titanium Alloys*, ASM International, 1987, p 2

Selected Binary Phase Diagrams

Low temperature α solubility limit for hydrogen in high purity titanium

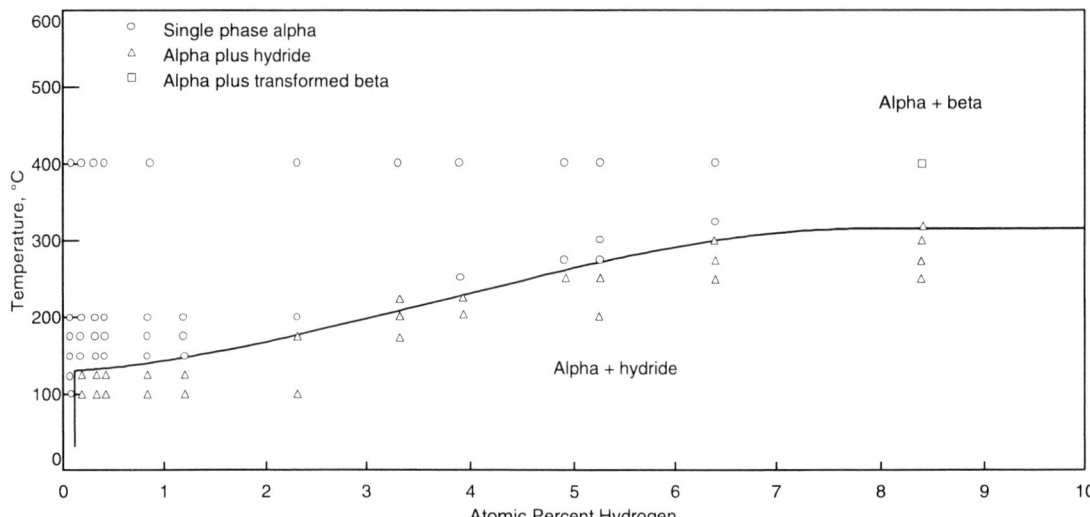

High-purity iodide titanium. The solubility of hydrogen in α titanium decreases from about 8 atomic pct at 300 °C to about 0.1 atomic pct at room temperature, the decrease in solubility being greatest down to 125 °C.
Source: G.A. Lenning, C.M. Craighead, and R.I. Jaffee, Constitution and Mechanical Properties of Titanium-Hydrogen Alloys, *Hydrogen Damage*, C.D. Beachem, Ed., American Society for Metals, 1977, p 100

Ti-O phase diagram

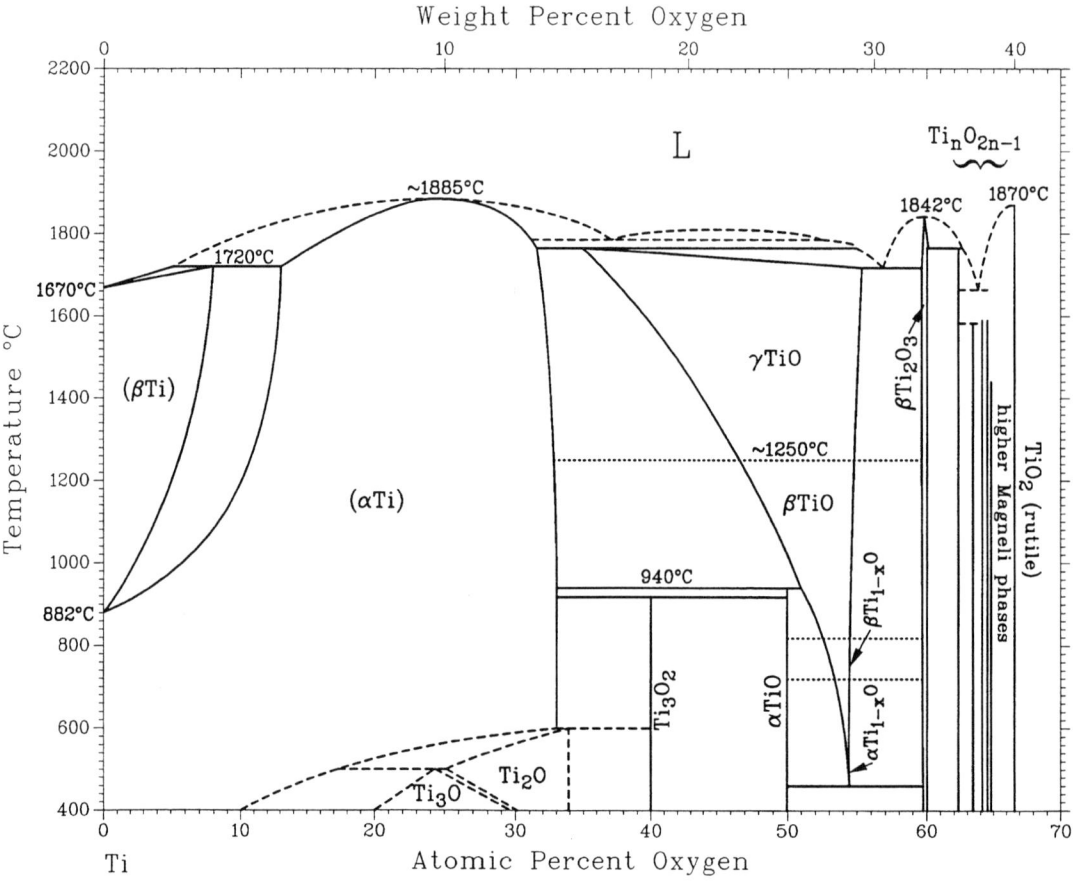

Source: J.L. Murray and H.A. Wriedt, 1987

Damping Properties

Internal Friction

A Q^{-1} value of 1.5×10^{-4} for strain amplitudes of 10^{-6} to 10^{-3} is a reported estimate of pure titanium internal friction at an unspecified temperature. Relaxation spectra of pure titanium (see figure) reveal a small peak in stress relaxation at about 600 °C (1100 °F) for small-grained specimens. This peak is found at higher temperatures as oxygen levels increase. The introduction of oxygen also results in the appearance of a very small additional relaxation peak at about 450 °C (850 °F). At temperatures well above the 450 °C peak, internal friction increases as oxygen levels increase.

Effect of oxygen on internal friction

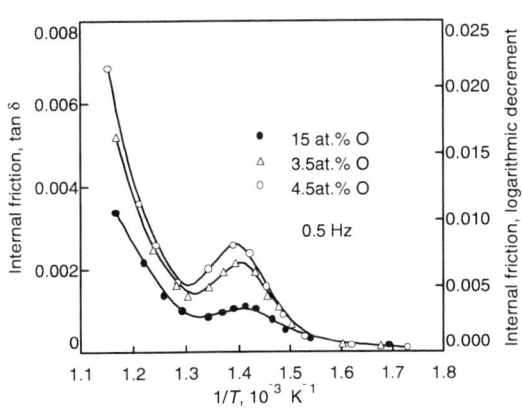

tan δ = logarithmic decrement/π.
Source: J.N. Pratt et al., *Acta Metall.*, Vol 2, 1954, p 203-208

Pure titanium: Internal friction

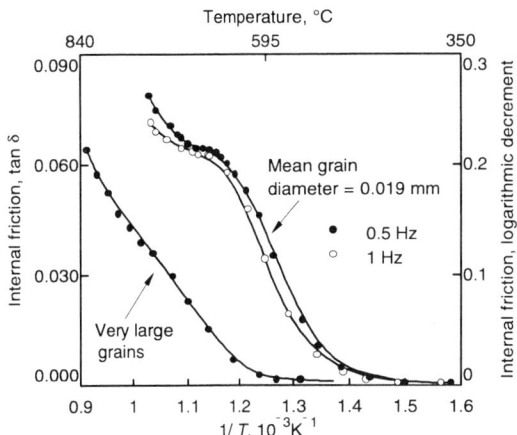

tan δ = logarithmic decrement/π.
Small-grained specimens (0.019 mm mean diam) were tested at vibration frequencies of 0.5 Hz an 1 Hz, while the large-grained specimen was tested at 0.5 Hz.
Source: J.N. Pratt et al., *Acta Metall.*, Vol 2, 1954, p 203-208

Elastic Properties

Because of the α(hcp) structure of titanium, elastic properties depend on the orientation of the titanium crystals and the texture of polycrystalline titanium. Interstitial impurities (e.g. oxygen and nitrogen) and residual stress can also have a slight influence on elastic constants.

Elastic constants of HCP titanium single crystals

Elastic stiffness (C constants)			Elastic compliances (S constants)		
C-Constant	10^6 psi	GPa	S-Constant	10^{-7} in.^{2}/lbf	10^{-12} m^2/N
C_{11}	22.33	153.97	S_{11}	0.688	9.98
C_{12}	12.47	85.98	S_{12}	−0.325	−4.71
C_{13}	9.74	67.16	S_{13}	−0.133	−1.93
C_{33}	26.54	183.0	S_{33}	0.476	6.90
C_{44}	6.71	46.27	S_{44}	1.475	21.4

Source: J.W. Flowers, Jr., K.C. O'Brien, and P.C. McEleney, "Elastic Constants of Alpha Titanium Single Crystals of 25 °C," *J. Less-Common Metals*, 1964, p 393-395, E.S. Fisher and D.J. Renken, "Single-Crystal Elastic Moduli and the HCP-BBC Transformations in Ti, Zr, and Hf," *Physical Review*, 135 (2A), 1964, p 482-494

Single-crystal titanium: Effect of temperature on elastic stiffness moduli (c-constants)

Temperature		c_{11}		c_{33}		c_{44}		$c_{66} = 1/2(c_{11}-c_{12})$		c_{13}		c_{12}	
°C	K	GPa	10^{12} dyne/cm²	GPa	10^{12} dyne/cm²	GPa	10^{12} dyne/cm²	GPa	10^{12} dyne/cm²	GPa	10^{12} dyne/cm²	GPa	10^{12} dyne/cm²
−269	4	176.1	1.761	190.5	1.905	50.8	0.508	44.6	0.446	68.3	0.683	86.9	0.869
−250	23	175.9	1.759	190.5	1.905	50.8	0.508	44.6	0.446	68.2	0.682	86.7	0.867
−200	73	174.9	1.749	189.4	1.894	50.5	0.505	43.9	0.439	68.0	0.680	87.1	0.871
−150	123	172.6	1.726	187.6	1.876	49.9	0.499	42.5	0.425	68.1	0.681	87.7	0.877
−100	173	169.9	1.699	185.7	1.857	49.0	0.490	40.5	0.405	68.4	0.684	88.9	0.889
−50	223	166.8	1.668	183.7	1.837	48.1	0.481	38.4	0.384	68.7	0.687	90.1	0.901
0	273	163.9	1.639	181.6	1.816	47.2	0.472	36.3	0.363	68.9	0.689	91.3	0.913
25	298	162.4	1.624	180.7	1.807	46.7	0.467	35.2	0.352	69.0	0.690	92.0	0.920
50	323	160.9	1.609	179.5	1.795	46.2	0.462	34.2	0.342	69.1	0.691	92.5	0.925
100	373	157.9	1.579	177.4	1.774	45.3	0.453	32.3	0.323	69.4	0.694	93.4	0.934
150	423	155.1	1.551	175.3	1.753	44.4	0.444	30.4	0.304	69.5	0.695	94.3	0.943
200	473	152.2	1.522	173.4	1.734	43.4	0.434	28.5	0.285	69.5	0.695	95.2	0.952
250	523	149.5	1.495	171.5	1.715	42.2	0.424	26.7	0.267	69.2	0.692	96.1	0.961
300	573	146.8	1.468	169.6	1.696	41.4	0.414	25.0	0.250	69.2	0.692	96.7	0.967
350	623	144.2	1.442	167.8	1.678	40.3	0.403	23.4	0.234	69.1	0.691	97.3	0.973
400	673	141.6	1.416	166.1	1.661	39.2	0.392	21.9	0.219	69.0	0.690	97.8	0978
450	723	139.2	1.392	164.4	1.644	38.1	0.381	20.5	0.205	69.2	0.692	98.3	0.983
500	773	136.8	1.368	162.7	1.627	37.0	0.370	19.1	0.191	68.8	0.688	98.5	0.985
550	823	134.5	1.345	161.0	1.610	35.9	0.359	17.8	0.178	68.8	0.688	98.8	0.988
600	873	132.2	1.322	159.3	1.593	34.8	0.348	16.6	0.166	68.8	0.688	99.1	0.991
650	923	129.9	1.299	157.6	1.576	33.7	0.337	15.4	0.154	68.8	0.688	99.2	0.992
700	973	127.6	1.276	156.0	1.560	32.6	0.326	14.2	0.142	...	...	99.3	0.993
750	1023	125.3	1.253	154.5	1.545	31.6	0.316	13.0	0.130	...	...	99.4	0.994
800	1073	123.1	1.231	152.9	1.529	30.7	0.307	11.8	0.118	...	...	99.6	0.996
810	1083	...	...	152.6	1.526	...	...	...	...	...	...	...	...
850	1123	121.0	1.210	...	...	29.7	0.297	10.7	0.107	...	...	99.6	0.996
880	1153	119.7	1.197	...	...	29.1	0.291	10.2	0.102	...	...	99.6	0.996
883	1156	...	...	150.4(a)	1.504(a)	29.1	0.291	10.0(a)	0.100(a)	68.8(a)	0.688(a)	99.6(a)	0.996(a)

(a) From extrapolated curve. Source: E.S. Fisher and J. Renken, *Physical Review*, Vol 135 (No. 2A), 1964, p A487

Elastic stiffness moduli vs pressure of single crystal Ti

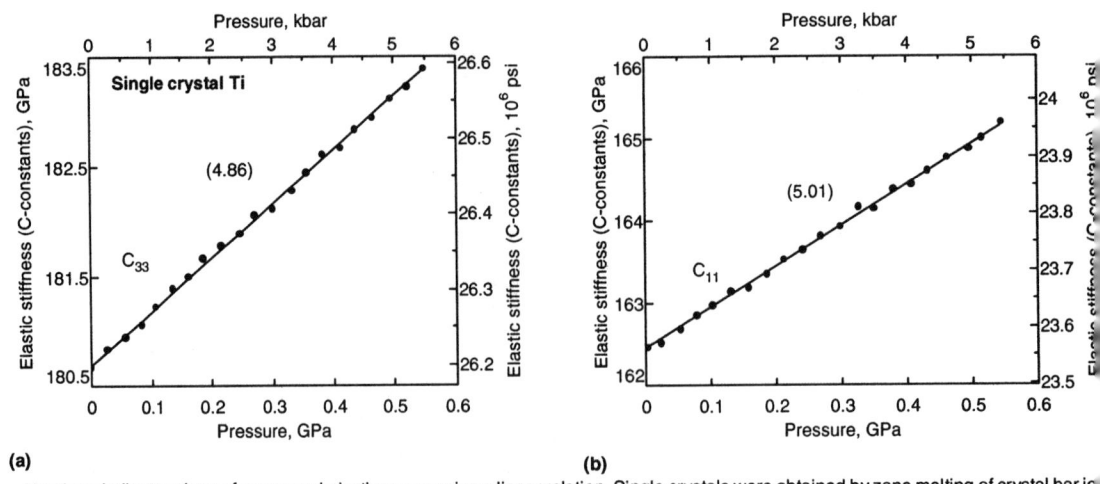

(a) (b)

Numbers indicate values of pressure derivatives assuming a linear relation. Single crystals were obtained by zone melting of crystal bar iodide titanium.
Source: E.S. Fisher and M.H. Manghnani, Effects of Changes in Volume and c/a Ratio on the Pressure Derivatives of the Elastic Moduli of HCP Ti and Zr, *J. Phys. Chem. Solids.*, Vol 32, 1971, p 657-667

(continued)

Elastic stiffness moduli vs pressure of single crystal Ti (continued)

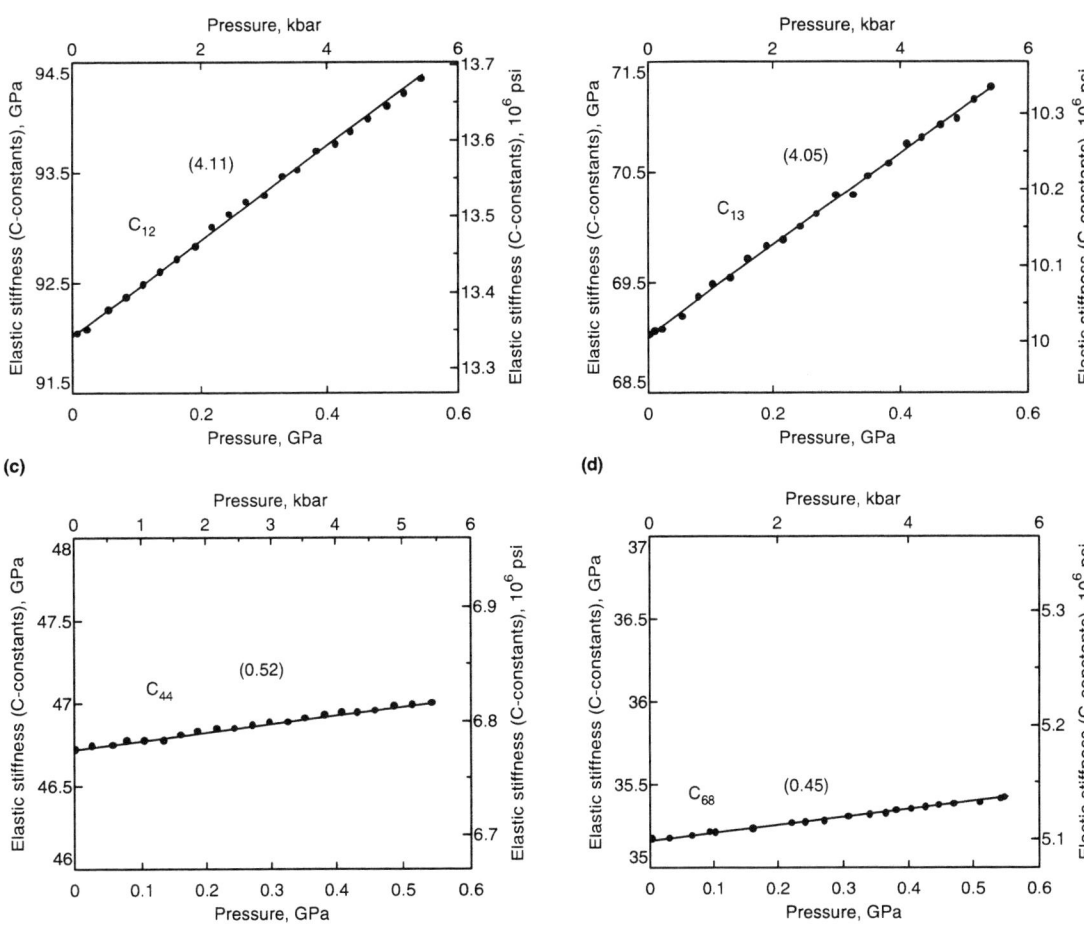

Numbers indicate values of pressure derivatives assuming a linear relation. Single crystals were obtained by zone melting of crystal bar iodide titanium..
Source: E.S. Fisher and M.H. Manghnani, Effects of Changes in Volume and c/a Ratio on the Pressure Derivatives of the Elastic Moduli of HCP Ti and Zr, *J. Phys. Chem. Solids.*, Vol 32, 1971, p 657-667

Young's Modulus

Typical values of Young's modulus (E) for pure titanium at room temperature is in the range of 100 to 110 GPa (15 to 16 × 10^6 psi). This range agrees with the evaluated elastic modulus of single titanium crystals in a direction perpendicular to the c-axis (see figure). However, a low elastic modulus of 85 GPa (12.3 × 10^6 psi) at 25 °C (298 K) is predicted from the following equation:

$$E \text{ (in GPa)} = 104.5 - 0.0645\,T$$

where T is the temperature in degrees kelvin (K). This equation is considered valid for high purity (99.9%) titanium below 500 K (W.H. Hill *et al.*, *Proc. ASTM*, Vol 61, 1961, p 890). For temperatures above 500 K (440 °F), the following equation is reported:

$$E \text{ (in GPa)} = 97 - 0.0486\,T$$

This equation for high purity titanium (99.9% Ti) predicts elastic moduli lower than that of commercial purity (<99.8%) titanium (see figure).

134 / Titanium Data Sheets

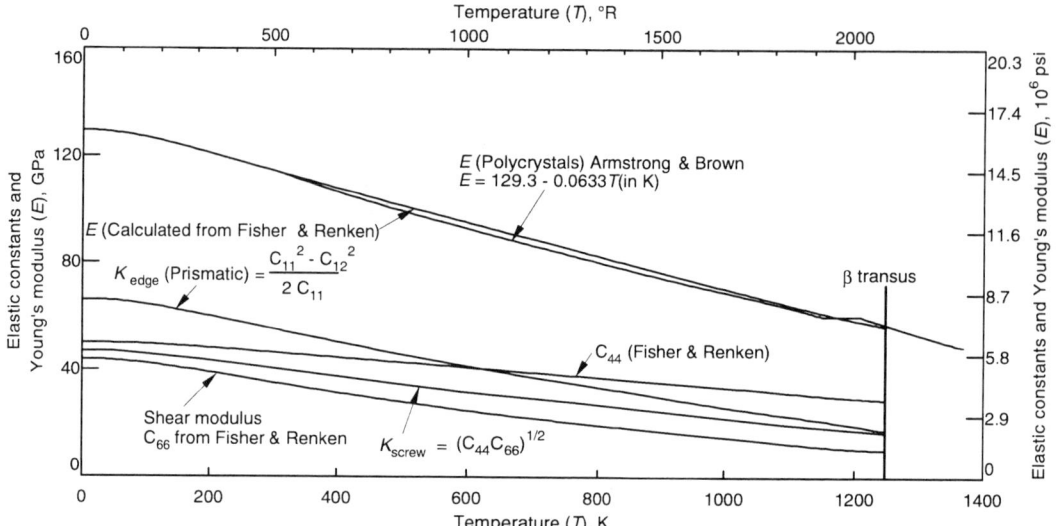

Commercially pure (>99.8%) titanium: Effect of temperature on various elastic constants

Elastic moduli for commercial purity titanium is higher than that of high-purity (99.9%) titanium, but interstitial impurities have less influence on temperature effects (H. Conrad et al., *Titanium Science and Technology*, Plenum Press, 1973, p 970)
Source: Data from: P.E. Armstrong and H.L. Brown, *Trans. AIME*, Vol 230, 1964, p 962; and E.S. Fisher and C.J. Renken, *Phys. Rev. A.*, Vol 135, No. 2A, 1964, p 482

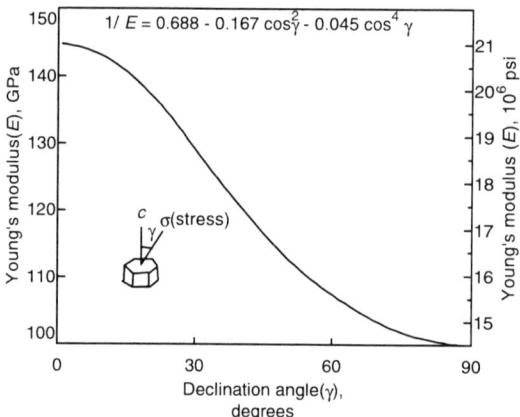

Variation of Young's modulus of single crystal titanium with orientation

Values were calculated from the compliance constants (S-constants). For example, when stress is applied parallel to the basal plane, then $E = 1/S = 100$ GPa (14.5×10^6 psi). For stress perpendicular to the basal plane, then $E = 1/S_{33} = 145$ GPa (21×10^6 psi).
Source: Reported in *Properties of Textured Titanium Alloys*, MCIC Report 74-20, Battelle, 1974, p 21

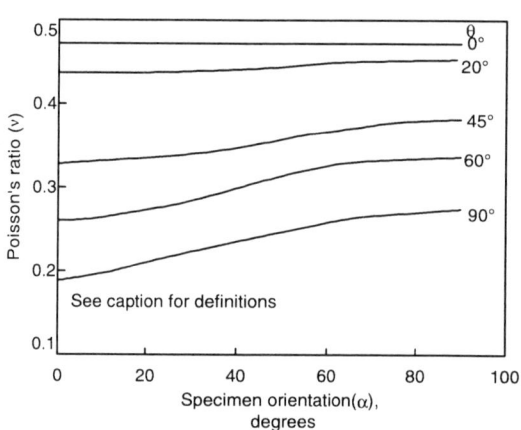

Theoretical variations of Poisson's ratio with orientation

Specimen orientation (α) is defined as the angle between the rolling direction and the specimen axis. The value of θ represents the angle between the basal pole and sheet normal.
Source: *Properties of Textured Titanium Alloys*, MCIC Report 74-20, Battelle, 1974, p 21

Shear Modulus

Poisson's Ratio. Like Young's modulus, the Poisson ratio of titanium depends on specimen orientation (see figure). Typical values for polycrystalline titanium range from 0.32 to 0.36.

Shear modulus (or modulus of rigidity) for pure titanium at room temperature ranges from about 42 to 45 GPa (6 to 6.5×10^6 psi) depending on the amount of interstitial impurities. In isotropic material, the shear modulus (G) is related to Young's modulus (E) as follows:

$$E = 2G(1 + \nu)$$

where ν is the poisson ratio. In textured material, this relation does not hold.

Bulk Modulus and Compressibility

Bulk Modulus. Bulk modulus (K) at room temperature is about 106 to 108 GPa (15.4 to 15.7 $\times 10^6$ psi) for pure titanium. The bulk modulus is related to Young's modulus as follows:

$$E = 3K(1 - 2\nu)$$

The bulk modulus also can be expressed as the reciprocal of compressibility in the elastic range. Therefore, a low bulk modulus of 90 GPa (13×10^6

Linear and volume compressibility of single crystal titanium vs pressure

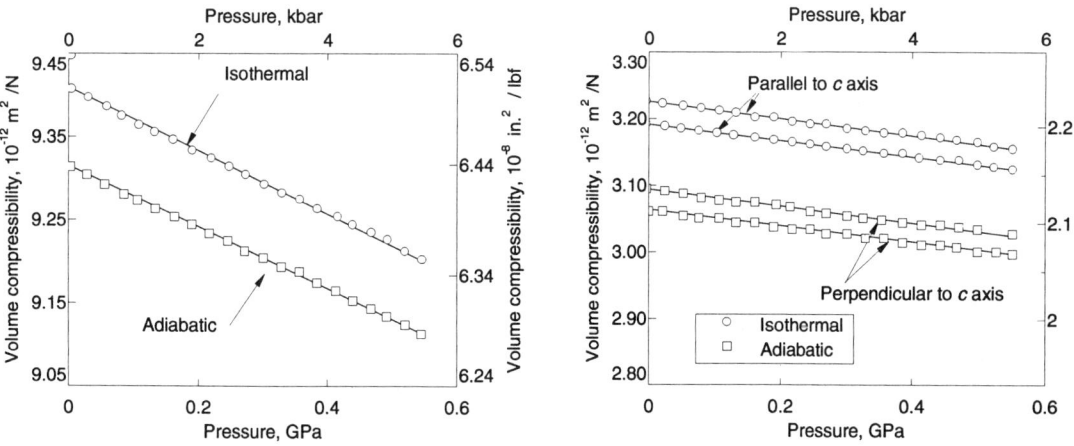

(a) Volume compressibility. (b) Linear compressibility. Material was a single crystal (zone refined from crystal iodide bar) titanium having an adiabatic bulk modulus of 107.3 GPa (15.56 × 10⁶ psi) and an isothermal bulk modulus of 106.3 GPa (15.42 × 10⁶ psi) at 25 °C (77 °F) and 1 bar (~1 atmosphere)
Source: E.S. Fisher and M.H. Manghani, Effects of Changes in Volume and c/a Ratio on the Pressure Derivatives of the Elastic Moduli of HCP Ti and Zr, *J. Phys. Chem. Solids*, Vol 32, 1971, p 657-667

Comparison of compression data for titanium

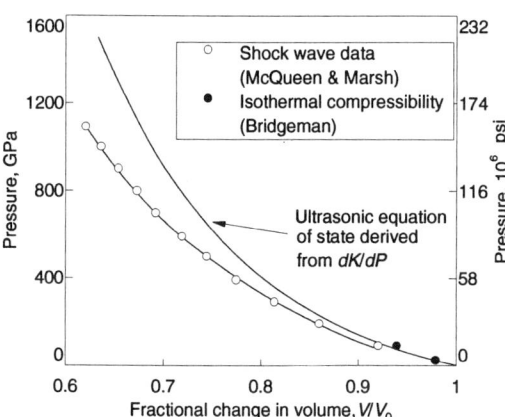

psi) for iodide titanium at room temperature and atmospheric pressure would correspond to a compressibility of about 1×10^{-11} m²/N (7.7×10^{-8} in.²/lbf). A low compressibility factor of 8.2×10^{-12} m²/N (0.8×10^{-6} cm²/kgf), has been reported from early work of Bridgman (*Proc. Am. Acad. Sci.*, Vol 64, 1929, p 51), while higher compressibilities are reported for iodide titanium (see figures).

There is a fairly good agreement between the isothermal compressibility data of Bridgman (*Proc. Am. Acad. Arts Sci.*, Vol 76, 1958, p 55) and the ultrasonic equation. There is a poor agreement between the latter and the shockwave data (*Phys. Rev.*, Vol 108, 1957, p 196) probably because of the phase change near 9 GPa (1.3 × 10⁶ psi). dK/dP = pressure derivative of the bulk modulus (K).
Source: E.S. Fisher and M.H. Manghani, Effects of Changes in Volume and c/a Ratio on the Pressure Derivatives of the Elastic Moduli of HCP Ti and Zr, *J. Phys., Chem. Solids*, Vol 32, 1971, p 657-667

Compressibility (κ) of single-crystal titanium

Temp., K	κ Perpendicular to c		κ Parallel to c		κ vol	
	10^{-12} m²/N	10^{-12} cm²/dyne	10^{-12} m²/N	10^{-12} cm²/dyne	10^{-12} m²/N	10^{-12} cm²/dyne
4	0.300	3.00	0.309	3.09	0.908	9.08
73	0.300	3.00	0.310	3.10	0.911	9.11
173	0.303	3.03	0.314	3.14	0.920	9.21
298	0.306	3.06	0.318	3.18	0.931	9.31
373	0.309	3.09	0.322	3.22	0.940	9.40
473	0.313	3.13	0.326	3.26	0.951	9.51
573	0.316	3.16	0.330	3.30	0.963	9.63
673	0.321	3.21	0.334	3.34	0.975	9.75
773	0.326	3.26	0.337	3.37	0.988	9.88
873	0.331	3.31	0.340	3.40	1.001	10.01
973	0.336(a)	3.36(a)	0.343(a)	3.43(a)	1.016(a)	10.16(a)
1073	0.342(a)	3.42(a)	0.345(a)	3.45(a)	1.029(a)	10.29(a)
1156	0.347(a)	3.47(a)	0.346(a)	3.46(a)	1.041(a)	10.41(a)

(a) Extrapolated from curves. Source: E.S. Fischer and C.J. Renken, *Physical Review*, Vol 135, No. 2A, 1964, p A482-A494

Electrical Properties

Resistivity

Resistivity values of 0.46 to 0.48 µΩ · m are usually obtained for iodide titanium at room temperature, although values as low as 0.42 µΩ · m have been obtained. The true resistivity of titanium has been assessed at 0.42 µΩ · m by W.J. Lepkowski and J.W. Holladay (TML Report No. 73 Battelle, 1957). Impurities increase resistivit (see the section "Physical Metallurgy of Titaniur Alloys" in this volume).

Effect of temperature on the resistivity of copper, silver, platinum and titanium

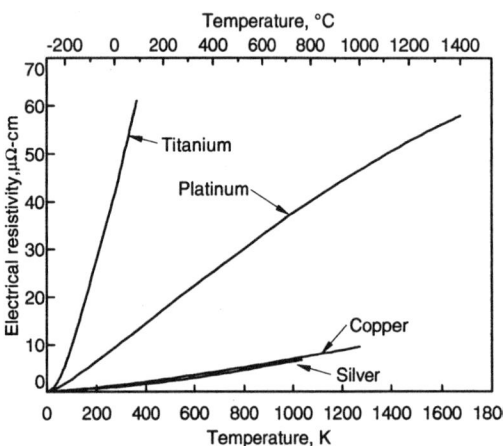

Source: J.K. Stanley, *Electrical and Magnetic Properties of Metals*, American Society for Metals, 1963

Change in resistivity with deformation

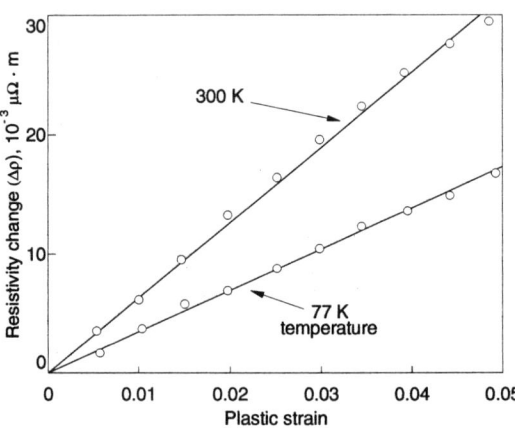

Source: F.F. Laurent'Yev, and V.N. Nikiforenko, "Influence of Plastic Deformation and Temperature on the Resistivity of Polycrystalline Titanium," *Fiz. Met. Metalloved.*, 52 (No. 6), 1981, p 1200-1204

Effect of temperature on resistivity on high-purity titanium

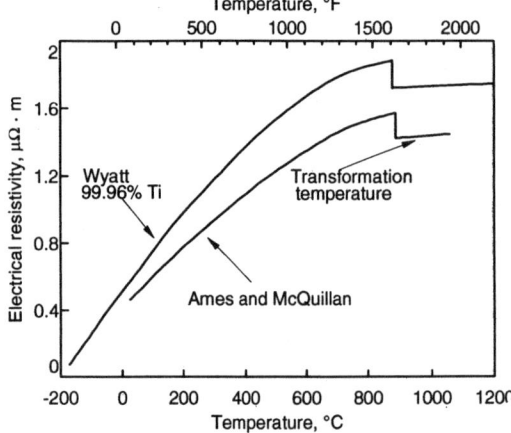

Source: J.L. Wyatt, Electrical Resistance of Titanium Metal, *J. Met.*, Vol 5, 1953, p 903-905; S.L. Ames and A.D. McQuillan, The Resistivity-Temperature Concentration Relationships in the System Niobium-Titanium, *Acta Metall.*, Vol 2, 1954, p 831

Pure titanium resistivity data

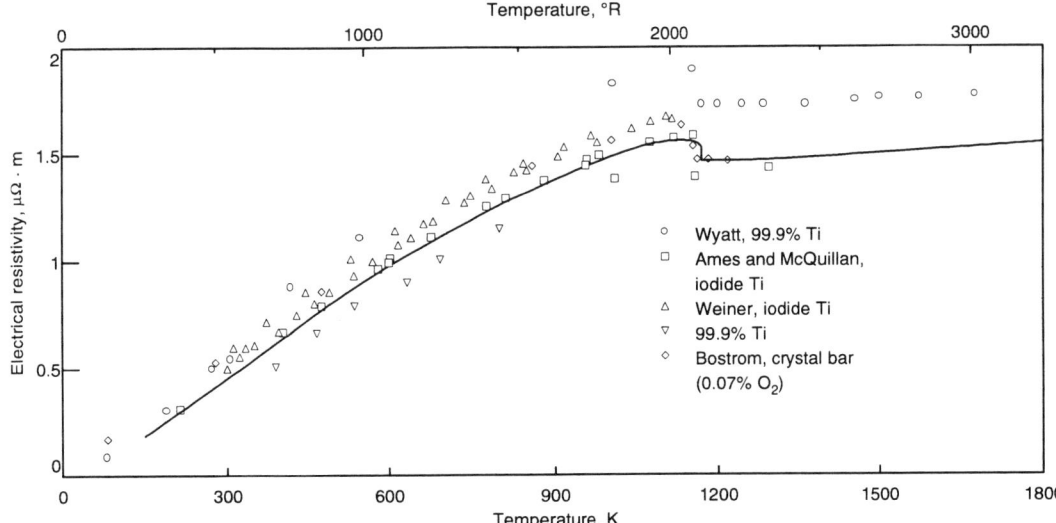

Source: *Thermophysical Properties of Solid Materials, Vol 1 - Elements,* WADA Technical Report 58-746, Wright Air Development Division, 1960

Relative resistance of a single crystal and a polycrystalline titanium sample

Temperature, K	Relative resistivity(a) Single crystal Parallel to c-axis	Relative resistivity(a) Single crystal Perpendicular to c-axis	Polycrystalline
4.2	...	...	0.0306
19.6	...	...	0.0309
77	0.1786	0.2086	0.15
196	0.6355	0.6450	0.67
273	1.000	1.000	1.000(b)
300	1.130	1.110	1.120
373	1.4997	1.4550	1.468
400	1.64	1.58	1.60
500	2.13	2.10	2.09
600	2.60	2.60	2.56
700	2.99	2.94	2.99
800	3.25	3.17	3.31
900	3.46	3.38	3.58
1000	3.59	3.58	3.80
1100	3.67	3.76	3.98
1200	3.43	3.61	3.705
1300	3.47	3.675	3.760
1400	3.525	3.75	3.835
1450	3.555	3.78	3.865

(a) Relative resistivity = resistivity at indicated temperature/resistivity at 273 K. (b) Electrical resistivity at 273 K (0 °C) is 42.67 ± 0.05 microhm-cm. Source: James K. Stanley, *Electrical and Magnetic Properties of Metals,* American Society for Metals, 1963, p 51

Pure titanium: Resistivity at cryogenic temperatures

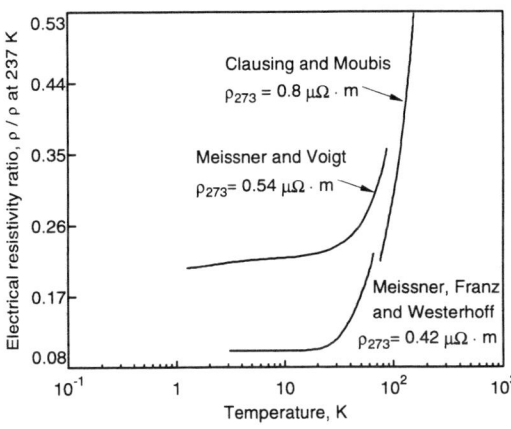

ρ_{273} = resistivity at 0 °C (273 K).
Source: R.B. Stewart and V.J. Johnson, Ed., *A Compendium of the Properties of Materials at Low Temperatures (Phase II),* WADD Technical Report 60-56 (Part IV), U.S. Air Force, 1961

Superconductivity

A state of superconductivity is stable only if temperature, magnetic field strength, and current density are all below the critical levels, or thresholds. The three critical parameters of temperature (T_c), magnetic field (H_c), and current density (J_c), are closely interdependent. For example, the Hc decreases with increasing temperature or current.

Critical Temperature of Superconductivity. A typical value for pure titanium is 0.40 ± 0.04 K (B.W. Roberts, Properties of Selected Superconducting Materials, 1978 Supplement, NBS Technical Note 983). Critical temperatures as high as 0.56 K have been reported (T.S. Smith and J.G. Daunt, Phy. Rev., Series 2, Vol 88, 1952, p 1172).

Critical Field Strength of Superconductivity. Pure metals are in the class of type I superconductors, which exhibit perfect diamagnetism up to some critical field (Hc) when temperature is below Tc. Values have ranged from 2×10^{-3} T (T.S. Smith *et al.,* Phy. Rev. Series 2, Vol 89, 1953, p 654) to 8.6×10^{-3} T for unannealed, cold-swaged wire (M.C. Steel and R.A. Hein, Phy. Rev., Series 2, Vol 92, 1953, p 243).

Critical current density of superconductivity depends on the degree of deformation and cooling rates during thermomechanical processing. References include E.W. Collings, Sourcebook of *Titanium Superconductivity,* and *Applied Superconductivity, Metallurgy, and Physics of Titanium Alloys* (Vol 1 and 2).

Thermoelectric Properties

Thermoelectric potentials of titanium-platinum couple

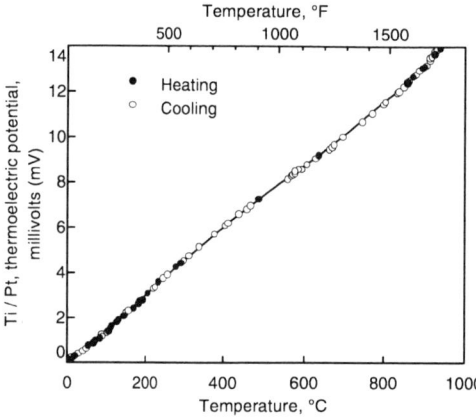

Source: H.W. Worner, Thermoelectric Properties of Titanium with Special Reference to the Allotropic Transformation, *Austral, J. Sci. Res.*, Vol 4 (No. 62), 1951

Magnetic Characteristics

Magnetic Properties

Magnetic Susceptibility. 3.17×10^{-6} ($\pm 0.03 \times 10^{-6}$) cm^3/g is the mass susceptibility (χ_m) established by Lepkowski and Holladay (TML Report No. 73, Battelle, 1957). This is in good agreement with more recent measurements (see tables). Additional information on the factors affecting magnetic susceptibility are discussed in the chapter "Physical Properties" of Section 1 in this Volume.

Magnetic Permeability. 1.00005 at 1600 A/m (20 Oe)

Magnetic susceptibility of polycrystalline titanium

Specimen purity, wt% Ti	Temperature, K	Mass of specimen, mg	Estimated Number of grains per specimen	Condition	Magnetic susceptibility, 10^{-6} cm^3/g			
					χ_1	χ_2	$\chi_1-\chi_2$	χ_{av}(a)
99.8	300	81	800	As cast	3.16_8			
99.8	300	84	800	As cast	3.16_8			
99.9	300	84	...	As received	3.16_9			
99.8	300	78	800	As cast	3.15_9	3.15_6	3.15_3	$\underline{3.15_6}$
99.8	300	73	800	As cast	3.19_8	3.12_9	3.12_7	$\underline{3.15_1}$
99.8	300	108	43,000	Recrystallized	3.19_3	3.14_9	3.13_3	$\underline{3.15_8}$
99.9	299	70	...	As received	3.20_9	3.12_1	3.11_0	$\underline{3.14_7}$
99.8	299	51	53,000	Recrystallized	3.25_6	3.13_5	3.10_0	$\underline{3.16_3}$
99.8	298	49	230,000	Recrystallized	3.35_3	3.08_5	3.07_8	$\underline{3.17_2}$

(a) Accuracy is limited to about three significant digits (as indicated by underlining). Source: E.W. Collings, Battelle, Columbus, Ohio

Magnetic susceptibility of single-crystal titanium

Magnetic susceptibility, 10^{-6} cm^3/g			
χ_1	χ_2	$\chi_1-\chi_2$	χ_{av}(a)
3.50_5	3.02_3	0.48_2	$\underline{3.18_4}$
3.47_3	3.02_0	0.45_3	$\underline{3.17_1}$
3.49_9	2.98_8	0.51_1	$\underline{3.15_8}$
3.52_7	3.00_6	0.52_1	$\underline{3.18_0}$

(a) Accuracy is limited to about three significant digits (as indicated by underlining). Source: E.W. Collings, Battelle, Columbus, Ohio

Magnetic Effects

Magnetoresistance Coefficient. 1.04×10^{-16} m²/A² (6.6×10^{-13} Oe⁻)

Hall Coefficient. Nonmetallic impurities greatly affect the Hall coefficient. Positive Hall coefficients for pure titanium have ranged from $+1 \times 10^{-4}$ nΩ cm/Oe to a high value of $+3 \times 10^{-4}$ n² cm/Oe reported by G.W. Scovil ("The Hall Effect in Titanium," *J. App. Phys.*, Vol 24, 1953, p 226). Negative Hall coefficients have been measured in specimens containing 0.34 wt% oxygen (S. Foner, *Phy. Rev.*, Series 2, Vol 91, 1953, p 447).

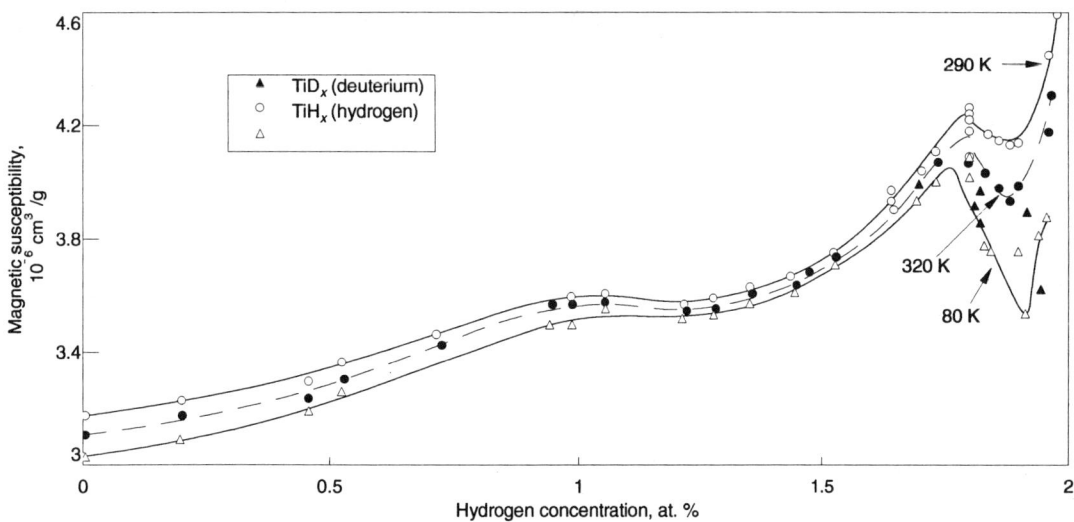

Magnetic susceptibility of titanium-hydrogen system

Source: W.J. Lepkowski and J.W. Holladay, The Physical Properties of Titanium and Titanium Alloys, TML Report No. 73, Battelle, 1957

Chemical/Corrosion Properties

The corrosion resistance of titanium depends on the formation of a protective oxide layer as described in "Technical Note 2: Corrosion" in this volume. The effectiveness of this layer seems to require the presence of either water or hydroxyl (even in trace amounts) during its formation. If titanium is exposed to a vigorously oxidizing environment in the complete absence of moisture, any surface film formed will not be protective and oxidation in depth may take place, often in the form of a violent exothermic reaction. Breakdown of the passive layer can also occur from dry oxidants, nonoxidizing aqueous environments as defined by a Pourbaix diagram, and pitting or crevice attack in near-neutral aqueous solutions (particularly in the presence of halides). Iron impurities, generally thought to detrimentally affect titanium corrosion, have little or no effect of practical significance at levels up to 0.2 wt% Fe (R.W. Schutz *et al.*, Effect of Solid Solution Iron on the Corrosion Behavior of Titanium, *Titanium Science and Technology*, Deutsche Gesellschaft für Metallkunde e.V., 1985, p 2617-2624.)

General corrosion becomes a concern in reducing acid environments, particularly as acid concentration and temperature increase (see figure). In strong and/or hot reducing acids (in the absence of inhibitors), the oxide film of titanium can

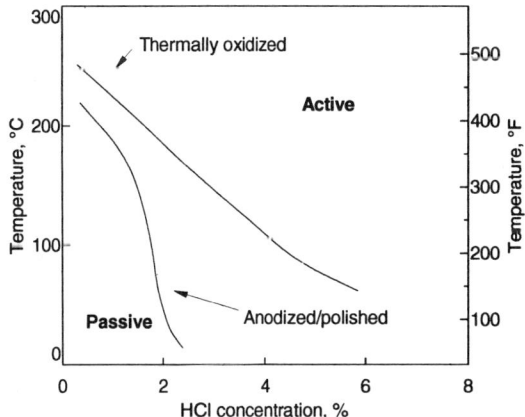

Variation in active and passive regions in HCl solution

Three types of titanium specimens were used; specimens polished with #400 emery paper, specimens anodized in a 2% phosphoric acid solution at 50V for 10 min. after polishing, and specimens oxidized in air at 600 °C for 10 min. after polishing. The thicknesses of all of these specimens were 1 mm.
Source: T. Fukuzuka *et al.*, On the Beneficial Effect of the Titanium Oxide Film Formed by Thermal Oxidation, *Titanium '80 Science and Technology*, The Metallurgical Society of AIME, 1980, p 2781-2792

Electrochemical Potentials

Electrochemical anodic and cathodic polarization testing is often used to supplement weight loss testing. Polarization testing can identify whether the alloy is truly fully passive or possibly metastable; this is often not discernible from weight loss tests alone.

Repassivation potentials generally decrease as iron content increases in titanium. For iron levels less than 0.3%, this reduction is considered to be insignificant, on the order of several tenths of a volt. In hot HCl solutions, for example, the useful passive region is reduced when iron levels exceed 0.3 wt% in base metal and 0.2 wt% in weldments. (Schutz et al., *Titanium Science and Technology*, 1985, p 2617-2624).

deteriorate and dissolve, and the unprotected metal is oxidized to the soluble trivalent ion ($Ti^{3+} + 3e^-$).

Corrosion for commercially pure titanium is covered in the article "CP and Modified Ti".

The steady-state electrode potential in flowing seawater is about 0.10 V vs SCE at 24 °C (75 °F).

Polarization curves in a solution of methyl alcohol and sulfuric acid ($CH_3OH + H_2SO_4$) were obtained for pure (iodide) titanium (see figure) and no significant differences were observed for Ti-75A (ASTM Grade 4) and Ti-6Al-4V. Without water additions, a short region of passivity can be detected, but a sharp increase of currents occurs at about –0.24V. Examination revealed attack in the form of pitting (see figure reference for details).

Pure titanium: Effect of water additions on potentiostatic polarization

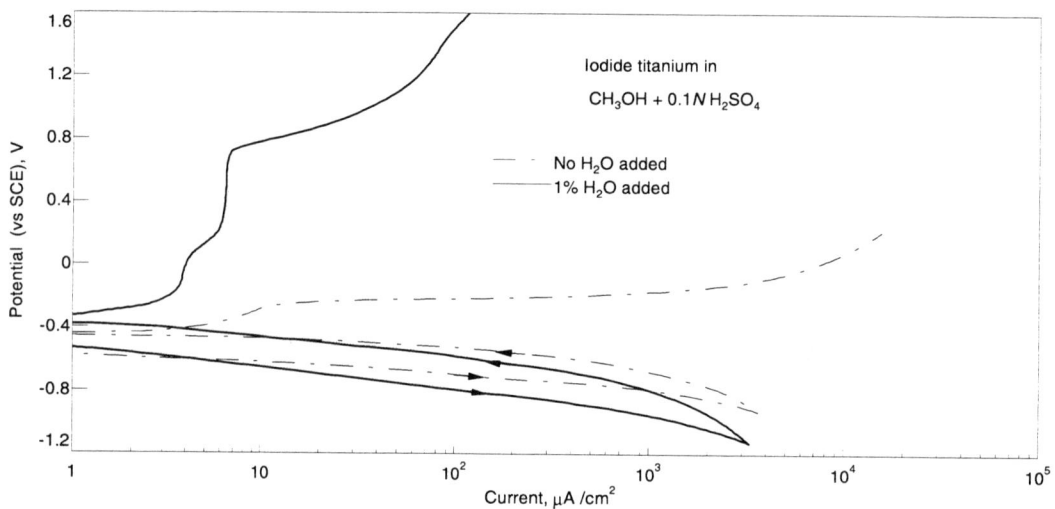

When 1% of water was added to the test solution, the titanium electrode apparently was in the passive state. The anodic current remained small and constant up to about +0.7 V. At higher potentials the current increased steadily, but not as sharply as in water-free solutions.
Source: F. Mansfeld, Pitting Caused by Chlorides or Sulfates in Organic Media, *Galvanic and Pitting Corrosion-Field and Laboratory Studies*, ASTM STP 576, ASTM, 1976, p 180-202

Oxidation

Oxidation of pure titanium does not differ significantly from that of ASTM commercially pure titanium grades 1 and 2 (UNS R50250 and R50400), and thin oxide films form in air temperatures between 315 and 650 °C (600 and 1200 °F). The oxide film is barely perceptible after exposure at 315 °C (600 °F), but it becomes darker and thicker with increasing temperature and time at temperature. The changes in surface color can be used as a rough guide of exposure temperature in air (see table), while time at temperature becomes a more significant factor for temperatures above 500 °C (see figure). Care must be used in estimating temperature due to the time factor mentioned, and the oxide color at a given temperature is strongly influenced by the cleanliness of the surface. A surface exposed to elevated temperature with, for example, oil on the surface, will have a different color than a clean surface.

The oxidation of pure titanium has also been studied at low oxygen pressures, so the first steps of oxidation can be evaluated by surface potential measurement, Auger analysis, or x-ray photoelectron microscopy (ESCA). The influence of oxidation on surface potentials is illustrated by typical pressure-temperature effects (see figure), while the growth of oxide layers have been evaluated by Auger analysis (see figure showing the effect of a sulfur monolayer on oxidation layers).

Effect of air temperature on titanium's appearance

| Air temperature | | Appearance of |
°C	°F	oxide film(a)
370	700	Straw yellow color
480	900	Blue color
650	1200	Dull gray

(a) Alloying elements and surface contaminants also influence color.

Surface potential vs time at various pressures

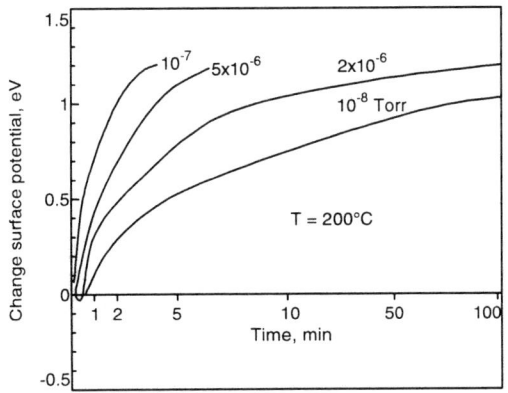
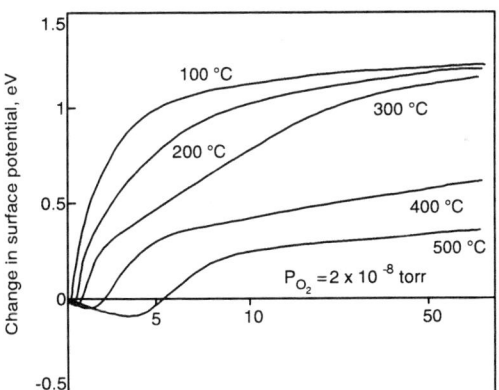

The adsorption of oxygen is related to surface potential measurements. Titanium sample had 40 ppm O and was surface cleaned by ion bombardment in a vacuum chamber.
Source: J.B. Bignolas et al., A Study of Oxygen Chemisorption on Titanium by Auger Analysis and Mirror Electron Microscopy, *Titanium '80 Science and Technology*, Metallurgical Society of AIME, 1980, 2829-2837

Oxidation kinetics of high-purity titanium

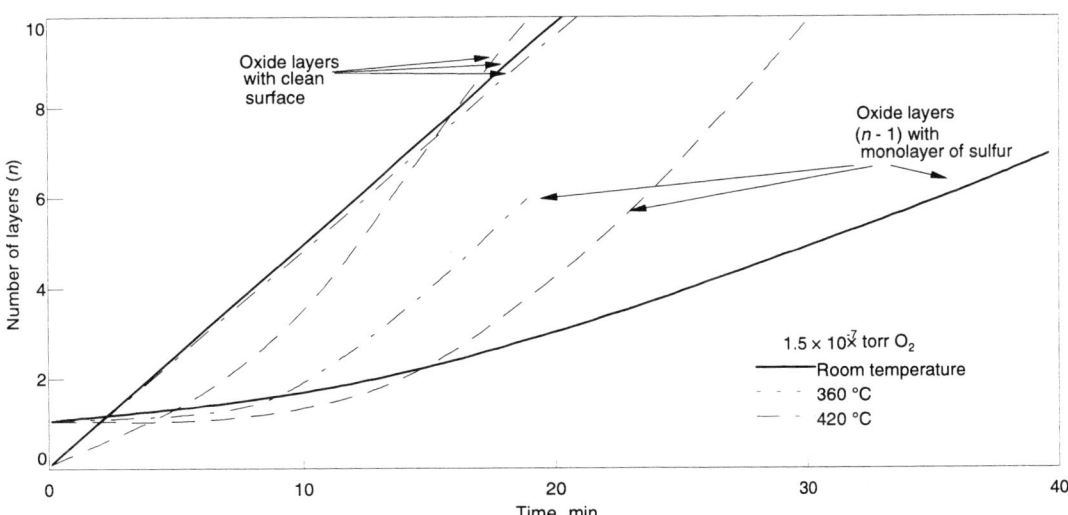

Samples were iodide titanium cleaned by successive treatments of argon ion bombardment (500 eV).
Source: J.P. Gspann et al., Oxidation Kinetics of High-Purity Titanium, Influence of a Monolayer of Chemisorped Sulfur, *Titanium '80 Science and Technology*, The Metallurgical Society of AIME, 1980, p 2839-2851

Crevice Corrosion

Crevice corrosion in boiling 6% NaCl solution

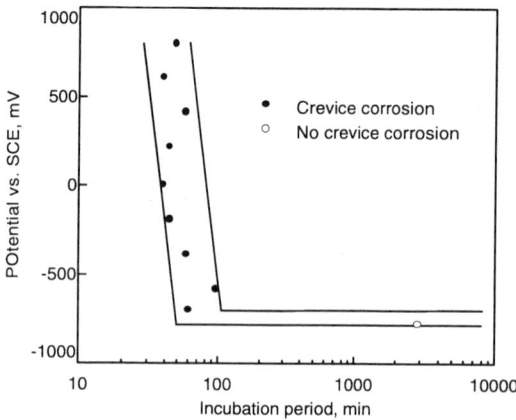

Specimens were machined to 15 or 30 mm (0.6 or 1.2 in.) squares with a 5.5 mm (0.215 in.) diam hole. Composition (wt%): 0.05 O, 0.004 C, 0.03 Fe, 0.014 N, 0.0003 H.
Source: M. Kobayashi et al., *Titanium '80 Science and Technology*, The Metallurgical Society of AIME, 1980, p 2613-2622

Effect of oxidation on crevice corrosion in NaCl solution

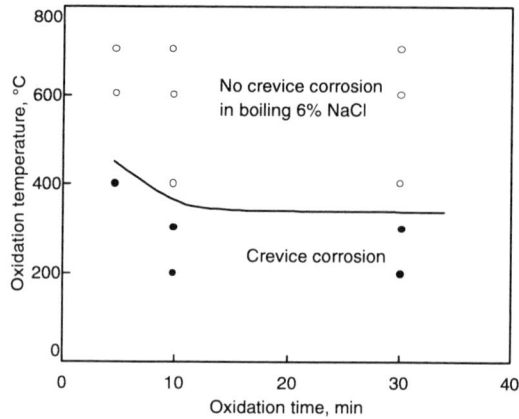

Specimens were machined to 15 or 30 mm (0.6 or 1.2 in.) squares with a 5.5 mm (0.215 in.) diam hole. Composition (wt%): 0.05 O, 0.004 C, 0.03 Fe, 0.014 N, 0.0003 H.
Source: M. Kobayashi et al., Study on Crevice Corrosion of Titanium, *Titanium '80 Science and Technology*, The Metallurgical Society of AIME, 1980, p 2613-2622

Hydrogen Damage

Titanium and its alloys suffer hydrogen damage primarily by hydride-phase formation. Uniaxial tensile properties of high-purity titanium are relatively unaffected by typical concentrations (<200 ppm) of hydrogen; however, the impact toughness can be impaired.

Commercially pure titanium is more sensitive to hydrogen than pure titanium is. The amount of hydrogen necessary to induce ductile-to-brittle transition behavior in commercially pure titanium is one-half the amount needed in pure titanium.

At room temperature, the solubility limit of hydrogen in α-titanium ranges from 20 to 200 ppm (wt%) depending on the impurity content. Hydrogen solubility in high-purity titanium is about 0.1 at.% (0.02 wt%) at room temperature (see the section "Phase Diagrams" in this datasheet).

Tensile properties, hardnesses, and impact strengths of three titanium alloys at two hydrogen levels

Alloy Addition, wt%	Hydrogen Atomic %	wt%	Treatment	Tensile strength, ksi	0.2% Offset yield strength, ksi	Elongation, %	Reduction in area, %	Room-temperature impact strength, in.-lb
0.06 N	0.07	0.0014	Vacuum annealed 775 °C 2 hr, furnace cooled	82	72	36	55	24
0.06 N	0.63	0.0133	700 °C 1 hr in argon, furnace cooled	77	63	30	57	4
0.18 N	0.09	0.0018	Vacuum annealed 775 °C 2 hr, furnace cooled	114	110	23	26	16
0.18 N	0.6	0.0127	700 °C 1 hr in argon, furnace cooled	122	117	25	23	3
2.55 Al	0.07	0.0015	Vacuum annealed 775 °C 2 hr, furnace cooled	93	78	26	47	41
2.55 Al	0.6	0.0129	900 °C 1 hr in argon, furnace cooled	96	82	23	39	7

Source: G.A. Lenning, C.M. Craighead, and R.I. Jaffee, Constitution and Mechanical Properties of Titanium-Hydrogen Alloys, *Hydrogen Damage*, C.D. Beachem, Ed., American Society for Metals, 1977, p 100

Stress-Corrosion Cracking

Pure titanium (and commercially pure ASTM grades 1, 2, 7, 11, and 12) is immune to stress-corrosion cracking (SCC), except in a few environments such as anhydrous methanol/halide solutions, nitrogen tetroxide (N_2O_4), red fuming HNO_3, and liquid or solid cadmium. The general SCC behavior of titanium and titanium alloys is described in "Technical Note 2: Corrosion" in this Volume and in the datasheet "CP and Modified Titanium."

Thermal Properties

Melting Point

At atmospheric pressure, pure titanium has an assessed melting temperature of 1670 ± 5 °C (3038 ± 9 °F) on the 1968 International Practical Temperature Scale (IPTS). This assessed value is based on past determinations (see table) and the demonstration by Rudy and Progulski (67Rud) that their samples were not contaminated during the melting experiments by checking for agreement of the βTi ↔ αTi transformation temperatures before and after the melting point determination.

Specific Heat

Electronic Specific Heat. 0.070 J/° kg (8.0 × 10^{-4} cal/°C · mol)

Effect of oxygen on the heat capacity (or specific heat) of pure titanium depends on the temperature range of the measurement (see table). Below about 165 K (–160 °F), titanium containing oxygen has a lower specific heat than pure titanium, but above about 165 K (–165 °F) the titanium-oxygen alloys have a higher specific heat than pure titanium. Consequently, specific heat measurements made on impure material will give values higher than the true specific heats at, or above, room temperature.

Experimental determinations of the melting point of Pure βTi

Reference(a)	Reported melting point, °C	Melting point (IPTS-68), °C
[51Han]	1720 ± 10	1723
[52Ade]	1700 ± 15	1703
[53May]	1680 ± 10	1683
[53Sch]	1660 ± 10	1663
[54Ori]	1672 ± 4	1675
[56Dea]	1660 ± 10	1663
[59Bic]	1667 ± 8	1670
[67Rud]	1668 ± 4	1671
[74Bcr]	1666 ± 4	1667
[77Cez]	1672 ± 5	1672
Assessed		1670 ± 5

(a) Full reference citations are given in *Phase Diagrams of Binary Titanium Alloys*, J. Murray, Ed., ASM International, 1987, p 1-5

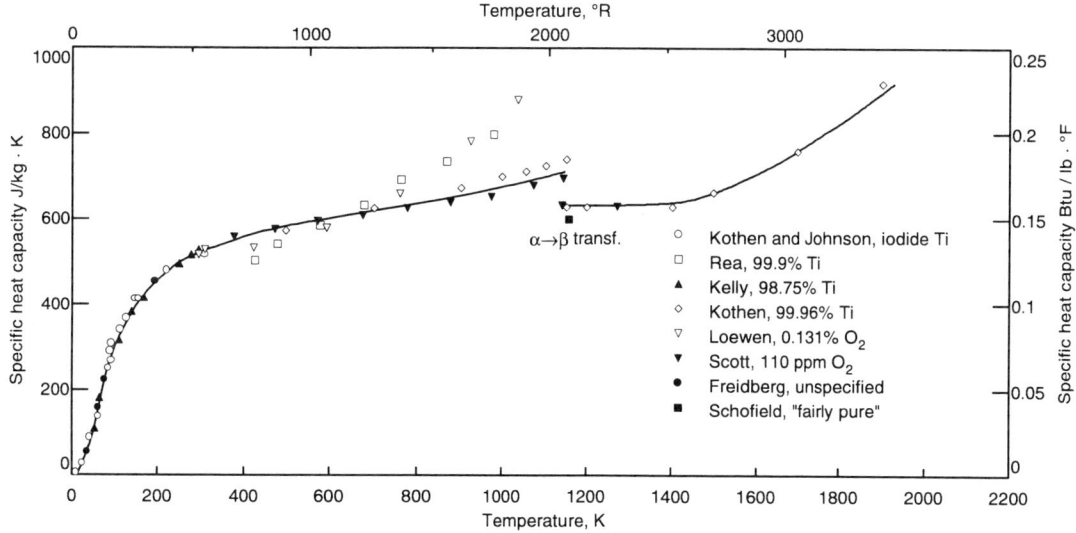

Pure titanium: Specific heat vs temperature

Source: *Thermophysical Properties of Solid Materials, Vol 1 - Elements*, WADC Technical Report 58-476, Wright Air Development Division, 1960

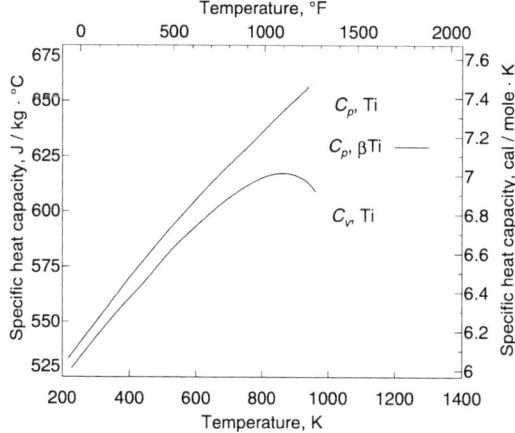

Heat capacity of pure titanium

Source: E.S. Fisher and J. Renken, *Phys. Rev.*, Vol 135, No. 2A, 1964, pA482-A494

Effect of oxygen on specific heat

K	°F	Specific heat (C_p), J/kg · °C (cal/mole · K)		
		99.95-99.99% pure titanium(a)	Ti-1.90 wt% O(b)	Ti-9.95 wt% O(b)
50	−370	101.1 (1.136)	91.78 (1.031)	60.8 (0.683)
75	−325	213.8 (2.402)	202.2 (2.271)	155.8 (1.750)
100	−280	(3.434)	(3.310)	(2.831)
125	−235	(4.155)	(4.074)	(3.780)
150	−190	(4.684)	(4.650)	(4.564)
175	−145	(5.043)	(5.067)	(5.141)
200	−100	(5.321)	(5.380)	(5.697)
225	−55	(5.539)	(5.627)	(6.101)
250	−10	(5.713)	(5.822)	(6.429)
275	+35	(5.864)	(5.991)	(6.703)
298.15	77	532.0 (5.976)	545.7 (6.130)	616.9 (6.930)

(a) Data from: M.H. Aven et al., *Phys. Rev*, Vol 102, 1956, p 1263. (b) Data from A.D. Mah et al., Thermodynamic Properties of Titanium-Oxygen Solutions and Compounds, United States Department of the Interior, Bureau of Mines, Report of Investigation 5316 (March, 1957)

Specific heat of titanium at elevated temperatures

Temperature		Specific heat (C_p)	
K	°F	J/kg · °C	cal/mole · K

Alpha titanium ($C_p = 669.0 - 0.037188T - 1.080 \times 10^7 T^{-2}$)(a)

300	80	547.7	6.153
350	170	578.3	6.496
400	260	597.4	6.711
450	350	610.0	6.852
500	440	618.4	6.947
550	530	624.1	7.011
600	620	628.0	7.055
650	710	630.7	7.085
700	800	632.4	7.104
750	890	633.4	7.115
800	980	633.9	7.121(b)
850	1070	633.9	7.121
900	1160	633.7	7.119
950	1250	633.2	7.113
1000	1340	632.5	7.105
1050	1430	631.6	7.095
1100	1520	630.6	7.084
1150	1610	629.5	7.071

Beta titanium

| 1150-1300 | 1610-1880 | 712 | 8.0 |

(a) Formula for specific heat (in J/kg · °C) at other temperatures (T, in K) between room temperature and the transformation temperature. (b) Anomalous maximum, which requires more work in the temperature range above 800 K (980 °F). Source: W.J. Lepkowski and J.W. Holladay, TML Report No. 73, Battelle, 1957, p 26-28

Pure Ti: Compilation of selected specific heats at low temperatures

Temperature		C_p		Temperature		C_p	
K	°R	J/kg·K	cal/lb·°F	K	°R	J/kg·K	cal/lb·°F
1	1.8	0.071	0.0043	70		189	11.4
2	3.6	0.146	0.0088	80		230	13.8
3	5.4	0.226	0.0136	90		267	16.1
4	7.2	0.317	0.0191	100	180	300	18.05
6	10.8	0.54	0.0325	120		352	21.2
8	14.4	0.84	0.0505	140		391	23.5
10	18	1.26	0.0760	160		422	25.4
15		3.3	0.20	180		446	26.8
20	36	7.0	0.42	200	360	465	28.0
25		13.4	0.807	220		480	28.9
30	54	24.5	1.475	240		493	29.7
40		57.1	3.44	260		504	30.35
50		99.2	5.97	280		514	30.95
60		146.7	8.835	300	540	522	31.44

References: Data listed in WADD Technical Report 60-56 Part II (V.J. Johnson, Ed.) from the following three sources: (1) M.H. Aven et al., *Phys. Rev.* Vol 102, 1956, p 1263; (2) C.W. Kothen and H.L. Johnston, *J. Am. Chem. Soc.*, Vol 75, 1953, p 1301; and (3) N.M. Wolcott, Conf. de Physique des Basses Temperatures, Paris, 1955

Latent Heat

The listed values of enthalpies (latent heats) of transformation and melting are from the *Metals Handbook*, Vol 2, 9th ed., 1979, and recent assessments given by the following:

- 79Jan: D. Stull and H. Prophet, Ed., *JANAF Thermochemical Tables*, 2nd ed., NSRDS-NBS 37, 1971, Supplements in *J. Phys. Chem. Ref Data*, Vol 3, p 311-480, Vol 4, p 1-175, Vol 7, p 793-940, Vol 11, p 695-940

- 84Des: P.D. Desai, "Thermodynamic Properties of Titanium," CINDAS Report 77, 1984

Values from the Metals Handbook are only estimates.

Latent Heat of Transformation:
87 ± kJ/kg (79Jan);
80 ± 8 kJ/kg (84Des);

Pure titanium: Specific heat at 1 to 10 K

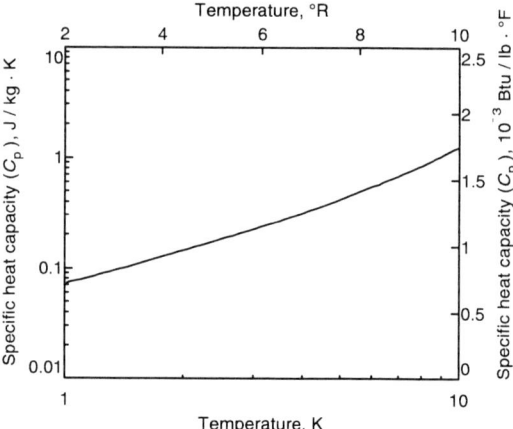

Source: V.J. Johnson, A Compendium of the Properties of Materials at Low Temperatures (Phase I), WADD Technical Report 60-56 Part II, Wright Air Development Division, 1960, p 4.141

92 kJ/kg (estimated value in *Metals Handbook*, Vol 2, 9th ed.)

Latent Heat of Fusion:
295 ± 3 kJ/kg (79Jan);
288 ± 10 kJ/kg (84Des);
440 kJ/kg (estimated value in *Metals Handbook*, Vol 2, 9th ed.)

Latent Heat of Vaporization. Estimated at 9.83 MJ/kg in *Metals Handbook*, 9th ed.

Thermal Expansion

The thermal coefficient of linear expansion is about 10.8 ppm/K at 900 to 1100 °C and 14.7 ppm/K at 100 to 1500 °C (see table) based on the work of Spreadborough (59Spr) and Schmitz-Pranghe (68Sch). These expressions lead to a lattice parameter of 0.32763 nm at room temperature. This value is comparable to values obtained by extrapolation of lattice parameters of quenched binary alloys (Ti-Cr and Ti-V) to 100% Ti.

Pure titanium: Coefficients of linear thermal expansion

Reference(a)	Coefficient $10^6 \frac{da}{a} dT$, °C^{-1}	$10^6 \frac{dc}{c} dT$, °C^{-1}	Temperature range, °C
[59Spr]	9.55 ± 0.5	10.65 ± 0.7	25 to 700
[53Mch]	11.0	8.8	25 to 225
[68Paw]	9.5	5.6	28 to 155
[53Ber]	11.03	13.37	25 to 700
[68Sch]	~6	~8	–200
	~10	~12.7	400
	~16	~18.7	882

(a) Full citations listed in *Phase Diagrams of Binary Titanium Alloys* (Murray, Ed), ASM International, 1987, p 2-4

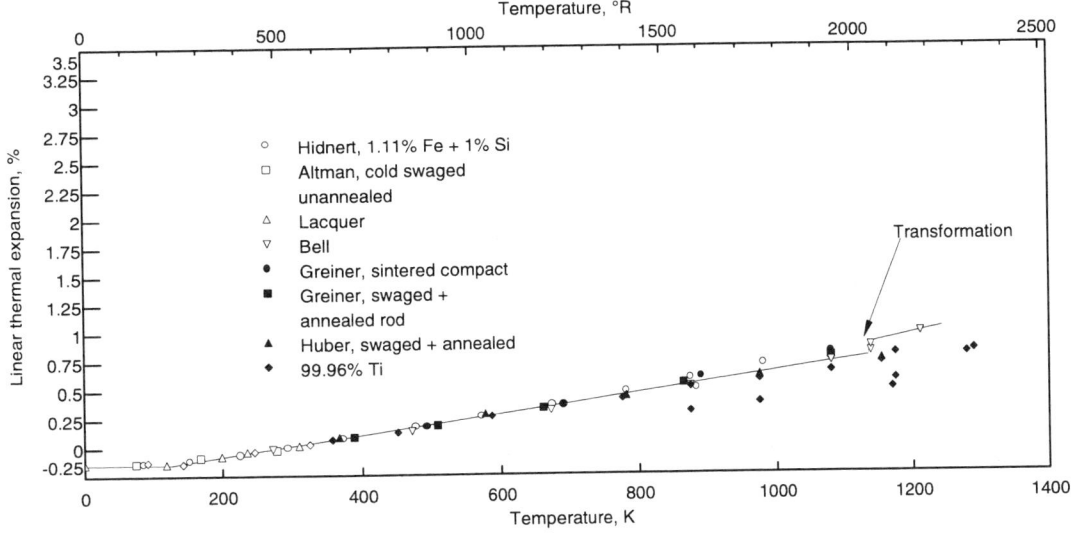

Pure Titanium: Linear thermal expansion

Source: *Thermophysical Properties of Solid Materials, Vol 1 - Elements*, WADC Technical Report 58-747, Wright Air Development Division, 1960

Transport Properties

Titanium has a low thermal conductivity compared to that of other metals, but its heat-transfer rate is greater than that of most copper-base alloys. The reason for this has to do with favorable surface film characteristics and lack of corrosion. The major factor in heat transfer relates to material thickness, corrosion resistance, and surface films, not the thermal conductivity of the metal. In this regard, titanium has the following advantages:

- Good strength (depending on impurity levels)
- Resistance to erosion and erosion-corrosion
- Very thin, conductive oxide surface film
- Hard, smooth surface that limits adhesion of foreign materials
- Surface promotes dropwise condensation

Thermal conductivity is affected by impurity content, and data on pure titanium are limited (see table). The value for very high purity titanium (99.99 wt% Ti) at –240 °C (–400 °F), is almost double that of 99.9 wt% Ti at the same temperature.

Thermal Diffusivity. 0.022 m^2/h at 570 °C (1057 °F), 0.02 m^2/h at 700 °C (1300 °F), and 0.0185 m^2/h at 840 °C (1540 °F) per McIntosh *et al.*, *Trans ASME*, Vol 76, 1954, p 407-410.

Thermal Conductivity

Thermal conductivity of various pure titanium samples

Sample description and data source footnote	Thermal conductivity, W/m·°C (Btu/ft·h·°F), at mean temperatures of:									
	−240 °C (−400 °F)	0 °C (32 °F)	20 °C (68 °F)	25 °C (77 °F)	50 °C (120 °F)	100 °C (212 °F)	150 °C (300 °F)	200 °C (390 °F)	250 °C (480 °F)	300 °C (570 °F)
99.99% Ti(a)	19.4 (11.2)	...	...	...	...	...	...	...	...	...
99.9% Ti(a)	11.4 (6.6)	...	...	...	...	...	...	...	...	...
Polycrystalline solid with unspecified impurity(b)	...	22.4 (12.95)	...	21.9 (12.65)	...	20.7 (11.97)	...	...	...	...
Normal commercial purity(c)	...	...	...	...	18.86 (10.90)	18.75 (10.84)	18.64 (10.77)	18.6 (10.75)	18.5 (10.7)	18.4 (10.6)
High-purity grade(c)	...	...	...	...	20.5 (11.85)	20.1 (11.6)	19.7 (11.4)	19.25 (11.1)	18.9 (10.9)	18.55 (10.7)
Very high purity(c)	...	...	20.0 (12.7)	...	...	...	...	...	...	...

(a) Mendelssohn and Rosenberg, *Proceeding of Physical Society* (London), Vol 65A, 1952, p 385. (b) C.Y. Ho et al., *Journal of Physical and Chemical Reference Data*, Vol 1, 1972, p 279-421. (c) E.W. Powell and R.P. Tye, *Journal of Less-Common Metals*, Vol 3, 1961, p 226-233

Pure Titanium: Thermal conductivity data

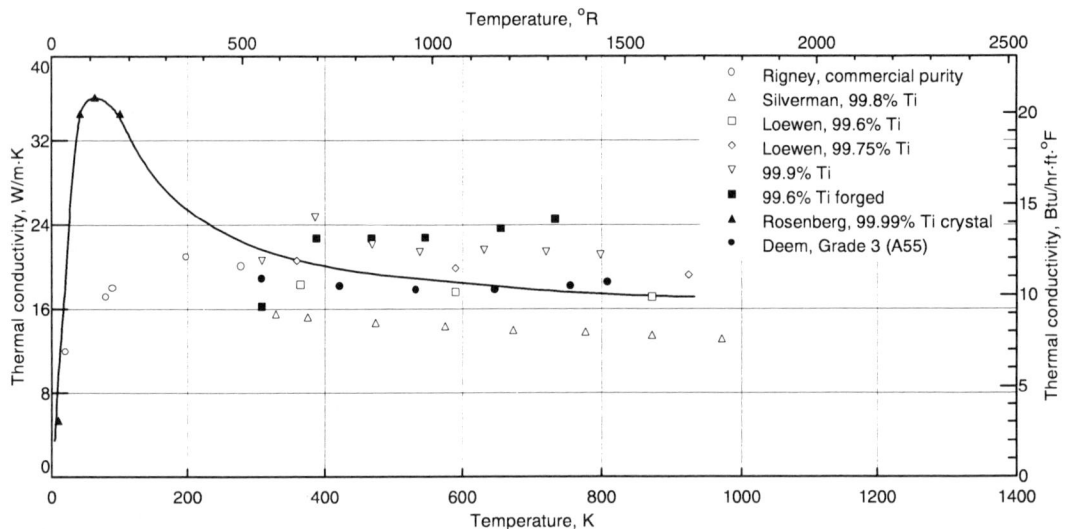

Source: *Thermophysical Properties of Solid Materials, Vol 1 - Elements*, WADC Technical Report 58-476, Wright Air Development Division, 1960

Mass Diffusivity

Pure titanium diffusion rates

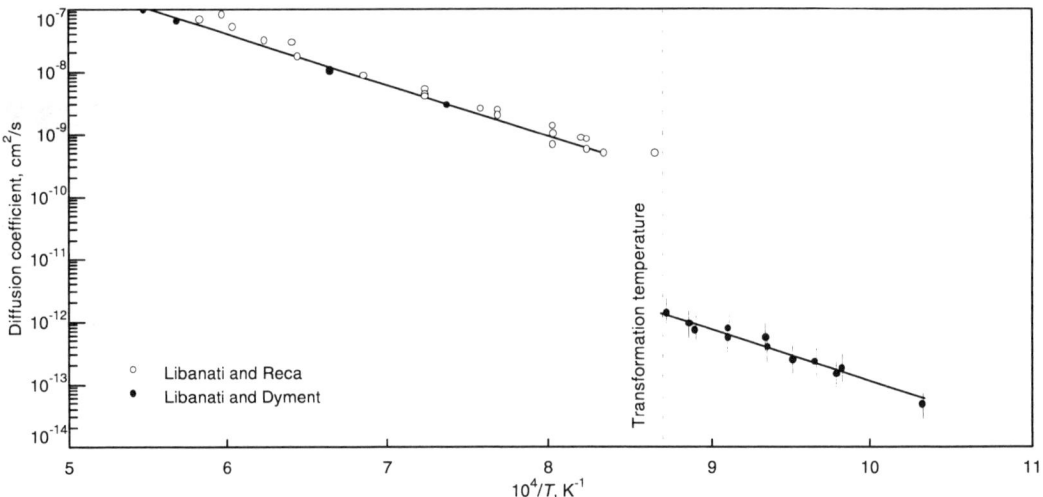

Diffusion of Ti^{44} in alpha and beta titanium (99.99 wt%).
Source: F. Dyment and C.M. Libanati, Self-Diffusion of Ti, Zr, and Hf in Their HCP Phases, and Diffusion of Nb^{95} in HPC Zr, *J. Mater. Sci.*, Vol 3, 1968, p 349-359.

Tensile Properties

Typical room-temperature tensile properties of pure titanium are:
- 235 MPa (34 ksi) ultimate tensile strength
- 140 MPa (20 ksi) tensile yield strength (0.2% offset)
- 50% elongation in 50 mm (2 in.)

Effect of Impurities. Interstitial impurities have a strong influence on tensile properties, which can be correlated to the hardening effect of the impurities (see figures).

Grain Size Effect. Effect of grain size on tensile yield strength obeys the Hall-Petch relation, but the effect is less pronounced for the more pure grades of titanium.

IMI 110 titanium sheet: Guaranteed mechanical properties

Minimum 0.2% yield strength	130 MPa (18.8 ksi)
Ultimate tensile strength	270-350 MPa (39-50 ksi)
Minimum elongation on 50 mm (2 in.)	30%
Maximum bend radius:,	
Sheet up to 1.83 mm (0.072 in.)	$1T$
Sheet up to 1.83-3.25 mm (0.128 in.)	$2T$

Pure Ti: Effect of impurities on tensile strength

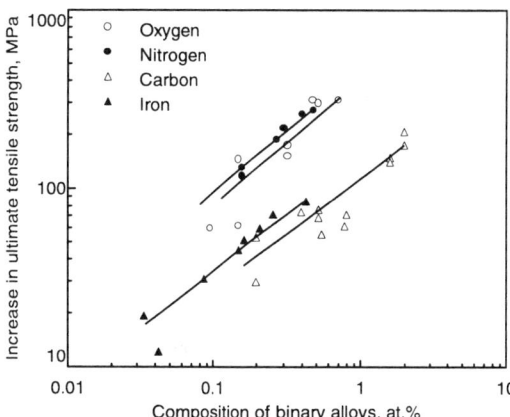

Titanium ingots having dilute carbon, nitrogen and oxygen impurity levels were cold rolled to 1 mm (0.040 in.) thick sheet, mechanically cleaned, 1000 °C vacuum annealed, cold rolled to 0.5 mm (0.020 in.), and finally were annealed for one hour at 700 °C in vacuum. This treatment gave an average equiaxed grain size of 0.020 to 0.025 mm.
Source: W.L. Finlay and J.A. Snyder, *J. Met.*, Vol 188, Feb 1950, p 280-282

Effect of Grain Size

Effect of grain size on room-temperature yield strength of three purities of titanium

Material	Impurities (by weight), ppm	Hall-Petch factors(a)			
		Friction stress (σ_i)(b)		Grain size constant (K)	
		kgf/mm^2	MPa	kgf/mm$^{3/2}$	MPa/$\sqrt{m}$
A70 (ASTM grade 4)	2830 O, 25 C, 250 N, 67 H, 3700 Fe	40	390	0.86	266
Battelle	330 O, 25 C, 140 N, 62 H, 90 Fe	18	175	0.32	99
Iodide	30 O, 18 C, 110 N, 18 H	14	137	0.47	145

(a) Factors in the Hall-Petch relation: yield stress = $\sigma_i + K\sqrt{d}$, where d is the average grain size. (b) Friction stress for a strain rate of 3×10^{-4}/s. Source: H. Conrad and R. Jones, *The Science, Technology and Application of Titanium*, Jaffee and Promise, Ed., Pergamon Press, 1970, p 493

Effect of Impurities

Pure titanium: Effect of nitrogen on tensile properties

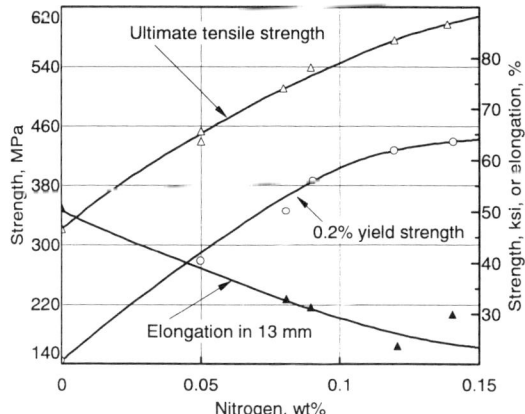

Source: W.L. Finlay and J.A. Synder, Effects of Three Interstitial Solutes (Nitrogen, Oxygen, and Carbon) on The Mechanical Properties of High-Purity, Alpha Titanium, *Trans. AIME*, Vol 188; *J. Met.*, Feb 1950, p 277-285

Effect of grain size on tensile yield strength

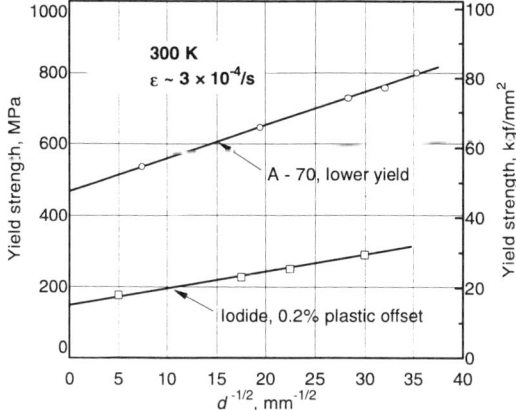

Room-temperature strength of ASTM grade 4 titanium (A-70) and iodide titanium (30 ppm O, 110 ppm N) tested at a strain rate of 3×10^{-4}/s.
Source: R.L. Jones and H. Conrad, *Trans. TMS/AIME*, Vol 245, 1969, p 779-789

Pure titanium: Effect of oxygen on tensile properties

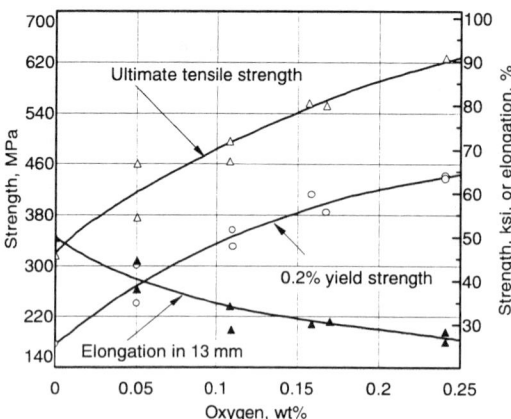

Source: W.L. Finlay and J.A. Synder, Effects of Three Interstitial Solutes (Nitrogen, Oxygen, and Carbon) on The Mechanical Properties of High-Purity, Alpha Titanium, *Trans. AIME,* Vol 188; *J. Met.,* Feb 1950, p 277-285

Pure titanium: Effect of carbon on tensile properties

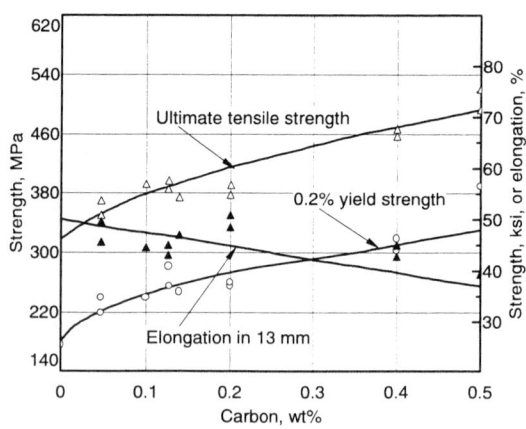

W.L. Finlay and J.A. Synder, Effects of Three Interstitial Solutes (Nitrogen, Oxygen, and Carbon) on The Mechanical Propeties of High-Purity, Alpha Titanium, *Trans. AIME* Vol 188; *J. Met.,* Feb 1950, p 277-285

Effect of Temperature

Typical elongation and ultimate tensile strength

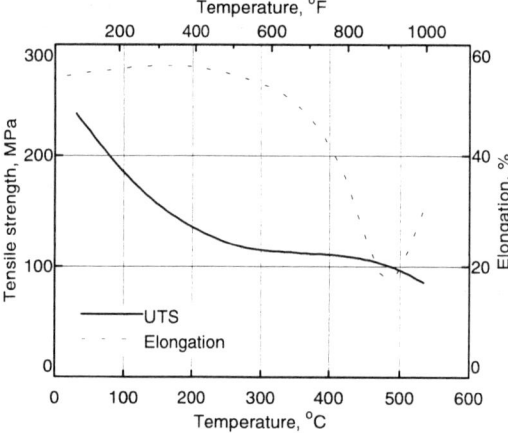

Source: *Metals Handbook,* 9th ed., Vol 2, p 815

Pure titanium: Effect of iron on tensile properties

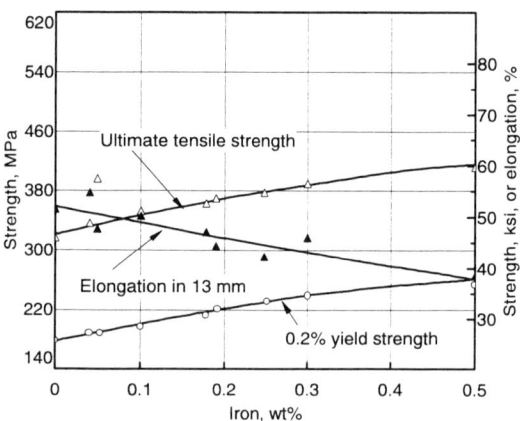

Source: W.L. Finlay and J.A. Synder, Effects of Three Interstitial Solutes (Nitrogen, Oxygen, and Carbon) on The Mechanical Properties of High-Purity, Alpha Titanium, *Trans. AIME,* Vol 188; *J. Met.,* Feb 1950, p 277-285

Effect of temperature on tensile yield strength and flow stress

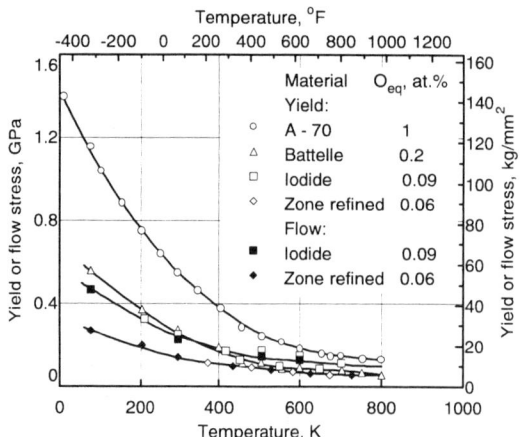

Grain sizes of 1 to 3 μm and strain rate of about 10^{-4}/s.
H. Conrad and R. Jones, *The Science, Technology and Application of Titanium,* Jaffee and Promisel, Ed., Pergamon, 1970, p 494

Correlation with Hardness

Pure Ti: Correlation of tensile properties with hardness

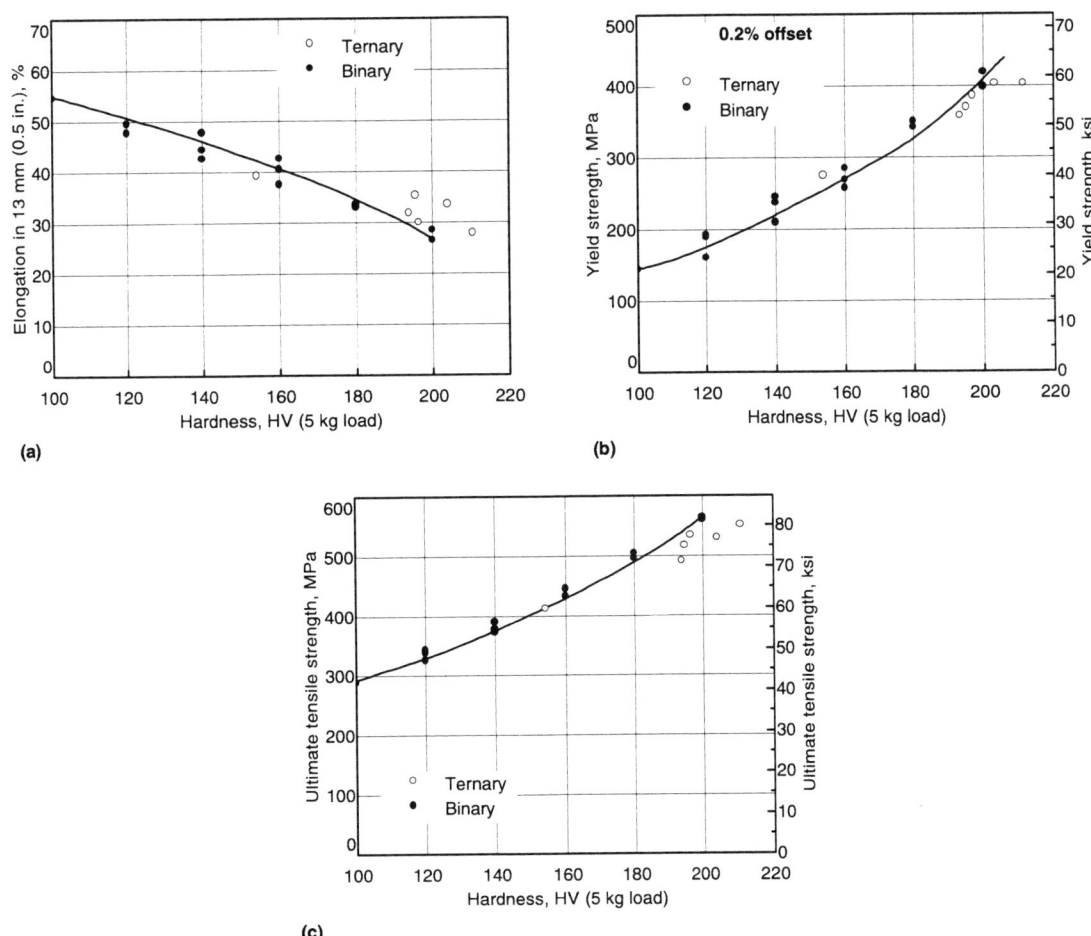

Impurity contents of binary alloys range from 0.04 at.% to 2.0 at.% (see the figure showing the increases in ultimate tensile strength for binary alloys). Annealed, high-purity titanium.
Source: W.L. Finlay and J.A. Synder, *J. Met.,* Vol 188, Feb 1950, p 280-282

Hardness

Pure titanium: Effect of impurities on Vickers hardness

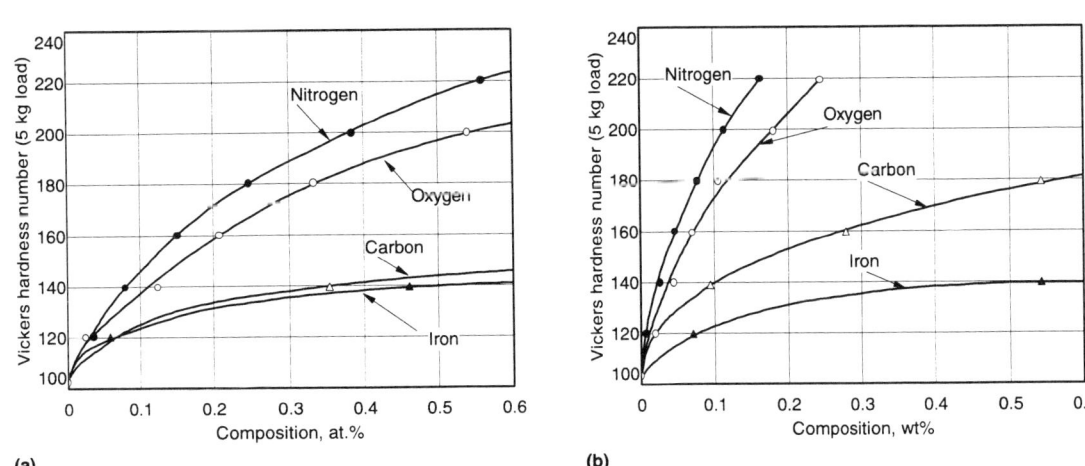

Titanium ingots having dilute carbon, nitrogen and oxygen impurity levels were cold rolled to 1 mm (0.040 in.) thick sheet, mechanically cleaned, 1000 °C vacuum annealed, cold rolled to 0.5 mm (0.020 in.), and finally were annealed for one hour at 700 °C in vacuum. This treatment gave an average equiaxed grain size of 0.020 to 0.025 mm.
Source: W. L. Finlay and J.A. Synder, *J. Met.*, Feb 1950, and *Trans. AIME*, Vol 188, p 280

Ingot melted from electrolytic titanium has a typical hardness of 70 to 74 HB. Ingot from iodide titanium has hardness of 65 to 72 HB.

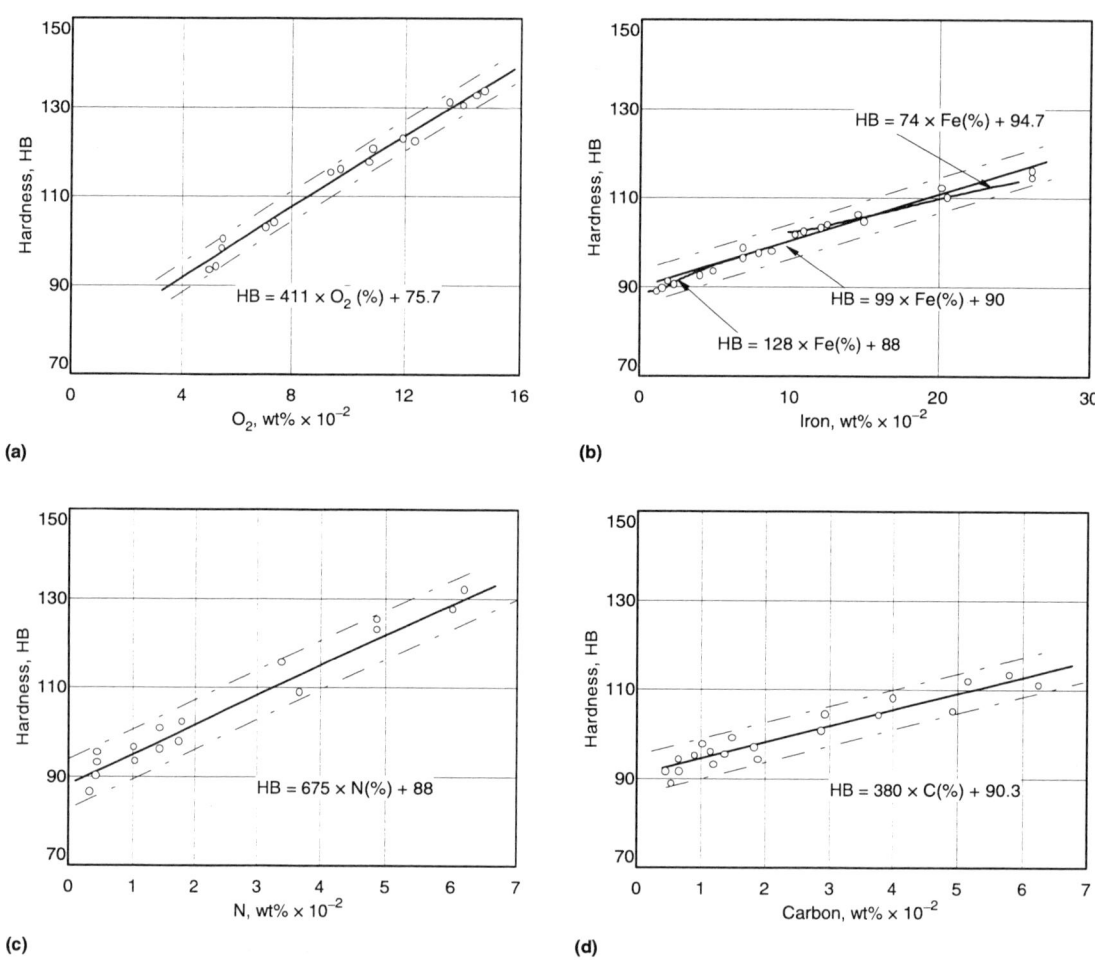

Pure titanium: Effect of impurities on Brinell hardness

Relation between Brinell hardness (HB) and (a) O_2, (b) iron, (c) nitrogen, and (d) carbon content.
Source: K. Shimasaki et al., Relation Between Brinell Hardness of Titanium and Impurities, *Titanium 80, Science and Technology*, AIME, 1980, p 1131-1136

Creep Properties

Creep Deformation

Diffusion creep of high purity (99.9%) titanium is similar to that of commercial purity titanium, and interstitial impurities have little influence on strength levels above 0.3 T_m (580 K, or 300 °C). At temperatures higher than 0.4 T_m (500 °C), titanium becomes very soft and undergoes dynamic recrystallization. Below 0.4 T_m, self-diffusion is not expected to play a role in mechanical properties.

Below 0.4 T_m. Between about 150 and 370 °C (300 and 700 °F) commercially pure titanium does not normally creep at stresses below the yield strength, although creep below 150 °C (300 °F) has been reported (*Metals Handbook*, 8th ed., Vol 1, p 537). This reversal in creep behavior may be associated with residual stress. Above 0.25 T_m (210 °C) and at stresses less than $3 \times 10^{-3} E$ (where E = Young's modulus), deformation contributes to the growth of voids, which results in transgranular creep fracture (Y. Krishnamohanrao et al., *Acta Metall.*, Vol 34, No. 9, 1986, p 1783-1806).

Above 0.4 T_m. At temperatures above 0.4 T_m (500 °C), creep can occur at relative low stresses on the order of $10^{-5} G$, where G is the shear modulus. Comprehensive data of creep behavior in this regime has been generated over a wide range of grain sizes by Malakondoiah and Rama Rao (see table). Creep mechanism maps (see figures) also have been generated for ranges of stress, temperature, and grain size. Over the range of temperature (550 to 815 °C) and grain size (34 to 443 μm) investigated, titanium exhibits a Newtonian viscous creep behavior indicating grain boundary creep to be the dominant mechanism. When the applied stress exceeds a transition-stress, a switch occurs from a viscous creep mechanism to a power-law, climb-creep mechanism with a stress exponent of 4.2 to 4.6.

Pure titanium: Creep behavior above $0.4\,T_m$ (500 °C)

Prior treatment of spring specimen(a)	Average grain size μm	Test temp K	Total duration of test h	Duration in transient creep h	Maximum applied stress MPa	Measured $d\varepsilon/d\sigma$ $(MPa \cdot s)^{-1}$
923 K/⅙ h	34	823	56.0	25.0	1.68	3.46×10^{-10}
1023 K/⅙ h	45	823	100.0	30.0	1.45	1.44×10^{-10}
1023 K/½ h	50	823	97.5	40.0	1.47	1.11×10^{-10}
1000 K/½ h + 25% RA + 1123 K/⅙ h	63	823	100.0	40.0	1.64	7.25×10^{-11}
1000 K/½ h + 25% RA + 1123 K/⅙ h	67	873	100.0	30.0	1.52	1.57×10^{-10}
1000 K/½ h + 25% RA + 1123 K/⅙ h	69	933	36.0	15.0	1.93	3.88×10^{-10}
1000 K/½ h + 25% RA + 1123 K/⅙ h	69	1000	23.5	10.0	1.55	1.45×10^{-9}
1000 K/½ h + 25% RA + 1123 K/⅙ h	69	1033	7.0	1.25	1.09	4.87×10^{-9}
1000 K/½ h + 25% RA + 1123 K/¼ h	75(b)	1048	8.5	...	1.51	5.54×10^{-9}
920 K/½ h + 16% RA + 1123 K/¼ h	84	1046	14.0	3.0	1.90	4.87×10^{-9}
1000 K/½ h + 25% RA + 1123 K/⅓ h	84	1042	8.25	2.5	1.37	3.79×10^{-9}
		1008	14.5	2.5	1.37	1.41×10^{-9}
		992	24.0	7.0	1.37	9.99×10^{-10}
920 K/½ h + 10% RA + 1123 K/½ h	86	1049	22.0	3.5	0.99	4.60×10^{-9}
1000 K/½ h + 15% RA + 1123 K/⅙ h	88	1000	47.7	15.0	1.27	1.16×10^{-9}
1000 K/½ h + 15% RA + 1123 K/⅙ h	91	1000	40.0	15.0	1.26	1.23×10^{-9}
920 K/½ h + 8% RA + 1123 K/½ h	110	1046	17.5	3.0	1.13	6.89×10^{-9}
1000 K/½ h + 15% RA + 1123 K/⅓ h	143	1000	30.5	10.0	1.28	1.13×10^{-9}
1000 K/½ h + 15% RA + 1123 K/½ h	195	1000	45.0	10.0	1.29	1.16×10^{-9}
1123 K/1 h	198	956	61.0	25.0	1.62	4.18×10^{-10}
1123 K/1 h	200	1058	12.0	1.0	0.98	4.41×10^{-9}
		1013	22.0	4.0	0.98	1.58×10^{-9}
1123 K/1 h	228	1058	13.0	6.0	1.39	3.21×10^{-9}
1123 K/1 h	238	1088	7.0	3.0	1.68	9.00×10^{-9}
1123 K/1 h	249	1010	23.0	9.0	1.67	1.45×10^{-9}
1100 K/1 h + 6% RA + 1143 K/2 h	276	1061	9.0	2.5	1.28	3.89×10^{-9}
		1032	11.5	1.5	1.28	1.61×10^{-9}
		1000	24.0	4.0	1.28	8.27×10^{-10}
1100 K/½ h + 15% RA + 1123 K/2 h	282	1000	24.0	7.5	1.28	1.05×10^{-9}
1100 K/½ h + 6% RA + 1123 K/1 h	315	1000	36.0	10.0	1.31	1.78×10^{-9}
1100 K/1 h + 6% RA + 1143 K/2 h	443	1000	40.0	10.0	1.25	1.67×10^{-9}

(a) To measure low strain rates ($<10^{-9}$/s) at stresses less than 2 MPa (300 psi) a sensitive spring geometry specimen was employed. RA = reduction of area by wire drawing. Initial material was 1.5 mm wire having an oxygen content of 2600 ppm. (b) During the test, grain size increased from 75 to 86 microns. Source: G. Malakondaiah and P. Rama Rao,"Creep of Titanium at Low Stresses, *Acta Metall.*,Vol 29,1981, p 1263-1275

Creep Mechanisms

Pure titanium: Creep mechanism map above $0.4\,T_m$

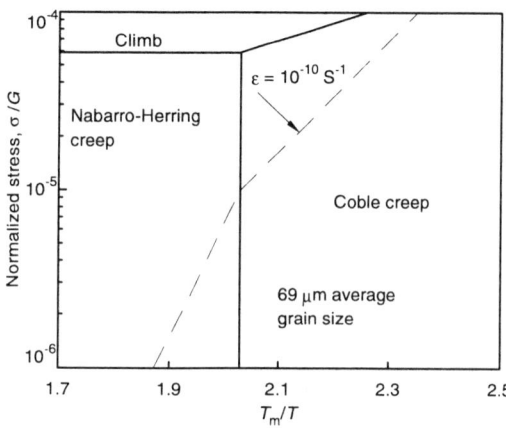

A transition from grain boundary diffusion controlled (Coble) creep to lattice diffusion controlled (Nabazzo-Herring) creep occurs at about 970 K ($0.5\,T_m$) for a grain size of ~65 μm. A similar transition is shown on the mechanism map for 69 μ grain size. Dashed line is a contour for a constant creep rate of 10^{-10}/s.
Map construction is based on an assumed room-temperature shear modulus of 49.5 GPa and a temperature coefficient of 25 MPa/K.
Source: G. Malakondaiah and P. Rama Rao, Creep of Alpha-Titanium at Low Stresses, *Acta Metall.*, Vol 29, 1981, p 1263-1275

Pure titanium: Effect of grain size on creep mechanism

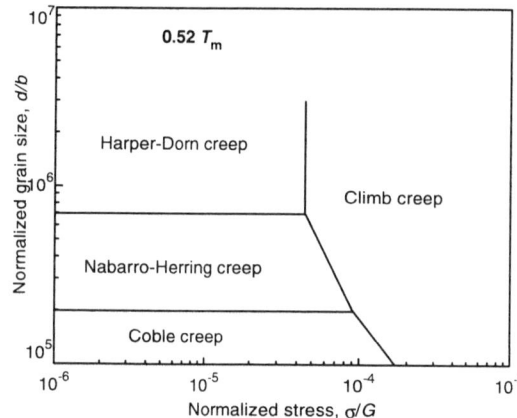

d = minimum grain diameter; b = Burgers vector.
Measured creep rates in the coarse grain size and intermediate temperature regime are found to be independent of grain size and are in good agreement with the predictions of the Harper-Dorn creep empirical relationship
Source: G. Malankondaiah and P. Rama Rao, *Acta Metall.*, Vol 29, 1981, p 1263-1275

Rupture

Stress-rupture characteristics of metals have been correlated by the following relation developed by Orr *et al.* (*Trans. Amer. Soc. Metals*, Vol 46, 1954, p 113):

$$\sigma = f[\,t_r \exp(H/RT)\,]$$

where σ is the rupture stress, t_r is the time to rupture, and H is a constant (which is considered to be the activation energy for creep and self-diffusion in the case of pure metals). Orr *et al* assigned an activation energy of 250 kJ/mole (60 kcal/mole) for commercial purity 75A (99.6% Ti), which is considered a good estimate for high purity titanium as well.

Fatigue Properties

Fatigue limits of unalloyed titanium depend on interstitial contents, but the ratio of fatigue limits and yield strength appears relatively constant at room temperature (see figure, Conrad *et al.*, Critical Review: Deformation and Fracture, *Titanium Science and Technology*, Plenum Press, 1973, p 996-1000). The ratio of fatigue limit and yield stress does show a temperature dependence (see figure, N.G. Turner and W.T. Roberts).

Unalloyed titanium: Effect of impurities on fatigue limits

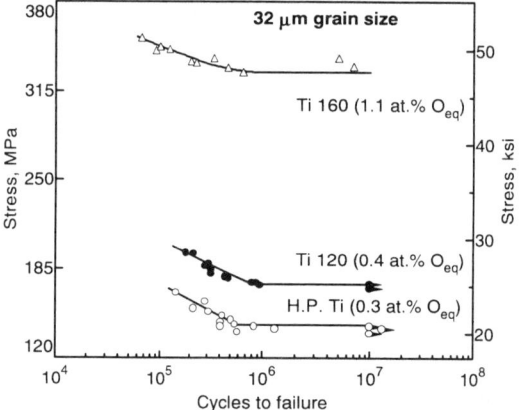

Source: N.G. Turner and W.T. Roberts, *Trans. AIME*, Vol 242, 1968, p 1223

Pure Ti: Yield stress/fatigue limit ratio vs temperature

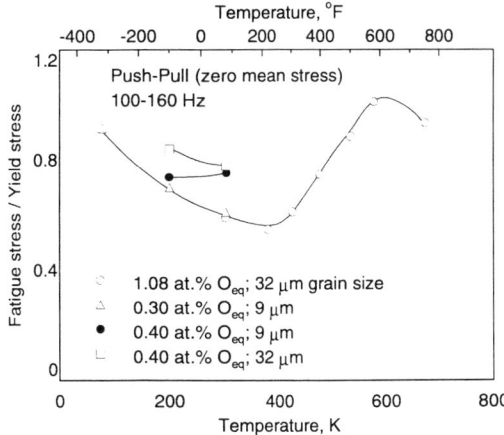

Source: N.G. Turner and W.T. Roberts, *J. Less-Common Met.*, Vol 16, 1968, p 37; and Trans. AIME, Vol 242, 1968, p 1223

Unalloyed titanium: Ratio of yield strength and fatigue limit

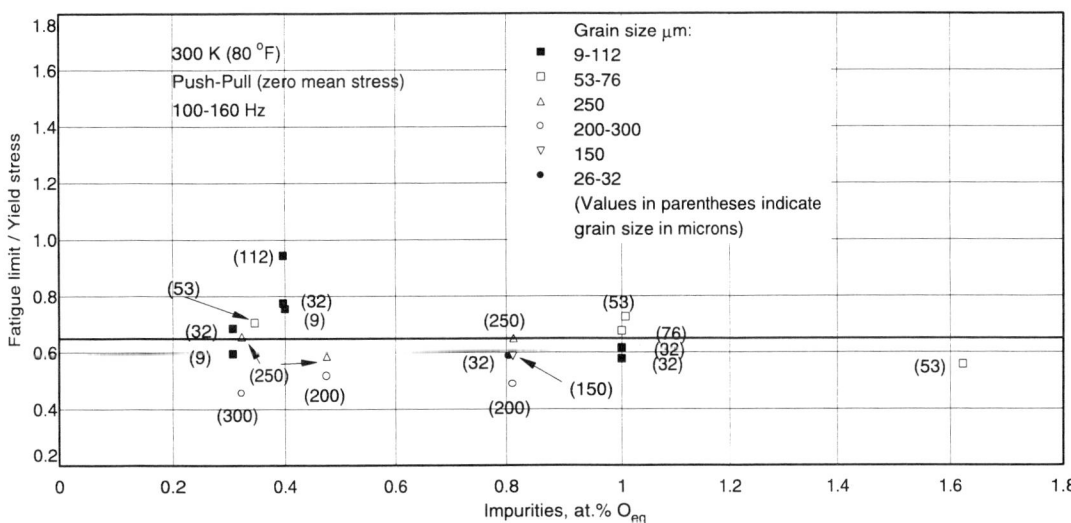

Source: Conrad *et al.*, *Titanium Science and Technology*, Plenum Press, 1973, p 998

Impact Strength

High-purity Ti: Effect of hydrogen and heat treatment on impact strength

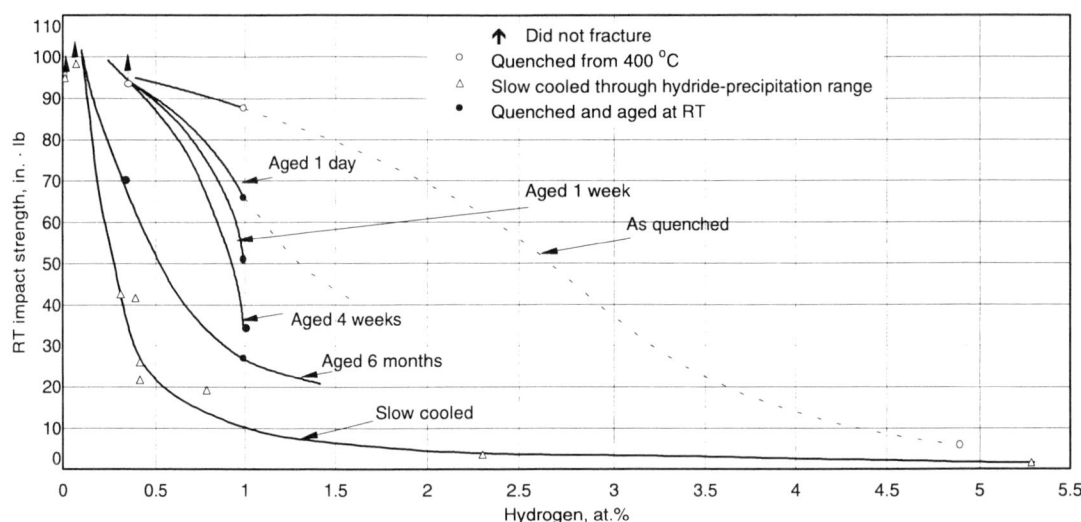

The beneficial effect on toughness of water quenching from α phase field occurs from the fine dispersion of hydrides formed during precipitation. Slow cooling precipitates plate-like hydrides.
Source: G.A. Lenning, C.M. Craighead, and R.I. Jaffee, Constitution and Mechanical Properties of Titanium-Hydrogen Alloys, *Hydrogen Damage* (C.D. Beachem, Ed.), American Society for Metals, 1977, p 100

High-purity Ti: Effect of hydrogen on impact strength

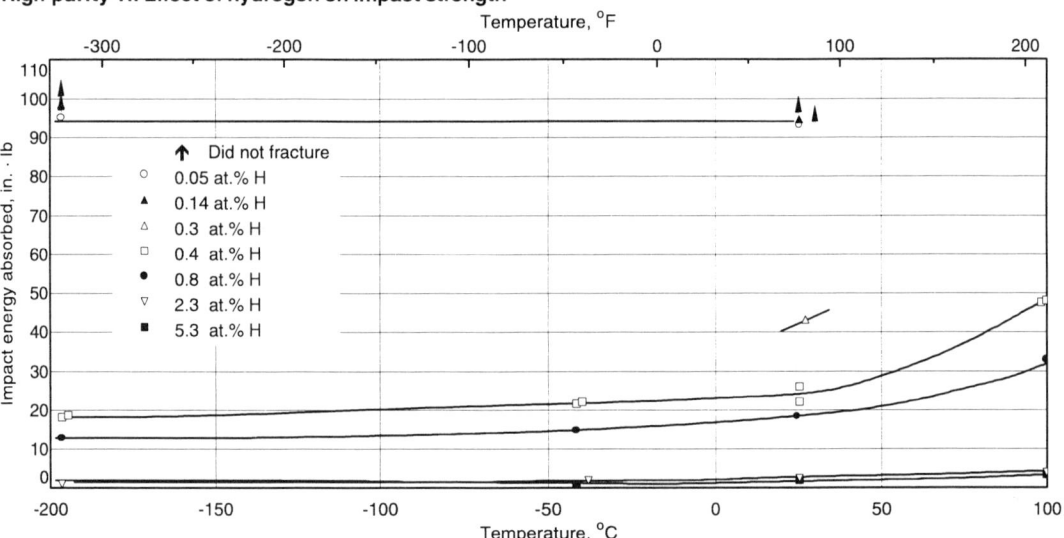

Effect of hydrogen on impact properties of high purity titanium. Specimens cooled through hydride-precipitation range at about 15 to 20 °C per min.
Source: G.A. Lenning, C.M. Craighead, and R.I. Jaffee, Constitution and Mechanical Properties of Titanium-Hydrogen Alloys, *Hydrogen Damage* (C.D. Beachem, Ed.), American Society for Metals, 1977, p 100

Results of impact and slow bend tests on Ti-H alloys

Hydrogen content, Atomic Pct	Energy absorbed in breaking, ft-lb	
	Impact tests 18.1 ft. per sec	Slow bend tests, 0.08 in. per Min
0.05	30	15.5
0.05	32	20
0.26	19.5	18.5
0.26	20	22
2.0	3.5	4.0
2.0		6.5

Source: G.A. Lenning, C.M. Craighead, and R.I. Jaffee, Constitution and Mechanical Properties of Titanium-Hydrogen Alloys, *Hydrogen Damage*. (C.D. Beachem, Ed.), American Society for Metals, 1977, p 100

Fracture Mechanisms

Pure (99.9%) titanium has only three fracture modes (see figure), all of which are ductile modes of fracture. Brittle fracture is absent at low temperatures and at high temperature-low stress conditions.

Below $0.3\,T_m$ fracture stress obeys the Hall-Petch relation, while fracture above $0.4\,T_m$ correlates with the following relation for stress rupture:

$$\sigma_r = f[\,t_r \exp(-H/RT)\,]$$

where σ_r is the rupture stress, t_r is the time to rupture, and H is a constant (which is considered to be the activation energy for creep and self-diffusion in the case of pure metals). This relation developed by Orr *et al.* (*Trans Amer. Soc. Metals*, Vol 46, 1954, p 113) identifies an activation energy of 250 kJ/mole (60 kcal/mole) for commercially pure (99.6% Ti) grade 75A. This activation energy is considered a close estimate for the activation energy of pure titanium.

Effect of Impurities. The reduction of ductility with high interstitial contents distinguishes the fracture behavior of high purity titanium (99.9%) from that of commercially pure grades. Rosi and Perkins (*Trans. Amer. Soc. Metals*, Vol 45, 1953, p 972), for example, observed a brittle (cleavage mode) failure of 99.7% Ti specimen [(25 μm grain size at 77 K (−195 °C)]. A recent review of other fracture tests also identifies a region of brittle fracture for commercially pure (mostly 99.7 wt%) titanium (see fracture mechanism map).

Iodide titanium: Fracture mechanism map

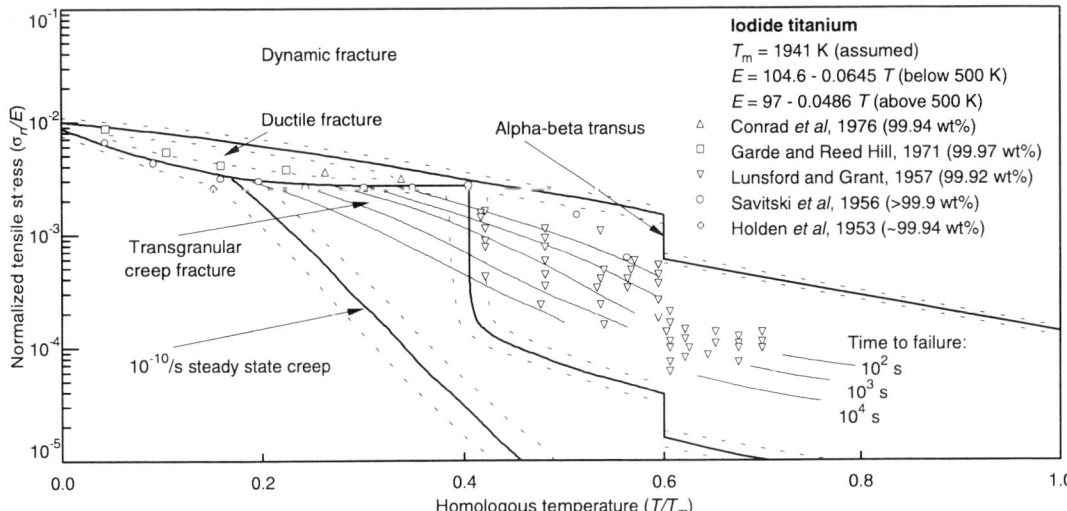

The left boundary for transgranular creep fracture is chosen at a steady state creep rate of $10^{-10}\,\text{s}^{-1}$, per Gandhi and Ashby, *Acta Metall.*, Vol 27, 1979, p 1565. Parameters normalized with Young's modulus (E) and melting temperature (T_m).
Source: P. Rama Rao *et al.*, Fracture Mechanism Maps for Titanium and its Alloys, *Acta Metall.*, Vol 34, 1986, p 1783-1806

CP titanium: Fracture mechanism map

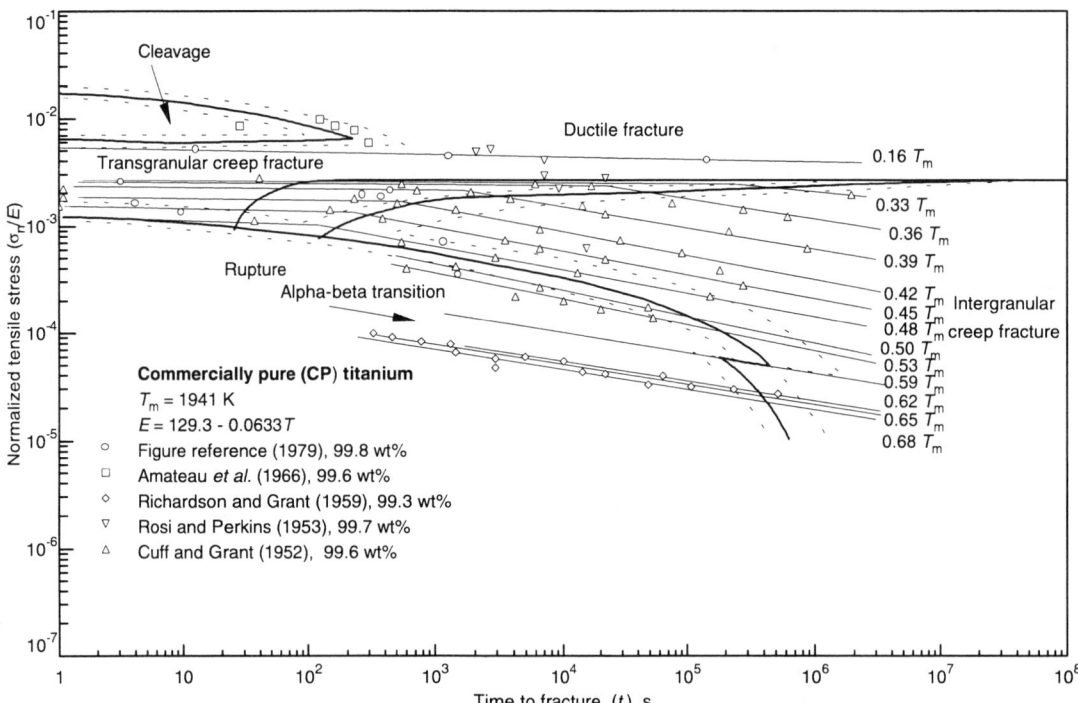

Full citations of data sources are given in the figure reference. The region of intergranular creep fracture is based on the data of Cuff and Grant (*Iron Age*, Vol 20, p 134), which attributes a sharp break in in stress-rupture plots to change in fracture mode (i.e. transgranular to intergranular).
Source: P. Rama Rao *et al.*, *Acta Metall.*, Vol 34, 1986, p 1783-1806

Fracture Stress

The fracture stress (σ_F) below 0.3 T_m obeys a modified Hall-Petch relation:

$$\sigma_F = \sigma_F + K_F\sqrt{d}$$

where σ_F is the friction stress, K_F is the grain size constant, and d is the average grain size. The value of K_F increases with interstitial content, especially above 500 K (400 °F) where dynamic strain aging becomes more pronounced with higher interstitial contents (see figure). The friction stress (σ_i) is a function of interstitial content, strain rate, temperature, and the amount of strain.

Elongation at fracture decreases with decreasing interstitial content but is relatively independent of grain size (H. Conrad *et al.*, Critical Review Deformation and Fracture, *Titanium Science and Technology*, Vol 2, Plenum Press, 1973, p 996).

Pure titanium: Hall-Petch coefficients for fracture stress (σ_F)

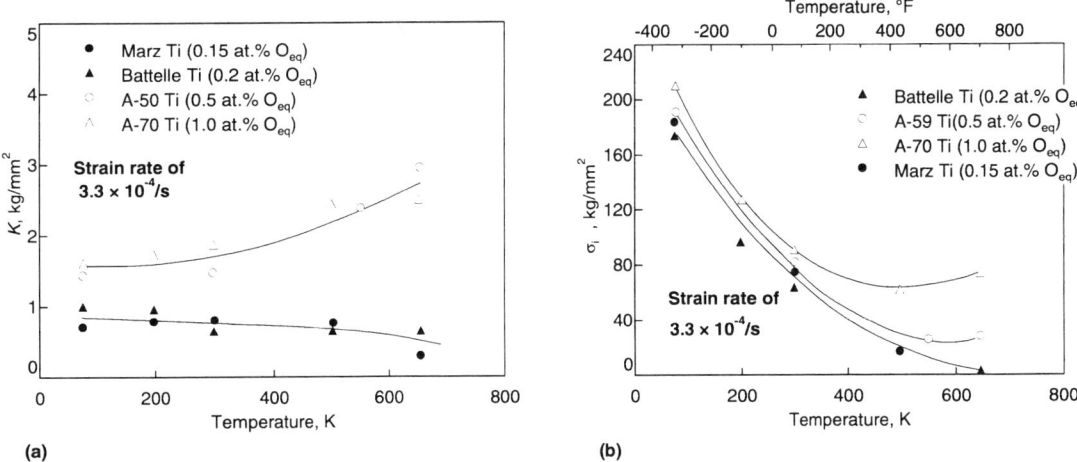

The results take into account the Bridgman correction factor for necking and show the effect of interstitial content on (a) the grain size constant (K_F) and (b) the friction stress (σ_i). $\sigma_F = \sigma_i + K_F\sqrt{d}$ where d is the average grain size.
Source: G. Sargent et al., Effect of Temperature, Interstitial Content and Grain Size on the Fracture Behavior of Titanium, *Titanium and Titanium Alloys* (Williams and Belov, Ed.), Plenum Press, 1982, p 679-689

Pure titanium: Effect of grain size and interstitial content on fracture stress

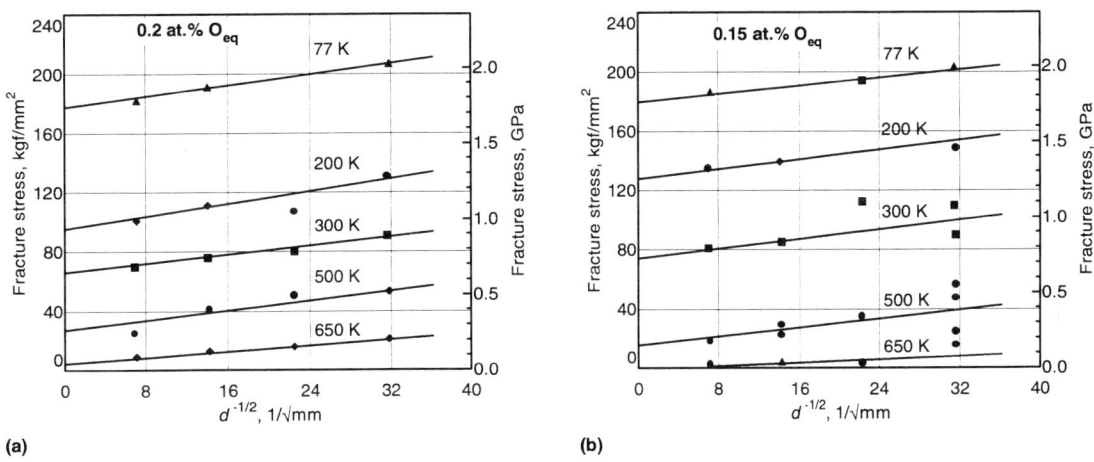

Strain rate was 3.3×10^{-4}/s.
Source: G. Sargent et al., Effect of Temperature, Interstitial Content and Grain Size on the Fracture Behavior of Titanium, *Titanium and Titanium Alloys*, Williams and Belov, Ed., Plenum Press, 1982, p 684

Effect of N, O, and C interstitials on elongation at fracture (300K)

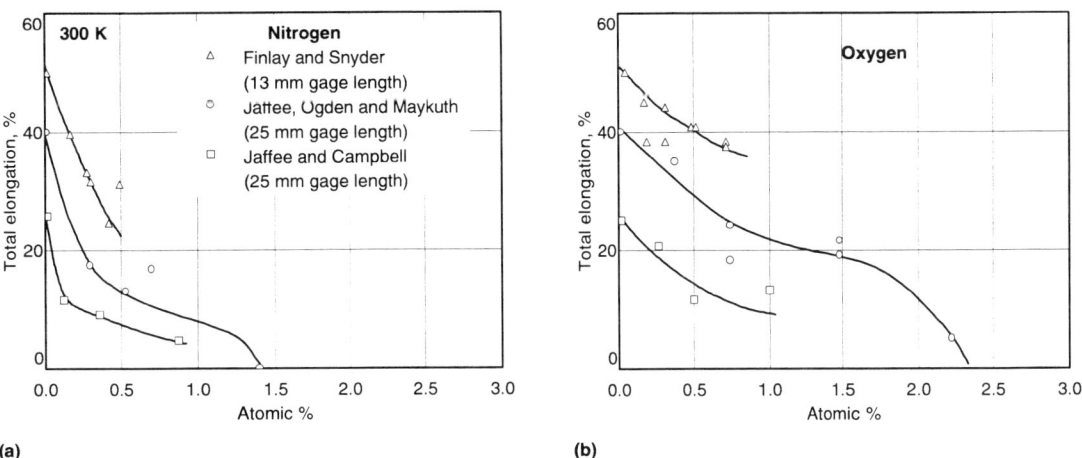

Source: H. Conrad et al., *Titanium Science and Technology*, Plenum Press, 1973, p 997

Effect of carbon on elongation at fracture (300K)

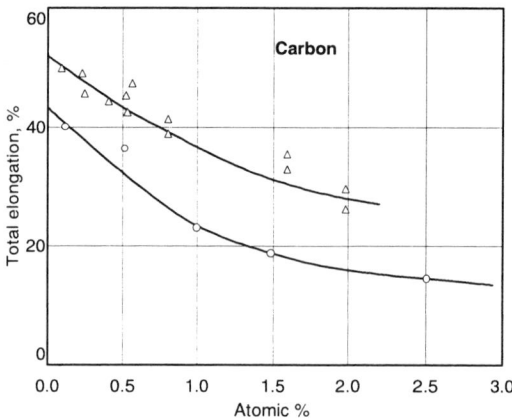

(c)

Source: H. Conrad et al., *Titanium Science and Technology*, Plenum Press, 1973, p 997

Plastic Deformation

Stress-strain plots from two different sources (see figures) indicate data scatter similar to that of ASTM grade 4 (A-70) titanium. Typically, the stress-strain behavior after yielding obeys a parabolic relation such that

$\sigma = \sigma(0) + h \sqrt{\varepsilon}$, where σ is the flow stress, $\sigma(0)$ is the intercept, and h is the strain hardening coefficient. However, a linear strain hardening relation has been observed for coarser (28 µm) grained specimens at larger (8%) strains (H. Conrad and R. Jones, *The Science, Technology and Application of Titanium*, Jaffee and Promisel, Ed., Pergamon Press, 1970, p 493).

Stress-strain curves of four purities of titanium

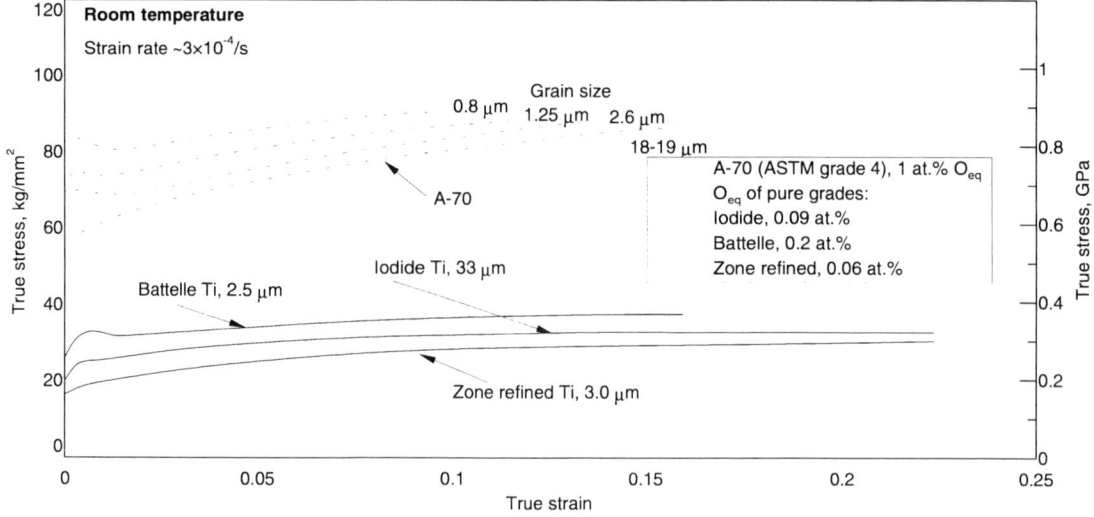

Oxygen equivalents (O_{eq}) calculated according to: N = 2 × O and C = 3/4 × O, in at.%. Analysis by weight in ppm: Zone refined (<30 O, 18 C, 110 N, 18 H); Iodide (30 O, 18 C, 110 N, 18 H); Battelle (330 O, 25 C, 140 N, 62 H, 90 Fe).

Source: H. Conrad and R. Jones, Effects of Interstitial Content and Grain Size on the Mechanical Behavior of Alpha Titanium Below $0.4 T_m$. *The Science, Technology and Application of Titanium* (Jaffee and Promise, Ed.), Pergamon Press, 1970, p 491

Stress-strain curves of zone-refined titanium

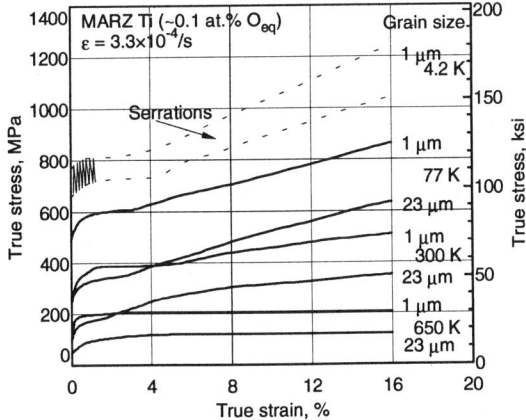

MARZ titanium, ~0.1 at.% O_{eq}, with indicated grain sizes (Gs).
Source: K. Okazaki and H. Conrad, *Acta Metall.*, Vol 21, 1973, p 1119

Strain Hardening

The strain hardening relation $\sigma = \sigma_0 + h\sqrt{\varepsilon}$ is affected by interstitial content, as shown by a comparison of σ_0 and h for iodide titanium and ASTM Grade 4 (see table). The effect of grain size on h for pure titanium is also less than that of ASTM Grade 4 (see table). Temperatures between room temperature and 325 °C (600 K) have little effect on the ratio of h with selected elastic constants (see figures).

Comparison of strain hardening parameters

Material	Grain size, μm	$\sigma(0)$(a) MPa	kgf/mm²	Strain hardening coefficient, h MPa	kgf/mm²
ASTM Grade 4 (A-70)	18 to 19	510	52	833	85
ASTM Grade 4 (A-70)	2.6	570	58	705	72
ASTM Grade 4 (A-70)	0.8	715	73	600	61
Iodide (30 ppm O, 110 ppm N)	3.3	215	22	343	35
Iodide (30 ppm O, 110 ppm N)	1.1	265	27	285	29

(a) $\sigma(0)$ = the maximum stress for zero plastic strain. Source: R.L. Jones and H. Conrad, Transactions of the Metallurgical Society of AIME, Vol 245, 1969, p 779-789

Pure titanium: Effect of grain size on the strain hardening coefficient (h)

Material/composition(a)	h_i MPa	h_i kg/mm²	K kg/mm^{3/2}
Iodide titanium (30 O, 18 C, 110 N)	440	45	0.5
Battelle titanium (330 O, 25 C, 140 N)	353	36	0.2
A-70 (ASTM grade 4)	872	89	0.8

(A) All compositions in ppm by weight. (b) Grain size effects on h modeled by the equation $h = h_i + K'\sqrt{d}$ where d is the average grain size. Source: H. Conrad and R. Jones, The Science, *Technology and Application of Titanium*, Jaffee and Promisel, Ed., Pergamon Press, 1970, p 495

Pure titanium: Temperature effect on strain hardening coefficient (h)

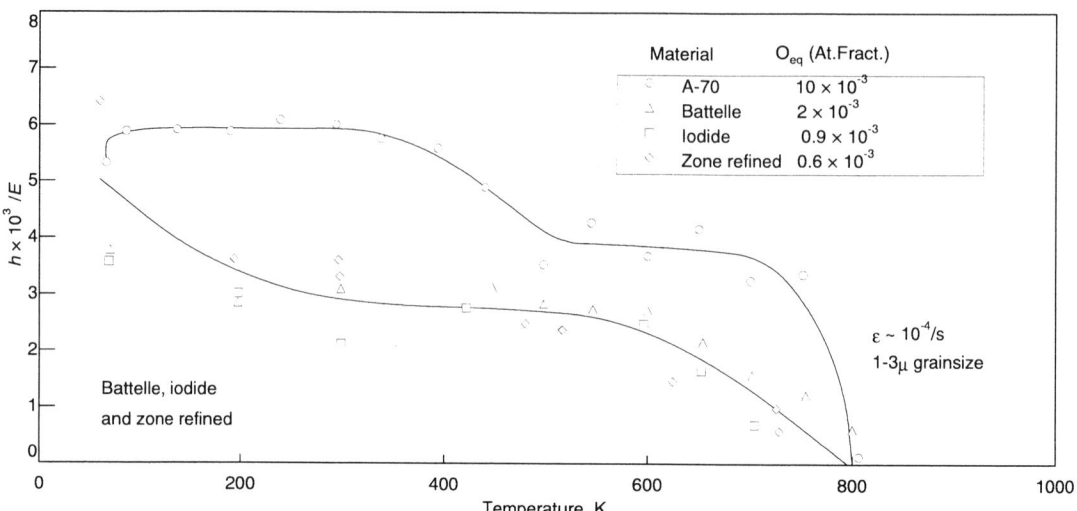

Material	O_{eq} (At.Fract.)
A-70	10×10^{-3}
Battelle	2×10^{-3}
Iodide	0.9×10^{-3}
Zone refined	0.6×10^{-3}

$\varepsilon \sim 10^{-4}$/s
1-3μ grainsize

Normalized to the modulus of elasticity (E).
Source: H. Conrad and R. Jones, *The Science, Technology and Application of Titanium*, Jaffee and Promisel, Ed., Pergamon Press, 1970, p 492

Pure titanium: Effect of temperature on h_i/G

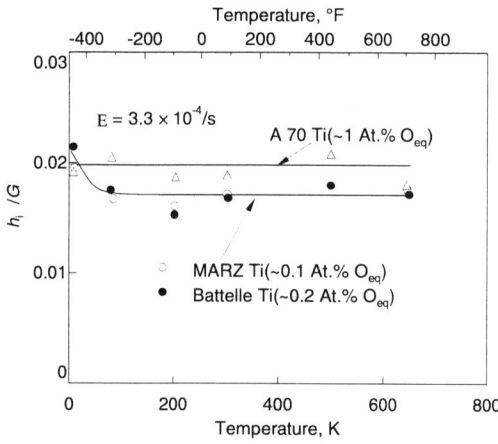

The strain hardening constant (h_i) is normalized with the shear modulus (G), where G is the C_{66} constant (taken from E.S. Fisher and C.J. Renkin, *Phys. Rev.*, Vol 135, 1964, p A482).
Source: K. Okasaki and H. Conrad, *Acta Metall.*, Vol 21, 1973, p 1121

Flow Stress

Pure titanium (99.9% Ti) has a flow stress comparable to medium-carbon (0.35% C) steel (see figures) and can be cold forged or formed. However, due to the tendency of titanium to stick, fret, and cold weld with steel tooling, cold working of titanium generally requires surface treatment and/or lubrication.

Surface Treatment. The sticking of titanium with steel tools during forming can be overcome with lubricants such as molybdenum disulfide. Physical vapor depositions (such as ion-plated copper) also provide lubrication during cold-working operations. For hot-working operations, lubricants include yttria ceramic coatings.

Pure titanium: Flow stress vs other materials

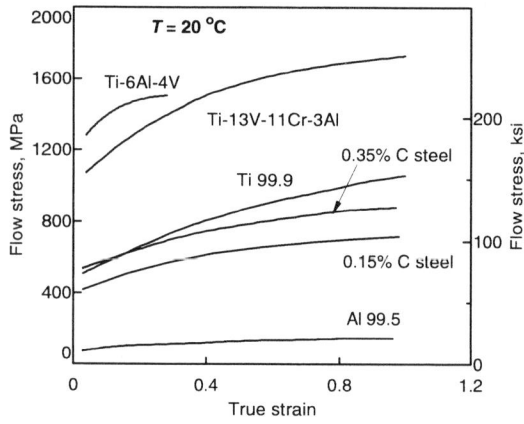

Titanium impurities (wt%) were 0.04 O_2, 0.01 N_2, 0.002 H_2, 0.04 Fe, and 0.010 C.
Source: H.W. Wagener and K.H. Tampe, Cold Impact Extrusion of Titanium, *Titanium Science and Technology*, Vol 1, DGM, 1985, p 578

Effect of Grain Size

Pure titanium: Effect of temperature on flow stress

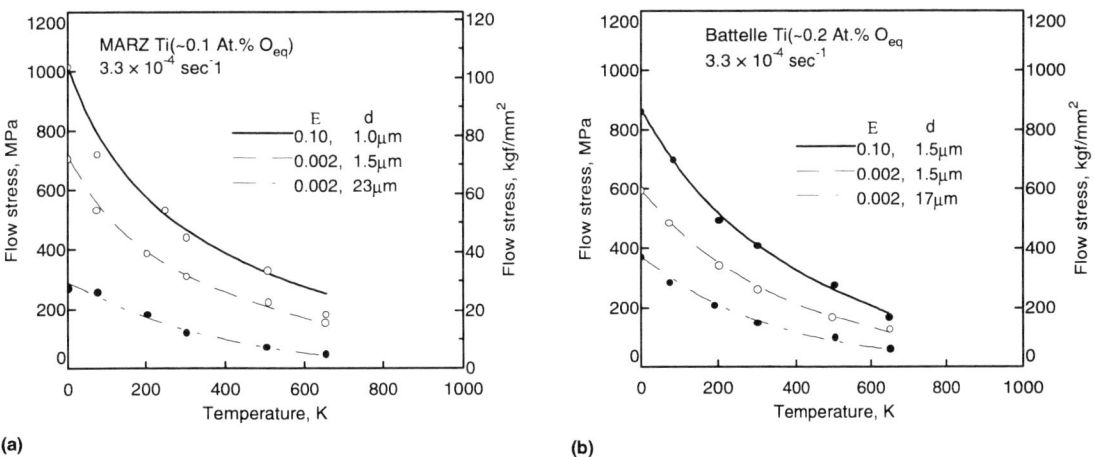

(a) MARZ titanium with 182 to 790 ppm O by weight and 0.11 at.% O_{eq}. (b) Battelle titanium with 330 to 569 ppm O by weight and 0.18 at.% O_{eq}. Both compositions tested at a strain rate of 3.3×10^{-4}/s.
Source: K. Okazaki and H. Conrad, *Acta Metall.*, Vol 21, 1973, p 1120

Effect of grain size (*d*) on flow stress

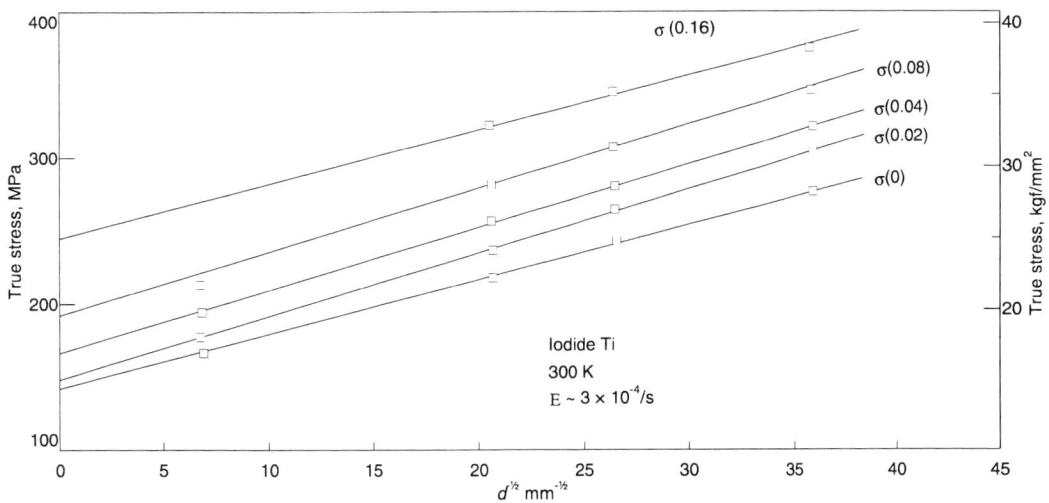

Room-temperature flow stress of iodide titanium (30 pm O, 110 pm N) for plastic strains of 0, 0.02, 0.04, 0.08 and 0.16.
Source: R.L. Jones and H. Conrad, *Trans. TMS/AIME*, Vol 245, 1969, p 779-789

Strain-Rate Sensitivity

Effect of impurities on strain rate sensitivity

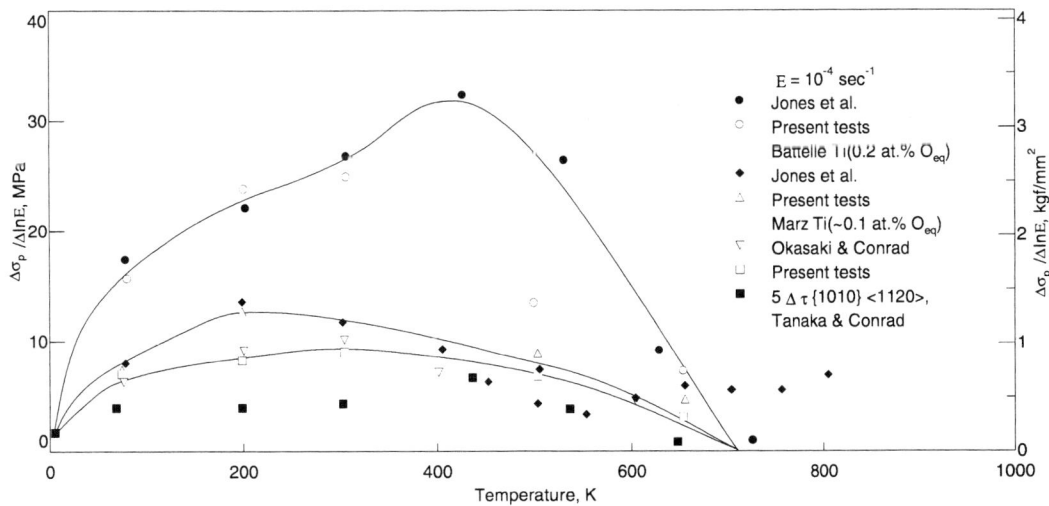

Source: K. Okazaki and H. Conrad, *Acta Metall.*, Vol 21, 1973, p 1122

Hall-Petch Parameters

Interstitial impurities increase both the Hall-Petch friction stress (σ_f) and the Hall-Petch grain size constant (K) at room temperature (see comparisons with ASTM Grade 4). As temperatures increase, interstitial impurities have less of an effect (see figure).

Effect of impurities on Hall-Petch parameters at room temperature

Material/composition(a)	Hall-Petch parameters(b)		Grain size constant (K) kgf/mm$^{3/2}$
	Friction stress (σ_f)		
	MPa	kgf/mm^2	
Iodide titanium (30 O, 18 C, 110 N)	137	14	0.47
Battelle titanium (330 O, 25 C, 140 N)	176	18	0.32
A-70 (ASTM grade 4)	390	40	0.86

(A) All compositions in ppm by weight. (b) Hall-Petch friction stress (σ_f) and grain size constant (K) used to model flow stress (σ) by the relation $\sigma = \sigma_i + K\sqrt{d}$, where d is the grain size. Source: H. Conrad and R. Jones, *The Science, Technology and Application of Titanium*, Jaffee and Promisel, Ed., Pergamon Press, 1970, p 495

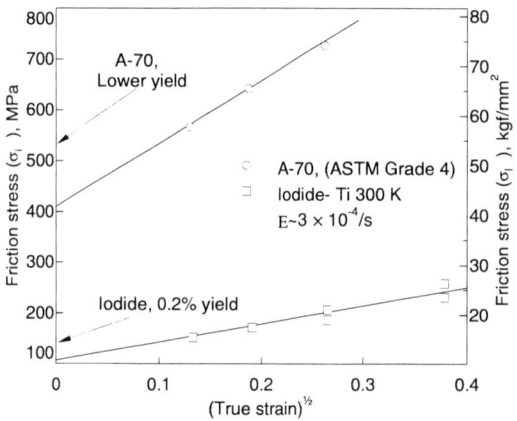

Hall-Petch friction stress of pure titanium and ASTM Grade 4

Room temperature data for Grade 4 and iodide titanium (30 ppm O, 110 ppm N) for various strains.
Source: R.L. Jones and H. Conrad, *Trans. TMS/AIME*, Vol 245, 1969, p 779-789

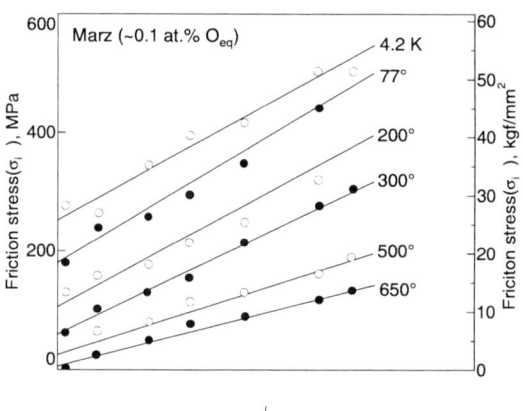

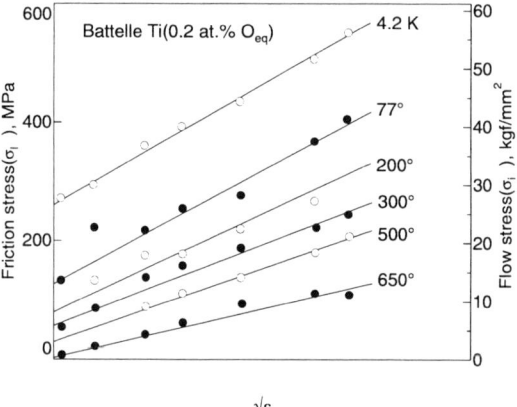

Effect of strain and temperature on friction stress

(a)
(b)
Strain rates were 3.3×10^{-4}/s.
Source: K. Okazaki and H. Conrad, *Acta Metall.*, Vol 21, 1973, p 1117-1129

Cold-Impact Extrusion

The main characteristics of extrusion can be described with force-travel diagrams (see figure). Cold-impact extrusion allows close dimensional control and the advantage of work hardening.

Pure titanium: Extrusion pressure vs other materials

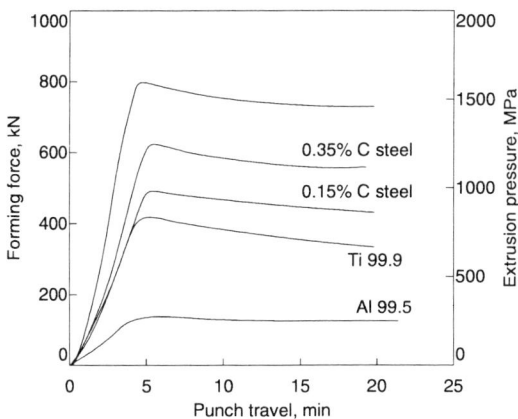

Forward rod extrusion with an extrusion strain of 0.85. Composition (wt%) of pure titanium was 99.9 Ti, 0.04 Fe, 0.04 O_2, 0.01 N_2, 0.002 H_2, and 0.010 C.
Source: H.W. Wagener and K.H. Tampe, Cold Impact Extrusion of Titanium, *Titanium Science and Technology*, Vol 1, DGM, 1985, p 577-584

Pure titanium: Force-travel diagrams of cold extrusion

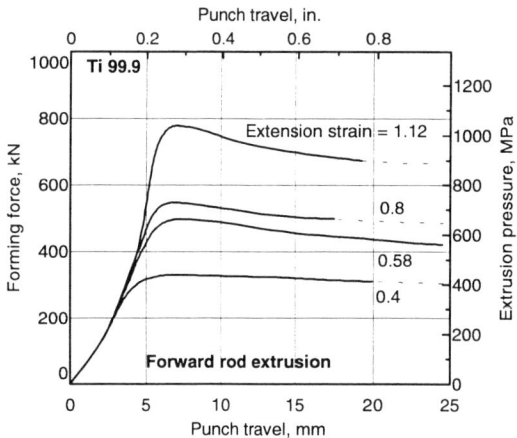

99.9% Ti with 0.04% O_2, 0.01% N_2, 0.002% H_2, 0.04% Fe, and 0.010% C
Source: H.W. Wagener and K.H. Tampe, Cold Impact Extrusion of Titanium, Titanium Science and Technology, Vol 1, DGM, 1985, p 577-584

Heat Treatment

In keeping with commonly observed behavior, increasing amount of cold work prior to annealing decreases the subsequent recrystallization temperature and grain size (see figure). Increasing the amount of cold work before annealing also decreases the spread in grain volume distribution (*Met. Trans.*, Vol 16A, May 1985, p 703-708). The reason for the decrease in grain size distribution is not completely clear.

Iodide Ti: Grain size vs annealing and deformation

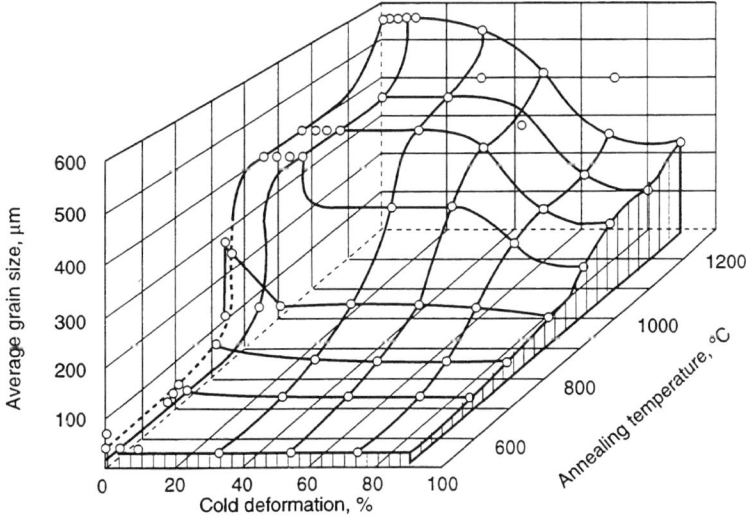

Influence of annealing temperature (1-h duration) and degree of deformation on the grain size of cold-deformed sheet-strips of iodide titanium.
Source: U. Zwicker, *Titan und Titanlegierungen*, Springer-Verlag, 1974, p 138

Commercially Pure and Modified Titanium

Commercially pure titanium has been available as mill products since 1950 and is used for applications that require moderate strength combined with good formability and corrosion resistance. Production was developed largely because of aerospace demands for a material lighter than steel and more heat resistant than aluminum alloys. However, commercially pure titanium is very useful when high corrosion resistance and good weldability are desired.

Commercially pure titanium is available in several grades, which have varying amounts of impurities such as carbon, hydrogen, iron, nitrogen, and oxygen. Some modified grades also contain small palladium additions (Ti-0.2 Pd) and nickel-molybdenum additions (Ti-0.3Mo-0.8Ni). These alloy additions allow improvements in corrosion resistance and/or strength.

Commercial purity titanium generally has more than 1000 ppm oxygen and iron, nitrogen, carbon, and silicon as principal impurities. Because small amounts of interstitial impurities greatly affect the mechanical properties of pure titanium, it is not convenient to distinguish between the various grades of unalloyed titanium on the basis of chemical analysis. Titanium mill products are more readily distinguished by mechanical properties. For example, the four ASTM grades of unalloyed titanium are grouped as follows:

ASTM grade	Minimum tensile strength		0.2% yield strength	
	MPa	ksi	MPa	ksi
Grade 1	240	35	170-310	25-45
Grade 2	345	50	275-450	40-65
Grade 3	440	64	380-550	55-80
Grade 4	550	80	480-655	70-95

Density. 4.51 g/cm^3 (0.16 3 lb/in.^{3})

Unalloyed Ti Grade I, R50250

Unalloyed titanium is available as four different ASTM grades, which are classified by their levels of impurities (primarily oxygen) and the resultant effect on strength and ductility. ASTM Grade 1 has the highest purity, lowest strength, and best room-temperature ductility and formability of the four ASTM unalloyed titanium grades.

ASTM titanium Grade 1 should be used where maximum formability is required and where low iron and interstitial contents might enhance corrosion resistance. It exhibits excellent corrosion resistance in highly oxidizing to mildly reducing environments, including chlorides. Grade 1 can be used in continuous service up to 425 °C (800 °F) and in intermittent service up to 540 °C (1000 °F). In addition, Grade 1 has good impact properties at low temperatures.

Chemistry

ASTM Grade 1 titanium has impurity limits of 0.18 O, 0.20 Fe, 0.03 N, and 0.10 C wt.% max. Equivalent compositions from other specifications are best determined by mechanical properties, because small variations in interstitial contents may raise yield strengths above maximum permitted values or lower ductility below minimum specifications.

Hydrogen content as low as 30 to 40 ppm can induce severe hydrogen embrittlement in commercially pure titanium (see the section "Hydrogen Damage" in this datasheet).

Product Forms and Condition

Unalloyed titanium Grade 1 is available in all wrought forms and has the best formability of the four ASTM grades. Like the other unalloyed titanium grades, Grade 1 can be satisfactorily welded, machined, cold worked, hot worked, and cast.

Unalloyed titanium typically has an annealed alpha structure in wrought, cast, and P/M forms. The yield strength of Grade 1 is comparable to that of fully annealed 304 stainless steel.

Applications

Typical uses for Grade 1 titanium include chemical, marine, and similar applications, heat exchangers, components for chemical processing and desalination equipment, condenser tubing, pickling baskets and anodes of various types. In the chemical and engineering industries, Grade 1 is an ideal material for a wide variety of chemical reactor vessels because of its resistance to attack by seawater, moist chlorine, moist metallic chlorides, chlorite and hypochlorite solutions, nitric and chromic acids. It lacks resistance to biofouling.

Unalloyed titanium grade 1 and equivalents: Specifications and compositions

Specification	Designation	Description	C	Fe	H	N	O	Si	OE	OT	Other
UNS	R50100		0.03 max	0.1 max	0.005 max	0.012 max	0.1 max				bal Ti
UNS	R50120		0.05	0.2	0.008	0.02	0.1				bal Ti
UNS	R50125		0.05	0.2	0.008	0.02	0.1-0.15				bal Ti
UNS	R50250		0.1	0.2	0.015	0.03	0.18				bal Ti
China											
GB 3620	TA-1		0.05	0.15	0.015	0.03	0.15	0.1			bal Ti
Europe											
AECMA prEN2525	P01	Sh Strp	0.08	0.2	0.0125	0.05	0.2		0.1	0.6	bal Ti
AECMA prEN3441	P01	Sh Strp Ann HR	0.08	0.2	0.0125	0.05	0.2		0.1	0.6	bal Ti
AECMA prEN3487	P01	Sh Strp Ann CR	0.08	0.2	0.0125	0.05	0.2		0.1	0.6	bal Ti
France											
AIR 9182	T-35	Sh Strp	0.08	0.12	0.01	0.05			0.04		bal Ti
Germany											
DIN 17850	3.7025	Plt Sh Strp Rod Wir Frg Ann	0.08	0.2	0.013	0.05	0.1				bal Ti
DIN 17850	Ti 1	Sh Strp Plt Rod Wir Frg Ann	0.08	0.2	0.013	0.05	0.1				bal Ti
DIN 17860	3.7025	Sh Strp	0.08	0.2	0.013	0.05	0.1				bal Ti
DIN 17862	3.7025	Rod	0.08	0.2	0.013	0.05	0.1				bal Ti
DIN 17863	3.7025	Wire	0.08	0.2	0.013	0.05	0.1				bal Ti
DIN 17864	3.7025	Frg	0.08	0.2	0.013	0.05	0.1				bal Ti
Japan											
JIS Class 1	Ti Class 1			0.2	0.015	0.05	0.15				bal Ti
JIS H4600	TP28H/C Class 1	HR CR Sh		0.2	0.013	0.05	0.15				bal Ti
JIS H4600	TR28H/C Class 1	HR CR Strp		0.2	0.013	0.05	0.15				bal Ti
JIS H4630	TTP28D/E Class 1	Smls pipe		0.2	0.015	0.05	0.15				bal Ti
JIS H4630	TTP28W/WD Class 1	As-weld/weld & drawn pipe		0.2	0.015	0.05	0.15				bal Ti
JIS H4631	TTH28D Class 1	Smls tube for heat exch		0.2	0.015	0.05	0.15				bal Ti
JIS H4631	TTH28W/WD Class 1	Weld tube for heat exch		0.2	0.015	0.05	0.15				bal Ti
JIS H4650	TB28C/H Class 1	HW CD Bar		0.2	0.015	0.05	0.15				bal Ti
JIS H4670	TW28 Class 1	Wire		0.2	0.015	0.05	0.15				bal Ti
Russia											
	Weld el		0.12	0.006	0.04	0.12	0.1				bal Ti
	Weld el		0.03	0.008	0.04	0.15					bal Ti
		All forms	0.1	0.3	0.015	0.04	0.15	0.15			bal Ti
			0.03	0.01	0.04	0.15	0.15				bal Ti
			0.05	0.3	0.01	0.04	0.15	0.15			bal Ti
			0.05	0.3	0.01	0.04	0.15	0.15			bal Ti
OST 1.90013-71	VT1-00	Sh Plt Strp Foil Rod Frg Ann	0.05	0.2	0.008	0.04	0.1	0.08		0.1	bal Ti
UK											
BS 2TA.1	2TA.1	Sh Strp HT		0.2	0.01						Ti 99.78 min; bal Ti
DTD 5013		Bar Bil		0.2	0.013						bal Ti
USA											
AMS 4951E	AMS 4951	Fill met gas-met W arc weld	0.08	0.2	0.005	0.05	0.18			0.6	bal Ti
ASME SB-265	Ti Grade 1	Sh Strp Plt Ann	0.1	0.2	0.015	0.03	0.18		0.1	0.4	bal Ti
ASME SB-381	F-1	Frg Ann	0.1	0.2	0.015	0.03	0.18		0.1	0.4	bal Ti
ASTM B265-79	Ti Grade 1	Sh Strp Plt Ann	0.1	0.2	0.015	0.03	0.18			0.4	bal Ti
ASTM B337-87	Ti Grade 1	Weld smls pipe Ann	0.1	0.2	0.015	0.03	0.18			0.4	bal Ti
ASTM B338-87	Ti Grade 1	Smls weld tube Exch Conds Ann	0.1	0.2	0.015	0.03	0.18			0.4	bal Ti
ASTM B348-87	Ti Grade 1	Bar Bil Ann	0.1	0.2	0.01-0.0125	0.03	0.18			0.4	bal Ti
ASTM B381-87	F-1	Frg Ann	0.1	0.2	0.015	0.03	0.18			0.4	bal Ti
ASTM F467-84a	Ti Grade 1	Nut	0.1	0.2	0.0125	0.05	0.18				bal Ti
ASTM F467M-84b	Ti Grade 1	Metric Nut	0.1	0.2	0.0125	0.05	0.18				bal Ti
ASTM F468-84a	Ti Grade 1	Bolt Screw Stud	0.1	0.2	0.0125	0.05	0.18				bal Ti
ASTM F468M-84b	Ti Grade 1	Metric Bolt Screw Stud	0.1	0.2	0.0125	0.05	0.18				bal Ti
ASTM F67-88	Ti Grade 1	Surg imp HW CW Frg Ann	0.1	0.2	0.0125-0.015	0.03	0.18				bal Ti
AWS A5.16-70	ERTi-1	Weld fill met	0.03	0.1	0.005	0.012	0.1				bal Ti
AWS A5.16-70	ERTi-2	Weld fill met	0.05	0.2	0.008	0.02	0.1				bal Ti
AWS A5.16-70	ERTi-3	Weld fill met	0.05	0.2	0.008	0.02	0.1-0.15				bal Ti
MIL T-81556A	CP-4	Ext Bar Shap Ann	0.08	0.2	0.015	0.05	0.15			0.3	bal Ti
MIL T-81915A		Invest Cast	0.08	0.2	0.015	0.05	0.2			0.6	bal Ti
MIL T-9046J	CP-4	Sh Strp Plt Ann	0.08	0.2	0.015	0.05	0.15			0.3	bal Ti

Unalloyed titanium grade 1 compositions: Producer specifications

Specification	Designation	Description	C	Fe	H	N	O	Si	OE	OT	Other
Germany											
Deutsche T Fuchs	Contimet 30 T2	Sh Strp Plt Bar Wir Frg Pip Frg	0.06	0.15	0.13	0.05	0.12				bal Ti
Japan											
Daido	DT1	Rod Bar Sh Strp Frg Ann	0.1	0.2	0.0125	0.03	0.18				bal Ti
Kobe	KS40	Sh Strp Tu Plt Wir Bar Pip Ann		0.1	0.01	0.03	0.1				bal Ti
Kobe	KS40LF	Low Fe grade		0.05	0.01	0.03	0.1				bal Ti
Kobe	KS40S	Ann		0.1	0.01	0.03	0.08				bal Ti
Kobe	KS50	Ann		0.15	0.01	0.03	0.15				bal Ti
Kobe	KS50LF	Low Fe grade		0.05	0.01	0.03	0.15				bal Ti
Sumitomo	ST-40										
UK											
Imp. Metal	IMI 110	Rod									Ti 99.8
Imp. Metal	IMI 115	All forms	0.1	0.2	0.013	0.03	0.15				bal Ti
USA											
Chase Ext.	CDX GR-1										
OREMET	Ti-1										
RMI	RMI 25	Chemical/marine/airframe apps	0.08	0.2	0.015	0.03	0.18				bal Ti
Tel.Rodney	A35										
TIMET	TIMETAL 35A	Ann	0.1 max	0.02 max	0.15 max	0.03 max	0.18 max				bal Ti
TMCA	Ti-1										

Unalloyed Ti Grade 2, R50400

Grade 2 titanium is the "workhorse" for industrial applications, having a guaranteed minimum yield strength of 275 MPa (40 ksi) and good ductility and formability. The yield strength of Grade 2 is comparable to those of annealed austenitic stainless steels, and it is used where excellent formability is required and where low interstitial contents might enhance corrosion resistance.

Grade 2 also has good impact properties at low temperatures and excellent resistance to erosion and to corrosion by seawater and marine atmospheres. Grade 2 can be used in continuous service up to 425 °C (800 °F) and in intermittent service up to 540 °C (1000 °F).

Chemistry

ASTM Grade 2 titanium has the same nitrogen content limits as ASTM Grade 1 (0.03% max), the same iron content limits as ASTM Grade 3 (0.30% max), and a maximum oxygen concentration of 0.25% that is approximately midway between the 0.18 to 0.40% range in the other three ASTM unalloyed titanium grades.

Effect of Impurities. The increased iron and oxygen concentrations of ASTM Grade 2 compared to ASTM Grade 1 impart additional tensile strength (345 vs 240 MPa, or 50 vs 35 ksi) and yield strength (275 vs 170 MPa, or 40 vs 25 ksi) to Grade 2 but at the expense of ductility (20% elongation for Grade 2 vs 24% elongation for Grade 1). Higher iron and interstitial contents also may degrade corrosion resistance relative to Grade 1.

Hydrogen content as low as 30 to 40 ppm can induce hydrogen embrittlement in CP titanium (see the section "Hydrogen Damage" in this datasheet).

Product Forms and Condition

Titanium Grade 2 is available in all wrought product forms. In cast form, ASTM Grade 2 constitutes about 5% of cast titanium products. Like other unalloyed titanium grades, Grade 2 can be welded, machined, cast, and cold worked.

Titanium Grade 2 typically has an annealed alpha structure in wrought, cast, and P/M forms. It is not heat treatable.

Applications

Typical uses for titanium Grade 2 include chemical, marine, and similar applications, airframe skin and nonstructural components, heat exchangers, cryogenic vessels, components for chemical processing and desalination equipment, condenser tubing, pickling baskets, anodes, shafting, pumps, vessels, and piping systems. Grade 2 offers high ductility for fabrication and moderate strength in service.

Aircraft applications include exhaust-pipe shrouds, fireproof bulkheads, gas-turbine bypass ducts, hot-air ducts, engine cowlings, formed brackets and skins for hot areas. Other aircraft applications include galley equipment, chemical toilets and floor supports under these areas.

Reaction vessels and heat exchangers are a major application of Grade 2 titanium because of its resistance to attack by seawater, moist chlorine, moist metallic chlorides, chlorite and hypochlorite solutions, nitric and chromic acids, organic acid, sulfides, and many industrial gaseous environments. Grade 2 titanium also has excellent resistance to deposit, impingement, and crevice attack even in highly polluted waters, and is therefore used extensively in tubular and plate-type heat exchangers for condensers, evaporators, and other components of marine vessels, power stations, oil refineries, offshore platforms, and water-purification plants.

Electrochemical Processing Equipment. The insulating property of the anodic film on titanium makes it an ideal and cost-efficient material for anodizing jigs and plating baskets. Other applications include high-efficiency heat-exchanger systems for electrolytes. A very thin coating of a precious metal such as platinum enables Grade 2 titanium anode to operate at high current density in many electrolytes. Consequently, non-consumable noble-metal coated Grade 2 titanium anodes are in demand for chlorine-production cells, electrodialysis plants, electroplating equipment, and cathodic protection of condensers, seagoing rigs, and jetties.

Most electrodeposits do not adhere well to commercial purity Grade 2 titanium. This characteristic has led to the widespread use of Grade 2 titanium for cathodes or starter-sheet blanks in many electrochemical metal-refining operations.

Unalloyed titanium grade 2 and equivalents: Specifications and compositions

Specification	Designation	Description	C	Fe	H	N	O	Si	OE	OT	Other
UNS	R50130		0.05	0.3	0.008	0.02	0.15-0.25				bal Ti
UNS	R50400		0.1	0.3	0.015	0.03	0.25				bal Ti
China											
GB 3620	TA-2		0.1 max	0.3 max	0.015 max	0.05 max	0.2 max	0.15 max			bal Ti
Europe											
AECMA prEN2518	Ti-PO2	Sh Strp Bar	0.08	0.2	0.01	0.06	0.25			0.6	bal Ti
AECMA prEN2526	Ti-PO2	Sh Strp	0.08 max	0.25 max	0.0125 max	0.05 max	0.25 max		0.1 max	0.6 max	bal Ti
AECMA prEN3378	Ti PO2	Wir	0.08 max	0.25 max	0.0125 max	0.05 max	0.25 max		0.1 max	0.6 max	bal Ti
AECMA prEN3442	Ti-PO2	Sh Strp Ann HR	0.08 max	0.25 max	0.0125 max	0.05 max	0.25 max		0.1 max	0.6 max	bal Ti
AECMA prEN3451	Ti-PO2	Frg NHT	0.08 max	0.25 max	0.0125 max	0.05 max	0.25 max		0.1 max	0.6 max	bal Ti
AECMA prEN3452	Ti-PO2	Frg Ann	0.08 max	0.25 max	0.0125 max	0.05 max	0.25 max		0.1 max	0.6 max	bal Ti
AECMA prEN3460	Ti-PO2	Bar Ann	0.08 max	0.25 max	0.0125 max	0.05 max	0.25 max		0.1 max	0.6 max	bal Ti
AECMA prEN3498	Ti-PO2	Sh Strp Ann CR	0.08 max	0.25 max	0.0125 max	0.05 max	0.25 max		0.1 max	0.6 max	bal Ti
France											
AIR 9182	T-35	Sh CR	0.08	0.12	0.015	0.05		0.04			Ti 99.69 min
AIR 9182	T-40	Sh	0.08	0.12	0.015	0.05		0.04			Ti 99.69 min
Germany											
DIN 17850	Ti II	Sh Strp Plt Rod Wir Frg Ann	0.08	0.25	0.013	0.06	0.2				bal Ti
DIN 17850	Ti III	Sh Strp Plt Rod Wir Frg Ann	0.1	0.3	0.013	0.06	0.25				bal Ti
DIN 17850	WL 3.7035	Plt Sh Strp Rod Wir Frg Ann	0.08	0.25	0.013	0.06	0.2				bal Ti
DIN 17850	WL 3.7055	Sh Plt Strp Rod Wir Frg Ann	0.1	0.3	0.013	0.06	0.25				bal Ti
DIN 17860	3.7035	Sh Strp	0.08 max	0.25 max	0.013 max	0.06 max	0.2 max				bal Ti
DIN 17862	3.7035	Rod	0.08 max	0.25 max	0.013 max	0.06 max	0.2 max				bal Ti
DIN 17863	3.7035	Wir	0.08 max	0.25 max	0.013 max	0.06 max	0.2 max				bal Ti
DIN 17864	3.7035	Frg	0.08 max	0.25 max	0.013 max	0.06 max	0.2 max				bal Ti
WL 3.7024		Sh Wir Ann	0.08	0.2	0.0125	0.05	0.2			0.6	bal Ti
WL 3.7034		Sh Bar Frg Wir Ann	0.08	0.25	0.0125	0.06	0.25			0.6	bal Ti
Japan											
	Class 2			0.25	0.015	0.05	0.2				bal Ti
JIS H4361	TTH 35D Class 2	Smls Tub		0.25	0.015	0.05	0.2				bal Ti
JIS H4600	TP 35 H/C Class 2	Sh HR CR		0.25	0.013	0.05	0.2				bal Ti
JIS H4600	TR 35 H/C Class 2	Strp HR CR		0.25	0.013	0.05	0.2				bal Ti
JIS H4630	TTP 35 D/E Class 2	Smls Pip		0.25	0.015	0.05	0.2				bal Ti
JIS H4630	TTP 35 W/WD Class 2	Weld Pip		0.25	0.015	0.05	0.2				bal Ti
JIS H4631	TTH 35 W/WD Class 2	Weld Tub		0.25	0.015	0.05	0.2				bal Ti
JIS H4650	TB 35 C/H Class 2	Bar Rod HW CD		0.25	0.015	0.05	0.2				bal Ti
JIS H4670	TW 35 Class 2	Wir		0.25	0.015	0.05	0.2				bal Ti
Russia											
OST 1.90000-76	VT1-O	Mult Forms Ann	0.07	0.3	0.01	0.04	0.2	0.1		0.3	bal Ti
OST 1.90060-72	VT1L	Cast	0.15	0.3	0.015	0.05	0.2	0.15		0.3	W 0.2; bal Ti
Spain											
UNE 38-711	L-7001	Sh Plt Strp Bar Wir Ext Ann	0.08	0.2	0.0125	0.05	0.2				bal Ti
UNE 38-712	L-7002	Sh Plt Strp Bar Wir Ext Ann	0.08	0.25	0.0125	0.05	0.25				bal Ti

(continued)

Unalloyed titanium grade 2 and equivalents: Specifications and compositions (continued)

Specification	Designation	Description	C	Fe	H	N	O	Si	OE	OT	Other
UK											
BS 2TA.2		Sh Strp HT		0.2	0.01						Ti 99.78 min
BS 2TA.3	2TA.3	Bar HT		0.2	0.01						Ti 99.78 min
BS 2TA.4	2TA.4	Frg HT		0.2	0.01						Ti 99.79 min
BS 2TA.5	2TA.5	Frg HT		0.2	0.01						Ti 99.78 min
DTD 5073		Tub	0.01 max	0.2 max	0.015 max						bal Ti
USA											
AMS 4902E		Sh Strp Plt Ann	0.08	0.3	0.015	0.05	0.2			0.3	bal Ti
AMS 4941C		Weld Tub Ann	0.1	0.2	0.015	0.05	0.25			0.15	bal Ti
ASM 4942C		Smls Tube Ann	0.1	0.3	0.015	0.03	0.25			0.3	bal Ti
ASME SB-265	Ti Grade 2	Sh Strp Plt Ann	0.1 max	0.3 max	0.015 max	0.03 max	0.25 max		0.1 max	0.4 max	bal Ti
ASME SB-381	F-2	Frg Ann	0.1 max	0.3 max	0.015 max	0.03 max	0.25 max		0.1 max	0.4 max	bal Ti
ASTM B 265	Ti Grade 2	Sh Strp Plt Ann	0.1	0.3	0.015	0.03	0.25			0.4	bal Ti
ASTM B 337	Ti Grade 2	Pip Ann	0.1	0.3	0.015	0.03	0.25			0.4	bal Ti
ASTM B 338	Ti Grade 2	Tube for heat exch/cond									bal Ti
ASTM B 348	Ti Grade 2	Bar Bil Ann	0.1	0.3	0.0125-0.01	0.03	0.25			0.4	bal Ti
ASTM B 367-87	Ti Grade 2	Cast	0.1 max	0.2 max	0.015 max	0.05 max	0.4 max		0.1 max	0.4 max	bal Ti
ASTM B 381	Ti Grade F-2	Frg Ann	0.1	0.3	0.015	0.03	0.25				bal Ti
ASTM F467-84	Ti Grade 2	Nut	0.1 max	0.3 max	0.0125 max	0.05 max	0.25 max				bal Ti
ASTM F467M-84a	Ti Grade 2	Nut Met	0.1 max	0.3 max	0.0125 max	0.05 max	0.25 max				bal Ti
ASTM F468-84	Ti Grade 2	Blt Scr Std	0.1 max	0.3 max	0.0125 max	0.05 max	0.25 max				bal Ti
ASTM F468M-84b	Ti Grade 2	Blt Scr Std Met	0.1 max	0.3 max	0.0125 max	0.05 max	0.25 max				bal Ti
ASTM F67	Ti Grade 2	Surg imp HW CW Frg Ann	0.1	0.3	0.015-0.0125	0.03	0.25				bal Ti
AWS A5.16-70	ERTi-4	Weld Fill Met	0.05	0.3	0.008	0.02	0.15-0.25				bal Ti
MIL T-81556A	Code CP-3	Ext Bar Shp Ann	0.08	0.3	0.015	0.05	0.2			0.3	bal Ti
MIL T-81915	Type I Comp A	Air/chem/marine apps Cast Ann	0.08	0.2	0.015	0.05	0.2			0.6	bal Ti
MIL T-9046J	Code CP-3	Sh Strp Plt Ann	0.08	0.3	0.015	0.05	0.2			0.3	bal Ti

Unalloyed titanium grade 2 compositions: Producer specifications

Specification	Designation	Description	C	Fe	H	N	O	Si	OE	OT	Other
France											
Ugine	UT35	Sh Plt Bar Frg Ann	0.08		0.0125	0.05	0.2				bal Ti
Ugine	UT40	Sh Plt Bar Frg Ann	0.08	0.25	0.0125	0.06	0.25				bal Ti
Germany											
Otto Fuchs	T3	Frg									Ti 99.5
Thyssen	Contimet 35	Sh Strp Plt Bar Wir Pip Ann	0.06	0.2	0.013	0.05	0.18				bal Ti
Thyssen	Contimet 35 D	Mult forms Ann	0.06	0.25	0.013	0.05	0.25				bal Ti
Japan											
Daido	DT 2	Rod Bar Sh Strp Frg Ann	0.1	0.3	0.0125	0.03	0.2				bal Ti
Kobe	KS60	Sh Strp Tub Plt Wir Bar Pi Ann		0.3	0.01	0.03	0.2				bal Ti
Kobe	KS60LF	Low Fe Ann		0.05	0.01	0.03	0.2				bal Ti
Nippon	T1X			0.5		0.1	0.2				bal Ti
Sumitomo	ST-50											
Sumitomo	ST6											
Toho	TIB		0.03	0.1	0.005	0.015	0.15				Ti 99.7 min
Toho	TIBLF	Low Fe	0.02	0.05	0.005	0.01	0.15				Ti 99.7 min
Toho	TIC		0.03	0.15	0.005	0.02	0.25				Ti 99.6 min
Toho	TICLF	Low Fe	0.02	0.05	0.005	0.01	0.25				Ti 99.6 min
UK											
Imp. Metal	IMI 125	Mult forms	0.1	0.2	0.013	0.03	0.2				bal Ti
Imp. Metal	IMI 130	Sh Bar	0.1	0.2	0.013	0.03	0.25				bal Ti
USA											
Chase Ext.	CDX GR-2											
OREMET	Ti-2											
RMI	RMI 40	Mult forms Ann	0.08	0.25	0.015	0.03	0.2				bal Ti
Tel.Rodney	A40											
TIMET	TIMETAL 50A	Ann	0.08 max	0.2 max	0.0125 max	0.05 max					bal Ti
TMCA	Ti 2											

Unalloyed Ti Grade 3, R50550

Grade 3 titanium is a general-purpose grade of commercially pure titanium that has excellent corrosion resistance in highly oxidizing to mildly reducing environments, including chlorides, and an excellent strength-to-weight ratio. Thus, like other titanium metals and alloys, Grade 3 bridges the design gap between aluminum and steel and provides many of the desirable properties of each. Grade 3 also has good impact toughness at low temperatures.

Chemistry

ASTM Grade 3 titanium has lower iron limits than ASTM Grade 4 (0.3 wt% vs 0.5 wt% max) and the second highest oxygen contents (0.35 wt%) of the four ASTM grades for unalloyed titanium. Only Grade 4 has higher strength levels than Grade 3.

Effect of Impurities. Excessive impurity levels may raise yield strength above maximum permitted values and decrease elongation or reduction in area below minimum values. Higher iron and interstitial contents may affect corrosion resistance.

Hydrogen content as low as 30 to 40 ppm can induce hydrogen embrittlement in commercially pure titanium (see the section "Hydrogen Damage" in this datasheet).

Product Forms and Condition

Like other unalloyed titanium grades, Grade 3 is available in all wrought product forms and can be satisfactorily welded, machined, and cast. Most forming operations can be carried out at room temperature but warm forming reduces springback and power requirements.

Titanium Grade 3 typically has an annealed alpha structure for wrought, cast, and P/M forms.

Applications

Grade 3 is used for nonstructural aircraft parts and for all types of applications requiring corrosion resistance. Typical uses for CP titanium include chemical and marine applications, airframe skin and nonstructural components, heat exchangers, cryogenic vessels, components for chemical processing and desalination equipment, condenser tubing, and pickling baskets.

Unalloyed titanium grade 3 and equivalents: Specifications and compositions

Specification	Designation	Description	C	Fe	H	N	O	Si	OE	OT	Other
UNS	R50550		0.1	0.3	0.015	0.05	0.35			0.4	bal Ti
France											
AIR 9182	T-50	Sh Ann	0.08	0.25	0.015	0.07		0.04			Ti 99.54 min
Germany											
DIN 17850	Ti IV	Sh Strp Plt Rod Wir Frg Ann	0.1	0.35	0.013	0.07	0.3				bal Ti
DIN 17850	WL 3.7065	Plt Sh Strp Rod Wir Frg Ann	0.1	0.35	0.013	0.07	0.3				bal Ti
DIN 17860	3.7055	Sh Strp	0.1 max	0.3 max	0.013 max	0.06 max	0.25 max				bal Ti
DIN 17862	3.7055	Rod	0.1 max	0.3 max	0.013 max	0.06 max	0.25 max				bal Ti
DIN 17863	3.7055	Wir	0.1 max	0.3 max	0.013 max	0.06 max	0.25 max				bal Ti
DIN 17864	3.7055	Frg	0.1 max	0.3 max	0.013 max	0.06 max	0.25 max				bal Ti
Japan											
JIS	Class 3			0.3	0.015	0.07	0.3				bal Ti
JIS H4600	TP 49 H/C Class 3	Sh HR CR		0.3	0.013	0.07	0.3				bal Ti
JIS H4600	TR 49 H/C Class 3	Strp HR CR		0.3	0.013	0.07	0.3				bal Ti
JIS H4630	TTP 49 D/E Class 3	Smls Pip Hot Ext CD		0.3	0.015	0.07	0.3				bal Ti
JIS H4630	TTP 49 W/WD Class 3	Weld Pip		0.3	0.015	0.07	0.3				bal Ti
JIS H4631	TTH 49 D Class 3	Smls Tub CD		0.3	0.015	0.07	0.3				bal Ti
JIS H4631	TTH 49 W/WD Class 3	Weld Tub		0.3	0.015	0.07	0.3				bal Ti
JIS H4650	TB 49 C/H Class 3	Bar HW CD		0.3	0.015	0.07	0.3				bal Ti
JIS H4670	TW 49 Class 3	Wir		0.3	0.015	0.07	0.3				bal Ti
UK											
BS 2TA.6		Sh Strp HT		0.2	0.01						Ti 99.78 min
BS 2TA.7		Bar HT		0.2	0.01						Ti 99.78 min
BS 2TA.8		Frg		0.2	0.01						Ti 99.79 min
BS 2TA.9		Frg HT		0.2	0.015						Ti 99.78 min
DTD 5023		Sh Strp		0.2 max	0.0125 max						bal Ti
DTD 5273		Bar		0.2 max	0.0125 max						bal Ti
DTD 5283		Frg		0.2 max	0.0125 max						bal Ti

(continued)

Unalloyed titanium grade 3 and equivalents: Specifications and compositions (continued)

Specification	Designation	Description	C	Fe	H	N	O	Si	OE	OT	Other
USA											
AMS 4900J		Sh Strp Plt Ann	0.08	0.3	0.015	0.05	0.3			0.3	bal Ti
AMS 4951E		Weld Wir	0.08 max	0.2 max	0.005 max	0.05 max	0.18 max		0.1 max	0.6 max	bal Ti
ASME SB-265	Grade 3	Sh Strp Plt Ann	0.1 max	0.3 max	0.015 max	0.05 max	0.35 max		0.1 max	0.4 max	bal Ti
ASME SB-381	F-3	Frg An	0.1 max	0.3 max	0.015 max	0.05 max	0.35 max		0.1 max	0.4 max	bal Ti
ASTM B 265	Grade 3	Sh Strp Plt Ann	0.1	0.3	0.015	0.05	0.35			0.4	bal Ti
ASTM B 337	Grade 3	Weld Smls Pip Ann	0.1	0.3	0.015	0.05	0.35			0.4	bal Ti
ASTM B 338	Grade 3	Smls Weld Tub Ann	0.1	0.3	0.015	0.05	0.35			0.4	bal Ti
ASTM B 348	Grade 3	Bar Bil Ann	0.1	0.3	0.0125	0.05	0.35			0.4	bal Ti
ASTM B 381	Grade F-3	Frg Ann	0.1	0.3	0.015	0.05	0.35			0.4	bal Ti
ASTM B 367-87	C-3	Cast	0.1 max	0.25 max	0.015 max	0.05 max	0.4 max		0.1 max	0.4 max	bal Ti
ASTM F 67	Grade 3	Surg Imp	0.1	0.3	0.015-0.0125	0.05	0.35				bal Ti
MIL T-81556A	Code CP-2	Ext Bar Shp Ann	0.08	0.3	0.015	0.05	0.3			0.3	bal Ti
MIL T-9046J	Code CP-2	Sh Strp Plt Ann	0.08	0.3	0.015	0.05	0.3			0.3	bal Ti

Unalloyed titanium grade 3 compositions: Producer specifications

Specification	Designation	Description	C	Fe	H	N	O	Si	OE	OT	Other
France											
Ugine	UT50	Sh Bar Frg Ann	0.08	0.25	0.0125	0.07	0.35				bal Ti
Germany											
Thyssen	Contimet 55	Mult Forms Ann	0.06	0.3	0.013	0.05	0.35				bal Ti
Titan	RT 20		0.1	0.35	0.013	0.07	0.3				bal Ti
Japan											
Daido	DT 3	Mult Forms Ann	0.1	0.3	0.0125	0.05	0.35				bal Ti
Kobe	KS70	Ann		0.3	0.01	0.05	0.3				bal Ti
Kobe	KS70LF	Low Fe Mult Forms Ann		0.05	0.01	0.05	0.3				bal Ti
Sumitomo	ST-70										
Toho	TID		0.05	0.2	0.01	0.04	0.3				Ti 99.4 min
UK											
Imp. Metal	IMI 130										
USA											
Chase Ext.	CDX GR-32										
OREMET	Ti-3										
RMI	RMI 55	Mult Forms Ann	0.08	0.25	0.015	0.05	0.3				bal Ti
Tel.Rodney	A55										
TIMET	TIMETAL 65A	Ann	0.1 max	0.2 max	0.015 max	0.05 max	0.35 max				bal Ti
TMCA	Ti 3										

Unalloyed Ti Grade 4, R50700

Grade 4 has the highest strength of the four ASTM unalloyed titanium grades in addition to good ductility and moderate formability. The benefits of strength and lightness of Grade 4 are retained at moderate temperatures. Its strength-to-weight ratio is higher than that of AISI type 301 stainless steel at temperatures up to 315 °C (600 °F). Grade 4 also has outstanding resistance to corrosion fatigue in salt water. The stress required to cause failure in several million cycles is 50% higher for this material than for K-Monel or AISI type 431 stainless steel.

Chemistry   ASTM Grade 4 has the highest oxygen (0.40 wt%) and iron (0.50 wt%) content of the four unalloyed titanium ASTM grades. The higher content of iron and interstitials may reduce corrosion resistance.

Hydrogen content as low as 30 to 40 ppm can induce hydrogen embrittlement in commercially pure titanium (see the section "Hydrogen Damage" in this datasheet).

Product Forms and Condition   Commercially pure Grade 4 is available in all wrought product forms and can be satisfactorily machined, cast, welded, and cold worked. Most forming operations are performed at room temperature but warm forming (150 to 425 °C, 300 to 800 °F) is often done to reduce springback and power requirements. Complex forms must be produced by warm forming.

Grade 4 typically has an annealed alpha structure in wrought, cast, and P/M forms.

Applications. Because Grade 4 has excellent resistance to corrosion and erosion applications, it is suitable for a wide range of chemical and marine applications, where it often can be used interchangeably with Grade 3. It can be used in continuous service at temperatures up to 425 °C (800 °F), and intermittent service to 540 °C (1000 °F).

Unalloyed titanium grade 4 and equivalents: Specifications and compositions

Specification	Designation	Description	C	Fe	H	N	O	Si	OE	OT	Other
UNS	R50700		0.1	0.5	0.015	0.05	0.4			0.4	bal Ti
China											
GB 3620	TA-3		0.1 max	0.4 max	0.015 max	0.05 max	0.3 max	0.15 max			bal Ti
Europe											
AECMA prEN2519	Ti-PO4	Bar Frg Sh Strp	0.08	0.35	0.01-0.0125	0.07	0.4			0.6	bal Ti
AECMA prEN2520	Ti-PO4	Frg	0.08 max	0.2 max	0.0125 max	0.07 max	0.4 max		0.1 max	0.6 max	bal Ti
AECMA prEN2527	Ti-PO4	Sh Strp	0.08 max	0.2 max	0.0125 max	0.07 max	0.4 max		0.1 max	0.6 max	bal Ti
AECMA prEN3443	Ti-PO4	Strp Sh Ann CR	0.08 max	0.2 max	0.0125 max	0.07 max	0.4 max		0.1 max	0.6 max	bal Ti
AECMA prEN3453	Ti-PO4	Frg NHT	0.08 max	0.2 max	0.0125 max	0.07 max	0.4 max		0.1 max	0.6 max	bal Ti
AECMA prEN3461	Ti-PO4	Bar Ann	0.08 max	0.2 max	0.0125 max	0.07 max	0.4 max		0.1 max	0.6 max	bal Ti
AECMA prEN3496	Ti-PO4	Frg Ann	0.08 max	0.2 max	0.0125 max	0.07 max	0.4 max		0.1 max	0.6 max	bal Ti
AECMA prEN3499	Ti-PO4	Sh Strp Ann CR	0.08 max	0.2 max	0.0125 max	0.07 max	0.4 max		0.1 max	0.6 max	bal Ti
France											
AIR 9182	T-60	Sh Ann	0.08	0.3	0.015	0.08		0.04			Ti 99.56 min
Germany											
DIN	3.7064	Sh Rod Bar Frg Ann	0.08	0.35	0.0125	0.07	0.4				bal Ti
DIN 17860	3.7065	Sh Strp	0.1 max	0.35 max	0.013 max	0.07 max	0.3 max				bal Ti
DIN 17862	3.7065	Rod	0.1 max	0.35 max	0.013 max	0.07 max	0.3 max				bal Ti
DIN 17863	3.7065	Wir	0.1 max	0.35 max	0.013 max	0.07 max	0.3 max				bal Ti
DIN 17864	3.7065	Frg	0.1 max	0.35 max	0.013 max	0.07 max	0.3 max				bal Ti
Spain											
UNE 38-714	L-7004	Mult Forms Ann	0.1	0.4	0.0125	0.07	0.4				bal Ti
UK											
BS 2TA6		Sh Strp	0.08 max	0.2 max	0.0125 max						bal Ti
BS 2TA7		Bar	0.08 max	0.2 max	0.0125 max						bal Ti
BS 2TA8		Frg	0.08 max	0.2 max	0.01 max						bal Ti
BS 2TA9		Frg		0.2 max	0.015 max						bal Ti
USA											
AMS 4901L		Sh Strp Plt Ann	0.08	0.5	0.015	0.05	0.4			0.3	bal Ti
AMS 4921F		Bar Wir Frg Bil Rng Ann	0.08	0.5	0.0125	0.05	0.4			0.3	bal Ti
ASTM B 265	Grade 4	Sh Plt Strp Ann	0.1	0.5	0.015	0.05	0.4			0.4	bal Ti
ASTM B 348	Grade 4	Bar Bil Ann	0.1	0.5	0.0125-0.01	0.05	0.4			0.4	bal Ti
ASTM B 367	Grade C-2	Cast	0.1	0.2	0.015	0.05	0.4			0.4	bal Ti
ASTM B 367	Grade C-3	Cast	0.1	0.25	0.015	0.05	0.4			0.4	bal Ti
ASTM B 381	Grade F-4	Frg Ann	0.1	0.5	0.015	0.05	0.4			0.4	bal Ti
ASTM F467-84	Grade 4	Nut	0.1 max	0.5 max	0.0125 max	0.07 max	0.4 max				bal Ti
ASTM F468-84	Grade 4	Blt Scrw Std	0.1 max	0.5 max	0.0125 max	0.07 max	0.4 max				bal Ti
ASTM F67	Grade 4	Sh Strp Bar HR CR Ann Frg	0.1	0.5	0.015-0.0125	0.05	0.4				bal Ti
MIL F-83142	Comp 1	Frg Ann	0.08	0.5	0.0125	0.05	0.4			0.3	bal Ti
MIL T-81556A	Code CP-1	Ext Bar Shp Ann	0.08	0.5	0.015	0.05	0.4			0.3	bal Ti
MIL T-9046J	Code CP-1	Sh Strp Plt Ann	0.08	0.5	0.015	0.05	0.4			0.3	bal Ti
MIL T-9047-G	SP-70	Bar	0.08 max	0.5 max	0.015 max	0.05 max	0.4 max			0.3 max	bal Ti
MIL T-9047G	Ti-CP-70	Bar Bil Ann	0.08	0.5	0.0125	0.05	0.4			0.3	Y 0.005; bal Ti

Unalloyed titanium grade 4 commercial equivalents: Compositions

Specification	Designation	Description	C	Fe	H	N	O	Si	OE	OT	Other
France											
Ugine	UT60	Bar Frg Sh Plt Ann	0.1	0.35	0.0125	0.07	0.4				bal Ti
Germany											
Otto Fuchs	T6	Frg									Ti 99
Japan											
Daido	DT 4	Bar Rod Shp Frg Ann	0.1	0.5	0.0125	0.05	0.5				bal Ti
Kobe	KS85	Sh Strp Plt Wir Bar Ann		0.4	0.01	0.05	0.4				bal Ti
Sumitomo	ST-80										
UK											
Imp. Metal	IMI 155	Sh	0.1	0.2	0.013	0.03	0.38				bal Ti
Imp. Metal	IMI 160	Rod Bar Bil Wir	0.1	0.2	0.017	0.05	0.4				bal Ti
USA											
Chase Ext.	CDX GR-4										
Crucible	A-70	Ann	0.05-0.15			0.07 max					bal Ti
OREMET	Ti-4										
RMI	RMI 70	Mult Forms Ann	0.08	0.5	0.015	0.05	0.4				bal Ti
Tel.Rodney	A40										
TIMET	Ti-75A	Ann	0.1 max	0.3 max	0.015 max		0.4 max				bal Ti
TIMET	TIMETAL 100A	Ann	0.01 max	0.3 max	0.01 max	0.05 max	0.4 max				bal Ti
TMCA	Ti 4										

Ti-0.2Pd, R52400 (Grade 7), R52250 (Grade 11)

The two Ti-0.2Pd ASTM grades (7 and 11) have better resistance to crevice corrosion at low pH and elevated temperatures than that of ASTM Grades 1, 2, and 12, and they are recommended for chemical-industry applications involving environments that are moderately reducing or that fluctuate between oxidizing and reducing. The palladium-containing alloys extend the range of titanium applications in hydrochloric, phosphoric, and sulfuric acid solutions. Their good fabricability, weldability, and strength are similar to those of corresponding grades of unalloyed titanium. Ti-0.2 Pd Grade 7 is comparable to Grade 2 in strength, while Grade 11 is comparable to unalloyed Grade 1 in strength.

Chemistry

A relatively small addition of palladium (0.15 to 0.20 wt%) to unalloyed titanium permits its use in stronger reducing media such as mild sulfuric and hydrochloric acids.

The higher oxygen content (0.25 wt%) and higher iron content (0.30 wt%) of the Grade 7 alloy results in lower ductility and cold formability but higher strength than Grade 11 which has a maximum oxygen content of 0.18 wt% and a maximum iron content of 0.20 wt%.

Hydrogen content as low as 30 to 40 ppm can induce hydrogen embrittlement in commercially pure titanium (see the section "Hydrogen Damage" in this datasheet).

Product Forms and Condition

Both Grade 7 and Grade 11 alloys are flat rolled products, extrusions, wires, tubing, and pipe. Ti-0.2Pd grades can be satisfactorily cast, welded, machined, and cold worked. Most forming operations are performed at room temperature, but warm forming (150 to 425 °C, or 300 to 800 °F) is sometimes employed.

Ti-0.2Pd products typically have an annealed alpha structure.

Applications

Ti-0.2Pd, Grade 7 and Grade 11 are used for chemical-industry equipment and for special corrosion applications. These alloys have excellent corrosion resistance for chemical processing applications. They are also used for storage applications involving media that are mildly reducing or that fluctuate between oxidizing and reducing. The palladium-containing alloys are also used where high cold formability in component fabrication is required, such as cold pressed plates for plate/frame heat exchangers and chlor-alkali anodes. ASTM Grades 7 and 11 can be used in continuous service up to 425 °C (800 °F) and in intermittent service up to 540 °C (1000 °F).

Ti-0.2Pd grades 7 and 11 and equivalents: Specifications and compositions

Specification	Designation	Description	C	Fe	H	N	O	Pd	Si	OT	Other
UNS	R52250	Grade 11	0.1	0.2	0.015	0.03	0.18	0.12-0.25			bal Ti
UNS	R52400	Grade 7	0.1	0.3	0.015	0.03	0.25	0.12-0.25			bal Ti
UNS	R52401	Filler	0.05	0.25	0.008	0.02	0.15	0.15-0.25			bal Ti
Germany											
DIN 17851	3.7225		0.06 max	0.15 max	0.0013 max	0.05 max	0.12 max	0.12-0.25		0.4 max	bal Ti
DIN 17851	3.7235		0.06 max	0.2 max	0.0013 max	0.05 max	0.18 max	0.12-0.25		0.4 max	bal Ti
DIN 17851	3.7255		0.06 max	0.25 max	0.0013 max	0.05 max	0.25 max	0.12-0.25		0.4 max	bal Ti
Japan											
JIS H 4635 type 11	TTP28PdD	Smls Pip CD		0.2 max	0.015 max	0.05 max	0.15 max	0.12-0.25			bal Ti
JIS H 4635 type 11	TTP28PdE	Smls Pip HE		0.2 max	0.015 max	0.05 max	0.15 max	0.12-0.25			bal Ti
JIS H 4635 type 11	TTP28PdW	Weld Pip		0.2 max	0.015 max	0.05 max	0.15 max	0.12-0.25			bal Ti
JIS H 4635 type 11	TTP28PdWD	Weld Pip CD		0.2 max	0.015 max	0.05 max	0.15 max	0.12-0.25			bal Ti
JIS H 4635 type 12	TTP35PdD	Smls Pip CD		0.25 max	0.015 max	0.05 max	0.2 max	0.12-0.25			bal Ti
JIS H 4635 type 12	TTP35PdE	Smls Pip HE		0.25 max	0.015 max	0.05 max	0.2 max	0.12-0.25			bal Ti
JIS H 4635 type 12	TTP35PdW	Weld Pip		0.25 max	0.015 max	0.05 max	0.2 max	0.12-0.25			bal Ti
JIS H 4635 type 12	TTP35PdWD	Weld Pip CD		0.25 max	0.015 max	0.05 max	0.2 max	0.12-0.25			bal Ti
JIS H 4635 type 13	TTP49PdD	Smls Pip CD		0.3 max	0.015 max	0.07 max	0.3 max	0.12-0.25			bal Ti
JIS H 4635 type 13	TTP49PdE	Smls Pip HE		0.3 max	0.015 max	0.07 max	0.3 max	0.12-0.25			bal Ti
JIS H 4635 type 13	TTP49PdW	Weld Pip		0.3 max	0.015 max	0.07 max	0.3 max	0.12-0.25			bal Ti
JIS H 4635 type 13	TTP49PdWD	Weld Pip CD		0.3 max	0.015 max	0.07 max	0.3 max	0.12-0.25			bal Ti
JIS H 4636 type 11	TTH28PdD	Smls Pip CD		0.2 max	0.015 max	0.05 max	0.15 max	0.12-0.25			bal Ti
JIS H 4636 type 11	TTH28PdW	Weld Pip		0.2 max	0.015 max	0.05 max	0.15 max	0.12-0.25			bal Ti
JIS H 4636 type 11	TTH28PdWD	Weld Pip CD		0.2 max	0.015 max	0.05 max	0.15 max	0.12-0.25			bal Ti
JIS H 4636 type 12	TTH35PdD	Smls Pip CD		0.25 max	0.015 max	0.05 max	0.2 max	0.12-0.25			bal Ti
JIS H 4636 type 12	TTH35PdW	Weld Pip		0.25 max	0.015 max	0.05 max	0.2 max	0.12-0.25			bal Ti
JIS H 4636 type 12	TTH35PdWD	Weld Pip CD		0.25 max	0.015 max	0.05 max	0.2 max	0.12-0.25			bal Ti
JIS H 4636 type 13	TTH49PdD	Smls Pip CD		0.3 max	0.015 max	0.07 max	0.3 max	0.12-0.25			bal Ti
JIS H 4636 type 13	TTH49PdW	Weld Pip		0.3 max	0.015 max	0.07 max	0.3 max	0.12-0.25			bal Ti
JIS H 4636 type 13	TTH49PdWD	Weld Pip CD		0.3 max	0.015 max	0.07 max	0.3 max	0.12-0.25			bal Ti
JIS H 4655 type 11	TB28PdC	Rod Bar CD		0.2 max	0.015 max	0.05 max	0.15 max	0.12-0.25			bal Ti
JIS H 4655 type 11	TB28PdH	Rod Bar HW		0.2 max	0.015 max	0.05 max	0.15 max	0.12-0.25			bal Ti
JIS H 4655 type 12	TB35PdC	Bar Rod CD		0.25 max	0.015 max	0.05 max	0.2 max	0.12-0.25			bal Ti
JIS H 4655 type 12	TB35PdH	Bar Rod HW		0.25 max	0.015 max	0.05 max	0.2 max	0.12-0.25			bal Ti
JIS H 4655 type 13	TB49PdC	Bar Rod CD		0.3 max	0.015 max	0.07 max	0.3 max	0.12-0.25			bal Ti
JIS H 4655 type 13	TB49PdH	Bar Rod HW		0.3 max	0.015 max	0.07 max	0.3 max	0.12-0.25			bal Ti
JIS H 4675 type 11	TW28Pd	Wir		0.2 max	0.015 max	0.05 max	0.15 max	0.12-0.25			bal Ti
JIS H 4675 type 12	TW35Pd	Wir		0.25 max	0.015 max	0.05 max	0.2 max	0.12-0.25			bal Ti
JIS H 4675 type 13	TW49Pd	Wir		0.3 max	0.015 max	0.07 max	0.25 max	0.12-0.25			bal Ti
Russia											
	4200		0.07	0.18	0.01	0.04	0.12	0.15-0.3	0.1	0.3	bal Ti
Spain											
UNE 38-715	L-7021	Sh Plt Strp Bar Wir Ext Ann	0.08	0.25	0.0125	0.05	0.25	0.12-0.25			bal Ti
USA											
ASTM B 265	Grade 11	Sh Plt Strp Ann	0.1	0.2	0.015	0.03	0.18	0.12-0.25		0.4	bal Ti
ASTM B 265	Grade 7	Sh Strp Plt Ann	0.1	0.3	0.015	0.03	0.25	0.12-0.25		0.4	bal Ti
ASTM B 337	Grade 11	Smls Weld Pip	0.1	0.2	0.015	0.03	0.18	0.12-0.25		0.4	bal Ti
ASTM B 337	Grade 7	Wld Smls Pip Ann	0.1	0.3	0.015	0.03	0.25	0.12-0.25		0.4	bal Ti
ASTM B 338	Grade 11	Smls Weld Tub Ann	0.1	0.2	0.015	0.03	0.18	0.12-0.25		0.4	bal Ti
ASTM B 338	Grade 7	Smls Weld Tub Ann	0.1	0.3	0.015	0.03	0.25	0.12-0.25		0.4	bal Ti
ASTM B 348	Grade 11	Bar Bil Ann	0.1	0.2	0.0125-0.01	0.03	0.18	0.12-0.25		0.4	bal Ti
ASTM B 348	Grade 7	Bar Bil Ann	0.1	0.3	0.0125	0.03	0.25	0.12-0.25		0.4	bal Ti
ASTM B 367	Grade Ti-Pd 7B	Cast	0.1	0.2	0.015	0.05	0.4	0.12		0.4	bal Ti
ASTM B 381	Grade F-11	Frg Ann	0.1	0.2	0.015	0.03	0.18	0.12-0.25		0.4	bal Ti
ASTM B 381	Grade F-7	Frg Ann	0.1	0.3	0.015	0.03	0.25	0.12-0.25		0.4	bal Ti
ASTM F467-84	Grade 7	Nut	0.1 max	0.3 max	0.0125 max	0.05 max	0.25 max	0.12-0.25			bal Ti
ASTM F467M-84a	Grade 7	Met Nut	0.1 max	0.3 max	0.0125 max	0.05 max	0.25 max	0.12-0.25			bal Ti
ASTM F468-84	Grade 7	Blt Scrw Std	0.1 max	0.3 max	0.0125 max	0.05 max	0.25 max	0.12-0.25			bal Ti
ASTM F468M-84b	Grade 7	Met Blt Scrw Std	0.1 max	0.3 max	0.0125 max	0.05 max	0.25 max	0.12-0.25			bal Ti
AWS A5.16-70	ERTi-0.2Pd	Weld Fill Met	0.05	0.25	0.008	0.02	0.15	0.15-0.25			bal Ti

Ti-0.2Pd grades 7 and 11 compositions: Producer specifications

Specification	Designation	Description	C	Fe	H	N	O	Pd	Si	OT	Other
France											
Ugine	UT35-02	Sh Plt Bar Frg Ann	0.08	0.2	0.015	0.05	0.2	0.2			bal Ti
Germany											
Deutsche T	Contimet Pd 02/30	Mult Forms Ann	0.06	0.15	0.013	0.05	0.12	0.15-0.25			bal Ti
Deutsche T	Contimet Pd 02/35	Mult Forms Ann	0.06	0.2	0.013	0.05	0.18	0.15-0.25			bal Ti
Deutsche T	Contimet Pd 02/35 D	Mult Forms Ann	0.06	0.25	0.013	0.05	0.25	0.15-0.25			bal Ti
Deutsche T	RT 12(Pd)	Sh Strp Bar Frg	0.08	0.2	0.013	0.05	0.1	0.15-0.25			bal Ti
Deutsche T	RT 15(Pd)		0.08	0.25	0.013	0.06	0.2	0.15-0.25			bal Ti
Deutsche T	RT 18(Pd)	Frg	0.1	0.3	0.013	0.06	0.25	0.15-0.25			bal Ti
Japan											
Kobe	KS40PdA	Mult Forms Ann		0.05	0.01	0.03	0.1	0.12-0.2			bal Ti
Kobe	KS40PdB	Mult Forms Ann		0.05	0.01	0.03	0.1	0.17-0.25			bal Ti
Kobe	KS50PdA	Mult Forms Ann		0.05	0.01	0.03	0.15	0.12-0.2			bal Ti
Kobe	KS50PdB	Mult Forms Ann		0.05	0.01	0.03	0.15	0.17-0.25			bal Ti
Kobe	KS70PdA	Mult Forms Ann		0.05	0.01	0.05	0.3	0.12-0.2			bal Ti
Kobe	KS70PdB	Mult Forms Ann		0.05	0.01	0.05	0.3	0.17-0.25			bal Ti
Sumitomo	ST-40P										
Sumitomo	ST-50P										
Sumitomo	ST-60P										
Toho	15PAT		0.02	0.05	0.005	0.01	0.1	0.15 min			bal Ti
Toho	15PBT		0.03	0.08	0.005	0.015	0.15	0.15 min			bal Ti
Toho	20PAT		0.02	0.05	0.005	0.01	0.1	0.2 min			bal Ti
Toho	20PBT		0.03	0.08	0.005	0.015	0.15	0.2 min			bal Ti
UK											
Imp. Metal	IMI 260	Sh						0.15			bal Ti
Imp. Metal	IMI 262	Mult Forms						0.15			bal Ti
USA											
Crucible	A-40 Pd										
OREMET	Ti-11										
OREMET	Ti-17										
RMI	RMI 0.2%Pd	Mult Forms Ann	0.08	0.3	0.015	0.03	0.2	0.2			bal Ti
TIMET	Ti-0.2Pd										
TIMET	TIMETAL 35A Pd										
TIMET	TIMETAL 50A Pd										
TMCA	Ti-7										
TMCA	Ti-11										

Ti-0.3Mo-0.8Ni, R53400

Ti-0.3Mo-0.8Ni (ASTM grade 12), introduced in 1974 for corrosion resistant applications, is considerably superior to unalloyed titanium in several respects. It exhibits better resistance to crevice corrosion in hot brines (similar to that of Ti-Pd but at much lower cost) and is more resistant than unalloyed Ti (but not Ti-0.2Pd) to corrosion by acids. It also offers significantly greater strength than unalloyed grades for use in high temperature, high pressure applications. This often permits the use of thinner wall sections in pressure vessels and piping, that often translates into cost advantages. Ti-0.3Mo-0.8Ni is less expensive than Ti-0.2Pd grades but does not offer the same crevice corrosion resistance at low pH (<3 pH). In near-neutral brines, crevice corrosion is similar to Ti-0.2Pd.

Chemistry — CP Grade 12 has allowable nitrogen, carbon, hydrogen, iron, and oxygen levels comparable to Grade 2 and Grade 7 except for a lower carbon content (0.08 wt% vs 0.10 wt% max). The titanium content in Grade 12 is lowered through the addition of two beta stabilizers, molybdenum and nickel.

Hydrogen content as low as 30 to 40 ppm can induce hydrogen embrittlement in commercially pure titanium (see the section "Hydrogen Damage" in this datasheet).

Product Forms and Condition — Grade 12 can be readily forged and can be cold worked on equipment used for stainless steels. It is available in all wrought forms and can be cast, welded, and machined.

Ti-0.3Mo-0.8Ni products typically have an annealed alpha structure. The tensile and yield strengths of Ti-0.3Mo-0.8Ni exceed those of either the Grade 2 alloy or the Grade 7 alloy. Compared to

Applications. Grade 12 is used in applications requiring moderate strength and enhanced corrosion resistance, such as equipment for chemical, marine, and other industries. Recommended environments for ASTM Grade 12 include seawater, brines, moist chlorine above 120 °C (250 °F), hot process streams containing chlorides where crevices may be pres-ent, oxidizing acids, dilute reducing acids, organic acids, and combinations of these with hot, brackish, or saline cooling waters. This material is used for equipment such as heat exchangers, pressure vessels, chlorine cells, salt evaporators, piping, pollution-control equipment, and other fabrications.

palladium-containing grades (ASTM Grade 11), Grade 12 has double the tensile and yield strengths of Grade 11.

Ti-0.3Mo-0.8Ni grade 12 and equivalents: Specifications and compositions

Specification	Designation	Description	C	Fe	H	Mo	N	Ni	O	OT	Other
UNS	R53400		0.08	0.3	0.015	0.2-0.4	0.03	0.6-0.9	0.25		bal Ti
USA											
ASTM B 265	Grade 12	Sh Strp Plt Ann	0.08	0.3	0.015	0.2-0.4	0.03	0.6-0.9	0.25	0.4	bal Ti
ASTM B 337	Grade 12	Smls Weld Pip Ann	0.08	0.3	0.015	0.2	0.03	0.6-0.9	0.25	0.3	bal Ti
ASTM B 338	Grade 12	Smls Weld Tub Ann	0.08	0.3	0.015	0.2-0.4	0.03	0.6-0.9	0.25		bal Ti
ASTM B 348	Grade 12	Bar Bill Ann	0.08	0.3	0.0125	0.2-0.4	0.03	0.6-0.9	0.25		bal Ti
ASTM B 381	Grade F-12	Frg Ann	0.08	0.3	0.015	0.2-0.4	0.03	0.6-0.9	0.25	0.3	bal Ti

Ti-0.3Mo-0.8Ni Grade 12 commercial equivalents: Compositions

Specification	Designation	Description	C	Fe	H	Mo	N	Ni	O	OT	Other
Germany											
Deutsche T	Contimet TiNiMo83	Ann	0.06	0.25	0.013	0.2-0.4	0.03	0.6-0.9	0.25		bal Ti
Japan											
Kobe	KSG12	Mult Forms Ann		0.3	0.01	0.2-0.4	0.03	0.6-0.9	0.25		bal Ti
Kobe	KSG12S	Soft Mult Forms Ann		0.3	0.01	0.2-0.4	0.03	0.6-0.9	0.2		bal Ti
USA											
OREMET	Ti-12											
TIMET	TIMETAL Code 12	Heat Exch	0.08 max	0.3 max	0.015 max	0.2-0.4	0.03 max	0.6-0.9	0.25 max		bal Ti
TMCA	Ti 12											

Phases and Structures (CP and Modified Ti, all grades)

Crystal Structure. The microstructures of commercially pure (CP) titanium and ASTM grades 7, 11, and 12 are typically 100% α-crystal structures at room temperature. As levels of impurities (primarily iron) increase, small but increasing amounts of spheroidal β are observed metallographically, usually at the grain boundaries. Larger amounts of spheroidal β are more likely in ASTM grade 12 (Ti-0.3Mo-0.8Ni) than in unalloyed titanium.

Beta transus temperatures are about 910 ± 15 °C (1675 ± 25 °F) for commercially pure titanium with 0.25 wt% O_2 max and 945 ± 15 °C (1735 ± 25 °F) with 0.40 wt% O_2 max (see table next page).

Lattice Parameters. Typical unit cell parameters for an α–crystal structure at 25 °C (77 °F) are:

$a = 0.2950$ nm

$c = 0.4683$ nm

Impurity elements (commonly oxygen, nitrogen, carbon, and iron) influence unit cell dimensions. The typical unit cell parameter for the β structure is 0.329 nm at 900 °C (1650 °F).

ASTM grade 2 titanium: Lattice parameters versus temperature

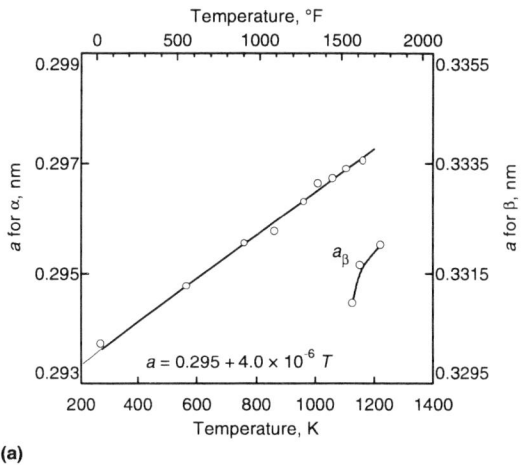

(a)

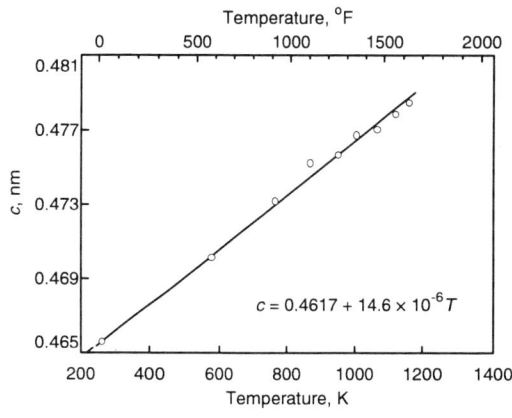

(b)

Unalloyed titanium with 0.25 Fe, 0.1 Si, 0.07 C, and 0.12 O in wt%.
Source: W. Szkliniarz and F. Grosman, "The Estimation of Refinement Possibility of Coarse-Grained Structures of Unalloyed Titanium by the Heat Treatment," presented at 6th World Conference on Titanium (France), 1988

Alpha Morphology. Annealed titanium may have an equiaxed or a platelike α morphology. Equiaxed α can be produced by recrystallization after extensive working in the α phase field. Platelet α occurs during cooling from the β phase. When the material is not fully recrystallized, the α will be elongated in the direction of working.

Grain Size

Rapid grain growth occurs when titanium is heated above the β transus. Grain size of ASTM grade 12 (Ti-0.3Mo-0.8Ni) is more stable than unalloyed titanium, because precipitate particles of the intermetallic compound $Ti_2(Ni,Mo,Fe)$ effectively retard grain growth.

CP and modified titanium grades: Typical transus temperature ±15 °C (±25 °F)

Designation	Typical β transus °C	°F	Typical α transus °C	°F
ASTM grade 1	888	1630	880	1620
ASTM grade 2	913	1675	890	1635
ASTM grade 3	920	1685	900	1650
ASTM grade 4	950	1740	905	1660
ASTM grade 7	913	1675	890	1635
ASTM grade 11	888	1630	880	1620
ASTM grade 12	890	1635	...	...

ASTM grade 2 titanium: Grain size versus annealing temperature

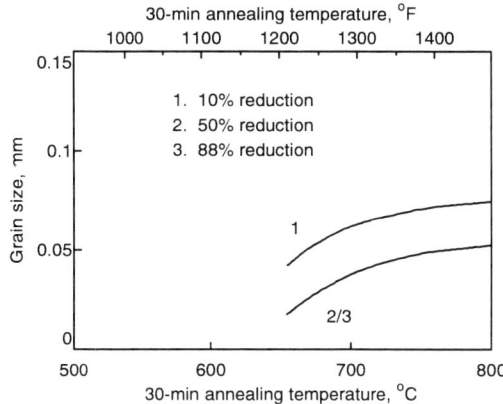

Source: "IMI Commercially Pure Titanium," IMI Titanium, Birmingham, UK, p 11

ASTM grade 3 titanium: Grain size versus annealing temperature

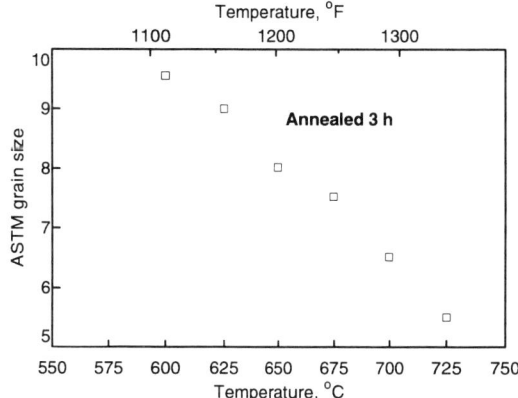

Annealing after isothermal working at less than 760 mm/min (30 in./min) with relatively low extrusion ratio. Extruded from billet preheated to an unspecified temperature in the range of 595 to 980 °C (1100 to 1800 °F).
Source: P.T. Finden, "Production of Seamless Titanium Tubing," presented at 6th World Conference on Titanium (France), 1988

Elastic Properties (all ASTM grades)

ASTM grade 4 titanium: Poisson's ratio versus test direction

ASTM grade 4 titanium: Young's modulus vs test direction

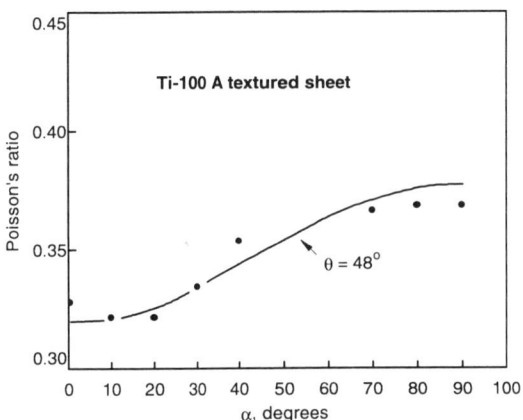

Source: F. Larson and A. Zarkades, *Properties of Textured Titanium Alloys*, MCIC-74-20, Battelle Columbus Laboratories, 1974

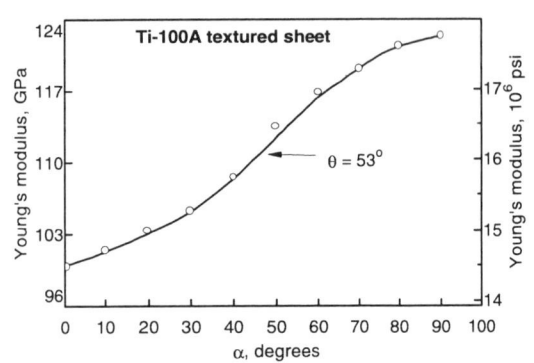

α is the angle between the rolling direction and the specimen axis. θ is the angle between the basal pole and the sheet normal.
Source: F. Larson and A. Zarkades, *Properties of Textured Titanium Alloys*, MCIC-74-20, Battelle Columbus Laboratories, 1974

CP and modified titanium: Typical room-temperature elastic properties

Designation	Tensile modulus(a) GPa	10^6 psi	Compressive modulus GPa	10^6 psi	Shear modulus in torsion GPa	10^6 psi	Poisson's ratio
ASTM grade 1 (0.18O, 0.2Fe, wt% max)	103-107	15-15.5	110	16.0	45	6.5	0.34-0.40
ASTM grade 2 (0.25O, 0.3Fe, wt% max)	103-107	15-15.5	110	16.0	45	6.5	0.34-0.40
ASTM grade 3 (0.35O, 0.3Fe, wt% max)	103-107	15-15.5	110	16.0	45	6.5	0.34-0.40
ASTM grade 4 (0.40O, 0.5Fe, wt% max)	103-107	15-15.5	110	16.0	45	6.5	0.34-0.40
ASTM grade 7 (grade 2 + 0.2Pd)	103-107	15-15.5	110	16.0	45	6.5	0.34-0.40
ASTM grade 11 (grade 1 + 0.2Pd)	103-107	15-15.5	110	16.0	45	6.5	0.34-0.40
ASTM grade 12 (Ti-0.3Mo-0.8Ni)	103-107	15-15.5	...	...	43	6.2	0.34-0.40

(a) Typical range only; textured material has wider range of values, depending on the test direction.

Effect of Temperature

ASTM grade 2 titanium: Tensile modulus at high temperature

Temperature °C	°F	Young's (tensile) modulus GPa	10^6 psi
27	80	117	17
77	170	100	14.5
200	390	90	13
300	575	79	11.5

Source: Beaton and Hewitt, *Physical Property Data for the Design Engineer*, Hemisphere Publishing, 1984

ASTM grade 4 titanium: Moduli vs temperature

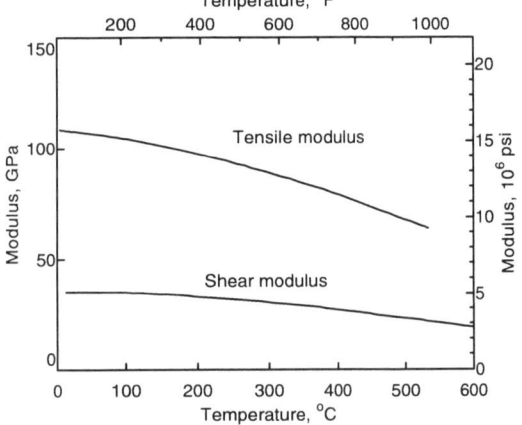

Source: *Metals Handbook*, 9th ed., Vol 3

CP titanium: Tensile modulus versus temperature

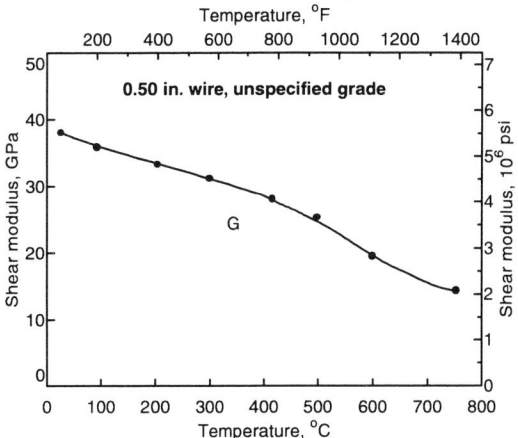

Source: *Aerospace Structural Metals Handbook*, Battelle Columbus Laboratories, Vol 4, 1963

CP titanium: Shear modulus versus temperature

Source: *Aerospace Structural Metals Handbook*

Compressive Tangent Modulus

CP grade 3 titanium: Compressive tangent modulus

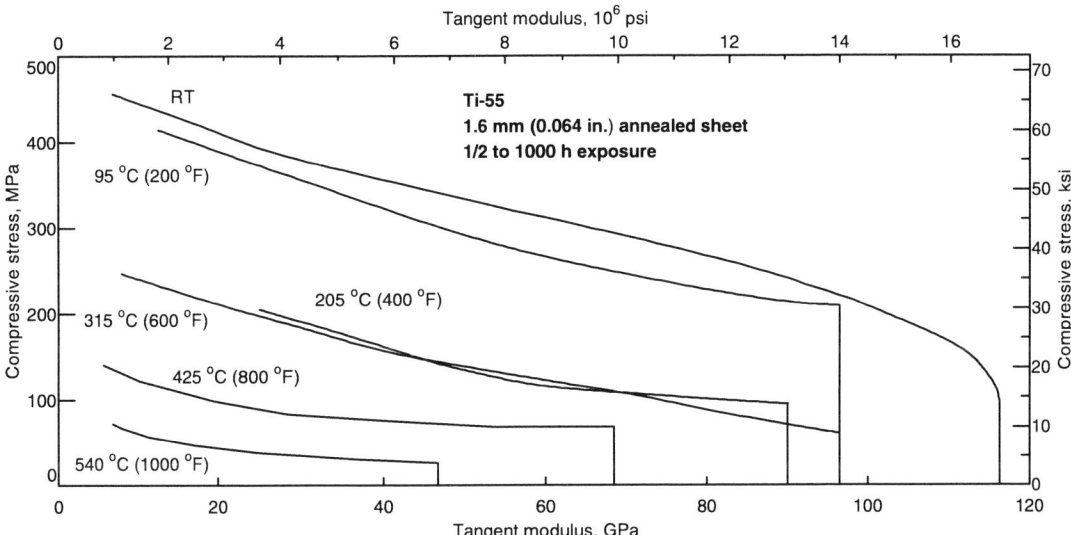

Source: *Aerospace Structural Metals Handbook*

Electrical Resistivity

Electrical resistivity decreases as the purity of titanium increases. With impurity contents of less than 0.18 wt% Fe and less than 0.20 wt% O, room-temperature resistivity (in microhm meters) can be estimated from the relation $\rho = 0.4231 + 0.3637$ (Fe, wt%) $+ 0.4540$ (O, wt%). Reported room-temperature values are

- Grade 2: 0.5 to 0.55 $\mu\Omega \cdot m$
- Grade 4: 0.6 $\mu\Omega \cdot m$
- Grades 7 and 11 (Ti-0.2 Pd): 0.55 $\mu\Omega \cdot m$
- Grade 12 (Ti-0.3Mo-0.8Ni): 0.52 $\mu\Omega \cdot m$

At room temperature, the resistivity of titanium is somewhat lower than that of 18Cr-8Ni stainless steel, but the resistivity of titanium increases more rapidly with temperature. Above 200 °C (400 °F), it has a higher resistivity than steel.

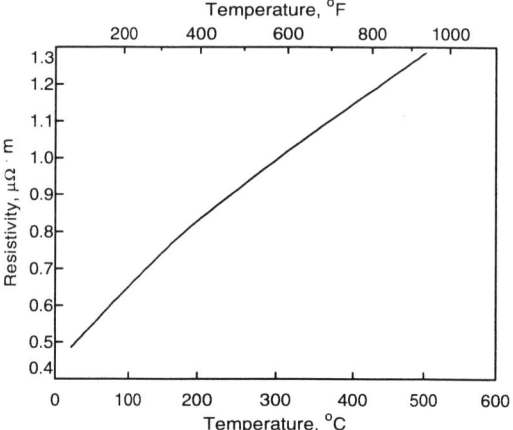

CP titanium: Electrical resistivity

Source: IMI Titanium

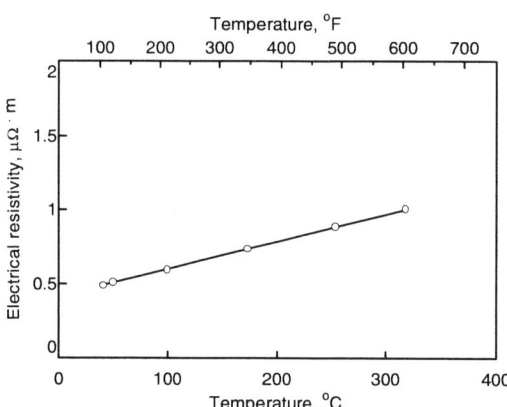

ASTM grade 2 titanium: Electrical resistivity

Source: Thermo Physical Research Lab Report 578 for RMI Titanium Company

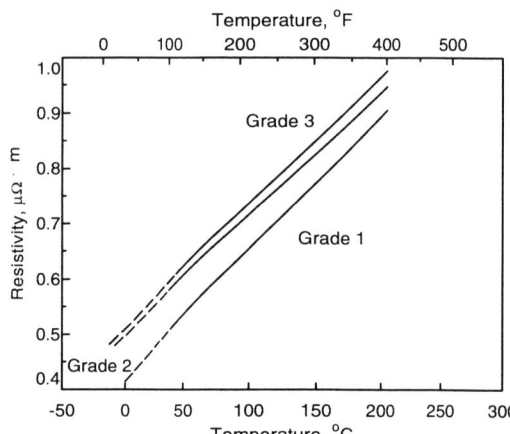

CP titanium: Resistivity vs temperature

Source: Corrosion data, RMI Titanium Company

Chemical Reactivity

The nature of the oxide film on titanium alloys basically remains unaltered in the presence of minor alloying constituents; thus, small additions (<2 to 3%) of most commercially used alloying elements or trace alloy impurities generally have little effect on the basic corrosion resistance of titanium in normally passive environments. For example, despite small differences in interstitial element (carbon, oxygen, and nitrogen) and iron content, all unalloyed grades of titanium possess the same useful range of resistance in environments in which corrosion rates are normally very low. However, under active conditions in which titanium exhibits significant general corrosion, certain alloying elements may accelerate corrosion. Increasing the alloy iron and sulfur content, for example, increases corrosion rates when corrosion rates exceed 0.13 mm/yr (5 mils/yr) (R.W. Schutz, J.S. Grauman, and J.A. Hall, Effect of Solid Solution Iron on the Corrosion Behavior of Titanium, in *Titanium, Science and Technology,* Proceedings of the Fifth International Conference on Titanium, Deutsche Gesellschaft für Metallkunde e.V., Germany, 1985, p 2617-2624; and L.C. Covington and R.W. Schutz, Effects of Iron on the Corrosion Resistance of Titanium, in *Industrial Applications of Titanium and Zirconium,* STP 728, ASTM, 1981, p 163-180).

Thus, minor variations in alloy chemistry may be of concern only under conditions in which the passivity of titanium is borderline, or when the metal is fully active. On the other hand, minor nickel and palladium additions are highly effective in enhancing the corrosion resistance of titanium alloys under reducing conditions. Weldments and castings generally exhibit corrosion resistance similar to that of their unwelded, wrought counterparts. However, under marginal or active conditions (for corrosion rates ≥0.10 mm, or 4 mils, per year), weldments may experience accelerated corrosion attack relative to the base metal, depending on the increasing impurity (iron, sulfur, or oxygen) content associated with a coarse, transformed β microstructure.

Titanium alloy combustion rates

Alloy	Combustion rate, $g/cm^2 \cdot s$	
	J_1	J_2
Ti (99.2%)	0.10	0.03
Ti-6Al-4V	0.10	0.05
Ti-5Al-2.5Sn	0.08	0.03
Ti-8Al-1Mo-1V	0.09	0.06
Ti-6Al-6V-2Sn	0.07	0.03
Ti-6Al-2Sn-4Zr-2Mo	0.10	0.04
Ti-6Al-2Sn-4Zr-6Mo	0.09	0.07
Ti-4Al-12Sn	0.07	0.04
Ti-12Mo-4Sn	0.11	0.05
Ti-0.24Fe-3.3Al-2.6V	0.10	0.06
Ti-11.5Mo-6Zr-4.5Sn	0.11	0.05
Ti-5.9Cr-4.2Mo-3.3V-0.07Fe (BetaIII)	0.10	0.04
Ti-5.8Al-2.0Mo-2.1Sn-1.9Cr-1.8Zr	0.07	0.05
Ti-3.3Al-8.1V-5.9Cr-4.2Mo-3.9Zr	0.09	0.07
Ti-4.9Al-5.0Sn-1.9Mo-1.9Zr-0.24Si	0.09	0.05
Ti-5Mo	0.12	0.07
Ti-11Mo	0.11	0.07
Ti-20Mo	0.10	0.06
Ti-30Mo	0.10	0.07
Ti-40Mo	0.10	0.06
Ti-2Sn	0.11	0.06
Ti-6Sn	0.10	0.05
	0.05	0.03
Ti-9Sn	0.10	0.05
	0.04	0.02
Ti-12Sn	0.10	0.05

Note: One goal of the study was to measure the combustion rates for a number of different alloys to see if the addition of certain alloying elements could reduce the rate of oxygen uptake. It is known, for example, that the addition of small amounts of silicon to iron or copper systems can markedly reduce their rates of oxygen absorption during combustion by forming a thin slag layer on the surface. The combustion rates for commercially pure titanium and a series of 23 alloys were measured in a gentle flow of oxygen (~65 cm^3/s) at atmospheric pressure. The combustion rates for these alloys during the first (J_1) and second (J_2) parts of the combustion reaction are presented. These results are the average from several specimens of each alloy. With only two exceptions, little variation in combustion rate was observed among the alloys studied.
Source: T. Strobridge *et al.*, "Titanium Combustion in Turbine Engines," FSS-RD-79-51, July 1979

Combustion

Although the ignition of titanium alloys in normal air is generally not of concern in typical mill product forms (except powders), ignition is possible in enriched oxygen atmospheres. Ignition is not easily achieved unless the oxide film is mechanically damaged and fresh metal surfaces are exposed. The problem increases as the material gets thinner.

Rapid, dangerous, exothermic halogenation reactions may occur with titanium in dry chlorine and bromine gas environments unless minimum water content (or oxygen content) maintains total alloy passivity. The critical water content depends on gas temperature and flow rate. Mechanical damage to metal surfaces exposing fresh metal facilitates reaction with dry chlorine, but thicker oxide films (thermal oxides) tend to retard initiation of the reaction. Rapid, pyrophoric reactions with titanium alloys are also possible in anhydrous N_2O_4 gas atmospheres, whereas the presence of 0.6 to 1.0 wt% nitric acid effectively inhibits metal attack.

Unalloyed Ti: Ignition/crack propagation limits in oxygen

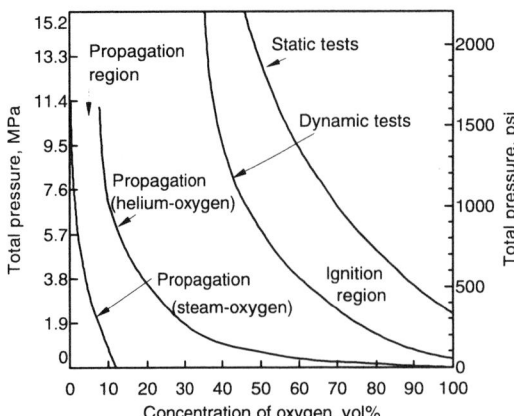

Source: J.D. Jackson, W.K. Boyd, and P.D. Miller, "Reactivity of Metals with Liquid and Gaseous Oxygen," DMIC Memorandum 163, Defense Materials Information Center, Battelle Memorial Institute, Jan 1963; and F.E. Littman and F.M. Church, "Reactions of Metals with Oxygen and Steam," Final Report AECU-4092, Stanford Research Institute to Union Carbide Nuclear Company, Feb 1959

Unalloyed Ti: Ignition limits in pure oxygen

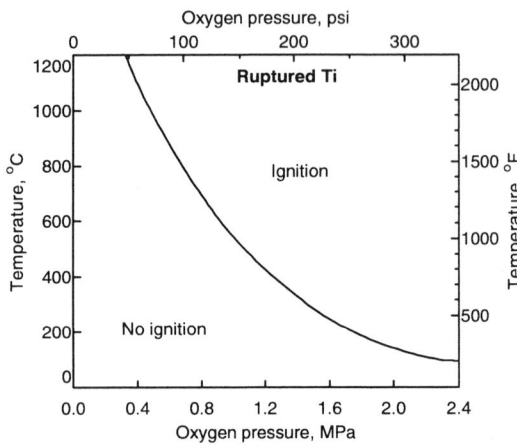

Source: F.E. Littman and F.M. Church, "Reactions of Metals with Oxygen and Steam," Final Report AECU-4092, Stanford Research Institute to Union Carbide Nuclear Company, Feb 1959

ASTM grade 4 Ti: Low-temperature ignition limits in pure O$_2$

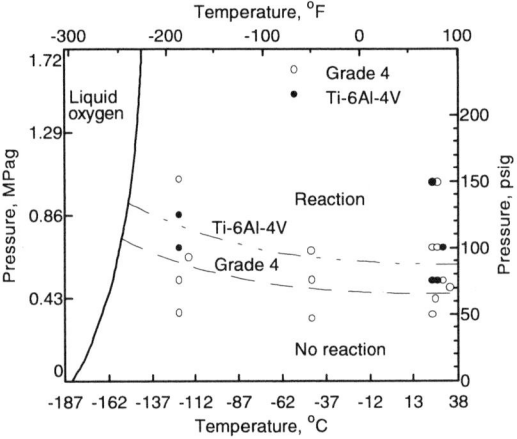

Source: R.L. Kane, The Corrosion of Titanium in *The Corrosion of Light Metals*, The Corrosion Monograph Series, John Wiley & Sons, 1967; and J.D. Jackson, W.K. Boyd, and P.D. Miller, "Reactivity of Metals with Liquid and Gaseous Oxygen," DMIC Memorandum 163, Defense Materials Information Center, Battelle Memorial Institute, Jan 1963

Ti reaction limits with red fuming HNO$_3$

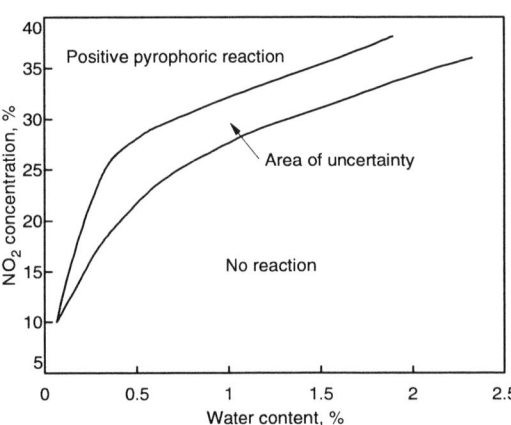

Fuming nitric acid containing less than 1.4 to 2.0% water or more than 6% NO$_2$ may cause a rapid impact-sensitive reaction to occur. Both water and NO are effective inhibitors.
Source: L.L. Gilbert and C.W. Funk, Explosions of Titanium and Fuming Nitric Acid Mixtures, *Met. Prog.*, Nov 1956, p 93-96

Passivation

Although titanium is the preferred metallic material for handling wet chlorine and bromine gas environments, a minimum water content (or oxygen content) in these cases is necessary to maintain total alloy passivity. The critical water content depends on gas temperature and flow rate. Titanium alloys cannot be fully passivated in liquid bromine, because of the extremely low solubility of water in this medium.

Repassivation potentials (E_p) are conservative measures of anodic pitting tendency because they represent minimum potentials below which pitting cannot be sustained. Unalloyed titanium exhibits the highest E_p value, which decreases as alloy aluminum content increases. Increasing iron content over the range of 0.02 to 0.20% results in a minor (several tenths of a volt) decrease in E_p values in unalloyed titanium (R.W. Schutz *et al.*, in *Titanium, Science and Technology*, 1985, p 2617-2624).

Like anodic pitting potentials, repassivation potentials are significantly lower in bromide and iodide media. Room-temperature E_p values of +1.2 and +0.95 V are measured for grades 2 and 5 titanium, respectively, whereas values of +0.9 V in dilute KBr solutions have been reported. Repassivation potentials for grades 2 and 5 titanium in dilute room-temperature iodide solutions have

Repassivation potentials of as-annealed titanium alloys in boiling chloride media

Alloy	Repassivation potentials, V(a)		
	5% NaCl (pH 3.5)	3% HCl	Saturated NaCl
Grade 1	...	...	+7.0
Grade 2	+6.7	+5.8	+5.7
Ti-6-4	+2.3	+1.7	...
Ti-550	+2.8	+2.3	...
Ti-6-2-4-6	+3.0	+2.4	...
Ti-3-8-6-4-4	+3.2	+2.6	...
Ti-8-8-2-3	+2.6	+2.4	...
Ti-15-5	+6.3	+5.6	...
Grade 12	...	...	+5.9
Grade 7	...	...	+5.6

(a) Measured versus Ag/AgCl reference electrode. Source: R.W. Schutz and J.S. Grauman, "Compositional Effects on Titanium Alloy Repassivation Potential in Chloride Media," paper presented at the International Conference on Localized Corrosion, Orlando, National Association of Corrosion Engineers, June 1987

Conditions to passivate Ti in pure chlorine gas

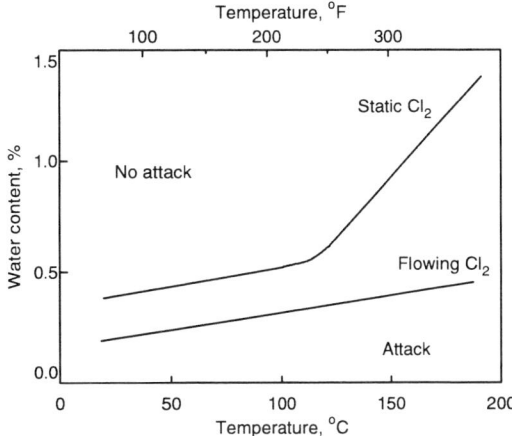

Source: H.B. Bomberger, in *Industrial Applications of Titanium and Zirconium: Third Conference*, STP 830, American Society for Testing and Materials, 1984, p 143-158

Passivation of unalloyed Ti in static chlorine gas

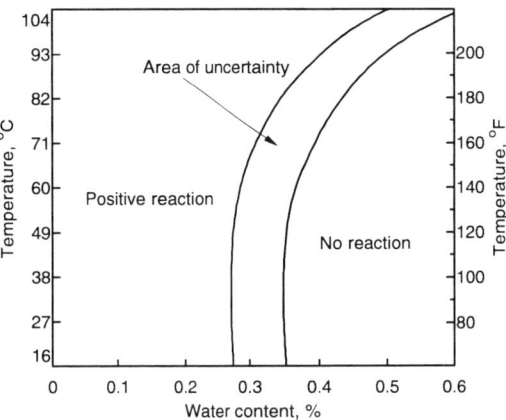

Source: E.E. Millaway and M.H. Klineman, Factors Affecting Water Content Needed to Passivate Titanium in Chlorine, *Corrosion*, Vol 23 (No. 4), 1972, p 88

been measured to be +1.8 and +1.5 V, respectively (H.J. Raetzer-Scheive, *Corrosion*, Vol 34 (No. 12), Dec 1978, p 437-442; and T.R. Beck, *J. Electrochem. Soc.*, Vol 120, 1973, p 1310).

Pitting **Anodic Breakdown Pitting.** Titanium exhibits relatively high anodic breakdown potentials (E_b) in aqueous solution compared to most engineering metals. This is the basis for its use as dimensionally stable anodes for chlor-alkali cells, anodes for recovery of metals or metal oxides from solutions, zinc and nickel plating anode baskets, aluminum anodizing racks, and platinum anode substrates for impressed cathodic protection systems. In sulfate and phosphate media, anodic pitting potentials of titanium alloys are typically in the range of +80 to +100 V (versus Ag/AgCl reference electrode). For this reason, dilute sulfuric and phosphoric acid solutions (and their salts) are typical electrolytes for anodizing titanium to grow protective surface oxides and/or produce colored surfaces.

In halide salt solutions, titanium alloys exhibit somewhat lower but yet reasonably high pitting potentials. Values of +9 to +10.5 V (versus Ag/AgCl) can be expected in room-temperature

Depassivation of Ti in various HCl concentrations

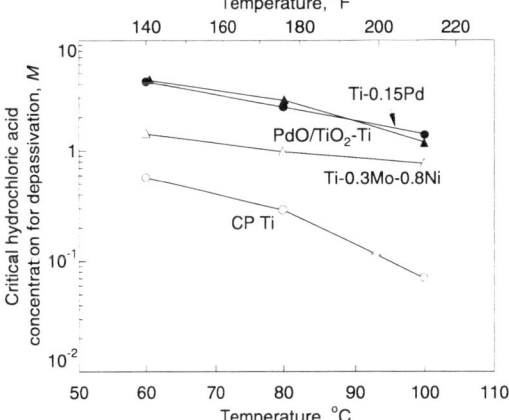

The effect of temperature on the critical hydrochloric acid concentration for the depassivation of four titanium materials in deaerated 1M sodium chloride solution is shown. For each material, the critical concentration decreases with increasing temperature.
Source: B. Satoh et al., The Crevice Corrosion Resistance of Some Titanium Materials, *Plat. Met. Rev.*, Vol 31, 1987, p 115-121

chloride solutions, decreasing to approximately +1.2 V at 175 to 250 °C (345 to 480 °F). These values depend on sample surface condition. For example, abraded or sand-blasted surfaces exhibit somewhat lower values than as-pickled surfaces.

Anodic pitting potential values are significantly lower in bromide solutions, and they decrease with increasing temperature. At room temperature, anodic pitting potentials of +0.90 to +1.4 V have been reported for titanium grades 2 and 5 (Ref 1, 2). One study has reported values for grade 1, 2, and 3 titanium ranging between +1.8 to +2.2 V in 1% NaBr (pH6) solution at room temperature, decreasing to +1.0 to +1.2 V at 100 °C (212 °F) (Ref 3). Thus, pitting of titanium alloys may be possible in pure bromide solutions at higher temperatures if highly oxidizing conditions prevail.

However, additions of various oxidizing anions may inhibit pitting in NaBr solutions by significantly raising anodic pitting potentials (Ref 2). Critical concentrations of these inhibitive anions have been determined, and the relative efficiency of inhibition decreases in the order $SO_4^{2-} > NO_3^- > CrO_4^{2-} > PO_4^{3-} > CO_3^{2-}$.

Studies in room-temperature iodide solutions have revealed anodic pitting potentials of +1.7 to +1.8 V, with little effect of acidification indicated (Ref 1, 4). Above 40 to 50 °C (100 to 120 °F), values near +0.5 V (versus SCE) are reported.

Anodic breakdown pitting potentials (E_b) for titanium alloys in chloride solutions

Alloy	Solution	pH	Temperature °C	°F	E_b, V(a)
Grade 2	1N NaCl	7	25	75	+11.0
Grade 5	1N NaCl	7	25	75	5.2
Grade 2	Saturated NaCl(b)	1,7	25	75	9.6
Grade 12	Saturated NaCl(b)	1,7	25	75	9.6
Grade 7	Saturated NaCl(b)	1,7	25	75	9.6
Grade 5	Saturated NaCl(b)	1,7	25	75	8.9
Grade 2	Saturated NaCl	1,7	95	200	5.0-6.5
Grade 12	Saturated NaCl	1,7	95	200	5.0-5.7
Grade 7	Saturated NaCl	1,7	95	200	5.2-7.0
Grade 5	Saturated NaCl	1,7	95	200	2.5-3.4
Grade 2	1N NaCl	7	125	255	~4.4
Grade 2	1N NaCl	7	150	300	~2.2
Grade 2	1N NaCl	7	175	345	~1.2
Grade 2	1N NaCl	7	200	390	~1.2
Grade 12	Seawater	8	245	475	2.3
Grade 12	O_2-saturated seawater	8	245	475	3.3
Grade 2	1N KCl + 0.2 M H_2SO_4	...	25	75	80.0

(a) Measured versus Ag/AgCl reference electrode. (b) Similar values were obtained in synthetic seawater (pH 8). Source: *Metals Handbook, Corrosion*, Vol 13, 9th ed., ASM International, 1987, p 688

Grade 2 titanium: Temperature vs E_b in dilute NaCl and NaBr

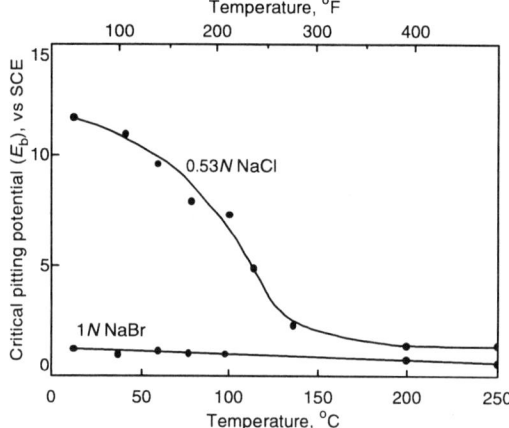

Source: T. Koizumi and S. Furuya, in *Titanium—Science and Technology*, Vol 4, Proceedings of the Second International Conference, Plenum Press, 1973, p 2383-2393

ASTM grade 2 Ti: Pitting potential in neutral chloride brine

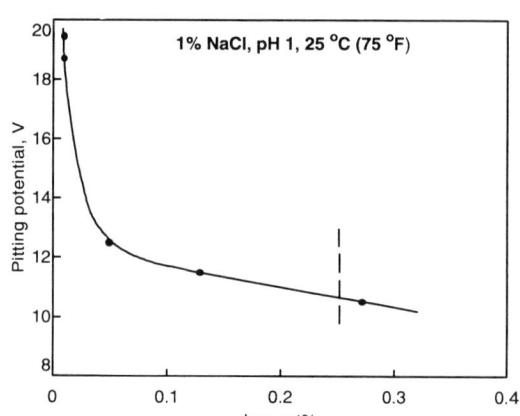

Source: L.C. Covington, Pitting Corrosion of Titanium Tubes in Hot Concentrated Brine Solutions, *Galvanic and Pitting Corrosion—Field and Laboratory Studies*, ASTM STP 576, American Society for Testing and Materials, 1976, p 150

CP Ti: Anode pitting potential vs iron

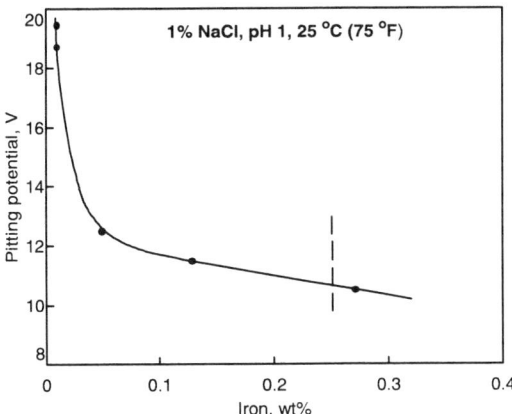

Iron content below 0.05% substantially increases the pitting potential, which is often used as an indication of the stability of the titanium oxide, with high voltage being desirable.

Electrochemical Potentials

CP Ti: Effect of Co ions on polarization

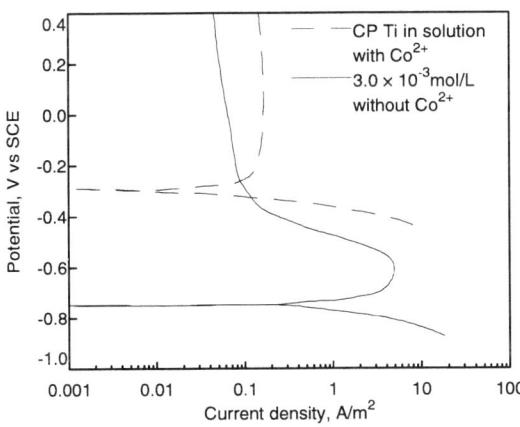

Polarization behavior of CP Ti in acidic NaCl solution (NaCl 4.27 mol/L, pH 0.5, boiling, sweep rate 0.02 V/min).
Source: ISIJ Int., Special Issue on Recent Advances on Titanium Technology, Vol 31(No. 8) 1991, p 903

CP and modified Ti: Polarization in NaCl solution

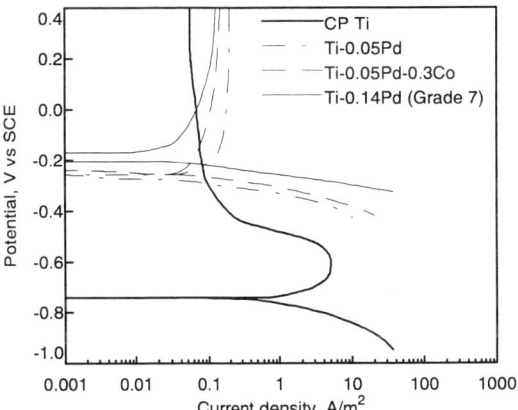

Ti-Pd-(Co) alloys in acidic NaCl solution (NaCl 4.27 mol/L, pH 0.5, boiling, sweep rate 0.92 V/min).
Source: ISIJ Int., Special Issue on Recent Advances on Titanium Technology, Vol 31 (No. 8), 1991, p 903

Ti-Pd alloys: Polarization in NaCl acidic solution

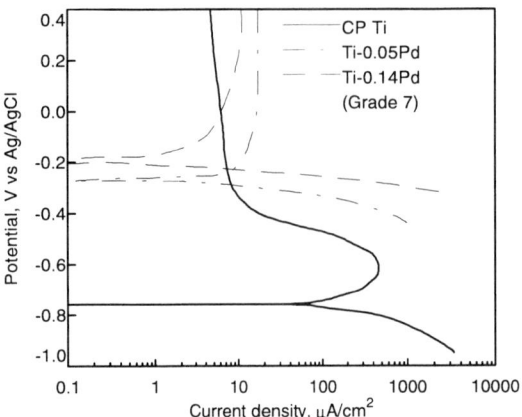

Acidic sodium chloride solution (NaCl, 250 g/L, pH 0.5, boiling).
Source: Y. Shida and S. Kitayama, Effect of Pd Additions on the Crevice Corrosion Resistance of Titanium, *Sixth World Conference on Titanium*, Les Editions de Physique, Paris, 1988, p 1729-1732

CP/modified Ti: Corrosion potentials in acidic NaCl solution

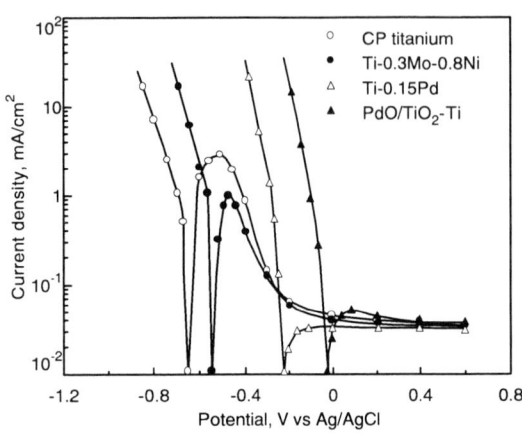

Polarization curves are shown for four titanium materials in boiling 1 M sodium chloride with 1 M hydrochloric acid. The corrosion potentials of both Ti-0.15 Pd and PdO/TiO$_2$-Ti, where no anodic peaks occur, are more noble than those of CP titanium and Ti-0.3Mo-0.8Ni.
Source: B. Satoh *et al.*, The Crevice Corrosion Resistance of Some Titanium Materials, *Plat. Met. Rev.*, Vol 31, 1987, p 115-121

CP/modified Ti: Corrosion potential in boiling NaCl solution

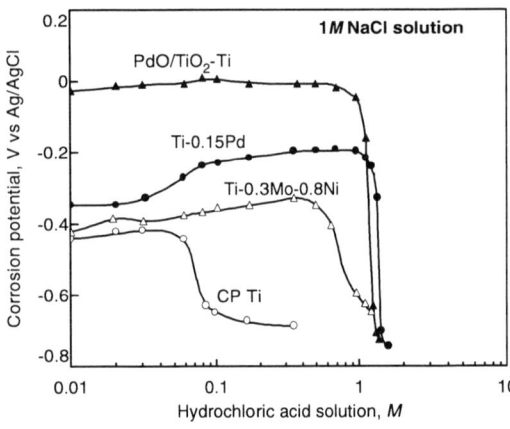

Source: B. Satoh *et al.*, The Crevice Corrosion Resistance of Some Titanium Materials, *Plat. Met. Rev.*, Vol 31, 1987, p 115-121

Anodic behavior of Ti in boiling 1M H$_2$SO$_4$

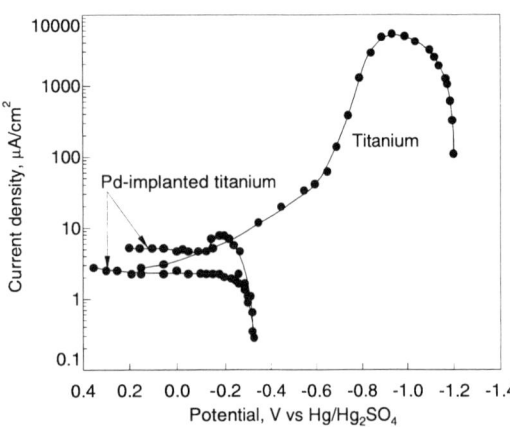

Source: E. McCafferty *et al.*, Effect of Laser Processing and Ion Implantation on Aqueous Corrosion, *Corrosion of Metals Processed by Directed Energy Beams*, AIME, 1982, p 6

ASTM grade 4 Ti: Polarization in CH₃OH with additions

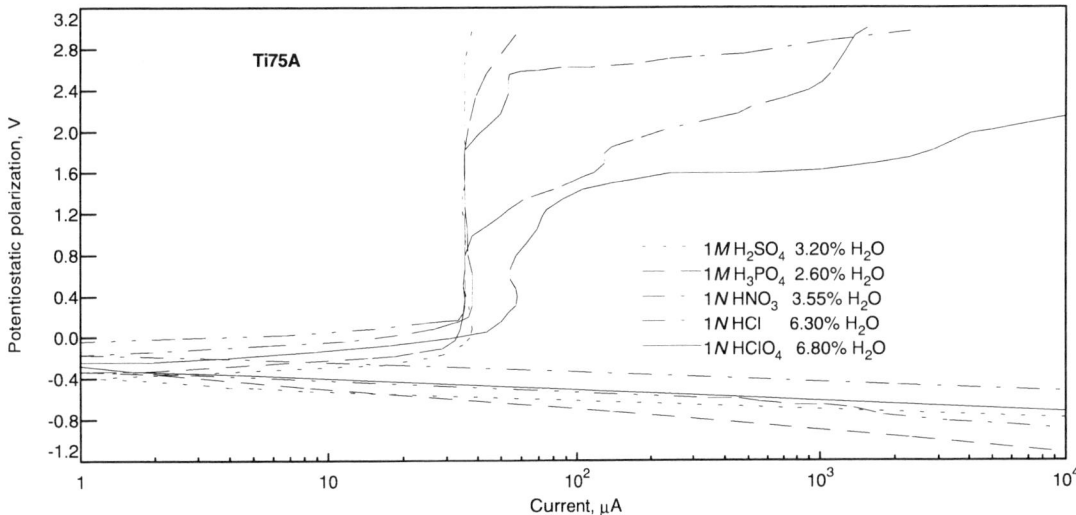

Source: F. Mansfield, Pitting Caused by Chlorides or Sulfates in Organic Media, *Galvanic and Pitting Corrosion—Field and Laboratory Studies*, ASTM STP 576, 1976, p 180-203

ASTM grade 4 Ti: Polarization in sulfuric acid solutions

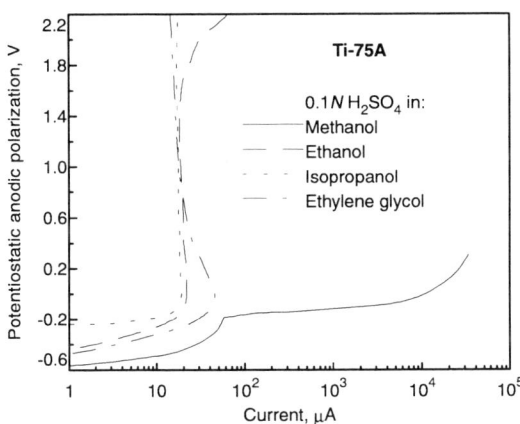

Source: F. Mansfield, Pitting Caused by Chlorides or Sulfates in Organic Media, *Galvanic and Pitting Corrosion—Field and Laboratory Studies*, ASTM STP 576, 1976, p 180-203

Potentials in boiling 1M H₂SO₄

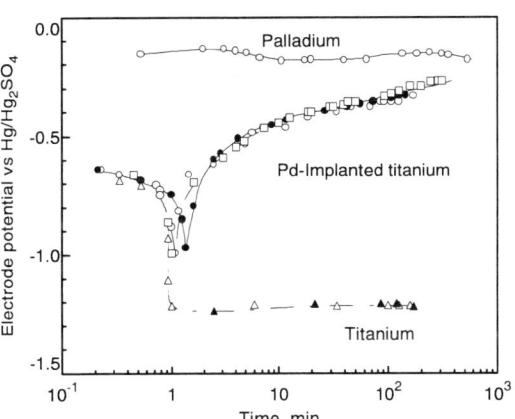

Source: E. McCafferty et al., Effect of Laser Processing and Ion Implantation on Aqueous Corrosion, *Corrosion of Metals Processed by Directed Energy Beams*, AIME, 1982

General Corrosion

Water and Seawater

Titanium and its alloys are fully resistant to water, all natural waters, and steam to temperatures in excess of 315 °C (600 °F) (*J. Inst. Met.*, Vol 89, 1960-1961, p 165). Slight weight gain is usually experienced in these benign environments, along with some surface discoloration at higher temperatures from finite passive film thickening. The typical contaminants encountered in natural water streams, such as iron and manganese oxides, sulfides, sulfates, carbonates, and chlorides, do not compromise the passivity of titanium. In media containing chloride levels greater than 1000 ppm (for example, seawater) at temperatures above 75 °C (165 °F), consideration should be given to possible crevice corrosion when tight crevices exist in service. Unalloyed titanium has provided more than 20 years of outstanding service in seawater for the chemical, oil refining, desalination, and power industries. Due to its immunity to ambient seawater corrosion, titanium is considered to be "technically correct" for many critical marine applications, including naval and offshore components.

Oxidizing Media

Titanium is generally highly resistant to oxidizing media and oxidizing acids over a wide range of concentrations and temperatures. Common chemicals in this category include chromic, nitric, perchloric, and hypochlorous acids and salts of these acids. Other oxidizing salts include thiosul-

Corrosion of titanium in ambient seawater

Alloy	Ocean depth m	Ocean depth ft	Corrosion rate mm/yr	Corrosion rate mils/yr
Unalloyed titanium	Shallow		8×10^{-7}	0.00003
Unalloyed titanium	720-2070	2360-6800	$<2.5 \times 10^{-4}$	<0.1
Unalloyed titanium	2-2070	6.5-6800	nil	
Unalloyed titanium	1720	5640	4×10^{-5}	0.0015
Ti-6Al-4V	2-2070	6.5-6800	$<2.5 \times 10^{-4}$	<0.01
Ti-6Al-4V	1720	5640	8×10^{-6}	0.0003
Ti-6Al-4V	1720	5640	$<1 \times 10^{-3}$	<0.04

Source: *Metals Handbook, Corrosion*, Vol 13, 9th ed., ASM International, 1987, p 677

Erosion-corrosion of grade 2 titanium in seawater at various locations

Location	Flow rate m/s	Flow rate ft/s	Type of test	Test duration, months	Erosion-corrosion rate mm/yr	Erosion-corrosion rate mils/yr
Brixham Sea	9.8	32	Model condenser	12	0.003	0.12
Kure Beach, NC	1	3.3	Ducting	54	7.5×10^{-7}	0.0003
Kure Beach, NC	8.5	28	Rotating disk	2	1.3×10^{-4}	0.005
Kure Beach, NC	9	29.5	Micarta wheel	2	2.8×10^{-4}	0.01
Kure Beach, NC	7.2	23.6(a)	Jet impingement	1	5×10^{-4}	0.02
Wrightsville Beach, NC	1.3	4.3	...	6	1×10^{-4}	0.004
Wrightsville Beach, NC	9	29.5	Micarta wheel	2	1.8×10^{-4}	0.007
Mediterranean Sea	7.2	23.6(a)	Jet impingement	0.5	0.5 mg/day	...
Dead Sea	7.2	23.6(a)	Jet impingement	0.5	0.2 mg/day	...

(a) Included air. Source: "Titanium Heat Exchangers for Service in Seawater, Brine and Other Natural Aqueous Environments: The Corrosion, Erosion and Galvanic Corrosion Characteristics of Titanium in Seawater, Polluted Inland Waters and in Brines," Titanium Information Bulletin. Imperial Metals Industries (Kynoch) Ltd., May 1970

Erosion-corrosion of titanium grade 2 in seawater containing suspended solids

Flow rate m/s	Flow rate ft/s	Seawater suspension	Test duration, h	Erosion-corrosion rate mm/yr	Erosion-corrosion rate mils/yr
7.2	23.6	No solids	10 000	nil	
2.0	6.5	40 g/L of 60-mesh sand	2000	0.0025	0.1
2.0	6.5	40 g/L of 10-mesh emery	2000	0.0125	0.5
3.5	11.5	1% 80-mesh emery	17.5	0.0037	0.15
4.1	13.5	4% 80-mesh emery	17.5	0.083	3.3
7.2	23.6	40% 80-mesh emery	1	1.5	60

Source: "Titanium Heat Exchangers for Service in Seawater, Brine and Other Natural Aqueous Environments: The Corrosion, Erosion and Galvanic Corrosion Characteristics of Titanium in Seawater, Polluted Inland Waters and in Brines," Titanium Information Bulletin, Imperial Metals Industries (Kynoch) Ltd., May 1970

fates, vanadates, permanganates, and molybdates. Corrosion rates at and below the boiling point of these aqueous salt solutions over the full range of concentration will typically be less than 0.03 mm/yr (1.2 mils/yr). Titanium is unique among the common engineering alloys in its immunity to general and pitting corrosion in oxidizing chloride environments. These comments also apply to bromine and iodine-containing media. Halide salts of oxidizing cationic species also enhance the passivity of titanium alloys such that negligible corrosion rates can be expected. Examples include $FeCl_3$, $CuCl_2$, and $NiCl_2$ solutions and their bromide counterparts.

Nitric Acid. Unalloyed titanium has been extensively used for handling and producing nitric acid in applications in which stainless steels have experienced significant uniform or intergranular attack.

In hot, very pure solutions or vapor condensates of nitric acid, significant uniform corrosion rates may occur, particularly as temperatures increase. However, as the impurity levels increase in hot HNO_3 solutions, the resistance of unalloyed titanium improves dramatically. In particular, relatively small amounts of certain dissolved metallic species, including Si^{4+}, Cr^{6+}, Fe^{3+}, Ti^{4+}, or various precious metal ions, can effectively inhibit the high-temperature corrosion of titanium in nitric acid (*Metals Handbook, Corrosion*, Vol 13, 9th ed., ASM International, 1987, p 677). The significant discrepancies and variations in titanium corrosion rates in hot HNO_3 media reported by investigators over the years appear to be the result of these inhibitive metal ion effects. Because titanium corrosion is inhibited by its own corrosion product (Ti^{4+}), the titanium surface area to acid volume ratio, the test duration, and the rate of solution replenishment will all be critical to the rate obtained. The container material and acid purity (chemistry) will also be influential. These factors are reviewed in more detail in D.E. Thomas, Titanium Alloy Corrosion Resistance in Nitric Acid Solutions, *Titanium 1986—Products and Applications*, Vol 1, Proceedings of the Technical Program from the 1986 International Conference, San Francisco, Titanium Development Association, 1986.

Fuming Nitric Acid. Titanium alloys exhibit good resistance to white fuming nitric acid. However, dangerous and violent pyrophoric reactions may occur with titanium alloys exposed to red fuming nitric acid or to nitrogen tetroxide (see the section on combustion above). Corrosion rate data in red fuming nitric acid for various alloys as a function of NO_2 and water content also can be found in N.D. Tomashov and P.M. Altovskii, *Corrosion and Protection of Titanium*, Government Scientific-Technical Publication of Machine-Building Literature, Russian translation, 1963.

Thickness of oxide on commercially pure titanium heated for 1/2 h in air (see also figure one page over)

Temperature		Measurable thickness	
°C	°F	mm	in.
315	600	None	
425	800	None	
540	1000	None	
650	1200	<0.005	<0.0002
705	1300	0.005	0.0002
760	1400	0.008	0.0003
815	1500	<0.025	<0.001
870	1600	<0.025	<0.001
925	1700	<0.05	<0.002
980	1800	0.05	0.002
1040	1900	0.10	0.004
1095	2000	0.36	0.014

Source: *Metals Handbook*, Vol 4, 9th ed., 1981, p 771

Corrosion of unalloyed titanium in solutions of oxidizing chlorine compounds

Reagent	Concentration, %	Temperature		Corrosion rate	
		°C	°F	mm/yr	mils/yr
Water saturated with chlorine	...	75	165	0.003	0.12
Water saturated with chlorine	...	88	190	0.002(a)	0.08
Water saturated with chlorine	...	97	207	0.07	2.8
NaOCl	6	25	77	nil	
ClO_2 + HOCl	15	43	110	nil	
ClO_2 + steam	5	100	212	0.005	0.2
$Ca(OCl)_2$	2	100	212	0.001	0.04
$Ca(OCl)_2$	6	100	212	0.001	0.04
$Ca(OCl)_2$	18	25	77	nil	
HOCl + ClO_2 + Cl_2	17	38	100	nil	

(a) Welded sample. Source: *Corrosion Resistance of Titanium*, Technical Handbook, Imperial Metals Industries (Kynoch) Ltd., Birmingham, UK

Corrosion of titanium grade 2 and type 304L stainless steel heated surfaces exposed to boiling 90% HNO_3

Metal temperature		Corrosion rate			
		Grade 2		Type 304L	
°C	°F	mm/yr	mils/yr	mm/yr	mils/yr
116	240	0.03-0.17	1.2-6.7	3.8-13.2	150-520
135	275	0.04-0.15	1.6-6	17.2-73.7	675-2900
154	310	0.03-0.06	1.2-2.4	18.3-73.7	720-2900

Source: T.F. Degnan, Materials for Handling Hydrofluoric, Nitric and Sulfuric Acids, in *Process Industries Corrosion*, National Association of Corrosion Engineers, 1975, p 229

Peroxides. Although peroxides are generally oxidizing, titanium alloys can experience general corrosion in aqueous peroxide solutions, depending on concentration, temperature, and pH. Corrosion rates are minimal in dilute near-neutral hydrogen peroxide solutions, but increase dramatically under alkaline conditions because of the formation of soluble titanium-peroxyl (complex) compounds. However, corrosion is effectively inhibited by small additions of calcium, strontium, or barium ions (U.S. patent 4,372,813). Sodium silicate and sodium hexametaphosphate additions have also been shown to reduce corrosion rates substantially (*Zashch Met.*, Vol 12, 1976, p 363-367). Significant attack may occur in highly concentrated (90%) H_2O_2 solutions.

Corrosion of various titanium alloys in boiling HNO_3 solutions
Test duration: 196 h

Alloy	Corrosion rate at indicated HNO_3 concentration					
	25%		45%		70%	
	mm/yr	mils/yr	mm/yr	mils/yr	mm/yr	mils/yr
Grade 1	0.15	6	0.39	15	0.08	3.1
Grade 7	0.17	6.7	0.38	14.9	0.07	2.8
Grade 12	0.18	7	0.27	10.6	0.06	2.4
Ti-6-2-1-0.8	0.39	15	0.73	28.7	0.21	8.3
Grade 9	0.18	7	0.54	21.3	0.10	4
Ti-550	0.83	32.6	1.14	44.9	0.30	12
Grade 5	0.67	26.4	0.86	33.8	0.02	0.8
Transage 207	8.0	315	15.6	614	0.95	37.4
Ti-6-2-4-6	4.3	170	5.7	224	0.78	30.7
Ti-10-2-3	0.48	18.9	1.2	47.2	0.07	2.8
Ti-3-8-6-4-4	1.13	44.5	3.6	141.7	1.46	57.5
Ti-5Ta	0.04	1.6	0.08	3.1	0.03	1.2

Source: Metals Handbook, Corrosion, Vol 13, 9th ed., ASM International, 1987, p 679

Effect of dissolved Ti^{4+} on the corrosion rate of unalloyed titanium in boiling HNO_3 solutions

Titanium ion added, mg/L	Corrosion rate			
	40% HNO_3		68% HNO_3	
	mm/yr	mils/yr	mm/yr	mils/yr
0	0.75	29.5	0.81	32
10	...	...	0.02	0.8
20	0.22	8.7	0.06	2.4
40	0.05	2	0.01	0.4
80	0.02	0.8	0.01	0.4

Source: S.H. Weiman, *Corrosion*, Vol 22, April 1966, p 98-106

Corrosion of unalloyed titanium in chromic acid solutions

Concentration of CrO_3, %	Temperature		Corrosion rate	
	°C	°F	mm/yr	mils/yr
10	Boiling		0.003	0.12
15	24	75	0.005	0.2
15	82	180	0.015	0.6
36.5	90	195	0.046	1.8
50	24	75	0.013	0.5
50	82	180	0.025	1.0

Source: *Corrosion Resistance of Titanium*, Technical Handbook, Imperial Metals Industries (Kynoch) Ltd., Birmingham, UK

General corrosion of grade 2 titanium in hydrogen peroxide solutions

Medium	pH(a)	Temperature		Corrosion rate	
		°C	°F	mm/yr	mils/yr
5% H_2O_2	1	23	73	0.064	2.5
5% H_2O_2	4.3	23	73	0.013	0.5
5% H_2O_2	1	66	150	0.152	6
5% H_2O_2	4.3	66	150	0.061	2.4
5% H_2O_2 + 500 ppm Ca^{2+}	1	66	150	nil	
20% H_2O_2	1	66	150	0.686	27
20% H_2O_2 + 500 ppm Ca^{2+}	1	66	150	nil	
10 g/L H_2O_2 + 20 g/L NaOH	...	60	140	55.9	2200
0.75 g/L H_2O_2	11	70	160	0.42	16.5
3.5 g/L H_2O_2 + 10 g/L NaOH + 10 g/L Na_2SiO_3 + 0.5 g/L Na_3PO_4	...	60	140	nil	

(a) Acid solutions were prepared with HCl additions. Source: Metals Handbook, Corrosion, Vol 13, 9th ed., ASM International, 1987, p 679

Unalloyed Ti: Growth of thermal oxide films in air

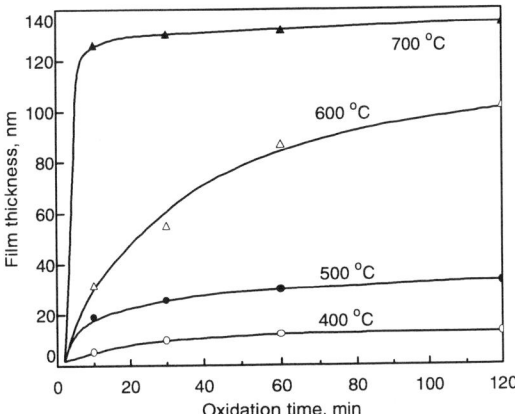

Source: T. Fukuzuka et al., On the Beneficial Effect of the Titanium Oxide Film Formed by Thermal Oxidation, in *Titanium '80—Science and Technology,* Vol 4, The Metallurgical Society, p 2783-2792

Corrosion of titanium alloys in boiling, uninhibited HNO₃

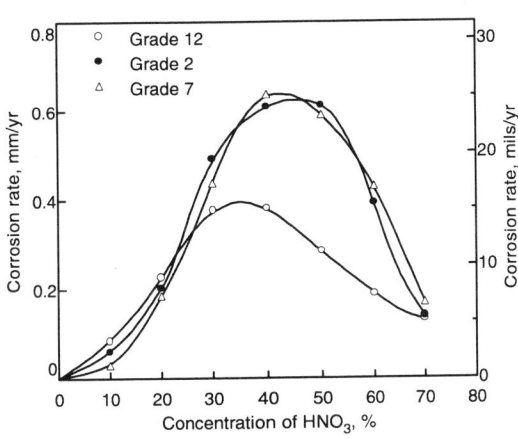

Acid solutions were refreshed every 24 h.
Source: *Metals Handbook, Corrosion,* Vol 13, 9th ed., ASM International, 1987, p 678

Ti: Effect of dissolved Ti^{3+} or Ti^{4+} ions on corrosion

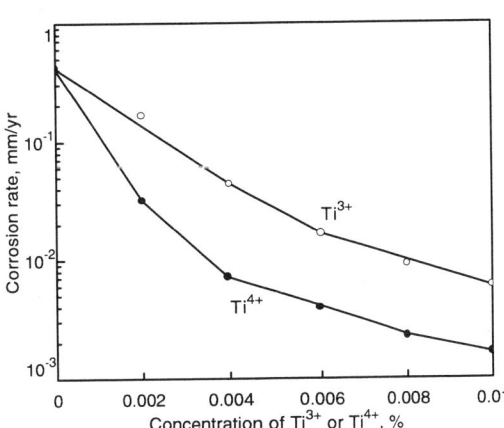

Corrosion was measured after 65 h in boiling 6N HNO₃ solutions.
Source: H. Satoh, F. Kamikubo, and K. Shimogori, Effect of Oxidizing Agents on Corrosion Resistance of CP Titanium in Nitric Acid Solution, in *Titanium—Science and Technology,* Proceedings of the Fifth International Conference on Titanium, Deutsche Gesellschaft für Metallkunde, e.V., 1985, p 2649-2655

Unalloyed Ti: Corrosion in high-temperature HNO₃ solutions

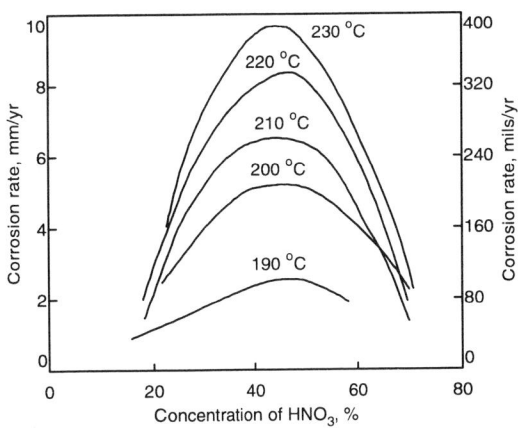

Source: *Corrosion Resistance of Titanium,* Technical Handbook, Imperial Metals Industries (Kynoch) Ltd., Birmingham, UK; and H. Keller and K. Risch, The Corrosion Behavior of Titanium in Nitric Acid at High Temperatures, *Werkst. Korros.,* Vol 9, 1964, p 741-743

CP Ti: Effect of dissolved $Cr_2O_7^{2-}$ ions on corrosion

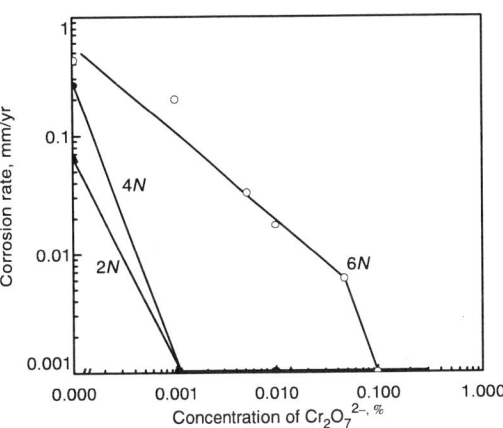

Corrosion was measured after 65 h in boiling HNO₃ solutions.
Source: H. Satoh, F. Kamikubo, and K. Shimogori, Effect of Oxidizing Agents on Corrosion Resistance of CP Titanium in Nitric Acid Solution, in *Titanium—Science and Technology,* Proceedings of the Fifth International Conference on Titanium, Deutsche Gesellschaft für Metallkunde, e.V., 1985, p 2649-2655

Reducing Acids

The corrosion resistance of titanium alloys in reducing acid media is very sensitive to acid concentration, temperature, background chemistry, and purity of the acid solution, in addition to titanium alloy composition. When the temperature and/or concentration of pure (uncontaminated) reducing acid solutions exceeds certain values, the protective oxide film of titanium may break down, which results in severe general corrosion. Included in this category are hydrochloric, sulfuric, hydrobromic, hydriodic, hydrofluoric, phosphoric, sulfamic, oxalic, and trichloroacetic acids. It should be noted that titanium-palladium alloys (e.g., ASTM grades 7 or 11) exhibit substantially greater resistance to dilute uninhibited reducing acids than any other commercial titanium alloy. Molybdenum also improves resistance to attack.

Inhibition of Reducing Acid Corrosion. Titanium alloys exhibit good resistance to most mildly reducing acid solutions whether they are inhibited or not. These environments include sulfurous acid, aqueous hydrogen sulfide solutions, boric acid, or carbonic acid. Near-nil corrosion rates can be expected over the full concentration range to temperatures well beyond their boiling points (*Metals Handbook, Corrosion*, Vol 13, 9th ed., ASM International, 1987, p 2679).

However, certain oxidizing species (cathodic depolarizers) in strongly reducing acids can effectively inhibit general corrosion, which expands the useful range of application of these alloys. For example, minute ferric ion concentrations extend the useful resistance of titanium grades, 2, 7, and 12 in HCl media (acid). Anodic protection is also an effective means of passivating and protecting titanium alloys in reducing acids. Generally, an increase in anodic potential will decrease the corrosion rate as long as the anodic pitting potential is not exceeded for titanium in the electrolyte.

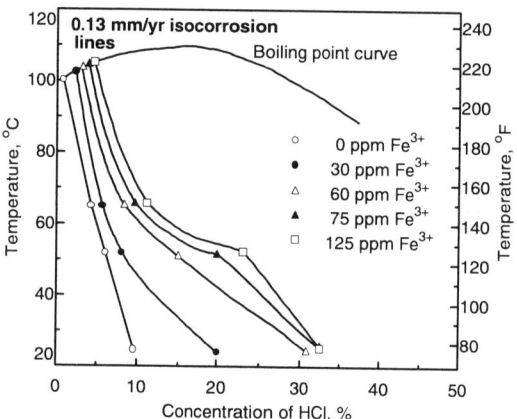

ASTM grade 2 Ti: Effect of Fe^{3+} on corrosion

In naturally aerated HCl solutions.
Source: *Metals Handbook, Corrosion*, Vol 13, 9th ed., ASM International, 1987, p 683

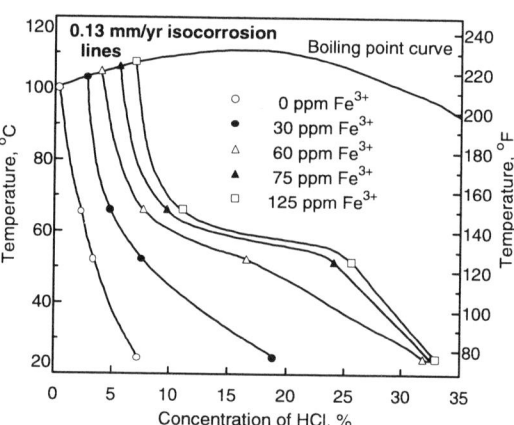

ASTM grade 12 Ti: Effect of Fe^{3+} on corrosion

In naturally aerated HCl solutions.
Source: *Metals Handbook, Corrosion*, Vol 13, 9th ed., ASM International, 1987, p 683

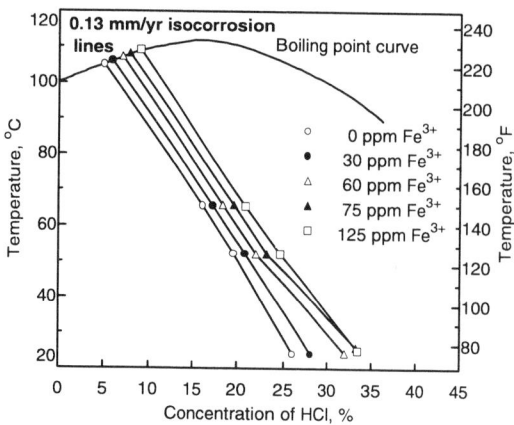

ASTM grade 7 Ti: Effect of Fe^{3+} on corrosion

In naturally aerated HCl solutions.
Source: *Metals Handbook, Corrosion*, Vol 13, 9th ed., ASM International, 1987, p 683

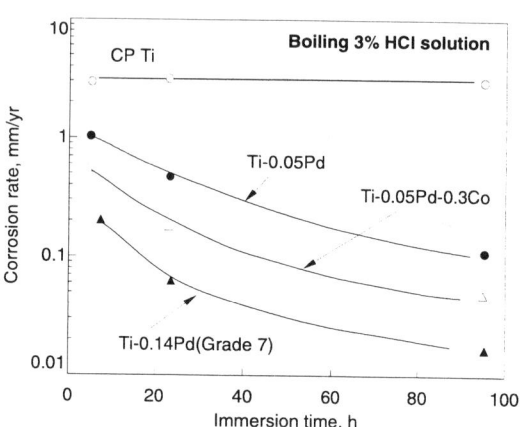

Modified Ti: Corrosion rates vs immersion time

Source: *ISIJ Int.*, Special Issue on Recent Advances on Titanium Technology, Vol 31 (No. 8), 1991, p 902

Ti-Pd alloys: Corrosion rate in boiling HCl solutions

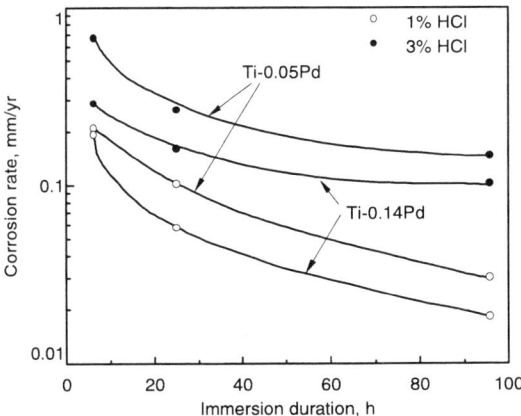

Source: Y. Shida and S. Kitayama, Effect of Pd Additions on the Crevice Corrosion Resistance of Titanium, *Sixth World Conference on Titanium*, Les Editions de Physique, Paris, 1988, p 1729-1732

Ti alloys: Corrosion rate vs HCl concentration (boiling)

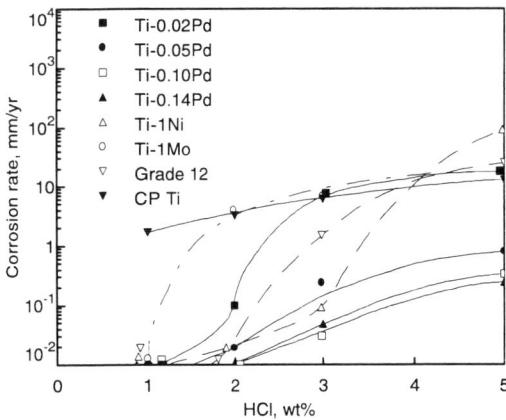

Source: Y. Shida and S. Kitayama, Effect of Pd Additions on The Crevice Corrosion Resistance of Titanium, *Sixth World Conference on Titanium*, Les Editions de Physique, Paris, 1988, p 1729-1732

Ti: Corrosion vs HCl concentration

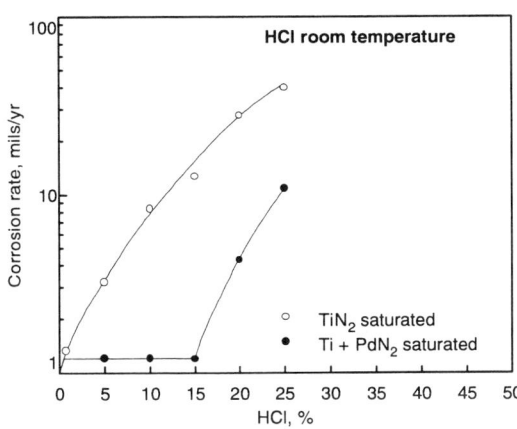

Source: H. Godard et al., *The Corrosion of Light Metals*, John Wiley & Sons, 1967

Modified Ti: Isocorrosion diagram in naturally aerated HCl

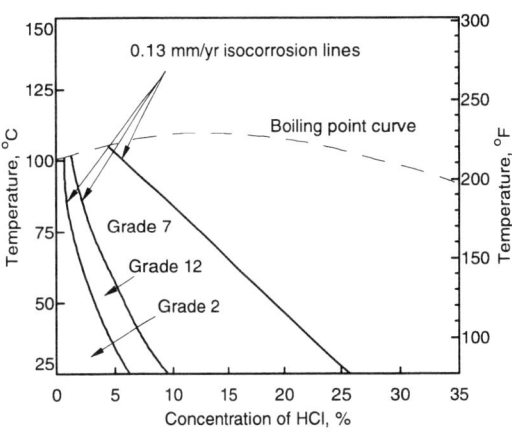

Source: *Metals Handbook, Corrosion*, Vol 13, 9th ed., ASM International, 1987, p 680

Modified Ti: Isocorrosion diagram in naturally aerated H_2SO_4

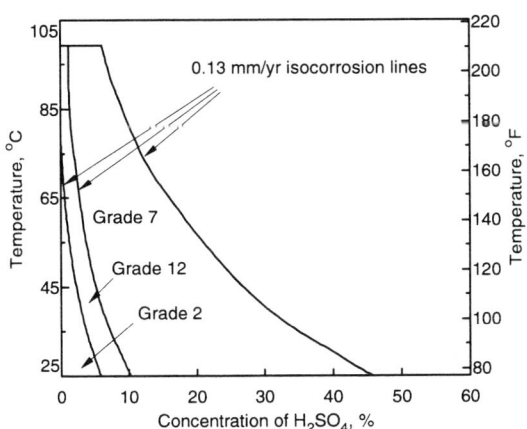

Source: *Metals Handbook, Corrosion*, Vol 13, 9th ed., ASM International, 1987, p 680

Ti and 0.2 Pd-Ti: Corrosion vs H₂SO₄ concentration

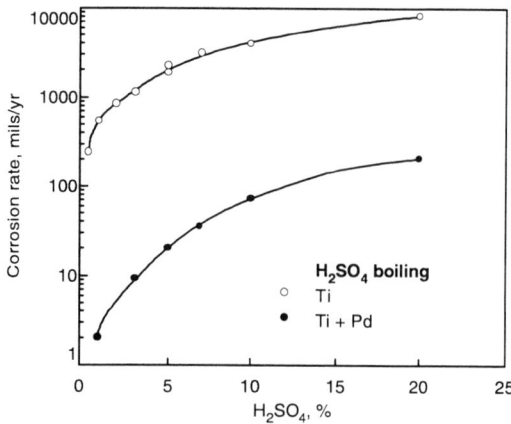

Source: H. Godard et al., *The Corrosion of Light Metals*, John Wiley & Sons, 1967, p 335

Modified Ti: Isocorrosion diagram in naturally aerated H₃PO₄

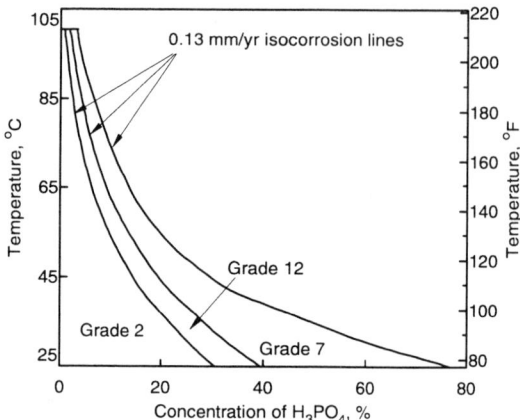

Source: *Metals Handbook, Corrosion,* Vol 13, 9th ed., ASM International, 1987, p 680

Effect of impressed anodic potentials on the corrosion of unalloyed titanium in hot reducing acids

Acid	Concentration, %	Temperature °C	°F	Applied potential, V versus SHE	Corrosion rate mm/yr	mils/yr	Reduction in corrosion rate
Sulfuric	40	60	140	+2.1	0.005	0.2	11 000×
	40	90	195	+1.4	0.07	2.8	896×
	40	114	237	+2.6	1.8	71	189×
	60	60	140	+1.7	0.035	1.4	662×
	60	90	195	+3.0	0.10	4	163×
Hydrochloric	37	60	140	+1.7	0.068	2.7	2080×
Phosphoric	60	60	140	+2.7	0.018	0.7	307×
	60	90	195	+2.0	0.5	20	100×
Formic	50	Boiling		+1.4	0.083	3.3	70×
Oxalic	25	Boiling		+1.6	0.25	10	350×
Sulfamic	20	90	195	+0.7	0.005	0.2	2710×

Source: *Corrosion Resistance of Titanium*, Technical Handbook, Imperial Metals Industries (Kynoch) Ltd., Birmingham, UK; J. Cotton, *Chem. Eng. Prog.*, Vol 66 (No. 10), 1970, p 57; and J. Cotton, *Chem. Ind.*, Vol 3, Jan 1958, p 68-69

Effect of certain multivalent metal ions on the corrosion of titanium in boiling reducing acids

Inhibiting ion	Concentration of inhibiting ion, ppm	Corrosion rate Boiling 5% HCl mm/yr	mils/yr	Boiling 10% H₂SO₄ mm/yr	mils/yr
Fe^{3+}	0	29	1142	>76.2	>3000
	100	0.025	1	0.208	8.2
	500	0.02	0.8	0.069	2.7
Cu^{2+}	0	29	1142	>76.2	>3000
	100	0.033	1.3	0.419	16.5
	500	nil		0.361	14.2
Mo^{6+}	0	29	1142	>76.2	>3000
	100	nil		0.001	0.04
	500	nil		nil	
Cr^{6+}	0	29	1142	>76.2	>3000
	100	nil		0.001	0.04
	500	nil		0.001	0.04
V^{5+}	0	29	1142	>76.2	>3000
	100	0.02	0.8	0.005	0.2
	500	0.008	0.3	0.005	0.2

Source: R.W. Schutz and L.C. Covington, Hydrometallurgical Applications of Titanium, in *Industrial Applications of Titanium and Zirconium: Third Conference*, STP 830, American Society for Testing and Materials, 1984, p 29-47

Corrosion resistance of CP Ti vs grade 12 (Ti-0.3Mo-0.8Ni) in HCl solutions

Test solution	Test time, h	Temperature °C	°F	Corrosion rate, mm/yr CP Ti	Ti-0.3Mo-0.8Ni
7.5% HCl	648	20	68	0.074	0.000
10% HCl	648	20	68	0.108	0.022
12.5% HCl	496	20	68	0.101	0.401
5% HCl	480	50	122	1.00	0.012
10% HCl	384	50	122	2.00	1.97
1% HCl	22	Boiling		0.013	0.026

Source: L. Zuochen *et al.*, The Corrosion Resistance and Application of Ti-0.3Mo-0.8Ni Alloy, *Titanium, Science and Technology*, Vol 4, G. Lütjering, J. Zwicker, and W. Bunk, Ed., Deutsche Gesselschaft für Metallkunde e.V., Germany, 1985, p 2611-2616

Corrosion resistance of CP Ti vs grade 12 in sulfuric acid solutions

Test solution	Test time, h	Temperature °C	°F	Corrosion rate, mm/yr CP Ti	Ti-0.3Mo-0.8Ni
7.5% H_2SO_4	648	20	68	0.120	0.006
10% H_2SO_4	648	20	68	0.130	0.032
15% H_2SO_4	496	20	68	0.158	0.469
4% H_2SO_4	480	50	122	1.51	0.078
5% H_2SO_4	480	50	122	1.67	0.17
1% H_2SO_4	24	Boiling		11.5	0.038

Source: L. Zuochen *et al.*, The Corrosion Resistance and Application of Ti-0.3Mo-0.8Ni Alloy, *Titanium, Science and Technology*, Vol 4, G. Lütjering, U. Zwicker, and W. Bunk, Ed., Deutsche Gesellschaft für Metallkunde, e.V., Germany, 1985, p 2611-2616

ASTM grade 12: Corrosion in naturally aerated HCl solutions

Concentration of HCl, %	Temperature °C	°F	Corrosion rate mm/yr	mils/yr
6	24	75	0.008	0.31
8	24	75	0.008	0.31
10	24	75	1.40	55
12	24	75	2.54	100
28.5	24	75	5.58	220
3	52	125	nil	
4	52	125	0.001	0.04
5.9	52	125	0.51	20
7	52	125	5.30	209
2.4	66	150	0.01	0.4
3.6	66	150	0.03	1.2
5.9	66	150	0.51	20
7	66	150	8.98	354
0.6	Boiling		0.025	1
1.7	Boiling		0.16	6.3
2	Boiling		0.51	20
2.5	Boiling		6.85	270

Source: *Metals Handbook, Corrosion*, Vol 13, 9th ed., ASM International, 1987, p 681

ASTM grade 7: Corrosion in naturally aerated HCl solutions

Concentration of HCl, %	Temperature °C	°F	Corrosion rate mm/yr	mils/yr
9	24	75	nil	
18	24	75	nil	
20	24	75	0.01	0.4
26.5	24	75	0.02	0.8
27	24	75	0.70	27.6
9	52	125	0.008	0.3
11.5	52	125	0.02	0.8
14.7	52	125	0.03	1.2
16.8	52	125	0.06	2.4
19	52	125	0.08	3.1
21.9	52	125	0.41	16.1
6	66	150	0.01	0.4
9.6	66	150	0.03	1.2
11.5	66	150	0.04	1.6
16.8	66	150	0.13	5
17	66	150	0.39	15.4
20.9	66	150	0.51	20
2	Boiling		0.025	1
3	Boiling		0.05	2
4	Boiling		0.10	4

(continued)

ASTM grade 7: Corrosion in naturally aerated HCl solutions (continued)

Concentration of HCl, %	Temperature °C	°F	Corrosion rate mm/yr	mils/yr
6	Boiling		0.23	9
9	Boiling		0.51	20
16.8	Boiling		2.97	117

Source: Metals Handbook, *Corrosion*, Vol 13, 9th ed., ASM International, 1987, p 681

ASTM grade 2: Corrosion in naturally aerated HCl solutions

Concentration of HCl, %	Temperature °C	°F	Corrosion rate mm/yr	mils/yr
5	24	75	nil	
6	24	75	0.07	2.8
8	24	75	0.2	8
9	24	75	0.25	1
17.3	24	75	0.51	2
26	24	75	2.59	102
1.4	52	125	0.02	0.8
5.8	52	125	0.51	2
6	52	125	0.68	26.8
7	52	125	1.27	50
11.5	52	125	3.07	121
1	66	150	0.01	0.4
1.5	66	150	0.02	0.8
1.7	66	150	0.13	5
2	66	150	0.61	24
3	66	150	1	40
4.7	66	150	7.08	279
0.05	Boiling		0.02	0.8
0.1	Boiling		0.1	4
0.2	Boiling		0.23	9
0.4	Boiling		0.53	21
0.5	Boiling		0.84	33.1
1	Boiling		1.83	72

Source: Metals Handbook, *Corrosion*, Vol 13, 9th ed., ASM International, 1987, p 681

Corrosion of titanium in naturally aerated HBr and HI solutions

Acid	Concentration, %	Temperature, °C (°F)	Alloy	Corrosion rate mm/yr	mils/yr
HBr	0.3	Boiling	Grade 2	nil	
HBr	0.6	Boiling	Grade 2	0.003	0.12
HBr	0.9	Boiling	Grade 12	0.008	0.32
HBr	3.0	Boiling	Grade 2	1.45	57
HBr	3.0	Boiling	Grade 12	0.013	0.5
HBr	3.0	Boiling	Grade 7	0.010	0.4
HBr	8.0	Boiling	Grade 7	0.094	3.7
HBr	40	25 (75)	Grade 2	nil	
HI	10	Boiling	Grade 2	nil	
HI	57	25 (75)	Grade 2	0.15	6

Source: Metals Handbook, *Corrosion*, Vol 13, 9th ed., ASM International, 1987, p 682

Alkaline Media

Titanium alloys are generally very resistant to alkaline media, including solutions of NaOH, KOH, Ca(OH)$_2$, Mg(OH)$_2$, and NH$_4$OH. Near-nil corrosion rates can be expected in boiling solutions of the latter three alkalies up to saturation. Although corrosion rates are relatively low in alkaline media, titanium alloys may experience excessive hydrogen pickup and eventual embrittlement under certain conditions.

Salt Solutions

Titanium alloys are highly resistant to practically all salt solutions over the pH range of 3 to 11 and to temperatures well in excess of boiling. Titanium withstands exposure to solutions of chlorides, bromides, iodides, sulfites, sulfates, borates, phosphates, cyanides, carbonates, bicarbonates, and ammonium compounds. Corrosion rates for titanium alloys in these various salt solutions are generally less than 0.03 mm/yr (1.2 mils/yr).

Oxidizing anionic salts such as nitrates, hypochlorites, chlorites, chlorates, perchlorates, molybdates, chromates, permanganates, and vanadates further extend titanium alloy passivity into stronger acidic and alkaline solutions. Similar

CP/modified Ti: Temperature guidelines in MgCl₂ solutions

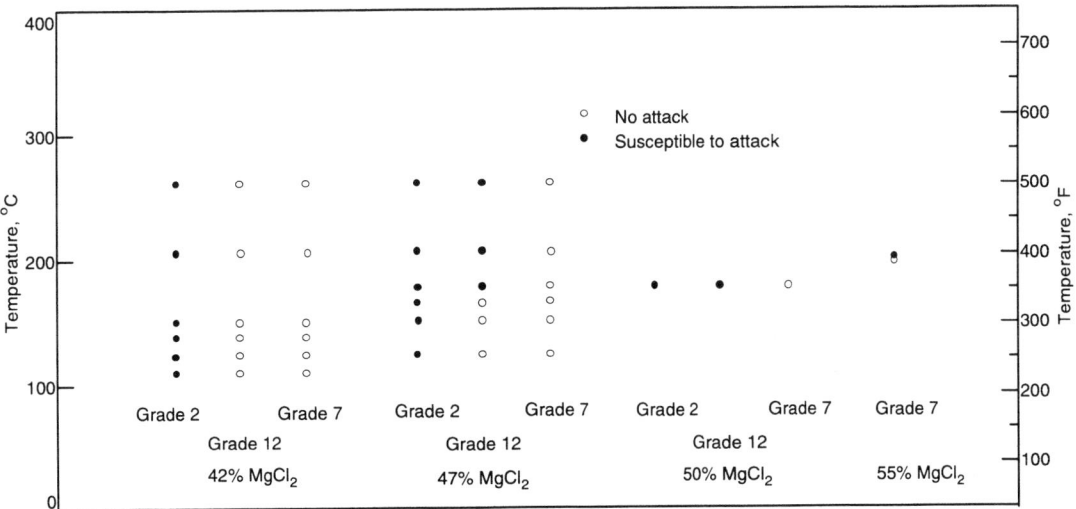

Source: R.S. Schutz and J.S. Grauman, Selection of Titanium Alloys for Concentrated Seawater, NaCl and MgCl₂ Brines, in *Titanium 1986—Titanium Products and Applications*, Titanium Development Association, 1986

CP/modified Ti: Temperature guidelines in NaCl solutions

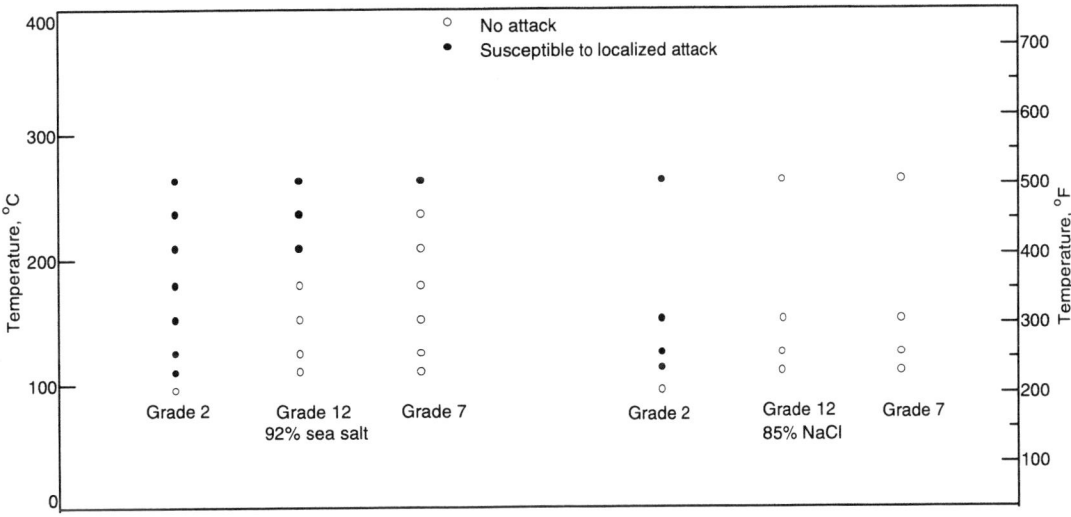

Temperature guidelines for avoiding localized attack of grades 2, 7, and 12 titanium in pressure-bled tests in concentrated sea salt and NaCl slurries in the absence of crevices.
Source: R.W. Schutz and J.S. Grauman, Selection of Titanium Alloys for Concentrated Seawater, NaCl and MgCl₂ Brines, in *Titanium 1986—Titanium Products and Applications*, Titanium Development Association, 1986

beneficial effects on the low pH (acidic) side can be expected from oxidizing cationic salts, such as ferric, cupric, and nickelous chlorides or sulfates. In fact, titanium is often the most practical metal for handling hot, oxidizing, acidic chloride conditions.

Titanium alloys are frequently selected because of their superior resistance to the chlorides typically found in many process streams, brines, and seawater. In hot chloride media, susceptibility to pitting is usually not an issue, but crevice corrosion may be possible, depending on pH and temperature. Special attention must be given to nonoxidizing acidic or hydrolyzable salt solutions as temperatures and concentrations increase. To avoid general or localized HCl attack resulting from salt hydrolysis, special concentration/temperature guidelines for titanium should be observed for concentrated $AlCl_3$, $ZnCl_2$, $MgCl_2$, and $CaCl_2$ solutions. Grades 12 and 7 and higher molybdenum-containing titanium alloys exhibit superior resistance to general and localized corrosion in these high-temperature acid salt solutions.

Titanium exhibits relatively high rates of attack in molten chloride salts, increasing with both temperature and the presence of oxygen. Aggressiveness of attack follows the order: KCl > NaCl > LiCl. High rates of metal dissolution have also been reported in fused sodium carbonate, sodium hydroxide, sodium peroxide, and sodium bisulfate (R.L. LaQue and H.R. Copson, *Corrosion Resistance of Metals and Alloys,* 2nd ed., ACS Monograph, Reinhold, 1963, p 646-661).

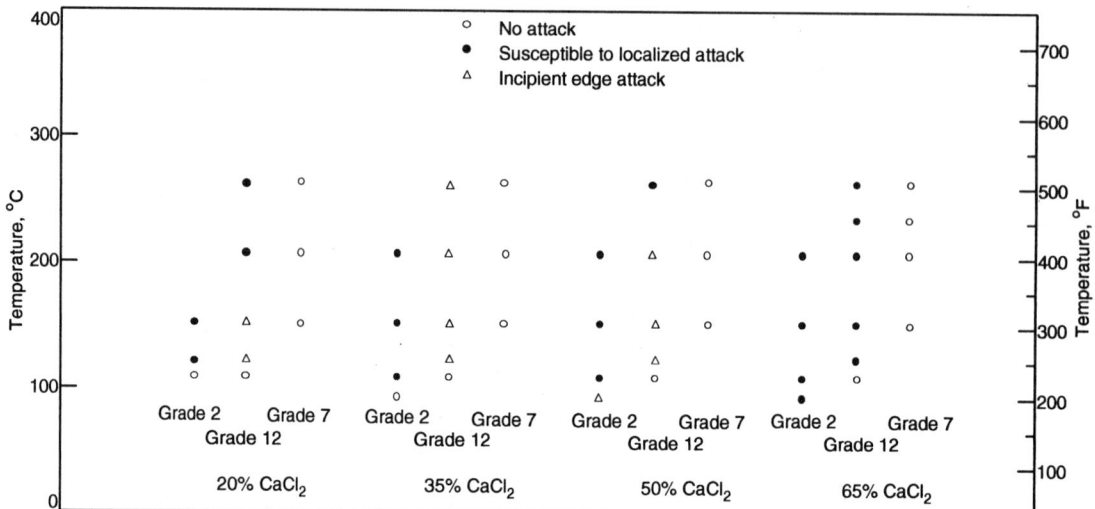

CP/modified Ti: Temperature guidelines in CaCl₂ solutions

Source: R.W. Schutz and J.S. Grauman, Selection of Titanium Alloys for Concentrated Seawater, NaCl and MgCl₂ Brines, in *Titanium 1986—Titanium Products and Applications*, Titanium Development Association, 1986

Corrosion of unalloyed titanium in highly alkaline solutions

Medium	Concentration, %	Temperature °C	Temperature °F	Corrosion rate mm/yr	Corrosion rate mils/yr
Ammonium hydroxide	28	26	79	0.002	0.08
Ammonium hydroxide	70	Boiling		nil	
Sodium carbonate	20	Boiling		nil	
Sodium hydroxide	28	25	75	0.003	0.12
Sodium hydroxide	10	Boiling		0.02	0.8
Sodium hydroxide	40	66	150	0.038	1.5
Sodium hydroxide	40	93	200	0.064	2.5
Sodium hydroxide	40	121	250	0.13	5
Sodium hydroxide	50	66	150	0.018	0.7
Sodium hydroxide	50-73	188	370	>1.1	>43.3
Sodium hydroxide	73	110	230	0.05	2
Sodium hydroxide	73	Boiling		0.13	5
Potassium hydroxide	10	Boiling		0.13	5
Potassium hydroxide	25	Boiling		0.3	12
Potassium hydroxide	50	25	75	0.010	0.4
Potassium hydroxide	50	Boiling		2.7	106

Source: N.D. Tomashov and P.M. Altovskii, *Corrosion and Protection of Titanium*, Government Scientific-Technical Publication of Machine-Building Literature (Russian translation),1963; L.C. Covington and R.W. Schutz,"Corrosion Resistance of Titanium," TIMET Corporation,1982; R.L. Kane,The Corrosion of Titanium, in *The Corrosion of Light Metals*,The Corrosion Monograph Series, John Wiley & Sons, 1967; *Corrosion Resistance of Titanium*,Technical Handbook,Imperial Metals Industries (Kynoch) Ltd., Birmingham,UK

Organic Compounds

Titanium alloys are highly resistant to most organic compounds, including alcohols, ketones, ethers, aldehydes, and hydrocarbons. The traces of moisture (ppm levels) normally present in industrial organic process streams are sufficient to maintain the protective oxide film of titanium. Totally anhydrous organic streams may prevent oxide film repair and should be avoided.

Chlorinated hydrocarbons generally do not pose any problems for most titanium alloys. If significant amounts of water are present, many chlorinated hydrocarbons may undergo hydrolysis to form HCl at higher temperatures. Performance will depend on the temperature and the extent of HCl formation and concentration in the aqueous phase.

Titanium alloys are generally very resistant to organic acids. Nil corrosion can be expected in concentrated solutions of very weak organic acids, such as adipic, hydroxyacetic, acetic, terephthalic, tannic, stearic, maleric, tartaric, benzoic, butyric, and succinic acids to temperatures in excess of 100 °C (212 °F). Corrosion rates can become significant in the stronger, nonaerated organic acids as acid concentration and temperature increase. These acids include formic, lactic, citric, trichloroacetic, and especially oxalic acid. Grades 7 and 12 exhibit much greater resistance to these acids than unalloyed titanium.

Solution aeration is often sufficient to inhibit corrosion of unalloyed titanium in the stronger organic acids, such as formic, lactic, and citric. Addition of oxidizing species may also effectively inhibit corrosion in the more aggressive organic acids, such as oxalic.

Corrosion of unalloyed titanium in organic media

Medium	Concentration, %	Temperature, °C (°F)	Corrosion rate mm/yr	mils/yr
Acetic anhydride	99-99.5	20 (70) to boiling	<0.13	<5
Adipic acid	0-67	204 (400)	<0.05	<2
Adipic acid + 20% glutaric acid +5% acetic acid	25	200 (390)	nil	
Aniline hydrochloride	5-20	35-100 (95-212)	<0.001	<0.04
Benzene + HCl + NaCl	Vapor + liquid	80 (175)	0.005	0.2
Carbon tetrachloride	99	Boiling	0.003	0.12
Carbon tetrachloride	100	Boiling	0.003	0.12
Chloroform	100	Boiling	nil	
Chloroform + water	50	Boiling	0.12	4.7
Cyclohexane + traces of formic acid	...	150 (300)	0.003	0.12
Ethyl alcohol	95	Boiling	0.013	0.5
Ethylene dichloride	100	Boiling	<0.13	<5
Formaldehyde	37	Boiling	<0.13	<5
Tetrachloroethylene	100	Boiling	nil	
Tetrachloroethylene + water	...	Boiling	0.13	5
Tetrachloroethane	100	Boiling	nil	
Trichloroethylene	99	Boiling	<0.13	<5

Source: N.D. Tomashov and P.M. Altovskii, *Corrosion and Protection of Titanium*, Government Scientific-Technical Publication of Machine-Building Literature (Russian translation), 1963; L.C. Covington and R.W. Schutz, "Corrosion Resistance of Titanium," TIMET Corporation, 1982; R.L. Kane, The Corrosion of Titanium, in *The Corrosion of Light Metals*, The Corrosion Monograph Series, John Wiley & Sons, 1967; R.L. LaQue and H.R. Copson, *Corrosion Resistance of Metals and Alloys*, 2nd ed., ACS Monograph, Reinhold, 1963, p 646-661

Corrosion of titanium in various organic acids

Acid	Concentration, %	Alloy	Temperature, °C (°F)	Corrosion rate mm/yr	mils/yr
Acetic	0-99.5	Grades 2, 7, 12	Boiling	nil	
Adipic	67	Grade 2	240 (465)	nil	
Citric, aerated	10-50	Grade 2	100 (212)	0.01	0.4
Citric	50	Grade 2	Boiling	0.35	13.8
Citric	50	Grades 7, 12	Boiling	0.01	0.4
Di- and mono-chloroacetic	100	Grade 2	Boiling	<0.013	<0.5
Formic, aerated	25-90	Grade 2	100 (212)	0.001	0.04
Formic, aerated	25	Grade 2	Boiling	2.4	94.5
Formic	45	Grade 2	Boiling	11.0	433
Formic	45	Grades 7, 12	Boiling	nil	
Formic	10	Grade 2	Boiling	nil	
Lactic, aerated	10	Grade 2	Boiling	0.014	0.55
Lactic	10	Grade 2	100 (212)	0.048	1.9
Lactic	25	Grade 2	Boiling	0.028	1.1
Lactic	85-100	Grade 2	Boiling	0.01	0.4
Oxalic	0.5	Grade 2	60 (140)	2.4	94.5
Oxalic	1	Grade 2	35 (95)	0.15	6
Oxalic	10	Grade 7	Boiling	32.3	1272
Stearic	100	Grade 2	180 (355)	0.003	0.12
Tartaric	10-50	Grade 2	100 (212)	≤0.013	<0.5
Terephthalic	77	Grade 2	225 (437)	nil	
Trichloroacetic	100	Grade 2	Boiling	14.6	575

Source: N.D. Tomashov and P.M. Altovskii, *Corrosion and Protection of Titanium*, Government Scientific-Technical Publication of Machine-Building Literature (Russian translation), 1963; L.C. Covington and R.W. Schutz, "Corrosion Resistance of Titanium," TIMET Corporation, 1982; *Corrosion Resistance of Titanium*, Technical Handbook, Imperial Metals Industries (Kynoch) Ltd., Birmingham, UK; D.W. Stough, F.W. Fink, and R.S. Peoples, "The Corrosion of Titanium," Report 57, Titanium Metallurgical Laboratory, Battelle Memorial Institute, 1956; R.L. LaQue and H.R. Copson, *Corrosion Resistance of Metals and Alloys*, 2nd ed., ACS Monograph, Reinhold, 1963, p 646-661

Hot Gases and Liquid Metals

Gases. The oxide film on titanium alloys provides an effective barrier against attack by most gases in wet or dry conditions, including oxygen, nitrogen, dry HCl, SO_2, NH_3, HCN, CO_2, CO, and H_2S. This protection extends to temperatures in excess of 150 °C (300 °F).

Titanium alloys experience no significant corrosion degradation in oxygen or sulfur-bearing gases below 300 °C (570 °F). However, if the protective oxide film is damaged, then a violent exothermic reaction can occur in oxygen. Excessive surface oxidation and eventual interstitial embrittlement may occur above 340 to 370 °C (645 to 700 °F) after prolonged, continuous exposure to air, depending on alloy composition. Embrittlement is the result of the enhanced diffusion rate of interstitial oxygen into the metal at higher temperatures such that time to failure will also depend on metal section thickness and state of stress.

Liquid Metals. Titanium exhibits good resistance to many liquid metals at moderately elevated temperatures. These metals include molten aluminum, sodium, potassium, sodium-potassium mixtures, magnesium, tin, and lead. In contrast, useful performance of titanium in molten lithium, bismuth, zinc, gallium, cadmium, and mercury is limited to relatively low temperatures. Liquid mercury below 150 °C (300 °F) does not appear to affect titanium unless wetting of freshly exposed (mechanically damaged) surfaces occurs (R.L. Kane, The Corrosion of Titanium, in *The Corrosion of Light Metals,* The Corrosion Monograph Series, John Wiley & Sons, 1967). Cracking from liquid/solid metal corrosion is discussed in the section "Stress-Corrosion Cracking".

Corrosion of unalloyed titanium in liquid metals

Liquid metal	Temperature °C	Temperature °F	Corrosion rate mm/yr	Corrosion rate mils/yr
Bismuth-lead	300	570	<0.1	<4
Bismuth-lead	600	1110	0.13-1.3	5-50
Gallium	400	750	0.1	4
Gallium	450	840	>1.0	>40
Lithium	850	1560	0.1-1.0	4-40
Magnesium	750	1380	0.1	4
Magnesium	850	1560	0.1-1.0	4-40
Lead	400	750	<0.13	<5
Lead	600-950	1110-1740	0.1-1.0	4-40
Mercury	150	300	<0.1	<4
Mercury	150-300	300-570	0.1-1.0	4-40
Sodium, potassium	600	1110	<0.1	<4
Sodium, potassium	800	1470	0.1-1.0	4-40
Sodium, potassium	600	1110	<0.1	<4
Tin	350	660	<0.1	<4
Tin	600	1110	0.1-1.0	4-40
Aluminum	750	1380	<0.1	<4
Aluminum	850	1560	>>0.1	>>4
Cadmium	500	930	>1.0	>40
Zinc	445	830	>>1.0	>>40

Source: *Metals Handbook,Corrosion,*Vol 13,9th ed., ASM International,1987, p 687

Unalloyed Ti: General Corrosion by Media

This compilation of general corrosion rate values is for unalloyed titanium (ASTM grades 1 to 4). These data should be used only as a guideline for alloy performance. Rates may vary depending on changes in medium chemistry temperature, length of exposure, and other factors. Also, total alloy suitability cannot be assumed from these values alone, because other forms of corrosion, such as localized attack, may be limiting. The text should be consulted to assess overall alloy suitability more thoroughly for a given set of environmental conditions. In complex, variable, and/or dynamic environments, *in situ* testing may provide more reliable data. In the following table, temperatures are given only in centigrade, and corrosion rates are reported only in millimeters per year.

Unalloyed Ti: General Corrosion

Medium	Concentration, %	Temperature, °C	Corrosion rate, mm/yr
Acetaldehyde	75	149	0.001
	100	149	nil
Acetate,*n*-propyl	...	87	nil
Acetin acid	5-99.7	124	nil
	33-vapor	Boiling	nil
	99	Boiling	0.003
	65	121	0.003
	58	130	0.381
	99.7	124	0.003
Acetic acid + 3% acetic anhydride	Glacial	204	1.02
Acetic acid + 1.5% acetic anhydride	Glacial	204	0.005
Acetic acid + 109 ppm Cl	31.2	Boiling	0.259
Acetic acid + 106 ppm Cl	62.0	Boiling	0.272
Acetic acid + 5% formic acid	58	Boiling	0.457
Acetic anhydride	100	21	0.025
	100	150	0.005
	99.5	Boiling	0.013
Adipic acid + 15-20% glutaric + 2% acetic acid	25	199	nil

(continued)

Unalloyed Ti: General Corrosion (continued)

Medium	Concentration, %	Temperature, °C	Corrosion rate, mm/yr
Adipic acid	67	240	nil
Adipylchloride and chlorobenzene solution	...	...	nil
Adiponitrile	Vapor	371	0.008
Aluminum chloride, aerated	10	100	0.002
	25	100	3.15
Aluminum chloride	10	100	0.002
	10	150	0.03
	25	60	nil
	25	100	6.55
Aluminum	Molten	677	164.6
Aluminum fluoride	Saturated	Room	nil
Aluminum nitrate	Saturated	Room	nil
Aluminum sulfate	Saturated	Room	nil
	10	80	0.05
	10	Boiling	0.12
Aluminum sulfate + 1% H_2SO_4	Saturated	Room	nil
Ammonium acid phosphate	10	Room	nil
Ammonium aluminum chloride	Molten	350-380	Very rapid attack
Ammonia, anhydrous	100	40	<0.127
Ammonia, steam water	...	222	11.2
Ammonium acetate	10	Room	nil
Ammonium bicarbonate	50	100	nil
Ammonium bisulfite, pH 2.05	Spent pulping liquor	71	0.015
Ammonium carbamate	50	100	nil
Ammonium chloride	Saturated	100	<0.013
Ammonium chlorate	300 g/L	50	0.003
Ammonium fluoride	10	Room	0.102
Ammonium hydroxide	28	Room	0.003
	28	100	nil
Ammonium nitrate	28	Boiling	nil
Ammonium nitrate + 1% nitric acid	28	Boiling	nil
Ammonium oxalate	Saturated	Room	nil
Ammonium perchlorate	20	88	nil
Ammonium sulfate	10	100	nil
Ammonium sulfate + 1% H_2SO_4	Saturated	Room	0.010
Aniline	100	Room	nil
Aniline + 2% $AlCl_3$	98	158	>1.27
Aniline hydrochloride	5	100	nil
	20	100	nil
Antimony trichloride	27	Room	nil
Aqua regia	3:1	Room	nil
	3:1	80	0.86
	3:1	Boiling	1.12
Arsenous oxide	Saturated	Room	nil
Barium carbonate	Saturated	Room	nil
Barium chloride	5	100	nil
	20	100	nil
	25	100	nil
Barium hydroxide	Saturated	Room	nil
Barium nitrate	10	Room	nil
Barium fluoride	Saturated	Room	nil
Benzaldehyde	100	Room	nil
Benzene (traces of HCl)	Vapor and liquid	80	0.005
	Liquid	50	0.025
Benzene	Liquid	Room	nil
Benzoric acid	Saturated	Room	nil
Bismuth	Molten	816	High
Bismuth/lead	Molten	300	Good resistance
Boric acid	Saturated	Room	nil
	10	Boiling	nil
Bromine	Liquid	30	Rapid attack
Bromine, moist	Vapor	30	<0.003
Bromine gas, dry	...	21	Dissolves rapidly
Bromine-water solution	...	Room	nil
Bromine in methyl alcohol	0.05	60	0.03 (cracking possible)
N-butyric acid	Undiluted	Room	nil
Calcium bisulfite	Cooking liquor	26	0.001
Calcium carbonate	Saturated	Boiling	nil
Calcium chloride	5	100	0.005
	10	100	0.007

(continued)

Unalloyed Ti: General Corrosion (continued)

Medium	Concentration, %	Temperature, °C	Corrosion rate, mm/yr
	20	100	0.015
	55	104	0.001
	60	149	<0.003
	62	154	0.406
	73	175	0.80
Calcium hydroxide	Saturated	Room	nil
	Saturated	Boiling	nil
Calcium hypochlorite	2	100	0.001
	6	100	0.001
	18	21	nil
	Saturated	21	nil
Carbon dioxide	100	...	Excellent
Carbon tetrachloride	99	Boiling	0.005
	Liquid	Boiling	nil
	Vapor	Boiling	nil
Carbon tetrachloride + 50% H_2O	50	25	0.005
Chlorine gas, wet	>0.7 H_2O	Room	nil
	>0.95 H_2O	140	nil
	>1.5 H_2O	200	nil
Chlorine saturated water	Saturated	97	nil
Chlorine gas, dry	<0.5 H_2O	Room	May react
Chlorine dioxide	5	82	<0.003
Chlorine dioxide + HOCl, H_2O + Cl_2	15	43	nil
Chlorine dioxide in steam	5	99	nil
Chlorine dioxide	10	70	0.03
Chlorine monoxide (moist)	Up to 15	43	nil
Chlorine trifluoride	100	30	Vigorous reaction
Chloracetic acid	30	82	<0.127
	100	Boiling	<0.127
Chlorosulfonic acid	100	Room	0.312
Chloroform	Vapor and liquid	Boiling	0.000
Chloroform + 50% H_2O	50	25	0.000
Chloropicrin	100	95	0.003
Chromic acid	10	Boiling	0.003
	15	24	0.006
	15	82	0.015
	50	24	0.013
	50	82	0.028
Chromic acid + 5% nitric acid	5	21	<0.003
Citric acid	10	100	0.009
	25	100	0.001
	50	60	0.000
	50	Boiling	0.127-1.27
	672	149	Corroded
Citric acid (aerated)	50	100	<0.127
Copper nitrate	Saturated	Room	nil
Copper sulfate	50	Boiling	nil
Copper sulfate + 2% H_2SO_4	Saturated	Room	0.018
Cupric carbonate + cupric hydroxide	Saturated	Ambient	nil
Cupric chloride	20	Boiling	nil
	40	Boiling	0.005
	55	118	0.003
Cupric cyanide	Saturated	Room	nil
Cuprous chloride	50	90	<0.003
Cyclohexane (plus traces of formic acid)	...	150	0.003
Cyclohexylamine	100	Room	nil
Dichloroacetic acid	100	Boiling	0.007
Dichlorobenzene + 4-5% HCl	...	179	0.102
Diethylene triamine	100	Room	nil
Ethyl alcohol	95	Boiling	0.013
	100	Room	nil
Ethylene dichloride	100	Boiling	0.005-0.127
Ethylene dichloride + 50% water	50	25	0.005
Ethylene diamine	100	Room	nil
Ferric chloride	10-20	Room	nil
	1-30	100	0.004
	10-40	Boiling	nil
	1-30	Boiling	nil
	50	150	0.003
Ferric chloride	10	Boiling	0.00
Ferric sulfate	10	Room	nil

(continued)

Unalloyed Ti: General Corrosion (continued)

Medium	Concentration, %	Temperature, °C	Corrosion rate, mm/yr
Ferrous chloride + 0.5% HCl	30	79	0.006
Ferrous sulfate	Saturated	Room	nil
Fluoboric acid	5-20	Elevated	Rapid attack
Fluorine, commercial	Gas-liquid	Gas-109	0.864
Fluorine, HF free	Liquid	−196	0.011
	Gas	−196	0.011
Fluorosilicic acid	10	Room	47.5
Formaldehyde	37	Boiling	nil
Formamide vapor	...	300	nil
Formic acid, aerated	10	100	0.005
	25	100	0.001
	50	100	0.001
	90	100	0.001
Formic acid, nonaerated	10	100	nil
	25	100	2.44
	50	Boiling	3.20
	90	100	3.00
Formic acid	9	50	<0.127
Furfural	100	Room	nil
Gluconic acid	50	Room	nil
Glycerin	...	Room	nil
Hydrogen chloride, gas	Air mixture	25-100	nil
Hydrochloric acid, aerated	1	60	0.004
	2	60	0.016
	5	60	1.07
	1	100	0.46
	5	35	0.01
	10	35	1.02
	20	35	4.45
Hydrochloric acid	0.1	Boiling	0.10
	1	Boiling	1.8
Hydrochloric acid + 4% FeCl$_3$ + 4% MgCl$_2$	19	82	0.51
Hydrochloric acid + 4% FeCl$_3$ + 4% MgCl$_2$ + Cl$_2$ saturated	19	82	0.46
Hydrochloric acid, chlorine saturated	5	190	<0.025
	10	190	28.5
Hydrochloric acid, +200 ppm Cl$_2$	36	25	0.432
Hydrochloric acid			
+1% HNO$_3$	5	40	nil
+1% HNO$_3$	5	95	0.091
+5% HNO$_3$	5	40	0.025
+5% HNO$_3$	5	95	0.030
+10% HNO$_3$	5	40	nil
+10% HNO$_3$	5	95	0.183
+3% HNO$_3$	8.5	80	0.051
+5% HNO$_3$	1	Boiling	0.074
Hydrochloric acid			
+2.5% NaClO$_3$	10.2	80	0.009
+5.0% NaClO$_3$	10.2	80	0.006
Hydrochloric acid			
+0.5% CrO$_3$	5	38	nil
+0.5% CrO$_3$	5	95	0.031
+1% CrO$_3$	5	38	0.018
+1% CrO$_3$	5	95	0.031
Hydrochloric acid			
+0.05% CuSO$_4$	5	38	0.040
+0.05% CuSO$_4$	5	93	0.091
+0.5% CuSO$_4$	5	38	0.091
+0.5% CuSO$_4$	5	93	0.061
+1% CuSO$_4$	5	38	0.031
+1% CuSO$_4$	5	93	0.091
+5% CuSO$_4$	5	38	0.020
+5% CuSO$_4$	5	93	0.061
+0.05% CuSO$_4$	5	Boiling	0.064
+0.5% CuSO$_4$	5	Boiling	0.084
Hydrochloric acid			
+0.05% CuSO$_4$	10	66	0.025
+0.20% CuSO$_4$	10	66	nil
+0.5% CuSO$_4$	10	66	0.023
+1% CuSO$_4$	10	66	0.023
+0.05% CuSO$_4$	10	Boiling	0.295
+0.5% CuSO$_4$	10	Boiling	0.290
Hydrochloric acid + 0.1% FeCl$_3$	5	Boiling	0.01

(continued)

Unalloyed Ti: General Corrosion (continued)

Medium	Concentration, %	Temperature, °C	Corrosion rate, mm/yr
Hydrochloric acid + 1 g/L Ti^{4+}	10	Boiling	0.000
Hydrochloric acid + 5.8 g/L Ti^{4+}	20	Boiling	0.000
Hydrochloric acid + 18% H_3PO_4 + 5% HNO_3	18	77	0.000
Hydrofluoric acid	1	26	127
Hydrofluoric acid, anhydrous	100	Room	0.127-1.27
Hydrofluoric-nitric acid 5 vol% HF-35 vol% HNO_3	...	25	452
Hydrofluoric-nitric acid 5 vol% HF-35 vol% HNO_3	...	35	571
Hydrogen peroxide	3	Room	<0.127
	6	Room	<0.127
	30	Room	<0.305
Hydrogen peroxide + 2% NaOH	1	60	55.9
Hydrogen peroxide			
pH 4	5	66	0.061
pH 1	5	66	0.152
pH 1	20	66	0.69
pH 11	0.08	70	0.42
Hydrogen sulfide (water saturated)	...	21	<0.003
Hydrogen sulfide, steam, and 0.077% mercaptans	7.65	93-110	nil
Hydroxy-acetic acid	...	40	0.003
Hypochlorous acid + ClO and Cl_2 gases	17	38	0.000
Iodine, dry or moist gas	...	25	0.1
Iodine in water + potassium iodide	...	Room	nil
Iodine in alcohol	Saturated	Room	Pitted
Lactic acid	10-85	100	<0.127
	10	Boiling	<0.127
Lead	...	816	Attacked
	...	324-593	Good
Lead acetate	Saturated	Room	nil
Linseed oil, boiled	...	Room	nil
Lithium, molten	...	316-482	nil
Lithium chloride	50	149	nil
Magnesium	Molten	760	Limited resistance
Magnesium chloride	5-20	100	<0.010
	5-40	Boiling	0.005
Magnesium hydroxide	Saturated	Room	nil
Magnesium sulfate	Saturated	Room	nil
Manganous chloride	5-20	100	nil
Maleic acid	18-20	35	0.002
Mercuric chloride	1	100	0.000
	5	100	0.011
	10	100	0.001
	Saturated	100	0.001
Mercuric cyanide	Saturated	Room	nil
Mercury	100	Up to 38	Satisfactory
	100	Room	nil
	...	371	3.03
Methyl alcohol	91	35	nil
	95	100	<0.01
Mercury + iron	...	371	0.079
Mercury + copper	...	371	0.063
Mercury + zirconium	...	371	0.033
Mercury + magnesium	...	371	0.083
Monochloracetic acid	30	80	0.02
	100	Boiling	0.013
Nickel chloride	5	100	0.004
	20	100	0.003
Nickel nitrate	50	Room	nil
Nitric acid, aerated	10	Room	nil
	30	Room	0.004
	40	Room	0.002
	50	Room	0.002
	60	Room	0.001
	70	Room	0.005
	10	40	0.003
	20	40	0.005
	30	50	0.015
	40	50	0.016
	50	60	0.937

(continued)

Unalloyed Ti: General Corrosion (continued)

Medium	Concentration, %	Temperature, °C	Corrosion rate, mm/yr
	60	60	0.040
	70	70	0.040
	40	200	0.610
	70	270	1.22
	20	290	0.305
Nitric acid	35	80	0.051-0.102
	70	80	0.025-0.076
	17	Boiling	0.076-0.102
	35	Boiling	0.127-0.508
	70	Boiling	0.064-0.900
Nitric acid, not refreshed	5-60	35	0.002-0.007
	5-60	60	0.01-0.02
	30-50	100	0.10-0.18
	5-20	100	0.02
	30-60	190	1.5-2.8
	70	270	1.2
	20	290	0.4
	70	290	1.1
Nitric acid, white fuming	Liquid or vapor	Room	nil
	...	82	0.152
	...	122	<0.127
	...	160	<0.127
Nitric acid, red fuming	<About 2% H_2O	Room	Ignition sensitive
	>About 2% H_2O	Room	Not ignition sensitive
Nitric acid	40	Boiling	0.63
+0.01% $K_2Cr_2O_7$	40	Boiling	0.01
+0.01% CrO_3	40	Boiling	0.01
+0.01% $FeCl_3$	40	Boiling	0.68
+1% $FeCl_3$	40	Boiling	0.14
+1% $NaClO_3$	40	Boiling	0.31
+1% $NaClO_3$	40	Boiling	0.02
+1% $Ce(SO_4)_2$	40	Boiling	0.10
+0.1% $K_2Cr_2O_7$	40	Boiling	0.016
Nitric acid, saturated with zirconyl nitrate	33-45	118	nil
Nitric acid + 15% zirconyl nitrate	65	127	nil
Nitric acid + 179 g/L $NaNO_3$ and 32 g/L NaCl	20.8	Boiling	0.127-0.295
Nitric acid + 170 g/L $NaNO_3$ and 2.9 g/L NaCl	27.4	Boiling	0.483-2.92
Oxalic acid	1	35	0.03
	5	35	0.13
	1	Boiling	107
	25	60	11.9
	Saturated	Room	0.508
Perchloroethylene + 50% H_2O	50	25	nil
Perchloryl fluoride + liquid ClO_3	100	30	0.002
Perchloryl fluoride + 1% H_2O	99	30	Liquid 0.290
	...	...	Vapor 0.003
Phenol	Saturated solution	25	0.102
Phosphoric acid	10-30	Room	0.020-0.051
	30-80	Room	0.051-0.762
	5.0	66	0.005
	6.0	66	0.117
	0.5	Boiling	0.094
	1.0	Boiling	0.266
	12	25	0.005
	20	25	0.076
	50	25	0.19
	9	52	0.03
	10	52	0.38
	5	Boiling	3.5
Phosphoric acid + 3% nitric acid	81	88	0.381
Phosphorus oxychloride	100	Room	0.004
Phosphorus trichloride	Saturated	Room	nil
Photographic emulsions	...	...	<0.127
Phthalic acid	Saturated	Room	nil
Potassium bromide	Saturated	Room	nil
Potassium chloride	Saturated	Room	nil
	Saturated	60	nil
Potassium dichromate	Saturated	Room	nil
Potassium ethylxanthate	10	Room	nil

(continued)

Unalloyed Ti: General Corrosion (continued)

Medium	Concentration, %	Temperature, °C	Corrosion rate, mm/yr
Potassium ferricyanide	Saturated	Room	nil
Potassium hydroxide + 13% potassium chloride	13	29	nil
Potassium hydroxide	50	29	0.010
	50	Boiling	<0.127
	25	Boiling	0.305
Potassium hydroxide	50	Boiling	2.74
	50 anhydrous	241-377	1.02-1.52
Potassium iodide	Saturated	Room	nil
Potassium permanganate	Saturated	Room	nil
Potassium perchlorate	20	Room	0.003
	0-30	50	0.003
Potassium sulfate	10	Room	nil
Potassium thiosulfate	1	Room	nil
Propionic acid	Vapor	190	Rapid attack
Pyrogallic acid	355 g/L	Room	nil
Salicylic acid	Saturated	Room	nil
Seawater	...	24	nil
Seawater, 4 1/2-year test	...	Ambient	nil
Sebacic acid	...	240	0.008
Silver nitrate	50	Room	nil
Sodium	100	To 1100 (593)	Good
Sodium acetate	Saturated	Room	nil
Sodium aluminate	25	Boiling	0.991
Sodium bifluoride	Saturated	Room	Rapid
Sodium bisulfate	Saturated	Room	nil
	10	66	1.83
Sodium bisulfite	10	Boiling	nil
	25	Boiling	nil
Sodium carbonate	25	Boiling	nil
Sodium chlorate	Saturated	Room	nil
Sodium chlorate + NaCl 80-250 g/L	0-72 g/L	40	0.003
Sodium chloride	Saturated	Room	nil
pH 7	23	Boiling	nil
pH 1.5	23	Boiling	nil
pH 1.2	23	Boiling	0.71
pH 1.2, some dissolved chlorine	23	Boiling	nil
Sodium citrate	Saturated	Room	nil
Sodium cyanide	Saturated	Room	nil
Sodium dichromate	Saturated	Room	nil
Sodium fluoride	Saturated	Room	0.008
pH 7	1	Boiling	0.001
pH 10	1	Boiling	0.001
pH 7	1	204	0.000
Sodium hydrosulfide + sodium sulfide and polysulfides	5-12	110	<0.003
Sodium hydroxide	5-10	21	0.001
	10	Boiling	0.021
	28	Room	0.003
	40	80	0.127
	50	57	0.013
	50	Boiling	0.051
	73	129	0.178
	50-73	188	>1.09
	50	38	0.023
Sodium hypochlorite	6	Room	nil
Sodium hypochlorite + 15% NaCl + 1% NaOH	1.5-4	66-93	0.030
Sodium nitrate	Saturated	Room	nil
Sodium perchlorate	900 g/L	50	0.003
Sodium phosphate	Saturated	Room	nil
Sodium silicate	25	Boiling	nil
Sodium sulfate	10-20	Boiling	nil
	Saturated	Room	nil
Sodium sulfide	10	Boiling	0.027
	Saturated	Room	nil
Sodium sulfite	Saturated	Boiling	nil
Sodium thiosulfate	25	Boiling	nil
Sodium thiosulfate + 20% acetic acid	20	Room	nil
Soils, corrosive	...	Ambient	nil
Stannic chloride	5	100	0.003
	24	Boiling	0.045
Stannic chloride, molten	100	66	nil

(continued)

Unalloyed Ti: General Corrosion (continued)

Medium	Concentration, %	Temperature, °C	Corrosion rate, mm/yr
Stannic chloride	100	35	nil
	Saturated	Room	nil
Steam + air	...	82	nil
Steam + 7.65% hydrogen sulfide	...	93-110	nil
Stearic acid, molten	100	180	0.003
Succinic acid	100	185	nil
	Saturated	Room	nil
Sulfanilic acid	Saturated	Room	nil
Sulfamic acid	3.75 g/L	Boiling	nil
	7.5 g/L	Boiling	2.74
Sulfamic acid + 0.375 g/L $FeCl_3$	7.5 g/L	Boiling	0.030
Sulfur, molten	100	240	nil
Sulfur monochloride	...	202	>1.09
Sulfur dioxide, dry	...	21	nil
Sulfur dioxide, water saturated	Near 100	Room	0.003
Sulfur dioxide gas + small amount SO_3 and approximately 3% O_2	18	316	0.006
Sulfuric acid, aerated	1	60	0.008
	3	60	0.013
	5	60	4.83
	10	35	1.27
	40	35	8.64
	75	35	1.07
	75	Room	10.8
	1	100	0.005
	3	100	23.4
	Concentrated	Room	1.57
	Concentrated	Boiling	5.38
	1	100	7.16
	3	100	21.1
Sulfuric acid	1	Boiling	17.8
	5	Boiling	25.4
Sulfuric acid + 0.25% $CuSO_4$	5	95	nil
	30	38	0.061
	30	95	0.088
Sulfuric acid + 0.5% $CuSO_4$	30	38	0.067
	30	95	0.823
Sulfuric acid + 1.0% $CuSO_4$	30	38	0.020
	30	95	0.884
Sulfuric acid + 0.5% CrO_3	5	95	nil
	30	95	nil
Sulfuric acid + 1.0% $CuSO_4$	30	Boiling	1.65
Sulfuric acid vapors	96	38	nil
	96	66	nil
	96	200-300	0.013
Sulfuric acid + 10% HNO_3	90	Room	0.457
Sulfuric acid + 50% HNO_3	50	Room	0.635
Sulfuric acid + 70% HNO_3	30	Room	0.102
Sulfuric acid + 90% HNO_3	10	Room	nil
Sulfuric acid + 90% HNO_3	10	60	0.011
Sulfuric acid + 95% HNO_3	5	60	0.005
Sulfuric acid + 50% HNO_3	50	60	0.399
Sulfuric acid + 20% HNO_3	80	60	1.59
Sulfuric acid saturated with chlorine	45	24	0.003
	62	16	0.002
	5, 10	190	<0.025
	82	50	>1.19
Sulfuric acid + 4 g/L Ti^{4+}	40	100	nil
Sulfurous acid	6	Room	nil
Tannic acid	25	100	nil
Tartaric acid	10-50	100	<0.127
	10	60	0.003
	25	60	0.003
	50	60	0.001
	10	100	0.003
	25	100	nil
	50	100	0.0121
Terephthalic acid	77	218	nil

(continued)

Unalloyed Ti: General Corrosion (continued)

Medium	Concentration, %	Temperature, °C	Corrosion rate, mm/yr
Tetrachloroethane, liquid and vapor	100	Boiling	0.001
Tetrachloroethylene + H_2O	...	Boiling	0.127
Tetrachloroethylene	100	Boiling	nil
Tetrachloroethylene, liquid and vapor	100	Boiling	0.001
Titanium tetrachloride	99.8	300	1.57
Trichloroacetic acid	100	Boiling	14.6
Trichloroethylene	99	Boiling	0.003-0.127
Trichloroethylene + 50% H_2O	50	25	0.001
Uranium chloride	Saturated	21-90	nil
Uranyl ammonium phosphate filtrate + 25% chloride + 0.5% fluoride + 1.4% ammonia + 2.4% uranium	20.9	165	<0.003
Uranyl nitrate containing 25.3 g/L Fe^{3+}, 6.9 g/L Cr^{3+}, 2.8 g/L Ni^{2+}, 4.0 M HNO_3 +1.0 M Cl	120 g/L	Boiling	nil
Uranyl sulfate + 3.1 M Li_2SO_4 + 100-200 ppm O_2	3.1 M	250	<0.020
Uranyl sulfate + 3.6 M Li_2SO_4, 50 psi oxygen	3.8 M	350	0.006-0.432
Urea + 32% ammonia + 20.5% H_2O, 19% CO_2	28	182	0.079
Water, degassed	...	316	nil
Water, river, saturated with chlorine	...	93	nil
X-ray developer solution	...	Room	nil
Zinc chloride	5	Boiling	nil
	20	104	nil
	50, 75	150	nil
	75	150	0.06
	75	200	Rapid pitting
	80	173	2.1
Zinc sulfate	Saturated	Room	nil

Source: *Metals Handbook, Corrosion*, Vol 13, 9th ed., ASM International, 1987, p 701

Modified Ti: General Corrosion by Media

This compilation of general corrosion rates is for ASTM grades 7 and 12 as specified. These data should be used only as a guideline for alloy performance. Rates may vary depending on changes in medium chemistry, temperature, length of exposure, and other factors. Total alloy suitability cannot be assumed from these values alone, because other forms of corrosion, such as localized attack, may be limiting. The text should be consulted to assess overall alloy suitability more thoroughly for a given set of environmental conditions. In complex, variable, and/or dynamic environments, *in situ* testing may provide more reliable data. In the following table, temperatures are given only in centigrade, and corrosion rates are reported only in millimeters per year.

Medium	Alloy	Concentration, %	Temperature, °C	Corrosion rate, mm/yr
Acetic acid + 5% formic acid	Grade 12	58	Boiling	nil
Ammonium hydroxide	Grade 12	30	Boiling	nil
Aluminum chloride	Grade 12	10	Boiling	nil
	Grade 7	10	100	<0.025
	Grade 7	25	100	0.025
Ammonium chloride	Grade 12	10	Boiling	nil
Aqua regia	Grade 7	3:1	Boiling	1.12
	Grade 12	3:1	Boiling	0.61
Calcium chloride	Grade 7	62	150	nil
	Grade 7	73	177	nil
Chlorine, wet	Grade 7	...	25	nil
Chromic acid	Grade 7	10	Boiling	nil
Citric acid	Grade 7	50	Boiling	0.025
	Grade 12	50	Boiling	0.013
Ferric chloride	Grade 7	10	Boiling	nil
	Grade 12	10	Boiling	nil
	Grade 7	30	Boiling	nil
Formic acid	Grade 7	45	Boiling	nil
	Grade 12	45, 50	Boiling	nil
	Grade 7	50	Boiling	0.01
	Grade 12	90	Boiling	0.56
	Grade 7	90	Boiling	0.056
Hydrochloric acid, deaerated	Grade 7	3	82	0.013
	Grade 7	5	82	0.051
	Grade 7	10	82	0.419
Hydrochloric acid	Grade 7	0.5	Boiling	nil
	Grade 7	1.0	Boiling	0.008
	Grade 7	1.5	Boiling	0.03
	Grade 7	5.0	Boiling	0.23
	Grade 12	0.5	Boiling	nil
	Grade 12	1.0	Boiling	0.04
	Grade 12	1.5	Boiling	0.25
Hydrochloric acid, hydrogen saturated	Grade 7	1-15	25	<0.025
	Grade 7	20	25	0.102
	Grade 7	5	70	0.076
	Grade 7	10	70	0.178
	Grade 7	15	70	0.33
	Grade 7	3	190	0.025
	Grade 7	5	190	0.102
	Grade 7	10	190	8.9
Hydrochloric acid, oxygen saturated	Grade 7	3, 5	190	0.127
	Grade 7	10	190	9.3
Hydrochloric acid, chlorine saturated	Grade 7	3, 5	190	<0.03
	Grade 7	10	190	29.0
Hydrochloric acid, aerated	Grade 7	1, 5	70	<0.03
	Grade 7	10	70	0.05
	Grade 7	15	70	0.15
Hydrochloric acid + 4% $FeCl_3$ + 4% $MgCl_2$	Grade 7	19	82	0.49
Hydrochloric acid + 4% $FeCl_3$ + 4% $MgCl_2$, chlorine saturated	Grade 7	19	82	0.46
Hydrochloric acid +5 g/L $FeCl_3$	Grade 7	10	Boiling	0.279
+16 g/L $FeCl_3$	Grade 7	10	Boiling	0.076
+16 g/L $CuCl_2$	Grade 7	10	Boiling	0.127

(continued)

Medium	Alloy	Concentration, %	Temperature, °C	Corrosion rate, mm/yr
Hydrochloric acid				
+2 g/L FeCl$_3$	Grade 12	4.2	91	0.058
+0.1% FeCl$_3$	Grade 7	5	Boiling	0.013
+0.1% FeCl$_3$	Grade 12	5	Boiling	0.020
Hydrochloric acid + 18% H$_3$PO$_4$ + 5% HNO$_3$	Grade 7	18	77	nil
Hydrogen peroxide				
pH 1	Grade 7	5	23	0.062
pH 4	Grade 7	5	23	0.010
pH 1	Grade 7	5	66	0.127
pH 4	Grade 7	5	66	0.046
–500 ppm Ca^{2+}, pH 1	Grade 7	5	66	nil
+500 ppm Ca^{2+}, pH 1	Grade 7	20	66	0.76
Hydrogen peroxide, pH 1 + 5% NaCl	Grade 7	20	66	0.008
Magnesium chloride	Grade 7	Saturated	Boiling	nil
Oxalic acid	Grade 7	1	Boiling	1.14
Phosphoric acid, naturally aerated	Grade 12	25	25	0.019
	Grade 12	30	25	0.056
	Grade 12	45	25	0.157
	Grade 12	8	52	0.02
	Grade 12	13	52	0.066
	Grade 12	15	52	0.52
	Grade 12	5	66	0.038
	Grade 12	7	66	0.15
	Grade 12	0.5	Boiling	0.071
	Grade 12	1.0	Boiling	0.14
	Grade 7	40	25	0.008
	Grade 7	60	25	0.07
	Grade 7	15	52	0.036
	Grade 7	23	52	0.15
	Grade 7	8	66	0.076
	Grade 7	15	66	0.104
	Grade 7	0.5	Boiling	0.050
	Grade 7	1.0	Boiling	0.107
	Grade 7	5.0	Boiling	0.228
Sodium fluoride				
pH 7	Grade 12	1	Boiling	0.001
pH 7	Grade 7	1	Boiling	0.002
Sodium sulfate, pH 1	Grade 7	10	Boiling	nil
Sulfamic acid	Grade 12	10	Boiling	11.6
	Grade 7	10	Boiling	0.37
Sulfuric acid, naturally aerated	Grade 12	9	24	0.003
	Grade 12	9.5	24	0.006
	Grade 12	10	24	0.38
	Grade 12	3.5	52	0.013
	Grade 12	3.75	52	1.73
	Grade 12	2.75	66	0.015
	Grade 12	3.0	66	1.65
	Grade 12	0.75	Boiling	0.003
	Grade 12	1.0	Boiling	0.91
	Grade 7	1.0	204	0.005
	Grade 7	2.0	204	nil
	Grade 12	1.0	204	0.91
Sulfuric acid, nitrogen saturated	Grade 7	5	70	0.15
	Grade 7	10	70	0.25
	Grade 7	1, 5	190	0.13
	Grade 7	10	190	1.50
Sulfuric acid, oxygen saturated	Grade 7	1-10	190	0.13
Sulfuric acid, chlorine saturated	Grade 7	10	190	0.051
	Grade 7	20	190	0.38
Sulfuric acid, nitrogen saturated	Grade 7	10	25	0.025
	Grade 7	40	25	0.23
Sulfuric acid, aerated	Grade 7	10	70	0.10
	Grade 7	40	70	0.94
Sulfuric acid +5 g/L Fe$_2$(SO$_4$)$_3$	Grade 7	10	Boiling	0.178
Sulfuric acid + 16 g/L Fe$_2$(SO$_4$)$_3$	Grade 7	10	Boiling	<0.03
Sulfuric acid + 16 g/L Fe$_2$(SO$_4$)$_3$	Grade 7	20	Boiling	0.15
Sulfuric acid + 15% CuSO$_4$	Grade 7	15	Boiling	0.64
Sulfuric acid + 1% CuSO$_4$	Grade 7	30	Boiling	1.75

(continued)

Medium	Alloy	Concentration, %	Temperature, °C	Corrosion rate, mm/yr
Sulfuric acid + 100 ppm Cu^{2+} + 1% thiourea (deaerated)	Grade 7	1	100	nil
Sulfuric acid + 100 ppm Cu^{2+} + 1% thiourea (deaerated)	Grade 12	1	100	0.23
Sulfuric acid + 1000 ppm Cl^-	Grade 7	15	49	0.015

Source: *Metals Handbook, Corrosion*, Vol 13, 9th ed., ASM International, 1987, p 705

Crevice Corrosion

The susceptibility of titanium to crevice corrosion should be considered when tight crevices exist in hot aqueous chloride, bromide, iodide, or sulfate solutions. Commercially pure titanium may also be susceptible to crevice corrosion when an environment contains more than 1000 ppm chloride at temperatures of about 75 °C or higher. Therefore, molybdenum-containing titanium alloys (>4 wt% Mo) with improved crevice corrosion resistance may be desirable for marine applications.

Crevice Corrosion in Chlorides

The pH-temperature limits for the crevice corrosion of titanium grades 2, 7, and 12 have been found to be applicable to most chloride salt solutions, including seawater over a wide range of chloride concentrations (>0.1%). The limits for grade 2 are applicable to all unalloyed grades as well. It should be cautioned that deviations from normal crevice corrosion guidelines can be expected in certain acidic salts that may hydrolyze to form HCl at high temperatures when highly concentrated. These salts include $MgCl_2$, $CaCl_2$, $ZnCl_2$, and $AlCl_3$. Grade 7 titanium is generally most resistant in these situations (*Titanium 1986 - Titanium Products and Applications*, Titanium Development Association, 1987).

Crevice attack of titanium alloys will generally not occur below 70 °C (160 °F) regardless of solution pH or chloride concentration, or when solution

Resistance of titanium ASTM grades 2, 7, and 12 to crevice corrosion in boiling salt solutions
Tight metal-to gasket crevices were tested.

Solution	pH	Alloy/resistance(a) Grade 2	Grade 12	Grade 7
Saturated $ZnCl_2$	3.0	F	R	R
10% $MgCl_2$	4.2	F	R	R
10% $CaCl_2$	3.0	F	R	R
10% KCl	3.0	F	R	R
Saturated NaCl	3.0	F	R	R
Saturated NaCl + Cl_2	1-2	F	F	R
10% NH_4Cl	4.1	F	R	R
10% $FeCl_3$	0.6	F	F	R
10% Na_2SO_4	2.0	F	R	R

(a) F, failed; R, resistant. Source: R.W. Schutz, J.A. Hall, and T.L. Wardlaw, "TI-CODE 12, An Improved Industrial Alloy," paper presented at the Japan Titanium Society 30th Anniversary International Symposium, Japan Titanium Society, Aug 1982

Results of crevice corrosion testing in various environments
Tests were conducted in 250 g/L, air-saturated NaCl; crevice, PTFE

Alloy	150 °C test				200 °C test			
	pH 2		pH 6		pH 2		pH 6	
	500 h	1000 h	500 h	1000 h	500 h	1000 h	500 h	1000 h
CP Ti	4/4	4/4	4/4	4/4	4/4	...	4/4	...
Ti-0.02Pd	1/4	2/4	2/4	3/4	1/4	4/8	2/4	3/8
Ti-0.05Pd	0/4	0/4	0/4	0/4	0/4	0/8	0/4	0/8
Ti-0.10Pd	0/4	0/4	0/4	0/4	0/4	0/8	0/4	0/8
Ti-0.14Pd	0/4	0/4	0/4	0/4	0/4	0/8	0/4	0/8
Ti-1Ni	4/4	...	2/4	...	4/4	...	2/4	...
Ti-1Mo	4/4	...	4/4	...	4/4	...	4/4	...
Grade 12	4/4	...	4/4	...	4/4	...	4/4	...

Note: 0/4 = no failures observed from 4 test specimens. Source: Y. Shida and S. Kitayama, Effect of Pd Additions on the Crevice Corrosion Resistance of Titanium, *Sixth World Conference on Titanium*, Les Editions de Physiques, Paris, 1988, p 1729-1732

Effect of crevice-forming material on crevice corrosion resistance of Ti-Pd alloys
Tests were conducted in 250 g/L, air-saturated NaCl for 1000 h.

Alloy	\multicolumn{3}{c}{150 °C test, pH 6, on:}			\multicolumn{2}{c}{200 °C test, pH 2, on:}
	PTFE	Asbestos	Neoprene rubber	PTFE	Asbestos
Ti-0.05Pd	0/4	0/2	0/2	0/4	0/2
Ti-0.14Pd (Grade 7)	0/4	0/2	0/2	0/4	0/2
CP Ti	4/4	2/2	1/2	4/4	2/2

Note: 0/4 = no failures observed for 4 test specimens. Source: Y. Shida and S. Kitayama, Effect of Pd Additions on the Crevice Corrosion Resistance of Titanium, *Sixth World Conference on Titanium*, Les Editions de Physiques, Paris, 1988, p 1729-1732

Crevice corrosion resistance of Ti-0.3Mo-0.8Ni alloy (ASTM grade 12) in chloride solutions

Test medium	Test temperature	Time, h	Corrosion condition	
			CP Ti	Ti-0.3Mo-0.8Ni
$MgCl_2 \cdot 6H_2O$	Boiling	96	F	R
44% NH_4Cl	Boiling	94	F	R
30% $CuCl_2$(a)	Boiling	94	F	R
30% $FeCl_3$(a)	Boiling	94	F	R

Note: Metal-to-tetrafluoride crevice samples used; F, failed (samples showed corrosion); R, resistant (samples showed no evidence of corrosion). (a) The bolts corroded away in testing.
Source: L. Zuochen *et al.*, The Corrosion Resistance and Application of Ti-0.3Mo-0.8Ni Alloy, *Titanium, Science and Technology*, Vol 4, G. Lütjering, U. Zwicker, and W. Bunk, Ed., Deutsche Gesellschaft für Metallkunde, e.V., Germany, 1985, p 2611-2616

Crevice corrosion resistance of Ti-0.3Mo-0.8Ni alloy in the saturated NaCl solution at different temperatures and acidities

Test medium	Acidity, pH	Temperature °C	°F	Time, h	Corrosion condition CP Ti	Ti-0.3Mo-0.8Ni
Saturated NaCl + trace HCl	1	Boiling		136	F	F
Saturated NaCl + trace HCl	3	Boiling		136	F	R
Saturated NaCl + trace HCl	5	Boiling		94	F	R
Saturated NaCl + trace HCl	7	Boiling		136	R	R
Saturated NaCl + trace HCl	2	65	145	624	R	R
Saturated NaCl + trace HCl	2	85	185	569	R	R
Saturated NaCl + trace HCl	2	95	200	177	R	R

Note: Metal-to-tetrafluoride crevice samples used; F, failed (samples showed corrosion); R, resistant (samples showed no evidence of corrosion). Source: L. Zuochen *et al.*, The Corrosion Resistance and Application of Ti-0.3Mo-0.8Ni Alloy, *Titanium, Science and Technology*, Vol 4, G. Lütjering, U. Zwicker, and W. Bunk, Ed., Deutsche Gesellschaft für Metallkunde, e.V., Germany, 1985, p 2611-2616

Time to initiation of crevice attack for titanium alloys in boiling chloride solutions

Alloy	Time to initiation of crevice attack, h, for:		
	6% NaCl (pH 6)	20% NaCl (pH 4)	42% $MgCl_2$
CP Ti	220	92	15
Ti-0.3Mo-0.8Ni	>720	>720	168
Ti-0.15Pd	>720	>720	>240
PdO/TiO_2-Ti	>720	>720	>240

Note: Ti/PTFE-creviced electrode. Source: B. Satoh *et al.*, The Crevice Corrosion Resistance of Some Titanium Materials, *Plat. Met. Rev.*, Vol 31, 1987, p 115-121

pH exceeds 10 regardless of temperature. ASTM grade 12 provides crevice corrosion resistance when brine pH falls between 3 and 11 to temperatures as high as 300 °C (570 °F). Grade 7 or 11 alloy extends this resistance to brine pH values as low as 0.6 to 0.7, depending on brine composition and temperature. The grade 7 alloy is considered to be the most crevice corrosion-resistant titanium alloy available commercially, and it is often preferred for hot, low-pH salt solutions.

ASTM Grade 12. The crevice corrosion resistance of grade 12 titanium has been extensively tested in concentrated near-neutral NaCl, NaCl-$MgCl_2$, and NH_4Cl brines at high temperatures. Studies were conducted to assess this alloy for potential application in hypersaline geothermal brine, high-level nuclear waste storage, oil refineries, and salt evaporator brine heaters. In all cases, crevice testing involving Teflon gasket-to-metal and metal-to-metal crevices for extended periods revealed no evidence of significant attack to temperatures as high as 250 °C (480 °F). In saturated NH_4Cl solution, no gasket-to-metal or under-salt-deposit attack was noted up to 177 °C (350 °F) and at pH 3 to 7. Similar performance is noted for grade 7 titanium under these conditions.

Effect of Chloride Concentration. Studies have addressed the effect of chloride concentration on crevice corrosion initiation on unalloyed titanium (*J. Jpn. Inst. Met.*, Vol 44, 1978, p 567-572). Threshold temperatures of 250, 200, and 150 °C (480, 390, and 300 °F) were indicated for chloride concentrations of 0.01, 0.1, and 1%, respectively, in neutral (pH 7) NaCl brine. Unpublished results suggest that 0.01% Cl^- at 90 °C (195 °F) and 0.10% Cl^- at 70 °C (160 °F) may be threshold conditions for crevice attack in aerated (pH 3 to 5) solutions given tight Teflon gasket-to-metal crevices.

Effect of Gasket Material. Teflon gasket-to-metal crevices are generally more susceptible to crack initiation than silicone rubber, neoprene rubber, asbestos, and polyvinyl chloride (PVC) gasket-to-metal crevices. Certain polymeric sealants, such as styrol-acrylic copolymer or methacrylate polymers, may significantly increase susceptibility to crevice attack, especially when the sealant

CP and modified Ti: Crevice corrosion limits

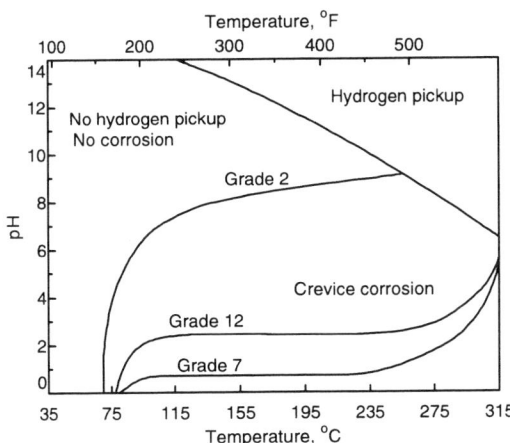

Temperature-pH limits for crevice corrosion in saturated NaCl brines. Grades 12 and 7 are more resistant than Grade 2.
Source: R.W. Schutz, *Mater. Perform.*, Vol 24 (No. 1), Jan 1985, p 39-47

Ti-Ni-Mo series alloys: Crevice corrosion

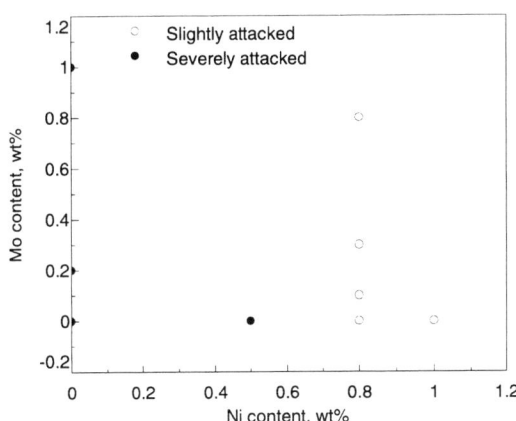

Measured in high-temperature NaCl solution, pH 6, NaCl 4.27 mol/L, 423 K for 500 h.
Source: *ISIJ Int.*, Special Issue on Recent Advances on Titanium Technology, Vol 31 (No. 8), 1991, p 900

Effect of Ni on crevice corrosion

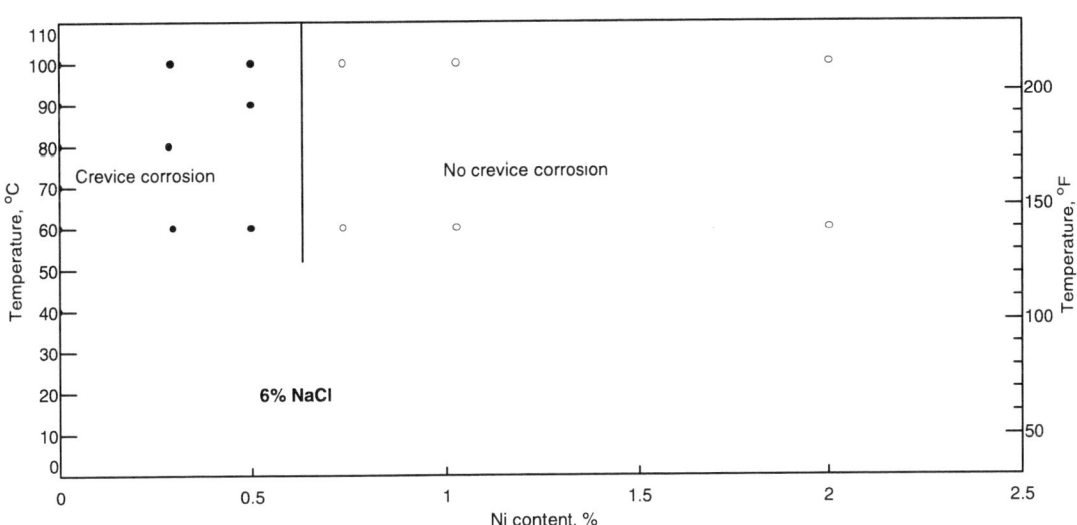

Source: M. Kobayashi, *Titanium '80, Science and Technology*, Vol 4, AIME, 1980, p 2620

Crevice Corrosion in Bromides and Sulfates

contains chloride salts. On the other hand, metal-to-metal crevices are generally least susceptible to attack. In addition, it has been found that the incubation time for crevice corrosion in NaCl brine may be reduced, and attack aggravated, by impressed anodic potential (M. Kobayashi et al., Study on Crevice Corrosion of Titanium, in *Titanium '80—Science and Technology*, Vol 4, The Metallurgical Society, 1980, p 2613-2622; and H. Satoh et al., Effect of Gasket Materials on Crevice Corrosion of Titanium, in *Titanium—Science and Technology*, Proceedings of the Fifth International Conference on Titanium, Deutsche Gesellschaft für Metallkunde e.V., 1985, p 2633-2639).

Unpublished test results suggest that the pH-temperature guidelines for crevice corrosion of titanium in saturated NaCl are applicable in saturated NaBr solutions. However, the rate of crevice attack is measurably lower than that in chlorides at corresponding pH and temperature values.

Crevice testing of unalloyed titanium in saturated Na_2SO_4 solutions revealed attack in neutral solutions at temperatures above 110 °C (230 °F) and only below pH 6 at 104 °C (220 °F), with no attack at 93 °C (200 °F) as low as pH 3. Teflon gasket-to-metal crevices were also exposed to a simulated acid sulfate galvanizing solution for 38 days at temperatures of 66, 82, and 100 °C (150, 180, and 212 °F). The pH 1.3 solution consisted of 100 g/L Na_2SO_4, 120 g/L $ZnSO_4$, 0.6 g/L Fe^{3+}, as $Fe_2(SO_4)_3$, and 0.008% Cl^-. Grade 2 titanium proved to be fully resistant below boiling (100 °C, or 212 °F). Grade 7 and 12 titanium alloys resisted crevice attack under all conditions tested. All results indicate that threshold temperatures for crevice corrosion in sulfate solution are measurably higher than those in chloride brines.

Hydrogen Damage

See also "High-Purity Ti" for additional information on hydrogen damage.

Hydrogen damage and embrittlement depends on the formation of hydrides, which have been observed in commercially pure titanium at concentrations as low as about 0.5 at.% H (100 ppm by weight) according to the phase diagram proposed by Lenning et al., (see figure). Hydrides have also been observed down to about 50 ppm and lower (0.25 at.%) in commercially pure titanium (E. Barta, R. Boyer, G. Narayaran, ISTFA Conference Proceedings, ASM International, p 387-396). However, small amounts of hydrides are not necessarily detrimental from an engineering standpoint depending on the particular test/application conditions. Severe embrittlement can occur in the commercial grades at hydrogen levels as low as 30 to 40 ppm (by weight) in the presence of a high residual stress or a stress riser, and elevated temperature. These conditions induce migration of the hydrogen to the stress riser, resulting in a much higher local concentration of hydrogen and the precipitation of hydrides.

The effect of this amount of hydrogen on tensile properties is negligible, but the effect on notch-bar impact properties is profound. The effect of hydrogen on mechanical properties is much greater in the presence of a notch. However, α and near-α titanium also show sensitivity to hydrogen levels for biaxial- and triaxial-stress properties such as bend ductility and cup drawing formability (Ref 1-6).

The effect of hydrogen on the notch sensitivity of α titanium, including high purity iodide titanium, commercial titanium, and α alloys, depends on purity to some extent. Commercially pure titanium is much more sensitive to hydrogen than is pure titanium. The amount of hydrogen necessary to induce ductile-to-brittle transition behavior in commercially pure titanium is less than one-half the amount needed in pure titanium. Pure α-titanium is relatively unaffected by small concentrations (<200 ppm) of hydrogen; however, above this content, the impact toughness is impaired.

Strain rate appears to have little effect on impact strength until higher hydrogen contents are reached. Low strain rate embrittlement is related to hydride formation caused by strain-enhanced precipitation, but embrittlement under impact is caused by hydride-phase formation after fabrication or heat treatment.

References

1. N.E. Paton and J.C. Williams, Effect of Hydrogen on Titanium and Its Alloys, in *Titanium and Titanium Alloys—Source Book*, American Society for Metals, 1982, p 185-207
2. R.R. Boyer and W.F. Spurr, Characteristics of Sustained-Load Cracking and Hydrogen Effects in Ti-6Al-4V, *Metall. Trans. A*, Vol 9A, Jan 1978, p 23-29
3. R. Bourcier and D. Koss, *Acta Metall.*, Vol 32 (No. 11), 1984, p 2091-2099
4. G.A. Lenning et al., Effect of Hydrogen on Alpha Titanium Alloys, *Trans. AIME*, Oct 1956, p 1235
5. C.M. Craighead et al., Hydrogen Embrittlement of Beta-Stabilized Titanium Alloys, *Trans. AIME*, Aug 1956, p 923
6. J.J. DeLuccia, "Electrolytic Hydrogen in Beta Titanium," Report NADC-76207-30, Air Vehicle Technical Department, Naval Air Development Center, June 1976

CP titanium Grade 3 (RC55): Solubility relation for hydrogen

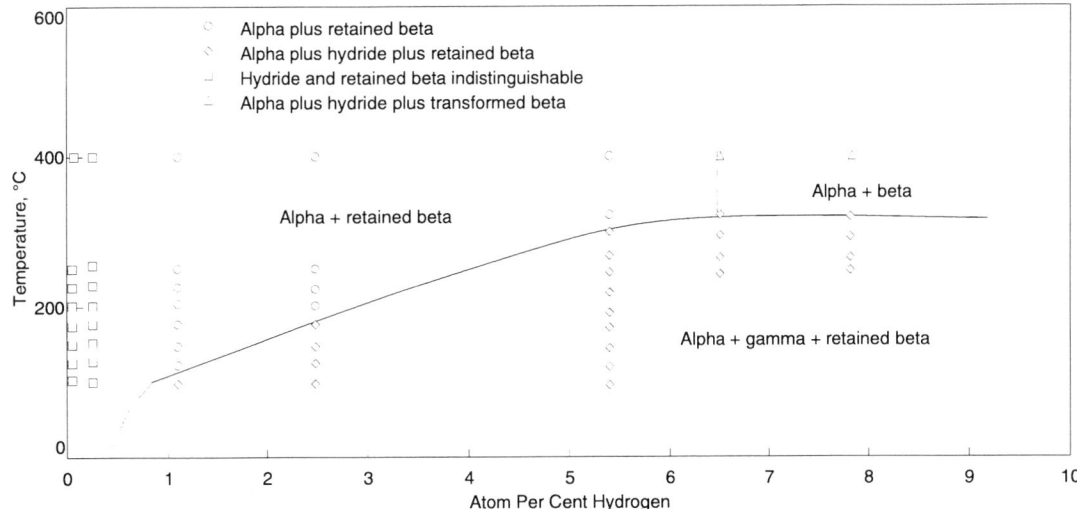

Low-temperature solubility relation for hydrogen in commercial (RC-55) titanium containing about 0.3 wt%. Fe. The presence of iron as an impurity in commercial titanium stabilizes a small amount of retained β phase to room temperature. Because hydrogen has a high solubility in β, the room-temperature solubility of hydrogen in commercial titanium is higher than in high purity titanium. However, this effect does not, apparently, eliminate the detrimental effects of hydrogen on notch toughness, perhaps because of embrittlement of the β phase.
Source: G.A. Lenning, C.M. Craighead, R.I. Jaffee, Constitution and Mechanical Properties of Titanium-Hydrogen Alloys, *Hydrogen Damage*, C.D. Beachem, Ed., American Society for Metals, 1977, p 100

Hydrogen absorption in grade 2 titanium after 96-h exposures in pure, dry hydrogen

Temperature		Hydrogen pressure		Freshly pickled	Hydrogen absorption, ppm/specimen surface preparation		
°C	°F	MPa	psi		Pickled 2-month exposure	Iron-contaminated	Anodized
149	300	Atmospheric		0	0	0	0
149	300	2.76	400	58	20	174	0
149	300	5.52	800	28	0	117	0
315	600	Atmospheric		0	0	0	0
315	600	2.76	400	2586	6911	5951	516
315	600	5.52	800	4480	10 550	13 500	10 000

Source: L.C. Covington, "Factors Affecting the Hydrogen Embrittlement of Titanium," Paper 59, presented at Corrosion/75, Toronto, Ontario, National Association of Corrosion Engineers, April 1975

Hydrogen absorption in unalloyed titanium after 14-day exposures

Sample condition	Hydrogen absorption, ppm		
	Dry hydrogen	Moist hydrogen	Hydrogen + CO_2
As-received	58	1	4
Iron-contaminated	780	21	125
Copper-contaminated	27	5	5
Platinum-contaminated	...	...	770
Anodized	9	2	3

Source: J.B. Cotton and J.G. Hines, Hydriding of Titanium Used in Chemical Plant and Protective Measures, in *The Science, Technology and Application of Titanium*, Pergamon Press, 1970, p 150-170

Gaseous Hydrogen

Absorption of hydrogen by titanium alloys in gaseous hydrogen is highly dependent on temperature, gas pressure, gas moisture or oxygen content, metal surface condition, alloy composition, and the nature of the surface oxide formed on the metal. In the absence of surface oxide film interferences, the hydrogen concentration in titanium is directly proportional to the square root of the hydrogen gas partial pressure. It has been shown that at least 2% H_2O content in 5.5-MPa (800-psi) hydrogen gas at 315 °C (600 °F) effectively retards hydrogen uptake in commercially pure titanium (L.C. Covington, "Factors Affecting the Hydrogen Embrittlement of Titanium," Paper 59, presented at Corrosion/75, Toronto, Ontario, National Association of Corrosion Engineers, April 1975).

Cathodic Hydrogen Uptake

Titanium may absorb hydrogen if there is an impressed cathodic current, or galvanic coupling to active metals, achieved by severe, continuous mechanical damage of titanium surfaces. In these situations, the factors of potential, current density, metal temperature, solution pH and chemistry, and alloy composition and surface condition all significantly influence the rate at which hydrogen is absorbed.

Unalloyed titanium may absorb hydrogen in near-neutral brines (seawater) at 25 and 100 °C (75 and 212 °F) when cathodic potentials are more active (negative) than –0.75 V versus SCE. Charging in this manner produced only thin, innocuous surface hydride films on unalloyed titanium, as long as potentials remain more noble than –1.0 V and temperatures are below 80 °C (175 °F). For example, grade 2 titanium galvanically coupled to aluminum alloy 6063 in a process water electrolyte at 65 °C (150 °F) generated surface hydrides only. However, accelerated hydriding of unalloyed titanium may occur at impressed potentials cathodic (negative) to –1.0 V even at room temperature. The potential threshold applies to α-β and β titanium alloys as well; however, significantly higher rates of hydrogen absorption and penetration can be expected at any given temperature under hydrogen charging conditions for these alloys. This results from the significantly higher solubility and diffusion coefficients for hydrogen in the β phase.

Increasing temperature and decreasing pH have both been shown to accelerate significantly the rate of hydrogen absorption of titanium alloys during cathodic charging (see Ref 1 to 6 below). These studies indicate that the rate of hydrogen absorption and surface hydride layer growth initially follow a parabolic rate law; this suggests that hydrogen diffusion is rate controlling at temperatures less than and equal to 100 °C (212 °F) (Ref 7). Linear rate behavior has been observed in long-term charging exposures (Ref 3). Dramatic increases in hydrogen absorption rate during cathodic charging are noted when electrolyte pH levels are reduced to 2 or below (Ref 1, 3, 4, and 8).

The presence of hydrogen recombination poisons, such as sulfide, arsenate, or antimony species, substantially increases hydrogen uptake during charging (Ref 4). This effect of sulfide has been observed in a few oil refinery heat exchangers in which grade 2 titanium tubes were galvanically coupled to carbon steel tubesheets and components in hot (≥80 °C, or 175 °F) sulfide-containing aqueous process streams. Severe hydriding and eventual embrittlement of the tubes occurred. Elimination of the detrimental couple with the active metal (steel) has been achieved by use of designs with more galvanically compatible passive alloys (including all-titanium design) or use of dielectric joints. In comparison, grade 2 titanium galvanically coupled to carbon steel in neutral sulfide-free electrolytes results in very minor increases in hydrogen uptake to temperatures as high as 120 °C (250 °F) (Ref 2, 8).

The surface condition of titanium also influences hydrogen absorption rates during cathodic charging. Studies consistently reveal that as-pickled and as-received surfaces are much less amenable to hydrogen uptake than abraded, vapor-blasted, or sandblasted surfaces. Furthermore, anodized and, particularly, thermally oxidized surfaces are highly effective barriers to hydrogen uptake in titanium during charging.

References

1. I.I. Phillips, P. Pool, and L.L. Shreir, Hydride Formation During Cathodic Polarization of Ti. II. Effect of Temperature and pH of Solution on Hydride Growth, *Corros. Sci.*, Vol 14, 1974, p 533-542
2. T. Fukuzuka, K. Shimogori, H. Satoh, and F. Kamikubo, "Corrosion Problems and Countermeasures in MSF Desalination Plant Using Titanium Tube," Kobe Steel Ltd., 1985
3. J. Lee and P. Chung, "A Study of Hydriding of Titanium in Seawater Under Cathodic Polarization," Paper 259, Presented at Corrosion/86, Houston, TX, National Association of Corrosion Engineers, Mar 1986
4. Z.A. Foroulis, Factors Influencing Absorption of Hydrogen in Titanium From Aqueous Electrolytic Solutions, in *Titanium '80—Science and Technology*, Vol 4, The Metallurgical Society, 1980, p 2705-2711
5. L.C. Covington and N.G. Feige, A Study of Factors Affecting the Hydrogen Uptake Efficiency of Titanium in Sodium Hydroxide Solutions, in *Localized Corrosion—Cause of Metal Failure*, STP 516, American Society for Testing and Materials, 1972, p 222-235
6. I. Phillips *et al.*, *Corros. Sci.*, Vol 12, 1972, p

Unalloyed Ti: Cathodic potentials for hydrogen absorption

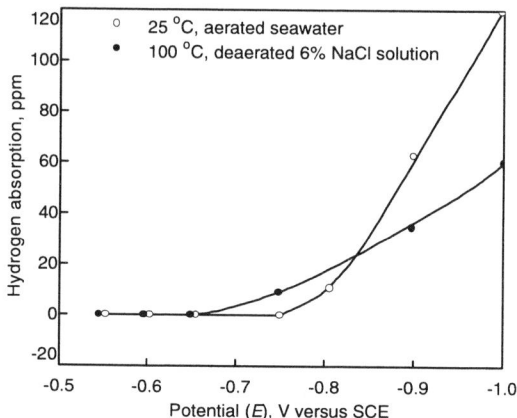

Source: "Get More Advantages by Applying Titanium Tubing Not Only for Power Plants but Also for Desalination Plants!!," Technical Brochure, Japan Titanium Society, May 1984; and H. Satoh, T. Fukuzuka, K. Shimogori, and H. Tanabe, "Hydrogen Pickup by Titanium Held Cathodic in Seawater," paper presented at the Second International Congress on Hydrogen in Metals, Paris, June 1977

855-866
7. R. Gruner, B. Streb, and E. Brauer, Hydrogen in Titanium, in *Titanium—Science and Technology*, Proceedings of the Fifth International Conference on Titanium, Deutsche Gesellschaft für Metallkunde e.V., 1985, p 2571
8. L.C. Covington, *Corrosion*, Vol 35 (No. 8), Aug 1979, p 378-382

Stress-Corrosion Cracking

Unalloyed and modified titanium (ASTM grades 1, 2, 7, 11, and 12) are immune to stress-corrosion cracking (SCC) except in a few specific environments. These environments include anhydrous methanol/halide solutions, nitrogen tetroxide (N_2O_4), red fuming HNO_3, liquid or solid cadmium, and liquid mercury

In addition, unalloyed titanium with higher interstitial contents may be susceptible to SCC in aqueous media. For example, Seagle *et al.* (The Influence of Composition and Heat Treatment on Aqueous Stress Corrosion of Titanium, *Applications Related Phenomena in Titanium Alloys*, STP 432, American Society for Testing and Materials, 1967, p 170) have studied SCC behavior of binary Ti-O alloys and showed that susceptibility occurred at oxygen contents between 0.2 and 0.4 wt%. This means that, although ASTM grades 1 and 2 fully resist aqueous SCC, certain higher interstitial chemistries (grades 3 and 4) may not (see table). For example, ASTM grade 2 (0.12 wt% O) has the same plane strain fracture toughness in air (K_{Ic} = 66 MPa$\sqrt{m}$ or 60 ksi$\sqrt{in.}$) as in ambient saltwater (K_{ISCC} = 66 MPa$\sqrt{m}$). On the other hand, ASTM grade 4 (0.38 wt% O) has a K_{Ic} value of 110 MPa$\sqrt{m}$ (100 ksi$\sqrt{in.}$) in air and a K_{ISCC} value of 38 MPa$\sqrt{m}$ (35 ksi$\sqrt{in.}$) in ambient saltwater.

Hot-Salt SCC. Unalloyed titanium with oxygen below about 0.20 to 0.25 wt% is considered immune to hot-salt SCC. Only limited published data on the influence of temperature on titanium alloy SCC behavior in aqueous media are available. However, most titanium alloys immune to ambient saltwater SCC, such as Grades 1, 2, 7, 12 and 9, remain so in halide solutions to temperatures in excess of 200 °C. For example, titanium Grades 2 and 12 exhibit resistance to SCC in slow strain rate tests in natural groundwaters and saturated NaCl brines to temperatures as high as 250 °C for the following studies: J.A. Ruppen, R.S. Glass, and M.A. Molecke, "Titanium Utilization in Long-Term Nuclear Waste Storage," in *Titanium for Energy and Industrial Applications*, AIME Publication, Warrendale, PA, 1981, p 355; and S.G. Pitman,

Unalloyed Ti: Toughness in air and 3.5% NaCl solution at 25 °C

Alloy	Thickness (in.)	Heat treatment(a)	Yield strength (ksi)	K_Q (ksi$\sqrt{in.}$)	K_{ISCC} or K_{SCC} (ksi$\sqrt{in.}$)
Ti-Gr. 2	...	αA-AC	...	60	60
Ti-Gr. 2 (0.060$_2$)	0.75	αA-AC	...	53	52
Ti-Gr. 3 (0.200$_2$)	0.75	αA-AC	...	72	68
Ti-Gr. 4 (0.400$_2$)	0.75	αA-AC	...	90	53
Ti-Gr. 4	0.50	αA-AC	83	123	33
	0.50	αA-WQ	85	128	34
	0.50	STA	82	113	39
	0.50	βA-WQ	76	105	70
	0.50	β-STA	77	96	48

This data was generated in ambient neutral 3.5% NaCl solution. It should be cautioned that these K_{ISCC} values are highly dependent on alloy composition, metallurgical conditions, and product form and thickness; and, therefore, may or may not be representative of alloy product materials commercially available.
Source: R. Schutz, Stress-Corrosion Cracking of Titanium Alloys, *Stress-Corrosion Cracking: Materials Performance and Evaluation*, ASM International, 1992. (a) A, anneal

Variation of SCC threshold for two grades of commercial-purity titanium in 3.5% NaCl solution

Alloy	Heat treatment	Phase structure	K_{IC} MPa$\sqrt{m}$	ksi$\sqrt{in.}$	K_{ISCC} MPa$\sqrt{m}$	ksi$\sqrt{in.}$
Ti-50A (0.12% O)	705 °C (1300 °F), AC	α	66	60	66	60
Ti-70A (0.38% O)	927 °C (1700 °F), WQ	α + α'	115	105	77	70
Ti-70A (0.38% O)	816 °C (1500 °F), WQ	α + (β + ω)	124	113	36	33

Note: Although SCC susceptibility at oxygen contents of 0.2 to 0.4% has been obtained on various grades of CP titanium, such data are complicated by differences in microstructure. For example, the effect of oxygen in CP titanium studied by Curtis *et al.* found that Ti-50A (0.12% O) was immune to SCC in 3.5% NaCl, whereas Ti-70A (0.38% O) was susceptible, as shown above. The reduction in K_{ISCC} produced by heat treatment was attributed to the formation of ω in the small β phase islands.
Source: M.J. Blackburn *et al.*, Stress Corrosion Cracking of Titanium Alloys, D1-82-1054, Boeing, 1970, p 179

"Enviromechanical Testing of Ti Grade 2 and Ti Grade 12 in Basalt Groundwater," *Industrial Applications of Ti and Zr: 3rd Conference*, ASTM STP 830, 1984, p 5-18.

Red Fuming Nitric Acid

The first reported observation of titanium SCC occurred in red fuming nitric acid. Cracking was observed for commercially pure titanium in a room-temperature environment containing 20% NO_2 (G.C. Kiefer, *Iron Age*, Vol 169, p 170). Later investigators indicated that SCC occurred in as little as 6.5% NO_2 with less than 0.7% H_2O. Intergranular cracking was observed on smooth specimens, which indicates SCC susceptibility in this medium. In addition to commercially pure titanium, cracking was observed on Ti-8Mn and Ti-6Al-4V, even in red fuming HNO_3 without free NO_2. Work on inhibitors showed that 1.5 to 2.0% H_2O completely inhibits SCC. The work on water as an inhibitor is especially interesting because it is of importance in a number of other environments that promote SCC.

Stress-Corrosion Cracking in Methanol

Unlike other engineering metals, titanium and zirconium alloys are unique in their strong susceptibility to SCC in methanol liquid and vapor. Methanol and NaBr are extremely corrosive to titanium, and in some cases, they promote intergranular SCC of smooth specimens. Methanol/HCl and methanol/H_2SO_4 mixtures also cause SCC of commercially pure titanium, once again on smooth specimens (*Corrosion*, Vol 22, 1966, p 29-31).

For titanium alloys not normally susceptible to aqueous SCC (such as ASTM grades 1 and 2), the SCC failure mode in methanol is primarily intergranular. This mode of cracking requires the presence of at least trace levels of halides or halogens, such as ≥0.3 ppm Cl^- or >0.00001N HCl for unalloyed titanium (*Corrosion*, Vol 25, 1969, p 87). Transgranular SCC is generally observed in titanium that is susceptible to SCC in aqueous solutions. It seems likely that anodic dissolution is the predominant SCC mechanism for intergranular failures. Titanium alloys known to be susceptible to aqueous SCC are also susceptible to methanol SCC, and the alloys most susceptible to seawater are also most susceptible to methanol.

Susceptibility to SCC measured as time to failure of smooth samples indicates that: increasing the halide content decreases time to failure, and water additions to a critical level decrease time to failure. Higher halide concentrations increase the critical level of water for maximum susceptibility to SCC, whereas water levels beyond the critical level reduce and can inhibit cracking susceptibility.

Anodic polarization increases the susceptibility of titanium to SCC in methanol/halide mixtures. On the other hand, cathodic polarization dramatically reduces SCC susceptibility. Potentials more negative than −250 mV versus Ag/AgCl prevent cracking in methanol (*Aust. Inst. Met.*, Vol 14 (No. 3), 1969, p 138).

Metal ion additions have also been examined and have been found to affect methanol SCC. In general, additions that have altered the cathodic reaction, such as palladium, chromium, iron, and gold, have accelerated cracking.

Other Alcohols and Organic Solvents. SCC susceptibility is primarily limited to methanol for most commercial titanium alloys, although a few highly susceptible alloys may also be affected by other low molecular weight alcohols. However, the addition of halogens, such as Cl_2, Br_2, or I_2, or other (non-oxygen containing) strong oxidizers (i.e., $FeCl_3$) to various anhydrous alcohols can induce SCC in all titanium alloys, even the unalloyed grades. Depending on the alloy and the oxidizer concentration, much higher water levels are required for SCC inhibition.

It is generally assumed that most commercial titanium alloys are resistant to SCC in most nonha-

CP Ti: Effect of Pd on SCC in methanol

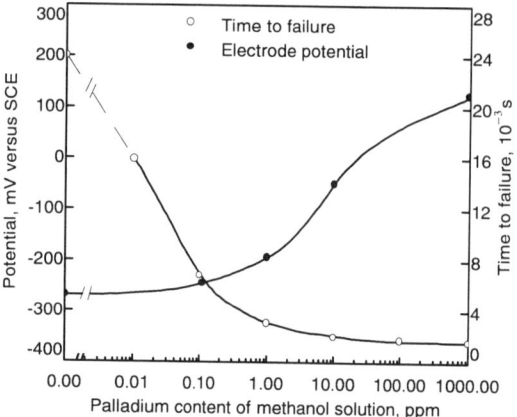

Source: *Metals Handbook, Corrosion*, Vol 13, 9th ed., ASM International, 1987, p 691

CP Ti: Effect of bromide and chlorides on SCC in methanol

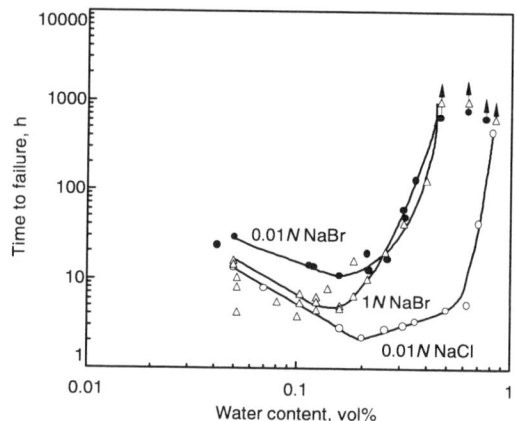

Cold rolled and annealed Ti, stressed to 75% of yield strength.
Source: *Metals Handbook, Corrosion*, Vol 13, 9th ed., ASM International, 1987, p 691

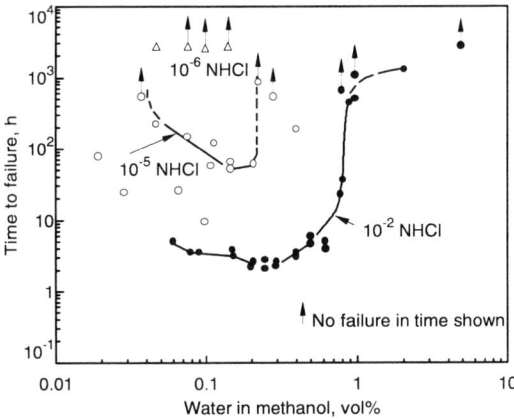

99.5 Ti: Effect of HCl and water on SCC in methanol

Cold rolled and annealed, 75% yield strength load.
Source: E.G. Haney and W.R. Wearmouth, Effect of 'Pure' Methanol on the Cracking of Titanium, *Corrosion*, Vol 25 (No. 2), 1969, p 87 and E.G. Haney and W.R. Wearmouth, Investigation of Stress-Corrosion Cracking of Titanium Alloys, Report No. 6, Research Grant NGR-39-008-014, NASA, May 1969

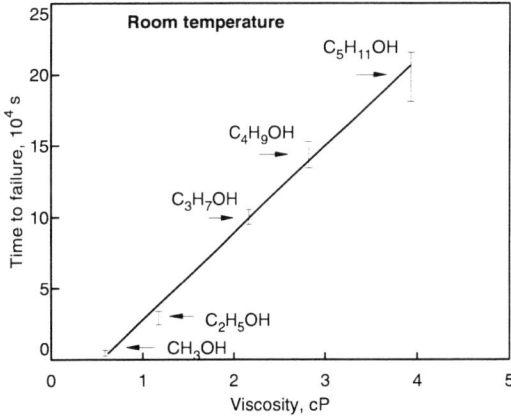

CP Ti: Time to failure vs viscosity in alcohol-iodine solutions

Source: A.J. Sedriks, Stress Corrosion Cracking in Alcohol-Iodine Solutions, *Corrosion*, Vol 25, 1969, p 207

logenated organic solvents. This includes ketones, aldehydes, ethers, hydrocarbons (alkanes, alkenes), and almost all alcohols (except for methanol). Worst-case testing involving anhydrous benzene and ethyl ether solvents containing 0.5% Br_2 additions revealed no SCC with unalloyed titanium (C.M. Chen, H.B. Kirkpatrick, and H.L. Gegel, *Proc. Int. Symp. Stress Corrosion Mechanisms in Titanium Alloys*, Jan 27-29, The Georgia Institute of Technology and NACE, 1971). Unalloyed and modified titanium are also resistant to SCC in chlorinated solvents and freons.

Liquid/Solid Metal Stress-Corrosion Cracking

Cadmium, mercury, zinc, and certain silver brazing alloys are known to cause cracking in titanium alloys. Susceptibility to cracking in cadmium (solid) occurs in commercially pure titanium with more than 0.2 wt% oxygen (ASTM grade 3), and liquid metal embrittlement (LME) from cadmium has been reported for grade 2 titanium. Penetration of the TiO_2 film by liquid cadmium is the critical factor.

Liquid mercury has produced embrittlement in ASTM grade 4 titanium. Exposures of stressed U-bend and C-ring samples of grades 2 and 12 and of Ti-3Al-8V-6Cr-4Zr-4Mo in liquid mercury reveal no cracking tendencies to temperatures as high as 230 °C (445 °F) (D.R. Klink and R.W. Schutz, Paper No. 63, presented at NACE Corrosion '92, Nashville, April 27-May 1, 1992). These static tests have shown that the surface oxide film of titanium is not wetted by liquid mercury, thereby precluding amalgamation and embrittlement. However, cracking failure may be induced when oxide films of stressed samples are severely mechanically damaged (heavily scraped or scratched) while immersed in the liquid metal. Enhanced oxide film barriers, achieved through anodizing or thermal oxidation, may offer improved resistance to liquid metal embrittlement in mercury, unless the film is physically compromised. The protective influence of a thermally nitrided, nonwetting titanium surface has similarly been demonstrated in mercury to temperatures as high as 370 °C (700 °F) (J.Y.N. Wang, Titanium and Titanium Alloys in Mercury—Some Observations on Corrosion and Inhibition, *Nucl. Sci. Engr.*, Vol 18, 1964, p 18-30).

Thermal Properties

CP and modified Ti: Liquidus temperatures

ASTM grade	Temperature °C	°F
Grade 1	1670 ± 5	3040 ± 10
Grade 2	1665 ± 5	3030 ± 10
Grade 3	1660 ± 10	3020 ± 20
Grade 4	1660 ± 10	3020 ± 20
Grade 7	1665 ± 5	3030 ± 10
Grade 11	1670 ± 5	3040 ± 10
Grade 12	~1660	~3020

Specific Heat

CP and modified Ti: Room-temperature specific heat

ASTM grade	Specific heat	
	J/kg · K	Btu/lb · °F
Grade 1	520	0.124
Grade 2	523	0.125
Grade 3	523	0.125
Grade 4	523-540	0.125-0.129
Grade 7	523	0.125
Grade 11	523	0.125
Grade 12	545-565	0.13-0.135

CP Ti: Effect of temperature on specific heat

Temperature		Specific heat	
°C	°F	J/kg · K	Btu/lb · °F
20	70	523	0.125
95	200	544	0.130
205	400	569	0.136
315	600	599	0.143
425	800	632	0.151
540	1000	670	0.160
650	1200	716	0.171
760	1400	766	0.183
845	1550	808	0.193

Source: M. Mote, R. Hooper, and P. Frost, "Report on the Engineering Properties of Commercial Titanium Alloys," TML Report 92, Titanium Metallurgical Laboratories, Battelle Memorial Institute, June 1958, p W-1

CP Ti: Specific heat vs temperature

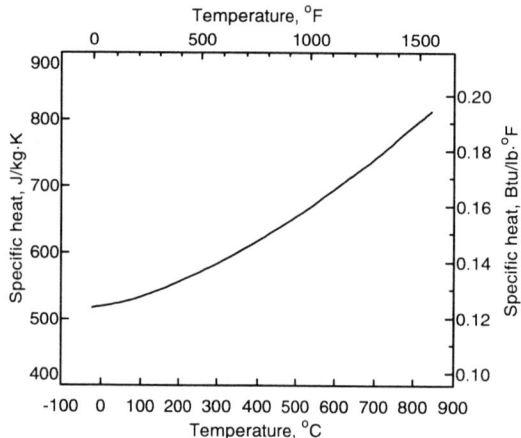

Source: MIL-HDBK-5

CP Ti: Specific heat vs temperature

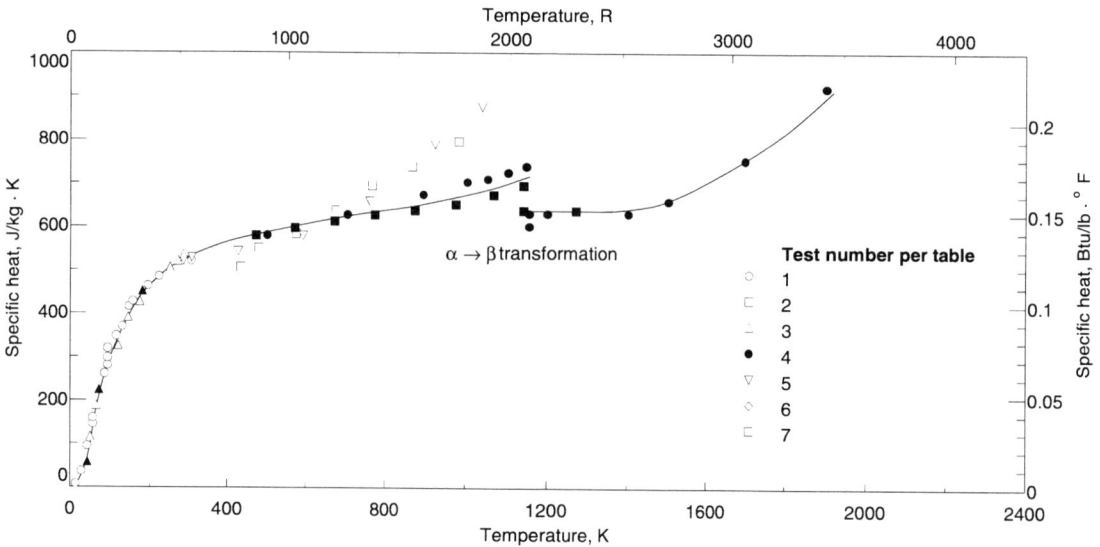

Source: A. Goldsmith, T.E. Waterman, and H.J. Hirshhorn, Ed., *Thermophysical Properties of Solid Materials*, Vol 1, Wright Air Development Division, Aug 1960. See table for description of test numbers.

Test number	Material/composition, wt%	Test method	Temperature K	Temperature R	Remarks
1	Iodide Ti: 0.0082 Mn, 0.007 Si, 0.0066 Al, 0.02 (N + Te + Pb + Cu)	Guarded specimen: Nernst-type vacuum calorimeter	16-305	28-550	Annealed in high vacuum at 800 °C (1470 °F)
2	99.9% pure, high-purity commercial grade	Comparative method: Measured rate of temperature increase in specimen compared with that in standard under same heating conditions	425-975	762-1752	Estimated accuracy, ± 18%
3	98.75% Ti: 98.75 Ti, 0.50 Si, 0.27 Fe, 0.15 V	Guarded specimen	54-295	97-532	Data corrected for impurities
4	99.96% Ti: 99.96 Ti, 0.009 Si, 0.0082 Mn, 0.0066 Al	Drop method: Copper block calorimeter	300-1900	536-3420	Estimated accuracy, ± 0.5%
5	99.75% Ti: 99.75 Ti, 0.131 O, 0.07 Fe, 0.06 C, 0.048 N, 0.0068 H_2	Drop method: Copper calorimeter	310-1035	560-1860	
6	Impurities in ppm: 320 C, 300 Fe, 110 O_2, 67 H, 10 Cu, 7.7 N_2	Guarded specimen	325-1275	582-2292	Run in vacuum
7	Not given	Not described here; refers to others	20-195	36-351	Tabular data from author's smoothed curve

Source: A. Goldsmith, T. Waterman, and H. Hirschhorn, Ed., *Thermophysical Properties of Solid Materials*, Vol 1, MacMillan, 1961, p I-T-6

Thermal Expansion

The coefficient of expansion is not greatly changed by elevated temperatures, heat treatment or impurities. However, the hexagonal crystal structure of α titanium causes some variation in coefficient of thermal expansion, depending on the direction of measurement. Expansion along the c-axis is about

Unalloyed Ti: Thermal expansion in crystallographic directions

Temperature °C	Temperature °F	Direction	Expansion(a), μm/m · K
20-400	68-750	Perpendicular to c-axis	10.2
20-700	68-1290	Perpendicular to c-axis	11.0
20-700	68-1290	Along c-axis	12.8

(a) Calculated from lattice parameters. Source: R. Berry and G. Raynor, *Research*, Vol 6, 1953, p 21S

CP and modified Ti: Typical thermal coefficients of linear expansion

Temperature range		Mean coefficient(a) for: ASTM grades 1, 2, 3, and 4		ASTM grade 12 (Ti-0.8Ni-0.3Mo)	
°C	°F	10^{-6}/°C	10^{-6}/°F	10^{-6}/°C	10^{-6}/°F
0-100	32-212	8.6	4.8	...	...
0-315	32-600	9.2	5.1	9.5	5.3
0-540	32-1000	9.7	5.4	...	...
0-650	32-1200	10	5.6	...	...
0-815	32-1500	10	5.6	...	...

(a) Mean coefficient between 0 °C (32 °F) and temperature indicated. Similar coefficients are expected for Ti-0.2Pd grades. Source: RMI Basic Design

ASTM grades 3 and 4: Variation of thermal coefficients of linear expansion

Temperature range		Mean coefficient(a) for: ASTM grade 3		ASTM grade 4	
°C	°F	10^{-6}/°C	10^{-6}/°F	10^{-6}/°C	10^{-6}/°F
20 to −240	68 to −400	...	...	1.26	0.7
−130	68 to −200	...	...	5.9	3.3
−75	68 to −100	7.9	4.4	7.0	3.9
−18	68 to 0	8.6	4.8	7.7	4.3
20 to 95	68 to 200	8.6	4.8	8.6	4.8
to 200	68 to 400	9.3	5.2	9.3	5.2
to 315	68 to 600	9.5	5.3	9.3	5.2
425	68 to 800	9.7	5.4	9.5	5.3
20 to 540	68 to 1000	9.9	5.5	9.7	5.4
to 650	to 1200	10.1	5.6	10.1	5.6
760	to 1400	10.2	5.7	10.1	5.6
870	to 1600	10.4	5.8	10.2	5.7
980	to 1800	10.6	5.9	10.8	6.0

(a) Mean coefficient between 20 °C (68 °F) and indicated temperature for Crucible A-55 and A-70 (ASTM grades 3 and 4, respectively). Source: Crucible Datasheet

20% greater than along the a-axis.

The coefficient of expansion for β titanium (from room temperature to 1000 °C, or 1830 °F) has been reported to be 10.9×10^{-6} per °C (5.6×10^{-6} per °F), which is lower than for α titanium (10.9×10^{-6} mm/mm · °C, or 6.1×10^{-6} in/in. · °F) for the same temperature range (W. Jepkowski and J. Holladay, "Report on the Physical Properties of Titanium and Titanium Alloys," TML 73, Battelle Memorial Institute, 1957, p 18).

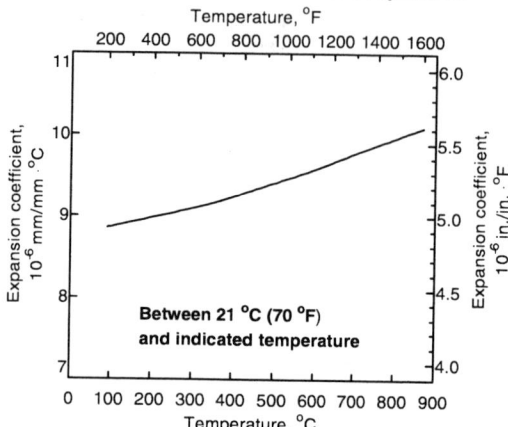

CP Ti: Thermal coefficients of linear expansion

Source: MIL-HDBK-5

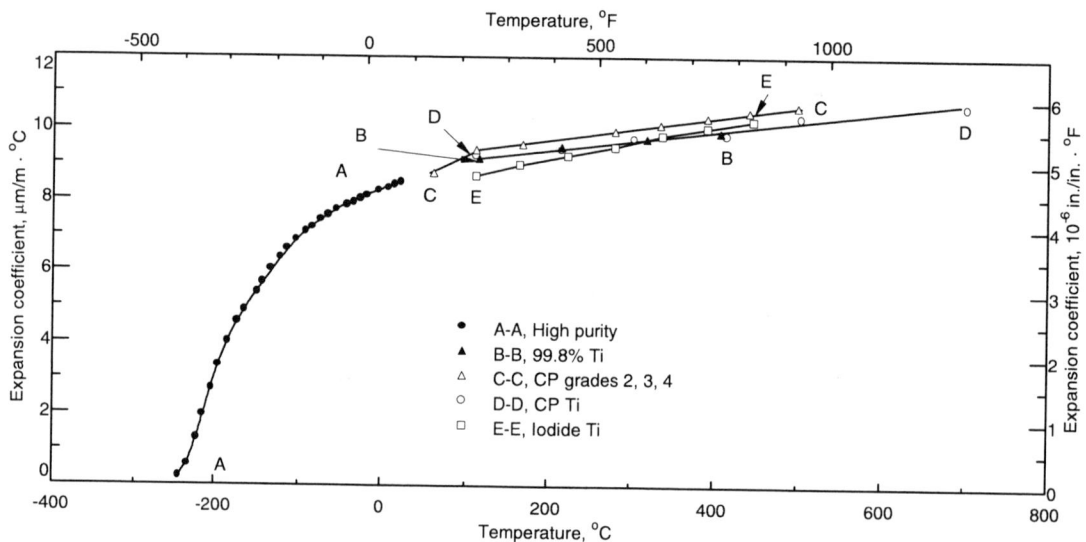

Unalloyed Ti: Effect of impurities on thermal expansion

- A-A, High purity
- B-B, 99.8% Ti
- C-C, CP grades 2, 3, 4
- D-D, CP Ti
- E-E, Iodide Ti

Curve A-A: Ultrahigh-purity titanium (W. Altman, T. Rubin, and H. Johnson, "Coefficient of Thermal Expansion of Solids at Low Temperatures. III. Thermal Expansion of Pure Metals with Data for Aluminum, Nickel, Titanium, and Zirconium," Technical Report TR-254-27, The Ohio State University, 1954). Curve B-B: 99.8% titanium, excluding O, N, C (E. Greiner and W. Ellis, "Thermal and Electrical Properties of Ductile Titanium," Met. Tech., Vol 15, 1948). Curve C-C: CP titanium, grades 2, 3, and 4 (Rem-Cru Titanium, Midland, PA). Curve D-D: CP Titanium (Imperial Chemical Industries, 1955). Curve E-E: Iodide titanium (H. Adenstedt, "Handbook on Titanium," WADC Technical Report 54-305, Part I, Aug 1954).
Source: W. Lepkowski and J. Holladay, "Report on the Physical Properties of Titanium and Titanium Alloys," TML 73, Battelle Memorial Institute, July 1957, p 19

Thermal Conductivity

CP and modified Ti: Thermal conductivity at room temperature

Ti grade	Typical RT thermal conductivity	
	W/m · K	Btu/ft · h · °F
ASTM grade 1 (0.18 O, 0.2 Fe wt% max)	22.0	12.7
ASTM grade 2 (0.25 O, 0.3 Fe wt% max)	21.8	12.6
ASTM grade 3 (0.35 O, 0.3 Fe wt% max)	21.8	12.6
ASTM grade 4 (0.40 O, 0.5 Fe wt% max)	17	9.8
ASTM grade 7 (grade 2 + 0.2 Pd)	16.4	9.5
ASTM grade 11 (grade 1 + 0.2 Pd)	...	...
ASTM grade 12 (Ti-0.3Mo-0.8Ni)	19	11.0

Source: S. Williams, "Report on Titanium: The Ninth Major Industrial Metal," Brundage, Story and Rose Research Reports, Series A-6, New York, 1965; and Alloys Digest, and with corrected values for Grades 1, 2, and 3 from Tom O'Connel, Timet

CP Ti: Thermal conductivity of various grades

ASTM grade	Alloy designation	−240 °C (−400 °F)	−185 °C (−300 °F)	−130 °C (−200 °F)	−75 °C (−100 °F)	0 °C (32 °F)	20 °C (68 °F)	95 °C (200 °F)	205 °C (400 °F)	315 °C (600 °F)	425 °C (800 °F)	535 °C (1000 °F)	650 °C (1200 °F)	760 °C (1400 °F)	815 °C (1500 °F)
CP Ti	...	12.9 (7.5)	17.6 (10.2)	20.0 (11.6)	20.9 (12.1)	19.9 (11.5)	...	...	...	...	...	...	...	...	...
CP Ti	...	...	...	...	...	...	...	15.2 (8.80)	14.7 (8.53)	14.3 (8.28)	13.9 (8.04)	13.5 (7.81)	13.1 (7.6)	...	...
Grade 3	A-55	...	...	...	20.4 (11.8)	...	19.9 (11.5)	18.6 (10.9)	18.0 (10.4)	18.1 (10.5)	18.1 (10.5)	18.5 (10.7)	19.5 (11.3)	20.9 (12.1)	21.6 (12.5)
Grade 3	A-55	...	...	...	...	...	...	19.3 (11.2)	18.8 (10.9)	18.3 (10.6)	17.8 (10.3)	17.1 (9.9)	...	...	...
Grade 3	RS-55	...	...	...	...	15.7 (9.1)	...	...	...	...	17.8	...	...	...	...
Grade 4	A-70	...	...	...	17.8 (10.3)	...	17.4 (10.1)	16.9 (9.8)	16.9 (9.8)	16.9 (9.8)	17.3 (10.0)	18.1 (10.5)	19.2 (11.1)	20.6 (11.9)	21.4 (12.4)
Grade 4	Ti-75A	...	...	...	...	17.0 (9.84)	...	...	...	...	...	...	...	...	...
Grade 4	Ti-75A	...	...	...	...	...	...	23.2 (13.4)	22.8 (13.2)	22.5 (13.0)	22.1 (12.8)	21.9 (12.7)	21.8 (12.6)	...	...
Grade 4	Ti-75A	...	...	...	...	...	...	20.2 (11.7)	19.9 (11.5)	19.5 (11.3)	19.5 (11.3)	19.2 (11.1)	19.0 (11.0)	...	...

Source: TML Report 73, Battelle, 1957

ASTM grade 2 Ti: Thermal conductivity and electrical resistivity at high temperatures

Temperature		Thermal conductivity(a)		Resistivity,
°C	°F	W/m · K	Btu/ft · h · °F	μΩ · cm
40	105	20.8	12.0	49.60
50	120	20.5	11.8	51.57
99	210	19.6	11.3	60.93
174	345	19.5	11.2	75.13
255	490	18.7	10.8	89.85
319	650	18.4	10.6	101.03

(a) Accurate within 5%, obtained by the Kohlrausch method. Source: R. Taylor and H. Groot, "Thermal Conductivity of Titanium and Titanium Alloys," Thermophysical Properties Research Laboratories, Purdue University, report to RMI, 1986

Unalloyed Ti: Comparison of thermal conductivity

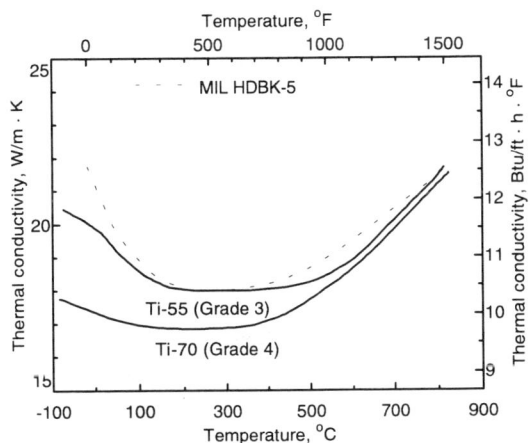

Source: *Aerospace Structural Metals Handbook*, Vol 14, Battelle, 1972

CP Ti: Thermal conductivity of various grades

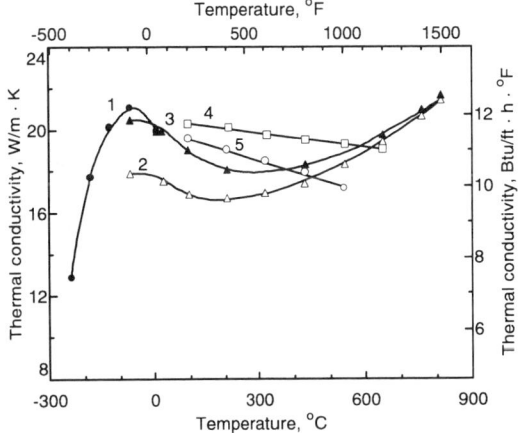

Curve 1: See original source for material composition. Curve 2: A-70 titanium, 99.0 wt% Ti, grade 4. Curve 3: A-55 titanium, 99.1 wt% Ti, grade 3. Curve 4: Ti-75A titanium, 99.0 wt% Ti, grade 4. Curve 5: A-55 titanium, 99.1% wt% Ti, grade 3.
Source: A. Goldsmith, T. Waterman, and H. Hirschhorn, Ed., MacMillan, 1961. Curve 1 from E. Rigney and L. Bockstahler, "The Thermal Conductivity of Titanium Between 20 and 273 Degrees Kelvin," *Phys. Rev.*, Vol 83, 1951, p 220; and G. Skinner, H. Johnston, and C. Beckett, *Titanium and Its Compounds*, H.L. Johnston Enterprises, Columbus, Ohio, 1954. Curves 4 and 5 from E. Loewen, "Thermal Properties of Titanium Alloys and Selected Tool Steels," *Trans. ASME*, Vol 78, 1956, p 667

Mechanical Properties

CP and Modified Ti: Minimum room-temperature tensile properties for various specifications

Designation	Chemical composition,% max				Tensile properties(a)				Minimum elongation, %
	C	O	N	Fe	Ultimate strength MPa	ksi	Yield strength MPa	ksi	
JIS Class 1	...	0.15	0.05	0.20	275-410	40-60	165(b)	24(b)	27
ASTM grade 1 (UNS R50250)	0.10	0.18	0.03	0.20	240	35	170-310	25-45	24
DIN 3.7025	0.08	0.10	0.05	0.20	295-410	43-60	175	25.5	30
GOST BT1-00	0.05	0.10	0.04	0.20	295	43	...	...	20
BS 19-27t/in.^{2}	...	...	...	0.20	285-410	41-60	195	28	25
JIS Class 2	...	0.20	0.05	0.25	343-510	50-74	215(b)	31(b)	23
ASTM grade 2 (UNS R50400)	0.10	0.25	0.03	0.30	343	50	275-410	40-60	20
DIN 3.7035	0.08	0.20	0.06	0.25	372	54	245	35.5	22
GOST BT1-0	0.07	0.20	0.04	0.30	390-540	57-78	...	...	20
BS 25-35t/in.^{2}	...	...	...	0.20	382-530	55-77	285	41	22
JIS Class 3	...	0.30	0.07	0.30	480-617	70-90	343(b)	50(b)	18
ASTM grade 3 (UNS R50500)	0.10	0.35	0.05	0.30	440	64	377-520	55-75	18
ASTM grade 4 (UNS R50700)	0.10	0.40	0.05	0.50	550	80	480	70	15
DIN 3.7055	0.10	0.25	0.06	0.30	460-590	67-85	323	47	18
ASTM grade 7 (UNS R52400)	0.10	0.25	0.03	0.30	343	50	275-410	40-60	20
ASTM grade 11 (UNS R52250)	0.10	0.18	0.03	0.20	240	35	170-310	25-45	24
ASTM grade 12 (UNS R53400)	0.10	0.25	0.03	0.30	480	70	380	55	12

(a) Unless a range is specified, all listed values are minimums. (b) Only for sheet, plate, and coil

Design Allowables

ASTM grades 7 and 11 have the same design mechanical properties as ASTM grades 2 and 1, respectively.

Unalloyed Ti: Design tensile properties for sheet, strip, and plate

ASTM grade	Designation(a)	Ultimate tensile strength (L-LT)(b) MPa	ksi	Tensile yield strength (L-LT)(b) MPa	ksi	Elongation(c),% L	LT
Grade 1 (0.18 O, 0.2 Fe wt% max)	CP4	240	35	170	25	24	24
Grade 2 (0.25 O, 0.3 Fe wt% max)	CP3	345	50	275	40	20	20
Grade 3 (0.35 O, 0.3 Fe wt% max)	CP2	445	65	380	55	18	18
Grade 4 (0.40 O, 0.5 Fe wt% max)	CP1	550	80	480	70	15	15

(a) Designation of corresponding ASTM grades in MIL-T-9046. (b) S-basis properties in both the longitudinal and long-transverse directions for thicknesses less than 25 mm (1 in.). (c) S-basis elongation for thicknesses of 0.635 to 25 mm (0.025 to 1 in.). Source: MIL-HDBK-5

CP Ti: S-basis tensile properties of annealed extruded bar and shapes

Property	Minimum longitudinal properties for extruded bars and shapes(a) less than 75 mm (3 in.) thick			
	Grade 1	Grade 2	Grade 3	Grade 4
Ultimate tensile strength, MPa (ksi)	275 (40)	345 (50)	445 (65)	550 (80)(b)(c)
Tensile yield strength, MPa (ksi)	205 (30)	275 (40)	380 (55)	480 (70)(b)(c)
Elongation,% for extrusions:,				
4.77 to 25 mm (0.188 to 1.000 in.) thick	25	20	18	15(b)(c)
25 to 50 mm (1.001 to 2.000 in.) thick	20	18	15	12
50 to 75 mm (2.001 to 3.000 in.) thick	18	15	12	10

(a) Extrusions per MIL-T-81556. (b) For rolled bar per MIL-T-9047, the minimums for ASTM grade 4 apply to sections up to 100 mm (4 in.) thick with a maximum 103 cm^2 (16 in.^{2}) cross section. (c) Minimums also apply to long-transverse direction in rolled rectangular bar 13 to 75 mm (0.5 to 3 in.) thick per MIL-T-9047. Source: MIL-HDBK-5, 1 Dec,1991

ASTM grade 4 Ti: S-basis design mechanical properties

Property	Rolled bar <100 mm (4 in.) thick(a)	Sheet, strip, plate <25 mm (1 in.) thick (b)
Ultimate tensile strength, MPa (ksi)	550 (80)(c)	550 (80) (L + LT)
Tensile yield strength, MPa (ksi)	480 (70)(c)	480 (70) (L + LT)
Elongation,%	15(c)	15 (d) (L + LT)
Reduction of area,%	30(c)	...
Compressive yield strength (L + LT), MPa (ksi)	...	480 (70)
Ultimate shear strength, MPa (ksi)	...	290 (42)
Ultimate bearing strength (e/D = 1.5), MPa (ksi)	...	825 (120)
Bearing yield strength (e/D = 1.5), MPa (ksi)	...	695 (101)

(a) Bar with a maximum cross section of 103 cm^2 (16 in.^{2}) per MIL-T-9047. (b) Longitudinal and long-transverse properties per MIL-T-9046 as indicated. (c) Applicable in ST direction for thickness above 75 mm (3 in.). (d) Applicable for thickness above 0.635 mm (0.025 in.). Source: MIL-HDBK-5

Ti-0.3Mo-0.8Ni: Minimum RT mechanical properties

Property	Value	Product form
Ultimate tensile strength, MPa (ksi)	480 (70)	All
Tensile yield strength (0.2%), MPa (ksi)	345 (50)	All
Elongation in 50 mm, or 2 in.,%	18	All
Reduction of area,%	25	Bar, billet, forgings
Bend radius (R/T)	2.0	T < 1.8 mm (0.070 in.)
	2.5	T = 1.8-4.7 mm (0.070-0.187 in.)
Flattening ability		
Distance between platens	9× Nominal wall thickness	Tubing
Flaring ability,%		
Expansion inside diameter	17	Tubing

Note: Minimum mechanical properties generally are used for material acceptance purposes and should not be construed as design allowables.

Effect of Impurities

Basically, oxygen and iron contents determine strength levels of commercially pure titanium. In higher strength grades, oxygen and iron are intentionally added to the residual amounts already in the sponge to provide extra strength. Conversely, carbon and nitrogen usually are kept to minimum residual levels to avoid embrittlement.

CP Ti: Effect of interstitials on weld ductility

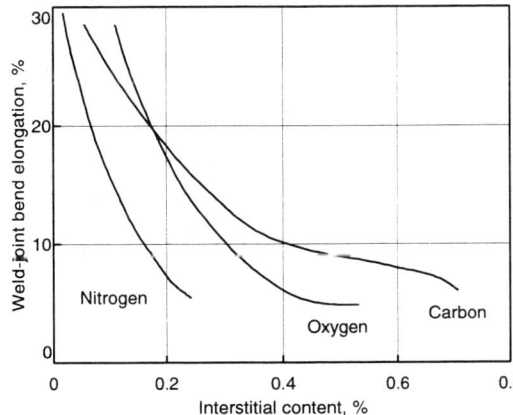

Source: *Metals Handbook*, Vol 3, 9th ed., American Society for Metals, 1980

CP Ti: Effect of interstitials on strength

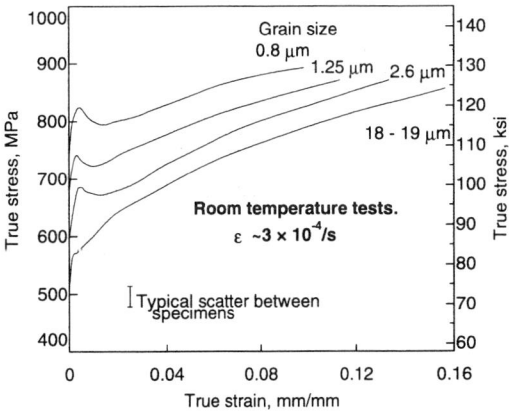

Source: *Metals Handbook*, Vol 3, 9th ed., American Society for Metals, 1980

Hardness

ASTM grade 2 Ti: Hardness comparison with other materials
Typical hardness values for condenser tube materials

Alloy	Hardness(a), HB	Hardness(b), HV
Titanium (ASTM grade 2)	180	145
Admiralty brass (annealed)	75	60
Aluminum brass	82	85
90-10 cupronickel (annealed)	65	75
70-30 cupronickel (annealed)	70	95
18-8 stainless steel (ST)	...	142
Type 304 stainless steel	180	...

(a) "CodeWeld Titanium Tubing," TIMET, Pittsburgh, p 7. (b) "Sumitomo Titanium," Sumitomo Metal Industries, Tokyo, p 14

CP Ti: Average Brinell and Vickers hardness at 20 °C (68 °F)

Specimen	Hardness, HB	Hardness, HV
JIS-1	105.5	110.4
JIS-2	121.0	127.1
JIS-3	166.0	182.5

Note: Each of six companies tested specimens and compared them to a standard specimen at four fixed points using JIS Z 2243, "Brinell Hardness Test," and JIS Z 2244, "Vickers Hardness Test," at temperatures of 10 °C, 20 °C, and 30 °C (50, 68, and 86 °F). Source: M. Nagai, K. Kashida, K. Ono, Y. Moriguchi, K. Murase, K. NOri, M. Koizui, N. Tsuruoka, T. Mimura, and S. Kashu, "Cooperative Examination of the Influence of Atmospheric Temperature in the Hardness of Titanium Mill Products," in *Titanium '80, Science and Technology*, H. Kimura and O. Izumi, Ed., TMS/AIME, 1980, p 1131

CP and modified Ti: Typical hardness of ASTM grades

ASTM grade	Condition	Hardness, HB	Hardness, HRB	Hardness, HK
Grade 1	Annealed	120	70	...
	unwelded sheet,	...	63.5	140
	single-bead weld	...	55.8	140
Grade 2	Annealed	200	80	...
	unwelded sheet,	...	80.6	165
	single-bead weld	...	83.1	175
Grade 3	Annealed	225	90	...
	unwelded sheet,	...	94.4	175
	single-bead weld	...	92.4	220
Grade 4	Annealed	265	100	...
	unwelded sheet,	...	23.4	215
	single-bead weld	...	21.2	240
Grade 7	Annealed	200	70-80	...
Grade 12	...	180-235	...	...

Source: *Metals Handbook, Properties and Selection: Stainless Steels, Tool Materials, and Special-Purpose Materials*, Vol 3, 9th ed., 1980, p 368-380; and *Metals Handbook, Properties and Selection: Nonferrous Alloys and Special Purpose Materials*, Vol 2, 10th ed., 1990, p 621

ASTM grade 2 Ti: Effect of processing on hardness

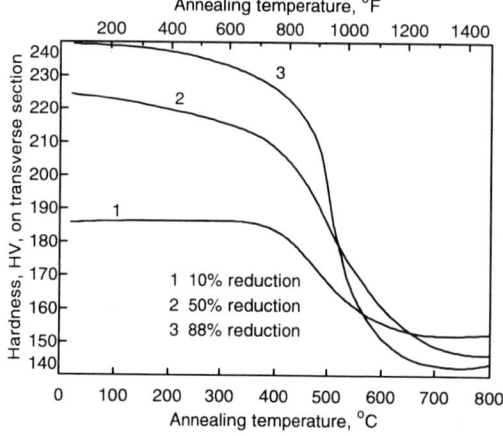

Effect of annealing for 30 min on cold worked specimens.
Source: "IMI Commercially Pure Titanium," IMI Titanium, Birmingham, UK, p 13

Knoop hardness vs indenter orientation for sheet material

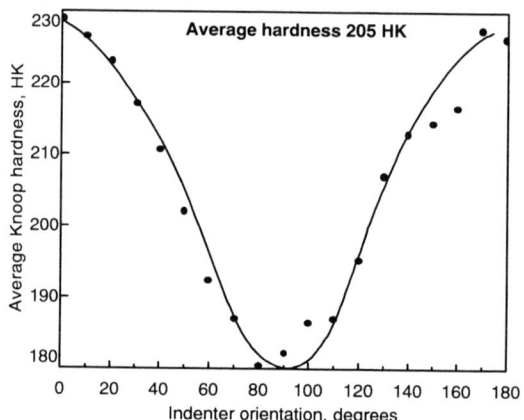

Deviation from longitudinal plane.
Source: A. Zarkades, "The Anisotropy of Knoop Hardness in Unalloyed Titanium Sheet," Army Materials Research Agency, AMRA-TR-67-04, January 1967; reported in *Properties of Textured Titanium Alloys*, F. Larson and A. Zarkades, Ed., MCIC-74-20, Metals and Ceramics Information Center, Battelle Columbus Laboratories,

Knoop hardness vs indenter orientation for sheet material

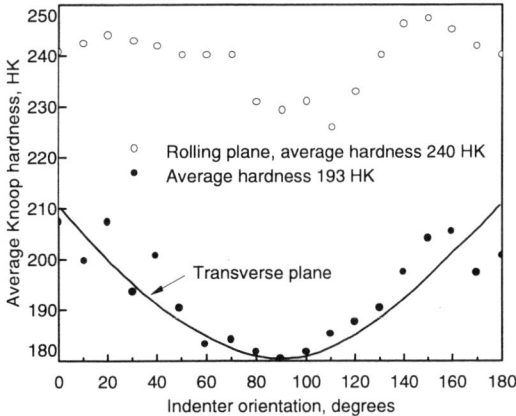

Deviation from transverse and rolling plane.
Source: A. Zarkades, "The Anisotropy of Knoop Hardness in Unalloyed Titanium Sheet," Army Materials Research Agency, AMRA-TR-67-04, January 1967; reported in *Properties of Textured Titanium Alloys*, F. Larson and A. Zarkades, Ed., MCIC-74-20, Metals and Ceramics Information Center, Battelle Columbus Laboratories, 1974, p 47

High-Temperature Strength

Design Allowables

CP grade 2 Ti: Typical mechanical strengths and ASME allowables

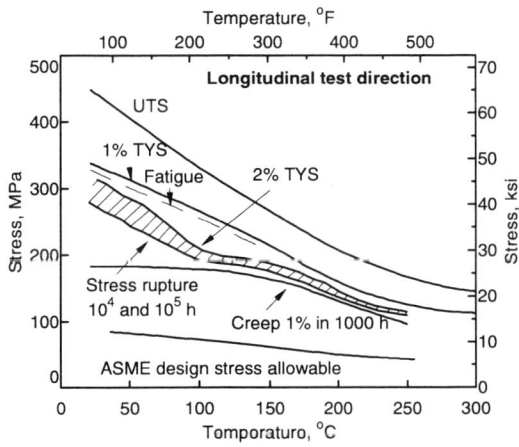

CP and modified Ti: Allowable stresses per ASME

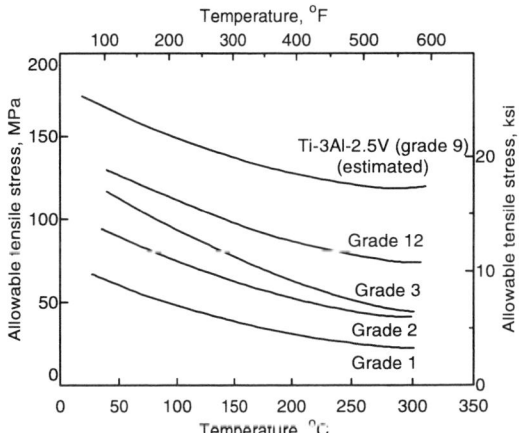

Relative Costs

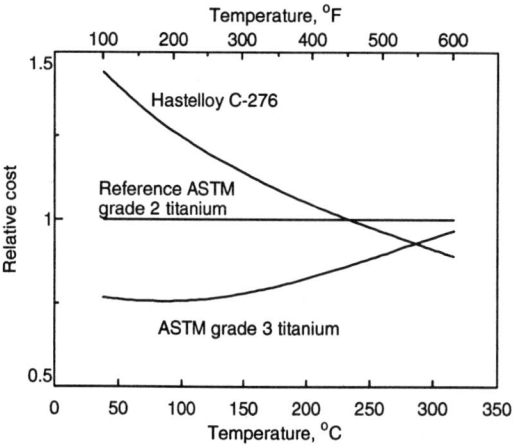

CP grade 2: Relative cost vs temperature

Source: P.A. Russo and J.D. Schöbel, "Mechanical Properties of Commercially Pure Titanium Containing Low Iron," presentation at ACHEMA 82, 1982

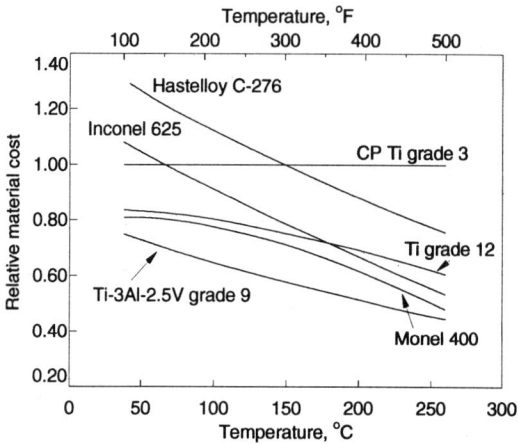

CP grade 3: Relative cost vs temperature

Source: *Industrial Applications of Titanium and Zirconium: Third Conference*, R.T. Webster and C.S. Young, Ed., STP 830, ASTM,

Typical Strengths vs Temperature

ASTM grade 3 Ti: Typical mechanical properties at various temperatures

Temperature		Ultimate tensile strength		Tensile yield strength(a)		Compressive yield strength		Shear strength		Ultimate bearing strength		Bearing yield strength(b)	
°C	°F	MPa	ksi	MPa	ksi	MPa	ksi	MPa	ksi	MPa	ksi	MPa	ksi
−253	−423	1210	175	...	...	...	...	...	...	...	...	...	...
−196	−320	1140	165	...	...	...	...	...	...	...	...	...	...
−54	−65	690	100	510	74	...	...	...	...	...	...	...	...
21	70	550	80	450	65	450	65	380	55	680	99	565	82
93	200	440	64	340	49	315	46	315	46	580	84	455	66
204	400	310	45	200	29	200	29	240	35	435	63	345	50
316	600	250	36	140	20	200	29	195	28	345	50	240	35
427	800	200	29	115	17	180	26	160	23	290	42	220	32
538	1000	165	24	90	13	140	20	110	16	180	26	145	21

(a) 0.2% offset. (b) 2% permanent set. Source: *Alloy Digest*, Datasheet on Crucible A-55, July 1971

ASTM grade 4: Typical mechanical properties vs temperature

Temperature		Ultimate tensile strength		Tensile yield strength(a)		Compressive yield strength		Shear strength		Ultimate bearing strength		Bearing yield strength(b)	
°C	°F	MPa	ksi	MPa	ksi	MPa	ksi	MPa	ksi	MPa	ksi	MPa	ksi
−253	−423	1275	185	...	...	...	...	...	...	...	...	...	...
−196	−320	1205	175	...	...	...	...	...	...	...	...	...	...
−54	−65	760	110	580	84	...	...	...	...	...	...	...	...
21	70	630	90	550	80	550	80	450	65	830	120	695	101
93	200	495	72	415	60	385	56	380	55	700	102	560	81
204	400	345	50	250	36	250	36	390	42	530	77	390	57
316	600	275	40	165	24	240	35	220	32	415	60	295	43
427	800	220	32	145	21	220	32	185	27	345	50	270	39
538	1000	185	27	110	16	165	24	140	20	215	31	170	26

(a) 0.2% offset. (b) 2% permanent set. Source: *Alloy Digest*, July 1971

ASTM grade 12: Typical tensile properties

Temperature		Ultimate tensile strength		Tensile yield strength (0.2%)		Elongation, %
°C	°F	MPa	ksi	MPa	ksi	
25	77	510	74	415	60	33
205	400	345	50	250	36	37
316	600	325	47	205	30	32

Source: *Metals Handbook*, 9th ed., Vol 3, 1980, p 381

CP and modified Ti: Typical UTS vs temperature

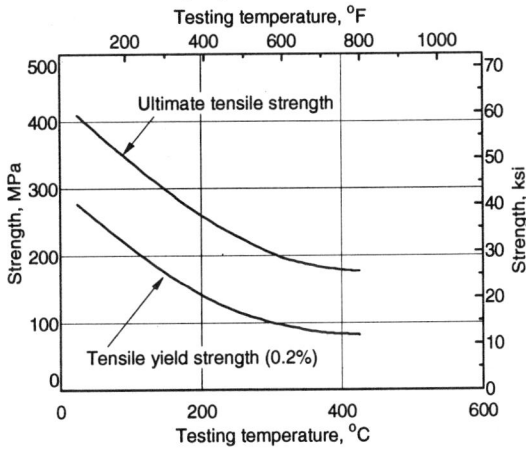

Source: *Industrial Applications of Titanium and Zirconium: Third Conference*, R.T. Webster and C.S. Young, Ed., STP 830, ASTM, 1984

99.2% Ti: Tensile properties vs temperature

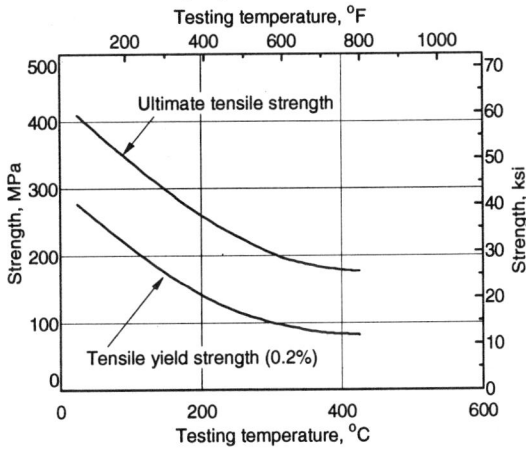

Source: *Metals Handbook, Properties and Selection*, Vol 1, 8th ed., American Society for Metals, 1961, p 537

Percent of Room-Temperature Strengths

CP and modified Ti: Percent drop in RT UTS

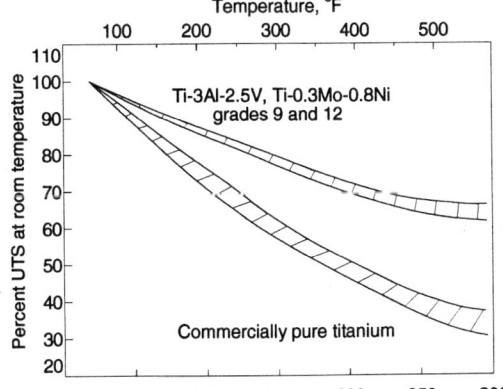

Source: *Industrial Applications of Titanium and Zirconium: Third Conference*, R.T. Webster and C.S. Young, Ed., STP 830, ASTM, 1984

ASTM grades 2, 7, and 12: Minimum tensile strength

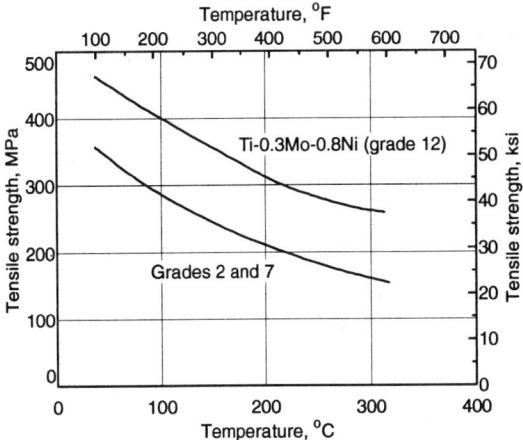

Source: *Metals Handbook*, Vol 3, 9th Ed., American Society for Metals, 1980, p 380

Low-iron grade 2 Ti: Tensile strength vs temperature

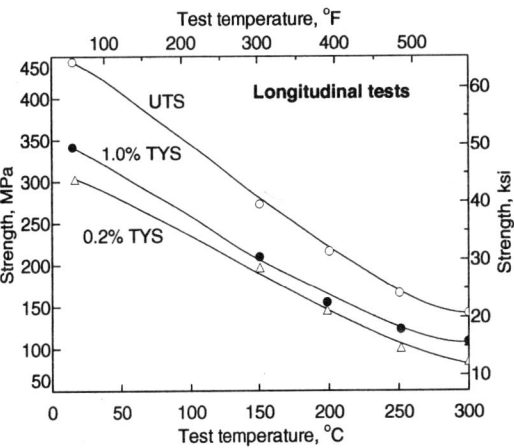

Commercially pure low-iron grade 2 titanium plate containing 0.03% iron or less were double vacuum melted by the consumable electrode arc melting process. The ingots were forged to slabs and rolled to 44.4 or 50.8 mm (1.7 or 2 in.) thick plate product. The plate product was annealed at 730 °C (1345 °F) for 30 min and air cooled. Source: P.A. Russo and J.D. Schöbel, "Mechanical Properties of Commercially Pure Titanium Containing Low Iron," presentation at ACHEMA 82, 1982

CP Ti: UTS vs temperature

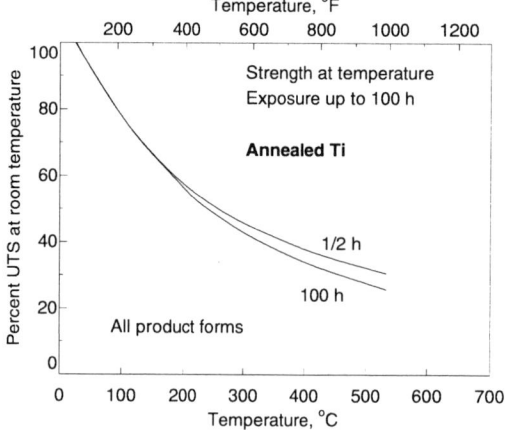

Source: MIL-HDBK-5

CP Ti: Tensile yield strength vs temperature

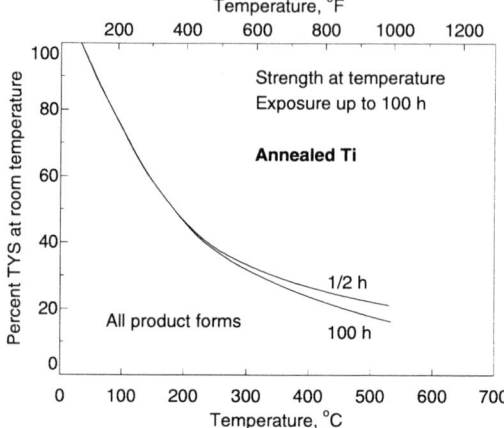

Source: MIL-HDBK-5

Low-iron CP Ti: Strength drop vs temperature

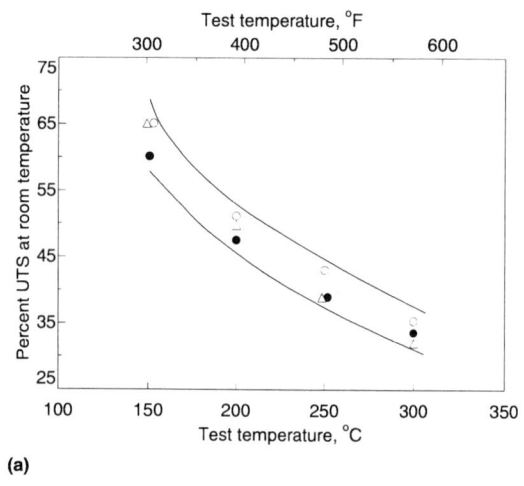
(a)

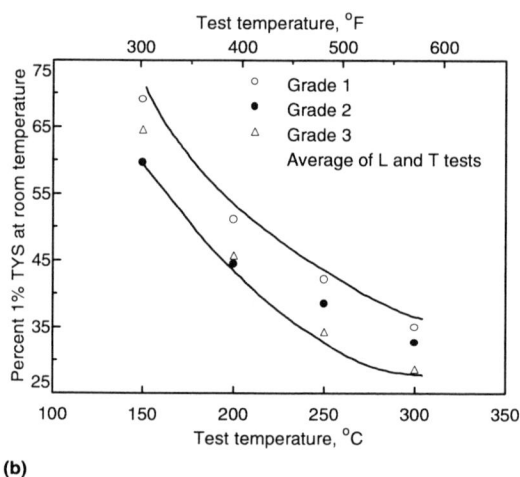
(b)

The 1% yield strength and ultimate strengths were 60 to 70% of room-temperature properties at 150 °C (300 °F), but were only 30 to 40% room-temperature properties at 300 °C (570 °F). Grade 1 maintains a higher percentage of room-temperature yield and ultimate strengths up to 300 °C (570 °F) than grades 2 and 3. This is probably due to the fact that strengthening due to interstitials diminishes with increasing temperature.
Three grades of commercially pure low-iron titanium plate containing 0.03% iron or less were double vacuum melted by the consumable electrode arc melting process. The ingots were forged to slabs and rolled to 44.4 or 50.8 mm (1.7 or 2 in.) thick plate product. The plate product was annealed at 730 °C (1345 °F) for 30 min and air cooled.
Source: P.A. Russo and J.D. Schöbel, "Mechanical Properties of Commercially Pure Titanium Containing low Iron," presentation at ACHEMA 82, 1982

Compressive Strength

Unalloyed Ti ASTM (grade 3 equivalent): Typical compressive and tensile strength vs temperature

Temperature		Compressive yield strength (0.2% offset)			Tensile yield strength (0.2% offset)		
°C	°F	MPa	ksi	% of RT values	MPa	ksi	% of RT values
21	70	448	65.0	100	448	65	100
95	200	313	45.5	70	336	48.75	75
205	400	201	29.2	45	201	29.25	45
315	600	197	28.6	44	134	19.5	30
425	800	179	26.0	40	116	16.9	26
540	1000	134	19.5	30	89	13	20

Note: Chemical composition: 0.02 wt% C, 0.20 wt% Fe, 0.005 wt% H, 0.01 wt% N, and 0.20 wt% O. Source: Crucible Data Sheet, Crucible Specialty Metals

Unalloyed Ti (ASTM grade 4 equivalent): Typical compressive and tensile strength vs temperature

Temperature		Compressive yield strength (0.2% offset)			Tensile yield strength (0.2% offset)		
°C	°F	MPa	ksi	% of RT values	MPa	ksi	% of RT values
20	70	552	80	100	552	80	100
95	200	386	56	70	413	60	75
205	400	248	36	45	248	36	45
315	600	243	35.2	44	165	24	30
425	800	221	32	40	143	20.8	26
540	1000	165	24	30	110	16	20

Chemical composition: 0.02 wt% C, 0.35 wt% Fe, 0.005 wt% H, 0.01 wt% N, and 0.35 wt% O.
Source: Crucible Data Sheet, Crucible Specialty Metals, Syracuse, NY

CP Ti: Compressive yield strength vs temperature

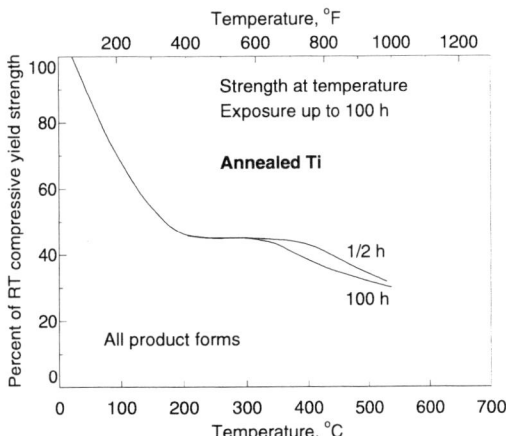

Source: MIL-HDBK-5

CP Ti: Compressive yield strength vs temperature

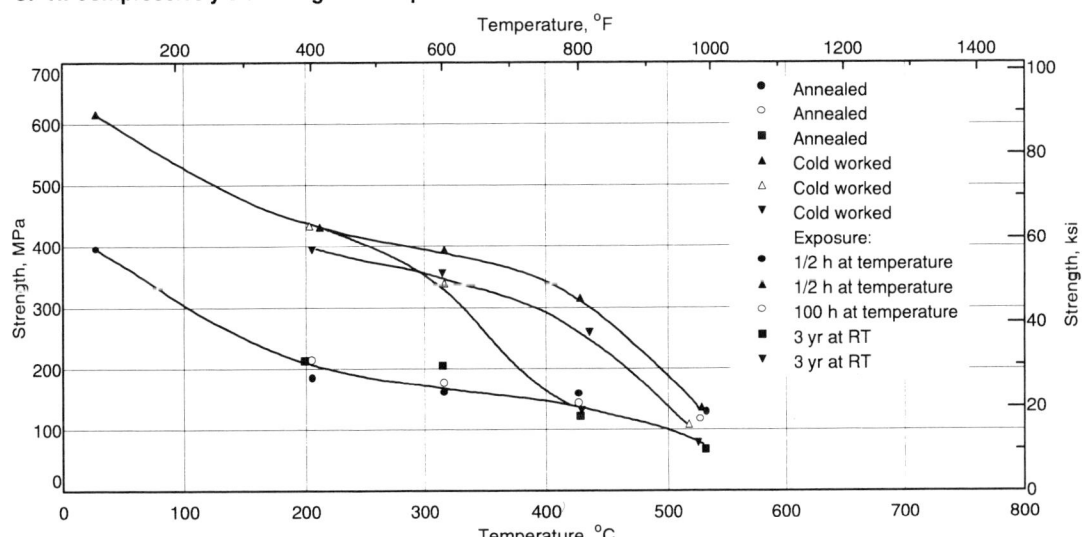

Ti-55 grade, 1.6 mm (0.064 in.) sheet.
Source: *Aerospace Structural Metals Handbook*, Vol 4, Code 3701, Battelle, 1981

Shear and Torsion Strength

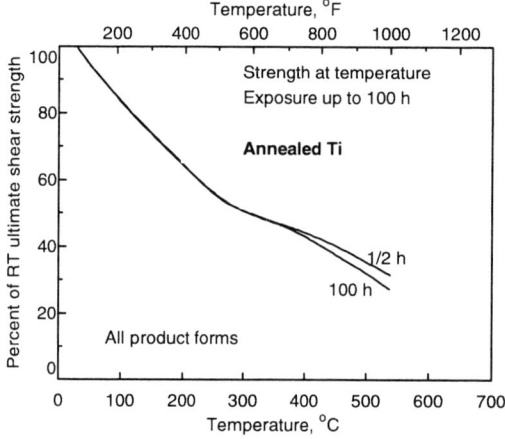

CP Ti: Ultimate shear strength vs temperature

Source: MIL-HDBK-5

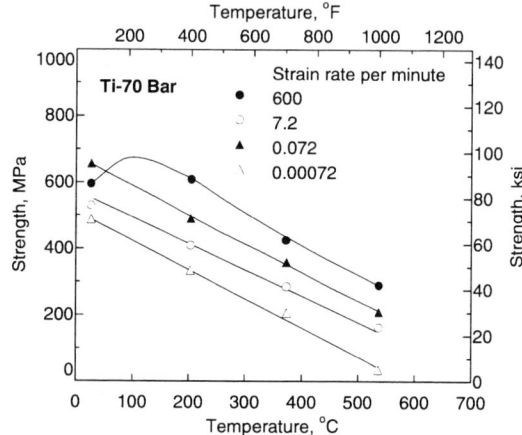

ASTM grade 4 Ti: Torsional strength vs temperature

Source: *Aerospace Structural Metals Handbook*, Vol 4, Code 3701, Battelle, 1981

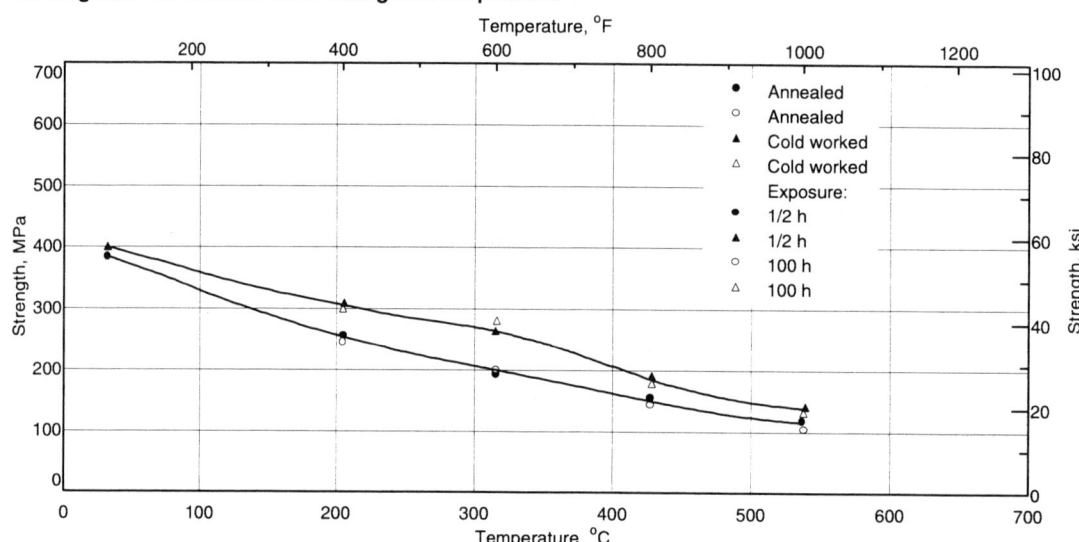

ASTM grade 3 Ti: Ultimate shear strength vs temperature

Ti-55 sheet, 4.75 mm (0.187 in.) thick.
Source: *Aerospace Structural Metals Handbook*, Vol 4, Code 3701, Battelle, 1981

Bearing Strength

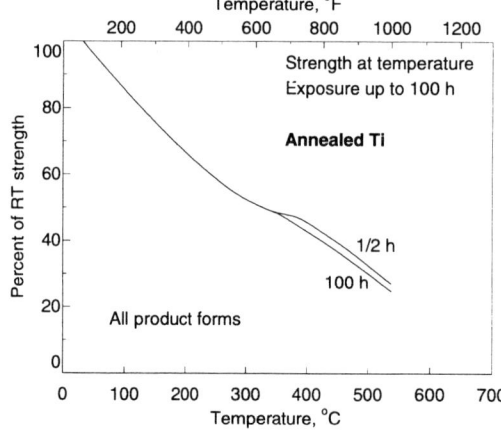

CP Ti: Ultimate bearing strength vs temperature

Source: MIL-HDBK-5

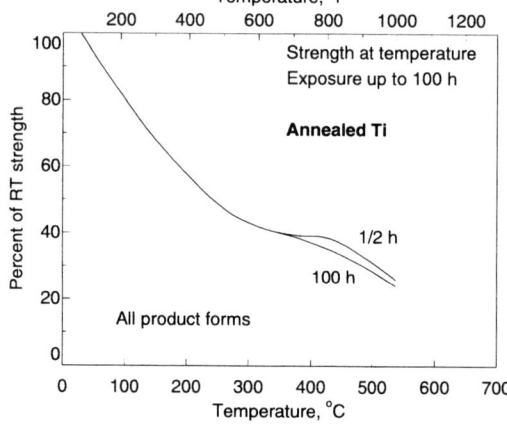

CP Ti: Bearing yield strength vs temperature

Source: MIL-HDBK-5

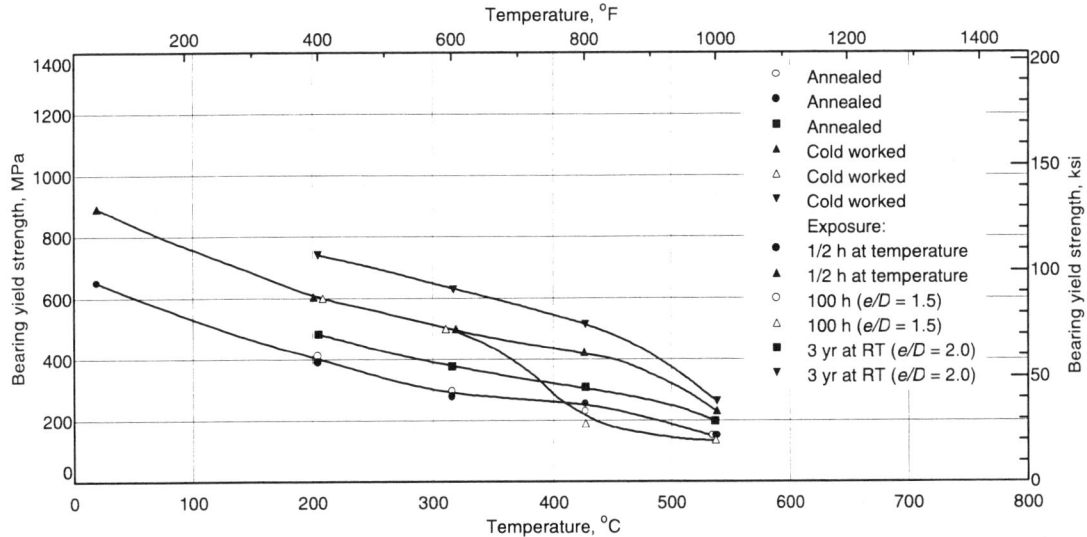

ASTM grade 3 Ti: Bearing yield strength vs temperature

Ti-55 sheet, 1.6 mm (0.064 in.) thick.
Source: *Aerospace Structural Metals Handbook*, Vol 4, Code 3701, Battelle, 1981

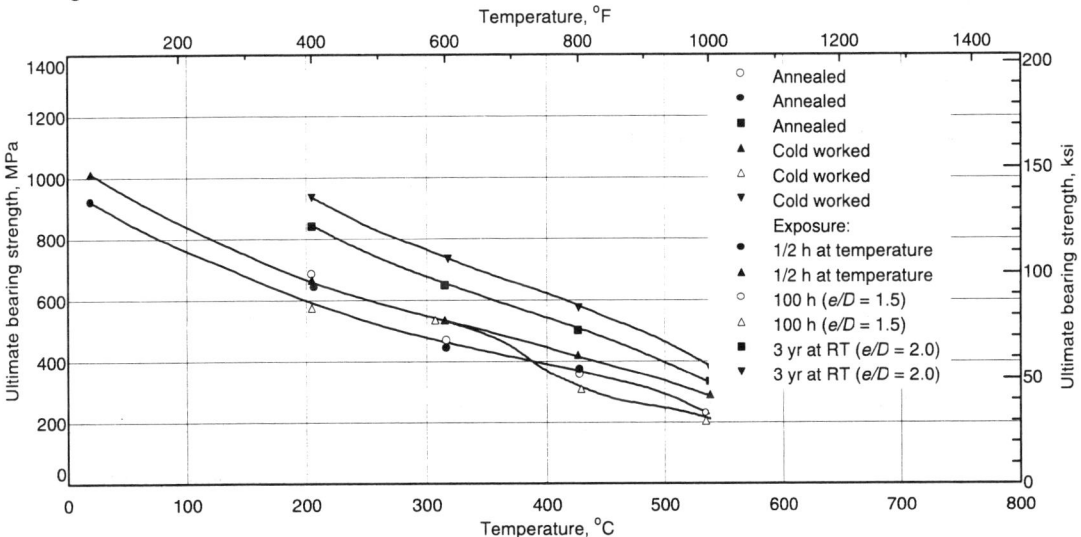

ASTM grade 3 Ti: Ultimate bearing strength vs temperature

Ti-55 sheet, 1.6 mm (0.064 in.) thick.
Source: *Aerospace Structural Metals Handbook*, Vol 4, Code 3701, Battelle, 1981

Creep Properties

Creep can occur, not only at elevated temperature, but also at ambient temperature in titanium as well as other metals. At ambient temperature, significant creep is encountered at relatively low percentages of 0.2% yield strength. As temperature increases, strain aging occurs and enhances creep performance. For instance, 1% creep will occur in 1000 h at room temperature for low-iron grade 2 at stresses less than 70% of the 0.2% yield strength. But at 250 °C (480 °F), the 1000 h/1% creep stress is very near the yield strength. Similar behavior exists for grades 1 and 3 commercially pure low-iron titanium.

Typical Creep Properties

Low-iron grade 2: Creep at 20 and 250 °C

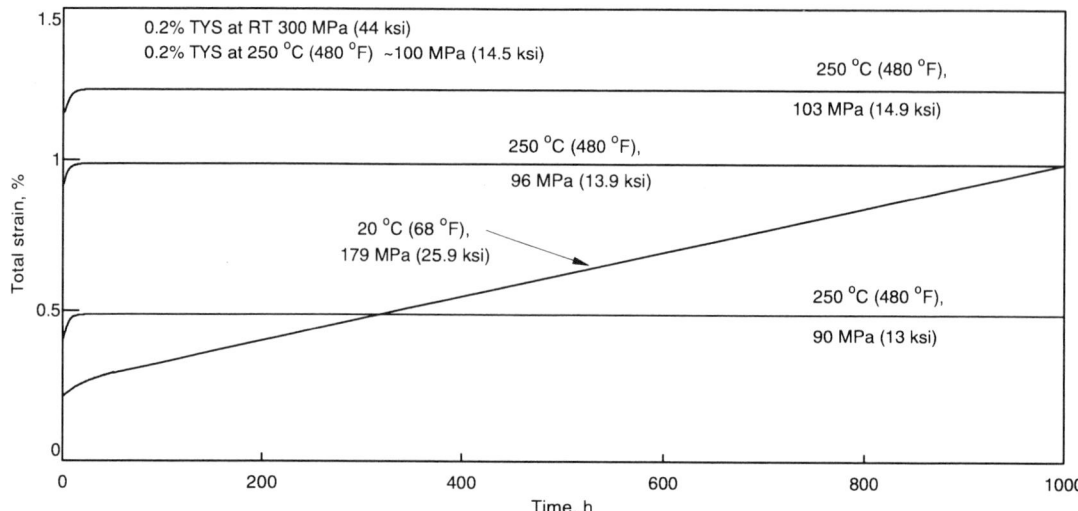

Strain aging is operative at 250 °C (480 °F) and results in a low secondary creep rate. The curve at 20 °C (68 °F) suggests an absence of strain aging.

Grade 2 commercially pure low-iron titanium plate containing 0.03 wt% iron. Production heats were double vacuum melted by the consumable electrode arc melting process. The ingots were forged to slabs and rolled to 44.4 or 50.8 mm (1.7 or 2 in.) thick plate product. The plate product was annealed at 730 °C (1345 °F) for 30 min and air cooled.

Source: P.A. Russo and J.D. Schöbel, "Mechanical Properties of Commercially Pure Titanium Containing Low Iron," presentation at ACHEMA 82, 1982

Creep and stress-rupture data for titanium alloys

Alloy	Stress to 1.0% creep in 1000 h at 250 °C (480 °F)		Stress to rupture in 1000 h at 250 °C (480 °F)	
	MPa	ksi	MPa	ksi
CP Grade 1	90	13	103	14
CP Grade 2	103	15	117	17
CP Grade 3	131	19	138	20
Ti-0.3Mo-0.8Ni, grade 12	221	32	297	43
Ti-3Al-2.5V grade 9	400	58	421	61

Source: *Industrial Applications of Titanium and Zirconium: Third Conference*, R.T. Webster and C.S. Young, Ed., STP 830, ASTM, 1984

ASTM Grade 12 Ti: Creep properties

Applied stress		Creep strain(a), %/h
MPa	ksi	
At 25 °C (77 °F)		
138	20.0	0
207	30.0	9.2×10^{-8}
290	42.0	3.6×10^{-5}
331	48.0	3.94×10^{-5}
At 150 °C (300 °F)		
108	15.6	0
179	26.0	7.2×10^{-6}
221	32.0	4.8×10^{-6}
At 315 °C (600 °F)		
48	7.0	3.3×10^{-6}
83	12.0	8.2×10^{-6}
103	15.0	1.4×10^{-5}
124	18.0	1.5×10^{-5}
138	20.0	1.86×10^{-5}

(a) Best fit linear rate over test period of 5000 to 10 000 h. Source: *Metals Handbook*, Vol 3, 9th ed., American Society for Metals, 1980, p 381

99.0% Ti: Tensile properties and creep stress vs temperature

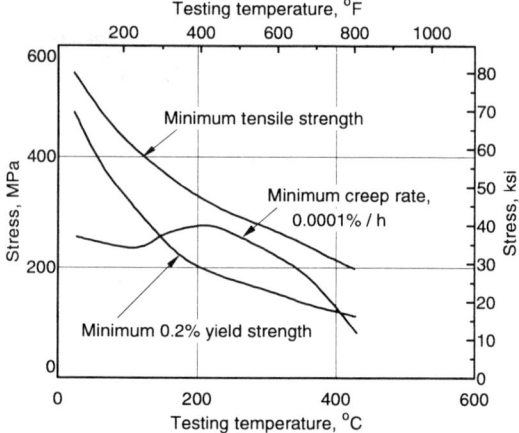

Source: *Metals Handbook*, Vol 1, *Properties and Selection,* 8th ed., American Society for Metals, p 537

Commercially Pure and Modified Ti / 235

Unalloyed Ti: Stress to rupture in 10^4 h

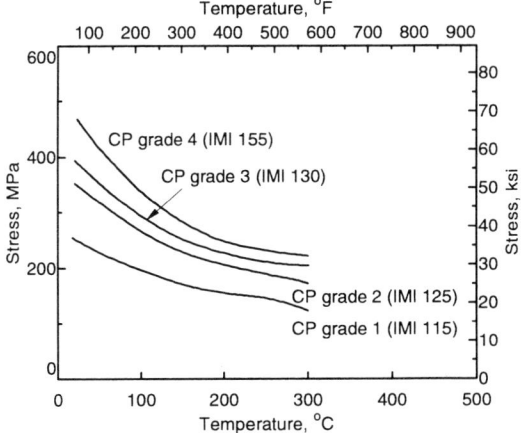

Based on Larson-Miller interpolation on tests extending 10^4 h at 250 °C (480 °F) and 1000 h at 350 °C (660 °F).
Source: "IMI Commercially Pure Titanium"

ASTM grade 3 Ti: Creep at 27 °C

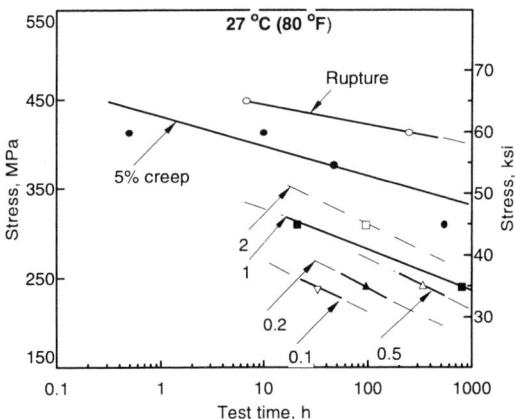

Mill annealed 99.0% Ti with a room-temperature yield strength of 380 MPa (55 ksi).
Source: *Metals Handbook*, Vol 1, *Properties and Selection*, 8th ed., p 537

ASTM Grade 3: Creep Data

ASTM grade 3 Ti: Creep at 425 °C

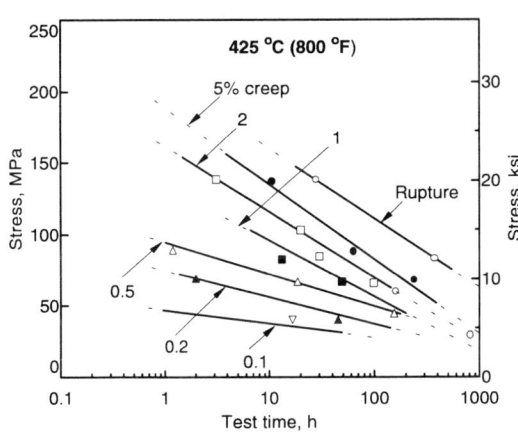

Mill annealed 99.0% Ti with a room-temperature yield strength of 380 MPa (55 ksi).
Source: *Metals Handbook*, Vol 1, *Properties and Selection*, 8th ed., p 537

ASTM grade 3 Ti: Creep at 540 °C

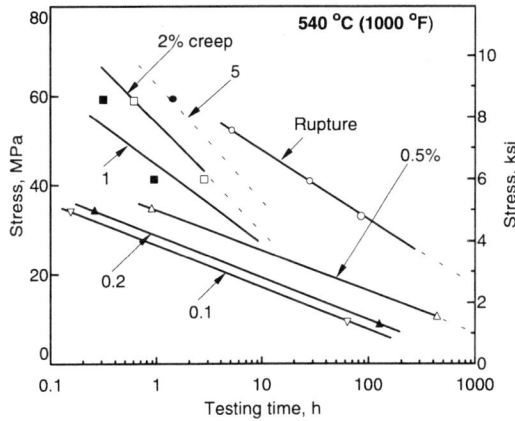

Mill annealed 99.0% Ti with a room-temperature yield strength of 380 MPa (55 ksi).
Source: *Metals Handbook*, Vol 1, *Properties and Selection*, 8th ed., p 537

ASTM Grade 4: Creep Data

ASTM grade 4 Ti: Creep at 25 °C

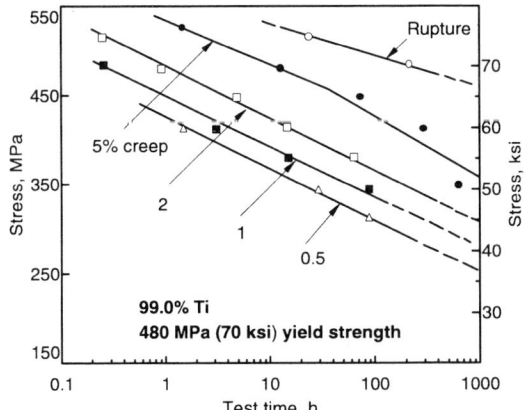

Mill annealed 99.0% Ti with a room-temperature yield strength of 480 MPa (70 ksi).
Source: *Metals Handbook*, Vol 1, *Properties and Selection*, 8th ed., p 537

ASTM grade 4 Ti: Creep at 370 °C

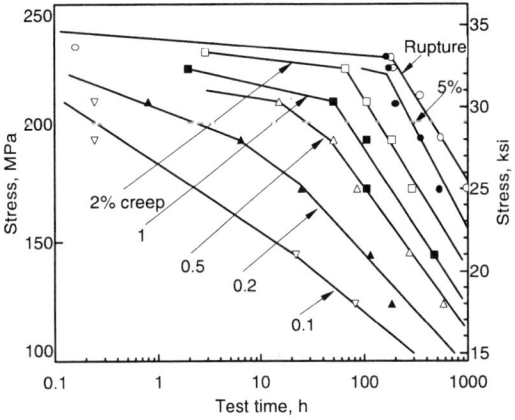

Mill annealed 99.0% Ti with a 480 MPa (70 ksi) yield strength at room temperature.
Source: *Metals Handbook*, Vol 1, *Properties and Selection*, 8th ed., p 537

ASTM grade 4: Creep at 425 °C

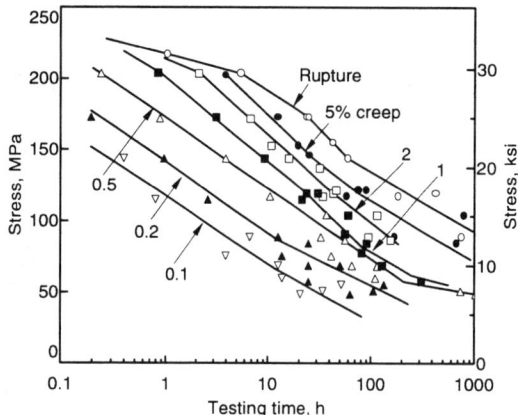

Mill annealed 99.0% Ti with a room-temperature yield strength of 480 MPa (70 ksi).
Source: *Metals Handbook*, Vol 1, *Properties and Selection*, 8th ed., p 537

ASTM grade 4: Creep at 540 °C

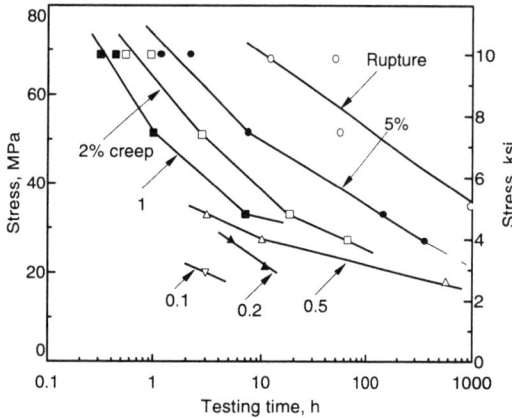

Mill annealed 99.0% Ti with a room-temperature yield strength of 480 MPa (70 ksi).
Source: *Metals Handbook*, Vol 1, *Properties and Selection*, 8th ed., p 537

Fatigue Properties

CP Ti: Fatigue strength at 10^7 cycles

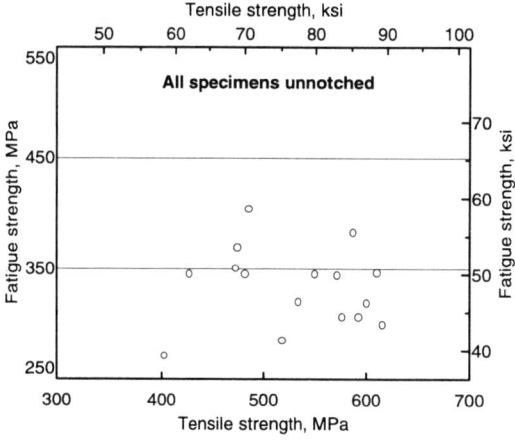

Source: *Metals Handbook*, Vol 1, *Properties and Selection*, 8th ed., American Society for Metals, 1961

Low-iron grade 2 Ti: Fatigue at 150 °C

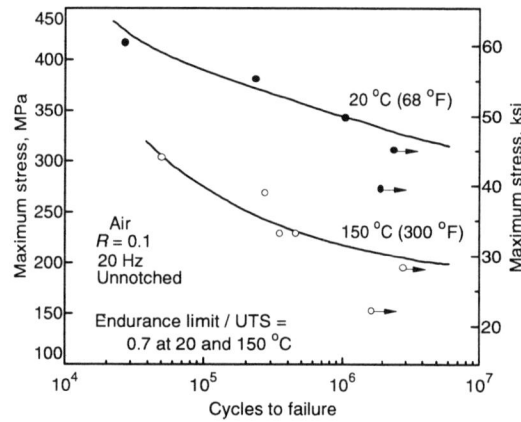

Commercially pure low-iron grade 2 titanium plates containing 0.03% iron or less were double vacuum melted by the consumable electrode arc melting process. The ingots were forged to slabs and rolled to 44.4 or 50.8 mm (1.7 or 2 in.) thick plate product. The plate product was annealed at 730 °C (1345 °F) for 30 min and air cooled.
Source: P.A. Russo and J.D. Schöbel, "Mechanical Properties of Commercially Pure Titanium Containing Low Iron," presentation at ACHEMA 82, 1982

ASTM grade 4: RT rotating and axial fatigue

Product form	Test method	Stress ratio (R)	Stress concentration	Fatigue strength at cycles					
				10^5		10^6		10^7	
				MPa	ksi	MPa	ksi	MPa	ksi
Bar	Rotating beam	-1	Smooth, $K = 1$	517	75	469	68	427	62
			Notched, $K = 2.7$	289	42	262	38	248	36
Sheet	Direct stress	0.6	Smooth, $K = 1$	...	...	...	...	538	78

Annealed Ti-70 sheet and bar. Source: *Aerospace Structural Metals Handbook*, Vol 4, Code 3701, Battelle, 1963

ASTM grade 3: Reverse bending fatigue

Temperature		Stress	Fatigue strength at cycles			
			10^6		10^7	
°C	°F	concentration	MPa	ksi	MPa	ksi
−191	−312	Smooth, $K_t = 1$	...	...	689	100
		Notched, $K_t = 2.7$	...	...	317	46
RT	RT	Smooth, $K_t = 1$	289	42	282	41
		Notched, $K_t = 2.7$	151	22	241	35
315	600	Smooth, $K_t = 1$	...	...	144	21

Annealed Ti-55 bar; $R = -1$. Source: *Aerospace Structural Metals Handbook*, Vol 4, Code 3701, Battelle, 1963

ASTM grade 3 Ti: RT rotating beam fatigue strength

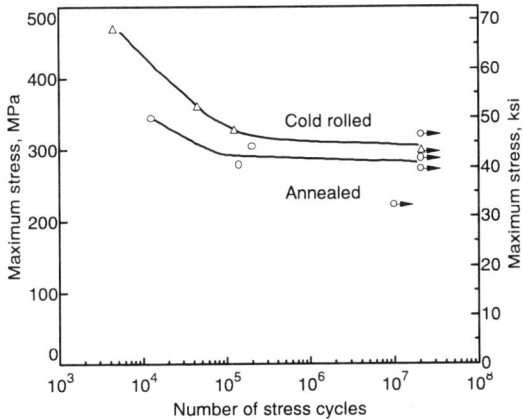

Source: *Metals Handbook*, Vol 3, *Properties and Selection of Stainless Steels, Tool Materials and Special Purpose Metals*, 9th ed., American Society for Metals, 1980, p 376

ASTM grade 4 Ti: Rotating-beam fatigue strength

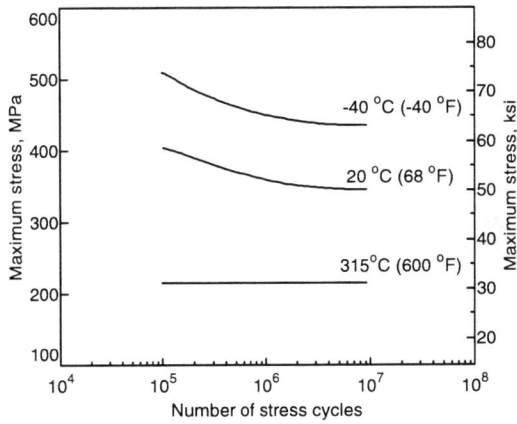

Data are for unnotched, polished specimens machined from annealed bar stock.
Source: *Metals Handbook*, Vol 3, 9th ed., American Society for Metals, 1980, p 378

Fracture Properties

Commercially pure titanium, like low-carbon steel, is considered ductile and tough. Because of the low strength and high toughness of commercially pure titanium, plane-strain fracture toughness tests are not meaningful unless specimen thickness is greater than about 150 mm (6 in.). This type of test is not practical for routine testing of commercially pure product. Unlike low-carbon steel, titanium does not exhibit a good correlation between absorbed impact energy and other more technically rigorous toughness tests such as dynamic tear energy and fracture toughness. Also titanium, unlike steel, does not exhibit a ductile/brittle transition temperature. However, the notched impact test does allow comparison within a given material type and can be used as a quality control tool.

CP Ti: Fracture toughness in air and 3.5% NaCl solution at 25 °C

Alloy	Thickness		Heat treatment(a)	Tensile yield strength		K_Q		K_{ISCC} or K_{SCC}	
	mm	in.		MPa	ksi	MPa√m	ksi √in.	MPa√m	ksi √in.
Ti, grade 2	...	...	αA, AC	...	...	66	60	66	60
Ti, grade 2 (0.06% O_2)	19	0.75	αA, AC	...	...	58	53	57	52
Ti, grade 3 (0.020% O_2)	19	0.75	αA, AC	...	...	79	72	74	68
Ti, grade 4 (0.40% O_2)	19	0.75	αA, AC	...	...	99	90	58	53
Ti, grade 4	13	0.50	αA, AC	572	83	135	123	36	33
			αA, WQ	586	85	140	128	37	34
			STA	565	82	124	113	43	39
			βA, WQ	524	76	115	105	77	70
			βSTA	530	77	105	96	52	48

(a) αA, alpha anneal; βA, beta anneal; βSTA, beta solution treated and annealed. Source: R. Schutz, Stress-Corrosion Cracking of Titanium Alloys, in *Stress-Corrosion Cracking: Materials Performance and Evaluation*, ASM International, 1992

CP Ti: Charpy V-notch impact toughness

Alloy	Absorbed energy at 20 °C (68 °F)(a)	
	J	ft · lb
Ti-3Al-2.5V	48	35
	48	35
	48	35
	54	40
Grade 1	302	223
	309	228
	305	225
	312	230
Grade 2	114	84
	148	109
	167	123
	171	126
Grade 3	30	22
	39	29
	43	32
	66	49

(a) Longitudinal test direction. Source: *Industrial Applications of Titanium and Zirconium: Third Conference*, R.T. Webster and C.S. Young, Ed., STP 830, ASTM, 1984

CP Ti: Charpy V-notch impact toughness vs yield strength

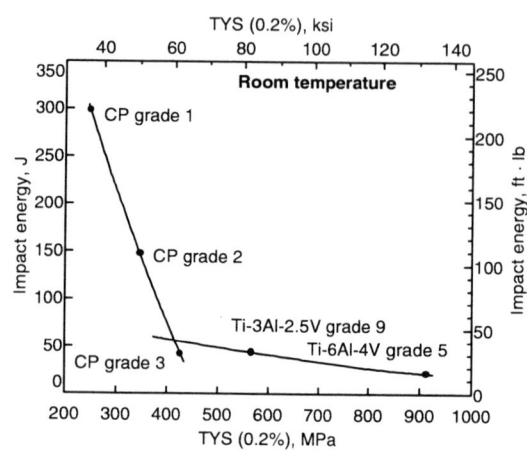

Source: *Industrial Applications of Titanium and Zirconium: Third Conference*, R.T. Webster and C.S. Young, Ed., STP 830, ASTM, 1984

CP Ti: Charpy V-notch impact toughness vs temperature

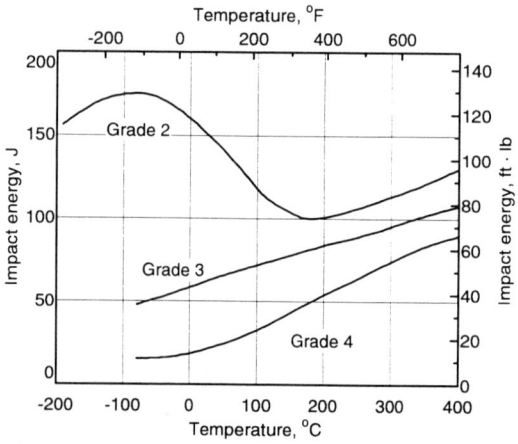

Source: *Metals Handbook*, Vol 3, 9th ed., American Society for Metals, 1980, p 375

Fracture Mechanism Maps

CP Ti: Fracture mechanisms plotted by stress and temperature

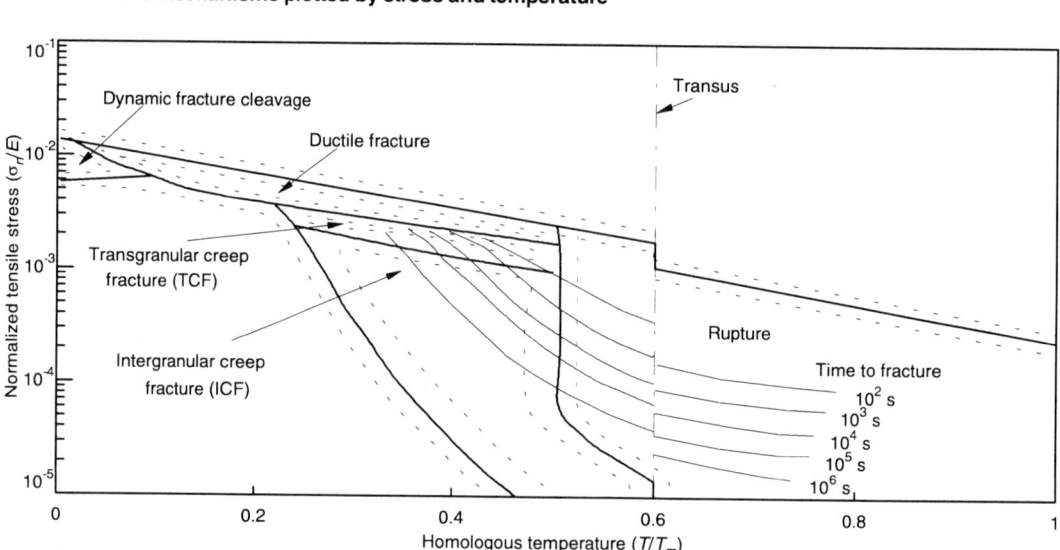

Although the TCF field in this figure appears very narrow compared to the intergranular field, this does not reflect the fact that TCF and rupture, and not ICF, dominate the high-temperature fracture behavior of CP titanium. The time-temperature diagram of fracture (see other figure) does not give rise to such an ambiguous impression.

Source: Y. Krishnamohanrao et al., *Fracture Mechanism Maps for Titanium and Its Alloys, Acta Metall.*, Vol 34(No. 9), 1986, p 1783-1806

CP Ti: Fracture mechanisms plotted by time and temperature

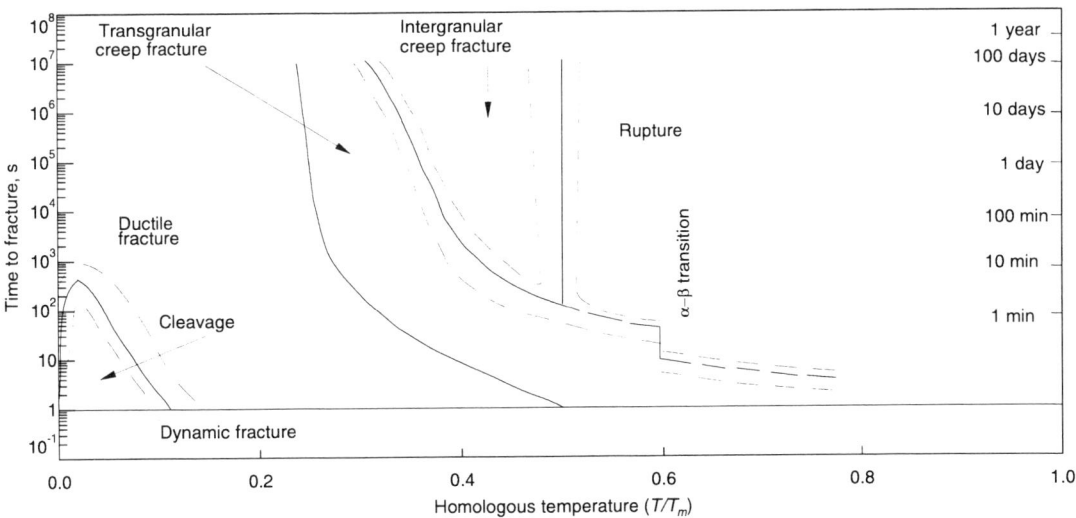

Unlike the stress-temperature diagrams of fracture mechanisms, which contain large ill-defined areas to which one cannot easily assign a fracture mode, this figure does not suffer from such ambiguities because it is more faithful to practical test conditions and data. Furthermore, the fracture fields in this figure are more reasonably proportional.
Source: Y. Krishnamohanrao et al., Fracture Mechanism Maps for Titanium and Its Alloys, Acta Metall., Vol 34(No 9), 1986, p 1783-1806

Stress Strain Curves

CP Ti: Room-temperature stress-strain curves

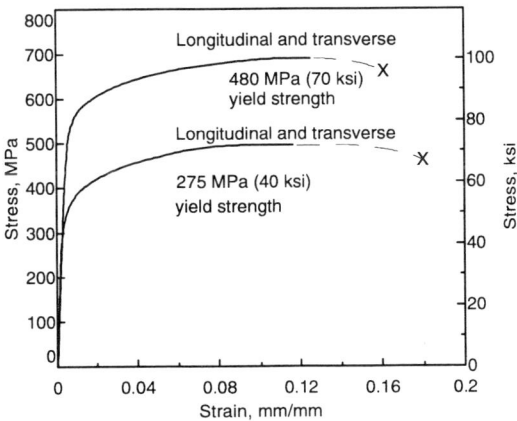

Source: MIL-HDBK 5, 1 Dec 1991

Grade 4 Ti: Effect of grain size on stress-strain

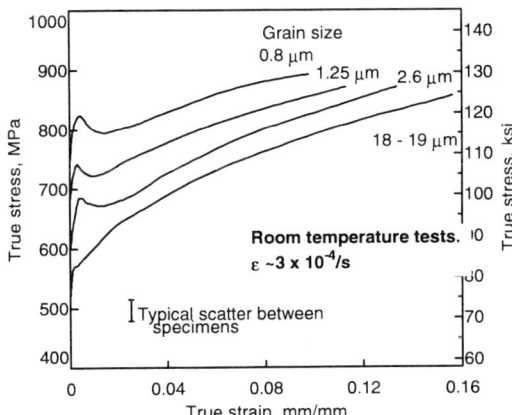

Room-temperature tests. $\varepsilon \sim 3 \times 10^{-4}$/s.
Source: H. Conrad and R. Jones, in *The Science, Technology and Application of Titanium*, R.I. Jaffee and N.E. Promisel, Ed., Pergamon Press, 1970, p 489-501

Textured Sheet

Grade 2 Ti: Engineering stress-strain curve for sheet

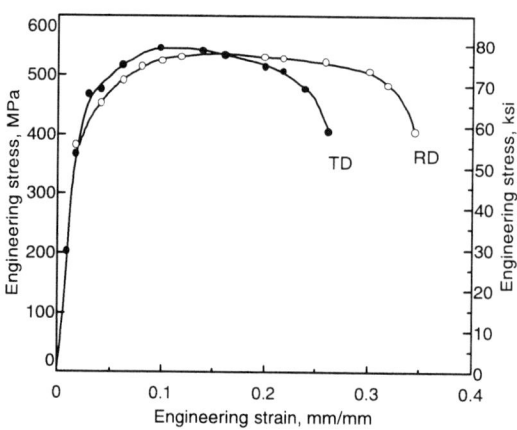

Source: L. Murugesh et al., Contractile Strain Ratios of Textured CP-Titanium (Grade II) Sheet, *J. Mater. Technol.*, Vol 7(No. 2), 1989, p 81-90

Grade 2 Ti: Stress-strain of textured sheet in rolling direction

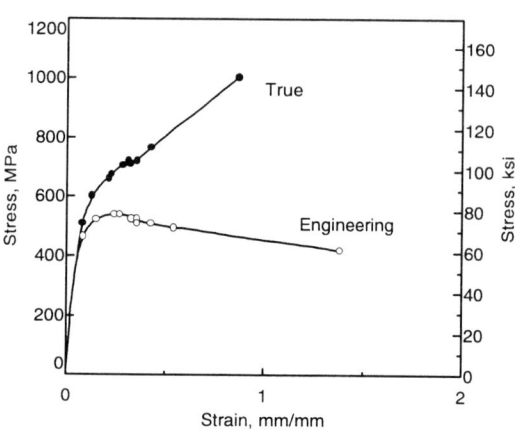

Source: L. Murugesh et al., *J. Mater. Shap. Technol.*, Vol 7(No. 2), 1989, p 86

Grade 2 Ti: Stress-strain of textured sheet in transverse direction

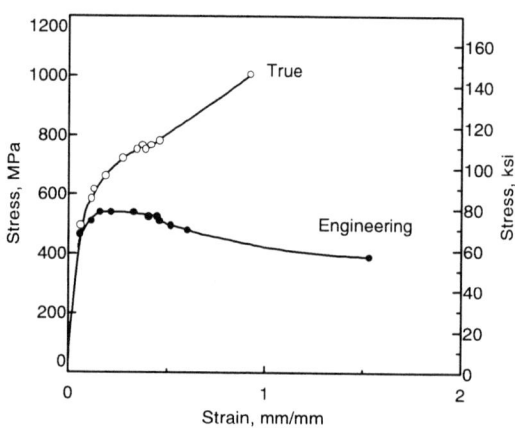

Data obtained from fractured grid elements.
Source: L. Murugesh et al., *J. Mater. Shap. Technol.*, Vol 7(No. 2), 1989, p 86

CP Ti: Effect of cross-head speed on stress-strain

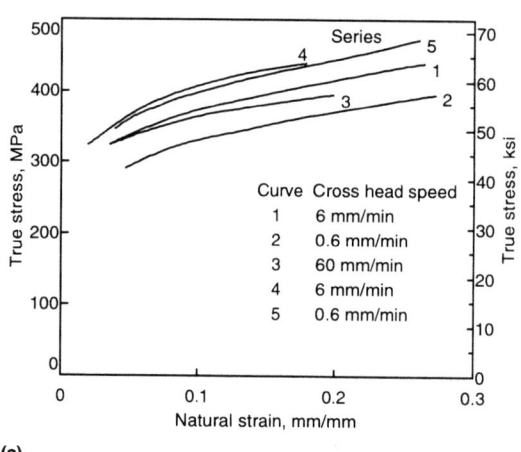
(a)

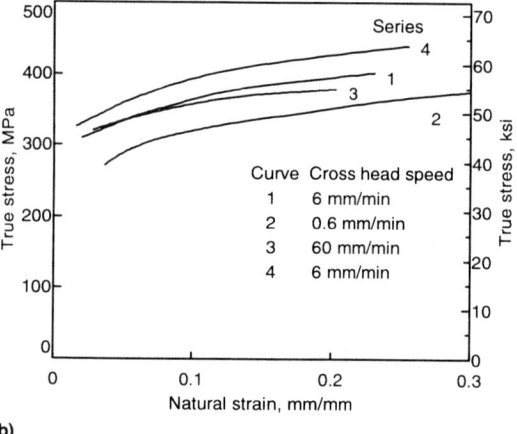
(b)

Tested at room temperature.
Source: *Titanium '80, Science and Technology*, H. Kimura and O. Izumi, Ed., TMS, 1980, p 1124

Commercially Pure and Modified Ti / 241

Effect of Temperature

CP Ti: Stress-strain curves at room and elevated temperature

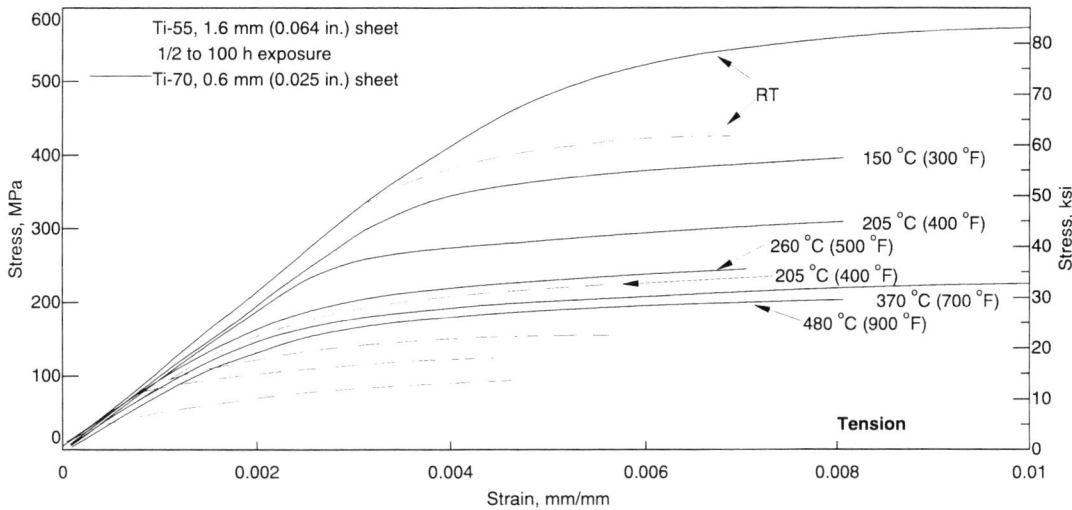

Source: *Aerospace Structural Metals Handbook*, Vol 4, Code 3701, Battelle Columbus Laboratories, 1963

Grade 2 equivalent: Typical stress-strain curves

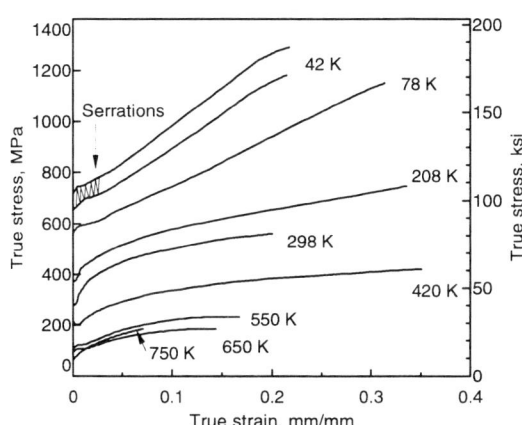

Strain rate 3.6×10^{-4}/s.
Ti-50A 0.5 at.% O_{eq} of 22 μm grain size.
Source: *Metall. Trans. A*, Vol 14, Dec 1983, p 2546

Grade 2 equivalent: True stress-strain curves

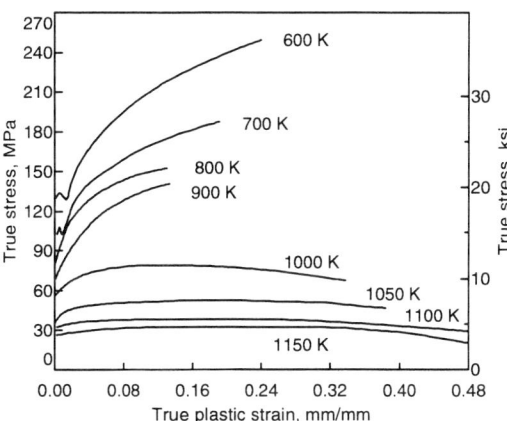

Strain rate of 3.3×10^{-2}/s.
Ti-50A 0.46 at.% O_{eq}.
Source: *Metall. Trans. A*, Vol 14, Dec 1973, p 2810

Grade 3 Ti: Typical compressive stress-strain curves

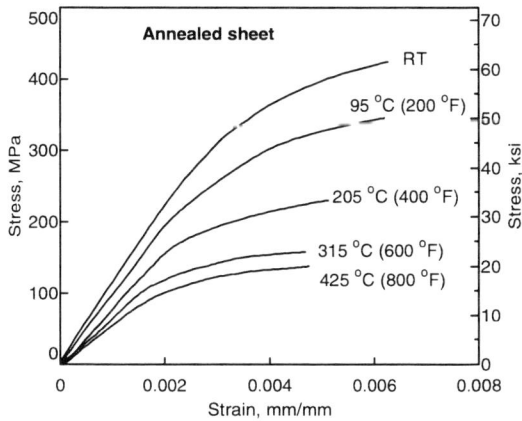

Chemical composition (wt%): 0.02 C, 0.20 Fe, 0.005 H, 0.01 N, and 0.20 O. Annealing performed at 705 °C (1300 °F), air cool. Source: Crucible Data Sheet, Crucible Specialty Metals

Grade 4 Ti: True stress-strain vs temperature and grain size

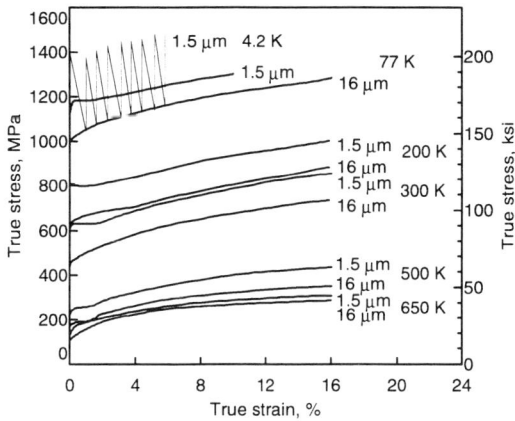

Strain rate 3.3×10^{-4}/s.
Ti-A70 (~1 at.% O_{eq}).
Source: *Acta Metall.*, Vol 21, Aug 1973, p 1117-1129

Plastic Deformation

Deformation Modes. In addition to $\langle 11\bar{2}0 \rangle$ slip on the basal plane, deformation twinning is another important crystallographic deformation mode in commercially pure titanium. Crystallographic deformation modes are stress-state dependent (*Metall. Trans. A*, Vol 12, 1981, p 853), and less deformation twinning is observed in uniaxial tension than in material strained with the same thickness strain under conditions of equibiaxial tension or plane-strain tension. Thus the influences of twinning are likely to be more prominent in press forming than in uniaxial tensile tests. The effects of twinning deformation on texture development and strain limits are discussed in subsequent sections on bulk working, forming properties, and forming methods.

Another effect of deformation twinning is to produce an effective refinement of the grain size. Because flow stress is related to grain size by the Hall-Petch relationship, and if the twin boundaries provide effective barriers to dislocation movement, then twinning during straining would be expected to enhance the rate of strain hardening. This has been referred to as the dynamic Petch effect (*Rev. on High-Temp. Mater.*, Vol 1, 1972, p 97).

Uniform and Post-Uniform Elongation. The ductility of a material exhibits convergence of the elongation and reduction of area values as temperature is increased to about 200 to 250 °C (390 to 480 °F). At higher temperatures, a divergence of these properties occurs. This is believed to be the result of a strain aging phenomenon that occurs due to the interaction of interstitial atoms and dislocations. This phenomenon is generally accompanied by reduced ductility and less uniform elongation, as suggested by the divergence of the tensile reduction of area and elongation curves.

Grade 2 Ti: Ductility vs temperature

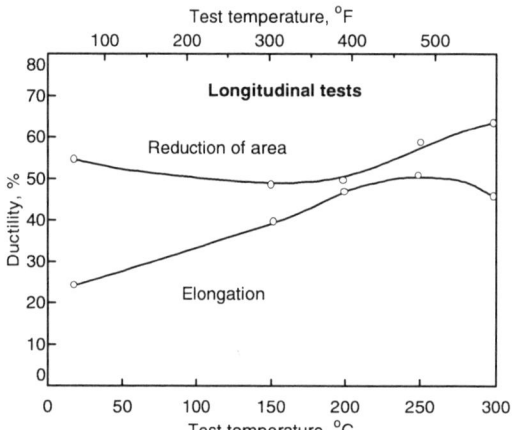

Source: P.A. Russo and J.D. Schöbel, "Mechanical Properties of Commercially Pure Titanium Containing Low Iron," presented at ACHEMA 82, 1982

Post-uniform elongation in α titanium

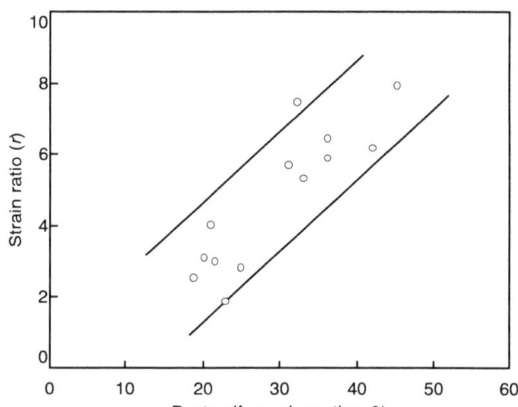

Post-uniform elongation is affected by strain-rate sensitivity (*m*) because a positive *m*-value extends the diffuse necking regime by delaying the onset of the local neck by virtue of strain-rate hardening. Specimens cut from a single sheet at different angles to a reference direction, giving r-values from 1.8 to 7.7 for a constant *m*-value (*m* = 0.026 ± 0.003). Here, the post-uniform elongation varies from about 20 to about 40%.
Source: R. Pearce, *Sheet Metal Forming*, IOP Publishing Ltd., 1991

Flow Stress

Grade 4 Ti: Flow stress vs grain size at room temperature

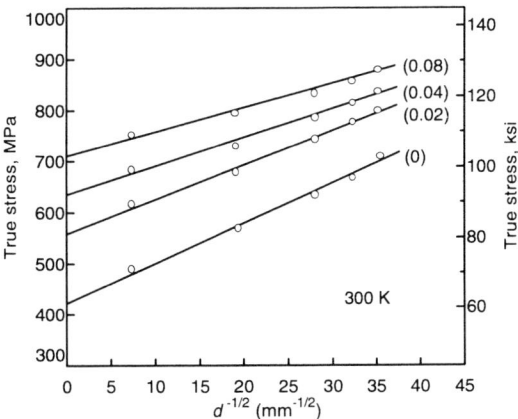

Strain rate of 3×10^{-4}/s.
Ti-A70. d is average grain diameter. Offset as indicated in parentheses. Source: *Trans. Metall. Soc. AIME*, Vol 245, Apr 1969, p 782

Grade 4 Ti: Flow stress vs dislocation density

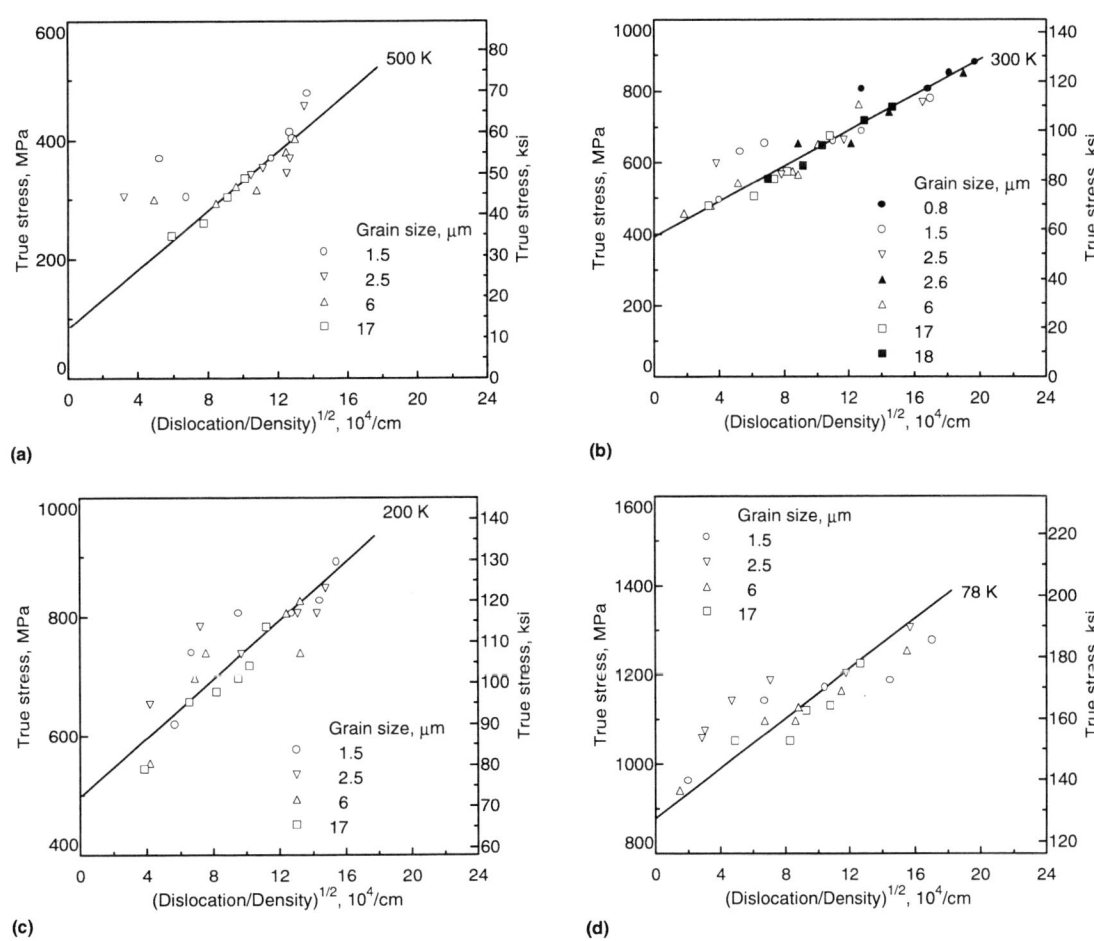

Strain rate of 3×10^{-4}/s.
A-70Ti (1.0 at.% O_{eq}).
Source: H. Conrad *et al.*, Critical Review Deformation and Fracture, *Titanium, Science and Technology*, R.I. Jaffee and H.M. Burte, Ed., Plenum Press, 1973, p 969-1005

Flow Stress vs Temperature

Grade 2 Ti: Flow stress vs temperature

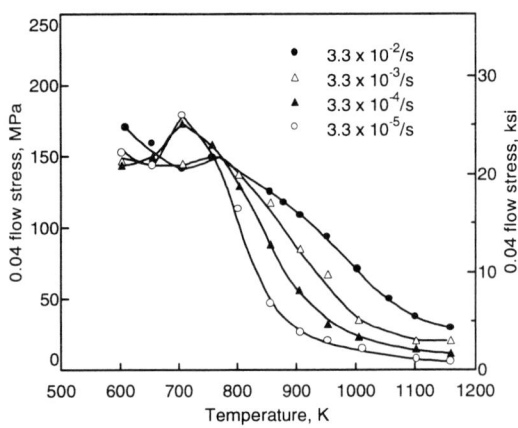

Effect of temperature on the flow stress at 4% strain of Ti-50A (0.46 at.% O_{eq}) at various strain rates. Grain size, 22 μm.
Source: *Metall. Trans. A*, Vol 4, Dec 1973, p 2810

Grade 2 Ti: Flow stress vs temperature

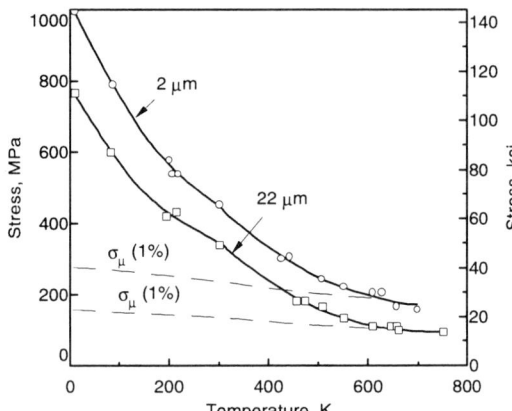

Strain rate of 3.3×10^{-4}/s. σ_μ is the athermal component of the flow stress.
Flow stress at 1% plastic strain vs temperature as a function of grain size for Ti-50A (0.5 at.% O_{eq}).
Source: *Metall. Trans. A*, Vol 14, Dec 1983, p 2545-2556

Strain Hardening

Unalloyed titanium work hardens relatively rapidly at room temperature as evidenced by the hard drawing of titanium wire (see figure). As a result, unalloyed titanium is better suited for stretch forming than deep drawing (especially when multiple draws are required). Plots of strain-hardening equations for annealed commercial purity (0.15 wt% O_2) sheet are shown in the accompanying figures for the following relations:

(1) Hollomon equation log true stress vs log true plastic strain ($\bar{\varepsilon}_p$); (2) Hollomon equation log true stress vs log true total strain ($\bar{\varepsilon}_t$); (3) Ludwik equation $\log(\sigma - \sigma_{lp})$ vs log true plastic strain; and (4) Ludwik equation $\log(\sigma - \sigma_{0.2})$ vs log true plastic strain. Also shown are curves of percentage elastic strain ($\bar{\varepsilon}_{el}$) in the total ($\bar{\varepsilon}_T$), and plastic ($\bar{\varepsilon}_p$) strain terms.

For annealed titanium, both the Hollomon and the Ludwik equations adequately describe the stress-strain curves at high strains (>0.025). Anomalies occur at lower stains where the elastic strains are significant. The strain-hardening exponent in the Ludwik equation is very dependent on the value of σ_o. For worked titanium, neither equation successfully describes the stress-strain curves. This is attributed to the very high elastic strain contribution at all strains at these high stress levels.

CP Ti: Work hardening of drawn wire

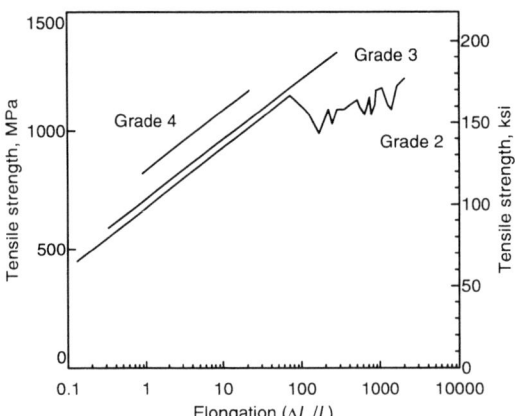

Hardening by drawing depends on surface preparation, but most processes do not allow multiple passes and are without practical interest. Source: *Titanium '80, Science and Technology*, H. Kimura and O. Izumi, Ed., TMS/AIME, 1980, p 2473

CP Ti: Strain-hardening equations (annealed)

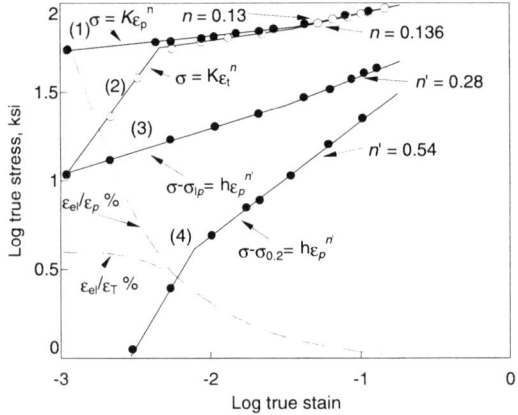

Annealed titanium (0.15 wt% O_2) sheet with 0.018 mm grain size tested at strain rate of 1.6×10^{-3}/s. Line numbers refer to discussion in text.
Source: A.W. Bowen and P.G. Partridge, Tensile Stress-Strain Curves of Cold-Worked Polycrystalline Titanium, *Titanium, Science and Technology*, R.I. Jaffee and H.M. Burte, Plenum Press, 1973, p 1029

CP Ti: Strain-hardening equations (cold worked)

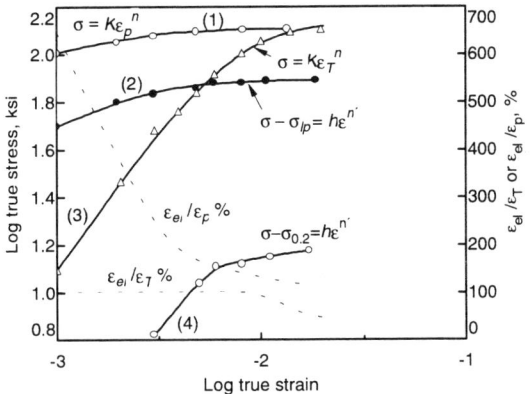

CP titanium (0.15 wt% O$_2$) reduced 50% at 20 °C. Grain size was 0.018 mm and strain rate was 1.6 × 10^{-3}/s. Line numbers refer to discussion in text.
Source: A.W. Bowen and P.G. Partridge, Tensile Stress-Strain Curves of Cold-Worked Polycrystalline Titanium, *Titanium, Science and Technology*, R.I. Jaffee and H.M. Burte, Ed., Plenum Press, 1973, p 1029

CP Ti: Effect of grain size on the strain-hardening coefficient (h)

Material/ composition(a)	Grain size parameters(b):		
	h_i MPa	kg/mm^2	K' kg/mm$^{3/2}$
Iodide titanium (30 O,18 C,110 N)	440	45	0.5
Battelle titanium (330 O, 25 C,140 N)	353	36	0.2
A-70 (ASTM grade 4)	872	89	0.8

(a) All compositions in ppm by weight. (b) Grain size effects on h modeled by the equation $h = h_i + K'\sqrt{d}$ where d is the average grain size. Source: H. Conrad and R. Jones, *The Science, Technology and Application of Titanium*, R.I. Jaffee and N.E. Promisel, Ed., Pergamon Press, 1970, p 495

CP Ti: Comparison of strain-hardening parameters

Material	Grain size, μm	Maximum stress(a) MPa	kgf/mm^2	Strain-hardening coefficient(h) MPa	kgf/mm^2
ASTM grade 4 (A-70)	18 to 19	510	52	833	85
ASTM grade 4 (A-70)	2.6	570	58	705	72
ASTM grade 4 (A-70)	0.8	715	73	600	61
Iodide (30 ppm O,110 ppm N)	3.3	215	22	343	35
Iodide (30 ppm O,110 ppm N)	1.1	265	27	285	29

(a) Maximum stress for zero plastic strain. Source: R.L. Jones and H. Conrad, *Trans. Metall. Soc. AIME*, Vol 245, 1969, p 779-789

Bulk Working

Forging

Forging of commercially pure titanium and modified titanium (ASTM grades 7, 11, and 12) presents little difficulty and can be performed by the full range of forging processes described in "Technical Note 4: Forging." Conventional equipment can be used to forge unalloyed titanium, and overall forging process technology is similar for all the ASTM grades of commercially pure and modified titanium. Time at elevated temperature is kept at a minimum to reduce oxidation and grain growth.

Recommended Die Temperatures. See "Technical Note 4: Forging."

Recommended Metal Temperature. Finish forging of commercially pure titanium is performed in the α + β phase field at a typical temperature of about 850 °C (1560 °F). The range of metal temperatures for α + β working varies from 815 °C (1500 °F) to about 870 or 900 °C (1600 or 1650 °F), where the upper limit depends on the β transus of the material. The upper limit of 900 °C (1650 °F) corresponds to ASTM grade 4, which has a β transus of 930 ± 15 °C (1700 ± 25 °F).

Higher forging temperatures may be permissible for initial roughing or cogging operations. For example, supra-transus β forging, at metal temperatures 30 to 55 °C (50 to 100 °F) above the β transus, may be used in early forging operations (e.g., open-die upsetting or preforming) to ease fabrication and assist in refining prior β grain size through dynamic recrystallization. Blocker temperatures range from 870 to 925 °C (1600 to 1700 °F).

Microstructure/Property Development. Microstructural objectives in forging grade 4 titanium are to achieve a refined macrostructure (prior β grain size) and α grain size suitable for the intended application. Care must be exercised in forging grade 4 to avoid thermal exposure conditions, e.g., very slow post-forging cooling, which may promote excessive grain boundary films, adversely affecting properties. The working history of grade 4 forgings is controlled to achieve a total of 50 to 75% cross-sectional area or upset reductions at sub-transus temperatures to achieve property objectives. Inadequate sub-transus working may lead to poor ductility.

Post-Forging Heat Treatment. Because the

CP Ti: Deformation resistance compared to steel

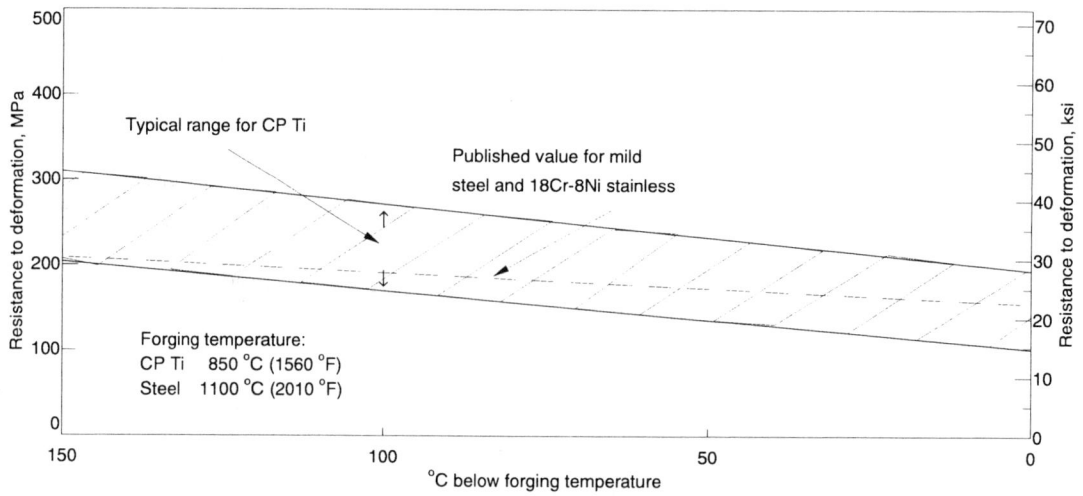

At 850 °C, the resistance to forming deformation of CP titanium is in the range 100 to 200 MPa (14 to 29 ksi), rising to 200 to 300 MPa (29 to 43 ksi) at 700 °C. Forging of commercially pure titanium is thus somewhat similar to that of low-carbon steel or to 18-8 Cr-Ni stainless steel, for which published values of resistance to deformation are around 140 to 160 MPa (20 to 23 ksi) for a forging temperature of 1100 °C (2010 °F), rising to a little over 200 MPa (29 ksi) at 950 °C (1740 °F).
Source: IMI Brochure

structure of commercially pure and modified titanium grades consists solely of α phase, the final heat treatment consists of a stress relief anneal at 650 to 760 °C (1200 to 1400 °F). Desired mechanical and corrosion properties in these alloys are achieved through control of elements such as oxygen and iron and/or additions of solid solution elements such as palladium, molybdenum, and nickel by alloy design and in ingot fabrication. Unalloyed titanium is not thermomechanically processed as are other titanium alloy classes in forging manufacturing. The objective of forging unalloyed titanium alloys is to produce the desired shape at the least cost.

Rolling

Effect of Work History on Grain Size. In keeping with commonly observed behavior, increasing amounts of cold work prior to annealing decrease the subsequent recrystallization temperature and grain size (see figure). A decrease in grain size distribution also occurs, but is not completely understood (*Metall. Trans. A*, Vol 16, p 707, 1985).

Rolling Textures. Commercially pure titanium sheet typically exhibits a classical α deformation structure (see figure), which is characterized by having a basal (0002) pole intensity on the sheet normal (SN), transverse direction (TD), and a great circle at about 27 to 30° from the SN. Annealing of cold or warm rolled sheet has only a slight sharpening effect on the (0002) poles. However, the ($10\bar{1}0$) poles rotate through an arbitrary angle of approximately 30° about the *c* axis, resulting in the so-called annealed α deformation texture (see figure).

In most cases, it is not necessary to distinguish between an annealed and a cold worked texture, because many properties are symmetrical about the *c* axis. Thus, a basal-pole figure is sufficient to define the crystallographic influence, and the above texture can be modified by either hot rolling (above 760 °C, or 1400 °F, but not above β transus) and/or alloying. Rolling textures are a combination of tension and compression textures, which form during {$10\bar{1}2$} and {$11\bar{2}1$} twinning reorientation of basal poles.

Twinning. The {$10\bar{1}2$} twinning rotation (about 85°) is such that the new basal pole orientation would be in or near the area labeled {$11\bar{2}2$}. The {$11\bar{2}1$} twinning rotation would be such that the basal poles move along a great circle toward the transverse direction, about 35°. From the compression stresses, {$11\bar{2}2$} twinning causes a rotation of the grains that have their basal poles near the sheet normal. This rotation, about 64°, occurs

Unalloyed Ti: Grain size vs temperature and deformation

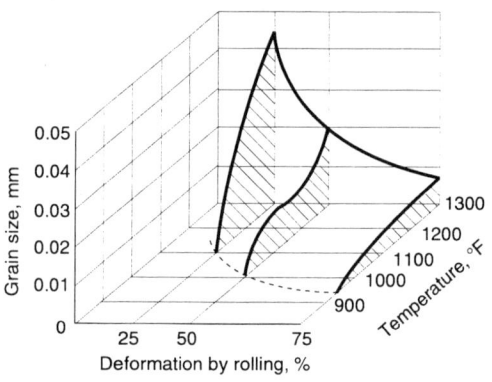

Effect of deformation and temperature on the recrystallized grain size of 0.56 mm (0.022 in.) thick unalloyed titanium.
Source: R.A. Wood, *Titanium Alloys Handbook*, MCIC HB-62, 1972

about the rolling direction toward the transverse direction. The {11$\bar{2}$2} twinning is considered to operate actively for thinning instead of slipping in the case of texture of small angle between the basal pole and the normal direction of the rolling plane (D.O. Hobson, *Trans. AIME*, Vol 242, 1968, p 1105).

Unalloyed Ti: Alpha deformation texture of sheet

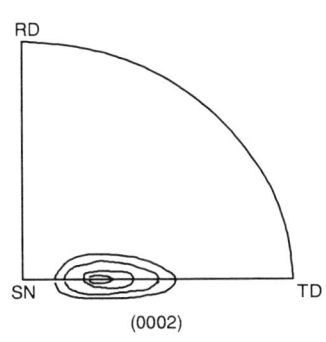

(a) (0002)

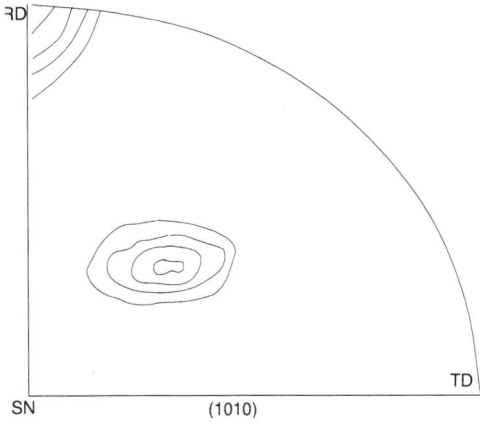

(b) (10$\bar{1}$0)

Along with the basal (0002) pole intensity shown, texture is further defined by the (10$\bar{1}$0) poles that lie near, or in, the rolling direction (RD).
Source: F. Larson and Z. Zarkades, *Properties of Textured Titanium Alloys*, MCIC 74-20, Battelle, June 1974

The slip direction of pure titanium is $\langle 11\bar{2}0 \rangle$, which lies on the basal lane. Because titanium pole figures have a strong [10$\bar{1}$0] component in the rolling direction, then tension in the rolling direction should cause $\langle 11\bar{2}0 \rangle$ duplex slip on {10$\bar{1}$0} and/or {10$\bar{1}$1}.

Because real textures of unalloyed titanium do not have basal pole peaks in the transverse direction (TD), an additional rotation back toward the sheet normal is required. The most likely mechanism would be basal slip. There is a possibility that {11$\bar{2}$1} twinning could also cause the rotation back toward the sheet normal, because the stress direction would be correct for second-order {11$\bar{2}$1} twinning within the {11$\bar{2}$2} twins. The rotation for this second-order twinning would be about 35° from the transverse direction toward the sheet normal; further rotation by slip would be required to orient the basal poles properly.

Unalloyed Ti: Annealed α deformation texture of sheet

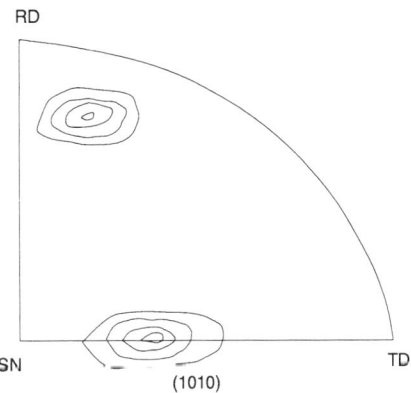

Source: F. Larson and A. Zarkades, *Properties of Textured Titanium Alloys*, MCIC-74-20, Battelle, June 1974

Forming Properties

As far as materials properties are concerned, the strain ratio (*r*-value) controls radial drawing, and some measure of ductility such as the strain-hardening exponent (*n*-value) or uniform or total elongation—the latter preferably—indicates stretching ability. The strain-rate sensitivity (*m*) factor is also important in the cold forming of sheet and in superplastic forming.

Typical forming properties of commercially pure titanium compare favorably with Ti-6Al-4V and the highly formable β alloy Ti-15-3 (see table). However, the table below does not provide an accurate comparison of stretching for CPTi and Ti-15-3. Boeing experience with ducts indicates much better stretch forming for CPTi than for Ti-15-3. Unalloyed titanium work hardens relatively rapidly, which is an advantage in stretch forming, but a limitation in severe (multiple-pass) drawing operations.

Titanium alloys: Room-temperature formability of annealed specimens (unspecified r values)

Alloy	Strength MPa	ksi	Gage mm	in.	Bend radius (min)	Stretch forming limit, %	Hydroform limit,% Stretch	Shrink
Ti-15-3	758	110	1	0.04	2.4T	...	12	...
			2	0.08	2.4T	28	20	...
CP Ti	275	40	1	0.04	3.0T	20	12	1.0
			2	0.08	3.0T	20	12	1.5
Ti-6-4	827	120	1	0.04	4.7T	7	8	1.0
			2	0.08	5.5T	7	8	...

Source: *Titanium, Science and Technology*, Vol 1, G. Lütjering, U. Zwicker, and W. Bunk, Ed., Deutsche Gesselschaft für Metallkunde, Germany, 1985, p 458

Bend Forming

Minimum bend radii at room temperature range from about 1T to 2T for 275 MPa (40 ksi) yield strength grade to about 2 or 3T for the highest strength grades of commercially pure titanium. Unlike tensile ductility, the bending ductility of commercially pure titanium can be adversely affected during warm working (100 to 400 °C) as compared to room-temperature working (see figure). Poor bendability during warm working can be explained by the lack of ductility in the thickness direction and the decrease of twinning at higher temperature.

Effect of Strain Ratio and Twinning. Titanium sheet with a high r-value has a texture in which the basal pole-direction is rather similar to the normal direction of the rolling plane. Consequently the slip along $\langle 11\bar{2}0 \rangle$ in the thickness direction is severely restrained in this material. However, $\{11\bar{2}2\}$ twinning is considered to operate actively for thinning instead of slipping in the case of the texture of small angle between the basal pole and the normal direction of the rolling plane. In addition to the contribution of twinning itself to thinning, this twinning operation causes a 64.4° reorientation of the basal plane, and it brings about the new texture, which is more favorable for thinning compared to the initial texture. Thus, once twinning generates this new texture, even material with a high r-value exhibits rather significant ductility in thinning. Twinning is particularly effective in both bending and biaxial stretching, which require a large reduction of thickness.

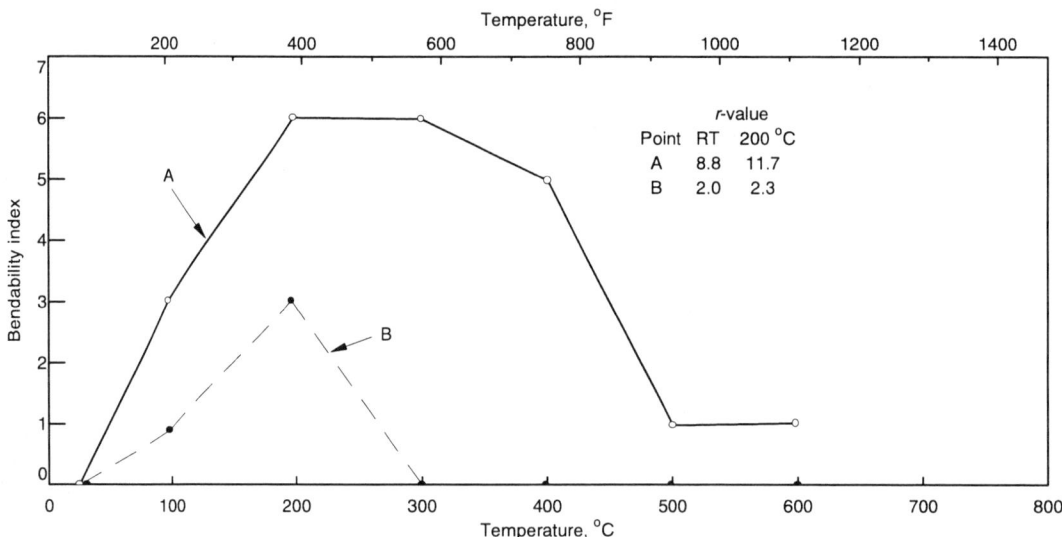

CP Ti: Warm working bendability

In this figure, bend index is applied to express the bendability of materials, and the larger the bend index value, the poorer the bendability of a given material. In the case of material A with a high r-value, the crack, accompanied by numerous deep surface pits, is observed at 200 and 400 °C (390 and 750 °F). Material B with a low r-value also exhibits a crack at 200 °C (390 °F), but its appearance is not as serious. Chemical composition (wt%): 0.006 C, 0.015 Fe, 0.0012 H, 0.004 N, and 0.072 O.
Source: N. Itoh et al., The Press Formability of Commercially Pure Titanium at Warm Working Temperature, in *Titanium '80, Science and Technology*, TMS/AIME, 1980, p 2523-2529

Strain-Hardening Exponent

The strain-hardening exponent (n) indicates the stretching ability and is thus sometimes termed the "stretchability index." Unalloyed titanium work hardens relatively rapidly. In general, the n value tends to be smaller in the transverse direction and greater in the rolling direction. This anisotropy of n is more distinct in the softer grades of titanium.

CP Ti: Strain-hardening exponent (n-value) at high and low strains

Series	Crosshead speed, mm/min	Strain rate, 10^{-3}/s	Orientation to rolling direction, degrees	Proof stress, MPa	Elongation, %	n-value at strain rate of: <0.15 ± 0.02	n-value at strain rate of: >0.15 ± 0.02
Air							
1	6	2.1	0	245	42	0.12	0.17
			45	260	44	0.11	0.12
			90	290	43	0.10	0.15
2	0.6	0.13	0	205	51	0.15	0.20
			45	220	51	0.13	0.20
			90	240	50	0.11	0.18
3	60	17	0	275	38	0.11	0.12
			45	280	33	0.08	0.09
			90	315	37	0.18	0.12
Water							
4	6	1.4	0	280	50	0.14	0.17
			45	275	48	0.11	0.15
			90	275	48	0.11	0.18
5	0.6	0.17	0	290	50	0.11	0.23

Note: Room-temperature tests, 0.81 mm (0.03 in.) thick sheet, 0.055 wt% O. Source: *Titanium '80, Science and Technology*, H. Kimura and O. Izumi, Ed., TMS/AIME, 1980, p 1121-1129

Grade 2 Ti: Strain-hardening exponent vs direction

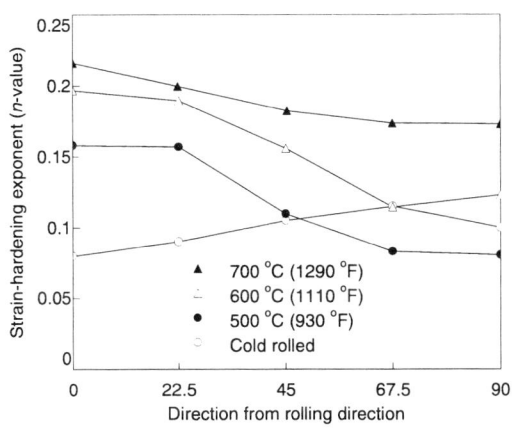

Cold rolled sheet (0.70 mm, or 0.02 in., thick), 0.1 wt% O, annealed 30 min at indicated temperature in argon.
Source: *Titanium '80, Science and Technology*, H. Kimura and O. Izumi, Ed., TMS/AIME, 1980, p 2534

Grade 2 Ti: Strain-hardening exponent vs annealing temperature

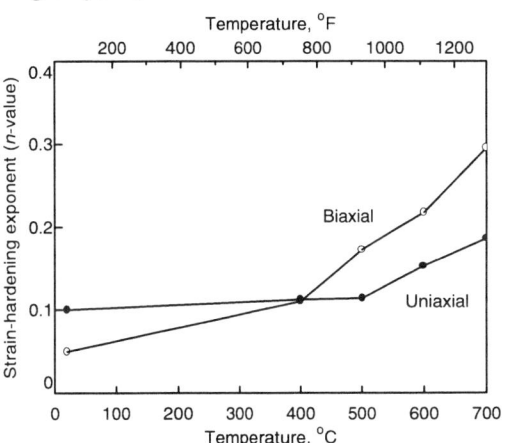

0.7 mm (0.02 in.) cold rolled sheet (0.1 wt% O).
Source: *Titanium '80, Science and Technology*, H. Kimura and O. Izumi, Ed., TMS/AIME, 1980, p 2534

CP Ti: Strain-hardening exponent vs tensile direction and anneal temperature

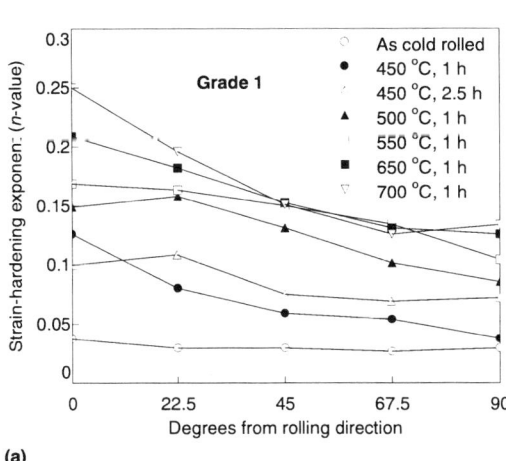

(a)

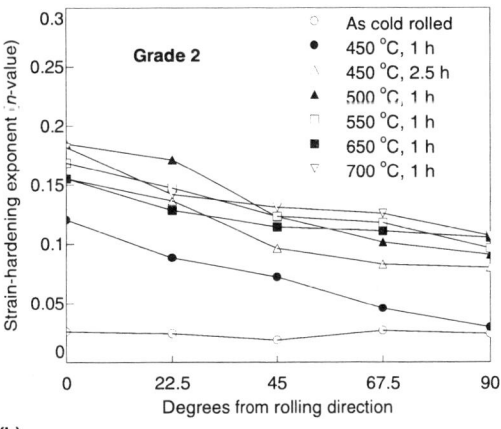

(b)

Effect of tensile direction on n-value of (a) titanium with 0.057 wt% O and (b) titanium with 0.1 wt% O.
Source: *Titanium '80, Science and Technology*, H. Kimura and O. Izumi, Ed., TMS/AIME, 1980, p 2541

Strain Ratio

Titanium has a relatively large strain ratio (*r*-value), which explains its superior drawability compared to steel or aluminum.

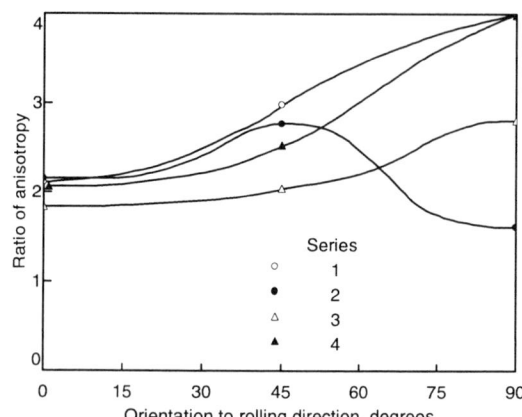

CP Ti: Typical variations in strain ratio (*r*)

See the previous table on strain-hardening exponents for a description of material series 1, 2, 3, and 4.
Source: *Titanium '80, Science and Technology*, H. Kimura and O. Izumi, Ed., TMS/AIME, 1980, p 1121-1129

Grade 1 Ti: Typical strain ratios (*r*)

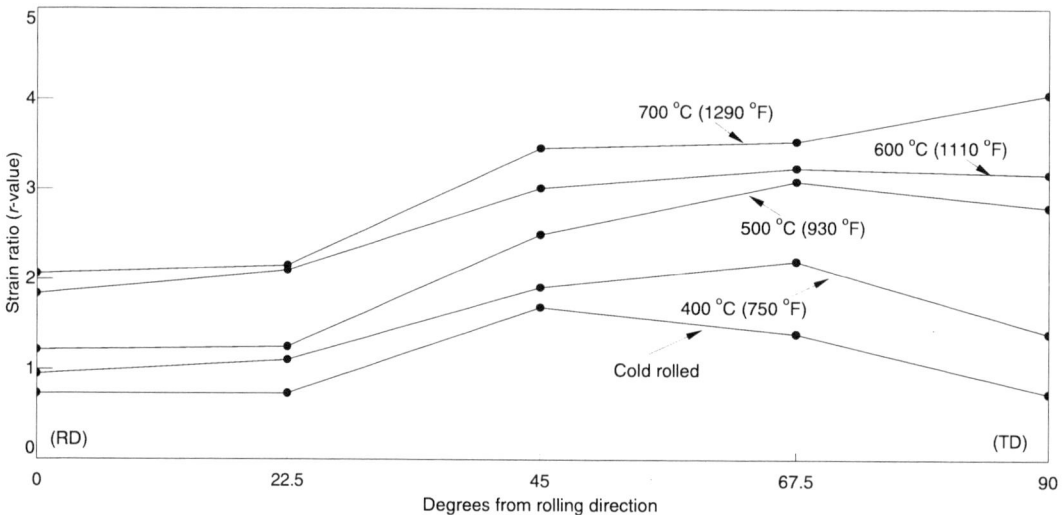

Cold rolled sheet (0.7 mm, or 0.02 in., thick), 0.01 wt% O, annealed 30 min at indicated temperature in argon.
Source: *Titanium '80, Science and Technology*, H. Kimura and O. Izumi, Ed., TMS/AIME, 1980, p 2534

CP Ti: Strain ratios vs temperature

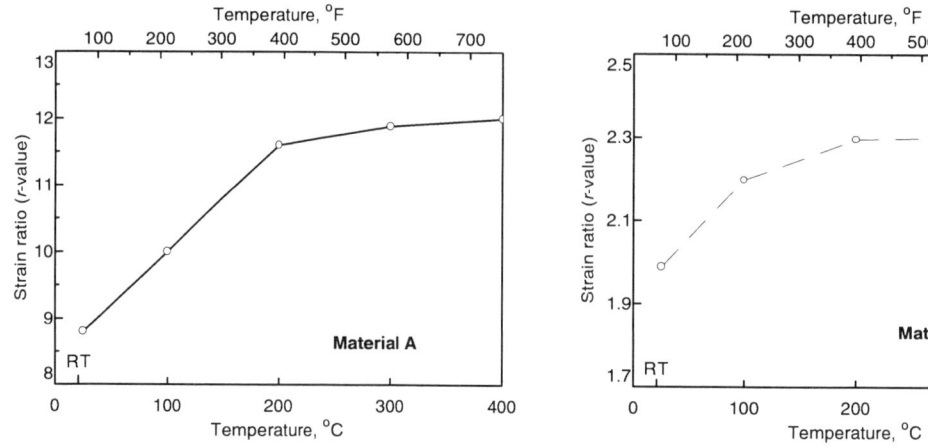

Dependence of *r*-values on temperature for a high *r*-value material and a low *r*-value material.
Source: *Titanium '80, Science and Technology*, H. Kimura and O. Izumi, Ed., TMS/AIME, 1980, p 2525

CP Ti: Strain ratio vs tensile direction

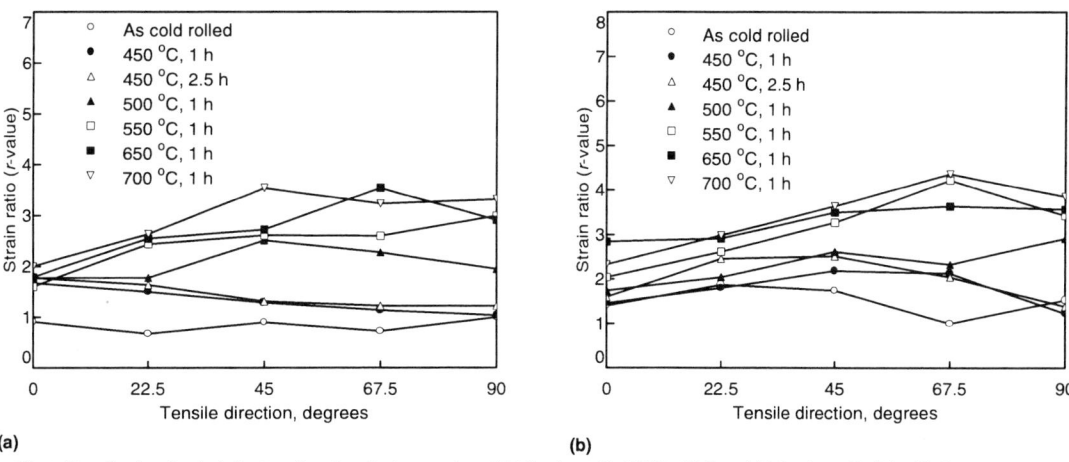

Effect of tensile direction (relative to rolling direction) on r-value of (a) titanium with 0.057 wt% O and (b) titanium with 0.1 wt% O.
Source: *Titanium '80, Science and Technology*, H. Kimura and O. Izumi, Ed., TMS/AIME, 1980, p 2542

CP Ti: Effect of rolling direction on strain ratio of annealed strip

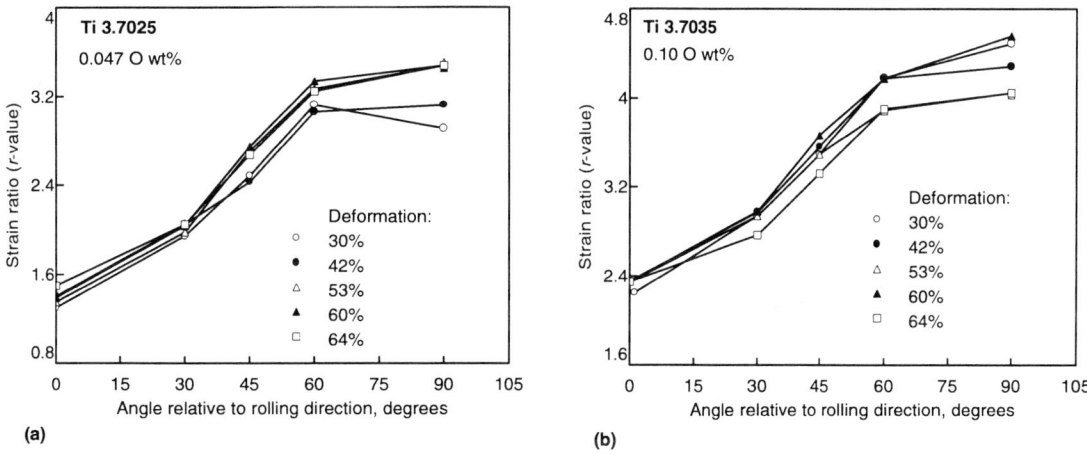

Deformed from 3 mm to 0.7 mm (0.1 to 0.02 in.) applying different degrees of deformation with or without an intermediate anneal. Final annealing time and temperature unspecified.
Source: *Titanium '80, Science and Technology*, H. Kimura and O. Izumi, Ed., TMS/AIME, 1980, p 2572

Strain-Rate Sensitivity

Grade 1 Ti: Elongation and strain-rate sensitivity at 20 °C

Grain size, μm	Direction, degrees	Uniform elongation,%	Total elongation,%	Strain-rate sensitivity exponent(a)
28	0	31	59(b)	0.018
	90	10	49(b)	0.028
19	0	18	46(c)	0.023
	90	8	40(c)	0.027
85	0	41	57(b)	0.027
	90	8	51(b)	0.026
114	0	...	53(b)	0.012
	90	12	40(b)	0.031
44	0	40	56(c)	0.020
	90	9	47(c)	0.024

Note: Mechanical properties measured in unaxial tension. Crosshead speed 5 mm/min. Initial strain rate 3.3×10^{-3}/s. (a) Crosshead speed change from 0.5 to 5 mm/min. (b) Gage length 20 mm (0.8 in.). (c) Gage length 25 mm (0.9 in.). Source: *Titanium, Science and Technology*, Vol 1, G. Lütjering, U. Zwicker, and W. Bunk, Ed., Deutsche Gesselschaft für Metallkunde, Germany, 1985, p 539-546

Grade 2 Ti: Strain-rate sensitivity index, m

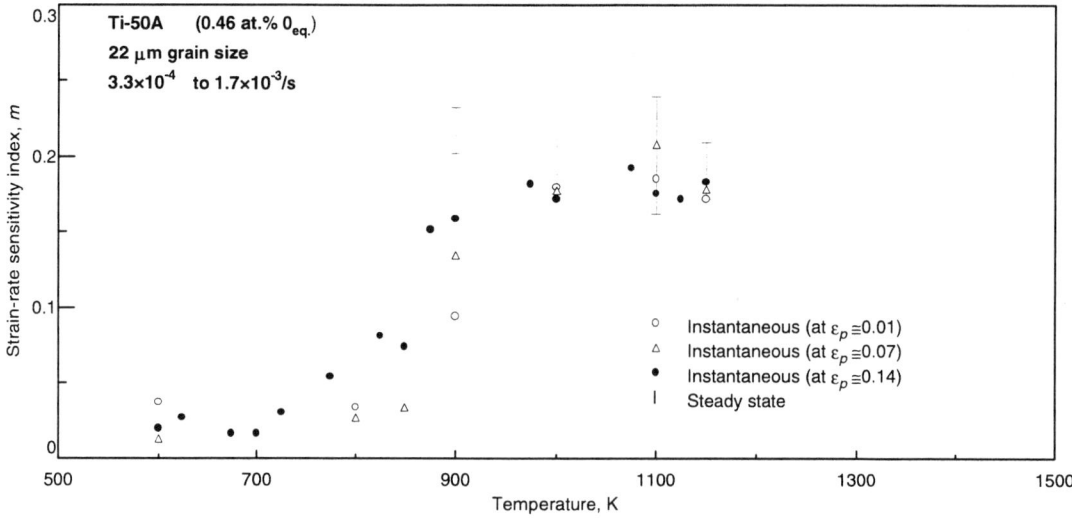

Source: *Metall. Trans. A*, Vol 4, Dec 1973, p 2811

Grade 1 Ti: Elongation and strain-rate sensitivity at 150 °C

Grain size, μm	Direction, degrees	Uniform elongation,%	Total elongation,%	Strain-rate sensitivity exponent(a)
28	0	28	57(b)	0.041
	90	9	66(b)	0.043
19	0	20	56(c)	0.041
	90	6	60(c)	0.042
85	0	24	56(b)	0.038
	90	9	55(b)	0.042
114	0	19	46(b)	0.050
	90	10	56(b)	0.055
44	0	...	...	0.034
	90	10	65+(c)	0.039

Note: Mechanical properties measured in uniaxial tension. Crosshead speed 5 mm/min. Initial strain rate 3.3×10^{-3}/s. (a) Crosshead speed change from 0.5 to 5 mm/min. (b) Gage length 20 mm (0.8 in.). (c) Gage length 25 mm (0.9 in.). Source: *Titanium, Science and Technology*, Vol 1, G. Lütjering, U. Zwicker, and W. Bunk, Ed., Deutsche Gesselschaft für Metallkunde, Germany, 1985, p 539-546

Forming Limit Diagrams

Grade 1 Ti: Effect of annealing temperature

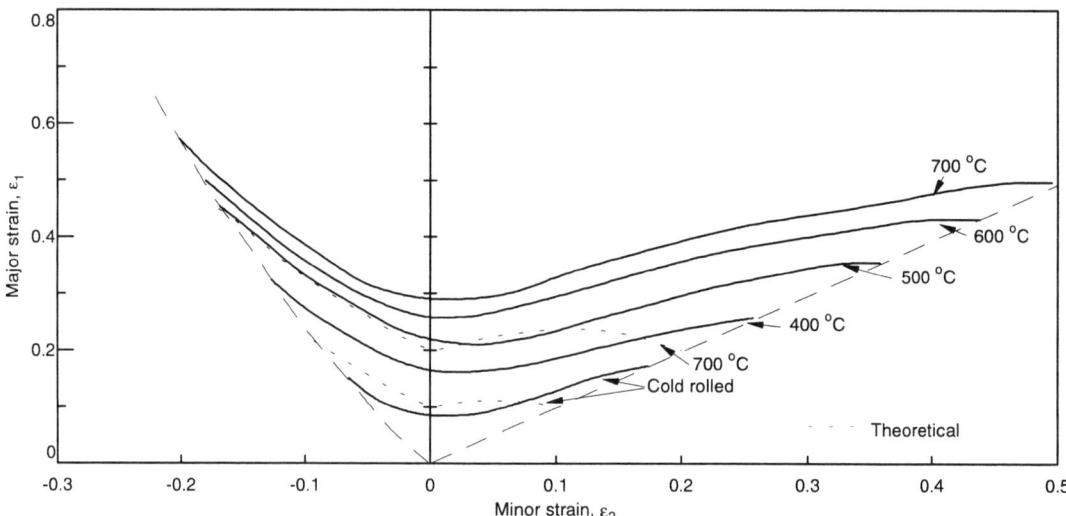

Forming limit diagram (FLD) for 0.70 mm (0.0275 in.) thick cold rolled sheet annealed 30 min in argon at indicated temperatures. Punch stretching and hydraulic bulging were used to obtain the FLD. The punch stretching was performed with a hemispherical punch of 60 mm (2.3 in.) and with varying the specimen width and the condition of lubrication.
Source: *Titanium '80, Science and Technology*, H. Kimura and O. Izumi, Ed., TMS/AIME, 1980, p 2535

CP Ti: Forming limit curve

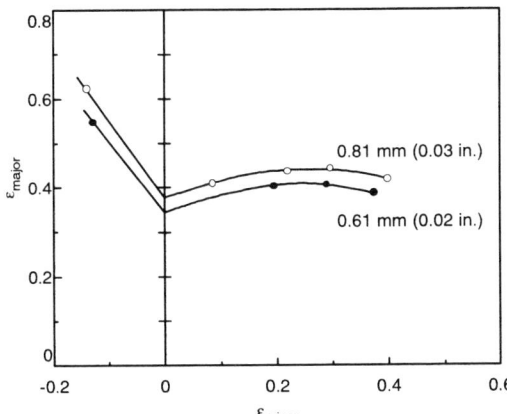

Four points on the forming limit curve were determined on the bulge test machine using one circular and three elliptical dies. The experiments with the circular die indicated that the formability of the material was lowest perpendicular to the rolling direction (fracture in rolling direction). On the basis of this, the rolling direction of the specimens for the experiments with the elliptical dies were aligned with the major axes of the dies before testing. In the tensile testing region, one point was calculated from the experiments perpendicular to the rolling direction. Composition: 0.055 wt% O.
Source: *Titanium '80, Science and Technology*, H. Kimura and O. Izumi, Ed., TMS/AIME, 1980, p 1129

CP Ti: Warm working FLD of high r-value material

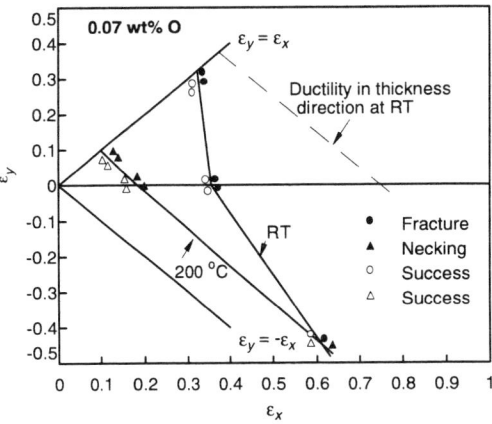

Forming limit diagram of material with an r-value of 8.8 at room temperature and 11.7 at 200 °C. See figure on next page for FLD with typical r value.
Source: *Titanium '80, Science and Technology*, H. Kimura and O. Izumi, Ed., TMS/AIME, 1980, p 2527

Warm Working Forming Limit Diagrams

At room temperature, the forming limit is determined by necking, because the strain at the thinning limit is larger than necking and is presumed to form the dotted lines shown in the forming limit diagram (FLD) in the accompanying figures. Under warm working conditions, forming limit strains are extremely small in the region of $\varepsilon_y > 0$ for strains in bending and biaxial stretching. Decrease of forming limit strain in this region is more remarkable in material A with a high r-value. On the other hand, forming limit strains in uniaxial stretching (tensile elongation) are rather large at the temperature of warm working condition and are also improved with an increase in r-value. These forming limit diagrams obtained at warm working temperature explain the inconsistency between large tensile elongation and poor bendability of titanium sheet, particularly, specimens with large r-values.

Effect of Strain Path

Forming limit diagrams for two grades of titanium sheet show a shift to higher limiting strains when the strain path involves a two-stage path such as equibiaxial tensile straining followed by uniaxial tensile straining, or uniaxial straining followed by biaxial straining (i.e., Path-I and Path-II in the figures, respectively). Thus, a higher forming limit without fracture can be achieved by selecting a proper strain path as opposed to a single-stage strain path (see next section). The increase in limit strains in the region around the plane-strain deformation region is especially remarkable.

CP Ti: Warm working FLD with a typical r-value

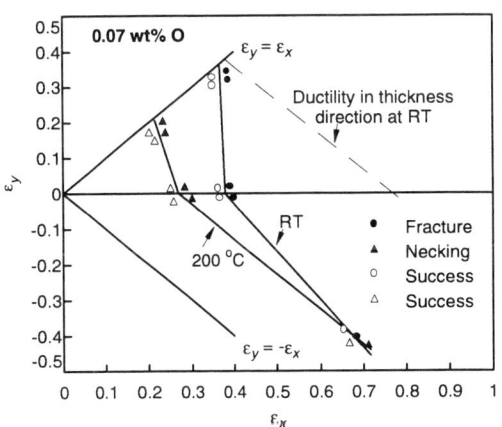

Forming limit diagram of material with an r-value of 2.0 at room temperature and 2.3 at 200 °C.
Source: *Titanium '80, Science and Technology*, H. Kimura and O. Izumi, Ed., TMS/AIME, 1980, p 2527

CP Ti: FLD for a two-stage strain path (Path I)

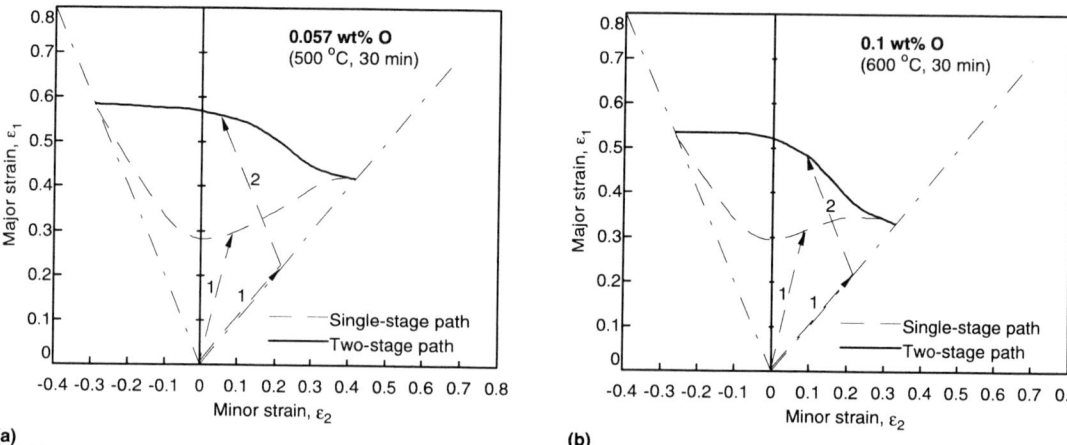

(a) (b)

Cold-rolled 0.7 mm (0.02 in.) sheet annealed as indicated. The two-stage strain path (1 and 2 in the figure) was a combination of equibiaxial tensile deformation followed by uniaxial tensile deformation. Circular blanks were bulged into cups of different heights (step 1) and specimens were cut from the cups for uniaxial elongation (step 2).
Source: S. Kohara, Forming Limit Curves of Titanium Sheet, in *Titanium, Science and Technology*, Vol 1, G. Lütjering, U. Zwicker, and W. Bunk, Ed., Deutsche Gesselschaft für Metallkunde, Germany, 1985, p 547-551

CP Ti: FLD for a two-stage strain path (Path II)

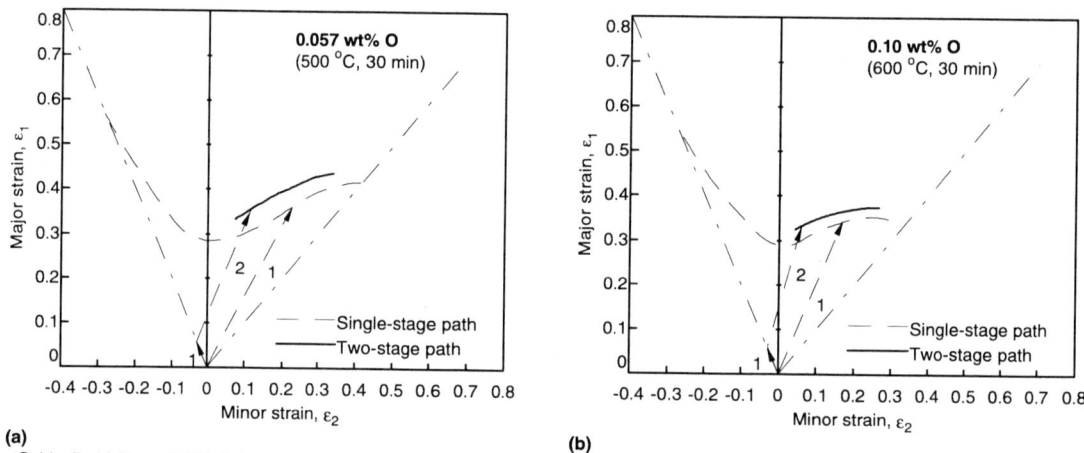

(a) (b)

Cold-rolled 0.7 mm (0.02 in.) sheet annealed as indicated. The two-stage strain path (1 and 2) was a combination of uniaxial tensile deformation (1) followed by biaxial tensile deformation (2).
Source: S. Kohara, Forming Limit Curves of Titanium Sheet, in *Titanium, Science and Technology*, Vol 1, G. Lütjering, U. Zwicker, and W. Bunk, Ed., Deutsche Gesselschaft für Metallkunde, Germany, 1985, p 547-551

Single-Stage Strain Path

CP Ti: FLD from single-path straining

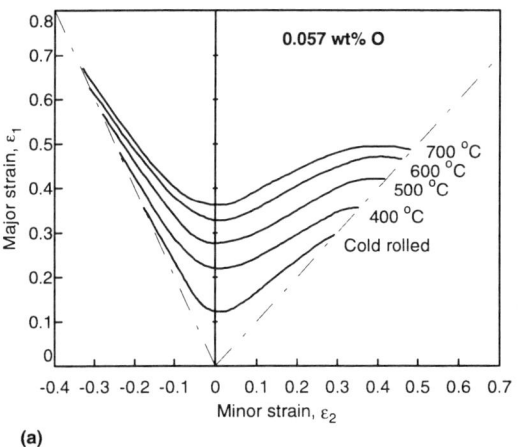

(a)

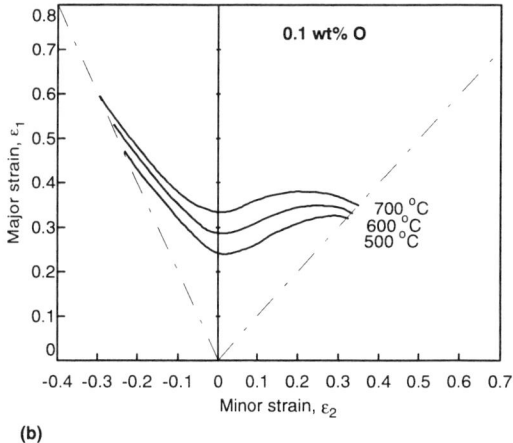

(b)

Cold-rolled sheet 0.7 mm (0.027 in.) thick annealed 30 min at indicated temperatures. The forming limit curves were determined by means of punch stretching, hydraulic bulging, and tensile testing. To produce various strain states between unidirectional tension and equibiaxial tension, rectangular blanks with various widths and three different lubricating conditions were adopted for punch stretching with a die of 100 mm (4 in.) diam and a punch of 60 mm (2.3 in.) diam, and the specimens with various widths were used for tensile testing. Furthermore, three types of elliptical dies with aspect ratios of 100:65, 100:75, and 100:85 mm and a circular die of 100 mm (4 in.) diam were used for hydraulic bulging.
Source: S. Kohara, Forming Limit Curves of Titanium Sheet, in *Titanium, Science and Technology*, Vol 1, G. Lütjering, U. Zwicker, and W. Bunk, Ed., Deutsche Gesellschaft für Metallkunde, Germany, 1985, p 547-551

Forming Methods

Depending on the severity of forming requirements and the capacity of equipment, titanium can be formed at room temperature or elevated temperature. When titanium parts are stretched at room temperature, a drop in compressive yield strength of up to 25% can be expected due to the Bauschinger effect. Titanium also requires a relatively large springback allowance during cold forming. Hot sizing is used to correct for variations in springback after cold forming.

CP Ti: Temperatures for specific forming operations

Forming method	Mild forming temperature		Severe forming temperature(a)	
	°C	°F	°C	°F
Drop hammer	200-315	400-600	480-540	900-1000
Hydropress	200-315	400-600	...	...
Brake forming	200-315	400-600	...	...
Spin forming	200-315	400-600	540-700	1000-1300
Drawing	200-315	400-600	480-540	900-1000
Matched die	200-315	400-600	...	...
Hydroform	200-315	400-600	...	...
Finish die	...	...	480-540	900-1000
Creep forming	...	...	480-540	900-1000

(a) Maximum time at temperature about 120 min. Temperatures should be held to a minimum to reduce scaling. Time at temperature is important because titanium is embrittled by oxygen above 540 °C (1000 °F) as a function of time and temperature. Generally, 2 h is maximum for 700 °C (1300 °F); 20 min is maximum for 870 °C (1600 °F). These are accumulated times to include heating times for single or multistage forming intermediate stress relief, and final stress relief. Source: *Titanium Alloys Handbook*, MCIC-HB-02, Battelle, 1972

Bending of Tubing

Round tubing of commercially pure titanium and alloy Ti-3Al-2.5V can be formed at room temperature in ordinary draw bending machines. When hot bending is required, the equipment is modified by adding some means of heating the tools. Minimum and preferred conditions for bending tubing of commercially pure titanium at room temperature and at elevated temperatures depend on the tube size (see table before "Heat Treatment" section). Tubing up to 175 mm (7 in.) in diameter ordinarily is bent at room temperature, while larger sizes are bent at temperatures of 175 to 205 °C (350 to 400 °C). In either case, bend radius is limited chiefly by tubing diameter, but maximum bend angle is affected by both diameter and wall thickness.

Commercially pure titanium deforms locally if tension is not applied evenly. Bending should be slow; rates of $\frac{1}{4}°$ to $4°$ per minute are suitable. A lubricant should be used.

Lubrication. Drawing oils are used as lubricants for forming commercially pure titanium tubing at room temperature. Grease with graphite is used as a lubricant for the hot bending of commercially pure titanium tubing, but is not recommended for temperatures above 315 °C (600 °F). Phosphate conversion coatings are sometimes used for hot bending of titanium tubing.

Suggested hot sizing conditions for unalloyed titanium: 480 to 540 °C (900 to 1000 °F) for 3 min with blank heated by contact with die.

Hot Forming. Most hot forming operations of titanium fall within the temperature ranges of 205 to 315 °C (400 to 600 °F). The time necessary to complete the hot forming steps may limit the temperature used.

Press Forming

Titanium and titanium alloys behave like stainless steels during press forming, except that the springback allowance is considerably greater (about 15° for cold bending). Titanium sheet is also more anisotropic than stainless steel. Therefore, sections being bent prior to stretching should be bent parallel to the grain. This practice permits more deformation during stretch forming, because maximum ductility occurs in the longitudinal direction. After cold stretching, hot sizing of some parts may be required due to variations in springback.

Press forming at room temperature usually is preferred if the press capacity is sufficient and if the bend radii are sufficiently large. Generally, the minimum bend radii decrease with increasing working temperature (see the discussions on bendability in the previous section on forming properties and tensile strength). However, no significant deterioration of bendability or stretchability at warm working temperature occurs in material that has been subjected to twinning deformation during prestraining under cold working conditions.

The effect of prestrain on bendability is shown for bending carried out at 200 °C (390 °F) after prestrain had been given by slight bending at both room temperature and –196 °C (–320 °F). Bendability under warm working conditions significantly improved by the prestrain at both room temperature and –196 °C (–320 °F). The prestrain at –196 °C (–320 °F) is more effective on the improvement of bendability at the warm working temperature than that at room temperature because of the larger activity of twinning at the lower temperature. These results prove that the deterioration of press formability of pure titanium sheet under warm working conditions is caused by the decrease in ductility in the thickness direction as a result of the small twinning potential at elevated

CP Ti: Effect of prestrain on bendability at 200 °C

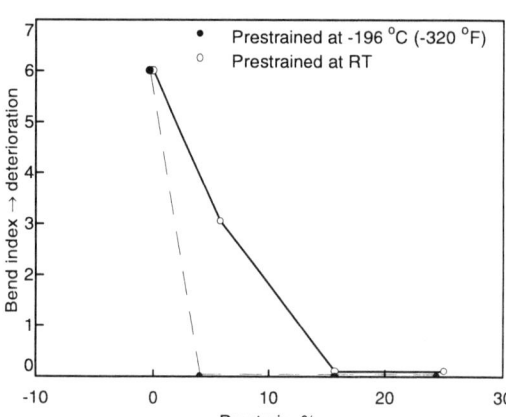

Effect of prestrain on bendability at 200 °C (390 °F) of commercially pure titanium sheet (0.072 wt% O) having a relatively high r-value of 8.8 at room temperature and 11.7 at 200 °C. A larger bending index value indicates poorer bendability.
Source: N. Itoh et al., The Press Formability of Commercially Pure Titanium at Warm Working Temperatures, Titanium '80, Science and Technology, H. Kimura and O. Izumi, Ed., TMS/AIME, 1980, p 2523-2529

CP Ti: Limiting pressing height for biaxial stretching

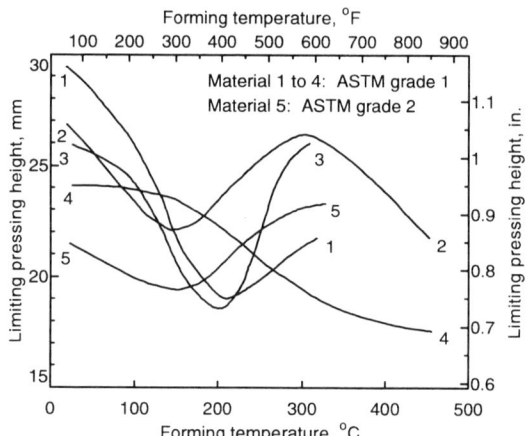

All materials were in the as-received annealed condition, except for material 3, which was lab annealed for a coarse grain size. Equibiaxial stretching was carried out over a hemispherical punch with a diameter of 50 mm (2 in.), the punch and the clamping plates being heated by cartridge resistance heaters. Graphite was used as a lubricant.
Source: W.T. Roberts and D.V. Wilson, Warm Forming of Titanium Sheet, Titanium, Science and Technology, Vol 1, G. Lütjering, U. Zwicker, and W. Bunk, Ed., Deutsche Gesellschaft für Metallkunde, Germany, 1985, p 541

temperature.

Biaxial Stretching. It is well established that room-temperature limit strains in biaxial stretching of many ductile sheet materials are related to the ratio of initial sheet thickness to initial grain size (*Metall. Trans. A*, Vol 12, 1981, p 1595). For unalloyed titanium skins, the stretch forming limit is about 20% for cold and hot conditions. Sections are normally stretch formed at 0.75 to 1 degree per minute. In both hot and cold stretching of skins, the elongation perpendicular to the stretch should not exceed one half of the stretch.

Like bending ductility, pressing performance during biaxial stretching degrades with a loss of deformation twinning at warm working temperatures (see figures). However, one curve shows some improvement in stretching performance in the range of 200 to 300 °C (390 to 570 °F). This temperature range appears to be in general agreement with the following brake forming temperatures recommended in the *Titanium Alloys Handbook* (MCIC-HB-02, Battelle, 1972, 3:72-93):

- Blank temperature: 200 to 315 °C (400 to 600 °F)
- Punch and die: 260 °C (500 °F)

Dwell times range around 30 s after the part is formed in heated dies.

CP Ti: Surface limit strains during biaxial stretching

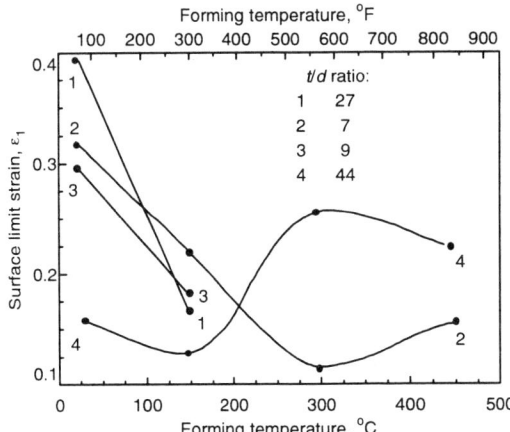

The low strain limit for material 4 is due to higher interstitial content of this material (grade 2 Ti) compared to the other materials (all grade 1 Ti).
Thickness (t) and grain size (d) ratio as indicated. The basal plane pole figures of all sheet consisted of two peaks on either side of the sheet normal on the normal-traverse direction diameter. Variations occurred in maximum peak intensity (× random) and ranged from 6 to 12, except for 28 for material 2.
Source: W.T. Roberts and D.V. Wilson, Warm Forming of Titanium Sheet, *Titanium Science and Technology*, Vol 1, G. Lütjering, U. Zwicker, and W. Bunk, Ed., Deutsche Gesselschaft für Metallkunde, Germany, 1985, p 541

Drawing

CP Ti: Drawability vs strain-hardening exponent

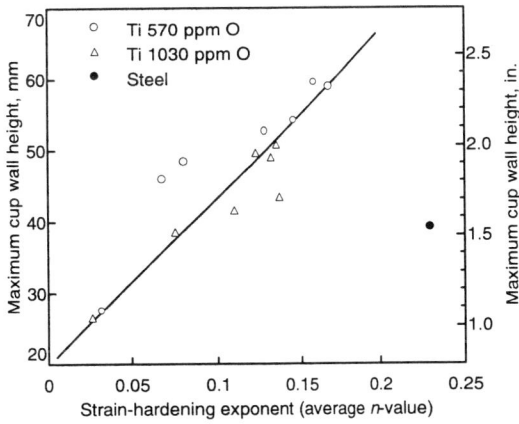

Effect of n-value on deep drawability of titanium annealed to various stages.
Source: *Titanium '80, Science and Technology*, H. Kimura and O. Izumi, Ed., TMS/AIME, 1980, p 2543

CP Ti: Drawability vs strain ratio (r)

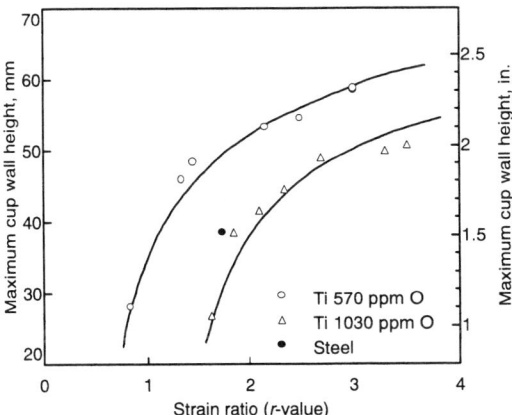

Effect of r-value on deep drawability of titanium annealed to various stages.
Source: *Titanium '80, Science and Technology*, H. Kimura and O. Izumi, Ed., TMS/AIME, 1900, p 2543

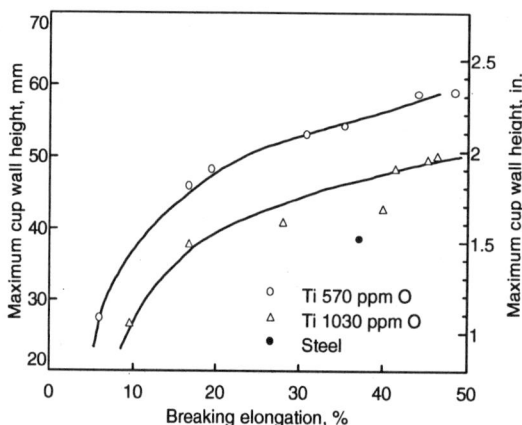

CP Ti: Drawability vs breaking elongation

Effect of elongation on deep drawability of titanium annealed to various stages.
Source: *Titanium '80, Science and Technology*, H. Kimura and O. Izumi, Ed., TMS/AIME, 1980, p 2543

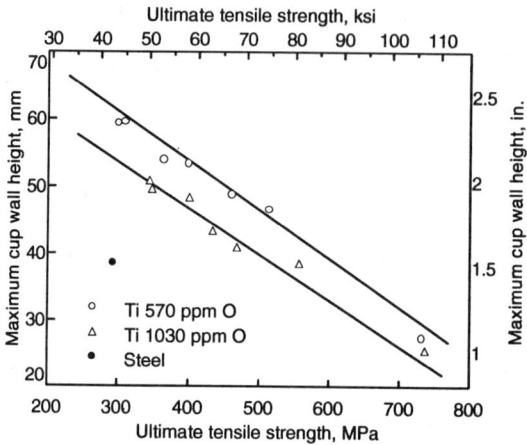

CP Ti: Drawability vs UTS

Effect of tensile strength on deep drawability of titanium annealed to various stages.
Source: *Titanium '80, Science and Technology*, H. Kimura and O. Izumi, Ed., TMS/AIME, 1980, p 2543

Tube Bending

Because commercially pure titanium tends to deform locally under tensile loads, tube bending speeds must be kept low; speeds of 1/4 to 4 degrees/min have been used to produce satisfactory parts. Bend quality can also be adversely affected by excessive wear of the mandrel and wiper die.

Grade 2 Ti: Bending limitations for tubing (see also "Bending of Tube")

Tube diameter		Wall thickness		Minimum bend radius		Preferred bend radius		Maximum bend angle(a), degrees	
								Minimum bend radius	Preferred bend radius
mm	in.	mm	in.	mm	in.	mm	in.		
Room-temperature bending									
38	1.5	0.40	0.016	57	2.25	75	3	90	120
		0.50	0.020	57	2.25	75	3	100	160
50	2	0.40	0.016	75	3	100	4	80	110
		0.40	0.016	75	3	100	4	100	150
64	2.5	0.40	0.016	95	3.75	125	5	70	100
		0.50	0.020	95	3.75	125	5	90	140
		0.89	0.035	95	3.75	125	5	110	180
Elevated-temperature bending(b)									
75	3	0.40	0.016	114	4.5	150	6	90	120
		0.50	0.020	114	4.5	150	6	110	160
		0.89	0.035	114	4.5	150	6	130	180
89	3.5	0.40	0.016	133	5.25	175	7	90	120
		0.50	0.020	133	5.25	175	7	110	160
		0.89	0.035	133	5.25	175	7	130	180
100	4	0.50	0.020	150	6	205	8	110	160
		0.89	0.035	150	6	205	8	120	180
114	4.5	0.50	0.020	170	6.75	230	9	130	140
		0.89	0.035	170	6.75	230	9	140	140
125	5	0.50	0.020	255	10	255	10	...	110
150	6	0.50	0.020	305	12	305	12	...	100

(a) Bend angle is predicated on a clamp section three times as long as the tube diameter and on maximum mandrel-ball support. (b) The best ductility for bending titanium is obtained between 175 and 200 °C (350 and 400 °F). Source: R. Woods, *Titanium Alloys Handbook*, MCIC-HB-02, Battelle, 1972

Heat Treatment

Because of the small amount of β stabilizer content in the highest purity grades of titanium, the temperature range where α and β phases are in equilibrium is very narrow. Practically no differences in mechanical properties (except perhaps for ductility) are found with acicular structures obtained by cooling from above the β transus and equiaxed structures obtained by thermal history below the β transus.

CP Ti: Recommended heat treatment conditions (all grades)

Heat treatment	Temperature °C	°F	Time	Cooling method
Stress relief	480 to 595	900 to 1100	15 min to 4 h	Air cool or step cool
Annealing	650 to 760	1200 to 1400	6 min to 2 h	Air cool

Grade 4 Ti: Stress relief recovery of yield strength after 1 to 3% stretching

Annealing time, min	Annealing temperature °C	°F	Compressive yield strength recovery, %
60	440	825	94
15	580	1075	94

Note: Tensile yield strength of 480 MPa (70 ksi). Source: R.A. Woods, *Titanium Alloys Handbook*, MCIC-HB-02, Battelle, 1972

Grade 3 Ti: Relief of residual stress vs time

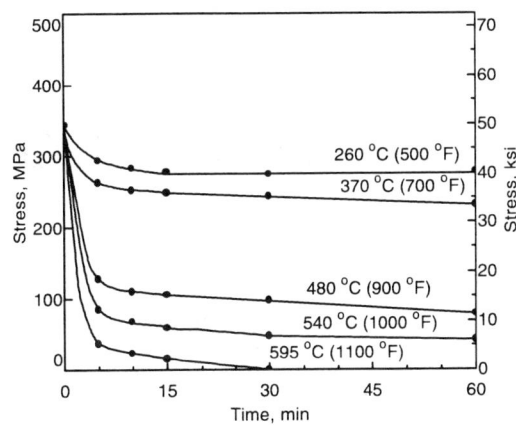

Source: R.A. Woods, *Titanium Alloys Handbook*, MCIC-HB-02, Battelle, 1972

Stress Relief

Stress relief treatments are designed to partially remove or largely overcome the effects of residual stresses, or to restore compressive yield strengths after stretch forming at room temperature.

In practice, where the degree of deformation is not always known, or where the reduction of residual stresses to a low level is desired, a treatment of 15 to 30 min at 540 to 595 °C (1000 to 1100 °F) is usually given for all unalloyed grades.

Annealing. For the complete removal of residual stresses or for obtaining optimum tensile ductility, higher temperatures than stress relief annealing temperatures are used. Commonly 2 h exposure at 705 °C (1300 °F) followed by air cooling is used for all unalloyed titanium grades. Higher temperatures are possible, but oxidation and contamination must be taken into account. Iron content can have significant effect on final grain size obtained during annealing (see figures).

CP Ti: Influence of annealing time and temperature on grain size

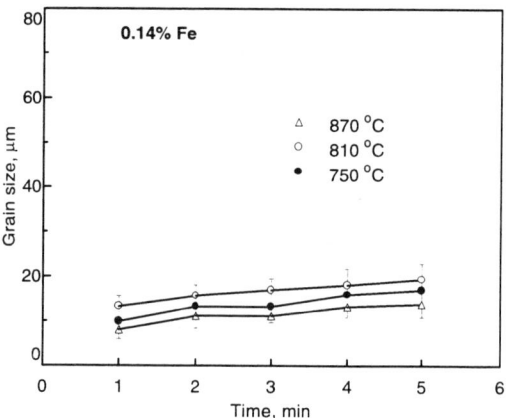

Source: K. Hülse *et al.*, in *Sixth World Conference on Titanium*, P. Lacombe, R. Tricot, and G. Beranger, Ed., Les Editions de Physiques, Paris, 1989, p 1679

CP Ti: Influence of annealing time and temperature on grain size

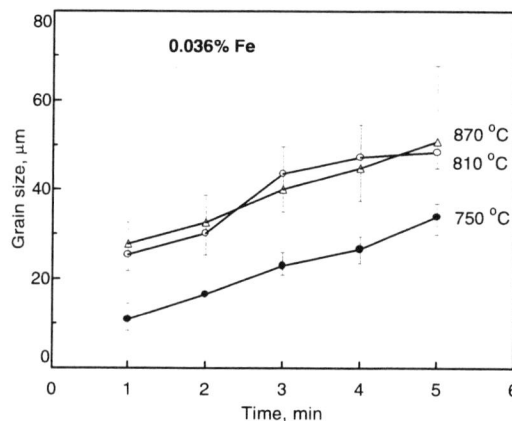

Source: K. Hülse *et al.*, in *Sixth World Conference on Titanium*, P. Lacombe, R. Tricot, and G. Beranger, Ed., Les Editions de Physiques, Paris, 1989, p 1679

Machining

Commercially pure titanium has machining characteristics similar to those of austenitic stainless steels. Titanium alloys, because of their higher hardness, are somewhat more difficult to machine, but the same general rules apply. Sharp tools, rigid setups, heavy feeds, slow speeds and an abundance of soluble oil coolant are needed for successful machining. Titanium requires low forces and demonstrates a complete absence of built-up edge. It can be machined to smooth surface finishes.

Sawing is best performed with high-speed friction saws running at a linear speed of 4000 to 4500 ft/min. The feed should be positive. Hack and band sawing are also possible. High-speed steel blades, heavy feeds, and slow speeds should be used. Surface scale and contaminated surfaces will result in excessive blade wear if not removed.

Alpha and Near-Alpha Alloys

Ti-3Al-2.5V

Common Name: Tubing Alloy, ASTM Grade 9
UNS Number: R56320

Ti-3Al-2.5V, which is intermediate in strength between unalloyed titanium and Ti-6Al-4V, has excellent cold formability required for production of seamless tubing, strip and foil. Like Ti-6Al-4V, Ti-3Al-2.5V has a high strength-to-weight ratio and is lighter than stainless steel. Ti-3Al-2.5V has 20 to 50% higher strength than unalloyed titanium at both room and elevated temperatures. It has comparable weldability, and is much more amenable to cold working than Ti-6Al-4V (which does not have good cold forming properties).

Chemistry and Density

With 3 wt% aluminum as an alpha stabilizer and 2.5 wt% vanadium as a beta stabilizer, Ti-3Al-2.5V is sometimes referred to as "half 6-4." High impurity levels may raise yield strength above maximum permitted values or decrease elongation or reduction in area below minimum values.

Density. 4.48 g/cm^3 (0.162 lb/in.^{3})

Product Forms

Ti-3Al-2.5V is available as foil, seamless tubing, pipe, forgings, and rolled products. Ti-3Al-2.5V was developed for tubing and foil applications. Seamless tubing made of Ti-3Al-2.5V is readily cold formed on the same type of conventional tube-bending equipment used for forming stainless steel. Cold worked and stress relieved tubing generally is not bent to radii less than 3 times the outer diameter in production shops, although radially textured tubing can be bent to 1.5. Relatively thin-wall tubing should be bent using tubing fillers or other inside-diameter constraints. Ti-3Al-2.5V tubing is readily welded by standard gas tungsten-arc welding with inert-gas shielding and by use of automatic welding tools with built-in inert-gas purge chambers.

Product Condition/Microstructure

Ti-3Al-2.5V is a near-alpha alpha-beta alloy that is generally used in the cold-worked and stress-relieved condition. Ti-3Al-2.5V can be heat treated to high strength, but it has very limited hardenability.

Applications

Ti-3Al-2.5V seamless tubing was originally developed for aircraft hydraulic and fuel systems and has a proven performance record in high-technology military aircraft, spacecraft, and commercial aircraft. The Lockheed C-5A was the first military production program in which Ti-3Al-2.5V tubing was employed. This tubing was also selected for the hydraulic system of the Concorde Supersonic Transport. Its first application in subsonic commercial aircraft was the Boeing 767. Since then, Ti-3Al-2.5V tubing has been chosen for most of the other commercial transports, commuter aircraft, and spacecraft. This alloy also can be readily rolled in strip and foil, the latter of which is used as the honeycomb layer between face sheets of Ti-6Al-4V sheet in sandwich structures.

Ti-3Al-2.5V is also employed, mostly in tubular form, in various nonaerospace applications such as sports equipment (golf-club shafts, tennis racquets, and bicycle frames), medical and dental implants, and expensive ballpoint-pen casings. In addition to its high strength-to-weight ratio, Ti-3Al-2.5V is being used in such applications because of its excellent torsion resistance (golf-club shafts and tennis racquets) and corrosion resistance (medical and dental products). Golf-club shafts of Ti-3Al-2.5V have been heat treated to tensile strengths of approximately 1140 MPa (165 ksi). Other sports products for which Ti-3Al-2.5V tubing is being investigated include ski poles, fishing poles, and tent stakes.

Use Limitations. The rotary flexure fatigue life of pressurized Ti-3Al-2.5V tubing is influenced by its crystallographic texture by residual stresses produced in straightening operations, surface roughness, and ovality. Flattening during bending operations reduces the impulse fatigue life of tubing as a result of the superposition of three additive stresses: residual stresses due to flattening, membrane stresses following pressurization, and bending stresses in the flattened tube wall. Overpressurization of tubing (auto-frettage) can decrease flattening, thus increasing the impulse fatigue life. Use of improper support assemblies may cause end fitting displacement with attendant installation stresses on the final system, outweighing the beneficial effect of overpressurization.

The reliability of tubing is adversely affected by cracking in service resulting from internal and surface irregularities. Production defects may be inclusions, separations in the tubing wall, or fissures at the inner and outer surfaces. Surface damage usually takes the form of chafing or denting.

Ti-3Al-2.5V: Specifications and compositions

Specification	Designation	Description	Al	C	Fe	H	N	O	V	OT	Other
UNS	R56320		2.5-3.5	0.05	0.25	0.013	0.02	0.12	2-3		bal Ti
UNS	R56321	Weld Fill Wir	2.5-3.5	0.04	0.25	0.005	0.012	0.1	2-3		bal Ti
China											
	Ti-3Al-2.5V		2.5-3.5	0.08 max	0.3 max	0.015 max	0.05 max	0.12 max	2-3		Si 0.15 max; bal Ti
Europe											
AECMA Ti-P69	prEN3120	Tub CW SR	2.5-3.5	0.05 max	0.3 max	0.015 max	0.02 max	0.12 max	2.5-3.5	0.4 max	Y 0.005 max; OE 0.1 max; bal Ti
Russia											
GOST	AK2		3					0.25-0.35	2.5		bal Ti
GOST	IMP-7	Powd	3		0.3	0.01	0.03	0.16	2		Si 0.6; bal Ti
USA											
AMS 4943D		Tub Ann	2.5-3.5	0.05	0.3	0.015	0.02	0.12	2-3	0.4	Y 0.005; bal Ti
AMS 4944D		Smls Tub CW SR	2.5-3.5	0.05 max	0.3 max	0.015 max	0.02 max	0.12 max	2-3	0.4 max	Y 0.005 max; OE 0.1 max; bal Ti
AMS 4944D		Tub CW SR	2.5-3.5	0.05	0.3	0.015	0.02	0.12	2-3	0.4	Y 0.005; bal Ti
AMS 4945		Smls Tub	2.5-3.5	0.05 max	0.3 max	0.015 max	0.02 max	0.12 max	2-3	0.4 max	Y 0.005 max; OE 0.1 max; bal Ti
ASTM B 337	Grade 9	Smls Weld Pip Ann	2.5-3.5	0.05	0.25	0.013	0.02	0.12	2-3	0.4	bal Ti
ASTM B 338	Grade 9	Smls Weld Tub CW SR	2.5-3.5	0.1	0.25	0.013	0.02	0.12	2-3	0.4	bal Ti
ASTM B 348	Grade 9	Bar Bil Ann	2.5-3.5	0.05	0.25	0.0125	0.02	0.12	2-3	0.4	bal Ti
ASTM B 381	Grade F-9	Frg Ann	2.5-3.5	0.05	0.25	0.015	0.02	0.12	2-3	0.4	bal Ti
ASTM B265-79		Sh Strp Plt	2.5-3.5	0.1 max	0.25 max	0.015 max	0.02 max	0.15 max	2-3		bal Ti
AWS A5.16-70	ERTi-3Al-2.5V-1	Weld Fill Met	2.5-3.5	0.04	0.25	0.005	0.012	0.1	2-3		bal Ti
AWS A5.16-70	ERTi-3Al-2.5V	Weld Fill Met	2.5-3.5	0.05	0.25	0.008	0.02	0.12	2-3		bal Ti
MIL T-9046J	Code AB-5	Sh Strp Plt Ann	2.5-3.5	0.05	0.3	0.015	0.02	0.12	2-3	0.4	bal Ti
MIL T-9047G	Ti-3Al-2.5V	Bar Bil Ann	2.5-3.5	0.05	0.3	0.015	0.02	0.12	2-3	0.4	Y 0.005; bal Ti

Ti-3Al-2.5V: Commercial compositions

Specification	Designation	Description	Al	C	Fe	H	N	O	V	OT	Other
Germany											
Deutsche T	Contimet AlV 32	Plt Bar Frg Pip Ann	2.5-3.5	0.05	0.3	0.015	0.04	0.12	2-3		bal Ti
Japan											
Kobe	KS3-2.5	Plt Sh Wir Bar Ann	2.5-3.5		0.3	0.0125	0.02	0.12	2-3		bal Ti
Sumitomo	SAT-325										
Toho	325AT		2.5-3.5	0.1	0.25	0.013	0.12	0.12	2-3		bal Ti
USA											
Cabot	Ti-3Al-2.5V										
Crucible	3Al-2.5V										
OREMET	Ti 3-25										
RMI	RMI 3Al-2.5V	Bar Tub Strp Ann	2.5-3.5	0.05	0.3	0.0125	0.02	0.12	2-3		bal Ti
RMI	RMI 3Al-2.5V	Bar Tub Strp CW SR	2.5-3.5	0.05	0.3	0.0125	0.02	0.12	2-3		bal Ti
TIMET	TIMETAL 3-2.5										
TMCA	Ti 325										

Phases and Structures

With a phase structure consisting mostly of α grains, with small amounts of β titanium in the matrix and grain boundaries, the major microstructural features of Ti-3Al-2.5V are the morphology of the α phase and the alignment (texture) of the α crystals. The structure is typically cold worked and partially recrystallized. However, transformation products can be achieved by heat treatment.

Beta Transus. 935 ± 15 °C (1715 ± 25 °F)

Alpha morphology can vary from 5 to 80% equiaxed alpha, depending on the amount of working and recrystallization. As a tubing material, large amounts of cold working produce an elongated structure of α grains, with the β phase strung out at the grain boundaries. Annealing recrystallizes the cold-worked structure to more rounded grains. Grain refinement during forging develops more slowly for Ti-3Al-2.5V than for commercial-purity titanium.

Ti-3Al-2.5V: Isothermal transformation

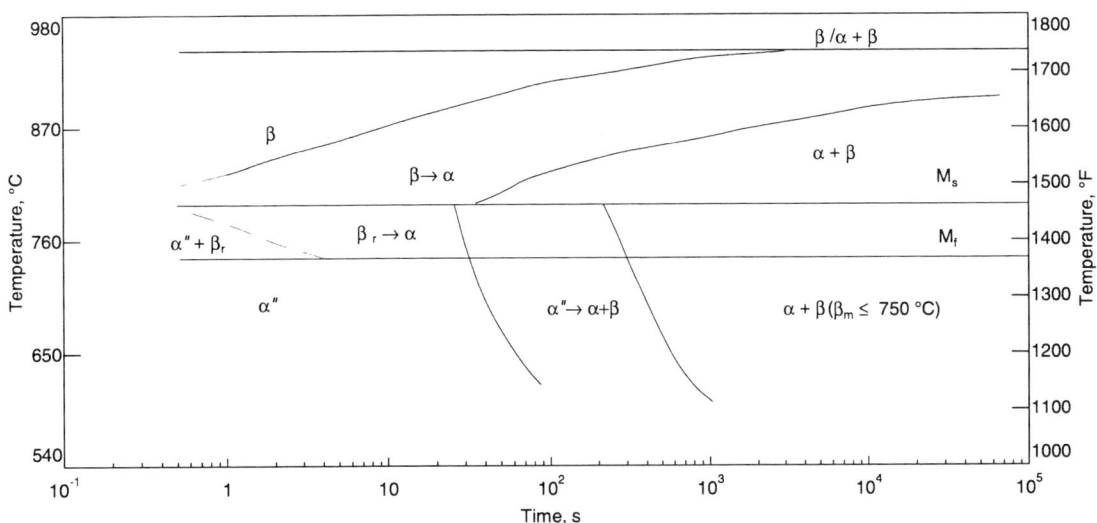

Composition: 3.1 wt% Al, 2.4 wt% V, 0.006 wt% C, 0.064 wt% Fe, 0.0035 wt% H, 0.0070 wt% N, 0.0795 wt% O, bal Ti.
Source: *Aerospace Structural Metals Handbook*, Code 3725, Battelle Columbus Laboratories, 1965, p 22

Ti-3Al-2.5V: Continuous cooling transformations

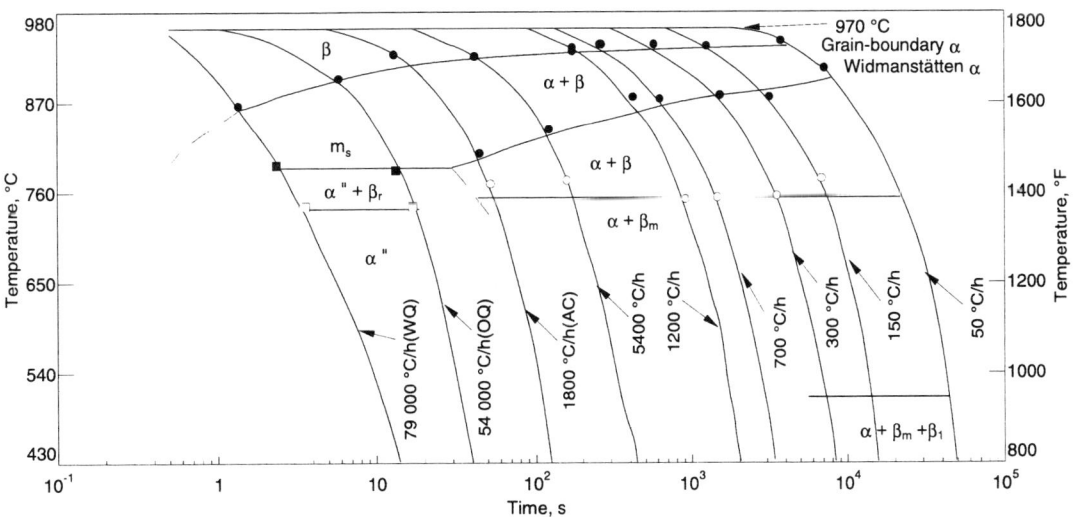

Composition: 3.1 wt% Al, 2.4 wt% V, 0.006 wt% C, 0.064 wt% Fe, 0.0035 wt% H, 0.0070 wt% N, 0.0795 wt% O, bal Ti.
Source: *Aerospace Structural Metals Handbook*, Code 3725, Battelle Columbus Laboratories, 1965, p 22

Transformation Products

On cooling to an isothermal temperature below the martensite start (M_s) point of 790 ± 5 °C (1454 ± 9 °F), first some α phase is formed above M_s and then the remaining, predominate portion of the β phase is transformed into a supersaturated hexagonal martensite (α″). Below M_s and above the martensite finish (M_f) temperature of 740 ± 5 °C (1364 ± 9 °F), there remains a residual β phase, which is probably transformed isothermally to α phase. The resulting structure for isothermal reaction is α + α″, where the α″ phase below 750 °C (1380 °F) decomposes *discontinuously* into a two-phase α + β structure and a metastable β phase enriched with β-stabilizing elements.

Transformation during continuous cooling at rates exceeding 22 °C/s (39.5 °F/s) results in a hexagonal martensite structure, while slower rates result in a structure of α titanium and a metastable β. At cooling rates slower than 5 °C/h (9.5 °F/h), a needlelike precipitate is formed in the metastable β phase when the temperature drops below 500 °C (930 °F). For a cooling rate of 50 °C/h (90 °F/h), the first α phase is nucleated at 935 °C (715 °F) in grain boundaries. The grain-boundary film grows at a rather moderate rate. At about 900 °C (1650 °F), Widmanstätten plates grow from grain boundaries and from nuclei within grains, and the growth rate increases markedly. The major portion of the transformation is terminated at about 750 °C (1380 °F), which corresponds well with the alteration from β to metastable β and appears independent of cooling rate.

Physical Properties

Ti-3Al-2.5V: Summary of typical physical properties

Beta transus	935 ± 15 °C (1715 ± 25 °F)
Melting (liquidus) point	1700 °C (3100 °F)
Density(a)	4.48 g/cm³ (0.162 lb/in.³)
Electrical resistivity(a)	1.27 μΩ · m
Magnetic permeability	Nonmagnetic
Thermal conductivity(a)	8.3 W/m · K (4.8 Btu/ft · h · °F)
Thermal coefficient of linear expansion (b)	9.61×10^{-6}/°C (5.34×10^{-6}/°F)

(a) Typical values at room temperature of about 20 to 25 °C (68 to 78 °F). (b) Mean coefficient from room temperature to 95 °C (200 °F)

Elastic Properties

Ti-3Al-2.5V: Elastic properties

Young's modulus	
At RT	95-105 GPa (14-15 × 10⁶ psi)
At 230 °C (450 °F)	75-85 GPa (11-12 × 10⁶ psi)
Shear modulus	
At RT	43-45 GPa (6.2-6.5 × 10⁶ psi)
Poisson's ratio	Typically 0.30

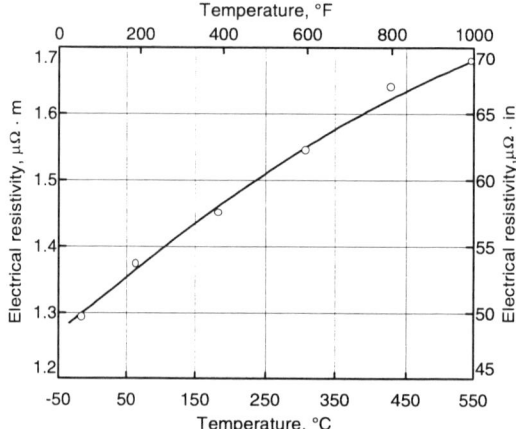

Ti-3Al-2.5V: Resistivity vs temperature

Source: *Aerospace Structural Metals Handbook*, Code 3725, Battelle Columbus Laboratories, 1965

Electrical Resistivity

Typical value at 20 °C (70 °F) is 1.27 μΩ · m (50 μΩ · in.).

Chemical/Corrosion Properties

Like titanium and its other alloys, successful use of Ti-3Al-2.5V can be expected in mildly reducing to highly oxidizing environments in which protective oxide films spontaneously form and remain stable. However, hot, concentrated, low-pH chloride salts corrode titanium, and warm or concentrated solutions of hydrochloric, phosphoric, and oxalic acids also are damaging. In general, all acidic solutions that are reducing in nature corrode titanium, unless they contain inhibitors. Strong oxidizers, including anhydrous red fuming nitric acid and 90% hydrogen peroxide, also cause attack. Ionizable fluoride compounds, such as sodium fluoride and hydrogen fluoride, activate the surface and can cause rapid corrosion. Dry chlorine gas is especially harmful.

For corrosion-resistant applications, Ti-3Al-2.5V usually is used in the annealed condition (as opposed to the cold-worked and stress-relieved condition typically used in aerospace applications). Weldments of Ti-3Al-2.5V also exhibit corrosion resistance similar to the base metal. This alloy contains so little alloy content and second phase that metallurgical instability and thermal response are not significant. Therefore, weldments and associated heat-affected zones (HAZ) generally do not experience corrosion limitations in welded components when normal passive conditions prevail for the base metal. However, under marginal or active conditions (for corrosion rates ≥0.10 mm/yr, or 4 mils/year), weldments may experience accelerated corrosion attack relative to the base metal, depending on alloy composition. The increasing impurity (iron, sulfur, or oxygen) content associated with the coarse, transformed β microstructure of weldments appears to be a factor (L.C. Covington and R.W. Schutz, Effects of Iron on the Corrosion Resistance of Titanium, in *Industrial Applications of Titanium and Zirconium*, STP 728, American Society for Testing and Materials, 1981, p 163-180).

Electrochemical Potentials

Pitting potentials of Ti-3Al-2.5V and CP titanium are estimated to be 9.2 and 9.6 V (SCE), respectively, in seawater at 50 °C (120 °F). Commercially pure titanium exhibits improved performance over Ti-3Al-2.5V under these conditions, which could make a difference in an anode application. However, pitting potential is better and at least 1 V (SCE) higher than stainless steel.

Electrode potentials of Ti-3Al-2.5V are more electropositive than Ti-6Al-4V in 3.5% NaCl solution. Consequently, it should be more corrosion resistant than Ti-6Al-4V.

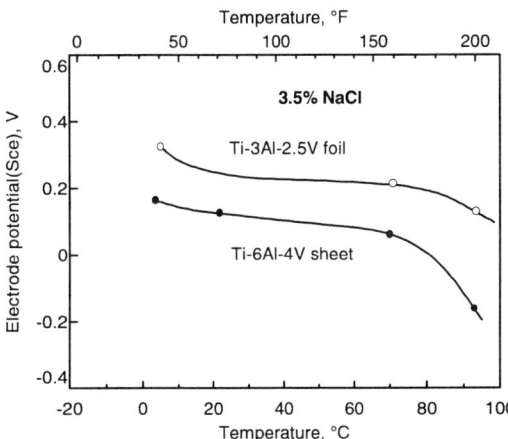

Ti-3Al-2.5V: Electrode potential in 3.5% NaCl

Source: S.D. Elrod and Y. Moji, Boeing Report No. FAA-SS-72-14, July 1972

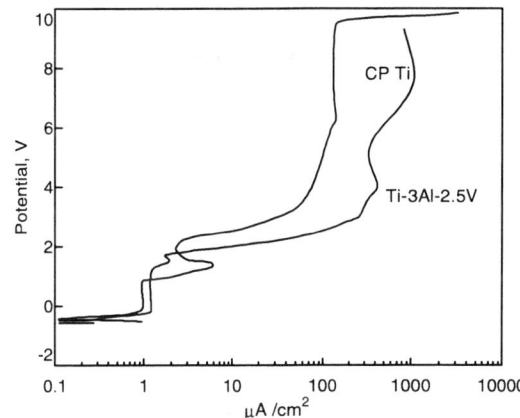

Ti-3Al-2.5V: Anodic polarization in seawater at 50 °C

Test conditions: Pickled sample; scan rate, 5 V/h
Source: Te-Lin Yau, Corrosion of Ti-3Al-2.5V in Seawater, in *Corrosion 89*, 1989

Crevice Corrosion

Although CP titanium has a higher pitting potential than Ti-3Al-2.5V in seawater, CP titanium may be susceptible to crevice corrosion in an environment that contains more than 1000 ppm chloride at temperatures of about 75 °C (168 °F). Therefore, titanium alloys with improved crevice corrosion resistance are desirable for marine applications.

In some cases, Ti-3Al-2.5V has better crevice corrosion resistance than CP titanium. For example, when anodic polarization tests were performed in seawater at 96 °C (205 °F), the passive region of Ti-3Al-2.5V was more stable than that at a CP Ti/PTFE gasket contact (see figure before "Thermal Properties"). Crevice corrosion specimens of Ti-3Al-2.5V have been tested in boiling seawater for 130 days with no detectable pitting or crevice corrosion.

General Corrosion

The general corrosion behavior of Ti-3Al-2.5V is similar to that of unalloyed titanium. The nature of the oxide film on titanium alloys basically remains unaltered in the presence of minor alloying constituents; thus, small additions (<2 to 3%) of most commercially used alloying elements or trace alloy impurities generally have little effect on the basic corrosion resistance of titanium in normally passive environments. However, under active conditions in which titanium exhibits significant general corrosion, certain alloying elements may accelerate corrosion.

Ti-3Al-2.5V: Comparative corrosion rate with unalloyed Ti

Environment	Temperature °C (°F)	Corrosion rate, mm/yr CP Ti	Ti 325
Acids			
HNO$_3$, 10%	Boiling	0.112	0.084
HNO$_3$, 40%	Boiling	0.620	0.709
HNO$_3$, 70%	Boiling	0.132	0.137
HNO$_3$, 65%	204 (400)	0.005	0.008
Aqua regia	RT	nil	0.015
Aqua regia	Boiling	1.12	1.30
Chromic, 30%	Boiling	0.010	0.053
HCl, 1%	Boiling	2.16	2.79
1% HCl to 0.2% FeCl$_3$	Boiling	<0.125	0.005
3% HCl (air agitated)	35 (95)	0.002	0.004
3% HCl (N$_2$ agitated)	35 (95)	0.147	0.126
5% HCl	Boiling	21.3	26.8
5% HCl + 0.2% FeCl$_3$	Boiling	<0.125	0.033
5% HCl (air agitated)	35 (95)	0.025	0.001
5% HCl (N$_2$ agitated)	35 (95)	0.298	0.185
5% H$_2$SO$_4$ (air agitated)	35 (95)	0.515	0.025
5% H$_2$SO$_4$ (N$_2$ agitated)	35 (95)	0.539	0.405
Alkalis			
50% NaOH	150 (300)	0.056	0.493
50% KOH	150 (300)	2.74	9.22
28% NH$_4$OH	150 (300)	nil	nil
Organic compounds			
100% acetic acid	Boiling	nil	nil
50% citric acid	Boiling	0.356	0.381
25% formic acid (air agitated)	35 (95)	0.001	<0.125
25% formic acid (N$_2$ agitated)	35 (95)	0.025	<0.125
50% formic acid	Boiling	11.0	5.08
100% methanol	Boiling	nil	nil
5% oxalic acid (air agitated)	35 (95)	0.690	0.525
5% oxalic acid (N$_2$ agitated)	35 (95)	1.04	0.560
50% urea	150 (300)	nil	0.2
Salts			
Seawater	Boiling	nil	nil
Simulated SO$_2$ scrubber solutions (up to 32 000 mg/L Cl$^-$ + 5 g/L fly ash + CaF$_2$ (Cl/F = 21) at pH 1)	Boiling	nil	nil

Source: Te-Lin Yau, Corrosion of Ti-3Al-2.5V in Seawater, in *Corrosion 89*, 1989

Ti-3Al-2.5V: Comparative corrosion rates with grade 2 titanium

Corrosion environment	Temperature, °C (°F)	Corrosion rate, mm/yr Ti grade 2	Ti-3Al-2.5V grade 9
Hydrochloric acid			
5 wt%, air agitation	35	0.025	0.001
5 wt%, nitrogen agitation	35	0.298	0.185
3 wt%, air agitation	35	0.002	0.004
3 wt%, nitrogen agitation	35	0.147	0.126
3 wt%, no agitation	88	3.54	3.10
1 wt%, no agitation	88	0.002	0.009
Sulfuric acid			
5 wt%, air agitation	35	0.515	0.025
5 wt%, nitrogen agitation	35	0.539	0.405
5 wt%, no agitation	88	12.9	16.6
Oxalic acid			
5 wt%, air agitation	35	0.690	0.525
5 wt%, nitrogen agitation	35	1.044	0.560
5 wt%, no agitation	88	18.24	25.0
Formic acid			
25 wt%, air agitation	35	0.001	<0.125
25 wt%, nitrogen agitation	35	0.025	<0.125
25 wt%, no agitation	88	0.001	<0.125
Mixed acids			
13 wt% H$_2$SO$_4$, 3.5 wt% HCl 1 wt% CuCl$_2$, 1 wt% FeCl$_3$	50	nil	nil
13 wt% H$_2$SO$_4$, 3.5 wt% HCl 1 wt% CuCl$_2$, 1 wt% FeCl$_3$	70	0.010	<0.100
13 wt% H$_2$SO$_4$, 3.5 wt% HCl 1 wt% CuCl$_2$, 1 wt% FeCl$_3$	Boiling	0.046	0.330

Source: R.T. Webster and C.S. Young, Ed., *Industrial Applications of Titanium and Zirconium: Third Conference*, STP 830, ASTM, 1984

Stress-Corrosion Cracking

Ti-3Al-2.5V is essentially immune to stress-corrosion cracking in boiling seawater and simulated sour-gas well brines at room temperature (Te-Lin Yau, Corrosion of Ti-3Al-2.5V in Seawater, in *Corrosion 89*, 1989). Like CP titanium, Ti-3Al-2.5V is also immune to hot-salt cracking.

Ti-3Al-2.5V: Corrosion rate vs Ti alloys

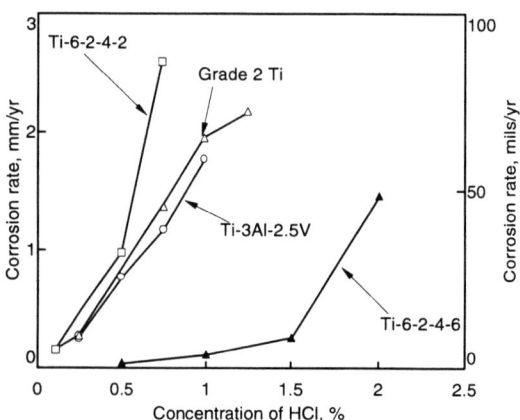

General corrosion in naturally aerated HCl solutions.
Source: *Metals Handbook, Corrosion*, Vol 13, 9th ed., ASM International, 1987

Crevice Corrosion

Ti-3Al-2.5V: Anodic polarization in seawater at 96 °C

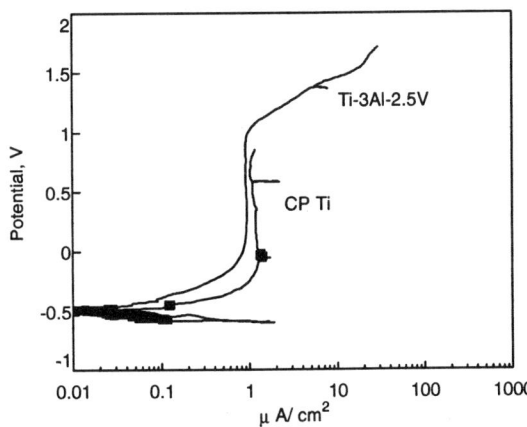

Ti-3Al-2.5V had a more stable passive region than CP titanium, and crevice corrosion occurred inconsistently at the CP Ti/PTFE-gasket contact.
Test conditions: Pickled sample; scan rate, 6 V/h.
Source: Te-Lin Yau, Corrosion of Ti-3Al-2.5V in Seawater, in *Corrosion 89*, 1989

Ti-3Al-2.5V: Crevice corrosion in saturated NaCl

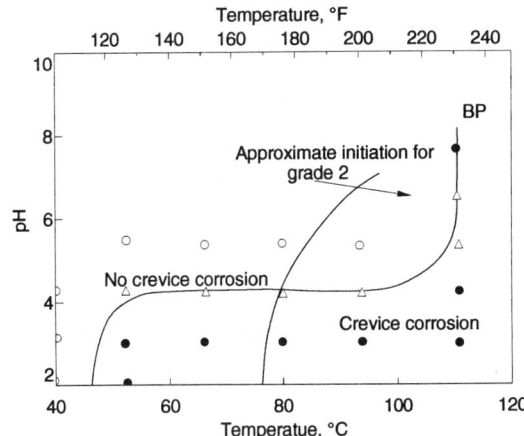

Source: R.T. Webster and C.S. Young, Ed., *Industrial Applications of Titanium and Zirconium: Third Conference*, STP 830, ASTM, 1984, p 133-142

Thermal Properties

Melting Point. 1700 °C (3100 °F)

Thermal Expansion

Ti-3Al-2.5V: Thermal coefficient of linear expansion for annealed alloy

Temperature range		Coefficient	
°C	°F	10^{-6}/K	10^{-6}/°F
20-95	70-200	9.61	5.34
20-205	70-400	9.67	5.37
20-315	70-600	9.86	5.48
20-425	70-800	9.92	5.51
20-540	70-1000	9.97	5.54

Source: C. Forney Jr. and J. Schemel, "Ti-3Al-2.5V Seamless Tubing Engineering Guide," 2nd ed., Sandvik Special Metals Corp., 1987, p 17

Thermal Conductivity

Ti-3Al-2.5V: Thermal conductivity

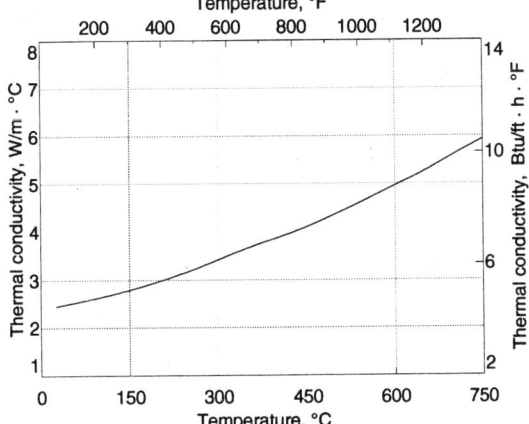

Source: *Aerospace Structural Metals Handbook*, Code 3725, Mechanical Properties Data Center, Battelle Columbus Laboratories, 1980, p 23

Ti-3Al-2.5V: Thermal conductivity at selected temperatures

Temperature		Thermal conductivity (calculated)	
°C	°F	W/m · K	Btu/ft · h · °F
22	72	8.3	4.8
95	200	9.2	5.3
205	400	10.7	6.2
315	600	11.8	6.8

Source: C. Forney, Jr. and J. Schemel, "Ti-3Al-2.5V Seamless Tubing Engineering Guide," 2nd ed., Sandvik Special Metals Corp., 1987, p 15

Mechanical Properties

Ti-3Al-2.5V: Tensile properties for tubing and pipe
All values are minimums unless otherwise indicated.

Specification	Condition	Ultimate tensile strength MPa	ksi	Tensile yield strength MPa	ksi	Elongation in 50 mm (2 in.), %
Seamless and welded pipe or tube (all outer diameters)						
ANSI/ASTM B337-78	Annealed pipe(a)	620	90	480	70	15
	CWSR(b) pipe	860	125	725	105	10
ANSI/ASTM B337-78	CWSR(b) tube	860	125	725	105	10
Seamless hydraulic tubing (all outside diameters, except as noted)						
Lockheed-Georgia STM 08-303C	CWSR(c) with OD >9.5 mm (3/8 in.)	860	125	655	95	10
	CWSR(c) with OD of 6.35 and 9.5 mm (1/4 and 3/8 in.)	825	120	620	90	10
GE-B50TF35-S7	Cold drawn, annealed(a)	790 max	115 max	515-620	75-90	15
Pratt & Whitney PWA 1260H	Cold drawn or cold reduced, annealed(d)	585-790	85-115	450	65	20
McDonnell Aircraft MM5-1205D	CWSR 70(e)	585-760	85-110	480	70	15
	CWSR 95	690-860	100-125	655	95	13
	CWSR 105	860-1035	125-150	725	105	10
Wolverine AFML-TR-76	CWSR(f)	860-1035	125-150	725	105	10
AMS 4943	Annealed	620	90(g)	515	75(g)	15(g)

(a) Anneal treatment unspecified. (b) Stress relief unspecified. (c) 315 °C (600 °F) minimum, 0.5 h minimum. (d) 705 ± 15 °C (1300 ± 25 °F) and 1 h minimum. (e) Stress relieved to minimum specified yield strength: 315 °C (600 °F) minimum. 0.5 h minimum, except rotary-straightened tubing which is stress relieved 370 °C (700 °F) minimum, 2 h minimum. (f) Stress relief based on strength level of cold-worked tube, with temperatures not less than 315 °C (600 °F). (g) S-basis. Source: *Aerospace Structural Metals Handbook*, Code 3725, Battelle Columbus Laboratories, 1980, p 13

Ti-3Al-2.5V: Minimum tensile properties of bar, sheet, strip, and foil

Specification	Condition	Ultimate tensile strength MPa	ksi	Tensile yield strength MPa	ksi	Elongation, %
MIL-T-9047, rolled or forged bar and reforging stock(a)	Annealed at 705-760 °C (1300-1400 °F) for 1 to 3 h, AC or slower	620	90(a)	515	75(a)	15(b)
GE-B50TF117-S3, sheet, strip, foil	Anneal treatment unspecified	620	90	515	75	20(c)
ASTECH-TRE 2311D, annealed(d) foil <0.15 mm (0.006 in.) thick	Foil width <13 mm (0.5 in.)	585	85	480	70	16(c)
	Foil width ≥ 13 mm (0.5 in.)	620	90	515	75	20(c)

(a) Properties apply in any grain direction and for products with cross-sectional area of 103 cm^2 (16 in.^{2}) or less and a distance between flats of 25 mm (1 in.) or less. (b) Elongation in 4D. (c) Elongation in 50 mm (2 in.). (d) Anneal temperature less than 925 °C (1700 °F). Source: *Aerospace Structural Metals Handbook*, Code 3725, Battelle Columbus Laboratories, 1980, p 13.

Hardness

Typical hardness of about 24 HRC has been reported with a range of 15 to 27 HRC. Hardness is highly dependent on annealing temperature.

Ti-3Al-2.5V: Effect of annealing temperature on hardness

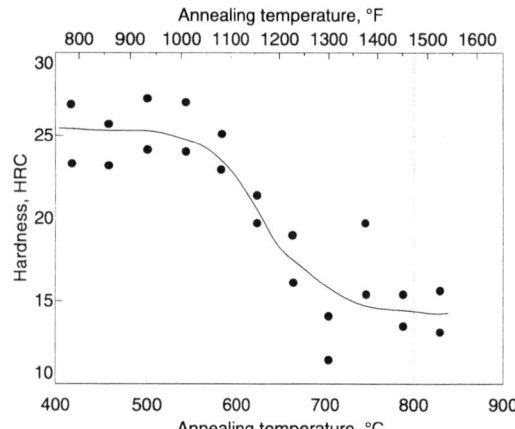

Alloy supplied as tubing 15.8 mm (0.625 in). OD by 0.9 mm (0.038 in.) wall, 50% cold worked (CW), annealed 2 h in vacuum, vacuum cooled to 425 °C (800 °F), air cooled. Source: American Society for Testing and Materials/American National Standards Institute Specification ANSI/ASTB B 337-78; reported in *Aerospace Structural Metals Handbook,* Code 3725, Mechanical Properties Data Center, Battelle Columbus Laboratories, 1980, p 21

Microindentation Hardness

When texture is developed in highly worked Ti-3Al-2.5V, the shape of the Knoop indenter results in anisotropic hardness readings. Vickers hardness typically ranges from 220 to 300 HV, depending on heat treatment.

Ti-3Al-2.5V: Effect of heat treatment on hardness

Heat treatment condition	Hardness, HV
As-quenched	
750 °C (1380 °F) 20 min, WQ	261
800 °C (1470 °F) 20 min, WQ	258
850 °C (1560 °F) 20 min, WQ	257
900 °C (1650 °F) 20 min, WQ	299
950 °C (1740 °F) 20 min, WQ	304
Tempered	
750 °C (1380 °F) 20 min, WQ, 500 °C (930 °F) 5 h, AC	263
800 °C (1470 °F) 20 min, WQ, 500 °C (930 °F) 5 h, AC	254
850 °C (1560 °F) 20 min, WQ, 500 °C (930 °F) 5 h, AC	278
900 °C (1650 °F) 20 min, WQ, 500 °C (930 °F) 5 h, AC	301
950 °C (1740 °F) 20 min, WQ, 500 °C (930 °F) 5 h, AC	292

Note: Commercially produced, annealed sheet 0.7 mm (0.03 in.) thick. Heat treatments were carried out in vacuum (10^{-4} to 10^{-5} torr). Source: J. Kaneko and M. Sugamata, "The Effect of Heat Treatment on the Days Drawability of Ti-3Al-2.5V Sheets," in *Titanium, Science and Technology*, Vol 1, G. Lütjering, U. Zwicker, and W. Bunk, Ed., Deutsche Gesellschaft für Metallkunde, e.V., 1985, p 563

Ti-3Al-2.5V: Knoop hardness vs annealing temperature

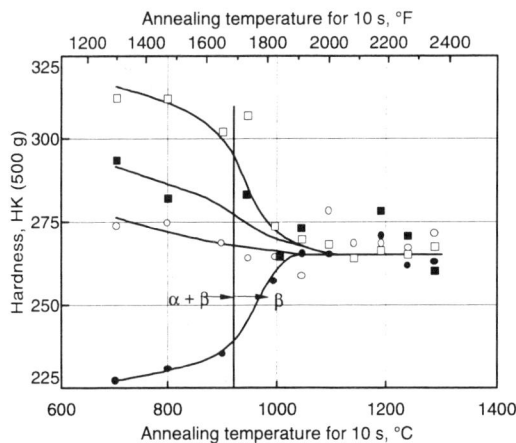

Alloy supplied as tubing 13 mm (0.5 in.) OD by 0.8 mm (0.033 in.) wall, 55% cold worked and partially recrystallized at 600 °C (1110 °F) for 1 h, annealed for 1 h, and air cooled. The different data points indicate the results of using various indenter orientations.
Source: T. Andersson and B. Lundquist, "Properties and Structure of Welded Joints in Ti-3Al-2.5V Hydraulic Tubing," *Welding J.*, Vol 53, 1974, p 314s; reported in *Aerospace Structural Metals Handbook*, Code 3725, Mechanical Properties Data Center, Battelle Columbus Laboratories, 1980, p 33

Ti-3Al-2.5V: Hardness vs homogenizing heat treatment/cooling

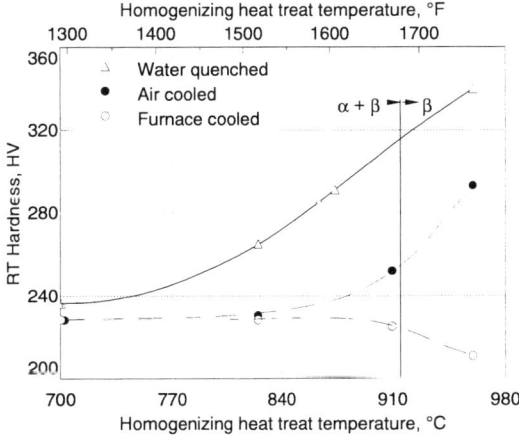

Effect of homogenizing heat treat temperature and cooling rate from that temperature on microhardness of small samples taken from forged billet.
As-forged hardness was about 210 HV.
Source: E. Baroch, B. McClanahan, and R. Cirtis, "Production of Extruded Tube Hollows for Titanium 3Al-2.5V Hydraulic Tubing," reported in *Aerospace Structural Metals Handbook*, Code 3725, Mechanical Properties Data Center, Battelle Columbus Laboratories, 1980, p 33

Typical Room-Temperature Strength

Typical ultimate shear strength for Ti-3Al-2.5V rivet wire is about 475 MPa (69 ksi) yielding a typical ultimate tensile strength of 640 MPa (93 ksi).

Ti-3Al-2.5V: Typical room-temperature tensile properties for solution treated plus aged specimens

Mill form	Solution treatment	Aging treatment	Tensile ultimate strength MPa	ksi	Tensile yield strength MPa	ksi	Elongation, %, in: 25 mm (1 in.)	50 mm (2 in.)
15.8 mm (0.625 in.) bar	1 h, 925 °C (1700 °F), WQ	6 h, 480 °C (900 °F), AC	827	120	723	105	15	...
1.7 mm (0.070 in.) sheet	¼ h, 910 °C (1675 °F), WQ	8 h, 510 °C (950 °F), AC	917	133	779	113	7	...
1.0 mm (0.040 in.) sheet	¼ h, 900 °C (1650 °F), WQ	2 h, 480 °C (900 °F), AC	792	115	675	98	...	14
0.73 mm (0.029 in.) wall tubing	¼ h, 870 °C (1600 °F), WQ	2 h, 480 °C (900 °F), VC	889	129	765	111	...	7
0.71 mm (0.028 in.) wall tubing	¼ h, 925 °C (1700 °F), WQ	6 h, 480 °C (900 °F), AC	910	132	827	120	...	11

Source: R. Wood and R. Favor, *Titanium Alloys Handbook*, MCIC-HB-02, Battelle, 1972

Ti-3Al-2.5V: Tensile properties of as-extruded and heat treated plate and weld metal

Condition ASTM B337 Grade 9b	Orientation unspecified	Ultimate tensile strength MPa	ksi	Tensile yield strength (0.2%) MPa	ksi	Elongation, %	Reduction of area, %
As extruded	T	737	107	717	104	19	60
	L	675	98	565	82	20	47
β annealed	T	593	86	510	74	14	24
	L	648	94	565	82	21	35
α + β annealed (I)	T	723	105	689	100	20	57
	L	655	95	544	79	21	47
α + β annealed (II)	T	730	106	717	104	20	60
	L	661	96	558	81	20	41
α annealed	T	723	105	710	103	20	58
	L	648	94	551	80	20	41
Weld metal	L	744	108	575	98	16	40

Ti-3Al-2.5V: RT tensile properties of cold worked

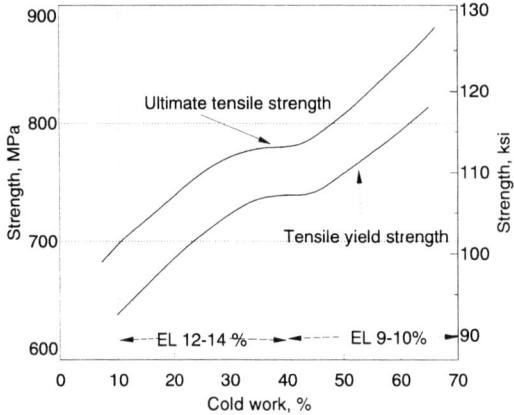

Effect of cold work on the room-temperature tensile properties of Ti-3Al-2.5V sheet.
Source: R. Favor and R. Wood, Ed., *Titanium Alloys Handbook*, MCIC-HB-02, Battelle Columbus Laboratories, 1972

Ti-3Al-2.5V: RT tensile properties vs cold work

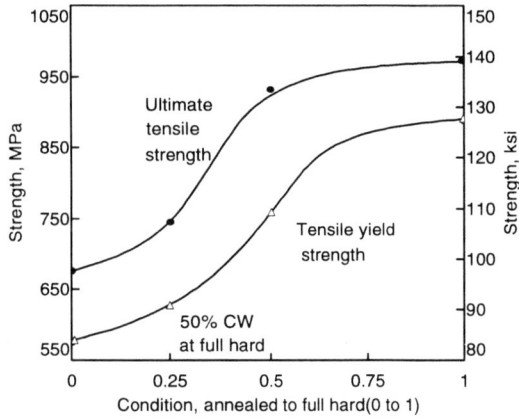

Source: *Aerospace Structural Metals Handbook*, Code 3725, Battelle Columbus Laboratories, 1980

Ti-3Al-2.5V: RT elongation vs cold work

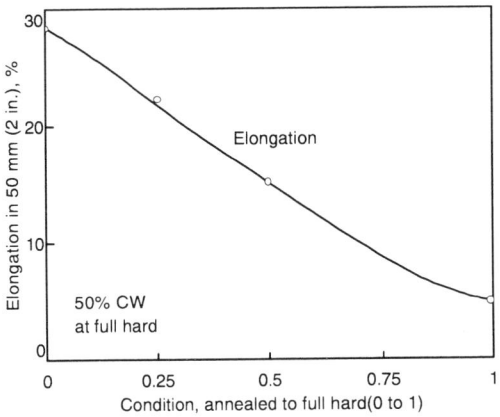

Source: *Aerospace Structural Metals Handbook*, 1980

Strength vs Temperature

Cryogenic Tensile Properties

Ti-3Al-2.5V: Typical smooth and notch tensile properties of annealed sheet and tubing

Test temperature		Ultimate tensile strength		Tensile yield strength		Elongation, %, in		Notched ultimate strength(a)		Notched/unnotched strength ratio
°C	°F	MPa	ksi	MPa	ksi	50 mm (2 in.)	25 mm (1 in.)	MPa	ksi	
Sheet annealed 1/2 h, 650 °C (1200 °F), AC										
RT		675	98	558	81	18	...	758	110	1.12
−195	−320	1144	166	1048	152	20	...	1255	182	1.10
−253	−423	1220	177	1186	172	2	...	1441	209	1.18
−268	−452	1310	190	1310	190	3	...	1482	215	1.13
Seamless tubing annealed 1 h, 650 °C (1200 °F), AC										
RT		703	102	551	80	...	19	758(b)	110(b)	1.12(b)
−195	−320	1179	171	986	143	...	20	1255(b)	182(b)	1.09(b)
−253	−423	1510	219	1386	201	...	2	1420(b)	206(b)	1.16(b)
−268	−452	1392	202	1365	198	...	4	...	...	...

(a) K_t = 6.3. (b) Data obtained on slit, flattened, and stress-relief annealed (½ h, 650 °C, 1200 °F, VC) tubing. Source: RMI Co., Zirconium Technology Corp., Titanium Metals Corporation of America, Mallory Sharon Metals Corp., and Garrett Corp., reported in *Titanium Alloys Handbook*, R. Wood and R. Favor, Ed., MCIC-HB-02, Battelle, 1972

Ti-3Al-2.5V: Smooth and notched tensile properties

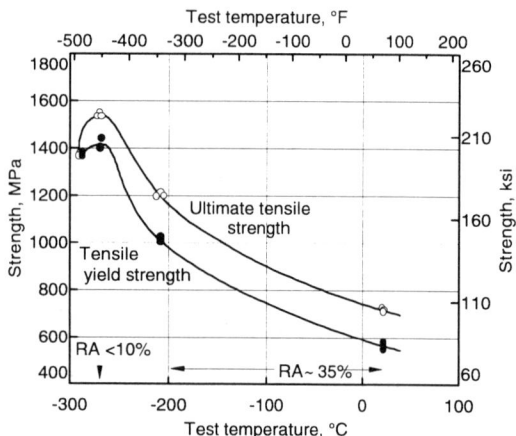

Indicated values of elongation are in 50 mm (2 in.). Material tested was seamless hydraulic tubing of 13 mm (0.5 in.) outside diameter and 1.3 mm (0.050 in.) wall. It was cold worked and stress relieved at 650 °C (1200 °F) for 0.5 h. Tubing was flattened for testing as parent metal, specimen welded with Ti-3Al-2.5V filler metal, and specimen welded with Ti-5Al-2.5Sn filler metal, as indicated.
Source: E. Swarens, Garrett Corp.; reported in *Aerospace Structural Metals Handbook*, Code 3725, Battelle Columbus Laboratories, 1980, p 27

Ti-3Al-2.5V: Cryogenic tensile properties of tubing

Test material was annealed tubing 13 mm (0.5 in.) OD by 1.3 mm (0.050 in.) wall.
Source: AiResearch Document 67-2009, March 25, 1969; reported in *Aerospace Structural Metals Handbook*, Code 3725, Battelle Columbus Laboratories, 1980, p 26

Ti-3Al-2.5V: Smooth and notched tensile properties of sheet

Sheet was annealed at either 650 °C (1200 °F) or 705 °C (1300 °F) for 0.5 h, as indicated. Elongation was less than 5% below −240 °C (−400 °F) and was about 20% above −184 °C (−300 °F).
Source: RMI Data Sheet, RMI Co., 1969; and H. Kessler, 1978; reported in *Aerospace Structural Metals Handbook*, Code 3725, J. Shannon, Jr., Ed., Battelle Columbus Laboratories, 1980, p 27

High-Temperature Strength

Ti-3Al-2.5V: Creep and creep rupture compared to CP titanium

Alloy	Stress to 1.0% creep in 1000 h at 250 °C (480 °F)		Stress to rupture in 1000 h at 250 °C (480 °F)	
	MPa	ksi	MPa	ksi
CP grade 1	90	13	103	15
CP grade 2	103	15	117	17
CP grade 3	131	19	138	20
Ti-3Al-2.5V (grade 9)	400	58	421	61
Ti-0.3Mo-0.8Ni (grade 12)	221	32	297	43

Source: "Ti-3Al-2.5V Seamless Tubing Engineering Guide," C. Forney, Jr. and J. Schemel, Sandvik Special Metals Corporation, Kennewick, WA, 1987, p 21

Ti-3Al-2.5V: Comparative ASME design stresses

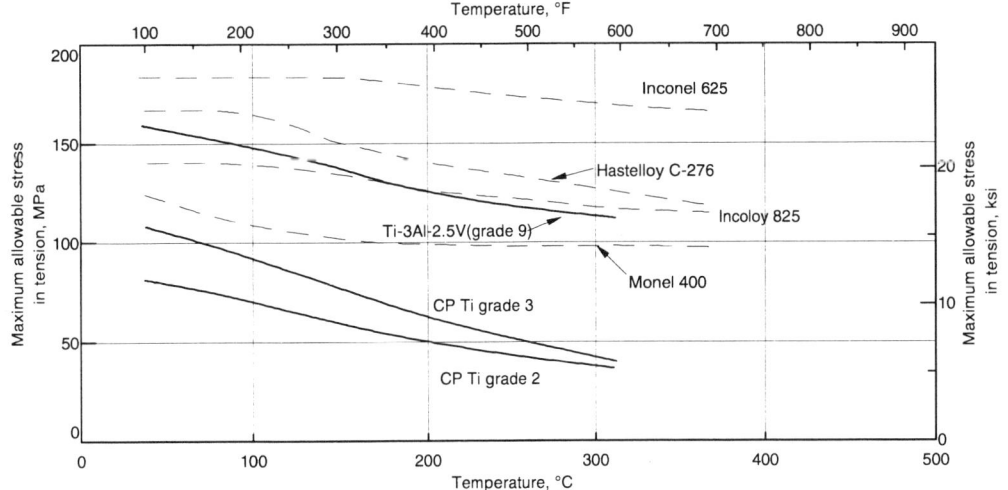

For comparative purposes only.
Source: From presentation at ACHEMA '82, Frankfurt, Germany, 1982 by P. Russo and J.D. Schöbel

Ti-3Al-2.5V: Ultimate tensile strength vs temperature of tubing

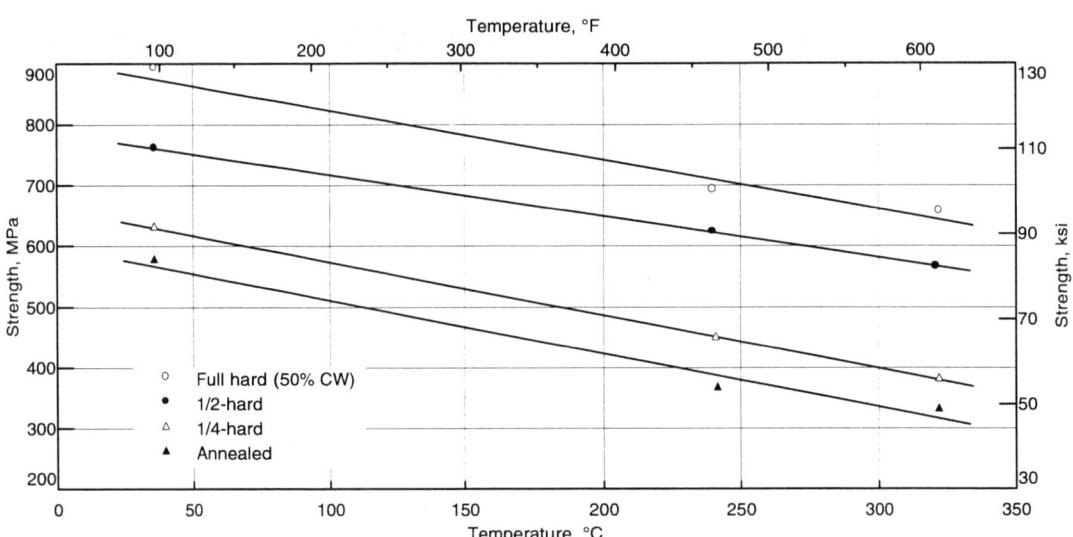

Test material was tubing annealed to full hard, 15.8 mm (0.625 in.) OD by 0.88 mm (0.035 in.) wall.
Source: Reactive Metals Inc., Feb 13, 1978; reported in *Aerospace Structural Metals Handbook*, Code 3725, Battelle Columbus Laboratories, 1980, p 26

Tensile yield strength vs temperature of tubing

Test material was tubing annealed to full hard, 15.8 mm (0.625 in.) OD by 0.88 mm (0.035 in.) wall.
Source: Reactive Metals Inc., Feb 13, 1978; reported in *Aerospace Structural Metals Handbook*, Code 3725, Battelle, Columbus Laboratories, 1980, p 26

Ti-3Al-2.5 V: Elongation vs temperature of tubing

Test material was tubing annealed to full hard, 15.8 mm (0.625 in.) OD by 0.88 mm (0.035 in.) wall.
Source: Reactive Metals Inc., Feb 13, 1978; reported in *Aerospace Structural Metals Handbook*, Code 3725, Battelle Columbus Laboratories, 1980, p 26

Ti-3Al-2.5V: Tensile properties

Test temperature(a)		Tensile yield strength 0.2%(b)		Ultimate tensile strength		Elongation, %	Reduction of area, %	TYS/UTS
°C	°F	MPa	ksi	MPa	ksi			
20	68	568.3	82.0	695.9	101	20	32	0.82
150	302	457.0	66.0	552.6	80	18	40	0.83
200	392	427.6	62.0	513.8	75	23	49	0.83
250	482	400.0	58.0	475.9	69	24	47	0.84
300	572	392.4	57.0	459.0	67	25	64	0.85

Note: Test Material from 75 mm OD by 50 mm (2 in.) ID pipe. Longitudinal direction. (a) Annealing cycle 730 °C (1345 °F), 1 h, air cool. (b) Average of two tests. Source: R.T. Webster and C.S. Young, Ed., *Industrial Applications of Titanium and Zirconium: Third Conference*, STP 830, ASTM, 1984

Fatigue Properties

Ti-3Al-2.5V: Smooth and notched bending fatigue

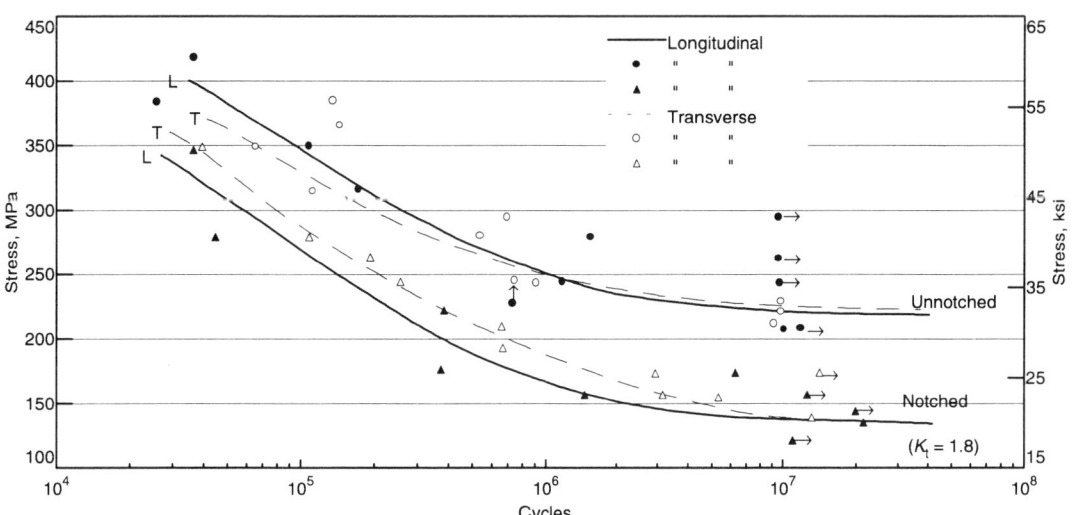

Specimens were annealed 1.0 mm (0.040 in.) sheet.
Source: R. Wood and R. Favor, Ed., *Titanium Alloys Handbook*, MCIC-HB-02, Battelle Columbus Laboratories, 1972

Ti-3Al-2.5V: Fatigue of plate and GTA weld metal

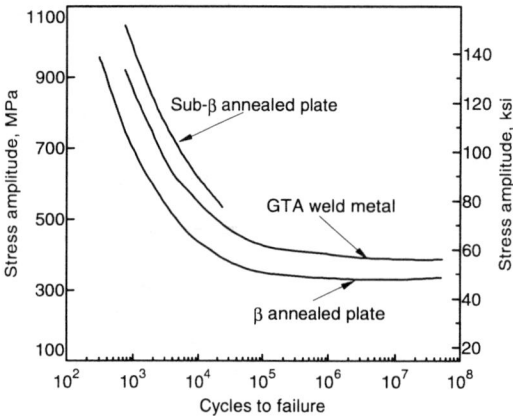

Strain-controlled low-cycle and load-controlled high-cycle fatigue tests were performed on β annealed, sub-β annealed and welded Ti-3Al-2.5V extruded plate. The axial low-cycle fatigue hourglass specimens had a minimum diameter of 6.35 mm (0.25 in.) and were tested at 2 cycles/min according to procedures outlined in ASTM E606. Rotating cantilever beam high-cycle fatigue specimens had a minimum diameter of 4.75 mm (0.187 in.) and were tested at a frequency of 6000 cycles/min.

Ti-3Al-2.5V: Bending fatigue strength of annealed sheet

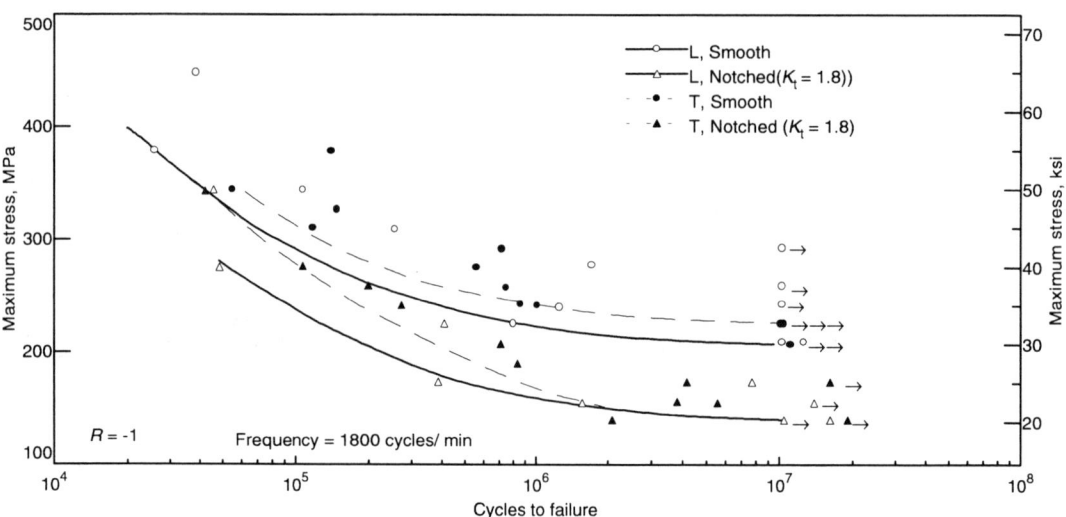

Test material was 1.0 mm (0.040 in.) sheet, annealed at 785 °C (1450 °F), 2 h, vacuum cooled; ultimate tensile strength (L and T) 538 MPa (78 ksi).
Source: Bridgeport Brass Co. Report 1000R436, M.O. 83025, Dec 21, 1962; reported in *Aerospace Structural Metals Handbook*, Code 3725, Battelle Columbus Laboratories, 1980, p 28

Ti-3Al-2.5V: Fatigue strength of annealed tubing

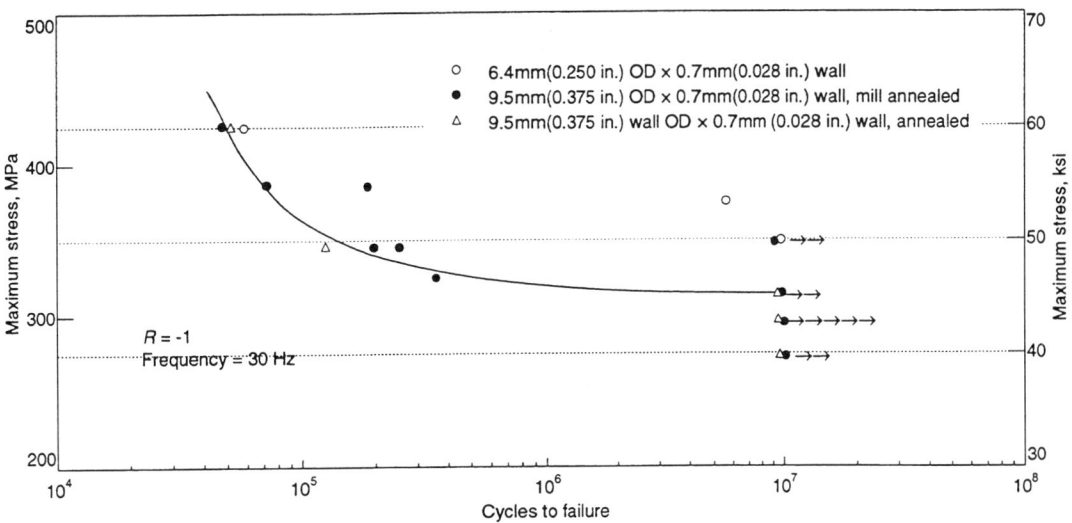

Source: Pratt & Whitney Aircraft Group Internal Report; June 1973; reported in *Aerospace Structural Metals Handbook*, Code 3725, Battelle Columbus Laboratories, 1980, p 28

Fracture Properties

Impact Toughness

Ti-3Al-2.5V: Charpy V-notch impact strength of extruded plate and welds

As-extruded α-β and α heat treated materials exhibit excellent impact toughness, about twice that of the β annealed plate.

As extruded	101 J (74 ft · lbf)
β annealed	44 J (32 ft · lbf)
α + β annealed(I)	82 J (60 ft · lbf)
α + β annealed(II)	87 J (64 ft · lbf)
α annealed	86 J (63 ft · lbf)
Weld metal	82 J (60 ft · lbf)

Note: TL orientation; test temperature, 0 °C (32 °F). Source: I. Caplan, "Ti-3Al-2.5V for Seawater Piping Applications," in *Industrial Applications of Titanium and Zirconium: Fourth Volume*, ASTM STP 917, C. Young and J. Durham, Ed., ASTM, Philadelphia, 1986, p 43

Ti-3Al-2.5V: Charpy V-notch impact strength of 25 mm (1 in.) extruded plate

Test temperature		As extruded		β anneal 955 °C (1755 °F), 30 min, AC		High-temperature α + β anneal 915 °C (1680 °F), 30 min, AC		High-temperature α + β anneal 915 °C (1680 °F) 30 min, WQ		Low-temperature α + β anneal 805 °C (1475 °F) 30 min, AC	
°C	°F	J	ft · lbf	J	ft · lbf	J	ft · lbf	J	ft · lbf	J	ft · lbf
93	200	...	...	48	36	118	87	123	91	116	86
RT		107	79	46	34	86	64	92	68	101	75
0	32	100	74	43	32	81	60	81	60	86	64
−62	−80	...	...	38	28	69	51	61	45	69	51

Source: *Aerospace Structural Metals Handbook*, Code 3725, Battelle Columbus Laboratories, 1980, p 20

Fracture Toughness

Seawater Stress Corrosion. Notched, deadweight loaded, cantilever beam specimens measuring 25 mm (1 in.) by 50 mm (2 in.) by 330 mm (13 in.) were used to evaluate the seawater stress-corrosion cracking performance of heat treated and welded plates. The specimens were step-loaded in seawater to a given stress intensity and held until failure occurred, or for a maximum of 1000 h. None

Ti-3Al-2.5V: Sustained load cracking of heat treated plate in seawater

Heat treatment/ condition	Threshold (K_{ISLC}) MPa√m	ksi √in.
β annealed(a)	75	68
Sub-β annealed	88	80

Note: Test duration 1000 h. (a) K_{max} in air = 81 MPa√m (74 ksi√in.). Source: I. Caplan, "Ti-3Al-2.5V for Seawater Piping Applications," in *Industrial Applications of Titanium and Zirconium: Fourth Volume*, C. Young and J. Durham, Ed., ASTM STP 917, ASTM, Philadelphia, 1986, p 43

Ti-3Al-2.5V: Fracture toughness of extrusions in several heat treated conditions compared to weld metal

Condition	Fracture toughness (J_{IC}) kJ/m²	in. · lb/in.²	Tear modulus(a)
As-extruded	40	230	7
β annealed	70	400	10
α + β annealed (near β transus)	93	530	24
α + β annealed (near α transus)	123	700	26
α annealed	100	570	31
Weld metal	151	860	27

Note: Chemical composition of extrusions: 2.71% Al, 0.011% C, 0.005% Cu, 0.191% Fe, 0.0014% H, 0.005% Mn, 0.013% N, 0.099% O, 0.015% Si, and 2.56% V. Weld metal composition: 0.033% H, 0.009% N, and 0.096% O. Fracture toughness was determined according to ASTM E813 using computer-interactive unloading compliance procedures. (a) Nondimensional. Source: I. Caplan, "Ti-3Al-2.5V for Seawater Piping Applications," in *Industrial Applications of Titanium and Zirconium: Fourth Volume*, C. Young and J. Durham, Ed., ASTM STP 917, ASTM, Philadelphia, 1986, p 43

of the materials displayed any stress-corrosion cracking susceptibility based on fractographic examination of failed specimens. However, the β and sub-β annealed material did exhibit time-dependent sustained load failures. The sustained load cracking threshold stress-intensity value in seawater (K_{ISLC}) was defined as the average of the minimum time-dependent failure and the maximum runout for a given material condition. The weld metal did not exhibit any time-dependent failure up to a maximum stress intensity of 123 MPa√m (112 ksi√in.).

Seamless Tubing

Ti-3Al-2.5V is used primarily as seamless tubing, which can exhibit variations in crystallographic orientation ranging from a radial texture to a circumferential texture (see figure). Texture variations of Ti-3Al-2.5V tubing provide a useful means of tailoring properties, and a radial texture has the characteristic of increasing both tensile yield strength and elongation (see figures).

Typical Properties

Extruded tube intermediates can be cold worked to a moderately high-strength ductile product (see table). For higher strengths and potential weight savings on aircraft hydraulic tubing, a seamless Ti-6Al-4V tubing product has been developed. Strength comparisons of Ti-6Al-4V and Ti-3Al-2.5V seamless tubing are provided below.

Ti-3Al-2.5V: Effect of tube reductions on texture and properties

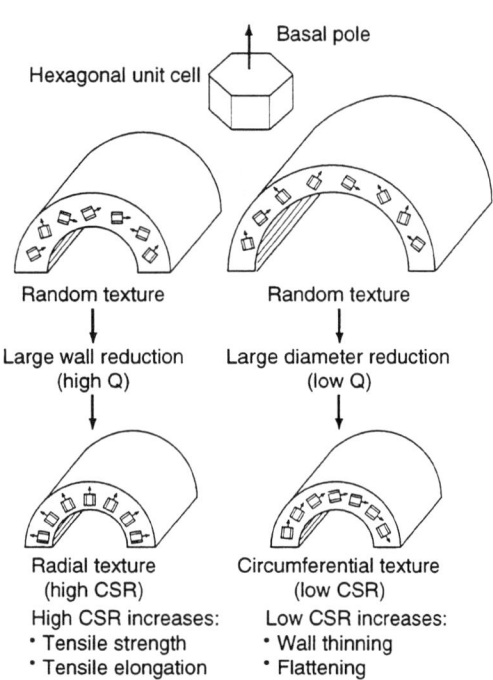

The contractile strain ratio (CSR) is the ratio of diametral to radial strain from a given stress. AMS 4945 specified a 1.3 minimum CSR for most tube sizes.. Low and high CSR both reduce fatigue strength, whereas a midrange CSR increases fatigue strength.

Ti-3Al-2.5V: Effect of texture on tensile properties

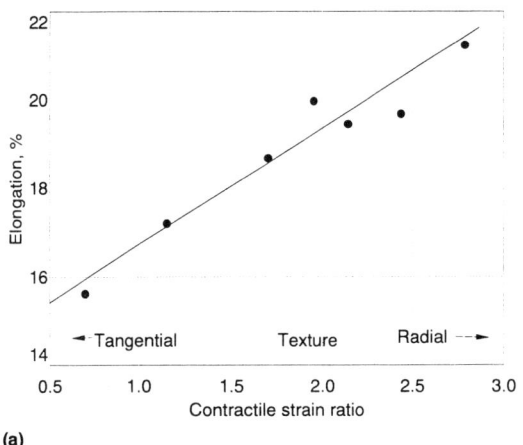
(a)

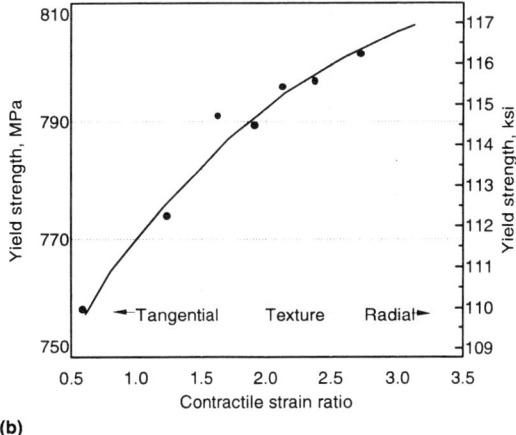
(b)

Data are an average of 71 samples.
Source: C.E. Forney, Jr. and J.H. Schemel, *Ti-3Al-2.5V Seamless Tubing Engineering Guide*, 2nd ed., Sandvik Special Metals, 1987

Ti-3Al-2.5V: Tensile properties of tubing

Condition	Basis	Ultimate tensile strength MPa	ksi	Tensile yield strength MPa	ksi	Elongation in 50 mm (2 in.), %
Cold worked	Typical	1034	150	896	130	7-11
	Full hard(a)		137		118	9
	Half hard(a)		126		98	12
Cold worked plus stress relieved	Typical(b)	909	132	792	115	19
	Minimum	861	125	723	105	16
Annealed	Typical(b)	648	94	579	84	29
	Minimum	620	90	517	75	15(c)

(a) Values reported in *Titanium Alloys Handbook*, MCIC-HB-02, 1972. (b) Typical values are an average from C.E. Forney, Jr., and J.H. Schemel, *Ti-3Al-2.5V Seamless Tubing Engineering Guide*, 2nd ed., Sandvik Special Metals, 1987. (c) 14% minimum for 6.35 and 9.5 mm (0.25 and 0.375 in.) OD sizes

Seamless tubing comparison

Alloy	System pressure MPa	ksi	Wall mm	in.	Theoretical burst(a) psi	Actual burst	lb/ft	Weight savings, %
6.35 mm (0.25 in.) OD								
Ti-3-2.5	34	5	0.55	0.022	23,900	...	0.0306	...
Ti-6-4	34	5	0.38	0.015	19,071	...	0.0213	30.04
	34	5	0.40	0.016	20,417	...	0.0226	26.14
Ti-3-2.5(b)	20	3	0.40	0.016	17,014	...	0.0229	...
Ti-6-4(b)	20	3	0.40	0.016	20,417	...	0.0226	1.31
13 mm (0.5 in.) OD								
Ti-3-2.5	34	5	1.09	0.043	23,317	...	0.1200	...
Ti-6-4	34	5	0.66	0.026	16,406	20,800(d)	0.0743	38.08
Ti-6-4	34	5	0.84	0.033	21,094	...	0.0930	22.50
Ti-3-2.5(b)	20	3	0.66	0.026	13,672	...	0.0753	...
Ti-6-4(c)	20	3	0.55	0.022	13,778	...	0.0634	15.8
19 mm (0.75 in.) OD								
Ti-3-2.5	34	5	1.65	0.065	23,511	...	0.2719	...
Ti-6-4	34	5	0.99	0.039	16,406	18,200(e)	0.1673	38.47
Ti-3-2.5	20	3	0.99	0.039	13,672	...	0.1693	...
Ti-6-4(c)	20	3	0.84	0.033	13,778	...	0.1427	15.71
25 mm (1 in.) OD								
Ti-3-2.5	34	5	2.2	0.088	23,900	...	0.4901	...
Ti-6-4	34	5	1.29	0.051	16,076	...	0.2919	40.44
Ti-6-4	34	5	1.67	0.066	21,094	...	0.3718	24.14
Ti-3-2.5(b)	20	3	1.29	0.051	13,397	...	0.2956	...
Ti-6-4(c)	20	3	1.09	0.043	13,452	...	0.2482	16.04

(a) Pressures calculated as: $P = \text{UTS} \times (\text{OD}^2 - \text{ID}^2)/(\text{OD}^2 + \text{ID}^2)$. (b) Minimum UTS of 860 MPa (125 ksi). (c) Minimum UTS of 1035 MPa (150 ksi). (d) Contractile strain ratio (CSR) 2.2. (e) Annealed to lower mechanical properties: 999 MPa (145 ksi) ultimate tensile strength, 862 MPa (125 ksi) yield strength CSR 1.08. Source: C. Forney, Sandvik Special Metals

Fabrication

Primary Working. Extrusion of billet to shapes such as tube hollows may be accomplished at temperatures of less than 870 °C (1650 °F). Extruded tube intermediates may be cold worked to final tube sizes with appropriate intermediate annealing treatments as required. Cleaning and pickling operations are included at appropriate intervals.

Mill products other than tubing can be produced following similar schedules. Typically, ingot breakdown may occur at temperatures above the β transus, major working of bloom and billet to intermediate section thickness at α-β temperatures (see table), and final primary fabrication steps at ambient temperatures with appropriate intermediate annealing.

Cold Working. Ti-3Al-2.5V offers excellent cold formability in combination with 30 to 50% higher tensile strengths than unalloyed titanium. Ti-3Al-2.5V can be cold worked 75 to 85% to result in moderately high strength and good ductility.

Machining is not a common practice on titanium alloy tubing, but hack sawing, abrasive wheel cut-off, and ordinary tube cutters are used in the cutting of titanium tubing. With the use of conventional machining techniques, titanium alloys are comparable to machining a good grade of stainless steel. In general, very sharp tools with a slightly larger rake angle and a very keen edge are suitable. Slower speed and heavier cuts are preferred because they maintain lower tool temperatures. Drilling of thin-walled titanium is not difficult as long as the drill is sharp. Thicker walled tube requires a heavy flood of coolant to remove heat and chips. General instructions for machining are given in "Technical Note 7: Machining."

Welding. Ti-3Al-2.5V has good weldability and, like all titanium alloys, is weldable by all methods except shielded arc welding and submerged arc welding (because no flux is permitted).

Filler Metal. Recommended filler metals are AWS Ti-9 and ERTi-9ELI.

Ti-3Al-2.5V: Seamless tubing production

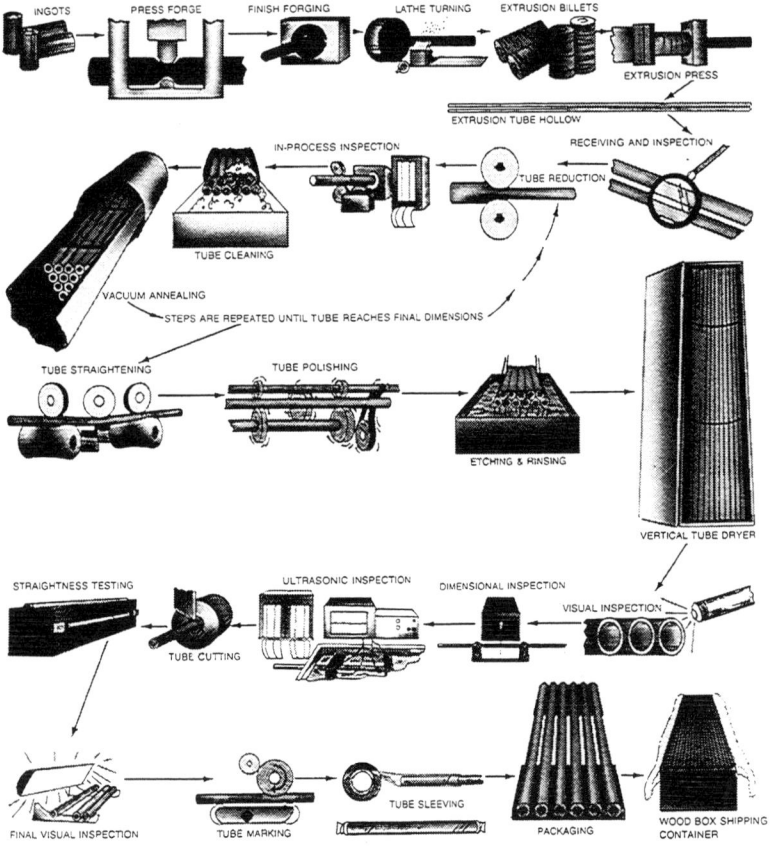

Source: Sandvik Special Metals

Ti-3Al-2.5V: Examples of primary working temperatures

Product form/condition	Working temperature
660 mm (26 in.) diam ingot, preheated to (2000 °F), 5 to 7 h, forged to 457 mm (18 in.) square billet	900 to 1095 °C (1650 to 2000 °F)
457 mm (18 in.) square billet forged to 355 mm (14 in.) square billet	900 to 1035 °C (1650 to 1900 °F) min
355 mm (14 in.) square billet forged to 250 mm (10 in.) square billet	900 to 980 °C (1650 to 1800 °F) min
250 mm (10 in.) square billet forged to 203 mm (8 in.) square billet	785 to 900 °C (1450 to 1650 °F) min
203 mm (8 in.) square billet forged to 150 mm (6 in.) diameter billet	900 °C (1650 °F)

Note: Intermediate conditioning operations as required. Source: R.A. Wood and R. Favor, *Titanium Alloys Handbook*, MCIC-HB-02, Battelle Columbus Laboratories, 1972

Ti-3Al-2.5V: Hot ductility

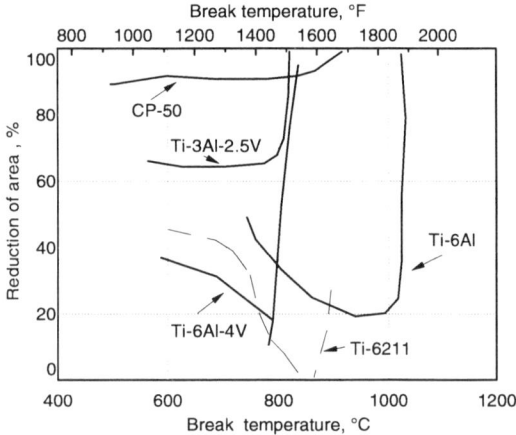

The on-cooling hot ductility behavior of titanium alloys with varying aluminum content. Ti-3Al-2.5V and unalloyed titanium do not exhibit a "hot ductility dip" as in other alloys containing more aluminum. Source: *Titanium, Science and Technology*, Vol 2, G. Lütjering, U. Zwicker, and W. Bunk, Ed., Deutsche Gesellschaft für Metallkunde, Germany

Ti-3Al-2.5V: Weldment property comparison

Alloy	Ti grade	Base metal			Weld(a)		
		Tensile yield strength (0.2%)		Elongation, %	Tensile yield strength (0.2%)		Elongation, %
		MPa	ksi		MPa	ksi	
CP Ti	1	221	32	52	297	43	42
CP Ti	3	386	56	24	441	64	23
Ti-0.3Mo-0.8Ni	12	372	54	25	469	68	12
Ti-3Al-2.5V	9	538	78	16	552	80	20
Ti-6Al-4V	5	883	128	12	814	118	10

(a) All weld metal tensile from 13 mm (0.5 in.) thick plate. Filled wire produced from same heats. Source: *Industrial Applications of Titanium and Zirconium: Third Conference*, R.T. Webster and C.S. Young, Ed., STP 830, ASTM, 1984

Forming

Bending

Both annealed and cold worked plus stress relieved conditioned tube can be readily formed at room temperature using the same dies and plug mandrels used in forming stainless steel tubing. Bend radii of 2.5 and 3.0 times the OD are typical for annealed and cold worked plus stress relieved tubing, respectively (see table). Spring back is about 15 to 25°. Tubing should be left in its protective sleeve or separator tray and handled carefully in bending to prevent surface damage that will reduce the fatigue life of the tube assembly.

Drawability of Ti-3Al-2.5V increases as the fraction of α phase increases, whereas low drawability occurs even after tempering of β quenched material. The best combination of strength and drawability is obtained by quenching from just below the β transus (see figure and table).

Ti-3Al-2.5V: Room-temperature formability
When smaller radii are required, Ti-3Al-2.5V can be heated to 200 to 300 °C (400 to 600 °F) in a proper atmosphere.

Product	Condition	Tensile yield strength		Elongation, %	Typical bend radius(a)	Flarability over 74° angle, %
		MPa	ksi			
Sheet	Annealed	558	81	18	2.5 R/t	...
	50% cold worked	758	110	10	...	...
Seamless tube	Annealed	558	81	20	2.5 R/OD	35-40(b)
	Cold worked, stress relieved	758	110	14	3.0 R/OD	20

(a) Higher or lower bend radii may be applicable depending on tube size and texture (see additional table and figure). (b) Tube 31.7 mm (1.25 in.) in diameter × 0.9 mm (0.035 in.) wall had been bent over a 50 mm (2 in.) radius, and tube 38 mm (1.5 in.) in diameter × 0.9 mm (0.035 in.) wall had been bent over a 57 mm (2.25 in.) radius, with wall thinning on the outside of the bend of less than 10%. Source: R.A. Wood and R.J. Favor, *Titanium Alloys Handbook*, MCIC-HB-02, Battelle Columbus Laboratories, 1972

Annealed Ti-3Al-2.5V: Bending limits vs. tube size

Tube OD	Room-temperature bending radius, R/OD
10 times wall thickness	1.2
10 to 25 times wall thickness	2.0
25 to 50 times wall thickness	2.75
50 to 60 times wall thickness	3.2

Source: C.E. Forney, Jr. and J.H. Schemel, *Ti-3Al-2.5V Seamless Tubing Engineering Guide*, 2nd ed., Sandvik Special Metals, 1987

Ti-3Al-2.5V: Effect of tube texture on bend radius

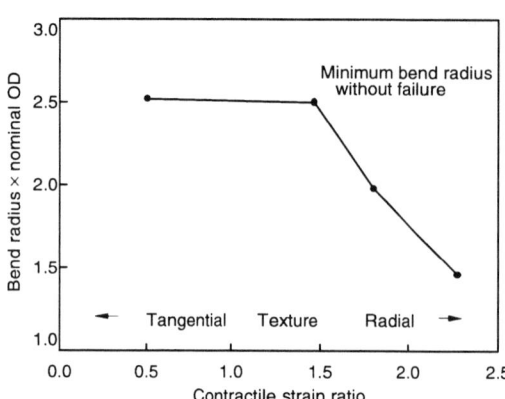

Cold worked and stress relieved tubing generally is not bent to radii less than 3 times the OD in production shops, although radially textured tubing can be bent to 1.5. Radial textures resist wall thinning and promote some shortening and wall thickening on the inside of the bend. Care must be taken to keep sufficient tensile force on the tube during bending to prevent buckling at the inside of the bend where compressive forces are developed.
Source: C.E. Forney, Jr. and J.H. Schemel, *Ti-3Al-2.5V Seamless Tubing Engineering Guide*, 2nd ed., Sandvik Special Metals, 1987

Ti-3Al-2.5V: Limiting drawing ratios for wall breakage

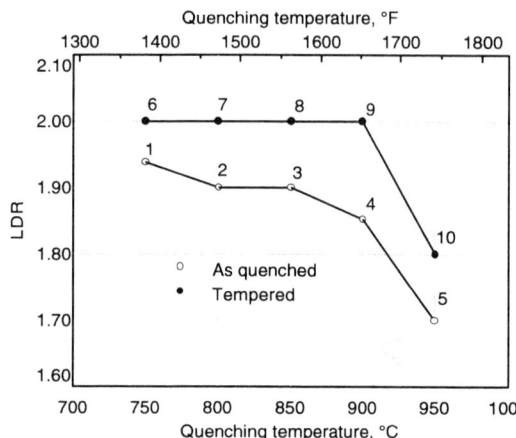

Limiting drawing ratio (LDR) for wall breakage of both quenched and tempered sheet (0.7 mm thick). Specimen numbers had the following average properties: (see table on next page).
Source: J. Kaneko and M. Sugamata, The Effects of Heat Treatment on the Deep Drawability of Ti-3Al-2.5V Sheets, *Titanium, Science and Technology*, Vol 2, G. Lütjering, U. Zwicker, and W. Bunk, Ed., Deutsche Gesellschaft für Metallkunde, Germany, 1985, p 563-568

Ti-3Al-2.5V: Properties of wall-breakage test specimens (previous figure)

Specimen No.	UTS, MPa	TYS, MPa	Elongation E_B %	Elongation E_u %	r
As-quenched					
1	575	508	24.2	10.0	3.08
2	587	490	25.4	11.6	2.67
3	600	470	24.8	11.8	2.21
4	771	578	9.3	4.0	1.91
5	827	590	5.5	3.8	0.56
Tempered					
6	593	528	22.4	9.4	3.01
7	605	506	21.2	8.8	3.14
8	611	537	21.3	8.9	2.96
9	753	621	8.1	3.4	2.90
10	820	627	5.6	3.9	0.91

Note: E_u = uniform elongation; E_B = breaking elongation. Source: J. Kaneko and M. Sugamata, The Effects of Heat Treatment on the Deep Drawability of Ti-3Al-2.5V Sheets, *Titanium, Science and Technology*, Vol 2, G. Lütjering, U. Zwicker, and W. Bunk, Ed., Deutsche Gesellschaft für Metallkunde, Germany, 1985, p 563-568

Ti-3Al-2.5V: Effect of heat treatment on strain ratio

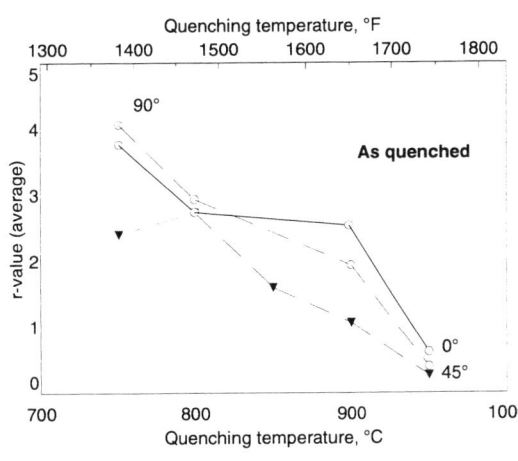 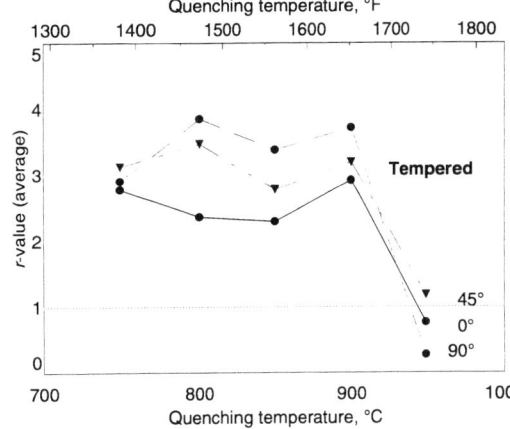

Change in r-value with quenching temperature for both 0.7 mm (0.02 in.) sheet.
Source: J. Kaneko and M. Sugamata, The Effects of Heat Treatment on the Deep Drawability of Ti-3Al-2.5V Sheets, *Titanium, Science and Technology*, Vol 2, G. Lütjering, U. Zwicker, and W. Bunk, Ed., Deutsche Gesellschaft für Metallkunde, Germany, 1985, p 563-568

Heat Treatment

Ti-3Al-2.5V is usually used in the annealed or in the cold worked plus stress relieved conditions. However, due to the small amount of β stabilizer present, a small age hardening response is possible from the solution heat treated condition. Only a small increase in strength is possible in thin sections by solution treating and aging. For example, 1.7 mm (0.070 in.) sheet solution treated 15 min at 910 °C (1675 °F) water quenched, and aged 8 h at 510 °C (950 °F) produces strengths only 138 to 172 MPa (20 to 25 ksi) greater than the annealed values and not much different from those achieved by cold working and stress relieving.

Ti-3Al-2.5V: Typical heat treatment conditions

Heat treatment	Temperature		Time, h	Cooling method
	°C	°F		
Minimum stress relief	315	600	0.5 min	AC
Typical stress relief(a)	370-650	700-1200	0.5 to 2 or 3	AC
Annealing(b)	650-760	1200-1400	0.5 to 2	AC
Solution treating	870-925	1600-1700	0.25 to 0.33	WQ
Aging	480-510	900-950	2 to 8	AC

(a) Heating above 540 °C (1000 °F) substantially reduced strength and hardness. (b) Heating to 700 °C (1300 °F) for 2 h appears to develop a fully annealed condition. There is no advantage in annealing above 800 °C (1475 °F), and annealing in the β phase field reduces impact strength.

Ti-3Al-2.5V: Effect of aging on strength

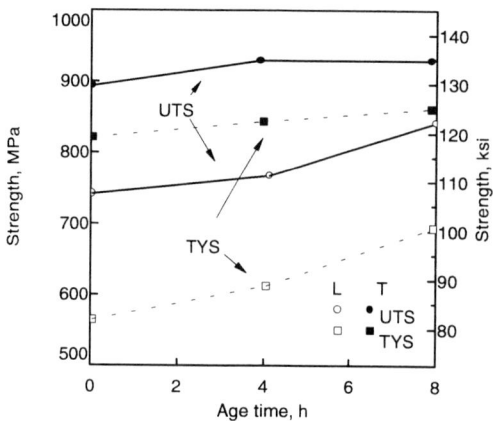

Effect of 540 °C (1000 °F) age on room-temperature tensile properties of 3.5 mm (0.140 in.) strip after solution treatment at 910 °C (1675 °F) for 15 min and water quench.
Source: *Aerospace Structural Metals Handbook*, Vol 4, Code 3725, Battelle Columbus Laboratories, 1980

Ti-3Al-2.5V: Tensile strength vs anneal temperature

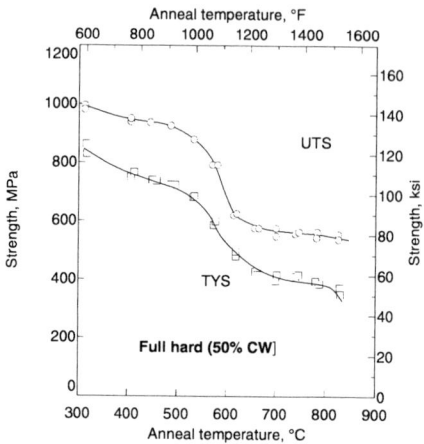

Effect of 2 h annealing (or stress relieving) temperature on room-temperature tensile properties of full hard (50% cold worked) tubing. With a room-temperature full hand strength of 999 MPa (145 ksi) UTS and 896 MPa (130 ksi) tensile yield strength. 15.8 mm (0.625 in.) OD × 0.96 mm (0.038 in.) wall 50% CW + anneal, 2 h in vacuum, vacuum cool to approximate 425 °C (800 °F), AC.
Source: *Aerospace Structural Metals Handbook*, Vol 4, Code 3725, Battelle Columbus Laboratories, 1980

Ti-5Al-2.5Sn

Common Name: Ti-5-2½ and Ti-5-2½ ELI
UNS Numbers: R54520/R54521

Chemistry and Density

Developed by Battelle for RemCru (later called Crucible Steel) as an intermediate-strength, weldable alloy, Ti-5Al-2.5Sn was first manufactured in 1950. Its primary use was in applications requiring moderate strength and excellent weldability. It was one of the first alloys to be developed commercially and is one of the few original alloys still in commercial use. Although it is still available from all producers, it is being replaced by Ti-6Al-4V in many applications.

As interstitial element content increases, both yield and tensile strengths increase and fracture toughness decreases. The extra low interstitial (ELI) grade of Ti-5Al-2.5Sn (UNS R54521) is especially well suited for service at cryogenic temperatures and exhibits an excellent combination of strength and toughness at −250 °C (−420 °F).

Density. 4.48 g/cm^3 (0.162 lb/in.^{3})

Product Forms

Ti-5Al-2.5Sn is available as bar, plate, sheet, strip, wire, forgings, and extrusions. The ELI grade is quite difficult to hot work into some product forms, particularly when converting from ignot to billet because of shear cracking, often referred to as strain-induced porosity. Ti-5Al-2.5Sn can be cast, machined, and welded.

Product Condition/ Microstructure

Ti-5Al-2.5Sn is a medium-strength, all-alpha titanium alloy. It has very high fracture toughness at both room temperature and elevated temperatures and is used only in the annealed condition.

Applications

Ti-5Al-2.5Sn is used for gas turbine engine castings and rings, rocket motor casings, aircraft forgings and extrusions, aerospace structural members in hot spots (near engines and leading edges of wings), ordnance equipment, chemical-processing equipment requiring elevated-temperature strength superior to that of unalloyed titanium and excellent weldability, and other applications demanding good weld fabricability, oxidation resistance, and intermediate strength at service temperatures up to 480 °C (900 °F).

Ti-5Al-2.5Sn ELI is employed for liquid hydrogen tankage and high-pressure vessels at temperatures below −195 °C (−320 °F), structural members for aircraft, and gas turbine parts. It is used in applications requiring ductility and toughness greater than those of the standard grade, although at some sacrifice in strength, particularly in hardware for service at cryogenic temperatures.

Precautions in Use. The elevated temperature stress-corrosion resistance of this alloy in the presence of solid salt is lower than those of other commonly used titanium alloys. Use of Ti-5Al-2.5Sn (like all titanium alloys) in contact with liquid oxygen, or in contact with gaseous oxygen at pressures above approximately 345 kPa (50 psi), constitutes severe fire and explosion hazard.

Ti-5Al-2.5Sn: Specifications and compositions

Specification	Designation	Description	Al	C	Fe	H	N	O	Sn	OT	Other
UNS	R54520		4-6	0.1	0.5	0.02	0.05	0.2	2-3		bal Ti
UNS	R54521	ELI	5						2.5		bal Ti
UNS	R54522	Weld Fill Met	4.7-5.6	0.05	0.4	0.008	0.03	0.12	2-3		bal Ti
UNS	R54523	ELI Weld Fill Met	4.7-5.6	0.04	0.25	0.005	0.012	0.1	2-3		bal Ti
China											
GB 3620	TA-7		4-6	0.1 max	0.3 max	0.015 max	0.05 max	0.2 max	2-3		Si 0.15 max; bal Ti
Germany											
	WL 3.7114		4.5-5.5	0.08	0.5	0.015-0.02	0.05	0.2	2-3	0.4	bal Ti
DIN 17851	Ti-5Al-2.5Sn	Sh Strp Plt Rod Wir	4-6	0.08	0.5	0.02	0.05	0.2	2-3		bal Ti
DIN 17851	WL 3.7115	Plt Sh Strp Ann	4-6	0.08	0.5	0.02	0.05	0.2	2-3		bal Ti
Russia											
GOST	VT5-1KT		4-5.5	0.05	0.2	0.008	0.04	0.12	2-3		Zr 0.2; Mn 0.1; Si 0.1; bal Ti
GOST 19807-74	VT5-1	Sh Plt Strp Rod Frg Ann	4-6	0.1	0.3	0.015	0.05	0.15	2-3	0.3	Zr 0.3; Si 0.15; bal Ti
Spain											
UNE 38-716	L-7101	Sh Strp Plt Bar Frg Ext	4.5-5.5	0.15	0.5	0.02	0.07	0.2	2-3		bal Ti

(continued)

Ti-5Al-2.5Sn: Specifications and compositions (continued)

Specification	Designation	Description	Al	C	Fe	H	N	O	Sn	OT	Other
UK											
BS TA14		Sh	4-6	0.08 max	0.5 max	0.0125 max			2-3		bal Ti
BS TA15		Bar	4-6	0.08 max	0.5 max	0.0125 max			2-3		bal Ti
BS TA16		Frg	4-6	0.08 max	0.5 max	0.0125 max			2-3		bal Ti
BS TA17		Frg	4-6		0.5 max	0.015 max			2-3		bal Ti
USA											
AMS 4909D		ELI Sh Strp Plt Ann	4.5-5.75	0.05	0.25	0.0125	0.035	0.12	2-3	0.3	Y 0.005; O + Fe = 0.32; bal Ti
AMS 4910J		Sh Strp Plt Ann	4.5-5.75	0.08	0.5	0.02	0.05	0.2	2-3	0.4	Y 0.005; bal Ti
AMS 4924D		ELI Bar Frg Rng Ann	4.7-5.6	0.05	0.25	0.0125	0.035	0.12	2-3	0.4	Y 0.005; O + Fe = 0.32; bal Ti
AMS 4926H		Bar Wir Bil Rng Ann	4-6	0.08	0.5	0.02	0.05	0.2	2-3	0.4	Y 0.005; bal Ti
AMS 4953D		Weld Fill Wir	4.5-5.75	0.08	0.5	0.015	0.05	0.175	2-3	0.4	Y 0.005; bal Ti
AMS 4966J		Frg Ann	4-6	0.08	0.5	0.02	0.05	0.2	2-3	0.4	Y 0.005; bal Ti
ASTM B 265	Grade 6	Sh Strp Plt Ann	4-6	0.1	0.5	0.02	0.05	0.2	2-3	0.4	bal Ti
ASTM B 348	Grade 6	Bar Bil Ann	4-6	0.1	0.5	0.0125	0.05	0.2	2-3	0.4	bal Ti
ASTM B 367	Grade C-6	Cast	4-6	0.1	0.5	0.015	0.05	0.2	2-3	0.4	bal Ti
ASTM B 381	Grade F-6	Frg Ann	4-6	0.1	0.5	0.02	0.05	0.3	2-3	0.4	bal Ti
AWS A5.16-70	ERTi-5Al-2.5Sn-1	ELI Weld Fill Met	4.7-5.6	0.04	0.25	0.005	0.012	0.1	2-3		bal Ti
AWS A5.16-70	ERTi-5Al-2.5Sn	Weld Fill Met	4.7-5.6	0.05	0.4	0.008	0.03	0.12	2-3		bal Ti
MIL F-83142A	Comp 2	Frg Ann	4.5-5.75	0.08	0.5	0.02	0.05	0.2	2-3	0.4	bal Ti
MIL F-83142A	Comp 2	Frg HT	4.5-5.75	0.08	0.5	0.02	0.05	0.2	2-3	0.4	bal Ti
MIL F-83142A	Comp 3	ELI Frg Ann	4.5-5.75	0.05	0.25	0.0125	0.035	0.12	2-3	0.3	bal Ti
MIL T-81556A	Code A-1	Ext Bar Shp Ann	4.5-5.75	0.08	0.5	0.02	0.05	0.2	2-3	0.4	bal Ti
MIL T-81556A	Code A-2	ELI Ext Bar Shp Ann	4.5-5.75	0.05	0.25	0.0125	0.035	0.12	2-3	0.3	bal Ti
MIL T-81915	Type II Comp A	Cast Ann	4.5-5.75	0.08	0.5	0.02	0.05	0.2	2-3	0.4	bal Ti
MIL T-9046J	Code A-1	Sh Strp Plt Ann	4.5-5.75	0.08	0.5	0.02	0.05	0.2	2-3	0.4	bal Ti
MIL T-9046J	Code A-2	ELI Sh Strp Plt Ann	4.5-5.75	0.05	0.25	0.0125	0.035	0.12	2-3	0.3	bal Ti
MIL T-9047G	Ti-5Al-2.5Sn	Bar Bil Ann	4.5-5.75	0.08	0.5	0.02	0.05	0.2	2-3	0.4	Y 0.005; bal Ti
MIL T-9047G	Ti-5Al-2.5Sn ELI	ELI Bar Bil Ann	4.5-5.75	0.05	0.25	0.0125	0.035	0.12	2-3	0.3	Y 0.005; bal Ti

(a) Maximum unless a range is specified

Ti-5Al-2.5Sn: Compositions

Specification	Designation	Description	Al	C	Fe	H	N	O	Sn	OT	Other
France											
Ugine	UTA5E	Sh Bar Ann	4.5-5.5	0.15	0.5	0.02	0.07	0.2	2-3		bal Ti
Ugine	UTA5EL	ELI Bar Ann	4.5-5.75	0.05	0.25	0.0125	0.035	0.12	2-3		bal Ti
Germany											
Deutsche T	Contimet AlSn 52	Sh Strp Plt Bar Frg Pip Ann	4.5-5.5	0.08	0.5	0.02	0.05	0.2	2-3		bal Ti
Deutsche T	Contimet AlSn 52 ELI	ELI Plt Bar Frg Pip Ann	4.7-5.6	0.06	0.15	0.013	0.05	0.12	2-3		bal Ti
Fuchs	TL52	Frg	5						2.5		bal Ti
Japan											
Kobe	KS5-2.5	Ann	4-6		0.5	0.02	0.05	0.2	2-3		bal Ti
Kobe	KS5-2.5ELI	ELI Ann	4.7-5.6		0.25	0.0125	0.035	0.12	2-3		bal Ti
MMA	5137										
Sumitomo	SAT-525										
Toho	525AT		4-6	0.03	0.5	0.02	0.05	0.3	2-3		bal Ti
USA											
OREMET	Ti 5-2.5										
RMI	RMI 5Al-2.5Sn	Mult Forms Ann	4-6	0.08	0.5	0.0175-0.02	0.05	0.2	2-3		
RMI	RMI 5Al-2.5Sn ELI	ELI Mult Forms Ann	4.7-5.75	0.08	0.25	0.0125-0.015	0.03	0.13	2-3		bal Ti
TIMET	TIMETAL 5-2.5	Ann	4-6	0.1 max	0.5 max	0.02 max	0.05 max	0.2 max	2-3		bal Ti
TIMET	TIMETAL 5-2.5 ELI	Ann	4.5-5.75	0.05 max	0.25 max	0.0125 max	0.035 max	0.12 max	2-3		bal Ti

(a) Maximum unless a range is specified

Physical Properties

The microstructure of Ti-5Al-2.5Sn is either acicular or equiaxed α, depending on prior processing. Acicular α is observed after thermal excursions above the β transus. Equiaxed alpha results from working the metal below the β transus, followed by annealing in the α field. There commonly is a very small amount of β in microstructures of Ti-5Al-2.5Sn that contains high iron. Equiaxed α is most frequently encountered in mill products of either standard Ti-5Al-2.5Sn or Ti-5Al-2.5Sn (ELI).

Beta Transus. The β phase transforms to α on cooling at 1040 to 1090 °C (1900 to 2000 °F).

Alpha Transus. On heating α to β at 955 to 985 °C (1750 to 1805 °F).

Beta Transus of ELI Grade. This occurs at 1010 ± 15 °C (1850 ± 25 °F).

Standard Ti-5Al-2.5Sn: Typical physical properties

Beta transus	1040 to 1090 °C	(1900 to 2000 °F)
Liquidus temperature	1590 ± 20 °C	(2895 ± 35 °F)
RT tensile modulus	110 to 125 GPa	(16 to 18 × 10^6 psi)
Density(a)	4.48 g/cm^3	(0.162 lb/in.^{3})
Electrical resistivity(a)	1.6 μΩ · m	
Magnetic permeability	Nonmagnetic	
Specific heat capacity(a)	530 J/kg · K	(0.127 Btu/lb · °F)
Thermal conductivity(a)	7.8 W/m · K	(4.5 Btu/ft · h · °F)
Thermal coefficient of linear expansion(b)	9.4 × 10^{-6}/°C	(5.2 × 10^{-6}/°F)

(a) Typical values at room temperature of about 20 to 25 °C (68 to 78 °F). (b) Mean coefficient from 0 to 100 °C (32 to 212 °F)

Ti-5Al-2.5Sn: Dynamic modulus of annealed sheet

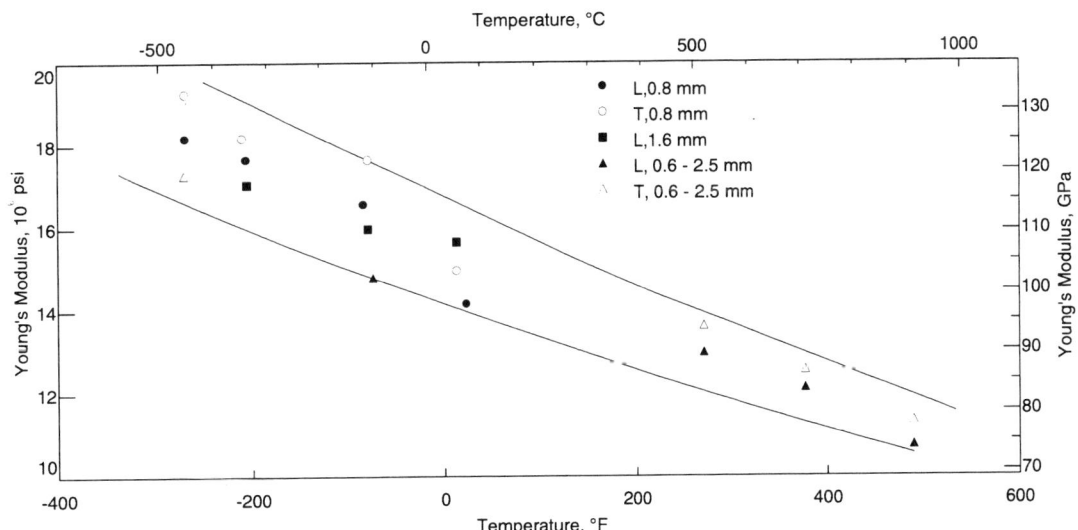

Source: *Aerospace Structural Metals Handbook*, Code 3706, Battelle Columbus Laboratories, 1965

Ti-5Al-2.5Sn: Static tensile modulus of annealed sheet

Source: *Aerospace Structural Metals Handbook*, Code 3706, Battelle Columbus Laboratories, 1965. More low-temperature modulus data is contained in the section "Low-Temperature Tensile Properties."

Electrical Resistivity

Ti-5Al-2.5Sn: Electrical resistivity

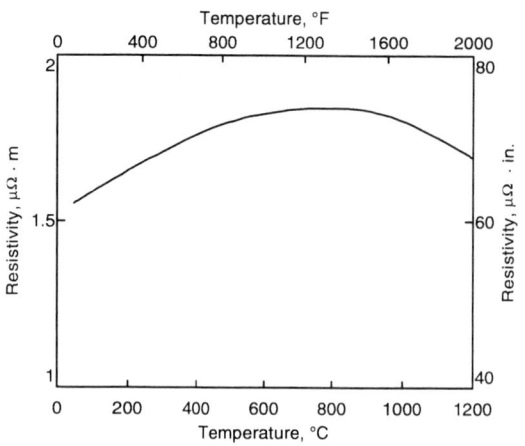

Source: *Aerospace Structural Metals Handbook*, Code 3706, Battelle Columbus Laboratories, 1965, p 4

Ti-5Al-2.5Sn: Electrical resistivity

Temperature		Resistivity	
°C	°F	μΩ · m	μΩ · in.
20	68	1.573	62
200	400	1.707	67
540	1000	1.860	73
760	1400	1.885	74

Optical Properties

Ti-5Al-2.5Sn: Emittance vs temperature

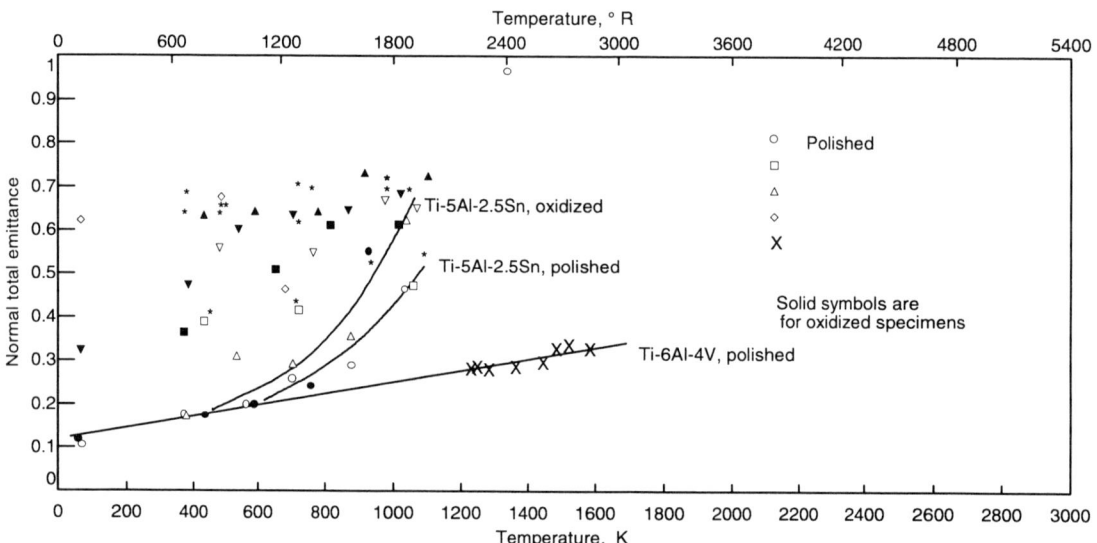

Source: O. Olson and J. Morris, WADC-TR-56-222, Part 2, Suppl. 1, reported in *Thermophysical Properties of High Temperature Solid Materials, Vol II, Nonferrous Alloys*, Y. Touloukian, Ed., Macmillan, 1967, p 1456

Ti-5Al-2.5Sn: Emissivity vs temperature

Source: Handbook of Thermophysical Properties of Solid Materials, Vol II, Alloys, A. Goldsmith, T. Waterman, and H. Hirschhorn, Ed., Macmillan, 1961, p 907

Chemical/Corrosion Properties

General Corrosion

Although few published data on the general corrosion rates of Ti-5Al-2.5Sn are available, successful use of this alloy, like other titanium alloys, can be expected in mildly reducing to highly oxidizing environments in which protective oxide films spontaneously form and remain stable. The nature of the oxide film on titanium alloys basically remains unaltered in the presence of minor alloying constituents. Consequently, small additions (<2 to 3%) of most commercially used alloying elements or trace alloy impurities generally have little effect on the basic corrosion resistance of titanium in normally passive environments. In boiling HCl concentrations below 0.5%, for example, Ti-5Al-2.5Sn is comparable to grade 2 titanium.

Like titanium, corrosion of the Ti-5Al-2.5Sn alloy is expected in hot, concentrated, low-pH chloride salts. Warm or concentrated solutions of hydrochloric, phosphoric, and oxalic acids also are damaging. In general, all acidic solutions that are reducing in nature corrode titanium, unless they contain inhibitors. Strong oxidizers, including anhydrous red fuming nitric acid and 90% hydrogen peroxide, also cause attack. Ionizable fluoride compounds, such as sodium fluoride and hydrogen fluoride, activate the surface and can cause rapid corrosion. Dry chlorine gas is especially harmful.

Ti-5Al-2.5Sn: Comparative corrosion in boiling HCl

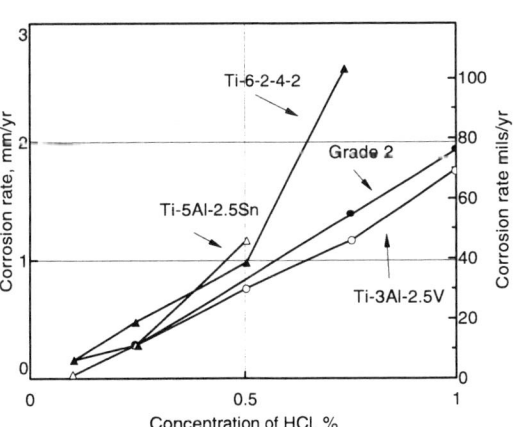

General corrosion of annealed titanium alloys in naturally aerated HCl solutions
Source: R. Schutz, Corrosion of Titanium and Titanium Alloys, in Metals Handbook, 9th ed., Vol 13, Corrosion, 1987, p 680

Ti-5Al-2.5Sn: Comparative oxidation at 550 °C in air

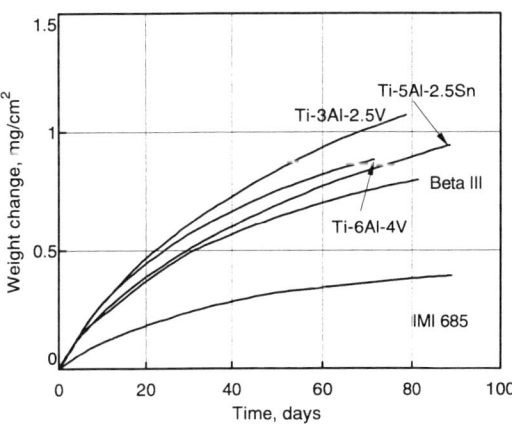

Source: C. Coddet et al., Oxidation of Titanium Base Alloys for Application in Turbines, Titanium '80, Science and Technology, Vol 4, H. Kimura and O. Izumi, Ed., TMS/AIME, 1980, p 2755-2764

Ti-5Al-2.5Sn: Comparative oxidation at 600 °C in air

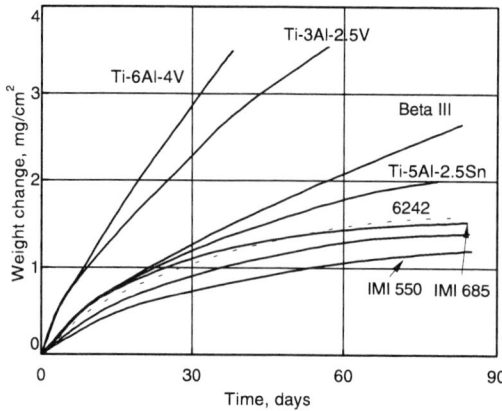

Source: C. Coddet et al., Oxidation of Titanium Base Alloys for Application in Turbines, *Titanium '80, Science and Technology*, Vol 4, H. Kimura and O. Izumi, Ed., TMS/AIME, 1980, p 2755-2764

Stress-Corrosion Cracking

Because tin and aluminum promote the formation of ordered Ti_3Al (α_2) structures, Ti-5Al-2.5Sn is one of the titanium alloys most susceptible to stress-corrosion cracking. Like step-cooled Ti-8Al-1Mo-1V, the Ti-5Al-2.5Sn alloy is susceptible to corrosion cracking in distilled water.

Ti-5Al-2.5Sn: Stress-corrosion cracking in distilled water

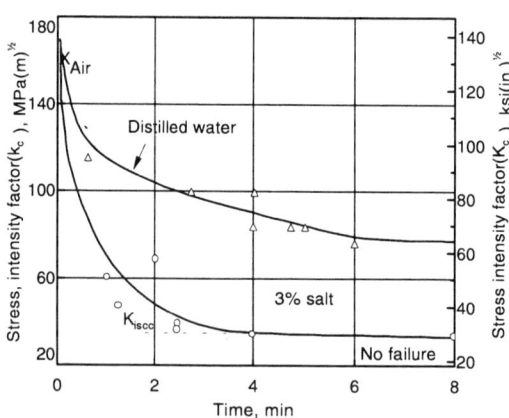

Source: T.L. Mackay and C.B. Gilpin, "Stress Corrosion Cracking of Titanium Alloys at Ambient Temperature in Aqueous Solutions," Missile & Space Systems Division, Astropower Laboratory, Douglas Aircraft Company, Report No. SM-49105-F1, June, 1967

Ti-5Al-2.5Sn: Environments known to promote stress-corrosion cracking

Medium	Temperature, °C (°F)	Other titanium alloys with known susceptibility
Oxidizers		
Nitric acid (red fuming)	RT	Ti, Ti-8Mn, Ti-6Al-4V
Organic compounds		
Methyl alcohol (anhydrous)	RT	Ti-6Al-4V, Gr. 2 Ti, Gr. 4 Ti, Ti-4Al-3Mo-1V, Ti-3Al-8V-6Cr-4Zr-4Mo, Ti-8Al-1Mo-1V, Ti-13V-11Cr-3Al
Methyl chloroform	370 (700)	Ti-8Al-1Mo-1V, Ti-6Al-4V, Ti-13V-11Cr-3Al
Ethyl alcohol (anhydrous)	RT	Ti-8Al-1Mo-1V
Trichloroethylene	370 (700), 620 (115), 815 (1500)	Ti-8Al-1Mo-1V
Trichlorofluoroethane	788 (1450)	Ti-8Al-1Mo-1V, Ti-6Al-4V, Ti-13V-11Cr-3Al
Chlorinated diphenyl	315-370 (600-700)	No alloy other than Ti-5Al-2.5Sn
Salts		
Hot chloride and other halide salts/residues	230-430 (445-805)	Most commercial alloys except grades 1, 2, 7, 11, 12, and Ti-3Al-2.5V
Seawater/NaCl solution	RT	Unalloyed Ti (with oxygen content >0.3%): Ti-2.5Al-1Mo-11Sn-5Zr-02Si (IMI-679), Ti-3Al-11Cr-13V, Ti-8Mn, Ti-6Al-4V, Ti-6Al-6V-2Sn, Ti-7Al-2Nb-1Ta, Ti-4Al-3Mo-1V, Ti-8Al-1Mo-1V, Ti-6Al-2Sn-4Zr-6Mo
Miscellaneous		
Distilled water	RT	Ti-8Al-1Mo-1V, Ti-11.5Mo-6Zr-4.5Sn (Beta III)
Ag metal + AgCl	232-480 (450-895)	7Al-4Mo
10% HCl	35-340 (94-645)	Ti-8Al-1Mo-1V

Source: R. Schutz, Stress-Corrosion Cracking of Titanium Alloys, in *Stress-Corrosion Cracking: Materials Performance and Evaluation*, ASM International, 1992

Ti-5Al-2.5Sn: Fracture toughness in air and 3.5% NaCl solution at 25 °C

Alloy	Thickness mm	Thickness in.	Heat treatment	Yield strength MPa	Yield strength ksi	K_{IC} or K_C MPa√m	K_{IC} or K_C ksi√in.	K_{ISCC} or K_{SCC} MPa√m	K_{ISCC} or K_{SCC} ksi√in.
Ti-5Al-2.5Sn	13	0.50	α anneal, AC	868	126	96	88	33	30
			α anneal, AC	855	124	113	103	29	27
			STA	889	129	91	83	25	23
			β anneal, WQ	868	126	130	119	40	37
Ti-5Al-2.5Sn (ELI)	9.6	0.38	α anneal, AC	827	120	107	98	45	41

Note: Data were generated in ambient neutral 3.5% NaCl solution. It should be cautioned that K_{ISCC} values are highly dependent on alloy composition, metallurgical condition, and product form and thickness. Therefore, they may or may not be representative of alloy product materials that are commercially available. Source: R. Schutz, Stress-Corrosion Cracking of Titanium Alloys, in *Stress-Corrosion Cracking: Materials Performance and Evaluation*, ASM International, 1992

294 / Alpha and Near-Alpha Alloys

Ti-5Al-2.5Sn: Stress intensity to failure

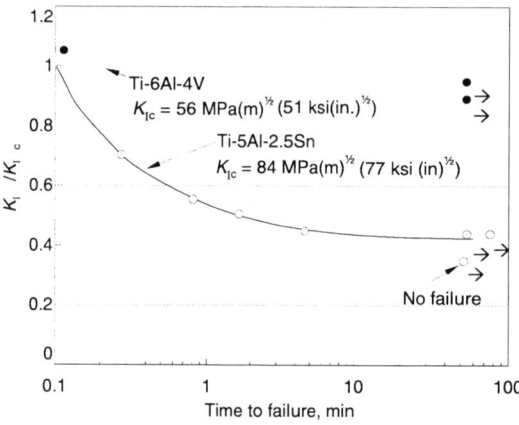

Variation of applied stress intensity (K_I) to critical stress intensity (K_{IC}) ratio with time to failure for martensite structures in Ti-6Al-4V and Ti-5Al-2.5Sn tested in 3.5% NaCl.
Both alloys were processed at 1095 °C (2000 °F), 15 min, water quenched.
Source: D. Fager and W. Spurr, *Trans. ASM*, Vol 61, 1968, p 283

Ti-5Al-2.5Sn: Hot-salt cracking

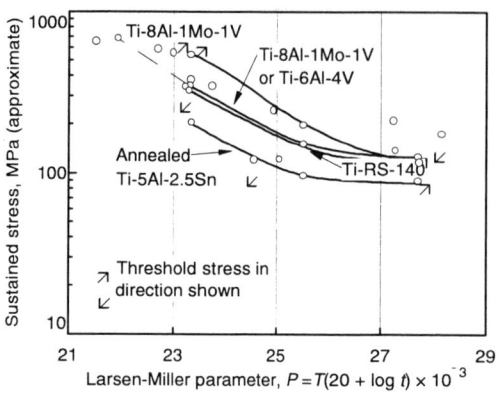

Larsen-Miller plot for hot-salt cracking of several annealed titanium alloys. T is temperature (°R) and t is exposure time (hours).
Source: R.V. Turley and C.H. Avery, "Elevated-Temperature Static and Dynamic Sea-Salt Stress Cracking of Titanium Alloys," *Stress-Corrosion Cracking of Titanium*, ASTM STP 397, ASTM, 1965, p 1

Hot-Salt Cracking

Ti-5Al-2.5Sn is one of the least resistant titanium alloys to hot-salt cracking. The presence of oxygen is necessary for hot-salt cracking to occur. At least one study has shown that cracking will not occur in Ti-5Al-2.5Sn when the environmental pressure is reduced below 10 μm (A.J. Hatch *et al.*, ASTM STP 397, 1965). Although the role of water (moisture) has not been clearly established, it appears that water is also a necessary environmental component in the cracking process. Chloride, bromide, and iodide salts have all been shown to produce similar cracking. Fluoride and hydroxide salts have not.

From a practical standpoint, hot-salt cracking appears to be a phenomenon that is restricted to the laboratory. No in-service failure has been attributed to hot-salt cracking. The likely reason for this is the critical relationship among environment, stress level, and alloy type. Unless all of the conditions are met simultaneously and for extended time, cracking will not occur.

Stress-Corrosion Cracking in Methanol

Stressed specimens of Ti-5Al-2.5Sn alloy are susceptible to stress-corrosion cracking in reagent-grade methanol liquid and in methanol vapor-air atmospheres after comparable exposure times. Unlike failure in oxygenated nitrogen tetroxide, the fracture surfaces in these environments were not visibly tarnished, and the failures were characterized by the formation of very few cracks. Cracks that formed in both the liquid and vapor were essentially intergranular. As with other susceptible titanium alloys, halides and halogens accelerate cracking and reduce K_{ISCC} values, whereas additions of water inhibit cracking.

Ti-5Al-2.5Sn: Effect of water and halogens on methanolic stress-corrosion cracking

Environment	Time to failure
CH_3OH (reagent grade 99.9 mol% pure)	8 days
CH_3OH (reagent grade, dried with CaO powder)	4 days
CH_3OH (reagent grade, redistilled)	2 days
CH_3OH vapor-dry air mixture	1-7 days
CH_3OH + 0.012% I_2	2 days
CH_3OH + 1.35% I_2	5 hours
CH_3OH + 2.0% Br_2	10 min
CH_3OH + 2.0% Br_2 + 2.5% H_2O	10 min
CH_3OH + 2.0% Br_2 + 10.0% H_2O	13 min
CH_3OH + 2.0% Br_2 + 45.0% H_2O	4 days
CH_3OH + 2.0% Br_2 + 50.0% H_2O	NF

U-bend specimens were annealed with a yield strength of 703 MPa (102 ksi) and a tensile strength of 772 MPa (112 ksi). Source: A.J. Sedriks, P.W. Slattery, and E.N. Pugh, "Stress Corrosion Cracking of Alpha Titanium in Nonaqueous Environments," Proceedings of Conference on Fundamental Aspects of Stress Corrosion Cracking, NACE, Sep 11-15, 1967, p 673

Thermal Properties

Heat Capacity

Ti-5Al-2.5Sn: Specific heat at various temperatures

Temperature		Heat	
°C	°F	J/kg · K	Btu/lb · °F
0(a)	32(a)	528(a)	0.126(a)
20(b)	70(b)	523(b)	0.125(b)
40	100	536	0.128
95	200	548	0.131
205	400	574	0.137
315	600	607	0.145
425(a)	800(a)	641(a)	0.153(a)
540	1000	674	0.161
650(a)	1200(a)	712(a)	0.170(a)
760	1400	754	0.180
845	1550	787	0.188

(a) Applied to both Ti-5Al-2.5Sn and Ti-5Al-2.5Sn ELI, reported in UNS No. R54521, Code Ti-91, *Alloy Digest*, May 1988; and "Rem-Cru Titanium Manual," Rem-Cru Titanium, reported in "Report on the Physical Properties of Titanium and Titanium Alloys," TML Report 73, W. Lepkowski and J. Holladay, Ed., Titanium Metallurgical Laboratories, Battelle Memorial Institute, 1957. (b) Applies only to Ti-5Al-2.5Sn (ELI), reported in RMI 5Al-2.5Sn ELI, Code Ti-75, *Alloy Digest*, Feb 1980

Ti-5Al-2.5Sn: Specific heat as a function of temperature

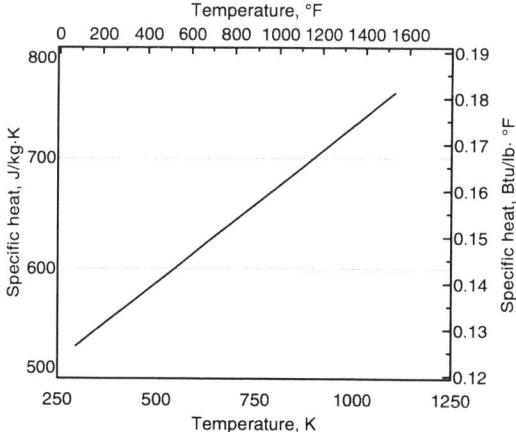

Source: *Metallic Materials and Elements for Aerospace Vehicle Structures*, MIL-HDBK-5E, Vol 2, June 1987, p 5.16

Thermal Expansion

Ti-5Al-2.5Sn and Ti-5Al-2.5Sn ELI: Thermal coefficient of linear expansion

Temperature		Coefficient of expansion	
°C	°F	10^{-6}/°F	10^{-6}/K
0-100	32-212	5.2	9.4
0-315	32-600	5.3	9.5
0-540	32-1000	5.3	9.5
		5.4(a)	9.7(a)
0-650	32-1200	5.4	9.7
		5.5(a)	9.9(a)
0-815	32-1500	5.6	10.1

(a) ELI, extra-low intensity. Source: TIMET datasheet

296 / Alpha and Near-Alpha Alloys

Ti-5Al-2.5Sn: Thermal expansion at low temperatures

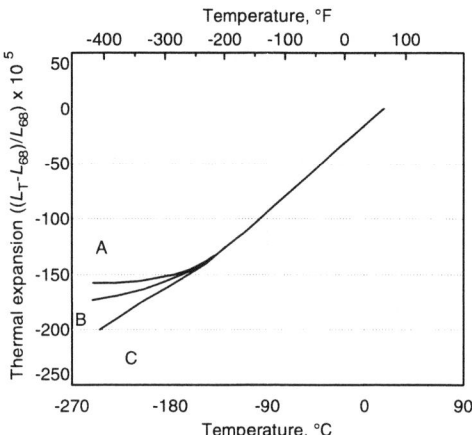

Curve A: 13 mm (0.500 in.) diameter bar, annealed. Curve B: 19 mm (0.750 in.) diameter bar, annealed. Curve C: 13 mm (0.500 in.) diameter bar, annealed.
Source: Curve A: Martin Marietta Company; Curve B: V. Arp et al., "Thermal Expansion of Some Engineering Materials from 20° to 293 °K," *Cryogenics*, Vol 2, 1962; Curve C: J. Belton et al., *Materials for Use at Liquid Hydrogen Temperatures*, ASTM Special Tech. Pub. 287, 1960, p 108; reported in *Cryogenic Materials Data Handbook (Revised)*, Vol 1, AFML-TR-64-280, Wright Patterson AFB, 1970, p 664

Ti-5Al-2.5Sn: Thermal expansion at room temperature/above

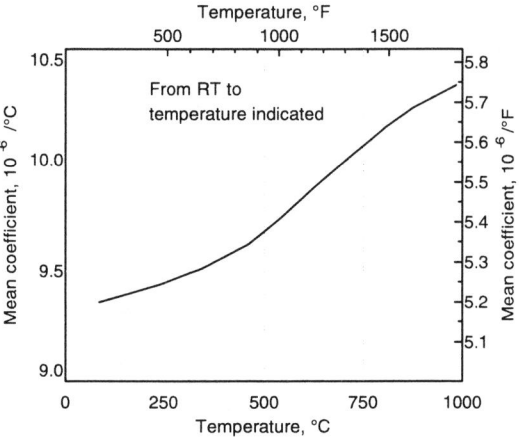

Source: *Aerospace Structural Metals Handbook*, Code 3706, Battelle Columbus Laboratories, 1965

Thermal Conductivity

Ti-5Al-2.5Sn: Thermal conductivity vs temperature

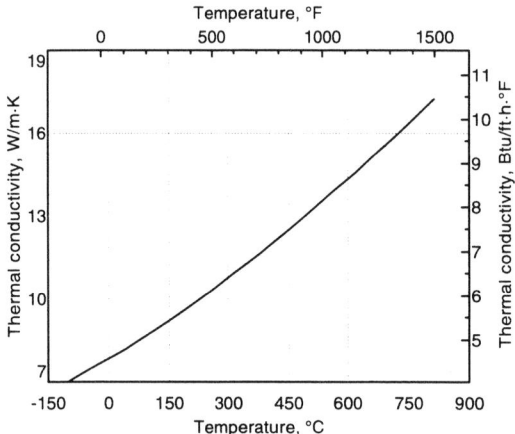

Source: *Aerospace Structural Metals Handbook*, Code 3706, Battelle Columbus Laboratories, 1965

Ti-5Al-2.5Sn: Thermal conductivity comparison

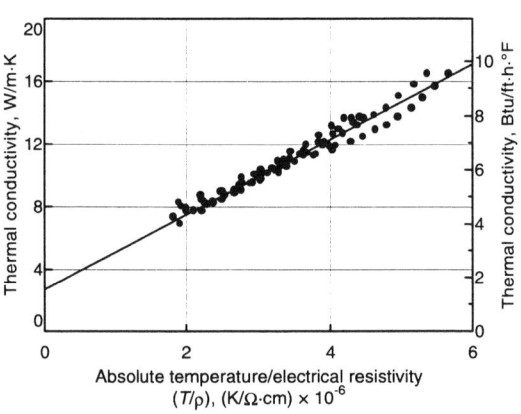

Thermal conductivity versus ratio of absolute temperature, T, to electrical resistivity.
Data are plotted for Ti-5Al-2.5Sn, Ti-6Al-4V, Ti-4Al-4Mn, Ti-2Al-2Mn, Ti-4Al-4Mo-2Sn-0.5Si, Ti-8Mn, Ti-5Al-1.4Cr-1.5Fe-1.2Mn, and Ti-2.1Cr-2.2Fe-2Mo. Values obtained by Powell and Tye were obtained at 50 °C (90 °F) intervals over the range 50 to 450 °C (120 to 840 °F). Previously reported values of Deem et al. (H. Deem, W. Wood, and C. Lucks, *Trans. Met. Soc. AIME*, Aug 1958, p 520) plotted at 55 °C (100 °F) intervals over the range 38 to 540 °C (100 to 1000 °F). Least-squares treatment gives an equation for the resulting line: $k = 2.43 \times 10^{-8} (T/\rho) + 0.027$. All values but one are within +7.5%.
Source: R. Powell and R. Tye, "The Thermal and Electrical Conductivity of Titanium and Its Alloys," *J. Less-Common Met.*, Vol 3, 1962, p 226

Ti-5Al-2.5Sn ELI: Thermal conductivity at various temperatures

Temperature		Thermal conductivity		Ref
°C	°F	Btu/ft · h · °F	W/m · K	
−75	−100	3.89	6.73	1
0	32	4.31	7.47	1
RT	RT	4.5	7.79	2
315	600	6.30	10.90	1
815	1500	10.48	18.13	1

References: (1) "UNS R 54521," Code 91, *Alloy Digest*, May 1988; (2) "Basic Design," RMI Titanium

Ti-5Al-2.5Sn: Experimental thermal conductivity

Temperature		Thermal conductivity	
°F	K	cal/(cm · s · °F)	W/m · K
100	311	0.0187	7.83
200	366	0.0200	8.37
300	422	0.0215	9.00
400	477	0.0229	9.59
500	533	0.0244	10.21
600	589	0.0260	10.88
700	644	0.0277	11.30
800	700	0.0294	12.31
900	755	0.0311	13.02
1000	811	0.0331	13.86

Note: The method used in making conductivity measurements was the steady-heat flow method. Specimens were protected by vacuum of ~5 × 10^{-5} mm Hg during measurements. The absolute error of experimental thermal conductivity values is estimated to be <5%. Specimens were mill annealed. Source: H. Deem, W. Wood, and C. Lucks, "The Relationship Between Electrical and Thermal Conductivities of Titanium Alloys," *Trans. Metall. Soc. AIME*, 1958, p 520

Mechanical Properties

Ti-5Al-2.5Sn: Typical tensile properties

Grade	Ultimate tensile strength		Tensile yield strength (0.2%)		Elongation in 50 mm (2 in.), %
	MPa	ksi	MPa	ksi	
Standard	861	125	827	120	15
ELI	779	113	717	104	17

Ti-5Al-2.5Sn: Tensile property specification minimums at room temperature

Specification	Min tensile strength		Min yield strength(a)		Min reduction of area, %
	MPa	ksi	MPa	ksi	
Plate, sheet, and strip					
AMS 4910 (Std)	827	120	779	113	...
ASTM B265, grade 6	827	120	793	115	...
MIL-T-9046					
Std grade, up to 1.5 in. thick	827	120	779	113	...
Std grade, 1.5 to 4.0 in. thick	793	115	758	110	...
ELI, all thicknesses	689	100	655	95	...
Billet, bar, forgings, and rings					
AMS 4924 (ELI)	689	100	620	90	20(b)
AMS 4926 and 4966 (Std)	793	115	758	110	25
ASTM B348, grade 6	827	120	793	115	25
ASTM B381, grade F-6	827	120	793	115	25
MIL-T-9047,					
Std grade	793	115	758	110	25
ELI	689	100	620	90	20(b)
Castings					
ASTM B367, grade C-6	793	115	724	105	...

(a) 0.2% offset. (b) 15% for material 2 to 4 in. thick. Source: *Metals Handbook*, Vol 3, 9th ed., American Society for Metals, 1980, p 383

Ti-5Al-2.5Sn ELI: Minimum tensile properties of bars, forgings, and rings

Nominal diameter or distance between parallel sides		Ultimate tensile strength		Yield Strength (0.2% offset)		Elongation in 4D, %			Reduction of area, %		
mm	in.	MPa	ksi	MPa	ksi	L	LT	ST	L	LT	ST
≤50	≤2.000	689	100	620	90	10	10	...	20	15	...
>50-100	>2.00-4.00	689	100	620	90	10	10	6	15	15	10

Source: AMS 4924, Jan 1985

Ti-5Al-2.5Sn ELI: Minimum tensile properties of sheet, strip, and plate

Nominal thickness		Ultimate tensile strength		Tensile yield strength (0.2% offset)		Elongation in 50 mm (2 in.) or 4D,
mm	in.	MPa	ksi	MPa	ksi	%
0.2-0.4	0.008-0.015	689	100	655	95	6
≥0.4-0.6	≥0.015-0.025	689	100	655	95	8
≥0.6-25	≥0.025-1.000	689	100	655	95	10

Source: AMS 4909, Oct 1984

Design Allowables

Ti-5Al-2.5Sn: Design tensile properties of annealed bar and forgings

Product form	Basis	Ultimate tensile strength(a)		Tensile yield strength(a)		Elongation(a), of area(a), %	Reduction %
		MPa	ksi	MPa	ksi		
Bar ≤75 mm (2.999 in.) thick(b)	A	793(c)(d)	115(c)(d)	758(c)(d)	110(c)(d)	10	25(d)
	B	868	126	827	120	...	...
Bar 76-100 m (3-4 in.) thick(b)	S	793	115	758	110	10 / 8(e)	25 / 20(e)
Forgings per AMS 4966	S	793(f)	115(f)	758(f)	110(f)	10(e)	25(e)

(a) S-basis. Limits apply to longitudinal and long transverse directions, except as noted. (b) Maximum of 105 cm^2 (16 in.^{2}) cross-sectional area per MIL-T-9047. (c) A values are higher than S values as follows: ultimate tensile strength, 806 MPa (117 ksi); tensile yield strength, 779 MPa (113 ksi) in the longitudinal direction. (d) S-basis in LT direction providing LT dimension is >75 mm (3 in.). (e) Minimum in the ST direction. (f) Applicable in LT or ST direction providing LT or ST dimension is ≥64 mm (2.5 in.). Source: MIL-HDBK-5, 1 Dec 1991

Ti-5Al-2.5Sn: S-basis tensile properties (longitudinal) of extruded and annealed bar and shapes Per MIL-T-81556, Comp. A-1

Thickness		Ultimate tensile strength		Tensile yield strength		Elongation, %
mm	in	MPa	ksi	MPa	ksi	
4.7-25	0.188-1.000	827	120	793	115	10
25-50	1.000-2.000	793	115	758	110	10
50-75	2.00-3.000	793	115	758	110	8
75-100	3.00-4.000	793	115	758	110	6

Source: MIL-HDBK-5, 1 Dec 1991

Ti-5Al-2.5Sn: Mechanical properties of annealed bar Per MIL-T-9047

	Thickness or diameter (a), mm (in.)		
	≤76.1 (≤2.999)		76.2-100 (3.000-4.000)(a)
Basis:	A	B	S
F_{tu}, MPa (ksi)			
L	793 (115)(b)	868 (126)	793 (115)
LT	793 (115)(c)	...	793 (115)
ST	...	...	793 (115)
F_{ty}, MPa (ksi)			
L	758 (110)(b)	827 (120)	758 (110)
LT	758 (110)(c)	...	758 (110)
ST	...	...	758 (110)
EL (d), %			
L	10	...	10
LT	10(c)	...	10
ST	...	...	8
RA (d), %			
L	25	...	25
LT	25(c)	...	25
ST	...	...	20

(a) Maximum of 105 cm^2 (16 in.^{2}) cross-sectional area. (b) A values are higher than S values as follows: ultimate tensile strength, 806 MPa (117 ksi); tensile yield strength, 779 MPa (113 ksi). (c) S-basis. Applicable providing LT dimension is >75 mm (3.000 in.). (d) S-basis. Source: MIL-HDBK 5, 1 Dec 1991

Ti-5Al-2.5Sn: Design mechanical properties of annealed sheet and strip
Per MIL-T-9046, Comp. A-1, and AMS 4910

	Per MIL-T-9046	Thickness, mm (in.)			
	<4.7 (<0.187)	0.4-2.0 (0.015-0.079)		2.0-4.7 (0.080-0.187)	
Basis:	S	A	B	A	B
F_{tu}, MPa (ksi)					
L	827 (120)	827 (120) (a)	882 (128)	827 (120) (b)	903 (131)
LT	827 (120)	827 (120) (a)	889 (129)	827 (120) (b)	910 (132)
F_{ty}, MPa (ksi)					
L	779 (113)	758 (110)	792 (115)	779 (113)	813 (118)
LT	779 (113)	779 (113)	813 (118)	779 (113) (c)	834 (121)
F_{cy}, MPa (ksi)					
L	792 (115)	792 (115)	827 (120)	779 (118)	848 (123)
LT	813 (118)	813 (118)	848 (123)	813 (118)	868 (126)
F_{su}, MPa (ksi)	517 (75)	517 (75)	551 (80)	517 (75)	565 (82)
F_{bru}, MPa (ksi), in:					
$e/D = 1.5$	1151 (167)	1151 (167)	1234 (179)	1151 (167)	1261 (183)
$e/D = 2.0$	1723 (250)	1723 (250)	1848 (268)	1723 (250)	1896 (275)
F_{bry}, MPa (ksi), in:					
$e/D = 1.5$	917 (133)	917 (133)	958 (139)	917 (133)	979 (142)
$e/D = 2.0$	1310 (190)	1310 (190)	958 (198)	1310 (190)	1399 (203)
EL (S-basis), %					
L	10	10 (d)	...	10	...
LT	10	10 (d)	...	10	...

(a) S basis. A-basis value is 848 MPa (123 ksi). (b) S-basis value is listed. A-basis value is 870 MPa (126 ksi). (c) S-basis value. (d) Thicknesses 0.635 mm (0.025 in.) and above. Source: MIL-HDBK-5, 1 Dec 1991

Ti-5Al-2.5Sn: Design mechanical properties of annealed plate
Per MIL-T-9046, Comp. A-1, and AMS 4910

	Thickness, mm (in.)			
	4.7-6.4 (0.188-0.250)		6.4-38 (0.25-1.500)	38-100 (1.50-4.000)
Basis:	A	B	S	S
F_{tu}, MPa (ksi)				
L	827 (120)(a)	930 (135)	827 (120)	793 (115)
LT	827 (120)(a)	944 (137)	827 (120)	793 (115)
F_{ty}, MPa (ksi)				
L	779 (113)(a)	848 (123)	779 (113)	758 (110)
LT	813 (113)(a)	862 (125)	779 (113)	758 (110)
F_{cy}, MPa (ksi)				
L	813 (118)	882 (128)	813 (118)	...
LT	813 (118)	896 (130)	813 (118)	...
F_{su}, MPa (ksi)	517 (75)	586 (85)	517 (75)	...
F_{bru}, MPa (ksi), in:				
$e/D = 1.5$	1151 (167)	1310 (190)	1151 (167)	...
$e/D = 2.0$	1723 (250)	1965 (285)	1723 (250)	...
F_{bry}, MPa (ksi), in:				
$e/D = 1.5$	917 (133)	1013 (147)	917 (133)	...
$e/D = 2.0$	1310 (190)	1448 (210)	1310 (190)	...
EL(a), %				
L	10	...	10	10
LT	10	...	10	10

(a) S-basis. Source: MIL-HDBK 5, 1 Dec 1991

Hardness

Ti-5Al-2.5Sn: Effect of low temperature on hardness

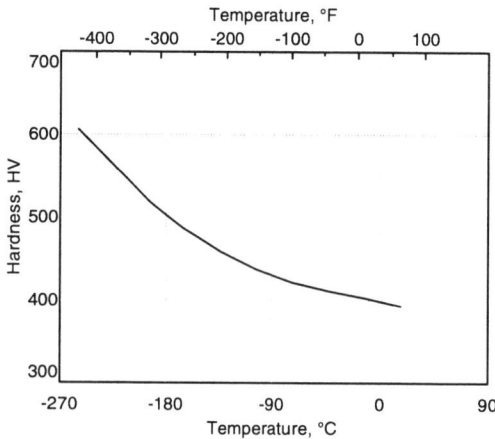

Specimens were made of normal interstitial material as-annealed 1.6 mm (0.064 in.) thick sheet.
Source: R. McGee, "The Mechanical Properties of Certain Aircraft Structural Metals at Very Low Temperatures," WADC-TR-58-336, Battelle Memorial Institute, Nov 1958; reported in *Cryogenic Materials Data Handbook*, AFML-TR-64-280, F. Schwartzberg, S. Osgood, R. Herzog, and M. Knight, Ed., Air Force Materials Laboratory, 1970, p 645

Ti-5Al-2.5Sn: Typical hardness at room temperature

Condition	Hardness Brinell, HB	Rockwell C, HRC	Knoop, HK
Annealed, Std O_2	...	30-36 (typical) 28 (min)	...
Annealed, ELI O_2	...	30-36	...
Unwelded ELI sheet	...	33.2	265
Single-bead weld of ELI sheet	...	28	310
Annealed bar	290	30-31	...
As cast	321 typical 335 max	...	...

Source: *Alloys Digest*, *Metals Handbook*, and *Aerospace Structural Metals Handbook*

Effects of Processing

Ti-5Al-2.5Sn: Tensile properties of annealed and cold drawn specimens

Property	Annealed	Cold drawn 15%
UTS, MPa (ksi)	982 (142.5)	1206 (175.0)
TYS (0.2% Offset), MPa (ksi)	879 (127.5)	1041 (151.0)
Proportional limit, MPa, (ksi)	844 (122.5)	517 (75.0)
EL in 50 mm (2 in.), %	17.0	10.0
RA, %	39.0	28.0
Elastic modulus, GPa (10^6 psi)	108 (15.7)	99 (14.4)
Type of fracture	Flat ½ cup	Flat ½ cup

Source: R. Wood and R. Favor, Ed., *Titanium Alloys Handbook*, MCIC-HB-02, Battelle, 1972

Ti-5Al-2.5Sn: Typical mechanical properties of as-received, stretched, and stress-relieved specimens

Specimen	Compressive yield strength (0.2% Offset) MPa	ksi	Tensile yield strength (0.2% Offset) MPa	ksi	Tensile ultimate strength MPa	ksi	Elongation, % in 50 mm (2 in.)	Percentage of as-received compressive yield strength
As received	895	129.9	839	121.7	908	131.8	19.8	100
1% stretch								
As stretched	605	87.8	866	125.6	909	131.9	20.0	67.6
Stress relieved	846	122.8	837	121.5	907	131.6	21.0	94.5
3% stretch								
As stretched	585	84.9	908	131.7	937	135.9	18.3	65.4
Stress relieved	837	121.5	832	120.7	917	133.0	17.5	93.5

Note: These are not design allowables. Source: R. Wood and R. Favor, Ed., *Titanium Alloys Handbook*, MCIC-HB-02, Battelle, 1972

Ti-5Al-2.5Sn: Effect of stretching on yield strengths

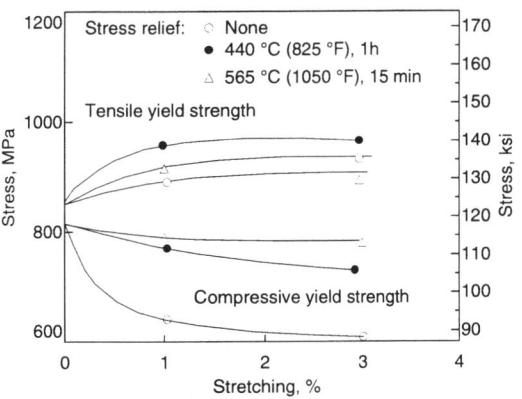

Specimens were 1.4 mm (0.057 in.) annealed sheet.
Source: Aerospace Structural Metals Handbook, Code 3706, Battelle Columbus Laboratories, 1965

Ti-5Al-2.5Sn: Yield strength vs rolling temperatures

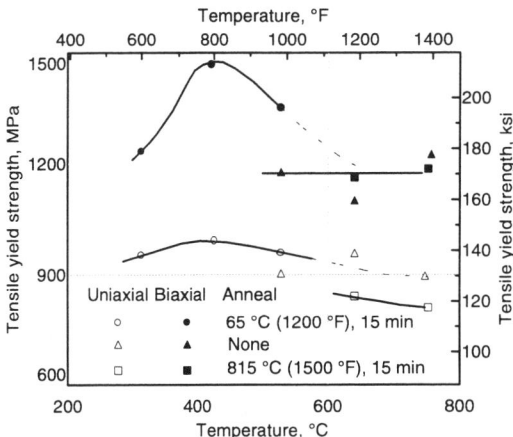

The uniaxial yield strength and the theoretical biaxial yield strength in a 2:1 stress field.
Source: *Aerospace Structural Metals Handbook*, Code 3706, Battelle Columbus Laboratories, 1965

Low-Temperature Tensile Properties

Ti-5Al-2.5Sn: Notch strength of annealed sheet

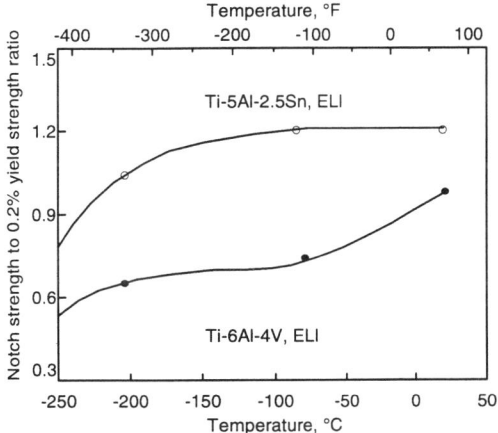

Specimens were 0.6 mm (0.025 in.) thick annealed sheet. Values are averages of several measurements in L and T directions. Specimen had a 60° notch on both sides. r less than 0.001 in.
Source: J. Shannon, Jr. and W. Brown, Jr., "A Review of Factors Influencing the Crack Tolerance of Titanium Alloys," in *Applications Related Phenomena in Titanium Alloys,* ASTM STP 432, ASTM, 1968, p 33

Ti-5Al-2.5Sn ELI: Sharp notch yield strength at −250 °C

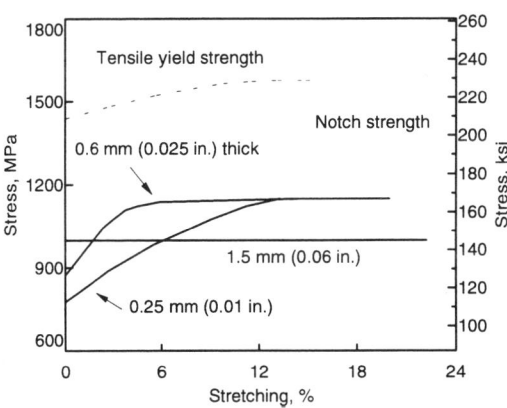

Specimens were annealed sheet heat treated at 815 °C (1500 °F), 2 h, FC; 60° notch on both sides with a 0.7 mil radius.
Source: *Aerospace Structural Metals Handbook*, Code 3706, Battelle Columbus Laboratories, 1965

Ti-5Al-2.5Sn: Low-temperature tensile properties compared with CP Ti and Ti-6Al-4V (continued)

Temperature		Ultimate tensile strength		Tensile yield strength		Elongation, %	Reduction of area, %	Notch tensile strength(a)		Young's modulus	
°C	°F	MPa	ksi	MPa	ksi			MPa	ksi	GPa	10⁶ psi
Ti-75A sheet, longitudinal orientation											
24	75	580	84.3	465	67.6	25	...	785	114	...	...
−78	−108	750	109	615	89.2	25	...	...	...	...	...
−196	−320	1050	152	940	136	18	...	1100	159	...	...
−253	−423	1280	186	1190	173	8	...	875	127	...	...
Ti-75A sheet, annealed, transverse orientation											
24	75	585	85.1	475	69.0	25	...	800	116	...	...
−78	−108	760	110	645	93.4	20	...	905	131	...	...
−196	−320	1060	153	965	140	14	...	1120	163	...	...

(continued)

Ti-5Al-2.5Sn: Low-temperature tensile properties compared with CP Ti and Ti-6Al-4V (continued)

Temperature °C	°F	Ultimate tensile strength MPa	ksi	Tensile yield strength MPa	ksi	Elongation, %	Reduction of area, %	Notch tensile strength(a) MPa	ksi	Young's modulus GPa	10⁶ psi
−253	−423	1340	194	1260	182	7	...	880	128	...	...
Ti-5Al-2.5Sn sheet, nominal interstitial annealed, longitudinal orientation											
24	75	850	123	795	115	16	...	1130	164	105	15.4
−78	−108	1080	156	1020	148	13	...	1310	190	115	16.6
−196	−320	1370	199	1300	188	14	...	1630	236	120	17.7
−253	−423	1700	246	1590	231	7	...	1430	208	130	18.5
Ti-5Al-2.5Sn sheet, nominal interstitial annealed, transverse orientation											
24	75	895	130	860	125	14	...	1170	170	...	...
−78	−108	1050	152	1020	148	12	...	1250	181	...	...
−196	−320	1430	208	1370	198	12	...	1630	236	...	...
−253	−423	1670	242	1610	234	6	...	1290	187	...	...
−268	−450	1590	231	...	...	1.5	...				
Ti-5Al-2.5Sn (ELI) sheet, annealed, longitudinal orientation											
24	75	800	116	740	107	16	...	1060	154	115	16.4
−78	−108	960	139	880	128	14	...	1190	173	125	18.0
−196	−320	1300	188	1210	175	16	...	1560	226	130	18.6
−253	−423	1570	228	1450	210	10	...	1670	242	130	19.2
Ti-5Al-2.5Sn (ELI) sheet, annealed, transverse orientation											
24	75	805	117	760	110	14	...	1100	159	110	16.0
−78	−108	950	138	895	130	12	...	1260	182	125	18.1
−196	−320	1300	188	1230	179	14	...	1570	228	130	18.9
−253	−423	1570	228	1480	214	8	...	1530	222	140	20.1
Ti-5Al-2.5Sn (ELI) sheet/weld weldment, annealed, EB weld											
24	75	815	118	785	114	...	...	...	...	...	...
−196	−320	1300	189	1210	176	...	...	...	...	...	...
−253	−423	1510	219	1380	200	...	...	...	...	...	...
Ti-5Al-2.5Sn (ELI) plate, annealed, longitudinal orientation											
24	75	765	111	705	102	33	43	...	...	...	...
−253	−423	1430	208	1390	202	17	32	...	...	...	...
Ti-5Al-2.5Sn (ELI) forgings, as forged, tangential orientation											
24	75	835	121	760	110	15	36	...	...	...	...
−78	−108	980	142	905	131	12	31	...	...	...	...
−196	−320	1260	182	1100	159	15	30	...	...	...	...
−253	−423	1420	206	1260	182	13	22	...	...	...	...
Ti-6Al-4V (ELI) sheet, annealed, longitudinal orientation											
24	75	960	139	890	129	12	...	1120	162	110	16.2
−78	−108	1160	168	1100	160	9	...	1220	177	115	16.6
−196	−320	1500	217	1420	206	10	...	1460	211	120	17.5
−253	−423	1770	256	1700	246	4	...	1500	217	130	18.6

(a) $K_t = 6.3$ for all three sheet forms; $K_t = 5$ to 8 for Ti-6Al-4V (ELI) forgings. (b) Recrystallization annealing treatment: 930 °C (1700 °F) 4 h, FC to 760 °C (1400 °F) in 3 h, cooled to 480 °C (900 °F) in ¾ h, AC. Source: *Metals Handbook, Properties and Selection: Stainless Steels, Tool Materials, and Special-Purpose Materials*, Vol 3, 9th ed., American Society for Metals, 1980

Ti-5Al-2.5Sn: Low-temperature tensile strength comparison

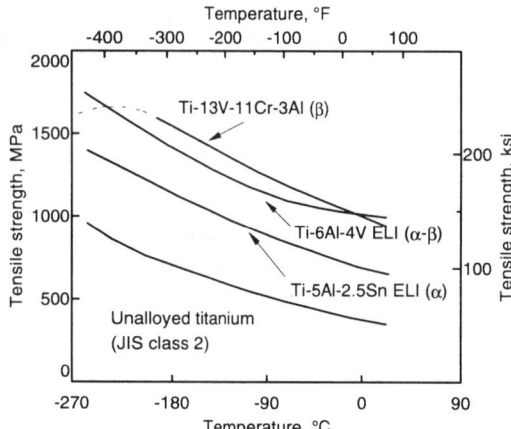

Source: Titanium Product Literature, Kobe Steel

Ti-5Al-2.5Sn: Low-temperature elongation comparison

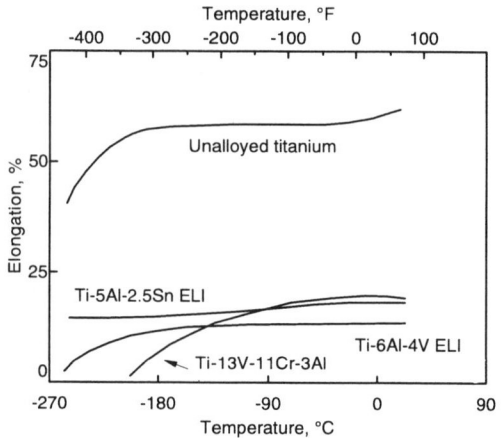

Source: Titanium Product Literature, Kobe Steel

Notched Strength

Ti-5Al-2.5Sn: Notch sensitivity at low temperatures

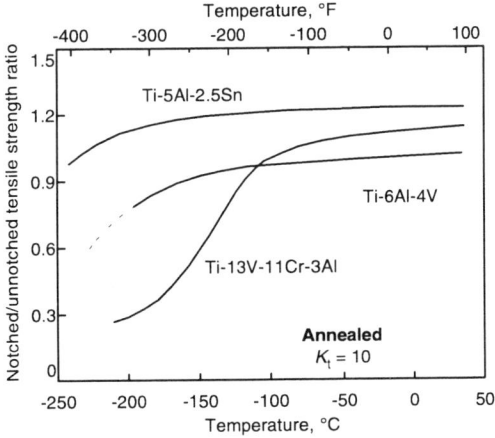

Source: F. Schwartzberg et al., *Cryogenic Materials Data Handbook*, AFML-TDR-64-280, 1964; reported in *Beta Titanium Alloys*, R. Wood, Ed., MCIC-72-11, Battelle Columbus Laboratories, 1972, p 17

Ti-5Al-2.5Sn: Notch strength of sheet vs thickness

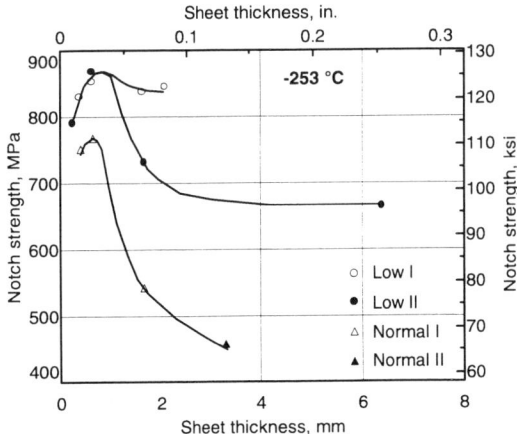

Low I material had a 0.2% tensile yield strength of 1392 MPa (202 ksi) and composition of 290 ppm C, 1200 ppm Fe, 8 ppm H, 40 ppm N, and 1030 ppm O. Low II had a yield strength of 1475 MPa (214 ksi) and composition of 280 ppm C, 1100 ppm Fe, 40 ppm H, 100 pm N, and 1000 ppm O. Normal I had a yield strength of 1544 MPa (224 ksi) and composition of 380 ppm C, 1600 ppm Fe, 8 ppm H, 90 ppm N, and 1800 ppm O. Normal II had a yield strength of 1661 MPa (241 ksi) and composition of 300 ppm C, 1700 ppm Fe, 70 ppm H, 140 ppm N, and 2000 ppm O. Alloy sheet was annealed at 815 °C (1500 °F) and furnace cooled. Data points are averages of at least four tests at –253 °C (–423 °F), in both longitudinal and transverse directions. Specimens had a 60° notch on both sides.
Source: J. Shannon, Jr. and W. Brown, Jr., "Effects of Several Production and Fabrication Variables on Sharp Notch Properties of 5Al-2.5Sn Titanium Alloy Sheet at Liquid Hydrogen Temperature," *Proc. ASTM*, Vol 63, 1963, p 809

High-Temperature Strength

Tensile Strengths vs Temperature

Ti-5Al-2.5Sn: Tensile strengths vs temperature

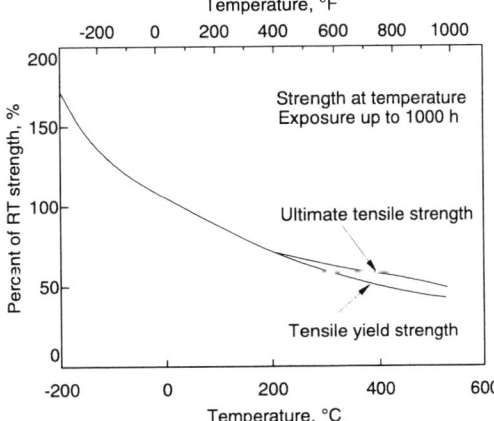

Specimens were annealed sheet.
Source: MIL-HDBK-5, 1 Dec 1991

Ti-5Al-2.5Sn: Ultimate tensile strength vs temperature

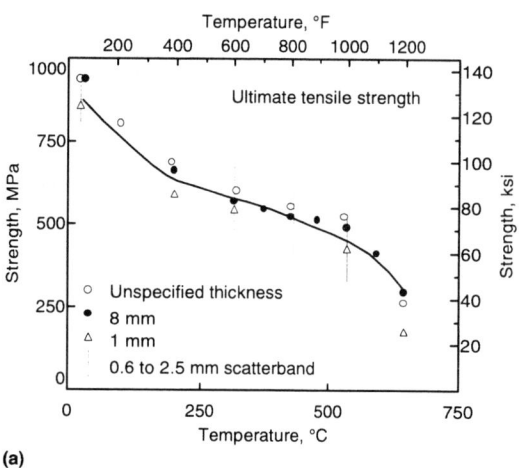

(a)

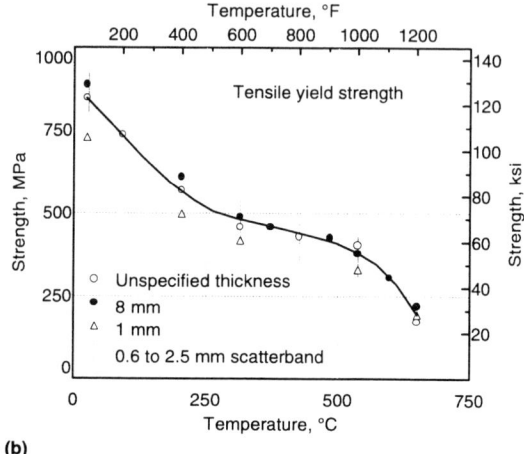

(b)

Specimens were annealed sheet.
Source: *Aerospace Structural Metals Handbook*, Code 3706, Battelle Columbus Laboratories, 1965

Ti-5Al-2.5Sn: Tensile strength vs temperature

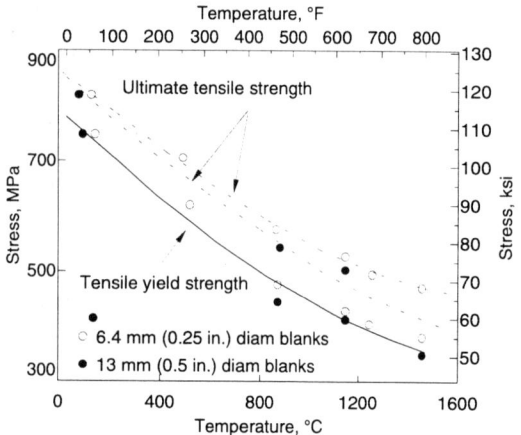

Specimens were annealed sheet.
Source: *Aerospace Structural Metals Handbook*, Code 3706, Battelle Columbus Laboratories, 1965

Ti-5Al-2.5Sn: Tensile elongation vs temperature

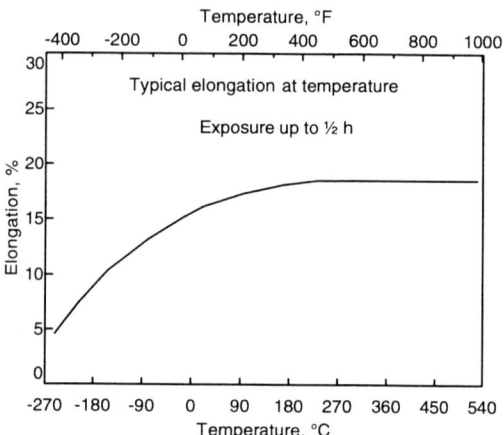

Blanks cast in graphite molds, annealed at 620 °C (1150 °F), 1 h, AC.
Source: *Aerospace Structural Metals Handbook*, Code 3706, Battelle Columbus Laboratories, 1965

Compressive and Shear Strengths

Ti-5Al-2.5Sn: Compressive yield and shear ultimate strengths vs temperature

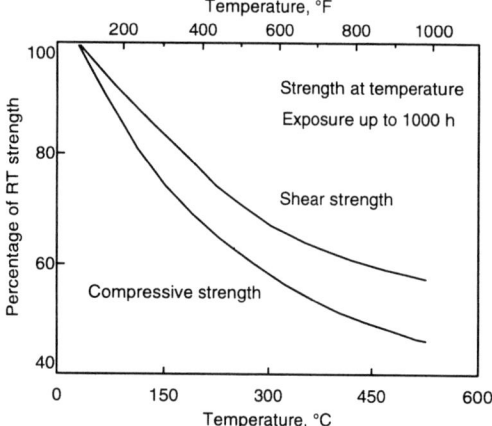

Annealed sheet.
Source: MIL-HDBK-5, 1 Dec 1991

Ti-5Al-2.5Sn: Compressive yield strength vs temperature

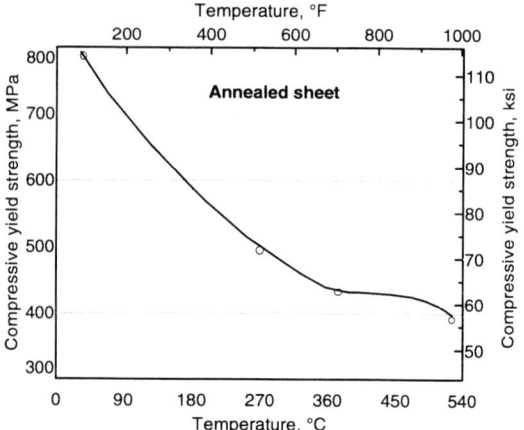

Source: *Aerospace Structural Metals Handbook*, Code 3706, Battelle Columbus Laboratories, 1965

Ti-5Al-2.5Sn: Shear strength vs temperature

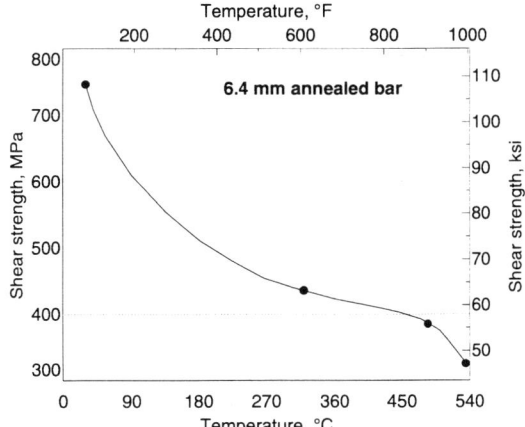

Source: *Aerospace Structural Metals Handbook*, Code 3706, Battelle Columbus Laboratories, 1965

Ti-5Al-2.5Sn: Bearing strengths vs temperature

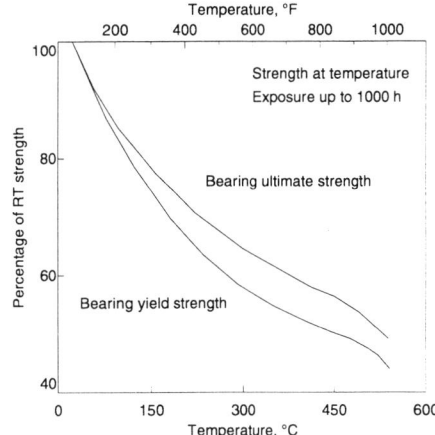

Annealed sheet. e/D ratio unspecified
Source: MIL-HDBK-5, 1 Dec, 1991

Bearing Strength

Ti-5Al-2.5Sn: Bearing strengths vs temperature

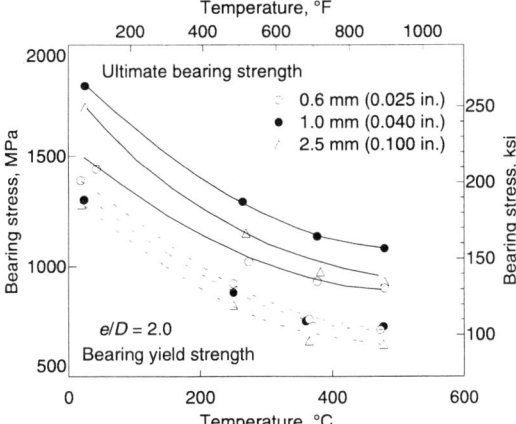

Annealed sheet.
Source: *Aerospace Structural Metals Handbook*, Code 3706, Battelle Columbus Laboratories, 1965

Creep Strength

Ti-5Al-2.5Sn: Creep strain at 125 and 150 °C

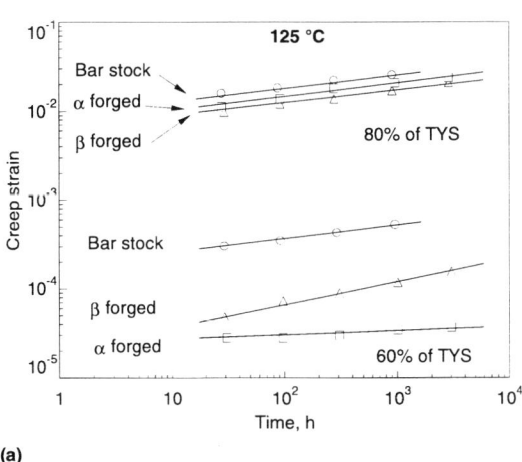
(a)

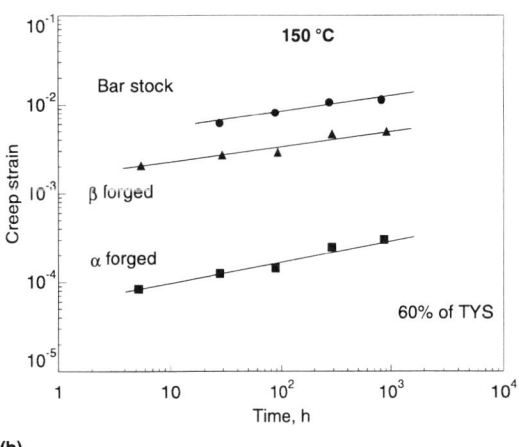
(b)

Log-log plots for tests at 125 °C (255 °F) (open symbols) and 150 °C (300 °F) (closed symbols).
Chemical composition: 5.13 wt% Al, 0.023 wt% C, 0.195 wt% Fe, 0.0012 wt% H, 0.02 wt% Mn, 0.020 wt% N, 2.68 wt% Sn. Bar stock was as-received hot-rolled round bar; forgings were forged at 955 °C (1750 °F), alpha forged, and at 1050 °C (1925 °F), beta forged, both air cooled. Creep tests were performed in conventional dead-weight loaded creep frames at stress levels, expressed as percentages of room-temperature yield stress. All tests were conducted in air.
Source: A. Thompson and B. Odegard, "Influence of Microstructure on Low Temperature Creep of Ti-5Al-2.5Sn," *Metall. Trans.*, Vol 4, 1973, p 899

Ti-5Al-2.5Sn: Typical creep properties

Temperature		Stress for total deformation in 1000 h of:					
		0.1%		0.2%		0.5%	
°C	°F	MPa	ksi	MPa	ksi	MPa	ksi
315	600	83	12	190	27	385	56
370	700	69	10	175	25	330	48
425	800	45	6.5	100	15	235	34
540	1000	2.8	0.4	5	0.7	23	3.3

Source: *Metals Handbook*, Vol 3, 9th ed., American Society for Metals, 1980, p 384

Ti-5Al-2.5Sn: Creep data for bar

Temperature		Stress		Time of test,	Deformation, %			Minimum creep rate,
°C	°F	MPa	ksi	h	Initial	Final	Total plastic	%/h
315	600	448	65	1002.0	0.500	0.544	0.044	0.000010
370	700	413	60	1039.7	0.505	0.744	0.239	0.000070
		379	55	793.0	0.442	0.566	0.124	0.000045
425	800	344	50	1010.3	0.438	1.317	0.879	0.00046
		310	45	486.0	0.383	0.792	0.409	0.00038
480	900	172	25	1000	0.232	1.577	1.345	0.00115
		138	20	650	0.187	0.610	0.423	0.00049
540	1000	69	10	1007.6	0.115	1.359	1.244	0.00096
		34	5	822.0	0.004	0.286	0.282	0.00022

Source: RMI data, reported in "Engineering Properties of Titanium Alloys," TML Report 92, Battelle Memorial Institute, 1959

Ti-5Al-2.5Sn: Creep data for sheet

Temperature		Stress		Time of test,	Deformation		Minimum
°C	°F	MPa	ksi	h	Initial, %	Total, %	Creep rate, %/h
315	600	434	63	1005.3	0.495	0.505	0.000028
		403	58.5	1067	0.455	0.515	0.000030
		372	54	1004.5	0.403	0.432	nil
370	700	395	57.4	1008.4	0.369	0.525	0.000065
		367	53.3	1125.5	0.414	0.556	0.000050
		339	49.2	1008.3	0.534	0.534	0.000040
		311	45.1	1000.9	0.339	0.442	0.000028
425	800	371	53.9	1062.7	0.426	3.93	0.00104
		344	50	1009.0	0.414	2.21	0.00075
		318	46.2	1083.2	0.420	1.67	0.00054
		291	42.3	1004.9	0.339	0.874	0.00038
480	900	351	51	68.3(a)	0.500	29.2(b)	0.065
		327	47.4	116.5(a)	0.400	43.5(b)	0.035
		302	43.8	240.9(a)	0.423	52.0(b)	0.025
		275	40	329.2(a)	0.374	55.0(b)	0.020
540	1000	281	40.8	3.5(a)	0.442	21.0(b)	...
		258	37.4	20.0(a)	0.286	45.1(b)	...
		234	34	44.0(a)	0.379	41.7(b)	...
		187	27.2	23.8(a)	0.370	45.1(b)	...
		103	15	783.7	0.179	11.5	...
		34	5	1000.3	0.056	0.751	...
		27	4	1003.1	0.048	0.554	...
		20	3	1002.2	0.034	0.481	...

Note: Alloy was annealed at 815 °C (1500 °F) for 30 min, air cooled. Ultimate tensile strength was (sheet) 896 MPa (130 ksi) and (bar) 930 MPa (135 ksi). (a) Ruptured in time indicated. (b) Percent elongation at failure. Source: RMI data, reported in "Engineering Properties of Titanium Alloys," TML Report 92, Battelle Memorial Institute, 1959

Ti-5Al-2.5Sn: Creep strength vs hydrogen content

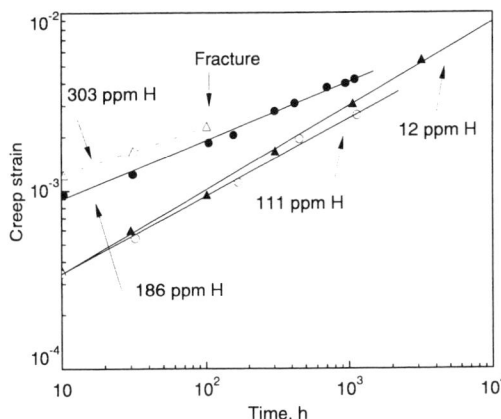

Specimens from alpha forged bar stock were vacuum annealed at 925 °C (1700 °F) then hydrogen charged to contents indicated in a Sieverts apparatus. Creep tests were performed at room temperature to 1000 h.

Ti-5Al-2.5Sn: Creep-rupture strength

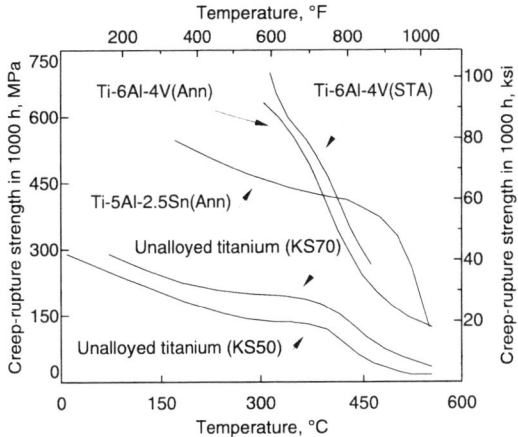

Annealed material compared to CP grade 2 and grade 3 titanium and to STA and annealed Ti-6Al-4V.
Source: Titanium product Literature, Kobe Steel

Ti-5Al-2.5Sn: Creep behavior of annealed sheet

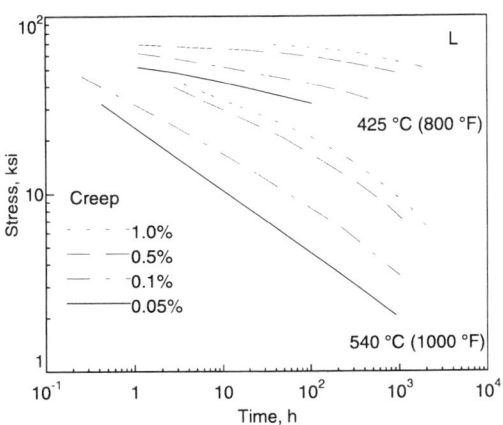

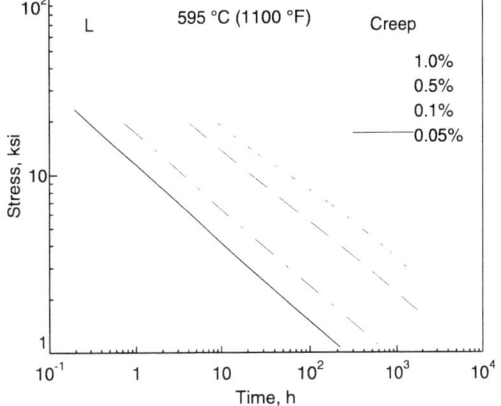

Alloy was used as 1.6 mm (0.064 in.) thick sheet.
Source: Aerospace Structural Metals Handbook, Battelle Columbus Laboratories, Code 3706, 1965

Fatigue Life

Room-Temperature Data

Ti-5Al-2.5Sn: Rotating bending fatigue strength at room temperature for smooth and notched annealed sheet

Stress concentration	Fatigue strength, MPa (ksi), for $R = -1$ at cycles:		
	10^5	10^6	10^7
Smooth			
$K = 1$	531 (77)	441 (64)	427 (62)
Notched			
$K = 2.4$	386 (56)	310 (45)	296 (43)
$K = 3.2$	275 (40)	209 (30)	186 (27)

Source: *Aerospace Structural Metals Handbook*, Vol 4, Code 3706, Battelle, 1965, p 3

308 / Alpha and Near-Alpha Alloys

Ti-5Al-2.5Sn: Fatigue endurance ratio comparison

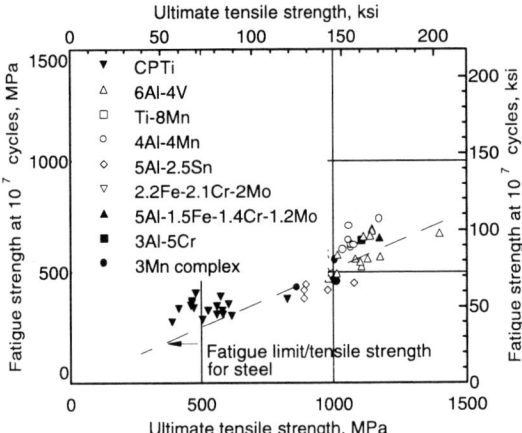

Source: TML Report No. 77, Battelle, 1957

Ti-5Al-2.5Sn: Rotating-beam fatigue strength

Effect of surface finish.
Source: *Metals Handbook, Properties and Selection: Stainless Steels, Tool Materials, and Special-Purpose Materials*, Vol 3, 9th ed., ASM, 1980

Ti-5Al-2.5Sn: Rotating-beam fatigue strength

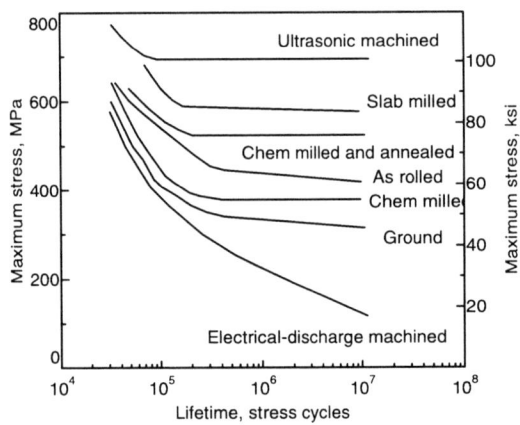

Effect of surface finish.
Source: *Metals Handbook, Properties and Selection: Stainless Steels, Tool Materials, and Special-Purpose Materials*, Vol 3, 9th ed., ASM, 1980

Ti-5Al-2.5Sn: Rotating-beam fatigue strength

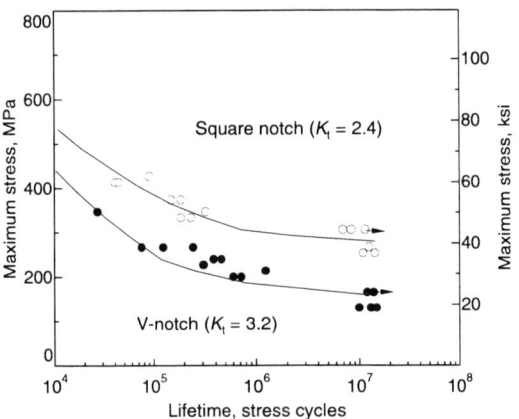

Notch fatigue strength for two notches.
Source: *Metals Handbook, Properties and Selection: Stainless Steels, Tool Materials, and Special-Purpose Materials*, Vol 3, 9th ed., ASM, 1980

Ti-5Al-2.5Sn: Constant life diagram of mill annealed sheet

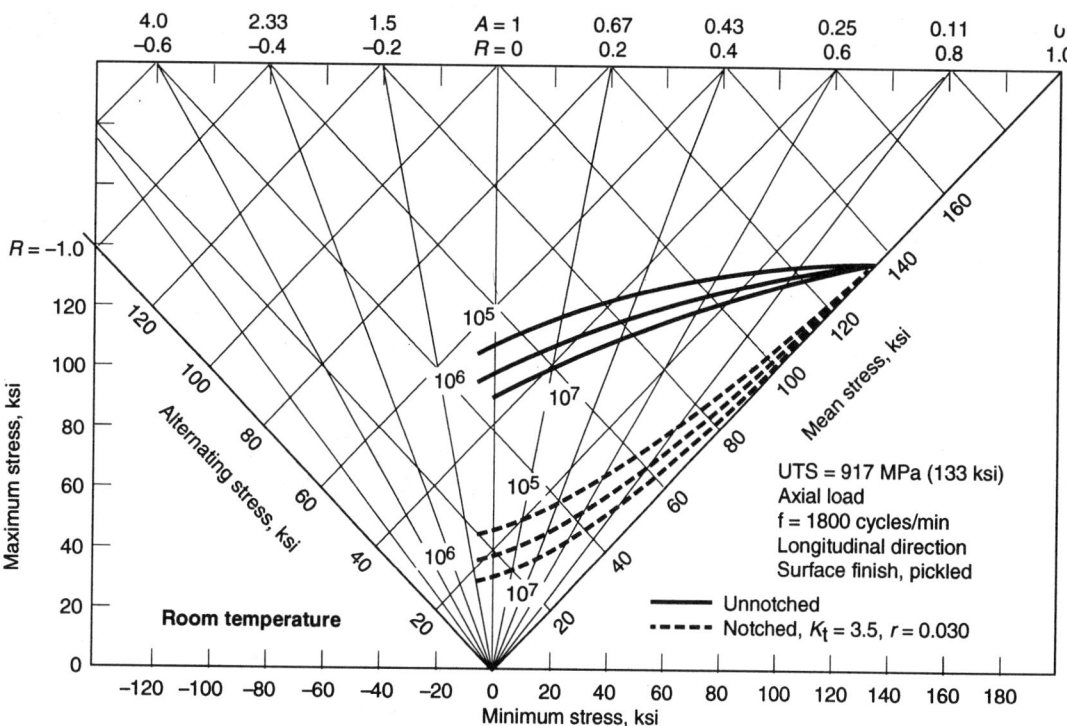

Source: MIL-HDBK-5, 1972

Low-Temperature Fatigue Data

Ti-5Al-2.5Sn ELI: Fatigue strength vs temperature

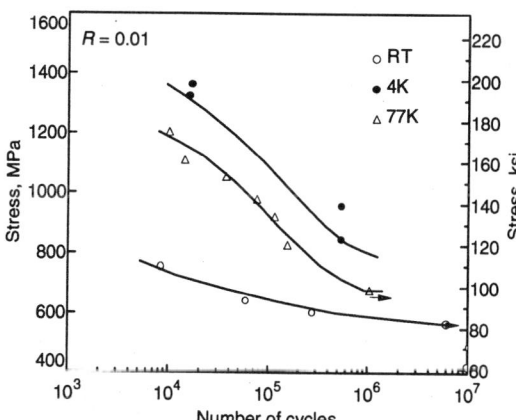

Source: Titanium Product Literature, Kobe Steel

Ti-5Al-2.5Sn: Low-temperature fatigue life of welded sheet

Alloy and condition	Stressing mode	Stress ratio, R	K_t	Fatigue strengths at 10^6 cycles:					
				24 °C (75 °F)		–196 °C (–320 °F)		–253 °C (–423 °F)	
				MPa	ksi	MPa	ksi	MPa	ksi
Ti-5Al-2.5Sn (ELI) sheet, annealed	Axial	0.01	1	495	72	815	118	760	110
			3.5	220	32	205	30	160	23
Ti-5Al-2.5Sn (ELI) sheet(a)	Axial	0.01	1	485	70	565	82	425	62
Ti-5Al-2.5Sn (ELI) bar, annealed(b)	Axial	0	1	760	110	985	143	925	134
Ti-6Al-4V (ELI) sheet(c)	Axial	0.01	1	505	73	675	98	895	130
			3.5	285	41	295	43	275	40
Ti-6Al-4V (ELI) sheet(a)	Axial	0.01	1	600	87	595	86	560	81
Ti-6Al-4V sheet, annealed	Flex	–1.0	1	345	50	550	80	530	77
			3.1	170	25	185	27	255	37

(a) Gas tungsten arc welded, base metal filler. (b) Cyclic frequency, 28 Hz. (c) STA: 900 °C (1650 °F) 5 min, WQ; 540 °C (1000 °F) 4 h, AC.
Source: *Metals Handbook, Properties and Selection: Stainless Steels, Tool Materials, and Special-Purpose Materials*, Vol 3, 9th ed., American Society for Metals, 1980

Fatigue Crack Growth

Ti-5Al-2.5Sn: Fatigue crack growth of annealed sheet at room temperature

Test environment	Fatigue crack growth, μm/cycle (μin./cycle), at:			
	$\Delta K = 5.5$ MPa $\sqrt{m}$ (5 ksi $\sqrt{in.}$), $R = 0.67$, and 55-58 Hz	$\Delta K = 11$ MPa $\sqrt{m}$ (10 ksi $\sqrt{in.}$), $R = 0.67$, and 55-58 Hz	$\Delta K = 22$ MPa $\sqrt{m}$ (20 ksi $\sqrt{in.}$), $R = 0.1$, and 30-50 Hz(a)	$\Delta K = 55$ MPa $\sqrt{m}$ (50 ksi $\sqrt{in.}$), $R = 0.1$, 30 Hz
L-T specimen orientation				
Dry argon	0.00075 (0.03)	0.00685 (0.27)	0.12-0.14 (4.77-5.56)	2.40 (94.7)
Lab air	0.0038 (0.15)	(2.13)	(11.6-11.7)	(124)
Distilled water	0.00635 (0.25)	(3.49)	(11.8)(b)	(124)
3.5% NaCl	0.0074 (0.29)	(7.97)	(23.5-30.2)	(157)
T-L specimen orientation				
Dry argon	...	(0.49)	(5.35-5.38)	(114)
Lab air	0.0038 (0.15)	(3.08)	(11.8-11.9)	(141)
Distilled water	0.009 (0.36)	(3.72)	(12.0-12.5)	(130)
3.5% NaCl	0.025 (0.98)	(14.6)	(24.5)(b)	(176)

(a) The higher measured values correspond to tests at 30 Hz. (b) 50 Hz. Source: J. Gallagher, *Damage Tolerant Design Handbook*, MCIC-HB-O1R, Battelle, 1983

Ti-5Al-2.5Sn: Fatigue crack growth rate compared to Ti-6Al-4V
Crack growth parameters per the relation $da/dN = C (\Delta K)^n$

Alloy and condition(a)	Orientation	Test temperature		C		n	Estimated range for ΔK	
		°C	°F	da/dN:mm/cycle ΔK:MPa $\sqrt{m}$	da/dN:in./cycle ΔK:ksi $\sqrt{in.}$		MPa $\sqrt{m}$	ksi $\sqrt{in.}$
Ti-5Al-2.5Sn (NI), annealed bar	T-S	24, –196, –269	75, –320, –452	5.1×10^{-11}	3.2×10^{-12}	4.8	14-30	13-27
Ti-5Al-2.5Sn (LI), annealed bar	T-L	24, –196, –269	75, –320, –452	4.9×10^{-10}	2.8×10^{-11}	4.0	10-60	9-54
Ti-6Al-4V (NI), annealed bar	T-L	24, –196, –269	75, –320, –452	3.1×10^{-12}	2.2×10^{-13}	6.0	14-30	13-27
Ti-6Al-4V (ELI), recrystallization annealed bar	T-L	24, –196, –269	75, –320, –452	1.9×10^{-13}	1.4×10^{-14}	7.0	10-20	9-18
		24, –196	75 –320	3.0×10^{-8}	1.6×10^{-9}	3.0	20-40	18-36

Note: Stress ratio: $R = 0.1$, at 20 to 28 Hz; compact specimens. (a) NI = normal interstitial, LI = low interstitial, ELI = extra low interstitial.
Source: R. L. Tobler and R.P. Reed, "Fatigue Crack Growth Resistance of Structural Alloys at Cryogenic Temperatures," in *Advances in Cryogenic Engineering*, K.D. Timmerhaus et al., Ed., Vol 24, Plenum Press, 1978, p 82-90

Ti-5Al-2.5Sn ELI: Crack growth at room temperature

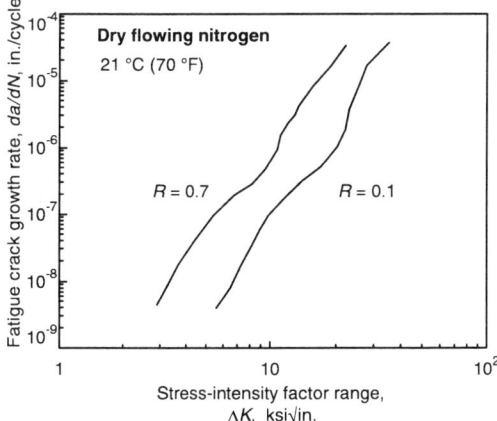

Source: D.E. Matejczyk et al., Fatigue Crack Retardation Following Overloads in Inconel 718, Ti-5Al-2.5Sn, and Haynes 188, Advanced Earth-to-Orbit Propulsion Technology 1986, Vol 2, NASA Conference Publication 2437, 1986, p 205-219

Ti-5Al-2.5Sn: Fatigue crack growth rates

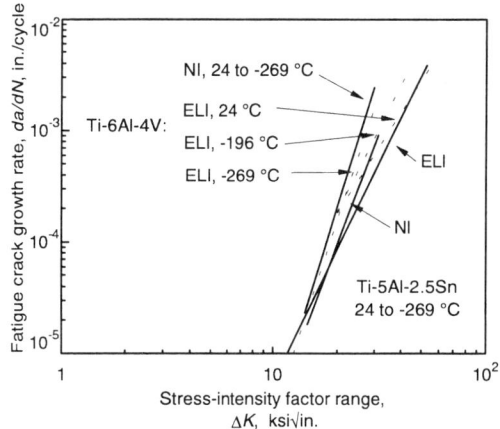

NI = normal interstitial content; ELI = extra-low interstitial content.
Source: R.L. Robler and R.P. Reed, in Advances in Cryogenic Engineering, Vol 24, K.D. Timmerhaus et al., Ed., Plenum Press, 1978, p 82-90

Fracture Properties

Ti-5Al-2.5Sn: Fracture mechanism map

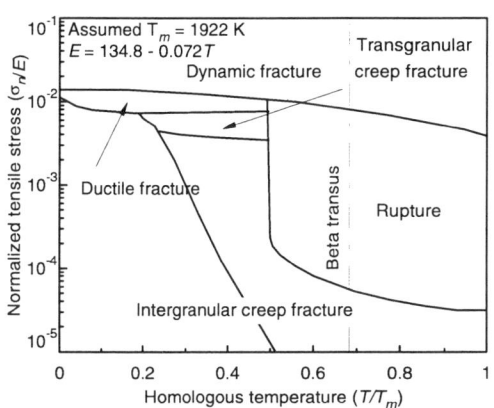

Source: Krishnamohanrao et al., Fracture Mechanism Maps for Titanium and Its Alloys, Acta Metall., Vol 34, 1986, p 1783-1806

Ti-5Al-2.5Sn: Time-to-fracture

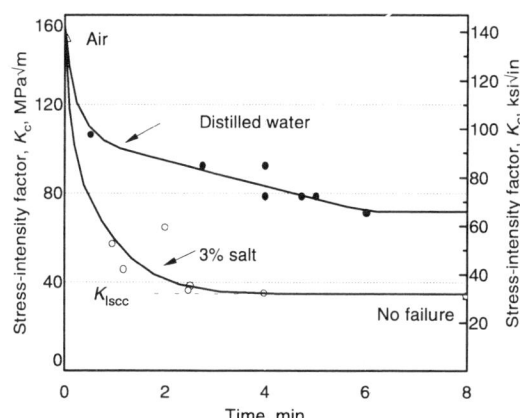

Effect of initial stress intensity on time-to-fracture at ambient temperature.
Source: R. Wood and R. Favor, Titanium Alloys Handbook, MCIC-HB-02, Battelle Columbus Laboratories, 1972

Low-Temperature Toughness (Standard and ELI)

Ti-5Al-2.5Sn: Fracture toughness

Heat treatment variable(a)	Test temperature		Stress intensity, K_{IC}		Specimen, orientation(b) and type(c)	Yield strength(d)	
	K	°F	MPa √m	ksi √in.		MPa	ksi
Standard grade							
Air cooled	295	72	71.4	65	LT-CT	876	127
	77	−320	53.8	49	LT-CT	1338	194
	20	−423	51.6	47	LT-B	1482	215
	20	−423	50.5	46	LS-B	...	...
Furnace cooled	295	72	65.9	60	LT-CT	882	128
	77	−320	57.1	52	LT-CT	1379	200
	20	−423	47.2	43	LT-B	1517	220
	20	−423	52.7	48	LS-B	...	...
ELI grade							
Air cooled	295	72	118.7	108(e)	LT-CT	703	102
	77	−320	111.0	101	LT-CT	1179	171
	20	−423	91.2	83	LT-B	1303	189
	20	−423	106.6	97	LS-B	...	...

(continued)

Ti-5Al-2.5Sn: Fracture toughness

Heat treatment variable(a)	Test temperature K	Test temperature °F	Stress intensity, K_{IC} MPa√m	Stress intensity, K_{IC} ksi√in.	Specimen, orientation(b) and type(c)	Yield strength(d) MPa	Yield strength(d) ksi
Furnace cooled	295	72	115.4	105(e)	LT-CT	682	99
	77	−320	82.4	75	LT-CT	1179	171
	20	−423	68.1	62	LT-B	1303	189
	20	−423	80.2	73	LS-B	...	...

(a) Air cooled or furnace cooled from annealing temperature. (b) Orientation notation per ASTM E399-74. (c) CT, compact tension specimen; B, bend specimen. (d) 0.2% offset. (e) Invalid toughness values (not 100% plane-strain conditions). Source: *Metals Handbook*, Vol 3, 9th ed., American Society for Metals, 1980, p 384

Ti-5Al-2.5Sn: Comparison of fracture toughness of two titanium alloys

Alloy and condition(a)	Form	Room-temperature yield strength MPa	Room-temperature yield strength ksi	Specimen design	Orientation	Fracture toughness, K_{IC} 24 °C (75 °F) MPa√m	Fracture toughness, K_{IC} 24 °C (75 °F) ksi√in.	−196 °C (−320 °F) MPa√m	−196 °C (−320 °F) ksi√in.	−253 °C (−423 °F) MPa√m	−253 °C (−423 °F) ksi√in.	−269 °C (−452 °F) MPa√m	−269 °C (−452 °F) ksi√in.
Ti-5Al-2.5Sn(NI), annealed	Plate	876	127	CT	L-T	71.8	65.4	53.4	48.6	...	...	...	...
		876	127	Bend	L-T	...	...	...	...	51.4	46.8	...	...
		876	127	Bend	L-S	...	...	...	...	50.2	45.7	...	...
	Bar	871	126	CT	T-S	77.2	70.3	42.1	38.3	...	...	42.0	38.2
Ti-5Al-2.5Sn(ELI) annealed	Plate	703	102	CT	L-T	...	...	111	101	...	...	...	...
		703	102	Bend	L-T	...	...	...	...	89.6	81.5	...	...
Ti-5Al-2.5Sn(ELI) as forged	Forging	760	110	CT	R-L	...	...	...	...	79.4	72.3	...	...
					R-C	...	...	...	...	58.5	53.2	...	...
Ti-5Al-2.5Sn(ELI)	Forging(b)	779	113	CT		...	...	...	...	54.4-75.3	49.5-68.5	...	...
Ti-6Al-4V (NI), annealed	Bar	942	136	CT	T-L	47.4	43.2	38.8	35.3	...	...	38.5	35.1
Ti-6Al-4V (ELI), as forged	Forging	830	120	CT	T-L	...	...	61.0	55.5	...	...	54.1	49.2
Ti-6Al-4V (ELI), RA	Forging	830	120	CT	M-L(c)	...	...	62.8	57.2	...	...	...	...
					M-R(c)	...	...	62.0	56.4	...	...	...	...
Ti-6Al-4V (ELI), RA, electron beam welded, SR	Forging	830	120	CT	M-R(c)	...	...	61.1(d)	55.6(d)	...	...	...	...
	Weldment	...	...	...	M-L(c)	...	...	56.9(d)	51.8(d)	...	...	...	...
					M-R(c)	...	...	57.1(e)	52.0(e)	...	...	...	...
					M-R(c)	...	...	51.0(f)	46.4(f)	...	...	...	...

(a) SR = stress relieved: 540 °C (1000 °F) 50 h, AC. FC = furnace cool. AC = air cool. NI = normal interstitial content. ELI = extra low interstitial content. RA = recrystallization annealed: 930 °C (1700 °F) 4 h, FC to 810 °C (1400 °F) 3 h, cooled to 480 °C (900 °F) in ¾ h, AC. (b) Range for 18 tests. (c) M-L and M-R are specific orientations in a spherical forging. (d) Fusion zone. (e) Heat affected zone. (f) Heat affected zone boundary. Source: *Metals Handbook, Properties and Selection: Stainless Steels, Tool Materials, and Special-Purpose Materials*, Vol 3, 9th ed., American Society for Metals, 1980

Ti-5Al-2.5Sn (ELI): Fracture toughness of 13 mm (0.50 in.) thick plate

Direction	Tensile yield strength MPa	Tensile yield strength ksi	K_{IC} MPa√m	K_{IC} ksi√in.
At −195 °C (−320 °F)				
LS	1206	175	73.0	67.0
	1206	175	76.0	69.8
	1206	175	67.0	60.8
TS	1199	174	58.0	52.8
	1199	174	51.0	46.4
	1199	174	56.5	51.5
LD	1206	175	60.6	55.2
	1206	175	63.0	58.0
	1206	175	60.0	55.0
TD	1199	174	60.6	55.2
	1199	174	64.0	58.4
	1199	174	59.5	54.2
At −252 °C (−423 °F)				
LS	1413	205	56.4	51.4
	1413	205	65.0	59.6
	1413	205	59.7	54.4

Ti-5Al-2.5Sn (ELI): Fracture toughness of 13 mm (0.50 in.) thick plate

Direction	Tensile yield strength		K_{IC}	
	MPa	ksi	MPa√m	ksi√in.
TS	1441	209	53.0	48.8
	1441	209	59.3	54.0
	1441	209	59.5	54.2
LD	1413	205	58.9	53.6
	1413	205	59.9	54.6
	1413	205	51.8	47.2
TD	1441	209	65.0	60.0
	1441	209	69.0	63.6
	1441	209	65.0	59.2

Note: Bend bar specimens 6.4 by 13 mm (¼ by ½ in.). Chemical composition: 5.0 wt% Al, 0.023 wt% C, 0.16 wt% Fe, 0.001 wt% H, 0.006 wt% Mn, 0.010 wt% N, 0.086 wt% O, and 2.6 wt% Sn. Plate was annealed by furnace cooling from 815 °C (1500 °F). The TS specimen orientation had a crack direction parallel to the rolling direction. Source: C. Carman and J. Katlin, "Plane Strain Fracture Toughness and Mechanical Properties of 5Al-2.5Sn ELI and Commercial Titanium Alloys at Room and Cryogenic Temperatures," in *Applications Related Phenomena in Titanium Alloys*, ASTM STP 432, ASTM, 1968, p 124

Ti-5Al-2.5Sn: Fracture toughness of 13 mm (0.5 in) thick plate

Direction	Tensile yield strength		K_{IC}	
	MPa	ksi	MPa√m	ksi√in.
At −195 °C (−320 °F)				
LS	1399	203	33.2	30.2
	1399	203	28.5	26.0
	1399	203	24.9	22.7
TS	1406	204	64.1	58.4
	1406	204	50.4	45.9
LD	1399	203	44.9	40.9
	1399	203	43.6	39.7
	1399	203	26.9	24.5
TD	1406	204	27.8	25.3
	1406	204	30.1	27.4
	1406	204	24.6	22.4
At −252 °C (−423 °F)				
LS	1606	233	21.5	19.6
	1606	233	29.4	26.8
	1606	233	30.1	27.4
TS	1634	237	58.3	53.1
	1634	237	30.2	27.5
LD	1606	233	52.7	48.0
	1606	233	34.8	31.7
	1606	233	21.3	19.4
TD	1634	237	24.0	21.9
	1634	237	25.6	23.3
	1634	237	23.3	21.2

Note: Bend bar specimens 6.4 by 13 mm (¼ by ½ in.). Chemical composition: 5.1 wt% Al, 0.023 wt% C, 0.34 wt% Fe, 0.017 wt% H, 0.006 wt% Mn, 0.015 wt% N, and 2.3 wt% Sn. Plate was annealed by furnace cooling from 815 °C (1500 °F). Source: C. Carman and J. Katlin, "Plane Strain Fracture Toughness and Mechanical Properties of 5Al-2.5Sn ELI and Commercial Titanium Alloys at Room and Cryogenic Temperatures," in *Applications Related Phenomena in Titanium Alloys*, ASTM STP 432, ASTM, 1968, p 124

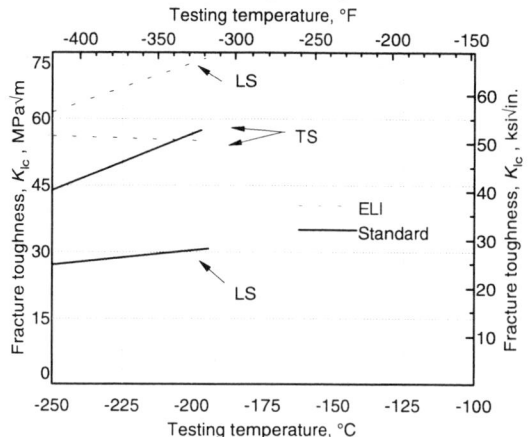

Ti-5Al-2.5Sn: Fracture toughness plate

LS and TS orientations.
Plate (13 mm, or 0.5 in.) thick. The TS specimen orientation has crack growth parallel to the rolling direction. All other crack orientations (LS, LD, TD,) are perpendicular to the rolling direction.
Source: C. Carman, "Influence of Purity on the Fracture Properties of High-Strength Aluminum, Titanium, and Steel," in *Fracture Toughness of High-Strength Materials: Theory and Practice*, ISI Publication 120, The Iron and Steel Institute, p 116

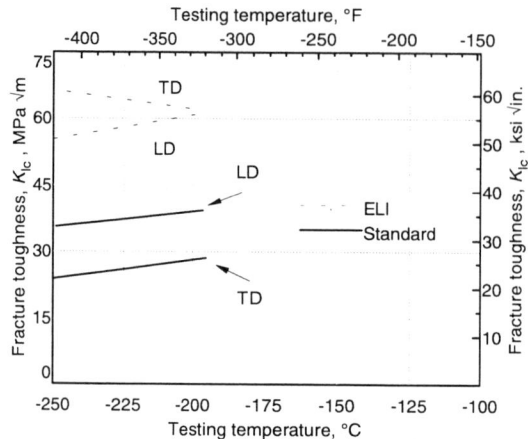

Ti-5Al-2.5Sn: Fracture toughness plate

LD and TD orientations.
Plate (13 mm, or 0.5 in.) thick. The TS specimen orientation has crack growth parallel to the rolling direction. All other crack orientations (LS, LD, TD) are perpendicular to the rolling direction.
Source: C. Carman, "Influence of Purity on the Fracture Properties of High-Strength Aluminum, Titanium, and Steel," in *Fracture Toughness of High-Strength Materials: Theory and Practice*, ISI Publication 120, The Iron and Steel Institute, p116

ELI Fracture Toughness

Ti-5Al-2.5Sn (ELI): Fracture toughness of 6.4 mm (0.25 in.) thick plate

Direction	Tensile yield strength		K_{IC}	
	MPa	ksi	MPa$\sqrt{m}$	ksi$\sqrt{in.}$
At −195 °C (−320 °F)				
LS	1172	170.5	68.1	62.0
	1172	170.5	67.4	61.4
	1172	170.5	69.4	63.2
TS	1199	174.5	54.1	49.3
	1199	174.5	62.8	57.2
	1199	174.5	63.7	58.0
LS	1172	170.5	67.0	61.0
	1172	170.5	70.9	64.6
	1172	170.5	70.8	64.5
	1172	170.5	74.6	67.9
TS	1199	174.5	66.9	60.9
	1199	174.5	72.5	66.0
	1199	174.5	64.4	58.6
	1199	174.5	70.3	64.0
At −252 °C (−423 °F)				
LS	1344	195	61.9	56.4
	1344	195	59.3	54.0
	1344	195	56.7	51.6
TS	1248	181	53.6	48.8
	1248	181	54.7	49.8
	1248	181	45.9	41.8

Note: Chemical composition: 5.0 wt% Al, 0.023 wt% C, 0.16 wt% Fe, 0.009 wt% H, 0.006 wt% Mn, 0.010 wt% N, 0.080 wt% O, and 2.6 wt% Sn. Plate was annealed by furnace cooling from 815 °C (1500 °F). Source: C. Carman and J. Katlin, "Plane Strain Fracture Toughness and Mechanical Properties of 5Al-2.5Sn ELI and Commercial Titanium Alloys at Room and Cryogenic Temperatures," in *Applications Related Phenomena in Titanium Alloys*, ASTM STP 432, ASTM, 1968, p 124

Ti-5Al-2.5Sn (ELI): Fracture toughness of 25 mm (1 in.) thick plate

Direction	Tensile yield strength		K_{IC}	
	MPa	ksi	MPa√m	ksi√in.
At –195 °C (–320 °F)				
LS	1213	176	62.5	56.9
	1213	176	57.2	52.1
	1213	176	50.9	46.4
TS	1213	176	50.5	46.0
	1213	176	55.7	50.7
LD	1213	176	65.4	59.5
	1213	176	67.3	61.3
	1213	176	58.2	53.0
TD	1213	176	70.5	64.2
	1213	176	81.9	74.6
	1213	176	66.3	60.4
At –252 °C (–423 °F)				
LS	1399	203	49.4	45.0
	1399	203	69.9	63.7
	1399	203	62.6	57.0
TS	1399	203	51.3	46.7
	1399	203	60.9	55.5
	1399	203	60.6	55.2
	1399	203	60.7	55.3
LD	1399	203	50.5	46.0
	1399	203	61.1	55.6
TD	1399	203	67.0	61.0
	1399	203	83.8	76.3
	1399	203	70.5	64.2

Note: Bend bar specimens. Chemical composition: 5.1 wt% Al, 0.026 wt% C, 0.14 wt% Fe, 0.003 wt% H, 0.004 wt% Mn, 0.101 wt% O, and 2.4 wt% Sn. Plate was annealed by furnace cooling from 815 °C (1500 °F). Source: C. Carman and J. Katlin, "Plane Strain Fracture Toughness and Mechanical Properties of 5Al-2.5Sn ELI and Commercial Titanium Alloys at Room and Cryogenic Temperatures," in *Applications Related Phenomena in Titanium Alloys*, ASTM STP 432, ASTM, 1968, p 124

Ti-5Al-2.5Sn ELI: Fracture strength of cracked cylinders

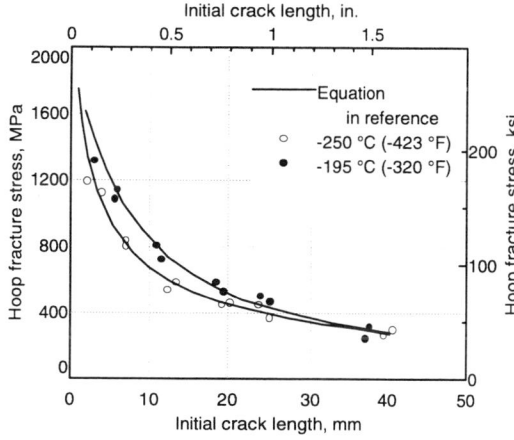

Source: T. Sullivan, "Texture Strengthening and Fracture Toughness of Titanium Alloy Sheet at Room and Cryogenic Temperatures," ASTM STP 432, ASTM, 1968

Ti-5Al-2.5Sn ELI: Fracture toughness at several temperatures

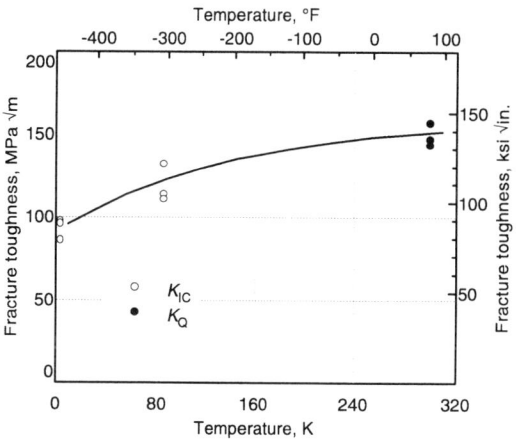

Alloy tested was extra-low interstitial grade with typical composition of 4.70 to 5.60 wt% Al, 0.25 wt% Fe max, 0.0125 wt% H max, 0.035 wt% N max, 0.12 wt% O max, and 2.0 to 3.0 wt% Sn.
Source: Titanium Product Literature, Kobe Steel

Forging

G.W. Kuhlman, ALCOA Forging Division

Ti-5Al-2.5Sn, an α alloy, is used commercially in the full range of forging products, e.g., open die forgings, closed die forgings, rings, etc., and is fabricated on all commercially available types of forging equipment. The forging process is used to fabricate useful engineering structures for elevated-temperature, cryogenic, and corrosion-resistant applications.

Ti-5Al-2.5Sn and its extra-low interstitial (ELI) grade are among the most difficult to forge of

Ti-5Al-2.5Sn: Forging process temperatures

Process	Metal temperature °C	°F
Conventional (subtransus) forging of Ti-5Al-2.5Sn	900-1010	1650-1850
Conventional forging of Ti-5Al-2.5Sn ELI	885-995	1625-1825
Supratransus forging(a)	β_T + 30 to 55 °C	β_T + 50 to 100 °F

Note: See "Technical Note 4: Forging" for recommended die temperatures. (a) Due to the high flow stresses, β forging may be used in early forging operations, including upsetting and open die forging. Typical transus is 1050 °C (1925 °F) for Ti-5Al-2.5Sn and 1035 °C (1895 °F) for Ti-5Al-2.5Sn ELI.

Ti-5Al-2.5Sn: Forging pressures at 870 °C

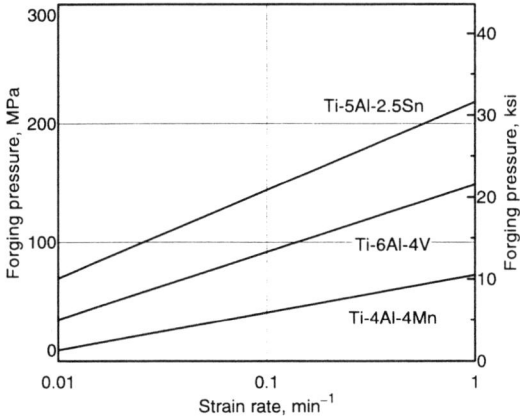

Forging pressures for a 10% upset reduction at 870 °C (1600 °F).

Ti-5Al-2.5Sn: Compressive stress-strain curves

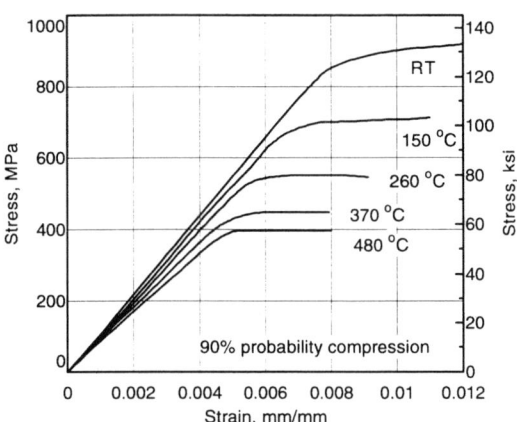

Annealed sheet, bar, and forgings.
Source: *Aerospace Structural Metals Handbook*, Vol 4, Code 3706, Battelle Columbus Laboratories, 1965

all titanium alloys. They are characterized by high unit pressures (flow stresses) and crack sensitivity in all types of forging processes, which may restrict the amount of forging reduction possible in a given forging step. Alpha phase predominates the final microstructure of Ti-5-2.5. Consequently, it is not thermomechanically processed in forging manufacture. Final thermal treatments consist of an anneal at 705 to 815 °C (1300 to 1500 °F), with in some cases a subtransus solution treatment at 1010 or 995 °C (1850 or 1825 °F) for the two alloy variants, followed by air cooling or faster quenches prior to annealing, to refine α grain size.

Forging Processes. The objective in forging Ti-5-2.5 is to achieve the final forging with least cost. Conventional subtransus (α + β) forging processes predominate in commercial forging. Due to its very high flow stresses and poor forgeability, supratransus (β) forging, at metal temperatures 30 to 55 °C (50 to 100 °F) above the transus, may be used in early forging operations, including upsetting and open die preforming. For successful fabrication of Ti-5-2.5 and to achieve desired final microstructure, total reductions, accumulated through subtransus forging in one or more steps, of 50 to 75% below the β transus, are used. This level of forging reduction, in combination with proper billet stock fabrication, achieves microstructural objectives of refined prior β grain size and fine α grain size.

Surface Treatment. Two important aspects of forging Ti-5-2.5 are precoats or other surface coating techniques used on billet stock and intermediate forging shapes during furnacing for forging operations and forging repair techniques. Ti-5-2.5 is particularly sensitive to the formation of excessive α case during reheating processes, which may lead to undue surface cracking in subsequent forging deformation. Thus, use of ceramic precoats and/or other methods of surface treatment prior to reheating are essential. Furthermore, the alloy is also quite sensitive to thermal stresses developed in crack repair processes, typically dry abrasive grinding techniques. Thus, Ti-5-2.5 is frequently heated to 425 to 590 °C (800 to 1100 °F) prior to repair, especially if surface cracking is severe.

Forming

Ti-5Al-2.5Sn is not readily formed into complex shapes as other alloys with similar room-temperature properties. Except for some forging operations, fabrication of Ti-5Al-2.5Sn is conducted at temperatures where the structure remains all α. Severe forming operations may be accomplished at temperatures up to 650 °C (1200 °F). Moderately severe forming can be done at 150 to 315 °C (300 to 600 °F), and simple forming may be done at room temperature. Most forming and welding operations are followed by an annealing treatment to relieve residual stresses imposed by the prior operation.

Ti-5Al-2.5Sn: Minimum bend radius

Temperature		Minimum bend, R/t	
°C	°F	Longitudinal(a)	Transverse(b)
21	70	7	6(c)
205	400	6	4.5
315	600	6	4.5
425	800	5.5	4
540	1000	5	3.5
650	1200	3.5	3
760	1400	2.5	2.5
815	1500	1.5	2

(a) Bend axis perpendicular to rolling direction. (b) Bend axis parallel to rod rolling direction. (c) Bend radius of 5.5 for joggling sheet under 2 mm (0.08 in.). Source: R.A. Wood and R.J. Favor, *Titanium Alloys Handbook*, MCIC HB-02, Battelle Columbus Laboratories, 1972

Ti-5Al-2.5Sn: Typical ranges of bend radii

Test temperature		Minimum bend radius, R/t	
°C	°F	Longitudinal	Transverse
RT		4.0-4.5	4.0-4.5
205	400	3.5-4.0	3.5-4.5
425	800	2.0-3.0	2.5-3.0
540	1000	1.5-2.5	2.0-2.5
650	1200	1.0-2.0	1.0-2.0

Source: R.A. Wood and R.J. Favor, *Titanium Alloys Handbook*, MCIC HB-02, Battelle Columbus Laboratories, 1972

Ti-5Al-2.5Sn: Forming temperatures

Method	Temperatures °C	°F	Comments
Hot sizing	650	1200	15 min for 0.8 mm (0.032 in.) sheet, 20 min for 1.6 mm (0.063 in.)
Brake forming	205-315	400-600	Mild forming
Drop hammer	540-705	1000-1300	Severe forming above 540 °C (1000 °F); maximum stretch of 12.6% at 480 °C (900 °F) in annealed condition
Stretching	...	...	Temperatures above 540 °C (1000 °F) are required for significant improvement in stretch formability
Drawing	540-705	1000-1300	...
Spinning or shear forming	650-760	1200-1400	Temperatures up to 870 °C (1600 °F) may be needed for spinning
Press forming			
Matched die	480-540	900-1000	Severe forming
Hydropress	205-315	400-600	Mild forming
Roll forming	...	...	...
Creep forming	540-705	1000-1300	...
Dimpling	870-980	1600-1800	...

Ti-5Al-2.5Sn: Springback from trapped rubber forming

Temperature(a)		Springback, degrees	
°C	°F	Stretch(b)	Shrink(b)
21	70	14	14
595	1100	18	14
650	1200	14	14
815	1500	12	12

(a) Some of the temperatures investigated are exceptionally high and probably impractical for production operations. (b) Increased pad pressure will decrease the springback on shrink flanges but has little effect on stretch flanges. Source: R.A. Wood and R.J. Favor, *Titanium Alloys Handbook*, MCIC HB-02, Battelle Columbus Laboratories, 1972

Ti-5Al-2.5Sn: Forming index vs temperature

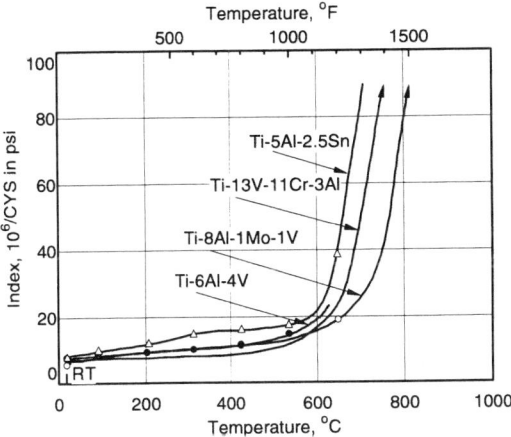

Optimum forming curves for rubber-press shrink flanges.
Source: R.A. Wood and R.J. Favor, *Titanium Alloys Handbook*, MCIC HB-02, Battelle Columbus Laboratories, 1972

Fabrication

Machining of titanium alloys is comparable to machining a good grade of stainless steel. In general, very sharp tools with a slightly larger rake angle and very keen edge are suitable. Slower speed and heavier cuts are preferred because they maintain lower tool temperatures and produce coarse chips, which are more difficult to ignite. Drilling of thin-walled titanium is not a problem as long as the drill is sharp. Thicker walled tube requires a heavy flood of coolant to remove heat and chips. A sample of turning parameters for Ti-5Al-2.5Sn is given below as a general example. For more detailed information, refer to "Technical Note 7: Machining."

Welding. Weldability of Ti-5Al-2.5Sn is good. Like all titanium alloys, it is weldable by all methods except shielded arc welding and submerged arc welding (because no flux is permitted).

Filler Metals. Recommended filler metals are AWS Ti-6 and ER Ti-6 ELI.

Ti-5Al-2.5Sn: Turning parameters for annealed material

Tool material	Tool geometry(a)	Depth of cut		Feed		Speed	
		mm	in.	mm/rev	in./rev	m/min	sfm
Rough turning							
Brazed carbide (C2)	A, E, F, G	2.5-6.35	0.10-0.25	0.25-0.38	0.010-0.015	42-55	140-180
Throwaway carbide (C2)	A, E, F, G	2.5-6.35	0.10-0.25	0.25-0.38	0.010-0.015	43-67	140-220
High-speed steel (M3, T5, T15)	B, D, E	2.5-6.35	0.10-0.25	0.25-0.38	0.010-0.015	110-290(b)	360-960(b)
Finish turning							
Brazed carbide (C3, C2)	A, B, C	0.635-2.5	0.025-0.10	0.13-0.25	0.005-0.010	50-65	165-215
Throwaway carbide (C3, C2)	A, B, C	0.635-2.5	0.025-0.10	0.13-0.25	0.005-0.010	67-76	220-250
High speed steel (M5, T5, T15)	C, E	0.635-2.5	0.025-0.10	0.13-0.25	0.005-0.010	146-290(b)	480-960(b)

(a) See accompanying table for tool geometry codes. (b) These high speeds would be lowered if higher feeds and deeper cuts are made with high-speed steel cutters. Source: R.A. Wood and R.J. Favor, *Titanium Alloys Handbook*, MCIC HB-02, Battelle Columbus Laboratories, 1972

Ti-5Al-2.5Sn: Tool geometry code

	Tool angles and nose radius for indicated tool geometry code						
	A	B	C	D	E	F	G
Back rake, degrees	−5	+5 to −5	0	0	0 to +5	0 to +10	+6 to +10
Side rake, degrees	−5	+6 to 0 0 to −6	5 or 6	15	+5 to +15	0 to +10	0 to +15
End relief, degrees	5	5-10	5	5	5-7	6-8	6.10
Side relief, degrees	5	5-10	5	5	5-7	6-8	6-10
End cutting edges, degrees	15-45	6-15	15 or 5	10-15	5-7	5-10	5-15
Side cutting edge (lead), degrees	15-45	5-20	15	15 to 45	15-20	0-30	0-45
Nose radius, mm (in.)	0.8-1.2 (1/32-3/64)	0.7-1.0 (0.03-0.04)	1.2 (3/64)	1.2 (3/64)	0.5-0.7 (0.02-0.03)	...	0.7-1.0 (0.03-0.04)

Source: R.A. Wood and R.J. Favor, *Titanium Alloys Handbook*, MCIC HB-02, Battelle Columbus Laboratories, 1972

Heat Treatment

Because Ti-5Al-2.5Sn is a single-phase α alloy, heat treatment is confined to stress relief or full annealing treatments (see table).

Temperatures above 870 °C (1600 °F) are seldom used for annealing Ti-5Al-2.5Sn, because excessive grain growth and oxidation can occur. Annealing in the preferred α temperature range imparts or restores optimum ductility and toughness.

Thermal stability measurements on Ti-5Al-2.5Sn made by comparing room-temperature properties before and after thermal exposure (stressed or unstressed) have indicated that this alloy is metallurgically stable under any conditions of stress, temperature, and time up to the annealing temperature. The only changes in properties due to thermal exposure that have been observed are believed to be traceable to the relief of residual stresses. For example, tensile and fatigue specimens prepared from butt-fusion-welded sheet were found to change in strength and ductility after a 500-h, 370 °C (700 °F) exposure.

Ti-5Al-2.5Sn: Recommended heat treatment conditions

Heat treatment	Temperature		Time, h	Cooling method
	°C	°F		
Stress relief	540-650	1000-1200	1/4 to 1	AC
Annealing	705-870	1300-1600	1/6 to 4	AC

Effect of Temperature

Ti-5Al-2.5Sn: Effect of annealing on tensile properties

Property	Annealed	Cold drawn 15%
UTS (0.2% offset), MPa (ksi)	982.5 (142.5)	1206.6 (175.0)
Tensile yield strength, MPa (ksi)	879.1 (127.5)	1041 (151.0)
Proportional limit, MPa (ksi)	844.6 (122.5)	517 (75.0)
Elongation in 50 mm (2 in.), %	17.0	10.0
RA, %	39.0	28.0
Elastic modulus, GPa (10^6 psi)	108.2 (15.7)	99.3 (14.4)
Type of fracture	flat 1/2 cup	flat 1/2 cup

Source: R.A Wood and R.J. Favor, *Titanium Alloys Handbook*, MCIC HB-02, Battelle Columbus Laboratories, 1972

Ti-5Al-2.5Sn: Residual stress relief

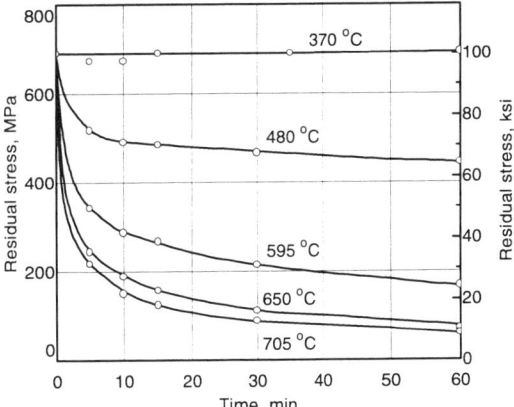

Source: R.A. Wood and R.J. Favor, *Titanium Alloys Handbook*, MCIC HB-02, Battelle Columbus Laboratories, 1972

Effect of Cooling Rate

Ti-5Al-2.5Sn: Effect of cooling from β anneal on tensile properties

Heating into the β field without subsequent working in the α field causes loss of ductility. Quenching reduces the loss.

Annealing treatment	Ultimate tensile strength MPa	ksi	Tensile yield strength MPa	ksi	Elongation in 4D, %	Reduction of area, %
1150 °C (2100 °F), water quench	1055	153.1	1052	152.6	12.5	25.0
1150 °C (2100 °F), air cool	1027	149.0	957	138.8	8.3	15.2
1150 °C (2100 °F), furnace cool	1046	151.8	973	141.2	7.5	13.9
1010 °C (1850 °F), water quench	1008	146.3	906	131.4	16.7	50.0
1010 °C (1850 °F), air cool	989	143.5	958	139.0	16.7	45.7
1010 °C (1850 °F), furnace cool	994	144.2	973	141.2	15.5	43.7
870 °C (1600 °F), air cool	954	138.4	908	131.8	16.7	50.3
650 °C (1200 °F), air cool	1006	146.0	987	143.2	16.7	43.4

Note: Annealed alloy extruded at 925 °C (1700 °F) (α-β extruded). Source: R.A. Wood and R.J. Favor, *Titanium Alloys Handbook*, MCIC HB-02, Battelle Columbus Laboratories, 1972

Ti-5Al-2.5Sn: Room-temperature properties after various treatments

Heat treated condition	Ultimate tensile strength MPa	ksi	Tensile yield strength (0.2%) MPa	ksi	K_{Ic} MPa√m	ksi√in.	K_{Iscc} MPa√m	ksi√in.
760 °C (1400 °F), 2 h, AC	953	138.2	890	129.2	79	72	28	26
900 °C (1650 °F), 10 h, AC	911	132.1	862	125.11	99	90	35	32
900 °C (1650 °F), 100 h, AC (in argon)	898	130.3	818	118.7	101	92	29	27
760 °C (1400 °F), 2 h, AC + 595 °C (1100 °F), 8 h, FC to 500 °C (930 °F), 120 h	964	139.8	899	130.5	50	46	23	21
900 °C (1650 °F), 1 h, WQ	933	135.4	856	124.2	113	103	29	27
1095 °C (2000 °F), 30 min, WQ	965	140.0	873	126.6	130	119	40	37

Source: R.A. Wood and R.J. Favor, *Titanium Alloys Handbook*, MCIC HB-02, Battelle Columbus Laboratories, 1972

Ti-6Al-2Nb-1Ta-0.8Mo

Common Name: Ti-621/0.8
UNS Number: R56210

Chemistry and Density

Ti-6Al-2Nb-1Ta-0.8Mo (Ti-621/0.8), which was developed around 1956, is used mainly in deep submersibles for the U.S. Navy. Other countries (Japan and France) are using Ti-6Al-4V for deep submersibles. Ti-621/0.8 has excellent fracture toughness in marine environments and resistance to salt-water stress-corrosion cracking. This alloy is forgeable and weldable and is intended for use as a structural alloy for marine applications.

Ti-621/0.8 is a modification of Ti-7Al-2Nb-1Ta (Ti-721) composition, which is itself a modification of the original Ti-8Al-2Nb-1Ta (Ti-821) alloy. The Ti-721 alloy was developed specifically to avoid weld cracking problems encountered in Ti-821 thick plate. Ti-621/0.8 was developed as a modification of Ti-721 to achieve resistance to stress-corrosion in salt water.
Density: 4.48 g/cm^3 (0.162 lb/in.^{3})

Effect of Impurities. For optimum toughness in deep submersibles, oxygen content should be kept below 0.10%, and other interstitials should be limited to minimum levels. Oxygen content influences the strength and toughness of this alloy. A modest but consistent increase in smooth tensile strength accompanies an increase in oxygen level from 0.058 to 0.122 wt.%.

Product Forms

Ti-621/0.8 is available as bar, plate, sheet, wire, extrusions and billet. The alloy has excellent weldability; the weld metal develops the same strength, ductility, and toughness as those of the base metal. Machinability of Ti-621/0.8 is similar to that of other titanium alloys. Hot working is normally preformed in the beta-phase region. On request, however, Ti-621/0.8 may be processed in the alpha-beta field, which results in improved strength at some sacrifice in toughness. Hot working in the alpha-beta field is difficult due to the alloy's relatively high cracking tendency.

Product Condition/ Microstructure

Ti-621/0.8 is a near-alpha titanium alloy for applications requiring high toughness and moderate strength. Ti-621/0.8, on the basis of fracture appearance, is considered resistant to seawater stress corrosion. However, sustained-load tests on precracked specimens indicate that the load-carrying ability of this alloy is reduced in seawater, although no evidence of stress-cracking has observed on the fracture surfaces of failed specimens.

Applications

Ti-621/0.8 is used for hulls of marine, hydrospace, and deep-submersible vehicles; for pressure vessels; and for other high-toughness applications.
Precautions in Use. Like most titanium alloys with alpha-beta microstructure, Ti-621/0.8 is susceptible to hydrogen embrittlement in hydrogenating solutions at room temperature, in air or reducing atmospheres at elevated temperatures, and even in pressurized hydrogen at cryogenic temperatures. Oxygen and nitrogen contamination can occur in air at elevated temperatures, and such contamination becomes more severe as exposure time and temperature increase. Ti-621/0.8 is susceptible to stress-corrosion cracking in hot salts (especially chlorides) and to accelerated crack propagation in aqueous solutions at ambient temperatures. The environments in which this alloy is to be used should be carefully controlled to prevent degradation of properties.

Ti-6Al-2Nb-1Ta-0.8Mo: Specifications and compositions

Specification	Designation	Description	Al	Fe	H	Mo	N	Nb	O	Ta	Other
UNS	R56210		6	...	...	0.8-8	...	2	...	1	bal Ti
USA											
AWS A5.16-70	ERTi-6Al-2Cb-1Ta-1Mo	Weld Fill Met	5.5-6.5	0.15	0.005	0.5-1.5	0.012	1.5-2.5	0.1	0.15-1.5	C 0.04; bal Ti
MIL T-9046J	Code A-3	Sh Strp Plt Ann	5.5-6.5	0.25	0.0125	0.5-1	0.03	1.5-2.5	0.1	0.5-1.5	C 0.05; OT 0.4; bal Ti

Ti-6Al-2Nb-1Ta-0.8Mo: Commercial compositions

Specification	Designation	Description	Al	Fe	H	Mo	N	Nb	O	Ta	Other
USA											
RMI	RMI 6Al-2Cb-1Ta-1Mo		5.5-6.5	0.25	0.0125	0.5-1	0.03	1.5-2.5	0.1	0.5-1.5	C 0.05; bal Ti
Timet	TIMETAL 6-2-1										

Physical Properties

Ti-6211: Summary of typical physical properties

Beta transus	1015 ± 15 °C (1860 ± 25 °F)
Melting (liquidus) point	~1650 °C (3000 °F)
Density(a)	4.48 g/cm^3 (0.162 lb/in.^{3})
Tensile modulus	120 GPa (17.5 × 10^6 psi)
Electrical resistivity(a)	1.6 μΩ · m
Magnetic permeability	Nonmagnetic
Specific heat capacity(a)	552 J/kg · K (0.132 Btu/lb · °F)
Thermal conductivity(a)	6.4 W/m · K (3.7 Btu/ft · h · °F)
Thermal coefficient of linear expansion(b)	9 × 10^{-6}/°C (5 × 10^{-6}/°F)

(a) Typical values at room temperature of about 20 to 25 °C (68 to 78 °F). (b) Mean coefficient from 0 to 640 °C (32 to 1200 °F)

Phases and Structures

The microstructure of Ti-6211 can be varied greatly by modifications in primary processing procedures and heat treatment, similar to Ti-6Al-4V. By suitable selection of working and annealing temperatures with respect to the β transus temperature, microstructures can be developed having equiaxed, platelet, or grain-boundary α in a transformed β matrix; both phases can be fine, medium, or coarse, and continuous or noncontinuous. The platelike α precipitates that nucleate and grow below the β transus produce a Widmanstätten structure. The plates often precipitate in colo-

Ti-6211: Effect of oxygen content on β transus temperature

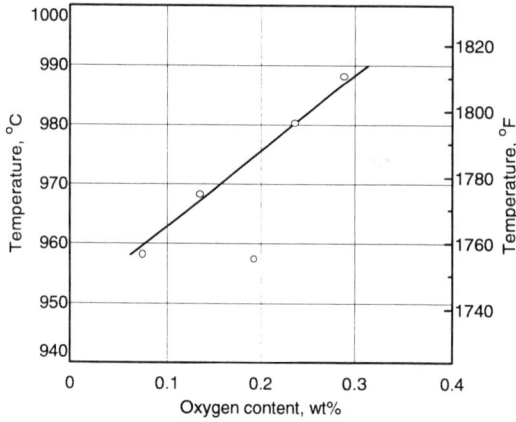

The β transus temperature increases in a linear manner with oxygen content at an approximate rate of 13 °C (23 °F) per 0.1 wt% oxygen. Sample β-annealed: heated to 1065 °C (1950 °F) for 2 h in a vacuum, followed by a moderate cooling rate in a helium atmosphere.
Source: M.A. Imam, B.B. Rath, and D.J. Gillespie, Effect of Oxygen on Microstructure and Properties of Ti-6Al-2Cb-1Ta-1Mo Alloy, *Titanium, Science and Technology*, G. Lütjering, U. Zwicker, and W. Bunk, Ed., Deutsche Gesellschaft fur Metallkunde, Germany, 1985, p 1514

Ti-6211: Effect of oxygen content on grain size

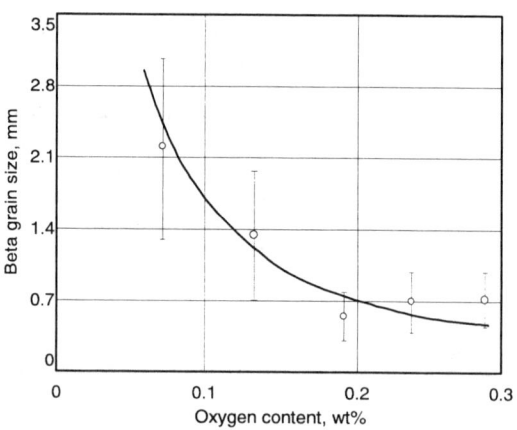

Grain size is reduced by a factor of three when the oxygen content is increased from 0.075 to 0.290%. The grain size is more sensitive to oxygen content in low levels up to about 0.2% oxygen, beyond which further addition of oxygen does not significantly alter the β grain size. Sample β-annealed: heated to 1065 °C (1950 °F) for 2 h in a vacuum, followed by a moderate cooling rate in a helium atmosphere.
Source: M.A. Imam, B.B. Rath, and D.J. Gillespie, Effect of Oxygen on Microstructure and Properties of Ti-6Al-2Cb-1Ta-1Mo Alloy, *Titanium, Science and Technology*, G. Lütjering, U. Zwicker, and W. Bunk, Ed., Deutsche Gesellschaft fur Metallkunde, Germany, 1985, p 1512

nies of the same crystallographic orientation, presumably because of autocatalytic nucleation. Martensite may form in quenched alloys with a platelike or lath morphology. The individual plates tend to have different crystallographic orientations, in contrast to the colony microstructure formed by nucleation and growth, and they often have an internal structure. In addition, Ti_3Al may precipitate in the α phase during aging at 500 °C (930 °F).

Ti-6211: Summary of heat treatment and microstructure of the Widmanstätten-type structure

Heat treatment	Microstructure
1. As-received	Widmanstätten α + β
2. Anneal: 950 °C, 6 h, AC + 700 °C, 2 h, AC	Widmanstätten α + β
3. Anneal: 900 °C, 6 h, AC + 700 °C, 2 h, AC	Widmanstätten α + β
4. Anneal: 1020 °C, 1 h; FC in 10 °C steps, holding 4 h at each step to 980 °C, AC + 700 °C, 2 h, AC	Coarse, blocky primary α in fine Widmanstätten α + β matrix
5. Anneal: 1050 °C, 2 h, AC + 700 °C, 2 h, AC	Fine Widmanstätten α + β
6. Anneal: 1050 °C, 2 h, AC + 950 °C, 6 h, AC + 700 °C, 2 h, AC	Widmanstätten α + β
7. Anneal: 1050 °C, 40 min, AC + 700 °C 2 h, AC	Fine Widmanstätten α + β (same as No. 5 except for prior β grain size)
8. Anneal: 1050 °C, 40 min, AC + 950 °C, 6 h, AC + 700 °C, 2 h, AC	Widmanstätten α + β (same as No. 6 except for prior β grain size)
9. Anneal: 1050 °C, 40 min, WQ + 800 °C, 1 h, WQ + 500 °C, 2 h, AC	Tempered martensite
10. Anneal: 1050 °C, 40 min, WQ + 700 °C, 2 h, AC	Tempered martensite
11. Anneal: 800 °C, 40 min, WQ + 500 °C, 2 h, AC	Widmanstätten α + β
12. Anneal: 950 °C, 40 min, WQ + 500 °C, 2 h, AC	Widmanstätten α + β + martensite

Source: *Metall. Trans. A*, Vol 15, 1984, p 1233

Electrical Properties

Ti-6211: Effect of oxygen content on RT resistivity

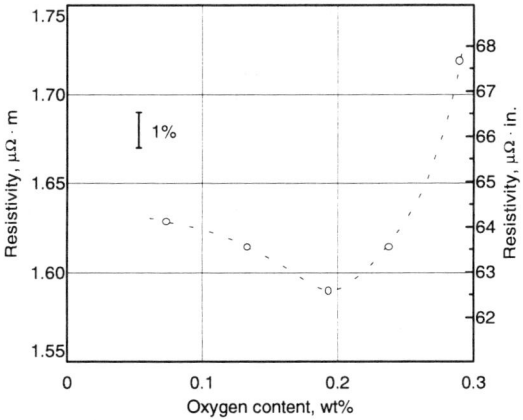

Source: M.A. Imam, B.B. Rath, and D.J. Gillespie, Effect of Oxygen on Microstructure and Properties of Ti-6Al-2Cb-1Ta-1Mo Alloy, *Titanium, Science and Technology*, G. Lütjering, U. Zwicker, and W. Bunk, Ed., Deutsche Gesellschaft fur Metallkunde, Germany, 1985

Stress-corrosion cracking threshold in seawater has been estimated at 77 to 90 MPa$\sqrt{m}$ (70 to 82 ksi $\sqrt{in}$) (*Aerospace Structural Metals Handbook*, Code 3720, Battelle Columbus Laboratories, June 1969).

Ti-6211: Corrosion rates in specific media

Medium	Concentration, %	Temperature, °C	Corrosion rate, mm/yr
Ferric chloride	10	Boiling	nil
Hydrochloric acid	0.5	Boiling	0.020
	1.0	Boiling	1.07
Hydrochloric acid +0.1% $FeCl_3$	5	Boiling	0.051

Mechanical Properties

Ti-6211: Minimum guaranteed mechanical properties

Product condition	Thickness mm	Thickness in.	Ultimate tensile strength MPa	Ultimate tensile strength ksi	Tensile yield strength MPa	Tensile yield strength ksi	Compressive yield strength MPa	Compressive yield strength ksi	Elongation, %	Reduction of area, %
Mill annealed plate per MIL-T-9046	≤70	≤2.750	710	103	655	95	...	...	10(a)	...
As-rolled plate (rolled above β transus)	≤25	≤1.0	827	120	758	110	813	118	10(b)	20
	25-75	1-3	793	115	758	100	744	108	10(b)	20
α-β processed and full annealed (c) sheet and plate	≤3.2	≤0.125	896	130	827	120	...	...	10(d)	...
	3.2-13	0.125-0.50	862	125	793	115	...	...	10(d)	...
	13-25	0.50-1.00	862	125	793	115	827	120	10(d)	20
	25-63.5	1.00-2.50	827	120	758	110	793	115	10(d)	20
	63.5-100	2.50-4.00	793	115	724	105	758	110	10(d)	20

(a) Minimum elongation in 50 mm (2 in.) or 4D. (b) Minimum elongation in 25 mm (1 in.). (c) Full anneal at 900 °C (1650 °F) for 1 h, AC. (d) Minimum elongation in 50 mm (2 in.). Source: *Aerospace Structural Metals Handbook*, Code 3720, Vol 4, Battelle Columbus Laboratories, 1969

Hardness

Rockwell Hardness. Typical Rockwell hardness is 30 HRC.

Ti-6211: Effect of oxygen content on hardness

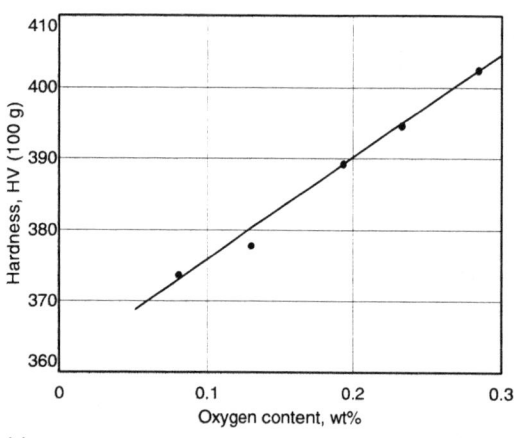
(a)

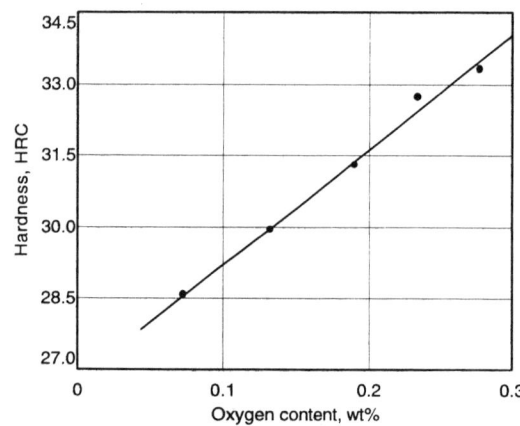
(b)

Note: Five Ti-6211 alloys with varying oxygen contents, ranging between 0.075 to 0.290 wt% (0.22 to 0.87 at.%), were prepared in 125-lb heats and fabricated by upset forging and hot rolling at 1065 °C (1945 °F) followed by an annealing treatment at 925 °C (495 °F) for 1 h and air cooling. Samples were then heat treated at 1065 °C (1945 °F) for 2 h and air cooled.
Source: M. Imam, B. Rath, and D. Gillespie, "Effect of Oxygen on Microstructure and Properties of Ti-6Al-2Cb-1Ta-1Mo Alloy," in *Titanium, Science and Technology*, G. Lütjering, U. Zwicker, and W. Bunk, Ed., Deutsche Gesellschaft fur Metallkunde, 1985, p 1511

Typical Tensile Strengths

Ti-6211: Tensile property variations of 25 mm (1 in.) rolled plate

Condition	Test direction	Ultimate tensile strength MPa	Ultimate tensile strength ksi	Tensile yield strength MPa	Tensile yield strength ksi	Elongation, %	Reduction of area, %
As-rolled	L	824.6	119.6	701.9	101.8	10.8	23.1
	T	848.0	123.0	723.9	105.0	12.5	33.6
α annealed: 870 °C (1600 °F), 1 h, AC	L	814.9	118.2	683.9	99.2	12.5	28.0
	T	843.9	122.4	740.5	107.4	12.5	31.9
α-β annealed, air cooled: 990 °C (1815 °F), 1 h, AC	L	814.9	118.2	673.6	97.7	12.5	31.3
	T	842.5	122.2	717.0	104.0	13.0	33.0
α-β solution treated, quenched: 990 °C (1815 °F), 1 h, AC	L	868.7	126.0	697.7	101.2	13.0	29.0
	T	889.4	129.0	747.4	108.4	11.8	30.7
α-β solution treated, aged: 990°C (1815 °F), 1 h, WQ + 595 °C (1100°F), 2 h, AC	L	875.6	127.0	750.8	108.9	10.5	20.8
	T	934.2	135.5	819.8	118.9	10.5	22.1
β annealed, air cooled 1035 °C (1900 °F), 1 h, AC	L	849.4	123.2	699.1	101.4	11.8	24.6
	T	849.4	123.2	712.9	103.4	12.2	27.7
β solution treated, quenched: 1035 °C (1900 °F), 1 hr, WQ	L	903.9	131.1	758.0	110.2	12.8	31.0
	T	883.2	128.1	744.6	108.0	8.8	23.8
β solution treated, aged: 1035 °C (1900 °F), 1 h, WQ + 595 °C (1100 °F), 2 h, AC	L	932.8	135.3	807.4	117.1	9.5	24.8
	T	933.5	135.4	801.1	116.2	9.5	21.2

Note: Each value average of two tests. Source: *Aerospace Structural Metals Handbook*, Code 3720, Vol 4, Battelle Columbus Laboratories, 1969

Ti-6211: Effect of oxygen content on tensile properties of 25 mm (1 in.) rolled plate

Condition	Test direction	Ultimate tensile strength MPa	Ultimate tensile strength ksi	Tensile yield strength MPa	Tensile yield strength ksi	Elongation in 25 mm (1 in.), %	Reduction of area, %
0.058 wt% oxygen							
1035 °C (1900 °F), 1 h, AC	L	849.4	123.2	699.1	101.4	11.8	24.6
	T	849.4	123.2	712.9	103.4	12.2	27.7
1035 °C (1900 °F), 1 h, WQ	L	903.9	131.1	759.8	110.2	12.8	31.0
	T	883.2	128.1	744.6	108.0	8.8	23.8
1035 °C (1900 °F), 1 h, WQ + 595 °C (1100 °F), 2 h, AC	L	932.8	135.3	807.4	117.1	9.5	24.8
0.122 wt% oxygen							
1065 °C (1950 °F), 1 h, AC	L	879.7	127.6	727.4	105.5	12.2	27.6
	T	885.9	128.5	732.9	106.3	12.5	26.5
1065 °C (1950 °F), 1 h, WQ	L	932.1	135.2	792.9	115.0	11.5	27.2
	T	941.8	136.6	803.9	116.6	10.8	19.4
1065 °C (1950 °F) 1 h, WQ + 595 °C (1100 °F), 2 h, AC	L	945.2	137.1	825.9	119.8	12.5	25.4
	T	958.3	139.0	834.9	121.1	9.2	16.0

Note: All values average of two tests. Source: *Aerospace Structural Metals Handbook*, Code 3720, Vol 4, Battelle Columbus Laboratories, 1969

Ti-6211: Tensile strengths vs solution treatment

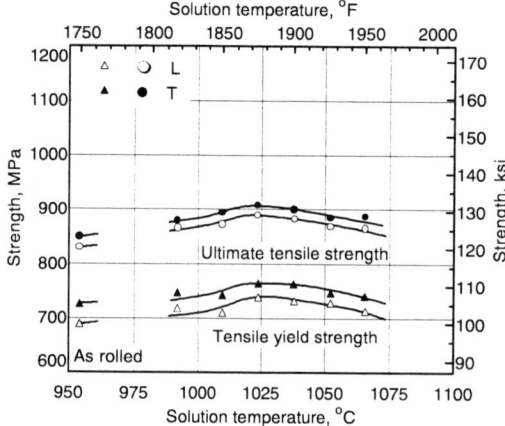

Quenched plate. 25 mm (1 in.) plate solution treated, 1 h, water quenched as 25 by 150 by 150 mm (1 by 6 by 6 in.) specimen blanks. Each data point is an average of two tests.
Source: *Aerospace Structural Metals Handbook*, Code 3720, Vol 4, Battelle Columbus Laboratories, 1969

Ti-6211: Tensile strengths vs solution treatment

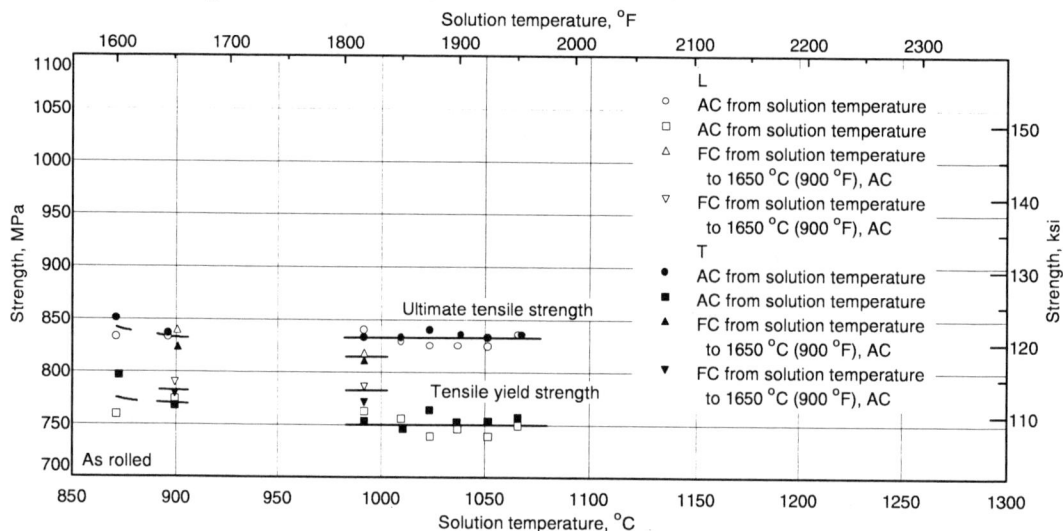

Air-cooled plate. Solution treated tensile properties of plate specimens air cooled or furnace cooled from various solution temperatures. Specimens were 25 mm (1 in.) plate solution treated 1 h, as 25 by 150 by 150 mm (1 by 6 by 6 in.) specimen blanks. Each data point is the average of two tests.
Source: *Aerospace Structural Metals Handbook*, Code 3720, Vol 4, Battelle Columbus Laboratories, 1969

Ti-6211: Effect of quench delay on tensile properties

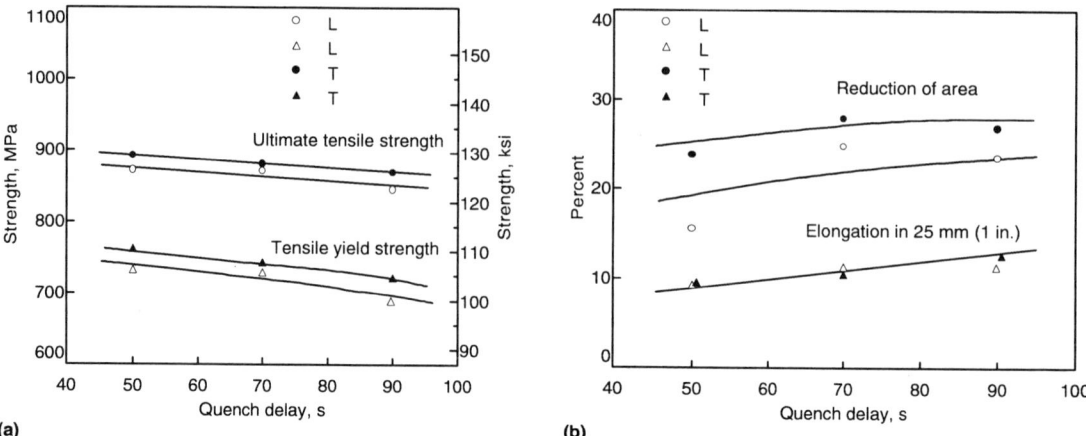

25 mm (1 in.) plate; 1095 °C (2000 °F), 1 h, delay (in air), water quenched as 25 by 150 by 150 mm (1 by 6 by 6 in.) specimen blanks. Each data point is an average of two tests.
Source: *Aerospace Structural Metals Handbook*, Code 3720, Vol 4, Battelle Columbus Laboratories, 1969

Ti-6211: Effect of aging temperature on tensile strengths

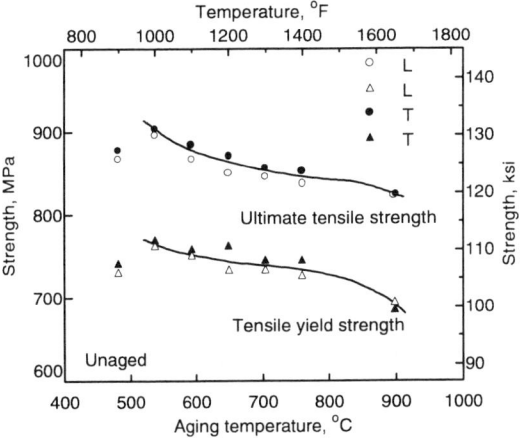

(a) 25 mm (1 in.) plate; 1095 °C (2000 °F), 1 h, 70 s delay (in air), water quench + age, 2 h, air cooled 25 by 150 by 150 mm (1 by 6 by 6 in.) specimen blanks. Each point is an average of two tests.
Source: *Aerospace Structural Metals Handbook*, Code 3720, Vol 4, Battelle Columbus Laboratories, 1969

Ti-6211: Effect of annealing temperature on yield strength

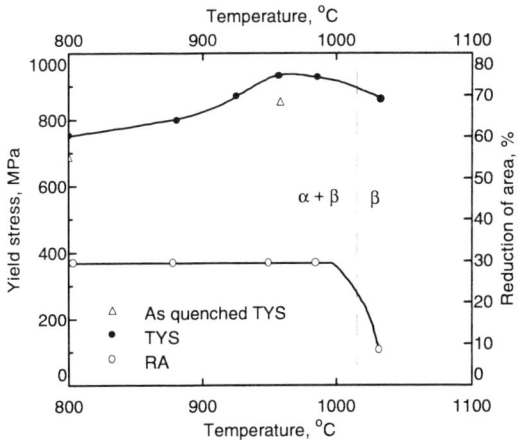

(b) Effect of annealing temperature on yield strength and ductility of Widmanstätten α + β material, annealed 40 min, water quenched + 500 °C (930 °F), 2 h, AC.
Source: *Metall. Trans. A*, Vol 15, June 1984, p 1236

Compressive Strengths

Compressive Yield Strength. Typical room-temperature compressive yield strength is 815 MPa (118 ksi).

Ti-6211: Compressive yield strength of as-rolled plate

Plate thickness		Location in plate thickness direction	Specimen direction	Compressive yield strength		Tensile yield strength		Reduction of area, %
mm	in.			MPa	ksi	MPa	ksi	
25	1	Midthickness	L	758	110	703	102	32
			T	765	111	696	101	23
63.5	2.5	Surface	L	731	106	668	97	28
			T	737	107	689	100	30
		Midthickness	L	737	107	662	96	30
			T	724	105	682	99	32
63.5	2.5	...	ST	710	103	662	96	38

Note: 12.8 mm (0.505 in.) diameter tensile specimens; each value average of two sets. Source: *Aerospace Structural Metals Handbook*, Code 3720, Vol 4, Battelle Columbus Laboratories, 1969

Ti-6211: Compressive yield strength of as-rolled plate

Specimen size, mm (in.)	Test direction	Compressive yield strength		Ultimate tensile strength		Elongation in 25 mm (in.), %
		MPa	ksi	MPa	ksi	
13 × 609 × 1524 (½ × 24 × 60)	L	...	...	934.2	135.5	10.0
	T	...	...	951.5	138.0	11.0
25 × 609 × 2438 (1 × 24 × 96)	L	779.1	113.0	834.2	121.0	13.0
	T	868.7	126.0	868.7	126.0	12.0
44 × 1650 (1¾ × 65) diam	L	...	...	827.3	120.0	11.0
	T	...	...	848.0	123.0	11.0
50 × 1016 × 1270 (2 × 40 × 50)	L	854.9	124.0	844.6	122.5	12.0
	T	786.0	114.0	841.1	122.0	14.0
60 × 914 × 1219 (2⅜ × 36 × 48)	L	786.0	114.0	841.1	122.0	14.0
	T	868.7	126.0	854.9	124.0	13.0
60 × 495 × 2845 (2⅜ × 19½ × 112)	L	847.3	122.9	834.2	121.0	13.5
	T	803.9	116.6	834.2	121.0	15.0
60 × 508 × 1905 (2⅜ × 20 × 75)	L	772.2	112.0	830.8	120.5	12.5
	T	823.9	119.5	837.7	121.5	13.5
75 × 1168 × 1168 (3 × 46 × 46)	L	767.4	111.3	809.4	117.4	13.5
	T	783.9	113.7	837.7	121.5	14.0
100 × 609 × 2438 (4 × 24 × 96)	L	...	...	792.9	115.0	13.0

Note: Not all from same heat. Center properties. Single tests at room temperature. Source: *Aerospace Structural Metals Handbook*, Code 3720, Vol 4, Battelle Columbus Laboratories, 1969

Ti-6211: Effect of tensile prestrain on compressive yield strength of plate

Tensile prestrain %	Stress relief treatment Temperature °C	°F	h	Compressive yield strength (0.01% offset) MPa	ksi	(0.2% offset) MPa	ksi
0.0	RT		...	599	87	758	110
1.28	RT		...	310	45	613	89
1.28	315	600	1	434	63	662	96
1.28	315	600	3	448	65	682	99
1.28	480	900	1	517	75	717	104

Note: 25 mm (1 in.) as-rolled plate. Source: *Aerospace Structural Metals Handbook*, Code 3720, Vol 4, Battelle Columbus Laboratories, 1969

High-Temperature Strength

Ti-6211: Tensile strengths vs temperature

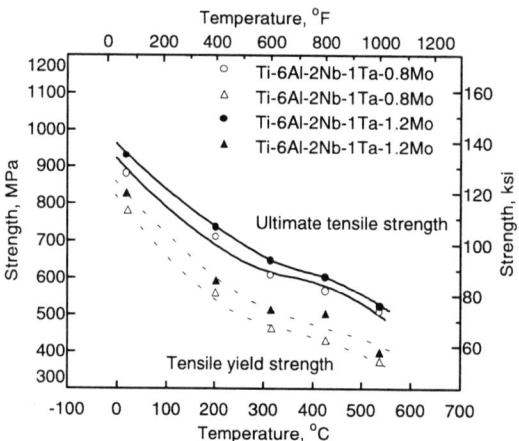

Full annealed forged pancake; 3:1 upset to 15.9 mm (⅝ in.) thickness; tangential, 900 °C (1650 °F), 1 h, air cooled. Each point is an average of two tests.
Source: Reactive Metals, Inc., Nov 25, 1968; *Aerospace Structural Metals Handbook*, Code 3720, Battelle Columbus Laboratories, 1969

Creep Strength

Ti-6211: Creep behavior

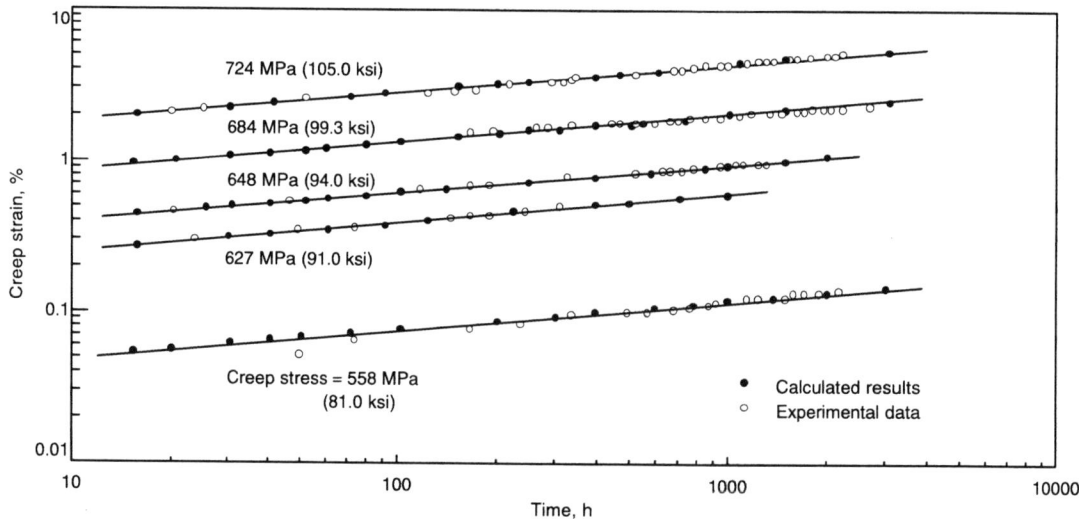

Data for room-temperature tension stress creep. Equation for calculation: $\varepsilon = t^{0.183} (\sigma/103.55)^{13.89}$
Source: H. Chu, "Room Temperature Creep and Stress Relaxation of a Titanium Alloy," *J. Mater.*, Vol 5, 1970, p 633

Ti-6211: Creep behavior of annealed hot rolled

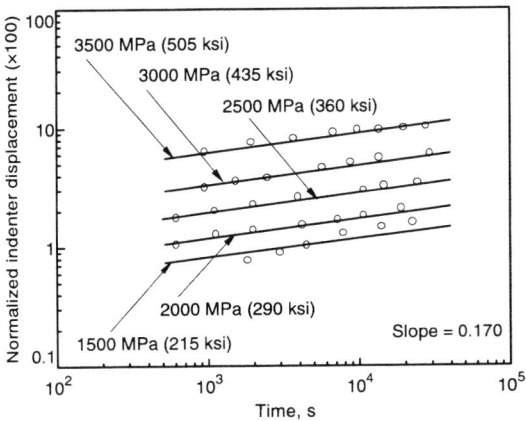

Chemical composition: 6.0 wt% Al, 0.02 wt% C, 1.95 wt% Nb, 0.0040 wt% H, 0.8 wt% Mo, 0.010 wt% N, 0.075 wt% O, and 0.88 wt% Ta. Alloy was hot rolled at 1065 °C (1950 °F), and annealed at 925 °C (1700 °F) for 1 h, air cooled. Specimens were cut to dimension 25 by 25 by 8 mm (1 by 1 by 0.3 in.). Creep was determined by impression test.
Source: H. Yu, M. Imam, and B. Rath, "Investigation of Creep Behavior of Ti-Alloy by Impression Test," in *Advances in Fracture Research*, S. Valluri, D. Taplin, P. Rao, J. Knott, and R. Dubey, Ed., Pergamon Press, 1985, p 3273

Ti-6211: Creep behavior at 180 °C

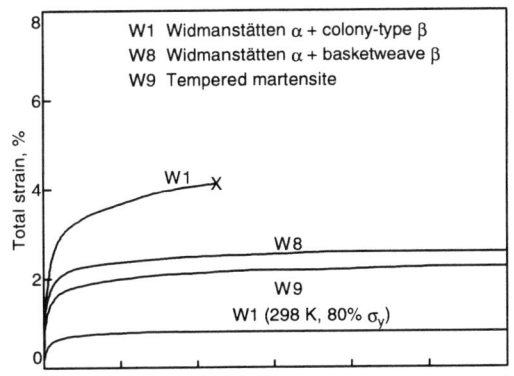

Chemical composition of alloys from two heats was 5.6 and 5.9% Al, 0.79 and 0.60% Mo, 2.16 and 2.0% Nb, and 0.95 and 0.93% Ta. Applied stress was 80% of yield strength at room temperature.
Source: W. Miller, Jr., R. Chen, and E. Starke, Jr., "Microstructure, Creep, and Tensile Deformation in Ti-6Al-2Nb-1Ta-0.8Mo," *Metall. Trans. A*, Vol 181, 1987, p 1451

Stress Relaxation

Ti-6211: Stress relaxation at 480 and 540 °C

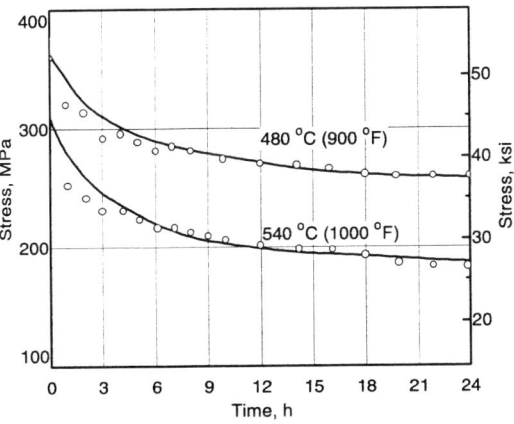

As-rolled 25 mm (1 in.) plate. Initial stress = 0.2% offset yield strength at the test temperature.
Source: *Aerospace Structural Metals Handbook*, Code 3720, Vol 4, Battelle Columbus Laboratories, 1969

Fatigue Properties

High- and low-cycle fatigue data for smooth and notched specimens of Ti-6211 tested in air and in salt water exhibited a decrease in the fatigue strength for notched specimens at all cyclic lives. The Severn River salt water, however, had no significant detrimental effect. The notch effect was much smaller in the low-cycle fatigue range (near 10^3 cycles to failure) than in the higher cyclic range. Fatigue crack growth rates were also unaffected by seawater under different cathodic potentials.

Ti-6211: Effect of seawater on high- and low-cycle fatigue

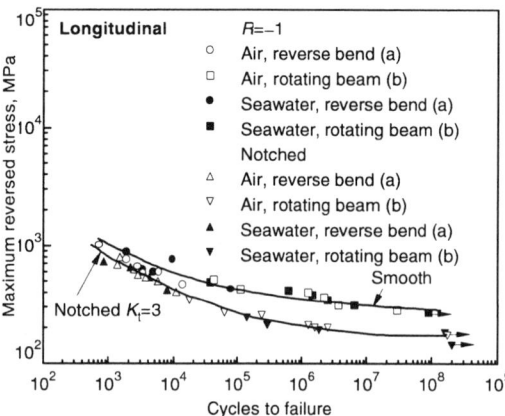

63.5 mm (2.5 in.) as-rolled plate. (a) Square wave load profile ½ to 1 cycles/min. (b) Sine wave load profile 1450 cycles/min.
Source: *Aerospace Structural Metals Handbook*, Code 3720, Vol 4, Battelle Columbus Laboratories, 1969; from R.C. Schwab and E.J. Czyryca, Naval Ship R&D Report 2854, May 1969, AD852521

Ti-6211: Fatigue crack growth in seawater

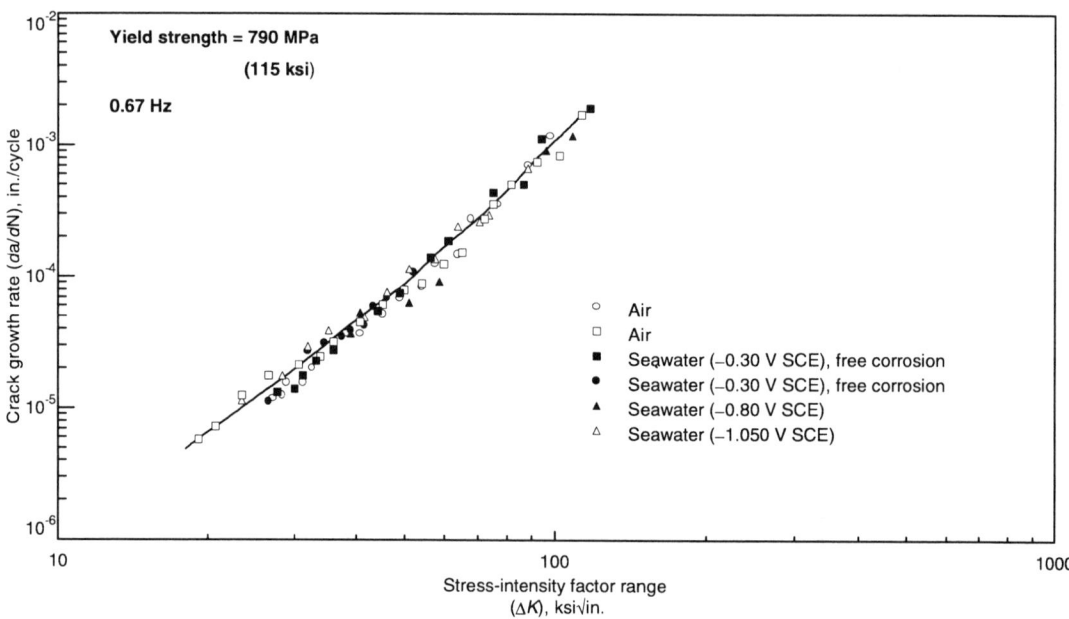

Source: W.R. Cares and T.W. Crooker, "Fatigue-Crack Growth of Ti-6Al-2Cb-1Ta-0.8 Mo Alloy in Air and Natural Sea Water Environments," NRL Memorandum Report 2617, June 1973, AD765318

Fracture Properties

Impact Toughness Minimum impact toughness values at 0 °C (32 °F) are approximately 28 to 34 J (21 to 25 ft · lbf) for standard Charpy V-notch specimens.

Ti-6211: Impact toughness vs temperature

As-rolled plate. Each data point is an average of two tests minimum.
Source: *Aerospace Structural Metals Handbook*, Code 3720, Vol 4, Battelle Columbus Laboratories, 1969

Ti-6211: Effect of heat treatment on impact strength of 25 mm (1 in.) rolled plate

Condition	Test direction	Tensile yield strength at RT		Charpy V-notch impact toughness at −62 °C (80 °F)		Drop weight tear energy at 0 °C (32 °F)	
		MPa	ksi	J	ft·lbf	J	ft·lbf
As-rolled	L	701.9	101.8	31.8	23.5	3072	2266(a)
	T	723.9	105.0	31.2	23.0	2909	2146(a)
α annealed: 870 °C (1600 °F), 1 h, AC	L	683.9	99.2	39.6	29.2	3312	2442
	T	740.5	107.4	43.6	32.2	3858	2846
α-β annealed, air cooled: 990 °C (1815 °F), 1 h, AC	L	673.6	97.7	44.0	32.5	3471	2560
	T	717.0	104.0	38.2	28.2	3072	2266
α-β solution treated, quenched: 990 °C (1815 °F), 1 h, WQ	L	697.7	101.2	37.7	27.8	3471	2560
	T	747.4	108.4	43.6	32.2	3312	2443
α-β solution treated, aged: 990 °C (1815 °F), 1 h, WQ + 595 °C (1100 °F), 2 h, AC	L	744.6	108.0	34.5	25.5	2828	2086
	T	819.8	118.9	39.3	29.0	2583	1905
β annealed, air cooled: 1035 °C (1900 °F), 1 h, AC	L	699.1	101.4	44.7	33.0	3549	2618
	T	712.9	103.4	38.2	28.2	2665	1966
β solution treated, quenched: 1035 °C (1900 °F), 1 h, WQ	L	759.8	110.2	41.3	30.5	3072	2266
	T	744.6	108.0	36.3	26.8	3072	2266
β solution treated, aged: 1035 °C (1900 °F), 1 h, WQ + 595 °C (1100 °F), 2 h, AC	L	807.4	117.1	32.5	24.0	2336	1723
	T	801.1	116.2	31.4	23.2	2909	2146

Note: All values average of two tests except drop weight tear values, which are individual results. (a) In a separate study on this same heat, the following results were obtained: 3232 J (2384 ft·lbf) and 2418 J (1784 ft·lbf) T direction. Source: *Aerospace Structural Metals Handbook*, Code 3720, Vol 4, Battelle Columbus Laboratories, 1969

Ti-6211: Effect of oxygen on impact strength of 25 mm (1 in.) rolled plate

Condition	Test direction	Tensile yield strength at RT		Charpy V-notch impact toughness at –62 °C (–80 °F)		Drop weight tear energy at 0 °C (32 °F)	
		MPa	ksi	J	ft·lbf	J	ft·lbf
0.058 wt% oxygen							
1035 °C (1900 °F), 1 h, AC	L	699	101.4	44.7	33.0	3549	2618
	T	712.9	103.4	38.2	28.2	2665	1966
1035 °C (1900 °F), 1 h, WQ	L	759.8	110.2	41.3	30.5	3072	2266
	T	744.6	108.0	36.3	26.8	3072	2266
1035 °C (1900 °F), 1 h, WQ + 595 °C (1100 °F), 2 h, AC	L	807.4	117.1	32.5	24.0	2336	1723
	T	801.1	116.2	31.4	23.2	2909	2146
0.122 wt% oxygen							
1065 °C (1950 °F), 1 h, AC	L	727.4	105.5	33.6	24.8	2418	1784
	T	732.9	106.3	34.5	25.5	...	...
1065 °C (1950 °F), 1 h, WQ	L	792.9	115.0	30.5	22.5	...	...
	T	803.9	116.6	32.5	24.0	...	...
1065 °C (1950 °F), 1 h, WQ + 595 °C (1100 °F), 2 h, AC	L	825.9	119.8	29.8	22.0	1590	1173
	T	834.9	121.1	31.4	23.2	...	...

Note: All values average of two tests, except drop weight tear values, which are individual results. Source: *Aerospace Structural Metals Handbook*, Code 3720, Vol 4, Battelle Columbus Laboratories, 1969

Ti-6211: Variation in room-temperature impact toughness for forged dome specimens

No.	Ultimate tensile strength		Tensile yield strength		Elongation in 25 mm (1 in.), %	Reduction of area, %	Charpy V-notch impact toughness at 0 °C (32 °F)	
	MPa	ksi	MPa	ksi			J	ft·lbf
1	841.8	122.1	740.5	107.4	10.5	27.2	...	...
2	845.6	122.6	746.0	108.2	12.5	29.9	28.4	21.0
3	845.6	122.6	747.4	108.4	12.0	30.5	...	...
4	860.4	124.8	783.2	113.6	17.0	30.5	...	...
5	896.3	130.0	786.0	114.0	9.5	22.3	29.8	22.0
6	892.2	129.4	813.6	118.0	17.5	38.8	31.8	23.5
7	897.7	130.2	817.7	118.6	17.5	37.2	31.8	23.5
8	934.9	135.6	853.5	123.8	15.0	37.5	36.6	27.0
9	842.5	133.4	842.5	122.2	15.0	37.2	29.8	22.0

Source: *Aerospace Structural Metals Handbook*, Code 3720, Vol 4, Battelle Columbus Laboratories, 1969

Fracture Toughness

Ti-6211: Sustained load cracking

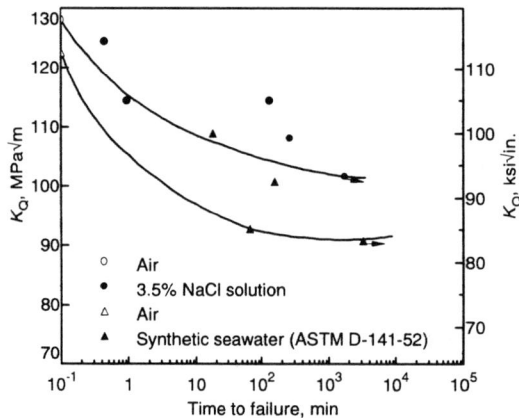

25 mm (1 in.) as-rolled plate. Specimens failed to meet the size requirements for plane-strain testing as defined by ASTM Committee E24. Therefore, the symbol K_Q is used rather than K_I.
Source: *Aerospace Structural Metals Handbook*, Code 3720, Vol 4, Battelle Columbus Laboratories, 1969

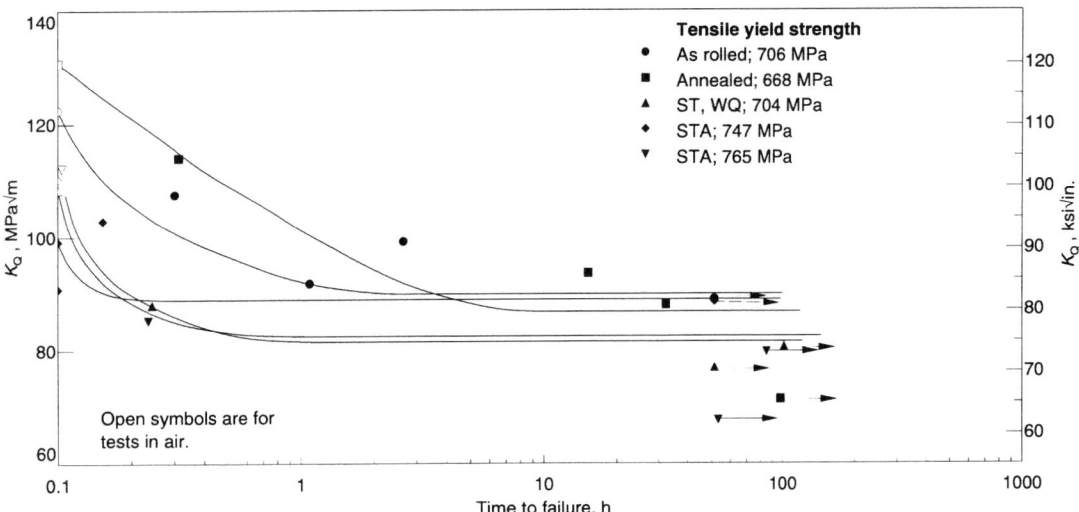

$$K_Q = \frac{4.12 M \sqrt{\frac{1}{\alpha^3} - \alpha^3}}{BD^{3/2}}$$

where M = moment at crack tip at rupture
B = specimen width (22 mm)
D = specimen depth (25 mm)
α = 1 – a/D
a = crack length

Sustained load seawater stress-corrosion behavior of fatigue cracked cantilever bend stress-corrosion specimen from 25 mm (1 in.) plate in several heat treated conditions as indicated.
Source: *Aerospace Structural Metals Handbook*, Code 3720, Vol 4, Battelle Columbus Laboratories, 1969

Ti-6211: Sustained load crack growth data

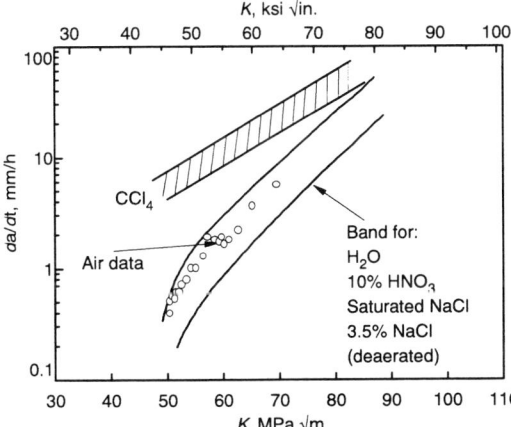

Ti-6211: Discontinuous crack growth in sodium chloride

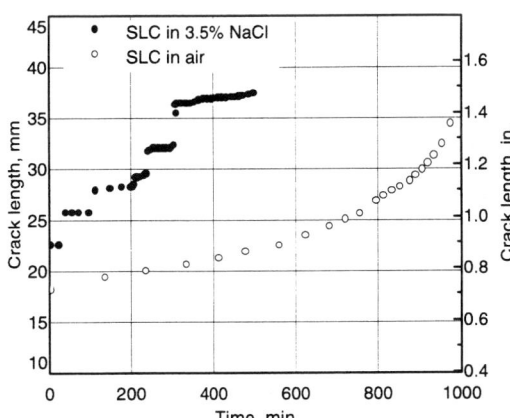

The alloy had an α β microstructure, had been rolled at 970 °C (1780 °F) and then heat treated at 705 °C (1300 °F) for 2 h, and then air cooled. The initial hydrogen content was approximately 30 ppm.
Source: C.L. Hoffmann, R.W. Rudy, Jr., and B.B. Rath, The Influence of Environmental Factors on Sustained Load Cracking in Ti-6211, in *Titanium, Science and Technology*, Vol 4, G. Lütjering, U. Zwicker, and W. Bunk, Ed., Gesellschaft fur Metallkunde, 1985, p 2495-2502

It has been proposed that sustained load cracking of titanium alloys in aqueous salt water environments results from the diffusion of hydrogen from the environment into the crack tip region. The increased hydrogen concentration in the hydrostatic stress field ahead of the crack tip results in hydride precipitation and subsequent cleavage fracture associated with the hydrides. Discontinuous crack growth occurs each time the crack moves out of the hydride-affected region and moves into hydride-free material where it is arrested.
Source: C.L. Hoffman, R.W. Rudy, Jr., and B.B. Rath, The Influence of Environmental Factors on Sustained Load Cracking in Ti-6211, in *Titanium, Science and Technology*, Vol 4, G. Lütjering, U. Zwicker, and W. Bunk, Ed., Deutsche Gesellschaft fur Metallkunde, 1985, p 2495-2502

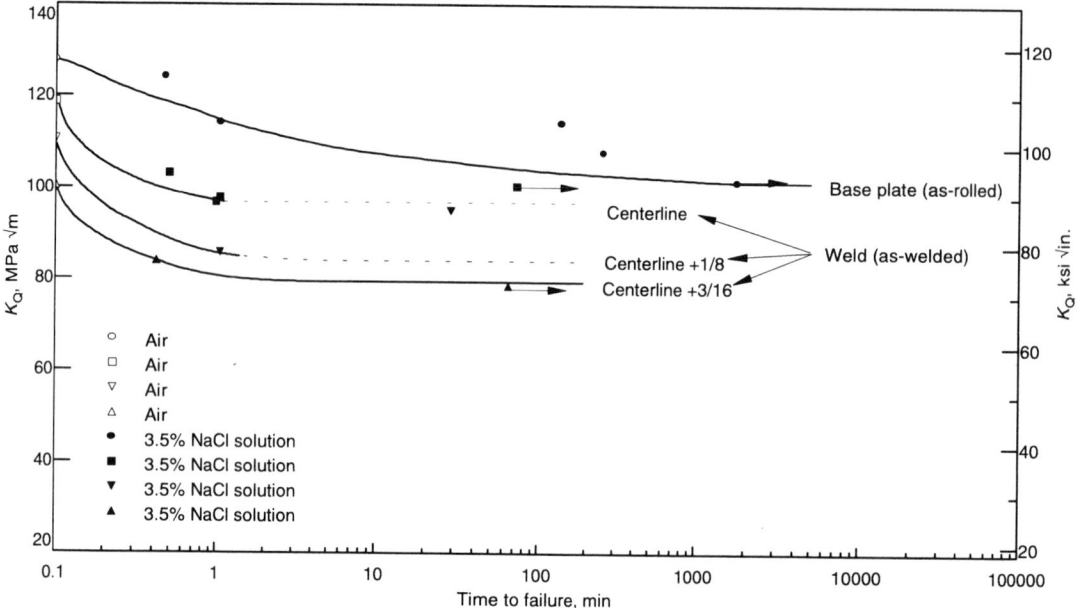

Fatigue cracked (and side-grooved) cantilever bend stress corrosion specimen of 25 mm (1 in.) rolled plate, longitudinal direction.

Plastic Deformation

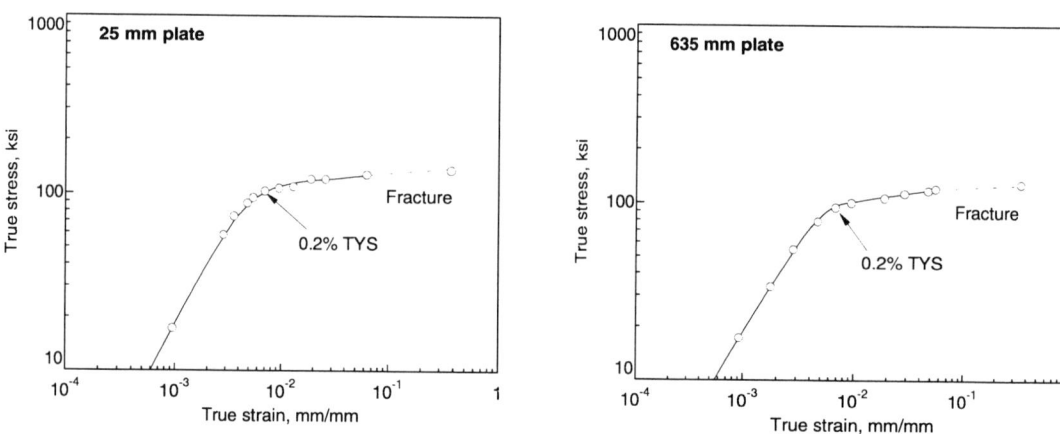

Longitudinal, mid-thickness stress-strain relations for as-rolled plate. Strain rate effects on the stresses required for plastic deformation should be similar to that of other alloys such as Ti-8Al-1Mo-1V.
Source: *Aerospace Structural Metals Handbook*, Vol 4, Code 3720, Battelle Columbus Laboratories, 1969

Forging

Ti-6211, an α alloy, is used commercially in the full range of forging product types and is produced on all types of forging equipment. As with other α titanium alloys, it is difficult to fabricate into forgings, exhibiting high flow stresses and crack sensitivity. However, the alloy has been successfully commercially produced in forgings for several applications including pressure vessels, reactor components, and armor where its excellent weldability is beneficial.

Ti-6211 is characterized by high unit pressures (flow stresses) and crack sensitivity in forging processes. The final microstructure of Ti-6211 is manipulated by thermomechanical processing in forging manufacture using combinations of sub- and/or supra-beta transus forging followed by thermal treatments. Final thermal treatments consist of an anneal, at 705 to 815 °C (1300 to 1500 °F) and for some applications, duplex annealing, a sub-transus solution treatment at 995 °C (1825 °F) followed by air cooling prior to anneal is used to refine the final structure and enhance strength or fracture-related properties.

Conventional Forging. The objectives in forging Ti-6211 are to obtain the final forging

Ti-6211: Hot ductility dip

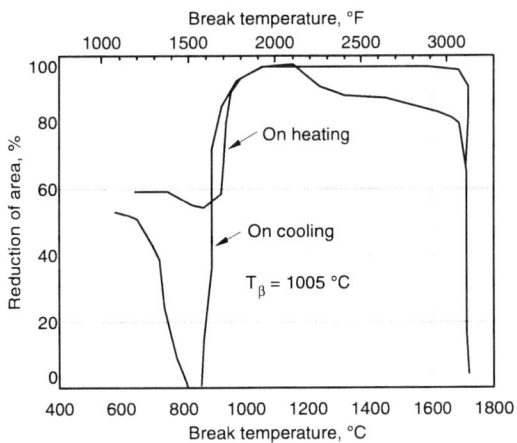
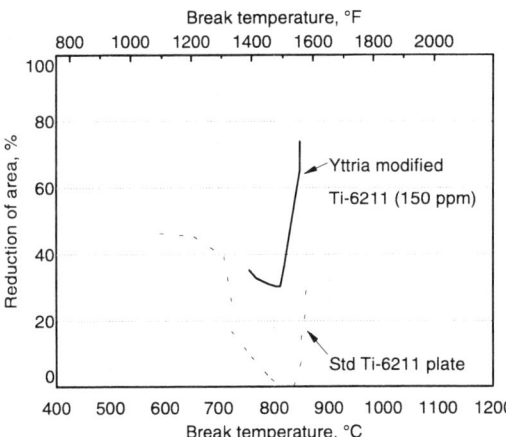

The severity of the ductility dip can be reduced by yttria additions; on-cooling curves for β-processed plate. The propensity for hot cracking can be related to hot ductility, which typically is determined by hydraulically applying a tensile load rapidly ($\dot{\varepsilon}$ = 160/min) during heating or cooling over a wide enough temperature range to identify minimum ductility regions.
Source: R.E. Lewis et al., The Elevated Temperature Ductility Dip Phenomenon in Alpha, Near-Alpha, and Alpha-Beta Alloys, *Titanium, Science and Technology*, Vol 2, G. Lütjering, U. Zwicker, and W. Bunk, Ed., Deutsche Gesellschaft für Metallkunde, Germany, 1985, p 895-902

shape and desired final microstructure at least cost. Conventional subtransus (α+β) forging processes predominate in commercial forging where optimum strength and ductility properties are desired. Conventional forging provides a final microstructure of predominately (80 to 90%) equiaxed α morphology. To achieve such structures and properties, subtransus reduction of 50 to 75%, accumulated through one or more forging steps, are required. Supra-transus (β) forging may be used in early forging operations, including upsetting and open die preforming, to reduce unit pressures and ease forging fabrication. High-temperature initial forging operations must be followed by sufficient subtransus reduction to achieve the desired equiaxed α structure. Conventionally forged Ti-6211 is then heat treated as noted above.

Supra-transus forging also may be used to achieve a transformed, elongated α structure for improved fracture-related properties. Successful β forging fabrication of Ti-6211 for this alternate microstructure includes heavy β reductions (e.g., 35 to 50%) in early forging stages followed by a low final subtransus reduction (e.g., 15 to 20%). Beta forging requires careful control of forging process conditions, particularly preheat times at temperature, to avoid excessive prior β grain growth. Beta forged Ti-6211 is then heated as noted above. Because of inherent variations in forging conditions, β forged alloy displays more final forging product variations than conventional forging.

Surface Treatment. As with other difficult-to-fabricate α alloys, precoats or other surface coating techniques used on billet stock and intermediate forging shapes during furnacing for forging operations are essential. Ti-6211 is sensitive to the formation of excessive α case during reheating processes, which may lead to undue surface cracking in subsequent forging deformation. The alloy is also sensitive to thermal stresses developed in crack repair processes, typically dry abrasive grinding techniques. Thus, Ti-6211 may be heated to 425 to 590 °C (800 to 1100 °F) prior to repair, especially if surface cracking is severe.

Ti-6211: Forging process temperatures

Process	Metal temperature	
	°C	°F
Conventional forging	940-995	1725-1825
Beta forging	1040-1120	1900-2050

Note: See "Technical Note 4: Forging" for recommended die temperatures

Other Fabrication

Machining of Ti-6211 is similar to Ti-5Al-2.5Sn.

Welding. Ti-6211 has excellent weldability. Like other titanium alloys, it is weldable by all methods except shielded arc welding and submerged arc welding (because no flux is permitted). Recommended filler metal is the same as the base metal.

Heat Treatment. Ti-6211 generally is used in the as-fabricated or fabricated plus annealed conditions (see table). A small increase in strength can be obtained by solution treating and aging, but at a sacrifice in ductility and toughness.

Ti-6211: Recommended heat treatment conditions

Heat treatment	Temperature °C	Temperature °F	Time, h	Cooling method
Stress relief	595-650	1100-1200	1/4 to 2	AC
Annealing	790-900	1450-1650	1 to 4	AC
Solution treating	1010	1850	1	WQ
Aging	620	1150	2	AC

Ti-6Al-2Sn-4Zr-2Mo-0.08Si

Common Name: Ti-6242S Ti-6242Si
UNS Number: R54620

Ti-6Al-2Sn-4Zr-2Mo-0.08Si (Ti-6242S or Ti-6242Si), developed in the late 1960s as an elevated-temperature alloy, has an outstanding combination of tensile strength, creep strength, toughness, and high-temperature stability for long-term applications at temperatures up to 425 °C (800 °F). Ti-6242S is one of the most creep-resistant titanium alloys and is recommended for use up to 565 °C (1050 °F). Proper heat treatment is important in allowing the alloy to develop its maximum creep resistance.

Chemistry and Density

The 6 percent aluminum addition in the Ti-6Al-2Sn-4Zr-2Mo composition is a potent alpha-phase stabilizer, while the 2 percent molybdenum addition represents only a moderate quantity of this potent beta-phase stabilizer. The tin and zirconium additions are solid-solution strengthening elements that are neutral with respect to phase stabilization. The net effect of this combination of alloying elements is the generation of a weakly beta-stabilized, alpha-beta alloy. Since it is weakly beta stabilized, the alloy is also properly described as a near-alpha, alpha-beta alloy.

The original composition of this alloy contained no silicon, but RMI introduced a nominal 0.08% silicon content which allowed the alloy to meet the creep requirements for its intended jet-engine applications. Before any major commercial applications were developed, all producers had added silicon to the original Ti-6242 composition.

Density. 4.54 g/cm^3 (0.164 lb/in.^{3})

Product Forms

Available mill forms include billet, bar, plate, sheet, strip, and extrusions. Cast Ti-6242S products constitute about 7% of cast titanium products. Some forming operations can be carried out at room temperature, and warm forming (425 to 705 °C, or 800 to 1300 °F) is employed when necessary. Ti-6242S has fair weldability. The molten weld metal and adjacent heated zones must be shielded from active gases (nitrogen, oxygen, and hydrogen).

Product Condition/Microstructure

Ti-6242S is sometimes described as a near-alpha or superalpha alloy, but in its normal heat treated condition this alloy has a structure better described as alpha-beta. Proper treatment is needed to develop good creep resistance. Limited hardening of Ti-6242S can be done by solution treating and aging.

Applications

Ti-6242S is used primarily for gas turbine components such as compressor blades, disks, and impellers, and also in sheet-metal form for engine afterburner structures and for various "hot" airframe skin applications, where high strength and toughness, excellent creep resistance, and stress stability at temperatures up to 565 °C (1050 °F) are required.

Ti-6Al-2Sn-4Zr-2Mo-0.08Si: Specifications and compositions

Specification	Designation	Description	Al	Fe	H	Mo	N	O	Sn	Zr	Other
UNS	R54620		6			2			2	4	bal Ti
UNS	R54621	Weld Fill Met	5.5-6.5	0.05	0.015	1.8-2.2	0.15	0.3	1.8-2.2	3.6-4.4	C 0.04; Cr 0.25; bal Ti
Germany											
	WL 3.7144		5.5-6.5	0.25	0.015	1.8-2.2	0.05	0.15	1.8-2.2	3.6-4.4	C 0.05; bal Ti
Spain											
UNE 38-718	L-7103	Sh Strp Plt Ann	5.5-6.5	0.25	0.015	1.8-2.2	0.05	0.12	1.8-2.2	3.6-4.4	C 0.05; OT 0.4; bal Ti
UNE 38-718	L-7103	Sh Strp Plt HT	5.5-6.5	0.25	0.015	1.8-2.2	0.05	0.12	1.8-2.2	3.6-4.4	C 0.05; OT 0.4; bal Ti
USA											
AMS 4919C		Sh Strp Plt	5.5-6.5	0.25 max	0.015 max	1.8-2.2	0.05 max	0.12 max	1.8-2.2	3.6-4.4	C 0.05 max; Si 0.06-0.1; Y 0.005 max; OE 0.1 max; OT 0.3 max; bal Ti
AMS 4919G		Sh Strp Plt DA	5.5-6.5	0.25	0.015	1.8-2.2	0.05	0.12	1.8-2.2	3.6-4.4	C 0.05; Si 0.1; Y 0.005; OT 0.3; bal Ti
AMS 4975E		Bar Wir Rng Bil STA	5.5-6.5	0.25	0.0125	1.8-2.2	0.05	0.15	1.8-2.2	3.6-4.4	C 0.05; Y 0.005; OT 0.3; Si 0.1; bal Ti
AMS 4975F		Bar Rng HT	5.5-6.5	0.1 max	0.0125 max	1.8-2.2	0.05 max	0.15 max	1.8-2.2	3.6-4.4	C 0.05 max; Si 0.06-0.1; Y 0.005 max; OE 0.1 max; OT 0.3 max; bal Ti
AMS 4976C		Frg STA	5.5-6.5	0.25	0.0125	1.8-2.2	0.05	0.15	1.8-2.2	3.6-4.4	C 0.05; Y 0.005; OT 0.3; Si 0.1; bal Ti

(continued)

Ti-6Al-2Sn-4Zr-2Mo-0.08Si: Specifications and compositions (continued)

Specification	Designation	Description	Composition, wt%									Other
			Al	Fe	H	Mo	N	O	Sn	Zr	
USA (continued)											
AMS 4976D		Frg HT	5.5-6.5	0.1 max	0.0125 max	1.8-2.2	0.05 max	0.15 max	1.8-2.2	3.6-4.4	C 0.05 max; Si 0.06-0.1; Y 0.005 max; OE 0.1 max; OT 0.3 max; bal Ti
MIL T-81556A	Code AB-4	Ext Bar Shp Ann	5.5-6.5	0.25	0.015	1.8-2.2	0.04	0.15	1.8-2.2	3.6-4.4	C 0.05; Y 0.005; Si 0.06-0.1; OT 0.3; bal Ti
MIL T-81556A	Code AB-4	Ext Bar Shp STA	5.5-6.5	0.25	0.015	1.8-2.2	0.04	0.15	1.8-2.2	3.6-4.4	C 0.05; Si 0.06-0.1; Y 0.005; OT 0.3; bal Ti
MIL T-81915	Type III Comp B	Cast Ann	5.5-6.5	0.35	0.015	1.5-2.5	0.05	0.12	1.5-2.5	3.6-4.4	C 0.08; OT 0.4; bal Ti
MIL T-9046J	Code AB-4	Sh Strp Plt DA	5.5-6.5	0.25	0.015	1.8-2.2	0.04	0.15	1.8-2.2	3.6-4.4	C 0.05; OT 0.3; bal Ti
MIL T-9046J	Code AB-4	Sh Strp Plt TA	5.5-6.5	0.25	0.015	1.8-2.2	0.04	0.15	1.8-2.2	3.6-4.4	C 0.05; OT 0.3; bal Ti
MIL T-9047G	Ti-6Al-2Sn-4Zr-2Mo	Bar Bil DA	5.5-6.5	0.25	0.015	1.8-2.2	0.04	0.15	1.8-2.2	3.6-4.4	C 0.05; Y 0.005; OT 0.3; bal Ti
MIL T-9047G	Ti-6Al-2Sn-4Zr-2Mo	Bar Bil STA	5.5-6.5	0.25	0.015	1.8-2.2	0.04	0.15	1.8-2.2	3.6-4.4	C 0.05; Y 0.005; OT 0.3; bal Ti

Ti-6Al-2Sn-4Zr-2Mo-0.08Si: Compositions

Specification	Designation	Description	Composition, wt%									Other
			Al	Fe	H	Mo	N	O	Sn	Zr	
France											
Ugine	UT6242	Bar Frg Ann	5.5-6.5			1.8-2.2			1.8-2.2	3.6-4.4	
Germany											
Deutsche T	Contimet AlSnZrMo 6-2-4-2	Plt Bar Frg Ann	5.5-6.5	0.25	0.015	1.8-2.2	0.05	0.15	1.8-2.2	3.6-4.4	C 0.05; Si 0.06-0.12; bal Ti
Deutsche T	Contimet AlSnZrMo 6-2-4-2	Plt Bar Frg STA	5.5-6.5	0.25	0.015	1.8-2.2	0.05	0.15	1.8-2.2	3.6-4.4	C 0.05; Si 0.06-0.12; bal Ti
Deutsche T	LT 24	Aged	5.5-6.5	0.25	0.015	1.8-2.2	0.05	0.12	1.8-2.2	3.6-4.4	C 0.05; bal Ti
Fuchs	TL62	Frg	6			2			2	4	bal Ti
Japan											
Kobe	KS6-2-4-2	Bar Frg STA	5.5-6.5	0.25	0.015	1.8-2.2	0.05	0.15	1.8-2.2	3.6-4.4	bal Ti
USA											
OREMET	Ti-6242										
RMI	RMI 6Al-2Sn-4Zr-2Mo-0.10Si	Bar Bil Plt Sh STA	5.5-6.5	0.25	0.01-0.0125	1.75-2.25	0.05	0.12	1.75-2.25	3.5-4.5	C 0.08; Si 0.1; bal Ti
TIMET	TIMETAL 6-2-4-2	Bar Bil Plt Sh STA	5.5-6.5	0.25	0.01-0.0125	1.75-2.25	0.05	0.12	1.75-2.25	3.5-4.5	C 0.08; Si 0.1; bal Ti

Phases and Structures

The structures of Ti-6Al-2Sn-4Zr-2Mo alloy are typically equiaxed α in a transformed β matrix, or a fully transformed structure that maximizes creep resistance. The equiaxed α grains found in sheet products tend to be smaller than those found in forgings, as with other alloys, and are present in greater proportion than in forgings. Primary α is typically about 80 to 90% of the structure in sheet products and can be somewhat lower than this in forged products, because the final forging temperature is normally higher than the final rolling temperature used for sheet. As in other near-α alloys, small amounts of residual β phase can be observed metallographically within the transformed β portion of the structure, typically between the acicular α grains of the transformed phase. Breakup of lamellar α into equiaxed α occurs during working (see figure).

Beta Transus. 995 ± 15 °C (1825 ± 25 °F)

Ti-6242: Effect of silicon on beta transus temperature

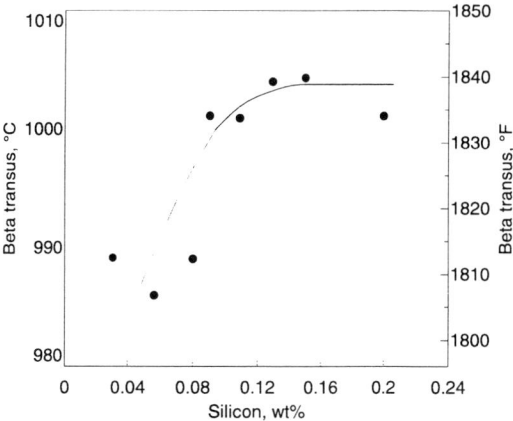

Source: *Aerospace Structural Metals Handbook*, Code 3718, Vol 4, Battelle Columbus Laboratories, June 1978

Ti-6242S: Alpha content vs solution temperature

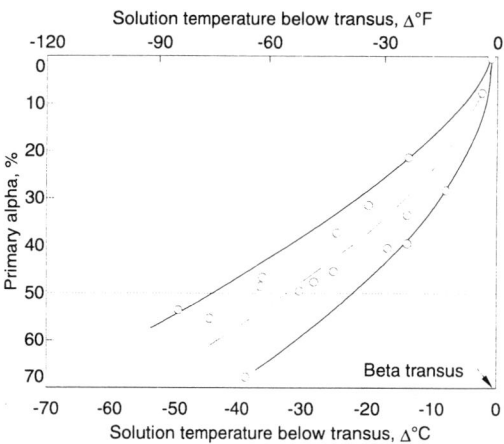

Solution treated, 1 h, AC + 595 °C (1100 °F), 8 h, AC.
Source: *Aerospace Structural Metals Handbook*, Code 3718, Vol 4, Battelle Columbus Laboratories, June 1978

Transformation Kinetics

Ti-6242: Time-temperature-transformation diagram

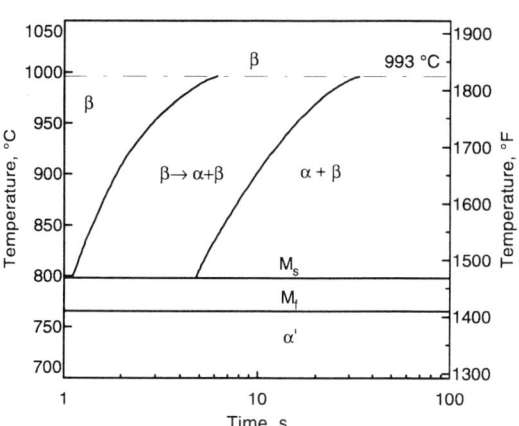

Source: H.W. Rosenberg, M.B. Vordahl, and D.B. Hunter, Project No. 48-8, Tech. Rep. No. 17, Timet, Jan 1966

Ti-6242: Breakup of lamellar α by shear strain

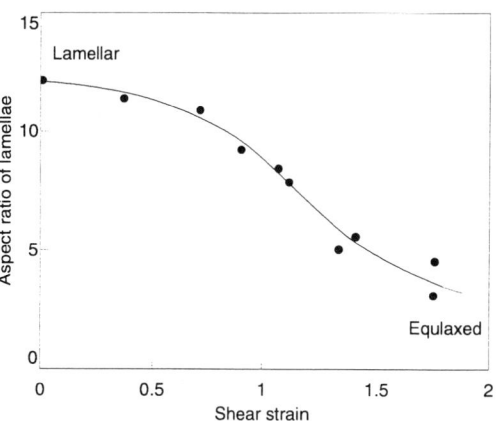

Lamellar breakup was done by torsional deformation at 925 °C (1700 °F), and the transformation from lamellar to equiaxed grains was evaluated by optical and electron microscopy. The average length-to-width aspect ratio of groups of 100 random lamella segments or equiaxed grains was calculated.
Source: G. Welsch, I. Weiss, D. Eylon, and F.H. Froes, Shear Deformation and Breakup of Lamellar Morphology in Ti-6Al-2Sn-4Zr-2Mo, in *Sixth World Conference on Titanium*, P. Lacombe, R. Tricot, and G. Beranger, Ed., Les Editions de Physique, Paris, 1989, p 1289-1293

Ti-6242: Comparison of CCT diagram and isothermal diagram

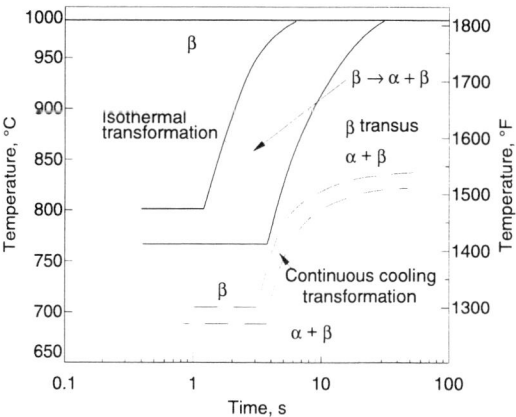

Source: D.R. Mitchell, Welding Evaluation of Ti-6Al-2Sn-4Zr-2Mo Sheet, TMCA Project BW-10-1, Final Report, June 1968, reported in *Aerospace Structural Metals Handbook*, Battelle Columbus Laboratories, June 1978

340 / Alpha and Near-Alpha Alloys

Ti-6242: Continuous cooling transformation diagram

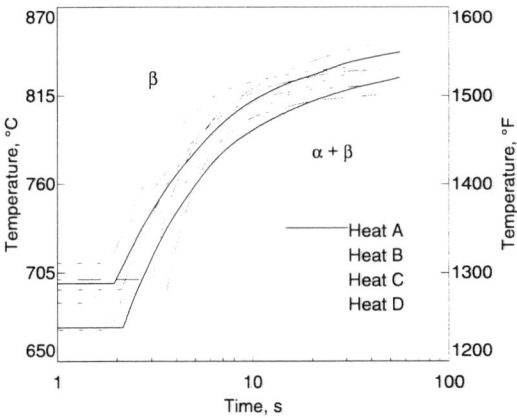

Effect of heat-to-heat variations, within specification, on continuous cooling transformation.
Source: D.R. Mitchell, Welding Evaluation of Ti-6Al-2Sn-4Zr-2Mo Sheet, TMCA Project BW-10-1, Final Report, June 1968, reported in *Aerospace Structural Metals Handbook*, Battelle Columbus Laboratories, June 1978

Physical Properties

Ti-6242: Summary of typical physical properties

Beta transus	995 ± 15 °C (1825 ± 25 °F)
Melting (liquidus) point	~1705 °C (3100 °F)
Density(a)	4.54 g/cm^3 (0.164 lb/in.^{3})
Electrical resistivity(a)	1.85 to 1.9 µΩ · m
Magnetic permeability	Nonmagnetic
Specific heat capacity(a)	460 J/kg · K (0.11 Btu/lb · °F)
Thermal conductivity(a)	7 W/m · K (4 Btu/ft · h · °F)
Thermal coefficient of linear expansion(b)	7.7×10^{-6}/°C (4.3×10^{-6}/°F)

(a) Typical values at room temperature of about 20 to 25 °C (68 to 78 °F). (b) Mean coefficient from 0 to 100 °C (32 to 212 °F)

Elastic Properties (Static)

Ti-6242: As-cast compressive modulus

Specimen No.	Tensile yield strength (0.2% offset)		Compressive modulus	
	MPa	ksi	GPa	10^6 psi
At room temperature				
1	993	144.3	119	17.3
2	898	130.3	118	17.1
3	910	132.1	116	16.9
At 200 °C (400 °F)				
4	675	98.0	107	15.5
5	615	89.3	114	16.4
6	670	97.2	112	16.3
At 425 °C (800 °F)				
7	546	79.2	103	14.9
8	520	75.4	101	14.7
9	535	77.6	96	13.9

Ti-6242: Effect of temperature on moduli

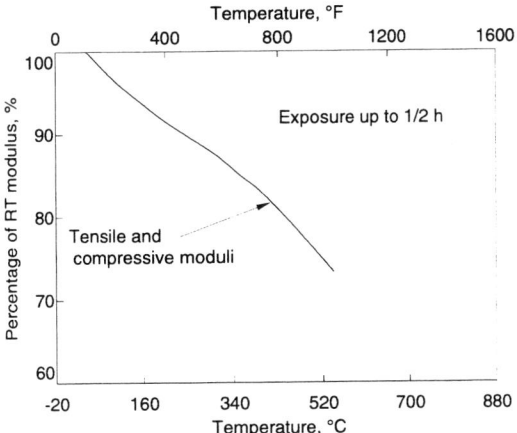

All product forms, duplex and triplex annealed.
Source: MIL-HDBK-5

Ti-6242: Cast compressive elastic modulus

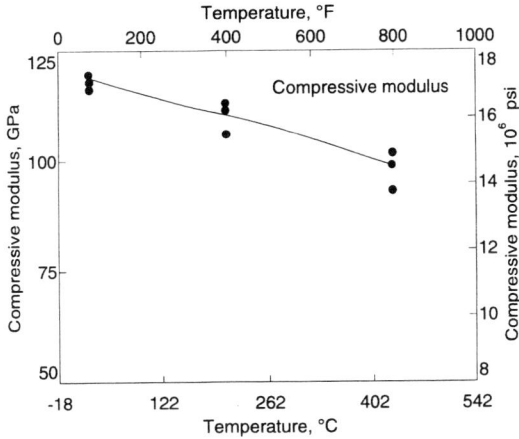

Alloy was supplied as cast wedges (tapered plate) 150 by 165 mm (6 by 6.5 in.). Chemical composition was 6.02 wt% Al, 0.018 wt% C, 0.010 wt% Fe, 0.0047 wt% H, 2.07 wt% Mo, 0.013 wt% N, 0.168 wt% O, 0.05 wt% Si, 2.04 wt% Sn, and 3.80 wt% Zr. Material was tested in the as-received, as-cast condition.
Source: O. Deel, "Engineering Data on New Aerospace Structural Materials," AFML-TR-77-190, Battelle Columbus Laboratories, 1977, p 29

Ti-6242: Static compressive modulus

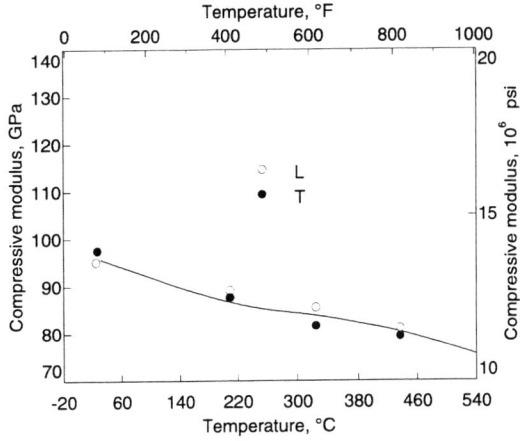

Specimens were 1 mm (0.040 in.) sheet duplex annealed at 900 °C (1650 °F), 30 min, AC + 790 °C (1450 °F), 15 min, AC. Each data point is an average of five tests.
Source: *Aerospace Structural Metals Handbook*, Code 3718, Vol 4, Battelle Columbus Laboratories, 1978, p 85

Ti-6242: Static Young's modulus (tensile)

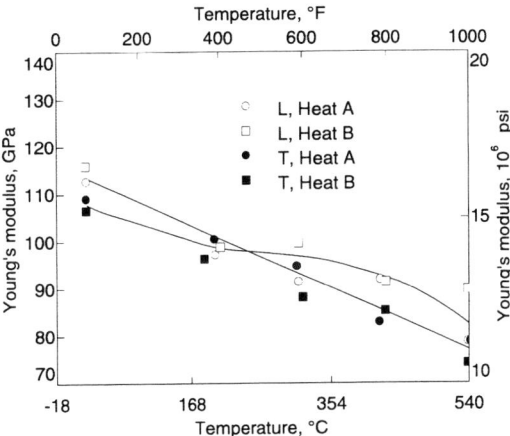

Specimens were 1 mm (0.040 in.) sheet, 30 min, duplex annealed 900 °C (1650 °F), ½ h, AC + 790 °C (1450 °F), 15 min, AC. Each point is an average of ten tests.
Source: *Aerospace Structural Metals Handbook*, Vol 4, Code 3718, Battelle Columbus Laboratories, 1978, p 85

Ti-6242: Poisson's ratio

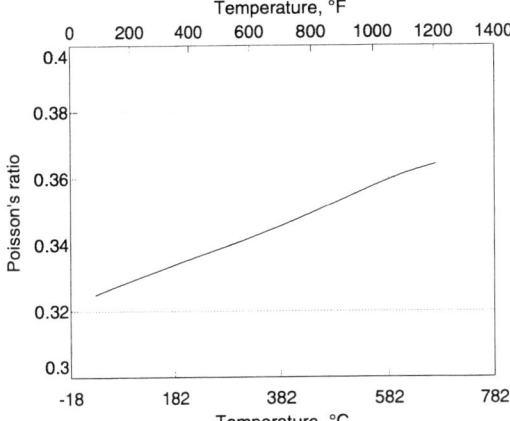

Specimens were pancake forgings, duplex annealed at 955 °C (1750 °F), 1 h, AC + 595 °C (1100 °F), 8 h, AC.
Source: *Aerospace Structural Metals Handbook*, Vol 4, Code 3718, Battelle Columbus Laboratories, 1978, p 85

Dynamic Modulus

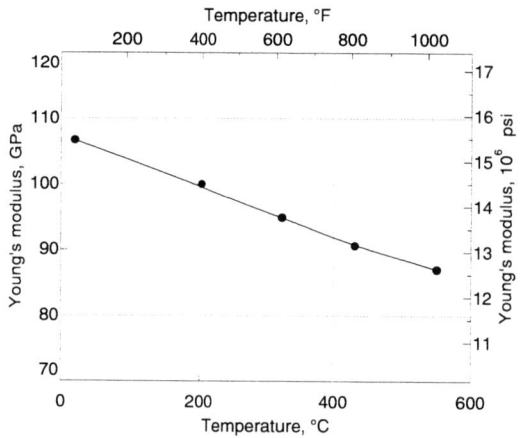

Ti-6242: Dynamic Young's modulus (tensile)

Specimens were 1 mm (0.040 in.) sheet duplex annealed at 900 °C (1650 °C), 30 min, AC + 790 °C (1450 °F), 15 min, AC. Average of longitudinal and transverse data.
Source: *Aerospace Structural Metals Handbook*, Vol 4, Code 3718, Battelle Columbus Laboratories, 1978, p 85

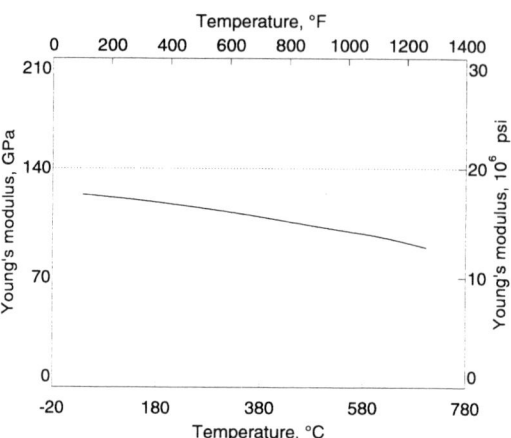

Ti-6242: Dynamic Young's modulus (tensile)

Specimens were duplex annealed pancake forgings, 955 °C (1750 °F), 1 h, AC + 595 °C (1100 °F), 8 h, AC.
Source: *Aerospace Structural Metals Handbook*, Vol 4, Code 3718, Battelle Columbus Laboratories, 1978, p 85

Electrical Resistivity

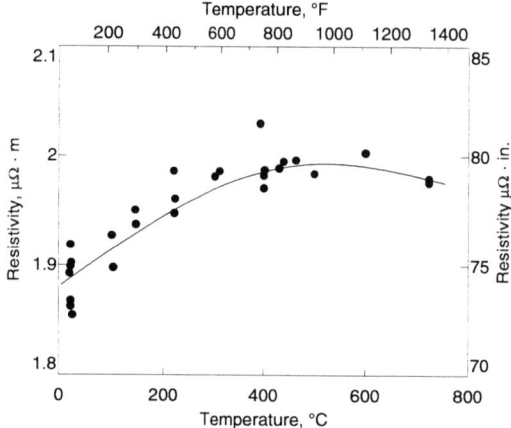

Ti-6242: Resistivity of duplex-annealed sheet

Specimens were 0.95 mm (0.038 in.) sheet, two heats, two specimens each heat, duplex annealed at 900 °C (1650 °F), 30 min, AC + 785 °C (1450 °F), 15 min, AC. Measurements were made parallel to the rolling direction.
Source: *Aerospace Structural Metals Handbook*, Vol 4, Code 3718, Battelle Columbus Laboratories, 1978, p 38

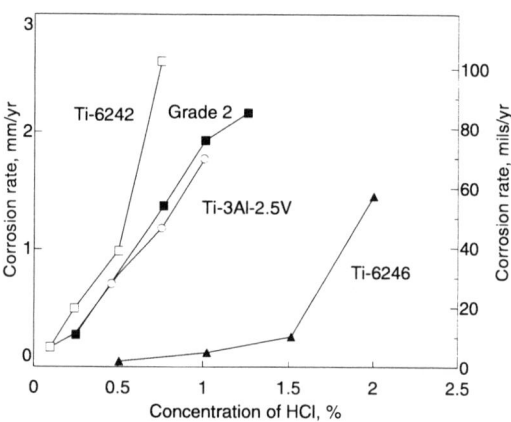

Ti-6242: Comparative corrosion in boiling HCl

The higher molybdenum content of Ti-6246 increases resistance in reducing environments compared to Ti-6242. General corrosion of annealed titanium alloys in naturally aerated HCl solutions.
Source: R. Schutz, "Corrosion of Titanium and Titanium Alloys," in *Metals Handbook, Corrosion*, Vol 13, 9th ed., 1987, p 680

General Corrosion

Resistance in Reducing Environments. The corrosion resistance of Ti-6242 in various media is not well documented, although it is probably comparable to other Ti-6Al-base alloys such as Ti-6Al-4V. The molybdenum content of Ti-6242 is not high enough to impart additional corrosion resistance in reducing environments (see figure above). The crevice corrosion resistance of Ti-6242 is probably less than grade 2 titanium, because crevice corrosion generally is associated with acidification from oxidant depletion.

Oxidation. A blue oxide film typically forms in about 6 to 10 h in exposures not exceeding 540 °C (1000 °F). Degradation of mechanical properties from oxidation at longer times and usual service temperatures has not been observed. In a strong oxidizing environment, resistance is probably comparable to grade 2 titanium or Ti-6Al-4V.

Stress-Corrosion Cracking

Under stress, Ti-6Al-2Sn-4Zr-2Mo has been shown to be subject to stress-corrosion cracking (SCC)—at room temperature in the presence of aqueous chloride solution and a preexisting crack (the so-called accelerated crack-growth type of salt-stress corrosion) and at elevated temperatures in the presence of a halogen salt (e.g., NaCl). The SCC susceptibility of Ti-6242 in hot salt appears to be less than that of Ti-8Al-1Mo-1V and Ti-6Al-4V (see figure). In ambient salt solution, the SCC threshold of Ti-6242 in the STA condition is comparable to mill annealed (MA) Ti-8Al-1Mo-1V (see table).

A substantial amount of laboratory testing on the hot-salt SCC behavior of Ti-6242 has been performed, but like other susceptible alloys (such as Ti-8Al-1Mo-1V), no failure in service has been attributed to hot-salt cracking. The likely reason for this is the critical relationship between stress and environment, which needs to peak simultaneously for extended periods of time for cracking to occur. Actual engine environmental conditions are unique and considered less conducive to stress corrosion than laboratory exposure conditions. Possible ameliorating engine conditions are high air velocities, high pressures, salt-air conditions, oil contamination, and/or unique operating cycles.

Triplex annealing appears to improve resistance to cracking slightly versus a duplex anneal.

Ambient Salt Solution Cracking

Ti-6242: Comparative toughness in air and stress-corrosion threshold in 3.5% NaCl solution at 25 °C

Alloy	Thickness mm	Thickness in.	Heat treatment	Tensile yield strength MPa	Tensile yield strength ksi	K_{Ic} or K_c MPa√m	K_{Ic} or K_c ksi√in.	K_{Iscc} or K_{scc} MPa√m	K_{Iscc} or K_{scc} ksi√in.
Ti-6Al-2Sn-4Zr-2Mo	13	0.50	STA	1048	152	58	53	29	27
Ti-6Al-2Sn-4Zr-6Mo	13	0.50	MA	1103	160	60	55	22	20
			DA	1034	150	88	80	49	45
	7	0.30	MA	965	140	57	52	28	26
			β STA	1172	170	89	81	49	45
Ti-8Al-1Mo-1V	1.3	0.05	MA	999	145	82	75	33	30
	1.3	0.05	DA	930	135	176	160	55	50
	13	0.5	MA	999	145	52	48	22	20
	13	0.5	DA	930	135	110	100	35	32
	13	0.5	MA, WQ	841	122	>110	>100	46	42
	13	0.5	βST, WQ	868	126	>110	>100	>110	>100
Ti-6Al-4V (standard grade)	13	0.50	MA	944	137	66	60	38	35
			DA	917	133	77	70	57	52
			STA	1103	160	51	47	27	25
			β STA	1068	155	77	70	49	45

Note: The data were generated in ambient neutral 3.5% NaCl solution. It should be cautioned that these K_{Iscc} values are highly dependent on alloy composition, metallurgical condition, and product form and thickness. Therefore, they may or may not be representative of alloy product materials commercially available. Source: R. Schutz, Stress-Corrosion Cracking of Titanium Alloys, in *Stress-Corrosion Cracking: Materials Performance and Evaluation*, ASM International, 1992

Ti-6242: Effect of heat treatment on saltwater SCC of sheet (precracked)

Heat treatment(a)	Applied net stress MPa	Applied net stress ksi	Time to rupture in 3.5% NaCl
730 °C (1350 °F), 8 h, FC	617	89.5	Broke on loading
	466	67.7	1 min
	308	44.7	7 min
	551	80	>73 h(a)
900 °C (1650 °F), 30 min, AC + 785 °C (1450 °F), 15 min, AC	586	85	>42 h(a)
	689	100	>43 h(a)
	758	110	>68 h(a)
	551	80	>48 h(a)
900 °C (1650 °F), 30 min, AC + 785 °C (1450 °F), 15 min, AC + 595 °C (1100 °F), 8 h, AC	586	85	>42 h(a)
	620	90	>24 h(a)
	689	100	>2 h(a)

(a) No failure, test discontinued. Source: R. Wood and R. Favor, *Titanium Alloys Handbook*, MCIC-HDBK-02, Battelle, 1972

Ti-6242: Comparative critical stress in 425 °C salt

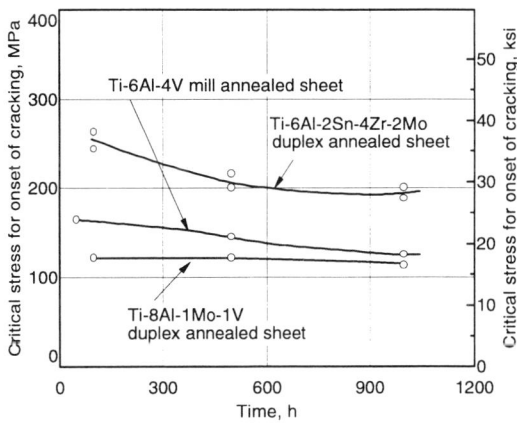

Crack-no crack limit stress.
Source: R. Wood and R. Favor, *Titanium Alloys Handbook*, MCIC-HDBK-02, Battelle, 1972

Ti-6242: Precracked tensile strength in air and in 3.5% NaCl
Center fatigue cracked samples

Sheet gage			Air (continuously loaded)		Notch tensile strength in: Saltwater (sustained load)			
					Failure in 1 to 2 h		Failure in <1 h	
mm	in.	Direction	MPa	ksi	MPa	ksi	MPa	ksi
Duplex anneal(a)								
11.4	0.45	T	...	...	758(c)	110(c)	(d)	(d)
2.0	0.080	L	875	127	696	101	(d)	(d)
2.0	0.080	T	827	120	696	101	744	108
3.1	0.125	L	889	129	765	111	(d)	(d)
3.1	0.125	T	827	120	696	101	744	108
Triplex anneal(b)								
0.4	0.015	T	...	...	689	100	(d)	(d)
2.0	0.080	L	882	128	...	...	434(e)	63(e)
2.0	0.080	T	855	124	600	87	(d)	(d)
3.1	0.125	L	827	120	496	72	551	80
3.1	0.125	T	744	108	...	...	365	53

(a) 900 °C (1650 °C), 30 min, AC + 785 °C (1450 °F), 15 min, AC. (b) 900 °C (1650 °F), 30 min, AC + 785 °C (1450 °F), 15 min, AC + 595 °C (1100 °F), 2 h, AC. (c) No failure in 68 h. (d) Failure stress not established. (e) Failed in 5 min. Source: R. Wood and R. Favor, *Titanium Alloys Handbook*, MCIC-HDBK-02, Battelle, 1972

Hot-Salt SCC Factors

Hot-salt SCC behavior of Ti-6242 in laboratory tests is influenced by several factors. Oxygen is necessary for hot-salt cracking to occur. At least one study has shown that cracking will not occur in the highly susceptible Ti-5Al-2.5Sn when the environmental pressure is reduced below 10 μm. Although the role of water (moisture) has not been clearly established, it appears that water is also a necessary environmental component in the cracking process (*Metals Handbook, Corrosion*, Vol 13, 9th ed., 1987, p 689).

Air velocity appears to have no effect on Ti-6242 cracking (see figure) and only a minor effect on Ti-8Al-1Mo-1V cracking (H.R. Gray and J.R. Johnston, "Hot-Salt Stress-Corrosion of a Titanium Alloy Under Simulated Turbine-Engine Compressor Environment," NASA Technical Note, TND-5510, Oct 1969).

Effect of Processing. There appears to be little difference between mill annealing and duplex annealing of Ti-6242 bar (see figure). Heavy forging can reduce the embrittlement and cracking thresholds. The cracking thresholds reported for sheet at 100-h exposure are generally lower than those reported for bar.

Ti-6242: 100-h crack threshold of NaCl-coated bar

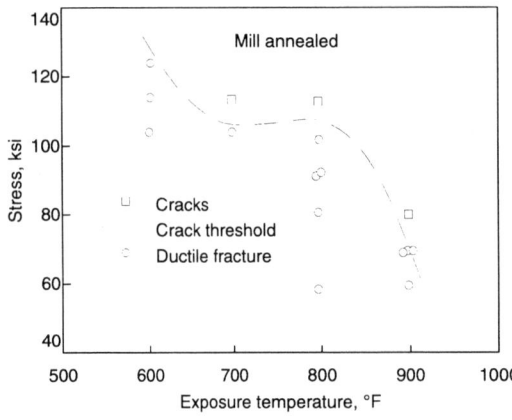
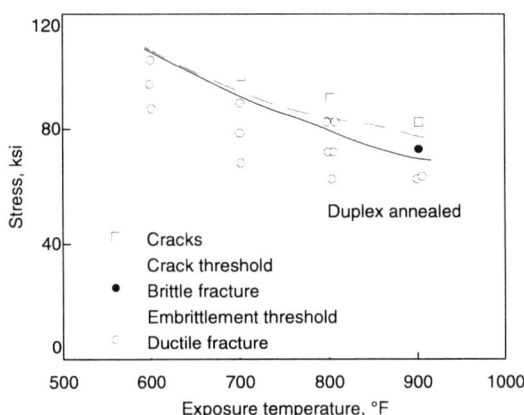

Little or no effect is noted between (a) mill annealed bar and (b) duplex annealed bar.
5 100-h embrittlement and crack threshold curves for NaCl-coated (aerosol deposit on hot surface) on machined tubular specimens exposed to elevated temperature with load in simulated aircraft turbine engine compressor environment and subsequently tensile tested at room temperature. Source: H.R. Gray, NASA TN D-6498, Nov 1971

Ti-6242: 100-h crack thresholds for NaCl-coated bar

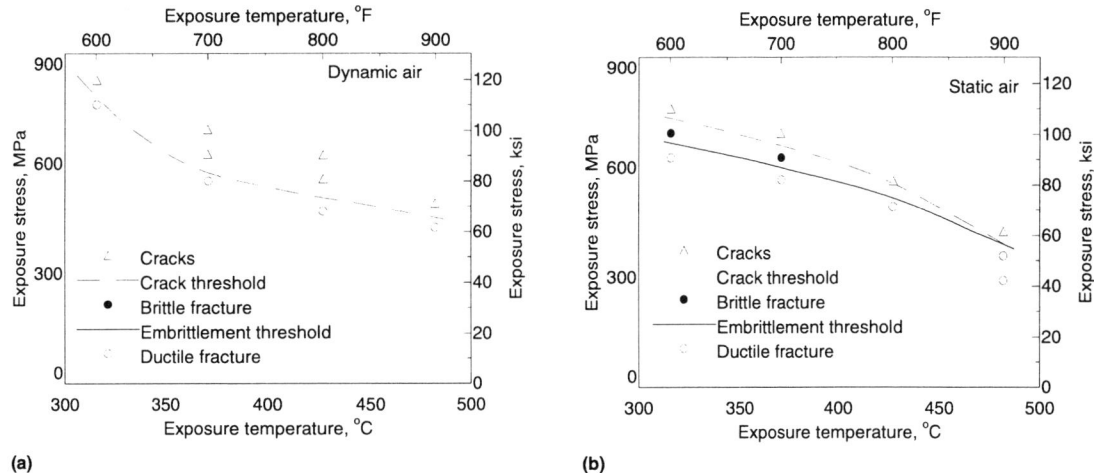

Little or no effect is noted between (a) dynamic air and (b) static air environment.
100-h embrittlement and crack threshold curves for NaCl-coated (aerosol deposited on chemically milled tubular specimens) (from mill annealed bar) exposed to elevated temperature with load in either a laboratory or simulated aircraft turbine engine compressor environment and subsequently tensile tested at room temperature.
Source: H.R. Gray, "Relative Susceptibility of Titanium Alloys to Hot-Salt Stress-Corrosion," NASA TN D-6498, Nov 1971

Ti-6242: Crack threshold of heavily forged bar

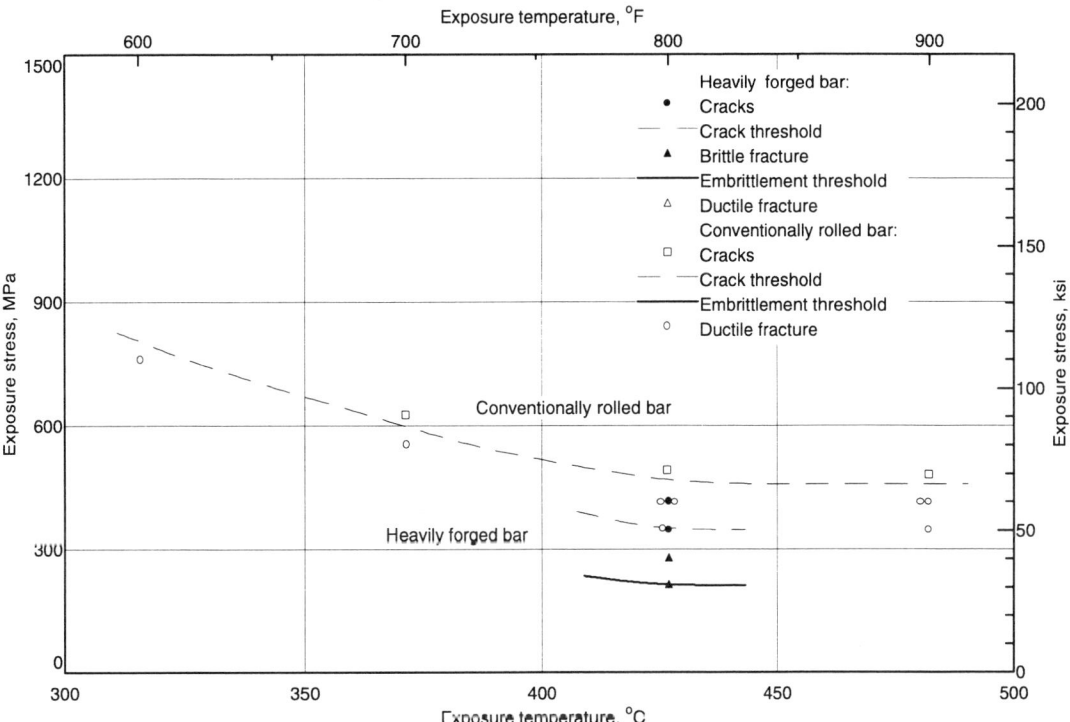

Effect of alloy processing on 100-h embrittlement and crack threshold curves for NaCl-coated (aerosol deposited on hot surface) chemically milled tubular specimens (from duplex annealed bar) exposed to elevated temperature with load in a laboratory environment and subsequently tensile tested at room temperature.
Source: *Aerospace Structural Metals Handbook*, Vol 4, Code 3718, June 1978, p 40

Ti-6242: 100-h hot-salt SCC

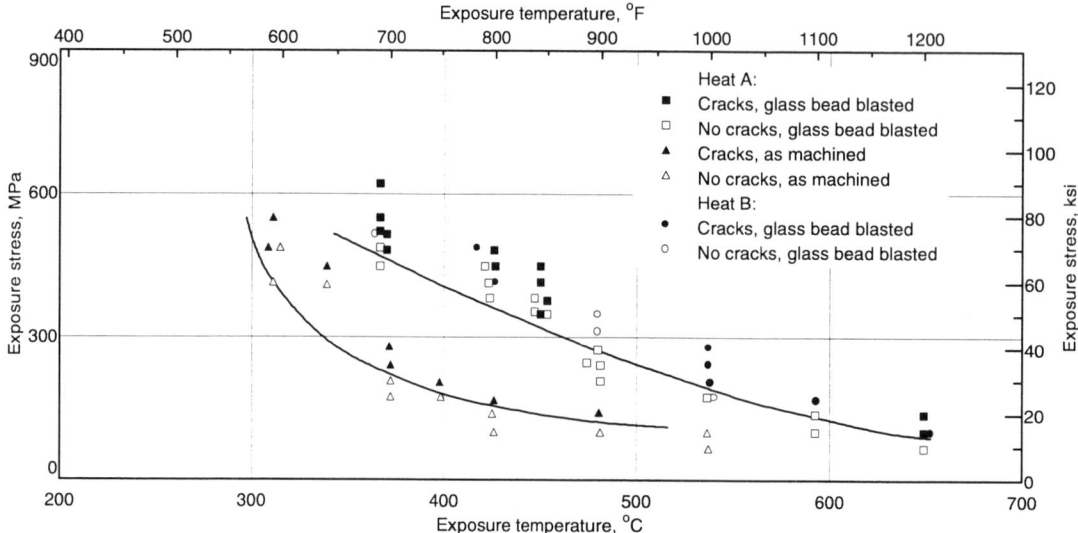

Hot-salt stress-corrosion cracking characteristics of duplex annealed compressor disc forging. Heat A: 975 °C (1790 °F), 1 h, AC + 595 °C (1100 °F), 8 h, AC; Heat B: 950 °C (1750 °F), 1 h, AC + 595 °C (1100 °F), 8 h, AC.
Source: *Aerospace Structural Metals Handbook*, Vol 4, Code 3718, June 1978, p 43

Thermal Properties

Melting Point. Approximately 1700 °C (3100 °F)

Thermal Coefficient of Linear Expansion. 7.7 µm/m · K (4.3 µin./in. · °F) at 205 °C (400 °F); 8.1 µm/m · K (4.5 µin./in. · °F) at 315 to 540 °C (600 to 1000 °F)

Specific Heat. 460 J/kg · K (0.110 Btu/lb · °F) at 100 °C (212 °F)

Thermal Conductivity. 7.1 W/m · K (4.1 Btu/ft · h · °F) at 100 °C (212 °F)

Specific Heat

Ti-6242: Specific heat vs temperature

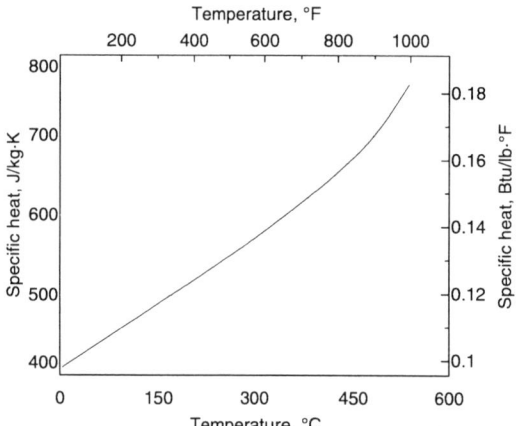

Source: MIL-HDBK-5E, Vol 2, June 1987, p 5.44

Ti-6242: Specific heat of duplex annealed bar

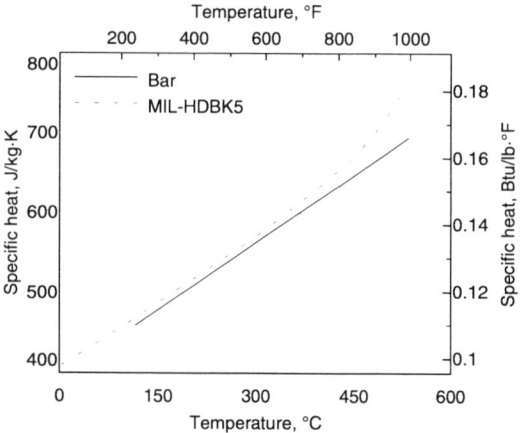

29 mm (1⅛ in.) diameter bar, annealed at 900 °C (1650 °F) for 1 h, air cooled, then at 595 °C (1100 °F) for 8 h, air cooled.
Source: H. Russell, "Physical Properties of Ti-6Al-2Sn-4Zr-2Mo," TMCA Case Study M-109, June 1967; reported in *Aerospace Structural Metals Handbook*, Vol 4, Code 3718, Battelle Columbus Laboratories, 1978, p 37

Ti-6242: Specific heat of duplex annealed sheet

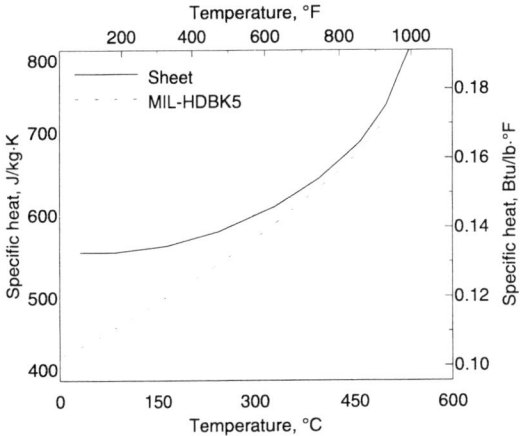

1.0 mm (0.040 in.) sheet, annealed at 900 °C (1650 °F) for 30 min, air cooled, then at 790 °C (1450 °F) for 15 min, air cooled.
Source: C. Dotson, "Mechanical and Thermal Properties of High-Temperature Titanium Alloys," AFML-TR-67-41, Apr 1967; reported in *Aerospace Structural Metals Handbook*, Vol 4, Code 3718, Battelle Columbus Laboratories, June 1978

Thermal Expansion

Ti-6242: Thermal coefficient of linear expansion of sheet

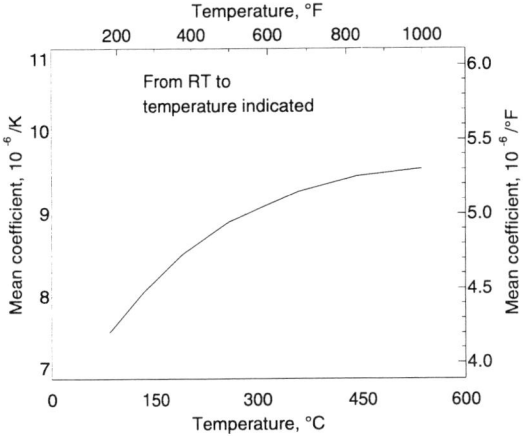

Alloy used as 1.0 mm (0.040 in.) thick sheet, heat treated at 900 °C (1650 °F) for 30 min, then air cooled, and heated at 790 °C (1450 °F) for 15 min, and air cooled. Values are averages of longitudinal and transverse directions.
Source: C. Dotson, "Mechanical and Thermal Properties of High-Temperature Titanium Alloys, AFML-TR-67-41, Apr 1967

Ti-6242: Thermal coefficient of linear expansion

Temperature		Coefficient of thermal expansion	
°C	°F	10^{-6}/K	10^{-6}/°F
0-100	32-212	7.7	4.3
0-315	32-600	8.1	4.5
0-540	32-1000	8.1	4.5

Source: RMI Titanium Basic Design

Ti-6242: Thermal coefficient of linear expansion of bar

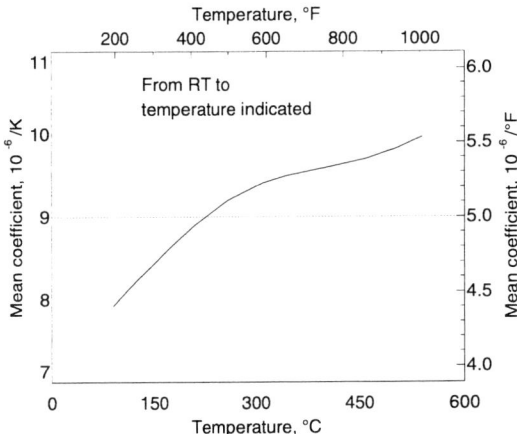

Alloy used as 29 mm (1⅛ in.) diameter bar, heat treated at 900 °C (1650 °F) for 1 h and air cooled, then at 595 °C (1100 °F) for 8 h, and air cooled. Measurements were taken in the longitudinal direction; values are averages of ascending and descending temperature values. Source: H. Russell, "Physical Properties of Ti-6Al-2Sn-4Zr-2Mo," TMCA Case Study M-109, June 1967

Ti-6242: Thermal coefficient of linear expansion of various product forms

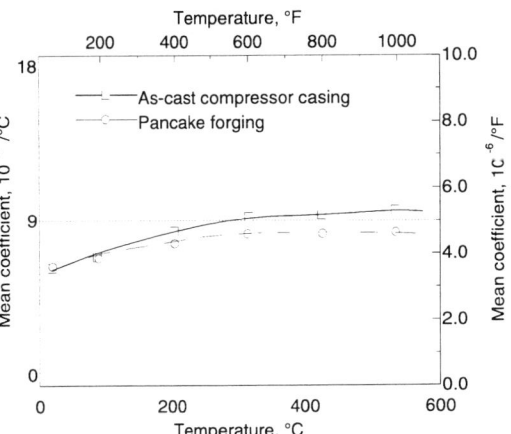

Specimens were an as-cast compressor casing and pancake forgings heat treated at 955 °C (1750 °F) for 1 h and air cooled, then at 595 °C (1100 °F) for 8 h and air cooled.
Source: *Aerospace Structural Metals Handbook*, Vol 4, Code 3718, Battelle Columbus Laboratories, 1978, p 37

Thermal Conductivity

Ti-6242: Thermal conductivity

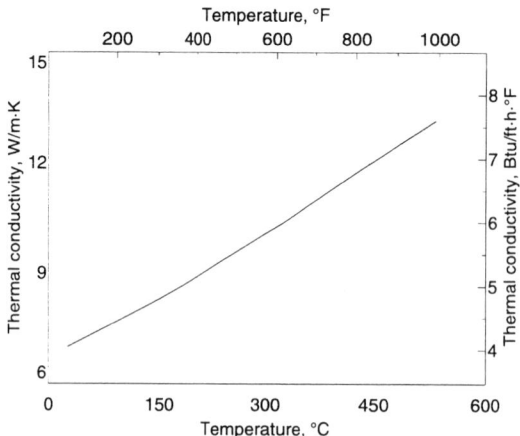

Source: MIL-HDBK-5E, Vol 2, June 1987, p 5.44

Ti-6242: Thermal conductivity of duplex annealed bar

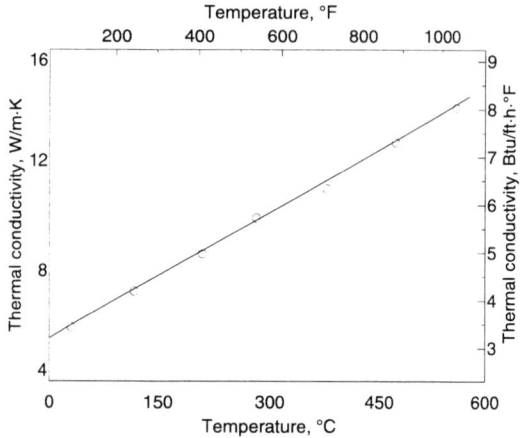

Measurements were taken parallel to the long direction of 54 mm (2 1/8 in.) diameter bar. Bar annealed at 900 °C (1650 °F) for 1 h, air cooled, then at 595 °C (1100 °F) for 8 h, air cooled.
Source: H. Russell, "Physical Properties of Ti-6Al-2Sn-4Zr-2Mo," TMCA Case Study M-109, June 1967; reported in *Aerospace Structural Metals Handbook*, Vol 4, Code 3718, Battelle Columbus Laboratories, 1978, p 36,37

Ti-6242: Thermal conductivity of duplex annealed sheet

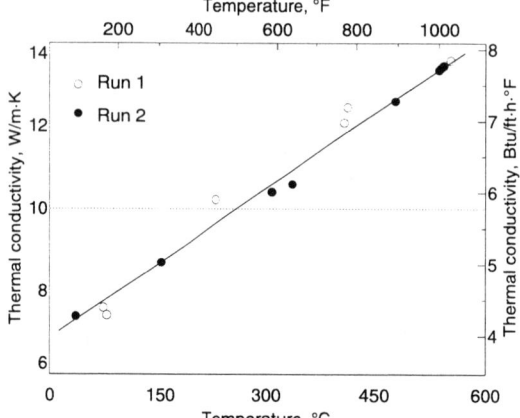

Measurements were taken parallel to the rolling direction. Alloy used as 1.0 mm (0.040 in.) sheet, annealed at 900 °C (1650 °F) for 30 min, air cooled, then at 790 °C (1450 °F) for 15 min, and air cooled.
Source: C. Dotson, "Mechanical and Thermal Properties of High-Temperature Titanium Alloys," AFML-TR-67-41, Apr 1967

Mechanical Properties

Design Allowables

Ti-6242: Design tensile and compressive strengths of duplex annealed sheet

Basis	Ultimate tensile strength (L-LT)(a) MPa	ksi	Tensile yield strength (L-LT)(a) MPa	ksi	Compressive yield strength (L-LT)(a) MPa	ksi
S-basis	958	139	889-903	129-131(b)	...	...
A-basis(c)	930	135	862	125	910	132
B-basis	986	143	924-937	134-136(b)	979	142

Note: Duplex annealed (DA) sheet thicknesses up to 4.75 mm (0.187 in.) per AMS 4919. (a) Longitudinal (L) and long transverse (LT) properties unless noted. (b) L and LT properties, respectively. (c) 8% elongation minimum for 0.635 to 1.57 mm (0.025 to 0.062 in.) thicknesses and 10% elongation for 1.57 to 4.75 mm (0.062 to 0.187 in.) thicknesses. Source: MIL-HDBK-5, Dec 1991

Ti-6242: Design bearing strengths of duplex annealed sheet

Thickness mm	in.	Basis	Ultimate bearing strength $e/D = 1.5$ MPa	ksi	$e/D = 2.0$ MPa	ksi	Bearing yield strength $e/D = 1.5$ MPa	ksi	$e/D = 2.0$ MPa	ksi
<1.17	<0.046	A	1344	195	1496	217	1179	171	1392	202
		B	1420	206	1586	230	1261	183	1496	217
1.19-2.36	0.047-0.093	A	1413	205	1675	243	1179	171	1392	202
		B	1496	217	1779	258	1261	183	1496	217
2.38-3.55	0.094-0.14	A	1475	214	1834	266	1179	171	1392	202
		B	1565	227	1944	282	1261	183	1496	217
3.58-4.75	0.141-0.187	A	1510	219	1923	279	1179	171	1392	202
		B	1599	232	2034	295	1261	183	1496	217

Note: Dry pin bearing values of duplex annealed (DA) sheet per AMS 4919. Source: MIL-HDBK-5, Dec 1991

Ti-6242: Design tensile properties of DA bar, forgings, and TA sheet

Product	Basis	Ultimate tensile strength MPa	ksi	Tensile yield strength MPa	ksi	Elongation(a), %	RA(a), %
Bar and forgings ≤75 mm (3 in.) thick							
STA (duplex annealed) bar(b) per AMS 4975	S	896(c)	130(b)	827(c)	120(c)	10(c)	25(c)
	A	951(d)	138(d)	862(d)	125(d)	...	...
	B	993(d)	144(d)	903(d)	131(d)	...	...
STA (duplex annealed) forgings(e) per AMS 4976	S	896	130	827	120	10(c)	25(c)
Triplex annealed (TA) sheet(f)							
Sheet ≤4.75 mm (0.187 in.) per MIL-T-9046	S	999(g)	145(g)	930(g)	135(g)	(g, h)	...

Note: Properties applicable in L, LT, and ST directions, except as noted. (a) S-basis. (b) Cross-sectional area less than 103 cm^2 (16 in.^{2}). (c) Applicable in LT and ST direction if transverse direction is greater than 63.5 mm (2.5 in.). (d) Longitudinal direction only. (e) Cross-sectional area less than 58 cm^2 (9 in.^{2}). (f) S-basis values are representative of test specimens excised from duplex annealed material and then treated to achieve a triplex annealed condition in the laboratory. (f) Specified only in L and LT directions. (g) 8% for 0.64 to 1.57 mm (0.025 to 0.062 in.) and 10% for thickness above 1.57 mm (0.062 in.). Source: MIL-HDBK-5, Dec 1991

Hardness

Typical room-temperature hardness is 32 to 36 HRC for duplex annealed Ti-6242.

Ti-6242: RT hardness vs high-temperature exposure

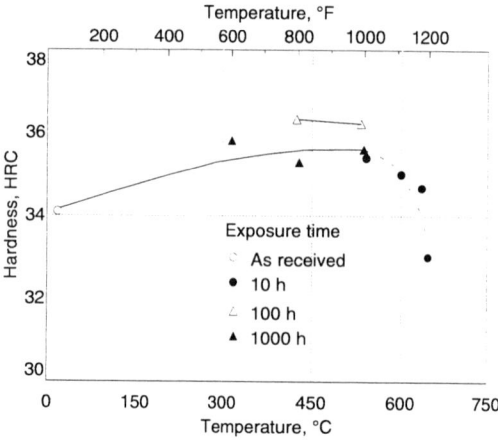

Duplex annealed sheet, 1.0 mm (0.040 in.) thick, heat treated for 30 min at 900 °C (1650 °F) and air cooled, then for 15 min at 790 °C (1450 °F), and air cooled.
Source: C. Dotson, "Mechanical and Thermal Properties of High-Temperature Titanium Alloys," AFML-TR-67-41, Apr 1967; reported in *Aerospace Structural Metals Handbook*, Vol 4, Code 3718, Battelle Columbus Laboratories, 1978, p 27

Ti-6242: Hardness of foil vs oxygen content

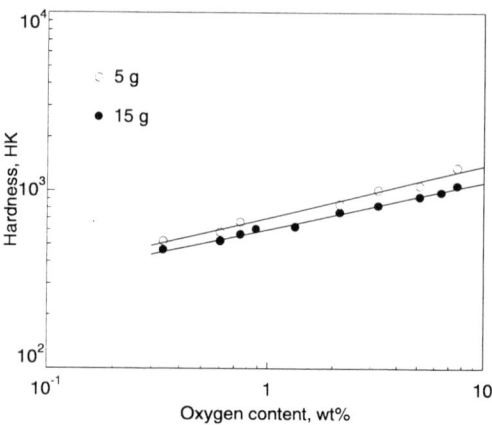

Specimens were foil, 80 μm (3.15 mils) thick. Composition was 5.80 wt% Al, 0.02 wt% C, 0.20 wt% Fe, 0.05 wt% Fe, 0.05 wt% H, 2.06 wt% Mo, 0.01 wt% N, 0.06 wt% Si, 1.87 wt% Sn, and 4.00 wt% Zr, <0.4 wt% others. Several foil specimens were oxidized in laboratory air at 855 °C (1570 °F) for various lengths of time and were subsequently homogenized by annealing for 120 h at 870 °C (1600 °F) in vacuum (7×10^{-7} torr). Hardness measurements were made using loads of 5 and 15 g. At least 15 Knoop indentations were made for each data point.
Source: R. Shenoy, J. Unnam, and R. Clark, "Oxidation and Embrittlement of Ti-6Al-2Sn-4Zr-2Mo Alloy," *Oxid. Met.*, Vol 26, 1986, p 105

Ti-6242: Hardness vs oxygen content

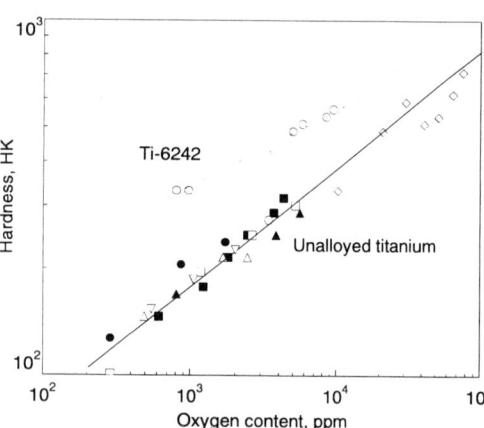

Bar stock, 28.58 mm (1.125 in.) in diameter. Composition: 6.4 wt% Al, 0.025 wt% C, 0.05 wt% Fe, 0.007 wt% H, 2.0 wt% Mo, 0.006 wt% N, 0.082 wt% O, 2.0 wt% Sn, and 4.2 wt% Zr. Disk specimens were machined to a diameter of 15 mm (0.60 in.) and a thickness of 3.2 mm (0.125 in.), cleaned with methanol, and oxidized for 5 to 1000 h at temperatures from 540 to 870 °C (1000 to 1600 °F). Oxidized specimens were metallographically prepared to present a chord surface with chord length approximately half the disk diameter. Hardness determinations were made with a Kentron microhardness tester using a Knoop indenter under a 1000-g load. Two to four traverses were made on each specimen.
Source: C. Shamblen and T. Redden, "Air Contamination and Embrittlement of Titanium Alloys," in *The Science, Technology and Application of Titanium*, R. Jaffee, Ed., Pergamon Press, 1970, p 199

Tensile Properties

Ti-6242: Effect of exposure on RT tensile strength

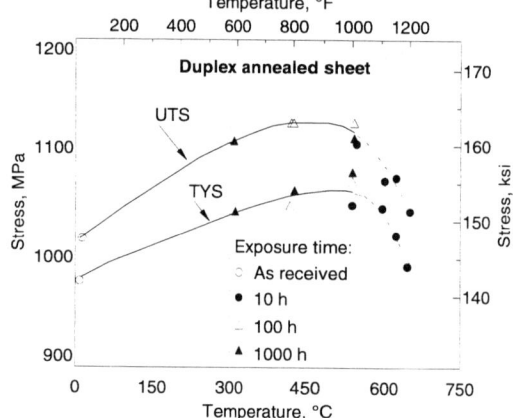

Specimens were 1 mm (0.040 in.) sheet duplex annealed at 900 °C (1650 °F), 30 min, AC + 785 °C (1450 °F), 15 min, AC. Longitudinal RT test results are reported. Exposure time as indicated.
Source: *Aerospace Structural Metals Handbook*, Vol 4, Code 3718, Battelle Columbus Laboratories, June 1978

Ti-6242: Typical room-temperature mechanical properties of sheet

Direction	Ultimate tensile strength		Tensile yield strength (0.2% offset)		Elongation, %	Compressive yield strength (0.2%)		Shear strength	
	MPa	ksi	MPa	ksi		MPa	ksi	MPa	ksi
Longitudinal	1010	146	990	144	13	1075	156.3	690	100
Transverse	1020	148	1010	146	12	1165	169	695	101

Source: *Metals Handbook*, Vol 3, 9th ed., American Society for Metals, 1987

Ti-6242: Effect of thermomechanical processing on tensile properties

Condition	Tensile yield strength		Ultimate tensile strength		Elongation, %	Reduction of area, %
	MPa	ksi	MPa	ksi		
Ti-6242						
α + β forged + STA(a)	1028	149	938	136	16	35
Ti-6242 + 0.25 wt% Si						
α + β forged + 965 °C (1770 °F), 1 h, AC + 595 °C (1100 °F), 8 h, AC	1007	146	1063	154	13	34
α + β forged + 965 °C (1770 °F), 1 h, OQ + 595 °C (1100 °F) 8 h, AC	1090	158	1187	172	12	20
β forged + 965 °C (1770 °F), 1 h, AC + 595 °C (1100 °F), 8 h, AC	952	138	1035	150	11	25
β forged + 965 °C (1770 °F), 1 h, AC + 595 °C (1100 °F), 8 h, AC	1042	151	1145	166	7	15

(a) STA = 985 °C (1810 °F) (T_β – 15 °C (27 °F)), air cool + 595 °C (1100 °F), 8 h, air cool. Source: M.M. Allen, AFML TR-71-78, Feb 1976

Ti-6242: Effect of silicon and exposure at 510 °C on tensile properties

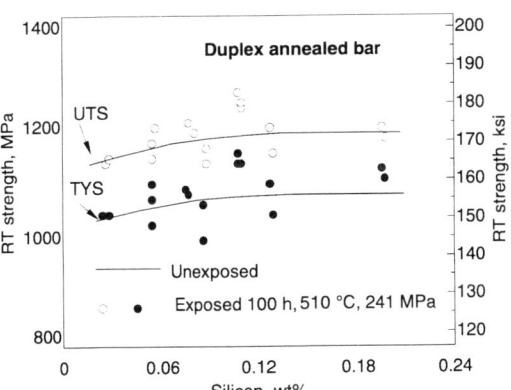

Specimens were of standard composition except for silicon. 15.8 mm (0.625 in.) bar. Heat treatment: 15 °C (25 °F) below β transus for 1 h, AC, 595 °C (1100 °F) for 8 h, AC.
Source: *Aerospace Structural Metals Handbook*, Vol 4, Code 3718, Battelle Columbus Laboratories, June 1978

Ti-6242: Effect of 1000-h exposure on RT strength

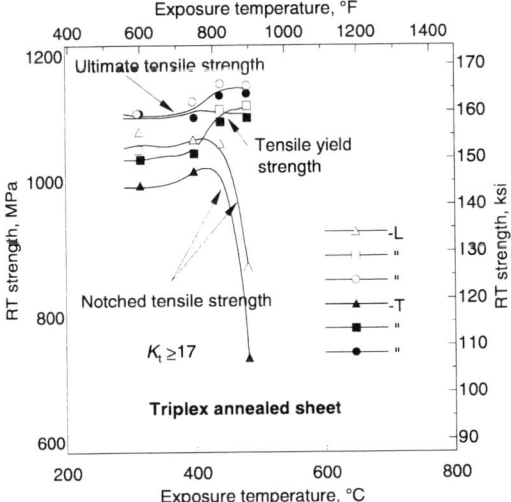

0.9 mm (0.036 in.) sheet triplex annealed at 900 °C (1650 °F), 30 min, AC + 785 °C (1450 °F), 15 min, AC + 595 °C (1100 °F), 2 h, AC. RT test results reported. All specimens were acid pickled after exposure prior to tensile tests.
Source: *Aerospace Structural Metals Handbook*, Vol 4, Code 3718, Battelle Columbus Laboratories, June 1978

352 / Alpha and Near-Alpha Alloys

Ti-6242: Effect of 1000-h exposure on RT strength

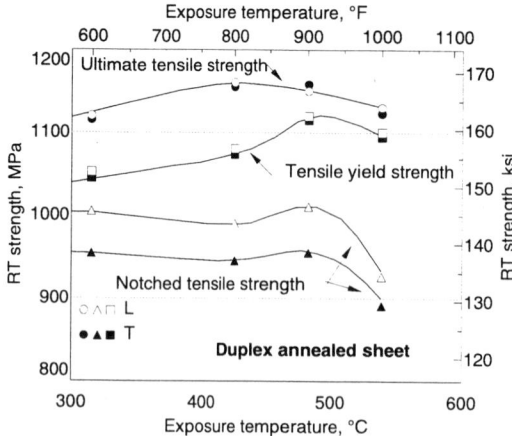

0.9 mm (0.036 in.) sheet duplex annealed at 900 °C (1650 °F), 30 min, AC + 785 °C (1450 °F), 15 min, AC. RT test results reported. All specimens were acid pickled after exposure prior to tensile tests.
Source: *Aerospace Structural Metals Handbook*, Vol 4, Code 3718, Battelle Columbus Laboratories, June 1978

Ti-6242: Effect of oxygen on cast tensile strength

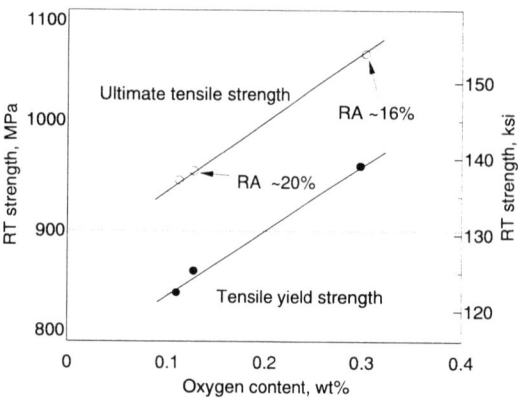

Specimens were consumable electrode skulls melted in water-cooled copper crucibles and cast into tungsten-lined ceramic molds. 6.4 mm (0.250 in.) diam tensile bars cast 0.25 to 0.4 mm (0.010 to 0.015 in.). Oversize and machined to final dimensions following heat treatment of 595 °C (1100 °F), 8 h, AC or HeC.
Source: *Aerospace Structural Metals Handbook*, Vol 4, Code 3718, Battelle Columbus Laboratories, June 1978

Cast Tensile Properties

Ti-6242: Room-temperature tensile properties of cast specimens

Heat treatment	Specimen type(a)	Ultimate tensile strength MPa	ksi	Tensile yield strength MPa	ksi	Elongation in 25 mm (1 in.), %	Reduction of area, %	Young's modulus GPa	10⁶psi
595 °C (1100 °F), 8 h, AC	(b)	948	137.5	853	123.7	10.5	21.4	...	...
		959	139.2	896	130.0	9.0	20.5	...	...
		968	140.4	853	123.7	10.0	18.2	...	...
955 °C (1750 °F), 1 h, AC + 595 °C (1100 °F), 8 h, AC	(b)	1010	146.5	894	129.7	8.0	21.5	127	18.4
		984	142.7	879	127.6	5.0	13.3	118	17.2
		993	144.1	892	129.4	4.0	8.1	117	17.1
1035 °C (1900 °F), 1 h, WQ + 595 °C (1100 °F), 8 h, AC	(b)	1272	184.5	1228	178.2	1.0	3.9	...	...
		1294	187.7	(c)	(c)	1.0	3.9	...	...
		1237	179.4	1237	179.4	1.0	3.1	...	...
1035 °C (1900 °F), 1 h, HeC + 595 °C (1100 °F), 4 h, HeC	(d), (e)	973	141.2	878	127.4	7.8	19.3	...	...
	(d), (f)	957-990	138.8-143.6	862-899	125.0-130.5	5.0-12.0	10.9-24.6	...	...

(a) Specimens were consumable electrode skulls melted in water-cooled copper crucibles and cast into tungsten-lined ceramic molds, unless otherwise indicated. (b) 6.4 mm (0.250 in.) diam tensile bars cast 0.25 to 0.38 mm (0.010 to 0.015 in.) oversize and machined to final dimensions following heat treatment. (c) Failed at less than 0.2% indicated strain. (d) 6.4 mm (0.250 in.) diam tensile bars machined from three thickest sections of heat treated, variable thickness step castings. (e) Average of nine values from three castings. (f) Range of nine values from three castings. Source: *Aerospace Structural Metals Handbook*, Vol 4, Code 3718, Battelle Columbus Laboratories, June 1978

High-Temperature Strength

Ti-6242: Typical tensile and compressive properties of sheet at high temperature

Temperature		Ultimate tensile strength		Tensile yield strength (0.2%)		Elongation, %	Compressive yield strength (0.2%)	
°C	°F	MPa	ksi	MPa	ksi		MPa	ksi
Longitudinal								
Room temperature		1010	146	990	144	3	1075	156.3
205	400	890	129	760	110	10.8	800	116
370	700	830	120	650	94	11.5	695	101
540	1000	710	103	570	83	17.8	635	92
Transverse								
Room temperature		1020	148	1010	146	2.7	1165	169
205	400	890	129	770	112	10.6	860	125
370	700	830	120	670	97	21.6	750	109
540	1000	720	104	590	86	16.5	695	101

Ti-6242: Hot tensile yield strength of bar at 510 °C vs Fe content

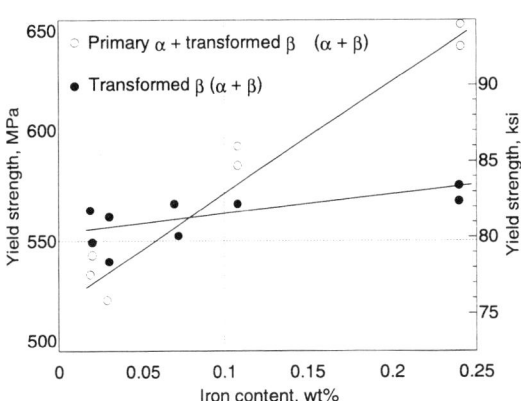

Source: S. Akem and S.R. Seagle, in *Titanium, Science and Technology*, Vol 4, 1985, p 2415

Ti-6242: Ultimate shear strength vs high-temperature exposure

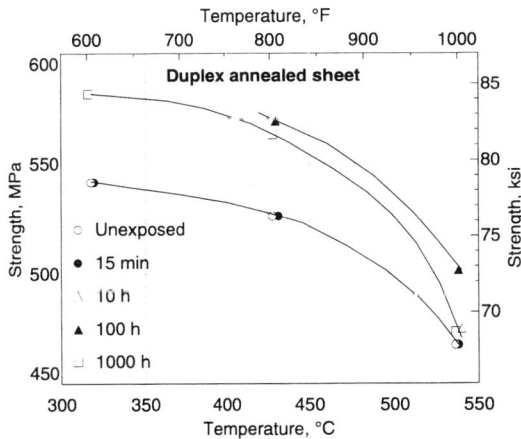

Sheet 1 mm (0.040 in.) thick was heat treated at 900 °C (1650 °F) for 30 min, air cooled, and at 785 °C (1450 °F) for 15 min, air cooled. Material was subsequently exposed to elevated temperatures as indicated. Each data point represents an average of two tests. Longitudinal test results are reported.
Source: C. Dotson, "Mechanical and Thermal Properties of High-Temperature Titanium Alloys," AFML-TR-67-41, 1967, and data from TIMET; reported in *Aerospace Structural Metals Handbook*, Vol 4, Code 3718, Battelle Columbus Laboratories, 1978, p 75

Ti-6242: Ultimate shear strength vs temperature

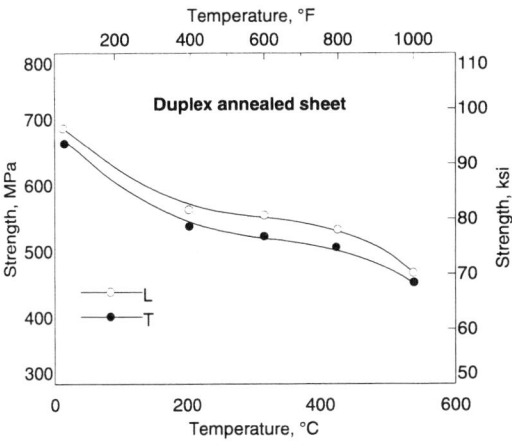

Sheet 1 mm (0.040 in.) thick was heat treated at 900 °C (1650 °F) for 30 min, air cooled, and at 785 °C (1450 °F) for 15 min, air cooled. Each data point represents an average of five tests.
Source: C. Dotson, "Mechanical and Thermal Properties of High-Temperature Titanium Alloys," AFML-TR-67-41, 1967, and data from TIMET; reported in *Aerospace Structural Metals Handbook*, Vol 4, Code 3718, Battelle Columbus Laboratories, 1978, p 75

Ti-6242: Design tensile strength vs temperature

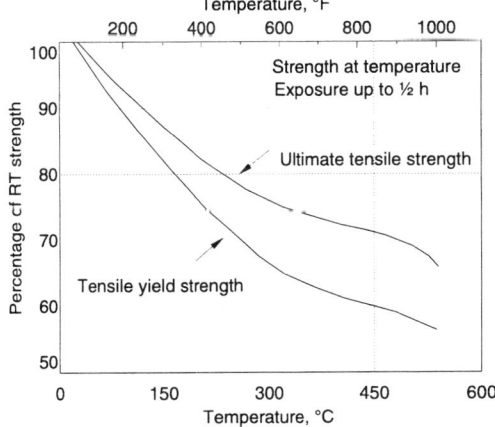

Design allowables for all product forms, duplex- and triplex annealed.
Source: MIL-HDBK 5, 1 Dec 1991

354 / Alpha and Near-Alpha Alloys

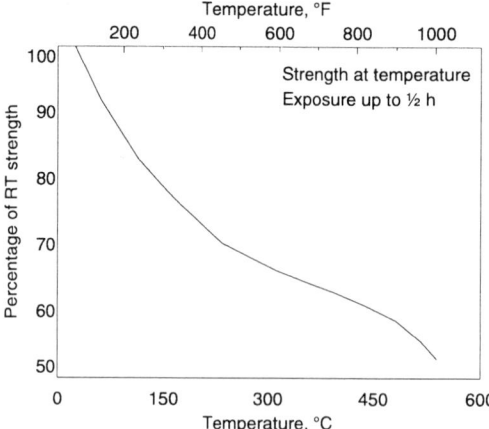

Ti-6242: Compressive strength vs temperature

Duplex annealed sheet.
Source: MIL-HDBK-5, 1 Dec 1991

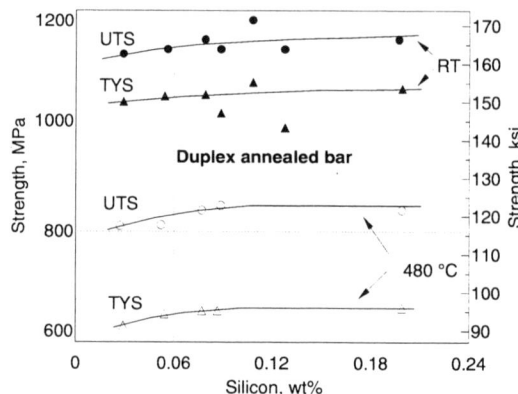

Ti-6242: Tensile properties of bar vs silicon content

Specimens were standard composition except for silicon. 15.8 mm (0.625 in.) bar heat treated 14 °C (25 °F) below the transus, 1 h, AC + 595 °C (1100 °F), 8 h, AC.
Source: *Aerospace Structural Metals Handbook*, Vol 4, Code 3718, Battelle Columbus Laboratories, June 1978

Cast Properties

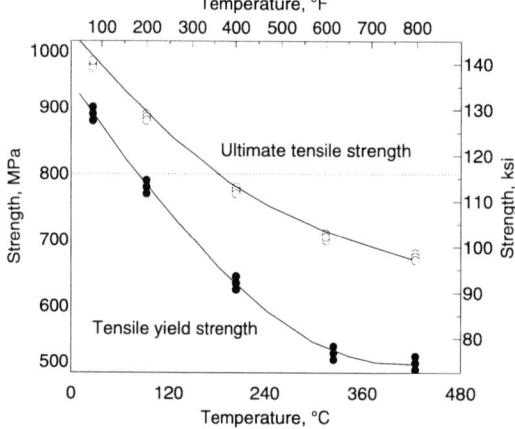

Ti-6242: Tensile strengths at high temperatures

Consumable electrode skulls melted in water-cooled copper crucibles and cast into tungsten-lined ceramic molds. 6.4 mm (0.250 in.) diam tensile bars machined from 31.7 mm (1.25 in.) thick sections of heat treated, variable thickness step castings. Specimens were treated at 1035 °C (1900 °F), 1 h, HeC + 595 °C (1100 °F), 4 h, HeC.
Source: *Aerospace Structural Metals Handbook*, Vol 4, Code 3718, Battelle Columbus Laboratories, June 1978

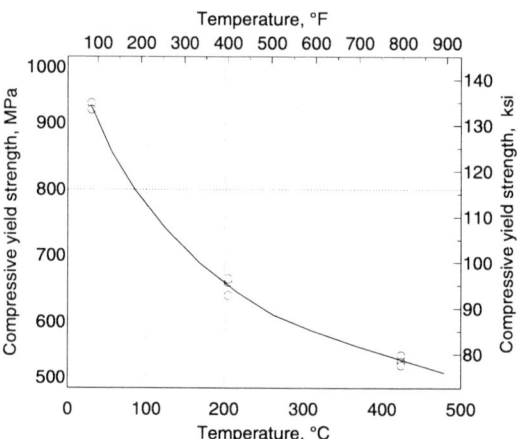

Ti-6242: Cast compressive yield strength

Alloy was supplied as cast wedges (tapered plates) of dimension 150 by 165 mm (6 by 6.5 in.). Chemical composition was 6.02 wt% Al, 0.018 wt% C, 0.010 wt% Fe, 0.0047 wt% H, 2.07 wt% Mo, 0.013 wt% N, 0.05 wt% Si, 2.04 wt% Sn and 3.80 wt% Zr. Material was tested in the as-received, as-cast condition with a compressive modulus of about 114 to 117 GPa (16.5 to 17×10^{-6} psi) at RT.
Source: O. Deel, "Engineering Data on New Aerospace Structural Materials," AFML-TR-77-198, Battelle Columbus Laboratories, 1977, p 29

Thermal Stability

Much information has been accumulated on the thermal stability of Ti-6242, due to the importance of this property in critical applications. The extent to which room-temperature properties will be affected by previous exposure will depend on the form and condition of the product as well as the time, temperature, and stress level of exposure. Data in the *Aerospace Structural Metals Handbook* (Vol 4, Code 3718) are extensive, although no systematic or firm generalizations can be made at this time. However, the following conclusions are suggested in the *Aerospace Structural Metals Handbook*.

Short-Time Stability (150 h). For sheet in both the duplex and triplex annealed conditions, 150-h exposure at 540 °C (1000 °F), 172 MPa (25 ksi) produced no deleterious effect on subsequent room-temperature tensile properties. This is also true for exposure at lower temperatures and higher stress levels—for example 480 °C (900 °F) at 310 MPa (45 ksi) and 425 °C (800 °F) at 448 MPa (65 ksi). Likewise, 100-h exposure to 540 °C (1000 °F) without load produces no loss in room-temperature sharp notch tensile strength of either duplex or triplex annealed sheet.

Duplex annealed rolled bar may or may not be affected by exposures of 150 h with load at temperatures up to 540 °C (1000 °F) depending on the exposure stress. At 172 MPa (25 ksi) stress level at 540 °C (1000 °F) no practical loss in room-temperature smooth tensile properties is observed, whereas at 206 MPa (30 ksi) a slight loss in ductility is observed (see figure). Bar rolled in the β region is unaffected by 150-h exposure at 540 °C (1000 °F), 206 MPa (30 ksi).

Forged bar in the recommended duplex annealed condition exposed 150 h up to 540 °C (1000 °F), 172 MPa (25 ksi) exhibits no loss in subsequent room-temperature smooth tensile properties. Contrasting results were obtained from forged compressor wheels. The hub location in one wheel exposed at 425 °C (800 °F), 482 MPa (70 ksi), 150 h showed losses in elongation and reduction of area. Other locations in the same forging, however, were essentially unaffected. A second wheel forging, water quenched from the solution anneal, exhibited losses in ductility after exposures at 480 °C (900 °F)/379 MPa (55 ksi), 540 °C (1000 °F)/241 MPa (35 ksi), or 595 °C (1100 °F)/103 MPa (15 ksi).

Long-Time Stability. It has been reported that 1000-h exposure up to 540 °C (1000 °F) without load has no deleterious effect on subsequent room-temperature tensile properties of duplex annealed sheet. In contrast, 1000-h exposure, with load produces a serious loss in ductility at 480 °C

Ti-6242: Ductility loss after high-temperature exposure

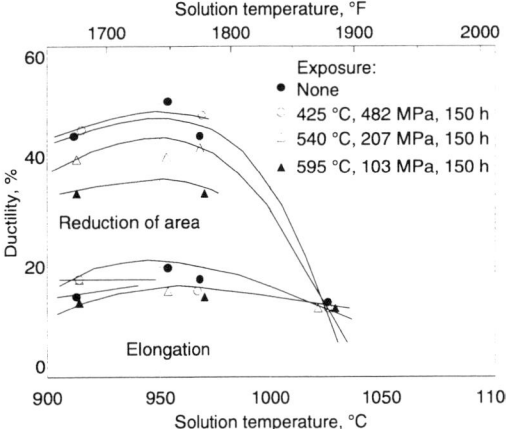

13 mm (0.5 in.) diameter round bar, α + β rolled and solution treated (ST) at indicated temperatures for indicated times + 595 °C (1100 °F), 8 h, AC.
Source: *Aerospace Structural Metals Handbook*, Vol 4, Code 3718, Battelle Columbus Laboratories, June 1978

Ti-6242: Effect of 150-h exposure on ductility of duplex annealed compressor hub forging

Location	150-h exposure at: Temperature °C	°F	Stress MPa	ksi	Creep in 150 h, %	Ultimate tensile strength MPa	ksi	Tensile yield strength MPa	ksi	Elongation, %	Reduction of area, %
Rim	None		None		...	965	140.0	893	129.6	13.5	28.6
Web	None		None		...	987	143.2	899	130.4	13.5	32.7
Hub	None		None		...	966	140.2	867	125.8	18.0	42.3
Rim	425	800	482	70	0.149	973	141.1	915	132.8	16.0	30.1
Web	425	800	482	70	0.160	963	139.7	902	130.9	16.0	28.1
Hub	425	800	482	70	0.148	955	138.5	899	130.5	15.0	28.8
Rim	540	1000	241	35	0.412	994	144.2	894	129.7	12.5	32.8
Web	540	1000	241	35	0.424	1001	145.2	921	133.6	17.0	29.4
Hub	540	1000	241	35	0.498	955	138.5	871	126.4	12.5	34.1

Note: Specimens were 470 mm (18.5 in.) diam compressor hub forgings duplex annealed at 955 °C (1750 °F), 1 h, AC + 595 °C (1100 °F), 8 h, AC tested in the radial direction. Source: *Aerospace Structural Metals Handbook*, Vol 4, Code 3718, Battelle Columbus Laboratories, June 1978

Ti-6242: Ductility loss after 150-h exposure of duplex annealed compressor wheel forging

150-h exposure at: Stress MPa	ksi	Temperature °C	°F	Plastic deformation(a), %	Subsequent RT tensile properties: Ultimate tensile strength MPa	ksi	Tensile yield strength MPa	ksi	Elongation, %	Reduction of area, %
Duplex annealed: 900 °C (1650 °F), 1 h, AC + 595 °C (1100 °F), 8 h, AC										
		None			1055	153.0	964	139.8	16.0	37.3
		None			1016	147.4	937	136.0	17.0	44.2
448	65	455	850	0.15	975	141.5	889	129.0	19.0	42.0
379	55	480	900	0.19	973	141.2	888	128.8	18.0	43.6
310	45	510	950	0.24	981	142.3	893	129.6	20.0	42.0
241	35	540	1000	0.30	983	142.7	891	129.3	19.0	47.6
103	15	595	1100	0.54	979	142.0	911	132.2	17.0	34.1
STA 955 °C (1750 °F), 1 h, WQ + 595 °C (1100 °F), 8 h, AC										
		None			1130	164.0	1016	147.4	14.0	41.9
		None			1150	166.9	1042	151.2	14.0	37.8
448	65	455	850	0.26	1136	164.8	1070	155.2	13.0	40.8
379	55	480	900	0.39	1139	165.2	1030	149.4	13.0	34.0
310	45	510	950	0.49	1135	164.6	1008	146.3	14.0	38.3
241	35	540	1000	0.81	1133	164.4	1032	149.7	11.0	31.5
103	15	595	1100	1.04	1110	161.0	1024	148.6	11.0	23.2

Ti-6242: Observed stability of ductility in rolled bar after 1000-h exposure

Temperature °C	Temperature °F	10000-h exposure at Stress MPa	10000-h exposure at Stress ksi	Deformation, %	Ultimate tensile strength MPa	Ultimate tensile strength ksi	Tensile yield strength MPa	Tensile yield strength ksi	Elongation in 25 mm (1 in.), %	Reduction of area, %
900 °C (1650 °F), 1 h, AC + 595 °C (1100 °F), 8 h, AC										
		None			944	137.0	832	120.7	15.5	31.2
425	800	379	55	0.11	1016	147.4	912	132.3	15.5	34.3
455	850	344	50	0.15	990	143.6	912	132.3	18.5	34.8
480	900	275	40	0.17	982	142.4	890	129.1	17.5	28.9
540	1000	103	15	0.12	979	142.1	904	131.2	15.0	31.6
955 °C (1750 °F), 1 h, AC + 595 °C (1100 °F), 24 h, AC										
		None			962	139.5	857	124.3	17.5	32.8
425	800	448	65	0.22	942	136.7	849	123.2	18.0	28.2
455	850	344	50	0.14	959	139.2	862	125.0	20.5	32.2
		379	55	0.18	941	136.5	846	122.7	18.0	30.2
480	900	275	40	0.22	942	136.6	850	123.3	19.0	30.2
		310	45	0.23	952	138.1	865	125.5	19.5	33.5
540	1000	138	20	0.17	957	138.9	888	128.9	16.5	30.8
1025 °C (1875 °F), 1 h, AC + 595 °C (1100 °F), 24 h, AC										
		None			954	138.4	834	121.0	15.0	21.2
425	800	448	65	0.19	955	138.5	869	126.1	13.5	23.4
455	850	344	55	0.14	962	139.6	867	125.8	14.5	21.9
480	900	310	45	0.16	964	139.8	873	126.7	14.5	23.4
540	1000	138	20	0.10	977	141.7	894	129.7	12.5	18.3

Note: Specimens were 57 mm (2.25 in.) diam rolled bar duplex annealed in full sections as indicated. Subsequent tensile properties were measured in the outside longitudinal direction and were tested after exposure without surface conditioning. Source: *Aerospace Structural Metals Handbook*, Vol 4, Code 3718, Battelle Columbus Laboratories, June 1978

Ti-6242: Loss of ductility after exposure at 425 °C
Effect of long-time exposure to elevated temperature with load on room-temperature tensile properties of duplex annealed pancake forgings

Heat	Temperature °C	Temperature °F	Stress MPa	Stress ksi	Time, h	Plastic deformation, %	Ultimate tensile strength MPa	Ultimate tensile strength ksi	Tensile yield strength MPa	Tensile yield strength ksi	Elongation, %	Reduction of area, %
A			Unexposed				993-1034(a)	144-150(a)	903-965(a)	131-140(a)	14-17(a)	42-51(a)
B			Unexposed				958-999(b)	139-145(b)	868-917(b)	126-133(b)	11-17(b)	33-52(b)
A	370	700	689	100	1500	1.681	1068	155	1061	154	14	44
B			551	80	1500	0.221	1013	147	979	142	18	39
B			413	60	1500	0.221	993	144	896	130	17	39
A	425	800	620	90	1500	1.444	1061	154	1048	152	14	28
A			586	85	1000	0.713	1061	154	1034	150	8	10
A			551	80	1000	0.433	1034	150	1013	147	15	44
A			517	75	500	0.256	1034	150	979	142	8	15
B			344	50	1500	9.158	986	143	910	132	16	39
B	480	900	344	50	1500	0.592	965	140	917	133	17	39
B			207	30	1500	0.173	1006	146	965	140	9	19
B	540	1000	207	30	1500	7.706	930	135	882	128	9	19
B			138	20	1500	2.051	1013	147	924	134	14	21

Note: Specimens were duplex annealed at 955 °C (1750 °F), 1 h, AC + 595 °C (1100 °F), 8 h, AC. (a) Range of values from six tests on two forgings. (b) Range of values from twelve tests on six forgings. Source: *Aerospace Structural Metals Handbook*, Vol 4, Code 3718, Battelle Columbus Laboratories, June 1978

(900 °F), 137 MPa (20 ksi) for duplex annealed sheet and at 540 °C (1000 °F), 69 MPa (10 ksi) for triplex annealed sheet. At 510 °C (950 °F), 103 MPa (15 ksi), no damage is done to triplex annealed sheet for 1000-h exposure. It appears that the triplex annealed condition is more stable than the duplex condition.

No change in smooth tensile properties at the exposure temperature is observed after 1000-h exposure without load up to 540 °C (1000 °F) and perhaps beyond. However, the influence of exposure to elevated temperature without load on the subsequent room-temperature sharp notch strength of sheet is a sensitive indicator of instability and reveals the onset of serious embrittlement at 480 °C (900 °F) for 1000-h exposure for both duplex and triplex annealed conditions (see the previous section "Notched Tensile Strength").

Bar in the recommended duplex annealed condition exposed 1000 h with load at temperatures up to 540 °C (1000 °F) exhibits no significant instability (see table). However, losses in ductility are observed for forgings in the recommended duplex annealed condition at exposure temperatures as low as 425 °C (800 °F).

Creep Properties

Studies on the creep behavior of Ti-6242-Si by Bania and Hall (see table) indicate the presence of two creep mechanisms, although the actual mechanisms are not clearly identified. This reinforces the argument that creep testing for design data generation or qualification testing must be performed under conditions (temperature and stress) similar to those anticipated in service. Otherwise, metallurgical changes that affect one mechanism and not another could be overlooked. Other conclusions of the study by Bania and Hall are as follows. Primary creep in Ti-6242-Si behaves in an irregular manner and is not easily correlated with microstructure. The steady-state creep rate correlates well with microstructure, indicating a direct proportionality with primary α content. Work by Thiehson et al. (Met. Trans. Vol 24A, Aug 1993, p 1819-1826) did show a correlation of both the steady-state and primary creep rates with microstructure. They also reported that Ni up to 0.1 wt.% had a deleterious effect on the creep rate, while Cr contents up about 0.3 wt.% had very little, if any, effect.

Much of the scatter in qualification testing can be attributed to varying contributions of erratic primary creep and microstructural variations. The fully transformed Ti-6242-Si β structure is superior to the conventional (19% equiaxed primary α) structure by roughly 20 °C (36 °F) or 70 MPa (10 ksi). Two creep mechanisms appear to be active: a high-activation energy (84 to 100 kcal/mol) process in the 495 to 565 °C (920 to 1045 °F) range and a low-activation energy (12 to 25 kcal/mol) process in the 455 to 480 °C (850 to 895 °F) range. Qualification creep testing should be conducted at temperatures typical of the intended use temperature and at stresses that minimize the impact of primary creep.

Ti-6242S: Creep test data

| Test temperature | | Stress | | Primary | Steady-state |
°C	°F	MPa	ksi	creep, %	creep, 10^4 %/h
0% primary α					
455	850	345	50	0.050	0.52
480	895	345	50	0.070	1.10
510	950	241	35	0.045	0.73
		276	40	0.083	1.07
		345	50	0.185	1.48
		414	60	0.182	10.7
540	1000	345	50	0.110	18.4
565	1045	345	50	0.10	137.0
7% primary α (average)					
510	950	241	35	0.06	1.4
		276	40	0.08	1.5
		345	50	0.135	4.2
		414	60	0.20	11.5
35% primary α (average)					
510	950	241	35	0.055	2.4
		276	40	0.085	3.0
		345	50	0.095	6.6
		414	60	0.190	23.0
19% primary α (average)					
455	850	276	40	0.06	0.36
		345	50	0.055	0.84
		414	60	0.095	2.05
480	895	276	40	0.045	0.65
		345	50	0.08	1.60
		414	60	0.11	2.9
495	920	276	40	0.08	1.7
		345	50	0.115	2.65
		414	60	0.14	6.50
510(a)	950(a)	241	35	0.075	1.30
		276	40	0.091	3.30
		345	50	0.116	5.85
		414	60	0.122	13.1
540	1000	276	40	0.11	21.2
		345	50	0.10	55.0
		414	60	0.20	174.0
565	1045	276	40	0.10	184.0
		345	50	0.20	349.0

(a) Average of eight tests; all other values represent an average of two tests. Source: P. Bania and J. Hall, Creep Studies of Ti-6242-Si Alloy, in *Titanium, Science and Technology*, Vol 4, Proc. of Fifth International Conference on Titanium, G. Lütjering, U. Zwicker, and W. Bunk, Ed., Deutsche Gesellschaft für Metallkunde, Germany, 1985, p 2371-2378

Ti-6242S: Effect of heat treatment on primary creep at 510 °C

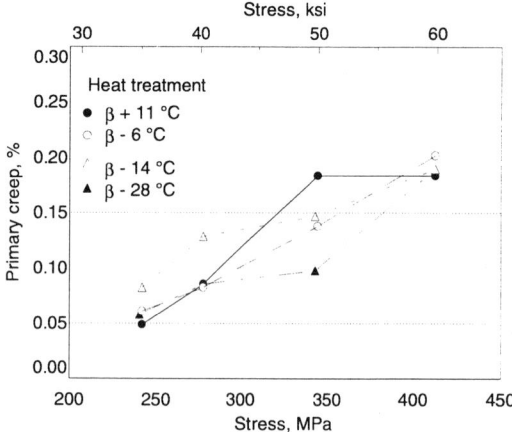

Primary creep vs applied stress for 510 °C (950 °F) creep tests of specimens solution treated at four temperatures, followed by air cooling and a stabilization age of 600 °C (1110 °F) for 8 h.
Source: P. Bania and J. Hall, Creep Studies of Ti-6242-Si Alloy, in *Titanium, Science and Technology*, Vol 4, G. Lütjering, U. Zwicker, and W. Bunk, Ed., Deutsche Gesellschaft für Metallkunde, Germany, 1985, p 2371-2378

Ti-6242S: Effect of heat treatment on steady state creep rates

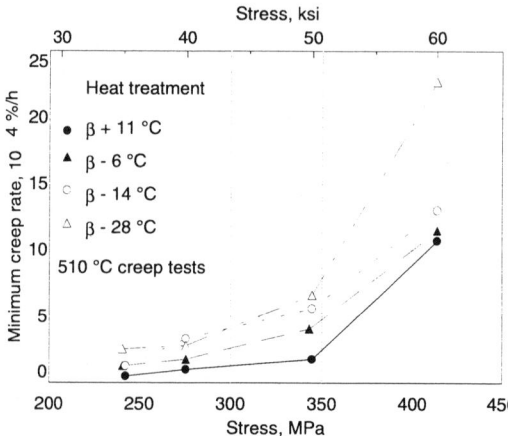

Steady-state (minimum) creep rate vs. applied stress for 510 °C (950 °F) creep tests of specimens solution treated at four temperatures, followed by air cooling and a stabilization age of 600 °C (1100 °F) for 8 h.
Source: P. Bania and J. Hall, Creep Studies of Ti-6242-Si Alloy, *Titanium, Science and Technology*, Vol 4, G. Lütjering, U. Zwicker, and W. Bunk, Ed., Deutsche Gesellschaft für Metallkunde, Germany, 1985, p 2371-2378

Creep Rates

Ti-6242S: Minimum creep rates at various temperatures

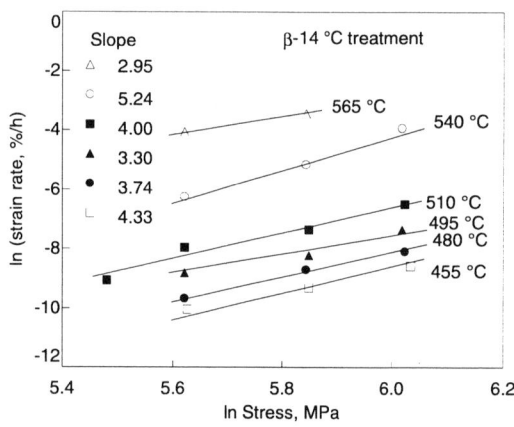

Plot of ln $\dot{\varepsilon}$ and ln (applied stress) for solution treated and stabilized Ti-6242 Si. Solution treatment was 14 °C (25 °F) below the β transus and for 1 h, followed by air cooling and aging at 600 °C (1110 °F) for 8 h.
Source: P. Bania and J. Hall, Creep Studies of Ti-6242-Si Alloy, *Titanium, Science and Technology*, Vol 4, G. Lütjering, U. Zwicker, and W. Bunk, Ed., Deutsche Gesellschaft für Metallkunde, Germany, 1985, p 2371-2378

Ti-6242S: Effect of primary alpha on creep rate

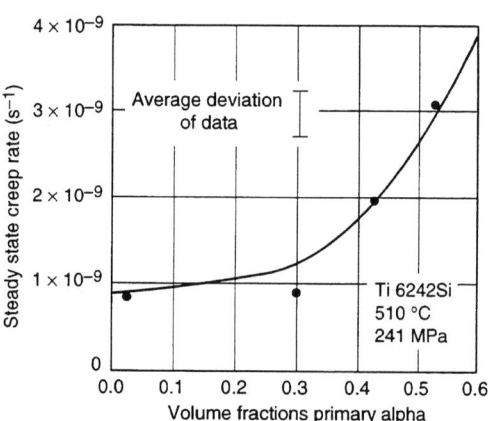

Source: *Met. Trans*, Vol 24A, August 1993, p 1819-1826

Ti-6242: Creep and creep rate data of castings

Stress		Temperature		Hours to indicated creep deformation:					Initial strain, %	Minimum creep rate, %/h
MPa	ksi	°C	°F	0.1%	0.2%	0.5%	1.0%	2.0%		
689	100	370	700	...	...	...	...	...	...	...
655	95	370	700	...	...	...	...	...	...	...
620	90	370	700	270	800	...	...	...	2.444	0.0002
551	80	370	700	55	340	4000(a)	...	...	1.455	0.00008
482	70	370	700	190	4600(a)	...	...	...	0.715	0.00001
620	90	480	900	0.02	0.05	0.17	0.47	1.5	6.430	0.92
551	80	480	900	0.9	3.0	20	66	154	0.963	0.010
413	60	480	900	4.5	22	232	680	1600(a)	0.452	0.0011
310	45	480	900	14	78	1030(a)	...	...	0.474	0.00024
241	35	480	900	165	975	4300(a)	...	...	0.018	0.00009

As-received cast wedges. (a) Estimated. Source: AFML-TR-77-198

Creep Rupture

Ti-6242: Creep and rupture properties of castings

Stress		Temperature		Hours to indicated creep deformation			Initial strain,	Rupture time,
MPa	ksi	°C	°F	0.1%	0.2%	0.5%	%	h
689	100	370	700	...	...	...	...	On loading
655	95	370	700	...	...	...	...	On loading
620	90	370	700	270	800	...	2.444	1132.8
551	80	370	700	55	340	4000(a)	1.455	502.7
482	70	370	700	190	4600(a)	...	0.715	1316.2
620	90	480	900	0.02	0.05	0.17	6.430	4.1
551	80	480	900	0.9	3.0	20	0.963	544.0
413	60	480	900	4.5	22	232	0.452	626.2
310	45	480	900	14	78	1030(a)	0.474	649.1
241	35	480	900	165	975	4300(a)	0.018	955.7

Note: Chemical composition was 6.02 wt% Al, 0.018 wt% C, 0.010 wt% Fe, 0.0047 wt% H, 2.07 wt% Mo, 0.013 wt% N, 0.168 wt% O, 0.05 wt% Si, 2.04 wt% Sn, and 3.80 wt% Zr. Alloy was tested in as-received condition, 150 by 165 mm (6 by 6.5 in.) cast wedges (tapered plate, 25 to 6.4 mm, or 1 to 0.25 in.). Source: O. Deel, "Engineering Data on New Aerospace Structural Materials," AFML-TR-77-198, Battelle Columbus Laboratories, 1977, p 25, 33. (a) Estimated

Effect of Silicon

Ti-6242: Effect of silicon on creep at 510 °C, 240 MPa

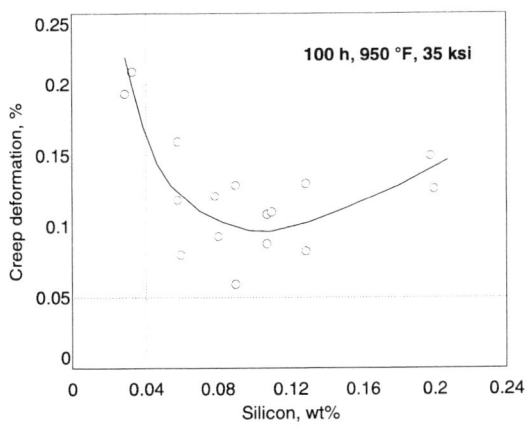

Specimens were standard composition except for silicon. 15.8 mm (0.625 in.) bar treated 14 °C (25 °F) below the transus, 1 h, AC + 595 °C(1100 °F), 8 h, AC.
Source: *Aerospace Structural Metals Handbook*, Vol 4, Code 3718, Battelle Columbus Laboratories, June 1978

Ti-6242: Larson-Miller plots of two Si levels

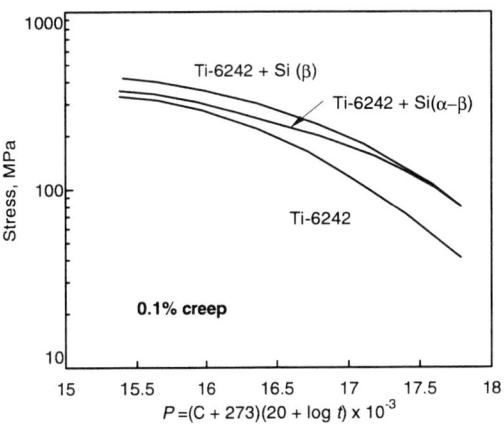

Condition for Ti-6242 without silicon is unspecified
Source: E. Huron and J. Miller, "Titanium Requirements for Current and Future Military Gas Turbine Engines," in *Space Age Metals Technology*, F. Froes and R. Cull, Ed., SAMPE, 1989, p 271

Effect of Nickel

Small concentrations of nickel appear to segregate uniformly into the beta phase and dramatically degrade the creep properties of Ti-6242S (*Met Trans*, Vol 24A, Aug 1993, p 1819-1826). The explanation for this is unclear.

Effect of Processing

Ti-6242S: Effect of solution annealing on creep at 540 °C, 170 MPa

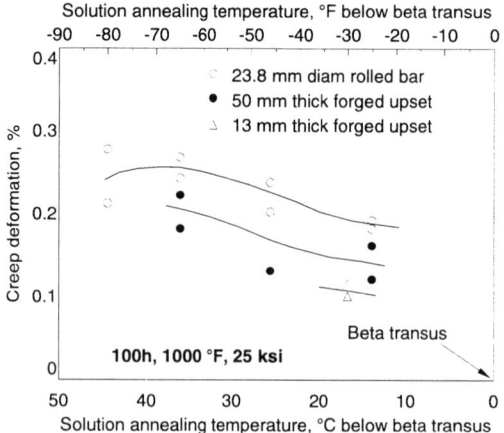

Effect of solution annealing temperature on 100 h, 540 °C (1000 °F), 170 MPa (25 ksi) creep deformation of duplex annealed Ti-6Al-2Sn-4Zr-2Mo-0.09Si rolled bar and forged upsets. Solution treated, 1 h, AC + 595 °C (1100 °F), 8 h, AC.
Source: *Aerospace Structural Metals Handbook*, Vol 4, Code 3718, Battelle Columbus Laboratories, June 1978

Ti-6242: Typical creep curves

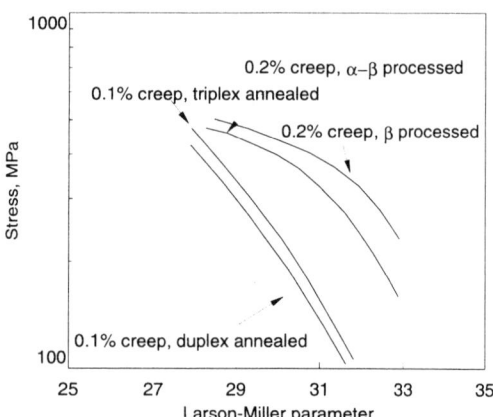

Larson-Miller parameter equals $10^{-3}T(20 + \log t)$, where T is temperature in °R and t is time in hours.
Source: *Metals Handbook*, Vol 3, 9th ed., 1980

Ti-6242: 0.1% creep data for heat treated sheet

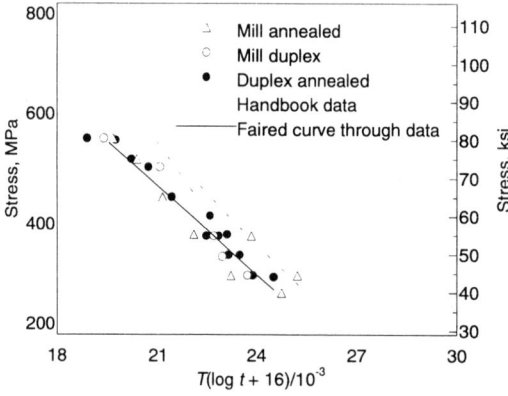

Mill annealed sheet of 0.8 mm (0.032 in.) thickness had a chemical composition of 6.0 wt% Al, 0.02 wt% C, 0.06 wt% Fe, 1.9 wt% Mo, 0.010 wt% N, 0.106 wt% O, 0.064 wt% Si, 2.0 wt% Sn, <0.005 wt% Y, and 4.0 wt% Zr. Duplex anneal was carried out at 900 °C (1650 °F) for 15 min, cooled by forced helium gas backfill, then at 785 °C (1450 °F) for 15 min, cooled by the same method. Creep tests were performed according to ASTM recommended practices.
Source: W. Ossa and R. Royster, "Material Characterization of Superplastically Formed Titanium (Ti-6Al-2Sn-4Zr-2Mo) Sheet," NASA Technical Paper 2674, National Aeronautics and Space Administration, 1987, p 26

Ti-6242: 0.1% creep data for superplastically formed sheet

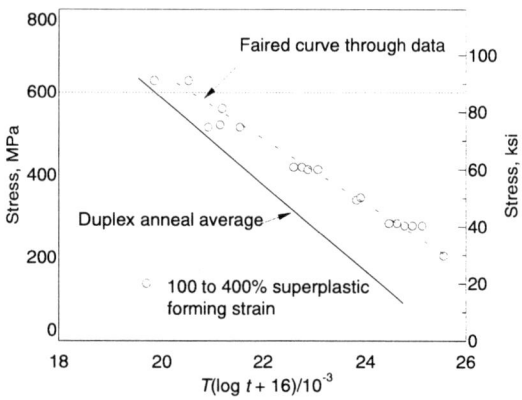

Mill annealed sheet of 0.8 mm (0.032 in.) thickness had a chemical composition of 6.0 wt% Al, 0.02 wt% C, 0.06 wt% Fe, 1.9 wt% Mo, 0.010 wt% N, 0.106 wt% O, 0.064 wt% Si, 2.0 wt% Sn, <0.005 wt% Y, and 4.0 wt% Zr. Duplex anneal was carried out at 900 °C (1650 °F) for 15 min, cooled by the same method.
Superplastically formed panels were produced by axially straining at 785 °C (1650 °F) for various times at a strain rate of 2×10^{-4}/s.
Creep tests were performed according to ASTM recommended practices.
Source: W. Ossa and R. Royster, "Material Characterization of Superplastically Formed Titanium (Ti-6Al-2Sn-4Zr-2Mo) Sheet," NASA Technical Paper 2674, National Aeronautics and Space Administration, 1987, p 26

Effect of Iron

Iron in the amounts of 0.02 to 0.24 wt% is detrimental to the creep resistance of α–β and β-annealed and stabilized Ti-6242S alloys. The detrimental effects of iron on creep properties are more pronounced in material that has undergone α–β anneal than in β-annealed structures. This is attributed to a change in the predominant creep mechanism, from grain boundary/interface sliding in α-β annealed structures to dislocation creep in β-annealed structures (S. Akem and S.R. Seagle, in *Titanium, Science and Technology*, Vol 4, G. Lütjering, U. Zwicker, and W. Bunk, Ed., Deutsche Gesellschaft für Metallkunde, Germany, 1985, p 2411-2418).

Ti-6242S: Effect of iron on creep rate

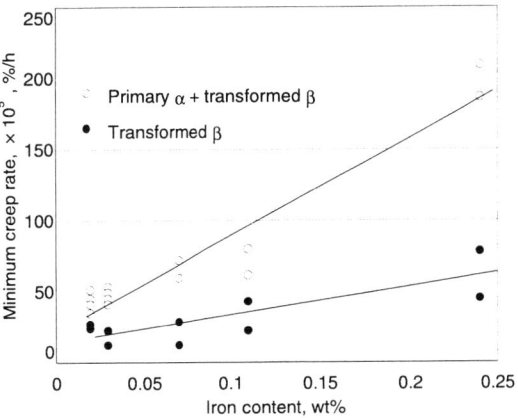

Specimens were tested at 510 °C (950 °F), 241 MPa (35 ksi) for 148 h.
Source: S. Akem and S.R. Seagle, The Detrimental Effect of Iron on Creep of Ti-6242S Alloy, *Titanium, Science and Technology*, Vol 14, G. Lütjering, U. Zwicker, and W. Bunk, Ed., Deutsche Gesellschaft für Metallkunde, Germany, 1985, p 2416

Fatigue Properties

Duplex Annealed Sheet

Ti-6242: RT fatigue properties

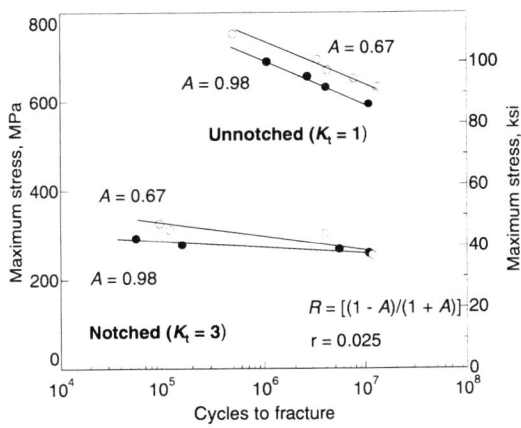

Specimens were 1 mm (0.040 in.) sheet duplex annealed at 900 °C (1650 °F), 30 min, AC + 785 °C (1450 °F), 15 min, AC. Axial fatigue, tension (sinusoidal). Surface: mill finish. Frequency: 2500 cycles/min. Test temperature, 21 °C (70 °F).
Source: AFML-TR-67-41; Apr 1967; reported in *Aerospace Structural Metals Handbook*, Vol 4, Code 3718, Battelle Columbus Laboratories, 1978, p 82

Ti-6242: Fatigue properties at 205 °C

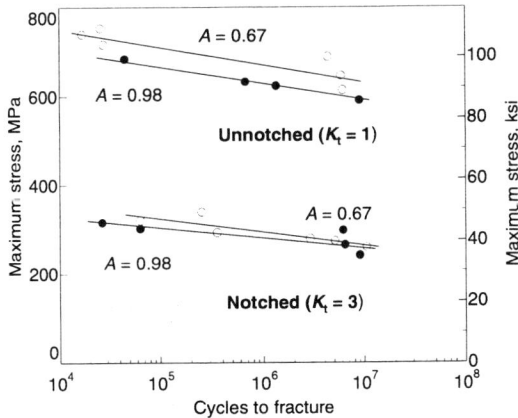

Specimens were 1 mm (0.040 in.) sheet duplex annealed at 900 °C (1650 °F), 30 min, AC + 785 °C (1450 °F), 15 min, AC. Axial fatigue, tension (sinusoidal). Surface: mill finish. Frequency: 2500 cycles/min. Test temperature, 205 °C (400 °F).
Source: AFML-TR-67-41, Apr 1967, reported in *Aerospace Structural Metals Handbook*, Vol 4, Code 3718, Battelle Columbus Laboratories, 1978, p 82

Ti-6242: Fatigue properties at 425 °C

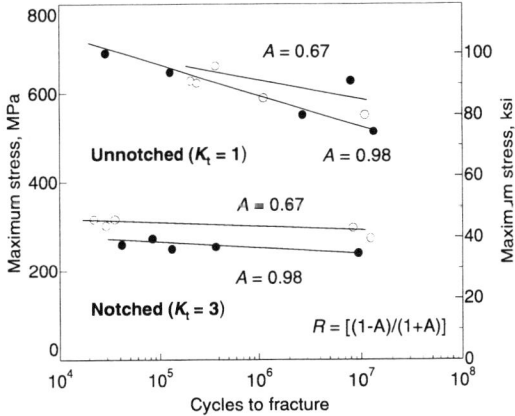

Specimens were 1 mm (0.040 in.) sheet duplex annealed at 900 °C (1650 °F), 30 min, AC + 785 °C (1450 °F), 15 min, AC. Axial fatigue: tension (sinusoidal). Surface: mill finish. Frequency: 2500 cycles/min. Test temperature, 425 °C (800 °F).
Source: AFML-TR-67-41; Apr 1967; reported in *Aerospace Structural Metals Handbook*, Vol 4, Code 3718, Battelle Columbus Laboratories, 1978, p 83

362 / Alpha and Near-Alpha Alloys

Duplex Annealed Bar

Ti-6242: RT and 480 °C fatigue properties

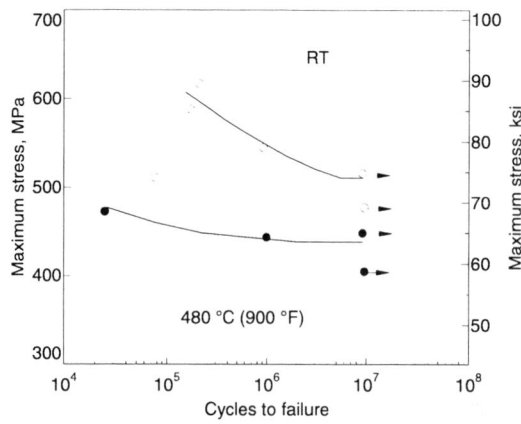

Specimens were 28.5 mm (1.125 in.) diameter bar duplex annealed at 970 °C (1775 °F), 1 h, AC + 595 °C (1100 °F), 8 h, AC. Ultimate tensile strength, 1006 MPa (146 ksi); tensile yield strength, 958 MPa (139 ksi); smooth, rotating beam tests.
Source: DMIC Data Sheet; Jan 1967, reported in *Aerospace Structural Metals Handbook*, Vol 4, Code 3718, Battelle Columbus Laboratories, 1978, p 83

Ti-6242: Fatigue properties at 480 °C

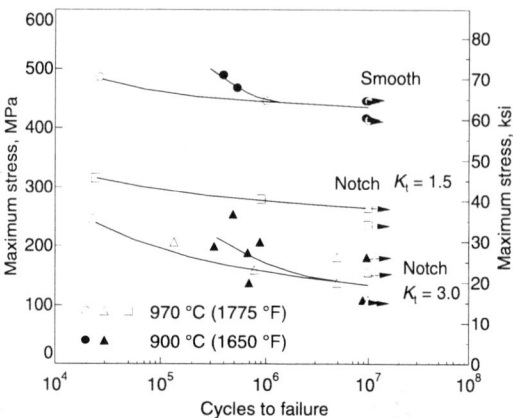

Specimens were 28.5 mm (1.125 in.) bar, duplex annealed as indicated. Rotating beam test results reported at 480 °C (900 °F).
Source: DMIC Data Sheet; Jan 1967; reported in *Aerospace Structural Metals Handbook*, Vol 4, Code 3718, Battelle Columbus Laboratories, 1978, p 84

Duplex Annealed Forgings

Ti-6242: Fatigue strength at 315 and 480 °C

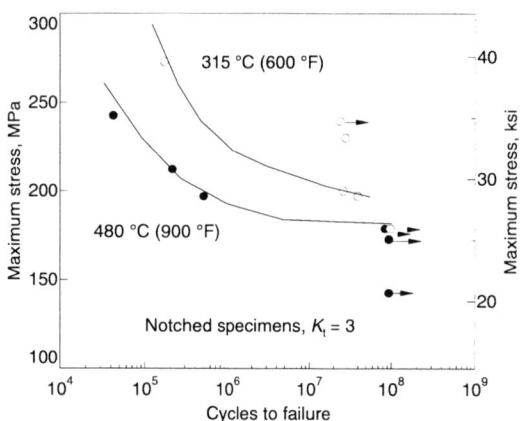

Specimens were compressor disk forgings duplex annealed at 955 °C (1750 °F), 1 h, AC + 595 °C (1100 °F), 8 h, AC. Specimens were stress relieved at 540 °C (1000 °F), 2 h in vacuum before testing. Reverse bending (cantilever); $R = -1$; frequency, 120 cycles/s.
Source: Pratt & Whitney Data; June 1972; reported in *Aerospace Structural Metals Handbook*, Vol 4, Code 3718, Battelle Columbus Laboratories, 1978, p 84

Ti-6242: High-frequency fatigue properties

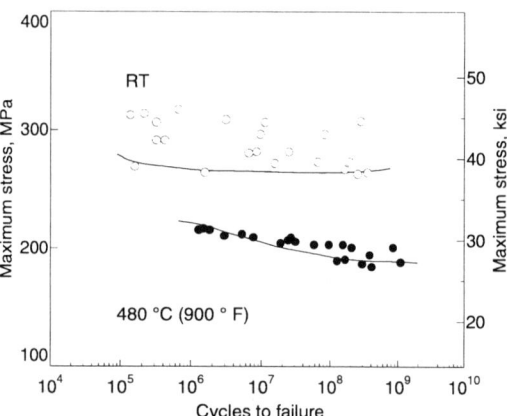

Specimens were 19 mm (0.75 in.) bar duplex annealed at 955 °C (1750 °F), 1 h, AC + 595 °C (1100 °F), 8 h, AC. Axial fatigue, $R = -1$. Frequency: 13.0 kHz at RT, 13.4 kHz at 480 °C (900 °F). Failure criterion: Crack grown to nearly half of specimen cross section.
Source: NASA TR-72618; July 1969; reported in *Aerospace Structural Metals Handbook*, Vol 4, Code 3718, Battelle Columbus Laboratories, 1978, p 84

Ti-6242: Fatigue strength at 200 °C

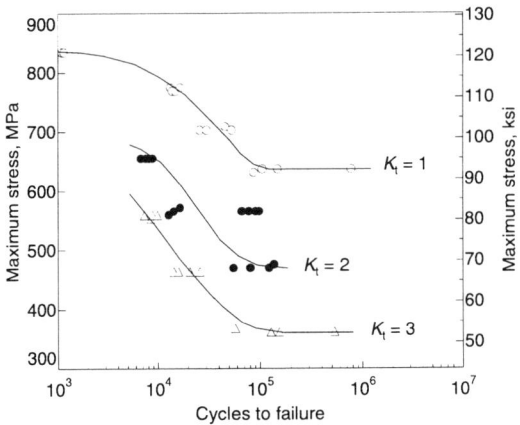

Specimens were compressor disk forgings (three forgings from three different heats) duplex annealed at 955 °C (1750 °F), 1 h, AC + 595 °C (1100 °F), 8 h, AC. Axial tension, $R = 0$; frequency, 1800 cycles/min.
Source: Pratt & Whitney Data; June 1972; reported in *Aerospace Structural Metals Handbook*, Vol 4, Code 3718, Battelle Columbus Laboratories, 1978, p 84

Ti-6242: Fatigue strength at 455 °C

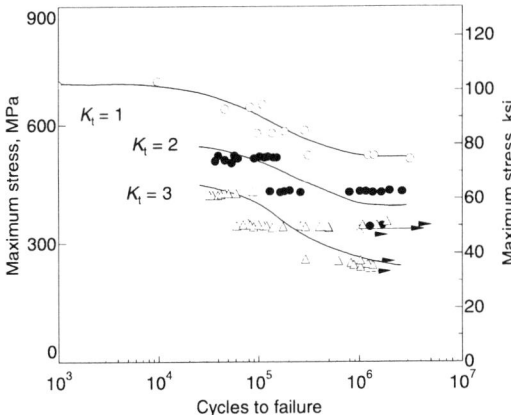

Compressor disk forgings from three different heats, treated at 955 °C (1750 °F) for 1 h, AC, 595 °C (1100 °F) for 8 h, AC. Axial tension, $R = 0$; frequency, 1800 cycles/min.
Source: Pratt & Whitney Data; June 1972; reported in *Aerospace Structural Metals Handbook*, Vol 4, Code 3718, Battelle Columbus Laboratories, 1978, p 84

Fracture Properties

Impact Toughness

Ti-6242: Effect of heat treatment on RT impact toughness of cast specimens

Heat treatment	Specimen type(a)	Charpy V-notch impact toughness J	ft · lbf
595 °C (1100 °F), 8 h, AC	(b)	30	22
		28	21
		28	21
955 °C (1750 °F), 1 h, AC + 595 °C (1100 °F), 8 h, AC	(b)	38	28
		34	25
		35	26
1035 °C (1900 °F), 1 h, WQ + 595 °C (1100 °F), 8 h, AC	(b)	8	6
		8	6
		8	6
1035 °C (1900 °F), 1 h, HeC + 595 °C (1100 °F), 4 h, HeC	(c) (d)	28	21
	(c) (e)	23-25	17-26

(a) Consumable electrode skull melted in water-cooled copper crucibles and cast into tungsten-lined ceramic molds unless otherwise indicated. (b) Standard Charpy-V bars cast 0.25 to 0.38 mm (0.010-0.015 in.) oversize and machined to final dimensions following heat treatment. (c) Standard Charpy-V bars machined from three thickest sections of heat-treated, variable thickness step castings. (d) Average of nine values from three castings. (e) Range of nine values from three castings. Source: *Aerospace Structural Metals Handbook*, Vol 4, Code 3718, Battelle Columbus Laboratories, June 1978

364 / Alpha and Near-Alpha Alloys

Ti-6242: Impact toughness vs silicon content at −40 °C

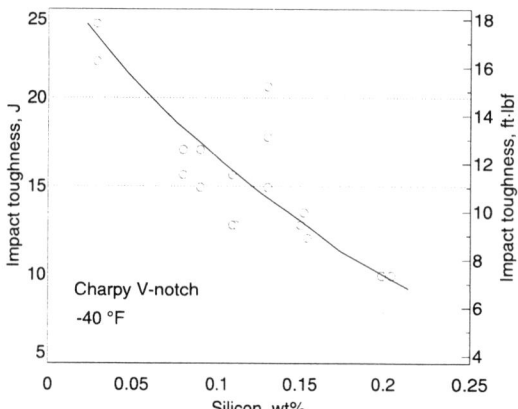

Effect of silicon content on −40 °C (−40 °F) Charpy-V-notch impact energy of duplex annealed 15.8 mm (⅝ in.) bar treated 15 °C (25 °F) below transus for 1 h, AC, 595 °C (1100 °F), for 8 h, AC.
Source: *Aerospace Structural Metals Handbook*, Vol 4, Code 3718, Battelle Columbus Laboratories, June 1978

Ti-6242: Effect of oxygen on cast impact toughness

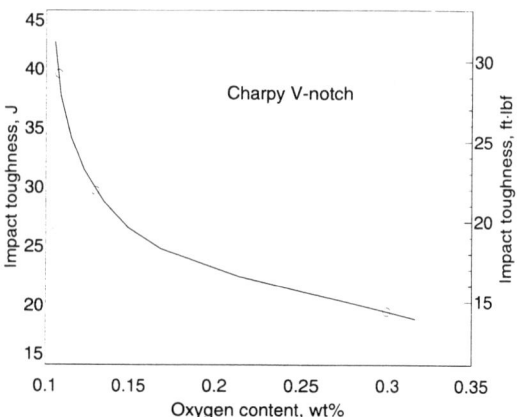

Effect of oxygen content on room-temperature Charpy V-notch impact energy for consumable electrode melted castings. Specimens were consumable electrode skulls melted in water-cooled copper crucibles and cast into tungsten-lined ceramic molds. Standard Charpy V-notch bars cast 0.25 to 0.38 mm (0.010 to 0.015 in.) oversize and machined to final dimensions following heat treatment of 595 °C (1100 °F), 8 h, AC or HeC.
Source: *Aerospace Structural Metals Handbook*, Vol 4, Code 3718, Battelle Columbus Laboratories, June 1978

Fracture Toughness

Ti-6242: Fracture toughness of plate

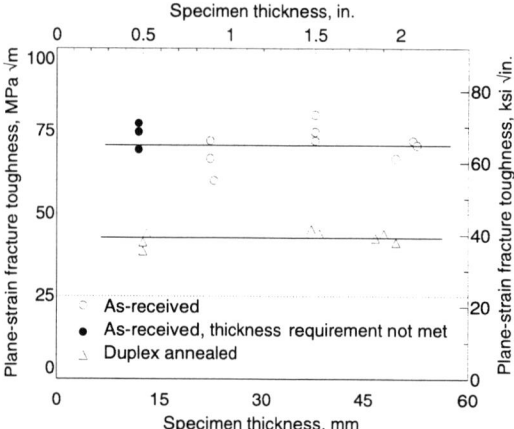

As-received 50 mm (2 in.) plate had a tensile yield strength of 855 MPa (124 ksi). Specimens were duplex annealed at 955 °C (1750 °F) for 30 min, in vacuum, water quenched, then at 455 °C (850 °F) for 17 h, and air cooled. Heat treated specimens had a tensile yield strength of 1041 MPa (151 ksi). Fracture toughness was determined by the three-point bend method, according to ASTM E399-74, except for 13 mm (0.5 in) thick specimens where the thickness requirement was not met. Specimen crack orientation was LT; specimens were symmetrical about the central plane of the plate.
Source: C. Carman, Frankford Arsenal data, 1970; reported in *Aerospace Structural Metals Handbook*, Vol 4, Code 3718, Battelle Columbus Laboratories, 1978, p 67

Ti-6242: Fracture toughness of forgings

Condition	Tensile yield strength MPa	ksi	Elongation, %	K_{Ic} MPa√m	ksi√in.
α + β forged, βST + aged	903	131	12.5	81	73
β forged, α + βST + aged	896	130	11.0	84	76

Source: J.C. Williams and E.A. Starke, Jr., The Role of Thermomechanical Processing in Tailoring the Properties of Aluminum and Titanium Alloys, *Deformation, Processing, and Structure*, American Society for Metals, 1984

Ti-6242S: Fracture toughness of forgings from various thermomechanical processing (TMP) routes

Property	TMP route 1(a)	TMP route 2(b)	TMP route 3(c)
Tensile yield strength, MPa (ksi)	937 (136)	903 (131)	917 (133)
Ultimate tensile strength, MPa (ksi)	979 (142)	986 (143)	1006 (146)
Elongation (in 4D), %	16	12	12
Reduction of area, %	34	24	26
Fracture toughness, MPa$\sqrt{m}$ (ksi$\sqrt{in.}$)	56 (51)	78 (71)	76 (69)
Creep at 510 °C, 241 MPa (950 °F, 35 ksi), h to 0.1%	117	447	430
Fatigue crack growth rate ($da/dn \times 10^{-6}$ in./cycle at ΔK 10 ksi$\sqrt{in.}$)	1.2	0.7	0.8

(a) TMP route 1 (baseline): α-β hot die forged plus sub-β transus duplex heat treatment. (b) TMP route 2: β hot die forged plus duplex heat treatment. (c) TMP route 3: β hot die forged plus direct age. Source: G. Kuhlman, T. Yu, A. Chakrabarti, and R. Pishko, "Mechanical Property Tailoring Titanium Alloys for Jet Engine Applications," in *Titanium 1986: Products and Applications*, Titanium Development Association, 1987, p 122

Ti-6242: Fracture toughness of duplex annealed bar

Alloy Condition	Tensile yield strength MPa	Tensile yield strength ksi	Crack plane orientation	Thickness mm	Thickness in.	Width mm	Width in.	Crack length mm	Crack length in.	K_Q MPa$\sqrt{m}$	K_Q ksi$\sqrt{in.}$
900 °C (1650 °F), 1 h, AC + 595 °C (1100 °F), 8 h, AC	855	124	LT	12.77	0.503	26.47	1.003	13.31	0.524	68(a,b)	62(a,b)
				12.80	0.504	25.42	1.001	12.54	0.494	69(a,b)	63(a,b)
				12.85	0.506	25.50	1.004	13.38	0.527	70(b)	64(b)
	848	123	TL	12.80	0.504	25.50	1.004	12.87	0.507	56(a)	51(a)
				12.75	0.502	25.42	1.001	13.28	0.523	74(b)	68(b)
				12.82	0.505	25.42	1.001	12.98	0.511	61(b)	56(b)
	848	123	TS	12.72	0.501	24.66	0.971	12.42	0.489	60(a)	55(a)
				12.72	0.501	25.65	1.010	12.98	0.511	65(a,b)	59(a,b)
				12.75	0.502	25.65	1.010	13.23	0.521	57(c)	52(c)
975 °C (1790 °F), 1 h, AC + 595 °C (1100 °F), 8 h, AC	820	119	LT	12.80	0.504	25.42	1.001	12.65	0.498	78(a,b)	71(a,b)
				12.80	0.504	25.42	1.001	14.45	0.569	89(b)	81(b)
				12.80	0.504	25.42	1.001	14.91	0.587	70(b)	64(b)
	848	123	TL	12.85	0.506	25.45	1.002	13.53	0.533	82(a,b)	75(a,b)
				12.85	0.506	25.42	1.001	14.09	0.555	77(b)	70(b)
				12.85	0.505	25.42	1.001	13.28	0.523	73(a,b)	67(a,b)
	848	123	TS	12.75	0.502	24.94	0.982	12.72	0.501	67(a,b)	61(a,b)
				12.75	0.502	24.66	0.971	13.66	0.538	71(a,b)	65(a,b)

Note: Specimens were duplex annealed bar 50 by 114 mm (2 by 4.5 in.). Fracture toughness was determined by the three-point bend method. (a) Excessive crack front curvature according to ASTM E-399-74. (b) Insufficient specimen thickness according to ASTM E-399-74. (c) Valid K_{Ic}. Source: C. Hickey, Jr., and T. DeSisto, "Mechanical Properties and Fracture Toughness of Ti-6Al-2Sn-4Zr-2Mo," reported in *Aerospace Structural Metals Handbook*, Vol 4, Code 3718, Battelle Columbus Laboratories, June 1987, p 67

Typical Stress-Strain Curves

Ti-6242: Tensile stress-strain curves

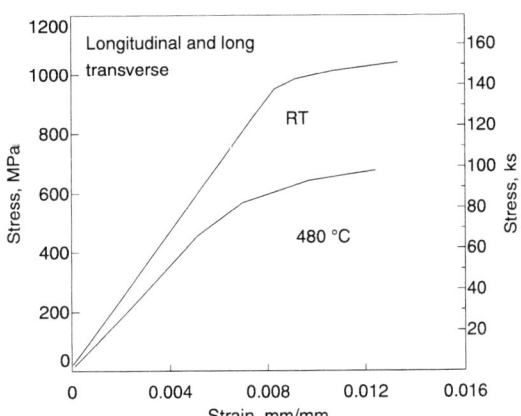

Duplex and triplex annealed 1.2 to 2.1 mm (0.048 to 0.085 in.) sheet after 1/2 h exposure.
Source: MIL-HDBK-5, 1 Dec 1991

Ti-6242: Compressive stress-strain curves

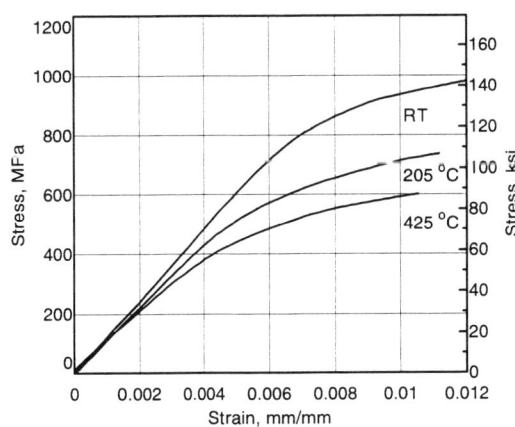

Specimens were cast wedges (tapered plates) and were tested in the as-received, as-cast condition.
Source: O. Deel, "Engineering Data on New Aerospace Structural Materials," AFML-TR-77-198, Battelle Columbus Laboratories, 1977, p 28

Ti-6242: Stress-strain curves up to 1010 °C

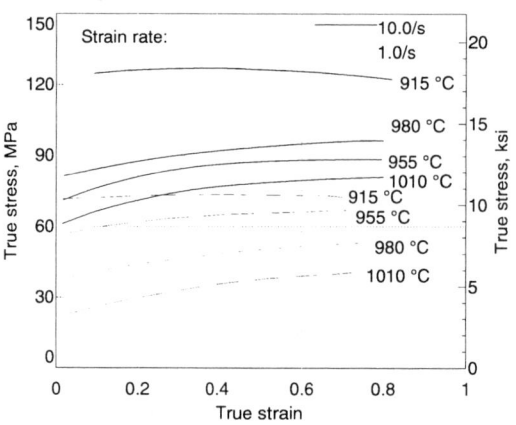

Source: G.D. Lahoti and T. Altan, AFML-TR-79-4156, Dec 1979

Ti-6242: Stress-strain curves up to 1010 °C

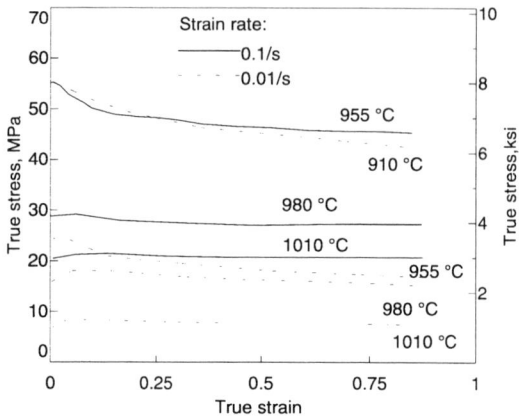

Source: G.D. Lahoti and T. Altan, "Research to Develop Process Models for Producing a Dual Property Titanium Alloy Compressor Disk," AFWAL-TR-80-4162, 1980

Flow Softening

From flow stress measurements using a mechanical press (Ref 1), measured load-stroke data were reduced to true stress-true strain data assuming frictionless, uniform deformation. This is a good assumption for α + β microstructure data, but somewhat approximate for β microstructure data, especially at 915 and 955 °C (1675 and 1750 °F). For both microstructures and strain rates, the stress-strain curves (see figures) decrease monotonically with strain (i.e., flow soften), an effect that is particularly pronounced for β microstructures. It is estimated that the error limits for α + β data are +5%, −2% for stresses and ±0.01 for the strains. The flow stress data for this microstructure are similar to those obtained previously in hydraulic and mechanical presses (Ref 2). The error limits for β microstructure are +10%, −2% for stresses, and ±0.01 for strains. However, it is thought that the basic trends in β stress-strain data are correct.

References

1. S.L. Semiatin, G.D. Lahoti, and T. Altan, Determination and Analysis of Flow Stress Data for Ti-6242 at Hot Working Temperatures, *Process Modeling Fundamentals and Applications to Metals*, American Society for Metals, 1980, p 387-408
2. J.R. Douglas and T. Altan, "A Study of Mechanics of Closed-Die Forging (Phase II)," Final Report on Contract No. DAAG46-71-C-0095, Battelle Columbus Laboratories, 1972

Ti-6242: Stress vs strain and strain rate at 915 °C

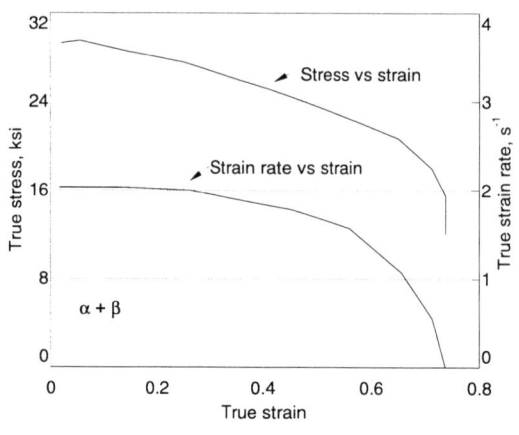

(a)

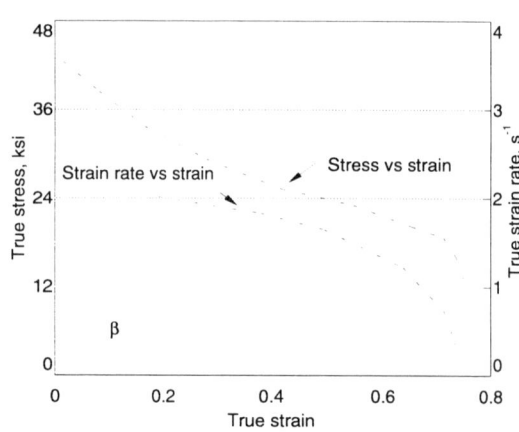

(b)

Source: S.L. Semiatin et al., in *Process Modeling Fundamentals and Applications to Metals*, American Society for Metals, 1980, p 387-408

Ti-6242: Softening of α + β and β structures

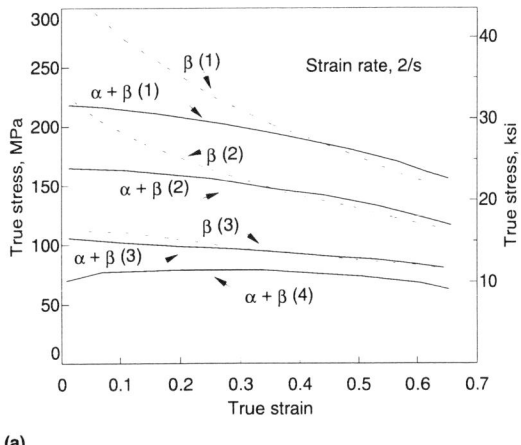

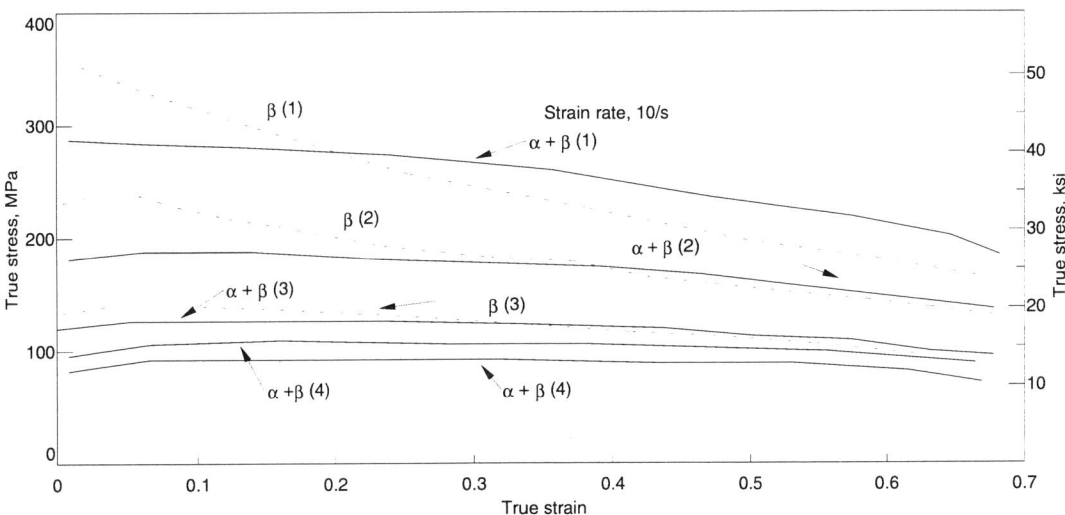

True stress-strain curves at various temperatures. (1) At 915 °C (1675 °F). (2) At 955 °C (1750 °F). (3) At 980 °C (1800 °F). (4) At 1010 °C (1850 °F), 1040 °C (1900 °F). The equilibrium microstructure of equiaxed α in a transformed β matrix deformed stably at all temperatures, but had decreasing stress-strain curves after an initial increase to a peak. In contrast, the nonequilibrium transformed β microstructure deformed unstably at temperatures in the α + β phase field in which it was tested. This instability gave rise to very sharply decreasing stress-strain curves. The flow softening of the equilibrium microstructure has been shown to be largely a result of adiabatic heating. The softening of the nonequilibrium microstructure results from adiabatic heating as well as dynamic recrystallization to produce the (softer) equilibrium microstructure. The development of a softer texture may also contribute to softening of the nonequilibrium microstructure.
Source: S.L. Semiatin et al., in *Process Modeling Fundamentals and Applications to Metals*, American Society for Metals, 1980, p 387-408

Flow Stress

Compression tests run on a mechanical press have been successful in obtaining high-strain-rate compression flow stress data for Ti-6242S (Semiatin et al., *Process Modeling Fundamentals and Applications to Metals*, American Society for Metals, 1980, p 387-480). Data were obtained at hot working temperatures for Ti-6242S having two different microstructures: equiaxed α in a transformed matrix, the α + β microstructure, and a microstructure consisting solely of transformed β, the β microstructure. The flow behavior of the two microstructures differed markedly and are summarized as follows:

The α + β microstructure deformed stably with monotonically increasing load-stroke curves. The flow behavior exhibited softening (see previous section), which could be attributed to adiabatic heating and a decreasing strain rate as the deformation proceeded, a characteristic of compression tests run in a mechanical press. Adjusting the data for this softening, the dependence of flow stress on strain was found to be weak, and hence, the flow stress is a function primarily of strain rate and temperature. This is a characteristic of metals that dynamically recover during hot working and has been observed in other α + β titanium alloys

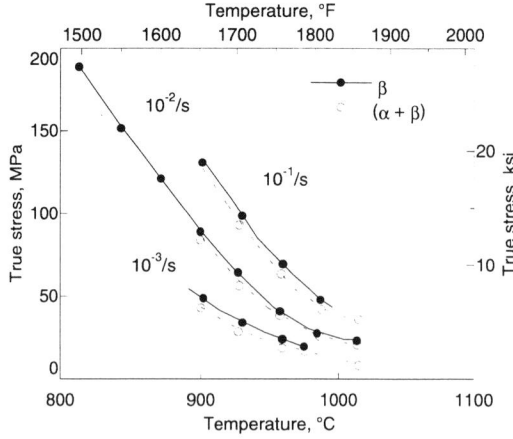

Temperature dependence of flow stress at different strain rates at ε = 0.6.
Source: G.D. Lahoti and T. Altan, "Research to Develop Process Models for Producing a Dual Property Titanium Alloy Compressor Disk," AFWAL-TR-80-4162, 1980

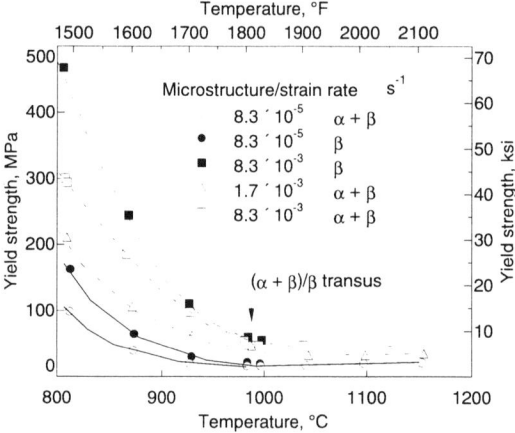

Temperature dependence of yield strength for Ti-6242Si alloy.
Source: G.D. Lahoti and T. Altan, AFML-TR-79-4156, Dec 1979

tested at much lower strain rates than those of the present tests. It appears, however, that flow stress is a function of strain-rate history, as evidenced by comparing strain-rate sensitivity data obtained through two different deformation schemes.

Beta Microstructure. The compressive deformation of β specimens gave rise to nonuniform, unstable flow characterized by decreasing load-stroke curves. The stress-strain curves for this microstructure exhibited an additional amount of flow softening in excess of that observed for the α + β microstructure. This increment has been attributed to breakup of the β microstructure and recrystallization to yield the equilibrium α phase at test temperatures below the β transus temperature. It is also believed that development of a softer texture may have caused some of the observed flow softening.

Selected References

- T. Altan and F.W. Boulger, Flow Stress of Metals and Its Application in Metal Forming Analyses, *J. Eng. Ind., Trans. ASME*, Vol 95, 1973, p 1009

- C.C. Chen, "Metallurgical Fundamentals to Ti-6Al-2Sn-4Zr-2Mo-0.1Si Alloy Forgings. I. Influence of Processing Variables on the Deformation Characteristics and the Structural Features of Ti-6242 Si Alloy Forgings," Report RD-77-110, Wyman-Gordon Company, North Grafton, MA, Oct 1977

- C.C. Chen, "Processing, Structure and Properties of Ti-6Al-2Sn-4Zr-2Mo-~0.1Si Alloy Forging," Subcontract Interim Report to Battelle Columbus Laboratories, under Air Force Contract AFML-F-33615-78-C-5025, Aug 1979

Strain-Rate Sensitivity

Ti-6242: Strain-rate sensitivity parameter

Micro-structure	Temperature °C	°F	Strain-rate sensitivity (m) at: (a)	(b)
α + β	915	1675	0.206	0.181
	955	1750	0.167	0.145
	980	1800	0.201	0.176
	1010	1850	0.259	0.145
	1040	1900	0.217	0.173
β	915	1675	0.134	0.153
	955	1750	0.121	0.156
	980	1800	0.190	0.153

(a) Tests at 2 and 10 s^{-1}; strain = 0.68. (b) Terminal portions of stress-strain data. Average of strain rate = 2 and 10 s^{-1} data. Source: S. Semiatin, G.D. Lahoti, and T. Altan, Determination and Analysis of Flow Stress Data for Ti-6242 at Hot Working Temperatures, *Process Modeling Fundamentals and Applications to Metals*, American Society for Metals, 1980, p 387-408

Determined from tests on nominal strain rates of 2 and 10 s^{-1}, the strain-rate sensitivity parameter, m, was found to show a non-negligible temperature dependence (see figures below). The dependence of m on strain is weak, except at small strains for the α + β microstructure. The m values determined from the terminal portions of stress-strain curves are of similar magnitude as these values. However, they are consistently lower for the α + β microstructure and approximately the same for the β microstructure when compared to measurements from the former method (see table). The lower values for the α + β microstructure can be rationalized from the viewpoint that the

Ti-6242: Strain-rate sensitivity

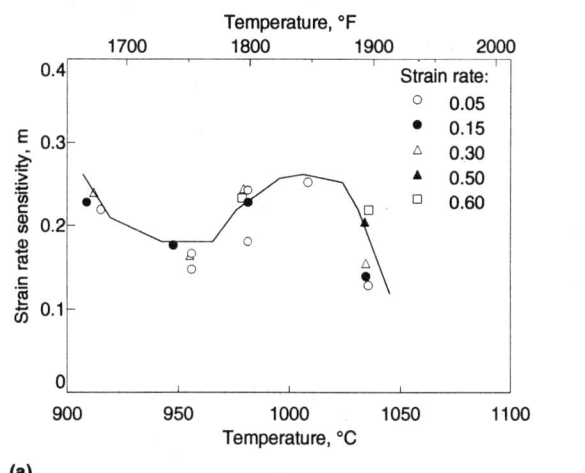
(a)

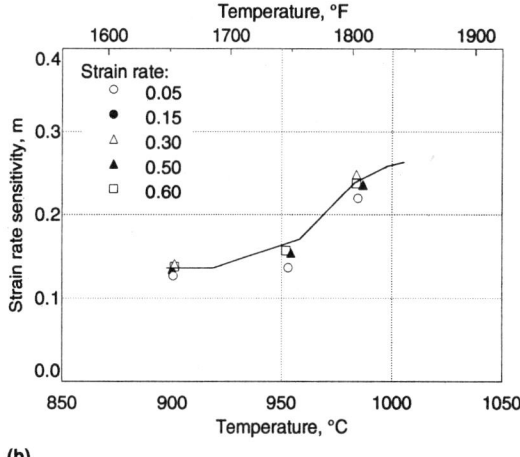
(b)

Strain-rate sensitivity parameter, m, for (a) α + β preform microstructure and (b) β preform microstructure.
Source: S. Semiatin, G.D. Lahoti, and T. Altan, Determination and Analysis of Flow Stress Data for Ti-6242 at Hot Working Temperatures, *Process Modeling Fundamentals and Applications to Metals*, American Society for Metals, 1980, p 387-408

Ti-6242: Strain-rate sensitivity

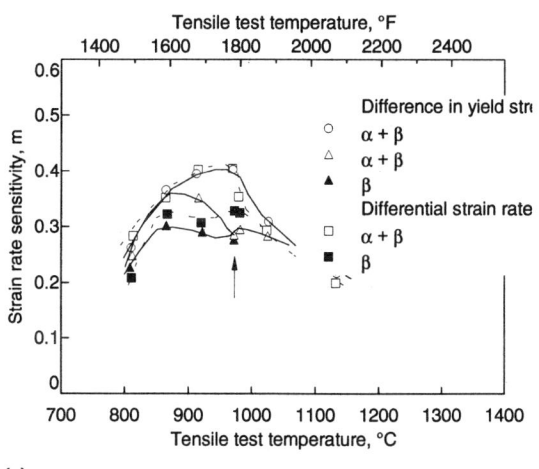
(a)

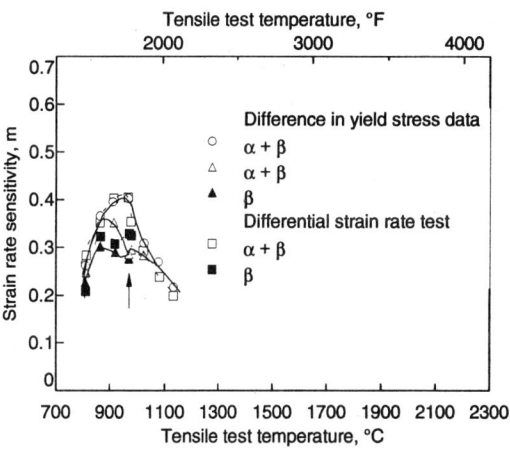
(b)

$\dot{\varepsilon}_1^{(m-1)}$ = 0.005 for all specimens; $\dot{\varepsilon}_2^{(m-1)}$ was 0.01 for differential strain rate test and 0.1 to 0.5 for others.
Source: C.C. Chen, "Metallurgical Fundamentals to Ti-6242 (0.1) Alloy Forging," Report RD-77-110, Wyman-Gordon, 1977

dislocation substructure changes at the end of a particular deformation are less than the substructure differences obtained in two separate tests at different nominal strain rates. It is these dislocation substructures and changes in substructure that determine the flow stress and rate sensitivity of the flow stress.

Selected References
- W.A. Backofen, *Deformation Processing*, Addison-Wesley, Reading, MA, 1972
- J.J. Jonas and M.J. Luton, Flow Softening at Elevated Temperatures, in *Advances in Deformation Processing*, Plenum Press, 1978, p 215
- J.J. Jonas and H.J. McQueen, Recovery and Recrystallization During High Temperature Deformation, in *Treatise on Materials Science and Technology, Vol 6; Plastic Deformation of Materials*, R.J. Arsenault, Ed., Academic Press, 1975, p 394
- C.M. Young and O.D. Sherby, Simulation of Extrusion Structures by Means of Torsion Testing for a High Strength Nickel-Base Alloy, Udimet 700, in *Metal Forming—Interrelation Between Theory and Practice*, A.L. Hoffmanner, Ed., Plenum Press, 1971, p 429
- A.J. Griest, A.M. Sabroff, and P.D. Frost, Effects of Strain Rate and Temperature on the Compressive Flow Stresses of Three Titanium Alloys, *Trans. ASM*, Vol 51, 1959, p 935

Forging

G.W. Kuhlman, ALCOA, Forging Division

Ti-6242 (and its modification Ti-6242S) is a near-alpha alloy produced commercially in all forging product types, although closed die forgings and rings predominate. The alloy is fabricated using all commercially available types of forging equipment. The silicon-modified version, Ti-6242S, due to its superior creep performance, predominates in commercial forging applications and is a preeminent titanium alloy for elevated-temperature service criteria, particularly rotating and static parts in turbine engines and heated airframe structures. For turbine engine disc forgings, hot die and isothermal forging technologies are important forging fabrication methods, producing near-net closed die forged product with significantly less machining.

Ti-6242S: Forging process temperatures

Process	Metal temperature	
	°C	°F
Conventional forging	900-975	1650-1790
Beta forging	1010-1065	1850-1950

Note: See "Technical Note 4: Forging" for recommended die temperatures.

Forging Process Technology

Ti-6242S is characterized by unit pressures (flow stresses), forgeability, and crack sensitivity similar to α–β titanium alloys such as Ti-6Al-4V (see Technical Note 4 and forging of Ti-6Al-4V elsewhere in this volume). The final microstructure of Ti-6242S forgings is developed by thermomechanical processing in forging manufacture tailored to achieve specific microstructural and mechanical property objectives. Thermomechanical processes use combinations of sub- and/or supra-beta transus forging followed by sub- or supra-transus thermal treatments to fulfill critical mechanical property criteria.

Final thermal treatments for forgings include solution treating and aging (stabilizing). Solution treating may be subtransus at 955 °C (1750 °F), followed by air cooling prior to aging, or supra-transus at β_t + 30 to 40 °C (50 to 75 °F), followed by air cooling or faster (e.g., fan cooling, water or oil quenching) cooling rates, prior to aging at 595 °C (1100 °F). Subtransus thermomechanical processes (forging and thermal treatment) for Ti-6242S forgings achieve equiaxed (50 to 75%) α in transformed β matrix microstructures to enhance strength, ductility, and particularly low-cycle fatigue properties. Supra-transus thermomechanical processes (forging and/or thermal treatments) achieve transformed, Widmanstätten and/or colony α microstructures that are exploited for the alloy to enhance creep and fracture-related properties (see Ref 1 in this section).

Conventional Forging. The objectives in forging Ti-6242S are to obtain the final forging shape and desired final microstructure at least cost. Conventional, subtransus (α + β) forging thermomechanical processes currently predominate in commercial forging manufacture. To achieve conventional equiaxed α structures, subtransus reductions of 50 to 75%, accumulated through one or more forging steps, are required. Supra-transus (β) forging for Ti-6242S may be used in early forging operations, including upsetting and open die preforming, to reduce unit pressures and ease forging fabrication. However, higher temperature initial forging operations must be followed by sufficiently subtransus reduction to achieve the desired predominately equiaxed α structure. Conventionally forged Ti-6242S is then subtransus heat treated as noted above.

Supra-transus thermomechanical processes for Ti-6242S are becoming commercially significant to achieve transformed Widmanstätten or colony α structures for improved creep and fracture-related properties. Successful β thermomechanical processes for Ti-6242S forgings include both β forging processes followed by subtransus heat treatment or conventional α + β forging followed by supra-transus heat treatment. The latter process provides a more uniform forging product and is currently preferred. Beta forging the alloy requires heavy β reductions (e.g., 35 to 50%) in early forging stages followed by a low final subtransus reduction (e.g., 15 to 25%). Beta forging Ti-6242S requires careful control of forging process conditions, particularly preheat times at temperature, to avoid excessive prior β grain growth. Beta forged Ti-6242S is then subtransus heat treated as noted above. Because of inherent variations in forging conditions, β forged alloy displays more final forging product variation than conventionally subtransus forged and heat treated equiaxed α structure.

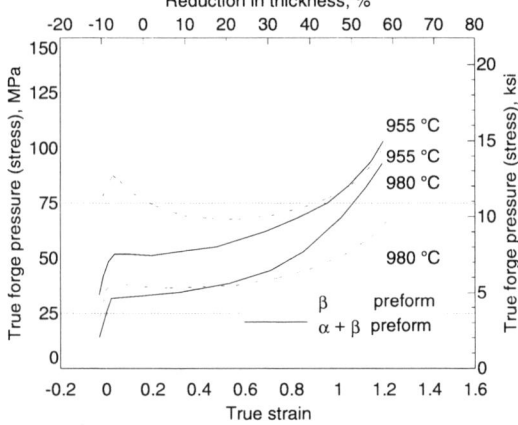

Ti-6242: Stress-strain for isothermal forging

True stress-strain curves for isothermal forging for α + β and β preforms at a strain rate of 5×10^{-3}/s (1 in./min).
Source: C. Chen, RD77-110, Wyman-Gordon, 1977

Ti-6242: Variation of forge pressure

Variation of forge pressure with reduction for forgings at various forge temperature-die temperature-ram rate combinations.
Source: G.D. Lahoti and T. Altan, AFML-TR-79-4156, Dec 1979

or conventionally forged and supra-transus heat treated product.

Hot die and/or isothermal forging techniques (see Ref 2 for details of these forging processes) are important commercial methods for fabrication of Ti-6242S rotating turbine engine discs to reduce final component cost (from less machining) and/or improve final component microstructural and property uniformity through improved control of forging process conditions. The axisymmetric shapes and designs of such engine components are very well suited to these forging methods. Isothermal forging of Ti-6242S discs is frequently accomplished in a single forging step from the bar or billet sock, under carefully controlled supra- or subtransus metal and die temperatures, levels of strain, and strain rate profiles. Hot die forging, where die temperature approaches but is not equivalent to metal temperature, is also used to reduce unit pressures, enhance forgeability, and produce more sophisticated final shapes in fewer forging operations.

References
1. C.C. Chen, U.S. patent 4,309,226, Jan 1982
2. G.W. Kuhlman, Forging Titanium Alloys, *Metals Handbook*, Vol 10, *Forging and Forming*, ASM International, 1988, p 267-287

Thermomechanical Effects

The effects of heat treatment alone, hot deformation, and the combination of hot deformation and subsequent heat treatment on the microstructures developed in α + β and β-preform samples of Ti-6242S were documented in AFWAL-TR-81-4130. The following conclusions were drawn.

Heat treatment of as-received α + β microstructure material below the transus temperature leads to minimal changes in the α grain size except at temperatures near the transus (≥954 °C). However, the morphology of the transformed β matrix of this microstructure is a function of heat treatment temperature over the observed range of 900 to 980 °C (1650 to 1800 °F), with higher temperatures leading to a refined matrix phase.

Subtransus heat treatment of as-received β microstructure material appears to lead to negligible changes in microstructure. It is believed that, although there may be α platelet reversion during heat treatment, the basic acicular or lamellar morphology remains the same during heat treatment and thus during cooling back to room temperature after heat treatment.

For the α + β microstructure, subtransus deformation or subtransus deformation plus subtransus heat treatment leads to a greater variation of α grain size than observed in the heat treatment studies. It was concluded that structures containing percentages of globular α close to the equilibrium amounts had been developed. This finding is similar to that for previously examined low-strain-rate compression samples and emphasizes the improvement in transformation kinetics brought about by hot work or residual hot work. This hot work also appears to be effective in speeding up the kinetics of the transformation of the β matrix phase when α + β compression samples are cooled after hot deformation.

As for low-strain-rate compression, subtransus/high-strain-rate compression of the β microstructure leads to grossly nonuniform flow, and at temperatures approaching the β transus, a phenomenon similar to dynamic recrystallization forms primary α. The temperatures at which this latter phenomenon occurs in high-rate deformation are slightly greater than those for low-rate deformation. This trend is analogous to that typically observed in single-phase materials. Heat treatment of high-strain-rate β microstructure compression samples after deformation leads to initial recrystallization in the areas of highest deformation followed by general recrystallization at long times, provided the heat treatment temperature is high enough. The lowest practical temperature for heat treatment of Ti-6242 with a β microstructure is about 915 °C (1680 °F).

The critical strains for die and preform design to make a dual microstructural/dual property disk using β preform material have been suggested by a series of low-strain-rate β microstructure compression tests at 900 °C (1650 °F), followed by heat treatment at 950 °C (1745 °F). The conclusions from these studies are as follows:

- At ε < 0.2, the as-forged microstructures show no distinct transformations. The transformed β structure is retained.
- At ε = 0.2 to 0.4, regions of intense localized shear develop, and these affect the post-deformation heat treatment response.
- At ε > 0.4, regions of dynamic recrystallization, which evolve to the equiaxed α + β structure during heat treatment, are first observed and increase in size with increasing strain.
- Strains ε > 1.0 are required for the equiaxed α + β structure to predominate after heat treatment.

Because complete transformation to the α + β microstructure does not occur even at ε = 1.0, it may be

Ti-6242S: Mechanical properties of alloy forgings

Forge/ heat treat conditions	Creep(a), h to 0.2% strain	Creep rupture time(b), h	K_{Ic} MPa√m	K_{Ic} ksi√in.	RT TYS MPa	RT TYS ksi	RT Elongation, %	480 °C (900 °F) TYS MPa	480 °C (900 °F) TYS ksi	480 °C (900 °F) Elongation, %	Post-creep TYS MPa	Post-creep TYS ksi	Post-creep Elongation, %
(α + β)/β	199	325	81	74	903	131	13	537	78	13	889	129	10
(α + β/β	169	...	~71(c)	~65(c)	924	134	10	565	82	11	...	...	...
(α + β)/(α + β)	56	...	~49(c)	~45(c)	951	138	14	593	86	15	...	...	...
(α + β)/β	19	...	~38(c)	~35(c)	1068	155	3	855	124	5	...	...	...
(α + β)/(α + β)	164	...	~55(c)	~50(c)	896	130	14	565	82	17	...	...	...
β/(α + β)	185	...	~77(c)	~70(c)	930	135	11	579	84	17	...	...	...
β/(α + β)	160	...	~82(c)	~75(c)	937	136	10	...	...	...	...	...	...
β/(α + β)	116	...	83	76	896	130	11	565	82	15	924	134	12
β/β	134	275	79	72	841	122	7	551	80	12	882	128	12

(a) Creep condition: 565 °C (1050 °F), 172 MPa (25 ksi). (b) Creep rupture: 565 °C (1050 °F), 344 MPa (50 ksi). (c) Estimated value. Source: G.D. Lahoti and T. Altan, AFML-TR-79-4156, Dec 1979

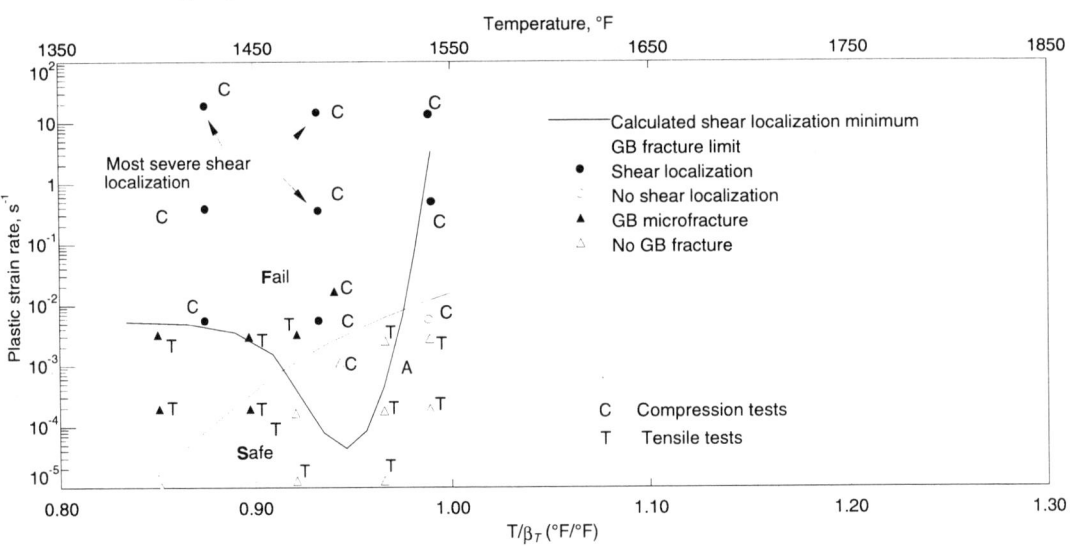

Ti-6242: Processing map

The safe region lies below the lines for shear localization and grain boundary (GB) fracture. Shear localization may be stress-state sensitive because it is more likely in compression than tension.
Ti-6242 preform with β microstructure. The region marked A close to the transus is recommended for high-strain-rate deformation.
Source: G.D. Lahoti and T. Altan, AFWAL-TR-81-4130, Oct 1981

that further investigation of disk processing at forging temperatures in the range 900 to 930 °C (1650 to 1705 °F) and heat treatment at temperatures in the range 950 to 970 °C (1740 to 1780 °F) would be worthwhile.

Deformation or heat treatment of the α + β and β microstructures above the transus temperature leads to development of transformed β microstructures of Widmanstätten α (slow cooling rate) or martensitic α (fast cooling rate).

Forming

Ti-6242 Sheet: Minimum bend radii of sheet

Sheet thickness		Test direction	Minimum bend radius(a)	
mm	in.		Duplex anneal	Triplex anneal
1.0	0.040	L	3.0	3.0
		T	3.0	3.0
2.0	0.080	L	3.0	3.1
		T	2.8	3.2
3.1	0.125	L	3.3	3.4
		T	3.5	3.4

(a) Minimum bend radius, r/t, for 105° after springback. Source: *Aerospace Structural Metals Handbook*, Vol 4, Code 3718, Battelle Columbus Laboratories, 1980

Ti-6242: Minimum bend radii

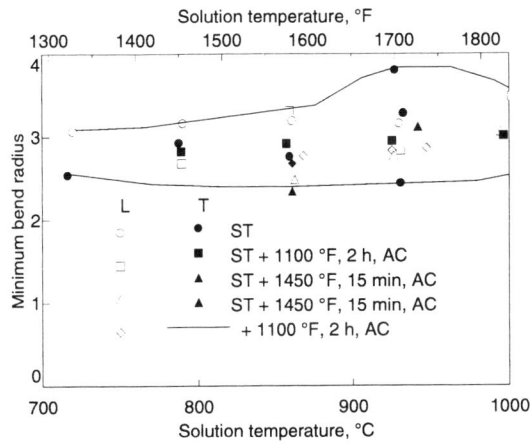

Effect of solution anneal temperature on room-temperature minimum bend radius of solution annealed, duplex annealed and triplex annealed 0.9 to 1.0 mm (0.036 to 0.040 in.) sheet.
Source: *Aerospace Structural Metals Handbook*, Vol 4, Code 3718, Battelle Columbus Laboratories, 1978

Heat Treatment

A variety of heat treatments for Ti-6242 are possible. A particular condition is usually selected on the basis of product type, part section size, and properties desired.

Annealing. Several different annealing treatments are available for Ti-6242. Choice depends on the product form and the section size of the product as well as on the properties desired (see tables). A general annealing treatment reported by a producer in the *Titanium Alloys Handbook* (Battelle, 1972) consists of a 1 to 8 h exposure at 705 to 845 °C (1300 to 1500 °F) followed by slow cooling to 565 °C (1050 °F) and subsequent air cooling. For bar and forged sections, solution annealing for 1 h at 900 to 955 °C (1650 and 1750 °F) plus stabilization annealing for 8 h at 595 °C (1100 °F), and air cooling are recommended. In this condition, the guaranteed room-temperature properties of 25 mm (1 in.) diameter bar are as follows:

Ultimate tensile strength	896 MPa (130 ksi)
Tensile yield strength (0.2%)	827 MPa (120 ksi)
Elongation	10%
Reduction of area	25%

Strengthening Heat Treatments. Ti-6242 may be heat treated to obtain higher uniaxial tensile strengths by conventional solution heat treatment and aging exposures. Because the β content of this α-β alloy is small, the response to such a strengthening heat treatment is not great. Furthermore, material in the solution treated and aged condition has lower creep strength than in the annealed-plus-stabilized condition.

Ti-6242: General types of heat treatment

Heat treatment	Temperature °C	Temperature °F	Time, h	Cooling method
Stress relief(a)	480-700	900-1300	0.25-4	AC or slow cool
Annealing	Varies by product form			
Solution treating	955-980	1750-1800	1	WQ
Aging	540-595	1000-1100	8	AC

(a) Stress relief varies depending on the portion of residual stress to be relieved. No data are available, and the given stress relief range covers the values specified in *Metals Handbook*, 9th ed. and *Titanium Alloys Handbook*, Battelle, 1972.

Ti-6242: Annealing treatments for sheet

Treatment	Temperature °C	Temperature °F	Duration, h	Cooling method
Duplex anneal				
1st stage	900	1650	0.5(a)	AC
2nd stage	785	1450	0.25	AC
Triplex anneal(b)				
1st stage	900	1650	2.5(a)	AC
2nd stage	785	1450	0.25	AC
3rd stage	595	1100	2	AC

(a) For sheet less than 3.1 mm (0.125 in.) in thickness, a shorter anneal time at 900 °C (1650 °F) is recommended. Ten minutes has been suggested. (b) The advantage of triplex anneal is higher uniaxial strength.

Ti-6242: Anneal treatments for bar and forgings

Treatment	Temperature °C	Temperature °F	Duration, h	Cooling method
Sections <63.5 mm (2.5 in.) diam				
Anneal	955	1750	1	AC
Stabilization	595	1100	8	AC
Sections >63.5 mm (2.5 in.) diam				
Anneal	900 or 955(a)	1650 or 1750(a)	1	AC
Stabilization	595	1100	8	AC

(a) The 900 °C (1650 °F) treatment, along with the 595 °C (1100 °F) stabilization anneal, provides somewhat higher tensile strengths at room and elevated temperatures, whereas the 955 °C (1750 °F) treatment combined with 595 °C (1100 °F) stabilization as above results in superior creep resistance at the higher temperatures and improved stability. Source: R.A. Wood and R.J. Favor, *Titanium Alloys Handbook*, MCIC-HB-02, Battelle Columbus Laboratories, 1972

Ti-6242: Duplex and triplex anneal strength

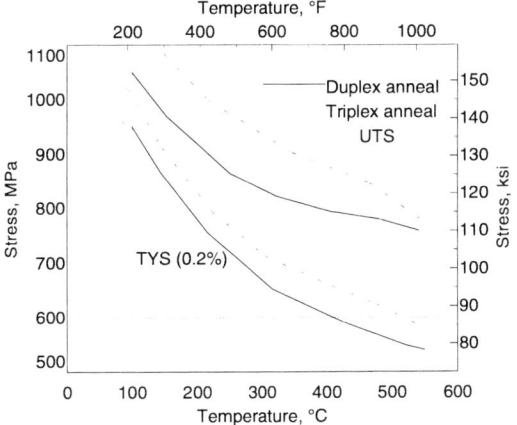

Elongation ductility is similar for both treatments.
Source: R.A. Wood and R.J. Favor, *Titanium Alloys Handbook*, MCIC-HB-02, Battelle Columbus Laboratories, 1972

Ti-6242: Beta fraction vs deformation

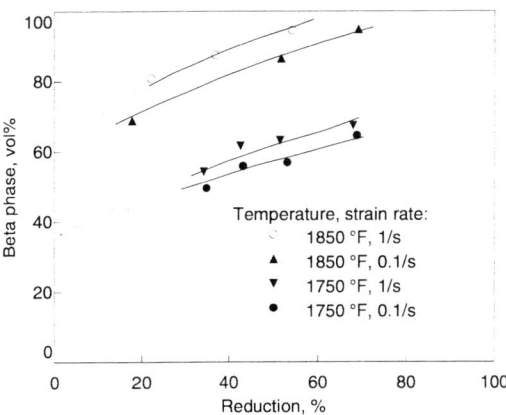

Plot of change in volume fraction of beta with deformation.
Source: G.D. Lahoti and T. Altan, "Research to Develop Process Models for Producing a Dual Property Titanium Alloy Compressor Disk," AFWAL-TR-80-4162, 1980

Ti-6242: Transformation hysteresis

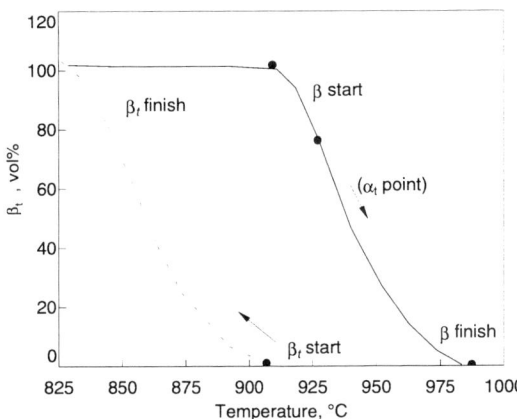

Source: G.D. Lahoti and T. Altan, "Research to Develop Process Models for Producing a Dual Property Titanium Alloy Compressor Disk," AFWAL-TR-80-4162, 1980

Ti-8Al-1Mo-1V

Common Name: Ti-811
UNS Number: R54810

Ti-8Al-1Mo-1V (Ti-811) was developed around 1954 for high-temperature gas turbine engine applications—specifically, compressor blades and wheels. It is now available from most titanium alloy producers. Ti-811 has the highest tensile modulus of all the commercial titanium alloys and exhibits good creep resistance at temperatures up to 455 °C (850 °F). Ti-811 has a room-temperature tensile strength similar to that of Ti-6Al-4V, but its elevated-temperature tensile strength and creep resistance are superior to those of other commonly available alpha and alpha+beta titanium alloys.

Chemistry and Density

The Ti-8Al-1Mo-1V alloy contains a relatively large amount of the alpha stabilizer, aluminum, and fairly small amounts of the beta stabilizers, molybdenum and vanadium (plus iron as an impurity). Although this alloy is metallurgically an alpha-beta alloy, the small amount of beta stabilizer in this grade (1Mo + 1V) permits only small amounts of the beta phase to become stabilized.

Density. 4.37 g/cm^3 (0.158 lb/in.^{3})

Product Forms

Ti-811 was developed for engine use, principally as forgings. Available forms include billet, bar, plate, sheet, and extrusions. Forming of sheet at room temperature is more difficult than for Ti-6Al-4V, and severe forming operations must be done hot. Ti-811 has good weldability like other alpha or near-alpha alloys. Weldments have similar strength but lower ductility in comparison with the base metal.

Product Condition/ Microstructure

Ti-811 is characterized as a near-alpha alloy with several alpha-alloy characteristics such as good creep strength and weldability. However, the alloy does have alpha-beta characteristics such as a mild degree of hardenability. Ti-811 is generally used in the annealed condition, where lamellar alpha morphology from transformed beta is produced by duplex and triplex annealing for enhanced creep resistance.

Applications

Ti-811 is used for airframe and turbine engine applications demanding short-term strength, long-term creep resistance, thermal stability, and stiffness. Ti-811 is predominantly an engine alloy and is available in three grades, including a "premium grade" (triple melted) and a "rotating grade," for use in rotating engine components.

Use Limitations. Like the alpha-beta alloys, Ti-811 is susceptible to hydrogen embrittlement in hydrogenating solutions at room temperature, in air or reducing atmospheres at elevated temperatures, and even in pressurized hydrogen at cryogenic temperatures. Oxygen and nitrogen contamination can occur in air at elevated temperatures and such contamination becomes more severe as exposure time and temperature increase. Ti-811 is susceptible to stress-corrosion cracking in hot salts (especially chlorides) and to accelerated crack propagation in aqueous solutions at ambient temperatures. The environment in which this alloy is to be used should be selected carefully to prevent material degradation.

Ti-8Al-1Mo-1V: Specifications and compositions

Specification	Designation	Description	Al	C	Fe	H	Mo	N	O	V	Other
UNS	R54810		8				1			1	bal Ti
China											
	Ti-8Al-1Mo-1V		7.5-8.5	0.1 max	0.3 max	0.015 max	0.75-1.25	0.04 max	0.15 max	0.75-1.25	Si 0.15 max; bal Ti
Spain											
UNE 38-717	L-7102	Sh Strp Plt Bar Ext Ann	7.35-8.35	0.08	0.3	0.015	0.75-1.25	0.05	0.12	0.75-1.25	OT 0.4; bal Ti
USA											
AMS 4915C		Sh Strp Plt Ann	7.35-8.35	0.08 max	0.3 max	0.015 max	0.75-1.25	0.05 max	0.12 max	0.75-1.25	OT 0.4; Y 0.005 max; OE 0.1 max; bal Ti
AMS 4915F		Sh Strp Plt Ann	7.35-8.35	0.08	0.3	0.015	0.75-1.25	0.05	0.12	0.75-1.25	OT 0.4; Y 0.005; bal Ti
AMS 4916E		Sh Strp Plt Dup Ann	7.35-8.35	0.08	0.3	0.015	0.75-1.25	0.05	0.12	0.75-1.25	OT 0.4; Y 0.005; bal Ti
AMS 4933A		Ext Rng SHT/Stab	7.35-8.35	0.08	0.3	0.015	0.75-1.25	0.05	0.12	0.75-1.25	OT 0.4; Y 0.005; bal Ti
AMS 4955B		Weld Fill Wir	7.35-8.35	0.08	0.3	0.01	0.75-1.25	0.05	0.12	0.75-1.25	OT 0.4; Y 0.005; bal Ti
AMS 4972C		Bar Wir Rng Bil SHT/Stab	7.35-8.35	0.08	0.3	0.015	0.75-1.25	0.05	0.12	0.75-1.25	OT 0.4; Y 0.005; bal Ti

(continued)

Ti-8Al-1Mo-1V: Specifications and compositions (continued)

Specification	Designation	Description	Al	C	Fe	H	Mo	N	O	V	Other
USA (continued)											
AMS 4973C		Frg Bil SHT/Stab	7.35-8.35	0.08	0.3	0.015	0.75-1.25	0.05	0.12	0.75-1.25	OT 0.4; Y 0.005; bal Ti
AWS A5.16-70	ERTi-8Al-1Mo-1V	Weld Fill Met	7.35-8.35	0.05	0.25	0.008	0.75-1.25	0.03	0.12	0.75-1.25	bal Ti
MIL F-83142A	Comp 5	Frg Ann	7.35-8.35	0.08	0.3	0.015	0.75-1.25	0.05	0.15	0.75-1.25	OT 0.4; bal Ti
MIL T-81556A	Code A-4		7.35-8.35	0.08	0.3	0.015	0.75-1.25	0.05	0.15	0.75-1.25	OT 0.4; bal Ti
MIL T-9046J	Code A-4	Sh Strp Plt Ann	7.35-8.35	0.08	0.3	0.015	0.75-1.25	0.05	0.15	0.75-1.25	OT 0.4; bal Ti
MIL T-9047G	Ti-8Al-1Mo-1V	Bar Bil Dup Ann	7.35-8.35	0.08	0.3	0.015	0.75-1.25	0.05	0.15	0.75-1.25	OT 0.4; Y 0.005; bal Ti
SAE J467	Ti-8-1-1		8	0.04 max	0.15 max		1	0.02 max		1	Si 0.07 max; Ni 0.008 max; bal Ti

OT, others total

Ti-8Al-1Mo-1V: Commercial compositions

Specification	Designation	Description	Al	C	Fe	H	Mo	N	O	V	Other
France											
Ugine	UTA8DV	Bar Frg DA	7.3-8.5	0.08	0.3	0.006-0.015	0.75-1.25	0.05	0.12	0.75-1.25	bal Ti
Germany											
Deutsche T	Contimet AlMoV 8-1-1	Plt Bar Frg Ann	7.5-8.5	0.08	0.3	0.015	0.75-1.25	0.05	0.12	0.75-1.25	bal Ti
Japan											
Kobe	KS8-1-1	Bar Frg STA	7.35-8.35		0.3	0.015	0.75-1.25	0.05	0.12	0.75-1.25	bal Ti
USA											
Chase Ext.	8Al-1Mo-1V										
OREMET	Ti-8-1-1										
RMI	RMI 8Al-1Mo-1V	Mult Forms DA	7.5-8.5	0.08		0.015	0.75-1.25	0.05	0.12	0.75-1.25	bal Ti
Timet	TIMETAL 8-1-1	Ann	7.35-8.35	0.08 max	0.3 max	0.015 max	0.75-1.25	0.05 max	0.12 max	0.75-1.25	bal Ti

Physical Properties

Ti-8Al-1Mo-1V: Summary of typical physical properties

Beta transus	~1040 °C (1900 °F)
Melting (liquidus) point	~1540 °C (2800 °F)
Density(a)	4.37 g/cm^3 (0.158 lb/in.^{3})
Electrical resistivity(a)	~1.97 μΩ · m
Magnetic permeability	Nonmagnetic
Specific heat capacity(a)	502 J/kg · K (0.120 Btu/lb · °F)
Thermal conductivity(a)	6 W/m · K (3.5 Btu/ft · h · °F)
Thermal coefficient of linear expansion(b)	8.5 × 10^{-6}/°C (4.7 × 10^{-6}/°F)

(a) Typical values at room temperature of about 20 to 25 °C (68 to 78 °F). (b) Mean coefficient from 0 to 100 °C (32 to 212 °F)

Ti-8Al-1Mo-1V: Retained β from diffraction data

Quenching temperature		Retained
°C	°F	β,%
1200	2200	<1
1150	2100	2
1095	2000	<1
1040	1900	2
980	1800	<1
925	1700	<2
870	1600	16
815	1500	14

Source: R.A. Wood and R.J. Favor, *Titanium Alloys Handbook*, MCIC-HB-02, Battelle Columbus Laboratories, 1972

Phases and Structures

Past work on the phase relationships of Ti-8Al-1Mo-1V (D.E. Austin, Boeing Document T6-314, 1965) indicates that the β phase transforms to martensite at temperatures in the α-β field from the transus down to about 900 °C (1650 °F). Below this temperature, the β phase is sufficiently enriched in molybdenum and vanadium to be retained (see table). The β phase decomposes during tempering below about 450 °C (840 °F). Like other α alloys, the structure is predominantly α, with small amounts of β.

The α phase may also undergo a metallurgical reaction, resulting in an ordered structure (DO_{19}-type superlattice) of the type found in binary titanium-aluminum alloys. It has been suggested that the presence of the ordered structure is responsible for the differences in mechanical properties between duplex and mill annealed material and is also responsible for the poor stress-corrosion resistance of this alloy.

Beta Transus. Approximately 1040 °C (1900 °F) with normal interstitial contents

Elastic Properties

Typical room-temperature elastic properties:
- Tensile modulus, 120 GPa (17.5×10^6 psi)
- Compressive modulus, 117 to 125 GPa (17 to 18×10^6 psi)
- Shear modulus, 46 GPa (6.7×10^6 psi)

Ti-8Al-1Mo-1V: Tensile modulus comparison

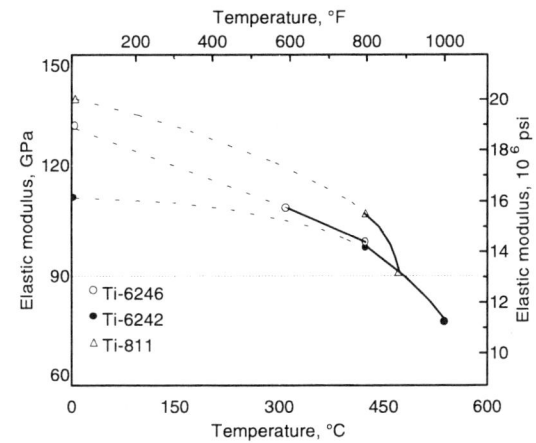

Ti-8Al-1Mo-1V: Tensile modulus of duplex annealed sheet

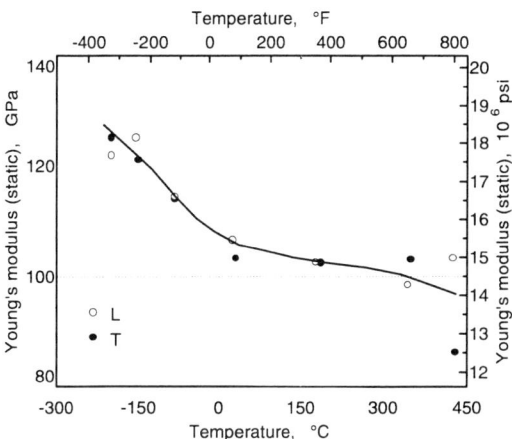

0.6 mm (0.025 in.) sheet duplex annealed at 1010 °C (1850 °F), 15 min, AC + 750 °C (1380 °F), 15 min, AC.
Source: *Aerospace Structural Metals Handbook*, Vol 4, Code 3709, Battelle Columbus Laboratories, 1980, p 20

Ti-8Al-1Mo-1V: Tensile modulus of bar and sheet

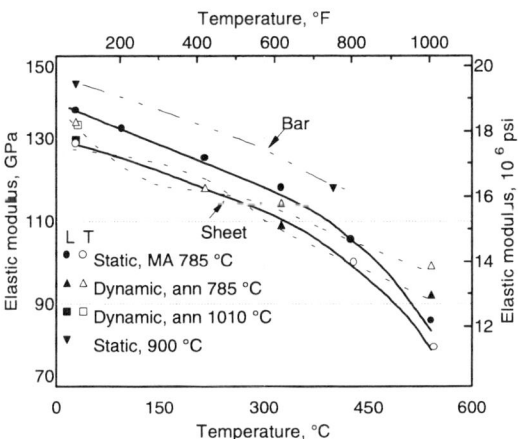

Sheet was 1.4 mm (0.056 in.) and 2.4 mm (0.096 in.) thick and annealed as indicated.
Source: *Aerospace Structural Metals Handbook*, Vol 4, Code 3709, Battelle Columbus Laboratories, 1966, p 20

Ti-8Al-1Mo-1V: Compressive modulus

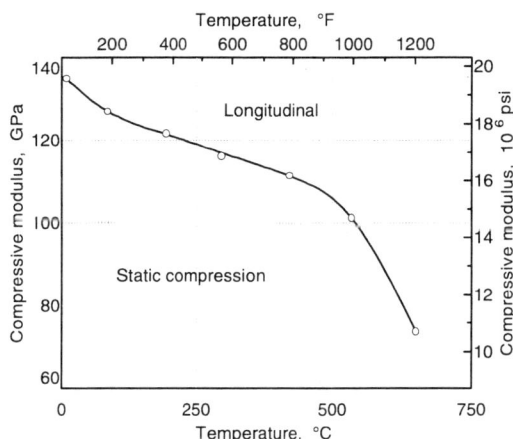

1.24 mm (0.049 in.) sheet mill annealed at 785 °C (1450 °F), 8 h.
Source: *Aerospace Structural Metals Handbook*, Vol 4, Code 3709, Battelle Columbus Laboratories, 1966, p 20

Ti-8Al-1Mo-1V: Compressive secant modulus

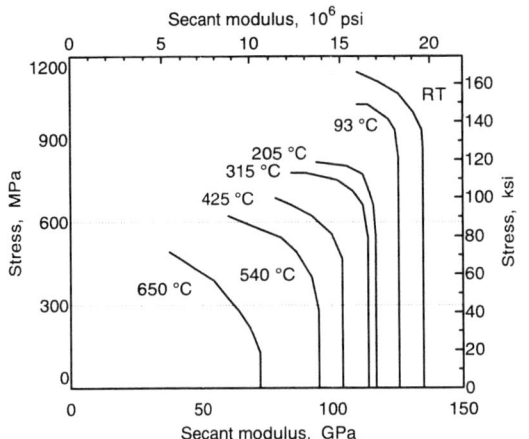

Ti-8Al-1Mo-1V: Compressive tangent modulus

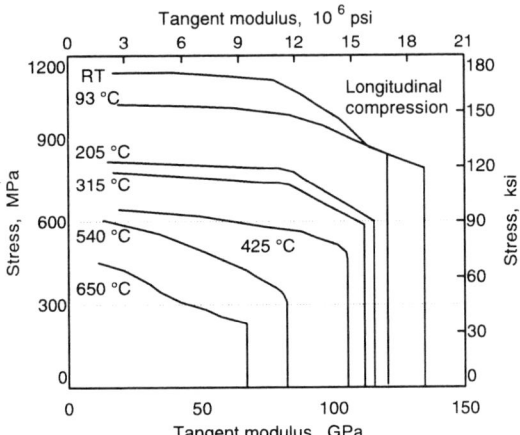

1.24 mm (0.049 in.) sheet mill annealed at 785 °C (1450 °F), 8 h.
Source: *Aerospace Structural Metals Handbook*, Vol 4, Code 3709, Battelle Columbus Laboratories, 1966, p 20

Electrical Resistivity

Ti-8Al-1Mo-1V: Electrical resistivity vs temperature

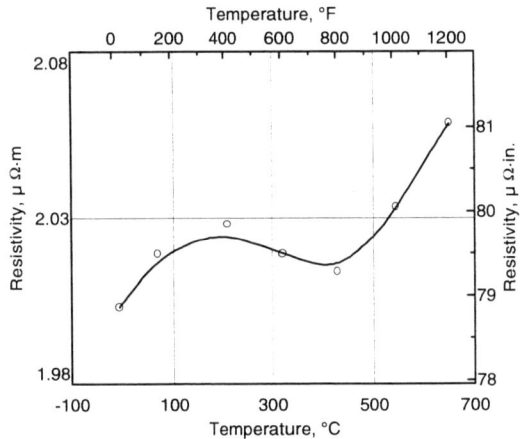

As-rolled bar.
Source: *Aerospace Structural Metals Handbook*, Vol 4, Code 3709, Battelle Columbus Laboratories, 1966, p 6

Ti-8Al-1Mo-1V: Repassivation time in neutral salt solution

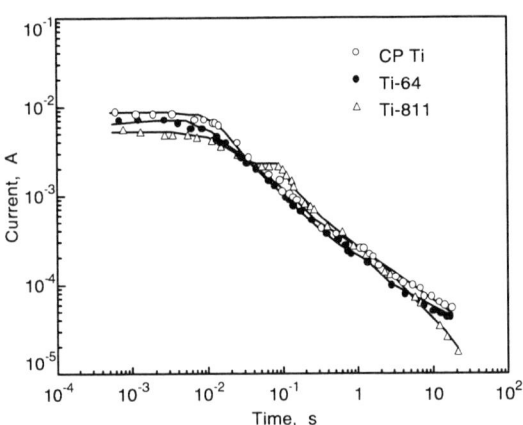

In 1 N NaCl at +200 mV.
Source: H.-J. Rätzer-Scheibe and H. Buhl, "Repassivation of Titanium and Titanium Alloys," *Titanium, Science and Technology*, Vol 4, Deutsche Gesselschaft für Metallkunde, 1985, p 2641

Corrosion Properties

Other than its high susceptibility to stress-corrosion cracking, few published data are available on the general or localized corrosion of Ti-8Al-1Mo-1V. One study, however, indicates repassivation behavior that is similar to that of CP Ti and Ti-6Al-4V in neutral salt solution (see above figure). Like other titanium alloys, successful application of Ti-811 can be expected in mildly reducing to highly oxidizing environments in which protective oxide films spontaneously form and remain stable. On the other hand, hot, concentrated, low-pH chloride salts corrode titanium. Warm or concentrated solutions of hydrochloric, phosphoric, and oxalic acids also are damaging. In general, all acidic solutions that are reducing in nature corrode titanium, unless they contain inhibitors. Strong oxidizers, including anhydrous red fuming nitric acid and 90% hydrogen peroxide, also cause attack. Ionizable fluoride compounds, such as sodium fluoride and hydrogen fluoride, activate the surface and can cause rapid corrosion. Dry chlorine gas is especially harmful.

Stress-Corrosion Cracking

Ti-8Al-1Mo-1V is one of the most susceptible titanium alloys to stress-corrosion cracking (SCC), which stems from the increased tendency to form the highly ordered Ti_3Al (α_2) phase when alumi-

Ti-8Al-1Mo-1V: SCC velocity in various environments

K_{Ic} and K_{Iscc} thresholds at 24 °C for double-cantilever beam specimens for 0.6M KCl solution, –500 mV

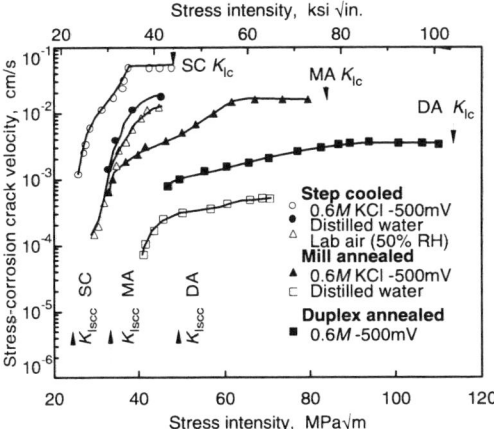

SC, step cooled; MA, mill annealed; DA, duplex annealed.
Source: M.J. Blackburn, J.A. Feeney, and T.R. Beck, "Stress Corrosion Cracking of Titanium Alloys," Boeing Report D1-82-1054, June 1970, p 92

Ti-8Al-1Mo-1V: Aqueous SCC vs chloride content

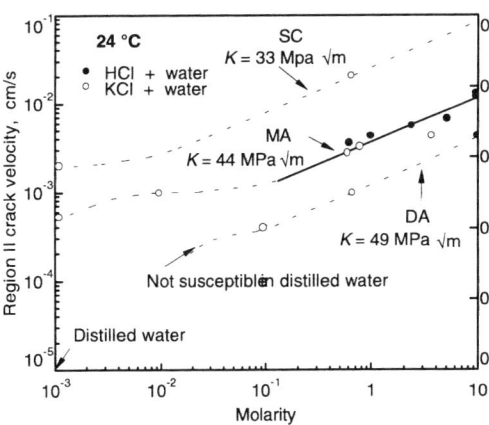

SC, step cooled; DA, duplex annealed; MA, mill annealed.
Source: T.R. Beck, M.J. Blackburn, W.H. Smyrl, and M.O. Speidel, "Stress-Corrosion Cracking of Titanium Alloys: Electrochemical Kinetics, SCC Studies with Ti:8-1-1, SCC and Polarization Curves in Molten Salts, Liquid Metal Embrittlement, and SCC Studies With Other Titanium Alloys," Contract NAS 7-489, Quarterly Progress Report 14, Dec 1969

num content exceeds 5 wt%. This low-ductility ordered phase forms in the 400 to 700 °C (750 to 1290 °F) temperature range and increases crack velocity and decreases K_{Iscc} as volume fraction increases. Oxygen levels above 0.20 to 0.25 wt% also promote SCC, again due to a transition from wavy to planar slip. The SCC fracture mode in Ti-811 and other susceptible α alloys is transgranular, with the fracture path highly oriented in unidirectionally processed (textured) material and more random in martensitic structures. References 1 to 7 in this section discuss the SCC behavior of Ti-811.

Ti-811 generally remains susceptible to SCC whether the structure is equiaxed α or martensitic (quenched from the β phase field). In general, SCC susceptibility of equiaxed α alloys diminishes with decreasing grain size, whereas step-cooled Ti-811 has the greatest susceptibility. Step-cooled Ti-811 has high susceptibility in that smooth specimens will produce cracking in ambient neutral salt solutions. In contrast, mill annealed Ti-811 has susceptibility in the presence of stress risers, such as a fatigue crack or machined notch. Step-cooled Ti-811 also exhibits SCC in distilled water.

Halide Solutions. Addition of halide ions such as Cl⁻, Br⁻, and I⁻ increases the susceptibility of Ti-811 and can induce susceptibility in other conditions that are immune to SCC in distilled water (see figures).

Other ionic species in solution can have a neutral or even an inhibitive effect on Ti-811 SCC if the alloy is not highly susceptible. These species include NO_3^-, SO_4^{-2}, F⁻, OH⁻, CrO_4^{-2}, and PO_4^{-3} (Ref 1, 2). The inhibitive influence diminishes as halide levels increase. Little influence of nonoxidizing cations such as Na⁺, K⁺, Li⁺, etc., is noted. However, oxidizing cations, such as Cu^{+2}, may raise K_{Iscc} values, depending on alloy heat treatment (Ref 3, 4).

Ti-8Al-1Mo-1V: Influence of halides on SCC

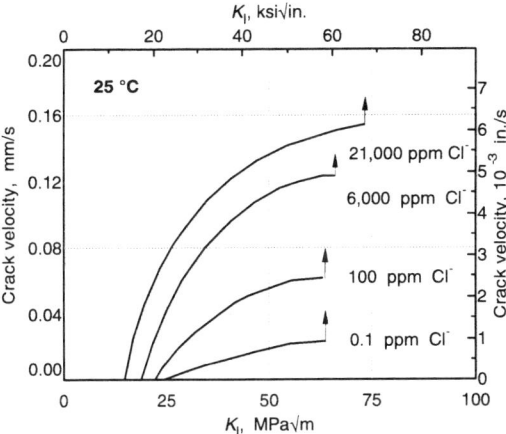

Source: J.D. Boyd, P.J. Moreland, W.K. Boyd, R.A. Wood, D.N. Williams, and R.I. Jaffee, "The Effect of Composition on the Mechanism of Stress-Corrosion Cracking of Titanium Alloys in N_2O_4, and Aqueous and Hot-Salt Environments," Contract NASr-100 (09), Aug 26, 1969

References

1. T.R. Beck, "Stress-Corrosion Cracking of Titanium Alloys. Preliminary Report on Ti-8Al-1Mo-1V Alloy and Proposed Electrochemical Mechanism," D1-82-0554, The Boeing Company, Seattle, July 1965
2. N.G. Feige and T. Murphy, "Fracture Behavior of Titanium Alloys in Aqueous Environments," Met. Eng. Quart., Vol 7(No. 1), 1967, p 53
3. M.J. Blackburn, J.A. Feeney, and T.R. Beck, "State-of-the-Art of Stress-Corrosion Cracking of Titanium Alloys," Part 4 of Monograph Review, The Boeing Company, Seattle, sponsored by the Advanced Research Projects Agency,

Ti-8Al-1Mo-1V: Environments known to produce SCC

Environment Medium	Temperature, °C	Other titanium alloys with known susceptibility
Organic compounds		
Methyl alcohol (anhydrous)	RT	Ti-6Al-4V, grade 2 Ti, Grade 4 Ti, Ti-4Al-3Mo-1V, Ti-3Al-8V-6Cr-4Zr-4Mo (Beta C), Ti-13V-11Cr-3Al, Ti-5Al-2.5Sn
Methyl chloroform	370	Ti-6Al-4V, Ti-5Al-2.5Sn, Ti-13V-11Cr-3Al
Ethyl alcohol (anhydrous)	RT	Ti-5Al-2.5Sn
Ethylene glycol	RT	No alloys other than Ti-8Al-1Mo-1V
Trichloroethylene	370, 620, 815	Ti-5Al-2.5Sn
Trichlorofluoroethane	788	Ti-5Al-2.5Sn, Ti-6Al-4V, Ti-13V-11Cr-3Al
Salts		
Hot salt: chloride and other halide salts/residues	230-430	Most commercial alloys except grades 1, 2, 7, 11, 12, and 9
Seawater/NaCl solution	RT	Unalloyed Ti (with oxygen content >0.3%) Ti-2.5Al-1Mo-11Sn-5Zr-0.2Si (IMI-679), Ti-3Al-11Cr-13V, Ti-5Al-2.5Sn, Ti-8Mn, Ti-6Al-4V, Ti-6Al-6V-2Sn, Ti-6Al-2Nb-1Ta, Ti-4Al-3Mo-1V, Ti-6Al-2Sn-4Zr-6Mo
Mercury (liquid)	370	Ti-13V-11Cr-3Al
Ag-5Al-2.5Mn (braze alloy)	340	Ti-6Al-4V
Miscellaneous		
Distilled water	RT	Ti-5Al-2.5Sn, Ti-11.5Mo-6Zr-4.5Sn (Beta III)
Chlorine gas	288	No alloys other than Ti-8Al-1Mo-1V
10% HCl	35, 340	Ti-5Al-2.5Sn
Molten chloride/bromide salts	300 - 500	No alloys other than Ti-8Al-1Mo-1V

Source: R. Schutz, Stress-Corrosion Cracking of Titanium Alloys, *Stress-Corrosion Cracking: Materials Performance and Evaluation*, ASM International, 1992

ARPA Order No. 878 and NAS7-489, June 1970
4. B.F. Brown, "Stress Corrosion Cracking in High Strength Steels and in Titanium and Aluminum Alloys," Naval Research Labs, Washington, DC, 1972
5. R.J.H. Wanhill, "Aqueous Stress Corrosion in Titanium Alloys," *Brit. Corrosion J.*, Vol 10(No. 2), 1975, p 69-78
6. R. Schutz, Stress-Corrosion Cracking of Titanium Alloys, in *Stress-Corrosion Cracking: Materials Performance and Evaluation*, ASM International, 1992
7. T.R. Beck, "Stress-Corrosion Cracking of Titanium Alloys. Preliminary Report on Ti-8Al-1Mo-1V Alloy and Proposed Electrochemical Mechanism," D1-82-0554, The Boeing Company, Seattle, July 1965

Aqueous K_{Iscc} Data

Because titanium alloys exhibit no Stage I type crack growth in neutral solutions (i.e., slowest crack velocities measured are approximately 10^{-3} cm/s), it may be concluded that true K_{Iscc} thresholds exist below which cracks will not propagate. This is not the case in highly acidic solutions, where both Stage I and II cracking behavior is observed. Increasing acidity generally reduces K_{Iscc} and increases Stage II cracking velocity (see figure for Ti-8Al-1Mo-1V). Increasing alkalinity appears to have no obvious or significant effect on SCC behavior relative to neutral conditions. As hydroxide concentrations exceed $1M$, increasing inhibition may be expected.

Temperature Effects. Only limited published data on the influence of temperature on titanium alloy SCC behavior in aqueous media are available. Data for Ti-811 suggest that K_{Iscc} values may be relatively unaffected by temperatures as high as 93 °C (200 °F) in neutral salt solutions. However, stress cracking velocity does increase sharply with increasing temperature.

Effect of Potential. In most SCC-susceptible titanium alloys, anodic or cathodic polarization tends to inhibit SCC and increase K_{Iscc} values. However, the anodic and cathodic polarization inhibition phenomenon is not as apparent in highly susceptible alloys, such as step-cooled Ti-811 (M.J. Blackburn et al., in Reference in previous section).

Ti-8Al-1Mo-1V: Fracture data in air and 3.5% NaCl solution at 25 °C

Alloy	Heat treatment	Yield strength MPa	ksi	K_{Ic} or K_C MPa√m	ksi√in.	K_{Iscc} or K_{scc} MPa√m	ksi√in.
Ti-8Al-1Mo-1V	Mill annealed	999	145	82	75	33	30
	Duplex annealed	930	135	176	160	55	50
	Mill annealed	999	145	52	48	22	20
	Duplex annealed	930	135	110	100	35	32
	Mill annealed, WQ	841	122	>110	>100	46	42
	βST, WQ	868	126	>110	>100	>110	>100
Ti-6Al-4V (standard grade)	Mill annealed	944	137	66	60	38	35
	Duplex annealed	917	133	77	70	57	52
	STA	1103	160	51	47	27	25
	β-STA	1068	155	77	70	49	45

Note: Data for 13 mm (0.5 in.) plate were generated in ambient neutral 3.5% NaCl solution. It should be cautioned that these K_{Iscc} values are highly dependent on alloy composition, metallurgical condition, and product form and thickness and, therefore, may not be representative of alloy product materials commercially available. Source: R. Schutz, Stress-Corrosion Cracking of Titanium Alloys, in *Stress-Corrosion Cracking: Materials Performance and Evaluation*, ASM International, 1992

Ti-8Al-1Mo-1V: Influence of acidity on SCC at 24 °C

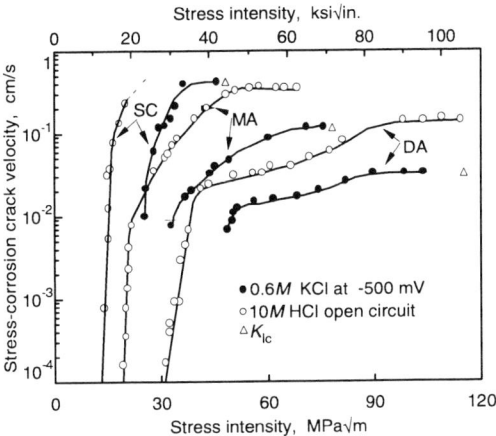

SC, step cooled; MA, mill annealed; DA, duplex annealed.
Source: T.R. Beck, M.J. Blackburn, W.H. Smyrl, and M.O. Speidel, "Stress-Corrosion Cracking of Titanium Alloys: Electrochemical Kinetics, SCC Studies with Ti: 8-1-1, SCC and Polarization Curves in Molten Salts, Liquid Metal Embrittlement, and SCC Studies with Other Titanium Alloys," Contract NAS 7-489, Quarterly Progress Report 14, Dec 1969

Ti-8Al-1Mo-1V: SCC initiation load vs potential at 24 °C

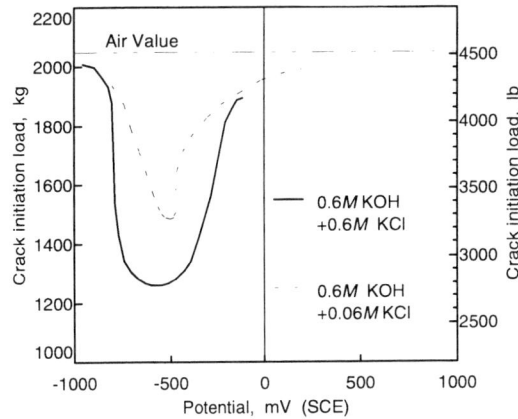

Source: M.J. Blackburn et al., Boeing Report D1-82-1054, June 1970

Ti-8Al-1Mo-1V: Crack initiation load vs potential at 24 °C

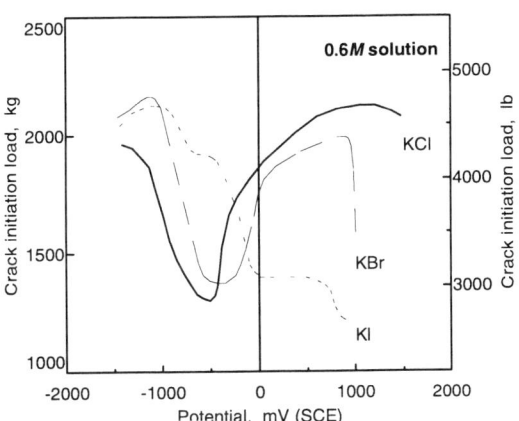

Duplex annealed SEN specimens.
Source: T.R. Beck and M.J. Blackburn, Stress-Corrosion Cracking of Titanium Alloys, AIAA J., Vol 6(No. 2), 1968, p 326

Ti-8Al-1Mo-1V: Effect of temperature on SCC velocity

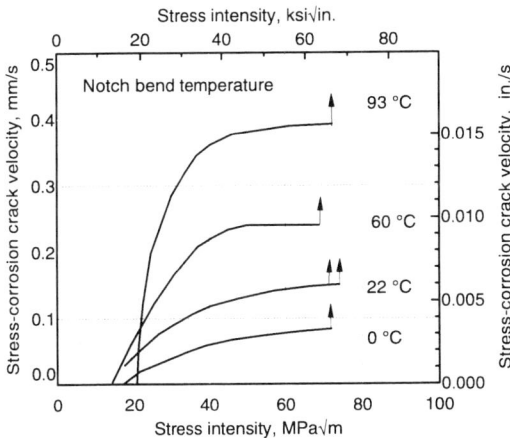

Source: M.J. Blackburn et al., Boeing Report D1-82-1054, June 1970

Aqueous SCC Velocity

The same factors that lower K_{Iscc} values generally increase SCC velocity in aqueous media. For example, higher acidity increases crack velocity and lowers K_{Iscc}. However, SCC velocity appears to be more temperature sensitive than K_{Iscc} values. Increasing potential also serves to increase stress cracking velocity in neutral halide solutions. This correlation is generally linear (see figure on next page). In highly acidic solutions, however, Stage II crack velocity is independent of applied potential. As a result, inhibition via cathodic polarization is not achievable in highly acidic solutions. Viscosity also has an effect.

Methanol/Water SCC

Ti-811 exhibits transgranular SCC in methanol like many other titanium alloys. Stress-corrosion cracking in pure methanol is difficult to observe for many susceptible titanium alloys, but it is more prone for Ti-811. The minimum level of water required for full SCC inhibition depends on alloy composition, metallurgical condition, temperature, halide level, acidity, and other species present. Alloys susceptible to SCC in distilled water, such as step-cooled Ti-811, cannot be inhibited regardless of water content. Increasing temperature increases Stage II cracking velocity and

Ti-8Al-1Mo-1V: Effect of temperature on SCC velocity

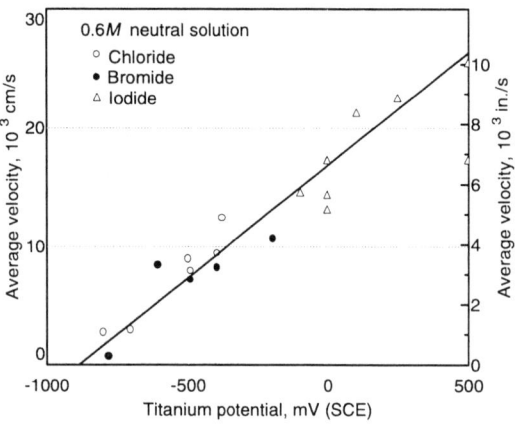

Mill annealed double cantilever beam specimens, 10M HCl, open circuit.
Source: M.J. Blackburn et al., Boeing Report D1-82-1054, June 1970

Ti-8Al-1Mo-1V: SCC velocity vs potential

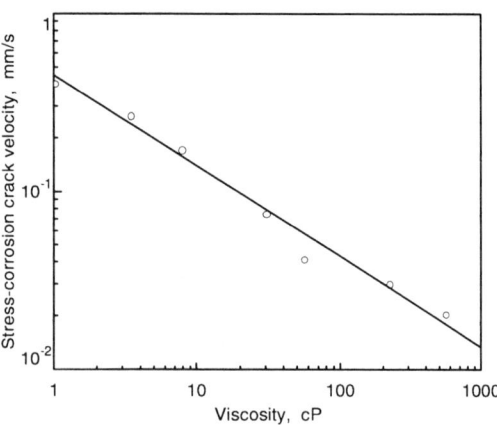

Relationship of stress-corrosion cracking plateau velocity to potential in 0.6M neutral halide solutions

Ti-8Al-1Mo-1V: SCC velocity vs viscosity at 24 °C

Variation of crack velocity with viscosity in 1M HCl water-glycerine solutions. SEN specimens; $K \cong 49$ MPa$\sqrt{m}$ (45 ksi$\sqrt{in.}$).
Source: J.A. Beavers, G.H. Koch, and W.E. Berry, *Corrosion of Metals in Marine Environments*, Metals and Ceramics Information Center, Battelle Columbus Laboratories, July 1986

Ti-8Al-1Mo-1V: SCC in methanol

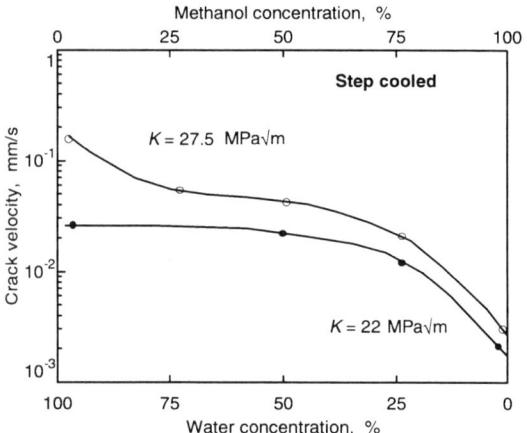

M.J. Blackburn et al., Boeing Report DI-82-1054, June 1970, p 131

appears to reduce K_{Iscc} values in pure methanol.

Methanol/Halide SCC

Like other titanium alloys, transgranular SCC of Ti-811 in methanol accelerates with halogen/halide additions. Water additions have an inhibitive effect, and numerous species have been found that inhibit transgranular (Stage II) stress cracking in methanol depending on halide level. These include nitrate and sulfate ions, NaF, and 0.1M concentrations of Al^{+3}, Zr^{+4}, Cd^+, and Sn^{+2} metallic ions (*Corrosion*, Vol 27, 1971, p 46-48). Investigators have also demonstrated that titanium alloy SCC in neutral methanol/halide solutions can be fully arrested by applied cathodic potentials of –1.0 to –1.5 V (SCE).

SCC in Other Alcohols

Although SCC susceptibility is limited primarily to methanol liquid and vapor for most commercial titanium alloys, a few highly susceptible alloys such as Ti-811 may also be affected by other low-molecular-weight alcohols. The highly susceptible Ti-811 alloy uniquely experiences cracking in non-halogen/halide-containing anhydrous ethanol and

Ti-8Al-1Mo-1V: Methanol/halide SCC vs heat treatment

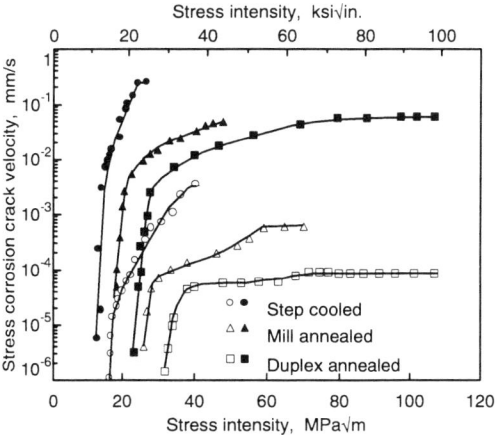

Crack velocity vs stress-intensity relationships for three heat treatments of double cantilever beam specimens tested in pure (Spectrograde) methanol and methanol-KI solutions at 24 °C. Open symbols tested in Spectrograde. Closed symbols tested in Spectrograde + 0.25M KI/methanol; –500 mV (SCE); double cantilever beam.
Source M.J. Blackburn, 1970.

Ti-8Al-1Mo-1V: Effect of temperature on SCC velocity

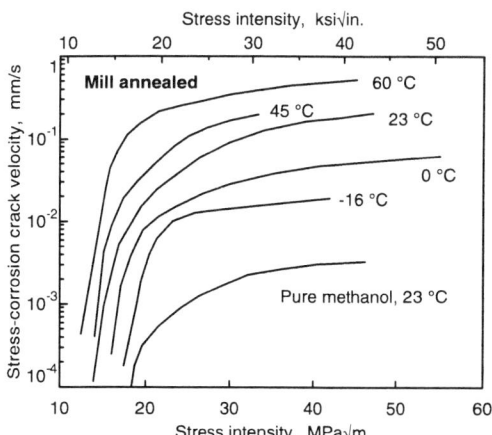

0.25M KI/methanol –500 mV(SCE); double cantilever beam.
Source: M.J. Blackburn et al., Boeing Report D1-82-1054, June 1970

Ti-8Al-1Mo-1V: Effect of NO_3^- on crack initiation at 24 °C

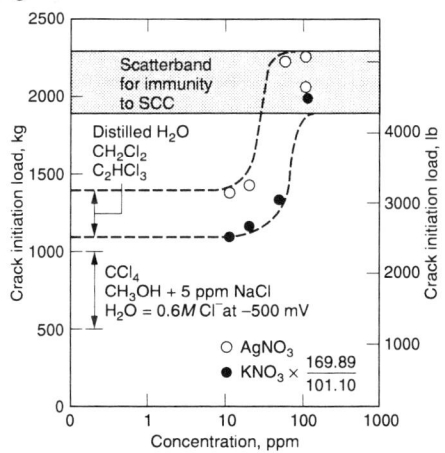

SEN specimens.
Source: M.J. Blackburn et al., Boeing Report D1-82-1054, June 1970

Ti-811: SCC velocity at 24 °C

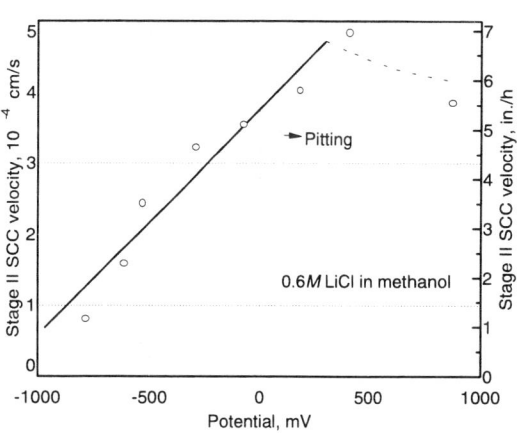

SEN specimens.
T.R. Beck and M.J. Blackburn, AIAA J., Vol 6(No. 2), 1968, p 326

ethylene glycol. This phenomenon appears to disappear in this alloy when alcohol carbon atom chain size exceeds three.

The addition of halogens, such as Cl_2, Br_2, or I_2, or other (nonoxygen-containing) strong oxidizers (i.e., $FeCl_3$) to various anhydrous alcohols also can induce SCC in all titanium alloys, even unalloyed grades. Depending on the alloy and on the oxidizer concentration, much higher water levels are required for SCC inhibition.

Halogenated/Hydrocarbons

Widespread use of titanium alloys in the aerospace industry has prompted considerable study of SCC in halogenated hydrocarbons common to aerospace processing. Carbon tetrachloride (CCl_4) SCC was first noted in Ti-811 (see selected references on next page). The threshold stress intensity was approximately the same as that observed for SCC in 3.5% NaCl. Crack velocities in CCl_4 were approximately ten times faster than velocities in methanol. Studies on dynamically loaded smooth specimens (category 3) also showed that Ti-5Al-2.5Sn was susceptible to SCC in CCl_4 at stresses approaching the tensile strength of the alloy.

Other hydrocarbons found to cause cracking in Ti-811 and Ti-5Al-2.5Sn are:

- Carbon tetrachloride
- Methylene chloride
- Methylene iodide
- Trichloroethylene

Ti-8Al-1Mo-1V: SCC in air, butane, and decane

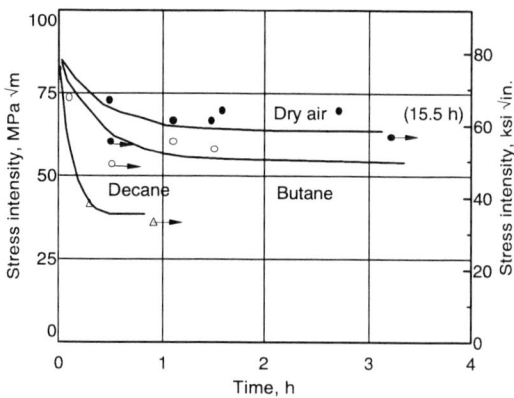

Effect of stress intensity on time to fracture of non-side-grooved 6 mm (0.25 in.) mill annealed specimens (848 MPa tensile yield strength) tested in dry air, n-butane, and n-decane.
Source: G. Sandoz, "Subcritical Crack Propagation in Ti-8Al-1Mo-1V Alloy in Organic Environments, Salt Water and Inert Environments," Proc. Conf. Fundamental Aspects of Stress Corrosion Cracking, NACE, Sept 11-15, 1967, p 684

Ti-8Al-1Mo-1V: K_{Iscc} in various alcohols

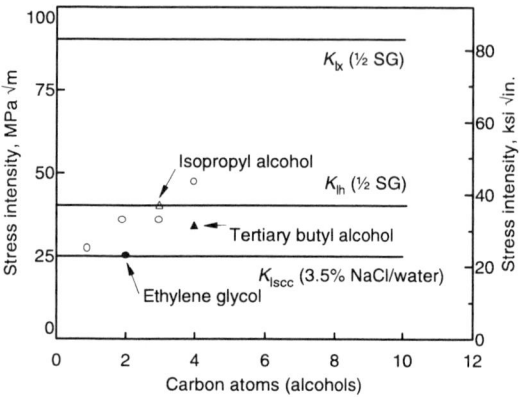

Apparent threshold stress intensity (K_{Iscc} in 360 min) for mill annealed specimens (848 MPa tensile yield strength) tested in alcohols with increasing number of carbon atoms.
Threshold stress intensity for crack growth and fracture of side-grooved ½ SG in. specimens (½) tested in normal alcohols containing increasing number of carbon atoms. Similar results with other alcohols and glycol as indicated.
Source: G. Sandoz, "Subcritical Crack Propagation in Ti-8Al-1Mo-1V Alloy in Organic Environments, Salt Water, and Inert Environments," Proc. Conf. Fundamental Aspects of Stress-Corrosion Cracking, NACE, 1969, p 684

Ti-8Al-1Mo-1V: K_{Iscc} in alkanes

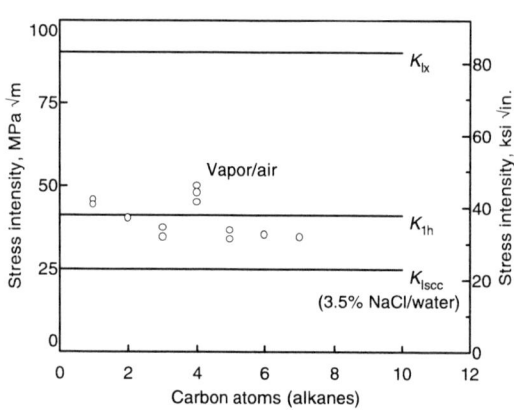

Apparent threshold stress intensity (K_{Iscc} in 360 min) for mill annealed specimens (848 MPa tensile yield strength) tested in normal alkanes with increasing number of carbon atoms.
Source: G. Sandoz, "Subcritical Crack Propagation in Ti-8Al-1Mo-1V Alloy in Organic Environments, Salt Water, and Inert Environments," Proc. Conf. Fundamental Aspects of Stress-Corrosion Cracking, NACE, 1969, p 684

Ti-8Al-1Mo-1V: SCC velocity in carbon tetrachloride

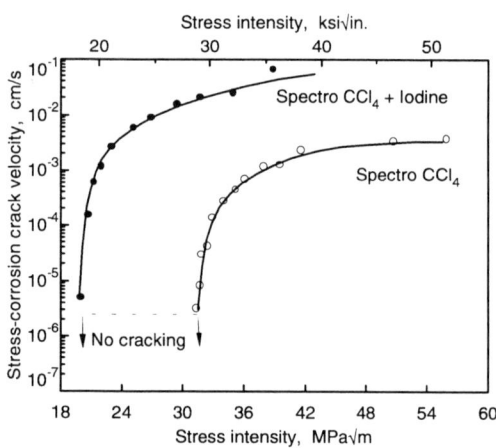

Mill annealed double cantilever beam specimens at 25 °C.
Source: M.J. Blackburn et al., Boeing Report D1-82-1054, June 1970

- Trichlorofluoromethane
- Trichlorofluoroethane
- Octafluorocyclobutane
- Freon

In most of these environments, precracked specimens are needed for susceptibility. No other titanium alloys were found to be similarly affected.

Selected References

- K.E. Weber, J.S. Fritzer, D.S. Cowgill, and W.C. Gillchriest, Similarities in Titanium Stress-Corrosion Cracking Processes in Salt Water and in Carbon Tetrachloride, in "Accelerated Crack Propagation of Titanium by Methanol, Halogenated Hydrocarbons, and Other Solutions," DMIC Memorandum 228, Defense Metals Information Center, Battelle Memorial Institute, March 1967, p 39

- H.R. Herrigel, Titanium U-Bends in Organic Liquids: Effect of Inhibitors, in "Accelerated Crack Propagation of Titanium by Methanol, Halogenated Hydrocarbons, and Other Solutions," DMIC Memorandum 228, Defense Metals Information Center, Battelle Memorial Institute, March 1967, p 16

- T.R. Beck, M.J. Blackburn, W.H. Smyrl, and M.O. Speidel, "Stress-Corrosion Cracking of Titanium Alloys: Electrochemical Kinetics, SCC Studies with Ti:8-1-1, SCC and Polarization Curves in Molten Salts, Liquid Metal Embrittlement, and SCC Studies With Other Titanium Alloys," Quarterly Progress Report 14, Contract NAS 7-489, Boeing Scientific Research Laboratories, Dec 1969

- T.R. Beck and M.J. Blackburn, Stress-Corrosion Cracking of Titanium Alloys, *AIAA J.*, Vol 6 (No. 2), 1968, p 326

- C.C. Seastrom and R.A. Gorski, The Influence of Fluorocarbon Solvents on Titanium Alloys, in "Accelerated Crack Propagation of Titanium by Methanol, Halogenated Hydrocarbons, and Other Solutions," DMIC Memorandum 228, Defense Metals Information Center, Battelle Memorial Institute, March 1967, p 20

Hot-Salt Cracking

Ti-811 is one of the least resistant titanium alloys to hot-salt cracking. Oxygen is necessary for hot-salt cracking to occur. Although the role of water (moisture) has not been clearly established, it appears that water is also a necessary environmental component in the cracking process (see selected references below).

Chloride, bromide, and iodide salts have all been shown to produce similar cracking. Fluoride and hydroxide salts have not. The cation associated with the salt has also been reported to affect cracking susceptibility. The severity of attack has been shown to increase as follows:

$MgCl_2 > SrCl_2 > CsCl > CaCl_2 > KCl > BaCl_2 > NaCl > LiCl$

Cracking is normally intergranular in nature, but it depends largely on alloy type. Alpha alloys exhibit both transgranular and intergranular fracture, depending on whether the material was annealed above or below the β transus, respectively. Alpha-beta alloys exhibit predominantly intergranular fracture.

From a practical standpoint, hot-salt cracking appears to be a phenomenon that is restricted to the laboratory. As indicated earlier, no in-service failure has been attributed to hot-salt cracking.

Ti-8Al-1Mo-1V: SCC velocity in halogenated hydrocarbons

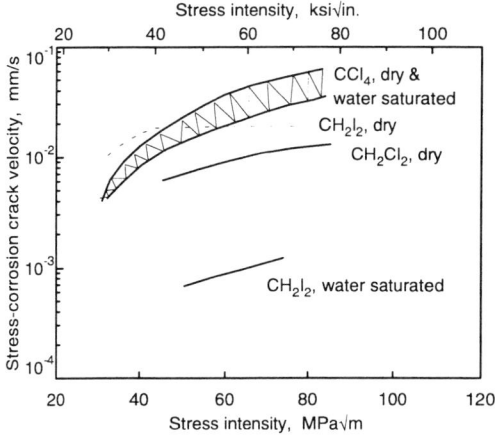

Ti-8Al-1Mo-1V: Hot-salt threshold stresses

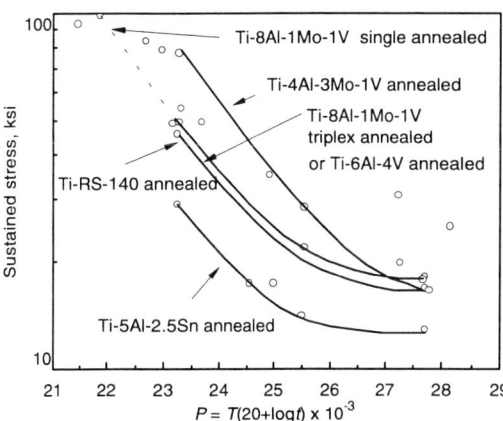

Source: R.V. Turley and C.H. Avery, Elevated-Temperature Static and Dynamic Sea-Salt Stress Cracking of Titanium Alloys, in *Stress-Corrosion Cracking of Titanium*, ASTM STP 397, ASTM, 1965

Ti-8Al-1Mo-1V: SCC velocity in fused salt at 375 °C

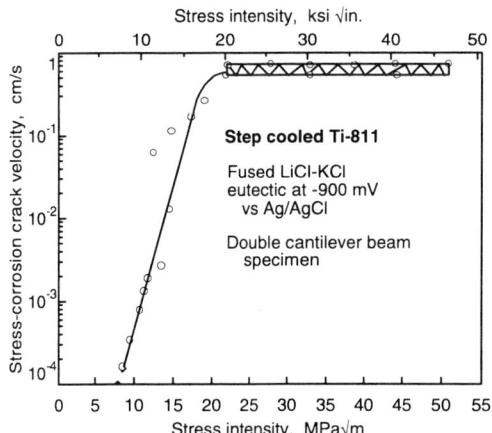

Source: M.J. Blackburn *et al.*, Boeing Report D1-82-1054, June 1970

388 / Alpha and Near-Alpha Alloys

The likely reason for this is the critical relationship among environment, stress level, and alloy type. Unless all of the conditions are met simultaneously and for extended time, cracking will not occur.

Selected References
- H.L. Logan, Studies of Hot-Salt Cracking of the Titanium-8% Al-1% Mo-1% V Alloy, in *Proceedings of Conference—Fundamental Aspects of Stress-Corrosion Cracking*, National Association of Corrosion Engineers, 1969, p 662
- H.L. Logan, M.J. McBee, G.M. Ugiansky, C.J. Bechtoldt, and B.T. Sanderson, *Stress-Corrosion Cracking of Titanium*, STP 397, American Society for Testing and Materials, 1965, p 215

Molten Salt

It would appear that Ti-811 is the only titanium alloy tested for SCC in molten salt environments. Cracking has been observed in pure chloride and bromide eutectic melts at temperatures between 300 and 500 °C (570 and 930 °F). In general, increasing temperature increases crack velocity. Cathodic protection has been observed to inhibit or stop cracking.

Nitrate salts below 125 °C (255 °F) do not induce cracking even when Cl⁻, Br⁻, or I⁻ anions are present. At higher temperatures in pure molten nitrates,

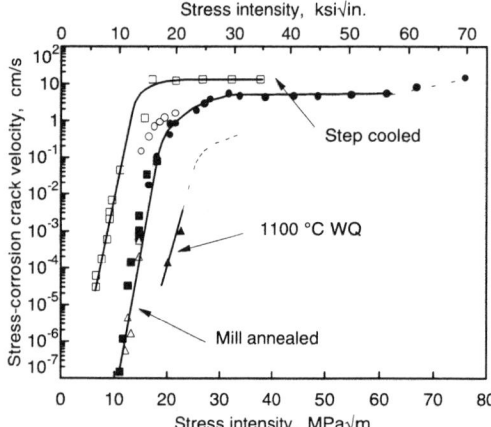

Ti-8Al-1Mo-1V: SCC velocity in mercury at 24 °C

Double cantilever beam specimens.
Source: M.J. Blackburn *et al.*, Boeing Report D1-82-1054, June 1970

cracking can occur only when halides are present (T.R. Beck *et al.*, Quarterly Progress Report 14, Boeing Scientific Research Laboratories, Dec 1969).

Metal Embrittlement

Several metals, both in liquid and solid form, have been found to induce cracking in contact with titanium alloys. Ti-811 has known susceptibility to cadmium, mercury, and silver brazing alloys.

Thermal Properties

Thermal Expansion

Specific Heat. At room temperature, 502 J/kg · K (0.120 Btu/lb · °F)

Melting Temperature. About 1540 °C (2800 °F)

Ti-8Al-1Mo-1V: Mean thermal coefficient of linear expansion

Temperature		Mean thermal coefficient of expansion	
°C	°F	10⁻⁶/°C	10⁻⁶/°F
0-100	32-212	8.5	4.7
0-315	32-600	9.0	5.0
0-540	32-1000	10.1	5.6
0-650	32-1200	10.3	5.7

Source: "Chase Extrusions," Chase Brass and Copper

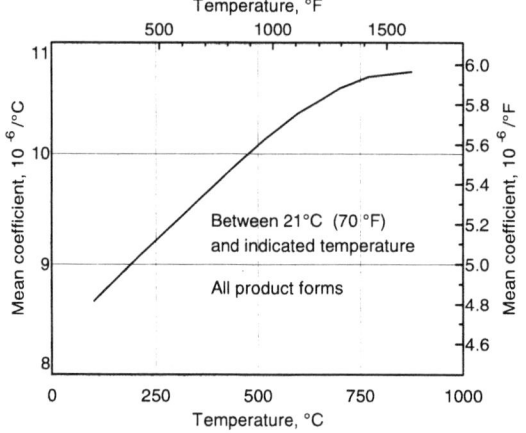

Ti-8Al-1Mo-1V: Thermal coefficient of linear expansion

Source: "Metallic Materials and Elements for Aerospace Vehicle Structures," MIL-HDBK-5E, Vol 2, 1987

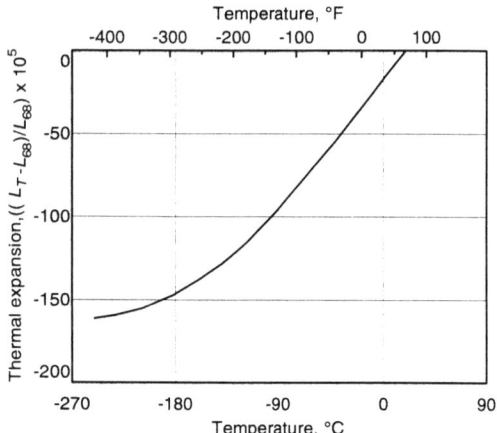

Ti-8Al-1Mo-1V: Thermal expansion at low temperature

Alloy used as 19 mm (0.750 in.) diameter rod, single annealed.
Source: Cryogenic Materials Data Handbook, Vol 1, AFML-TR-64-280, Wright Patterson AFB, Ohio, revised 1970

Ti-8Al-1Mo-1V: Coefficient of linear expansion

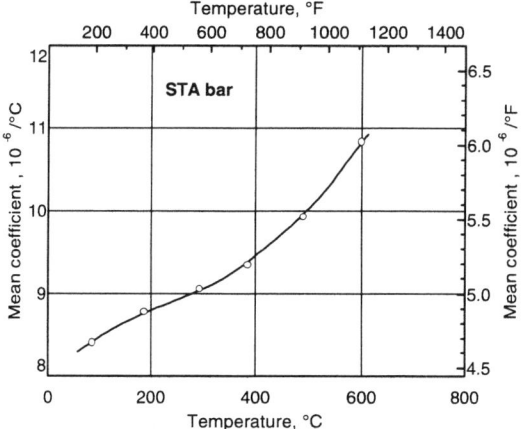

Alloy used as bar, heat treated at 900 °C (1650 °F) for 1 h and air cooled, then aged at 595 °C (1100 °F) for 24 h and air cooled.
Source: *Aerospace Structural Metals Handbook*, Vol 4, Code 3709, Battelle Columbus Laboratories, 1966

Thermal Conductivity

Ti-8Al-1Mo-1V: Thermal conductivity

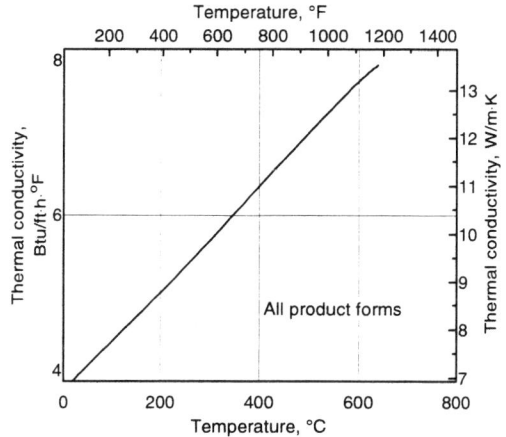

Source: "Metallic Materials and Elements for Aerospace Vehicle Structures," MIL-HDBK-5E, Vol 2, 1987

Ti-8Al-1Mo-1V: Thermal conductivity

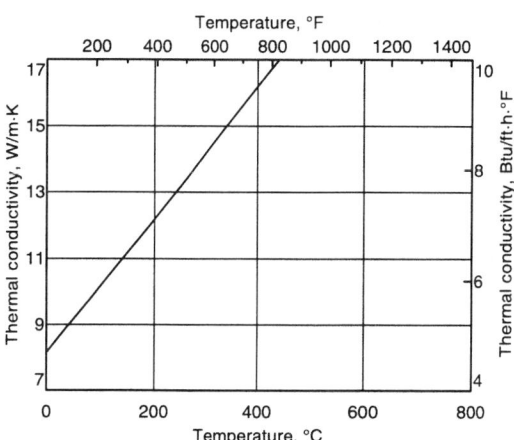

Source: *Aerospace Structural Metals Handbook*, Vol 4, Code 3709, Battelle Columbus Laboratories, 1966

Mechanical Properties

Mechanical properties are also included in "Heat Treatment" section

Design Allowables

Ti-8Al-1Mo-1V: S-basis tensile properties of plate, bar, and forgings

Product thickness		Ultimate tensile strength(a)		Tensile yield strength(a)		Elongation(a),
mm	in.	MPa	ksi	MPa	ksi	%
Single-annealed plate per MIL-T-9046						
4.7-13	0.1875-0.500	999	145	930	135	10
13-25	0.50-1.000	965	140	896	130	10
25-64	1.00-2.500	896	130	827	120	10
64-100	2.50-4.000	827(b)	120(b)	758(b)	110(b)	8(b)
Duplex-annealed plate per MIL-T-9046						
4.7-13	0.1875-0.500	896	130	827	120	10
13-25	0.50-1.000	896	130	827	120	10
25-50	1.00-2.000	862	125	793	115	10
50-100	2.00-4.000	827	120	758	110	8

(continued)

Ti-8Al-1Mo-1V: S-basis tensile properties of plate, bar, and forgings (continued)

Product thickness		Ultimate tensile strength(a)		Tensile yield strength(a)		Elongation(a),
mm	in.	MPa	ksi	MPa	ksi	%
Single-annealed bar per MIL-T-9047						
<64(c)	<2.500(c)	896(d)	130(d)	827	120(d)	10(d)
64-100(c)	2.50-4.000(c)	827(b)	120(b)	758(b)	110(b)	10(d)(e)
Solution treated and stabilized forging (AMS 4973)						
<63	<2.499	896(f)	130(f)	827(f)	120(f)	10(f)
64-100	2.500-4.000	827(g)	120(g)	758(g)	110(g)	10(g)

(a) Applicable in the longitudinal and long transverse directions, except as noted. (b) Applicable in ST direction, providing ST dimension is greater than 75 mm (3.00 in.). (c) Maximum of 103 cm^2 (16 in.^{2}) area. (d) Applicable in LT dimension if it is greater than 75 mm (3.0 in.). (e) ST elongation is 8% if ST dimension is greater than 75 mm (3.0 in.). (f) Applicable in LT dimension if it is greater than 63.5 mm (2.5 in.). (g) Applicable in both LT and ST direction. Source: MIL-HDBK-5, Dec 1991

Ti-8Al-1Mo-1V: S-basis mechanical properties of sheet

Thickness,	Single annealed	Duplex annealed	
mm (in.)	≤4.7 (≤0.1875)	0.4-0.6 (0.015-0.024)	0.6-4.7 (0.025-0.1875)
F_{tu}, MPa (ksi):			
L	999 (145)	930 (135)	930 (135)
LT	999 (145)	930 (135)	930 (135)
ST	...	...	...
F_{ty}, MPa (ksi):			
L	930 (135)	827 (120)	827 (120)
LT	930 (135)	827 (120)	827 (120)
ST	...	...	...
F_{cy}, MPa (ksi):			
L	993 (144)	868 (126)	868 (126)
LT	1027 (149)	868 (126)	868 (126)
ST	...	...	...
F_{su}, MPa (ksi):	641 (93)	579 (84)	519 (84)
F_{bru}, MPa (ksi):			
($e/D = 1.5$)	1648 (239)	1537 (223)	1537 (223)
($e/D = 2.0$)	2027 (294)	1854 (269)	1854 (269)
F_{bry}, MPa (ksi):			
($e/D = 1.5$)	1351 (196)	1199 (174)	1199 (174)
($e/D = 2.0$)	1475 (214)	1317 (191)	1317 (191)
EL, %			
L	(a)	8	10
LT	(a)	8	10
ST	...	...	...

(a) 0.2 to 0.3 mm (0.008 to 0.014 in.) thickness, 6%; 0.4 to 0.6 mm (0.015 to 0.024 in.) thickness, 8%; >0.6 mm (>0.025 in.) thickness, 10%. Source: MIL-HDBK-5 per AMS 4915 and MIL-T-9046, Comp A-4

Hardness

Ti-8Al-1Mo-1V: Effect of hydrogen content on hardness

Processing/condition	Thickness		Hydrogen, ppm	Hardness, HRC
	mm	in.		
As received, mill annealed	17	0.673	14-21	36
As received, mill annealed	17	0.673	14-21	36
Annealed at 870 °C (1600 °F) in vacuum, 8 h, furnace cooled	17	0.673	5	33
Annealed at 870 °C (1600 °F) in vacuum, 8 h, furnace cooled (twice)	17	0.673	7	33
Annealed at 870 °C (1600 °F), hydrogenated, 8 h, furnace cooled	17	0.673	39	33
Annealed at 1065 °C (1945 °F) in vacuum, 4 h, furnace cooled	15	0.597	21	~30

Note: Alloy was received in the hot rolled and mill annealed condition. Chemical composition was 7.92 wt% Al, 0.03 wt% C, 0.15 wt% Fe, H as indicated, 0.98 wt% Mo, 0.01 wt% N, 0.11 wt% O, and 1.01 wt% V. Source: D. Meyn, "Effect of Hydrogen on Fracture and Inert-Environment Sustained Load Cracking Resistance of Alpha-Beta Titanium Alloys," *Metall. Trans.*, Vol 5, 1974, p 2405

Ti-8Al-1Mo-1V: Effect of solution temperature on hardness

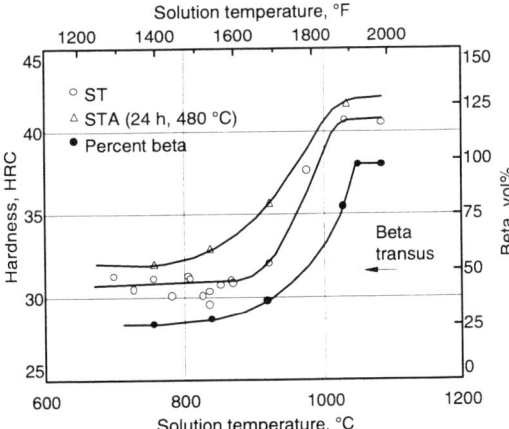

Alloy was supplied in the form of 64 mm (2.5 in.) square bar stock. Chemical composition was 7.6 wt% Al, 0.022 wt% C, 0.06 wt% Fe, 0.005 wt% H, 1.1 wt% Mo, 0.008 wt% N, 0.09 wt% O, and 1.1 wt% V. Hardness measurements were made on 13 mm (0.5 in.) cubes in the plane normal to the rolling direction. Cube surfaces had been ground at least 1.3 mm (0.050 in.) and mechanically polished. Hardness was also determined for Jominy bar that had been sectioned along the center line and surface ground. Specimens were solution treated in air for 1 h, followed by brine quench.
Source: P. Fopiano and C. Hickey, "The Effect of Heat Treatment on the Mechanical Properties of the Alloy Ti-8Al-1Mo-1V," in *Titanium, Science and Technology*, R. Jaffee and H. Burte, Ed., 1973, p 2009

Ti-8Al-1Mo-1V: Typical room-temperature hardness

Product condition	Hardness Rockwell C	Knoop
Duplex annealed	35 HRC	...
Unwelded sheet	36 HRC	325
Single-bead weld	35.2 HRC	345

Source: *Metals Handbook*, Vol 2, 10th ed., 1990, p 617-621

Typical Tensile Properties

Ti-8Al-1Mo-1V: Typical transverse tensile strengths

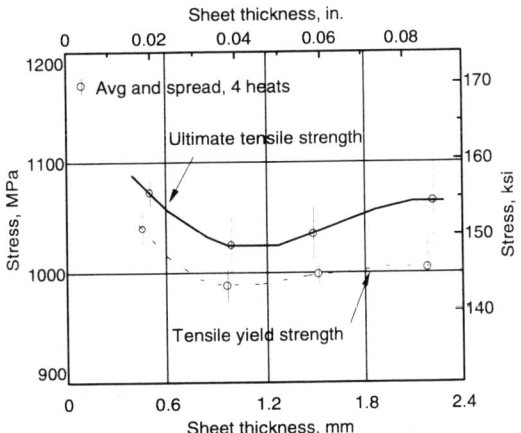

Data were from four heats of sheet mill annealed at 785 °C (1450 °F), 4 h.
Source: *Aerospace Structural Metals Handbook*, Vol 4, Code 3709, Battelle Columbus Laboratories, 1966

Ti-8Al-1Mo-1V: Typical longitudinal tensile strengths

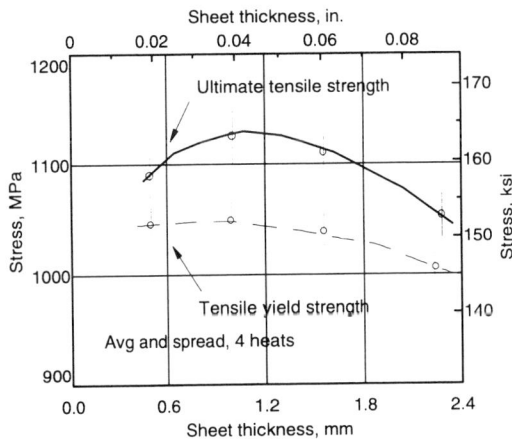

Data were from four heats of sheet mill annealed at 785 °C (1450 °F), 4 h.
Source: *Aerospace Structural Metals Handbook*, Vol 4, Code 3709, Battelle Columbus Laboratories, 1966

Ti-8Al-1Mo-1V: Effect of thermal, stressed exposure on room-temperature tensile properties

Rolling temperature		Heat treatment	Exposure conditions				Time, h	Ultimate tensile strength		Tensile properties Tensile yield strength (0.2% offset)		Elongation, %	Reduction of area, %
°C	°F		Temperature		Stress								
			°C	°F	MPa	ksi		MPa	ksi	MPa	ksi		
17 mm (¹¹⁄₁₆ in.) diameter bar													
950	1750	760 °C (1400 °F), 24 h, AC	No exposure (control)					937	136	910	132	18	47
			540	1000	103	15	200	1013	147	986	143	15	38
1010	1850	900 °C (1650 °F), 1 h, AC + 595 °C (1100 °F), 24 h	No exposure (control)					1048	152	1020	148	18	40
			540	1000	103	15	150	1061	154	1006	146	19	40
			540	1000	207	30	150	1075	156	986	143	24	40
1065	1950	900 °C (1650 °F), 1 h, AC + 595 °C (1100 °F), 24 h	No exposure (control)					1144	166	1034	150	17	26
			540	1000	103	15	150	1158	168	1020	148	15	21
			540	1000	207	30	150	1158	168	1006	146	16	18
1.6 mm (0.063 in.) sheet													
1010	1850	925 °C (1700 °F), ½ h, AC + 760 °C (1400 °F), 2 h	No exposure (control)					1013	147	910	132	15	...
			425	800	448	65	150	1013	147	924	134	6	...
			540	1000	172	25	150	1068	155	1013	147	19	...
1.0 mm (0.040 in.) sheet (solution treated and aged)													
1010	1850	980 °C (1800 °F), 5 min, WQ + 540 °C (1000 °F), 8 h	No exposure (control)					1179	171	1068	155	5	...
			540	1000	172	25	150	1186	172	1068	155	10	...
1010	1850	980 °C (1800 °F), 5 min, WQ + 650 °C (1200 °F), 8 h	No exposure (control)					1130	164	1048	152	7	...
			540	1000	172	25	150	1186	172	1089	158	6	...

Source: R. Wood and R. Favor, *Titanium Alloys Handbook*, MCIC-HB-02, Battelle, 1972

High-Temperature Strength

Tensile Strength vs Temperature

Ti-8Al-1Mo-1V: Typical tensile properties of beta annealed specimens

Test temperature		Ultimate tensile strength		Tensile yield strength (0.2% offset)		Elongation in 50 mm (2 in.), %	Reduction of area, %
°C	°F	MPa	ksi	MPa	ksi		
	RT	1020	148	896	130	13	19
93	200	951	138	793	115	13	20
205	400	855	124	668	97	12	21
315	600	779	113	586	85	11	23
425	800	724	105	551	80	10	26
540	1000	655	95	503	73	14	30

Note: 13 mm (1/2 in.) round specimens, rolled from 1065 °C (1950 °F), annealed 1 h, at 1065 °C (1950 °F), air cooled, reheated 8 h at 595 °C (1100 °F). Source: *Alloy Digest*, Jan 1962

Ti-8Al-1Mo-1V: Typical tensile properties of alpha-beta annealed specimens

Test temperature		Ultimate tensile strength		Tensile yield strength (0.2% offset)		Elongation in 50 mm (2 in.), %	Reduction of area, %
°C	°F	MPa	ksi	MPa	ksi		
	RT	999	145	896	130	19.0	48
93	200	910	132	758	110	20.0	49
205	400	813	118	634	92	20.5	50
315	600	744	108	551	80	20.0	52
425	800	689	100	517	75	20.0	55
540	1000	593	86	448	65	28.0	70

Note: Rolled in beta field, annealed 1 h at 900 °C (1650 °F), air cooled, and reannealed 8 h at 595 °C (1100 °F). Source: *Alloy Digest*, Jan 1962

Ti-8Al-1Mo-1V: Tensile strength vs temperature of sheet

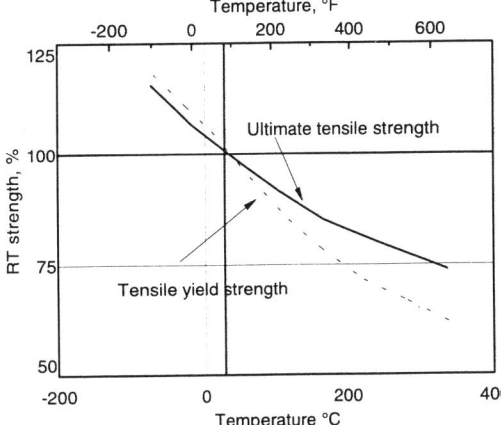

Duplex annealed sheet.
Source: MIL-HDBK-5, 1 Dec 1991

Ti-8Al-1Mo-1V: Tensile strength vs temperature of sheet

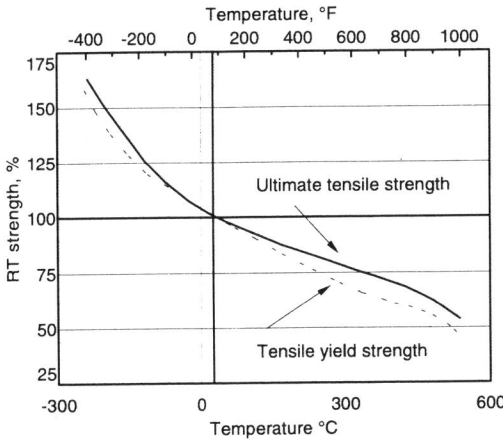

Single annealed sheet.
Source: MIL-HDBK-5, 1 Dec 1991

Ti-8Al-1Mo-1V: Tensile strength vs temperature of sheet

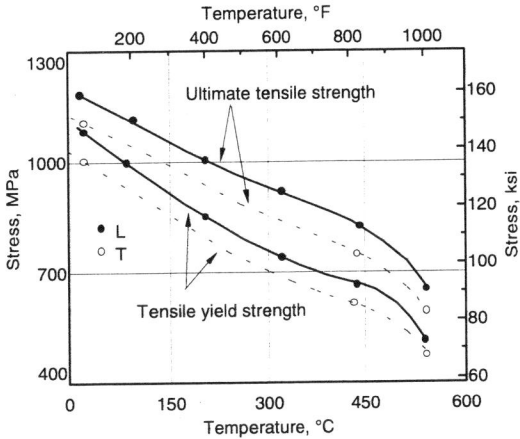

1.4 mm (0.056 in.) sheet mill annealed, 785 °C (1450 °F), 8 h.
Source: *Aerospace Structural Metals Handbook*, Vol 4, Code 3709, Battelle Columbus Laboratories, 1966

Ti-8Al-1Mo-1V: Tensile strength vs temperature of sheet

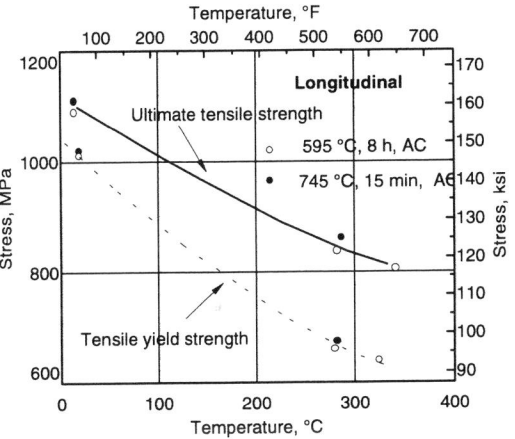

Effect of test temperature on two heats of triplex annealed sheet.
Source: *Aerospace Structural Metals Handbook*, Vol 4, Code 3709, Battelle Columbus Laboratories, 1966

Bearing, Compressive and Shear Strength

Ti-8Al-1Mo-1V: Bearing strength vs temperature of sheet

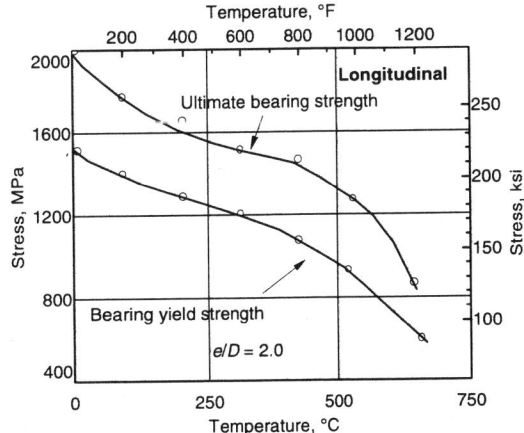

1.2 mm (0.049 in.) sheet mill annealed, 785 °C (1450 °F), 8 h.
Source: *Aerospace Structural Metals Handbook*, Vol 4, Code 3709, Battelle Columbus Laboratories, 1966

Ti-8Al-1Mo-1V: Compressive yield strength vs temperature

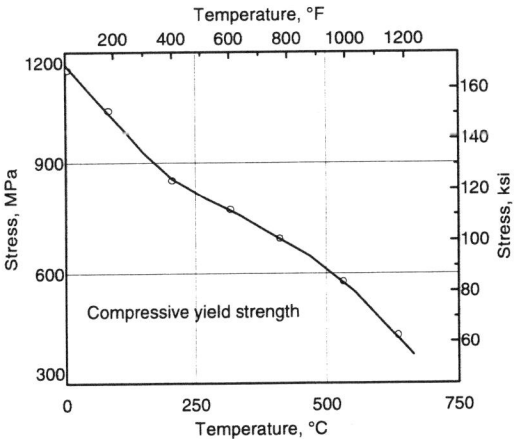

1.2 mm (0.049 in.) sheet mill annealed, 785 °C (1450 °F), 8 h.
Source: *Aerospace Structural Metals Handbook*, Vol 4, Code 3709, Battelle Columbus Laboratories, 1966

394 / Alpha and Near-Alpha Alloys

Ti-8Al-1Mo-1V: Typical shear strength of bar

Duplex anneal	Test temperature °C	°F	Shear strength(a) MPa	ksi	Ratio of shear/UTS
900 °C (1650 °F), 1 h, AC + 595 °C (1100 °F), 8 h, AC	RT		671	97.3	0.67
	315	600	497	72.1	0.67
	425	800	471	68.3	0.68
	540	1000	449	65.2	0.75
980 °C (1800 °F), 1 h, AC + 595 °C (1100 °F), 8 h, AC	RT		674	97.8	0.69
	315	600	496	72.0	0.68
	425	800	467	67.8	0.69
	540	1000	444	64.4	0.77
1065 °C (1950 °F), 1 h, AC + 595 °C (1100 °F), 8 h, AC	RT		641	93.0	0.63
	315	600	477	69.2	0.62
	425	800	450	65.3	0.61
	540	1000	433	62.9	0.65

Note: 13 mm (1/2 in.) round bar rolled from 1065 °C (1950 °F). (a) 9.5 mm (3/8 in.) diameter shear pins. Source: *Alloy Digest*, Jan 1962

Ti-8Al-1Mo-1V: Ultimate shear strength vs temperature

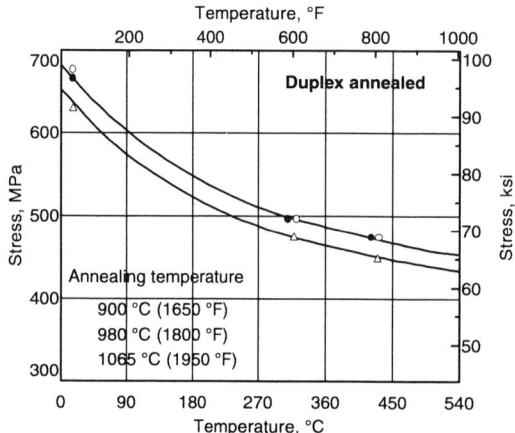

13 mm (0.5 in.) diam bar, annealed, 1 h, AC + 595 °C (1100 °F), 8 h, AC.
Source: *Aerospace Structural Metals Handbook*, Vol 4, Code 3709, Battelle Columbus Laboratories, 1966

Creep Properties

Ti-8Al-1Mo-1V: Creep strength at 315 and 480 °C

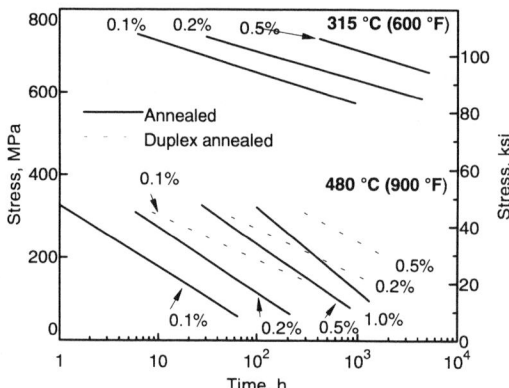

Mill annealed 1.3 mm (0.050 in.) thick sheet was heated at 790 °C (1450 °F) for 8 h; duplex annealed 1.3 mm (0.050 in.) thick sheet was heated at 1010 °C (1850 °F) for 5 min, air cooled and held at 745 °C (1375 °F) for 15 min, air cooled.
Source: *Aerospace Structural Metals Handbook*, Vol 4, Code 3709, Battelle Columbus Laboratories, 1966

Ti-8Al-1Mo-1V: Stress for 0.1% creep

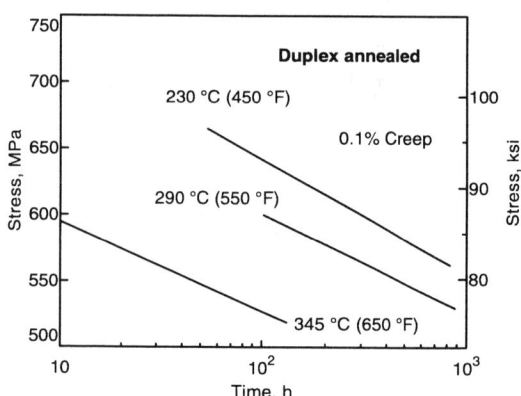

Sheet 1.3 mm (0.050 in.) thick was heated at 790 °C (1450 °F) for 8 h, furnace cooled, then held at 790 °C (1450 °F) for 15 min, air cooled
Source: *Aerospace Structural Metals Handbook*, Vol 4, Code 3709, Battelle Columbus Laboratories, 1966

Ti-8Al-1Mo-1V: Stress-rupture data at 540 °C (1000 °F)

Rolled 13 mm (1.2 in.) bar stock as a function of solution annealing temperature. A constant stabilizing anneal of 595 °C (1100 °F) for 24 h was used after the indicated solution anneal

Solution annealing temperature		Stress		Rupture time,
°C	°F	MPa	ksi	h
900	1650	551	80	4.0
		482	70	17.5
		400	58	65
		344	50	160
980	1800	537	78	15
		482	70	40
		413	60	80
		344	50	202
1065	1950	537	78	20
		413	60	150
		331	48	504

Source: *Alloy Digest*, Jan 1962

Ti-8Al-1Mo-1V: Stress rupture at 540 °C (1000 °F)

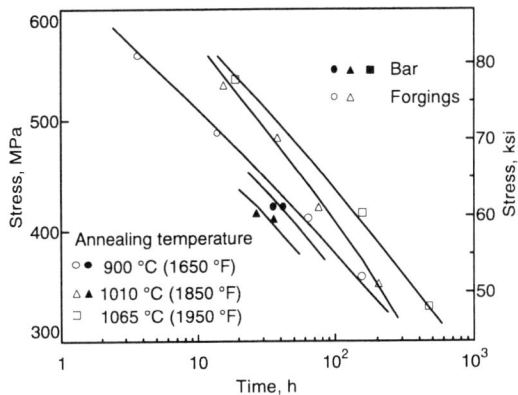

Bar 13 mm (0.5 in.) in diameter was heated at temperatures indicated, then at 595 °C (1100 °F) for 24 h. Forgings 38 mm (1.5 in.) thick were heated at temperatures indicated and at 595 °C (1100 °F) for 8 h.
Source: *Aerospace Structural Metals Handbook*, Vol 4, Code 3709, Battelle Columbus Laboratories, 1966

Fatigue Properties

Ti-8Al-1Mo-1V: Typical rotating beam fatigue of rolled barstock

Condition	Stress MPa	ksi	Cycles to failure
Simplex anneal(a) with a	724	105	45,000 and 55,000
937 MPa (136 ksi) UTS	689	100	50,000
and 9903 MPa (131 ksi)	655	95	200,000
TYS	620	90	140,000
Duplex anneal(b) with a	724	105	85,000
1013 MPa (147 ksi) UTS	689	100	140,000
and 951 MPa (138 ksi)	655	95	200,000 and 300,000
TYS	620	90	1,100,000 and 3,000,000

(a) 760 °C (1400 °F), 24 h, AC. (b) 980 °C (1800 °F), 4 h, AC + 540 °C (1000 °F), 24 h, AC. Source: *Alloy Digest*, Jan 1962

Unnotched Fatigue Life

Ti-8Al-1Mo-1V: Best-fit S/N curves for unnotched sheet at RT

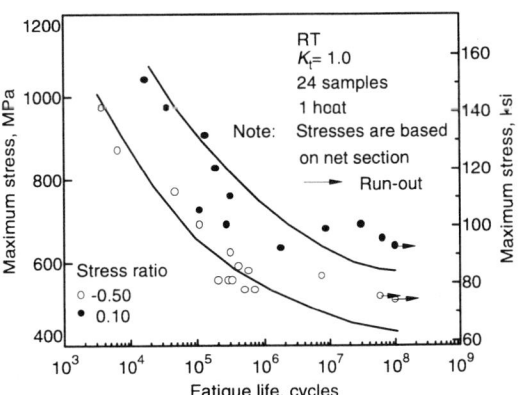

See table. *Caution*: The equivalent stress model may provide unrealistic life predictions for stress ratios beyond those represented.
Source: MIL-HDBK-5, Dec 1991

Test conditions for best-fit S/N curves

Test Material Common Name:	Ti-8Al-1Mo-1V
Specification Designation:	N/A
Composition:	N/A
Product form:	1.3 mm (0.050 in.) sheet
Heat treatment:	Duplex annealed
Condition/microstructure:	N/A
Hardness:	N/A
Modulus of elasticity (avg at RT):	N/A
RT tensile strength/elongation:	1014.9 MPa (147.2 ksi)
RT yield strength:	934.9 MPa (135.6 ksi)
Test Parameters	
Test temperature:	RT, 200 °C (400 °F) and 345 °C (675 °F)
Test environment:	Air
Failure criterion:	Fracture
Loading condition:	Axial (see figures for R ratio)
Specimen orientation:	Long-transverse direction
Specimen geometry:	Unnotched, 19 mm (0.75 in.) net width
Surface:	HNO_3/HF pickled
Gauge length:	N/A
Frequency:	1800 cycles/min
Strain rate:	N/A
Waveform:	N/A
Test specifications:	N/A
Remarks/Reference:	MIL-HDBK-5, Dec 1991

Ti-8Al-1Mo-1V: Best-fit S/N curves at 200 °C

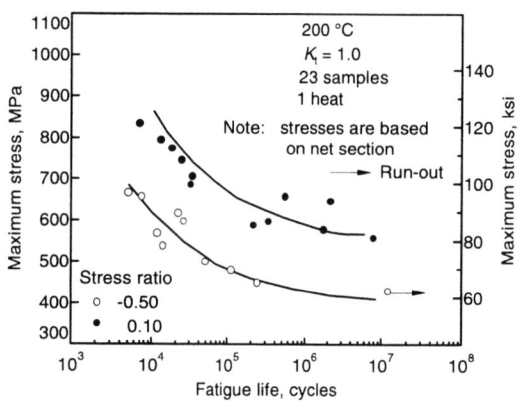

See table on previous page for test conditions. UTS at 200 °C (400 °F) was 825 MPa (119.5 ksi). *Caution:* The equivalent stress model may provide unrealistic life predictions for stress ratios beyond those represented.
Source: MIL-HDBK-5, Dec 1991

Ti-8Al-1Mo-1V: Best-fit S/N curves at 345 °C

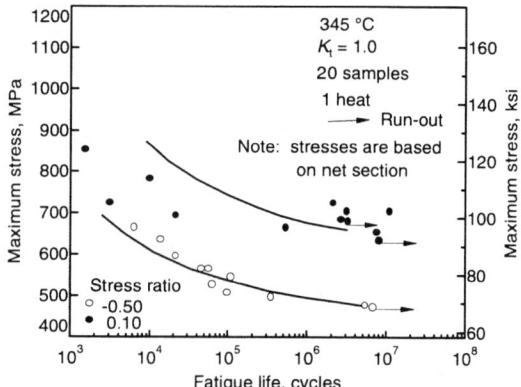

See table on previous page for test conditions. UTS at 345 °C (650 °F) was 760 MPa (110.2 ksi). *Caution:* The equivalent stress model may provide unrealistic life predictions for stress ratios beyond those represented.
Source: MIL-HDBK-5, Dec 1991

Notched Fatigue Life

Ti-8Al-1Mo-1V: Fatigue crack growth data
Duplex annealed 1.27 mm (0.05 in.) specimens

Design mean stress		Fatigue life, flights	Crack initiation(a)		Crack growth(b)	
MPa	ksi		Period, flights	Percentage of total life, flights	Period, flights	Percentage of total life, flights
Accelerated tests at room temperature(c)						
172	25	137 158	127 000	93	10 160	7
207	30	18 540	13 300	72	5 240	28
241	35	7 472	5 300	71	2 170	29
172	25	105 988	78 000	74	27 990	26
207	30	57 290	40 300	70	16 990	30
241	35	22 290	12 000	54	10 290	46
Accelerated tests at 560 K(c)						
172	25	36 243	25 500	70	10 740	30
207	30	12 498				
241	35	4 580	2 850	62	1 730	38
Real-time tests(d)						
172	25	19 014	12 000	63	7 010	37
207	30	10 420	8 100	78	2 320	22
241	35	5 093				

(a) For cracks extending 1 mm (0.04 in.) from the notch. (b) For cracks from 1 mm (0.04 in.) long until failure. (c) One test at each design stress. (d) Median value from test. Source: L.A. Imig, "Crack Growth in Ti-8Al-1Mo-1V with Real-Time and Accelerated Flight-by-Flight Loading," *Fatigue Crack Growth under Spectrum Loads*, ASTM STP 595, ASTM, 1976, p 251-264

Ti-8Al-1Mo-1V: Best-fit S/N curves for notched sheet at RT

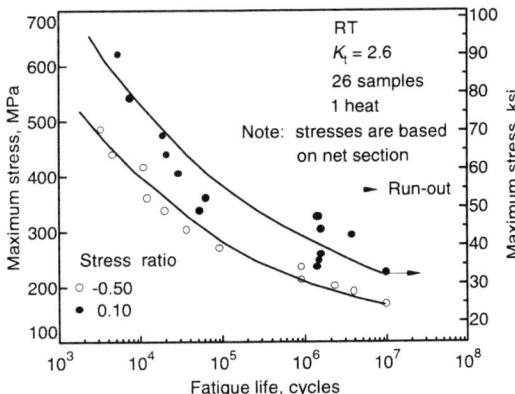

See table for test conditions. *Caution*: The equivalent stress model may provide unrealistic life predictions for stress ratios beyond those represented.
Source: MIL-HDBK-5, Dec 1991

Ti-8Al-1Mo-1V: Best-fit S/N curves at 200 °C

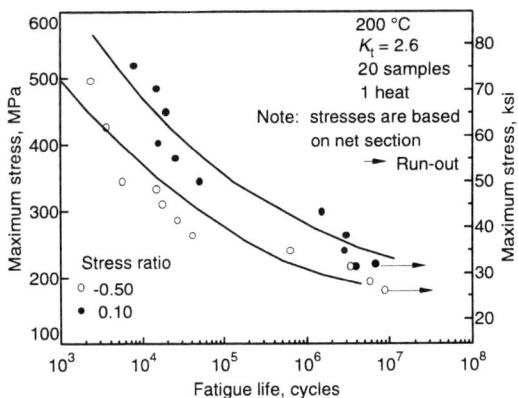

See table for test conditions. *Caution*: The equivalent stress model may provide unrealistic life predictions for stress ratios beyond those represented.
Source: MIL-HDBK-5, Dec 1991

Ti-8Al-1Mo-1V: Best-fit S/N curves at 345 °C

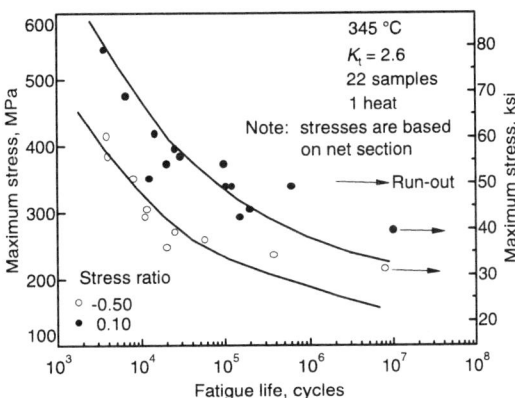

See table. *Caution*: The equivalent stress model may provide unrealistic life predictions for stress ratios beyond those represented.
Source: MIL-HDBK-5, Dec 1991

Test conditions for best-fit SN curves

Test Material Common Name:	Ti-8Al-1Mo-1V
Specification Designation:	N/A
Composition:	N/A
Product form:	1.33 mm (0.050 in.) sheet
Heat treatment:	Duplex annealed
RT tensile strength/elongation:	1014.9 MPa (147.2 ksi)
RT yield strength:	934.9 MPa (135.6 ksi)
Test Parameters	
Test temperature:	RT, 200 °C (400 °F), and 345 °C (675 °F)
Test environment:	Air
Failure criterion:	Fracture
Loading condition:	Axial (see figures for *R* ratios)
Specimen orientation:	Long transverse grain direction
Specimen geometry:	Notched, hole type, K_t = 2.6; 38 mm (1.500 in.) gross width; 31.7 mm (1.250 in.) net width; 6.4 mm (0.250 in.) diameter hole
Surface:	HNO_3/HF pickled
Gauge length:	N/A
Frequency:	1800 cycles/min
Remarks/Reference:	MIL-HDBK-5, Dec 1991

Ti-8Al-1Mo-1V: Notched fatigue at low temperatures

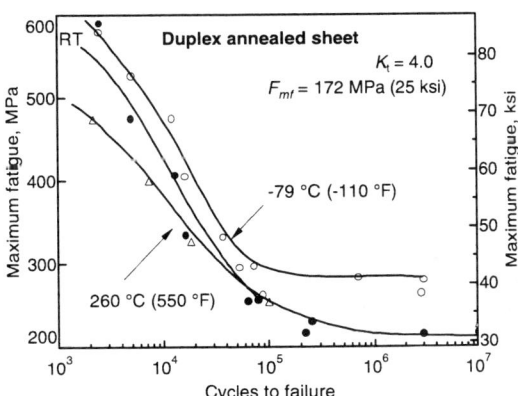

1 mm (0.04 in.) sheet duplex annealed 1010 °C (1850 °F), 5 min, AC + 745 °C (1375 °F), 15 min, AC.
Source: *Aerospace Structural Metals Handbook*, Vol 4, Code 3709, Battelle Columbus Laboratories, 1966

Ti-8Al-1Mo-1V: Axial load sharp notch fatigue

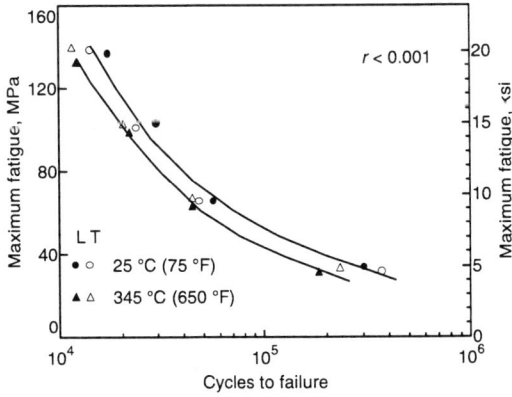

Duplex annealed 0.635 mm (0.025 in.) sheet, 1010 °C (1850 °F), 5 min, AC + 745 °C (1375 °F), 15 min, AC. Notched on both sides after annealing with 60° notch.
Source: *Aerospace Structural Metals Handbook*, Vol 4, Code 3709, Battelle Columbus Laboratories, 1966

Implant Material Fatigue

Ti-8Al-1Mo-1V: Fatigue in salt solution

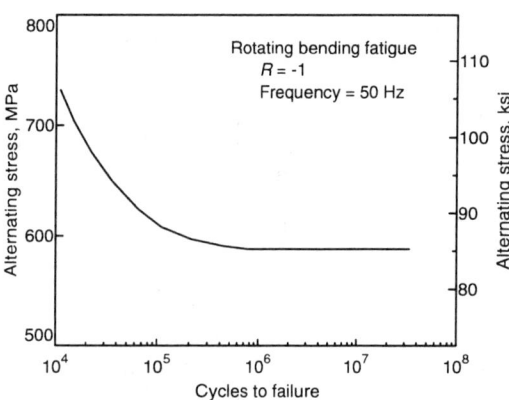

Source: M. Levy, et al., The Corrosion Behavior of Titanium Alloys in Chloride Solutions: Materials for Surgical Implants, in *Titanium, Science and Technology*, R.I. Jaffee and H.M. Burte, Ed., 1973, p 2459-2474

Fatigue-Crack Growth

Forged Fan Blades

Fatigue-crack growth rate of Ti-811 at ambient temperatures in both room air and 3.5% NaCl solution for compact-tension specimens made from forged first-stage-turbine fan blades are shown in the accompanying figures. Environment/lower frequency had little effect at low and high ΔK levels, but caused a significant increase in crack-growth rate at intermediate ΔK levels. Comparison of the data with results for conventionally rolled Ti-811 material revealed that the fatigue-crack growth rates for forged specimens were nearly the same as those for the conventionally rolled specimens.

Ti-8Al-1Mo-1V: Crack growth in fan blade specimens

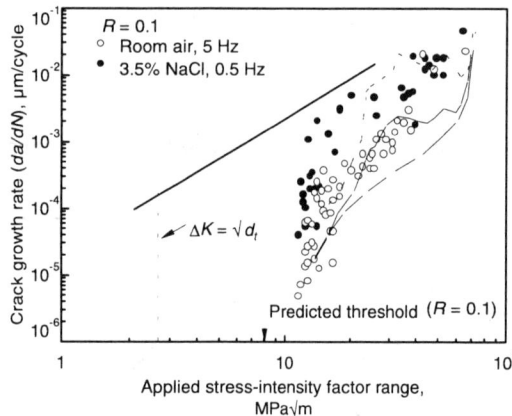

Specimens were cut from blades to simulate suspected in-service fracture path.
Source: Reported in MCIC-81-42 from W.H. Cullen and F.R. Stonesifer, "Fatigue-Crack-Growth Analysis of Titanium Gas-Turbine Fan Blades," NAVAIR, NRL-MR-3378, Oct 1976, ADA031836

Environmental Effects

Ti-8Al-1Mo-1V: Crack growth vs environment

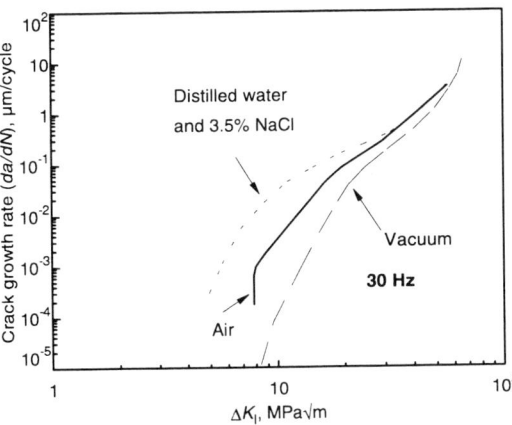

Source: H. Doker and D. Munz, Influence of Environment on the Fatigue Crack Propagation of Two Titanium Alloys, in *The Influence of Environment on Fatigue*, Institution of Mechanical Engineers, London, 1977, p 123-130

Ti-8Al-1Mo-1V: Crack growth vs CP titanium

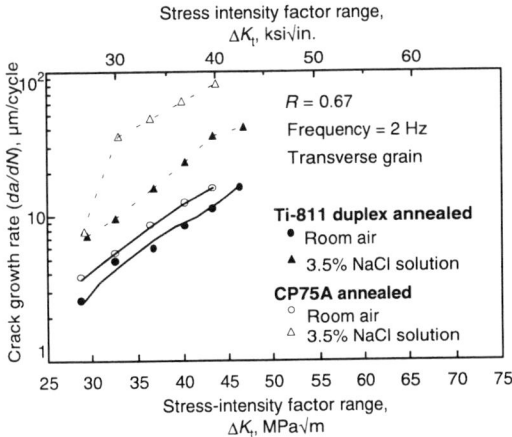

Source: M.J. Blackburn et al., Boeing Report D1-82-1054, June 1970

Ti-8Al-1Mo-1V: Microstructure and corrosion fatigue

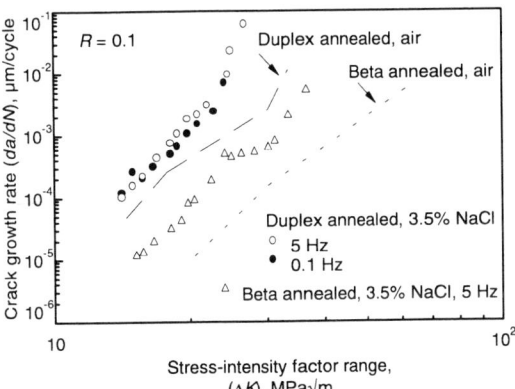

Corrosion-fatigue crack growth rates in 3.5% NaCl solution for duplex annealed and beta annealed microstructures show the expected improvement from transformed structures of beta treatment.
Source: G.R. Yoder, L.A. Cooley, and T.W. Crooker, "Improvement of Environmental Crack Propagation Resistance in Ti-8Al-1Mo-1V Through Microstructural Modification," Final Report No. NRL-MR-3955, Mar 1979, ADA069084

Effect of Frequency

Ti-8Al-1Mo-1V: Effect of frequency on da/dN vs ΔK

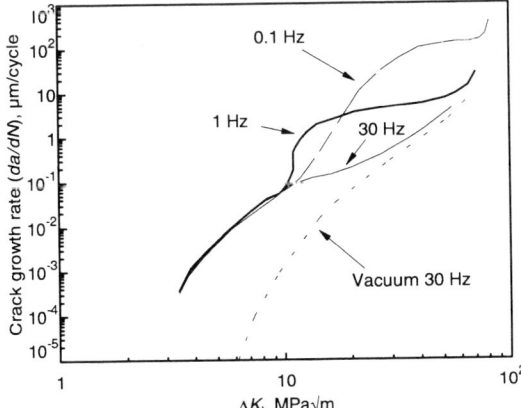

Effect of frequency on da/dN versus ΔK curves in 3.5% sodium chloride solution. All fatigue tests were run at $R = 0.1$ with sinusoidal waveform.
Source: H. Doker and D. Munz, Influence of Environment on the Fatigue Crack Propagation of Two Titanium Alloys, in *The Influence of Environment on Fatigue*, Institution of Mechanical Engineers, London, 1977, p 123-130

Fracture Properties

Impact Toughness

Ti-8Al-1Mo-1V: Effects of rolling temperature on Charpy impact toughness

Hot working temperature		Test temperature		Impact toughness	
°C	°F	°C	°F	J	ft · lbf
1010	1850		−40	11.5	8.5
		RT		12.9	9.5
		93	200	16.9	12.5
		205	400	25.1	18.5
		315	600	39.3	29.0
1035	1950		−40	16.9	12.5
		RT		23.0	17.0
		93	200	37.3	27.5
		205	400	46.7	34.5
		315	600	59.6	44.0

Source: *Alloy Digest*, Jan 1962

Fracture Properties

An acicular α from β fabrication or heat treatment improves fracture toughness in air but has less of an effect on fracture toughness in salt water. Toughness in salt water is directionally sensitive (see figure).

Ti-8Al-1Mo-1V: Stress-corrosion susceptibility

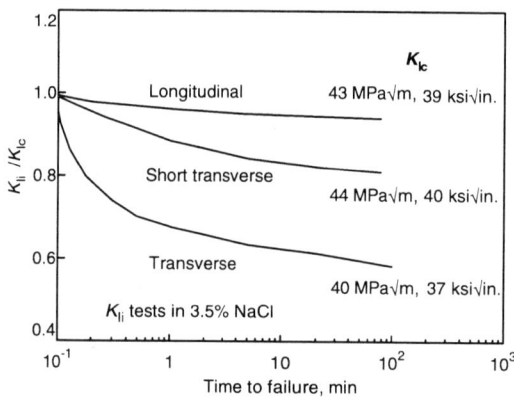

Stress-corrosion susceptibility as a function of specimen orientation in 13 mm (0.5 in.) annealed plate. Three-point-loaded notched bend specimens.
Source: R. Wood and R. Favor, *Titanium Alloys Handbook*, MCIC-HB-02, Battelle Columbus Laboratories, 1972

Ti-8Al-1Mo-1V: SCC resistance for bending or torsion

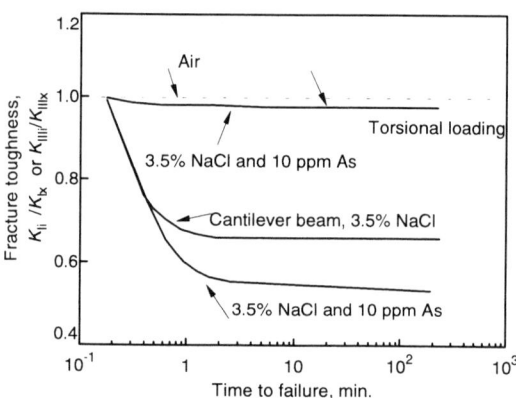

Arsenic additions were used to facilitate hydrogen entry and were carried out at −500 mV (SCE). The lower resistance in Mode I to stress-corrosion cracking was taken as an indication that hydrogen played an important role in the cracking process because in Mode I, a state of hydrostatic tensile stress exists ahead of the crack tip, which was thought to promote the concentration of hydrogen.
Source: M. Fontana and R. Staehle, Ed., *Advances in Corrosion Science and Technology*, Vol 7, Plenum Press, 1980

Ti-8Al-1Mo-1V: Time to failure compared to Ti-6Al-4V at 24 °C

Ti-8Al-1Mo-1V: K_{Ix}, K_{IH}, and K_{Iscc} vs hydrogen

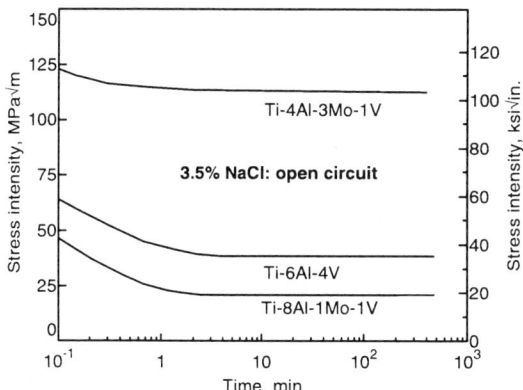

Mill annealed specimens.
Source: T.L. MacKay, C.B. Gilpin, and N.A. Tiner, "Stress-Corrosion Cracking of Titanium Alloys at Ambient Temperatures in Aqueous Solutions," Contract NAS 7-488, Report SM-49105-F1, McDonnell Douglas Corp., July 1967; P. Finden, "Comparative Data—Titanium Alloy Screening Tests," D6-24541-TN, Boeing Company

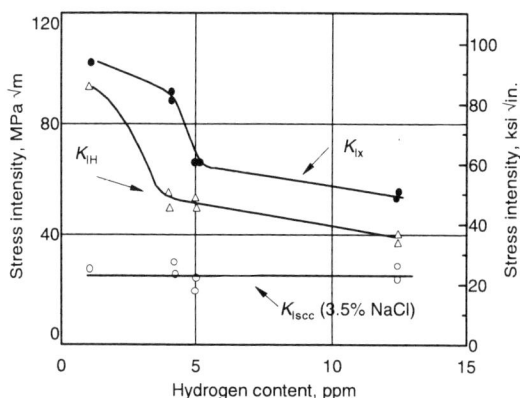

Notched cantilever beam specimens at room temperature.
Source: M.J. Blackburn et al., Boeing Report D1-82-1054, June 1970

Ti-8Al-1Mo-1V: Typical toughness at room temperature

| Thickness | | Heat | Yield strength | | K_{Ic} or K_c | | K_{Iscc} or K_{scc} | |
mm	in.	treatment	MPa	ksi	MPa√m	ksi√in.	MPa√m	ksi√in.
1.3	0.05	Mill annealed	999	145	82	75	33	30
		Duplex annealed	930	135	176	160	55	50
13	0.50	Mill annealed	999	145	52	48	22	20
		Duplex annealed	930	135	110	100	35	32
		Mill annealed, WQ	841	122	>110	>100	46	42
		βST, WQ	868	126	>110	>100	>110	>100

Source: R.W. Schutz, Stress Corrosion of Titanium Alloys, in *Stress Corrosion*, ASM International, 1992

Ti-8Al-1Mo-1V: Plane-stress toughness-(K_c)

| Condition | Ultimate tensile strength(a) | | Tensile yield strength(a) | | Elongation(a), % | Fracture toughness, +(K_c) | |
	MPa	ksi	MPa	ksi		MPa√m	ksi√in.
Mill annealed	999	145	930	135	8-10	151	138
Duplex annealed	930	135	862	125	8-10	274	250

(a) Guaranteed minimums. Source: R. Wood and R. Favor, *Titanium Alloys Handbook*, MCIC-HB-02, Battelle Columbus Laboratories, 1972

Ti-8Al-1Mo-1V: Effect of heat treatment on impact toughness

| Material/condition | Tensile yield strength | | Impact toughness | | | |
| | | | Air | | Saltwater | |
	MPa	ksi	MPa√m	ksi√in.	MPa√m	ksi√in.
Annealed	937	136	70	64	26	24
β fabricated	958	139	107	98	25	23
β heat treated	924	134	104	95	32	29
Annealed plus exposed 48 h at 550-600 °C (1020-1110 °F)	965	140	36	33	16	15
Mill annealed	972	141	50	46	19	18

Source: R. Wood and R. Favor, *Titanium Alloys Handbook*, MCIC-HB-02, Battelle Columbus Laboratories, 1972

Notch Strength

Ti-8Al-1Mo-1V: Transverse notch strength vs temperature

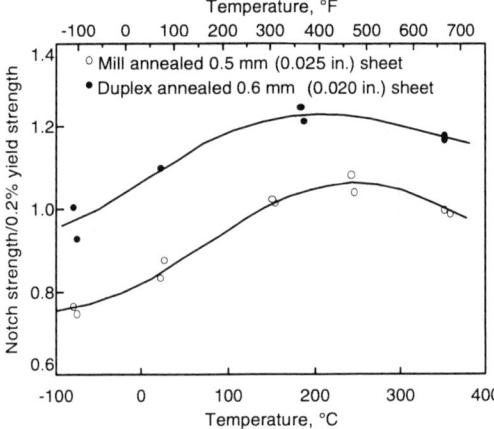

Source: J.L. Shannon and W.F. Brown, A Review of Factors Influencing the Crack Tolerance of Titanium Alloys, in *Applications Related Phenomena in Titanium Alloys*, ASTM STP 432, 1968, p 33-63

Stress-Strain Curves

Tensile

Ti-811: Tensile stress-strain curves for sheet

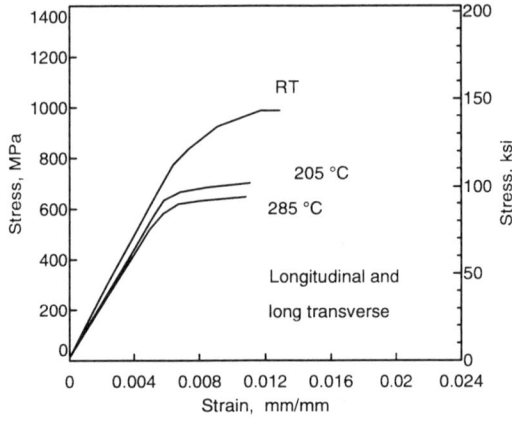

Duplex annealed after ½-h exposure.
Source: MIL-HDBK-5, 1 Dec 1991

Ti-811: Tensile stress-strain curves for MA sheet

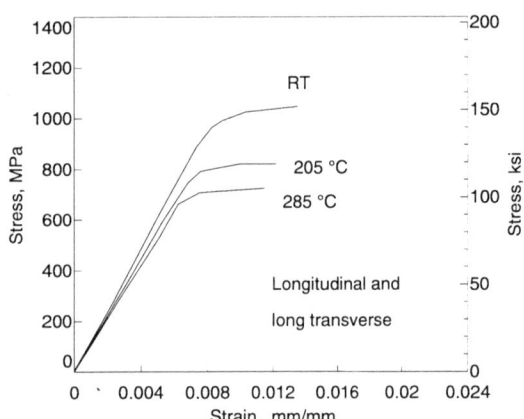

Single annealed after ½-h exposure.
Source: MIL-HDBK-5, 1 Dec 1991

Compressive

Ti-811: Compressive stress-strain curves for sheet

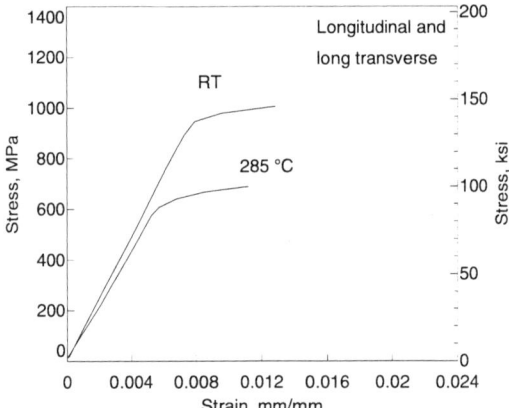

Duplex annealed after ½-h exposure.
Source: MIL-HDBK-5, 1 Dec 1991

Ti-811: Compressive stress-strain for MA sheet

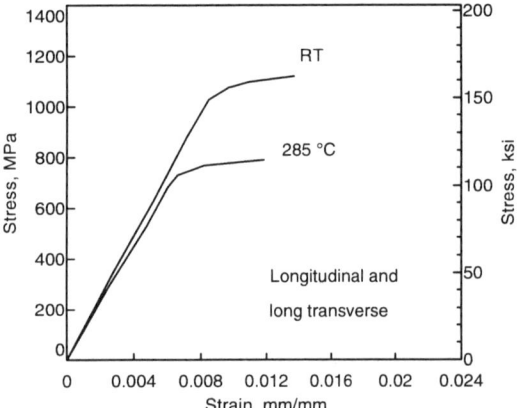

Single annealed after ½-h exposure.
Source: MIL-HDBK-5, 1 Dec 1991

Forging

G.W. Kuhlman, ALCOA, Forging Division

Ti-811 is an α alloy that is commercially produced in all forging product types, although closed die forgings and rings predominate. The alloy is produced on all types of commercially available forging equipment; however, because the major commercial application for the alloy is high-volume turbine engine compressor blades and vanes, rapid-strain-rate hammers, mechanical presses, and/or screw presses are the primary types of forging equipment used to forge this alloy. On this equipment, the alloy is forged to precise airfoil shapes requiring minimal final machining. Ti-811 is generally subtransus forged.

As an α alloy, Ti-811 is very difficult to fabricate, exhibiting very high unit pressures (flow stresses), among the highest of commonly forged titanium alloys, crack sensitivity, and relatively poor forgeability. For intended compressor blade and vane applications, desired final microstructure consists of predominately equiaxed α (e.g., 80 to 90%) with a small amount of transformed β. Thus, the primary forging thermomechanical process for Ti-811 involves conventional forging, in one or more operations, to achieve overall α/β reductions of at least 50 to 75%, followed by solution treating and stabilizing or duplex annealing. The thermal treatment process includes subtransus solution treatment at 995 °C (1825 °F), followed by air cooling or faster quenching and stabilization at 565 to 595 °C (1050 to 1100 °F). The aluminum equivalent in Ti-811 is high enough that excessively slow cooling after forging and/or solution treatment must be avoided to preclude short-range ordering (α_2) that may lead to very poor ductility and toughness.

The alloy is very sensitive to the formation of excessive α case in the course of furnacing prior to forging operations; consequently, ceramic precoats or other surface treatment methods to retard case formation are essential, and reheat furnace times are limited to only that necessary to reach desired metal temperatures. Ti-811 blades and vanes commonly are forged on rapid-strain-rate equipment, and thus, metal temperatures are adjusted to preclude inadvertent overheating of the alloy due to the adiabatic heating typical of such processes.

Ti-811: Forging process temperatures

| | Metal temperature | |
Process	°C	°F
Conventional forging	900-1020	1650-1870

Note: See "Technical Note 4: Forging" for recommended die temperatures.

Ti-811: Forging pressure

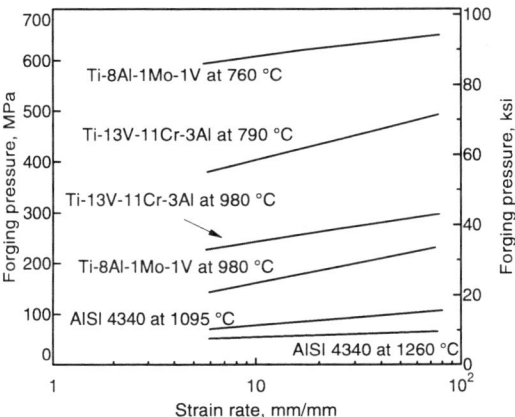

Forging pressures at 10% upset reduction with 4340 steel presented for comparison

Forming

Forming practices for Ti-811 are similar to those of other α-rich alloys. The β transus of this alloy is higher than Ti-6Al-4V, and this permits higher hot work temperatures. Finish working temperatures above the β transus (approximately 1035 °C, or 1900 °F) produce coarser grain size than is characteristic of finishing below the β transus. Sheet forming is more difficult than in Ti-6Al-4V, and for severe operations, forming temperatures between 715 and 745 °C (1325 and 1375 °F) are required. Duplex annealed Ti-811 can be hot formed in a broad temperature range without reduction in properties provided the material after forming is heat treated to 785 °C (1450 °F) for 15 min followed by quick cooling.

Stretching

The formability limits for a formed section or extrusion to be stretch formed depend on the ductility and buckling limits of the material. The formability index for splitting limits depends on the conventional strain, ε, for a 50 mm (2.0 in.) gage length. The formability index for elastic buckling is a function of the ratio of tensile modulus to tensile yield (E_T/S_{ty}).

The index governing the optimum forming temperature will largely depend on the material thickness (t) and section height (H). For small values of t, the ratio H/t becomes large, thereby placing this H/t value in the elastic buckling region, which is a function of (E_T/S_{ty}). For large values of t, the conventional strain for a 50 mm (2.0 in.) gage length is the formability index.

Ti-811: Forming temperatures

Method	Temperature °C	Temperature °F	Comments
Hot sizing	785	1450	15-min exposure, ceramic dies
Brake forming	730	1350	Severe forming
Drop hammer	730	1350	Severe
Stretching	...	...	Temperatures above 540 °C (1000 °F) are required before significant improvement in stretchability is obtained
Drawing	785	1450	...
Spinning	...	...	Better formability starts at 540 °C (1000 °F) and increases rapidly at 650 °C (1200 °F)
Press forming			
Hydropress	730	1350	Severe
Matched die	785	1450	Severe
Roll forming	540	1000	1T bend radius after 20 roll passes with rolls heated to 95 to 260 °C (205 to 500 °F). Alpha case removed per pass
Creep forming	785	1450	
Joggling	...	...	...
Dimpling	400	750	0.56 to 16.0 mm (0.022 to 0.63 in.) sheet; lower temperatures are permissible, but springback is greater and more erratic

Ti-811: Linear-stretch limit curves

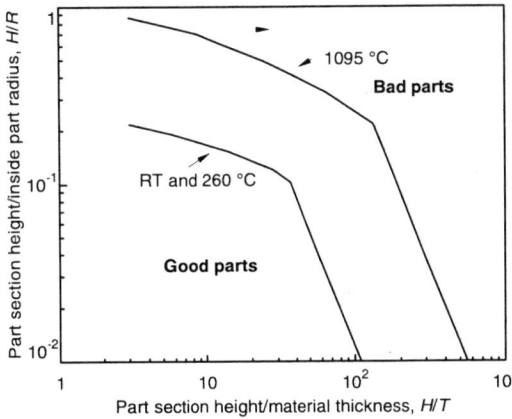

Heel-in (inboard) angle and channel section.
Source: R.A. Wood and R.J. Favor, *Titanium Alloys Handbook*, MCIC-HB-02, Battelle Columbus Laboratories, 1972

Ti-811: Linear stretch limit curves

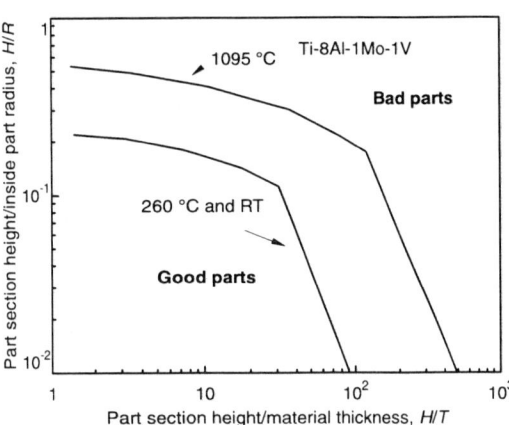

Heel-in-hat section.
Source: R.A. Wood and R.J. Favor, *Titanium Alloys Handbook*, MCIC-HB-02, Battelle Columbus Laboratories, 1972

Ti-811: Analytical extension of deep drawing limit

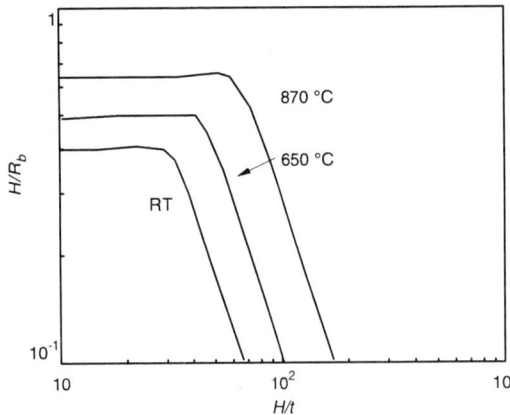

The important dimensions are the cup depth H, the blank diameter D_b, and the inside cup diameter, D. The material thickness and the draw radius are also important parameters, but do not enter into the formability limits directly.
H = height of cup after drawing; R_b = radius of blank.
Source: R.A. Wood and R.J. Favor, *Titanium Alloys Handbook*, MCIC-HB-02, Battelle Columbus Laboratories, 1972

Ti-811: Elastic buckling limit in spinning

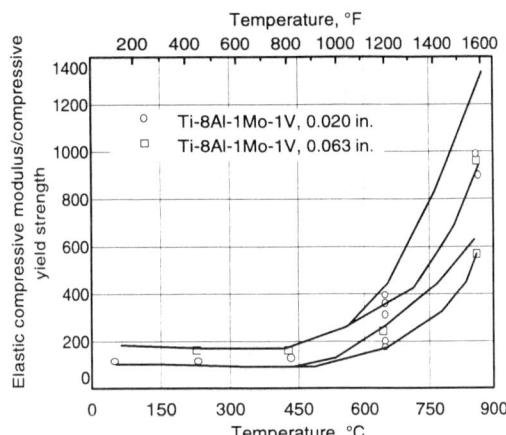

Source: R.A. Wood and R.J. Favor, *Titanium Alloys Handbook*, MCIC-HB-02, Battelle Columbus Laboratories, 1972

Spinning and Drawing

Deep Drawing Forming Limits. The forming limit for deep drawing is established by buckling in the flange area and by fracture or splitting in the cup wall area. The limits have been found to be related to the ratio of compressive modulus and compressive yield strength for buckling and the tensile yield strength: compressive yield strength ratio for splitting.

Spinning. The effect of temperature on the ratio of compression modulus/compression yield strength controls elastic buckling. The change toward better formability starts around 540 °C (1000 °F) and increases rapidly around 760 °C (1400 °F). The latter temperature is approximately the highest temperature that can be used without degrading the properties of the alloys. The total time required for forming may also influence the choice of spinning temperature.

Ti-811: Analytical extension of spinning limit

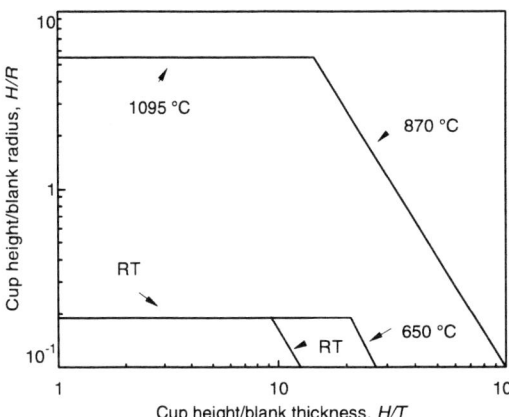

Within the envelope, good parts can be made; failures will occur by plastic buckling if the height-to-radius ratio (H/R) becomes too large, and failure by elastic buckling will occur if the height-to-thickness (H/T) becomes too large. The position of the curves will vary according to the properties of the material and the forming temperature.
Source: R.A. Wood and R.J. Favor, *Titanium Alloys Handbook*, MCIC-HB-02, Battelle Columbus Laboratories, 1972

Beading

Trapped rubber forming is often used for forming beaded panels. The forming limits for beaded panels are determined by failures resulting from splitting or from buckling. Consequently, success or failure depends on the ratio of the bead radius to the thickness of the material, R/T, or on the spacing of the beads, R/L. The limits for several titanium alloys made into beaded panels by trapped rubber forming (see table) indicates that increasing the forming pressure increases the limiting R/T ratios and that increasing the forming temperature permits closer beads in sheet of a particular gage.

Ti-811: Beading limits
Limits on forming beaded panels by the trapped rubber process with a pressure of 20 MPa (3000 psi)

Critical ratio, L/T	Temperature, °C (°F)	Insufficient pressure limits, R/L				Temperature, °C (°F)	Splitting limits, R/T					
		Bead(a)	For R/T ratios of:				Bead(a)	For R/L ratios of:				
			2	5	15			0.01	0.03	0.06	0.10	0.15
147	RT	R/L	0.093	0.123	0.178	RT	R/T	52	47	44	...	...
		L/T	21.5	40.6	84.3		L/T	5200	1568	733		
121	540 (1000)	R/L	0.107	0.140	0.195	RT	R/L	52	47	44	...	...
		L/T	18.7	35.7	77.0		L/T	5200	1568	733		

(a) R = bead radius; T = sheet thickness; L = bead spacing. Source: R.A. Wood and R.J. Favor, *Titanium Alloys Handbook*, MCIC-HB-02, Battelle Columbus Laboratories, 1972

Dimpling

Ti-811: Dimpling limits

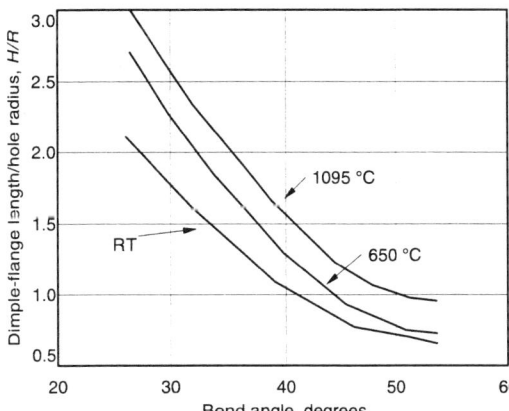

Good parts can be formed for conditions under the curves, whereas split parts can be expected for conditions above the curves. The major failure in dimpling is caused by simple tension.
Theoretical relationship between ratio H/R and bend angle for dimpling.
Source: R.A. Wood and R.J. Favor, *Titanium Alloys Handbook*, MCIC-HB-02, Battelle Columbus Laboratories, 1972

Ti-811: Beading limits

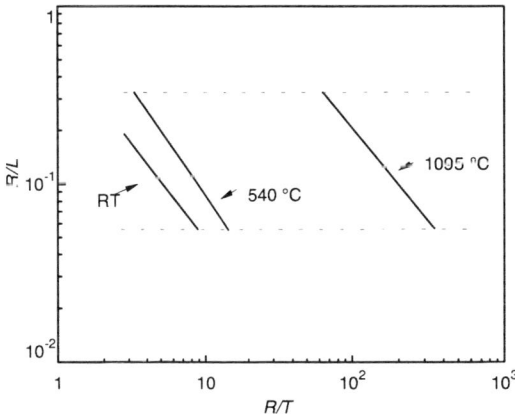

Although the limits apply to beaded panels, they can be used with caution as guides to forming other types of parts with drop hammers.
Limits for forming beaded panels with a drop hammer.
Source: R.A. Wood and R.J. Favor, *Titanium Alloys Handbook*, MCIC-HB-02, Battelle Columbus Laboratories, 1972

Ti-811: Dimpling limits
Radial splitting at edge of hole

Material	Condition	Dimpling temperature, °C (°F)	Dimpling limits, H/R, for various bend angles, α, above and below standard bend angle (Standard)				
			30°	35°	40°	45°	50°
Ti-6Al-4V	Mill annealed	RT	2.00	1.50	1.17	0.92	0.74
Ti-8Al-1Mo-1V	Duplex annealed	RT	1.88	1.42	1.08	0.82	0.70
Ti-8Al-1Mo-1V	Duplex annealed	650 (1200)	2.30	1.72	1.30	1.00	0.85

Source: R.A. Wood and R.J. Favor, *Titanium Alloys Handbook*, MCIC-HB-02, Battelle Columbus Laboratories, 1972

Heat Treatment

Like Ti-6Al-2Sn-4Zr-2Mo, a variety of heat treatment conditions are possible with Ti-811. A particular condition is usually selected on the basis of part section size, fabrication history, service environment, and the mechanical properties desired. For example, a thick section product for engine use may be most desirable in the duplex annealed condition having maximum creep strength, whereas a sheet may be heat treated to produce a formable condition.

Stress relief annealing treatment should consist of thermal exposures that do not disturb the metallurgical stability of the alloy. Annealing data have been generated between 595 and 785 °C (1100 and 1450 °F). Complete relaxation occurs in about 2 h at 595 °C (1100 °F) to about 15 to 20 min at 785 °C (1450 °F). Differences in cooling from the stress-relief annealing temperatures, particularly if temperatures on the high side of the range are used, may significantly affect mechanical properties. When stress relief annealing at 760 to 785 °C (1400 to 1450 °F), it is possible to control the an-

Ti-811: Typical heat treatments

Heat treatment	Temperature °C	Temperature °F	Time, h	Cooling method
Stress relief	600-700	1100-1300	0.25-4	Air or slow cool
Mill annealing	760-790	1400-1450	1-8	Air or furnace cool
Solution treating	980-1010	1800-1850	1	Oil or water quench
Aging	565-595	1050-1100	...	Air cool
Duplex anneal for thick sections(a)				
1st stage	900-1010	1650-1850	...	Air cool
stabilization	600-745	1100-1375	...	Air cool
Duplex anneal for sheet				
1st stage (mill anneal)	760-790	1400-1450	1-8	Air or furnace cool
2nd stage	600-790	1100-1450	...	Air cool

(a) Treatment for good creep strength

Ti-811: Volume fraction of α and β

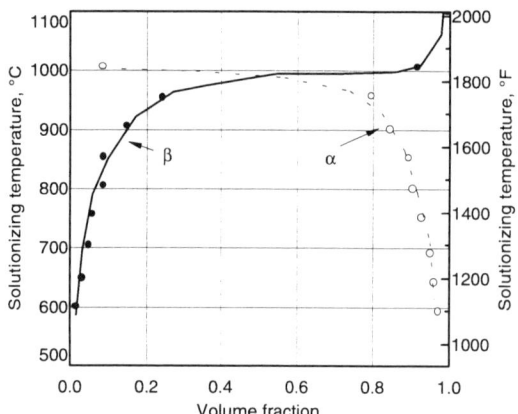

Source: *Titanium '80, Science and Technology*, H. Kimura and O. Izumi, Ed., TMS/AIME, 1980, p 1224

Ti-811: Stress relief

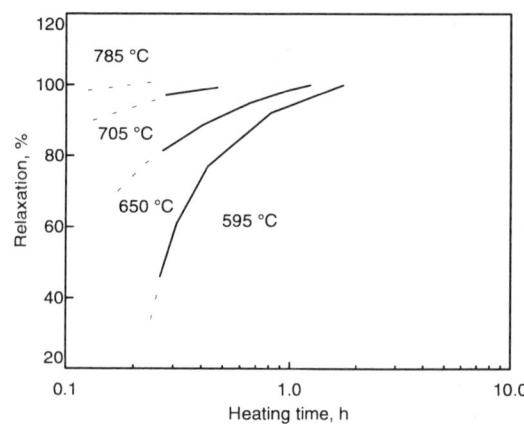

1.4 mm (0.056 in.) sheet mill annealed at 785 °C (1450 °F) for 8 h, then cold worked.
Source: R.A. Wood and R.J. Favor *Titanium Alloys Handbook*, MCIC-HB-02, Battelle Columbus Laboratories, 1972

Ti-811: Effect of cooling on properties

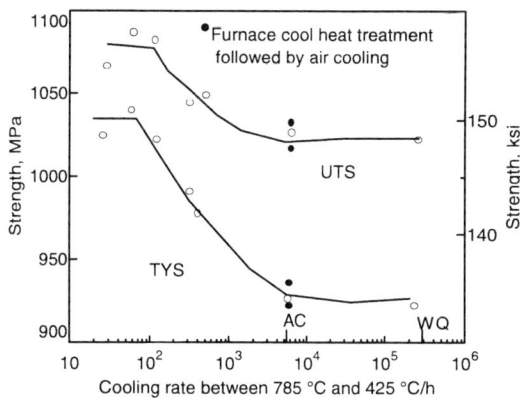
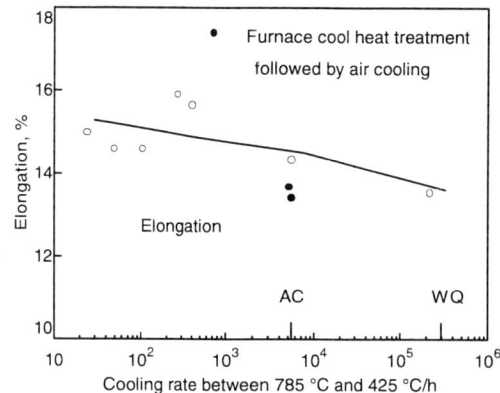

Sheet cooled from 785 °C (1450 °F) after heating 8 h

nealed properties by thermal exposure time and terminal cooling rate in the same way that is used by the mill in supplying the two conditions—mill and duplex annealed.

Reannealing. In the shop, forming, machining, or joining operations may require reannealing to regain the properties inherent in the as-received metal. This may be a step beyond stress relief annealing, although it may be accomplished in much the same way. In the above cases, longer exposure time followed by slow cooling at about 55°C/h (100 °F/h) from 785 to 480 °C (1450 to 900 °F) results in the mill-annealed condition. A 20-min exposure followed by faster cooling (785 to 480 °C, or 1450 to 900 °F in less than 1 h, i.e., air cooling is quite satisfactory) results in the duplex-annealed condition. Intermediate cooling rates result in mechanical properties that are intermediate to those of mill and duplex-annealed material.

Strengthening Heat Treatments. Ti-811 may be strengthened by solution heat treatment followed by aging. Solution treatment is performed high in the α-β field at 900 to 1010 °C (1650 to 1850 °F), followed by water quenching. Usually aging is accomplished at intermediate temperatures in the 480 to 650 °C (900 to 1200 °F) aging temperature range. Strengthening is limited to smaller sections and is not widely used.

Machining and Welding

Machining. Like other titanium alloys, machining of Ti-811 is comparable to machining a good grade of stainless steel. In general, very sharp tools with a slightly larger rake angle and very keen edge work quite well. Slower speed and heavier cuts are preferred, because they keep tool temperatures down and produce coarse chips. Drilling of thin-walled titanium is not much of a problem as long as the drill is sharp. Thicker walled tube requires a heavy flood of coolant to remove heat and chips. General guidelines are given in "Technical Note 7: Machining".

Welding. Ti 811 has fair weldability and requires the use of a filler metal that is the same as the base metal.

Ti-811: Turning parameters for annealed material

Tool material	Tool geometry(a)	Depth of cut mm	in.	Feed mm/rev	in./rev	Speed m/min	sfm
Rough turning							
Brazed carbide (C2)	A, E, F	2.5-6.35	0.10-0.25	0.25-0.38	0.010-0.015	21-42	70-140
Throwaway carbide (C2)	A, E, F	2.5-6.35	0.10-0.25	0.25-0.38	0.010-0.015	46-60	150-200
High-speed steel (M, T5, T15)	B, D, E, F	2.5-6.35	0.10-0.25	0.25-0.38	0.010-0.015	73-220(b)	240-720(b)
Finish turning							
Brazed carbide (C3, C2)	A, C	0.635-2.5	0.025-0.10	0.13-0.25	0.005-0.010	27-47	90-155
Throwaway carbide (C3, C2)	A, C	0.635-2.5	0.025-0.10	0.13-0.25	0.005-0.010	50-56	165-185
High-speed steel (M3, T5, T15)	C, E, F	0.635-2.5	0.025-0.10	0.13-0.25	0.005-0.010	73-183(b)	240-600(b)

(a) See accompanying table for tool geometry codes. (b) These high speeds would be lowered if higher feeds and deeper cuts are made with high-speed steel cutters. Source: R.A. Wood and R.J. Favor, *Titanium Alloys Handbook*, MCIC-HB-02, Battelle Columbus Laboratories, 1972

Ti-811: Tool geometry data

	Tool angles and nose radius for indicated tool geometry code					
	A	B	C	D	E	F
Back rake, °	−5	+5 to −5	0	0	0 to +5	0 to +10
Side rake, °	−5 0 to −6	+6 to 0	5 or 6	15	+5 to +15	0 to +10
End relief, °	5	5-10	5	5	5-7	6-8
Side relief, °	5	5-10	5	5	5-7	6-8
End cutting edges, °	15-45	6-15	15 or 5	10 to 15	5-7	5-10
Side cutting edge (lead), °	15-45	5-20	15	15-45	15-20	0-30
Nose radius, mm (in.)	0.08-1.2 (1/32-3/64)	0.75-1.0 (0.03-0.04)	1.2 (3.64)	1.2 (3.64)	0.5-0.75 (0.02-0.03)	...

Source: R.A. Wood and R.J. Favor, *Titanium Alloys Handbook*, MCIC-HB-02, Battelle Columbus Laboratories, 1972

Ti-11
Ti-6Al-2Sn-1.5Zr-1Mo-0.35Bi-0.1Si

No longer produced, Ti-11 was developed by TIMET for improved creep resistance and stability. Ti-11 was not commercially marketed because qualification costs exceeded the technical benefit.

Ti-11: Creep behavior comparison

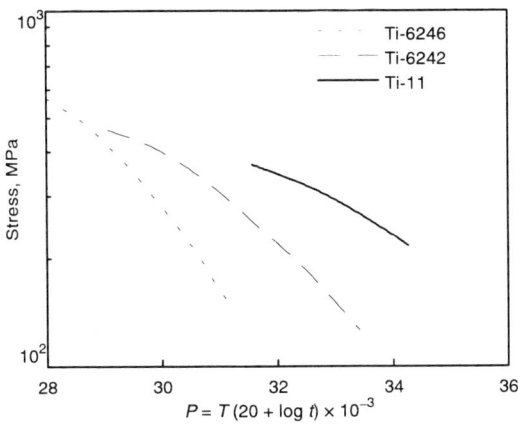

Larson-Miller comparison of creep behavior for beta forged and annealed Ti-11 Ti-6Al-2Sn-4Zr-2Mo, and Ti-6Al-2Sn-4Zr-6Mo. 0.2% creep deformation. Annealed, AC condition for Ti-11.

Ti-11: Summary of typical physical properties

Density(a)	4.5 g/cm^3 (0.162 lb/in.^{3})
Beta transus	~ 980 – 1015 °C (1800 – 1860 °F)
Magnetic permeability	Nonmagnetic

(a) Typical values at room temperature of about 20 to 25 °C (68 to 78 °F)

Ti-11: Typical Tensile properties of 25 mm (1 in.) thick beta forged pancakes

Process cooling(a)	Annealing treatments	Test temperature °C	°F	Tensile yield strength MPa	ksi	Ultimate tensile strength MPa	ksi	Elongation %	Reduction of area, %
AC	1065 °C (1950 °F), 15 min, AC + 705 °C (1300 °F), 1 h, AC		RT	848	123	937	136	16	29
OQ	815 °C (1500 °F), 1 h, OQ + 595 °C (1100 °F), 8 h, AC	540	1000	434	63	565	82	18	40
			RT	951	138	1048	152	15	26
OQ	705 °C (1300 °F), 1 h, OQ + 595 °C (1100 °F), 8 h, AC	540	1000	517	75	675	98	16	40
			RT	944	137	1041	151	19	32
WQ	705 °C (1300 °F), 1 h, AC		RT	930	135	1034	150	16	34
		540	1000	524	76	682	99		

(a) Cooling method directly off the forging press
Source: *Titanium Science and Technology*, (Jaffe and Burte, ed), Vol 4, 1973, p 2219-2225

TIMETAL® 1100

Ti-6Al-2.75Sn-4Zr-0.4Mo-0.45Si
Ti-1100
UNS No.: Unassigned

Tom O'Connell, TIMET

Ti-1100 is a near-alpha alloy developed for elevated-temperature use up to 600 °C (1100 °F). It was developed to be used primarily in the beta-processed (beta-worked or beta-annealed) condition. Ti-1100 offers the highest combination of strength, creep resistance, fracture toughness, and stability of any commercially available titanium alloy. It is also recommended for castings.

Effects of Alloying and Impurities. The effects of tin, iron, oxygen, silicon, zirconium, molybdenum, and aluminum on creep, strength, and stability of Ti-1100 have been determined. The alloy development program began with the screening of over 250 compositions of button (250-g) heats. These studies identified compositions that were scaled to 45-kg (100-lb) heats to provide forged product for evaluation. The most promising of these alloys were then scaled to several 815-kg (1800-lb) heats for melting and conversion studies as well as thermomechanical processing (TMP) studies. This successful progression culminated with the production and evaluation of a production-sized 3630-kg (8000-lb) ingot. The outcome of this alloy development study was a composition consisting of Ti-6Al-2.75Sn-4Zr-0.4Mo-0.45Si-0.07O_2-0.02Fe(max).

This alloy is clearly a modification of the Ti-6242-Si alloy that is so widely used today. Although the chemistry differences would appear to be subtle, they are quite dramatic in their effect on creep response, as indicated below:

- **Silicon:** Creep resistance is significantly enhanced up to 0.5% silicon, but beyond that point post-creep ductility (stability) is compromised with no further creep enhancement.

- **Tin:** A similar relationship exists for tin, with stability sacrificed above the 3% level.

- **Iron:** Iron demonstrates a strong effect on time to 0.2% creep strain at the 510 °C (950 °F), 410 MPa (59 ksi) test condition, necessitating iron levels well below those typically encountered in the Ti-6242-Si alloy.

- **Aluminum:** The aluminum level in the new alloy was kept at 6% due to stability problems at higher levels and strength problems at lower levels.

- **Zirconium:** Zirconium was kept high to promote a uniform distribution of silicides in light of the high silicon level. Thus, the chemistry of this alloy was optimized not only for creep strength, but also for stability, strength, and uniformity.

Sponge and Melting Practice. Ti-1100, due to its extremely low iron limit, requires a select grade of titanium sponge. However, sponge containing roughly 100 ppm iron (0.01%) has been produced on a commercial basis, and no problems exist concerning raw material supply. In terms of melting, the high silicon content of this alloy calls for special controls during vacuum arc remelting, especially on the third and final melt. However, Ti-550 (Ti-4Al-2Sn-4Mo-0.5Si) has a comparable silicon content, and this alloy has been successfully melted for several years.

Product Forms. Ti-1100 has been processed successfully to billet, bar, sheet, and weld wire. Forgings have been produced using isothermal and warm die methods, and foil has been produced for use in metal matrix composites.

Investment castings have been produced. The lack of a quench requirement from the solution treatment temperature may enhance the producibility of castings. No data are available on P/M products.

Product Condition. The two standard conditions recommended for the alloy are (1) beta processed ($T > 1065$ °C, 1950 °F) and annealed ($T = 595$ °C, 1100 °F) and (2) alpha-beta processed; beta annealed ($T > 1065$ °C, 1950 °F) plus anneal ($T = 595$ °C, 1100 °F).

The alloy has only a slight response to cooling rate or section size from the solution treatment (or processing) temperature. Very rapid quenches increase strength and decrease elevated-temperature creep resistance. Ti-1100 generally is used in the beta-processed or beta heat-treated condition, but it is provided in an equiaxed alpha-beta condition for the product forms to enhance processibility.

Applications. Ti-1100 is designed for applications requiring excellent creep strength or fracture properties at temperatures up to approximately 600 °C (1100 °F). High-pressure compressor disks, low-pressure turbine blades, and automotive valves are typical examples.

Ti-1100: Typical composition range

	Composition, wt%								
	Al	Sn	Zr	Fe	Mo	Si	O_2	N_2	C
Minimum	5.7	2.4	3.5	...	0.35	0.35	...	...	...
Maximum	6.3	3.0	4.5	0.02	0.50	0.50	0.09	0.03	0.04
Nominal	6.0	2.7	4.0	...	0.40	0.45	0.07	...	...

Physical Properties

Phases and Structures. Typical microstructures for Ti-1100 include equiaxed α-β for billet and sheet stock. It also transforms to a Widmanstätten or colony α + β structure depending on cooling rate. The effects of cooling rate on the transformed β structure are as follows: alpha-beta processing with a normal cooling rate results in equiaxed primary α plus transformed β with a colony structure plus silicides; beta processing with rapid cooling results in a Widmanstätten structure, whereas slower cooling after β processing results in a colony structure.

In addition to α and β phases, various silicides exist for both α + β or β processed conditions. The silicide solvus has been measured at between 1030 and 1065 °C (1885 and 1950 °F). The α-2 solvus is approximately 740 °C (1365 °F). The β transus is nominally 1015 °C (1860 °F).

Ti-1100: Summary of typical physical properties

Ti_3Al (α_2) solvus	~740 °C (1365 °F)
Beta transus (nominal)	1015 °C (1860 °F)
Silicide solvus	1030 to 1065 °C (1885 to 1950 °F)
Calculated liquidus point	1637 °C (2978 °F)
Density(a)	4.5 g/cm^3 (0.163 lb/in.^{3})
Modulus of elasticity(a)	107 to 117 GPa (15.5 to 17 × 10^6 psi)
Electrical resistivity(a)	1.8 μΩ · m
Magnetic permeability	Nonmagnetic
Specific heat capacity(a)	545 J/kg · K (0.13 Btu/lb · °F)
Thermal conductivity(a)	7 W/m · K (4 Btu/ft · h · °F)
Coefficient of linear expansion	8.5 × 10^{-6}/°C (4.7 × 10^{-6}/°F)
Calculated solidus	1615 °C (2939 °F)

(a) Typical values at room temperature of about 20 to 25 °C (68 to 78 °F)

Ti-1100: Electrical resistivity vs temperature

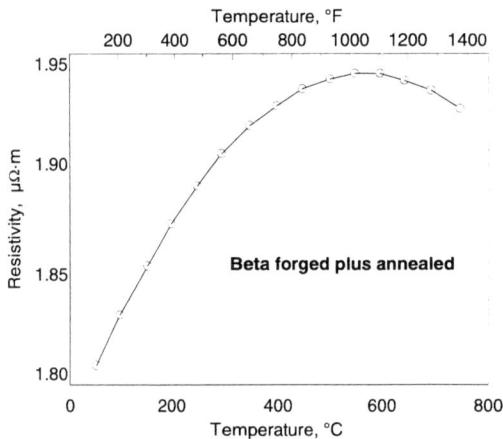

Resistivity (R) for beta forged plus annealed material between 25 and 750 °C (77 and 1380 °F) has been determined to fit the expression: $R(10^{-8} \Omega \cdot m) = 178 + 0.057 T + 5 \times 10^{-5} T^2$

Ti-1100: Specific heat vs temperature

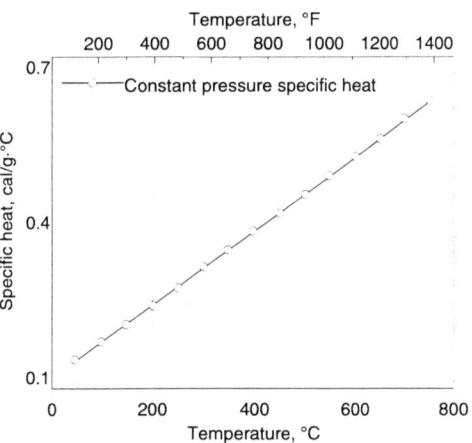

The specific heat (C_p) for beta-processed material between 25 and 750 °C (77 and 1380 °F) can be expressed as: C_p (cal/g · °C) = 0.117 + 6.7 × 10^{-4} (T)

Ti-1100: Thermal coefficient of linear expansion

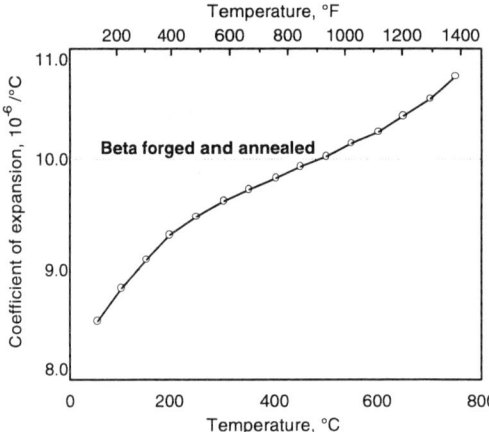

The thermal coefficient of linear expansion (α) between 25 and 750 °C (77 and 1380 °F) for beta-processed material is given by: α (ppm/°C) = 8.12 + 8.17 × 10^{-3} T − 1.37 × 10^{-5} T^2 + 10^{-8} T^3

Ti-1100: Thermal conductivity

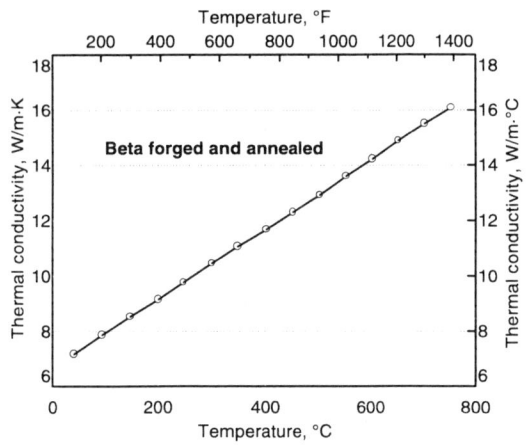

Thermal conductivity (Q) between 25 and 750 °C (77 and 1380 °F) for beta-processed material follows the equation: Q (W/m · °C) = 6.62 + 1.27 × 10^{-2} T

Mechanical Properties

Creep Properties. It has been determined that beta processing greatly improves the creep resistance of Ti-1100 and that the quench rate from the beta forging or annealing temperature will subtly affect the creep resistance. Faster cooling (i.e., oil quench versus air cool) will improve the high-stress, low-temperature portion of the Larson-Miller plot at the expense of the low-stress, high-temperature portion of the plot. Alpha-beta processed material has higher strength and ductility at low temperatures, but has decreased strength at high temperatures (600 °C, or 1110 °F).

Ti-1100: Typical 0.2% creep of beta forged material

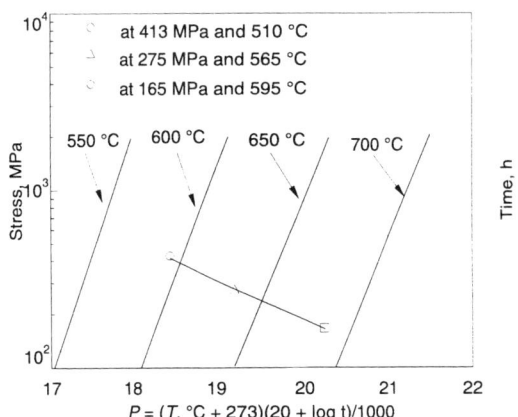

Ti-1100: Yield and tensile strength vs temperature

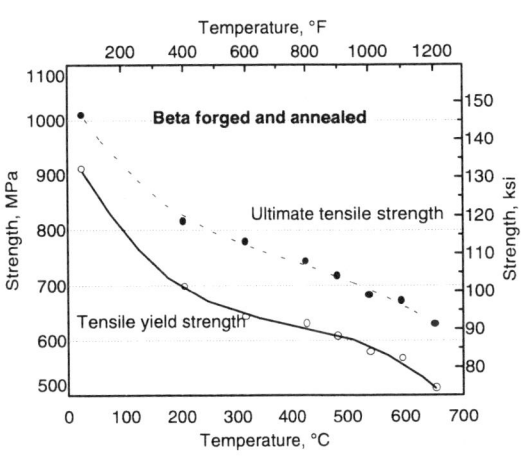

Ti-1100: Tensile ductility vs temperature

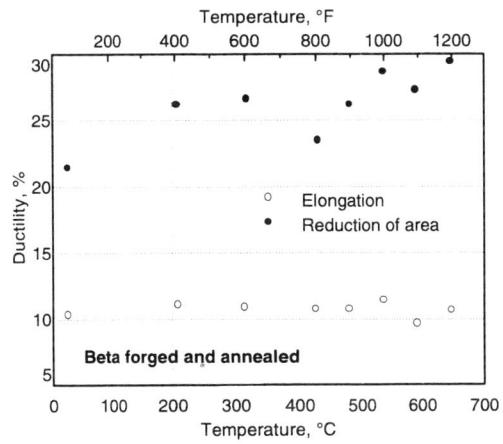

Fatigue Properties

Fatigue Crack Growth. Room temperature data are shown. High-temperature crack growth is reported in *Met. Trans.*, Vol 24 A, p 1321.

Ti-1100: Room-temperature fatigue crack growth

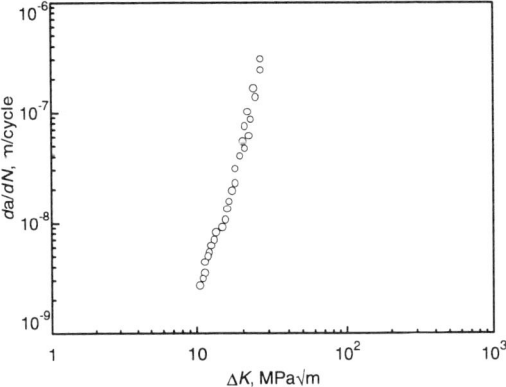

Ti-1100 demonstrates excellent crack growth resistance compared to other conventional alloys from RT to 600 °C (1110 °F). Isothermally forged plus annealed. Tested at 23 °C (75 °F); 20 Hz; $R = 0.1$

Ti-1100: Fatigue strength at 10^7 cycles

Temperature			Fatigue strength(a)	
°C	°F	K_t	MPa	ksi
22	71	1.0	655	95
		3.0	250	36
480	895	1.0	500	72
		3.0	235	34

(a) Beta forged and annealed; tested at 30 Hz; $R = 0.1$

Fracture Properties

Ti-1100: Fracture toughness of beta forged and annealed material

Heat treatment	As processed MPa√m	ksi√in.	Exposed at 650 °C (1200 °F) for 300 h MPa√m	ksi√in.
1095 °C (2000 °F), FAC + 595 °C (1100 °F), 8 h	62.9	57.2	43.5	39.6
1095 °C (2000 °F), OQ + 595 °C (1100 °F), 8 h	63.7	57.9	40.2	36.6
1150 °C (2100 °F), FAC + 705 °C (1300 °F), 8 h	53.5	48.7	45.7	41.6
1095 °C (2000 °F), FAC + 995 °C (1825 °F), 1 h + 595 °C (1100 °F), 8 h	64.1	58.3	53.2	48.4
1095 °C (2000 °F), FAC + 1095 °C (2000 °F), 0.5 h + 595 °C (1100 °F), 8 h	71.0	64.6	48.3	43.9
995 °C (1825 °F), FAC + 995 °C (1825 °F), 1 h + 595 °C (1100 °F), 8 h	39.4	35.8	30.3	27.5
995 °C (1825 °F), FAC + 1095 °C (2000 °F), 0.5 h + 595 °C (1100 °F), 8 h	75.9	69.0	44.4	40.4

Note: As expected, the alpha-beta heat treated material has the lowest fracture toughness.

Processing

Casting. In a casting study, the strength of Ti-1100 was found to be equivalent to cast Ti-6Al-2Sn-4Zr-2Mo at 540 °C (1000 °F), but was stronger at 595 °C (1100 °F). Ti-1100 is somewhat weaker in thick cross sections than thin ones and exhibits a significant creep advantage relative to Ti-6Al-2Sn-4Zr-2Mo. It also has higher low-cycle fatigue strength at 550 °C (1022 °F) than Ti-6Al-2Sn-4Zr-2Mo.

Forming. Ti-1100 possesses limited cold formability and behaves similarly to Ti-6Al-2Sn-4Zr-2Mo in cold and hot forming. Although the α + β window is relatively small, the alloy has demonstrated superplasticity in a simulated manufacturing environment.

Machining. Ti-1100 machines essentially the same as Ti-6Al-2Sn-4Zr-2Mo.

Joining. Welding and brazing of Ti-1100 is similar to Ti-6Al-2Sn-4Zr-2Mo.

Rolling characteristics and texture formation are similar to Ti-6Al-2Sn-4Zr-2Mo.

Surface Treatments. Although material-specific surface treatments have not been fully explored, Ti-1100 should behave essentially the same as Ti-6Al-2Sn-4Zr-2Mo.

Forging

Ti-1100 may be hammer forged or press forged using isothermal, warm die, or conventional die methods. The resulting properties will vary depending on the effective cooling rate and strain rate of the deformation process. Finer structures will result in higher tensile strength at the expense of creep strength at high temperatures. Forging below the β transus followed by a beta anneal to obtain the appropriate microstructure generally is not recommended. However, early forging operations (e.g., preform block) may be conducted in the subtransus field with the finish forging conducted in the β field.

As with other difficult to fabricate near-α alloys, precoats or other surface coating techniques are essential on billet stock and intermediate forging shapes during furnacing for forging operations. Ti-1100 may be sensitive to the excessive formation of a case during reheating processes, which may lead to undue surface cracking in forging deformation. As with other α alloys, care must be exercised in the use of dry abrasive grinding techniques used for crack repair.

Recommended Forging Temperatures. The recommended beta forging range is 1090 to 1120 °C (1990 to 2050 °F). Conventional forging is not recommended for this alloy. Die temperatures are listed in "Technical Note 4: Forging."

To achieve desired elevated-temperature and creep performance characteristics, Ti-1100 is designed to be beta processed, creating a transformed, Widmanstätten α-type microstructure, with minimum grain boundary α. To date, thermomechanical processing work with the alloy suggests that β forging, followed by an appropriate post-forging cooling process based on section size, and final stabilization thermal treatment provides optimum properties.

Subtransus forging and beta heat treatment is not currently recommended because Ti-1100, as a near-alpha alloy, is characterized by high unit pressures. However, hot working above the β transus is not cumulative; thus, if multiple forging steps are required (e.g., preform, block, and finish), early forging operations may be conducted subtransus, with the finish forging being conducted above the transus with a sufficiently high level of work.

Supra-transus, beta forging of Ti-1100 significantly reduces unit pressure requirements and crack sensitivity and is conducted from a temperature above the silicide solvus—1040 °C (1905 °F)—to avoid excessive silicide formation. Typically, beta forging reductions of 50 to 75% are recommended for Ti-1100. Low levels of deformation above the β transus should be avoided.

Post-Forging Treatment. The post-forging cooling rate is not highly critical, and generally an air cool is sufficient. However, for thicker section forgings, fan cooling or oil quenching may be required to achieve final part mechanical properties. Final stabilization age thermal treatments may be adjusted to modify final strength properties. Stabilization treatments are generally in the range of 500 to 650 °C (930 to 1200 °F).

IMI 230

Ti-2.5Cu

As a binary alloy containing 2.5 wt% copper, IMI 230 combines the formability and weldability of unalloyed titanium with improved mechanical properties particularly at elevated temperatures. This alloy can be used at temperatures up to 350 °C (660 °F) and is used in the annealed condition as sheet, forgings and extrusions for fabricating components such as bypass ducts of gas turbine engines. Aging treatment raises room-temperature tensile properties by about 25%, and almost doubles the elevated-temperature properties (e.g., creep at 200 °C).

Product Forms and Condition. IMI 230 has a structure consisting mainly of a supersaturated solid solution of copper in alpha (close-packed hexagon) titanium in the solution-treated state. This structure is amenable to an age-hardening reaction similar to that in the conventional Al-Cu-Mg type of alloy. The aging treatment causes precipitation of a finely divided compound, Ti_2Cu, giving the usual strain-hardening effect. Available forms include billet, bar, rod, wire, extruded sections, and sheet. Sheet, extrusions, and bar for machining are supplied annealed, and solution treated (suitable for aging).

Physical Properties

IMI 230: Summary of typical physical properties

Beta transus	895 ± 10 °C (1645 ± 20 °F)
Melting (liquidus point)	Not Available
Density(a)	4.56 g/cm³ (0.165 lbf/in.³)
Electrical resistivity(a)	~0.65 μΩ · m
Magnetic permeability	Nonmagnetic
Specific heat capacity(a)	Not Available
Thermal conductivity(a)	13 W/m · K (7.5 Btu/ft · h · °F)
Thermal coefficient of linear expansion(b)	9.02×10^{-6}/°C (5×10^{-6}/°F)

(a) Typical values at room temperature of about 20 to 25 °C (68 to 78 °F). (b) Mean coefficient from room temperature to 100 °C (212 °F).

IMI 230: Thermal expansion

Temperature range		Mean thermal coefficient of linear expansion	
°C	°F	10^{-6}/°C	10^{-6}/°F
20-100	68-212	9.02	5.01
20-200	68-390	8.73	4.85
20-300	68-570	9.10	5.05
20-400	68-750	9.29	5.16
20-500	68-930	9.47	5.26
20-600	68-1110	9.46	5.25
20-700	68-1290	9.62	5.34
20-800	68-1470	9.39	5.22

The linear dimensional change on aging is between nil and 0.2%. The thermal expansion coefficient is similar to other titanium alloys.

IMI 230: Young's modulus

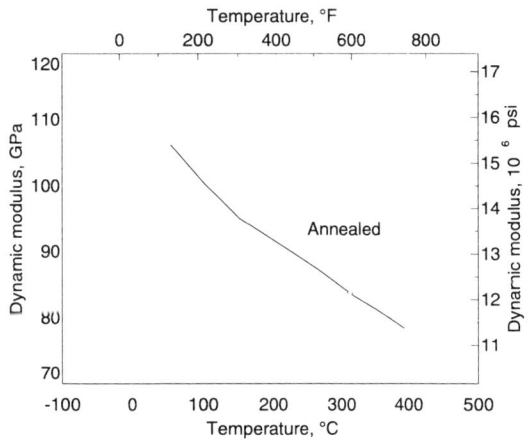

IMI 230: Electrical resistivity

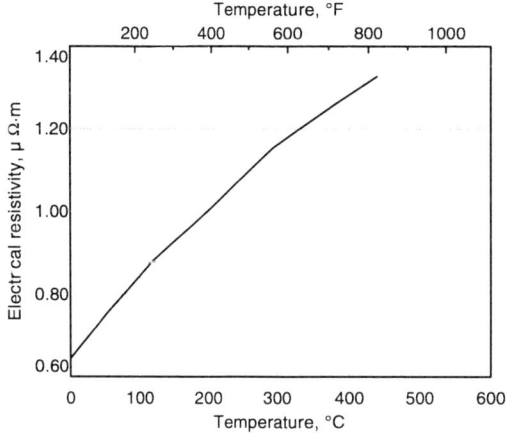

IMI 230: Thermal conductivity

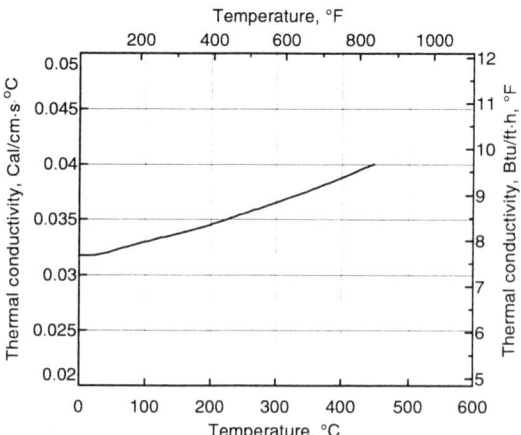

Mechanical Properties

IMI 230: Specification properties

Condition and form	Specification	Bend radius	Ultimate tensile strength MPa	ksi	Tensile yield strength (min) MPa	ksi	Minimum elongation, %	Minimum RA, %
Annealed sheet	BS TA 21	2T(max)(a)	540-700	78-101	460	66.7	18(b)	...
STA sheet	BS TA 52	2T(max)(a)	690-920	100-133	550	80	10(b)	...
Annealed bar and forging stock	BS TA 22 and 23	...	540-770	78-111	400	58	16(c)	35
STA bar and forging stock	BS TA 53 and 54	...	650-880	94-127	525	76	10(c)	25

(a) Bend radius from 0.5 to 3 mm (0.02 to 0.12 in.) only. (b) Elongation on 50 mm (2 in.) with thickness above 0.9 mm (0.035 in.). (c) Elongation on 5D.

IMI 230: Typical room-temperature properties

IMI 230 is relatively free from directionality apart from yield strength which is about 30 to 80 MPa (4 to 12 ksi) higher in the transverse direction.

Condition and form	Thickness mm	in.	Bend radius	Ultimate tensile strength MPa	ksi	0.1%-0.2% tensile yield strength(a) MPa	ksi	Elongation, %	RA, %
Annealed sheet	1.3	0.05	...	620	90	480	70	24	...
Solution treated sheet (transverse)	...	...	1.5T	620	90	510-530	74-77(b)	24	...
STA sheet (c)	1.3	0.05	...	770	112	585	85	22	...
STA sheet (transverse)	...	...	2T	770	112	650-660	94-96(b)	24	...
Annealed bar	13	0.5	...	655	95	480	70	27	...
Solution treated bar	...	...	...	630	91	500(d)	72	27	45
STA bar(c)	13	0.5	...	793	115	620	90	22	...
STA bar	...	...	...	740	107	580	84(d)	22	41
Annealed extrusion	...	...	...	630	91	500	72.5	30	40
STA extrusion	...	...	...	790	115	670	97	28	30

(a) 0.1% yield stress unless noted. (b) 0.1 and 0.2% yield stress, respectively. (c) STA treatment: 850 °C (1560 °F), plus 24 h at 400 °C (750 °F), 8 h at 475 °C (885 °F), air cooled. (d) 0.2% yield stress

IMI 230: Typical creep strength

Condition	Exposure temperature		Stress to produce 0.1% plastic strain in 300-h exposure	
	°C	°F	MPa	ksi
Annealed	100	212	255	37
	215	420	235	34
	300	575	215	31
Solution treated and aged	400	750	185	27

IMI 230: Typical fatigue properties

	Tensile strength		Fatigue strength 10^7 cycles(a)		Fatigue ratio
	MPa	ksi	MPa	ksi	
Rod, direct stress					
Annealed	640	93	±280	±40	0.43
Aged	790	115	±470	±68	0.58
Aged, notched $K_t = 3.3$	...	...	±200	±29	...
Rod, rotating bend					
Annealed	600	87	±370	±53.5	0.62
Aged	790	115	±450	±65	0.57
Rod, rotating bend at 400 °C					
Annealed	...	...	±150	±21.75	...
Aged	...	...	±290	±42	...
Sheet, reversed bend					
Annealed	565	82	±390	±56.5	0.68
Aged	770	112	±490	±71	0.64
Sheet, direct stress					
Aged	760	110	0-570	0-82.5	...

(a) A ± sign implies a stress ratio of $R = -1$

IMI 230: Tensile properties vs temperature

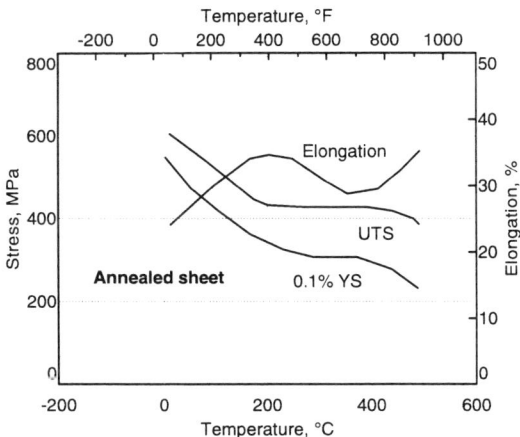
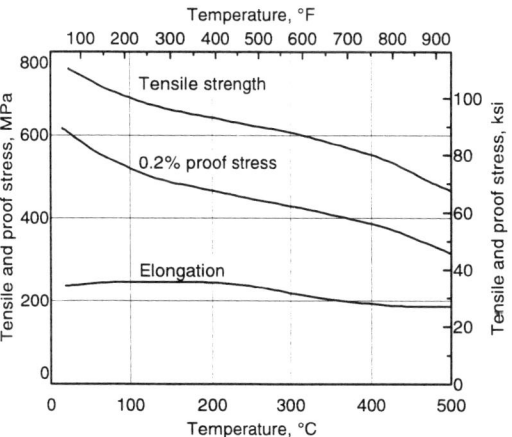

Processing

Forging

IMI Titanium 230 is very easy to forge. A certain amount of forging in the alpha+beta field is required to develop optimum properties. The ideal forging preheating temperature is 800 to 820 °C (1470 to 1500 °F), though a preheating temperature of 850 °C (1560 °F) is commonly used. It may, on occasion, be permissible to go as high as 875 °C (1600 °F) for initial roughing operations, provided that a reduction of at least 2-to-1 or 4-to-1 is subsequently carried out at the lower temperature. IMI 230 is an easier alloy to work than the well-known Ti-6Al-4V alloy at their recommended forging temperature. Preheating time should be kept as short as possible consistent with uniform heating; as a rough guide, a preheating time of ½ h per 25 mm (1 in.) of section thickness should be allowed.

Forming

Although forming operations are often performed prior to aging, IMI 230 may in fact be cold worked at any stage after solution treatment. The cold working of sheet after partial aging (24 h at 400 °C, or 750 °F) or after duplex aging gives material which, when fully aged, has higher yield and tensile strength and only slightly lower ductility than if it had been cold worked in the solution-treated condition and then fully aged (see table). Warm forming temperatures of 350 °C (660 °F) or below are preferable, since they do not interfere with subsequent aging at 400 to 475 °C (750 to 890 °F).

IMI 230: Effect of cold working on aged properties

Sheet	0.1% proof stress MPa	0.1% proof stress ksi	Tensile strength MPa	Tensile strength ksi	Elongation on 25 mm (in.) %
1. Solution treated (ST)	550	80	630	91	21
2. ST + 24 h/400 °C	580	84	690	100	21
3. ST + duplex age	685	99	820	119	16
4. As 2 + 20% cold work + 8 h/475 °C	810	117.5	900	130	10
5. As 2 + 30% cold work + 8 h/475 °C	800	116	900	130	10
6. As 2 + 40% cold work + 8 h/475 °C	780	113	880	127.5	9
7. As 3 + 30% cold work	895	130	955	138.5	8
8. As 3 + 30% cold work + 6 h/475 °C	700	101.5	865	125	8

Heat Treatment

IMI 230: Recommended heat treatments

Treatment	Temperature °C	Temperature °F	Duration	Cooling method
Stress relief	600	1110	1 hour	Air cool
Annealing	675-785	1250-1450(a)	0.5 to 2 hours(a)	Air cool
Solution heat treat	850	1560	0.5 hour	Forced air
Aging (first stage)(b)	400	750	8 hours	Air cool
Aging (second stage)(b)	475	885	8 hours	Air cool

(a) Annealing from 675 to 700 °C (1250 to 1300 °F) is frequently used for full annealing. Duration depends on product thickness. (b) Two-stage aging is recommended.

Joining

Welding. IMI 230 can be joined by fusion, resistance, flash-butt, and pressure welding. Fusion welds can be made by both argon-arc and electron-beam welding. With adequate control of welding techniques, welds of 100% strength can be obtained, with only a slight loss in tensile or bend ductility. If the alloy is to be used in the annealed condition, then welding should be followed by stress relieving for ½ h at 600 °C (1110 °F).

Brazing. It is possible to make brazed joints, but difficulty arises owing to the formation of brittle intermetallic phases between the titanium and the filler metal.

IMI 417

IMI 417 is the general engineering version of the IMI 834 near-alpha alloy. The two alloys have identical composition specifications and are ideal for high-temperature (600 °C, max) fatigue-sensitive applications. Major uses include cast or wrought parts for turbine and internal combustion engines. Product manufacture may differ for IMI 417 applications versus the aerospace IMI 834 alloy.

Like IMI 834, the carbon addition in IMI 417 widens the a+b phase field and thus allows solution treatment high in the a+b field for a combination of excellent creep resistance, fatigue strength, and tensile strength/ductility. A target primary alpha content of 12 to 15% is recommended for solution treatment of IMI 417. This is equivalent to a heat treatment temperature of 1020 to 1025 °C (1870 to 1880 °F) for a typical transus approach curve. Current experience indicates that for sections exceeding 15 mm, highest tensile strength and creep resistance are obtained by heat treatment for 2 h at temperature followed by oil-quenching and aging at 700 °C for 2 h, air cool. For optimum ductility after heat treatment, aging at 625 °C for 2 h, air cool, is recommended. For thinner sections, air cooling or equivalent inert gas quenching after vacuum heat treatment is adequate.

IMI 417: Typical composition range (wt%) and density

	Al	Sn	Zr	Nb	Mo	Si	C	Fe	O_2	N_2	H_2
Minimum	5.5	3.0	3.0	0.5	0.25	0.20	0.04	...	0.075	...	...
Maximum	6.1	5.0	5.0	1.0	0.75	0.6	0.08	0.05	0.150	0.03	0.006
Nominal	5.8	4.0	3.5	0.7	0.5	0.35	0.06	...	0.10	...	...

Composition and density are identical to IMI 834.

IMI 417: Effect of primary alpha content and aging on tensile properties

Metallurgical condition		Room-temperature tensile properties				100-h total plastic strain, %, at:		
Alpha phase, %	Aging temperature, °C	0.2% yield stress MPa	Tensile strength MPa	Elongation (5D) %	Reduction in area, %	600 °C and 125 MPa	600 °C and 150 MPa	700 °C and 50 MPa
7.5	625	943	1092	15	32	...	0.130	...
	700	957	1086	12	22	...	0.072	...
	625	949	1086	14	31	0.110	0.136	0.193
15	700	945	1079	12	23	0.054	0.082	0.151
	750	942	1058	12	17	0.054	...	0.146

Refer to IMI 834 data for applicable properties of IMI 417.

IMI 679
Ti-11Sn-5Zr-2.25Al-1Mo-0.25Si

IMI 679 was introduced in 1961 as a high-temperature alloy for jet engine components, but has been superseded by other alloys such as Ti-6242S (which was introduced in 1974 as an improvement of IMI 679). IMI 679 has a maximum use temperature of about 450 °C (840 °F) and appears to be metallurgically stable up to 455 °C (850 °F). For comparable products in the annealed condition, the strength of IMI 679 from room temperature to 540 °C (1000 °F) exceeds that of Ti-6Al-4V and Ti-8Al-1Mo-1V and is about equal to Ti-6Al-2Sn-4Zr-2Mo. Its creep strength is superior to Ti-8Al-1Mo-1V and Ti-6Al-4V at all temperatures, but inferior to Ti-6Al-2Sn-4Zr-2Mo at temperatures above 480 °C (900 °F). At elevated temperatures, this alloy is less fatigue resistant than Ti-8Al-1Mo-1V and Ti-6Al-2Sn-4Zr-2Mo. Forgeability and machinability of this alloy are comparable to Ti-8Al-1Mo-1V. Welding of IMI 679 is not recommended.

IMI 679: Typical composition range (wt%) and density

	Al	Sn	Zr	Mo	Si	Fe	C	O_2	N_2	H_2	Ti
Minimum	2.0	10.5	4.0	0.8	0.10	...	...	...	...	...	...
Maximum	2.5	11.5	6.0	1.2	0.50	0.20	...	0.20	...	0.125	...
Nominal	2.25	11.0	5.0	1.0	0.25	...	...	...	...	...	bal

Density of IMI 679 is 4.84 g/cm^3 (0.175 lb/in.^{3}).

Phases and Structures

The combination of low-aluminum, medium-zirconium, and high-tin strengthens and stabilizes the alpha phase. Considerable strengthening at all temperatures is derived from the active eutectoid compound Ti$_x$Si$_y$. The alloy may be classified as both a weakly stabilized, martensitic alloy and an active eutectoid. It displays the isothermal transformation characteristics of two-phase titanium alloys.

Ti-Si diagram

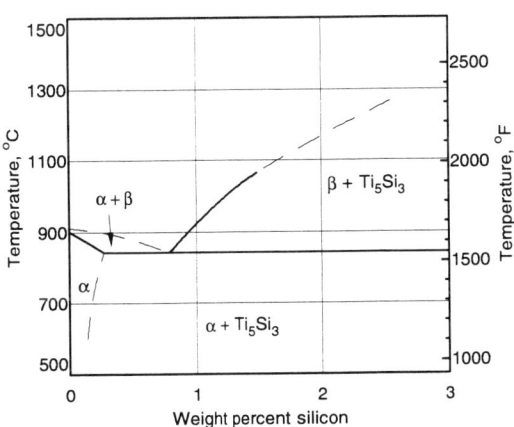

Source: *Aerospace Structural Metals Handbook*, Vol 4, Code 3711, Battelle Columbus Laboratories, 1969

IMI 679: Time-temperature transformation diagram

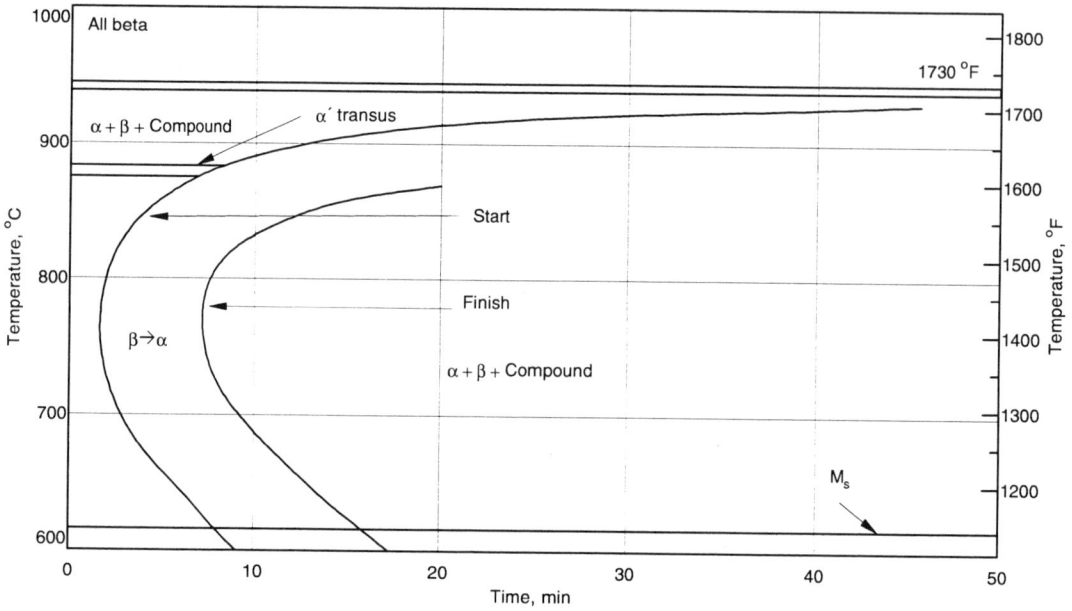

Source: *Aerospace Structural Metals Handbook*, Vol 4, Code 3711, Battelle Columbus Laboratories, 1969

Physical Properties

IMI 679: Summary of typical physical properties

Alpha prime transus	870-890 °C (1600-1630 °F)
Beta transus	950 ± 10 °C (1740 ± 20 °F)
Melting (liquidus point)	...
Density(a)	4.84 g/cm^3 (0.175 lbf/in.^{3})
Electrical resistivity(a)	1.6 µΩ · m
Magnetic permeability	Nonmagnetic
Specific heat capacity(a)	500 J/kg · K (0.12 Btu/lb · °F)
Thermal conductivity	8.3 W/m · K (4.8 Btu/ft · h · °F)
Thermal coefficient of linear expansion(b)	8.2 × 10^{-6}/ °C (4.6 × 10^{-6}/ °F)

(a) Typical values at room temperature of about 20 to 25 °C (68 to 78 °F). (b) Mean coefficient from room temperature to 100 °C (212 °F).

IMI 679: Static compressive modulus of STA bar

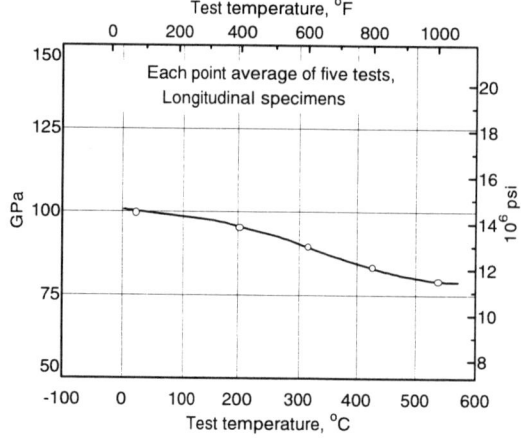

STA treatment: 900 °C (1650 °F) for 2 h, AC, plus 500 °C (930 °F) for 24 h, AC.
Source: *Aerospace Structural Metals Handbook*, Vol 4, Code 3711, Battelle Columbus Laboratories, 1969

IMI 679: Static tensile modulus of STA bar

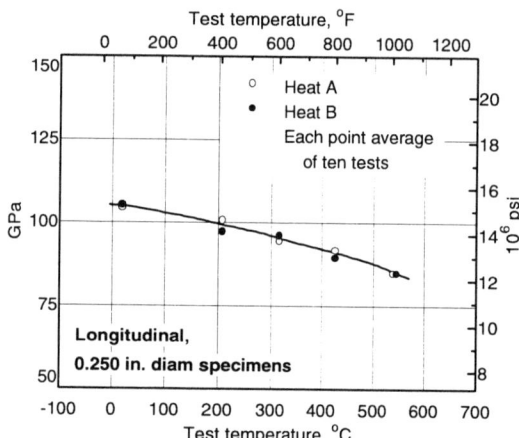

The dynamic modulus was similar for specimens excited to resonance longitudinally.
Rolled bar, 900 °C (1650 °F) for 2 h, AC, 500 °C (930 °F), 24 h AC.
Source: *Aerospace Structural Metals Handbook*, Vol 4, Code 3711, Battelle Columbus Laboratories, 1969

IMI 679: Electrical resistivity

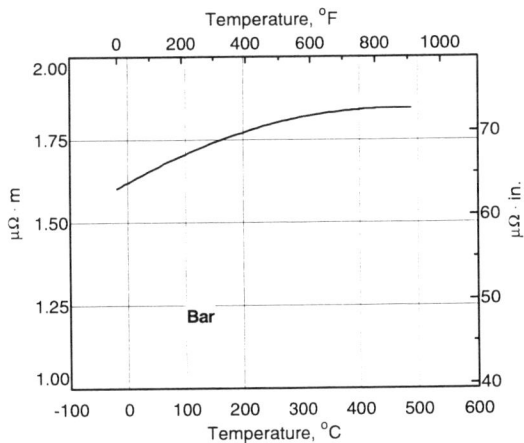

Source: *Aerospace Structural Metals Handbook*, Vol 4, Code 3711, Battelle Columbus Laboratories, 1969

IMI 679: Specific heat

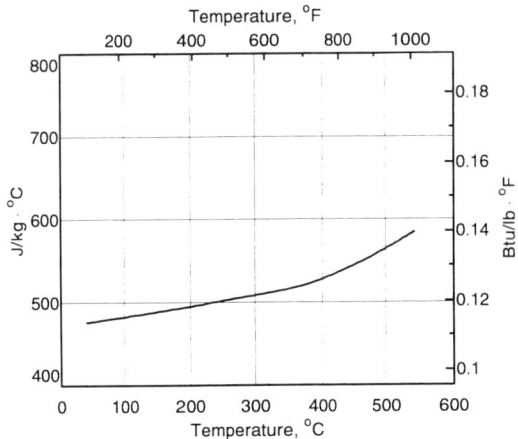

STA bar.
Source: *Aerospace Structural Metals Handbook*, Vol 4, Code 3711, Battelle Columbus Laboratories, 1969

IMI 679: Thermal conductivity

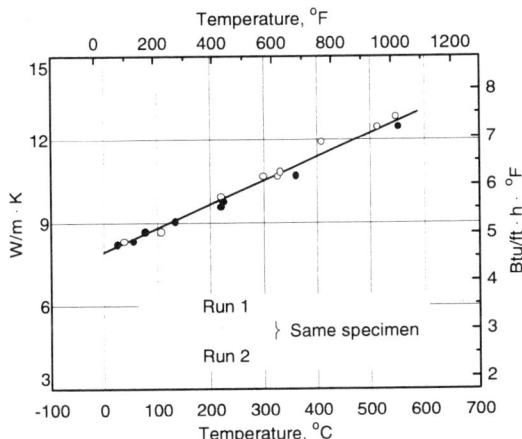

Measurement in short transverse direction of STA bar.
Source: *Aerospace Structural Metals Handbook*, Vol 4, Code 3711, Battelle Columbus Laboratories, 1969

IMI 679: Thermal coefficient of linear expansion

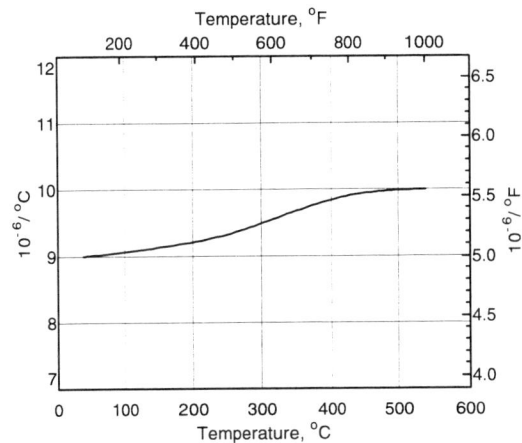

Mean coefficient from room temperature to temperature indicated.
Measurement in long direction of STA bar.
Source: *Aerospace Structural Metals Handbook*, Vol 4, Code 3711, Battelle Columbus Laboratories, 1969

Tensile Properties

IMI 679: Minimum tensile properties at room temperature

Condition	Ruling section		Minimum ultimate tensile strength		Minimum tensile yield strength		Minimum elongation	RA
	mm	in.	MPa	ksi	MPa	ksi	%	%
Oil quenched and aged: 900 °C/OQ, 500 °C/24 h/AC	50(a)	2(a)	1110	161	970	140	8	25
Air cooled and aged: 900 °C/AC, 500 °C/24 h/AC	75(b)	3(b)	1030(c)	149(c)	880(c)	127(c)	8	30

(a) Bar, forging stock, and forgings per BSTA 18, 19, and 20, respectively. (b) Bar, forging stock, and forgings per BSTA 25, 26, and 27, respectively. (c) Slightly lower values are quoted in AMS 4974 and MIL-T-9047D.

IMI 679: Tensile strength

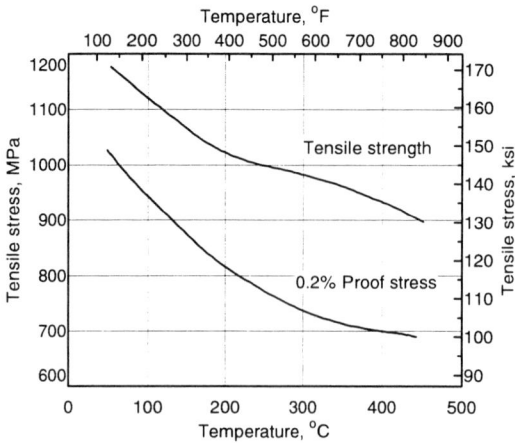

Oil-quenched and aged IMI 679.
Source: IMI Titanium

IMI 679: Tensile properties vs temperature

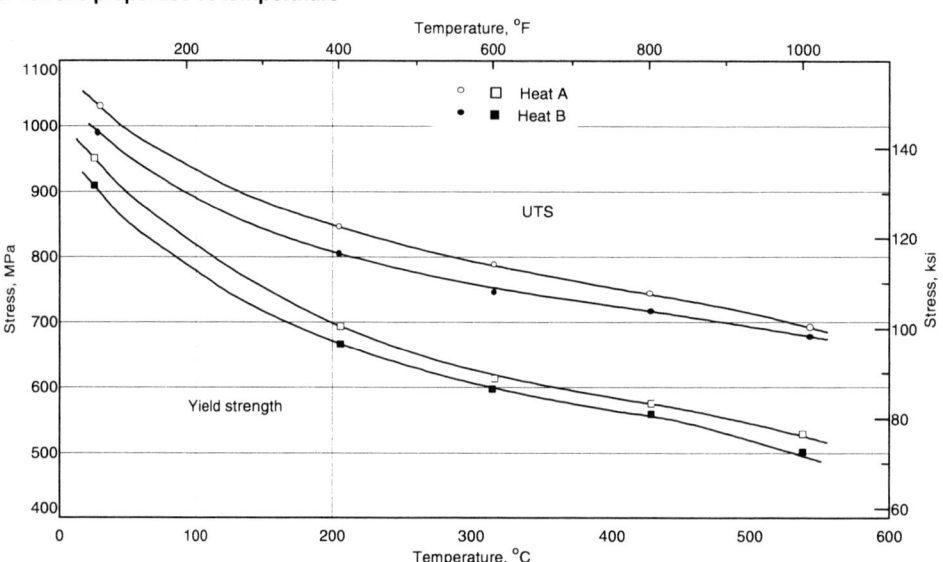

(a)

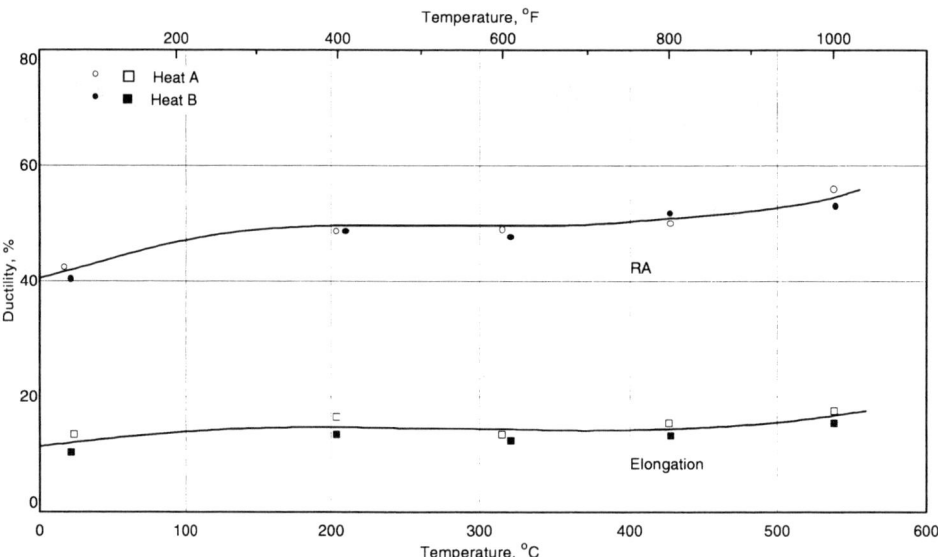

(b)

IMI 679: Tensile properties as a function of thermal exposure

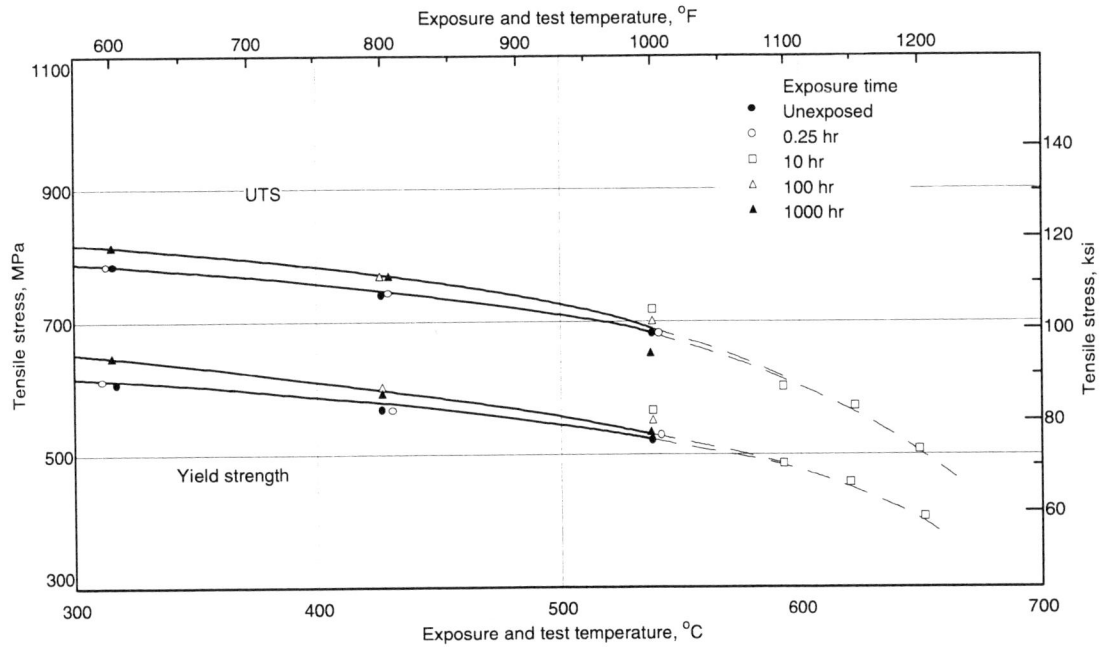

(a)

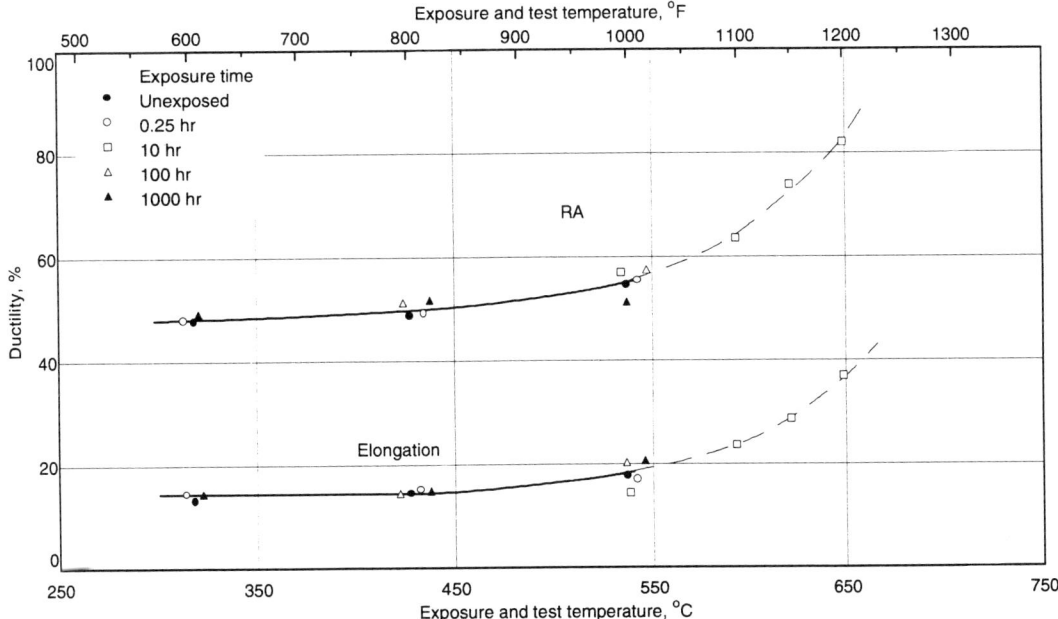

(b)

Solution treated, air cooled, and aged bar: 900 °C (1650 °F) for 2 h, AC, plus 500 °C (930 °F) for 24 h, AC. Each point an average of ten tests on 6.35 mm (0.25 in.) specimen.
Source: *Aerospace Structural Metals Handbook*, Vol 4, Code 3711, Battelle Columbus Laboratories, 1969

Creep Properties

IMI 679: Minimum creep rate

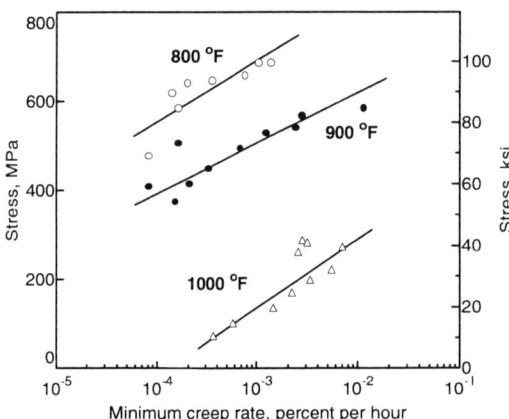

Solution treated, air cooled, and aged bar. Duplicate tests up to 540 °C (1000 °F), single tests above 540 °C (1000 °F).
Source: *Aerospace Structural Metals Handbook*, Vol 4, Code 3711, Battelle Columbus Laboratories, 1969

IMI 679: Creep deformation at 425 °C

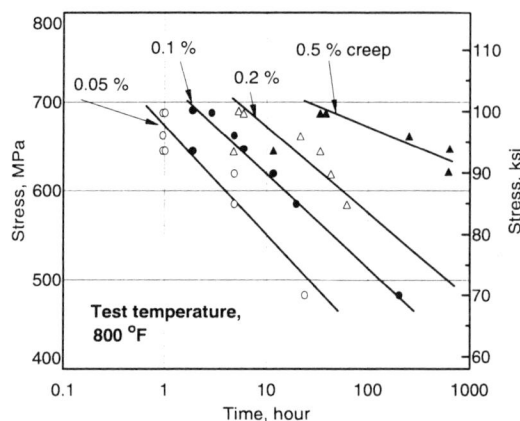

6.35 mm (0.25 in.) diam specimens from 13 × 28.5 mm (0.5 × 1 1/8 in.) bar solution treated and aged as follows: 900 °C (1650 °F) for 2 h, AC, plus 500 °C (930 °F) for 24 h, AC.
Source: *Aerospace Structural Metals Handbook*, Vol 4, Code 3711, Battelle Columbus Laboratories, 1969

IMI 679: Creep deformation at 480 °C

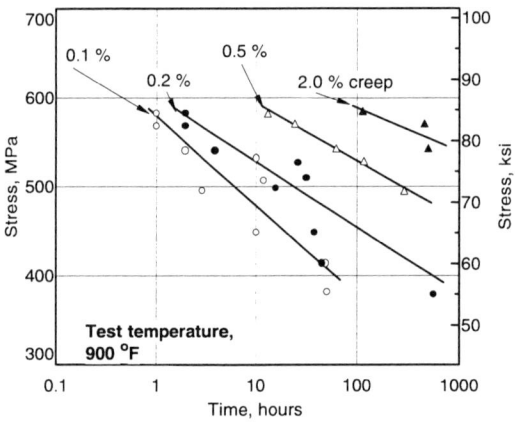

Solution treated, air cooled, and aged bar.
Source: *Aerospace Structural Metals Handbook*, Vol 4, Code 3711, Battelle Columbus Laboratories, 1969

IMI 679: Creep deformation at 540 °C

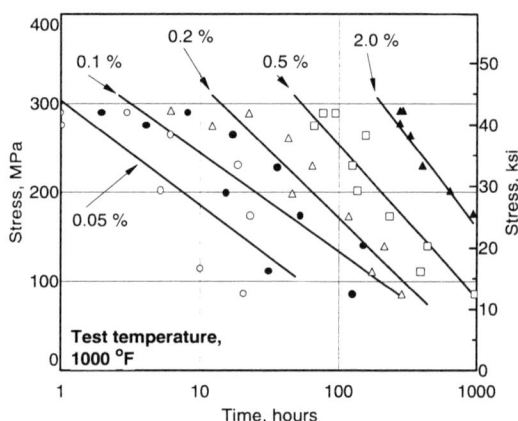

Solution treated, air cooled, and aged bar.
Source: *Aerospace Structural Metals Handbook*, Vol 4, Code 3711, Battelle Columbus Laboratories, 1969

Fatigue Properties

IMI 679: Axial fatigue of STA bar

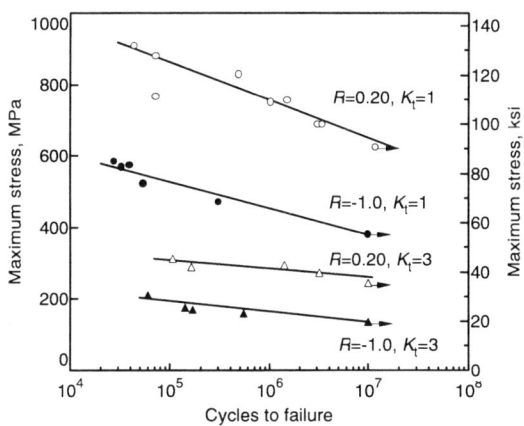

Room-temperature tests at 2500 cycles/min of longitudinal specimens from solution treated, air cooled and aged bar.
Source: *Aerospace Structural Metals Handbook*, Vol 4, Code 3711, Battelle Columbus Laboratories, 1969

IMI 679: Reversed bending fatigue

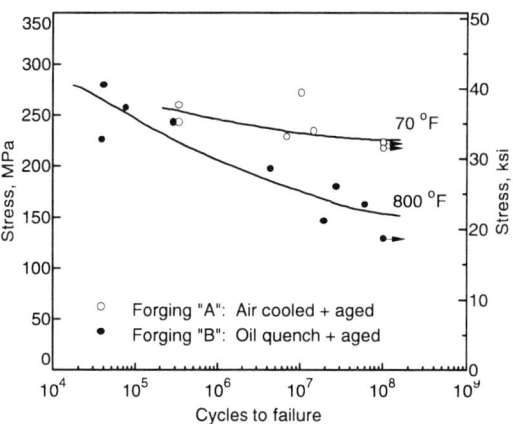

Mild-notch ($K_t = 3$) fatigue strength of forgings at 21 and 425 °C (70 and 800 °F) and a frequency of 1800 cycles/min. Specimens were from mid-radius (forging "A") or web (forging "B") location, radial direction. Treatment temperatures were standard: 900 °C (1650 °F) solution, AC or OQ, plus 500 °C (930 °F) age for 24 h, AC.
Source: *Aerospace Structural Metals Handbook*, Vol 4, Code 3711, Battelle Columbus Laboratories, 1969

IMI 679: Axial fatigue at 200 °C

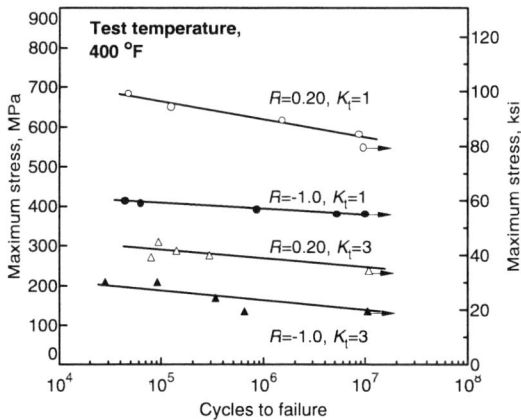

Axial load smooth and mild-notch fatigue properties for solution treated, air cooled, and aged bar. Longitudinal polish (5 µin., rms), frequency of 2500 cycles/min.
Source: *Aerospace Structural Metals Handbook*, Vol 4, Code 3711, Battelle Columbus Laboratories, 1969

Axial fatigue at 425 °C

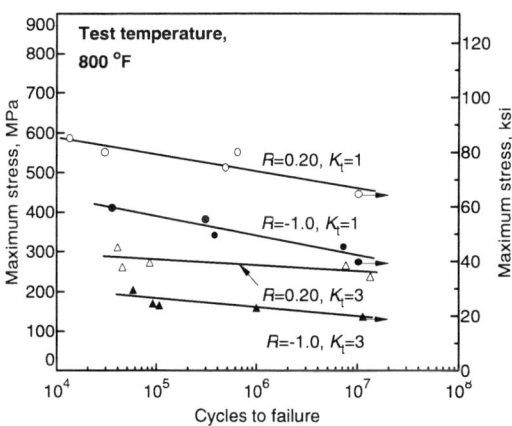

Axial load smooth and mild-notch fatigue properties for solution treated, air cooled, and aged bar. 2500 cycles/min test frequency.
Source: *Aerospace Structural Metals Handbook*, Vol 4, Code 3711, Battelle Columbus Laboratories, 1969

IMI 679: Constant-life diagram (10^7 cycles, axial stress)

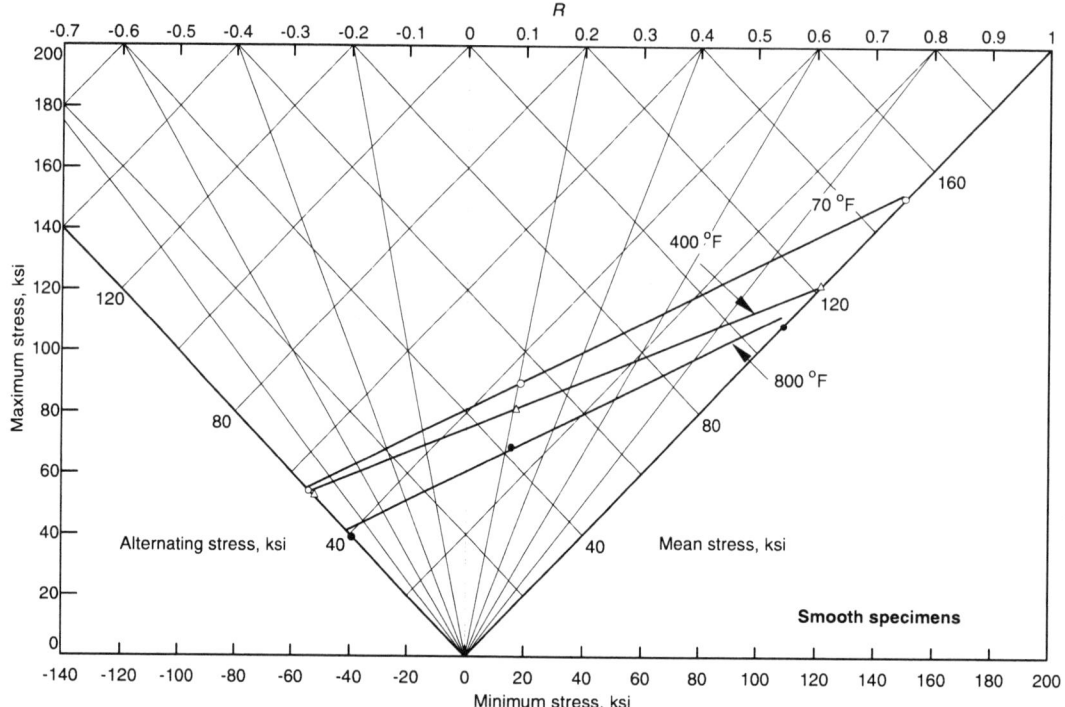

Constant-life fatigue diagram for 13 × 28.5 mm (0.5 × 1⅛ in.) rolled bar heat treated as follows: 900 °C (1650 °F) for 2 h, AC, plus 500 °C (930 °F) for 24 h, AC. Longitudinal smooth specimens with 5 μ in. (rms) longitudinal polish. Axial tension sinusoid at 2500 cycles/min.
Source: *Aerospace Structural Metals Handbook*, Vol 4, Code 3711, Battelle Columbus Laboratories, 1969

Plastic Deformation

IMI 679: Typical stress-strain curve (RT)

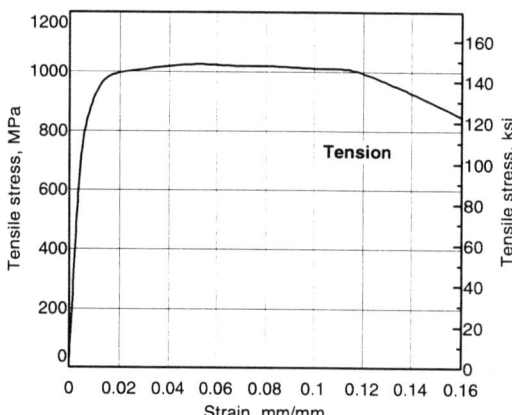

Typical room-temperature stress-strain curve in tension from a large ring forging. Solution treated, AC, and aged.
Source: *Aerospace Structural Metals Handbook*, Vol 4, Code 3711, Battelle Columbus Laboratories, 1969

IMI 679: Typical compressive stress-strain

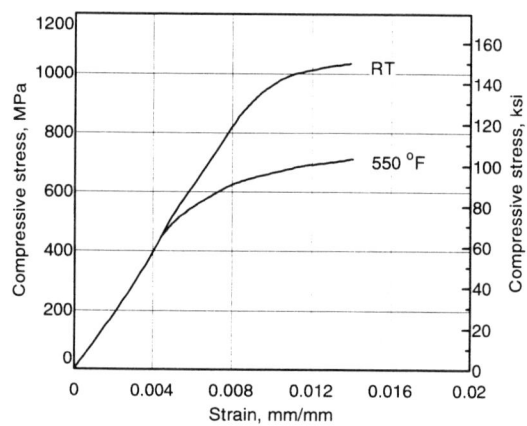

Solution treated, AC, and aged large ring forging.
Source: *Aerospace Structural Metals Handbook*, Vol 4, Code 3711, Battelle Columbus Laboratories, 1969

IMI 679: Compressive yield strength of bar

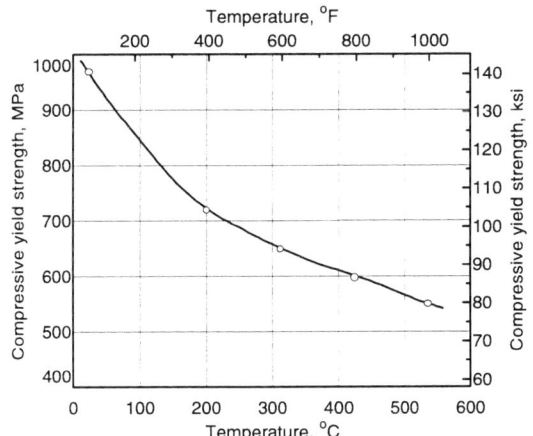

Each point is an average of five tests on a 12 mm (0.475 in.) diam specimen from STA bar.
Source: *Aerospace Structural Metals Handbook*, Vol 4, Code 3711, Battelle Columbus Laboratories, 1969

IMI 679: Flow stress vs temperature

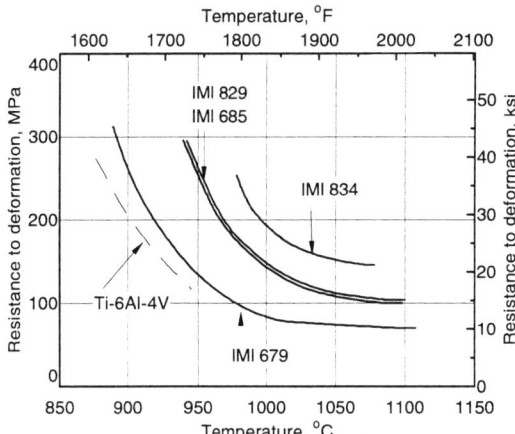

IMI 679 flow stress at a strain rate of 16/S.
Source: IMI Titanium

Processing

Forging. To develop optimum mechanical properties, hot working should be restricted to the alpha+beta+compound field and a maximum temperature of 925 °C (1700 °F) is recommended. The normal preheating temperature for forging is therefore 900 to 920 °C (1650 to 1690 °F).

Heat Treatment. Solution treatment at 900 °C (1650 °F), followed by air cooling and aging 24 hours at 500 °C (930 °F), has been found to produce the best combination of creep strength and ductility. Faster cooling from solution-treatment temperature by oil quenching significantly increases the tensile strength while slightly decreasing ductility. Thin sections, such as gas-turbine compressor blades, are usually air cooled and aged; thicker sections, such as discs and spacer rings, are more usually oil quenched and aged. Above 450 °C (840 °F), creep strength of oil-quenched material is slightly lower than that of air-cooled material.

IMI 679: Resistance to deformation

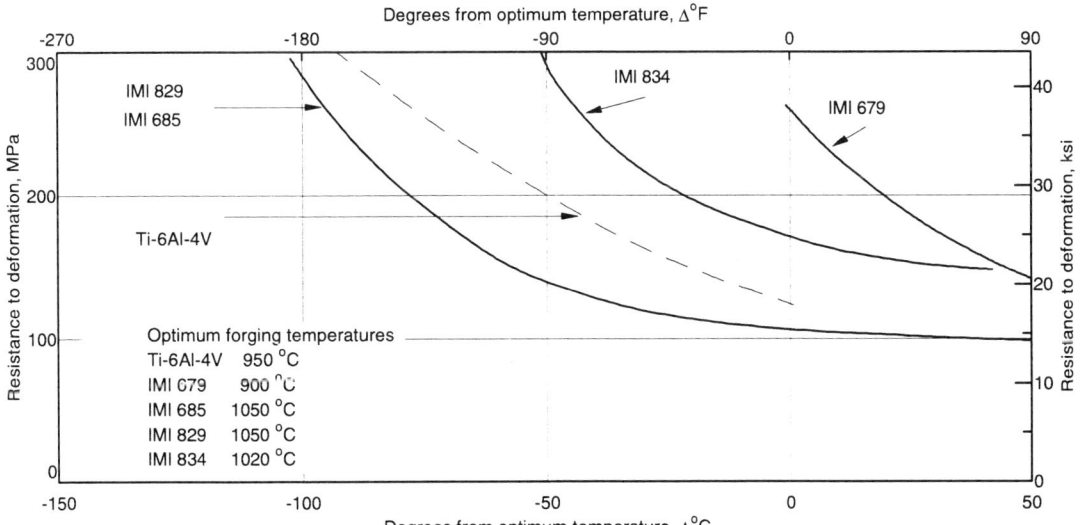

IMI 679: Recommended heat treatments

Treatment	Temperature		Duration	Cooling method
	°C	°F		
Stress relief	480-510	900-950	5 to 10 hours	Air cool
Solution treatment	900	1650	1 to 2 hours	Air cool or oil quench
Aging	500	930	24 h	Air cool

IMI 685
Ti-6Al-5Zr-0.5Mo-0.25Si

IMI 685 is a titanium alloy specifically developed to meet aerospace-engine requirements. It was the first of the near-alpha alloys, and represented a significant step forward in combining good creep strength, weldability, and ease of working. Forging temperatures above the beta transus can be used.

IMI 685: Typical composition range (wt%) and density

	Al	Zr	Mo	Si	Fe	C	O_2	N_2	H_2	Ti
Minimum	5.7	4.5	0.25	0.15	...	...	...	...	...	...
Maximum	6.3	6.0	0.75	0.35	0.05	0.08	0.20	0.03	0.01	...
Nominal	6	5	0.50	0.25	...	...	...	...	...	bal

Density of IMI 685 is 4.45 g/cm^3 (0.161 lb/in.^{3}).

Physical Properties

IMI 685: Summary of typical physical properties

Beta transus	1020 °C (1870 °F)
Melting (liquidus point)	Not Available
Density(a)	4.45 g/cm^3 (0.161 lbf/in.^{3})
Electrical resistivity(a)	1.67 μΩ · m
Magnetic permeability	Nonmagnetic
Specific heat capacity(a)	Not Available
Thermal conductivity(a)	4.2 W/m · K (2.4 Btu/ft · h · °F)
Thermal coefficient of linear expansion(b)	9.8 × 10^{-6}/°C (5.4 × 10^{-6}/°F)

(a) Typical values at room temperature of about 20 to 25 °C (68 to 78 °F). (b) Mean coefficient from room temperature to 100 °C (212 °F)

IMI 685: Young's modulus (dynamic)

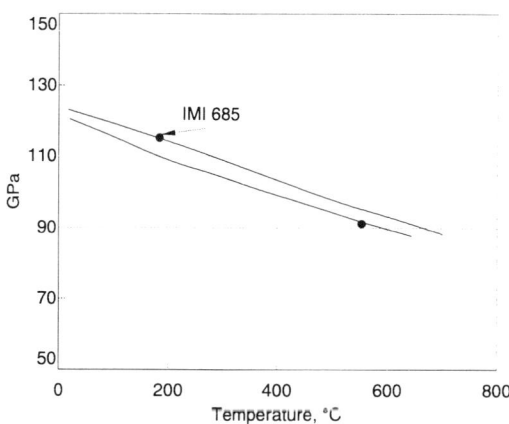

IMI 685: Electrical resistivity

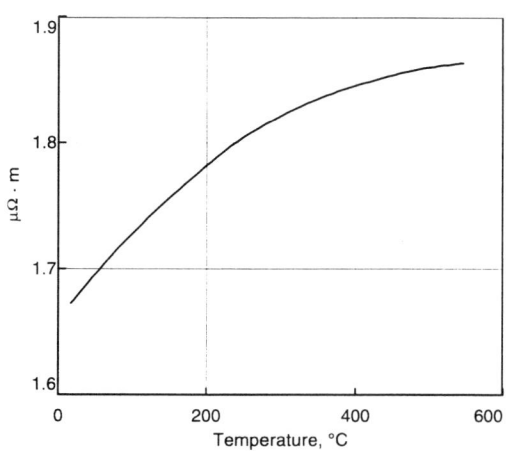

IMI 685: Thermal conductivity

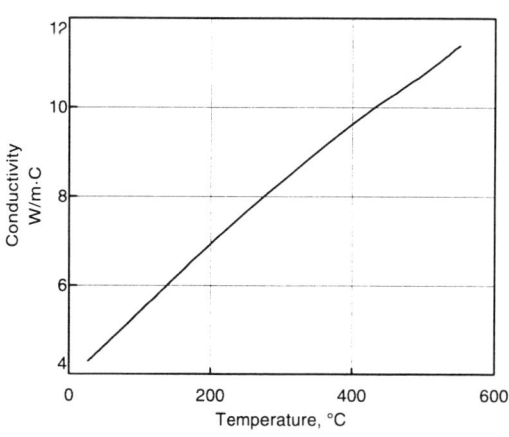

Mechanical Properties

IMI 685: Typical room-temperature tensile properties

Condition or form	Ultimate tensile strength		Tensile yield strength (0.2%)		Elongation on 5D	RA
	MPa	ksi	MPa	ksi	%	%
Rod	1020	148	914	132.5	11	22
Billet	1030	149	900	130.5	10	20
Aged 24 h at 550 °C	1010	146.5	901	130.6	10.5	18.5
Aged 48 h at 550 °C	1030	149	910	132	11.5	19
Specification minimum	990	143.5	850	123	6	15

IMI 685: Effect of cooling rate on room-temperature tensile properties after aging

Heat treatment	0.2% yield stress		Tensile strength		Elongation on 5D	Reduction in area	Notched-tensile ratio
	MPa	ksi	MPa	ksi	%	%	$K_t = 3$
1050 °C, OQ within 15 s	924	134	1060	153	10	22	1.64
1050 °C, delay 30 s, OQ	881	128	1030	149	10	26	1.62
1050 °C, delay 60 s, OQ	873	126.5	1000	145	10	25	1.62
1050 °C, AC	858	124	983	142.5	10	20	1.65

High-Temperature Strength

IMI 685 rod: Typical tensile properties vs temperature

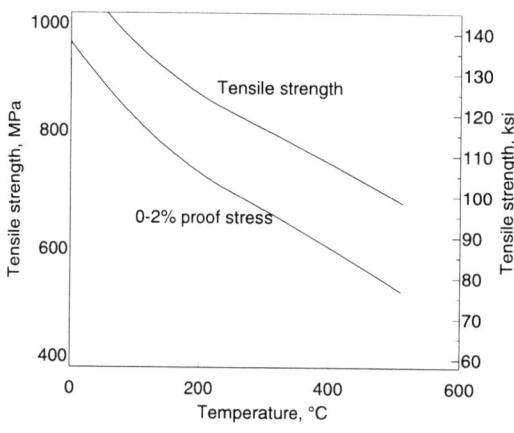
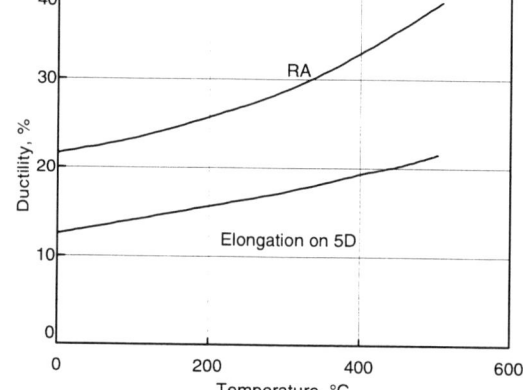

Beta heat treated rod

IMI 685: Typical and design stresses for 0.1% total plastic strain in 100 h

Typical stresses measured on samples from forged compressor disc

The heat treatment section at the end of this datasheet includes the effect of cooling rate on creep resistance.

Fatigue Properties

IMI 685: Alternating direct-stress fatigue

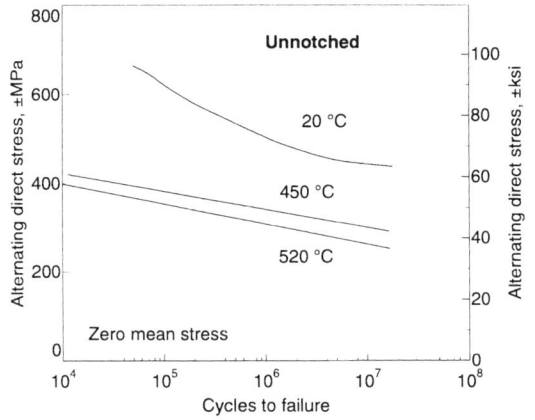

IMI 685: Direct-stress fatigue at 450 and 520 °C

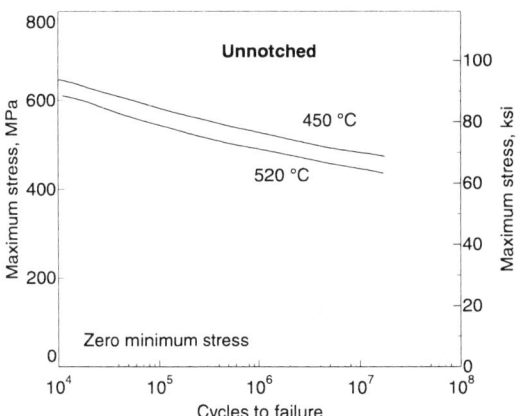

IMI 685: Direct-stress fatigue of forged disc

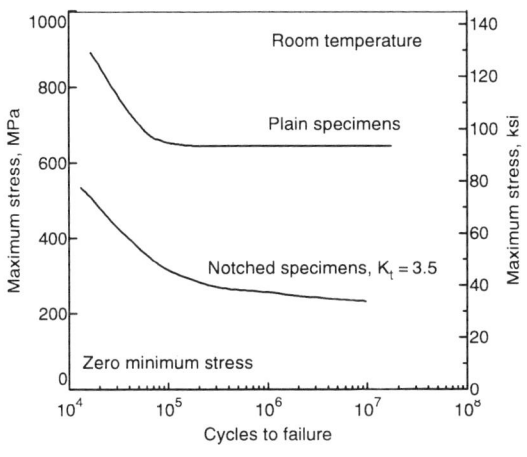

IMI 685: Direct-stress fatigue of forged disc at 475 °C

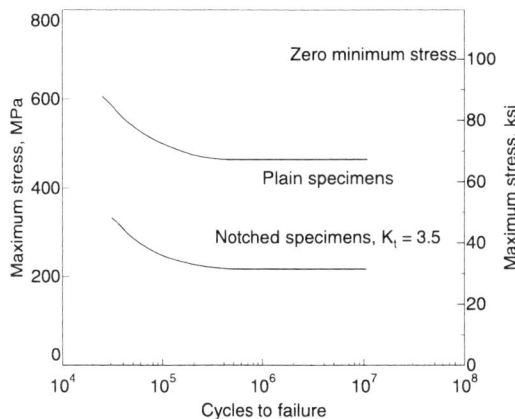

Processing and Heat Treatment

Forging routes for IMI 685 should aim to provide a material which on subsequent treatment in the beta field recrystallizes to a fine uniform beta grain structure of average grain sizes less than 1.5 mm.

Preheating in the beta field at temperatures up

IMI 685: Recommended STA heat treatment for optimum creep resistance

Treatment	Temperature °C	°F	Duration	Cooling method
Solution treatment	1050	1920	30 min per 25 mm (1 in.) of thickness	Oil quench(a)
Aging	550	1020	24 hours	Air cool

(a) The transfer time between furnace and oil bath should not be prolonged. A maximum of 15 to 30 s, depending on part mass, is recommended.

IMI 685: Effect of cooling rate on creep resistance and post-creep tensile properties after aging

Heat treatment	Total plastic strain % in 100 h, 310 MPa at 520 °C	0.2 yield stress MPa	ksi	Tensile strength MPa	ksi	Elongation on 5D %	Reduction in area %
1050 °C OQ within 15 s	0.041	920	133	1010	146.5	13	20
1050 °C, delay 30 s, OQ	0.053	920	133	1000	145	11	19
1050 °C, delay 60 s, OQ	0.055	890	129	990	143.5	11	19

to 20-30 °C (35 to 55 °F) above the beta transus is permissible provided sufficient work is introduced per heat to avoid undesirable grain coarsening effects. If small reductions per heat are likely, alpha-beta forging from 980 to 1000 °C (1795 to 1830 °F) is recommended.

Preheat times in the beta field should be the minimum necessary for uniform soaking of the material prior to forging.

Welding. IMI 685 can be joined by the processes normally used in the fabrication of titanium, including argon-arc, electron-beam, and friction welding. Material should be fully heat treated prior to welding, and also given a suitable post-weld treatment (for example 4-8 h at 550 °C). The weld zone of material treated in this way will have similar structure, and hence similar properties, to those of the parent metal.

IMI 829
Ti-5Al-3.5Sn-3.0Zr-1Nb-0.3Si

Common Name: Ti-5331S

IMI 829 is a near-alpha titanium alloy of medium strength and high-temperature capability up to ~540 °C (1000 °F). IMI 829 develops its properties from a combination of solid-solution strengthening and "beta" heat treatment which produces an acicular transformed structure. IMI 829 is lean in beta stabilizers in order to give high creep resistance. It therefore has limited hardenability; good property levels being achieved in sections up to ~75 mm (3 in.) thick.

Product Forms and Condition. IMI 829 is available in the form of bar, billet, plate, sheet, wire, and castings. IMI 829 is fully weldable using any of the titanium welding techniques, MIG, TIG, EB, etc.

The standard condition of use is beta solution treated, 1050 °C (1922 °F)/½ h oil quench plus aged at 625 °C (1157 °F)/2 h air cool). Instead of quench, air cooling is recommended below 30 mm diameter.

In this condition, the alloy possesses a microstructure of acicular transformed beta with a grain size typically ~0.5 mm (0.02 in.). The alpha/beta to beta transus temperature for IMI 829 is 1015 °C ± 10 °C (1860 °F ± 20 °F).

Applications. The major use for IMI 829 is as discs and blades in aeroengine compressors.

IMI 829: Typical composition range (wt%) and density

	Al	Sn	Zr	Nb	Mo	Si	O_2	N_2	H_2
Minimum	5.2	3.0	2.5	0.7	0.20	0.20	0.09	...	...
Maximum	5.7	4.0	3.5	1.3	0.35	0.50	0.15	0.03	0.0060
Nominal	5.6	3.5	3.0	1.0	0.25	0.30	0.115	...	...

Density of IMI 829 is 4.54 g/cm^3 (0.164 lb/in.^{3}).

Physical Properties

IMI 829: Summary of typical physical properties

Beta transus	1015 ± 10 °C (1860 ± 20 °F)
Melting (liquidus point)	Not Available
Density(a)	4.54 g/cm^3 (0.164 lbf/in.^{3})
Electrical resistivity(a)	Not Available
Magnetic permeability	Nonmagnetic
Specific heat(a)	516 J/kg · K (0.123 Btu/lb · °F)
Thermal conductivity(a)	6.9 W/m · K (4.0 Btu/ft · h · °F)
Thermal coefficient of linear expansion(b)	9.45 × 10^{-6}/°C (5.24 × 10^{-6}/°F)

(a) Typical values at room temperature of about 20 to 25 °C (68 to 78 °F). (b) Mean coefficient from room temperature to 200 °C (390 °F)

IMI 829: Young's modulus (dynamic)

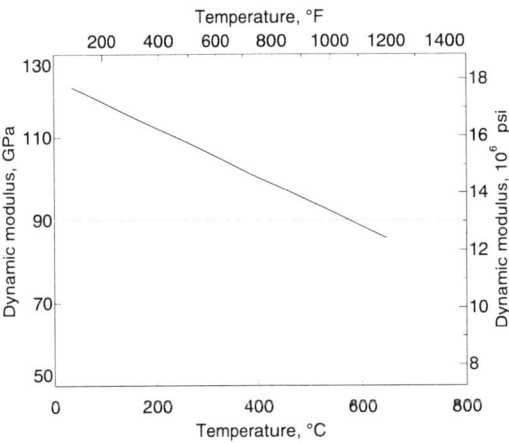

The modulus of IMI 829 is typical of other high-aluminum, near-alpha alloys and is therefore somewhat higher than alpha-beta alloys. Source: IMI Titanium "High-Temperature Alloys" brochure

Thermal Properties

IMI 829: Specific heat

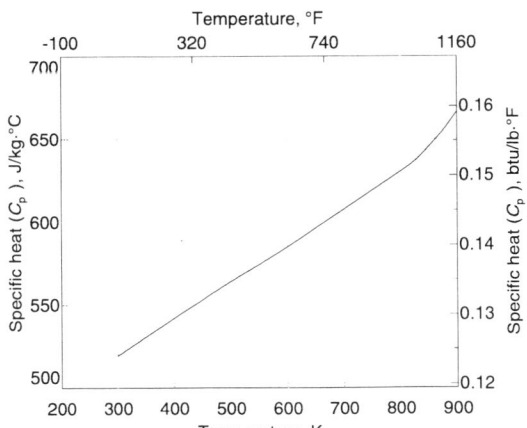

Heat-treated bar.
Source: IMI Titanium "High Temperature Alloys" brochure

IMI 829: Thermal coefficient of linear expansion

Temperature range		Mean coefficient of thermal expansion	
°C	°F	$10^{-6}/°C$	$10^{-6}/°F$
20-200	68-390	9.45	5.24
20-400	68-750	9.77	5.43
20-600	68-1110	9.98	5.54
20-800	68-1470	10.34	5.74
20-1000	68-1830	10.39	5.77

Heat-treated bar. Thermal expansion is typical of other near-alpha titanium alloys. Source: IMI Titanium "High-Temperature Alloys" brochure

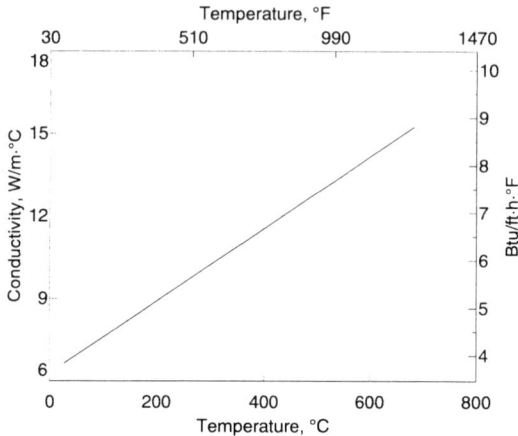

IMI 829: Thermal conductivity

Heat-treated disc material, 75 mm (3 in.) ruling section.
Source: IMI Titanium "High-Temperature Alloys" brochure

Mechanical Properties

Hardness of heat treated IMI 829 is typically 320 HV (20 kg) or 32 HRC.

Notch tensile ratio is typically 1.6 ($K_t = 3$).

Fracture toughness is typically ~75 MPa$\sqrt{m}$ (68 ksi$\sqrt{in.}$).

IMI 829: Room-temperature tensile properties

Property	Typical	Minimum
0.2% PS MPa (ksi)	860 (125)	820 (119)
U.T.S. MPa (ksi)	950 (142)	930 (135)
El (5D)%	11	9
Reduction in area %	19	15

Discs, ~75 mm (3 in.) ruling section, heat treated

High-Temperature Strength

IMI 829 is regarded as having good creep performance up to around 550 °C (1020 °F) and somewhat higher for short-time applications. At 540 °C (1000 °F) a total plastic strain of less than 0.1% in 100 hours is achieved under a stress of about 300 MPa (43.5 ksi).

IMI 829: Typical tensile properties vs temperature

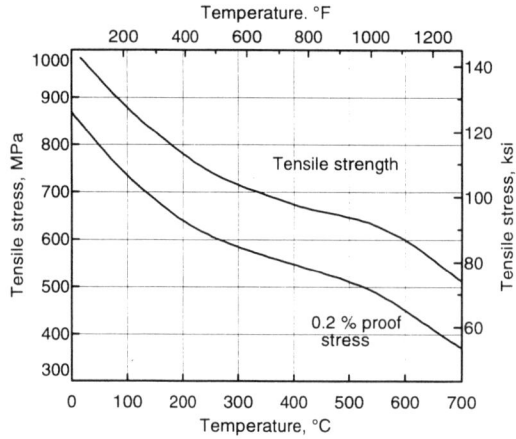

Heat-treated disc forging ~75 mm (3 in.) ruling section.
Source: IMI Titanium "High-Temperature Alloys" brochure

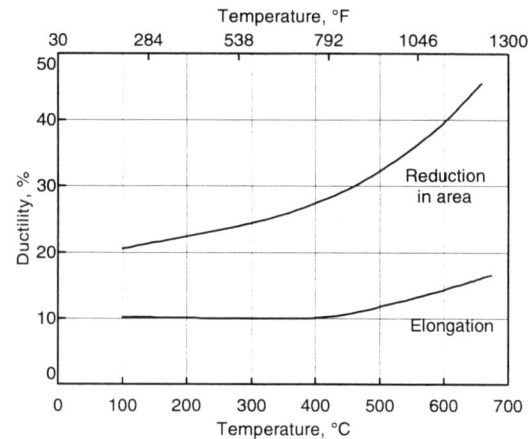

IMI 829: Typical 0.1% creep strength

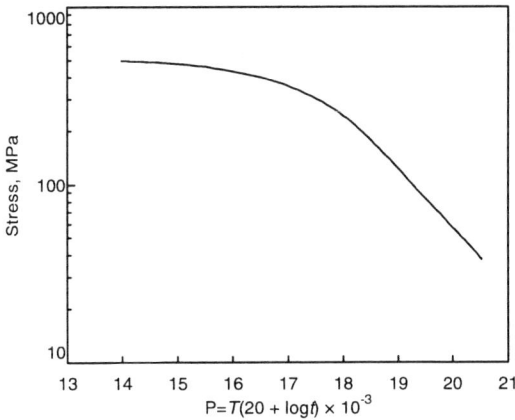

Disc forging ~75 mm (3 in.) ruling section.
T in °C, t in hours

Fatigue Properties

Typical low-cycle fatigue properties are shown at 20 °C (68 °F), 300 °C (572 °F), and 540 °C (1000 °F) (see figure). Crack propagation resistance is superior to that of typical alpha-beta titanium alloys.

IMI 829: Typical plain section fatigue strength

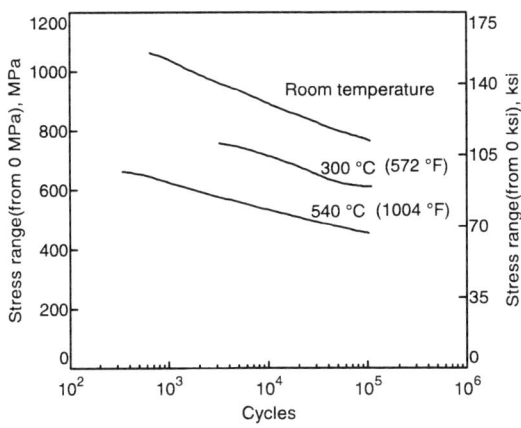

Heat-treated disc forging ~75 mm (3 in.) ruling section.
Source: IMI Titanium "High-Temperature Alloys" brochure

IMI 829: Typical fatigue crack growth rate

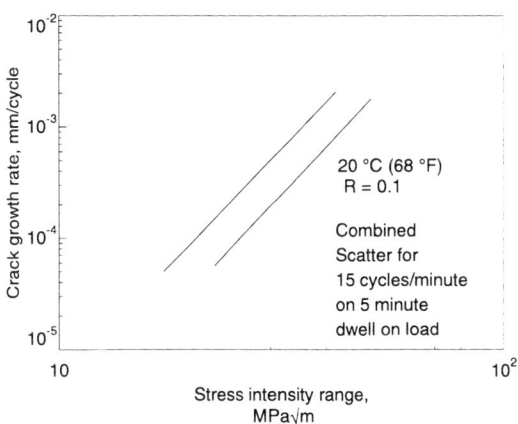

Heat-treated disc forging ~75 mm (3 in.) ruling section.
Source: IMI Titanium "High -Temperature Alloys" brochure

IMI 829: Flow stress

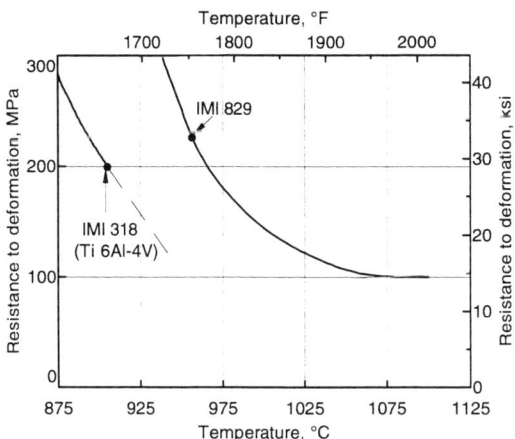

Heat-treated bar, plastometer test at a strain rate of 15/s.
Source: IMI Titanium "High-Temperature Alloys" brochure

Plastic Deformation

Flow stress of IMI 829 around its typical forging temperature is shown (see figure). IMI 829 is stiffer than alpha-beta alloys such as Ti-6Al-4V, and flow stress rises more rapidly as temperature falls.

Processing

Casting. IMI 829 is an excellent casting alloy because it is designed to be used in the microstructural condition analogous to casting (i.e. beta heat treated). Cast properties therefore approach, or are better than those, of wrought products.

Forging. IMI 829 is readily forgeable by conventional hammer, press, or isothermal forging. Its flow stress is a little higher than most other titanium alloys and forging temperature must not be allowed to drop excessively. Typical forging temperature is 1000 °C (1830 °F).

Forming. IMI 829 can be superplastically formed although it requires a relatively high temperature to achieve the required two-phase microstructure (at about 975 °C, or 1785 °F).

Heat Treatment. Beta solution treatment followed by aging (see table).

Welding. IMI 829 is weldable using all of the normal techniques used for titanium welding. Properties of welds are comparable with those of the parent metal.

IMI 829: Recommended heat treatments

Treatment	Temperature		Duration	Cooling method
	°C	°F		
Solution treatment	1050	1920	30 min	Oil quench(a)
Aging	625	1160	2 hours	Air cool

(a) For sections less than 30 mm (1.25 in.), air cooling is recommended.

IMI 834
Ti-5.8Al-4Sn-3.5Zr-0.7Nb-0.5Mo-0.35Si

IMI 834 is a near-alpha titanium alloy of medium strength (typically 1050 MPa, or 152 ksi) and temperature capability up to about 600 °C (1110 °F) combined with good fatigue resistance. The alloy derives its properties from solid-solution strengthening, and heat treatment high in the alpha + beta phase field. The addition of carbon facilitates treatment by widening the heat treatment window (see figure). IMI 834 has a low beta stabilizer content and therefore has limited hardenability. It retains a good level of properties in sections up to around 75 mm (3 in.) diameter, with small reductions in strength in larger sections.

Product Forms and Condition. IMI 834 is available in the form of bar, billet, plate, sheet, wire, and castings. IMI 834 is weldable using all of the established titanium welding techniques. It is normally alpha + beta solution treated (15% α) and aged. Microstructural characterization of IMI 834 is discussed in *Met. Trans.*, Vol 24A, June 1993, p 1273-1280.

Applications. The major use for IMI 834 is compressor discs and blades in the aerospace industry. General purpose use is intended for IMI 417.

IMI 834: Typical composition range (wt%) and density

	Al	Sn	Zr	Nb	Mo	Si	C	Fe	O_2	N_2	H_2
Minimum	5.5	3.0	3.0	0.5	0.25	0.20	0.04	...	0.075	...	...
Maximum	6.1	5.0	5.0	1.0	0.75	0.60	0.08	0.05	0.150	0.03	0.006
Nominal	5.8	4.0	3.5	0.7	0.5	0.35	0.06	...	0.10	...	...

Density of IMI 834 is 4.55 g/cm^3 (0.164 lb/in.^{3}).

Physical Properties

IMI 834: Summary of typical physical properties

Beta transus	1045 ± 10 °C (1915 ± 20 °F)
Melting (liquidus point)	Not Available
Density(a)	4.55 g/cm^3 (0.164 lbf/in.^{3})
Electrical resistivity(a)	Not Available
Magnetic permeability	Nonmagnetic
Specific heat capacity(a)	Not Available
Thermal conductivity(a)	Not Available
Thermal coefficient of linear expansion(b)	10.6 × 10^{-6}/°C (5.9 × 10^{-6}/°F)

(a) Typical values at room temperature of about 20 to 25 °C (68 to 78 °F). (b) Mean coefficient from room temperature to 200 °C (390 °F)

IMI 834: Thermal coefficient of linear expansion

Temperature range		Mean coefficient of thermal expansion	
°C	°F	10^{-6}/°C	10^{-6}/°F
20-200	68-392	10.6	5.9
20-400	68-752	10.9	6.1
20-600	68-1112	11.0	6.1
20-800	68-1472	11.2	6.2
20-1000	68-1832	11.3	6.3

The thermal expansion coefficient of IMI 834 is typical of other titanium alloys. Heat treated bar

IMI 834: Beta approach curve

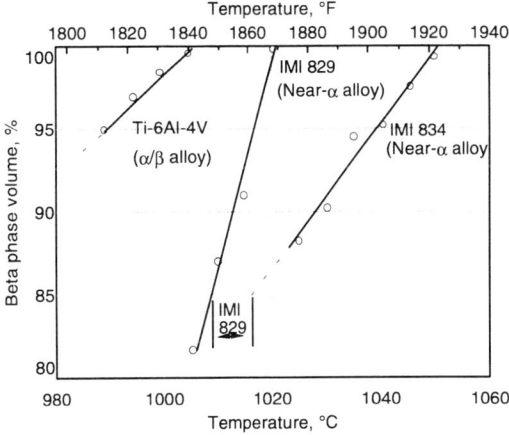

Beta transus approach curves of IMI 834, IMI 829, and Ti-6Al-4V

Mechanical Properties

Hardness of heat treated IMI 834 is typically 350 HV (20 kg load) or about 35 HRC.

Notch tensile ratio is typically 1.45 ($K_t = 3$).

Impact Strength. Typical Charpy (U-notch) impact strength is 15 J (11 ft · lbf) at room temperature.

Fracture toughness of IMI 834 is typically 45 MPa$\sqrt{m}$ (40 ksi$\sqrt{in.}$) in heat treated discs.

IMI 834: Minimum tensile properties
Typical UTS is 1050 MPa (152 ksi).

Property	Room-temperature minimum
0.2% PS MPa (ksi)	910 (132)
U.T.S. MPa (ksi)	1030 (149)
Elongation (in 5D), %	6
Reduction in area, %	15

Heat treated discs

IMI 834: Young's modulus (dynamic)

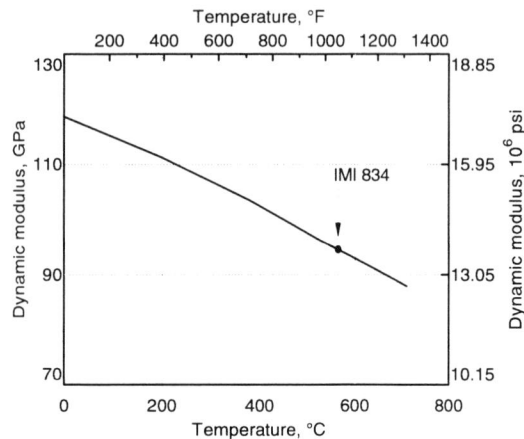

The dynamic modulus of IMI 834 is typical of other near-alpha titanium alloys. Heat treated bar.
Source: IMI Titanium "High-Temperature Alloys" brochure

High-Temperature Strength

IMI 834 has useful strength up to 600 °C (1110 °F). IMI 834 is regarded as having long term creep performance up to around 600 °C (1110 °F) and good short term performance up to significantly higher temperatures. Typically, the alloy gives less than 0.1% total plastic strain in 100 hours at 600 °C (1110 °F) under a stress of 150 MPa (21.8 ksi).

IMI 834: Typical tensile properties

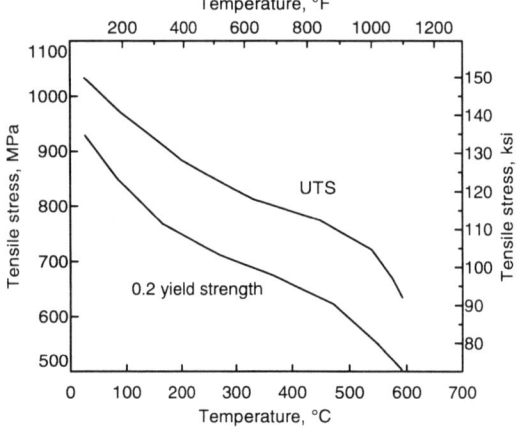

(a) Heat treated discs.
Source: IMI Titanium "High-Temperature Alloys" brochure

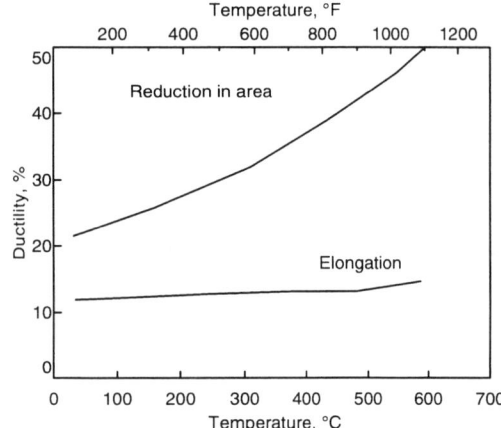

(b)

IMI 834: 0.2% creep strain conditions

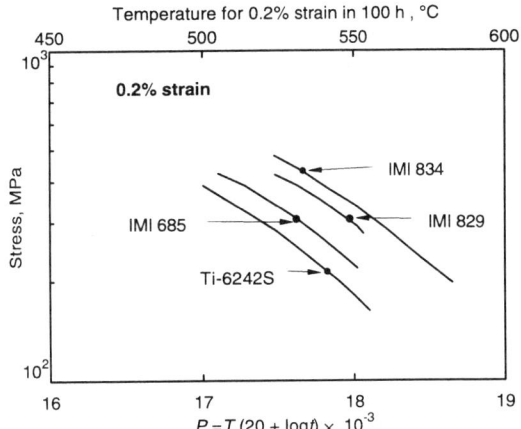

Heat treated discs or bars.
Source: IMI Titanium "High-Temperature Alloys" brochure

IMI 834: Stress rupture properties

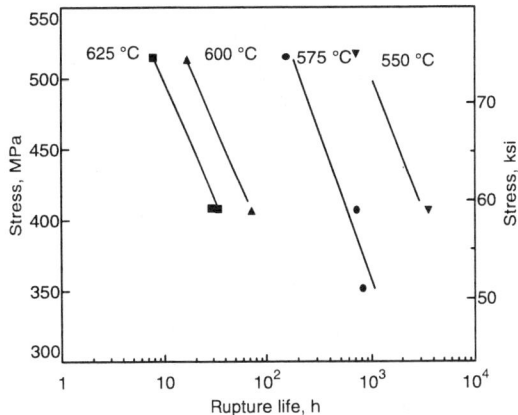

Heat treated bar

Fatigue Properties

Low-Cycle Fatigue

Cast IMI 834: Fatigue strength at 10^5 cycles

Condition	Fatigue strength at 10^5 cycles MPa (ksi)
Cast, alpha+beta HIP, plus ½ h at 1070 °C, OQ plus 2 h at 700 °C	700 ± 50 (101.6 ± 7.26)
Cast, beta HIP, plus 2 h at 700 °C	500 ± 50 (72.6 ± 7.26)
Wrought 50 mm (2 in.) diam bar	800 (116.1)

Direct stress, zero minimum ($R = 0$)

IMI 834: Low-cycle fatigue ($R = 0$)

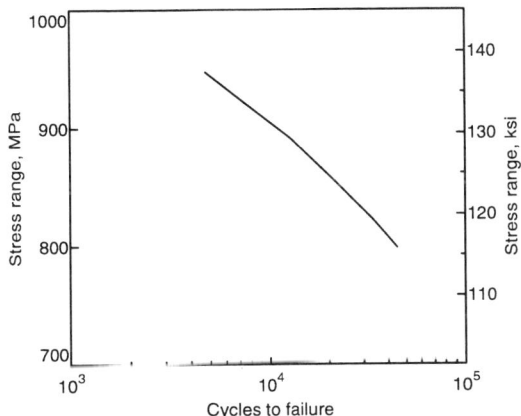

Unnotched specimens from heat treated bar, direct (axial) stress, room-temperature tests.
Source: IMI Titanium "High-Temperature Alloys" brochure

IMI 834: Elevated-temperature low-cycle fatigue

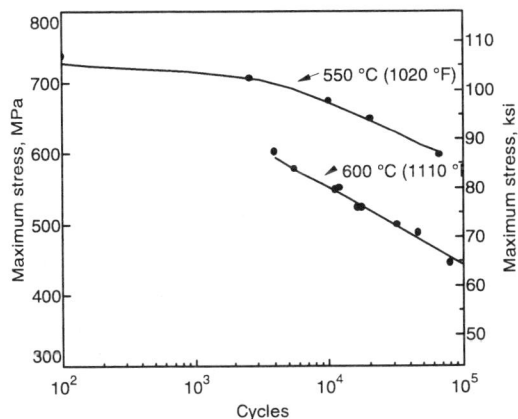

Unnotched specimens from heat treated bars, direct (axial) stress, zero minimum stress ($R = 0$)

High-Cycle Fatigue

Cast IMI 834: Fatigue strength at 10^7 cycles

Condition	Fatigue strength at 10^7 cycles MPa (ksi)
Cast, alpha+beta HIP, plus ½ h at 1070 °C, AC plus 2 h at 700 °C	500 ± 25 (72.6 ± 3.6)
Cast, beta HIP, plus 2 h at 700 °C	400 ± 25 (58.1 ± 3.6)
Wrought, 50 mm (2 in.) diam bar	500 (79.8)

Unnotched specimens, direct stress, zero minimum ($R = 0$)

Cast IMI 834: Notched fatigue strength

Condition	K_t	Fatigue strength at 10^7 cycles MPa (ksi)
Cast, alpha+beta HIP, plus ½ at 1070 °C, AC, plus 2 h at 700 °C	3.0	250 ± 25 (36.3 ± 3.6)
Wrought, 50 mm (2 in.) Ø bar	2.0	340 (49.3)

Direct stress, zero minimum ($R = 0$)

IMI 834: High-cycle fatigue properties ($R = 0$)

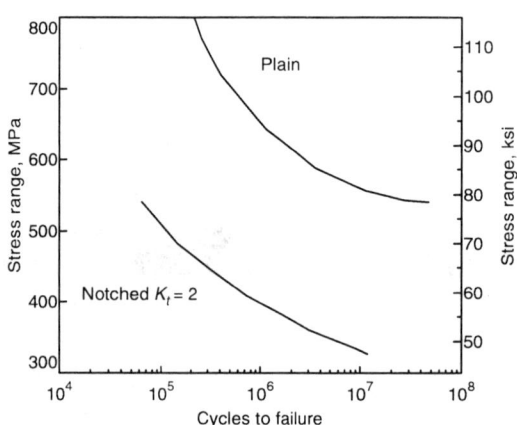

Heat treated bar, direct (axial) stress at room temperature.
Source: IMI Titanium "High-Temperature Alloys" brochure

Crack Propagation

IMI 834: Crack propagation ($R = 0$)

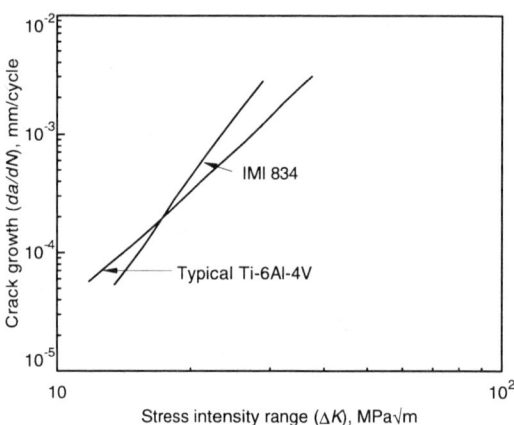

Heat treated bar, longitudinal crack direction room-temperature tests.
Source: IMI Titanium "High-Temperature Alloys" brochure

Processing

Casting

IMI 834 can be cast using the normal techniques developed for titanium alloys. Typical tensile properties of cast IMI 834 at room temperature and at 600 °C (1110 °F) are given in tables. Cast IMI 834 gives lower tensile ductility than the alpha-beta wrought product but gives better creep performance.

Cast IMI 834: Room-temperature tensile properties

Bar condition	Yield strength (0.2%) MPa	ksi	Ultimate tensile strength MPa	ksi	Elongation %	Reduction in area %
Cast + (α+β) HIP + 1070 °C AC + 2 h 700 °C	944	137.0	1071	155.4	5	7
	966	140.2	1072	155.6	5	9
Cast + β HIP + 2 h 700 °C	898	130.3	1040	147.2	6	10
	901	130.8	1025	148.8	4	9
Wrought (15% alpha) OQ + 2 h 700 °C 50 mm bar (2 in.)	950	137.9	1070	155.2	13	23

Cast IMI 834: Tensile properties at 600 °C

Bar condition	Yield strength (0.2%) MPa	ksi	Ultimate tensile strength MPa	ksi	Elongation, %	Reduction in area, %
Cast + (α+β) HIP + 1070 °C AC + 2 h 700 °C	526	76.3	663	96.2	6	16
	515	74.7	669	97.1	10	29
Cast + β HIP + 2 h 700 °C	467	67.8	566	82.1	6	16
	472	68.5	575	83.5	7	16
Wrought (15% alpha) OQ + 2 h 700 °C 50 mm (2 in.) bar	518	75.2	682	99.0	23	52

Forging

IMI 834 is readily forgeable using conventional hammer, press, or isothermal techniques. Typical forging temperature is around 1010 °C (1850 °F). IMI 834 is stiffer than most other titanium alloys, but it has good forgeability at its recommended forging temperature.

IMI 834: Flow stress

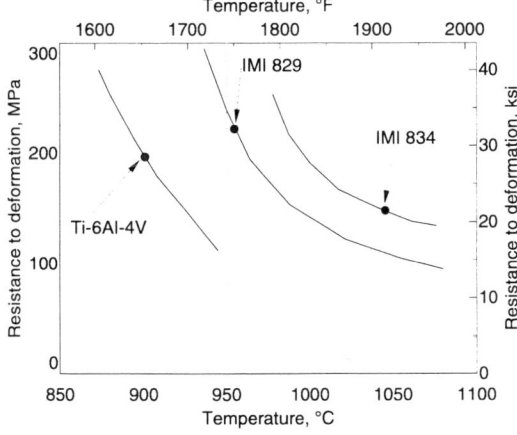

As-rolled bar tested with a plastometer up to 1100 °C (2010 °F) at a strain rate of 15/s.
Source: IMI Titanium "High-Temperature Alloys" brochure

Forming

IMI 834 has very limited cold formability, but good hot formability. It can be produced in sheet and plate form. Typical sheet properties are shown (see table). Superplastic forming is also possible at about 990 °C (1814 °F).

IMI 834: Properties of 2 mm sheet

Material condition(a)	Orientation	Yield strength (0.2%) MPa	ksi	Ultimate tensile strength MPa	ksi	Elongation (50 mm), %	Creep strain(b), %
Room-temperature properties							
Rolled + *Annealed (800 °C)	L	996	144.6	1114	161.7	11.5	...
	T	1014	147.2	1120	162.5	12	...
*1025 °C (α/β) AC + 2 h 700 °C	L	998	144.8	1145	166.2	11.5	...
	T	1009	146.4	1111	161.2	11	...
*1060 °C (β) AC + 2 h 700 °C	L	947	137.4	1098	159.4	6	...
	T	963	139.8	1103	160.1	6	...
High-temperature (600 °C) properties							
Rolled + *Annealed (800 °C)	L	473	68.7	671	97.4	18	...
	T	510	74.0	720	104.5	14	...
*1025 °C (α/β) AC + 2 h 700 °C	L	518	75.2	702	100.6	16	0.213
	T	546	79.2	728	105.7	18	0.247
*1060 °C (β) AC + 2 h 700 °C	L	554	80.4	716	103.9	12	0.055
	T	532	77.2	729	105.8	12	0.064

(a) An asterisk * indicates a heating duration of 30 minutes. (b) Total plastic strain after exposure of 150 MPa (21.8 ksi) at 600 °C (1110 °F) for 100 hours

Heat Treatment

IMI 834: Recommended heat treatments

Treatment	Temperature °C	°F	Duration	Cooling method
Solution treatment (15% alpha)	1015 ± 5	1860 ± 9	2 hours	Oil quench(a)
Aging	700	1290	2 hours	Air cool

(a) For sections less than about 15 mm (0.6 in.), air cooling is recommended.

IMI 834: Typical tensile properties after recommended heat treatment (STA)

	Room temperature	600 °C	1110 °F
0.2% MPa (ksi)	950 (138)	520	(75.4)
U.T.S. MPa (ksi)	1050 (152)	650	(94.3)
Elongation (in 5D), %	12	20	
Reduction in area, %	20	50	

Ti-5Al-6Sn-2Zr-1Mo-0.25Si

Common Name: Ti-5621S
UNS: Unassigned

Compiled by P. Russo, RMI Titanium Company

Ti-5621S is a semicommercial alloy developed by RMI Titanium Company in the mid-1960s to extend the use of titanium-base alloys to 540 °C (1000 °F). The β-lean α+β alloy combines a well-selected alpha base with small additions of the β-stabilizers molybdenum and silicon to optimize creep resistance and thermal stability. Ti-5621S has good tensile and stress-rupture properties up to 540 °C (1000 °F), combined with excellent creep. It can be machined and formed at room temperature or warm forming temperatures of 425 to 540 °C (800 to 1000 °F) and can be welded.

Effects of Impurities and Alloying. Exceeding impurity limits may decrease the ductility of the alloy below required minimums due to increases in yield strength. As for all α+β alloys, excessive aluminum, oxygen, and nitrogen can reduce ductility and fracture toughness. These elements, together with excessive tin and zirconium, can reduce the thermal stability of Ti-5621S and other high-temperature titanium alloys.

Product Forms and Conditions. Ti-5621S has been produced in standard wrought product forms, such as forged billets and bars, and as flat products, such as sheet and plate. The alloy is available only by special order. It cannot be strengthened by thermal treatment.

Applications. Ti-5621S is a semicommercial alloy intended for high-temperature applications up to 540 °C (1000 °F) where creep is the limiting concern and where good elevated-temperature strength and thermal stability are required. Applications include jet engine components.

Ti-5621S: Typical composition range

	Composition, wt%									
	Al	Sn	Zr	Mo	Si	O	N	C	Fe	Ti
Min, wt%	4.5	5.0	1.5	0.5	0.15	...	...	...	...	...
Max, wt%	5.5	7.0	2.5	0.95	0.35	0.15	0.03	0.05	0.3	bal

Physical Properties

Phases and Structures. At room temperature, Ti-5621S consists primarily of α phase (hcp), with a small amount of β phase (bcc). Heating to higher temperatures increases the amount of β phase in the alloy until, at temperatures exceeding 1010 °C (1850 °F), the alloy transforms to all β phase. The alloy also contains a small amount of $(Ti,Zr)_5Si_3$. Microstructures resulting from cooling through the β-transus temperature typically consist of packets of α platelets separated by films of β phase.

Elastic Modulus. (see figures)
Poisson's Ratio: 0.326
Chemical/Corrosion Properties. Although actual data are not available, Ti-5621S is expected to have general corrosion behavior similar to other near-α and lean-β α + β alloys.

Ti-5621S: Summary of typical physical properties

Beta transus	1010 ± 14 °C (1850 ± 25 °F)
Melting (liquidus point)	Not Available
Density(a)	4.51 g/cm³ (0.163 lbf/in.³)
Electrical resistivity(a)	1.7 μΩ · m
Magnetic permeability	Nonmagnetic
Specific heat capacity(a)	Not Available
Thermal conductivity	Not Available
Thermal coefficient of linear expansion	Not Available

(a) Typical values at room temperature of about 20 to 25 °C (68 to 78 °F)

Ti-5621S: Effect of temperature on tensile modulus

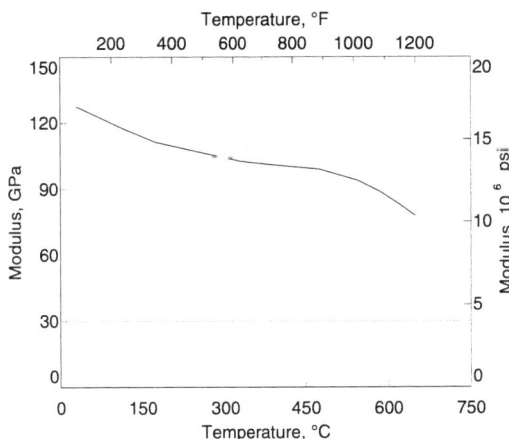

Beta forged heat treated at 980 °C (1800 °F), 1 h, air cooled + 590 °C (1100 °F), 2 h, air cooled.
Source: RMI Titanium Company

446 / Alpha and Near-Alpha Alloys

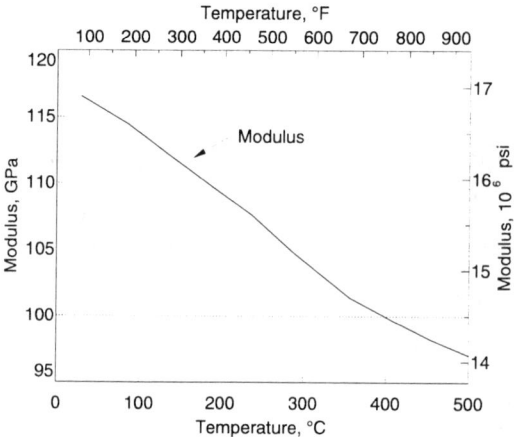

Ti-5621S: Effect of temperature on compressive modulus

38 mm (1.5 in.) thick pancake forging, β forged, α + β annealed and aged.
Source: RMI Titanium Company

Mechanical Properties

Ti-5621S: Typical tensile properties of forgings

Forge process and test temperature	Anneal	Yield strength MPa	Yield strength ksi	Ultimate tensile strength MPa	Ultimate tensile strength ksi	Elongation, %	Reduction of area, %
22 °C (72 °F)							
α + β	α + β	993	144	1096	159	16	40
β	α + β	860	125	1000	145	12	25
480 °C (900 °F)							
α + β	α + β	593	86	745	108	18	48
β	α + β	515	75	690	100	16	36

Source: RMI Titanium Company

High-Temperature Strength

Ti-5621S: Stress-rupture properties

Property	RT	315 (600)	425 (800)	510 (950)
Stress rupture (radial), MPa (ksi) at:				
100 h	NA	745 (108)	631 (91.5)	593 (86)
1000 h	NA	738 (107)	627 (91)	545 (79)

Temperature, °C (°F)

Source: Battelle Memorial Institute

Ti-5621S: Effect of temperature on tensile properties

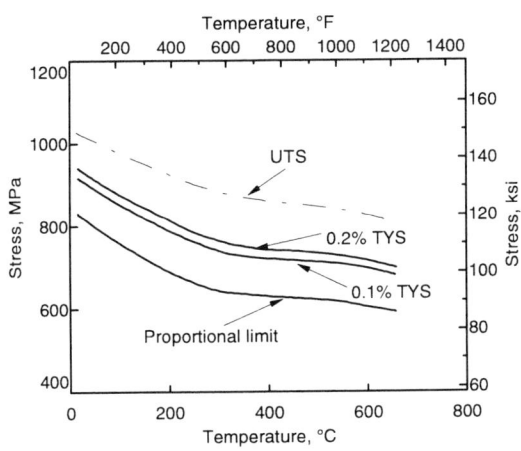
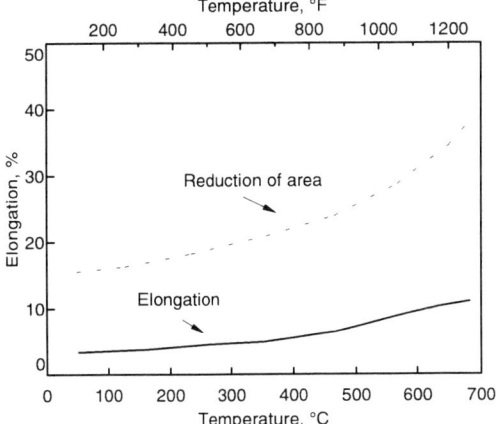

Beta forged, heat treated at 980 °C (1800 °F), 1 h, air cooled + 590 °C (1100 °F), 2 h, air cooled.
Source: RMI Titanium Company

Ti-5621S: Larson-Miller plot of creep

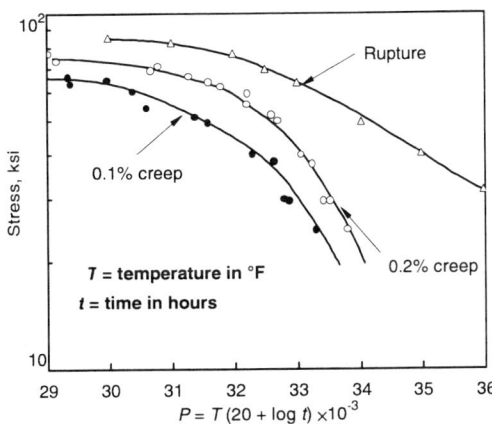

Pancake forging, β forged, α + β annealed and aged.
Source: RMI Titanium Company

Fatigue Properties

Ti-5621S: Fatigue data

Fatigue (Radial)	Temperature, °C (°F)		
	RT	205 (400)	370 (700)
Unnotched, $R = 0.1$			
10^3 cycles, MPa (ksi)	1035 (150)	965 (140)	854 (124)
10^5 cycles, MPa (ksi)	786 (114)	760 (110)	710 (103)
10^7 cycles, MPa (ksi)	590 (85)	565 (82)	537 (78)
Notched ($K_t = 3.0$), $R = 0.1$			
10^3 cycles, MPa (ksi)	760 (110)	730 (106)	690 (100)
10^5 cycles, MPa (ksi)	372 (54)	338 (49)	303 (44)
10^7 cycles, MPa (ksi)	262 (38)	262 (38)	262 (38)

38 mm (1.5 in.) thick pancake forging, β forged, α + β annealed and aged. Source: Battelle Memorial Institute

Ti-5621S: Smooth fatigue results

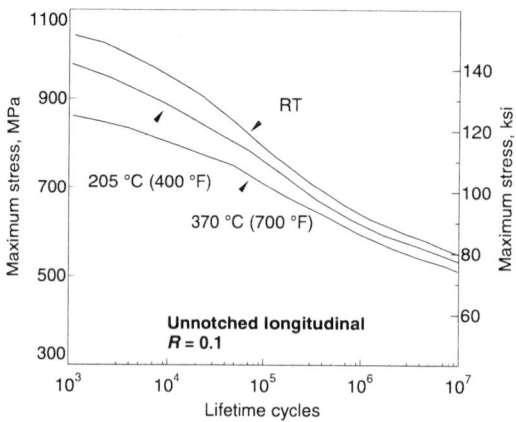

38 mm (1.5 in.) thick pancake forging, β forged, α + β annealed and aged.
Source: RMI Titanium Company

Ti-5621S: Notched fatigue results

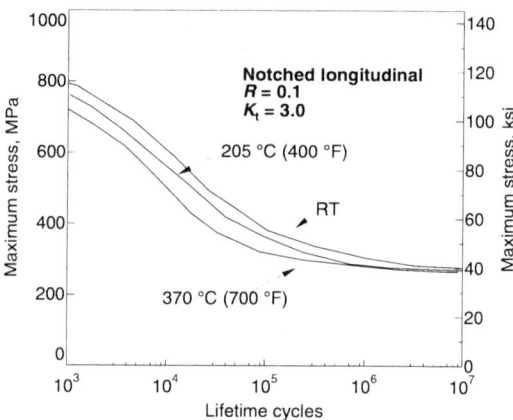

38 mm (1.5 in.) thick pancake forging, β forged, α + β annealed and aged.
Source: RMI Titanium Company

Ti-5621S: Fatigue crack growth in air and vacuum

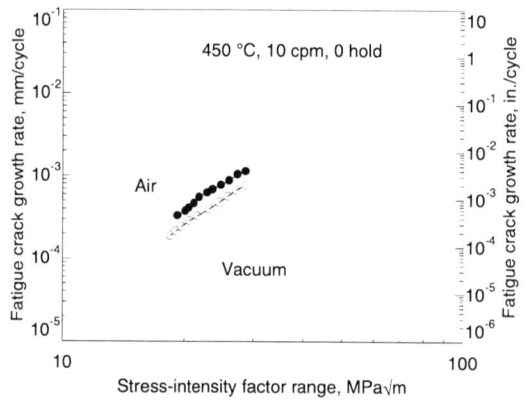

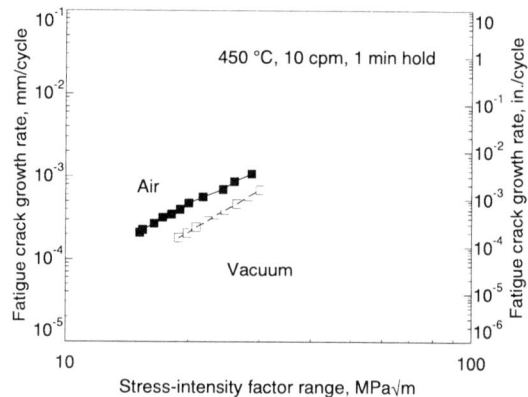

Stress ratio, 0.05; 2.5 mm (0.1 in.) sheet; hot rolled at 955 °C (1750 °F); annealed at 975 °C (1785 °F), 1 h, air cooled + 590 °C (1100 °F), 2 h, air cooled.
Source: H.H. Smith, P.S. Kullen, and D.J. Michel, Fatigue Crack Propagation Behavior of Titanium Alloys 6242S and 5621S at Elevated Temperature, *Metall. Trans. A*, Vol 19A, April 1988, p 881-885

Fracture Properties

Ti-5621S: Charpy impact data

Test temperature		Absorbed energy	
°C	°F	N · m	lbf · ft
22	72	28.9	21.3
−40	−40	23.3	17.2
−70	−100	19.0	14.0

Average of three tests. 38 mm (1.5 in.) thick pancake forging, β forged, α + β annealed and aged. Source: Battelle Memorial Institute

Ti-5621S: Fracture toughness

Heat treatment(a)	Fracture toughness		Ultimate tensile strength		Yield strength		Elongation, %	Reduction of area, %
	MPa√m	ksi√in.	MPa	ksi	MPa	ksi		
A (annealed)	106.7	97.0	1020	148	900	130	12.7	22.5
A + 480 °C (900 °F), 1000 h	100.1	91.0	1027	149	900	130	10.5	20.1

Four-point loading; pancake forging. (a) A, 1015 °C (1860 °F), 1 h, air cooled + 590 °C (1100 °F), 2 h, air cooled. Source: RMI Titanium Company

Plastic Deformation

Ti-5621S: Stress-strain curves

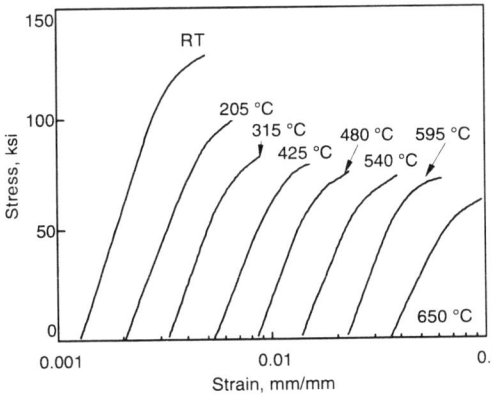

Beta forged, heat treated at 980 °C (1800 °F), 1 h, air cooled + 590 °C (1100 °F), 2 h, air cooled.
Source: RMI Titanium Company

Processing

Forging

To optimize creep properties, Ti-5621S can be β forged, followed by α + β annealing. Alpha-beta forging must be controlled carefully to minimize surface cracking.

Machining

Like many titanium alloys, Ti-5621S has a tendency to seize and therefore requires sharp tools, correct tool angles, heavy feeds, and slow speeds, as well as rigid tool supports and adequate coolant. Typical tool geometries are provided in "Technical Note 7: Machining" in this Volume. Carbides are used on heavier, faster cuts, while high-speed tools are used on lighter, slower cuts. Most drills and taps are made of high-speed steels.

Cutting feeds and speeds of alloy Ti-5621S are similar to those of Ti-6242S or Ti-6Al-4V. In turning operations, the typical range of feeds and speeds are:

- Roughing cuts at 0.4 to 0.75 mm/rev (0.015 to 0.03 in./rev) and 30 to 45 m/min (100 to 150 sfm)
- Finishing cuts at 0.25 mm/rev (0.01 in./rev) and 90 to 150 m/min (300 to 500 sfm)

In grinding operations, Ti-5621S requires many of the same precautions against surface damage as other titanium alloys (see "Technical Note 7: Machining" in this Volume).

Alpha-Beta Alloys

Ti-5Al-2Sn-2Zr-4Mo-4Cr

Common Name: Ti-17
UNS Number: R58650

Ti-5Al-2Sn-2Zr-4Cr-4Mo (Ti-17) is a high-strength, deep hardenable, forging alloy that was developed primarily for gas turbine engine components, such as disks for fan and compressor stages. Ti-17 has strength properties superior to those of Ti-6Al-4V, and also exhibits higher creep resistance at intermediate temperatures.

Product Conditions/ Microstructure

Ti-17 can be heat treated to yield strengths of 1030 to 1170 MPa (150 to 170 ksi). It is more ductile than Ti-6Al-6V-2Sn, and it is superior to Ti-6Al-4V in creep behavior. With hardenability characteristics comparable to those of some beta type alloys, Ti-17 is lower in density and higher in modulus and creep strength than the beta alloys.

Chemistry and Density

Ti-17 may be classified as a "beta-rich" alpha-beta alloy, because it has a beta-stabilizer (Mo + Cr) content of 8%.

Density. 4.65 g/cm^3 (0.168 lb/in.^{3})

Product Forms

Ingot, billet, forgings

Product Conditions/ Microstructure

Ti-17 can be processed in either the beta or alpha plus beta region, and subsequent heat treatment depends on processing history. Special ingot melting conditions are required, particularly during the final melt, to minimize segregation of beta stabilizers (primarily chromium) during solidification. Excessive segregation of beta stabilizers can cause "beta flecks" during forging or upon heat treatment, which constitute microregions of subnormal fracture toughness and ductility. Both forging and heat treating practices must be controlled carefully to minimize the effects of microsegregation (beta flecks).

Applications

Ti-17 is used for heavy-section forgings up to 150 mm (6 in. thick) for gas turbine engine components and other elevated-temperature applications demanding high tensile strength and good fracture toughness. It is used only by General Electric.

Ti-5Al-2Sn-2Zr-4Mo-4Cr: Specifications and Compositions

Specification	Designation	Description	Al	Cr	Fe	H	Mo	N	Sn	Zr	Other
UNS	R58650		4.5-5.5	3.5-4.5	0.3 max	0.0125 max	3.5-4.5	0.04 max	1.5-2.5	1.5-2.5	Mn 0.1 max; Cu 0.1 max; O 0.08-0.13; C 0.05 max; OT 0.3 max; OE 0.1 max; Y 0.005 max; bal Ti
USA											
AMS 4995		Bil STA	4.5-5.5	3.5-4.5	0.3	0.0125	3.5-4.5	0.04	1.5-2.5	1.5-2.5	Mn 0.1 max; Cu 0.1 max; O 0.08-0.13; C 0.05; OT 0.3; Y 0.005; bal Ti
AMS 4997		Powd	4.5-5.5	3.5-4.5	0.3	0.0125	3.5-4.5	0.04	1.5-2.5	1.5-2.5	Mn 0.1 max; Cu 0.1 max; O 0.08-0.12; C 0.05; OT 0.3; Y 0.005; bal Ti

Ti-5Al-2Sn-2Zr-4Mo-4Cr: Commercial Compositions

Specification	Designation	Description	Al	Cr	Fe	H	Mo	N	Sn	Zr	Other
Japan											
Kobe	KS5-2-2-4-4	Bar Frg STA	4.5-5.5	3.5-4.5	0.3	0.0125	3.5-4.5	0.04	1.5-2.5	1.5-2.5	O 0.08-0.13; bal Ti
USA											
OROMET	Ti-17										
TIMET	TIMETAL 17										

Phases and Structures

Various types of phase transformations can be achieved by the decomposition of β during continuous quenching and isothermal treatment of Ti-17. At high isothermal temperatures or low cooling rates, the transformation structure has a characteristic Widmanstätten or basketweave appearance, with a thick α layer initially nucleated along grain boundaries. The prior β phase transforms to a Widmanstätten α + β mixture by nucleation and growth according to the classical Burger orientation. At temperatures above 500 °C (930 °F), the α forms first on the grain boundaries and then grows into the β matrix. Below 500 °C (930 °F), α precipitates and grows throughout the β matrix as well as in the β grain boundaries. At moderate cooling rates, the fineness of the α structure increases and is fairly uniform over the matrix, with some preference for α formation at prior β grain boundaries as cooling rates increase.

Beta Transus. 890 °C (1635 °F)

Ti-17: Continuous cooling transformation diagram

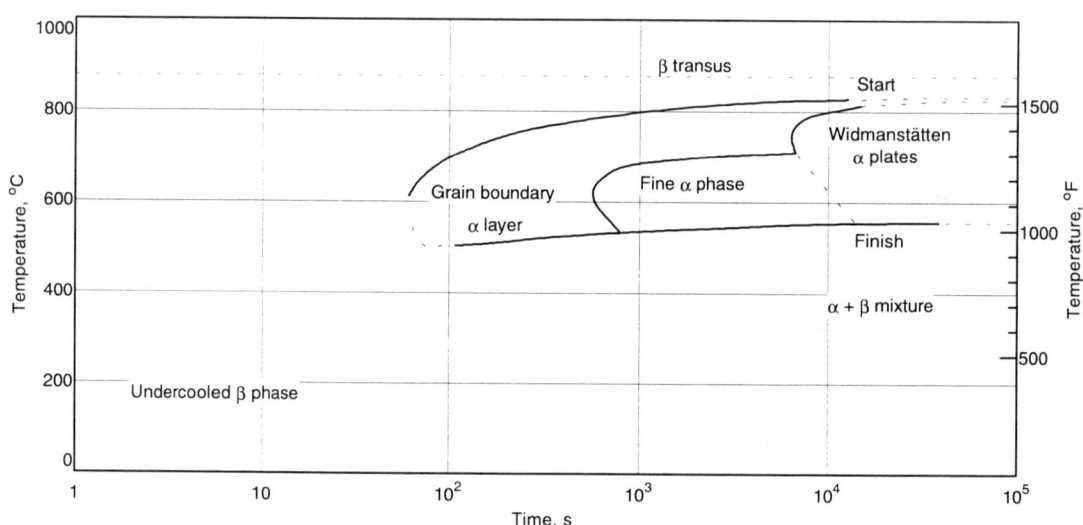

Solution treated at 930 °C (1700 °F) for 30 min.
Source: J. Béchet and B. Hocheid, Decomposition of the Beta-Phase in Titanium Alloy Ti-17, *Titanium Science and Technology*, G. Lütjering, U. Zwicker, and W. Bunk, Ed., Deutsche Gesellschaft fur Metallkunde, Germany, 1985, p 1617

Ti-17: Isothermal transformation diagram

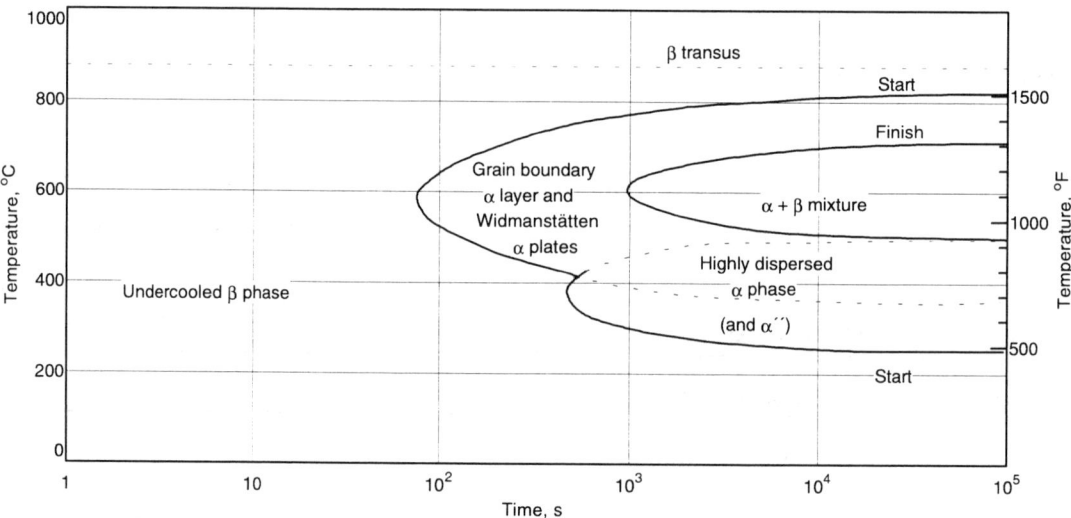

Solution treated at 930 °C (1700 °F).
Source: J. Béchet and B. Hocheid, Decomposition of the Beta-Phase in Titanium Alloy Ti-17, *Titanium Science and Technology*, G. Lütjering, U. Zwicker, and W. Bunk, Ed., Deutsche Gesellschaft fur Metallkunde, Germany, 1985, p 1615

Physical Properties

Ti-17: Summary of typical physical properties

Beta transus	890 ± 10 °C (1635 ± 20 °F)
Melting (liquidus) point	Not Available
Density(a)	4.65 g/cm^3 (0.168 lb/in.^{3})
Electrical resistivity	Not Available
Magnetic permeability	Nonmagnetic
Specific heat capacity	Not Available
Thermal conductivity	Not Available
Thermal coefficient of linear expansion(b)	9.7 × 10^{-6}/ °C (5.4 × 10^{-6}/ °F)

(a) Typical values at room temperature of about 20 to 25 °C (68 to 78 °F). (b) Mean coefficient from room temperature to 400 °C (750 °F)

Corrosion Properties. No data are available on the corrosion of Ti-17, but it should be susceptible to stress-corrosion cracking because its aluminum content is above 3 wt%. Corrosion of Ti-17 may be similar to Ti-6Al-2Sn-4Zr-2Mo given the comparable compositions.

Ti-17: Dynamic modulus of elasticity

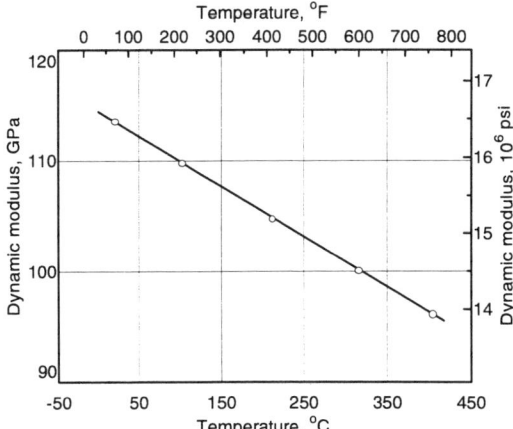

Specimens were heat treated at 855 °C (1575 °F), 4 h, air cooled + 800 °C (1475 °F), 4 h, fan air cooled, 620 °C (1150 °F), 8 h, air cooled.
Source: *Aerospace Structural Metals Handbook*, Vol 4, Code 3724,

Mechanical Prorperties

Typical Hardness: 39 to 40 HRC

Tensile Properties

Ti-17 retains a high fraction of its strength at elevated temperatures. Its notched tensile to tensile ratio is also stable at elevated temperatures.

Ti-17: Typical STA tensile properties

Temperature		Tensile yield strength		Ultimate tensile strength		Elongation, %	Reduction of area, %
°C	°F	MPa	ksi	MPa	ksi		
24	75	1035-1075	150-170	1105-1240	160-180	8-15	20-45
93	200	930-1000	135-145	1035-1105	150-160	8-15	30-45
205	400	795-860	115-125	930-1000	135-145	8-15	30-45
315	600	760-825	110-120	930-1000	135-145	8-15	30-45
370	700	700-760	100-110	860-930	125-135	8-15	30-45

Source: *Beta Titanium Alloys in the 1980's*, R.R. Boyer and H.W. Rosenberg, Ed., TMS/AIME, 1984, p 436

Ti-17: Typical STA notch tensile properties

Temperature		Notched tensile strength(a)		NTS/UTS
°C	°F	MPa	ksi	
24	75	1380-1515	200-220	1.3
93	200	1450-1515	210-220	1.4
205	400	1380-1450	200-210	1.4
315	600	1275-1345	185-195	1.4
370	700	1275-1345	185-195	1.4

(a) K_t = 4.0. Source: *Beta Titanium Alloys in the 1980's*, R.R. Boyer and H.W. Rosenberg, Ed., TMS/AIME, 1984, p 436

Ti-17: Effect of temperature on radial tensile properties of spool forgings

Test temperature		Ultimate tensile strength		Tensile yield strength		Elongation in 50 mm (2 in.), %	Reduction of area, %
°C	°F	MPa	ksi	MPa	ksi		
24	75	1185	172	1144	166	8	26
		1179	171	1117	162	12	32
		1124	163	1068	155	11	39
93	200	1075	156	875	127	12	36
205	400	993	144	841	122	14	46
315	600	985	143	813	118	12	46
		944	137	793	115	11	49
370	700	917	133	744	108	13	47

Note: Specimens were heat treated at 845 °C (1550 °F), 4 h, AC + 800 °C (1475 °F), 4 h, FAC, 620 °C (1150 °F), 8 h, AC. Source: *Aerospace Structural Metals Handbook*, Vol 4, Code 3724, Battelle Columbus Laboratories, 1976

Ti-17: Effect of temperature on notched tensile strength of spool forgings

Test temperature		Ultimate tensile strength		Notched tensile strength(a)		NTS/UTS
°C	°F	MPa	ksi	MPa	ksi	
24	75	1144	166	1468	213	1.28
93	200	1075	156	1482	215	1.38
205	400	972	141	1413	205	1.45
315	600	958	139	1310	190	1.36
370	700	910	132	1303	189	1.44

Note: Specimens were heat treated at 845 °C (1550 °F), 4 h, AC + 800 °C (1475 °F), 4 h, FAC, 620 °C (1150 °F), 8 h, AC. Data are an average of three tests. Notch radius = 0.12 mm (0.0047 in.). (a) K_t = 4. Source: *Aerospace Structural Metals Handbook*, Vol 4, Code 3724, Battelle Columbus Laboratories, 1976

Effect of Heat Treatment

Ti-17: Effect of aging temperature on tensile yield strength

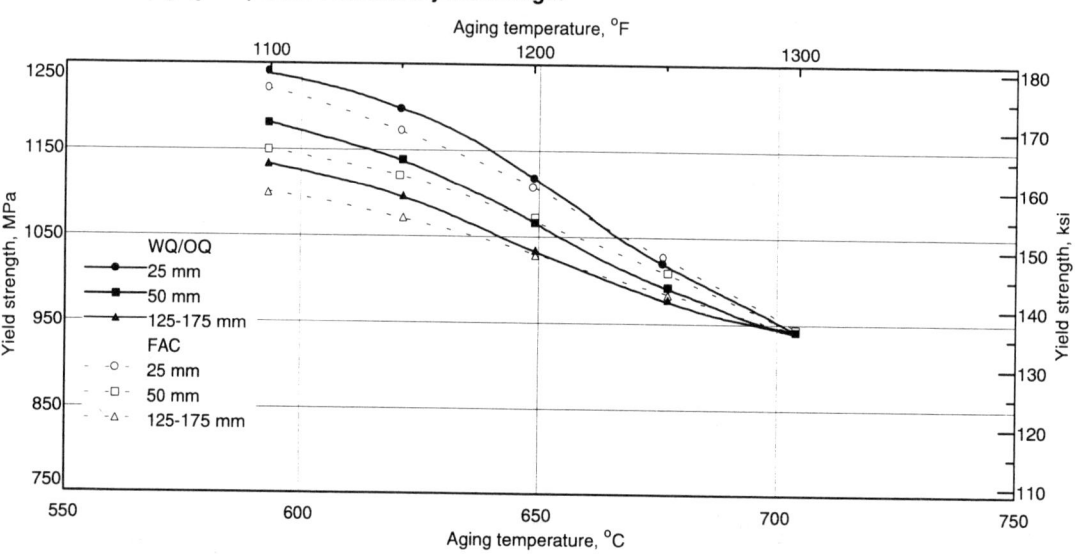

Alpha-beta processed disks heat treated at 855 °C (1575 °F), 4 h, AC, 800 °C (1475 °F), 4 h, water quench (WQ) or oil quench (OQ) or fan air cool (FAC).
Source: *Aerospace Structural Metals Handbook*, Vol 4, Code 3724, Battelle Columbus Laboratories, 1976

Ti-17: Effect of solution temperatures on tensile strengths

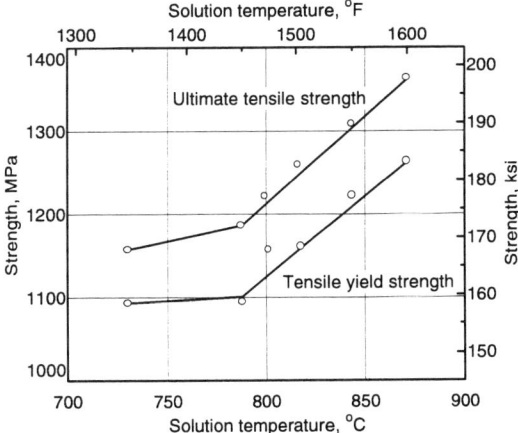

Alpha-beta processed disk forgings heat treated at 855 °C (1575 °F), 4 h, AC + solution treated, 4 h, WQ, 620 °C (1150 °F), 8 h, AC.
Source: *Aerospace Structural Metals Handbook*, Vol 4, Code 3724, Battelle Columbus Laboratories, 1976

Ti-17: Effect of solution treatment on aged strength

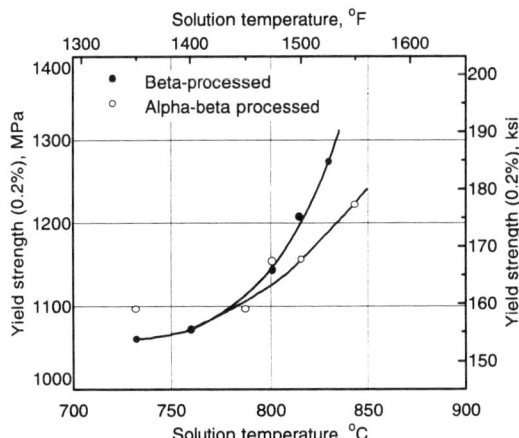

Solution treated at indicated temperature for 4 h, WQ, and aged at 635 °C (1175 °F) for 8 h.
Source: *Beta Titanium Alloys in the 1980's*, R.R. Boyer and H.W. Rosenberg, Ed., TMS/AIME, 1984, p 244

Creep Properties

Ti-17: Creep-rupture curves for forgings

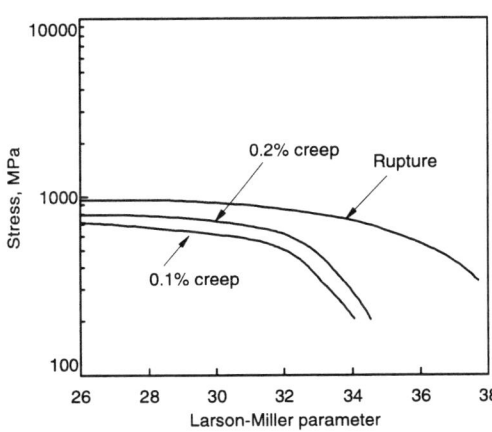

Larson-Miller parameter equals $10^{-3} T$ $(25 + \log t)$, where T is temperature in °R and t is time in hours.
Source: *Metals Handbook*, Vol 3, 9th ed., American Society for Metals, 1980

Ti-17: Creep behavior comparison

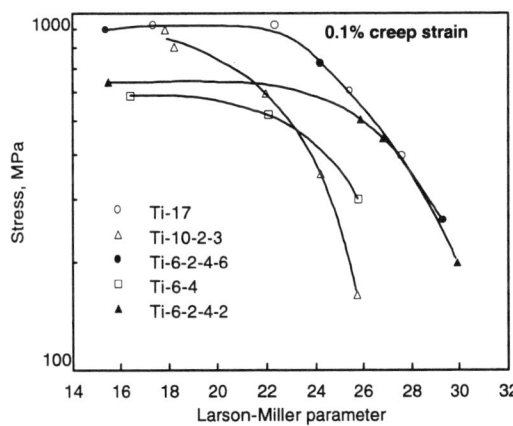

Chemical composition (wt%): 4.98 Al, <0.005 C, 4.00 Cr, <0.005 Cu, 0.059 Fe, 0.0010 H, <0.005 Mn, 3.96 Mo, 0.0036 N, 0.081 O, 1.98 Sn, <0.0010 Y, and 1.96 Zr. Forged disks were double solution treated at 835 °C (1525 °F) for 4 h and at 810 °C (1490 °F) for 4 h, and aged at 610 °C (1130 °F) for 8 h.
Source: T. Matsumoto and T. Nishimura, Effects of Forging and Heat Treatment Conditions on Mechanical Properties of a High-Strength Titanium Alloy, Ti-17, in *Sixth World Conference on Titanium*, P. Lacombe, R. Tricot, and G. Beranger, Ed., Les Editions de Physique, Paris, 1989, p 1319

Ti-17: Typical STA creep properties

Temperature		Stress		
°C	°F	MPa	ksi	Time to 0.2%, h
205	400	793	115	2200
		814	118	400
315	600	690	100	1000
		724	105	500
		745	108	125
425	800	241	35	150
		310	45	75
		345	50	75
		414	60	30

Source: *Beta Titanium Alloys in the 1980's*, R.R. Boyer and H.W. Rosenberg, Ed., TMS/AIME, 1984, p 439

Ti-17: Typical STA stress-rupture properties

Temperature		Stress		
°C	°F	MPa	ksi	Time to rupture, h
205	400	945	137	>1000
		965	140	800
		983	142.5	600
		1000	145	0.01
315	600	896	130	>1000
		917	133	0.01
		931	135	0.01
425	800	690	100	>500
		758	110	25
		793	115	10

Source: *Beta Titanium Alloys in the 1980's*, R.R. Boyer and H.W. Rosenberg, Ed., TMS/AIME, 1984, p 439

Ti-17: Creep-rupture of α – β processed forging

Temperature		Stress		Time,
°C	°F	MPa	ksi	h
205	400	999	145	0.01
		982	142.5	>671.9
		979	142	793.6
315	600	965	140	0.1
		948	137.5	>670.5
425	800	896	130	>721.5
		793	115	8
480	900	482	70	>140
510	950	482	70	~16

Note: Spool forgings were heat treated at 845 °C (1550 °F) for 4 h, air cooled, then 800 °C (1475 °F) for 4 h, furnace air cooled, and aged at 620 °C (1150 °F) for 8 h, air cooled. Source: *Aerospace Structural Metals Handbook*, Vol 4, Code 3724, Battelle Columbus Laboratories, 1976, p 7

Fatigue Properties

Ti-17: Typical STA high-cycle fatigue (unnotched)
Load control, A = 0.95

Temperature		Maximum stress		Alternating stress		Cycles to failure
°C	°F	MPa	ksi	MPa	ksi	
24	75	965	140	470	68.2	21,000
		827	120	403	58.5	75,000
		758	110	370	53.6	7,000,000
		724	105	353	51.2	6,000,000
315	600	758	110	370	53.6	37,000
		690	100	336	48.7	50,000
		676	98	329	47.7	88,000
		655	95	319	46.3	15,000,000
		621	90	43.8	302	12,000,000

Source: *Beta Titanium Alloys in the 1980's*, R.R. Boyer and H.W. Rosenberg, Ed., TMS/AIME, 1984, p 438.

Ti-17: Typical STA low-cycle fatigue (unnotched)
Strain control, A = 1.0

Temperature		Strain,%			Cycles to failure
°C	°F	Plastic	Elastic	Total	
24	75	0.38	1.49	1.87	3,180
		0.10	1.365	1.465	5,040
		0.03	1.23	1.26	9,650
		0.02	1.135	1.155	15,400
		0.01	1.04	1.05	25,700
		0.025	0.97	0.995	60,600
315	600	0.23	1.30	1.54	3,600
		0.133	1.20	1.34	5,500
		0.044	0.982	1.03	>12,700
		0.055	0.92	0.98	>56,300
		0.017	0.9075	0.93	>86,000
		0.045	0.88	0.93	>16,000

Source: *Beta Titanium Alloys in the 1980's*, R.R. Boyer and H.W. Rosenberg, Ed., TMS/AIME, 1984, p 437

Ti-17: Axial fatigue of STA disk forgings

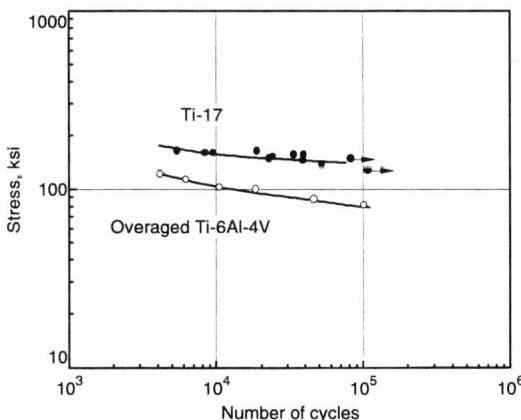

Ti-17 heat treatment: 860 °C (1575 °F), 4 h, AC, 800 °C (1475 °F), 4 h, FAC, 620 °C (1150 °F), 8 h, AC. Axial loaded, $R = 0$, $K_t = 1$; frequency, 20 cycles/min.
Source: *Aerospace Structural Metals Handbook*, Vol 4, Code 3724, Battelle Columbus Laboratories, 1976

Ti-17: Typical STA low-cycle fatigue (unnotched)
Load control, A = 1.0

Temperature		Stress range		Cycles to failure
°C	°F	MPa	ksi	
24	75	1103	160	5,000
		1069	155	10,000
		1000	145	25,000
		931	135	35,000
		896	130	170,000
		827	120	290,000
315	600	896	130	5,000
		862	125	6,000
		841	122	7,000
		834	121	13,000
		827	120	64,000

Source: *Beta Titanium Alloys in the 1980's*, R.R. Boyer and H.W. Rosenberg, Ed., TMS/AIME, 1984, p 437

Fatigue Crack Growth

Ti-17: Fatigue crack growth at room temperature

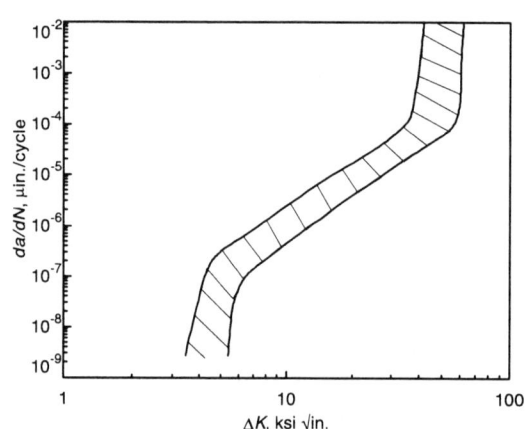

Alpha-beta processed spool forgings were heat treated at 860 °C (1575 °F), 4 h, AC + 800 °C (1475 °F), 4 h, FAC + 620 °C (1150 °F), 8 h, AC. Tensile yield strength, 1075 MPa (156 ksi); B = 1; W/B = 2; L-R orientation.
Source: *Aerospace Structural Metals Handbook*, Vol 4, Code 3724, Battelle Columbus Laboratories, 1976

Fracture Properties

Ti-17: Plane-strain fracture toughness at room temperature STA

Yield strength		K_{IC}	
MPa	ksi	MPa$\sqrt{m}$	ksi $\sqrt{in.}$
Alpha-beta processed			
1172	170	33	30
1103	160	40	36
1034	150	50	45
Beta-processed			
1172	170	53	48
1103	160	65	59
1034	150	88	80

Source: *Beta Titanium Alloys in the 1980's*, R.R. Boyer and H.W. Rosenberg, Ed., TMS/AIME, 1984, p 438

Ti-17: Effect of reduction ratio on fracture toughness of disk forgings

Reduction ratio	Tensile yield strength(a)		Fracture toughness(b) (K_{IC})	
	MPa	ksi	MPa√m	ksi √in.
Alpha-beta forged + STA(c)				
2:1	1150	167	41.5	37.8
3:1	1145	166	36	32.9
4:1	1165	169	37.2	33.9
Beta forged + STA(d)				
2:1	1117	162	68.8	62.2
3:1	1103	160	61	55.5
4:1	1110	161	55.2	50.2

(a) Average of two tests. (b) 25 mm (1 in.) thick compact tension specimen. (c) 845 °C (1550 °F) for 4 h, FAC, 800 °C (1475 °F) for 4 h, FAC; and 620 °C (1150 °F) for 8 h, AC. (d) 800 °C (1475 °F) 4 h, FAC; 620 °C (1150 °F) 8 h, AC. Source: *Aerospace Structural Metals Handbook*, Vol 4, Code 3724, Battelle Columbus Laboratories, 1976

Ti-17: Fracture toughness vs yield strength (aged)

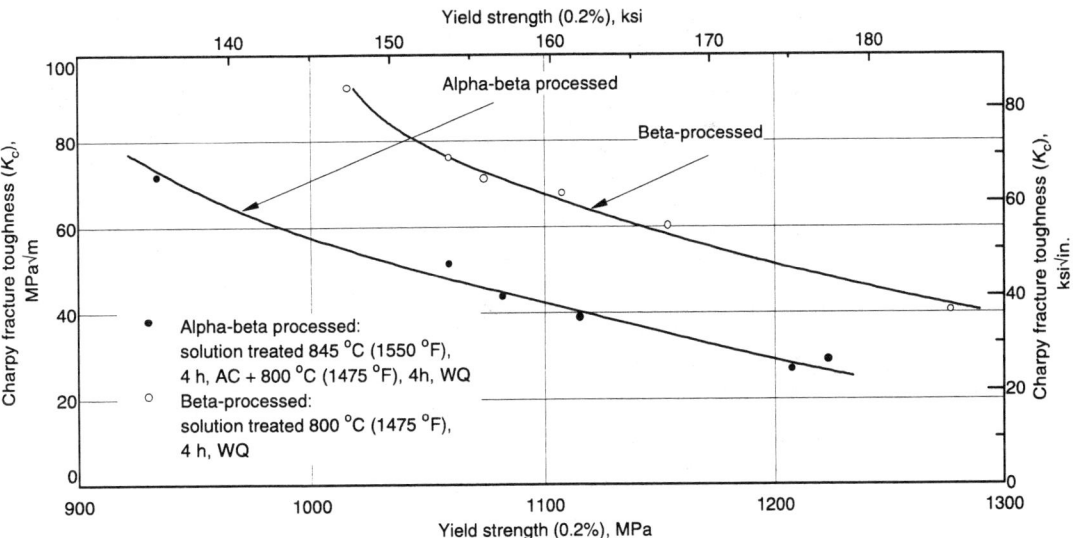

Source: *Beta Titanium Alloys in the 1980's*, R.R. Boyer and H.W. Rosenberg, Ed., TMS/AIME, 1984, p 245. Aged to strength (8 h with temperatures from 900 to 1300 °F)

Ti-17: Effect of solution temperature on toughness

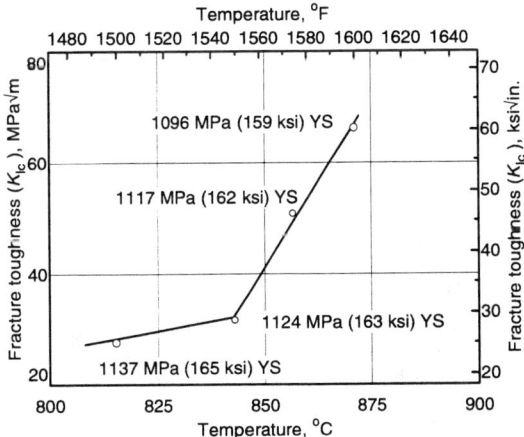

25 mm (1 in.) thick compact tension specimen from 457 mm (18 in.) diam × 50 mm (2 in.) thick disk forging. Indicated solution temperature plus 785 °C (1450 °F), 4 h, WQ + 620 °C (1150 °F), 8 h.
Source: *Beta Titanium Alloys in the 1980's*, R.R. Boyer and H.W. Rosenberg, Ed., TMS/AIME, 1984, p 246

Forging

G.W. Kuhlman, ALCOA, Forging Division

Ti-17 is a high-strength, highly beta-stabilized, α-β (near-beta) alloy whose primary commercial application is turbine engine rotating components. It can be fabricated into all forging product types, although closed die forgings and rings predominate. Ti-17 is commercially fabricated on all types of forging equipment. Turbine engine disks are frequently produced using hot die or isothermal forging techniques, resulting in near-net closed die forgings with reduced final machining.

Ti-17 is a highly forgeable alloy with lower unit pressures (flow stresses), improved forgeability, and less crack sensitivity than the α-β alloy Ti-6Al-4V. The final microstructure of Ti-17 forgings is developed by thermomechanical processing in forging manufacture tailored to achieve specific microstructural and mechanical-property objectives. Thermomechanical processes use combinations of subtransus and/or supra-transus forging followed by subtransus thermal treatments to fulfill critical mechanical-property criteria.

Final thermal treatments for Ti-17 forgings include two- or three-step practices of single or two-step solution treatments followed by quenching and aging. Solution treatment is subtransus, at 800 °C (1475 °F), followed by water quench or fan air cool for thin sections. For forgings fabricated conventionally, a solution anneal at 855 °C (1575 °F), followed by an air cool, may be used to improve toughness and creep properties. Aging treatment is conducted at 620 °C (1150 °F). Subtransus thermomechanical processes (forging and thermal treatment) for Ti-17 forgings achieve equiaxed (20 to 30%) α in transformed β matrix microstructures that enhance strength, ductility, and particularly low-cycle fatigue properties. Supra-transus thermomechanical processes (beta forging followed by subtransus thermal treatments) achieve transformed, Widmanstätten α microstructures that enhance creep and/or fracture-related properties (T.K. Redden, Ref 1).

Conventional Forging. The objectives in forging Ti-17 are to obtain the final forging shape and desired final microstructure at least cost. Conventional subtransus (α + β) forging thermomechanical processes are most widely used in commercial engine disk forging manufacture. To achieve conventional equiaxed α structures, subtransus reduction of 50 to 75%, accumulated through one or more forging steps, are required. Supra-transus (β) forging for Ti-17 may be used in early forging operations, including upsetting and open die preforming, to reduce unit pressures and ease forging fabrication. However, higher temperature initial forging operations must be followed by sufficient subtransus reduction to achieve the desired predominately equiaxed α structure. Conventionally forged Ti-17 is then subtransus solution treated, quenched, and aged as noted above.

Supra-transus thermomechanical processes for Ti-17 are used for selected disk applications to achieve transformed, Widmanstätten α structures for improved creep and fracture-related properties. Successful β thermomechanical processes for Ti-17 forgings include controlled β forging processes followed by subtransus solution treatment and aging. The β forging thermomechanical processes are particularly well suited to isothermal or hot die forging technology. Beta forging requires subtransus reduction (e.g., 20 to 50%) in early forging (blocker die) stages followed by a controlled, single β forging step, that achieves 30 to 50% reductions. Beta forging Ti-17 requires careful control of forging process conditions, particularly preheat times at temperature, to avoid excessive prior β grain growth. Beta forged Ti-17 is then subtransus heat treated as noted above. Because of inherent variations in forging conditions, β forged Ti-17 may exhibit more final forging product variation than conventionally subtransus forged and heat treated Ti-17 forged product.

Hot die and/or isothermal forging techniques are important commercial methods for fabrication of Ti-17 rotating turbine engine disks to reduce final component cost (from less machining) and/or improve final component microstructural and property uniformity through improved control of forging process conditions. The axisymmetric shapes and designs of such engine components are very well suited to these forging methods. Isothermal forging of Ti-17 disks is frequently accomplished in a single forging step from bar or billet stock, under carefully controlled supra- or subtransus metal and die temperatures, levels of strain, and strain-rate profiles. Hot die forging, where die temperature approaches but is not equivalent to metal temperature, is also used to reduce unit pressures, enhance forgeability, and produce more sophisticated final shapes in fewer forging operations. With either subtransus or supra-transus forging via both of these "hot die" processes and controlled post-forging cooling rates, desired tensile strength, fracture toughness, and creep properties can be achieved in Ti-17 using direct aging, thus eliminating the solution treatment processes (G.W. Kuhlman, Ref 2).

References

1. T.K. Redden, Processing and Properties of Ti-17 Alloy for Aircraft and Turbine Applications, *Beta Titanium in the 1980's*, R.R. Boyer and H.W. Rosenberg, Ed., TMS/AIME, 1984, p 239-254
2. G.W. Kuhlman, et al., "Mechanical Property Tailoring Titanium Alloys for Jet Engine Applications," *Proc. 1986 Int. Conf. Titanium Products and Applications*, Titanium Development Association, 1987, p 122-153

Ti-17: Forging process temperatures

Process	Metal temperature	
	°C	°F
Conventional forging	800-845	1480-1550
Beta forging	915-940	1675-1725

Note: See "Technical Note 4: Forging" for recommended die temperatures.

Heat Treatment

The recommended heat treatment for Ti-17 depends on process history. For α-β processed material, the heat treatment consists of a double solution treatment followed by aging. The first solution treatment should be done at 815 to 860 °C (1500 to 1575 °F) for 4 h, followed by rapid air cooling. The higher temperature (855 °C, or 1575 °F) produces higher toughness as a result of an increased amount of acicular or, which precipitates both during cooling and subsequent heat treatment. The second solution treatment is done at 800 °C (1475 °F) and nucleates additional acicular α and produces a β matrix, that is responsive to subsequent aging. Fan air cooling may be used from the second solution treatment for sections up to 75 mm (3 in.) thick, although more consistent and slightly higher strengths are achieved with a water quench. Heat treatment of β-processed material includes only a single 800 °C (1475 °F) 4 h solution treatment, because acicular α is already nucleated or precipitated during cooling from the forging temperature. An aging treatment of 620 to 650 °C (1150 to 1200 °F) for 8 h is recommended for both the α-β and β processed material.

Ti-17: Recommended heat treatments

Treatment	Temperature		Duration, h	Cooling method
	°C	°F		
Double solution treat and age				
ST1	860	1575	2	AC
ST2	800	1475	4	WQ
Age	620	1150	8	AC
Solution treat and age				
ST	800	1475	4	AC
Age	635	1175	8	AC
Stress relief				
Before machining	550	1020	4	AC
Other(a)	480-650	900-1200	1-4	AC or slow cool

Note: Rationales behind these treatments are described by T.K. Redden in *Beta Titanium Alloys in the 1980's*, R.R. Boyer and H.W. Rosenberg, Ed., TMS/AIME, 1984, p 239-254, and p 435. (a) From *Metals Handbook*, 9th Ed., Vol 4, p 764

Ti-6Al-2Sn-4Zr-6Mo

Common Name: Ti-6246
UNS Number: R56260

Ti-6Al-2Sn-4Zr-6Mo (Ti-6246) is a heat-treatable alpha-beta alloy designed to combine the long-term, elevated-temperature strength properties of Ti-6Al-2Sn-4Zr-2Mo-0.08Si (Ti-6242S) with much-improved short-term strength properties of a fully hardened alpha-beta alloy. It is used for forgings in intermediate-temperature sections of gas turbine engines, particularly in compressor disks and fan blades. This alloy is used at lower temperatures than Ti-6242S, but should be considered for long-term load-carrying applications at temperatures up to 400 °C (750 °F) and short-term load-carrying applications at temperatures up to 540 °C (1000 °F).

Chemistry and Density

Ti-6246 is a solid-solution-strengthened alloy that responds to heat treatment as a result of the beta-stabilizing effect of its 6% molybdenum content. Silicon additions (0.08 wt%) improve creep resistance. As for all alpha-beta alloys, excessive amounts of aluminum, oxygen, and nitrogen can decrease ductility and fracture toughness.

Density. 4.65 g/cm^3 (0.168 lb/in.^{3})

Product Forms

Ti-6246 is produced by all U.S. melters as billets and bars for forging stock. It has also been produced and evaluated in sheet and plate form.

Product Condition/Microstructure

Special ingot melting practices must be employed, particularly during final melting, to minimize microsegregation of the beta-stabilizing element, molybdenum, which could result in "beta flecks" (see Technical Note 1). Forging and heat treating practices require special controls to minimize beta flecks, which could result in microregions of high strength and low fracture toughness. Beta flecks are less of a problem for Ti-6246 than for Ti-17.

Applications

Ti-6246 is used for forgings in intermediate-temperature sections of gas turbine engines, particularly for compressor disks and fan blades and also for seals and airframe components. Ti-6246 is also under evaluation for deep, sour-well applications.

Ti-6Al-2Sn-4Zr-6Mo: Specifications and compositions

Specification	Designation	Description	Al	Fe	H	Mo	N	O	Sn	Zr	Other
UNS	R56260		6			6			2	4	bal Ti
USA											
AMS 4981B		Bar Wir Frg Bil STA	5.5-6.5	0.15	0.0125	5.5-6.5	0.04	0.15	1.75-2.25	3.5-4.5	C 0.04; OT 0.4; Y 0.005; bal Ti
MIL F-83142A	Comp 11	Frg Ann	5.5-6.5	0.15	0.0125	5.5-6.5	0.04	0.15	1.75-2.25	3.6-4.4	C 0.04; bal Ti
MIL F-83142A	Comp 11	Frg HT	5.5-6.5	0.15	0.0125	5.5-6.5	0.04	0.15	1.75-2.25	3.6-4.4	C 0.04; bal Ti
MIL T-9047G	Ti-6Al-2Sn-4Zr-6Mo	Bar Bil DA	5.5-6.5	0.15	0.0125	5.5-6.5	0.04	0.15	1.75-2.25	3.6-4.4	C 0.04; OT 0.4; Y 0.005; bal Ti

Ti-6Al-2Sn-4Zr-6Mo: Commercial compositions

Specification	Designation	Description	Al	Fe	H	Mo	N	O	Sn	Zr	Other
Japan											
Kobe	KS6-2-4-6	Bar Frg STA	5.5-6.5	0.15	0.0125	5.5-6.5	0.04	0.15	1.75-2.25	3.5-4.5	bal Ti
USA											
Astro	Ti-6Al-2Sn-4Zr-6Mo	Bar	5.5-6.5	0.15 max	0.0125	5.5-6.5	0.04 max	0.15 max	1.8-2.2	3.6-4.4	C 0.1 max; bal Ti
Howmet											

(continued)

Ti-6Al-2Sn-4Zr-6Mo: Commercial compositions (continued)

Specification	Designation	Description	Al	Fe	H	Mo	N	O	Sn	Zr	Other
USA (continued)											
Martin Mar Oremet RMI Tel. AllVac	Ti-6246 6Al-2Sn-4Zr-6Mo	Bar Bil STA	5.5-6.5	0.15	0.0125	5.5-6.5	0.04	0.15	1.75-2.25	3.5-4.5	C 0.04; bal Ti
Timet	TIMETAL 6-2-4-6	DA	5.5-6.5	0.15 max	0.0125 max	5.5-6.5	0.04 max	0.15 max	1.75-2.25	3.5-4.5	C 0.04 max; bal Ti

Phases and Structures

The microstructure of Ti-6246 is typically equiaxed primary α in a transformed β matrix; this can vary, depending on processing and heat treatment history. A microstructure with an optimum combination of strength, ductility, and toughness contains about 10% equiaxed α (primary α) plus a transformed β matrix with relatively coarse secondary α and aged β.

Beta Transus: 935 °C (1715 °F). The 1020 °C transus in figure is suspect.

Transformation Products

Ti-6246: Continuous cooling transformation and aging diagram

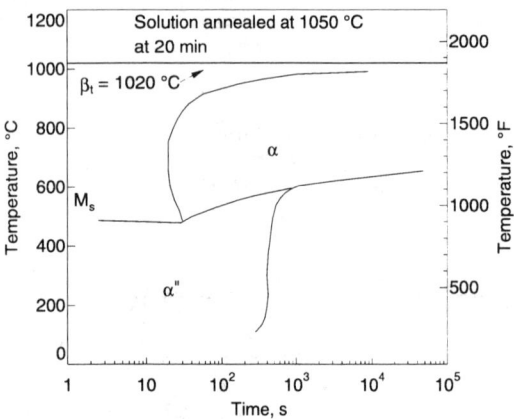

Source: W.W. Cias, "Phase Transformation Kinetics, Microstructures, and Hardenability of the Ti-6Al-2Sn-4Zr-6Mo Titanium Alloy," Rp-27-71-02, Climax Molybdenum, 2 March 1972

Ti-6246: Continuous cooling transformation diagram

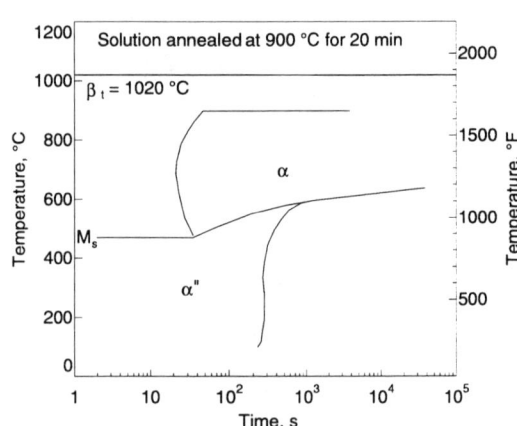

Source: W.W. Cias, "Phase Transformation Kinetics, Microstructures, and Hardenability of the Ti-6Al-2Sn-4Zr-6Mo Titanium Alloy," Rp-27-71-02, Climax Molybdenum, 2 March 1972

Physical Properties

Ti-6246: Summary of typical physical properties

Beta transus	935 °C (1715 °F)
Melting range	1595 to 1675 °C (2900 to 3050 °F)
Density(a)	4.65 g/cm^3 (0.168 lb/in.^{3})
Electrical resistivity(a)	1.9 to 2.05 μΩ · m
Magnetic permeability	Nonmagnetic
Specific heat capacity	500 J/kg · K (0.12 Btu/lb · °F)
Thermal conductivity(a)	7.7 W/m · K (4.4 Btu/ft · h · °F)
Thermal coefficient of linear expansion(b)	9×10^{-6}/°C (5×10^{-6}/°F)

(a) Typical values at room temperature of about 20 to 25 °C (68 to 78 °F). (b) Mean coefficient from room temperature to 100 °C (212 °F)

Damping Characteristics

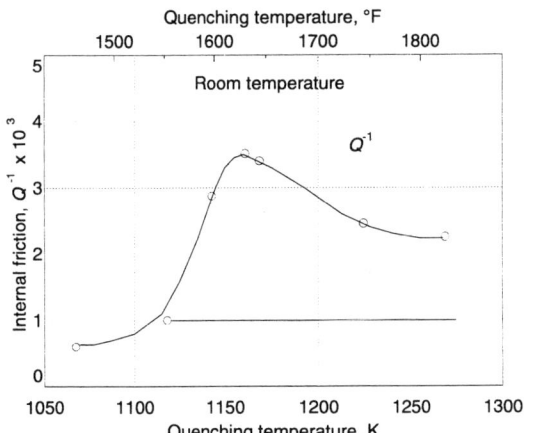

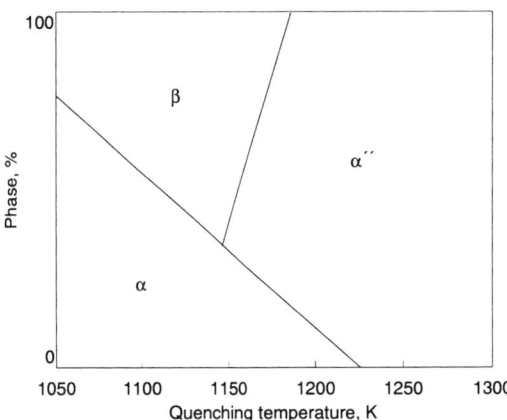

Source: K. Sugimoto et al., "Influence of Composition and Microstructure on Damping Capacity in Some Ti-Mo Based Alloys," presented at 6th World Conference on Titanium (France), 1988

Elastic Properties

Typical tensile modulus at room temperature is 114 GPa (16.5×10^6 psi) in the solution treated and aged condition.

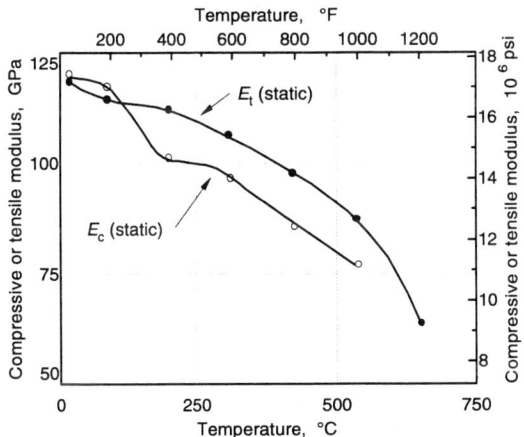

Triplex annealing: 855 °C (1575 °F), 15 min, forced air cooled + 730 °C (1350 °F), 15 min, air cooled + 595 °C (1100 °F), 2 h, air cooled.
Source: Aerospace Structural Metals Handbook, Vol 4, Code 3714, 1972

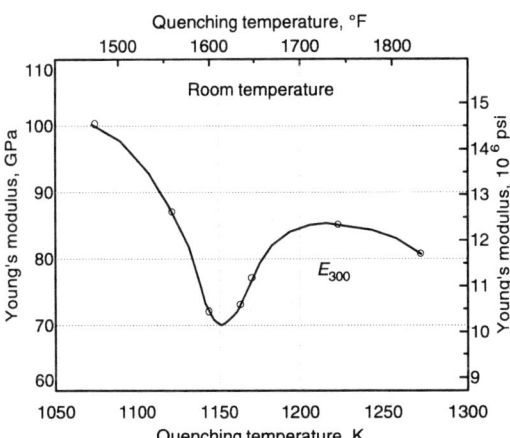

Source: K. Sugimoto et al., "Influence of Composition and Microstructure on Damping Capacity in Some Ti-Mo Based Alloys," presented at 6th World Conference on Titanium (France), 1988

Ti-6246: Tensile modulus of elasticity

Temperature		Modulus of elasticity		Yield strength (0.2%)	
°C	°F	GPa	10^6 psi	MPa	ksi
20	70	130	18.9	1165	169.0
315	600	107	15.5	834	121.0
425	800	100	14.5	787	114.2

Strain rate was 0.005 in./in./min.

Resistivity

Ti-6246: Electrical resistivity vs aging time

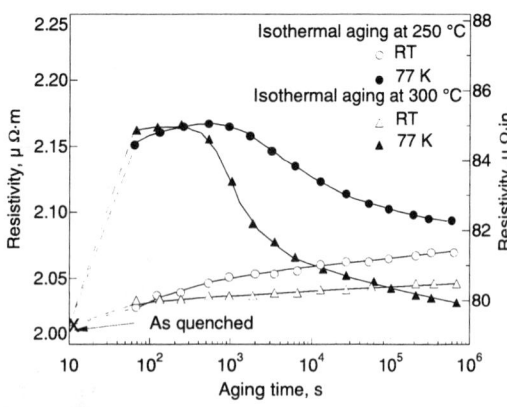

Measured using dc four-prove method and 2 × 2 × 50 mm (0.08 × 0.08 × 2 in.) specimens. Composition (wt%): 6.14Al, 0.0023H, 6.08Mo, 0.0019N, 2.03Sn, 4.17Zr. Heat treated in evacuated silica capsules: 900 °C (1650 °F), 2 h, water quenched, then aged at 250 or 300 °C (480 or 570 °F) for times indicated.
Source: K. Sugimoto et al., Aging Behavior of α″ Martensite Formed in a Quenched Ti-6Al-2Sn-4Zr-6Mo Alloy, *Titanium, Science and Technology*, Vol 3, G. Lütjering, U. Zwicker, and W. Bunk, Ed., Deutsche Gesellschaft für Metallkunde, Germany, 1984, p 1583

Ti-6246: Unaged resistivity vs quenching temperature

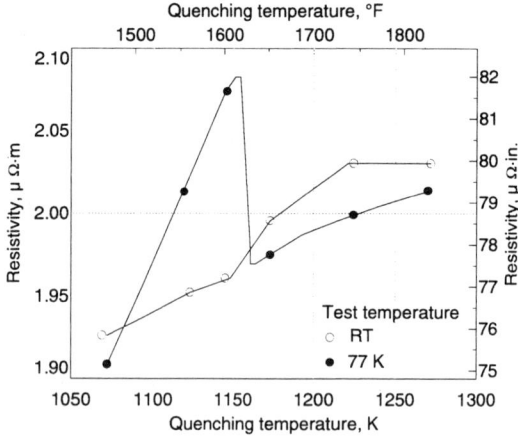

Source: K. Sugimoto et al., Aging Behavior of α″ Martensite Formed in a Quenched Ti-6Al-2Sn-4Zr-6Mo Alloy, *Titanium, Science and Technology*, Vol 3, G. Lütjering, U. Zwicker, and W. Bunk, Ed., Deutsche Gesellschaft für Metallkunde, Vol 3, Germany, 1984, p 1583

Ti-6246: Electrical resistivity vs aging time

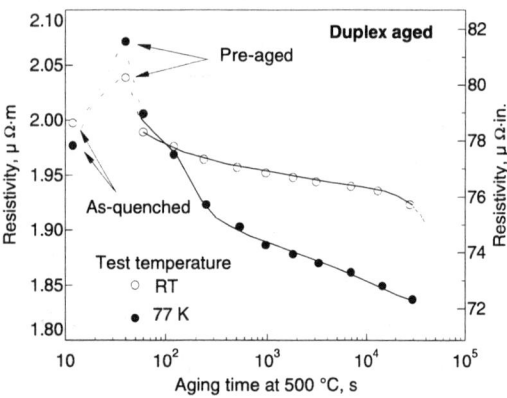

Heat treated at 900 °C (1650 °F), 2 h, water quenched, aged at 250 °C (480 °F), 170 h, then aged at 500 °C (930 °F) for times indicated.
Source: K. Sugimoto et al., Aging Behavior of α″ Martensite Formed in a Quenched Ti-6Al-2Sn-4Zr-6Mo Alloy, *Titanium, Science and Technology*, Vol 3, G. Lütjering, U. Zwicker, and W. Bunk, Ed., Deutsche Gesellschaft für Metallkunde, Vol 3, Germany, 1984, p 1583

Ti-6246: Electrical resistivity vs aging temperature

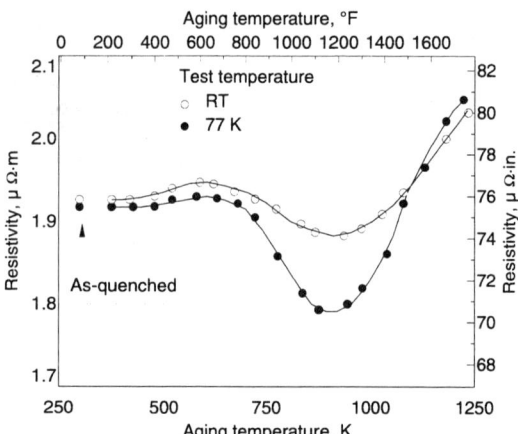

Heat treated at 800 °C (1470 °F), 2 h, water quenched, then aged for 0.5 h at temperatures indicated.
Source: K. Sugimoto et al., Aging Behavior of α″ Martensite Formed in a Quenched Ti-6Al-2Sn-4Zr-6Mo Alloy, *Titanium, Science and Technology*, Vol 3, G. Lütjering, U. Zwicker, and W. Bunk, Ed., Deutsche Gesellschaft für Metallkunde, Vol 3, Germany, 1984, p 1583

Ti-6246: Electrical resistivity vs aging temperature

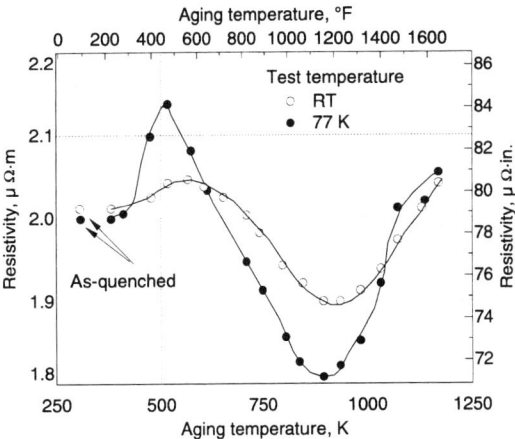

Heat treated at 900 °C (1650 °F), 2 h, water quenched, then aged for 30 min at temperatures indicated.
Source: K. Sugimoto et al., Aging Behavior of α″ Martensite Formed in a Quenched Ti-6Al-2Sn-4Zr-6Mo Alloy, *Titanium, Science and Technology*, Vol 3, G. Lütjering, U. Zwicker, and W. Bunk, Ed., Deutsche Gesellschaft für Metallkunde, Vol 3, Germany, 1984, p 1583

Chemical/Corrosion Properties

General Corrosion

Molybdenum additions greater than 4 wt% improve the corrosion resistance of titanium alloys in reducing media, and this effect is evidenced by the general corrosion of Ti-6246 in HCl solutions. The increase in reducing environment resistance is achieved, however, at the expense of oxidizing environment resistance.

Because Ti-6246 is less resistant to oxidizing media than CP Ti, it is expected that pitting resistance would likewise suffer. This is indeed the situation observed for repassivation potentials, which represent conservative measures of anodic pitting below which pitting cannot be sustained.

Ti-6246: Corrosion in naturally aerated HCl solutions

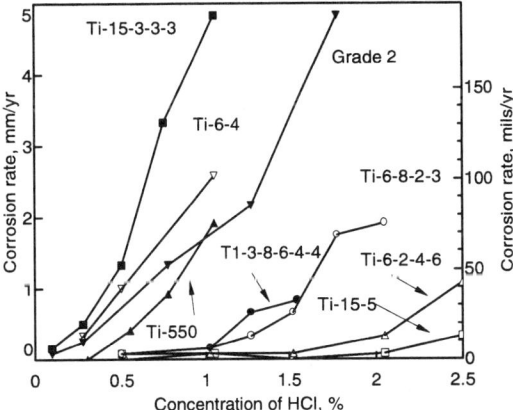

Source: *Metals Handbook, Corrosion*, Vol 13, 9th ed., ASM International, 1987, p 682

Ti-6246: Repassivation potential

Repassivation potentials in boiling chloride solutions for as-annealed alloys and unalloyed titanium

Alloy	Repassivation potential, V(a)		
	5% NaCl (pH 3.5)	3% HCl	Saturated NaCl
Grade 1	...	...	+7.0
Grade 2	+6.7	+5.8	+5.7
Ti-6-4	+2.3	+1.7	...
Ti-6-2-4-6	+3.0	+2.4	...
Beta C	+3.2	+2.6	...
Ti-8-8-2-3	+2.6	+2.4	...
Ti-15-5	+6.3	+5.6	...

(a) Measured versus Ag/AgCl reference electrode. Source: Data from R. Schutz and J. Grauman, reported in "Corrosion of Titanium and Titanium Alloys," R. Schutz and D. Thomas, in *Metals Handbook, Corrosion*, Vol 13, 9th ed., ASM International, 1987, p 669

Ti-6246: Corrosion rates in specific media

Medium	Concentration, %	Temperature, °C	Corrosion rate, mm/yr
Ferric chloride	10	Boiling	0.06
Formic acid	50	Boiling	0.62
Hydrochloric acid	0.5	Boiling	nil
	1.0	Boiling	0.03
Hydrochloric acid, aerated	pH 1	Boiling	0.01
Hydrochloric acid + 0.1% $FeCl_3$	5	Boiling	0.068

These data should be used only as a guideline for alloy performance. Rates may vary depending on changes in medium chemistry, temperature, length of exposure, and other factors. Total alloy suitability cannot be assumed from these values alone, because other forms of corrosion, such as localized attack, may be limiting. In complex, variable, and/or dynamic environments, *in situ* testing may provide more reliable data. Source: *Metals Handbook, Corrosion*, Vol 13, 9th ed., 1987

Crevice Corrosion

In contrast to the anodic breakdown associated with pitting, crevice corrosion is usually the result of acidification in the crevice region by oxidant depletion. Therefore, Ti-6246 should be very resistant to crevice corrosion due to its reducing environment resistance from molybdenum. In test results reported by Dees, the alloy did not crevice corrode in saturated salt brines at any of the pH levels tested (see figure). Crevice corrosion testing in salt brines above boiling has been pursued by Dees.

Stress-Corrosion Cracking

Resistance to stress-corrosion cracking in salt water is reported to be better after β forging than α-β forging (Wyman-Gordon Co., Project EM-06-1, Dec 1968). Duplex annealing also improves cracking resistance in salt water.

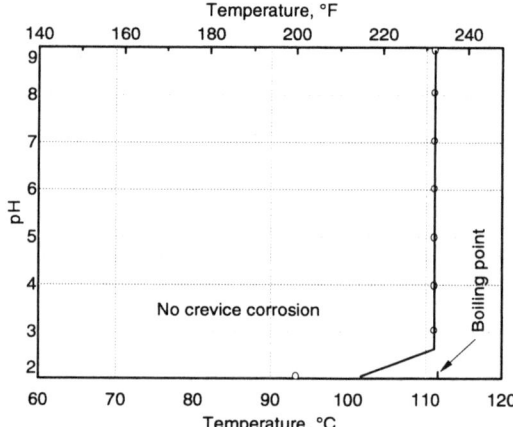

Ti-6246: Crevice corrosion in saturated brine

No crevice corrosion occurred.
Source: D. Dees, "Crevice Corrosion of High-Strength Titanium Alloys in Saturated Brine," *Industrial Applications of Titanium and Zirconium*, ASTM STP 830, 1984, p 133-142

Ti-6246: Fracture toughness in air and 3.5% NaCl solution at 25 °C

Alloy	Thickness mm	Thickness in.	Heat treatment	Yield strength MPa	Yield strength ksi	Toughness(a) K_{Ic} or K_c MPa√m	Toughness(a) K_{Ic} or K_c ksi√in.	K_{Iscc} or K_{scc} MPa√m	K_{Iscc} or K_{scc} ksi√in.
Ti-6246	13	0.50	Mill annealed	1103	160	60	55	22	20
			Duplex annealed	1034	150	88	80	49	45
	7	0.30	Mill annealed	965	140	57	52	28	26

Source: R. Schutz, Stress-Corrosion Cracking of Titanium Alloys, in *Stress Corrosion Cracking: Materials Performance and Evaluation*, ASM International, 1992. Note (a): Listed values of toughness in air are much higher than typical values in "Fracture Toughness" section.

Thermal Properties

Specific Heat. 500 J/kg · K (0.12 Btu/lb · °F) at 20 °C (70 °F).

Melting Range. 1595 to 1675 °C (2900 to 3050 °F).

Thermal Expansion

Ti-6246: Thermal coefficient of linear expansion

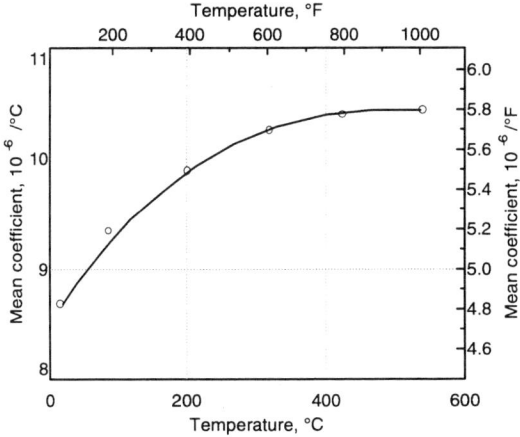

Alloy used as forgings. Solution treated at 870 °C (1600 °F) for 1 h, then water quenched, and aged at 595 °C (1100 °F) for 8 h, then air cooled.
Source: TIMET data reported in *Aerospace Structural Metals Handbook*, Code 3714, Vol 4, Battelle Columbus Laboratories, 1972

Ti-6246: Mean thermal coefficient of linear expansion

Temperature		Mean coefficient	
°C	°F	10^{-6}/K	10^{-6}/°F
0-100	32-212	9.4(a)	5.2(a)
0-315	32-600	10.3	5.7
0-540	32-1000	10.4	5.8
0-650	32-1200	10.4	5.8
0-815	32-1500	10.6	5.9

(a) A low coefficient of 8.6×10^{-6}/°C (4.8×10^{-6}/°F) is reported in "Corrosion Data," RMI Titanium Co. Source: "Quality Products and Services," TIMET, p 13

Ti-6246: Thermal coefficient of linear expansion
Solution treated and aged specimens

Temperature		Average coefficient	
°C	°F	10^{-6}/K	10^{-6}/°F
20	68	8.6	4.8
93	200	9.4	5.2
205	400	9.9	5.5
315	600	10.3	5.7
425	800	10.4	5.8
540	1000	10.4	5.8

Source: *Metals Handbook, Properties and Selection: Stainless Steels, Tool Materials, and Special-Purpose Materials*, Vol 3, 9th ed., American Society for Metals, 1980, p 395

Ti-6246: Thermal coefficient of linear expansion

Temperature		Coefficient	
°C	°F	10^{-6}/K	10^{-6}/°F
20-100	70-212	9.0	5.0
20-205	70-400	9.2	5.1
20-315	70-600	9.4	5.2
20-425	70-800	9.5	5.3
20-540	70-1000	9.5	5.3

Source: *Metals Handbook*, Vol 2, 10th ed., 1990, p 620

Thermal Conductivity

Ti-6246: Thermal conductivity

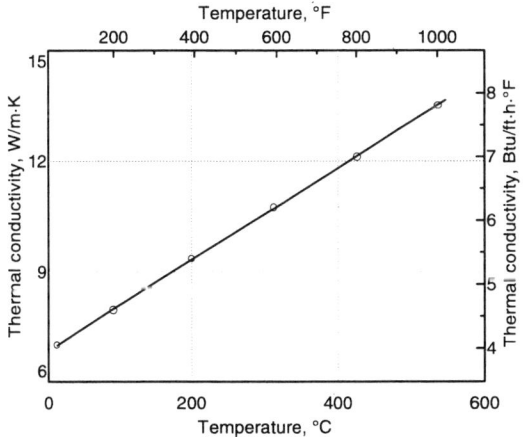

Alloy used as forgings in STA condition. Heated at 870 °C (1600 °F) for 1 h, water quenched, and held at 595 °C (1100 °F) for 8 h, then air cooled.
Source: "Metallurgical and Mechanical Properties of an Advanced High Strength Titanium Alloy," Titanium Metals Corp. of America, Publication EP 4-69, 1969; reported in *Aerospace Structural Metals Handbook*, Code 3714, Vol 4, Battelle Columbus Laboratories, 1972

Ti-6246: Thermal conductivity of STA specimens

Temperature		Conductivity	
°C	°F	W/m · K	Btu/ft · h · °F
20	68	7.7	4.4
93	200	7.9	4.6
205	400	9.3	5.4
315	600	10.7	6.2
425	800	12.1	7.0
540	1000	13.5	7.8

Source: *Metals Handbook, Properties and Selection: Stainless Steels, Tool Materials, and Special-Purpose Materials*, Vol 3, 9th ed., American Society for Metals, 1980, p 396

Mechanical Properties

Ti-6246: Producer guaranteed tensile properties of forgings

Thickness		Ultimate tensile strength		Tensile yield strength		Elongation(a),	RA,
mm	in.	MPa	ksi	MPa	ksi	%	%
STA forgings(b)							
<75	<3	1170	170	1100	160	6	12
75 to 100	3 to 4	1100	160	1035	150	6	12
DA forgings(c)							
<50	<2	1100	160	1035	150	10	20
50 to 100	2 to 4	1035	150	965	140	8	20

(a) Transverse elongation in 25 mm (1 in.). (b) Solution treated and aged: 870 °C (1600 °F) for 1 h, WQ; plus 595 °C (1100 °F) for 8 h, AC. (c) Duplex annealed: 870 °C (1600 °F) for 1 h, AC; plus 540 °C (1000 °F) for 8 h, AC. Source: TIMET EP 4-695M

Hardness

Rockwell Hardness. Typical room-temperature hardness in the STA condition ranges from 36 to 42 HRC. As-forged hardness is reported to be around 33 to 38 HRC.

Ti-6246: Rockwell hardness of different forging and treatment conditions

Forging conditions	Heat treatment	Hardness, HRC
α-β forge (β_t–100 °F), AC + finish (β_t–25 °F), AC	885 °C (1625 °F), 1 h, AC + 595 °C (1100 °F), 8 h, AC	39.3
α-β forge (β_t–100 °F), AC + finish (β_t–100 °F), AC	885 °C (1625 °F), 1 h, AC + 705 °C (1300 °F), 1 h, AC	38.4
β forge (β_t+75 °F), AC	885 °C (1625 °F), 1 h, AC + 595 °C (1100 °F), AC	39.9

Source: AFML-TP–78–68

Ti-6246: Compressive yield strength of 1.5 mm (0.06 in.) sheet

Condition	Compressive yield strength	
	MPa	ksi
Duplex annealed: 870 °C (1600 °F) for 15 min, AC; plus 700 °C (1300 °F) for 15 min, AC	1270	184
Triplex annealed: same as above plus 595 °C (1100 °F) for 2 h, AC	1255	182

Source: *Aerospace Structural Metals Handbook*, Code 3714, Vol 4, Battelle Columbus Laboratories, 1972

Ti-6246: Effect of aging on Vickers hardness

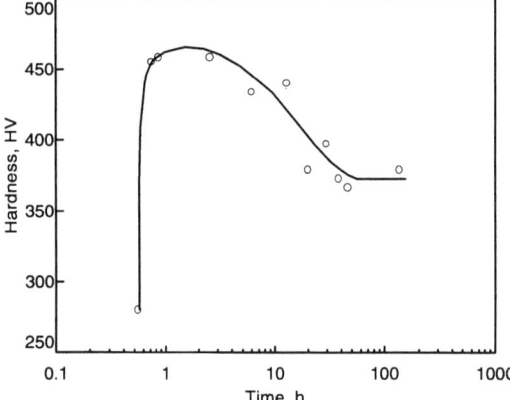

Ti-6246 in the as-quenched condition exhibits low hardness (283 HV), but a short aging period of 15 min increases hardness considerably (462 HV). Specimens were machined from alloy bar stock that had been extruded to 95% reduction in area at 940 °C (1725 °F). Coupons were annealed at 950 °C (1740 °F) for 1 h, water quenched, and aged at 550 °C (1020 °F) for times as indicated.
Source: M. Mendiratta and J. Roberson, "Tensile Properties to 550 °C and Microstructures in Quenched and Aged Ti-6Al-2Sn-4Zr-6Mo Alloy," *Metall. Trans. A*, Vol 6, 1975, p 940

Ti-6246: Effect of aging on Vickers hardness

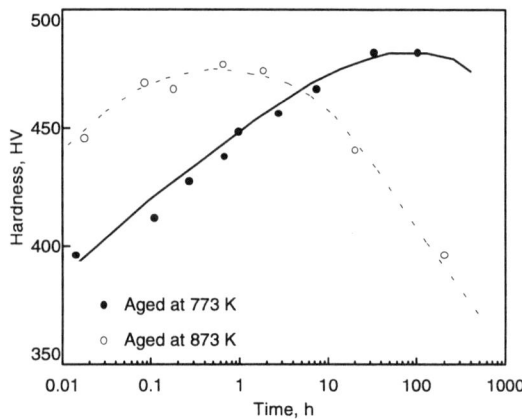

Isothermal aging curves for alloy aged at 773 and 873 K after water quenching from the β field. Chemical composition: 5.48 wt% Al, 0.072 wt% Fe, 6.35 wt% Mo, 0.004 wt% N, 0.083 wt% O, 1.94 wt% Sn, and 4.00 wt% Zr. Beta transus temperature was 1211 K. Alloy used was in the form of flat bar stock previously warm worked in the α+β field. Hardness determinations were obtained on electro-polished specimens using a Zwick diamond pyramid hardness tester at a load of 10 kg.
Source: M. Young, E. Levine, and H. Margolin, "The Aging Behavior of Orthorhombic Martensite in Ti-6-2-4-6," *Metall. Trans.*, Vol 5, 1974, p 1891

Typical Tensile Properties

Ti-6246: Typical variations in tensile properties with heat treatment/condition

Condition	Tensile yield strength MPa	ksi	Ultimate tensile strength MPa	ksi	Elongation, %	Reduction of area, %
10 to 20% primary α + STA(a)	1118	162	1214	176	13	37
10 to 20% primary α + STOA(b)	1021	148	1090	158	16	42
40 to 50% primary α + STA(a)	1152	167	1242	180	14	42
40 to 50% primary α + STOA(b)	1070	155	1145	166	14	41
β forged + STA(c)	1049	152	1201	174	6.5	13

(a) STA = 885 °C (1630 °F), 1 h, air cool + 595 °C (1100 °F), 8 h, air cool. (b) STOA = 885 °C (1630 °F), 1 h, air cool + 705 °C (1300 °F), 1 h, air cool. (c) STA = 985 °C (1810 °F), (β_t – 15 °C), air cool + 595 °C (1100 °F), 8 h, air cool. Source: R.B. Sparks and J.R. Long, AFML-TR-73-301, Feb 1974

Ti-6246: Effect of creep exposure on tensile properties

Thickness mm	in.	Creep exposure Temperature °C	°F	Stress MPa	ksi	Time, h	Creep deformation, %	F_{tu} MPa	ksi	F_{ty} MPa	ksi	Elongation in 25 mm (1 in.), %	Reduction of area, %
Duplex annealed(a)													
75	3	...	...	...	...	...	...	1190	172.6	1079	156.6	16.5	43.3
		425	800	585	85	50	0.102	1157	167.8	1063	154.2	14.0	40.5
50	2	...	...	...	...	...	...	1164	168.8	1066	154.7	14.0	41.9
		425	800	585	85	50	0.136	1121	162.7	1023	148.4	17.0	41.8
25	1	...	...	...	...	...	...	1200	174.1	1087	157.7	14.0	39.4
		425	800	585	85	50	0.098	1215	176.2	1106	160.5	14.0	35.0
Solution treated and aged(b)													
75	3	...	...	...	...	...	...	1385	200.9	1305	189.3	7.0	27.5
		425	800	585	85	50	0.071	1348	195.6	1284	186.3	7.0	20.4
50	2	...	...	...	...	...	...	1393	202.1	1312	190.3	7.0	18.3
		425	800	585	85	50	0.106	1371	198.9	1285	186.4	5.0	17.5
25	1	...	...	...	...	...	...	1576	228.6	1468	212.9	2.5	8.2
		425	800	585	85	50	0.165	1570	227.8	1461	212.0	3.5	10.1

Note: α-β forged discs, 150 mm (6 in.) diam at various thicknesses indicated. (a) Duplex anneal of 910 °C (1675 °F) for 1 h, AC; plus 595 °C (1100 °F) for 4 h, AC. (b) Solution treated and aged: 910 °C (1675 °F) for 1 h, OQ; plus 595 °C (1100 °F) for 4 h, AC. Source: *Aerospace Structural Metals Handbook*, Code 3714, Vol 4, Battelle Columbus Laboratories, 1972

High-Temperature Strength

Ti-6246: Tensile strength of DA sheet

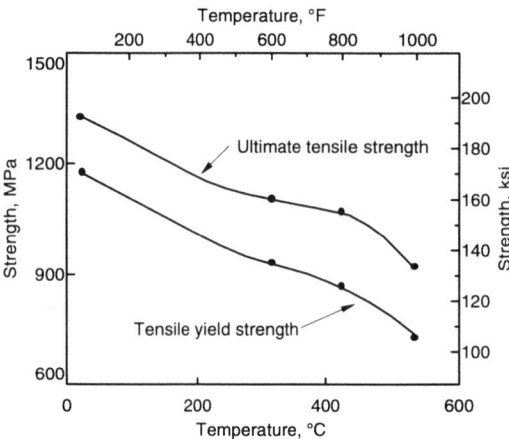

Duplex annealed 870 °C (1600 °F) 15 min, AC, 700 °C (1300 °F), 15 min, AC.
Source: *Aerospace Structural Metals Handbook*, Code 3714, Vol 4, Battelle Columbus Laboratories, 1972

Ti-6246: Tensile strength of TA sheet

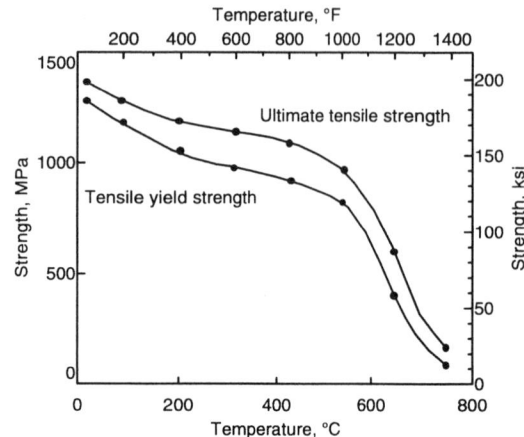

Triplex annealed sheet, 855 ° C (1575 °F), 15 min, FAC; 730 °C (1350 °F) 15 min, AC; 595 °C (1100 °F), 2 h, AC.
Source: TIMET, Nov 1971; reported in *Aerospace Structural Metals Handbook*, Code 3714, Vol 4, Battelle Columbus Laboratories, 1972

Ti-6246: Compressive strength vs temperature

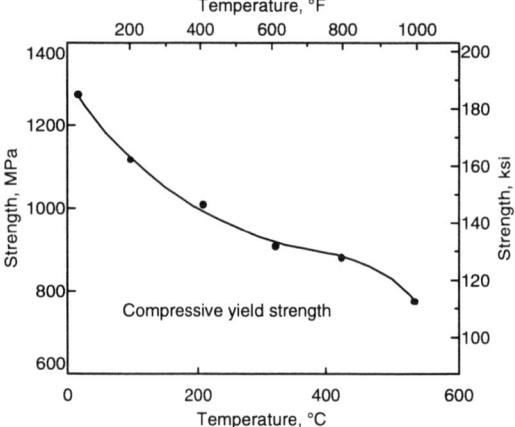

Triplex annealed sheet.
Source: TIMET, Nov 1971; reported in *Aerospace Structural Metals Handbook*, Code 3714, Vol 4, Battelle Columbus Laboratories, 1972

Creep Properties

Ti-6246: Effect of forging on creep behavior

Creep exposure					Time to 0.1% creep, h	Time to 0.2% creep, h	Total plastic deformation, %
Temperature		Stress					
°F	°C	MPa	ksi	Time, h			
α − β forged at 880 °C (1620 °F)							
1380	750	620	90	1228	270	840	0.251
1470	800	613	89	150	...	...	0.101
		517	75	504	105	420	0.217
		586	85	241	40	210	0.234
		586	85	150	...	...	0.180
1830	1000	138	20	51	6	27	0.246
		207	30	72	2	8	0.760
		207	30	150	...	...	0.820

(continued)

Ti-6246: Effect of forging on creep behavior (continued)

Temperature		Creep exposure			Time to 0.1% creep, h	Time to 0.2% creep, h	Total plastic deformation, %
°F	°C	Stress MPa	ksi	Time, h			

β forged at 1010 °C (1850 °F)

°F	°C	MPa	ksi	Time, h	Time to 0.1% creep, h	Time to 0.2% creep, h	Total plastic deformation, %
1380	750	620	90	336	75	264	0.203
		613	89	150	...	...	0.091
1470	800	517	75	290	35	250	0.223
		586	85	313	25	165	0.252
1830	1000	138	20	120	4	22	0.427
		207	30	72	1	6	0.613

Source: M. Greenlee and W. Heil, "Evaluation of Alpha-Beta and Beta Forged Disc Forgings in Ti-6Al-2Sn-4Zr-6Mo," Wyman-Gordon Co. Report; reported in *Aerospace Structural Metals Handbook*, Code 3714, Vol 4, Battelle Columbus Laboratories, 1972

Ti-6246: Creep behavior comparison

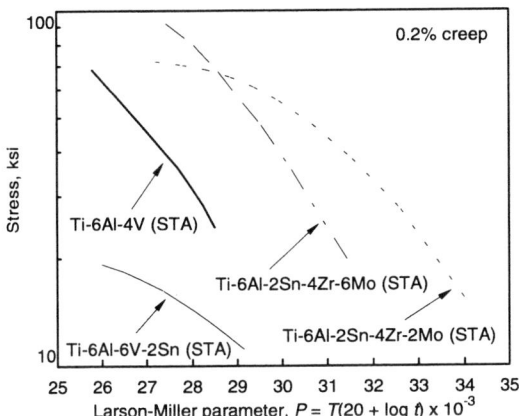

All alloys STA.
Source: MIL-HDBK-697A, p 37

Ti-6246: Typical creep curves

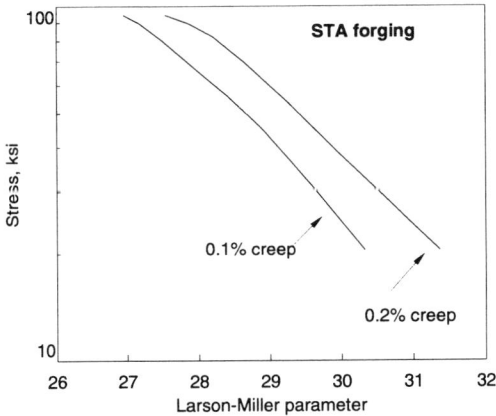

Larson-Miller parameter equals $10^{-3} T (20 + \log t)$, where T is temperature in °R and t is time in hours.
Source: *Metals Handbook, Properties and Selection: Stainless Steels, Tool Materials and Special-Purpose Materials*, Vol 3, 9th ed., American Society for Metals, 1980, from TIMET data reported in *Titanium Alloys Handbook*, R. Wood and R. Favor, Ed., Battelle, MCIC-HB-02, 1972

Ti-6246: Creep stress for 0.2% strain

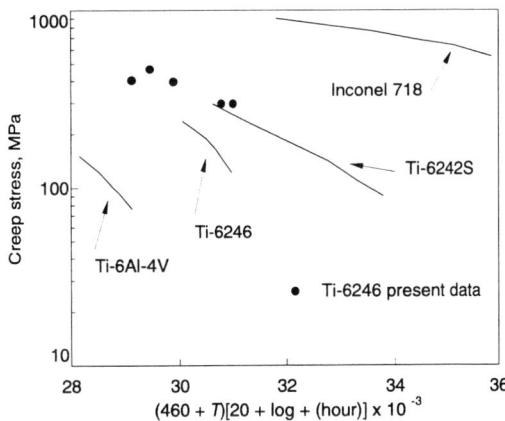

Curves represent previously reported data (*J. Metall.*, Nov 1984). Data points are for β-processed impeller forgings (heat treated at β transus temperature minus 40 °C (72 °F) for 2 h, fan cooled, then held at 595 °C (1100 °F) for 8 h, air cooled). Room-temperature ultimate tensile strength was 1117 MPa (162 ksi). Forging dimensions were 22.8 cm (9 in.) diameter by 13.7 cm (5.4 in.) thick. Forging history was α-β preform, β finish. Creep exposure was at 450 and 500 °C (750 and 930 °F).
Source: A. Chakrabarti, M. Burn, D. Fournier, and G. Kuhlman, "Structure and Mechanical Property Optimization Through Thermomechanical Processing in Ti-6-4 and Ti-6-2-4-6 Alloys," in *Sixth World Conference on Titanium*, Vol 2, P. Lacombe, R. Tricot, and B. Beranger, Ed., Les Editions de Physiques, Paris, 1989, p 1339

Fatigue Properties

High-Cycle Fatigue

Ti-6246: Room-temperature axial fatigue strength at 10^7 cycles

Heat treatment	Axial strength at $K_t = 1$		Fatigue strength 10^7 cycles for: $K_t = 3.8$	
	MPa	ksi	MPa	ksi
870 °C (1600 °F), 1 h, AC + 595 °C (1100 °F), 8 h, AC	793	115	380	55
910 °C (1675 °F), 1 h, AC + 595 °C (1100 °F), 8 h, AC	825	120	345	50

Note: 25 mm (1 in.) round duplex annealed forgings. Source: *Aerospace Structural Metals Handbook*, Code 3714, Vol 4, Battelle Columbus Laboratories, 1972

Ti-6246: Fatigue and tensile data for various microstructural conditions

Condition	Tensile yield strength MPa	ksi	Ultimate tensile strength MPa	ksi	Elongation, %	Reduction of area, %	Stress at 10^7 cycles Smooth MPa	ksi	Notched MPa	ksi
10% equiaxed primary α + annealed(a)	1020	148	1109	161	15	37	620	90	289	42
10% equiaxed primary α + STA(b)	1116	162	1213	176	13	37	620	90	248	36
50% equiaxed primary α + annealed	1061	154	1130	164	13	34	620	90	282	41
50% equiaxed primary α + STA	1151	167	1240	180	14	42	675	98	262	40
50% equiaxed primary α + STOA(c)	1068	155	1144	166	14	41	620	90	262	38
50% elongated primary α + STA	1096	159	1206	175	10	23	751	109	276	40
20% elongated primary α + STA	1109	161	1206	175	11	26	620	90	282	41
β forged + STA	1047	152	1199	174	7	13	675	98	262	38

(a) Annealed = 705 °C (1300 °F), 1 h, AC. (b) STA = 885 °C (1630 °F), 1 h, AC + 595 °C (1100 °F), 8 h, AC. (c) STOA = 885 °C (1630 °F), 1 h, AC + 705 °C (1300 °F), 1 h, AC. Source: J.C. Williams and E.A. Starke, in *Deformation, Processing, and Structure*, G. Krauss, Ed., American Society for Metals, 1984, p 332

Low-Cycle Fatigue

Low-cycle fatigue (LCF) behavior of Ti-6246 has been studied to determine the effect of microstructure on cyclic deformation and LCF initiation (Mahajan and Margolin, *Metall. Trans. A*, Vol 13, 1982, p 257-268). Widmanstätten + grain boundary α and equiaxed structures of different α particle sizes were produced in smooth bar specimens of a Ti-6246 alloy, heat treated to produce a 0.2% yield stress of about 1100 MPa (159 ksi). Specimens were cycled at room temperature under total strain control.

At low strains for both Widmanstätten + grain boundary and equiaxed α structures, crack initiation took place at α-β interfaces and in the aged β matrix. In Widmanstätten + grain boundary α structures, profuse extrusion formation was noted as well. At higher strains, cracking was more predominant at slip bands within α.

In Widmanstätten + grain boundary α structures, Widmanstätten α and grain boundary α particles provided sites at which ready crack formation and link-up could take place, thus leading to much longer surface cracks in the Widmanstätten + grain boundary α than in equiaxed α structures for given cycling conditions. Beta grain size played an indirect role in development of fatigue cracks. Larger β grains permitted longer Widmanstätten and grain boundary α particles to form. These longer particles provided longer paths where crack growth could take place preferentially and longer surface cracks could develop.

At larger plastic strains, Widmanstätten α colonies at large angles to the crack propagation direction served to produce multiple cracking along the α-β interfaces and to slow or change the direction of crack propagation at both surface and interior locations. Coarse α particles, which have a small aspect ratio, are favorable for multiple slip and associated multiple cracking at the crack tip.

Ti-6246: Low-cycle fatigue

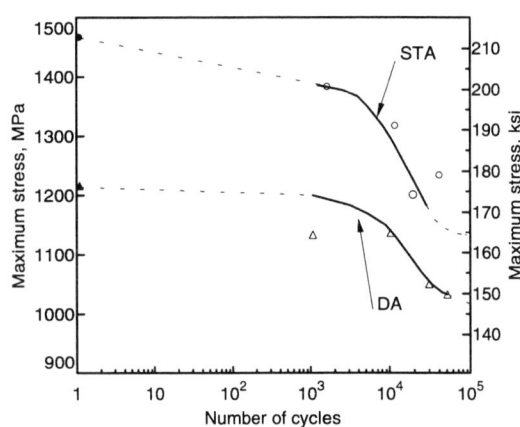

STA condition: 1 h at 870 °C (1600 °F), water quench, age 8 h at 595 °C (1100 °F) and air cool. DA (duplex annealed) condition: 15 min at 870 °C, air cool, then 8 h at 540 °C (1000 °F) and air cool. All fatigue tests conducted at a stress ratio of R = 0.1. Open symbols indicate fatigue tests; solid symbols, tension tests.
Source: *Aerospace Structural Metals Handbook*, Code 3714, Vol 4, Battelle Columbus Laboratories, 1972

Fatigue Crack Growth

A Ti-6246 alloy containing 68 ppm H with basal texture was tested to determine the influence of dwell time at maximum tensile stress on the fatigue crack growth rates (see table). All of the fatigue tests were conducted using displacement-controlled constant stress intensity (K), in air (relative humidity not specified), at room temperature. The Ti-6246 alloy exhibited a nominal two- to threefold increase in the total fatigue crack growth rate as a result of the 10-min dwell at $\Delta K = 38.5$ MPa$\sqrt{m}$ (35 ksi$\sqrt{in.}$), but there was little effect on the fatigue crack growth at the lower values of ΔK. The small changes in the fatigue crack growth rate were due to the crack advance by cleavage during the dwell periods. The cleavage fracture was the result of hydrogen embrittlement.

Ti-6246: Fatigue crack growth vs dwell time
Basal transverse textured titanium alloys tested at 21 °C ± 1 °C; R = 0.01; TL orientation; 0.3 Hz

Fatigue crack growth (ΔK)		Dwell time, min	da/dN prior to dwell, μm/cycle	da/dt during dwell, μm/min	da/dN after dwell, μm/cycle	Total da/dN(a), μm/cycle
MPa$\sqrt{m}$	ksi$\sqrt{in.}$					
38.46	35.0	10	6.25	44.5	4.62	20.1
27.80	25.3	45	1.38	7.29	1.28	2.87
23.52	21.4	45	0.9	1.0	0.9	1.01

(a) Includes crack advance during the dwell time. Source: *Metall. Trans. A*, Vol 14, 1983, p 2179

Fracture Properties

Ti-6246: Impact toughness

Forging temperature		Solution temperature(a)		Cooling from solution	Aging temperature(b)		Charpy V-notch impact toughness		Ultimate tensile strength	
°C	°F	°C	°F		°C	°F	J	ft·lbf	MPa	ksi
885	1625	830	1525	Air cool	540	1000	12.2	9	1241	180
					595	1100	13.5	10	1255	182
				Oil quench	540	1000	10.1	7.5	1296	188
					595	1100	14.9	11	1268	184
		870	1600	Air cool	540	1000	11.5	8.5	1310	190
					595	1100	9.5	7	1248	181
				Oil quench	540	1000	8.1	6	1489	216
					595	1100	8.1	6	1324	192
915	1675	830	1525	Air cool	540	1000	10.8	8	1193	173
					595	1100	10.8	8	1165	169
				Oil quench	540	1000	10.8	8	1337	194
					595	1100	12.9	9.5	1241	180
		870	1600	Air cool	540	1000	12.2	9	1255	182
					595	1100	12.9	9.5	1275	185
				Oil quench	540	1000	9.5	7	1461	212
					595	1100	8.8	6.5	1350	197

Note: 45 mm (1.75 in.) thick upset forgings. (a) 1 h at temperature. (b) 8 h at temperature. Source: *Aerospace Structural Metals Handbook*, Code 3714, Vol 4, Battelle Columbus Laboratories, 1972

Fracture Toughness

Ti-6246: Fracture toughness of forgings

Condition	Tensile yield strength		Ultimate tensile strength		Elongation, %	K_{IC} (K_Q)	
	MPa	ksi	MPa	ksi		MPa$\sqrt{m}$	ksi$\sqrt{in.}$
α + β forged + STA(a) (10% primary α)	1116	162	1213	176	13	34	31
α + β forged + STA(a) (50% primary α)	1150	166	1240	180	14	26	23
α + β forged + annealed(b) (50% primary α)	1061	154	1130	164	13	26	23
β forged + STA(a)	1047	152	1199	174	7	57	52

(a) 885 °C (1630 °F), 1 h, AC + 595 °C (1100 °F), AC. (b) 705 °C (1300 °F), 1 h, AC. Source: J.C. Williams and E.A. Starke, in *Deformation, Processing, and Structure*, American Society for Metals, 1984

Ti-6246: Fracture toughness of forgings of several forging and heat treatment conditions and section thicknesses

Forging conditions	Heat treatment conditions	Section thickness mm	Section thickness in.	Ultimate tensile strength MPa	Ultimate tensile strength ksi	Fracture toughness K_{Ic} MPa√m	Fracture toughness K_{Ic} ksi√in.
885 °C (1625 °F), AC	870 °C (1600 °F), 2 h, AC + 595 °C (1100 °F), 8 h, AC	50	2	1144	166	36.7	33.4
				...	...	33.5	30.5
900 °C (1650 °F), AC	900 °C (1650 °F), 1 h, WQ + 650 °C (1200 °F), 8 h, AC	75	3	1303	189	20.9	19.1
	900 °C (1650 °F), 1 h, AC + 650 °C (1200 °F), 8 h, AC	75	3	1172	170	26.6	24.2
885 °C (1625 °F), AC	915 °C (1675 °F), 1 h, AC + 595 °C (1100 °F), 8 h, AC	50	2	1158	168	50.4	45.9
980 °C (1800 °F), WQ		50	2	1158	168	65.8	59.9
885 °C (1625 °F), AC	915 °C (1675 °F), AC + 525 °C (975 °F), 8 h, AC	25	1	1220	177	32.6	29.7
980 °C (1800 °F), WQ	95 °C (1675 °F), 1 h, AC + 525 °C (975 °F), 8 h, AC	25	1	1220	177	46.0	41.9
980 °C (1800 °F), WQ	915 °C (1675 °F), 1 h, AC + 595 °C (1100 °F), 8 h, AC	25	1	1186	172	47.5	43.3
900 °C (1650 °F), AC	915 °C (1675 °F), 1 h, AC + 845 °C (1550 °F), 8 h, OQ + 595 °C (1100 °F), 8 h, AC	75	3	1220	177	39.0	35.5
		54	2.125	1255	182	36.3	33.1
980 °C (1800 °F), AC	845 °C (1550 °F), 1 h, OQ + 595 °C (1100 °F), 8 h, AC	75	3	1186	172	37.2	33.9
		54	2.125	1268	184	35.4	32.2
		38	1.5	1296	188	27.7	25.2
885 °C (1625 °F), AC	915 °C (1675 °F), 1 h, AC + 595 °C (1100 °F), 4 h, AC	75	3	1186	172	33.7	30.7
		50	2	1165	169	33.3	30.3
		25	1	1234	179	29.9	27.3

Source: W. Heil; reported in *Aerospace Structural Metals Handbook*, Code 3714, Vol 4, Battelle Columbus Laboratories, 1972

Ti-6246: Fracture toughness of STA forgings of two forging conditions and specimen locations

	Fracture toughness (K_{Ic}) for material:			
	α + β forged at 880 °C (1620 °F)		β forged at 1010 °C (1850 °F)	
Specimen location	MPa√m	ksi√in.	MPa√m	ksi√in.
Center tangential	28.28	25.74	...	...
Outside tangential	24.47	22.27	21.81	19.85
	26.49	24.11	...	...
Center diametral	...	...	21.45	19.52

Note: K_{Ic} values determined with precracked three-point notched bend specimens. Heat treatment was at 870 °C (1600 °F) for 1 h, water quench, then at 595 °C (1100 °F) for 8 h, air cool. Source: M. Greenlee and W. Heil, Wyman-Gordon Co. data, 1968; reported in *Aerospace Structural Metals Handbook*, Code 3714, Vol 4, Battelle Columbus Laboratories, 1972

Ti-6246: Fracture toughness vs yield strength

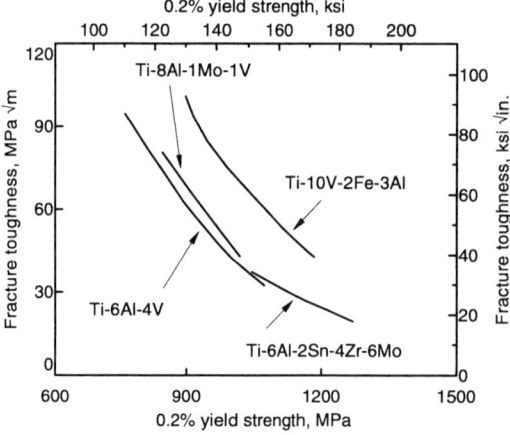

Fracture toughness is in part dependent on microstructure and is higher in the presence of an acicular structure.
Source: "Titanium," Kobe Steel

Forging

G.W. Kuhlman, ALCOA, Forging Division

Ti-6246 is a high-strength, highly beta-stabilized, α-β alloy whose primary commercial application is turbine engine rotating components. The alloy can be fabricated into all forging product types, although closed die forgings and rings predominate. Ti-6246 is commercially fabricated on all types of forging equipment. Turbine engine disks are frequently produced using hot die or isothermal forging techniques, resulting in near-net closed die forgings with reduced final machining.

Ti-6246 is a reasonably forgeable alloy with unit pressures (flow stresses), forgeability, and crack sensitivity similar to the α-β alloy Ti-6Al-4V. The final microstructure of forgings is developed by thermomechanical processing in forging manufacture tailored to achieve specific microstructural and mechanical-property objectives. Thermomechanical processes use combinations of subtransus and/or supra-transus forging followed by subtransus thermal treatments to fulfill critical mechanical-property criteria.

Final thermal treatments for forgings include two-step practices of solution treatments followed by quenching and aging. Solution treatment is subtransus at 870 to 900 °C (1600 to 1650 °F), followed by water or oil quenching and/or fan air cooling for thin sections. Aging is conducted at 535 to 620 °C (1000 to 1150 °F). Subtransus thermomechanical processes (forging and thermal treatment) for forgings achieve equiaxed α (20 to 40%) in transformed β matrix microstructures that enhance strength, ductility, and particularly low-cycle fatigue properties. Supra-transus thermomechanical processes (β forging followed by subtransus thermal treatments) achieve transformed, Widmanstätten α microstructures that enhance creep and fracture-related properties such as fatigue-crack growth resistance (Chakrabarti, Ref 1).

Conventional Forging. The objectives in forging Ti-6246 are to obtain the final forging shape and desired final microstructure at least cost. Conventional, subtransus (α + β) forging thermomechanical processes are currently most widely used in commercial engine disk forging manufacture. To achieve conventional equiaxed α structures, subtransus reductions of 50 to 75%, accumulated through one or more forging steps, are required. Supra-transus (β) forging may be used in early forging operations, including upsetting and open die preforming, to reduce unit pressures and ease forging fabrication. However, higher temperature initial forging operations must be followed by sufficient subtransus reduction to achieve the desired predominately equiaxed α structure. Conventionally forged Ti-6246 is then subtransus solution treated, quenched, and aged as noted above.

Recently developed (Chakrabarti, Ref 2) alternative subtransus or supra-transus thermomechanical processes for Ti-6246 forgings are now used for selected disk applications to achieve superior combinations of low-cycle fatigue and fatigue-crack growth resistance. Alternative subtransus processes involve solution treatment of conventionally forged product in close proximity to the transus, e.g., β_t–5 to 25 °C (10–45 °F), followed by oil or water quenching followed by aging at 425 to 650 °C (800 to 1200 °F).

Beta Thermomechanical Processes. Successful practices for forgings include controlled β forging processes followed by the above noted subtransus solution treatment and age. The β forging thermomechanical processes are particularly well suited to isothermal or hot die forging technology.

Ti-6246: Forging process temperatures

Process	Metal temperature °C	°F
Conventional forging	845-915	1550-1675
Beta forging	955-1010	1750-1850

Note: See "Technical Note 4: Forging" for recommended die temperatures.

Ti-6246: Effect of TMP options

Although the introduction of acicular α from TMP options can improve fracture-related properties of Ti-6246, the benefit of α morphology manipulation is not as great as in Ti 6Al 4V.

Forge	Heat treat	TYS MPa	UTS MPa	EL, %	RA, %	K_{Ic} MPa$\sqrt{m}$	da/dN(a) 10^{-8} m/cycle	LCF life(b) 10^3 cycles
α/β	STA	1082	1192	12	29	31	2.48	91
α/β	Duplex STA	1054	1262	7	11	51	2.03	81
α/β	Duplex βA	1037	1183	8	10	67	1.52	25
β	STA	1072	1183	12	17	74	1.78	37
α/β	HSTA(c)	1089	1176	11	21	58	1.90	66

(a) da/dN at ΔK of 11 MPa$\sqrt{m}$, with R = 0.05 and F = 20 cpm in air. (b) Low cycle strain controlled, R = 0.0, F = 20 cpm, triangular waveform in air. (c) HSTA - high temperature solution treatment and age. Source: L.J. Bartlo, H.B. Bomberger, and S.R. Seagle, "Deep-Hardenable Titanium Alloy," AFML-TR-73-122, AFML, 1973 May for the data shown here. See also U.S. Patent 4,975,125 (Chakrabarti, et al., Dec 4, 1990) for alternative TMP.

Beta forging requires subtransus reductions (e.g., 20 to 50%) in early forging (blocker die) stages followed by a controlled, single β forging step, that achieves 30 to 50% reductions. Beta forging requires careful control of forging process conditions, particularly preheat times at temperature, to avoid excessive prior β grain growth. Beta forged Ti-6246 is then subtransus heat treated as noted above.

Hot die and/or isothermal forging techniques are important commercial methods for fabrication of Ti-6246 rotating turbine engine disks to reduce final component cost (from less machining) and/or improve final component microstructural and property uniformity through improved control of forging process conditions. The axisymmetric shapes and designs of such engine components are very well suited to these forging methods. Isothermal forging of Ti-6246 disks is frequently accomplished in a single forging step from bar or billet stock, under carefully controlled supra- or subtransus metal and die temperatures, levels of strain, and strain-rate profiles. Hot die forging, where die temperature approaches but is not equivalent to metal temperature, is also used to reduce unit pressures, enhance forgeability, and produce more sophisticated final shapes in fewer forging operations.

References

1. A.K. Chakrabarti et al., "Microstructure and Mechanical Property Optimization Through Thermomechanical Processing in Ti-6-4 and Ti-6-2-4-6 Alloys," *6th World Conference on Titanium*, Vol 3, P. Lacombe, R. Tricot, and G. Beranger, Ed., Les Editions de Physiques, Paris, 1988, p 1339-1344
2. A.K. Chakrabarti et al., U.S. patent 4,975,125, Dec 4, 1990

Other Fabrication Methods

Forming. Ti-6246 may be formed similar to Ti-6Al-4V alloy, although the reported bend properties are somewhat inferior. The room-temperature minimum bend radius, for Ti-6246 ranges between 3.5 and 6.0 for solution treated or duplex annealed sheet. Hot forming and sizing of sheet may be accomplished in the 595 to 705 °C (1100 to 1300 °F) range using the usual titanium forming techniques. If hot forming is performed in the 595 to 705 °C (1100 to 1300 °F) range, stress relief annealing would not ordinarily be required; limited cold forming is possible. Depending on property requirements, stress relief in the 595 to 705 °C (1100 to 1300 °F) range is satisfactory.

Superplasticity. Flow stresses and strain rate sensitivity of α + β preforms indicate superplastic behavior (High-Temperature Deformation of Ti-6246, *Titanium, Science and Technology*, Vol 2, DGM, 1985, p 745-752).

Machinability of Ti-6246 in the as-forged condition is similar to that of annealed Ti-6Al-4V and Ti-6Al-6V-2Sn. In the solution treated and aged condition, it is similar to that of Ti-6Al-6V-2Sn in the same type of heat treated condition.

Welding. Ti-6246 is very difficult to weld. Recommended filler metal is the same as the base alloy.

Heat Treatment

Ti-6246 may be used in a number of heat treated conditions, which can be categorized as anneals or solution treatment and aging (see tables). As previously described, optimum combinations of strength, ductility, and toughness in forgings are obtained by superimposing the heat treatments on processing schedules, which result in a microstructure having about 10 vol% equiaxed primary α and a relatively coarse transformed β matrix. If relatively high fabrication temperatures are used, solution heat treatment on the low side of the range can be used. If moderate α-β fabrication temperatures are used, double solution treatments—the first at a high temperature, the second at about 845 °C (1550 °F)—should result in the desirable structure.

Ti-6246: Typical heat treatment conditions

Heat treatment	Temperature °C	Temperature °F	Time, h	Cooling method
Stress relief	595-705	1100-1300	0.25-4	Air or slow cool
Solution treating(a)	815-925	1500-1700	1	Water or oil quench
Aging(b)	580-605	1075-1125	4-8	AC
Overaging	>650	>1200	...	...

(a) See separate table for specific temperatures by product form. (b) The most commonly used aging temperature is 595 °C (1100 °F).

Ti-6246: Annealing treatments

Treatment	Temperature °C	Temperature °F	Cooling method
Solution anneal (SA)	815-925	1500-1700	AC
Duplex anneal			
First stage (SA)	815-925	1500-1700	AC
Second stage (age)	540-730	1000-1350	AC
Triplex anneal			
SA stage	815-925	1500-1700	AC
First age(a)	540-730(a)	1000-1350(a)	AC
Second age(a)	540-730(a)	1000-1350(a)	AC

(a) First aging higher than the second

Ti-6246: Hardenability

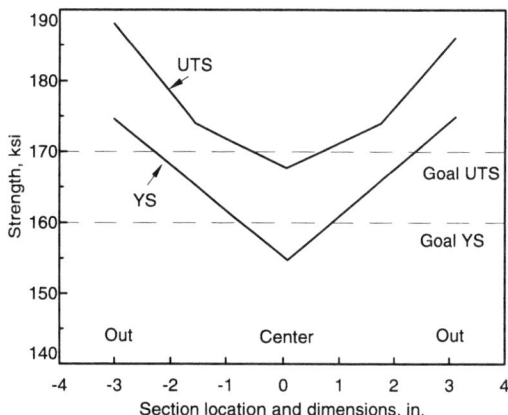

Variation of room-temperature tensile properties across a 150 mm (6 in.) section; solution treated 1 h at 870 °C (1600 °F), WQ + aged 8 h, 595 °C (1100 °F), AC.
Source: R.A. Wood and R.J. Favor, *Titanium Alloys Handbook*, MCIC-HB-02, Battelle Columbus Laboratories, 1972

Ti-6246: Solution treatment conditions

Product	Temperature °C	Temperature °F	Duration, h	Cooling method
Common	870	1600	Up to 1	WQ or OQ
Sheet	870	1600	0.25	Quench(a)
Forgings	845-900	1550-1650	Up to 1	WQ or OQ

(a) Solution treatment of sheet may be followed by a stabilization exposure of 0.25 h at 720 to 730 °C (1325 to 1350 °F) with an air cool prior to aging. Sheet may be age hardened to optimum properties in as little as 2 h at 595 °C (1100 °F) (air cooled).

Ti-6246: Effect of aging on transverse tensile properties of sheet

Time, h	Aging treatment(a) Temperature °C	Aging treatment(a) Temperature °F	UTS MPa	UTS ksi	TYS MPa	TYS ksi	Elongation, %
1	540	1000	1648	239	1455	211	3
16			1579	229	1420	206	4
1	595	1100	1489	216	1379	200	6
2			1517	220	1386	201	3
16			1461	212	1365	198	6
1	650	1200	1406	204	1337	194	7
16			1296	188	1261	183	9
1	705	1300	1255	182	1220	177	6
16			1158	168	1110	161	12

(a) Solution treatment for 0.25 h, 870 °C (1600 °F), air cooled. Source: R.A. Wood and R.J. Favor, *Titanium Alloys Handbook*, MCIC-HB-02, Battelle Columbus Laboratories, 1972

Ti-6Al-4V

Ti64, 6Al-4V, 6-4
UNS Number: R56400 (normal interstitial grade); R56401 (extra-low interstitial grade); R56402 (filler metal)

Introduction

Ti-6Al-4V presently is the most widely used titanium alloy, accounting for more than 50% of all titanium tonnage in the world. To date, no other titanium alloy threatens its dominant position. The aerospace industry accounts for more than 80% of this usage. The next largest application of Ti-6Al-4V is medical prostheses, which accounts for 3% of the market. The automotive, marine, and chemical industries also use small amounts of Ti-6Al-4V (see the section "Applications" in this introduction).

Chemistry

Effects of Impurities and Alloying. Ti-6Al-4V is produced in a number of formulations. Depending on the application, the oxygen content may vary from 0.08 to more than 0.2% (by weight), the nitrogen content may be adjusted up to 0.05%, the aluminum content may reach 6.75%, and the vanadium content may reach 4.5%. The higher the content of these elements, particularly oxygen and nitrogen, the higher the strength. Conversely, lower additions of oxygen, nitrogen, and aluminum will improve the ductility, fracture toughness, stress-corrosion resistance, and resistance against crack growth.

ELI Grade. Ti-6Al-4V is available in ELI (extra-low interstitial) grades with high damage-tolerance properties, especially at cryogenic temperatures. The principal compositional characteristics are low oxygen and iron contents.

Ti-6Al-4V-Pd is a grade that has palladium additions (about 0.2 wt% Pd) for enhanced corrosion resistance. Sumitomo Titanium has produced this grade.

Product Forms

Ti-6Al-4V is available in wrought, cast, and powder metallurgy (P/M) forms, with wrought products accounting for more than 95% of the market. The properties of these various product forms will vary depending on their interstitial contents and thermal-mechanical processing. Processing methods and characteristics of Ti-6Al-4V are discussed in a separate section entitled "Processing."

Wrought Product Forms. Ti-6Al-4V is available in a wide range of wrought product forms (see Table).

The aircraft industry uses all wrought product forms. Forgings are used to fabricate various attachment fittings, and sheet and plate are used to fabricate numerous clips, brackets, skins, bulkheads, etc. Extrusions are not used extensively, but are used for parts such as wing chords and other parts with long, constant cross-sections. Wire is used to produce the numerous fasteners found on wings. Ti-6Al-4V tubing has been used for components such as torque tubes. In missile and space applications, Ti-6Al-4V has been used for rocket engine and motor cases, pressure vessels, wings, and generally in applications where weight is critical.

Castings. Ti-6Al-4V of the same chemistry as for wrought materials has excellent casting characteristics. However, the high reactivity of titanium in the molten state requires suitable casting technology and has limited the number of titanium foundries. In general terms, the mechanical and fatigue properties of castings will be slightly lower

Ti-6Al-4V: Wrought products

Product	Size and weight ranges	Price comparison(a)
Ingot	3200 to 13,600 kg (7000 to 30,000 lb)	...
Billet	Normally 100 mm (4 in.) diam to about 355 mm (14 in.) diam or square. Billets up to 5000 lb have been sold, but this is not necessarily the upper limit.	...
Bar	Cross-sections up to 0.4 × 0.4 m (16 × 16 in.)	...
Die forging	From <0.5 kg to >1300 kg (<1 lb to >3000 lb)	Ti, $30/lb; Al, $10/lb; stainless steel, $8/lb
Plate	Typical dimensions: Thickness: 5 to 75 mm (0.1875 to 3 in.); Width: 915 and 1220 mm (36 and 48 in.); Length: 1.8, 2.4, and 3 m (72, 96, and 120 in.)	...
Sheet	Typical dimensions: Thickness: 0.4 to 4.75 mm (0.016 to 0.187 in.); Width: 915 and 1220 mm (36 and 48 in.); Length: 1.8, 2.4, and 3 m (72, 96, and 120 in.)	Ti, $16/lb; stainless steel, $3/lb; Al, $2-4/lb; Inco 718, $10/lb
Tube	Specialty item	...
Forged block	Available in a wide range of sizes, with maximum size related to ingot size and the amount of work that can be imparted to the forged block	Ti, $8/lb; stainless steel and Al, $2.50-3/lb
Extrusion	From circle sizes of about 25 to 760 mm (1 to 30 in.) diam. Minimum thickness of about 3 mm (1/8 in.) for small circle sizes, and about 13 mm (1/2 in.) for large circle sizes	Ti, $13-15/lb; 300 series stainless steel, $3-4/lb; 15-5PH, $4-5/lb; 13-8PH, $9-12/lb; Al, $2-4/lb
Wire	Typically manufactured in sizes ranging from 0.28 to 12.2 mm (0.011 to 0.480 in.) diam	1/4 in. wire: Ti, $26/lb; A283, $6/lb; stainless steel, $7.50/lb; 8740, $1/lb; Al 7075, $2.30/lb

(a) Due to its lower density, 1 lb of titanium is approximately 1.7 to 1.8 more material by volume than 1 lb of steel or nickel-base alloy.

than for the wrought product, but fracture toughness, stress-corrosion resistance, and crack growth resistance will be comparable to that of annealed wrought Ti-6Al-4V.

Ti-6Al-4V castings are about two to three times the cost of superalloy castings. The cost effectiveness depends on the size, complexity, and the number of castings. Major application is in aerospace and marine use. Other industrial applications include well-logging hardware for the petroleum industry, special automotive parts, boat deck hardware, and medical implants.

P/M Products. The major reason for using the P/M products is to produce near-net shapes. Most of the titanium P/M effort has been with Ti-6Al-4V because it is the most widely used alloy having a large data base for comparison.

The two general approaches to titanium P/M are the blended elements (BE) method and the prealloyed (PA) approach. Blended elemental powders cost $6 to $30/lb, depending on the chloride impurity content. Chloride content ranges from 10 to 2000 ppm; powders with low amounts of chloride are more expensive. High-chloride powders cannot be used if good fatigue strength is needed.

Blended elemental P/M parts of Ti-6Al-4V are currently in production for aerospace and nonaerospace applications where full wrought properties are not required and where there is economic advantage to this approach. (See the section "Applications" for examples). The PA approach, however, has been less successful in establishing a commercial market. Prealloyed powders are not cold compactable, and their cost is high ($60 to $100/lb).

Product Condition/Microstructure

Wrought Ti-6Al-4V is most commonly used in the mill-annealed condition, where it has a good combination of strength, toughness, ductility, and fatigue. Its minimum yield strength may vary from 760 to 895 MPa (110 to 130 ksi), depending on processing, heat treatment, section size, and chemistry (primarily oxygen).

Almost all titanium castings are hot isostatically pressed (HIP'ed) to heal internal porosity not linked to the surface. This minimizes the amount of weld repair, improves the consistency of mechanical properties, and enhances the fatigue performance. Ti-6Al-4V castings are generally used in the ($\alpha + \beta$)-annealed condition, although some special heat treatments can be used to enhance the performance of the castings in comparison to the β anneal.

Annealed Condition. Although Ti-6Al-4V is commonly used in the mill-annealed condition, other annealing treatments are also utilized. For example, annealing just above the beta transus, or annealing high in the $\alpha + \beta$ phase field, creates a Widmanstätten or lamellar $\alpha + \beta$ microstructure with good fracture toughness, stress-corrosion resistance, and crack growth resistance, and creep resistance. Recrystallization annealing of wrought alloy improves tensile ductility and fatigue performance.

Solution Treated, Quenched and Aged Ti-6Al-4V Alloy. Solution-treated and quenched alloys may either have an acicular α'-martensite structure (quenched from above β-transus) or mixed $\alpha' + \alpha$ microstructure (quenched from 900-1000 °C) or mixed $\alpha'' + \alpha$ microstructure (quenched from 800-900 °C), of which the latter is exceptionally soft and ductile. They serve as starting conditions for subsequent aging treatments. Quenched components contain high residual stresses which may not be fully relieved upon aging at low temperatures. Such components may distort during machining. Ti-6Al-4V has excellent hardenability in sections up to about 25 mm (1 in.) thick; strengths as high as 1140 MPa (165 ksi) may be achieved at aging temperatures between 300 and 600 °C.

Applications

Designed primarily for high strength at low to moderate temperatures, Ti-6Al-4V has a high specific strength (strength/density), stability at temperatures up to 400 °C (750 °F), and good corrosion resistance. Cost continues to be an inhibitive factor for its use in industries where weight and corrosion are not critical considerations.

Aerospace Applications. Ti-6Al-4V was developed in the 1950s and initially used for compressor blades in gas turbine engines. Today, wrought Ti-6Al-4V is used extensively for turbine engine and airframe applications. Engine components include blades, discs, and wheels. Wrought forms are used for airframe components. In addition, the superplastic characteristics of fine-grained, equiaxed Ti-6Al-4V is being used increasingly for aerospace applications. It also has good diffusion-bonding characteristics, which, when combined with superplastic forming, enables the fabrication of very complex structures. Significant amounts of superplastically formed and diffusion-bonded structures are used today, particularly for military aircraft.

Aerospace casting applications include the range from major structural components weighing more than 135 kg (300 lb) each to small switch guards weighing less than 30 g (1 oz).

Ti-6Al-4V castings are used extensively for large, complex housings in the turbine engine industry. They are used in a variety of airframe applications, including cargo-handling equipment, flow diverters, torque tubes for brakes, and helicopter rotor hubs. In missile and space applications, they are used for wings, missile bodies, optical sensor housings, and ordnance. Also, Ti-6Al-4V castings are used to attach the main external fuel tanks to the Space Shuttle and the boosters to the external tanks.

Surgical Implants. Wrought Ti-6Al-4V is a useful material for surgical implants because of its low modulus, good tensile and fatigue strength, and biological compatibility. It is used for bone screws and for partial and total hip, knee, elbow, jaw, finger, and shoulder replacement joints. Where fatigue properties are not an issue, the cast alloy also has had minor use as an implant product.

Automotive Applications. In the automotive industry, wrought Ti-6Al-4V is used in special applications in high-performance and racing cars where weight is critical, usually in reciprocating and rotating parts, such as valves, valve springs, connecting rods, and rocker arms. It also has been used for drive shafts and suspension springs. Cast Ti-6Al-4V also has had minor use in automotive applications.

Marine applications of wrought Ti-6Al-4V include armaments, sonar equipment, deep-submergence applications, hydrofoils, and capsules for telephone-cable repeater stations. Casting applications include water-jet inducers for hydrofoil propulsion and seawater ball valves for nuclear submarines.

P/M Applications. The BE method produces a product with less than full density that can be as strong as wrought material, but that generally has lower ductility, toughness, and fatigue strength. Process modifications can improve these latter properties, even making them comparable to wrought, but they increase costs.

The BE approach has found a niche for the production of near-net-shape components or of low-cost preforms for subsequent processing, such as forging. Applications include sidewinder missile housing, missile fins, connecting rods, turbine blade preforms, hex stock preforms for fittings, nuts, mirror hubs, and lens housings.

High cost has thus far limited potential applications of PA technology to, for the most part, the manufacture of critical aerospace components. A number of demonstration parts are now flying in the F-15 and the F-18 airplanes, but none is made on a production basis. The increased demand for titanium aluminides in higher-temperature applications is creating interest in PA technology of P/M titanium.

Ti-6Al-4V and equivalents: specifications and compositions

Specification	Designation	Description	Al	C	Fe	H	N	O	V	OT	Other
UNS	R56400	Weld Wir	5.5-6.75	0.1	0.4	0.015	0.05	0.2	3.5-4.5		bal Ti
UNS	R56401		6						4		bal Ti
UNS	R56402	Fill Met	5.5-6.75	0.04	0.15	0.005	0.012	0.1	3.5-4.5		bal Ti
Europe											
AECMA prEN2517	Ti-P63	Sh Strp Plt Bar Ann	5.5-6.75	0.08	0.3	0.01	0.05	0.2	3.5-4.5	0.4	bal Ti
AECMA prEN2530		Bar Ann	5.5-6.75	0.08 max	0.3 max	0.0125 max	0.05 max	0.2 max	3.5-4.5	0.4 max	OE 0.1 max; bal Ti
AECMA prEN2531		Frg Ann	5.5-6.75	0.08 max	0.3 max	0.0125 max	0.05 max	0.2 max	3.5-4.5	0.4 max	OE 0.1 max; bal Ti
AECMA prEN3310		Frg NHT	5.5-6.75	0.08 max	0.3 max	0.0125 max	0.05 max	0.2 max	3.5-4.5	0.4 max	OE 0.1 max; bal Ti
AECMA prEN3311		Bar Ann	5.5-6.75	0.08 max	0.3 max	0.0125 max	0.05 max	0.2 max	3.5-4.5	0.4 max	OE 0.1 max; bal Ti
AECMA prEN3312		Frg Ann	5.5-6.75	0.08 max	0.3 max	0.0125 max	0.05 max	0.2 max	3.5-4.5	0.4 max	OE 0.1 max; bal Ti
AECMA prEN3313		Frg NHT	5.5-6.75	0.08 max	0.3 max	0.0125 max	0.05 max	0.2 max	3.5-4.5	0.4 max	OE 0.1 max; bal Ti
AECMA prEN3314		Bar STA	5.5-6.75	0.08 max	0.3 max	0.0125 max	0.05 max	0.2 max	3.5-4.5	0.4 max	OE 0.1 max; bal Ti
AECMA prEN3315		Frg STA	5.5-6.75	0.08 max	0.3 max	0.0125 max	0.05 max	0.2 max	3.5-4.5	0.4 max	OE 0.1 max; bal Ti
AECMA prEN3352		Inv Cast Ann HIP	5.5-6.75	0.1 max	0.3 max	0.015 max	0.05 max	0.22 max	3.5-4.5	0.4 max	OE 0.1 max; bal Ti
AECMA prEN3353		Bar Wir STA	5.5-6.75	0.08 max	0.3 max	0.0125 max	0.05 max	0.2 max	3.5-4.5	0.4 max	OE 0.1 max; bal Ti
AECMA prEN3354		Sh Ann	5.5-6.75	0.08 max	0.3 max	0.0125 max	0.05 max	0.2 max	3.5-4.5	0.4 max	OE 0.1 max; bal Ti
AECMA prEN3355		Ext Ann	5.5-6.75	0.08 max	0.3 max	0.0125 max	0.05 max	0.2 max	3.5-4.5	0.4 max	OE 0.1 max; bal Ti
AECMA prEN3456		Sh Strp Ann	5.5-6.75	0.08 max	0.3 max	0.0125 max	0.05 max	0.2 max	3.5-4.5	0.4 max	OE 0.1 max; bal Ti
AECMA prEN3457		Frg NHT	5.5-6.75	0.08 max	0.3 max	0.0125 max	0.05 max	0.2 max	3.5-4.5	0.4 max	OE 0.1 max; bal Ti
AECMA prEN3458		Bar Wir Ann	5.5-6.75	0.08 max	0.3 max	0.0125 max	0.05 max	0.2 max	3.5-4.5	0.4 max	OE 0.1 max; bal Ti
AECMA prEN3464		Plt Ann	5.5-6.75	0.08 max	0.3 max	0.0125 max	0.05 max	0.2 max	3.5-4.5	0.4 max	OE 0.1 max; bal Ti
AECMA prEN3467		Remelt NHT	5.5-6.75	0.08 max	0.3 max	0.0125 max	0.05 max	0.2 max	3.5-4.5	0.4 max	OE 0.1 max; bal Ti
France											
AIR 9183	T-A6V	Bar Rod Frg	5.5-7	0.08	0.25	0.012	0.07	0.2	3.5-4.5		bal Ti
AIR 9184	T-A6V	Blt	5.5-7	0.08 max	0.25 max	0.12 max	0.07 max	0.2 max	3.5-4.5		bal Ti
Germany											
DIN	3.7164	Sh Strp Plt Bar Frg Ann	5.5-6.75	0.08	0.3	0.0125-0.015	0.05	0.2	3.5-4.5	0.4	bal Ti
DIN	3.7264	Cast Ann	5.5-6.75	0.1	0.3	0.015	0.05	0.2	3.5-4.5	0.4	bal Ti
DIN 17850	3.7165	Plt Sh Strp Rod Wir Ann	5.5-6.75	0.08	0.3	0.015	0.05	0.2	3.5-4.5		bal Ti

(continued)

Ti-6Al-4V and equivalents: specifications and compositions (continued)

Specification	Designation	Description	Al	C	Fe	H	N	O	V	OT	Other
DIN 17851	3.7165	Sh Plt Strp Rod Wir Ann	5.5-6.75	0.08	0.3	0.015	0.05	0.2	3.5-4.5		bal Ti
DIN 17860	3.7615	Sh Strp	5.5-6.75	0.2 max	0.3 max	0.015 max	0.05 max		3.5-4.5		bal Ti
DIN 17862	3.7615	Rod	5.5-6.75	0.08 max	0.3 max	0.015 max	0.05 max	0.2 max	3.5-4.5		bal Ti
DIN 17864	3.7615	Frg	5.5-6.75	0.08 max	0.3 max	0.015 max	0.05 max	0.2 max	3.5-4.5		bal Ti
Russia											
GOST 19807-74	VT6S	Sh Plt Strp Foil Rod Ann	5.3-6.8	0.08	0.25	0.007	0.05	0.015	3.5-4.5	0.3	Zr 0.3; Si 0.15; bal Ti
OST 1.90000-70	VT6	Sh Plt Strp Foil Rod Frg Ann	5.5-7	0.1	0.3	0.015	0.05	0.2	4.2-6	0.3	Si 0.15; bal Ti
OST 1.90060-72	VT6L	Cast	5-6.5	0.1	0.3	0.015	0.05	0.15	3.5-4.5	0.3	Zr 0.3; Si 0.15; W 0.2; bal Ti
Spain											
UNE 38-723	L-7301	Sh Plt Strp Bar Ex Ann	5.5-6.75	0.1	0.3	0.125	0.05	0.2	3.5-4.5	0.4	bal Ti
UNE 38-723	L-7301	Sh Plt Strp Bar Ex HT	5.5-6.75	0.1	0.3	0.125	0.05	0.2	3.5-4.5	0.4	bal Ti
UK											
BS 2TA.10		Sh Strp HT	5.5-6.75		0.3	0.01					V. 3.5-4.5; Ti 88.18 max; O+N=0.25
BS 2TA.11		Bar	5.5-6.75		0.3	0.01	0.05	0.2			V. 3.5-4.5; Ti 88.18 max;
BS 2TA.12		Frg	5.5-6.75		0.3	0.01	0.05	0.2	3.5-4.5		Ti 88.19 max;
BS 2TA.13		Frg HT	5.5-6.75		0.3	0.01		0.2	3.5-4.5		Ti 88.18 max;
BS 2TA.28		Wir Frg HT Quen	5.5-6.75		0.3	0.01	0.05	0.2	3-5		Ti 88.19 max;
BS 3531 Part 2		Srg Imp	5.5-6.75	0.08 max	0.3 max	0.015 max		0.2 max	3.5-4.5		bal Ti
BS TA.56		Plt to 100 mm HT	5.5-6.75		0.3				3.5-4.5		Ti 88.2 max; O+N=0.25
BS TA.59		Sh Strp	5.5-6.75	0.08 max	0.3 max	0.0125 max			3.5-4.5		N+O=0.25; bal Ti
DTD 5303		Bar Ann	5.5-6.75	0.2 max	0.3 max	0.0125 max	0.05 max		3.5-4.5		bal Ti
DTD 5313		Frg Ann	5.5-6.75		0.3 max	0.01 max	0.05 max	0.2 max	3.5-4.5		bal Ti
DTD 5323		Frg Ann	5.5-6.75		0.3 max	0.015 max	0.05 max	0.2 max	3.5-4.5		bal Ti
DTD 5363		Cast	5.5-6.75		0.3 max	0.15 max	0.05 max	0.25 max	3.5-4.5		N+O=0.27; bal Ti
USA											
AMS 4905A		ELI Plt	5.6-6.3	0.05 max	0.25 max	0.0125 max	0.03 max	0.12 max	3.6-4.4	0.4 max	Y 0.005 max; OE 0.1 max; bal Fe
AMS 4905A		Plt Beta Ann	5.6-6.3	0.05	0.25	0.0125	0.03	0.12	3.6-4.4	0.4	Y 0.005; bal Ti
AMS 4906		Sh Strp	5.5-6.75	0.08 max	0.3 max	0.0125 max	0.05 max	0.2 max	3.5-4.5	0.4 max	Y 0.005 max; bal Ti
AMS 4907D		ELI Sh Strp Plt Ann	5.5-6.5	0.08	0.25	0.0125	0.05	0.13	3.5-4.5	0.3	Y 0.005; bal Ti
AMS 4911F		Sh Strp Plt Ann	5.5-6.75	0.08	0.3	0.015	0.05	0.2	3.5-4.5	0.4	bal Ti
AMS 4920		Frg Ann	5.5-6.75	0.1	0.3	0.0125	0.05	0.2	3.5-4.5	0.4	Y 0.005; bal Ti
AMS 4928K		Bar Wir Frg Bil Rng Ann	5.5-6.75	0.1	0.3	0.0125	0.05	0.2	3.5-4.5	0.4	bal Ti
AMS 4930C		ELI Bar Wir Frg Bil Rng Ann	5.5-6.5	0.08	0.25	0.0125	0.05	0.13	3.5-4.5	0.4	Y 0.005; bal Ti
AMS 4931		ELI Bar Frg Bil Rng	5.5-6.5	0.08	0.25	0.0125	0.03	0.13	3.5-4.5	0.4	Y 0.005; bal Ti
AMS 4934A		Ex Rng STA	5.5-6.75	0.1	0.3	0.0125	0.05	0.2	3.5-4.5	0.4	Y 0.005; bal Ti
AMS 4935E		Ex Rng Ann	5.5-6.75	0.1	0.3	0.0125	0.05	0.2	3.5-4.5	0.4	Y 0.005; bal Ti
AMS 4954D		Fill met gas-met/W-arc weld	5.5-6.75	0.05	0.3	0.015	0.03	0.18	3.5-4.5	0.4	Y 0.005; bal Ti
AMS 4956B		ELI Fill Met Wir	5.5-6.75	0.03	0.15	0.005	0.012	0.08	3.5-4.5	0.1	Y 0.005; bal Ti
AMS 4965E		Bar Frg Rng STA/Mach Press ves	5.5-6.75	0.08	0.3	0.0125	0.05	0.2	3.5-4.5	0.4	Y 0.005; bal Ti
AMS 4967F		Bar Frg Rng Mach/STA Press ves	5.5-6.75	0.08	0.3	0.0125	0.05	0.2	3.5-4.5	0.4	Y 0.005; bal Ti
AMS 4985A		Cast Ann	5.5-6.75	0.1	0.3	0.015	0.05	0.2	3.5-4.5	0.4	Y 0.005; bal Ti
AMS 4991A		Cast Ann	5.5-6.75	0.1	0.3	0.015	0.05	0.2	3.5-4.5	0.4	Y 0.005; bal Ti
AMS 4993A		Powd Sint Nuts	5.5-6.75	0.1	0.3	0.01	0.05	0.3	3.5-4.5	0.4	Si 0.05; Na 0.15; Cl 0.15; bal Ti

(continued)

Ti-6Al-4V and equivalents: specifications and compositions (continued)

Specification	Designation	Description	Al	C	Fe	H	N	O	V	OT	Other
AMS 4996		Bill Powd Ann	5.5-6.75	0.1	0.3	0.0125	0.04	0.13-0.19	3.5-4.5	0.2	Mo 0.1 max; Sn 0.1 max; Zr 0.1 max; Mn 0.1 max; Cu 0.1 max; Y 0.001; bal Ti
AMS 4996		ELI Bil	5.5-6.75	0.1 max	0.3 max	0.0125 max	0.04 max	0.13-0.19	3.5-4.5	0.2 max	Y 0.001 max; OE 0.1 max; bal Ti
AMS 4998		ELI Powd	5.5-6.75	0.1 max	0.3 max	0.0125 max	0.04 max	0.13-0.19		0.2 max	Y 0.001 max; OE 0.1 max; bal Ti
AMS 4998		Powd	5.5-6.75	0.1	0.3	0.012	0.04	0.13-0.18	3.5-4.5	0.2	Mo 0.1 max; Sn 0.1 max; Zr 0.1 max; Mn 0.1 max; Cu 0.1 max; Y 0.001; bal Ti
ASTM B 265	Grade 5	Sh Strp Plt Ann	5.5-6.75	0.1	0.4	0.015	0.05	0.2	3.5-4.5	0.4	bal Ti
ASTM B 348	Grade 5	Bar Bil Ann	5.5-6.75	0.1	0.4	0.0125	0.05	0.2	3.5-4.5	0.4	bal Ti
ASTM B 367	Grade C-5	Cast	5.5-6.75	0.1	0.4	0.015	0.05	0.25	3.5-4.5	0.4	bal Ti
ASTM B 381	Grade F-5	Frg Ann	5.5-6.75	0.1	0.4	0.0125	0.05	0.2	3.5-4.5	0.4	bal Ti
ASTM F136		ELI Wrought Ann for Surg Imp	5.5-6.5	0.08	0.25	0.012	0.05	0.13	3.5-4.5		bal Ti
ASTM F467-84	Grade 5	Blt Scr Std	5.5-6.75	0.1 max	0.4 max	0.0125 max	0.05 max	0.2 max	3.5-4.5		bal Ti
ASTM F468-84		Blt Scr Std	5.5-6.75	0.1 max	0.4 max	0.0125 max	0.05 max	0.2 max	3.5-4.5		bal Ti
AWS A5.16-70	ERTi-6Al-4V	Weld fill met	5.5-6.75	0.05	0.25	0.008	0.02	0.15	3.5-4.5		bal Ti
AWS A5.16-70	ERTi-6Al-4V-1	ELI Fill Met Wir Rod	5.5-6.75	0.04	0.15	0.005	0.012	0.1	3.5-4.5		bal Ti
MIL A-46077D		Weld armor plt Ann	5.5-6.5	0.04	0.25	0.0125	0.02	0.14	3.5-4.5	0.4	bal Ti
MIL F-83142A	Comp 6	Frg Ann	5.5-6.75	0.08	0.3	0.015	0.05	0.2	3.5-4.5	0.4	bal Ti
MIL F-83142A	Comp 6	Frg HT	5.5-6.75	0.08	0.3	0.015	0.05	0.2	3.5-4.5	0.4	bal Ti
MIL F-83142A	Comp 7	ELI Frg Ann	5.5-6.5	0.08	0.2-0.25	0.0125	0.05	0.13	3.5-4.5	0.3	bal Ti
MIL F-83142A	Comp 7	ELI Frg HT	5.5-6.5	0.08	0.25	0.0125	0.05	0.13	3.5-4.5	0.3	bal Ti
MIL T-81556A	Code AB-1	Ex Bar Shp Ann	5.5-6.75	0.08	0.3	0.0125	0.05	0.2	3.5-4.5	0.4	bal Ti
MIL T-81556A	Code AB-1	EX Bar Shp STA	5.5-6.75	0.08	0.3	0.0125	0.05	0.2	3.5-4.5	0.4	bal Ti
MIL T-81556A	Code AB-2	ELI Ext Bar Ann	5.5-6.5	0.08	0.25	0.0125	0.05	0.13	3.5-4.5	0.3	bal Ti
MIL T-81915	Type III Comp A	Cast Ann	5.5-6.75	0.08	0.3	0.015	0.05	0.2	3.5-4.5	0.4	bal Ti
MIL T-9046J	Code AB-1	Sh Strp Plt Ann	5.5-6.75	0.08	0.3	0.0125	0.05	0.2	3.5-4.5	0.4	bal Ti
MIL T-9046J	Code AB-1	Sh Strp Plt STA	5.5-6.75	0.08	0.3	0.0125	0.05	0.2	3.5-4.5	0.4	bal Ti
MIL T-9046J	Code AB-2	ELI Sh Strp Plt Ann	5.5-6.5	0.08	0.25	0.0125	0.05	0.13	3.5-4.5		bal Ti
MIL T-9047G		Bar Bil STA	5.5-6.75	0.08	0.3	0.015	0.05	0.2	3.5-4.5	0.4	Y 0.005; bal Ti
MIL T-9047G		ELI Bar Bil Ann	5.5-6.5	0.08	0.25	0.0125	0.05	0.13	3.5-4.5	0.3	Y 0.005; bal Ti
MIL T-9047G	MIL-T-9047G	Bar Bil Ann	5.5-6.75	0.08	0.3	0.015	0.05	0.2	3.5-4.5	0.4	Y 0.005; bal Ti
SAE J467		ELI	6.18	0.023	0.22	0.008	0.026	0.097			bal Ti

Ti-6Al-4V commercial equivalents: compositions

Specification	Designation	Description	Al	C	Fe	H	N	O	V	OT	Other
France											
Ugine	UTA6V	Sh Strp Plt Bar Frg Ann	5.5-6.75	0.08	0.3	0.015	0.07	0.2	3.5-4.5		bal Ti
Ugine	UTA6V	Sh Strp Plt Bar Frg STA	5.5-6.75	0.08	0.3	0.015	0.07	0.2	3.5-4.5		bal Ti
Germany											
Deutsche T.	LT 31	Ann	5.5 6.5	0.08	0.25	0.013	0.07	0.2	3.5-4.5		bal Ti
Fuchs	TL64	Frg	6						4		bal Ti
Fuchs	TL64 ELI	ELI Frg	6						4		bal Ti
Thyssen	Contimet AlV 64	Plt Bar Frg Ann	5.5-6.75	0.1	0.3	0.015	0.05	0.2	3.5-4.5		bal Ti
Thyssen	Contimet AlV 64	Plt Bar Frg STA	5.5-6.75	0.1	0.3	0.015	0.05	0.2	3.5-4.5		bal Ti
Thyssen	Contimet AlV 64 ELI	ELI Plt Bar Frg Pip Ann	5.5-6.75	0.06	0.15	0.013	0.05	0.13	3.5-4.5		bal Ti
Japan											
Daido	DAT 5	Rod Bar Rng Frg Ann	5.5-6.75	0.1	0.3	0.015		0.05	3.5-4.5		bal Ti
Daido	DAT 5	Rod Bar Rng Frg STA	5.5-6.75	0.1	0.3	0.015	0.05	0.2	3.5-4.5		bal Ti
Daido	DT 5	Rod Bar Frg Rng STA	5.5-6.75	0.1	0.3	0.015	0.05	0.2	3.5-4.5		bal Ti

(continued)

Ti-6Al-4V commercial equivalents: compositions (continued)

Specification	Designation	Description	Al	C	Fe	H	N	O	V	OT	Other
Kobe	KS6-4	Plt Sh Wir Bar Ann	5.5-6.75		0.3	0.0125	0.05	0.2	3.5-4.5		bal Ti
Kobe	KS6-4ELI	ELI Plt Sh Ann	5.5-6.5		0.25	0.0125	0.05	0.13	3.5-4.5		bal Ti
Toho	64AT		5.5-6.75	0.03	0.4	0.0125	0.05	0.2	3-5		bal Ti
Toho	64AT	STA	5.5-6.75	0.03	0.4	0.0125	0.05	0.2	3-5		bal Ti
UK											
IMI	IMI 318	Sh Rod Bar Bil Wir Plt Ex	6						4		bal Ti
USA											
Oremet	Ti-6Al-4V										
RMI	6Al-4V-ELI	ELI Bar Bil Ex Plt Sh Strp Ann	5.5-6.5	0.08	0.25	0.01-0.015		0.13	3.5-4.5		bal Ti
RMI	6Al-4V	Bar Bil Ex Plt Sh Strp Wir Ann	5.6-6.75	0.08	0.25	0.01-0.015	0.05	0.2	3.5-4.5		bal Ti
RMI	6Al-4V	Bar Bil Ex Plt Sh Strp Wir STA	5.6-6.75	0.08	0.25	0.01-0.015	0.05	0.2	3.5-4.5		bal Ti
Tel.Allvac	Allvac 6-4	Ann	6	0.18					4		bal Ti
TIMET	TIMETAL 6-4		5.5-6.75	0.1 max	0.4 max	0.015 max	0.05 max	0.2 max	3.5-4.5		bal Ti
TIMET	TIMETAL 6-4 ELI	ELI Ann	5.5-6.5	0.08 max	0.25 max	0.0125 max	0.05 max	0.13 max	3.5-4.5		bal Ti
TIMET	TIMETAL 6-4 STA	Ing Bil Bar Plt Sh Str STA	5.5-6.75	0.1	0.4	0.015	0.05	0.2	3.5-4.5		bal Ti

Phases and Structures

As an alpha beta alloy, Ti-6Al-4V may have different volume fractions of alpha and beta phases, depending on heat treatment and interstitial (primarily oxygen) content. Beta is stable at room temperature only if it is enriched with more than 15 wt.% vanadium. Such enrichment is obtained when the alloy is slow cooled or annealed below about 750 °C (1400 °F). Slow cooled Ti-6Al-4V contains up to about 90 vol% of the alpha phase.

In addition, Ti-6Al-4V can acquire a large variety of microstructures with different geometrical arrangements of the alpha and beta phases, depending on the particular thermomechanical treatment. These different alpha "morphologies" and microstructures can be roughly classified into three different categories: lamellar, equiaxed, or a mixture of both (bimodal).

Lamellar structures can be readily controlled by heat treatment. Slow cooling into the two-phase region from above the β transus leads to nucleation and growth of the α-phase in plate form starting from β-grain boundaries. The resulting lamellar structure is fairly coarse and is often referred to as plate-like alpha. Air cooling results in a fine needle-like alpha phase referred to as acicular alpha. Certain intermediate cooling rates develop Widmanstätten structures. Water-quenching from the β-phase field followed by annealing in the (α + β)-phase region leads to a much finer lamellar structure. Quenching from temperatures greater than 900 °C (1650 °F) results in a needle-like hcp martensite (α′), while quenching from the 750 to 900 °C (1380 to 1650 °F) temperature range produces an orthorhombic martensite (α″).

Equiaxed microstructures are obtained by extensive (>75% reduction) mechanical working the material in the (α + β)-phase field, where the breakup of lamellar alpha into equiaxed alpha depends on the exact deformation procedure (e.g., see figure). Subsequent annealing at about 700 °C (1300 °F) produces the so-called "mill-annealed" microstructure, which gives microstructure that is very dependent upon previous working. A more reproducible equiaxed structure is obtained by a recrystallization anneal of 4 h at 925 °C (1700 °F) followed by slow cooling. The resulting structure is fairly coarse with an α-grain size of about 15-20 μm.

Bimodal type microstructures consist of isolated primary α-grains in a transformed beta matrix. These microstructures are best obtained by a 1 h anneal at 955 °C (1750 °F) followed by water quenching (or more commonly an air cool) and aging at 600 °C (1100 °F). The resulting primary α-grain size is usually about 15-20 μm in such "solution treated and aged" microstructures. Aging below 650 °C (1200 °F) can also produce precipitates of alpha in previously quenched beta.

Interface Phase. An FCC crystal structure, often called the "interface phase," is frequently observed in thin foils for electron microscopy at lamellar boundaries between alpha phase and beta phase. The interpretation of the interface phase is still controversial. It has been reported to be a phenomenon of thin foil preparation while others claim that it occurs in bulk material as well.

Typical lattice parameters of alpha phase in slow cooled or aged Ti-6Al-4V alloy are $a = 0.2925 \pm 0.0002$ nm, $c = 0.4670 \pm 0.0005$ nm. The lattice parameters vary only slightly as a function of heat treatment because the composition of alpha is relatively constant. The room-temperature

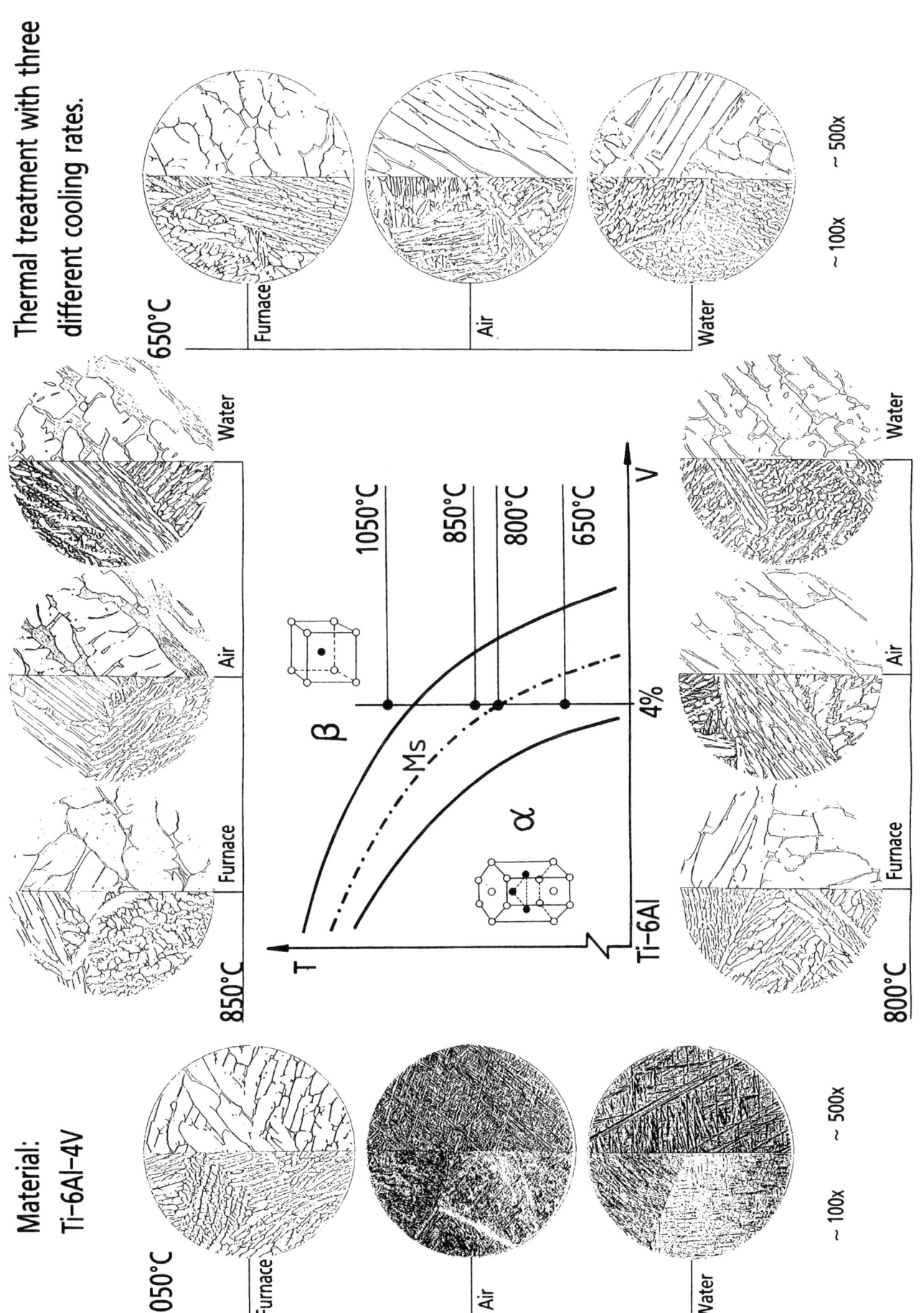

Thermal treatment with three different cooling rates.

Material: Ti-6Al-4V

Ti-6Al-4V: Lattice parameters after quenching from various temperatures

Quench temperature (°C)	Phase α			Phase β
	a nm	c nm	c/a	a nm
950	0.29313	0.46813	0.15969	...
900	0.29320	0.46798	0.15962	...
850	0.29288	0.46750	0.15963	...
800	0.29281	0.46729	0.15960	...
750	0.29259	0.46711	0.15966	...
725	0.29261	0.46706	0.15962	0.32530
700	0.29255	0.46709	0.15966	0.32510
650	0.29243	0.46706	0.15972	0.32295
600	0.29254	0.46711	0.15969	0.22250
550	0.29254	0.46716	0.15970	0.32160
500	0.29245	0.46718	0.15974	0.32145
450	0.29246	0.46705	0.15970	0.32150
400	0.29240	0.46718	0.15977	0.32120

Source: R. Castro and L. Seraphin, *Memoires Scientifique Rev. Metallurg.*, Vol LXIII, No.12, 1966, p 1036

Ti-6Al-4V: Beta transus vs oxygen content

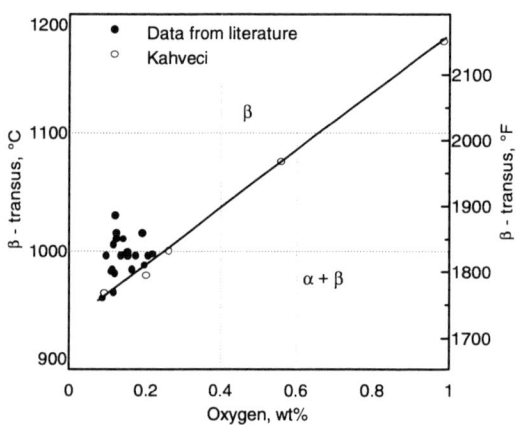

Source: A. Kahveci, Thesis, Case Western Reserve University

Ti-6Al-4V: Lattice parameter of β phase

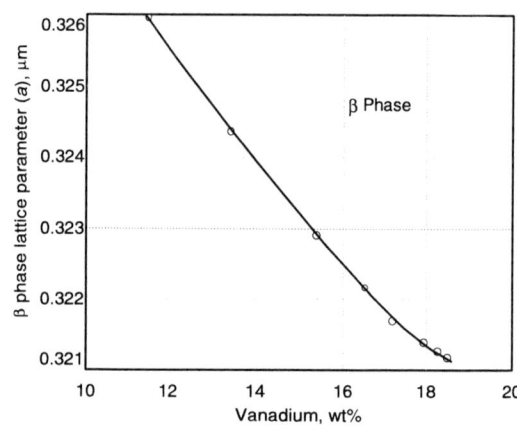

Variation in lattice parameter beta phase in Ti-6Al-4V alloy after quenching from various heat treatment temperatures.
Source: R. Castro and L. Seraphin, *Memoires Scientifique Rev. Metall.*, Vol 63, No. 12, 1966, p 1036

Ti-6Al-4V: Fraction of phase constituents after quenching

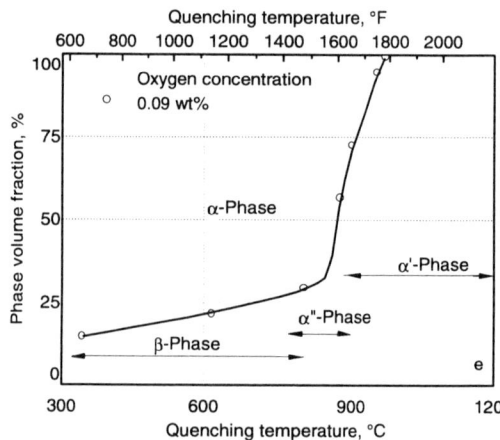

lattice parameter of the beta phase in furnace cooled Ti-6Al-4V has been measured as $a = 0.319$ nm ± 0.001 nm (G. Welsch et al., *Met. Trans. A*, Vol 8A, 1977, p 169-177). Increasing vanadium concentrations decrease the lattice parameter of beta, while interstitial elements increase the lattice parameters of alpha by occupying a fraction of the octahedral interstitial sites. For oxygen concentrations less than 6 at.%, oxygen increases the lattice parameters of alpha as follows (S. Anderson et al., *Acta Chem. Scand.*, Vol 11, 1957, p 1641):

$$a = a_0 + 7 \times 10^{-5} \text{ nm/at.\% O}$$

$$c = c_0 + 36 \times 10^{-5} \text{ nm/at.\% O}$$

Transformation Structures

Hexagonal close packed martensite (α') is obtained by quenching from above 900 °C (1650 °F) and has an acicular or sometimes fine-lamellar microstructure. It is related crystallographically to the alpha phase and has similar lattice parameters as the alpha phase.

Orthorhombic martensite (α'') is a rather soft martensite that forms during quenching of beta phase with 10 ± 2 wt% vanadium. This occurs when Ti-6Al-4V is quenched from temperatures between 750 and 900 °C (1380 and 1650 °F). The α'' martensite can also form as a stress-induced product by straining metastable beta.

Omega (ω) **Precipitation.** Oxygen suppresses omega formation, and it does not occur in Ti-6Al-4V alloy of commercial purity. If the β phase is highly enriched with vanadium (over 15 wt%), ω-precipitates might occur during low-tempera-

Ti₃Al Precipitation

ture aging (200 to 350 °C) or during cooling through the same temperature range. However, no such precipitation has been reported in Ti-6Al-4V.

The formation of Ti₃Al (α_2) has been experimentally verified in Ti-6Al-4V containing less than 0.2 wt% oxygen (*Met. Trans.*, Vol 8A, 1977, p 169-177), and it occurs in Ti-6Al-4V at aging temperatures from 500 to 600 °C (930 to 1100 °F) when oxygen concentrations (still within specification limits for Ti-6Al-4V) are increased. Oxygen is known to restrict the solubility limit of aluminum in the alpha phase of titanium, thus enhancing the likelihood of Ti₃Al formation. Vanadium also restricts aluminum solubility in alpha titanium. However, no quantitative dependency on oxygen concentration has been established for the $\alpha/(\alpha + \alpha_2)$ solvus line.

The Ti₃Al precipitates are known to promote coarse planar glide on $\{10\bar{1}0\}$ prismatic planes, and one should be able to produce a predominance of either planar pyramidal glide or coarse planar prismatic glide in the α-phase of a Ti-6Al-4V alloy with sufficient oxygen by the choice of aging treatment. At high aging temperatures, ordering should not be expected because oxygen would have a high jump frequency. (G. Welsch and W. Bunk, *Met. Trans. A*, Vol 13A, 1982, p 889-899). On the other hand, an alloy with a very low oxygen concentration (ELI grade) should exhibit only a predominance of prismatic slip, regardless of aging treatment.

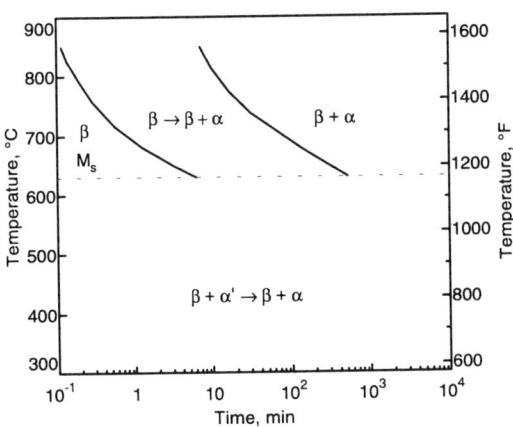

Ti-6Al-4V: Time-temperature-transformation diagram

Solution annealed at 1020 °C (1868 °F), and quenched directly to reaction temperatures.
Source: L.E. Tanner, "Time-Temperature-Transformation Diagrams of the Titanium Sheet-Rolling-Program Alloys," DMIC Report 46G, Battelle Memorial Institute, Columbus, OH, October 1959

Damping Characteristics

G.E. Welsch, Case Western Reserve University
Y.T. Lee, Korea Institute of Machinery and Materials

The solution treatment and quenching condition of Ti-6Al-4V determines its starting (high) value of damping capacity. Aging causes a rapid decrease in the damping capacity. Changes in damping capacity occur at aging temperatures as low as 200 °C (400 °F), which is well below the temperature at which interstitial or substitutional elements can diffuse in alpha or beta titanium by thermal activation. Although the damping capacity decreases rapidly upon aging, it is slightly sensitive to oxygen content.

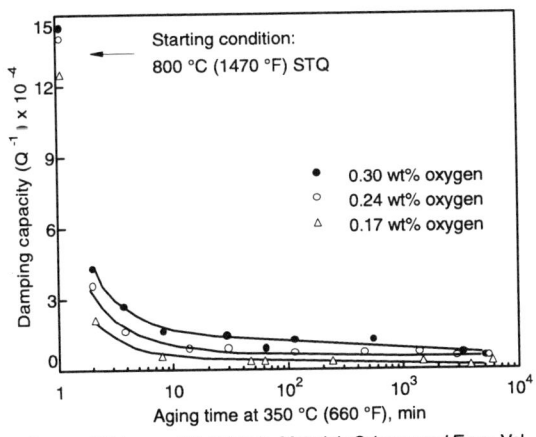

Ti-6Al-4V: Damping capacity with different oxygen contents

Source: Y.T. Lee and G. Welsch, *Materials Science and Engr.*, Vol A128, 1990, p 77-89

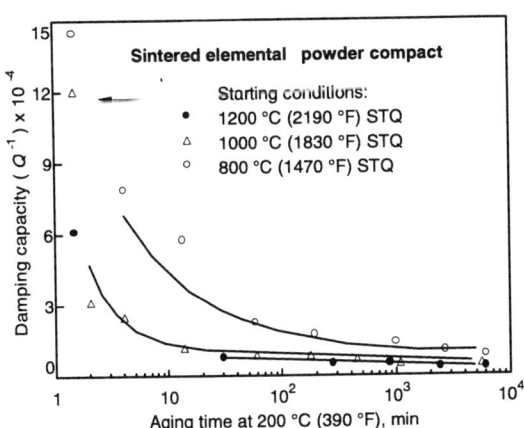

Ti-6Al-4V: Damping capacity (Q^{-1}) as a function of aging time

Source: Y.T. Lee and G. Welsch, *Materials Science and Engr.*, Vol A128, 1990, p 77-89

Ti-6Al-4V: Attenuation and ultrasonic noise level for rolled bar in several heat treated conditions

Heat treatment	Test direction	Attenuation (dB per mm sound-travel)
2 h 700 °C (1300 °F)/AC	L	0.11
	LT	0.09
10 h 955 °C (1750 °F)/AC + 2 h 700 °C (1300 °F)/AC	L	0.12
	LT	0.08
1 h 1010 °C (1850 °F)/AC + 2 h 700 °C (1300 °F)/AC	L	0.16
	LT	0.15

Source: "The Role of Ultrasonic Testing for the Production of Titanium Mill Products," K. Rudinger, A. Ismer, D. Fischer, and W. Schula, in *Titanium and Titanium Alloys*, J.C. Williams and A. Belov, Ed., Plenum Press, NY, 1973, p 1231

Ti-6Al-4V: Damping capacity as a function of aging

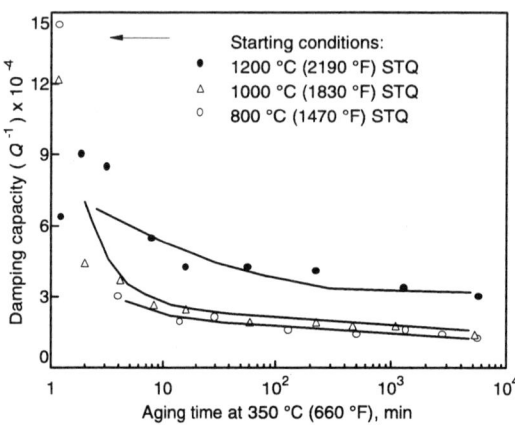

Source: Y.T. Lee and G. Welsch, *Materials Science and Engr.*, Vol A128, 1990, p 77-89

Ti-6Al-4V: Damping as a function of aging

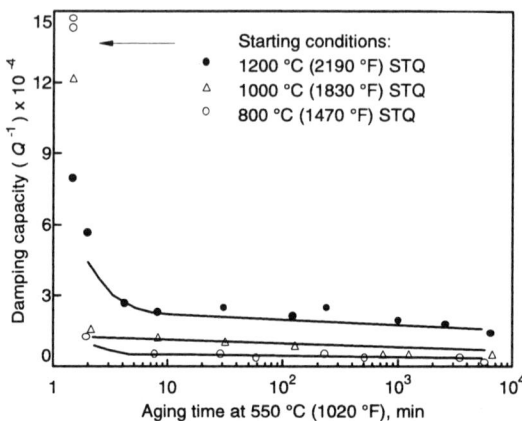

Source: Y.T. Lee and G. Welsch, *Materials Science and Engr.*, Vol A128, 1990, p 77-89

Ti-6Al-4V: Damping capacity, influenced by composition and temperature

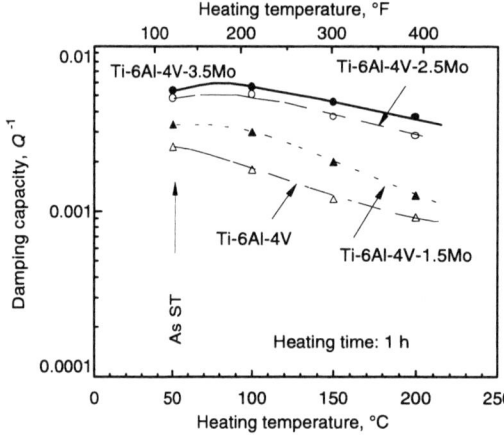

Test specimens 2 mm × 10 mm × 90 mm were fabricated from forged bar.
Source: Y. Ito, Y. Moriguchi, N. Nagai, A. Hiromoto, Y. Takeda, T. Daikoku, and S. Ueda, "Effects of Alloying Elements and Heat Treatment on the Internal Friction of Ti-Al-V Base Alloys," in *Titanium '80*, H. Kimura and O. Izumi, Ed., p 593

Ti-6Al-4V: Attenuation for shear waves at 10 MHz at several temperatures

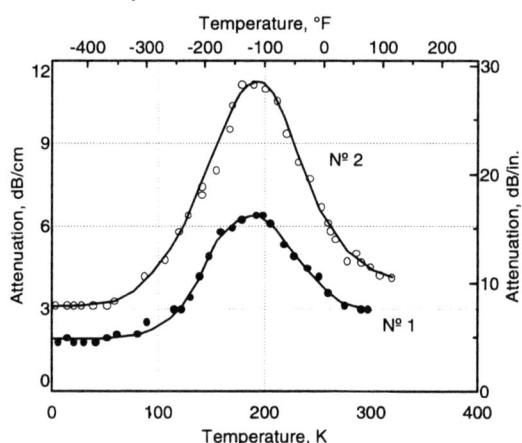

Unannealed samples.
Source: W. Mason and J. Wehr, "Internal Friction and Ultrasonic Yield Stress of the Alloy Ti-6Al-4V," *J. Phys. Chem. Solids 31*, 1970, 1925

Elastic Properties

Young's Modulus

Young's modulus of Ti-6Al-4V is about mid-range for titanium alloys but relatively low compared to other high-strength materials (see table). Tensile strains up to about 0.8% can be accommodated elastically.

Like other titanium alloys, Ti-6Al-4V has a wider range of elastic moduli as compared to other alloy systems. The literature reports values of 100 to 130 GPa (14.5 to 19×10^6 psi) for the Young's modulus of Ti-6Al-4V. In a multi-phase alloy like Ti-6Al-4V, the value of Young's modulus is determined by the moduli of specific phase and their volume fraction. Heat treatment of Ti-6Al-4V thus has an effect on its modulus. More significantly in commercial practice, variations in Young's modulus also accompany variations in texture because of the strong crystallographic anisotropy of the elastic constants of alpha titanium. In general, it has been found that interstitial and substitutional alpha-stabilizing solutes increase the modulus, whereas beta-stabilizing solutes decrease it. Small variations (up to 3%) may occur due to variations in impurity levels (especially oxygen) and alloy concentration. A low value is obtained after solution treatment from about 800 °C (1470 °F). Aging causes the modulus to recover from a low value to a high value.

Ti-6Al-4V: Temperature dependence of Young's modulus

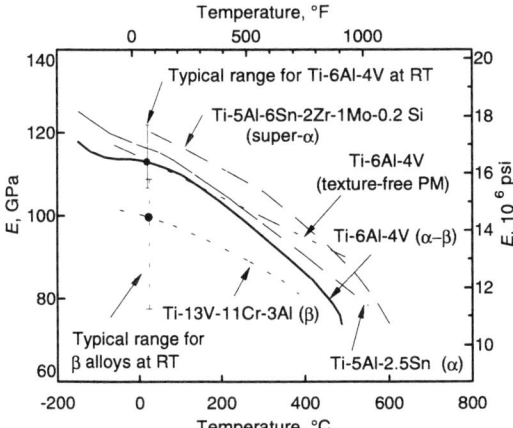

Range of values for an alloy may be due to variations in composition, structure, texture, heat treatment, and testing variables.
Source: *Titanium and Titanium Alloys*, MIL HDBK-697A, 1974, p 36 and texture-free P/M data from G. Welsch, Weller, and Y.T. Lee

Ti-6Al-4V: Elastic modulus for aged sheet from eight heats

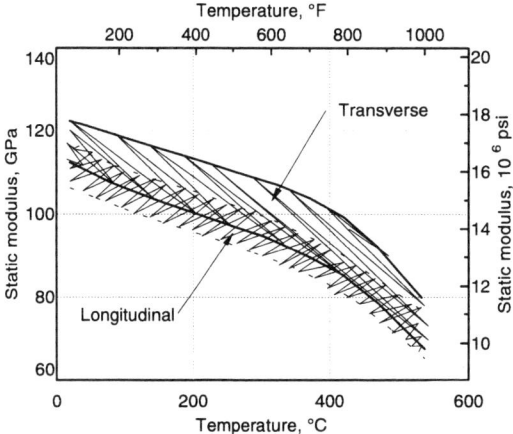

Sheet of 0.5 to 3.2 mm (0.020 to 0.125) thickness was heat treated at 925 °C (1700 °F) for 3 to 20 min, water quenched, and aged at 480 to 510 °C (900-950 °F) for 4 h, air cool.
Source: Data from Lockheed-Georgia Co., 1962, reported in *Aerospace Structural Metals Handbook*, Vol 4, Code 3707, p 26

Ti-6Al-4V: Young's modulus as a function of specimen orientation in textured material

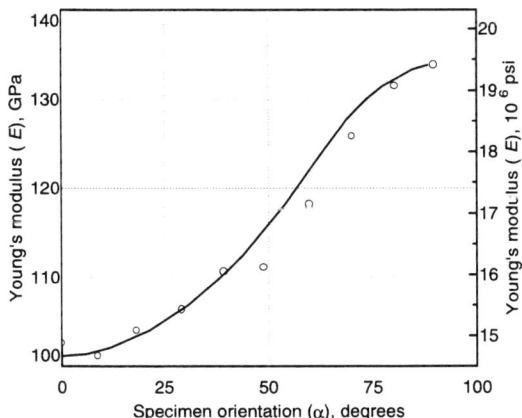

Young's modulus as a function of specimen orientation, α, in textured rolled sheet at room temperature. α is the angle between the rolling direction and the specimen axis.
Source: F. Larson and A. Zarkades, Properties of Textured Titanium Alloys, MCIC Report MCIC-74-20, Metals and Ceramics Information Center, Battelle Memorial Institute, Columbus, OH, June 1974, p 16-23

Ti-6Al-4V: Young's modulus as a function of oxygen content and thermal treatment

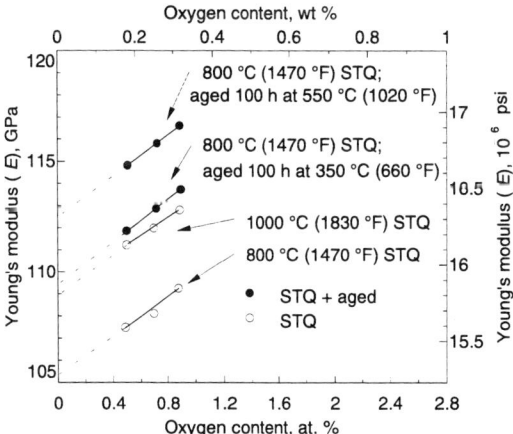

Equiaxed microstructure. STQ = solution treated and quenched.
Source: Y.T. Lee and G. Welsch, Young's Modulus and Damping Capacity of Ti-6Al-4V Alloy as a Function of Heat-Treatment and Oxygen Concentration, *Materials Science and Engineering*, Vol A128, 1990, p 77-89

Ti-6Al-4V: Typical room temperature values of texture-free material

Alloy	Young's modulus, E GPa	psi × 10⁶	Shear modulus GPa	psi × 10⁶	Poisson's ratio, ν	Ref
Ti-6Al-4V	105-116	15.2-16.8	41-45	5.9-6.5	0.26-0.36	1, 2
Duraluminum	72	10.4	26-27	3.8-3.9	0.345	3
Nickel alloys	200-222	29.0-32.2	76-85	11.0-12.3	0.31-0.32	3
Stainless steels	190-215	27.6-31.2	74-83	10.7-12.0	0.27-0.30	3

References: (1) "Young's Modulus and Damping of Ti-6Al-4V as a Function of Heat Treatment and Oxygen Concentration," *Materials Science and Engineering*, A128, 1990, p 77-89. (2) "Elastic Moduli and Tensile and Physical Properties of PM Ti-6Al-4V Alloy," *Metall. Trans. A*, March 1991 issue. (3) *Smithells Metals Reference Book*, Butterworths

Ti-6Al-4V: Young's modulus as a function of aging time

Aging time, min	800 °C (1470 °F) STQ GPa	10⁶ psi	1000 °C (1830 °F) STQ GPa	10⁶ psi	1200 °C (2190 °F) STQ GPa	10⁶ psi
At 200 °C (390 °F)						
0	108.2	15.69	112.4	16.30	113.1	16.40
2	...	...	112.5	16.32	...	...
4	109.3	15.85	112.7	16.35	...	...
15	109.7	15.91	112.9	16.37	...	...
30	...	...	...	...	113.9	16.52
60	110.4	16.01	113.2	16.42	...	...
200	...	...	113.5	16.46	...	...
280	111.0	16.10	...	...	114.4	16.59
420	...	...	113.8	16.51	...	...
1000	111.8	16.22	...	...	114.6	16.62
1200	...	...	113.9	16.52	...	...
2400	...	...	114.1	16.55	...	...
2700	112.3	16.29	...	...	114.9	16.66
6000	112.5	16.32	114.2	16.56	115.0	16.68
At 350 °C (660 °F)						
0	108.1	15.68	112.0	16.24	113.2	16.42
2	110.5	16.03	113.3	16.43	113.9	16.52
4	111.0	16.10	113.6	16.48	114.1	16.55
8	...	...	115.1	16.69	...	...
15	112.3	16.29	115.8	16.80	116.2	16.85
30	112.5	16.32	...	...	...	...
60	...	...	116.8	16.94	117.2	17.00
140	113.4	16.45	...	...	...	...
240	113.9	16.52	117.3	17.01	117.7	17.07
500	114.3	16.58	...	...	...	...
1000	...	...	118.3	17.16	...	...
1200	...	...	...	...	118.1	17.13
1400	115.0	16.68	...	...	...	...
3000	115.7	16.78	...	...	...	...
6000	116.2	16.85	119.2	17.29	118.3	17.16
At 550 °C (1020 °F)						
0	108.1	15.68	112.0	16.24	113.2	16.42
2	113.7	16.49	114.7	16.64	114.4	16.59
4	...	...	...	...	114.7	16.64
8	115.8	16.80	115.2	16.71	114.8	16.65
30	116.6	16.91	116.1	16.84	115.5	16.75
60	116.7	16.93	...	...	...	...
120	117.0	16.97	116.5	16.90	116.0	16.82
240	117.2	17.00	...	...	116.3	16.87
500	117.3	17.01	...	...	...	...
800	...	...	116.9	16.95	...	...
1000	117.4	17.03	...	...	116.6	16.91
1200	...	...	117.0	16.97	...	...
3000	...	...	...	...	117.1	16.98
3500	117.7	17.07	...	...	...	...
6000	117.9	17.10	117.3	17.01	117.4	17.03

Note: All specimens contained 0.24 wt% oxygen. Source: Y.T. Lee and G. Welsch, Young's Modulus and Damping Capacity of Ti-6Al-4V Alloy as a Function of Heat-Treatment and Oxygen Concentration, *Materials Science and Engineering*, Vol A128, 1990, p 77-89

Ti-6Al-4V: Young's modulus as a function of deformation

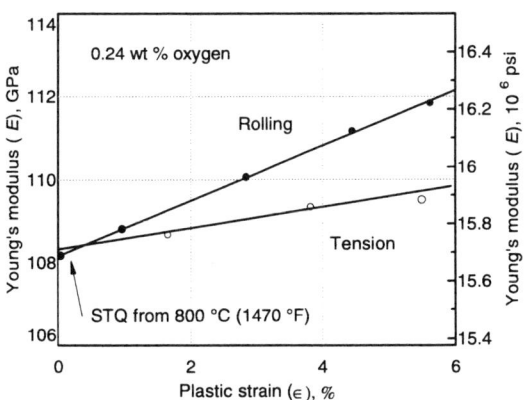

Plastic deformation results in transformation into orthorhombic martensite (α″), which increases the modulus. Texture-free elemental powder compacts sintered at 1260 °C (2300 °F) to more than 99% of theoretical density, then solution heat treated for 1 h at 700 to 1200 °C (1290 to 2190 °F), water quenched, and aged at 200 to 550 °C (390 to 1020 °F). Modulus values were determined by the resonance bar technique at room temperature.
Source: Y.T. Lee and G. Welsch, Young's Modulus and Damping Capacity of Ti-6Al-4V Alloy as a Function of Heat-Treatment and Oxygen Concentration, *Materials Science and Engineering*, Vol A128, 1990, p 77-89

Ti-6Al-4V: Young's modulus at cryogenic temperatures

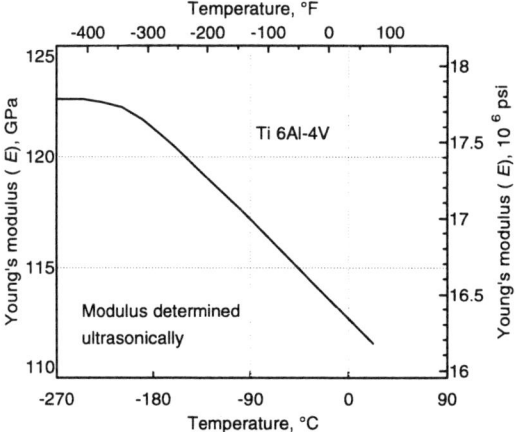

Source: M.J. Donachie, Jr., Ed., *Titanium: A Technical Guide*, ASM International, Materials Park, OH, 1988, p 202

Ti-6Al-4V: Young's modulus as a function of quenching temperature

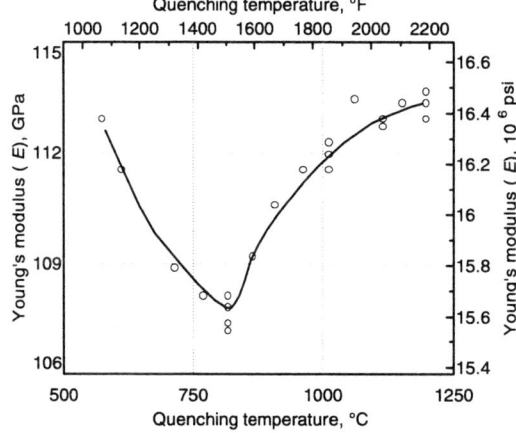

Polycrystalline alloy of equiaxed grain structure. Isotropic values for untextured alloy evaluated from longitudinal resonance vibrations at room temperature. The modulus minimum corresponds to a maximum of retained beta.
Source: Y.T. Lee and G. Welsch, Young's Modulus and Damping Capacity of Ti-6Al-4V Alloy as a Function of Heat-Treatment and Oxygen Concentration, *Materials Science and Engineering*, Vol A128, 1990, p 77-89

Compressive Moduli

The compressive elastic modulus of Ti-6Al-4V closely mimics the material's tensile elastic modulus (Young's modulus), where in all cases (all mill forms, gages and heat treatment conditions), the compressive modulus slightly exceeds that of the material's tensile modulus. A typical room temperature value is generally around 113 GPa (16.4×10^6 psi). This value provides essentially the same stiffness/weight ratio as those for aluminum and ferrous alloys.

Typical compressive elastic moduli and related properties for various materials

Material	Compressive elastic modulus, E_c		Density		E_c/density	
	GPa	10^6 psi	kg/in.³	lb/in.³	M	in.
Ti-6Al-4V	113	16.4	4.430	0.160	2.6×10^6	1.03×10^8
7075-T6	73	10.6	2.795	0.101	2.66×10^6	1.05×10^8
301 SS 1/4 hard	193	28	7.916	0.286	2.49×10^6	0.98×10^6

Ti-6Al-4V: Compressive modulus for highly textured sheet at room temperature

Thickness		Grain	Young's modulus		Compressive yield strength		Compressive elastic modulus	
mm	in.	direction	GPa	10^6 psi	MPa	ksi	GPa	10^6 psi
1.3	0.050	L	102	14.8	871.5	126.4	108	15.6
		T	134	19.4	1355	196.5	136	19.7
		45°	110	15.9	988.1	143.3	113	16.4
1.5	0.060	L	102	14.8	928.8	134.7	108	15.7
		T	130	18.8	1327	192.5	137	19.9
		45°	110	16.0	1001	145.2	120	17.4
2.0	0.080	L	...	...	921.2	133.6	...	...
		T	...	...	1188	172.3	...	...
		45°	...	...	988.1	143.3	...	...
3.2	0.125	L	101	14.7	924.6	134.1	105	15.3
		T	122	17.7	1138	165.1	118	17.1
		45°	112	16.2	986.7	143.1	114	16.6
3.8	0.150	L	102	14.8	896.4	130.0	110	15.9
		T	130	18.9	1223	177.4	134	19.4
		45°	105	15.3	1033	149.8	124	18.0

Source: Boeing Co. data from F. Parkinson, 1972, reported in *Properties of Textured Titanium Alloys*, F. Larson and A. Zarkades, MCIC-74-20, Battelle Columbus Laboratories, Columbus, 1974, p 39

Ti-6Al-4V: Compressive elastic modulus (E_c) for sheet at room and elevated temperatures

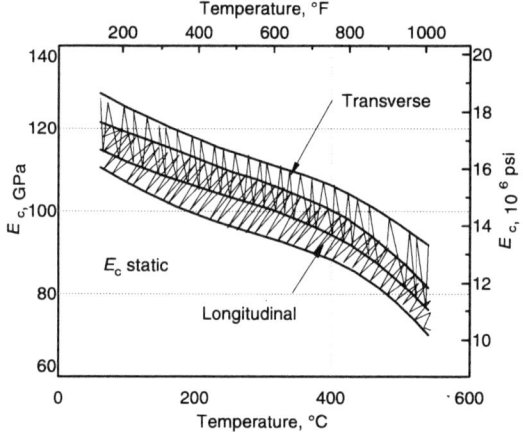

Alloy sheet rolled from 8 different heats, one producer. Sheet thickness varied from 0.5 mm (0.020 in.) to 3.2 mm (0.125 in.). All specimens heat treated at 925 °C (1700 °F) for 3 to 20 min, water quenched, and aged at 482-510 °C (900-950 °F) for 4 h, air cool. Source: "Properties and Processing, Ti-6Al-4V", TIMET Corp., 1986

Ti-6Al-4V: Typical compressive tangent-modulus curves for annealed extrusion at room and elevated temperatures

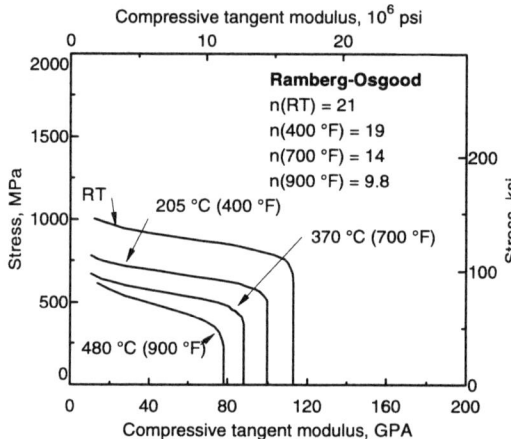

Specimens were exposed to temperature indicated for 1/2 h before testing began. Testing was done in the longitudinal direction.
Source: MIL-HDBK 5E, 1987, p 5-68

Ti-6Al-4V: Compressive modulus of heat treated sheet

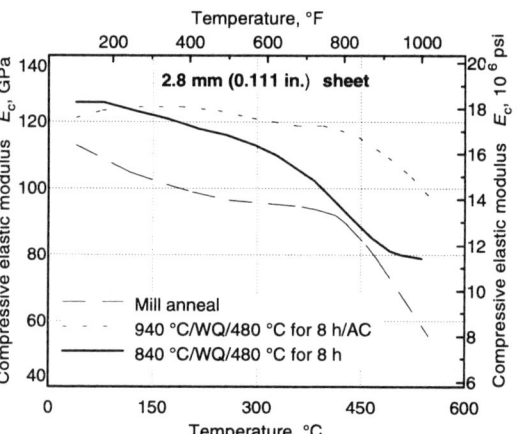

Source: "Report on the Engineering Properties of Commercial Titanium Alloys," M. Mote, R. Harper, and P. Frost, TML Report 92, Battelle Memorial Institute, Columbus, 1958

Shear Moduli and Poisson's Ratio

Poisson's Ratio. Room temperature values of Poisson's ratio range between 0.26 and 0.36 in texture-free Ti-6Al-4V alloy. The variation depends on the phases that have been established during alloy heat treatments. α + β microstructures have a Poisson's ratio around 0.27. Martensitic microstructures (α', α'') have a Poisson ratio around 0.35. The Poisson's ratio for α + β microstructures is virtually independent of temperature between the temperature range from 995 °C to 600 °C (−1760 °F to 1100 °F). There is also a direct correlation with strength for annealed material. As with other properties, it is also texture dependent.

Shear modulus values for Ti-6Al-4V are strongly affected by texture, prior heat treatments, and temperature. For isotropic material, the shear modulus (G) is related to Young's modulus by the relation $G = E/2(\nu - 1)$, where ν is Poisson's ratio.

Ti-6Al-4V: Elastic moduli at several temperatures

Temperature	Young's modulus (E)		Shear modulus (G)		Poisson's ratio(a),
°C	GPa	10^6 psi	GPa	10^6 psi	v
−196	114.9	16.6	46.6	6.75	0.23
0	105.0	15.2	42.8	6.2	0.23
200	94.7	13.7	38.5	5.6	0.23
400	84.1	12.2	34.2	4.95	0.23
600	74.2	10.75	30.0	4.35	0.24
800	62.8	9.1	...	...	...

(a) Poisson ratio was computed from $v = (E/2G) - 1$ (applicable to quasi-isotropic materials), although a weak undetermined texture was reported. Material/test parameters: Alloy in the form of mill annealed (alpha + beta) 9 mm rod had composition in weight % of Al: 5.8, V: 4.5. The resonant bar method was used to determine elastic moduli. Source: "Dynamic Elastic and Damping Properties of Some Practical Ti-Base Alloys," E. Torok and J. Simpson, in *Titanium '80*, p 601

Ti-6Al-4V: Room temperature elastic properties of texture-free blended/elemental sintered alloy in several heat treated conditions

Condition	Young's modulus, E GPa	Poisson's ratio	Shear modulus, G GPa
RT	114.0	0.272	44.8
550 °C STQ	113.0	0.274	44.3
600 °C STQ	111.3	0.275	43.6
700 °C STQ	109.0	0.270	42.9
750 °C STQ	108.2	0.267	42.7
800 °C STQ	107.9	0.265	42.6
850 °C STQ	109.1	0.318	41.4
900 °C STQ	110.4	0.370	40.3
950 °C STQ	111.6	0.360	41.0
1000 °C STQ	111.8	0.352	41.3
1050 °C STQ	113.0	0.351	41.8
1100 °C STQ	112.8	0.350	41.8
1150 °C STQ	113.0	0.351	41.8
1200 °C STQ	113.3	0.352	41.9

Alloy as sintered had composition in weight % of Al: 6.2, C: 0.02, Cl:0.12, Fe: 0.18, N: 0.016, Na: 0.10, O: 0.24, V: 4.1. Materials were solution treated at temperatures indicated and water quenched. RT = as sintered, slow cooled. Source: Personal communication, Y.T. Lee, OLR, Institut für Werkstoff-Forschung, Cologne, 1989

Physical Properties

Electrical Resistivity

Ti-6Al-4V is an alloy with relatively low electrical conductivity or high resistivity compared to other metals. Much of the resistivity is due to the high alloy content in solid solution. Thus, even at low temperatures (20 K), the resistivity is substantial. This translates into low thermal conductivity relative to other metals. The resistivity ranges from about 1.3 microhm meters at 20 K to 1.9 microhm meters at 1000 K. This indicates the residual resistivity from alloy elements, impurity elements and lattice imperfections outweighs the thermal contribution to resistivity over the entire temperature range of 4 to 1200 K. The absolute value of resistivity depends on the distribution of alloy elements and on the degree of cold work. Therefore, aging heat treatments tend to lower the resistivity of solution treated and quenched alloys. Resistivity change can be used to study the temperature and time dependence of aging.

Ti-6Al-4V: Poisson's ratio for textured sheet

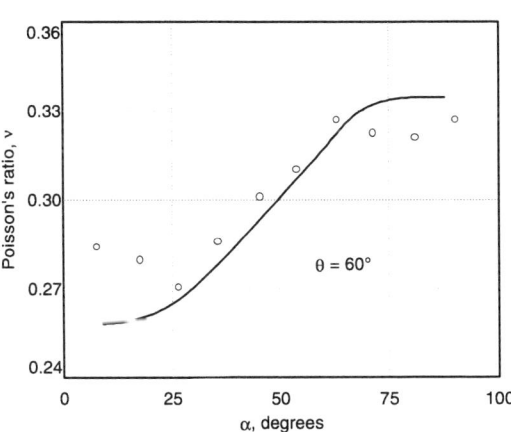

The data are presented along with a solid line which is derived from a constant stress model discussed by Larson and Zarkades (see Source). α is the angle between the rolling direction and specimen axis. θ is the angle between the basal pole of α-titanium and the sheet normal.
Source: F. Larson and A. Zarkades, "Properties of Textured Titanium Alloys," MCIC Report MCIC 74-20, Metals and Ceramics Information Center, Battelle Columbus Laboratories, Columbus, June 1974, p 16-23

498 / Alpha-Beta Alloys

Ti-6Al-4V and Ti-6Al-4V ELI: Electrical resistivity

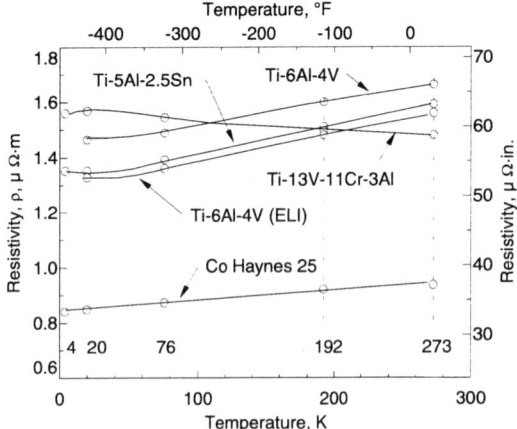

Source: A. Clark, G. Childes, and G. Wallace, "Electrical Resistivity of Some Engineering Alloys at Low Temperatures," *Cryogenics*, Vol 10, 1970, p 295

Ti-6Al-4V and Ti-6Al-4V ELI: Electrical resistivity

Temperature K	Electrical resistivity (microhm meter)	
	Ti-6Al-4V	Ti-6Al-4V ELI
19.65	1.469	1.331
75.75	1.501	1.366
192.4	1.610	1.490
273.	1.675	1.563

Composition of Ti-6Al-4V in weight % was Al: 6.2, Fe: 0.13, V: 4.0, C, H, N present but amount not reported, Ti: rem. Alloy was used in annealed condition; average grain size was 0.027 mm. Average grain size for Ti-6Al-4V ELI was 0.015 mm and Rockwell C hardness was 43. Source: A. Clark, G. Childes, and G. Wallace, "Electrical Resistivity of Some Engineering Alloys at Low Temperatures," *Cryogenics*, Vol 10, 1970, p 295

Magnetic Susceptibility

Ti-6Al-4V: Magnetic susceptibility

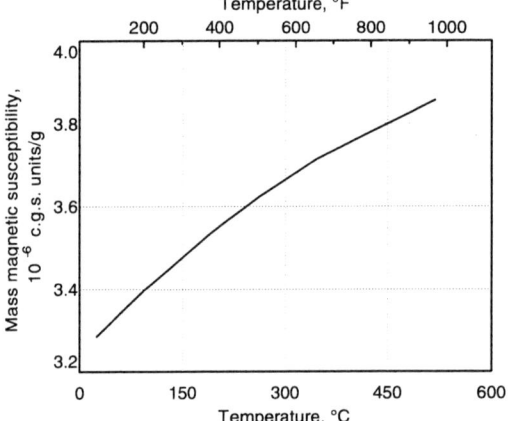

Like other titanium alloys, Ti-6-4 is virtually non-magnetic, as shown by the very low values for mass magnetic susceptibility.

Ti-6Al-4V: Electrical resistivity

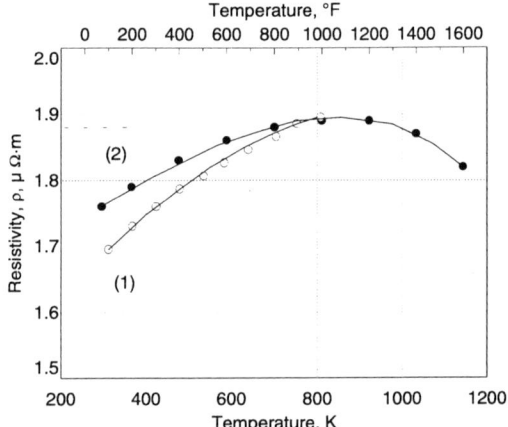

Source: W. Lepkowski and J. Holladay, "Report on the Physical Properties of Titanium and Titanium Alloys," TML 73, Battelle Memorial Institute, Columbus, 1957, p 44 and H. Deem et al., *Trans. Metall. Soc. AIME*, Vol 212, 1958, p 520

Surface Properties

Coefficient of Friction. From theoretical calculations, metals with low theoretical tensile and shear strengths exhibit higher coefficients of friction (μ) than higher-strength materials. Within the class of hexagonal close-packed (hcp) structures, titanium has relatively low values for these properties. Consequently, it is expected that titanium would exhibit high frictional values, which has been demonstrated for titanium sliding against itself in air (μ = 60, *ASLE Trans.*, Vol 27, 1981, p 15-23). Addition of a lubricant, such as polyperfluoroalkylether (PFPE) or kerosene oil, reduces the coefficient of friction and wear damage somewhat, compared with unlubricated conditions, although reaction of titanium surfaces with these lubricants (generally under high-temperature conditions) can reduce lubricant performance.

Ti-6Al-4V: Mechanical and electrochemical properties of the oxide layer of pure Ti and Ti-6Al-4V with pure tantalum and Ti-6Al-4V

	Titanium	TiAl6V4	Tantalum	CoCrMo
Friction coefficient	0.34	0.28	0.3	0.14
Shear resistance, MPa	3.7	0.67	1.09	4.9
Repassivation time, msec	172	104	96	77
Corrosion current density (μA/cm^2)	0.14	0.17	0.22	0.1

All materials are manufactured in accordance to the ISO 5836/1-6 standards for implant materials. Source: Tümmler and Thull, Surface Properties of Titanium and Its Alloys—Mechanical and Electrochemical Investigation, *Titanium Science and Technology*, Vol 2, 1984, p 1335-1342

Ti-6Al-4V: Reduced coefficients of friction for PFPE-lubricated and unlubricated Ti-6Al-4V in air environment

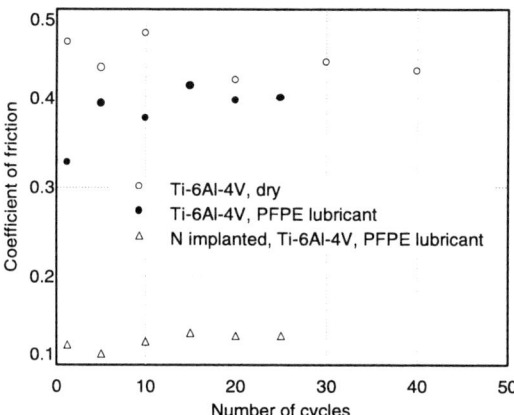

Sliding material, WC-Co; sliding speed, 1 mm/s (0.04 in./s); load, 0.5 kgf.
Source: Independent Research and Development Brochure, Project D-81R, Martin Marietta Astronautics Group, 1990; and C.E. Snyder, Jr., and R.E. Dolle, Jr., Development of Polyperfluoroalkylethers as High Temperature Lubricants and Hydraulic Fluids, *ASME Trans.*, Vol 19 (No. 3), 1975, p 171-180

Surface energy vs Young's modulus at 0.9 T_m

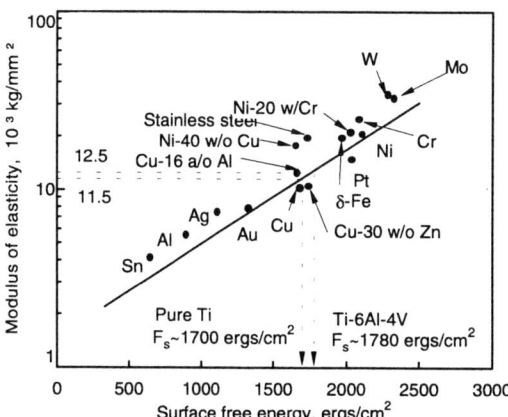

Alloy used as drawn wire, 0.0127 cm (0.005 in.) in diameter, and annealed for 20 h near the test temperature, in helium.
Source: J. Roth and P. Sappayak, "The Surface and Grain Boundary Free Energies of Pure Titanium and the Titanium Alloy Ti-6Al-4V", *J. Mat. Sci. Eng.* 35, 1978, p 187; After L.E. Murr, Interfacial Phenomena in Metals and Alloys, Addison-Wesley, Reading, Mass., 1975

Ti-6Al-4V: Surface and grain boundary free energy

Temperature °C	°F	Surface free energy, (F_s) ergs/cm^2	Grain boundary free energy, ergs/cm^2	γ_{gb}/F_s
1137.3	2080	2095	810	0.39
1176.7	2150	2110	815	0.39
1221.1	2230	2010	815	0.40
1285.0	2345	2050	830	0.40

T.A. Roth, Elevated Temperature Values of the Surface and Grain Boundary Free Energies for Pure Titanium and Ti-6Al-4V, InterAmerican Conference on Materials Technology, Southwest Research Institute, 1978

Other Properties

Ti-6Al-4V: Sound velocity of shock wave

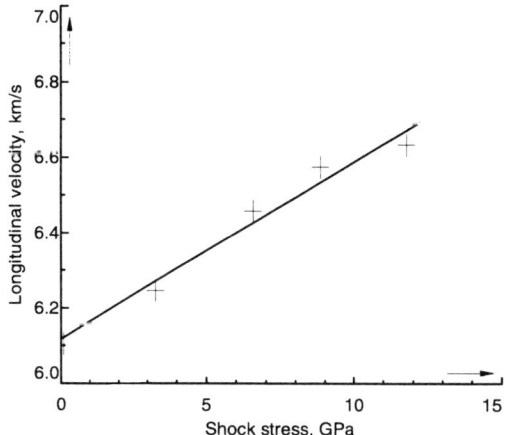

Variation of the longitudinal elastic wave velocity with shock stress in shock loaded Ti-6Al-4V specimens.
Source: Z. Rosenberg and Y. Heybar, "Determination of Changes in the Sound Velocity in Shock Loaded Ti-6Al-4V with In-Material Manganin Gauges," *J. Phys. D: Appl. Phys.*, Vol 16, 1983, L193-L194

Ti-6Al-4V: Sound velocity after quenching

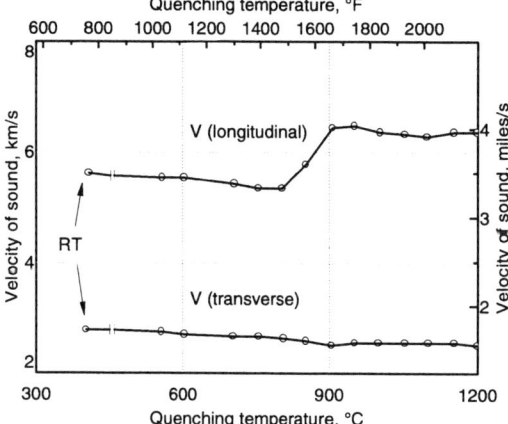

Source: Y.T. Lee

Ti-6Al-4V: Thermoelectric power coupled with Cu

Temperature K	Thermoelectric power of Ti-6Al-4V-Cu couple, microvolts/K	Absolute thermoelectric power of Cu	Absolute thermoelectric power of Ti-6Al-4V
17.18	0.58	+0.35	−0.23
19.91	0.72	+0.45	−0.27
23.35	1.06	+0.55	−0.51
42.37	2.87	+1.05	−1.82
65.26	3.74	+1.40	−2.34
78.76	4.00	+1.40	−2.60
82.59	4.08	+1.40	−2.68
100.6	4.20	+1.30	−2.90
141.3	4.55	+1.10	−3.45
196.1	5.43	+1.40	−4.03
204.1	5.45	+1.40	−4.05
276.3	6.57	+1.70	−4.87
278.0	6.62	+1.70	−4.92

The negative sign for absolute thermoelectric power of Ti-6Al-4V indicates that electrical conduction is primarily by electrons rather than holes. Source: W. Tyler and A. Wilson, Jr., "Thermal Conductivity, Electrical Resistivity, and Thermoelectric Powers of Titanium Alloy RC-130B," reported in *The Physical Properties of Titanium and Titanium Alloys*, W. Lepkowski and J. Holladay, Ed., Battelle, 1957, p 53

Ti-6Al-4V: Reflectivity as a function of the electrode potential for titanium and Ti-6Al-4V

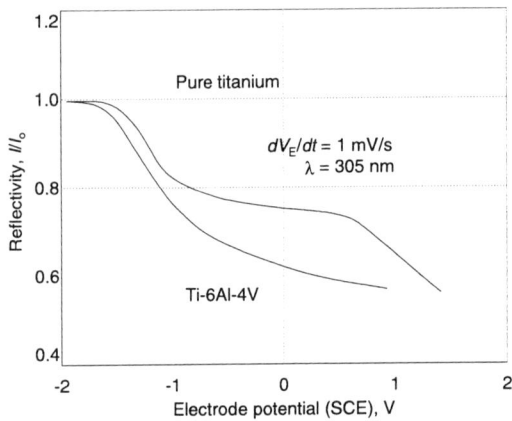

Source: Tümmler and Thull, *op cit*, Titanium Science and Technology, Vol 2, 1985, p 1335-1338

Corrosion/Chemical Properties

Although not as corrosion-resistant as commercially pure titanium alloys, Ti-6Al-4V has excellent corrosion resistance compared to other alloy systems. The exceptional corrosion resistance is due primarily to oxide-film formation. At low temperatures, a thin protective oxide film is present on Ti-6Al-4V in normal atmospheric environments. When penetrated or broken, the oxide layer is quickly reformed, ensuring the protection. The oxide film consists primarily of TiO_2 rutile and grows by inward diffusion of oxygen.

Oxidation

The oxidation behavior of Ti-6Al-4V alloy is similar to that of unalloyed titanium. The reaction rate laws range from logarithmic (at 300-500 °C) to parabolic (500-750 °C) to linear (above 750 °C) with increasing temperatures of oxidation (see figure). Above about 400 °C (750 °F), oxygen begins to be dissolved from the oxide into the underlying metal substrate. This causes surface embrittlement of the alloy. Above about 500 °C (930 °F), the oxide grows into a multi-layered, porous, non-protective scale and causes embrittlement.

Thin surface layers of TiO_2, less than 100 nm thick, can be generated at room temperature by anodizing. This avoids oxygen-embrittlement of the alloys. Such treatment is useful on screws and fasteners or any component subjected to a rubbing action.

Ti-6Al-4V: Oxidation as a function of temperature

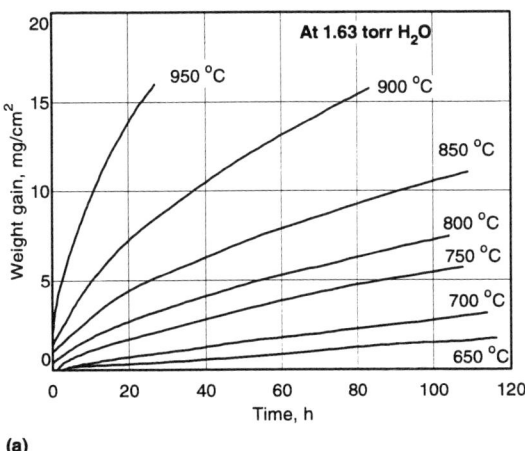
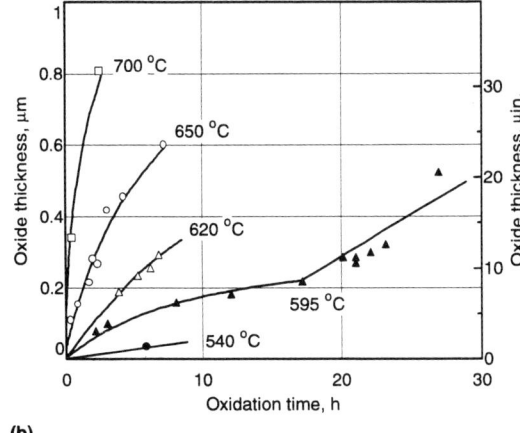

(a) (b)

Oxidation in terms of (a) weight gain and (b) oxide thickness. The oxide forms multilayered structures and spalls off easily. In addition, oxygen is readily taken up into solution in the surface layer of the alloy and may reach 33 at.%. It is the cause of the surface alpha-case with its severe embrittlement.
Source: P. Motte et al., A Comparative Study of the Oxidation with Water Vapor of Pure Titanium and Ti-6Al-4V, Oxid. Met., Vol 10, 1976, p 113-125

General Corrosion

Ti-6Al-4V is highly resistant to natural environments and a great many aqueous chemicals up to at least their boiling temperatures. It is resistant to general corrosion by sea water and brine, oxidizing acids, aqueous chloride solutions, wet chlorine gas and sodium hypochlorite at typical product-operating temperatures. It is, however, vulnerable to reducing acids such as hydrofluoric, hydrochloric, sulfuric, oxalic, formic, and phosphoric acids.

Ti-6Al-4V: General corrosion rates in specific media

Medium	Concentration, wt%	Temperature °F	Corrosion rate, mils/yr	Medium	Concentration, wt%	Temperature °F	Corrosion rate, mils/yr
Acids				**Acids**			
Hydrochloric	2	100	nil-1.2	Sulfuric-nitric	10-90	160	19.0-23.0
Hydrochloric	5	100	2.4-7.2	**Acids (vapors)**			
Hydrochloric	10	100	20.0-24.0	Hydrochloric	37	100	328.0-408.0
Hydrochloric	30	100	208.0-253.0	Nitric	70	150	nil
Nitric	65	Boiling	3.0-5.0	Nitric	70	200	2.0
Nitric (white fuming)	90	180	6.0	Sulfuric	96	100	nil
Phosphoric	10	Room	0.8-2.0	Sulfuric	96	150	nil
Phosphoric	10	170	132.0	Sulfuric	96	200-300	0.4-0.6
Phosphoric	85	Room	16.0-24.0	**Alkalis**			
Sulfuric	2	100	15.6-21.6	Sodium hydroxide	25	Boiling	1.8-2.0
Sulfuric	10	100	38.4-39.6	**Chlorides**			
Acids (inhibited)				Aluminum chloride	25	Boiling	780.0-840.0
5% HCl + 1% CuSO$_4$	...	100	0.6	Barium chloride	25	Boiling	nil
5% HCl + 1% CrO$_3$	...	100	nil	Calcium chloride	28	208	2.7-2.9
10% HCl + 1% CuSO$_4$	...	150	3.0	Cupric chloride	40	Boiling	0.2
10% HCl + 5% CuSO$_4$	...	150	4.0	Ferric chloride	20	Room	nil
10% HCl + 1% CrO$_3$	...	150	nil-0.4	Ferric chloride	30	200	nil
10% HCl + 5% CrO$_3$	...	150	nil-0.2	Magnesium chloride	5	Boiling	nil
Acids (mixed)				Magnesium chloride	20	Boiling	nil
Aqua regia (3 parts HCl, 1 part HNO$_3$)	...	Room	2.0	Magnesium chloride	40	Boiling	nil
				Mercuric chloride	Sat.	200	4.0
Sulfuric-nitric	90-10	Room	18.0	Nickel chloride	20	200	nil
Sulfuric-nitric	90-10	160	295.0-298.0	Seawater	...	Room	nil
Sulfuric-nitric	70-30	Room	22.0-25.0	Sodium chloride	20	Room	nil
Sulfuric-nitric	70-30	160	269.0	Stannic chloride	100	Molten	nil
Sulfuric-nitric	50-50	Room	20.0-25.0	**Gases**			
Sulfuric-nitric	50-50	160	175.0-179.0	Sulfur dioxide (dry)	...	Room	nil
Sulfuric-nitric	30-70	Room	4.0	**Organic chemicals: acids**			
Sulfuric-nitric	30-70	160	92.0-95.0	Formic	50	Boiling	7.92
Sulfuric-nitric	10-90	Room	nil	Oxalic	1	Room	12.0-13.0

Source: D.W. Stough et al., "The Corrosion of Titanium," TML Report No.57, Titanium Metallurgical Laboratory, Battelle Memorial Institute, 29 Oct 1956, p 109-111

Ti-6Al-4V: General corrosion of Ti-6Al-4V in the presence of rocket propellants and oxidizers

Propellant/oxidizer	Corrosion resistance	Penetration rate, mils/yr	Ti-6Al-4V Compatibility Decomposition of propellant	Shock sensitivity	Temperature, °C
Fluorine, gas	Good	<5	No	No	250
Fluorine, liquid	Good	<5	No	No	−320(a)
Fluorine, liquid	Poor	>50	Extensive	Yes	−310
Chlorine trifluoride (ClF_3), liquid	Poor	>50	Extensive	Yes	75
Perchloryl fluoride (ClO_3F), gas	Excellent	<1	No	No	85
Perchloryl fluoride (ClO_3F), liquid	Excellent	<1	No	No	85(a)
Perchloryl fluoride (ClO_3F), gas + 1% H_2O	Excellent	<1	No	No	85
Perchloryl fluoride (ClO_3F), gas + 1% H_2O	Poor	>50	Extensive	Yes	160
Perchloryl fluoride (ClO_3F), liquid + 1% H_2O	Fair	5 to 50	Some	No	85
25% ClO_3F + 75% ClF_3, liquid	Poor	>50	Extensive	Yes	85
50% ClO_3F + 50% N_2F_4, liquid or gas	Excellent	<1	No	No	−109
Trichlorotrifluoroethane (CCl_2F-$CClF_2$, Freon PCA), gas	Excellent	<1	No	No	75
Dibromotetrafluoroethane ($CBrF_2$-CBr-F_2, Freon 114B2), gas	Excellent	<1	No	No	75
Hydrazine (N_2H_4), liquid	Excellent	<1	No	No	160
Unsymmetrical dimethyl hydrazine (UDMH), gas or liquid	Excellent	<1	No	No	160
50% N_2H_4 + 50% UDMH, gas or liquid	Excellent	<1	No	No	160
(50% N_2H_4 + 50% UDMH) + 3% H_2O, liquid	Excellent	<1	No	No	>60

(a) Impact ignition exhibited but flame does not propagate. Source: "Compatibility of Materials with Rocket Propellants and Oxidizers," W. Boyd, W. Berry, E. White, DMIC-201, 1965, reported in *Ti-6Al-4V Handbook*, D. Maykuth, R. Monroe, R. Favor, and D. Moon, Ed., Battelle Memorial Institute, Columbus, OH, 1971, p 1-8

Ti-6Al-4V: General corrosion in naturally aerated HCl solution

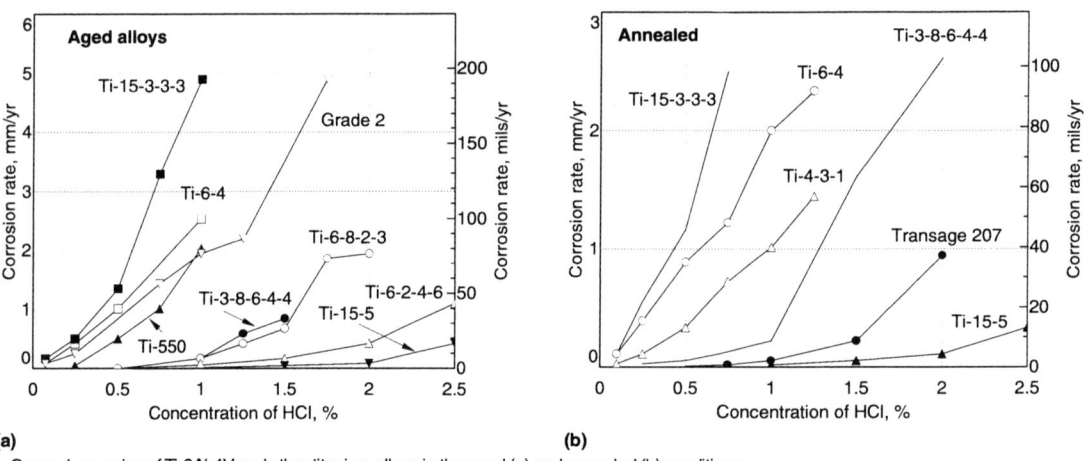

General corrosion of Ti-6Al-4V and other titanium alloys in the aged (a) and annealed (b) conditions.
Source: *Metals Handbook*, Vol 13, 9th ed., *Corrosion*, ASM International, 1987

Galvanic Corrosion Rates

Ti-6Al-4V is generally immune to galvanic corrosion in contact with other structural materials and is compatible with nickel alloys and stainless steels. However, Ti-6Al-4V can accelerate galvanic attack when in contact with less noble metals such as aluminum, carbon steel, and magnesium alloys.

Ti-6Al-4V: Galvanic corrosion rates for Ti-6Al-4V coupled with other materials after 1 month in artificial seawater

Material coupled with Ti-6Al-4V	Ti-6Al-4V Corrosion appearance	Hydrogen content(b) ppm	Materials coupled to 6-4				Corrosion appearance
			Weight loss		Corrosion rate		
			mg	grains	mm/yr	in./yr	
NiAl Bronze	NVC(a)	22.2	8.2	0.13	0.012	0.00047	Dealuminization
Naval Brass	NVC	17.6	11.4	0.18	0.016	0.00063	Dezincification
60/40 Brass	NVC	20.4	17.4	0.26	0.025	0.00098	Dezincification
Steel (304)	NVC	20.5	1.3	0.02	0.002	0.00008	NVC
Steel (316)	NVC	11.6	1.2	0.02	0.002	0.00008	NVC
Fe-9Ni	NVC	16.2	87.8	1.35	0.143	0.0056	General(c)
Fe	NVC	23.2	84.3	1.30	0.137	0.0054	General
Al (1100)	NVC	17.1	19.6	0.30	0.093	0.0037	Pitting
Al (7075)	NVC	12.4	19.3	0.29	0.092	0.0036	Pitting
Al (6061)	NVC	20.6	25.2	0.39	0.120	0.0047	Pitting
Zn	NVC	16.2	180.5	2.79	0.323	0.013	General

(a) NVC: no visible corrosion. (b) Hydrogen content of Ti-6Al-4V before testing is 11.0 to 23.5 ppm. (c) General: general corrosion. Composition of Ti-6Al-4V in (wt%) was Al: 6.4, C: 0.008, Fe: 0.22, H: 0.0038, N: 0.0028, O: 0.15, and V: 4.2. Bar 153 mm in diameter and 1.1 m long was machined and annealed after rolling in an alpha-beta range from beta heat treated forging. Source: "A Titanium Alloy for Marine Properller Shaft," T. Nishimura, Y. Tsumoori, H. Yuguchi, and H. Satoh, in *Titanium Science and Technology*, G. Lütjering, U. Zwicker, and W. Bunk, Ed., DGM, 1985, p 1127

Cavitation/Erosion Rates

Ti-6Al-4V: Comparison of corrosion-erosion and cavitation rates of various alloys in seawater

Material	Corrosion-erosion rate(a)		Cavitation rate(b)	
	mm/yr	Mils/yr	mm/yr	in./yr
Ti-6Al-4V	0.025	1	20	0.8
AM355	0.064	2.5	34	1.35
Hastalloy C	0.076	3	15	0.6
Inconel 718	0.102	4	13	0.5
K Monel	0.241	9.5	27	1.05

(a) 30 day exposure, seawater at 90 knots, 45 degrees impingement angle. (b) Double amplitude 0.025 mm (0.001 in.), 22,000 Hz, 8 h exposures, seawater. Source: A. Hohman, and W. Kennedy, Materials Protection 2, 1963, p 56-58

Ti-6Al-4V: Rain erosion behavior

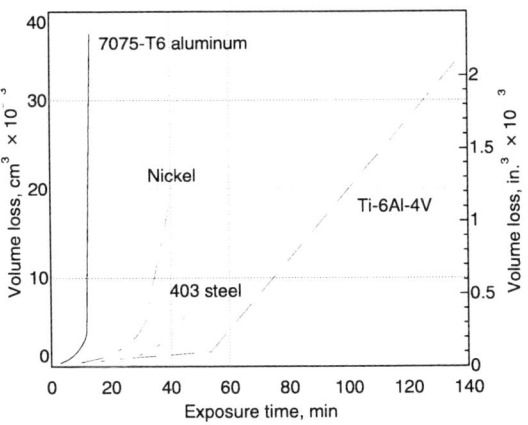

Rain erosion behavior of Ti-6Al-4V at 341 m/s (1120 ft/s) droplet velocity compared with other ductile metallics.
Source: Data from W. Adler and R. Syhnal, Bell Aerospace, 1974, reported in *Properties and Processing Ti-6Al-4V*, Timet, Pittsburgh

Crevice Corrosion

Crevice corrosion of Ti alloys in hot 5% NaCl and saturated NaCl brines

Alloy	90 °C pH 1	pH 2	pH 3	121 °C (250 °F) pH 5[a]	177 °C (350 °F) pH 3	232 °C (450 °F) pH3[a]	pH 5[a]
Ti-Grade 2	C[b]	C[b]	C,C[b]	C	C[a],C[c],C[d], C, C	N	N
Ti-6-4	...	...	C,C[b]	C	C[a], C[c], C[d], (C), (C)	N	N
Ti-4-3-1	...	...	...	C	N[a]	N	N
Ti-550	(C)[b]	(C)[b]	(C)[b]	N	N[a], C[c], N[d], (N), (N)	N	N
Ti-6-2-4-6	C[b]	C[b]	N(N)[b]	N	N[a], N[c], N[d], (N), (N)	N	N
Transage 210	(C)[b]	(C)[b]	...	...	...	...	...
Transage 207	...	...	...	N	N[a]	N	N
Beta-C	C[b]	N[b]	N,(N)[b]	...	N[c], N[d], (N), (N)	...	...
Ti-8-8-2-3	...	...	(N)[b]	...	(N), (N)	...	...
Ti-15-5	N[b]	N[b]	(N)[b]	N	N[a], N[c], N[d], (N), (N)	N	N

Note: C/N = crevice attack/no crevice attack. ( ) = aged specimens. [a] Saturated NaCl brine. [b] Aerated brine. [c] O_2 saturated brine. [d] 0.1% $NaCl_2$ brine additions. Annealed condition = 732 °C (1350 °F), 3 h, FC + 566 °C (1050 °F), AC. STA condition = 954 °C (1210 °F), 15 min, WQ + 538 °C (1000 °F), 4 h, AC. Source: "Fundamental Corrosion Characterization of High Strength Titanium Alloys," R. Schutz and J. Grauman, in *Industrial Applications of Titanium and Zirconium: Fourth Volume*, ASTM STP 917, C. Young and J. Durham, Ed., ASTM, Philadelphia, PA, 1986, p 130

Hydrogen Damage

Hydrogen damage of titanium and titanium alloys is manifested as a loss of ductility (embrittlement) and/or a reduction in the stress-intensity threshold for crack propagation. The damage is caused by hydrides, which form as hydrogen diffuses into the material during exposure with either gaseous or cathodic hydrogen. Because the phenomenon depends on both hydrogen diffusion and hydride formation, there may be a peak in hydrogen embrittlement as a function of temperature (see figure). The marked decreases in crack-growth susceptibility shown in the graph are attributed to an increased difficulty in hydride nucleation at higher temperatures. The exact level of hydrogen at which a separate hydride phase is formed depends on the composition of the alloy and the previous metallurgical history.

Relative Hydrogen Susceptibility. Alloy Ti-6Al-4V has moderate sensitivity to hydrogen damage, as illustrated in the disk-pressure test results for several materials (see next page) per the disk-pressure method ASTM STP 543.

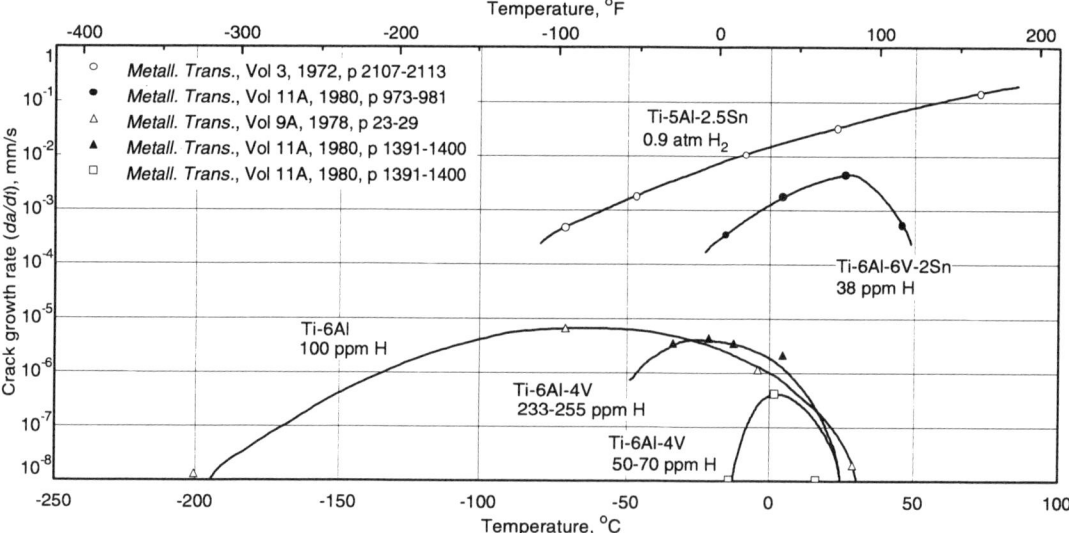

Ti-6Al-4V: Hydrogen embrittlement susceptibility vs temperature

Ti-6Al with a continuous α phase and Ti-6Al-4V exhibit a maximum in crack growth susceptibility near 0 °C (32 °F), whereas Ti-6Al-6V-2Sn with continuous β phase exhibits a maximum near 50 °C (120 °F). No maximum was observed for Ti-5Al-2.5Sn at the temperatures tested. All crack-growth rates correspond to an applied stress intensity near 50 MPa$\sqrt{m}$ (45 ksi$\sqrt{in.}$).
Source: N.R. Moody and J.E. Costa, A Review of Microstructural Effects on Hydrogen-Induced Sustained Load Cracking in Structural Titanium Alloys, in *Proc. Symp. Property/Structure Relationships in Titanium Alloys and Titanium Aluminides*, The Metallurgical Society, 1991

Ti-6Al-4V: Hydrogen susceptibility vs other metals

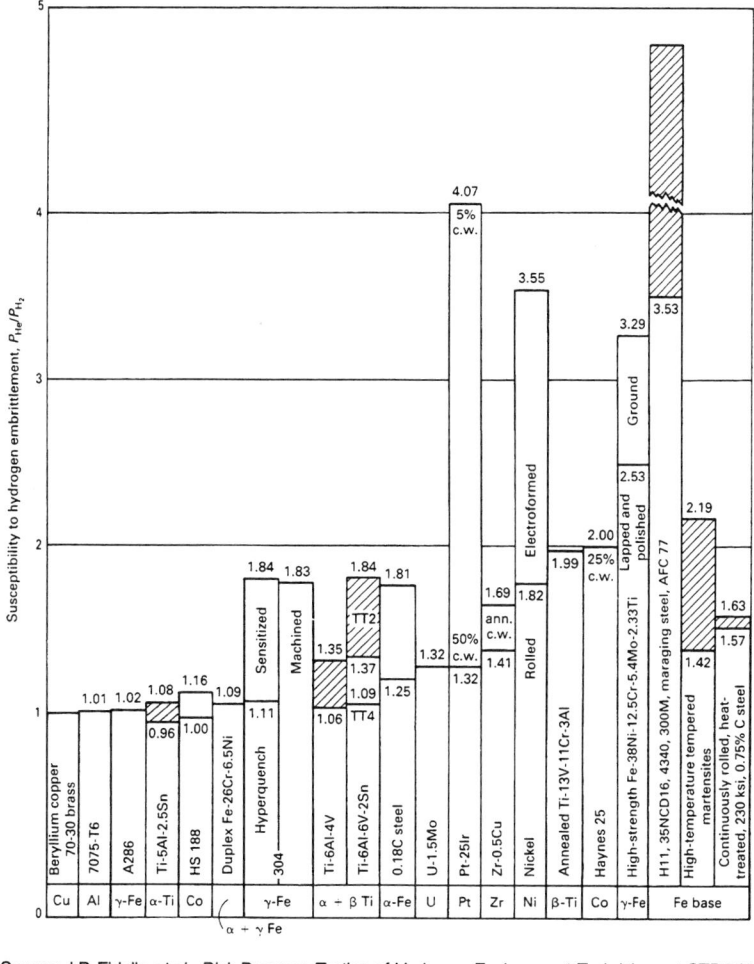

Source: J.P. Fidelle et al., *Disk Pressure Testing of Hydrogen Environment Embrittlement*, STP 543, American Society for Testing and Materials, 1974, p 221-253

Ti-6Al-4V: Effect of microstructure on hydrogen embrittlement

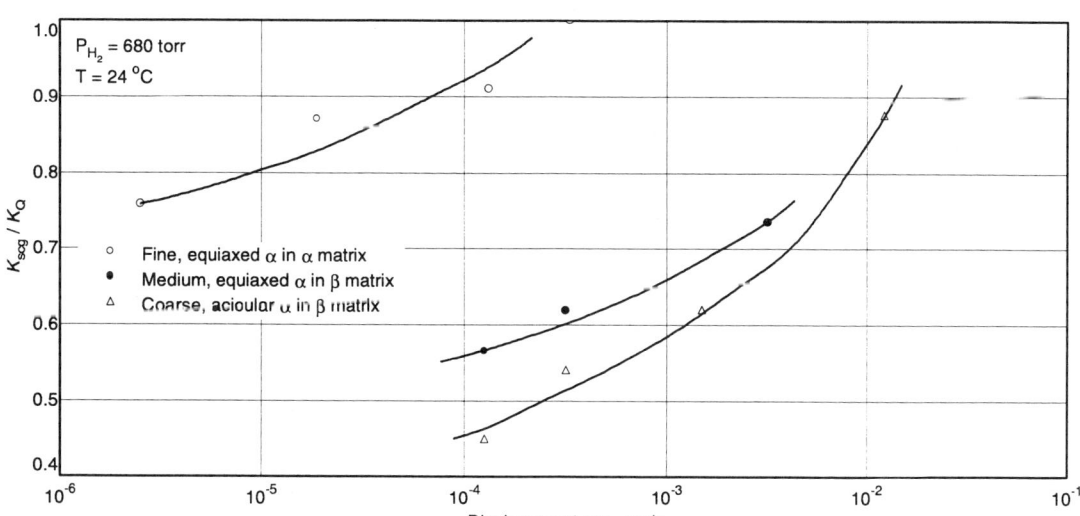

K_{scg} is the stress intensity for the beginning of measurable subcritical crack growth for embrittled Ti-6Al-4V. K_Q is the stress intensity for failure for unembrittled Ti-6Al-4V. Those microstructures that contained a continuous α-phase matrix with a fine dispersed β-phase in their boundaries (not shown in the figure) are the least embrittled, while those containing a continuous β-phase throughout are the most hydrogen embrittled.
Source: H. Nelson, D. Williams, and J. Stein, Environmental Hydrogen Embrittlement of an Alpha-Beta Titanium Alloy: Effect of Microstructure, *Metall. Trans.*, Vol 3, 1972, p 469

Annealed Condition

Annealed Ti-6Al-4V: Hydrogen effects on sustained load fracture

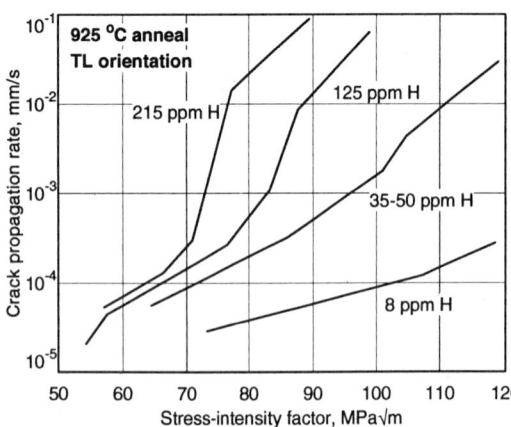

Source: D.A. Meyn, Effect of Hydrogen Content on Inert Environment Sustained Load Crack Growth Propagation Mechanisms of Ti-6Al-4V, in *Environmental Degradation of Engineering Materials in Hydrogen*, M.R. Louthan, Jr., et al., Ed., VPI Press, 1981, p 383-392

Annealed Ti-6Al-4V: Effect of hydrogen on fracture strength

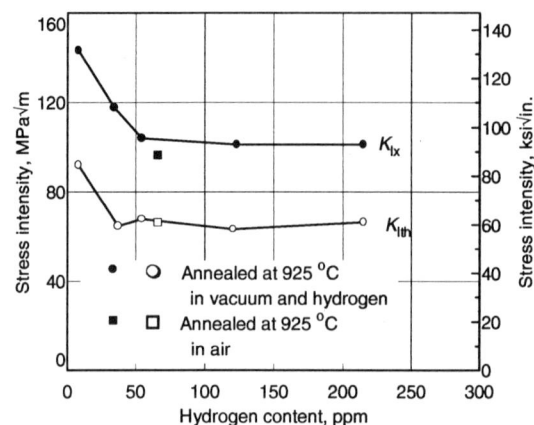

Fracture toughness K_{Ix} and sustained-load crack propagation threshold (K_{Ith}) vs hydrogen constant illustrate embrittlement from interstitial hydrogen.

Material/test parameters: Hot-rolled, mill-annealed material had composition (wt%) 6.5 Al, 0.02 C, 0.18 Fe, 0.012 N, 0.18 O, and 4.2 V.

Source: D. Meyn, Effect of Hydrogen in Fracture and inert-Environment Sustained Load Cracking Resistance of Alpha-Beta Titanium Alloys. *Metall. Trans.*, Vol 5, 1974, p 2405

Solution Treated and Aged Condition

STA Ti-6Al-4V: Average crack-growth rate with varying internal hydrogen

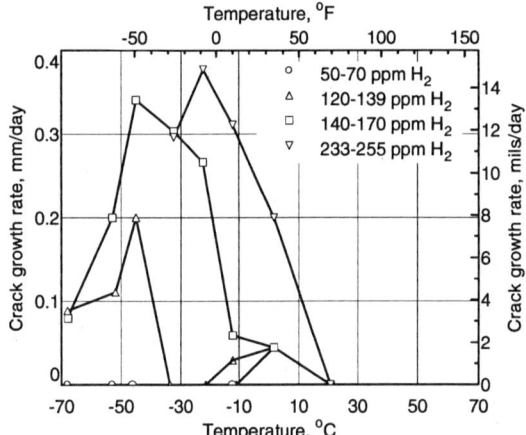

Plate material had a composition (wt%) of 6.5 Al, 0.0028 N, 0.1782 O, and 3.92 V. Blanks were pickled, hydrogenated, then solution treated at 940 °C (1725 °F) for 30 min, water quenched, aged at 538 °C (1000 °F) for 4 h and machined into a single-edge cracked plate for crackline bending.

Source: R. Boyer and W. Spurr, Characteristics of Sustained-Load Cracking and Hydrogen Effects in Ti-6Al-4V, *Metall. Trans.*, Vol 9A, 1978, p 23

STA Ti-6Al-4V: Crack growth rate variations as a function of external hydrogen pressure

Hydrogen pressure P_{H_2}		Crack growth rate (da/dt)		Step width	
KPa	psi	mm/s	in./s	μm	mils
90.6	13.1	0.1	0.0039	3.25	0.128
90.6	13.1	0.09	0.0035	3.5	0.138
75	10.9	0.08	0.0031	3.5	0.138
60.8	8.8	0.036	0.0014	4.25	0.167
48.8	7.1	0.0475	0.0019	4.25	0.167
6.1	0.9	0.0072	0.0003	5.25	0.207

Note: Crack growth rates were the result of a constant stress intensity near 76 MPa$\sqrt{m}$ at 24 °C. The material used for this investigation was commercially produced Ti-6Al-4V alloy in the form of 0.32 cm and 1.27 cm thick plates. The material was machined into specimens and solution-treated at 1040 °C for 40 min, stabilized at 700 °C for 1 h and at 590 °C for 1 h in vacuum, and then air-cooled to produce a microstructure of acicular alpha-phase platelets in a continuous beta-phase matrix. Source: "A Film-Rupture Model of Hydrogen-Induced Slow Crack Growth in Acicular Alpha-Beta Titanium," H. Nelson, *Met. Trans. A*, 7A, 1976, p 621

Stress-Corrosion Cracking

Stress-corrosion cracking (SCC) mechanisms fall into two general categories: anodic-assisted cracking and cathodic-assisted cracking. Anodic SCC (active path corrosion) involves the dissolution of metal during the initiation and propagation of cracks. Cathodic SCC (embrittlement by corrosion product hydrogen) involves the deposition of hydrogen at cathodic sites on the metal surface or on the walls of a fissure or crack and its subsequent absorption into the metal lattice. The close interaction of these two mechanisms accounts for the diversity of SCC phenomena.

Ti-6Al-4V: Environments and temperatures known to promote SCC of Ti-6Al-4V

Environment	SCC threshold temperature
Nitric acid, RFNA	Room
Nitrogen tetroxide (no excess NO)	30-75 °C (85-165 °F)
Methanol	Room
Methyl chloroform, inhibited	370 °C (700 °F)
Trichlorofluoroethane	788 °C (1450 °F)
Chloride salts	288-426 °C (550-800 °F)
Cadmium	Room
Mercury	Room
Ag-5Al-2.5Mn	340 °C (650 °F)
Seawater	Ambient

Source: Adapted from "Corrosion of Titanium and Titanium Alloys," R. Schutz and D. Thomas, in *Metals Handbook Ninth Edition, Volume 13: Corrosion*, ASM International, Materials Park, OH, 1987, p 690

Ti-6Al-4V: Stress intensity thresholds of SCC (K_{Iscc}) for Ti-6Al-4V in various media

Medium	Alloy condition	K_{Iscc} ksi√in.	Reference
Freon TF ($C_2Cl_3F_3$)	STA	33	1
Freon TF ($C_2Cl_3F_3$)	Mill annealed	52	2
Freon MF (CCl_3F)	STA	23	3
Freon MF (CCl_3F)	Mill annealed	25	2
Deionized water	Mill annealed	73.4	4
3.5% NaCl, pH = 4.5	Mill annealed	52.7	4
3.5% NaCl, pH = 9.0	Mill annealed	48.0	4
3.5% NaCl, pH = 11.0	Mill annealed	42.2	4
3.5% NaCl + 0.5% FeCl, pH = 2.2	Mill annealed	48.9	4
3.5% NaCl + 1.0% $AlCl_3$, pH = 7.0	Mill annealed	46.1	4
3.5% NaCl + 0.1% Dupanol (wetting agent)	Mill annealed	45.8	4
Kerosene	Mill annealed	70.5	4
Dodecyl alcohol	Mill annealed	66.7	4
Methyl alcohol	Mill annealed	70.3	4
Mercury	Mill annealed	22 (average)	4

Source: (1) S.V. Glorioso, *Lunar Module Pressure Vessel Operating Criteria*, Specification SE-V-0024, NASA/MSC, Oct. 1968. (2) C.C. Seastrom and R.A. Gorski, "The Influence of Fluorocarbon Solvents on Titanium Alloys," *Accelerated Crack Propagation of Titanium by Methanol Halogenated Hydrocarbons, and Other Solutions*, DMIC Memorandum 228, March 6, 1967, p 20. (3) C.F. Tiffany and J.N. Masters, *Investigation of the Flaw Growth Characteristics of Ti-6Al-4V Titanium Used in Apollo Spacecraft Pressure Vessels*, CR-65586, NASA, March 1967. (4) "Studies of the Mechanism of Crack Propagation in Salt Water Environments of Candidate Supersonic Transport Titanium Alloy Materials," D. Williams, R. Wood, E. White, W. Boyd, and H. Ogden, FAA Report SST-66-1, 1966

Electrochemical Effects

The halide ions (chloride, bromide, and iodide) are SCC agents unique for titanium alloys in aqueous solutions at room temperature. The crack initiation load and velocity are controlled by the applied potential, as illustrated in the figure for Ti-6Al-4V in a KCl solution. The width of the critical potential range and the potential for maximum susceptibility varies with the anion.

Crack propagation can be halted by switching the potential to either the anodic or cathodic protection zone. The potential of titanium alloys in 3.5% sodium chloride and seawater of about −800 mV versus SCE is the potential at which SCC susceptibility reached a maximum (M.J. Blackburn, W.H. Smyrl, and J.A. Sweeney, Titanium Alloys, in *Stress Corrosion Cracking in High Strength Steels and in Titanium and in Aluminum Alloys*, B.F. Brown, Ed., Naval Research Laboratory, 1972, p 246-363).

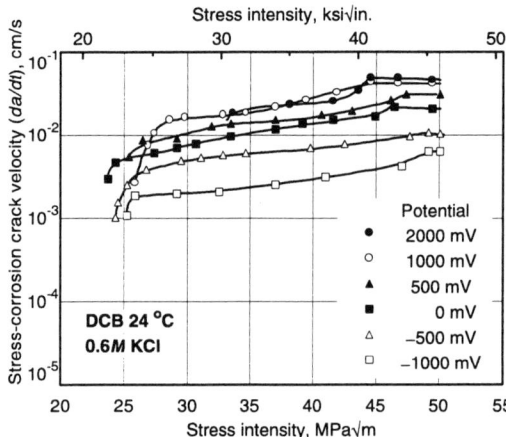

Ti-6Al-4V: Stress-corrosion crack velocity in KCl solution

Effect of stress intensity and potential on stress-corrosion crack velocity in Ti-6Al-4V tested in 0.6M KCl solution. Note that K_{Iscc} exhibits a minimum value between 0 and −500 mV. The region II stress-corrosion crack velocity increases in a similar manner to iodide solutions. Note, however, that the 2000 and 1000 mV curves are rather similar. DCB = double cantilever beam.
Source: *Corrosion Fatigue: Chemistry, Mechanics and Microstructure*, Vol 2, National Association of Corrosion Engineers, 1972

SCC in Salt Water

SCC-resistance for titanium alloys in 3.5% saltwater

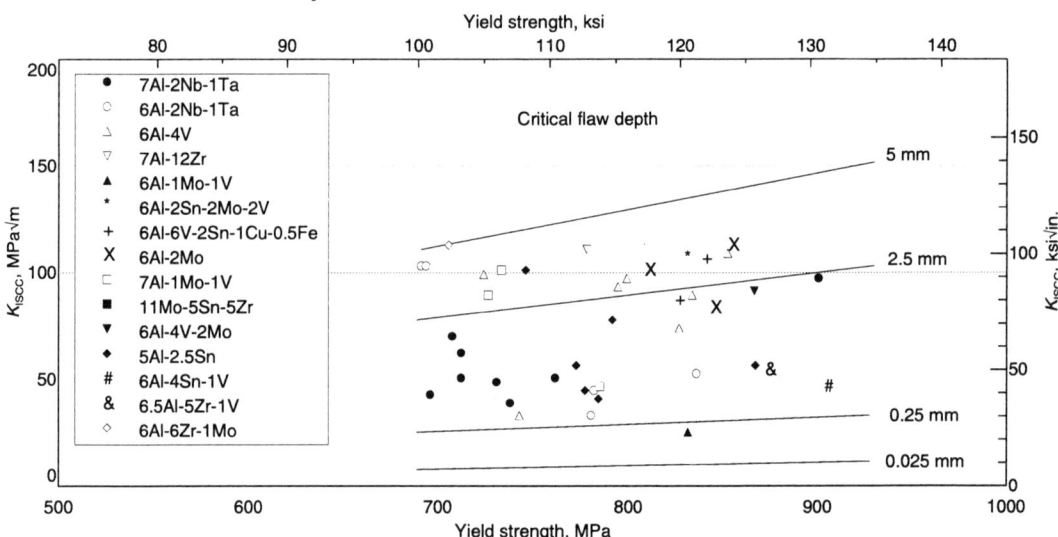

The critical flaw depth is obtained from the relation $0.22 \cdot YS \cdot \sqrt{a} = K_{Iscc}$. This expression assumes that the applied stress is at the yield stress level and that a surface crack of length $2c$ and of depth a is present, where a is much less than c. For a given yield stress and flaw size, the value of K_{Iscc} must exceed that given by this relation if failure is to be avoided.
Source: M. Schumacher, Ed., *Seawater Corrosion Handbook*, Noyes Data Corp., 1979

Ti-6Al-4V: SCC-thresholds in salt water

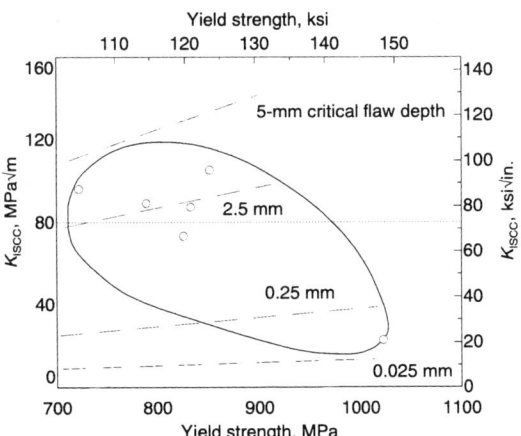

K_{Iscc} threshold of various lots of Ti-6Al-4V at various strength levels in salt water. The oval encloses the known K_{Iscc} data for this alloy in salt water.
Source: B.F. Brown, *Stress Corrosion Cracking Control Measures*, National Bureau of Standards, Monograph 156, U.S. Department of Commerce, June 1977, p 36

Ti-6Al-4V: Stress intensities for cracking in salt water

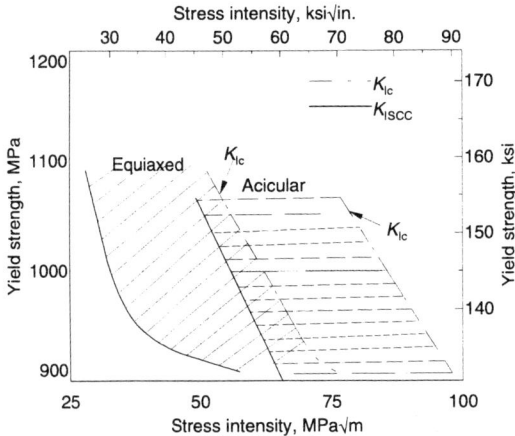

Relationship between yield stress and stress intensity for either unstable fast fracture (K_{Ic}) or stress-corrosion cracking (K_{Iscc}) in salt water. A variety of microstructures for each type were tested.
Source: A.W. Thompson and I.M. Bernstein, The Role of Metallurgical Variables in Hydrogen-Assisted Environmental Fracture, in *Corrosion Science and Technology*, Vol 7, M.G. Fontana and R.W. Staehle, Ed., Plenum Press, 1980, p 111

Ti-6Al-4V: Stress-intensity thresholds and ultimate tensile strength of plate in several heat-treated conditions

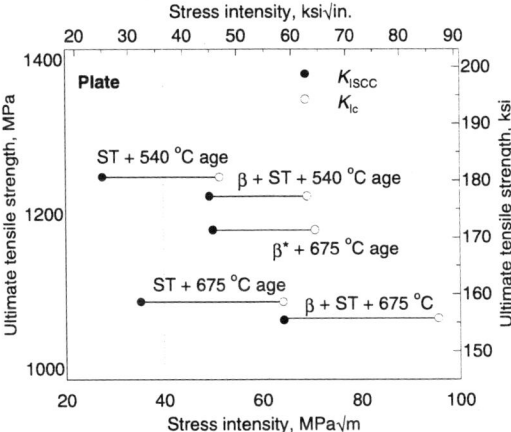

Fracture toughness (K_{Ic}) and stress-corrosion resistance (K_{Iscc}) for the same heat treatment are connected by a line. The heat treatments are identified as follows: β = β anneal (AC), β^* = β quench (WQ), and ST = $\alpha - \beta$ solution treatment (940 °C - WQ). Aging was conducted at the temperature shown. Material with composition (wt%) of 6.3 Al, 0.02 C, 0.17 Fe, 0.007 H, 0.006 N, 0.11 O, and 4.1 V was used as 12.7 mm (0.5 in.) thick plate. Three notched-bend specimens per data point were loaded in hydraulic testing apparatus, immersed in 3.5% NaCl solution (pH = 6.5), and held to failure or at least 6 h.
Source: R. Curtis and W. Spurr, Effect of Microstructure on the Fracture Properties of Titanium Alloys in Air and Salt Solution, *Trans. ASM*, Vol 61, 1968, p 115

Ti-6Al-4V: SCC of annealed and aged sheet in salt water

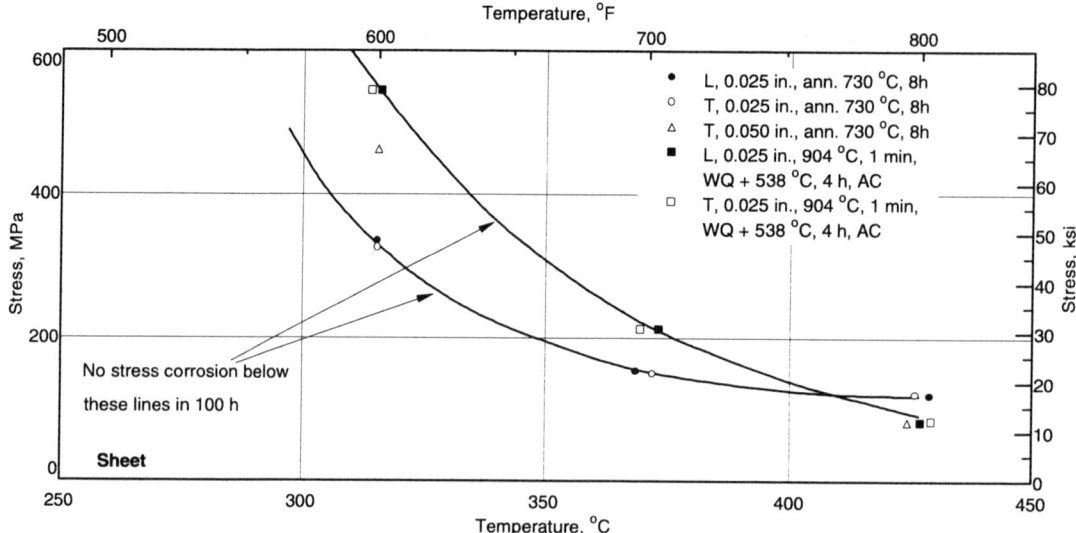

Alloy sheets of 0.64 and 0.13 mm (0.025 and 0.050 in.) thicknesses were heat treated as indicated. Smooth specimens 6.4 mm (0.25 in.) wide were coated with 1.6 mm (1/16 in.) thick layer of ASTM synthetic sea salt for 100 h. L = longitudinal, T = transverse.
Source: Data from Douglas Aircraft Co., "Chloride Stress-Corrosion Susceptibility of High Strength Stainless Steel, Titanium Alloy, and Superalloy Sheet," ML-TDR-64-44, 1964, reported in *Aerospace Structural Metals Handbook*, Code 3707, p 10

Ti-6Al-4V: SCC behavior of smooth P/M specimens in salt water

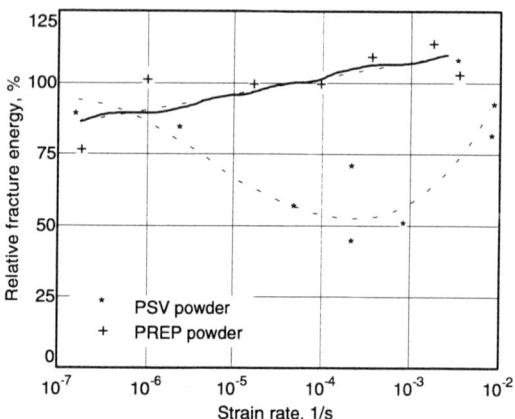

Dashed line represents powder sprayed under vacuum (PSV powder). Solid line represents plasma rotating electrode process (PREP powder). The experiments were performed in 3.5% sodium chloride solution at 25 °C (77 °F).
Source: H. Buhl, Stress Corrosion Cracking Behavior of I/M and P/M Ti6Al4V Dependent upon the Strain Rate, in *Titanium Science and Technology*, Vol 4, G. Lütjering et al., Ed., Deutsche Gesellschaft für Metallkunde e.V., West Germany, 1985, p 2549-2556; Fracture energy relative to that in $Mg(ClO_4)_2$

Ti-6Al-4V: SCC behavior of specimens produced by ingot metallurgy (I/M)

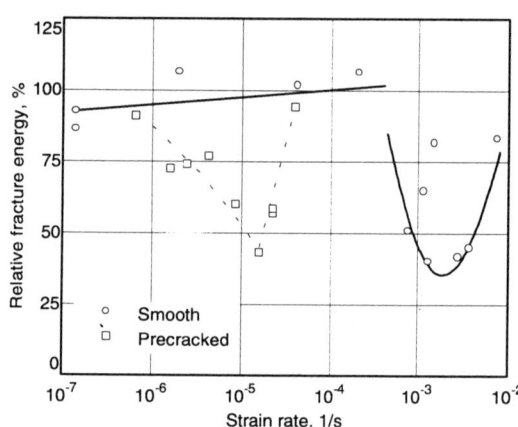

Stress-corrosion cracking behavior of I/M Ti-6Al-4V in the smooth (solid line) and precracked (dashed line) conditions in 3.5% sodium chloride. The strain rates represent overall deformation rates and do not reflect the much higher strain rates at the crack front.
Source: H. Buhl, Stress Corrosion Cracking Behavior of I/M and P/M Ti6Al4V Dependent upon the Strain Rate, in *Titanium Science and Technology*, Vol 4, G. Lütjering et al., Ed., Deutsche Gesellschaft für Metallkunde e.V., West Germany, 1985, p 2549-2556

SCC in Hot Salts

In the 260 to 425 °C (500 to 800 °F) range, the stresses to produce salt cracking are well below the 0.1% creep design criterion. At about 450 °C (840 °F), the creep and hot salt curves again merge. Ti-6Al-4V has shown hot-salt SCC susceptibility in lab tests, but there have been no failures in service ascribed to this phenomenon.

Ti-6Al-4V: Hot-salt SCC

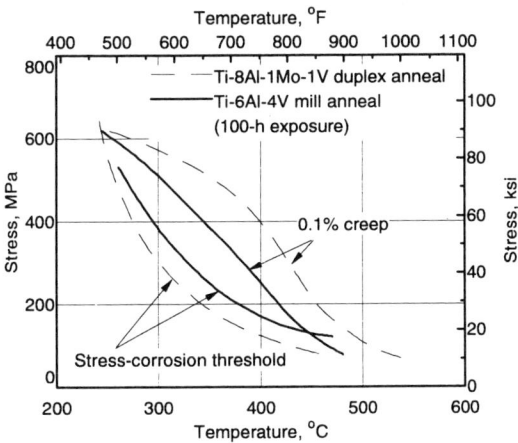

Source: V.C. Petersen, Hot-Salt Stress Corrosion of Titanium, in *Source Book on Titanium and Titanium Alloys*, M.J. Donachie, Jr., American Society for Metals, 1982, p 167-174

Ti-6Al-4V: Hot-salt SCC of sheet

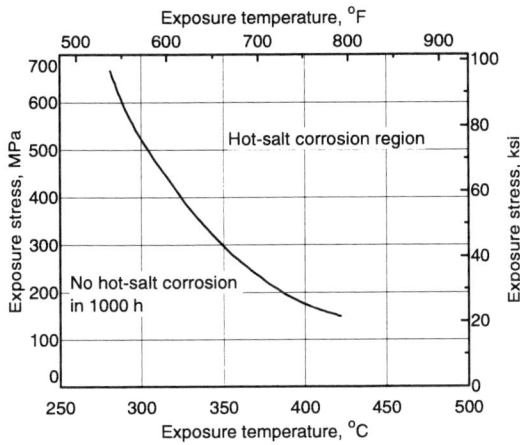

Sheet was mill annealed at 790 °C (1450 °F) for 15 min, air cool. Salt used in testing was dry NaCl.
Source: "RMI 6Al-4V," RMI Co., Niles, OH

Ti-6Al-4V: Stress-temperature thresholds for stress-corrosion cracking of several titanium alloys in hot salt

Alloy	100 hour threshold stress to produce cracking at:																	
	290 °C (550 °F)		315 °C (600 °F)		345 °C (650 °F)		370 °C (700 °F)		400 °C (750 °F)		425 °C (800 °F)		455 °C (850 °F)		480 °C (900 °F)		510 °C (950 °F)	
	MPa	ksi	MPa	ksi	MPa	ksi	MPa	ksi	MPa	ksi	MPa	ksi	MPa	ksi	MPa	ksi	MPa	ksi
Ti-5Al-2.5Sn annealed	195	28	205	30	...	...	105	15	...	...	70-140	10-20	...	...	...	...	...	...
Ti-8Al-1Mo-1V aged	...	...	...	...	...	...	...	...	170	25	...	...	140	20	...	...	105	...
annealed	170	25	380	55	...	...	160	23	...	...	125	18	...	...	...	...	...	...
Ti-6Al-4V aged	...	...	655	95	450	65	170	25	205	30	85	12	105	15	...	...	...	...
annealed	345	50	345	50	...	...	150	22	...	...	125-165	18-24	...	...	...	...	...	...

Source: MIL-HDBK-697A, 1974, p 100

SCC in Methanol

Titanium alloys susceptible to saltwater and/or aqueous SCC also are most susceptible to methanol. Anodic polarization increases the susceptibility of titanium to SCC in methanol/halide mixtures. On the other hand, cathodic polarization dramatically reduces SCC susceptibility. Potentials more negative than –250 mV versus Ag/AgCl prevent cracking in methanol [B.S. Hickman, J.C. Williams, and H.L. Marcus, Transgranular and Intergranular Stress-Corrosion Cracking of Titanium Alloys, *Aust. Inst. Met.*, Vol 14(No.3), 1969, p 138].

Potentials

Ti-6Al-4V: Cathodic polarization in salt water

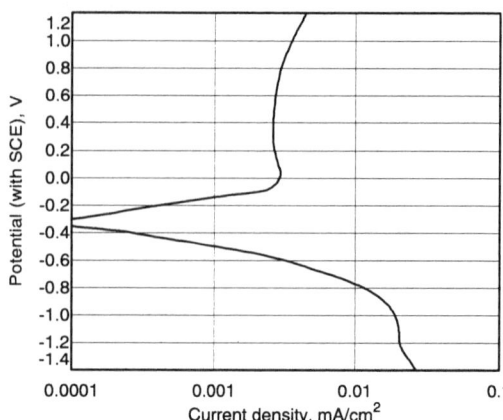

Cathodic polarization curve with a saturated calomel electrode (SCE) is shown for Ti-6Al-4V at room temperature in 3.5% synthetic seawater. Polarization characteristics of this alloy are similar to those of unalloyed titanium. Solution used was ASTM synthetic seawater with pH = 8. Pickled surface scan rate was 0.5 mV/s. Source: "Properties and Processing of Ti-6Al-4V," Timet, Pittsburgh

ELI Ti-6Al-4V: Anodic polarization in deaerated 1N sulfuric acid

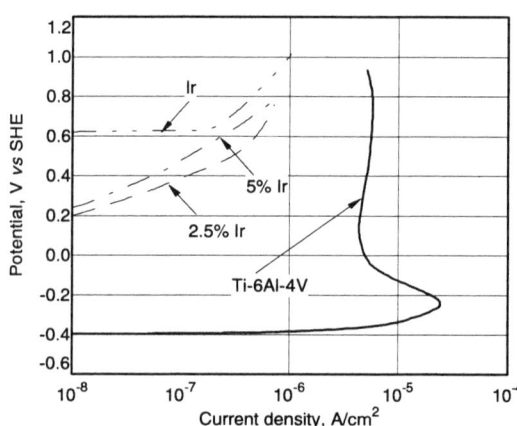

Anodic polarization with standard hydrogen electrode (SHE) also is shown for pure iridium and iridium-implanted Ti-6Al-4V. Annealed ELI grade alloy was mechanically polished and subjected to ion implantation with Ir^{++} ions at 244 keV and fluence of 0.74×10^{16} and 1.48×10^{16} ions/cm^2. Peak concentrations occurred at a depth of about 30 nm.
Source: R. Buchanan et al., Iridium Ion Implantation of Surgical Titanium Alloy: Corrosion Inhibition and Change Injection Effects, in *Ion Implantation and Plasma Assisted Processes*, R. Hochman et al., Ed., ASM International, 1988, p 53

Repassivation/Pitting

Ti-6Al-4V: Repassivation potentials in boiling chloride solutions for as-annealed and unalloyed titanium

Alloy	Repassivation potential, V(a)		
	5% NaCl (pH 3.5)	3% HCl	Saturated NaCl
Grade 1	...	...	+7.0
Grade 2	+6.7	+5.8	+5.7
Ti-6-4	+2.3	+1.7	...
Ti-550	+2.8	+2.3	...
Ti-6-2-4-6	+3.0	+2.4	...
Ti-3-8-6-4-4	+3.2	+2.6	...
Ti-8-8-2-3	+2.6	+2.4	...
Ti-15-5	+6.3	+5.6	...
Grade 12	...	...	+5.9
Grade 7	...	...	+5.6

(a) Measured versus Ag/AgCl reference electrode. Source: Data from R. Schutz and J. Grauman, 1987, reported in "Corrosion of Titanium and Titanium Alloys," R. Schutz and D. Thomas, in *Metals Handbook, Ninth Edition, Volume 13: Corrosion*, ASM International, Materials Park, OH, 1987, p 669

Ti-6Al-4V: Characteristic potentials of pitting corrosion of titanium and Ti-6Al-4V at 25 °C in different solutions

Metal or alloy	Solution	Potential of pit repassivation	Potential of pit nucleation
Ti	1N NaBr	1170 mV	1425-1745 mV
Ti-6Al-4V	1N NaBr	950 mV	1330 mV
Ti-6Al-4V	1N NaI	1450-1580 mV	1690 mV
Ti	1N NaCl	8.75 V	11-11.4 V
Ti-6Al-4V*	1N NaCl	2.84 V	...
Ti-6Al-4V	1N NaCl	2.15 V	5.2-5.7 V

*F.A. Posey, E.G. Bohlman, *Desalination*, Vol 3, 1976, p 269

The current potential curves were determined potentiodynamically using a Wenking potentiostat (model 68 FRO.5). A platinum counterelectrode and a Ag/AgCl reference electrode were used together with a potentiostat. The electrolyte was air saturated and was kept at a constant temperature of 25 °C.
The Ti-6Al-4V specimens were mechanically polished with 600 grit SiC papers, followed by polishing with alumina paste (mean grit size 0.1 μm). The electrodes had a disc surface of approximately 2.3 cm^2 area. Source: H.J. Raetzer-Scheibe, "The Relationship Between Repassivation Behavior and Pitting Corrosion for Ti and Ti-6Al-4V," *Corrosion*, Vol 34, No. 12, December 1978, p 437-442

Ti-6Al-4V: Combustion rates of titanium alloys

Alloy	J_1 g/cm$^2 \cdot$s	J_2 g/cm$^2 \cdot$s
Ti (99.2%)	0.10	0.03
Ti-6Al-4V	0.10	0.05
Ti-5Al-2.5Sn	0.08	0.03
Ti-8Al-1Mo-1V	0.09	0.06
Ti-6Al-6V-2Sn	0.07	0.03
Ti-6Al-2Sn-4Zr-2Mo	0.10	0.04
Ti-6Al-2Sn-4Zr-6Mo	0.09	0.07
Ti-4Al-12Sn	0.07	0.04
Ti-12Mo-4Sn	0.11	0.05
Ti-.24Fe-3.3Al-2.6V	0.10	0.06
Ti-11.5Mo-6Zr-4.5Sn	0.11	0.05
Ti-5.9Cr-4.2Mo-3.3V-0.7Fe	0.10	0.04
Ti-5.8Al-2.0Mo-2.1Sn-1.9Cr-1.8Zr	0.07	0.05
Ti-3.3Al-8.1V-5.9Cr-4.2Mo-3.9Zr	0.09	0.07
Ti-4.9Al-5.0Sn-1.9Mo-1.9Zr-.24Si	0.09	0.05

J_1, J_2 = combustion rates during first and second parts of reaction. Source: "Titanium Combustion in Turbine Engines," T. Strobridge, J. Noulder, and A. Clark, FAA-RD-79-51, Federal Aviation Administration, 1979, p 59

Thermal Properties

As a lightweight structural material, Ti-6Al-4V is useful at temperatures below 400 °C (750 °F). Melting range, specific heat, and thermal conductivity are important in casting and heat-treatment processing. Thermal conductivity of Ti-6Al-4V is less than that of pure titanium, while its thermal expansion and specific heat are close to those of pure titanium. The liquidus temperature of Ti-6Al-4V is below the melting point (1688 °C, or 3070 °F) of pure titanium.

Specific Heat

The specific heats of Ti-6Al-4V at constant pressure (C_p) and at constant volume (C_v) are nearly identical (within 3%). Therefore, no distinction is made in this data tabulation. The specific heat of Ti-6Al-4V depends only weakly on composition and heat treatment, and its absolute value is close to that of pure titanium. Its temperature dependence is also like that of pure titanium. Specific heat is needed to calculate the change in enthalpy and the increase in entropy upon heating.

Ti-6Al-4V: Specific heat and Debye temperature vs quenching temperature

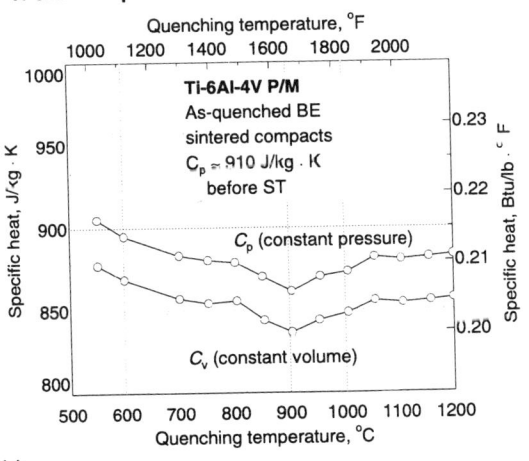
(a)

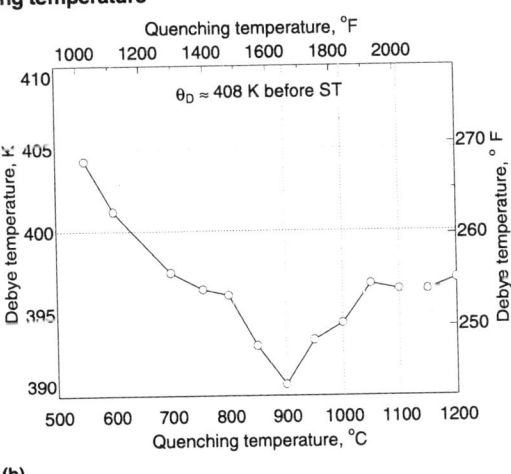
(b)

Ti-6Al-4V specimens were made by powder metallurgy processing with 0.17, 0.24, and 0.30 wt% oxygen; blended elemental (BE) powder compacts were sintered at 1260 °C (2300 °F) to over 99% of theoretical density and had an initial equiaxed grain structure free of texture. The specimens were solution-heat-treated at indicated temperatures for one hour each and then water quenched.
Source: Y.T. Lee, Ph.D. thesis, Case Western Reserve University

Ti-6Al-4V: Specific heat vs temperature

Temperature		Specific heat	
°C	°F	J/kg·K	Btu/lb·°F
20	68	580	0.14
205	400	610	0.15
425	800	670	0.16
650	1200	760	0.18
870	1600	930	0.22

Source: *Metals Handbook*, Vol 3, 9th ed.

Ti-6Al-4V: Specific heat vs temperature

Temperature		Specific heat	
K	°F	J/kg·K	Btu/lb·°F
273	32	565	0.135
366	200	565	0.135
477	400	574	0.137
588	600	603	0.144
699	877	649	0.155
810	1000	699	0.167
922	1200	770	0.184
1033	1400	858	0.205
1144	1600	959	0.229

Source: M. Mots, R. Hooper, and P. Frost, "Report on the Engineering Properties of Commercial Titanium Alloys," TML Report 92, Titanium Metallurgical Laboratory, Battelle Memorial Institute, Columbus, Ohio, June 1958

Ti-6Al-4V: Specific heat vs temperature

Temperature		Specific heat	
K	°F	J/kg·K	Btu/lb·°F
33	−400	71	0.017
61	−350	167.5	0.040
83	−310	251	0.060
105	−270	335	0.087
144	−200	414.5	0.099
228	−50	502	0.127
255	0	527.5	0.126
366	200	569	0.136
478	400	594.5	0.142
588	600	620	0.148
700	800	641	0.153
810	1000	661.5	0.158
866	1100	670	0.160
922	1200	687	0.162

3.2 mm (0.125 in.) sheet, heat treated at 925 °C (1700 °F) for 20 min, water quenched, aged at 480 °C (900 °F) for 4 h. Source: M. McGann and S. Mathews, "Determination of Design Data for Heat Treated Titanium Alloy Sheet," Vol 2a, ASD-TDR-62-355, May 1962, reported in *Aerospace Structural Metals Handbook*, Mechanical Properties Data Center, Battelle Columbus Laboratories, Columbus, Ohio, Code 3707, p 9

Ti-6Al-4V: Specific heat vs temperature

Temperature		Specific heat	
K	°F	J/kg·K	cal/g·K
20	−423	8.2	0.00196
25	−414	16.0	0.00383
30	−405	27.0	0.00640
40	−387	58.4	0.01394
50	−369	99.5	0.02377
60	−351	144.2	0.03443
70	−333	187.6	0.04481
80	−315	228.9	0.05466
90	−297	266.9	0.06375
100	−279	301.0	0.07190
120	−243	356.9	0.08524
140	−207	398.5	0.09518
160	−171	432.5	0.1033
180	−135	458.0	0.1094
200	−99	478.1	0.1142
220	−63	494.5	0.1181
240	−27	507.9	0.1213
260	9	519.6	0.1241
280	45	530.0	0.1263
300	81	526.3	0.1257

Alloy from the Mallory-Sharon Metals Corporation (now RMI), solution treated at 925 °C (1700 °F) for 20 min, oil quenched, then aged at 480 °C (900 °F) for 4 h, air cooled. Composition in weight % was Al: 5.89, C: 0.02, Fe: 0.15, H: 50 ppm, N: 0.015, V: 3.87. Approximately 40 measurements were made in the range 21 K to 300 K, with temperature increments of 2 to 7 degrees. Four temperature regulating baths were used to cover the desired temperature range: liquid hydrogen, liquid nitrogen, solid carbon dioxide-ethanol, and ice water. Measurements within a given bath were taken to overlap measurements from the adjacent temperature range. Source: W. Ziegler, J. Mullins, and S. Hwa, "Specific Heat and Thermal Conductivity of Four Commercial Titanium Alloys," Advances in Cryogenic Engineering, Vol 8, Plenum Press, New York, 1963, p 268

Heat Transfer

Thermal Conductivity. Ti-6Al-4V has a room-temperature thermal conductivity of about 7 W/m · K (4 Btu/ft · h · °F), which is lower than that of pure titanium (22 W/m · K) and high-alloy steel (12 to 30 W/m · K). Thermal conductivity is related to the heat capacity and the mean free path of electrons and phonons. At absolute zero, thermal conductivity is zero.

Thermal Diffusivity. The thermal diffusivity of Ti-6Al-4V is reported to be 0.027 cm^2/s (0.0042 in.^{2}/s) at 27 °C (80 °F) (J. Hust and A. Clark, *Cryogenics,* Vol 13, 1973, p 325). Diffusivity is the quantity of heat passing through a unit area per unit time divided by the product of the specific heat, density, and temperature gradient.

Ti-6Al-4V: Thermal conductivity

Temperature		Conductivity	
°C	°F	W/m · K	Btu/ft · h · °F
Mill annealed			
20	68	6.6	3.8
93	200	7.3	4.2
205	400	9.1	5.3
315	600	10.6	6.1
425	800	12.6	7.3
540	1000	14.6	8.4
650	1200	17.5	10.1
Solution treated and aged			
20	68	6.8	3.9
93	200	7.5	4.3
205	400	8.5	4.9
425	800	10.9	6.3
540	1000	12.6	7.3
650	1200	14.1	8.1

Source: *Metals Handbook,* Vol 3, 9th ed

Ti-6Al-4V: Thermal conductivity vs temperature

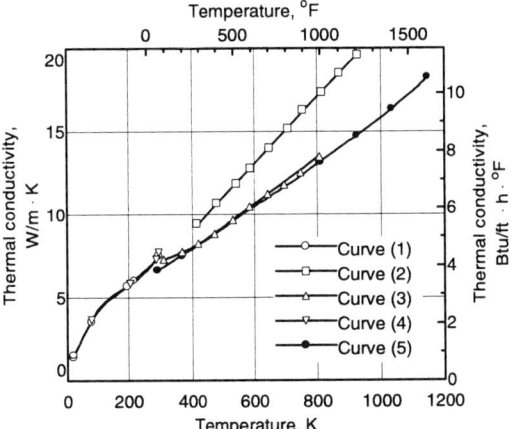

Curve (1) Sample: 5.89 Al, 0.2 C, 0.15 Fe, 0.015 N, 3.87 V. Specimen: 4 in. × 0.375 in. × 0.125 in. Solution heat-treated at 1200 K for 20 min and aged at 755 K for 4 h. Temperature range: 22 to 294 K. Curve (2) Sample: 5.89 Al, 0.2 C, 0.15 Fe, 0.015 N, 3.87 V. Specimen: 1.0 in × 0.5 × 0.125 in. Solution heat-treated at 1200 K for 20 min. and aged at 755 K for 4 h. Temperature range 422 to 922 K. Curve (3) Sample: Nominal composition. Mill annealed. Temperature range: 311 to 811 K. Curve (4) Sample: 5.89 Al, 0.02 C, 0.15 Fe, 50 H (ppm), 0.015 N, 3.87 V; solution heat-treated at 1700 °C for 20 min; oil quenched; aged at 900 °C for 4 h and air cooled. Curve (5) Sample: Nominal composition, no condition given. As reported *Thermophysical Properties of Matter,* Vol 1: Thermal Conductivity, Metallic Elements and Alloys, Y.S. Touloukran, *et al.,* Ed., Plenum Press, New York, 1970, p 1073-1076

Ti-6Al-4V: Normal spectral emissivity vs wavelength

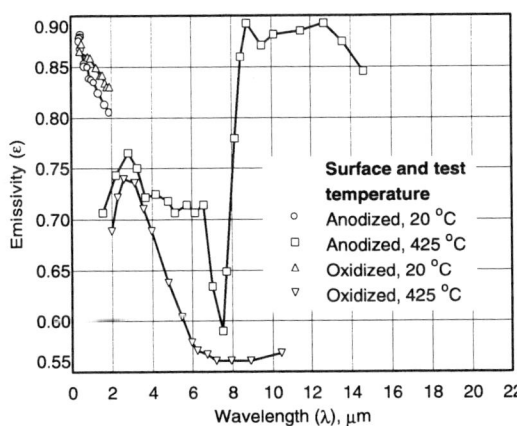

Source: *Thermophysical Properties of Selected Aerospace Materials, Part I: Thermal Radiative Properties,* Y.S. Touloukian and C.Y. Ho, Ed., Thermophysical and Electronic Properties Information Center, CINDAS, Purdue University, 1976, p 185-187

Ti-6Al-4V: Total emittance vs. temperature

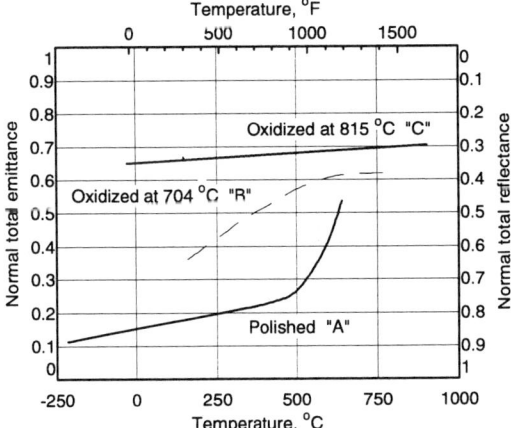

Source: "RMI Ti-6Al-4V," RMI Co., Niles, OH

516 / Alpha-Beta Alloys

Transition Temperatures

Beta transus temperatures of Ti-6Al-4V range from 955 to 1010 °C (1750 to 1850 °F) depending on alloying (see Technical Note B) and interstitial impurities, particularly oxygen.

Standard Ti-6Al-4V: 995 ± 15 °C (1825 ± 25 °F)

ELI grade: 975 ± 15 °C (1790 ± 25 °F)

Liquidus Temperature. Aluminum and vanadium alloying decreases the melting temperature such that the liquidus temperature of Ti-6Al-4V is below the melting point of pure titanium. Reported values range from 1635 to 1670 °C (2975 to 3040 °F). The nominal value is about 1650 to 1660 °C (3000 to 3020 °F), with a variance of ±15 °C (±25 °F).

Solidus Temperature. Nominal value of 1605 ± 10 °C (2920 ± 20 °F), with values as high as 1635 °C (2975 °F) reported (W. Lepkowski and J. Holladay, *The Physical Properties of Titanium and Titanium Alloys*, Battelle, 1956, p 36).

Martensite-start Temperature. 800 °C (1470 °F)

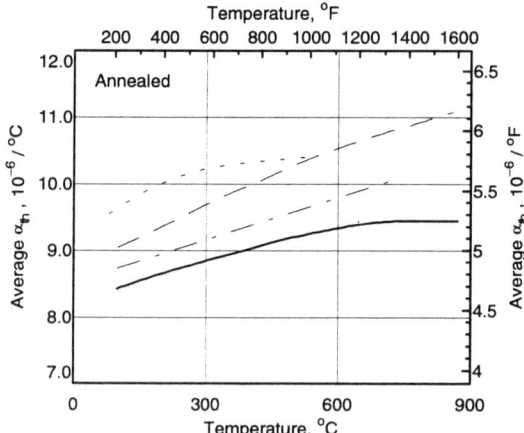

Ti-6Al-4V annealed sheet: Variation in linear thermal expansion coefficients

Average coefficients from room temperature to the indicated temperatures.
Source: *Aerospace Structural Metals Handbook*, Vol 4, Belfour Stulen Inc., 1972

Thermal Expansion

Thermal Expansion Coefficients. The average linear thermal expansion coefficient is isotropic in a texture-free alloy and anisotropic in a textured alloy. Different expansion coefficients are measured along the longitudinal and transverse directions of a rolled sheet because of the anisotropy of the α phase. The thermal expansion along the c-axis is up to 20% larger than perpendicular to the c-axis. The volume thermal expansion coefficient, β_{th}, is related to the linear isotropic expansion coefficient, α_{th}, by: $\beta_{th} \approx 3\,\alpha_{th}$.

Ti-6Al-4V: Typical thermal expansion coefficients

Temperature interval		Linear expansion coefficient	
°C	°F	10^{-6}/°C	10^{-6}/°F
0-100	32-212	9.0	5.0
20-100	70-212	8.6	4.8
20-200	70-400	9.0	5.0
0-315	32-600	9.5	5.3
20-315	70-600	9.2	5.1
20-425	70-800	9.4	5.2
0-540	32-1000	10.1	5.6
20-540	70-1000	9.5	5.3
20-650	70-1200	9.7	5.4
0-650	0-1200	10.6	5.9
0-815	0-1500	11.0	6.1

Source: RMI Co., and *Metals Handbook*, Vol 2, 10th ed.

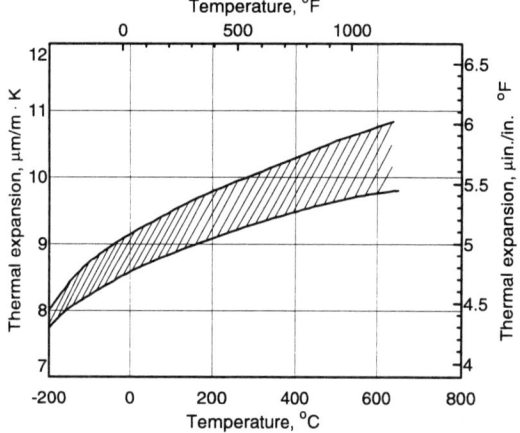

Ti-6Al-4V: Mean coefficient of thermal expansion

Expansion data are for room temperature to indicated temperature for both mill-annealed and solution treated and aged stock.
Source: *Metals Handbook*, Vol 3, 9th ed., *Properties and Selection: Stainless Steels, Tool Materials, and Special-Purpose Metals*, American Society for Metals, 1980

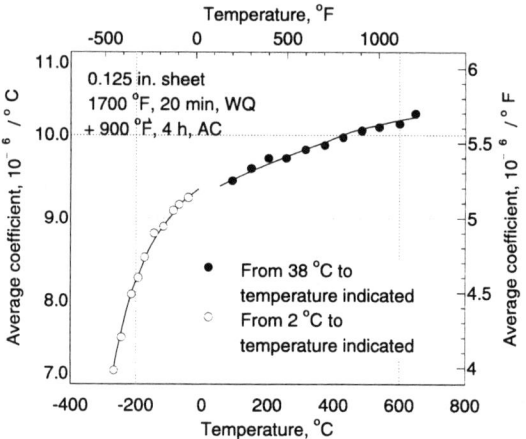

Ti-6Al-4V: Linear thermal expansion coefficient

0.125 in. sheet
1700 °F, 20 min, WQ
+ 900 °F, 4 h, AC

● From 38 °C to temperature indicated
○ From 2 °C to temperature indicated

Aged sheet as indicated.
Source: *Aerospace Structural Metals Handbook*, Vol 4

Design Mechanical Properties

Ti-6Al-4V: Mechanical properties of various metallic alloys

Property	2024-T3	7075-T6	Ti-6Al-4V	4130 Steel
Strength(a) and density				
F_{tu}, MPa (ksi)	441 (64)	537 (78)	924 (134)	1379 (200)
F_{ty}, MPa (ksi)	324 (47)	482 (70)	868 (126)	1213 (176)
F_{cy}, MPa (ksi)	269 (39)	475 (69)	917 (133)	1248 (181)
Density, ρ, g/cm^3 (lb/in.^{3})	2.795 (0.101)	2.795 (0.101)	4.428 (0.160)	7.833 (0.283)
Specific strengths				
F_{tu}/ρ, m (in.) $\times 10^3$	16.1 (634)	19.6 (772)	21.3 (838)	17.9 (707)
F_{ty}/ρ, m (in.) $\times 10^3$	11.8 (465)	17.6 (693)	20.0 (788)	15.8 (622)
F_{cy}/ρ, m (in.) $\times 10^3$	9.8 (386)	17.3 (683)	21.1 (831)	16.2 (640)

(a) A or S values, for sheet gage 1.5 mm (0.060 in.) at room temperature. Design values obtained from MIL-HDBK 5

Sheet and Plate

Ti-6Al-4V: Design tensile properties for annealed sheet and plate per MIL-T-9046 and AMS 4911

Thickness		Basis	Ultimate tensile strength (L-LT)(a)		Tensile yield strength (L-LT)(a)		S-basis elongation, %	
mm	in.		MPa	ksi	MPa	ksi	L	LT
<4.75	<0.1875	A	924	134	868	126	8	8
		B	958	139	903	131	...	...
4.75-50	0.1875-2.000	A	896	130(b)	827-848	120-123(c)	10	10
		B	930-951	135-138(c)	862-903	125-131(c)	...	...
50-100	2.00-4.000	S	896	130	827	120	10	10

(a) Longitudinal (L) and long-transverse (LT) values unless noted. (b) Actual A values are higher than specification values as follows: 131 ksi in L direction and 132 ksi in LT direction. (c) L and LT values, respectively. Source: MIL-HDBK 5, 1992

Ti-6Al-4V: Design compressive and shear strengths of annealed sheet and plate

Thickness		Basis	Compressive yield strength				Ultimate shear strength	
			L direction		LT direction			
mm	in.		MPa	ksi	MPa	ksi	MPa	ksi
<4.75	<0.1875	A	917	133	930	135	600	87
		B	951	138	972	141	620	90
4.75-50	0.1875-2.000	A	855	124	896	130	544	79
		B	889	129	979	142	579	84
50-100	2.00-4.000	S	855	124	896	130	544	79

Source: MIL-T-9046, AMS 4911, MIL-HDBK 5, 1992

Ti-6Al-4V: Design bearing strengths of annealed sheet and plate

Thickness		Basis	Ultimate bearing strength(a)				Bearing yield strength(a)			
			e/D = 1.5		e/D = 2.0		e/D = 1.5		e/D = 2.0	
mm	in.		MPa	ksi	MPa	ksi	MPa	ksi	MPa	ksi
<4.75	<0.1875	A	1468	213	1875	272	1179	171	1434	208
		B	1523	221	1951	283	1227	178	1496	217
4.75-50	0.1875-2.000	A	1420	206	1792	260	1130	164	1337	194
		B	1475	214	1903	276	1234	179	1461	212
50-100	2.00-4.000	S	1420	206	1792	260	1130	164	1337	194

(a) Dry pin bearing values. Source: MIL-T-9046, AMS 4911, and MIL-HDBK 5, 1992

Annealed Bar

Ti-6Al-4V: S-basis tensile strengths and A-basis ductility of annealed bar

Thickness		S-basis Ultimate tensile strength (L-LT)(a)		S-basis Tensile yield strength (L-LT)(a)		A-basis Elongation, %		A-basis RA, %	
mm	in.	MPa	ksi	MPa	ksi	L	LT	L	LT
Per AMS 4928									
<13	<0.500	930(b)	135(b)	861	125(b)	10(f)	10(b)(f)	25(f)	20(b)(f)
13-25	0.500-1.000	930	135	861	125	10	10(b)	25	20(b)
25-50	1.000-2.000	930	135	861	125	10	10(b)	25	20(b)
50-75	2.000-3.000	896	130	827	120	10	10(b)	25	20(b)
75-100	3.000-4.000	896(c)	130(c)	827(c)	120(c)	10	10	25	20
100-125	4.000-5.000	896(c)	130(c)	827(c)	120(c)	10	10	20	20
125-150	5.000-6.000	896(c)	130(c)	...	...	10	10	20	20
Per MIL-T-9047(d)									
<13	<0.500	896(e)	130(e)	827(e)	120(e)	10(f)	10(e)(f)	25(f)	25(e)(f)
13-25	0.500-1.000	896	130	827	120	10	10(e)	25	25(e)
25-50	1.000-2.000	896	130	827	120	10	10(e)	25	25(e)
50-75	2.000-3.000	896	130	827	120	10	10(e)	25	25(e)
75-100	3.000-4.000	896(c)	130(c)	827(c)	120(c)	10	10	25	25
100-125	4.000-5.000	896(c)	130(c)	827(c)	120(c)	10	10	20	20
125-150	5.000-6.000	896(c)	130(c)	...	...	10	10	20	20

(a) S-basis in longitudinal (L) and long-transverse (LT) direction, except as noted. (b) Applicable in LT direction providing LT or ST dimension is greater than 63.5 mm (2.5 in.). (c) LT direction only. (d) Bar with cross-sectional area less than 3.1 m^2 (48 in.^{2}). (e) Applicable in LT direction provided LT dimension is greater than 75 mm (3 in.). (f) S-basis ductility. Source: MIL-HDBK 5, Table 5.4.1.0 (c1)

Ti-6Al-4V: A- and B-basis tensile strengths of annealed bar
Per AMS 4928 and MIL-T-9047. See above table for A-basis ductility

Property	Annealed bar with a thickness, mm (in.), of:					
	13-25 (0.500-1.000)	25-50 (1.00-2.000)	50-75 (2.00-3.000)	75-100 (3.00-4.000)	100-125 (4.00-5.000)	125-150 (5.00-6.000)
A-basis values						
F_{tu}, MPa (ksi)						
L	944 (137)	924 (134)	910 (132)	896 (130)	882 (128)	862 (125)
LT	965 (140)	958 (139)	951 (138)	937 (136)	930 (135)	924 (134)
F_{ty}, MPa (ksi)						
L	889 (129)	868 (126)	848 (123)	827 (120)	806 (117)	786 (114)
LT	889 (129)	875 (127)	862 (125)	848 (123)	834 (121)	...
B-basis values						
F_{tu}, MPa (ksi)						
L	979 (142)	965 (140)	951 (138)	930 (135)	917 (133)	903 (131)
T	993 (144)	986 (143)	979 (142)	972 (141)	896(a) (130)(a)	951 (138)
F_{ty}, MPa (ksi)						
L	924 (134)	903 (131)	882 (128)	862 (125)	841 (122)	820 (119)
T	924 (134)	910 (132)	903 (131)	889 (129)	875 (127)	862 (125)

(a) MIL-T-9047 has a B-basis value of 958 MPA (139 ksi) instead of 895 (130 ksi). Source: MIL-HDBK 5, 1 Dec 1991

Ti-6Al-4V: Design compressive and shear strengths of annealed bar

Thickness		Basis	Compressive yield strength L direction		LT direction		Ultimate shear strength	
mm	in.		MPa	ksi	MPa	ksi	MPa	ksi
Per AMS 4928								
<13	<0.500	S	903	131	...	...	572	83
13-25	0.50-1.000	A	903	131	...	...	572	83
		B	965	140	...	...	600	87
25-50	1.00-2.000	A	903	131	...	...	565	82
		B	944	137	...	...	593	86
Per MIL-T-9047(a)								
<13	<0.500	S	855	124	...	...	551	80
13-25	0.500-1.000	A	855	124	...	...	551	80
		B	951	138	...	...	600	87
25-50	1.00-2.000	A	855	124	...	...	551	80
		B	930	135	...	...	593	86

(a) Bar with cross-sectional area less than 3.1 m^2 (48 in.^{2}). Source: MIL-HDBK 5

Ti-6Al-4V: Design bearing strengths of annealed bar

Thickness			Ultimate bearing strength				Bearing yield strength			
			$e/D = 1.5$		$e/D = 2.0$		$e/D = 1.5$		$e/D = 2.0$	
mm	in.	Basis	MPa	ksi	MPa	ksi	MPa	ksi	MPa	ksi
Per AMS 4928										
<13	<0.500	S	1392	202	1765	256	1248	181	1468	213
13-25	0.50-1.000	A	1392	202	1765	256	1248	181	1468	213
		B	1475	214	1869	271	1337	194	1579	229
25-50	1.000-2.000	A	1386	201	1751	254	1248	181	1468	213
		B	1455	211	1841	267	1310	190	1544	224
Per MIL-T-9047(a)										
<13	<0.500	S	1337	194	1682	244	1172	170	1358	197
13-25	0.500-1.000	A	1337	194	1682	244	1172	170	1358	197
		B	1461	212	1834	266	1310	190	1517	220
25-50	1.000-2.000	A	1337	194	1682	244	1172	170	1358	197
		B	1441	209	1806	262	1282	186	1482	215

(a) Bar with cross-sectional area less than 3.1 m^2 (48 in.^{2}). Source: MIL-HDBK 5

Solution Treated and Aged Bar

Shear Strength. A minimum ultimate shear strength of 634 MPa (92 ksi) is specified in MIL-T-9047 for rectangular bar less than 13 mm (0.5 in) thick and with a width of 13 to 200 mm (0.5 to 8 in.).

Ti-6Al-4V: Design (S-basis) tensile properties of STA rectangular bar

Thickness		Ultimate tensile strength (L-LT)		Tensile yield strength (L-LT)		Elongation, %		RA, %	
mm	in.	MPa	ksi	MPa	ksi	L	LT	L	LT
<13	<0.500	1103	160	1034	150	10	10	25	25
13-25	0.50-1.000	1034-1068	150-155	965-999	140-145	10	10	20	20
25-40	1.00-1.500	999-1034	145-150	930-965	135-140	10	10	20	20
40-50	1.50-2.000	965-999	140-145	896-930	130-135	10	10	20	20
50-75	2.00-3.000	930	135	862	125	10	10	20	20
75-100	3.00-4.000	896	130	827	120	8	8	15	15

Note: Per MIL-T-9047. Source: MIL-HDBK 5, 1 Dec 1991

Ti-6Al-4V: Design (S-basis) tensile properties of STA bar (round, square, hexagon)

Thickness		Ultimate tensile strength (L-LT)(a)		Tensile yield strength (L-LT)(a)		Elongation, %		RA, %	
mm	in.	MPa	ksi	MPa	ksi	L	LT	L	LT
<13	<0.500	1135	165	1070	155	10	10	20	20
13-25	0.50-1.000	1105	160	1035	150	10	10	20	20
25-40	1.00-1.500	1070	155	1000	145	10	10	20	20
40-50	1.50-2.000	1035	150	965	140	10	10	20	20
50-75	2.00-3.000	965	140	895	130	10	10	20	20

Note: Per MIL-T-9047. (a) Longitudinal (L) and long-transverse (LT) values unless noted. Source: MIL-HDBK 5, 1 Dec 1991

Extrusions and Forgings

Ti-6Al-4V: Design mechanical properties of annealed extrusions
Per AMS 4935

	Thickness or diameter, mm (in.)			
	Basis			
	<50 (<2.000)		50-75 (2.00-3.000)	
Mechanical properties	A	B	A	B
F_{tu}, MPa (ksi)				
L	896 (130)(a)	944 (137)	896 (130)(b)	930 (135)
LT(c)	896 (130)(a)	958 (139)	896 (130)(b)	958 (139)
F_{ty}, MPa (ksi)				
L	827 (120)	855 (124)	813 (118)	841 (122)
LT(c)	827 (120)(a)	882 (128)	827 (120)	862 (125)
F_{cy}, MPa (ksi)				
L	882 (128)	910 (132)	855 (124)	882 (128)
LT(d)	882 (128)	945 (137)	868 (126)	903 (131)
F_{su}, MPa (ksi)	579 (84)	944 (89)	579 (84)	606 (88)
F_{bru}(d), MPa (ksi)				
(e/D = 1.5)	1489 (216)	1572 (228)	1489 (216)	1544 (224)
(e/D = 2.0)	1848 (268)	1944 (282)	1848 (268)	1916 (278)
F_{bry}(d), MPa (ksi)				
(e/D = 1.5)	1255 (182)	1296 (188)	1234 (179)	1275 (185)
(e/D = 2.0)	1489 (216)	1537 (223)	1461 (212)	1517 (220)
EL(e), %				
L	10	...	10	...
LT(c)	8	...	8	...
RA(e), %				
L	20	...	20	...
LT(c)	15	...	15	...

(a) A values are higher than specification values as follows: F_{tu}(L) and (LT) = 132 ksi and F_{ty}(LT) = 121 ksi. (b) A values are higher than specification values as follows: F_{tu}(L) = 132 ksi and F_{tu}(LT) = 136 ksi. (c) Applicable, providing LT dimension is ≥63.5 mm (2.500 in.). (d) Bearing values are "dry pin" values. (e) S basis. Source: MIL-HDBK 5, 1 Dec 1991

Ti-6Al-4V: Design tensile properties of STA extrusions

Thickness			Ultimate tensile strength (L-LT)(a)		Tensile yield strength (L-LT)(a)		Elongation(b), %		RA(b), %	
mm	in.	Basis	MPa	ksi	MPa	ksi	L	LT	L	LT
<13	<0.500	A	1068	155	951	138	6	6(a)	12	12(a)
		B	1124	163	1013	147	...	...	...	...
13-19	0.50-0.750	A	1041	151	951	138	6	6(a)	12	12(a)
		B	1082	157	986-999	143-145	...	...	...	...
19-25	0.75-1.000	A	1013	147	917	133	6	6(a)	12	12(a)
		B	1055-1068	153-155	965-979	140-142	...	...	...	...
25-50	1.00-2.000	S	965	140	896	130	6	6	12	12
50-75	2.00-3.000	S	896	130	827	120	6	6	12	12

Note: Per AMS 4934. (a) Applicable in LT direction if LT dimension is greater than 63.5 mm (2.5 in.). (b) S basis. Source: MIL-HDBK 5, 1 Dec 1991

Ti-6Al-4V: Design compressive and shear strengths of STA extrusions

Thickness			Compressive yield strength				Ultimate shear strength	
			L direction		LT direction(a)			
mm	in.	Basis	MPa	ksi	MPa	ksi	MPa	ksi
<13	<0.500	A	1013	147	1013	147	648	94
		B	1082	157	1082	157	682	99
13-19	0.50-0.750	A	1013	147	1013	147	634	92
		B	1055	153	1068	155	662	96
19-25	0.75-1.000	A	979	142	958	139	613	89
		B	1034	150	1048	152	641	93
25-50	1.00-2.000	S	958	139	958	139	586	85
50-75	2.00-3.000	S	882	128	882	128	544	79

Note: Per AMS 4934. (a) Applicable in LT direction if LT dimension is greater than 63.5 mm (2.5 in.). Source: MIL-HDBK 5

Ti-6Al-4V: Design bearing strengths of STA extrusions

Thickness			Ultimate bearing strength(a)				Bearing yield strength(a)			
			$e/D = 1.5$		$e/D = 2.0$		$e/D = 1.5$		$e/D = 2.0$	
mm	in.	Basis	MPa	ksi	MPa	ksi	MPa	ksi	MPa	ksi
<13	<0.50	A	1675	243	2144	311	1434	208	1668	242
		B	1765	256	2254	327	1530	222	1772	257
13-19	0.50-0.750	A	1634	237	2089	303	1434	208	1668	242
		B	1696	246	2172	315	1489	216	1723	250
19-25	0.75-1.000	A	1592	231	2034	295	1386	201	1606	233
		B	1654	240	2116	307	1461	212	1689	245
25-50	1.00-2.000	S	1517	220	1937	281	1351	196	1572	228
50-75	2.00-3.000	S	1406	204	1799	261	1255	182	1448	210

Note: Per AMS 4934. (a) Dry pin bearing values. Source: MIL-HDBK 5

Ti-6Al-4V: S-basis mechanical properties of α–β processed and annealed die forgings

Mechanical properties	Thickness, mm (in.)		
	≤50 (≤2.000)	50-100 (2.00-4.000)	100-150 (4.00-6.000)
F_{tu}, MPa (ksi)			
L	930 (135)	896 (130)	896 (130)
LT	930 (135)(a)	896 (130)(a)	896 (130)
ST	...	896 (130)(a)	896 (130)
F_{ty}, MPa (ksi)			
L	862 (125)	827 (120)	827 (120)
LT	862 (125)(a)	827 (120)(a)	827 (120)
ST	...	827 (120)(a)	827 (120)
F_{cy}, MPa (ksi)			
L	...	848 (123)	848 (123)
LT	...	882 (128)	882 (128)
ST	...	...	...
F_{su}, MPa (ksi)	...	544 (79)	544 (79)
F_{bru}, MPa (ksi)			
($e/D = 1.5$)	...	1399 (203)	1399 (203)
($e/D = 2.0$)	...	1772 (257)	1772 (257)
F_{bry}, MPa (ksi)			
($e/D = 1.5$)	...	1179 (171)	1179 (171)
($e/D = 2.0$)	...	1386 (201)	1386 (201)
EL, %			
L	10	10	10
LT	10(a)	10(a)	10
ST	...	10(a)	8
RA, %			
L	25	25	20
LT	20(a)	20(a)	20
ST	...	15(a)	15

Note: Per AMS 4928. (a) Applicable providing LT or ST dimension is ≥63.5 mm (2.500 in.). Source: MIL-HDBK 5, 1 Dec 1991

Typical Strengths

See next section for typical tensile properties

Bearing strength of Ti-6Al-4V is a function of its heat treatment condition and varies directly with tensile strength. It has approximately the same bearing strength as ¼-hard austenitic stainless steel, but up to twice that of aluminum alloys. Ti-6Al-4V is slightly weaker than Ti-6Al-2Sn-4Zr-2Mo. Ti-6Al-4V offers acceptable bearing strength design properties up to 540 °C (1000 °F), though creep could become a factor at these temperatures.

Compressive yield strengths of Ti-6Al-4V vary with mill product forms, gages, and heat treatment conditions. In all cases, it slightly exceeds the corresponding tensile yield strengths. This makes Ti-6Al-4V an ideal candidate in structural design considerations. By comparison, austenitic stainless steels exhibit compression strengths of about only two-thirds of their tensile yield strengths for all conditions from annealed to full hard. Compression and tension yield strengths for Ti-6Al-4V vary directly as a function of temperature up to 540 °C (1000 °F). This applies to both the annealed and solution heat treated and aged (STA) conditions.

In all forming operations, Ti-6Al-4V is susceptible to the Bauschinger effect. The Bauschinger effect is most pronounced at room temperature. Plastic deformation (1 to 5% tensile elongation) at room temperature always introduces a significant loss in compressive yield strength, regardless of the initial heat treatment or strength of the alloys. At 2% tensile strain, for instance, the compressive yield strength of Ti-6Al-4V drops to less than half that for solution treated material. Increasing the deformation temperature reduces the Bauschinger effect; subsequent full thermal stress relieving completely removes it.

Temperatures as low as the aging temperature remove most of the Bauschinger effect in solution treated Ti-6Al-4V and most titanium alloys. Heating or plastic deformation at temperatures above the normal aging temperature for solution treated Ti-6Al-4V causes overaging to occur, and as a result, all mechanical properties decrease.

For highly textured Ti-6Al-4V sheet, the compressive yield strength and compressive elastic modulus are functions of grain direction and are independent of sheet thickness.

Ultimate shear strength of Ti-6Al-4V is generally 60% of the tensile strength. This applies to all wrought and cast forms and includes both annealed and solution treated and aged conditions. No significant change has been noted in this relationship up to 300 °C (600 °F).

Depending on heat treatment condition, shear strengths range from 480 to 700 MPa (70 to 100 ksi). These properties compare well to those of low-alloy steels of similar strength classes. For example, 4130 quench and tempered steel with a tensile strength of 860 MPa (125 ksi) exhibits a shear strength of 520 MPa (75 ksi). Comparatively, Ti-6Al-4V in the mill annealed condition has a tensile strength of 925 MPa (134 ksi) and a shear strength of 545 MPa (79 ksi). For 4130 steel with a tensile strength of 1100 MPa (160 ksi), shear strength is 640 MPa (93 ksi). In the STA condition, Ti-6Al-4V has a tensile strength of 1100 MPa (160 ksi) and shear strength of 660 MPa (96 ksi).

Ti-6Al-4V: S-basis design mechanical properties of α–β or β processed and annealed die forgings

Mechanical properties	Thickness, mm (in.) ≤50 (≤2.000)	50-150 (2.00-6.000)
F_{tu}, MPa (ksi)		
L	896 (130)	896 (130)
LT	896 (130)(a)	896 (130)(a)
ST	...	896 (130)(a)
F_{ty}, MPa (ksi)		
L	827 (120)	827 (120)
LT	827 (120)(a)	827 (120)(a)
ST	...	827 (120)(a)
F_{cy}, MPa (ksi)		
L	...	848 (123)
LT	...	882 (128)
ST	...	...
F_{su}, MPa (ksi)	...	544 (79)
F_{bru}, MPa (ksi)		
(e/D = 1.5)	...	1399 (203)
(e/D = 2.0)	...	1772 (257)
F_{bry}, MPa (ksi)		
(e/D = 1.5)	...	1179 (171)
(e/D = 2.0)	...	1386 (201)
EL, %		
L	8	8
LT	8(a)	8(a)
ST	...	8(a)
RA, %		
L	15	15
LT	15(a)	15(a)
ST	...	15(a)

Note: Per AMS 4920. (a) Applicable providing LT or ST dimension is ≥63.5 mm (2.500 in.). Source: MIL-HDBK 5

Ti-6Al-4V: Double shear strength vs room-temperature UTS

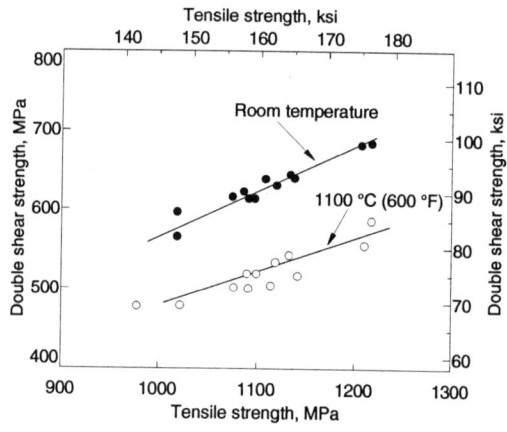

Source: M. Mote, R. Hooper, and P. Frost, Ed., *The Engineering Properties of Commercial Titanium Alloys*, TML-92, Battelle Memorial Institute, 1958, p G-19

Ti-6Al-4V: Typical room-temperature strengths compared to other materials

Material	Ultimate bearing strength		Compressive yield strength		Ultimate shear strength	
	MPa	ksi	MPa	ksi	MPa	ksi
Ti-6Al-4V	1380-2070	200-300	825-895	120-130	480-690	70-100
Alloy steel (4130 or 4xxx)	1380-2755	200-400	415-1725	60-250	480-690	70-100
2xxx aluminum	...	...	...	...	205-275	30-40
7xxx aluminum	620-825	90-120	380-480	55-70	205-345	30-50
Stainless steel	690-1930	100-280	275-620	40-90	...	...

Ti-6Al-4V: Bearing/tensile strength ratios for sheet in several heat treated conditions

			Ratio, %			
			BUS/UTS		BYS/TYS	
Condition	CYS/TYS	SUS/UTS	$e/D = 1.5$	$e/D = 2.0$	$e/D = 1.5$	$e/D = 2.0$
Mill annealed						
n	21	22	22	22	22	22
Average	104.8	67.8	167.1	214.1	145.1	175.9
Ratio	103.5	66.0	163.1	209.5	141.8	169.8
940 °C (1725 °F), 10 min, WQ + 540 °C (1000 °F), 4 h						
n	6	6	6	6	6	6
Average	107.5	62.8	155.0	198.0	147.0	175.1
Ratio	103.6	61.0	151.2	190.4	138.8	163.4
940 °C (1725 °F), 10 min, WQ + 675 °C (1250 °F), 4 h						
n	12	12	12	12	12	12
Average	104.5	65.4	158.4	205.1	141.8	167.5
Ratio	101.5	64.0	154.9	199.4	137.4	162.8
955 °C (1750 °F), 10 min, AC + 675 °C (1250 °F), 4 h						
n	16	16	16	16	16	16
Average	106.4	64.7	163.9	211.3	145.4	174.3
Ratio	104.8	63.8	162.3	207.6	142.6	171.4

Note: n = number of ratios of paired measurements. Average = average of ratios. Ratio = lower 95% confidence interval estimate of average. Testing direction was the same within paired measurements. Approximately half of meaurements were longitudinal; half were long-transverse. Source: D. Maykuth, R. Monroe, R. Favor, and D. Moon, *Ti-6Al-4V Handbook*, Battelle Memorial Institute, 1971, p IV-8

Ti-6Al-4V: Bearing/tensile strength ratios for plate in several heat treated conditions

			Ratio, %			
			BUS/UTS		BYS/TYS	
Condition	CYS/TYS	SUS/UTS	$e/D = 1.5$	$e/D = 2.0$	$e/D = 1.5$	$e/D = 2.0$
Mill annealed						
n	6	6	6	6	6	6
Average	112.5	70.4	161.9	206.3	153.9	184.1
Ratio	110.6	66.3	159.5	202.3	145.9	177.8
940 °C (1725 °F), 10 min, WQ + 540 °C (1000 °F), 4 h						
n	15	15	11	14	11	14
Average	110.6	65.8	157.8	200.6	156.6	182.2
Ratio	108.9	64.1	153.6	198.1	150.8	178.4
940 °C (1725 °F), 10 min, WQ + 675 °C (1250 °F), 4 h						
n	22	18	16	17	16	17
Average	111.2	65.6	163.7	204.2	156.7	181.2
Ratio	109.7	64.3	160.7	202.8	153.4	179.0

Note: n = number of ratios of paired measurements. Average = average of ratios. Ratio = lower 95% confidence interval estimate of average. Testing direction was the same within paired measurements. Approximately half of meaurements were longitudinal; half were long-transverse. Source: D. Maykuth, R. Monroe, R. Favor, and D. Moon, *Ti-6Al-4V Handbook*, Battelle Memorial Institute, 1971, p IV-9

Ti-6Al-4V: Compression properties of highly textured, continuously rolled sheet

Thickness		Grain	Compressive yield strength		Compressive modulus	
mm	in.	direction	MPa	ksi	GPa	10⁶ psi
1.3	0.050	L	871	126.4	107	15.7
		T	1355	196.5	136	19.7
		45°	988	143.3	113	16.4
1.5	0.060	L	928	134.7	108	15.7
		T	1327	192.5	137	19.9
		45°	1001	145.2	120	17.4
2.0	0.080	L	921	133.6	...	...
		T	1188	172.3	...	...
		45°	988	143.3	...	...
3.2	0.125	L	924	134.1	105	15.3
		T	1138	165.1	118	17.1
		45°	986	143.1	114	16.6
3.8	0.150	L	896	130.0	109	15.9
		T	1223	177.4	133	19.4
		45°	1033	149.8	124	18.0

Source: F. Larson and A. Zarkades, in *Properties of Textured Titanium Alloys*, MCIC-74-20, Battelle, 1974, p 39

Typical Room-Temperature Tensile Properties

Room-temperature tensile properties are affected by heat treatment (microstructure), composition (oxygen content), and texture (primarily in sheet). Tensile strength can be changed by more than 200 MPa (30 ksi) by heat treatment and about 70 to 100 MPa (10 to 15 ksi) by oxygen content. Textured sheet can also exhibit variations on the order of 200 MPa (30 ksi) as a function of direction.

Weldments normally fail in the base metal. Generally weld strength is higher and ductility is lower. The exception to this is material that is welded in the β annealed/solution treated and aged condition followed by stress relieving.

Ti-6Al-4V: Minimum and typical tensile properties

Alloy	Condition	Ultimate tensile strength(a)		Tensile yield strength 0.2% offset(a)		Elongation, %
		MPa	ksi	MPa	ksi	
6Al-4V	Annealed	900-993	130-144	830-924	120-134	14
	Solution treated + aged	1172	170	1103	160	10
6Al-4V (low oxygen)	Annealed	830-896	120-130	760-827	110-120	15

(a) Lower values are minimums and upper values are averages. Source: Metals Handbook, *Properties and Selection: Nonferrous Alloys and Special-Purpose Metals*, Vol 2, 10th ed., ASM International, 1990

Effect of Oxygen

Ti-6Al-4V: Tensile properties of annealed castings vs oxygen

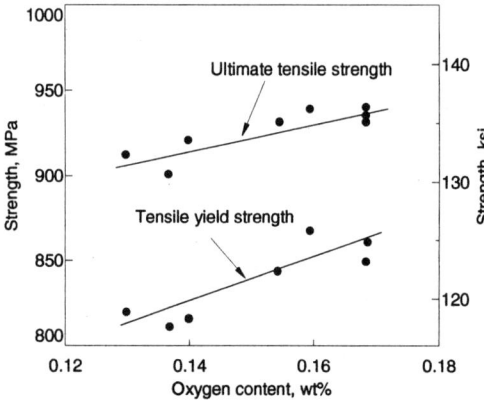

Test specimens 6.4 mm (0.252 in.) in diameter were cast to size and polished with grit paper. Specimens were annealed in argon at 705 °C (1300 °F) for 2 h, air cooled. Each data point represents an average of at least four tests.
Source: Data from Precision Castparts Corp., Portland, OR, D. Coney and M. Lasker, 1969, reported in *Aerospace Structural Metals Handbook*, Code 3801, p 14

Ti-6Al-4V: Tensile strength compared to other materials

Material	Tensile strength	
	MPa	ksi
Ti-6Al-4V	895-1250	130-180
Titanium alloys	240-1500	35-215
Alloy steel	100-2300	15-335
Aluminum	70-700	10-100
Copper	170-1500	25-220

Ti-6Al-4V: Effect of oxygen content on tensile properties

Condition	Oxygen content	Tensile yield strength MPa	ksi	Ultimate tensile strength MPa	ksi	Elongation, %	Reduction of area, %
BA	Low(a)	773	112	856	124	11	23
	Standard(b)	883	128	994	144	11	20
STOA	Low	904	131	973	141	16	47
	Standard	945	137	1042	151	14	33
RA	Low	711	103	876	127	12	36
	Standard(c)	1055	153	1076	156	13	...
BQ	Low	863	125	932	135	6	6
	Standard	980	142	1076	156	10	20

(a) 0.12 wt% oxygen. (b) 0.19 wt% oxygen. (c) 0.20 wt% oxygen (plate). Source: J.C. Williams and E.A. Starke, The Role of Thermomechanical Processing in Tailoring the Properties of Aluminum and Titanium Alloys, in *Deformation, Processing and Structure*, G. Krauss, Ed., American Society for Metals, 1984, p 301-349

Effect of Texture

Ti-6Al-4V: Effect of basal texture on tensile properties

Test direction	Tensile yield strength MPa	ksi	Ultimate tensile strength MPa	ksi	Elongation, %	Reduction in area, %	E GPa	10^6 psi
Longitudinal(a)	883	128	918	133	13	32	110	16.0
Transverse(b)	1063	154	1132	164	16	33	146	21.2

(a) Normal to high density of basal poles. (b) Parallel to high density of basal poles. Source: J.C. Williams and E.A. Starke, The Role of Thermomechanical Processing in Tailoring the Properties of Aluminum and Titanium Alloys, in *Deformation, Processing and Structure*, G. Krauss, Ed., American Society for Metals, 1984, p 301-349

Ti-6Al-4V: Tension properties of highly textured, continuously rolled sheet

Thickness mm	in.	Grain direction	Ultimate tensile strength MPa	ksi	Tensile yield strength MPa	ksi	Elongation in 50 mm (2 in.), %
1.3	0.050	L	994	144.2	865	125.5	10.0
		T	1092	158.5	1070	155.2	10.0
		45°	920	133.5	900	130.6	10.8
1.5	0.060	L	1048	152.0	927	134.5	8.2
		T	1127	163.5	1117	162.0	9.3
		45°	946	137.3	936	135.8	11.7
2.0	0.080	L	1008	146.3	892	129.4	8.8
		T	1057	153.3	1017	147.5	8.5
		45°	938	136.1	895	129.8	10.5
3.2	0.125	L	990	143.7	889	129.0	10.8
		T	1040	150.9	982	142.5	10.0
		45°	962	139.5	909	131.9	12.8
3.8	0.150	L	944	136.9	859	124.6	12.7
		T	1026	148.8	1015	147.3	11.3
		45°	919	133.3	902	130.9	14.7

Source: A. Zarkades and F. Larson, *Properties of Textured Titanium Alloys*, MCIC-74-20, Battelle, 1974

Effect of Processing

Ti-6Al-4V: Effect of thermomechanical processing on tensile properties

Condition	Tensile yield strength MPa	ksi	Ultimate tensile strength MPa	ksi	Elongation, %	Reduction of area, %
α + β forged + recrystallization annealed(a)	711	103	876	127	12.4	36
α + β forged + mill annealed (minimum values)	828	120	897	130	10.0	25
α + β forged + STA (aged 4 h at 594 °C, or 1095 °F)	876	127	938	136	15.2	34
α + β forged + STOA (aged 24 h at 594 °C, or 1095 °F)	904	131	973	141	15.5	47
β forged + AC (BA) + 705 °C, or 1300 °F/2 h/AC	773	112	856	124	11.2	23
β forged + WQ (BQ) + 705 °C, or 1300 °F/2 h/AC	863	125	932	135	5.9	6
α + β forged DA (870 °C, or 1600 °F/2 h/ AC + 705 °C, or 1300 °F/2 h/AC)	856	124	911	132	15.3	47

(a) 925 °C (1700 °F)/4 h/ cool at 50 °C (90 °F)/h to 760 °C (1400 °F)/air cool. Source: J.C. Williams and E.A. Starke, The Role of Thermomechanical Processing in Tailoring the Properties of Aluminum and Titanium Alloys, in *Deformation, Processing and Structure*, G. Krauss, Ed., American Society for Metals, 1984

Ti-6Al-4V: Room-temperature tensile properties as a function of heat treatment

Heat treatment/condition	Micro-structure	Test condition Strain rate, s^{-1}	Direction	Tensile yield strength (0.2% offset) MPa	ksi	Ultimate tensile strength MPa	ksi	Elonga-tion, %	Reduc-tion of area, %	Ref
Cross rolled at 955 °C (1750 °F) and mill annealed	Large elongated α plates + β	1.30×10^{-4}	RD	880	127	952	138	12.7	36.2	4
Cast + hot isostatically pressed, 925 °C (1700 °F), 103 MPa (15 ksi) 4 h	Lamellar α + β	...	...	813	118	917	133	8	13	3
As-cast	Lamellar α + β	...	...	834	121	884	128	4	13	3
Annealed	...	...	...	834	121	906	131	17.5	33	7
1050 °C (1920 °F), 1 h, FC to 700 °C (1290 °F), AC + 625 °C (1155 °F), 4.5 h, AC	...	...	...	876	127	917	133	15.0	...	7
1050 °C (1920 °F), 1 h, AC + 625 °C (1155 °F), 4.5 h, AC	...	...	...	913	132	982	142	12.0	...	7
1050 °C (1920 °F) 1 h, AC	Plate martensite + β layers	...	...	1080	156	1120	162	...	...	10
1050 °C (1920 °F), 0.25 h, WQ + 800 °C (1470 °F), 1 h, WQ	Lamellar	1.00×10^{-3}	RD	1056	153	1278	185	...	15	1
1050 °C (1920 °F), 0.25 h, WQ + 800 °C (1470 °F), 1 h, WQ	Fine lamellar	8.40×10^{-4}	TD	1040	151	...	...	...	20	2
1050 °C (1920 °F), quenched	...	...	...	1070	155	1170	169	11.5	...	5
1050 °C (1920 °F), quenched and aged	...	...	...	1085	157	1140	165	3.5	...	5
1050 °C (1920 °F), AC and aged	...	...	...	945	137	1020	148	2.5	...	5
1050 °C (1920 °F), AC	...	...	...	940	136.3	1050	152	10.5	...	5
1040 °C (1900 °F), 1 h, AC to RT, 730 °C (1345 °F), 4 h, AC to RT	Acicular α + β	1.30×10^{-4}	RD	860	124	956	139	8.8	21.5	4
1040 °C (1900 °F), 0.5 h, WQ + 700 °C (1290 °F), 2 h, AC	...	1.67×10^{-4}	...	862	125	931	135	5.9	6	8
1010 °C (1850 °F), 1 h, WQ	Martensite	...	...	1004	145	1155	167	5	10	6
1040 °C (1900 °F), 1 h, WQ	Martensite	...	...	882	128	1296	188	...	...	12
1040 °C (1900 °F), 1 h, WQ	Martensite	...	...	848	123	1400	203	...	...	12
1040 °C (1900 °F), 0.4 h, 925 °C (1700 °F), 46.5 h, WQ + 760 °C (1400 °F), 1.5 h, WQ	Widmanstätten α	...	...	744	108	1255	182	...	...	12
850 °C (1560 °F), 1 h, WQ + 540 °C (1000 °F), 0.5 h, WQ	...	...	...	940	136.3	...	...	...	...	13
850 °C (1560 °F), 1 h, WQ + 425 °C (800 °F), 0.5 h, WQ	...	...	...	1000	145	...	...	...	...	13
850 °C (1560 °F), 1 h, WQ	...	...	...	906	131	...	...	...	...	13
850 °C (1560 °F), 0.5, WQ + 500 °C (930 °F), 8 h, AC	...	...	...	938	136	1025	148	16.0	28	7
850 °C (1560 °F), quenched	...	...	...	830	120	990	143	16.5	...	5
850 °C (1560 °F), quenched and aged	...	...	...	935	135	1010	146.5	13.0	...	5
850 °C (1560 °F), AC and aged	...	...	...	960	139	1030	149	13.5	...	5
850 °C (1560 °F), AC	...	...	...	920	133	1010	146.5	14.0	...	5
845 °C (1550 °F), 24 h, WQ + 540 °C (1000 °F), 24 h	...	...	...	993	144	1055	153	19	43	6
845 °C (1550 °F), 24 h, WQ + 480 °C (900 °F), 24 h	...	...	...	966	140	1062	154	18	44	6
845 °C (1550 °F), 24 h, WQ	...	...	...	772	112	938	136	25	55	6
845 °C (1550 °F), 24 h, WQ	...	...	...	772	112	938	136	25	55	6
845 °C (1550 °F), 1 h, WQ + 540 °C (1000 °F), 24 h	...	...	...	1014	147	1028	149	20	52	6
845 °C (1550 °F), 1 h, WQ + 480 °C (900 °F), 24 h	...	...	...	1069	155	1138	165	14	47	6
845 °C (1550 °F), 1 h, WQ	...	...	...	710	103	1000	145	26	55	6
845 °C (1550 °F), 1 h, FC (28 °C, 50 °F)/h	...	...	...	897	130	959	139	20	49	6
845 °C (1550 °F), 1 h, FC + 540 °C (1000 °F), 24 h	...	...	...	938	136	938	136	20	53	6
845 °C (1550 °F), 10 h, FC + 480 °C (900 °F), 24 h	...	...	...	945	137	945	137	19	51	6

Note: AC = air cooled, WC = water quenched, FC = furnace cooled, RD = rolling direction, RT = room temperature, TD = transverse direction.
References. 1. C. Muller, A. Gysler, and G. Lutjering, *Strength of Metals and Alloys*, Vol 2, H.J. McQueen, J.P. Bailon, J.I. Jonas, and M.G. Akben, Ed., p 1175-1180. 2. S. Adachi, L. Wagner, and G. Lutjering, *Strength of Metals and Alloys*, Vol 1, H.J. McQueen, J.P. Bailon, J.I. Jonas, and M.G. Akben, Ed., p 2117-2122. 3. D.J. Chronister, S.W. Scott, D.R. Stickle, D. Eylong, and F.H. Froes, *J. Metals*, Sept 1986, p 51-54. 4. D. Eylon and C.M. Pierce, *Metall. Trans. A. Vol 7A, 1976, p 111-121.* 5. W. Hall and Hammond, *Titanium and Titanium Alloys*, Vol 1, J.C. Williams and A.F. Belov, Ed., 1976, p 601-613. 6. R.G. Sherman and H.D. Kessler, *Trans. ASM 48*, 1956, p 657-676. 7. A.W. Bowen and C.A. Stubbington, *Titanium and Titanium Alloys*, Vol 3, J.C. Williams and A.F. Belov, Ed., 1976, p 1989-2001. 8. A.W. Thompson, J.C. Williams, J.D. Frandsen, and J.C. Chesnutt, *Titanium and Titanium Alloys*, Vol 1, J.C. Williams and A.F. Belov, 1976, p 691-704. 9. A.W. Bowen, *Titanium Science and Technology*, Vol 2, R.I. Jaffee and H.M. Burte, Ed., 1973, p 1271-1281. 10. I.W. Hall and C. Hammond, *Titanium Science and Technology*, Vol 2, R.I. Jaffee and H.M Burte, Ed., 1973, p 1365-1376. 11. G.R. Yoder, L.A. Cooley, and T.W. Crooker, *Metall. Trans. A*, Vol 8, 1977, p 1737-1743. 12. H. Margolin and Y. Mahajan, *Metall. Trans. A, Vol 9, 1978, p 781-791.* 13. K. Bose, N.C. Birla and D.B. Goel, *Trans. Indian Inst. Metals*, Vol 36 (No. 3), 1983, p 181-188

(continued)

Ti-6Al-4V: Room-temperature tensile properties as a function of heat treatment (continued)

Heat treatment/condition	Micro-structure	Test condition Strain rate, s^{-1}	Direction	Tensile yield strength (0.2% offset) MPa	ksi	Ultimate tensile strength MPa	ksi	Elonga-tion, %	Reduc-tion of area, %	Ref
845 °C (1550 °F), 1 h, AC + 540 °C (900 °F), 24 h	...	...	...	969	140	986	143	23	54	6
845 °C (1550 °F), 1 h, AC + 480 °C (900 °F), 24 h	...	...	...	959	139	1041	151	20	50	6
845 °C (1550 °F), 1 h, AC	...	...	...	852	123	993	144	21	50	6
845 °C (1550 °F), 0.75 h, WQ	...	...	...	672	97.5	1014	147	22	55	6
845 °C (1550 °F), 0.5 h, WQ	...	...	...	669	97	1007	146	21	55	6
845 °C (1550 °F), 0.25 h, WQ	...	...	...	690	100	1014	147	21	56	6
815 °C (1500 °F), 1 h, WQ	...	...	...	697	101	973	141	22	53	6
800 °C (1470 °F), 15 h, WQ	Equiaxed	1.00×10^{-3}	RD	1030	149	1404	203	...	59	1
788 °C (1450 °F), 1 h, WQ	...	...	...	790	114	993	144	22	47	6
788 °C (1450 °F), 1 h, AC + 1040 °C (1910 °F), 0.5 h, AC + 732 °C (1350 °F), 2 h, AC	...	...	LT, L	892	129	960	139.2	16	23	11
788 °C (1450 °F), 1 h, AC + 1040 °C (1910 °F), 0.5 h, AC + 732 °C (1350 °F), 2 h, AC	...	...	TL, T	869	126	958	138	11	16	11
788 °C (1450 °F), 1 h, AC	...	...	LT, L	948	137	986	143	15	26	11
788 °C (1450 °F), 1 h, AC	...	...	TL, T	1007	146	1034	150	14	29	11
760 °C (1400 °F), 1 h, WQ	...	...	...	855	124	931	135	20	56	6
750 °C (1380 °F), 6 h, AC	Equiaxed α + 10% untransformed β	...	...	750	108	1020	148	...	...	10
750 °C (1380 °F), quenched	...	...	...	970	140	1050	152	12.5	...	5
750 °C (1380 °F), quenched and aged	...	...	...	975	141	1030	149	12.0	...	5
750 °C (1380 °F), AC and aged	...	...	...	985	142	1030	149	13.0	...	5
750 °C (1380 °F), AC	...	...	...	950	137.8	1020	147.9	13.5	...	5
704 °C (1300 °F), 1 h, WQ	...	...	...	948	137.5	959	139	21	53	6
702 °C (1290 °F), 2 h, AC	...	1.67×10^{-4}	...	774	112	853	123	11.2	23	8

Note: AC = air cooled, WC = water quenched, FC = furnace cooled, RD = rolling direction, RT = room temperature, TD = transverse direction.
References. 1. C. Muller, A. Gysler, and G. Lutjering, *Strength of Metals and Alloys,* Vol 2, H.J. McQueen, J.P. Bailon, J.I. Jonas, and M.G. Akben, Ed., p 1175-1180. 2. S. Adachi, L. Wagner, and G. Lutjering, *Strength of Metals and Alloys,* Vol 1, H.J. McQueen, J.P. Bailon, J.I. Jonas, and M.G. Akben, Ed., p 2117-2122. 3. D.J. Chronister, S.W. Scott, D.R. Stickle, D. Eylong, and F.H. Froes, *J. Metals,* Sept 1986, p 51-54. 4. D. Eylon and C.M. Pierce, *Metall. Trans. A, Vol 7A, 1976, p 111-121.* 5. *W. Hall and Hammond, Titanium and Titanium Alloys,* Vol 1, J.C. Williams and A.F. Belov, Ed., 1976, p 601-613. 6. R.G. Sherman and H.D. Kessler, *Trans. ASM 48,* 1956, p 657-676. 7. A.W. Bowen and C.A. Stubbington, *Titanium and Titanium Alloys,* Vol 3, J.C. Williams and A.F. Belov, Ed., 1976, p 1989-2001. 8. A.W. Thompson, J.C. Williams, J.D. Frandsen, and J.C. Chesnutt, *Titanium and Titanium Alloys,* Vol 1, J.C. Williams and A.F. Belov, 1976, p 691-704. 9. A.W. Bowen, *Titanium Science and Technology,* Vol 2, R.I. Jaffee and H.M. Burte, Ed., 1973, p 1271-1281. 10. I.W. Hall and C. Hammond, *Titanium Science and Technology,* Vol 2, R.I. Jaffee and H.M Burte, Ed., 1973, p 1365-1376. 11. G.R. Yoder, L.A. Cooley, and T.W. Crooker, *Metall. Trans. A, Vol 8, 1977, p 1737-1743.* 12. *H. Margolin and Y. Mahajan, Metall. Trans. A, Vol 9, 1978, p 781-791.* 13. *K. Bose, N.C. Birla and D.B. Goel, Trans. Indian Inst. Metals,* Vol 36 (No. 3), 1983, p 181-188

Ti-6Al-4V: Tensile properties of sheet treated by hydrogenation

Hydrogenation-de-hydrogenation can provide significant strengthening through microstructural refinement during dehydrogenation.

H_2, wt.%	Transformation Temperature °C	°F	Time, h	Dehydrogenation Temperature °C	°F	Time, h	UTS MPa	ksi	TYS MPa	ksi	Elongation in 38 mm (1.5 in.), %	Reduction of area %
0.00	590	1095	4	700	1290	6	990	143	930	135	12.7	32.1
0.43	540	1000	2	650	1200	6	1140	165	1040	151	9.6	22.5
0.46	540	1000	2	760	1400	5	1060	153	960	139	12.3	28.4
0.43	590	1095	2	650	1200	6	1140	165	1030	149	3.8	9.2
0.61	540	1000	16	700	1290	6	1100	159	1010	146	11.8	32.2
0.62	590	1095	4	700	1290	6	1100	159	1010	146	12.2	35.3
0.62	565	1050	8	675	1240	8	1140	165	1050	152	10.7	28.5
0.85	540	1000	16	700	1290	6	1140	165	1100	159	11.6	34.7
0.86	565	1050	8	675	1240	8	1220	177	1180	171	9.8	26.2
0.85	590	1095	4	700	1290	6	1150	166	1120	162	11.6	34.3
0.94	540	1000	16	700	1290	6	1130	164	1100	160	11.8	32.4
0.95	565	1050	8	675	1240	8	1220	177	1180	171	9.4	27.1
0.98	590	1095	4	650	1200	6	1230	178	1170	169	3.6	6.8
0.97	590	1095	4	700	1290	6	1150	166	1120	162	12.6	45.4
0.97	590	1095	4	760	1400	3.5	1080	156	1040	151	14.2	47.7
0.98	590(a)	1095(a)	2	650	1200	6	1250	186	1170	169	4.0	7.6
0.99	590(a)	1095(a)	2	700	1290	6	1160	168	1110	161	14.0	37.0
0.99	590(a)	1095(a)	2	760	1400	3.5	1080	156	1030	149	14.0	43.4
0.96	700	1290	8	650	1200	6	940	136	820	120	9.4	12.4
0.95	590	1095	4	675	1240	8	1210	175	1170	169	8.6	16.4
0.96	700	1290	8	700	1290	6	940	136	830	120	9.8	12.6
0.97	700	1290	8	760	1400	3.5	940	136	840	122	10.3	13.8
1.25	590	1095	8	650	1200	6	1220	177	1180	171	1.8	3.5
1.25	590(a)	1095(a)	8	760	1400	3	1060	165	1020	148	14.0	47.1
1.25	590(a)	1095(a)	8	650	1200	6	1230	178	1200	174	2.1	2.3
As-received, mill annealed							1030	149	1000	145	11.1	29.8
Typical annealed							930	135	900	130	12.4	...
Typical STA							1220	177	1130	164	6.4	...

Note: Sheet 3.3 mm (0.13 in.) thick had a composition of 6.4 to 6.5 wt.% Al, 0.13 to 0.18 wt.% Fe, 0.015 to 0.022 wt.% H, 0.012 to 0.013 wt.% N, 0.13 to 0.14 wt.% O, and 4.2 to 4.4 wt.% V. Specimens were machined, hydrogenated, heat treated at 870 °C (1590 °F) for 0.5 h, then dehydrogenated. Testing was carried out on specimens 15 by 200 mm (0.6 by 7.8 in.), having a reduced gage section 38 mm (1.5 in.) long by 9.5 mm (0.37 in.) wide by 3.3 mm (0.13 in.) thick, with an Instron machine and cross head speed 0.002 cm/s. Values reported are an average of at least two tests. (a) Indicates specimens air cooled to room temperature between beta treatment and transformation. Source: W. Kerr, The Effect of Hydrogen as a Temporary Alloying Element on the Microstructure and Tensile Properties of Ti-6Al-4V, *Metall. Trans. A*, Vol 16, 1985, p 1077

High-Temperature Strength

Typical Values

Ti-6Al-4V was initially developed for aerospace applications with temperatures that did not exceed 350 °C (660 °F). Other more creep-resistant titanium alloys have been developed for temperatures up to about 600 °C (1100 °F).

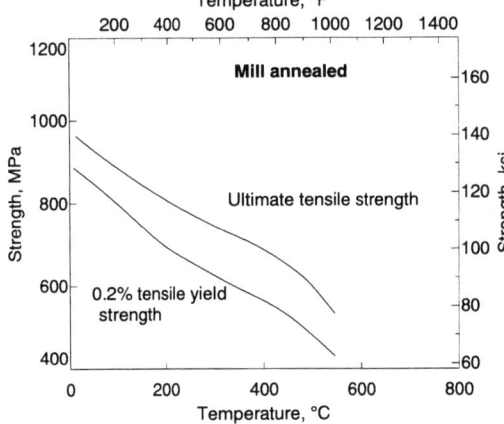

Ti-6Al-4V: Typical tensile strengths vs temperature

Source: *Metals Handbook, Properties and Selection: Stainless Steels, Tool Materials, and Special-Purpose Materials*, Vol 3, 9th ed., American Society for Metals, 1980

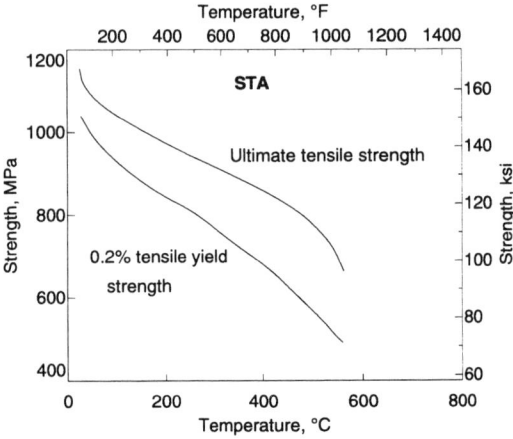

Ti-6Al-4V: STA tensile strengths vs temperature

STA condition—1 h at 955 °C (1750 °F), water quench, age 4 h at 525 °C (975 °F) and air cool.
Source: *Metals Handbook, Properties and Selection: Stainless Steels, Tool Materials, and Special-Purpose Materials*, Vol 3, 9th ed., American Society for Metals, 1980

Ti-6Al-4V: Ultimate bearing strength of aged sheet

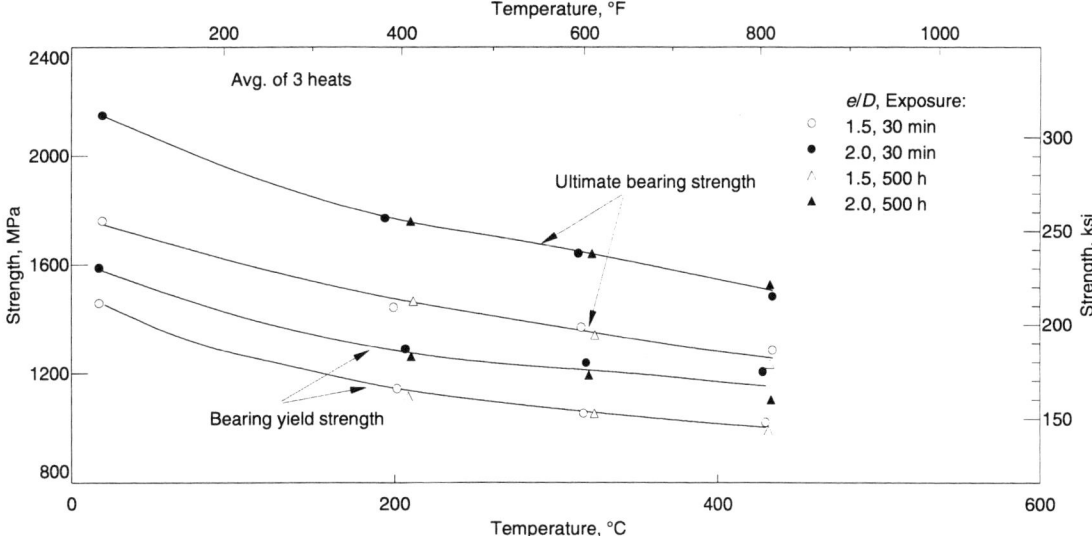

Source: Aerospace Structural Metals Handbook, Vol 4, Batelle

Ti-6Al-4V: STA compressive yield strength vs temperature

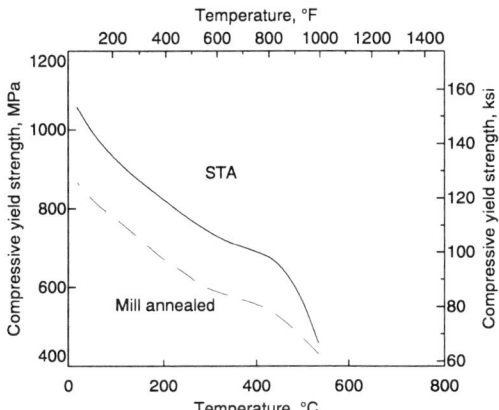

Sheet 1.6 mm (0.063 in.) thick and bar 25 mm (1 in.) diameter or less was tested. STA = heat treatment at 925 °C (1700 °F) for 20 min, water quench, then at 540 °C (1000 °F) for 4 h, air cool.
Source: "Facts About Titanium," RMI Co.

Ti-6Al-4V: Compressive yield/shear strength vs temperature

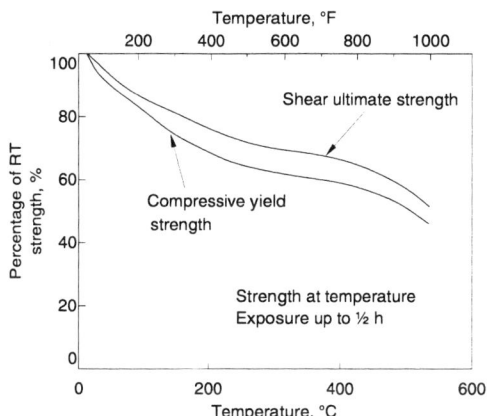

All product forms, annealed.
Source: MIL-HDBK 5, 1 Dec 1991

Annealed Design Values

Ti-6Al-4V: Tensile strength vs temperature

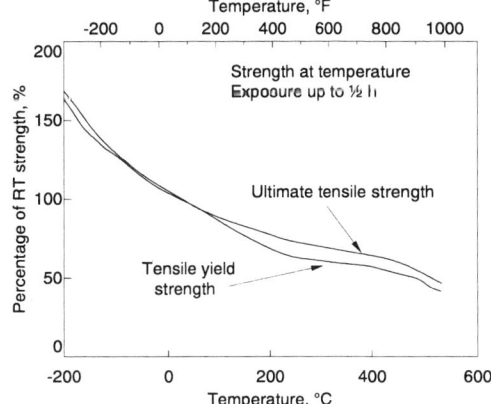

All product forms, annealed.
Source: MIL-HDBK 5, 1 Dec 1991

Ti-6Al-4V: Annealed bearing strengths vs temperature

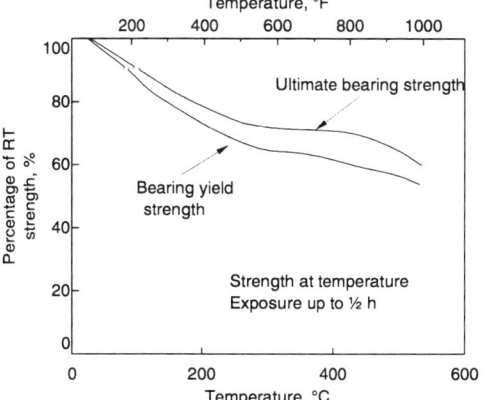

All product forms, annealed.
Source: MIL-HDBK 5, 1 Dec 1991

STA Design Values

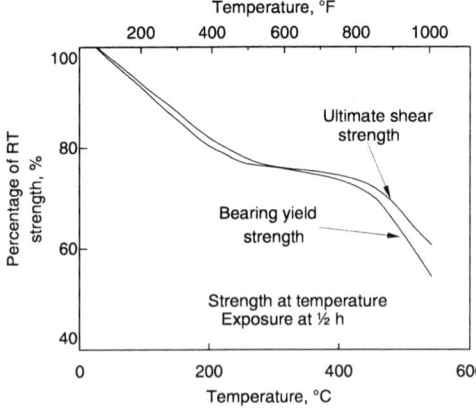

Ti-6Al-4V: STA bearing strength vs temperature

All product forms, solution treated and aged.
Source: MIL-HDBK 5, 1 Dec 1991

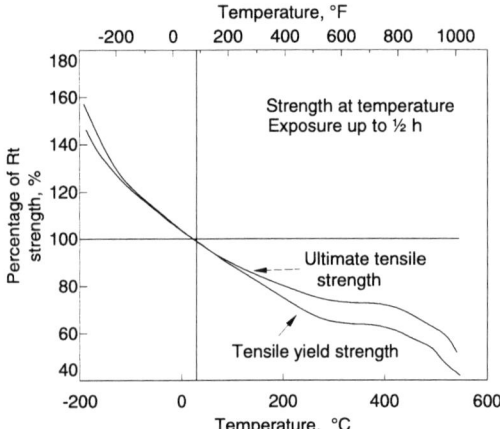

Ti-6Al-4V: STA tensile strength vs temperature

All product forms, solution treated and aged.
Source: MIL-HDBK 5, 1 Dec 1991

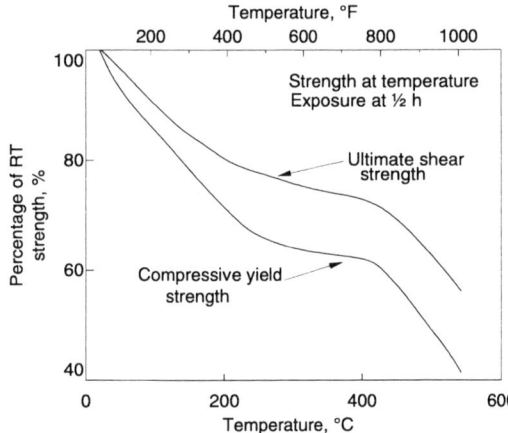

Ti-6Al-4V: STA compressive yield/shear strength vs temperature

All product forms, solution treated and aged.
Source: MIL-HDBK 5, 1 Dec 1991

Creep Properties

Ti-6Al-4V is used up to about 300 or 350 °C (570-660 °F). In general, the near α alloys have the best creep resistance, followed by the α + β alloys, and lastly the β alloys. Optimum creep resistance can be obtained by altering an equiaxed structure to an acicular structure, from β processing or by heat treatment high in the α + β phase field. Like Ti-6242, steady-state (minimum) creep rates for Ti-6Al-4V should be correlated to the amount of primary alpha. (See datasheet for Ti-6Al-2Sn-4Zr-2Mo.)

Ti-6Al-4V: Comparison of temperatures for various structural materials where a stress of 400 MPa (58 ksi) will result in a creep strain of 0.2% in 100 h

Material	Temperature °C	°F
Ti-6Al-4V	385	725
Ti-6Al-2Sn-4Zr-2Mo (α-β processed)	445	835
Ti-6Al-2Sn-4Zr-2Mo (β processed)	520	970
Ti-1100 (β processed)	565	1050
Aluminum (6061)	100	210
IN-718	705	1300

Source: T. O'Connell, Timet

Ti-6Al-4V: 0.2% creep comparison for various alloys

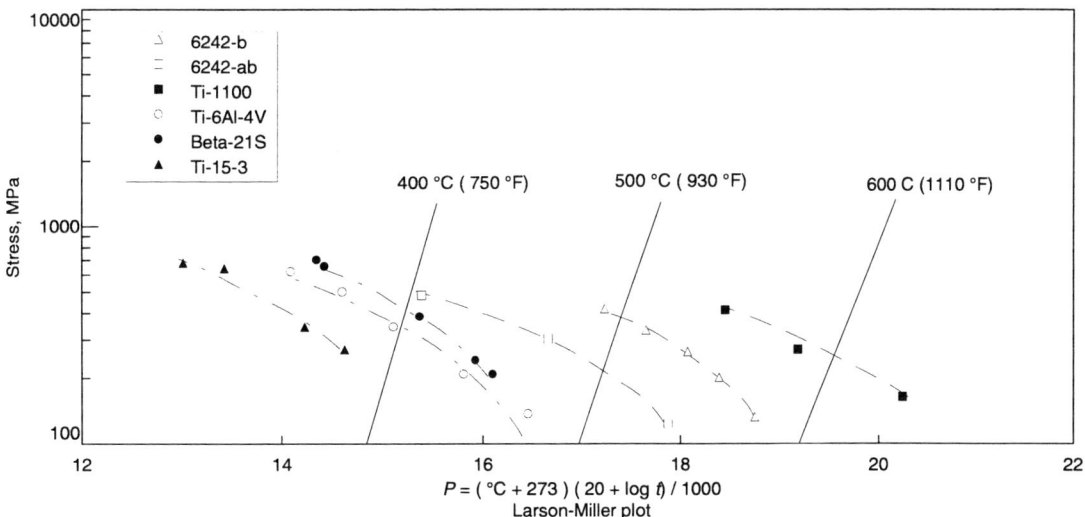

Source: T. O'Connell, Timet

Ti-6Al-4V: Effect of β processing on creep strength

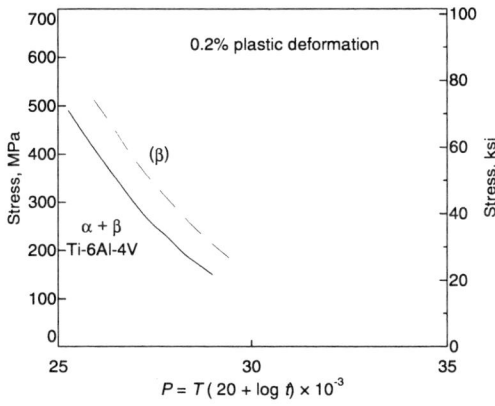

Source: R.A. Wood and R.J. Favor, *Titanium Alloys Handbook*, MCIC HB-02, 1972, Section 1-4:72, p 8

Ti-6Al-4V: 0.2% creep of aged sheet

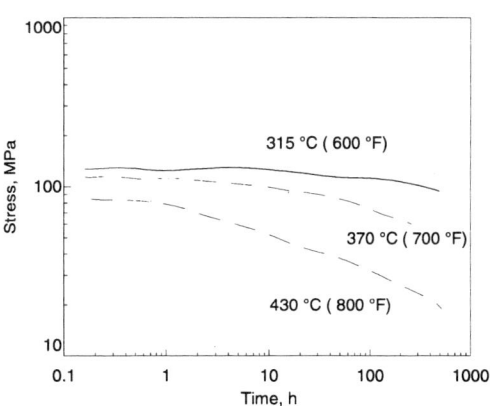

1.6 mm sheet, ST at 925 °C/WQ/aged 480-510 °C, 4 h, AC.
Source: *Aerospace Structural Metals Handbook*, Vol 4, Code 3707, Battelle

Annealed Ti-6Al-4V: Creep at 400 °C (750 °F)

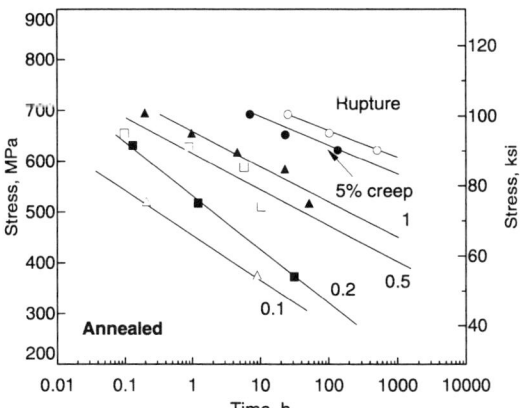

Source: *Metals Handbook, Properties and Selection*, Vol 1, 8th ed., American Society for Metals, 1961

Annealed Ti-6Al-4V: Creep at 455 °C (850 °F)

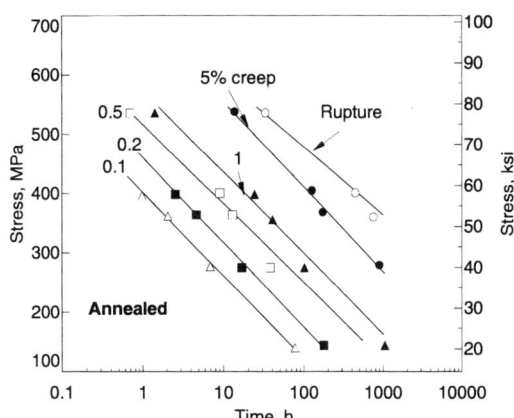

Source: *Metals Handbook, Properties and Selection*, Vol 1, 8th ed., American Society for Metals, 1961

Ti-6Al-4V (STA): Creep at 400 °C

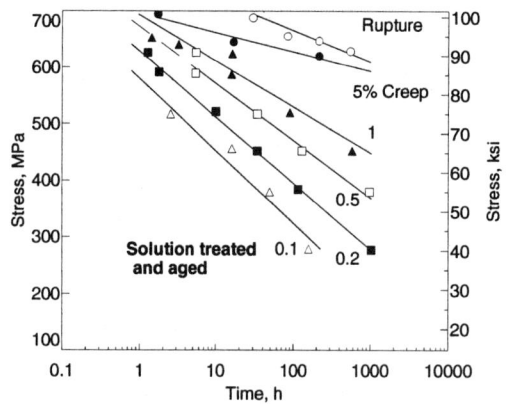

Source: *Metals Handbook, Properties and Selection*, Vol 1, 8th ed., American Society for Metals, 1961

Ti-6Al-4V (STA): Creep at 455 °C

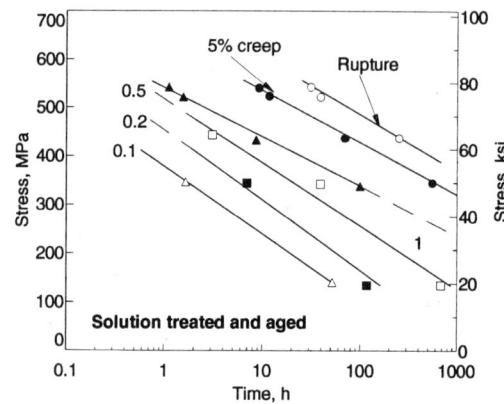

Source: *Metals Handbook, Properties and Selection*, Vol 1, 8th ed., American Society for Metals, 1961

Ti-6Al-4V: Creep strength of cast material

Test temperature		Stress		Plastic strain on loading, %	Test duration, h	Time, h, to reach creep of:		
°C	°F	MPa	ksi			0.1%	0.2%	1.0%
455	850	276	40.0	0	611.2	2.0	9.6	610.0
425	800	276	40.0	0	500.0	15.0	60.0	...
425	800	345	60.0	0	297.5	3.5	11.0	291.5
400	750	448	65.0	0.7	251.4	7.5	22.0	...
370	700	414	60.0	0.3	500	240.0	...	...
315	600	517	75.0	2.04	330.9	0.02	0.04	0.1
260	500	534	77.5	2.1	307.9	0.01	0.02	0.1
205	400	552	80.0	0.56	138.0	0.1	0.13	1.5
205	400	531	77.0	0.8	18.2	0.02	0.04	0.16
175	350	517	75.0	0.01	1006.0	0.4	2.2	...
150	300	517	75.0	...	500	0.25	1.2	...
150	300	517	75.0	...	500	1.7	12.2	...
120	250	517	75.0	0.0	1006.1	9.8	160.0	...

Note: Specimens from hubs of centrifugal compressor impellers that were cast, HIPed (2 h at 900 °C, or 1650 °F), and 103.5 MPa, 15.0 ksi, and aged 1.5 h at 675 °C (1250 °F). Specimen blanks approximately 5.72 by 0.95 by 0.96 cm (2.25 by 0.37 by 0.37 in.) in section size, with the long axis oriented tangential to the hub section, were machined to standard-type creep specimens 3.81 mm (0.150 in.) in diameter. The specimens were lathe turned and then polished with 320-grit emery paper. The creep-rupture tests were performed at 120 to 455 °C (250 to 850 °F) using dead-load-type creep frames in air over a stress range of 276 to 552 MPa (40 to 80 ksi). The microstructure consisted of transformed β grains with discontinuous grain-boundary α and colonies of transformed β which contained packets of parallel-oriented α platelets separated by a thin layer of aged β. Source: A. Chakrabarti and E. Nichols, Creep Behavior of Cast Ti-6Al-4V Alloy, *Titanium '80: Science and Technology*, Proceedings of the 4th International Conference on Titanium, Kyoto, Japan, 19-22 May 1980, Vol 2, H. Kimura and O. Izumi Ti-6Al-4V: , Ed., TMS/AIME, 1980, p 1081-1096

Creep Stability

Ti-6Al-4V: Typical as-exposed creep properties

Test condition	Testing time, h	Ultimate tensile strength		Tensile yield strength (0.2% offset)		Elongation, %	Reduction of area, %
		MPa	ksi	MPa	ksi		
Annealed							
Unstressed, 21 °C (70 °F)	...	925	134	855	124	20	42
Stressed, 345 MPa (50 ksi), 345 °C (650 °F)	16	980	142	860	125	18	49
	100	1070	155	915	133	15	44
	300	1025	149	915	133	18	40
	1000	1025	149	895	130	13	41
Stressed, 345 MPa (50 ksi), 400 °C (750 °F)	16	1005	146	895	130	16	43
	100	960	139	885	128	16	43
	300	1020	148	915	133	20	43
	1000	1015	147	895	130	17	45
Stressed, 345 MPa (50 ksi), 455 °C (850 °F)	16	995	144	885	128	17	39
	100	940	136	850	123	16	48
	300	985	143	915	133	17	34
	1000	1075	156	970	141	15	30
Solution treated and aged							
Unstressed	...	1145	166	1055	153	18	57
Stressed, 310 MPa (45 ksi), 425 °C (800 °F)	150	1180	171	1040	151	16	55

Source: Timet Brochure. Under strain at high temperature, embrittlement may occur from the formation of alpha-2 intermetallic.

General Fatigue Behavior

When evaluating fatigue behavior, comparisons should account for differences in yield strength, grain size, and microstructure. There is evidence regarding Ti-6Al-4V which indicates that superior high-cycle (10^7 cycles) smooth-bar fatigue is obtained when the slip length is small (Ref 1-5). Small slip lengths accompany a fine-grain equiaxed material or by quenching from the β-phase field to produce fine, acicular α′. There is general agreement that the Widmanstätten or colony α + β microstructure has decidedly poorer fatigue strength. In the coarser, equiaxed microstructure the fatigue strength is significantly lower, but it is still better than in the colony microstructure (Ref 6). In general, all microstructural parameters that increase yield strength and/or reduce slip length should improve HCF strength. However, variations in texture, test method (axial versus bending), test conditions (load ratio, frequency), and surface-preparation methods may make comparisons difficult.

References
1. M. Peters, A. Gysler, and G. Lütjering, in *Titanium '80, Science and Technology*, edited by H. Kimura and O. Izumi, Vol 1, TMS-AIME, 1981, p 1777.
2. J.J. Lucas, in *Titanium Science and Technology*, edited by R.I. Jaffee and H.M. Burte, Plenum Press, 1973, p 2081.
3. C.A. Stubbington, 1976 AGARD Conferences Proceedings (No. 185), Reprinted in *Titanium and Titanium Alloys Source Book*, American Society for Metals, 1982, p 140-158.
4. C.A. Stubbington and A.W. Bowen, *J. Mater. Sci.*, Vol 9, 1974, p 941.
5. J.C. Williams and G. Lütjering, in *Titanium '80, Science and Technology*, edited by H. Kimura and O. Izumi, Vol 1, TMS-AIME, p 671.
6. J.C. Chesnutt, A.W. Thompson and J.C. Williams, AFML-TR-78-68, 1978.

Aged Ti-6Al-4V: HCF strength in air for structures with a basal texture

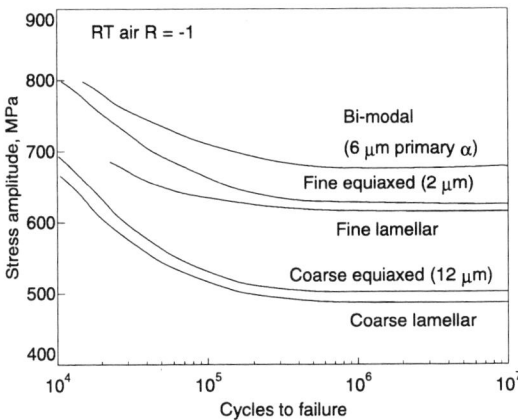

The bimodal microstructure exhibited the highest HCF strength in air, because the basal planes of the strongly textured α grains are separated from each other. The aggressive effect of laboratory air (about 50% relative humidity) is thought to be due to hydrogen, which is most damaging along the basal plane.
Material/Test Parameters: Solution annealed at 800 °C (1470 °F) for 1 h, water quenched, and aged at 500 °C (930 °F) for 24 h.
Source: G. Lütjering and A. Gysler, *Titanium Science and Technology*, Vol 4, Deutsche Gesellschaft für Metallkunde e.V., 1985, p 2068

Ti-6Al-4V: Effect of α grain size on crack-initiation stress

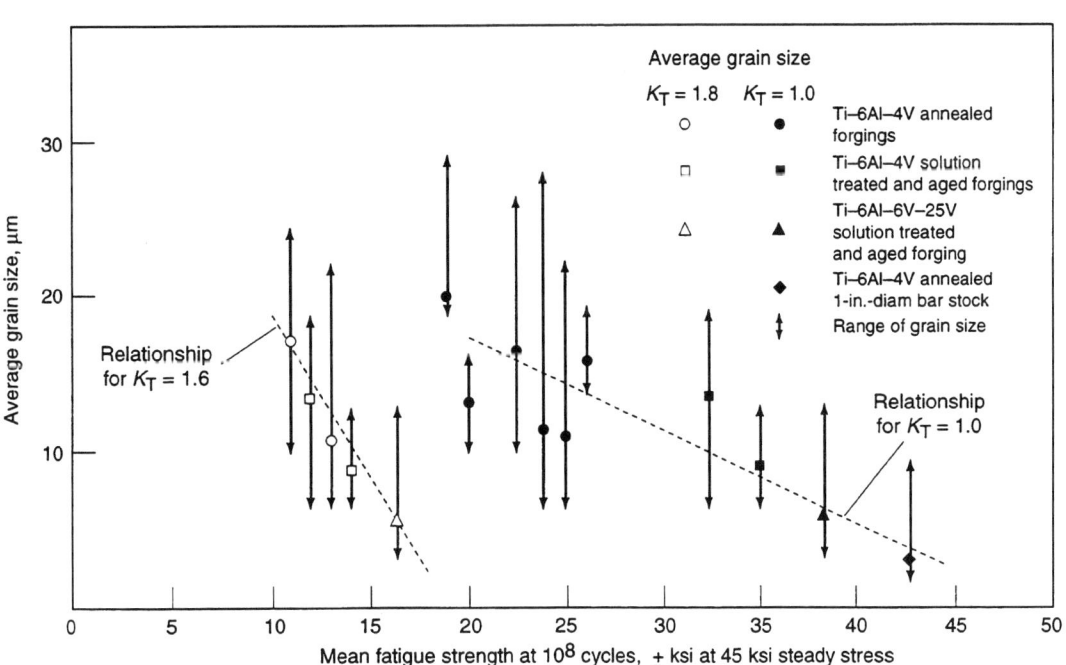

Source: J.J. Lucas and P.P. Konieczny, Relationships Between α Grain Size and Crack-Initiation Fatigue Strength of Ti-6Al-4V, *Metall. Trans.*, Vol 2, 1971, p 911-912

Low-Cycle Fatigue

Low-cycle fatigue (LCF) is the regime characterized by a high maximum stress in a cyclic loading situation. It is also characterized by the existence of significant plastic deformation during the fatigue cycle, at least on a localized scale. Low cycle fatigue failure often occurs with a cyclic life of less than about 10^4 to 10^6 cycles.

The LCF life of Ti-6Al-4V is quite sensitive to heat treatment and microstructural details. The tables illustrate the wide range of fatigue properties obtainable with this alloy. As expected, the variations in fatigue properties are accompanied by variations in other properties, making compromises and optimization for specific uses feasible with alloys like Ti-6Al-4V. Crack initiation, known to consume a large portion of total LCF life, is reported (Gilmore and Imam, see table) to be very sensitive to microstructure, with an $\alpha + \beta$ anneal providing the most resistance. However, an STOA treatment of an $\alpha + \beta$ forging of a beta quenched billet offers promise but with hints of some short life tests (Chakrabarti et al., see table).

Ti-6Al-4V: Fatigue crack initiation vs heat treatment

Heat treatment	Strain range(a)	Cycles to crack initiation(b)		
		Mean	Minimum	Standard deviation
α-β anneal:	±0.006 torsional strain	84,370	73,690	9,875
800 °C (1472 °F) for 3 h, furnace cool (FC) to 600 °C (1112 °F), followed by air cool (AC) to room temperature	±0.012 torsional strain	9,697	9,110	386
Recrystallization anneal:	±0.006 torsional strain	52,840	40,920	10,637
928 °C (1702 °F) for 4 h, FC to 760 °C (1400 °F) at 180 °C/h, FC to 482 °C (900 °F) at 372 °C (702 °F)/h, AC	±0.012 torsional strain	6,232	4,560	1,025
β anneal	±0.006 torsional strain	42,720	35,240	5,684
	±0.012 torsional strain	963	705	184

(a) Sinusoidal strain amplitude with a frequency of 28.6 Hz for ±0.006 and 0.2 Hz for ±0.012 strain. (b) Based on four specimens for each heat treatment with fatigue life taken as the cycles to initiation of a circumferential crack as indicated by an increase in axial elongation.
Source: C.M. Gilmore and M.A. Imam, *Titanium and Titanium Alloys*, J.C. Williams and A.F. Belov, Ed., Plenum Press, 1982, p 637

LCF and fracture toughness of Ti-6Al-4V pancake forgings

Prior stock treatment	Forging condition (soaking condition)	Heat treatments	Number of cycles to failure(a)	Yield strength MPa	ksi	Ultimate tensile strength MPa	ksi	Fracture toughness MPa√m	ksi√in.
As received	$T_\beta - 100$ °C for 0.5 h/Press/OQ	965 °C for 0.5 h/OQ + 705 °C for 2 h/AC	22,325(b)	1038	150	1071	155	40.5	37
			18,808(b)	1038	150	1071	155	40.5	37
As received	$T_\beta - 100$ °C for 0.5 h/Press/OQ	801 °C for 1 h/OQ + 500 °C for 24 h/AC	13,934(b)	1113	161	1126	163.3	33.3	30.3
			16,769(b)	1113	161	1126	163.3	33.3	30.3
Beta quenched	$T_\beta - 165$ °C for 0.5 h/Press/OQ	975 °C for 0.5 h/FAC + 801 °C for 1 h/OQ + 500 °C for 24 h/AC	32,581(c)	1087	157	1124	163	52.8	48.1
			17,960(d)	1087	157	1124	163	52.8	48.1

(a) Tested with a stress ratio of $R = 0.05$, a frequency of 20 cycles/min, and a maximum stress of 880 MPa (127.7 ksi). (b) Failed in gauge. (c) Run-out. (d) Failed at interface of radius and uniform section. *Material parameters:* Bar stock of 7.6 cm diameter had the composition in weight % of Al:6.1, C:0.04, Fe:0.23, H:0.0061, N:0.036, O:0.187, V:4.1, Y:<0.0050. *Reference:* "Microstructure and Mechanical Property Optimization Through Thermomechanical Processing in Ti-6-4 and Ti-6-2-4-6 Alloys," A. Chakrabarti, M. Burn, D. Fournier, and G. Kuhlman, in *Sixth World Conference on Titanium*, 1989, p 1339

Strain Life

Ti-6Al-4V extruded rod: LCF from shear strain

With high shear strains, unstable microstructures, such as those provided by quenching from the solution treatment, provide greater LCF life when compared to annealed or stable microstructures. However, such unstable microstructures are not recommended for cyclic loading applications.

Thermal treatment	Mean life(a)	Minimum life	Standard deviation
α-β Annealed	944	429	443
843 °C (1550 °F) + WQ	2497	1837	717
900 °C (1650 °F) + WQ	9616	8917	758
927 °C (1700 °F) + WQ	2223	2142	605
β solution – 1065 °C (1950 °F) + WQ	2396	1633	487

(a) Results based on four specimens (6.4 mm diam extruded rod) for a shear strain of ±0.02 at 0.2 Hz. The α-β anneal involved holding at 800 °C (1470 °F) for 3 h followed by furnace cooling to 600 °C (1110 °F). Source: M.A. Imam and C.M. Gilmore, Fatigue and Microstructural Properties of Quenched Ti-6Al-4V, *Metall. Trans. A*, Vol 14A, 1983, p 233-240

Stress-Controlled LCF

Ti-6Al-4V: Low-cycle fatigue of heat-treated bar

This figure illustrates the effect of tensile strength on stress controlled LCF of Ti-6Al-4V where the high strength STA condition gives the greatest life. Although strain controlled test results are not presented, the conclusions may be different with strong correlation expected with true ductility.

Ti-6Al-4V: Bar tensile properties for LCF figure

Heat treatment	0.2% proof stress ksi	Tensile strength ksi	Elongation on 5D, %
900 °C WQ + 500 °C (STA 1)	146	163	15
960 °C WQ + 700 °C (STA 2)	141	156	14
Annealed 700 °C	141	149	14

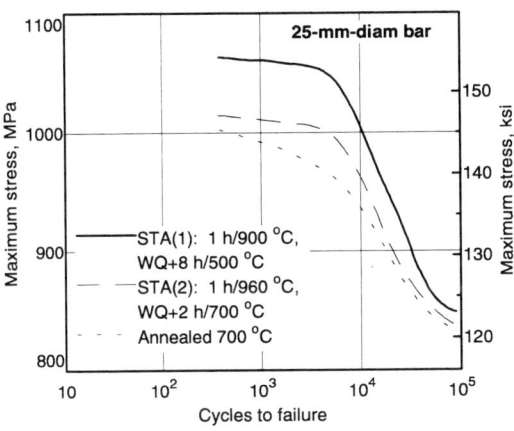

Material/Test Parameters: See table for material condition. Testing was performed at constant maximum load, zero minimum, frequency = 10 cycles/min.
Source: Data from J.R.B. Gilbert, IMI Titanium Ltd., 1988

Cast and P/M

Castings and powder metallurgy (P/M) products are being applied with increasing frequency, primarily for their net shape manufacturing capabilities. There is an LCF penalty for castings and P/M products (see figures), however, these debits can be reduced significantly by hot isostatic pressing (HIP) and heat treatment processing as well as improved PM processing. A compilation from 260 °C LCF tests (see figure) gives results for polished, notched, cast test bars of Ti-6Al-4V with no HIP consolidation.

Cast and P/M

Ti-6Al-4V: LCF properties of cast and wrought STA alloy

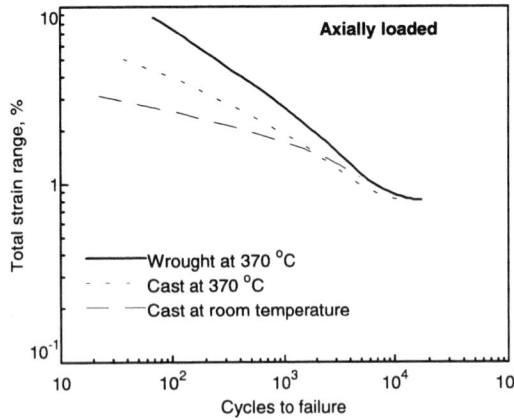

Specimens tested at room temperature were subjected to 20 cycles/min, while the elevated-temperature tests used a testing frequency of 5 cycles/min.
Material/Test Parameters: Cast specimens taken from thick sections.
Source: Data from U. Hellmann and T. Tsareff, General Motors Corp., Detroit Diesel Allison Division, 1971, reported in *Titanium Alloys Handbook*, R. Wood and R. Favor, MCIC-HB-02, Battelle, reprinted 1985, p 2:72-35

Ti-6Al-4V: LCF of PREP-HIP P/M components

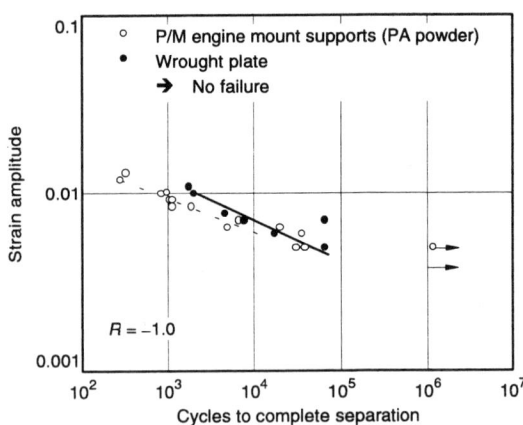

Source: A.S. Sheinker et al., Evaluation and Application of Prealloyed Titanium P/M Parts for Airframe Structures, *Int. J. Powder*, Vol 23 (No. 3), 1987, p 171-176

Ti-6Al-4V: Low-cycle axial fatigue for notched (K_t = 3.5) annealed castings (without HIP)

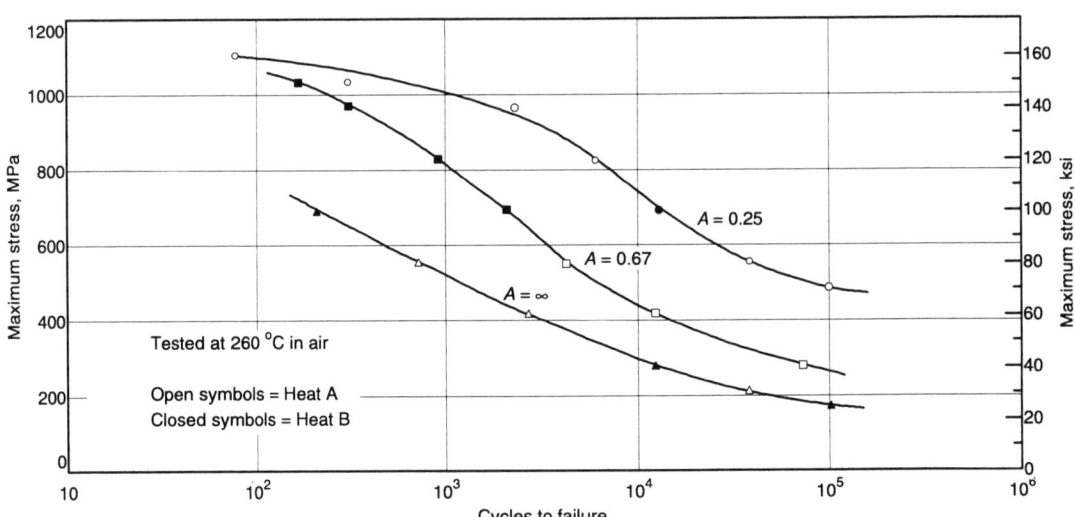

Material/Test Parameters: Cast-to-size specimens were annealed at 700 °C (1300 °F) for 2 h, air cooled, then polished with wire and diamond paste, and finish machined. Test frequency was 300 cycles/min.
Source: Data from R. Dalal, AVCO Corp., reported in *Aerospace Structural Metals Handbook*, Battelle, Code 3801, p 18

Fatigue Limits and Endurance Ratios

Fatigue limits (or endurance limits) represent the value of stress below which a material can presumably endure an infinite number of cycles. For many variable-amplitude loading conditions, fatigue limits may be observed in the regime of 10^7 cycles or more. Fatigue limits generally are influenced by surface conditions because 80 to 90% of fatigue life in the high-cycle regime (about 10^4 cycles or more) involves nucleation of fatigue cracks at the surface. Fatigue limits are also influenced by microstructure (see figures for fatigue strength in air and vacuum). In some cases, stress-controlled fatigue limits are not observed. These cases are generally attributed to periodic overstrains and the absence of interstitial hardening as with very low oxygen levels (see figure on effect of yield strength). Also, the absence of a fatigue limit in this alloy is often associated with subsurface initiations, especially at cryogenic temperatures (see figure).

Ti-6Al-4V: Effect of microstructure on fatigue strength in vacuum

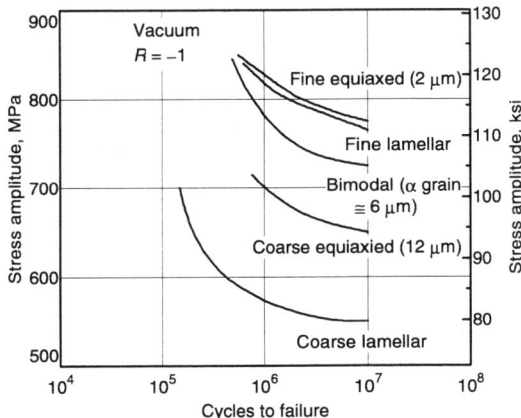

Grain size of primary α shown. The prior β grain size, which limits the length of individual lamellae of bimodal and lamellar microstructures, was 6 to 10 μm for the bimodal microstructure and 300-600 μm for the fine lamellar structure.
Material/Test Parameters: Annealed at 800 °C (1470 °F) for 1 h, water quenched, and aged at 500 °C (930 °F) for 24 h.
Source: G. Lütjering and A. Gysler, *Titanium Science and Technology*, Vol 4, Deutsche Gesellschaft für Metallkunde e.V., 1985, p 2068

Ti-6Al-4V: Effect of microstructure on fatigue strength in air

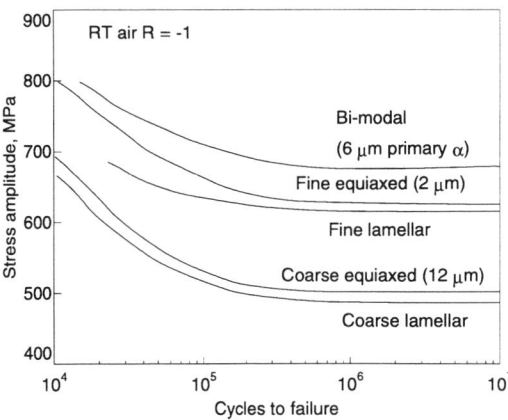

Grain size of primary alpha shown. Annealed at 800 °C (1470 °F) for 1 h, WQ, aged at 500 °C for 24 h.
Source: G. Lütjering and A. Gysler, *Titanium Science and Technology*, Vol 4, DGM, 1985, p 2068

Ti-6Al-4V ELI: Fatigue strength at cryogenic temperatures

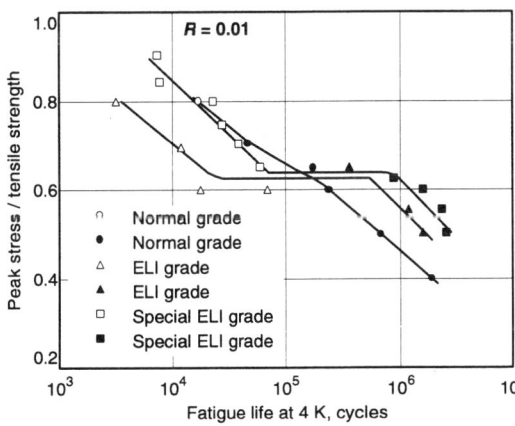

Open symbols indicate fracture from surface; closed symbols indicate internal fatigue initiation near surface.
Material/Test Parameters: Testing was performed at 4.2 K in a liquid-helium-cooled servohydraulic testing apparatus with a sinusoidal cyclic load. Specimens were taken from α-β forged bars (70-mm, 2.75-in., square) that were annealed at 700 °C (1290 °F), for 2 h.
Source: Y. Ito *et al.*, Cryogenic Properties of Extra-Low Oxygen Ti-6Al-4V Alloy, in *6th World Conf. Titanium*, 1989, p 87

Ti-6Al-4V: Effect of yield strength (YS) on fatigue strength

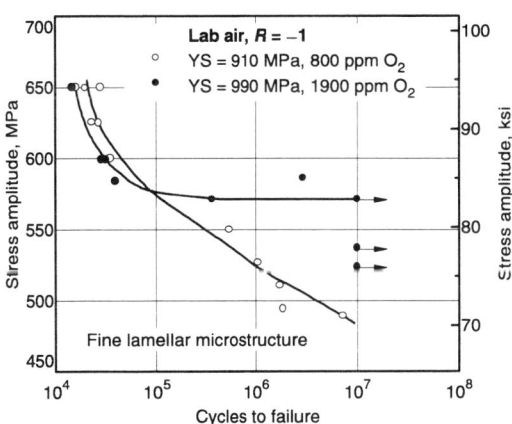

Decreasing the oxygen from the typical value of 0.19 wt% to 0.08 wt% lowers the yield strength and thus the fatigue strength. Oxygen levels influence the mechanisms of precipitation and hardening, which improves resistance to dislocation motion (and thus increases yield and fatigue strength).
Material/Test Parameters: Water quenched from 800 °C (1470 °F) and aged at 500 °C (930 °F) for 4 h.
Source: E.A. Starke, Jr. and G. Lütjering, Cyclic Plastic Deformation and Microstructure, in *Fatigue and Microstructure*, American Society for Metals, 1979, p 237

Endurance Ratio

Fatigue Strength and Fatigue Limits. Fatigue limits may be related to tensile strength, although the fatigue limit-to-tensile strength ratio of titanium alloys may reveal more scatter than quenched and tempered low-alloy steels (see figure, *Metals Handbook*). For alloy Ti-6Al-4V, extensive tensile and smooth-bar fatigue data are presented for different alloy conditions by Sparks and Long and summarized by Williams and Starke (see table). Using their data, regression analysis has been performed to see if a correlation exists between 10^7-cycle fatigue strength and yield or tensile strength. In both cases the coefficient of correlation was smaller than 0.1, indicating that essentially no correlation exists. This tends to point out an important difference between Ti alloys and steels—namely, that the effects of microstructure and strength can be offsetting factors so that no change in fatigue performance might be observed even when strength is increased. Thus, fatigue strength may not correlate with tensile strength alone.

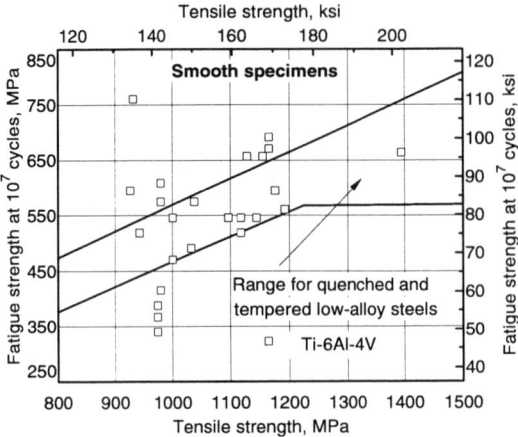

Ti-6Al-4V: Scatter of fatigue strength vs. tensile strength

Source: *Metals Handbook*, Vol 1, 8th ed., *Properties and Selection of Metals*, American Society for Metals, 1961, p 529

Ti-6Al-4V: Fatigue and tensile data for various microstructural conditions

Condition	Yield strength MPa	Yield strength ksi	Tensile strength MPa	Tensile strength ksi	Elongation. %	Reduction in area, %	Stress (smooth) at 10^7 cycles MPa	Stress (smooth) at 10^7 cycles ksi	Stress (notched) at 10^7 cycles MPa	Stress (notched) at 10^7 cycles ksi
10% equiaxed primary α + ann(a)	971	141	1068	155	14	35	537	78	214	31
40% equiaxed primary α + ann	930	135	1013	147	15	41	579	84	255	37
10% equiaxed primary α + STOA(b)	978	142	1061	154	15	41	489	71	220	32
10% equiaxed primary α + ann(c)	958	138	1040	151	14	37	606	88	262	38
50% elongated primary α + ann(c)	923	134	1020	148	13	32	620	90	227	33
β forge + ann	882	128	992	144	11	20	565	82	220	32
β forge/water quench + ann	951	138	1054	153	10	21	606	88	186	27
β forge + STOA	978	142	1075	156	10	20	586	85	220	32
10% equiaxed primary α + ann(d)	882	128	985	143	13	33	620	90	214	31

(a) ann = 705 °C/2 h/AC. (b) STOA = 955 °C/1 h/WQ + 705 °C/2 h/AC. (c) Water quenched off forging press. (d) Low-oxygen material. Source: R.B. Sparks and J.R. Long, AFML-TR-73-301, February 1974. Data summary reported by J.C. Williams and E.A. Starke, in *Deformation, Processing, and Structure*, American Society for Metals, 1984, p 326

Variation of Endurance Ratio

The spread in the endurance ratio (fatigue limit/ultimate tensile strength) is also documented in ASTM STP 459 (see figure). Endurance ratios varied from 0.42 to 0.62 for the unnotched condition. Several of the data points which make up the low side of the band represent slow cooling rates for the beta phase field. The data point for the 1350 °F treatment on the low side of the band was from material containing coarse plate-like alpha, which is further evidence that the coarse plate-like alpha structures lower the endurance ratio of the Ti-6Al-4V alloy. Therefore, heavy sections of Ti-6Al-4V, which contain coarse plate-like or even coarse equiaxed alpha, might be expected to have endurance ratios of 0.4 to 0.45. Fine grained alpha-beta structures or structures produced by water quenching or quenching and aging can be expected to have higher endurance ratios (e.g., between 0.55 and 0.62). However, this trend does not appear to hold true for the notched condition as evidenced by the scatter present in the lower band. The notched endurance ratios varied between 0.17 and 0.3.

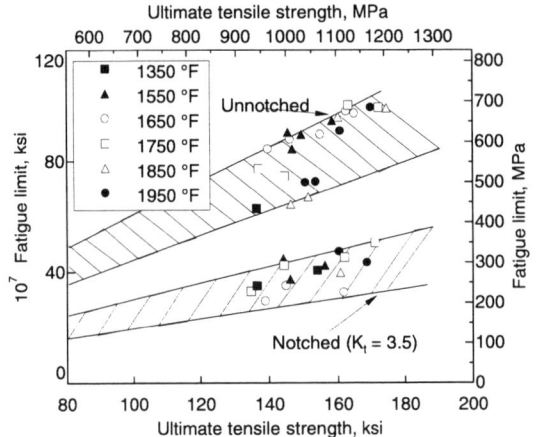

Ti-6Al-4V: Variation in RT endurance ratio

6.35 mm (0.25 in.) specimens cut from as-rolled bar, solution treated at indicated temperatures, and cooled at various rates (furnace, air, water quench). Rotating-beam fatigue at 8000 rpm. Fatigue limits at 10^7 cycles determined by highest stress amplitude at which three specimens ran 10^7 cycles without failure.
Source: L.J. Bartlo, Effect of Microstructure on Fatigue of Ti-6Al-4V Bar, ASTM STP 459, 1969

Surface and Texture Effects on Fatigue

Effect of Residual Stress

Surface residual stress is a predominant factor influencing fatigue, and the residual stress effect is most pronounced in the infinite-life stress range of the endurance-limit regime (10^7 cycles or more). Residual stress is even a more potent indicator than surface roughness in this regime, although residual stress and surface roughness are closely related in many instances. This accounts for the traditional correlation between fatigue strength and surface roughness within reasonable ranges of 2.5 to 5 μm (100 to 200 μin., arithmetic average). However, the correlation between fatigue limits and surface roughness may not be as strong as residual stress.

Surface residual stress is important, but the effect is not simple because of the combined influences of residual stress, cold worked structure and the surface roughness on HCF strength. Surface effects on components of HCF are as follows:

Surface effect	Crack nucleation	Crack propagation
Surface roughness	Accelerates	No effect
Cold work	Retards	Accelerates
Residual compressive stress	Minor or no effect	Retards

Surface roughness and cold work are often associated with the process of inducing residual stress. Because HCF is often largely dominated by fatigue crack nucleation and small crack growth to an observable size, applications of general rules of thumb regarding residual stresses must be exercised with care.

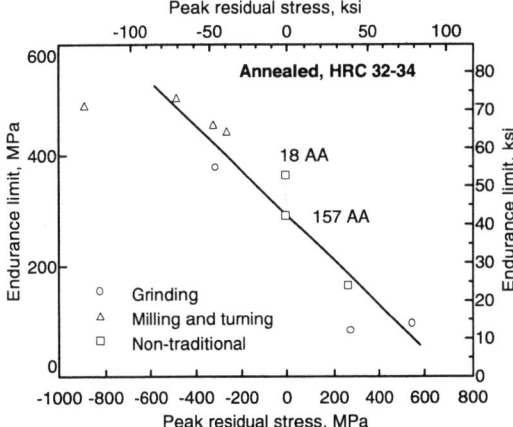

Ti-6Al-4V: Correlation between endurance limit and peak residual stress

Source: W.P. Koster, Effect of Residual Stress on Fatigue of Structural Alloys, Practical Applications of Residual Stress Technology (Clayton Rudd, Ed.), ASM International, 1991

Effect of Texture

There are two major texture considerations. There is a microstructural texture in which directly observable microstructural features are aligned on a scale large compared to the size of the individual features. Elongated beta grains in the rolling or forging direction is a typical example of microstructure texture.

There is also a crystallographic texture in which most of the alpha phase grains are aligned such that a unique direction in the hexagonal unit cell of alpha titanium is oriented close to the same direction relative to some physical attribute of the titanium sample, such as the rolling direction for sheet or plate. An example of the effect of crystallographic texture is shown (see figure from Zarkades and Larson), where fatigue life in the various test directions in a plate product are large.

Crystallographic texture is seldom reduced by sub-beta transus heat treatments, and the texture of the α phase in equiaxed and bi-modal microstructures can be a factor on fatigue limits. In a vacuum environment (see figure), the application of stress in the transverse direction resulted in higher HCF strength values as compared to tests in rolling direction. This can be directly correlated to the higher yield stress value observed when the stress is applied perpendicular to the basal planes. The opposite ranking was observed when the tests were performed in laboratory air (see figure), which is explained by hydrogen damage on basal planes. If no shear or normal stresses are acting on the basal planes (tests in rolling direction), then the highest fatigue strength values are observed.

Ti-6Al-4V: Effect of test orientation on fatigue

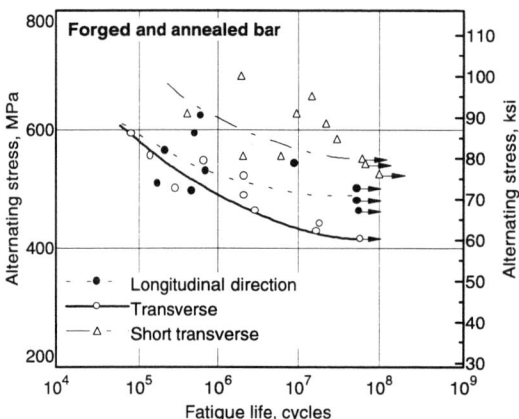

Although much scatter is evident, a large difference in endurance limit is shown for different test orientations of forged bar (57-mm, or 2.25-in., thick) having a mild and varying texture.
Source: A. Bower, The Effect of Testing Direction on the Fatigue and Tensile Properties of a Ti-6Al-4V Bar, *2nd Int. Conf. Titanium*, reported in *Properties of Textured Titanium Alloys*, F. Larson and A. Zarkades, Ed., MCIC-74-20, Battelle Columbus Laboratories, 1974, p 67

Ti-6Al-4V: Influence of texture and test direction

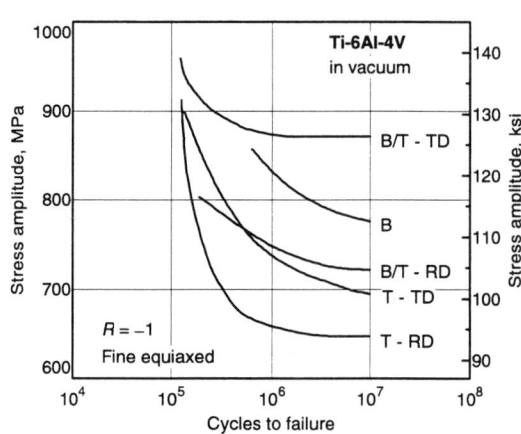

(a)

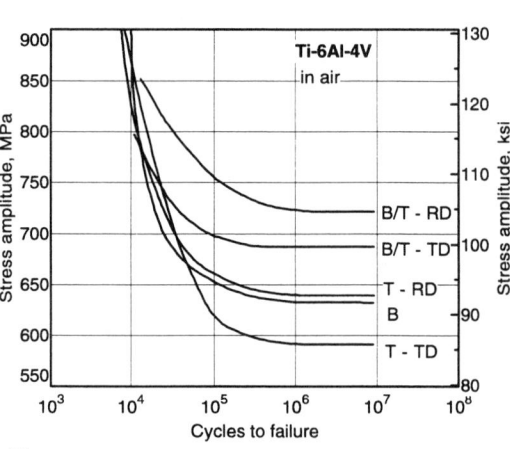

(b)

B, basal; T, transverse texture; B/T, basal-transverse texture; RD, test in rolling direction; TD, transverse test direction.
Material/Test Parameters: Fine equiaxed Ti-6Al-4V, $R = -1$.
Source: A.W. Bowen, *Titanium 80, Science and Technology*, AIME, 1980, p 947

Effect of Surface Treatment

Shot peening and surface finish interact in a complex way to influence HCF. Starting with an undisturbed electropolished surface of a fine lamellar microstructure, shot peening will enhance room-temperature fatigue strength whereas additional electropolishing adds to this enhancement. But, stress relief of the shot peened specimens reduces fatigue strength to levels below the baseline electropolished samples. Adding an electropolish to the stress relieved samples restores most of the benefit. Tests at 500 °C (see two-part figure, Gray et al.) indicate that the effect of shot peening is negative without restoration of the original electropolished surface. Similar effects are seen with fine, equiaxed, microstructures (see figure, Wagner et al).

Ti-6Al-4V: Effect of shot peening and electrolytic polishing

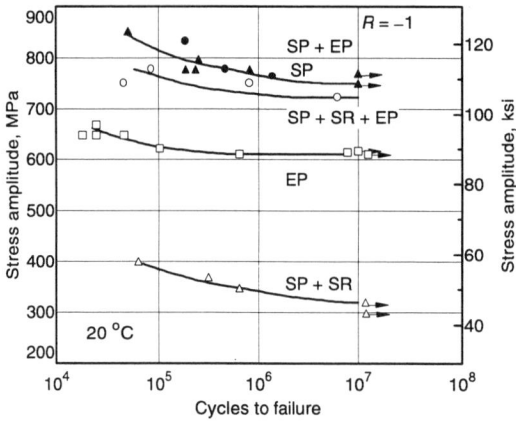

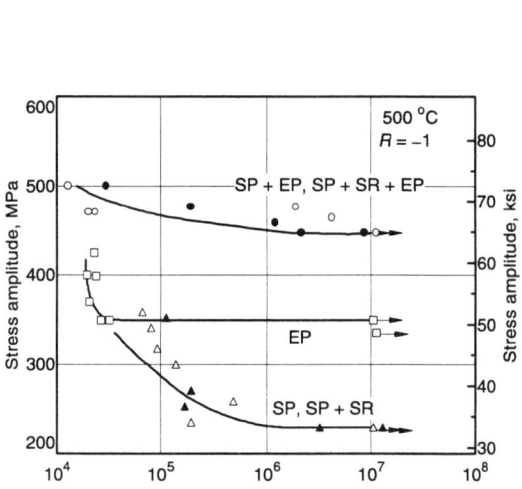

At (a) 20 °C (68 °F) and (b) 500 °C (930 °F). EP, electrolytically polished; SP, shot peened; SR, stress relieved.
Material/Test Parameters: Alloy with a fine lamellar (β-quenched) microstructure was machined into blanks 7 × 7 mm (0.28 × 0.28 in.), annealed at 800 °C (1470 °F) for 1 h and quenched, then heat treated at 600 °C (1110 °F) for 24 h. Specimens were electrolytically polished to remove a layer about 100 μm thick. Shot peening was performed using an Almen intensity of I = 0.28 A (mm). Fatigue tests were performed on smooth hour-glass-shaped specimens in rotating-beam loading ($R = -1$).
Source: H. Gray et al., Influence of Surface Treatment on the Fatigue Behavior of Ti-Alloys at Room and Elevated Temperatures, in *6th World Conf. Titanium*, 1989, p 1895

Ti-6Al-4V: Effect of shot peening on fatigue strength

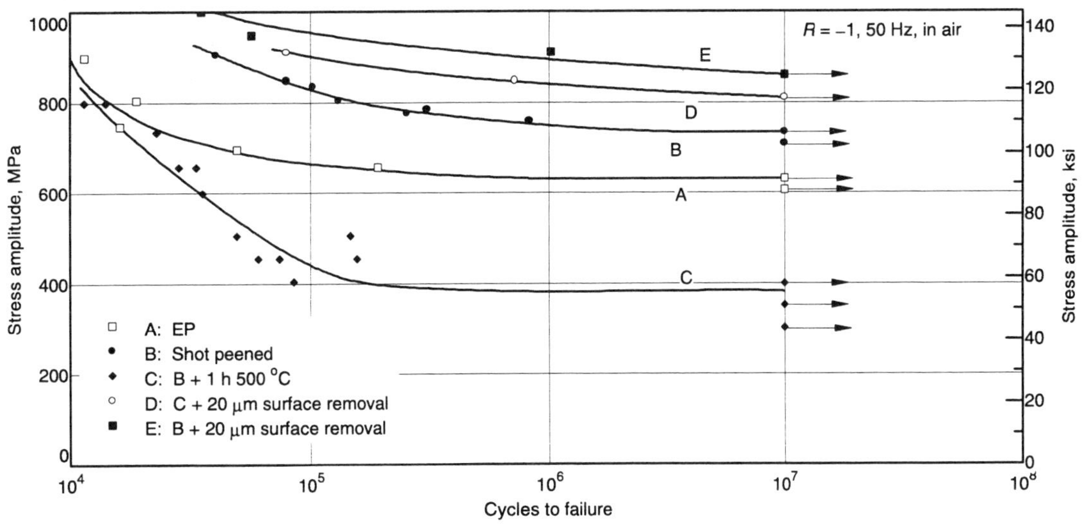

- A: EP
- B: Shot peened
- C: B + 1 h 500 °C
- D: C + 20 μm surface removal
- E: B + 20 μm surface removal

EP, electrolytically polished.
Material/Test Parameters: Test material had a fine equiaxed microstructure and a yield strength of 1100 MPa (160 ksi). Shot peening was performed with S230 steel balls (0.6-mm diam) and an Almen intensity of 15A mm/100. Fatigue testing was done on hourglass-shaped specimens with 3.8-mm (0.15-in.) gage diameter in air on a rotating-beam apparatus ($R = -1$, $f = 50$ Hz).
Source: L. Wagner et al., Influence of Surface Treatment on Fatigue Strength of Ti-6Al-4V," in *Titanium Science and Technology*, Deutsche Gesellschaft für Metallkunde e.V., 1985, p 2147

Fretting Fatigue

Titanium has notoriously poor wear resistance when there is sliding contact with itself and other materials. The resultant fretting has a strong effect on fatigue strengths and limits.

Ti-6Al-4V: Fretting fatigue in shot-peened and coated conditions

Fatigue specimen surface treatment	Normal pressure MPa	Fretting fatigue strength 3×10^6 cycles			
		20 °C		400 °C	
		MPa	ksi	MPa	ksi
None (as received)	35	220	32	220	32
Shot peened	35	270	39	240	35
Shot peened + Cu-Ni-In	35	320	46	245	36
None	140	215	31	190	28

Cyclic tensile load on fretting pad. Source: "Fretting Fatigue in High Temperature Oxidizing Gases," D. Taylor, in *Fretting Fatigue*, R. Waterhouse, Ed., Applied Science Publishers, Ltd., London, 1981, p 177

Ti-6Al-4V: Fretting fatigue at room temperature and 350 °C (660 °F) for alloy in polished and shot-peened conditions

Specimen treatment	Temperature °C	°F	Fatigue strength Unfretted MPa	ksi	After 10^7 cycles Fretted MPa	ksi	Percent reduction of basic room-temperature properties Unfretted	Fretted
Plain polished	20	68	650	95	140	20	0	78
Plain polished	350	660	580	84	140	20	11	78
Shot-peened Almen A7	20	68	600	87	340	49	7	48
Shot-peened Almen A7	350	660	540	78	310	45	17	52

Cyclic tensile load on fretting pad. 93 MPa fretting stress. Source: "Fretting Fatigue in High Temperature Oxidizing Gases," D. Taylor, in *Fretting Fatigue*, R. Waterhouse, Ed., Applied Science Publishers, Ltd., London, 1981, p 177

Influence of Mean Stress

There are indications that the Ti-6Al-4V alloy exhibits an anomalous mean stress dependence of HCF strength if the material was forged in the ($\alpha + \beta$) phase field in contrast to a normal mean stress dependence if the material was β forged or β heat treated (Ref 1-3). The results of an investigation on a fine lamellar structure and a bi-modal structure with a pronounced mixed B/T type of texture tested in RD and TD are shown (see figure, Influence of mean stress). It can be seen that the fine lamellar and the bi-modal structure tested in RD exhibited a normal mean stress dependence of HCF strength whereas the bi-modal structure tested in TD showed the anomalous mean stress dependence, i.e., much lower fatigue strength values with increasing mean stress. No reasonable explanation for this effect is given.

References

1. R.K. Steele and A.J. McEvily, *Eng. Fracture Mech.*, 8 (1976), p 31.
2. J. Broichhausen and H. van Kann, "Titanium Science and Technology," Plenum Press (1973), 1785.
3. A. Atrens, M. Müller, H. Meyer, G. Faber, and M.O. Speidel, "Corrosion Fatigue of Steam Turbine Blade Material," Pergamon Press (1983), p 4-50.

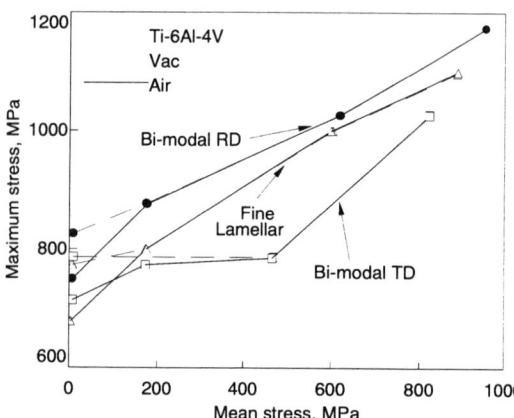

Ti-6Al-4V: Influence of mean stress on HCF strength (10^7 cycles)

Source: G. Lütjering and A. Gysler, *Titanium Science and Technology*, Vol 4, Deutsche Gesellschaft für Metallkunde e.V., 1985, p 2072

Effect of Processing

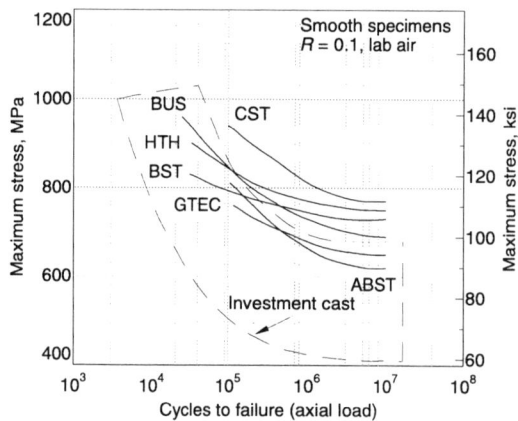

Ti-6Al-4V: Fatigue of investment castings after treatments

ABST, α-β solution treatment; BST, β solution treatment; BUS, broken-up structure; CST, constitutional solution treatment; GTEC, Garrett treatment (long-time, low-temperature anneal); HTH, high-temperature hydrogenation.
Material/Test Parameters: 5-Hz triangular wave form.
Source: D. Eylon and R. Boyer, "Titanium Alloy Net-Shape Technologies," in *Proc. Int. Conf. Titanium and Aluminum*, Paris, Feb 1990

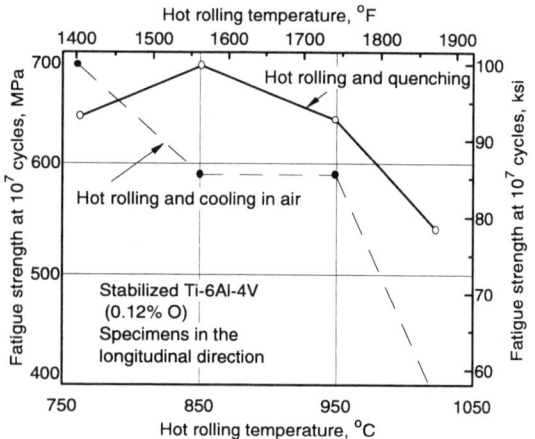

Ti-6Al-4V: Effect of rolling temperature on HCF strength

Material/Test Parameters: Material was 22 mm (0.865 in.) in diameter and was produced by 65% hot rolling at indicated temperatures, followed by annealing at 700 °C (1300 °F) for 1 h, furnace cooled to 500 °C (930 °F), held for 12 h, and air cooled.

Effect of Thermomechanical Processing

Ti-6Al-4V: Effect of working temperatures on HCF strength

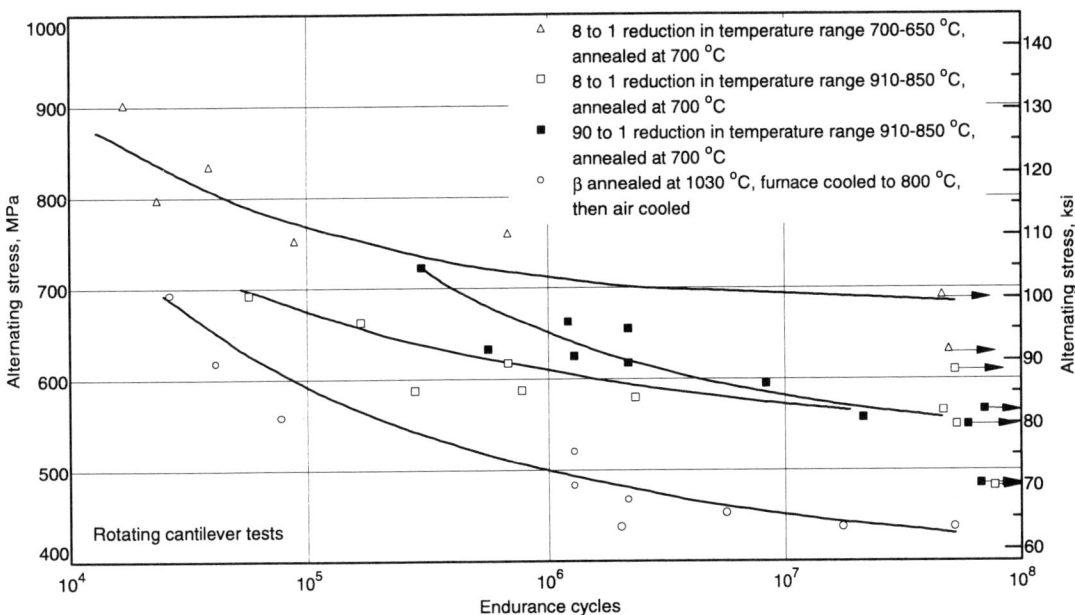

Material/Test Parameters: Ti-6Al-4V bar worked at a high and low temperature in the α-β field are compared with β-annealed material.
Source: C.A. Stubbington and A.W. Bowen, Improvements in the Fatigue Life of Ti-6Al-4V Through Microstructural Control, *J. Mat. Sci.*, Vol 9, 1974, p 941-947

Ti-6Al-4V: Effect of forging and heat treatment on HCF strength (see table on next page for treatments)

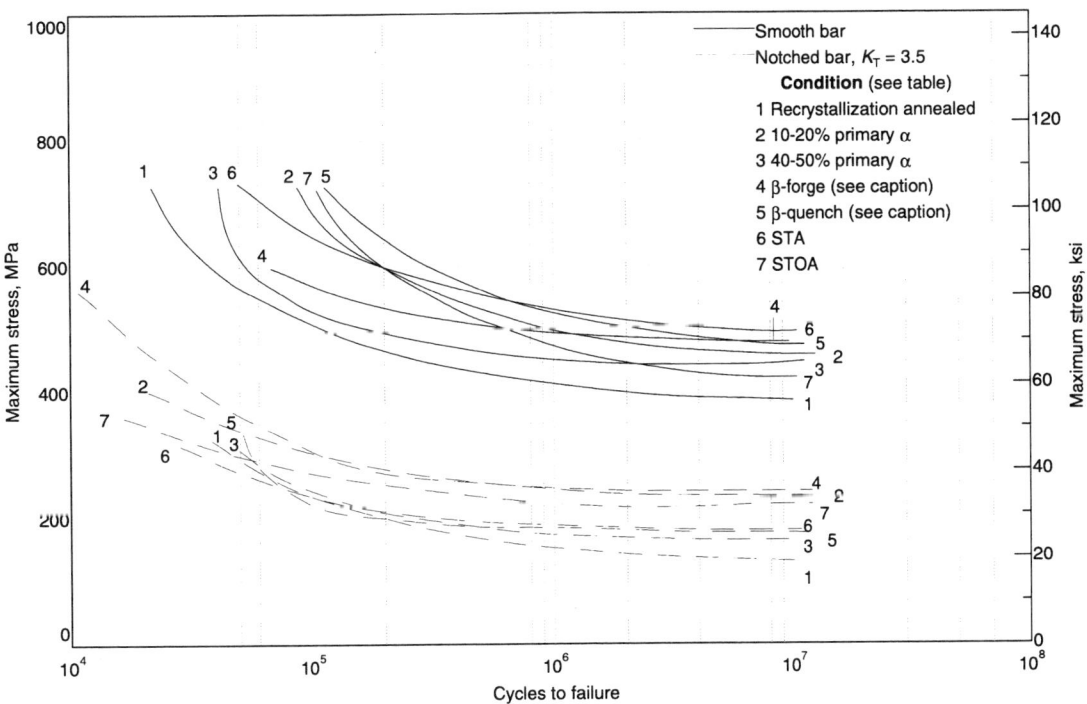

Ti-6Al-4V: Caption table for bottom figure on previous page

Condition	Fabrication	Heat treatment
1	α-β forge ($\beta_t - 75$ °F)/AC α-β finish ($\beta_t - 100$ °F)/AC	925 °C (1700 °F)/4 h/Cool at 50 °C/h (90 °F/h) to 760 °C (1400 °F)/AC
2	α-β forge ($\beta_t - 50$ °F)/AC α-β finish ($\beta_t - 25$ °F)	955 °C (1750 °F)/1 h/AC + 700 °C (1300 °F)/2 h/AC
3	α-β forge ($\beta_t - 75$ °F)/AC α-β finish ($\beta_t - 100$ °F)/AC	870 °C (1600 °F)/2 h/AC + 700 °C (1300 °F)/2 h/AC
4 (β-forge)	β forge ($\beta_t + 75$ °F)/AC β finish ($\beta_t + 75$ °F)/WQ	700 °C (1300 °F)/2 h/AC
5 (β-quench)	β forge ($\beta_t + 75$ °F)/AC β finish ($\beta_t + 75$ °F)/AC	1040 °C (1900 °F)/30 min/WQ + 700 °C (1300 °F)/2 h/AC
6	α-β forge ($\beta_t - 75$ °F)/AC α-β finish ($\beta_t - 75$ °F)/AC	955 °C (1750 °F)/1 h/WQ + 595 °C (1100 °F)/4 h/AC
7	α-β forge ($\beta_t - 75$ °F)/AC α-β finish ($\beta_t - 75$ °F)/AC	955 °C (1750 °F)/1 h/WQ + 595 °C (1100 °F)/24 h/AC

Source: J.C. Chesnutt et al., "Influence of Metallurgical Factors on Fatigue Crack Growth Rate in Alpha-Beta Titanium Alloys", AFML-TR-78-68, 1978

Effect of Heat Treatment on Fatigue

Annealing

High-cycle fatigue strength is lowered during annealing due to the coarsening of grain sizes. During α-β annealing, for example, a longer annealing time decreases HCF strength because of the increase in the grain size of equiaxed α. Annealing above the β transus reduces fatigue strength still further.

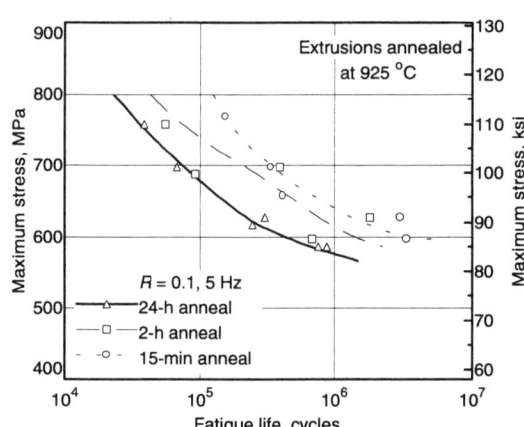

Ti-6Al-4V: Effect of annealing time

Material/Test Parameters: Extrusions had a composition (wt%) of 6.7 Al, 0.01 C, 0.18 Fe, 0.0044 H, 0.013 N, 0.164 O, and 4.1 N. Mill-annealed extrusions as 76-mm (3-in.) diam cylinders were annealed at 925 °C (1700 °F) for times indicated, followed by slow cooling at 50 °C/h (90 °F/h).
Source: I. Weiss et al., Recovery, Recrystallization, and Mechanical Properties of Ti-6Al-4V Alloy, in Proc. 8th Int. Conf. Strength of Metals and Alloys, H. McQueen et al., Ed., Pergamon Press, 1985, p 1073

Ti-6Al-4V: Effect of annealing temperature on fatigue strength

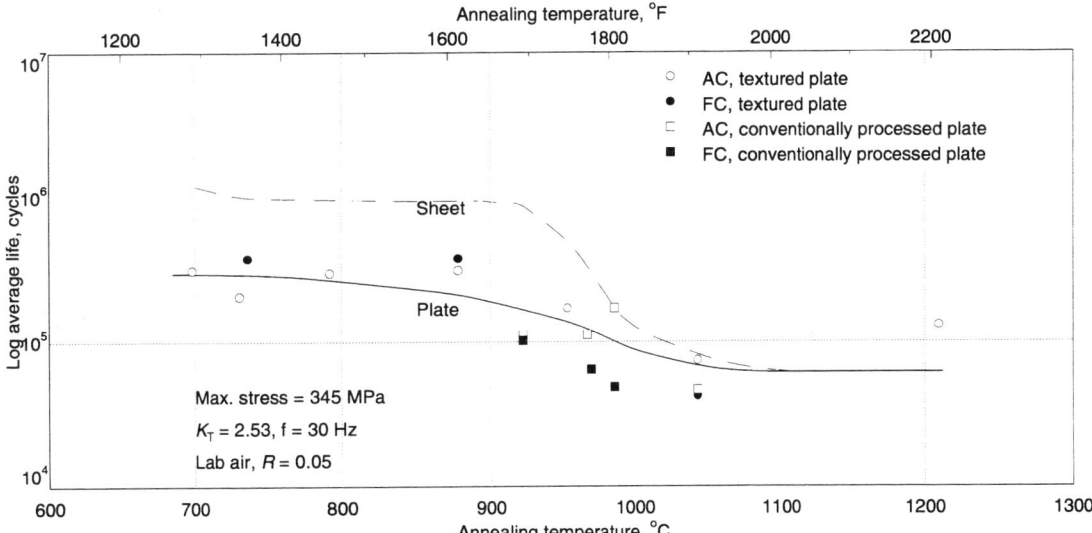

Comparison between sheet and plate shows the difference due to the finer grain sizes of sheet (however, the comparison is only general because gage, composition, and rolling practices are different for sheet).
Material/Test Parameters: Time and temperature for all anneals below 1040 °C (1900 °F) was 2 h. Above 1040 °C (1900 °F), annealing time was 20 min. All specimens having annealing temperatures above 870 °C (1600 °F) were heated to 730 °C (1350 °F) for 2 h and air cooled.
Source: R. Boyer et al., The Effects of Thermal Processing Variations on the Properties of Ti-6Al-4V, in *Microstructure, Fracture Toughness, and Fatigue Crack Growth Rate in Titanium Alloys*, A. Chakrabarti and J.C. Chesnutt, Ed., The Metallurgical Society, 1987, p 149

Effect of Cooling

Fatigue strength is improved by rapid cooling from either the α-β region or from above the β transus. Fast cooling leads to the martensitic formation of α', which improves fatigue strength. Water quenching without further aging may result in a low HCF strength if the retained β phase is unstable against stress-induced martensitic transformation.

Ti-6Al-4V: Effect of cooling from α-β region on HCF strength

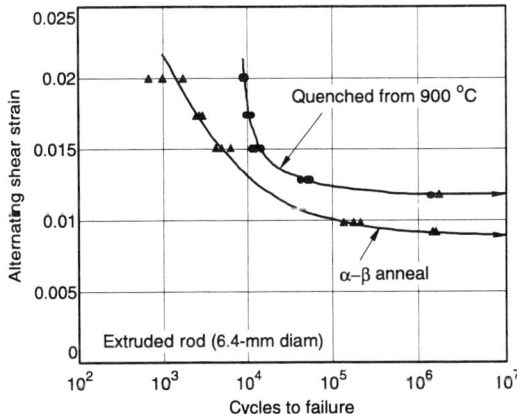

Material/Test Parameters: The α-β anneal alloy was annealed at 800 °C (1470 °F) for 3 h, furnace cooled to 600 °C (1110 °F), and vacuum cooled to room temperature.
Source: M.A. Imam and C.M. Gilmore, Fatigue and Microstructural Properties of Quenched Ti-6Al-4V, *Metall. Trans. A*, Vol 14A, 1983, p 233-240

Ti-6Al-4V: Effect of cooling rates from β region

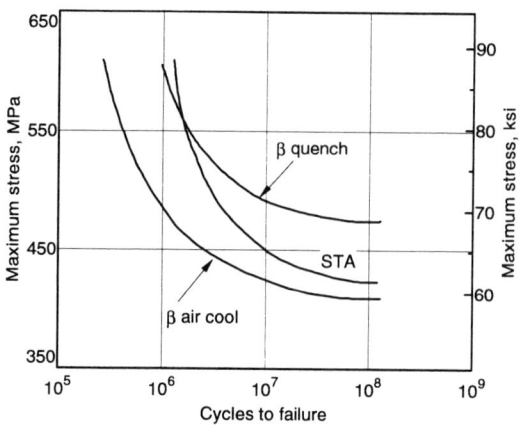

HCF comparison for three conditions for general comparison only.

Ti-6Al-4V: Effect of cooling rate from solution anneal on aged HCF strength

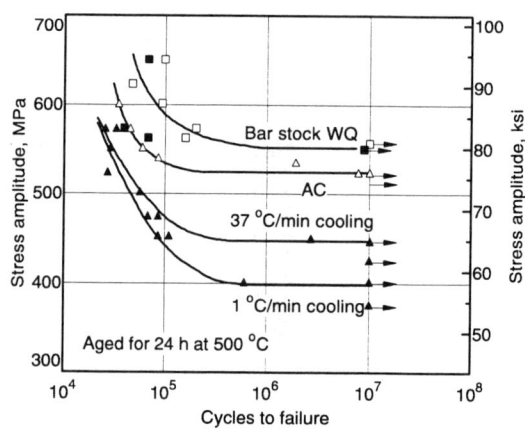

Material/Test Parameters: Rotating-beam fatigue tests were performed on hourglass-shaped specimens with gage diameter of 3.8 mm (0.15 in.), electrolytically polished surface, f = 50 Hz, at room temperature in the longitudinal direction. See table for alloy condition.
Source: R. Jaffee et al., The Effect of Cooling Rate From the Solution Anneal on the Structure and Properties of Ti-6Al-4V, in 6th World Conf. Titanium, 1989, p 1501

Ti-6Al-4V: Material condition for HCF strength (see above figure)

Treatment	Primary α, vol%	Primary α, 0.2% μm	Yield strength MPa	ksi	Reduction in area, %
965 °C/WQ/aged	35	8	1035	150	50
965 °C/AC/800 °C for 1 h/AC/aged	35	8	985	143	47
965 °C/37 °C per min/800 °C for 1 h/AC/aged	45	10	975	141	34
965 °C/1 °C per min/800 °C for 1 h/AC/aged	75	12	955	139	38

STA Condition

Solution-treated and aged (STA) material has good fatigue strength but not as good as that of the fine equiaxed or β quenched materials. Age hardening results in the strengthening of the β phase by the precipitation of small α grains and/or the strengthening of the α phase by Ti_3Al precipitates. The degree of age hardening depends on the solution-anneal temperature and cooling rates.

Ti-6Al-4V: Influence of age hardening on HCF strength

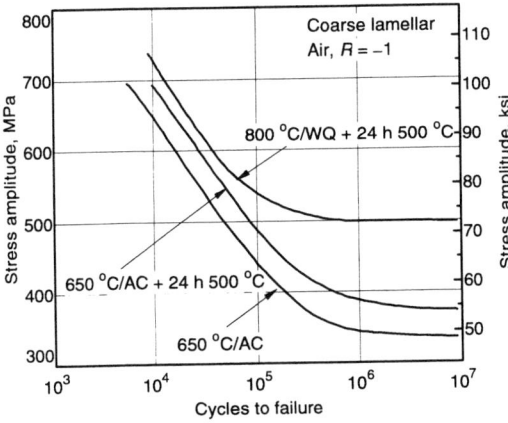

Water quenching from 800 °C (1470 °F) without further aging resulted in a low HCF strength value of 290 MPa (42 ksi) because (in addition to the absence of age hardening) the β phase is unstable against stress-induced martensitic transformation, resulting in crack nucleation at low applied stresses. Annealing at 650 °C (1200 °F) instead of 800 °C (1470 °F) leads to stabilization of the β phase due to the higher vanadium content, but the effect of subsequent age hardening at 500 °C (930 °F) is smaller for the β phase and also for the α phase. This is due to the lower vacancy concentration after the 650 °C (1200 °F) anneal as compared to the 800 °C (1470 °F) treatment and its effect on Ti_3Al precipitation.

Source: G. Lütjering and A. Gysler, Critical Review of Fatigue, in *Titanium Science and Technology*, Deutsche Gesellschaft für Metallkunde e.V., 1985, p 2069

Ti-6Al-4V: HCF strength of age-hardened bar

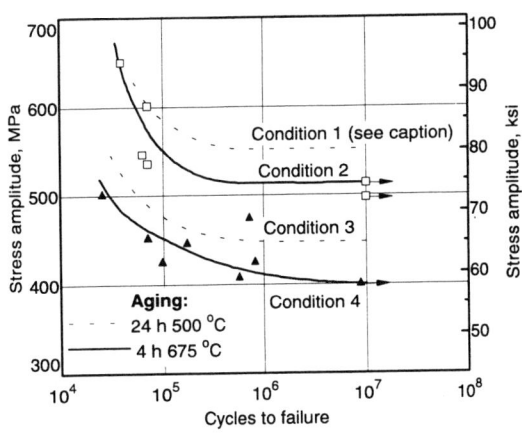

Material/Test Parameters: Rotating-beam fatigue tests were performed on hourglass-shaped specimens with 3.8-mm (0.15-in.) gage diameter, electrolytically polished surface, f = 50 Hz, at room temperature in the longitudinal direction. See table below for conditions.

Source: R. Jaffee *et al.*, The Effect of Cooling Rate From the Solution Anneal on the Structure and Properties of Ti-6Al-4V, in *6th World Conf. Titanium*, 1989, p 1501

Ti-6Al-4V: HCF strength of age-hardened bar (see above figure)

Condition	Solution treatment
1	1 h 955 °C/WQ + 1 h 800 °C/WQ
2	1 h 955 °C/WQ + 1 h 800 °C/AC + 4 h 675 °C/AC
3	1 h 965 °C/37 °C/min to 800 °C + 1 h 800 °C/AC
4	1 h 965 °C/37 °C/min to 800 °C + 1 h 800 °C/AC + 2 h 650 °C/AC

Constant-Life Fatigue Diagrams

Ti-6Al-4V: Constant-life diagram for (α + β) annealed bar

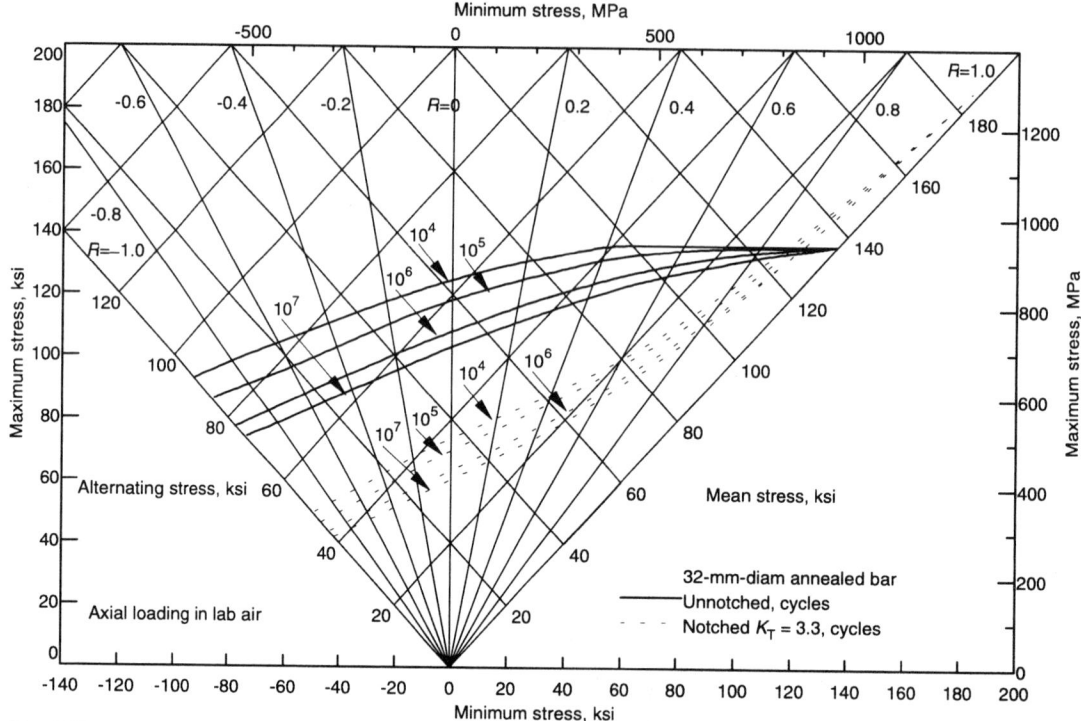

Material/Test Parameters: Unnotched specimen had a 5.15-mm (0.203-in.) diameter, a tensile strength of 940 MPa (136.5 ksi), and was polished longitudinally with 240, 400, and 600 emery belts. Notched specimen had a gross diameter of 8.4 mm (0.331 in.), a net diameter of 6.4 mm (0.252 in.), and was machined into a V-groove followed by polishing notch root with 600-grit slurry and rotating copper wire. Test frequency: 1750 cycles/min.
Source: R. Wood and R. Favor, *Titanium Alloys Handbook*, MCIC-HB-02, Battelle Columbus Laboratories, p 5-4:72-23

Ti-6Al-4V: Constant-life diagram for STA sheet

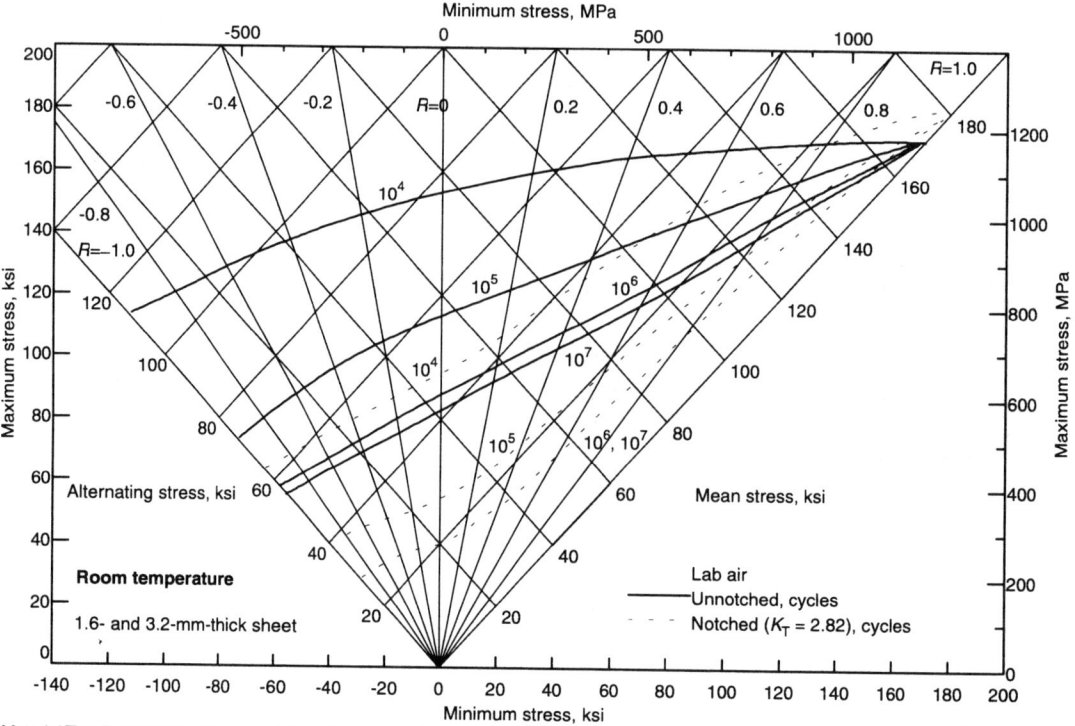

Material/Test Parameters: *Unnotched:* As-rolled surface, but edges machined and hand polished with No. 1 and 00 grit emery paper, cleaned with methyl ethyl ketone. *Notched:* Surface and edges prepared as above; 1.587-mm (0.0625-in.) diam hole drilled and reamed. Test frequency, 1500-2200 cycles/min.
Source: R. Wood and R. Favor, *Titanium Alloys Handbook*, MCIC-HB-02, Battelle Columbus Laboratories, 1972, reprinted 1985, p 5-4:72-27, 28, 29

Ti-6Al-4V: Constant-life diagram of extrusions at room temperature (RT)

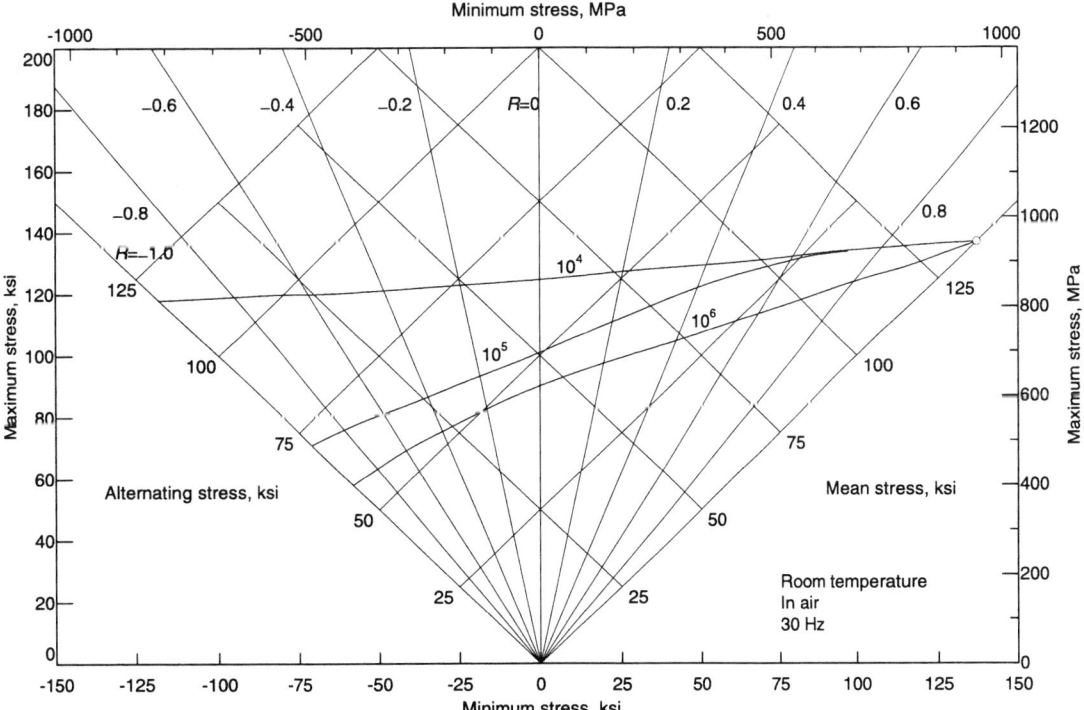

Material/Test Parameters: Specimens were from 7.6- and 14-mm (0.300- and 0.560-in.) thick extrusions with a tensile strength of 985 MPa (143 ksi) and an RMS surface roughness of 1.6 µm (63 µin.). Grain direction was longitudinal.
Source: R. Wood and R. Favor, *Titanium Alloys Handbook*, MCIC-HB-02, Battelle Columbus Laboratories, p 5-4:72-24

Duplex Annealed Sheet

Ti-6Al-4V: Constant-life diagram of unnotched duplex-annealed ELI sheet

Material/Test Parameters: Duplex anneal (DA) sheet = 910 °C (1675 °F) for 10 min/AC + 730 °C (1350 °F) for 4 h/AC. Sheet composition (wt%): 0.11 O, 6.0 Al, 4.0 V, 0.19 Fe, 0.01 N, 0.02 C, and 0.0034 H.
Source: R.R. Boyer and R. Bajoraitis, "Standardization of Ti-6Al-4V Processing Conditions," AFML-TR-78-131, 1978

550 / Alpha-Beta Alloys

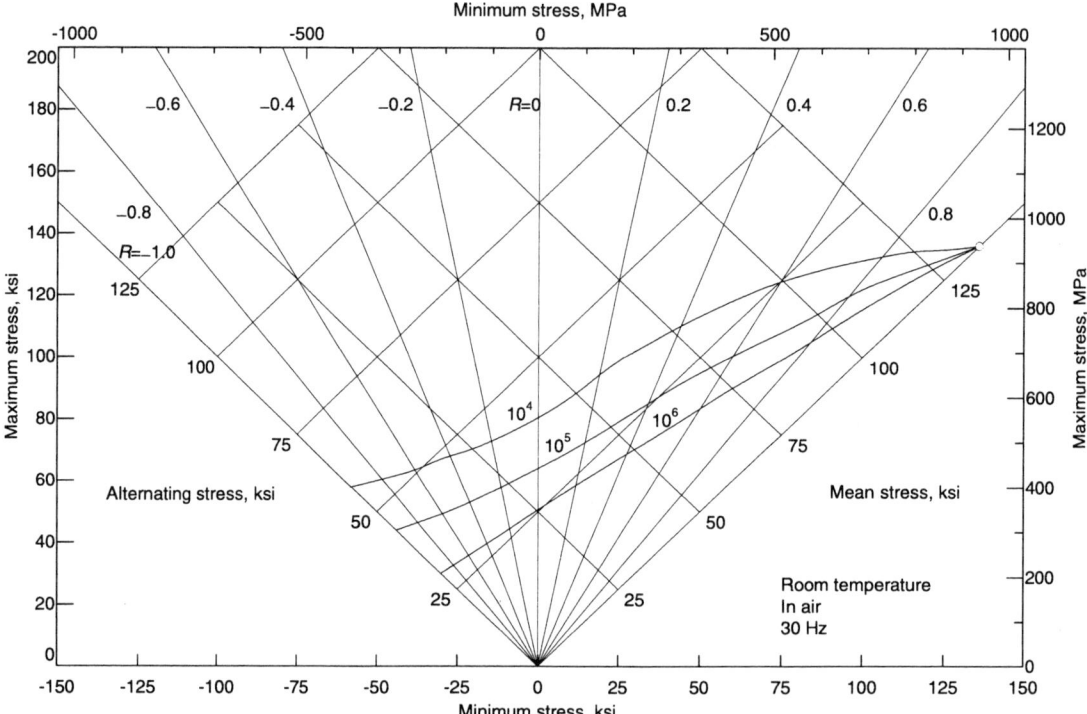

Ti-6Al-4V: Constant-life diagram of DA and notched (K_T = 2.53) ELI sheet

Material/Test Parameters: Duplex anneal (DA) sheet = 910 °C (1675 °F) for 10 min/AC + 730 °C (1350 °F) for 4 h/AC. Sheet composition (wt%): 0.11 O, 6.0 Al, 4.0 V, 0.19 Fe, 0.01 N, 0.02 C, and 0.0034 H
Source: R.R. Boyer and R. Bajoraitis, "Standardization of Ti-6Al-4V Processing Conditions," AFML-TR-78-131, 1978

Beta Annealed Plate

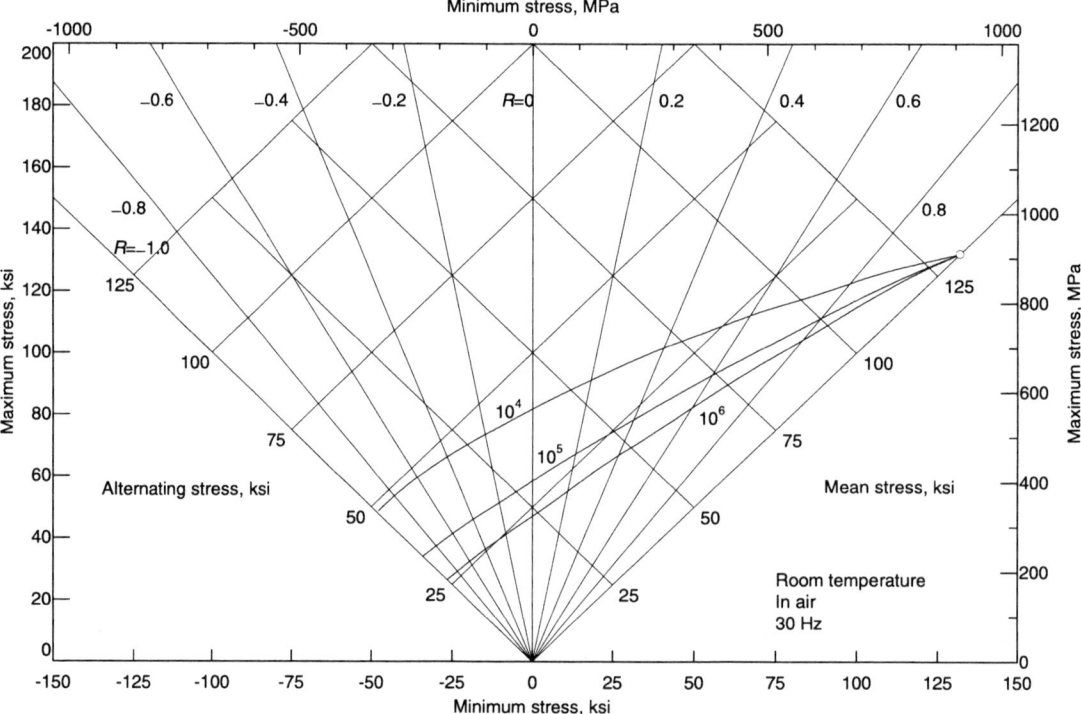

Ti-6Al-4V: Constant-life diagram of notched (K_T = 2.53) β-annealed ELI plate

Material/Test Parameters: Beta anneal (BA) = 1000 °C (1845 °F) for 20 min/AC + 730 °C (1350 °F) for 2 h/AC. Material composition (wt%): 0.10 O, 6.1 Al, 4.0 V, 0.15 Fe, 0.01 N, 0.02 C, and 0.048 H.
Source: R.R. Boyer and R. Bajoraitis, "Standardization of Ti-6Al-4V Processing Conditions," AFML-TR-78-131, 1978

Ti-6Al-4V: Constant-life diagram for unnotched β-annealed ELI plate

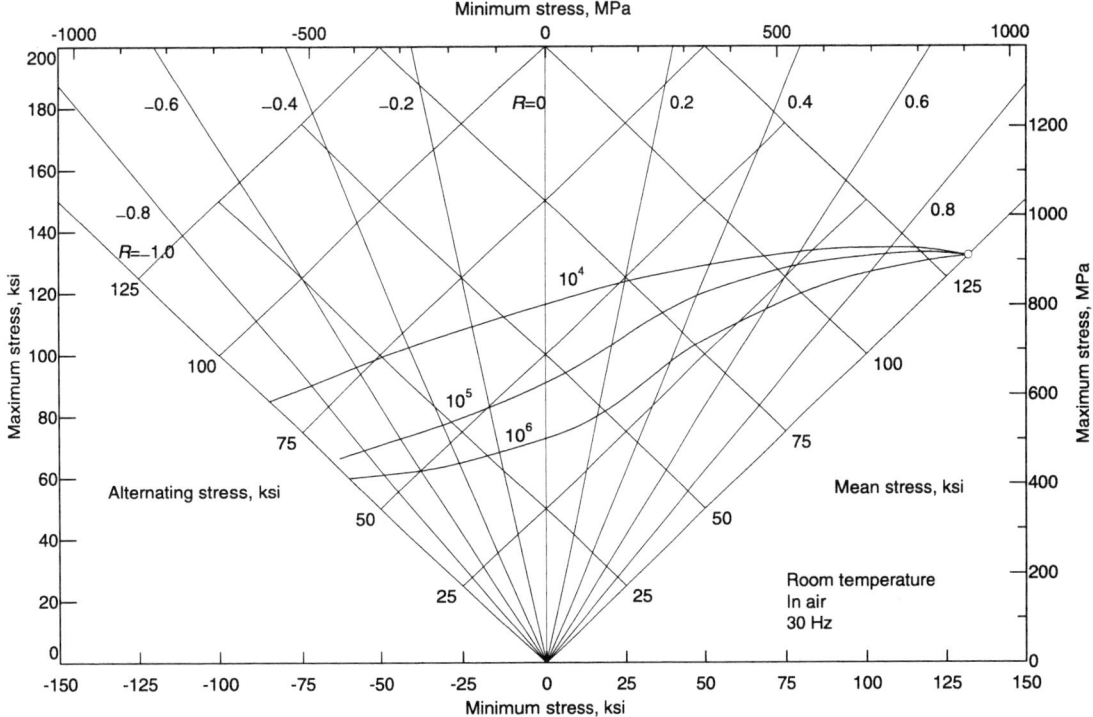

Material/Test Parameters: Beta anneal (BA) = 1000 °C (1845 °F) for 20 min/AC + 730 °C (1350 °F) for 2 h/AC. Material composition (wt%): 0.10 O, 6.4 Al, 4.0 V, 0.15 Fe, 0.01 N, 0.02 C, and 0.048 H.
Source: R.R. Boyer and R. Bajoraitis, "Standardization of Ti-6Al-4V Processing Conditions," AFML-TR-78-131, 1978

At 315 °C   **Ti-6Al-4V: Notched axial fatigue of extrusions at 315 °C (400 °F)**

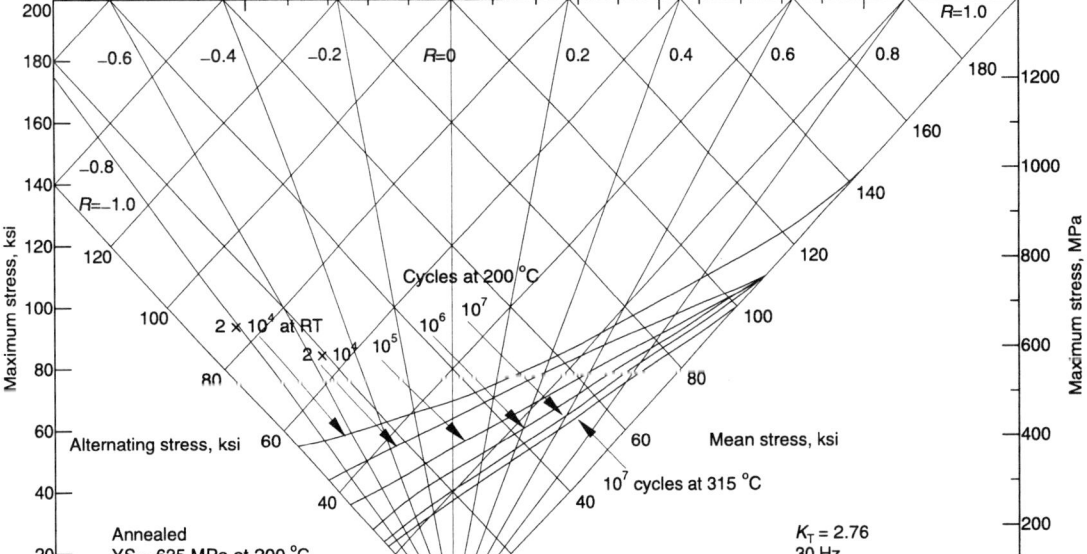

Material/Test Parameters: Grain direction was longitudinal, and specimens had an RMS surface roughness of 1.6 μm (63 μin.). Tensile strength at 200 °C (400 °F) was 770 MPa (112 ksi).
Source: R. Wood and R. Favor, *Titanium Alloys Handbook*, MCIC-HB-02, Battelle Columbus Laboratories, p 5-4:72-25

Ti-6Al-4V: Typical constant-life diagram for unnotched STA sheet at 315 °C (600 °F)

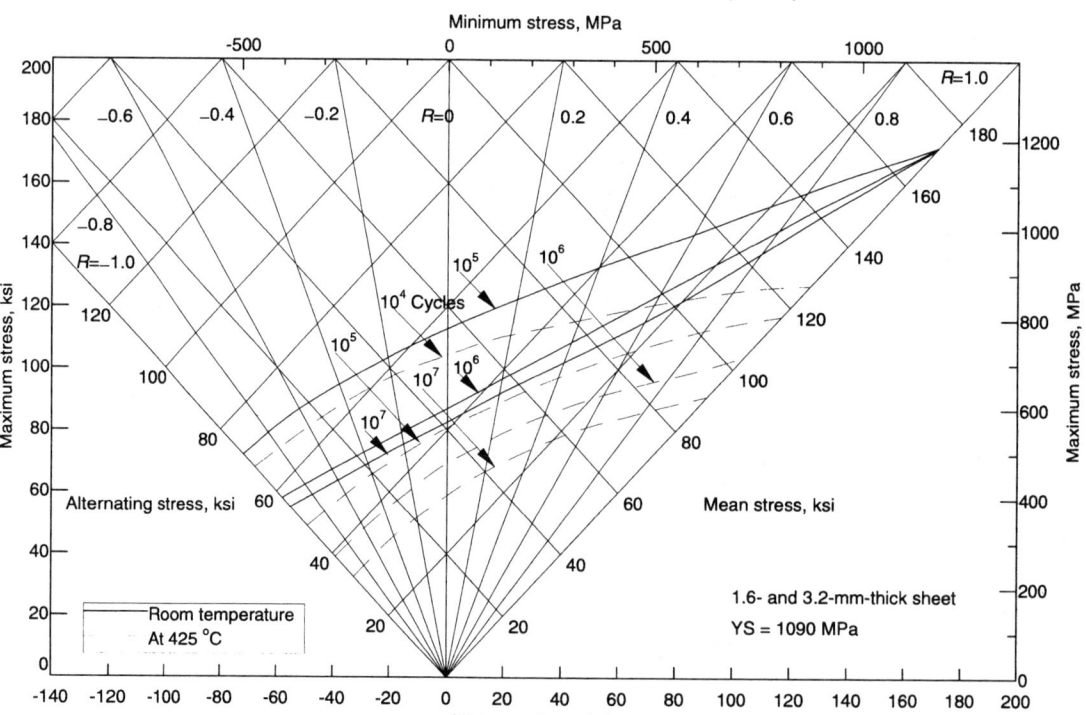

Material/Test Parameters: As-rolled surfaces, but edges machined and hand polished with No. 1 and 00 grit emery paper, cleaned with methyl ethyl ketone. Axial loading in air; test frequency, 1500-2200 cycles/min.
Source: R. Wood and R. Favor, *Titanium Alloys Handbook*, MCIC-HB-02, Battelle Columbus Laboratories, 1972, reprinted 1985, p 5-4:72-27, 28, 29

Ti-6Al-4V: Typical constant-life diagram for unnotched STA sheet at 425 °C (800 °F)

Material/Test Parameters: As-rolled surfaces, but edges machined and hand polished with No. 1 and 00 grit emery paper, cleaned with methyl ethyl ketone. Axial loading in air; test frequency, 1500-2200 cycles/min.
Source: R. Wood and R. Favor, *Titanium Alloys Handbook*, MCIC-HB-02, Battelle Columbus Laboratories, 1972, reprinted 1985, p 5-4:72-27, 28, 29

Ti-6Al-4V: Notched vs. unnotched fatigue of STA sheet at 425 °C (800 °F)

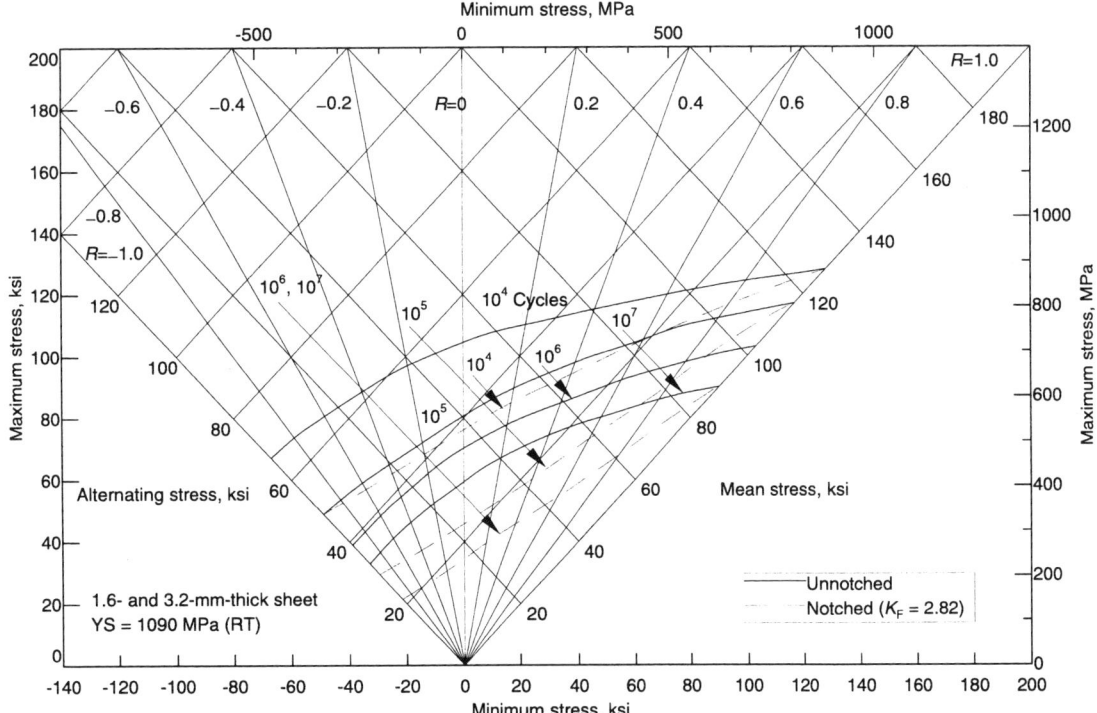

Material/Test Parameters: *Unnotched:* As-rolled surface, but edges machined and hand polished with No. 1 and 00 grit emery paper, cleaned with methyl ethyl ketone. *Notched:* Surface and edges prepared as above; 1.587-mm (0.0625-in.) diam hole drilled and reamed. Test frequency, 1500-2200 cycles/min.
Source: R. Wood and R. Favor, *Titanium Alloys Handbook*, MCIC-HB-02, Battelle Columbus Laboratories, 1972, reprinted 1985, p 5-4:72-27, 28, 29

Unnotched Fatigue Strength

Ti-6Al-4V: Axial fatigue of cast and wrought forms (R = 0.1, unnotched)

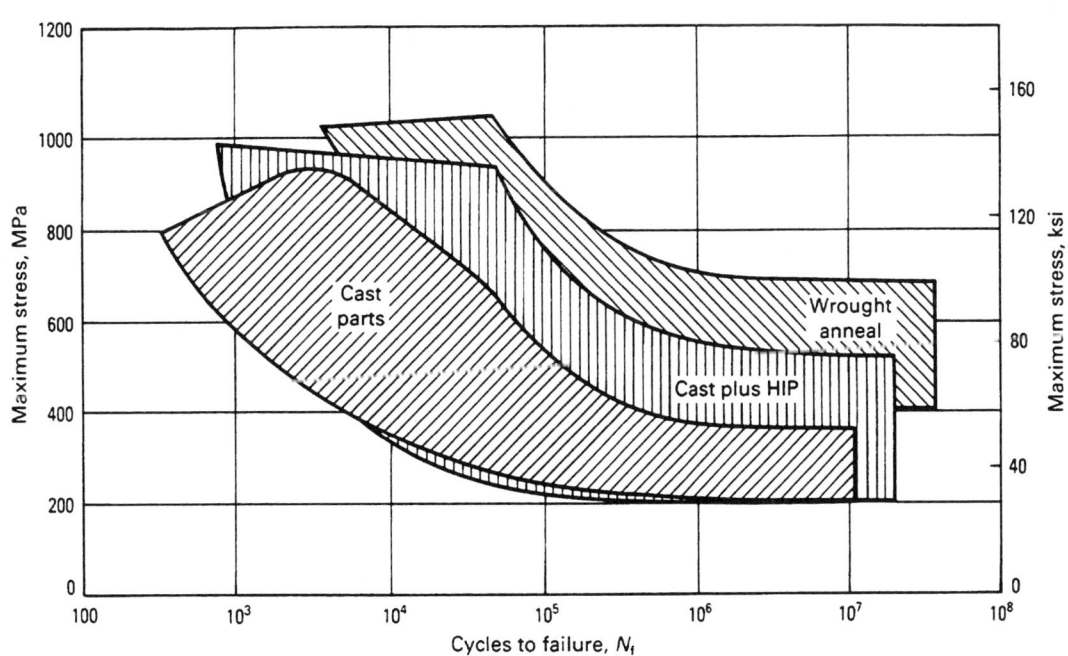

Source: D. Eylon and R. Boyer, "Titanium Alloy Net-Shape Technologies," in *Proc. Int. Conf. Titanium and Aluminum*, Paris, Feb 1990

Plate

Ti-6Al-4V: Fatigue life of unnotched β-annealed plate in aqueous 3.5% NaCl

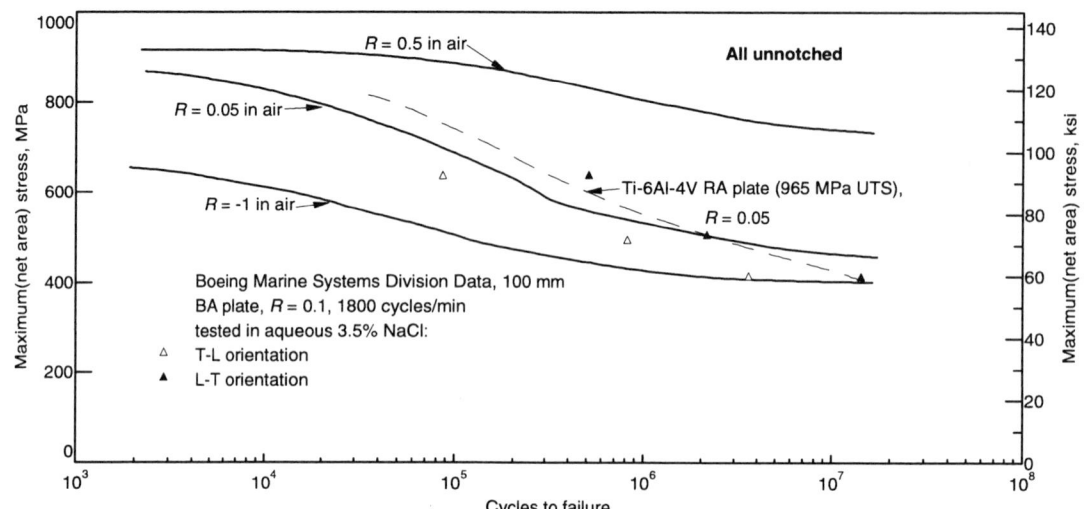

Material/Test Parameters: Beta anneal (BA) = 1000 °C (1845 °F) for 20 min/AC + 730 °C (1350 °F) for 2 h/AC. Material composition (wt%): 0.10 O, 6.1 Al, 4.0 V, 0.15 Fe, 0.01 N, 0.02 C, and 0.048 H.
Source: Reported in R.R. Boyer and R. Bajoraitis, "Standardization of Ti-6Al-4V Processing Conditions," AFML-TR-78-131, 1978; recrystallization annealed (RA) data from "B-1 Airframe Fatigue Design Properties Manual," Technical Report NA-72-1088, Rockwell International, 1975

Ti-6Al-4V: Axial fatigue of unnotched STA plate

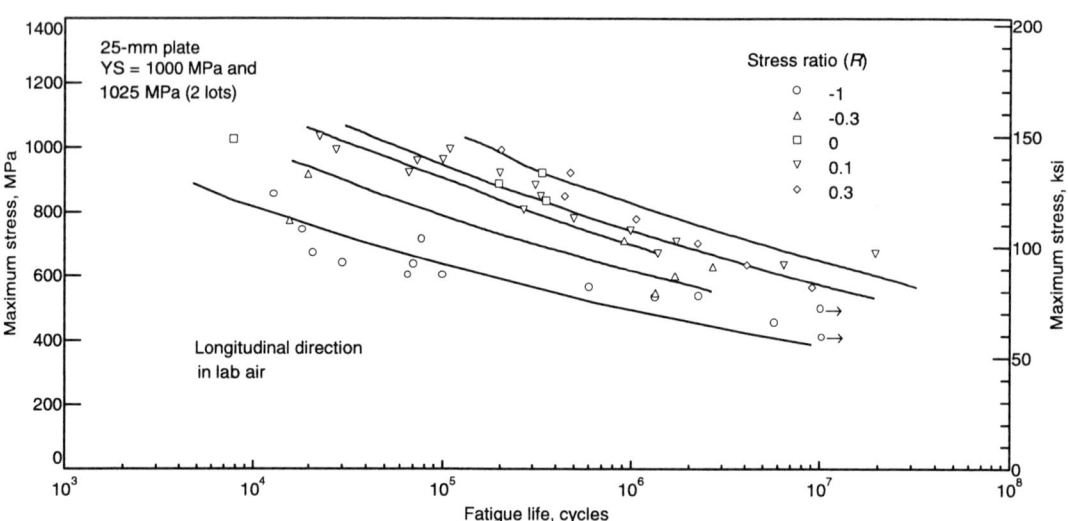

Material/Test Parameters: Specimens were longitudinally polished with No. 000 emery paper, removing all circumferential marks. Test frequency, 30 to 300 Hz.
Source: Data from A. Sommer and G. Martin, North American Rockwell, TR-69-161, 1969; and A. Marrocco, Grumann Aircraft, "Fatigue Characteristics of Ti-6Al-4V and Ti-6Al-6V-2Sn Sheet and Plate," 1968, reported in *MIL-HDBK-5E*, U.S. Dept. of Defense, 1987, p 5-88b

Ti-6Al-4V: Axial fatigue of unnotched STA plate

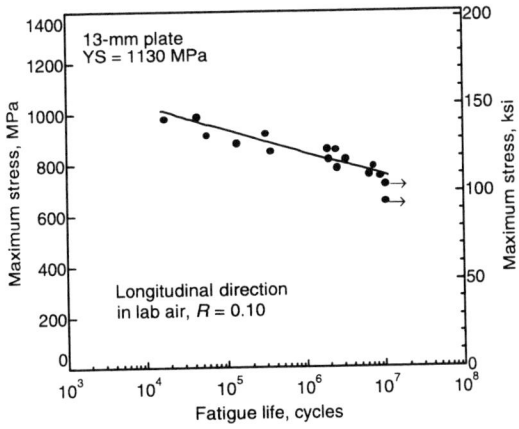

Machined RMS surface roughness of 1.6 μm (63 μin.).
Data from A. Marrocco, Grumann Aircraft, "Fatigue Characteristics of Ti-6Al-4V and Ti-6Al-6V-2Sn Sheet and Plate," 1968, reported in *MIL-HDBK-5E*, U.S. Dept. of Defense, 1987, p 5-88c

Sheet

Ti-6Al-4V: Axial fatigue of unnotched STA sheet

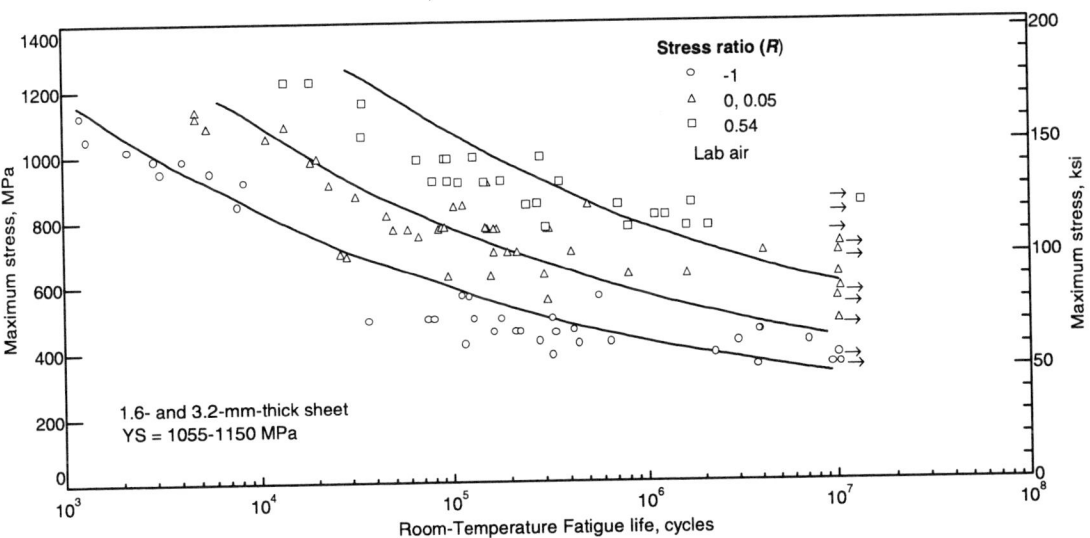

Conservative best fit per MIL-HDBK 5.
Material/Test Parameters: Longitudinal test direction. Machined specimens were cleaned with methyl ethyl ketone. Edges polished with No. 1 and 00 grit emery paper, recleaned with methyl ethyl ketone. Test frequency not specified.
Source: Data from C. Hickey, Jr., Watertown Arsenal, 1962, and from North American Aviation Report, 1960, reported in *MIL-HDBK-5E*, U.S. Dept. of Defense, 1987, p 5-84

Ti-6Al-4V: Fatigue life of annealed unnotched sheet

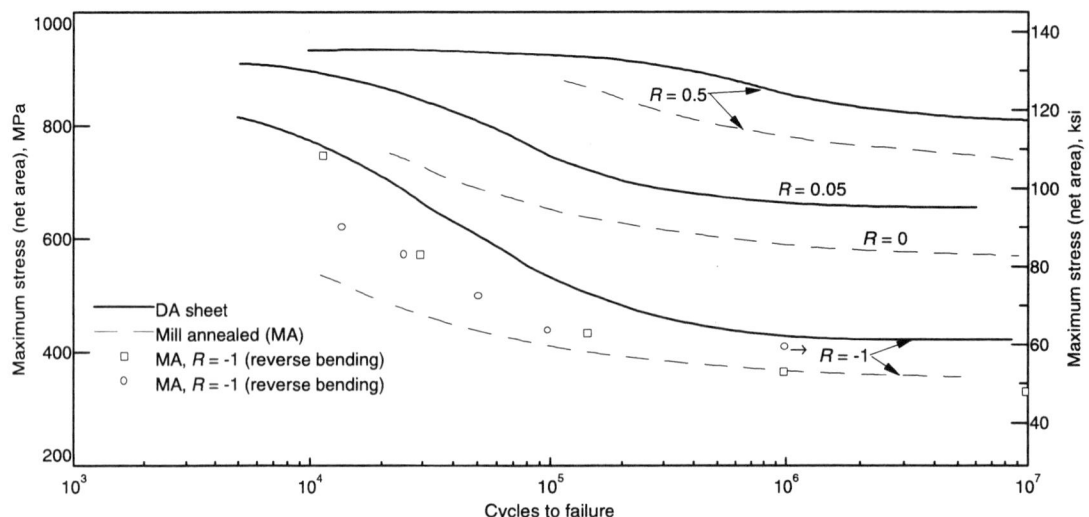

Material/Test Parameters: Axial test except as indicated. Duplex anneal (DA) sheet = 910 °C (1675 °F) for 10 min/AC + 730 °C (1350 °F) for 4 h/AC. Sheet composition (wt%): 0.11 O, 6.0 Al, 4.0 V, 0.19 Fe, 0.01 N, 0.02 C, and 0.0034 H.
Source: R.R. Boyer and R. Bajoraitis, "Standardization of Ti-6Al-4V Processing Conditions," AFML-TR-78-131, 1978

Ti-6Al-4V: Axial fatigue of unnotched STA sheet at 200 and 315 °C (400 and 600 °F)

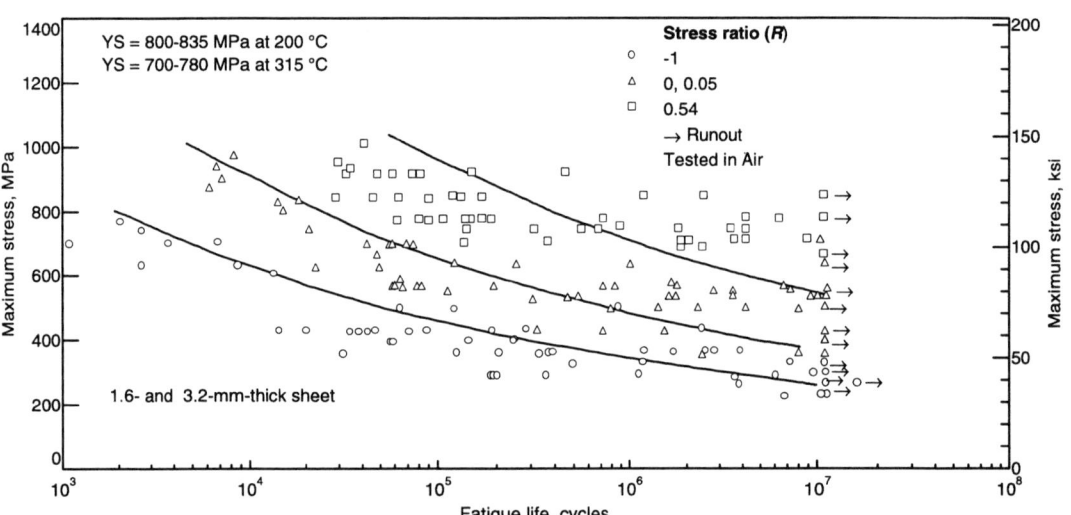

Conservative best-fit per MIL-HDBK 5.
Material/Test Parameters: Longitudinal test direction. Machined specimens were cleaned with methyl ethyl ketone. Edges polished with No. 1 and 00 grit emery paper, recleaned with methyl ethyl ketone. Test frequency not specified.
Source: Data from C. Hickey, Jr., Watertown Arsenal, 1962, and from North American Aviation Report, 1960, reported in *MIL-HDBK-5E*, U.S. Dept. of Defense, 1987, p 5-86

Ti-6Al-4V: Axial fatigue of unnotched sheet at 425 and 480 °C (800 and 900 °F)

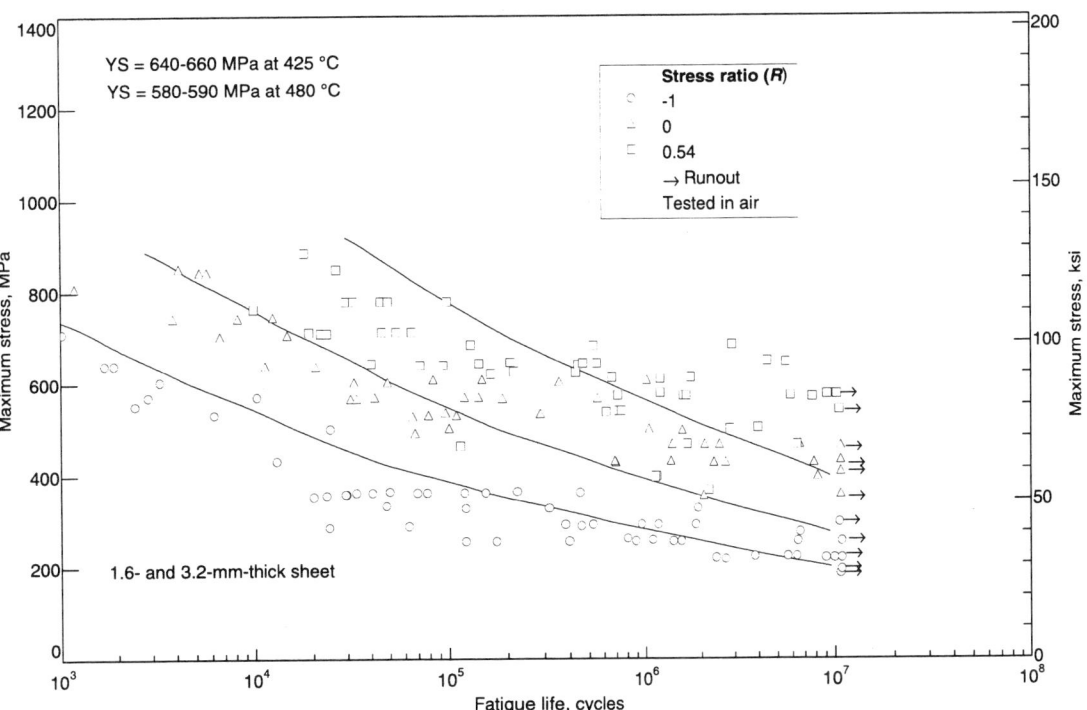

Conservative best-fit per MIL-HDBK 5.
Material/Test Parameters: Longitudinal test direction. Machined specimens were cleaned with methyl ethyl ketone. Edges polished with No. 1 and 00 grit emery paper, recleaned with methyl ethyl ketone. Test frequency not specified.
Source: Data from Lockheed-Georgia Co., "Determination of Design Data for Heat Treated Titanium Sheet," 1962, reported in *MIL-HDBK-5E*, U.S. Dept. of Defense, 1987, p 5-88

Strain Life

Initial cyclic strain hardening has been found to occur in Ti-6Al-4V material forged and heat-treated below the beta-transus (*Trans ASM*, 1972, p 263-270). This initial cyclic strain hardening, caused by an increase in dislocation density, was followed by subsequent strain softening.

In the beta-annealed material, initial strain hardening has not been observed (see figure). Softening, which is often associated with localized strains, can be enhanced for material with large slip length. In the beta-annealed (BA) titanium microstructure condition, the alpha platelets in a single colony tend to act as a single slip unit especially in the absence of a significant amount of beta phase between the alpha platelets as shown by Hack and Leverant (*Met Trans*, 13A, 1982, p 1729-1738).

Ti-6Al-4V: Wrought and P/M strain life fatigue

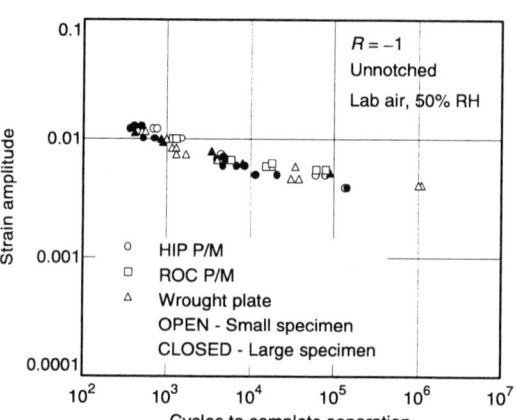

P/M bars consolidated from plasma rotating electrode process (PREP) powder with consolidation by hot isostatic pressing (HIP) and rapid omnidirectional compaction (ROC). Oxygen for P/M bar and wrought plate was near average for standard Ti-6Al-4V. The fatigue lives of the large specimens from the HIP P/M bars were equal to those of the I/M plate. The fatigue lives of the large HIP P/M specimens were all shorter than those of the small P/M specimens (both HIP and ROC), except for one specimen tested at the lowest strain amplitude, 0.004, where they were nearly equal, suggesting a size effect in the HIP P/M materials.
Source: *SAMPE Quarterly*, Oct 1988, p 15-19

Ti-6Al-4V: LCF at room temperature

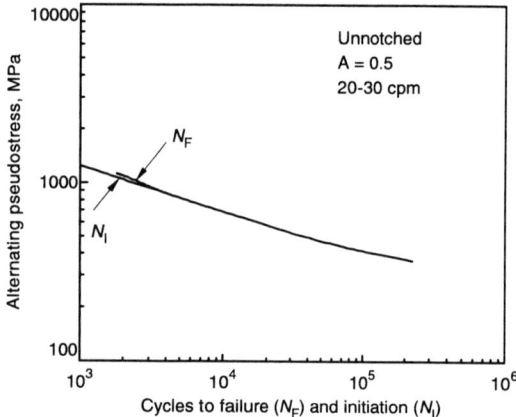

Axial-axial longitudinal strain control for forgings up to 150 mm (6 in.) thick; smaller size forgings may actually demonstrate higher pseudostress levels. Tangential orientation, test elastic modulus of 118.6 GPa (17.2×10^6 psi).
Source: C. Shamblen, GE Aircraft

Ti-6Al-4V: Beta annealed strain life fatigue

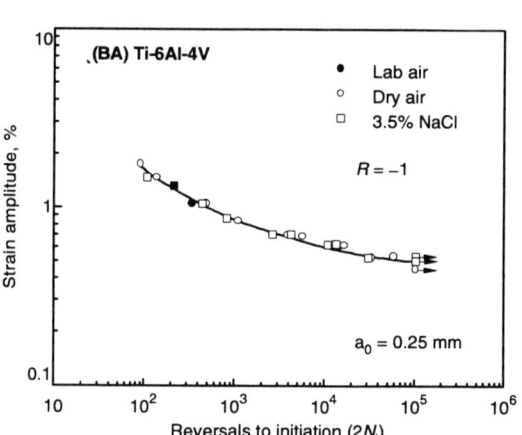

6.35 mm diam hourglass specimens from 22.2 mm plate. Cyclic softening occurred in the beta-annealed Ti-6Al-4V alloy at higher strain amplitudes. Both the tensile stress and compressive stress amplitudes decreased as a function of cycling. The percentage decrease in compressive stress amplitude was used to measure cyclic softening occurring and thus allowing the effects of "load-shedding" and cyclic softening to be separated in the measurements of maximum tensile stress. The onset of a 0.250 mm deep fatigue crack initiation was defined as the number of cycles when the tensile stress amplitude showed a 2% greater decrease than the compressive stress amplitude.
Source: *Corrosion Cracking*, American Society for Metals, 1986, p 157-165

Ti-6Al-4V: LCF at 315 °C (600 °F)

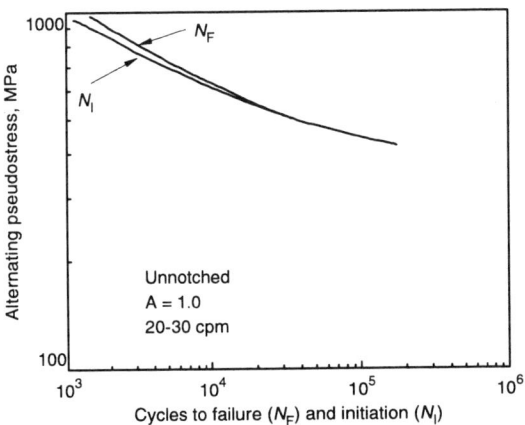

Axial-axial longitudinal strain control for forgings up to 150 mm (6 in.) section thickness; smaller size forgings would demonstrate higher pseudostress levels. Tangential orientation, elastic modulus 103 GPa (15×10^6 psi) at test temperature.
Source: C. Shamblen, GE Aircraft Engines

Ti-6Al-4V: Cyclic softening in beta annealed condition

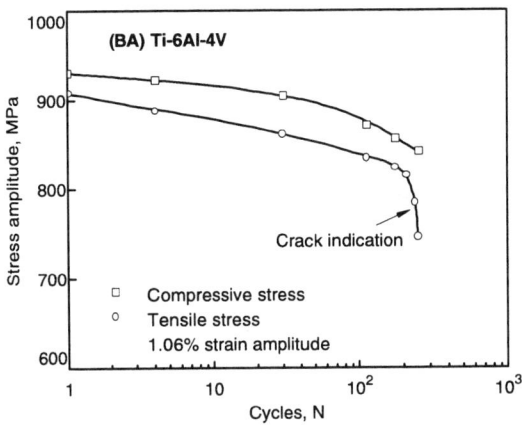

Typical cyclic stress response curves in beta annealed (BA) Ti-6Al-4V. The significant softening is attributed to the alpha colony size, which was quite large (about 100 μm).
Source: *Corrosion Cracking*, American Society for Metals, 1986, p 157-165

Notched Fatigue Strength

Ti-6Al-4V: Notch effects on wrought, cast, and P/M forms

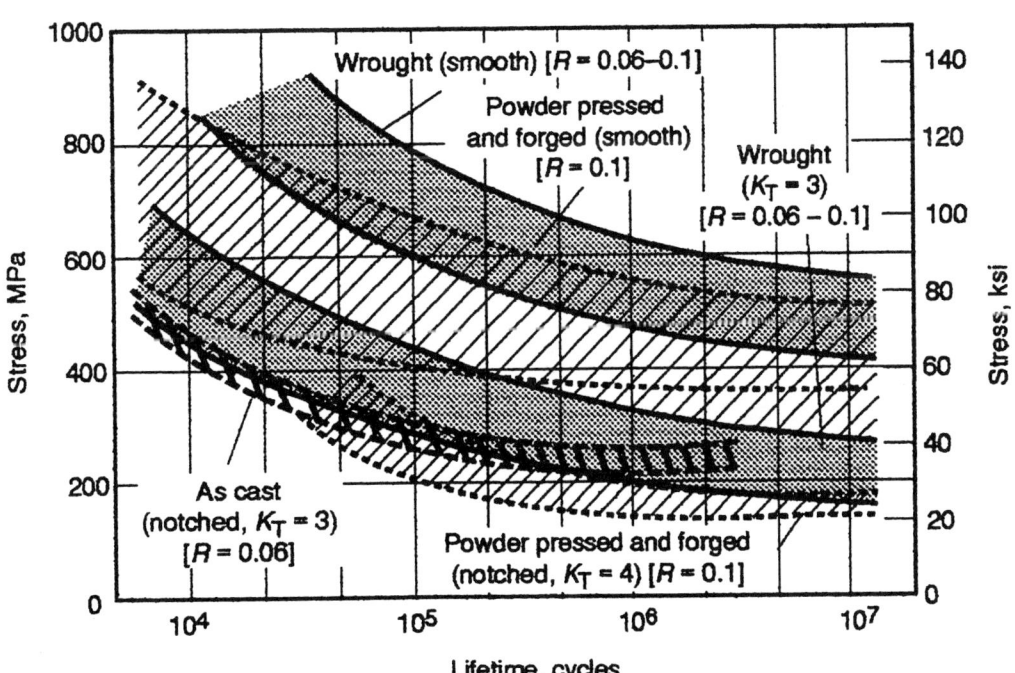

Source: *Titanium and Titanium Alloys*, MIL-HDBK-697A, U.S. Dept. of Defense, 1974, p 46

Annealed Ti-6Al-4V: Smooth axial fatigue

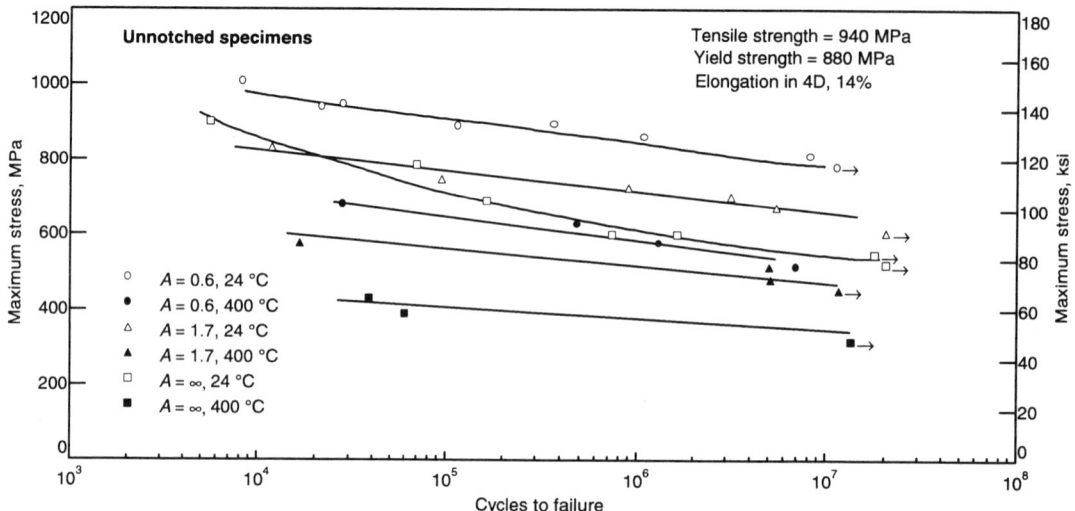

See figure below for condition. Source: *Metals Handbook*, Vol 1, 8th ed., *Properties and Selection of Metals*, American Society for Metals, 1961, p 530

Annealed Ti-6Al-4V: Notched axial fatigue

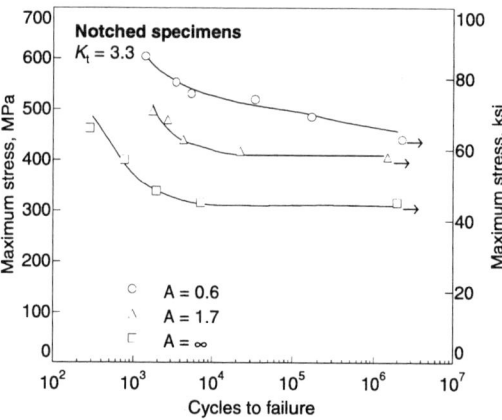

Annealed sheet, UTS 136 ksi; TYS, 128 ksi; Elongation (in 4D), 14%.
Source: *Metals Handbook*, Vol 1, 8th ed., *Properties and Selection of Metals*, American Society for Metals, 1961, p 530

STA Ti-6Al-4V: (a) Notched axial fatigue

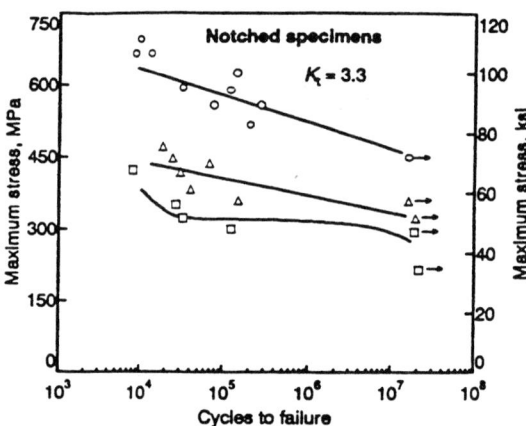

STA, solution treated and aged. See part (b) for conditions.
Source: *Metals Handbook*, Vol 1, 8th ed., *Properties and Selection of Metals*, American Society for Metals, 1961, p 530

STA Ti-6Al-4V: Smooth axial fatigue

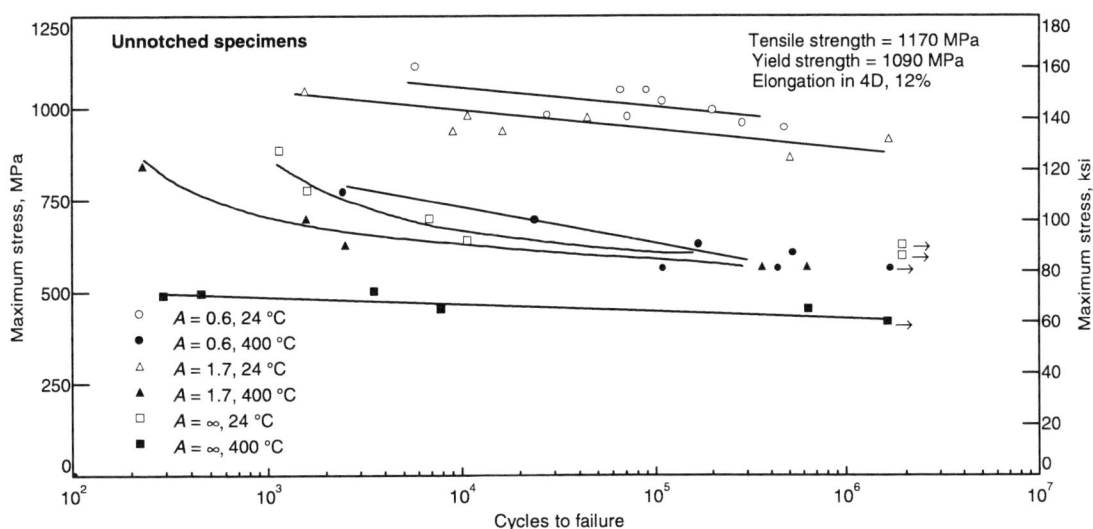

STA, solution treated and aged sheet
Source: *Metals Handbook*, Vol 1, 8th ed., *Properties and Selection of Metals*, American Society for Metals, 1961, p 530

Plate **Ti-6Al-4V: Fatigue of notched (K_t = 2.53) β-annealed plate**

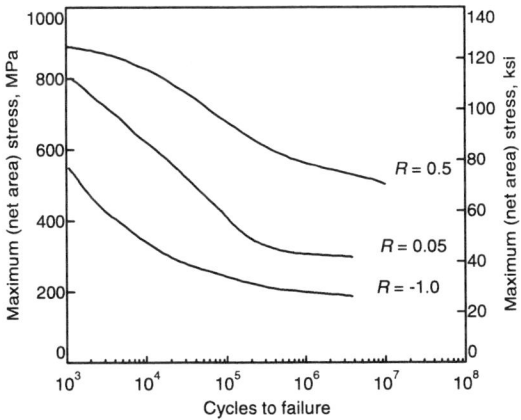

Stresses based on net section.
Material/Test Parameters: Beta anneal (BA) = 1000 °C (1845 °F) for 20 min/AC + 730 °C (1350 °F) for 2 h/AC. Material composition (wt%): 0.10 O, 6.1 Al, 4.0 V, 0.15 Fe, 0.01 N, 0.02 C, and 0.048 H. UTS of 805 MPa (130 ksi).
Source: R.R. Boyer and R. Bajoraitis, "Standardization of Ti-6Al-4V Processing Conditions," AFML-TR-78-131, 1978

Ti-6Al-4V: Axial fatigue of notched STA plate

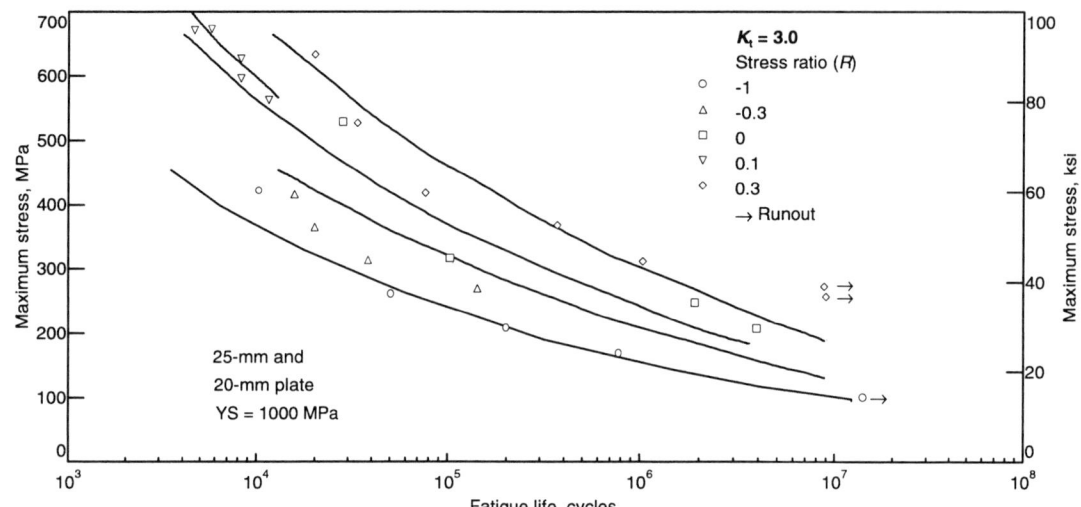

Note: Stresses are based on net section. Conservative best-fit curve from MIL-HDBK.
Material/Test Parameters: Longitudinal test direction in laboratory air with a test frequency of 30-300 Hz.
Source: Data from A. Sommer and G. Martin, North American Rockwell, TR-69-161, 1969; and M. Sargent, General Dynamics, "Fatigue Characteristics of Ti-6Al-4V Plate and Forgings," 1965, reported in *MIL-HDBK-5E*, U.S. Dept. of Defense, 1987, p 5-88d

Bar and Extrusions

Ti-6Al-4V: Axial notched fatigue in longitudinal direction of bar

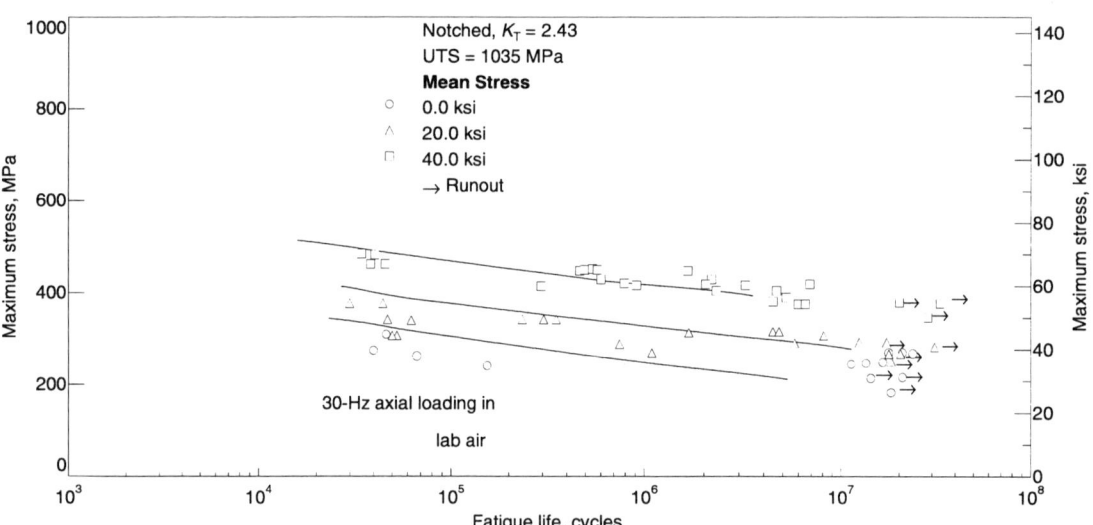

Note: Stresses are based on net section.
Material/Test Parameters: Specimens were taken from 25-mm (1-in.) diam annealed bar and had a 2.5-μm (100-μin.) machined surface roughness (R_{RMS}).
Source: Data from Sikorsky Aircraft, "Fatigue Evaluation of Ti-6Al-4V Bar Stock," 1970, reported in *MIL-HDBK-5E*, U.S. Dept. of Defense, 1987, p 5-71

Ti-6Al-4V: Axial fatigue of annealed extrusions (YS = 875 MPa), smooth and notched

Material/Test Parameters: Unnotched and notched specimens were from 7.5- and 14-mm (0.300- and 0.560-in.) thick extrusions with a tensile strength of 985 MPa (143 ksi) and an RMS surface roughness of 1.6 μm (63 μin.). Testing was performed in longitudinal direction at room temperature.
Source: Data from R. Brockett and J. Gottbrath, Lockheed-California Co., AFML-TR-67-189, 1967, reported in *MIL-HDBK-5E*, U.S. Dept. of Defense, 1987, p 5-72

Sheet

Ti-6Al-4V: Axial fatigue of notched (K_t =2.8) STA sheet at 425 and 480 °C (800 and 900 °F)

Note: Stresses are based on net section.
Material/Test Parameters: Machined specimens were cleaned with methyl ethyl ketone. Edges polished with No. 1 and 00 grit emery paper, recleaned with methyl ethyl ketone. Test frequency, 1500-2200 cycles/min.
Source: Data from Lockheed-Georgia Co., "Determination of Design Data for Heat Treated Titanium Sheet," 1962, reported in *MIL-HDBK-5E*, U.S. Dept. of Defense, 1987, p 5-88a

Ti-6Al-4V: Axial fatigue of notched (K_t =2.8) STA sheet at 200 and 315 °C (400 and 600 °F)

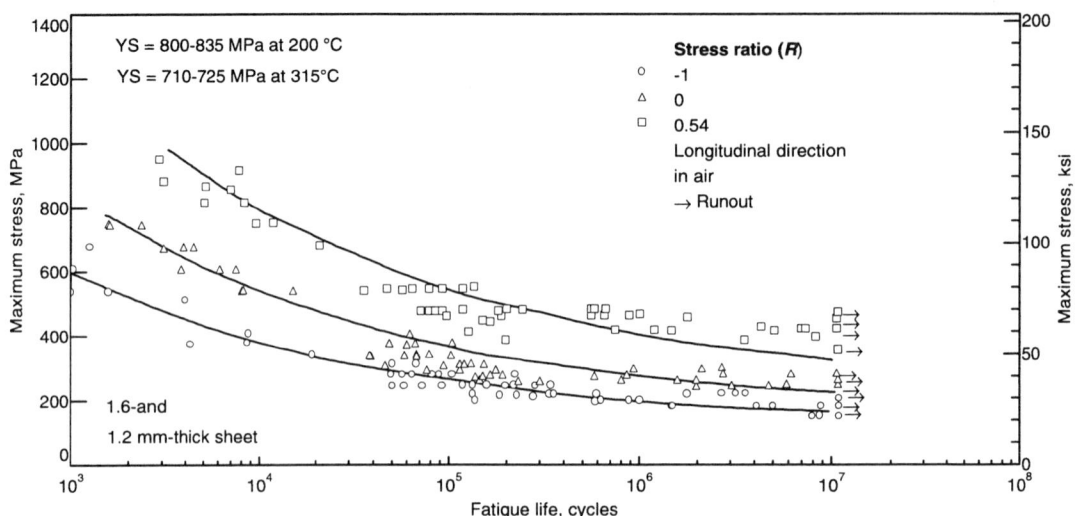

Note: Stresses are based on net section.
Material/Test Parameters: Machined specimens were cleaned with methyl ethyl ketone. Edges polished with No. 1 and 00 grit emery paper, recleaned with methyl ethyl ketone. Test frequency, 1500-2200 cycles/min.
Source: Data from Lockheed-Georgia Co., "Determination of Design Data for Heat Treated Titanium Sheet," 1962, reported in *MIL-HDBK-5E*, U.S. Dept. of Defense, 1987, p 5-87

Ti-6Al-4V: Longitudinal axial fatigue of notched (K_t = 2.8) STA sheet

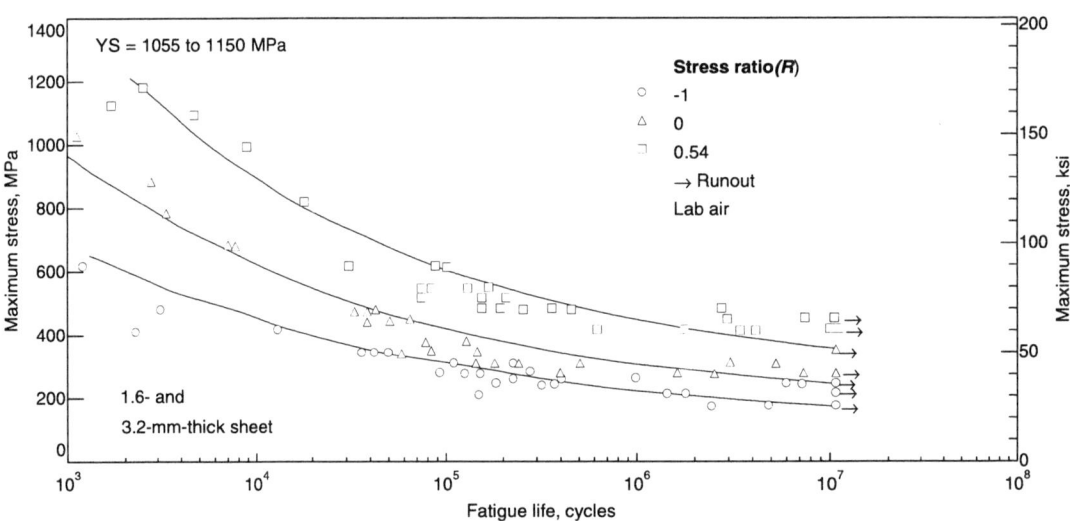

Note: Stresses are based on net section. UTS 1145 to 1220 MPa (166 to 177 ksi).
Material/Test Parameters: Machined specimens were cleaned with methyl ethyl ketone. Edges polished with No. 1 and 00 grit emery paper, recleaned with methyl ethyl ketone. Test frequency, 1500-2200 cycles/min.
Source: Data from Lockheed-Georgia Co., "Determination of Design Data for Heat Treated Titanium Sheet," 1962, reported in *MIL-HDBK-5E*, U.S. Dept. of Defense, 1987, p 5-85

Ti-6Al-4V: Notched fatigue life of ($K_t = 2.53$) annealed sheet

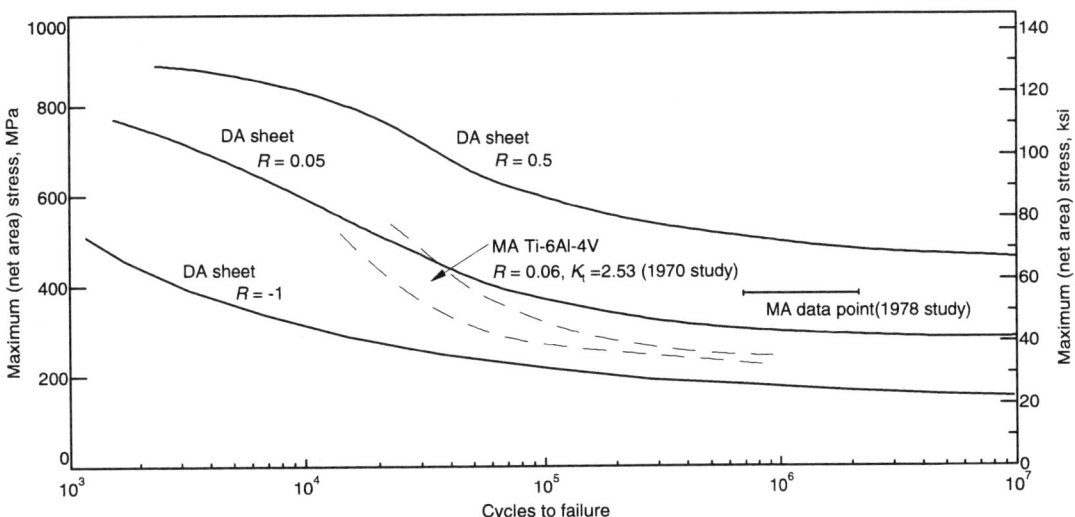

Material/Test Parameters: Duplex anneal (DA) sheet = 910 °C (1675 °F) for 10 min/AC + 730 °C (1350 °F) for 4 h/AC. Sheet composition (wt%): 0.11 O, 6.0 Al, 4.0 V, 0.19 Fe, 0.01 N, 0.02 C, and 0.0034 H.
Source: R.R. Boyer and R. Bajoraitis, "Standardization of Ti-6Al-4V Processing Conditions," AFML-TR-78-131, 1978

Cast and P/M Fatigue

Ti-6Al-4V: Fatigue of smooth and notched castings compared

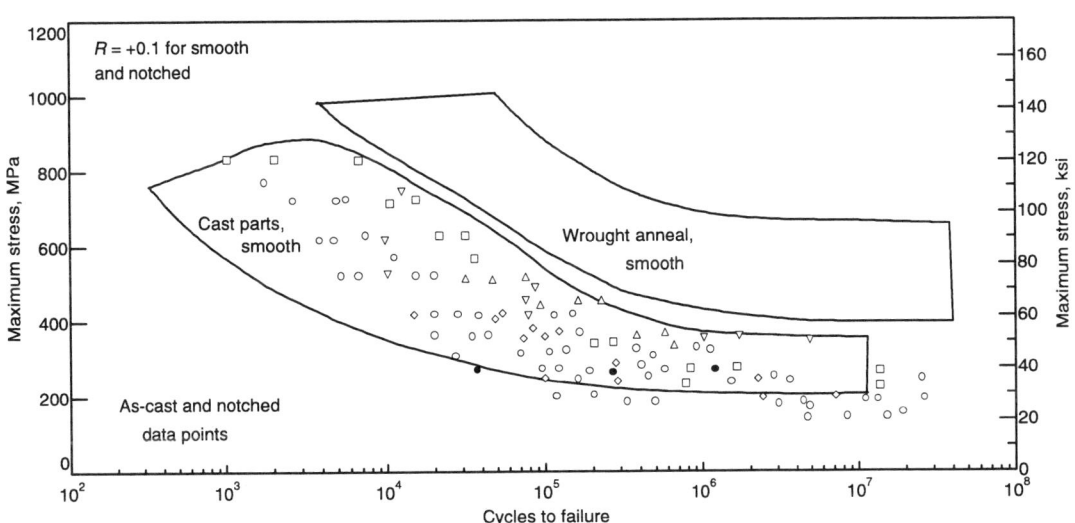

Data points for notched castings are replotted here.
Source: *Metals Handbook*, Vol 3, 9th ed., *Properties and Selection: Stainless Steels, Tool Materials and Special-Purpose Metals*, American Society for Metals, 1980, p 411. Most data points are based on $K_t = 3.0$, but some are for smooth specimens ($K_t = 1$). Each symbol represents data from a different source.

Ti-6Al-4V: Smooth fatigue of castings compared to wrought Ti-6Al-4V

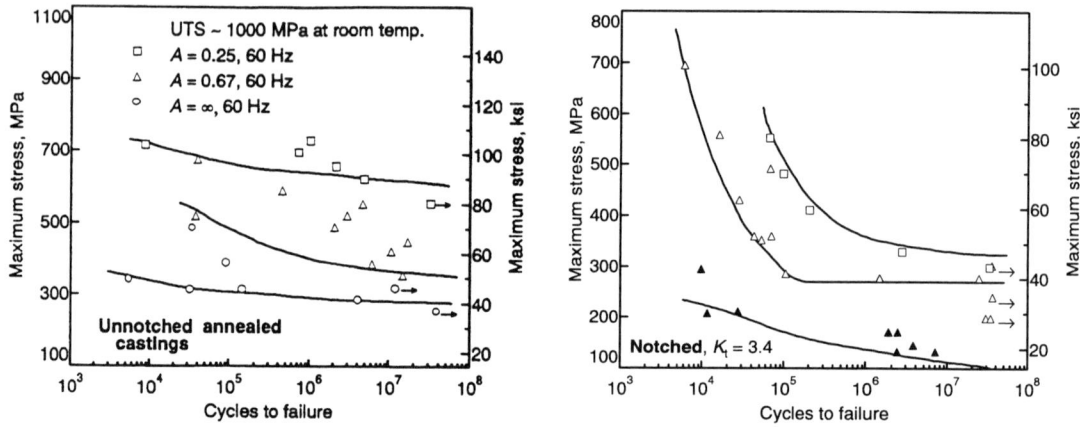

Source: Data from Titech International, Inc., Research Report 114, 1979, reported in *Titanium Alloys Handbook*, MCIC-HB-02, Battelle Columbus Laboratories, 1972, reprinted 1985, p 2-72:37

Ti-6Al-4V: (a) Smooth and (b) notched axial fatigue at 260 °C (500 °F)

Material/Test Parameters: Data were obtained from two heats of annealed castings having room-temperature ultimate tensile strengths of 997 and 1000 MPa (144.6 and 145.1 ksi). Cast-to-size specimens were annealed at 700 °C (1300 °F) for 2 h and air cooled. Longitudinal polish of smooth specimen had an RMS roughness of less than 0.1 μm (4 μin.).

Source: Data from R. Dalal, AVCO Corp., reported in *Aerospace Structural Metals Handbook*, Battelle Columbus Laboratories, Code 3801, p 19

P/M Prealloyed Ti-6Al-4V P/M parts: (a) Smooth and (b) notched fatigue

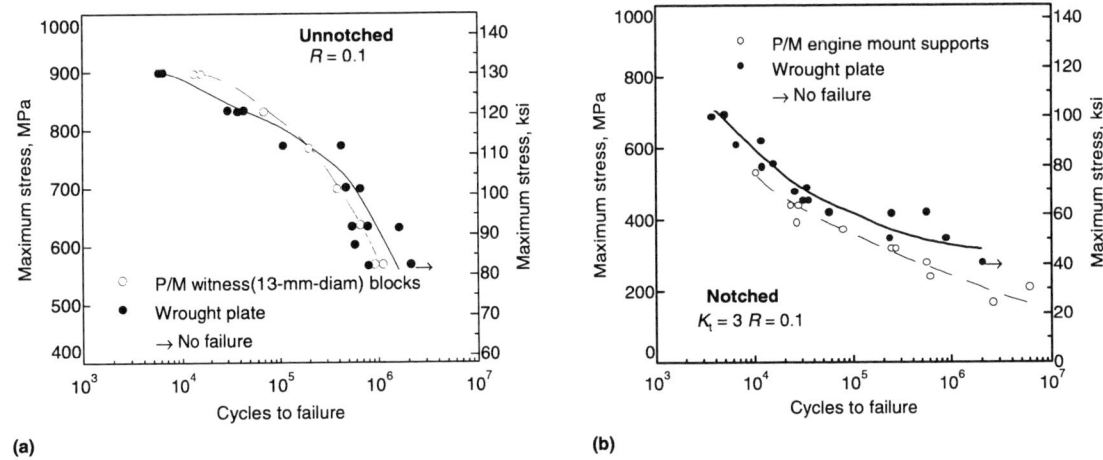

Source: A.S. Sheinker et al., Evaluation and Application of Prealloyed Titanium P/M Parts for Airframe Structures, *Int. J. Powder*, Vol 23 (No. 3), 1987, p 171-176

Ti-6Al-4V: Fatigue bands of I/M and P/M products

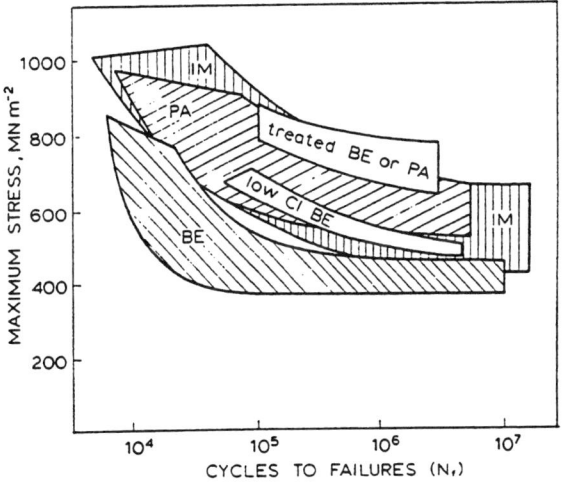

Note: I/M, ingot metallurgy; PA, prealloyed P/M products; BE, blended elemental P/M products.
Source: F.H. Froes and D. Eylon, Powder Metallurgy of Ti Alloy, *Int. Mat. Rev.*, Vol 35 (No. 3), 1990

Ti-6Al-4V: Axial fatigue of notched ($K_t = 2.16$) BE powder specimens

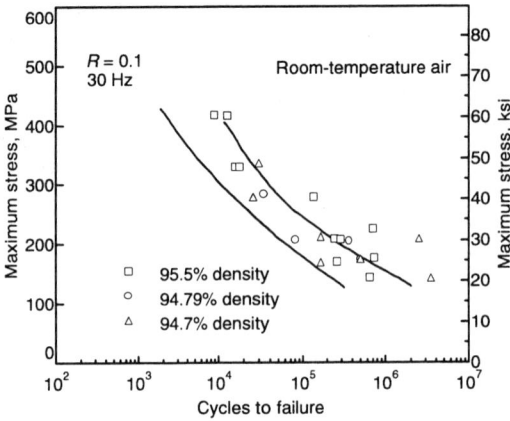

Note: Combined middle-density cycles to failure results.
Material/Test Parameters: Pressed and sintered without post-sinter compaction.
Source: J.A. Miller and G. Brodi, "Consolidation of Blended Elemental Ti-6Al-4V Powder to Near Net Shapes," AFML-TR-79-4028, 1979

Ti-6Al-4V: Axial fatigue of notched ($K_t = 2.16$) BE powder specimens

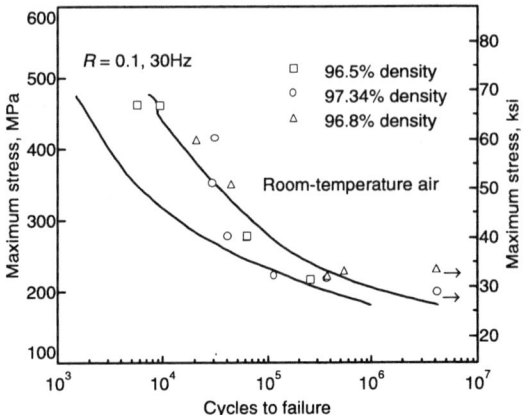

Note: Combined high-density cycles to failure results.
Material/Test Parameters: Pressed and sintered with no post-sinter compaction.
Source: J.A. Miller and G. Brodi, "Consolidation of Blended Elemental Ti-6Al-4V Powder to Near Net Shapes," AFML-TR-79-4028, 1979

Ti-6Al-4V: Axial fatigue of notched ($K_t = 2.16$) BE powder specimens with porosity

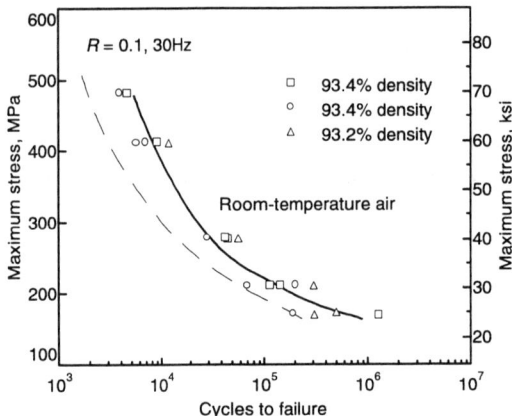

Note: Combined cycles-to-failure results for low-density P/M.
Material/Test Parameters: Pressed and sintered without post-sinter compaction.
Source: J.A. Miller and G. Brodi, "Consolidation of Blended Elemental Ti-6Al-4V Powder to Near Net Shapes," AFML-TR-79-4028, 1979

Corrosion Fatigue

Ti-6Al-4V: Fatigue strength in pure water

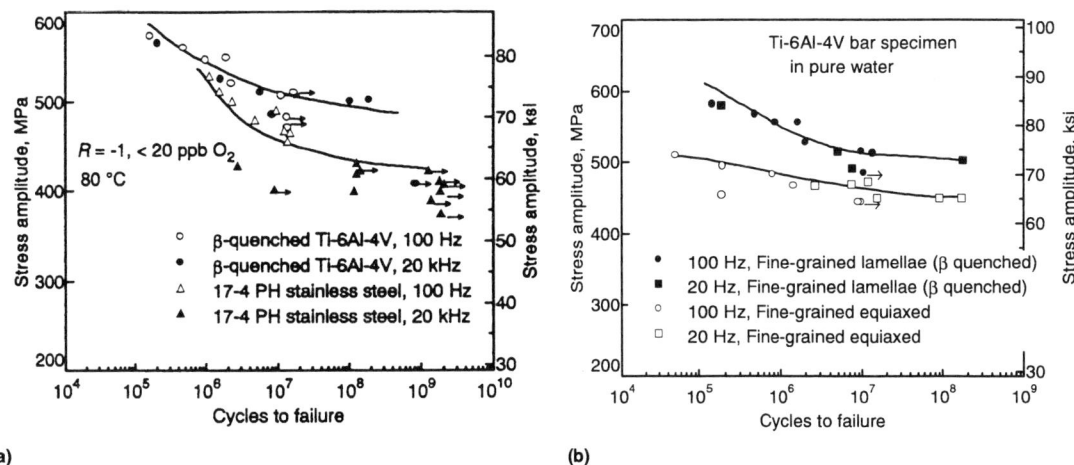

(a) β-quenched material compared to a stainless steel in water with dissolved O_2. (b) Fine-grained equiaxed microstructure compared to a fine-grained transformed structure.
Source: Adapted from L.D. Roth and L.E. Willertz, in *Environment Sensitive Fracture: Evaluation and Comparison of Test Methods*, STP 821, E.N. Pugh and G.M. Ugiansky, Ed., American Society for Testing and Materials, 1984, p 497; and L.E. Willertz et al., High and Low Frequency Corrosion Fatigue of Some Steam Turbine Blade Alloys, in *Ultrasonic Fatigue*, J.M. Wells et al., Ed., AIME, 1982, p 333-348

Ti-6Al-4V: Effect of texture and environment on fatigue

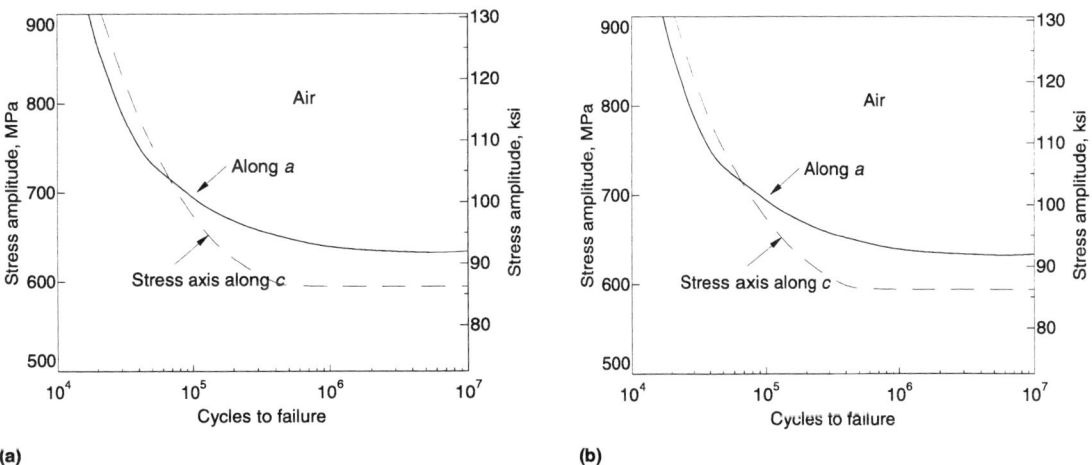

The S-N curves for Ti-6Al-4V processed in the α-β field show the effects of texture and environment on fatigue strength in (a) air and (b) 3.5% NaCl. These data show that testing in an aqueous 3.5% NaCl solution reduces fatigue strength when the stress axis is along [0001].
Source: J.C. Williams and E.A. Starke, Jr., The Role of Thermomechanical Processing in Tailoring the Properties of Aluminum and Titanium Alloys, in *Deformation, Processing, and Structure*, G. Krauss, Ed., American Society for Metals, 1984, p 335

Ti-6Al-4V: Fatigue in air and water of castings with varying oxygen content

○ 0.09% oxygen, air
● 0.09% oxygen, Severn River water
□ 0.11% oxygen, air
■ 0.11% oxygen, Severn River water
▽ 0.18% oxygen, air
▼ 0.18% oxygen, Severn River water
▲ 0.11% oxygen, air, R.R. Moore specimens
△ 0.15% oxygen, air, R.R. Moore specimens

Material/Test Parameters: Rotating cantilever beam notched test specimens were machined to 10.7 mm (0.420 in.) square by 12.5 mm (5 in.) long, containing a 0.005-mm (0.0002-in.) root radius notch.
Source: A. Morton and I. Lane, Jr., Titanium Castings for Marine Propellers, in *Titanium Science and Technology*, R. Jaffee and H. Burte, Ed., Plenum Press, 1973, p 119

Ti-6Al-4V: Fatigue of castings and weldments in water and air

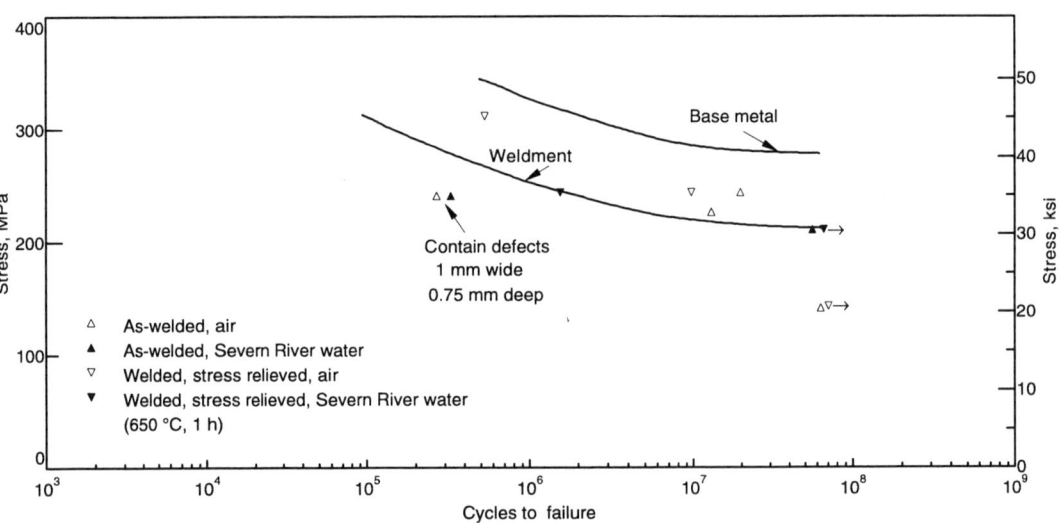

△ As-welded, air
▲ As-welded, Severn River water
▽ Welded, stress relieved, air
▼ Welded, stress relieved, Severn River water
(650 °C, 1 h)

Material/Test Parameters: Composition (wt%) of castings was 6.02 Al, 0.029 C, 0.11 Fe, 0.0035 H, 0.015 N, 0.11 O, and 4.12 V. Weldments were made by gas metal arc welding with Ti-6Al-4V filler metal containing 0.08% O. Rotating cantilever beam test specimens were machined.
Source: A. Morton and I. Lane, Jr., Titanium Castings for Marine Propellers, in *Titanium Science and Technology*, R. Jaffee and H. Burte, Ed., Plenum Press, 1973, p 119

Compared to Stainless Steel

Ti-6Al-4V corrosion fatigue: Compared with type 403 stainless steel

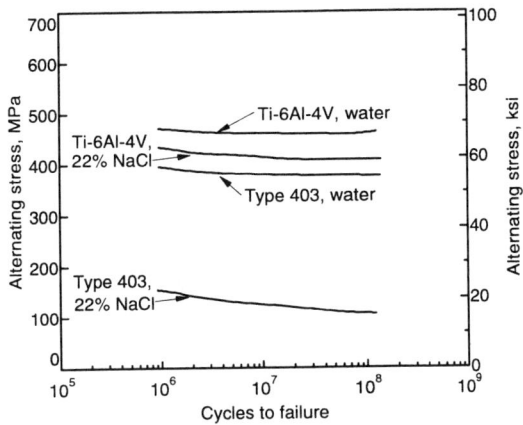

Material/Test Parameters: Corrosion fatigue strength comparison of Ti-6Al-4V and type 403 stainless steel in 22% NaCl solution at 80 °C (176 °F), with <20 ppb O_2, pH 4

Ti-6Al-4V: Weibull plot for 30 specimens in water

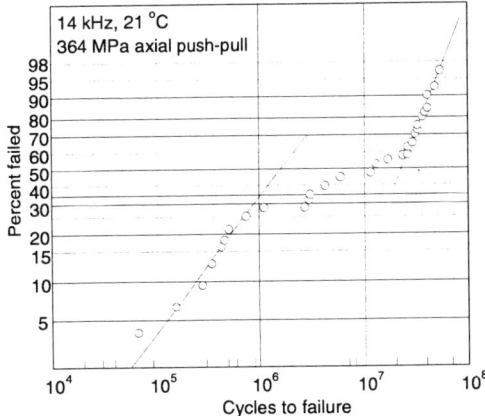

Annealed Ti-6Al-4V in water cooling bath.
Source: A. Thiruvengadam, Corrosion Fatigue at High Frequencies and High Hydrostatic Pressures, in *Stress Corrosion Cracking of Metals—A State of the Art*, STP 518, ASTM, 1972, p 139-154

Ti-6Al-4V: Fatigue compared with a stainless steel

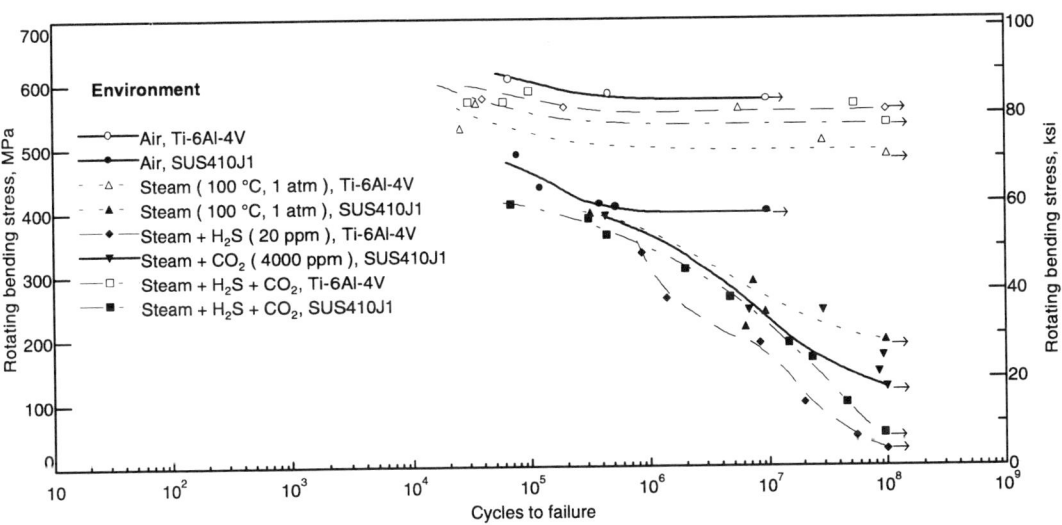

Comparison of the effect of industrial gas in steam on the fatigue strength of Ti-6Al-4V and an 11.5-14% Cr stainless steel (SUS410J1)

Ti-6Al-4V: Fatigue compared with stainless steel

[Bar chart comparing fatigue strength (MPa, at 10⁹ cycles) of 403 stainless steel, 17-4 pH stainless steel, and Ti-6Al-4V across various environments: Pure water; 100 × limit specified; 25% Na₂SO₄ < 20 ppb O₂; < 200 ppb O₂; Air saturated 22% NaCl + Na₂SO₄ + oxides; NaCl (saturated); Na₂SiO₂ (28%) + SiO₂ + Fe₃O₄. L = low, H = high.]

Asterisk (*) denotes extrapolation from 10^8 cycles. Fatigue strength of Ti-6Al-4V in the last environment (denoted by *) is meaningless because the material dissolves. H, high. L, low.
Source: L.E. Willertz et al., in *Ultrasonic Fatigue*, J.M. Wells et al., Ed., TMS-AIME, 1982, p 333

Fatigue Crack Growth in Air

Because fatigue crack propagation (FCP) resistance generally is improved with microstructures containing increased amounts of transformed β morphology, the slowest crack growth rates are frequently observed in products such as castings and β annealed or β processed parts. Age hardening reduces FCP resistance due to lower intrinsic ductility associated with increased strength of the STA condition.

Microstructure morphology can have significant effects on FCP resistance in inert environments, and it must be cautioned that laboratory air is not inert, even for room-temperature FCP testing.

Ti-6Al-4V has slower crack growth rates than aluminum and somewhat faster rates than steel. In aggressive environments such as seawater, the comparative advantage of Ti-6Al-4V improves because seawater has more of an effect on the FCP rates of steel.

Ti-6Al-4V: FCP data compared with aluminum and steel

Alloy and condition	da/dN(a) μm/cycle	μin./cycle	Test frequency cycles/min
Ti-6Al-4V, mill annealed	0.5	20	60
Ti-6Al-4V, β annealed	0.2	8	60
7075-T7351	2.5	100	360
HP-9-4-30(b)	0.1	4	360
300M steel	0.13	5	360

(a) All data are for a stress-intensity range (ΔK) of 22 MPa$\sqrt{m}$ (20 ksi$\sqrt{in.}$) and a stress ratio of $R = 0.3$ tested in low-humidity air. (b) High fracture toughness steel having a quenched and tempered yield strength of 1515 to 1655 MPa (220 to 240 ksi). Source: Fracture Mechanics Evaluation of B-1 Materials AFML-TR-76-137, Oct 1976

Ti-6Al-4V: Beta-quenched FCP rates vs. other heat treatments

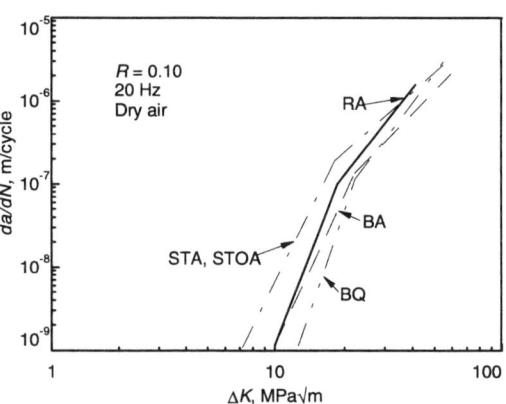

Material/Test Parameters: The five microstructures were produced from pancake forgings to obtain material that was uniformly and very weakly textured. Microstructures selected were either β forged and annealed (BA) or water quenched (BQ), or were forged in the α + β region and heat treated to produce microstructures that were recrystallization annealed (RA), solution treated and aged (STA) at 590 °C (1095 °F), or solution treated and overaged (STOA). These microstructures were characterized in some detail with transmission microscopy.
Cited: A.W. Thompson et al., The Effect of Microstructure on Fatigue Crack Propagation Rate in Ti-6Al-4V, in *Titanium and Titanium Alloys, Scientific and Technological Aspects*; Vol 1, J.C. Williams and A.F. Belov, Ed.

Ti-6Al-4V: Scatterbands for FCP rates of cast and β-annealed wrought products

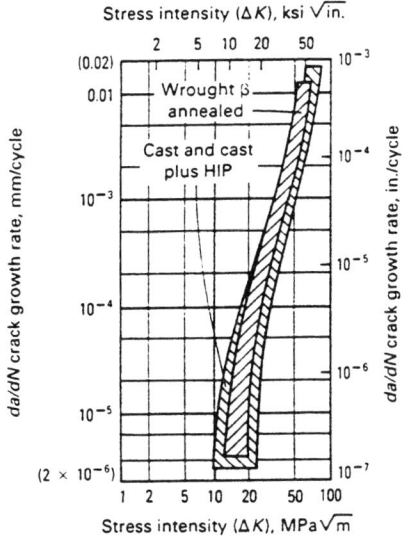

Source: *Metals Handbook*, Vol 2, 10th ed., *Properties and Selection: Nonferrous Alloys and Special-Purpose Materials*, ASM International, 1990, p 641

Ti-6Al-4V: Influence of α morphology on FCP resistance at room temperature

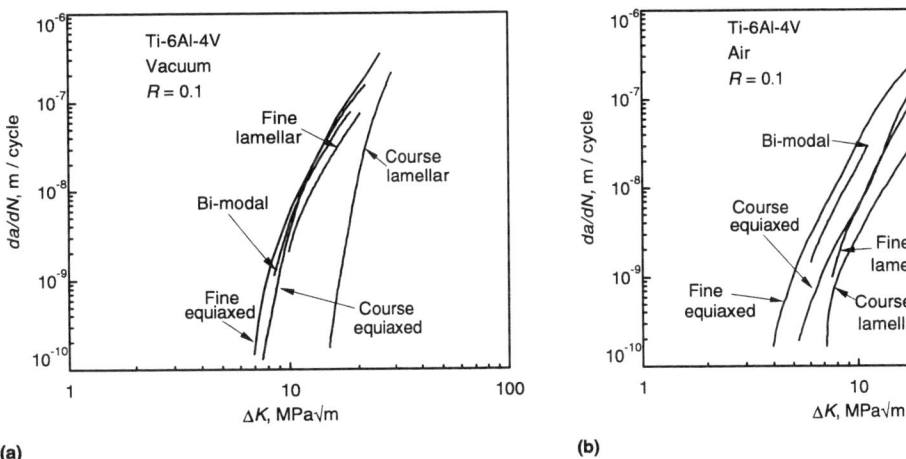

(a) In vacuum. (b) In air.
Material/Test Parameters: The equiaxed and bi-modal microstructures had a mixed basal/transverse texture after rolling. The fine equiaxed structure had an average α grain size of 2 μm, and the coarse equiaxed structure had an average α grain size of 12 μm.
Source: G Lütjering and A. Gysler, *Titanium Science and Technology*, Vol 4, Deutsche Gesellschaft für Metallkunde e.V., 1985, p 2077

Effect of Texture. Texture and specimen orientation are considered to be more important for crack nucleation than for crack propagation. However, crystallographic orientation can have significant effects when aggressive environments are involved (see the next section "Crack Growth with Corrosion"). Smaller effects of texture and test orientation are observed in laboratory air.

Effect of α-β Processing

Thermal processing or working in the α + β field generally produces an equiaxed microstructure which has less FCP resistance than transformed β structures. Significant amounts of scatter can occur because of variations in microstructure (grain size, morphology), texture, and strength levels. Besides lot-to-lot variations, mill annealed products may also have significant heat-to-heat variations (see figure for mill annealed plate). Improved FCP resistance at somewhat reduced levels of strength can be obtained by recrystallization annealing. Compared to the mill annealed condition, recrystallization annealing softens the alpha phase, thereby increasing its capacity for strain energy absorption at the tip of a propagating crack. As a result, the material has an increased resistance to crack propagation with monotonic loading (intrinsic fracture toughness) and with fatigue loading (FCP resistance). A secondary, beneficial effect of recrystallization anneal is a measurable microstructural coarsening which serves to retard crack growth under certain conditions.

Ti-6Al-4V: Scatter of FCP data for mill-annealed plate

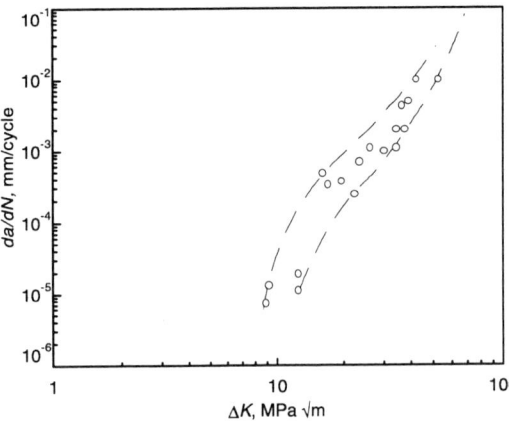

This is one of the more extreme cases of FCP data scatter, which can arise for the mill-annealed condition due to inconsistencies of microstructure, texture, and strength. Data are for six heats of mill-annealed Ti-6Al-4V.
Material/Test Parameters: Room-temperature air, $R = 0.1$, 10-Hz frequency, and T-L test orientation of 25-mm (1-in.) plate.
Source: *Titanium: A Technical Guide*, ASM International, 1988, p 180

Ti-6Al-4V: Effect of recrystallization annealing on FCP in plate

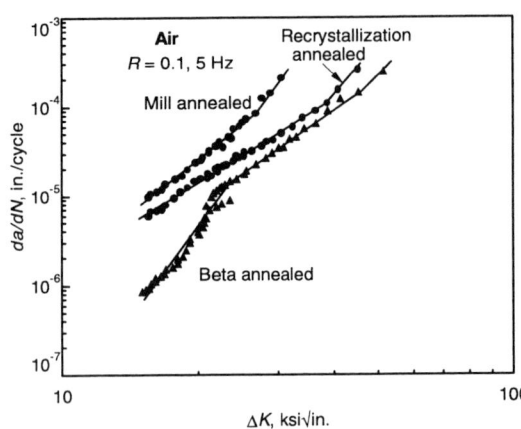

Material/Test Parameters: Room-temperature air, $R = 0.1$, 5-Hz frequency, and haversine wave form.
Source: G.R. Yoder et al., "Effects of Microstructure and Frequency on Corrosion-Fatigue Crack Growth in Ti-7Al-1Mo-1V and Ti-6Al-4V, in *Corrosion Fatigue: Mechanics, Metallurgy, Electrochemistry, and Engineering*, STP 801, T.W. Crooker and B.N. Leis, Ed., American Society for Testing and Materials, 1983, p 159-174

FCP Resistance of Transformed β

Like fracture toughness, FCP resistance improves when a microstructure of transformed β is obtained during β annealing or β solution treatment. In β annealed products, FCP rates below the transition point (ΔK_t) are related to the average Widmanstätten packet size (see figure). Interstitial oxygen (acting as an α stabilizer) and prior β grain size have an indirect influence on FCP rates by affecting the size of the average Widmanstätten packets that form upon cooling from above the β transus. The slowest FCP rates are obtained by slow (furnace) cooling from above the β transus.

Beta-annealed Ti-6Al-4V: Effect of Widmanstätten packet size on FCP rates

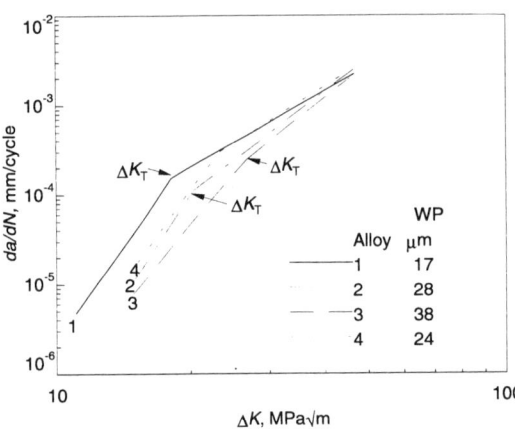

Each plot shows a significant change in slope at a transition point (ΔK_t) at which the reversed plastic zone appears to attain the average Widmanstätten packet (WP) size.
Material/Test Parameters: Room-temperature tests in laboratory air at $R = 0.1$ and a frequency of 5 Hz with haversine wave form.
Source: *Metall Trans A*, Vol 9A (1978), p 1413-1420

Ti-6Al-4V forgings: FCP rates for various R ratios

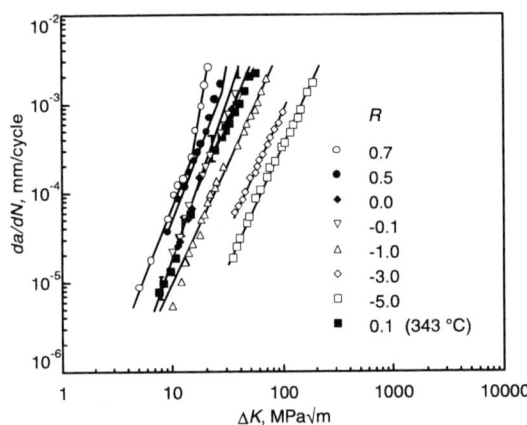

The data for $R = 0.1$ is an average of 10 tests. STOA.
Material/Test Parameters: Pancake forgings had a composition (wt%) of 6.3-6.43 Al, 0.02-0.033 C, 0.10-0.18 Fe, 0.0050-0.0062 H, 0.013-0.015 N, 0.172-0.183 O, and 4.28-4.3 V. Forged at 970 °C (1775 °F) in the α-β range, solution treated at 955 °C (1750 °F) for 1 h, water quenched, then aged at 700 °C (1300 °F) for 2 h. Specimen dimensions were 15.2 × 5.1 × 0.2 cm (6 × 2 × 0.08 in.) with a 0.76-cm (0.3-in.) slot (electric discharge machined) in center, normal to loading direction.
Source: A. Yuen et al., Correlations Between Fracture Surface Appearance and Fracture Mechanics Parameters for Stage II Fatigue Crack Propagation in Ti-6Al-4V, *Metall. Trans.*, Vol 5, 1974, p 1833

Ti-6Al-4V: Effect of microstructure and R ratio on FCP

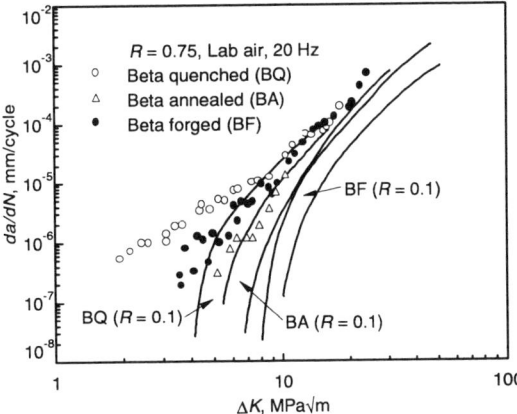

Source: J.C. Chesnutt, Fatigue Crack Propagation in Titanium Alloys, *Titanium Science and Technology*, Vol 4, Deutsche Gesellschaft für Metallkunde e.V., 1985, p 2227-2233

Beta-annealed Ti-6Al-4V: FCP rates with different oxygen contents

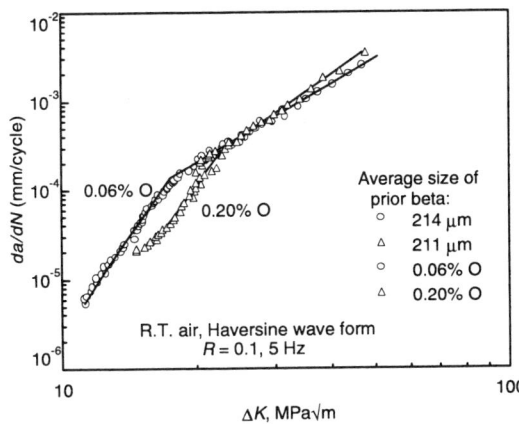

The slower crack growth rates for 0.20% O are associated with a larger Widmanstätten packet (WP) size (24 μm) vs a WP size of 17 μm for the specimens with 0.06% O.
Source: G.R. Yoder *et al.*, Fatigue Crack Propagation Resistance of Beta Annealed Ti-6Al-4V Alloys of Different Interstitial Oxygen Contents, *Metall Trans A*, Vol 9A, 1978, p 1413-1420

Crack Growth and Corrosion

The test environment can have a very strong influence on crack growth rate depending on the environment, alloy microstructure, alloy texture, and testing parameters (such as cycling frequency, dwell-time, stress-intensities, and electrodynamical potentials). Air is a more aggressive environment than a vacuum, and high-oxygen steam is slightly more detrimental than low-oxygen steam. Aqueous halide solutions are known contributors to stress-corrosion cracking (SCC), which in turn affects FCP rates of titanium alloys.

Ti-6Al-4V: Effect of oxygen of steam on near-threshold fatigue crack growth rates in Ti-6Al-4V

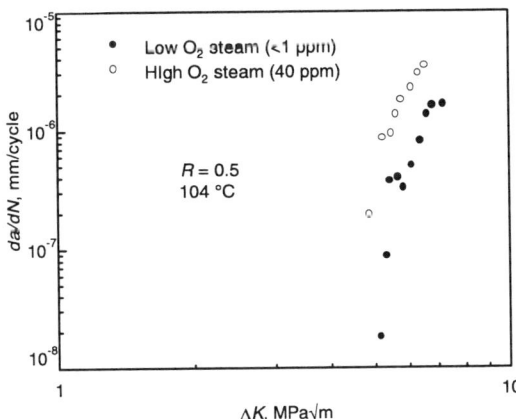

Source: P.K. Liaw, J. Anello, and J.K. Donald, Effects of Corrosive Environments on Near-Threshold Fatigue Crack Growth Behavior of Ti-6Al-4V, *Eng. Fract. Mech.*, Vol 19 (No. 6), 1984, p 1047-1056

Ti-6Al-4V: Effect of texture on near-threshold FCP in NaCl solution

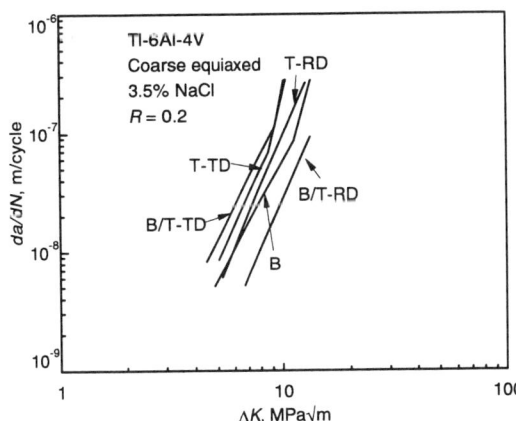

B, basal texture; B/T, basal/transverse texture; T, transverse texture; RD, rolling direction of test; TD, transverse direction of test.
Source: G Lütjering and A. Gysler, *Titanium Science and Technology*, Vol 4, Deutsche Gesellschaft für Metallkunde e.V., 1985, p 2078

Ti-6Al-4V: Crack-growth rates for continuously rolled textured sheet in air and salt water

Stress-intensity range		Environment	Grain direction	Crack growth ($\Delta 2a/\Delta N$)			
				μm/cycle		10^{-6} in./cycle	
MPa$\sqrt{m}$	ksi$\sqrt{in.}$			Range	Avg	Range	Avg
55	50	Air	Longitudinal	0.43-0.71	0.5	17-28	20
			Transverse	0.43-0.74	0.53	17-29	21
55	50	3.5% NaCl	Longitudinal	0.53-1.4	0.8	21-56	32
			Transverse	1.6-6.2	3.7	64-245	146
55	50	Distilled H_2O	Longitudinal	0.5-1.3	0.79	20-53	31
			Transverse	1.3-4.0	2.7	53-158	105
66	60	Air	Longitudinal	0.7-1.1	0.84	27-44	33
			Transverse	0.7-1.1	0.84	27-44	33
66	60	3.5% NaCl	Longitudinal	0.94-2.1	1.5	37-84	60
			Transverse	3.6-7.1	5.8	142-280	228
66	60	Distilled H_2O	Longitudinal	0.84-1.9	1.3	33-74	53
			Transverse	2.1-4.6	3.2	83-182	125

Material/test parameters: Preponderance of basal poles parallel to the transverse direction. Tested at $R = 0.67$ and 2 Hz. The specimen was continuously rolled sheet having a thickness of 1.3 mm (0.050 in.) and an area of 305×910 mm (12×36 in.). The maximum gross-area stress was 170 MPa (25 ksi). Source: Data from F. Parkinson, Boeing Co., 1972, reported in *Properties of Textured Titanium Alloys*, F. Larson and A. Zarkades, Ed., MCIC-74-20, Battelle Columbus Laboratories, 1974, p 70

Ti-6Al-4V: Effect of test direction and texture on FCP rates
Texture has little effect in a totally inert environment

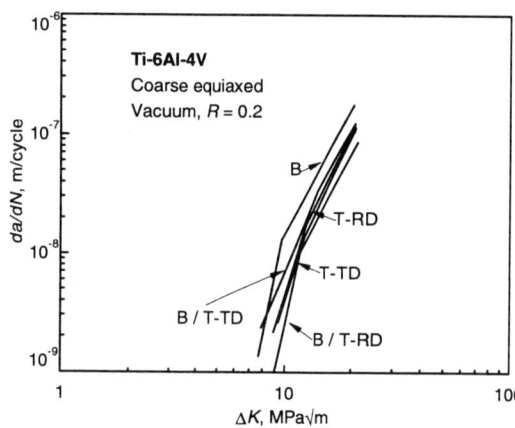

B, basal texture; B/T, basal/transverse texture; T, transverse texture; RD, rolling direction of test; TD, transverse direction of test.
Source: G Lütjering and A. Gysler, *Titanium Science and Technology*, Vol 4, DGM, e.V., 1985, p 2078

Aqueous Halide Solutions

Ti-6Al-4V: Typical FCP rates in salt water

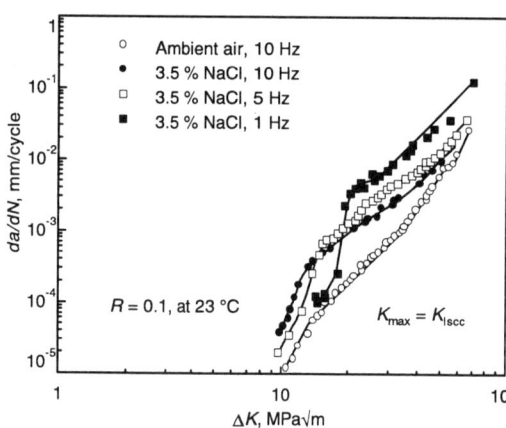

Mill annealed sheet.
Source: D.B. Dawson and R.M. Pelloux, Corrosion Fatigue Crack Growth of Titanium Alloys in Aqueous Environments, *Metall. Trans. A*, Vol 5A, 1974, p 723-731

Ti-6Al-4V: FCP behavior in salt water

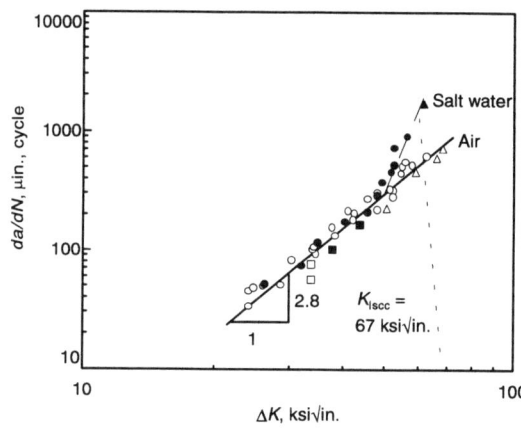

Salt water has little effect below the K_{Iscc} threshold.
Material/Test Parameters: Stress ratio, test frequency, and microstructure unspecified.
Source: N.G. Feige and R.L. Kane, Service Experience With Titanium Structures, in *Proceedings of the 26th Conference*, National Association of Corrosion Engineers, 1970, p 194-199

Mill-annealed Ti-6Al-4V: FCP rates in iodide solutions and distilled water

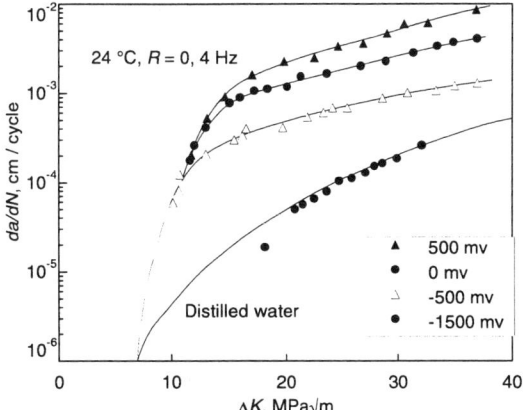

Electrode potentials must be closely monitored during testing. At a potential of −1500 mV, FCP rates in iodide solution are equal to those in distilled water.
Source: *Corrosion Fatigue: Chemistry, Mechanics and Microstructure*, Vol 2, National Association of Corrosion Engineers, 1972

Mill-annealed Ti-6Al-4V: Effect of testing frequency and ΔK on FCP rates in salt water

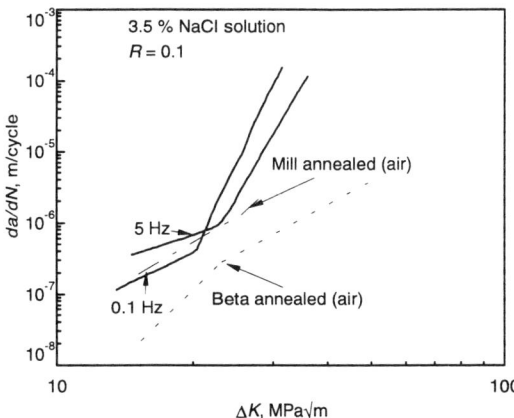

Source: G.R. Yoder et al., "Effects of Microstructure and Frequency on Corrosion-Fatigue Crack Growth in Ti-8Al-1Mo-1V and Ti-6Al-4V, in *Corrosion Fatigue: Mechanics, Metallurgy, Electrochemistry, and Engineering*, STP 801, T.W. Crooker and B.N. Leis, Ed., American Society for Testing and Materials, 1983, p 159-174

Effect of Test Frequency

Beta-annealed Ti-6Al-4V: Effect of testing frequency and ΔK on FCP rates in salt water

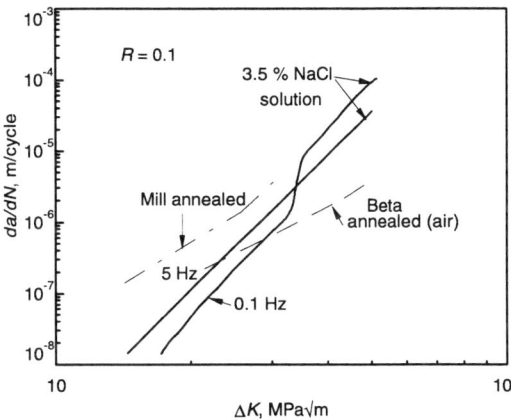

Source: G.R. Yoder et al., "Effects of Microstructure and Frequency on Corrosion-Fatigue Crack Growth in Ti-8Al-1Mo-1V and Ti-6Al-4V, in *Corrosion Fatigue: Mechanics, Metallurgy, Electrochemistry, and Engineering*, STP 801, T.W. Crooker and B.N. Leis, Ed., American Society for Testing and Materials, 1983, p 159-174

Recrystallization-annealed Ti-6Al-4V: Effect of dwell loading and hydrogen on FCP rates

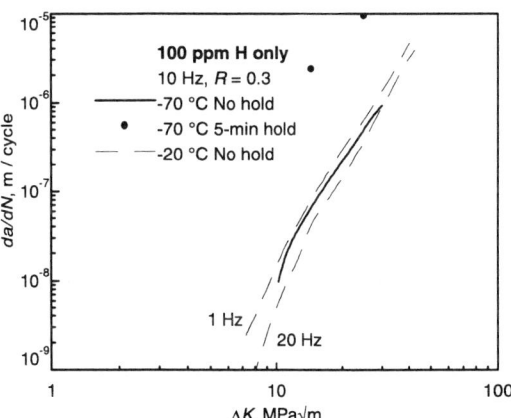

The RA microstructure, at 100 ppm H_2, shows a significant acceleration with a 5 minute hold time at maximum load when tested at −70 °C. There is even a slight increase in growth rate comparing 1 Hz and 20 Hz test frequencies at −20 °C. The acceleration, in both cases, is attributed to migration of hydrogen to the crack tip during the hold and precipitation of hydrides. This results in quasicleavage fracture of the α phase.
Source: J.C. Chesnutt and N.E. Paton, Hold Time Effects on Fatigue Crack Propagation in Ti-6Al and Ti-6Al-4V, in *Microstructure, Fracture Toughness, and Fatigue Crack Growth Rate in Titanium Alloys*, A.K. Chakrabarti and J.C. Chesnutt, Ed., The Metallurgical Society, 1987

Holding the specimen at the maximum load for some time period (or dwell) during the cycling can have a very significant effect on the crack growth rate. The effect is related to migration of hydrogen to the crack and the formation of hydrides; hence factors such as test temperature, microstructure and texture are important. As the fracture mechanism involves quasi-cleavage of the α-phase, which occurs on a plane near the basal plane, this phenomenon is strongly influenced by crystallographic texture. When testing a moderately textured plate where the direction of crack propagation was normal to the principal orientation of the basal planes, no acceleration was observed under any conditions, whereas a significant hold time effect was observed when the crack was propagating along the basal planes (J.C. Chesnutt and N.E. Paton, in figure reference above).

There are three conditions which all must be satisfied in order to observe acceleration of growth rate due to dwell effects: (1) hydrogen contents >100 ppm, (2) a temperature below room temperature (which is dependent on hydrogen level), and (3) a significant hold time in the tensile portion of the loading cycle. Stubbington and Pearson, however, demonstrated that an acceleration could be observed at temperatures up to 60 °C with 200 ppm hydrogen using 45 minute hold times (*Engr. Fracture Mechanics*, Vol 10, 1978, p 723-756).

Cyclic load frequency is an important variable that influences corrosion fatigue for most material, environment, and stress intensity conditions. The effect of frequency (or dwell-time effects) is directly related to the time dependence of the mass transport and chemical reaction steps required for environmental cracking.

Generally, the rate of environmental cracking above that produced in a vacuum increases with decreasing frequency. For titanium alloys and many other materials, however, the effects of frequency also depend on stress-intensity levels (see figures). At low ΔK's the lower frequency grows at a slower rate as repassivation of the surface can reduce the effect of the environment, and at the slower frequency, there is more time for repassivation. The crossover occurs at a ΔK level associated with the onset of cyclic stress corrosion cracking (SCC), and has been referred to as ΔK_{SCC} (see Dawson, D.B. and Pelloux, R.M., *Met. Trans.*, Vol 5, No. 3, March 1974, p 723-731). Above ΔK_{SCC} the effects of frequency are consistent with a cyclic SCC mechanism. With the lower frequency the crack has a longer residence time at the higher ΔK's than at the lower frequencies, resulting in more stress corrosion cracking and faster growth rates. This has been interpreted by Stubbington and Pearson (*Engr. Fracture Mechanics*, 1978, ibid) that above a hydrogen embrittlement mechanism dominates, and da/dN is related to hydrogen mobility to the plastically deformed region near the crack tip. The lower frequency then exhibits higher crack growth rates as there is more time for hydrogenation embrittlement to occur during each cycle.

Effect of test frequency on ΔK_{SCC}, the transition stress-intensity factor range for cyclic stress-corrosion cracking

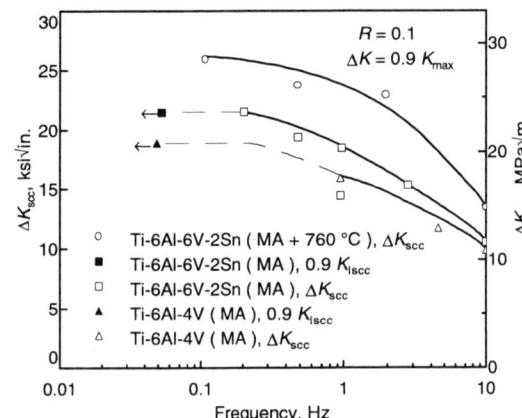

The ΔK level for which $K_{max} = K_{ISCC}$ is shown for reference. ΔK_{SCC} is lower than K_{ISCC} (the stress-corrosion cracking threshold under sustained loading), and it decreases with increasing frequency due to repeated rupture of the passive film at the crack tip.
Source: D.B. Dawson and R.M. Pelloux, Corrosion Fatigue Cracking of Titanium Alloys in Aqueous and Methanolic Environments, in *1972 Tri-Service Conference on Corrosion*, MCIC 73-19, M.M. Jacobson and A. Gallaccio, Ed., Battelle Columbus Laboratories, 1972, p 77-94

Impact Toughness

Impact properties are important in that they can be related to the residual strength or fracture toughness of a material. Also, impact tests are less expensive to conduct than compact-tension fracture toughness tests. Frequently, they are run as a screening test to evaluate alloys, heat treatment procedures, and other variables to select tougher materials prior to fracture toughness testing. The correlation between impact energy and fracture toughness, however, is not precisely proportional. Typical room-temperature impact strength of Ti-6Al-4V is about 20 J (15 ft · lbf) for Izod specimens and about 24 J (18 ft · lbf) for standard grade Charpy specimens.

Ti-6Al-4V: Izod impact strength vs temperature

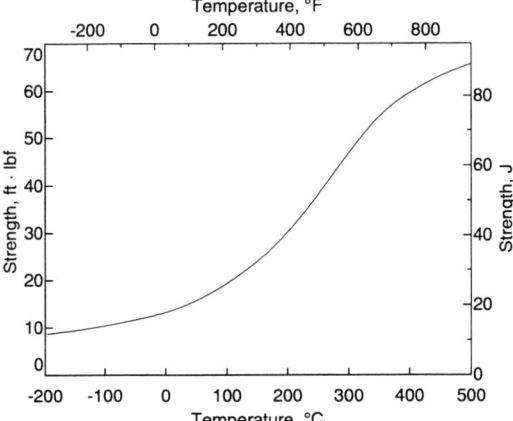

Source: "Titanium Alloy IMI 318," IMI Titanium, Ltd.

Ti-6Al-4V: Charpy impact strength of annealed bar vs temperature

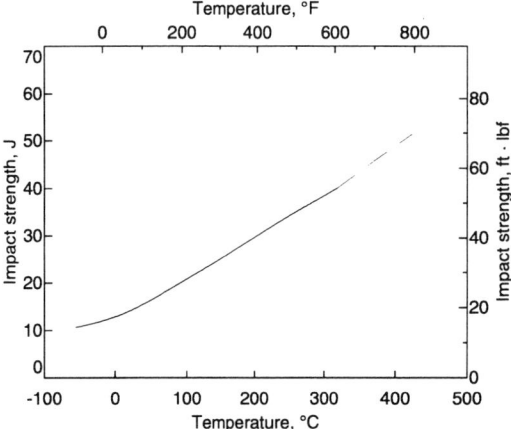

At –40 °C (–40 °F), annealed bars exhibit average Charpy V-notch impact energy of 24 J (18 ft · 1bf).
Alloy annealing conditions: holding at 730 °C (1350 °F) for 0.5 to 2 h, depending on section size, air cool. For maximum ductility, furnace cool at a rate of less than 2 °C (50 °F)/min to 540 °C (1000 °F), then air cool.
Source: Crucible Specialty Metals, Colt Industries

Ti-6Al-4V: Charpy impact strength of sheet

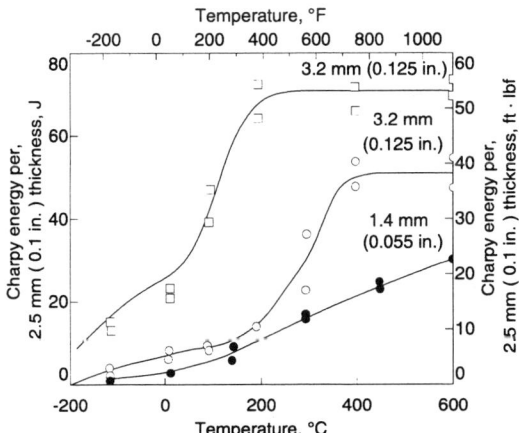

Test results from various heats of annealed sheet. Transverse specimen orientation.

Ti-6Al-4V: Impact strength of textured plate

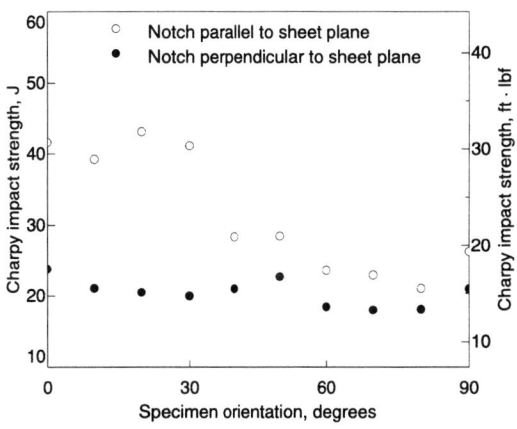

The highest impact strengths occur in a prism pole specimen (notch length parallel to c-axis) with the crack propagating perpendicular to the c crystal axis.
Specimen orientation refers to the angle between the specimen axis and the rolling direction. Plate thickness was 13 mm (0.525 in.) with 0.2% yield strengths of 840 and 925 MPa (122 and 135 ksi) in the L and T directions, respectively.
Source: F. Larson and A. Zarkades, Properties of Textured Titanium Alloys, MCIC 74-20, Battelle, 1974, p 55

Ti-6Al-4V: Energy to propagate and fracture

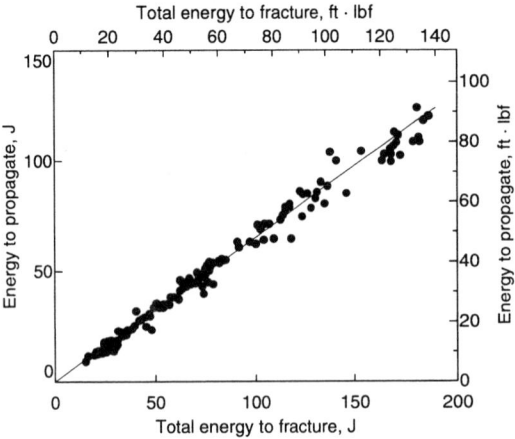

Relationship between total energy-to-fracture and energy-to-propagate in Kahn-type tensile-tear test.

Ti-6Al-4V: Charpy energy per 2.5 mm

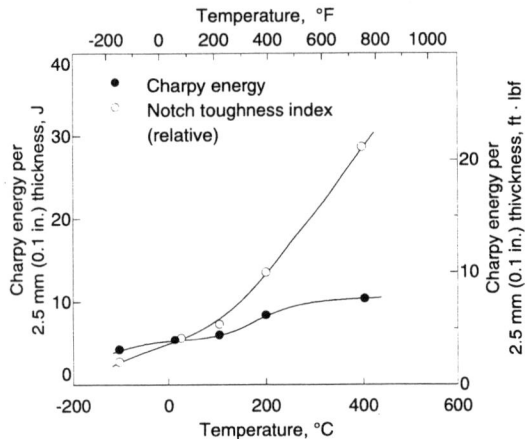

The material was Ti-6Al-4V 1.6 mm (0.063 in.). Note that the notch toughness index increases more rapidly with temperature than Charpy energy due to the decrease in yield strength with increasing temperature.

Ti-6Al-4V: Charpy impact strength and ultimate tensile strength compared to titanium alloys

Alloy	Ultimate tensile strength		Charpy impact toughness	
	MPa	ksi	J	ft · lbf
Unalloyed titanium	345	50	54	40
Unalloyed titanium	550	80	27	20
Ti-5Al-2.5Sn	825	120	25	19
Ti-8Al-1Mo-1V	895	130	27	20
Ti-6Al-6V-2Sn	1035	150	16	12
Ti-3Al-8V-6Cr-4Mo-4Zr	1240	180	9.5	7

Source: "The Titanium Industry in the mid-1970's," MCIC-75-26, Battelle, 1975, p 59

Ti-6Al-4V: Charpy impact strength vs temperature for 15.9 mm (5/8 in.) rounds in various heat treated conditions

Temperature		Charpy V-notch impact toughness, for indicated heat treatment: J (ft · lbf)		
°C	°F	705 °C (1300 °F), 2 h, mill annealed	845 °C (1550 °F), 1 h, water quenched, 480 °C (900 °F), 8 h, air cooled	940 °C (1725 °F), 1 h, water quenched, 480 °C (900 °F), 8 h, air cooled
−73	−100	18.9 (14.0), 16.2 (12.0)	20.3 (15.0), 18.9 (14.0)	18.3 (13.5), 15.6 (11.5)
17	0	24.4 (18.0), 25.1 (18.5)	21.7 (16.0), 24.4 (18.0)	18.9 (14.0), 23.0 (17.0)
RT	RT	24.4 (18.0), 23.0 (17.0)	20.3 (15.0), 24.4 (18.0)	20.3 (15.0), 21.7 (16.0)
93	200	28.4 (21.0), 31.8 (23.5)	28.4 (21.0), 28.4 (21.0)	25.7 (19.0), 23.0 (17.0)
150	300	41.3 (30.5), 47.5 (35.0)	36.6 (27.0), 28.4 (21.0)	37.9 (28.0), 28.4 (21.0)
205	400	69.1 (51.0), 45.4 (33.5)	42.0 (31.0), 50.8 (37.5)	43.3 (32.0), 51.5 (38.0)
260	500	66.4 (49.0), 84.1 (62.0)	47.4 (35.0), 46.1 (34.0)	56.9 (42.0), 60.3 (44.5)
315	600	70.5 (52.0), 90.1 (66.5)	64.4 (47.5), 52.8 (39.0)	71.2 (52.5), 65.0 (48.0)
370	700	103.7 (76.5), 93.5 (69.0)	69.8 (51.5), 94.9 (70.0)	56.2 (41.5), 62.3 (46.0)

Source: M. Mote, R. Hooper, and P. Frost, *Engineering Properties of Commercial Titanium Alloys*, TML Report 92, Battelle, 1958, p G-6

Charpy impact strength for castings in several heat treated conditions

Heat treatment	Charpy V-notch impact toughness	
	J	ft · lbf
Conventional anneal at 845 °C (1550 °F), 2 h in vacuum/argon fan cool	34.7	25.6
	33.7	24.9
	33.6	24.8
Below beta solution treatment and age, consisting of a solution treatment at 985 °C (1810 °F), 1 h in vacuum, an inert gas fan cool at a rate equivalent to air cooling, and an age at 540 °C (1010 °F), 8 h in vacuum/GFC	30.6	22.6
	28.7	21.2
	28.2	20.8
Above beta solution treatment and age, consisting of 1015 °C (1860 °F), 1 h in vacuum/GFC + 540 °C (1000 °F), 8 h in vacuum/GFC	27.1	20.0
	26.0	19.2
	28.7	21.2
Below beta cyclic exposure involving six cycles of 980 °C (1800 °F), 10 min in vacuum/GFC to 540 °C (1000 °F) + 540 °C (1000 °F), 10 min in vacuum, heat to 980 °C (1800 °F), followed by a gas fan cool to 21 °C (70 °F)	40.5	29.9
	29.9	22.1
	35.5	26.2
Nontraditional, proprietary thermal treatment process known as CST-I	24.9	18.4
	23.8	17.6
	23.8	17.6

Note: Materials from three heats were centrifugally cast into preheated MonoShell investment molds. Composition range was 5.9 to 6.1 wt.% Al, 0.001 to 0.003 wt.% H, 0.01 to 0.02 wt.% N, 0.16 to 0.19 wt.% O, and 3.8 to 4.1 wt.% V. Source: R.J. Smickley, and L.E. Dardi, "The Thermal Processing Response of HIP'ed Investment Cast Ti-6Al-4V Alloy," *Titanium Net Shape Technologies*, F.H. Froes and D. Eylon, Ed., The Metallurgical Society of AIME, 1984, p 201-209

Fracture Toughness

The fracture toughness (K_{Ic}) of Ti-6Al-4V is higher than that of aluminum alloys, but lower than steels. In very general terms, fracture toughness increases as the amount of transformed β (lamellar α/β) structure increases, with β annealing providing the highest fracture toughness. The exception to this is the recrystallization annealed structure, which contains no transformed β, but has a fracture toughness almost as good as β annealed material. Also, coarser microstructures generally provide higher toughness values. The ELI grade alloy is used for fracture-critical applications, due to its superior toughness.

Ti-6Al-4V: Fracture toughness scatter bands

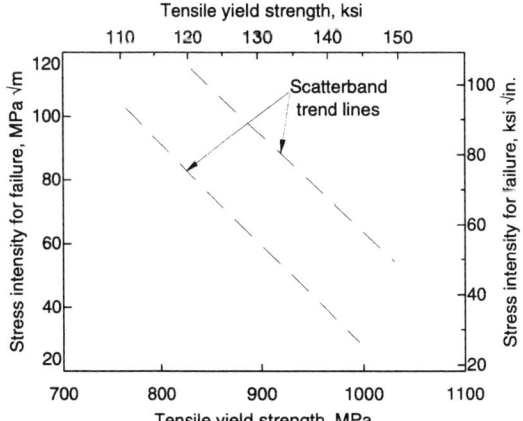

Annealed specimens (bar, plate, and forgings) tested were all within specification limitations. A large number of specimens from numerous heats were tested by compact tension or four-point bending tests.. Source: *Titanium and Titanium Alloys*, MIL-HDBK-697A, 1974, p 43

Ti-6Al-4V: Range of yield strength and fracture toughness

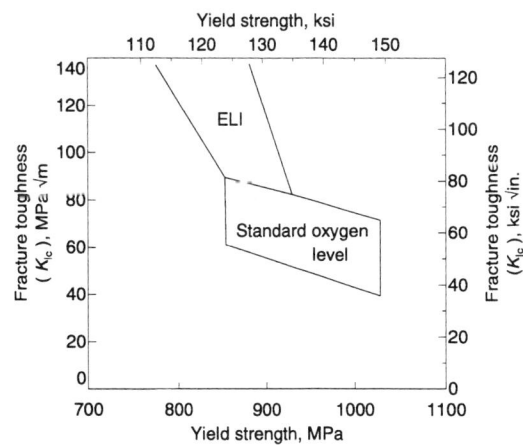

Ti-6Al-4V: Fracture toughness vs yield strength

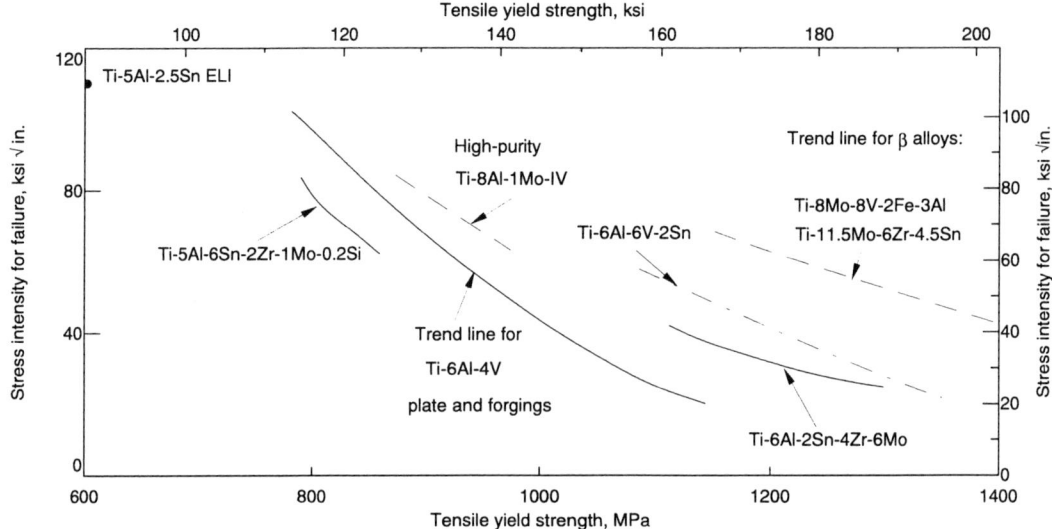

Alloys in the annealed condition were Ti-5Al-2.5Sn, Ti-8Al-1Mo-1V, and alloys in the solution heat treated and aged conditions were Ti-6Al-2Sn-4Zr-6Mo, Ti-6Al-6V-2Sn, and β alloys Ti-11.5Mo-6Zr-4.5Sn and Ti-8Mo-8V-2Fe-3Al. Ti-6Al-4V trend line represents both annealed and STA conditions.
Source: *Titanium and Titanium Alloys*, MIL-HDBK-697A, 1974, p 44

Effects of Processing

The fracture toughness of Ti-6Al-4V ranges from about 33 to more than 110 MPa√m (30 to 100 ksi√in.), depending on oxygen content and heat treatment. At similar strengths, β processed material has a fracture toughness on the order of 50% greater than α-β processed material.

Effect of Welding. Generally, the fracture toughness (and crack-growth resistance) of Ti-6Al-4V welds depends on both fusion zone microstructure and interstitial content. The martensitic microstructure characteristic of electron beam and laser welds with high depth-to-width ratios is characterized by toughnesses below that of the mill annealed base metal. Welds produced at lower cooling rates exhibit increasing toughness levels that are comparable or superior to that of the mill annealed base metal.

Obata *et al.*, studied the fracture toughness of electron beam welds as a function of filler metal thickness and stress relief temperature. The thickness of the filler metal was shown to have a strong

Typical fracture toughness of several alloys

Alloy	Fracture toughness MPa√m	ksi√in.
Ti-6Al-4V (annealed)	65	60
Ti-6Al-4V (β annealed)	90	80
7075 aluminum (T73)	35	32
4140 steel(a)	115	105
4340 steel(a)	115	105
15-5 PH (a)	80	75

(a) Hardened to 1240 to 1380 MPa (180 to 200 ksi)

Ti-6Al-4V: Oxygen content/thermal treatment vs fracture toughness

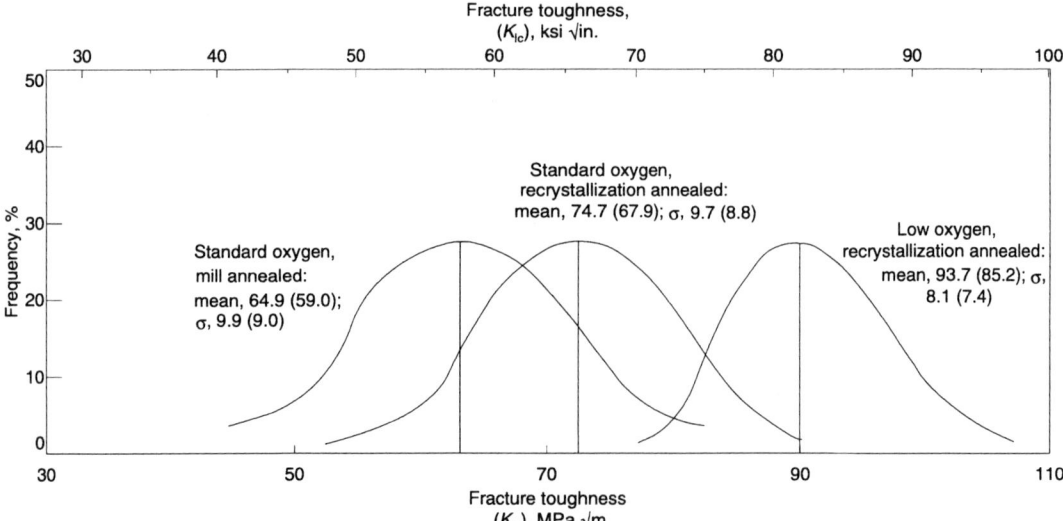

Source: G.W. Kuhlman and F.R. Billman, Selecting Processing Options for High-Fracture Toughness Titanium Airframe Forgings, *Met. Prog.*, March 1987

influence on the toughness of the weld metal, but no effect in the heat-affected zone (HAZ). Furthermore, the welding sequence did not have a significant effect on the toughness of the weld metal, but heat treatment after welding, rather than prior to welding, was beneficial in the HAZ.

Ti-6Al-4V: (ELI): Annealing temperature vs fracture toughness of plate

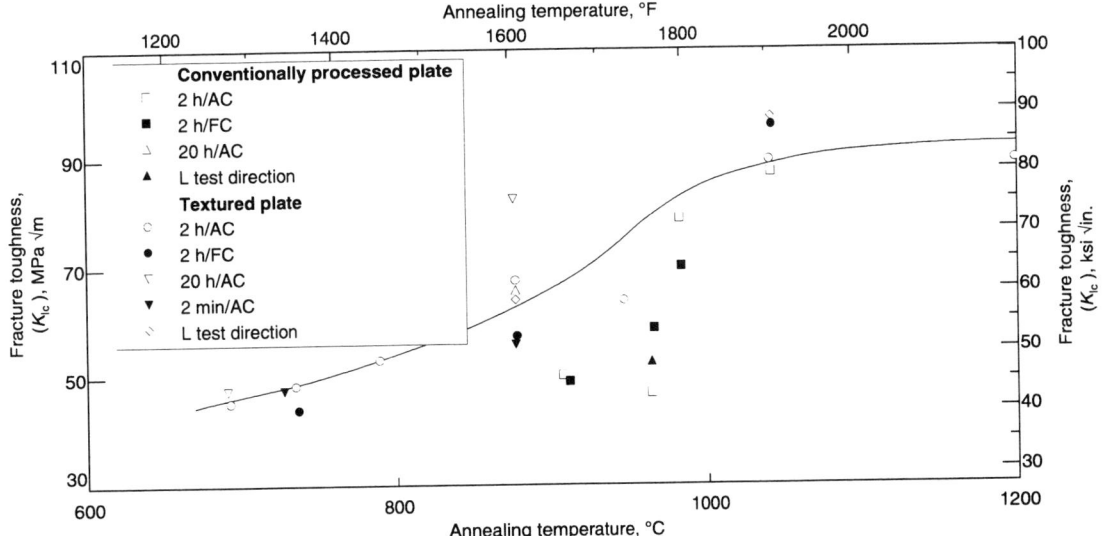

Material description	Oxygen content, wt%	Heat treatment	Grain direction	Tensile strength MPa	Tensile strength ksi	Yield strength MPa	Yield strength ksi	Elongation, %	Reduction of area, %
Conventional plate	0.13	MA	L	986	143	910	132	13	22
		MA	T	1034	150	972	141	15	33
		BA	T	951	138	876	127	11	26
Textured plate	0.126	As rolled	L	945	137	869	126	15	27
		As rolled	T	1034	150	979	142	15	27
		BA	T	931	135	855	124	9	17

Note: Composition 6.3% Al, 0.026% C, 0.07% Fe, 0.005% H, 0.015% N, 0.13% O, 4.10% V. Conventional plate (Air Force plate, 30.5 mm or 1.2 in. thick) was conventionally a-b cross rolled and mill annealed; textured plate (31.75 mm or 1.25 in. thick) was unidirectionally rolled to produce a moderately transverse basal plane (TD) texture. Specimens were annealed 2 min to 20 h at 690 to 1040 °C (1275 to 1900 °F). R.R. Boyer, R. Bajoraitis, and W.F. Spurr, The Effects of Thermal Processing Variations on the Properties of Ti-6Al-4V, in Microstructure, Fracture Toughness, and Fatigue Crack Growth Rate in Titanium Alloys, Proceedings of the 1987 TMS-AIME annual symposia on Effect of Microstructure on Fracture Toughness and Fatigue Crack Growth Rate in Titanium Alloys, Denver, A.K. Chakrabarti and J.C. Chesnutt, Ed., The Metallurgical Society, 1987, p 149-170

Ti-6Al-4V: Fracture toughness vs rolling temperature tensile strength

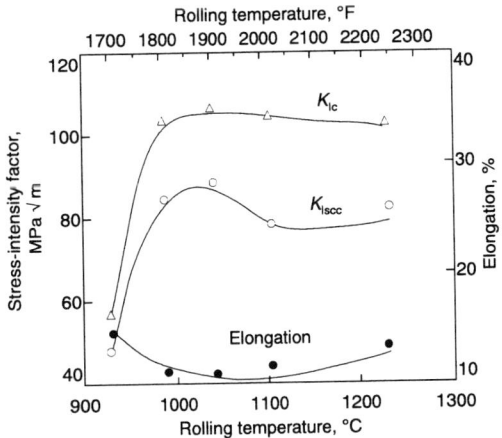

Solution treated and aged, 940 °C (1725 °F), and water quenched plus 8 h at 675 °C (1250 °F) and air cooled, after 30% reduction by rolling.
Source: R.E. Curtis and W.F. Spurr, Effect of Microstructure on the Fracture Properties of Titanium Alloys in Air and Salt Solution, *Trans. ASM*, Vol 61, p 115-127

Ti-6Al-4V: Fracture toughness vs solution temperature on tensile/fracture strength

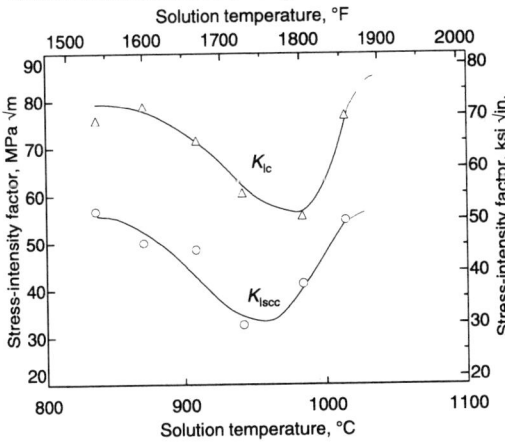

Specimens for tensile and fracture tests were aged 4 h at 675 °C (1250 °F) and air cooled after solution treatment.
Source: R.E. Curtis and W.F. Spurr, Effect of Microstructure on the Fracture Properties of Titanium Alloys in Air and Salt Solution, *Trans. ASM*, Vol 61, p 115-127

Ti-6Al-4V: Fracture toughness of powder compacts compared to corresponding wrought alloys

Alloy	Powder compact Fracture toughness (K_{Ic}) MPa√m	ksi√in.	Yield strength MPa	ksi	Wrought alloy Fracture toughness (K_{Ic}) MPa√m	ksi√in.	Yield strength MPa	ksi
Ti-10V-2Fe-3Al	51	46	854	124	60	55	1100	160
Ti-6Al-6V-2Sn	44	40	910	132	65	59	965	140
Ti-6Al-2Sn-4Zr-6Mo	31	28	1068	155	33	30	1100	160
Ti-6Al-4V*	35	32	900	130	66	60	860	125

Note: Compositions of cold isopressed and vacuum sintered billets were as follows. Ti-10-2-3: 2.98 wt.% Al, 0.010 wt.% C, 2.03 wt.% Fe, 0.0008 wt.% H, 0.011 wt.% N, 0.13 wt.% O, and 10.25 wt.% V. Ti-6-6-2: 6.00 wt.% Al, 0.019 wt.% C, 0.55 wt.% Cu, 0.48 wt.% Fe, 0.0016 wt.% H, 0.0045 wt.% N, 0.27 wt.% O, 1.65 wt.% Sn, and 5.31 wt.% V. Ti-6-2-4-6: 6.03 wt.% Al, 0.024 wt.% C, 0.060 wt.% Fe, 0.0009 wt.% H, 5.60 wt.% Mo, 0.0055 wt.% N, 0.032 wt.% O, 2.20 wt.% Sn, and 3.60 wt.% Zr. Data for Ti-6-4 powder compact were taken from P. Andersen et al., *Modern Developments in Powder Metallurgy*, Vol 13, 1981. Data for wrought alloys taken from AMS specifications, except Ti-10-2-3 (TIMET data). Source: J. Smugeresky and N. Moody, Properties of High Strength Blended Elemental Powder Metallurgy Titanium Alloys, in *Titanium Net Shape Technologies*, F. Froes and D. Eylon, Ed., 1984, p 131

Ti-6Al-4V (ELI): Fracture toughness of recrystallization annealed forgings

Direction	Yield strength MPa	ksi	Fracture toughness MPa√m	ksi√in.
LT	841	122	91	83
TL	890	129	92	84

Note: Specimens were recrystallization annealed at 925 °C (1700 °F), 4 h FC to 760 °C (1400 °F), AC. Room-temperature data from section 38.5 mm (1.5 in.) thick. Oxygen in 0.10 to 0.13% range. Source: R. Judy Jr., and B. Goode, "Stress Corrosion Cracking Characteristics of Alloys of Titanium in Salt Water," Interim Report 6564, Naval Research Laboratory, Washington DC, July 1967, as reported in *Damage Tolerant Design Handbook*, Vol 1, MCIC-HB-01, Metals and Ceramics Information Center, Battelle, Columbus, 1972

Ti-6Al-4V: Effect of heat treatment on fracture toughness, strength, and elongation

Heat treatment	Temperature °C	°F	Fracture toughness (K_{Ic}) MPa√m	ksi√in.	Tensile yield strength (0.2%) MPa	ksi	Ultimate tensile strength MPa	ksi	Elongation, %
Air cooled	750	1380	67.5	61.4	950	138	1020	148	13.5
	850	1560	69.0	62.8	920	133	1010	146	14.0
	950	1740	73.0	66.4	935	136	1020	148	13.5
	1050	1920	78.0	71.0	940	136	1050	152	10.5
Air cooled and aged	750	1380	58.5	53.2	985	143	1030	149	13.0
	850	1560	55.5	50.5	960	139	1030	149	13.0
	950	1740	85.0	77.4	965	140	1030	149	13.5
	1050	1920	105.0	95.6	945	137	1020	148	2.5
Quenched	750	1380	49.0	44.6	970	141	1050	152	12.5
	850	1560	69.5	63.2	830	120	990	144	16.5
	950	1740	59.0	53.7	1050	152	1150	167	12.0
	1050	1920	59.0	53.7	1070	155	1170	170	11.5
Quenched and aged	750	1380	69.0	62.8	975	141	1030	149	12.0
	850	1560	78.0	71.0	935	136	1010	146	13.0
	950	1740	59.5	54.1	1060	154	1125	163	12.0
	1050	1920	68.0	61.9	1085	157	1140	165	3.5

Note: Composition: 6.35% Al, 0.05% C, 0.07% Fe, 45 ppm H, 0.01% N, 0.18% O, and 3.78% V. β/α + β transus, 995 ± 15 °C (1825 ± 27 °F). Source: I.W. Hall and C. Hammond, Fracture Toughness, Strength and Microstructure in Alpha + Beta Titanium Alloys, *Titanium and Titanium Alloys*, Vol 1, J.C. Williams and A.F. Belov, Ed., Plenum Press, 1976, p 601-613

Ti-6Al-4V: Effects of overaging on fracture toughness and tensile strength

Type of processing	Heat treatment	Fracture toughness (K_{Ic}) MPa√m	ksi√in.	Tensile strength MPa	ksi
Alpha-beta	940 °C (1725 °F), WQ + aged 4 h at 540 °C (1000 °F), AC	54	49	1195	173
Alpha-beta	940 °C (1725 °F), WQ + overaged 4 h at 675 °C (1250 °F), AC	66	60	1080	157
Beta	940 °C (1725 °F), WQ + aged 4 h at 540 °C (1000 °F), AC	91	83	1160	168
Beta	940 °C (1725 °F), WQ + overaged 4 h at 675 °C (1250 °F), AC	101	92	1055	153

Source: R.A. Wood and R.J. Favor, *Titanium Alloys Handbook*, MCIC, Battelle, Columbus, 1972

Ti-6Al-4V: Fracture toughness

Condition/heat treatment	Product form	Thicknesses mm	Thicknesses in.	L-T Mean	L-T Standard deviation	T-L Mean	T-L Standard deviation
Alpha-beta forged, mill annealed	Forging	57	2.25	...	...	38.9 (35.4)	2.9 (2.7)
Annealed	Forging	75	3.00	92.7 (84.4)	1.9 (1.8)	91.6 (83.4)	10.9 (9.9)
	Extrusion	100	4.00	...	...	102.5 (93.3)	2.5 (2.3)
	Billet	100	6.0	87.4 (79.6)	10.5 (9.6)	...	...
Annealed 540 °C (1000 °F) 2 h, AC	Billet	58	2.30	55.9 (50.9)	0.65 (0.6)	...	...
Annealed 705 °C (1300 °F) 4 h, AC	Forging	58	2.30	63.8 (58.1)	1.3 (1.2)	68.3 (62.2)	3.3 (3.0)
Annealed 745 °C (1375 °F) 3 h, AC	Plate	70	2.75	66.3 (60.4)	6.0 (5.5)	...	...
As received	Forged bar	25-89	1.00-3.50	62.7 (57.1)	11.4 (10.4)	60.3 (54.9)	11.8 (10.8)
Forged, beta forged reheated to 1065 °C (1950 °F) drawn to size, annealed 705 °C (1300 °F)	Forged bar	57	2.25	...	...	46.8 (42.6)	4.7 (4.3)
Beta forged, mill annealed 705 °C (1300 °F) 2 h, AC	Forging	50	2.00	77.5 (70.6)	5.4 (4.9)	78 (71.0)	0.43 (0.4)
β processed-mill annealed	Plate	75	3.00	104.2 (94.9)	5.2 (4.8)	...	...
Diffusion bond annealed	Billet	25-89	1.00-3.50	74.9 (68.2)	10.6 (9.7)	70.5 (64.2)	12.9 (11.8)
Mill annealed	Plate	25-38	1.00-1.50	61.1 (55.6)	1.4 (1.3)	...	...
Mill annealed	Plate	31-50	1.25-2.00	...	...	110.5 (100.6)	7.4 (6.8)
	Extrusion	45-100	1.80-4.00	91.7 (83.5)	3.4 (3.1)	96.1 (87.5)	4.5 (4.1)
Mill annealed 705 °C (1300 °F) 2 h, AC	Forging	50	2.00	52.4 (47.7)	3.2 (2.9)	54.4 (49.5)	4.3 (3.9)
	Billet	58	2.30	92.3 (84.0)	3.7 (3.4)	...	...
Recrystallize anneal	Plate	25-63	1.00-2.50	90.9 (82.8)	8.5 (7.8)	88.8 (80.8)	11.8 (10.8)
	Forging	30-170	1.20-6.70	91.8 (83.6)	6.0 (5.5)	92.2 (83.9)	7.6 (6.9)
STA	Plate	15	0.62	...	...	46.6 (42.6)	2.2 (2.0)
925 °C (1700 °F) 6 h, AC, 760 °C (1400 °F) 6 h, AC	Forging	35	1.40	83.4 (75.9)	4.6 (4.2)	89.2 (81.2)	6.3 (5.8)
955 °C (1750 °F) 1 h, WQ, 540 °C (1000 °F) 4 h	Forging	75	3.00	...	...	87.1 (79.3)	5.4 (4.9)
955 °C (1750 °F) 1 h, FC to 595 °C (1100 °F), AC	Plate	38	1.50	...	...	100.5 (91.5)	2.3 (2.1)
955 °C (1750 °F) 1 h, FC to RT	Plate	38	1.50	78.6 (71.8)	3.5 (3.2)	100.6 (91.6)	1.4 (1.3)
955 °C (1750 °F) 2 h, WQ, 540 °C (1000 °F) 2 h, AC, 705 °C (1300 °F) 2 h, AC, STA	Plate	15	0.62	45.5 (41.4)	2.5 (2.3)	...	...

Fracture toughness (K_{Ic}), MPa√m (ksi√in.)

Source: *Damage Tolerant Design Handbook*, J. Gallagher, Battelle, 1983. FC, fan cooled.

Weldments

Ti-6Al-4V: Effects of specimen type, orientation, and test temperature on fracture toughness of electron beam welds

Specimen type and orientation(a)	Specimen thickness mm	Specimen thickness in.	Test temperature °C	Test temperature °F	K_Q MPa√m	K_Q ksi√in.	K_{Ic} MPa√m	K_{Ic} ksi√in.
BM/T-L	50.8	2.0	RT		115.9	105.5	115.9	105.5
BM/L-T	50.8	2.0	RT		123.1	112.0	119.3	108.6
BM/S-L	50.8	2.0	RT		96.7	88.0	...	...
HAZ/T-L	50.8	2.0	RT		131.6	119.8	...	...
HAZ/L-T	50.8	2.0	RT		129.5	117.9	...	...
HAZ/S-L	50.8	2.0	RT		97.9	89.1	...	...
WM/T-L	50.8	2.0	RT		90.8	82.6	...	...
BM-T-L	50.8	2.0	−54	−65	96.6	87.9	96.6	87.9
HAZ/T-L	50.8	2.0	−54	−65	118.7	108.0	...	...
WM/T-L	50.8	2.0	−54	−65	102.2	93.0	...	...
BM/T-L	50.8	2.0	79	175	128.6	117.0	...	...
HAZ/T-L	50.8	2.0	79	175	149.4	136.0	...	...
WM/T-L	50.8	2.0	79	175	122.2	111.2	...	...
WM/T-L	25.4	1.0	RT		63.1	57.4	...	...

Note: Compact tension specimens were tested to ASTM E399; values shown are averages of three specimens per condition. All specimens were stress relieved for 5 h at 705 °C (1300 °F) prior to testing. (a) BM, base metal; HAZ, heat-affected zone; WM, weld metal; T-L and S-L indicate base-metal orientations. (b) Most WM and HAZ tests were invalid because of insufficient crack length or specimen thickness. Occasional high values in HAZ resulted from unstable crack propagation at large angles from desired path. Source: J.G. Bjeletich, "Development of Engineering Data on Thick Section Electron Beam Welded Titanium," AFML-TR-73-197, U.S. Air Force Materials Laboratory, 1973

Ti-6Al-4V: Fracture toughness of electron beam welds as a function of processing order and stress-relief temperature

Location of notch	Filler-metal thickness mm	in.	Welding position(a)	No. of pass	Fracture toughness (K_Q)(b)(c) STA→EBW→SR for SR temperature of: 545 °C (1015 °F) MPa√m	ksi√in.	645 °C (1195 °F) MPa√m	ksi√in.	EBW→STA MPa√m	ksi√in.
Weld metal	None		H	1	...	...	54.9	50.0	71.0*	64.6*
				2	69.5	63.2	71.0	64.6	71.0*	64.6*
			F	1	65.7	59.8	60.2	54.8	71.0*	64.6*
				2	59.2	53.9	65.7	59.8	71.0*	64.6*
	0.5	0.020	H	1	76.0*	69.2*	90.2*	82.1*	...	...
				2	74.4	67.7	76.3*	69.4*	...	...
			F	1	86.5*	78.7*	74.7*	68.0*	...	...
				2	89.0*	81.0*	87.7*	79.8*	...	...
	1.0	0.039	H	1	84.7*	77.1*	...	...	90.2*	82.1*
				2	87.4*	79.5*	89.6*	81.5*	90.2*	82.1*
			F	1	83.4*	75.9*	88.1*	80.2*	90.2*	82.1*
				2	90.2*	82.1*	87.1*	79.3*	90.2*	82.1*
Heat-affected zone	None		H	1	58.0	52.8	49.0	44.6	69.8	63.5
				2	60.8	55.3	55.2	50.2	69.8	63.5
			F	1	61.4	55.9	58.0	52.8	69.8	63.5
				2	60.8	55.3	54.6	49.7	69.8	63.5
	0.5	0.020	H	1	54.6	49.7	59.2	53.9	...	...
				2	67.0	61.0	58.0	52.8	...	...
			F	1	...	...	58.6	53.3	...	...
				2	61.1	55.6	51.2	46.6	...	...
	1.0	0.039	H	1	57.7	52.5	58.9	53.6	76.9*	70.0*
				2	65.7*	59.8*	52.4	47.7	76.9*	70.0*
			F	1	67.0*	61.0*	52.7	48.0	76.9*	70.0*
				2	65.4*	59.5*	51.5	46.9	76.9*	70.0*

Note: Composition, 6.17% Al, 0.017% C, 0.117% Fe, 0.0058% H, 0.0024% N, 0.149% O, and 4.04% V. Focal length, 350 mm (13.8 in.) for horizontal specimens, 400 mm (15.7 in.) for flat specimens. Work distance, 300 mm (11.8 in.). Accelerating voltage, 50 kV. Beam current, 150 mA. Welding speed, 300 mm/min (11.8 in./min). (a) H, horizontal; F, flat. (b) STA, solution treated and aged; EBW, electron beam welded; SR, stress relief. (c) Asterisks indicate invalid K_{Ic}. Source: Y. Obata et al., Fracture Behavior in Electron Beam Welds of Ti-6Al-4V Alloy, *Titanium Science and Technology*, G. Lütjering, U. Zwicker, and W. Bunk, Ed., Deutsche Gesellschaft für Metallkunde eV, 1985, p 807-813

Ti-6Al-4V: Fracture toughness of electron beam welds

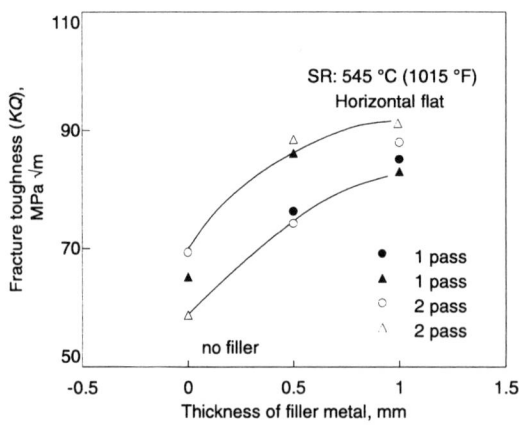
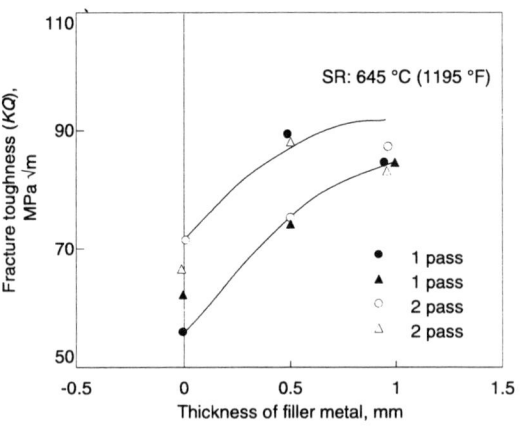

Effect of filler metal thickness and stress relief temperature on fracture toughness.
Composition: 6.17% Al, 0.017% C, 0.117 % Fe, 0.0058% H, 0.0024%, N, 0.149% O, and 4.04% V. Focal length, 350 mm (13.8 in.) for horizontal specimens, 400 mm (15.7 in.) for flat specimens. Work distance, 300 mm (11.8 in.). Accelerating voltage, 50 kV. Beam current, 150 mA. Welding speed, 300 mm/min (11.8 in./min).
Source: Y. Obata et al., Fracture Behavior in Electron Beam Welds of Ti-6Al-4V Alloy, *Titanium Science and Technology*, G. Lütjering, U. Zwicker, and W. Bunk, Ed., Deutsche Gesellschaft für Metallkunde eV, Germany, 1985, p 807-813

Effect of Temperature

Ti-6Al-4V: Fracture mechanism map

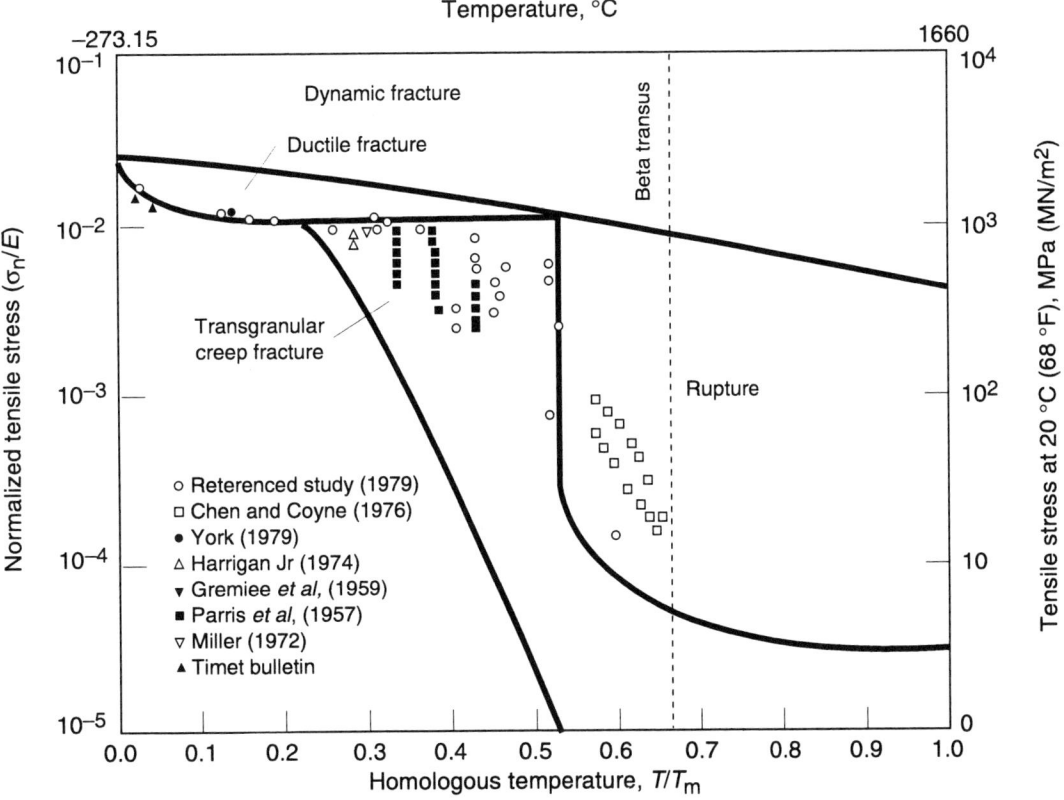

Normalizing parameters: T_m = 1933 K and E(GPa) = 122.7 − 0.056T (in K).
Source: Krishnamohanrao et al., Fracture Mechanism Maps for Titanium and Its Alloys, *Acta Metall.*, Vol 34, No. 9, 1986, p 1783-1806

Ti-6Al-4V: Fracture toughness vs oxygen content/temperature

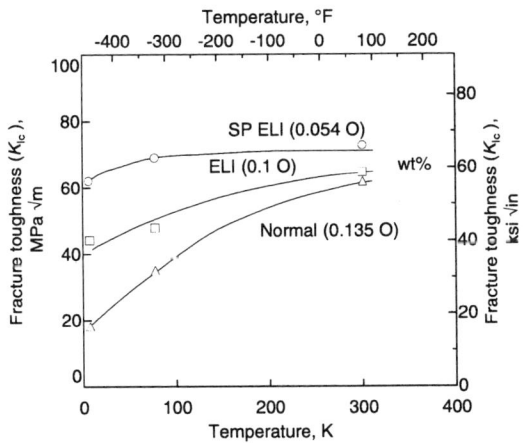

Ti-6Al-4V/Ti-6Al-4V(ELI): Low-temperature fracture toughness

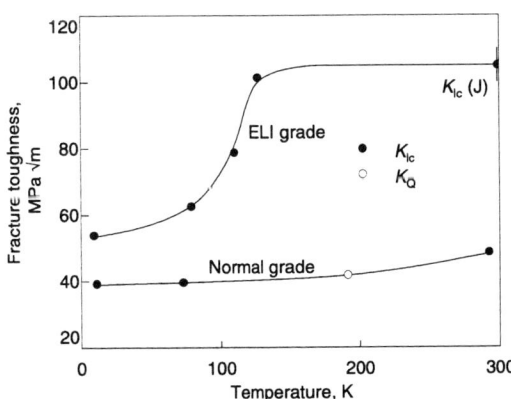

Source: C. Fowlkes and R. Tobler, "Fracture Testing and Results for Ti-6Al-4V Alloy at Liquid Helium Temperature," *Eng. Fract. Mech.*, Vol 8, 1976, p 487, reported in *Materials at Low Temperatures*, R. Reed and A. Clark, Ed., American Society for Metals, 1983, p 407

Ti-6Al-4V (ELI): Fracture toughness

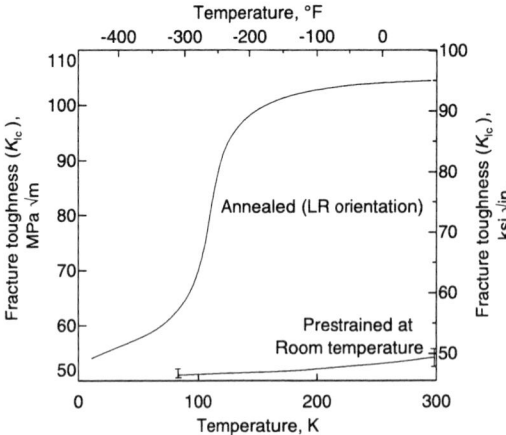

Effect of test temperature and prestraining in biaxial tension on recrystallization annealed alloy. Composition, 5.91% Al, 0.018% C, 0.103% Fe, 52 ppm H, 0.014% N, 0.110% O, and 3.95% V. R.L. Tobler, Low Temperature Fracture Behavior of a Ti-6Al-4V Alloy and Its Electron Beam Welds, in *Toughness and Fracture Behavior of Titanium*, STP 651, American Society for Testing and Materials, 1978, p 267-294

Ti-6Al-4V: Fracture toughness of electron beam welds

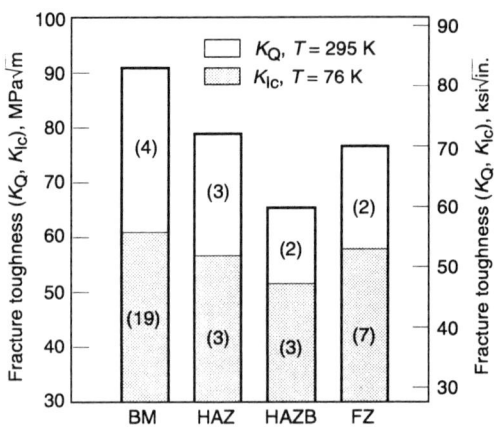

Effect of test temperature and test location. Number of tests performed are indicated in parentheses. Source: R.L. Tobler, Low Temperature Fracture Behavior of a Ti-6Al-4V Alloy and Its Electron Beam Welds, in *Toughness and Fracture Behavior of Titanium*, STP 651, American Society for Testing and Materials, 1978, p 267-294

Hydrogen Embrittlement

There are two basic types of hydrogen embrittlement: gaseous hydrogen embrittlement, which basically assumes an infinite source of external hydrogen, and embrittlement involving internal hydrogen contained within the material. The mechanisms involved are similar, but the resulting fracture topography can be different.

Gaseous Hydrogen Embrittlement. In a report on the effects of hydrogen pressure, stress intensity, and temperature on crack-growth rate under gaseous hydrogen embrittlement conditions in a lamellar microstructure, Nelson (*Metall. Trans. A*, Vol 7, 1976, p 621-627) proposes that the rate of growth depends on migration of the hydrogen ahead of the crack tip and on saturation of the β phase with hydrogen to form hydrides. Propagation of the crack occurs when the hydride grows to a thickness that is not capable of withstanding the applied load. Thus, a reduction of hydrogen reduces the growth rate by prolonging the time required for the film to grow to a given thickness; an increase in the stress intensity increases the growth rate by reducing the film thickness that must be achieved to cause rupture of the hydride. A reduction in temperature, however, reduces the growth rate by slowing the diffusion rate of the hydrogen. Gaseous hydrogen embrittlement results in a brittle, terraced fracture.

Microstructure is also an important factor in this type of embrittlement. The rate of hydrogen diffusion is higher by several orders of magnitude in the β phase than in the α phase. Therefore, microstructures with the more continuous β phase, such as β annealed material, are more severely embrittled than those with discontinuous β, such as the fine equiaxed α in the α matrix (Nelson *et al.*, *Metall. Trans.*, Vol 3, 1972, p 469-475). Transport of the hydrogen to the critical location is much more rapid with continuous β; consequently, the effect is more severe.

Embrittlement from Internal Hydrogen. Hydrogen effects involving the presence of internal hydrogen are similar, but it may take longer for these effects to be observed because of the longer transport distances involved. The hydrogen must diffuse to the stress raiser from within the matrix, because the infinite source from the atmosphere is not available. Meyn (*Metall. Trans.*, Vol 5, 1974, p 2405-2414) shows a decrease in fracture toughness due to hydrogen in concentrations of up to 50 ppm. The rate of sustained load cracking increases with increasing hydrogen concentrations above this level. Additionally, hydrogen contents up to 50 ppm affect the sustained load cracking, whereas higher hydrogen levels do not, as reported by Meyn.

Another study, Boyer and Spurr, (*Metall. Trans. A*, Vol 9, 1978, p 23) demonstrates the effect of hydrogen and temperature on sustained load cracking (see figure). The higher the hydrogen content, the higher the temperature for the onset of sustained load cracking. The only specimens that did not exhibit any significant crack growth at temperatures down to –68 °C (–90 °F) were those with hydrogen contents of 122 ppm or lower. Crack growth was occurring at stress intensities as low as about 12 MPa√m (10.9 ksi√in.) at –68 °C, whereas material with hydrogen contents of about 50 to 60 ppm did not exhibit crack growth at stress intensities of about 50 MPa√m (45.5 ksi√in.).

Crack-growth rate is also a function of hydrogen content and temperature. The crack-growth rate increases with temperature until it reaches a peak, and then the rate decreases. Prior to reaching the peak (from the high-temperature side), hydride nucleation becomes easier, but the diffusion rate decreases. As long as nucleation is the rate-controlling process, crack-growth rate increases with decreasing temperature. When the temperature gets low enough for hydrogen diffusion to be-

Ti-6Al-4V: Variation in crack-growth rate and step width as a function of hydrogen pressure

Hydrogen pressure		Crack-growth rate (da/dt)		Step width	
kPa	psi	m/s	ft/s	m	ft
90	13.1	1×10^{-4}	3.3×10^{-4}	3.25	10.7
90.6	13.1	9×10^{-5}	3.0×10^{-4}	3.5	11.5
75	10.9	8×10^{-5}	2.6×10^{-4}	3.5	11.5
60.8	8.8	3.6×10^{-5}	1.2×10^{-4}	4.25	13.9
48.8	7.1	4.75×10^{-5}	1.56×10^{-4}	4.25	13.9
6.1	0.9	7.2×10^{-6}	2.4×10^{-5}	5.25	17.2

Note: DT specimens. Constant stress intensity of 76 MPa$\sqrt{m}$ (69.2 ksi$\sqrt{in.}$) in Stage II. Test temperature, 24 °C (75 °F). Source: H.G. Nelson, A Film-Rupture Model of Hydrogen-Induced, Slow Crack-Growth in Acicular Alpha-Beta Titanium, *Metall. Trans.* A, Vol 7 (No. 5), May 1976, p 621-627

Ti-6Al-4V: Gaseous hydrogen embrittlement

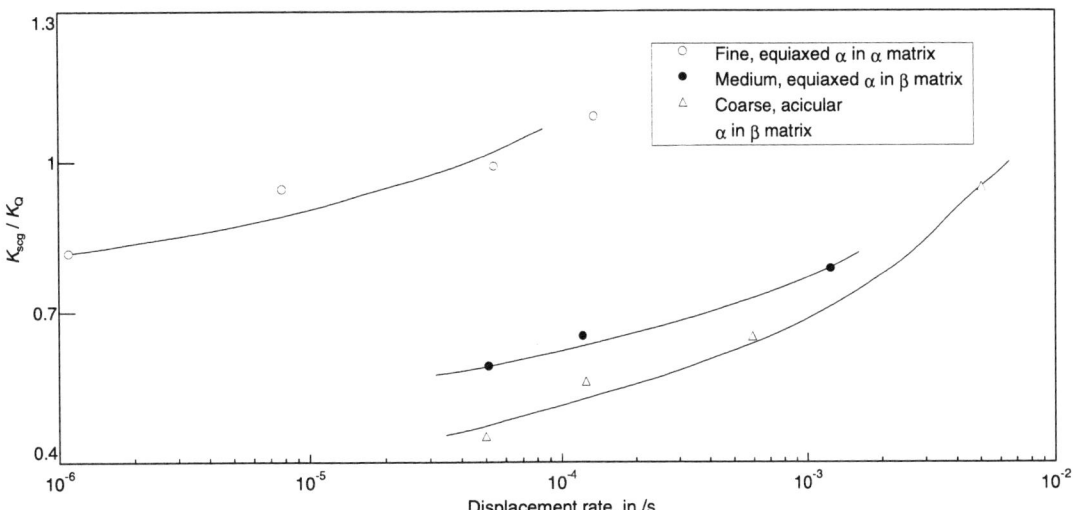

Effect of test displacement rate on hydrogen embrittlement for three microstructures.
Hydrogen pressure, 90.7 kPa (680 torr). Test temperature, 24 °C (75 °F) K_{scg}, stress intensity for subcritical crack growth.
Source: H.G. Nelson, D.P. Williams, and J.E. Stein. Environmental Hydrogen Embrittlement of an Alpha-Beta Titanium Alloy: Effect of Microstructure, *Metall. Trans.*, Vol 3 (No.2), Feb 1972, p 469-475

Ti-6Al-4V: Step width vs applied stress intensity

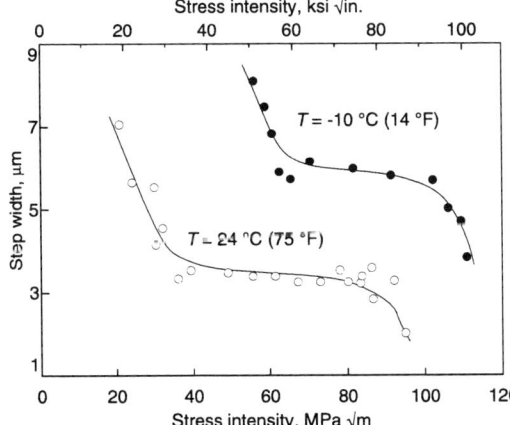

Double cantilever beam specimens. Hydrogen pressure, 90.6 kPa (13.1 psi).
Source: H.G. Nelson, A Film-Rupture Model of Hydrogen-Induced, Slow Crack-Growth in Acicular Alpha-Beta Titanium, *Metall. Trans.* A, Vol 7 (No. 5), May 1976, p 621-627

Ti-6Al-4V: Sustained load cracking behavior

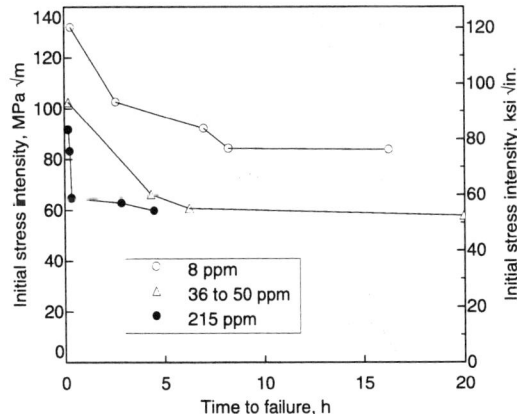

Effect of hydrogen content on sustained load cracking of precracked cantilever beam, dead weight loaded specimens.
Source: D.A. Meyn, Effect of Hydrogen on Fracture and Inert-Environment Sustained Load Cracking Resistance of Alpha-Beta Titanium Alloys, *Metall. Trans.*, Vol 5 (No. 11), Nov 1974, p 2405-2414

come the rate-controlling process, the growth rate decreases with further reductions in temperature.

Boyer and Spurr (*Metall. Trans. A*, Vol 9, 1978, p 23) also report a significant effect of texture on sustained load cracking. When an attempt was made to test specimens in which cracks would have to grow in a direction normal to the basal planes, the cracks would rotate 90° and would not grow in the intended orientation.

Ti-6Al-4V: Effects of hydrogen content

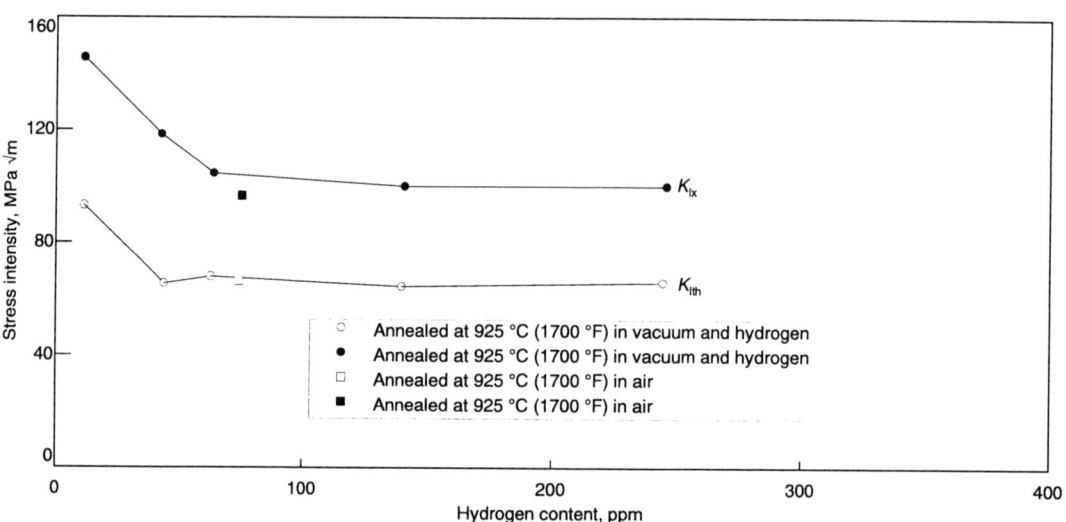

Fracture toughness (K_{Ix}) vs hydrogen content sustained load crack propagation threshold (K_{Ith}).
Source: D.A. Meyn, Effect of Hydrogen on Fracture and Inert-Environment Sustained Load Cracking Resistance of Alpha-Beta Titanium Alloys, *Metall. Trans.*, Vol 5 (No. 11), Nov 1974, p 2405-2414

Ti-6Al-4V: Crack length as a function of time

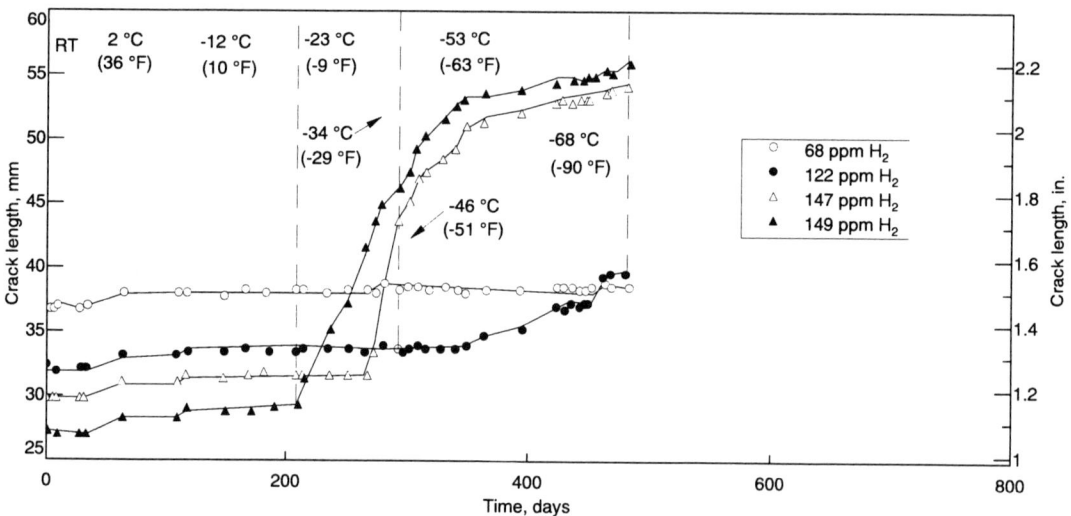

At various test temperatures for hydrogen contents from 68 to 149 ppm.
Source: R.R. Boyer and W.F. Spurr, Characterization of Sustained Load Cracking and Hydrogen Effects in Ti-6Al-4V, *Metall. Trans. A*, Vol 9, Jan 1978

Ti-6Al-4V: Crack length as a function of time

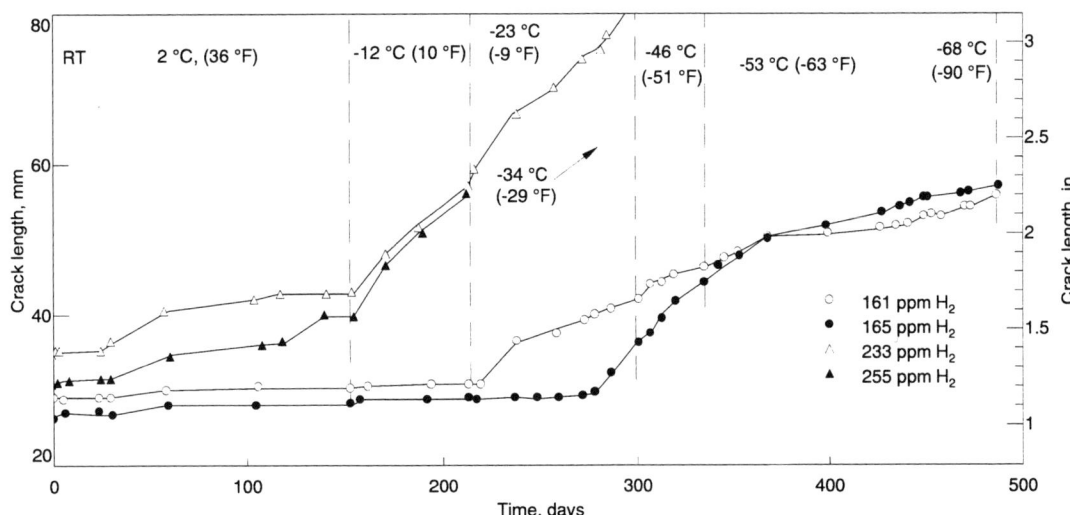

At various test temperatures for hydrogen contents from 161 to 255 ppm.
Solid symbols indicate termination of test.
Source: R.R. Boyer and W.F. Spurr, Characterization of Sustained Load Cracking and Hydrogen Effects in Ti-6Al-4V, *Metall. Trans. A*, Vol 9, Jan 1978

Ti-6Al-4V: Average maximum crack-growth rate vs temperature

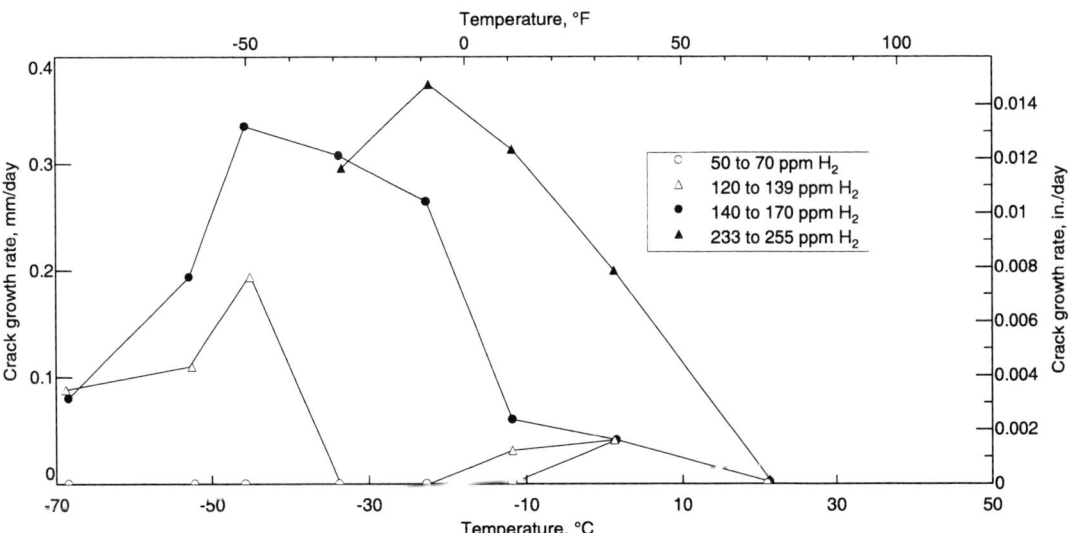

These results ignore stress-intensity effects, but they demonstrate the competition between diffusion of hydrogen and nucleation of hydrides. General shape of curve confirmed by W.J. Pardee and N.E. Paton in *Metall. Trans. A*, Vol 11, 1980, p 1391-1400.
Source: R.R. Boyer and W.F. Spurr, Characterization of Sustained Load Cracking and Hydrogen Effects in Ti-6Al-4V, *Metall. Trans. A*, Vol 9, Jan 1978

Plastic Deformation

Ti-6Al-4V: Tensile elongation vs temperature

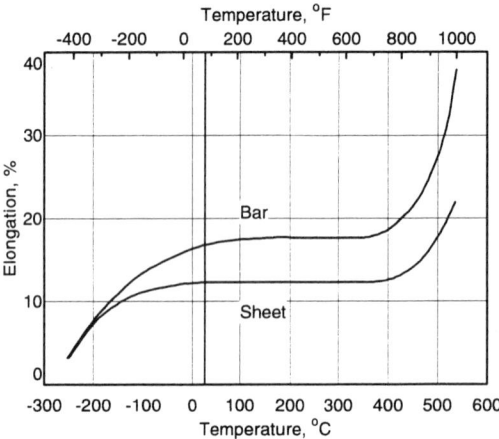

Typical elongation at temperature of annealed sheet and bar after 1/2-h exposure.
Source: MIL-HDBK 5, 1 Dec 1991

Ti-6Al-4V: Hot ductility dip

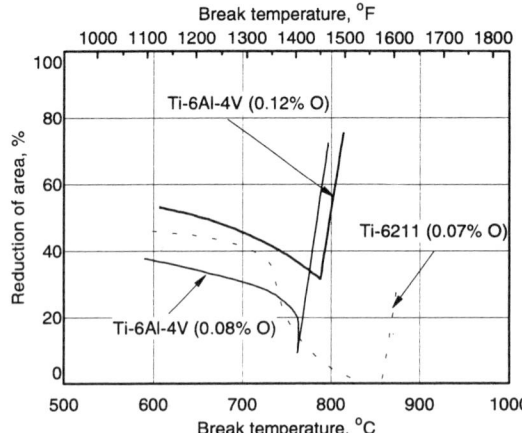

Source: *Titanium, Science and Technology*, Vol 2, G. Lütjering, U. Zwicker, and W. Bunk, Ed., Deutsche Gesellschaft für Metallkunde, Germany, 1985, p 901

Compressive Stress-Strain

Ti-6Al-4V: Compressive stress-strain

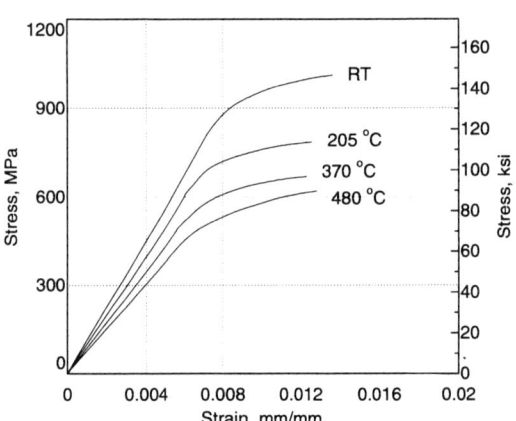

Annealed extrusions, longitudinal direction, after 1/2-h exposure.
Source: MIL-HDBK 5, 1 Dec 1991

Ti-6Al-4V: Compressive stress-strain

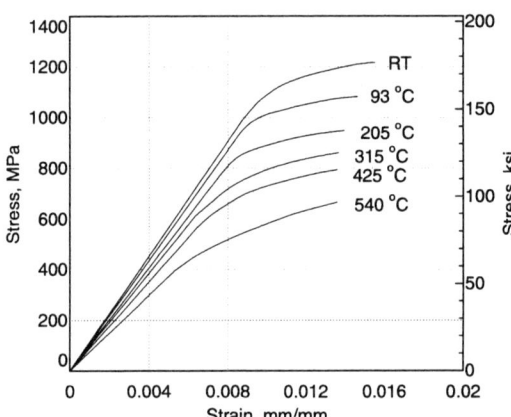

Sheet, solution treated and aged, longitudinal direction, after 1/2-h exposure.
Source: MIL-HDBK 5, 1 Dec 1991

Tensile Stress-Strain

Ti-6Al-4V: Annealed tensile stress-strain

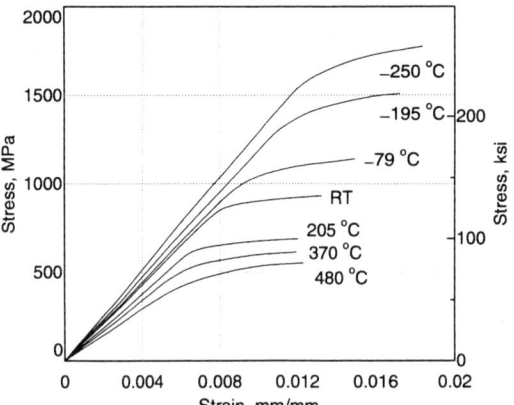

Annealed extrusions. Static strain rate, after 1/2-h exposure.
Source: MIL-HDBK 5, 1 Dec 1991

Ti-6Al-4V: STA tensile stress-strain

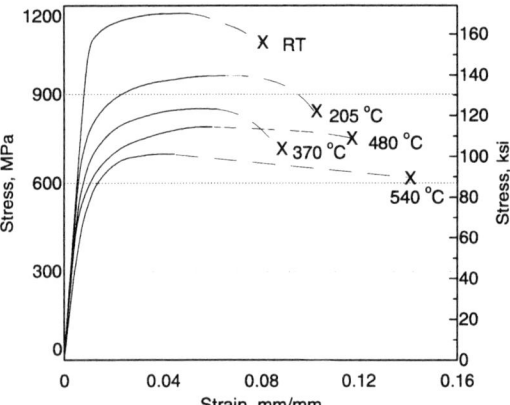

All product forms, solution treated and aged, longitudinal direction, after 1/2-h exposure.
Source: MIL-HDBK 5, 1 Dec 1991

Flow Stress

Effect of Temperature

Ti-6Al-4V: Flow stress vs temperature

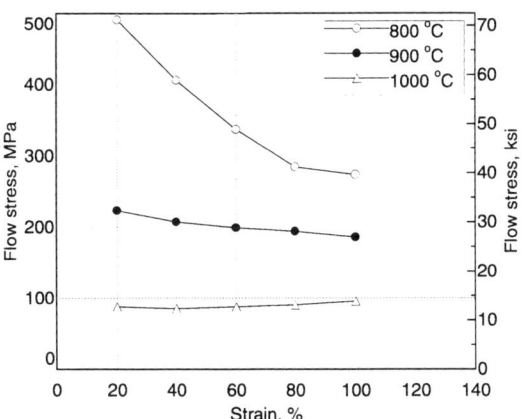

Flow stress at a strain rate of 10/s with a starting microstructure of about 50% α in a transformed β matrix.
Source: G.W. Kuhlman, ALCOA, Forging Division

Ti-6Al-4V: Tensile stress-strain

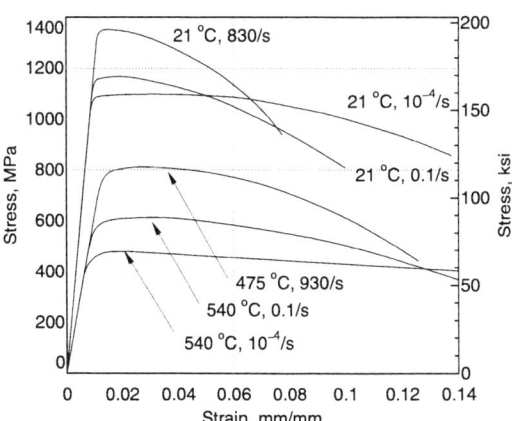

Solution treated and aged rod.
Source: D.L. McLellan and T.W. Eichenberger, "Constitutive Equation Development (COED), Vol 1, Technical Summary, SAMSO-TR-68-320, July 1968, p 80

Ti-6Al-4V: Flow stress vs temperature

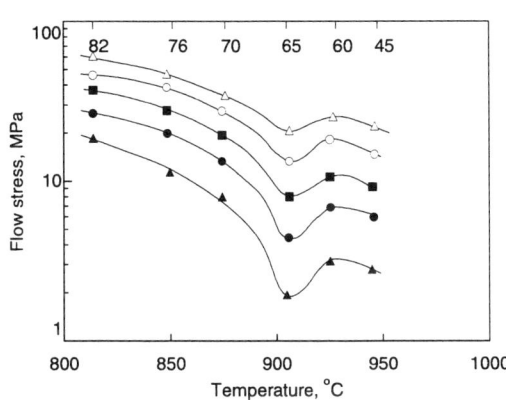

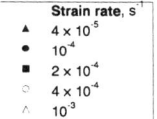

Source: C.H. Hamilton, Superplasticity in Titanium Alloys, *Superplastic Forming*, ASM International, 1985, p 13-22

Ti-6Al-4V: Effect of microstructure, strain rate, and temperature on flow stress

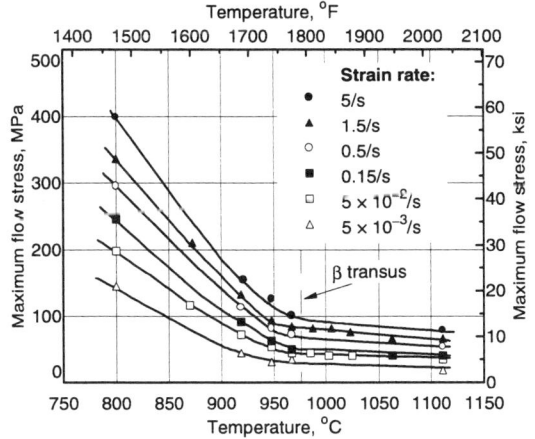

(a)

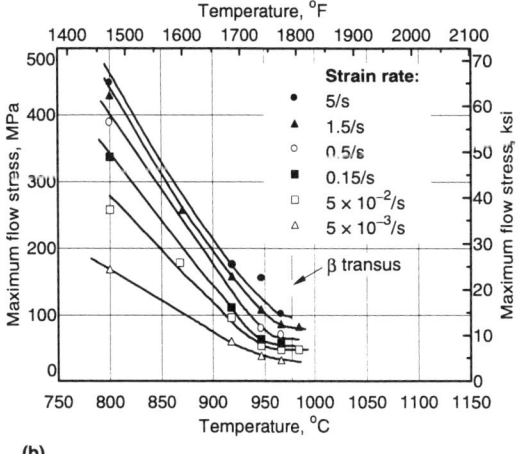

(b)

6.08% Al, 4.03% V, 0.14% Fe, and 0.13% O. (a) Equiaxed initial microstructure. (b) Lamellar initial microstructure.
Source: J.G. Malcor, Mechanical and Microstructural Behavior of Ti-6Al-4V Alloy in the Hot Working Range, *Titanium, Science and Technology*, G. Lütjering, U. Zwicker, and W. Bunk, Ed., Deutsche Gesellschaft für Metallkunde, Germany, 1985, p 1495-1502

Effect of Strain Rate

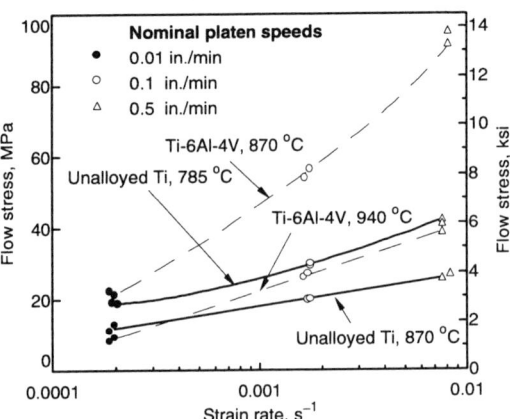

Ti-6Al-4V: Flow stress vs strain rate

Effect of strain rate and temperature on deformation resistance of unalloyed titanium and Ti-6Al-4V for 25% compression. Average strain rate over 0.25% strain.
Source: Trans. ASM, Vol 51, 1959, p 941

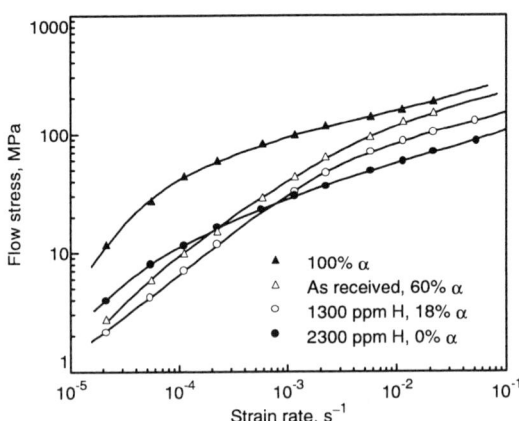

Ti-6Al-4V: Flow stress at 870 °C

Effect of α phase content on the flow stress vs strain rate characteristics of the Ti-6Al-4V at 870 °C (1600 °F).
Source: C.H. Hamilton, Superplasticity in Titanium Alloys, *Superplastic Forming*, ASM International, 1985, p 13-22

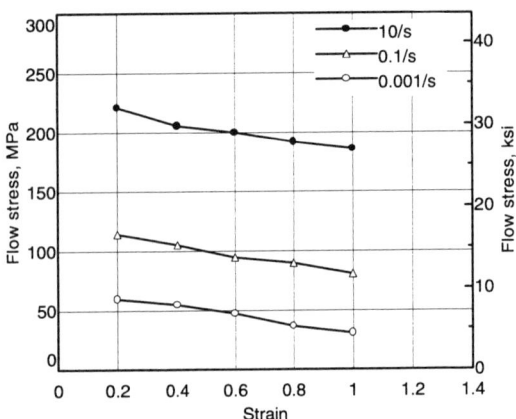

Ti-6Al-4V: Effect of strain rate on flow stress at 900 °C

Flow stress at 900 °C (1650 °F) with a starting microstructure of about 50% α in a transformed β matrix.
Source: G.W. Kuhlman, ALCOA, Forging Division

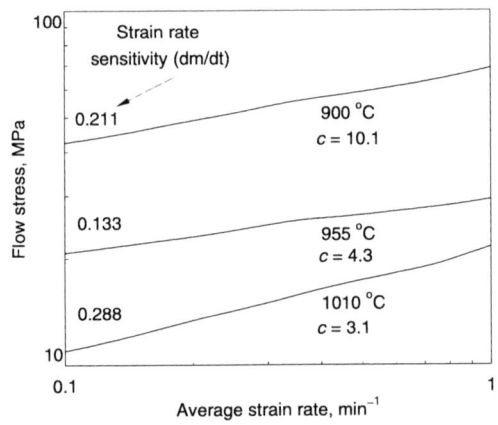

Ti-6Al-4V: Effect of strain rate on flow stress

1.5 to 1.6% strain.
Source: G.W. Kuhlman

Forging

G.W. Kuhlman, ALCOA Forging Division

Ti-6Al-4V is supplied in several forging types, including open die (or hand) forgings, rings, closed die forgings, and precision forgings. Hand and closed die forgings and rings range in size from a few ounces to 4000 or 5000 lb and are suitable for subsequent processing (machining) by the user. Among titanium alloys that are forged, Ti-6Al-4V is intermediate in forgeability. It is more difficult to forge, as measured by flow stress and crack sensitivity, than most ferrous alloys (and all aluminum alloys), but is less difficult to forge than most nickel- and cobalt-base superalloys.

Ti-6Al-4V may be forged using either conventional (sub-β transus) or β (supra-β transus) forging techniques. Both techniques are used in combination with annealing and solution treating and annealing or aging thermal treatments. Other thermal treatments such as recrystallized annealing and/or β annealing or β solution treatment and annealing may be combined with conventional forging to achieve tailored properties. Conventional forging of Ti-6Al-4V predominates commercially because it achieves an equiaxed α in a transformed β matrix microstructure that is preferred for many applications. Beta forging of Ti-6Al-4V creates an acicular α microstructure that is preferred for service conditions where fracture-related and/or creep properties are highly critical.

Resistance to deformation is reduced by increasing metal temperatures; however, in commercial practice, reheat furnace capabilities and uniformity constrain the upper limit of metal temperature in conventional forging to ensure that the beta transus is not exceeded to achieve the desired equiaxed α microstructure. Flow stresses of Ti-6Al-4V are significantly reduced in beta forging, with metal temperatures above the beta transus. However, this forging technique results in transformed, Widmanstätten α microstructures that may not be suitable for all applications. Ti-6Al-4V is also strain-rate-sensitive in forging deformation processes. In rapid-strain-rate (≥10/s) processes (e.g., hammers, mechanical presses, etc.), flow stresses are 2 to 3 times greater than slower strain rate (≈0.1/s) processes (e.g., hydraulic presses). Hot die and isothermal forging capture further reduction in flow stress by forging at very slow strain rates (≈0.001/s), achieving highly refined, precision forging shapes through the enhanced deformation achievable.

Open Die Forging. Metal temperatures at the high end of the ranges specified for conventional and beta forging are preferred because the workpieces tend to cool during the forging process. This usually involves multiple squeezes and/or blows and therefore relatively long times on the press equipment.

Closed Die Forging. In rapid-strain-rate forging equipment, e.g., hammers, mechanical and/or screw presses, high-energy-rate forming equipment, etc., the Ti-6Al-4V workpiece may be adiabatically heated by the rapid rate of deformation. Thus workpiece temperatures at the lower end of the ranges (see table) are used. For slower strain rate equipment, e.g., hydraulic presses, metal temperatures at the high end of the range are preferred. With hot die or isothermal forging, metal temperatures at the middle of the range typically are used.

Heating Methods. Ti-6Al-4V is heated for forging using a variety of commercially available furnace equipment including induction heating, resistance heating, electric (radiant), oil, and natural gas. Forging reheat furnaces include batch, rotary, and continuous (walking beam) furnaces designed for the temperatures required. The alloy may be heated to required metal temperatures for forging at any convenient heating rate. However, Ti-6Al-4V has a low coefficient of thermal conductivity. Therefore, commercial forging preheat practices allow a hold of 20 to 30 min/in. of ruling section thickness (bar or forging) to ensure the workpieces reach the desired temperature throughout. Ti-6Al-4V is heated in ambient to oxidizing atmospheres to retard hydrogen pickup from furnace products of combustion. Heating of forgings in this type of atmosphere creates a surface layer, termed α case, from diffusion of oxygen and nitrogen, that must be removed from finished forgings by pickling (or chemical milling) in appropriately constituted and controlled solutions of hydrofluoric and nitric acids.

Forging Equipment. Ti-6Al-4V forgings are produced on all types of forging equipment including hammers (standard and counter-blow), upsetters, mechanical presses, screw presses, HERF machines, hydraulic presses, ring mills, etc. Selection of the type of forging equipment is predicated on the forging type, forging design and tolerances, forging volume, etc. Hammers and small-to-intermediate (3 to 10,000 ton) hydraulic presses are preferred for open die forging. Hammers, mechanical presses, and screw presses are best suited to the manufacture of small-to-intermediate size closed die forgings. Mechanical and screw presses are uniquely suited to the manufacture of certain types of precision forgings, including turbine engine blades, prosthetic devices, etc. Large closed die forgings are predominantly manufactured on hydraulic presses (up to 65,000 ton). Hot die and isothermal forging is restricted to small-to-intermediate (up to 10,000 ton) hydraulic presses that possess required speed and pressure controls and programmable press operation. Annular shapes are best produced on vertical or horizontal ring mills, although rings are manufactured by mandrel forging in open die presses or hammers.

Workpiece lubrication in forging of Ti-6Al-4V combines workpiece precoats and die lubricants to suit the forging equipment and forging type. Precoats, proprietary combinations of a variety of ceramic compounds and other additives, are applied prior to furnacing to retard the reaction with the furnace atmosphere, act as parting agents, act as insulators, and/or assist in lubrication. Precoats are applied by spraying, dipping, or other techniques. With standard die materials in closed die forging, die lubricants, applied by spraying, are used that are proprietary graphite compounds in water or mineral oil carriers with other additives. Die lubricants are formulated to with-

Forging type/ equipment	Die temperature	
	°C	°F
Open die forging	150-260	300-500
Ring rolling	95-260	200-500
Closed die forging		
Hammers	95-260	200-500
Upsetters	150-260	300-500
Mechanical/screw presses	150-315	300-600
Orbital forging	150-315	300-600
Spin forging	95-315	200-600
Roll forging	95-260	200-500
Hydraulic presses	315-485	600-900
Hot die forging	705-890	1300-1600
Isothermal forging	925-980	1700-1800

Ti-6Al-4V: Recommended forging temperatures

Alloy	Beta transus		Forging process	Metal temperature	
	°C	°F		°C	°F
Ti-6Al-4V	995	1825	Conventional	870-980	1600-1800
			Beta forging	900-1065	1850-1950
Ti-6Al-4V ELI	975	1790	Conventional	870-950	1600-1740
			Beta forging	990-1045	1815-1915

stand high temperatures and pressures in the forging process and reduce friction and wear between the workpiece and the tools. Frequently, in closed die forging in standard die materials, insulative blankets of suitable materials are used to retard cooling of the pieces by the dies. In hot die and isothermal forging, die lubricants, other than a parting agent such as boron nitride, are typically not used.

Forging Procedures

Working procedures in the manufacture of Ti-6Al-4V open die and closed die forgings are designed to impart, at selected metal temperatures, sufficient deformation to the workpiece to achieve, after appropriate thermal treatments, microstructural and mechanical-property objectives. Forging processes begin with billet and/or bar fabricated from ingot by the metal producer using primary working processes that refine the grain size and begin microstructural development. Deformation from subsequent open die or closed die forging processes add to or modify the work imparted in forging stock fabrication to achieve overall deformation objectives. Forging deformation objectives differ for the two major forging processes. Working histories are therefore designed to achieve both microstructural and forging shape/design requirements.

Conventional Forging. In forging Ti-6Al-4V subtransus, deformation through one or more forging operations is cumulative. This forging process does not modify grain size, but modifies microstructural features, in particular α morphology, through dynamic recrystallization and through residual strain that is exploited in the final thermal treatments. Deformation imparted in each operation (e.g., preforming, block forging, and finish forging, etc.) is additive and controlled to achieve the final forging shape with the fewest operations. Although total reduction in cross-sectional area is not always a reasonable measure of necessary forging reduction, in conventional open die forging of billet/bar stock, reductions of ≥50% are preferred. In conventional closed die and precision forging, total forging reduction is not measured, but may approach several hundred percent in areas. Conventional forging working histories are used in combination with annealing, solution treating and aging or annealing, recrystallization annealing, and/or beta annealing thermal treatment processes.

Beta Forging. Heating the workpiece above the β transus statically recrystallizes Ti-6Al-4V. Thus, deformation in beta forging is not cumulative. In this process, the workpiece macrostructure (grain size) and microstructure are modified; however, when the alloy is heated supra-transus, grain coarsening is likely. Preheating and working histories for this forging process are therefore carefully designed and controlled. In a multiple-step beta forging process, early operations (e.g., preforming), including forging stock working, are conducted subtransus. Supra-transus working is limited to a single operation (e.g., blocking or first finishing) with dies designed to impart a high level of deformation (e.g., ≥75%). Supra-transus working is followed by subtransus deformation (e.g., finish forging) in the range of 10 to 25% reduction. The final subtransus reduction is critical to development of reasonable ductility in β forged components. Beta forging working histories are used in combination with annealing, solution treating and annealing (STAN), or solution treating and aging (STA).

Post-Forging Cooling. In commercial practice, post-forging cooling rate for most forging processes is not a critical process element other than to ensure that the forgings cool uniformly to prevent undue distortion and rapidly enough to avoid undesirable microstructures such as coarse or blocky α. Generally, most Ti-6Al-4V forgings are therefore air cooled. For some thermomechanical processes (TMP), post-forging cooling rates may be an important element of the overall TMP process. For example, rapid post-forging cooling using quenching techniques (water, oil, etc.) may be used for both conventional and beta forging processes when fine transformation products and/or reduced grain boundary α are preferred microstructural features. Selection of the post-forging cooling rate may therefore depend on the TMP, microstructure, and mechanical-property objectives for individual forging shapes.

Subsequent Processing Procedures. The manufacturing processes for most Ti-6Al-4V forgings are multiple-step processes consisting of forging stock preparation (cutting), one or more forging operations (frequently several operations in two or more sets of forging dies), intermediate cleaning and repairing processes, final thermal treatments and final cleaning, repairing, and inspection processes. Forgings may then be machined by all commercially available machining techniques. Forging stock is cut by all commercially available techniques including band sawing, abrasive sawing, shearing, and flame cutting. Ti-6Al-4V is crack sensitive in most forging processes; thus, intermediate cleaning (HF-HNO$_3$ pickling) and repairing (abrasive grinding or other techniques) processes may be necessary to ensure satisfactory forging quality.

Prior to shipment to users, heat treated forgings are cleaned (HF-HNO$_3$ chem mill) to remove α case, repaired if necessary, inspected dimensionally, and nondestructive evaluation inspected for internal quality (ultrasonic inspection for engine disks) and in the case of precision forgings, surface quality (liquid-penetrant inspection). In some cases, precision forgings are chemically milled, machined, and/or surface treated (conversion processes, painting, etc.) prior to shipment as finished components.

Thermomechanical Processing

For Ti-6Al-4V, forging is a critical element of thermomechanical processing used to achieve selected microstructure and therefore mechanical-property objectives. Thermomechanical processing of forgings may incorporate several possible final thermal treatments depending on microstructure and mechanical-property objectives, such as annealing (A), solution treating, quenching (water, oil or air) and annealing or aging (STA or STAN), recrystallization annealing (RA), and beta annealing (BSTAN or BA).

Conventional Forging Combined with Annealing and/or Solution Treating and Aging or Annealing. Conventional forging plus A, STA, or STAN TMP sequences for standard composition Ti-6Al-4V are the most widely used commercially in all open die, closed die, ring, and precision forging products. This TMP achieves an excellent combination of yield and tensile strengths, ductility (elongation and reduction of area), and high- (HCF) and low- (LCF) cycle smooth and notched fatigue properties. Fracture-related properties including fracture toughness (K_{Ic}) and fatigue-crack growth resistance (FCGR) and creep resistance are good, but inferior to other TMP routes. The STA heat treatment provides the highest strength possible in Ti-6Al-4V forgings. The STAN thermal treatment in combination with conventional forging is used to improve forging product uniformity.

Conventional Forging Combined with Recrystallization Annealing. Conventional forging plus recrystallized annealing, usually in combination with the Ti-6-4 ELI composition, achieves superior fracture-related properties, including K_{Ic} and FCGR, excellent HCF and LCF properties and good ductility and creep resistance. However, this is at a sacrifice in yield and tensile strengths. This TMP process also provides a uniform and consistent forging product, because the RA thermal treatment prevails over forging history in determining the final microstructure. This TMP route has been applied successfully to all Ti-6Al-4V products.

Conventional Forging Combined with Beta Annealing. Conventional forging of either standard or ELI compositions with BA or BSTAN achieves the best fracture-related properties (e.g., K_{Ic} and FCGR) and creep properties possible in Ti-6Al-4V. However, there is a sacrifice in yield and tensile strengths, ductility and HCF and LCF

Ti-6Al-4V: Forging properties (minimum RT values)

Composition	Forging process	Heat treatment	Tensile yield strength MPa	ksi	Ultimate tensile strength MPa	ksi	Elongation, %	Reduction of area, %	Fracture toughness MPa√in.	ksi√in.
Standard(a)	α + β	MA	827-862	120-125	896-931	130-135	10	20-25	44-45	40-50
Standard(a)	α + β	RA	793-827	115-120	862-896	125-130	10	20-25	60-71	55-65
Special(b)	α + β	RA	758-793	110-115	827-862	120-125	10	20-25	77-88	70-80
Standard(a)	α + β	βA	793-862	115-125	931-1000	135-145	6-8	10-15	82-93	75-85
Special(b)	α + β	βA	758-827	110-120	827-862	120-125	6-8	10-15	88-99	80-90
Standard(a)	β preform, α + β block and finish	MA	813-848	118-123	896-931	130-135	10	20-25	66-77	60-70
Standard(a)	β preform, β block, α + β finish	MA	813-848	118-123	862-896	125-130	8-10	15-20	71-88	65-80
Special(b)	β preform, β block, α + β finish	Duplex STAN	758-793	110-115	827-862	120-125	6-8	10-15	82-93	75-85

(a) Standard specimen: 5.5-6.75 Al, 0.10 C, 0.30 Fe, 0.015 H, 0.05 N, 0.20 O, 3.5-4.5 V, .40 other. (b) Special composition: 5.5-6.2 Al, 0.08 C, 0.25 Fe, 0.0125 H, 0.05 N, 0.13 O, 3.5-4.5 V, .30 other. Source: G.W. Kuhlman and F.R. Billman, Selecting Processing Options for High Fracture Toughness Titanium Airframe Forgings, *Met. Prog.*, Mar 1977, p 39-49

Ti-6Al-4V: Effect of commercially available TMP routes

Composition(a)	Forging process(b)	Heat treatment(c)	Strength	K_{Ic}	Smooth HCF	Notched HCF	FCGR	LCF	Product cost	Product uniformity
Std	α/β	MA			Baseline				100	100
CMG	α/β	MA	(+)	(−)	(+)	(=)	(=)	(≥)	105	100
Std	α/β	RA	(=)	(+)	(=)	(=)	(=)	(=)	105	120
ELI	α/β	RA	(−)	(++)	(−)	(≥)	(+)	(=)	115	130
Std	β1	MA	(−)	(+)	(−)	(≥)	(+)	(−)	95	90
Std	β2	MA	(−)	(+)	(− −)	(+)	(+)	(−)	90	85
ELI	β2	MA	(−)	(++)	(− −)	(+)	(++)	(− −)	95	95
Std	βHDF	MA	(=)	(+)	(−)	(+)	(+)	(−)	110	110
ELI	βHDF	MA	(− −)	(++)	(− −)	(+)	(++)	(− −)	115	110
Std	α/β	βA	(−)	(++)	(−)	(+)	(++)	(−)	105	130
ELI	α/β	αA	(− −)	(++)	(− −)	(+)	(+++)	(− −)	115	150
Std	β2	Duplex STAN	(=)	(+)	(≤)	(+)	(+)	(−)	102	110
ELI	β2	Duplex STAN	(−)	(++)	(−)	(+)	(++)	(−)	110	115
Std	α/β	Duplex STA	(+)	(−)	(+)	(=)	(=)	(+)	110	120
CMG	α/β	Duplex STA	(++)	(−)	(+)	(=)	(=)	(++)	115	120

LEGEND: (a) Std = standard composition; ELI = controlled composition; CMG = high interstitial composition. (b) α/β = α + β working; β1 = beta preforming only; β2 = beta preform block; βHDF = beta hot die forged. (c) MA = mill anneal; RA = recrystallized anneal; BA = beta anneal; STA/STAN = solution treated and aged/annealed; HSTA = high-temperature solution treatment and aged. Source: G.W. Kuhlman, A Critical Appraisal of Thermomechanical Processing of Structural Titanium Alloys, *Microstructure / Property Relationships in Titanium Aluminides and Alloys*, R.R. Boyer and Y.W. Kim, Ed., TMS/AIME, 1993

Ti-6Al-4V: Typical microstructure after working and annealing

lamellar structure following "beta-processing"

Rolled 30% 955°C — Rolling Direction → Annealed 925°C 2 hr

Forged 70% 955°C — Forging Direction → Annealed 925°C 2 hr

Extruded 140% 955°C — Extrusion Direction → Annealed 925°C 2 hr

Source: I. Weiss, Wright State University

properties. This TMP route also achieves the most consistent and uniform forging product possible, due to the preeminence of the beta heat treatment over the conventional working history. However, subtransus forging is an essential element to proper execution of this TMP route. Closed die and precision Ti-6Al-4V forgings predominate in the application of this TMP route, although open die forgings are produced.

Beta Forging Combined with Annealing or Solution Treating plus Aging or Annealing. In combination with either the standard or ELI composition variant, beta forging with A, STA, or STAN TMP routes are used in closed die forgings to achieve superior fracture-related properties (e.g., K_{Ic} and FCGR) and creep resistance. There is a necessary trade-off in yield and tensile strengths, ductility, and HCF and LCF properties. The working history in this TMP route dominates thermal treatments in determining final microstructure. Furthermore, there may be variations in the specific beta forging process conditions from forging-to-forging in a lot and from lot-to-lot that result in a less consistent final forging product than other TMP routes. However, forging shape sophistication may be enhanced, and forging total cost may be reduced due to the enhanced fabricability with beta forging.

Forming

Ti-6Al-4V typically is formed by hot forming above 540 °C (1000 °F). Normal production hot forming is conducted usually at 650 °C (1200 °F), but hot forming temperatures can be defined between 540 and 760 °C (1000 and 1400 °F), or even higher for superplastic forming. At 760 °C (1400 °F), stresses are self-relieved.

When forming temperatures above 650 °C (1200 °F) are used, forming procedures must be done in a vacuum or protective atmosphere, or must be followed by removal of the alpha case (e.g., by chemical milling). Superplastic forming of Ti-6Al-4V is performed at 870 to 925 °C (1600 to 1700 °F). Cold forming of Ti-6Al-4V is generally not advised because of cracking and excessive springback problems. However, when cold formed, hot sizing is usually used. Ti-6Al-4V is moderately formable compared to other titanium alloys.

As with all other titanium alloys, forming of Ti-6Al-4V usually involves overcoming and contending with galling tendencies, potential impurity embrittlement, low shrink capabilities, and notch sensitivity. Because titanium alloys are stronger than ferrous, aluminum, and copper alloys, they

Ti-6Al-4V: Forming temperatures
Normal production hot forming usually is done at 650 °C (1200 °F).

Method	Temperature(a) °C	Temperature(a) °F	Comments
Hot sizing	650	1200	3 to 15 min for 0.8 mm (0.032 in.) sheet; 3 to 20 min for 1.6 mm (0.063 in.) sheet
Brake forming	540-650	1000-1200	...
Drop hammer	540-785	1000-1450	Temperatures up to 870 °C (1600 °F) may be needed
Superplastic	870-925	1600-1700	...
Drawing	540-785	1000-1450	Temperatures up to 870 °C (1600 °F) may be needed
Spinning	540-785	1000-1450	Temperatures up to 870 °C (1600 °F) may be needed
Hydropress	205-315	400-600	Mild forming
Press forming	480-540	900-1000	Hydroform and finish die
Matched die	480-785	900-1450	Temperatures up to 870 °C (1600 °F) may be needed
Creep forming	540-650	1000-1200	...

(a) Temperatures should be held to a minimum to reduce scaling. Time at temperature is important, because the titanium surface is embrittled by oxygen above 540 °C (1000 °F) as a function of time and temperature. Generally, 2 h is maximum for 705 °C (1300 °F); 20 min is maximum for 870 °C (1600 °F). These are accumulated times to include heating times for single or multistage forming, intermediate stress relief, and final stress relief.

Ti-6Al-4V: Formability comparison

Process	Ti-5Al-2.5Sn	Ti-6Al-4V	Ti-13V-11Cr-3Al
Brake press (min radius at RT)	3.5T	4.5T	1.5T
Drop hammer (max stretch) at 455-510 °C (850-950 °F)	13%	13%	16%
Hydropress (trapped rubber) at 315-370 °C (600-700 °F):			
Stretch, max	5%	5%	10%
Shrink, max	3%	4%	6%
Joggle-depth ratio at:			
RT	4	4.5	1.25
315-370 °C (600-700 °F)	4.5	3.0	1.00
Maximum stretch wrap at RT	8%	3.5%	5.5%(a)
Maximum skin stretch at 455-510 °C (850-950 °F)	12.5%	17%	13.5%(a)

(a) Solution treated condition. Source: MIL-HDBK 697A, June 1974

Ti-6Al-4V: Typical alpha case after hot forming

Temperature °C	Temperature °F	Time, h	Amount of removal necessary per side mm	Amount of removal necessary per side in.
540-595	1000-1100	4	0.013	0.0005
		12	0.025	0.001
595-650	1100-1200	2	0.013	0.0005
		6	0.025	0.001
650-705	1200-1300	1	0.013	0.0005
		2	0.025	0.001
		12	0.051	0.002
705-760	1300-1400	1	0.025	0.001
		2	0.051	0.002

require higher forming pressures, which must be closely controlled over a much smaller workability range.

Surface Treatment. Sheet forming and bending of titanium alloys must be accomplished with sheet that is free of α case. If the sheet has α case on the surface, it will crack when bent. Therefore, all residual α case must be removed with chemical cleaning before bending procedures can be used.

Production practice of hot forming titanium alloys usually requires the sheet to be coated by painting or dipping before hot forming. The coatings provide lubrication, make scale easier to remove, and inhibit oxygen contamination.

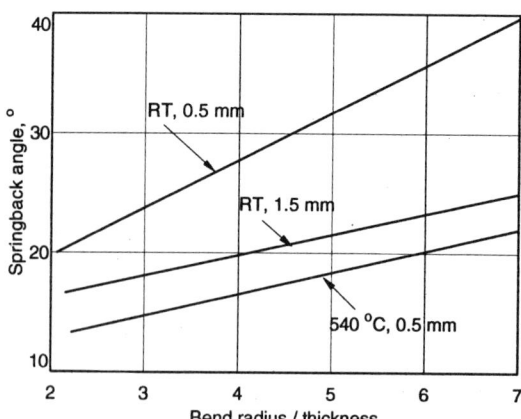

Ti-6Al-4V: Springback of brake formed sheet

Source: R.A. Wood and R.J. Favor, *Titanium Alloys Handbook*, MCIC HB-02, Battelle Columbus Laboratories, 1972

Springback

Ti-6Al-4V: Effect of hot sizing conditions on springback

Forming temperature		Time,	Springback angle, °		
°C	°F	min	Mill annealed	ST(a)	STA(b)
540	1000	20	...	3.5	5.0
		40	...	3.0	4.2
		100	...	2.0	3.5
		150	...	1.2	...
		210	...	0.75	3.0
565	1050	30	1.5	...	...
595	1100	40	1.1	...	1.0
		60	...	...	1.0
620	1150	30	0.6	...	...
650	1200	20	...	...	0
		30	0	...	...
675	1250	30	0	...	...

(a) Water quenched from 940 °C (1725 °F). (b) Water quenched from 940 °C (1725 °F) and aged 4 h at 540 °C (1000 °F). Source: R.A. Wood and R.J. Favor, *Titanium Alloys Handbook*, MCIC HB 02, Battelle Columbus Laboratories, 1972

Ti-6Al-4V: Springback from trapped rubber forming

Temperature(a)		Springback, °	
°C	°F	Stretch(b)	Shrink(b)
20	70	11-15	12-13
595	1100	12-14	13-14
650	1200	12-13	11-12
815	1500	6-8	4-7
		9-11(c)	10-11(c)

(a) Some of the temperatures investigated are exceptionally high and probably impractical for production operations. (b) Increased pad pressure will decrease the springback on shrink flanges, but has little effect on stretch flanges. (c) Two sets of data at 815 °C. Source: R.A. Wood and R.J. Favor, *Titanium Alloys Handbook*, MCIC HB-02, Battelle Columbus Laboratories, 1972

Bending

Bending properties of Ti-6Al-4V sheet vary with processing temperature. For production operations on mill annealed Ti-6Al-4V sheet, the room-temperature bend radius is essentially $6t$. This bend radius will usually cover both longitudinal and long-transverse grain (sheet rolling) directions. For special applications, minimum bend radii can be as low as $4.5t$ in the transverse direction and $5t$ in the longitudinal direction when formed with care. In nearly all cases, cold forming of production parts requires stress relieving.

Ti-6Al-4V: Minimum inside radii for sheet

Forming temperature		Inside bend radius (multiples of sheet gauge)	
°C	°F	Production	Occasional
	RT	6t	5t
540	1000	4t	3t
650	1200	3.5t	2.5t
760	1400	3t	2t

Note: Mill annealed sheet. Source: J. Woodward, Rohr Industries

Ti-6Al-4V: Minimum bend radius

Temperature		Minimum bend, R/t	
°C	°F	Longitudinal(a)	Transverse(b)
21	70	4.5-6	5-7
205	400	3.5-4	3.5-6.5
315	600	3.5	3-5.5
425	800	3-3.5	2.5-5
540	1000	2.5-3	2.5-4
650	1200	2-2.5	2-3
760	1400	1.5	1.5-2
815	1500	1	1

(a) Bend axis perpendicular to rolling direction. (b) Bend axis parallel to rolling direction. Source: R.A. Wood and R.J. Favor, *Titanium Alloys Handbook*, MCIC HB-02, Battelle Columbus Laboratories, 1972

Ti-6Al-4V: Roll bending of heel-out channel

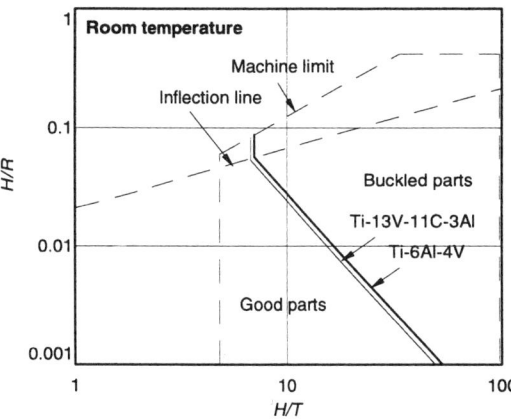

Transverse buckling and wrinkling, respectively, are the common modes of failure in bending heel-out and heel-in channels. The principal parameters are the bend radius, R; channel height, H; web width, W; and material thickness, T.
Source: R.A. Wood and R.J. Favor, *Titanium Alloys Handbook*, MCIC HB-02, Battelle Columbus Laboratories, 1972

Forming Limits

Ti-6Al-4V: Limits for manual spinning
Formability limits for manual spinning of flat-bottom cylindrical cups at room temperature

Thickness			Limiting ratio	
mm	in.	Blank diameter/ sheet thickness	Blank diameter/ cup diameter	Cup height/ cup diameter
0.5	0.020	25	1.3	0.22
		50	1.3	0.22
		100	1.2	0.14
		150	1.2	0.14
		200	1.1	0.07
1.6	0.063	25	1.3	0.22
		50	1.2	0.14
3.1	0.125	25	1.2	0.14
		50	1.1	0.07

Note: These formability limits for manual spinning at room temperature are expected to hold for relatively small forces and limited amounts of thinning. The data show that spinnability is favored by smaller ratios of blank diameter to sheet thickness. Ti-6Al-4V will not withstand very severe deformation at room temperature. For example, the limit for an 80 mm (3.125 in.) diam, 3.1 mm (0.125 in.) thick blank of Ti-6Al-4V appears to be a flat cup 66 mm (2.6 in.) in diameter and 9.2 mm (0.365 in.) high. The term cup diameter refers to the inside diameter; the cup height is based on the outside dimension. Source: R.A. Wood and R.J. Favor, *Titanium Alloys Handbook*, MCIC HB-02, Battelle Columbus Laboratories, 1972

602 / Alpha-Beta Alloys

Ti-6Al-4V: Stretch forming limits for sheet

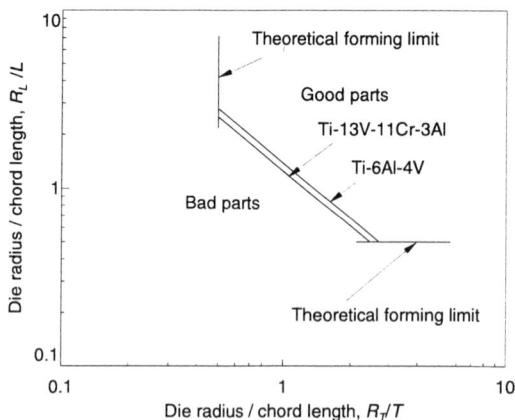

Forming limit curve for double contouring relatively large sheet metal parts. The parameters that determine the curve are the longitudinal radius (R_L), transverse radius (R_T), longitudinal chord length (L), and transverse chord length (T).
Source: R.A. Wood and R.J. Favor, *Titanium Alloys Handbook*, MCIC HB-02, Battelle Columbus Laboratories, 1972

Ti-6Al-4V: Stretch limits

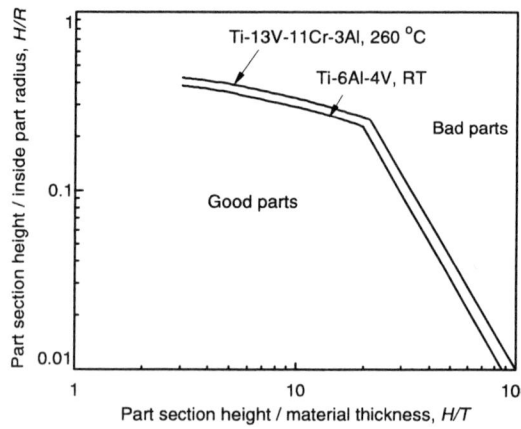

Composite limit curves for linear-stretch heel-in (inboard) angle and channel sections.
The formability index for splitting limits depends on the conventional strain, ε, for a 50 mm (2.0 in.) gage length. The formability index for elastic buckling is a function of the ratio of tensile modulus to tensile yield (E_T/S_{ty}). The index governing the optimum forming temperature will largely depend on the material thickness (t). For small values of t, the ratio h/t becomes large, thereby placing this h/t value in the elastic buckling region, which is a function of (E_T/S_{ty}). For large values of t, the conventional strain for a 50 mm (2.0 in.) gage length is the formability index.
Source: R.A. Wood and R.J. Favor, *Titanium Alloys Handbook*, MCIC HB-02, Battelle Columbus Laboratories, 1972

Ti-6Al-4V: Stretch limits

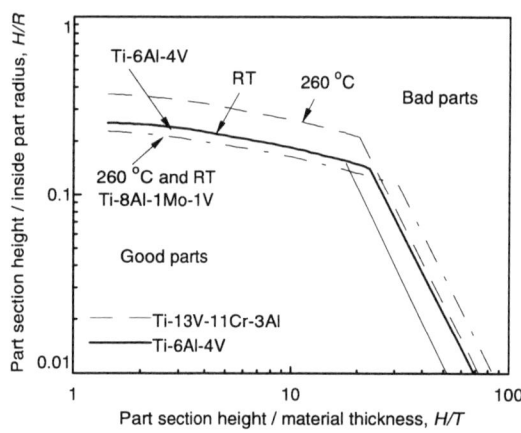

Composite of optimum linear-stretch heel-in hat-section-limit curves.
Source: R.A. Wood and R.J. Favor, *Titanium Alloys Handbook*, MCIC HB-02, Battelle Columbus Laboratories, 1972

Ti-6Al-4V: Forming limit diagram for sheet

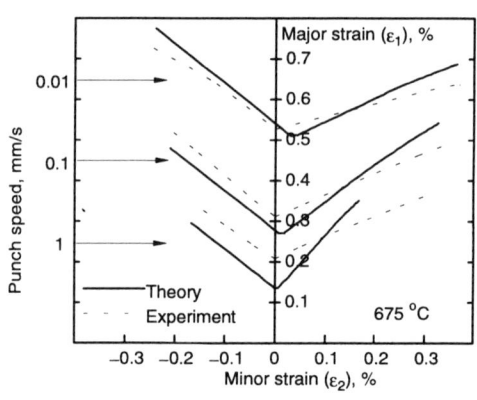

Comparison of forming limit curves from Marciniak theory and experimental diagrams versus punch speed for annealed sheet.
Source: *Titanium, Science and Technology*, Vol 2, G. Lütjering, U. Zwicker, and W. Bunk, Ed., Deutsche Gesellschaft für Metallkunde, Germany, 1985, p 456

Superplastic Forming

Superplastic forming performance is affected by several factors (see "Technical Note 5A: Superplastic Forming of Titanium Alloys"), and most superplastic procedures are based on Ti-6Al-4V sheet. In addition, the potential cost and weight savings associated with superplastic forming of thin Ti-6Al-4V sheet structures has led to increased interest in the forming of other product forms such as bar and extruded sections. However, compared with thin sheet, extruded Ti-6Al-4V alloy has a coarser and less desirable (transformed β) microstructure, which is not very superplastic (see figure on right).

Forming Temperatures. Superplastic forming of Ti-6Al-4V is performed at 870 to 925 °C (1600 to 1700 °F).

Post-forming tensile properties of superplastic Ti-6Al-4V sheet (M. Cope, D. Evetts, and N. Ridley, *Materials Science and Technology*, Vol 3, 1987, p 455-461) may exhibit some changes in tensile properties, depending on the deformation conditions. Room-temperature strength tends to increase with increasing strain rate above the optimum value for superplastic flow, probably due to the increased contribution to deformation from dislocation creep leading to a higher dislocation density. Superplastic flow results in the development of new, complex textures rather than texture reduction. This is in agreement with previous work on hexagonal materials.

Ti-6Al-4V: Strain-rate sensitivity

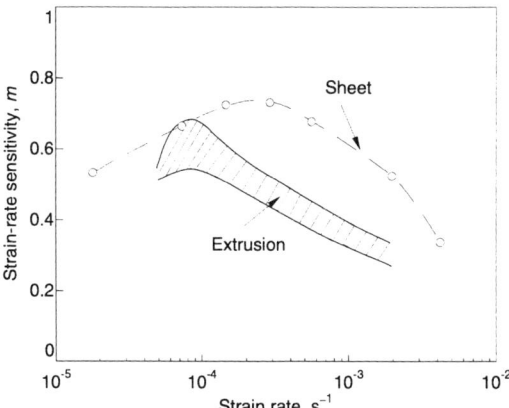

A conventionally extruded U-channel section, tested under conditions that produced superplasticity in thin sheet, exhibited enhanced plasticity, but the deformation in the gage length was nonuniform, and in most positions in the extrusion, was anisotropic with flow stresses a factor 2 greater than for sheet. Changes in processing to produce more equiaxed microstructure were required. Source: D.V. Dunford and P.G. Partridge, Deformation of Extruded Titanium Alloys under Superplastic Conditions, Sixth World Conference on Titanium, P. Lacombe, R. Tricot, G. Beranger, Ed., Les Editions de Physique, Paris, 1989, p 1197

Ti-6Al-4V: Strain-rate sensitivity

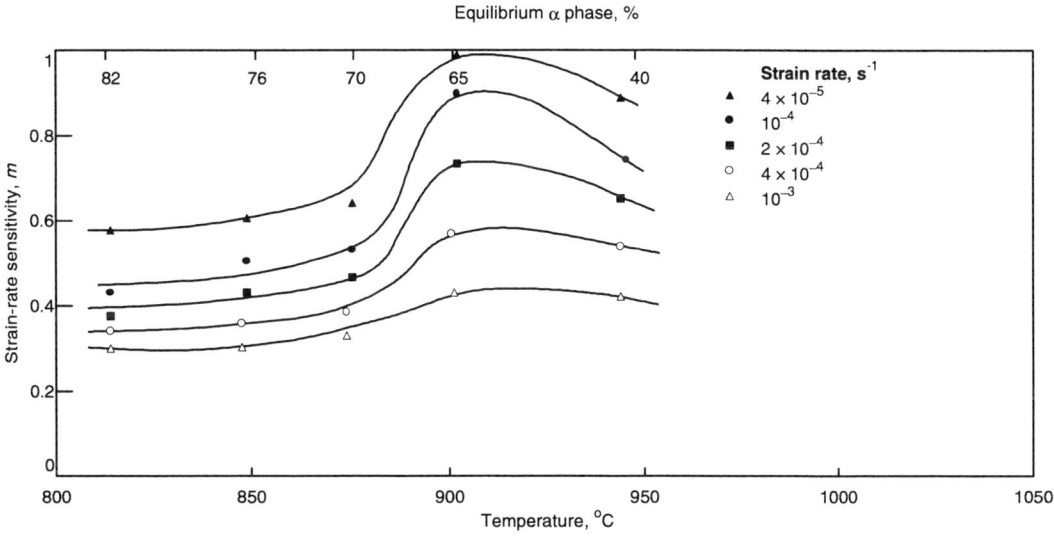

Source: C.H. Hamilton, Superplasticity in Titanium Alloys, *Superplastic Forming*, ASM International, 1985, p 13-22

Machining

Based on a rating system in which AISI B1112 steel has a machinability rating of 100, the machinability of Ti-6Al-4V in the annealed condition is 22 (see table). This means that Ti-6Al-4V can be machined at only 22% of the speed used for B1112 steel for the same tool life with the same feeds and depths of cut. Ti-6Al-4V is rated as slightly lower than 4340 steel heat treated to a hardness of 40 HRC, but slightly higher than 4340 steel heat treated to a hardness of 52 HRC.

For optimum machining of titanium alloys, low cutting speeds and heavy feed rates should be used to remove contaminated surfaces or for rough machining. A large volume of nonchlorinated cutting

Ti-6Al-4V: Machinability comparison/ratings

Alloy	Condition(a)	Rating(b)
2017 aluminum	T4	300
Leaded brass	...	200
B1112, resulfurized steel	HR	100
1020 carbon steel	CD	70
4340 alloy steel	A	45
CP Ti	A	40
Type 302 stainless steel	A	35
Ti-5Al-2.5Sn	A	30
A4340 alloy steel	Q&T(40HRC)	25
Ti-8Mn	A	25
Ti-6Al-4V	A	22
Ti-8Al-1Mo-1V	A	22
Ti-6Al-6V-2Sn	A	20
Ti-6Al-4V	HT	18
Ti-6Al-6V-2Sn	HT	16
Ti-13V-11Cr-3Al	A	16
Ti-13V-11Cr-3Al	HT	12
A4340 alloy steel	Q&T(52HRC)	10
HS25 (cobalt base)	A	10
Rene' 41 (nickel base)	HT	6

(a) T4, solution heat treated and artificially aged; HR, hot rolled; A, annealed; HT, solution treated and aged; CD, cold drawn; Q&T, quenched and tempered. (b) Based on AISI B1112 steel as 100

fluid should be used. Chlorinated fluids are often more efficient and may be used if adequate care is taken for their complete removal. Sharp tools should be used and replaced at the first sign of wear. While the tool and work are in moving contact, feed should never be stopped. Additionally, rigid setups should be used.

Titanium chips are highly combustible. Extreme care should be taken to avoid possible fire hazards. For a more complete discussion of the machining of titanium alloys, see "Technical Note 7: Machining."

Forming Examples

Deep Drawing

Dome-Shaped Part (200 mm diam). A rubber diaphragm press was used for drawing 200 mm (8 in.) diam hemispheres from 400 mm (16 in.) diam blanks sheared from Ti-8Al-1Mo-1V and Ti-6Al-4V sheet 1.27 mm (0.050 in.) thick. Before forming, the edges of the blanks were deburred on a grinding wheel and smoothed with 400-grit abrasive paper; then the blanks were wiped clean with methylene ketone.

A disk of 1018 steel 400 mm (16 in.) in diameter, 3.2 mm (1/8 in.) thick and with a 4.9 mm (0.1935 in.) diam center hole, was used as an overlay to protect the diaphragm and to prevent wrinkles in the workpiece. Using a wax lubricant, the hemispheres were drawn in one operation with a ball-end punch 200 mm (8 in.) in diameter. The punch rose against the titanium blank, pressing the steel overlay up against the diaphragm under hydraulic pressure of 13.8 MPa (2000 psi). The pressure was gradually increased to 95 MPa (14,000 psi) max.

After forming, the hemisphere was separated from the overlay by a blast of air through the center hole. The workpiece was cleaned in an alkaline bath, rinsed, pickled in an acid bath, rinsed and dried in an oven. The hemispheres were annealed for 1 h at 705 °C (1300 °F) in air, followed by cooling in air.

Dome-Shaped Part (400 mm diam). Hemispheres 400 mm (16 in.) in diameter were deep drawn in a hydraulic press (punch-and-diaphragm type) having a capacity for blanks 660 mm (26 in.) in diameter and 300 mm (12 in.) depth of draw, from blanks of alloy Ti-6Al-4V, 1.27 mm (0.050 in.) thick sheared to 660 mm (26 in.) diam. The edges were deburred by grinding, then smoothed with 400-grit paper; the blanks were wiped clean with methylethyl ketone.

A disk of 1018 steel 660 mm (26 in.) in diameter and 3.2 mm (1/8 in.) thick, with a 4.9 mm (0.1935 in.) diam center hole was used as an overlay to protect the diaphragm and to prevent wrinkles in the workpiece. Using a wax lubricant, the punch initially applied 2000 psi pressure, which was gradually increased to 12,000 psi, to form the hemisphere, which had a flange radius of 12.7 mm (1/2 in.).

The workpiece was separated from the overlay by a blast of air in the center hole, then cleaned in an alkaline bath, rinsed, pickled in an acid bath, rinsed, and dried in an oven. The workpiece was annealed for 1 h at 705 °C (1300 °F) in an air atmosphere furnace, and cooled in air; cleaning, rinsing, pickling and drying were then repeated. The overlays were wiped clean with methylethyl ketone and replaced on the workpiece for further forming in the same setup.

Using 90 MPa (13,000 psi) to press the punch deeper into the work, the ball end formed a short cylinder as an extension of the hemisphere, thus completing the draw. The completed workpiece was again separated from the overlay and cleaned and annealed as before.

Tube Bending

Ti-6Al-4V: Minimum bend radii and springback for 120° bends in annealed seamless tubing

Tube OD		Wall thickness		Minimum bend radius		Spring- back
mm	in.	mm	in.	mm	in.	
12.7	$\frac{1}{2}$	0.40	0.016	64	2.5	8°
		0.71	0.028	50	2	14°
15.9	$\frac{5}{8}$	0.40	0.016	64(a)	2.5(a)	8°
		0.89	0.035	50	2	12°
25	1	0.58	0.023	100(a)	4(a)	8°
		1.44	0.057	89	3.5	14°
38	$1\frac{1}{2}$	0.89	0.035	125(a)	5(a)	12°
		2.16	0.085	100(a)	4(a)	12°

(a) Using a pressure-die boost cylinder on the pressure-die carriage

Power Spinning

Most titanium alloys are difficult to form by power spinning. Ti-6Al-4V and Ti-13V-11Cr-3Al and some grades of commercially pure titanium are the most responsive to forming by this method. Conditions used in production power spinning are described in the examples that follow.

Conical part from Ti-6Al-4V sheet, 2.0 mm (0.080 in.) thick was spun at room temperature, using a roller feed of 12.7 mm (1/2 in.) per minute and mandrel speed of 150 rpm, to a 50% thickness reduction in one pass. The same alloy, 3.86 mm (0.152 in.) thick, was power spun at 425 °C (800 °F), using a roller feed of 50 mm (2 in.) per min and mandrel speed of 400 rpm. The workpiece was 144 mm (5.7 in.) in diameter and thickness was reduced 38% in one pass.

Conical part from Ti-13V-11Cr-3Al sheet, 3.3 mm (0.130 in.) thick, was power spun at room temperature, using a roller feed of 89 mm (3.5 in.) per min and mandrel speed of 280 rpm. The 244 mm (9.6 in.) diam part was reduced 53% in one pass.

Cylinders 355 mm (14 in.) in diameter were power spun at room temperature from Ti-13V-11Cr-3Al rolled ring blanks, in two passes. A roller feed of 100 mm (4 in.) per min and mandrel speed of 200 rpm were used. Annealing at 760 °C (1400 °F) was required between passes.

In the first pass, the cylinder was spun to a diameter of 359.1 mm (14.138 in.), reducing the thickness by 51% from 8.38 to 3.73 mm (0.301 to

0.147 in.). The second pass reduced the thickness to 1.9 mm (0.075 in.)—a reduction of 49%. All cylinders that were power spun at room temperature had to be sized to minimize out-of-roundness.

Ti-6Al-4V: Typical power spinning temperatures

Alloy	Spin temperature °C	°F
CP Ti	425-540	800-1000
Ti-5Al-2.5Sn	650-760	1200-1400
Ti-6Al-4V	595-650	1100-1200
Ti-13V-11Cr-3Al	870-980	1600-1800

Ti-6Al-4V: Lubricants used for power spinning

Operating temperature °C	°F	Lubricant
Commercially pure titanium		
Up to 205	Up to 400	Heavy drawing oils, graphited greases, colloidal graphite
Commercially pure titanium and titanium alloys		
205-425	400-800	Bentonite greases with graphite, colloidal graphite, or molybdenum disulfide
425-980	800-1800	Bentonite greases with graphite, colloidal graphite, powdered mica

Heat Treatment

R.R. Boyer, Boeing Commercial Airplanes

Ti-6Al-4V is most commonly used in the fully annealed condition, often referred to as mill annealed. Ti-6Al-4V has limited hardenability, but is sometimes used in the solution treated and aged (STA) condition. Using STA, strengths as high as 1110 MPa (160 ksi) can be achieved in section thicknesses up to 12.7 to 19 mm (0.5 to 0.75 in.). Damage tolerance properties are optimized (fracture toughness is maximized and the crack growth rate is minimized) by β annealing (BA), which provides a totally transformed β structure. This, however, results in reduced ductility (although the material would still have at least 5% elongation) and reduced fatigue strength. Beta annealing should not be used on sheet gages due to the possible incidence of a single prior β grain across the entire thickness. Improvements in fatigue performance can be achieved by water quenching from above the β transus temperature of 995 °C (1825 °F). This refines the transformed structure.

Recrystallization annealing (RA) also increases damage tolerant properties. Fracture toughness and crack growth resistance is typically not quite as good as BA, but fatigue performance and ductility are improved. To maximize damage tolerance properties, the extra-low-interstitial (ELI) grade, which contains lower oxygen, should be used with both the recrystallization anneal and the β anneal.

Another option for achieving improved damage tolerance over mill annealing is a duplex anneal, although this is not a common heat treatment. The first anneal of a duplex anneal is high in the α–β phase field. On cooling, the β at temperature transforms to a lamellar α-β structure, which improves damage tolerance properties. The higher the first anneal temperature, the higher the fracture toughness and crack growth resistance; fatigue strength diminishes until the anneal temperature exceeds the β transus.

The high-temperature anneal is followed by a full anneal or mill anneal. This heat treatment is very similar to RA, except the cooling rate from the first anneal temperature is faster, resulting in the transformed structure. The RA treatment uses a slower cooling rate, resulting in a predominantly equiaxed structure. There are many variations of these basic heat treatments to tailor the resulting properties to the application.

Thermomechanical Processing Patent. Although several thermomechanical processing schedules are described in the section "Forging," a recent patent (Patent No. 5,118,363, ALCOA, Pittsburgh, PA) has been issued for improving mechanical properties of Ti-6Al-4V forgings. The processing steps include heating slightly above the β transus temperature to form the β phase, followed by rapid cooling; reheating the billet (which now has a fine β transformed structure) to 855 to 925 °C (1570 to 1700 °F); forging to obtain a reduction ratio of about 3:1; cooling and solution treating to form primary α particles; and aging and cooling.

Furnaces and Atmospheres. Heat treatment should be accomplished in atmospheres free of reducing gases and other contaminants that might produce excessive hydrogen pick up. Gas-fired furnaces should have oxidizing flames, and there should be no flame impingement on the part.

Furnaces used above 650 °C (1200 °F) that have contained endothermic or dissociated ammonia atmospheres should be equipped to prevent leakage, purged, and tested for hydrogen pickup prior to heat treating the first load of titanium parts. The use of inert gas or vacuum atmospheres minimizes contamination when properly controlled. Molten salt and fluidized beds should not be used. See MIL-H-81200 for further guidelines.

Selected References

- C.R. Brooks, *Heat Treatment, Structure and Properties of Nonferrous Alloys*, American Society for Metals, 1982, p 361-376

- J.A. Burger and D.K. Hanink, Heat Treating Titanium and Its Alloys, *Met. Prog.*, Vol 91 (No. 6), June 1957, p 70-75
- *ASM Trans*, 1956, p 657-676
- W. Herman et al., Heat Treating of Titanium and Titanium Alloys, *Metals Handbook*, Vol 4, 9th ed., *Heat Treating*, American Society for Metals, 1981, p 763-774
- R.A. Wood and R.J. Favor, *Titanium Alloy Handbook*, MCIC-HB-02, Battelle Memorial Institute, 1972

Property Data References

- **Forgings:** G.W. Kuhlman and A.K. Chakrabarti, Room Temperature Fatigue Crack Propagation in Beta Titanium Alloys, *Microstructure, Fracture Toughness and Fatigue Crack Growth Rate in Titanium Alloys*, A.K. Chakrabarti and J.C. Chesnutt, Ed., TMS, 1987, p 3-15
- **Std grade plate:** R.R. Boyer, R. Bajoraitis, and W.F. Spurr, The Effects of Thermal Processing Variations on the Properties of Ti-6Al-4V, *Microstructure, Fracture Toughness and Fatigue Crack Growth Rate in Titanium Alloys*, A.K. Chakrabarti and J.C. Chesnutt, Ed., TMS, 1987, p 149-170
- **ELI forged pancakes:** J.C. Chesnutt, A.W. Thompson, and J.C. Williams, "Influence of Metallurgical Factors on the Fatigue Crack Growth Rate in Alpha-Beta Titanium Alloys," AFML-TR-78-68, May 1978
- **Std grade plate:** R.R. Ferguson and R.C. Berryman, "Fracture Mechanics Evaluation on B-1 Materials," Vol 1, AFML-TR-76-137, Oct 1976
- **Std grade sheet:** R.R. Boyer and R. Bajoraitis, "Standardization of Ti-6Al-4V Processing Conditions," AFML-TR-78-131, Sep 1978
- **ELI sheet:** R.R. Boyer and R. Bajoraitis, "Standardization of Ti-6Al-4V Processing Conditions," AFML-TR-78-131, Sep 1978
- **STD and ELI grade forgings:** G.W. Kuhlman and F.R. Billman, Selecting Processing Options for High Fracture Toughness Titanium Airframe Forgings, *Met. Prog.*, Mar 1977
- **12.7 mm (0.5 in.) std grade plate:** C. Ouichi, H. Suenaga, T. Otoh, and T. Sawumura, Effect of Hot Rolling Variables on the Mechanical Properties of Ti-6%Al-4%V, *Titanium Science and Technology*, G. Lütjering, U. Zwicker, and W. Bunk, Ed., Deutsche Gesellschaft für Metallkunde, eV, Germany, 1985, p 483-490
- **15.9 mm (5/8 in.) diam bar, std grade:** L.J. Bartlo, "The Effect of Microstructure on the Fatigue Properties of Ti-6Al-4V," RMI Research and Development Report 493, Apr 1967
- **56.9 mm (2.24 in.) thick ELI bar:** C.A. Stubbington and S. Pearson, Effect of Dwell on the Growth of Fatigue Cracks in Ti-6Al-4V, *Eng. Fract. Mechan.*, Vol 10 (No. 4), Apr 1978, p 723-756

Ti-6Al-4V: Heat treatments

Heat treatment	Procedure	Comments
Stress relief annealing	2 to 4 h 595 °C (1100 °F), air cool	Relieves residual stresses from welding, forming, etc. This cycle only provides a partial stress relief. A full anneal must be used for full stress relief. Low strength, good ductility
Full annealing (mill annealing)	2 h, 735 ± 15 °C (1350 ± 25 °F), air cool	Most common heat treatment, has good overall properties combinations. Low strength, good ductility
Annealing for continuously rolled sheet	5 min 870 °C (1600 °F), rapid furnace cooled, plus 5 min 595 °C (1100 °F), air cool	Approximately the same strength and ductility as above, but with improved fracture toughness
Recrystallization annealing	4 or more h, 925 °C (1700 °F), furnace cool to 760 °C (1400 °F) at 55 °C (100 °F)/h (or slower), cool to 480 °C (900 °F) at 370 °C (670 °F)/h (or faster), air cool	Usually used for ELI material. Strength comparable to above conditions, but improved damage tolerance (fracture toughness, stress-corrosion resistance, reduced crack growth rates). Strength is lower for ELI material
Duplex annealing	10 min, 940 °C (1725 °F), air cool, plus 4 h 675 °C (1250 °F), air cool	Improved damage tolerance
Beta annealing	30 min, 1035 °C (1900 °F), air cool, plus 2 h, 730 °C (1350 °F), air cool	Used to maximize damage tolerance properties. These properties are attained with a slight loss of ductility and a significant fatigue debit. A preliminary teatment followed by other treatments such as annealing
Beta STA	30 min, 1035 °C (1900 °F), water quench plus 4 h, 510 to 675 °C (950 to 1250 °F)	...
Beta STOA	30 min, 1035 °C (1900 °F), water quench plus 2 to 4 h, 675 to 730 °C (1250 to 1350 °F)	Similar to beta anneal with improvement in fatigue performance may be preferable to beta anneal because of improved fatigue properties, but at a cost in damage tolerant properties
Solution heat treatment	10 min, 940 °C (1725 °F), water quench	Used as an intermediate step for forming material ultimately to be used in the STA condition. Not to be used as a final condition due to instability
Solution heat treatment and aging	10 min, 940 °C (1725 °F), water quench, plus 4 h, 510 to 540 °C (950 to 1000 °F), air cool	Highest strength condition, but less ductility, stress-corrosion resistance, and fracture toughness than annealed
Solution heat treatment and overaging	10 min, 940 °C (1725 °F), water quench, plus 4 h, 675 °C (1250 °F), air cool	Intermediate strength between annealed and STA, but improved ductility and damage tolerance properties compared to STA

Source: R. Boyer, Boeing Commercial Airplane

Ti-6Al-4V: Mechanical properties vs heat treatment

Heat treatment	Ultimate tensile strength		Tensile yield strength		Elongation, %	Reduction of area, %	K_{Ic}	
	MPa	ksi	MPa	ksi			MPa√m	ksi√in.
Mill anneal	1020	148	951	138	...	...	43(a)	39(a)
	930	135	869	126	15	...	...	...
	930	135	860	125	10	20	55	50
STOA								
ELI	1000	145	895	130	12	30	60	55
STD	1000	145	938	136	...	...	47	43
Recrystallize anneal								
ELI	930	135	860	125	10	25	83	75
STD	1000	145	931	135	...	...	56	51
Beta anneal								
ELI	860	125	795	115	8	20	99	90
STD	990	144	910	132	11	20	95	86
Duplex anneal								
ELI, forged	892	129	814	118	12	36	124(b)	112(b)
STD, plate	931	135	903	131	16	...	...	...
ELI, sheet	945	137	895	130	15	...	176(c)	160(c)
STD, bar	1014	147	903	131	17	47	...	...
ELI, bar	934	136	832	121	13	...	90	82
Beta STOA								
ELI	938	136	860	125	11	18	134(b)	122(b)
STD	972	141	900	130	9	13	99	90
Beta STA, STD	1172	170	1069	155	8	15	...	...
Solution treat and age	1270	184	1181	171	16	41.5	...	...
	1186	172	1069	155	16	56	...	...
Solution treat	1117	162	951	138	17	60	...	...

(a) Estimated. (b) K_{Ic} invalid. (c) K_c

Stress Relief

The effect of time and temperature on stress relieving at elevated temperatures is shown in the nomograph for Ti-6Al-4V. At 595 °C (1100 °F), it takes 50 hours to achieve a full stress relief; a 50% stress relief would take about 1 hour at 595 °C (1100 °F) or 5 hours at 540 °C (1000 °F). Such a diagram is extremely valuable in selection of thermal treatments to reduce residual stress levels. Stress relaxation data can often be fitted to a time-temperature parameter such as the Larson-Miller parameter.

Ti-6Al-4V: Stress relief vs time

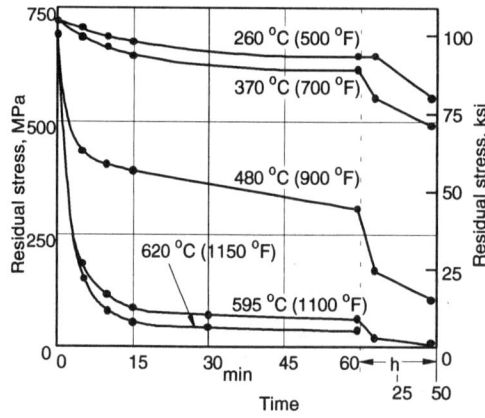

Source: *Metals Handbook*, Vol 4, 9th ed., 1981

Ti-6Al-4V: Stress relief nomograph

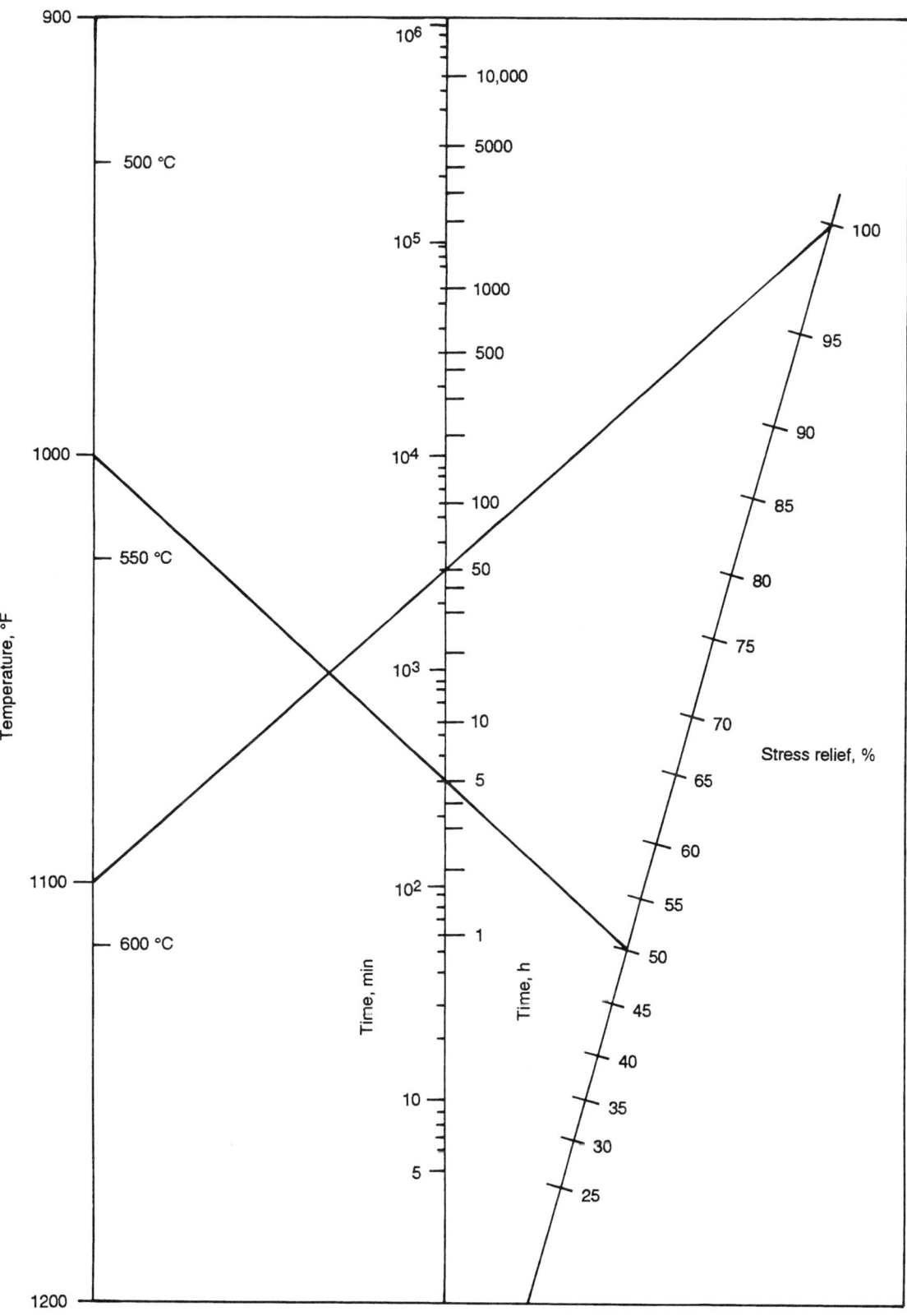

Relationship of time, temperature, and percent of stress relief for Ti-6Al-4V. Two examples (50% and 100% relief) shown.
Source: Maykuth, 1968

Annealed Properties

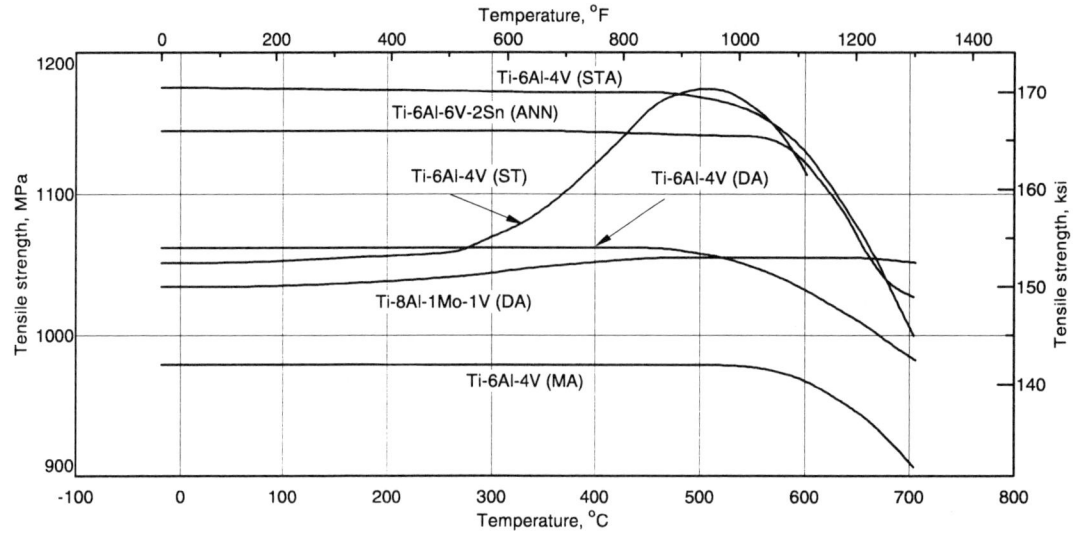

Ti-6Al-4V: Effect of temperature on tensile strength (1/2 h exposure)

Ti-6Al-4V: Comparison of heat treated and mill annealed sheet

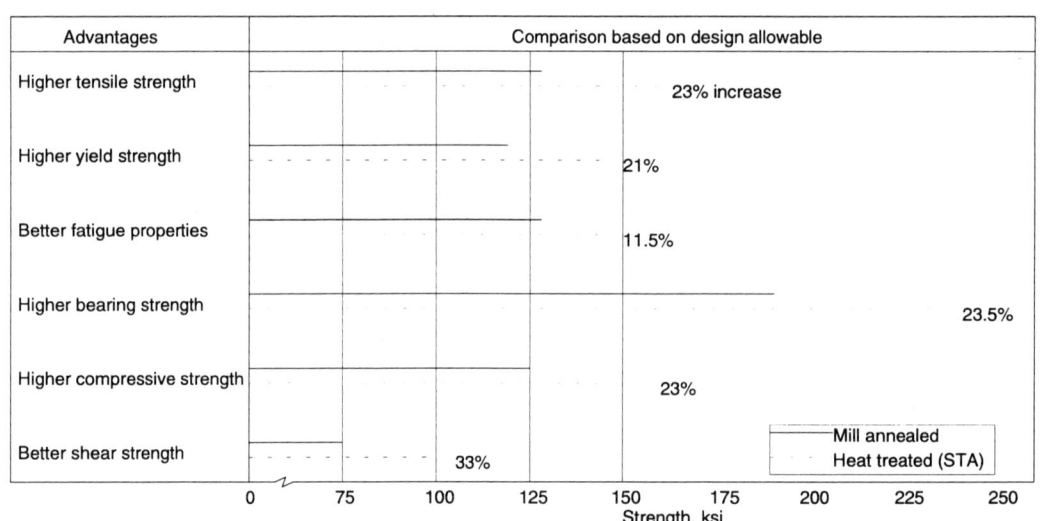

Source: J.C. Chang, Forming Ti-6Al-4V Sheet Metal in Four Heat Treated Conditions, *The Science, Technology and Application of Titanium*, R.I. Jaffee and N.E. Promisel, Ed., Pergamon Press, 1970, p 1053-1063

STA Properties

Ti-6Al-4V: Variation of tensile properties of bar stock with solution treating temperature

Solution-treating temperature		Room-temperature tensile properties(a)				Elongation in 4D,
		Tensile strength		Yield strength(b)		
°C	°F	MPa	ksi	MPa	ksi	%
845	1550	1025	149	980	142	18
870	1600	1060	154	985	143	17
900	1650	1095	159	995	144	16
925	1700	1110	161	1000	145	16
940	1725	1140	165	1055	153	16

(a) Properties determined on 13 mm (0.5 in.) bar after solution treating, quenching and aging. Aging treatment: 8 h at 480 °C (900 °F), air cool. (b) At 0.2% offset

Ti-6Al-4V: Effect of heating rate and time at 955 °C on growth

Mill heat(a)	Heating rate °C/min	°F/min	Holding time(b), h	Net growth(c), %
A	3.3	6	0	0.27
B	3.3	6	0	0.22
A	3.3	6	1	0.60
B	3.3	6	1	0.49
A	3.3	6	2	1.00
B	3.3	6	2	0.90(d)
B	10	18	1	0.32
B(e)	10	18	1	0.35

Note: 50 mm (2 in.) specimens were taken in the longitudinal direction (except where otherwise indicated) from material annealed 2 h at 705 °C (1300 °F) and air cooled. No growth was observed in specimens tested during annealing. (a) Beta transus temperatures (determined metallographically) were 990 °C (1810 °F) for heat A and 1015 °C (1860 °F) for heat B. (b) All specimens water quenched after holding for time indicated. (c) As determined by Leitz-Wetzler dilatometer. (d) Calculated from curve. (e) Specimen taken in transverse direction

Ti-6Al-4V: Section size vs STA tensile properties

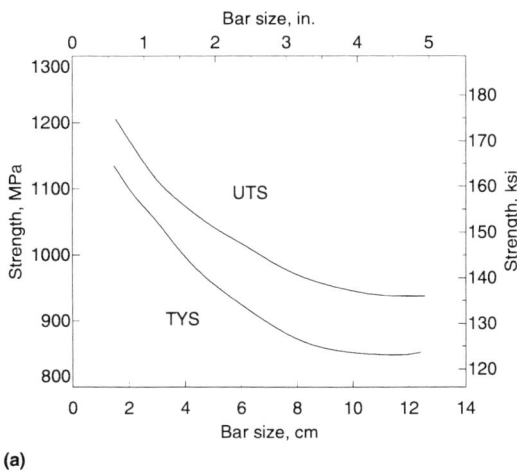

(a)

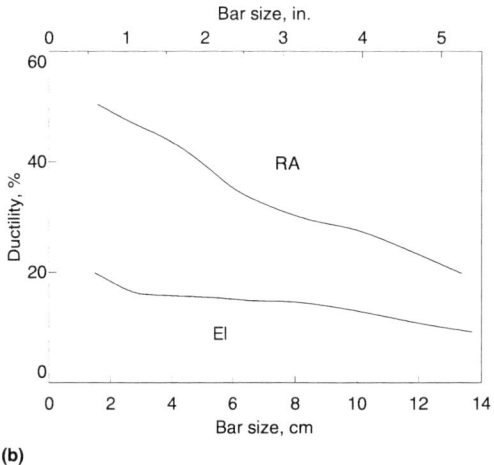

(b)

Beta Heat Treated Microstructures

The microstructures achieved from supratransus solution treatment depend on the degree of homogenization during solution treatment and the cooling rate from the β phase field. Quenching suppresses the diffusion-controlled β to α transformation and leads to a martensitic transformation, where the β phase becomes needlelike α' martensite. This α' martensite also can be produced by quenching from solution treatment below the β transus. However, the volume fraction of α' would not be as large as the volume fraction of α' in β-quenched material.

A slower cooling rate (e.g., air or furnace cooling) permits diffusion-controlled partitioning between the α and β as the temperature falls below the β transus and the nucleation of α begins. The microstructure is lamellar, in which broad α and fine β lamellae alternate to form packets. The size of the lamella packets depends on prior β grain size and cooling rate. The fineness of the lamellae ranges from a fine acicular structure to a more coarse lamellar structure depending on the cooling

Ti-6Al-4V: Effects of quench delay on tensile properties of bar

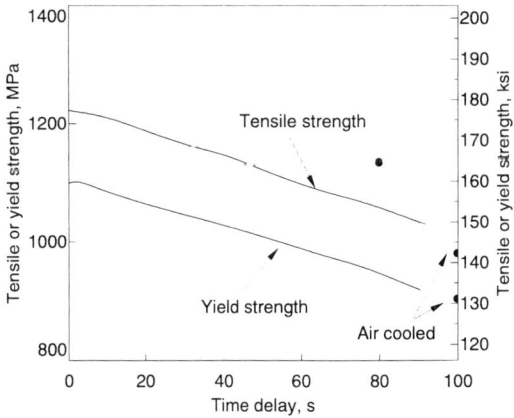

Bar, 13 mm (1/2 in.) in diameter, was solution treated 1 h at 955 °C (1750 °F), water quenched, aged 6 h at 480 °C (900 °F) and air cooled.
Source: *Metals Handbook*, Vol 4, 9th ed., 1981, p 768

rate *through* the transus.

The degree of homogenization or microstructural changes depends on self-diffusion and diffusion of alloy elements. The rates at which microstructural changes occur depend on temperature and time. For example, an approximate assessment of the extent of possible microstructural change can be made on the basis of the magnitude of the titanium self-diffusion coefficient. At 500 °C (930 °F), $D_{\alpha Ti} \approx 10^{-19}$ m^2/s and $D_{\beta Ti} \approx 10^{-18}$ m^2/s. At 1000 °C (1830 °F), $D_{\alpha Ti} \approx 10^{-15}$ m^2/s and $D_{\beta Ti} \approx 10^{-13}$ m^2/s [K. Ouchi *et al.*, 1980, and K. Inouye *et al.*, 1980, cited below]. Thus, the average diffusion distances at 500 °C (930 °F) for 50 h are approximately 0.8 and 0.9 μm in the α and β phase, respectively. However, for 1 hour at 1000 °C (1830 °F), they are ~4 μm in α and 40 μm in the β phase. This illustrates that homogenization of the microstructure occurs rapidly above the β transus, whereas microstructural changes are sluggish at temperatures well below the transus temperature.

Selected Reference
- K. Inouye, Y. Iijima, and K. Hirano, Interdiffusion in Ti-V Alloys, p 569-576; and K. Ouchi, Y. Iijima, and K. Hirano, Interdiffusion in Ti-Al Systems, p 559-568; both in *Titanium '80, Science and Technology*, H. Kimura and O. Izumi, Ed., TMS/AIME, 1980

Ti-6Al-4V: Heat treatment cycles of transformed β

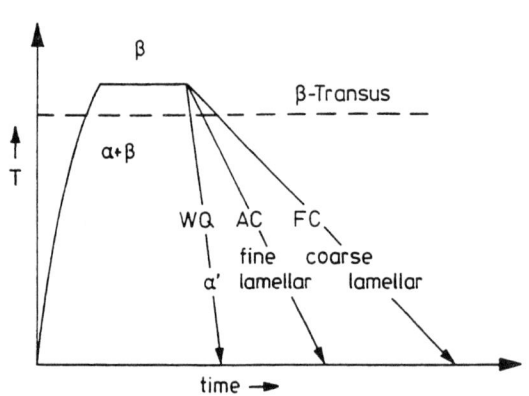

Cooling rate through the transus determines the fineness of transformed structures (α', acicular, or coarse lamellar).

Ti-6Al-4V: α' martensite

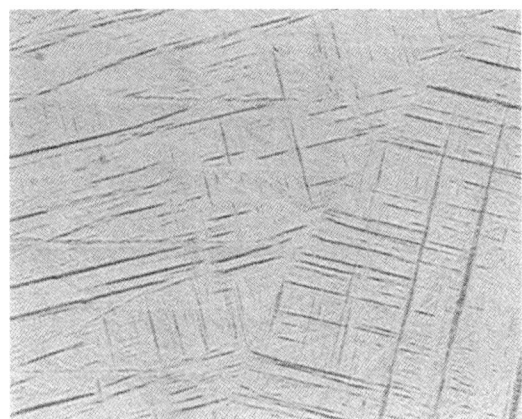

As-quenched, 200X

Ti-6Al-4V: Fine lamellar

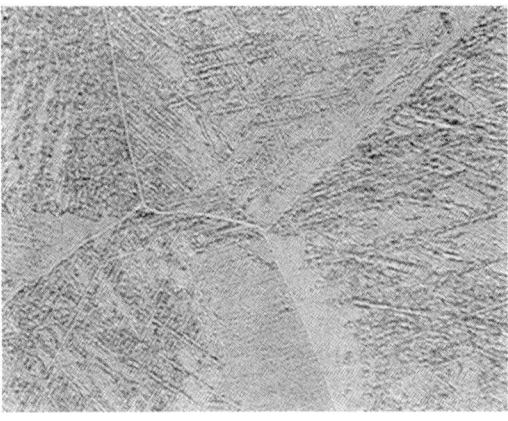

Fast air cool, 200X

Ti-6Al-4V: Coarse lamellar

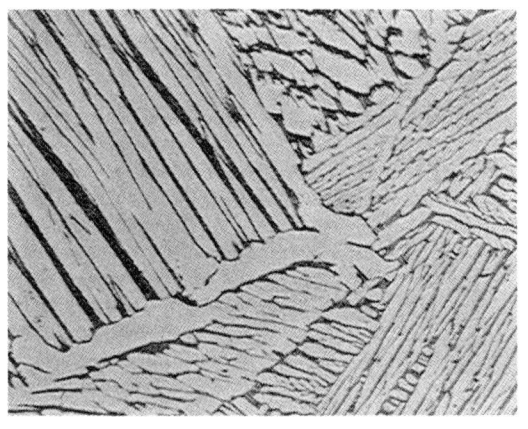

Slow (furnace cool), 200X. Photos courtesy of J.K. Gregory, GKSS, Germany

Ti-6Al-4V: Transformed structure from slow cooling

A broad lamellar or plate-like α develops during slow cooling, and most of the microstructural development occurs just below the transus. 90 to 95 vol% α phase.
Optical micrograph after furnace cooling from 1050 °C (1920 °F). Prior β grain boundaries are outlined by ribbons of grain boundary α. 50×.
Source: J. Hemptenmacher, DLR, Cologne, Germany

Ti-6Al-4V: Microstructure after slow cooling from just above the transus

Structure with relatively small prior β grain size but broad α lamellae because of the slow cooling rate. Slow furnace cool (~100 °C/h) from 1010 °C (1850 °F). 100×.
Source: G. Welsch, Case Western Reserve University

Alpha + Beta Annealed Microstructures

Completely different microstructures are obtained by hot working in the α + β phase field and then recrystallizing at subtransus temperatures. The presence of both phases during hot-working prevents the coarsening which occurs when only β grains are present. If the subsequent recrystallization anneal (RA) is performed below the martensite start temperature, a fine microstructure consisting of α and β grains (see figures on next page) forms. Above the martensite start temperature, the fine β grains transform in the same way as described earlier, and the microstructure consists of α grains and fine lamellar regions ("duplex" microstructure). The duplex annealing (DA) temperature determines the relative amounts of α and fine lamellae (transformed β) according to the lever rule. A final aging treatment of 2-24 h at 450-600 °C (840-1110 °F) brings out the α_2 phase.

During α + β annealing, the microstructure at temperature consists of primary α and some volume fraction of β (which becomes more substantial at the high-temperature end of the α + β phase field). Primary α, by definition, is not transformed during the anneal, and it therefore retains the morphology (equiaxed, lamellar, acicular) from prior thermomechanical processing. In contrast, the transformation of the remaining volume fraction of β depends on the cooling rates from the annealing temperature. Quenching produces martensites (β to α' or α''), whereas air or furnace cooling induces transformation of β into secondary α lamellae. The term "transformed β" usually refers to the lamellar packets of secondary α.

Heat treatments below the transus do not change the lamellar primary α from prior β treatments (though a slow cooling rate can coarsen it), but subtransus treatments can affect microstructural details in respective phases. Slow cooling from between 900 and 1000 °C (1650 and 1830 °F) usually produces a microstructure of fully recrystallized α lamellae (or equiaxed grains depending on prior working) and 10 to 20 vol% vanadium enriched β phase.

The recrystallization annealed condition combines high fracture toughness, good hot formability, and a somewhat higher strength than β-processed material. The mill annealed condition (see figure below) also falls into this category; however, the processing history is less tightly controlled, and therefore, the microstructure and mechanical properties are less reproducible. Depending on the degree of prior working, mixed microstructures between lamellar and equiaxed may be obtained.

Ti-6Al-4V: Typical microstructures

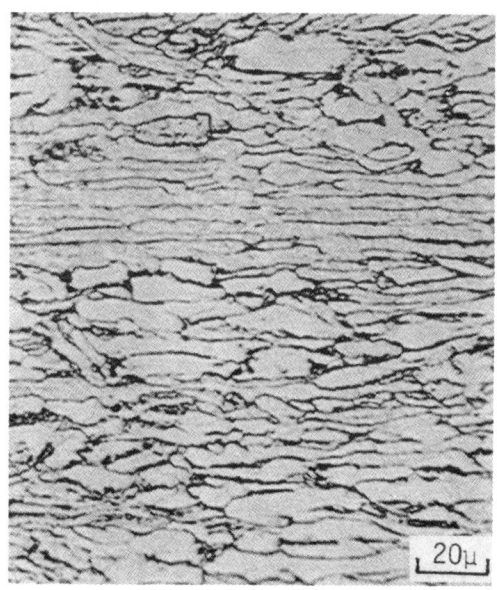
(a)

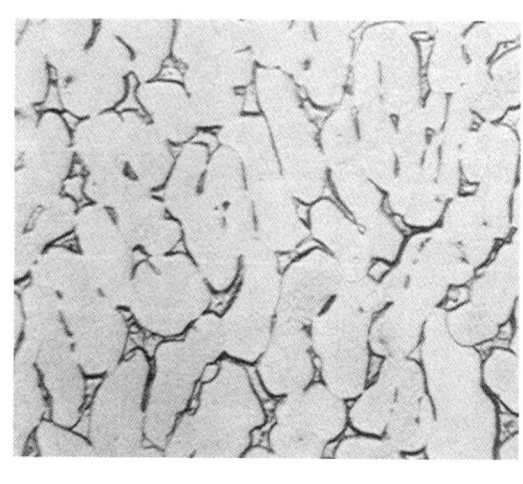
(b)

(a) Transverse view of mill annealed textured plate. (b) Conventionally processed plate (recrystallization annealed). Mill annealed treatment: 690 °C (1275 °F) for 2 h, AC. Recrystallization anneal: 980 °C (1795 °F) for 30 min, FC. 500×.
Source: R. Boyer, "Standardization of Ti-6Al-4V Processing Conditions," Report AF33615-75-C-5176, AFML, Dec 1976

Slow Cool from Transus

Ti-6Al-4V: Equiaxed microstructure obtained from prior forging below the transus

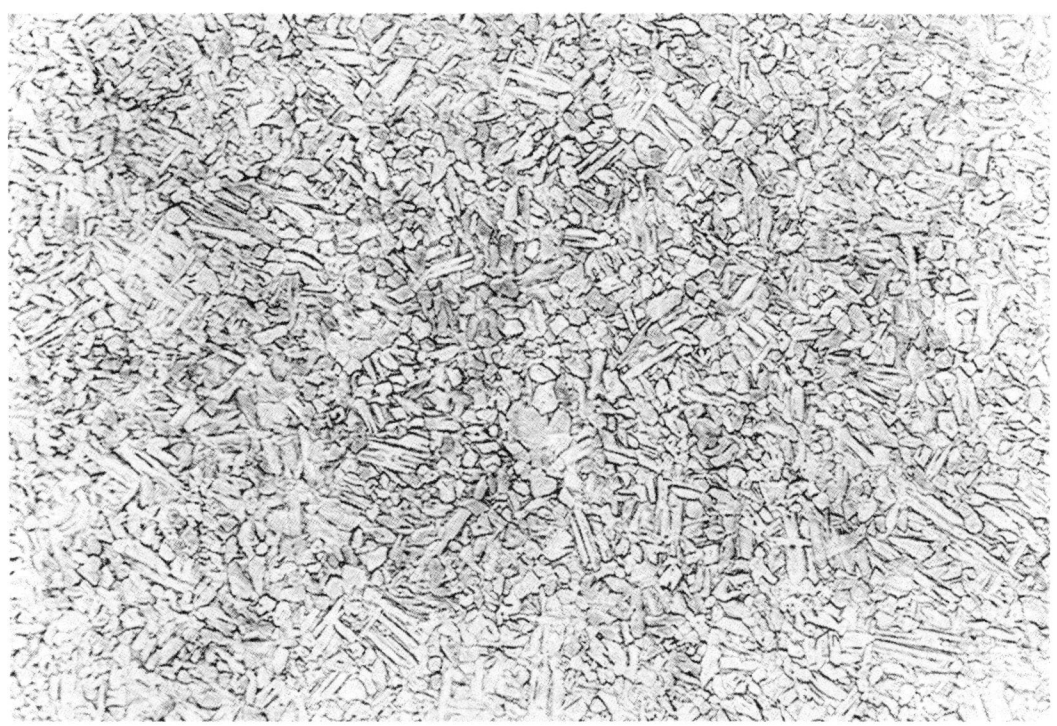

Mixed α morphology after 1000 °C (1830 °F) anneal. Annealed near the transus (1000 °C, or 1830 °F) for 30 min, cooled to 800 °C (1470 °F) in 5 h, and water quenched. On slow cooling, a recrystallized equiaxed grain structure is obtained with vanadium-enriched β (dark) between α grains (light). 100×.
Source: G. Welsch, Case Western Reserve University

Anneal after β Treatment

Ti-6Al-4V: Beta treated and α–β annealed structure

Beta treated at 1050 °C (1920 °F), air cooled to room temperature, annealed at 800 °C (1470 °F), and water quenched.
The fine lamellar α (light) is from intermediate (air) cooling through the transus. Subsequent annealing at 800 °C (1470 °F) causes little coarsening of the prior α lamellae, but it permits some vanadium enrichment in the interlamellar β phase regions. On quenching, the latter transform to α″ martensite. 100×.
Source: G. Welsch, Case Western Reserve University

Processing cycles for fine microstructures in α + β alloys

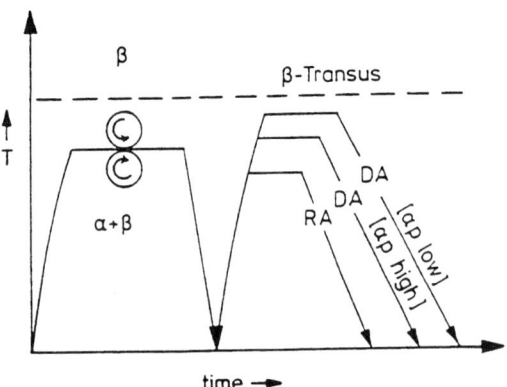

Although not indicated here, recrystallization anneal has the slower cooling rate. Corresponding photos courtesy of J.K. Gregory, GKSS, Germany.

Ti-6Al-4V: Fine equiaxed structure

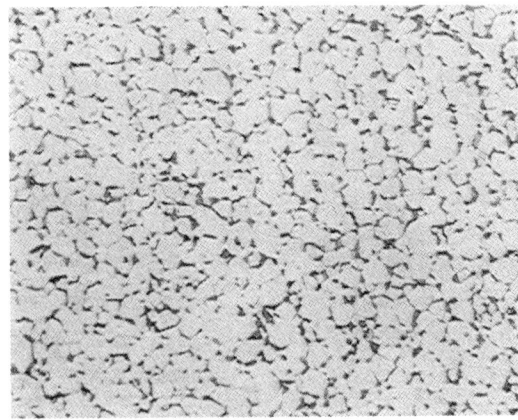

200×

Ti-6Al-4V: Duplex microstructure

40% primary alpha (α_p), 200×

Ti-6Al-4V: Duplex microstructure

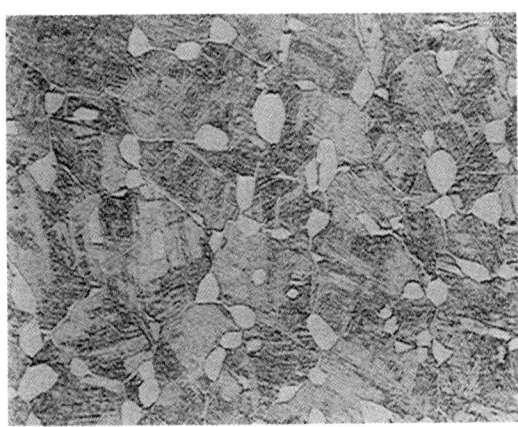

20% primary alpha, 200×

Effect of Working

Ti-6Al-4V: Effect of working on fine α lamellae

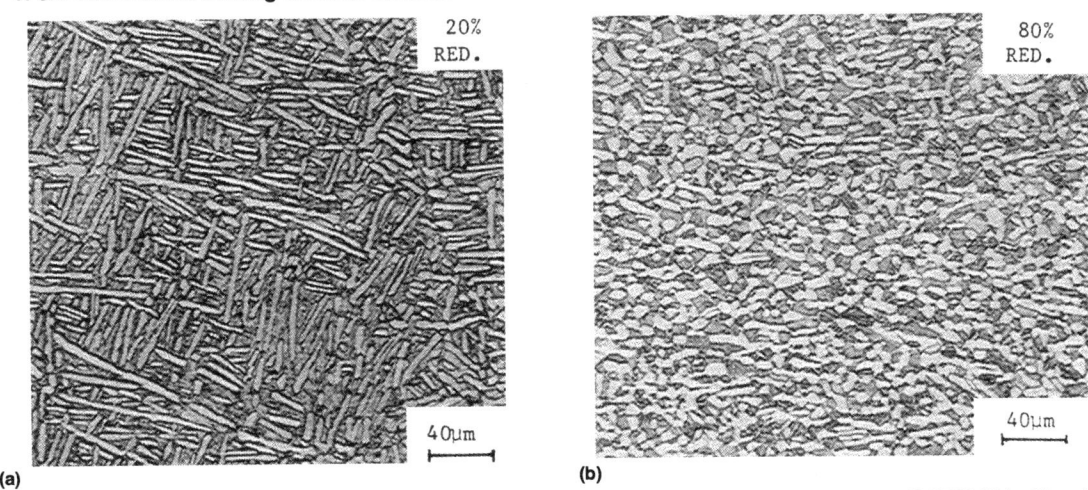

Initial fine lamellar structure broken up with a forging reduction of 20% (a) and 80% (b). After forging, annealed at 925 °C (1700 °F) for 2 h and quenched. The microstructure consists of primary α (light) and dark regions of transformed β (secondary α or α' martensite).
Source: I. Weiss et al., Metall. Trans. A, Vol 17, 1986, p 1935-1947

Ti-6Al-4V: Effect of working on coarse α lamellae

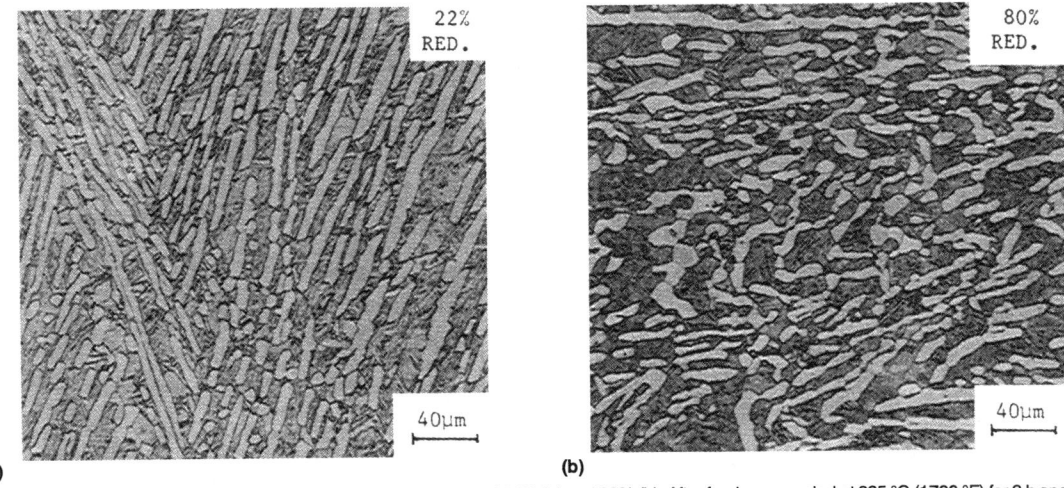

Initial coarse lamellar structure broken up by a forging reduction of 22% (a) and 80% (b). After forging, annealed at 925 °C (1700 °F) for 2 h and quenched. The microstructure consists of primary α (light) and dark regions of transformed β (secondary α or α' martensite).
Source: I. Weiss et al., Metall. Trans. A, Vol 17, 1986, p 1935-1947

Quenched and Aged Microstructures

As-quenched structures vary from 100 vol% of α' martensite for β quenched material to mixtures of primary α with α' or α" depending on the treatment temperature (see figure). The microstructures for an α + β quench are essentially the same as α + β anneals, except for the fact that the β phase regions transform to martensites (α' or α") instead of secondary α. The martensites decompose into fine α after aging.

Hexagonal Martensite (α')

Quenching from above 900 °C (1650 °F) causes the transformation of β into either fine secondary α or needlelike α' martensite. The volume fraction of secondary α and α' depend on the quench temperature and cooling rate. According to the Burgers orientation relation, up to twelve orientations of the hexagonal α' laths form within a given β grain.

Ti-6Al-4V: Quenching and aging processes schematic

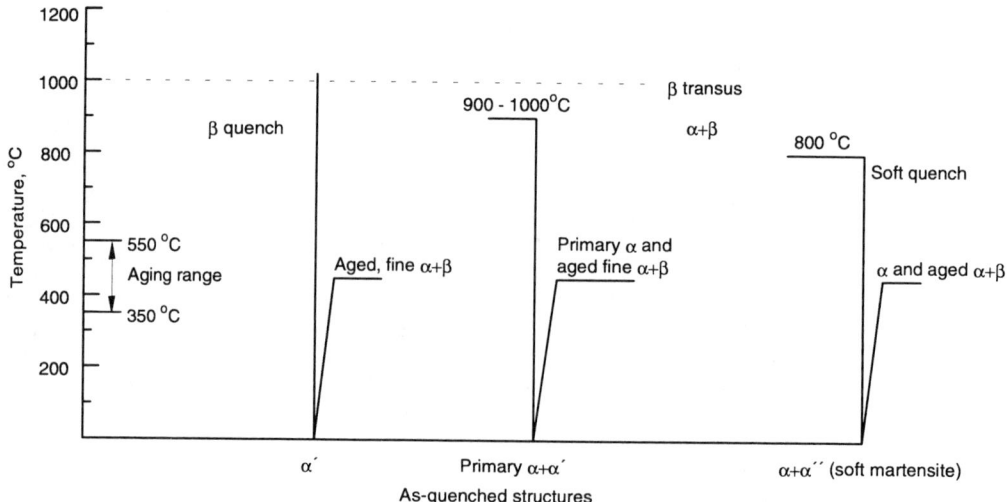

Lower solution temperatures allow higher amounts of solute, which generally favors the formation of orthorhombic (α'') martensite upon quenching.

Ti-6Al-4V: Primary α and transformed β

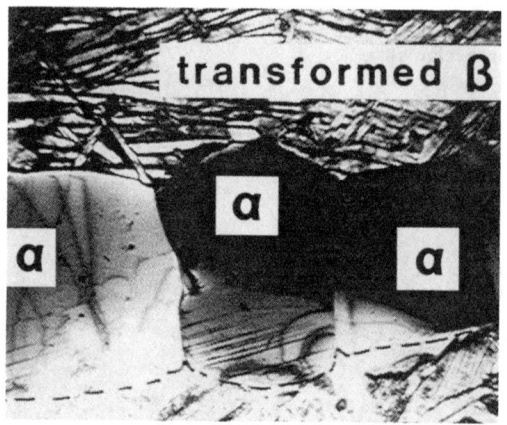

A broad primary α lamellae recrystallized into a string of α grains and is surrounded by transformed β. No effort is made to distinguish between α' and secondary α in the transformed region. TEM micrograph, 4300×.
Source: G. Welsch, Case Western Reserve University

Ti-6Al-4V: Needlelike α' and prior primary α

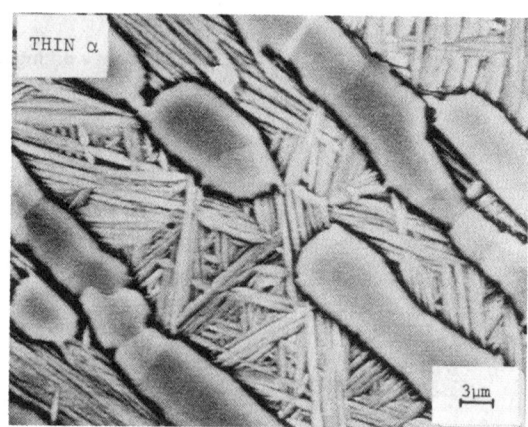

The primary α lamellae segments are divided into recrystallized grains. Channeling contrast SEM.
Source: G. Welsch, Case Western Reserve University

Soft α''

Quenching of Ti-6Al-4V from ~ 800 °C (1470 °F) produces the soft orthorhombic martensite (α'') during the transformation of β phase regions. Because the vanadium enrichment is not sufficient to stabilize the bcc β structure to room temperature, a martensitic transformation to the α'' martensite takes place. Thin martensite is mechanically soft and gives the microstructure high ductility and the lowest yield strength and Young's modulus values at room temperature. However, this soft microstructure can be aged to high strengths.

The soft α'' martensite undergoes noticeable effects after aging. This soft α'' martensite is replaced by a mixture of β phase with fine α precipitates. The fine precipitate structures are obtained at low aging temperatures between 150 and 450 °C (300 and 840 °F), but long aging times (days or weeks) are needed to obtain the highest strength levels. At higher aging temperatures up to 600 °C (1110 °F), the precipitates become noticeably larger and can be identified as individual α grains or islands within the β phase.

Examples of the microstructural changes at aging temperatures of 350 and 550 °C (660 and 1020 °F) are shown. These changes are only observable at very high magnifications using transmission electron microscopy (TEM), but they nevertheless have a significant influence on the mechanical properties of the alloy.

Ti-6Al-4V: Equiaxed α (dark) and α″ (light)

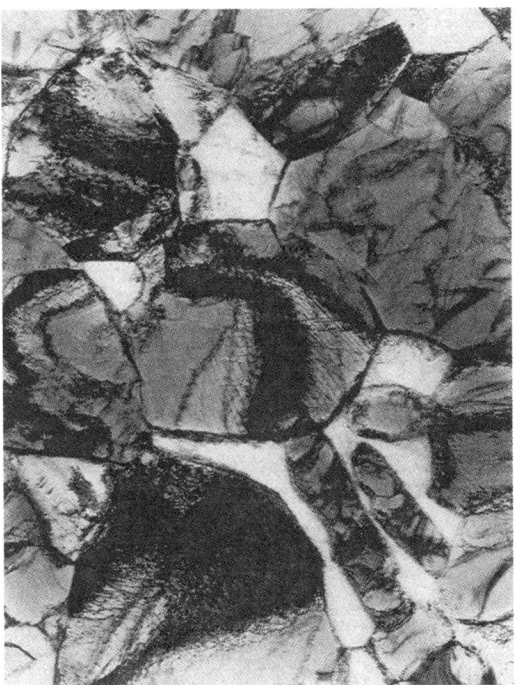

TEM Source: G. Welsch, Case Western Reserve University

Ti-6Al-4V: Aged region of prior α″

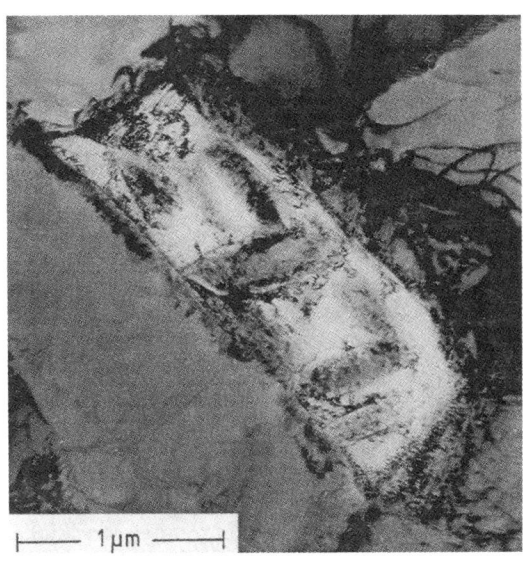

Aged at 550 °C for 67 h. The region of prior α″ has decomposed into β phase with islands of secondary α embedded in it.
Source: G. Welsch, Case Western Reserve University

Ti-6Al-4V: Aged region of prior α″

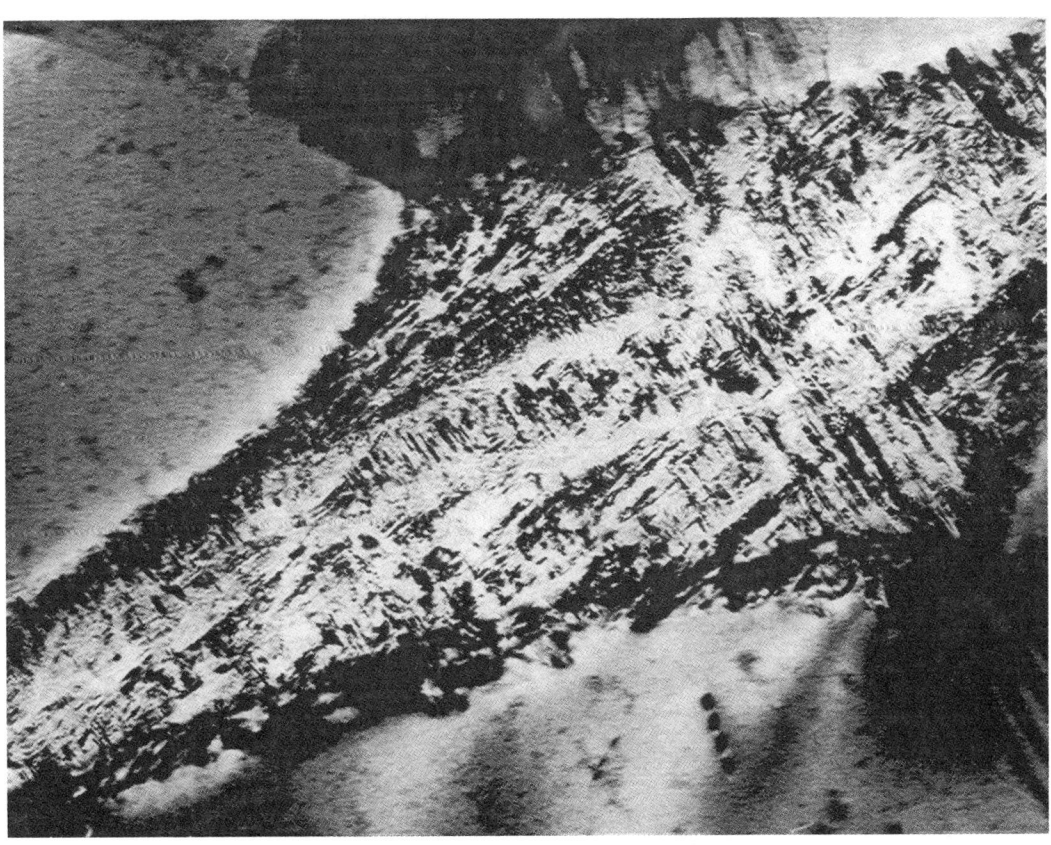

Region of α″ between primary alpha grains decomposed into β (dark) plus fine secondary α (light) after aging at 350 °C (660 °F) for 67 h. TEM

Welding

W.A. Baeslack III, Department of Welding Engineering, Ohio State University

Compared to most structural titanium alloys, Ti-6Al-4V is considered to be highly weldable. This weldability can be attributed to two principal factors. First, the α' martensite that forms in Ti-6Al-4V is not as hard and brittle as that exhibited by more heavily β-stabilized α-β alloys such as Ti-6Al-6V-2Sn. Secondly, Ti-6Al-4V exhibits relatively low hardenability, which allows the formation of higher proportions of the more desirable Widmanstätten α plus retained β microstructure even at relatively high weld cooling rates.

Weldability. In many alloy systems, weldability is determined by the capability of the alloy to produce a weld that is free of discontinuities or defects. Due to its single-phase mode of solidification (i.e, absence of low-melting-point eutectics), Ti-6Al-4V is highly resistant to solidification- and liquidation-related cracking. However, the occurrence of solid-state cracking and the formation of porosity can be encountered during welding. Fortunately, these defects are not metallurgically inherent in Ti-6Al-4V, but rather originate from readily correctable deficiencies in preweld cleaning of the workpieces and shielding of the weld zone from atmospheric contamination.

The weldability of an alloy is also dependent on its capability to produce a joint that exhibits acceptable mechanical properties either in the as-welded condition or following suitable postweld heat treatment. The weldability of structural titanium alloys, such as Ti-6Al-4V depends primarily on this second criterion. Weldment properties of titanium alloys depend on the structural characteristics of each weld region, which in turn depend on the specific thermal cycle(s) imposed during welding and on subsequent postweld heat treatment.

Welding Specifications. The American Welding Society has published two specifications that relate to welding of titanium. AWS Specification 5.16, entitled "Specifications for Titanium and Titanium Alloy Bare Welding Rods and Electrodes," provides requirements for use with GTAW and GMAW processes. Although the basic specification principally covers filler metal chemical composition requirements, an appendix presents additional general information on welding of titanium, including toughness data for welds in Ti-6Al-4V. Compositions of Ti-6Al-4V and Ti-6Al-4V extra-low interstitial (ELI) filler metals are specified by this document (see table).

AWS Specification D10.6, entitled "Recommended Practices for Gas Tungsten-Arc Welding of Titanium Pipe and Tubing," is specific to single-phase α alloys and does not cover Ti-6Al-4V. However, it does provide useful general information that is pertinent to welding of Ti-6Al-4V.

Numerous additional specifications developed by the government and by industry discuss the procedural aspects of the welding of Ti-6Al-4V and set forth minimum weld-quality and mechanical-property requirements. These specifications may be quite general or very specific, describing in detail the fabrication or weld repair of an individual component.

Ti-6Al-4V: Recommended filler metal specifications

ERTi-6Al-4V	5.5-6.75 Al, 3.5-4.5 V, 0.05 C max, 0.15 O max, 0.008 H max, 0.020 N max
ERTi-6Al-4V-1	5.5-6.75 Al, 3.5-4.5 V, 0.04 C max, 0.10 O max, 0.005 H max, 0.012 N max

Source: American Welding Society Specification A5.16, "Titanium and Titanium Alloy Bare Welding Rods and Electrodes"

Selected References

- M.J. Donachie Jr., *Titanium—A Technical Guide*, American Society for Metals, 1988
- *American Welding Society Handbook, Metals and their Weldability*, Vol 4, 7th ed., American Welding Society, 1982
- *Metals Handbook, Welding, Brazing and Soldering*, Vol 6, 9th ed., American Society for Metals, 1983
- W.A. Baeslack III, Advances in Titanium Alloy Welding Metallurgy, *J. Met.*, Vol 35 (No. 5), 1984, p 46-58
- *Welding Technology for the Aerospace Industry*, American Welding Society, 1981

Welding Processes

Ti-6Al-4V may be welded by a wide variety of conventional fusion and solid-state processes, although its chemical reactivity typically requires special procedures and precautions. In the United

States, fusion welding of titanium is performed principally in inert-gas-shielded arc and high-energy beam welding processes. To date, the commercial joining of titanium alloys using flux-based processes (e.g., submerged-arc, flux-cored, and electroslag welding) has not been shown to be competitive due to requirements for costly, high-purity, fluoride-based fluxes.

Weld Design Criteria and Limitations. Designers of Ti-6Al-4V welded structures must consider both welding process applicability and the physical characteristics and mechanical properties of welds. From a welding process standpoint, an efficient design will incorporate a welding process that is optimally suited to a particular material thickness and joint configuration. Such process suitability is also based on component size and shape (e.g., whether the component will fit in an available electron beam welding chamber, or whether the entire part can be suitably protected from the atmosphere during GTAW or PAW), the cost of producing the weld (including both capital equipment and operating costs), and postweld processing requirements.

The design of Ti-6Al-4V welded structures also is influenced by the characteristics and mechanical properties of a joint produced by a particular welding process. Physical characteristics of the weld, including undercuts or underfills in fusion welds, upsets in solid-state welds, and weld distortion are important considerations, because these characteristics influence not only the physical dimensions of the component but also joint mechanical properties. Obviously, the mechanical properties of the joint, as influenced by the integrity and metallurgical structure of the weld zone, are the principal considerations in weld design. In this regard, low fusion zone ductility and toughness in rapidly cooled welds and poor axial fatigue behavior in welds containing defects are important characteristics of Ti-6Al-4V welds that may influence the design of the welded component.

Special Welding Precautions and Requirements. Production of high-quality welds of Ti-6Al-4V requires meticulous preweld cleaning of the workpiece and consumables and adequate gas shielding to prevent atmospheric contamination. Mill annealed titanium sheet and plate generally are supplied in a condition that requires only the removal of the surface grease, oil, and dirt. Such cleaning normally is accomplished by wiping with or dipping in a nonchlorinated solvent such as acetone or methylethyl ketone. Materials that exhibit a light oxide scale as a result of heat treatment below about 600 °C (1110 °F), or which contain entrapped oil from machining operations, should be pickled for 5 to 10 min in a solution of 30 to 40% nitric acid and 4 to 5% hydrofluoric acid in water at a temperature between 20 and 70 °C (68 and 160 °F). Heavier oxide scales resulting from high-temperature heat treatment or thermomechanical processing may require cleaning in a molten salt bath or mechanical removal, followed by pickling. Pickled pieces should be rinsed in water, dried, and either directly welded or wrapped in clean plastic. All handling should be performed with clean, white gloves. Weld contamination also can result from filler metal that has become dirty due to inadequate storage or careless handling prior to or during the welding operation. Degreasing the filler rod or wire is generally sufficient to eliminate contamination problems originating from these sources.

Effective shielding of the weld zone from the atmosphere during welding is extremely important in ensuring maximum weld ductility and toughness and in reducing the potential for solid-state weld cracking. Optimum shielding conditions are provided by welding within a permanent or collapsible clear plastic enclosure that has been evacuated and purged with argon (dew point, –24 °C or –11 °F). Alternatively, localized trailing and backing shields can be used, particularly in automatic welding of simple joint geometries (e.g., automatic butt welding of sheet). Use of localized shielding for manual welding of complex parts, although feasible, requires specially designed shielding fixtures and generally is not recommended. Fortunately, evidence of titanium weld contamination is readily apparent by discoloration of the weld surface, with a change in color from bright silver to straw, to magenta, to blue, with a gradual increase in the level of contamination. Severe contamination is associated with a white or gray, powdery-appearing weld surface and is often accompanied by solid-state cracking in the weld metal. In many high-performance applications, a bright silver color is required, although a light straw color may be accepted in less-critical applications.

Because Ti-6Al-4V welds commonly are used in fatigue-critical applications, stress relief generally is required following welding. Specific stress relief temperatures and times depend on the base-metal microstructure (see table). Stress relief should be performed on components that are free of surface contamination and in an inert or vacuum environment (or in air if the resulting alpha case is subsequently removed, usually by etching or chemical mill).

Ti-6Al-4V: Stress-relief times and temperatures for welds

Base metal condition	Stress-relief temperature °C	°F	Stress-relief time(a), h
Annealed	480	900	20
	540	1000	2-4
	595	1100	0.5-1
Solution heat treated	490	915	15
	540	1000	4
Aged	480	900	15
	540	1000	5

Note: Surface contamination should be removed from surface prior to stress-relief heat treatment. Heat treatment should be performed in inert gas or vacuum environment. (a) Followed by gas fan cool. Source: *Metals Handbook, Welding, Brazing and Soldering*, Vol 6, 9th ed., American Society for Metals, 1983

GTAW

Gas tungsten-arc welding (GTAW), in which the heat of welding is provided by an arc maintained between a nonconsumable tungsten electrode and the workpiece, is the most widely used process for joining titanium sheet. In general, GTA welds can be produced manually (in complex joint geometries or for weld repair) and automatically in sheet up to about 32 mm (0.125 in.) in thickness without joint preparation or filler metal additions. For welding of greater thicknesses, suitable joint preparation and the cold or hot wire addition of a filler metal are required. Although conventional GTAW equipment can be used for welding of Ti-6Al-4V, the heated metal in the weld zone must be shielded from the atmosphere to prevent pickup of oxygen and nitrogen, which can be deleterious to weld mechanical properties. Such protection is best provided by welding within a rigid or collapsible plastic chamber that has been evacuated and purged with argon, although local shielding is also suitable for many welding operations.

Ti-6Al-4V: Typical parameters and conditions for gas tungsten-arc welding

Material thickness(a), in.	Tungsten electrode diameter, in.	Filler rod diameter, in.	Nozzle size ID, in.	Shielding gas flow, ft^3/h	Welding current(b), A	Number of passes	Travel speed(c), in./min
Square-groove and fillet welds							
0.024	1/16	...	3/8	18	20-35	1	6
0.063	1/16	...	5/8	18	85-140	1	6
0.093	3/32	1/16	5/8	25	170-215	1	8
0.125	3/32	1/16	5/8	25	190-235	1	8
0.188	3/32	1/8	5/8	25	220-280	2	8
V-groove and fillet welds							
0.25	1/8	1/8	5/8	30	275-320	2	8
0.375	1/8	1/8	3/4	35	300-350	2	6
0.50	1/8	5/32	3/4	40	325-425	3	6

Note: Tungsten used for the electrode; first choice 2% thoriated EWTh2, second choice 1% thoriated EWTh1. Use filler metal one or two grades lower in strength than the base metal. Adequate gas shielding is essential not only for the arc, but for heated material also. Backing gas is recommended at all times. A trailing gas shield is also recommended. Argon is preferred. For higher heat input, on thicker material use argon-helium mixture. Without backing or chill bar, decrease current 20%. (a) Or fillet size. (b) Direct current electrode negative. (c) Per pass. Source: *Metals Handbook, Welding, Brazing, and Soldering*, Vol 6, 9th ed., 1983

GMAW

Gas metal-arc welding (GMAW), in which the heat of welding is provided by an arc maintained between a consumable electrode and the workpiece, is more cost-effective than GTAW for the welding of titanium plate over 12.7 mm (0.5 in.) in thickness. However, the application of GMAW to titanium has been restricted by poor arc stability, which promotes appreciable spatter during welding, thereby reducing weld quality and overall process efficiency.

Ti-6Al-4V: Parameters and conditions for gas metal-arc welding

Material thickness(a), in.	Wire feed speed(b), in./min	Shielding-gas flow(c), ft^3/h	Welding current, A	Arc voltage(d), V	Travel rate, in./min
0.125	200-225	36	250-260	20	15
0.250	300-320	36	300-320	30	15
0.500	375-400	36	340-360	40	15
0.625	400-425	50	350-370	45	15
1.00(e)	380	36	320-350	36	20-23
2.00(e)	550	50	...	33	25

(a) Groove welds produced in flat position. (b) 0.062 in. diam filler wire. (c) 75% Ar + 25% He. (d) DCEP. (e) Multipass weld. Source: *AWS Handbook, Metals and Their Weldability*, Vol 4, 7th ed., American Welding Society, Miami, Florida, 1982, and M.J. Donachie, Jr., *Titanium—A Technical Guide*, American Society for Metals, 1988

Plasma Arc

Plasma-arc welding (PAW) is an extension of the GTAW process in which the arc plasma is constricted by a nozzle, thereby increasing its temperature and energy density as compared with the more diffuse GTAW arc. This higher energy density provides greater penetration capabilities, allowing the production of full-penetration, keyhole square-butt welds in thicknesses up to 12.7 mm (0.5 in.).

Welding Speed of PAW vs GTAW. Three metals used for aerospace components were welded by plasma-arc and gas tungsten-arc processes to compare times required for making welds. Samples welded were 6.4 mm (1/4 in.) thick type 410 stainless, 6.4 mm (1/4 in.) thick maraging steel (18% Ni), and 2.4 mm (0.095 in.) thick titanium alloy Ti-6Al-4V. Welding conditions and results are given in the accompanying table. Plasma-arc welding took a fifth to a tenth as long to complete 255 cm (100 in.) of weld as gas tungsten-arc welding, partly because fewer passes were required and partly because travel speed was greater.

Ti-6Al-4V: Typical parameters and conditions for plasma-arc welding

Thickness, in.	Travel speed, in./min	Current (DCEN), A	Arc voltage, V	Shielding gas	Gas flow, ft³/h Orifice gas	Shielding gas	Joint type
0.125	20	185	21	Argon	8	60	Square butt
0.187	13	175	25	Argon	18	60	Square butt
0.390	10	225	38	75He-25Ar	32	60	Square butt
0.500	10	270	36	50He-50Ar	27	60	Square butt
0.600	7	250	39	50He-50Ar	30	60	V-groove(a)

Note: Backing gas and trailing shield required; keyhole technique used with orifice-to-work distance of 0.187 in.

Ti-6Al-4V: Plasma-arc welding vs gas tungsten-arc welding
Comparison of the two processes for making 255 cm (100 in.) welds in three aerospace metals

Condition	410 stainless steel PAW	GTAW	Maraging steel, 18% Ni PAW	GTAW	Ti-6Al-4V PAW	GTAW
Thickness of metal, in.	0.250	0.250	0.250	0.250	0.095	0.095
No. of passes	1	2	1	3	2(a)	5(b)
Current, A	240	170-200	240-260	180-200	90-175	120-175
Travel speed, in./min	12	4	12-13	4	15	6
Time per 100 in. of weld, min	8.3	50	7.4	75	13.4	83.5

(a) One keyhole welding pass and one filler metal pass on the outside of a rocket motor case. (b) One root pass without filler metal and two passes with filler metal on the outside of a rocket motor case. The root pass was back gouged, and two filler passes were made on the inside. Source: *Metals Handbook*, Vol 6, 8th ed., p 145

Electron Beam

Electron beam welding (EBW) involves melting of the base metals to be joined by the impingement of a focused beam of high-energy electrons. This process has been used widely in the aerospace industry for producing high-quality welds in titanium plate ranging from 6.4 to more than 76 mm (0.25 to more than 3 in.) in thickness. Advantages of EB welding include high depth-of-penetration, minimum joint preparation, a narrow weld heat-affected zone (HAZ), low distortion, and excellent weld cleanness due to the fact that welding is performed in a high vacuum. Disadvantages of the EB process include high capital equipment cost and high weld cooling rates; the latter promotes formation of undesirable martensite in the weld zone. Typical welding parameters used in electron beam welding of Ti-6Al-4V are shown in the accompanying table.

Example: Welding a Heat Treated Assembly for Close Dimensional Control. Design of the complex fuel injection system of a lunar module landing engine was based on the use of electron beam welding for joining components of various subassemblies. Electron beam welds provided strength, freedom from distortion, and freedom from burn-through of thin walls. A typical subassembly was the manifold (see figure). It consisted of two conically shaped parts and was assembled by fitting the upper chambered portion into the solid lower portion, which served as the cover of the combustion chamber. Circular welds A and B sealed the subassembly, providing a conical space between the parts for smooth fuel distribution and cooling of the cover. Other details of fuel and oxygen flow, as well as other welds, have been omitted for simplicity. Enlarged details A and B show that the weld at B was the more critical, because of its proximity to 60 drilled holes (0.76 mm, or 0.030 in., from the centerline of the weld joint).

Material for this application was Ti-6Al-4V, solution treated and aged before welding. Postweld heat treatment was not possible because it would have caused excessive distortion, even though several areas were finish machined after welding.

The joints were machined with a 0.05 mm (0.002 in.) maximum diametral clearance. Also, to facilitate fit, the corner edge of the male part was chamfered, as shown in the accompanying figure.

Before assembly, joint surfaces were cleaned by acid pickling and ultrasonic cleaning. Just before welding, the joints were wiped with acetone.

After assembly, the workpiece was clamped in a fixture mounted on a turntable in the vacuum chamber. The beam was aligned for weld B while the chamber was being evacuated. Four tack welds were placed at 90° intervals, between holes. The part was moved sidewise, the beam was aligned for weld A, and eight equally spaced tack welds were deposited. Weld A was then completed in one revolution, using beam oscillation and power downslope control. Machine settings and other conditions for these operations are given in the table below the figure.

For the final weld at B, the vacuum was released and the beam realigned. For making this difficult weld, part of the clamping fixture was removed to facilitate the positioning of copper inserts in the holes drilled close to the joint and to fit a copper ring over the inserts. This precaution was necessary to prevent burn-though in areas where the holes were within 0.76 mm (0.030 in.) of the joint. A gear change was also required in the turntable drive to maintain a high welding speed on the smaller circle. After pumpdown, the weld was completed in a single pass, using the same oscillation and slope techniques that had been used for weld A. However, as shown in the table below the figure, other machine settings were different. Despite the copper chills, it was necessary to use lower amperage for better heat control. Weld penetration was

Mechanical Properties of EB Weld Produced in 90 mm (3.5 in.) Thick Ti-6Al-4V ELI Plate. Transverse weld-oriented round tensile specimens, ASTM E8 type, 35 mm (1.3 in.) gage length sectioned from the top (1/4), center (1/2) and bottom (3/4) of the weld thickness were tested in the as-welded condition and after PWHT at 650 °C (1200 °F)/5 h. All specimens failed in the base metal at tensile strengths from 855 to 860 MPa (124 to 125 ksi). All-weld-metal tensile specimens were tested using a JISZ2201 No. 4 specimen design, 14 mm (0.5 in.) gage length. Bend specimens 20 mm (0.78 in.) wide and 10 mm (0.4 in.) thick were made to JISZ3122, No. 3. Charpy impact specimens were made to JISZ220Z, No. 4 and tested at 0 °C (see table on next page for test results).

Ti-6Al-4V: Typical parameters and conditions for electron beam welding

Material thickness, in.	Accelerating voltage, kV	Beam current, mA	Travel rate, in./min
0.125	20	95	30
0.20	28	170	98
0.66	45	275	45
1.0	36.5	375	44
2.0	45	450	26
3.5	90.0	500	20

Source: *Metals and Their Weldability*, AWS Handbook, Vol 4, 7th ed., American Welding Society, 1982; R.H. Witt, J.G. Maciora, and H.P. Ellison, Sliding Seal Electron Beam Welding of Titanium, *Source Book on Electron Beam and Laser Welding*, M.M. Schwartz, Ed., American Society for Metals, 1981; and R. Sasano, K. Fuchigami, T. Toyohara and G. Takano, *Mitsubishi Tech. Rev.*, 1983, p 237

Ti-6Al-4V: Comparison of two joints welded by EBW

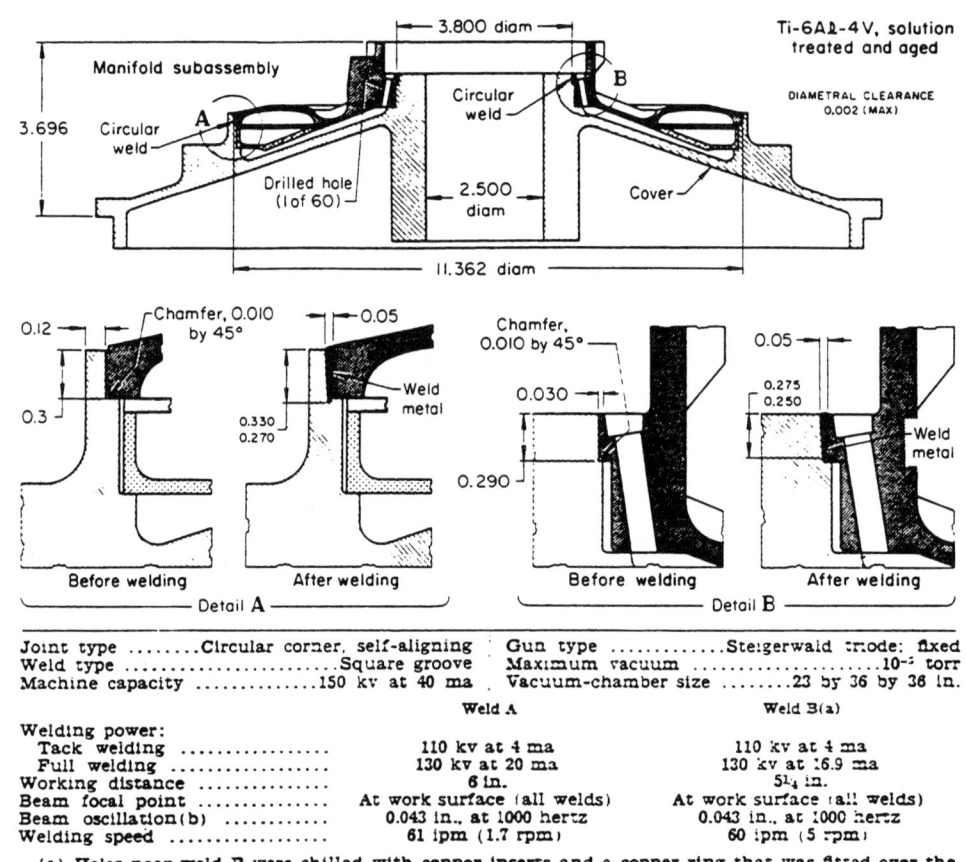

Manifold subassembly of a rocket engine fuel injector showing the two major joints. Weld shown in detail B was especially critical because of proximity to drilled holes.

Ti-6Al-4V: Electron beam weldment properties from 90 mm plate

Specimen type	Joint tensile strength MPa	ksi	Weld metal proof strength MPa	ksi	Weld metal tensile strength MPa	ksi	Elongation, %	Side bend	Face bend	Root bend	Impact toughness J	ft · lbf
1/4t-AW	856	124.2	882	128.0	1010	146.5	14	P	P	P	53.3	39.3
1/2t-AW	856	124.2	867	125.8	1000	145.1	12.5	P	P	P	39.3	29.0
3/4t-AW	858	124.5	858	124.5	990	143.7	12	P	P	P	37.0	27.3
1/4t-HT	866	125.7	897	130.2	980	142.2	13	P	P	P	23.9	17.7
1/2t-HT	853	123.7	909	131.9	980	142.2	13	P	P	P	23.9	17.7
3/4t-HT	862	125.0	897	130.2	980	142.2	13	P	P	P	22.9	16.9

Note: AW, as welded; HT, postweld heat treated; P, good to $8tR \times 180°$. Alloy chemistry: 6.25% Al, 4.15% V, 0.119% O, 0.0025% N, 0.012% C, 0.0025% H, and 0.2% Fe. Base metal properties: 0.2% proof stress; 834 MPa (121 ksi), UTS, 905 MPa (131.3 ksi); elongation, 14%. Welding parameters: 90 kV, 500 mA, 320 mm work-to-material distance, 410 mm/min, single-pass, no oscillation. Source: R. Sasano, K. Fuchigami, T. Toyohara, and G. Takano, "Fundamental Study of Electron Beam Welding of Heavy Thickness 6Al-4V Titanium Alloy," *Mitsubishi Heavy Ind. Tech. Rev.*, Vol 20 (No. 3), 1983, p 237-243

Laser Beam

During the past decade, laser beam welding (LBW) has become increasingly competitive as a joining process for titanium sheet and thin plate. In laser welding, melting of the workpiece is produced by the impingement of a high-intensity, coherent beam of light. Because the laser beam can be readily transmitted through air, the LBW process offers significant practical advantages over conventional electron beam welding, which requires welding in a high vacuum. However, the LBW process is more limited than EBW from the standpoint of joining thick titanium plate, with 15 kW required to produce a full penetration weld in 12.7 mm (0.5 in.) thick Ti-6Al-4V. As expected, considering similarities in the heat sources, the metallurgical and mechanical properties of laser and electron beam welds are quite comparable. Typical parameters for laser beam welding are given (see table).

Ti-6Al-4V: Typical parameters and conditions for laser beam welding

Material thickness, in.	Laser beam power(a), kW	Travel rate, in./min
0.140	5.5	60-70
0.230	5.5	50-60
0.50	13.0	NA

(a) CO_2 laser, continuous power. Source: E.B. Breinan, C.M. Banas, and M.A. Greenfield, Laser Welding—The Present State-of-the-Art, *Source Book on Electron Beam and Laser Welding*, M.M. Schwartz, Ed., American Society for Metals, 1981, p 247; and P. Denny and E.A. Metzbower, Laser Beam Welding of Titanium, *Weld. J.*, Vol 68 (No. 8), 1989, p 342s

Resistance Welding

Resistance welding is a process in which heat is produced by resistance to the flow of high electrical current (5 to 10 kA) across the interface between two contacting surfaces. At sufficiently high current levels, melting is induced at the interface and a weld fusion zone (i.e., weld nugget) is created. At lower current levels, welding of the faying surfaces occurs entirely by a solid-state deformation or diffusion welding mechanism. Resistance welding is most commonly used to join sheet and can produce either discrete, individual spot welds, or continuous overlapping seam welds. Due to similarities in thermal conductivity, the resistance welding pa-

Ti-6Al-4V: Typical parameters and conditions for resistance spot welding

Material thickness, in.	Welding current, kA	Electrode force(a), lb	Weld time, cycles (60 Hz)
0.035	5.5	600	7
0.062	10.6	1500	10
0.093	12.5	2400	16

(a) RWMA class 2 copper alloy electrodes, 3 in. face radius. Source: *Metals and Their Weldability, AWS Handbook*, Vol 4, 7th ed., American Welding Society

Ti-6Al-4V: Typical parameters for conventional resistance spot welding
Results of early parametric study on spot welding of Ti-6Al-4V sheet

Sheet thickness, in.	Electrode force, lb	Weld time, cycles	Weld current, A	Tension shear strength, lb	Weld diam, in.	Tension shear strength, psi
0.035	600	7	5,500	1720	0.255	34,400
0.062	1500	10	10,600	5000	0.359	50,000
0.070	1700	12	11,500	6350	0.391	52,916
0.093	2400	16	12,500	8400	0.431	57,534

Note: Alloy chemistry: 6.1-6.2% Al, 4.1% V, 0.12-0.16% Fe, 0.009% N, 0.009-0.012% H; oxygen analysis not provided, although was likely somewhat higher than current (circa 1990s) alloy. Sheet provided in mill annealed condition. Welds in 0.035 in. thick sheet produced using 30 kV · A, 60 cycle single-phase 30 in. rocker arm type of spot welding machine. Welds in 0.062 in. thickness and greater produced in 600 kV · A machine. Welds were produced with 5/8 in. diam class 2 copper alloy electrodes, 3 in. spherical radius. Source: R.K. Nolen, J.F. Rudy, H. Schwartzbart, and H.D. Kessler, Spot Welding of Ti-6Al-4V Alloy, *Weld. J.*, Apr 1958, p 129s

Diffusion Weld

Diffusion welding is a solid-state joining process that is finding increased application for the fabrication of complex Ti-6Al-4V components. Diffusion welding (or bonding) of titanium involves four basic steps: (1) development of intimate physical contact between the faying surfaces through deformation of asperity peaks (i.e, surface roughness) by yielding and creep at low pressures and high temperatures; (2) formation of a metallic bond; (3) diffusion across the faying surface; and (4) grain growth across the original weld interface. Prerequisites for accomplishing these steps include a clean and smooth surface combined with a low applied pressure to promote steps 1 and 2 and moderate-to-high temperatures to promote steps 3 and 4. Titanium alloys are particularly well suited for diffusion welding because titanium exhibits a high solubility for oxygen, which promotes dissolution of surface oxides at elevated temperatures. In addition, the yield and creep strengths of titanium are low at the temperatures used for welding, thereby enhancing stages 1 and 2 of the diffusion welding process. Diffusion welding of Ti-6Al-4V generally is performed at temperatures ranging from about 900 to 950 °C (1650 to 1740 °F), pressures ranging from 1.38 to 13.8 MPa (200 to 2000 psi), and for times ranging from about 1 to 6 h. In combination with superplastic forming, diffusion welding has been used to produce a wide variety of complex Ti-6Al-4V shapes such as honeycomb panels for aerospace applications.

Special Methods

A wide variety of additional, more specialized welding processes have been used to join Ti-6Al-4V, including flash-butt welding, continuous-drive and inertia friction welding, ultrasonic welding, and explosive welding. In general, these processes effectively produce defect-free joints, with properties that depend primarily on the cooling rates from high temperatures and the characteristics of the resulting transformed β microstructure.

Production of a Titanium Ring by Flash Welding. The ring shown in the accompanying figure was used after machining as a jet engine inlet case seal. The workpiece was an AMS 4935 (titanium alloy Ti-6Al-4V) extrusion with a hardness of 34 HRC, tensile strength of 999 MPa (145 ksi), and 14% elongation. The welded joint had the same hardness, but a tensile strength of 1047 MPa (151.9 ksi) and elongation of 12.5%. Thus, the joint efficiency was 105%. The welded area was about 322 mm^2 (0.5 in.^{2}).

The joint was made in a 750 kV · A, single-phase 440 V flash welding machine with a maximum secondary output rating of 7500 A at 16.9 V. Hydraulically actuated clamps 75 mm (3 in.) in length and made of high-conductivity copper (RWMA class 2), held the roll-formed ring during welding. The movable platen was also hydraulically actuated.

The extruded section was cut to length, heated to 675 °C (1250 °F), and then roll formed into a circle in a horizontal three-roll forming machine. To ensure maximum current flow, the contact surfaces of the workpiece at the clamping dies were polished before the ring was placed in the welding machine. After welding, the weld upset was removed, the weld area was ground manually to the same dimensions as the adjoining area, and the joint was inspected for alignment and for defects. Then, the welded rings were heat treated in accordance with AMS 4928.

The joint was proof tested by placing the welded ring in a hydraulically actuated expander and applying uniform pressure to all sections of the ring simultaneously. The circumference of the ring was increased about 50 mm (2 in.), or 1¼% during this operation, which also made the ring round and sized the inside diameter to specification.

One ring from each production lot was used for destructive tests on the base metal and the welded joint. The tests were conducted according to AMS 7948, which covers flash welding of titanium alloys.

Ti-6Al-4V: Flash welded ring

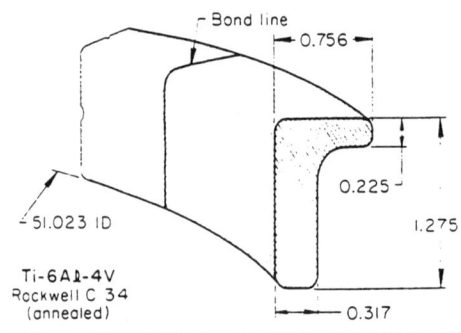

```
Initial die opening .......................2⅛ in.
Final die opening ........................¾ in.
Metal lost ..............................1⅜ in.
Preheat time ...........................10 sec
Flashing time ..........................10 sec
```

Weldment Microstructures

W.A. Baeslack III, Ohio State University

Mechanical properties of composite weld structures in titanium alloys depend on the structural characteristics of each weld region, which in turn depend on the specific thermal cycle(s) imposed during welding and on subsequent postweld heat treatment. The key microstructural features from the welding process are (1) prior β grain size in the fusion zone and (2) the manner in which the high-temperature β phase transforms during cooling. The weld microstructure and mechanical properties may also be influenced by postweld heat treatment, with specific postweld heat treatment effects depending on heat treatment time and temperature and on the as-welded microstructure.

Prior β Grain Size in Fusion Zone. The weld fusion zone in titanium alloys is characterized by coarse, columnar prior β grains that originate during weld solidification. The size and morphology of these grains depend on the nature of the heat flow that occurs during weld solidification. Beta grain size depends primarily on the weld energy input, with higher energy promoting larger grain sizes.

Due to epitaxial grain growth, the fusion zone β grain size may also depend on the β grain size in the near heat-affected zone directly adjacent to the fusion line. This latter effect of base metal grain size is most significant in the welding of extremely coarse-grain cast or beta annealed alloys (see figure). Because weld mechanical properties, particularly ductility, can be degraded by a coarse prior β grain size, it is important to maintain as fine a grain structure as possible by minimizing the weld energy input.

Appreciable β grain growth in the near heat-affected zone occurs directly adjacent to the weld fusion line, where peak temperatures range from between the alloy solidus down to the β transus (approximately 995 °C or 1825 °F) for Ti-6Al-4V. As in the fusion zone, the extent of this growth increases with energy input into the weld zone. Consequently, this region can vary markedly in width, being almost unresolvable in electron beam and laser welds and yet being several β grains wide in gas tungsten-arc welds (see figure). Farther from the fusion line, temperatures below the β transus are encountered, which promotes transformation of α phase in the mill annealed microstructure to various proportions of the high-temperature β phase. The presence of even small quantities of α phase at peak temperatures in the weld thermal cycle is sufficient to prevent β grain growth, thereby contributing to the improved ductility of this region compared to the coarser grain fusion zone and near heat-affected zone.

Ti-6Al-4V: Fusion zone in β-processed forging

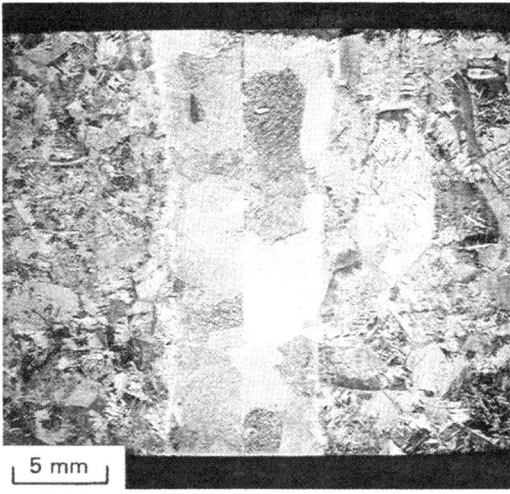

Light macrograph of electron beam weld produced in 25 mm (1.0 in.) thick β-processed forging.
Source: Courtesy of M. Scott, The Welding Institute, Cambridge

Ti-6Al-4V: GTAW weldment

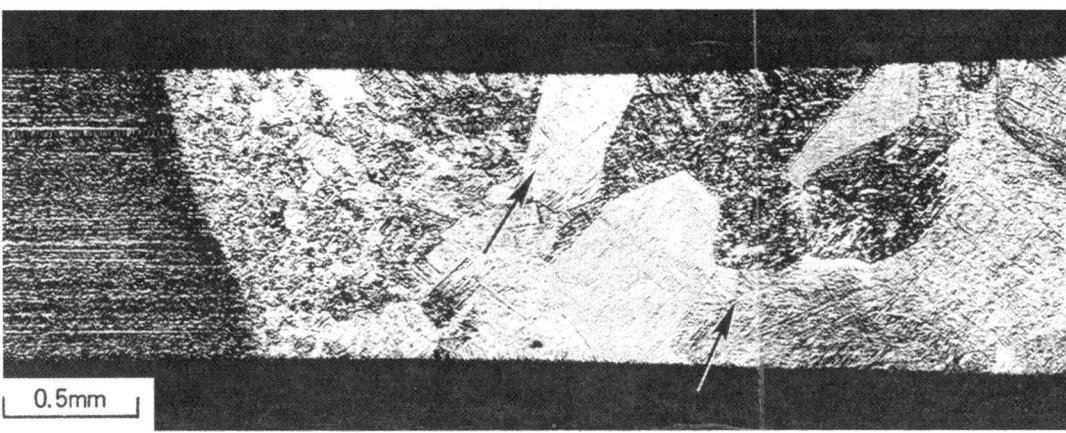

Light macrograph of gas tungsten-arc weld produced in 1.2 mm (0.04 in.) thick sheet, 90 A, 10 V, 3.4 mm/s travel rate, argon shielding. Arrows indicate approximate fusion line.
Source: Courtesy of M. Scott, The Welding Institute, Cambridge

Transformation Products

In addition to prior β grain size, weld zone mechanical properties in Ti-6Al-4V are significantly influenced by the manner in which the high-temperature, bcc β phase transforms on cooling to the low-temperature cph phase. Characteristics of this transformed β microstructure depend primarily on the cooling rate from above the β transus temperature, which is correspondingly influenced by the type of welding process, process parameters, and other welding conditions (such as workpiece geometry and fixturing).

In this context, Ti-6Al-4V is considered to be highly weldable relative to most structural titanium alloys. This weldability can be attributed to two principal factors. First, the α′ martensite that forms in Ti-6Al-4V is not as hard and brittle as that exhibited by more heavily β-stabilized α-β alloys, such as Ti-6Al-6V-2Sn. Secondly, Ti-6Al-4V exhibits relatively low hardenability, which allows the formation of higher proportions of the more desirable Widmanstätten α plus retained β microstructure even at relatively high weld cooling rates.

In the fusion zone and near heat-affected zone regions, high cooling rates associated with low energy input welding processes such as laser, electron beam, and resistance welding (100 to 10,000 °C/s, or 180 to 18,000 °F/s), promote transformation of β to α′ martensite. This extremely fine, acicular transformation product exhibits high strength and hardness, but relatively low ductility and toughness. At the lower cooling rates associated with gas tungsten-arc or plasma-arc welding (10 to 100 °C/s, or 18 to 180 °F/s), a coarser structure of Widmanstätten α plus retained β, or a mixture of this structure and α′, is produced, which exhibits yield and tensile strengths superior to those of the mill annealed base metal and ductility and toughness greater than those of an entirely martensitic microstructure. In the far heat-affected zone, microstructures are comprised of primary α phase originating from the base metal microstructure in a matrix of transformed β.

Postweld heat treatments at relatively low temperatures from 450 to 600 °C (750 to 1110 °F), which are commonly used for stress relief, age the martensitic structure and can further increase strength and decrease ductility and toughness. At higher postweld heat treatment temperatures, which are practically limited to about 730 °C (1345 °F) for large fabricated structures, aging of the martensite and subsequent coarsening of the resultant α phase promote softening of the fusion zone with concurrent increases in ductility and toughness. Microstructures comprised of Widmanstätten α plus retained β, which are produced at lower weld cooling rates, are influenced to a lesser extent by postweld heat treatment at these low to moderate temperatures. However, welds exhibiting these microstructures are still normally postweld heat treated to relieve residual stresses.

Fusion Zone β

The size and morphology of β grains depends on the nature of heat flow. Under simple, uniaxial heat flow, such as occurs in a spot weld, the β grains nucleate epitaxially on β grains in the base metal substrates and solidify preferentially in a direction parallel to the maximum temperature gradient (i.e. parallel to the welding electrodes) until they impinge at a horizontal weld centerline (see

Ti-6Al-4V: Prior β from two-dimensional heat flow

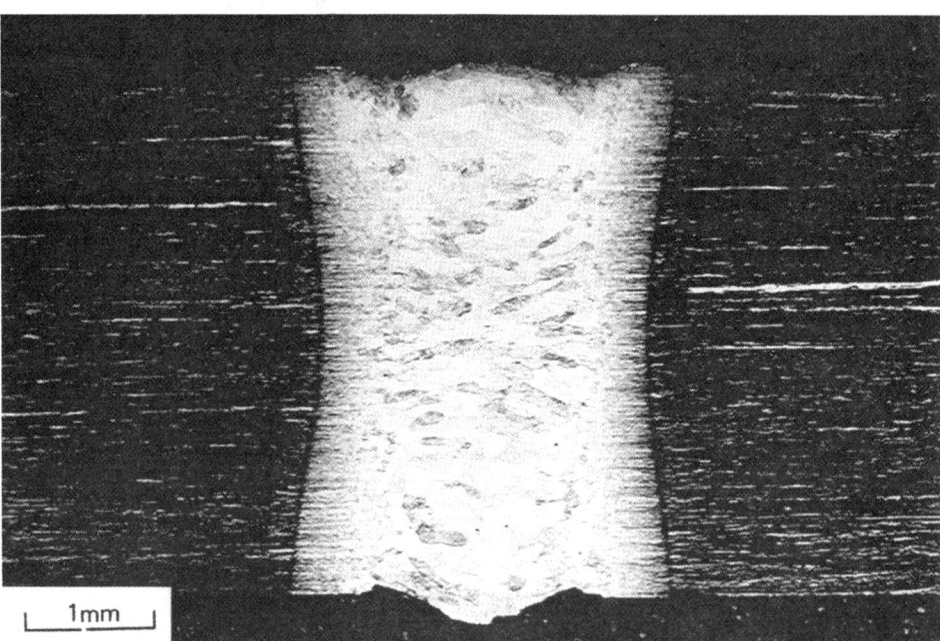

Light macrographs of laser welds produced in 4.0 mm (0.15 in.) thick Ti-6Al-4V continuous-wave, CO_2 laser operated at 5.0 kW; 33 mm/s travel rate.
Source: C.J. Dawes, "Welding of Titanium Alloy by Laser," Welding Institute Final Report for Royal Aircraft Establishment, 1980

figure on next page). Under two-dimensional heat flow conditions characteristic of full-penetration plasma-arc, laser beam, and electron beam welds, the columnar β grains solidify inward from the base metal in a direction nearly parallel to the sheet or plate surface, ultimately impinging to form a vertical grain boundary at the weld centerline (see figure on previous page). Finally, three-dimensional or mixed two-dimensional/three-dimensional heat flow conditions, such as those present in single-pass and multi-pass tungsten-arc and gas metal-arc weldments, promote the formation of more complex, multidirectional β grain morphologies.

Ti-6Al-4V: Prior β from uniaxial heating

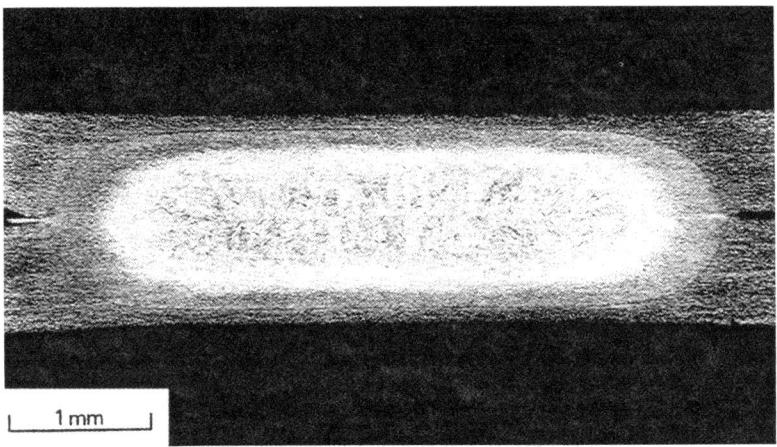

Light macrograph of capacitor discharge spot weld in 1.2 mm (0.04 in.) thick sheet.

Weldment Fatigue

Consistent with its excellent weldability, the mechanical properties of Ti-6Al-4V welds generally compare extremely well with those of the mill annealed base material. The high cooling rates encountered in the fusion and HAZ regions of Ti-6Al-4V weldments promote the formation of fine martensitic or Widmanstätten α microstructures. These microstructures are stronger than the equiaxed α + β microstructures exhibited by the mill annealed base metal, and consequently, fracture of a transverse-weld-oriented tensile specimen occurs in the unaffected base metal (i.e., joint efficiency of 100%). Conversely, the ductility of the weld zone is generally less than that of the mill annealed base metal, which is consistent with the inherently lower ductility of the β-processed versus α-β processed titanium. Although the lower weld zone ductility is not revealed by transverse weld oriented tensile testing, it can be measured by testing longitudinal weld-oriented and all-weld metal tensile or bend specimens. The strength and ductility of weldments produced in Ti-6Al-4V that has been solution heat treated and aged to high strength levels are influenced more by the weld zone structure and properties. The completely martensitic microstructures exhibited in electron beam and laser beam welds with high depth-to-width ratios are sufficiently strong to promote base metal fracture in the solution heat treated and aged base metal. However, the softer Widmanstätten plus retained β microstructures exhibited by the more slowly cooled GTAW and PAW welds may promote preferential fracture in the weld zone at lower macroscopic elongation.

In the absence of weld defects and discontinuities, the generally higher yield and tensile strengths of the weld zone versus the mill annealed base metal promote equivalent axial fatigue properties, with transverse weld-oriented specimens failing in the unaffected base metal adjacent to the weld zone. In practice, lower axial fatigue properties commonly are observed, particularly in welds made with high energy density. This behavior is attributed to the presence of fine porosity, often unresolvable by radiography, and of other defects such as undercuts or underfills in fusion welds, or lack of bonding regions at the weld interface in solid-state diffusion or friction welds. As with tensile properties, the axial fatigue behavior of welds produced in solution heat treated and aged base metals depends on both the weld integrity and the strength of the weld zone.

Poor axial fatigue properties also can limit the application of lap joints produced in titanium using resistance, diffusion, and ultrasonic welding. In the case of resistance welds, the mechanical notch at the perimeter of the weld nugget, the presence of a brittle martensitic microstructure, and residual stresses can promote early fatigue-crack initiation and low fatigue life. Although the metallurgical structure and residual stresses can be modified through process control, the presence of a notch is inherent to a lap joint and subject to less modification.

Ti-6Al-4V: Fatigue curves for resistance spot welds

Comparison of standard as-welded and postweld processed resistance spot welds in 2.3 mm (0.090 in.) thickness. Material was mill annealed. Each stress is the log average of six specimens.

A conventional 400 kV · A frequency converter spot welding machine was used with phase shift control. Plot compares data for welds made in conventional manner with those using: (1) lower weld current density by increasing weld force, increasing electrode radius and drilling a hole in center of electrode, (2) preheating the weld to reduce temperature gradients, (3) postweld coining to reduce internal stresses, and (4) shot peening to create surface compressive stresses. Tension-tension fatigue tests were conducted on lap specimens at a speed of 1800 cycles/min with a stress ratio of 0.06. Two-spot overlap fatigue coupon, 25 mm (1 in.) wide, 25 mm (1 in.) center-to-center spot spacing.
Source: R.D. Beemer, High Fatigue Life Results from Titanium Spot Welding Innovations, *Weld. J.*, Mar 1970, p 89s

The fracture toughness of Ti-6Al-4V welds depends on both the fusion zone microstructure and interstitial content. The martensitic microstructure characteristic of high depth-to-width ratio electron beam and laser welds is characterized by toughness below that of the mill annealed base metal. Measuring the toughness of discrete weld regions is often difficult, because the crack has a strong tendency to propagate out of that region into the base metal. Although this effect can be reduced by stress relief prior to fatigue precracking, the actual crack path should always be determined and considered in evaluations of toughness and fatigue-crack growth rate data in welded specimens. Welds produced at lower cooling rates exhibit increasing toughness levels comparable or superior to that of the mill annealed base metal. Fracture toughness is also markedly influenced by the interstitial content of the weld metal. The use of extra-low interstitial (ELI) grades of Ti-6Al-4V base metal and consumables are recommended for fracture toughness critical applications.

Finally, fatigue-crack growth behavior of the weld zone has been shown to be superior to that of the mill annealed base metal. This observation parallels the higher fatigue-crack propagation resistance of β annealed versus α-β processed Ti-6Al-4V and is attributed to the greater crack path tortuosity induced by coarse prior β grain structure. As indicated above, fatigue-crack propagation rates can be significantly influenced by residual stresses, and this effect should be considered in evaluations of crack propagation rate data.

LBW Fatigue

Ti-6Al-4V: S/N curves for laser beam weld (weld reinforcement intact)

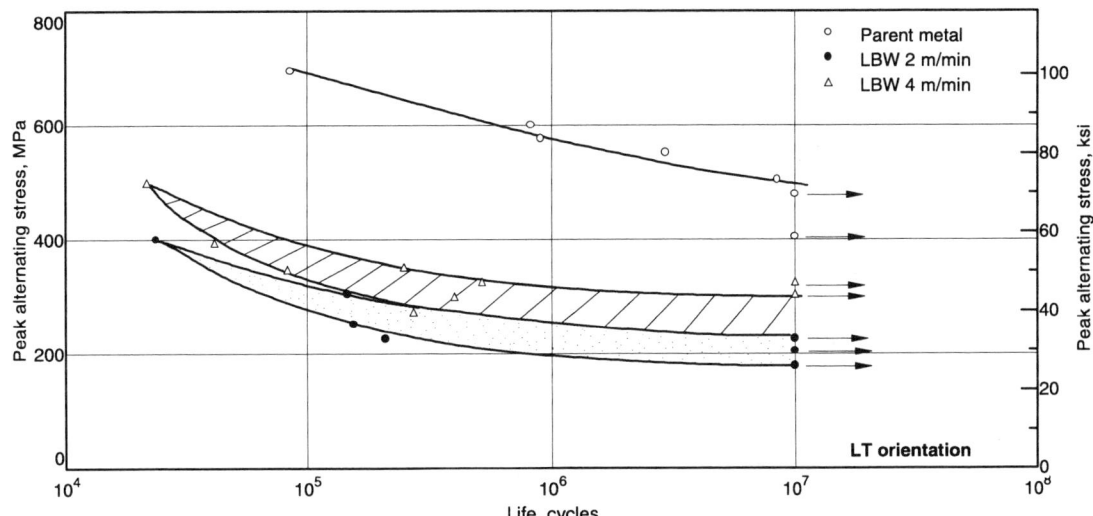

Laser beam welds in 4 mm (0.15 in.) sheet produced at 2 and 4 m/min. Tested in as-welded condition. Fracture initiated at weld undercut. Base metal properties are provided for comparative purposes.
Mill annealed 4 mm (0.15 in.) thick Ti-6Al-4V sheet. Welding parameters: 5 kW continuous CO_2 laser, welds produced at 2 and 4 direction normal to sheet rolling direction. 150 mm focal length, spot diameter of 0.2 mm. Fatigue testing: Standard, transverse w specimen, details not provided. $R = 0.1$. Laser welded specimens tested in as-welded condition and after removal of undercut and
Source: T.S. Baker and P.G. Partridge, "Fatigue Properties of 4 mm Thick Laser Welded Ti-6Al-4V Sheet," *Fatigue '84*, C.J. Beev neering Materials Advisory Services Ltd., London, 1984

Ti-6Al-4V: S/N curves for laser beam weld (weld reinforcement removed)

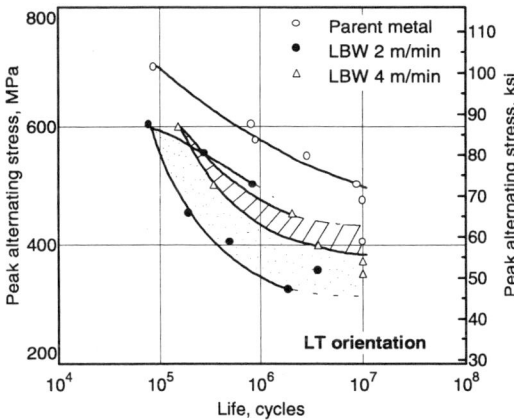

Laser beam welds in 4 mm (0.15 in.) sheet produced at 2 and 4 m/min. Tested following removal of weld reinforcement and undercut. Fracture initiated at pores in fusion zone. Base metal properties are provided for comparative purposes. $R = 0.1$
Source: T.S. Baker and P.G. Partridge, "Fatigue Properties of 4 mm Thick Laser Welded Ti-6Al-4V Sheet," *Fatigue '84*, C.J. Beevers, Ed., Engineering Materials Advisory Services Ld., London, 1984

EBW Fatigue

Ti-6Al-4V: S/N curve for 25 mm thick base plate

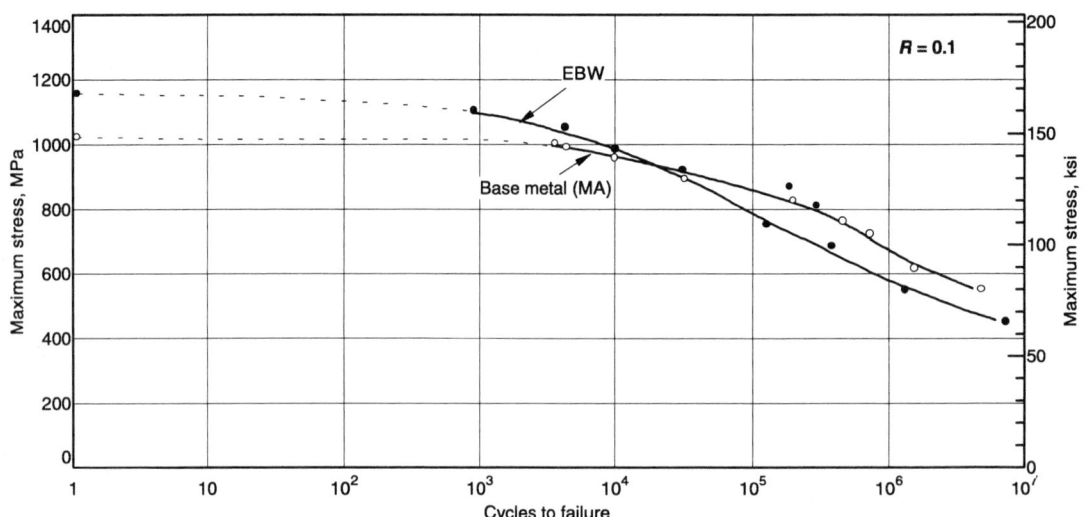

Axial fatigue data for transverse weld-oriented electron beam welded specimens in 25 mm (1 in.) thick plate.
Note that the higher maximum stress of the welded specimens at low cycles to failure resulted from the higher strength of the weld metal in the reduced region of the fatigue specimen. Of the ten welded specimens, eight failed outside the minimum section in the base metal. Specimen design: Round specimen 648 mm (6.0 in.) in length, 57 mm (2.25 in.) tapered gage section from 19 to 6.4 mm (0.75 to 0.25 in.) at center. 16 rms finish, weld at center ±0.25 mm (±0.010 in.). Alloy chemistry: 6.48% Al, 4.29% V, 0.15% O_2, 0.016% N, 0.023% C, 0.008% H, 0.13% Fe. Welding parameters: Penetration pass: 36.5 kV, 375 A, 44 in./min, 178 mm (7 in.) gun-to-work distance; wash (cover) pass –25 kV, 200 mA, 44 in./min, 7 in. gun-to-work distance.
Source: J.G. Bjeletich, "Development of Engineering Data on Thick-Section Electron Beam Welded Titanium," AFML-TR-73-197, 1973

Crack Growth

Ti-6Al-4V: PAW crack growth

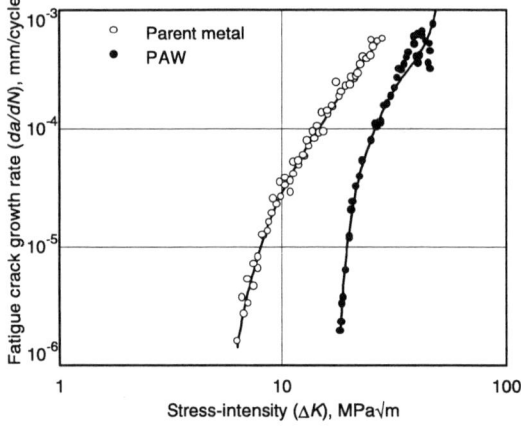

Fatigue crack growth rate curves for plasma-arc welded and parent metal Ti-6Al-4V sheet 4 mm (0.15 in.) in thickness. LT orientation, $R = 0.1$.
Welding parameters: 100 A, 1.3 mm/s; nonkeyhole type weld. Tensile specimens with transverse-oriented welds all failed in the unaffected base metal. Fatigue crack propagation was evaluated using SEN three-point bending specimens, $R = 0.1$. Welds were not stress relieved prior to fatigue testing; weld reinforcement and undercut were removed prior to testing. Note appreciably greater fatigue crack propagation rate in base metal than in weld fusion zone.
Source: T.S. Baker, "Fatigue Crack Propagation and Fracture Toughness of Plasma Arc Welded Ti-6Al-4V," Royal Aircraft Establishment, TR 85066, July, 1985

Ti-6Al-4V: Crack growth of various weldments

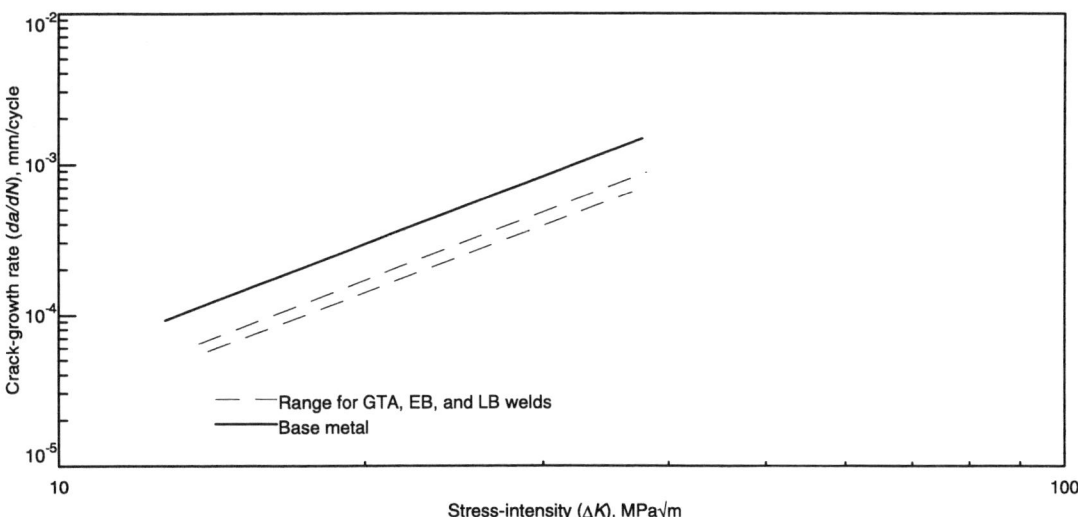

Mill annealed 3.18 mm (0.125 in.) sheet. Welding parameters: Full-penetration butt welds; details of welding parameters not provided. All welds were stress-relieved at 575 °C (1070 °F) 1 h. Tensile testing performed on transverse weld oriented specimens.

Ti-6Al-4V ELI: Crack growth rates

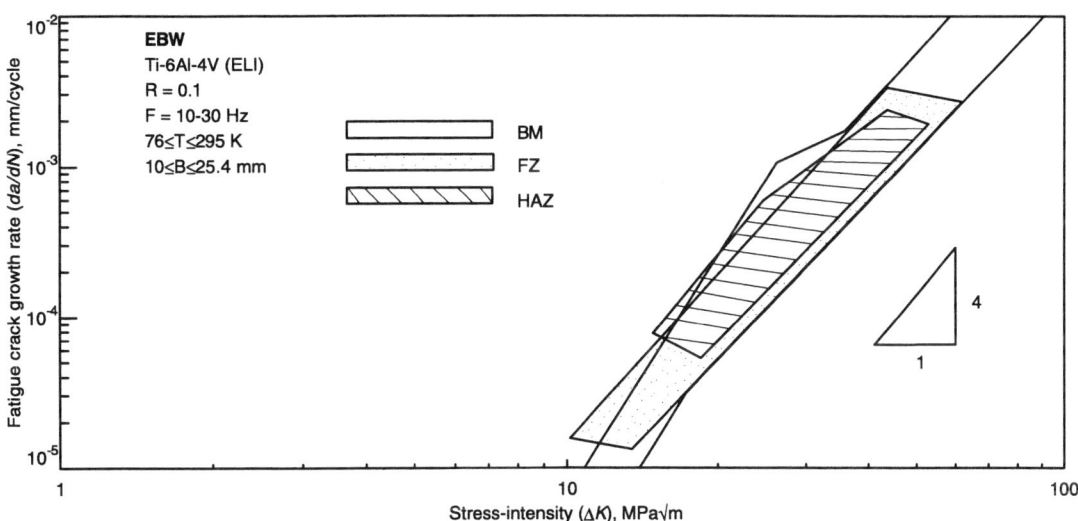

Base metal condition: ELI Ti-6Al-4V was in form of 1 m sphere with 50 mm (2 in.) wall thickness, recrystallization annealed at 925 °C (1700 °F), 4 h, furnace cooled to 760 °C (1400 °F) in 3 h, cooled to 480 °C (900 °F) in 45 min and cooled to room temperature. Base metal chemistry: 5.91% Al, 3.94% V, 0.103% Fe, 0.110% O, 0.018% C, 0.014% N and 5 ppm H. Electron beam welds were produced on arc-shaped bars machined from sphere, square-groove joints, self-backing, self-aligning joint preparation. Welding was performed using a tacking pass (115 kV, 15 mA, 4.8 mm/s), penetration pass (150 kV, 45 mA, 4.8 mm/s, focus) and a cosmetic pass (120 kV, 25 mA, 3.5 mm/s). All welds were produced at sharp focus. Welds were stress-relief heat treated at 540 °C (1000 °F), 5 h.
Source: R.L. Tobler, "Low Temperature Fracture Behavior of a Ti-6Al-4V Alloy and its Electron Beam Welds," *Toughness and Fracture Behavior of Titanium*, ASTM STP 651, ASTM, 1978, p 267-294

Casting

Unlike many other alloy systems, cast titanium products do not necessarily have inferior mechanical properties compared to wrought titanium products. In fact, Ti-6Al-4V casting properties generally are comparable to their wrought counterparts (though generally somewhat lower in the annealed condition), and in some instances, properties such as creep resistance and resistance to crack propagation are superior.

Ti-6Al-4V accounts for 85% of all titanium castings produced worldwide. Ti-6Al-4V castings continue to gain ground in market share against Ti-

6Al-4V wrought products because of their economical overall cost for complex parts, even with the additional cost of hot isostatic pressing, which is typically necessary to attain maximum properties. Suppliers of titanium castings must contend with the high reactivity of titanium at elevated temperatures, the rapid solidification of the liquid metal, and the absence of customary eutectics. Special, and often expensive, methods of melting, moldmaking, and surface cleaning may be required to maintain product integrity (see "Technical Note 3: Casting" for general information on the casting of titanium alloys).

Mechanical Properties. The mechanical properties of castings are influenced by the cooling rates of the casting process, subsequent heat treatments, oxygen or hydrogen impurity levels, and subsequent HIP procedures. Mechanical properties also depend on whether the test specimen is separately cast or machined from a casting. The effect of casting and pouring methods on mechanical properties can be attributed to their effects on cooling rates. For example, the difference between rammed graphite and investment castings (see table) is influenced by the more massive graphite mold, which would make a better heat sink than a ceramic die. Centrifugal or static pouring can also affect cooling rates. Statically poured castings will generally have lower strength, because the heated molds result in lower cooling rates and a coarser structure.

Differences in mechanical properties are due to casting geometry and the size and location of test specimens. Large thick castings will tend to have low cooling rates that will produce large β grains and in some cases large α plate colonies. This kind

Ti-6Al-4V: Wrought and cast property comparison

Material condition	Heat treatment	Tensile yield strength MPa	ksi	Ultimate tensile strength MPa	ksi	Elongation, %	Reduction of area, %	Impact toughness J	ft·lbf	K_{Ic} MPa√m	ksi√in.
Wrought											
Mill annealed	...	875	127	965	140	13	25	...	...	...	...
Annealed	...	875	127	970	141	16	27	27	20	52	47
Beta annealed	...	860	125	955	139	9	21	...	...	91	83
Cast											
As cast	...	895	130	1000	145	8	16	...	...	107	97
Annealed	...	825	120	930	135	12	22	17.5	24	103	94
Annealed	730 or 845 °C anneal	825-855	120-124	895-930	130-135	6-10	10-15	...	...	...	...
α-β solution + aged	955 °C, 1 h, cool + 620 °C, 2 h	855-900	124-131	935-970	136-141	5-8	6-14	...	...	...	...
β solution + aged	1025 °C, 1 h, cool + 620 °C, 2 h	860-925	125-134	965-1025	140-149	5-8	10-14	...	...	...	...
Cast HIP	...	870	130	1000	145	8	16	...	...	109	99

Ti-6Al-4V: Effect of casting method on tensile properties

Specimen	Tensile yield strength (0.2% offset) MPa	ksi	Ultimate tensile strength MPa	ksi	Elongation, %	Reduction of area, %
25 mm diam bar cast in rammed graphite						
0.18 oxygen	874	126.7	1010	146.5	11	17
0.11 oxygen	893	129.5	1008	146.2	9	16
0.09 oxygen	818	118.6	940	136.4	11	22
19 × 50 × 200 mm bar (0.15% O), investment cast	834	121.0	917	133.0	7	12
Cast propeller (0.17% O)	840	121.9	968	140.4	9	16

(a) 13.8 mm (0.505 in.) diameter specimen tested at a strain rate of 0.002/s to yielding. Source: A. Morton and I. Lane, Jr., Titanium Castings for Marine Propellers, *Titanium Science and Technology*, Vol 1, R. Jaffe and H. Burte, Ed., Plenum Press, 1973, p 119-130

Ti-6Al-4V: Tensile properties of specimens

Specimen type	Ultimate tensile strength MPa	ksi	Tensile yield strength MPa	ksi	Elongation, %	Reduction of area, %
Specimen 1(a)						
Separately cast	907	131.6	821	119.1	10.6	20.8
Cast on part	886	128.5	810	117.5	10.1	22.2
Machined from casting	851	123.4	780	113.2	8.6	15.6
Specimen 2(b)						
Separately cast(c)	829	120.3	740	107.3	...	...
Cast on part	821	119.1	738	107.1	...	...
Machined from casting(d)	792	114.9	724	105.0	...	...

(a) HIP + annealed casting described in *Titanium: A Technical Guide*, ASM International, 1988, p 112. (b) Data from a single casting source. (c) From 312 tests. (d) From 229 tests

of microstructure is beneficial for fracture toughness, creep resistance, and fatigue crack propagation resistance and detrimental to low- and high-cycle fatigue strength and tensile elongation properties.

Thermal and Thermomechanical Treatments. Although solution treated and aged castings are produced for enhanced tensile or fatigue strength, most Ti-6Al-4V castings are supplied in the annealed condition. Other processing options (see table) are also used to improve the fatigue performance of Ti-6Al-4V castings (see figure).

Ti-6Al-4V: Effect of thermal treatments on tensile properties

Method(a)	Solution treatment(b)	Annealing or aging treatment	Tensile yield strength MPa	ksi	Ultimate tensile strength MPa	ksi	Elongation, %	Reduction of area, %
Broken up structure (BUS)	1040 °C (1900 °F), 1/2 h	845 (1550 °F), 24 h	938	136	1041	151	8	12
Garrett treatment (GTEC)	1050 °C (1925 °F), 1/2 h	845 °C (1550 °F), 1/2 h, 705 °C (1300 °F), 2 h	938	136	1027	149	8	11
Beta solution + age	1040 °C (1900 °F), 1/2 h, GFC	540 °C (1000 °F), 8 h	931	135	1055	153	9	15
Alpha-beta solution + age	955 °C (1750 °F), 1 h, GFC	540 °C (1000 °F), 8 h	931	135	1020	148	8	12
Cast and HIP	...	...	870	126	960	139	10	18

(a) All methods include HIP prior to heat treatment. (b) GFC, gas fan cooled. Source: *ASM Handbook*, Vol 2, 1990, p 641

Ti-6Al-4V: Effect of hydrogen treatments on cast tensile properties

Method(a)	Solution treatment	Hydrogenation temperature °C	°F	Intermediate treatment °C	°F	Dehydrogenation temperature °C	°F	Ultimate tensile strength MPa	ksi	Tensile yield strength MPa	ksi	Elongation, %
Cast + HIP	...	...	...	...	...	...	...	960	139	870	126	10
Hydrovac process	...	650	1200	870(b)	1600(c)	760	1400	...	...	...	...	...
Thermochemical treatment	1040 °C (1900 °F), 1/2 h	595	1100	Cool to RT		760	1400	1125	163	1055	153	6
Conventional solution treatment	...	870	1600	No intermediate step (continuous process)		815	1500	1055	153	985	143	8
High-temperature hydrogenation	...	900	1650	Cool to room temperature		705	1300	1103	160	1055	153	8

(a) All methods include HIP. (b) Glass encapsulated prior to heat treatment. Source: *ASM Handbook*, Vol 2, 1990, p 639, 641

Ti-6Al-4V: Smooth axial fatigue of casting

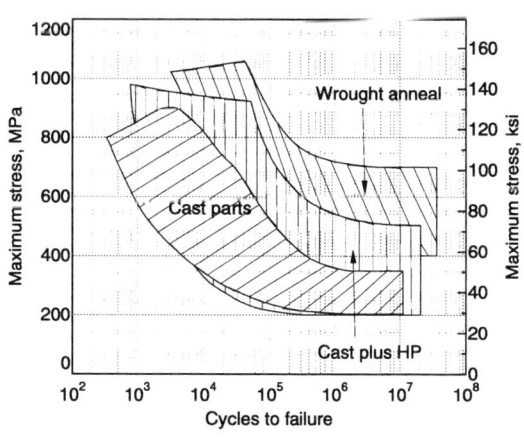
(a)

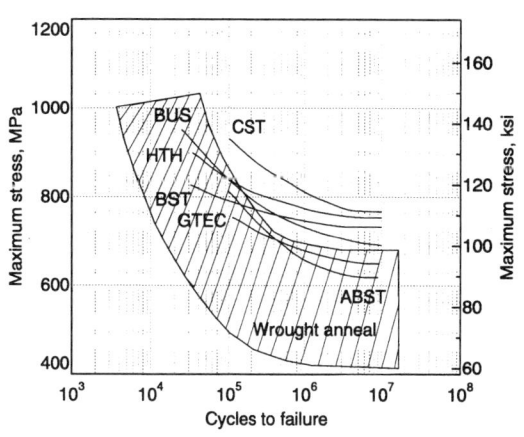
(b)

(a) Room-temperature fatigue after HIP. (b) Fatigue after HIP and various thermal and hydrogen treatments. R = 0.1. BUS, broken up structure; HTH, high-temperature hydrogenation; CST, constitutional solution treatment; GTEC, Garrett treatment; ABST, alpha-beta solution treatment and aging; BST, beta solution treatment and aging.
Source: *ASM Handbook*, Vol 2, 1990, p 642

Ti-6Al-4V: Tensile property distribution of HIP and annealed cast test bars

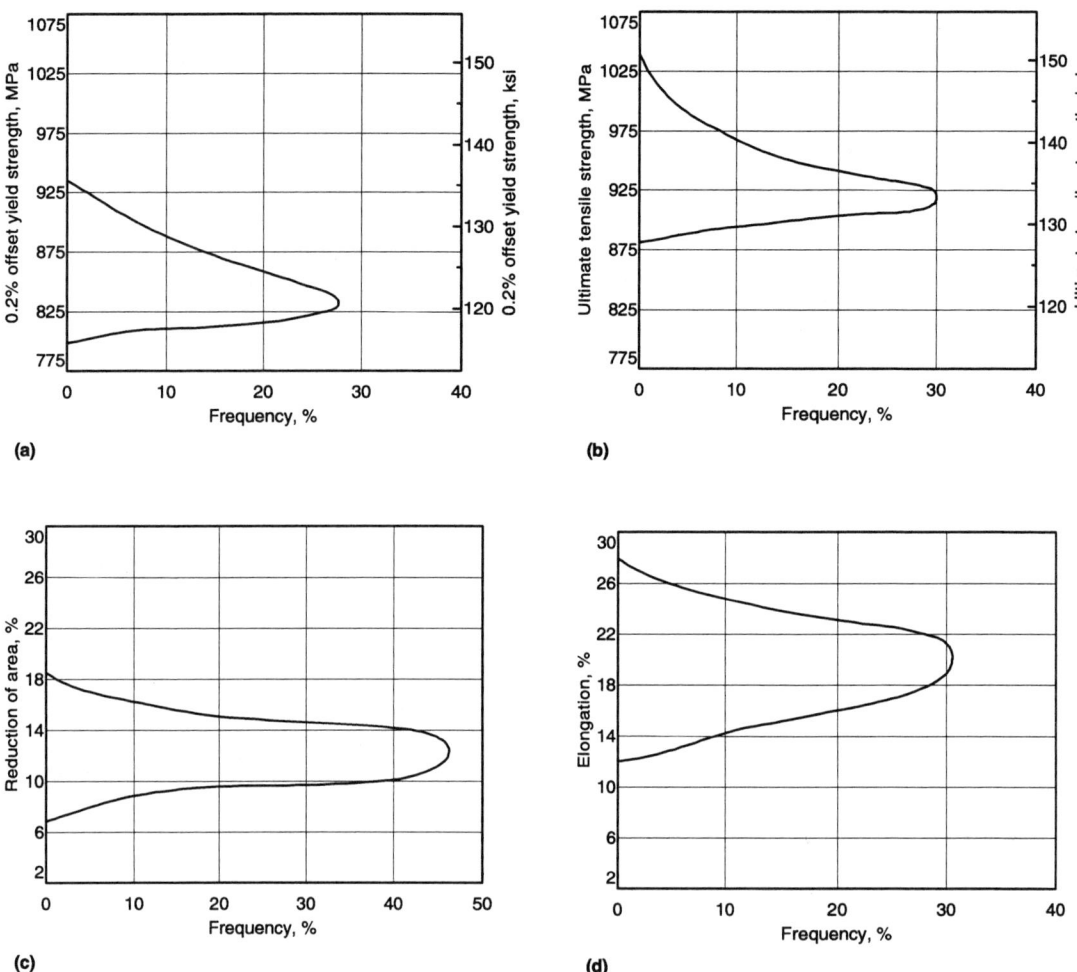

Sample size of 500 heats with oxygen from 0.16 to 0.20 wt%.
Source: TiTech International, Inc.

Ti-6Al-6V-2Sn

Common Name: Ti-662
UNS Number: R56620

Ti-6Al-6V-2Sn was developed at New York University on a U.S. Army contract as a higher-strength version of Ti-6Al-4V. It is a corrosion-resistant, high-strength alloy which offers an ultimate tensile strength of 1200 MPa (175 ksi) in the heat treated condition in sizes up to 25 mm (1 in.) diameter. This grade is used in applications requiring high strength-to-weight ratios at temperatures up to 315 °C (600 °F).

Chemistry and Density

Ti-662 contains a total of about 1% (Cu + Fe) in approximately equal proportions, which give it much-improved heat treatability. Its nominal 6% aluminum content stabilizes the alpha phase and increases the hot workability range by raising the beta transus temperature to approximately 945 °C (1735 °F). Cooling from above this temperature with little concurrent or subsequent deformation generally results in inferior ductility. As a neutral stabilizer, the 2% tin strengthens both the alpha and beta phases, and in combination with the aluminum, provides better room- and elevated-temperature strength properties than those of Ti-6Al-4V and other lower-alloy alpha-beta compositions. Beta stabilization is accomplished by nominal additions of 6% vanadium, 0.5% copper, and 0.5% iron. Acting together, these elements permit heat treatment of the alloy to high strength levels by solution treatment and aging.

Density. 4.54 g/cm^3 (0.164 lb/in.^{3})

Alloy Segregation. Ingot composition must be controlled within specified limits, and special melting practices, particularly for the final melt, are required to minimize segregation during solidification. Excessive macrosegregation results in "beta flecks," which are harder, less-ductile areas after heat treatment. Detrimental effects of beta flecks have not been demonstrated for this alloy.

Exceeding Composition Limits. As for all alpha-beta alloys, excessive amounts of aluminum, oxygen, and nitrogen can decrease ductility and fracture toughness. Excessive amounts of beta stabilizers (molybdenum and vanadium) affect the stability of the alloy and increase its heat treatability, therefore making control of properties more difficult. Excessive impurity levels may raise yield strength above maximum permitted values or decrease elongation or reduction in area below minimum values.

Product Forms

Ti-662 is produced by all U.S. titanium melters as bar and billet for forging stock. Plate, sheet, wire, and extrusions are also available.

Product Conditions/Microstructure

In forged sections and plate up to 25 mm (1 in.) thick, solution-treated-and-aged material has a guaranteed minimum ultimate tensile strength of 1170 MPa (170 ksi). For forged sections between 75 and 100 mm (3 and 4 in.) thick, the corresponding ultimate tensile strength is 1035 MPa (150 ksi). The response to heat treatment may vary from heat to heat and the correct aging temperature is best determined by tests on the heat in question. Cooling from above the beta transus with concurrent or subsequent deformation generally results in inferior ductility.

Annealing treatments at temperatures of 640 to 790 °C (1200 to 1450 °F) are applied to produce maximum stability at temperatures up to 450 °C (600 °F). The strengthening response to the precipitation-hardening reaction is dependent on the ability to retain the beta phase during quenching from the solution temperature, and this alloy is sufficiently beta stabilized to attain heat treated properties through section thicknesses up to 100 mm (4 in.).

Applications

Ti-662 is used in applications requiring high strength at temperatures up to 315 °C (600 °F) in the forms of sheet, light-gage plate, extrusions, and small forgings. This alloy is used for airframe structures where strength higher than that of Ti-6Al-4V is required. Usage is generally limited to secondary structures because the attractiveness of higher strength efficiency is minimized by lower fracture toughness and fatigue properties. Ti-662 is used for aircraft structural members, centrifuge parts, and rocket-engine parts.

Limitations in Use. As is characteristic of other titanium alloys, exposure to stress at elevated temperature produces changes in the retained mechanical properties. The stress and temperature limits below which these changes will not occur have not been established for this alloy. Structural applications should be based on a knowledge of the low toughness characterizing the higher-strength conditions of this alloy and the limited toughness of welds. Particular attention should be given to the influence of aggressive environments in the presence of cracks. Such environments include aqueous solutions of chlorides and possibly certain organic solvents such as methanol.

638 / Alpha-Beta Alloys

Ti-662: Equivalent specifications

Specification UNS	Designation R56620	Description	Al 5.5	Cu	Fe	H	N	O	Sn 2	V 5.5	Other bal Ti(a)
China											
GB 3620	TC-10		5.5-6.5	0.5	0.5	0.015 max	0.04 max	0.2 max	1.5-2.5	5.5-6.5	C 0.1 max; Si 0.15 max; bal Ti
Europe											
AECMA Ti-P64	prEN3316	Sh Strp Ann	5-6	0.35-1	0.35-1	0.0125 max	0.04 max	0.2 max	1.5-2.5	5-6	C 0.05 max; OT 0.4 max; OE 0.1 max; bal Ti
AECMA Ti-P64	prEN3317	Plt Ann	5-6	0.35-1	0.35-1	0.0125 max	0.04 max	0.2 max	1.5-2.5	5-6	C 0.05 max; OT 0.4 max; OE 0.1 max; bal Ti
AECMA Ti-P64	prEN3318	Frg NHT	5-6	0.35-1	0.35-1	0.0125 max	0.04 max	0.2 max	1.5-2.5	5-6	C 0.05 max; OT 0.4 max; OE 0.1 max; bal Ti
AECMA Ti-P64	prEN3319	Bar Ann	5-6	0.35-1	0.35-1	0.0125 max	0.04 max	0.2 max	1.5-2.5	5-6	C 0.05 max; OT 0.4 max; OE 0.1 max; bal Ti
AECMA Ti-P64	prEN3320	Frg Ann	5-6	0.35-1	0.35-1	0.0125 max	0.04 max	0.2 max	1.5-2.5	5-6	C 0.05 max; OT 0.4 max; OE 0.1 max; bal Ti
Germany											
WL 3.7174		Sh Strp Plt Bar Frg Ann	5-6	0.35-1	0.35-1	0.0125-0.015	0.04	0.2	1.5-2.5	5-6	C 0.05; OT 0.4; bal Ti
WL 3.7174		Sh Strp Plt Bar Frg STA	5-6	0.35-1	0.35-1	0.0125-0.015	0.04	0.2	1.5-2.5	5-6	C 0.05; OT 0.4; bal Ti
Spain											
UNE 38-725	L-7303	Sh Strp Plt Bar Ext Ann	5-6	0.35-1	0.35-1	0.0125	0.04	0.2	1.5-2.5	5-6	C 0.05; OT 0.4; bal Ti
UNE 38-725	L-7303	Sh Strp Plt Bar Ext HT	5-6	0.35-1	0.35-1	0.0125	0.04	0.2	1.5-2.5	5-6	C 0.05; OT 0.4; bal Ti
USA											
AMS 4918F		Sh Strp Plt Ann	5-6	0.35-1	0.35-1	0.015	0.04	0.2	1.5-2.5	5-6	C 0.05; OT 0.4; Y 0.005; bal Ti
AMS 4936B		Ext Rng Ann	5-6	0.35-1	0.35-1	0.015	0.04	0.2	1.5-2.5	5-6	C 0.05; OT 0.4; Y 0.005; bal Ti
AMS 4936B		Ext Rng STA	5-6	0.35-1	0.35-1	0.015	0.04	0.2	1.5-2.5	5-6	C 0.05; OT 0.4; Y 0.005; bal Ti
AMS 4936C		Beta Ext Ann Rng Flsh Wld	5-6	0.35-1	0.35-1	0.015 max	0.04 max	0.2 max	1.5-2.5	5-6	C 0.05 max; OT 0.4 max; Y 0.005 max; OE 0.1 max; bal Ti
AMS 4971C		Bar Frg Wir Rng Bil Ann	5-6	0.35-1	0.35-1	0.015	0.04	0.2	1.5-2.5	5-6	C 0.05; OT 0.4; Y 0.005; bal Ti
AMS 4978B		Bar Wir Frg Bil Rng Ann	5-6	0.35-1	0.35-1	0.015	0.04	0.2	1.5-2.5	5-6	C 0.05; OT 0.4; Y 0.005; bal Ti
AMS 4978C		Bar Frg Rng Ann	5-6	0.35-1	0.35-1	0.015 max	0.04 max	0.2 max	1.5-2.5	5-6	C 0.05 max; OT 0.4 max; Y 0.005 max; OE 0.1 max; bal Ti
AMS 4979B			5-6	0.35-1	0.35-1	0.015	0.04	0.2	1.5-2.5	5-6	C 0.05; OT 0.4; Y 0.005; bal Ti
MIL F-83142A	Comp 8	Frg Ann	5-6	0.35-1	0.35-1	0.015	0.04	0.2	1.5-2.5	5-6	C 0.05; OT 0.3; bal Ti
MIL F-83142A	Comp 8	Frg HT	5-6	0.35-1	0.35-1	0.015	0.04	0.2	1.5-2.5	5-6	C 0.05; OT 0.3; bal Ti
MIL T-81556A	Code AB-3	Ext Bar Shp Ann	5-6	0.35-1	0.35-1	0.015	0.04	0.2	1.5-2.5	5-6	C 0.05; OT 0.3; bal Ti
MIL T-81556A	Code AB-3	Ext Bar Shp STA	5-6	0.35-1	0.35-1	0.015	0.04	0.2	1.5-2.5	5-6	C 0.05; OT 0.3; bal Ti
MIL T-9046J	Code AB-3	Sh Strp Plt Ann	5-6		0.35-1	0.015	0.04	0.2	1.5-2.5	5-6	C 0.05; OT 0.3; bal Ti
MIL T-9046J	Code AB-3	Sh Strp Plt ST	5-6		0.35-1	0.015	0.04	0.2	1.5-2.5	5-6	C 0.05; OT 0.3; bal Ti
MIL T-9046J	Code AB-3	Sh Strp Plt STA	5-6		0.35-1	0.015	0.04	0.2	1.5-2.5	5-6	C 0.05; OT 0.3; bal Ti
MIL T-9047G	Ti-6Al-6V-2Sn	Bar Bil Ann	5-6	0.35-1	0.35-1	0.015	0.04	0.2	1.5-2.5	5-6	C 0.05; OT 0.3; Y 0.005; bal Ti

(a) OT, others total; OE, others each; single values are maximums.

(continued)

Ti-662: Equivalent specifications (continued)

Specification UNS	Designation R56620	Description	Al 5.5	Cu	Fe	H	N	O	Sn 2	V 5.5	Other bal Ti(a)
MIL T-9047G	Ti-6Al-6V-2Sn	Bar Bil STA	5-6	0.35-1	0.35-1	0.015	0.04	0.2	1.5-2.5	5-6	C 0.05; OT 0.3; Y 0.005; bal Ti
USA (continued)											
SAE J467	Ti662		5.5 (nom)	0.7 (nom)	0.7 (nom)	...	0.02 max		2 (nom)	5.5 (nom)	C 0.02 max; Ni 0.006 max; Si 0.1 max; bal Ti

(a) OT, others total; OE, others each; single values are maximums.

Ti-662: Commercial compositions

Specification	Designation	Description	Al	Cu	Fe	H	N	O	Sn	V	Other
France											
Ugine	UT662	Sh Plt Frg Ann	5-6	0.35-1	0.35-1	0.015	0.04	0.2	1.5-2.5	5-6	bal Ti
Ugine	UT662	Sh Plt Frg QA	5-6	0.35-1	0.35-1	0.015	0.04	0.2	1.5-2.5	5-6	bal Ti
Germany											
Deutsche T	Contimet AlVSn 6-6-2	Plt Bar Frg Pip Ann	5-6	0.35-1	0.35-1	0.015	0.04	0.2	1.5-2.5	5-6	C 0.05; bal Ti
Deutsche T	Contimet AlVSn 6-6-2	Plt Bar Frg Pip STA	5-6	0.35-1	0.35-1	0.015	0.04	0.2	1.5-2.5	5-6	C 0.05; bal Ti
Deutsche T	LT 33	Frg Aged	5-6	0.35-1	0.35-1	0.015	0.04	0.2	1.5-2.5	5-6	C 0.05; bal Ti
Deutsche T	LT 33	Frg Ann	5-6	0.35-1	0.35-1	0.015	0.04	0.2	1.5-2.5	5-6	C 0.05; bal Ti
Japan											
Kobe	KS6-6-2	Plt Sh Ann	5-6	0.35-1	0.35-1	0.0125	0.04	0.2		5-6	bal Ti
Kobe	KS6-6-2	Plt Sh STA	5-6	0.35-1	0.35-1	0.0125	0.04	0.2		5-6	bal Ti
Sumitomo	Ti-6Al-6V-2Sn		5-6		0.35-1	0.015	0.04	0.12-0.2	1.5-2.5	5-6	C 0.05; bal Ti
Toho	662AT	STA									
USA											
OREMET	Ti 6-6-2										
RMI	RMI 6Al-6V-2Sn	Mult Forms Ann	5-6	0.35-1	0.35-1	0.0125-0.015	0.04	0.2	1.5-2.5	5-6	C 0.08; bal Ti
RMI	RMI 6Al-6V-2Sn	Mult Forms STA	5-6	0.35-1	0.35-1	0.0125-0.015	0.04	0.2	1.5-2.5	5-6	C 0.08; bal Ti
Timet	TIMETAL 6-6-2	Ann	5-6		0.35-1	0.015	0.05 max	0.2 max	1.5-2.5	5-6	C 0.05 max; bal Ti
Timet	TIMETAL 6-6-2 STA	Bil Bar Plt Sh Str STA	5-6	0.35-1	0.35-1	0.015	0.04	0.2	1.5-2.5	5-6	C 0.05; bal Ti

Single values are maximums.

Phases and Structures

Alloy Ti-662 is normally processed in the α + β two-phase field, resulting in primary equiaxed α and some β. For example, annealing treatments (~760 °C or 1400 °F) moderately low in the two-phase α + β field after normal α + β processing result in microstructures with a high volume percentage of primary α with stabilized β at the equiaxed α grain boundaries. If the processing involves less exposure time or less working in the α + β region and is subsequently annealed at approximately 760 °C (1400 °F), the primary α grains appear more elongated, and the volume percentage is high. Both structures develop acceptable mechanical properties.

Totally transformed β structures are often considered unacceptable, although acicular products do have advantages. Annealing temperatures and cooling rates determine the presence and the coarseness of secondary α (transformed β). For solution treatments up to 825 °C (1515 °F), β is sufficiently enriched with vanadium to prevent decomposition into martensitic α. At temperatures above 900 °C (1650 °F), β decomposes completely to martensitic α. Between these two temperatures, partial transformation of β occurs (see the isothermal TTT diagram after quenching from 850 °C or 1560 °F). From above the β transus, the M_s temperature is about 420 °C (790 °F).

640 / Alpha-Beta Alloys

Crystal Structure **Beta Transus:** 945 ± 10 °C (1733 ± 20 °F) to 955 ± 5 °C (1750 ± 10 °F)

Transformation Products

Ti-662: Time-temperature transformations from 850 °C (1560 °F)

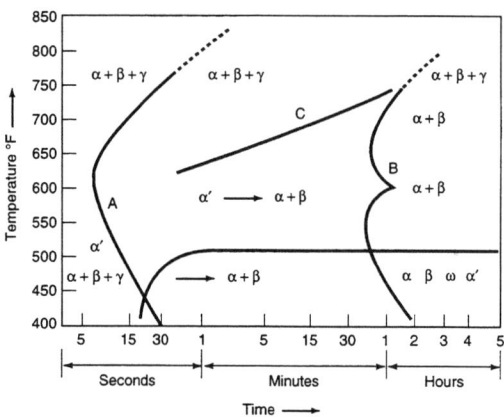

Dilatometric tests indicated M_s temperature of 640 °C (1185 °F), and X-ray measurements indicated that α′ martensite formed when isothermal holds were stopped by quenching before line A. Beyond line B, β is sufficiently enriched with vanadium to prevent martensitic transformation. Measurements indicated the disappearance of $Ti_3Al(γ)$ beyond line C. 25 mm (1 in.) diam specimens solution treated at 850 °C (1560 °F) for 1 h. Composition (wt%): 5.5 V, 5.65 Al, 2.35 Sn, 0.5 Cu, 0.62 Fe.
Source: B. Hocheid et al., Isothermal Transformation of Ti-6Al-6V-2Sn Alloy After Preheating in the α-β Range, *Titanium Science and Technology*, R.I. Jaffee and H.M. Burte, Ed., TMS-AIME, 1973, p 1609-1619

Ti-662: Isothermal transformation diagram

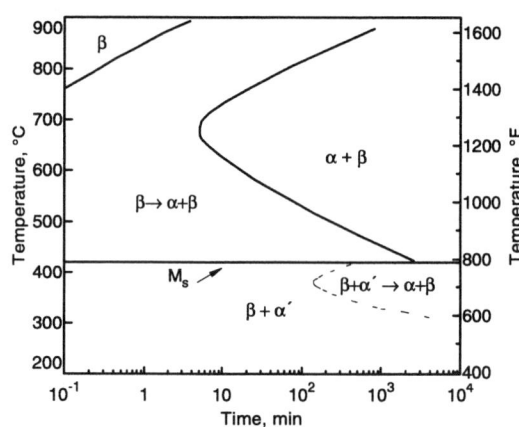

Quenched from β field to temperature indicated.
Source: *Titanium Alloy Handbook*, MCIC-HB-02, Battelle Columbus Laboratories, 1972

Ti-662: Phase transformation diagram

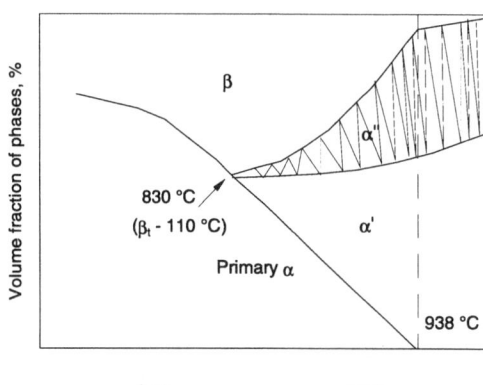

Source: Y. Murakami et al., Phase Transformation and Heat Treatment in Ti Alloys, *Titanium Science and Technology*, G. Lütjering, U. Zwicker, and W. Bunk, Ed., Deutsche Gesellschaft für Metallkunde, Germany, 1985, p 1405

Physical Properties

Ti-6Al-6V-2Sn: Summary of typical physical properties

Beta transus	945 ± 15 °C (1735 ± 25 °F)
Melting (liquidus) point	1650 to 1700 °C (3000 to 3100 °F)
Density(a)	4.54 g/cm^3 (0.164 lb/in.^{3})
Electrical resistivity(a)	1.57 μΩ · m
Magnetic permeability	1.00 (nonmagnetic)
Specific heat capacity(a)	635 J/kg · K (0.155 Btu/lb · °F)
Thermal conductivity(a)	5.5 W/m · K
Thermal coefficient of linear expansion(b)	~9 × 10^{-6}/°C (~5 × 10^{-6}/°F)

(a) Typical values at room temperature of about 20 to 25 °C (68 to 78 °F). (b) Mean coefficient from room temperature to 100 °C (212 °F)

Ti-662: Electrical resistivity vs temperature

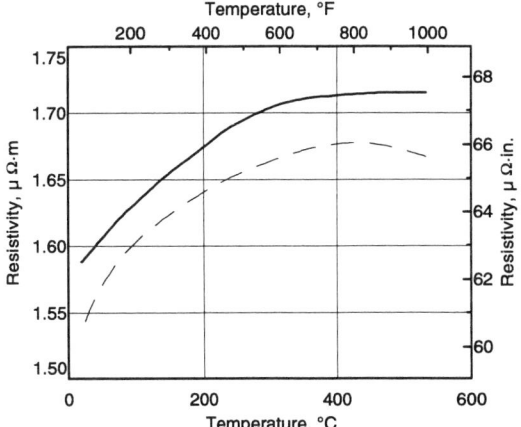

Solid line: 900 °C (1650 °F), 1 h, water quenched + 565 °C (1050 °F), 4 h, air cooled. Dashed line: 705 °C (1300 °F), 2 h, air cooled.
Source: *Aerospace Structural Metals Handbook*, Code 3715, Vol 4, Battelle Columbus Laboratories, 1975, p 9

Elastic Properties

Typical tensile modulus at room temperature is 110 GPa (16 × 10^6 psi) in the annealed condition.

Ti-662: Typical room-temperature tangent modulus

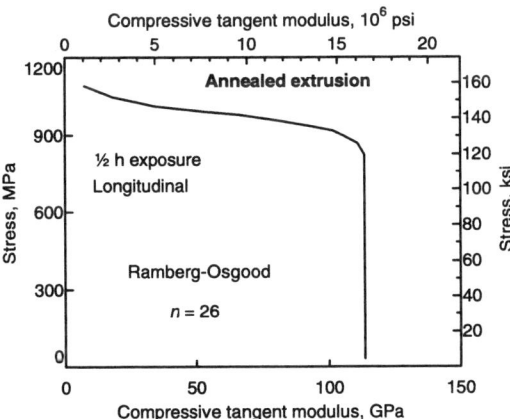

Source: MIL-HDBK-5E

Chemical / Corrosion Properties

Ti-6Al-6V-2Sn has less corrosion resistance in reducing media than several other titanium alloys. It also is one of the least resistant titanium alloys to crevice corrosion in salt solution.

642 / Alpha-Beta Alloys

Ti-662: Corrosion comparison in HCl solutions

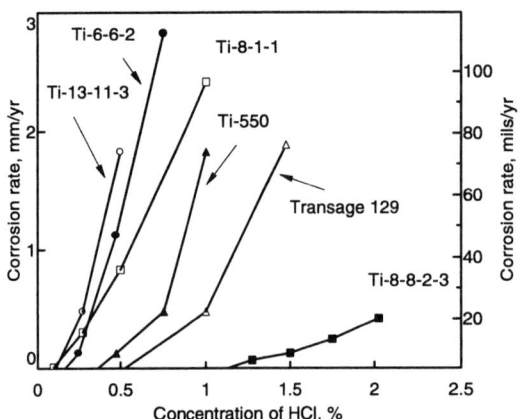

General corrosion of annealed titanium alloys in naturally aerated HCl solutions.
Source: *Metals Handbook, Corrosion*, Vol 13, 9th ed., ASM International, 1987, p 680

Ti-662: Crevice corrosion in saturated brine

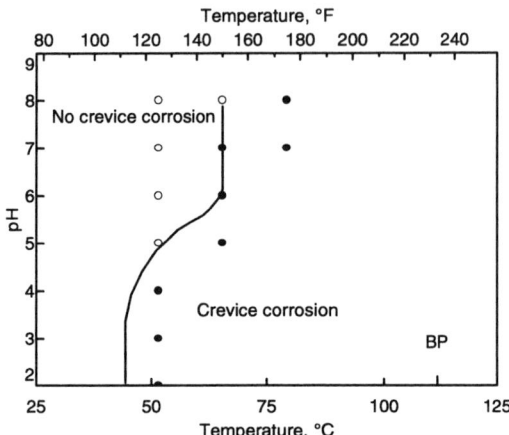

Source: D. Dees, "Crevice Corrosion of High-Strength Titanium Alloys in Saturated Brine," *Industrial Applications of Titanium and Zirconium*, ASTM STP 830, 1984, p 133-142

Stress-Corrosion Cracking

Ti-662: Fracture toughness in air and 3.5% NaCl solution at 25 °C

Heat treatment	Tensile yield strength		K_{Ic} or K_c		K_{Iscc} or K_{scc}	
	MPa	ksi	MPa√m	ksi√in.	MPa√m	ksi√in.
Mill annealed	1082	157	66	60	22	20
Duplex annealed	1006	146	88	80	27	25
Solution treated and aged	1172	170	49	45	33	30
β solution treated and aged	1048	152	77	70	49	45

Source: R. Schutz, Stress-Corrosion Cracking of Titanium Alloys, in *Stress-Corrosion Cracking: Materials Performance and Evaluation*, ASM International, 1992

Ti-662: Delayed failure in salt water

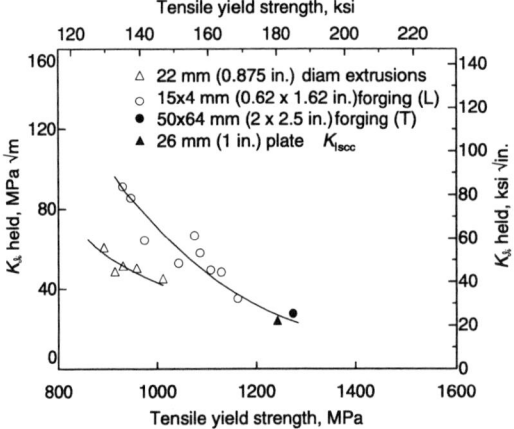

Cracked notch bend specimens exposed to 3.5% NaCl. Specimens were heat treated at 855 to 885 °C (1575 to 1625 °F), WQ or AC + 595 to 760 °C (1100 to 1400 °F), AC and annealed at 705 °C (1300 °F), 2 h, AC.
Source: *Aerospace Structural Metals Handbook*, Code 3715, Vol 4, Battelle Columbus Laboratories, Dec 1975

Thermal Properties

Liquidus Temperature. 1650 °C (3000 °F), reported in *Metals Handbook*, Vol 3, 9th ed., 1978.
Solidus Temperature. 1627 °C (2940 °F).

Melting Point. 1700 °C (3100 °F), reported in "RMI Titanium Basic Design."

Specific Heat

Ti-662: Specific heat

Temperature		Specific heat	
°C	°F	J/kg·K	Btu/lb·°F
93	200	670	0.160
150	300	674	0.161
205	400	682	0.163
260	500	687	0.164
315	600	691	0.165
370	700	699	0.167
425	800	703	0.168
480	900	712	0.170

Annealed condition. Source: *Metals Handbook*, Vol 3, 9th ed., American Society for Metals, 1978, p 392

Ti-662: Specific heat vs temperature

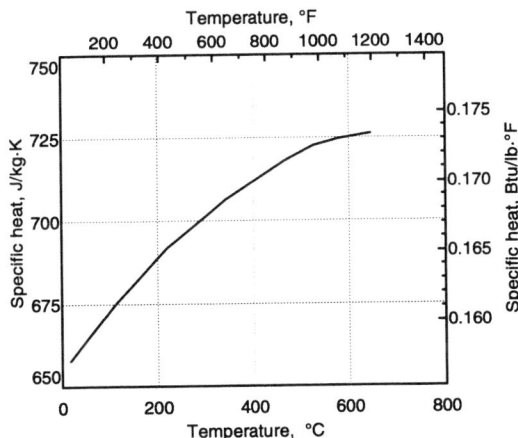

Source: *Metallic Materials and Elements for Aerospace Vehicle Structures*, MIL-HDBK 5E, June 1987

Ti-662: Specific heat vs temperature

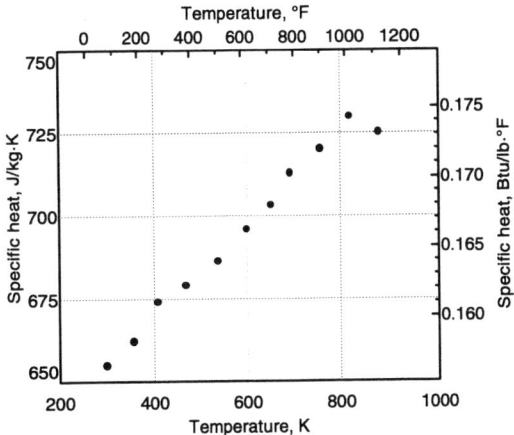

Source: *Aerospace Structural Metals Handbook*, Vol 4, Code 3715, Battelle Columbus Laboratories, Dec 1975, p 9

Thermal Expansion

Ti-662: Thermal coefficient of linear expansion at select temperatures

°C	°F	10^{-6}/K	10^{-6}/°F
93	200	9.0	5.0
205	400	9.2	5.1
315	600	9.3	5.2
425	800	9.5	5.3

Source: *Metals Handbook*, Vol 3, 9th ed., American Society for Metals, 1978, p 392

Ti-662: Thermal coefficient of linear expansion

Temperature range		Average coefficient	
°C	°F	10^{-6}/K	10^{-6}/°F
0–100	32–212	9.0	5.0
0–315	32–600	9.3	5.2
0–540	32–1000	9.5	5.3

Ti-662: Thermal coefficient of linear expansion

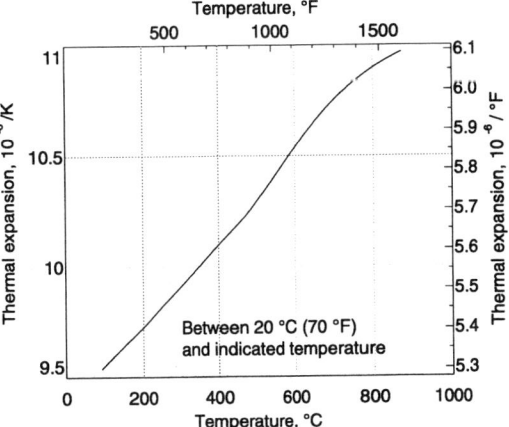

Between 20 °C (70 °F) and indicated temperature

Source: *Metallic Materials and Elements for Aerospace Vehicle Structures*, MIL-HDBK 5E, Vol 2, 1987

Thermal Conductivity

Ti-662: Thermal conductivity

Temperature		Conductivity	
°C	°F	W/m·K	Btu/ft·h·°F
93	200	6.6	3.8
205	400	8.12	4.7
315	600	9.86	5.7
425	800	11.9	6.9

Annealed condition. Source: *Metals Handbook*, Vol 3, 9th ed., American Society for Metals, 1978, p 392

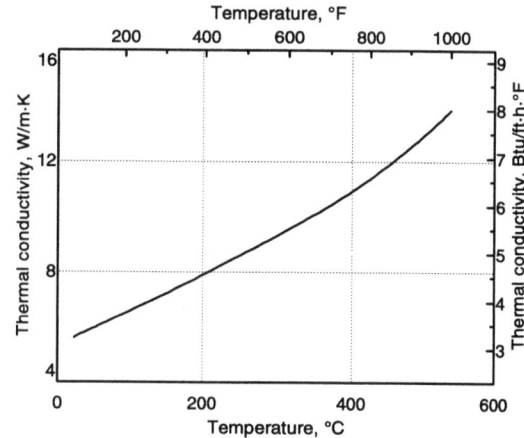

Ti-662: Thermal conductivity vs temperature

Source: *Metallic Materials and Elements for Aerospace Vehicle Structures*, MIL-HDBK 5E, Vol 2, June 1987

Mechanical Properties

See also "Heat Treatment" and "Forging" for mechanical properties.

Design Allowables

Ti-662: Design bearing strengths

Thickness		Ultimate bearing strength				Bearing yield strength			
		e/D = 1.5		e/D = 2.0		e/D = 1.5		e/D = 2.0	
mm	in.	MPa	ksi	MPa	ksi	MPa	ksi	MPa	ksi
Annealed sheet, strip, and plate (AMS 4918, MIL-T-9046)									
4.75-13.0	0.1875-0.500	1627	236	2027	294	1330	193	1482	215
13-25	0.50-1.00	1661	241	2089	303	1351	196	1537	223
25-38	1.0-1.50	1703	247	2151	312	1372	199	1613	234
38-50	1.5-2.00	1723	250	2185	317	1392	202	1654	240
STA sheet, strip, and plate (MIL-T-9406)									
4.75-38	0.1875-1.500	1820	264	2234	324	1634	237	1834	266
Extruded bar and shape, annealed (MIL-T-81556)									
<50	≤2.00	1503-1580	218-229(a)	1847-1937	268-281(a)	1351-140	196-203(a)	1565-1620	227-235(a)

(a) A- and B-basis values, respectively. Source: MIL-HDBK 5, Dec 1991

Ti-662: Design compression and shear strengths

Thickness		Compressive yield strength				Ultimate shear strength	
		L direction		LT direction			
mm	in.	MPa	ksi	MPa	ksi	MPa	ksi
Annealed sheet, strip, and plate (MIL-T-9046, AMS 4918)							
4.75-13	0.1875-0.50	958	139	1040	151	627	91
13-25	0.50-1.00	979	142	1013	147	641	93
25-38	1.00-1.50	1006	146	972	141	655	95
38-50	1.50-2.00	1020	148	937	136	655	95
STA sheet, strip, and plate (MIL-T-9406)							
4.75-38	0.1875-1.50	1172	170	1172	170	696	101
Extruded bar and shapes (MIL-T-81556), annealed							
<50	<2.0	944-993(a)	137-144(a)	937-979(a)	136-142(a)	641-668(a)	93-97(a)
50-75	2.0-3.0	965	140	965	140	...	...
75-100	3.0-4.0	930	135	930	135	...	...
STA extruded bar and shapes (MIL-T-81556)							
4.7-13	0.188-0.50	1137	165	1137	165	...	...,
13-38	0.50-1.50	1103	160	1103	160	...	...
38-64	1.5-2.50	1068	155	1068	155	...	...
64-100	2.5-4.00	999	145	999	145	...	...

Note: All minimums are S-basis values unless otherwise noted. (a) A- and B-basis values, respectively. Source: MIL-HDBK 5, 1 Dec 1991

Ti-662: Design tensile properties of extruded bar and shapes

Thickness			Ultimate tensile strength (L-LT)(a)		Tensile yield strength (L-LT)(a)		Elongation(b), %		Reduction of area(b), %	
mm	in.	Basis	MPa	ksi	MPa	ksi	L	LT	L	LT
Annealed extrusions										
<50.0	<2.00	A	980-972(c)	142-141(c)	890-882(c)	129-128(c)	10	8	20	15
		B	1020	148	930	135	...	...	...	...
50-75	2.0-3.00	S	999	145	930	135	10	8	20	15
75-100	3.0-4.00	S	965	140	896	130	10	8	20	15
STA extrusions										
4.78-13	0.188-0.500	S	1172	170	1103	160	8	6	15	12
13-38	5.0-1.500	S	1137	165	1068	155	8	6	15	12
38-63	1.50-2.50	S	1103	160	1034	150	8	6	15	12
63-100	2.5-4.00	S	1034	150	965	140	8	6	15	12

(a) Listed values are for longitudinal (L) and long-transverse (LT) direction unless otherwise noted. (b) S-basis. (c) Values in L and LT direction, respectively. Source: MIL-HDBK 5, 1 Dec 1991

Ti-662: Design tensile properties of bar and forgings

Thickness			Ultimate tensile strength (L-LT)(a)		Tensile yield strength (L-LT)(a)		Elongation, %		Reduction of area, %	
mm	in.	Basis	MPa	ksi	MPa	ksi	L	LT	L	LT
Air-cooled annealed (b) bar (AMS 4978)										
<38	<1.5	A	993-1013	144-147	903-937	131-136	8	10	20	15
		B	1035-1048	150-152	951-972	138-141	...	...	...	...
38-75	1.5-3.0	A	958-986	139-143	868-903	126-131	10	8(c)	20	15(c)
		B	999-1020	145-148	910-937	132-136	...	...	...	...
75-100	3.0-4.0	A	937-965	136-140	848-875	123-127	10	8(c)	15	15(c)
		B	979-999	142-145	889-910	129-132	...	...	...	...
Annealed forgings (AMS 4978)										
<50	<2.0	S	1035	150(e)	965	140(e)	10	8	20	15
50-100	2.0-4.0	S	1000	145(f)	930	135(f)	10	8	20	15
STA bar and forgings(d)										
<25	≤1.00	S	1205	175(e)	1103	160(e)	8	8	20	15
25-50	1.0-2.0	S	1172	170(e)	1068	155(e)	8	8	20	15
50-75	2.0-3.0	S	1068	155(f)	1000	145(f)	8	6(c)	20	15(c)
75-100	3.0-40	S	1035	150(f)	965	140(f)	8	6(c)	20	15(c)

(a) Listed values are for longitudinal (L) and long-transverse (L-T) values, respectively. LT values applicable providing LT or ST dimension is greater than 63.5 mm (2.5 in.). (b) 700 to 730 °C (1300 to 1350 °F) for 1 to 3 h, AC to room temperature. (c) Applicable to ST direction. (d) Per AMS 4971 and 4979. (e) L and LT direction. (f) L, LT, and ST direction. Source: MIL-HDBK 5, 1 Dec 1991

Ti-662: Design tensile properties for sheet, strip, and plate

Thickness		Ultimate tensile strength(a)		Tensile yield strength(a)		Elongation(b), %	
mm	in.	MPa	ksi	MPa	ksi	L	LT
Annealed per MIL-T-9046 and AMS 4918							
<4.7	<0.1875	1068(c)	155(c)	999(d)	145(d)	10(e)	8(f)
4.7-13	0.1875-0.50	1034	150	965	140	10	8
13-50	0.50-2.00	1034	150	965	140	10	8
50-100	2.00-4.00	999	145	930	135	8	6
Solution treated and aged per MIL-T-9406							
<4.7		1172	170	1103	160	8	6
4.7-38	0.1875-1.50	1172	170	1103	160	8	8
38-63	1.50-2.5	1103	160	1034	150	6	6
63-100	2.5-4.0	1034	150	965	140	6	6

(a) S-basis values for longitudinal (L) and long-transverse (LT) directions except where specified. (b) S-basis. (c) A-basis value in L and LT direction; B-basis values are 1100 MPa (160 ksi) in L direction and 1035 MPa (150 ksi) in LT direction. (d) A-basis value in L and LT direction; B-basis values are 1050 MPa (152 ksi) in L and 1060 MPa (154 ksi) in LT direction. (e) Longitudinal ≤0.6 mm (0.025 in.) is 8%. (f) Long-transverse <0.6 mm (0.025 in.) is 6%. Source: MIL-HDBK 5, 1 Dec 1991

Hardness

Ti-662: Typical hardness at room temperature

Material condition	Rockwell C hardness, HRC	Knoop hardness, HK
Unwelded sheet(a)	34	350
Single-bead weld(a)	46.8	420
Mill annealed sheet(b)	36	...
Weldments with heat treatments of(c):		
760 °C (1400 °F) for 4 h, air cooled	38	...
870 °C (1600 °F) for 4 h, air cooled	34	...
925 °C (1700 °F) for 4 h, air cooled	33	...
25 mm (1 in.) bar annealed(d)		
2 h at 690 to 720 °C (1275 to 1325 °F)	34	...
Mill annealed bar(d)	38	...

(a) From *Metals Handbook*, Vol 3, 9th ed., American Society for Metals, 1978, p 368. (b) From *Metall. Trans.*, Vol 5, 1974, p 2405. (c) Full-penetration weldments were produced on 3.5-mm (0.1 in.) thick sheet using the gas tungsten-arc welding process (205 A, 12 V DCSP, 0.1 cm/s). After welding, coupons were heat treated in vacuum. Specimens were machined so that the weld fusion zone was oriented along the longitudinal axis. (d) From *Alloy Digest*, Code 61 and 79, Sept 1980. Source: *Scripta Metall.*, Vol 13, 1979, p 1125

Ti-662: Variations in Rockwell C hardness

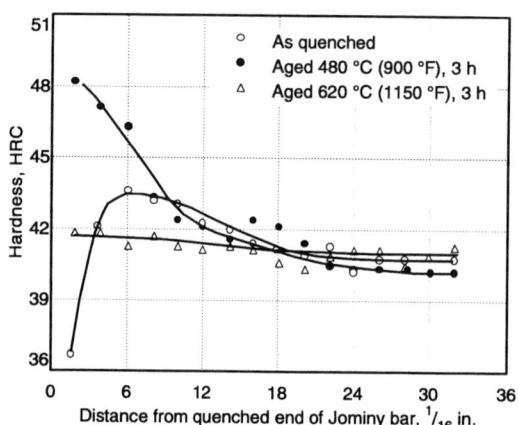

Chemical composition (wt%): 5.08 Al, 0.66 Cu, 0.55 Fe, 0.002 H, 0.02 N, 0.172 O, 1.78 Sn, and 5.41 V. Jominy bars with diameter of 28.55 mm (1.125 in.) and length of 125 mm (5 in.) were trepanned longitudinally from a position approximately halfway from the center of a 150 mm (6 in.) diameter forging. Bars were solution treated at 870 °C (1600 °C) for 2 h, followed by a conventional Jominy end quench. Aged specimens were held at indicated temperatures for 3 h and air cooled. Hardness measurements were performed along the length of the bar on which a surface flat had been ground at the center line of the bar.
Source: C. Hickey and P. Fopiano, Some Observations on the Hardenability of Ti-6Al-6V-2Sn, *Metall. Trans.*, Vol 1, 1970, p 1775

Ti-662: Knoop hardness after oxidation

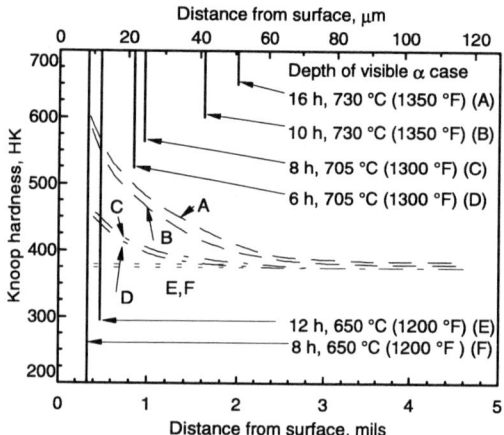

Source: A. Marrocco, "Investigation of Depth of Oxide Contamination on Titanium as a Function of Thermal Exposure," Grumman Aircraft Engineering Corporation, Bethpage, New York, Sept 1970, reported in *Titanium Alloys Handbook*, R. Wood and R. Favor, Ed., MCIC-HB-02, Metals and Ceramics Information Center, Battelle Columbus Laboratories, 1972

Ti-662: Variation in Vickers hardness after aging

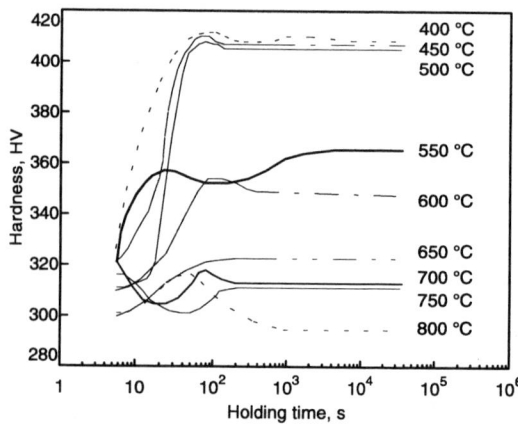

Alloy used as 25 mm (1 in.) diameter forged bar. Chemical composition (wt%): 5.65 Al, 0.2215 C, 0.0050 Cr, 0.5 Cu, 0.62 Fe, 0.0010 H, 0.0050 Mg, 0.0100 Mo, 0.0110 N, 0.0050 Ni, 0.0400 Si, 2.35 Sn, and 5.5 V. Beta transus temperature was determined to be 935 to 945 °C (1725 to 1730 °F). Specimens were solution treated at 850 °C (1560 °F) in salt baths and quenched in ice water. Surfaces were machined to remove contaminants.
Source: B. Hocheid, C. Fontalirand, C. Beauvais, C. Roux, and J. Fidelle, "Isothermal Transformations of Ti-6Al-6V-2Sn Alloy after Preheating in the Alpha-Beta Range," in *Titanium and Titanium Alloys*, J. Williams and A. Belov, Ed., Plenum Press, 1976, p 1609

Typical Strengths

Ti-662: Typical RT bearing and shear strengths

Product form	Condition	Ultimate shear strength		Bearing yield strength(a)		Ultimate bearing strength(a)	
		MPa	ksi	MPa	ksi	MPa	ksi
125 × 150 mm (5 × 6 in.) Forging	STA(b)	724	105	2330	338	2550	370
Plate	Annealed(c)	668	97	1579	229	1980	287
	STA(b)	765	111	1806	262	2250	326
Sheet	Annealed	710	103	1648	239	2060	299
	STA	758	110	1806	262	2250	326

(a) $e/D = 2$. (b) 870 °C (1600 °F) for 1 h, WQ, plus 595 °C (1100 °F) for 4 h, AC. (c) Annealed at 730 °C (1350 °F). Source: *Aerospace Structural Metals Handbook*, Vol 4, Code 3715, Battelle Columbus Laboratories, Dec 1975

Ti-662: Typical tensile properties of plate, bar, and forgings

Product form	Heat treatment(a)	Ultimate tensile strength		Tensile yield strength		Elongation(b), %	Reduction of area, %
		MPa	ksi	MPa	ksi		
50 mm (2 in.) plate	A	1110	161	1082	157	10	22
	B	1089	158	1034	150	13	26
	C	1027	149	958	139	15	28
38 mm (1.5 in.) bar	A	1082	157	1055	153	14	44
Large-die forging	A	1041	151	993	144	18	44
	B	999	145	951	138	19	45

(a) Heat treatments: A: mill annealed. B: 910 °C (1675 °F) 2 h, AC + 870 °C (1600 °F) 1 h, FC. 925 °C (1700 °F), 4 h, AC + 760 °C (1400 °F), 1 h, AC. (b) Elongation in 25 mm (1 in.). Source: *Aerospace Structural Metals Handbook*, Vol 4, Code 3715, Battelle Columbus Laboratories, Dec 1975

Ti-662: Notched strength of STA forgings

Specimen location(a)	Direction	Ultimate tensile strength		Notched ($K_t = 3.9$) strength(b)	
		MPa	ksi	MPa	ksi
Edge	L	1261	183	1213	176
	LT	1241	180	1179	171
	ST	1310	190	1268	184
Midradius	LT	1213	176	1199	174
	ST	1227	178	1179	171
Center	L	1275	185	1365	198

(a) 6.3 mm (0.25 in.) diam specimens taken from 125 × 150 mm (5 × 6 in.) forged section and then heat treated as follows: 870 °C (1600 °F) for 1 h, WQ, plus 595 °C (1100 °F) for 4 h. (b) 60° circumferential notch to a depth of 4.47 mm (0.176 in.) with a notch radius of 0.013 mm (0.005 in.). Source: *Aerospace Structural Metals Handbook*, Vol 4, Code 3715, Battelle Columbus Laboratories, 1975, p 23

Ti-662: Tensile properties of investment cast specimens

Condition	Ultimate tensile strength		Tensile yield strength		Elongation, %	Reduction of area, %
	MPa	ksi	MPa	ksi		
As cast + 595 °C (1100 °F), 1 h, AC	1061	154	972	141	8	18
870 °C (1600 °F), 1 h, WQ, 650 °C (1200 °F), 1 h, AC	1151	167	1089	158	4	10
980 °C (1800 °F), 1 h, WQ, 595 °C (1100 °F), 4 h, AC	1303	189	(a)	(a)	1.2	3

(a) Specimen broke before 0.2% yield. Source: *Aerospace Structural Metals Handbook*, Vol 4, Code 3715, Battelle Columbus Laboratories, 1975

Ti-662: Tensile strengths vs aging temperature

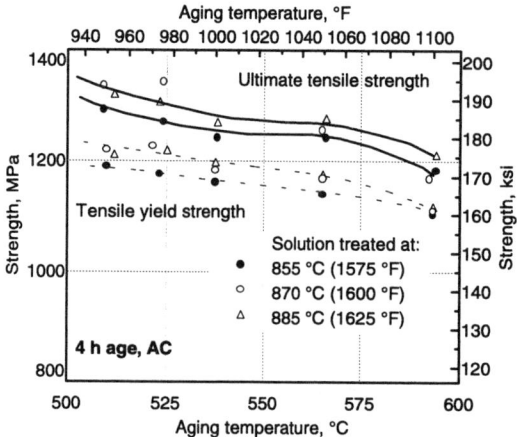

Beta forged at 1095 °C (2000 °F) followed by a 1 h solution treatment at indicated temperatures, WQ.
Source: *Aerospace Structural Metals Handbook*, Vol 4, Code 3715, Battelle Columbus Laboratories, 1975

Ti-662: Strength vs thickness

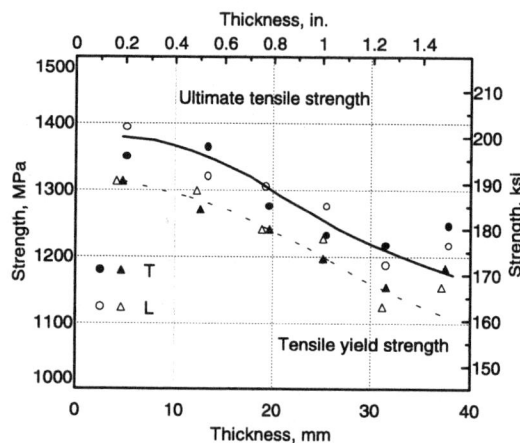

Aged tensile strength was determined on solution treated and aged plate, 845 to 900 °C (1550 to 1650 °F), WQ + 595 °C (1100 °F), 4h, AC.

Ti-662: Effect of solution treatment on aged strength

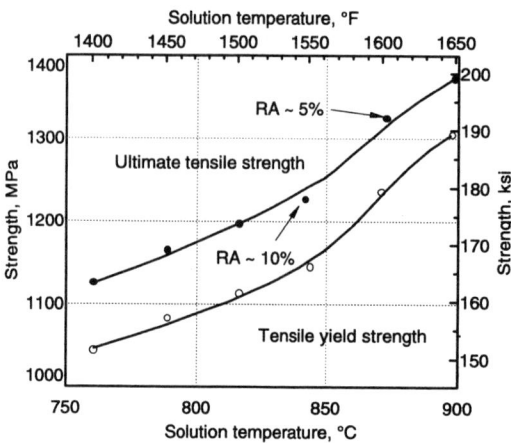

56 cm (22 in.) wide extrude panels, aged 4 h at 540 °C (1000 °F). The thickest portion of the panel was 18 mm (0.71 in.).
Source: *Aerospace Structural Metals Handbook*, Vol 4, Code 3715, Battelle Columbus Laboratories, 1975

Ti-662: Strength of cracked specimen

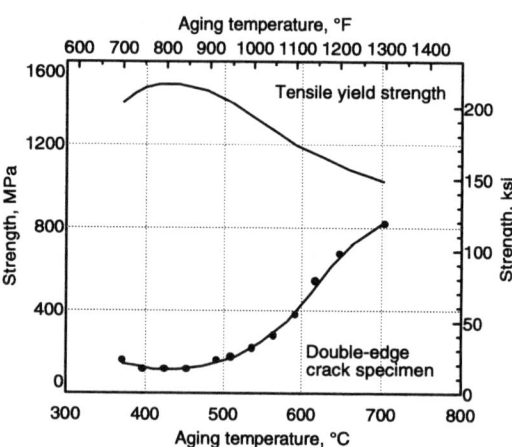

Specimens were 3 mm (0.125 in.) double edge crack sheet treated at 845 °C (1550 °F), 30 min, WQ + aging 4 h. Composition: 0.16 O_2, 0.025 N_2, 0.007 H_2, 0.026 C, and 0.072 Fe. Tests were conducted in the longitudinal direction.
Source: *Aerospace Structural Metals Handbook*, Vol 4, Code 3715, Battelle Columbus Laboratories, 1975

High-Temperature Strength

Tensile Strengths

Ti-662: Design tensile strengths of plate

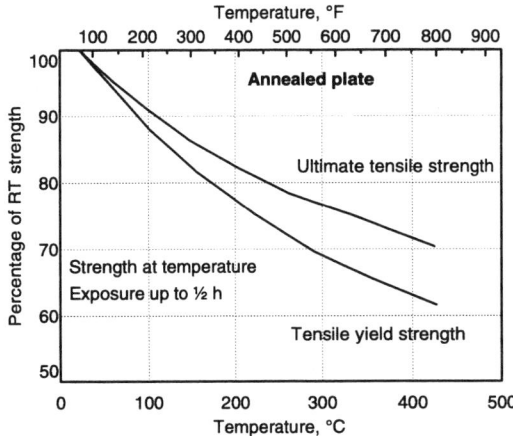

(a)

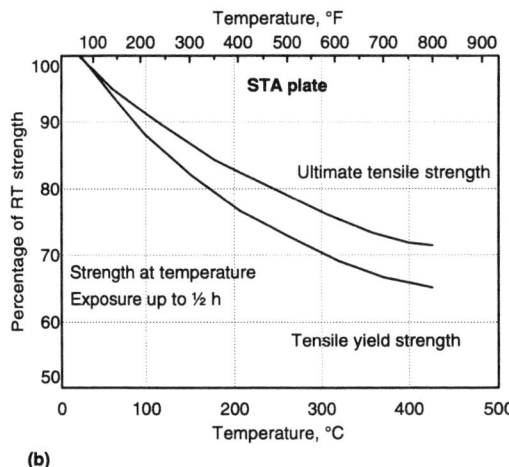

(b)

(a) Annealed. (b) Solution treated and aged.
Source: MIL-HDBK 5, 1 Dec 1991.

Ti-662: Tensile strengths of STA specimens

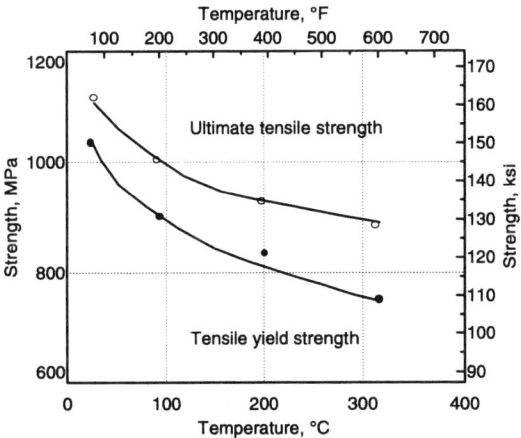

Metals Handbook, Vol 3, 9th ed., American Society for Metals, 1978

Ti-662: Effect of temperature on tensile strength of sheet

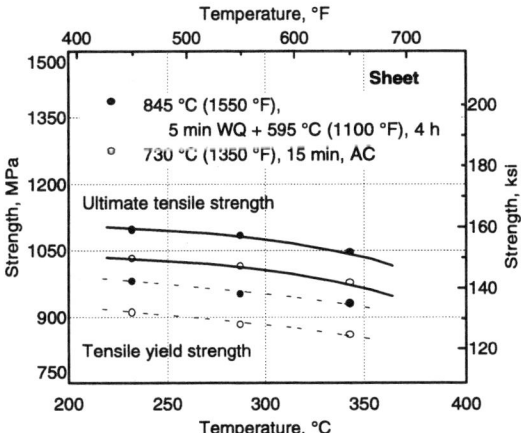

Specimens were 0.3 mm (0.120 in.) sheet heat treated as indicated.
Source: *Aerospace Structural Metals Handbook*, Vol 4, Code 3715, Battelle Columbus Laboratories, 1975, p 31

Ti-662: Effect of temperature on tensile strengths

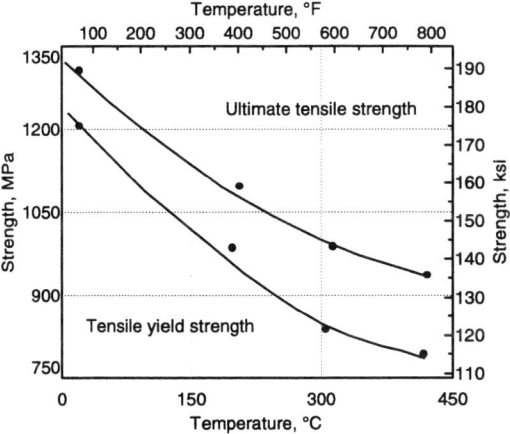

Specimen was a 33 cm (13 in.) diam by 3.1 mm (1.25 in.) thick hammer forged disk tested in the tangential direction. Specimens were forged at 925 °C (1700 °F), WQ + 595 °C (1100 °F), 4 h.
Source: *Aerospace Structural Metals Handbook*, Vol 4, Code 3715, Battelle Columbus Laboratories, 1975, p 27

Compressive Yield and Ultimate Shear Strengths

Ti-662: Design allowables for ultimate shear and compressive yield strengths vs temperature

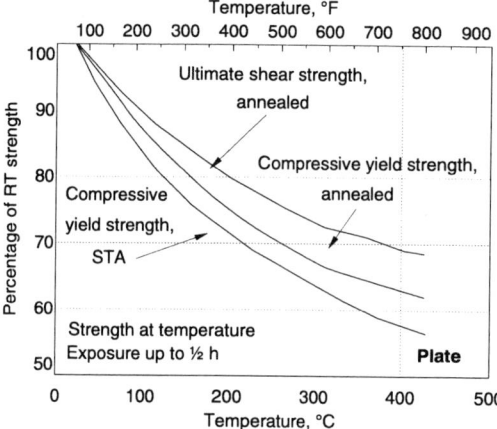

Source: MIL-HDBK 5, Dec 1991

Ti-662: Effect of temperature on shear and compressive strength

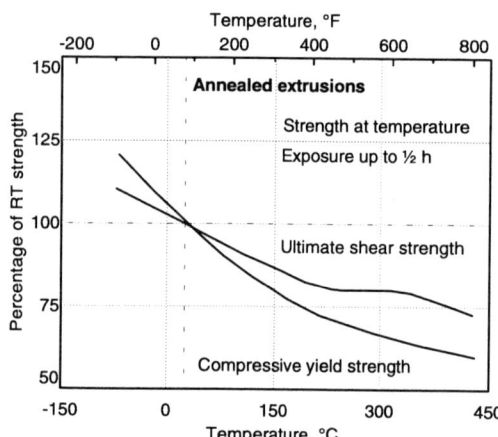

Source: MIL-HDBK 5, 1 Dec 1991

Ti-662: Compressive yield strength vs temperature

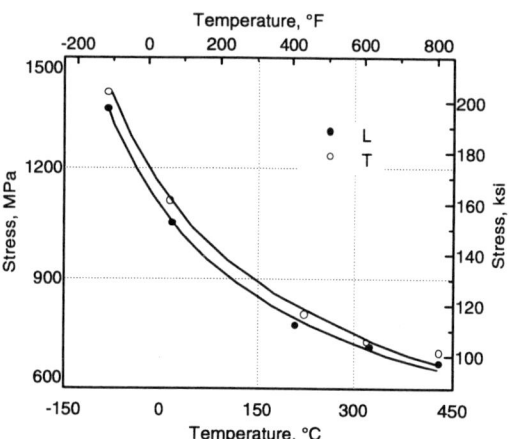

Specimens were extrusion annealed at 705 °C (1300 °F), 40 to 60 min, AC.
Source: *Aerospace Structural Metals Handbook*, Vol 4, Code 3715, Battelle Columbus Laboratories, 1975

Ti-662: Compressive yield strength of STA forging

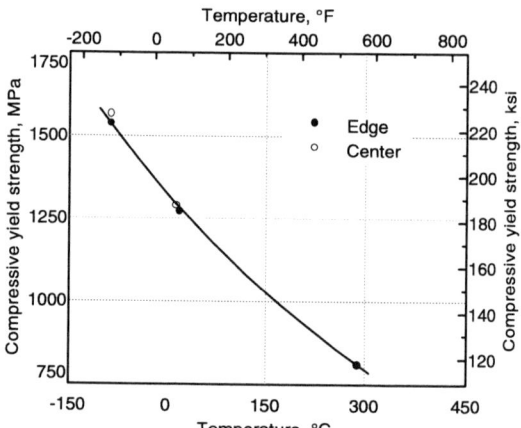

Effect of test temperature and specimen location on compressive yield strength of solution treated and aged specimens. Specimens were taken from 125 by 150 mm (5 by 6 in.) forgings and heat treated as follows: 870 °C (1650 °F), 1 h + WQ + 595 °C (1100 °F), 4 h.
Source: *Aerospace Structural Metals Handbook*, Vol 4, Code 3715, Battelle Columbus Laboratories, 1975, p 32

Ti-662: Compressive yield strength vs temperature

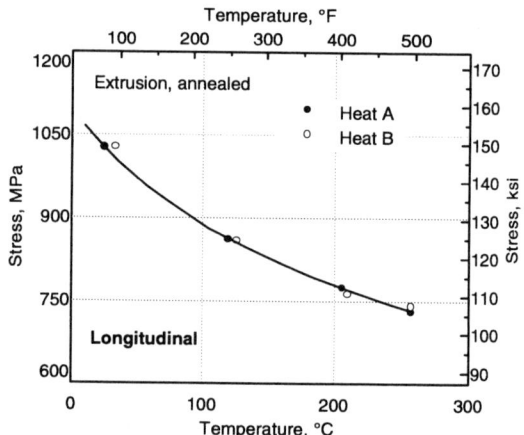

Source: *Aerospace Structural Metals Handbook*, Vol 4, Code 3715, Battelle Columbus Laboratories, 1975

Ti-662: Ultimate shear strength vs temperature

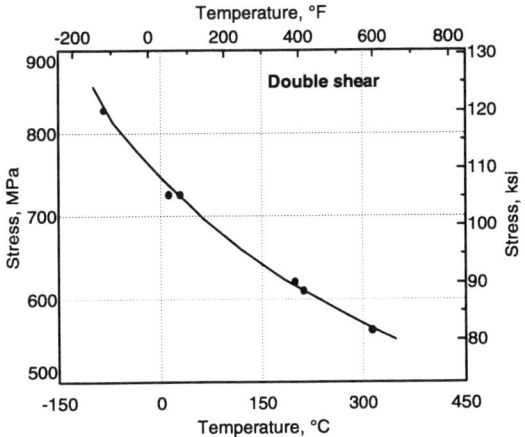

Effect of test temperature on double shear strength of extrusions annealed at 705 °C (1300 °F), 40 to 60 min, AC.
Source: *Aerospace Structural Metals Handbook*, Vol 4, Code 3715, Battelle Columbus Laboratories, 1975, p 33

Bearing Strength

Ti-662: Design bearing strength vs temperature

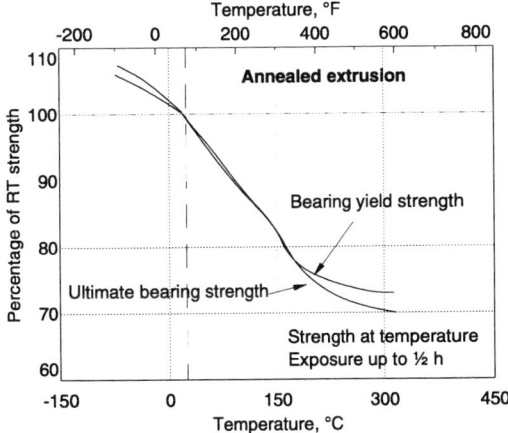

Source: MIL-HDBK 5, Dec 1991. e/D ratio unspecified.

Ti-662: Design bearing strength vs temperature

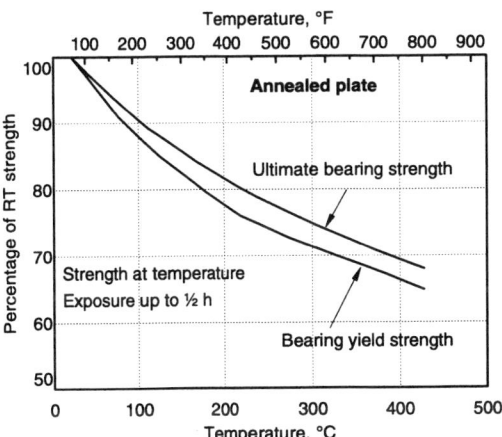

Source: MIL-HDBK 5, Dec 1991. e/D ratio unspecified.

Ti-662: Bearing strength vs temperature

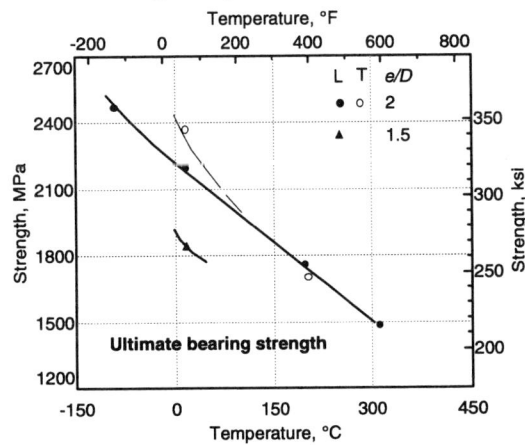

(a) Ultimate bearing strength. (b) Bearing yield strength.
Effect of test temperature on extrusions, annealed at 705 °C (1300 °F), 40 to 60 min, AC.
Source: *Aerospace Structural Metals Handbook*, Vol 4, Code 3715, Battelle Columbus Laboratories, 1975

Creep Properties

Ti-662: 0.2% creep strain curves

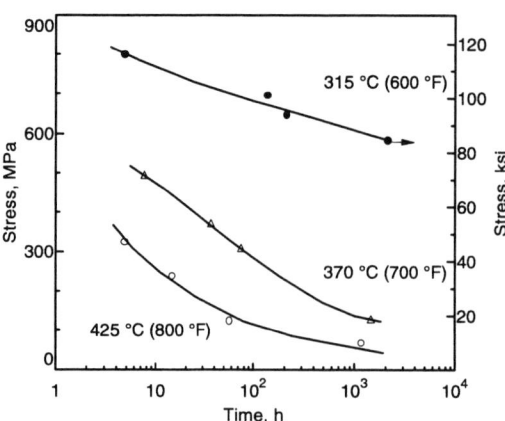

Bar, solution treated and aged at 885 °C (1625 °F), 1 h, WQ + 605 °C (1125 °F), 4 h, AC.
Source: *Aerospace Structural Metals Handbook*, Vol 4, Code 3715, Battelle Columbus Laboratories, 1975, p 37

Ti-662: 0.2% creep comparison

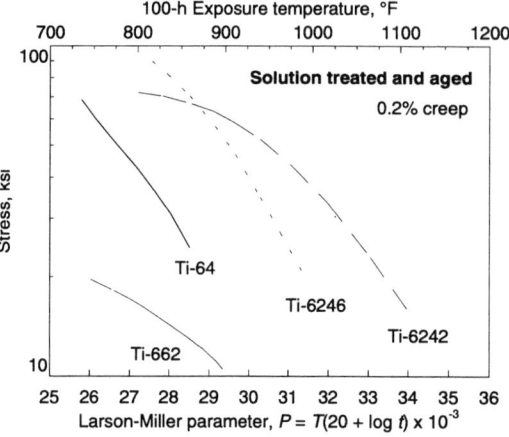

Results reported after 100-h exposure at temperature.
Source: *Military Handbook: Titanium and Titanium Alloys*, MIL-HDBK 697A, 1974, p 37

Ti-662: Creep deformation of STA sheet

Exposure	Stress MPa	ksi	Total creep, %
Condition 1(a)			
315 °C (600 °F), 150 h	413	60	0.07
	482	70	0.08
	689	100	0.21
455 °C (850 °F), 47 h	172	25	0.44
Condition 2(b)			
285 °C (550 °F), 150 h	827	120	0.12
315 °C (600 °F), 150 h	344	50	0.08
345 °C (650 °F), 150 h	241	35	0.10
Condition 3(c)			
285 °C (550 °F), 150 h	758	110	0.13
345 °C (650 °F), 307 h	241	35	0.11
455 °C (850 °F), 22 h	207	30	0.55
455 °C (850 °F), 24 h	220	32	0.54

(a) Condition 1: 885 °C (1625 °F) for 0.25 h, WQ, plus 565 °C (1050 °F) for 4 h, AC. (b) Condition 2: 915 °C (1675 °F) for 0.25 h, WQ, plus 595 °C (1100 °F) for 4 h, AC. (c) Condition 3: 845 °C (1550 °F) for 0.25 h, WQ, plus 595 °C (1100 °F) for 4 h, AC. Source: *Aerospace Structural Metals Handbook*, Vol 4, Code 3715, Battelle Columbus Laboratories, Dec 1975

Ti-662: Creep deformation of annealed extrusion

Exposure temperature °C	°F	Exposure time, h	Stress MPa	ksi	Total creep strain, %
315	600(a)	100	800	116	0.5
			820	119	1.0
315	600(a)	500	772	112	0.5
			800	116	1.0
425	800(b)	100	840	122	2.0
425	800(b)	500	820	119	2.0

Note: Specimens were annealed at 700 °C (1300 °F) for 40 to 60 min, AC. (a) Yield strength at 315 °C was 655 MPa (95 ksi). (b) Yield strength at 425 °C was 620 MPa (90 ksi). Source: *Aerospace Structural Metals Handbook*, Vol 4, Code 3715, Battelle Columbus Laboratories, Dec 1975

Ti-662: Creep stability of annealed or aged bar

Temperature		Creep conditions		Time, h	Total strain, %	Tensile properties after exposure:				Elongation in 25 mm (1 in.), %	Reduction of area, %
		Stress				Ultimate tensile strength		Tensile yield strength (0.2% offset)			
°C	°F	MPa	ksi			MPa	ksi	MPa	ksi		
Annealed bar(a)											
			No exposure		...	1110	161	1055	153	17	51
1110	600	606	88	149	0.450	1330	193	1110	161	14	32
1110	600	103	15	955	0.176	1324	192	1117	162	12	23
1290	700	275	40	148	0.318	1324	192	1158	168	11	23
1290	700	138	20	630	0.120	1337	194	1213	176	11	24
1470	800	103	15	144	0.300	1261	183	1206	175	12	30
1470	800	69	10	357	0.176	1282	186	1227	178	13	30
Aged bar(b)											
			No exposure		...	1289	187	1255	182	10	31
1110	600	862	125	143	0.749	1337	194	1324	192	9	4
1110	600	689	100	191	0.217	1296	188	1227	178	11	32
1290	700	448	65	143	0.481	1379	200	1268	184	8	14
1290	700	120	17.5	1144	0.173	1282	186	1241	180	14	26
1470	800	310	45	142	0.765	1337	194	1248	181	6	10
1470	800	138	20	461	0.365	1344	195	1296	188	6	11

(a) 1 h, 760 °C (1400 °F), AC. (b) 1 h, 885 °C (1625 °F), WQ + 4 h, 605 °C (1125 °F), AC. Source: R. Wood and R. Favor, *Titanium Alloys Handbook*, MCIC-HB 02, Metals and Ceramics Information Center, Columbus, Ohio, 1985

Low-Cycle Fatigue

Ti-662: Strain cycling for annealed bar

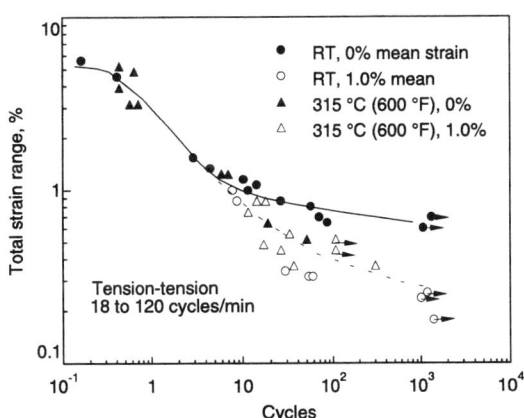

Specimens were 25 mm (1 in.) diam bar vacuum annealed at 705 °C (1300 °F), 2 h, FC.
Source: *Aerospace Structural Metals Handbook*, Vol 4, Code 3715, Battelle Columbus Laboratories, 1975

Ti-662: Low-cycle axial fatigue

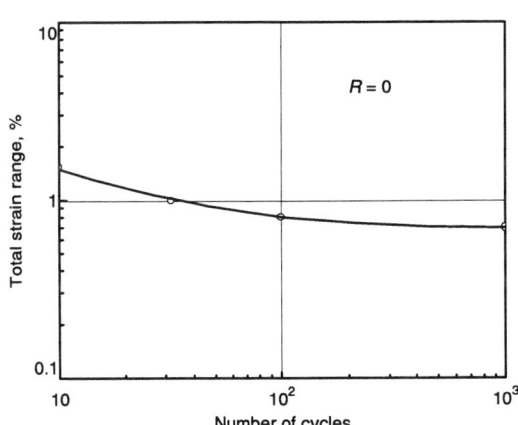

Stock annealed 2 h at 700 °C (1300 °F) and furnace cooled.
Source: *Metals Handbook*, Vol 3, 9th ed., American Society for Metals, 1978

High-Cycle Fatigue

Ti-662: Axial fatigue strength of notched specimens ($R = -1$)

Tensile strengths, MPa (ksi):		Test conditions(a), Notch factor (K_t)	Fatigue strength, MPa (ksi), at:		
UTS	TYS		10^5 cycles	10^6 cycles	10^7 cycles
RT test temperature					
1100 (160)	1055 (153)	3.4	186 (27)	145 (21)	138 (20)
		5.7	131 (19)	82 (12)	76 (11)
		10.0	138 (20)	69 (10)	...
315 °C (600 °F) test temperature					
875 (127)	717 (104)	3.4	144 (21)	138 (20)	131 (19)
		5.7	110 (16)	76 (11)	69 (10)
		10.0	...	69 (10)	62 (9)

(a) Specimen from 25 mm (1 in.) bar; specimen vacuum annealed at 700 °C (1300 °F) for 2 h, FC. 60° notch with a radius of 0.025, 0.13, and 0.38 mm (0.001, 0.005, and 0.015 in.). Source: *Aerospace Structural Metals Handbook*, Vol 4, Code 3715, Battelle Columbus Laboratories, Dec 1975

Ti-662: Axial fatigue strength of STA forging ($R = 0.1$)

Material condition	Test condition	Fatigue strength, MPa (ksi), at:			
		10^4 cycles	10^5 cycles	10^6 cycles	10^7 cycles
0.16% O_2, 0.66% Fe	Smooth, RT	1068 (155)	896 (130)	744 (108)	...
	Smooth, 285 °C (550 °F)	827 (120)	724 (105)	641 (93)	551 (80)
	$K_t = 3.9$, RT	537 (78)	289 (42)	207 (30)	172 (25)
0.10% O_2, 1.0% Fe	Smooth, RT	1068 (155)	896 (130)	724 (105)	620 (90)
	$K_t = 3.9$, RT	537 (78)	310 (45)	241 (35)	172 (25)

Note: Tests were conducted on 125 × 150 mm (5 × 6 in.) forged section with specimens heat treated as follows: 870 °C (1600 °F) for 1 h, WQ; plus 595 °C (1100 °F) for 4 h. Source: AFML-TR-65-206

Ti-662: Axial fatigue strength of extrusions ($R = 0.1$)

Material condition	Test condition	Fatigue strength, MPa (ksi), at:			
		10^4 cycles	10^5 cycles	10^6 cycles	10^7 cycles
Mill annealed, yield strength 945 to 993 MPa (137 to 144 ksi)	Smooth, RT	965 (140)	827 (120)	731 (106)	689 (100)
	$K_t = 4$, RT	482 (70)	289 (42)	220 (32)	220 (32)
T-extrusion, STA(a), L direction	Smooth, RT	...	724 (105)	655 (95)	620 (90)
	$K_t = 3.3$, RT	...	310 (45)	207 (30)	200 (29)

(a) 845 °C (1550 °F), WQ; plus 565 °C (1050 °F) for 4 h, AC. Source: *Aerospace Structural Metals Handbook*, Vol 4, Code 3715, Battelle Columbus Laboratories, Dec 1975

Ti-662: RT axial fatigue strength of annealed plate ($R = 0.1$)

Product form/condition	Test conditions	Fatigue strength, MPa (ksi), at:			
		10^4 cycles	10^5 cycles	10^6 cycles	10^7 cycles
25 mm (1 in.), mill annealed, 1055 MPa (153 ksi) yield strength	Smooth	1034 (150)(a)	848 (123)	689 (100)	606 (88)
0.18% O_2, 3.2 mm (1.25 in.), annealed(b)	Smooth, L and T direction	...	...	758 (110)	593 (86)
	$K_t = 3.5$, L and T direction	...	241 (35)	193 (28)	179 (26)
0.11% O_2, 25 mm (1 in.), annealed(b)	Smooth, L and T direction	...	...	793 (115)	620 (90)
	$K_t = 3.5$, L and T direction	...	275 (40)	310 (45)	207 (30)

(a) Extrapolated value. (b) 730 °C (1350 °F) annealed for 8 h, AC. Source: *Aerospace Structural Metals Handbook*, Vol 4, Code 3715, Battelle Columbus Laboratories, Dec 1975

Ti-662: Axial RT fatigue strength of STA plate

Product form/condition	Test conditions	Fatigue strength, MPa (ksi), at:			
		10^4 cycles	10^5 cycles	10^6 cycles	10^7 cycles
25 mm (1 in.) STA plate(a), 1180 MPa (171 ksi) yield strength	$R = 0.1$, Smooth	1068 (155)(b)	951 (138)	827 (120)	744 (108)
	$R = -1$, Smooth	882 (128)	620 (90)	551 (80)	386 (56)
0.18% O_2, 3.2 mm (1.25 in.), STA condition(c)	$R = 0.1$, Smooth, L direction	...	...	917 (133)	724 (105)
	Smooth, T direction	...	...	827 (120)	655 (95)
	$K_t = 3.5$, L direction	...	275 (40)	220 (32)	193 (28)
	$K_t = 3.5$, T direction	...	207 (30)	193 (28)	172 (25)
0.11% O_2, 25 mm (1 in.), STA condition(d)	Smooth, L and T direction	...	...	...	793 (115)
	$K_t = 3.5$, L and T direction	...	344 (50)	296 (43)	275 (40)
0.18% O_2, 50 mm (2 in.), STA condition(e)	Smooth, L and T direction	...	...	...	655 (95)
	$K_t = 3.5$, L and T direction	...	207 (30)	193 (28)	172 (25)

(a) 870 °C (1600 °F), WQ, plus 565 to 595 °C (1050 to 1100 °F) for 4 h, AC. (b) Extrapolated value. (c) 730 °C (1350 °F), 8 h, AC. (d) 845 °C (1550 °F), 1 h, WQ + 650 °C (1200 °F), 4 h. (e) 885 °C (1625 °F), WQ + 565 °C (1050 °F), 4 h. Source: *Aerospace Structural Metals Handbook*, Vol 4, Code 3715, Battelle Columbus Laboratories, Dec 1975

Ti-662: Axial fatigue strength of sheet

RT tensile strengths, MPa (ksi):		Test	Fatigue strength, MPa (ksi), at:			
UTS	TYS	conditions(a)	10^4 cycles	10^5 cycles	10^6 cycles	10^7 cycles
Mill annealed sheet(b)						
...	1090 (158)	$R=0.1$, smooth, RT	1070(c) (155)	860 (125)	758 (110)	710 (103)
		$R=0.1$, $K_t=4$, RT	358 (52)	220 (32)	193 (28)	193 (28)
		$R=0.25$, smooth, RT	...	862 (125)	793 (115)	758 (110)
		$R=0.1$, smooth, 450 °F	882 (128)	786 (114)	710 (103)	668 (97)
		$R=-0.1$, $K_t=4$, 450 °F	372 (54)	193 (28)	165 (24)	165 (24)
2.5 mm (0.1 in.) sheet, annealed						
1130 (164)	1070 (155)	$R=0.1$, smooth, RT	1103 (160)	862 (125)	758 (110)	703 (102)
		$R=0.1$, smooth, 450 °F	895 (130)	772 (112)	689 (100)	655 (95)
		$R=0.1$, $K_t=4.2$, RT	344 (50)	207 (30)	193 (28)	193 (28)
		$R=0.1$, $K_t=4.2$, 450 °F	365 (53)	193 (28)	165 (24)	165 (24)
3.2 mm (0.125 in.) STA sheet						
1248 (181)	1220 (177)	$R=0.1$, smooth, RT	...	827 (120)	786 (114)	786 (114)
		$R=0.1$, $K_t=4.2$, RT	310 (45)	213 (31)	207 (30)	207 (30)

(a) Conditions include load type/stress ratio/notch factor (K_t)/ test temperature/etc. (b) Sheet 0.5, 2.5, and 3.2 mm (0.02, 0.10, and 0.125 in.) thick, annealed at 700 to 760 °C (1300 to 1400 °F). (c) Extrapolated. Source: *Aerospace Structural Metals Handbook*, Vol 4, Code 3715, Battelle Columbus Laboratories, Dec 1975

Ti-662: RT axial fatigue strength of forgings

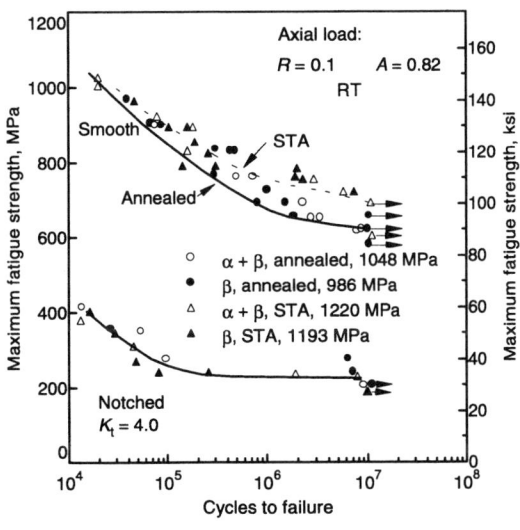

Smooth and notched fatigue strength at room temperature for α + β and for β processed forging. Beta forging involved beta block forging followed by α-β finish forging. Heat treatments were as follows: annealed at 705 to 760 °C (1300 to 1400 °F), 2 h, AC; solution treated and aged at 855 °C (1575 °F), 1 h, WQ + 565 °C (1050 °F), 4 h, AC. Source: *Aerospace Structural Metals Handbook*, Vol 4, Code 3715, Battelle Columbus Laboratories, 1975

Ti-662: Typical axial fatigue strength

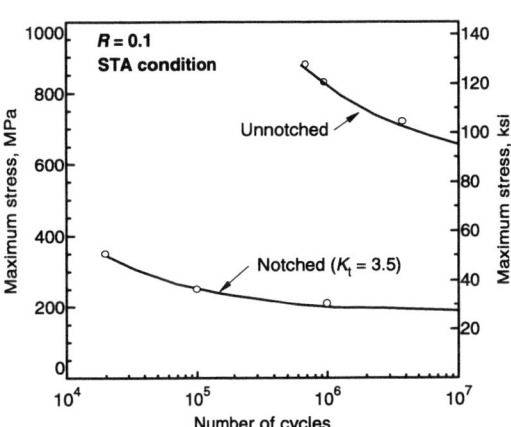

Source: *Metals Handbook*, Vol 3, 9th ed., American Society for Metals, 1978

Constant Lifetime Diagrams

Ti-662: RT smooth axial fatigue of mill annealed plate

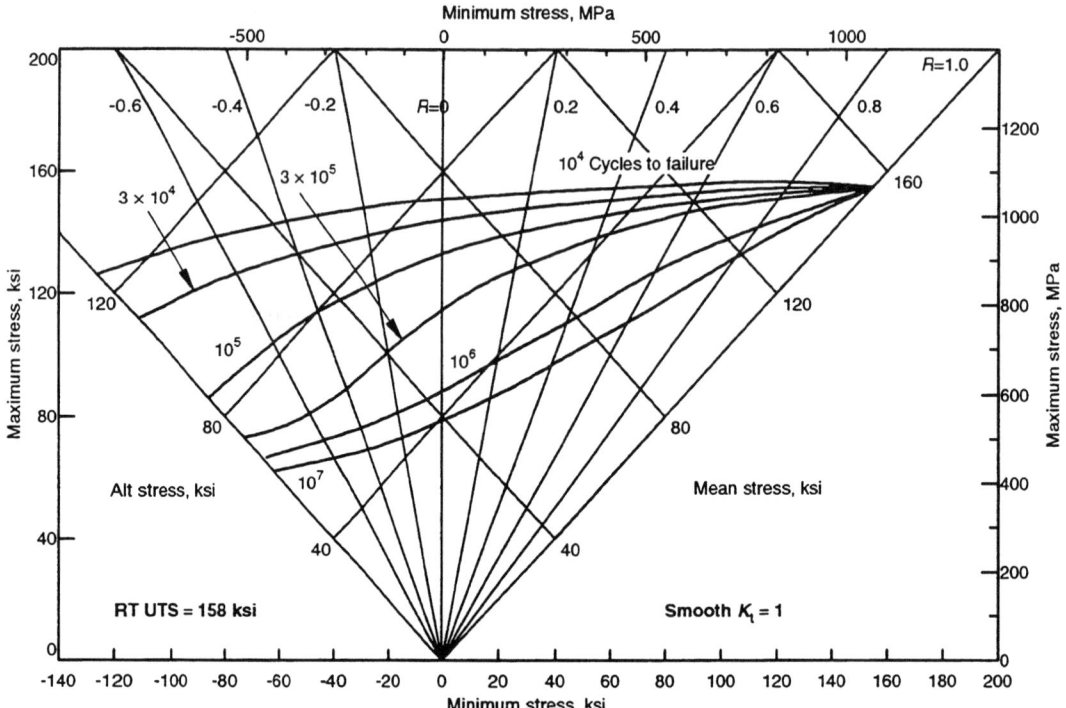

Source: *Aerospace Structural Metals Handbook*, Vol 4, Code 3715, Battelle Columbus Laboratories, 1975

Ti-662: RT notched axial fatigue of mill annealed plate

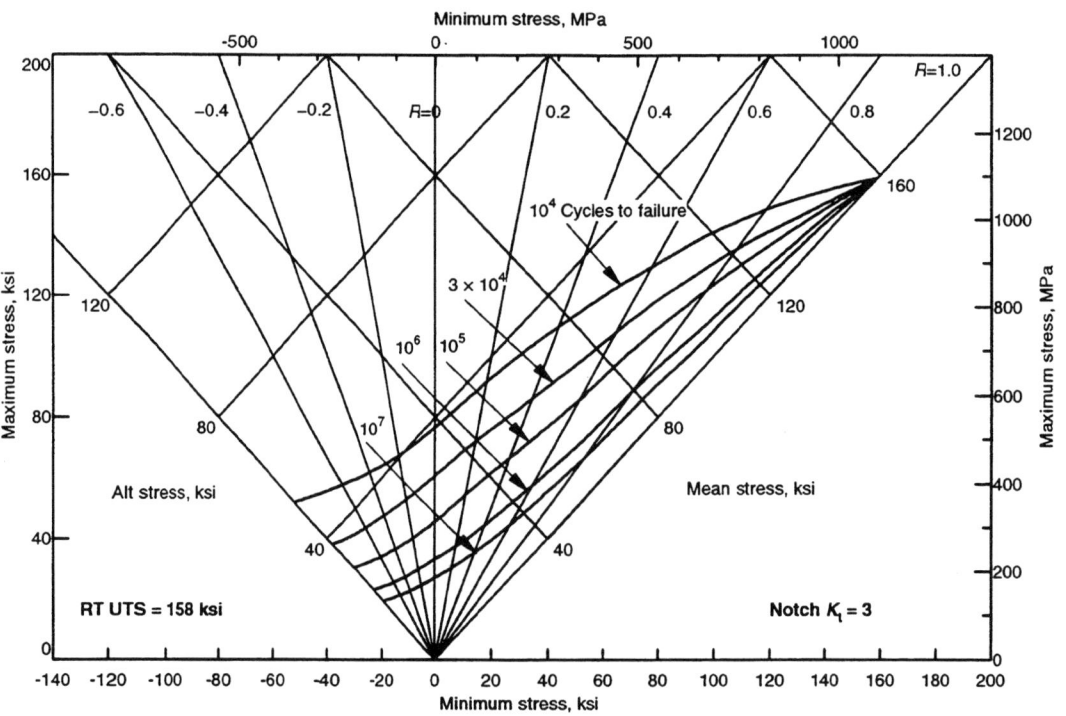

Source: *Aerospace Structural Metals Handbook*, Vol 4, Code 3715, Battelle Columbus Laboratories, 1975

Ti-662: RT smooth axial fatigue of STA plate

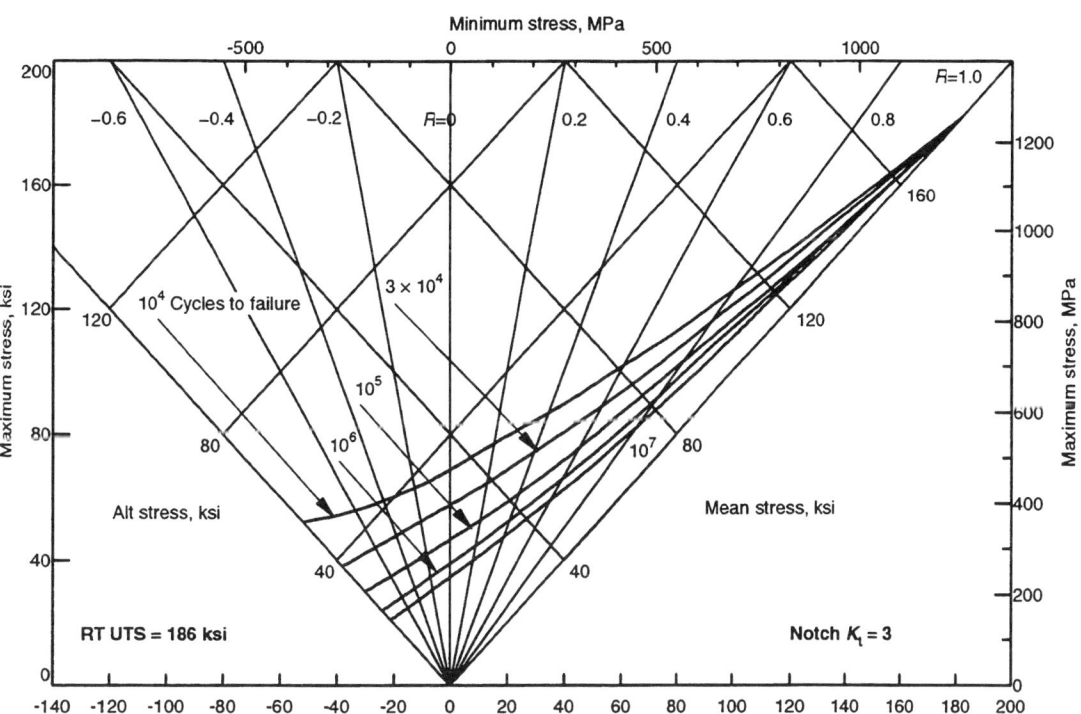

Source: *Aerospace Structural Metals Handbook*, Vol 4, Code 3715, Battelle Columbus Laboratories, 1975

Ti-662: RT notched axial fatigue of STA plate

Source: *Aerospace Structural Metals Handbook*, Vol 4, Code 3715, Battelle Columbus Laboratories, 1975

Fatigue Crack Propagation

Ti-662: Average crack growth rates

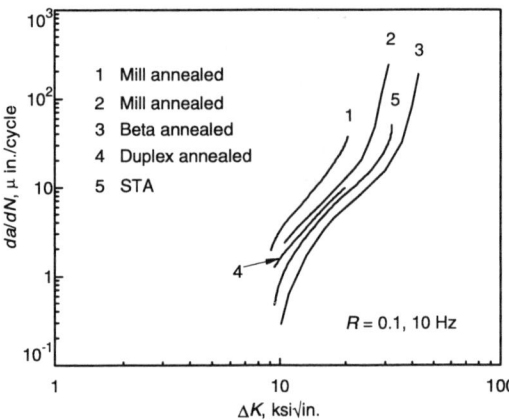

Line No.	Treatment	Yield strength MPa	ksi
1	Mill anneal	1095	159
2	Mill anneal	1124	163
3	1010 °C (1850 °F) in vacuum	965	140
4	925 °C (1700 °F) + 760 °C (1400 °F)	1041	151
5	915 °C (1675 °F), WQ, 595 °C (1100 °F)	1193	173

Fatigue crack growth rates at room temperature, tested in laboratory air at 50 to 70% relative humidity. See table for treatments and yield strengths.
Source: *Aerospace Structural Metals Handbook*, Vol 4, Code 3715, Battelle Columbus Laboratories, 1975

Ti-662: Crack growth rates for annealed plate

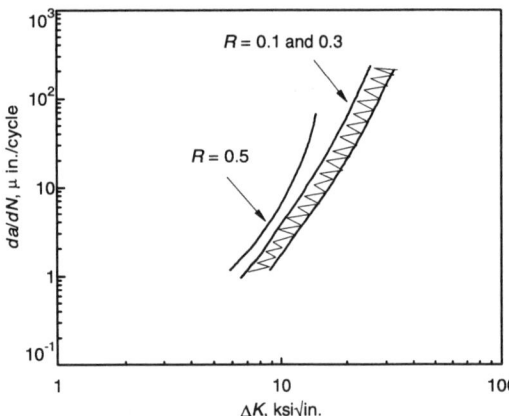

13 mm (0.5 in.) mill annealed plate was tested at room temperature in air at 50 to 70% relative humidity.
Source: *Aerospace Structural Metals Handbook*, Vol 4, Code 3715, Battelle Columbus Laboratories, 1975

Ti-662: Crack growth of β annealed plate

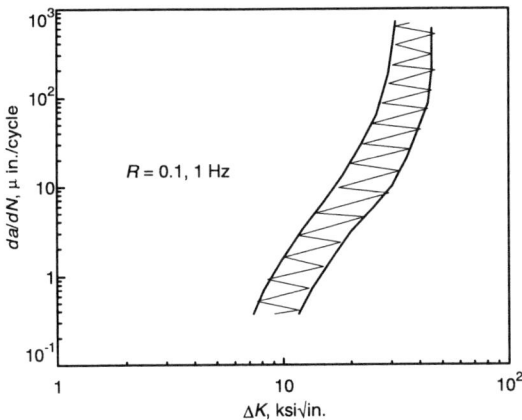

13 mm (0.5 in.) β annealed plate was tested at room temperature in air and 3.5% NaCl.
Source: *Aerospace Structural Metals Handbook*, Vol 4, Code 3715, Battelle Columbus Laboratories, 1975

Ti-662: Crack growth rates at –54 °C for STA specimens

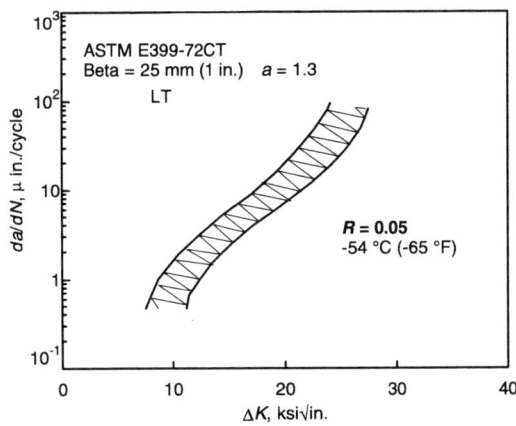

96 mm (3.8 in.) square forged bar heat treated at 870 °C (1600 °F), 30 min, WQ + 540 °C (1000 °F), 6 h.
Source: *Aerospace Structural Metals Handbook*, Vol 4, Code 3715, Battelle Columbus Laboratories, 1975

Ti-662: Crack growth in simulated body environments

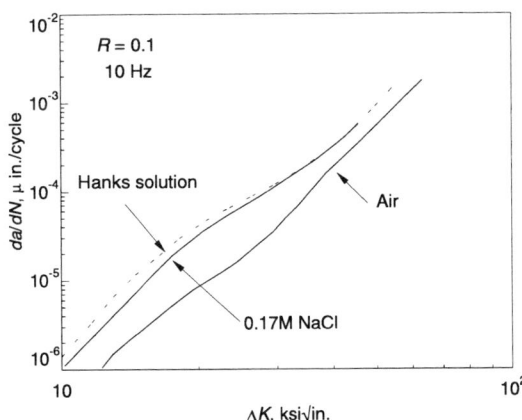

Annealed sheet at room temperature. Tensile yield strength, 986 MPa (143 ksi).
Source: *Aerospace Structural Metals Handbook*, Vol 4, Code 3715, Battelle Columbus Laboratories, 1975

Ti-662: Crack growth range at several temperatures

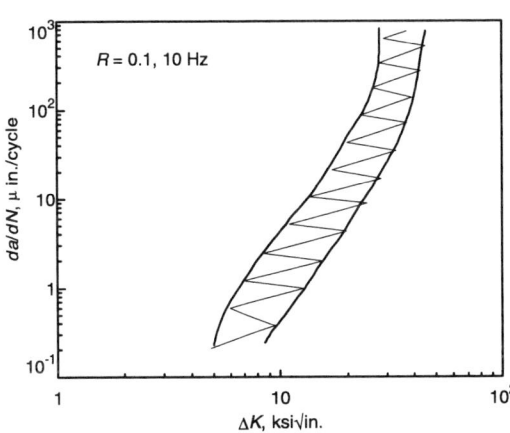

13 mm (0.5 in.) mill annealed plate tested at –62 to 82 °C (–80 to 180 °F).
Source: *Aerospace Structural Metals Handbook*, Vol 4, Code 3715, Battelle Columbus Laboratories, 1975

Fracture Properties

Import Toughness

Ti-662: Impact toughness of annealed extrusions

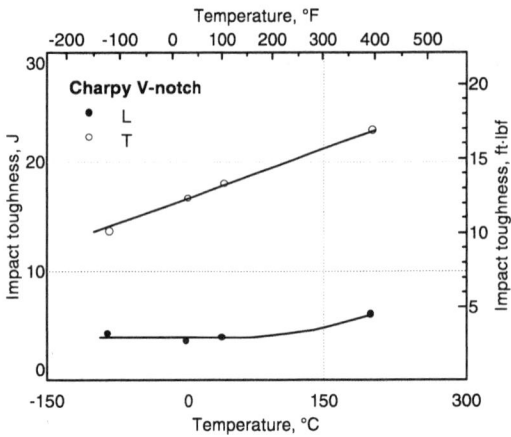

Extrusions were annealed at 705 °C (1300 °F), 40 to 60 min, AC. RT yield strength ~930 MPa (135 ksi).
Source: *Aerospace Structural Metals Handbook*, Vol 4, Code 3715, Battelle Columbus Laboratories, 1975

Ti-662: Impact toughness of STA bar

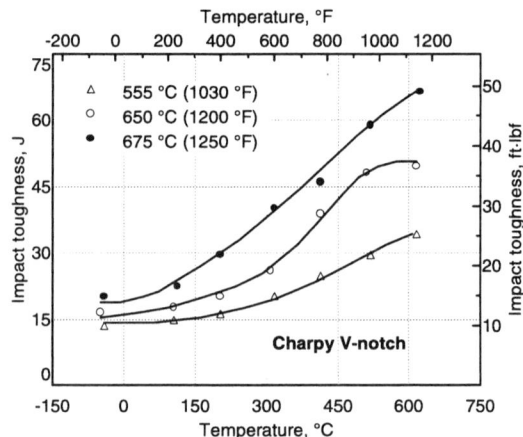

Effect of temperature on impact toughness of bar heat treated at 885 °C (1630 °F), 90 min, WQ and aged 4 h, at temperature indicated.
Source: *Aerospace Structural Metals Handbook*, Vol 4, Code 3715, Battelle Columbus Laboratories, 1975

Fracture Toughness

Ti-662: Fracture toughness/yield strength

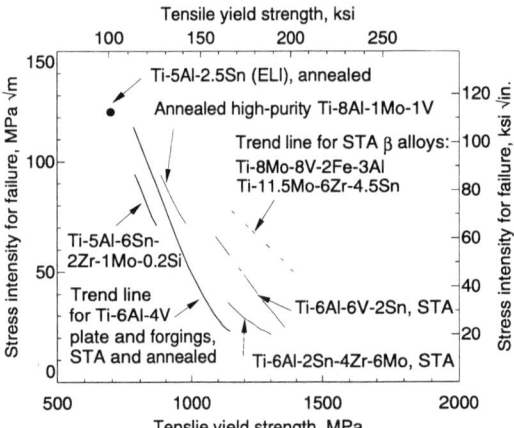

Source: *Titanium and Titanium Alloys*, MIL-HDBK 697A, 1974

Ti-662: Impact toughness of plate

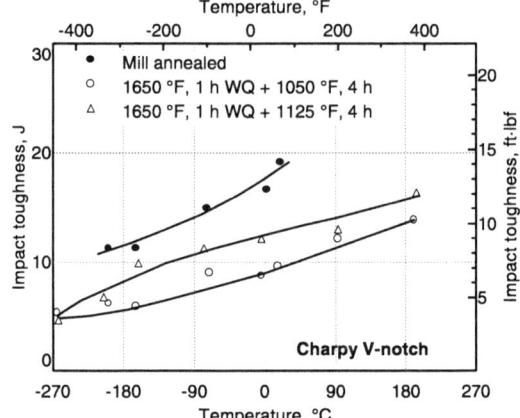

25 mm (1 in.) plate composition: 0.081 O_2, 0.018 N_2, 0.006 H_2, 0.015 C, 0.59 Fe. Approximate RT yield strengths: A, 999 MPa (145 ksi); B, 1241 MPa (180 ksi); C, 1172 MPa (170 ksi).
Source: *Aerospace Structural Metals Handbook*, Vol 4, Code 3715, Battelle Columbus Laboratories, 1975

Ti-662: Fracture toughness vs temperature

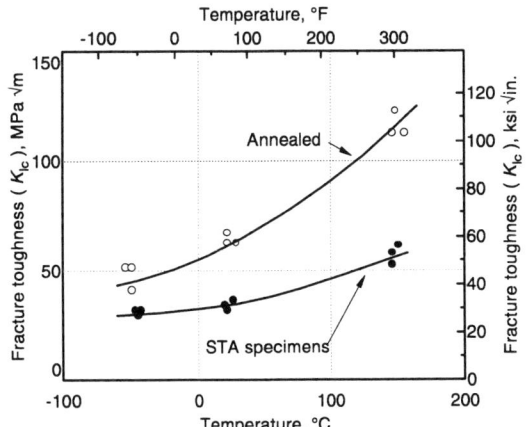

96 mm (3.8 in.) square forged bar. Plane-strain fracture toughness determined as per ASTM E399-72. Compact tension tests, LT direction with standard specimen B = 25 mm (1 in.), a = 33 mm (1.3 in.). Solution treated and aged specimens were heat treated at 870 °C (1600 °F), 30 min, WQ + 540 °C (1000 °C), 6 h, AC.
Source: *Aerospace Structural Metals Handbook*, Vol 4, Code 3715, Battelle Columbus Laboratories, 1975

Ti-662: Influence of yield strength on fracture toughness

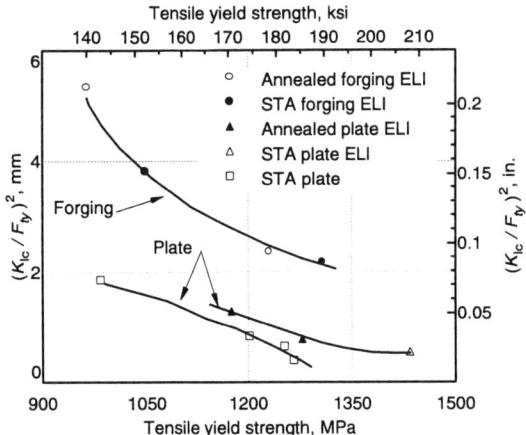

Source: *Aerospace Structural Metals Handbook*, Vol 4, Code 3715, Battelle Columbus Laboratories, 1975

Ti-662: RT fracture toughness of plate

Condition(a)	Direction	Fracture toughness (b) (K_{Ic}) MPa√m	ksi√in.
870 °C (1600 °F), 1 h, water quench, 565 °C (1050 °F), 4 h, air cool	L-T	32.7 ± 0.55	29.8 ± 0.5
900 °C (1650 °F), 1 h, water quench	L-T	37.3 ± 3.8	34.0 ± 3.5

(a) 25 mm (1 in.) plate. (b) Mean ± standard deviation. Source: J. Gallagher, *Damage Tolerant Design Handbook*, Vol 1, Battelle Columbus Laboratories, 1983

Ti-662: RT fracture toughness of plate, forging, and billet

Thickness mm	in.	Condition	Fracture toughness(b) (K_{Ic}) MPa√m	ksi√in.	Direction
Plate					
9.6	0.38	STOA at 925 °C (1700 °F), 1 h, water quench, 760 °C (1400 °F), 1 h, air cool	47.1 ± 1.3	42.9 ± 1.2	L-T
			50.6 ± 3.4	46.1 ± 3.1	T-L
13	0.50	β anneal at 985 °C (1810 °F), 1 h, argon cool	59.6 ± 2.2	54.3 ± 2.0	T-L
13	0.50	Duplex anneal	71.5 ± 2.2	65.1 ± 2.0	T-L
13	0.50	Mill anneal	38.4 ± 5.7	35.0 ± 5.2	T-L
15.7	0.62	β anneal + STOA at 980 °C (1800 °F), 30 min, air cool, 855 °C (1575 °F), 30 min, water quench, 565 °C (1050 °F), 8 h, air cool	55.0 ± 1.9	50.1 ± 1.8	L-T
32	1.25	STA at 915 °C (1675 °F), 15 min, water quench, 595 °C (1100 °F), 4 h	37.4 ± 4.1	34.1 ± 3.8	T-L
Forging					
96.5	3.80	...	64.4 ± 2.9	58.6 ± 2.7	L-T
96.5	3.80	STA at 870 °C (1600 °F), 30 min, water quench, 540 °C (1000 °F), 6 h, air cool	33.8 ± 0.7	30.8 ± 0.7	L-T
Billet					
55.8	2.20	...	57.4 ± 7.0	52.3 ± 6.4	L-T
55.8	2.20	Mill anneal at 540 °C (1000 °F), 2 h, air cool	62.7 ± 2.4	57.1 ± 2.2	L-T
305	12.00	...	69.0 ± 7.6	62.8 ± 6.9	L-T
		...	62.6 ± 4.0	57.0 ± 3.7	T-L

Source: J. Gallagher, *Damage Tolerant Design Handbook*, Vol 1, Battelle Columbus Laboratories, 1983

Ti-662: RT fracture toughness of plate and forgings

Product form and specimen	Condition	Yield stress MPa	ksi	Apparent K_{Ic} MPa√m	ksi√in.
25 mm (1 in.) plate; (three-point bend B = 1 in., a = 1 in., W = 2 in.)(a)	1550 °F, 30 min, WQ + 900 °F, 4 h, AC	1289	187	21	19
	1550 °F 30 min, WQ + 1000 °F, 4 h, AC	1261	183	27	25
	1550 °F, 30 min, WQ + 1100 °F, 4 h, AC	1193	173	34	31
	1550 °F, 30 min, WQ + 1300 °F, 4 h, AC	999	145	43	39
114 × 114 mm (4.5 × 4.5 in.) forging (double edge crack B = 0.5 in., 2a = 1 in., W = 3 in.)(b)	1650 °F, 1 h, WQ + 1050 °F, 4 h, AC	1027	149	66	60
25 mm (1 in.) ELI plate; (three-point bend B = 0.25 in., a = 0.2 in., W = 0.5 in.)(c)	1600 °F, 1 h, WQ + 1050 °F, 4 h, AC	1234	179	33	30
	1650 °F, 1 h, WQ + 1125 °F, 4 h, AC	1179	171	37	34
75 × 228 mm (3 × 9 in.) forging (center crack B = 1 in., 2a = 3 in., W = 9 in.)(d)	Annealed at 1300 °F, 2 h, AC	979	142	61	56
	1575 °F, 1 h, WQ + 1200 °F, 4 h, AC	1186(e)	172(e)	30(e)	28(e)
		1310(e)	190(e)	30(e)	28(e)

(a) J. Strawley, M. Jones, and W. Brown, Jr., "Determination of Plane Strain Fracture Toughness," *Mater. Res. Stand.*, Vol 7, 1967, p 262. (b) R. Bubsey, NASA Lewis Research Center. (c) T. DeSisto and C. Hickey, Jr., "Low Temperature Mechanical Properties and Fracture Toughness of Ti-6Al-6V-2Sn," *Proc. ASTM*, Vol 65, 1965, p 641. (d) J. Shannon, Jr., and W. Brown, Jr., "Thick Section Fracture Toughness," AFML-TDR-64-236, 1964, reported in "A Review of Factors Influencing the Crack Tolerance of Titanium Alloys," in *Applications Related Phenomena in Titanium Alloys*, ASTM STP 432, ASTM, 1968, p 33. (e) At –80 °C (–110 °F)

Plastic Deformation

Ti-6Al-6V-2Sn: Effect of thermal history on hot ductility

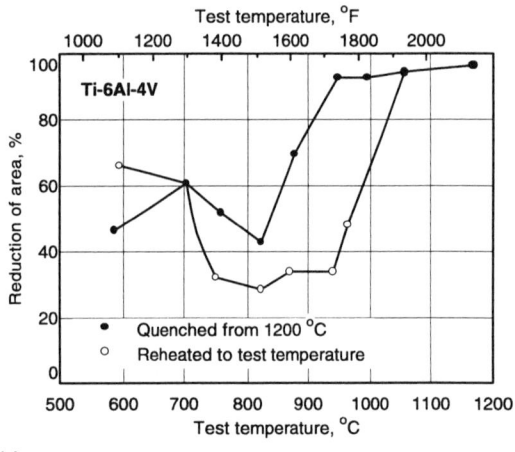
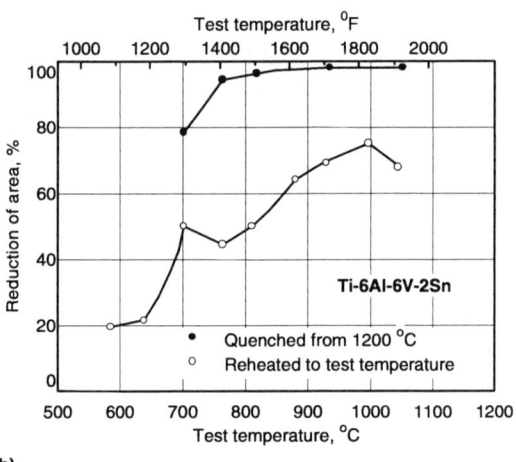

In Ti-6Al-4V, hot ductility is improved, and a ductility trough becomes narrow when rapidly cooled from the β region. In the more hardenable Ti-6Al-6V-2Sn alloy, excellent ductility is obtained down to 700 °C (1290 °F) in the sensitized mode. The reheated mode produces poorer ductility than the sensitized mode, and ductility improves more in alloys that exhibit more retardation of the β to α transformation on cooling. Sensitized specimens were quenched from 1200 °C (2190 °F) to the test temperature. Strain rate was 5/s.

Source: H.G. Suzuki *et al.*, Effect of Phase Transformation on the Hot Workability of Ti-6Al-6V-2Sn, Ti-5Al-2.5Sn and Other Alloys, *Sixth World Conference on Titanium*, P. Lacombe, R. Tricot, and G. Beranger, Ed., Les Editions de Physique, Paris, 1988, p 1427-1432

Ti-6Al-6V-2Sn: True stress-strain curves

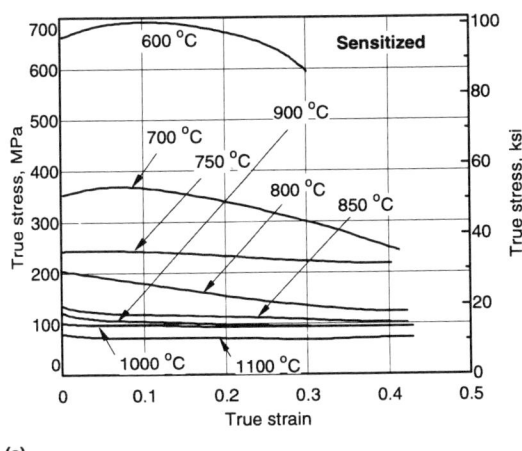
(a)

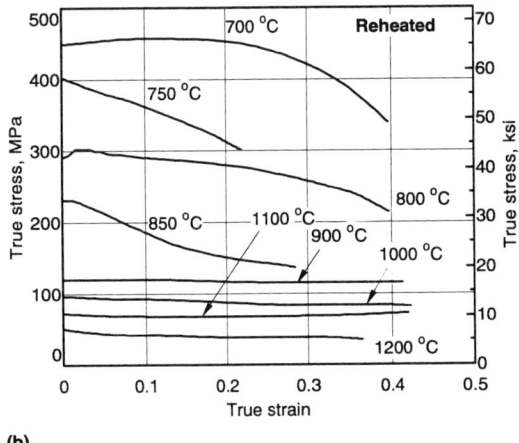
(b)

In the sensitized mode, smooth stress-strain curves are shown above 750 °C (1380 °F), and work hardening occurs below 665 °C (1220 °F). At 850 °C (1560 °F), for example, the stress level of the reheated material is almost twice that of the sensitized material at low strain. The sensitized mode involved quenching from 1200 °C (2190 °F) to the test temperature. The reheated mode involved heating to the test temperature in 60 s.
Source: H.G. Suzuki et al., Effect of Phase Transformation on the Hot Workability of Ti-6Al-6V-2Sn, Ti-5Al-2.5Sn and Other Alloys, *Sixth World Conference on Titanium*, P. Lacombe, R. Tricot, and G. Beranger, Ed., Les Editions de Physique, Paris, 1988, p 1427-1432

Ti-6Al-6V-2Sn: Tensile stress strain

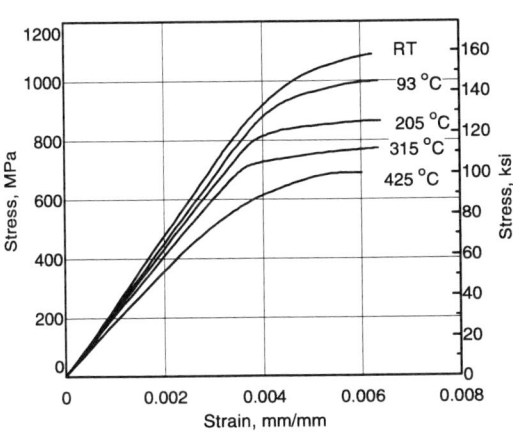

Elevated temperature tension stress-strain curves for sheet annealed 4 h at 760 °C (1400 °F).
Source: *Aerospace Structural Metals Handbook*, Vol 4, Code 3715, Battelle Columbus Laboratories, 1975, p 25

Ti-6Al-6V-2Sn: Tensile stress-strain

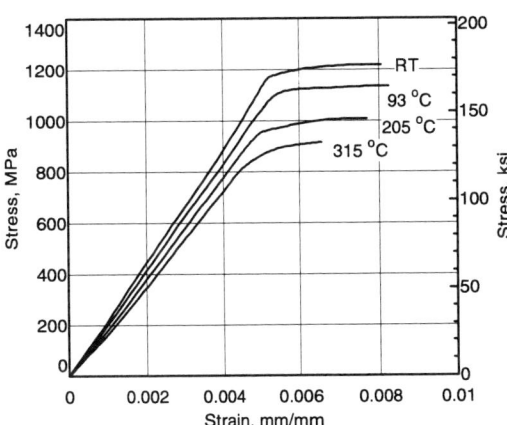

Elevated temperature tension stress-strain curves for aged bar, heat treated at 870 °C (1600 °F), 1 h, WQ + 565 °C (1050 °F), 4 h.
Source: *Aerospace Structural Metals Handbook*, Vol 4, Code 3715, Battelle Columbus Laboratories, 1975, p 25

Forging

G.W. Kuhlman, ALCOA, Forging Division

Ti-6Al-6V-2Sn is a high-strength, highly beta-stabilized, deep-hardenable α-β alloy whose primary commercial application is in aerospace structural components. The alloy can be fabricated into all forging product types, although closed-die forgings predominate.

Ti-6Al-6V-2Sn is a reasonably forgeable alloy with lower unit pressures (flow stresses), improved forgeability, and less crack sensitivity than

Ti-6Al-6V-2Sn: Forging process temperatures

Process	Metal temperature	
	°C	°F
Conventional forging	845-915	1550-1675
Supra-transus forging	(a)	(a)

Note: See "Technical Note 4: Forging" for recommended die temperatures. (a) Supra-transus may be used in early forging operations, but it must be followed by sufficient subtransus reduction.

Ti-6Al-6V-2Sn: Effect of thermomechanical processing on properties

Microstructural observations for thermomechanical processing options of Ti-662 suggested that α morphology is the key microstructural feature affected by the TMP route. However, less significant α grain size modification was realized than observed for alloy Ti-6Al-6V. Cost and product uniformity implications parallel those observed for Ti-6Al-4V.

Alloy	TMP option	Direction	Tensile yield strength MPa	Tensile yield strength ksi	Ultimate tensile strength MPa	Ultimate tensile strength ksi	Elongation, %	Reduction of area, %	K_{Ic} MPa√m	K_{Ic} ksi√in.
Ti-6Al-6V-2Sn Std	α+β forge/MA	L	1094	158	1164	169	18	31	39	35
		T	1049	152	1128	163	15	24	...	...
	α+β forge/RA	L	1041	151	1110	161	17	33	50	45
		T	1028	149	1095	159	16	29	...	...
Ti-6Al-6V-2Sn ELI	α+β forge/RA	L	1022	148.2	1089	158	19	37	74	67
		T	993	144	1068	155	15	29	68	62
Ti-6Al-6V-2Sn Std	β preform/MA	L	1032	150	1094	158.6	11	22	58	52
		T	1021	148.0	1090	158.1	12	23	59	53
	β preform/block/MA	L	1024	148.5	1110	161	9	19	71	64
		T	973	141	1076	156	10	22	69	63

the α-β alloy Ti-6Al-4V (see figure). The final microstructure of Ti-6Al-6V-2Sn forgings is developed by conventional thermomechanical processing in forging manufacture. Thermomechanical processes for the alloy use a combination of subtransus forging followed by subtransus thermal treatments to fulfill mechanical-property criteria.

Final thermal treatments for forgings include annealing (A), solution treatment and annealing (STAN), and solution treatment and aging (STA), with final thermal treatment selected based on strength requirements. Ti-6Al-6V-2Sn forgings may be supplied in an annealed condition to facilitate machining and subsequently solution treated and aged to optimum strength levels.

Annealing is conducted at 705 to 760 °C (1300 to 1400 °F). Solution treatment is subtransus, at 845 to 900 °C (1550 to 1650 °F), followed by water quenching. Forgings may then be annealed or aged. Aging is conducted at 510 to 620 °C (950 to 1150 °F) depending on strength mechanical-property objectives for the STA condition. Subtransus thermomechanical processes (forging and thermal treatment) for Ti-6Al-6V-2Sn forgings achieve equiaxed α in transformed β matrix microstructures that enhance strength, ductility, and high-cycle fatigue properties. Annealed microstructures consist of 40 to 80% α, whereas solution treated and aged microstructures are 10 to 20% equiaxed α.

Deformation objectives in forging Ti-6Al-6V-2Sn are to obtain the final forging shape and desired final microstructure at least cost with the conventional subtransus (α + β) forging thermomechanical processes most widely used. To achieve desired equiaxed α structures, subtransus reductions of 50 to 75%, accumulated through one or more forging steps, are required. Supra-transus

Ti-6Al-6V-2Sn: Flow stress comparison

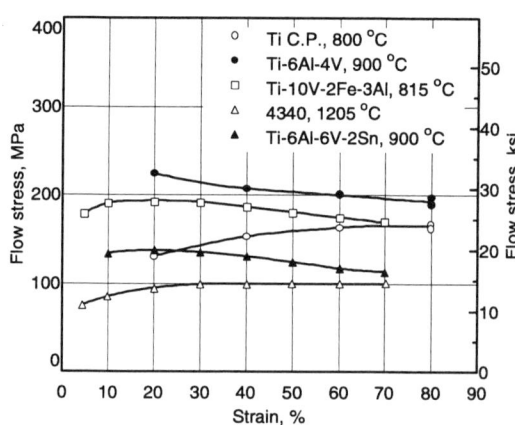

Flow stress of commonly forged titanium alloys at 10/s strain rate compared to 4340 alloy steel at 27/s strain rate.

(β) forging for Ti-6Al-6V-2Sn may be used in early forging operations, including upsetting and open die preforming, to reduce unit pressures and ease forging fabrication. However, higher temperature initial forging operations must be followed by sufficient subtransus reduction to achieve the desired predominately equiaxed α structure. Conventionally forged Ti-6Al-6V-2Sn is then subtransus solution treated, quenched and aged as noted above.

Due to the high iron and copper content (up to 1.0%) of Ti-6Al-6V-2Sn, it may be susceptible to the formation of β flecks, or small fully transformed areas resulting from excessive microsegregation. To avoid excessive β flecking, which may adversely affect ductility and toughness, metal temperature in forging and solution treatment are targeted at the low end of the temperature ranges cited above.

Fabrication

Forming. Ti-6Al-6V-2Sn is readily formable in the annealed condition. For sheet or plate, it is generally used in the annealed condition, although the alloy is capable of heat treatment to higher strength levels with some loss of toughness. When Ti-6Al-6V-2Sn sheet and plate are hot formed at any temperature over 540 °C (1000 °F) and air cooled, the material should be stabilized by reheating to 540 °C (1000 °F) followed by air cooling.

Welding. Ti-6Al-6V-2Sn is difficult to weld. Like all titanium alloys, it is weldable by all methods except shielded arc welding and submerged arc welding (because no flux is permitted).

The ductility of Ti-6Al-6V-2Sn weldments is poor unless a postweld annealing treatment is used. Solution heat treatment followed by water quenching does not improve weld ductility, and subsequent aging of such conditioned material results in weld metal embrittlement. Solution of the weld metal ductility problem is possible in some applications by using an annealing treatment. The treatment of 4-h exposure at 725 °C (1340 °F), followed by air cooling has been recommended.

Machining of titanium alloys is comparable to machining a good grade of stainless steel. In general, very sharp tools with a slightly larger rake angle and a very keen edge work quite well. Slower speed and heavier cuts are preferred because they keep tool temperatures down and produce coarse chips, which are more difficult to ignite.

Drilling of thin-walled titanium is not much of a problem as long as the drill is sharp. Thicker walls require a heavy flood of coolant to remove heat and chips. General information on the machining of titanium alloys is covered in "Technical Note 7: Machining."

Ti-6Al-6V-2Sn: Temperatures for hot forming

Alloy	Forming temperature °C	°F
CP Ti (all grades)	480-705	900-1300
Alpha and near-alpha alloys		
Ti-8Al-1V-1Mo	790 ± 15	1450 ± 25
Ti-5Al-2.5Sn	620-815	1150-1500
Alpha-beta alloys		
Ti-6Al-6V-2Sn	790 ± 15	1450 ± 25
Beta alloy		
Ti-13V-11Cr-3Al	605-790	1125-1450

Note: Annealed or solution treated material. Source: "Fabrication Practices for Titanium and Titanium Alloys," Lockheed Corporate Process Specification LCP70-1099, Revision B, Lockheed-California Company, Oct 1983

Heat Treatment

Annealing. Ti-6Al-6V-2Sn is one of the strongest titanium grades available in the annealed condition, which consists of about 2 to 8 h of exposure at 705 to 760 °C (1300 to 1400 °F), followed by air cooling or furnace cooling. This alloy is so highly beta stabilized that annealing should ideally be terminated by slow cooling from the annealing temperature to an intermediate temperature. Slow cooling, such as furnace cooling, produces maximum annealed strength. Air cooling may be used from annealing temperatures below 760 °C (1400 °F), but strength will generally be lowered. Annealing at temperatures higher than 760 °C (1400 °F) is also possible (see table).

Solution heat treatments vary for some products. The typical 885 °C (1625 °F) temperature is a workable solution temperature for a wide range of products and applications. About 30% primary α, balance β phase, is found in the microstructure after this treatment. Water quenching is the standard method of terminating solution heat treatment, although fast air cooling achieved by forced air stream may be satisfactory for thin-section material because the β phase in this alloy is fairly stable. Solution treatment above the transus results in a severe loss of ductility (see figure next page).

Ti-6Al-6V-2Sn: Stress relief and annealing treatments

Heat treatment	Temperature °C	°F	Time, h	Cooling method
Typical stress relief range	480-650	900-1200	1-4	Air or slow cool
50 to 90% relaxation of residual stress	595	1100	2	Air cool
Typical anneal	700-760	1300-1400	2-8	Air or slow cool
Extended anneal range	700-815	1300-1500	0.75-4	Air or slow cool(a)

(a) Annealing at the higher temperature, followed by a furnace cool to 595 °C (1100 °F) then air cooling to room temperature is recommended.

Ti-6Al-6V-2Sn: Solution treatment and aging

Treatment	Temperature °C	°F	Duration	Cooling method
Solution treatments				
Typical for most products	885 ± 15	1625 ± 25	60 min	WQ
Sheet <3.22 mm (0.125 in.) thick	830-870	1525-1600	5 to 15 min	WQ
Sheet >3.2 mm (0.125 in.) thick	845-885	1550-1625	30 min	WQ
Bar, forging, extrusions	845-900	1550-1650	60 min	WQ
Aging				
Typical	540-620	1000-1150	4-8 h	AC
Low aging temperature	480-540	900-1000	...	AC
Overage	595-650	1100-1200	...	AC
Flat rolled products	565-620	1050-1150	4 h	AC
Bar, forging, extrusions	510-595	950-1100	4-12 h	AC

Ti-6Al-6V-2Sn: Effect of 565 °C (1050 °F) aging on tensile properties

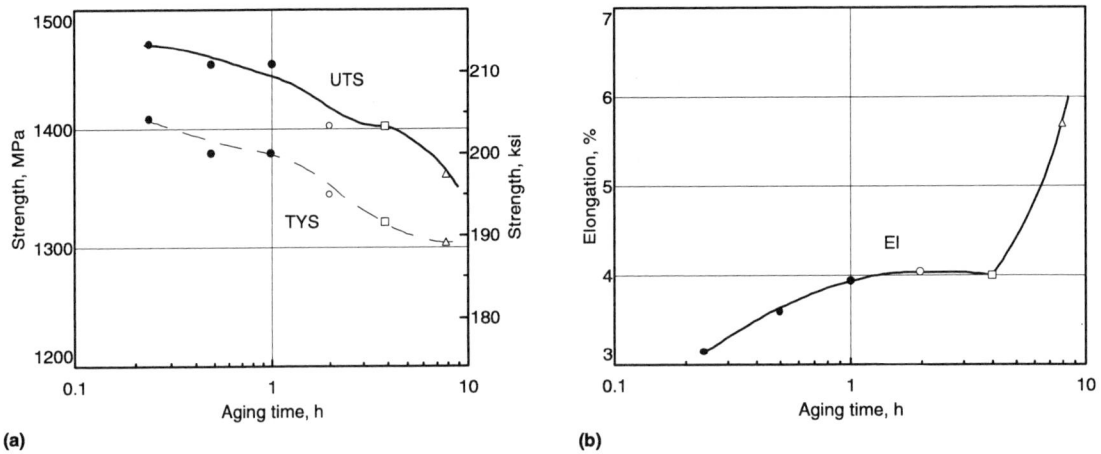

1.5 mm (0.060 in.) sheet solution treated 15 min at 885 °C (1625 °F), water quenched.
Source: R.A. Wood and R.J. Favor, *Titanium Alloys Handbook*, MCIC HB-02, Battelle Columbus Laboratories, 1972

Ti-6Al-6V-2Sn: Effect of solution treatment on tensile properties

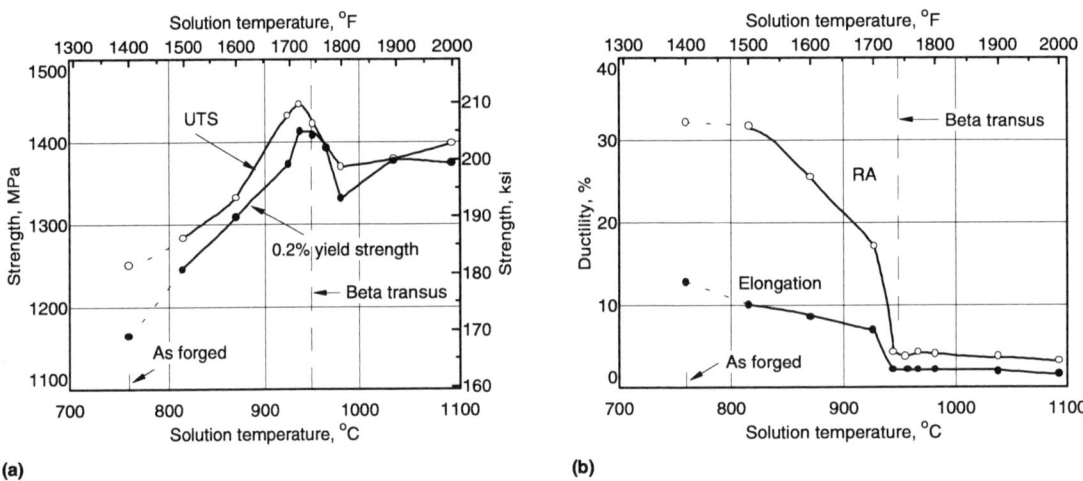

Effect of solution heat treatment temperature (1 h exposure terminated by water quenching) on the aged tensile properties of die forgings aged 3 h, 585 °C (1085 °F), air cooled.
Source: R.A. Wood and R.J. Favor, *Titanium Alloys Handbook*, MCIC HB-02, Battelle Columbus Laboratories, 1972

Ti-7Al-4Mo

UNS Number: R56740

Ti-7Al-4Mo is a heat-treatable alpha-beta alloy with roughly 10% higher strength than Ti-6Al-4V. In the annealed condition, Ti-7Al-4Mo also offers a considerable improvement in creep resistance at temperatures up to 480°C (900 °F) over Ti-6Al-4V. Short-term high-temperature tensile properties are improved as well. Ti-7Al-4Mo is used up to 455 °C (850 °F), although it offers good stability under stress at temperatures at least as high as 480 °C (900 °F) for prolonged loading times.

Chemistry and Density

The molybdenum addition, in stabilizing the beta phase, improves heat treatment and strengthens beta by solid solution. Compared with Ti-6Al-4V, both the alpha and beta phases are more highly alloyed and consequently stronger. The replacement of vanadium with molybdenum and the addition of 1% aluminum are the reasons why the elevated-temperature properties of this alloy are better than those of Ti-6Al-4V. This improvement reflects itself in increased resistance to deformation at forging temperatures in comparison with Ti-6Al-4V, although this difference may be small.

Density. 4.48 g/cm^3 (0.162 lb/in.^{3}).

Product Forms

Billet, bar, plate, wire, extrusions are available. Ti-7Al-4Mo is used primarily in the form of light and intermediate forgings and extrusions. Welding is not recommended.

Product Condition/Microstructure

Ti-7Al-4Mo is an alpha-beta alloy with more beta in the annealed condition than Ti-6Al-4V because of the much lower solubility of molybdenum in alpha. The strength of Ti-7Al-4Mo is about 100 MPa (15 ksi) higher than Ti-6Al-4V in the annealed condition and about 170 MPa (25 ksi) in the solution treated and aged condition. The depth of hardening is similar to that of Ti-6Al-4V. Guaranteed minimum tensile properties for reforge and upset tests are 10% higher than those of Ti-6Al-4V.

Applications

This alloy has limited use today, being used primarily for horns on ultrasonic welding equipment. Given the alloys combined high-temperature strength and stability, bar, forgings, and forging stock are in limited current use for jet engine disks, compressor blades and spacers.

Ti-7Al-4Mo: Equivalent specifications

Specification	Designation	Description	Al	C	Fe	H	Mo	N	O	OT	Other bal Ti
UNS	R56740		7				4				
USA											
AMS 4970E		Frg Bar Wir Bil STA	6.5-7.3	0.1	0.3	0.013	3.5-4.5	0.05	0.2	0.4	Y 0.005; bal Ti
MIL F-83142A	Comp 9	Frg HT	6.5-7.3	0.1	0.3	0.013	3.5-4.5	0.05	0.2	0.4	Y 0.005; bal Ti
MIL T-9047G	Ti-7Al-4Mo	Bar Bil STA Ann	6.5-7.3	0.1	0.3	0.013	3.5-4.5	0.05	0.2	0.4	Y 0.005; bal Ti

Ti-7Al-4Mo: Commercial compositions

Specification	Designation	Description	Al	C	Fe	H	Mo	N	O	OT	Other
France											
Ugine	UTA7D	Frg Quen Aged	6.5-7.3	0.08	0.25	0.0125	3.5-4.5	0.5	0.2		bal Ti
USA											
RMI	Ti-7Al-4Mo	Bil	6.5-7.5	0.08 max	0.25 max	0.01	3.5-4.5	0.05 max			bal Ti
Timet	TIMETAL 7-4	Ann	6.5-7.3	0.1 max	0.3 max	0.13 max	3.5-4.5	0.05 max	0.2 max		bal Ti

Physical Properties

Ti-7Al-4Mo: Summary of typical physical properties

Beta transus	1005 ± 15 °C (1840 ± 25 °F)
Melting (liquidus) point	~1650 °C (3000 °F)
Density(a)	4.48 g/cm^3 (0.162 lb/in.^{3})
Electrical resistivity(a)	1.7 µΩ · m
Magnetic permeability	Nonmagnetic
Specific heat capacity(a)	515 J/kg · K (0.123 Btu/lb · °F)
Thermal conductivity(a)	6.1 W/m · K (3.5 Btu/ft · h · °F)
Thermal coefficient of linear expansion(b)	9.7 x 10^{-6}/°C (5.4 x 10^{-6}/°F)

(a) Typical values at room temperature of about 20 to 25 °C (68 to 78 °F). (b) Mean coefficient from room temperature to 455 °C (850 °F).

Elastic Properties

Ti-7Al-4Mo: Typical RT elastic moduli

Condition	Poisson's ratio	Tensile modulus		Compressive modulus		Shear modulus	
		GPa	10^{6}psi	GPa	10^{6}psi	GPa	10^{6}psi
Annealed	0.32	111	16.2	111	16.2	42	6.1
STA	0.32	116	16.9	116	16.9	44	6.4

Source: MIL-HDBK 5, Dec 1991

Ti-7Al-4Mo: Static tensile modulus of bar

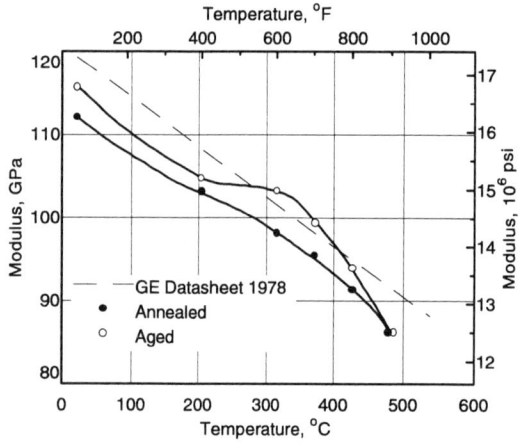

Source: *Aerospace Structural Metals Handbook*, Vol 4, Code 3708, Battelle Columbus Laboratories, 1972

Ti-7Al-4Mo: Dynamic tensile modulus of bar

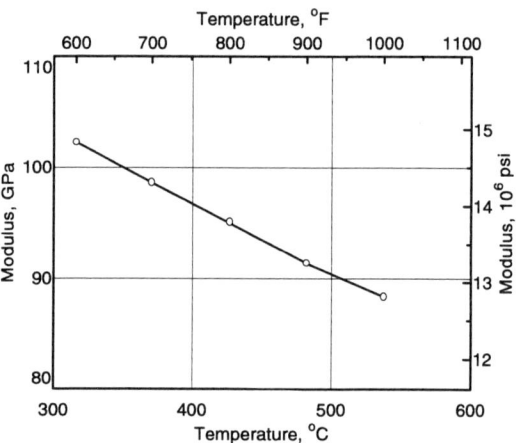

Bar forged at 980 to 1010 °C (1800 to 1850 °F) + 785 °C (1450 °F), 1 h, FC to 565 °C (1050 °F), AC + 565 °C (1050 °F), 24 h, AC.
Source: *Aerospace Structural Metals Handbook*, Vol 4, 1972

Electrical Resistivity

Ti-7Al-4Mo: Electrical resistivity

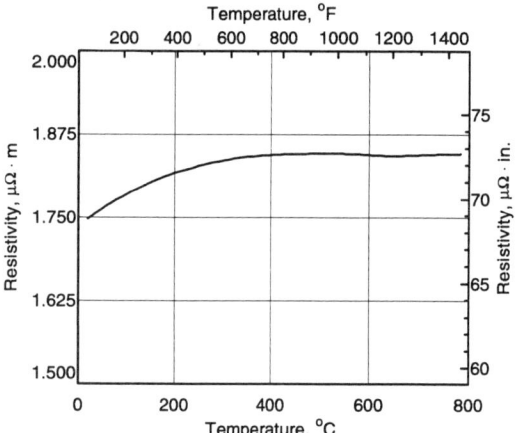

Source: *Aerospace Structural Metals Handbook*, Vol 4, Code 3708, Battelle Columbus Laboratories, 1972, p 9

Ti-7Al-4Mo: Dynamic shear modulus of bar

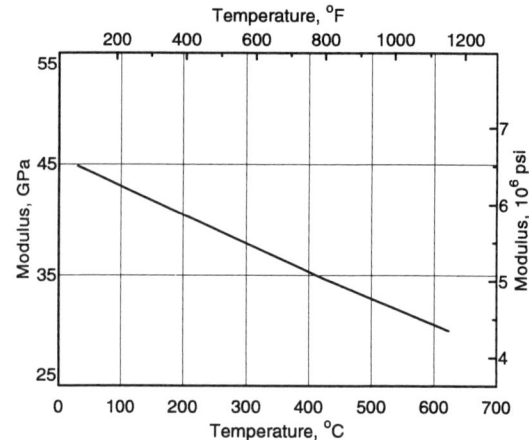

Source: *Aerospace Structural Metals Handbook*, Vol 4, Code 3708, Battelle Columbus Laboratories, 1972

Corrosion

General corrosion data on Ti-7Al-4Mo are limited, although molybdenum additions may increase corrosion resistance in reducing (nonoxidizing) environments at the expense of less resistance in oxidizing environments. Molybdenum also is usually beneficial for stress-corrosion cracking (SCC) resistance, but SCC thresholds in salt solutions are comparable to Ti-6Al-4V with similar yield strengths (see table). Hot-salt SCC of a Ti-7Al-4Mo power plant component has been observed in conjunction with silver coatings (*Metals Handbook, Corrosion*, Vol 13, 9th ed., 1987, p 1039-1040).

Erosion-Corrosion. Extensive erosion-corrosion testing of Ti-6Al-4V, Ti-5Al-2.5Sn, and Ti-7Al-4Mo alloys has been conducted in high-velocity wet steam environments for use in low-pressure steam turbine blading in power plants. These alloys demonstrated superior resistance to type 403 stainless steel (12 to 13% Cr steel) in operating turbines and in water droplet erosion and water jet impingement tests. (R.A. Wood, "Status of Titanium Blading for Low Pressure Steam Turbines," EPRI AF-445, Final Report, Electric Power Research Institute, Feb 1977). Single-shot water jet impingement testing has shown that annealed Ti-7Al-4Mo alloy is significantly more erosion resistant than 12% Cr steel, type 303 stainless steel, or Stellite alloy 6 at jet velocities of 610 and 915 m/s (2000 and 3000 ft/s).

Ti-7Al-4Mo: Fracture toughness in air and 3.5% NaCl solution at 25 °C

Alloy	Thickness mm	Thickness in.	Heat treatment	Tensile yield strength MPa	Tensile yield strength ksi	K_{Ic} or K_c MPa√m	K_{Ic} or K_c ksi√in.	K_{Iscc} or K_{scc} MPa√m	K_{Iscc} or K_{scc} ksi√in.
Ti-7Al-4Mo	13	0.50	Mill annealed	993	144	80	73	34	31
			STA	1151	167	39	36	28	26
Ti-6Al-4V (standard grade)	15	0.06	Mill annealed	944	137	165	150	121	110
			Duplex annealed	917	133	165	150	121	110
			STA	1103	160	104	95	71	65
			β STA	1068	155	104	95	71	65
	13	0.50	Mill annealed	944	137	66	60	38	35
			Duplex annealed	917	133	77	70	57	52
			STA	1103	160	51	47	27	25
			β STA	1068	155	77	70	49	45

Source: R. Schutz, "Stress Corrosion Cracking of Titanium Alloys," in *Stress-Corrosion Cracking*, ASM International, 1992

Thermal Properties

Melting (Liquidus) Temperature. ~1650 °C (~3000 °F)

Specific Heat. At 20 °C (68 °F): 515 J/kg · K (0.123 Btu/lb · °F) (*Metals Handbook*, Vol 3, 9th ed., 1978, p 393).

Ti-7Al-4Mo: Specific heat

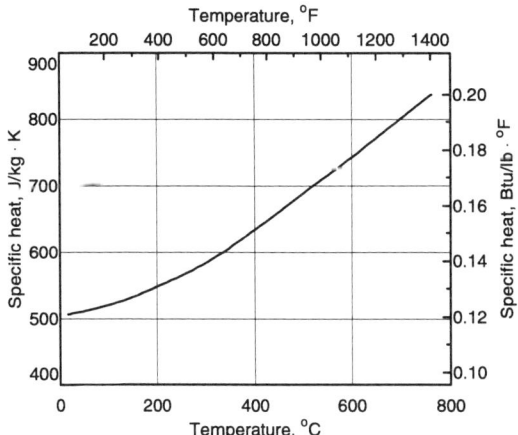

Source: MIL-HDBK 5, Sept 1976

670 / Alpha-Beta Alloys

Thermal Expansion

Ti-7Al-4Mo: Thermal coefficient of linear expansion

Temperature		Average coefficient	
°C	°F	10^{-6}/K	10^{-6}/°F
20-100	70-212	9.0	5.0
20-205	70-400	9.2	5.1
20-315	70-600	9.4	5.2
20-425	70-800	9.7	5.4
20-540	70-1000	10.1	5.6
20-650	70-1200	10.4	5.8
20-815	70-1500	11.2	6.2

Source: *Metals Handbook*, Vol 2, 10th ed., ASM International, 1990, p 620

Ti-7Al-4Mo: Thermal coefficient of linear expansion

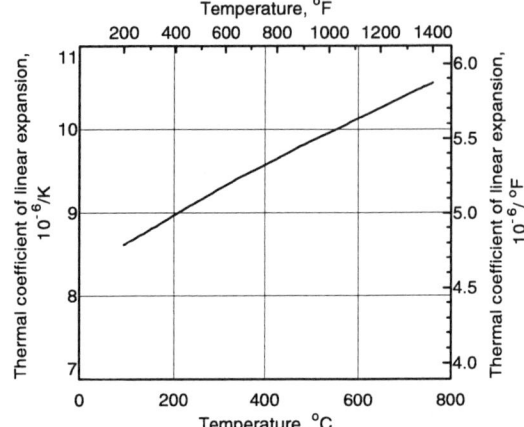

Source: MIL-HDBK 5, Sept 1976

Thermal Conductivity

Ti-7Al-4Mo: Thermal conductivity

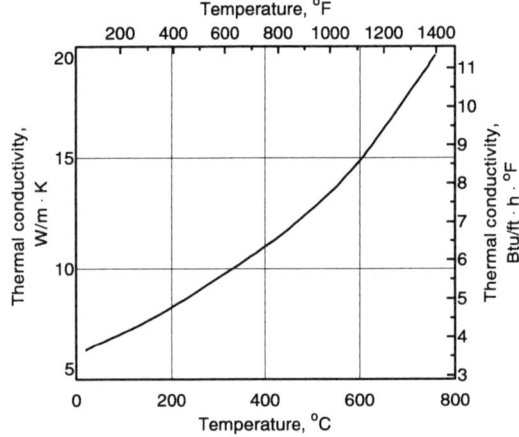

MIL-HDBK 5, Sept 1976

Ti-7Al-4Mo: Thermal conductivity

Temperature		Conductivity	
°C	°F	W/m·K	Btu/ft·h·°F
20	68	6.1	3.5
315	600	10.4	6.0
427	800	12.1	7.0
540	1000	13.8	8.0

Source: *Metals Handbook*, Vol 3, 9th ed., Americal Society for Metals, 1978

Mechanical Properties

Ti-7Al-4Mo: Design tensile properties for extrusions

Thickness		Ultimate tensile strength(a)		Tensile yield strength(a)		Elongation (a), %
mm	in.	MPa	ksi	MPa	ksi	
Annealed						
≤50	≤2.0	1000	145	930	135	10
50-100	2.01-4.0	965	140	896	130	10
≤13	≤0.5	1172	170	1103	160	6
STA						
13-25	0.51-1.0	1103	160	1034	150	6
25-50	1.01-2.0	1034	150	965	140	6
50-64	2.01-2.5	1000	145	930	135	6
64-100	2.51-4.0	965	140	896	130	6

(a) Data are for S-basis values in the longitudinal (L) direction for extended bar, rod, and special-shaped section per MIL-T-81556. Source: MIL-HDBK 5, Dec 1991

Ti-7Al-4Mo: Design tensile properties for MIL-T-9047 bar

Thickness		Ultimate tensile strength (L-LT)(a)		Tensile yield strength (L-LT)(a)		Elongation, %	
mm	in.	MPa	ksi	MPa	ksi	L	LT
Annealed							
≤50	≤2.00(b)	1000	145	930	135	10	10
50-75	2.0-3.00(b)	965(c)	140(c)	896(c)	130(c)	10	10(c)
STA							
≤25	≤1.00	1172	170	1103	160	8	8
25-50	1.0-2.00	1103	160	1034	150	8	8
50-100	2.0-4.00	1034(c)	150(c)	965(c)	140(c)	8	8(c)

(a) Data are S-basis values in the longitudinal (L) and long-transverse (LT) direction unless otherwise noted. (b) Maximum of 6500 mm^2 (10 in.^{2}) cross-sectional area. (c) Applicable in ST direction. Source: MIL-HDBK 5, Dec 1991

Hardness

Reported typical hardness value range is 32 to 38 HRC for solution treated and aged material.

Typical Tensile Properties

Typical room-temperature tensile properties are higher than the guaranteed minimums (see table), which vary with section size. Lower strengths occur in the annealed condition, or when STA material has been worked in the β phase field prior to heat treatment.

Ti-7Al-4Mo: Vickers hardness

Heat treatment	Ultimate tensile strength		Hardness, HV
	MPa	ksi	
845 °C (1550 °F) 1 h, WQ, 480 °C (900 °F) 16 h, AC	1148	166.5	400
	1147	166.4	360
	1138	165.1	388
	1135	164.7	376
	1128	163.3	373
	1139	165.2	365
	1146	166.3	349
	1168	169.5	366
	1196	173.5	386
870 °C (1600 °F) 1 h, WQ, 480 °C (900 °F) 16 h, AC	1165	169.0	348
	1168	169.5	347
	1148	166.6	355
	1145	166.1	348
	1150	166.8	348
	1153	167.2	356
	1174	170.3	360
	1198	173.8	367
	1248	181.0	364

Note: Measurements were made from consecutive 9.5 mm (0.375 in.) locations on a standard Jominy bar. Source: "High Strength Titanium Alloy Forgings, Boeing Airplane Company, Nov 1956, reported in *The Engineering Properties of Commercial Titanium Alloys*, M. Mote, R. Hooper, and P. Frost, Ed., TML Report 92, Battelle Memorial Laboratory, 1958

Ti-7Al-4Mo: Minimum RT tensile properties in STA condition

Thickness as rolled or forged		Thickness as heat treated		Tensile properties (guaranteed minimum)							
				Ultimate tensile strength		Tensile yield strength (0.2% offset)		Elongation in 4D,%		Reduction of area, %	
mm	in.	mm	in.	MPa	ksi	MPa	ksi	L	T	L	T
≤13	≤0.5	≤13	≤0.5	1240	180	1135	165	8	...	20	...
>13-25	>0.5	>13-25	>0.5-1	1170	170	1103	160	8	...	20	...
>25-50	>1-2	>25-50	>1-2	1105	160	1035	150	8	6	20	12
>50-100	>2-4	>50-100	>2-4	1035	150	965	140	8	6	20	10
>100(a)	>4(a)	25	1	1170	170	1105	160	...	8	...	15
>100(b)	>4(b)	22	0.875 square	1170	170	1105	160	8	...	20	...
>100-150	>4-6	≤25	≤1	1105	160	1035	150	6	4	15	12

Note: Specimens were heat treated. (a) Upset forged to 25 mm (1 in.) maximum using 3 to 1 ratio. (b) Reforged to 22 mm (0.875 in.) square. Source: R. Wood, *Titanium Alloys Handbook*, MCIC-HB-02, Battelle Columbus Laboratories, 1972

672 / Alpha-Beta Alloys

Ti-7Al-4Mo: RT tensile strength of aged bar

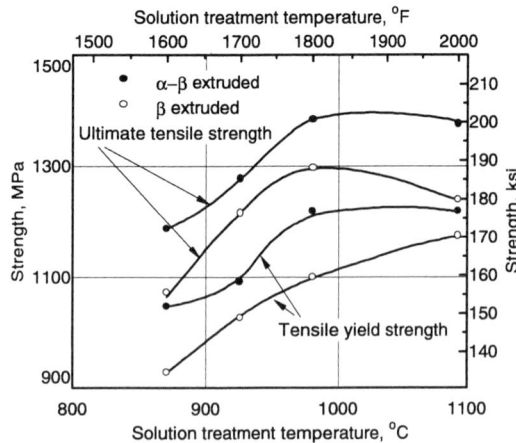

Bar aged at 480 °C, 8 h.
Source: *Aerospace Structural Metals Handbook*, Vol 4, Code 3708, Battelle Columbus Laboratories, 1967

Ti-7Al-4Mo: Strength of unaged bar

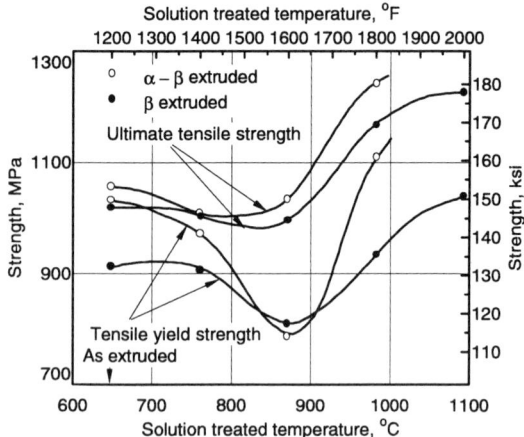

Source: *Aerospace Structural Metals Handbook*, Vol 4, Code 3708, Battelle Columbus Laboratories, 1967, p 7

High-Temperature Strength

The typical aged tensile strengths at elevated temperature are high percentages of the room-temperature strength, as shown in the table below.

Thermal Stability. Ti-7Al-4Mo may be used up to 455 °C (850 °F) in the STA condition. Thermal stability is good if annealing includes sufficient exposure and if the aging temperature is just above the proposed service temperature.

Strength	Percentage of RT strengths at:		
	315 °C (600 °F)	425 °C (800 °F)	540 °C (1000 °F)
Ultimate tensile strength	88	82	70
Yield strength (0.2% offset)	72	66	56

Tensile and Comprehensive

Ti-7Al-4Mo: STA strength vs temperature

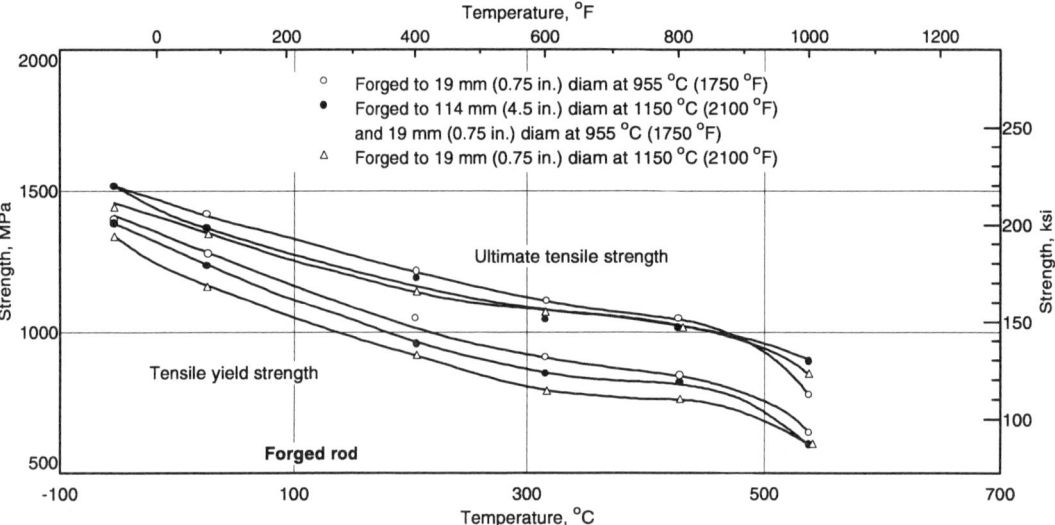

Forged as indicated from 190 mm (7.5 in.) diam and then heat treated 1 h at 955 °C (1750 °F), WQ, and aged 4 h at 595 °C (1100 °F).
Source: *Aerospace Structural Metals Handbook*, Vol 4, Code 3708, Battelle Columbus Laboratories, 1967, p 14

Ti-7Al-4Mo: STA tensile strength

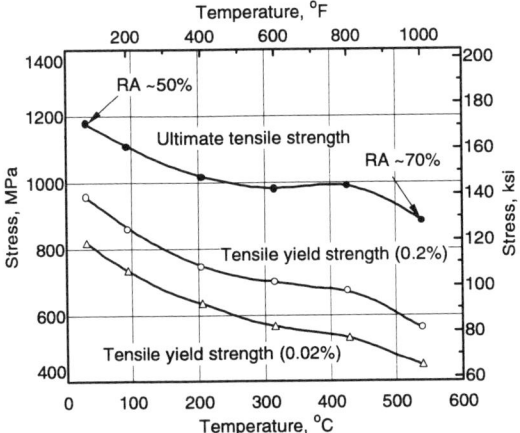

Source: R. Wood and R. Favor, *Titanium Alloys Handbook*, MCIC-HB-02, Battelle Columbus Laboratories, 1972

Ti-7Al-4Mo: Annealed compressive properties

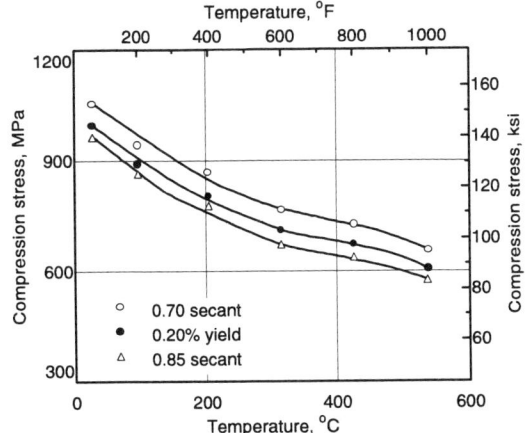

Compressive modulus ranged from about 110 GPa (16×10^6 psi) at RT to about 70 GPa (10×10^6 psi) at 540 °C (1000 °F).
Source: R. Wood and R. Favor, *Titanium Alloys Handbook*, MCIC-HB-02, Battelle Columbus Laboratories, 1972

Bearing and Shear Strengths

Ti-7Al-4Mo: Annealed bearing strength

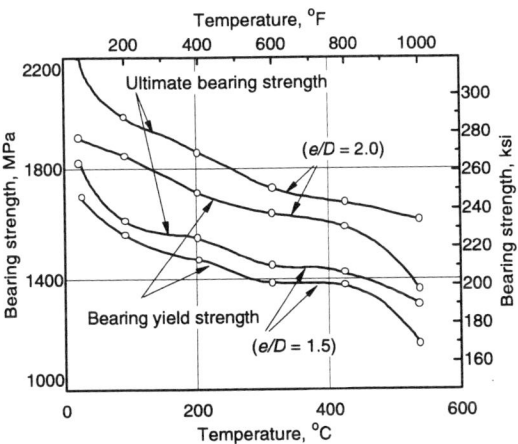

Source: R. Wood and R. Favor, *Titanium Alloys Handbook*, MCIC-HB-02, Battelle Columbus Laboratories, 1972

Ti-7Al-4Mo: Annealed ultimate shear strength

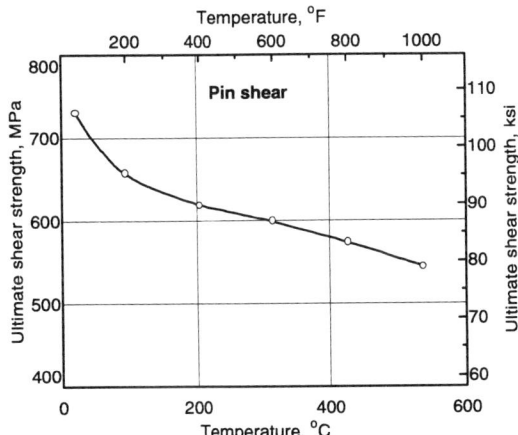

Source: R. Wood and R. Favor, *Titanium Alloys Handbook*, MCIC-HB-02, Battelle Columbus Laboratories, 1972

Creep Properties

The creep strength of solution treated plus aged Ti-7Al-4Mo alloy may be altered by fabrication and heat treatment procedure, in that β processing or heat treatment improves creep resistance. Typical creep strengths after 150-h exposure for bar solution heat treated at 955 °C (1750 °F) and aged at 595 °C (1100 °F) (8 h) are:

Exposure stress		Exposure temperature		Total strain, %
MPa	ksi	°C	°F	
345	50	455	850	0.18
205	30	540	1000	0.72
275	40	540	1000	1.94

Ti-7Al-4Mo: Creep and creep-rupture curves

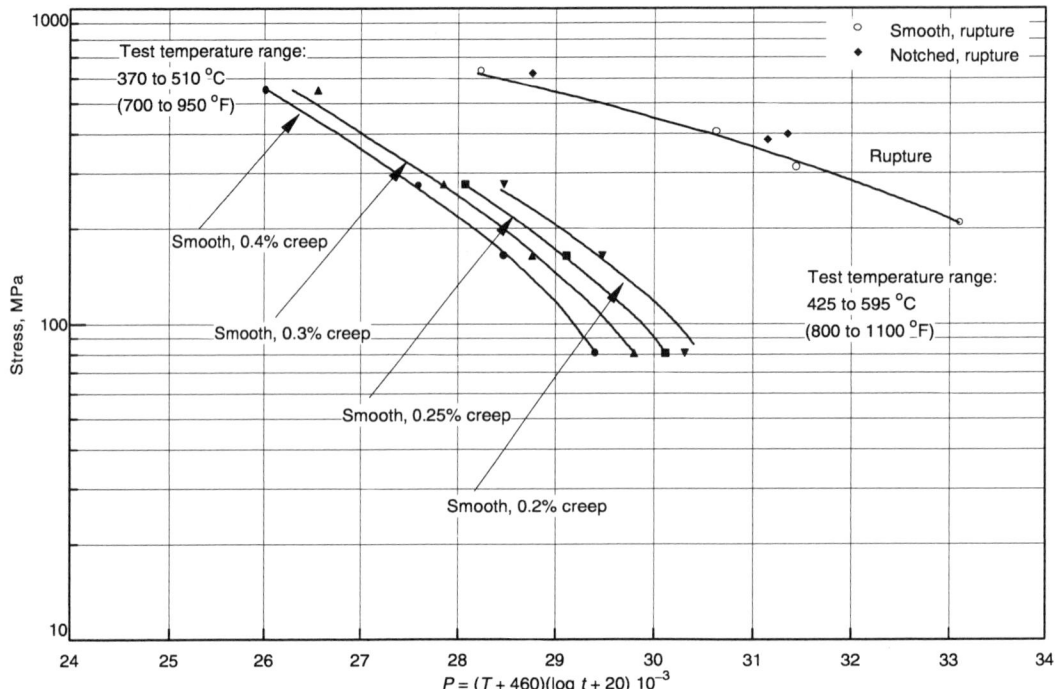

Aged compressor blade forgings forged at 925 to 955 °C (1700 to 1750 °F), ST 850 °C (1560 °F), 30 min argon, AC + age at 550 °C (1020 °F), 24 h, AC.
Source: *Aerospace Structural Metals Handbook*, Vol 4, Code 3708, Battelle Columbus Laboratories, Mar 1967

Ti-7Al-4Mo: Rupture stress of STA bar

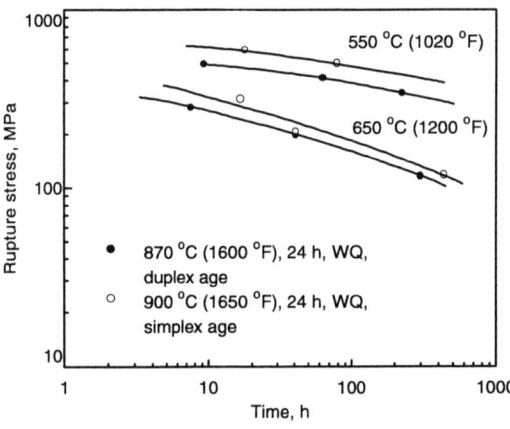

Duplex age was 48 h at 550 °C (1020 °F), AC, plus 48 h at 650 °C (1200 °F), AC. Simplex age was 48 h at 550 °C (1020 °F), AC.
Source: *Aerospace Structural Metals Handbook*, Vol 4, Code 3708, Battelle Columbus Laboratories, Mar 1967

Ti-7Al-4Mo: Minimum creep rate curves of STA bar

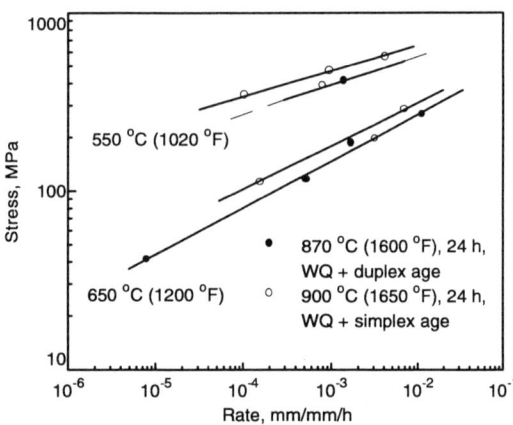

Duplex age was 48 h at 550 °C (1020 °F), AC, plus 48 h at 650 °C (1200 °F), AC. Simplex age was 48 h at 550 °C (1020 °F), AC.
Source: *Aerospace Structural Metals Handbook*, Vol 4, Code 3708, Battelle Columbus Laboratories, Mar 1967

Fatigue and Fracture

Fracture Toughness. Ti-7Al-4Mo in the STA condition achieves higher strength than Ti-6Al-4V at the expense of fracture toughness. Typical fracture toughness in the STA condition is 39 MPa$\sqrt{m}$ (36 ksi$\sqrt{in.}$) for a 13 mm (0.5 in.) specimen with a yield strength of 1151 MPa (167 ksi). Mill annealed Ti-7Al-4Mo has a toughness of about 80 MPa$\sqrt{m}$ (73 ksi$\sqrt{in.}$) for a yield strength of 993 MPa (144 ksi).

Ti-7Al-4Mo: Typical fatigue strengths at 10^7 cycles

Product condition	Test condition	RT fatigue strength	
		MPa	ksi
Annealed bar, 13 to 19 mm (0.5 to 0.75 in.)	Smooth specimen, bending load, $R = -1$	668	97
	Notched ($K_t = 3.9$), bending load, $R = -1$	200	29
STA bar(a)	Round specimen, bending	617	89.5
	Round specimen, torsional loading	310	45

(a) Heat treatment consisted of 1 h at 785 °C (1450 °F) in argon, furnace cool to 565 °C (1050 °F), AC, plus aging 2 h at 550 °C (1020 °F).
Source: *Aerospace Structural Metals Handbook*, Vol 4, Code 3708, Battelle Columbus Laboratories, 1967

Ti-7Al-4Mo: Charpy impact toughness of bar

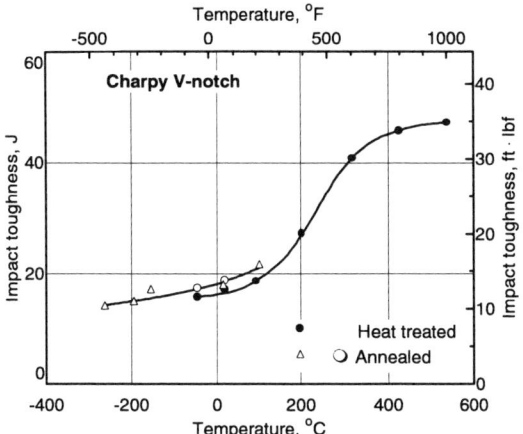

Source: *Aerospace Structural Metals Handbook*, Vol 4, Code 3708, Battelle Columbus Laboratories, Mar 1967

Ti-7Al-4Mo: Low-cycle fatigue at RT

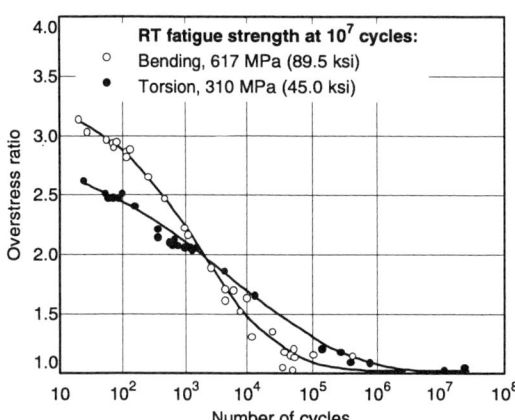

Low-cycle fatigue expressed in terms of overstress ratio = applied stress/fatigue strength at 10^7 cycles. Round specimens from STA bar solution treated at 785 °C (1450 °F), 1 h, argon, FC at 166 °C/h (300 °F/h) max to 565 °C (1050 °F), AC + 550 °C (1020 °F), 2 h, AC.
Source: *Aerospace Structural Metals Handbook*, Vol 4, Code 3708, Battelle Columbus Laboratories, 1967

Ti-7Al-4Mo: Stress range diagram for 1000 h

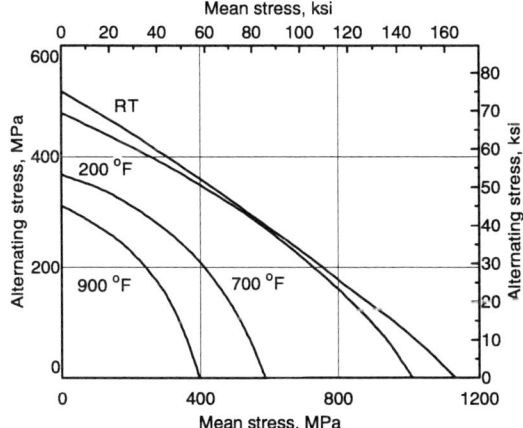

For calculation purposes, 1 h = 1.08×10^5 cycles for a mean stress of 482 MPa (70 ksi), or 1 h = 6×10^5 cycles for a mean stress of zero. Annealed bar at room temperature to 480 °C (900 °F)
Source: *Aerospace Structural Metals Handbook*, Vol 4, Code 3708, Battelle Columbus Laboratories, Mar 1967

Ti-7Al-4Mo: Stress range diagram at 250 h

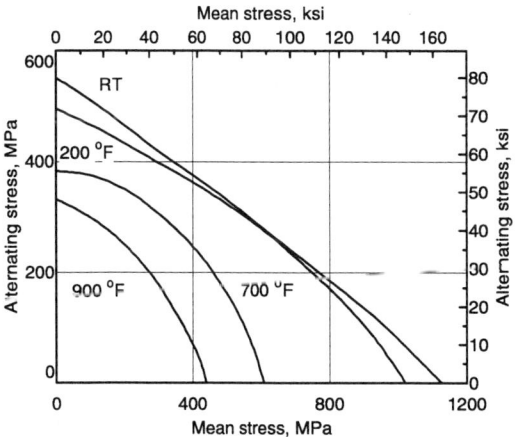

For calculation purposes, 1 h = 1.08×10^5 cycles for a mean stress of 482 MPa (70 ksi), or 1 h = 6×10^5 cycles for a mean stress of zero. Annealed bar at room temperature to 480 °C (900 °F).
Source: *Aerospace Structural Metals Handbook*, Vol 4, Code 3708, Battelle Columbus Laboratories, Mar 1967

Fabrication

Machining of titanium alloys is comparable to machining a good grade of stainless steel. In general, very sharp tools with a slightly larger rake angle and a very keen edge work quite well. Slower speed and heavier cuts are preferred because they keep tool temperatures down and produce coarse chips. Drilling of thin-walled titanium is not much of a problem as long as the drill is sharp. Thicker walls require a heavy flood of coolant to remove heat and chips. More information on machining can be found in "Technical Note 7: Machining."

Welding. Weldability of Ti-7Al-4Mo is fair. Like all titanium alloys, Ti-7Al-4Mo is weldable by all methods except shielded arc welding and submerged arc welding (because no flux is permitted). Recommended filler metal is the same as the base metal. Welding is not recommended.

Heat Treatment. Ti-7Al-4Mo is used in the annealed or solution treated and aged condition. Typical treatments are shown below (see table).

Ti-7Al-4Mo: Heat treatment conditions

Heat treatment	Temperature °C	°F	Time, h	Cooling method
Stress relief	480-700	900-1300	1-8	Air or slow cool
Annealing range	700-790	1300-1450	1-8	Air cool
Recommended anneal	790	1450	1	Furnace cool to 565 °C (1050 °F), then air cool
Solution treatment range	870-980	1600-1800	0.5 to 1.5	WQ
Recommended solution treatment	925-955	1700-1750	1	WQ
Aging range (min to max)	510-620	950-1150	Up to 24	AC
Typical age	565	1050	4-8	AC

Forging

Ti-7Al-4Mo is used primarily as a forging alloy for sonic horns (and once turbine engine components). It can be fabricated into all forging product types, although closed die forgings predominate. Ti-7Al-4Mo is commercially fabricated on all types of forging equipment.

This titanium alloy is reasonably forgeable, with higher unit pressures (flow stresses), less forgeability, and higher crack sensitivity than the α-β alloy Ti-6Al-4V. The final microstructure of Ti-7Al-4Mo forgings is developed by conventional thermomechanical processing in forging manufacture. Thermomechanical processes use a combination of subtransus forging followed by subtransus thermal treatments to fulfill mechanical-property criteria.

Final thermal treatments for forgings include annealing and solution treatment and aging, with final thermal treatment selected based on strength requirements. Forgings may be supplied in an annealed condition to facilitate machining and subsequently solution treated and aged to optimum strength levels.

Annealing is conducted at 815 °C (1500 °F), followed by furnace cooling to 565 °C (1050 °F) and then air cooling. Solution treatment is subtransus at 925 to 955 °C (1700 to 1750 °F), followed by water quenching. Forgings may then be annealed or aged. Aging is typically at 535 to 620 °C (1000 to 1150 °F), depending on strength mechanical-property objectives for the STA condition. Subtransus thermomechanical processes (forging and thermal treatment) for forgings produce equiaxed α in transformed β matrix microstructures that enhance strength, ductility, and high-cycle fatigue properties. Annealed and solution treated and annealed microstructures consist of 40 to 80% α, whereas solution treated and aged microstructures are 10 to 20% equiaxed α.

Deformation objectives in forging Ti-7Al-4Mo are to obtain the final forging shape and de-

Ti-7Al-4Mo: Forging process temperatures

Process	Metal temperature °C	°F
Conventional forging	900-985	1650-1800
Supra-transus forging	(a)	(a)

Note: See "Technical Note 4: Forging" for recommended die temperatures. (a) Beta forging can be performed in early forging operations if it is followed by substantial subtransus working.

Ti-7Al-4Mo: Tensile stress-strain curves for bar

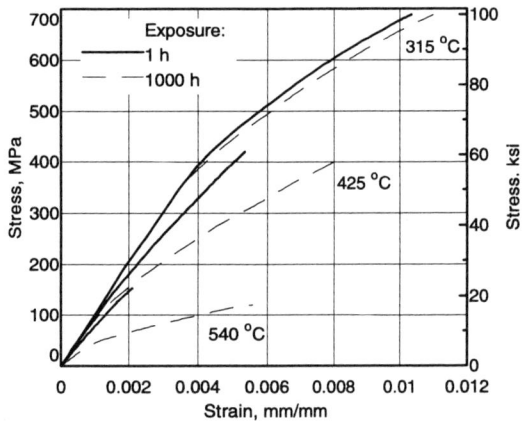

Isochronous stress-strain curves in tension at 315 to 540 °C (600 to 1000 °F) for bar. Specimens were forged at 980 to 1010 °C (1800 to 1850 °F) + 785 °C (1450 °F), 1 h, FC to 565 °C (1050 °F), 24 h, AC. Source: *Aerospace Structural Metals Handbook*, Vol 4, Code 3708, Battelle Columbus Laboratories, 1967, p 20

sired final microstructure at least cost with the conventional subtransus (α + β) forging thermomechanical processes most widely used. To achieve desired equiaxed α structures, subtransus reduction of 50 to 75%, accumulated through one or more forging steps, are required. Supra-transus (β) forging for Ti-7Al-4Mo may be used in early forging operations, including upsetting and open die preforming, to reduce unit pressures and ease forging fabrication. However, higher temperature initial forging operations must be followed by sufficient subtransus reduction to achieve the desired predominately equiaxed α structure.

TIMETAL® 62S

Ti-6Al-1.7Fe-0.1Si
62S
UNS Number: Unassigned

T. O'Connell, TIMET

Because iron is used as a stabilizer in lieu of more expensive elements, alloy 62S has a lower formulation cost than most titanium alloys, yet the properties and processing characteristics of 62S are equivalent to or better than those of Ti-6Al-4V. The combination of reasonable cost and excellent mechanical properties makes 62S a practical substitute for other engineering materials in numerous industrial applications that require low weight and high corrosion resistance. The microstructural response of 62S to heat treatment is quite similar to that of Ti-6Al-4V. Alloy 62S has a relatively high modulus-to-density ratio.

Chemistry Formulation. From alloy formulations used to assess the acceptable limits of alloying elements, the aluminum level of 6% (nominal) appeared optimum, based on evidence of poor strength at low aluminum levels and poor ductility at higher levels. Silicon was also felt to be optimized at 0.1 wt% because higher levels contribute to melting difficulties (hence cost). Thus, iron and oxygen contents were selected for further study.

The chemistries melted and processed to study the effects of iron and oxygen supported the following conclusions:

- On average, an increase of about 0.07 wt% oxygen is equivalent to (or provides) about a 60 MPa increase in strength (i.e., about 8.5 MPa per 0.01 wt% oxygen).

- On average, a 1 wt% change in iron content resulted in only about a 40 MPa increase in strength.

- For all heat treated conditions, the combination of high iron (2.4%) and high oxygen (0.25%) resulted in unacceptable post-creep ductility.

- Although annealing treatment had only a minor effect on creep properties (700 °C anneal was worse than 790 °C), the solution treated and aged condition provided substantially better properties than both annealed conditions.

- Post-creep ductility was maximized by the 790 °C (1455 °F) anneal, low oxygen, and in general low iron.

Product Forms. Ingot is available in 710, 815 or 865 mm (28 to 34 in.) diameter in masses ranging from 3180 through 6365 kg. Bloom is a semifinished form forged from above the β transus. Forging billet and bar are available as rounds, squares, or rectangles. Sheet and plate are also available. Plate is available in thickness from 4.8 to 102 mm (0.2 to 4 in.) in widths up to 3.05 m (10 ft) and lengths up to 10.67 m (35 ft). Not all these maxima are available simultaneously. The distinction between plate and sheet is made at 4.8 mm (3/16 in.). The standard sheet thickness minimum is 0.41 mm (0.016 in.). Sheet widths are available up to 1220 mm (48 in.). Cut lengths beyond 4880 mm (192 in.) are not standard. Finish grinding on both sides is standard procedure.

Product Conditions. 62S is typically processed to plate or billet either in the β or α-β temperature fields. Beta processing is used to improve yield and reduce the cost of processing in cases where the β structure is acceptable such as industrial applications (in which the corrosion resistance and low density are the primary reasons for use) and automotive applications (in which improved creep strength may be used). Alpha-beta processing is used for applications such as armor for improved ballistic response.

62S: Typical composition range

	Composition, wt%					
	Al	Fe	Si	O_2	N_2	H_2
Minimum	5.5	1.3	0.07	0.15	...	...
Maximum	6.5	2.0	0.13	0.20	...	...
Nominal	6.0	1.65	0.10	0.18	...	...

Physical Properties

Phases and Structures. The microstructural response of 62S to heat treatment is quite similar to that of Ti-6Al-4V. The transformed β microstructure is typically a colony structure after air cooling, but can be Widmanstätten for more rapid cooling. Alpha-beta processing results in a structure consisting of primary α with transformed β. The transformed β structure varies with the cooling rate.

62S: Summary of typical physical properties

Beta transus	1000 to 1025 °C (1840 to 1880 °F)
Liquidus (calculated)	1611 °C (2932 °F)
Solidus (calculated)	1555 °C (2831 °F)
Density(a)	4.44 g/cm^3 (0.160 lb/in.^{3})
Electrical resistivity(a)	1.62 μΩ · m
Magnetic permeability	Nonmagnetic
Specific heat capacity(a)	565 J/kg · K (0.135 Btu/lb · °F)
Thermal conductivity(a)	7.75 W/m · K (4.48 Btu/ft · h · °F)
Thermal coefficient of linear expansion(b)	8.35 × 10^{-6}/°C (4.6 × 10^{-6}/°F)

(a) Typical values at room temperature of about 20 to 25 °C (68 to 78 °F). (b) Mean coefficient from room temperature to 50 °C (120 °F)

62S: Elastic properties at room temperature

Modulus of elasticity	127 GPa (18.4 × 10^6 psi)
Shear modulus	48.7 GPa (7.06 × 10^6 psi)
Poisson's ratio	0.301
Bulk modulus	105.6 GPa (15.3 × 10^6 psi)

62S: General corrosion rates for α-β processed sheet

Solution/condition	Corrosion rate, mm/year
0.25% HCl, boiling	1.19
0.5% HCl, boiling	2.3475
1.0% HCl, boiling	8.35
1.0% HCl, 65 °C (145 °F)	1.15
3.0% HCl, 65 °C (145 °F)	4.675
5.0% HCl, 65 °C (145 °F)	11.45
Seawater, pH 1.5, boiling	5.45
Seawater, pH 3.0, boiling	0.00375
Seawater, pH 3.5, boiling	0.000
50 vol% acetic acid, 50 vol% formic acid, boiling	1.775
10 vol% acetic acid, 10 vol% formic acid, boiling	3.625

Temperature Effects

The electrical resistivity (R) of beta-processed material between 25 and 815 °C (77 and 1500 °F) has been determined to fit the expression:

$$R(10^{-8}\ \Omega \cdot m) = 160 + 8 \times 10^{-2}\ T - 6.3 \times 10^{-5}\ T^2$$

62S: Electrical resistivity vs temperature

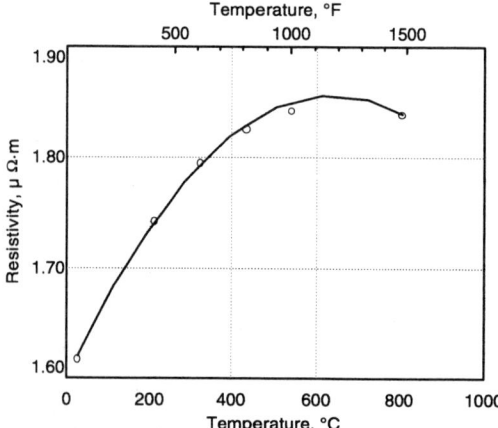

Resistivity (R) for beta-processed material between 25 and 815 °C (77 and 1500 °F) has been determined to fit the expression:
$R(10^{-8}\ \Omega \cdot m) = 160 + 8 \times 10^{-2}\ T - 6.3 \times 10^{-5}\ T^2$
82.5 mm (3.25 in.) plate, beta forged, and heat treated at 970 °C (1780 °F), 2 h, force air cool plus 790 °C (1455 °F), 2 h, air cool

62S: Specific heat vs temperature

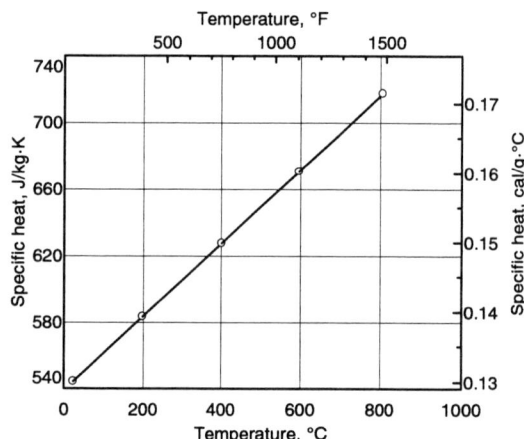

The specific heat (C_p) for beta-processed material between 25 and 815 °C (77 and 1500 °F) can be expressed as:
$C_p\ (cal/g \cdot °C) = 0.130 + 5.26 \times 10^{-5}\ T$
82.5 mm (3.25 in.) plate, beta forged, and heat treated at 970 °C (1780 °F), 2 h, force air cool plus 790 °C (1455 °F), 2 h, air cool

62S: Thermal coefficient of linear expansion

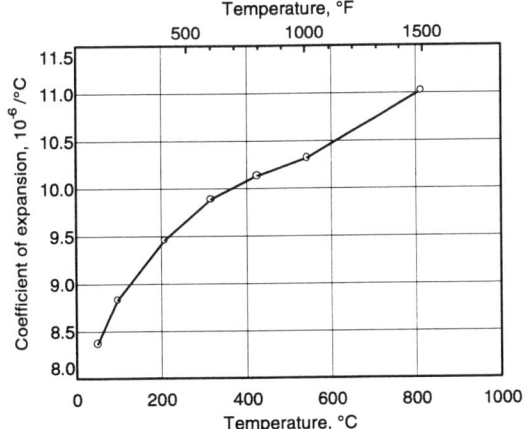

The thermal coefficient of linear expansion α for beta-processed material is given by:
$$\alpha\,(\text{ppm}/°C) = 7.89 + 1.1 \times 10^{-2}\,T - 1.78 \times 10^{-5}\,T^2 + 1.14 \times 10^{-8}\,T^3$$
82.5 mm (3.25 in.) plate, beta forged, and heat treated at 970 °C (1780 °F), 2 h, force air cool plus 790 °C (1455 °F), 2 h, air cool

62S: Thermal conductivity

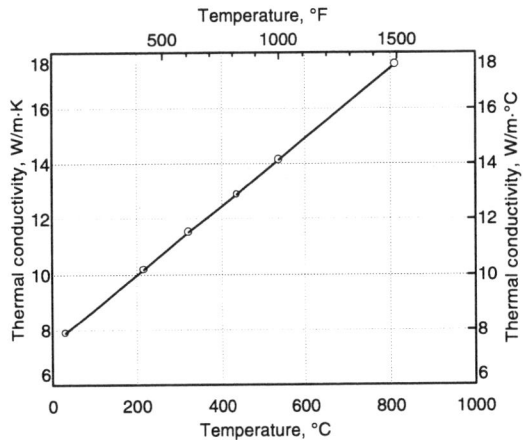

Thermal conductivity (Q) for beta-processed material follows the equation:
$$Q\,(\text{W/m} \cdot °C) = 7.54 + 1.24 \times 10^{-2}\,T$$
82.5 mm (3.25 in.) plate, beta forged, and heat treated at 970 °C (1780 °F), 2 h, force air cool plus 790 °C (1455 °F), 2 h, air cool

Mechanical Properties

In both the beta- and alpha-beta processed conditions, the tensile properties of 62S are comparable to those of Ti-6Al-4V. Creep properties and Larson-Miller plots are also expected to be essentially the same as similarly processed Ti-6Al-4V.

62S: Minimum tensile properties

Alloy	Ultimate tensile strength		Tensile yield strength		Elongation, %
	MPa	ksi	MPa	ksi	
62S	930	135	895	130	10
Ti-6Al-4V	895	130	825	120	10

Note: Annealed plate and forgings to 75 mm (3 in.) thick. 62S and Ti-6Al-4V α-β processed plus annealed at 700 to 790 °C (1290 to 1455 °F) for 2 h, air cooled

62S: Typical tensile properties of bar

Processing	Test temperature		Tensile yield strength		Ultimate tensile strength		Elongation, %	Reduction of area, %
	°C	°F	MPa	ksi	MPa	ksi		
α-β processed bar	24	75	991	143	1052	152	16.5	37.3
			991	143	1043	151	15.5	36.8
	300	570	667	96	798	115	17.5	44.5
			646	93	807	117	17.0	41.1
β-processed bar	24	75	996	144	1024	148	14.5	26.6
			1000	145	1028	149	15.5	29.2
	300	570	625	90	741	107	17.0	49.8
			641	93	747	108	15.5	48.8

Note: 45-mm (1.7-in.) bar α-β and β-processed plus annealed at 700 °C (1290 °F) for 2 h, air cooled

High-Temperature Strength

62S: High-temperature and post-creep mechanical properties

Alloys were double melted as 35-kg (77-lb) ingots then beta-processed (forged and rolled above the β transus) to 1.2-cm (0.47-in.) diam rod and subsequently heat treated as indicated.

Composition, wt%		Tensile properties at:						Creep(a) Post-creep tensile properties			Reduction of area, %
		RT Yield strength		Reduction of area, %	480 °C (900 °F) Yield strength		Reduction of area, %		Yield strength		
Fe	O	MPa	ksi		MPa	ksi		Time 0.2% creep, h	MPa	ksi	
STA condition(b)											
2.4	0.25	1180	171	7	635	92	70	500	...	...	...
2.0	0.24	1055	153	19	595	86	56	740	1080	156	9
1.4	0.24	1040	150	17	570	82	52	500	1050	152	8
2.3	0.18	1115	161	8	610	88	71	330	1140	165	6
1.9	0.17	1005	145	19	580	84	72	780	1010	146	18
1.4	0.17	980	142	24	540	78	57	690	1000	145	17
Annealed(c)											
2.4	0.25	1100	159	26	595	86	73	25	...	...	...
2.0	0.24	1055	153	30	575	83	71	13	1060	153	9
1.4	0.24	1050	152	32	550	79	64	22	1040	151	12
2.3	0.18	1050	152	26	580	84	70	12	1030	149	8
1.9	0.17	1013	147	33	600	87	68	17	1020	148	5
1.4	0.17	980	142	29	540	78	66	26	985	143	16

Note: (a) Creep test run at 480 °C (895 °F), 579 MPa, 84 ksi for approximately 500 h. (b) Solution treated for 1 h at 60 °C (110 °F) below the β transus followed by water quenching and aging at 540 °C (1000 °F), 8 h. (c) Annealed 700 °C (1290 °F) for 2 h. Source: A.J. Hutt, R.E. Adams, W.M. Parris, and P.J. Bania, "A New Low Cost Titanium Alloy," 7th Int. Titanium Conf., San Diego, July, 1992.

62S: Property comparisons from alloy development

Alloy	Temperature		Ultimate tensile strength		Tensile yield strength		Reduction of area, %	Elongation, %
	°C	°F	MPa	ksi	MPa	ksi		
Ti-6Al-4V	24	75	989	143	950	137	37.2	13.5
	150	300	858	124	795	115	53.0	16.5
	300	570	715	103	652	94	58.1	15.0
	480	895	650	94	558	81	60.4	18.5
Ti-3Al-1.5Cr-1.5Fe	24	75	863	125	793	115	41.5	17.5
	150	300	744	108	625	90	54.6	23.0
	300	570	610	88	479	69	64.0	21.0
	480	895	491	71	407	59	83.0	27.0
Ti-6Al-2Fe	24	75	1047	152	990	143	30.6	15.5
	150	300	922	133	815	118	39.9	15.0
	300	570	793	115	643	93	39.7	15.0
	480	895	650	94	547(a)	79(a)	63.7	21.0
Ti-6Al-2Fe-0.1Si	24	75	1059	153	1024	148	31.3	14.5
	150	300	950	137	838	121	36.0	15.0
	300	570	816	118	668	97	37.4	14.0
	480	895	661	96	562(b)	81(b)	63.9	23.0
Ti-6Al-2Fe-0.02Y	24	75	1019	147	987	143	31.1	15.0
	150	300	901	130	791	114	38.1	15.5
	300	570	775	112	626	90	46.8	15.5
	480	895	655	93	559	81	66.2	21.0
Ti-6Al-1Fe-1Cr	24	75	1015	147	969	140	29.1	14.5
	150	300	907	131	793	115	38.9	15.0
	300	570	769	111	636	92	40.0	14.5
	480	895	675	98	566	82	57.7	18.5
Ti-8Al-2Fe	24	75	1164	169	1120	162	5.8	4.0
	150	300	1072	155	973	141	10.6	5.0
	300	570	972	141	816	118	28.3	3.5
	480	895	808	117	687	99	42.8	19.5

Note: 1.2-cm (0.42-in.) diam bar beta rolled and annealed at 700 °C (1290 °F) for 2 h, AC. (a) 0.2% creep deformation observed after 172 h at 480 °C (895 °F) and 579 MPa (84 ksi). (b) 0.2% creep deformation observed after 331 h at 480 °C (895 °F) and 579 MPa (84 ksi). Source: A.J. Hutt, R.E. Adams, W.M. Parris, and P.J. Bania, "A New Low Cost Titanium Alloy," 7th Int. Titanium Conf., San Diego, July, 1992

Fatigue and Fracture

62S: Fracture toughness

Condition	Yield strength		Fracture toughness (K_{Ic})	
	MPa	ksi	MPa√m	ksi√in.
Alpha-beta forged + recrystallized annealed	960	139	45	41
Beta forged + recrystallized annealed	950	138	59	53
Beta forged + mill annealed	950	138	55	50

Source: A.J. Hutt, R.E. Adams, W.M. Parris, and P.J. Bania, "A New Low Cost Titanium Alloy," 7th Int. Titanium Conf., San Diego, July, 1992

62S: Crack growth at room temperature

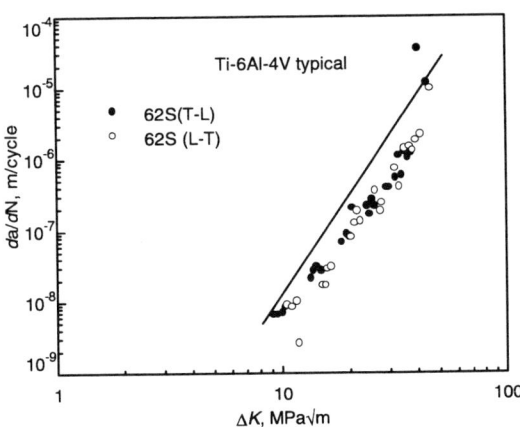

Beta-processed plus annealed specimens were tested in both the T-L and L-T directions at 25 °C (77 °F); $R = 0.1$; frequency, 20 Hz. Beta-processed plus annealed specimens were tested in both the T-L and L-T directions at 25 °C (77 °F); $R = 0.1$; frequency, 20 Hz.

Plastic Deformation

62S: Stress versus axial strain at 24 °C (75 °F)

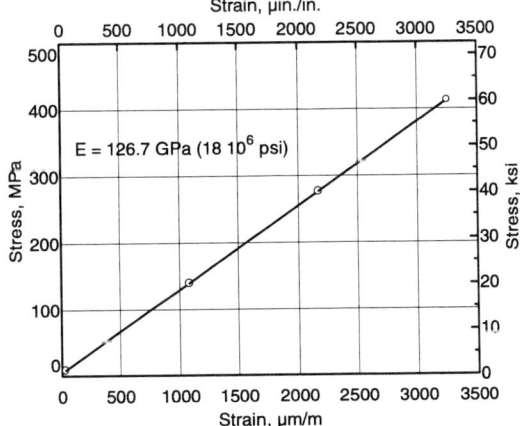

Alpha-beta cross forged 44.5 mm (1.7 in.) thick plate with 790 °C (1455 °F), 2 h anneal

62S: Stress versus diametral strain at 24 °C (75 °F)

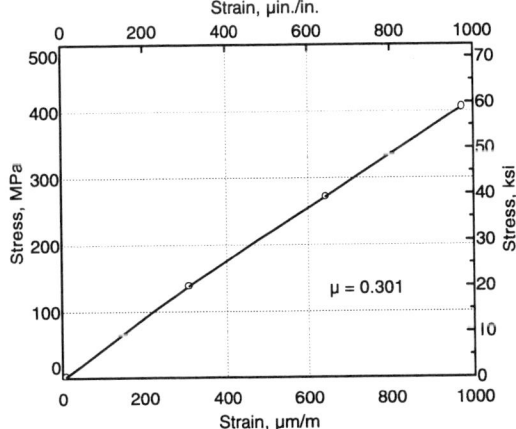

Alpha-beta cross forged 44.5 mm (1.7 in.) thick plate with 790 °C (1455 °F), 2 h anneal. Poisson's ratio (ν) is 0.301 for the axial and diametral strains.

Processing

Bulk working, forming, machining, and welding of 62S should be similar to Ti-6Al-4V, although no data are available. Heat treatments are summarized in the accompanying table.

62S: Typical heat treatments

Treatment	Temperature °C	°F	Duration, h	Cooling method
Stress relief(a)	480-650	900-1200	1 to 4	Air cool
Mill anneal	700-790	1290-1455	2	Air cool
Recrystallization anneal	970	1780	1	Furnace cool to 760 °C (1400 °F), hold 2 h, fan air cool
Solution treatment	β_t–60 °C	β_t–110 °F	1	Water quench
Age	540	1000	8	Air cool

(a) Stress relief same as Ti-6Al-4V

Ti-4.5Al-3V-2Mo-2Fe

Common Name: SP-700
UNS Number: Unassigned

SP-700 is a β-rich α-β titanium alloy designed to offer superplastic formability properties that are superior to those of Ti-6Al-4V. The low flow stress of the alloy, together with its fine microstructure, results in excellent superplastic formability at a temperature level of 700 °C (1290 °F), which provides the origin of the name SP-700.

The fine microstructure results in an excellent combination of mechanical properties. The alloy exhibits excellent heat treatability, cold formability, and hot forgeability.

Product Forms. SP-700 is available in all mill product forms (plate, sheet, round bar, etc.), as well as in cast and powder metallurgy (P/M) forms. The alloy can be fusion and spot welded.

Product Conditions. SP-700 is specified in either the annealed or the solution-treated and aged condition. The alloy consists of a very fine microstructure in all heat treatment conditions—for example, the primary α grains are typically smaller than 3 μm in the recrystallization annealed condition. It can be hardened to 450 HV or higher by solution treatment in the α + β region followed by short-time aging.

Applications. SP-700 is superplastically formed into such components as aerospace parts, metal wood heads, and metal balloons. Other uses include working tools, automobile parts, wrist watch casings, and mountain-climbing equipment.

Physical Properties

Internal Friction. A Q^{-1} value of 9.3×10^{-5} was measured at a vibration frequency of 400 Hz using the flexural vibration method.

Young's modulus for material in the mill-annealed condition at room temperature is 110 GPa (16×10^6 psi).

Poisson's ratio for material in the mill-annealed condition at room temperature is 0.33.

Magnetic Susceptibility. 3.51×10^{-6} cm^3/g

Magnetic Permeability. 1.00020

SP-700: Composition requirements

Element	Wt%
Aluminum	4.0-5.0
Vanadium	2.5-3.5
Molybdenum	1.8-2.2
Iron	1.7-2.3
Oxygen	0.15 max
Carbon	0.08 max
Nitrogen	0.05 max
Hydrogen	0.01 max
Yttrium	0.005 max
Other	
Each	0.10 max
Total	0.40 max
Titanium	bal

SP-700: Summary of typical physical properties

Beta transus	900 ± 5 °C (1650 ± 9 °F)
Melting (liquidus) point	1593 ± 5 °C (2900 ± 9 °F)
Density(a)	4.54 g/cm^3 (0.164 lb/in.^{3})
Electrical resistivity(a)	1.64 μΩ · m
Magnetic permeability	1.0020
Specific heat capacity (a)	495 J/kg · K (0.12 Btu/lb · °F)
Thermal conductivity(a)	7 W/m · K (4 Btu/ft · h · °F)
Thermal coefficient of linear expansion(b)	7.7×10^{-6}/°C (13.9×10^{-6}/°F)

(a) Typical values at room temperature of about 20 to 25 °C (68 to 78 °F). (b) Mean coefficient from room temperature to 100 °C (212 °F)

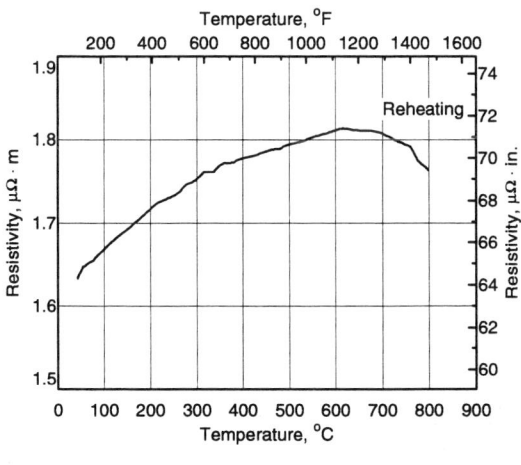

SP-700: Effect of temperature on resistivity

Corrosion Properties

General Corrosion. The corrosion resistance of SP-700 depends on the formation of a protective oxide layer, such as commercially pure titanium. SP-700 resists corrosion under a salt environment and has slightly higher corrosion resistance in hot or concentrated solutions of reducing acids such as hydrochloric and sulfuric acid than pure titanium and Ti-6Al-4V (see table). Corrosion resistance in acid solutions depends on concentration and temperature (see table).

Crevice Corrosion. SP-700 resists crevice corrosion in boiling 5% NaCl aqueous solution.

Hot Corrosion. SP-700 has excellent resistance to hot salt cracking. A specimen with a salt coating 15.5 g/m^2 (0.05 oz/ft^2) did not crack under a stress of 245 MPa (35.5 ksi) at 350 °C (660 °F).

SP-700: Effect of boiling environment on general corrosion rate

Alloy	General corrosion rate, mm/yr (in./yr), in:		
	25% NaCl	6% HCl	6% H_2SO_4
SP-700	0 (0)	51.4 (2.0)	45.9 (1.8)
CP Grade 2 Ti	0 (0)	77.3 (3.0)	55.9 (2.2)
Ti-6Al-4V	0 (0)	68.4 (2.7)	56.3 (2.22)

General corrosion rates were determined from weight loss after a 24-hr exposure.

SP-700: Effect of H_2SO_4 concentration on general corrosion rate

Temperature	General corrosion rate, mm/yr (in./yr), in:	
	2% H_2SO_4	6% H_2SO_4
20 °C (70 °F)	0 (0)	0.04 (0.0016)
Boiling	20.49 (0.80)	45.87 (1.80)

General corrosion rates were determined from weight loss after a 24-hr exposure.

Thermal Properties

The heat capacity of SP-700 is slightly lower than that of pure titanium and Ti-6Al-4V (see figure).

The thermal coefficient of linear expansion of SP-700 is about 7.7×10^{-6}/°C at 25 to 100 °C (75 to 212 °F) and 8.9×10^{-6}/°C at 25 to 600 °C (75 to 1110 °F) (see table).

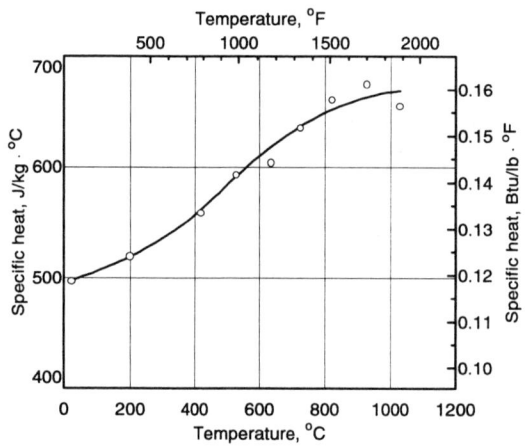

SP-700: Specific heat

Laser flashing was performed on a specimen 10 mm (0.4 in.) in diameter and 2 mm (0.08 in.) thick.

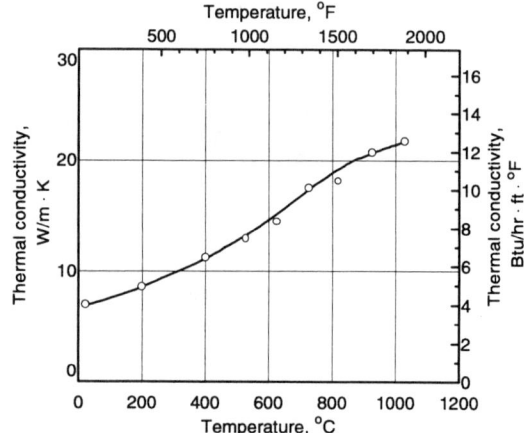

SP-700: Thermal conductivity

Laser flashing was performed on a specimen 10 mm (0.4 in.) in diameter and 2 mm (0.08 in.) thick.

SP-700: Coefficients of linear expansion

Temperature range		Coefficient of linear expansion	
°C	°F	10^{-6}/°C	10^{-6}/°F
25-100	75-212	7.71	4.3
25-200	75-390	8.15	4.5
25-300	75-570	8.53	4.7
25-400	75-750	8.82	4.9
25-500	75-930	8.95	4.97
25-600	75-1110	8.90	4.94

Differential expansion was measured on a specimen 5 mm (0.2 in.) in diameter and 20 mm (0.8 in.) long.

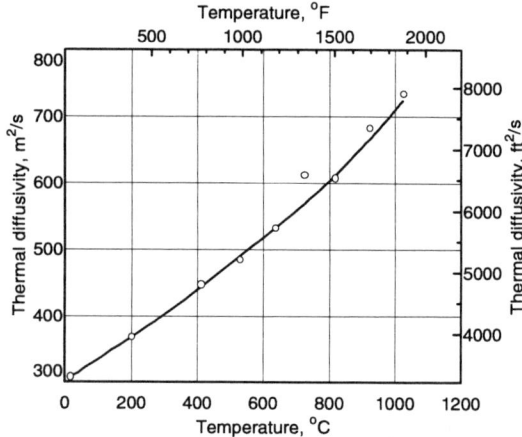

SP-700: Thermal diffusivity

Laser flashing was performed on a specimen 10 mm (0.4 in.) in diameter and 2 mm (0.08 in.) thick.

Mechanical Properties

See also "Heat Treatment" for tensile data.

Hardness. Mill-annealed SP-700 has a hardness of 300 to 330 HV. Recrystallization-annealed SP-700 has a hardness of 280 to 320 HV. Solution treated and aged SP-700 offers a wide range of hardness, from 350 to 510 HV, depending on solution treating and aging conditions.

Minimums of annealed SP-700 sheet:

Ultimate tensile strength	960 MPa (140 ksi)
Tensile yield strength	900 MPa (130 ksi)
Elongation in 0.50-2.00 mm (0.02-0.078 in.)	8%
Elongation in 2.01-5.00 mm (0.079-0.2 in.)	10%

SP-700: Typical tensile properties of sheet

Thickness			0.2% Yield strength		Tensile strength		Elongation,
mm	in.	Direction(a)	MPa	ksi	MPa	ksi	%
0.8	0.03	L	1023	148	1073	156	10.4
		T	1023	148	1073	156	10.2
2.0	0.08	L	953	138	1014	147	13.2
		T	924	134	996	144	15.0
3.0	0.12	L	910	132	1020	148	18.5
		T	1009	146	1042	151	19.7
3.8	0.15	L	949	137	1025	149	22.8
		T	929	135	1015	147	21.0

(a) L, longitudinal; T, transverse. Mill-annealed sheet. Tensile testing was performed on rectangular specimens with a 6 mm (0.24 in.) width and a 25 mm (1 in.) gage length.

SP-700: Comparison of annealed tensile properties among product forms

Product form	0.2% Yield strength		Tensile strength		Elongation,
	MPa	ksi	MPa	ksi	%
Plate	990	144	1028	149	16.8
Sheet	949	138	1025	148	22.8
Bar	936	136	1007	146	18.4

Mill-annealed material. Tensile testing was performed on 6.25 mm (0.25 in.) diam round specimens with a 25 mm (1 in.) gage length for a 15 mm (0.6 in.) thick plate and a 22 mm (0.9 in.) bar, and on rectangular specimens with a 6 mm (0.24 in.) width and a 25 mm (1 in.) gage length for a 3.8 mm (0.15 in.) thick sheet.

SP-700: Properties of heat-treated 15 mm (0.6 in.) plate

Alloy	Heat treatment	0.2% Yield strength		Tensile strength		Elongation,	Reduction of area,
		MPa	ksi	MPa	ksi	%	%
SP-700	Mill annealing (720 °C/1 hr/AC)	972	141	1023	148	19.0	61.9
	Recrystallization annealing (800 °C/1 hr/AC)	917	133	966	140	20.8	61.6
	ST (WQ) and aged (850 °C/1 hr/WQ + 560 °C/6 hr/AC)	1240	180	1377	200	11.6	28.0
	ST (AC) and aged (850 °C/1 hr/AC + 510 °C/6 hr/AC)	1114	162	1213	176	14.4	39.6
Ti-6Al-4V	Mill annealing (720 °C/1 hr/AC)	945	137	1003	145	19.6	38.0
	STA (955 °C/1 hr/WQ + 538 °C/6 hr/AC)	1129	164	1205	175	10.5	31.4

Tensile testing was performed on 6.25 mm (0.25 in.) diam round specimens with a 25 mm (1 in.) gage length.

SP-700: Effect of temperature on tensile properties

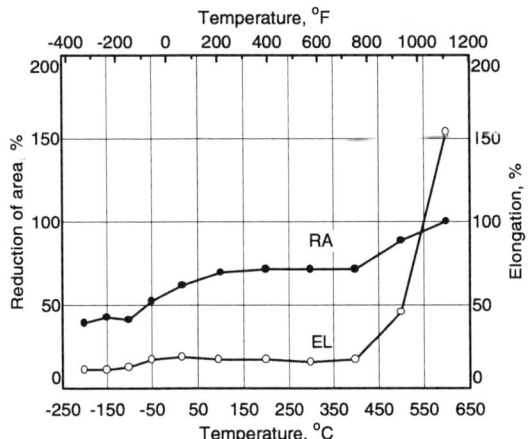
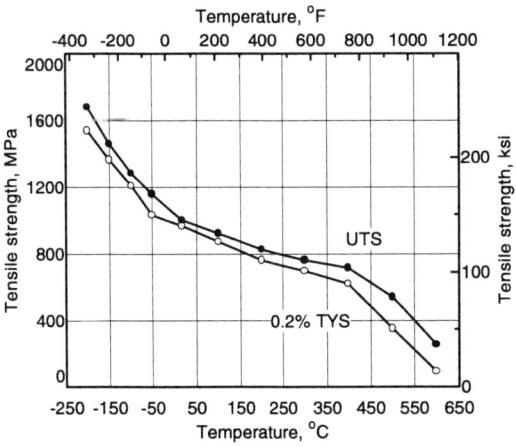

(a) (b)
15 mm (0.6 in.) thick mill-annealed plate. Tensile testing was performed on 6.25 mm (0.25 in.) diam round specimens with a 25 mm (1 in.) gage length.

Fatigue and Fracture Properties

Fatigue Life Data

The fatigue strength of SP-700 is higher than that of Ti-6Al-4V because of its fine microstructure (see figures).

SP-700: Fatigue strength, compared with Ti-6Al-4V

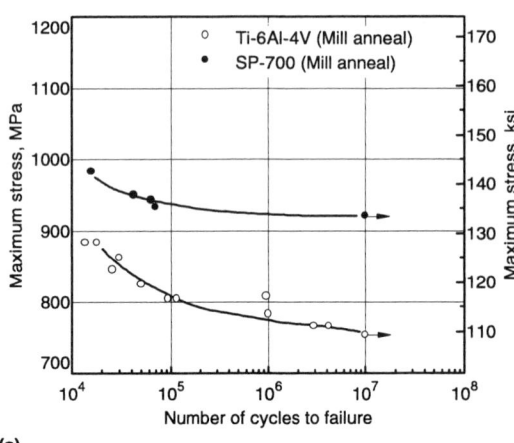
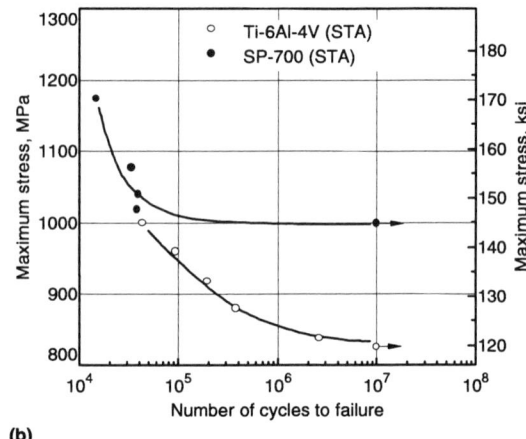

(a) Mill annealing. (b) Solution treating and aging. 20 mm (0.8 in.) diam bar. Fatigue testing was performed on 7 mm (0.28 in.) diam round specimens. The loading wave-form was sinusoidal, and all tests were performed at a frequency of 10 Hz and a stress ratio of $R = 0$. Surface roughness: $R_a = 0.1 \mu m$

Fatigue Crack Propagation

SP-700: Fatigue crack propagation behavior

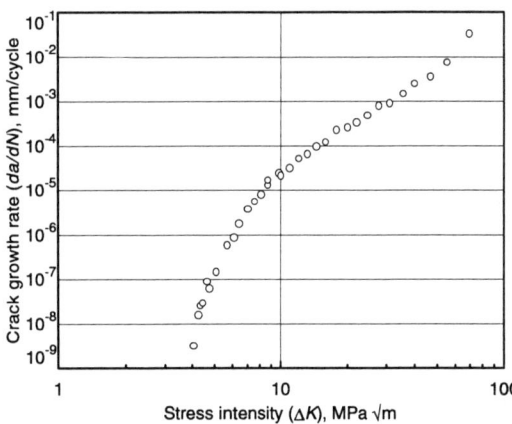

SP-700: Fatigue strength ($R = -1$)

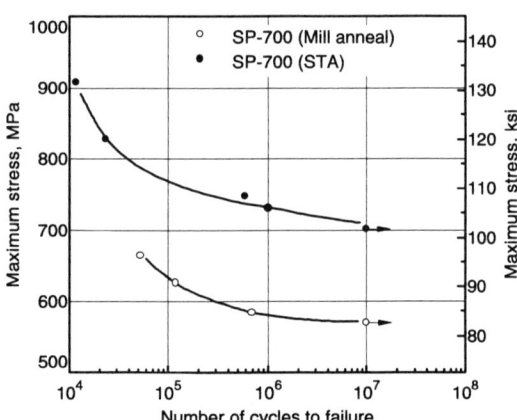

15 mm (0.6 in.) thick mill-annealed plate. Fatigue testing was performed on a 12.5 mm (0.5 in.) thick compact specimen. The loading wave-form was sinusoidal, and all tests were performed at a frequency of 10 Hz and a stress ratio of $R = 0.05$.

20 mm (0.8 in.) diam bar. Fatigue testing was performed on 7 mm (0.28 in.) diam round specimens. The loading wave-form was sinusoidal, and all tests were performed at a frequency of 10 Hz and a stress ratio of $R = -1$. Surface roughness: $R_a = 0.1 \mu m$

Fracture Properties

Fracture Toughness. SP-700 has excellent fracture toughness compared to Ti-6Al-4V because of its superior ductility (see figure).

Impact Toughness. (see figure)

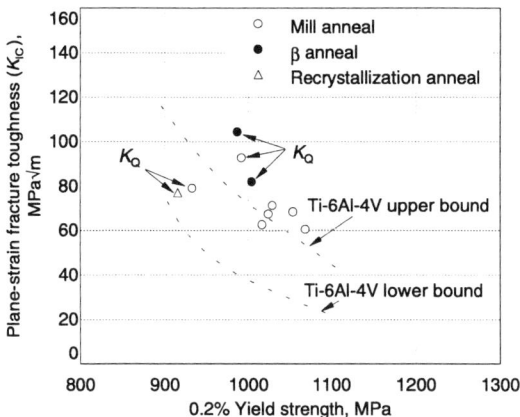

SP-700: Fracture toughness compared with Ti-6Al-4V

15 mm (0.6 in.) thick plate. SP-700 was annealed; Ti-6Al-4V was either annealed or solution treated and aged. Fracture toughness testing was performed on 12.5 mm (0.5 in.) thick compact specimens. K_Q denotes invalid values that do not meet the validity requirements in terms of limitations on P_{max}/P_Q and specimen size.

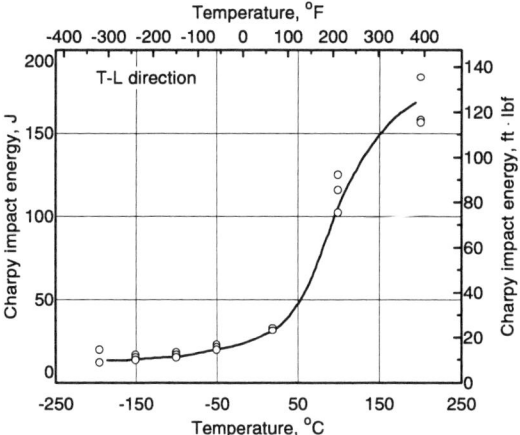

SP-700: Effect of temperature on impact strength

15 mm (0.6 in.) thick mill-annealed plate. Impact testing was performed on Charpy specimens with a 2 mm (0.08 in.) V-notch.

Plastic Deformation

Flow Stresses SP-700 produces a much lower flow stress than Ti-6Al-4V at temperatures between 500 and 900 °C (930 and 1650 °F) (see figures).

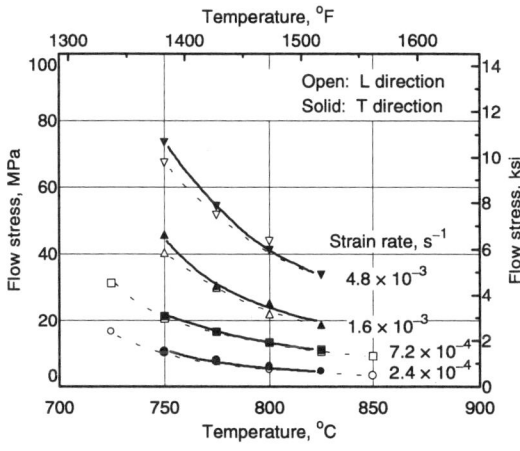

SP-700: Effect of temperature on flow stress

3 mm (0.12 in.) thick mill-annealed sheet. High-temperature tensile testing was performed on 5 mm (0.2 in.) wide rectangular specimens with a 5 mm (0.2 in.) gage length.

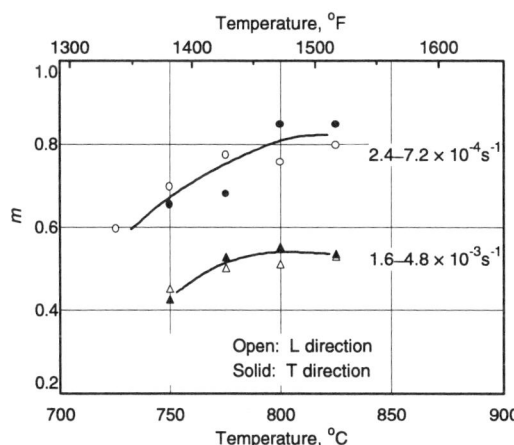

SP-700: Strain-rate sensitivity index, m

The m values were determined from the log strain rate versus log flow stress diagram.

Forging **Forgeability.** SP-700 exhibits much higher resistance to hot deformation cracking than Ti-6Al-4V (see figure).

690 / Alpha-Beta Alloys

SP-700: Forgeability, compared with Ti-6Al-4V

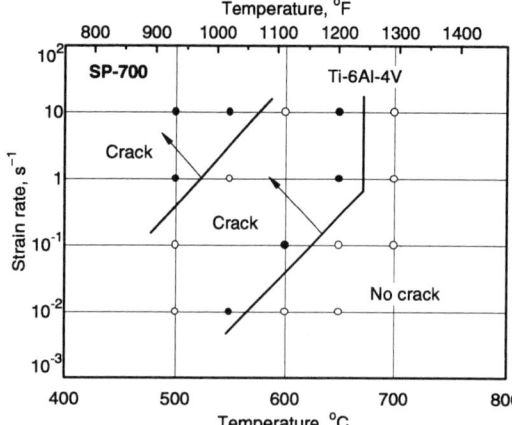

Hot compression testing was performed on specimens 6 mm (0.24 in.) in diameter and 10 mm (0.4 in.) long with a circumferential notch. Open circles denote samples that did not exhibit cracking at 70% height reduction. Closed circles denote samples that showed cracking.

SP-700: Cold formability versus Ti-6Al-4V

Alloy	Direction(a)	Bend factor (R/t)	Cold rolling reduction limit, %
SP-700	L	2.1	69
	T	2.1	58
Ti-6Al-4V	L, T	4	20

(a) L, longitudinal; T, transverse. SP-700 was recrystallization annealed; Ti-6Al-4V was mill annealed. Bend factors were determined by bending tests on specimens 20 mm (0.8 in.) wide, 140 mm (5.5 in.) long, and 4 mm (0.16 in.) thick. Maximum reduction limits were obtained by cold rolling tests.

Forming

Cold Formability. SP-700 has much better cold formability in comparison with Ti-6Al-4V (see table).

Superplastic Formability. SP-700 shows excellent formability at 775 °C (1425 °F), more than 100 °C (180 °F) lower than Ti-6Al-4V (see figure).

SP-700: Superplastic formability, compared with Ti-6Al-4V

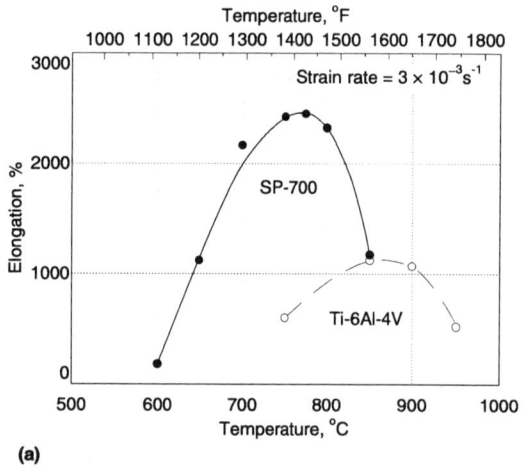
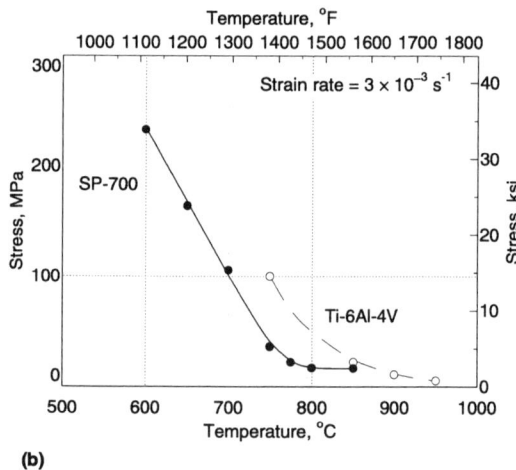

Data for Ti-6Al-4V are not necessarily representative, as elongations up to 2000% have been reported for this material.
13.5 mm (0.53 in.) diam mill-annealed bar. Tension testing was performed on 5 mm (0.2 in.) diam round specimens with a 6 mm (0.24 in.) gage length.

Heat Treatment

SP-700 is specified either in the annealed condition or in the fully heat-treated condition.

Annealing Temperatures. Mill annealing requires 0.5 to 2 h at 650 to 750 °C (1200 to 1380 °F) followed by furnace or air cooling. Recrystallization annealing (see table) provides maximum cold working capability. The highest volume fraction of β phase, 40%, was retained for the annealing at 800 °C (1470 °F).

Solution Heat Treatment Temperatures. Solution treating requires 0.5 to 2 h at 800 to 850 °C (1470 to 1560 °F), followed either by water quenching to obtain very high strength or by air cooling to obtain high strength with good ductility.

Aging Temperatures. Aging requires 1 to 6 h at 450 to 600 °C (840 to 1110 °F), followed by air cooling (see figures). A relatively short aging period is enough to achieve an excellent combination of strength and ductility (see figure).

SP-700: Recommended heat treatments

Treatment	Temperature °C	Temperature °F	Duration, hr	Cooling method
Mill anneal	650-750	1200-1380	0.5-2	FC or AC
Recrystallization anneal	750-850	1380-1560	0.5-2	FC or AC
Solution treatment	800-850	1470-1560	0.5-2	WQ or AC
Aging	450-600	840-1110	1-6	AC

Aging Effects

SP-700: Effect of aging temperature on hardness and tensile properties

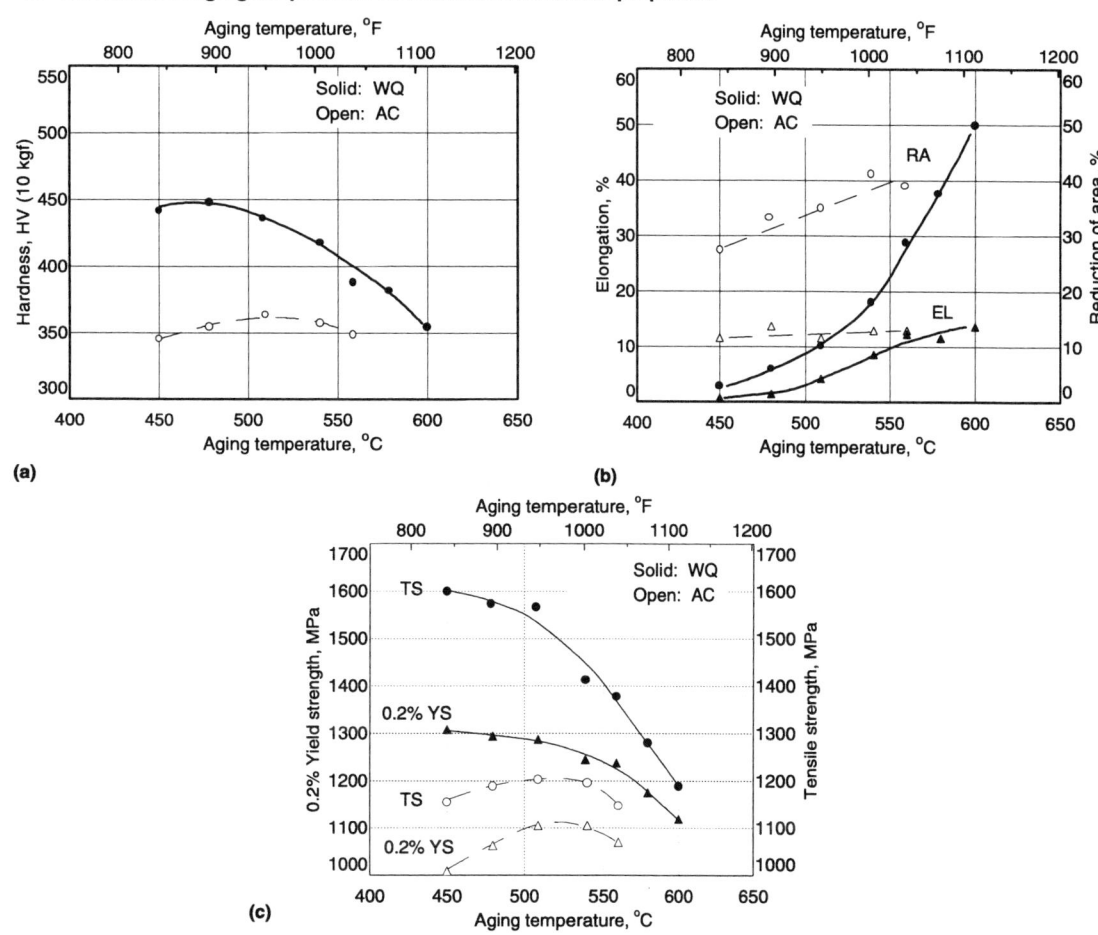

A 15 mm (0.6 in.) thick plate was solution treated at 850 °C (1560 °F) for 1 hr, followed by either water quenching or air cooling. Aging was done at temperatures from 450 to 600 °C (840 to 1110 °F) for 1 hr, followed by air cooling.

SP-700: Effect of aging time on tensile properties

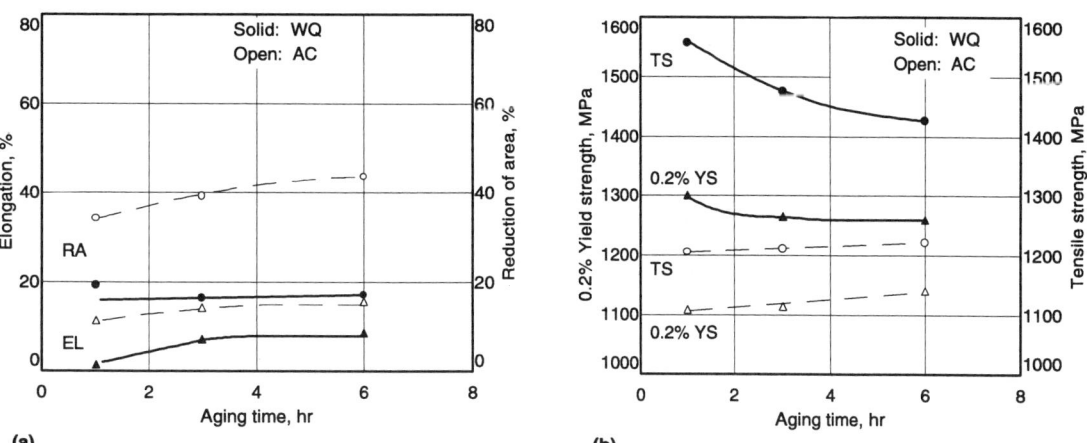

A 15 mm (0.6 in.) thick plate was solution treated at 850 °C (1560 °F) for 1 hr, followed by either water quenching or air cooling. Aging was done at 510 °C (950 °F) for 1 to 6 hr, followed by air cooling.

692 / Alpha-Beta Alloys

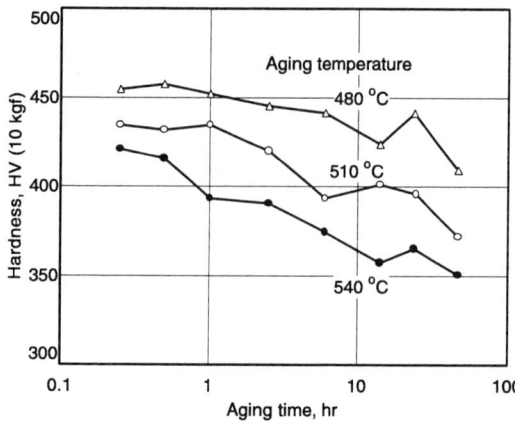

SP-700: Aging response

Solution treatment was performed at 850 °C (1560 °F) for 1 hr, followed by water quenching. Aging was done at 480 to 540 °C (895 to 1005 °F) for 0.25 to 48 hr, followed by air cooling.

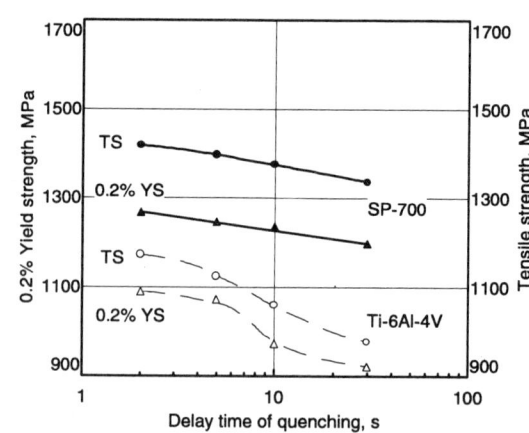

SP-700: Effect of delay time of quenching on strength

Solution treatment was done on 12.5 mm (0.5 in.) thick plates, followed by a quench delay ranging from 2 to 30 s before water quenching. SP-700: Solution treated at 850 °C (1560 °F) for 1 hr, followed by water quenching; aged at 510 °C (950 °F) for 6 hr, followed by air cooling. Ti-6Al-4V: Solution treated at 950 °C (1740 °F) for 1 hr, followed by water quenching; aged at 538 °C (1000 °F) for 6 hr, followed by air cooling.

Quench Effects

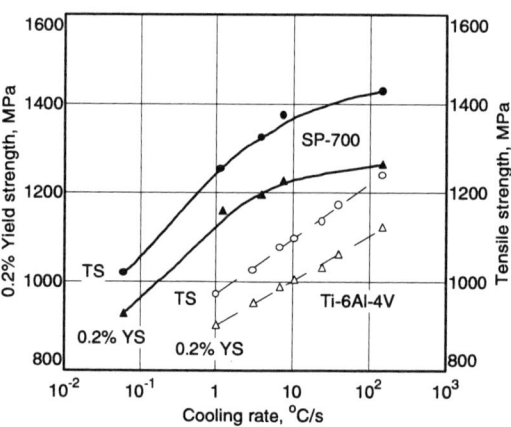

SP-700: Effect of cooling rate on strength

Solution treatment was done on 12.5 mm (0.5 in.) thick plates, followed by cooling at a rate that varied from 0.06 to 150 °C/s (0.1 to 270 °F/s) for SP-700 and from 1 to 150 °C/s (2 to 270 °F/s) for Ti-6Al-4V. SP-700: Solution treated at 850 °C (1560 °F) for 1 hr, followed by water quenching; aged at 510 °C (950 °F) for 6 hr, followed by air cooling. Ti-6Al-4V: Solution treated at 950 °C (1740 °F) for 1 hr, followed by water quenching; aged at 538 °C (1000 °F) for 6 hr, followed by air cooling.

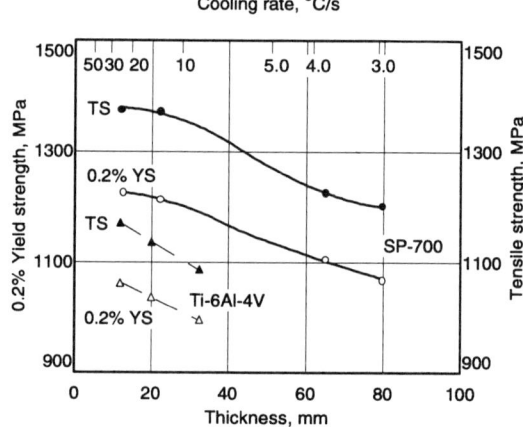

SP-700: Effect of quenched section size on strength

Solution treatment was done on plates with thicknesses ranging from 12.5 to 80 mm (0.5 to 3 in.) for SP-700 and from 12.5 to 32 mm (0.5 to 1.3 in.) for Ti-6Al-4V. SP-700: Solution treated at 850 °C (1560 °F) for 1 hr, followed by water quenching; aged at 510 °C (950 °F) for 6 hr, followed by air cooling. Ti-6Al-4V: Solution treated at 950 °C (1740 °F) for 1 hr, followed by water quenching; aged at 538 °C (1000 °F) for 6 hr, followed by air cooling.

Hardenability

SP-700 has better hardenability than Ti-6Al-4V. SP-700 is also less sensitive to section size on quenching. The greater stability of the β phase of SP-700 compared to Ti-6Al-4V provides greater flexibility during heat treatment. SP-700 is relatively insensitive to quench delays of up to 30 s. It is also less sensitive than Ti-6Al-4V with regard to cooling rate down to about 1 °C/s (2 °F/s) and relatively insensitive to cooling rate in the range of about 7 to 150 °C/s (13 to 270 °F/s).

IMI 367
Ti-6Al-7Nb

IMI 367 is a high-strength titanium alloy with excellent biocompatibility for surgical implants. The alloy was developed specifically for the manufacture of femoral component items for hip prostheses.

Hot forging procedures and mechanical working practices are the same as the standard interstitial Ti-6Al-4V. IMI 367 is available as rod and bar (8 to 100 mm, diam) and rectangular bar (25 × 75 mm). Heat treatment is a 1-hr anneal at 700 °C (1290 °F), air cool.

IMI 367: Typical composition range (wt%) and density

	Al	Nb	Ta	Fe	C	O_2	N_2	H_2	Ti
Minimum	5.50	6.50	...	...	...	...	...	...	...
Maximum	6.50	7.50	0.50	0.25	0.08	0.20	0.05	0.009	...
Nominal	6.0	7.0	...	...	...	...	...	...	bal

Density of IMI 367 is 4.52 g/cm^3 (0.163 lb/in.^{3}).

IMI 367: Summary of typical physical properties

Beta transus	1010 ± 15 °C (1850 ± 30 °F)
Melting (liquidus point)	Not Available
Density(a)	4.52 g/cm^3 (0.163 lbf/in.^{3})
Elastic modulus	105 GPa (15.2 × 10^6 psi)
Electrical resistivity(a)	Not Available
Magnetic permeability	Nonmagnetic
Specific heat capacity(a)	...
Thermal conductivity	...
Thermal coefficient of linear expansion	...

(a) Typical values at room temperature of about 20 to 25 °C (68 to 78 °F).

IMI 367: Mechanical properties

Property	
0.2% yield stress	800 MPa (min) to 900 MPa (typ)
Tensile strength	900 MPa (min) to 1000 MPa (typ)
Elongation on 5D	10% (min) to 12% (typ)
Reduction of area	25% (min) to 35% (typ)
Fatigue	The 10^7 cycle fatigue strength of IMI 367 in rotating bending is ± 500 MPa (± 72.5 ksi) which is in the range of Ti-6Al-4V

IMI 550

Ti-4Al-4Mo-2Sn-0.5Si
Trade names: IMI 550 (Previously Hylite 50)

IMI 550 is an alpha-beta titanium alloy of medium strength with a typical ultimate tensile strength of 1100 MPa (159 ksi) and temperature capability up to about 400 °C (750 °F). The alloy derives its properties from solid solution strengthening and age hardening. Because of the significant level of beta stabilizer content, IMI 550 exhibits a useful aging response in sections up to 150 mm (6 in.) thick.

Product Forms and Condition. IMI 550 is available in the form of bar, billet, plate, sheet and castings. It is a relatively easy alloy to forge compared with other titanium alloys. With careful control of welding parameters, it is considered to be weldable. IMI 550 is normally used in the solution treated and aged (STA) condition, which possesses a microstructure of primary alpha, transformed beta, and silicide.

Applications. The major use for IMI 550 is in the aerospace industry, both as airframe and aeroengine components. Typical components are flap-tracks and engine compressor discs.

IMI 550: Typical composition range (wt%) and density

	Al	Sn	Fe	Mo	Si	$O_2 + 2N_2$	H_2
Minimum	3.0	1.5	...	3.0	0.3	...	...
Maximum	5.0	2.5	0.2	5.0	0.7	0.27	0.0125
Nominal	4	2	...	4	0.5	...	...

Density of IMI 550 is 4.60 g/cm^3 (0.166 lb/in.^{3}).

Physical Properties

IMI 550: Summary of typical physical properties

Beta transus	975 ± 10 °C (1787 ± 18 °F)
Melting (liquidus point)	Not Available
Density(a)	4.60 g/cm^3 (0.166 lbf/in.^{3})
Electrical resistivity(a)	1.59 μΩ · m
Magnetic permeability	Nonmagnetic
Specific heat capacity(a)	Not Available
Thermal conductivity(a)	7.5 W/m · K (4.35 Btu/ft · h · °F)
Thermal coefficient of linear expansion(b)	8.8 × 10^{-6} °C (4.9 × 10^{-6} °F)

(a) Typical values at room temperature of about 20 to 25 °C (68 to 78 °F). (b) Mean coefficient from room temperature to 100 °C (212 °F)

IMI 550: Young's modulus (dynamic)

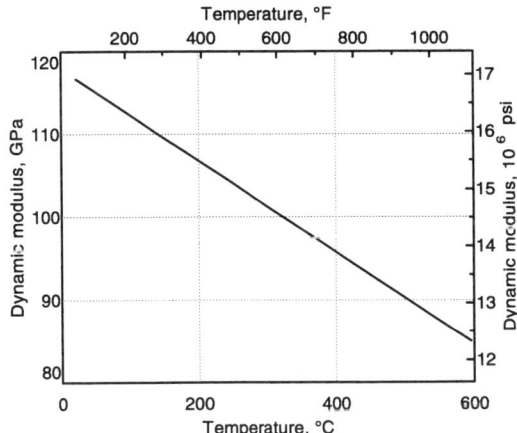

Heat treated bar.
Source: IMI Titanium 550 brochure

IMI 550: Electrical resistivity

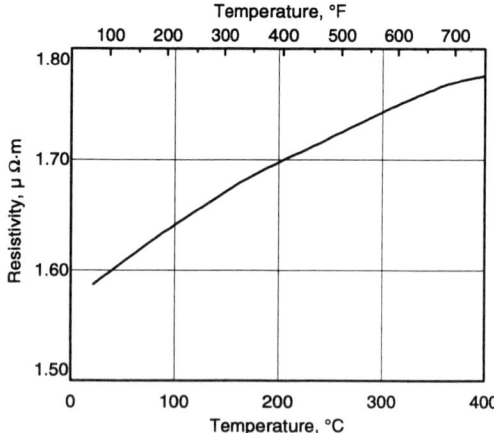

Heat treated bar.
Source: IMI Titanium 550 brochure

IMI 550: Thermal conductivity

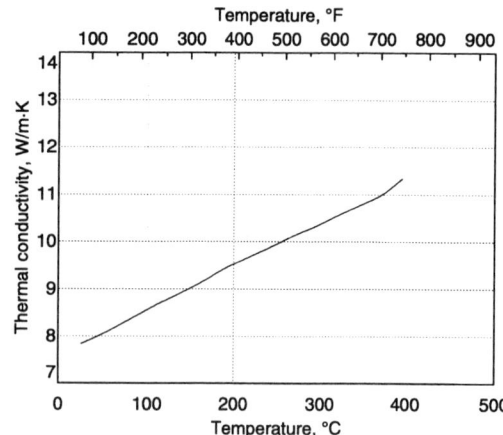

Heat treated bar.
Source: IMI Titanium 550 brochure

IMI 550: Thermal coefficient of linear expansion

Temperature range		Mean coefficient of thermal expansion	
°C	°F	$10^{-6}/$°C	$10^{-6}/$°F
20-100	68-210	8.8	4.9
20-200	68-390	9.0	5.0
20-300	68-570	9.2	5.1
20-400	68-750	9.3	5.2
20-500	68-930	9.7	5.4
20-600	68-1110	10.1	5.6

Heat treated bar

Mechanical Properties

Hardness of heat treated IMI 550 is typically 360 HV (50 kg).

Notch tensile ratio is typically 1.5 ($K_t = 3$).

Fracture toughness of IMI 550 is typically 60 MPa$\sqrt{m}$ (55 ksi$\sqrt{in.}$).

IMI 550: U-notch Charpy impact values

Temperature		Energy absorbed	
°C	°F	J	ft · lbf
−60	−76	19	14
−40	−40	22	16
−20	−4	22	16
+20	+68	23	17

Heat treated bar. Source: IMI Titanium 550 brochure

IMI 550 Bar: Minimum tensile properties
Typical UTS in 75 mm section is 1100 MPa (160 ksi).

	Diameter	
	Less than 25 mm (1 in.)	Less than 100 mm (4 in.)
0.2% yield strength, MPa (ksi)	960 (139)	920 (133)
Ultimate tensile strength, MPa (ksi)	1100 (160)	1050 (152)
Elongation (in 5D)	9 (9)	9 (9)
Reduction in area, %	25 (25)	20 (20)

High-Temperature Strength

Tensile properties up to 600 °C (1110 °F) are shown (see figure). IMI 550 is regarded as having useful creep performance up to about 400 °C (750 °F) giving less than 0.1% total plastic strain in 100 hours at 465 MPa (67 ksi).

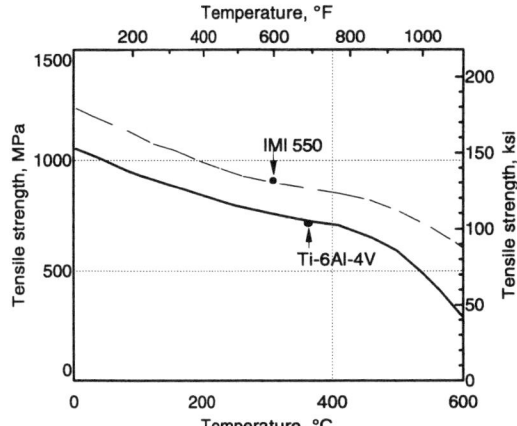

IMI 550: Tensile strength vs temperature

Heat treated bar.
Source: IMI Titanium "Medium Temperature Alloys" brochure

IMI 550: Tensile properties up to 600 °C

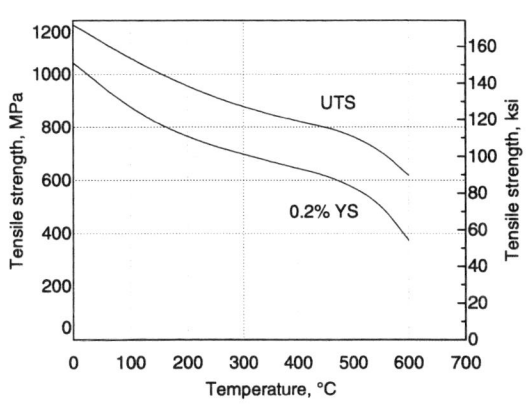

(a)
Heat treated bar.
Source: IMI Titanium 550 brochure

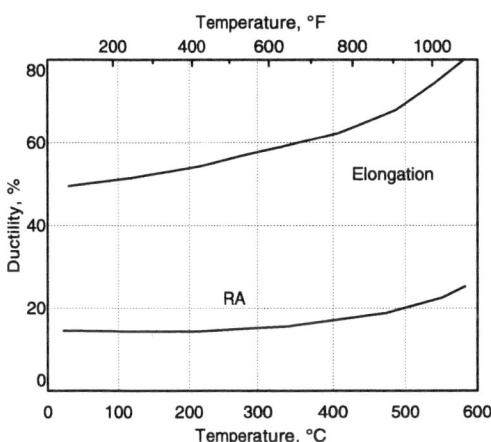

(b)

IMI 550: Creep properties at 100 and 300 °C

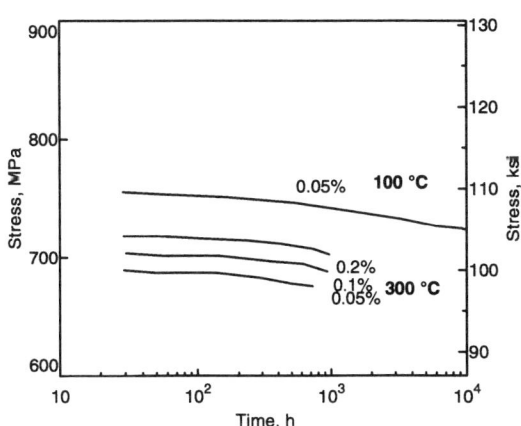

Heat treated bar.
Source: IMI Titanium 550 brochure

IMI 550: Flow stress below 900 °C

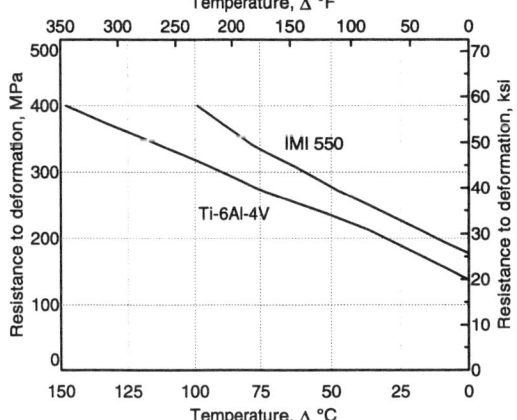

Heat treated bar tested with plastometer at a strain rate of 15/s.
Ti-64 flow stress below 950 °C compared.

698 / Alpha-Beta Alloys

IMI 550: Creep properties at 400 to 500 °C

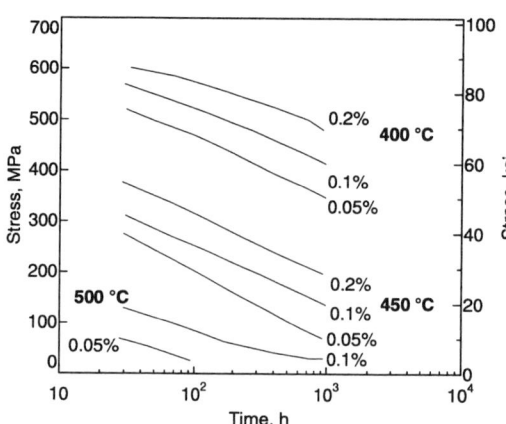

Source: IMI Titanium 550 brochure

IMI 550: Creep rupture

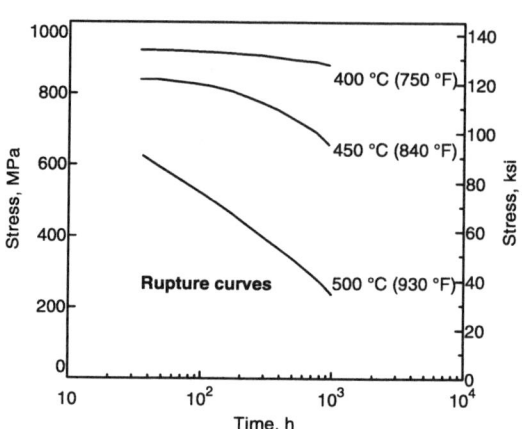

Fatigue Properties

IMI 550: Low-cycle fatigue properties

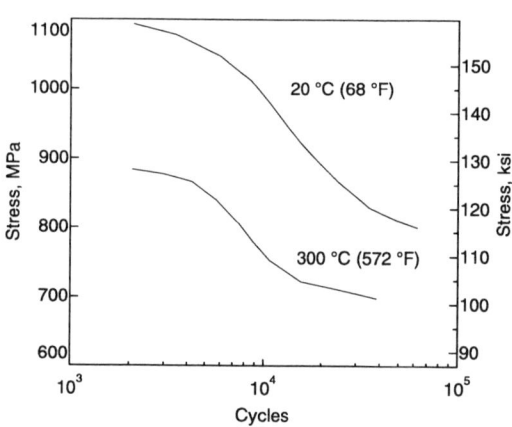

Heat treated bar; tests at 20 °C (68 °F) and 300 °C (572 °F); smooth specimens; axial loading; zero minimum stress.
Source: IMI Titanium 550 brochure

IMI 550: High-cycle fatigue

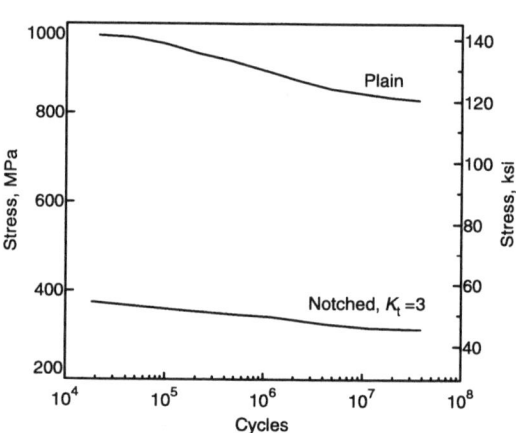

Heat treated bar; room temperature, axial loading, zero minimum stress.
Source: IMI Titanium 550 brochure

IMI 550: Rotating bending fatigue

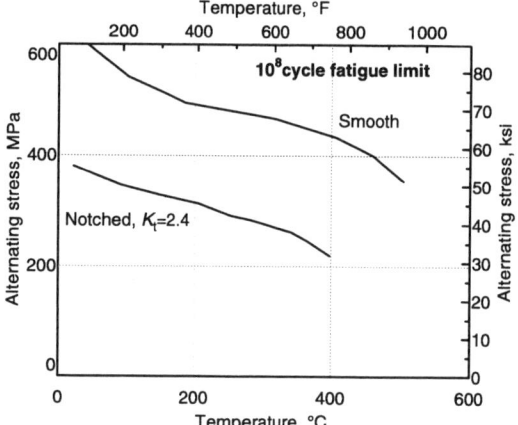

Heat treated bar.
Source: IMI Titanium 550 brochure

Processing

Casting

IMI 550 can be cast using the normal techniques developed for titanium alloys. It has similar flow and "castability" performance to that of Ti-6Al-4V (IMI 318).

Mechanical property levels of cast IMI 550 are generally equivalent, or superior, to the wrought product and therefore superior to cast Ti-6Al-4V. Tensile ductility and low-cycle fatigue performance are inferior to those of the wrought product as is the case with cast Ti-6Al-4V.

Bulk Working

Extrusion. IMI 550 can be alpha beta or beta extruded, and typical properties after such extrusions are shown (see table). Property levels in the alpha beta extruded condition are similar to those of the forged or rolled forms. Beta extruded materials exhibit lower ductility, but better fracture toughness than the equivalent alpha beta extruded product.

Forging. IMI 550 is readily forgeable by conventional hammer, press, or isothermal forging and is one of the easier titanium alloys to forge. Typical forging temperature is 900 °C (1650 °F).

Flow stress of IMI 550, around its typical forging temperature, is slightly stiffer than Ti-6Al-4V at their respective forging temperatures. The flow stress of IMI 550 increases slightly more rapidly than Ti-6Al-4V as temperature falls. IMI 550 is, however, regarded as a very easy alloy to forge.

IMI 550: Typical mechanical properties of extrusions

Section shape and extrusion type(a)	Testpiece(a) direction	0.2% Proof stress MPa	ksi	Tensile strength MPa	ksi	Elongation, %	Reduction in area, %	Fracture toughness MPa√m	ksi√in.
Extruded in α + β field	L	957	139	1107	161	14	40	53.6 T-L	48.7
	T	1016	147	1127	163	14.5	40	46.3 L-T	42.0
Extruded in β field	L	998	145	1183	172	12.5	24.5	62.9 T-L	57.1
	T	1009	146	1236	179	10.5	16	64.7 L-T	58.8

(a) Testpieces taken from heat treated web sections

Forming

IMI 550 can, in general, only be formed hot and it can be treated in the same way as other high-strength alpha-beta alloys such as Ti-6Al-4V. Plate and sheet can be produced by conventional rolling techniques. IMI 550 is also very amenable to superplastic forming; its fine grain size giving good flow characteristics and high m values (about 0.8). Typical tensile properties of superplastically formed IMI 550 sheet are shown (see table).

IMI 550: Tensile properties of superplastically formed sheet

Condition	Approximate SPF strain %	0.2% Yield stress MPa	ksi	Tensile strength MPa	ksi	Elongation (in 50 mm), %
SPF	0(a)	908	132	1097	159	12
SPF	100	937	136	1109	161	7
SPF	100	965	140	1123	163	5
SPF + STA	100	1126	163	1338	194	7

(a) Testpieces taken from the flange were not deformed in the superplastic formed (SPF) condition but were exposed to the SPF heat cycle. STA, solution treated and aged. Material/Test conditions: 2 mm (0.08 in.) sheet, annealed prior to forming, strain rate 2×10^{-4}/s. Source: IMI Titanium "Medium Temperature Alloys" brochure

Heat Treatment

The recommended heat treatment (see table) for IMI 550 consists of alpha-beta solution treatment followed by aging, and stress relieving if required during manufacture. Typical properties after such treatments are shown.

IMI 550: Tensile properties after various heat treatments

Treatment	0.2% Proof stress		Tensile strength		Elongation (in 5D) %	Reduction in area, %
	MPa	ksi	MPa	ksi		
Solution treated 900 °C/1/2 h/AC	930	135	1080	157	12	40
Fully aged, 900 °C/1/2 h/AC + 500 °C/24 h/AC	1070	155	1200	174	14	42
Stress-relieved 650 °C/2 h/AC	1020	148	1130	164	12	42

IMI 550: Recommended heat treatments

Treatment	Temperature		Duration	Cooling method
	°C	°F		
Alpha-beta solution treatment	900	1625	1 hour	Air cool
Aging	500	930	24 hours	Air cool
Stress relief	650	1200	2 hours	Air cool

Welding

IMI 550 is weldable using controlled electron beam or laser welding techniques. The limiting factor is fracture toughness of the weld due to high cooling rates obtained after welding. The use of slower welding speeds to produce lower cooling rates gives adequate weld properties. Electron beam welded compressor discs are in service in military aeroengines.

IMI 551
Ti-4Al-4Mo-4Sn-0.5Si

IMI Titanium 551 is an alpha + beta alloy containing 4Al-4Mo-4Sn-0.5Si, and belongs to the same alloy group as IMI 550; however, the higher alloying content of IMI 551 gives increased strength at room temperature, while still preserving good forging characteristics. IMI 551 is one of the strongest of the commercially available titanium alloys, with room-temperature strengths ranging from 1250 to 1400 MPa (181 to 203 ksi), and useful creep properties up to 400 °C (750 °F). The alloy is not normally regarded as weldable.

Applications. Primarily intended for airframe structural forgings and machined parts such as undercarriage components, mounting brackets, and pump casings where the strength is required at low weight; and for gas-turbine engine components. This alloy is also suitable for general engineering and chemical applications such as steam-turbine blades, axial and radial compressor parts, connecting rods, and other high-speed rotating and reciprocating components. Relevant British standard specifications for IMI 551 are listed (see table).

IMI Titanium 551: British Standard Specifications

Limiting ruling section	Bar for machining	Forging stock	Forgings
Up to 25 mm (1 in.)	TA 38	TA 39	
Over 25 and up to 75 mm (1 to 3 in.)	TA 40	TA 41	TA 42

IMI 551: Typical composition range (wt%) and density

	Al	Sn	Mo	Si	Fe	C	O_2	N_2	H_2
Minimum	3.0	3.0	3.0	0.3	...	0.05	...	...	...
Maximum	5.0	5.0	5.0	0.7	0.20	0.20	0.25	0.05	0.125(a)
Nominal	4.0	4.0	4.0	0.5	...	...	...	...	...

(a) 0.15 wt% maximum of H_2 in forgings. Typical density is 4.62 g/cm^3 (0.167 lbf/in.^{3}).

Physical Properties

IMI 551: Summary of typical physical properties

Beta transus	1050 ± 15 °C (1920 ± 30 °F)
Melting (liquidus point)	Not available
Density(a)	4.62 g/cm^3 (0.167 lbf/in.^{3})
Electrical resistivity(a)	1.7 μΩ · m
Magnetic permeability	Nonmagnetic
Specific heat capacity(a)	370 J/kg · K (0.088 Btu/lb · °F)
Thermal conductivity	7 W/m · K (4 Btu/ft · h · °F)
Thermal coefficient of linear expansion(b)	8.4 × 10^{-6}/°C (4.7 × 10^{-6}/°F)

(a) Typical values at room temperature of about 20 to 25 °C (68 to 78 °F). (b) Mean coefficient from room temperature to 100 °C (212 °F)

IMI 551: Elastic properties

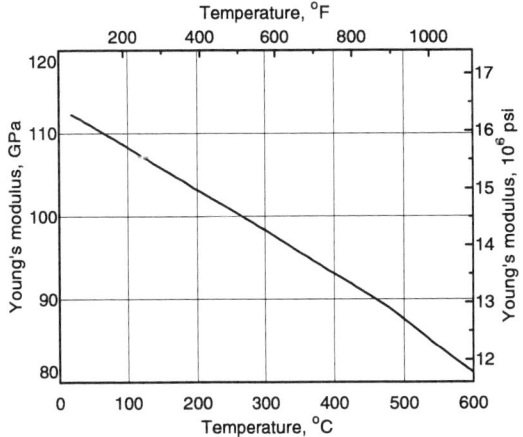

A curve of the dynamic elastic modulus of IMI 551, measured at a low level of strain, shown below. Torsional testing gives a shear modulus of 43 GPa (6.2 × 10^6 psi), which leads to a value for Poisson's ratio of 0.30.

IMI 551: Electrical resistivity

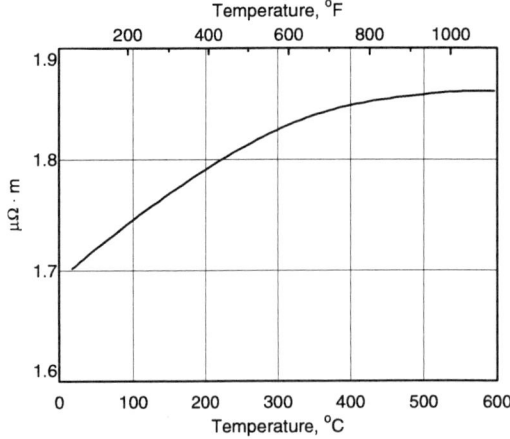

The relatively high alloy content leads to a slightly higher resistivity than other titanium alloys.

Thermal Properties

IMI 551: Thermal coefficient of linear expansion

Temperature range		Mean coefficient of thermal expansion	
°C	°F	10^{-6}/°C	10^{-6}/°F
20-100	68-212	8.4	4.7
20-200	68-390	9.0	5.0
20-300	68-570	9.3	5.1
20-400	68-750	9.5	5.2
20-500	68-930	9.6	5.3
20-600	68-1110	9.7	5.4

IMI 551: Specific heat

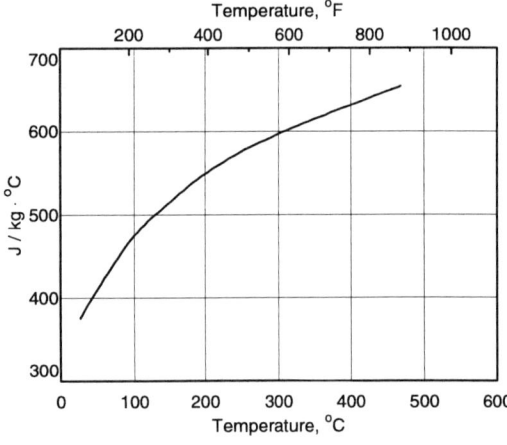

IMI 551: Thermal conductivity

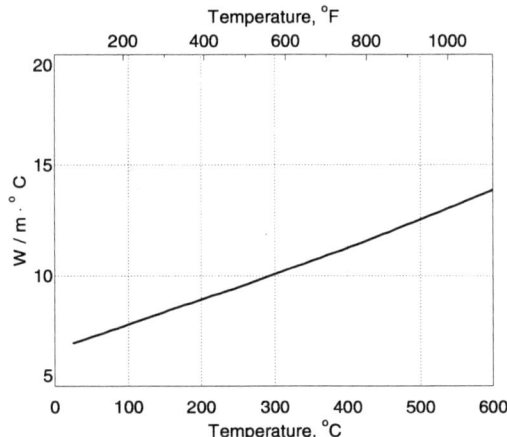

The relatively high alloying content leads to a thermal conductivity slightly lower than that of many titanium alloys.

Mechanical Properties

IMI 551 rod and bar: Guaranteed properties (BS TA 38-41)

Limiting ruling section	0.2% yield stress (min.)		Tensile strength		Elongation on 5D (min.) %	Reduction in area (min.) %
	MPa	ksi	MPa	ksi		
Up to 25 mm (1 in.)	1095	159	1250-1420	181-206	8	20
Over 25 and up to 75 mm (1 to 3 in.)	1065	154	1205-1375	175-199	8	20

IMI 551: Typical tensile properties for various section sizes

Product	Sample	0.2% yield stress		Tensile strength		Elongation on 5D	Reduction in area
		MPa	ksi	MPa	ksi	%	%
175 mm (7 in.) diam billet	Transverse, 3:1 upset, heat treated	1140	165	1300	188.5	9	25
100 to 125 mm (4 to 5 in.) square billet	Transverse, 3:1 upset, heat treated	1200	174	1310	190	10	30
75 mm (3 in.), square bar heat treated	Longitudinal, surface	1210	175	1320	191	13	40
	Longitudinal, center	1150	167	1260	183	16	37
25 to 50 mm (1 to 2 in.) rod	Longitudinal, center	1140	165	1300	188.5	12	40
16 mm (5/8 in.) rod	Center	1340	194	1480	214.5	10	28

IMI 551: Elevated-temperature tensile properties

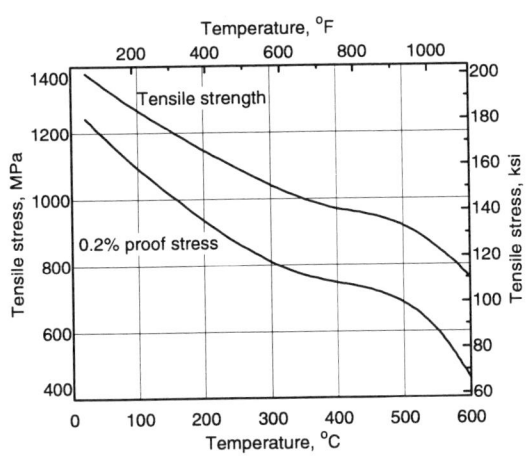

(a)

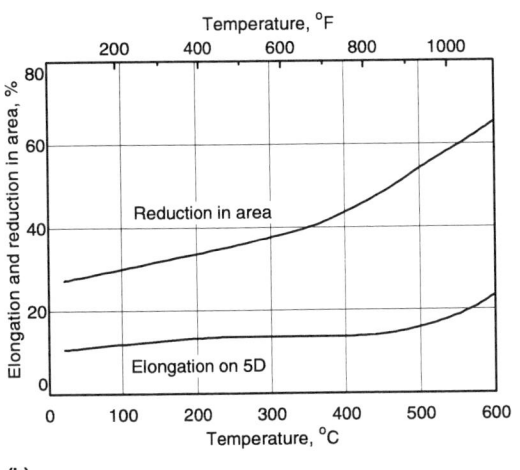

(b)

IMI 551: Creep properties

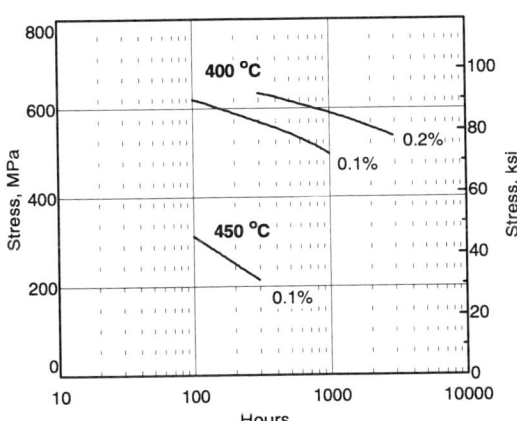

25 mm (1 in.) diam rod

IMI 551: Charpy impact strength

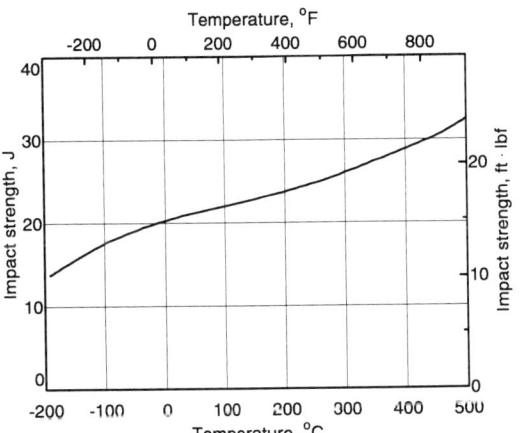

Fatigue

IMI 551: Rotating-bending fatigue properties

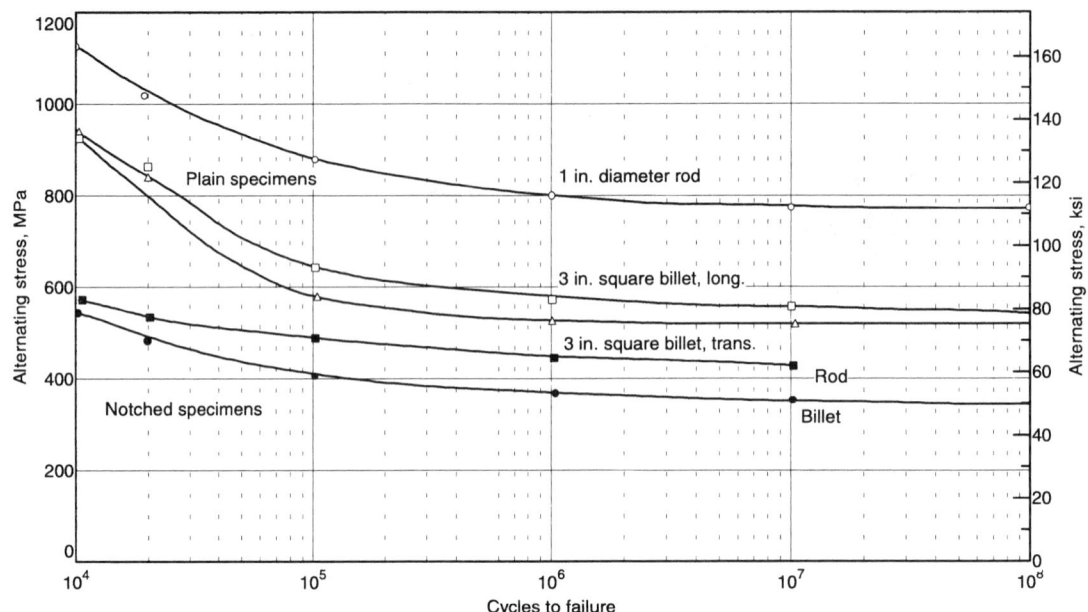

Rotating-bending fatigue properties of 25 mm (1 in.) diameter rod and 75 mm (3 in.) square billet in IMI 551 (notched specimens, K_t = 3.2). The results give a ratio of fatigue strength to tensile strength of between 0.45 and 0.55.

Processing

Forging. IMI 551 has a higher beta transus temperature than most other titanium alloys. This permits a fairly high forging temperature to be used so that, despite the high room-temperature strength of the alloy, it is only marginally more difficult to forge than Ti-6Al-4V.

To obtain good properties in a finished component, the alloy should be given at least a 4:1 forging reduction in the alpha + beta phase field. So as to ensure that no part of the forging exceeds the beta transus temperature as a result of internal work, it is recommended that the initial forging preheat temperature should not exceed 950 °C (1740 °F). The preheat temperature should be reduced to 930 °C (1700 °F) for subsequent forging operations to obtain the optimum combination of strength and ductility.

Heat Treatment. The recommended heat treatment for IMI 551 consists of solution treatment at 900 °C (1650 °F) for 1 hour per 25 mm of section thickness (1 h/in.), followed by air cooling and aging for 24 hours at 500 °C (930 °F) and again air cooling.

When small-diameter sections of 16 mm (5/8 in.) or less are air cooled, or when thicker sections are water or oil quenched, higher tensile strengths can be developed on aging, but only at the expense of ductility and creep strength. It is therefore better to slow down the cooling rate of thin sections (for example, in a box filled with refractory material) to avoid the undesirable effects of an excessive cooling rate (see table). Various solution-treatment temperatures have been suggested for IMI 551, but 900 °C followed by air cooling has been found to give the best strength, while avoiding the low ductility brought about by oil or water quenching.

Typical tensile properties after various heat treatments of small rod, 16 mm diameter (5/8 in.)

Treatment	0.2% yield stress MPa	ksi	Tensile strength MPa	ksi	Elongation on 5D %	Reduction in area %
1 h 900 °C, WQ, 24 h 500 °C, AC	1390	201.5	1650	239	3	9
1 h 900 °C, AC, 24 h 500 °C, AC	1310	190	1450	210	10	28
1 h 900 °C, cool in vermiculite, 24 h 500 °C, AC	1240	180	1410	204.5	13	47

Corona 5
Ti-4.5Al-5Mo-1.5Cr

UNS Number: Unassigned

Compiled by Elihu Bradley

Corona 5 is an experimental alloy. It is a versatile alloy capable of being processed to conditions covering a broad range of strength/fracture toughness combinations. Typical toughness values are well above the minimum required levels for fracture-controlled titanium alloy parts and offer potential cost savings in terms of quality assurance through a reduction in toughness verification testing frequency. At current design minimum requirements for fracture toughness, Corona 5 can be produced to significantly greater strength levels than Ti-6Al-4V is capable of achieving without an acceptable reduction in toughness.

Chemistry. Corona 5 requires less restricted chemistry control (particularly oxygen) and can be processed at lower temperatures than Ti-6Al-4V.

Product Forms. Corona 5 has been produced for experimental Navy programs in production-sized ingots and has been converted to forgings and flat rolled plate. Sheet and rod also have been produced in limited quantities.

Applications. Corona 5 was intended for structural applications, especially for fracture-critical components used primarily in the aerospace industry.

Physical Properties

Phases and Structures. Corona 5 is an α-β alloy with a β transus of approximately 925 °C (1700 °F). Its microstructure is characterized by the α phase within the β matrix. In the development of Corona 5, both α and β stabilizers were studied. Aluminum was selected as the α stabilizer, and the 4.5% level was chosen to avoid problems with stress-corrosion resistance, which might be lowered at high aluminum contents. Molybdenum and chromium were used as strong β stabilizers. The composition allows about 50% α phase to be easily precipitated at temperatures within 100 °C (200 °F) of the β transus in discrete lenticular form rather than in pockets or colonies.

The α morphology strongly influences mechanical properties. Two basic processing operations are available: (1) beta processing, carried out completely above the β transus or in which finish processing is completed below the β transus, but at a high enough temperature so that very little α phase is present, or (2) α-β processing carried out below the β transus temperature in the presence of the α phase. Subsequent annealing below the β transus within about 175 °C (300 °F) of the transus temperature results in a distribution of primary α, which is related to the processing sequence and annealing temperature. With β-processed material, lenticular α morphology is achieved and maintained. With α-β processing, the primary α-β becomes globular during subsequent heat treatment.

The changes in the morphology of α from lenticular to globular is a direct result of the prior deformation of the α. The strain energy in the α after α-β working causes it to recrystallize and relax to the lower surface energy globular form. The rate at which lenticular α transforms to globular α is a function of annealing temperature and time and the amount of work the α has received. Lightly worked α remains lenticular longer than heavily worked α (see figure on next page).

The strength of Corona 5 is virtually unaffected by the shape of the primary α, but other properties such as fracture toughness and elevated-temperature flow characteristics are strongly influenced. High fracture toughness is associated with α having a high aspect ratio (i.e., lenticular), whereas lower fracture toughness at the same strength level corresponded to α with a low aspect ratio (i.e., globular). Optimum superplastic forming and diffusion bonding are found in material with a globular α microstructure.

Corona 5: Summary of typical physical properties

Density at 20 °C (68 °F)	4.54 g/cm^3 (0.164 lb/in.^{3})
Melting point (approximate)	1750 °C (3180 °F)
Beta transus	925 °C (1700 °F)
Modulus of elasticity	114 GPa (16.6 × 10^6 psi)

Source: Corona 5 Alloy, *Alloy Digest*, May 1978

Chemical/Corrosion Properties

Corona 5 has excellent resistance to general corrosion and stress corrosion. It also has excellent corrosion resistance in saline and physiological environments (see figure on next page).

Corona 5: Alpha morphology vs annealing time and temperature

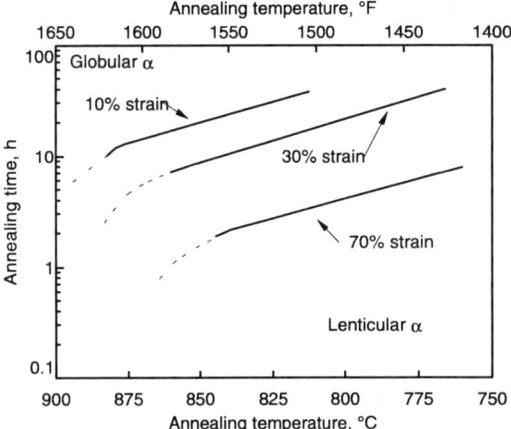

After various amounts of deformation in the α-β field.
Source: F.H. Froes et al., Synthesis of Corona 5 (Ti-4.5Al-5Mo-1.5Cr), JOM, May 1980

Corona 5: Anodic polarization curves

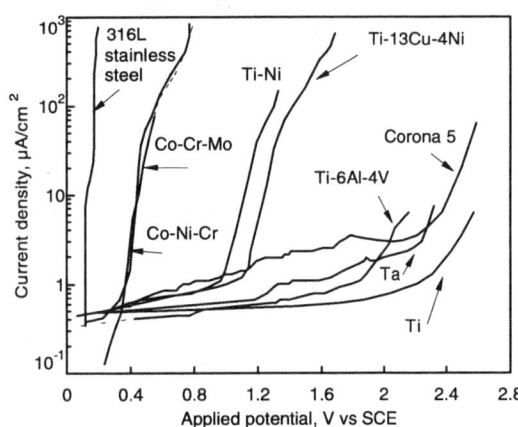

Potentiostatic measurements were made in Hanks' solution. The solution was in a closed vessel and was not aerated or deaerated.
Source: A.C. Fraker et al., Surface Preparation and Corrosion of Titanium Alloys for Surgical Implants

Mechanical Properties

Corona 5: Mechanical properties for fracture-critical applications

Property	Corona 5	Ti-6Al-4V
Ultimate tensile strength (UTS), MPa (ksi)	895 (130)	895 (130)
Tensile yield strength, MPa (ksi)	825 (120)	795 (115)
Elongation, %	12.0	12.0
Reduction of area, %	25.0	25.0
Fracture toughness MPa√m (K_{Ic}, ksi√in.)	110 (100)	88 (80)
Modulus of elasticity GPa (10^6 psi)	110 (16.0)	110 (16.0)
Density, g/cm³ (lb/in.³)	4.5 (0.164)	4.4 (0.160)
Critical crack length design parameter: ($2 a_c$), mm (in.)	21.5 (0.848)	13.7 (0.542)

$$2a_c = \frac{2}{\pi}\left\{\frac{K_{Ic}}{\frac{2}{3} \times \text{UTS}}\right\}^2$$

Source: F.H. Froes et al., Synthesis of Corona 5 (Ti-4.5Al-5Mo-1.5Cr), JOM, May 1980

Corona 5: Elevated-temperature tensile strength

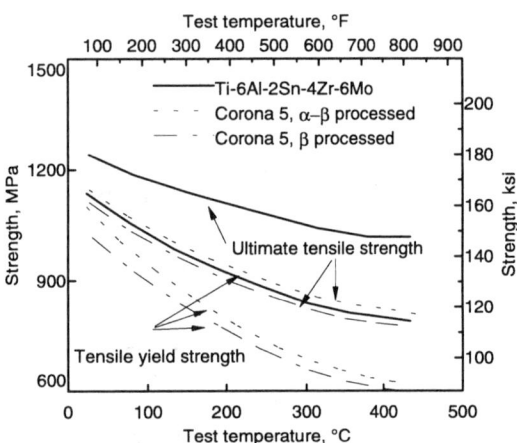

Elevated-temperature tensile properties of Corona 5 and Ti-6Al-2Sn-4Zr-6Mo.
Source: F.H. Froes et al., Synthesis of Corona 5 (Ti-4.5Al-5Mo-1.5Cr), JOM, May 1980

Corona 5: Weld and base-metal property comparison

	Corona 5			Ti-6Al-4V	
Property	Base metal(a)	Weldment(b) 705 °C (1300 °F), 4 h	Weldment(b) 775 °C (1425 °F), 4 h	Base metal (mill annealed)	Weldment(b) 675 °C (1250 °F) 30 min
Ultimate tensile strength, MPa (ksi)	1137 (165)	1220 (177)	1220 (177)	979 (142)	951 (138)
Tensile yield strength, MPa (ksi)	1055 (153)	1179 (171)	1144 (166)	910 (132)	910 (132)
Elongation, %	21	2-3	3-4	14	8
Toughness (K_Q), MPa√m (ksi√in.)	81 (74)	57 (52)	67 (61)	...	41-44 (38-40)

(a) 915 °C (1675 °F), 30 min, FC, 845 °C (1550 °F), 4 h. (b) Longitudinal fusion zone with post-weld heat treatment as indicated. Source: F.H. Froes et al., Synthesis of Corona 5 (Ti-4.5Al-5Mo-1.5Cr), JOM, May 1980

Corona 5: Effect of processing on tensile properties

Temperature °C (°F)	Time, h	Direction(a)	Ultimate tensile strength MPa	Ultimate tensile strength ksi	Tensile yield strength MPa	Tensile yield strength ksi	Elongation, %	Reduction of area, %	Modulus of elasticity GPa	Modulus of elasticity 10^6 psi
High-temperature processing route										
970 (1780)	0.25(b)	L	865	125.6	777	112.7	16.0	30.4	103	15.0
		T	898	130.3	811	117.7	14.0	30.1	110	16.0
855 (1570)	4.0(c)	L	800	116.1	719	104.3	21.5	62.3	109	15.8
		T	822	119.3	756	109.7	18.5	39.7	113	16.4
830 (1525)	4.0(c)	L	854	124.0	761	110.4	19.5	53.5	111	16.1
		T	861	125.0	812	117.9	19.0	55.2	111	16.1
830 (1525)	16.0(c)	L	803	116.6	695	100.8	20.5	67.7	99	14.4
		T	830	120.5	765	111.0	19.0	49.5	107	15.5
Low-temperature processing route										
970 (1780)	0.25(b)	L	794	115.3	722	104.8	16.0	34.2	105	15.2
		T	814	118.1	738	107.1	19.0	35.9	103	15.0
855 (1570)	4.0(c)	L	773	112.2	741	107.5	24.0	64.9	99	14.3
		T	835	121.2	810	117.5	22.5	60.1	101	14.7
	4.0(c)	L	785	114.0	760	110.3	22.5	69.7	103	15.0
		T	830	120.5	801	116.2	20.0	59.1	110	16.0
830 (1525)	16.0(c)	L	777	112.8	727	105.5	24.0	61.0	114	16.6
		T	828	120.2	798	115.8	20.5	60.6	113	16.4

(a) L, longitudinal; T, transverse. (b) Air cooled and stabilize annealed 4 h, 705 °C (1300 °F), air cooled. (c) Furnace cooled 38 °C (100 °F)/h to 540 °C (1000 °F), air cooled. Source: J.C. Williams et al., *Development of High Fracture Toughness Titanium Alloys*, ASTM Special Publication 651

Corona 5: Mechanical properties of HIP'd compacts

HIP cycle(a)	Type(b)	Ultimate tensile strength MPa	Ultimate tensile strength ksi	Tensile yield strength MPa	Tensile yield strength ksi	Elongation, %	Reduction of area, %	Toughness (K_Q) MPa√m	Toughness (K_Q) ksi√in.
955 °C (1750 °F)	Beta	985	143	917	133	8.5	15	~71	~65
900 °C (1650 °F)	Alpha-beta	999	145	930	135	19	50	~71	~65

(a) Followed by an 845 °C (1550 °F), 8 h, AC + 705 °C (1300 °F), 4 h, AC. (b) Beta transus, ~945 °C (~1735 °F). Source: F.H. Froes et al., Synthesis of Corona 5 (Ti-4.5Al-5Mo-1.5Cr), *JOM*, May 1980

Corona 5: Thermal stability of tensile properties after creep exposure

Processing	Creep exposure	Plastic strain, %	Tensile yield strength MPa	Tensile yield strength ksi	Ultimate tensile strength MPa	Ultimate tensile strength ksi	Elongation, %	Reduction of area, %
Alpha-beta	As-processed, RT	...	962	139.6	1148	166.6	33.0	27.3
	370 °C (700 °F), 100 h, 689 MPa (100 ksi)	3.2	1121	162.6	1154	167.4	7.0	19.7
Beta	As-processed, RT	...	1046	151.7	1114	161.6	14.0	30.6
	370 °C (700 °F), 100 h, 689 MPa (100 ksi)	0.7	1076	156.1	1116	161.9	11.0	23.8
Ti-6Al-4V(a)	As-processed, RT	...	1220	177	1282	186	8	39
	370 °C (700 °F), 100 h, 689 MPa (100 ksi)	0.11	1179	171	1255	182	6	32

(a) Data from DMIC Report 46, Nov 1958, p 13. Ti-6Al-4V annealed 925 °C (1700 °F), 20 min, WQ, aged 540 °C (1000 °F), 4 h, AC. Source: F.H. Froes et al., Synthesis of Corona 5 (Ti-4.5Al-5Mo-1.5Cr), *JOM*, May 1980

Corona 5: Elevated temperature ductility

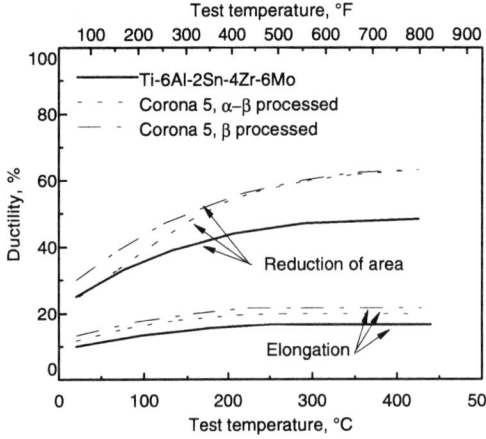

β processed material is more ductile, which is not normally true.
Source: F.H. Froes et al., Synthesis of Corona 5 (Ti-4.5Al-5Mo-1.5Cr), JOM, May 1980

Fatigue Life

Corona 5: Fatigue comparison of mill annealed alloy

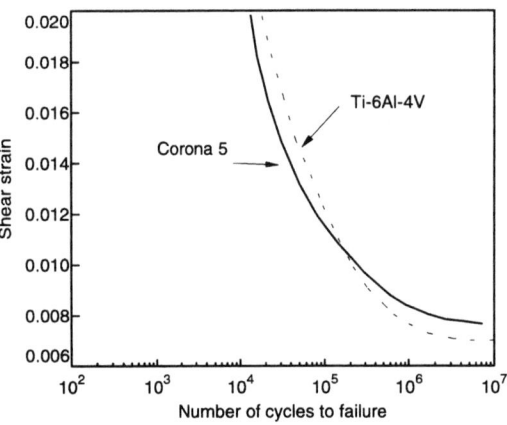

Source: M.A. Iman et al., Fatigue Crack Growth in a Ti-4.5Al-5Mo-1.5Cr Alloy with Metastable β Phase, Fracture Mechanics, Vol 21

Corona 5: Smooth and notched fatigue behavior of material for post-superplastic forming applications

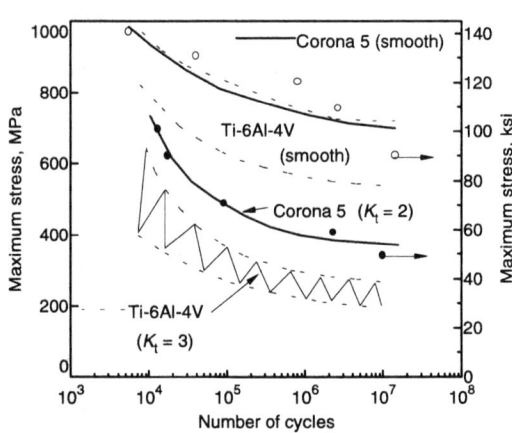

Room-temperature axial fatigue; $R = 0.05$; longitudinal data.
Source: M.A. Iman et al., Fatigue Crack Growth in a Ti-4.5Al-5Mo-1.5Cr Alloy with Metastable β− Phase, Fracture Mechanics, Vol 21

Corona 5: Fatigue life for powder

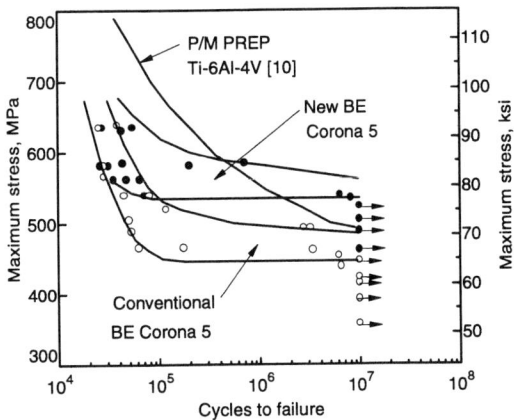

Smooth axial fatigue; $R = 0.1$; 80 Hz.
Source: F.H. Froes et al., Effect of Microstructure Strength and Oxygen Content on Fatigue Crack Growth Rate of Ti-4.5Al-5.0Mo-1.5Cr (Corona 5), Metall. Trans. A, Vol 15, Jan 1984

Corona 5: Fatigue of shear-strained material in air

Anneal temperature		Mean life, cycles	Minimum life, cycles	Standard deviation, cycles
°C	°F			
870	1600	21,041	18,006	1919
900	1650	24,738	20,536	2018
915	1680	50,808	40,981	4646
955	1750	17,489	15,484	1584

Note: Axial load, 450 g. Cycled at a shear strain of ± 0.02 at 0.2 Hz in air. Source: Luckey and Kubli, Titanium Alloys in Surgical Implants, ASTM Publication 04-796000-54

Corona 5: Corrosion fatigue for possible implant application

Anneal temperature		Mean life, cycles	Minimum life, cycles	Standard deviation, cycles
°C	°F			
760	1400	10,218	10,165	75
815	1500	22,408	20,111	2018
845	1550	22,780	20,356	2934
870	1600	20,650	14,714	8340
915	1680	30,332	26,483	5443

Note: Cycled at a shear strain of ± 0.020 at 1 Hz. Source: Luckey and Kubli, Titanium Alloys in Surgical Implants, ASTM Publication 04-796000-54

Fatigue Crack Growth Rates

Corona 5: Fatigue crack growth rate comparison

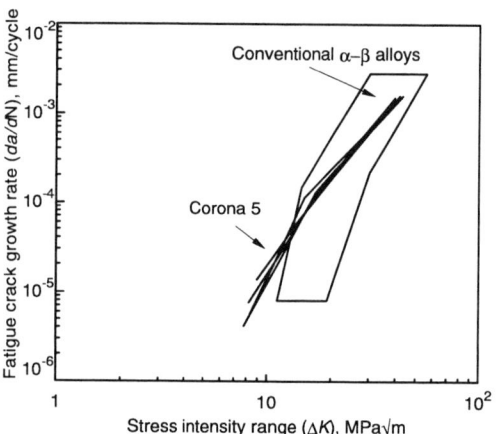

Room-temperature data in air; $R = 0.10$; 5 Hz.
Source: M.A. Iman et al., On Fatigue Crack Growth in a Ti-4.5Al-5Mo-1.5Cr Alloy with Metastable β-Phase, *Fracture Mechanics*, Vol 21

Corona 5: Fatigue crack growth

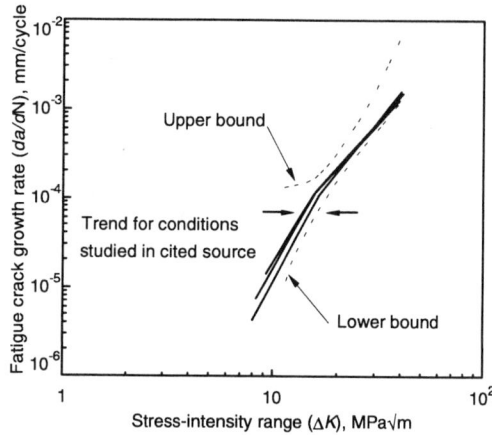

Room-temperature data in air; $R = 0.10$; 5 Hz.
Source: M.A. Iman, et al., On Fatigue Crack Growth in a Ti-4.5Al-5Mo-1.5Cr Alloy with Metastable β-Phase, *Fracture Mechanics*, Vol 21

Corona 5: Fatigue crack growth

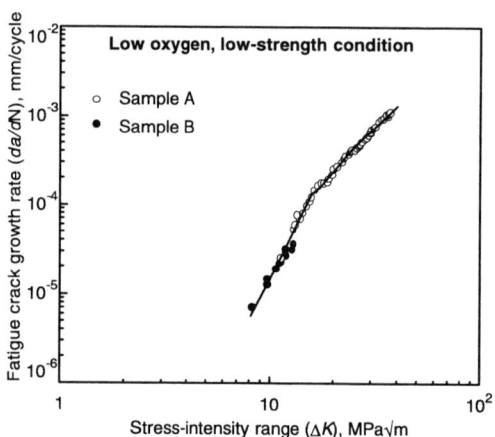

825 MPa (120 ksi) UTS. Room-temperature data; $R = 0.10$; 5 Hz.
Source: M.A. Iman et al., On Fatigue Crack Growth in a Ti-4.5Al-5Mo-1.5Cr Alloy with Metastable β-Phase, *Fracture Mechanics*, Vol 21

Corona 5: Fatigue crack growth

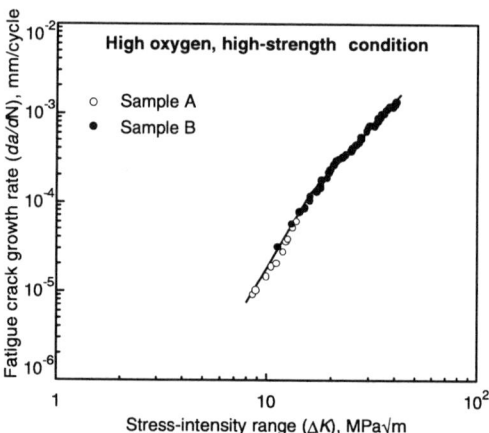

1100 MPa (160 ksi) UTS. Room-temperature data; $R = 0.10$; 5 Hz.
Source: M.A. Iman, et al., On Fatigue Crack Growth in a Ti-4.5Al-5Mo-1.5Cr Alloy with Metastable β-Phase, *Fracture Mechanics*, Vol 21

Corona 5: Fatigue crack growth

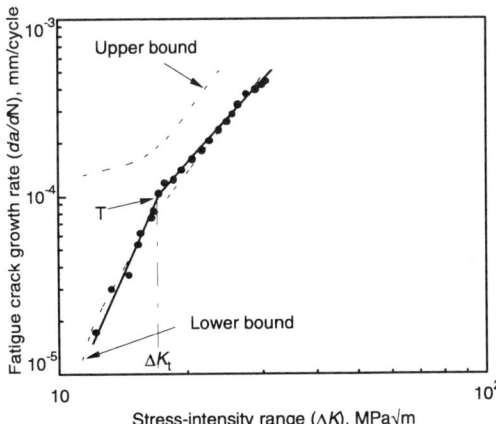

The effect of helium cooled from 900 °C (1650 °F) vs the upper bound (UB) and lower bound (LB) of different heat treatments. A transition point (T) is noted along with the transition stress intensity range ΔK_T.
Source: M.A. Iman et al., On Fatigue Crack Growth in a Ti-4.5Al-5Mo-1.5Cr Alloy with Metastable β-Phase, *Fracture Mechanics*, Vol 21

Corona 5: Fatigue crack growth

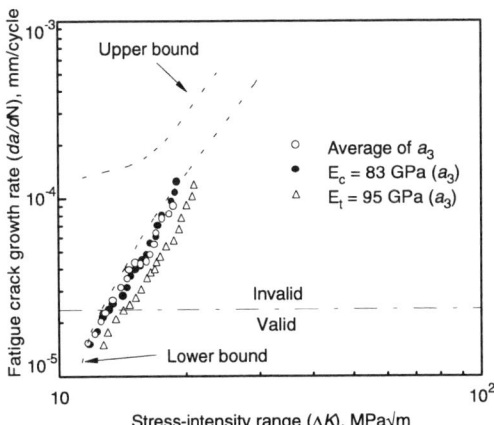

The effect of water quenching from 900 °C (1650 °F) to the upper bound (UB) and lower bound (LB). No transition was observed within the valid region of testing.
Source: M.A. Iman et al., On Fatigue Crack Growth in a Ti-4.5Al-5Mo-1.5Cr Alloy with Metastable β-Phase, *Fracture Mechanics*, Vol 21

Fracture Toughness

Corona 5: Fracture toughness comparison

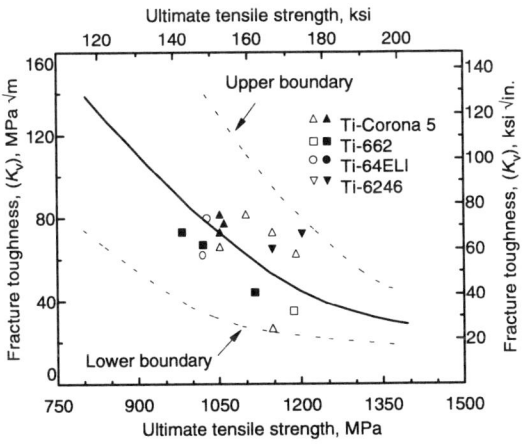

Open symbols represent postweld heat treatments of 705 °C (1300 °F) and below; solid symbols are 705 °C (1300 °F) and higher
Source: W.A. Baeslack et al., Weldability of a High Toughness Titanium Alloy—A Metallurgical Evaluation

Corona 5: Fracture toughness vs strength

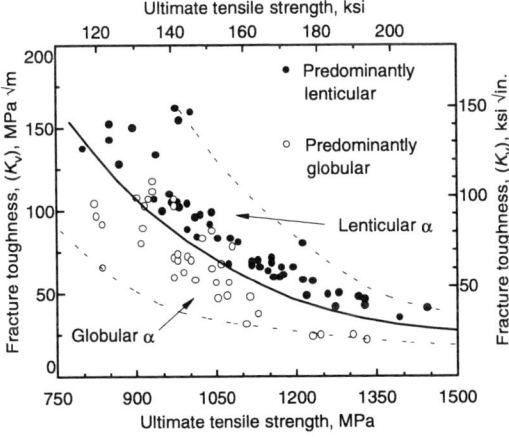

Source: F.H. Froes et al., Synthesis of Corona 5, *JOM*, May 1980

Corona 5: Fracture toughness comparison of base and welded materials

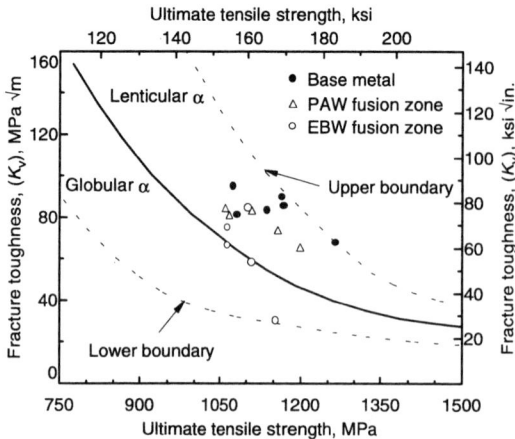

Source: W.A. Baeslack et al., Weldability of a High Toughness Titanium Alloy—A Metallurgical Evaluation

Corona 5: Fracture toughness of plate

Processing route	Fracture toughness		Tensile yield strength		Ultimate tensile strength	
	MPa√m	ksi√in.	MPa	ksi	MPa	ksi
Beta anneal	129	117	775	112	865	126
	156	142	860	125	950	138
Beta processed	130	118	920	134	980	142
	145	132	760	110	855	124
	160	145	890	129	1000	145
Alpha-beta processed	111	101	905	131	935	136

Source: F.H. Froes et al., Synthesis of Corona 5 (Ti-4.5Al-5Mo-1.5Cr), JOM, May 1980

Ti-6-22-22S
Ti-6Al-2Sn-2Zr-2Mo-2Cr-0.25Si

UNS Number: Unassigned

Compiled by P. Russo, RMI Titanium Co., and R. Boyer, Boeing

Ti-6-22-22S was developed by RMI Titanium Co., with additional development funded through an Air Force Contract in the early 1970s. The alloy was conceived to provide high strength in heavy sections with good fracture toughness and to retain that strength up to moderate temperatures through the addition of silicon. The lack of any production applications precluded further development at that time. Interest in this alloy for fighter aircraft applications, because of its strength advantage over Ti-6Al-4V and good damage tolerance properties, has been revived. A strong effort is underway to develop thermomechanical processing procedures to optimize the strength, toughness, and crack growth rate properties of Ti-6-22-22S in sheet, plate, and forged forms.

Some data report a relatively high elastic modulus, which could be important for certain applications. Sheet can be formed at room temperature and has excellent superplastic forming characteristics.

Effects of Impurities and Alloying. Exceeding impurity limits may result in decreasing the ductility and fracture toughness below required minimums due to the associated increase in strength. As with other α-β titanium alloys, excessive aluminum, oxygen, and nitrogen can reduce ductility and fracture toughness. High amounts of the β stabilizers, chromium and molybdenum, may result in higher strength than desired.

Product Forms. Ti-6-22-22S has been produced in standard wrought product forms such as sheet, plate, bar, and forgings. Sheet exhibits excellent superplastic forming characteristics.

Product Condition. Ti-6-22-22S can be used in the annealed and heat treated conditions; solution treatment and aging can provide significant strengthening. The main emphasis at present, except for sheet, is on a triplex heat treatment involving a β solution treatment with a controlled cooling rate followed by an α-β solution treatment followed by aging to maximize damage-tolerant properties. Sheet should be used in the α-β processed condition.

Applications. There are no production applications for Ti-6-22-22S at this time, but it is bill-of-material for the aft fuselage of the F-22 ATF fighter. The primary interest in this alloy lies in its improved damage tolerance properties with respect to strength in relation to Ti-6Al-4V.

Specifications and Compositions. The only specifications for Ti-6-22-22S to date are those written by Lockheed/Boeing/General Dynamics for the ATF fighter. The composition limits are established as follows (except Si content may be reduced):

	Composition, wt%										
	Al	Sn	Zr	Mo	Cr	Si	Fe	O	C	N	H
Minimum	5.25	1.75	1.75	1.75	1.75	0.20	...	...	...	...	...
Maximum	6.25	2.25	2.25	2.25	2.25	0.27	0.15	0.13	0.04	0.03	125 ppm

Selected References

- H.R. Phelps and J.R. Wood, "Correlation of Mechanical Properties and Microstructures of Ti-6Al-2Sn-2Zr-2Mo-2Cr-0.25Si Titanium Alloy," Proc. 7th Int. Titanium Conf., San Diego, TMS/AIME, June 1992, to be published

- R.R. Boyer and A.E. Caddey, "The Properties of Ti-6Al-2Sn-2Zr-2Mo-2Cr Sheet," Proc. Int. Titanium Conf., San Diego, TMS/AIME, June 1992, to be published

- R.C. Bliss, "Evaluation of Ti-6Al-2Sn-2Zr-2Cr-2Mo-0.23 Si Sheet," Proc. 7th Int. Titanium Conf., San Diego, TMS/AIME, June 1992, to be published

- G.W. Kuhlman et al., "Characterization of Ti-6-22-22S: A High-Strength Alpha-Beta Titanium Alloy for Fracture Critical Applications," Proc. 7th Int. Titanium Conf., San Diego, TMS/AIME, June 1992, to be published

- "Mechanical-Property Data: Ti-6Al-2Zr-2Sn-2Mo-2Cr Alloy Solution Treated and Aged Plate," F33615-72-C-1280, Battelle Columbus Laboratories, Apr 1973

- A.K. Chakrabarti et al., "TMP Conditions-Microstructure-Mechanical Property Relationship in Ti-6-22-22S Alloy," Proc. 7th Int. Titanium Conf., San Diego, TMS/AIME, June 1992, to be published

- G.W. Kuhlman, Beta Processed Ti-6-22-22S Aging Studies, Alcoa Report, Mar 1992

- L.J. Bartlo, H.B. Bomberger, and S.R. Seagle, Deep Hardenable Titanium Alloy, AFML-TR-73-122, Battelle Columbus Laboratories, May 1973

- O.L. Deel, P.E. Ruff, and H. Mindlin, "Engineering Data on New Aerospace Structural Materials," AFML-TR-75-97, Battelle Columbus Laboratories, June 1975

Physical Properties

Phases and Structures

Triplex heat treatment results in a coarse lamellar α structure in a transformed β matrix. Cooling rates from the solution treatments must be controlled within a given window to provide desired strengths. The retained β contains a fine acicular α precipitate due to aging. Very fine, submicron-size silicides have been observed in this alloy.

Sheet can be used in the annealed or solution treated and aged condition. An air cool from the solution treatment temperature provides adequate heat treatment response. The annealed condition consists basically of equiaxed α with intergranular β (see figure). Material in the solution treated and aged condition is very similar to that of Ti-6Al-4V, with equiaxed α in a β matrix. With solution treatments below about 850 °C (1560 °F), β phase at temperature will be retained upon cooling (for sheet gages with an air cool), which provides a strength minimum in the solution treated condition. Solution treating at temperatures above this results in increased amounts of martensite formation as the temperature is increased and higher strengths.

Elastic Properties

Young's Modulus. The high modulus reported by Battelle (see table) has never been explained and has not been duplicated. The F-22 program is now working with a modulus of 113.8 GPa (16.5×10^6 psi).

Poisson's ratio. 0.33

Ti-6-22-22S: Summary of typical physical properties

Beta transus	960 ± 15 °C (1760 ± 25 °F)
Melting (liquidus) point	Not available
Density(a)	4.65 g/cm^3 (0.164 lb/in.^{3})
Electrical resistivity	Not available
Magnetic permeability(a)	Nonmagnetic
Specific heat capacity	Not available
Thermal conductivity	Not available
Thermal coefficient of linear expansion(b)	9.2×10^{-6}/°C (5.1×10^{-6}/°F)

(a) Typical values at room temperature of about 20 to 25 °C (68 to 78 °F). (b) Mean coefficient from room temperature to 425 °C (800 °F)

Ti-6-22-22S: Mill annealed microstructure

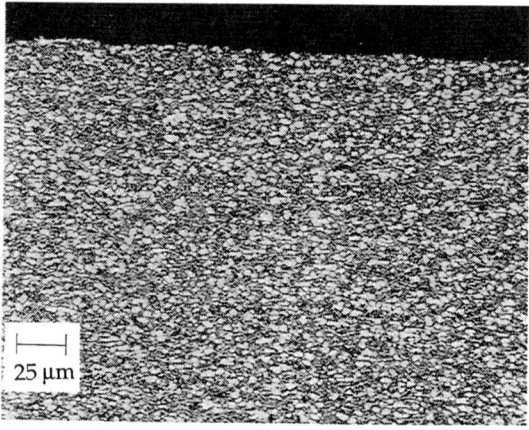

1.2 mm (0.050 in.) sheet annealed at 730 °C (1350 °F), 15 min, AC

Ti-6-22-22S: Elastic properties of forgings

Young's modulus	
Conventionally processed	122×10^3 MPa (17.7×10^3 ksi)
Beta processed (a)	115×10^3 MPa (16.8×10^3 ksi)
Bulk modulus	117×10^3 MPa (17.0×10^3 ksi)
Shear modulus	46×10^3 MPa (6.7×10^3 ksi)
Poisson's ratio	0.33

(a) Conventional subtransus forging with a triplex heat treatment consisting of β treat/fan cool/ α + β treat/fan or air cool/and aging

Ti-6-22-22S: Variation in Young's modulus

Test temperature		Longitudinal		Transverse	
°C	°F	GPa	10^6 psi	GPa	10^6 psi
Room temperature(a)		110	15.9	112	16.2
Room temperature(b)		123	17.9	122	17.8
200	400(b)	110	15.9	112	16.2
315	600(b)	107	15.6	110	16.0
425	800(b)	99	14.4	100	14.6

(a) Beta processed. From G.W. Kuhlman et al., "Characterization of Ti-6-22-22S: A High-Strength Alpha-Beta Titanium Alloy for Fracture Critical Applications," Proc. 7th Int. Titanium Conf., San Diego, TMS/AIME, June 1992, to be published. (b) From "Mechanical-Property Data: Ti-6Al-2Zr-2Sn-2Mo-2Cr Alloy," AFML, Battelle Columbus Laboratories, Apr 1973

Corrosion

Stress-Corrosion Cracking. Boeing has reported the stress-corrosion threshold for this alloy to be about 55 MPa$\sqrt{m}$ (50 ksi$\sqrt{in.}$) in an aqueous 3.5% NaCl solution. Previous work (Battelle, Apr 1973) reported a value of 80% of the tensile yield strength.

Tensile Properties

Because Ti-6-22-22S is an age-hardenable alloy, a range of tensile properties is attainable. The alloy can be heat treated in sections up to 75 to 100 mm (3 to 4 in.) thick. Tensile properties will depend on processing history and on the solution treatment and aging temperature (although strength is not very sensitive to aging temperature over a fairly wide temperature range).

Ti-6-22-22S: Typical mechanical properties for α–β processed STA products

Product	Ultimate tensile strength MPa	ksi	Tensile yield strength MPa	ksi	Elongation, %	Reduction of area, %
Sheet(a)	1331	193	1193	173	7.5	...
Plate(b)	1204	175	1131	164	12.0	35
Billet(b)	1200	174	1089	158	15.0	41

Source: G.A. Bella, RMI Titanium Co., 8 May 1991. (a) Subtransus ($\beta_T - 50$ °F) solution treatment with 540 °C (1000 °F) age, 8 h. (b) Supratransus and subtransus treat plus aging

Ti-6-22-22S: Typical mechanical properties for β-processed STA products

Product	Ultimate tensile strength MPa	ksi	Tensile yield strength MPa	ksi	Elongation, %	Reduction of area, %
50 mm (2 in.) plate	1138	165	1020	148	10	17
100 mm (4 in.) plate	1103	160	979	142	10	15
150 mm (6 in.) plate	1076	156	958	139	10	15

Source: J.R. Wood, RMI Titanium Co., 15 Aug 1991. $\beta_T - 28$ °C (50 °F) for 1 h, AC/540 °C (1000 °F) age, 8 h, AC

Ti-6-22-22S: Typical mechanical properties for α–β processed mill annealed products

Product	Ultimate tensile strength MPa	ksi	Tensile yield strength MPa	ksi	Elongation, %	Reduction of area, %
Sheet	1103	160	1034	150	10	...
Plate	1076	156	1014	147	13	28

Source: G.A. Bella, RMI Titanium Co., 8 May 1991. Mill anneal 730 °C (1350 °F), 2 h, AC

Plate and Forgings

The F-22 program has established a minimum tensile strength of 1035 MPa (150 ksi), and it is felt that the strength should be controlled within the range of 1070-1137 MPa (155 to 165 ksi) to meet the minimum fracture toughness requirement. This has resulted in a cooling rate window (see figure). The effect of slower cooling rates is a coarser lamellar α and lower strength. Oxygen content has the expected influence on strength and toughness.

Sheet

Ti-6-22-22S: Determination of acceptable processing window

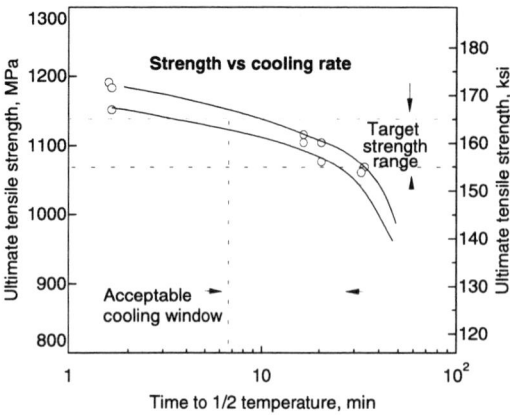

Beta processing improves fracture and crack growth resistance, and this study identified through-transus beta forging followed by α-β STA as the optimum TMP route.
Beta-processed plate α-β solution treated and aged at 540 °C (1000 °F), 8 h. Time to 1/2 temperature refers to time from ST to 1/2 ST temperature.
Source: H.R. Phelps and J.R. Wood, "Correlation of Mechanical Properties and Microstructures of Ti-6Al-2Sn-2Zr-2Mo-2Cr-0.25Si Titanium Alloy," Proc. 7th Int. Titanium Conf., San Diego, TMS/AIME, June 1992, to be published

Ti-6-22-22S: Effect of oxygen content on tensile strengths

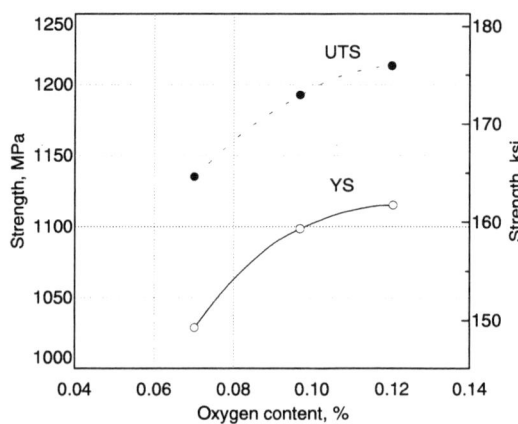

Forged pancakes α-β forged + α-β solution annealed + 540 °C (1000 °F), 8 h, aged, AC.
Source: A.K. Chakrabarti et al., "TMP Conditions-Microstructure-Mechanical Property Relationship in Ti-6-22-22S Alloy," Proc. 7th Int. Titanium Conf., San Diego, TMS/AIME, June 1992, to be published

The solution treat temperature has a very strong effect on the mechanical properties of Ti-6-22-22S (see figures). The yield strength minimum is associated with the retention of the maximum amount of β. The effect of superplastic forming temperature is similar to that of solution treat temperature, and superplastic formed parts can be aged to higher strengths (see section on superplastic forming).

Ti-6-22-22S: Strength and ductility vs solution treating temperature

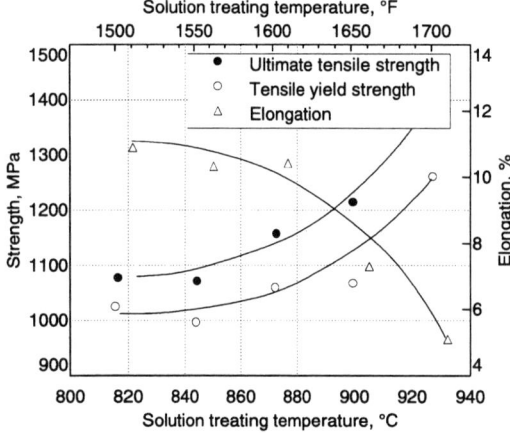

1.2 mm (0.050 in.) sheet solution treated 30 min, AC, no aging.
Source: R.R. Boyer and A.E. Caddey, "The Properties of Ti-6Al-2Sn-2Zr-2Mo-2Cr Sheet," Proc. 7th Int. Titanium Conf., San Diego, TMS/AIME, June 1992, to be published

Ti-6-22-22S: Effect of solution temperature on tensile properties

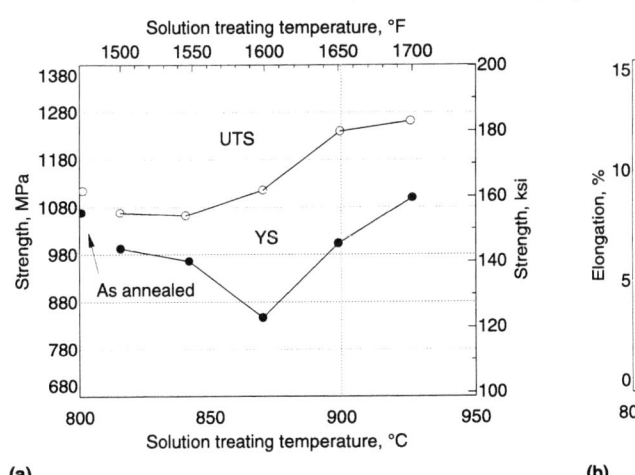
(a)

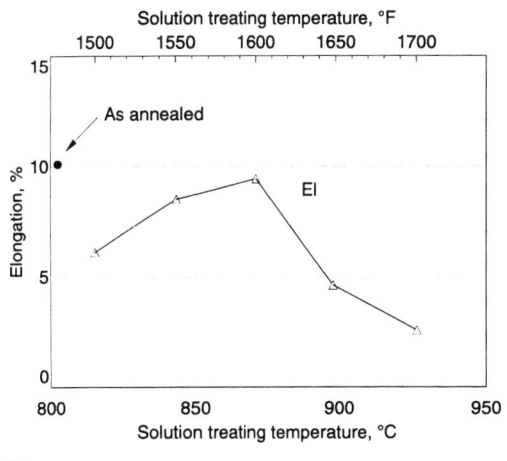
(b)

1.6 mm (0.063 in.) thick sheet heat treated as indicated.
Source: R.C. Bliss, "Evaluation of Ti-6Al-2Sn-2Zr-2Cr-2Mo-0.23Si Sheet," Proc. 7th Int. Titanium Conf., San Diego, TMS/AIME, June 1992, to be published

Ti-6-22-22S: Yield strength vs aging temperature

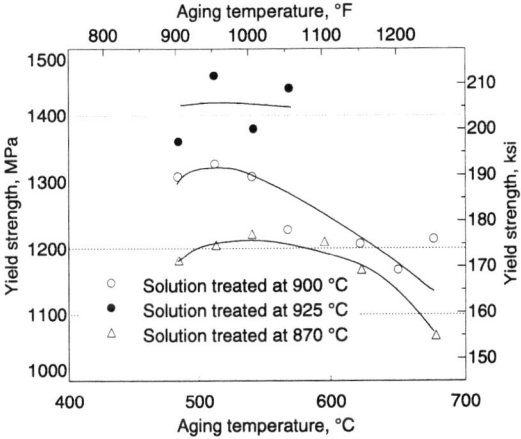

1.2 mm (0.050 in.) sheet heat treated as indicated.
Source: R.R. Boyer and A.E. Caddey, "The Properties of Ti-6Al-2Sn-2Zr-2Mo-2Cr Sheet," Proc. 7th Int. Titanium Conf., San Diego, TMS/AIME, June 1992, to be published

High-Temperature Strength

Ti-6-22-22S: Effect of temperature on tensile, compressive, and shear properties
Room-temperature ratio of tensile strength and compressive and shear strength should be similar for different heat treatments and product forms.

Properties	RT	205 °C (400 °F)	315 °C (600 °F)	425 °C (800 °F)
Tension				
Ultimate tensile strength				
Longitudinal, MPa (ksi)	1160 (168.3)	1002 (145.3)	958 (139.0)	910 (132.0)
Transverse, MPa (ksi)	1163 (168.7)	1006 (146.0)	963 (139.7)	910 (132.0)
Tensile yield strength				
Longitudinal, MPa (ksi)	1073 (155.6)	799 (116.0)	737 (107.0)	697 (101.2)
Transverse, MPa (ksi)	1079 (156.6)	825 (119.7)	749 (108.7)	717 (104.0)
Elongation in 25 mm (1 in.)				
Longitudinal, %	18.0	19.5	18.5	21.3
Transverse, %	17.7	19.7	18.2	21.0
Reduction of area				
Longitudinal, %	24.8	33.2	34.9	42.1
Transverse, %	26.2	33.7	33.3	41.4
Young's modulus				
Longitudinal, GPa (10^6 psi)	123 (17.9)	109 (15.9)	107 (15.6)	99 (14.4)
Transverse, GPa (10^6 psi)	122 (17.8)	111 (16.2)	110 (16.0)	100 (14.6)
Compression				
Compressive yield strength				
Longitudinal, MPa (ksi)	1170 (169.7)	884 (128.3)	772 (112.0)	728 (105.7)
Transverse, MPa (ksi)	1195 (173.3)	891 (129.3)	793 (115.0)	733 (106.3)
Compressive modulus				
Longitudinal, GPa (10^6 psi)	125 (18.1)	115 (16.7)	109 (15.8)	100 (14.6)
Transverse, GPa (10^6 psi)	127 (18.5)	112 (16.3)	109 (15.8)	100 (14.6)
Ultimate shear strength				
Longitudinal, MPa (ksi)	746 (108.3)	...	...	...
Transverse, MPa (ksi)	744 (108.0)	...	...	...

Note: Specimens were 38 mm (1.5 in.) thick plate in the STA condition heat treated at 950 °C (1740 °F), 1 h, AC + 540 °C (1000 °F), 8 h.
Source: "Mechanical-Property Data: Ti-6Al-2Zr-2Sn-2Mo-2Cr Alloy Solution Treated and Aged Plate," F33615-72-C-1280, Battelle Columbus Laboratories, Apr 1973

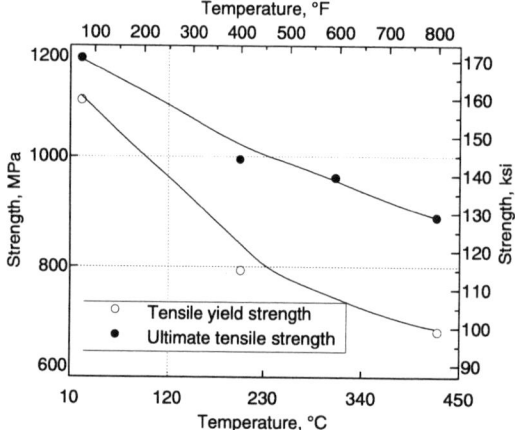

Ti-6-22-22S: High-temperature tensile strength of STA billet

Source: G.A. Bella, RMI Titanium Co., 8 May 1991

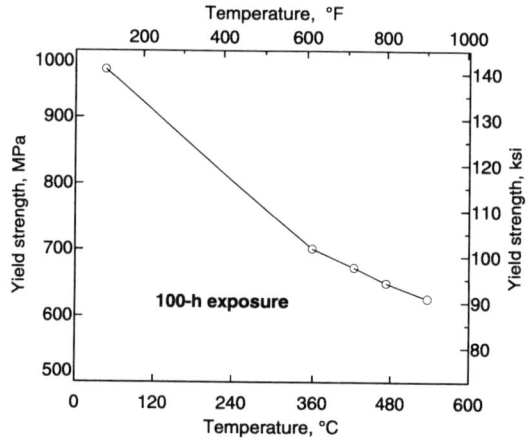

Ti-6-22-22S: High-temperature tensile strength

Forgings heat treated at β_t + 30 °C (50 °F), 30 min, FC + β_t – 40 °C (70 °F), 1 h, AC + 540 °C (1000 °F), 8 h.
Source: G.W. Kuhlman et al., "Characterization of Ti-6-22-22S: A High-Strength Alpha-Beta Titanium Alloy for Fracture Critical Applications," Proc. 7th Int. Titanium Conf., San Diego, TMS/AIME, June 1992, to be published

High-Temperature Strength

Creep Strength/ Creep Rupture

The limited creep data available on Ti-6-22-22S are presented below. In general, the creep properties are similar to those of Ti-6Al-2Sn-4Zr-6Mo and are superior to Ti-6Al-4V.

Ti-6-22-22S: Stress-rupture and creep properties for STA billet

Property	Temperature		
	205 °C (400 °F)	315 °C (600 °F)	425 °C (800 °F)
Stress to rupture			
Stress, MPa (ksi)	979 (142)	910 (132)	841 (122)
Time, h	100	100	100
Creep			
Stress, MPa (ksi)	841 (122)	827 (120)	572 (83)
Time, h	100	100	100
Creep, %	0.2	0.2	0.2

Source: O.L. Deel, P.E. Ruff, and H. Mindlin, "Engineering Data on New Aerospace Structural Materials," AFML-TR-75-97, Battelle Columbus Laboratories, June 1975

Ti-6-22-22S: Creep properties of β solution treated and aged forgings

Temperature		Stress		Time, h		
°C	°F	MPa	ksi	0.1%	0.2%	Rupture
370	700	415	60	225	687	687
425	800	345	50	190	606	606
480	900	240	35	69	251	251

Note: Forgings were processed at 30 °C (50 °F) above the β transus temperature, 30 min, fan cooled + 40 °C (70 °F) below the β transus, 1 h, AC + 540 °C (1000 °F), 8 h. Source: G.W. Kuhlman et al., "Characterization of Ti-6-22-22S: High-Strength Alpha-Beta Titanium Alloy for Fracture Critical Applications," Proc. 7th Int. Titanium Conf., San Diego, TMS/AIME, June 1992, to be published

Ti-6-22-22S: Larson-Miller creep curves

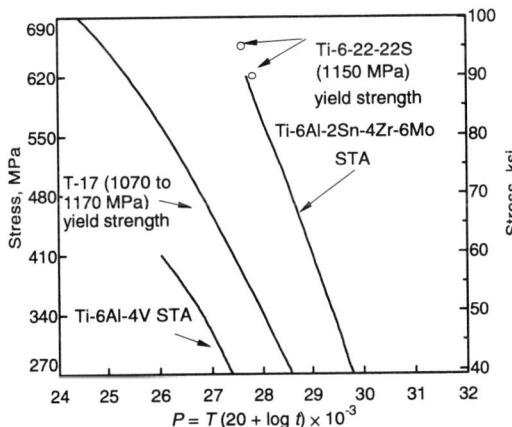

Larson-Miller creep curves at 0.2% deformation for specimens heat treated at 950 °C (1740 °F, 1 h, AC + 540 °C (1000 °F), 8 h.
Source: L.J. Bartlo, H.B. Bomberger, and S.R. Seagle, "Deep Hardenable Titanium Alloy," AFML-TR-73-122, Battelle Columbus Laboratories, May 1973

Ti-6-22-22S: Creep of STA plate

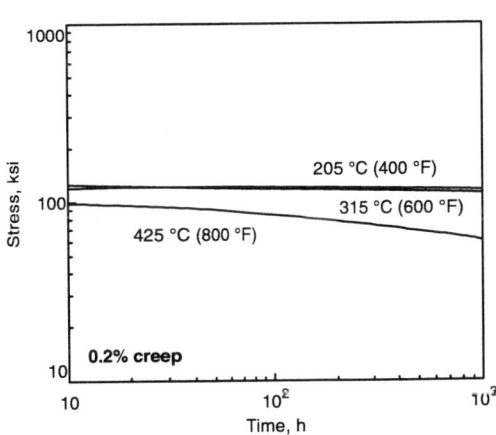

38 mm (1.5 in.)(thick plate; 950 °C (1740 °C), 1 h, AC + 540 °C (1000 °F), 8 h, AC.
Source: "Mechanical Property Data: Ti-6Al-2Zr-2Sn-2Mo-2Cr Alloy Solution Treated and Aged Plate," F33615-72-C-1280, Battelle Columbus Laboratories, Apr 1973

Ti-6-22-22S: Creep and stress rupture of forged billet

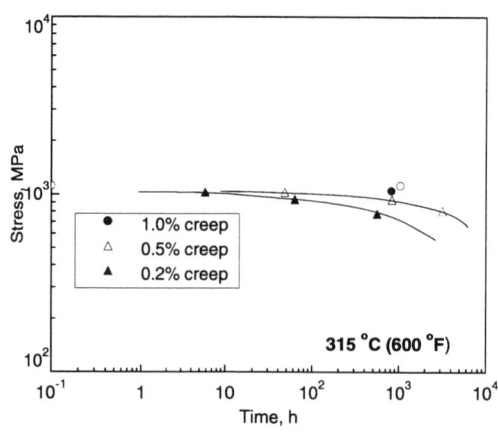
(a)

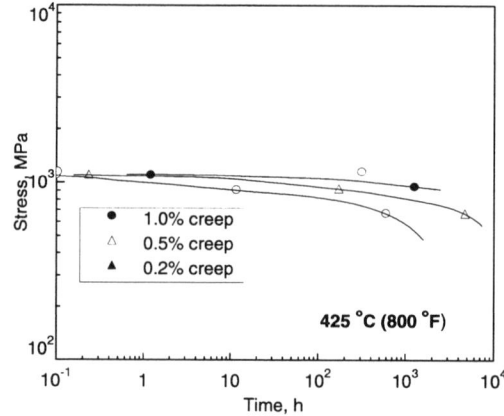
(b)

Duplex annealed; test direction, long transverse. Source: O.L. Deel, P.E. Ruff, and H. Mindlin, "Engineering Data on New Aerospace Structural Materials," AFML-TR-75-97, Battelle Columbus Laboratories, June 1975

High- and Low-Cycle Fatigue

The axial fatigue data on duplex annealed (DA) forged billet given below represent data generated in the early 1970s during initial alloy development/characterization and should be representative of α-β processed material. There is some doubt about the 38 mm (1.5 in.) plate data. The last two curves are representative of the β-processed material being studied today.

Ti-6-22-22S: Transverse axial fatigue of STA plate

		Temperature	
Fatigue	RT	205 °C (400 °F)	315 °C (600 °F)
Unnotched, $R = 0.1$			
10^3 cycles, MPa (ksi)	1158 (168)	1034 (150)	924 (134)
10^5 cycles, MPa (ksi)	930 (135)	848 (123)	799 (116)
10^7 cycles, MPa (ksi)	517 (75)	517 (75)	517 (75)
Notched, $K_t = 3.0$, $R = 0.1$			
10^3 cycles, MPa (ksi)	868 (126)	703 (102)	620 (90)
10^5 cycles, MPa (ksi)	413 (60)	379 (55)	344 (50)
10^7 cycles, MPa (ksi)	289 (42)	255 (37)	255 (37)

Note: 38 mm (1.5 in.) plate, 950 °C (1740 °F), 1 h, AC + 540 °C (1000 °F) for 8 h. Source: "Mechanical-Property Data of Ti-6Al-2Sn-2Zr-2Mo-2Cr," AFML, Battelle Columbus Laboratories, Apr 1973

DA Forged Billet

Ti-6-22-22S: Unnotched axial fatigue of DA forged billet

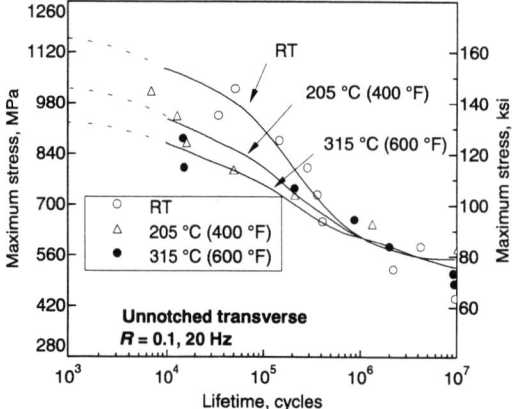

Ti-6-22-22S: Notched axial fatigue of DA forged billet

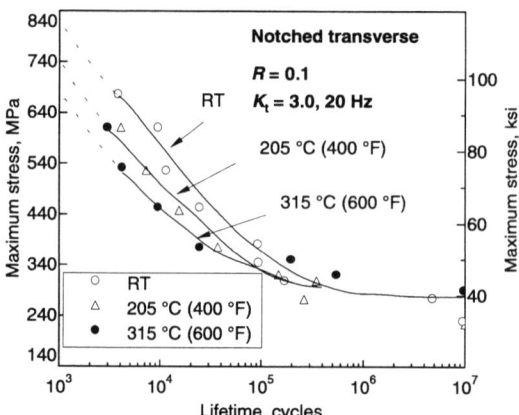

See also accompanying tables on next page.

Ti-6-22-22S: Unnotched axial fatigue of DA forged billet ($R = 0.1$)

RT		400 °F		600 °F	
ksi	cycles	ksi	cycles	ksi	cycles
145	52,730	145	6,400	135	(b)
135	37,730	135	12,900	125	15,400
125	159,300	125	15,800	115	14,700
115	303,270	115	47,900	105	218,300
105	392,790	105	212,400	95	836,600
95	429,580	95	1,277,700	95	1,912,100
85	4,527,700	85	10,130,900(a)	75	9,789,300
75	2,268,600			70	13,808,600(a)
65	10,003,500(a)				

(a) Did not fail. (b) Failed on loading. Source: O.L. Deel, P.E. Ruff, and H. Mindlin, "Engineering Data on New Aerospace Structural Materials," AFML-TR-75-97, Battelle Columbus Laboratories, June 1975

Ti-6-22-22S: Notched axial fatigue of DA forged billet ($R = 0.1$, $K_t = 3.0$)

RT		400 °F		600 °F	
ksi	cycles	ksi	cycles	ksi	cycles
95	3,600	85	3,700	85	2,900
85	8,600	75	6,850	75	4,000
75	11,400	65	14,700	65	8,600
65	23,400	55	33,300	55	22,500
55	89,100	47.5	141,200	50	194,600
50	89,900	45	417,400	47.5	527,800
45	153,200	40	237,000	45	10,084,900(a)
40	5,069,900	35	17,270,800(a)		
35	11,645,200(a)				

(a) Did not fail. Source: O.L. Deel, P.E. Ruff, and H. Mindlin, "Engineering Data on New Aerospace Structural Materials," AFML-TR-75-97, Battelle Columbus Laboratories, June 1975

STA Plate

Ti-6-22-22S: Fatigue behavior of unnotched STA plate

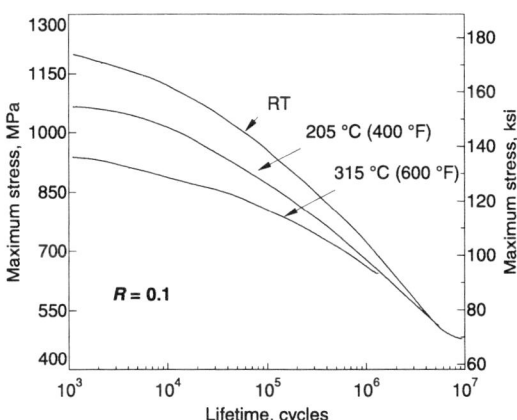

38 mm (1.5 in.) thick plate heat treated at 950 °C (1740 °F), 1 h, AC + 540 °C (1000 °F), 8 h, AC; test direction, transverse; $R = 0.1$.
Source: "Mechanical Property Data: Ti-6Al-2Zr-2Sn-2Mo-2Cr Alloy Solution Treated and Aged Plate," F33615-72-C-1280, Battelle Columbus Laboratories, Apr 1973

Ti-6-22-22S: Fatigue of notched STA plate

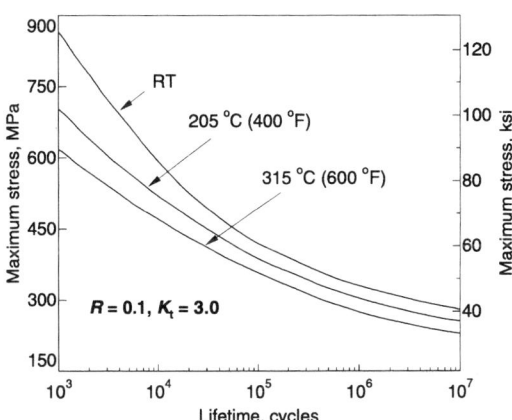

38 mm (1.5 in.) thick plate heat treated at 950 °C (1740 °F), 1 h, AC + 540 °C (1000 °F), 8 h, AC; test direction, transverse; $R = 0.1$; $K_t = 3.0$.
Source: "Mechanical Property Data: Ti-6Al-2Zr-2Sn-2Mo-2Cr Alloy Solution Treated and Aged Plate," F33615-72-C1280, Battelle Columbus Laboratories, Apr 1973

Beta-Processed Material

Ti-6-22-22S: Smooth high-cycle fatigue

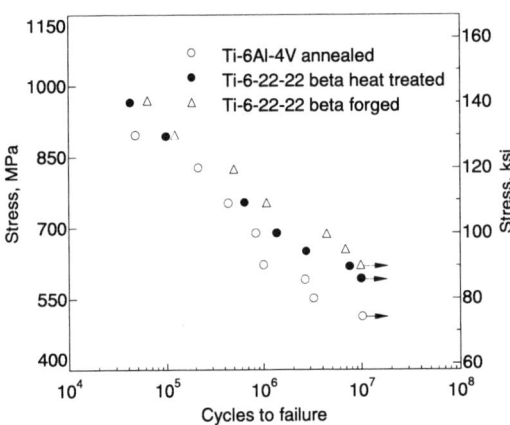

Close-die forgings; processed as noted. $R = 0.1$; $K_t = 1.0$, 30 Hz.
Source: G.W. Kuhlman et al., "Characterization of Ti-6-22-22S: A High-Strength Alpha-Beta Titanium Alloy for Fracture Critical Applications," Proc. 7th Int. Titanium Conf., San Diego, TMS/AIME, June 1992, to be published

Ti-6-22-22S: Notched high-cycle fatigue

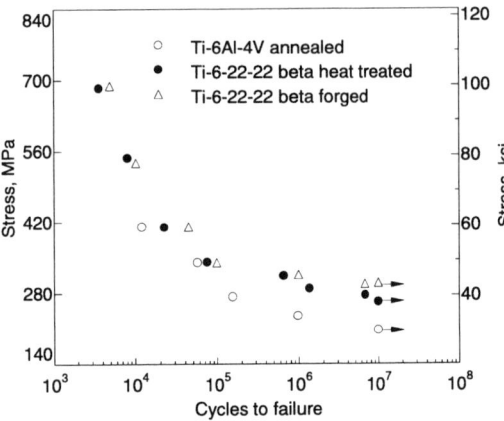

Close-die forgings; processed as noted. $R = 0.1$, $K_t = 3.0$, 30 Hz.
Source: "Characterization of Ti-6-22-22S: A High-Strength Alpha-Beta Titanium Alloy for Fracture Critical Applications," Proc. 7th Int. Titanium Conf., San Diego, TMS/AIME, June 1992, to be published

Fatigue-Crack Propagation

The first figures represent early data on α-β processed STA plate, whereas the remaining figures are for β-processed material. The crack propagation rates of the latter are similar to that of β-annealed ELI Ti-6Al-4V. It is readily apparent that the rapid cooling rates, which refine the transformed β structure, detract from fatigue-crack growth resistance. The effect of thermomechanical processing is also illustrated. Basically, the data indicate that the lamellar α-β structure, i.e., β-processed material, provides the slowest crack growth rates.

Billet

Ti-6-22-22S: Fatigue crack growth rate in forged billet

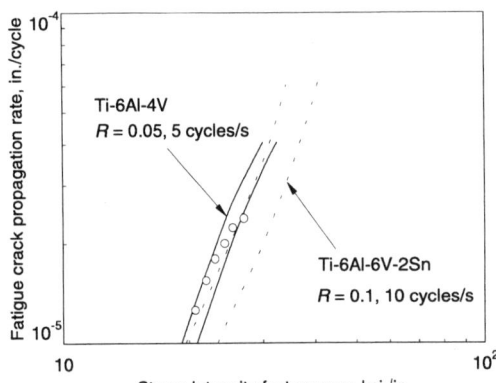

150 mm (6 in.) diam billet; 950 °C (1740 °F), 1 h, water quenched + 540 °C (1000 °F), 8 h, AC; test direction, L-S; yield strength, 1083 MPa (157 ksi); $R = 0.044$; 25 Hz.
Source: L.J. Bartlo, H.B. Bomberger, and S.R. Seagle, "Deep Hardenable Titanium Alloy," AFML-TR-73-122, Battelle Columbus Laboratories, May 1973

STA Plate

Ti-6-22-22S: Fatigue cracking in 3.5% NaCl of STA plate

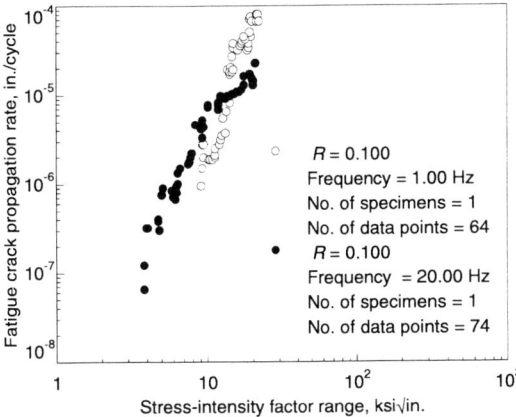

16 mm (5/8 in.) thick plate; test direction, longitudinal transverse; environment, 20 °C (70 °F), 3.5% NaCl; yield strength, 1083 MPa (157 ksi); specimen, 3.8 mm (0.15 in.) thick.
Source: *Damage Tolerant Design Handbook,* Part 2, Metals and Ceramic Information Center, Battelle Columbus Laboratories, Jan 1975

Ti-6-22-22S: Fatigue cracking in air of STA plate

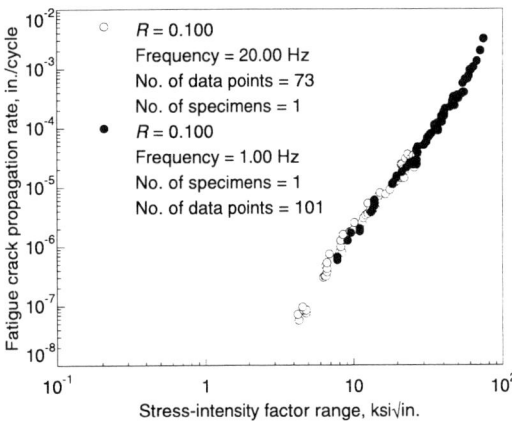

16 mm (5/8 in.) thick plate; test direction, longitudinal transverse; environment, 20 °C (70 °F), 95% relative humidity; yield strength, 1083 MPa (157 ksi); specimen, 3.8 mm (0.15 in.) thick.
Source: *Damage Tolerant Design Handbook,* Part 2, Metals and Ceramic Information Center, Battelle Columbus Laboratories, Jan 1975

Ti-6-22-22S: Fatigue cracking in 3.5% NaCl of STA plate

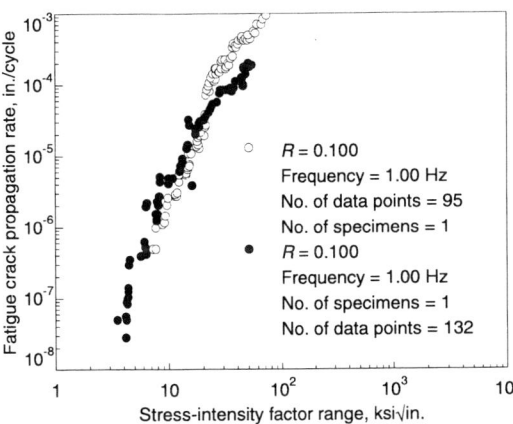

16 mm (5/8 in.) thick plate; test direction, longitudinal transverse; environment, 20 °C (70 °F), 3.5% NaCl; yield strength, 1083 MPa (157 ksi); specimen, 3.8 mm (0.15 in.) thick.
Source: *Damage Tolerant Design Handbook,* Part 2, Metals and Ceramic Information Center, Battelle Columbus Laboratories, Jan 1975

Ti-6-22-22S: Fatigue cracking in 3.5% NaCl of STA plate

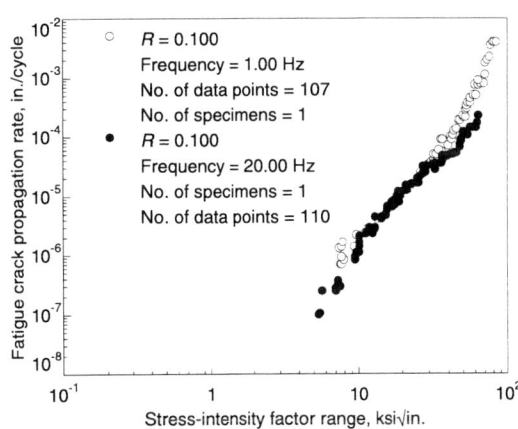

16 mm (5/8 in.) thick plate; test direction, longitudinal transverse; environment, 20 °C (70 °F), 3.5% NaCl; yield strength, 1083 MPa (157 ksi); specimen, 3.8 mm (0.15 in.) thick.
Source: *Damage Tolerant Design Handbook,* Part 2, Metals and Ceramic Information Center, Battelle Columbus Laboratories, Jan 1975

Beta-Processed Condition

Ti-6-22-22S: Fatigue crack growth rate of forgings

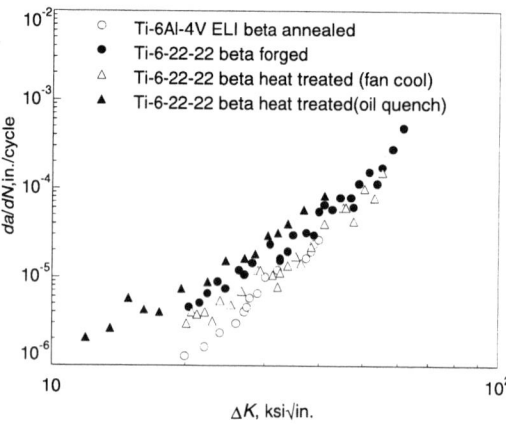

Forged pancakes processed as indicated. $R = 0.1$, 20 Hz, lab air; Ti-6Al-4V ELI, $R = 0.01$.
Source: G.W. Kuhlman et al., "Characterization of Ti-6-22-22S: A High-Strength Alpha-Beta Titanium Alloy for Fracture Critical Applications," Proc. 7th Int. Titanium Conf., San Diego, TMS/AIME, June 1992, to be published

Ti-6-22-22S: Fatigue crack growth rate comparison

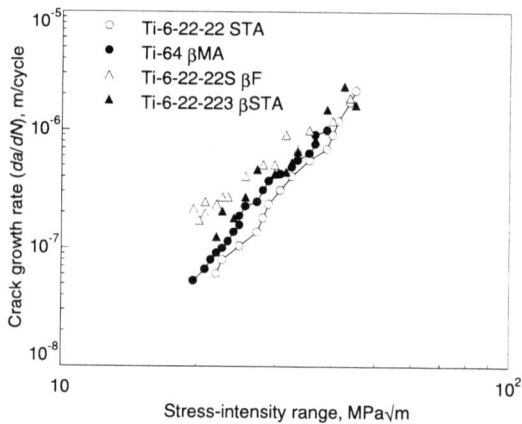

$R = 0.01$, 10 Hz, lab air.
Source: A.K. Chakrabarti, R. Pishko, V.M. Sample, and G.W. Kuhlman, "TMP Conditions-Microstructure-Mechanical Property Relation in Ti-6-22-22S Alloy," Proc. 7th Int. Titanium Conf., San Diego, TMS/AIME, June 1992, to be published

Ti-6-22-22S: Fatigue crack growth rate vs applied stress intensity of forgings

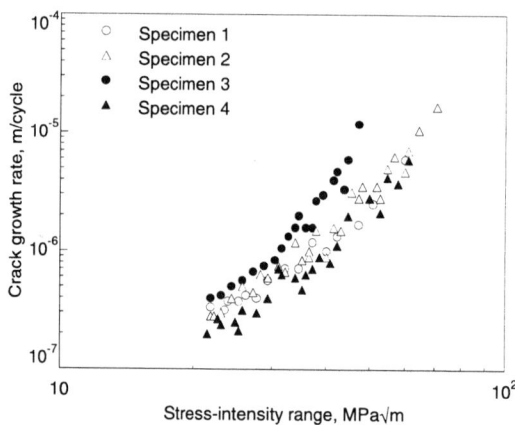

Forged pancakes, triplex beta heat treated with varying cooling rates.
Source: A.K. Chakrabarti, R. Pishko, V.M. Sample, and G.W. Kuhlman, "TMP Conditions-Microstructure-Mechanical Property Relation in Ti-6-22-22S Alloy," Proc. 7th Int. Titanium Conf., San Diego, TMS/AIME, June 1992, to be published

Ti-6-22-22S: Fatigue crack growth rate of plate

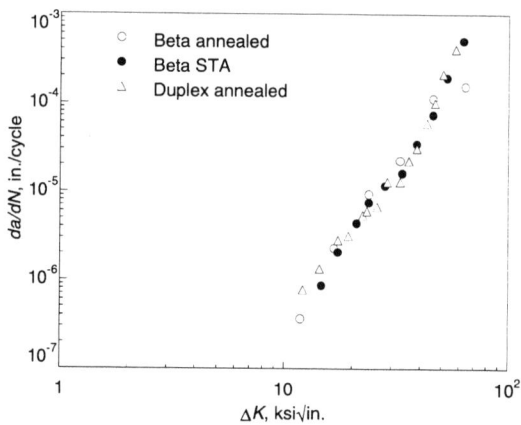

Specimens were α-β rolled 14 mm (0.58 in.) plate in three conditions: (1) β annealed at 35 °C (65 °F) above the β transus, 1 h, AC; (2) β solution treated and aged at 35 °C (65 °F) above the β transus, 1 h, AC + 540 °C (1000 °F), 8 h; or (3) duplex heat treated by β annealing + 30 °C (50 °F) below the β transus, 1 h, AC + 540 °C (1000 °F), 8 h. Beta annealed: tensile strength, 930 MPa (135 ksi); K_q, 100 MPa√m (91 ksi√in.). Beta solution treated and aged: tensile strength, 1027 MPa (149 ksi); K_q, 81 MPa√m (74 ksi√in.). Duplex annealed: tensile strength, 1034 MPa (150 ksi); K_q, 71 MPa√m (65 ksi√in.). $R = 0.1$, lab air.
Source: H.R. Phelps and J.R. Wood, "Correlation of Mechanical Properties and Microstructures of Ti-6Al-2Sn-2Zr-2Mo-2Cr-0.25Si Titanium Alloy," Proc. 7th Int. Titanium Conf., San Diego, TMS/AIME, June 1992, to be published

3.5% NaCl

Ti-6-22-22S: Fatigue crack growth rate comparison in 3.5% NaCl

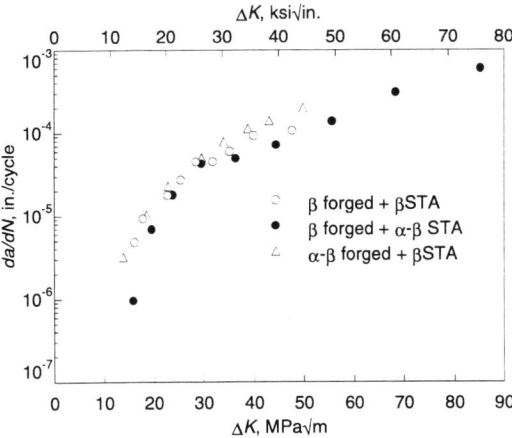

Specimens were α-β rolled 14 mm (0.58 in.) plate in three conditions: (1) β annealed at 35 °C (65 °F) above the β transus, 1 h, AC; (2) β solution treated and aged at 35 °C (65 °F) above the β transus, 1 h, AC + 540 °C (1000 °F), 8 h; or (3) duplex heat treated by β annealing + 30 °C (50 °F) below the β transus, 1 h, AC + 540 °C (1000 °F), 8 h. Specimens were tested in the L-T direction, $R = 0.1$, 2 Hz. Source: H.R. Phelps and J.R. Wood, "Correlation of Mechanical Properties and Microstructures of Ti-6Al-2Sn-2Zr-2Mo-2Cr-0.25Si Titanium Alloy," Proc. 7th Int. Titanium Conf., San Diego, TMS/AIME, June 1992, to be published

Fracture Properties

The fracture properties of Ti-6-22-22S, as for other α + β titanium alloys, are quite dependent on strength and microstructure. In general, the β-processed/α-β annealed and α-β-processed/β annealed conditions produce better toughness than conditions that result in fine α + β microstructures. The effects of thermomechanical processing and oxygen content on fracture toughness are shown below. Similar to the results for fatigue crack propagation rates, the data show that the transformed β structure provides the maximum fracture toughness. It also illustrates an unexplained drop in fracture toughness as the aging temperature is increased. Similar behavior has been observed for plain-stress or mixed mode fracture toughness in sheet, as shown below. It is speculated that this drop in toughness is related to an ordering reaction in the alpha and/or silicide precipitation.

Ti-6-22-22S: Fracture toughness of sheet

Aging temperature		Ultimate tensile strength		Tensile yield strength		Elongation, %	Toughness (K_{app})	
°C	°F	MPa	ksi	MPa	ksi		MPa√m	ksi√in.
480	900	1275	185	1160	168	11.7	165	150
565	1050	1240	180	1170	170	11.2	109	99
675	1250	1140	165	1078	155	10.4	112	102

Note: Tensile properties are the average of six values each, and toughness values are the averages of two tests each. 1.2 mm (0.05 in.) sheet was solution treated at 900 °C for 30 min, aged 8 h. Source: R.R. Boyer and A.E. Caddey, "The Properties of Ti-6Al-2Sn-2Zr-2Mo-2Cr Sheet," Proc. 7th Int. Titanium Conf., San Diego, TMS/AIME, June 1992, to be published

Ti-6-22-22S: Fracture toughness and impact toughness

Direction	Ultimate tensile strength		Elongation in 25 mm (1 in.), %	Charpy V-notch impact toughness		Fracture toughness (K_{Ic})	
	MPa	ksi		J	ft·lbf	MPa√m	ksi√in.
Longitudinal	1160	168.3	18.0	18.8	13.9	...	...
Transverse	1163	168.7	17.7	22.1	16.3	...	...
L-T	...	...	...	...	...	96	88
T-L	...	...	...	...	...	102	93

38 mm (1.5 in.) thick plate, STA condition. Source: AFML, Apr 1973

Ti-6-22-22S: Typical fracture toughness of β-processed STA products

Product	Ultimate tensile strength		Yield strength		Elongation, %	Reduction of area, %	Fracture toughness (K_{Ic})	
	MPa	ksi	MPa	ksi			MPa√m	ksi√in.
50 mm (2 in.) plate	1138	165	1020	148	10	17	85	77
100 mm (4 in.) plate	1103	160	979	147	10	15	89	81
150 mm (6 in.) plate	1076	156	958	139	10	15	98	89

Source: J.R. Wood, RMI Titanium Co., 15 Aug 1991

Ti-6-22-22S: Fracture toughness of α + β processed STA products

Product	Ultimate tensile strength		Tensile yield strength		Elongation, %	Reduction of area, %	Fracture toughness (K_{Ic})	
	MPa	ksi	MPa	ksi			MPa√m	ksi√in.
50 mm (2 in.) plate	1207	175	1131	164	12	35	67	61
150 mm (6 in.) billet	1200	174	1089	158	15	41	65	72

Source: J.R. Wood, RMI Titanium Co., 15 Aug 1991

Ti-6-22-22S: Effect of oxygen content on K_{Ic}

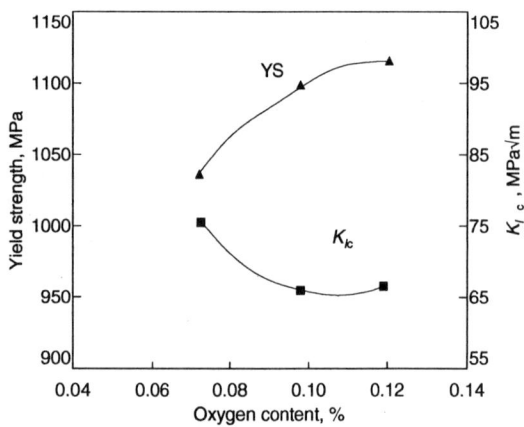

Specimens were forged pancakes α-β forged + α-β solution treated + 540 °C (1000 °F), 8 h, aged, AC.
Source: A.K. Chakrabarti et al., "TMP Conditions-Microstructure-Mechanical Property Relationship in Ti-6-22-22S Alloy," Proc. 7th Int. Titanium Conf., San Diego, TMS/AIME, June 1992, to be published

Plastic Deformation

Strain Hardening

The m-values, indicators of the superplasticity of material, from 790 to 925 °C (1450 to 1700 °F) at two strain rates are illustrated.

Flow Stress

The flow stress over the temperature range of 790 to 925 °C (1450 to 1700 °F) over the range of strain rates from 8×10^{-5} to 2×10^{-4} is illustrated.

Ti-6-22-22S: Typical m-values

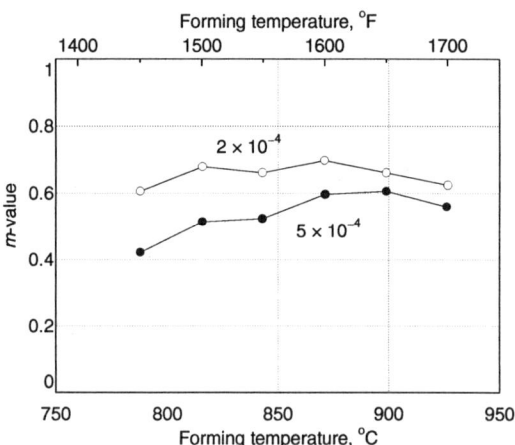

Data are shown for two different strain rates. 2.5 mm (0.10 in.) sheet, as annealed, 75% total strain using a step-strain-rate tensile test.
Source: R.C. Bliss, "Evaluation of Ti-6-22-22S Sheet," Proc. 7th Int.

Ti-6-22-22S: Flow stress

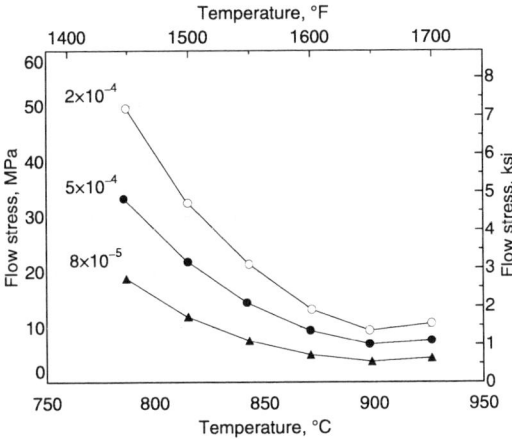

2.5 mm (0.10 in.) sheet, as annealed, step-strain-rate tensile tests under argon.
Source: RMI Titanium Co., unpublished data

Stress-Strain Curves

Ti-6-22-22S: Tensile stress-strain curves for STA plate

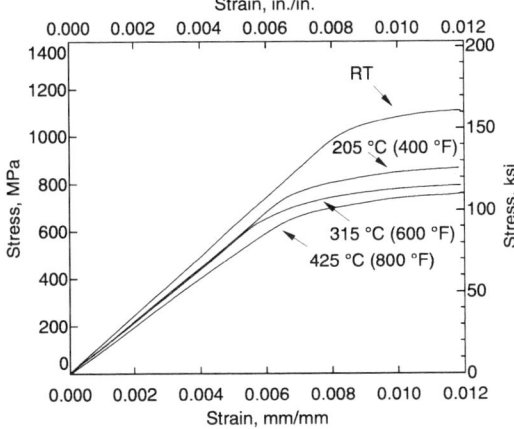

Test direction, longitudinal.
Source: O.L. Deel, P.E. Ruff, and H. Mindlin, "Engineering Data on New Aerospace Structural Materials," AFML-TR-73-114, Battelle Columbus Laboratories, June 1973

Ti-6-22-22S: Compressive stress-strain for STA plate

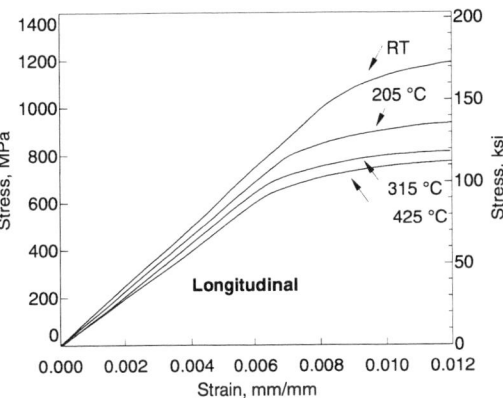

Test direction, longitudinal.
Source: O.L. Deel, P.E. Ruff, and H. Mindlin, "Engineering Data on New Aerospace Structural Materials," AFML-TR-73-114, Battelle Columbus Laboratories, June 1973

Ti-6-22-22S: Compressive stress-strain for STA plate

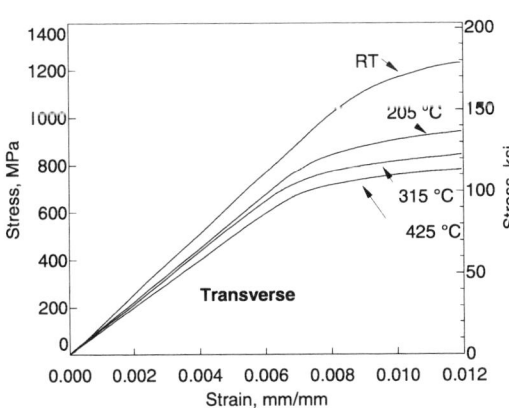

Test direction, transverse.
Source: O.L. Deel, P.E. Ruff, and H. Mindlin, "Engineering Data on New Aerospace Structural Materials," AFML-TR-73-114, Battelle Columbus Laboratories, June 1973

Ti-6-22-22S: Tensile stress-strain curves for STA plate

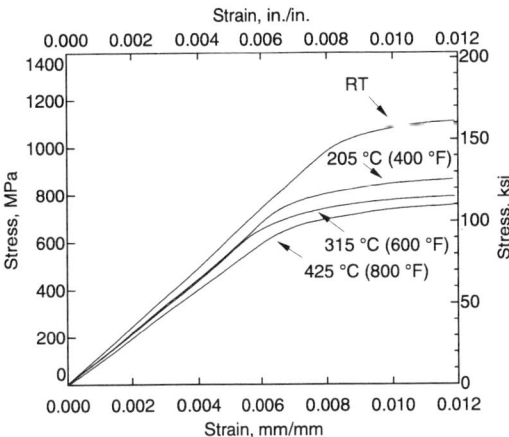

Test direction, transverse.
Source: O.L. Deel, P.E. Ruff, and H. Mindlin, "Engineering Data on New Aerospace Structural Materials," AFML-TR-73-114, Battelle Columbus Laboratories, June 1973

Forging

Ti-6-22-22S is a high-strength, highly β-stabilized, α-β alloy whose primary commercial application is aerospace and airframe structural and turbine engine components. The alloy can be fabricated into all forging product types, although closed die and precision forgings predominate. Ti-6-22-22S is commercially fabricated on all types of forging equipment at the process temperatures indicated (see table).

Ti-6-22-22S is a reasonably forgeable alloy with comparable unit pressures (flow stresses), forgeability, and crack sensitivity to the α-β alloy Ti-6Al-4V. The final microstructure of Ti-6-22-22S forgings is developed by thermomechanical processing in forging manufacture tailored to achieve specific microstructural and mechanical property objectives (see table). Thermomechanical processes for the alloy use combinations of conventional (subtransus) and/or β (supra-transus) forging followed by subtransus and/or supra-transus thermal treatments to fulfill critical mechanical property criteria.

Ti-6-22-22S: Recommended forging process temperatures

Process parameter	Temperature °C	Temperature °F
Metal temperature		
Conventional forging	865-925	1590-1700
Beta forging(a)	980-1015	1800-1860
Die temperatures	See "Technical Note 4: Forging"	

(a) Beta transus, 960 °C (1760 °F)

Ti-6-22-22S: Forging TMP conditions and mechanical properties

Preform forging	Finish forging	Heat treatments	Tensile yield strength MPa (ksi)	Ultimate tensile strength MPa (ksi)	Elongation, %	Reduction of area, %	Fracture toughness (K_{Ic}) MPa√m (ksi√in.)
α-β	β finish 28 °C (50 °F) above β transus	$β_t$ − 22 °C (40 °F), 1 h, FAC + 540 °C (1000 °F), 8 h, AC	993 (144)	1110 (161)	11	21	90.4 (82)
α-β	β finish 50 °C (90 °F) above β transus	$β_t$ − 22 °C (40 °F), 1 h, FAC + 540 °C (1000 °F), 8 h, AC	1000 (145)	1117 (162)	12	21	87.6 (79)
β preform	α-β finish (25%)	$β_t$ − 22 °C (40 °F), 1 h, FAC + 540 °C (1000 °F), 8 h, AC	1034 (150)	1124 (163)	12	25	68.0 (62)
β preform	α-β finish (50%)	$β_t$ − 22 °C (40 °F), 1 h, FAC + 540 °C (1000 °F), 8 h, AC	1027 (149)	1117 (162)	12	25	58.4 (53)
α-β preform	α-β finish (50%)	$β_t$ + 28 °C (50 °F), 1/2 h, FAC + 540 °C (1000 °F), 8 h, AC	958 (140)	1110 (161)	10	15	75.1 (68)
α-β preform	α-β finish (50%)	$β_t$ + 28 °C (50 °F), 1/2 h, FAC + $β_t$ − 50 °C (90 °F), 1 h, AC + 540 °C (1000 °F), 8 h, AC	972 (141)	1096 (159)	10	17	85.1 (77)

Note: Heat treatment sequence temperature, time, cooling method. Source: A.K. Chakrabarti, R. Pishko, V.M. Sample, and G.W. Kuhlman, "TMP Conditions-Microstructure-Mechanical Property Relation in Ti-6-22-22S Alloy," Proc. 7th Int. Titanium Conf., San Diego, TMS/AIME, June 1992, to be published

Thermomechanical Processes

Conventional Forging. The objectives in forging Ti-6-22-22S are to obtain the final forging shape and desired final microstructure at the least cost. Conventional, subtransus (α + β) forging thermomechanical processes followed by triplex β heat treatment are the most widely used in commercial forging manufacture. To achieve conventional equiaxed α structures in preparation for final β heat treatment, subtransus reductions of at least 50 to 75%, accumulated through one or more forging steps, are required. Supra-transus β forging for Ti-6-22-22S may be used in early operations such as upsetting and preforming, to ease fabrication. However, higher temperature initial forging operations must be followed by sufficient subtransus reduction to achieve the desired predominately equiaxed α structure prior to heat treatment. Conventionally forged Ti-6-22-22S is then subtransus solution treated, cooled, and aged, or preferably, for optimum fracture-related properties, triplex β heat treated as noted above.

Supra-transus β forging thermomechanical processes for Ti-6-22-22S are used for applications to achieve transformed, Widmanstätten α structure desired for improved toughness and fracture-related properties. Successful β forging thermomechanical processes for Ti-6-22-22S forgings include controlled β forgings followed by subtransus solution treatment and aging as noted above. Beta forging the alloy may entail supra-transus forging in all stages, but the preferred process is subtransus reduction (e.g., 20 to 50%) in early forging (preform or blocker die) stages followed by a controlled, single β forging step, that achieves 30 to 50% reduction. Beta forging Ti-6-22-22S requires careful control of forging process conditions, particularly preheat times at temperature, to avoid excessive prior β grain growth. Because of inherent variations in forging conditions, β forged Ti-6-22-22S may exhibit more final forging product variation than conventionally subtransus forged and heat treated or conventionally subtransus forged and triplex β heat treated Ti-6-22-22S forged product.

Hot die and/or isothermal forging techniques may be an important commercial method for fabrication of Ti-6-22-22S rotating turbine engine discs to reduce final component cost (from less machining) and/or improve final component microstructural and property uniformity through improved control of forging process conditions. The axisymmetric shapes and designs of such engine components are well suited to these forging methods. Isothermal forging of Ti-6-22-22S discs frequently is accomplished in a single forging step from bar or billet stock, under carefully controlled supra- or subtransus metal and die temperatures, levels of strain, and strain rate profiles. Hot die forging, where die temperature approaches but is not equivalent to metal temperature, also is used to reduce unit pressures, enhance forgeability, and produce more sophisticated final shapes in fewer forging operations.

Final thermal treatments for Ti-6-22-22S forgings include two- or three-step practices of single or two-step (duplex) solution treatments followed by controlled cooling and aging (stabilizing). Subtransus thermal treatments, used in combination with conventional and/or β forging processes, are done at 30 °C (50 °F) below the β transus temperature, followed by air or fan cooling. Aging (stabilization) is at 480 to 540 °C (900 to 1000 °F). For conventional subtransus forgings, a triplex β heat treatment has been developed (G.W. Kuhlman *et al.*, "Characterization of Ti-6-22-22S: A High-Strength Alpha-Beta Titanium Alloy for Fracture Critical Applications," Proc. 7th Int. Titanium Conf., San Diego, TMS/AIME, June 1991, to be published) consisting of processing 30 to 40 °C (50 to 75 °F) above the β transus temperature, followed by fan cooling plus 30 to 50 °C (50 to 90 °F) below the β transus, followed by air or fan cooling and aging (stabilization) at 480 to 540 °C (900 to 1000 °F). Subtransus thermomechanical processes (forging and thermal treatment) for Ti-6-22-22S forgings achieve equiaxed α (10 to 30%) in transformed β matrix microstructures that enhances strength, ductility, and low-cycle fatigue properties. Supra-transus thermomechanical processes (β forging followed by subtransus thermal treatments or subtransus forging followed by supratransus thermal treatments) achieve either transformed Widmanstätten α microstructure (β forged, subtransus heat treatment) or colony α (subtransus forged, triplex β heat treated) that enhances creep and particularly fracture-related properties such as fatigue crack growth resistance. The latter thermomechanical process is preferred for current large structural airframe forging applications due to superior fatigue crack growth resistance and enhanced final forging product uniformity over β forging.

Other Fabrication Methods

Rolling. The rolling schedule for Ti-6-22-22S is very much like that for Ti-6Al-4V. Sheet is rolled below the β transus and has a very small grain size.

Forming is similar to Ti-6Al-4V with slightly higher pressure required for hot forming due to higher strength of Ti-6-22-22S. Superplastic forming is similar to Ti-6Al-4V with same pressure and strain rates.

Machining characteristics are very similar to those of Ti-6Al-4V, and preliminary investigations indicate that Ti-6-22-22S may machine slightly easier than Ti-6Al-4V.

Superplastic Forming

Ti-6-22-22S: Effect of post superplastic forming aging on 1.2 mm (0.05 in.) sheet

Aging temperature		Ultimate tensile strength		Tensile yield strength		Elongation, %
°C	°F	MPa	ksi	MPa	ksi	
Formed at 800 °C (1470 °F)						
510	950	1247	180	1110	160	11.0
540	1000	1197	173	1069	155	9.2
565	1045	1216	176	1111	161	10.0
No age		1024	148	949	137	10.0
Formed at 885 °C (1620 °F)						
510	950	1349	195	1271	184	7.5
565	1045	1431	207	1282	186	6.0
No age		1236	179	1051	152	8.0

Note: Aging time was 8 h. Source: R.R. Boyer and A.E. Caddey, "The Properties of Ti-6Al-2Sn-2Zr-2Mo-2Cr Sheet," Proc. 7th Int. Titanium Conf., San Diego, TMS/AIME, June 1992, to be published

Ti-6-22-22S: Effect of superplastic forming temperature on tensile properties of sheet

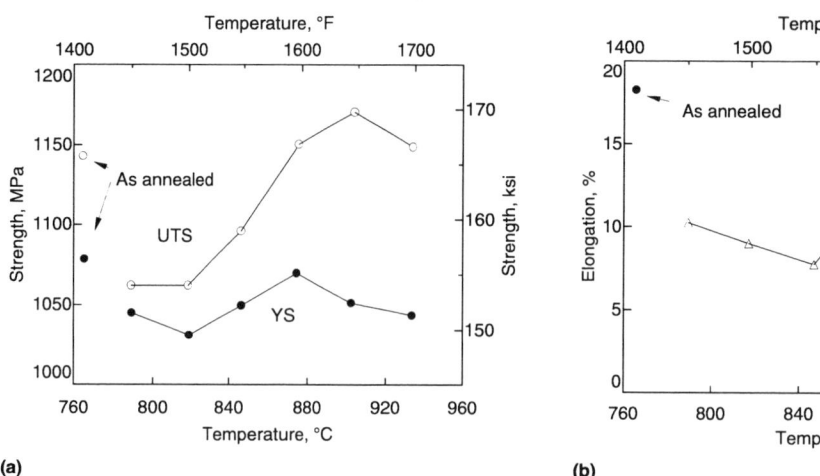

(a)   (b)

As-formed properties at 75% strain.
Source: R.C. Bliss, "Evaluation of Ti-6Al-2Sn-2Zr-2Cr-2Mo-0.23Si Sheet," Proc. 7th Int. Titanium Conf., San Diego, TMS/AIME, June 1992, to be published

Heat Treatment

See "Forging" section for additional discussion of heat treatment.

Effect of Aging

The accompanying table on aging illustrates a virtual independence of strength on aging temperature, with fracture toughness decreasing steadily as aging temperature increases. This toughness trend is opposite of what would be expected, and this anomaly cannot be explained at this time.

Ti-6-22-22S: Effect of aging temperature on tensile properties

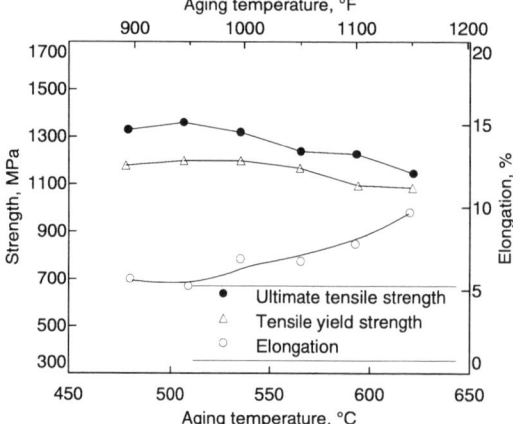

1.6 mm (0.062 in.) thick STA sheet, solution treated at 870 °C (1600 °F).
Source: G.A. Bella, RMI Titanium Co., 8 May 1991

Ti-6-22-22S: Effect of aging temperature on tensile properties

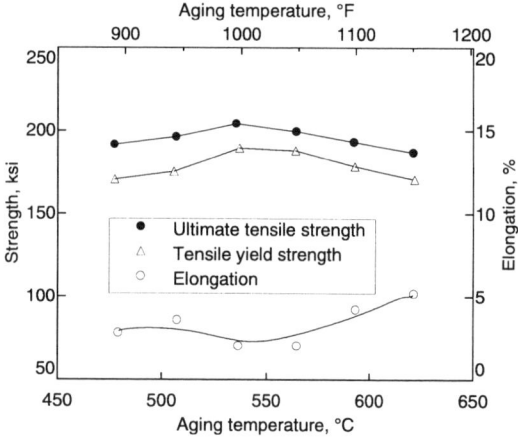

1.6 mm (0.062 in.) thick STA sheet, solution treated at 925 °C (1700 °F).
Source: G.A. Bella, RMI Titanium Co., 8 May 1991

Ti-6-22-22S: Mechanical properties after aging

Aging temperature °C	°F	Direction	Tensile yield strength, MPa (ksi)	Ultimate tensile strength, MPa (ksi)	Elongation, %	Reduction of area, %	Fracture toughness (K_{Ic}), MPa√m (ksi√in.)
425	800	L	1048 (152.0)	1206 (175.0)	7.0	7.8	88 (80.3)(a)
		ST	980 (142.2)	1121 (162.7)	6.2	10.6	
480	900	L	1096 (159.0)	1223 (177.4)	10.0	12.3	87 (79.8)(a)
		ST	1028 (149.1)	1197 (173.6)	10.1	16.6	
510	950	L	1096 (159.0)	1224 (177.6)	8.5	9.7	85 (77.4)(a)
		ST	1011 (146.7)	1138 (165.1)	10.1	16.5	
540	1000	L	1106 (160.4)	1218 (176.7)	9.0	9.7	83 (76.2)(a)
		ST	1024 (148.6)	1139 (165.3)	8.6	13.8	
565	1050	L	1061 (154.0)	1162 (168.6)	10.0	12.7	74 (67.8)(a)
		ST	1042 (151.1)	1155 (167.5)	13.3	36.3	
595	1100	L	1117 (162.0)	1205 (174.8)	8.0	9.3	62 (56.3)(a)
		ST	1049 (152.1)	1157 (167.8)	10.9	21.5	
620	1150	L	1113 (161.5)	1183 (171.6)	6.5	7.4	49 (45.3)(a)
		ST	1030 (149.5)	1102 (159.9)	6.2	11.1	

Note: 50 mm (2 in.) thick α-β rolled plate heat treated as follows: $β_t$ plus 28 °C (50 °F), fan cool, reheat to 40 °C (70 °F) below $β_t$, air cool, + age as indicated for 8 h, air cool. (a) T-L direction. Source: G.W. Kuhlman, Beta Processed Ti-6-22-22S Aging Studies, Alcoa Report, Mar 1992

Ti-6-22-22S: Effect of solution treating temperature on tensile properties

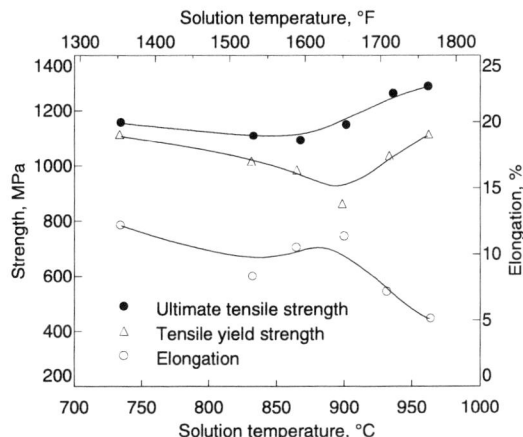

1.6 mm (0.062 in.) thick sheet.
Source: G.A. Bella, RMI Titanium Co., 8 May 1991

Ti-6-22-22S: Typical heat treatments

Treatment	Temperature °C	°F	Time/Cooling
Mill anneal	730	1350	2 h/AC
Sheet solution treatment	$β_T - 28$	$β_T - 50$	30 min/AC
Plate/billet solution treatment			
First stage (supratransus)	$β_T + 28$	$β_T + 50$	1 h/AC
Second stage (subtransus)	$β_T - 28$	$β_T - 50$	1 h/AC
Beta processed plate solution treat	$β_T - 28$	$β_T - 50$	1 h/AC
Age (all product forms)	540	1000	8 h/AC

732 / Alpha-Beta Alloys

Effect of Solution Treatment and Cooling

Ti-6-22-22S: Cooling rate/microstructure correlation

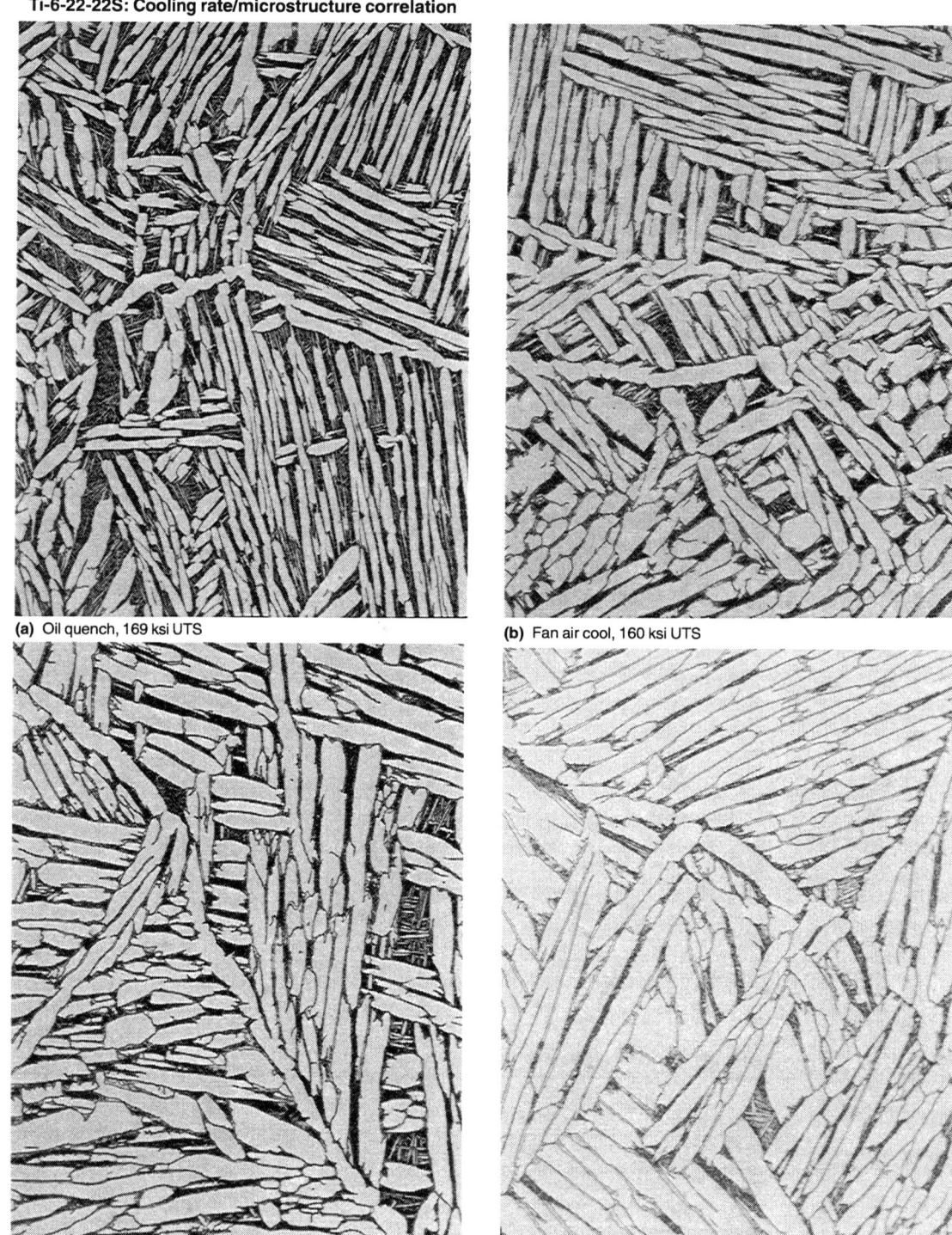

(a) Oil quench, 169 ksi UTS
(b) Fan air cool, 160 ksi UTS
(c) Slow oven cool, 154 ksi UTS
(d) ST in vacuum, 140 ksi UTS

150 mm (6 in.) β forged billet machined to 100 mm (4 in.) thick with α + β solution treatment. Times to half temperature for the various cooling rates were: (a) 100 s, (b) 1000 s, (c) 2000 s, and (d) 2880 s.
Source: H.R. Phelps and J.R. Wood, "Correlation of Mechanical Properties and Microstructures of Ti-6Al-2Sn-2Zr-2Mo-2Cr-0.25Si Titanium Alloy," Proc. 7th Int. Titanium Conf., June 1992, to be published. x 810

Ti-4Al-3Mo-1V

Ti-431

This alpha-beta alloy is considered a sheet alloy that was also available in plate thicknesses. Ti-4Al-3Mo-1V may be used in the annealed, solution treated, or solution treated plus aged conditions.

Ti-4Al-3Mo-1V: Summary of typical physical properties

Density(a)	4.5 g/cm^3 (0.163 lb/in.^{3})
Modulus of elasticity (aged)	117 GPa (17 × 10^6 psi)
Magnetic permeability	Nonmagnetic

(a) Typical values at room temperature of about 20 to 25 °C (68 to 78 °F). See also Alloy Digest, Ti-26

Ti-4Al-3Mo-1V: Minimum mechanical properties

Condition	Tensile yield strength (0.2%)		Ultimate tensile strength		Elongation in 50 mm (2 in.), %	Bend radius
	MPa	ksi	MPa	ksi		
Solution treated	895	130	...	...	10	3.5-4.0T
STA	1100	160	1275	185	3-5	...
Annealed	860	125	860	125	10	3.5-4.0T

Ti-4Al-3Mo-1V: Creep strength of aged sheet

Temperature		Time, h	Stress		Creep deformation, %
°C	°F		MPa	ksi	
1.6 mm (0.063 in.) gage					
315	600	200	689	100	0.18
		200	551	80	0.09
425	800	200	448	65	0.50
		200	241	35	0.15
2.3 mm (0.090 in.) gage					
315	600	200	689	100	0.22
		200	551	80	0.12
425	800	200	448	65	0.75
		200	241	35	0.17

Ti-4Al-3Mo-1V: Tensile strength of aged sheet

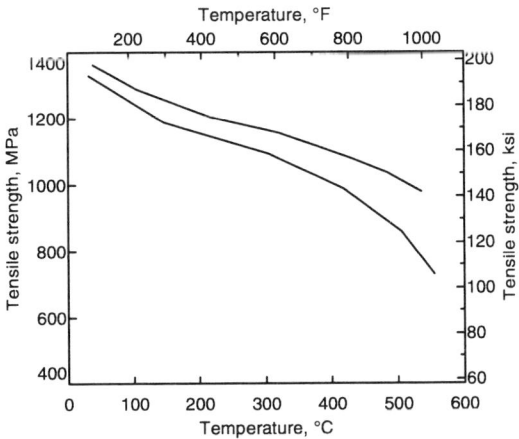

Ti-4Al-3Mo-1V: Bearing and shear of aged sheet

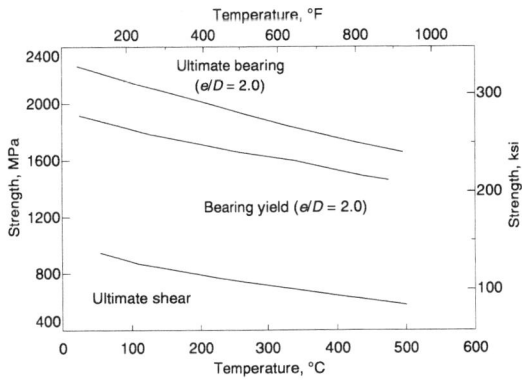

Ti-4Al-3Mo-1V: Axial fatigue

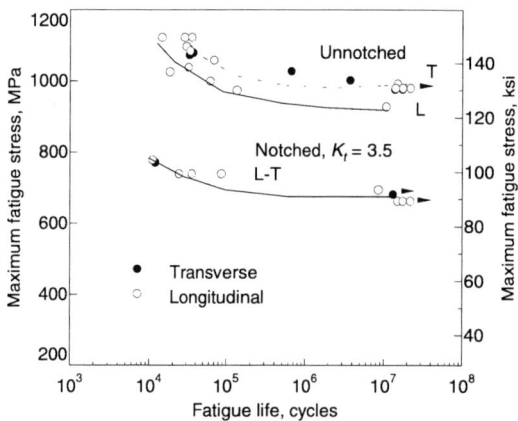

Notched and unnotched tension-tension fatigue properties of Ti-4Al-3Mo-1V sheet, solution treated and aged. Room-temperature ultimate tensile strengths were 1340 and 1350 MPa (194 and 196 ksi) in the longitudinal and transverse directions, respectively. Yield strengths (0.2%) were 1117 and 1170 MPa (162 and 170 ksi), respectively, and elongation was 6 and 5%, respectively. $R = 0.4$.

Ti-5Al-1.5Fe-1.4Cr-1.2Mo
Ti-155A

Also known as Ti-155A, this alloy was dropped from production.

Ti-5Al-1.5Fe-1.4Cr-1.2Mo: Summary of typical physical properties

Beta transus	985 to 1007 °C	(1805 to 1845 °F)
Thermal coefficient of linear expansion	10.3×10^{-6}/°C	(5.71×10^{-6}/°F)
Young's modulus	114 GPa	(16.5×10^6 psi)
Poisson's ratio	0.327	
Shear modulus	42.9 GPa	(6.22×10^6 psi)

Ti-5Al-1.5Fe-1.4Cr-1.2Mo: Typical tensile properties

Condition	Tensile yield strength MPa	ksi	Ultimate tensile strength MPa	ksi	Elongation, %	Reduction of area, %
Mill annealed	1020	148	1061	154	14	42
880 °C (1625 °F), 1 h, WQ + 540 °C (1000 °F), 8 h, AC	1172	170	1227	178	11	40

Ti-5Al-2.5Fe

Trade Name: Tikrutan LT 35
DIN 3.7110

Prof. Dr. Jürgen Breme

Materials used for permanent implants in the human body must exhibit corrosion resistance, biocompatibility, amenability to osseointegration, and biofunctionality (a high ratio of fatigue strength to Young's modulus). These requirements are met by Ti-5Al-2.5Fe, which does not contain any toxic or allergenic constituents or decomposition products. Potential applications for this alloy include total hip endoprostheses, knee joints, spinal implants, dental implants, and all types of joint prostheses, as well as bone nails, screws, and plates.

Chemistry/Density

4.45 g/cm^3 (0.160 lb/in.^{3})

Ti-5Al-2.5Fe: Chemical composition of annealed sheet or bar

	Composition, wt.%					
	Al	Fe	O	H	N	C
Minimum	3.0	2.0	...	...	...	...
Maximum	5.0	3.0	0.2	0.015	0.05	0.08

Source: Titanium Tikrutan Catalogue, Deutsche Titan, 1991

Product Forms

The formability of Ti-5Al-2.5Fe is poor at room temperature. Good formability (even superplastic deformation) is achieved in the α + β phase field at 850 to 950 °C (1560 to 1740 °F), as well as good formability in the β phase field at temperatures above 950 °C (1740 °F).

Weldability. Like other titanium alloys, Ti-5Al-2.5Fe can be welded, but preferably by the more sophisticated methods such as gas tungsten-arc and electron beam welding. Other precautionary procedures such as preheating, postheating, control of interpass temperature, and control of environmental conditions should be used.

Machinability. Ti-5Al-2.5Fe is readily machinable, but only under well-controlled conditions in terms of requirements for rigid machine tools, tool materials, tool geometry, and specially prepared cutting fluids. Like other titanium alloys, reduced rates of speed, feed, and depth of cut should be used.

Wrought Forms. Hot rolled strip, sheet, or plate can be manufactured upon request. Hot rolled, extruded, or drawn round bar is available, as well as square and flat bar with a maximum cross-sectional width-to-thickness ratio of 5:1.

Castability. Ti-5Al-2.5Fe exhibits similar behavior to other α + β alloys.

Cast Forms. Cast implant devices, such as sockets of hip prosthesis, are readily produced from Ti-5Al-2.5V.

Special Products. Special implant devices that are porous sintered to facilitate tissue ingrowth, e.g., heart pacemaker electrodes and dental implants, are manufactured upon request.

Product Condition and Microstructure

Deformation in the α + β phase field and annealing at 850 °C (1560 °F) produces a fine-grained α + β microstructure. Age hardening is also possible.

Phases and Structures

Crystal Structure

Beta Transus. The α + β transus is 950 °C (1740 °F) for an alloy composition of 5.15 wt.% Al, 2.45 wt.% Fe, 0.14 wt.% O, 0.09 wt.% N, 0.01 wt.% Si, and 0.025 wt.% H.

738 / Alpha-Beta Alloys

Ti-5Al-2.5Fe: Effect of temperature on α and β volume fraction

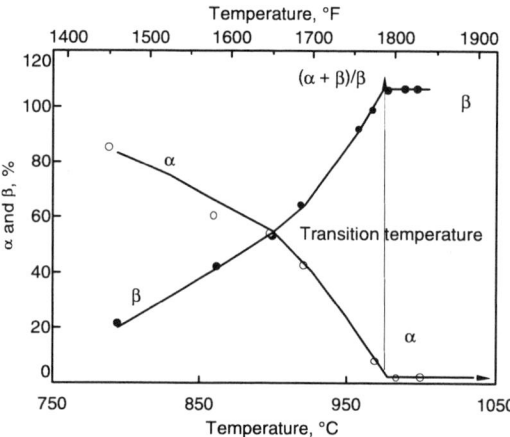

Composition: 4.8 Al wt.%, 2.2 Fe wt.%, and 0.12 wt.% O.
U. Zwicker, J. Breme, and K. Nigge, Optimizing of the Microstructure of the Implant Alloy TiAl5Fe2.5 by Microprobe Analysis, *Microchim. Acta Suppl.*, Vol 11, 1985, p 333-341

Ti-5Al-2.5Fe: Al and Fe contents of α and β phases

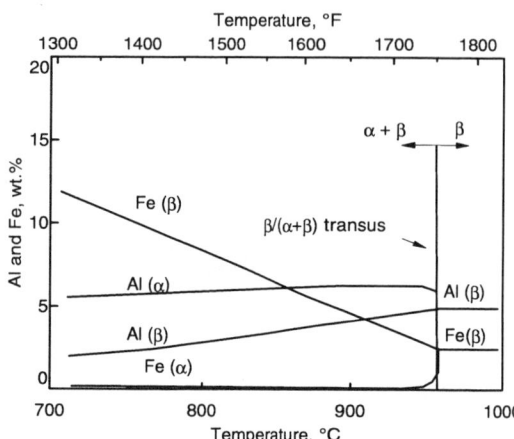

Alloy composition: 4.8 wt.% Al, 2.2 wt.% Fe, and 0.12 wt.% O.
U. Zwicker, J. Breme, and K. Nigge, Optimizing of the Microstructure of the Implant Alloy TiAl5Fe2.5 by Microprobe Analysis, *Microchim. Acta Suppl.*, Vol 11, 1985, p 333-341

Ti-5Al-2.5Fe: 5 wt.% Al section in the Ti-Al-Fe system

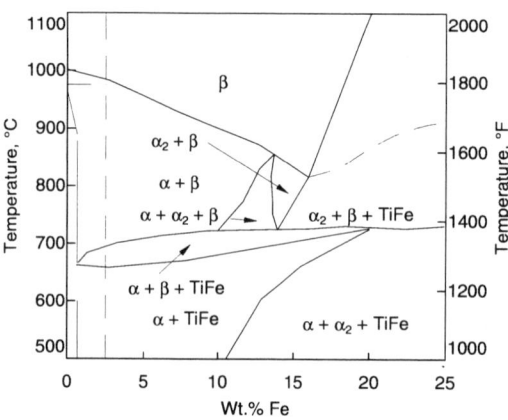

M.A. Volkava and J.J. Kornilov, *Inst. Metall. Akad. Nauk S.S.R.*, 1973, p 77-80

Ti-5Al-2.5Fe: TTT diagram

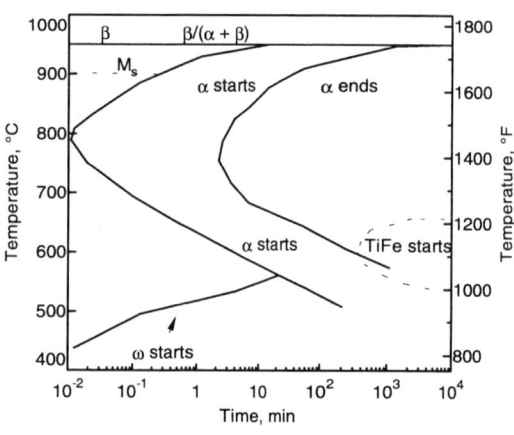

J. Breme and W. Schade, Phase Transformation in TiAl5Fe2.5 Alloy, *Titanium, Science and Technology*, Vol 3, G. Lütjering, U. Zwicker, and W. Bunk, Ed., Deutsche Gesellschaft für Metallkunde, e.V., 1985, p 1487-1494

Physical Properties

Elastic constants

Young's modulus in tension for Ti-5Al-2.5Fe sheet rolled at 850 °C (1560 °F) is 105 to 120 GPa (15 to 17 × 10^6 psi).

Electrical Properties

Dielectric Constants: $\varepsilon_R = 25$ and $\varepsilon_i = 122$ (Source: H.P. Thümmler and R. Thull, Surface Properties of Titanium and Its Alloys—Mechanical and Electrochemical Investigation, in *Titanium, Science and Technology*, Vol 2, G. Lütjering, U. Zwicker, and W. Bunk, Ed., Deutsche Gesellschaft für Metallkunde, e.V., 1985, p 1335-1342)

Ti-5Al-2.5Fe: Effects of aging on electrical conductivity

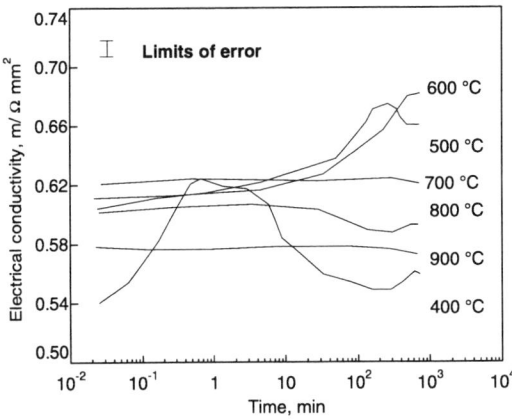

Isothermally aged after annealing 10 min at 1000 °C (1830 °F), and water quenching.
J. Breme and W. Schade, Phase Transformation in TiAl5Fe2.5 Alloy, in *Titanium, Science and Technology*, Vol 3, G. Lütjering, U. Zwicker, and W. Bunk, Ed., Deutsche Gesellschaft für Metallkunde, e.V., 1985, p 1487-1494

Ti-5Al-2.5Fe: Current density-potential curves in NaCl solution

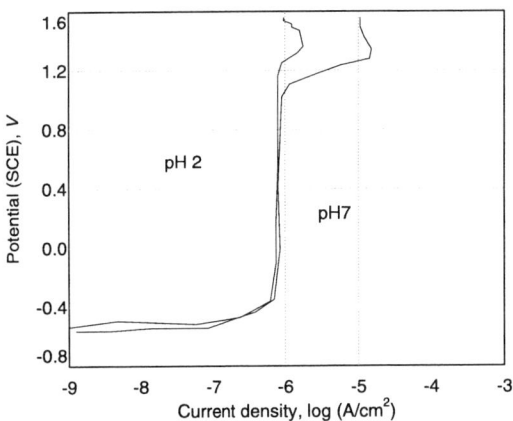

Buffered to pH 2 by HCl-citrate.
J. Geis-Gerstorfer *et al.*, Elektrochemische Untersuchungen zum Einfluß des pH-Wertes auf die Korrosionsbeständigkeit von Implantatlegierungen, *Z. Zahnärztl. Implantalogie*, Vol 4, 1988, p 31-36

Electrochemical Properties

Current density is 7.8 µA/cm² in 0.9 NaCl and 3.7 µA/cm² in albumin (protein) solution (5 g per 100 mL) (Source: H.P. Thümmler and R. Thull, Surface Properties of Titanium and Its Alloys—Mechanical and Electromechanical Investigation in *Titanium, Science and Technology*, Vol 2, G. Lütjering, U. Zwicker, and W. Bunk, Ed., Deutsche Gesellschaft für Metallkunde, e.V., 1985, p 1335-1342)

Thermal Properties

Thermal Coefficient of Expansion. At 200 to 1000 °C (390 to 1830 °F), 11.9×10^{-6}/K is an average value for the coefficient of Ti-5Al-2.5Fe. (Source: J. Breme, V. Wadewitz, and K. Burger, Verbund Titanlegierung/Al_2O_3 fur Dentale Implantate, Entwicklung Geeigneter Legierungen, Verbundwerkstoffe 1988, Deutsche Gesellschaft fur Metallkunde, 1988, p 123-132)

Mechanical Properties

Hardness

After undergoing deformation at 850 °C (1560 °F), Ti-5Al-2.5Fe has a typical Vickers hardness of 320 to 340 HV.

Tensile Properties

Ti-5Al-2.5Fe: Tensile properties of centrifugally cast specimens

Tensile yield strength (0.2%)		Ultimate tensile strength		Elongation at fracture, %	Reduction of area, %
MPa	ksi	MPa	ksi		
820	120	900	130	6	18

Note: Specimens had 14.5 mm (0.57 in.) diameter. Source: U. Zwicker and J. Breme, Evaluation of Centrifugally Cast TiAl5Fe2.5 Alloy for Implant Material, in *Titanium, Science and Technology*, Vol 1, G. Lütjering, U. Zwicker, and W. Bunk, Ed., Deutsche Gesellschaft für Metallkunde, e.V. 1985, p 171-178

Ti-5Al-2.5Fe Effects of aging on hardness

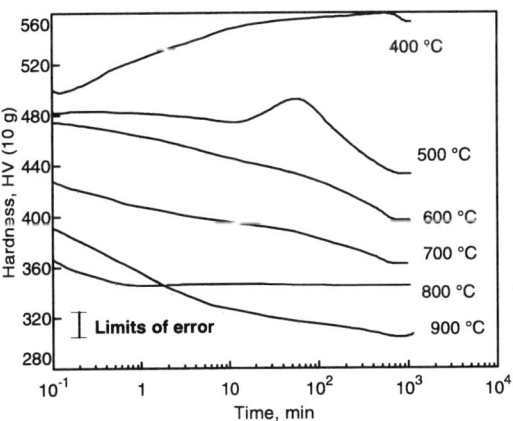

Isothermally aged after annealing 10 min at 1000 °C (1830 °F), and water quenching.
J. Breme and W. Schade, Phase Transformation in TiAl5Fe2.5 Alloy, *Titanium, Science and Technology*, Vol 3, G. Lütjering, U. Zwicker, and W. Bunk, Ed., Deutsche Gesellschaft für Metallkunde, e.V., 1985

Ti-5Al-2.5Fe: Tensile properties of annealed sheet at room temperature

Thickness mm	Tensile yield strength		Minimum tensile strength		Minimum elongation at fracture, %	Minimum reduction of area, %
	MPa	ksi	MPa	ksi		
0.2 to 2	780	115	860	125	8	...
2 to 6	780	115	860	125	10	...
≤80	780	115	860	125	10	25
≤160	780	115	860	125	8	20

Source: Titanium Tikrutan Catalogue, Deutsche Titan, 1991

Ti-5Al-2.5Fe: Tensile properties of aged bar at room temperature

Heat treatment	Tensile yield strength (0.2%)		Ultimate tensile strength		Elongation at fracture, %	Reduction of area, %
	MPa	ksi	MPa	ksi		
10 min at 1000 °C (1830 °F)/WQ	1173	170	1314	190	2	7
10 min at 1000 °C (1830 °F)/WQ + 65 h at 600 °C (1110 °F)	1035	150	1079	156	6	11.7
10 min at 1000 °C (1830 °F)/WQ + + 65 h at 700 °C (1290 °F)	955	140	1018	145	14	21
1 h at 900 °C (1650 °F)/WQ + 2 h at 700 °C (1650 °F)	920	135	1050	152	15.5	38

Source: U. Zwicker, Metallkundliche Untersuchungen an der Implantatlegierung Ti5AlFe2.5, Z. Metallkunde, Vol 77, 1986, p 714-720

Ti-5Al-2.5Fe: Effect of rolling temperature on tensile properties

Rolling temperature		Tensile yield strength (0.2%)		Ultimate tensile strength		Elongation at fracture, %	Reduction of area, %
°C	°F	MPa	ksi	MPa	ksi		
850	1560	853	123	1006	146	16	42
800	1470	820	119	1035	150	17.5	43
600	1110	1103	160	1273	184	8	22

Source: U. Zwicker, Metallkundliche Untersuchungen an der Implantatlegierung Ti5AlFe2.5, Z. Metallkunde, Vol 77, 1986, p 714-720

Ti-5Al-2.5Fe: Effect of orientation on tensile properties of sheet rolled at 850 °C (1560 °F)

Position of the sample with respect to the rolling direction	Tensile yield strength (0.2%)		Ultimate tensile strength		Elongation at fracture, %	Reduction of area, %	Young's modulus	
	MPa	ksi	MPa	ksi			GPa	10^6 psi
Parallel	784	113	967	140	13	35	109	15.8
45°	876	127	913	132	12	30	116	16.8
Perpendicular	926	134	1015	147	12.5	32.5	124	17.9

Source: U. Zwicker and J. Breme, Investigations of the Friction Behavior of Oxidized Ti-Al5Fe2.5 Surface Layers of Implant Material, J. Less-Common Met., Vol 100, 1984, p 371-375

Ti-5Al-2.5Fe: Tensile strengths at elevated temperatures

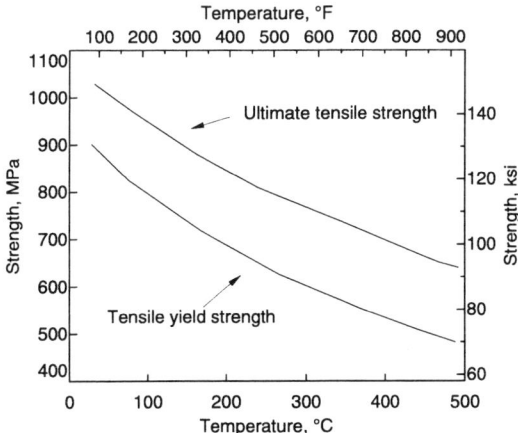

Private communication, Deutsche Titan

Ti-5Al-2.5Fe: RT mechanical properties

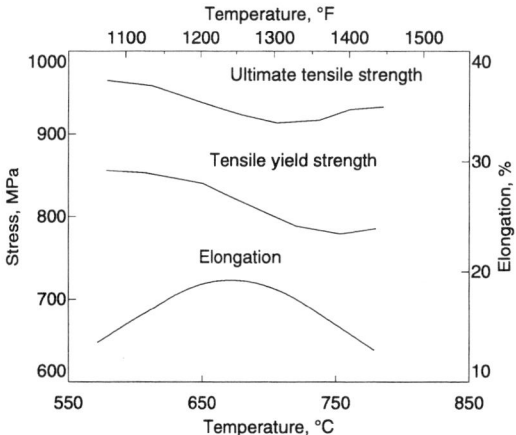

Effect of annealing temperature on RT properties.
Source: *Titanium, Science and Technology*, Vol 2, 1985, p 1393-1400

Fatigue and Fracture Properties

Ti-5Al-2.5Fe: Tension-tension fatigue of bar (R = 0.1)

Condition	Stress concentration factor (K_t)	Fatigue strength ($\geq 10^7$ cycles) MPa	ksi
Annealed	1.0	725	105
1 h at 900 °C (1650 °F)/WQ + 2 h at 700 °C (1290 °F)	3.6	300	43.5
Cast (centrifugally)	1.0	425	61.6
Cast, hot isostatically pressed	1.0	450	65.2
	3.6	300	43.5

Source: K.H. Borowy and K.H. Kramer, On the Properties of a New Titanium Alloy (Ti-Al5Fe2.5) as Implant Material, *Titanium, Science and Technology*, Vol 2, G. Lütjering, U. Zwicker, and W. Bunk, Ed., Deutsche Gesellschaft für Metallkunde, e.V., 1985, p 1381-1386

Ti-5Al-2.5Fe: High-cycle fatigue of forged and cast material

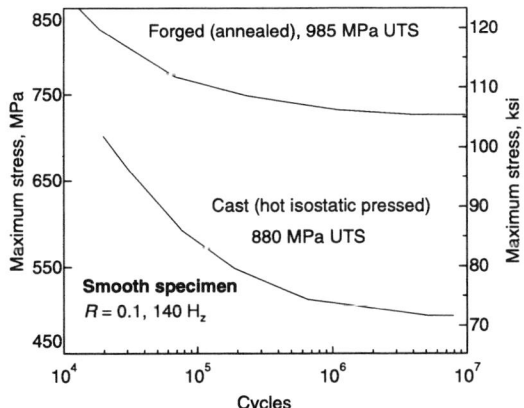

Tensile yield strengths of 900 MPa (130 ksi) for annealed forging and 820 MPa (119 ksi) for HIP casting.
K.H. Borowy and K.H. Kramer, On the Properties of a New Titanium Alloy (TiAl5Fe2.5) as Implant Material, *Titanium, Science and Technology*, Vol 2, G. Lütjering, U. Zwicker, and W. Bunk, Ed., Deutsche Gesellschaft für Metallkunde, e.V., 1985, p 1381-1386; and private communication, Deutsche Titan

Ti-5Al-2.5Fe: High-cycle notched fatigue

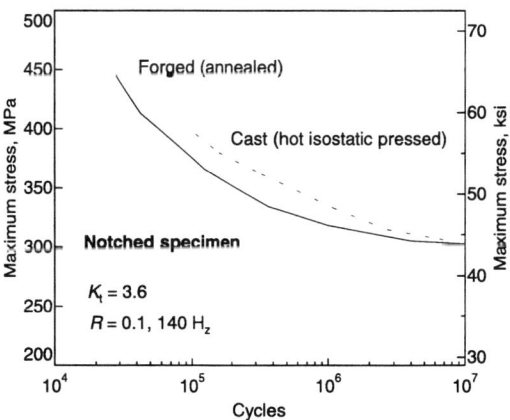

Annealed forgings with ultimate tensile strength of 985 MPa (143 ksi) and a yield strength of 900 MPa (130 ksi). HIP casting had ultimate tensile strength of 880 MPa (127 ksi) and a yield strength of 820 MPa (119 ksi).
Source: K.H. Borowy and K.H. Kramer, On the Properties of a New Titanium Alloy (Ti-Al5Fe2.5) as Implant Material, *Titanium, Science and Technology*, Vol 2, G. Lütjering, U. Zwicker, and W. Bunk, Ed., Deutsche Gesellschaft für Metallkunde, e.V., 1985, p 1381-1386

Ti-5Al-2.5Fe: Rotating bending fatigue strength

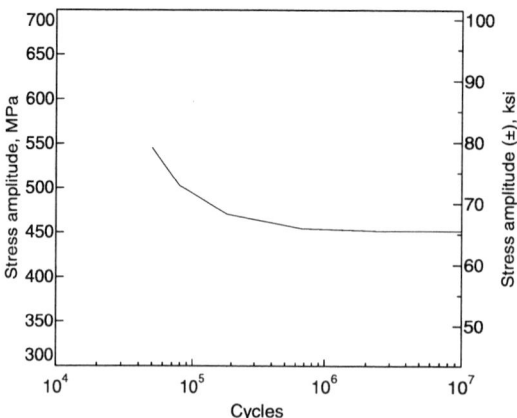

Smooth specimens; annealed; $R = -1$.
K.H. Borowy and K.H. Kramer, On the Properties of a New Titanium Alloy (TiAl5Fe2.5) as Implant Material, *Titanium, Science and Technology*, Vol 2, G. Lütjering, U. Zwicker, and W. Bunk, Ed., Deutsche Gesellschaft für Metallkunde, e.V., 1985, p 1381-1386; and private communication, Deutsche Titan

Ti-5Al-2.5Fe: Rotating fatigue strength

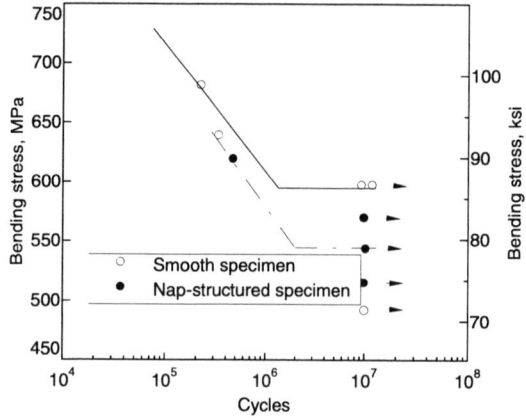

NAP-structured specimens have hemispherical cavities on the surface.
M. Merget and F. Aldinger, Influence of Technological Parameters on the Fatigue Strength of Ti5Al2.5Fe—A New Material for Endoprostheses, *Titanium, Science and Technology*, Vol 2, G. Lütjering, U. Zwicker, and W. Bunk, Ed., Deutsche Gesellschaft für Metallkunde, e.V., 1985, p 1393-1400

Corrosion Fatigue

Ti-5Al-2.5Fe: Rotating fatigue in 0.9% NaCl

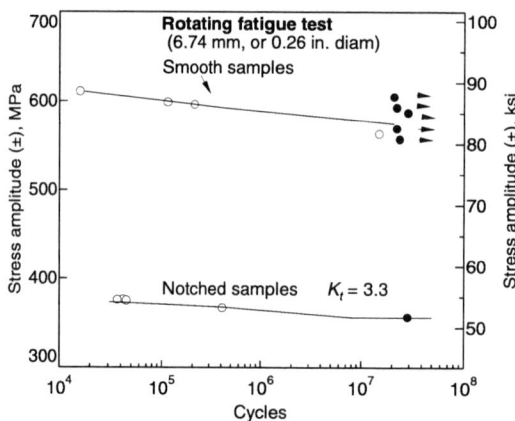

U. Zwicker *et al.*, Mechanical Properties and Tissue Reactions of a Titanium Alloy for Implant Material, *Titanium '80, Science and Technology*, H. Kimura and O. Izumi, Ed., The Metallurgical Society of AIME, 1980, p 505-514

Ti-5Al-2.5Fe: Tension-tension fatigue in 0.9% NaCl

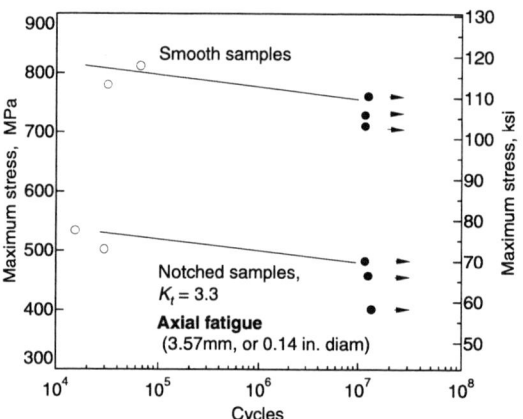

U. Zwicker *et al.*, Mechanical Properties and Tissue Reactions of a Titanium Alloy for Implant Material, *Titanium '80, Science and Technology*, H. Kimura and O. Izumi, Ed., The Metallurgical Society of AIME, 1980, p 505-514

Ti-5Al-2.5Fe: Wöhler curve of hip prostheses

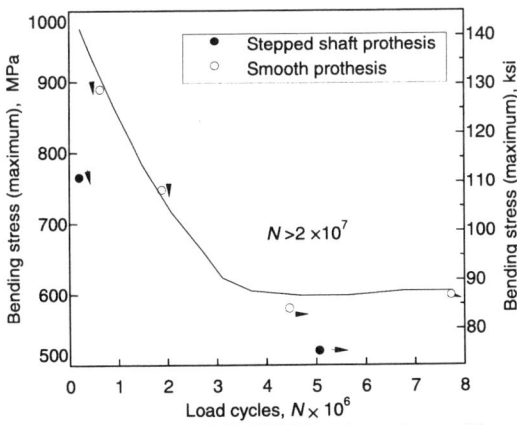

Tests in 0.9% NaCl at 39 °C (102 °F) under service conditions, stepped shaft prosthesis with lacunae (notch effect).
Forged Ti-5Al-2.5Fe with ultimate tensile strength of 990 MPa (145 ksi) and a tensile yield strength of 900 MPa (130 ksi).
J. Breme and G. Heimke, Corrosion Fatigue Test of TiAl5Fe2.5 Hip Implant Under High Stresses, *Titanium, Science and Technology*, Vol 2, 1985, p 1351-1357

Toughness

Ti-5Al-2.5Fe: Fracture toughness

Condition	Fracture toughness (K_{Ic})	
	MPa√m	ksi√in.
Annealed	38	34.5
1 h at 900 °C (1650 °F) WQ + 2 h at 700 °C (1290 °F)	56	51

Source: K.H. Borowy and K.H. Kramer, On the Properties of a New Titanium Alloy (Ti-Al5Fe2.5) as Implant Material, *Titanium, Science and Technology*, Vol 2, G. Lütjering, U. Zwicker, and W. Bunk, Ed., Deutsche Gesellschaft für Metallkunde, e.V., 1985, p 1381-1386

Plastic Deformation

Extrusion. Ti-5Al-2.5Fe exhibits better extrudability compared to Ti-6Al-4V in the α + β and the β phase fields due to lower flow stresses.

Forging. Ti-5Al-2.5Fe exhibits forgeability

Ti-5Al-2.5Fe: Flow stress at 700 and 850 °C

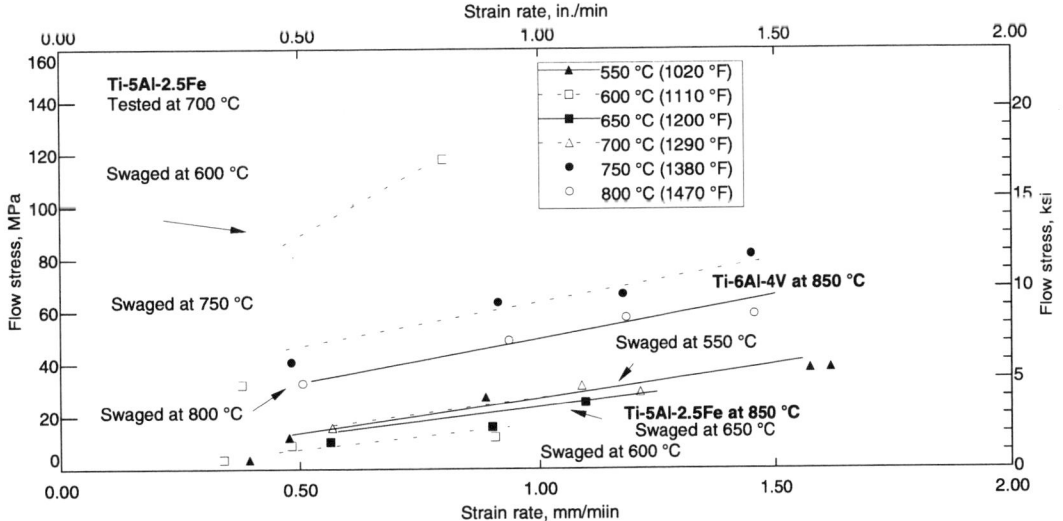

Effects of swaging temperature and strain rate.
U. Zwicker et al., Investigation on the Superplastic Behavior of TiAl6V4 and TiAl5Fe2.5 (Rod Material), *Advanced Technology of Plasticity*, Springer, 1987, p 363-372

compared to Ti-6Al-4V in the α + β and the β phase fields due to lower flow stresses.

Forming. This alloy exhibits better formability compared to Ti-6Al-4V in the α + β and the β phase fields due to lower flow stress.

Ti-5Al-2.5Fe: Stress strain curves of porous sintered specimens

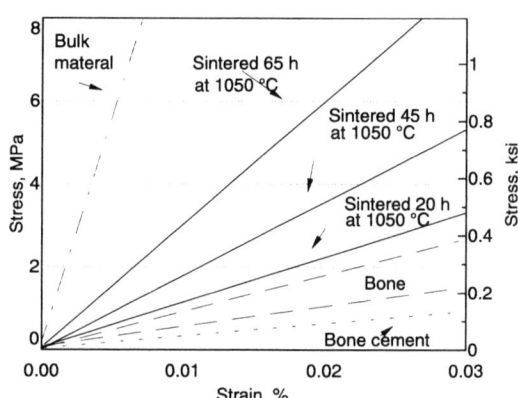

Effects of sintering parameters on bulk Ti-5Al-2.5Fe, bone, and bone cement.
U. Zwicker, Metallkundliche Untersuchungen an der Implantatlegierung Ti5AlFe2.5, *Z. Metallkunde*, Vol 77, 1986, p 714-720

Hot Ductility

Ti-5Al-2.5Fe: Superplastic behavior at 850 °C

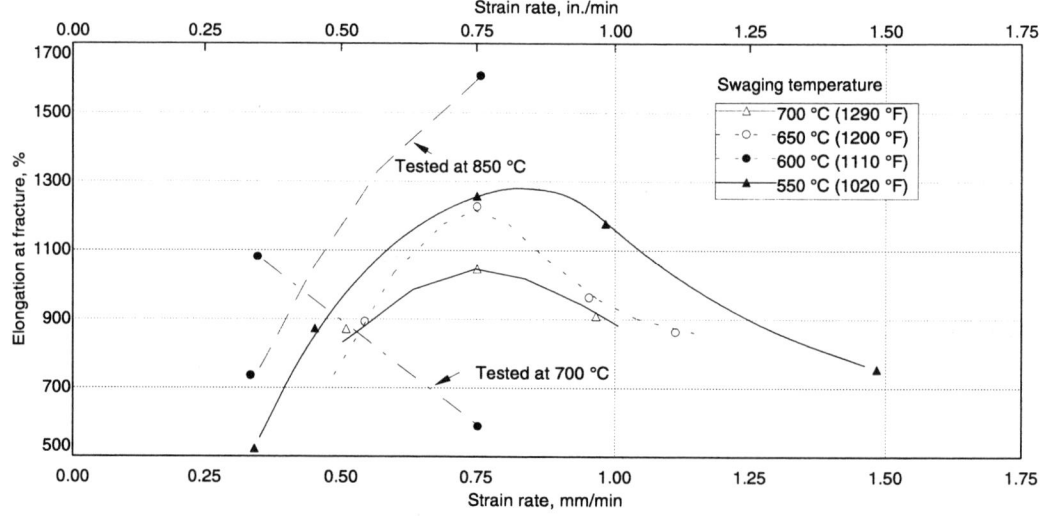

U. Zwicker *et al.*, Investigation on the Superplastic Behavior of TiAl6V4 and TiAl5Fe2.5 (Rod Material), *Advanced Technology of Plasticity*, Springer, 1987, p 363-372. Hot rolled at 800 °C and swaged as indicated. Test at 700 °C shown for comparison

Ti-5Al-2.5Fe: Swaging temperature vs ductility

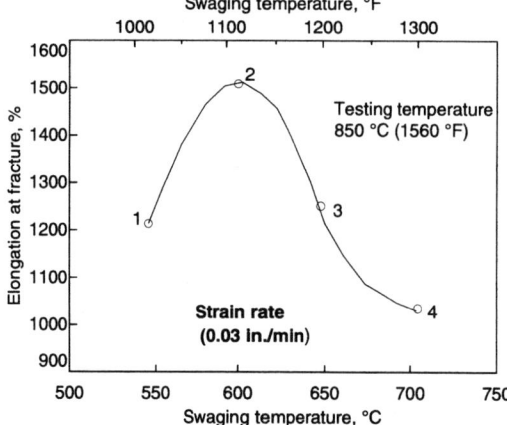

Source: U. Zwicker *et al.*, Investigation on the Superplastic Behavior of TiAl6V4 and TiAl5Fe2.5 (Rod Material), *Advanced Technology of Plasticity*, Springer, 1987, p 363-372

Net Shaping

Casting. Castability of Ti-5Al-2.5Fe is similar to other alloys. Titanium investment casting is possible with the use of special molding materials that exhibit a limited reaction with the molten material, such as graphite artificial resin, thermodynamic stable oxides or oxide binders (ThO_2, Y_2O_3, CaO), and high-melting metals/oxide binders (tungsten and molybdenum). Centrifugal casting is possible with permanent molds of copper, which do not react due to a high heat conductivity.

Powder Metallurgy. A special application of Ti-5Al-2.5Fe is the porous sintering of implant devices. Due to sufficient pore size (>50μm), securing of the implants is provided by tissue ingrowth. Examples for such devices are heart pacemaker electrodes and dental implants.

Ti-5Al-2.5Fe: Effects of porosity produced by cold compression

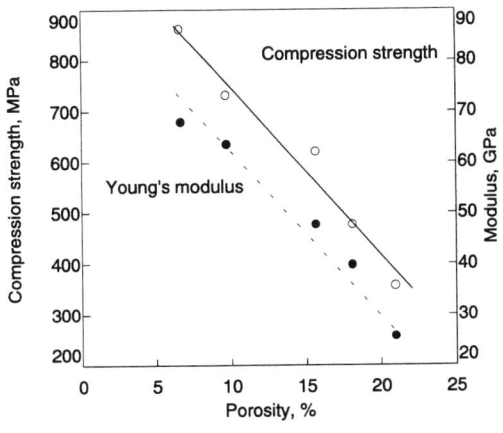

HDH powder (90 to 200 μm) sintered 60 min at 1100 °C (2010 °F). J. Breme *et al.*, Optimierung der mechanischen Eigenschaften poröser Sinterkörper aus der Implantatlegierung TiAl5Fe2.5 mit dem Ziel der Herstellung von Zahnimplantaten, *Z. Zahnörztl. Implantologie*, Vol 3, 1987, p 41-46

Ti-5Al2.5Fe: Effect of sintering time at 1100 °C (2010 °F)

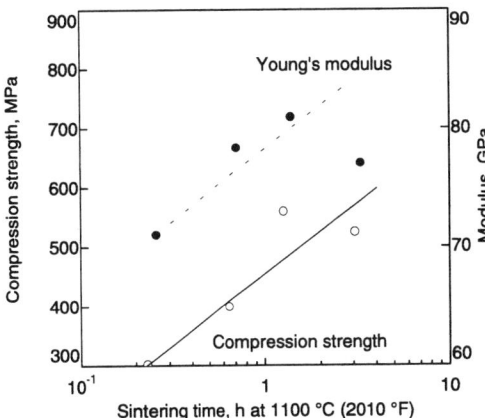

REP powder (300 to 450 μm). J. Breme *et al.*, Optimierung der mechanischen Eigenschaften poröser Sinterkörper aus der Implantatlegierung TiAl5Fe2.5 mit dem Ziel der Herstellung von Zahnimplantaten, *Z. Zahnörztl. Implantologie*, Vol 3, 1987, p 41-46

Treatments

Heat Treatment

Annealing at 850 °C (1560 °F)
Solution treatment at 800 to 920 °C (1470 to 1690 °F), or in the β phase field at 1000 °C (1830 °F) aging at 400 to 700 °C (750 to 1290 °F)
Aging at 400 to 700 °C (750 to 1290 °F)

Surface Treatments

Ti-5Al-2.5Fe: Surface hardening by oxidation

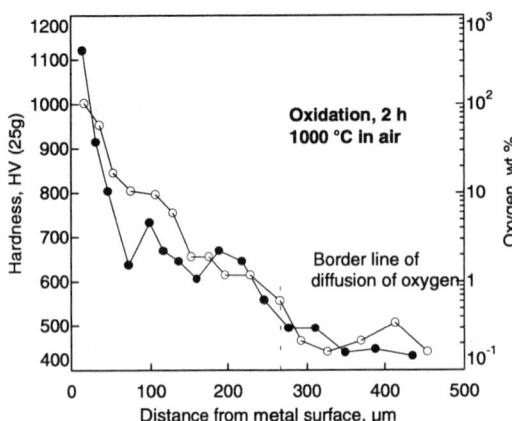

Oxidation occurred after an annealing of 2 h at 1000 °C (1830 °F), and water quenching.
U. Zwicker et al., Abrasive Properties of Oxide Layers on TiAl5Fe2.5 in Contact with High Density Polyethylene, *Titanium, Science and Technology*, Vol 2, G. Lütjering, U. Zwicker, and W. Bunk, Ed., Deutsche Gesellschaft für Metallkunde, e.V., 1985, p 1343-1350

Ti-5Al-2.5Fe: Surface laser gas nitriding

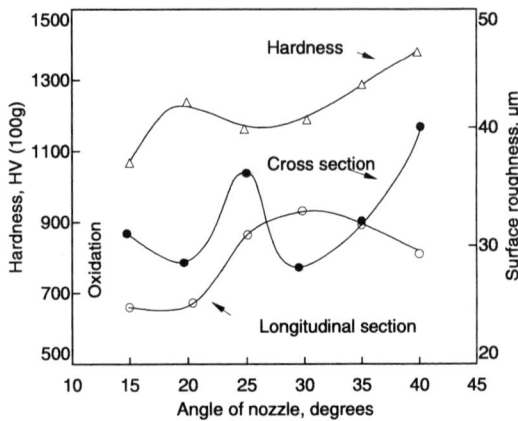

Influence of the angle incidence.
S.Z. Lee and H.W. Bergmann, Laser Surface Alloying of Titanium and Titanium Alloys, *Sixth World Conference on Titanium*, P. Lacombe, R. Tricot, and G. Beranger, Ed., Les Editions de Physique, Paris, 1988, p 1811-1816

Ti-5Al-2.5Fe: Friction behavior

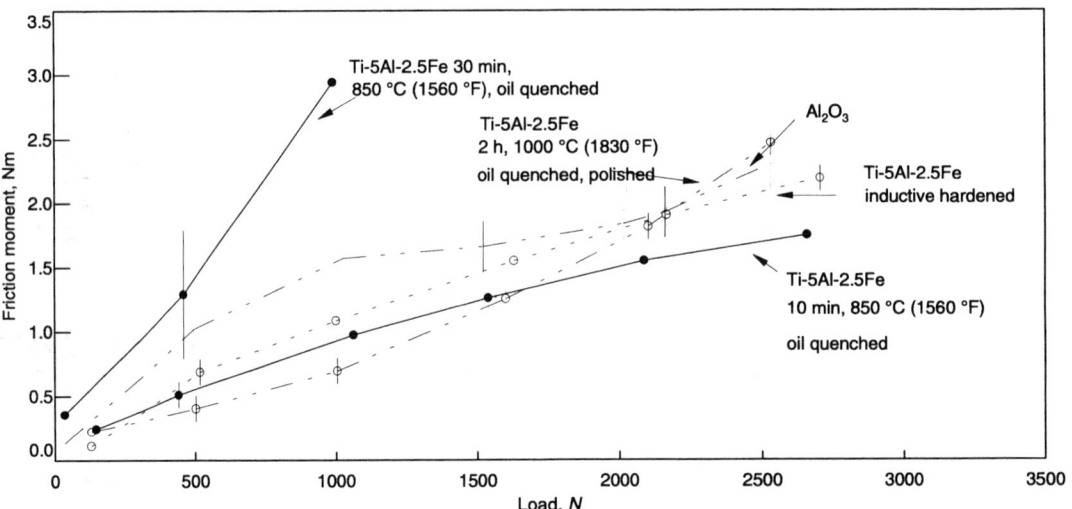

Friction behavior of Ti-5Al-2.5Fe hip prosthesis heads compared to an Al_2O_3 head.
U. Zwicker et al., Abrasive Properties of Oxide Layers on TiAl5Fe2.5 in Contact with High Density Polyethylene, *Titanium, Science and Technology*, Vol 2, G. Lütjering, U. Zwicker, and W. Bunk, Ed., Deutsche Gesellschaft für Metallkunde, e.V., 1985, p 1343-1350

Ti-5Al-5Sn-2Zr-2Mo-0.25Si

Common Name: Ti-5522-S
UNS Number: R54560

Reviewed by P. Russo, RMI Titanium Company

This semicommercial alloy was developed by RMI in a program sponsored by the Air Force Materials Laboratory in the early 1970s for use in aircraft components subjected to prolonged exposure near 500 °C (930 °F). It is a beta-lean α-β alloy with relatively low content of β-stabilizing elements. Ti-5522-S exhibits good tensile, stress rupture, and creep properties at elevated temperature into the 425 to 540 °C (800 to 1000 °F) range. It can be welded, formed at room temperatures, or warm formed temperatures 540 to 700 °C (800 to 1300 °F).

Effects of Impurities and Alloying. Exceeding impurity limits may result in raising yield strength above maximum permitted or in lower elongation or reduction in area below minimum. As for all α-β alloys, excessive aluminum, oxygen, and nitrogen can reduce ductility and fracture toughness. Excessive beta stabilizers (for example, molybdenum or vanadium) affect the stability of the alloy and increase its heat treatability, therefore making it more difficult to control properties.

Product Forms and Conditions. Alloy Ti-5522-S is available in standard wrought product form as a forged billet or bar. In addition, it is produced as a special wrought product in the form of plate and sheet. The alloy cannot be hardened or strengthened by any thermal treatment. After cold or warm forming, it may need stress relief or a full anneal.

Applications. Alloy Ti-5522-S is a semicommercial product for high-temperature applications in the range of 425 to 540 °C (800 to 1000 °F). The alloy was developed for high creep strength and elevated-temperature stability in applications such as jet-engine components.

Specifications. Ti-5Al-5Sn-2Zr-2Mo-0.25Si is listed in government specifications MIL-T-9046, MIL-T-9047, and MIL-T-81556.

Ti-5522S: Typical composition

	Al	Sn	Zr	Mo	Si	Fe	C	N	O	Ti
Min	4.5	4.5	1.75	1.75	0.20	...	...	...	...	...
Max	5.5	5.5	2.25	2.25	0.30	0.15	0.04	0.03	0.13	
Aim	5	5	2	2	0.25	...	...	...	...	bal

Physical Properties

Phases and Structures. Annealed Ti-5522-S has a hexagonal closed-packed crystal structure with a small amount of beta phase at room temperature. Microstructures resulting from cooling through the β-transus temperature typically consist of packets of α-platelets, similarly aligned and crystallographically oriented, that are separated by films of β-phase. Beta annealing results in a long α-platelet structure while β-working results in a shorter α-platelet structure.

Elastic properties:

- Young's modulus (tension): 114 GPa (16.5×10^6 psi)
- Poisson's ratio: 0.326

Corrosion/Chemical Properties. The corrosion-resisting characteristics of Ti-5Al-5Sn-2Zr-2Mo-0.25Si are comparable to unalloyed titanium and to other near-alpha and α-β titanium alloys. This alloy is unaffected after 1000 h in the standard ASTM salt-spray test.

Ti-5522-S: Summary of typical physical properties

Beta transus	980 ± 15 °C (1800 ± 25 °F)
Melting (liquidus point)	~1700 °C (3100 °F)
Density(a)	4.51 g/cm³ (0.163 lbf/in.³)
Electrical resistivity(a)	1.71 μΩ·m
Magnetic permeability	Nonmagnetic
Specific heat capacity(a)	Not available
Thermal conductivity(a)	Not available
Thermal coefficient of linear expansion(b)	10.2×10^{-6}/°C (5.7×10^{-6}/°F)

(a) Typical values at room temperature of about 20 to 25 °C (68 to 78 °F). (b) Mean coefficient from room temperature to 815 °C (1500 °F)

Mechanical Properties

Hardness. 32 to 38 HRC (at room temperature).

Impact Strength. Charpy V-notch at room temperature: 22 J (16 ft · lbf).

Plane-Strain Fracture Toughness. 82.4 MPa$\sqrt{m}$ (75 ksi$\sqrt{in.}$) for forging with yield strength of 890 MPa (129 ksi).

Creep Properties:
- Minimum creep strength: 345 MPa (50 ksi) for 0.1% creep in 100 h at 510 °C (950 °F).
- Minimum rupture stress: 1170 MPa (170 ksi) for rupture in 5 h at room temperature, notched specimen.

Tensile Properties

Room temperature, minimum values (no heat treatment specified): tensile strength, 900 MPa (130 ksi); yield strength, 830 MPa (120 ksi); elongation, 10%; reduction in area, 25%. At 535 °C (1000 °F), minimum values: tensile strength, 689 MPa (100 ksi); yield strength, 517 MPa (75 ksi); elongation, 15%; reduction in area, 35%. Average or typical values (see table).

Ti-5Al-5Sn-2Zr-2Mo-0.25Si: Typical tensile properties

Condition	Temperature °C	°F	Ultimate tensile strength MPa	ksi	Yield strength MPa	ksi	Elongation, %	Reduction in area, %
Mill annealed	RT		965	140	868	126	12	35
	315.5	600	813	118	586	85	18	41
	426.6	800	745	108	558	81	20	50
	537.7	1000	730	106	552	80	20	50
975 °C (1785 °F)	RT		1048	152	965	140	13	...
1/2 h, AC + 595	315.5	600	793	115	565	82	15	...
°C (1100 °F) 2	426.6	800	780	113	530	77	17	...
h, AC	537.7	1000	690	100	503	73	19	...

Source: M.J. Donachie, Jr., *Titanium: A Technical Guide*, ASM, 1988, p 449; and "Basic Design," RMI Titanium, RMI Company, Niles, OH

Fatigue Properties

Although an equiaxed alpha morphology provides the best resistance to fatigue crack initiation, an acicular morphology produced by a beta treatment may be desired for creep resistance. Low-cycle fatigue strength of beta processed Ti-5522-S depends on temperature, frequency and the size of alpha platelets (see figure).

Ti-5522-S: Low-cycle fatigue strength

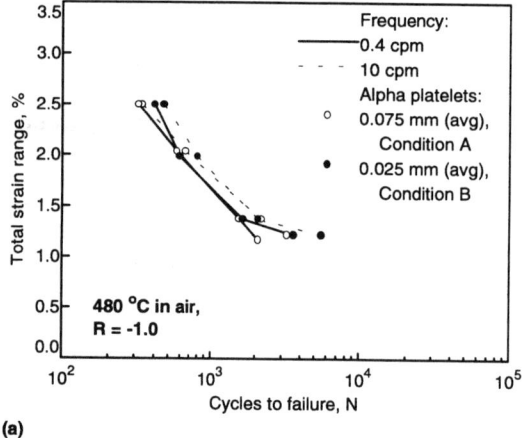
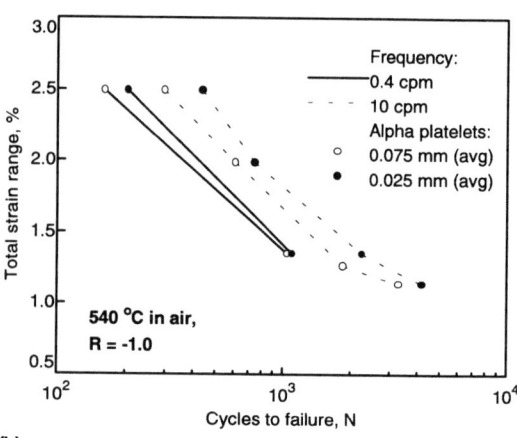

Condition A and B tested at (a) 480 °C (900 °F) and (b) 540 °C (1000 °F).
Condition A (0.075 mm average platelet length) was produced by alpha-beta working (6:1 extrusion, 3:1 swaging) followed by a beta anneal and a stabilization treatment.
Condition B (0.025 mm platelet length) was produced by beta working (6:1 extrusion, 3:1 swaging) and a stabilization treatment.
Source: D. Eylon, T. Bartel, and M. Rosenblum, "High Temperature Low Cycle Fatigue in Beta Processed Ti-5Al-5Sn-2Zr-2Mo-0.25Si," *Met. Trans. A*, 11A, August 1980, p 1361

Processing

Forging. Ti-5Al-5Sn-2Zr-2Mo-0.25Si should be forged at the lowest possible temperature to minimize surface contamination by oxygen. Heat to and start forging at 925 to 955 °C (1700 to 1750 °F), finish at 790 to 815 °C (1450 to 1500 °F).

Forming. Most forming operations can be done at room temperature but warm (425 to 700 °C, or 800 to 1300 °F) forming is sometimes employed.

Bend radius recommendations:

- For $t < 1.8$ mm (<0.070 in.); $4.5\,t$
- For $t \geq 1.8$ mm ($\geq$0.070 in.); $5.0\,t$

Heat Treatment. The alloy is usually given a full anneal with a reheat (see table). A beta anneal enhances creep resistance and toughness.

Machining. Like many titanium alloys, Ti-5Al-5Sn-2Zr-2Mo-0.25Si has a seizing tendency and therefore requires sharp tools, correct tool angles, heavy feeds and slow speeds; also rigid tool supports and adequate coolant. Typical tool geometries are provided in "Technical Note 7: Machining" in this Volume.

Cutting feeds and speeds of alloy Ti-5Al-5Sn-2Zr-2Mo-0.25Si are similar to those of Ti-6242S and Ti-6Al-4V. In turning operations, the typical range of feeds and speeds are:

- Roughing cuts at 0.4 to 0.75 mm/rev (0.015 to 0.03 in./rev) and 30 to 45 m/min (100 to 150 sfm)
- Finishing cuts at 0.25 mm/rev (0.01 in./rev) and 90 to 150 m/min (300 to 500 sfm)

Grinding. In grinding operations, alloy Ti-5Al-5Sn-2Zr-2Mo-0.25Si requires many of the same precautions against surface damage as other titanium alloys (see "Technical Note 7: Machining" in this Volume).

Welding. Alloy Ti-5Al-5Sn-2Zr-2Mo-0.25Si can be welded readily by inert-gas shielded arc welding, using it as the filler metal. Oxyacetylene welding and other forms of welding using active gases, electrode coatings, or fluxes are not recommended because the gases tend to embrittle the titanium and make it impossible to produce ductile welds.

Ti-5522-S: Typical heat treatments

Heat treatment	Heat				Reheat			
	Temperature		Time	Cooling method	Temperature		Time	Cooling method
	°C	°F			°C	°F		
Stress-relief anneal	595-650	1100-1200	2	Air	...	...	...	...
Full anneal	955	1750	1	Air	595	1100	2	Air
Beta anneal	1015	1860	1	Air	595	1100	2	Air

Source: "RMI 5Al-5Sn-2Zr-2Mo-Si," Ti-78, *Engineering Alloys Digest*, July 1980

Ti-6.4Al-1.2Fe RMI Low-Cost Alloy

The RMI VM (virgin material) alloy was designed as a low-cost alternative to the industry standard Ti-6Al-4V alloy. Iron, a low-cost beta stabilizer, was substituted for vanadium, which is substantially more expensive. The amount of iron added to the Ti-6.4Al base was selected to provide beta stability similar to Ti-6Al-4V. The alloy can achieve strength similar to Ti-6Al-4V at ambient temperatures.

Low-cost alloy: Typical tensile data

Heat treatment	0.2% YS MPa	ksi	UTS MPa	ksi	Elongation, %	RA, %
α + β anneal	862	125	965	140	20	40
β anneal	862	125	965	140	10	20

0.62 in. diameter bar at room temperature. Source: RMI Titanium Company

Ti-2Fe-2Cr-2Mo

Not produced since 1960, Ti-2Fe-2Cr-2Mo was developed by Timet in the mid-1950s as a commercial sheet alloy. It was capable of being heat treated to increase strength and hardness, but was replaced by Ti-6Al-4V. Applications requiring a high strength-to-weight ratio at room temperature and moderately elevated service temperatures (e.g., jet engine compressor components) were typical. Its corrosion resistance is essentially the same as that of unalloyed titanium.

The usual hot working range is 620 to 950 °C (1150 to 1750 °F) with the preferred range between 785 to 900 °C (1450 to 1650 °F). Up to 90% reduction during one heating may be accomplished without danger of cracking. The recommended heat treatment for this alloy is an isothermal anneal at 595 to 650 °C (1100 to 1200 °F) followed by air cooling.

Ti-2Fe-2Cr-2Mo: Summary of typical physical properties

Melting (liquidus) point	1650 °C	(3000 °F)
Density(a)	4.67 g/cm^3	(0.169 lb/in.^{3})
Magnetic permeability	Nonmagnetic	
Thermal coefficient of linear expansion(b)	$9 \times 10^{-6}/°C$ ($5 \times 10^{-6}/°F$)	

(a) Typical values at room temperature of about 20 to 25 °C (68 to 78 °F). (b) Mean coefficient from 95 to 315 °C (200 to 600 °F)

Ti-2Fe-2Cr-2Mo: Typical tensile properties

Test temperature		Ultimate tensile strength		Tensile yield strength (0.2%)		Elongation in 50 mm (2 in.), %	Reduction of area, %
°C	°F	MPa	ksi	MPa	ksi		
	RT	917	133	882	128	26	55
150	300	751	109	620	90	29	58
260	500	634	92	475	69	30	61
370	700	551	80	386	56	25	60
480	900	393	57	330	48	40	83

Ti-8Mn

Common Name: 8Mn
Trade Names: No longer commercial; RMI 8Mn, Rem-Cru C-110M, MST 9M, Republic RS-110A
UNS Number: R56080

Compiled by E. Bradley, Metallurgical Consulting Services

Although Ti-8Mn is no longer made in production quantities for commercial applications, this α-β type sheet and plate titanium alloy was originally developed for its excellent formability and intermediate strength by Rem-Cru and subsequently produced by RMI, Timet, and Republic. Ti-8Mn was one of the first titanium material used in airplane bodies. This alloy has good elevated-temperature strength and stability up to about 315 °C (600 °F) and is used in the annealed condition only. Heat treatment is not recommended. It has excellent formability; severe forming is accomplished at 260 to 540 °C (500 to 1000 °F). Used extensively in the F9 fighter, the major application of this alloy was for the tail section of the F8 fighter. Ti-8Mn has been replaced by Ti-6Al-4V. Additionally, Ti-8Mn is not recommended for welding.

Ti-8Mn: Chemical composition and equivalent specifications

Specification	Product form	Composition, wt%						Ti
		Mn	C	H	N	Fe	Other elements	
AMS 4908 MIL SPEC MIL-T-9046	Sheet, plate, strip	6.0-9.0	0.2 max	0.015 max	0.07 max	0.3 max	0.8 max	Bal

Physical Properties

Phases and Structures. Ti-8Mn is an α–β titanium alloy containing both the α and β phases at room temperature. The 8% manganese stabilizes a considerable amount of the β phase at room temperature. The alloy structure is entirely β at temperatures above 800 °C (1475 °F).

The α phase of this alloy is similar to that of unalloyed titanium. The β phase is the high-temperature phase of titanium, but a considerable amount is stabilized to room temperature by the manganese content. The β phase composition of the Ti-8Mn alloy may be varied by solution treatment above the β transus and below the transus in the α-β region, resulting in the β phase containing 11 and 15% Mn, respectively.

The microstructure consists of equiaxed grains of α and β phases, the volume fraction of which is almost equal following the standard anneal. The stabilized β remains untransformed after very long periods in the temperature range up to 375 °C (700 °F) due to the sluggish eutectoid reaction. The eutectoid temperature is 675 °C (1248 °F). It should be noted that Ti-8Mn behaves as a superplastic material at temperatures below the β transus in the α-β region. (J.C. Williams et al., The Effect of Omega Phase on the Mechanical Properties of Titanium Alloys, Metall. Trans., Vol 2, July 1971).

Ti-8Mn: Summary of typical physical properties

Beta transus	800 ± 15 °C	(1475 ± 25 °F)
Melting (liquidus point)	~1565 °C	(~2850 °F)
Density(a)	4.7 g/cm^3	(0.17 lbf/in.^{3})
Electrical resistivity(a)	0.92 μΩ · m	
Magnetic permeability(a)	Nonmagnetic	
Specific heat capacity(a)	494 J/kg · K	(0.118 Btu/lb · °F)
Thermal conductivity(a)	11 W/m · K	(6.3 Btu/ft · h · °F)
Thermal coefficient of linear expansion(b)	8.6 × 10^{-6}/°C	(4.8 × 10^{-6}/°F)

(a) Typical values at room temperature of about 20 to 25 °C (68 to 78 °F). (b) Mean coefficient from room temperature to 100 °C (212 °F)

Elastic Properties

Young's modulus of elasticity (Tension): 113 GPa (16.4 × 10⁶ psi)

Shear modulus of elasticity (Torsion): 48 GPa (7.0 × 10⁶ psi)

Ti-8Mn: Elastic modulus at low temperature

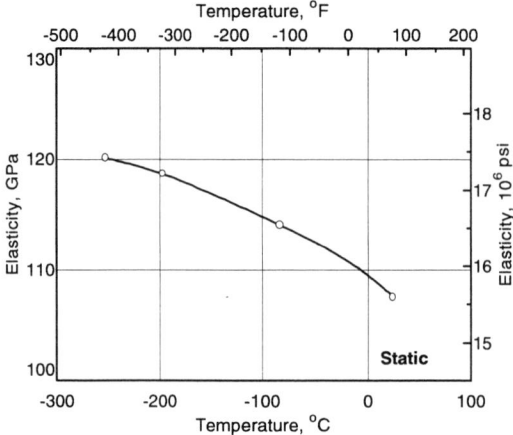

1.6 mm (0.064 in.) annealed sheet tested in the longitudinal direction.
Source: *Aerospace Structural Metals Handbook*, Vol 4, Code 3712, Battelle Columbus Laboratories, 1963

Ti-8Mn: Elastic modulus at elevated temperatures

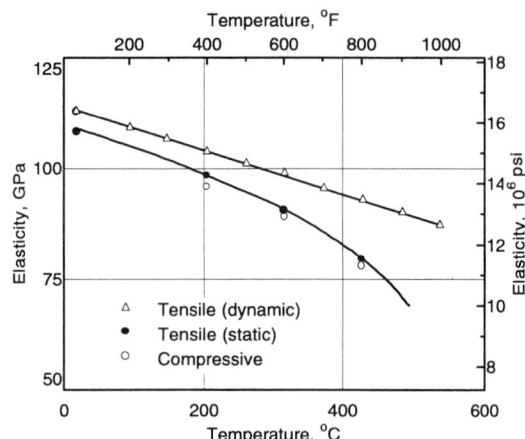

1.7 mm (0.070 in.) sheet after ½ to 100 h of exposure at indicated temperature.
Source: *Aerospace Structural Metals Handbook*, Vol 4, Code 3712, Battelle Columbus Laboratories, 1963

Ti-8Mn: Tangent modulus curves

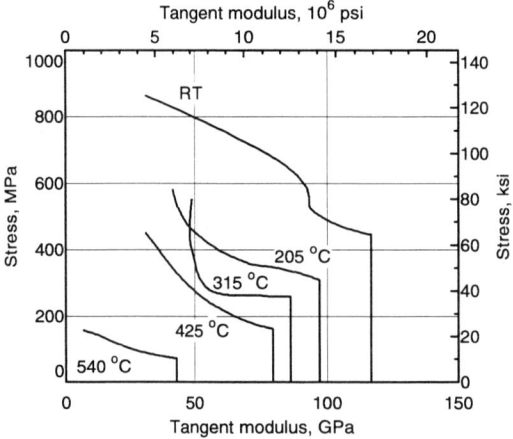

Source: *Aerospace Structural Metals Handbook*, Vol 4, Code 3712, Battelle Columbus Laboratories, 1963

Ti-8Mn: Electrical resistivity

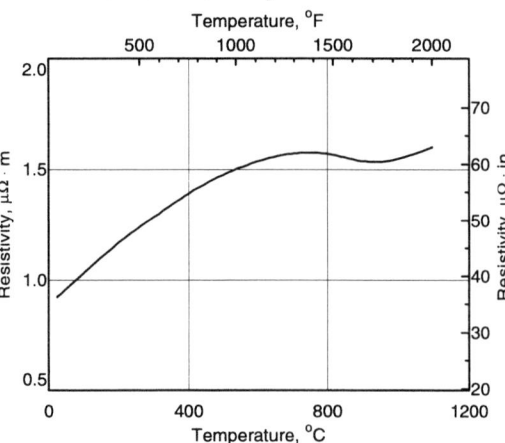

Source: *Aerospace Structural Metals Handbook*, Vol 4, Code 3712, Battelle Columbus Laboratories, 1963, p 2

Electrical Resistivity

Ti-8Mn: Electrical resistivity

Temperature °C	°F	Resistivity, µΩ·cm
38	100	93
93	200	100
150	300	108
205	400	115
260	500	122
315	600	128
370	700	135
425	800	141
480	900	146
540	1000	151

Note: Experimental values of metals in the mill annealed condition. Electrical resistivity measurements were made using the voltage-drop method. Source: H. Dean, W. Wood, and C. Pucks, "The Relationship Between Electrical and Thermal Conductivities of Titanium Alloys," *Trans. TMS/AIME*, Aug 1958, p 520

Ti-8Mn: Electrical resistivity

Temperature °C	°F	Resistivity, µΩ·cm
20	68	90.7
93	200	101.6
205	400	117.3
315	600	131.0
425	800	141.3
540	1000	149
650	1200	154.7
760	1400	156.2
870	1600	153.9
980	1800	155.3
1095	2000	157.0

Source: W. Lepkowski and J. Holladay, "Report on the Physical Properties of Titanium and Titanium Alloys," TML Report No. 73, Titanium Metallurgical Laboratory, Battelle Memorial Institute, Columbus, 1957, p 44

Corrosion Properties

The general corrosion resistance of Ti-8Mn is excellent, being similar to that of commercially pure titanium. Corrosion resistance depends on the formation of a protective oxide layer. General corrosion becomes a concern in reducing acid environments, particularly as acid concentration and temperature increase. In strong and/or hot acids, the protective oxide film can deteriorate and dissolve. Oxidation resistance is similar to that of pure titanium. Oxide films form in air at temperatures of 315 °C (600 °F and above). The oxide film is barely perceptible after exposure at 315 °C (600 °F), but becomes darker and thicker with increasing temperature and time at temperature.

Hydrogen Embrittlement

The accompanying figures show a slight influence of hydrogen content on tensile strength, but a significant degradation in ductility with high hydrogen contents. Creep-rupture strength is also strongly influenced by hydrogen content.

Ti-8Mn: Hydrogen content vs tensile ductility of annealed sheet

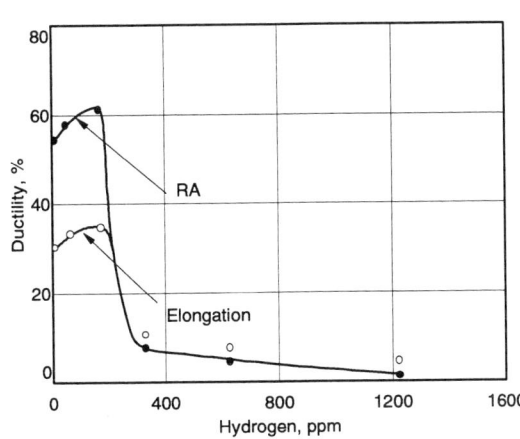
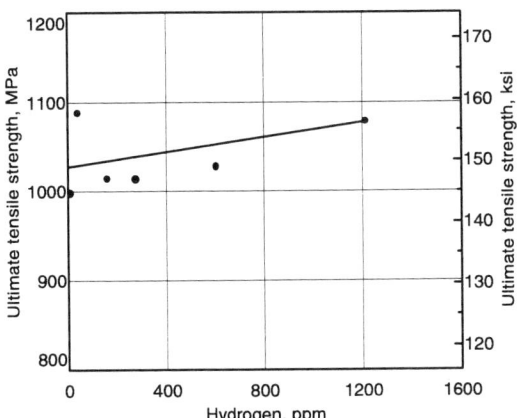

Source: *Aerospace Structural Metals Handbook*, Vol 4, Code 3712, Battelle Columbus Laboratories, 1963

Ti-8Mn: Hydrogen content vs creep-rupture of notched sheet

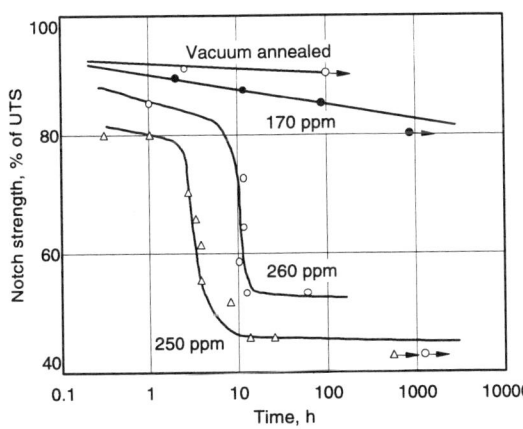

Annealed sheet; K_t = 5.5. RT test
Source: *Aerospace Structural Metals Handbook*, Vol 4, Code 3712, Battelle Columbus Laboratories, 1963

Ti-8Mn: Test temperature vs tensile ductility of sheet with varying hydrogen contents

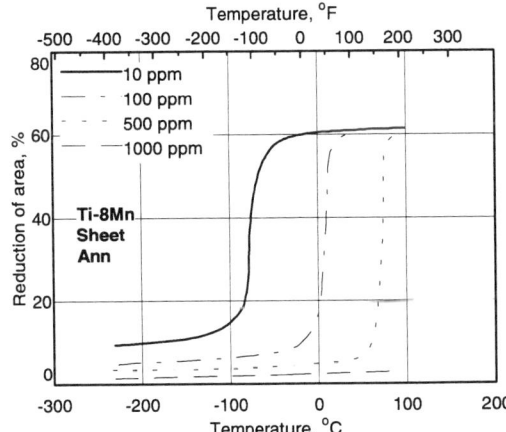

Source: *Aerospace Structural Metals Handbook*, Vol 4, Code 3712, Battelle Columbus Laboratories, 1963

Stress-Corrosion Cracking

The stress-corrosion crack velocity depends on the heat treatment used. Ti-8Mn has slower crack velocities than Ti-13V-11Cr-3Al or Ti-11.5Mo-6Zr-4.5Sn.

758 / Alpha-Beta Alloys

Ti-8Mn: Crack velocity vs stress intensity in 0.6 *M* KCl

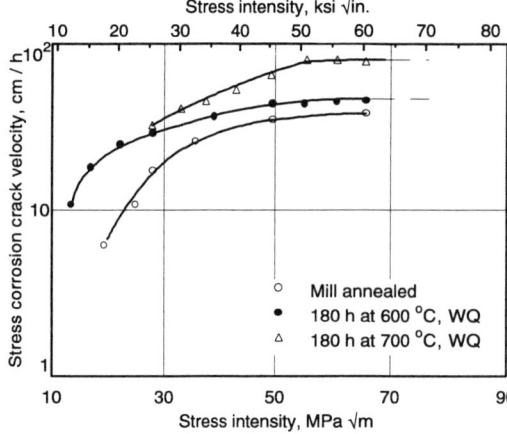

Crack velocity vs stress intensity of SEN specimens in 0.6*M* KCl at −500 mV and 23 °C (74 °F)

Thermal Properties

Ti-8Mn: Specific heat

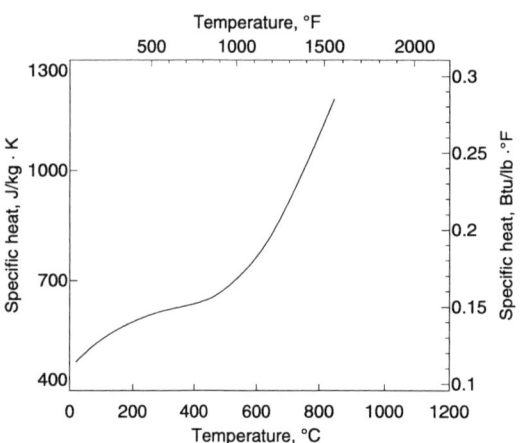

Source: *Aerospace Structural Metals Handbook*, Vol 4, Code 3712, Battelle Columbus Laboratories, 1963

Ti-8Mn: Thermal conductivity

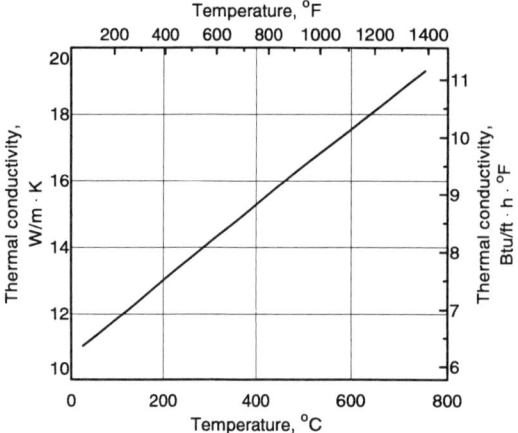

Source: *Aerospace Structural Metals Handbook*, Vol 4, Code 3712, Battelle Columbus Laboratories, 1963

Ti-8Mn: Thermal coefficient of linear expansion

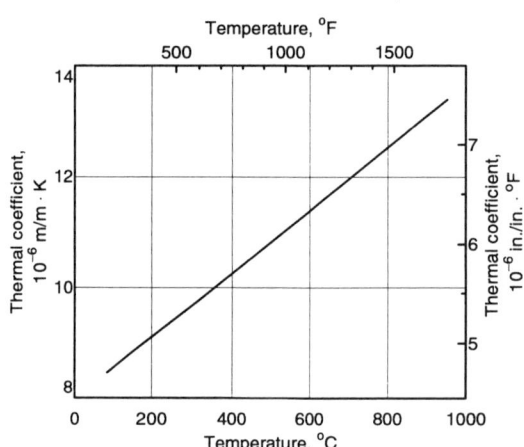

Mean coefficient of thermal expansion from RT to temperature indicated.
Source: *Aerospace Structural Metals Handbook*, Vol 4, Code 3712, Battelle Columbus Laboratories, 1963

Ti-8Mn: Thermal coefficient of linear expansion

Temperature			
°C	°F	m/m · K	10^{-6} in./in. · °F
0-100	32-212	8.6	4.8
0-315	32-600	9.7	5.4
0-540	32-1000	10.8	6.0
0-650	32-1200	11.7	6.5
0-815	32-1500	12.6	7.0

Source: "Basic Design," RMI Titanium

Ti-8Mn: Thermal conductivity

Temperature			
°C	°F	W/m · K	Btu/ft · h · °F
20	68	10.9	6.3
93	200	11.6	6.7
205	400	12.8	7.4
315	600	14.0	8.1
425	800	15.2	8.8
540	1000	16.9	9.8
650	1200	17.8	10.3
760	1400	19.5	11.3

Source: W. Lepkowski and J. Holladay, "Report on the Physical Properties of Titanium and Titanium Alloys," TML Report No. 73, Titanium Metallurgical Laboratory, Battelle Memorial Institute, Columbus, 1957, p 25

Mechanical Properties

Ti-8Mn: Minimum mechanical properties of annealed sheet per AMS 4908

Ultimate tensile strength, MPa (ksi)	827 (120)
Tensile yield strength, MPa (ksi)	758 (110)
Elongation, %	10

Ti-8Mn: Typical room-temperature properties

	Nominal	Minimum
Ultimate tensile strength, MPa (ksi)	1000 (145)	827 (120)
Tensile yield strength, MPa (ksi)	930 (135)	758 (110)
Elongation in 4D or 50 mm (2 in.), %	15	...
Reduction of area, %	30	...
Bend radius, T	3	...
Hardness (annealed), HRC	33-36	...

Ti-8Mn: Tensile properties

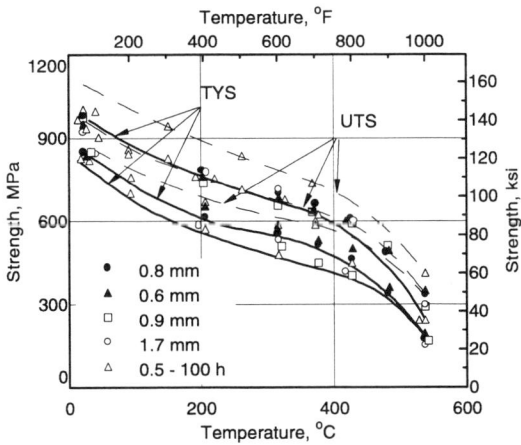

Effect of test temperature on tensile properties of annealed sheet. Source: *Aerospace Structural Metals Handbook*, Vol 4, Code 3712, Battelle Columbus Laboratories, 1963

Ti-8Mn: Compressive yield strength

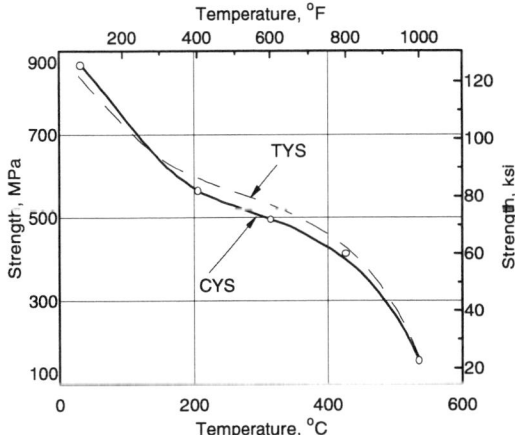

Effect of test temperature on compressive yield strength of annealed 1.7 mm (0.070 in.) sheet after 0.5 to 100 h of exposure. Source: *Aerospace Structural Metals Handbook*, Vol 4, Code 3712, Battelle Columbus Laboratories, 1963

Bearing and Shear Strength

Ti-8Mn: Bearing strength

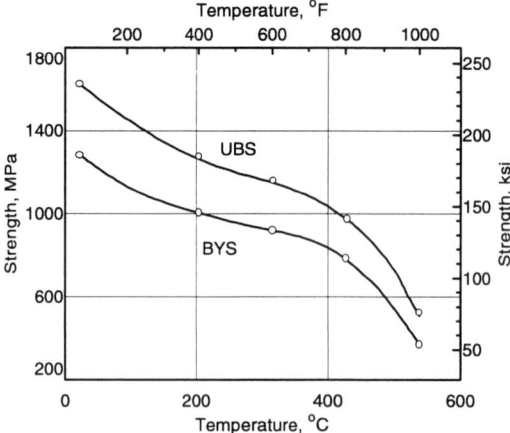

Effect of test temperature on bearing properties of annealed 1.7 mm (0.070 in.) sheet, $e/D = 2.0$, after 0.5 to 100 h of exposure.
Source: *Aerospace Structural Metals Handbook*, Vol 4, Code 3712, Battelle Columbus Laboratories, 1963

Ti-8Mn: Shear strength

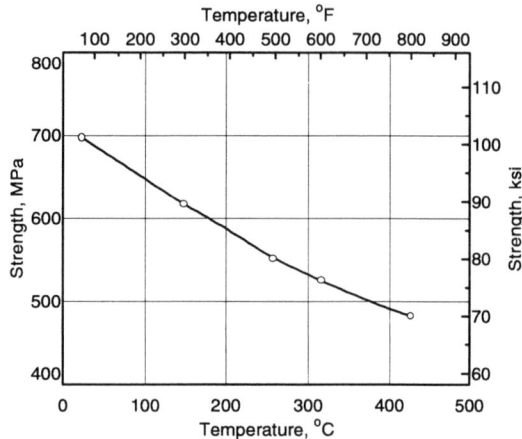

Effect of test temperature on shear strength of annealed 4.7 mm ($^3/_{16}$ in.) sheet.
Source: *Aerospace Structural Metals Handbook*, Vol 4, Code 3712, Battelle Columbus Laboratories, 1963

Creep

Ti-8Mn: Creep at 370 and 425 °C (700 and 800 °F)

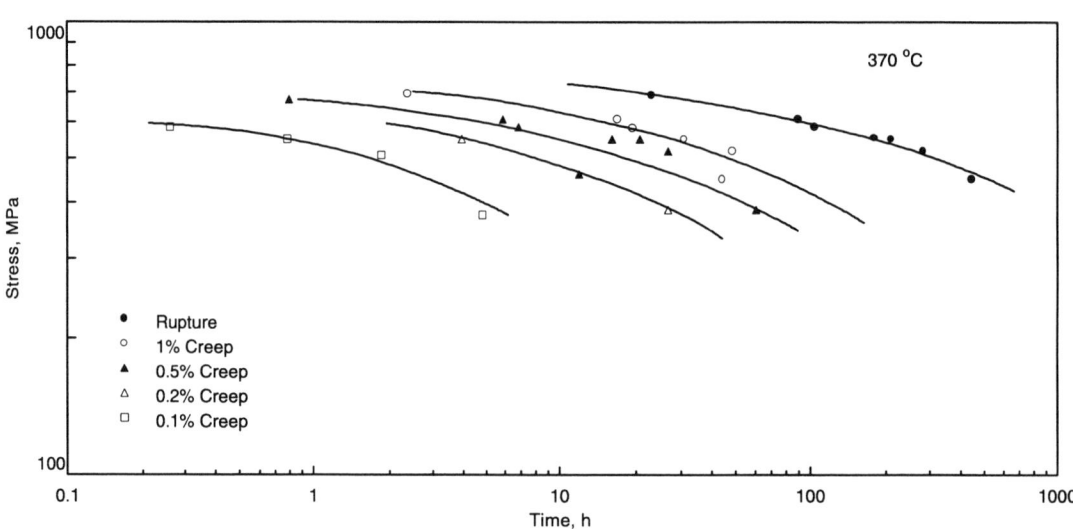

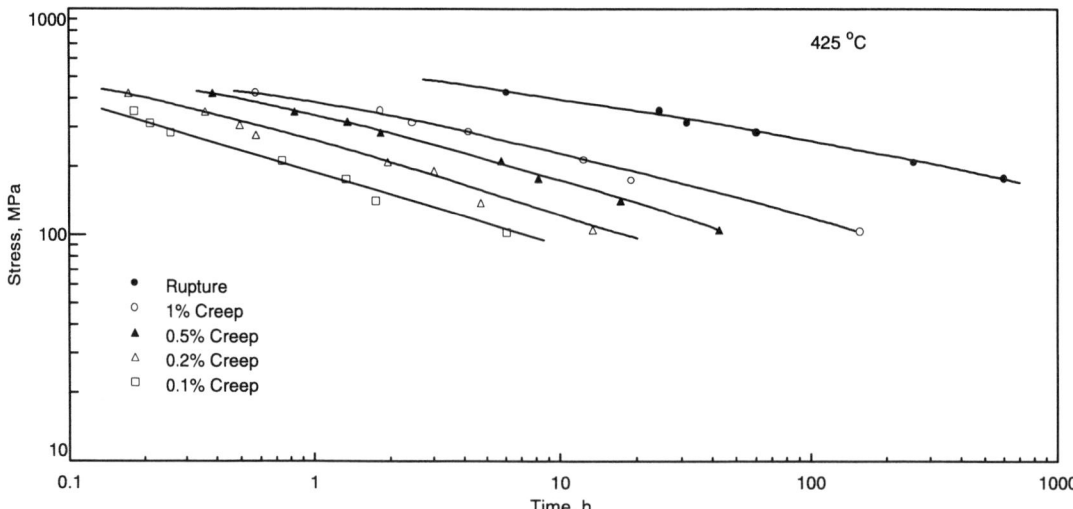

Creep and creep-rupture curves for annealed sheet.

Fatigue

Ti-8Mn: Cyclic stress-strain

Room-temperature test, material forged, extruded, swaged to 20.8 mm (0.8 in.) diam, heat treated 700 °C, 6 h, water quenched

Stress amplitude		Strain amplitude, %	Cycles to failure
MPa	ksi		
1050.0	152.3	2.0000	237.6
1012.5	146.8	1.5000	412.5
867.5	125.8	1.0000	670.5
940.0	136.3	1.2000	757.0

Total strain control, $R = -1$, electropolished surface. Source: Y. Saleh and H. Margolin, *Metall. Trans. A*, Vol 11, 1980, p 1295-1302

Ti-8Mn: Cyclic stress-strain

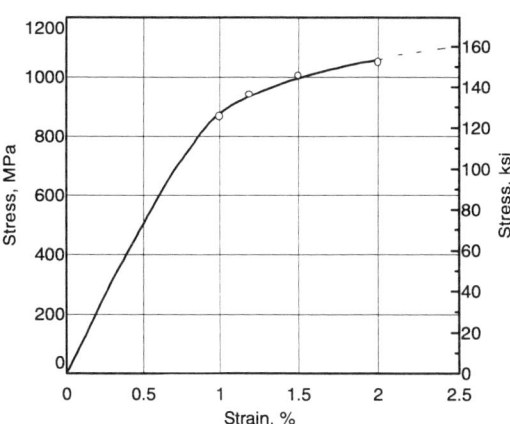

Ti-8Mn: Tension-tension fatigue of annealed sheet

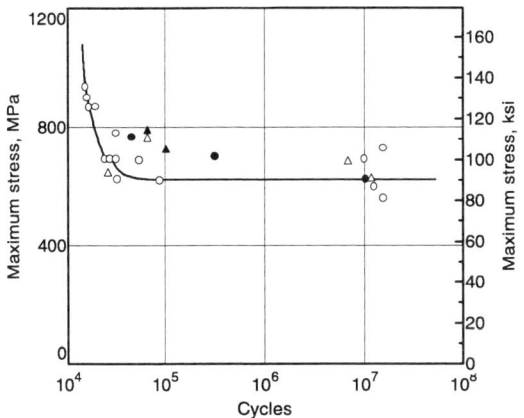

$S_{min} = 0.25\ S_{max}$.
Source: "An Evaluation of the Fatigue Properties of Titanium and Titanium Alloys," Titanium Metallurgical Laboratory, Battelle Memorial Institute

Ti-8Mn: Crack propagation

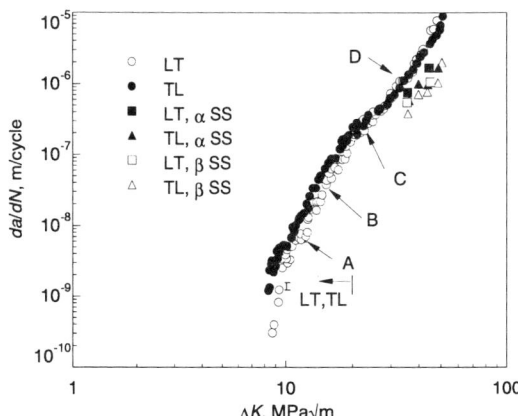

Specimens were tested in the LT and TL directions. Arrows indicates end of stage I. Letters indicate ΔK values where SEM photomicrographs were taken. SS refers to crack propagation rates determined from striation spacings.
Source: H. Margolin *et al.*, The Role of Alpha and Beta in Fatigue Crack Propagation of Ti-Mn Alloys, *Metall. Trans. A*, Vol 15, Jan 1984

Rotating Beam

Ti-8Mn: Rotating beam reversed fatigue tests

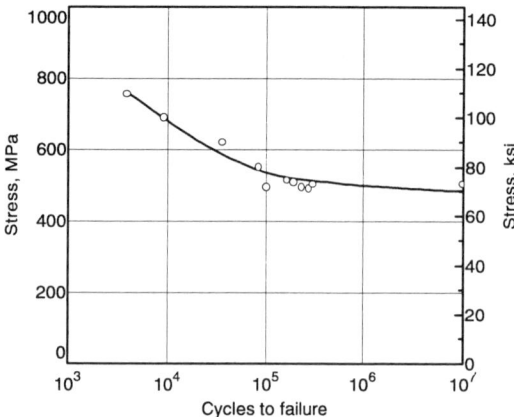

Unnotched 15.8 mm (⁵⁄₈ in.) diam polished bar tested at room temperature and 10,000 cycles/min. UTS, 917 MPa (133 ksi); TYS, 841 MPa (122 ksi); 56% RA; 42-44 HRC.
Source: "An Evaluation of the Fatigue Properties of Titanium and Titanium Alloys," Titanium Metallurgical Laboratory, Battelle Memorial Institute

Ti-8Mn: Rotating beam fatigue with 19 ppm H

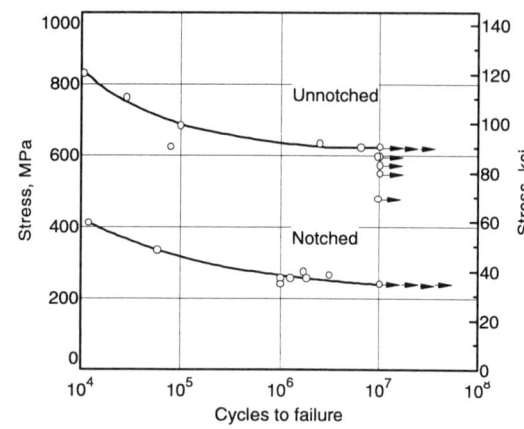

15.8 mm (⁵⁄₈ in.) diam polished bar tested at room temperature and 10,000 cycles/min. Notched specimens: V, $r = 0.1$ mm (0.005 in.), $K_t = 3.0$. Heat treatment: 19 ppm H, vacuum annealed 6 h at 820 °C (1510 °F). UTS, 930 MPa (135 ksi); TYS, 868 MPa (126 ksi); 33% El; 59% RA; 304 HV (10 g).
Source: "An Evaluation of the Fatigue Properties of Titanium and Titanium Alloys," Titanium Metallurgical Laboratory, Battelle Memorial Institute

Ti-8Mn: Rotating beam fatigue with 368 ppm H

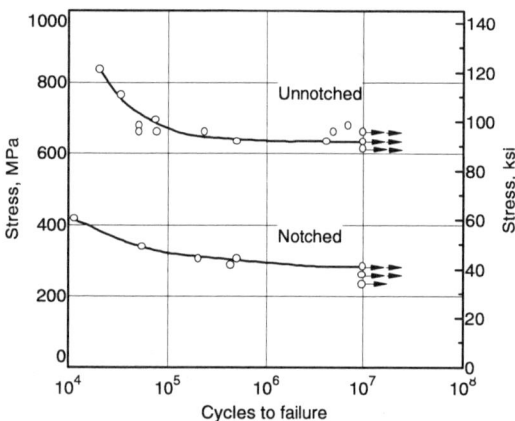

15.8 mm (⁵⁄₈ in.) diam polished bar tested at room temperature and 10,000 cycles/min. Notched specimens: V, $r = 0.1$ mm (0.005 in.), $K_t = 3.0$. Heat treatment: 368 ppm H, hydrogenated at 21 h at 820 °C (1510 °F). UTS, 945 MPa (137 ksi); TYS, 890 MPa (129 ksi); 16% El; 16% RA; 302 HV.
Source: "An Evaluation of the Fatigue Properties of Titanium and Titanium Alloys," Titanium Metallurgical Laboratory, Battelle Memorial Institute

Ti-8Mn: Effect of temperature on unnotched rotating beam fatigue

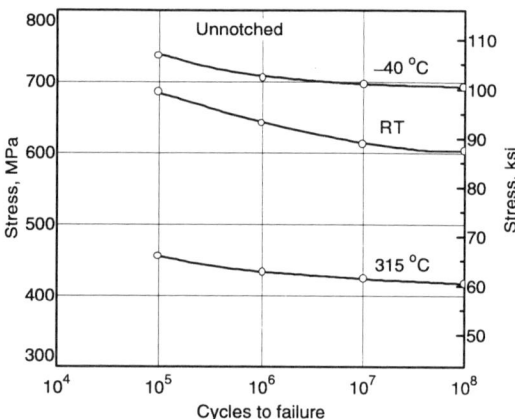

Machine finished, hand polished annealed rod tested at 5000 cycles/min at room temperature and 4000 cycles at –40 °C (–40 °F) and 315 °C (600 °F). Room temperature UTS, 958 MPa (139 ksi); TYS, 868 MPa (126 ksi); El, 16%.
Source: "An Evaluation of the Fatigue Properties of Titanium and Titanium Alloys," Titanium Metallurgical Laboratory, Battelle Memorial Institute

Plastic Deformation

Ti-8Mn: Stress-strain curves for α and β phases

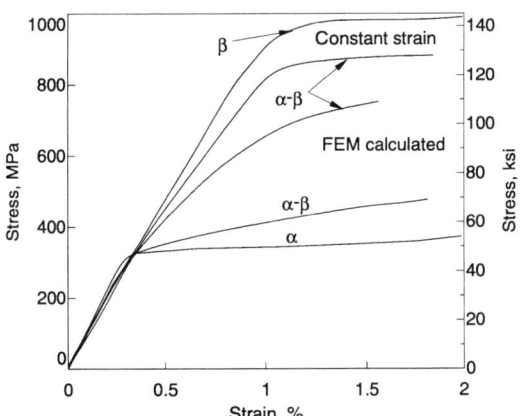

Source: H. Margolin et al., Calculations of Stress-Strain Curves and Stress and Strain Distribution for an Alpha-Beta Ti-8Mn Alloy, Mater. Sci. Eng., Vol 34, 1978, p 203-211

Ti-8Mn: Stress-strain comparison

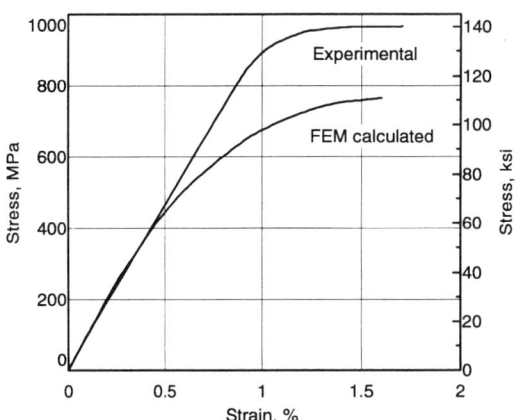

Source: H. Margolin et al., Calculations of Stress-Strain Curves and Stress and Strain Distribution for an Alpha-Beta Ti-8Mn Alloy, Mater. Sci. Eng., Vol 34, 1978, p 203-211

Beta and Near-Beta Alloys

Ti-11.5Mo-6Zr-4.5Sn

Common Name: Beta III
UNS No. R58030

Beta III was developed in the 1960s by Crucible Steel. This alloy was intended to supplement Ti-13V-11Cr-3Al. It has excellent cold workability, heat treatability and mechanical properties, but it is very difficult to melt without molybdenum segregation. Crucible stopped making Beta III when they decreased their participation in the titanium market.

Chemistry. The chemistry balance ultimately selected for Beta III (11.5 Mo, 6 Zr, 4.5 Sn wt%) is a solute-rich composition developed by a semi-empirical balancing of desired properties. Molybdenum is a strong beta stabilizing element completely soluble in beta titanium at elevated temperatures, and the nominal composition of Beta III contains enough of this element by itself to stabilize the beta phase to room temperature. Zirconium and tin, often called neutral stabilizing additions to titanium, augment the beta phase stabilization in the quantities used in the Beta III alloy. Both zirconium and tin strengthen the alpha and beta phases of titanium and are soluble in both phases. Molybdenum, zirconium, and tin were combined in the Beta III formulation in quantities which produce very sluggish beta-phase reaction kinetics.

At the same time, the amount of beta stabilizers was limited by cost and density considerations. Also, the uniformity of tensile ductility in the solution treated condition decreased with a more highly stabilized beta phase.

Density. 5.06 g/cm^3 (0.183 lb/in.^{3})

Product Forms. Limited availability at present in all mill product forms.

Applications. Aircraft fasteners, especially rivets, and sheet metal parts where cold formability and strength potential can be used to greatest advantage. Commercial applications have included springs and orthodontic appliances. Possible use in plate and forging applications where high strength, deep hardenability, and resistance to stress corrosion are required and somewhat lower ductility is acceptable.

Product Condition. Beta III is solution treated above and below the beta transus depending on the desired properties. Brief solution treatment above the transus is sometimes used when maximum cold formability or deep hardenability is sought, but from the mill the Beta III alloy is usually solution treated slightly below the transus temperature. Solution treatment of worked material below or at the beta transus preserves a high dislocation density, which in turn results in a fine alpha dispersion upon subsequent aging. This condition generally gives the best combination of strength and ductility. The microstructure of Beta III is equiaxed beta when solution annealed above the beta transus, while a mixture of equiaxed alpha in a beta matrix is present when solution annealing is performed in the α + β phase field.

Ti-11.5Mo-6Zr-4.5Sn (Beta III): Specifications and compositions

Specification	Designation	Description	C	Fe	H	Mo	N	O	Sn	Zr	Other
UNS	R58030		0.1	0.35	0.02	10-13	0.05	0.18	3.75-5.25	4.5-7.5	
Spain											
UNE 38-730	L-7702	Sh Str Bar Frg Tube HT	0.1	0.35	0.02	10-13	0.05	0.18	3.75-5.25	4.5-7.5	OT 0.4; bal Ti
USA											
AMS 4980B		Bar Wir SHT	0.1	0.35	0.015	10-13	0.05	0.18	3.75-5.25	4.5-7.5	OT 0.4; Y 0.005; bal Ti
AMS 4980B		Bar Wir STA	0.1	0.35	0.015	10-13	0.05	0.18	3.75-5.25	4.5-7.5	OT 0.4; Y 0.005; bal Ti
ASTM B265	Grade 10	Sh Plt Str ST	0.1	0.35	0.02	10-13	0.05	0.18	3.75-5.25	4.5-7.5	OT 0.4; bal Ti
ASTM B337	Grade 10	Pip ST	0.1	0.35	0.02	10-13	0.05	0.18	3.75-5.25	4.5-7.5	OT 0.4; bal Ti
ASTM B338	Grade 10	Tube Heat Ex/Con SHT	0.1	0.35	0.02	10-13	0.05	0.18	3.75-5.25	4.5-7	OT 0.4; bal Ti
ASTM B348(10)-87		Bar	0.1 max	0.35 max	0.02 max	10-13	0.05 max	0.18 max	3.75-5.25	4.5-7.5	OT 0.4 max; OE 0.1 max; bal Ti
ASTM B348(10)-87		Bil	0.1 max	0.35 max	0.015 max	10-13	0.05 max	0.18 max	3.75-5.25	4.5-7.5	OT 0.4 max; OE 0.1 max; bal Ti
MIL F-83142A	Comp 13	Frg Ann	0.1	0.35	0.02	10-13	0.05	0.18	3.75-5.25	4.5-7.5	OT 0.4; bal Ti
MIL F-83142A	Comp 13	Frg HT	0.1	0.35	0.02	10-13	0.05	0.18	3.75-5.25	4.5-7.5	OT 0.4; bal Ti
MIL T-9046J	Code B-2	ST	0.1	0.35	0.02	10-13	0.05	0.18	3.75-5.25	4.5-7.5	OT 0.4; bal Ti
MIL T-9047G	Ti-4.5Sn-6Zr-11.5Mo	Bar Bil SHT	0.1	0.35	0.02	10-13	0.05	0.18	3.75-5.25	4.5-7.5	OT 0.4; Y 0.005; bal Ti

(a) Maximum unless a range is specified

Ti-11.5Mo-6Zr-4.5Sn (Beta III): Commercial compositions

Specification	Designation	Description	Composition, wt% (nominal content only)								
			C	Fe	H	Mo	N	O	Sn	Zr	Other
France											
Ugine	TD12ZrE	Bar Sh Quen	...	...	...	11.5	...	...	4.5	6	bal Ti
Ugine	TD12ZrE	Bar Sh Quen Aged	...	...	...	11.5	...	...	4.5	6	bal Ti
USA											
Crucible	Beta III	Ann	...	...	...	11.5	...	...	4.5	6	bal Ti
Crucible	Beta III	HT	...	...	...	11.5	...	...	4.5	6	bal Ti
Oremet	Ti Beta 3										

Phases and Structures

Crystal Structure

Beta III solution annealed above the transus is completely β, which becomes enriched with molybdenum in some regions during aging through alloy partitioning as alpha-phase precipitation occurs. As with other β alloys this enrichment affects the lattice parameter and makes the β phase more stable at lower temperatures. Consequently, β decomposition can be sluggish in Beta III, as in other β alloys.

Beta Transus. About 760 °C (1400 °F) at nominal molybdenum concentrations.

Beta III: Alloy element partitioning data

Heat treatment	Phase composition, wt%								Vol %, α phase	Lattice parameter, β phase, nm
	Mo		Zr		Sn		Ti			
	β	α	β	α	β	α	β	α		
900 °C (1650 °F)/25 h	11.37(a)	...	5.19(a)	...	4.13(a)	...	80.31(a)	...	0	0.3284
705 °C (1300 °F)/25 h + 760 °C (1400 °F)/25 h	12.7	2.3	5.2	5.0	4.1	4.6	79.6	88.5	12.5	0.3282
705 °C (1300 °F)/25 h + 705 °C (1300 °F)/25 h	14.8	1.9	5.3	4.8	3.9	4.7	77.0	89.3	26.0	0.3279
705 °C (1300 °F)/25 h + 650 °C (1200 °F)/100 h	14.9	1.3	5.7	4.6	3.8	4.9	74.8	91.2	35.0	0.3274
705 °C (1300 °F)/25 h + 595 °C (1100 °F)/100 h	17.2	1.2	5.7	4.6	3.6	4.7	72.0	91.2	44.0	0.3268
705 °C (1300 °F)/25 h + 540 °C (1000 °F)/100 h	19.4	1.4	5.9	4.5	3.5	4.8	69.6	90.4	53.0	0.3264

(a) By chemical analysis. Source: F.H. Froes, J.M. Capenos, and M.G.H. Wells, Alloy Partitioning in Beta III and Effect on Aging Characteristics, *Titanium Science and Technology*, R.I. Jaffee, and H.M. Burte, Ed., TMS/AIME, 1973, p 1621-1633

Transformation Products

Beta III forms athermal ω during quenching and isothermal ω during aging at low temperatures. It is generally recognized that ω phase formation leads to ductility losses, although proper control of ω phase volume fraction can lead to high strength and reasonable ductility.

Detailed microstructural work (F.H. Froes *et al.*, *Metall. Trans. A*, Vol 11, 1980, p 21-31) has shown the following:

- Supertransus solution treatment should be avoided, because it removes the heterogeneous nucleation sites, which accelerate the rate of formation of α-phase relative to that of ω phase

- The ω phase forms on aging at temperatures up to 480 °C (900 °F), although at 480 °C, it has limited time of stability

- In recovered and incompletely recrystallized material, heterogeneous nucleation of α phase occurs at dislocations and subboundaries, whereas ω phase is uniformly nucleated in the matrix

It was also found that the hardening due to ω phase formation exhibits very little overaging response as long as ω phase is present. The incubation time for ω phase formation is substantially lengthened in directly aged samples compared to quenched and aged samples. There is a critical α phase particle size and spacing above which only very limited hardening is observed. Increasing oxygen content accelerates the kinetics of α phase formation and retards those of ω phase formation. Cold working prior to aging in the ω phase range accelerates formation of ω phase, possibly from the

Beta III: Ti-Mo section of the phase diagram

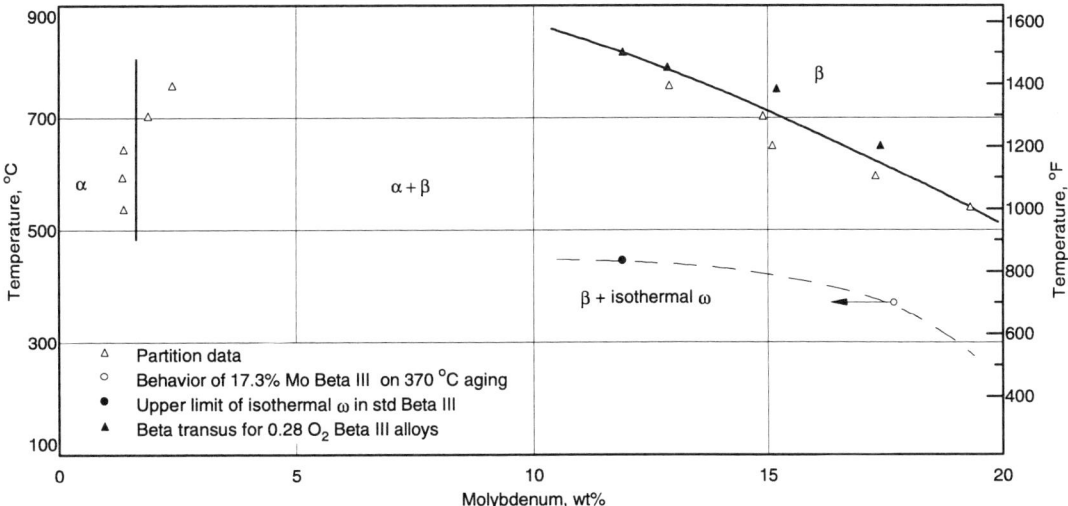

Source: F.H. Froes, J.M. Capenos, and M.G.H. Wells, Alloy Partitioning in Beta III and Effect on Aging Characteristics, *Titanium Science and Technology*, R.I. Jaffee and H.M. Burte, TMS/AIME, 1973, p 1621-1633

Beta III: Variation of β lattice parameter with Mo content

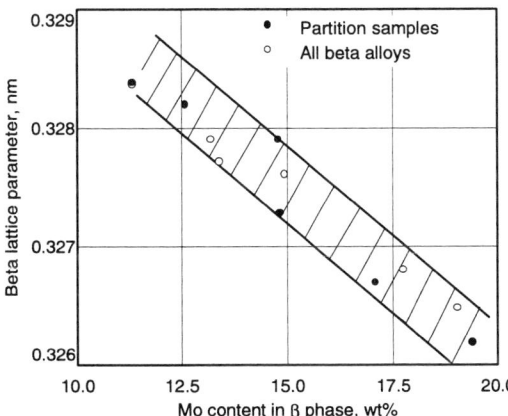

Source: F.H. Froes, J.M. Capenos, and M.G.H. Wells, Alloy Partitioning in Beta III and Effect on Aging Characteristics, *Titanium Science and Technology*, R.I. Jaffee and H.M. Burte, Ed., TMS/AIME, 1973, p 1621-1633

Beta III: ITT diagram

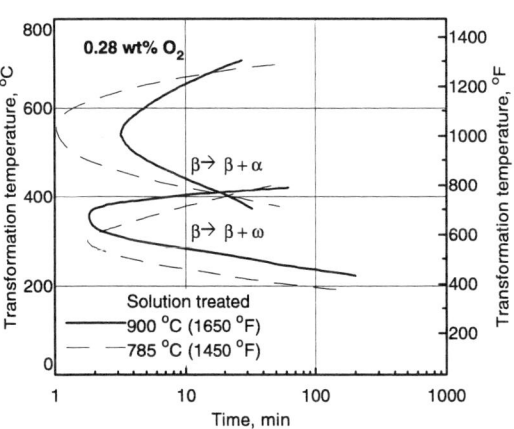

Solution annealed above and below the β transus. Actual temperatures are indicated on the diagram.
Source: *Metall. Trans. A*, Vol 11, Dec 1980, p 26

Beta III: ITT diagram

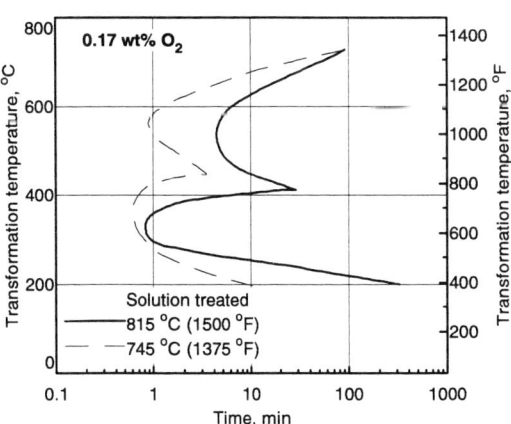

Solution annealed above and below the β transus. Actual temperatures are indicated on the diagram.
Source: *Metall. Trans. A*, Vol 11, Dec 1980, p 26

creation of vacancies during cold working.

Duplex (low-high) aging has no advantage with regard to strengthening potential in subtransus solution treated material. In supertransus solution treated material, the potential advantage of duplex aging is much greater. Duplex aging at low and then at high temperatures results in very fine α phase particles, possibly due to an *in situ* ω to α transformation. Further studies on the effect of quench history and varying amounts of cold work on the kinetic balance between ω and α phase formation are warranted, because they suggest additional ways of controlling ω phase formation.

Physical Properties

Beta III: Summary of typical physical properties

Beta transus	760 °C (1400 °F)
Melting (liquidus) point	1690 °C (3075 °F)
Density(a)	5.06 g/cm^3 (0.183 lb/in.^{3})
Electrical resistivity(a)	1.56 μΩ · m
Magnetic permeability	1.00 (nonmagnetic)
Thermal conductivity(a)	6.275 W/m · K (3.625 Btu/ft · h · °F)
Thermal coefficient of linear expansion(b)	7.6 × 10^{-6}/°C (4.2 × 10^{-6}/°F)

(a) Typical values at room temperature of about 20 to 25 °C (68 to 78 °F). (b) Mean coefficient from room temperature to 100 °C (212 °F)

Elastic Properties

The elastic (Young's) modulus of Beta III alloy has been reported to be as low as 70 GPa (1 × 10^7 psi) for solution annealed and quenched tensile samples. Values up to 110 GPa (16 × 10^6 psi) have been reported for aged specimens. Young's modulus decreases with increasing test temperature to about 98 GPa (14.2 × 10^6 psi) at 200 °C (400 °F), and 83 GPa (12.1 × 10^6 psi) at 425 °C (800 °F). Compressive modulus values range almost 14 GPa (2 × 10^6 psi) higher than tensile modulus values at corresponding temperatures. (R.A. Wood, *Beta Titanium Alloys*, MCIC 72-11, Battelle, 1972, p 125).

Beta III: Room-temperature elastic properties

	Tensile modulus		Compressive modulus(a)		Poisson's
Material/Condition	GPa	10^6 psi	GPa	10^6 psi	ratio
STQ bar(b)					
Unaged	68	9.9	76.5-80	11.1-11.7	...
Aged 8 h at 480 °C (900 °F)	102	14.8	110-111.5	16.0-16.2	...
Aged 8 h at 565 °C (1050 °F)	108	15.7	106	15.4	...
STQ plate(c)					
Unaged	79	11.5	81-83	11.8-12.0	...
Aged 8 h at 510 °C (950 °F)	109	15.8	115-117	16.7-17.0	...
Aged 8 h at 540 °C (1000 °F)	110	15.9	109	15.8	...
STQ sheet(d)					
Unaged	...	...	...	...	0.359-0.368
Aged 8 h at 480 °C (900 °F)	...	...	...	...	0.312-0.313
Aged 8 h at 590 °C (1100 °F)	...	...	...	...	0.325-0.335

(a) Compression specimens were 13 mm (0.5 in.) diam by 32 mm (1.25 in.) in height. (b) 13 mm (0.5 in.) bar solution treated at 770 °C (1420 °F) and water quenched. (c) 13 mm (0.5 in.) plate hot rolled from 28 mm (1.1 in.), solution treated at 730 °C (1350 °F) for 15 min, 15-s delay, water quenched. (d) 1.6 mm (0.063 in.) sheet, solution treated at 730 °C (1350 °F). Source: *Aerospace Structural Metals Handbook*

Beta III: Compressive elastic modulus of plate

Aging temperature		Modulus of elasticity		Compressive modulus(a)	
°C	°F	GPa	10^6 psi	GPa	10^6 psi
Unaged		79	11.5	82	12.0
				81	11.8
510	950	108	15.8	115	16.7
				117	17.0
540	1000	109	15.9	108	15.8
				108	15.8

Note: Specimens were 13 mm (0.5 in.) plate hot rolled to 30 mm (1.1 in.) to 13 mm (0.5 in.) + 730 °C (1350 °F), 15 min, 15-s delay, WQ + age, 8 h. (a) Compression specimens were 13 mm (0.5 in.) diameter by 32 mm (1.250 in.) height. Source: *Aerospace Structural Metals Handbook*, Vol 4, Code 3722, Battelle Columbus Laboratories, 1972

Beta III: Poisson's ratio for sheet

Condition	Poisson's ratio, μ
785 °C (1450 °F), WQ	0.382
730 °C (1350 °F), WQ	0.368, 0.359
730 °C (1350 °F), WQ + 480 °C (900 °F), 8 h, AC	0.312, 0.313
730 °C (1350 °F), WQ + 595 °C (1100 °F), 8 h, AC	0.335, 0.325

Note: All specimens were longitudinal sheet 1.6 mm (0.063 in.). Source: *Aerospace Structural Metals Handbook*, Vol 4, Code 3722, Battelle Columbus Laboratories, 1972

Beta III: Tensile modulus of bar vs temperature

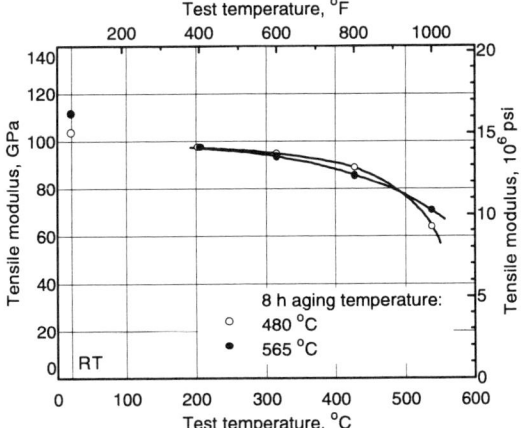

13 mm (0.5 in) bar solution treated at 770 °C (1420 °F) for 15 min, WQ, and aged as indicated.
Source: *Aerospace Structural Metals Handbook*, Vol 4, Code 3722, Battelle Columbus Laboratories, 1972

Beta III: Compressive modulus of sheet vs temperature

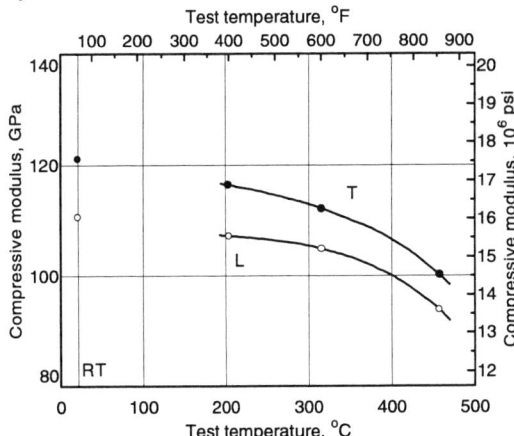

1.6 mm (0.063 in.) sheet solution treated and aged 8 h at 510 °C (950 °F).
Source: *Aerospace Structural Metals Handbook*, Vol 4, Code 3722, Battelle Columbus Laboratories, 1972

Beta III: Tensile modulus of sheet vs temperature

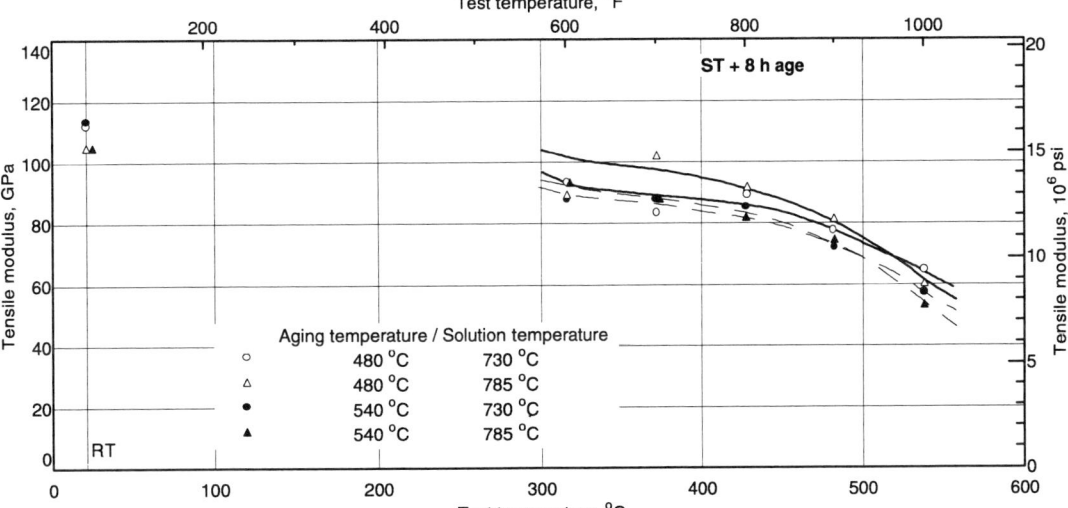

1.7 mm (0.067 in.) sheet, 46% cold reduced, plus solution treated (ST) for 5 min and aged as indicated.
Source: *Aerospace Structural Metals Handbook*, Vol 4, Code 3722, Battelle Columbus Laboratories, 1972

Beta III: Compressive tangent modulus of sheet

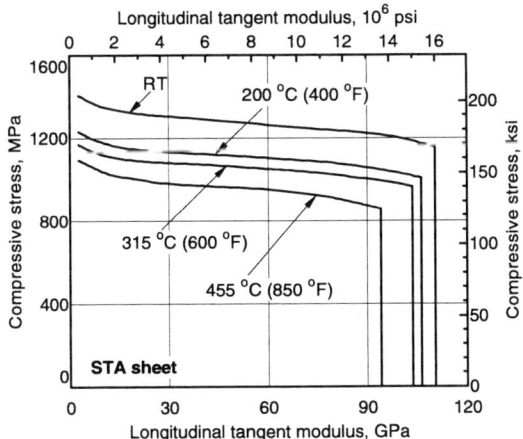

Solution treated 1.6 mm (0.063 in.) sheet aged at 510 °C (950 °F) for 8 h, AC.
Source: *Aerospace Structural Metals Handbook*, Vol 4, Code 3722, Battelle Columbus Laboratories, 1972

Beta III: Transverse compressive tangent modulus

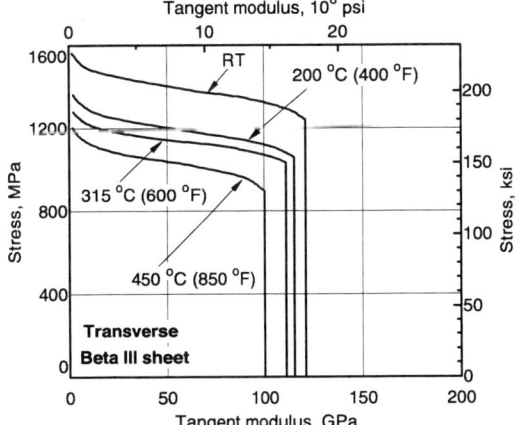

Source: R.A. Wood and R.J. Favor, *Titanium Alloys Handbook*, MCIC-HB-02, Battelle, 1972

772 / Beta and Near-Beta Alloys

Electrical Properties

Room-temperature resistivity is about 1.5 $\mu\Omega \cdot m$.

Chemical/Corrosion Properties

The high molybdenum content of Beta III alloy imparts excellent corrosion resistance under reducing conditions and should produce excellent crevice corrosion resistance (because crevice corrosion is generally associated with acidification from oxidant depletion in the crevice region). At room temperature, the general corrosion of Beta III alloy in HCl of concentrations up to about 20% is acceptable, whereas for unalloyed titanium, the HCl concentration limit is about 5%. A similar advantage was observed in boiling H_2SO_4 (see table).

The enhanced corrosion resistance in reducing environments from molybdenum is achieved, however, at the expense of corrosion resistance in oxidizing conditions. Therefore, pitting and repassivation potential are expected to be lower than most other titanium alloys, although no data are available.

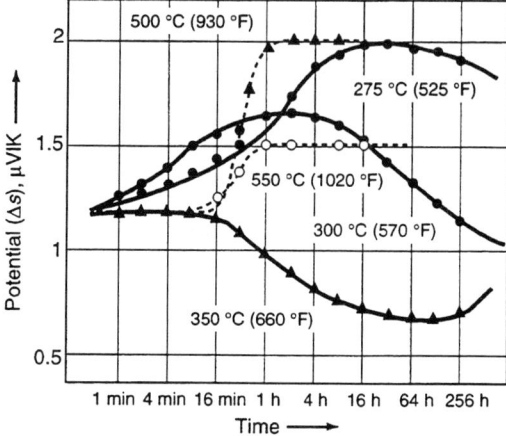

Beta III: Thermoelectric potential vs aging time

Source: R. Borelly and J. Merlin, "Application des Measures de Pouvoir Thermoelectrique a la Characterization Metallurgique des Alliages de Titane," *J. Less-Common Met.*, Vol 69, 1980, p 49

Beta III: Corrosion rates in hydrochloric and sulfuric acids

Medium		Corrosion rate, mils per year, at:			
		RT		Boiling	
Acid	Wt%	A70(a) 500-h test	Beta III(b) 1000-h test	A70(a) 48-h test	Beta III(b) 48-h test
HCl	2	0.034	0.062	297.0	4.28
	3	0.104	0.062	569.0	15.5
	5	7.45	0.078	1393	104.3
	10	26.55	0.164	3155	676.5
	15	...	...	5205	3093
	20	72.0	8.69	...	...
	30	316.5	17.55	...	...
H_2SO_4	1	...	...	456.5	1.80
	3	...	...	1182	15.05
	5	...	...	1830	38.89
	10	...	...	2833	135.1
	15	...	...	4055	241.0

(a) Unalloyed titanium grade A70, mill annealed sheet 2.2 mm (0.087 in.) thick. (b) Beta III titanium alloy, solution treated 760 °C (1400 °F), WQ, sheet, 1.0 mm (0.040 in.) thick. Source: R.A. Wood, *Beta Titanium Alloys*, MCIC 72-11, Battelle, 1972, p 164

Stress-Corrosion Cracking

The high molybdenum content of Beta III is also probably responsible for its relatively good resistance to stress-corrosion cracking in aqueous-chloride and hot-salt environments. Beta alloys solution treated to contain 100% beta stabilized by molybdenum, vanadium, niobium, or tantalum are immune to aqueous stress-corrosion cracking. In the solution treated and aged condition, however, Beta III has exhibited stress-corrosion cracking susceptibility in distilled water (B.F. Brown, "Stress Corrosion Cracking in High Strength Steels and in Titanium and Aluminum Alloys," Naval Research Labs, 1972). Addition of halide ions such as Cl^-, Br^-, and I^- increase susceptibility in Beta III and can induce susceptibility in other alloys that are immune to stress-corrosion cracking in distilled water.

Studies of uncracked bend specimens of Beta III aged sheet (8 h, 510 °C, or 950 °F) in 3.5% NaCl solution showed a total lack of stress-corrosion reaction under stress equal to 80% of the tensile yield stress. (AFML-TR-70-252, Oct 1970). On the other hand, various degrees of immunity or susceptibility have been reported for Beta III materials where precracked specimens were exposed under stress to salt water (see table). In general, the susceptibility of titanium alloys to stress-corrosion cracking in aqueous media is influenced by the type and concentration of species in solution, the pH, temperature, and viscosity of the solution, and the metal potential in the solution. See "Technical Note 2: Corrosion" for general information.

Effect of Potential. In most stress-corrosion cracking susceptible titanium alloys, anodic or cathodic polarization tends to inhibit stress-corrosion cracking and increase K_{Iscc} values (see figure).

Increasing potential also increases cracking velocity in neutral halide solutions, but not in highly acidic solutions. In highly acidic solution, stage II crack velocity becomes independent of applied potential.

Hot-Salt Stress-Corrosion Cracking. Beta III is highly resistant to hot-salt stress-corrosion cracking, and hot-salt cracking only occurs at stresses high enough to cause substantial creep deformation (see figure).

Stress-Corrosion Cracking in Methanol. Titanium is susceptible to stress-corrosion cracking in methanol, but no instances of Beta III susceptibility have been obtained for this compilation. For general information on stress-corrosion cracking in methanol, see "Commercially Pure and Modified Ti" or "Technical Note 2: Corrosion" in this Volume.

Beta III: Fracture toughness in air and 3.5% NaCl solution at 25 °C

Product form	Processing/heat treatment	Yield strength MPa	Yield strength ksi	K_{Ic} MPa√m	K_{Ic} ksi √in.	K_{Iscc} MPa √m	K_{Iscc} ksi √in.
13 mm (0.5 in.) plate	β STA	1034	150	71	65	27	25
Spar forging	1 h at 720 °C (1325 °F), WQ, + 8 h, at 510 °C (950 °F), AC	1172–1275	170–185	65–69	59–63	60–69	55–63
Unspecified	Solution treated 8 h at 595 °C (1100 °F), AC	1034	150	82	75	77	70
Unspecified	Solution treated	689	100	128(a)	117(a)	103	94
25 mm (1 in.) plate	Nonrecrystallized, processed from slab at 1035 °C (1900 °F)	1255	182	>55	>50	58	53
13 mm (0.5 in.) plate	Recrystallized, processed at 925 °C (1700 °F) with 2 reheats	1324	192	58	53	26	24

(a) K_Q. Source: R.A. Wood, *Beta Titanium Alloys*, Battelle, 1972, p 165; and R. Schutz, Stress-Corrosion Cracking of Titanium Alloys, in *Stress-Corrosion Cracking: Material Performance and Evaluation*, ASM International, 1992

Beta III: Effect of potential on aqueous SCC

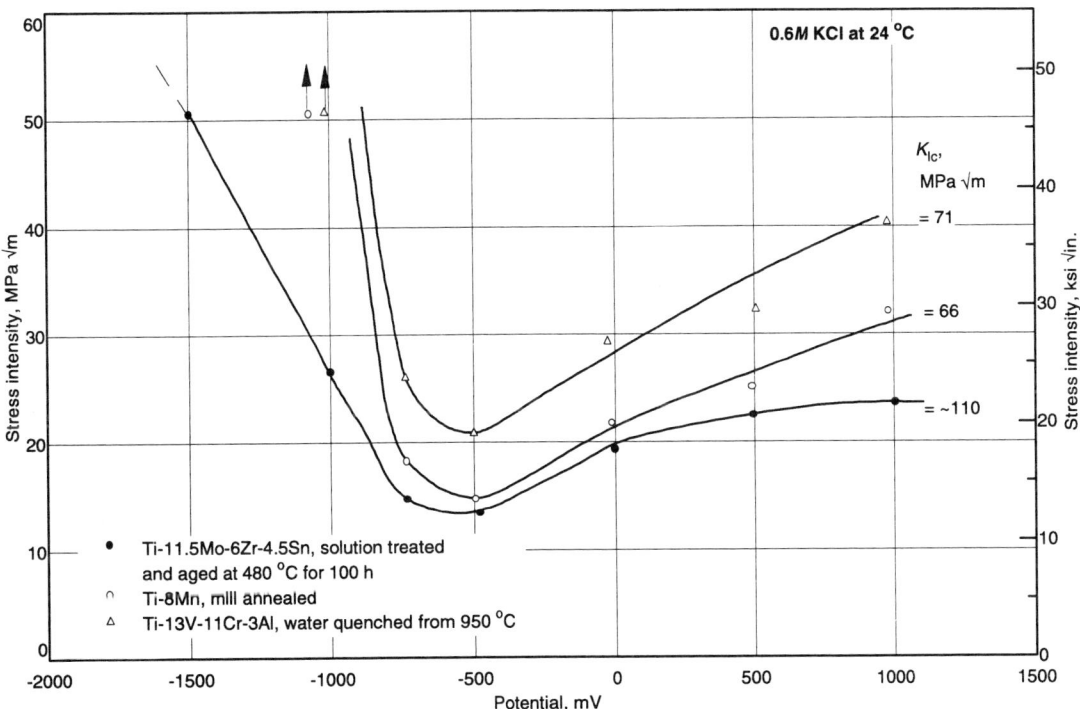

Within a narrow potential region (typically centered around −500 mV SCE), K_{Iscc} reached a minimum value that depends on the alloy and its metallurgical condition.
Source: T.R. Beck, M.J. Blackburn, W.H. Smyrl, and M.O. Speidel, "Stress-Corrosion Cracking of Titanium Alloys: Electrochemical Kinetics, SCC Studies With Ti: 8-1-1, SCC and Polarization Curves in Molten Salts, Liquid Metal Embrittlement, and SCC Studies With Other Titanium Alloys," Quarterly Progress Report 14, Contract NAS 7-489, Boeing Scientific Research Laboratories, Dec 1969

Beta III: Crack velocity vs KCl molarity at 24 °C

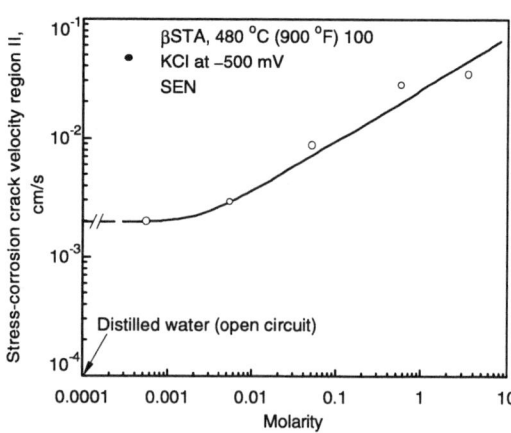

Source: M.J. Blackburn, J.A. Feeny, and T.R. Beck, "Stress-Corrosion Cracking of Titanium Alloys," Boeing Report DI-82-1054, June 1970, p 96

Beta III: Hot-salt SCC threshold for bar

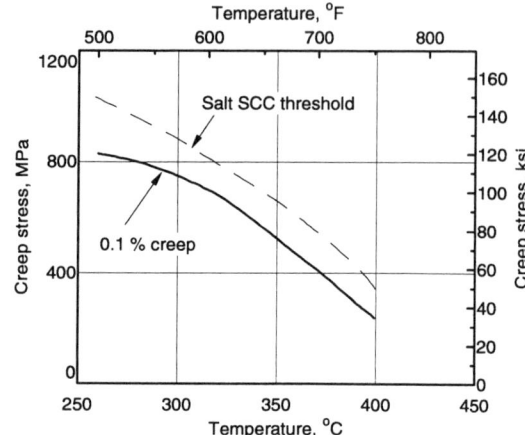

100-h tensile creep.
Source: *J. Metals*, Vol 23, Apr 1971, p 40-47

Thermal Properties

Liquidus. ~1690 °C (3075 °F)
Solidus. ~1573 °C (2863 °F)
Thermal Conductivity. At room temperature, 6.275 W/m · K (0.015 cal/cm · s · °C)

Thermal Expansion

Thermal Coefficient of Linear Expansion. At room temperature, $8 \times 10^{-6}/°C$ ($4.4 \times 10^{-6}/°F$)

Beta III: Thermal coefficient of linear expansion

Temperature		Mean thermal coefficient	
°C	°F	µm/m · K	µin./in. · °F
20-95	68-200	7.6	4.2
20-205	68-400	8.1	4.5
20-315	68-600	8.5	4.7
20-425	68-800	8.7	4.8
20-540	68-1000	8.7	4.8

Note: Specimens were in the STA condition. Source: R.R. Boyer and H.W. Rosenberg, Ed., *Beta Titanium Alloys in the 1980's*, AIME, 1984

Beta III: Thermal expansion

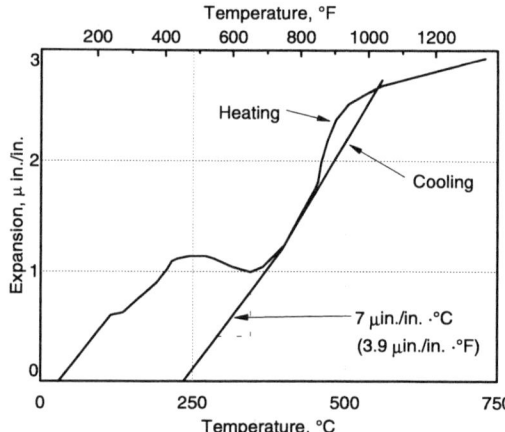

Thermal expansion behavior of Beta III alloy in the 1000 °C (1830 °F) solution annealed and water quenched condition. The irregularities in the heating curve reportedly are related to the formation of ω phase. Reheating of the same specimen after slow cooling results in a heating curve without irregularities like the cooling curve shown. The value of 7 µin./in. · °C (3.9 µin./in. · °F) obtained from this work compares reasonably well with other work.
Source: A.M. Adair and J.A. Roberson, AFML-TR-70-277, 1971

Beta III: Thermal coefficient of linear expansion

Temperature		Mean thermal coefficient for alloys processed at:			
		Aged 8 h, 540 °C (1000 °F)		730 °C (1350 °F) solution annealed	
°C	°F	µm/m · K	µin./in. · °F	µm/m · K	µin./in. · °F
21-93	70-200	7.6	4.2	7.6	4.2
21-205	70-400	18.9	4.5	19.3	4.6
21-315	70-600	19.7	4.7	...	...
21-425	70-800	20.1	4.8	...	...
21-540	70-1000	20.1	4.8	...	...

Source: V.C. Peterson et al., AFML-TR-69-171, 1969

Mechanical Properties

Beta III: Minimum tensile properties

Condition	Diameter or thickness mm	Diameter or thickness in.	Ultimate tensile strength MPa	Ultimate tensile strength ksi	Tensile yield strength MPa	Tensile yield strength ksi	Elongation in 50 mm (2 in.) or 4D, %	Reduction of area, %
Annealed	Unspecified(a)		690	100	620	90	...	...
As-quenched bar and wire for fastener stock(b)	<41.3	<1.625	760	110	620	90	15	50
	41.3-75	1.625-3.00	690	100	620	90	15	50
As-quenched bar and wire(c) per AMS 4980	<41.3	<1.625	...	...	...	...	15	50
	41.3-75	1.625-3.00	...	...	...	...	15	50
STA bar and wire per AMS 4980(c)(d)	<41.3	<1.625	1240	180	1205	175	8	22
	41.3-75	1.625-3.00	1240	180	1170	170	4	10
STA fastener stock(b)(e) per AMS 4977	<41.3	<1.625	930	135	895	130	12	40

(a) *Metals Handbook*, Vol 2, 10th ed., p 622. (b) 690 to 730 °C (1275 to 1350 °F) for 15 min, WQ. (c) 15 min at 705 to 785 °C (1300 to 1450 °F), WQ. (d) Aged 8 h at 485 to 505 °C (910 to 940 °F), AC. (e) Aged 2 h minimum at 565 to 595 °C (1050 to 1100 °F), AC

Hardness

Beta III: Effect of aging on Vickers hardness

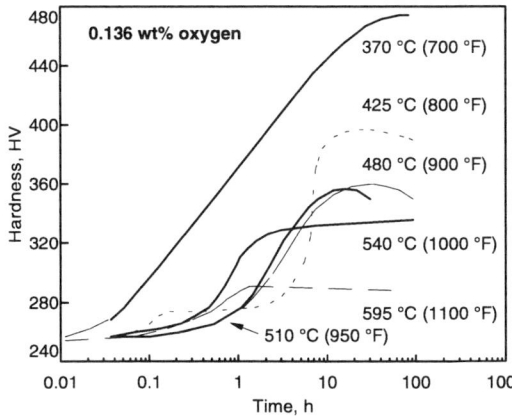

Solution treated at 785 °C (1450 °F) (above the β transus) and aged as indicated. Material was obtained in the form of sheet or plate that was fabricated by hot working and finishing above the β transus, but at a temperature where recovery and recrystallization were relatively sluggish. Extensive warm work was retained in the material in the as-rolled condition.
Source: J.C. Williams, F. Froes, and S. Fujishiro, "Microstructure and Properties of the Alloy Ti-11.5Mo-6Zr-4.5Sn (Beta III)," in *Titanium and Titanium Alloys*, J. Williams and A. Belov, Ed., 1982, p 1421

Beta III: Effect of aging on Rockwell hardness

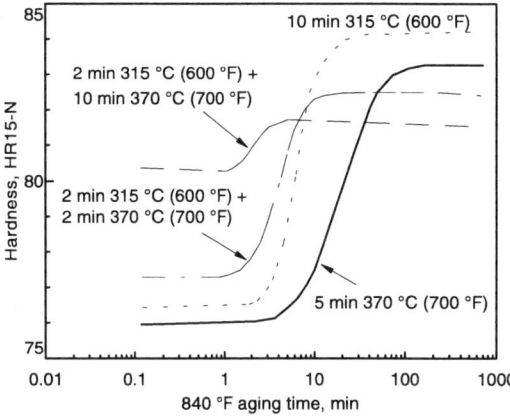

Source: L. Rosales, K. Ono, A. Somer and L.A. Lee, "Microstructures and Mechanical Properties of Thermomechanically Treated High-Strength Beta-Titanium Alloys," North American Rockwell Corporation Report NA-72-232, Feb 1972; reported in *Beta Titanium Alloys*, R. Wood, Ed., MCIC-72-11, 1972, p 118

Beta III: Effect of oxygen content on hardness

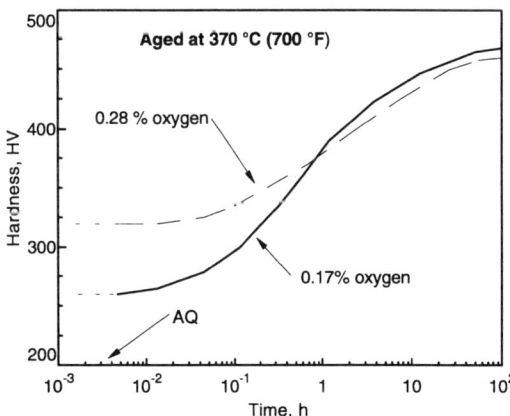

Variation of hardness with aging time at 360 °C (700 °F) for Beta III alloy containing two oxygen levels and solution treated above the β transus.
Source: F. Froes, C. Yolton, J. Capenos, M. Wells, and J.C. Williams, "The Relationship Between Microstructure and Age Hardening Response in the Metastable Beta Titanium Alloy Ti-11.5Mo-6Zr-4.5Sn (Beta III)," *Metall. Trans. A*, Vol 11, 1980, p 21

Beta III: Effect of cold work and aging time on hardness

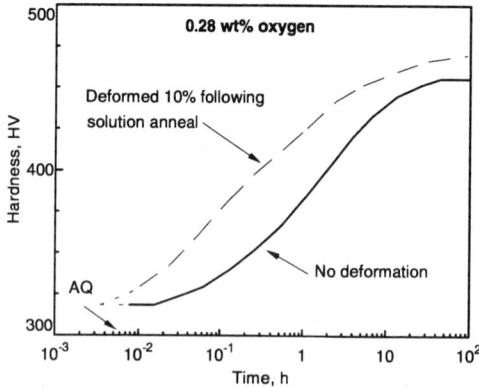

Variation of hardness with aging time at 370 °C (700 °F) for alloy quenched from 900 °C (1650 °F). Several specimens of high oxygen content alloy were solution treated above the beta transus, quenched, cold worked 10% and aged for various times. Data indicate that cold work prior to aging accelerated the aging response. Source: F. Froes, C. Yolton, J. Campenos, M. Wells, and J.C. Williams, "The Relationship Between Microstructure and Age Hardening Response in the Metastable Beta Alloy Ti-11.5Mo-6Zr-4.5Sn (Beta III)," *Metall. Trans. A*, 1980, p 21

Typical Tensile Properties

Beta III: Typical mechanical properties

Aging temperature		Ultimate tensile strength		Tensile yield strength(a)		Elongation(b), %	Reduction of area, %
°C	°F	MPa	ksi	MPa	ksi		
Rivet wire(c)							
As solution treated		993	144	792	115	24	65
480	900	1365	198	1269	184	15	36
510	950	1303	189	1213	176	18	38
540	1000	1186	172	1124	163	21	44
565	1050	1089	158	1034	150	25	56
590	1100	986	143	945	137	27	65
Bar, 13.6 mm (0.522 in.) diam(c)							
As solution treated		855	124	752	109	21	72
480	900	1386	201	1317	191	11	33
540	1000	1165	169	1096	159	17	63
590	1100	1041	151	1007	146	17	67
Plate 13 and 25 mm (0.5 and 1.0 in.) thick(c)(d)							
As solution treated		896	130	827	120	22	62
480	900	1351	196	1262	183	3	6.4
510	950	1289	187	1200	174	5	10.7
540	1000	1255	182	1179	171	4.8	12.3
590	1100	1041	151	979	142	11	24
Sheet 1.6 mm (0.063 in.) thick							
Solution treated, 720 °C (1325 °F)							
Air cooled		972	141	882	128	17	45
Water quenched		841	122	738	107	20	52
480	900	1413	205	1317	191	7	29
540	1000	1158	168	1089	158	8	45
Solution treated, 770 °C (1425 °F)							
Air cooled		896	130	834	121	18	45
Water quenched		827	120	745	108	21	48
480	900	1310	190	1234	179	6	35
540	1000	1138	165	1062	154	8	42
Lab foil specimen(e)							
As solution treated, 760 °C (1400 °F)							
0.010 in. thick		1000	145	958	139	8.0	...
0.005 in. thick		979	142	924	134	8.5	...
0.002 in. thick		1014	147	958	139	6.5	...
480	900						
0.010 in. thick		1282	186	1248	181	6.7	...
0.005 in. thick		1510	219	1413	205	4.5	...
0.002 in. thick		1586	230	1538	223	2.0	...
540	1000						
0.010 in. thick		1158	168	1082	157	6.5	...
0.005 in. thick		1262	183	1186	172	8.2	...
0.002 in. thick		1344	195	1276	185	4.0	...

(a) 0.2% offset. (b) In 2 in. or $4d$ where d is diameter of reduced section of tensile test specimen. (c) Solution treated 730 to 790 °C (1350 to 1450 °F), water quenched, aged 8 h. (d) Longitudinal properties. (e) Solution treated, descaled and pickled, aged 8 h. Source: *Metals Handbook*, Vol 3, 9th ed., 1980, p 405

Beta III: Tensile properties of seamless tubing

Condition	Ultimate tensile strength(a) MPa	ksi	Tensile yield strength MPa	ksi	Elongation in 50 mm (2 in.), %	Reduction of area, %
38 mm (1.5 in.) OD, 4.4 mm (0.173 in.) wall						
As extruded(b)	799	116.0	687	99.7	16.0	75.6
	854	123.9	738	107.1	18.0	72.0
As extruded + 510 °C (950 °F), 8 h	1381(c)	200.4(c)	1270	184.3	8.0	18.4
	1376(c)	199.6(c)	1264	183.4	8.0	22.6
32 mm (1.26 in.) OD, 3 mm (0.120 in.) wall						
Mill annealed(d)	991	143.8	924	134.1	13.2	42.0
Mill annealed + pickled	988	143.4	935	135.6	13.0	47.5
Mill annealed + 730 °C (1350 °F), WQ	905	131.3	791	114.8	13.5	50.9
Mill annealed + 730 °C (1350 °F), WQ + 510 °C (950 °F), 8 h	1278	185.4	1215	176.3	7.5	26.7
Mill annealed + 730 °C (1350 °F), WQ + 540 °C (1000 °F), 8 h	1197	173.7	1136	164.8	9.2	36.5
22.4 mm (0.884 in.) OD, 1.3 mm (0.05 in.) wall						
Mill annealed(d)	1066	154.7	972	141.0	8.5	16.6
Mill annealed + pickled	1041	151.0	953	138.2	8.5	24.7
Mill annealed + 730 °C (1350 °F), WQ	809	117.4	697	101.2	20.8	36.0
Mill annealed + 730 °C (1350 °F), WQ + 510 °C (950 °F), 8 h	1244	180.4	1165	169.0	6.0	12.3
Mill annealed + 730 °C (1350 °F), WQ + 540 °C (1000 °F), 8 h	1170	169.8	1101	159.8	7.2	15.6
13 mm (0.5 in.) OD, 0.6 mm (0.024 in.) wall						
As finish cold drawn(e)	1117	162	827	120	7.0	...
Cold drawn + 785 °C (1450 °F), WQ	841	122	765	111	18.0	...
Cold drawn + 730 °C (1350 °F), rapid AC + 510 °C (950 °F), 8 h	1303	189	1261	183	5.5	...
Cold drawn + 730 °C (1350 °F), rapid AC + 540 °C (1000 °F), 8 h	1268	184	1193	173	7.5	...
Cold drawn + 730 °C (1350 °F), rapid AC + 565 °C (1050 °F), 8 h	1158	168	1034	150	9.5	...

(a) Specimens machined from longitudinal sections of tubing subtended by 19 mm (0.75 in.) cords. Specimen uniform section 13 mm (0.5 in.) wide; gage length 50 mm (2 in.). (b) 71 mm (2.8 in.) OD by 32 mm (1.260 in.) ID hollow billets canned in mild steel, heated to 815 °C (1500 °F), extruded to tube hollow (reduction 7.2 to 1) and water quenched. (c) Machined round specimen; uniform section 3 mm (0.125 in.) diameter; gage length 13 mm (0.5 in.). (d) Tube hollows cold reduced to finished tubing in a single pass with tube reducers, followed by vacuum anneal of 730 °C (1350 °F), 1/2 h, rapid cool + 620 °C (1150 °F), 4 h, furnace cool to 530 °C (900 °F). (e) 26 mm (1.02 in.) diameter bar gun drilled and cold drawn to finished tubing. Source: AFML-TR-69-171

Beta III: Tensile properties of 50 mm (2 in.) as-quenched plate

Solution temperature °C	°F	Test direction	Ultimate tensile strength MPa	ksi	Tensile yield strength MPa	ksi	Elongation in 25 mm (1 in.), %	Reduction of area, %
760	1400	L	853	123.7	733	106.4	22.3	63.8
		T	826	119.8	694	100.7	25.0	65.9
730	1350	T	944	137.0	884	128.3	11.0	33.4

Note: Beta transus ~745 °C (~1375 °F). Source: AFML-TR-69-171

Beta III: Tensile properties of vacuum-arc melted plate

Heat treatment	Test direction	Ultimate tensile strength MPa	ksi	Tensile yield strength MPa	ksi	Elongation in 25 mm (1 in.), %	Reduction of area, %
Double vacuum-arc melted plate(a)							
760 °C (1400 °F), 10 min, AC	L	923	140.3	923	133.9	14.0	37.6
	T	950	137.9	888	128.8	8.5	29.2
760 °C (1400 °F), 10 min, WQ	L	881	127.8	788	114.3	18.0	51.9
	T	917	133.0	821	119.1	12.5	45.0
760 °C (1400 °F), 10 min, WQ + 510 °C (950 °F), 8 h	L	1309	189.9	1215	176.2	6.5	7.5
	T	1352	196.2	1294	187.7	2.5	12.5
760 °C (1400 °F), 10 min, WQ + 565 °C (1050 °F), 8 h	L	1012	164.7	950	160.9	8.5	16.7
Triple vacuum-arc melted plate(a)							
760 °C (1400 °F), 10 min, AC	L	1012	146.8	950	137.8	11.0	29.5
760 °C (1400 °F), 10 min, WQ	L	908	131.7	784	137.8	15.0	37.7
	T	914	132.6	792	114.9	15.0	41.5
760 °C (1400 °F), 10 min, WQ + 510 °C (950 °F), 8 h	L	1333	193.4	1237	179.4	5.5	9.0
	T	1352	196.2	1275	185.0	4.0	10.5
760 °C (1400 °F), 10 min, WQ + 565 °C (1050 °F), 8 h	L	1169	169.6	1085	157.4	6.7	8.6
	T	1187	172.2	1120	162.5	7.0	14.5

Note: (a) 13 mm (0.5 in.) plate hot rolled at 760 °C (1400 °F) and heat treated as indicated. Source: AFML-TR-69-171 (III)

Beta III: Typical tensile properties of wire (see also "Fastener/Spring" section)

Condition	Ultimate tensile strength MPa	ksi	Tensile yield strength (0.2% offset) MPa	ksi	Elongation in 4D, %	Reduction of area, %
2.3 mm (0.090 in.) diam						
Cold drawn 67%	1245	181	770	112	15	80
Cold drawn 67% + 455 °C (850 °F) 4 h	1745	253	1640	238	10	57
Cold drawn 67% + 480 °C (900 °F) 4 h	1585	230	1530	222	10	63
1.3 mm (0.050 in.) diam						
Cold drawn 68%	1310	190	910	132	10	84
Cold drawn 68% + 455 °C (850 °F) 4 h	1750	254	1645	239	7.5	62
Cold drawn 68% + 480 °C (900 °F) 4 h	1545	224	1435	208	7.5	66
0.76 mm (0.030 in.) diam						
Cold drawn 83%	1460	212	...	...	10	81
Cold drawn 83% + 455 °C (850 °F) 4 h	1985	288	...	...	7.5	54
Cold drawn 83% + 480 °C (900 °F) 4 h	1840	267	...	...	10	55
0.38 mm (0.015 in.) diam						
Cold drawn 87%	1550	225	...	...	...	76
Cold drawn 87% + 455 °C (850 °F) 4 h	2205	320	...	...	...	...
Cold drawn 87% + 480 °C (900 °F) 4 h	2040	296	...	...	...	55

Source: R.R. Boyer and H.W. Rosenberg, Ed., *Beta Titanium Alloys in the 1980's*, AIME, 1984

Beta III: Tensile properties of bar
Properties are for 13 to 38 mm (0.5 to 1.5 in.) diameter bar solution treated at 705 to 790 °C (1350 to 1450 °F), water quenched and aged as indicated.

Condition	Ultimate tensile strength MPa	ksi	Tensile yield strength (0.2% offset) MPa	ksi	Elongation in 4D, %	Reduction of area, %
As ST	855	124	750	109	21	72
ST + age 480 °C (900 °F) 8 h	1385	201	1315	191	11	33
ST + age 540 °C (1000 °F) 8 h	1165	169	1095	159	17	63
ST + age 595 °C (1100 °F) 8 h	1040	151	1005	146	17	67

Source: R.R. Boyer and H.W. Rosenberg, Ed., *Beta Titanium Alloys in the 1980's*, AIME, 1984

Beta III: Tensile properties of sheet and foil

Condition(a)	Ultimate tensile strength MPa	ksi	Tensile yield strength (0.2% offset) MPa	ksi	Elongation in 50 mm (2 in.), %	Reduction of area, %
1.6 mm (0.063 in.) thick						
ST 720 °C (1325 °F), WQ	840	122	735	107	20	52
ST 720 °C (1325 °F), AC	970	141	880	128	17	45
ST 720 °C (1325 °F) + age 480 °C (900 °F) 8 h	1410	205	1315	191	7	29
ST 720 °C (1325 °F) + age 540 °C (1000 °F) 8 h	1160	168	1090	158	8	45
ST 775 °C (1425 °F), WQ	825	120	745	108	21	48
ST 775 °C (1425 °F) AC	895	130	835	121	18	45
ST 775 °C (1425 °F) + age 480 °C (900 °F) 8 h	1310	190	1235	179	6	35
ST 775 °C (1425 °F) + age 540 °C (1000 °F) 8 h	1135	165	1060	154	8	42
0.3 mm (0.012 in.) thick						
Mill ST	1005	146	925	134	12	...
0.13 mm (0.005 in.) thick						
Mill ST	990	144	915	133	12	...
ST 790 °C (1450 °F) + age 510 °C (950 °F) 8 h	1400	203	1315	191	3	...
ST 790 °C (1450 °F) + age 540 °C (1000 °F) 8 h	1180	171	1150	167	4	...

Source: R.R. Boyer and H.W. Rosenberg, Ed., *Beta Titanium Alloys in the 1980's*, AIME, 1984

Compressive Strength

Beta III: Compressive yield strength

Product form	Material condition	Test condition	Compressive yield strength MPa	ksi	Tensile yield strength MPa	ksi
13 mm (0.5 in.) bar	As-quenched(a)	13 × 31.75 mm (0.5 × 1.25 in.) compression specimens	798	115.7(b)	689	99.9
	ST(a) at 8 h at 480 °C (900 °F)		1370	198.7(b)	1293	187.6
	ST(a) + 8 h at 540 °C (1000 °F)		1122	162.8	1073	155.6
13 mm (0.5 in.) plate	As-quenched(c)	Compression specimens same as above	854	123.85(b)	728	105.6
	ST(c) + 8 h at 510 °C (950 °F)		1284	186.2(b)	1210	175.5
	ST(c) + 8 h at 540 °C (1000 °F)		1170	169.75(b)	1124	163.0
1.6 mm (0.063 in.) sheet	ST + 8 h at 510 °C (950 °F)	Longitudinal(d)	1344	195	1206	175
		Transverse(d)	1455	211	1275	185

(a) Solution treated at 770 °C (1420 °F), WQ. (b) Average of two test results. (c) Solution treated 15 min at 730 °C (1350 °F), 15-s delay, WQ. (d) Compression specimens had four 90° V-notches 0.25 mm (0.010 in.) deep. Source: AFML-TR-70-252 and AFML-TR-69-171 (III)

Beta III: Compressive yield strength vs temperature

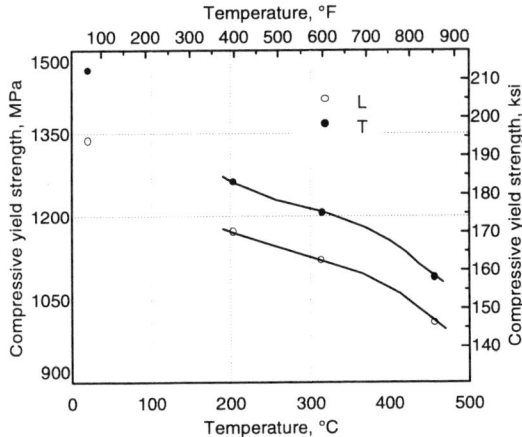

Source: *Aerospace Structural Metals Handbook*, Vol 4, Code 3722, Battelle Columbus Laboratories, 1972

Shear Strength

Ultimate shear strength of STA sheet was 805 MPa (117 ksi) in the longitudinal direction and 815 MPa (118 ksi) in the transverse direction.

Beta III: Typical double shear strength and notched tensile strength of STA bar

Condition(a)	Ultimate tensile strength		Elongation in 4D, %	Double shear strength		Notched tensile strength(K_t = 6.1)	
	MPa	ksi		MPa	ksi	MPa	ksi
Solution treated	896	130	20	627	91	...	...
ST + 480 °C (900 °F) 8 h	1337	194	12	799	116	1461	212
ST + 510 °C (950 °F) 8 h	1303	189	12	779	113	1544	224
ST + 540 °C (1000 °F) 8 h	1186	172	14	737	107	1592	231
ST + 565 °C (1050 °F) 8 h	1103	160	17	724	105	1510	219

(a) Bar 25 to 38 mm (1 to 1.5 in.) in diameter taken from four heats were solution treated at 705 to 730 °C (1300 to 1350 °F), water quenched, and aged as indicated. Source: V. Petersen, F. Froes, and R. Malone, "Metallurgical Characteristics and Mechanical Properties of Beta III, a Heat-Treatable Beta Titanium Alloy," in *Titanium Science and Technology*, R. Jaffee and H. Burte, Ed., 1973, p 1969

Beta III: Double shear strength and tensile properties of STA rod

Condition(a)	Ultimate tensile strength		Tensile yield strength (0.2% offset)		Elongation in 4D, %	Reduction of area, %	Double shear strength	
	MPa	ksi	MPa	ksi			MPa	ksi
As ST	990	144	790	115	24	65	625	91
ST + age 480 °C (900 °F) 8 h	1365	198	1270	184	15	36	790	115
ST + age 510 °C (950 °F) 8 h	1300	189	1215	176	18	38	780	113
ST + age 540 °C (1000 °F) 8 h	1185	172	1125	163	21	44	750	109
ST + age 565 °C (1050 °F) 8 h	1090	158	1035	150	25	56	725	105
ST + age 595 °C (1100 °F) 8 h	985	143	945	137	27	65	655	95

(a) Mill processed rod was solution treated at 690 to 730 °C (1275 to 1350 °F), water quenched, and aged as indicated. Source: J. Beckman and C. Yolton, "Beta III (Ti-11.5Mo-6Zr-4.5Sn)," in *Beta Titanium Alloys in the 80's*, R. Boyer and H. Rosenberg, Ed., TMS/AIME, 1984, p 401

High-Temperature Strength

Beta III: Elevated temperature tensile properties of plate

Aging temperature(a)		Test temperature(b)		Specimen direction	Ultimate tensile strength		Tensile yield strength (0.2% offset)		Elongation in 25 mm (1 in.), %	Reduction in area, %
°C	°F	°C	°F		MPa	ksi	MPa	ksi		
510	950	RT		L	1299	188.4	1210	175.5	4.0	9.6
				T	1409	204.4	1354	196.4	4.0	9.7
540	1000	RT		L	1210	175.6	1124	163.0	5.0	11.0
				T	1301	188.8	1253	181.8	3.0	9.7
510	950	205	400	L	1131	164.1	992	143.9	7.0	16.6
				T	1179	171.1	1086	157.5	5.0	17.7
540	1000	205	400	L	1064	154.3	933	135.4	7.0	14.6
				T	1104	160.2	1024	148.5	9.0	22.1
510	950	315	600	L	1078	156.4	939	136.2	7.0	16.4
				T	1135	164.7	1055	153.0	4.0	17.3
540	1000	315	600	L	985	142.9	851	123.4	7.0	20.2
				T	1042	151.1	958	139.0	6.0	19.6
510	950	425	800	L	1030	149.5	893	129.5	10.0	23.4
				T	1079	156.6	975	141.5	6.0	18.7
540	1000	425	800	L	967	140.3	828	120.2	10.0	35.1
				T	1006	145.9	902	130.8	9.0	25.2
510	950	540	1000	L	824	119.5	587	85.2	30.0	82.6
				T	798	115.8	573	83.2	34.0	65.5
540	1000	...		L	803	116.5	598	86.8	21.0	39.4
				T	839	121.7	617	89.5	17.0	48.3

(a) 13 mm (0.5 in.) panels were laboratory solution treated for 15 min at 730 °C (1350 °F) and water quenched after a 15-s delay and aged 8 h at temperature indicated. (b) All specimens held 15 min at test temperature before loading. Source: R.A. Wood, *Beta Titanium Alloys*, MCIC 72-11, 1972

Beta III: Elevated temperature tensile properties of hand mill sheet

Aging (8-h) temperature(a)		Test temperature(b)		Ultimate tensile strength(c)		Tensile yield strength (0.2% offset)		Elongation in 50 mm (2 in.), %	Reduction of area, %	Elastic modulus	
°C	°F	°C	°F	MPa	ksi	MPa	ksi			GPa	10⁶ psi
510	900	RT		1397	202.7	1350	195.8	3.7	18.5	111	16.2
540	1000	...		1201	174.3	1117	162.1	6.0	37.1	113	16.4
510	900	315	600	1082	157.0	1007	146.1	4.5	28.5	92	13.4
540	1000	315	600	958	139.0	897	130.2	5.0	44.9	87	12.6
510	900	370	700	1038	150.6	959	139.1	5.0	41.2	83	12.0
540	1000	370	700	981	142.3	911	132.2	5.5	46.7	88	12.7
510	900	425	800	1009	146.4	852	123.6	8.0	42.8	89	12.9
540	1000	425	800	970	140.7	872	126.5	5.0	38.5	84	12.2
510	900	480	900	857	124.4	624	90.5	14.0	46.6	78	11.3
540	1000	480	900	853	123.8	740	107.4	6.5	67.7	72	10.4
510	900	540	1000	666	96.6	382	55.4	28.5	89.1	64	9.3
540	1000	540	1000	693	100.6	439	63.7	28.5	92.8	57	8.3

(a) 1.7 mm (0.067 in.) sheet cold reduced 46% and solution treated 5 min at 730 °C (1350 °F) and air cooled. (b) Tensile coupons held 15 min at test temperatures prior to loading. (c) Tensile properties are average of duplicate longitudinal tests. Source: R.A. Wood, *Beta Titanium Alloys*, MCIC-72-11, 1972

Beta III: Tensile strength vs temperature

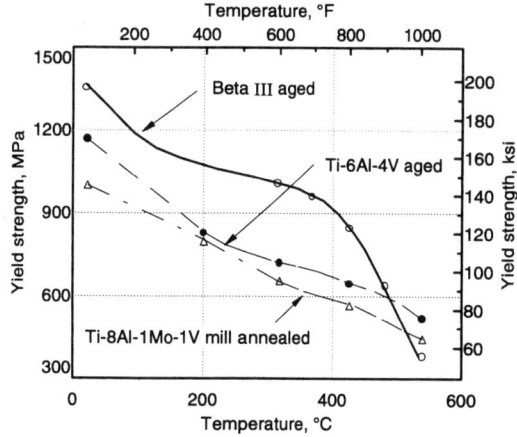

The strength of Beta III is outstanding up to 425 °C (800 °F).
Source: *Alloy Digest*, Code Ti-59, 1970

Beta III: Tensile and yield strengths of plate

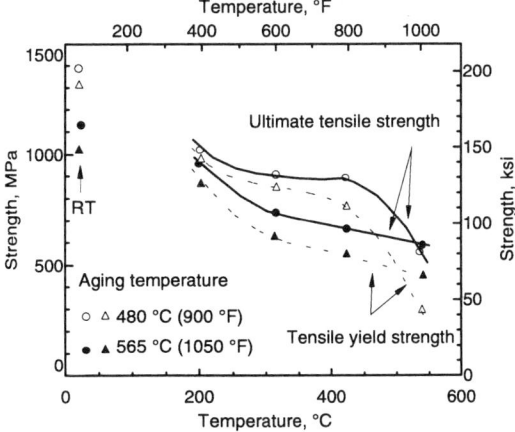

13 mm (0.5 in.) diam bar, solution treated 5 min at 770 °C (1420 °F), WQ, + 8 h age.
Source: *Aerospace Structural Metals Handbook*, Vol 4, Code 3722, Battelle Columbus Laboratories, 1972

Beta III: Tensile strength of STA plate

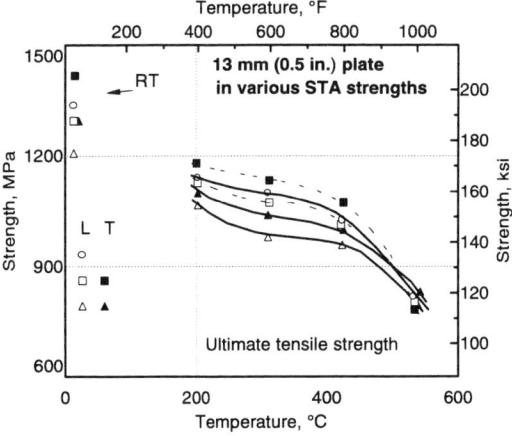

Source: AFML-TR-69-171-(III), June 1969; *Aerospace Structural Metals Handbook*, Code 3722, Battelle Columbus Laboratories, 1972

Beta III: Yield strength of STA plate vs temperature

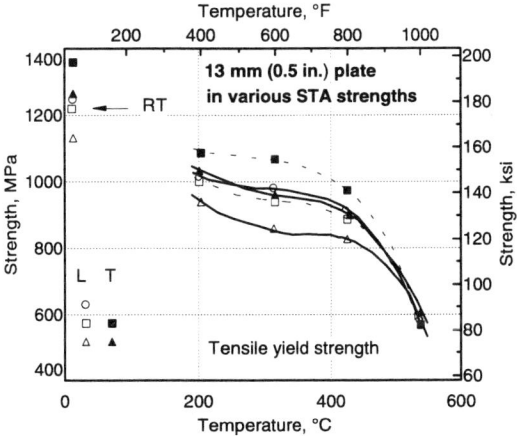

Source: AFML-TR-69-171-(III), June 1969; *Aerospace Structural Metals Handbook*, Vol 4, Code 3722, Battelle Columbus Laboratories, 1972

Beta III: Tensile strength of STA sheet

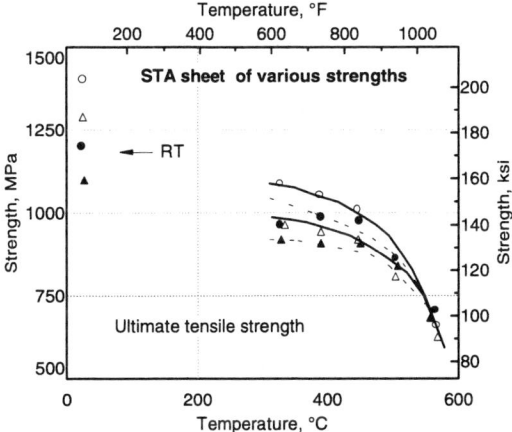

Source: AFML-TR-69-171-(III), June 1969; *Aerospace Structural Metals Handbook*, Vol 4, Code 3722, Battelle Columbus Laboratories, 1972

Beta III: Yield strength of sheet

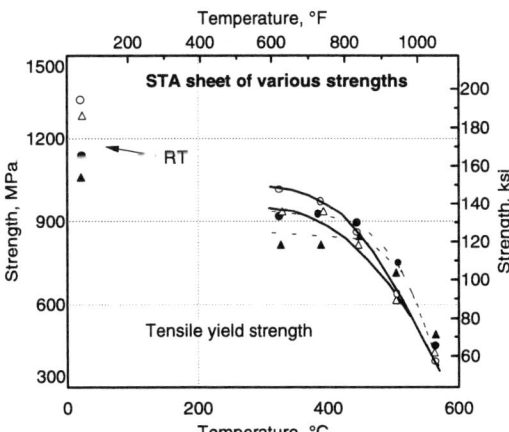

Source: AFML-TR-69-171-(III), June 1969; *Aerospace Structural Metals Handbook*, Vol 4, Code 3722, Battelle Columbus Laboratories, 1972

Creep Properties

Beta III: Creep properties

Temperature		Creep stress		Plastic creep strain(a),
°C	°F	MPa	ksi	%
260	500	895	130	0.07
315	600	795	115	0.16
370	700	240	35	0.05
		515	75	0.17
425	800	70	10	0.04
		160	23	0.20

(a) For 100 h at temperature under load. Source: *Metals Handbook*, Vol 3, 9th ed., 1980

Beta III: Creep strength comparison

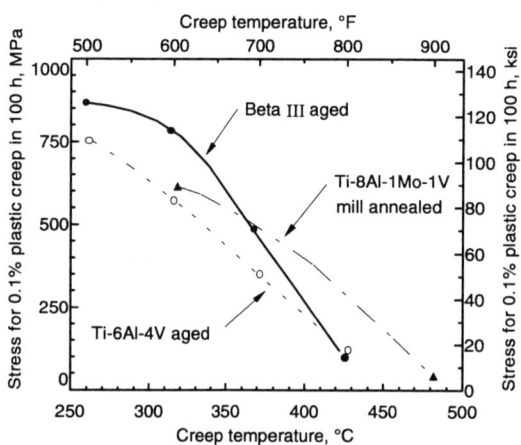

Source: V. Petersen, F. Froes, and R. Malone, "Metallurgical Characteristics and Mechanical Properties of Beta III, A Heat-Treatable Beta Titanium Alloy," in *Titanium Science and Technology*, R. Jaffee and H. Burte, Ed., 1973, p 1969

Beta III: Creep and rupture behavior of STA sheet

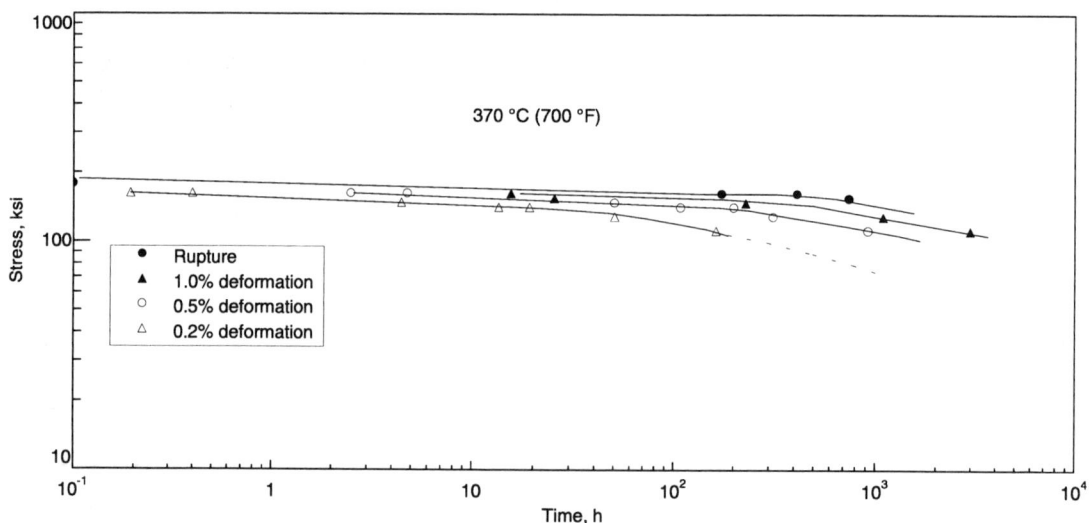

Sheet of 1.6 mm (0.063 in.) thickness was solution heat treated and aged at 510 °C (950 °F) for 8 h.
Source: O. Deel and H. Mindlin, "Engineering Data on New and Emerging Structural Materials," AFML-TR-70-252, 1970; reported in *Beta Titanium Alloys*, R. Wood, Ed., MCIC-72-11, 1972, p 156

Beta III: Elevated temperature stability for bar, sheet and plate

	Creep temperature		Creep stress		Plastic creep,	RT properties after creep:				Reduction of area,
						Ultimate tensile strength		Tensile yield strength (0.2% offset)		
Condition(a)	°C	°F	MPa	ksi	%	MPa	ksi	MPa	ksi	%
Aged 8 h 510 °C (950 °F) (high strength)			As aged		None	1317	191	1248	181	29.1
	260	500	896	130	0.13	1372	199	1324	192	22.4
	370	700	517	75	0.16	1358	197	1310	190	22.1
Aged 8 h 540 °C (1000 °F) (medium strength)			As aged		None	1179	171	1117	162	35.0
	260	500	827	120	0.16	1186	172	1151	167	32.2
	370	700	482	70	0.12	1234	179	1165	169	33.6

(a) Solution treated material was aged as indicated, pickled, creep loaded for 100 h, and tensile tested without removing creep scale. Source: V. Petersen, F. Froes, and R. Malone, "Metallurgical Characteristics and Mechanical Properties of Beta III, a Heat-Treatable Beta Titanium Alloy," in *Titanium Science and Technology*, R. Jaffee and H. Burte, Ed., 1973, p 1969

Beta III: Creep stability for STOA rod

Temperature °F	Exposure data			Plastic creep, %	RT tensile properties after exposure:								
	Time, h	Stress			Tensile strength		Yield strength (0.2% offset)		Elongation in 4D, %	Reduction of area, %	Double shear strength	
		MPa	ksi		MPa	ksi	MPa	ksi			MPa	ksi
...	...	...	...	...	1011	146.7	974	141.3	25.5	66.1	643	93.3
450	2000	275	40	0.01	1017	147.5	948	137.6	...	72.7	...	...
450	4000	275	40	0.15	1026	148.8	1013	146.9	20.0	73.6	...	...
450	4000	620	90	0.11	1054	152.9	1025	148.7	22.0	71.4	...	...
550	2000	...	...	...	1022	148.3	1008	146.2	16.0	66.7	624	90.5
550	4000	...	...	...	1023	148.4	1009	146.3	20.0	69.9	662	96.1
550	8000	...	...	...	...	...	...	...	...	...	654	94.9
550	2000	482	70	0.33	1019	147.9	1004	145.6	24.0	70.1	649	94.1

Note: Chemical composition was 0.03 wt% C, 0.03 wt% Fe, 0.0087 wt% H, 11.37 wt% Mo, 0.017 wt% N, 0.28 wt% O, 4.13 wt% Sn, 5.19 wt% Zr. High oxygen content resulted in a β transus temperature of 815 °C (1500 °F) (about 84 °C, or 150 °F higher than usual). Rod of 6.4 mm (0.25 in.) diameter was mill solution treated at 690 to 705 °C (1275 to 1300 °F), water quenched, then overaged for 8 h at 595 °C (1100 °F) before creep exposure. Temper strain was not removed before tensile testing. Source: F. Froes, J. Capenos, and M. Wells, "Alloy Partitioning and Effect on Aging Characteristics," in *Titanium Science and Technology*, R. Jaffee and H. Burte, Ed., 1973, p 1621

Beta III: Creep stability of STA bar

Aging temperature		Creep data					Plastic creep, %	RT tensile properties after creep exposure						RA loss after creep, %
		Temperature		Stress		Time, h		Ultimate tensile strength		Tensile yield strength (0.2% offset)		Elongation in 25 mm (1 in.), %	Reduction of area, %	
°C	°F	°C	°F	MPa	ksi			MPa	ksi	MPa	ksi			
480	900			As aged				1376	199.6	1293	187.6	9.6	27.2	...
		260	500	862	125	100	0.11	1409	204.4	1328	192.7	12.0	26.5	2.6
		315	600	724	105	100	0.05	...	...	...	...	...	...	...
		425	800	103	15	100	0.26	1462	212.1	1372	199.1	8.0	19.6	27.5
		260	500	862	125	500	0.00	1450	210.4	1378	199.9	6.0	21.2	20.0
		370	700	551	80	500	0.61	1490	216.2	1439	208.7	7.0	16.7	38.6
510	950			As aged				1335	193.7	1248	181.0	8.0	34.3	...
		260	500	862	125	100	0.00	1379	200.1	1310	190.0	5.0	16.4	52.2
		260	500	965	140	100	0.36	1357	196.9	1326	192.4	11.0	32.5	5.2
		370	700	413	60	100	0.16	1317	191.1	1259	182.7	10.0	29.5	14.0
		370	700	620	90	100	0.30	1395	202.4	1319	191.3	8.0	25.1	26.6
565	1050			As aged				1179	171.0	1130	163.9	11.0	40.6	...
		260	500	758	110	100	0.41	1117	162.1	1097	159.1	14.0	54.1	0.0
		260	500	758	110	500	0.00	1122	162.8	1077	156.3	11.0	49.7	0.0
		315	600	620	90	100	0.18	1191	172.8	1139	165.3	12.0	41.3	0.0
		425	800	103	15	100	0.02	1219	176.9	1139	165.3	12.0	38.9	4.2
		370	700	413	60	500	0.07	1182	171.5	1110	161.0	13.0	43.5	0.0

Note: Bar of 13 mm (0.5 in.) diameter was solution treated at 770 °C (1425 °F) in a continuous furnace, water quenched and aged at temperatures indicated for 8 h. Tensile testing after exposure was performed without removal of surface oxidation. Source: V. Petersen, J. Guernsey, and R. Buehl, "Manufacturing Procedures for a New, High-Strength Titanium Alloy Having Superior Formability," AFML-TR-69-171, 1969; R. Wood, reported in *Beta Titanium Alloys*, MCIC-72-11, 1972, p 155

Beta III: Creep stability of STA plate

Treatment	Creep data Temperature °C	°F	Stress MPa	ksi	Time, h	Plastic creep, %	Ultimate tensile strength MPa	ksi	Tensile yield strength (0.2% offset) MPa	ksi	Elongation in 50 mm (2 in.), %	RT tensile properties after creep exposure: Reduction of area, %	RA loss after creep, %
ST 730 °C (1350 °F), WQ + 8 h 510 °C (950 °F)			As aged				1299	188.4	1210	175.5	4.0	9.6	...
	260	500	827	120	100	0.08	1292	187.4	1215	176.2	5.0	7.8	19.0
			896	130	100	0.05	1343	194.4	1279	185.5	3.0	7.0	27.0
			724	105	500	0.05	1275	184.9	1199	174.0	4.0	6.3	34.0
	315	600	827	120	100	0.13	1340	194.4	1268	183.9	3.0	6.3	34.0
			896	130	100	0.13	1379	200.0	1322	191.8	2.0	2.7	72.0
	370	700	448	65	100	0.13	1364	197.9	1270	184.3	3.0	4.0	58.0
			344	50	500	0.17	1336	193.8	1250	181.3	2.0	5.6	42.0
	425	800	69	10	100	0.12	1333	193.4	1235	179.2	4.0	5.6	42.0
			138	20	100	0.17	1347	195.4	1258	182.5	3.0	7.8	19.0
ST 730 °C (1350 °F), WQ + 8 h 540 °C (1000 °F)			As aged				1211	175.7	1124	163.0	5.0	11.0	...
	260	500	758	110	100	0.04	1230	178.4	1149	166.7	4.0	7.0	36.0
			827	120	500	0.11	1264	183.4	1199	174.0	4.0	11.4	0.0
			896	130	100	0.08	1254	181.9	1201	174.3	1.0	4.0	64.0
	315	600	689	100	100	0.10	1261	182.9	1177	170.7	5.0	7.8	29.0
			827	120	100	0.08	1260	182.8	1209	175.4	4.0	9.3	15.0
	370	700	448	65	100	0.11	1257	182.3	1161	168.5	4.0	8.7	21.0
			379	55	500	0.17	1268	183.9	1175	170.4	5.0	10.0	9.1
			655	95	100	0.20	1253	181.8	1180	171.2	4.0	6.3	43.0
	425	800	69	10	100	0.12	1271	184.4	1164	168.9	3.0	7.8	29.0
			138	20	100	0.25	1264	183.4	1170	169.8	4.0	12.4	0.0

Note: Plate of 13 mm (0.5 in.) thickness was heat treated as indicated. Tensile testing after exposure was performed without removal of surface oxidation. Source: V. Petersen, J. Guernsey, and R. Buehl, "Manufacturing Procedures for a New High-Strength Beta Titanium Alloy Having Superior Formability," AFML-TR-69-171, 1969; R. Wood, reported in *Beta Titanium Alloys*, MCIC-72-11, 1972, p 153

Fatigue Properties

Beta III: Smooth axial fatigue of STA plate

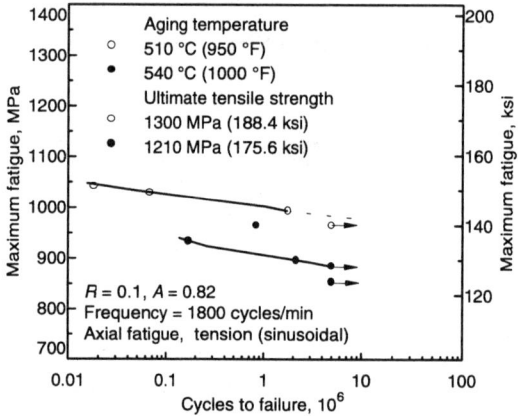

13 mm (0.5 in.) plate hot rolled + 730 °C (1350 °F), 15 min, 15-s delay, WQ + age, 8 h.
Source: AFML-TR-69-171 (III), 1969

Beta III: Axial fatigue of STA sheet

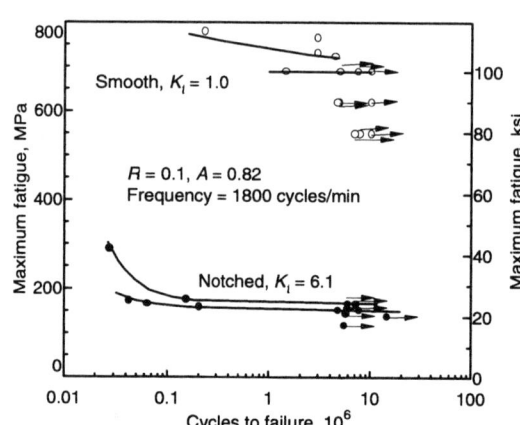

1.7 mm (0.067 in.) sheet, acid pickled.
Source: AFML-TR-69-171

Beta III: Axial and rotating beam fatigue strength

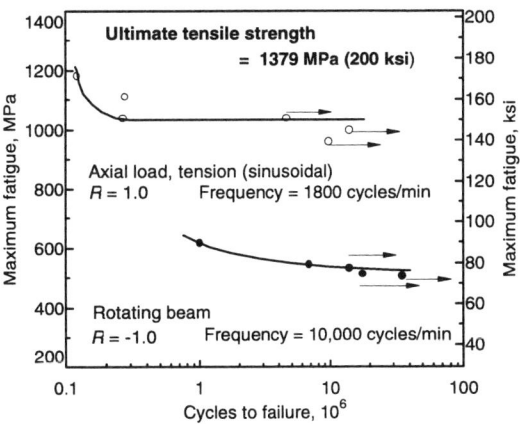

13 mm (0.5 in.) diameter bar hot rolled + 770 °C (1420 °F), WQ + 480 °C (900 °F), 8 h.
Source: AFML-TR-69-171 (III)

Fracture Properties

Beta III: Effect of aging on toughness

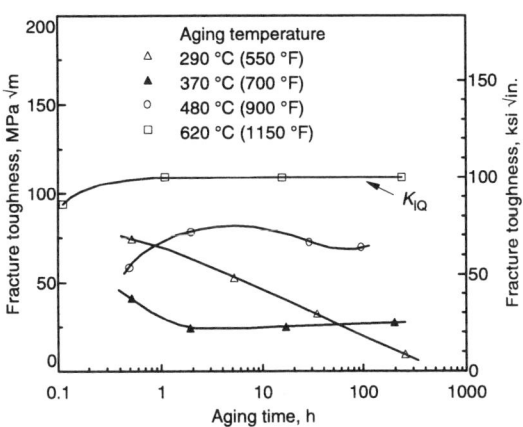

Plate (16 mm, or 0.640 in., thick) and bar (13 mm, or 0.5 in., diameter) were solution treated above the β transus, water quenched, and aged at temperatures indicated in neutral salt baths or air circulating furnace with temperature control to ±10 °C. Yield strengths varied from 65 to 1103 MPa (140 to 160 ksi). Tests were performed on standard Charpy specimens in three-point bending to determine plane-strain fracture toughness.
Source: J. Feeney and M. Blackburn, "Effect of Microstructure on the Strength, Toughness, and Stress-Corrosion Cracking Susceptibility of a Metastable Beta Titanium Alloy (Ti-11.5Mo-6Zr-4.5Sn)," Metall. Trans., Vol 1, 1970, p 3300

Beta III: Fracture toughness comparison

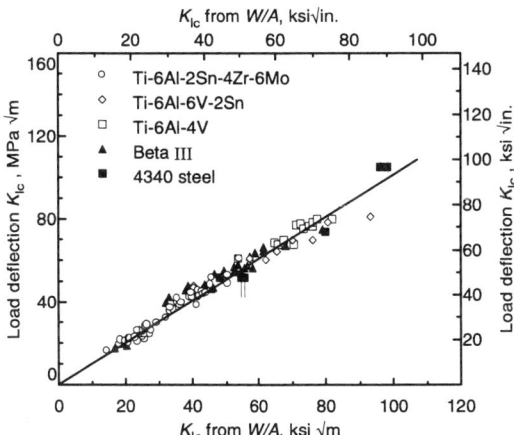

K_{Ic} determined from slow bend load-deflection curves vs K_{Ic} calculated from W/A values for precracked specimens tested in slow bend.
A Physmet SB-750 tester was used for slow-bend tests with crosshead speed of 2.5 mm/min (0.1 in./min) and load deflection curves were recorded along with energy values. Precracked Charpy specimens with notch roots approximately 1.5 mm (0.060 in.) deep were used.
Source: T. Ronald, J.A. Hall, and C. Pierce, "Usefulness of Precracked Charpy Specimens for Fracture Toughness Screening Tests of Titanium Alloys," Metall. Trans., Vol 3, 1972, p 813

Beta III: Fracture toughness of plate

Condition(a)	Transverse tensile strength		Transverse yield strength (0.2% offset)		Transverse elongation in 4D, %	Reduction of area, %	Fracture toughness			
							K_{Ic}		K_{Iscc}(b)	
	MPa	ksi	MPa	ksi			MPa√m	ksi√in.	MPa√m	ksi√in.
Solution treated	860	125	760	110	20	58	163	148	...	...
480 °C (900 °F) 8 h	1460	212	1380	200	3	5	57	52	26	24
	1355(d)	197	1280	186	(c)	11	...	...	58	53(d)
510 °C (950 °F) 8 h	1405	204	1335	194	3	6	66	60	...	...
595 °C (1100 °F) 8 h	1170	170	1145	162	6	20	95	87	...	...

(a) 13 mm (0.5 in.) plate was mill rerolled from 25 mm (1 in.) plate with two reheatings to 925 °C (1700 °F). Structure recrystallized. Plate solution treated 15 min at 720 to 790 °C (1325 to 1450 °F), water quenched and aged as indicated. (b) Estimated. Step loaded in increments of 5.5 MPa√m (5 ksi√in.) with 20 min at each stress level. Specimens immersed in 3.5% NaCl. (c) Fracture outside gage marks. (d) 25 mm (1 in.) plate rolled to size from slab in one heating from 1040 °C (1900 °F). Nonrecrystallized structure. Source: R.R. Boyer and H.W. Rosenberg, Ed., *Beta Titanium Alloys in the 1980's*, AIME, 1984

Beta III: Fracture toughness vs processing for extrusions

Extrusion temperature		Cooling rate from extrusion operation	Intermediate solution treatment(a)	Tensile yield strength		Ultimate tensile strength		Fracture toughness (K_{Ic})	
°C	°F			MPa	ksi	MPa	ksi	MPa√m	ksi√in.
980	1800	Air cool	Yes	1151	167	1186	172	60	55
760	1400	Air cool	Yes	1165	169	1213	176	58	53
815	1500	Air cool	Yes	1165	169	1199	174	57	52
980	1800	Water quench	Yes	1137	165	1255	182	56	51
815	1500	Water quench	Yes	1144	166	1248	181	54	49
980	1800	Water quench	None	1165	169	1282	186	52	48
760	1400	Water quench	None	1282	186	1392	202	45	41

Note: Chemical composition was 0.0244 wt% C, 0.11 wt% Fe, 0.0055 wt% H, 10.95 wt% Mo, 0.0080 wt% N, 0.1515 wt% O, 4.48 wt% Sn, 5.86 wt% Zr. Alloy specimens were heat treated as indicated, then aged at 480 °C (900 °F) for 8 h, air cooled. Fracture toughness was determined from Charpy V-notch precracked specimens in slow, three-point loading bend tests. (a) Solution treatment: 0.5 h at 760 °C (1400 °F) water quench. Source: R. Wood, *Beta Titanium Alloys*, MCIC-72-11, 1972

Beta III: Fracture toughness of forgings, plate, and extrusions vs heat treatment conditions

Processing and heat treatment	Yield strength		Fracture toughness					
			K_Q		K_{Ic}		K_f(a)	
	MPa	ksi	MPa√m	ksi√in.	MPa√m	ksi√in.	MPa√m	ksi√in.
Navaho forging, 100 mm (4 in.) section thickness								
900-980 °C (1650-1800 °F) forging, 1 h, 720 °C (1325 °F), WQ + 8 h, 510 °C (950 °F), AC	1255-1261	182-183	...	...	65-69	59-63	...	...
Hot rolled plate, 50 mm (2 in.) section thickness								
730-745 °C (1350-1375 °F) solution treatment, WQ + aging: 8 h, 480 °C (900 °F), AC	1399	203	54	49	...	...	56	51
8 h, 540 °C (1000 °F), AC	1158	168	71	65	...	...	73	67
8 h, 595 °C (1100 °F), AC	1089	158	77	70	...	...	81	74
Hot rolled plate, 16 mm (0.640 in.) thick								
1 h, 815 °C (500 °F), WQ + aging: 30 min, 285 °C (550 °F), AC	1041	151	...	...	71	65	...	...
5 h, 285 °C (550 °F), AC	1068	155	...	...	51	47	...	...
32 h, 285 °C (550 °F), AC	1096(d)	159(d)	...	...	32	29	...	...
250 h, 285 °C (550 °F)	1124(d)	163(d)	...	...	11	10	...	...
30 min, 320 °C (700 °F), AC	993	144	...	...	40	37	...	...
2 h, 320 °C (700 °F), AC	1082(d)	157(d)	...	...	23	21	...	...
15 h, 320 °C (700 °F), AC	1379(d)	200(d)	...	...	23	21	...	...
200 h, 320 °C (700 °F), AC	1461(d)	212(d)	...	...	28	26	...	...
1 h, 870 °C (1600 °F), WQ + aging: 30 min, 480 °C (900 °F), AC	896	130	...	...	56	51	...	...
2 h, 480 °C (900 °F), AC	910	132	...	...	72	66	...	...
8 h, 480 °C (900 °F), AC	1151	167	...	...	55	50	...	...
32-36 h, 480 °C (900 °F), AC	1055	153	...	...	73-74	67-68	...	...
100 h, 480 °C (900 °F), AC	1020	148	...	...	78	71	...	...
6 min, 620 °C (1150 °F), AC	799	116	93	85	...	...	...	...
1 h, 620 °C (1150 °F), AC	862	125	108	99	...	...	...	...
16 h, 620 °C (1150 °F), AC	806	117	108	99	...	...	...	...
250 h, 620 °C (1150 °F), AC	799	116	108	99	...	...	...	...
Hot rolled 925 °C (1700 °F) plate, 13 mm (0.50 in.) thick								
15 min, 720-790 °C (1325-1450 °F), WQ + aging: No aging	765	111	162	148	...	...	...	...
8 h, 480 °C (900 °F), AC	1386	201	...	...	58	53	...	...
8 h, 510 °C (950 °F), AC	1337	194	...	...	66	60	...	...
8 h, 595 °C (1100 °F), AC	1117	162	...	...	95	87	...	...

(continued)

Beta III: Fracture toughness of forgings, plate, and extrusions vs heat treatment conditions (continued)

Processing and heat treatment	Yield strength MPa	ksi	K_Q MPa√m	ksi√in.	K_{Ic} MPa√m	ksi√in.	K_f(a) MPa√m	ksi√in.
Extrusions; 10.9 to 1 reductions on 75 mm (2.95 in.) diameter billet								
980 °C (1800 °F) extrusion, AC + STA(b)	1151	167	...	...	59	54	...	...
980 °C (1800 °F) extrusion, WQ + STA(b)	1137	165	...	...	56	51	...	...
980 °C (1800 °F) extrusion, WQ + A(c)	1165	169	...	...	52	48	...	...
815 °C (1500 °F) extrusion, AC + STA(b)	1165	169	...	...	57	52	...	...
815 °C (1500 °F) extrusion, WQ + STA(b)	1144	166	...	...	54	49	...	...
760 °C (1400 °F) extrusion, AC + STA(b)	1165	169	...	...	58	53	...	...
760 °C (1400 °F) extrusion, WQ + A(c)	1282	186	...	...	45	41	...	...

(a) K_f = stress intensity for catastrophic failure. (b) STA = 30 min at 760 °C (1400 °F), WQ + 8 h, 480 °C (900 °F), AC. (c) A = 8 h at 480 °C (900 °F), AC. (d) Compressive yield strength. Source: R. Wood, *Beta Titanium Alloys*, MCIC-72-11, 1972

Stress-Strain Curves

Beta III: Tension stress-strain curves for sheet

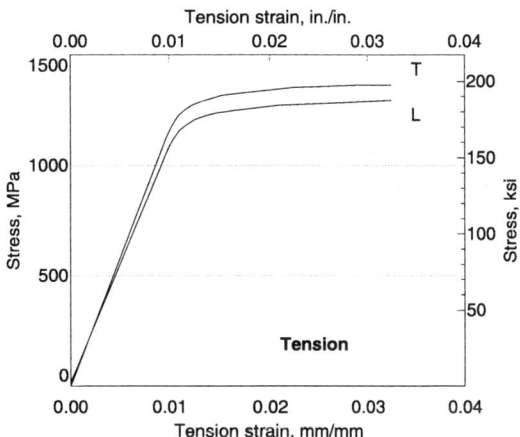

1.6 mm (0.063 in.) sheet solution treated at 510 °C (950 °F), 8 h, AC. Source: *Aerospace Structural Metals Handbook*, Vol 4, Code 3722, Battelle Columbus Laboratories, 1972, p 11

Beta III: Compressive stress-strain curves for sheet

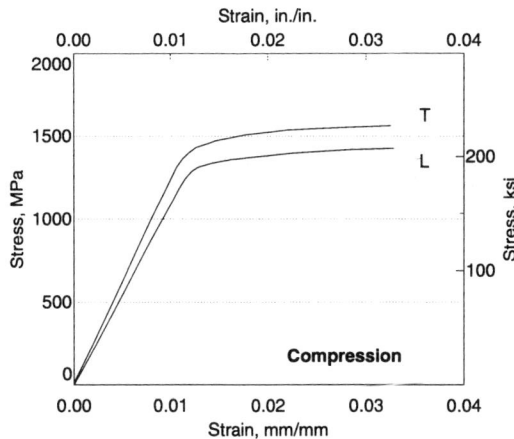

1.6 mm (0.063 in.) sheet solution treated at 510 °C (950 °F), 8 h, AC. Source: *Aerospace Structural Metals Handbook*, Vol 4, Code 3722, Battelle Columbus Laboratories, 1972, p 30

Tensile Curves

Beta III: Effect of temperature on tensile behavior of sheet

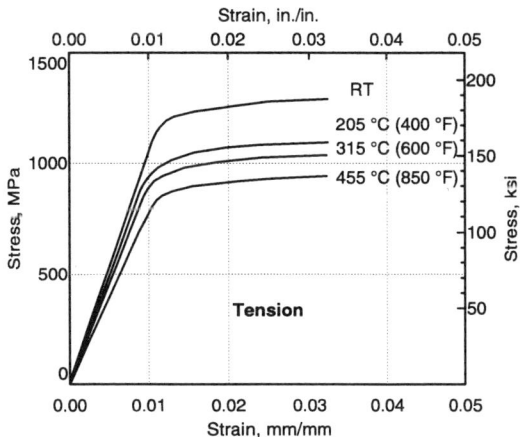

1.6 mm (0.063 in.) sheet solution treated and aged at 510 °C (950 °F), 8 h, AC. Longitudinal stress-strain at indicated test temperature. Source: *Aerospace Structural Metals Handbook*, Vol 4, Code 3722, Battelle Columbus Laboratories, 1972, p 33

Beta III: Effect of temperature on tensile behavior of sheet

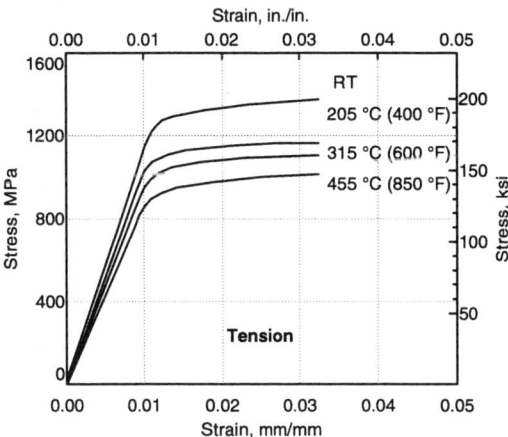

1.6 mm (0.063 in.) sheet solution treated and aged at 510 °C (950 °F), 8 h, AC. Transverse stress-strain at indicated test temperature. Source: *Aerospace Structural Metals Handbook*, Vol 4, Code 3722, Battelle Columbus Laboratories, 1972, p 33

Compressive Curves

Beta III: Effect of temperature on compressive behavior of sheet

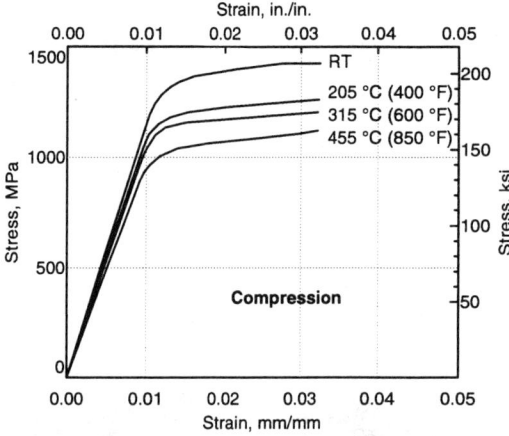

1.6 mm (0.063 in.) sheet, solution treated and aged at 510 °C (950 °F), 8 h, AC. Longitudinal compressive strain at indicated test temperature.
Source: *Aerospace Structural Metals Handbook*, Vol 4, Code 3722, Battelle Columbus Laboratories, 1972, p 35

Beta III: Effect of temperature on compressive behavior of sheet

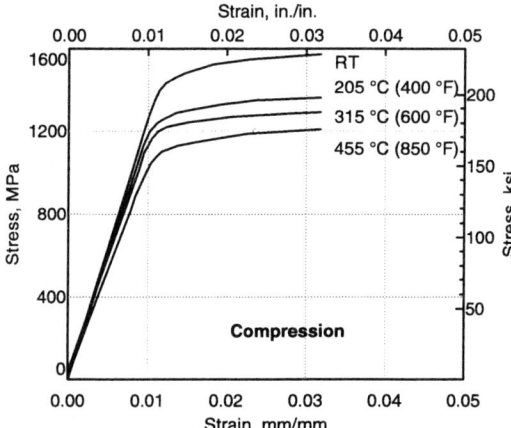

1.6 mm (0.063 in.) sheet solution treated and aged at 510 °C (950 °F), 8 h, AC. Transverse stress-strain at indicated test temperature.
Source: *Aerospace Structural Metals Handbook*, Vol 4, Code 3722, Battelle Columbus Laboratories, 1972, p 35

Forging

G.W. Kuhlman, ALCOA, Forging Division

Commercially important metastable β and near-β alloys respond well to thermomechanical processing (TMP), and several complex thermomechanical processing routes have been reduced to commercial practice. With this alloy class, thermomechanical processing is focused on optimal combinations of high strength, good fracture toughness, and ductility. These alloys possess superior high-cycle fatigue (HCF) properties due to their refined microstructures. Several of these alloys have been successfully direct aged, thereby producing even finer microstructures and improved smooth and notched fatigue properties.

Supra-transus forging processes prevail in this class of materials, except for Ti-10V-2Fe-3Al in which a combination of supra- and subtransus working is used to achieve desired properties through α phase manipulation and control. None of the other β and near-β alloys respond to thermomechanical processing to improve fracture-related properties because α morphology cannot be modified to the extent possible in α + β alloys. Ti-10V-2Fe-3Al may offer the broadest range of engineering properties for this class of alloys, which are ideal for use in structural applications where durability is the critical design criterion.

Beta III is a very high strength, deep hardening, metastable β alloy whose primary commercial applications in forgings are aerospace structural components, corrosion-resistant applications, and prosthetic devices. The alloy can be fabricated into all forging product types, although closed die forgings predominate. Beta III is commercially fabricated on all types of forging equipment.

Beta III is a highly forgeable alloy (when forged above the β transus), with comparable unit pressures (flow stresses), improved forgeability, and less crack sensitivity in forging than the α-β alloy Ti-6Al-4V. Flow stresses and unit pressures exceed that of the near-β alloy Ti-10V-2Fe-3Al (see figure). The desired final microstructure from Beta III forging processing is transformed β with a fine recrystallized prior β grain size, in preparation for final thermal treatments. Thus, Beta III is typically forged above the transus through one or more forging operations. Reheating for subsequent forging operations recrystallizes the alloy from prior hot working refining prior β grain size. Beta III may be subtransus (α + β) forged in final stages, with a significant increase in unit pressure requirements, to accomplish further recrystallization during heat treatment.

Final thermal treatments for Beta III forging include solution treating and aging. Forgings may be supplied in the solution treated condition and/or fully aged. In the solution treated condition, Beta III has lower strengths, but much higher ductility and toughness than in the solution treated and aged condition. Solution treatment is conducted at 690 to 730 °C (1275 to 1350 °F), followed by water quenching. Aging is conducted at 565 to 595 °C (1050 to 1100 °F).

Beta forging working histories for Beta III require imparting enough hot work to reach final macrostructure and microstructure objectives. Generally, reductions in any given forging process are 30 to 50% to achieve desired dynamic and static recrystallization. Very low levels of β reduction are not recommended. Although Beta III is cold worked in other products (sheet), cold working is not used for forgings.

Beta III, as with all β alloys, has a higher affinity for hydrogen than other alloy classes. Although Beta III forms less α case from heating operations than other alloy classes, therefore requiring less metal removal in chemical pickling (milling processes), control of chemical removal processes is essential to preclude excessive hydrogen pickup.

Beta III: Forging process temperatures

Process	Metal temperature °C	°F
Beta forging	700-955	1300-1750

Note: See "Technical Note 4: Forging" for recommended die temperatures.

Beta III: Extrusion breakthrough pressure

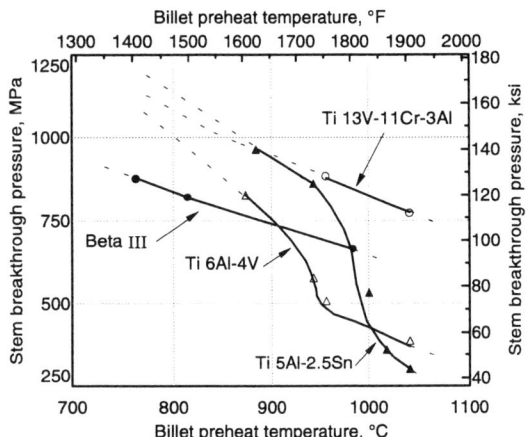

Relationship between stem breakthrough pressure and billet preheat temperature for axisymmetric extrusion of the titanium alloys indicated. Reduction ratio 10.9:1 for Beta III; all others 10.0:1.

Beta III: Flow stress vs strain rate

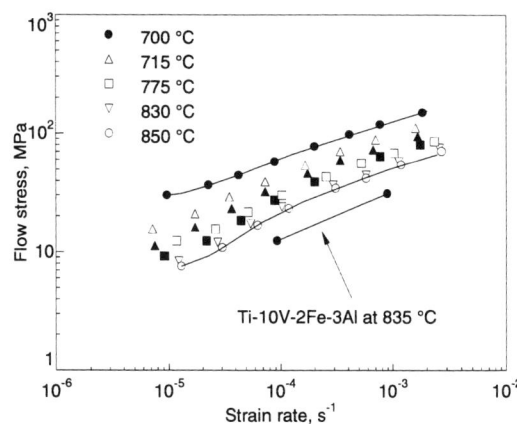

Source: Superplastic Deformation of β-Ti Alloys, *Mater. Sci. Eng.*, Vol 86, 1987, p 159-177

Beta III: Tensile properties of STA forgings at room temperature

Specimen location(a)	Tensile yield strength (0.2%) MPa	ksi	Ultimate tensile strength MPa	ksi	Elongation, %	Reduction of area, %
Panel, T (thin)	1213	176	1296	188	5.0	17
Panel, L (thin)	1275	185	1330	193	5.0	12
Rib, L (thin)	1268	184	1323	192	4.5	10
Center, surface, L (thick)	1261	183	1268	184	2.5	5
Midradius, center, L (thick)	1255	182	1303	189	2.5	6
Center, L (thick)	1255	182	1310	190	4.0	11
Center, short T (thick)	1199	174	1261	183	3.0	10
Center, T (thick)	1172	170	1241	180	5.0	14
Center, end, T (medium)	1193	173	1268	184	5.5	15
Center, end, T (medium)	1199	174	1268	184	4.5	16
Center, end, L (medium)	1193	173	1268	184	4.0	11

Note: Specimens were solution treated and aged: 1 h, 715 °C (1325 °F), water quenched + 8 h, 510 °C (950 °F), air cool. (a) L = longitudinal; T = transverse with respect to grain direction.

Forming

All of the β titanium alloys are highly cold formable (see table), but Beta III has the lowest yield strength combined with excellent ductility of any of the beta compositions. Cold workability is particularly advantageous for rivets and other fasteners. Manufacturing is facilitated by the good formability of Beta III.

Warm forming of Beta III parts can be accomplished to eliminate the springback that occurs during some cold forming operations. Solution treated material may be formed at temperatures as low as 315 °C (600 °F) where yield strength may be as low as 550 to 620 MPa (80 to 90 ksi) and elongation as high as 25%. On the other hand, temperatures of 510 to 540 °C (950 to 1000 °F) were found to minimize springback in hot sizing (see figure). Hot draw forming at 200 °C (400 °F) was found to result in considerable springback. Forming at 510 °C (950 °F) appears to be required for obtaining a precise shape. Either cold or warm forming followed by a hot-sizing operation, which can be a part of the aging heat treatment, appears as an acceptable method of processing Beta III parts.

Beta III: Formability comparison in the annealed condition
Typical tensile and bend properties and optimum formability

Alloy	Tensile yield strength		Tensile elongation, %	Minimum bend radius, height, (R/t)	Olsen cup Headability		D_f/D_i
	MPa	ksi			mm	in.	
Unalloyed Ti	379	55	25	2.0	7.3	0.290	1.87(b)
Ti-8Mn	827	120	15	2.5	7.3	0.290	...
Ti-5Al-6Sn-2Zr-1Mo-0.25Si	862	125	10	...	5.2	0.205	...
Ti-6Al-4V	862	125	15	4.0	4.5	0.180	1.31
Ti-8Mo-8V-2Fe-3Al	827	120	20	2.5	...	...	2.70
Beta III	724	105	25	2.0	8.4	0.330	2.50
Ti-3Al-8V-6Cr-4Zr-4Mo	862	125	20	...	8.4	0.380	2.23
Ti-13V-11Cr-3Al	896	130	15	2.5	6.6	0.260	...

(a) Determined by cold upsetting a specimen having length equal to twice the diameter and measuring the ratio of diameter at fracture (D_f) to initial diameter (D_i). (b) Test discontinued prior to fracture. Source: R.A. Wood, *Beta Titanium Alloys*, MCIC 72-11, Battelle Columbus Laboratories, 1972

Beta III: Springback after hot sizing

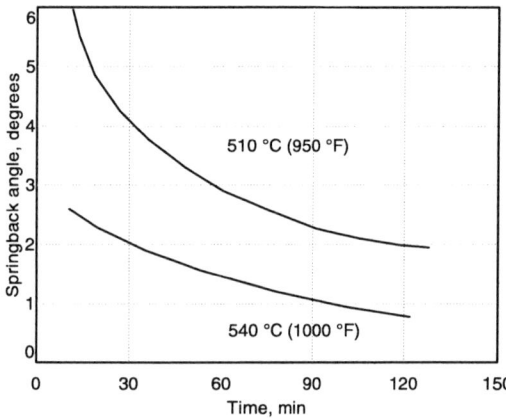

Source: R.A. Wood, *Beta Titanium Alloys*, MCIC 72-11, Battelle Columbus Laboratories, 1972

Sheet

Rolling

The overall cold rollability of Beta III is at least as good as that of commercially pure unalloyed titanium. The cold rolling limit for Beta III, as determined by the onset of edge cracking, is reported to be somewhat greater than 90% reduction. The resistance to rolling is about the same for Beta III and unalloyed titanium (see figure). Because of this capability and ease for cold work, thin strip and foils and other sophisticated mill products can be produced in Beta III relatively inexpensively. Foil has been made down to 0.025 mm (0.001 in.) thick at a width of 635 mm (25 in.).

Beta III: Cold rollability

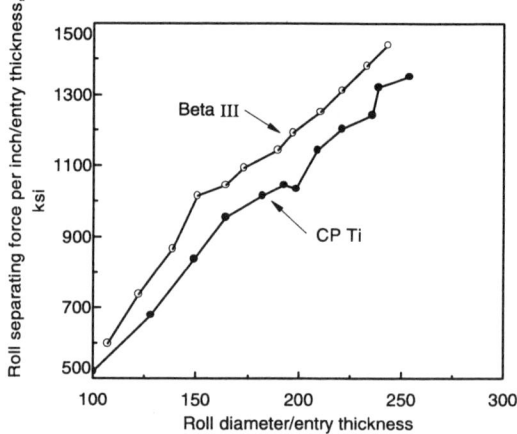

Source: R.A. Wood, *Beta Titanium Alloys*, MCIC 72-11, Battelle Columbus Laboratories, 1972

Sheet Formability

Beta III: Sheet formability comparison

Alloy	Olsen cup height mm	Olsen cup height in.	Transverse bend radius (R/t)	Tensile yield strength MPa	Tensile yield strength ksi	Hardness, HRC
Transage 129, solution treated, fan air cooled	7.1	0.281	2	734	105	27.3
Ti-70A (commercial purity)	7.3	0.290	2	483(a)	70(a)	26
Ti-13V-11Cr-3Al, solution treated	6.3	0.25	3	827(a)	120(a)	32-36
Beta III						
Solution treated, water quenched	8.4	0.330	2	738	107	...
Solution treated, air cooled	6.3	0.25	3.5	883	128	...
Ti-6Al-4V, mill annealed	3.0	0.12	4.5	827	120	36

(a) Guaranteed minimum

Beta III: Sheet formability comparison

Alloy	Condition	Olsen cup height mm	Olsen cup height in.	Minimum transverse bend radius(a), R/t
Beta III	ST, water quench	7.6	.30	2.0
	ST, air cool	6.4	.25	3.5
Ti-6Al-4V	Mill anneal	3.0	.12	4.0
Ti-13V-11Cr-3Al	Mill ST	6.4	.25	2.5
Ti-8Mn	Mill anneal	7.4	.29	2.5
Type 301 stainless steel	Half hard	8.1	.32	2.5

(a) R, bend radius; t, sheet thickness. Source: *Beta Titanium Alloys in the 1980s*, R.R. Boyer and H.W. Rosenberg, Ed., TMS/AIME, 1984

Beta III: Bend ductility and springback of sheet at room temperature

Test direction(a)	Minimum bend radius(b)	Springback angle, °
1.72 mm (0.068 in.) sheet		
Longitudinal	3.7	27
Transverse	2.8	27
Diagonal	2.6	25
0.5 mm (0.020 in.) sheet		
Longitudinal	3.0	22
Transverse	2.5	20
Diagonal	2.0	18

(a) Bend axis with respect to rolling direction. Longitudinal specimen has bend axis parallel to rolling direction. (b) Minimum bend radius, T, where $T = R/t$; R, bend radius; t, sheet thickness. Source: R.A. Wood, *Beta Titanium Alloys*, MCIC 72-11, Battelle Columbus Laboratories, 1972

Beta III: Bend and Olsen cup data for sheet

Sheet thickness mm	Sheet thickness in.	Condition	T-bend(a) ratio, min	Olsen cup height(b) mm	Olsen cup height(b) in.
1.7	0.067	ST 730 °C (1350 °F), AC	3.5	6.3-6.8	0.25-0.27
		ST 730 °C (1350 °F), AC	1.7	7.1-7.8	0.28-0.31
		ST 785 °C (1450 °F), AC	3.0	6.6-7.1	0.26-0.28
		ST 785 °C (1450 °F), WQ	1.7	7.1-7.6	0.28-0.30
0.5	0.020	ST 770 °C (1425 °F), AC	3.5	5.3	0.21
		ST 730 °C (1350 °F), WQ	1.7	6.8	0.27

(a) Transverse coupons were bent through 105° around a mandrel to determine minimum value, which is reported as bend radius/sheet thickness. Bends were examined at 30×. (b) Olsen Cup test conditions: Clamp pressure, 1500 lb; test speed, 22 mm/min (1 in./min); ball indenter (7/8 in.) diameter; die, 25 mm (1 in.) inside diameter. Source: R.A. Wood, *Beta Titanium Alloys*, MCIC 72-11, Battelle Columbus Laboratories, 1972

Sheet Weldments

The weldability of Beta III sheet and plate is considered to be very good. Welds have good ductility with strength comparable to solution treated and aged material. Like all titanium alloys, Beta III is weldable by all methods, except shielded arc and submerged arc welding (because no flux is permitted). Ductility is better if aging is done after welding.

Beta III: Mechanical properties of sheet weldments

Specimen orientation	Preweld aging treatment Temperature °C	°F	Time, h	Postweld aging treatment Temperature °C	°F	Time, h	Ultimate tensile strength MPa	ksi	Tensile yield strength MPa	ksi	Elongation in 50 mm (2 in.), %	Reduction of area, %	Location of tensile break(a)	Minimum bend radius(T), bend ratio(b), R/t
Longitudinal	As ST	...	...	...	...	...	734	106.5	615	89.3	12.0	49.4	...	3.0
	510	950	8	...	...	...	773	112.1	686	99.6	20.0	48.0	...	...
	595	1100	8	...	...	...	781	113.3	692	100.4	17.0	56.1	...	7.0
Transverse	As ST	...	...	...	...	...	753	109.2	657	95.3	2.5	20.6	W	3.0
	510	950	8	...	...	...	803	116.5	735	106.6	6.8	54.0	W	4.6
	595	1100	8	...	...	...	766	111.2	689	100.0	6.3	53.2	W	4.3
Transverse	As ST	...	...	510	950	8	1155	167.6	1137	164.9	1.0	9.1	W	...
				510	950	8	1310	190.0	1104	160.1	4.0	24.7	P	...
				540	1000	8	1035	150.2	993	144.0	3.0	26.9	W	...
				540+	1000+	8	1085	157.4	1043	151.3	2.0	15.9	W	...
				565	1050	8	960	139.3	904	131.1	3.0	25.4	W	7.0

Note: Fusion butt welded. (a) Location of tensile break: W, weld nugget; P, parent metal. (b) Welded sheet bent through 105° around a mandrel. Bends were examined at 30×. R, bend radius; t, thickness. Source: R.A. Wood, *Beta Titanium Alloys*, MCIC 11-72, Battelle Columbus Laboratories, 1972

Fasteners/Springs

Because good cold workability is particularly advantageous for rivets and other fasteners, Beta III in rod and wire form has found use as fasteners and is a candidate for certain spring applications. Beta III rivets can be reliably gun driven in either the solution treated or overaged (565 to 595 °C, or 1050 to 1100 °F) conditions. Great versatility of fastener design is possible with Beta III alloy because of the wide range of strengths and degree of formability available with the material.

Beta III: Typical tensile and shear properties of rod for fasteners

Condition(a)	Ultimate tensile strength MPa	ksi	Tensile yield strength (0.2% offset) MPa	ksi	Elongation in 4D, %	Reduction of area, %	Double shear strength MPa	ksi	Cold head ratio(b), D_f/D_i
As ST	990	144	790	115	24	65	625	91	2.10
ST + age 480 °C (900 °F), 8 h	1365	198	1270	184	15	36	790	115	1.20
ST + age 510 °C (950 °F), 8 h	1300	189	1215	176	18	38	780	113	...
ST + age 540 °C (1000 °F), 8 h	1185	172	1125	163	21	44	750	109	1.40
ST + age 565 °C (1050 °F), 8 h	1090	158	1035	150	25	56	725	105	...
ST + age 595 °C (1100 °F), 8 h	985	143	945	137	27	65	655	95	1.77

(a) Mill processed rod solution treated at 690 to 730 °C (1275 to 1350 °F), water quenched, and aged as indicated. (b) Determined by cold upsetting a specimen with the initial length equal to twice the diameter and measuring the ratio of diameter at fracture to initial diameter. Source: *Beta Titanium Alloys in the 1980's*, R.R. Boyer and H.W. Rosenberg, Ed., TMS/AIME, 1984

Beta III: Tensile properties of wire at room temperature (see also "Mechanical Properties")
Solution heat treated at 1350 °F and water quenched

Processing and heat treatment	Ultimate tensile strength MPa	ksi	Tensile yield strength (0.2% offset) MPa	ksi	Elongation in 4D, of area, %	Reduction, %
1.6 mm (0.063 in.) diameter						
Solution annealed	979	142	703	102	29	83
Solution annealed + 8 h, 480 °C (900 °F)	1406	204	1289	187	13	...
Solution annealed + 8 h, 510 °C (950 °F)	1344	195	1268	184	15	...
3.5 mm (0.140 in.) diameter						
Cold drawn 50%	1110	161	572	83	18	70
Cold drawn + 8 h, 480 °C (900 °F)	1455	211	1386	201	20	45
2.3 mm (0.092 in.) diameter						
Cold drawn 79%	1310	190	896	130	12	60
Cold drawn + 8 h, 425 °C (800 °F)	1860	270	1820	264	7	34
Cold drawn + 8 h, 370 °C (700 °F)	1889	274	1834	266	3	10

Source: R.A. Wood, *Beta Titanium Alloys*, MCIC 11-72, Battelle Columbus Laboratories, 1972

Beta III: Mechanical properties of seamless tubing

Tube size, mm (in.)		Condition	Ultimate tensile strength(c)		Tensile yield strength (0.2% offset)		Elongation in 50 mm (2 in.), %	Reduction of area, %	Flattening(e) test, (OD)f/t	Flaring(f) test, %
OD	Wall		MPa	ksi	MPa	ksi				
(1.5)	(0.173)	As extruded(a)	799	116.0	687	99.7	16.0	75.6	3.7	...
			854	123.9	738	107.1	18.0	72.0	...	...
		As extruded + age 8 h 510 °C (950 °F)	1381(d)	200.4(d)	1270	184.3	8.0	18.4	...	...
			1376(d)	199.6(d)	1264	183.4	8.0	22.6	...	...
(1.26)	(0.120)	MA(b)	991	143.8	924	134.1	13.2	42.0	9.0	13.5
		MA + pickled	988	143.4	935	135.6	13.0	47.5	8.1	12.0
		MA + ST 730 °C (1350 °F), WQ	905	131.3	791	114.8	13.5	50.9	7.0	32.3
		Ma + ST 730 °C (1350 °F), WQ + age 8 h 510 °C (950 °F)	1278	185.4	1215	176.3	7.5	26.7	...	...
		MA + ST 730 °C (1350 °F), WQ + age 8 h 540 °C (1000 °F)	1197	173.7	1136	164.8	9.2	36.5	...	...
(0.884)	(0.050)	MA(b)	1066	154.7	972	141.0	8.5	16.6	11.8	16.3
		MA + pickled	1041	151.0	953	138.2	8.5	24.7	12.2	15.7
		MA + ST 730 °C (1350 °F) + ST 730 °C (1350 °F), WQ	809	117.4	697	101.2	20.8	36.0	6.6	27.8
		MA + ST 730 °C (1350 °F), WQ + age 8 h 510 °C (950 °F)	1244	180.4	1165	169.0	6.0	12.3	...	...
		MA + ST 730 °C (1350 °F), WQ + age 8 h 540 °C (1000 °F)	1170	169.8	1101	159.8	7.2	15.6	...	...

(a) Hollow billets, 78 mm (2.8 in.) OD by 32 mm (1.260 in.) ID, were canned in mild steel, heated to 815 °C (1500 °F), extruded to tube hollow (reduction ratio 7.2:1) and water quenched. (b) Processing of tube hollows to size was done in a single cold pass with tube reducers and was followed by a vacuum mill anneal: 30 min 730 °C (1350 °F) fast cool + 4 h at 620 °C (1150 °F) + FC to 480 °C (900 °F). (c) Tensile specimen machined from a longitudinal section of tubing subtended by a 19 mm (0.75 in.) chord. Gage length, 50 mm (2 in.). Reduced section, 13 mm (0.5 in.). (d) Machined round specimen. Gage length, 13 mm (0.5 in.). Reduced section, 3.1 mm (0.125 in.). (e) Flattening performance is expressed as the ratio of the height (outside diameter) at failure/wall thickness. (f) Flaring performance is expressed as the percentage increase in outside diameter at the point of failure. Source: R.A. Wood, *Beta Titanium Alloys*, MCIC 11-72, Battelle Columbus Laboratories, 1972

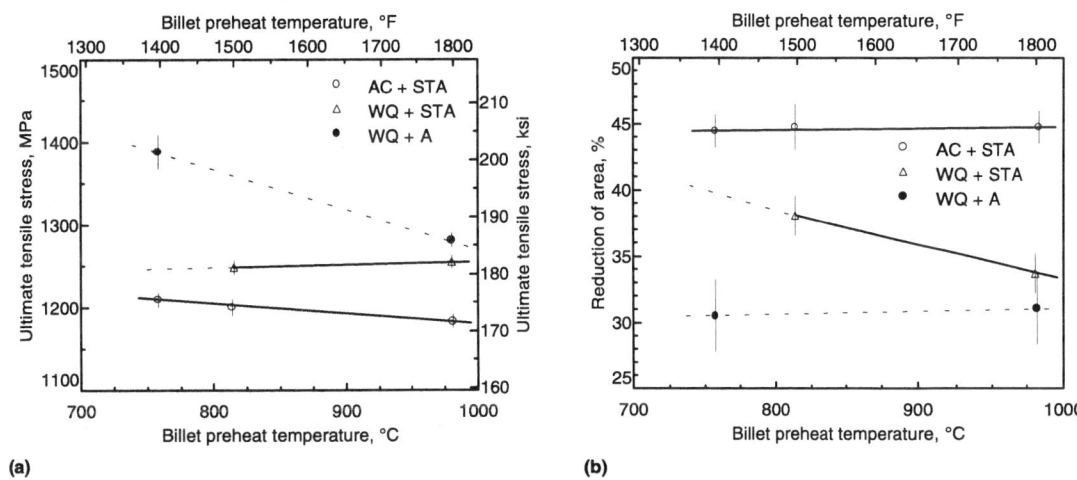

Effect of billet preheat treatment on the room-temperature properties of Beta III extrusions. AC, air cooled after extruding; WQ, water quenched after extruding; STA, 30 min at 760 °C (1400 °F), water quench, plus 8 h at 480 °C (900 °F), air cool.
Source: R.A. Wood, *Beta Titanium Alloys*, MCIC 72-11, Battelle Columbus Laboratories, 1972

Heat Treatment

Beta III may be supplied in either the solution treated (highly formable) or the solution heat treated plus aged condition (high strength or moderately high strength, depending on treatment selected). Heat treatment may consist of simply aging the material from the solution treated condition supplied, or re-solution treatment prior to aging may be preferred to achieve maximum ductility and toughness in heavily worked areas of the aged part. The best combination of properties is obtained after solution treatment near the β transus.

Stress relief annealing for Beta III may consist of re-solution heat treatment or an aging heat treatment. Re-solution treatment may be accomplished in as short a time as 1 to 2 min at 715 to 730 °C (1325 to 1350 °F), followed by either water quenching or air cooling. No special treatment is used as a stress relief annealing treatment.

Annealing of Beta III is also the same as solution treating. The recommended annealing temperature range for Beta III alloys is 705 to 760 °C (1300 to 1400 °F), generally for short times and usually followed by either water quenching or air cooling.

Beta III: Solution treatments

Heat treatment	Temperature(a) °C	°F	Time, min	Cooling method(b)
Typical ST	730-785	1350-1450	5	AC or WQ
Low ST for rod, wire, etc.	690-730	1275-1350	5	AC or WQ
High ST for thicker section	815-870	1500-1600	5(c)	AC or WQ

(a) The best combination of properties is obtained by solution treatment near the beta transus. (b) Either air cool (AC) or water quench (WQ) might allow the same aging response depending on section thickness. (c) Exposure is usually short, but may be longer for thicker sections.

Aging

The recommended aging heat treatment for producing the high strength condition in Beta III alloy is 8 h at 480 °C (900 °F) followed by air cooling. An overaged condition may be achieved by exposure for 8 h at 595 °C (1100 °F) and air cooling.

Precipitation of the α phase in 8-h exposure in the 480 to 595 °C (900 to 1100 °F) temperature range results in tensile strengths in the range 1380 to 930 MPa (200 to 135 ksi), respectively.

Beta III: Aging vs tensile properties of sheet

Room-temperature tensile properties of 1.7 mm (0.067 in.) thick hand sheet as solution treated and after various aging treatments

8-h aging temperature °C	°F	Specimen direction	Ultimate tensile strength(a) MPa	ksi	Tensile yield strength (0.2%) MPa	ksi	Elongation in 50 mm (2 in.), %	Reduction of area, %
Solution treated 3 min, 730 °C (1350 °F) and air cooled								
None		L	955	138.5	900	130.6	10.8	56.2
		T	925	134.2	901	130.7	10.0	52.4
480	900	L	1397	202.7	1350	195.8	3.7	18.2
		T	1418	205.7	1397	202.6	2.7	9.3
510	950	L	1311	190.2	1208	175.3	7.3	33.2
		T	1383	200.6	1297	188.1	5.5	23.9
540	1000	L	1201	174.3	1117	162.1	6.0	37.2
		T	1184	171.8	1100	159.6	7.0	37.2
595	1100	L	966	140.1	903	131.0	11.0	62.4
Solution treated 3 min 730 °C (1350 °F) and water quenched								
None		L	837	121.4	753	109.2	19.2	61.2
		T	809	117.4	718	104.2	29.7	66.2
480	900	L	1370	198.7	1337	194.0	4.0	26.4
		T	1419	205.8	1407	204.1	2.8	11.3
510	950	L	1348	195.6	1247	180.9	5.5	26.5
		T	1354	196.4	1279	185.6	5.5	30.1
540	1000	L	1237	179.4	1173	170.1	5.0	40.5
		T	1288	186.9	1223	177.4	5.0	29.8

(a) Tensile properties are average of duplicate tests. Source: R.A. Wood, *Beta Titanium Alloys*, MCIC 11-72, Battelle Columbus Laboratories, 1972

Beta III: Effect of pre-age and cold work on age time

Pre-age, min, at: 315 °C (600 °F)	370 °C (700 °F)	Cold work, %	Time to full strength, min
0	0	0	1440
2	10	0	2
2	2	0	10-20
2	2	10	7
2	2	20	3
10	0	0	20
0	5	0	100

Note: Effects pre-aging and cold work on aging time at 450 °C (840 °F) to reach full strength. Source: H.W. Rosenberg, Property Scatter in Beta Titanium: Some Problems and Solutions, *Beta Titanium Alloys in the 1980's*, TMS/AIME, 1984

Beta III: Aging vs tensile properties of sheet
Room-temperature tensile properties of 13 mm (0.5 in.) thick plate as rolled, solution treated, and aged

Solution treatment temperature		Quench delay, s	8-h aging temperature		Ultimate tensile strength		Tensile yield strength (0.2%)		Elongation in 4D, %	Reduction of area, %	Charpy V-notch impact toughness(a)	
°C	°F	s	°C	°F	MPa	ksi	MPa	ksi	%	%	J	ft·lbf
As hot rolled(a)			...		992	143.9	921	133.6	20.0	67.2	10.8	8
785(b)	1450(b)	15	...		922	133.7	862	125.1	20.0	69.5	20.3	15
			480	900	1359	197.2	1281	185.9	4.0	8.2	10.8	8
			540	1000	1297	188.1	1207	175.1	4.0	13.7	...	...
			595	1100	1086	157.6	1026	148.8	10.0	22.6	14.9	11
730(b)	1350(b)	15	...		834	121.0	728	105.6	34.0	68.3	35.2	26
			480	900	1343	194.9	1248	181.1	2.0	8.2	...	...
			510	950	1288	186.9	1200	174.1	5.0	10.7	10.8	8
			540	1000	1219	176.9	1137	165.0	6.0	11.5	10.8	8
			565	1050	1143	165.8	1057	153.3	8.0	13.4	...	...
			595	1100	1090	158.2	1023	148.4	8.0	29.7	13.5	10
730(b)	1350(b)	30	...		949	137.7	906	131.4	20.0	56.6	23.0	17
			480	900	1397	202.6	1277	185.3	4.0	8.1	6.7	5
			540	1000	1255	182.0	1158	168.0	6.0	14.8	12.2	9
			595	1100	1036	150.3	979	142.1	13.0	26.7	...	...

(a) At −40 °C (−40 °F). (b) Plate 27.9 mm (1.1 in.) thick was hot rolled to 13 mm (0.5 in.) thickness using furnace temperature 925 °C (1700 °F), reheat temperature of 925 °C (1700 °F), final pass exit temperature 895 °C (1640 °F). Plate straightened warm without reheating. (c) Panels were laboratory solution treated for 15 min at the indicated temperature and water quenched with the indicated delay time. Source: R.A. Wood, *Beta Titanium Alloys*, MCIC 11-72, Battelle Columbus Laboratories, 1972

Beta III: Effect of pre-aging and cold work

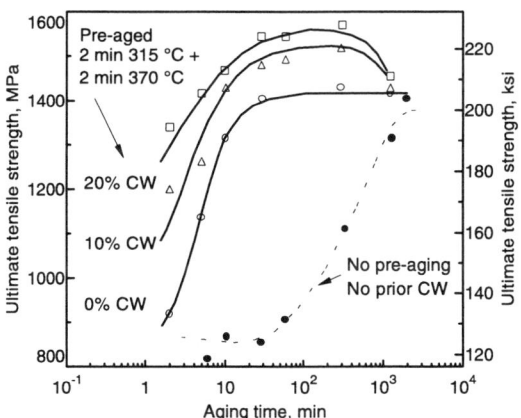

Effect of pre-aging heat treatments with and without a prior cold work on the tensile strength of aged (450 °C, 840 °F) specimens. Solution heat treated 5 min at 730 °C (1350 °F) and water quenched prior to cold working or pre-aging.
Source: R.A. Wood, *Beta Titanium Alloys*, MCIC 72-11, Battelle Columbus Laboratories, 1972

Beta III: Tensile strength vs age temperature

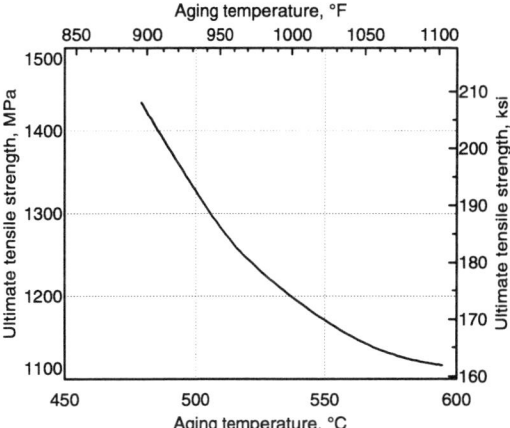

Effect of 8-h aging temperature on tensile strength of specimens solution treated at 730 to 745 °C (1350 to 1375 °F).
Source: R.A. Wood, *Beta Titanium Alloys*, MCIC 72-11, Battelle Columbus Laboratories, 1972

Beta III: Effect of aging temperature and cold work on tensile properties

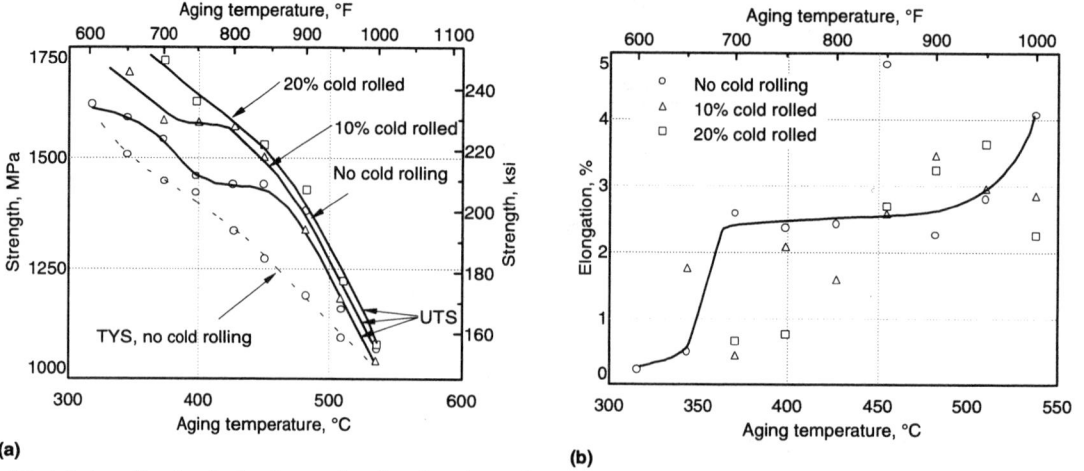

(a) (b)
Effect of prior cold work on the tensile properties of aged specimens. Solution heat treated 5 min at 730 °C (1350 °F) and water quenched prior to cold working or aging. Aging times not given except that they are peak aging times.

Ti-3Al-8V-6Cr-4Mo-4Zr (Beta C)

Common Name: Beta C™, 38-6-44
UNS Number: R58640

Ti-3Al-8V-6Cr-4Mo-4Zr (Beta C) is a commercial alloy developed by RMI in the mid-to-late 1960s. It has similar characteristics to Ti-13V-11Cr-3Al (13-11-3), but is easier to melt and shows less segregation. Beta C was developed as an improvement of 13-11-3 which had some melting problems due to a high chromium content. Due to its significant molybdenum content, Beta C exhibits superior resistance to reducing acids and chloride crevice corrosion compared to other high-strength titanium alloys. Currently, Beta C holds a small amount (much less than 1%) of the production market.

Chemistry and Density

Beta C is formulated by depressing the beta transus with the beta isomorphous elements, molybdenum and vanadium, and the sluggish beta eutectoid element, chromium. It is slightly more beta-stabilized than Ti-11.5Mo-6Zr-4.5Sn (Beta III) and less beta-stabilized than Ti-13V-11Cr-3Al.

Density. 4.82 g/cm^3 (0.174 lb/in.^{3})

Product Condition/Microstructure

Beta C is cold rollable and drawable, and is used mainly as bar and wire material for aircraft springs; it has also been explored as a spring material for automotive applications. It constitutes less than 1% of titanium products. Beta C can be heated to high levels above 1380 MPa, 200 ksi—by aging between 480 and 595 °C (900 and 1100 °F).

Large variations in tensile strength can be obtained by varying the aging temperature and time. A portion of the beta phase transforms to a finely dispersed alpha during aging. Also, it does not grain-coarsen as rapidly as other beta alloys when heat treated or worked at temperatures above the beta transus.

Applications

Beta C is used in fasteners, springs, torsion bars, and in foil form for making cores for sandwich structures. It is also used for tubulars and casings in oil, gas, and geothermal wells.

Use Limitations. Beta C, like other beta titanium alloys, is highly susceptible to hydrogen pickup and rapid hydrogen diffusion during heating, pickling, and chemical milling. However, because of the much higher solubility of hydrogen in the beta phase than in the alpha phase of titanium, this alloy has a higher tolerance for hydrogen than the alpha or alpha-beta alloys.

Beta C can be welded in the solution-treated condition; however, welding is not recommended after solution treating and aging. Care is necessary in pickling to minimize hydrogen absorption.

Ti-3Al-8V-6Cr-4Mo-4Zr (Beta C): Specifications

Specification	Designation	Description	Al	Cr	Fe	Mo	N	O	V	Zr	Other
	UNS R58640		3	6		4			8	4	bal Ti
USA											
AMS 4957		Bar Wir CD	3-4	5.5-6.5	0.3	3.5-4.5	0.03	0.12	7.5-8.5	3.5-4.5	H 0.03; C 0.05; OT 0.4; Y 0.005; bal Ti
AMS 4958		Bar Rod STA	3-4	5.5-6.5	0.3	3.5-4.5	0.03	0.12	7.5-8.5	3.5-4.5	H 0.03; C 0.05; OT 0.4; Y 0.005; bal Ti
MIL T-9046J	Code B-3	Sh Strp Plt SHT	3-4	5.5-6.5	0.3	3.5-4.5	0.03	0.12	7.5-8.5	3.5-4.5	H 0.02; C 0.05; OT 0.4; bal Ti
MIL T-9046J	Code B-3	Sh Strp Plt STA	3-4	5.5-6.5	0.3	3.5-4.5	0.03	0.12	7.5-8.5	3.5-4.5	H 0.02; C 0.05; OT 0.4; bal Ti
MIL T-9047G	Ti-3Al-8V-6Cr-4Mo-4Zr	Bar Bil STA	3-4	5.5-6.5	0.3	3.5-4.5	0.03	0.12	7.5-8.5	3.5-4.5	H 0.02; C 0.05; OT 0.4; Y 0.005; bal Ti
MIL T-9047G	Ti-3Al-8V-6Cr-4Mo-4Zr	Bar Bil SHT	3-4	5.5-6.5	0.3	3.5-4.5	0.03	0.12	7.5-8.5	3.5-4.5	H 0.02; C 0.05; OT 0.4; Y 0.005; bal Ti

Ti-3Al-8V-6Cr-4Mo-4Zr (Beta C): Commercial compositions

Specification	Designation	Description	Al	Cr	Fe	Mo	N	O	V	Zr	Other
USA											
Astro	Ti-3Al-8V-6Cr-4Zr-4Mo	Bar Sprg Pip	3-4	5.5-6.5		3.5-4.5	0.03 max	0.14 max	7.5-8.5	3.5-4.5	C 0.05 max; bal Ti
Oremet	Ti-38-6-44										
RMI	3Al-8V-6Cr-4Zr-4Mo	Sh Plt Bar Bil Wir Ex	3-4	5.5-6.5	0.3	3.5-4.5	0.03	0.14	7.5-8.5	3.5-4.5	C 0.05; bal Ti
Teledyne	Tel-Ti-3Al-8V-6Cr-4Mo-4Zr		3-4	5.5-6.5	0.3	3.5-4.5	0.03	0.14	7.5-8.5	3.5-4.5	
Timet	TIMETAL 3-8-6-4-4	Ing Bil STA	3-4	5.5-6.5	0.3	3.5-4.5	0.03	0.14	7.5-8.5	3.5-4.5	H 0.02; C 0.05; bal Ti

Phases and Structures

As a solute-rich β alloy, precipitation of α within the solute-lean β regions (β') of Beta C is slow. Prior cold work accelerates the formation of intragranular α and also reduces the extent of grain boundary α. Peak aging occurs at around 480 °C (900 °F), and smaller quantities of α (in the form of coarse precipitates) are found at higher temperatures. Type 2 α occurs during certain aging treatments. Recrystallization occurs after short times above the β transus, although β grain growth is not a problem. The possibility of a second phase responsible for inhibiting grain growth above the β transus has been suggested (R.A. Wood and R.J. Favor, *Titanium Alloys Handbook*, MCIC-HB-02, Battelle Columbus Laboratories, 1972, Section 1-12, p 72-1).

Beta C: Effect of aging temperature

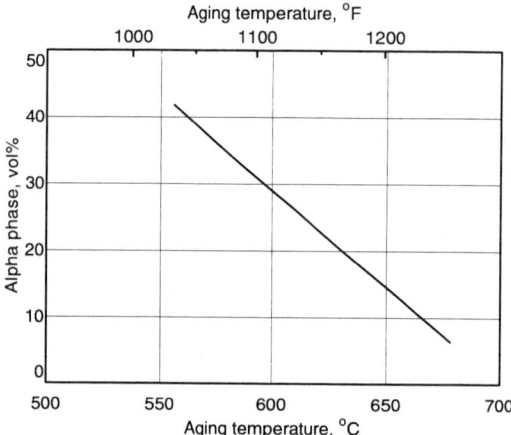

Effect of aging temperature on amount of α phase precipitation in 5 mm (1.25 in.) thick plate solution heat treated at 925 °C (1700 °F) and aged.
Source: R.A. Wood and R.J. Favor, *Titanium Alloys Handbook*, MCIC-HB-02, Battelle Columbus Laboratories, 1972, Section 1-12, p 72-1

Beta C: Variation of β lattice parameter

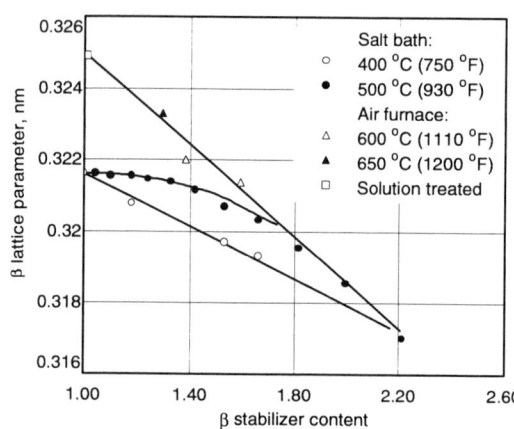

Variation of β lattice parameter with β stabilizing alloying element content (normalized to unity at zero volume fraction of α).
Source: G.H. Isaac and C. Hammond, The Formation of Type 2 α Phase in Ti-3Al-8V-6Cr-4Zr-4Mo, in *Titanium, Science and Technology*, G. Lütjering, U. Zwicker, and W. Bunk, Ed., Deutsche Gesellschaft für Metallkunde eV, Germany, 1985, p 1608

Beta Transus. 730 °C (1350 °F). The previously published transus temperature of 795 °C (1460 °F) is too high.

Beta C: Lattice parameters of the α and β phases in solution heat treated and aged plate

Aging temperature		Lattice spacing, Å (±0.004 Å)		
		Alpha phase		Beta phase
°C	°F	a	c	a_o
565	1050	2.948	4.680	3.218
620	1150	2.936	4.682	3.226
675	1250	2.920	4.684	3.229

Note: 32 mm (1.25 in.) plate was solution treated at 925 °C (1700 °F) and aged 8 h. Source: R.A. Wood and R.J. Favor, *Titanium Alloys Handbook*, MCIC-HB-02, Battelle Columbus Laboratories, 1972, Section 1-12, p 72-1

Transformations

Beta C: Transformation diagram

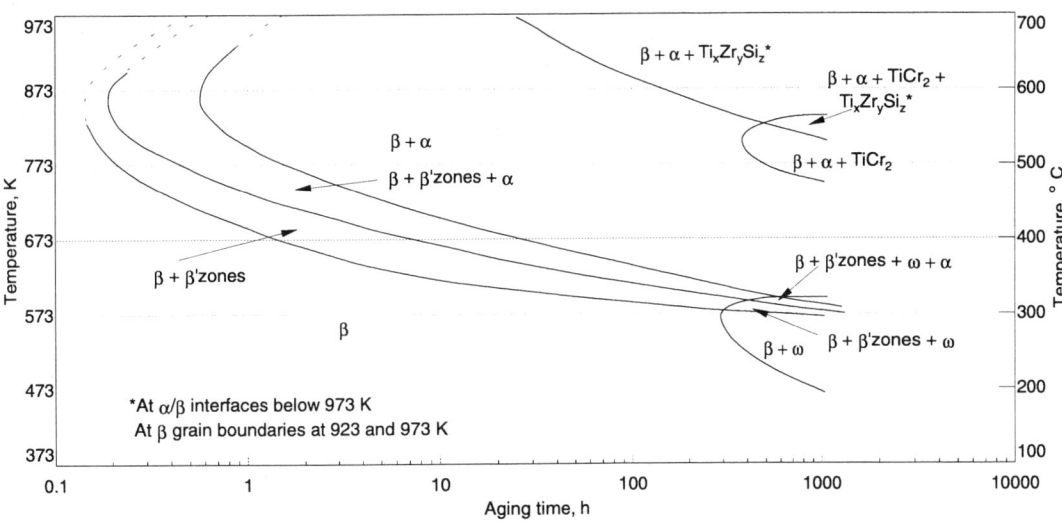

Solution treated, air quenched and aged RMI 38644. Zones were probably β' but was not confirmed.
Source: T.J. Headley and H.J. Rack, *Metall. Trans. A*, Vol 10, 1979, p 909

Physical Properties

Elastic Properties

Annealed or solution treated Beta C has a tensile modulus of between 78 and 91 GPa (11.4 and 13.2×10^6 psi). Overaging produces slightly higher tensile strength and elastic modulus (up to 96.5 GPa, or 14×10^6 psi), whereas full age hardening results in moduli in the range of 100 GPa (15×10^6 psi) and up to 124 GPa (18×10^6 psi).

Beta C: Summary of typical physical properties

Beta transus	730 ± 15 °C (1350 ± 25 °F)
Melting (liquidus) point	~1650 °C (3000 °F)
Density(a)	4.82 g/cm³ (0.174 lb/in.³)
Electrical resistivity(a)	1.6 μΩ · m
Magnetic permeability	Nonmagnetic
Specific heat capacity(a)	515 J/kg · K (0.123 Btu/lb · °F)
Thermal conductivity(a)	6.2 W/m · K (3.6 Btu/ft · h · °F)
Thermal coefficient of linear expansion(b)	8.3×10^{-6}/°C (4.6×10^{-6}/°F)

(a) Typical values at room temperature of about 20 to 25 °C (68 to 78 °F). (b) Mean coefficient from room temperature to 100 °C (212 °F)

Beta C: Modulus of elasticity vs tensile yield strength

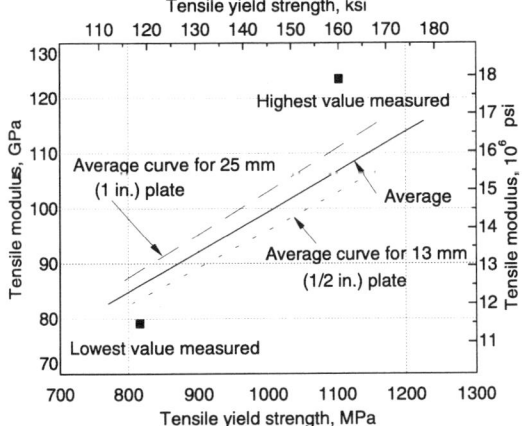

Measured from plate in several annealed and aged conditions.
Source: R.A. Wood, *Beta Titanium Alloys*, MCIC 72-11, Battelle Columbus Laboratories, 1972, p 180

Beta C: Tensile and compressive moduli vs temperature

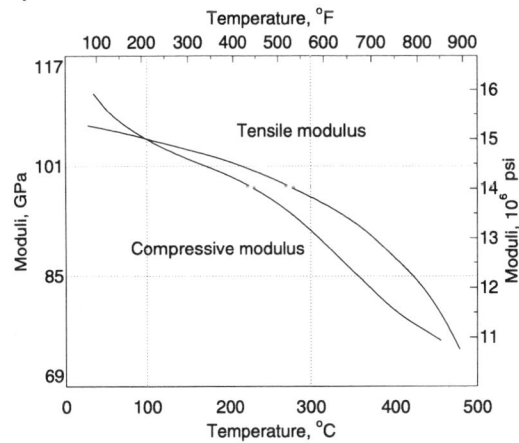

Specimens were solution treated and aged forged billets, 15 min at 815 °C (1500 °F), AC + 12 h at 565 °C (1050 °F), AC.
Source: R.A. Wood, *Beta Titanium Alloys*, MCIC 72-11, Battelle Columbus Laboratories, 1972

Beta C: Transverse tangent modulus

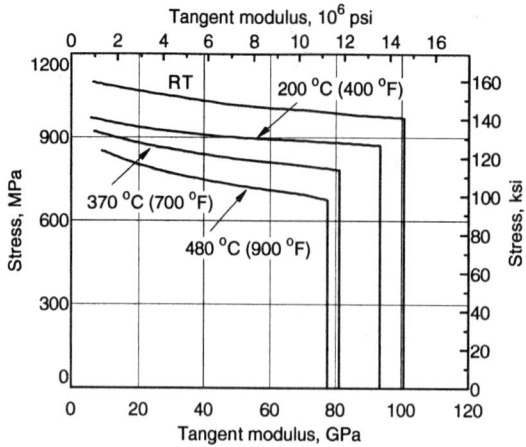

Specimens were 150 mm (6 in.) square billets solution treated at 815 °C (1500 °F) for 15 min, AC, aged at 565 °C (1050 °F) for 2 h, AC.
Source: *Aerospace Structural Metals Handbook*, Vol 4, Code 3723, Battelle Columbus Laboratories, 1972

Beta C: Longitudinal tangent modulus

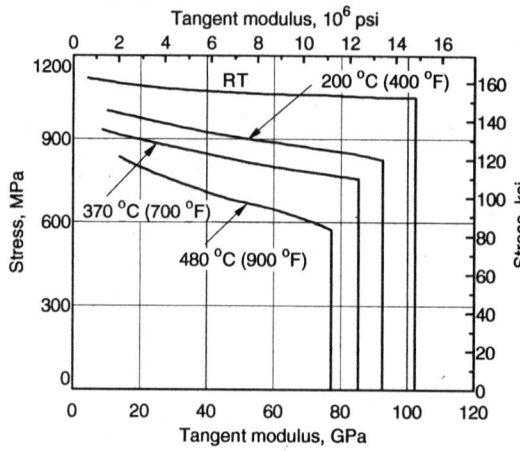

Specimens were 150 mm (6 in.) square billets, solution treated at 815 °C (1500 °F) 15 min, AC, aged at 565 °C (1050 °F) for 2 h, AC.
Source: *Aerospace Structural Metals Handbook*, Vol 4, Code 3723, Battelle Columbus Laboratories, 1972

Beta C: Compressive strength and modulus at high temperatures

Temperature		Test direction	Compressive modulus		Compressive yield strength	
°C	°F		GPa	10⁶ psi	MPa	ksi
RT	RT	L	102	14.8	1110	161
		T	101	14.7	1070	155
200	400	L	93	13.5	965	140
		T	94	13.7	945	137
370	700	L	85	12.4	895	130
		T	82	11.9	890	129
480	900	L	77	11.2	780	113
		T	78.5	11.4	800	116

Note: Full section tests on 150 mm (6 in.) diam forgings; solution treated at 815 °C (1500 °F), AC, aged at 565 °C (1050 °F) for 4 h, AC. Source: D.H. Wilson and C.M. Esler, in *Beta Titanium in the 1980's*, R.R. Boyer and H.W. Rosenberg, Ed., TMS/AIME, 1984, p 470

Electrical Resistivity

Beta C: Electrical resistivity

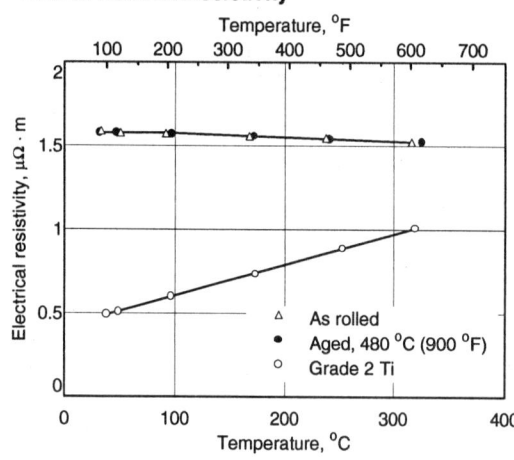

Beta C: Shear modulus of bar after aging

Aging treatment	Shear modulus	
	GPa	10⁶ psi
480 °C (900 °F), 12 h, AC	41.0	5.95
480 °C (900 °F), 24 h, AC	43.3	6.28
495 °C (925 °F), 8 h, AC	40.7	5.91
510 °C (950 °F), 8 h, AC	41.6	6.04

Note: Modulus determined statistically from torsion data on 6.66 mm (2.625 in.) diam bar. Treatment was solution annealed at 925 °C (1700 °F) for 30 min, AC, and 815 °C (1500 °F) for 30 min, AC, prior to aging. Source: *Aerospace Structural Metals Handbook*, Vol 4, Code 3723, Battelle Columbus Laboratories, 1972

Chemical/Corrosion Properties

General Corrosion

Molybdenum additions improve the corrosion resistance of titanium alloys in reducing media, and this effect is evidenced by the general corrosion rates of Beta C in reducing media such as hydrochloric and sulfuric acid (see figures). This increase in reducing environment resistance is achieved, however, at the expense of corrosion resistance in oxidizing environments such as nitric acid (see figures). Oxidizing agents such as ferric chloride ($FeCl_3$) have a similar adverse effect on Beta C corrosion in sulfuric acid.

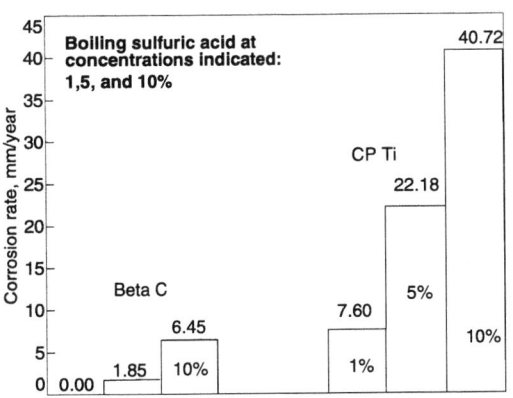

Beta C and CP Ti: Corrosion in boiling sulfuric acid

No corrosion was observed for Beta C at 1% concentration.
Source: D.E. Thomas, S. Ankem, W.D. Goodin, and S.R. Seagle, in *Industrial Applications of Titanium and Zirconium: Fourth Volume*, C.S. Young and J.C. Durham, Ed., ASTM STP 917, ASTM, 1986, p 144-163

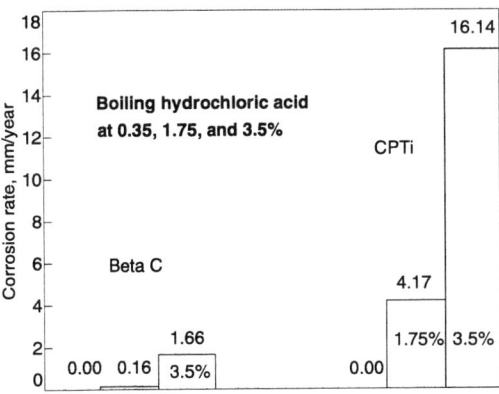

Beta C and CP Ti: Corrosion in boiling HCl

No corrosion was observed for Beta C at 0.35% concentration.
Source: D.E. Thomas, S. Ankem, W.D. Goodin, and S.R. Seagle, in *Industrial Applications of Titanium and Zirconium: Fourth Volume*, C.S. Young and J.C. Durham, Ed., ASTM STP 917, ASTM, 1986, p 144-163

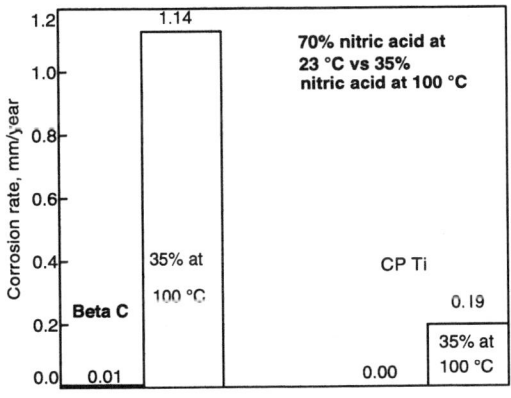

Beta C and CP Ti: Corrosion in nitric acid

No corrosion was observed for Beta C at 70% concentration.
Source: D.E. Thomas, S. Ankem, W.D. Goodin, and S.R. Seagle, in *Industrial Applications of Titanium and Zirconium: Fourth Volume*, C.S. Young and J.C. Durham, Ed., ASTM STP 917, ASTM, 1986, p 144-163

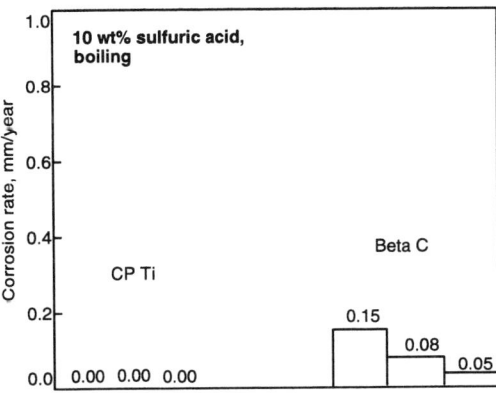

Beta C and CP Ti: Inhibition by $FeCl_3$ in sulfuric acid

Ferric chloride as an oxidizing agent is an effective inhibitor of sulfuric acid attack of CP Ti, but adversely affects Beta C. Tested at 1, 10, and 50 g/L of ferric chloride.
Source: D.E. Thomas, S. Ankem, W.D. Goodin, and S.R. Seagle, in *Industrial Applications of Titanium and Zirconium: Fourth Volume*, C.S. Young and J.C. Durham, Ed., ASTM STP 917, ASTM, 1986, p 144-163

Beta C: Corrosion rates in specific media

Medium	Concentration, %	Temperature, °C	Corrosion rate, mm/yr
Ferric chloride	10	Boiling	nil
Formic acid	50	Boiling	0.98
Hydrochloric acid	0.5	Boiling	0.003
	1.0	Boiling	0.058
	1.5	Boiling	0.26
Hydrochloric acid, aerated	pH 1	Boiling	nil
Sulfuric acid,			
naturally aerated	1	Boiling	nil
	5	Boiling	1.85
Sulfuric acid + 3% $Fe_2(SO_4)_3$	50	Boiling	<0.03
Sulfuric acid + 1 g/L $FeCl_3$	10	Boiling	0.15
Sulfuric acid + 50 g/L $FeCl_3$	10	Boiling	0.05

These data should be used only as a guideline for alloy performance. Rates may vary depending on changes in medium chemistry, temperature, length of exposure, and other factors. Total alloy suitability cannot be assumed from these values alone, because other forms of corrosion, such as localized attack, may be limiting. In complex, variable, and/or dynamic environments, *in situ* testing may provide more reliable data. Source: *Metals Handbook*, Vol 13, 9th ed., 1987

Pitting

Because Beta C is less resistant to oxidizing media than CP Ti, it is expected that resistance to pitting would likewise suffer. For example, Beta C has pitting potentials around 1.5 to 2.0 V (SCE) in saturated sodium chloride at 80 °C (180 °F). This is a significantly lower potential than CP Ti, but it is greater than the pitting potentials of most titanium alloys. A similar trend is observed for repassivation potentials (see table), which represent conservative measures of anodic pitting tendency because they are the minimum potentials below which pitting cannot be sustained.

Crevice Corrosion

In contrast to the anodic breakdown associated with pitting, crevice corrosion is usually the result of acidification in the crevice region by oxidant depletion. Therefore, Beta C is expected to outperform CP Ti in terms of crevice corrosion resistance (see table).

Beta C: Repassivation potential
Repassivation potentials in boiling chloride solutions for as-annealed alloys and unalloyed titanium

	Repassivation potential, V(a)		
Alloy	5% NaCl (pH 3.5)	3% HCl	Saturated NaCl
Grade 1	...	...	+7.0
Grade 2	+6.7	+5.8	+5.7
Ti-6-4	+2.3	+1.7	...
Ti-6-2-4-6	+3.0	+2.4	...
Beta C	+3.2	+2.6	...
Ti-15-5	+6.3	+5.6	...
Grade 12	...	...	+5.9
Grade 7	...	...	+5.6

(a) Measured versus Ag/AgCl reference electrode. Source: R. Schutz and D. Thomas, "Corrosion of Titanium and Titanium Alloys," in *Metals Handbook*, Vol 13, 9th ed., 1987, p 669

Beta C: Crevice corrosion compared to CP Ti
Crevice corrosion was measured in 250 g/L NaCl, 15 psi H_2S, 1 g/L sulfur, pH 3.0 after 720 h

Temperature			Grade 12	
°C	°F	CP Ti	(Ti-0.3Mo-0.8Ni)	Beta C
25	78	No	No	No
50	120	No	No	No
75	165	Yes	No	No
100	210	Yes	...	...
200	480	...	Yes	No
250	570	...	Yes	No
300	572	...	Yes	No

Note: Artificial crevice produced by a flat PTFE washer clamped to the titanium surface with a Beta C nut and bolt arrangement. The ratio of crevice to uncreviced area was approximately 1 to 10. Source: D.E. Thomas, S. Ankem, W.D. Goodin, and S.R. Seagle, in *Industrial Applications of Titanium and Zirconium: Fourth Volume*, C.S. Young and J.C Durham, Ed., ASTM STP 917, ASTM, 1986

Stress-Corrosion Cracking

Aqueous Media

In β titanium alloys, the β phase may be susceptible to either transgranular or intergranular stress-corrosion cracking (SCC), depending on alloy composition and microstructure. Intergranular cracking has only been observed in a few aged β alloys, particularly with fine α precipitates formed at lower aging temperatures. Beta phase stabilized by either molybdenum, vanadium, niobium, or tantalum is immune to SCC, except for a β phase stabilized by the eutectoid elements manganese and chromium. Although β + ω phase structures appear to be highly resistant, the precipitation of compounds, such as $TiCr_2$ in Beta C can be expected to degrade cracking resistance.

Only limited published data on the influence of temperature on titanium alloy SCC behavior in aqueous media are available. Data for Beta C suggests that K_{Iscc} values may be relatively unaffected by temperatures as high as 93 °C (200 °F) in neutral salt solutions. However, a few titanium alloys exhibiting negligible SCC tendencies in near ambient saltwater may exhibit well-defined temperature thresholds for SCC susceptibility in high-temperature brines. Beta C, for example (STA condition), resists SCC in near-neutral NaCl brines up to ~180 °C (355 °F), but can exhibit stress cracking during C-ring and slow strain-rate testing above this temperature. The presence of hydrogen sulfide and/or sulfur appears to impart little effect.

In this case, the SCC threshold temperatures can be raised significantly (by 15 to 80 °C, or 27 to 144 °F) by microstructural refinement and/or minor (≤0.1%) additions of palladium (R.W. Schutz, M. Xiao, and T.A. Bednarowicz, "Stress Corrosion Behavior of the Ti-3Al-8V-6Cr-4Zr-4Mo Titanium Under Deep Sour Gas Well Conditions," Paper No. 51, presented at NACE Corrosion '92, Nashville, Apr 27-May 1, 1992).

In most SCC-susceptible titanium alloys, anodic or cathodic polarization tends to inhibit SCC and increase K_{Iscc} values. Increasing potential also serves to increase stress cracking velocity in neutral halide solutions. In highly acidic solutions, however, stage II crack velocity becomes independent of applied potential. As a result, inhibition via cathodic polarization is not achievable in highly acidic solutions.

Beta C: Fracture toughness in air and 3.5% NaCl solution at 25 °C

Thickness		Yield strength		K_{Ic} or K_c		K_{Iscc} or K_{scc}	
mm	in.	MPa	ksi	MPa√m	ksi √in.	MPa√m	ksi √in.
13	0.50	1330	193	55	50	40	47
14	0.56	1260	183	38	35	38	35
>6	>0.25	1135	165	52	48	52	48
>6	>0.25	985	143	56	51	56	51

Note: Specimens were β solution treated and aged. The data were generated in ambient neutral 3.5% NaCl solution. These K_{Iscc} values are highly dependent on alloy composition, metallurgical condition, and product form and thickness and, therefore, may or may not be representative of commercially available products. Source: R. Schutz, Stress-Corrosion Cracking of Titanium Alloys, in *Stress-Corrosion Cracking: Materials Performance and Evaluation*, ASM International, 1992

Methanol and Other Alcohols

Unlike other engineering metals, titanium and zirconium alloys are unique in their strong susceptibility to stress-corrosion cracking (SCC) in methanol liquid and vapor. Stress-corrosion cracking susceptibility is primarily limited to methanol alcohol for most commercial titanium alloys including Beta C. Titanium alloys that are susceptible to SCC in aqueous solutions tend to exhibit intergranular stage I cracking and transgranular stage II cracking, although there are exceptions. Exceptions to this behavior include Beta III and fine-grained Beta C, which exhibit intergranular

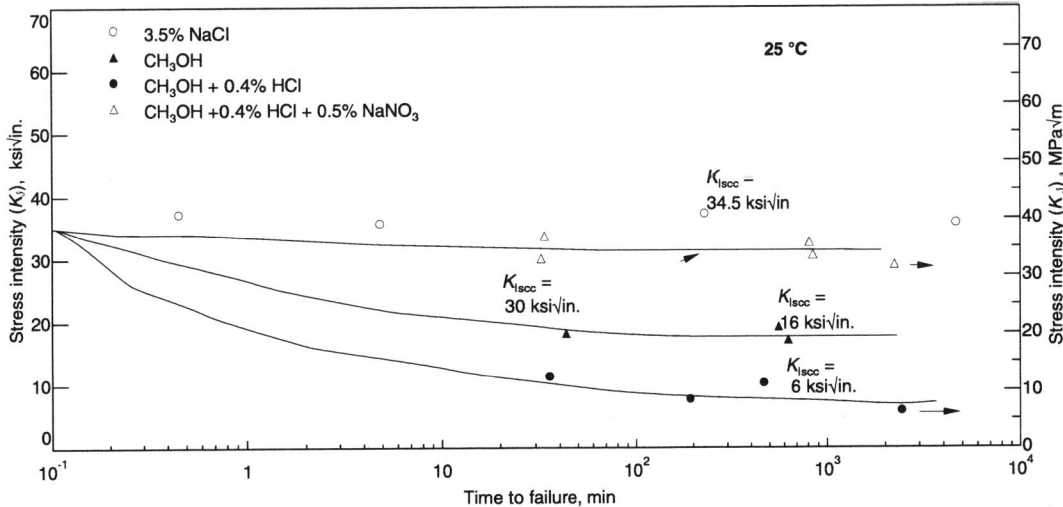

Beta C: Effect of acid and nitrates on K_{Iscc} in methanol

Source: W.F. Czyrklis and M. Levy, "Stress Corrosion Susceptibility of β Titanium Alloy 38-6-44," *Corrosion*, Vol 32 (No. 3), Mar 1976, p 99-102

stage II cracking. On the other hand, coarser grained structures of aged Beta C predominantly exhibit transgranular stages I and II cracking. Grain size and phase morphology are highly influential factors in β alloy SCC behavior.

Stress-corrosion cracking is generally difficult to observe in absolutely pure methanol, but becomes favored at HCl levels above 10^{-6} M. The minimum level of water required for full SCC inhibition depends on alloy composition, metallurgical condition, temperature, halide level, acidity, and other species present.

The SCC inhibition in neutral halide-containing methanol solutions can be achieved with 3 to 5.0 wt% water additions for Beta C. Studies on Beta C in methanol-water solutions, however, suggest that the minimum water level required for SCC inhibition may actually diminish with increasing temperature. (R.W. Schutz and M. Xiao, "Stress Corrosion Cracking Resistance of Beta C Titanium Springs in Methanol Solutions," paper presented at OMAE-Calgary '92, 11th Offshore Mechanics & Arctic Engineers Conference, Calgary, Canada, June 7-11, 1992). Numerous species also inhibit transgranular (stage II) stress cracking in methanol depending on halide level. These include nitrate and sulfate ions, NaF, and 0.1 M concentrations of Al^{+3}, Zr^{+4}, Cd^+, and Sn^{+2} metallic ions.

Hot-Salt Stress-Corrosion Cracking

Although no actual hot service component failures of titanium alloys have been assigned to hot-salt SCC to date, laboratory studies indicate that hot-salt SCC may occur on highly stressed titanium alloys with halide-containing surface residues after exposure in the 210 to 510 °C (410 to 950 °F) range.

Although some test results (see tables) indicate no cracking of Beta C up to about 200 °C (400 °F), Beta C has exhibited stress cracking during C-ring and slow strain-rate testing above 180 °C (355 °F). The SCC susceptibility of Beta C in hot brines has well-defined temperature thresholds for some of the sweet and sour brine concentrations expected in downhole tubular applications. A substantial, rigorous stress corrosion and crevice corrosion database for Pd-enhanced Beta C has been generated for worst-case sour gas well brine up to 260 °C (R. Schutz, RMI Titanium Co.).

Beta C: H_2S exposure without cracking

Heat treatment	Yield stress MPa	ksi	Stress level MPa	ksi	Test results(a)
870 °C (1600 °F), 1 h, AC	896	130	503	73	No cracking
			669	97	No cracking
785 °C (1450 °F), 1 h, AC + 480 °C (900 °F), 8 h, AC	1448	210	883	128	No cracking
			1082	157	No cracking
785 °C (1450 °F), 1 h, AC + 595 °C (1100 °F), 8 h, AC	965	140	627	91	No cracking
			724	105	No cracking

(a) After 350-h exposure. Source: D.E. Thomas, S. Ankem, W.D. Goodin, and S.R. Seagle, in *Industrial Applications of Titanium and Zirconium: Fourth Volume*, C.S. Young and J.C. Durham, Ed., ASTM STP 917, ASTM, 1986

Beta C: H_2S exposure without cracking

Chloride, %	Environmental conditions								Results(a)
	H_2S MPa	psi	CO_2 MPa	psi	Temperature °C	°F	Pressure(b) kPa	psi	
2	6.9	1	68.9	10	23	75	6895	1000	
2	6.9	1	68.9	10	23	75	6895	1000	
2	6.9	1	68.9	10	65	150	6895	1000	No environmental
2	20.9	3	20.9	3	23	75	103	15	cracking at a
3	103	15	0	0	23	75	103	15	stress level of
2	68.9	10	137.9	200	150	300	6895	1000	1205 MPa (175 ksi) after
15	68.9	10	137.9	200	150	300	6895	1000	exposure for
2	137.9	200	137.9	200	150	300	6895	1000	14 days
15	137.9	200	137.9	200	150	300	6895	1000	
15	137.9	200	137.9	200	200	400	6895	1000	

Note: Beta C solution treated and aged to a yield strength of 1205 (175 ksi). (a) C-ring samples tested with and without steel galvanic couples. (b) Total pressure made up with argon and indicated gases. Source: D.E. Thomas, S. Ankem, W.D. Goodin, and S.R. Seagle, in *Industrial Applications of Titanium and Zirconium: Fourth Volume*, C.S. Young and J.C. Durham, Ed., ASTM STP 917, ASTM, 1986

Beta C: SCC threshold ranges in sour gas

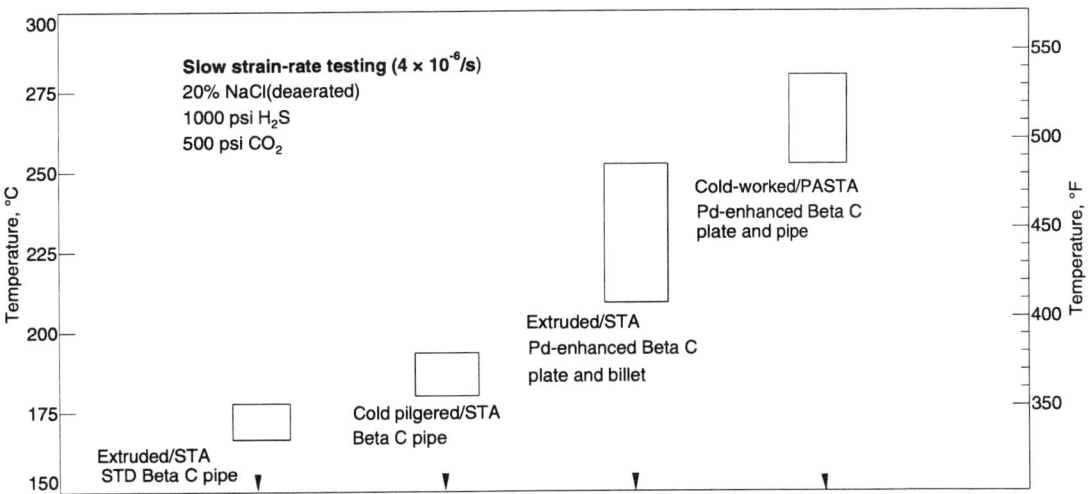

Specimens were solution treated and aged (STA) or proprietary (RMI) solution treated and aged (PASTA).
Source: R.W. Schutz, M. Xiao, and T.A. Bednarowicz, "Stress Corrosion Behavior of Ti-3Al-8V-6Cr-4Zr-4Mo Alloy Under Deep Sour Gas Well Conditions," Paper No. 51, Corrosion '92, Nashville, 1992

Thermal Properties

Solidus Temperature. 1555 °C (2830 °F)
Liquidus Temperature. 1650 °C (3000 °F)

Specific Heat. 515 J/kg · K (0.123 Btu/lb · °F) at RT

Thermal Expansion

Beta C: Thermal coefficient of linear expansion

Temperature range		Mean coefficient, μm/m · K, on:	
°C	°F	Heating	Cooling
20-38	68-100	8.44	8.44
20-66	68-150	8.57	9.11
20-93	68-200	8.66	9.20
20-121	68-250	8.80	9.25
20-149	68-300	8.89	9.31
20-177	68-350	8.96	9.38
20-205	68-400	9.05	9.43
20-230	68-450	9.16	9.49
20-260	68-500	9.25	9.54
20-290	68-550	9.34	9.59
20-315	68-600	9.41	9.65
20-345	68-650	9.49	9.70
20-370	68-700	9.54	9.74
20-400	68-750	9.58	9.79
20-425	68-800	9.61	9.83
20-455	68-850	9.65	9.88
20-480	68-900	9.68	9.92

Note: Alloy used was forgings, solution treated and aged 12 h at 565 °C (1045 °F). Source: D. Thomas, S. Ankem, W.D. Goodin, and S.R. Seagle, "Beta C: An Emerging Titanium Alloy for the Industrial Marketplace," Research Report No. 643, RMI Co., Nov 1984

Beta C: Thermal coefficient of linear expansion

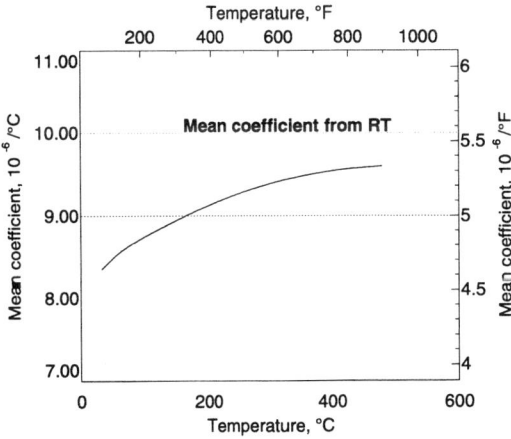

Alloy used was 150 mm (6 in.) square billet, heat treated at 815 °C (1500 °F) for 15 min, then air cooled, and aged at 565 °C (1050 °F) for 2 h, then air cooled.
Source: *Aerospace Structural Metals Handbook*, Vol 4, Code 3723, Battelle Columbus Laboratories, 1972, p 6

Thermal Conductivity

Beta C: Thermal conductivity and electrical resistivity

Specimen condition	Temperature °C	Temperature °F	Conductivity, W/m·K	Resistivity, μΩ·cm
As rolled	36	96	6.27	157.45
	52	125	6.54	156.92
	96	205	7.30	155.66
	170	340	9.02	154.15
	239	460	10.3	153.38
	317	600	11.9	151.99
Aged 480 °C (900 °F)	34	93	6.12	157.20
	50	122	6.36	156.66
	98	208	7.29	155.34
	174	345	9.07	153.84
	243	470	10.3	153.19
	326	618	12.1	151.97

Note: Thermal conductivity and electrical resistivity were determined using the Kohlrausch apparatus. Thermal conductivity values accurate within 5% are obtained by the Kohlrausch method. Source: R. Taylor and H. Groot, "Thermal Conductivity of Titanium and Titanium Alloys," Thermophysical Properties Research Laboratories, Purdue University, 1986

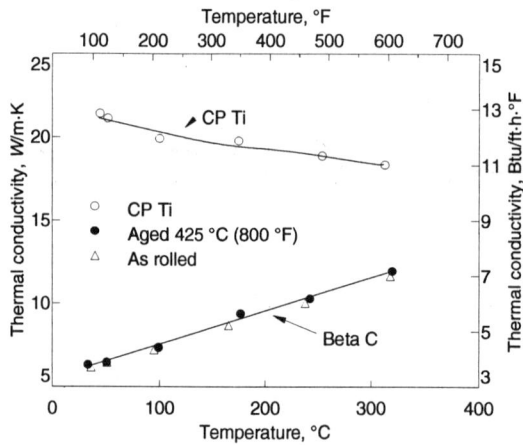

Beta C: Thermal conductivity

Source: R. Taylor and H. Groot, "Thermal Conductivity of Titanium and Titanium Alloys," Thermophysical Properties Research Laboratories, Purdue University, 1986

Mechanical Properties

Beta C: Guaranteed minimum tensile properties of bar and wire

Form and condition	Thickness mm	Thickness in.	Ultimate tensile strength MPa	Ultimate tensile strength ksi	Tensile yield strength MPa	Tensile yield strength ksi	Elongation in 4D, %	Reduction of area, %
Cold drawn bar and wire per AMS 4957	≤4.75	≤0.187	1310-1445	190-210	...	...	10	20
	4.75-9.5	0.187-0.375	1275-1415	185-205	...	...	10	20
	9.5-1.58	0.375-0.625	1240-1380	180-200	...	...	8	20
Bar and wire, solution heat treated, centerless ground, and aged per AMS 4958	Unspecified		1240	180	...	...	8	20

Beta C: Minimum tensile properties in solution annealed condition

Product(a)	Thickness mm	Thickness in.	Ultimate tensile strength MPa	Ultimate tensile strength ksi	Tensile yield strength (0.2% offset) MPa	Tensile yield strength (0.2% offset) ksi	Elongation in 50 mm (2 in.) or 4D, %	Reduction of area, %	Bend test TR
Sheet and strip	≤1.7	≤0.070	860	125	825	120	8.0	...	3.5
							6.0(b)	...	
	>1.7-4.7	>0.070-0.187	860	125	825	120	8.0	...	4.0
Plate	≤50	≤2	860	125	825	120	10.0	...	...
	>50-100	>2-4	825	120	790	115	8.0(L)	...	...
							6.0(T)		
							(ST)3.0		
Bar or test forged fillet	13-38	0.50-1.50	860	125	825	120	10	30	...
	>38-75	>1.50-3.00	825	120	790	115	10(L)	25(L)	...
							8(T)	20(T)	
	>75-228(c)	>3.00-9.00(c)	860	125	825	120	10	25	...

(a) Solution annealed condition: 815 °C (1500 °F), 15 to 30 min at temperature plus air cool. (b) Gauges under 0.7 mm (0.030 in.). (c) Results based on test forged samples 3:1 minimum upset. Source: *Beta Titanium Alloys in the 1980's*, R.R. Boyer and H.W. Rosenberg, Ed., TMS/AIME, 1984

Beta C: Minimum tensile properties in STA condition

Product(a)	Thickness mm	Thickness in.	Ultimate tensile strength MPa	Ultimate tensile strength ksi	Tensile yield strength (0.2% offset) MPa	Tensile yield strength (0.2% offset) ksi	Elongation in 50 mm (2 in.) or 4D, %	Reduction of area, %
Sheet and strip	All gauges		1240	180	1170	170	6	...
Plate	≤50	≤2	1240	180	1170	170	8	...
	>50-100	>2-4	1240	180	1170	170	8(L)	...
							6(T)	...
Bar and billet	13-38	0.5-1.5	1310	190	1240	180	8(b)	15(b)
	>38-75	>1.5-3	1240	180	1170	170	8(L)(b)	15(b)
							6(T)(b)	15(b)
	>75-150	>3-6	1170	170	1105	160	6(L)(b)	5(T)
	>75-228	>3-9(c)	1240	180	1170	170	10(b)	20(b)

(a) Solution anneal: 815 to 925 °C (1500 to 1700 °F) for 1 h, age: 455 to 540 °C (850 to 1000 °F) for 24 h maximum total plus air cool. (b) Higher ductilities may be obtained at lower strength levels. (c) Results based on test forged samples 3:1 minimum upset. Source: *Beta Titanium Alloys in the 1980's*, R.R. Boyer and H.W. Rosenberg, Ed., TMS/AIME, 1984

Hardness

Beta C: Rockwell hardness with aging at 480 and 510 °C

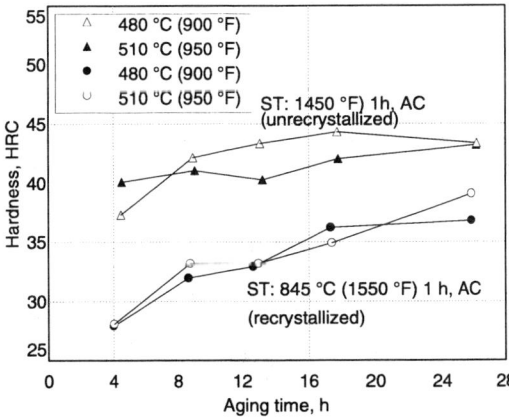

Residual strain from cold/warm working accelerates the aging process, as evidenced in the unrecrystallized structure of the 785 °C (1450 °F) solution treatment.
Average Rockwell hardness versus aging time (five tests) for 75 mm (3 in.) diameter bar for 480 °C (900 °F) and 510 °C (950 °F) ages.
Source: G.A. Bella et al., Effects of Processing on Microstructure and Properties of Ti-3Al-8V-6Cr-4Mo-4Zr (Beta C™), in *Microstructure and Property Relationships in Titanium Aluminides and Alloys*, Y-W. Kim and R.R. Boyer, Ed., TMS/AIME, 1991, p 493-510

Beta C: Rockwell hardness with aging at 540 and 565 °C

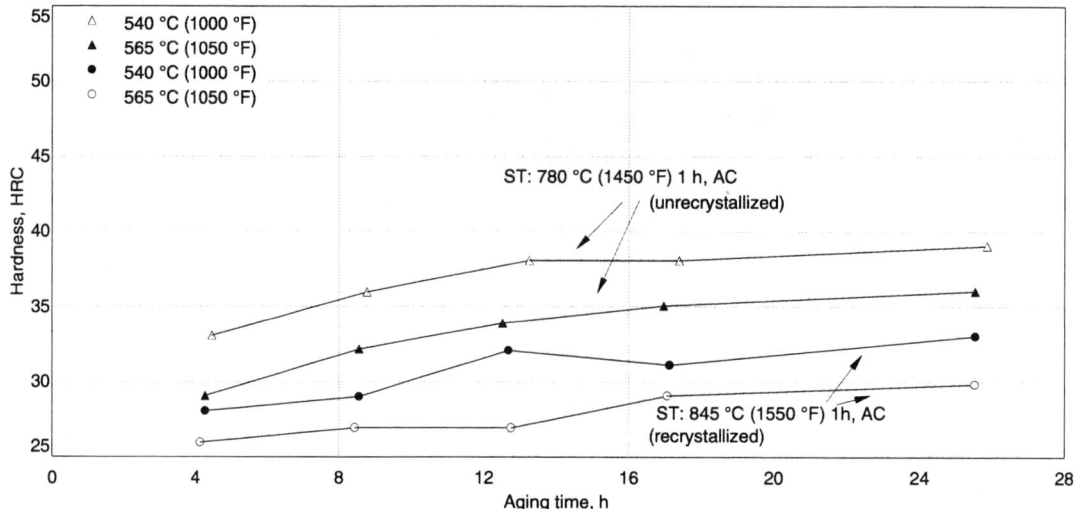

Average Rockwell hardness versus aging time (five tests) for 75 mm (3 in.) diameter bar for 540 °C (1000 °F) and 565 °C (1050 °F) ages.
Source: G.A. Bella et al., Effects of Processing on Microstructure and Properties of Ti-3Al-8V-6Cr-4Mo-4Zr (Beta C™), in *Microstructure and Property Relationships in Titanium Aluminides and Alloys*, Y-W. Kim and R.R. Boyer, Ed., TMS/AIME, 1991, p 500

Beta C: Knoop hardness of weldments

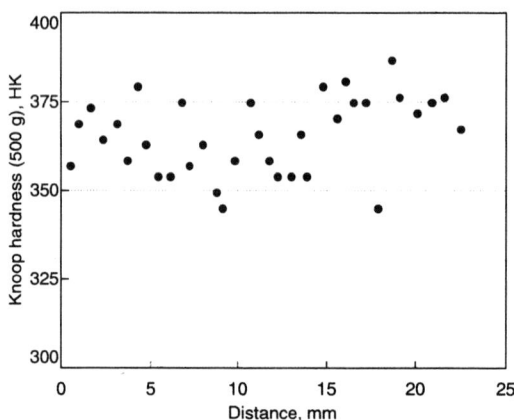

Hardness traverse from centerline of TIG butt welds in 1.6 mm (0.065 in.) Beta C sheet solution treated, welded, and aged at 540 °C (1000 °F) 8 h.
Source: R. Kaneko and C. Woods, "Low-Temperature Forming of Beta Titanium Alloys," NASA Contractor Report 3706, NASA, 1983, p 125

Beta C: Vickers hardness with aging at 500 °C

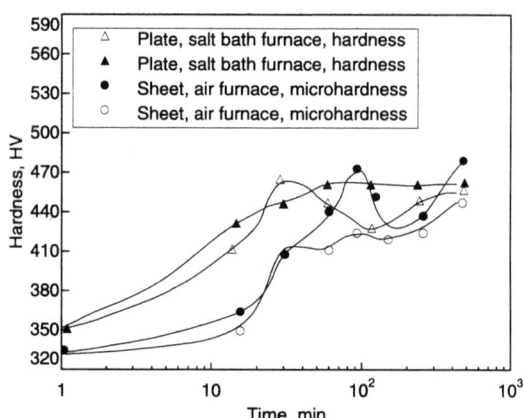

The double peak aging response has been observed in this figure in terms of the different amounts of ω formation following different aging (air/salt bath) treatments. This explanation would not appear to apply to the present work because such double peaks were found both in air and salt bath aged material, and no ω phase was detected. The double peaks in the present work may arise as a result of small changes in the balance of hardness between α precipitation and recovery.
Alloy was supplied as 1 mm (0.04 in.) thick sheet and as 7 mm (0.275 in.) thick plate. Specimens were encapsulated in argon-filled silica tubes and β solution treated at 900 °C (1650 °F) for 60 min to obtain an equiaxed grain size of about 150 μm and water quenched (tubes being broken on water contact). Samples were cold rolled to 75% reduction in thickness and aged in an air furnace (sheet) or salt bath (plate) at 200-650 °C (390 to 1200 °F) for up to 1500 min, then water quenched. Values represent an average of five tests. Error ±5.
Source: G. Isaac and C. Hammond, "The Formation of Type 2 Alpha Phase in Ti-3Al-8V-6Cr-4Zr-4Mo," in *Titanium, Science and Technology*, G. Lütjering, U. Zwicker, and W. Bunk, Ed., Deutsche Gesellschaft für Metallkunde eV, Germany, 1985, p 1605

Beta C: Effect of aging temperature on hardness

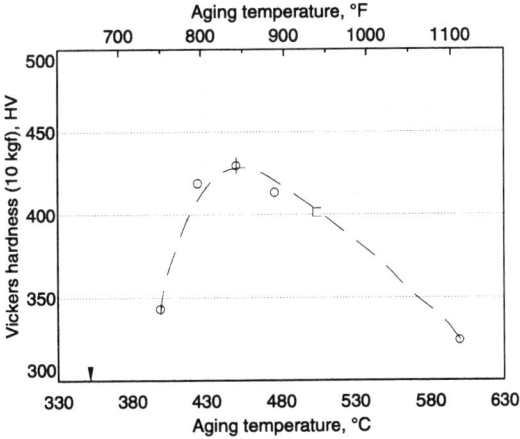

Blocks 5 × 20 × 60 mm (0.2 × 0.8 × 2.3 in.) were cut from hot rolled plate and solution treated at 750 °C (1380 °F) 30 min, WQ, plus 20 h age at indicated temperatures. Specimens were mounted in resin. Testing surfaces were planed 1 mm (0.04 in.) to remove oxide scale and the contaminated layer and were polished and buffed to smooth surfaces. Vickers hardness measurements were carried out at 10 kgf.
Source: Y. Shida and Y. Sugimoto, "Water Jet Erosion Behavior of Titanium Alloys," in *Sixth World Conference on Titanium*, P. Lacombe, R. Tricot, and G. Beranger, Ed., Les Editions de Physique, Paris, 1989, p 1935

Typical Compressive Strength

Beta C: Typical compressive properties at RT of several heats

Condition	Test direction	Prestrain direction	Compressive yield strength MPa	ksi	Compressive modulus (E_c) GPa	10^6 psi
STA	L	...	1234	179	107	15.5
			1241	180	101	14.7
			1234(a)	179(a)	104(b)	15.1(a)
ST + CW + A	L	L	1241	180	102	14.9
			1255	182	103	15.0
			1248(a)	181(a)	102(a)	14.9(b)
STA	T	...	1303	189	107	15.5
			1261	183	111	16.2
			1282(a)	186(a)	109(a)	15.8(a)
ST + CW + A	T	L	1324	192	100	14.6
			1344	195	107	15.5
			1330(b)	193(a)	103(a)	15.0(a)
ST + Weld + A	T	...	Specimen bent		...	...
			Specimen bent		...	...
			1248	181	113	16.5
			1234	179	117	17.0
			1227	178	115	16.7
			1234(a)	179(a)	115(a)	16.7(a)
ST + braze (Beta C/Ti-6-4) + A	T	...	1055	153	100	14.6
			1061	154	105	15.2
			1061(a)	154(a)	102(a)	14.9(a)
ST + braze cycle + A	T	...	1220	177	108	15.6
			1234	179	107	15.5
			1227(a)	178(a)	107(a)	15.5(a)
ST + CW + braze cycle + A	T	L	1317	191	105	15.3
			1261	183	108	15.6
			1289(a)	187(a)	106(a)	15.4(a)

(a) Average of all specimens tested in a given heat. Source: R.S. Kaneko and C.A. Woods, "Low-Temperature Forming of Beta Titanium Alloys," NASA Contractor Report 3706, NASA, 1983

Beta C: Compressive properties at −55 °C (−70 °F) of several heats

Condition(a)	Test direction	Prestrain direction	Compressive yield strength MPa	ksi	Compressive modulus (E_c) GPa	10^6 psi
STA	T	...	1392	202	107	15.6
			1365	198	108	15.7
			1379(a)	200(a)	107.5(a)	15.6(a)
ST + CW + A	T	L	Specimen bent		...	...
			Specimen bent		...	...
ST + braze (Beta C/Ti-6-4) + A	T	...	Specimen bent		...	...
			1055	153	102	14.9
			1151	167	106	15.4
			1103(a)	160(a)	104(a)	15.1(a)
ST + braze cycle + A	T	...	1289	187	104	15.1
			1248	181	119	17.3
			1268(a)	184(a)	111(a)	16.2(a)
ST + CW + braze cycle + A	T	L	1324	192	114	16.5
			1303	189	113	16.4
			1310(a)	190(a)	113(a)	16.4(a)

(a) Average of all specimens tested in a given heat. Source: R.S. Kaneko and C.A. Woods, "Low-Temperature Forming of Beta Titanium Alloys," NASA Contractor Report 3706, NASA, 1983

Typical Shear Strength

Beta C: Typical ultimate shear and tensile strength of wire at RT

Size mm	in.	Condition(a)	Ultimate tensile strength MPa	ksi	Elongation in 50 mm (2 in.), %	Double shear strength MPa	ksi
7.9	0.312	815 °C (1500 °F), 15 min, AC	896	130	16.0	634	92.0
		SA + age 425 °C (800 °F), 6 h, AC	1503	218	8.0	862	125.0
		SA + age 480 °C (900 °F), 6 h, AC	1448	210	6.7	848	123.0
13	0.500	SA + 30% cold drawn + age 480 °C (900 °F), 6 h, AC	1643	238.3	5.0	834	121
		SA + 30% cold drawn + age 510 °C (950 °F), 6 h, AC	1572	228	5.0	862	125
		SA + 30% cold drawn + age 540 °C (1000 °F), 4 h, AC	1388	201.4	10.0	834	121
7.6	0.300	SA + 39% cold drawn + age 480 °C (900 °F), 6 h, AC	1606	233	8.0	951	138
6	0.238	SA + 59% cold drawn + age 480 °C (900 °F), 6 h, AC	1675	243	7.0	924	134

(a) SA, solution anneal: 815 °C (1500 °F), 15 min, AC. Source: "RMI Basic Design," RMI Co.

Beta C: Typical ultimate tensile and shear strength of wire and bar at RT

Diameter mm	in.	Condition	Tensile yield strength MPa	ksi	Ultimate tensile strength MPa	ksi	Elongation, %	Reduction of area, %	Shear strength(a) MPa	ksi
15.8	0.625	815 °C (1500 °F), 15 min, AC	853	123.7	891	129.2	18	52	644	93.5
13	0.500	815 °C (1500 °F), 15 min, AC	869	126.1	887	128.7	15	40	644	93.5
		815 °C (1500 °F), 15 min, AC	853	123.7	869	126.1	16	42	660	95.7
		815 °C (1500 °F), 15 min, AC + 510 °C (950 °F), 6 h, AC	1210	175.5	1337	194.0	9	17	774	112.3
		815 °C (1500 °F), 15 min, AC + 510 °C (950 °F), 6 h, AC	1241	180.0	1365	198.0	9	23	760	110.2
		815 °C (1500 °F), 15 min, + 540 °C (1000 °F), 6 h, AC	1139	165.2	1230	178.5	10	23	...	...
		815 °C (1500 °F), 15 min, AC + 540 °C (1000 °F), 6 h, AC	1128	163.7	1229	178.3	10	23	...	...
		815 °C (1500 °F), 15 min, AC + 565 °C (1050 °F), 6 h, AC	1121	162.7	1221	177.1	13	33	762	110.5
		815 °C (1500 °F), 15 min, AC + 565 °C (1050 °F), 6 h, AC	1084	157.3	1186	172.0	14	38	777	112.7
9.5	0.375	815 °C (1500 °F), 15 min, AC + 565 °C (1050 °F), 6 h, AC	1079	156.5	1162	168.6	18	47	...	...
		815 °C (1500 °F), 15 min, AC + (1050 °F), 6 h, AC	1073	155.6	1149	166.7	18	50	...	...
7.9	0.312	815 °C (1500 °F), 15 min, AC	893	129.6	901	130.7	16	49	634	92.0
		815 °C (1500 °F), 15 min, AC	877	127.2	892	129.4	16	49	629	91.3
		815 °C (1500 °F), 15 min, AC + 425 °C (800 °F), 6 h, AC	1420	206.0	1503	218.0	8	17	862	125
		815 °C (1500 °F), 15 min, AC + 480 °C (900 °F), 6 h, AC	1289	187.0	1448	210.0	7	21	820	119
		815 °C (1500 °F), 15 min, AC + 540 °C (1000 °F), 6 h, AC	1130	164.0	1234	179.0	15	38	758	110
		815 °C (1500 °F), 15 min, AC + 565 °C (1050 °F), 6 h, AC	1048	152.0	1117	162.0	14	49	710	103
		815 °C (1500 °F), 15 min, AC + 675 °C (1250 °F), 6 h, AC	889	129.0	930	135.0	17	48	627	91
6.3	0.248	815 °C (1500 °F), 15 min, AC	942	136.7	972	141.0	13	54	617	89.6
		815 °C (1500 °F), 15 min, AC	952	138.1	977	141.7	15	53	620	89.9

(continued)

Beta C: Typical ultimate tensile and shear strength of wire and bar at RT (continued)

Diameter		Condition	Tensile yield strength		Ultimate tensile strength		Elongation, %	Reduction of area, %	Shear strength(a)	
mm	in.		MPa	ksi	MPa	ksi			MPa	ksi
4.7	0.187	815 °C (1500 °F), 15 min, AC + 565 °C (1050 °F), 6 h, AC	995	144.4	1102	159.9	13	46	726	105.3
		815 °C (1500 °F), 15 min, AC + 565 °C (1050 °F), 6 h, AC	1003	145.5	1102	159.9	17	49	735	106.6
		815 °C (1500 °F), 15 min, WQ	862	125.0	874	126.8	27	50	646	93.8
			866	125.7	879	127.5	27	53	644	93.5
		815 °C (1500 °F), 15 min, WQ + 565 °C (1050 °F), 6 h, AC	1044	151.4	1115	161.8	17	43	744	108.0
			1053	152.7	1117	162.0	20	45	751	109.0
		815 °C (1500 °F), 15 min, WQ + 675 °C (1250 °F), 6 h, AC	868	125.9	900	130.6	12	26	657	95.4
			874	126.8	910	132.0	20	28	667	96.8

(a) Based on double-shear specimens. Source: Data from RMI Co., in *Beta Titanium Alloys*, R. Wood, Ed., MCIC-72-11, Battelle Columbus Laboratories, 1972, p 188

Typical Tensile Properties

See "Mechanical Properties" for minimum tensile properties.

Beta C: Typical RT tensile properties.

Product	Condition	Test direction	Ultimate tensile strength		Tensile yield strength		Elongation, %	Reduction of area, %
			MPa	ksi	MPa	ksi		
Strip, 0.5 mm (0.020 in.)	SA	L	896	130	855	124	16	...
	STA	L	1338	194	1241	180	7	...
Sheet, 1.2 mm (0.050 in.)	SA	L	896	130	883	128	10	...
		T	931	135	917	133	6	...
	STA	L	1372	199	1276	185	8	...
		T	1441	209	1344	195	5	...
Plate, 12 mm (0.500 in.)	SA	L	924	134	896	130	14	37
		T	924	134	910	132	6	21
	STA	L	1276	185	1179	171	11	13
		T	1296	188	1207	175	9	16
Pipe, 75 mm OD × 50 in. ID (2.95 in. OD, 2.00 in. ID)	STA	L	1151	167	1014	147	11	25
Fastener stock, 4.7 mm (0.187 in.) diam	SA	L	876	127	862	125	27	50
	STA	L	1117	162	1048	152	18	44
Spring wire, 5.8 mm (0.229 in.) diam	CW + age	L	1469	213	...	...	12	28
Billet, 150 mm (6 in.) diam	STA	L-edge	1193	173	1145	166	9	18
		L-midradius	1207	175	1151	167	9	17
		L-center	1220	177	1158	168	9	16

Note: STA, solution treated and aged; SA, solution annealed. Source: C.S. Young and J.C. Durham, *Industrial Applications of Titanium and Zirconium: Fourth Volume*, ASTM STP 917, 1986, p 154

Beta C: Sharp notch tensile strength of sheet

Condition	Prestrain direction	Test direction	Notch radius		K_t	Notch tensile strength		NTS/TYS
			mm	in.		MPa	ksi	
At room temperature								
STA	...	L	0.10	0.005	8.4	648	94	0.53
			0.07	0.003	10.6	620	90	0.51
ST + CW + A	L	L	0.06	0.0025	11.6	606	88	0.49
					11.6	593	86	0.48
STA	...	T	0.07	0.003	10.6	627	91	0.51
					10.6	579	84	0.47
At −53 °C (−65 °F)								
STA	...	L	0.03	0.0012	16	641	93	0.45
ST + CW + A	L	L	0.06	0.0025	11.5	531	77	0.38
			0.10	0.005	8.4	572	83	0.40
At 315 °C (600 °F)								
STA	...	L	0.03	0.0012	16	875	127	0.91
					16	841	122	0.87
ST + CW + A	L	L	0.10	0.005	8.3	765	111	0.78
			0.15	0.006	7.7	765	111	0.78
STA	...	T	0.03	0.0012	16	793	115	0.81
					16	737	107	0.75

Note: Sheet of 1.65 mm (0.065 in.) thickness was aged at 540 °C (1000 °F) for 8 h. Tests were conducted according to ASTM E338.
Source: R. Kaneko and C. Woods, "Low-Temperature Forming of Beta Titanium Alloys," NASA Contractor Report 3706, NASA, 1983, p 73

Effect of Heat Treatment

Beta C: Effect of heat treatment on tensile properties

Heat treatment (designation)	Modulus of elasticity		Tensile yield strength (0.2% offset)		Ultimate tensile strength		Elongation, %	Reduction of area, %
	GPa	10⁶ psi	MPa	ksi	MPa	ksi		
800 °C (1470 °F), 1/2 h, AC, (as-SHT 800)	88	12.7	895	130	895	130	22	48
925 °C (1700 °F), 1/2 h, AC, (as-SHT 927)	88	12.7	850	123	850	123	25	62
as-SHT 800 + 535 °C (995 °F), 8 h, (LO-simplex)	104	15	1225	177	1320	191	8	15
as-SHT 800 + 425 °C (795 °F), 4 h, + 560 °C (1040 °F), 8 h, (LO-duplex)	104	15	1220	176	1300	188	10	13
as-SHT 927 + 530 °C (985 °F), 16 h, (HI-simplex)	103	14.9	1140	165	1220	177	12	21
as-SHT 927 + 455 °C (850 °F), 4 h, + 555 °C (1025 °F), 16 h, (HI-duplex)	105	15.2	1075	156	1180	171	14	23

Source: H.E. Krugmann and J.K. Gregory, Microstructure and Crack Propagation in Ti-3Al-8V-6Cr-4Mo-4Zr, *Microstructure and Property Relationships in Titanium Aluminides and Alloys*, Y-W. Kim and R.R. Boyer, Ed., TMS/AIME, 1991, p 549-560

Beta C: Yield strength vs elongation (3 in. diam bar)

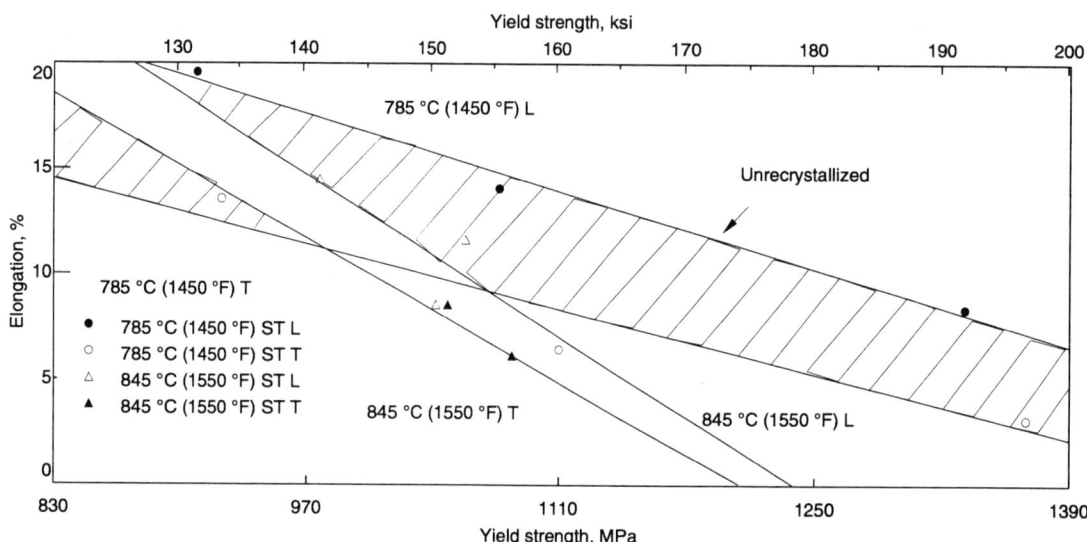

Source: G. Bella et al., Effects of Processing on Microstructure and Properties of Ti-3Al-8V-6Cr-4Mo-4Zr (Beta C™), in *Microstructure and Property Relationships in Titanium Aluminides and Alloys*, Y-W. Kim and R.R. Boyer, TMS/AIME, 1991, p 493-510. Aged 24 hours at three temperatures from 900 °F up to 1125 °F or 1025 °F max (maximum age temperatures for 1450 °F and 1550 °F ST, respectively)

Beta C: Yield strength vs reduction of area (3 in. diam bar)

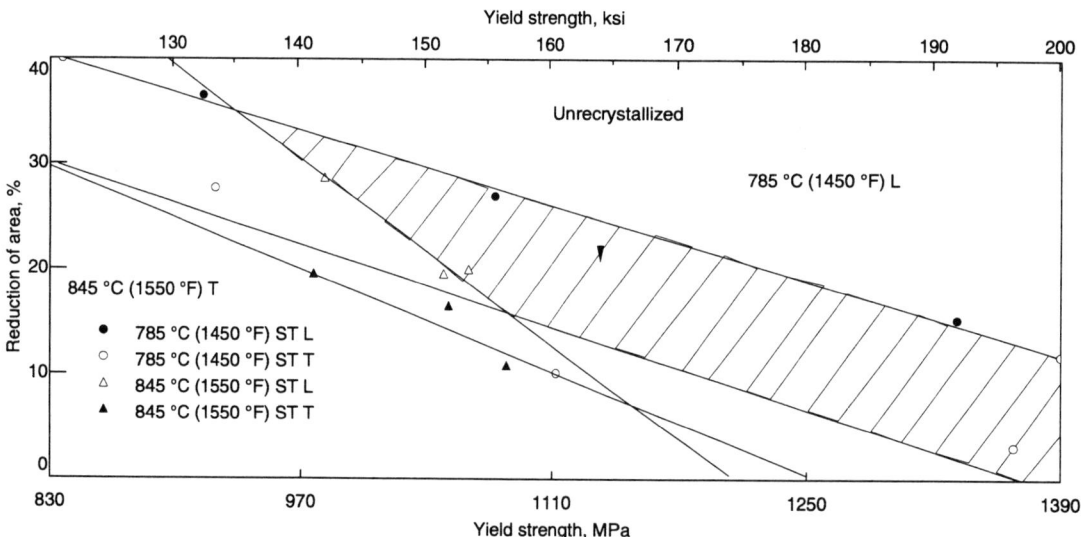

Source: G. Bella et al., Effects of Processing on Microstructure and Properties of Ti-3Al-8V-6Cr-4Mo-4Zr (Beta C™), in *Microstructure and Property Relationships in Titanium Aluminides and Alloys*, Y-W. Kim and R.R. Boyer, Ed., TMS/AIME, 1991, p 493-510

Beta C: Effect of aging time on ultimate tensile strength

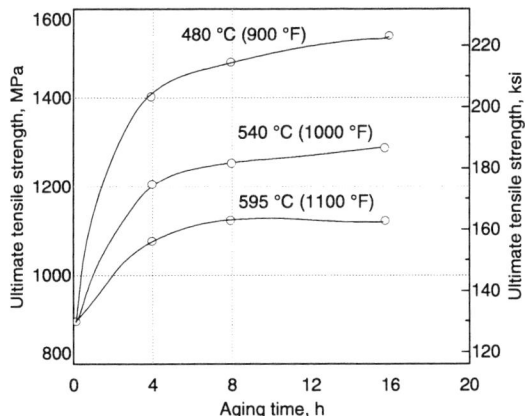

Pancake upset forged to 16 mm (0.625 in.) from 64 mm (2.5 in.) at 815 °C (1500 °F). Treatment: 815 °C (1500 °F) for 30 min and air cooled.
Source: *Beta Titanium Alloys in the 1980's*, R.R. Boyer and H.W. Rosenberg, Ed., TMS/AIME, 1984

Beta C: Effect of aging time on ultimate tensile strength

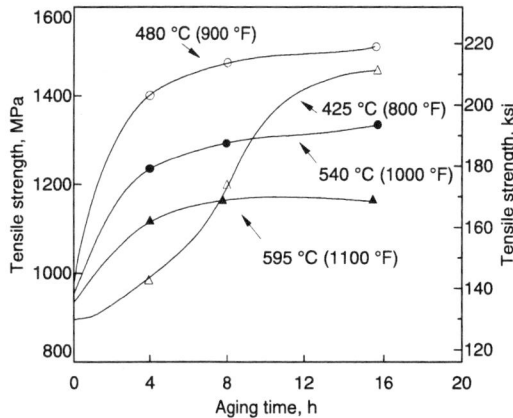

Forgings upset to 16 mm (0.625 in.) from 64 mm (2.5 in.) at 815 °C (1500 °F), followed by solution annealing at 815 °C.
Source: *Metals Handbook*, Vol 13, 9th ed., 1980

Effect of Cold Work

Cold working of Beta C prior to aging has the capability of increasing tensile properties while maintaining good ductility. Other mechanical properties such as double shear strength and fatigue are also improved.

Beta C: As-cold drawn tensile properties of fastener stock

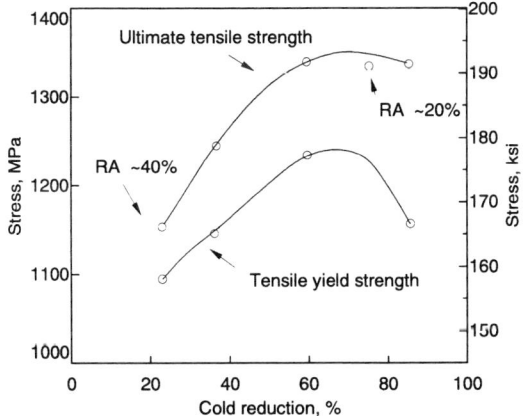

Source: *Beta Titanium Alloys in the 1980's*, R.R. Boyer and H.W. Rosenberg, Ed., TMS/AIME, 1984. See also top table on next page.

Beta C: Cold drawn and aged tensile properties

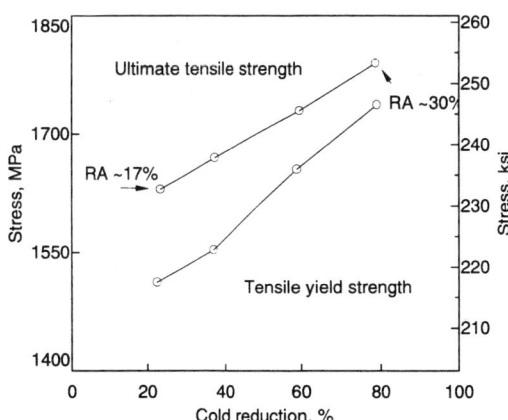

Cold drawn and aged 6 h at 480 °C (900 °F).
Source: *Beta Titanium Alloys in the 1980's*, R.R. Boyer and H.W. Rosenberg, Ed., TMS/AIME, 1984

Beta C: Effect of aging on cold drawn fastener stock

Size		Condition	Ultimate tensile strength		Yield strength		Elongation in 50mm (2 in.), %	Reduction of area, %	Double shear strength	
mm	in.		MPa	ksi	MPa	ksi			MPa	ksi
8.4	0.330	CD 22.5% + age 480 °C (900 °F), 6 h, AC	1579	229	1475	214	5	16	...	
7.6	0.299	CD 37.2% + age 480 °C (900 °F), 6 h, + AC	1606	233	1510	219	8	15	951	138
6.0	0.238	CD 59.3% + age 480 °C (900 °F), 6 h, + AC	1675	243	1613	234	7	21	938	134
4.5	0.176	CD 78.0% + age 480 °C (900 °F), 6 h, + AC	1730	251	1682	244	7	32	876	127

Source: *Beta Titanium Alloys in the 1980's*, R.R. Boyer and H.W. Rosenberg, Ed., TMS/AIME, 1984

Beta C: Effect of cold work on fastener stock

Size		Cold reduction, %	Ultimate tensile strength		Tensile yield strength		Elongation in 50 mm (2 in.), %	Reduction of area, %	Headability D_f/D_t	Double shear strength	
mm	in.		MPa	ksi	MPa	ksi				MPa	ksi
Solution annealed 815 °C (1500 °F), 15 min, AC											
8.4	0.330	22.5	1137	164.9	1084	157.2	13.0	43.3	1.14	696	101
7.6	0.299	37.2	1230	178.5	1141	165.6	10.6	47.0	1.75	731	106
6.0	0.238	59.5	1323	191.9	1220	177.0	8.0	44.4	1.68	744	108
4.7	0.188	75.0	1320	191.5	934	135.5	6.2	23.0	1.63	703	102
3.3	0.132	87.6	1317	191.0	1140	165.4	8.0	41.0	1.85	...	...
Solution annealed 925 °C (1700 °F), 30 min, WQ											
8.1	0.320	20.0	948	137.6	912	132.3	18.0	51.0	1.79	662	96
7.2	0.284	37.7	1093	158.5	955	152.8	14.0	47.0	1.79	717	104
6.0	0.238	56.2	1086	157.6	1086	139.2	13.3	43.5	1.50	689	100

Source: *Beta Titanium Alloys in the 1980's*, R.R. Boyer and H.W. Rosenberg, Ed., TMS/AIME, 1984

Beta C: Effect of aging on cold drawn fastener stock

Size		Condition	Ultimate tensile strength		Tensile yield strength		Elongation in 50 mm (2 in.), %	Reduction of area, %	Double shear strength	
mm	in.		MPa	ksi	MPa	ksi			MPa	ksi
8.4	0.330	As CD 22.5%	1064	164.3	1084	157.2	13.0	43.3	...	...
		CD + age 425 °C (800 °F), + 6 h, + AC	1666	241.7	1568	227.5	5.0	17.2	...	...
		CD + age 425 °C (800 °F), + 12 h, + AC	1700	246.4	1612	233.8	1.0	11.5	...	...
		CD + age 480 °C (900 °F) + 6 h + AC	1581	229.3	1480	214.7	5.0	16.3	...	...
		CD + age 480 °C (900 °F) + 12 h + AC	1534	222.5	1412	204.9	7.0	18.2	...	...
		CD + age 540 °C (1000 °F) + 6 h + AC	1313	190.5	1222	177.3	12.0	17.9	...	...
12.9	0.510	As CD 30%	1153	167.8	1103	160.0	8.0	25.0	689	100
		CD + age 480 °C (900 °F) + 6 h + AC	1643	238.3	1529	221.8	5.0	6.0	834	121
		CD + age 510 °C (950 °F) + 6 h + AC	1572	228.0	1451	210.5	5.0	6.0	862	125
		CD + age 540 °C (1000 °F) + 4 h + AC	1388	201.4	1275	185.0	10.0	24.0	834	121

Source: *Beta Titanium Alloys in the 1980's*, R.R. Boyer and H.W. Rosenberg, Ed., TMS/AIME, 1984

High-Temperature Strength

The excellent crevice corrosion resistance of Beta C, coupled with its excellent elevated-temperature strength, provides a useful combination of properties for industrial applications such as seawater piping. Beta C could age in service at temperatures greater than 350 °C (660 °F) and therefore is not recommended for use above 350 °C (660 °F).

Beta C: Yield strength vs corrosion resistance comparison

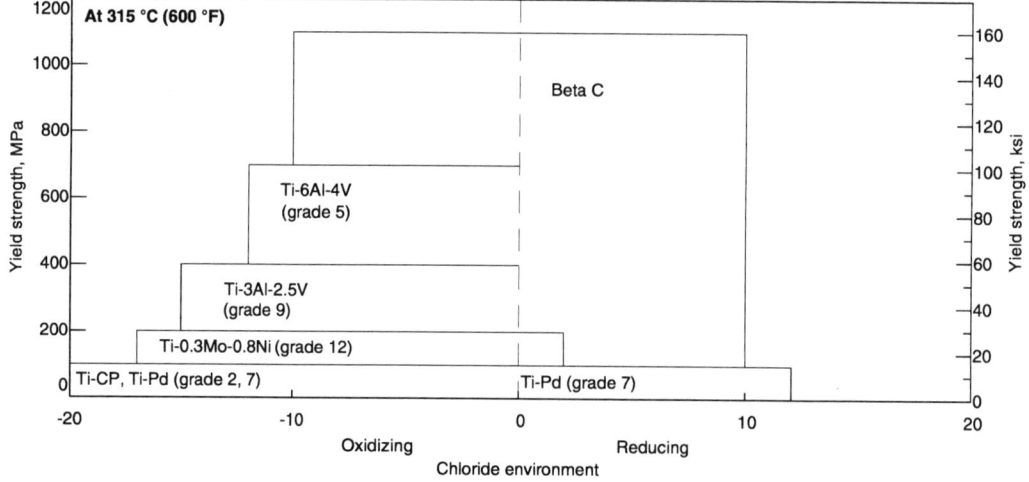

Source: C.S. Young and J.C. Durham, *Industrial Applications of Titanium and Zirconium: Fourth Volume*, ASTM STP 917, ASTM, 1986, p 159

Tensile Strengths

Beta C: Elevated-temperature tensile properties of bar

Test temperature		Ultimate tensile strength		Tensile yield strength (2% offset)		Elongation in 50 mm (2 in.), %	Reduction of area, %
°C	°F	MPa	ksi	MPa	ksi		
815 °C (1500 °F), solution anneal + age 565 °C (1050 °F), 6 h, AC							
RT		1200	174	1103	160	14	35
93	200	1034	150	889	129	17	40
205	400	1027	149	841	122	16	38
315	600	1082	157	903	131	12	30
425	800	938	136	758	110	17	40
Solution annealed 925 °C (1700 °F), 30 min, AC							
RT		886	128.5	835	121.2	15.0	37.2
93	200	772	112.0	724	105.0	21.0	47.0
205	400	724	105.0	648	94.0	20.0	50.1
315	600	669	97.0	600	87.0	22.0	49.3
425	800	703	102.0	593	86.0	23.5	48.7

Note: Specimens were 13 mm (0.5 in.) diameter bar. Source: C.S. Young and J.C. Durham, *Industrial Applications of Titanium and Zirconium: Fourth Volume*, ASTM STP 917, 1986, p 154

Beta C: Short-time elevated-temperature tensile properties of ST material

Test temperature		Ultimate tensile strength		Tensile yield strength (0.2% offset)		Elongation in 50 mm (2 in.), %	Reduction of area, %
°C	°F	MPa	ksi	MPa	ksi		
21	70	886	128.5	835	121.2	15.0	37.2
93	200	772	112	724	105	21.0	47.0
205	400	724	105	648	94	20.0	50.1
315	600	668	97	600	87	22.0	49.3
425	800	703	102	593	86	23.5	48.7

Note: Specimens were solution treated 925 °C (1700 °F) for 30 min, AC. Source: *Alloy Digest*, 1987

Beta C: Effect of temperature on tensile properties of ST bar stock

Test temperature		Ultimate tensile strength		Tensile yield strength (0.2% offset)		Elongation in 50 mm (2 in.), %	Reduction of area, %
°C	°F	MPa	ksi	MPa	ksi		
RT		886	128.5	835	121.2	15.0	37.2
93	200	772	112.0	724	105.0	21.0	47.0
205	400	724	105.0	648	94.0	20.0	50.1
315	600	668	97.0	600	87.0	22.0	49.3
425	800	703	102.0	593	86.0	23.5	48.7

Note: Specimens were solution annealed 30 min at 925 °C (1700 °F), AC. Source: *Beta Titanium Alloys in the 1980's*, R.R. Boyer and H.W. Rosenberg, Ed., TMS/AIME, 1984

Beta C: Tensile properties of STA bar

Test temperature		Ultimate tensile strength		Tensile yield strength		Elongation in 50 mm (2 in.), %	Reduction of area, %
°C	°F	MPa	ksi	MPa	ksi		
RT		1199	174	1103	160	14	35
93	200	1034	150	889	129	17	40
205	400	1027	149	841	122	16	38
315	600	1082	157	903	131	12	30
425	800	937	136	758	110	17	40
480	900	944	137	799	116	25	70
540	1000	744	108	386	56	31	83

Note: Specimens were 13 mm (0.5 in.) bar 815 °C (1500 °F) solution annealed + aged at 565 °C (1050 °F), 6 h, AC. Source: *Beta Titanium Alloys in the 1980's*, R.R. Boyer and H.W. Rosenberg, Ed., TMS/AIME, 1984

Beta C: Tensile properties of tube at high temperatures

Test temperature		Ultimate tensile strength		Tensile yield strength		Elongation in 50 mm (2 in.),
°C	°F	MPa	ksi	MPa	ksi	%
19 mm (0.750 in.) OD × 1 mm (0.042 in.) wall						
RT		875	127	841	122	20
230	450	731	107	710	103	34
345	650	710	103	627	91	31
13 mm (0.540 in.) OD × 0.7 mm (0.030 in.) wall						
RT		889	129	862	125	18
230	450	724	105	689	100	28
345	650	682	99	606	88	30
9.5 mm (0.375 in.) OD × 0.6 mm (0.025 in.) wall						
RT		944	137	896	130	14
230	450	744	108	689	100	16
345	650	703	102	620	90	21

Note: Data are average of two tests. Specimens were solution annealed at 815 °C (1500 °F), 90 min, vacuum cooled. Source: *Aerospace Structural Metals Handbook*, Vol 4, Code 3723, Battelle Columbus Laboratories, 1981

Beta C: Tensile strength vs temperature of STA bar

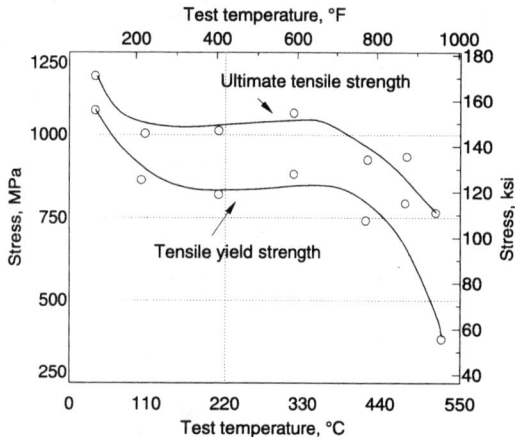

13 mm (0.5 in.) bar, solution annealed at 815 °C (1500 °F) + 565 °C (1050 °F), 6 h, AC.
Source: *Beta Titanium Alloys in the 1980's*, R.R. Boyer and H.W. Rosenberg, Ed., TMS/AIME, 1984

Beta C: Strength vs temperature of STA billet

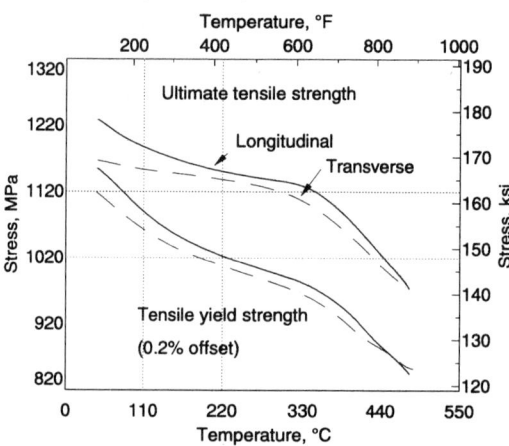

Full section tests were performed on 150 mm (6 in.) diam billets heat treated at 815 °C (1500 °F), 15 min, AC + aged at 565 °C (1050 °F), 4 h, AC.
Source: *Beta Titanium Alloys in the 1980's*, R.R. Boyer and H.W. Rosenberg, Ed., TMS/AIME, 1984

Compressive Strength

Beta C: Effect of temperature on compressive properties

Test temperature		Compressive yield strength				Compressive modulus			
		Longitudinal		Transverse		Longitudinal		Transverse	
°C	°F	MPa	ksi	MPa	ksi	GPa	10^6 psi	GPa	10^6 psi
RT		1110	161	1068	155	102	14.8	101	14.7
205	400	965	140	944	137	93	13.5	94	13.7
370	700	896	130	889	129	85	12.4	82	11.9
480	900	781	113.3	799	116	77	11.2	78	11.4

Note: Specimens were 150 mm (6 in.) diam forgings heat treated at 815 °C (1500 °F), 15 min, AC + age 565 °C (1050 °F), 4 h, AC. Full section tests. Source: *Beta Titanium Alloys in the 1980's*, R.R. Boyer and H.W. Rosenberg, Ed., TMS/AIME, 1984

Beta C: Typical compressive properties at 315 °C

Condition	Prestrain direction	Compressive yield strength(a)		Compressive modulus(a)	
		MPa	ksi	GPa	10^6 psi
STA	...	1110	161	107	15.5
ST + CW + A	L	1193	173	111	16.2
ST + braze (Beta C/Ti-6-4) + A	...	786(b)	114(b)	102	14.9
ST + braze cycle + A	...	999	145	108	15.7
ST + CW + braze cycle + A	L	1006	146	109	15.9

(a) Average of two transverse tests. (b) Permanent bending noted. Source: R.S. Kaneko and C.A. Woods, "Low-Temperature Forming of Beta Titanium Alloys," NASA Contractor Report 3706, NASA, 1983

Beta C: Compressive yield strength vs temperature

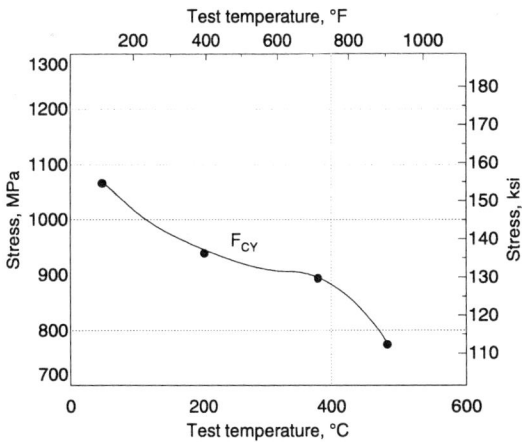

Full section tests were performed on 150 mm (6 in.) diam forgings solution treated and aged at 815 °C (1500 °F), 15 min, AC + 565 °C (1050 °F), 4 h, AC.
Source: *Aerospace Structural Metals Handbook*, Vol 4, Code 3723, Battelle Columbus Laboratories, 1981

Creep Properties

Beta C: Creep behavior of STA billet at 480 °C

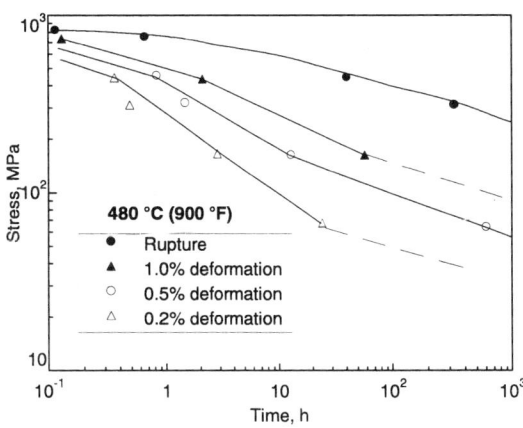

Forged billet was solution treated at 815 °C (1500 °F) for 15 min, air cooled, then aged at 565 °C (1050 °F) for 12 h, air cooled. Transverse specimens were tested for times and temperatures indicated.
Source: O. Deel and H. Mindlin, "Engineering Data on New and Emerging Structural Materials," AFML-TR-70-252, 1970, reported in *Beta Titanium Alloys*, R. Wood, MCIC-72-11, Battelle Columbus Laboratories, 1972, p 199

Beta C: Creep behavior of STA billet at 370 °C

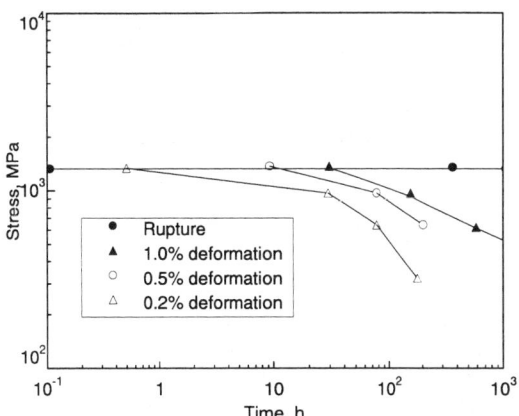

Forged billet was solution treated at 815 °C (1500 °F) for 15 min, air cooled, then aged at 565 °C (1050 °F) for 12 h, air cooled. Transverse specimens were tested for times and temperatures indicated.
Source: O. Deel and H. Mindlin, "Engineering Data on New and Emerging Structural Materials," AFML-TR-70-252, 1970, reported in *Beta Titanium Alloys*, R. Wood, MCIC-72-11, Battelle Columbus Laboratories, 1972, p 199

Beta C: Creep of STA sheet at 315 °C

Estimated aging temperature		315 °C (600 °F) creep exposure		Deformation in 96 h, %
°C	°F	Stress MPa	ksi	
480-510	900-950	793	115	0.41
920-565	975-1050	448	65	0.08
565-620	1050-1150	379	55	0.06

Source: RMI Co., reported in *Beta Titanium Alloys*, R. Wood, MCIC-72-11, Battelle Columbus Laboratories, 1972, p 202

Beta C: Creep of STA bar

Temperature		Stress		Creep exposure data Time, h, for: 0.1% deformation	Time, h, for: 0.2% deformation	Time, h, for: Total deformation	Total deformation, %
°C	°F	MPa	ksi				
315	600	551	80	800	...	1107	0.126
370	700	517	75	50	155	190	0.241
370	700	517	75	74	185	212	0.234
425	800	207	30	32	59	72	0.250
425	800	207	30	16	38	69	0.333

Note: Bar 13 mm (0.5 in.) thick was solution annealed at 815 °C (1500 °F), then aged at 565 °C (1050 °F) for 6 h. Source: RMI Co., reported in *Beta Titanium Alloys*, R. Wood, MCIC-72-11, Battelle Columbus Laboratories, 1972, p 200

Creep Stability

Beta C: Room-temperature shear strength of fastener stock before exposure

Material condition	Ultimate tensile strength		Tensile yield strength		Elongation, %	Reduction of area, %	Double shear strength	
	MPa	ksi	MPa	ksi			MPa	ksi
815 °C (1500 °F), 15 min, AC	897	130.1	885	128.4	16.0	49.0	632	91.7
+ aged 425 °C (800 °F), 6 h, AC	1503	218.0	1419	205.9	8.0	17.4	862	125.0
+ aged 480 °C (900 °F), 6 h, AC	1450	210.4	1290	187.1	6.7	20.8	826	119.8
+ aged 540 °C (1000 °F), 6 h, AC	1234	179.0	1130	164.0	15.3	38.7	762	110.5
+ aged 565 °C (1050 °F), 6 h, AC	1119	162.3	1055	153.0	14.0	49.5	710	103.0
+ aged 675 °C (1250 °F), 6 h, AC	933	135.4	892	129.4	17.0	48.6	631	91.5

Note: Specimens were 7.9 mm (0.312 in.) fastener stock heat treated as indicated. Source: *Beta Titanium Alloys in the 1980's*, R.R. Boyer and H.W. Rosenberg, Ed., TMS/AIME, 1984

Beta C: Room-temperature shear strength after exposure (properties prior to exposure above)

Condition before exposure	Ultimate tensile strength		Tensile yield strength		Elongation in 50 mm (2 in.), %	Reduction of area, %	Shear strength	
	MPa	ksi	MPa	ksi			MPa	ksi
815 °C (1500 °F), 15 min, AC	937	136	896	130	15	39	655	95
	937	136	889	129	15	44	655	95
815 °C (1500 °F), 15 min, AC	1565	227	1448	210	7	10	862	125
425 °C (800 °F), 6 h, AC							862	125
815 °C (1500 °F), 15 min, AC	1461	212	1365	198	8	17	855	124
480 °C (900 °F), 6 h, AC	1441	209	1358	197	9	16	855	124
815 °C (1500 °F), 15 min, AC	1186	172	1082	157	15	41	758	110
540 °C (1000 °F), 6 h, AC	1193	173	1096	159	17	37	758	110
815 °C (1500 °F), 15 min, AC	1103	160	1027	149	20	42	724	105
565 °C (1050 °F), 6 h, AC	1110	161	1013	147	20	44	724	105
815 °C (1500 °F), 15 min, AC	1013	147	958	139	21	45	689	100
620 °C (1150 °F), 6 h, AC	1027	149	951	138	20	46	682	99
815 °C (900 °F), 15 min, AC	924	134	875	127	20	46	627	91
675 °C (1250 °F), 6 h, AC	951	138	903	131	13	39	634	92

Note: Alloy was used as 7.9 mm (0.312 in.) diameter bar, exposed to 285 °C (550 °F) temperature for 500 h. Prior heat treatment was as indicated. Source: RMI, reported in *Aerospace Structural Metals Handbook*, Vol 4, Code 3723, Battelle Columbus Laboratories, 1975, p 14

Beta C: Creep stability of STA and cold worked rod after exposure

Thermal exposure cycle	Tensile yield strength		Ultimate tensile strength		Elongation, %	Reduction of area, %
	MPa	ksi	MPa	ksi		
As cold drawn 22.5%						
Unexposed	1082	157	1137	165	13	43
260 °C (500 °F), 275 MPa (40 ksi), 100 h	1041	151	1130	164	12	50
As 815 °C (1500 °F) solution annealed						
Unexposed	882	128	896	130	16	49
260 °C (500 °F), 0 MPa, 25 h	889	129	903	131	19	47
260 °C (500 °F), 0 MPa, 100 h	882	128	903	131	16	49
260 °C (500 °F), 275 MPa (40 ksi), 25 h	889	129	910	132	19	48
260 °C (500 °F), 275 MPa (40 ksi), 100 h	882	128	917	133	18	50
260 °C (500 °F), 275 MPa (40 ksi), 200 h	875	127	896	130	18	48
315 °C (600 °F), 275 MPa (40 ksi), 100 h	917	133	951	138	23	48

Note: Rod 8.4 mm (0.330 in.) in diameter was solution annealed at 815 °C (1500 °F), or cold drawn, as indicated. Source: RMI Co., reported in *Beta Titanium Alloys*, R. Wood, MCIC-72-11, Battelle Columbus Laboratories, 1972, p 200

Fatigue Properties

Beta C: Fatigue life of shot peened wire

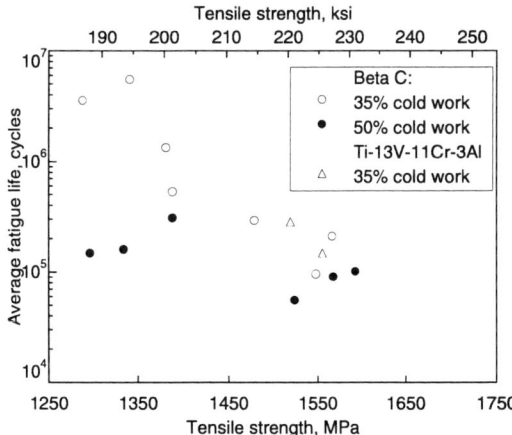

Shot peening is a critical parameter and shot peen intensities of at least 0.016 to 0.018 A should be used. Higher intensities would provide additional improvement in fatigue life, but a higher intensity callout could limit the number of shot peening sources available due to equipment limitations.

The effects of cold work and tensile strength on the average fatigue life of 10 mm (0.4 in.) diam shot peened to 0.016 to 0.018 A. Each data point represents the log average of six tests. Ti-13V-1Cr-3Al data points included for comparison. 1034 MPa (150 ksi) maximum stress, $R = 0.1$, 30 Hz.
Source: *Beta Titanium Alloys in the 1980's*, R.R. Boyer and H.W. Rosenberg, Ed., TMS/AIME, 1984

Beta C: Fatigue life of recrystallized wire

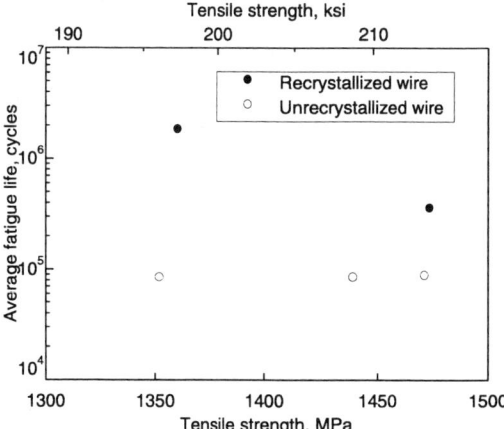

Control of grain size is desired, and the wire should be recrystallized during solution treatment.
Specimens were 9 mm (0.35 in.) diam wire cold worked 35%. Shot peen intensity of 0.016 to 0.018 A, 1034 MPa (150 ksi) maximum stress, $R = 0.1$, 30 Hz.
Source: *Beta Titanium Alloys in the 1980's*, R.R. Boyer and H.W. Rosenberg, Ed., TMS/AIME, 1984

Beta C: Fatigue life at high temperatures

Test condition(a)	Fatigue life, MPa (ksi), at:		
	RT	205 °C (400 °F)	370 °C (700 °F)
Unnotched			
10^3 cycles	1144 (166.0)	1089 (158.0)	1020 (148.0)
10^5 cycles	855 (124.0)	731 (106.0)	634 (92.0)
10^7 cycles	600 (87.0)	551 (80.0)	372 (54.0)
Notched(b)			
10^3 cycles	827 (120.0)	717 (104.0)	634 (92.0)
10^5 cycles	303 (44.0)	248 (36.0)	275 (40.0)
10^7 cycles	275 (40.0)	207 (30.0)	234 (34.0)

(a) Axial fatigue of transverse specimens from 150 mm (6 in.) diam STA forging treated 15 min at 815 °C (1500 °F), AC, plus aging at 565 °C (1050 °F) for 12 h, AC. $R = 0.1$. (b) $K_t = 3.0$. Source: *Beta Titanium Alloys in the 1980's*, R.R. Boyer and H.W. Rosenberg, Ed., TMS/AIME, 1984

Beta C: Axial fatigue at high temperature

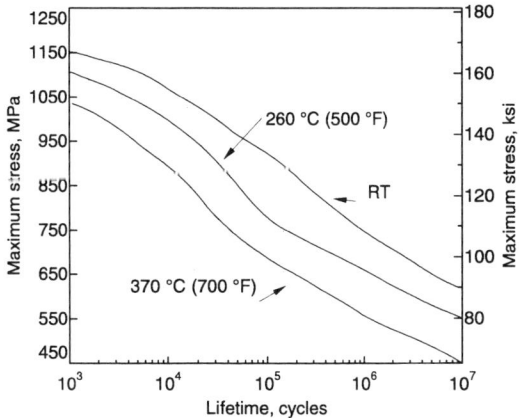

Unnotched specimens ($R = 0.1$) from 150 mm (6 in.) diam forging, 815 °C (1500 °F) for 15 min, AC, plus aged 12 h at 565 °C (1050 °F), AC.
Source: *Beta Titanium Alloys in the 1980's*, R.R. Boyer and H.W. Rosenberg, Ed., TMS/AIME, 1984

Beta C: Notched fatigue strength at high temperature

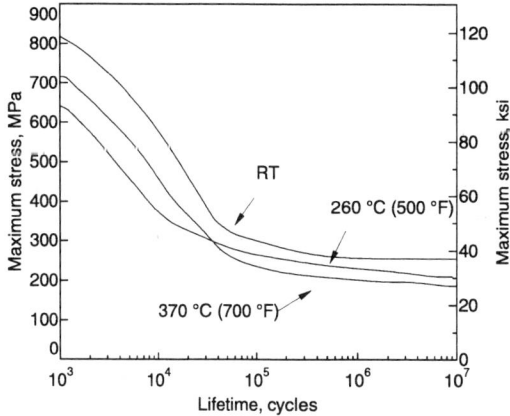

Axial fatigue of notched specimens ($R = 0.1$, $K_t = 3.0$) from 150 mm (6 in.) diam STA forgings.
Source: *Beta Titanium Alloys in the 1980's*, R.R. Boyer and H.W. Rosenberg, Ed., TMS/AIME, 1984

Fatigue Crack Growth

Crack growth rates in the accompanying figures were determined for Beta C in various conditions (see table). Because Beta C is an attractive alloy for highly corrosive environments such as in sour wells, the effect of aggressive environments on mechanical behavior also is of interest. For the test results presented here (see figures), no noticeable acceleration in crack growth rates was found when going from air to a saltwater environment, or when the frequency was reduced. Because the differences in da/dN–ΔK behavior are insignificant, data are presented as single scatterbands.

A slight tendency toward faster growth rates was observed for LO-Simplex and LO-Duplex (see figure below) as opposed to HI-Simplex and HI-Duplex, presumably as a consequence of the lower ductilities. No effect of the duplex versus the simplex aging treatment was detected. A significant difference, however, was found between aged and unaged material. The value of ΔK_{th} is roughly 3 MPa$\sqrt{m}$ (2.7 ksi$\sqrt{in.}$) for aged material under all testing conditions, as opposed to values of 4 to 5 MPa$\sqrt{m}$ (3.6 to 4.5 ksi$\sqrt{in.}$) for as-SHT material. Correcting for crack closure (ΔK_{eff}) brings the data into accord and reduces $\Delta K_{th,eff}$ to ≈2 MPa$\sqrt{m}$ (1.8 ksi$\sqrt{in.}$), suggesting that the difference between as-SHT and aged material may not be present at high R ratios.

Beta C: Material condition in crack growth tests

Heat treatment(a) (designation)	Tensile yield strength(0.2% offset)		Ultimate tensile strength		Elongation, %	Reduction of area, %
	MPa	ksi	MPa	ksi		
800 °C, 30 min, AC (as-SHT 800)	895	130	895	130	22	48
925 °C, 30 min, AC (as-SHT 927)	850	125	850	125	25	62
as-SHT 800 + 535 °C, (8 h) (LO-simplex)	1225	177	1320	190	8	15
as-SHT 800 + 425 °C, (4 h) (LO-duplex)	1220	176	1300	188	10	13
as-SHT 927 + 530 °C (16 h) (HI-simplex)	1140	165	1220	176	12	21
as-SHT 927 + 455 °C (4 h) + 555 (16 h) (HI-duplex)	1075	156	1180	171	14	23

(a) The grain sizes after solution heat treating at 800 and 925 °C were 45 and 160 μm, respectively. The 800 °C SHT did not fully recrystallize the as-hot worked structure, and left approximately 20 vol. % unrecrystallized. Almost no unrecrystallized grains were present after SHT at 925 °C. For both SHT, the 4 hr cycle promotes a somewhat more homogeneous α distribution. Source: H.E. Krugmann and J.K. Gregory, Microstructure and Crack Propagation in Ti-3Al-8V-6Cr-4Mo-4Zr, in *Microstructure and Property Relationships in Titanium Aluminides and Alloys*, Y-W. Kim and R.R. Boyer, Ed., TMS/AIME, 1991, p 551

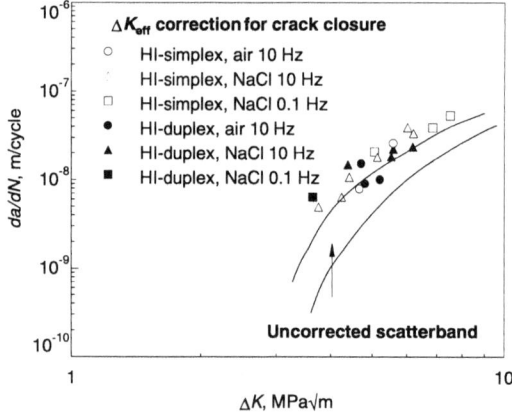

Beta C: Crack growth with high-temperature ST

Source: H.E. Krugmann and J.K. Gregory, Microstructure and Crack Propagation in Ti-3Al-8V-6Cr-4Mo-4Zr, in *Microstructure and Property Relationships in Titanium Aluminides and Alloys*, Y-W. Kim and R.R. Boyer, Ed., TMS/AIME, 1991, p 549-560

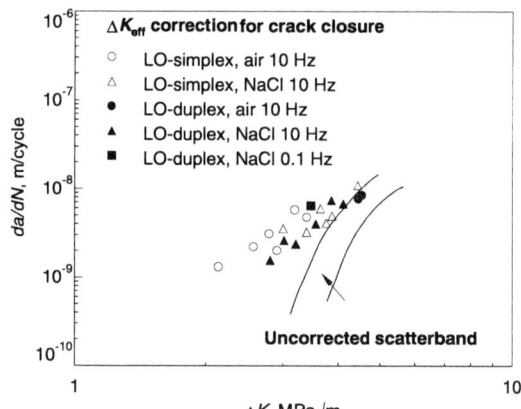

Beta C: Crack growth with low-temperature ST

Source: H.E. Krugmann and J.K. Gregory, Microstructure and Crack Propagation in Ti-3Al-8V-6Cr-4Mo-4Zr, in *Microstructure and Property Relationships in Titanium Aluminides and Alloys*, Y-W. Kim and R.R. Boyer, Ed., TMS/AIME, 1991, p 549-560

Fracture Properties

Beta C: Fracture toughness of bar

Condition	Test direction	Fracture toughness (K_{Ic}) MPa√m	ksi √in.
785 °C (1450 °F), 1 h, AC	C-R	53.7	48.9
+ 550 °C (1025 °F), 24 h, AC	C-R	55.7	50.7
	R-L	53.3	48.5
	R-L	56.7	51.6
840 °C (1550 °F), 1 h, AC	C-R	55.1	50.1
+ 480 °C (900 °F), 24 h, AC	C-R	69.2	63.0
	R-L	57.6	52.4
	R-L	55.2	50.2

Note: Specimens were from 75 mm (3 in.) bar. Source: G. Bella et al., Effects of Processing on Microstructure and Properties of Ti-3Al-8V-6Cr-4Mo-4Zr (Beta C™), in *Microstructure and Property Relationships in Titanium Aluminides and Alloys*, Y-M. Kim and R.R. Boyer, Ed., TMS/AIME, 1991, p 493-510

Beta C: Crack growth in solution treated condition

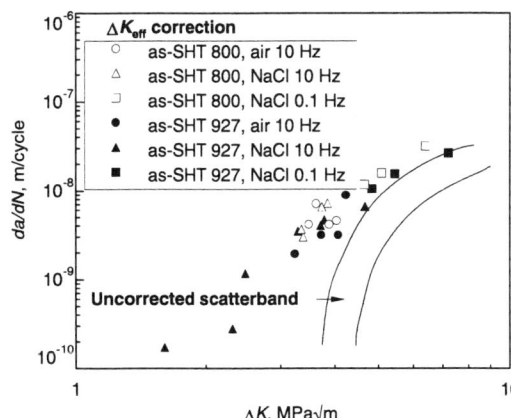

Source: H.E. Krugmann and J.K. Gregory, Microstructure and Crack Propagation in Ti-3Al-8V-6Cr-4Mo-4Zr, in *Microstructure and Property Relationships in Titanium Aluminides and Alloys*, Y-W. Kim and R.R. Boyer, Ed., 1991, p 549-560

Beta C: Fracture toughness of STA billet

Treatment	Test direction	Fracture toughness (K_{Ic}) MPa√m	ksi √in.	Ultimate tensile strength MPa	ksi	Tensile yield strength (2% offset) MPa	ksi	Elongation, %	Reduction of area, %
Water quench	L	96.7	88.0	1189	172.5	1125	163.2	9.5	19.6
	T	62	56.4	1188	172.4	1145	166.0	3.0	5.6
Air cool	L	89.9	81.8	1208	175.2	1150	166.8	9.2	17.4
	T	63.3	57.6	1242	180.2	1184	171.7	3.5	6.6

Note: Specimens were 150 mm (6 in.) billet solution treated 815 °C (1500 °F), 15 min, cooled (WQ,AC), then aged 12 h at 565 °C (1050 °F), AC. Source: RMI Co., reported in *Industrial Applications of Titanium and Zirconium: Fourth Volume*, C.S. Young and J.C. Durham, Ed., ASTM STP 917, 1986, ASTM, p 155

Beta C: Fracture toughness of billet, forging, and plate

Product form/ specimen location	Heat treatment	Direction	Tensile yield strength MPa	ksi	Fracture toughness K_Q MPa√m	ksi √in.	K_{Ic} MPa√m	ksi √in.
Billet								
150 mm (6 in.) diam	15 min, 815 °C (1500 °F), AC	L	1151	167	...	...	90(a)	82(a)
(midradius specimens)	+ 12 h, 565 °C (1050 °F), AC	T	1186	172	...	...	64	58
	15 min, 815 °C (1500 °F), WQ	L	1124	163	...	...	97	88
	+ 12 h, 565 °C (1050 °F), AC	T	1144	166	...	...	61	56
150 mm (6 in.) (location unknown)	Annealed(b) + aged:							
	8 h, 510 °C (950 °F), AC	T	1330	193	...	...	55(c)	50(c)
	plus exposed(d)	T	...	...	...	...	54	49
	8 h, 565 °C (1050 °F), AC	T	...	...	...	...	69	63
	plus exposed(d)	T	...	...	...	...	70	64
	8 h, 620 °C (1150 °F), AC	T	...	...	...	...	81	74
	plus exposed(d)	T	...	...	...	...	80	73
150 × 150 mm (6 × 6 in.)								
Surface specimens	15 min, 815 °C (1500 °F), AC	(c)	1151	167	...	...	60-66(e)	55-60(e)
Center specimens	+ 12 h, 565 °C (1050 °F), AC	(c)	1151	167	...	...	57-71	52-65
Navaho spar forging								
100 × 150 mm (4 × 6 in.)	15 min, 815 °C (1500 °F), AC							
(forging center specimens)	+ 12 h, 565 °C (1050 °F), AC	L	1158	168	...	...	58-59(e)	53-54(e)
Plate								
32 mm (1.25 in.)	925 °C (1700 °F), annealed, AC							
(center specimens)	+ 8 h, 565 °C (1050 °F), AC	RW	1137	165	53(f)	48(f)	...	...
	925 °C (1700 °F), annealed, AC							
	+ 8 h, 675 °C (1250 °F), AC	RW	862	125	56(f)	51(f)	...	...
19 mm (0.75 in.)	1 h, 815 °C (1500 °F), AC							
(center specimens)	+ 4 h, 525 °C (975 °F), AC	(c)	1206	175	...	...	90(c)	82(c)

(a) Four-point loaded, slow-bend test. (b) Heat treatment details not given. (c) Test details not given. (d) 1000-h exposure at 285 °C (550 °F) under 172 MPa (25 ksi) load, cooled to room temperature and tested. (e) Slow-bend tests. (f) Compact-tension tests. Source: *Beta Titanium Alloys*, R. Wood, MCIC-72-11, Battelle Columbus Laboratories, 1972

Stress-Strain Curves

Beta C: Typical tensile stress-strain

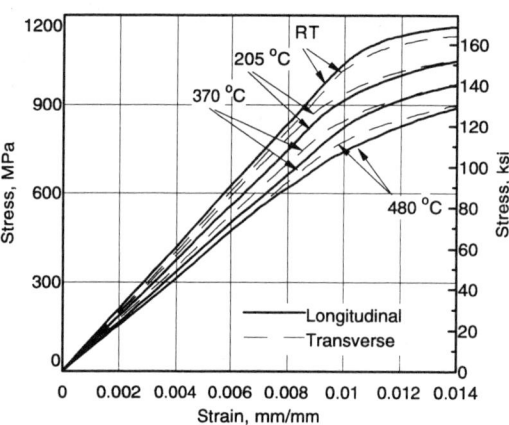

Tension stress-strain curves for solution treated and aged forged billet. Specimens cut from a 150 × 150 mm (6 × 6 in.) billet heat treated for 15 min at 815 °C (1500 °F) air cooled, plus 12 h, 565 °C (1050 °F) air cooled.

Beta C: Longitudinal stress-strain

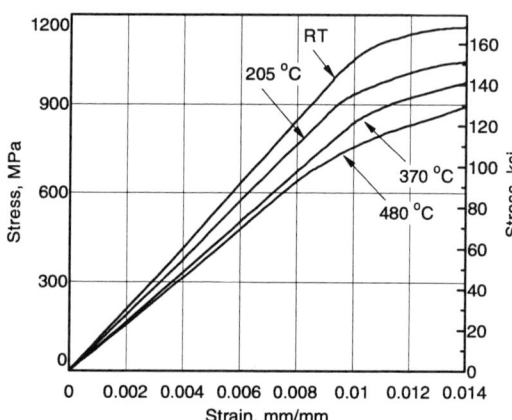

150 mm (6 in.) square billet heat treated at 815 °C (1500 °F), 15 min, AC + 565 °C (1050 °F), AC.
Source: *Aerospace Structural Metals Handbook*, Vol 4, Code 3723, Battelle Columbus Laboratories, 1981

Beta C: Compressive stress-strain

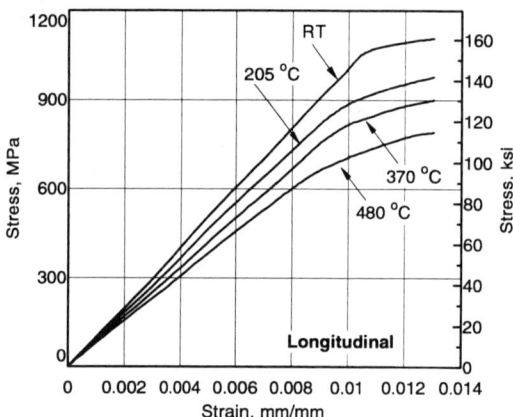

150 mm (6 in.) square billet heat treated at 815 °C (1500 °F), 15 min, AC + 565 °C (1050 °F), AC.
Source: *Aerospace Structural Metals Handbook*, Vol 4, Code 3723, Battelle Columbus Laboratories, 1981

Beta C: Compressive stress-strain

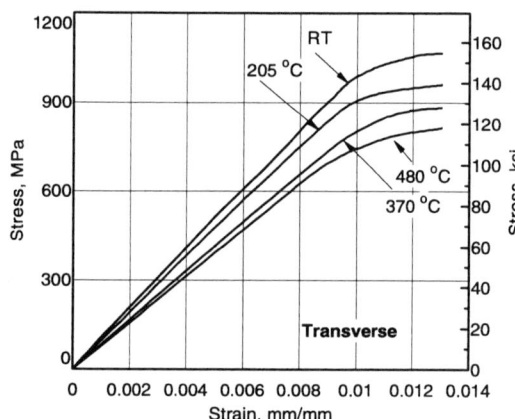

150 mm (6 in.) square billet heat treated at 815 °C (1500 °F), 15 min, AC + 565 °C (1050 °F), AC.
Source: *Aerospace Structural Metals Handbook*, Vol 4, Code 3723, Battelle Columbus Laboratories, 1981

Beta C: Evaluation of stretch-formed specimens

Section height		Stretch	Stretch,	Strain rate,	
mm	in.	direction	%	min^{-1}	Remarks
38	1.5	Longitudinal	8	0.013	Good part
50	2.0	Transverse	14	0.023	Good part
89	3.5	Longitudinal	28	0.046	Good part, except that horizontal flange had slight wave and vertical flange drew into die block due to die block separations

Source: R. Kaneko and C.A. Woods, "Low-Temperature Forming of Beta Titanium Alloys," NASA Report 3706, 1983

Forging

G.W. Kuhlman, ALCOA, Forging Division

Commercially important metastable β and near-β alloys respond well to thermomechanical processing (TMP), and several complex thermomechanical processing routes have been reduced to commercial practice. With this alloy class, thermomechanical processing is focused on optimal combinations of high strength, good fracture toughness, and ductility. These alloys possess superior high-cycle fatigue properties due to the refined nature of their microstructure. Several of these alloys have been successfully direct aged, thereby producing even finer microstructures and improved smooth and notched fatigue properties.

Supra-transus forging processes prevail in this class of materials (except for Ti-10V-2Fe-3Al in which a combination of supra- and subtransus working can achieve desired properties through α phase manipulation and control). None of these alloys respond to thermomechanical processing to improve fracture-related properties, because α morphology cannot be modified to the extent possible in α + β alloys. All of these alloys are ideal for use in structural applications in which durability is the critical design criterion.

Beta C is a very high strength, deep hardening, metastable β alloy whose primary commercial applications in forgings are corrosion-resistant applications in a wide variety of media and aerospace and automotive applications. It can be fabricated into all forging product types, although closed die forgings predominate.

Beta C is a highly forgeable alloy (when forged above the β transus), with lower unit pressures (flow stresses), improved forgeability, and less crack sensitivity in forging than the α-β alloy Ti-6Al-4V. Flow stresses and unit pressures exceed that of the near-β alloy Ti-10V-2Fe-3Al (see figure). Beta C is thermomechanically processed in forging manufacture to achieve the desired final microstructure of fine transformed β, limited grain boundary films, with a fine recrystallized prior β grain size, in preparation for final thermal treatments. The highly refined microstructures of Beta C forgings are responsible for its excellent corrosion, strength, and fatigue properties.

Beta C is forged above the transus through one or more forging operations. Reheating for subsequent forging operations recrystallize the alloy from prior hot work refining prior β grain size. Beta C is not subtransus (α + β) forged, because no microstructural advantages are gained and there is a significant increase in unit pressure requirements.

Final thermal treatments for Beta C forgings include annealing or solution annealing and aging. Forgings may be supplied in the annealed or solution annealed (ST) condition and/or fully aged (STA). In the ST condition, Beta C has lower strength, but much higher ductility and toughness than in the STA condition (not recommended for high-temperature use). Solution treatment is conducted at 815 to 925 °C (1500 to 1700 °F), followed by air cooling. Aging is conducted at 455 to 540 °C (850 to 1000 °F). For thick section Beta C forgings, it has been reported (Gurganus, Ref 1) that a three-step heat treatment process improves the overall combination of strength, ductility, toughness, and fatigue. The process studied was solution annealing at 920 °C (1685 °F), air cooling or faster plus-resolution annealing at 820 °C (1525 °F), air cool or faster, and aging at 280 °C (535 °F). Silicides (silicon is a tramp element from master alloys), which may adversely affect ductility and toughness, may form at grain boundaries in Beta C under certain supra-transus annealing process conditions. Recent work (Ankem, Ref 2) has suggested that solution annealing above the silicide solvus (980 °C or 1800 °F), followed by rapid quenches and then above-mentioned thermal treatments may be used to reduce continuous grain boundary silicides and improve properties.

Beta forging working histories for Beta C require imparting enough hot work to reach final macrostructure and microstructure objectives. Generally, reductions in any given forging process are 30 to 50% to achieve desired dynamic and static recrystallization. Very low levels of β reduction are not recommended. Although Beta C is highly cold workable and is processed in other product forms such as sheet and plate, cold working is generally not used in forging manufacture. Beta C may be successfully isothermal or hot die forged. However, it has lower strain-rate sensitivity than α-β and near-β alloys, and thus, unit pressure reductions and forging shape sophistication improvements through these technologies are modest in comparison to other titanium alloys.

Beta C, as with all β alloys, has a higher affinity for hydrogen than other titanium alloy classes. Al-

Beta C: Forging process temperatures

Process	Metal temperature	
	°C	°F
Beta forge	815-980	1500-1800
Preferred range(a)	$\beta_t + 85$ °C	$\beta_t + 150$ °F

Note: See "Technical Note 4. Forging" for recommended die temperatures. (a) Final billet conversion and final forging temperatures about 85 °C (150 °F) above the transus are preferred over higher temperatures to reduce grain growth.

Beta C: Effect of TMP on properties

Due to the fine microstructural features of aged β alloys, their resistance to crack growth is generally inferior to α + β alloys.

Forging route	Heat treatment	Tensile yield strength		Ultimate tensile strength		Elongation, %	Reduction of area, %	Fracture toughness (K_{Ic})		Critical crack length		ΔK_{th}(b)		Smooth fatigue stress(c)	
		MPa	ksi	MPa	ksi			MPa√m	ksi √in.	mm	in.	MPa√m	ksi √in.	MPa	ksi
$\beta_t + 85$ °C	535 °C, 8 h	1123	163	1179	171	6	9	53	48	2.5	0.10	...	...	790	114
$\beta_t + 85$ °C	910 °C, AC + 535 °C, 8 h	1199	174	1262	183	8	14	48	43	1.8	0.071	<4.4	<4	800	116
$\beta_t + 195$ °C	910 °C, AC + 535 °C, 8 h	1190	172	1258	182	3	7	50	45	1.9	0.075	<4.4	<4	700	114
$\beta_t + 85$ °C	815 °C, AC + 565 °C, 8 h	1151	167	1192	173	11	22	57	52	3.0	0.12	4.4	4	780	113

(a) Critical crack length, ≈ 1.1 $(K_{Ic}/YS)^2$. (b) ΔK_{th} is the threshold stress-intensity in fatigue crack growth rate tests. (c) Smooth fatigue stress at 10^7 cycles, tests conducted at R = 0.1 to 0.3, F = 30 to 125 Hz. Source: G.W. Kuhlman, "A Critical Appraisal of Thermomechanical Processing (TMP) of Structural Titanium Alloys"

though Beta C forms less α case from heating operations than other alloy classes, therefore requiring less metal removal in chemical pickling (milling processes), control of chemical removal processes is essential to preclude excessive hydrogen pickup.

References
1. T.B. Gurganus, "Improvement of Reliability and the Mechanical Properties of Titanium Alloy Forgings," AFML-TR-75-311, Air Force Materials Laboratory, Dec 1975
2. S. Ankem et al., "Silicide Formation in Ti-3Al-8V-6Cr-4Zr-4Mo," *Metall. Trans. A*, Vol 18, Dec 1987, p 2015-2025

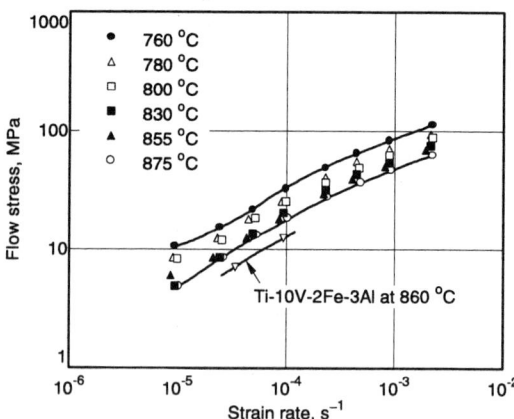

Beta C: Flow stress

Source: *Mater. Sci. Eng.*, Vol 86, 1987, p 159-177

Beta C: Forging tensile properties
Room-temperature tensile properties of forging solution heat treated 15 min, 815 °C (1500 °F), air cooled, plus aged 12 h, 565 °C (1050 °F), air cooled

Sample location and orientation	Tensile yield strength (0.2%)		Ultimate tensile strength		Elongation, %	Reduction of area, %
	MPa	ksi	MPa	ksi		
Panel (thin section), long transverse	1089	158	1179	171	10.0	23
Panel (thin section), longitudinal	1165	169	1241	180	4.5	14
Rib (thin section), longitudinal	1193	173	1268	184	5.5	12
Center (thick section near surface), longitudinal	1186	172	1241	180	7.5	15
Center (thick section mid radius), longitudinal	1158	168	1241	180	8.0	14
Center (thick section center), longitudinal	1089	158	1103	160	8.0	20
Center (thick section), short transverse	1027	149	1061	154	5.0	16
Center (thick section), long transverse	1034	150	1068	155	4.0	17

Source: R.A. Wood, *Beta Titanium Alloys*, MCIC-72-11, Battelle Columbus Laboratories, 1972

Forming

Total cold reductions of 60 to 70% are common for Beta C, and the capacity for cold work enhances the manufacture of seamless tubing, rod, wire, strip, and foil products. The material is markedly strengthened by cold work to about 60% but further cold work has little additional effect. As in

Beta C: Aged tensile properties vs cold work

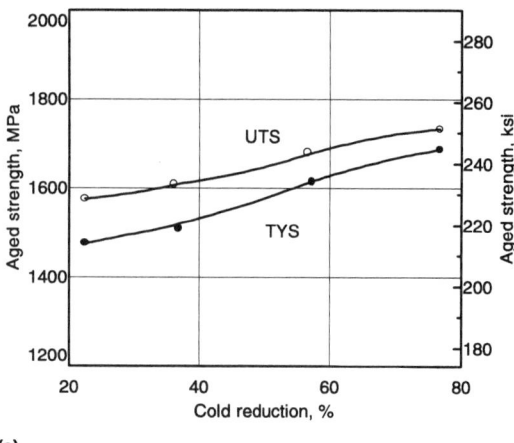
(a)

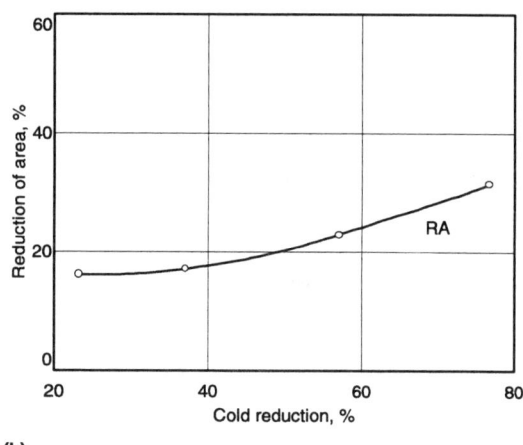
(b)

Effect of increasing amounts of cold work on the room-temperature tensile properties of aged alloy cold drawn as indicated plus aged for 6 h at 480 °C (900 °F) and air cooled.
Source: R.A. Wood, *Beta Titanium Alloys*, MCIC-72-11, Battelle Columbus Laboratories, 1972

other β titanium alloys, prior cold work accelerates the aging reaction, and a fine, uniformly dispersed α precipitate may be obtained. Limited information is available on warm working of Beta C (see table), but good results would be expected in the 230 to 345 °C (450 to 650 °F) range. Data available suggest that aged material would not be very workable below about 540 °C (1000 °F).

Beta C: Minimum bend radii
Cold press-brake bending of solution treated β alloys

	Minimum bend radius as a function of sheet thickness, t	
Alloy	$t <$ 1.75 mm (0.069 in.)	1.75 mm (0.069 in.) $< t <$ 4.76 mm (0.1875 in.)
Ti-13V-11Cr-3Al	3.0	3.5
Ti-11.5Mo-6Zr-4.5Sn	3.0	3.0
Beta C	3.5	4.0
Ti-8Mo-8V-2Fe-3Al	3.5	3.5

Source: Military Standard MIL-T-9046J, US Government Printing Office

Beta C: Warm brake bending of 1.65 mm (0.065 in.) sheet

Bend axis direction	Bend radius	Forming temperature °C	Forming temperature °F	Average free bend angle, °	Average springback, °	Visual examination at 15×
Longitudinal	5.0t	175	350	90	15	Acceptable
	4.0t	175	350	92	14	Apparent separations
	3.0t	175	350	88	12	Apparent crack
	2.0t	175	350	90	5	Cracked
	2.0t	215	420	90	7	Cracked
Transverse	3.0t	175	350	89	11	Acceptable
	2.0t	175	350	87	8	Acceptable
	2.0t	215	420	90	8	Acceptable

Source: R. Kaneko and C.A. Woods, "Low-Temperature Forming of Beta Titanium Alloys," NASA Report 3706, 1983

Beta C: Room-temperature brake bending of 1.65 mm (0.065 in.) sheet

Bend axis direction	Bend radius	Average free bend angle, °	Average springback, °	Visual examination at 20×
Longitudinal		40	17	Acceptable
	6.0t	87	30	Acceptable
		121	28	Acceptable
		39	7	Apparent metal separation
		45	11	Apparent metal separation
	5.0t	94	23	Apparent metal separation
		132	24	Apparent metal separation
	4.0t	90	15	Apparent metal separation
	3.5t	(a)	...	...
Transverse		43	13	Acceptable
	4.0t	90	16	Acceptable
		120	33	Acceptable
	3.5t	90	...	Apparent metal separation

(a) Specimen fractured at bend before reaching 90°. Source: R. Kaneko and C.A. Woods, "Low-Temperature Forming of Beta Titanium Alloys," NASA Report 3706, 1983

Beta C: Tensile properties of bar and rod as a function of cold work and heat treatment

Diameter mm	Diameter in.	Condition	Ultimate tensile strength MPa	Ultimate tensile strength ksi	Tensile yield strength MPa	Tensile yield strength ksi	Elongation, %	Reduction of area, %	Shear strength(b) MPa	Shear strength(b) ksi	Cold head ratio(c) D_f/D_i
9.5	0.375	15 min, 815 °C (1500 °F), AC	875	127	862	125	16	41	655	95	2.04
8.4	0.330	Cold drawn 22.5%	1137	165	1082	157	13	43	696	101	1.14
		Cold drawn 22.5% + 6 h, 425 °C (800 °F), AC	1668	242	1565	227	5	17	...	...	...
		Cold drawn 22.5% + 6 h, 480 °C (900 °F), AC	1579	229	1475	214	5	16	...	...	...
		Cold drawn 22.5% + 6 h, 540 °C (1000 °F), AC	1310	190	1220	177	12	18	...	...	...
7.6	0.299	Cold drawn 37.2%	1234	179	...	...	...	47	...	...	1.75
		Cold drawn 37.2% + 480 °C (900 °F) age(a)	1606	233	1510	219	8	15	951	138	...
6.0	0.238	Cold drawn 59.5%	1324	192	...	...	...	44	...	...	1.68
		Cold drawn 59.5% + 480 °C (900 °F) age(a)	1675	243	1613	234	7	21	924	134	...
4.7	0.188	Cold drawn 75.0%	1317	191	...	...	...	23	...	...	1.63
4.4	0.176	Cold drawn 78.0% + 480 °C (900 °F) age(a)	1730	251	1682	244	7	32	875	127	...
3.3	0.132	Cold drawn 87.6%	1317	191	...	...	...	41	...	...	1.85

(a) Probably 6-h aging at 480 °C (900 °F). (b) Based on double shear specimens. (c) D_f = final diameter; D_i = initial diameter in upset specimens where D_i = one half specimen length. Source: R.A. Wood, *Beta Titanium Alloys*, MCIC-72-11, Battelle Columbus Laboratories, 1972

Beta C: Tensile properties of annealed 7.9 mm (0.312 in.) diam rod
Room temperature tensile, shear and headability properties as a function of solution annealing treatment

Condition(a)	Tensile yield strength MPa	ksi	Ultimate tensile strength MPa	ksi	Elongation, %	Reduction of area, %	Shear strength MPa	ksi	Headability(b), D_f/D_i
760 °C (1400 °F), AC	875	127	903	131	15	57	655	95	1.94
760 °C (1400 °F), WQ	875	127	910	132	21	58	648	94	2.02
815 °C (1500 °F), AC	875	127	910	132	16	56	662	96	2.02
815 °C (1500 °F), WQ	875	127	910	132	19	58	655	95	2.04
870 °C (1600 °F), AC	868	126	903	131	17	58	662	96	2.13
870 °C (1600 °F), WQ	862	125	903	131	19	56	648	94	2.14
925 °C (1700 °F), AC	834	121	862	125	17	58	648	94	...
925 °C (1700 °F), WQ	827	120	868	126	20	58	648	94	...

(a) 30 min at temperature. (b) D_f = final diameter, D_i = initial diameter in upset specimens where D_i = one half specimen length. (c) Using double shear specimens. Source: R. A. Wood, *Beta Titanium Alloys*, MCIC-72-11, Battelle Columbus Laboratories, 1972

Heat Treatment

Beta C is capable of achieving many strength/ductility combinations, depending on processing history and heat treatment. A wide range of aging heat treatments may be used for Beta C to achieve a preferred strength level and associated mechanical properties.

Full annealing may be used to alleviate undesirable residual stresses in Beta C, and in certain cases where the material is to be used in the aged or overaged condition, stress relief annealing may be accomplished simultaneously with the aging or overaging heat treatment. Annealing for Beta C is the same as solution treating.

Solution Treatment. Some care should be exercised in matching the solution annealing treatment with the mill product form being used, its processing history, and the mechanical properties expected. A typical recommended solution annealing treatment for Beta C is 30 min at 815 °C (1500 °F) (terminated by either water quenching or air cooling). However, good properties (high ductility in the solution annealed condition and combinations of good strength and ductility in the aged condition) also may be obtained by solution annealing at temperatures up to 925 °C (1700 °F). The higher temperatures (e.g., 925 °C) are generally favored for thicker section products such as plate and bar, whereas the lower temperatures (e.g., 1500 to 1550 °F) may be used for products such as wire and sheet. Solution annealing temperature has a significant effect on the ductility of plate, but little effect on the properties of small-diameter rod.

Aging temperatures range from 455 to 540 °C (850 to 1000 °F). Typical aging times range from 6 to 12 h at the aging temperature, although 12 to 24 h exposure may be used to age material to maximum strength at aging temperatures of 455 to 465 °C (850 to 875 °F). Aging is terminated by air cooling.

Beta C: Effect of aging time on room-temperature tensile properties

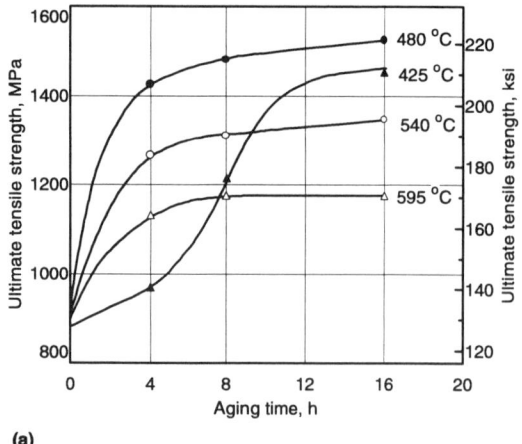
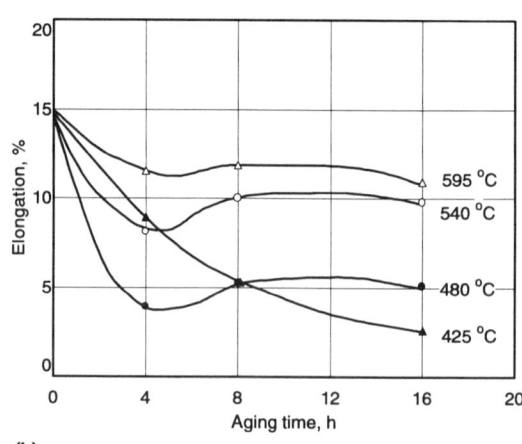

Forgings upset to 15.8 mm (0.625 in.) from 63.5 mm (2.5 in.) at 815 °C (1500 °F) followed by 815 °C (1500 °F) solution annealing.
Source: R.A. Wood and R.J. Favor, *Titanium Alloys Handbook*, MCIC-HB-02, Battelle Columbus Laboratories, 1972

Beta C: Effect of aging and cold work (6 hour age)

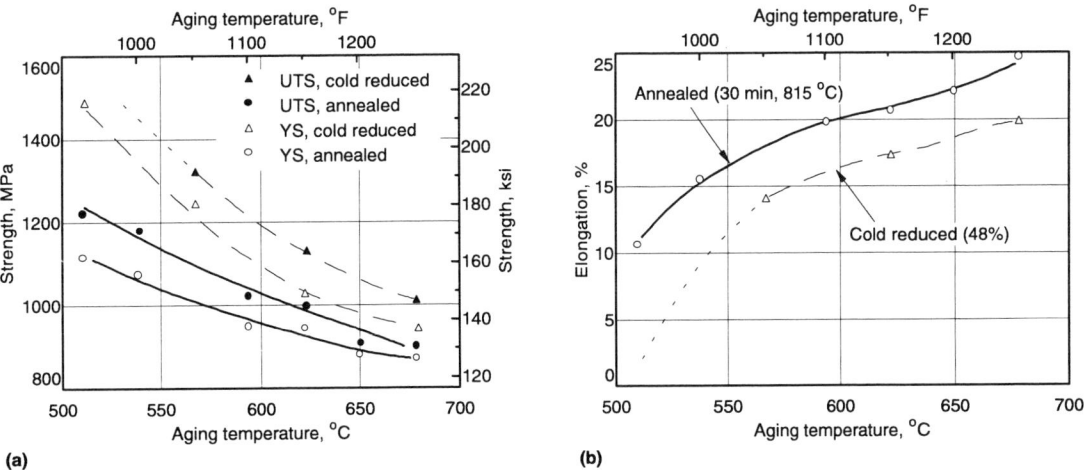

Effect of aging temperature variation and of cold work plus aging temperature on the tensile properties of tubing. 19 mm (0.75 in.) OD × 1.0 mm (0.042 in.) wall. Aged for 6 h.
Source: R.A. Wood, *Beta Titanium Alloys*, MCIC-72-11, Battelle Columbus Laboratories, 1972

Ti-10V-2Fe-3Al

Common Name: Ti-10-2-3
UNS Number: Unassigned

Ti-10-2-3 is a high-strength titanium-base alloy. Metallurgically it is a near-beta alloy, and it is capable of attaining a wide variety of strength levels depending on selection of heat treatment. Major advantages of this alloy are its excellent forgeability; its high toughness in air and saltwater environments; and its high hardenability, which provides good properties in sections up to 125 mm (5 in.) thick. It is used in the aerospace industry for applications up to 315 °C (600 °F).

A major advantage of Ti-10-2-3 over commercially available alpha-beta compositions of similar strength levels is its toughness in air and salt water environments. This near-beta alloy was developed primarily for high-strength and toughness applications at temperatures up to 315 °C (600 °F) and tensile strengths of 1240 MPa (180 ksi) in order to provide weight savings over steels in airframe forging applications. Of special interest for high-strength forgings for aircraft, it is being used for components by much of the aerospace industry.

Chemistry and Density

Ti-10-2-3 has a near-beta composition and is slightly more beta stabilized than Ti-11.5Mo-6Zr-4.5Sn (Beta III).
Density. 4.65 g/cm^3 (0.168 lb/in.^{3})

Product Forms

Ti-10-2-3 has the best hot-die forgeability of any commercial titanium alloy and is often used for near-net-shape forging applications. Mill products are billet, bar, and plate.

Product Condition/Microstructure

Developed for use in the aerospace industry, Ti-10-2-3 combines many of the advantages of the metastable beta titanium alloys without sacrificing certain inherent alpha-beta characteristics. It shows excellent hardenability in section sizes up to 125 mm (5 in.), but also demonstrates good short-transverse ductility. In the solution-treated and aged condition, this alloy maintains greater than 80% of its room-temperature strength at 315 °C (600 °F) and has creep-stability characteristics similar to those of the alpha-beta alloys at this temperature.

Applications

Ti-10-2-3 is used at temperatures up to 315 °C (600 °F) where medium-to-high strength and high toughness are required in bar, plate, or forged sections up to 125 mm (5 in.) thick. It can be heat treated over a wide strength-toughness range, allowing the tailoring of properties. It is employed for applications requiring uniformity of tensile properties at surface and center locations. Specific applications include aerospace airframes hot-die and conventional forgings, and other forged parts in a wide variety of components. The major user, Boeing, uses the alloy up to 260 °C (500 °F).

Ti-10V-2Fe-3Al: Specifications and compositions

Specification	Designation	Description	Al	C	Fe	H	N	O	V	Y	Other
USA											
AMS 4986		Frg STOA	2.6-3.4	0.05	1.6-2.2	0.015	0.05	0.13	9-11	0.005	OT 0.3; bal Ti
AMS 4983A		Frg STA	2.6-3.4	0.05	1.6-2.2	0.015	0.05	0.13	9-11	0.005	OT 0.3; bal Ti
AMS 4984		Frg STA	2.6-3.4	0.05	1.6-2.2	0.015	0.05	0.13	9-11	0.005	OT 0.3; bal Ti
AMS 4987		Frg STOA	2.6-3.4	0.05	1.6-2.2	0.015	0.05	0.13	9-11	0.005	OT 0.3; bal Ti

Ti-10V-2Fe-3Al: Commercial compositions

Specification	Designation	Description	Al	C	Fe	H	N	O	V	Y	Other
Japan											
Kobe	KS10-2-3	Bar Frg STA	2.6-3.4		1.6-2.2	0.015	0.05	0.13	9-11		bal Ti
USA											
Timet	TIMETAL 10-2-3	Frg	2.6-3.4		1.6-2.2	0.015	0.05	0.13	9-11		bal Ti

Phases and Structures

As a solute-lean β alloy, the microstructure of Ti-10V-2Fe-3Al typically has a bimodal (equiaxed and lamellar) α phase in a β matrix. The precise microstructural characteristics depend on the deformation and heat treatment history of the alloy. Coarse, globular α, which has little effect on strength but has desirable effects on ductility, is produced during solution treatment in the α + β phase field through recrystallization of some of the lamellar α as a result of α/β work. The remnant acicular alpha, which detracts from ductility, is necessary to meet fracture toughness requirements. Ti-10V-2Fe-3Al generally is not solution treated above the β transus because ductility and toughness are lower than that of α/β solution treated material. However, extreme overaging of β quenched material can restore ductility and toughness to acceptable levels.

Beta Transus. 800 °C (1475 °F) is often reported as the typical transus, with a range of 790 to 805 °C (1450 to 1480 °F).

Ti-10V-2Fe-3Al: Lattice parameters

Alpha phase
$a = 0.293595$ nm
$c = 0.467454$ nm
Beta phase
$a = 0.3238$ nm

Note: Determined by X-ray diffraction. Source: AFML-TR-78-114

Ti-10V-2Fe-3Al: Effect of solution temperature on primary α content

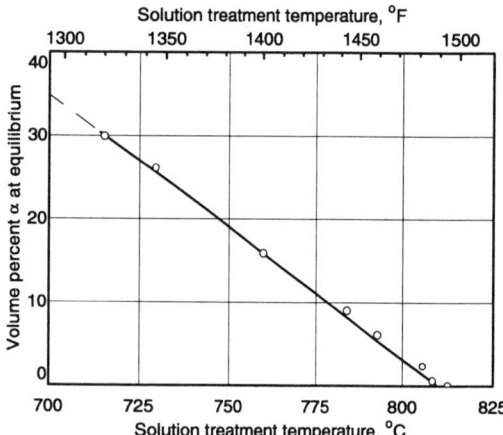

Source: Duerig et al., Stress Assisted Transformation in Ti-10V-2Fe-3Al, Metall. Trans. A, Vol 11, Dec 1980, p 1987-1998

Grain Structure

Recrystallization and grain structure of Ti-10V-2Fe-3Al are influenced greatly by minor modifications in thermomechanical processing. High amounts of deformation and high temperatures enhance dynamic recrystallization and the formation of equiaxed β, whereas high volume fractions of α retard recrystallization. High deformation also reduces the aspect ratio of α platelets. Forgings are normally beta forged followed by α-β forging, with about 10 or 15 to 25% reductions (for high-strength condition) to break up grain-boundary α and recrystallize some of the primary α to a globular shape for improved ductility.

Transformation Products

Detailed microstructural work (Duerig et al., Metall. Trans. A, Vol 11, 1980, p 1987-1998) has revealed a number of transformations and transformation products that may occur in Ti-10V-2Fe-3Al. They are summarized below.

Athermal ω. Like all solute-lean β alloys, athermal ω appears upon rapid quenching of solution treated Ti-10V-2Fe-3Al. Although these fine nondescript particles have no discernible effect on mechanical properties, they affect subsequent aging behavior by catalyzing isothermal ω phase formations.

Isothermal ω occurs at aging temperatures (below 450 °C, or 840 °F) in β quenched Ti-10V-2Fe-3Al. Isothermal ω particles formed at any temperature are very uniform and have been associated with displacive growth of athermal ω. Three morphologies of ω exist—nondescript, ellipsoidal, and cuboidal. As aging continues, misfit increases and a transition from ellipsoidal to a cuboidal morphology is observed. Increased amounts of hydrogen limit the amount of athermal ω formation during quenching (J.E. Costa et al., Titanium Science and Technology, 1985, p 2480).

Uniform α of both the Burger's and non-Burger's variety occurs at aging temperatures above 400 °C (750 °F) (see TTT and ITT diagrams). This type of α appears to nucleate on particles of

Ti-10V-2Fe-3Al: Isothermal grain growth for β

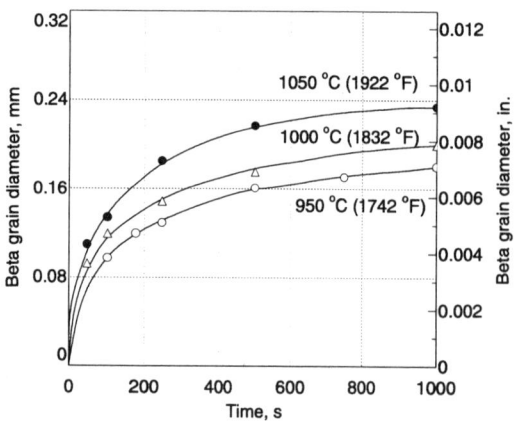

Source: J.R. Toran and R.R. Biederman, Phase Transformation Study of Ti-10V-2Fe-3Al, Titanium '80 Science and Technology, H. Kimura and O. Izumi, Ed., TMS/AIME, 1980, p 1494

isothermal ω. With continued aging, α changes from a blocky morphology into fine "stubby" plates that are uniformly distributed throughout the β matrix. The fine nature of these dispersions allows very high strengths to be achieved, even in solution treated specimens containing as much as 25 vol% of primary α.

Sympathetic plate α of both the Burger's and non-Burger's type occurs above 400 °C (750 °F) (see TTT diagram on next page). These precipi-

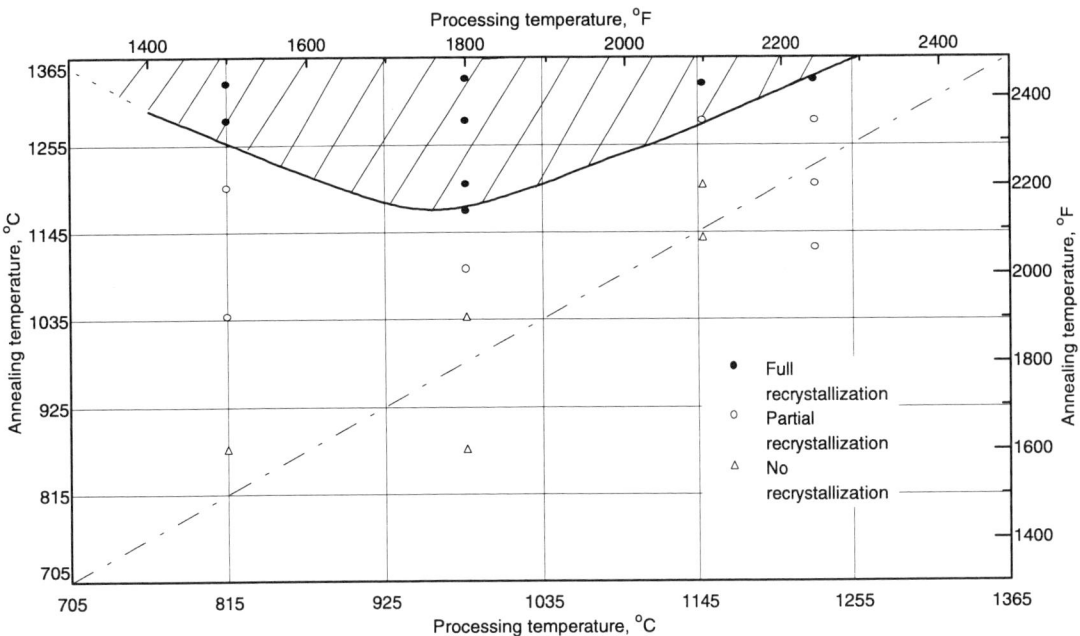

Ti-10V-2Fe-3Al: Recrystallization regions with 30% reduction

Effect of processing and annealing temperatures on the microstructure of specimens formed to 30% reduction. True strain, 36%. Annealing time, 1 h.
Source: I. Weiss and F.H. Froes, The "Processing Window" for the Near-Beta Ti-10V-2Fe-3Al Alloy, *Titanium, Science and Technology*, G. Lütjering, U. Zwicker, and W. Bunk, Ed., Deutsche Gesellschaft für Metallkunde, e.V., Germany, 1985, p 504

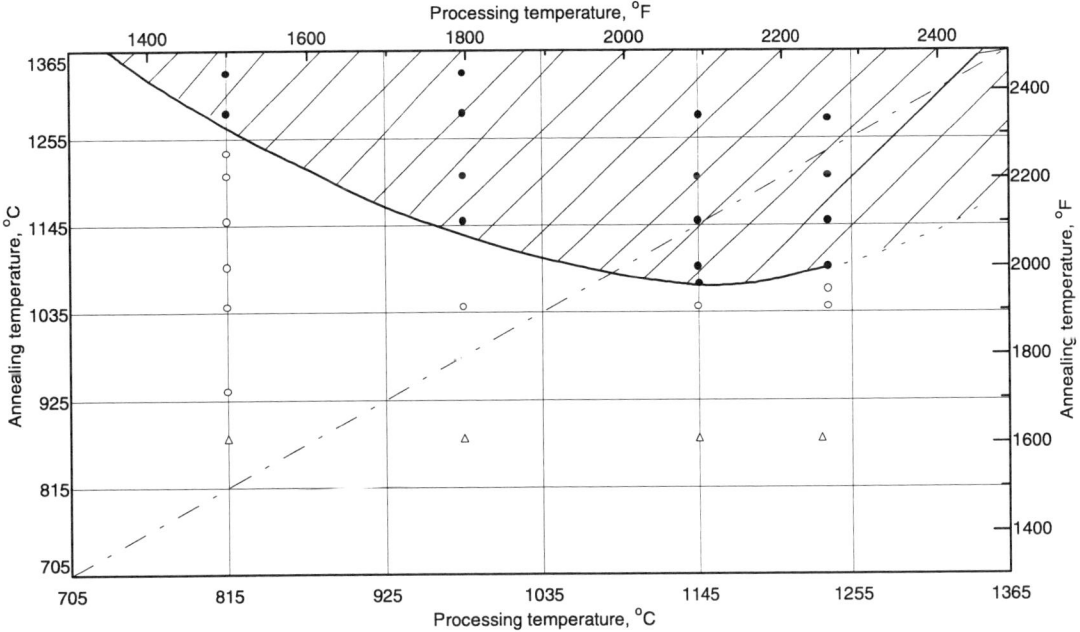

Ti-10V-2Fe-3Al: Recrystallization regions with 65% reduction

Effect of processing and annealing temperatures on the microstructure of specimens forged to 65% reduction. True strain 109%; 30-50% reduction. Annealing time, 1 h.
Source: I. Weiss and F.H. Froes, The "Processing Window" for the Near-Beta Ti-10V-2Fe-3Al Alloy, *Titanium, Science and Technology*, G. Lütjering, U. Zwicker, and W. Bunk, Ed., Deutsche Gesellschaft für Metallkunde, e.V., Germany, 1985, p 504

tates appear as large plates with very high aspect ratios. Although these coarser plates do not strengthen as efficiently as the fine uniform dispersions, very high strengths are nonetheless achievable in β solution treated material.

Grain-boundary α first forms during α-β solution treatment. It is thickened with aging, and the thickness increases with aging temperature.

Orthorhombic martensite (α″) is found only as a stress-induced transformation product in Ti-10V-2Fe-3Al. The transformation can be inhibited by the precipitation of α and the associated chemical stabilization of β. The formation of α″ decreases tensile yield strength by plastically accommodating relatively small elastic strains.

Inclusions rich in titanium, phosphorus, sulfur, and silicon can occur. Chemical microsegregation (or beta flecks), which is found in other β alloys and even some α + β alloys, is due to iron segregation.

Ti-10V-2Fe-3Al: TTT diagram for β-solution treated alloy

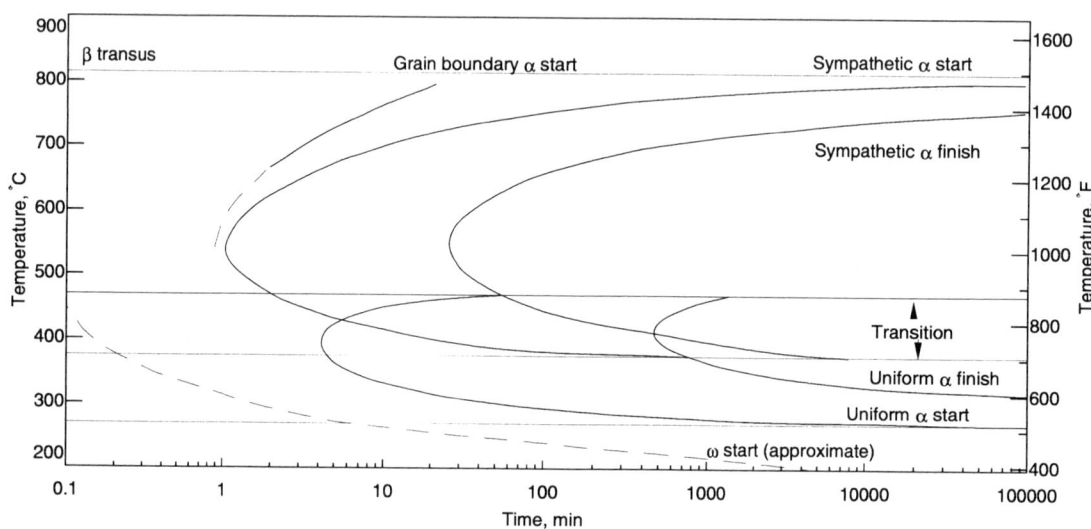

Qualitative diagram illustrating the competition between nucleation regimes.
Source: Duerig et al., Metall. Trans. A, 1980, p 1987-1998

Ti-10V-2Fe-3Al: ITT diagram

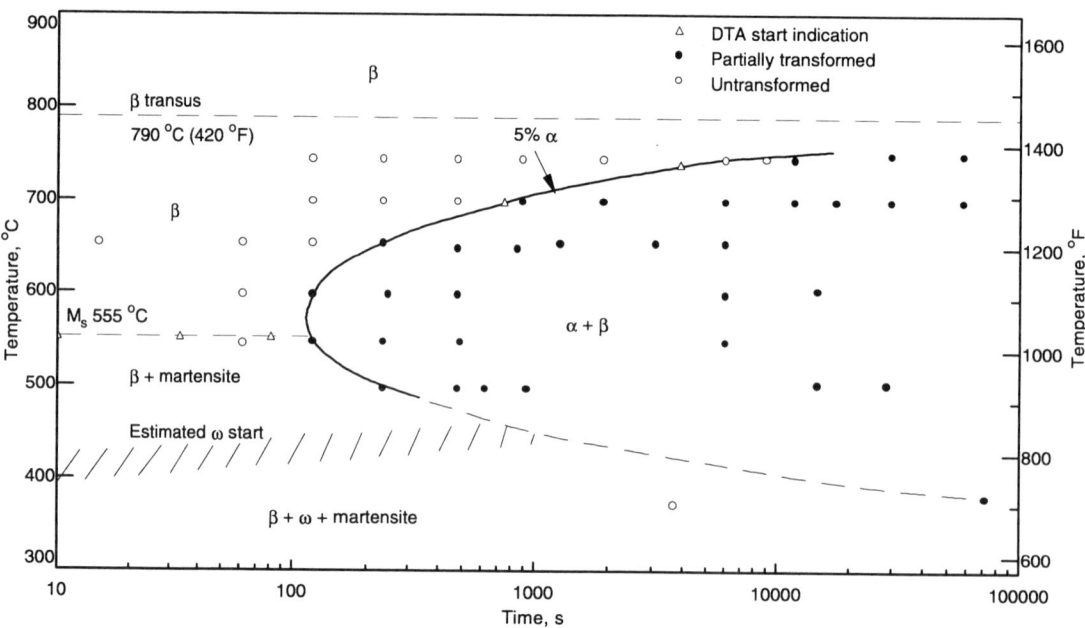

Solution treated at 860 °C (1580 °F) for 240 s DTA, differential thermal analysis.
Source: J.R. Toran and R.R. Biederman, Phase Transformation Study of Ti-10V-2Fe-3Al, Titanium '80 Science and Technology, H. Kimura and O. Izumi, Ed., TMS/AIME, 1980, p 1494

Physical Properties

Ti-10V-2Fe-3Al: Summary of typical physical properties

Beta transus	790-805 °C (1450-1480 °F)
Melting (liquidus) point	Not available
Density(a)	4.65 g/cm^3 (0.168 lb/in.^{3})
Electrical resistivity(a)	Not available
Magnetic permeability(a)	Nonmagnetic
Specific heat capacity(a)	Not available
Thermal conductivity(a)	Not available
Thermal coefficient of linear expansion(b)	9.7×10^{-6}/°C (5.4×10^{-6}/°F)

(a) Typical values at room temperature of about 20 to 25 °C (68 to 78 °F). (b) Mean coefficient from room temperature to 100 °C (212 °F)

Elastic Properties

Ti-10V-2Fe-3Al: Room-temperature elastic properties

Material condition	Tensile modulus GPa	Tensile modulus 10^6 psi	Compressive modulus GPa	Compressive modulus 10^6 psi
As forged, different conditions	83-103	12-15	...	...
Solution treated and aged	103-110	15-16	107-114	15.5-16.5
Solution treated and overaged	96.5-107	14-15.5	...	...

Source: R. Boyer and H. Rosenberg, Ed., *Beta Titanium Alloys in the 1980's*, TMS/AIME, 1984, p 443; AFML-TR-78-114

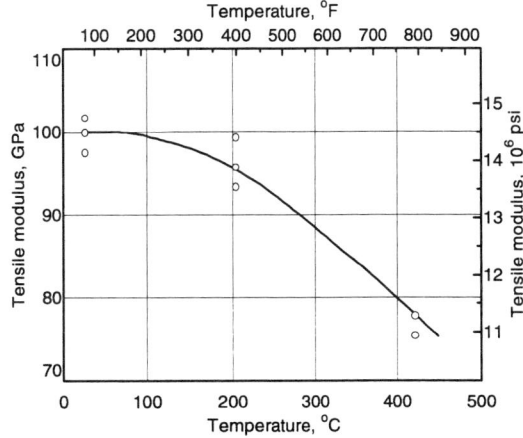

Ti-10V-2Fe-3Al: Tensile modulus vs temperature

Round bar, solution treated and overaged.
Source: *Metals Handbook, Properties and Selection: Stainless Steels, Tool Materials, and Special-Purpose Materials*, Vol 3, 9th ed., American Society for Metals, 1980

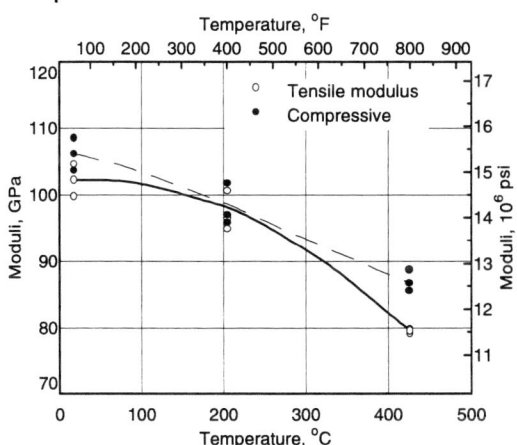

Ti-10V-2Fe-3Al: Tensile and compressive moduli vs temperature

Solution treated and overaged bar.
Source: D. Deel, AFML-TR-77-198, 1977

Modulus in Torsion. 41 GPa (6×10^6 psi)

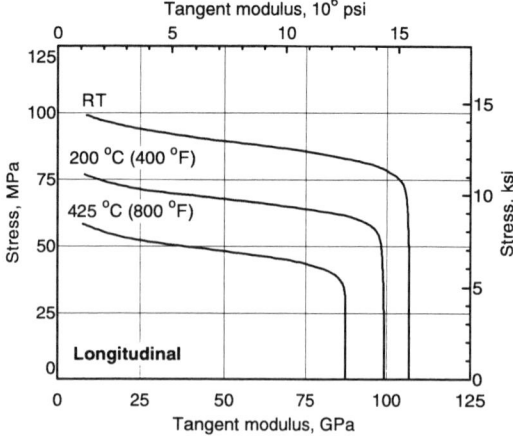

Ti-10V-2Fe-3Al: Compressive tangent modulus curves (see also "Stress-Strain Curves")

Solution treated and overaged round bar tested in the longitudinal direction.
Source: O. Deel, "Engineering Data on New Aerospace Structural Materials," AFML-TR-77-198, Air Force Materials Laboratory, 1977, p 99

Corrosion

Ti-10V-2Fe-3Al: Corrosion rates in specific media

Medium	Concentration, %	Temperature, °C	Corrosion rate, mm/yr
Ferric chloride	10	Boiling	nil
Hydrochloric acid	0.5	Boiling	1.10
Hydrochloric acid +0.1% $FeCl_3$	5	Boiling	0.008

These data should be used only as a guideline for alloy performance. Rates may vary depending on changes in medium chemistry, temperature, length of exposure, and other factors. Total alloy suitability cannot be assumed from these values alone, because other forms of corrosion, such as localized attack, may be limiting. The text should be consulted to assess overall alloy suitability more thoroughly for a given set of environmental conditions. In complex, variable, and/or dynamic environments, *in situ* testing may provide more reliable data.

Mechanical Properties

Ti-10V-2Fe-3Al is capable of being heat treated to provide a wide range of properties. There are four AMS specifications (AMS 4984, 4986, 4987, and 4983A) covering strengths ranging from 140 to 180 ksi, with the lower strengths being utilized when higher fracture toughness is required.

Design Allowables

Ti-10V-2Fe-3Al: Design tensile properties of forgings

Thickness		Ultimate tensile strength (L-LT)(a)		Tensile yield strength (L-LT)(a)		Elongation, %		Reduction of area(a), %	
mm	in.	MPa	ksi	MPa	ksi	L	LT	L	LT
Solution treated and aged(b) hand forgings per AMS 4986									
<75	<3.00	1103	160(c)	1000	145(c)	6	6(c)	10	10(c)
75-100	3.00-4.00	1103	160	1000	145	6	6	10	10
Conventional solution treated and aged(b) die forgings									
<25	<1.00(e)	1240	180(c)	1103	160(c)	4	4(c)	...	...
<75	<3.00(f)	1193	173(d)	1103	160(d)	4	4(d)	...	...

(a) S-basis values applicable in both longitudinal (L) and long transverse (LT) directions, except as noted. (b) Aged at 510 to 535 °C (950 to 1000 °F). (c) Applicable in LT direction providing LT dimension is greater than 63.5 mm (2.5 in.). (d) Applicable in LT and ST directions providing LT or ST dimension is greater than 63.5 mm (2.5 in.). (e) Per AMS 4983. (f) Per AMS 4984. Source: MIL-HDBK 5

Ti-10V-2Fe-3Al: S-basis design bearing strengths of forgings

Thickness		Ultimate bearing strength(a)				Bearing yield strength(a)			
		e/D = 1.5		e/D = 2.0		e/D = 1.5		e/D = 2.0	
mm	in.	MPa	ksi	MPa	ksi	MPa	ksi	MPa	ksi
Hand forgings per AMS 4986									
<75	<3.00	1660	241	2020	293	1503	218	1690	245
Die forgings per AMS 4983 and 4984									
<25	<1.00	1680	244	2035	295	1565	227	1800	261
<75	<3.00	1613	234	1958	284	1565	227	1800	261

(a) Dry pin bearing values for solution treated and aged material aged at 480 to 510 °C (900 to 950 °F). Source: MIL-HDBK 5

Ti-10V-2Fe-3Al: S-basis design compressive and shear strengths of forgings

Thickness		Compressive yield strength				Ultimate shear strength	
		L direction		LT direction			
mm	in.	MPa	ksi	MPa	ksi	MPa	ksi
Solution treated and aged(a) hand forgings per AMS 4986							
<75	<3.00	1062	154	...	...	669	97
Conventional solution treated and aged(a) die forgings							
<25	<1.00	1158	168	1145	166	695	101
<75	<3.00	1158	168	1145	166(b)	669	97

(a) Aged at 480 to 510 °C (900 to 950 °F). (b) Applicable in ST direction. Source: MIL-HDBK 5

Hardness

The following data demonstrate the effect of forging parameters, aging conditions and hydrogen content on the hardness of Ti-10V-2Fe-3Al. Also included is a plot of hardness as a function of tensile strength. The very high hardnesses obtained by aging at temperatures of 375 °C (700 °F) or below are due to the precipitation of omega at these low temperatures. Hydrogen is a β stabilizer, so high H_2 concentrations stabilize the β, making the aging reaction more sluggish.

Typical Vickers Hardness. 300 to 470 HV
Typical Rockwell C Hardness. 32 to 41 HRC

Ti-10V-2Fe-3Al: Effect of forging conditions on Rockwell hardness

Forging temperature		Forging speed		Hardness,	Calculated flow stress	
°C	°F	mm/min	in./min	HRC	MPa	ksi
954	1750	0.75	0.03	32.2	18	2.6
871	1600	0.75	0.03	31.6	30	4.3
788	1450	0.75	0.03	32.1	35	5.1
760	1400	0.75	0.03	33.0	50	7.2
732	1350	0.75	0.03	34.3	67	9.8
954	1750	75	3.00	32.1	74	10.7
704	1300	0.75	0.03	33.7	88	12.7
871	1600	75	3.00	31.8	110	16.0
677	1250	0.75	0.03	34.1	109	15.9
788	1450	75	3.00	34.5	133	19.3
643	1190	0.75	0.03	33.8	139	20.2
760	1400	75	3.00	35.6	165	23.9
732	1350	75	3.00	35.9	199	28.8
704	1300	75	3.00	35.9	233	33.8
677	1250	75	3.00	37.0	269	39.0
643	1190	75	3.00	40.7	312	45.2

Note: Hardness measurements for alloy forged isothermally to 0.50 mm/mm (in./in.) (nominal) at various conditions; mean grain diameter, 255 μm. Composition, 2.95% Al, 0.008% C, 1.90% Fe, 0.0051% H, 0.020% N, 0.116% O, 10.10% V, 0.08% other total. Material was obtained from Titanium Metals Corporation of America from a single heat of 7.5 cm (3 in.) diam bar. Hardness was determined for each ring specimen forged to 0.50 in./in. (nominal) at each temperature and speed indicated. Measurements were made at random locations with at least ten measurements per specimen. Surfaces were prepared by grinding and etching; final polishing was done with 400-grit silicon carbide paper. Source: I. Martorell, "Effects of Isothermal Forging Conditions on the Properties and Microstructures of Ti-10V-2Fe-3Al," AFML-TR-78-114, Air Force Materials Laboratory, Wright Patterson AFB, Dec 1978, p 88

Ti-10V-2Fe-3Al: Effect of aging on Vickers hardness

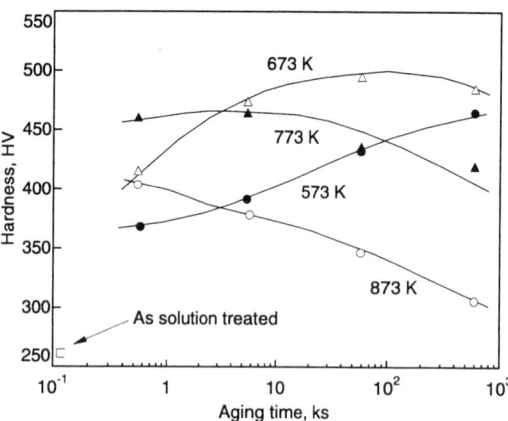

Source: I. Maeda and H. Flower, The Effect of Aluminum on the Phase Transformations in Ti-15%V, in *Sixth World Conference on Titanium*, P. Lacombe, R. Tricot, and G. Beranger, Ed., Les Editions de Physique, Paris, 1989, p 1589

Ti-10V-2Fe-3Al: Effect of aging on Vickers hardness

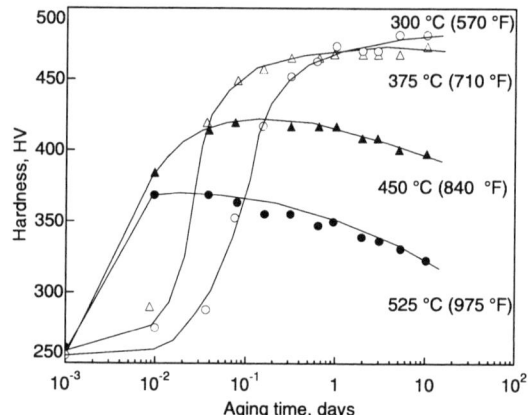

Hardness as a function of aging time at various temperatures for solution treated alloy processed at 780 °C (1435 °F), 2 h, water quenched.

Alloy was supplied in the form of round bars, 18 mm (0.7 in.) in diameter, α + β forged. Chemical composition: 3.20 wt% Al, 0.041 wt% C, 1.98 wt% Fe, 0.0125 wt% N, 0.1480 wt% O, and 10.24 wt% V. Microstructure in the as-received condition was fully recrystallized, with globular α in a β matrix. The transformation temperature, microscopically determined, was 830 + 10 °C (1525 + 18 °F). Specimens were solution treated below the β transus temperature, at 780 °C (1435 °F) for 2 h and water quenched. Isothermal aging treatments were performed in salt baths or in an air furnace on cylindrical blanks 14 mm (0.55 in.) in diameter. Age hardening response of the alloy was observed by means of Vickers hardness measurements.
Source: M. Champagnac and A. Vassel, Influence of Microstructure on the Tensile and Fracture Toughness Properties of Ti-10V-2Fe-3Al Alloy, in *Designing with Titanium*, The Institute of Metals, London, 1986, p 261

Ti-10V-2Fe-3Al: UTS vs hardness

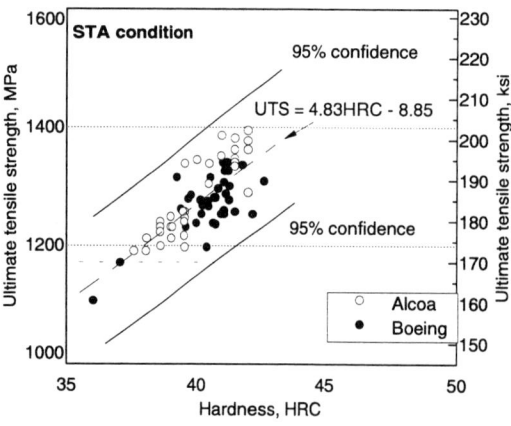

UTS (in ksi) = 4.83 HRC - 8.85

Ti-10V-2Fe-3Al: Knoop hardness vs aging time

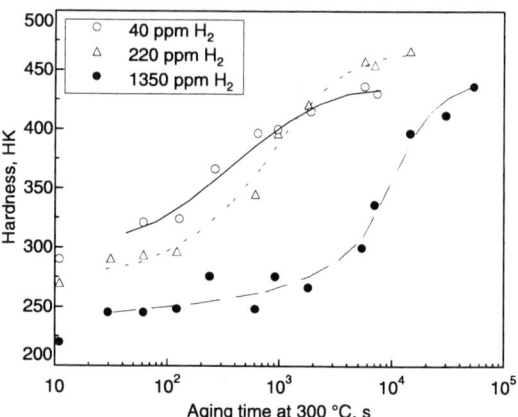

Samples were aged at 300 °C (570 °F) for indicated times to form isothermal ω. Initial hardness increases are difficult to interpret, but latter changes are attributed to isothermal ω.

Room-Temperature Tensile Properties

Almost all product forms are capable of being heat treated over the strength range cited above, except for castings (Ti-10V-2Fe-3Al is not really considered a casting alloy) and conventional, high Cl, blended elemental powder compacts. The alloy is capable of being heat treated to 173 ksi in section thicknesses up to 3 inches thick with uniform properties. Uniform properties can be obtained for heavier section thicknesses at lower strength. This alloy demonstrates a high strain-rate sensitivity of the tensile strength as demonstrated below. The expected strength ductility trend is observed. The amount of α/β work, following primary forging in the β-phase field, has a strong influence on the ductility through its effect on the morphology of the primary α phase.

Ti-10V-2Fe-3Al: Typical tensile properties of various product forms

Product form	Ultimate tensile strength MPa	ksi	Tensile yield strength MPa	ksi	Elongation, %	Reduction of area, %	Plane-strain fracture toughness MPa √m	ksi √in.
High-strength condition								
Isothermal forgings	1300-1380	188-200	1200-1255	174-182	3-6	5-13	29	26
Conventional forgings	1230-1350	178-196	1145-1280	166-185	4-10	5-28	44-60	40-54
Pancake forgings	1275-1310	185-190	1150-1160	167-168	5-8	5-29	47	43
Extrusions	1240	180	1170	169	4	...	...	...
P/M High strength								
Prealloyed, HIP	1310	190	1205	175	9	13	...	...
Prealloyed, HIP + isothermal forge	1345-1400	195-203	1240-1305	180-189	6-8	15-28	28	25
P/S (0.19 wt% Cl max)	1195	173	1110	161	3.5	...	...	...
P/S + HIP	1228-1275	177-185	1185-1245	172-180	7-9	...	28-29	25-26
Reduced strength condition								
Isothermal forgings	1060-1100	154-159	985-1060	143-154	8-12	22-32	70	64
Pancake forgings	965	140	930	135	16	50	100	91
Extrusions	1110-1170	161-169	1000-1105	145-160	6-7	10-18	45-48	41-44
P/M Prealloyed HIP + isothermal forge	1125-1145	163-166	1050-1090	152-158	13-15	37-45	55	50
P/M P/S + HIP	1120-1160	162-168	1070-1105	155-160	9-10	...	32	29
Castings	1105-1130	160-164	1010-1030	146-149	6-10	6-15	...	...

Source: R. Boyer, D. Eylon, and F. Froes, Comparative Evaluation of Ti-10V-2Fe-3Al Cast, P/M and Wrought Product Forms, *Titanium, Science and Technology*, Vol 2, G. Lütjering, U. Zwicker, and W. Bunk, Ed., Deutsche Gesellschaft für Metallkunde e.V., Germany, 1985, p 1307

Ti-10V-2Fe-3Al: Typical room-temperature tensile properties of an airframe forging

Condition(a)	Ultimate tensile strength MPa	ksi	Tensile yield strength (0.2% offset) MPa	ksi	Elongation(b), %	Reduction of area, %
Longitudinal direction, 15 mm (0.6 in.) section thickness						
STA	1275	185	1200	174	11	25
STOA	980	152	940	136	22	56
Transverse direction, 15 mm (0.6 in.) section thickness						
STA	1260	183	1200	174	9	20
STOA	950	138	895	130	21	56
Transverse direction, 56 mm (2.2 in.) section thickness						
STA	1270	184	1195	173	7	33
STOA	970	141	910	132	21	56
Short transverse direction, 56 mm (2.2 in.) section thickness						
STA	1280	186	1200	174	8	21
STOA	950	138	890	129	19	55

(a) STA condition: Solution treat 1 h at 760 °C (1400 °F), water quench and age 8 h at 510 °C (950 °F). STOA condition: Solution treat 1 h at 730 °C (1350 °F), air cool and age 8 h at 580 °C (1075 °F). (b) In 50 mm or 2 in. Source: *Metals Handbook, Properties and Selection: Stainless Steels, Tool Materials, and Special-Purpose Materials*, Vol 13, 9th ed., American Society for Metals, 1980

Ti-10V-2Fe-3Al: Longitudinal tensile properties of heavy sections

Square specimens were heat treated as follows: as forged + 535 °C (1000 °F), 8 h, AC

Section size		Location	Ultimate tensile strength		Tensile yield strength (0.2% offset)		Elongation (4D), %	Reduction of area, %
mm	in.		MPa	ksi	MPa	ksi		
25	1	Center	1268	184.0	1255	182.0	14.0	57.7
50	2	Outside	1136	164.8	1100	159.5	16.0	50.1
		Center	1086	157.5	1053	152.7	18.0	56.3
75	3	Outside	1057	153.4	1009	146.4	16.0	44.1
		Center	1037	150.5	1009	146.4	17.0	50.6
100	4	Outside	1149	166.7	1068	154.9	9.0	22.6
		Midradius	1032	149.7	991	143.8	17.0	45.4
		Center	1010	146.5	966	140.1	15.0	47.9
125	5	Outside	978	141.9	924	134.0	16.0	42.2
		Midradius	982	142.5	934	135.5	19.0	48.7
		Center	957	138.8	916	132.9	19.0	60.1

Source: E. Bohanek, Deep Hardenable Titanium Alloys for Large Airframe Elements, *Titanium, Science and Technology*, Vol 3, R.I. Jaffee and H.M. Burte, Ed., 1973, p 1993

Ti-10V-2Fe-3Al: Typical tensile properties of forgings

Condition	Ultimate tensile strength		Tensile yield strength		Elongation, %	Reduction of area, %	Plane-strain fracture toughness	
	MPa	ksi	MPa	ksi			MPa√m	ksi√in.
STA: 1 h, 750-765 °C (1385-1410 °F), WQ + 8 h, 480-510 °C (900-950 °F), AC	1240-1380	180-200	1158-1268	168-184	4-12	10-30	46-61.5	42-56
STOA: 1 h, 730 °C (1350 °F), AC + 8 h, 580-595 °C (1075-1100 °F), AC	965-1035	140-150	895-965	130-140	20	45	102	93
BAOA: 1 h, 815 °C (1500 °F), AC + 8 h, 620 °C (1150 °F), AC	1000	145	930	135	17	46	110	100

Note: Maximum oxygen 0.13%, with final processing in the α-β field. Source: R. Boyer and H. Rosenberg, Ed., *Beta Titanium Alloys in the 1980's*, TMS/AIME, 1984, p 446

Effect of Microstructure

The strength-ductility-toughness relationship is dependent on the microstructure (and the processing which provides the microstructural variations). In very general terms, globular primary α provides higher ductility at a given strength (see next two pages). The ductility of a microstructure with no primary α will be minimum, ductility will then improve with increasing amounts of primary α, and then decrease again as the amount of primary α is increased. The formation of ω will provide high strengths, but poor ductility, often nil.

Effect of Hydrogen

The addition of hydrogen can strongly influence tensile properties in Ti-10-2-3. The type and magnitude of hydrogen effects, for a given microstructure, is a function of both hydrogen concentration and thermal processing prior to and/or subsequent to hydrogen introduction. Hydrogen can display two separate types of effects in Ti-10-2-3; the first is as an intrinsic embrittling agent, akin to the hydrogen embrittlement observed in other metals. The intrinsic embrittling effect is generally observed as decreased ductility and a reduction in tensile strength. The second hydrogen effect results from hydrogen being a powerful beta stabilizing element. Changes in tensile properties may be a direct result of hydrogen-induced changes in microstructure. Generally, hydrogen introduced into the material after final thermomechanical processing results in an intrinsic effect. Where there is thermomechanical processing subsequent to the introduction of hydrogen, any changes in mechanical properties are likely to

Ti-10V-2Fe-3Al: Effect of strain rate on tensile strength

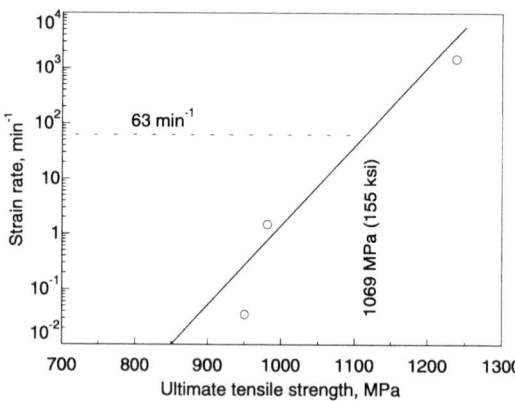

Ti-10-2-3 tensile strength is quite strain rate sensitive, as are all β alloys.
Nominal β transus: 810 °C (1490 °F). Using conventional forging techniques, the pancake forgings were about 229 mm (9 in.) in diameter and 25.4 to 32 mm (1 to 1.3 in.) thick from starting stocks of 114 mm (4.5 in.) and 152 mm (6 in.) diameter billets. All pancake forgings were initially β forged at the β transus plus 10 to 24 °C (50 to 75 °F) to produce a 50 to 70% thickness reduction. The amount of α/β forging reduction at the β transus minus 10 to 25 °C (50 to 75 °F) varied from about 2 to 58%. The heat treatments were conducted within a ±5 °C (9 °F) furnace tolerance. The tensile specimens had a 6.3 mm (0.25 in.) diameter, 25.4 mm (1 in.) gage length, and were tested in accordance with ASTM E-8.
Source: R.R. Boyer and G.W. Kuhlman, Processing Properties Relationships of Ti-10V-2Fe-3Al, *Metall. Trans. A*, Vol 18, 1987, p 2095

Ti-10V-2Fe-3Al: Effect of percentage and morphology of on α ductility at constant yield stress

Globular α is beneficial to ductility compared to elongated α.

| Primary α | | Secondary α Aging temperature | | Strain to | Reduction of area, |
Morphology	Vol %	°C	°F	fracture	%
Globular	10	500	260	0.50	39
Elongated	10	500	260	0.30	26
Globular	30	350	175	0.20	18
Elongated	30	370	185	0.11	11

All specimens had a 0.2% yield strength of 1250 MPa (180 ksi). Good heat treatment practice was used to avoid atmospheric contamination. Tensile tests were performed on electropolished and etched cylindrical specimens with a diameter of 6.4 mm (0.25 in.) and a 32 mm (1.25 in.) gage length using an Instron testing machine with a clip-on extensometer. The tests were carried out in the L-direction of the plate at a strain rate of 5.5×10^{-4}/s. Source: G.T. Terlinde, T.W. Duerig, and J.C. Williams, Microstructure, Tensile Deformation, and Fracture in Aged Ti-10V-2Fe-3Al, *Metall. Trans. A*, Vol 14, 1983, p 2101

Ti-10V-2Fe-3Al: Effect of microstructure on tensile properties

Microstructure	Heat treatment	Tensile yield strength MPa	ksi	Ultimate tensile strength MPa	ksi	Uniform elongation, %	Elongation to failure, %	Reduction of area, %
20% primary α + β + athermal ω	730 °C (1345 °F) (48 h) + WQ	741	107	862	125	9.7	18.6	35
β + athermal ω	850 °C (1560 °F) (2 h) + WQ	262	38	878	127	15.7	21.8	32
α + β + isothermal ω	700 °C (1290 °F) (300 min) + WQ + 250 °C (480 °F) (6000 min)	1218	176	1266	183	0.26	0.58	2.25
β + isothermal ω	850 °C (1560 °F) (2 h) + WQ + 250 °C (480 °F) (10^4 min)	Brittle, no yield		...	...	0	0	0
20% primary α + β + α (uniform)	720 °C (1330 °F) (100 min) + WQ + 370 °C (700 °F) (1000 min)	1240	180	1430	207	2.7	8.9	16
β + α (uniform)	850 °C (1560 °F) (100 min) + WQ + 370 °C (700 °F) (1000 min)	Brittle, no yield		...	...	0	0	0
20% primary α + β + α (sympathetic)	730 °C (1345 °F) + WQ + 500 °C (930 °F) (60 min)	1063	154	1106	160	4.6	17.5	58
β + α (sympathetic)	850 °C (1560 °F) (100 min) + WQ + 500 °C (930 °F) (240 min)	1225	177	1243	180	2.3	8.7	14

0.15 wt% O. Tensile testing was performed on an Instron machine using a clip-on extensometer. The strain rate was 0.00055/s, and the tensile specimen gage sections were 0.640 cm (0.25 in.) in diam and 3.2 cm (1.3 in.) in length. Specimens were pulled with the rolling direction parallel to the tensile axis. Source: T.W. Duerig, G.T. Terlinde, and J.C. Williams, Phase Transformations and Tensile Properties of Ti-10V-2Fe-3Al, *Metall. Trans. A*, Vol 11, 1980, p 1987

Ti-10V-2Fe-3Al: Tensile properties of selected microstructures

Heat treatment	Microstructure	Tensile yield strength (0.2% offset) MPa	ksi	Ultimate tensile strength MPa	ksi	Elongation, %	Strain to fracture
Primary α							
725 °C (1330 °F) 20 h, WQ + 500 °C (930 °F) 1 h (salt)	30% primary α + large secondary α	1063	154	1106	160	17.7	0.99
725 °C (1330 °F) 100 min, WQ + 370 °C (700 °F) 10^3 min	30% primary α + small secondary α	1246	181	1419	206	7.6	0.19
780 °C (1435 °F) 3 h, WQ + 500 °C (930 °F) 1 h (salt)	10% primary α + large secondary α	1202	174	1247	181	10.3	0.63
780 °C (1435 °F) 3 h, WQ + 500 °C (930 °F) 1 h (air)	10% primary α + small secondary α	1445	209	1544	224	2.4	0.09
850 °C (1560 °F) 2 h, WQ + 500 °C (930 °F) 4 h (salt)	0% primary α + grain boundary α + large secondary α	1250	181	1308	190	3.9	0.16
Elongated primary α							
700 °C (1290 °F) 8 h, WQ + 200 °C (390 °F) 6800 min	35% primary α + ω	1218	176	1266	184	0.5	0.02
850 °C (1560 °F) 2 h, WQ + 500 °C (930 °F) 4 h (salt)	0% primary α + large secondary α + grain boundary α	1182	171	1265	183	3.8	0.17
760 °C (1400 °F) 75 min, WQ + 500 °C (930 °F) 1 h (salt)	~10% primary α + large secondary α	1298	188	1381	200	4.6	0.23
700 °C (1290 °F) 75 min, WQ + 350 °C (660 °F) 10^3 min	~30% primary α + small secondary α	1239	180	1395	202	3.9	0.11

be a combination of the two separate hydrogen effects and difficult to consider independently.

An interesting example in Ti-10-2-3 is hydrogen's effect on beta annealed and water quenched microstructure. This microstructure exhibits a stress-induced martensitic transformation at very low stress magnitude. With the addition of hydrogen prior to heat treating, the triggering stress, defined as the stress required to initiate martensitic transformation, increases (see yield strength figure). At very high hydrogen concentrations, the triggering stress disappears and the martensitic transformation changes from stress-induced to strain-induced. With the addition of hydrogen, UTS suffers an initially steep decrease that approaches an asymptote at high concentration (see two-part figure). Ductility, as measured by reduction in area, RA, of the tensile cross-section, initially increases with hydrogen but then decreases precipitously. The initial increase in RA is understood to be the result of hydrogen-induced microstructural changes, when the sudden decrease at high concentrations is the result of the domination of the intrinsic hydrogen effect to decrease ductility.

Ti-10V-2Fe-3Al: Yield stress vs ductility

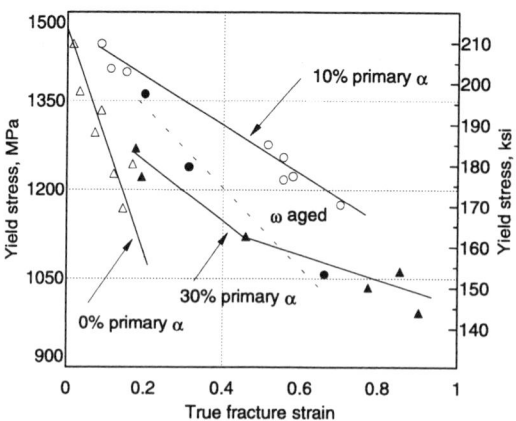

Source: T.W. Duerig, et al., *Metall. Trans. A*, Vol 11, 1983, p 1987

Ti-10V-2Fe-3Al: Ductility vs α morphology

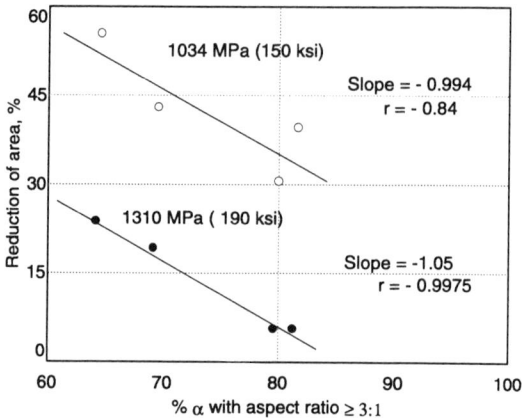

Source: *Metall. Trans. A*, Vol 18, Dec 1987, p 2100

Ti-10V-2Fe-3Al: Effect of H$_2$ concentration on yield strength

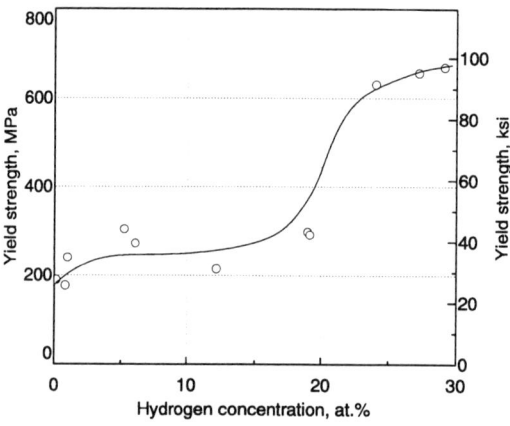

The base hydrogen level was 0.2 at.% (40 ppm). Additional hydrogen was introduced into the test material with the use of a Sievert's apparatus at temperatures ranging from 765 to 800 °C (1400 to 1470 °F). No appreciable difference in β grain size was detected due to varying the hydrogen charging temperature within this range. Hydrogen levels greater than 29 at.% were obtained by this method (although not for each microstructure). Sievert's charging was done prior to heat treating to establish the desired microstructure.
Source: J. Costa, D. Banerjee, and J.C. Williams, The Effect of Hydrogen on Microstructure and Properties of Ti-10V-2Fe-3Al, in *Titanium, Science and Technology*, Vol 4, G. Lütjering, U. Zwicker, and W. Bunk, Ed., Deutsche Gesellschaft für Metallkunde e.V., Germany, 1984, p 2479

Ti-10V-2Fe-3Al: Effect of H$_2$ concentration on tensile properties of β ST material

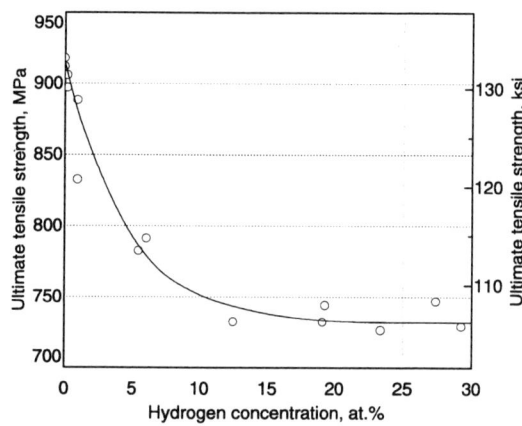

(a) 1 at.% H = 0.003 wt% H

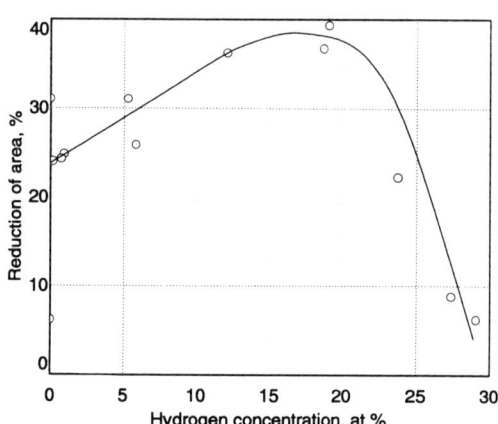

(b) 1 at.% H = 0.003 wt% H

Tensile tests were performed on a screw-driven Instron universal test machine at room temperature. The tensile specimen had a 6.40 mm (0.25 in.) diameter and a gage length of 25.4 mm (1 in.). The initial strain rate was 3.3×10^{-5}/s. Specimen extension was measured with a clip gage extensometer.
Source: J.E. Costa, J.C. Williams, and A.W. Thompson, *Metall. Trans. A*, Vol 18, Aug 1987, p 1421

High-Temperature Strength

Tensile Strength

Ti-10V-2Fe-3Al: Yield strength compared to Ti-6Al-4V

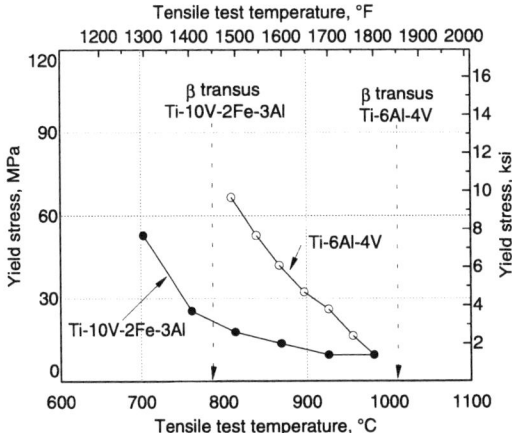

Strain rate, 0.005/min.
Source: R. Boyer and H. Rosenberg, Ed., *Beta Titanium Alloys in the 1980's*, TMS/AIME, 1984, p 264

Ti-10V-2Fe-3Al: Yield strength vs temperature

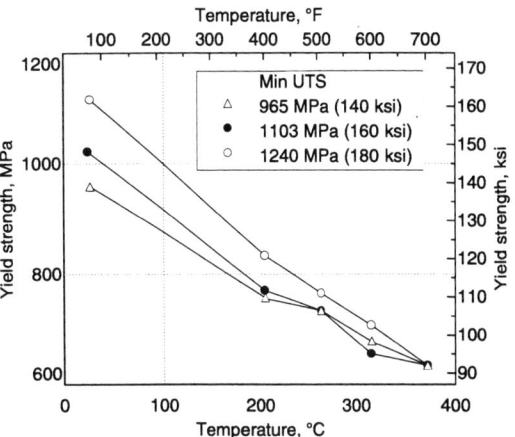

Ti-10V-2Fe-3Al elevated temperature yield strength for forgings processed to three strength levels.
Source: G.W. Kuhlman, Alcoa Green Letter No. 224, August 1987

Ti-10V-2Fe-3Al: Effect of temperature on longitudinal tensile properties of STOA 75 mm (3 in.) diam round bar

	Ultimate tensile strength		Tensile yield strength (0.2% offset)		Elongation mm, or %	Reduction of area, %	Tensile modulus,	
	MPa	ksi	MPa	ksi			GPa	10^6 psi
Room temperature								
1	977	141.7	949	137.7	18	60.5	101	14.7
2	978	141.8	951	137.9	18	63.5	99	14.4
3	972	141.0	950	137.8	19	63.5	103	15.0
Average	975	141.5	949	137.7	18.3	62.5	101	14.7
205 °C (400 °F)								
4	835	121.1	740	107.4	23	67.3	100	14.5
5	826	119.9	731	106.0	20	68.5	96	13.9
6	817	118.5	729	105.7	21	65.9	94	13.7
Average	826	119.8	733	106.4	21.3	65.6	97	14.0
425 °C (800 °F)								
7	665	96.5	542	78.6	21	79.6	79	11.5
8	670	97.3	546	79.1	24	79.0	79	11.5
9	674	97.8	545	79.0	22	79.9	77	11.3
Average	670	97.2	544	78.9	22.3	79.5	78	11.4

Heat treatment: 760 °C (1400 °F) for 1 h, furnace cool + 565 °C (1050 °F) for 8 h, air cool. Source: O. Deel, "Engineering Data on New Aerospace Structural Materials," Air Force Materials Laboratory, Wright Patterson AFB, AFML-TR-77-198, 1977

Ti-10V-2Fe-3Al: Effect of temperature on tensile properties of forgings

Temperature		Treatment	Ultimate tensile strength		Tensile yield strength		Elongation, %	Reduction of area, %
°C	°F		MPa	ksi	MPa	ksi		
23	RT	STA(a)	1276	185	1200	174	9	27
		STOA(b)	972	141	896	131	20	55
205	400	STA(a)	1117	162	1048	152	13	32
		STOA(b)	800	116	683	99	21	58
315	600	STA(a)	1103	160	979	142	13	42
		STOA(b)	738	107	600	87	22	63

(a) STA: 750 to 765 °C (1385 to 1410 °F) 1 h, WQ 480 to 495 °C (900 to 950 °F), 8 h, AC (b) STOA: 730 °C (1350 °F) 1 h, AC 580 to 595 °C (1075 to 1100 °F), 8 h, AC. Source: C.C. Chen and C.P. Gure, "Forgeability, Structures and Properties of Hot-Die Pressed Ti-10V-2Fe-3Al Thin Section Forgings," Wyman Gordon Report RD 74-120, Nov 1974; reported in *Beta Titanium Alloys for the 1980's*, R. Boyer and H. Rosenberg, Ed., TMS/AIME, 1984, p 441

Ti-10V-2Fe-3Al: Tensile strength of STOA bar

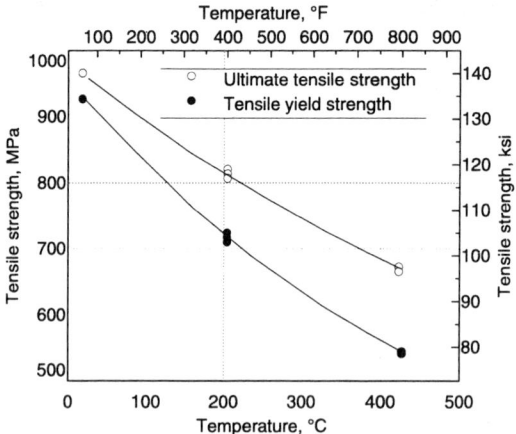

Alloy in the form of 75 mm (3 in.) round bar was heat treated at 760 °C (1400 °F) for 1 h, furnace cooled, then at 565 °C (1050 °F) for 8 h, and air cooled. Tensile tests were performed in the longitudinal direction.

Ti-10V-2Fe-3Al: Bearing strength vs temperature

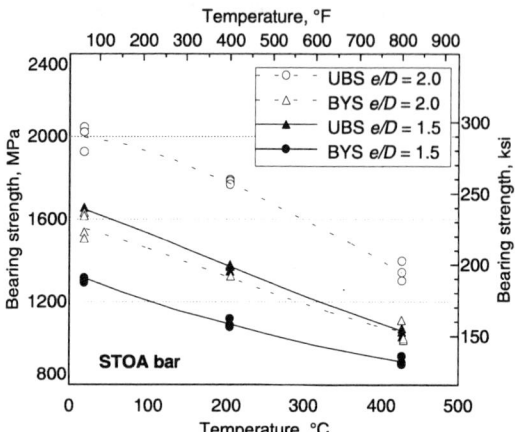

Source: O. Deel, "Engineering Data on New Aerospace Structural Materials," Air Force Materials Laboratory, AFML-TR-198, Wright Patterson AFB, 1977

Bearing, Compression, and Shear Strength

Compression strength at temperatures up to about 1800 °F are used to determine forging flow stresses. The compression, as well as bearing and shear strengths to a more useful application range, 800 °F are also indicated.

Ti-10V-2Fe-3Al: Compressive yield strength vs temperature

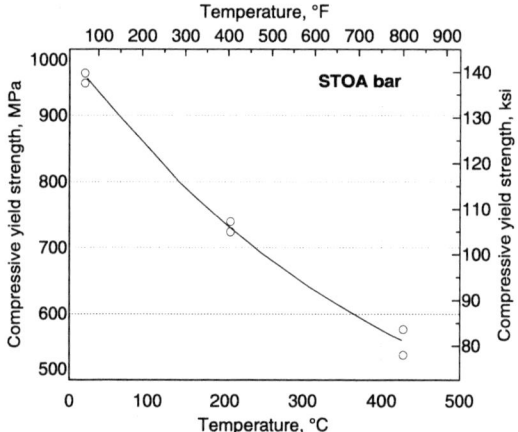

Source: O. Deel, "Engineering Data on New Aerospace Structural Materials," AFML-TR-77-198, Wright Patterson AFB, 1977

Ti-10V-2Fe-3Al: Shear strength vs temperature

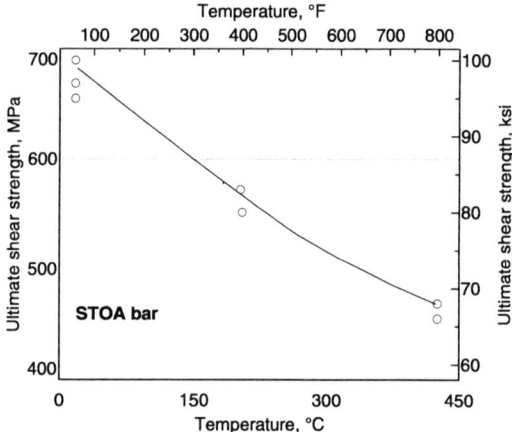

Source: O. Deel, "Engineering Data on New Aerospace Structural Materials," AFML-TR-77-198, Wright Patterson AFB, 1977

Ti-10V-2Fe-3Al: Effect of temperature on bearing strength

Specimen No.	Ultimate bearing strength				Bearing yield strength			
	e/D = 1.5		e/D = 2.0		e/D = 1.5		e/D = 2.0	
	MPa	ksi	MPa	ksi	MPa	ksi	MPa	ksi
Room temperature								
1	1655	240.0	2027	294.0	1324	192.0	1510	219.0
2	1655	240.0	1930	280.0	1296	188.0	1544	224.0
3	1641	238.0	2048	297.0	1310	190.0	1627	236.0
Average	1650	239.3	2001	290.3	1310	190.0	1560	226.3
205 °C (400 °F)								
4	1358	197.0	1772	257.0	1089	158.0	1324	192.0
5	1372	199.0	1778	258.0	1075	156.0	1324	192.0
6	1379	200.0	1792	260.0	1124	163.0	1324	192.0
Average	1370	198.7	1781	258.3	1096	159.0	1324	192.0
425 °C (800 °F)								
7	1069	155.0	1296	188.0	937	136.0	1020	148.0
8	1048	152.0	1392	202.0	896	130.0	1117	162.0
9	1048	152.0	1344	195.0	903	131.0	1034	150.0
Average	1055	153.0	1344	195.0	912	132.3	1057	153.3

Note: 75 mm (3 in.) diam STOA bar. Heat treatment: 760 °C (1400 °F) for 1 h, furnace cool + 565 °C (1050 °F) for 8 h, air cool. Source: O. Deel, "Engineering Data on New Aerospace Structural Materials," Air Force Materials Laboratory, Wright Patterson AFB, AFML-TR-77-198, 1977

Ti-10V-2Fe-3Al: Effect of temperature on longitudinal ultimate shear strength

Specimen No.	Ultimate shear strength	
	MPa	ksi
Room temperature		
1	682	99.0
2	658	95.5
3	670	97.2
Average	670	97.2
205 °C (400 °F)		
4	562	81.6
5	571	82.9
6	567	82.3
Average	567	82.3
425 °C (800 °F)		
7	457	66.3
8	469	68.1
9	459	66.6
Average	467	67.0

Note: 75 mm (3 in.) diam STOA bar. Heat treatment: 760 °C (1400 °F) for 1 h, furnace cool + 565 °C (1050 °F) for 8 h, air cool. Source: O. Deel, "Engineering Data on New Aerospace Structural Materials," Air Force Materials Laboratory, Wright Patterson AFB, AFML-TR-77-198, 1977

Creep Properties

Ti-10V-2Fe-3Al, as with other β alloys, would not be expected to have very good creep resistance in comparison to alloys such as Ti-6Al-4V, and particularly Ti-6Al-2Sn-4Zr-2Mo. The creep resistance of the lower strength conditions have been studied and are reported here. It can be seen that the higher strength condition is slightly more creep resistant. One might assume that the 173 and 180 ksi conditions would be more creep resistant than shown for the lower strength conditions.

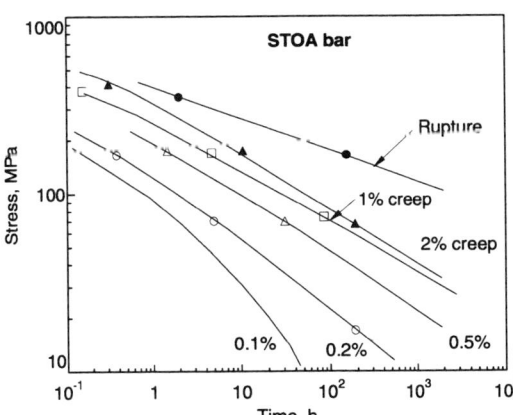

Ti-10V-2Fe-3Al: Creep properties at 480 °C (900 °F)

Source: O. Deel, "Engineering Data on New Aerospace Structural Materials," AFML-TR-77-198, Wright Patterson AFB, 1977

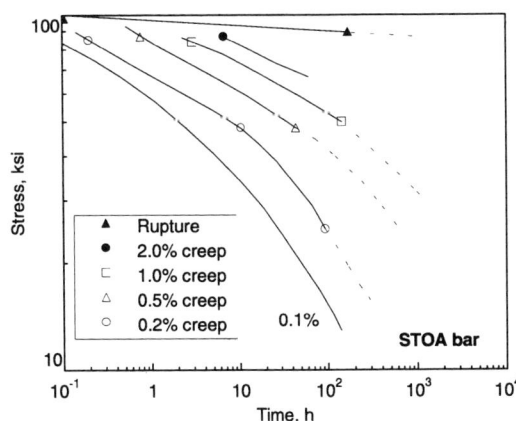

Ti-10V-2Fe-3Al: Creep properties at 370 °C (700 °F)

Source: O. Deel, "Engineering Data on New Aerospace Structural Materials," AFML-TR-77-198, Wright Patterson AFB, 1977

Ti-10V-2Fe-3Al: Creep and rupture properties of STOA

Exposure stress		Hours to indicated creep deformation, %:					Initial strain, %	Rupture time, h	Elongation (in 50 mm, or 2 in.), %	Reduction of area, %	Minimum creep rate, %
MPa	ksi	0.1	0.2	0.5	1.0	2.0					
At 370 °C (700 °F)											
724	105	...	...	...	...	...	...	On loading	12.9	68.3	...
620	90	0.1	0.2	0.8	3.0	7.0	1.228	169.6	25.2	67.1	0.074
345	50	1.5	4.0	47	152	930	0.543	1605.2(a)	2.935	...	0.00053
172	25	30	95	1165	4500(b)	...	0.098	1320.2(a)	0.616	...	0.00015
At 480 °C (900 °F)											
379	55	0.01	0.03	0.08	0.15	0.32	0.807	2.0	34.3	89.0	6.0
172	25	0.15	0.4	1.65	4.6	10.5	0.313	131.4	59.5	94.3	0.13
69	10	2.2	5.6	27	82	210	0.068	2306.2	161.0	98.0	0.008(a)
17	2.5	28	173	960	2500(b)	...	0.024	1319.9(a)	0.641	...	0.00033

(a) Test discontinued. (b) Estimated. Source: O. Deel, "Engineering Data on New Aerospace Structural Materials," Air Force Materials Laboratory, Wright Patterson AFB, AFML-TR-77-198, 1977

Ti-10V-2Fe-3Al: 0.1% creep behavior

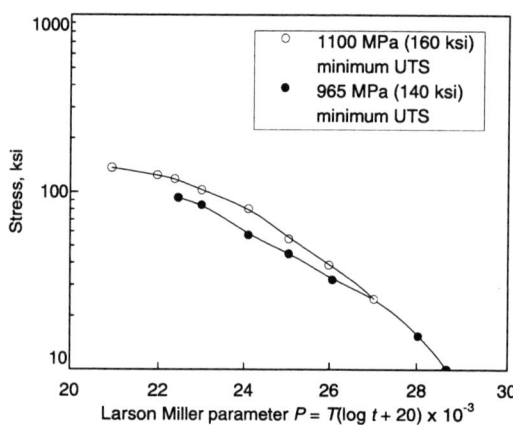

Source: G.W. Kuhlman et al., Sixth World Conference on Titanium, P. Lacombe, R. Tricot, and G. Beranger, Ed., Les Editions de Physique, Paris, 1989, p 1269-1275

Ti-10V-2Fe-3Al: 0.2% creep data

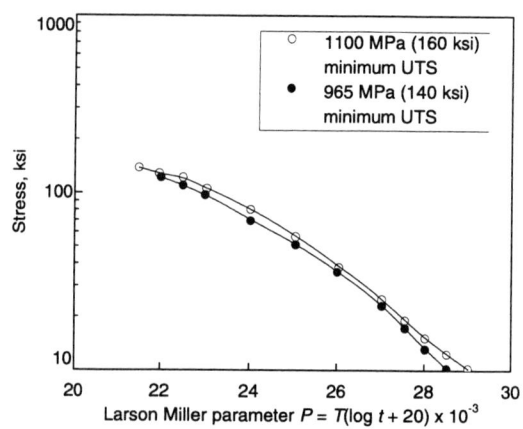

Source: G.W. Kuhlman, Alcoa Green Letter No. 224

Ti-10V-2Fe-3Al: 1.0% creep data

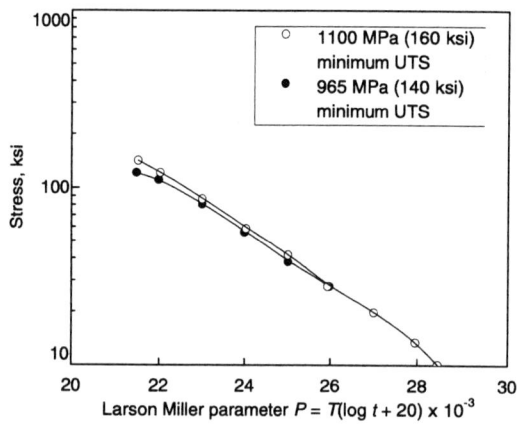

Source: G.W. Kuhlman, Alcoa Green Letter No. 224

Fatigue (Smooth)

Care must be taken in analyzing this fatigue data and comparing to a given set of conditions as there are so many variables, including specimen geometry, surface finish, R-ratio, and loading condition factors such as load controlled or strain controlled frequency, and wave form.

The high-cycle fatigue strength is a function of the tensile strength as one might expect. Generally the S-N curves are quite flat. Direct aging (with no solution treatment, which can be used over a limited thickness range) has a pronounced advantage over the solution treated and aged condition. This is attributed to two factors, the minimization of grain boundary α and the precipitation of a finer, more uniform dispersion of aged α when using a direct age. A primary α grain size effect has been recently reported (see bottom right figure). The effect of test temperature on fatigue properties is also illustrated. Again, one might expect the higher strength condition to have a lower fatigue debit as a function temperature than the lower strength conditions.

Ti-10V-2Fe-3Al: Effect of temperature on axial fatigue

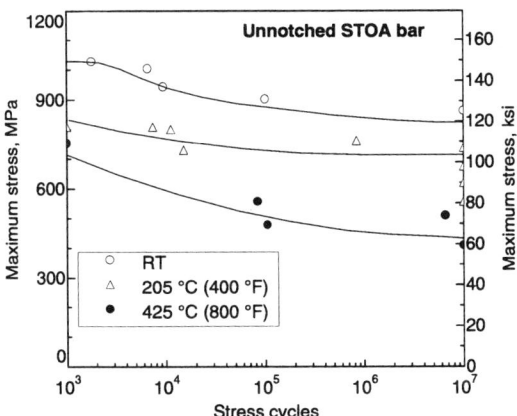

Axial fatigue of Ti-10V-2Fe-3Al bar stock in the STOA (solution treated and overaged) condition. Specimens were taken from round bars 75 mm (3 in.) in diameter that had been solution treated 1 h at 760 °C (1400 °F), furnace cooled, overaged 8 h at 565 °C (1050 °F), and air cooled. Tests were conducted at a stress ratio of $R = 0.1$ and a frequency of 20 Hz.
Source: O. Deel, "Engineering Data on New Aerospace Structural Materials," Air Force Materials Laboratory, AFML-TR-77-198, Wright Patterson AFB, 1977

Ti-10V-2Fe-3Al: Comparison of smooth fatigue strengths

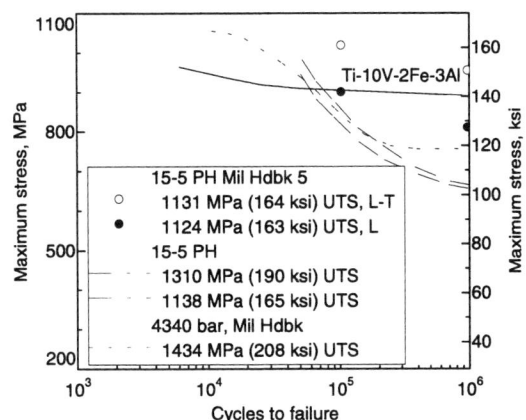

Source: J. of Metals, March, 1980

Ti-10V-2Fe-3Al: Comparison of smooth fatigue strengths

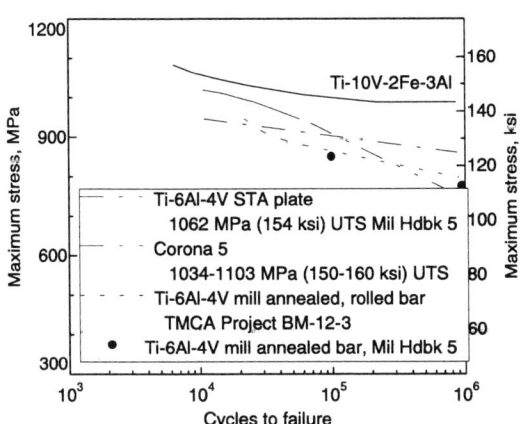

Source: J. of Metals, March, 1980

Ti-10V-2Fe-3Al: Fatigue endurance and grain size

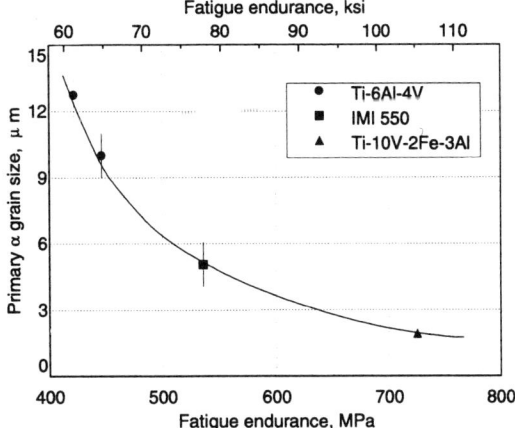

Source: D.P. Davies, 7th World Conf on Titanium

Low-Cycle Fatigue

Ti-10V-2Fe-3Al: Fatigue of smooth specimens (1190 MPa UTS)

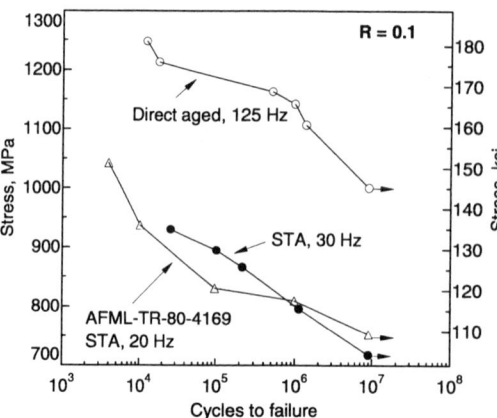

Ti-10V-2Fe-3Al: Solution treated and aged (STA) specimens were taken from β hot die forgings, solution treated at 30 °C (54 °F) below β transus temperature, water quenched, and aged to a strength level of 1190 MPa (175 ksi). Ti-10V-2Fe-3Al direct aged specimens were β hot die forged, post-forge cooled at a rate of 5 °C/s (9 °F/s), and aged to 1190 MPa (172 ksi). Fatigue tests for STA specimens were performed on specimens 3 mm (0.125 in.) in diameter with $K_t = 1$, $R = 0.1$, and frequency of 30 Hz, low stress ground. Fatigue test for direct aged specimens were performed on 3 mm (0.125 in.) diam samples with $K_t = 1$, $R = 0.1$, and frequency of 125 Hz; surfaces were low stress ground and electropolished.
Source: G. Kuhlman, A. Chakrabarti, T. Yu, R. Pishko, and G. Terlinde, LCF, Fracture Toughness, and Fatigue/Fatigue Crack Propagation Resistance Optimization in Ti-10V-2Fe-3Al Alloy Through Microstructural Modification, in *Microstructure, Fracture Toughness, and Fatigue Crack Growth Rate in Titanium Alloys*, A. Chakrabarti and J.C. Chesnutt, Ed., TMS/AIME, 1987, p 171

Ti-10V-2Fe-3Al: Fatigue of smooth specimens (965 MPa UTS)

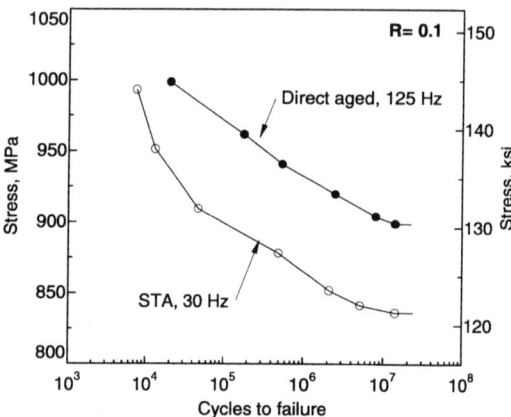

Ti-10V-2Fe-3Al solution treated and aged (STA) specimens were taken from β hot die forgings, solution treated at 30 °C (54 °F) below the β transus temperature, water quenched, and aged to a strength level of 965 MPa (140 ksi). Ti-10V-2Fe-3Al direct aged specimens were β hot die forged, post-forge cooled at a ratio of 5 °C/s (9 °F/s), and aged to the desired strength level.
Source: G. Kuhlman, A. Chakrabarti, T. Yu, R. Pishko, and G. Terlinde, LCF, Fracture Toughness, and Fatigue/Fatigue Crack Propagation Resistance Optimization in Ti-10V-2Fe-3Al Alloy Through Microstructural Modification, in *Microstructure, Fracture Toughness, and Fatigue Crack Growth Rate in Titanium Alloys*, A. Chakrabarti and J.C. Chesnutt, Ed., TMS/AIME, 1987, p 171

Ti-10V-2Fe-3Al: Fatigue of smooth specimens (1100 MPa UTS)

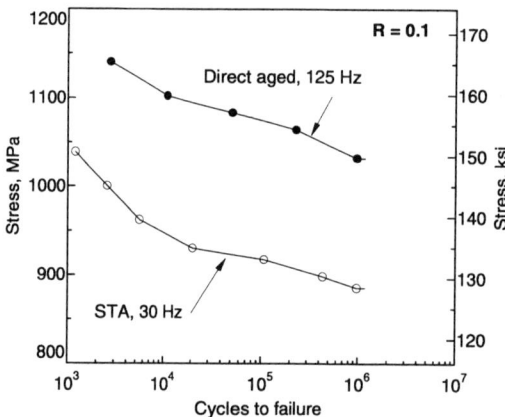

Ti-10V-2Fe-3Al solution treated and aged (STA) specimens were taken from β hot die forgings, solution treated at 30 °C (54 °F) below the β transus temperature, water quenched, and aged to a strength level of 1100 MPa (160 ksi). Ti-10V-2Fe-3Al direct aged specimens were β hot die forged, post-forge cooled at a rate of 5 °C/s (9 °F/s), and aged to the desired strength level.
Source: G. Kuhlman, A. Chakrabarti, T. Yu, R. Pishko, and G. Terlinde, LCF, Fracture Toughness, and Fatigue/Fatigue Crack Propagation Resistance Optimization in Ti-10V-2Fe-3Al Alloy Through Microstructural Modification, in *Microstructure, Fracture Toughness, and Fatigue Crack Growth Rate in Titanium Alloys*, A. Chakrabarti and J.C. Chesnutt, Ed., TMS/AIME, 1987, p 171

Ti-10V-2Fe-3Al: S/N data at two mean stress levels

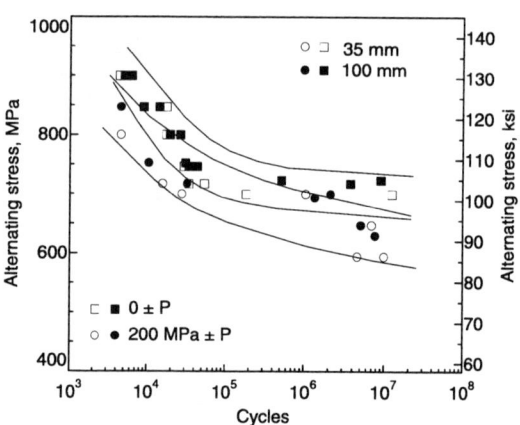

The fatigue endurance limit is influenced by the position in the billet, i.e. superior fatigue endurance values were obtained from the outer portion more heavily worked area of the billet ring, although the effect was considered negligible.
Source: D.P. Davies, Effect of Heat Treatment on the Mechanical Properties of Ti-10V-2Fe-3Al for Dynamically Critical Helicopter Components, 7th World Conf on Titanium

Ti-10V-2Fe-3Al: RT axial fatigue of STA forgings

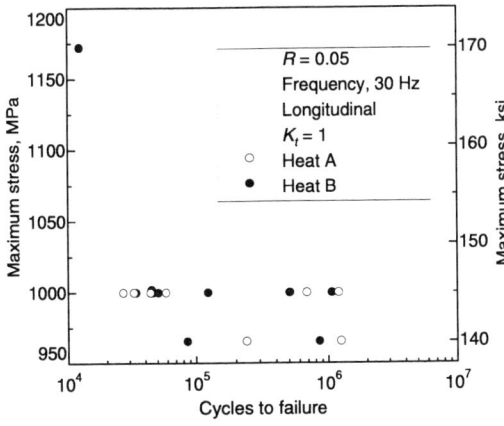

Boeing 747 lower link fitting, β forged with α-β (≤20%) finish; 775 °C (1435 °F), 2 h, AC + 770 °C (1425 °F), 2 h, WQ + 510 °C (950 °F), 8 h, AC.
Source: *Aerospace Structural Metals Handbook*, Vol 4, Code 3726, Battelle, 1972

Ti-10V-2Fe-3Al: LCF under strain control

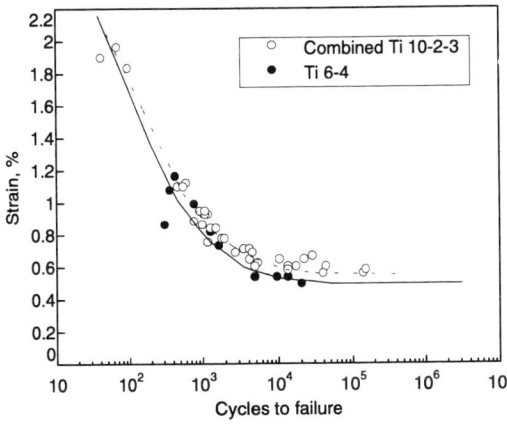

All of the forging heat treat combinations tested in this study cyclically softened. Most of the stress reduction occurred early in the test. For relatively short lives in low-cycle fatigue, the load never completely stabilizes.
Ti-10V-2Fe-3Al forgings were processed under four different conditions to an average yield strength of 1103 ± 12 MPa. Low-cycle fatigue testing was performed on a closed loop hydraulic MTS Systems machine according to ASTM E606, "Standard Recommended Practice for Constant Amplitude Low Cycle Fatigue Testing." $R = -1$, and constant strain rate was 0.01/s.
Source: R. Carey, R. Boyer, and H. Rosenberg, Fatigue Properties of Ti-10V-2Fe-3Al, in *Titanium, Science and Technology*, Vol 2, G. Lütjering, U. Zwicker, and W. Bunk, Ed., Deutsche Gesellschaft für Metallkunde, e.V., Germany, 1985, p 1261

Ti-10V-2Fe-3Al: LCF under load control

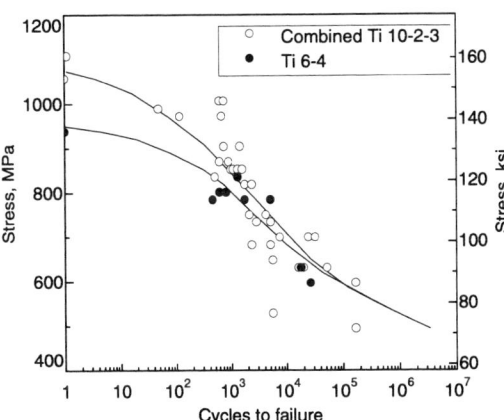

Ti-10V-2Fe-3Al forgings were processed under four different conditions to an average yield strength of 1103 ± 12 MPa. Low-cycle fatigue testing was performed on a closed loop hydraulic MTS Systems machine according to ASTM E606, "Standard Recommended Practice for Constant Amplitude Low Cycle Fatigue Testing." $R = -1$.
Source: R. Carey, R. Boyer, and H. Rosenberg, Fatigue Properties of Ti-10V-2Fe-3Al, in *Titanium, Science and Technology*, Vol 2, G. Lütjering, U. Zwicker, and W. Bunk, Ed., Deutsche Gesellschaft für Metallkunde, e.V., Germany, 1985, p 1261

Ti-10V-2Fe-3Al: Notched and smooth fatigue vs Ti-6Al-4V

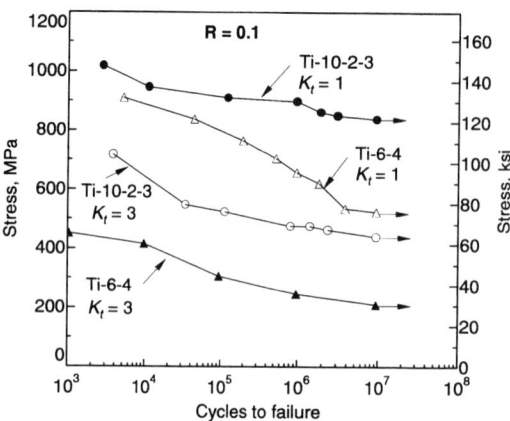

S-N curves for Ti-10V-2Fe-3Al (strength level, 965 MPa, or 140 ksi) and Ti-6Al-4V (strength level, 896 MPa, or 130 ksi). Ti-10V-2Fe-3Al specimens were taken from β hot die forgings, solution treated at 30 °C (54 °F) below the β transus temperature, water quenched, and aged to a strength level of 965 MPa (140 ksi). Ti-6Al-4V isothermal forgings were annealed to a minimum strength level of 896 MPa (130 ksi) with an actual ultimate tensile strength of 1000 MPa (145 ksi). Fatigue tests were performed on specimens 3 mm (0.125 in.) in diameter with K_t = 1 or 3, R = 0.1, and a frequency of 30 Hz, low stress ground.
Source: G. Kuhlman, A. Chakrabarti, T. Yu, R. Pishko, and G. Terlinde, LCF, Fracture Toughness, and Fatigue/Fatigue Crack Propagation Resistance Optimization in Ti-10V-2Fe-3Al Alloy Through Microstructural Modification, in *Microstructure, Fracture Toughness, and Fatigue Crack Growth Rate in Titanium Alloys*, A. Chakrabarti and J.C. Chestnutt, Ed., TMS/AIME, 1987, p 171

Ti-10V-2Fe-3Al: Fatigue of STA notched (K_t = 3) specimens

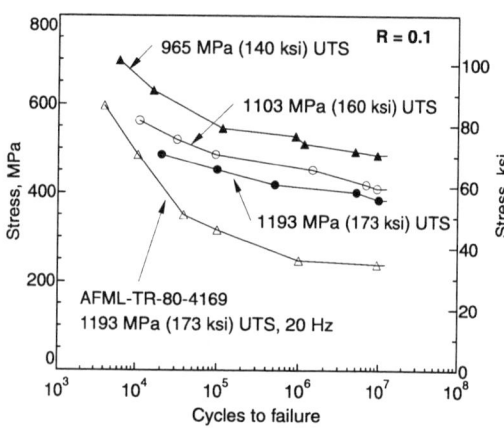

Ti-10V-2Fe-3Al solution treated and aged (STA) specimens were taken from β hot die forgings, solution treated at 30 °C (54 °F) below the β transus temperature, water quenched, and aged to a strength level of 965 MPa (140 ksi). Fatigue tests for STA specimens were performed on specimens 3 mm (0.125 in.) in diameter with K_t = 3, R = 0.1, and a frequency of 30 Hz, low stress ground.
Source: G. Kuhlman, A. Chakrabarti, T. Yu, R. Pishko, and G. Terlinde, LCF, Fracture Toughness, and Fatigue/Fatigue Crack Propagation Resistance Optimization in Ti-10V-2Fe-3Al Alloy Through Microstructural Modification, in *Microstructure, Fracture Toughness and Fatigue Crack Growth Rate in Titanium Alloys*, A. Chakrabarti and J.C. Chesnutt, Ed., TMS/AIME, 1987, p 171

Ti-10V-2Fe-3Al: Smooth and notched fatigue of STA forgings

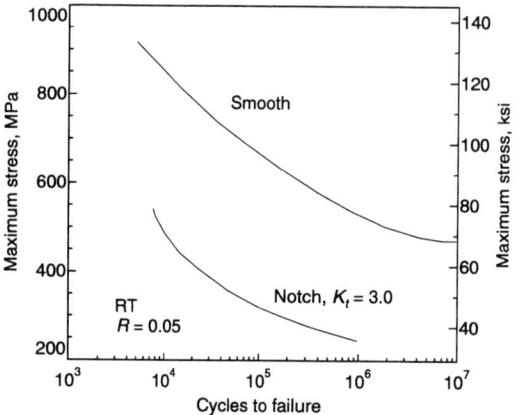

Heat treatment: 815 °C (1500 °F), 1 h, AC + 620 °C (1150 °F), 8 h, AC.
Source: *Aerospace Structural Metals Handbook*, Vol 14, Code 3726, Battelle, 1972

Ti-10V-2Fe-3Al: Notched fatigue of STA forging

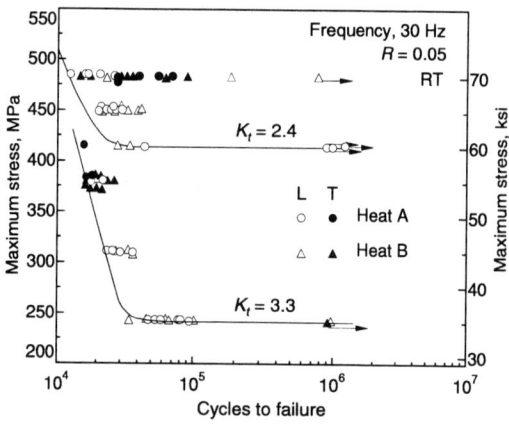

Boeing 747 lower link fitting; β forged with α-β (≤20%) finish; 775 °C (1435 °F), 2 h, AC + 770 °C (1425 °F), 2 h, WQ + 510 °C (950 °F), 8 h, AC.
Source: *Aerospace Structural Metals Handbook*, Vol 14, Code 3726, Battelle, 1972

Ti-10V-2Fe-3Al: Fatigue with single-hole notch

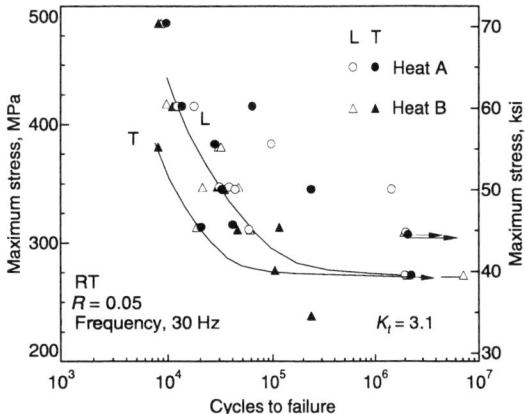

Boeing 747 lower link fitting; β forged with α-β (≤20%) finish; 775 °C (1435 °F), 2 h, AC + 770 °C (1425 °F), 2 h, WQ + 510 °C (950 °F), 8 h, AC.
Source: *Aerospace Structural Metals Handbook*, Vol 14, Code 3726, Battelle, 1972

Ti-10V-2Fe-3Al: Fatigue with double-hole notch

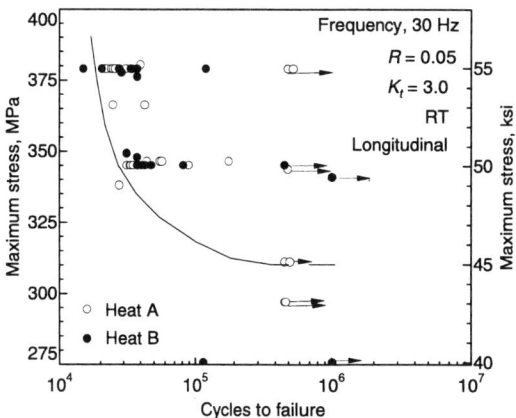

Boeing 747 lower link fitting; β forged with α-β (≤20%) finish; 775 °C (1435 °F), 2 h, AC + 1425 °F), 2 h, WQ + 510 °C (950 °F), 8 h, AC.
Source: *Aerospace Structural Metals Handbook*, Vol 14, Code 3726, Battelle, 1972

Ti-10V-2Fe-3Al: Notched fatigue performance of forgings

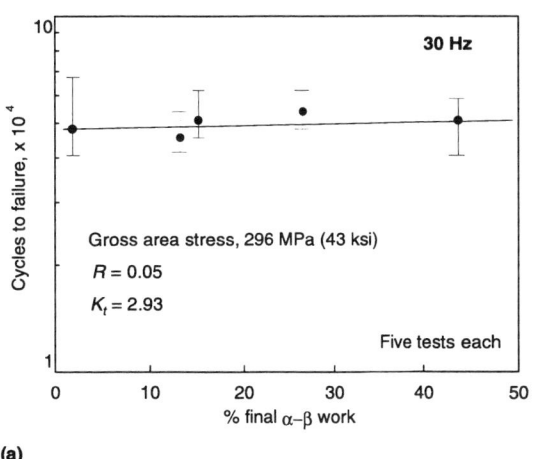
(a)

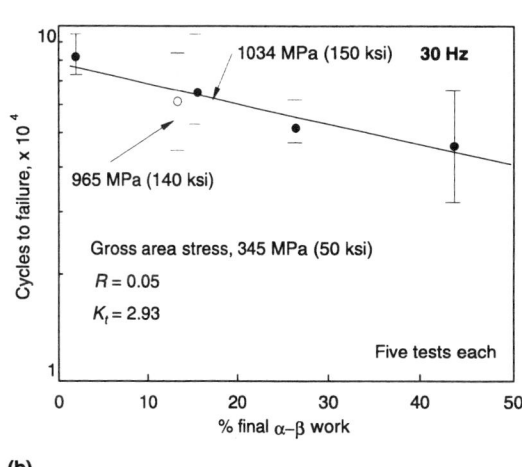
(b)

Notched fatigue (cycles to failure) of pancake forgings vs amount of work at (a) strength level of 1310 MPa (190 ksi) and (b) strength level of 865 MPa (140 ksi) and 1034 MPa (150 ksi). Log average lives and scatterband indicated.
β transus temperature was 810 °C (1490 °F). Pancake forgings were produced by β forging at temperatures 10 to 25 °C (18 to 45 °F) above the β transus to produce 50 to 70% thickness reduction. Additional reduction of 2 to 58% was accomplished by forging in the α-β range (10 to 25 °C, 18 to 45 °F, below the β transus temperature).
Source: R. Boyer and G. Kuhlman, Processing Properties Relationships of Ti-10V-2Fe-3Al, *Metall. Trans. A*, Vol 18, 1987, p 2095

High-Cycle Notched Fatigue

Room Temperature

The notched fatigue strength at a $K_t = 3$ decreases as the strength level increases. (At all strength levels it is superior to that of Ti-6Al-4V). The drop in fatigue strength as the strength is increased is attributed to increased notch sensitivity at the higher strength levels, leading to earlier crack initiation. Data from several notch geometries are presented. The effect of notch geometry is shown. A round and a flat specimen, with K_t's of 2.4 and 2.5 respectively, show a much larger difference in properties than can be ascribed to the difference in K_t. Microstructure has virtually no effect on the fatigue strength at the high strength level (190 ksi), but does have an influence at lower tensile strength levels for a K_t of 2.93. Higher amounts of α/β work, which result in a more equiaxed primary α, appears to have a negative effect on fatigue strength for the lower strength conditions (140 and 150 ksi).

Elevated Temperature

Ti-10V-2Fe-3Al: Fatigue of notched STOA bar

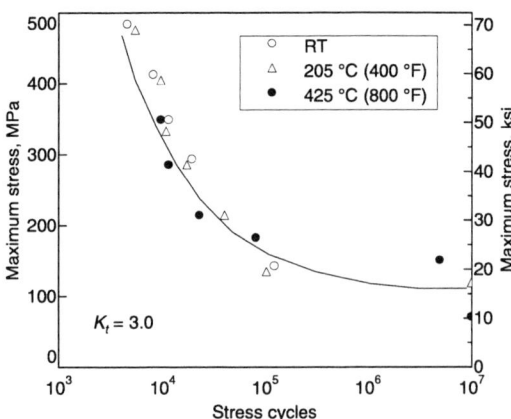

Specimens were taken from round bar 75 mm (3 in.) in diameter, solution treated at 760 °C (1400 °F) for 1 h, furnace cooled, overaged at 565 °C (1050 °F) for 8 h, and air cooled. Fatigue testing performed at $R = 0.1$ and a frequency of 20 Hz.
Source: O. Deel, "Engineering Data on New Aerospace Structural Materials," Air Force Materials Laboratory, AFML-TR-77-198, Wright Patterson AFB, 1977

Ti-10V-2Fe-3Al: Smooth and notched fatigue at RT

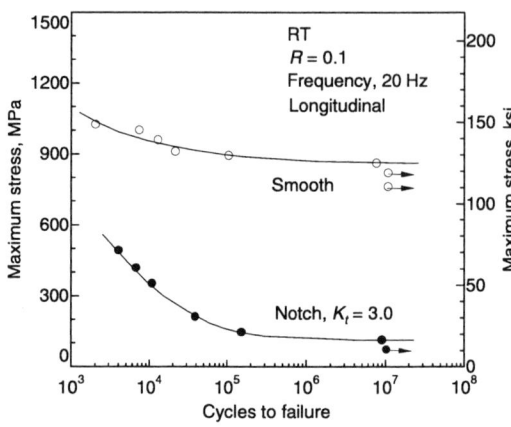

Solution treated and aged 75 mm (3 in.) diam bar treated at 760 °C (1400 °F), 1 h, FC + 565 °C (1050 °F), 8 h, AC.
Source: O. Deel, "Engineering Data on New Aerospace Structural Materials," Air Force Materials Laboratory, AFML-TR-77-198, Wright Patterson AFB, 1977

Ti-10V-2Fe-3Al: Smooth and notched fatigue at 200 °C

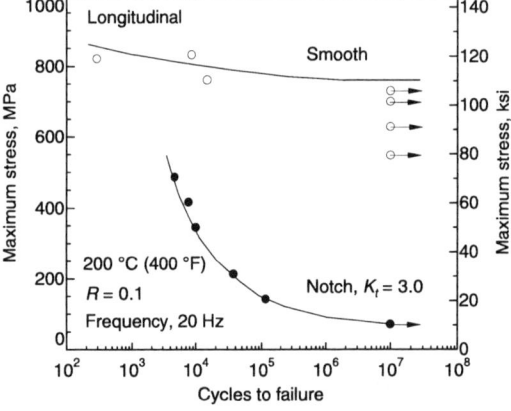

Solution treated and aged 75 mm (3 in.) diam bar treated at 760 °C (1400 °F), 1 h, FC + 565 °C (1050 °F), 8 h, AC.
Source: O. Deel, "Engineering Data on New Aerospace Structural Materials," Air Force Materials Laboratory, AFML-TR-77-198, Wright Patterson AFB, 1977

Ti-10V-2Fe-3Al: Smooth and notched fatigue at 425 °C

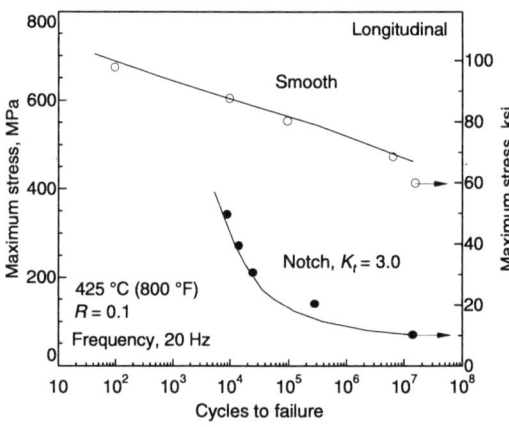

Solution treated and aged 75 mm (3 in.) diam bar treated at 760 °C (1400 °F), 1 h, FC + 565 °C (1050 °F), 8 h, AC.
Source: O. Deel, "Engineering Data on New Aerospace Structural Materials," Air Force Materials Laboratory, AFML-TR-77-198, Wright Patterson AFB, 1977

High-Cycle Fatigue: P/M and Cast

Early work with powder metallurgy compacts indicated a debit in comparison to wrought forgings. Thermomechanical processing can be seen to have an influence. Of the two compaction techniques studied for pre-alloyed powder, the compaction technique does not appear to be important. The blended elemental compact fatigue performance could also be improved by thermomechanical processing and/or the use of low Cl powder.

Ti-10V-2Fe-3Al: Fatigue in notched specimens for several product forms in high-strength and low-strength conditions

Product form	Ultimate tensile strength MPa	Ultimate tensile strength ksi	Log average fatigue life
Cast and wrought			
Isothermal forgings	1300-1380	188-200	20 200
	1060-1100	154-159	25 700
Conventional forgings	1230-1350	178-196	50 000
Pancake forgings	1260-1345	183-195	50 000
43% α/β work	1050-1070	152-155	50 000
2% α/β work	1050	152	83 300
Extrusions	1105-1175	160-170	32 500
P/M			
Prealloyed HIP	1345-1415	195-205	16 900
Isothermally forged	1125-1145	163-166	53 300

Note: Test frequency was 30 Hz and $R = 0.05$, with $K_t = 2.93$ and tests performed at stress level of 345 MPa (50 ksi). Source: R. Boyer, D. Eylon, and F. Froes, Comparative Evaluation of Ti-10V-2Fe-3Al Cast, P/M and Wrought Product Forms, *Titanium, Science and Technology*, Vol 2, G. Lütjering, U. Zwicker, and W. Bunk, Ed., Deutsche Gesellschaft für Metallkunde, e.V. Germany, 1985, p 1307

Ti-10V-2Fe-3Al: Effect of notch geometry on fatigue strength

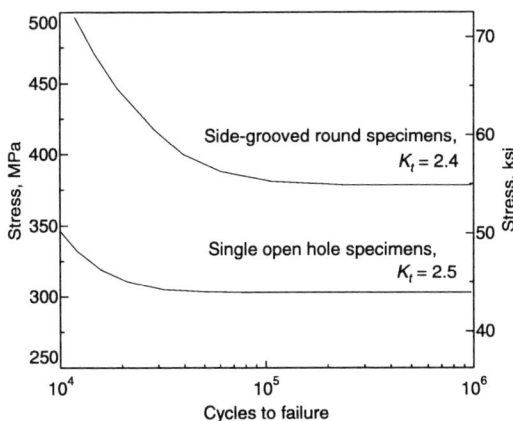

Forgings were heat treated to a strength level of 1241 MPa (180 ksi). Flat sheet-type specimens with holes drilled to a notch factor $K_t = 2.5$ and round side-grooved specimens with $K_t = 2.4$ were used. Heat treating and machining sequences were the same in both cases. Source: R. Carey, R. Boyer, and H. Rosenberg, Fatigue Properties of Ti-10V-2Fe-3Al, in *Titanium, Science and Technology*, Vol 2, G. Lütjering, U. Zwicker, and W. Bunk, Ed., Deutsche Gesellschaft für Metallkunde, e.V., Germany, 1985, p 1261

Ti-10V-2Fe-3Al: Fatigue of cast and wrought specimens

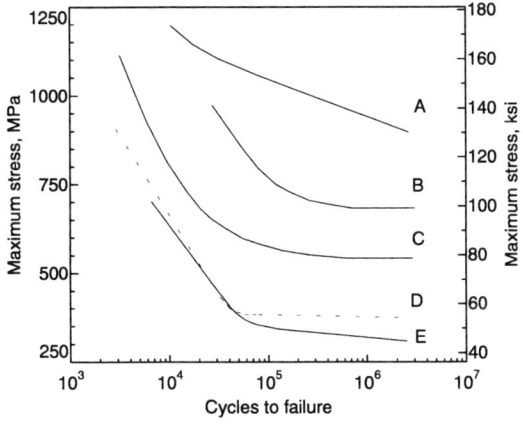

Smooth specimen fatigue for (A) cast and wrought plus isothermal forge, ultimate tensile strength = 1300 to 1380 MPa (188 to 200 ksi); (B) prealloyed P/M HIP plus isothermal forge, ultimate tensile strength = 1345 to 1400 MPa (195 to 203 ksi); (C) prealloyed HIP, ultimate tensile strength = 1310 MPa (190 ksi); (D) P/S + HIP, ultimate tensile strength 1228 to 1275 MPa (178 to 185 ksi); (E) P/S, ultimate tensile strength = 1195 MPa (173 ksi). Curves represent data lower limits. Test frequency was 5 Hz and $R = 0.1$.
Source: R. Boyer, D. Eylon, and F. Froes, Comparative Evaluation of Ti-10V-2Fe-3Al Cast, P/M and Wrought Product Forms, in *Titanium, Science and Technology*, Vol 2, G. Lutjering, U. Zwicker, and W. Bunk, Ed., Deutsche Gesellschaft für Metallkunde, e.V., Germany, 1985, p 307

Ti-10V-2Fe-3Al: Fatigue in powder compacts

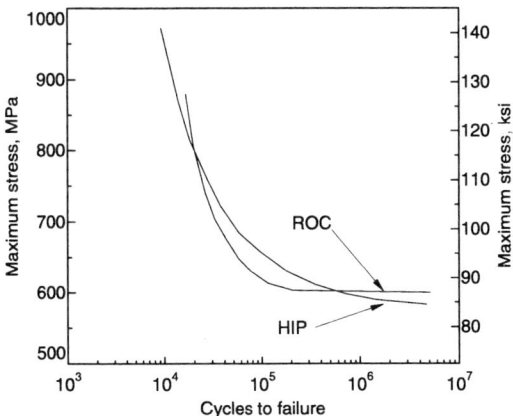

S-N curve for powder compacts consolidated by rapid omnidirectional compaction (ROC) by hot isostatic pressing. Both specimens were heat treated.
Chemical composition of the alloy was 3.0 wt% Al, 0.065 wt% C, 2.1 wt% Fe, 0.0063 wt% H, 0.0093 wt% N, 0.1485 wt% O, 9.2 wt% V, and 0.006 wt% W. Processing parameters for consolidation (ROC) were 775 °C (1425 °F) at 830 MPa (120 ksi), 1/2-s dwell, air cool. Processing parameters for HIP were 790 °C (1450 °F) at 103 MPa (15 ksi) for 20 h. Heat treatment was carried out at 745 °C (1365 °F) for 1 h, water quench, and 490 °C (915 °F) for 8 h, air cool. Fatigue tests were performed at room temperature on a servo-hydraulic MTS machine. Constant load triangular waveform cycling was done at $R = 0.1$ and a frequency of 5 Hz.
Source: Y. Mahajan, D. Eylon, C. Kelto, and F. Froes, Evaluation of Ti-10V-2Fe-3Al Powder Compacts Produced by the ROC Method, *Metal Powder Rep.*, Oct 1986, p 749

Fatigue Crack Growth

Using conventional processing and heat treatments, the crack growth rate (da/dN) of this alloy is essentially independent of microstructure, strength level and test environment, and, in air, is similar to that of mill annealed Ti-6Al-4V. Aging to produce the omega phase significantly reduced da/dN, but, as mentioned previously, this is not a practical microstructure to use

Ti-10V-2Fe-3Al: Crack growth in two aged conditions

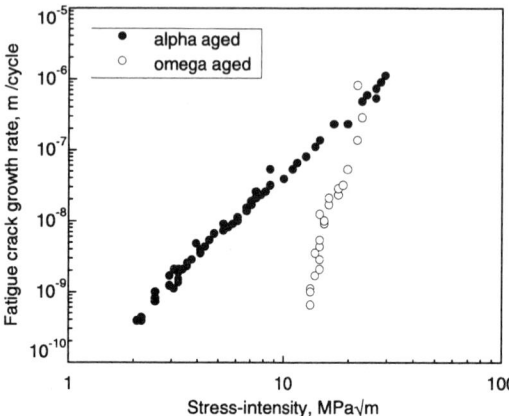

Source: T.W. Duerig and J.C. Williams, Overview: Microstructure and Properties of Beta Titanium Alloys, in *Beta Titanium Alloys in the 1980's*, R. Boyer and H. Rosenberg, Ed., TMS/AIME, 1984, p 44

Ti-10V-2Fe-3Al: Crack growth in air and 3.5% NaCl

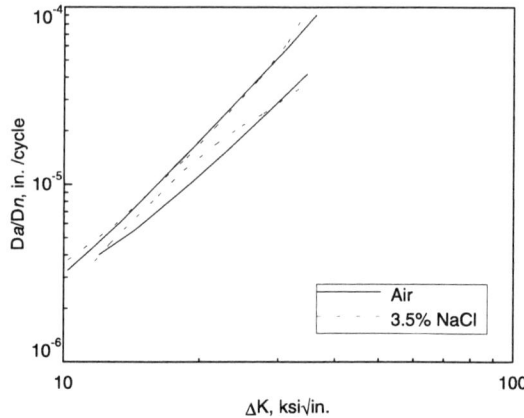

Superimposed air and 3.5% NaCl fatigue crack propagation rate scatterbands for Ti-10V-2Fe-3Al, R = 0.05, frequency 1-30 Hz, various orientations.
Source: R. Boyer, WestTech, 1981

Ti-10V-2Fe-3Al: FCG with low aspect ratio of primary α

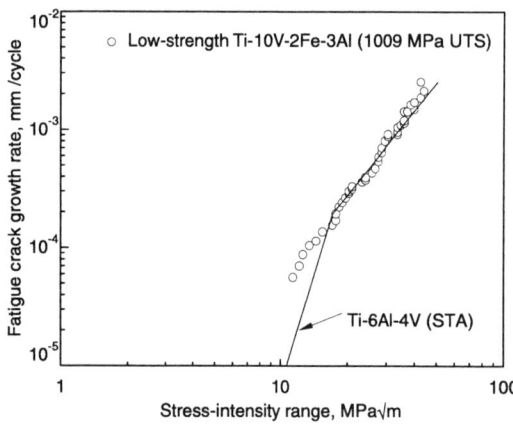

Chemical composition of the alloy was 3.2 wt% Al, 0.03 wt% C, 1.8 wt% Fe, 0.005 wt% H, 0.01 wt% N, 0.087 wt% O, and 9.7 wt% V. Material was β forged then α-β worked to effect an additional 55% reduction, followed by heat treatment at 750 °C (1380 °F) for 2 h, and water quench, aged at 550 °C (1020 °F) for 8 h. Ultimate tensile strength was 1009 MPa (146 ksi). Tests were performed in air at room temperature, with Haversine waveform and R = 0.10.
Source: G. Yoder, L. Cooley, and R. Boyer, Microstructure/Crack Tolerance Aspects of Notched Fatigue Life in Ti-10V-2Fe-3Al Alloy, in *Microstructure, Fracture Toughness and Fatigue Crack Growth Rate in Titanium Alloys*, A. Chakrabarti and J.C. Chesnutt, Ed., TMS/AIME, 1987, p 209

Ti-10V-2Fe-3Al: FCG with high aspect ratio of primary α

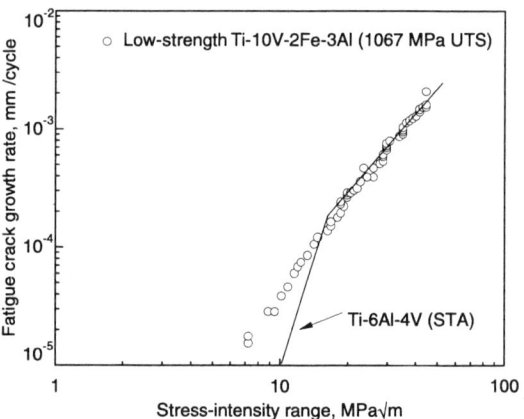

Chemical composition of the alloy was 3.2 wt% Al, 0.03 wt% C, 1.8 wt% Fe, 0.05 wt% H, 0.01 wt% N, 0.087 wt% O, and 9.7 wt% V. Material was β forged then α-β worked to effect an additional 2% reduction, followed by heat treatment at 750 °C (1380 °F) for 2 h, and water quench, aged at 550 °C (1020 °F) for 8 h. Ultimate tensile strength was 1067 MPa (154 ksi). Tests were performed in air at room temperature, with Haversine waveform and R = 0.10.
Source: G. Yoder, L. Cooley, and R. Boyer, Microstructure/Crack Tolerance Aspects of Notched Fatigue Life in Ti-10V-2Fe-3Al Alloy, in *Microstructure, Fracture Toughness and Fatigue Crack Growth Rate in Titanium Alloys*, A. Chakrabarti and J.C. Chesnutt, Ed., TMS/AIME, 1987, p 209

Ti-10V-2Fe-3Al: FCG with high aspect ratio of primary α

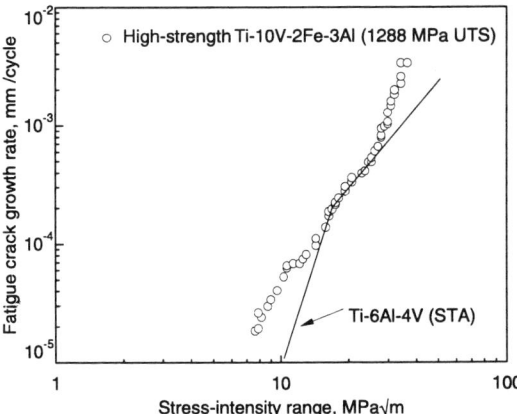

Chemical composition of the alloy was 3.2 wt% Al, 0.03 wt% C, 1.8 wt% Fe, 0.05 wt% H, 0.01 wt% N, 0.087 wt% O, and 9.7 wt% V. Material was β forged then α-β worked to effect an additional 2% reduction, followed by heat treatment at 750 °C (1380 °F) for 2 h, and water quench, aged at 495 °C (255 °F) for 8 h. Ultimate tensile strength was 1288 MPa (187 ksi). Tests were performed in air at room temperature, with Haversine waveform and $R = 0.10$.
Source: G. Yoder, L. Cooley, and R. Boyer, Microstructure/Crack Tolerance Aspects of Notched Fatigue Life in Ti-10V-2Fe-3Al Alloy, in *Microstructure, Fracture Toughness and Fatigue Crack Growth Rate in Titanium Alloys*, A. Chakrabarti and J.C. Chesnutt, Ed., TMS/AIME, 1987, p 209

Ti-10V-2Fe-3Al: FCG with low aspect ratio of primary α

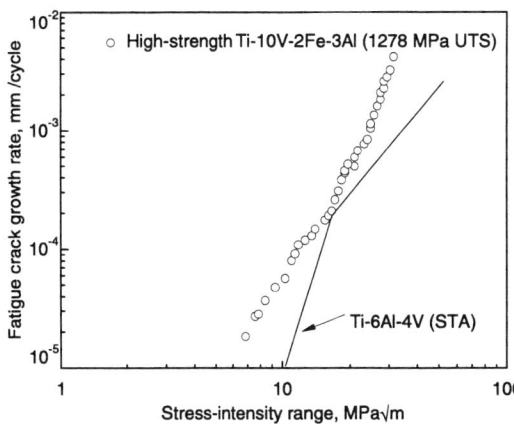

Chemical composition was 3.2 wt% Al, 0.03 wt% C, 1.8 wt% Fe, 0.005 wt% H, 0.01 wt% N, 0.087 wt% O, and 9.7 wt% V. Material was β forged then α-β worked to effect an additional 55% reduction, followed by heat treatment at 750 °C (1380 °F) for 2 h, and water quench, aged at 495 °C (255 °F) for 8 h. Ultimate tensile strength was 1278 MPa (185 ksi). Tests were performed in air at room temperature, with Haversine waveform and $R = 0.10$.
Source: G. Yoder, L. Cooley, and R. Boyer, Microstructure/Crack Tolerance Aspects of Notched Fatigue Life in Ti-10V-2Fe-3Al Alloy, in *Microstructure, Fracture Toughness and Fatigue Crack Growth Rate in Titanium Alloys*, A. Chakrabarti and J.C. Chesnutt, Ed., TMS/AIME, 1987, p 209

Ti-10V-2Fe-3Al: FCG in STA and direct age conditions

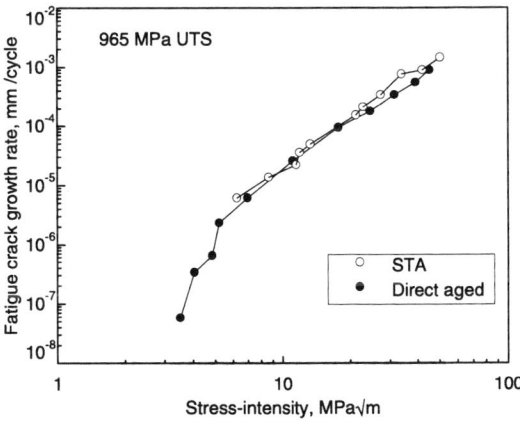

Ti-10V-2Fe-3Al STA specimens were taken from β hot die forgings solution treated at 30 °C (54 °F) below β transus temperature, water quenched, and aged to a strength level of 965 MPa (140 ksi). Ti-10V-2Fe-3Al direct aged specimens were β hot die forged, post-forge cooled at a rate of 5 °C/s (9 °F/s), and aged to the desired strength level. Fatigue crack propagation tests for STA specimens were performed on specimens 6 mm (0.25 in) thick and 37 mm (1.25 in.) in length and width with $R = 0.1$ and a frequency of 30 Hz, compact tension. Fatigue crack propagation tests for direct aged specimens were performed according to ASTM E606 on 6 mm (0.25 in.) diameter specimens, low stress ground, triangular waveform, 20 cycles/min, $R = 0$, and $A = 1.0$, with a frequency of 50 Hz, constant strain.
Source: G. Kuhlman, A. Chakrabarti, T. Yu, R. Pishko, and G. Terlinde, LCF, Fracture Toughness and Fatigue/Fatigue Crack Propagation Resistance Optimization in Ti-10V-2Fe-3Al Alloy Through Microstructural Modification, in *Microstructure, Fracture Toughness, and Fatigue Crack Growth Rate in Titanium Alloys*, A. Chakrabarti and J.C. Chesnutt, Ed., TMS/AIME, 1987, p 171

Ti-10V-2Fe-3Al: FCG in direct age condition

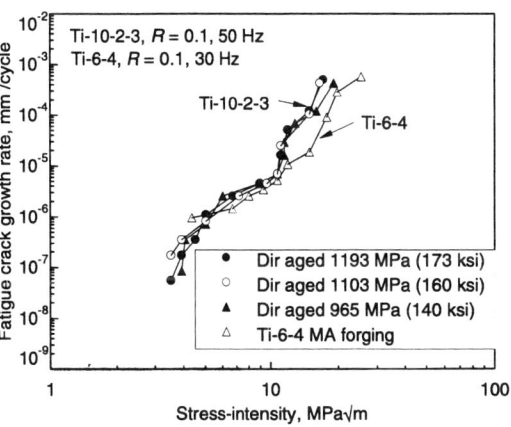

Ti-10V-2Fe-3Al direct aged specimens were β hot die forged, post-forge cooled at a rate of 5 °C/s (9 °F/s), and aged to the desired strength level. Fatigue crack propagation tests for direct aged specimens were performed according to ASTM E606 on 6 mm (0.25 in.) diam specimens, low stress ground, triangular waveform, 20 cycles/min, $R = 0$, $A = 1.0$, constant strain.
Source: G. Kuhlman, A. Chakrabarti, T. Yu, R. Pishko, and G. Terlinde, LCF, Fracture Toughness, and Fatigue/Fatigue Crack Propagation Resistance Optimization in Ti-10V-2Fe-3Al Alloy Through Microstructural Modification, in *Microstructure Fracture Toughness, and Fatigue Crack Growth Rate in Titanium Alloys*, A. Chakrabarti and J.C. Chesnutt, Ed., TMS/AIME, 1987, p 171

Fracture Toughness

The fracture toughness is strongly dependent on the tensile strength and the microstructure as reported by several authors. The processing, in terms of the amount of α/β work affects the toughness by modification of the morphology of the primary α. Higher amounts of α/β work, following primary working in the β-phase field, changes the primary α to a more globular morphology, which improves ductility at the expense of toughness. There would also appear to be an optimum amount of primary α to achieve maximum toughness (a 10% volume fraction of elongated primary α had significantly higher fracture toughness than 30 vol.%). There seems to be a lot of variation in the toughness reported for powder compacts. There is some evidence that the fracture toughness is related to the volume fraction of defects in P/M products.

Ti-10V-2Fe-3Al: Room temperature Charpy impact toughness of STOA bar

Direction	Impact toughness	
	J	ft · lb
Longitudinal	35.9	26.5
	40.7	30.0
	40.7	30.0
Average	39.1	28.9
Transverse	27.8	20.5
	26.5	19.5
	23.1	17.0
Average	25.8	19.0

Source: AFML-TR-78-114

Ti-10V-2Fe-3Al: Fracture toughness for several product forms

Product form	Ultimate tensile strength		Tensile yield strength		Elongation, %	Plane-strain fracture toughness	
	MPa	ksi	MPa	ksi		MPa√m	ksi√in.
High strength condition							
Isothermal forgings	1300-1380	188-200	1200-1255	174-182	3-6	29	26
Conventional forgings	1230-1350	178-196	1145-1280	166-186	4-10	44-60	40-54
Pancake forgings	1275-1310	185-190	1150-1160	167-168	5-8	47	43
Extrusions	1240	179	1170	169	4	...	...
P/M high strength							
Prealloyed, HIP	1310	190	1205	175	9	...	...
Prealloyed, HIP + isothermal forge	1345-1400	195-203	1240-1305	179-189	6-8	28	25
P/S	1195	173	1110	161	3.5	...	...
P/S + HIP	1228-1275	178-185	1185-1245	172-180	7-9	28-29	25-26
Reduced strength condition							
Isothermal forgings	1060-1100	153-159	985-1060	143-153	8-12	70	64
Pancake forgings	965	140	930	135	16	100	91
Extrusions	1110-1170	161-169	1000-1105	145-160	6-7	45-48	41-44
P/M Prealloyed HIP + isothermal forge	1125-1145	163-166	1050-1090	152-158	13-15	55	50
P/M, P/S + HIP	1120-1160	162-168	1070-1105	155-160	9-10	32	29
Castings	1105-1130	160-164	1010-1030	146-149	6-10	...	...
AMS specification (forgings)							
AMS 4984	1190	173	1100	160	4 (in 4D)	44	40
AMS 4986	1100	160	1000	145	6 (in 4D)	60	55
AMS 4987	965	140	895	130	8 (in 4D)	88	80

Source: R. Boyer, D. Eylon, and F. Froes, Comparative Evaluation of Ti-10V-2Fe-3Al Cast, P/M, and Wrought Product Forms, *Titanium Science and Technology*, Vol 2, G. Lütjering, U. Zwicker, and W. Bunk, Ed., Deutsche Gesellschaft für Metallkunde e.V., Germany, 1985, p 1307

Ti-10V-2Fe-3Al: Typical α/β forged room-temperature tensile properties and fracture toughness of forgings

Forging thickness		Orientation/ location	Ultimate yield strength (0.2% offset)		Ultimate tensile strength		Elongation, %	Reduction of area, %	Plane-strain fracture toughness	
mm	in.		MPa	ksi	MPa	ksi			MPa√m	ksi√in.
75	3	L/S, MC, C	1256-1263	182-183	1318-1325	191-192	9-11	32-34	40.2	36.58
		LT/S, MS, C	1270-1283	184-186	1332-1339	193-194	8-9	20-30	39.9	36.26
		ST/S, MS, C	1214-1311	176-190	1283-1380	186-200	5-9	12-34	43.2	39.26
		Range	1256-1311	182-190	1283-1380	186-200	5-11	12-34	39~43	36~39
50	2	L/S, MS, C	1249-1256	181-182	1325-1311	190-192	8-11	27-35	35.4	32.22
		LT/S, MS, C	1270-1325	184-192	1346-1394	195-202	5-8	12-27	35.1	31.88
		ST/S, MS, C	1173-1194	170-173	1194-1242	173-180	13-14	46-59	...	...
		Range	1173-1325	170-192	1228-1394	178-202	5-14	12-59	~35	~32
25	1	L/S, MS, C	1256-1283	182-186	1339-1342	194-196	5-9	10-25	30.1	27.32
		LT/S, MS, C,	1221-1241	177-176	1270-1270	184-184	10-13	36-48	30.9	28.05
		Range	1214-1256	176-186	1270-1352	184-196	5-13	10-48	30~31	27-28

Note: L, longitudinal; LT, long transverse; ST, short transverse; S, surface; MS, midsurface; C, center. α + β forging was conducted at 760 °C (1400 °F) with about 60% deformation, followed by hand forging. The alloy was double solution treated and aged. The first solution treatment was performed close to, but below, the beta transus (788 to 802 °C, or 1450 to 1480 °F), followed by a slow cool. The second solution treatment took place at a temperature lower than the first, followed by water quench. Source: C. Chen and R. Boyer, Practical Considerations for Manufacturing High-Strength Ti-10V-2Fe-3Al Alloy Forgings, *J. Metals*, July 1979, p 33

Ti-10V-2Fe-3Al: Fracture toughness vs UTS

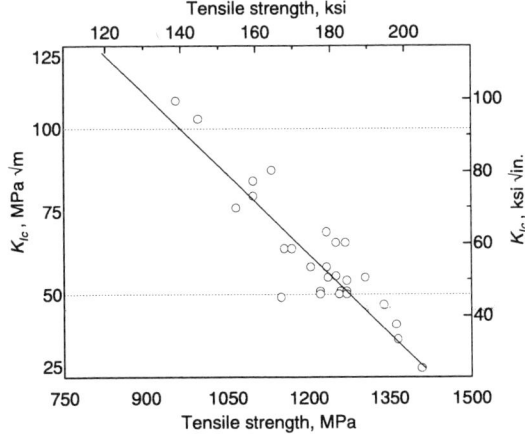

Testing performed in air on compact tension specimens according to ASTM E399.
Source: G. Kuhlman, "Alcoa Titanium Alloy Ti-10V-2Fe-3Al Forgings, Alcoa Green Letter No. 224," Aluminum Company of American, Forging Division, Cleveland, Aug 1987. With permission

Ti-10V-2Fe-3Al: Fracture toughness vs yield strength

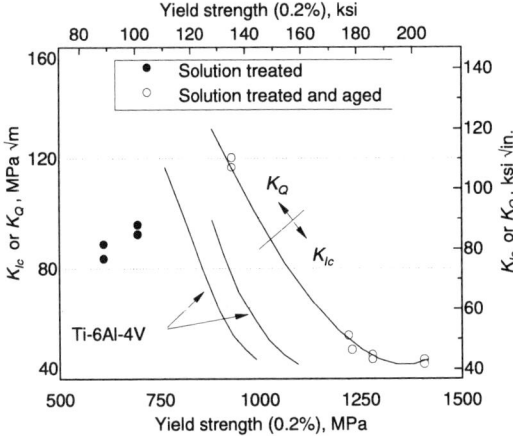

Source: *The Sumitomo Search*, No. 35, Nov 1987, p 21-28

Stress Corrosion Resistance. The stress corrosion threshold has been reported to be at least 80% of K_{Ic} except when it is stressed in the short transverse direction, where it is 70% of K_{Ic}.

Ti-10V-2Fe-3Al: Plane-strain fracture toughness vs UTS

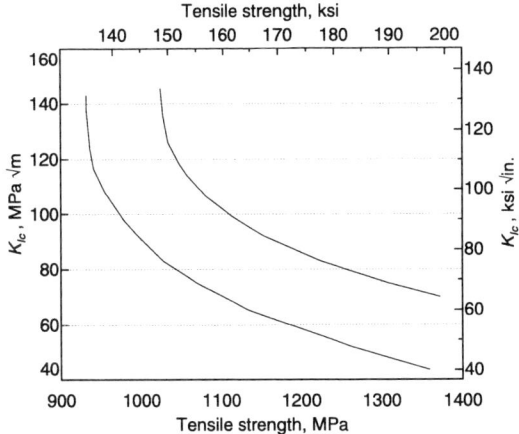

Data represent a composite of fracture toughness values for β forged die forgings, β forged block forgings, and β forged plus α-β forged die forgings of Ti-10V-2Fe-3Al.
Source: *Metals Handbook, Properties and Selection: Stainless Steels, Tool Materials, and Special-Purpose Materials*, Vol 3, 9th ed., American Society for Metals, 1980

Effect of Microstructure and Processing

Ti-10V-2Fe-3Al: Fracture toughness of forgings with different aspect ratios of primary α

Aspect ratio(a)	Tensile yield strength (0.2% offset)		Ultimate tensile strength		Elongation, %	Reduction of area, %	Plane-strain fracture toughness	
	MPa	ksi	MPa	ksi			MPa√m	ksi√in.
HAR	1149	166.6	1288	186.8	3.0	3.3	46.8	42.6
LAR	1190	172.6	1278	185.4	7.0	12.8	37.0	33.7
HAR	1002	145.4	1067	154.8	9.0	24.3	86.1	78.4
LAR	990	143.6	1009	146.4	15.0	50.7	67.0	61.0

Note: Alloy was forged above the β transus for thickness reduction of 50 to 70%, followed by α-β range forging and heat treatment to indicated strengths. (a) HAR, high aspect ratio; LAR, low aspect ratio of primary alpha. Source: G. Yoder, L. Cooley, and R. Boyer, Microstructure/Crack Tolerance Aspects of Notched Fatigue Life in Ti-10V-2Fe-3Al Alloy, *Microstructure, Fracture Toughness, and Fatigue Crack Growth Rate in Titanium Alloys*, A. Chakrabarti and J.C. Chesnutt, Ed., TMS/AIME, 1987, p 209

Ti-10V-2Fe-3Al: Fracture toughness/microstructure for forgings

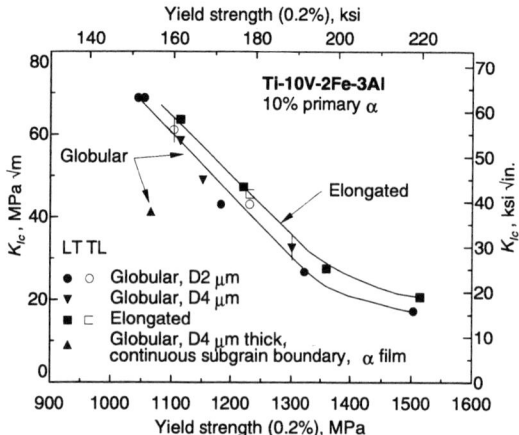

Source: G. Terlinde, H.-J. Rathjen, and K.H. Schwalbe, Microstructure and Fracture Toughness of the Aged β-Ti Alloy Ti-10V-2Fe-3Al, Metall. Trans. A, Vol 19, 1988, p 1037

Ti-10V-2Fe-3Al: Effect of elongated α on toughness

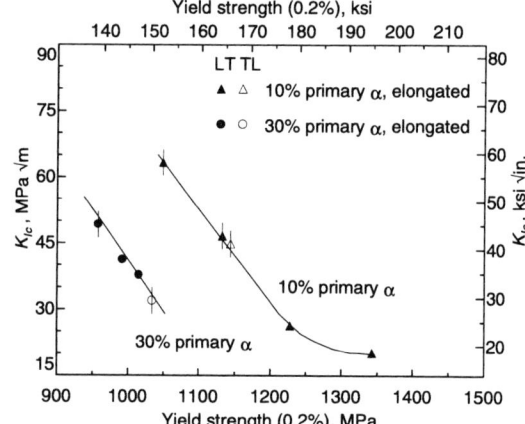

Source: Metall. Trans. A, Vol 19, 1988, p 1041

Ti-10V-2Fe-3Al: Effect of α morphology on toughness/ductility

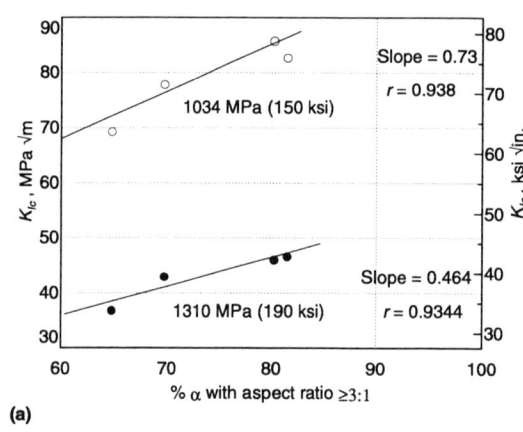

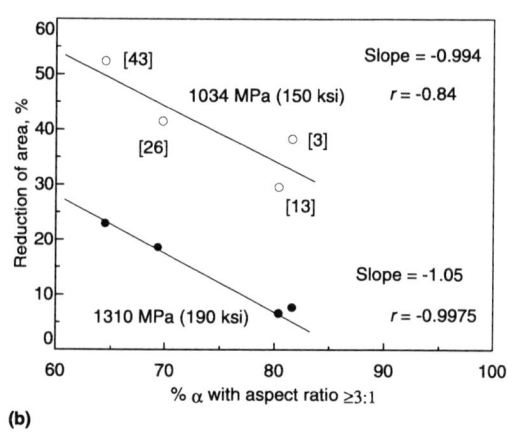

(a) Toughness. (b) Reduction of area. Numbers in brackets indicate amount of final α-β work.
Source: R. Boyer and G. Kuhlman, Processing Properties Relationships of Ti-10V-2Fe-3Al, Metall. Trans. A, Vol 18, 1987, p 2095-2103

Effect of Processing

Ti-10V-2Fe-3Al: Fracture toughness of forgings vs final working

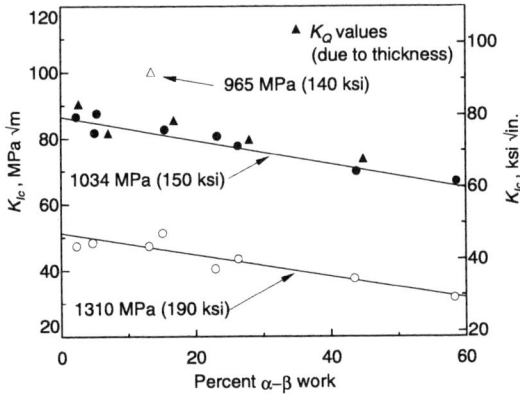

Alloy from two heats. No difference in behavior was observed for the two heats. Material was β forged at the β transus temperature plus 10 to 25 °C (18 to 45 °F) for a 50-70% reduction, followed by α-β range forging at the β transus temperature minus 10 to 25 °C (18 to 45 °F). Specimens were solution treated at 750 °C (1380 °F) for 2 h, followed by aging for 8 h at 495 °C (920 °F) for the high-strength condition, and as indicated for the low-strength condition.
Source: R. Boyer and G. Kuhlman, Processing Properties Relationships of Ti-10V-2Fe-3Al, *Metall. Trans. A*, Vol 18, 1987, p 2095

Ti-10V-2Fe-3Al: Fracture toughness vs forging/heat treatment

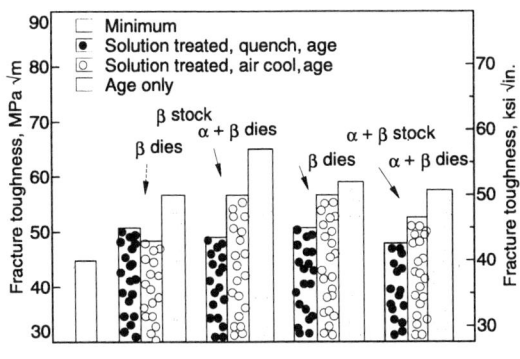

Source: R. Pishko, T. Yu, and G. Kuhlman, Precision Forging of Titanium Alloy, in *Titanium 1986 Products and Applications*, Titanium Development Association, 1987, p 376

Ti-10V-2Fe-3Al: Toughness vs defect content

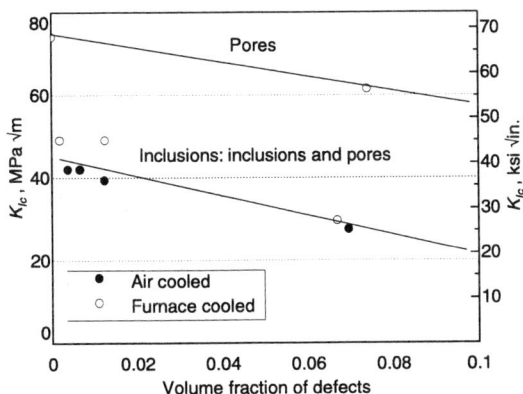

Source: N. Moody et al., The Role of Inclusion and Pore Content on the Fracture Toughness of Powder Processed Blended Elemental Ti-10V-2Fe-3Al, in *Microstructure, Fracture Toughness, and Fatigue Crack Growth Rate in Titanium Alloys*, A. Chakrabarti and J.C. Chesnutt, Ed., TMS/AIME, 1987, p 83

Ti-10V-2Fe-3Al: Toughness from conventional and hot die forging

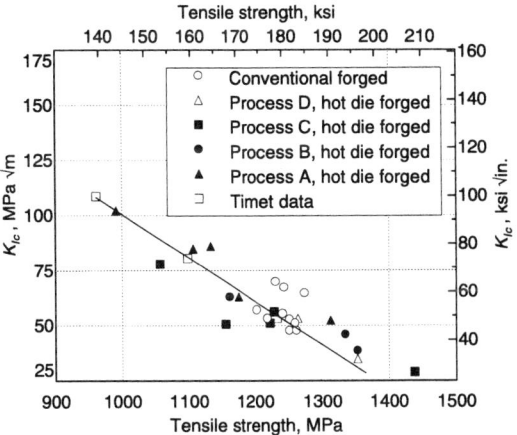

Source: R. Pishko, T. Yu, and G. Kuhlman, Precision Forging of Titanium Alloy, in *Titanium 1986 Products and Applications*, Titanium Development Association, 1987, p 401

Ti-10V-2Fe-3Al: Comparison of fracture toughness of powder compacts vs wrought alloys

| | Powder compact | | | | Wrought alloy | | | |
| | Plane-strain fracture toughness | | Tensile yield strength | | Plane-strain fracture toughness | | Tensile yield strength | |
Alloy	MPa√m	ksi√in.	MPa	ksi	MPa√m	ksi√in.	MPa	ksi
Ti-10V-2Fe-3Al	51	46	854	124	60	55	1100	160
Ti-6Al-6V-2Sn	44	40	910	132	65	59	965	140
Ti-6Al-2Sn-4Zr-6Mo	31	28	1068	155	33	30	1100	160
Ti-6Al-4V	35	32	900	130	66	60	860	125

Note: Compositions of cold isopressed and vacuum sintered billets were as follows. Ti-10-2-3: 2.98 wt% Al, 0.010 wt% C, 2.03 wt% Fe, 0.0008 wt% H, 0.011 wt% N, 0.13 wt% O, 10.25 wt% V. Ti-6-6-2: 6.00 wt% Al, 0.019 wt% C, 0.55 wt% Cu, 0.48 wt% Fe, 0.0016 wt% H, 0.0045 wt% N, 0.27 wt% O, 1.65 wt% Sn, 5.31 wt% V. Ti-6-2-4-6: 6.03 wt% Al, 0.024 wt% C, 0.060 wt% Fe, 0.0009 wt% H, 5.60 wt% Mo, 0.0055 wt% N, 0.32 wt% O, 2.20 wt% Sn, 3.60 wt% Zr. Source: Data for Ti-6-4 powder compact as taken from P. Anderson et al., *Modern Developments in Powder Metallurgy*, Vol 13, 1981. Data for wrought alloys taken from AMS specifications, except Ti-10-2-3 (TIMET data). J. Smugeresky and N. Moody, Properties of High Strength Blended Elemental Powder Metallurgy Titanium Alloys, *Titanium Net Shape Technologies*, F. Froes and D. Eylon, Ed., TMS/AIME, 1984, p 131

Ti-10V-2Fe-3Al: Fracture toughness of powder compacts

Condition(a)	Defects	Volume fraction of defects	Tensile yield strength MPa	ksi	Ultimate tensile strength MPa	ksi	Reduction of area, %	Strain to fracture, %	Plane-strain fracture toughness MPa√m	ksi√in.
~0.15 wt% Cl, 0.38 wt% O										
As sintered, FC	Inclusions, pores	0.067	883	128	966	140	2.1	2.1	29.7	32.6
β annealed, AC	Inclusions, pores	0.069	945	137	961	139	0.9	0.9	27.8	30.5
~0.15 wt% Cl, 0.13 wt% O										
As sintered, FC	Inclusions, pores	0.014	852	123	928	134	12.0	13.0	47.7	52.4
β annealed, AC	Inclusions, pores	0.012	1033	150	1083	157	10.7	11.3	39.5	43.4
HIP (60 MPa, 1000 °C), 1 h, AC	Inclusions	0.0055	977	142	1067	154	21.4	24.0	41.7	45.8
HIP (207 MPa, 750 °C) 850 °C, 1 h, FC	Inclusions	0.0027	928	134	1027	149	21.3	23.9	48.3	53.1(b)
<0.001 wt% Cl(c)										
As sintered, FC	Pores	0.075	786	114	888	128	5.6	5.8	62.9	69.1
HIP (207 MPa, 750 °C), 850 °C, 1 h, FC	None	...	996	144	1102	160	29.0	34.3	75.5(b)	82.9

Note: Fracture toughness values were determined with 15.24 mm (0.6 in.) thick compact tension specimens and with 10.2 mm (0.4 in.) square cross section three-point bend specimens. Both types of specimens were precracked and tested in accordance with ASTM E399 at a rate of 1.5 MPa√m/s (1.3 ksi√in./s). (a) Chlorine content of starting titanium powder. (b) One fracture toughness test. Source: N. Moody, W. Garrison, Jr., J. Smugeresky, and J. Costa, The Role of Inclusion and Pore Content on the Fracture Toughness of Powder Processed Blended Elemental Ti-10V-2Fe-3Al, *Microstructure, Fracture Toughness, and Fatigue Crack Growth Rate in Titanium Alloys*, A. Chakrabarti and J.C. Chesnutt, Ed., TMS/AIME, 1987, p 83

Stress-Strain Curves

Typical

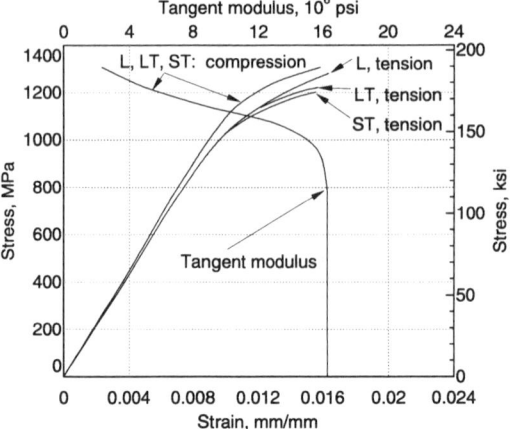

Ti-10V-2Fe-3Al: Typical stress-strain curves

Solution treated and aged die forging 78 to 84 mm (3.1 to 3.3 in.) thick section aged at 480 to 510 °C (900 to 950 °F).
Source: MIL-HDBK 5, 1 Dec 1991

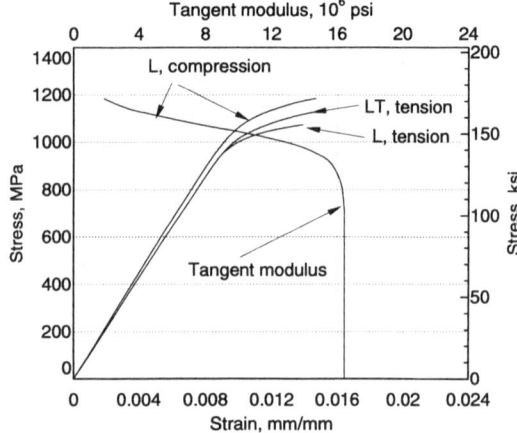

Ti-10V-2Fe-3Al: Typical stress-strain curves

Solution treated and aged, hand forging; 75 mm (3.0 in.) thick.
Source: MIL-HDBK 5, 1 Dec 1991

Effect of Temperature

Ti-10V-2Fe-3Al: Tensile stress-strain

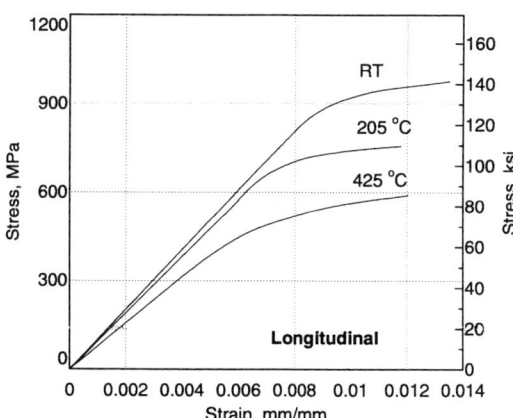

Solution treated and overaged Ti-10V-2Fe-3Al round bar; 0.16 wt% O max, 0.05 wt% N max.
Source: O. Deel, "Engineering Data on New Aerospace Structural Materials," AFML-TR-77-198, Battelle Columbus Laboratories, 1977, p 97

Ti-10V-2Fe-3Al: Compression stress-strain curves

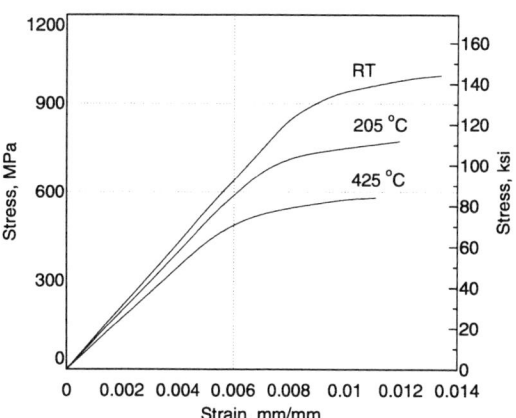

Solution treated and overaged round bar tested in the longitudinal direction.
Source: O. Deel, "Engineering Data on New Aerospace Structural Materials," AFML-TR-77-198, Battelle Columbus Laboratories, 1977, p 98

Effect of Microstructure

Ti-10V-2Fe-3Al: Effect of alpha fraction on true stress/true strain curves on unaged material

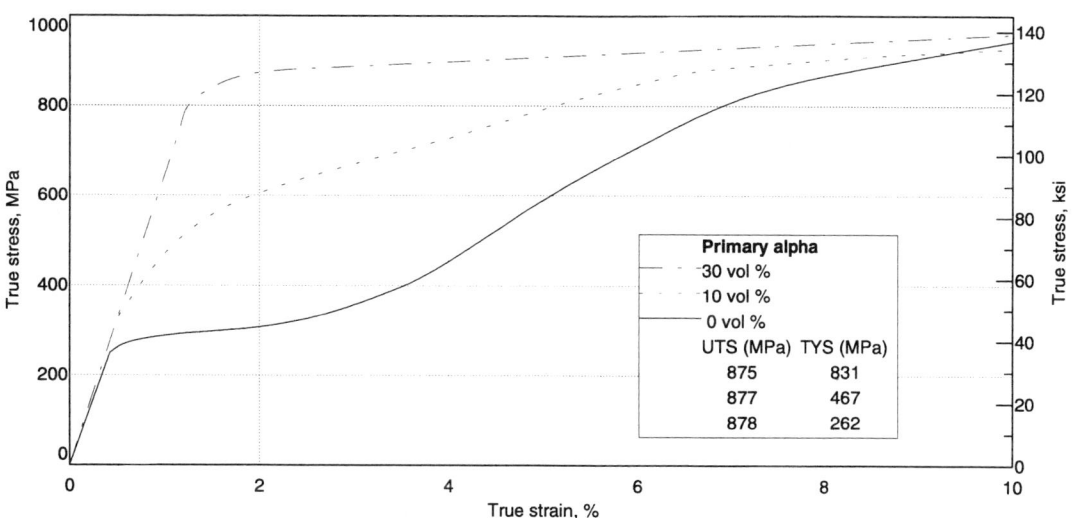

Increasing the amount of α increases the yield strength of unaged Ti-10V-2Fe-3Al, but does not affect the ultimate tensile strength.
The β transus was measured as 805 ± 3 °C (1480 °F), somewhat high compared to other heats. This probably reflects the oxygen (0.15 wt%) content, which is on the high side of the normal range. Treatments above 600 °C (1110 °F) were done by vacuum encapsulating specimens wrapped with tantalum foil. Below 600 °C, treatments were performed in a liquid nitrate salt bath. The strain rate was 0.00055 s^{-1}, and the tensile specimens were pulled with the rolling direction parallel to the tensile axis.
Source: T.W. Duerig, G.T. Terlinde, and J.C. Williams, Phase Transformations and Tensile Properties of Ti-10V-2Fe-3Al, *Metall. Trans. A*, Vol 11, Dec 1980, p 1987

Ti-10V-2Fe-3Al: Stress-strain at 790 °C for β and α + β processed material

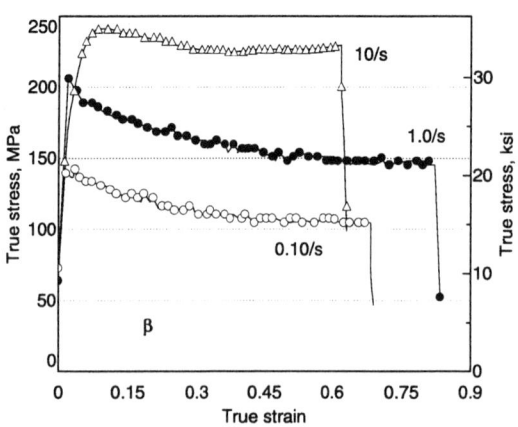
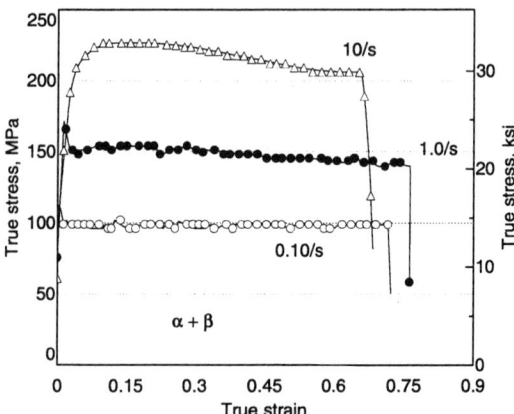

(a) Beta structure. (b) Alpha + beta structure.
Source: G.W. Kuhlman et al., *Sixth World Conference on Titanium*, P. Lacombe, R. Tricot, and G. Beranger, Ed., Les Editions de Physique, Paris, 1989, p 1269-1275

Flow Stress

Ti-10V-2Fe-3Al: Flow stress

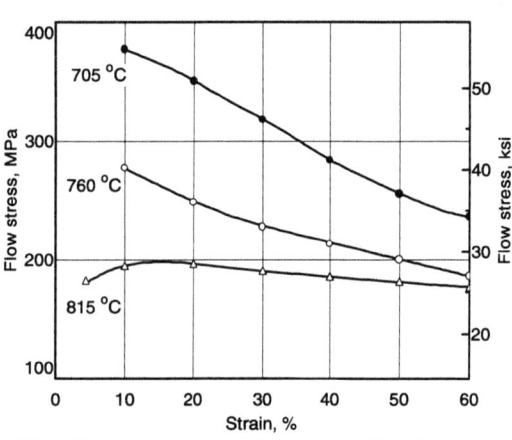

Effect of forging temperature on flow stress at 10/s strain rate.
Source: *Metals Handbook*, Vol 14, 9th ed., 1988

Ti-10V-2Fe-3Al: Flow stress

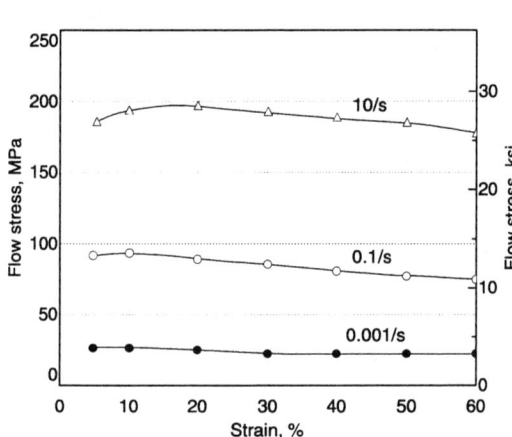

Effect of strain rate at 815 °C (1500 °F).
Source: *Metals Handbook*, Vol 14, 9th ed., 1988

Ti-10V-2Fe-3Al: Effect of microstructure on flow stress

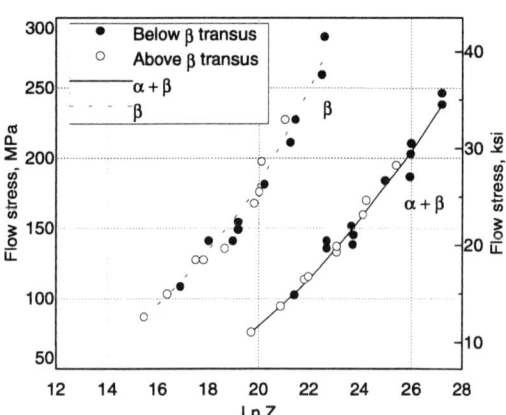

Source: G.W. Kuhlman et al., *Sixth World Conference on Titanium*, P. Lacombe, R. Tricot, and G. Beranger, Ed., Les Editions de Physique, Paris, 1989, p 1269-1275. LnZ is the temperature compensated strain rate defined by C.D. Zener and J.H. Hollaman, *J. Applied Physics*, Vol 15, 1944, p 22-32.

Effect of Strain Rate

Ti-10V-2Fe-3Al: Flow stress comparison at 815 °C

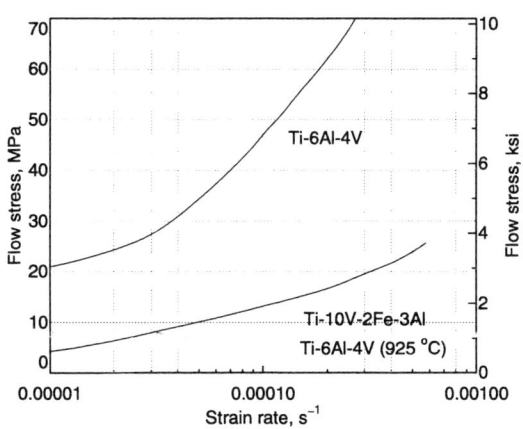

Source: C.C. Chen, "On the Forgeability of Hot Die Processed Ti-10V-2Fe-3Al Alloy Forgings," Wyman Gordon Report RD 75-118, Nov 1975. Top curve is for 6-4 at 815 °C

Ti-10V-2Fe-3Al: Flow stress vs strain rate comparison

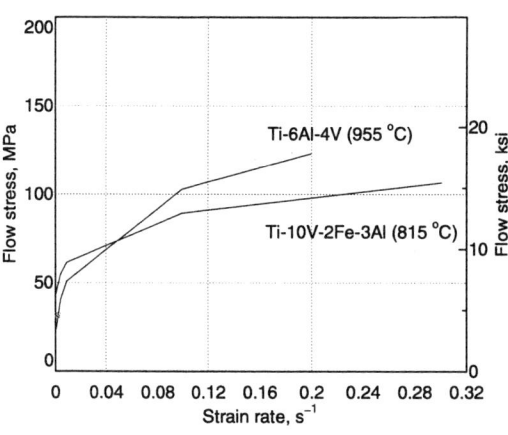

Ti-10V-2Fe-3Al: Flow stress vs temperature compensated strain rate (LnZ)

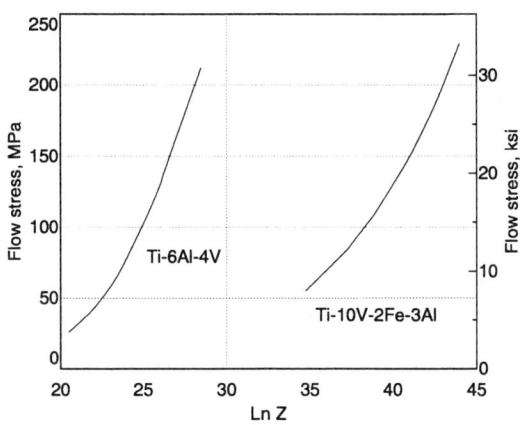

LnZ defined in *J. Applied Physics*, Vol 15, 1944, p 22-32.
Source: G.W. Kuhlman, et al, Sixth International

Ti-10V-2Fe-3Al: Flow stress vs strain rate

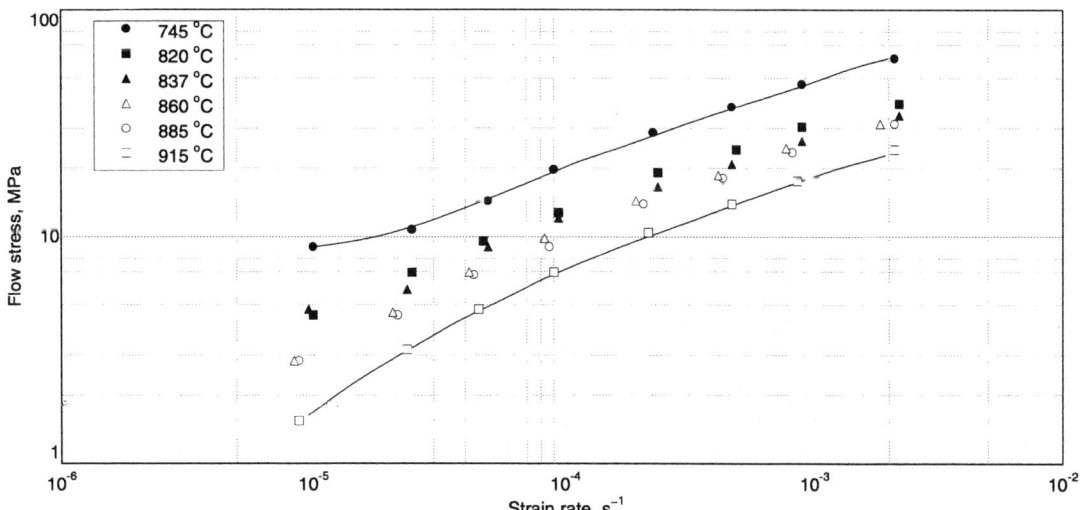

Source: G. Morgan and C. Hammond, Superplastic Deformation Properties of β-Ti Alloys, *Mater. Sci. Eng.*, Vol 86, 1987, p 159-177

Forging

G.W. Kuhlman, ALCOA, Forging Division

Ti-10V-2Fe-3Al is a deep-hardening, metastable, near β alloy that may be thermomechanically processed to a range of strength levels combined with excellent fracture toughness, whose primary commercial applications are forgings for aerospace components. The alloy can be fabricated into all forging product types, although closed die and precision forgings predominate. Ti-10V-2Fe-3Al is commercially fabricated on all types of forging equipment, although hydraulic presses are preferred for some products.

Ti-10V-2Fe-3Al is an extremely forgeable alloy, among the most forgeable of all titanium alloys, with lower unit pressures (flow stresses), improved forgeability, and significantly less crack sensitivity in forging than the α-β alloy Ti-6Al-4V. Ti-10V-2Fe-3Al forgings are supplied at three major strength/fracture toughness combination levels: 1190 MPa (173 ksi) ultimate tensile strength and 44 MPa√m (40 ksi√in.) fracture toughness (K_{Ic}); 1100 MPa (160 ksi) ultimate tensile strength and 60 MPa√m (55 ksi√in.) fracture toughness (K_{Ic}); and 965 MPa (140 ksi) ultimate tensile strength and 88 MPa√m (80 ksi√in.) fracture toughness (K_{Ic}). Forging thermomechanical processes have been commercially established to provide Ti-10V-2Fe-3Al products in a range of forging types to meet these properties.

The preferred forging process to meet the above mechanical-property criteria is controlled β forging followed by controlled α/β forging. This, in combination with final thermal treatment, provides the optimum combination of strength, ductility, toughness, fatigue, and fracture-related properties. Ti-10V-2Fe-3Al may be conventionally α + β forged and thermally treated. With such conventional processes, the alloy achieves high strength and fatigue properties and superior ductility, but poor toughness and fracture-related properties.

Ti-10V-2Fe-3Al: Forging process temperatures

Process	Metal temperature	
	°C	°F
Conventional forge	700-785	1300-1450
Beta forge	815-870	1500-1600

Note: See "Technical Note 4: Forging" for recommended die temperatures.

Final thermal treatments for forgings include two- and/or three-step subtransus solution treating, quenching, and aging. Forgings may be supplied in the annealed or solution and aged (STA) condition and/or overaged (STOA) conditions. In the STA condition, Ti-10V-2Fe-3Al has lower strengths, higher ductility, and toughness than in the STA condition. Solution treatment is conducted at 745 to 765 °C (1370 to 1410 °F) followed by air cooling or faster for very thin section precision forgings, with water and/or other quenchants (including polymers) for thicker section forgings. Aging is conducted at 480 to 525 °C (900 to 975 °F) for the highest strength; 510 to 540 °C (950 to 1000 °F) for the intermediate strength; and 565 to 620 °C (1050 to 1100 °F) for the lower strength. Some work (Ref 1, Kuhlman, and Ref 2, Chen) has found that, for Ti-10V-2Fe-3Al forgings meeting the highest strength requirement, a three-step heat treatment process improves the overall combination of strength, ductility, toughness, and fatigue. The process uses a solution anneal at 770 to 775 °C (1420 to 1430 °F), followed by air cool or faster prior to the above-noted solution treatment. Furthermore, other work (Ref 3, Kuhlman) has suggested that, when hot die or isothermal forging technologies are used on precision forgings with controlled post-forging cooling (quenches in appropriate media), direct aging at the above-noted aging temperatures achieves target mechanical properties and even further refined structures with enhanced high-cycle fatigue properties.

Processing

Beta forging working histories for Ti-10V-2Fe-3Al require imparting enough controlled levels of supra-transus and subtransus hot work to reach final macrostructure and microstructure objectives. Specifically, optimum β forging processes involve subtransus reductions in early preform stages followed by heavy β reductions (30 to 50%) in intermediate (blocker or first finishing) stages, followed by controlled final α + β work (10 to 25%). Low levels of β reduction in blocking are not recommended. High levels of α + β work in finish are also not recommended, because fracture toughness properties will be affected. The high-strength (1190 MPa, or 173 ksi) version of the alloy is particularly sensitive to the β forging process. For the lower strength version of the alloy (965 MPa, or 140 ksi), mechanical-property objectives have been met with forgings where all process steps are conducted supra-transus; however, reductions of 30 to 50% in each step are preferred.

Ti-10V-2Fe-3Al is suited for and successfully fabricated via isothermal or hot die forging into net or near-net precision forging shapes. The relatively low metal temperature permits use of less costly nickel- and cobalt-base superalloys as die materials. The alloy exhibits strain-rate sensitivity, and thus, unit pressure reductions and forging shape sophistication improvements through these technologies are realized. Furthermore, these technologies may provide improved control over forging process conditions including temperatures, levels of strain, and strain-rate histories, resulting in more uniform final forging products. Precision forgings of Ti-10V-2Fe-3Al are also manufactured on more conventional types of forging equipment.

Ti-10V-2Fe-3Al, as with all β alloys, has a higher affinity for hydrogen than other alloy classes. Although Ti-10V-2Fe-3Al forms less α case from heating operations than other alloy classes, therefore requiring less metal removal in chemical pickling (milling processes), control of chemical removal processes is essential to preclude excessive hydrogen pickup.

References

1. G.W. Kuhlman et al., Isothermal Forging of Beta and Near-Beta Alloys, *Beta Titanium Alloys in the 1980's*, Boyer and Rosenberg, Ed., TMS/AIME, 1984, p 255-280
2. C.C. Chen et al., Practical Considerations for Manufacturing High-Strength Ti-10V-2Fe-3Al Alloy Forgings, *J. Met.*, Vol 31(No. 7), July 1979, p 33-39
3. G.W. Kuhlman et al., LCF, Fracture Toughness, Fatigue/Fatigue Crack Propagation Resistance Optimization in Ti-10V-2Fe-3Al Alloy Through Microstructural Control, *Microstructure, Fracture Toughness and Fatigue Crack Growth Rates in Titanium Alloys*, Chakrabarti and Chesnutt, Ed., TMS/AIME, 1987, p 171-191

Thermomechanical Processing Effects

Unlike other β alloys, thermomechanical processing (TMP) of Ti-10V-2Fe-3Al achieves desired final microstructure through manipulation of α phase morphology. Microstructural objectives range from fully transformed, aged β structures to controlled amounts of elongated primary α in an aged β matrix, characterized by extremely fine secondary (aged) α. The latter microstructure is preferred for most aerospace applications (specifications) and forms the basis for most commercial use of the alloy in forgings. The highly refined microstructures achieved in Ti-10V-2Fe-3Al forgings are responsible for its superior strength and high-cycle fatigue properties and excellent fracture tough-

Ti-10V-2Fe-3Al: Mechanical properties of forgings

Heat treatment(a)	Tensile yield strength MPa	ksi	Ultimate tensile strength MPa	ksi	Elongation, %	Reduction of area, %	Fracture toughness (K_{Ic}) MPa√m	ksi√in.	Critical crack length(b) mm	in.	ΔK_{th}(c) MPa√m	ksi√in.	Smooth fatigue stress(d) MPa	ksi
STA	1100	160	1195	173	4	...	44	40	1.8	0.07	4.0	3.6	895	130
STA	1000	145	1100	160	6	15	60	54	4.1	0.16	4.1	3.7	860	125
STOA	895	130	965	140	8	20	88	80	10.7	0.42	4.3	3.9	830	120

(a) STA/STOA: single or duplex solution treated and aged or overaged. Forging process is β hot die forge or β block/α + β finish per AMS 4983, 4986, and 4987. (b) Critical crack length, ≈ 1.1 (K_{Ic}/YS)². (c) ΔK_{th} is the threshold stress-intensity in fatigue crack growth rate tests. (d) Smooth fatigue stress at 10^7 cycles, tests conducted at $R = 0.1$ to 0.3, $F = 30$ to 125 Hz.

Ti-10V-2Fe-3Al: Thermomechanical processing route for forgings

Forging process(a) Block	Finish	Solution heat treatment(a)	Quenching	Aging temperature(c), °C	Tensile yield strength MPa	ksi	Ultimate tensile strength MPa	ksi	Elongation, %	Reduction of area, %	Fracture toughness (K_{Ic}) MPa√m	ksi√in.
Conventional closed die forgings												
α/β	α/β	Single	WQ	510	1296	188	1365	198	10	28	25	22
β	α/β	Single	WQ	510	1158	168	1269	184	6	18	32	29
α/β	β	Single	WQ	510	1179	171	1317	191	3	6	46	42
β	β	Single	WQ	510	1268	184	1413	205	2	5	43	39
β	α/β	Duplex	AC/WQ	495	1213	176	1268	184	7	10	51	46
β	α/β	Single	WQ	540	1048	152	1144	166	11	24	73	66
β	α/β	Single	WQ	580	990	143	1049	152	14	33	99	90
Hot die/isothermal forging												
β(d)	β(e)	Single	WQ	485	1158	168	1269	184	7	15	51	46
β(d)	β(e)	Single	WQ	485	1220	177	1316	191	8	19	46	42
β(d)	β(e)	None	(f)	485	1172	170	1269	184	8	23	55	50
β(d)	α/β(e)	Single	WQ	485	1193	173	1289	187	8	16	46	42
β(d)	α/β(e)	Single	AC	485	1144	166	1254	182	9	21	55	50
β(d)	α/β(e)	None	(f)	485	1144	166	1260	183	9	20	64	58
α/β(d)	β(e)	Single	WQ	485	1199	174	1330	193	10	26	48	43
α/β(d)	β(e)	Single	AC	485	1158	168	1248	181	13	28	55	50
α/β(d)	β(e)	None	(f)	485	1207	175	1351	196	11	22	57	52
α/β(d)	α/β(e)	Single	WQ	485	1227	178	1365	198	9	21	48	43
α/β(d)	α/β(e)	Single	AC	485	1193	173	1303	189	13	30	52	47
α/β(d)	α/β(e)	None	(f)	485	1240	180	1372	199	13	34	56	51
β(d)	β(e)	Single	WQ	540	1034	150	1131	164	9	31	66	60
β(d)	β(e)	None	(f)	540	1069	155	1138	165	7	27	68	62
β(d)	β(e)	Single	WQ	590	938	136	1027	149	16	53	90	82
β(d)	β(e)	None	(f)	590	958	139	1006	146	16	52	89	81

(a) Block is the initial process; finish is the final process in closed die forging. (b) Single: Solution treatment at β transus 30 °C plus quench. Duplex: Solution anneal at β transus 25 °C, AC prior to solution treatment. (c) 8-h aging treatment. (d) Metal temperature. (e) Die temperature. (f) Controlled cooling rate plus direct aging

ness and low-cycle fatigue properties.

In finalizing commercial thermomechanical processing routes and property capabilities of Ti-10V-2Fe-3Al, a range of thermomechanical processing options including warm and hot die forging process conditions, thermal treatment processes, etc., have been evaluated (Ref 1-5). Much of this work attempted to optimize strength-ductility-toughness property combinations for each strength level, but also to enhance fatigue-crack growth resistance and high-cycle fatigue strength (see table). Thermomechanical processing requirements to meet these high-strength minimum properties, including fracture toughness and ductility, proved to be a significant challenge, because these properties are at the upper limit of the alloy capability. Initially, four possible forging process combinations and two-step heat treatments were evaluated. β forging processes enhanced toughness, but at a sacrifice in ductility, whereas α + β forging processes enhanced ductility and reduced toughness. Thermomechanical processing for high-strength closed die forgings emerged from extensive studies (Ref 1, 4) that combined controlled final α + β working with a three-step heat treatment. The first step solution anneal achieved the desired combination of equiaxed primary α for ductility and elongated α for toughness, as well as enhanced forging structural uniformity. This thermomechanical processing, when combined with final solution treatment and aging, consistently achieved the desired strength-ductility-toughness combinations.

Forging history in the thermomechanical processing of the low and intermediate strength levels for Ti-10V-2Fe-3Al is less critical than for the high-strength version. However, optimal property combinations are achieved with similar forging approaches. Subsequently, some work (Ref 2) has shown that, for the intermediate strength level, working history (e.g., level of α + β working) has a significant impact on fracture toughness and notched fatigue properties.

References

1. G.W. Kuhlman and T.B. Gurganus, *Met. Prog.*, Vol 118 (No. 2), July 1980, p 30-35
2. R.R. Boyer and G.W. Kuhlman, *Metall. Trans. A*, Vol 18, Dec 1987, p 2095-2103
3. G.W. Kuhlman and R. Pishko, *Titanium, Science and Technology*, Vol 1, G. Lütjering, U. Zwicker, and W. Bunk, Ed., Deutsche Gesellschaft für Metallkunde, Germany, 1985, p 469-476
4. C.C. Chen and R.R. Boyer, *J. Met.*, July 1979
5. A.K. Chakrabarti, M. Brun, D. Fournier, and G.W. Kuhlman, *Proceedings Sixth International Conference on Titanium*, Vol 3, P. Lacombe, R. Tricot, and G. Beranger, Ed., Les Editions de Physique, Paris, 1988, p 1339-1344

Ti-10V-2Fe-3Al: Effect of α/β work on toughness

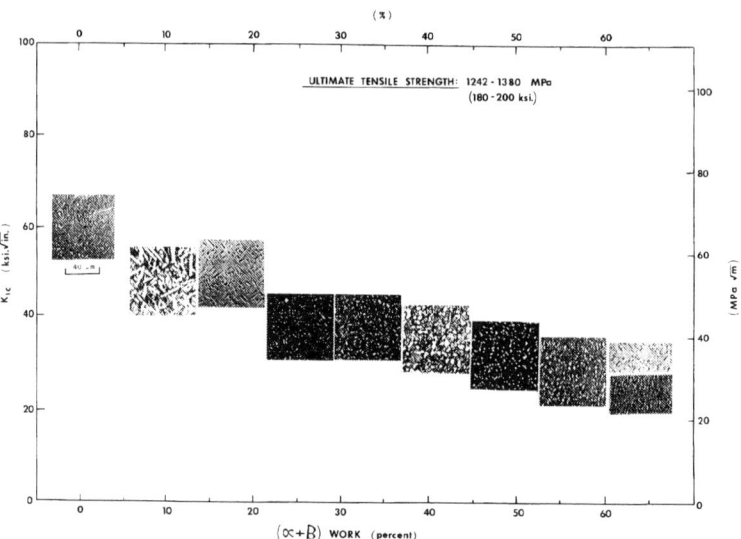

Heat Treatment

Ti-10V-2Fe-3Al is used in the solution treated and aged condition. In general, aged strength can be increased by increasing the solution temperature or decreasing the aging temperature. There are applications in which it may be desirable to create a stable microstructure that will not change greatly with service exposure at intermediate temperatures. One way to achieve this is to solution treat, followed by furnace cooling to an overaging temperature, overaging, and air cooling. A final long-time (48 h) age at 480 °C (900 °F) would further stabilize the alloy. The aging temperature should be at least 475 °C (885 °F) to avoid the formation of ω phase.

Ti-10V-2Fe-3Al: Typical heat treatments

Heat treatment	Temperature °C	Temperature °F	Time, h	Cooling method
Stress relief	657-700(a)	1250-1300(a)	0.5-2	Air or slow cool
Beta	1500	815	...	...
Solution treating range	730-775	1325-1425	1	WQ
Aging range	480-620(b)	900-1100(b)	8	AC
Overage	580-620	1075-1150	8	AC

(a) Aging at 480 °C (900 °F) and above also effectively stress relieves. (b) Aging temperature should be at least 475 °C (885 °F) to avoid formation of ω phase. For high to intermediate strength, the most useful aging temperatures are 480 to 540 °C (900 to 1000 °F), respectively.

Ti-10V-2Fe-3Al: Heat treating schedules for forgings

Solution treat and age	750 to 765 °C (1385 to 1410 °F), 1 h, WQ
	480 to 495 °C (900 to 950 °F), 8 , AC
Solution treat and overage	730 °C (1350 °F), 1 h, AC
	580 to 595 °C (1075 to 1100 °F), 8 h, AC
Stabilize	765 °C (1400 °F), 1 h, FC to 565 °C (1050 °F)
	565 °C (1050 °F), 8 h, AC
	480 °C (900 °F), 48 h, AC (optional)
Beta anneal and overage	815 °C (1500 °F), 1 h, AC
	620 °C (1150 °F), 8 h, AC

Source: *Beta Titanium Alloys in the 80's*, R.R. Boyer and H.W. Rosenberg, Ed., TMS/AIME, 1984

Ti-10V-2Fe-3Al: Effect of solution treatment on tensile properties

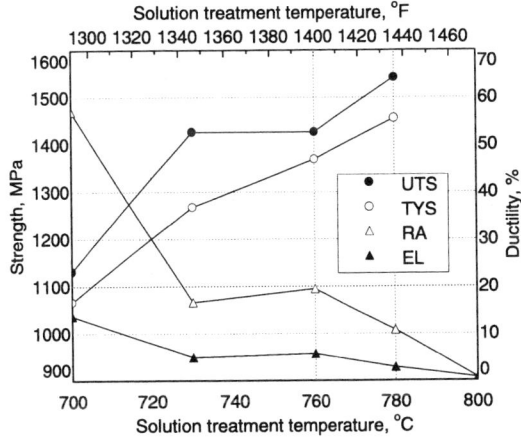

Typical properties after solution treating for 8 h at various temperatures and aging at 500 °C (930 °F) for 60 min.
Source: *Beta Titanium Alloys in the 80's*, R.R. Boyer and H.W. Rosenberg, Ed., TMS/AIME, 1984, p 25

Ti-10V-2Fe-3Al: Effect of aging on UTS

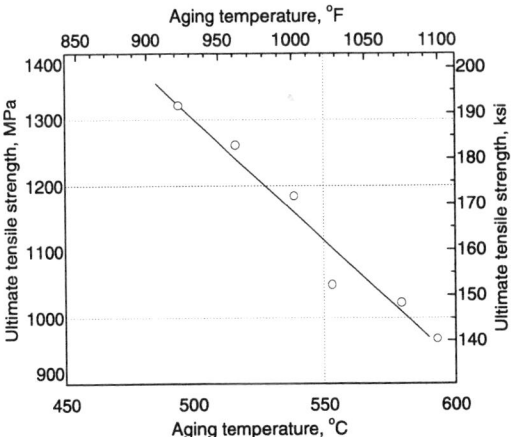

All specimens were solution treated at 750 °C (1385 °F) followed by water quenching. Aging time was 8 h.
Source: *Metall. Trans. A.*, Vol 18, 1987, p 2095

Ti-10V-2Fe-3Al: Heat treatment per AMS specifications

Specification	Solution treatment(a) Temperature below transus Δ °C	Δ °F	Aging temperature(b) °C	°F	Min UTS MPa	ksi
AMS 4984	15-40 (c)	60-100	480-510	900-950	1190	173
AMS 4986	15-40 (c)	60-100	510-540	950-1000	1100	160
AMS 4987	15-40 (c)	60-100	565-620	1050-1150	965	140
AMS 4983A	...	...	480-510	900-950	1240	180

(a) Hold for a minimum of 30 min. (b) Hold within ± 5 °C (10 °F) for not less than 8 h. (c) Conversions for Δ °C per specification

Ti-13V-11Cr-3Al

Common Name: Ti-13-11-3
UNS Number: R58010

Ti-13V-11Cr-3Al, developed by Rem Cru (later called Crucible Steel) in the mid-1950s, was the first alloy to stabilize the beta phase of titanium to room temperature. Thus Ti-13-11-3 for titanium alloys is analogous to austenitic 18-8 stainless steel for iron alloys. Its primary use over the last decade has been for springs.

Ti-13-11-3, also known as B120VCA, is suitable for operation in the range from –54 to +315 °C (–65 to +600 °F) and higher in certain uses. It possesses good ductility for ease of fabrication. When aged, it can be hardened to extremely high strength levels. Its high strength and low density make Ti-13-11-3 one of the most efficient structural materials available. It was used for airborne structures that must sustain temperatures up to 650 °C (1200 °F) for short periods of time and for lightweight pressure vessels that operate at temperatures from –54 to +315 °C (–65 to +600 °F) and is also used as a high-strength fastener material having cold headability and shear strengths over 825 MPa (120 ksi).

Chemistry and Density

As a solute-rich beta alloy, Ti-13-11-3 contains relatively large amounts of beta-stabilizer elements and relatively small amounts of alpha-stabilizer elements. At the nominal composition, Ti-13V-3Al base plus 11 percent chromium and low oxygen content, the alloy is hypoeutectoidal. The high vanadium content contributes to the stabilization of the beta phase, but it does not contribute to the titanium-chromium eutectoid relationship.

Density. 4.82 g/cm^3 (0.174 lb/in.^{3})

Product Forms

Although it is used in limited quantities today, Ti-13-11-3 is still used for some airframe sheet-metal applications and for springs. Wrought forms include billet, bar, plate, sheet, and wire.

Product Condition/Microstructure

Ti-13V-11Cr-3Al is heat treatable to high strength; is cold rollable; and can be solution treated without distortion because it can be air cooled. It is highly ductile in the solution-treated condition and can be severely cold worked without intermittent annealing. It has little directionality in sheet and strip, low scratch sensitivity, and a relatively high tolerance for hydrogen. The final combination of strength and ductility is controlled over a wide range by selection of heat treating temperature and time.

Applications

For several years, Ti-13-11-3 was the only beta titanium alloy of commercial significance, until the advent of alloys such as Ti-10-2-3, Ti-15-3, and Beta C.

Ti-13-11-3 is used for missile applications such as solid rocket motor cases where extremely high strengths are required for short periods of time, and for other structural applications in advanced manned and unmanned airborne systems. It is also used for springs for airframe applications, solid rocket pressure chambers, airframe components, welded pressure vessels, fasteners, and bonded and brazed honeycomb cover sheets.

Ti-13V-11Cr-3Al: Specifications

Specification	Designation	Description	Al	C	Cr	Fe	H	N	O	V	Other
	UNS R58010		3		11					13	bal Ti
Russia											
	IMP-10		3		11					13	bal Ti
Spain											
UNE 38-729	L-7701	Sh Str Plt Wir Bar Ann	2.5-3.5	0.05	10-12	0.35	0.02	0.05	0.18	12.5-14.5	OT 0.4; bal Ti
UNE 38-729	L-7701	Sh Strp Plt Wir Bar HT	2.5-3.5	0.05	10-12	0.35	0.02	0.05	0.18	12.5-14.5	OT 0.4; bal Ti
USA											
AMS 4917D		Sh Str Plt SHT	2.5-3.5	0.05	10-12	0.35	0.025	0.05	0.17	12.5-14.5	OT 0.4; bal Ti
AMS 4917D		Sh Str Plt STA	2.5-3.5	0.05	10-12	0.35	0.025	0.05	0.17	12.5-14.5	OT 0.4; bal Ti
AMS 4959B			2.5-3.5	0.05	10-12	0.35	0.03	0.05	0.17	12.5-14.5	OT 0.4; Y 0.005; bal Ti
AWS A5.16-70	ERTi-13V-11Cr-3Al	Wir Rod	2.5-3.5	0.05	10-12	0.25	0.008	0.03	0.12	12.5-14.5	bal Ti
MIL F-83142A	Comp 12	Frg Ann	2.5-3.5	0.05	10-12	0.35	0.025	0.05	0.17	12.5-14.5	OT 0.4; bal Ti
MIL F-83142A	Comp 12	Frg HT	2.5-3.5	0.05	10-12	0.35	0.025	0.05	0.17	12.5-14.5	OT 0.4; bal Ti
MIL T-9046J	Code B-1	Sh Str Plt SHT	2.5-3.5	0.05	10-12	0.15-0.35	0.025	0.05	0.17	12.5-14.5	OT 0.4; bal Ti

(continued)

Ti-13V-11Cr-3Al: Specifications (continued)

Specification	Designation	Description	Composition, wt%								Other
			Al	C	Cr	Fe	H	N	O	V	
USA (continued)											
MIL T-9046J	Code B-1	Sh Str Plt STA	2.5-3.5	0.05	10-12	0.15-0.35	0.025	0.05	0.17	12.5-14.5	OT 0.4; bal Ti
MIL T-9047G	Ti-13V-11Cr-3Al	Bar Bil SHT	2.5-3.5	0.05	10-12	0.35	0.025	0.05	0.17	12.5-14.5	OT 0.4; Y 0.005; bal Ti
MIL T-9047G	Ti-13V-11Cr-3Al	Bar Bil STA	2.5-3.5	0.05	10-12	0.35	0.025	0.05	0.17	12.5-14.5	OT 0.4; Y 0.005; bal Ti

Ti-13V-11Cr-3Al: Commercial compositions

Specification	Designation	Description	Composition,%								Other
			Al	C	Cr	Fe	H	N	O	V	
Japan											
Kobe	KS13-11-3	Bar Frg SHT	2.5-3.5		10-12	0.35	0.025	0.05	0.17	12.5-14.5	bal Ti
Kobe	KS13-11-3	Bar Frg STA	2.5-3.5		10-12	0.35	0.025	0.05	0.17	12.5-14.5	bal Ti
USA											
Astro	Ti-13V-11Cr-3Al	Bar Bil Sprg	2.5-3.5	0.05 max	10-12	0.35 max	0.025	0.05 max	0.17	12.5-14.5	bal Ti
RMI	13V-11Cr-3Al										
Teledyne	Tel-Ti-13V-11Cr-3Al										
Timet	TIMETAL 13-11-3	Ing Plt Sh Str STA	2.5-3.5	0.05	10-12	0.35	0.025	0.05	0.17	12.5-14.5	bal Ti

Phases and Structures

Strengthening of Ti-13V-11Cr-3Al occurs from the precipitation of $TiCr_2$ and α in solute-lean β regions (β'). As might be expected, long periods at solution treatment temperatures result in undesirable grain growth and the associated breakdown of favorable nucleation sites for strengthening precipitates.

Beta Transus. At nominal alloying concentrations and commercial-grade oxygen contents of 0.15 wt% O_2, Ti-13V-11Cr-3Al has a β transus of about 700 °C (1300 °F). For lower oxygen concentrations (0.05 wt%), the β transus is lowered to about 650 °C (1220 °F).

Ti-13V-11Cr-3Al: Phase diagram with variable chromium content

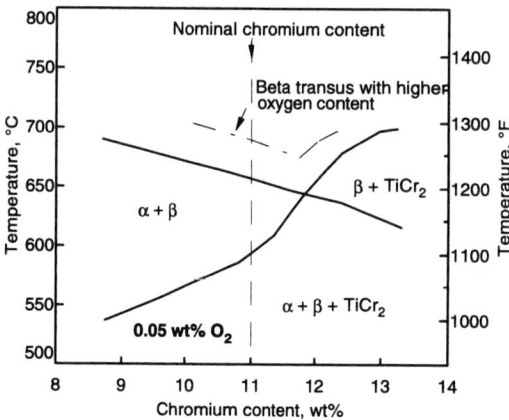

Source: R.A. Wood and R.J. Favor, *Titanium Alloys Handbook*, MCIC-HB 02, Battelle Columbus Laboratories, 1972

Ti-13V-11Cr-3Al: Phase diagram with variable aluminum content

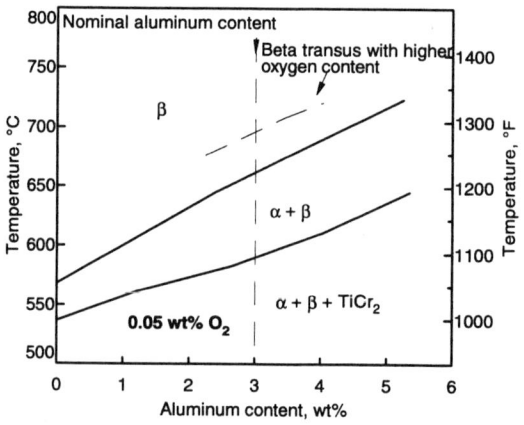

Source: R.A. Wood and R.J. Favor, *Titanium Alloys Handbook*, MCIC-HB 02, Battelle Columbus Laboratories, 1972

Transformation Products

Normal air cooling from above the β transus retains a room-temperature β structure that is metastable. However, because the decomposition of metastable β is so sluggish in this solute-rich β al-

Ti-13V-11Cr-3Al: Grain size

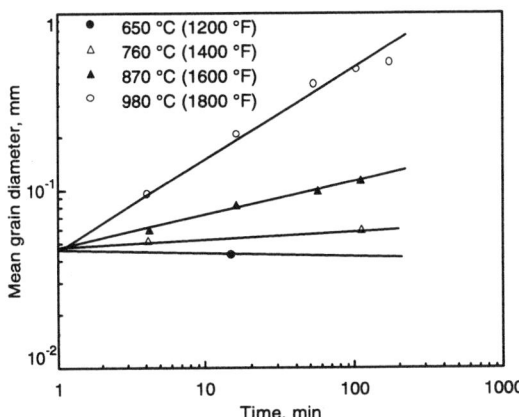

Source: R.A. Wood and R.J. Favor, *Titanium Alloys Handbook*, MCIC-HB 02, Battelle Columbus Laboratories, 1972

Ti-13V-11Cr-3Al: TTT diagram

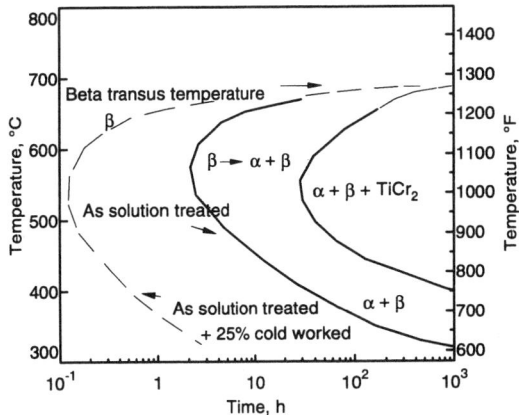

Source: R.A. Wood and R.J. Favor, *Titanium Alloys Handbook*, MCIC-HB 02, Battelle Columbus Laboratories, 1972

loy, decomposition of β below about 315 °C (600 °F) is essentially nonexistent for practical purposes. Deformation can accelerate the decomposition of β, however (see the TTT diagram).

As a solute-rich β alloy, Ti-13V-11Cr-3Al is also more susceptible to the formation of grain boundary α during aging. However, prior deformation can enhance the formation of intragranular α and reduce the extent of grain boundary α. Like other solute-rich β alloys, athermal ω does not form in Ti-13V-11Cr-3Al during quenching. Conditions for the formation of isothermal ω have not been defined.

Physical Properties

Ti-13V-11Cr-3Al: Summary of typical physical properties

Beta transus	~700 °C (1300 °F)
Melting (liquidus) point	Not available
Density(a)	4.82 g/cm³ (0.174 lb/in.³)
Electrical resistivity(a)	1.4 μΩ · m (STA condition)
Magnetic permeability	Nonmagnetic
Specific heat capacity(a)	545 J/kg · K (0.13 Btu/lb · °F)
Thermal conductivity(a)	6.9 W/m · K (4.0 Btu/ft · h · °F)
Thermal coefficient of linear expansion(b)	9.67×10^{-6}/°C (5.37×10^{-6}/°F)

(a) Typical values at room temperature of about 20 to 25 °C (68 to 78 °F). (b) Mean coefficient from room temperature to 95 °C (200 °F)

Ti-13V-11Cr-3Al: Elastic modulus in tension

Temperature		Tensile modulus(a)			
		Solution treated material		Aged material	
°C	°F	GPa	10⁶ psi	GPa	10⁶ psi
−54	−65	102	14.8	112	16.2
21	70	100	14.5(b)	107(b)	15.5(b)
21	70	101	14.7	110	16.0
205	400	96.5	14.0	107	15.5
315	600	91.0	13.2	103	15.0
425	800	85.5	12.4	99.3	14.4
540	1000	80.0	11.6	94.5	13.7

(a) Typical shear modulus is 436 GPa (6.2×10^6 psi) at room temperature. (b) Design modulus from MIL-HDBK 5. Source: *Mater. Eng.*, Dec 1987, p 112; and *Metals Handbook, Properties and Selection: Stainless Steels, Tool Materials, and Special Purpose Materials*, Vol 3, 9th ed., American Society for Metals, 1980. Additional data on elastic properties are contained in MIL-HDBK 5 and the *Aerospace Structural Metals Handbook*.

Electrical Resistivity

Ti-13V-11Cr-3Al: Electrical resistivity

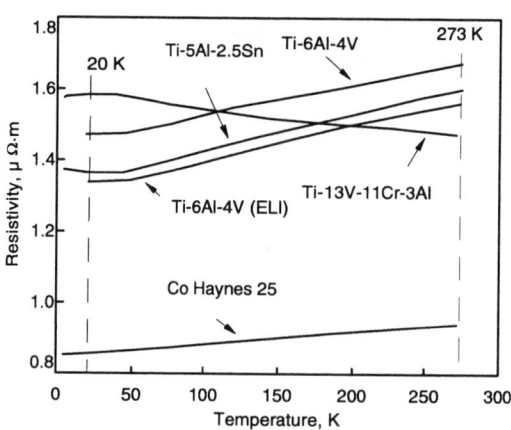

Source: *Cryogenics*, Vol 10, 1970, p 295

Ti-13V-11Cr-3Al: Electrical resistivity

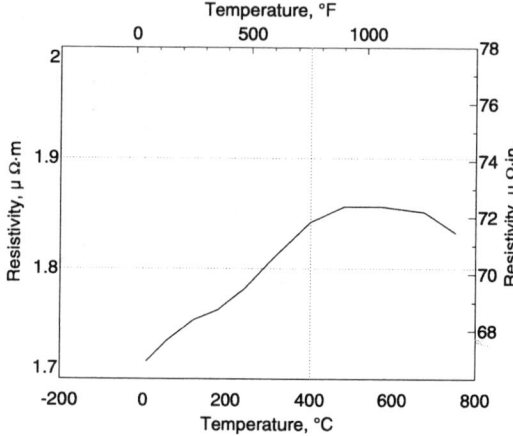

Source: *Aircraft Designer's Handbook for Titanium and Titanium Alloys*, AFML-TR-67-142, Mar 1967, reported in R. Wood, *Beta Titanium Alloys*, MCIC 72-11, Battelle Columbus Laboratories, 1972, p 34

Chemical/Corrosion Properties

The chemical reactivity of Ti-13V-11Cr-3Al is fairly typical of other titanium alloys.

In either the solution treated or aged condition, it is corrosion resistant to seawater, salt and other natural environments, oxidizing media, inhibited reducing acids, alkalies, and metallic chlorides at room temperature. In salt-spray tests, aged Ti-13V-11Cr-3Al exhibits no pitting and experiences no general corrosion or degradation in mechanical properties. Its corrosion resistance in reducing environments appears to be less than other titanium alloys (see figure). In hot air, however, Ti-13V-11Cr-3Al does not appear to discolor and scale as badly as other titanium alloys at temperatures 260 to 315 °C (500 to 600 °F). Little difference among Ti-13-11-3 and other alloys is noted in terms of discoloration and scaling at higher temperatures.

Ti-13V-11Cr-3Al: Corrosion comparison in HCl

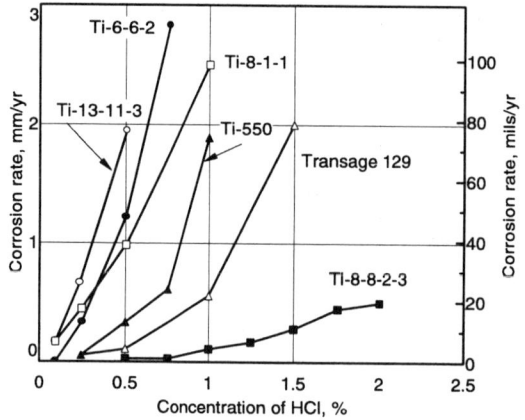

General corrosion of annealed titanium alloys in naturally aerated HCl solutions.
Source: *Metals Handbook, Corrosion*, 9th ed., Vol 13, ASM International, 1987, p 680

Ti-13V-11Cr-3Al: Depth of scale after solution treatment

Solution heat treatment	Hardness(a), HV	Maximum scale depth(b)	
		mm	in.
1 h 870 °C (1600 °F), AC	306	0.03	0.0013
2 h 870 °C (1600 °F), AC	318	0.05	0.0022
4 h 870 °C (1600 °F), AC	349	0.08	0.0033
1 h 925 °C (1700 °F), AC	304	0.06	0.0024
2 h 925 °C (1700 °F), AC	347	0.04	0.0017
4 h 925 °C (1700 °F), AC	296	0.09	0.0038
1 h 980 °C (1800 °F), AC	366	0.06	0.0025
1 h 980 °C (1800 °F), FC	344	0.10	0.0047

(a) Average, base metal away from scale, 20-kg load. (b) Determined by microscopic examination of cross section. Source: R.A. Wood, *Beta Titanium Alloys*, MCIC 72-11, Battelle Columbus Laboratories, 1972

Stress-Corrosion Cracking

Like most titanium alloys, Ti-13V-11Cr-3Al can be fairly resistant to aqueous halide stress-corrosion cracking (SCC) when used in its standard metallurgical condition. However, step-cooled Ti-13V-11Cr-3Al is highly susceptible to SCC in neutral salt solutions, in that loading of smooth samples can show cracking susceptibility. Typically, a stress riser (fatigue crack or notch) is needed to initiate SCC susceptibility of titanium alloys.

In β titanium alloys, the β phase may be susceptible to either transgranular or intergranular SCC in aqueous media, depending on alloy composition and microstructure. Intergranular cracking has only been observed in a few aged β alloys, particularly with fine α precipitates formed at lower aging temperatures. Transgranular cleavage of β phase is known to occur in solution treated Ti-13V-11Cr-3Al (R.J.H. Wanhill, Aqueous Stress Corrosion in Titanium Alloys, *Brit. Corrosion J.*, Vol 10(No. 2), 1975, p 69-78). Stress-corrosion cracking susceptibility in this case is mitigated by decreasing the grain size and mean free path of the β phase.

Ti-13V-11Cr-3Al: Environments known to promote cracking

	Environment		Other titanium alloys with
	Temperature		
Medium	°C	°F	known susceptibility
Organic compounds			
Methyl alcohol (anhydrous)	RT		Ti-6Al-4V, Gr. 2 Ti, Gr. 4 Ti, Ti-4Al-3Mo-1V, Ti-8V-3Al-6Cr-4Zr-4Mo (Beta C), Ti-8Al-1Mo-1V, Ti-5Al-2.5Sn
Methyl chloroform	370	700	Ti-8Al-1Mo-1V, Ti-6Al-4V, Ti-5Al-2.5Sn
Trichlorofluoroethane	790	1455	Ti-8Al-1Mo-1V, Ti-5Al-2.5Sn, Ti-6Al-4V
Salts			
Chloride and other halide salts/residues	230-430	445-805	Most commercial alloys except grade 1, 2, 7, 11, 12, and 9
Seawater/NaCl solution(a)	RT		Unalloyed Ti (with oxygen content >0.3%) Ti-2.5Al-1Mo-11Sn-5Zr-0.2Si (IMI-679), Ti-5Al-2.5Sn, Ti-8Mn, Ti-6Al-4V, Ti-6Al-6V-2Sn, Ti-6Al-2Nb-1Ta, Ti-4Al-3Mo-1V, Ti-8Al-1Mo-1V, Ti-6Al-2Sn-4Zr-6Mo
Metal embrittlement			
Cadmium (solid + liquid)	25-600	75-1110	Ti-8Mn, grade 2, Ti-6Al-4V
Mercury (liquid)	370	700	Gr. 4 Ti, Ti-6Al-4V, Ti-8Al-1Mo-1V

Source: R. Schutz, Stress-Corrosion Cracking of Titanium Alloys, in *Stress-Corrosion Cracking: Materials Performance and Evaluation*, ASM International, 1992. (a) Smooth sample susceptibility for step cooled Ti-13-11-3

Aqueous Media

Like other susceptible titanium alloys, the SCC of Ti-13V-11Cr-3Al in aqueous media is also influenced by the type and concentration of species in solution, the pH, temperature, and viscosity of the solution, and the metal potential in the solution. The general effect of these environmental factors is discussed in "Technical Note 2: Corrosion" in this Volume. Like most SCC-susceptible titanium alloys, anodic or cathodic polarization tends to inhibit SCC and increase K_{Iscc} values for Ti-13V-11Cr-3Al (see figure). Increasing potential also increases cracking velocity in neutral halide solutions, but not in highly acidic solutions. Cracking (K_{Iscc}) thresholds in 3.5% NaCl solutions at room temperature range from about 28 to 44 MPa$\sqrt{m}$ (26 to 40 ksi$\sqrt{in.}$).

Ti-13V-11Cr-3Al: Fracture toughness in air and 3.5% NaCl solution at 25 °C

Thickness		Yield strength		K_{Ic} or K_t		K_{Iscc} or K_{scc}	
mm	in.	MPa	ksi	MPa$\sqrt{m}$	ksi$\sqrt{in.}$	MPa$\sqrt{m}$	ksi$\sqrt{in.}$
3.3	0.13	882(ST)	128	97	89	28	26
13	0.50	827(ST)	120	>110	>100	38	35
		1103(STA)	160	77	70	33	30
5	0.20	1055(STA)	153	71	65	35	32

Specimens were ST or STA as shown. Data were generated in ambient neutral 3.5% NaCl solution. It should be cautioned that these K_{Iscc} values are highly dependent on alloy composition, metallurgical condition, and product form and thickness and, therefore, may or may not be representative of alloy commercially available product materials. Source: R. Schutz, Stress-Corrosion Cracking of Titanium Alloys, in *Stress-Corrosion Cracking: Materials Performance and Evaluation*, ASM International, 1992

Ti-13V-11Cr-3Al: Effect of potential on aqueous SCC

Within a narrow potential region (typically centered around –500 mV SCE), K_{Iscc} reaches a minimum value that depends on the alloy and its metallurgical condition.
Source: T.R. Beck, M.J. Blackburn, W.H. Smyrl, and M.P. Speidel, "Stress-Corrosion Cracking of Titanium Alloys: Electrochemical Kinetics, SCC Studies With Ti: 8-1-1, SCC and Polarization Curves in Molten Salts, Liquid Metal Embrittlement, and SCC Studies With Other Titanium Alloys," Quarterly Progress Report 14, Contract NAS 7-489, Boeing Scientific Research Laboratories, Dec 1969

Methanol and Other Alcohols

Unlike other engineering metals, titanium and zirconium alloys are unique in their strong susceptibility to SCC in methanol liquids and vapor. Transgranular SCC generally is observed in those titanium alloys that are susceptible to SCC in aqueous solutions. However, intergranular failure in methanol is observed primarily in Ti-13V-11Cr-3Al. This fracture mode is also commonly observed in almost all titanium alloys during stage I cracking at low stress levels. Intergranular cracking in methanol generally involves anodic dissolution and requires little or no stress to propagate. Application of stress accelerates cracking, but this mode of propagation remains independent of stress level.

For both intergranular and transgranular SCC in methanol, halogen/halide additions accelerate cracking, whereas water has an inhibitive effect. Stress-corrosion cracking is generally difficult to observe in absolutely pure methanol, but becomes favored at HCl levels above 10^{-6} M. The minimum level of water required for full SCC inhibition (about 2 to 3 wt% for Ti-13-11-3) depends on alloy composition, metallurgical condition, temperature, halide level, acidity, and other species present.

Other Alcohols. The SCC susceptibility is limited primarily to methanol for most commercial titanium alloys such as Ti-13V-11Cr-3Al. However, the addition of halogens, such Cl_2, Br_2, or I_2, or other (nonoxygen containing) strong oxidizers (i.e., $FeCl_3$) to various anhydrous alcohols can induce SCC in all titanium alloys, even the unalloyed grades. Depending on the alloy and on the oxidizer concentration, much higher water levels are required for SCC inhibition.

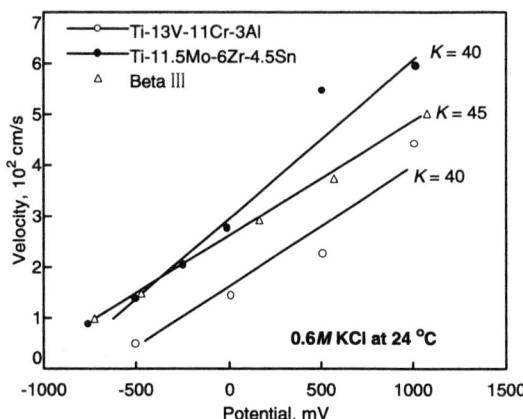

Ti-13V-11Cr-3Al: Effect of potential on crack velocity

Although the increase of potential serves to increase cracking velocity in this neutral halide solution, stage II crack velocity becomes independent of applied potential in highly acidic solutions. As a result, inhibition via cathodic polarization is not achievable in highly acidic solutions.
Source: R. Schutz, Stress-Corrosion Cracking of Titanium Alloys, in *Stress Corrosion Cracking: Materials Performance and Evaluation*, ASM International, 1992

Hot-Salt Cracking

Ti-13V-11Cr-3Al has intermediate susceptibility to hot-salt SCC, but the alloy is not considered quite as susceptible to salt corrosion as Ti-6Al-4V or Ti-8Al-1Mo-1V. Early NASA experiments (NASA TN D-2011, Dec 1963) indicate that most salt-exposed material in the aged condition will crack with a 285 °C (550 °F), 689 MPa (100 ksi) exposure during a 4000- to 6000-h run. Other Lockheed salt tests at 260 °C (500 °F) have indicated a pronounced stress-corrosion susceptibility of both annealed and aged Ti-13V-11Cr-3Al when exposed in tests simulating various fastening techniques (e.g., rivet, screw, spot, and fusion welds). Some susceptibility also was indicated on material worked by bending. Although the stress levels that promote stress corrosion are undefined in these tests, it is apparent that operations resulting in high residual stresses promote susceptibility to the salt-cracking phenomenon at 260 °C (500 °F) in less than 2000 h, and in most fastener applications in less than 100 h (R. Woods, *Beta Titanium Alloys*, MCIC 72-11, Battelle Columbus Laboratories, 1972, p 67).

Thermal Properties

Heat Capacity

Specific Heat. 545 J/kg · K (0.13 Btu/lb · °F) at room temperature.

Ti-13V-11Cr-3Al: Specific heat at low temperatures

Temperature, K	Specific heat	
	J/kg · K	Btu/lb · °F
20	12	0.00291
25	22	0.00528
30	35	0.00839
40	69	0.01663
50	112	0.02680
60	157	0.03754
70	200	0.04780
80	240	0.05737
90	276	0.06607
100	309	0.07381
120	361	0.08632
140	400	0.09555
160	430	0.1029
180	454	0.1085
200	473	0.1131
220	488	0.1167
240	500	0.1196
260	510	0.1219
280	518	0.1239
300	526	0.1258

Note: Specimens were solution treated at 785 °C (1450 °F) for 20 min, air cooled, then aged at 480 °C (900 °F) for 60 h, air cooled. Composition: 3.5 wt% Al, 0.04 wt% C, 10.4 wt% Cr, 0.25 wt% Fe, 114 ppm H, and 13.9 wt% V. Approximately 50 measurements were made in the temperature range 21 to 300 °K with temperature increments of 2 to 7°. Four temperature regulating baths were used to cover the desired temperature range: liquid hydrogen, liquid nitrogen, solid carbon dioxide/ethanol, and ice water. Measurements within a given bath were taken to overlap measurements from the adjacent temperature range. Source: W. Ziegler, J. Millins, and S. Hwa, Specific Heat and Thermal Conductivity of Four Commercial Titanium Alloys, *Advances in Cryogenic Engineering*, Vol 8, Plenum Press, 1963, p 268

Ti-13V-11Cr-3Al: Specific heat

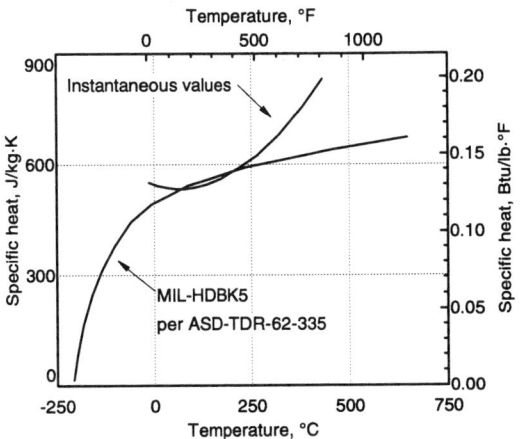

Source: R. Wood and H. Ogden, "The All-Beta Titanium Alloy (Ti-13V-11Cr-3Al)," DMIC Report 110, Defense Metals Information Center, Apr 1959

Ti-13V-11Cr-3Al: Instantaneous specific heat

Temperature		Specific heat	
°C	°F	J/kg · K	Btu/lb · °F
0	32	548	0.131
38	100	536	0.128
93	200	531	0.127
205	400	573	0.137
315	600	669	0.160
425	800	833	0.199

Source: R.A Wood and H.R. Ogden, "The All-Beta Titanium Alloy (Ti-13V-11Cr-3Al)," Defense Metals Information Center, DMIC Report 110, 17 Apr 1959

Ti-13V-11Cr-3Al: Specific heat

Temperature		Specific heat	
°C	°F	J/kg · K	Btu/lb · °F
93	200	532	0.127
205	400	557	0.133
315	600	590	0.141
425	800	615	0.147
540	1000	645	0.154
650	1200	678	0.162

Source: H.W. Rosenberg, Ti-13V-11Cr-3Al Data Sheet, *Beta Titanium Alloys in the 80's*, R.R. Boyer and H.W. Rosenberg, Ed., TMS/AIME, 1984, p 397-400

Thermal Expansion

Ti-13V-11Cr-3Al: Thermal coefficient of linear expansion

Typical values for thermal expansion between 21 °C (70 °F) and temperature indicated

Temperature		Thermal coefficient	
°C	°F	10^{-6}/°C	10^{-6}/°F
93	200	9.67	5.37
205	400	9.88	5.49
315	600	9.99	5.55
425	800	10.10	5.61
540	1000	10.24	5.69
650	1200	10.44	5.80

Source: H. Rosenberg, Ti-13V-11Cr-3Al Data Sheet, in *Beta Titanium Alloys in the 1980's*, R. Boyer and H. Rosenberg, Ed., TMS/AIME, p 397

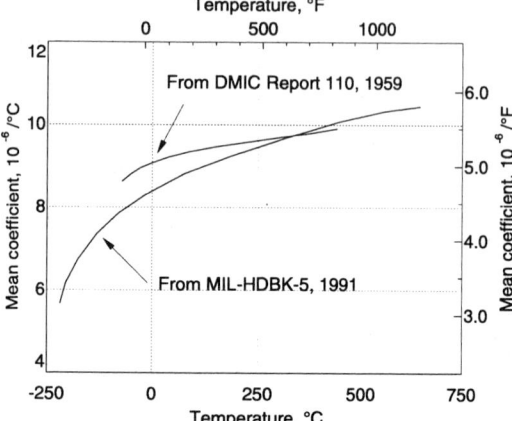

Ti-13V-11Cr-3Al: Mean thermal coefficient of linear

Mean coefficient from room temperature to indicated temperature

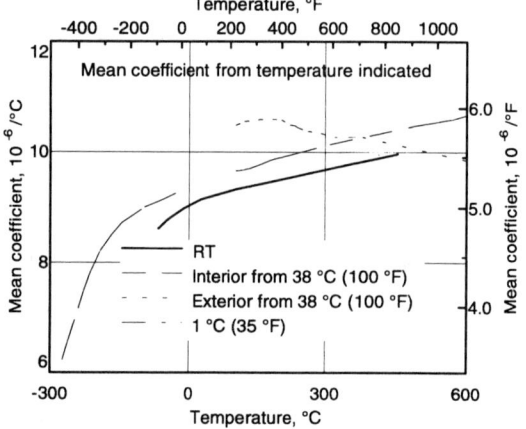

Ti-13V-11Cr-3Al: Thermal coefficient of linear expansion

Source: *Aerospace Structural Metals Handbook*, Vol 4, Code 3713, Battelle Columbus Laboratories, 1972, p 6

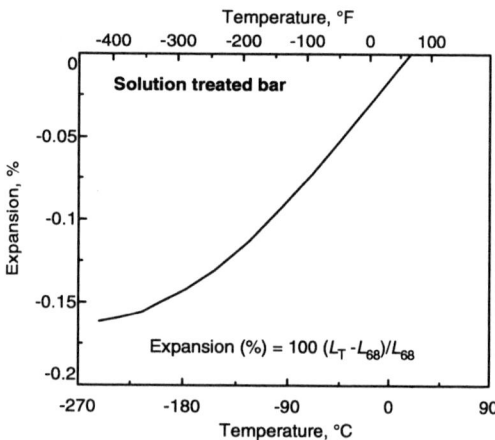

Ti-13V-11Cr-3Al: Thermal expansion

19 mm (0.750 in.) diam bar solution treated 770 °C (1425 °F), AC. Source: V. Arp *et al.*, Thermal Expansion of Some Engineering Materials from 20° to 293 °K, *Cryogenics*, Vol 2(No. 4), June 1962, reported in *Cryogenics Materials Data Handbook*, Vol 1, F. Schwartzberg, R. Herzog, S. Osgood, and M. Knight, AFML-TR-64-280, 1970, p 739

Thermal Conductivity

Room-Temperature Value. 6.9 W/m · K (4.0 Btu/ft · h · °F).

Ti-13V-11Cr-3Al: Thermal conductivity at various temperatures

Temperature		Conductivity	
°C	°F	W/m · K	Btu/ft · h · °F
93	200	9.69	5.60
205	400	12.02	6.95
315	600	14.53	8.40
425	800	17.12	9.90
540	1000	19.81	11.45
650	1200	22.57	13.05

Source: H. Rosenberg, Ti-13V-11Cr-3Al Data Sheet, in *Beta Titanium Alloys in the 1980's*, R. Boyer and H. Rosenberg, Ed., TMS/AIME, 1984, p 397

Ti-13V-11Cr-3Al: Thermal conductivity at low temperatures

Temperature, K	Thermal conductivity	
	W/m · K	Btu/ft · h · °F
296.65	8.02	4.63
283.96	7.67	4.43
223.90	6.23	3.60
212.78	5.90	3.41
82.06	2.94	1.69
25.37	1.11	0.64
25.34	1.09	0.63
23.91	0.92	0.53
23.89	0.92	0.53
24.96	1.08	0.62
24.96	1.09	0.63

Source: *Advances in Cryogenic Engineering*, Vol 8, K. Timmerhaus, Ed., Plenum Press, 1963, p 272

Ti-13V-11Cr-3Al: Thermal conductivity

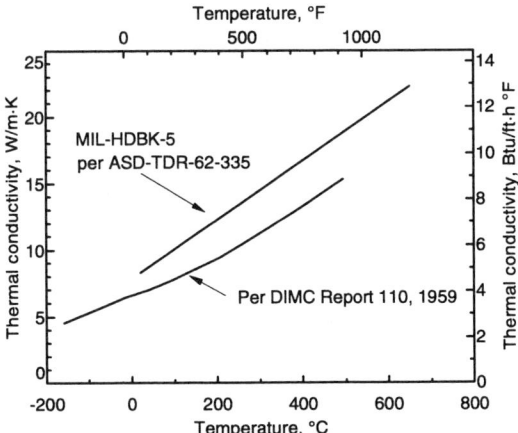

Mechanical Properties

Design Allowables

Ti-13V-11Cr-3Al: S-basis design tensile properties

Condition	Thickness mm	Thickness in.	Ultimate tensile strength(a) MPa	Ultimate tensile strength(a) ksi	Tensile yield strength(a) MPa	Tensile yield strength(a) ksi	Elongation(a), %	Reduction of area(a), %
Annealed sheet, strip, and plate per MIL-T-9046	0.3-1.24	0.012-0.049	910	132	870	126	8	...
	1.25-100	0.050-4.00	862(b)	125(b)	827(b)	120(b)	10(b)	...
STA sheet, strip, and plate per MIL-T-9046	≤100	≤4.00	1172(b)	170(b)	1103(b)	160(b)	3-4(c)	...
Annealed bar per MIL-T-9047	≤175	≤7.00	862(d)	125(d)	827(d)	120(d)	10(d)	25(d)
STA bar per MIL-T-9047 with cross sections ≤103 cm^2 (16 in.^{2})	≤100	≤4.00	1172(d)	170(d)	1103(d)	160(d)	2-6(e)	5-10(f)

(a) Applicable in longitudinal (L) and long-transverse (LT) directions except as noted. (b) Also applicable in short-transverse direction. (c) 4% elongation for thickness greater than 0.635 mm (0.025 in.). (d) Applicable in LT or ST direction if LT or ST dimension is greater than 75 mm (3 in.). (e) 2% elongation in the LT and ST directions. (f) 5% reduction of area in LT and ST directions. Source: MIL-HDBK 5, 1 Dec 1991

Ti-13V-11Cr-3Al: S-basis bearing, compressive, and shear strengths of sheet and plate

Property	Test condition	MPa	(ksi)
Annealed 1.25 to 100 mm (0.50 to 4.00 in.) thick			
Compressive yield strength, MPa (ksi)	L, LT, ST	827	(120)
Ultimate shear strength, MPa (ksi)		634	(92)
Ultimate bearing strength, MPa (ksi)	e/D = 1.5	1427	(207)
	e/D = 2.0	1861	(270)
Bearing yield strength, MPa (ksi)	e/D = 1.5	1165	(169)
	e/D = 2.0	1379	(200)
STA ≤100 mm (≤4.00 in.) thick			
Compressive yield strength, MPa (ksi)	L, LT, ST	1117	(162)
Ultimate shear strength, MPa (ksi)	...	724	(105)
Ultimate bearing strength, MPa (ksi)	e/D = 1.5	1710	(248)
	e/D = 2.0	2158	(313)
Bearing yield strength, MPa (ksi)	e/D = 1.5	1496	(217)
	e/D = 2.0	1703	(247)

Source: MIL-T-9046 and MIL-HDBK 5

Hardness

Ti-13V-11Cr-3Al: Knoop and Rockwell hardness

Condition	Knoop hardness	Rockwell C hardness
Unwelded sheet, 965 MPa (140 ksi) UTS	300	30.6
Single-bead weld, 950 MPa (138 ksi) UTS	320	30.1

Source: *Metals Handbook, Properties and Selection: Stainless Steels, Tool Materials, and Special-Purpose Materials*, Vol 3, 9th ed., American Society for Metals, 1980, p 368

Ti-13V-11Cr-3Al: Vickers hardness of weldments

Post-weld heat treatment	Hardness(a), HV		
	Base metal	HAZ	Weld zone
As welded	280	278	253
Weld + 2 h 315 °C (600 °F)	257	287	315
Weld + 4 h 315 °C (600 °F)	287	291	368
Weld + 8 h 315 °C (600 °F)	265	281	383

Note: Spot welding characteristics were investigated by hardness testing a cross section of the weld nugget. The hardness survey indicated that decomposition takes place after short-term thermal exposure at 315 °C (600 °F). Increase in hardness is confined to the weld metal. (a) 150-g load. Source: DMIC Report 110, 1959

Ti-13V-11Cr-3Al: Rockwell hardness vs aging time

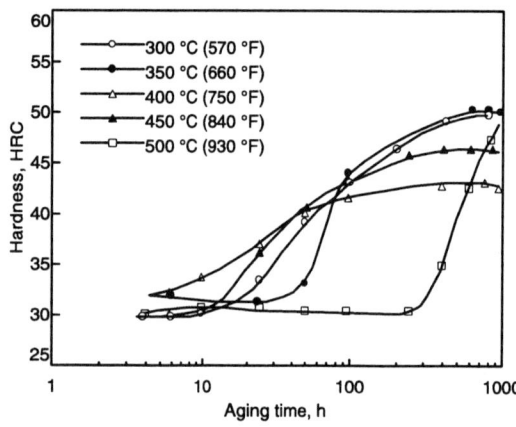

The rapid increase in hardness that follows the initial stages of aging occurs at times corresponding to the appearance of the α phase. Strips from sheet were vacuum annealed at 850 °C (1560 °F) for about 4 h and cold rolled to a thickness of 0.4 mm (0.015 in.). Coupons were prepared from rolled material and solution treated at 800 °C (1470 °F) for 90 min in purified helium, then quenched in water, oil or air. Several coupons were solution treated at 900 °C (1650 °F) and water quenched. Aging was done in salt baths held at 250 to 500 °C (480 to 930 °F) to 1000 h.
Source: G.H. Narayanan and T. Archbold, Decomposition of the Metastable Beta Phase in the All-Beta Alloy Ti-13V-11Cr-3Al, *Metall. Trans.*, Vol 1, 1970, p 2281-2290

Ti-13V-11Cr-3Al: Rockwell hardness vs reduction of area at room temperature

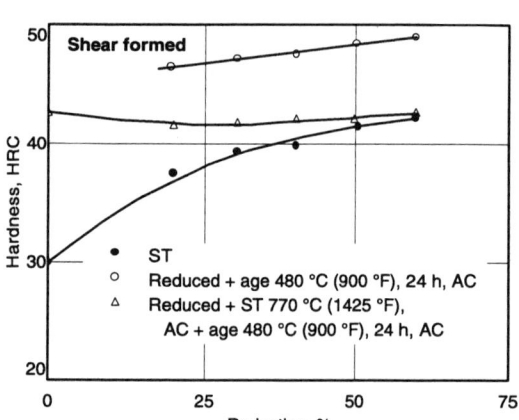

Source: F. Jacobs, "Mechanical Properties of Materials Fabricated by Shear Forming," ASD TDR-62-830, Feb 1963, reported in *Aerospace Structural Metals Handbook, Vol 4*, Code 3713, Battelle Columbus Laboratories, 1972, p 5

Ti-13V-11Cr-3Al: Hardness at low temperatures

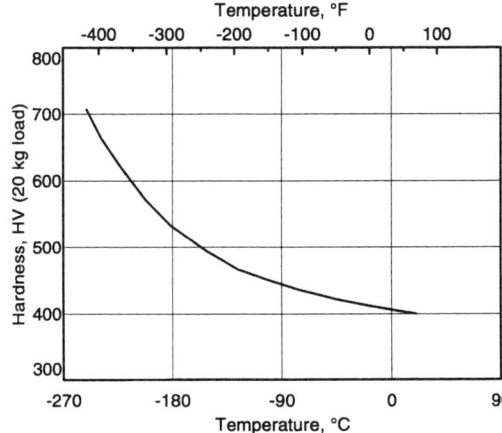

1.5 mm (0.060 in.) sheet solution treated and aged.
Source: L. Rice, "The Evaluation of the Effects of Very Low Temperatures on the Properties of Aircraft and Missile Metals," WADD TR 60-254, Feb 1960, reported in *Cryogenic Materials Data Handbook*, Vol 1, F. Schwartzberg, R. Herzog, S. Osgood, and M. Knight, Ed., AFML-TR-64-280, 1970, p 736

Typical Tensile Properties

See also "Heat Treatment" for tensile properties.

Ti-13V-11Cr-3Al: Typical RT tensile properties of bar, rod, and wire

Diameter			Ultimate tensile strength		Tensile yield strength (0.2% offset)		Elongation,
mm	in.	Condition	MPa	ksi	MPa	ksi	%
Annealed or annealed + aged							
19	0.750	64 h, 480 °C (900 °F) + 1 h, 565 °C (1050 °F)	1289	187	1213	176	5.0
14	0.575	Annealed	993	144	986	143	22.5
		48 h, 480 °C (900 °F)	1365	198	1206	175	8.0
		72 h, 480 °C (900 °F)	1461	212	1296	188	6.0
6.5	0.257	Annealed	1034	150	993	144	23.3
		48 h, 480 °C (900 °F)	1420	206	1289	187	10.0
		72 h, 480 °C (900 °F)	1475	214	1344	195	6.7
Cold worked or cold worked + aged							
1.5	0.062	Cold drawn, 92%	1751	254	...	...	4.0
		92% + 24 h, 370 °C (700 °F)	1999	290	...	...	8.0
		92% + 24 h, 425 °C (800 °F)	2151	312	...	...	4.0

Source: R.A. Wood, *Beta Titanium Alloys*, MCIC-72-11, Battelle Columbus Laboratories, 1972

Ti-13V-11Cr-3Al: RT tensile properties of forged and heat treated bars

Bar size		480 °C (900 °F) aging treatment, h	Direction	Ultimate tensile strength		Tensile yield strength (0.2% offset)		Elongation,	Reduction of area,
mm	in.			MPa	ksi	MPa	ksi	%	%
150	6	48	L	1264	183.3	1153	167.2	8.0	13.7
			T	1297	188.1	1169	169.6	8.0	8.5
100	4	48	L	1474	213.9	1388	201.3	4.0	11.6
			T	1396	202.5	1307	189.6	4.0	12.4
75	3	48	L	1478	214.4	1369	198.6	6.0	9.3
			T	1462	212.1	1364	197.9	3.0	6.2
		30	L	1438	208.6	1341	194.6	6.5	8.5
			T	1407	204.1	1288	186.9	5.0	12.0
50	2	30	L	1368	198.5	1241	180.0	7.0	10.0
			T	1340	194.4	1232	178.7	5.0	12.0
30	1.2	20	L	1424	206.6	1290	187.2	10.0	15.8

Center of bar samples. Bars aged directly from forging operation. Source: R.A. Wood, *Beta Titanium Alloys*, MCIC-72-11, Battelle Columbus Laboratories, 1972, p 38

Ti-13V-11Cr-3Al: Typical tensile strength after aging

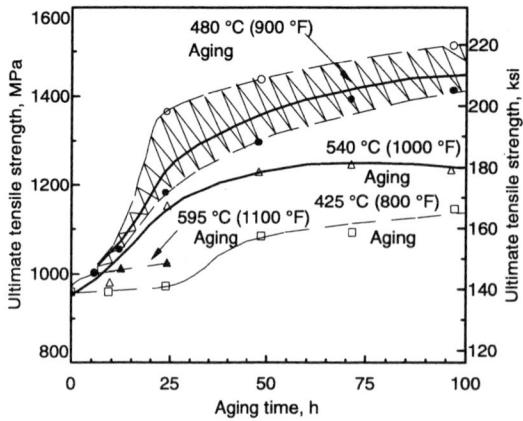

Effect of aging time on the tensile strength of longitudinal sheet.
Source: R.A. Wood, *Beta Titanium Alloys*, MCIC-72-11, Battelle Columbus Laboratories, 1972

Effect of Cold Work

Ti-13V-11Cr-3Al: Effect of cold work on tensile properties of STA sheet

Cold reduction, %	Ultimate tensile strength		Tensile yield strength (0.2%)		Elongation in 50 mm (2 in.), %	Reduction of area, %
	MPa	ksi	MPa	ksi		
0	924	134	903	131	25	50
10	1013	147	951	138	17	42
20	1117	162	1013	147	12	36
30	1206	175	1103	160	8	32
40	1289	187	1193	173	6	28
50	1372	199	1268	184	6	25
60	1434	208	1337	194	5	22
70	1489	216	1399	203	4	22
80	1537	223	1482	215	2	16

Source: *Alloy Digest*, Code Ti-28

Ti-13V-11Cr-3Al: As-drawn tensile properties

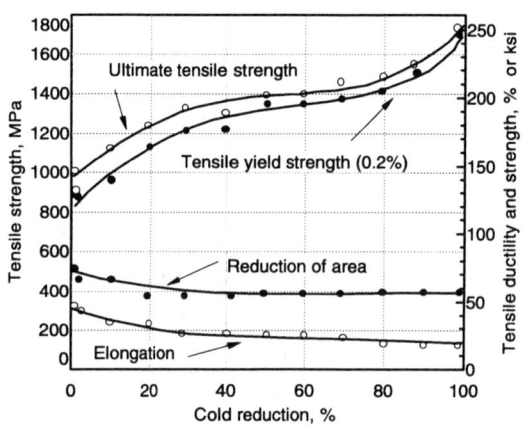

Wire size not reported.
Source: R.A. Wood, *Beta Titanium Alloys*, MCIC-72-11, Battelle Columbus Laboratories, 1972

Ti-13V-11Cr-3Al: Tensile strengths of annealed sheet vs cold work

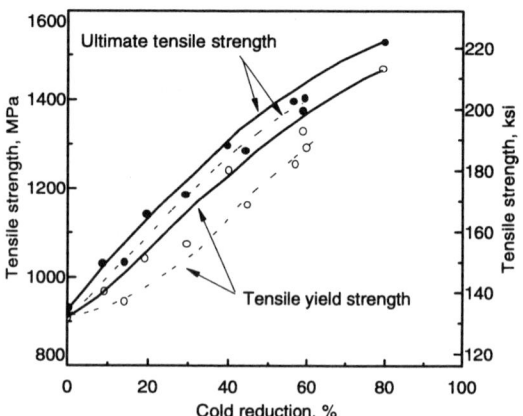

Source: R.A. Wood, *Beta Titanium Alloys*, MCIC-72-11, Battelle Columbus Laboratories, 1972

Ti-13V-11Cr-3Al: Aged UTS after reduction

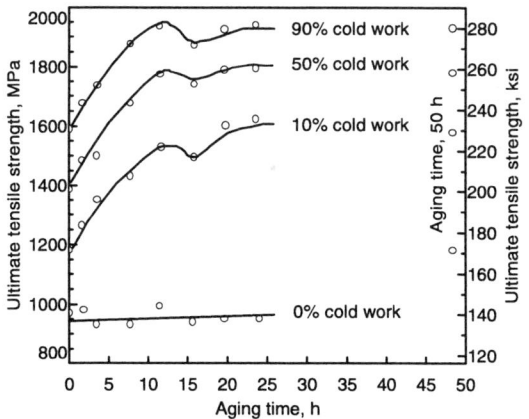

Wire size not reported, aged at 425 °C (800 °F).
Source: R.A. Wood, *Beta Titanium Alloys*, MCIC-72-11, Battelle Columbus Laboratories, 1972

Ti-13V-11Cr-3Al: Aged tensile strengths of strained sheet

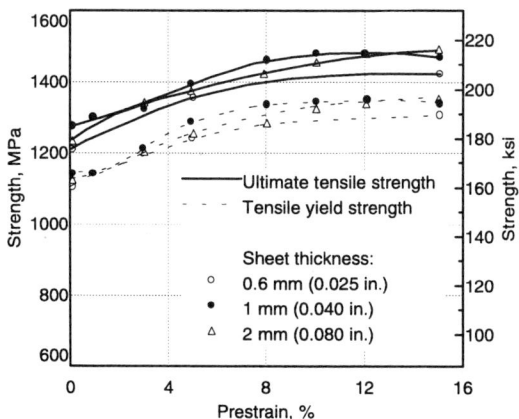

Effect of straining after solution treatment on the subsequent aged properties of sheet aged 72 h at 480 °C (900 °F).
Source: R.A. Wood, *Beta Titanium Alloys*, MCIC-72-11, Battelle Columbus Laboratories, 1972

Notched Strength

Ti-13V-11Cr-3Al: Notched tensile properties of cold worked sheet

Cold work %	−65 °F Tests			Room-temperature tests		
	Notched strength MPa	ksi	Notched/ unnotched ratio	Notched strength MPa	ksi	Notched/ unnotched ratio
None	1558	226	1.03	1199	174	1.24
10	1523	221	1.19	1255	182	1.20
20	1634	237	1.25	1344	195	1.16
40	1710	248	1.18	1482	215	1.08
60	1710	248	1.09	1586	230	1.12
80	1668	242	1.01	1586	230	1.02

Note: Solution treated sheet was cold worked as indicated. Source: "The All-Beta Alloy Ti-13V-11Cr-3Al," DMIC 110, R. Wood and H. Ogden, Ed., Battelle Columbus Laboratories, 1959, p 79

Ti-13V-11Cr-3Al: Notch strength ratio of sheet after cold rolling

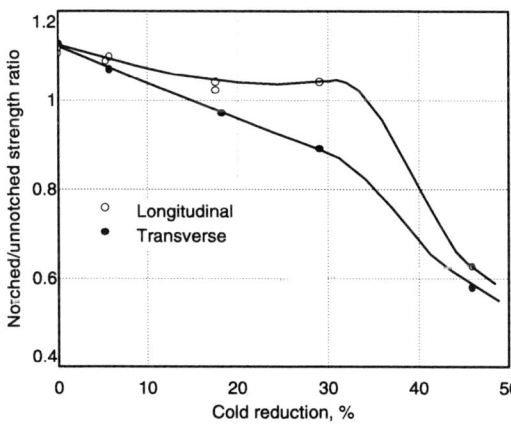

Sheet 1.6 mm (0.063 in.) thick was solution treated. Notch radius was less than 0.0178 mm (0.0007 in.) for 60° notch on two sides.
Source: A. Repko and W. Brown, Jr., Influence of Cold Rolling and Aging on Sharp Notch Properties of Beta Titanium Sheet, *Proc. ASTM*, Vol 62, 1962, p 869; also in *A Review of Factors Influencing the Crack Tolerance of Titanium Alloys*, J. Shannon, Jr., and W. Brown, Jr., Ed., ASTM STP 432, ASTM, 1968, p 33

Ti-13V-11Cr-3Al: RT notch strength of ST sheet

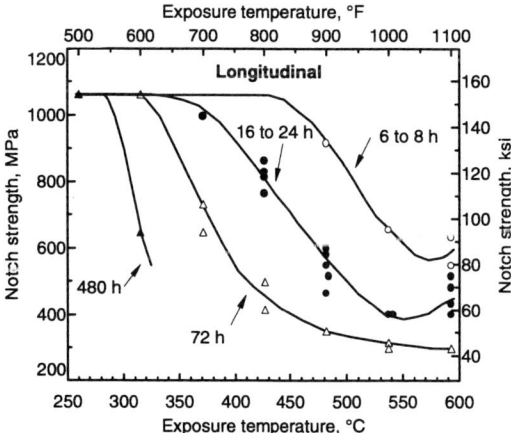

Sheet 1.3 mm (0.050 in.) thick was solution heat treated, then exposed at the elevated temperatures indicated.
Source: A. Repko and W. Brown, Jr., *A Review of Factors Influencing the Crack Tolerance of Titanium Alloys*, J. Shannon, Jr., and W. Brown, Jr., Ed., ASTM STP 432, ASTM, 1968, p 33

High-Temperature Strength

Tensile Strength

Ti-13V-11Cr-3Al: Tensile properties vs temperature of aged specimens

Test temperature		Ultimate tensile strength		Tensile yield strength (0.2%)		Elongation in 50 mm (2 in.), %	Modulus of elasticity	
°C	°F	MPa	ksi	MPa	ksi		GPa	10^6 psi
21	70	1406	204	1275	185	6	107	15.6
93	200	1320	191.5	1158	168	7	106	15.4
205	400	1255	182	1041	151	8	102	14.8
315	600	1220	177	986	143	10	98	14.3
425	800	1068	155	937	136	10	95	13.8
540	1000	786	114	672	97.5	12	92	13.4

Source: *Alloy Digest*, Code Ti-27

Ti-13V-11Cr-3Al: Tensile properties vs temperature of annealed sheet and bar

Test temperature		Ultimate tensile strength		Tensile yield strength (0.2%)		Elongation in 50 mm (2 in.), %	Modulus of elasticity	
°C	°F	MPa	ksi	MPa	ksi		GPa	10^6 psi
21	70	920	133.5	903	131	21	101	14.7
93	200	862	125	817	118.5	21	98	14.3
205	400	841	122	765	111	23	96	14.0
315	600	793	115	668	97	23	91	13.2
425	800	851	123.5	717	104	18	85	12.4
540	1000	696	101	655	95	34	80	11.6

Source: *Alloy Digest*, Code Ti-27

Ti-13V-11Cr-3Al: Typical tensile properties vs temperature

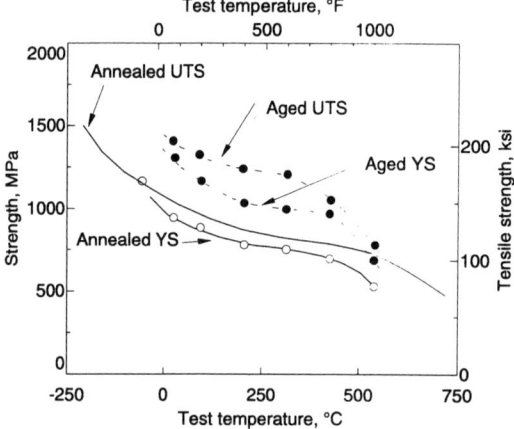

Annealed/solution heat treated and solution heat treated + 480 °C (900 °F) aged conditions of sheet.
Source: R.A. Wood, *Beta Titanium Alloys*, MCIC-72-11, Battelle Columbus Laboratories, 1972

Ti-13V-11Cr-3Al: Strength vs temperature of STA sheet (design allowables)

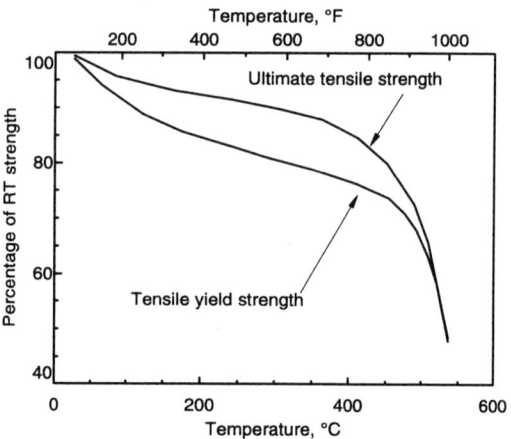

Strength at temperature after exposure up to 1/2 h.
Source: MIL-HDBK 5, 1 Dec 1991

Ti-13V-11Cr-3Al: Strength vs temperature of annealed sheet (design allowables)

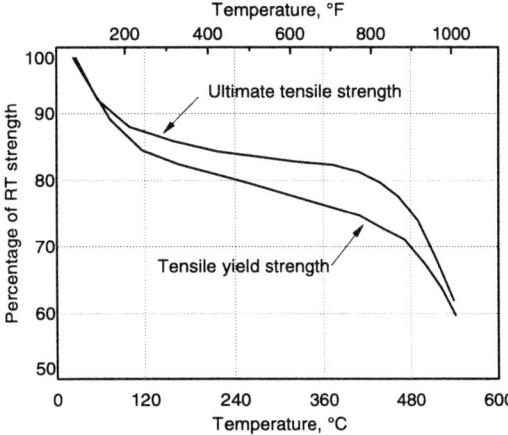

Strength at temperature after exposure up to 1/2 h.
Source: MIL-HDBK 5, 1 Dec 1991

Bearing Strength

Ti-13V-11Cr-3Al: STA high-temperature bearing strength

Bearing strength	\multicolumn{10}{c}{Strength, MPa (ksi), at:}									
	26 °C	(80 °F)	93 °C	(200 °F)	205 °C	(400 °F)	315 °C	(600 °F)	425 °C	(800 °F)	540 °C	(1000 °F)
e/d = 1.5, L												
Ultimate	1634	(237)	1496	(217)	1413	(205)	1386	(201)	1344	(195)	1179	(171)
Yield	1324	(192)	1193	(173)	1179	(171)	1144	(166)	1137	(165)	951	(138)
e/d = 1/5, T												
Ultimate	1606	(233)	1503	(218)	1420	(206)	1386	(201)	1330	(193)	1137	(165)
Yield	1310	(190)	1227	(178)	1179	(171)	1165	(169)	1144	(166)	937	(136)
e/d = 2.0, L												
Ultimate	2123	(308)	1930	(280)	1840	(267)	1806	(262)	1772	(257)	1510	(219)
Yield	1544	(224)	1386	(201)	1413	(205)	1392	(202)	1351	(196)	1144	(166)
e/d = 2.0, T												
Ultimate	2096	(304)	1944	(282)	1848	(268)	1806	(262)	1751	(254)	1551	(225)
Yield	1572	(228)	1448	(210)	1420	(206)	1420	(206)	1358	(197)	1137	(165)

Note: STA sheet 1.6 mm (0.063 in.) thick; pin diameter, d, was 6 mm (0.250 in.). Source: R.A. Wood, *Beta Titanium Alloys*, MCIC-72-11, Battelle Columbus Laboratories, 1972

Ti-13V-11Cr-3Al: Bearing yield strength vs temperature of STA sheet

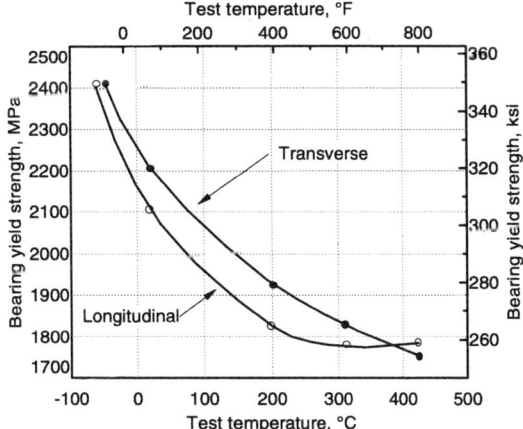

Effect of test temperature on the bearing yield strength of sheet solution heat treated + aged 72 h at 425 °C (800 °F). 1.2 mm (0.048 in.); pin diameter, d, was 6 mm (0.250 in.). e/D = 2.0.
Source: R.A. Wood, *Beta Titanium Alloys*, MCIC-72-11, Battelle Columbus Laboratories, 1972

Compressive and Shear Strengths

Ti-13V-11Cr-3Al: Bearing yield strength of an-

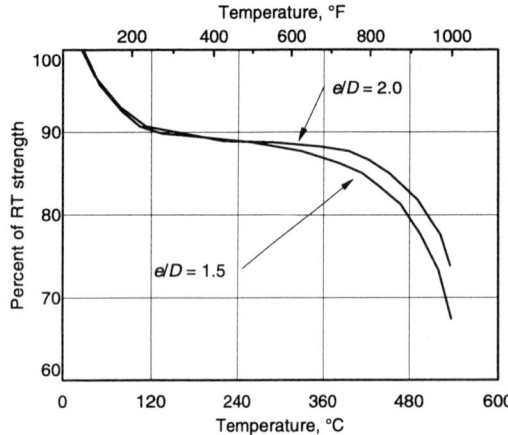

Strength at temperature after exposure up to 1/2 h.
Source: MIL-HDBK 5, 1 Dec 1991

Ti-13V-11Cr-3Al: High-temperature compressive strength

Sheet thickness			Test temperature		Compressive yield strength (0.2%)	
mm	in.	Condition	°C	°F	MPa	ksi
1.6	0.063	Annealed	26	80	903	131
			93	200	813	118
			205	400	751	109
			315	600	717	104
			425	800	689	100
			540	1000	620	90
1	0.040	Aged 50 h, 425 °C (800 °F)	23	74	993	144
			315	600	765	111
			425	800	710	103
0.6	0.025	Aged 100 h, 425 °C (800 °F)	23	74	1068	155
			315	600	841	122
			425	800	765	111

Source: R.A. Wood, *Beta Titanium Alloys*, MCIC-72-11, Battelle Columbus Laboratories, 1972

Ti-13V-11Cr-3Al: Shear strength of annealed bar and sheet

Form	Ultimate shear strength MPa (ksi), at:											
	−53 °C	(−65 °F)	24 °C	(75 °F)	93 °C	(200 °F)	205 °C	(400 °F)	315 °C	(600 °F)	425 °C	(800 °F)	540 °C	(1000 °F)
13 mm (0.500 in.) bar	834	(121)	758	(110)	717	(104)	662	(96)	634	(92)	606	(88)	…	…
1.6 mm (0.063 in.) sheet	…	…	723	(105)	703	(102)	655	(95)	620	(90)	600	(87)	510	(74)

Source: R.A. Wood, *Beta Titanium Alloys*, MCIC-72-11, Battelle Columbus Laboratories, 1972

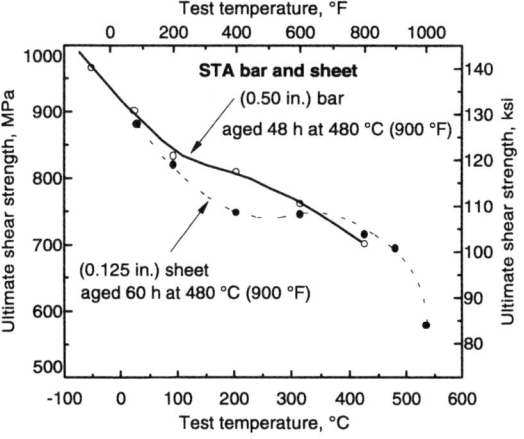

Source: R.A. Wood, *Beta Titanium Alloys*, MCIC-72-11, Battelle Columbus Laboratories, 1972

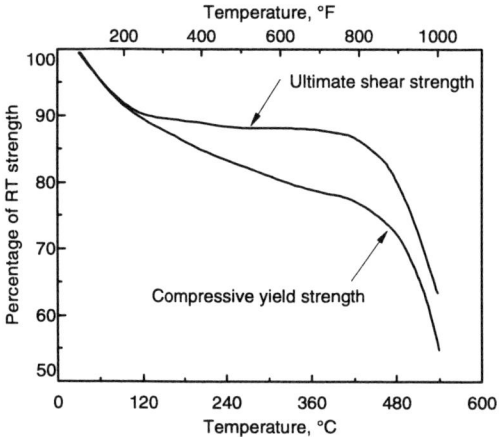

Strength at temperature after exposure up to 1/2 h.
Source: MIL-HDBK 5, 1 Dec 1991

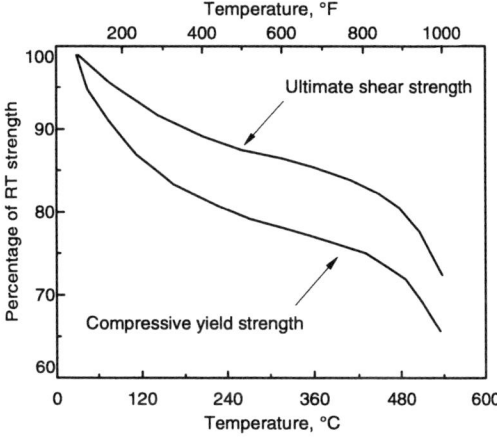

Design allowables for annealed sheet at temperature after exposure up to 1/2 h.
Source: MIL-HDBK 5, 1 Dec 1991

Creep Properties

Creep Deformation

Ti-13V-11Cr-3Al: Creep data for annealed sheet

Creep-exposure conditions				Test duration, h	Total plastic deformation, %
Temperature		Stress			
°C	°F	MPa	ksi		
150	300	413	60	500	−0.3(a)
		413	60	500	−0.2(a)
205	400	620	90	474	0.15
		689	100	496	0.19
260	500	275	40	500	−0.2(a)
		275	40	500	−0.4(a)
		551	80	547	0.02
		620	90	305	0.15
315	600	207	30	500	0.2
		207	30	500	0.8

(a) Negative strain measurements. Source: R.A. Wood, *Beta Titanium Alloys*, MCIC-72-11, Battelle Columbus Laboratories, 1972

Ti-13V-11Cr-3Al: Creep data for STA sheet

Creep test temperature		Creep stress		Test duration,	Minimum creep rate,	Total plastic deformation,
°C	°F	MPa	ksi	h	%/h	%
26	80	1034	150	500	0.000012	0.016
93	200	965	140	500	0.000010	0.031
205	400	896	130	500	0.000024	0.075
260	500	724	105	1502	...	0.050
300	575	448	65	1503	...	0.070
		689	100	1507	...	0.100
315	600	655	95	1502	0.000018	0.110
		689	100	712	...	0.160
		758	110	500	0.000040	0.123
330	625	448	65	100	...	0.000
425	800	448	65	500	0.003000	3.020
		551	80	10	...	0.150

Note: Sheet 0.9 mm (0.036 in.) thick aged for 96 h at 480 °C (900 °F). Source: R.A. Wood, *Beta Titanium Alloys*, MCIC-72-11, Battelle Columbus Laboratories, 1972

Ti-13V-11Cr-3Al: Creep at 425 °C for annealed sheet

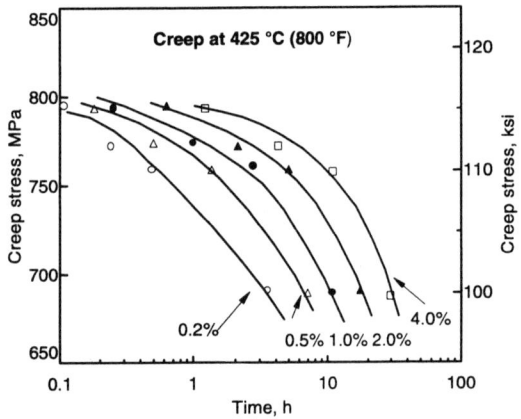

Source: A. Blatherwick and A. Cers, "Fatigue, Creep, and Stress-Rupture Properties of Ti-13V-11Cr-3Al Titanium Alloy (B-120VCA)," AFML-TR-66-293, 1966

Ti-13V-11Cr-3Al: Creep strain of STA sheet

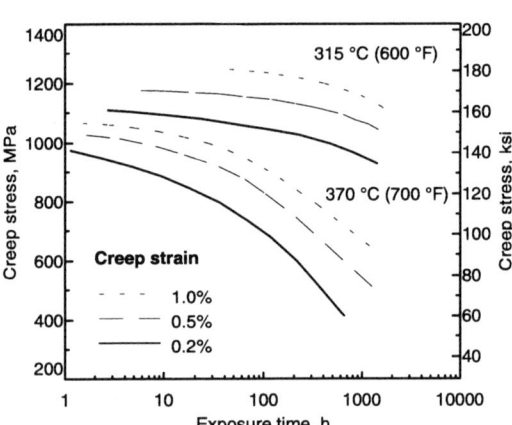

1.6 mm (0.064 in.) sheet aged for 50 h at 470 °C (875 °F).
Source: R.A. Wood, *Beta Titanium Alloys*, MCIC-72-11, Battelle Columbus Laboratories, 1972

Creep Stability

Ti-13V-11Cr-3Al: Creep stability for annealed sheet

Annealed Ti-13V-11Cr-3Al has good creep resistance at high stress levels up to 260 °C (500 °F), but would not ordinarily be used above 315 °C (600 °F) because of instability.

Temperature		Creep-exposure conditions Stress		Test duration,	Total plastic deformation,	RT elongation after exposure,	
°C	°F	MPa	ksi	h	%	%	Remarks
150	300	413	60	500	−0.3(a)	24	Stable
		413	60	500	−0.2(a)	23	Stable
205	400	620	90	474	0.15	20	Stable
		689	100	496	0.19	23	Stable
260	500	275	40	500	−0.2(a)	25	Stable
		275	40	500	−0.4(a)	23	Stable
		551	80	547	0.02	18	Stable
		620	90	305	0.15	20	Stable
315	600	207	30	500	0.2	1.2	Unstable
		207	30	500	0.8	1.5	Unstable

(a) Negative strain measurements. Source: R. Wood and H. Ogden, "The All-Beta Titanium Alloy Ti-13V-11Cr-3Al," DMIC Report 110, Battelle Columbus Laboratories, 1959, p 109

Ti-13V-11Cr-3Al: Creep stability of STA sheet

Creep exposure of 1000 h at 315 °C (600 °F) and 690 MPa (100 ksi)

Aging treatment	Prestrain(a), %	Creep strain, %	Minimum creep rate, %/h	Room-temperature elongation,% After exposure	No exposure(b)
480 °C (900 °F)/36 h	0	0.079	0.000018	6	5.3
	0	0.084	0.000020	5	5.3
	8	0.097	0.000025	...	5.7
480 °C (900 °F)/12 h	0	0.114	0.000057	7	8.3
	0	0.109	0.000054	7	8.3
	8	0.127	0.000055	5	5.0

Note: Sheet 1.6 mm (0.063 in.) thick was solution treated and aged at 480 °C (900 °F) for 12 or 36 h, as indicated. (a) Nominal amount of stretch prior to aging to simulate forming strains. (b) Average properties. Source: R. Kaneko and C. Woods, "Low-Temperature Forming of Beta Titanium Alloys," NASA Contractor Report 3706, NASA, 1983, p 15

Fatigue Properties

Room-Temperature S/N Curves

Ti-13V-11Cr-3Al: Rotating beam fatigue of bar at room temperature

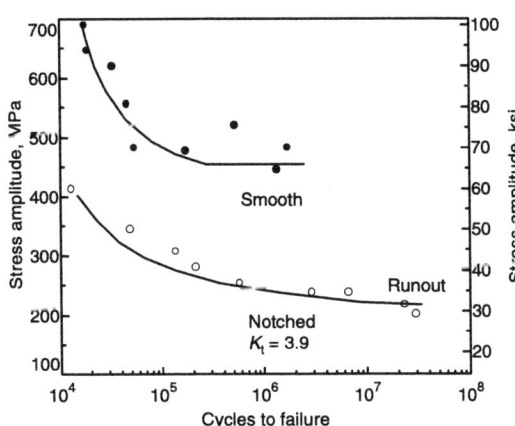

17.5 mm (0.69 in.) diam bar solution treated and aged.
Source: R.A. Wood, Beta Titanium Alloys, MCIC-72-11, Battelle Columbus Laboratories, 1972

Ti-13V-11Cr-3Al: Tension-tension fatigue of wire at room temperature

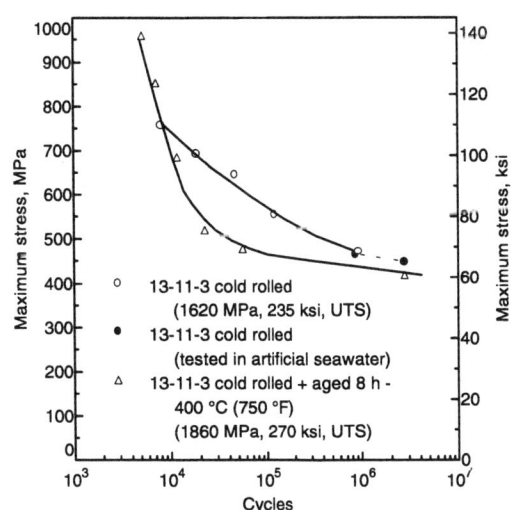

1570 cycles.min, $R = 0.95$, $A = 0.025$.
Source: R.A. Wood, Beta Titanium Alloys, MCIC-72-11, Battelle Columbus Laboratories, 1972, p 59

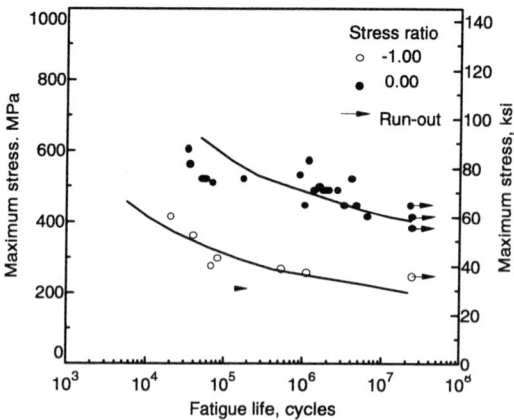

Ti-13V-11Cr-3Al: RT axial fatigue of annealed smooth specimens in air

Specimens from annealed sheet, longitudinal direction, with an ultimate tensile strength of 955 MPa (138.5 ksi) and tensile yield strength of 915 MPa (132.8 ksi). Stresses are based on net section, 3600 cycles/min. Unnotched 7 mm (0.30 in.) wide, specimens as machined, edges polished with emery paper.
Source: MIL-HDBK 5, 1 Dec 1991

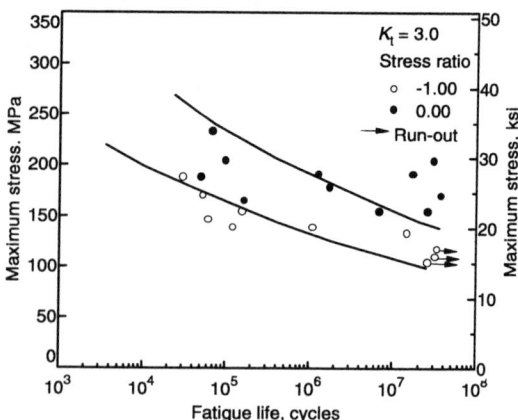

Ti-13V-11Cr-3Al: RT axial fatigue of annealed notched specimens in air

Longitudinal specimens from 1.09 mm (0.043 in.) annealed sheet, with an ultimate tensile strength of 955 MPa (138.5 ksi) and tensile yield strength of 915 MPa (132.8 ksi). Stresses are based on net section 3600 cycles/min. Specimens had 11 mm (0.448 in.) gross width, 7 mm (0.300 in.) net width, 0.5 mm (0.022 in.) root radius and 60° flank angle, and were as machined, edges polished with emery paper.

High-Temperature S/N Curves

Ti-13V-11Cr-3Al: High-temperature fatigue of STA bar in air

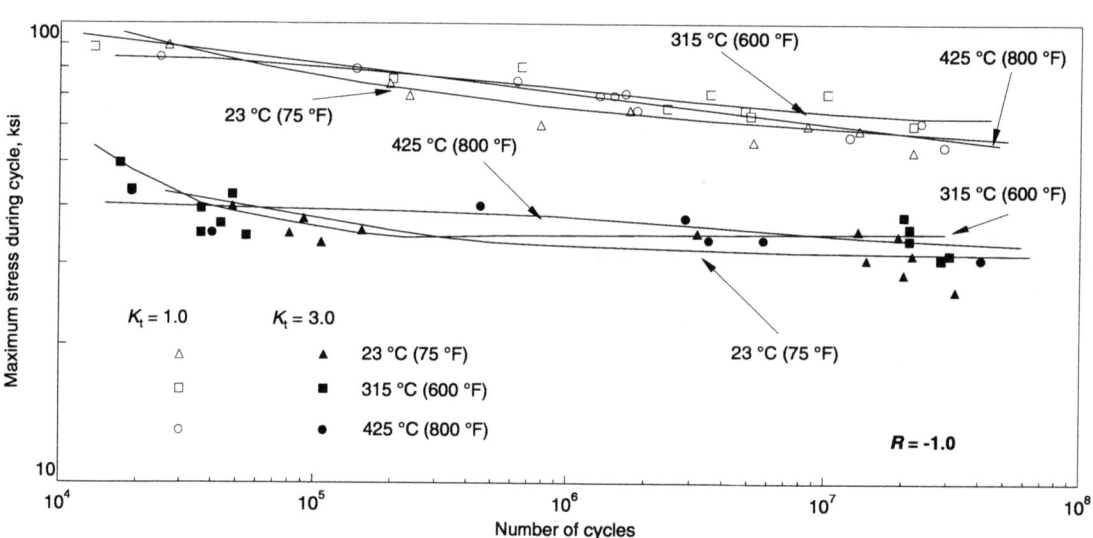

Bar aged 72 h at 480 °C (900 °F) to 1185 to 1345 MPa (172 to 195 ksi) yield strength.
Source: R.A. Wood, *Beta Titanium Alloys*, MCIC-72-11, Battelle Columbus Laboratories, 1972

Ti-13V-11Cr-3Al: Axial fatigue at 315 °C of annealed sheet

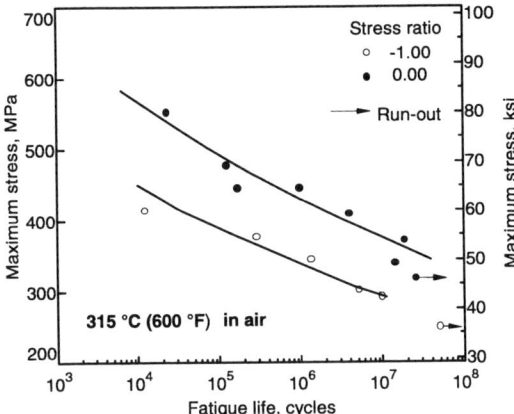

Smooth 1.09 mm (0.043 in.) longitudinal specimen from annealed sheet, with an ultimate tensile strength of 800 MPa (116 ksi) and a tensile yield strength of 707 MPa (102.61 ksi). 3600 cycles./min. Unnotched specimen, 7 mm (0.300 in.) wide, as machined, edges polished with emery paper.
Source: MIL-HDBK 5, 1 Dec 1991

Ti-13V-11Cr-3Al: Axial fatigue at 425 °C of annealed sheet

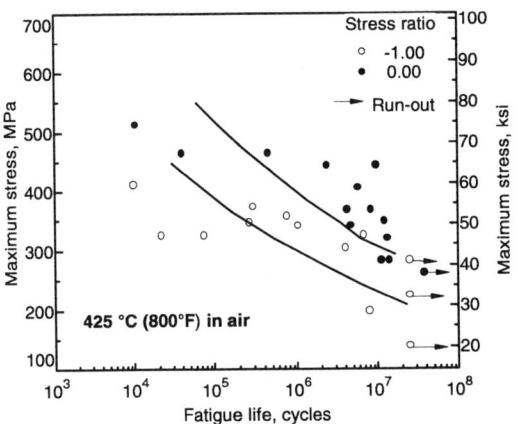

Smooth 1.09 mm (0.043 in.) longitudinal specimens from annealed sheet, with an ultimate tensile strength of 798 MPa (115.8 ksi) and a tensile yield strength of 700 MPa (98.61 ksi). 3600 cycles/min. Unnotched specimen, 7 mm (0.300 in.) wide, as machined, edges polished with emery paper.
Source: MIL-HDBK 5, 1 Dec 1991

Constant-Life Diagrams

Ti-13V-11Cr-3Al: Constant-life axial fatigue at RT

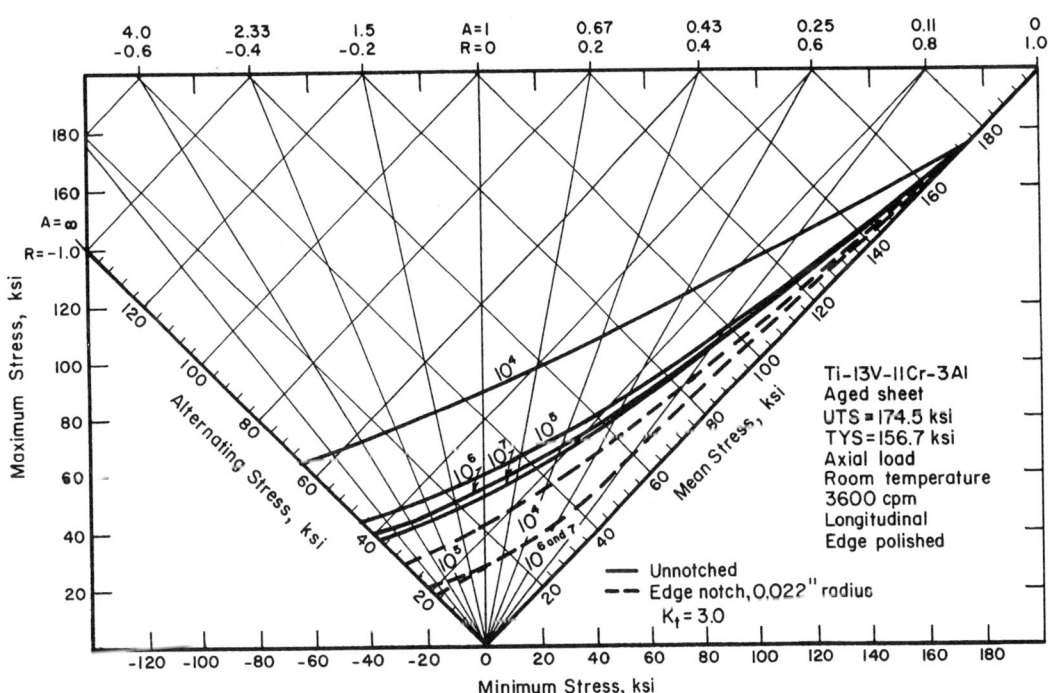

Solution treated and aged sheet. UTS, 1203 MPa (174.5 ksi); TYS, 1080 MPa (156.7 ksi); 3600 cycles/min, longitudinal, edge polished.
Source: R.A. Wood, *Beta Titanium Alloys*, MCIC-72-11, Battelle Columbus Laboratories, 1972, p 57

Fracture Properties

Ti-13V-11Cr-3Al: Charpy impact toughness

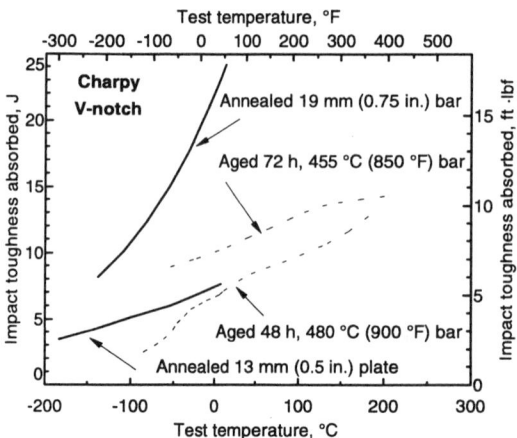

Source: R.A. Wood, *Beta Titanium Alloys*, MCIC-72-11, Battelle Columbus Laboratories, 1972

Ti-13V-11Cr-3Al: Plane-strain fracture toughness

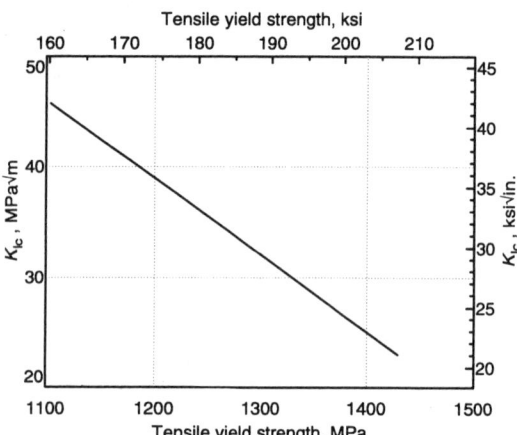

Aging condition not specified.
Source: R.A. Wood, *Beta Titanium Alloys*, MCIC-72-11, Battelle Columbus Laboratories, 1972

Ti-13V-11Cr-3Al: K_c fracture toughness vs yield strength

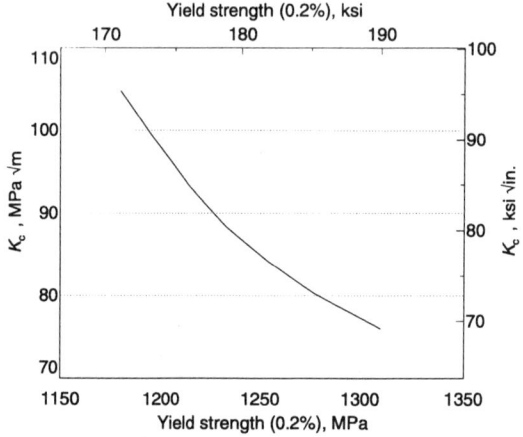

1.6 mm (0.063 in.) sheet aged at 480 °C (900 °F).
Source: A. Repko and W.F. Brown, Jr., "Influence of Cold Rolling and Aging on Sharp-Notch Properties of Beta Titanium Sheet," NASA Lewis Paper E-1274, 1961, reported in *Beta Titanium Alloys*, R.A. Wood, MCIC-72-11, Battelle Columbus Laboratories, 1972

Ti-13V-11Cr-3Al: K_c fracture toughness after working and aging

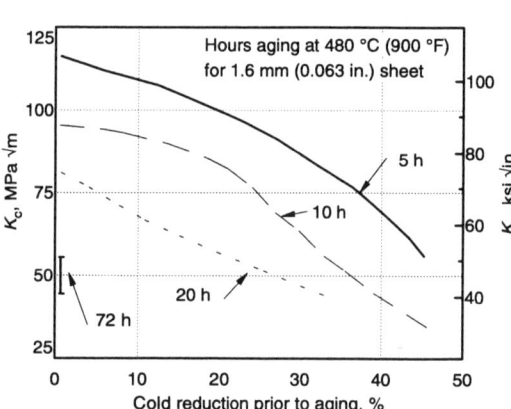

Source: R.A. Wood, *Beta Titanium Alloys*, MCIC-72-11, Battelle Columbus Laboratories, 1972

Ti-13V-11Cr-3Al: K_c fracture toughness vs sheet thickness

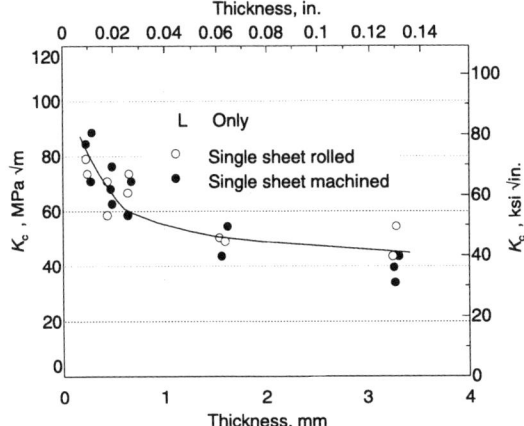

Specimens were aged at 480 °C (900 °F) for 72 h to achieve yield strength of 1172 to 1310 MPa (170 to 190 ksi).
Source: A. Repko, M. Jones, and W.F. Brown, Jr., "Influence of Sheet Thickness on the Sharp Edge Notch Properties of a Beta Titanium Alloy at Room and Low Temperatures," NASA Lewis Paper E-1274, 1961; reported in Beta Titanium Alloys, R. Wood, MCIC-72-11, Battelle Columbus Laboratories, 1972

Ti-13V-11Cr-3Al: K_c fracture toughness vs sheet thickness

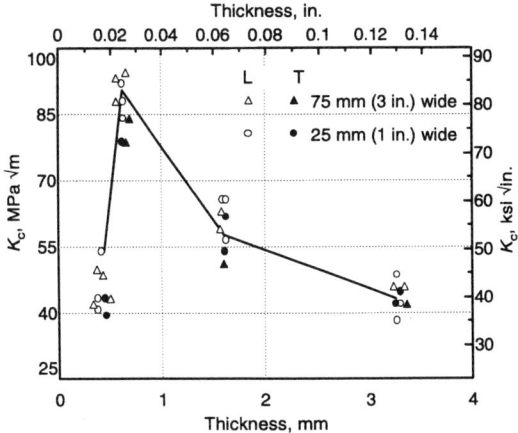

Specimens were aged 72 h at 480 °C (900 °F) to 1172 to 1310 MPa (170 to 190 ksi) yield strength.
Source: R.A. Wood, Beta Titanium Alloys, MCIC-72-11, Battelle Columbus Laboratories, 1972

Deformation

Ti-13-11-3: Tensile ductility

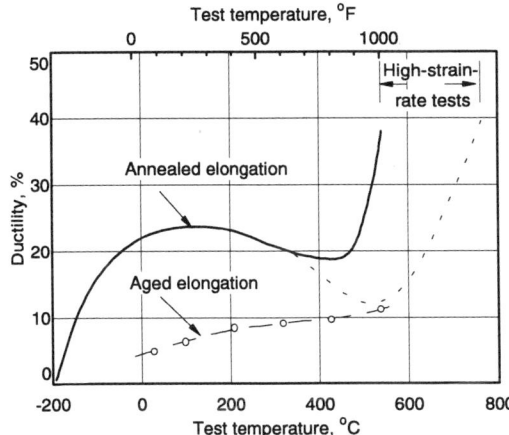

Tensile ductility of sheet vs test temperature.
Annealed solution heat treated and solution heat treated plus 480 °C (900 °F) aged conditions.

Ti-13-11-3: Elongation

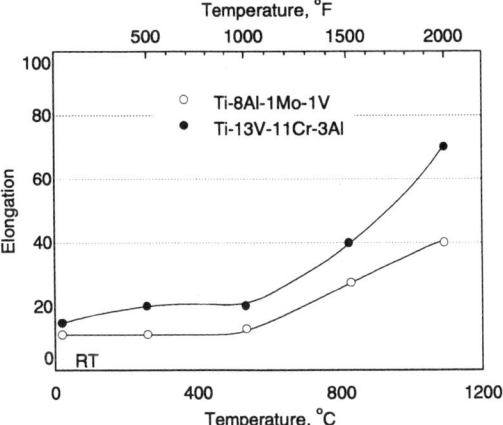

Source: R.A. Wood and R.J. Favor, Titanium Alloys Handbook, MCIC-HB-02, Battelle Columbus Laboratories, 1972

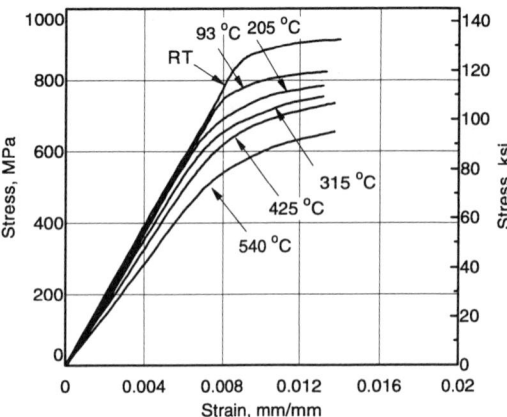

Ti-13-11-3: Typical tensile stress-strain curves for annealed sheet

Longitudinal and long-transverse specimens of annealed sheet after 1/2 h exposure at temperature.
Source: MIL-HDBK 5, 1 Dec 1991

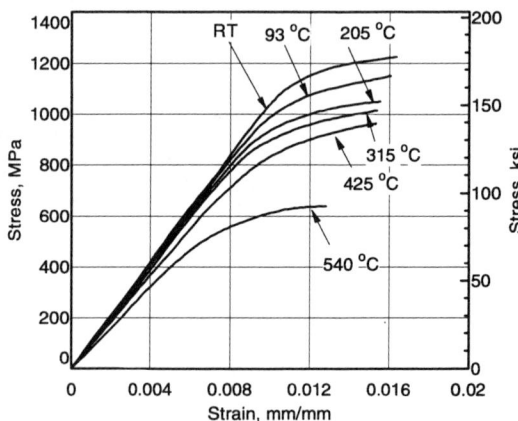

Ti-13-11-3: Typical tensile stress-strain curves for STA sheet

Solution treated and aged sheet (longitudinal and long-transverse) after 1/2 h exposure at temperature.
Source: MIL-HDBK 5, 1 Dec 1991

Forging

G.W. Kuhlman, ALCOA, Forging Division

Commercially important metastable β and near-β alloys respond well to thermomechanical processing (TMP) and several complex thermomechanical processing routes have been simplified for commercial practice. With this alloy class, thermomechanical processing is focused on optimal combinations of high strength, good fracture toughness, and ductility. All of these alloys possess superior high-cycle fatigue properties due to the refined nature of their microstructures. Several of these alloys have been successfully direct aged, thereby achieving even finer microstructures and improved smooth and notched fatigue properties.

Supra-transus forging processes prevail in this class of materials, except for Ti-10V-2Fe-3Al, in which the combination of supra- and subtransus working is used to produce desired properties through α phase manipulation and control. With the exception of Ti-10V-2Fe-3Al none of these alloys respond to thermomechanical processing to improve fracture-related properties, because α morphology cannot be modified to the extent possible in α + β alloys. Ti-10V-2Fe-3Al may offer the broadest range of engineering properties for this class of alloys. All of these alloys are ideal for use in structural applications where durability is the critical design criterion.

Ti-13-11-3 is a very high strength, metastable β alloy whose primary commercial applications in forgings are aerospace components, pressure vessels, and corrosion-resistant applications. The alloy can be fabricated into all forging product types, although closed die forgings predominate. Ti-13-11-3 is commercially fabricated on all types of forging equipment.

Ti-13-11-3 is a moderately forgeable alloy (when forged above the β transus), with higher unit pressures (flow stresses), improved forgeability, and less crack sensitivity in forging than the α-β alloy Ti-6Al-4V. Due to the high alloying content of Ti-13-11-3, its flow stresses are among the highest of commonly forged titanium alloys, more than double that of the near-β alloy Ti-10V-2Fe-3Al. The desired final microstructure from Ti-13-11-3 forging processing is transformed β with a fine recrystallized prior β grain size in preparation for final thermal treatments. Thus, Ti-13-11-3 is typically forged above its transus through one or more forging operations. Reheating for subsequent forging operations recrystallizes the alloy, thus refining prior beta grain size. Ti-13-11-3 may be subtransus (α + β) forged in final stages, with a significant increase in unit pressure requirements, to accomplish further recrystallization during heat treatment.

Final thermal treatments for Ti-13-11-3 forging include solution treatment annealing and aging. Forgings may be supplied in the solution treatment annealed (ST) condition and/or fully aged (STA). In the solution treatment annealed

Ti-13-11-3: Forging process temperatures

Process	Metal temperature	
	°C	°F
Beta forge	650-955	1200-1750

Note: See "Technical Note 4: Forging" for recommended die temperatures.

condition, Ti-13-11-3 has lower strengths but much higher ductility and toughness than in the STA condition. Solution treatment is conducted at 775 °C (1425 °F), followed by air cooling. Aging is conducted at 425 to 480 °C (800 to 900 °F).

Beta forging working histories for Ti-13-11-3 require imparting enough hot work to reach final macrostructure and microstructure objectives. Generally, reductions in any given forging process are 30 to 50% to achieve desired dynamic and static recrystallization. Very low levels of β reduction are not recommended. Although Ti-13-11-3 is cold worked in other product forms (sheet), cold working is not used for forgings.

Surface Treatment. Ti-13-11-3, as with all β alloys, has a higher affinity for hydrogen than other alloy classes. Although Ti-13-11-3 forms less α case from heating operations than other alloy classes, therefore requiring less metal removal in chemical pickling (milling processes), control of chemical removal processes is essential to preclude excessive hydrogen pick up.

Ti-13-11-3: Relative forging pressure comparison

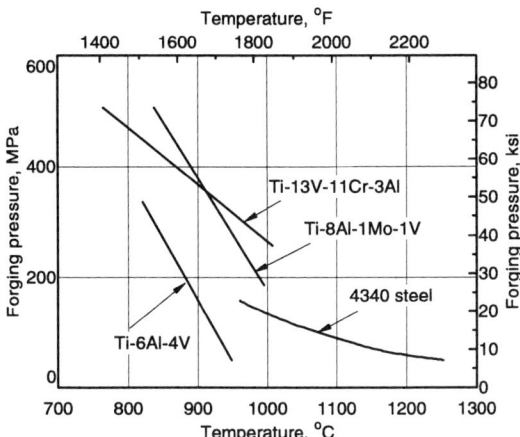

Effect of forging temperature on forging pressure.
Source: A. Sabroff, F. Boulger, and H. Henning, *Forging Materials and Practices*, Reinhold, 1968

Ti-13-11-3: Flow stress comparison at 900 °C

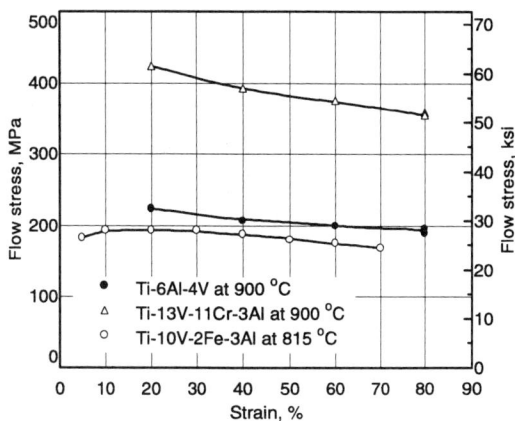

Flow stress of commonly forged titanium alloys at 10/s strain rate.

Formability

Ti-13-11-3 normally is fabricated to flat rolled products in the β-phase temperature field. However, the final fabrication of sheet by rolling to finish gages is often done cold to obtain improved flatness and gage uniformity. Similarly, in the production of rod and wire products aimed at spring manufacture, initial fabrication at elevated temperatures may be followed by cold working to improve the finished surface and the mechanical properties of the finished product.

Although requiring higher work forces, Ti-13-11-3 in the solution treated condition is more amenable to cold forming than in any other high-strength titanium alloys. It also has good cold heading properties. In very severe cold forming such as spinning or deep drawing, intermediate anneals may be advisable.

Forming by all conventional methods is possible.

Although the uniform elongation of Ti-13-11-3 at room temperature is fairly low, bend ductility is excellent. Flow stresses at strain rates typical of stretch forming are high. Thus, although such operations as stretch forming at room temperature may be difficult even on fully annealed material, operations involving bending are easily performed.

Ti-13-11-3: Forging pressures

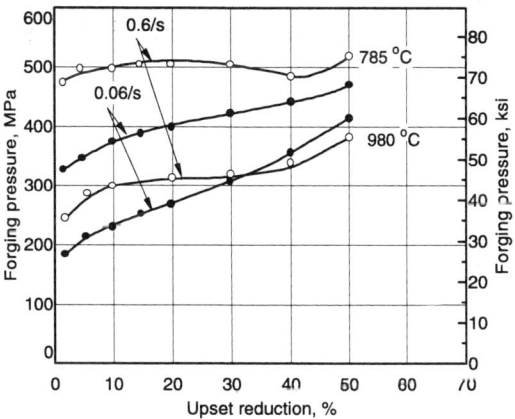

Source: A. Sabroff, F. Boulger, and H. Henning, *Forging Materials and Practices*, Reinhold, 1968

With respect to the effect of warm working after solution treatment on the subsequent aged properties of Ti-13-11-3, there is little difference between warm working and cold working. Both operations accelerate aging response.

892 / Beta and Near-Beta Alloys

Ti-13-11-3: Formability comparison

Material	Olsen cup height mm	Olsen cup height in.	Transverse bend radius (R/t)	Tensile yield strength MPa	Tensile yield strength ksi	Hardness, HRC
Ti-70A (commercial purity)	7.37	0.290	2	483(a)	70(a)	26
Ti-13-11-3, solution treated	6.35	0.25	3	827(a)	120(a)	32-36
Beta III						
Solution treated, water quenched	8.38	0.330	2	738	107	...
Solution treated, air cooled	6.35	0.25	3.5	883	128	...
Ti-6Al-4V, mill annealed	3.05	0.12	4.5	827(a)	120(a)	36

(a) Guaranteed minimum

Ti-13-11-3: Hot forming temperatures for annealed or solution treated material

Alloy	Forming temperature °C	Forming temperature °F
CP Ti (all grades)	480-705	900-1300
α and near-α alloys		
Ti-8Al-1V-1Mo	790 ± 15	1450 ± 25
Ti-5Al-2.5Sn	620-815	1150-1500
α-β alloys		
Ti-6Al-6V-2Sn	790 ± 15	1450 ± 25
β alloy		
Ti-13V-11Cr-3Al	605-790	1125-1450

Source: "Fabrication Practices for Titanium and Titanium Alloys," Lockheed Corporate Process Specification LCP70-1099, Revision B, Lockheed-California Company, Oct 1983

Ti-13-11-3: Stretching and bending strain limit

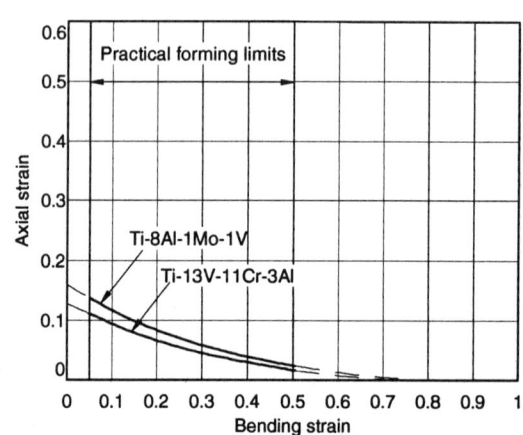

Composite of bead forming limits. These curves indicate that a part with a stretching strain of 0.1 mm/mm should have a bending strain of less than 0.1 mm/mm.
Source: R.A. Wood and R.J Favor, *Titanium Alloys Handbook*, MCIC-HB-02, Battelle Columbus Laboratories, 1972

Bending and Stretching Limits

Bending Limits

Ti-13-11-3: Heel-in bending limit

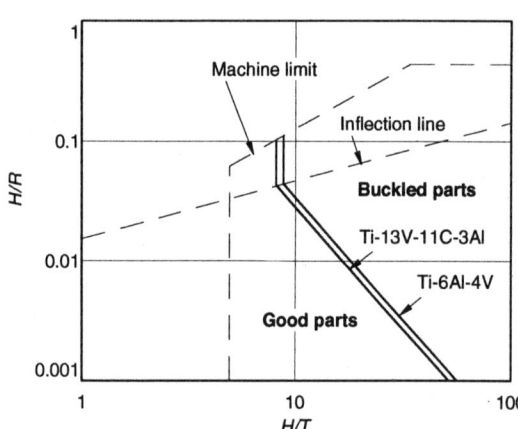

Transverse buckling and wrinkling, respectively, are the common modes of failure in bending heel-out and heel-in channels. The principal parameters are the bend radius, R; the channel height, H; the web width, W; and the material thickness, T.
Source: R.A. Wood and R.J. Favor, *Titanium Alloys Handbook*, MCIC-HB-02, Battelle Columbus Laboratories, 1972

Ti-13-11-3: Bend radius vs pad pressure

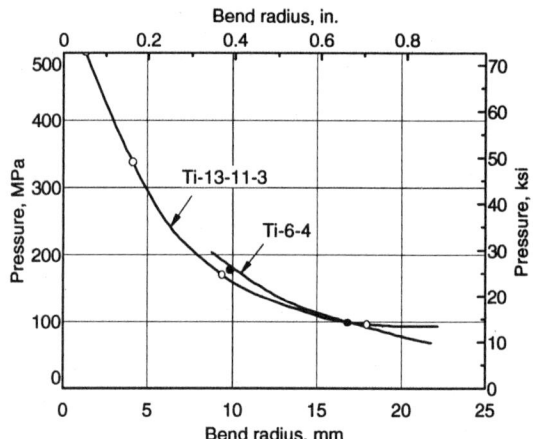

Effect of pad pressure on radii formed in 1.60 mm (0.063 in.) thick titanium alloy sheets at room temperature.

Ti-13-11-3: Bend radii comparison of annealed or solution treated material

	Minimum bend radius as a function of sheet thickness, t	
Alloy	$t <$ 1.75 mm (0.069 in.)	1.75 mm (0.069 in.) $< t <$ 4.76 mm (0.1875 in.)
CP titanium		
ASTM grade 1	2.5	3.0
ASTM grade 2	2.0	2.5
ASTM grade 3	2.0	2.5
ASTM grade 4	1.5	2.0
α alloys		
Ti-5Al-2.5Sn	4.0	4.5
Ti-5Al-2.5Sn ELI	4.0	4.5
Ti-6Al-2Nb-1Ta-0.8Mo	...	...
Ti-8Al-1Mo-1V	4.5(a)	5.0(b)
α-β alloys		
Ti-6Al-4V	4.5	5.0
Ti-6Al-4V ELI	4.5	5.0
Ti-6Al-6V-2Sn	4.0	4.5
Ti-6Al-2Sn-4Zr-2Mo	4.5	5.0
Ti-3Al-2.5V	2.5	3.0
Ti-8Mn	6.0	7.0
β alloys		
Ti-13V-11Cr-3Al	3.0	3.5
Ti-11.5Mo-6Zr-4.5Sn	3.0	3.0
Ti-3Al-8V-6Cr-4Mo-4Zr	3.5	4.0
Ti-8Mo-8V-2Fe-3Al	3.5	3.5

(a) 4.0 in transverse direction. (b) 4.5 in transverse direction.
Source: Military Standard MIL-T-9046J, US Government Printing Office

Stretching Limits

Ti-13-11-3: Stretch limits of heel-out sections

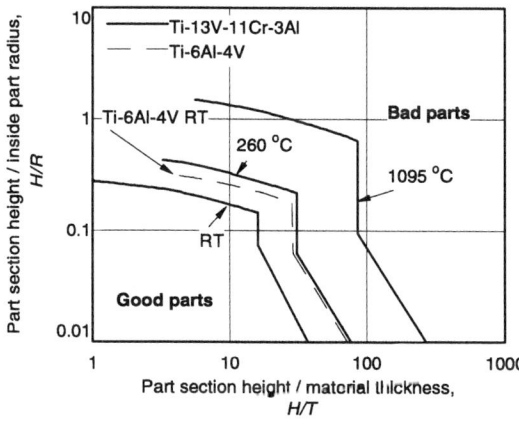

Linear stretch heel-out angle section (outboard) test results at elevated temperatures.
Source: R.A. Wood and R.J. Favor, *Titanium Alloys Handbook*, MCIC-HB-02, Battelle Columbus Laboratories, 1972

Ti-13-11-3: Inboard stretch limits

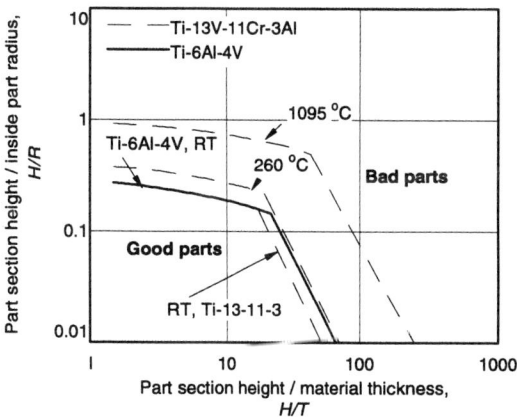

The formability limits for a formed section or extrusion to be stretch formed inboard depend on the ductility and buckling limits of the material. The index governing the optimum forming temperature will largely depend on the material thickness, t.
Source: R.A. Wood and R.J. Favor, *Titanium Alloys Handbook*, MCIC-HB-02, Battelle Columbus Laboratories, 1972

Ti-13-11-3: Stretch limit

Material	Condition	Maximum stretch(a) at 480 °C
Ti-13V-11Cr-3Al	Solution treated	15.8
Ti-8Mn	Annealed	15.8
Ti-5Al-2.5Sn	Annealed	12.6
Ti-6Al-4V	Annealed	12.6
Ti-3.25Mn-2.25Al	Annealed	15.8

(a) Percent stretch = $(L_1 - L_0)/L_0 \times 100$, where L_1 = stretched length; and L_0 = original length. Source: R.A. Wood and R.J. Favor, *Titanium Alloys Handbook*, MCIC-HB-02, Battelle Columbus Laboratories, 1972

Ti-13-11-3: Inboard stretching limits

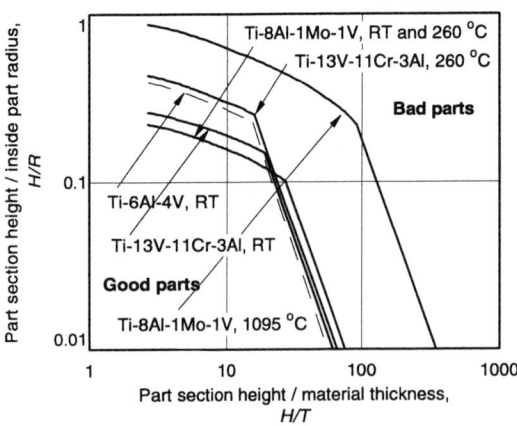

Composite limit curves for titanium linear-stretch heel-in (inboard) angle and channel sections at various temperatures.
Source: R.A. Wood and R.J. Favor, *Titanium Alloys Handbook*, MCIC-HB-02, Battelle Columbus Laboratories, 1972

Ti-13-11-3: Deep drawing limits

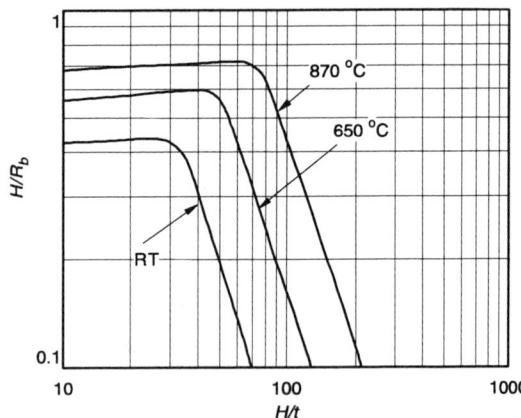

The important dimensions are the cup depth, H; the blank diameter, D_B; and the inside cup diameter, D_D. The material thickness and the draw radius are also important parameters, but do not enter into the formability limits directly.
Analytical extension of deep-draw-limit curve. R_b = blank radius.
Source: R.A. Wood and R.J. Favor, *Titanium Alloys Handbook*, MCIC-HB-02, Battelle Columbus Laboratories, 1972

Spinning, Beading, and Dimpling

Spinning Limits

Ti-13-11-3: Spinning limits

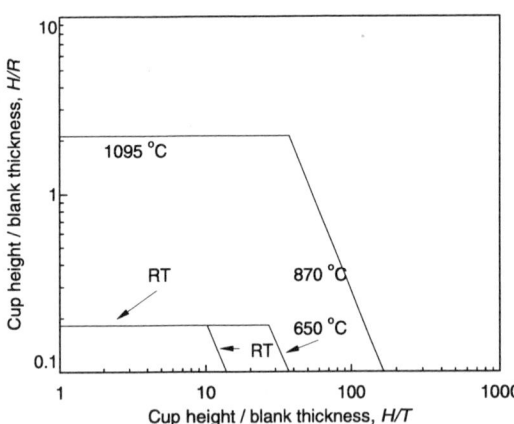

Plastic buckling, if the height-to-radius ratio (H/R) becomes too large and failure by elastic buckling will occur if the height-to-thickness (H/T) becomes too large. The position of the curves will vary according to the properties of the material and the forming temperature.
Analytical extension of spinning-limit curve.
Source: R.A. Wood and R.J. Favor, *Titanium Alloys Handbook*, MCIC-HB-02, Battelle Columbus Laboratories, 1972

Ti-13-11-3: Elastic buckling limit in spinning

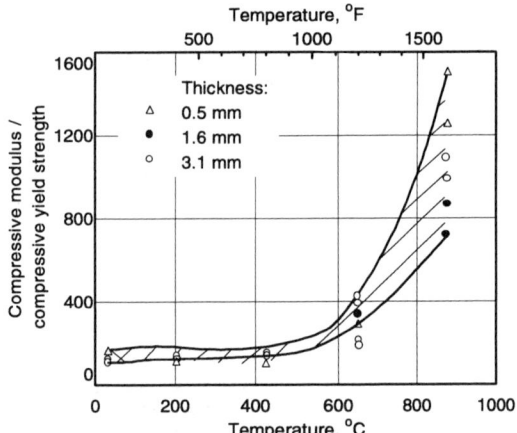

The change toward better formability starts around 540 °C (1000 °F) and increases rapidly around 760 °C (1400 °F). The latter temperature is approximately the highest temperature that can be used without degrading the properties of the alloys. The total time required for forming may also influence the choice of spinning temperature.
Effect of temperature on elastic buckling limit in spinning.
Source: R.A. Wood and R.J. Favor, *Titanium Alloys Handbook*, MCIC-HB-02, Battelle Columbus Laboratories, 1972

Ti-13-11-3: Manual spinning

Thickness		Blank diameter/	Limiting ratio for alloys indicated(a)			
			Blank diameter/ cup diameter		Cup height/ cup diameter	
mm	in.	sheet thickness	AMS 4911	AMS 4917	AMS 4911	AMS 4917
0.5	0.020	25	1.3	1.2	0.22	0.14
		50	1.3	1.2	0.22	0.14
		100	1.2	1.2	0.14	0.14
		150	1.2	1.1	0.14	0.07
		200	1.1	...	0.07	...
1.6	0.063	25	1.3	1.2	0.22	0.14
		50	1.2	1.2	0.14	0.14
3.1	0.125	25	1.2	1.2	0.14	0.14
		50	1.1	...	0.07	...

(a) Alloys AMS 4911 and AMS 4917 for Ti-6Al-4V and Ti-13V-11Cr-3Al, respectively. The term cup diameter refers to the inside diameter; the cup height is based on the outside dimension. Source: R.A. Wood and R.J. Favor, *Titanium Alloys Handbook*, MCIC-HB-02, Battelle Columbus Laboratories, 1972

Trapped rubber forming is often used for forming beaded panels. The forming limits for beaded panels are determined by failures resulting from splitting or from buckling. Consequently, success or failure depends on the ratio of the bead radius to the thickness of the material (R/T), or on the spacing of the beads (R/L). Increasing the forming pressure increases the limiting R/T ratios, and increasing the forming temperature permits closer beads in sheets of a particular gage.

Ti-13-11-3: Dimpling limit comparison to Ti-811

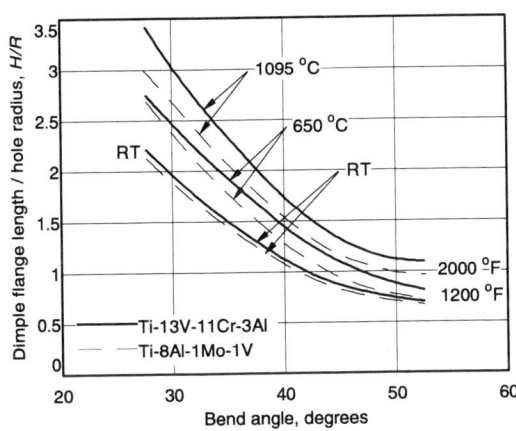

Good parts can be formed for conditions under the curves, whereas split parts can be expected for conditions above the curves. The major failure in dimpling is caused by simple tension.
Theoretical relationship between ratio H/R and bend angle for the dimpling of titanium alloys.
Source: R.A. Wood and R.J. Favor, *Titanium Alloys Handbook*, MCIC-HB-02, Battelle Columbus Laboratories, 1972

Ti-13-11-3: Beading limits

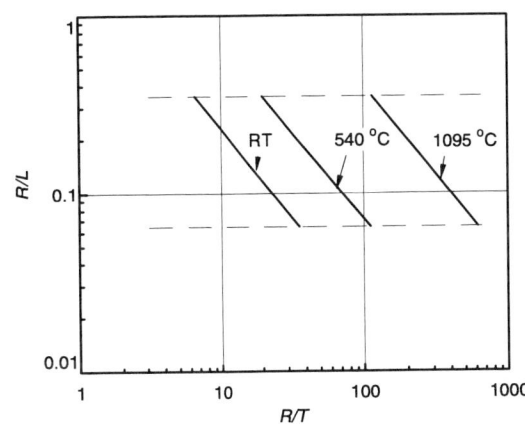

Although the limits apply to beaded panels, they can be used with caution as guides to forming other types of parts with drop hammers.
Limits for forming beaded panels with a drop hammer.
Source: R.A. Wood and R.J. Favor, *Titanium Alloys Handbook*, MCIC-HB-02, Battelle Columbus Laboratories, 1972

Dimpling Limits

Dimpling limits

Material	Condition	Dimpling temperature		Dimpling limits, H/R, for various bend angles above and below standard bend angle (standard)				
		°C	(°F)	30°	35°	40°	45°	50°
Ti-6Al-4V	Mill annealed	RT		2.00	1.5	1.17	0.92	0.74
Ti-13V-11Cr-3Al	Aged 480 °C	RT		1.58	1.17	0.91	0.73	0.60
Ti-8Al-1Mo-1V	Duplex annealed	RT		1.88	1.42	1.08	0.82	0.70
Ti-13V-11Cr-3Al	Solution annealed	650	(1200)	2.58	1.95	1.48	1.15	0.96
Ti-8Al-1Mo-1V	Duplex annealed	650	(1200)	2.30	1.72	1.30	1.00	0.85

Note: Dimpling limits for radial splitting at edge of hole. Bend angles above and below the standard 40° angle are given. Other conditions of heat treatment and dimpling at elevated temperatures would necessitate the use of other dimpling limits. Source: R.A. Wood and R.J. Favor, *Titanium Alloys Handbook*, MCIC-HB 02, Battelle Columbus Laboratories, 1972

Heat Treatment

See also "Forging" for heat treatment description.

High strength can be achieved by solution treating and aging of Ti-13V-11Cr-3Al. Strength and ductility combinations from aging and the rate of aging depend on the processing history of the metal being heat treated. Optimum aged properties are obtained when the prior history of the metal is such that it creates a favorable nucleation distribution. Therefore, some residual strain energy should promote aging response. Cold working or warm working can be used to achieve the residual strain required. Residual strain energy accelerates the aging reaction and imparts somewhat better ductility for some strength levels.

Solution Treatment. Within the broad solution treatment range of 705 to 1035 °C (1300 to 1900 °F), there is little change in aging response. However, long periods at solution temperatures degrade ductility, presumably from grain growth and the breakdown of nucleation site distribution. Water quenching from the solution heat treatment temperature does not offer a significant advantage over air cooling, except where it might aid in removing heat from thick sections.

Weldment Stress Relief. Much of the available data on stress relief annealing pertains to weldments. Fusion weldments in 6.4 mm (0.250 in.) plate are reported to be stress relieved to zero residual stress levels by any of the following treatments:

4 h at 480 °C (900 °F), AC
1 h at 540 °C (1000 °F), AC
<30 min at 595 °C (1100 °F), AC
<30 min at 650 °C (1200 °F), AC

A 285 °C (550 °F) preheat on material to be welded may also reduce residual tensile stresses in the weldment. It has been reported that 15 min 760 °C (1400 °F) or 5 min 980 °C (1800 °F) treatments resulted in weld embrittlement (R.A. Wood, *Beta Titanium Alloys*, Battelle Columbus Laboratories, 1972, p 26).

Ti-13-11-3: Aging response variations

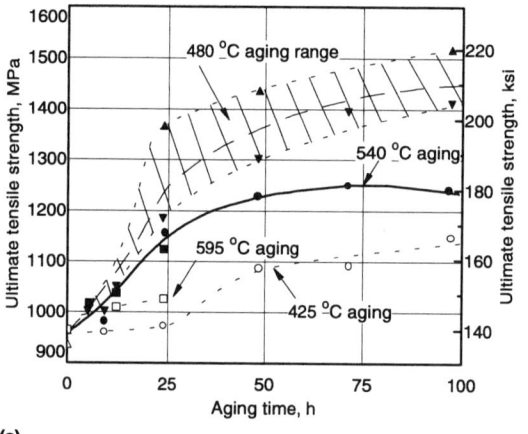
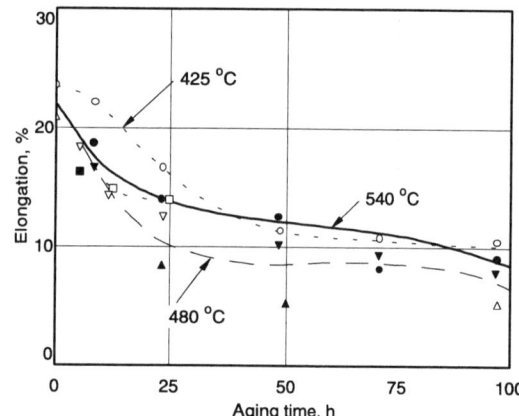

(a) (b)

Effect of aging time on the longitudinal tensile properties of sheet.
Source: R.A. Wood, *Beta Titanium Alloys*, MCIC-72-11, Battelle Columbus Laboratories, 1972

Ti-13-11-3: Solution treatment and aging

Heat treatment	Temperature °C	°F	Time, h	Cooling method
Typical solution treatment	>760	>1400	0.25-1	AC or WQ
Narrow solution treating range	775-800	1425-1475	0.25-1	AC or WC
Broad ST range	700-1040	1300-1900	0.25-1	AC or WQ
Aging range	425-540	800-1000	...	AC
Typical age	425-510	800-950	20-100	AC

Ti-13-11-3: Grain size at solution temperatures

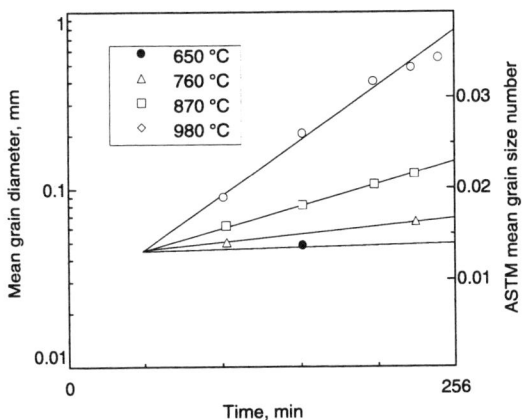

Source: R.A. Wood, *Beta Titanium Alloys*, MCIC 72-11, Battelle Columbus Laboratories, 1972

Ti-13-11-3: Effect of solution temperature on hardness

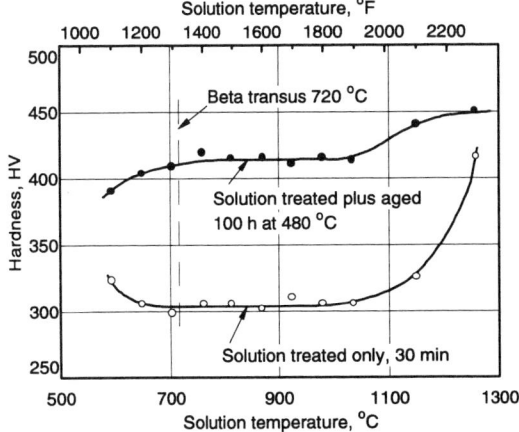

Effect of solution temperature on the annealed and annealed plus 480 °C (900 °F) aged Vickers hardness. Hardnesses shown are averages of five impressions, using a 5-kg load.
Source: R.A. Wood, *Beta Titanium Alloys*, MCIC-72-11, Battelle Columbus Laboratories, 1972

Ti-13-11-3: Effect of working on aging

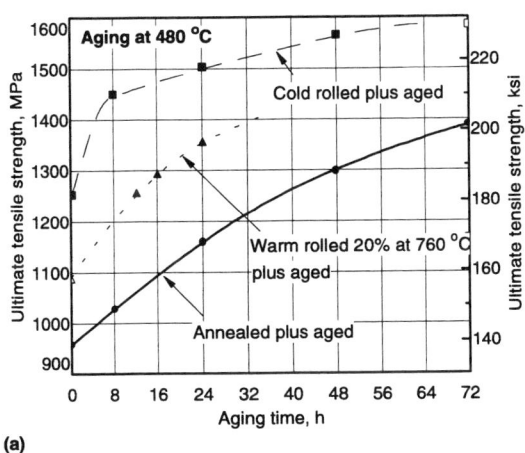
(a)

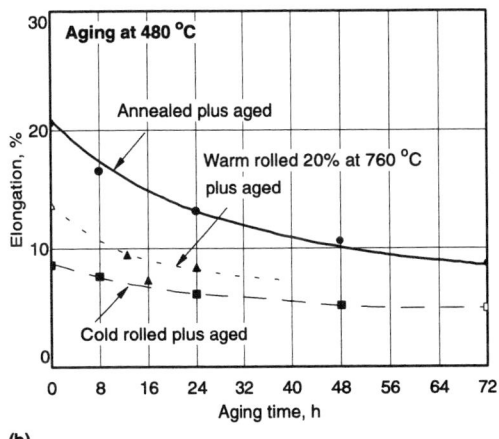
(b)

Effect of warm and cold rolling on longitudinal aged tensile properties.
Source: R.A. Wood, *Beta Titanium Alloys*, MCIC-72-11, Battelle Columbus Laboratories, 1972

Ti-13-11-3: Stress relief and annealing treatments

Treatment	Temperature °C	Temperature °F	Duration, min	Cooling method
Interstage anneal of sheet after severe deformation	730-760	1350-1400	...	AC
Stress relief of ST stock	540(a)	1000(a)	15	AC
Typical stress relief	700-785(b)	1300-1450(b)	5-15	AC
Typical anneal	Same as solution treatment			AC or WQ

(a) Stress relief if aging is not planned; stress relief can be accomplished during 480 °C (900 °F) aging. (b) Stress relief for material other than weldments

Ti-15V-3Cr-3Al-3Sn

Common Name: Ti-15-3
UNS Number: Unassigned

Ti-15-3 was developed during the 1970's on an Air Force contract and was later scaled up to produce titanium strip. It is a solute-rich beta titanium alloy developed primarily to lower the cost of titanium sheet metal parts by reducing processing cost through the capability of being strip producible and its excellent room-temperature formability characteristics. It can also be aged to a wide range of strength levels to meet a variety of applications. Although originally developed as a sheet alloy, it has expanded into other areas such as fasteners, foil, plate, tubing, castings and forgings.

Chemistry and Density

Ti-15-3 is formulated by depressing the beta transus with vanadium and chromium additions. It is less beta-stabilized than Ti-13V-11Cr-3Al.

Density. 4.76 g/cm^3 (0.172 lb/in.^{3})

Product Forms

Ingot, billet, plate, sheet, strip, seamless tube, castings, and welded tube.

Product Conditions/Microstructure

The alloy can be directly aged after forming. However, strength will vary depending upon the amount of cold work in the part. Heating times prior to hot forming should be minimized in order to prevent appreciable aging prior to forming.

Applications

Ti-15-3 is used primarily in sheet metal applications since it is strip-producible, age-hardenable, and highly cold-formable. It is used in a variety of airframe applications, in many cases replacing hot-formed Ti-6Al-4V. Ti-15-3 can also be produced as foil, is an excellent casting alloy, and has also been evaluated for aerospace tankage applications, high-strength hydraulic tubing and fasteners.

Airframe Structures. Ti-15-3 possesses good potential for lowering the manufacturing costs of titanium airframe structures. Studies on its formability led to use as the lower half of the A-10 fuselage frame. Production costs are lower than those for Ti-6-4. Ti-15-3 welded tubing is used for pneumatic ducting, and Ti-15-3 sheet is formed into hemispheres and welded to fabricate fire extinguisher bottles on the Boeing 777. Other potential applications for this material are as seamless tubing, wire, rivets, and foil for honeycomb structures. High-strength castings are in use.

Use Limitations. Ti-15-3, like other beta titanium alloys, is highly susceptible to hydrogen pickup and rapid hydrogen diffusion during heating, pickling, and chemical milling. However, because of the much higher solubility of hydrogen in the beta phase than in the alpha phase of titanium, this alloy has a higher tolerance to hydrogen embrittlement than the alpha or alpha-beta alloys.

Ti-15-3 can be welded in the solution-treated condition; however, welding is not recommended after solution treating and aging. Care is necessary in pickling to minimize hydrogen absorption.

Ti-15V-3Cr-3Al-3Sn: Specifications and Compositions

Specification	Designation	Description	Al	Cr	Fe	H	N	O	Sn	V	Other
USA											
AMS 4914		Sh Strp SHT	2.5-3.5	2.5-3.5	0.25	0.015	0.05	0.13	2.5-3.5	14-16	C 0.05; OT 0.4; bal Ti
AMS 4914		Sh Strp STA	2.5-3.5	2.5-3.5	0.25	0.015	0.05	0.13	2.5-3.5	14-16	C 0.05; OT 0.4; bal Ti

Ti-15V-3Al-3Cr-3Sn: Commercial compositions

Specification	Designation	Description	Composition, wt%									
			Al	Cr	Fe	H	N	O	Sn	V	Other
Japan											
Kobe	KS15-3-3-3	Plt Sh SHT	2.5-3.5	2.5-3.5	0.25	0.015	0.05	0.13	2.5-3.5	14-16	bal Ti
Kobe	KS15-3-3-3	Plt Sh STA	2.5-3.5	2.5-3.5	0.25	0.015	0.05	0.13	2.5-3.5	14-16	bal Ti
USA											
Timet	TIMETAL 15-3	Strp Plt Sh Frg SHT	2.5-3.5	2.5-3.5	0.25	0.015	0.05	0.13	2.5-3.5	14-16	bal Ti
Timet	TIMETAL 15-3	Strp Plt Sh Frg STA	2.5-3.5	2.5-3.5	0.25	0.015	0.05	0.13	2.5-3.5	14-16	bal Ti

Phases and Structures

Ti-15-3 can retain an all-beta structure with sufficiently rapid cooling (e.g., air cooling of a 6.5-mm thick section from the β field). Subsequent aging produces a fine α phase, which is very difficult to resolve optically as it is extremely fine.

Beta Transus. 750 to 770 °C (1385 to 1415 °F)

Grain Structure

Ti-15-3: Grain size after full recrystallization of cold rolled strip

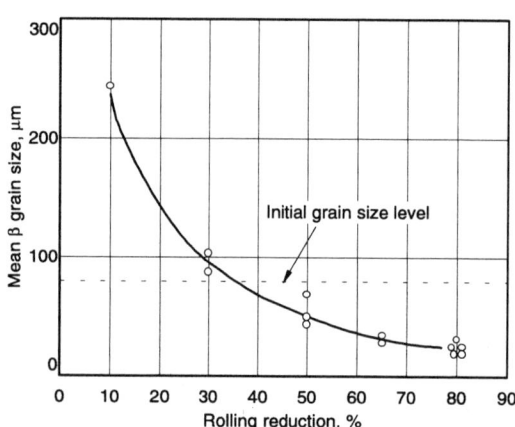

Source: H. Ohyama and Y. Ashida, Estimation of Recrystallized Grain Size Under Continuous Annealing of Cold-Rolled β Titanium Alloy Strip, *ISIJ Int.*, Special Issue on Recent Advances on Titanium Technology, Vol 31 (No. 8), 1991, p 800

Ti-15-3: Annealing time vs recrystallized grain size

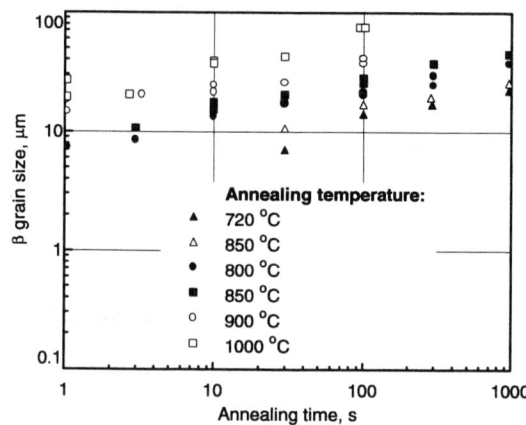

80% cold rolled. Unrecrystallized regions were present for annealing times of 1 s and for several samples with grain sizes less than 20 μm.
Source: H. Ohyama and Y. Ashida, Estimation of Recrystallized Grain Size Under Continuous Annealing of Cold-Rolled β Titanium Alloy Strip, *ISIJ Int.*, Special Issue on Recent Advances on Titanium Technology, Vol 31 (No. 8), 1991, p 801

Ti-15-3: Measured vs calculated grain size

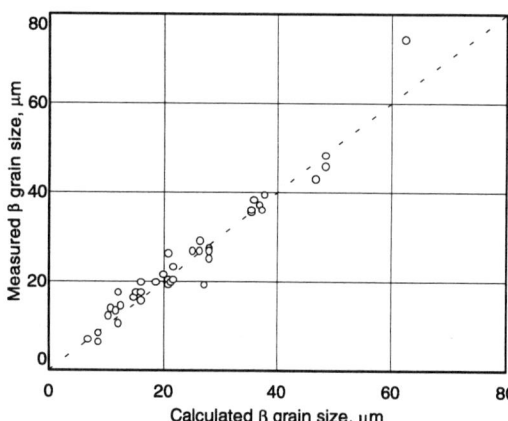

$D = 0.80 \times 10^4 \, t^{0.24} \exp(-1.50 \times 10^4 / RT)$, where D = grain size in μm; T = annealing temperature, K; t = annealing time, seconds.
Source: H. Ohyama and Y. Ashida, Estimation of Recrystallized Grain Size Under Continuous Annealing of Cold-Rolled β Titanium Alloy Strip, *ISIJ Int.*, Vol 31 (No. 8), 1991, p 802

Transformation Products

As a solute-rich β alloy, Ti-15-3 undergoes a phase separation into a solute-rich phase (β) and a solute-lean phase (β′) prior to the formation of uniform, needle-like α within the β′. This uniformly dispersed α provides strengthening, but its formation requires longer aging times because of slower nucleation kinetics (compared to solute-lean β alloys). The refinement of the size and spacing of α precipitates can be achieved by minimizing the amount of recovery and recrystallization after deformation. Stored energy (deformation) alters the aging kinetics of uniform α and generally produces a finer precipitate.

An accelerated rate of intragranular α formation also reduces the extent of grain boundary α formation. Like other solute-rich β alloys, Ti-15-3 is also susceptible to the formation of grain boundary α. The tendency to form grain boundary α is more pronounced because the nucleation kinetics of grain boundary α is not as solute sensitive as the kinetics of homogeneous α precipitation.

Physical Properties

Ti-15-3: Summary of typical physical properties

Beta transus	750 to 770 °C (1385 to 1415 °F)
Density(a)	4.7 g/cm^3 (0.172 lb/in.^{3})
Electrical resistivity(a)	1.4 μΩ · m
Magnetic permeability	Nonmagnetic
Specific heat capacity(a)	500 J/kg · K (0.12 Btu/lb · °F)
Thermal conductivity(a)	8.08 W/m · K (4.67 Btu/ft · h · °F)
Thermal coefficient of linear expansion(b)	8.5 × 10^{-6}/°C (4.7 × 10^{-6}/°F)

(a) Typical values at room temperature of about 20 to 25 °C (68 to 78 °F). (b) Mean coefficient from room temperature to 95 °C (200 °F)

Ti-15-3: Electrical resistivity

Temperature		Resistivity, μΩ · m
°C	°F	
25	77	1.48
260	50	1.55
540	1000	1.60

Ti-15-3: Room-temperature tensile modulus

Condition	Heat treatment	Young's modulus	
		GPa	10^6 psi
Solution treated	785 °C (1450 °F), 10 min, AC	81-84	11.8-12.2
Aged	540 °C (1000 °F), 8 h, AC	102-105	14.8-15.2

Source: P.J. Bania et al., Development and Properties of Ti-15V-3Cr-3Sn-3Al, in *Beta Titanium Alloys in the 1980's*, R.R. Boyer and H.W. Rosenberg, Ed., TMS/AIME, 1984, p 217

Corrosion Properties

Like other titanium alloys, successful application of Ti-15-3 can be expected in mildly reducing to highly oxidizing environments in which protective oxide films spontaneously form and remain stable. On the other hand, hot, concentrated, low-pH chloride salts corrode titanium; warm or concentrated solutions of hydrochloric, phosphoric and oxalic acids also are damaging. In general, all acidic solutions that are reducing in nature corrode titanium, unless they contain inhibitors. Strong oxidizers, including anhydrous red fuming nitric acid and 90% hydrogen peroxide, also cause attack. Ionizable fluoride compounds, such as sodium fluoride and hydrogen fluoride, activate the surface and can cause rapid corrosion. Dry chlorine gas is especially harmful.

Weldments and Castings. Few published data on the corrosion resistance of weldments and castings are available for most β alloys. Under marginal or active conditions (for corrosion rates ≥0.10 mm/year or 4 mils/year) weldments may experience accelerated corrosion attack relative to the base metal.

Hydrogen Tolerance

Ti-15-3 has a high tolerance to hydrogen up to about 4000 to 5000 ppm. At hydrogen levels of up to 2000 pm, tests indicate no effect on solution annealed strength or ductility. Cup tests on solution annealed material showed no effect up to 4000 ppm. Similarly, tests have shown no effect on aged tensile properties up to 2000 ppm when the hydrogen was added after the material was fully aged. However, when hydrogen was added before the age cycle, aged strengths were reduced by approximately 275 MPa (40 ksi) at 1000 ppm and an additional 138 MPa (20 ksi) at 2400 ppm. Base hydrogen was approximately 130 ppm. In no case was the ductility reduced, even up to 2400 ppm. The strength reduction arises from the fact that hydrogen is a potent β stabilizer and acts to suppress the aging response. Nonetheless, Ti-15-3 is very hydrogen tolerant and is well within the safe range at the typical specification upper limit of 150 ppm.

902 / Beta and Near-Beta Alloys

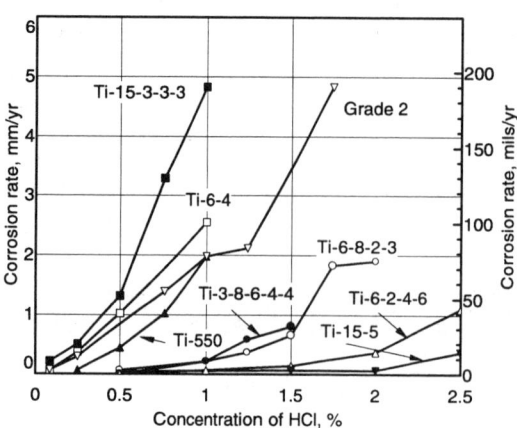

Aged Ti-15-3: Corrosion comparison in boiling HCl

General corrosion of aged titanium alloys in naturally aerated boiling HCl solutions.
Source: *Metals Handbook, Corrosion*, Vol 13, 9th ed., ASM International, 1987

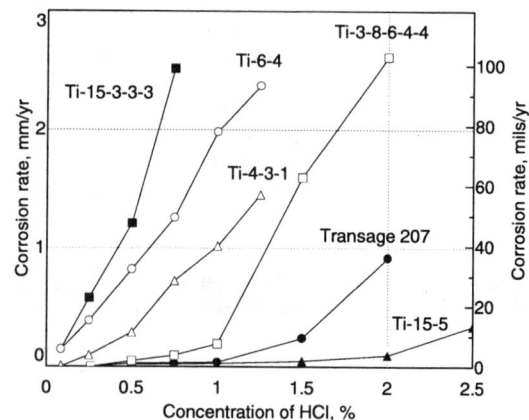

Annealed Ti-15-3: Corrosion comparison in boiling HCl

General corrosion of annealed titanium alloys in naturally aerated boiling HCl solutions.
Source: *Metals Handbook, Corrosion*, Vol 13, 9th ed., ASM International, 1987

Stress-Corrosion Cracking

Ti-15-3 is expected to be less susceptible to stress-corrosion cracking (SCC) than Ti-6Al-4V due to its high vanadium content and lower aluminum content. A β phase stabilized by vanadium (or molybdenum, niobium, or tantalum) is immune to SCC in aqueous media. (R. Schutz, Stress-Corrosion Cracking of Titanium Alloys, in *Stress-Corrosion Cracking: Materials Performance and Evaluation*, ASM International, 1992).

Thermal Properties

Heat Capacity

Specific Heat. At room temperature: 500 J/kg · K (0.12 Btu/lb · °F)

Ti-15-3: Specific heat

Specific heat		Temperature	
J/kg · K	Btu/lb · °F	°C	°F
508	0.121	25	77
574	0.137	200	392
649	0.155	400	752
724	0.173	600	1112
784	0.187	800	1400

Source: H.W. Rosenberg, Ti-15-3 Property Data, in *Beta Titanium Alloys in the 1980's*, R.R. Boyer and H.W. Rosenberg, Ed., TMS/AIME, 1984, p 411

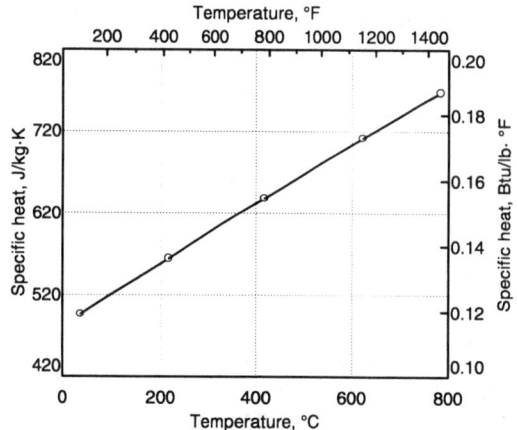

Ti-15-3: Specific heat

Alloy aged at 540 °C (1000 °F) for 8 h.
Source: G. Lenning, J. Hall, M. Rosenblum and W. Trepel, "Cold-Formable Titanium Sheet," AFWAL-TR-82-4187, Materials Laboratory, Wright Patterson AFB, Ohio, Dec 1982, p 162

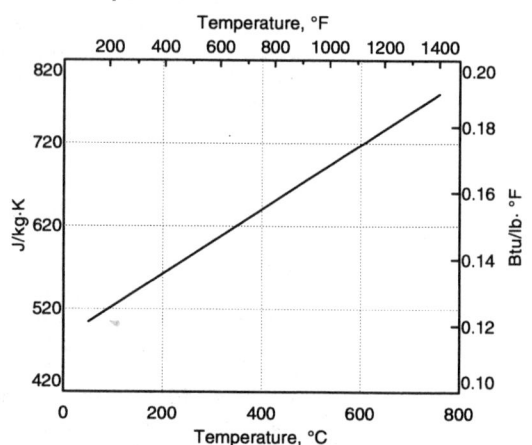

Ti-15-3: Specific heat

Source: MIL-HDBK 5, Dec 1991

Thermal Expansion

Ti-15-3: Thermal coefficient of linear expansion

Temperature		L/L_o, %		Average coefficient $10^{-6}/°C$	
°C	°F	L	T	L	T
23	73	0	0	...	...
100	212	0.0647	0.0653	8.41	8.48
200	390	0.1582	0.1537	8.95	8.68
300	570	0.2540	0.2514	9.16	9.07
400	750	0.3558	0.3559	9.43	9.43
500	930	0.4622	0.4614	9.68	9.67
600	1110	0.5763	0.5701	9.88	9.99
700	1290	0.7054	0.7085	10.42	10.48
800	1470	0.8046	0.8080	10.93	10.96

Source: H.W. Rosenberg, Ti-15-3 Property Data, in *Beta Titanium Alloys in the 1980's*, R.R. Boyer and H.W. Rosenberg, Ed., TMS/AIME, 1984, p 410

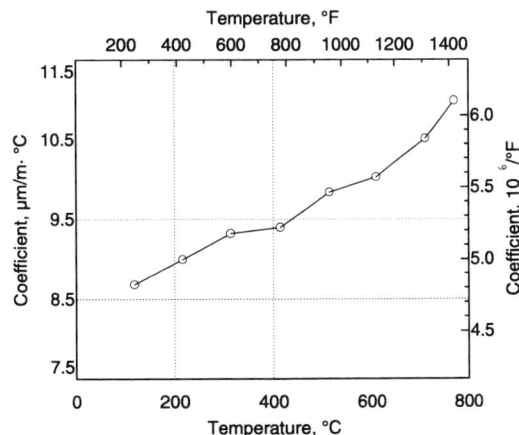

Ti-15-3: Thermal coefficient of linear expansion

Alloy used was aged at 540 °C (1000 °F) for 8 h.
Source: G. Lenning, J. Hall, M. Rosenblum, and W. Trepel, "Cold Formable Titanium Sheet," AFML-TR-4187, Materials Laboratory, Wright Patterson AFB, Ohio, Dec 1982, p 162

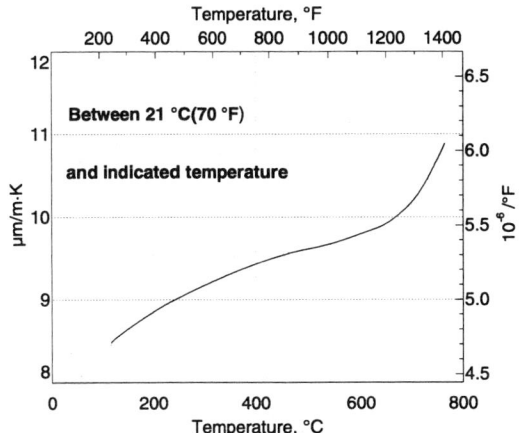

Ti-15-3: Thermal coefficient of linear expansion

Source: MIL-HDBK 5, Dec 1991

Thermal Conductivity

Ti-15-3: Thermal conductivity

Temperature		Conductivity	
°C	°F	Btu · in./h · ft² · °F	W/m · K
25	75	56.0	8.08
260	500	83.1	11.99
538	1000	115.4	16.64
760	1400	137.2	19.79

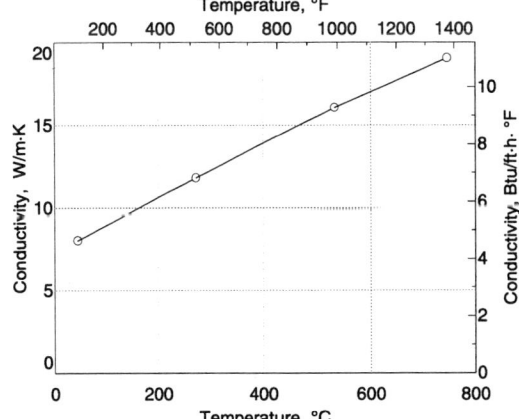

Ti-15-3: Thermal conductivity

Alloy aged at 540 °C (1000 °F) for 8 h.
Source: G. Lenning, J. Hall, M. Rosenblum, and W. Trepel, "Cold Formable Titanium Sheet," AFML-TR-82-4187, Materials Laboratory, Wright Patterson AFB, Ohio, Dec 1982, p 164

Ti-15-3: Thermal conductivity

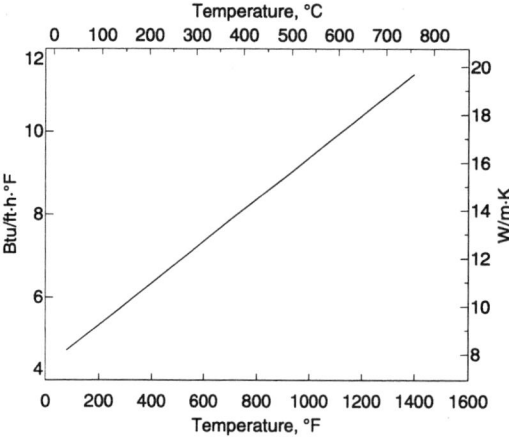

Source: MIL-HDBK 5, Dec 1991

Mechanical Properties

Ti-15-3: Design allowables of sheet
S-basis values for sheet ≤3.175 mm (0.125 in.), per AMS 4914

Condition	Solution treated(a)	Solution treated and aged at 540 °C (1000 °F)
F_{tu}, MPa (ksi)		
L	675 (98)	972 (141)
LT	703 (102)	99 (145)
F_{ty}, MPa (ksi)		
L	662 (96)	937 (136)
LT	689 (100)	965 (140)
F_{cy}, MPa (ksi)		
L	689 (100)	958 (139)
LT	724 (105)	993 (144)
F_{su}, MPa (ksi)	524 (76)	627 (91)
F_{bru}(b), MPa (ksi)		
e/D = 1.5	1165 (169)	1489 (216)
e/D = 2.0	1517 (220)	1896 (275)
F_{bry}(b), MPa (ksi)		
e/D = 1.5	930 (135)	1399 (203)
e/D = 2.0	1027 (149)	1586 (230)
EL, %		
LT	12	7
Tensile modulus, GPa (10^6 psi)		
L	...	105 (15.2)
LT	...	109.5 (15.9)
Compressive modulus, GPa (10^6 psi)		
L	...	105.5 (15.3)
LT	...	110 (16.0)

(a) Solution treated material should not be subjected to service temperatures above approximately 205 °C (400 °F). Solution treated material, which has been cold worked, should not be exposed to service temperatures above approximately 150 °C (300 °F). Long-time exposures above these temperatures could result in low ductility. (b) Bearing values are "dry pin" values. Source: MIL-HDBK 5, Dec 1991

Ti-15-3: Effect of cold rolling on tensile properties

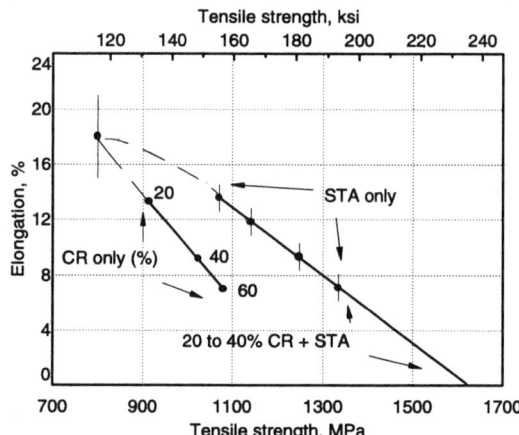

Relationship between strength and ductility showing effect of cold rolling before aging.
Source: *Beta Titanium Alloys in the 1980's*, R.R. Boyer and H.W. Rosenberg, Ed., TMS/AIME, 1984

Hardness

Ti-15-3: Effect of warm work-extrusion on Rockwell hardness

Product form	Room-temperature hardness, HRA
63 mm (2.5 in.) OD × 11 mm (0.430 in.) wall tube, as reduced from 86 mm (3.4 in.) OD × 15 mm (0.600 in.) wall tube	62-64
44 mm (1.75 in.) OD × 7.6 mm (0.300 in.) wall tube, as reduced from 86 mm (3.4 in.) OD × 15 mm (0.600 in.) wall tube	66-67
63 mm (2.5 in.) OD × 11 mm (0.430 in.) wall tube, 840 °C (1545 °F) continuous anneal	61-62
44 mm (1.75 in.) OD × 7.6 mm (0.300 in.) wall tube, 800 °C (1475 °F) continuous anneal	62

Note: Holed extrusion billets (machined from as-quenched forging) were coated with lubricant, induction heated to 595 to 980 °C (1100-1800 °F), transferred automatically to the press and extruded over a mandrel. The extrusion process was essentially isothermal and was characterized by a slow extrusion speed (<30 in./min) with relatively low extrusion ratios (<20). Extruded tube was cold reduced as indicated.
Source: P. Finden, Production of Seamless Titanium Alloy Tubing, in *Sixth World Conference on Titanium*, P. Lacombe, R. Tricot, and G. Beranger, Ed., Les Editions de Physique, Paris, 1989, p 1251

Ti-15-3: Effect of aging time on hardness

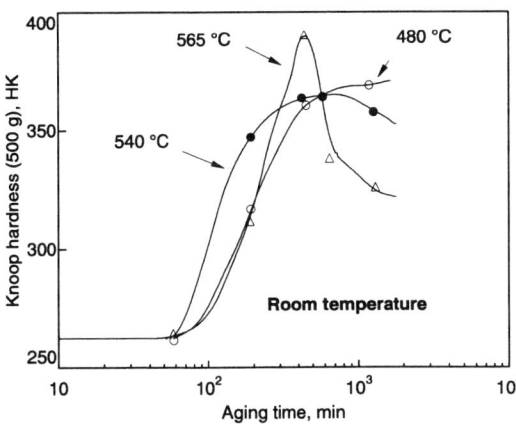

Alloy in the as-received condition was fabricated by press forging at 1090 °C (1995 °F) followed by hot rolling at 925 °C (1700 °F) to 38 mm (1.5 in.) plate thickness. Chemical composition (wt%) was 2.91 Al, 3.06 Cr, 0.128 Fe, 0.009 N, 0.122 O, 3.03 Sn, and 14.79 V. Beta transus temperature was between 750 and 770 °C (1380 and 1420 °F). Samples were isothermally annealed at 480, 540, and 565 °C for times from 30 min to 20 h.
Source: M. Imam and B. Rath, Transformation-Strengthened Beta Titanium Alloys, in *Sixth World Conference on Titanium*, P. Lacombe, R. Tricot, and G. Beranger, Ed., Les Editions de Physique, Paris, 1989, p 1513

Ti-15-3: Effect of aging on Vickers hardness

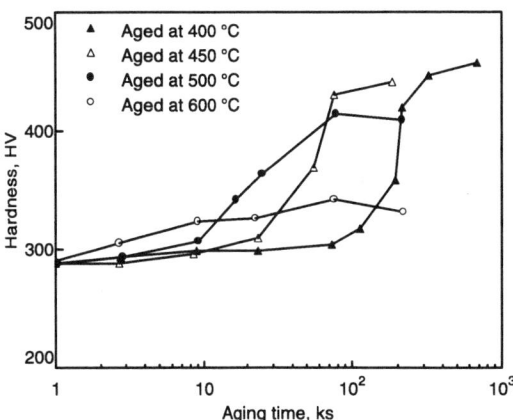

Chemical composition (wt%) was 3.37 Al, 0.004 C, 3.36 Cr, 0.17 Fe, 0.0061 H, 0.008 N, 0.14 O, 3.36 Sn, and 15.1 V. Alloy was supplied as plate 12 mm (0.47 in.) thick which had been hot rolled at 900 °C (1650 °F). Solution heat treatment was carried out at 850 °C (1560 °F) for 3.6 ks (kiloseconds) in inert atmosphere and air cooled.
Source: N. Niwa, K. Ito, H. Takatori, and H. Sakuyama, Influence of Heat Treatment on Microstructures and Mechanical Properties of Ti-15-3 Alloy, in *Sixth World Conference on Titanium*, P. Lacombe, R. Tricot, and G. Beranger, Ed., Les Editions de Physique, Paris, 1989, p 1507

Ti-15-3: Effect of heat treatment on Knoop hardness
Hardness drops below peak values when samples are annealed at higher temperatures.

Heat treatment	Hardness, HK (500 g)
Quenched from 900 °C (1650 °F)	268
Quenched + 300 °C (570 °F), 2.5 h	474
Quenched + 345 °C (650 °F), 4 h	481
Quenched + 565 °C (1045 °F), 3 h	303

Sheet Mechanical Properties

Aged Sheet

Tensile Properties. Ti-15-3 can be aged to a tensile strength of at least 1310 MPa (190 ksi) while guaranteeing ductility in excess of 5% (see table). Between aging temperatures of 510 and 540 °C (950 and 1000 °F), fully aged strength is a linear function of aging temperature. Aging at 455 to 480 °C (850 to 900 °F) would be recommended only in those situations requiring the highest possible strength and where ductility is of lesser importance.

For the aged condition, cold deformation of up to at least 40% does not significantly affect the relationship between aged strength and ductility (see figure).

Ti-15-3: Tensile properties vs aging temperature

		14-h aging temperature					
		510 °C (950 °F)		525 °C (975 °F)		540 °C (1000 °F)	
		T	L	T	L	T	L
Ultimate tensile strength, MPa (ksi)	Mean(a)	1335 (193.6)	1313 (190.5)	1225 (177.6)	1205 (174.7)	1114 (161.6)	1096 (159.0)
	0.99% point	1311 (190.1)	1276 (185.0)	1202 (174.3)	1169 (169.6)	1090 (158.1)	1059 (153.6)
Tensile yield strength, MPa (ksi)	Mean(a)	1245 (180.5)	1222 (177.2)	1126 (163.3)	1105 (160.2)	1009 (146.3)	987 (143.2)
	0.99% point	1190 (172.6)	1161 (168.4)	1075 (155.9)	1047 (151.9)	954 (138.4)	927 (134.4)
Elongation	Mean(a)	7.8		10.2		12.6	
	0.66% point	5.7		8.2		10.6	

(a) Means and percentage points calculated by regression technique. Regressed data from four lots. Gages: 0.89 to 1.78 mm (0.035 to 0.070 in.). Source: *Beta Titanium Alloys in the 1980's*, R.R. Boyer and H.W. Rosenberg, Ed., TMS/AIME, 1984

Ti-15-3: Typical aging curves for sheet

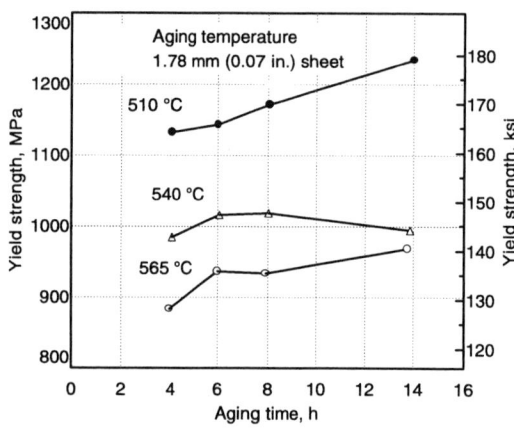

Source: *Beta Titanium Alloys in the 1980's*, R.R. Boyer and H.W. Rosenberg, Ed., TMS/AIME, 1984

Ti-15-3: Effect of prior cold work on aging

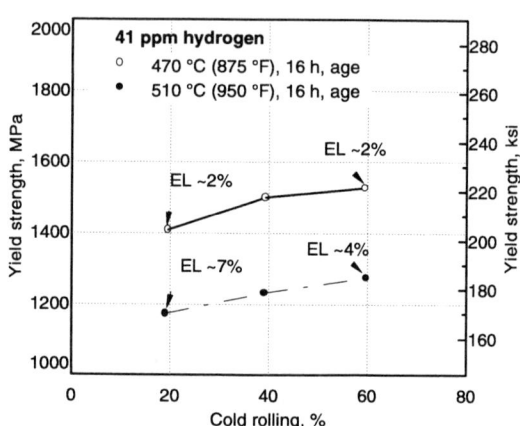

Source: *Beta Titanium Alloys in the 1980's*, R.R. Boyer and H.W. Rosenberg, Ed., TMS/AIME, 1984

Annealed Tensile Properties

Typical variations of tensile properties in the annealed condition (see table) are from test direction and lot-to-lot differences, the latter being larger. Sample location within a coil is of little consequence. Test direction differences typically are small, but can amount to 30 MPa (4 ksi); the lot-to-lot difference can be up to about twice as much. These values are small, however, when compared with unalloyed titanium strip or α-β alloy sheet.

Cold deformation over the range of 20 to 60% has an approximately linear effect on strength.

Annealed Ti-15-3: Typical tensile properties

	Average	Standard deviation
Ultimate tensile strength, MPa (ksi)	787 (114.1)	23 (3.4)
Tensile yield strength, MPa (ksi)	773 (112.1)	25 (3.6)
Elongation, %	21.5	2.7

Source: *Beta Titanium Alloys in the 1980's*, R.R. Boyer and H.W. Rosenberg, Ed., TMS/AIME, 1984, p 412

Ti-15-3: Effect of hydrogen on tensile properties

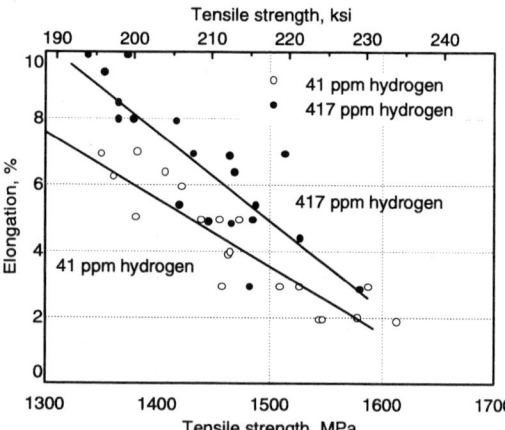

Cold worked 20 to 60%, aged at 480 to 510 °C (900 to 950 °F), 4 h, 16 h. Hydrogen addition has little effect on cold worked and aged properties.
Source: *Beta Titanium Alloys in the 1980's*, R.R. Boyer and H.W. Rosenberg, Ed., TMS/AIME, 1984

Ti-15-3: Effects of cold rolling on annealed tensile properties

Cold work, %	Ultimate tensile strength		Tensile yield strength		Elongation, %
	MPa	ksi	MPa	ksi	
0	789	114.4	763	110.7	16.0
20	893	129.5	851	123.4	13.1
40	996	144.5	938	136.1	10.2
60	1100	159.5	1025	148.8	7.2

Source: *Beta Titanium Alloys in the 1980's*, R. R. Boyer and H.W. Rosenberg, Ed., TMS/AIME, 1984, p 414

Cast Tensile Properties

A study of tensile properties from various castings and suppliers (see figures) does not show any significant trends regarding strength as a function of supplier. In general, thicker material exhibits a slightly lower average tensile strength. Ductility, however, does exhibit significant variation from supplier to supplier and as a function of thickness. It should be noted that a material with a higher oxygen content was not associated with higher strength or lower ductility. The ductility of castings from Suppliers A and E is significantly lower than that of the other suppliers for the thinner gages (≤12.7 mm, or 0.5 in.). It is interesting to note that Suppliers D and E used material from the same starting billet. The observed property variation in this case is attributed to casting practice.

Ti-15-3: Variation in tensile strength

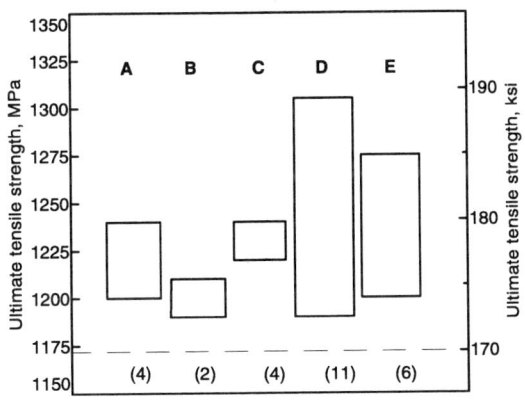 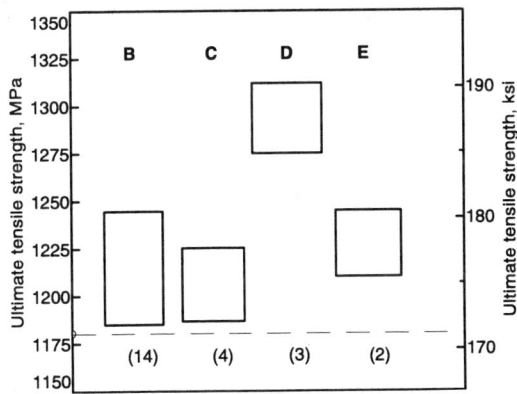

(a) t < 12.7 mm (0.5 in.)   (b) t > 12.7 mm (0.5 in.)

Castings were supplied by TiTech, Howmet, Tiline, and Precision Castparts from the United States, and from Tital in Germany. The compositions were within the requirements of AMS 4914 for Ti-15-3 sheet except for oxygen. The oxygen content ranged as high as 0.1645 wt% (the specification maximum is 0.13%). Sources are identified randomly as Sources A through E.

The starting material was selected with a range of section sizes and consisted of round cast test bars from 12.7 to 25.4 mm (1/2 to 1 in.) in diam, rectangular cast test bars from 25.4 × 25.4 mm to 25.4 × 152.4 mm (1 × 1 to 1 × 6 in.) in size, cast test plate from 3.8 to 25.4 mm (0.15 to 1 in.) in thickness, and complex castings with gages ranging from about 2 mm (0.08 in.) to 50 mm (2 in.) in thickness. The castings were all hot isostatically pressed at either 895 °C, 103.4 MPa, 2 h (1650 °F, 15 ksi, 2 h) or 955 °C, 103.4 MPa, 2 h (1750 °C, 15 ksi, 2 h), followed by direct aging at 525 °C (975 °F) for 12 h. The castings obtained by the University of Dayton were solution treated at 955 °C (1750 °F), 1 h, after HIP and prior to aging. Numbers in parentheses are the number of tests for each source.

Source: R. Boyer et al., *Microstructural/Property Relationships in Titanium Aluminides and Alloys*, The Minerals, Metals, and Materials Society, 1991, p 511-520

Cast Ti-15-3: Variation in tensile elongation

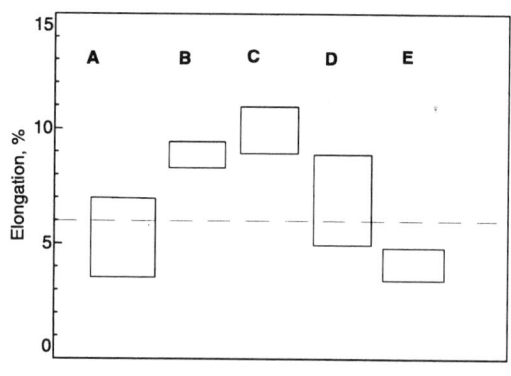
(a) $t \leq 12.7$ mm (0.5 in.)

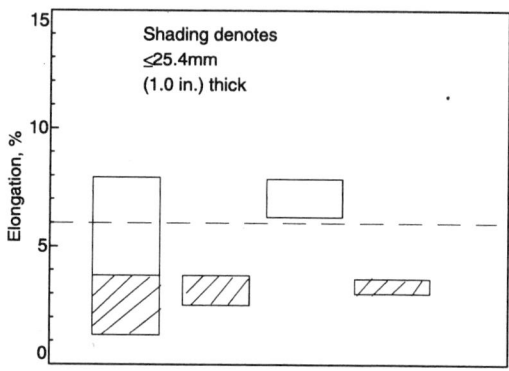
(b) $t > 12.7$ mm (0.5 in.)

Letters denote supplier designation. Each bar represents data from a given source. Castings were supplied by TiTech, Howmet, Tiline, and Precision Castparts from the United States and from Tital in Germany. The compositions were within the requirements of AMS 4914 for Ti-15-3 sheet except for oxygen. The oxygen content ranged as high as 0.1645 wt% (the specification maximum is 0.13%). Sources are identified randomly as Sources A through E.

Given the association between low ductility and thicker cross sections, the likely factors that influence ductility are β grain size and the thickness of grain boundary α. When comparing β grain sizes of high- and low-ductility specimens from a given supplier, the thicker material, with the lower elongation, had a consistently larger grain size. However, if data for castings of various shapes and sizes from all suppliers are considered together, no clear trend is apparent.

Source: R. Boyer et al., *Microstructural/Property Relationships in Titanium Aluminides and Alloys*, The Minerals, Metals, and Materials Society, 1991, p 511-520

High-Temperature Strength

Ti-15-3: Typical STA compressive yield strength

Temperature			Compressive yield strength (0.2%)(a)	
°C	°F	Direction	MPa	ksi
−51	−60	L	1253 ± 34	181.7 ± 4.9
		T	1292 ± 43	187.4 ± 6.2
24	75	L	1092 ± 69	158.4 ± 10.0
		T	1165 ± 32	169.0 ± 11.9
205	400	L	961 ± 46	139.4 ± 6.6
		T	962 ± 32	139.5 ± 4.6

Note: Solution treated and aged 8 h at 540 °C (1000 °F). (a) Standard deviations are for two or three samples. Source: *Beta Titanium Alloys in the 1980's*, R.R. Boyer and H.W. Rosenberg, Ed., TMS/AIME, 1984

Ti-15-3: High-temperature tensile properties

	Direction	RT	204 °C (400 °F)	425 °C (800 °F)
Ultimate tensile strength, MPa (ksi)	L	1250 (181.3)	1135 (164.6)	999 (144.9)
	T	1245 (180.6)	1147 (166.3)	1010 (146.5)
Tensile yield strength, MPa (ksi)	L	1127 (163.4)	980 (142.2)	837 (121.4)
	T	1146 (166.2)	996 (144.5)	854 (123.8)
Elongation, %	L	8.7	7.8	13.0
	T	8.2	7.3	11.3
Modulus of elasticity, GPa (10^3 ksi)	L	99.3 (14.4)	92.4 (13.4)	84.8 (12.3)
	T	98.6 (14.3)	93.8 (13.6)	82.7 (12.1)

Note: Sheet aged 8 h at 495 °C (925 °F). Average of three tests. Source: *Beta Titanium Alloys in the 1980's*, R.R. Boyer and H.W. Rosenberg, Ed., TMS/AIME, 1984, p 415

Ti-15-3: Yield strength vs temperature

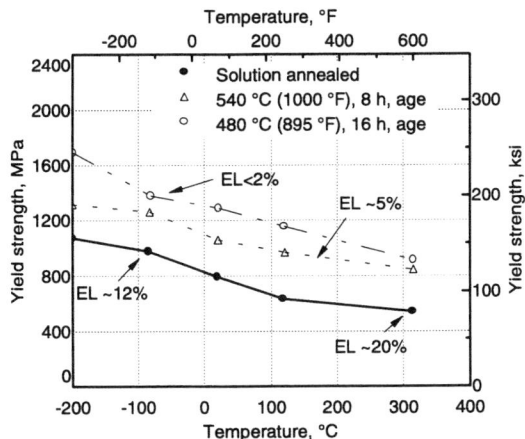

Data points represent averages of three tests of 1.27 mm (0.05 in.) gage material tested in the longitudinal direction.
Source: *Beta Titanium Alloys in the 1980's*, R.R. Boyer and H.W. Rosenberg, Ed., TMS/AIME, 1984

Ti-15-3: Notch/smooth strength ratio

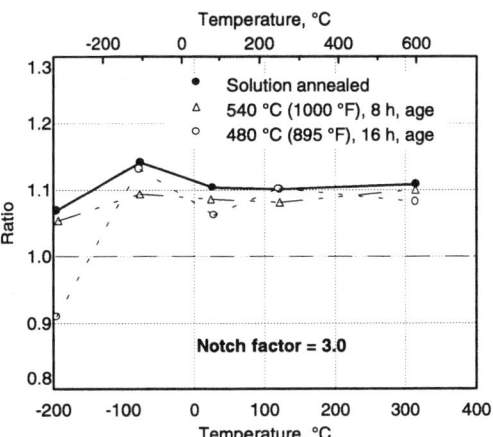

Notch/smooth strength ratio as a function of temperature for 1.27 mm (0.05 in.) gage sheet material.
Source: *Beta Titanium Alloys in the 1980's*, R.R. Boyer and H.W. Rosenberg, Ed., TMS/AIME, 1984

Ti-15-3: Bearing strength at high temperatures

Temperature			Bearing ultimate strength				Bearing yield strength			
			e/D = 1.5		e/D = 2.0		e/D = 1.5		e/D = 2.0	
°C	°F	Dir.	MPa	ksi	MPa	ksi	MPa	ksi	MPa	ksi
24	RT	L	1987	288.2	2231	323.6	1689	244.9	1830	265.4
		T	1878	272.3	2323	336.9	1743	252.8	1989	288.5
205	400	L	1705	247.3	2057	298.4	1545	224.1	1762	255.6
		T	1746	253.2	2062	299.1	1558	225.9	1795	260.3
425	800	L	1667	241.7	2006	290.9	1460	211.8	1615	234.2
		T	1651	239.5	2049	297.1	1431	207.5	1644	238.5

Note: Aged 8 h at 495 °C (925 °F). Source: *Beta Titanium Alloys in the 1980's*, R.R. Boyer and H.W. Rosenberg, Ed., TMS/AIME, 1984

Ti-15-3: Ultimate shear strength at high temperatures

Temperature		L		T	
°C	°F	MPa	ksi	MPa	ksi
24	RT	784	113.7	799	115.9
205	400	711	103.1	705	102.3
425	800	636	92.3	637	92.4

Note: Aged 8 h at 495 °C (925 °F). Source: *Beta Titanium Alloys in the 1980's*, R.R. Boyer and H.W. Rosenberg, Ed., TMS/AIME, 1984

Creep Properties

Ti-15-3 follows a stress exponential creep law at 430 °C (800 °F). Product aged 8 h at 495 °C (925 °F) fits the following equation:

$$\dot{\varepsilon} = 0.00010865 \exp\frac{95.087\sigma}{RT}$$

where $\dot{\varepsilon}$ is the minimum creep strain rate in %/h, σ is the applied stress in ksi, T is absolute temperature in K, and R is the gas constant. This relationship removed 95.1% of the variance in $\dot{\varepsilon}$.

Ti-15-3: Transverse creep at 425 °C (800 °F)

Stress		Time to indicated creep, h				
MPa	ksi	0.1%	0.2%	0.5%	1.0%	2.0%
827	120	0.03	0.10	0.35	0.8	2.4
724	105	0.10	0.25	1.0	2.5	7.0
483	70	0.20	0.9	5.6	19.4	75
207	30	4.2	13.5	87	400(a)	...
138	20	36	117	520(a)	...	...
103	15	170	535	2500(a)	...	...

(a) Extrapolated. Source: *Beta Titanium Alloys in the 1980's*, R.R. Boyer and H.W. Rosenberg, Ed., TMS/AIME, 1984

Ti-15-3: Creep at 205 °C (400 °F)

Sheet	Direction	Stress MPa	Stress ksi	Plastic creep strain, %	Time, h
A	L	827	120	0.18	579
A	L	862	125	0.11	1008
A	T	827	120	0.02	1001
A	T	862	125	0.14	1004
B	L	827	120	0.07	1002
B	L	862	125	0.08	1003
B	T	827	120	0.05	1001
B	T	862	125	...	...

Note: Sheet was aged at 540 °C (1000 °F). Source: *Beta Titanium Alloys in the 1980's*, R.R. Boyer and H.W. Rosenberg, Ed., TMS/AIME, 1984

Ti-15-3: Creep at 600 °F and 100 ksi

	Prestrain(a), %	Total creep strain in 10^3 h, %	Minimum creep rate, %/h
Aging treatment			
Aged at 455 °C (850 °F), 16 h	0	0.324	0.00016
	0	0.293	0.00015
	8	0.364	0.00018
Aged at 480 °C (900 °F), 12 h	0	0.293	0.00017
	0	0.320	0.00020
	8	0.377	0.00020

Note: Sheet 1.6 mm (0.063 in.) thick was solution treated and aged as indicated. (a) Nominal amount of stretch prior to aging to simulate forming strains. Source: R. Kaneko and C. Woods, "Low Temperature Forming of Beta Titanium Alloys," NASA Contractor Report 3706, NASA, 1983, p 15

Fatigue Properties

Ti-15-3: Smooth and notched fatigue

Temperature °C	Temperature °F	Runout stress(a) Smooth MPa	Smooth ksi	Notched(b) MPa	Notched ksi
−51	−60	724	105	207	30
24	75	655-758	95-110	207-241	30-35
205	400	655-690	95-100	221-241	32-35

(a) Runout >10^7 cycles, $R = 0.1$, maximum stress shown. (b) $K_t = 3$. Source: *Beta Titanium Alloys in the 1980's*, R.R. Boyer and H.W. Rosenberg, Ed., TMS/AIME, 1984, p 419

Fatigue Crack Growth

Ti-15-3 exhibits crack growth characteristics much like mill annealed Ti-6Al-4V, although Ti-15-3 is not as sensitive to environments such as salt water.

Ti-15-3: Crack growth at $\Delta K = 22$ MPa$\sqrt{m}$ (20 ksi$\sqrt{in.}$)

Evaluating the data at $\Delta K = 22$ MPa$\sqrt{m}$ (20 ksi$\sqrt{in.}$) shows da/dN increases slightly as sheet gage increases. The combined salt water plus frequency effect is just at the "detection limit" statistically.

		da/dN at $\Delta K = 22$ MPa$\sqrt{m}$ (20 ksi$\sqrt{in.}$): 10^{-6} in.	10^{-6} mm
Environmental effect	Air at 20 Hz	9.12	232
	Salt at 5 Hz	9.80	249
Gage effect	1.3 mm (0.050 in.)	8.58	218
	2.5 mm (0.100 in.)	10.33	262

Note: Test error for these data was estimated to be 12×10^{-6} mm/cycle (0.49×10^{-6} in./cycle). Source: *Beta Titanium Alloys in the 1980's*, R.R. Boyer and H.W. Rosenberg, Ed., TMS/AIME, 1984, p 419

Ti-15-3: Crack growth in air and salt solution

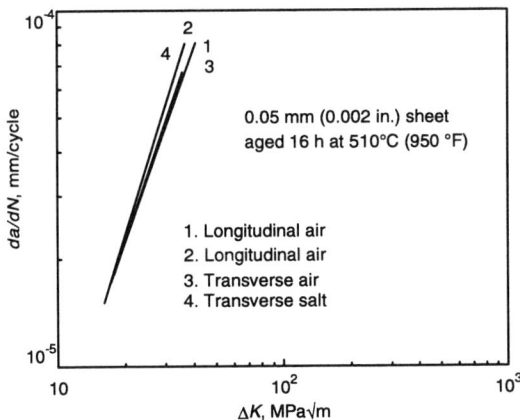

Source: *Beta Titanium Alloys in the 1980's*, R.R. Boyer and H.W. Rosenberg, Ed., TMS/AIME, 1984, p 420

Ti-15-3: Crack growth data for sheet

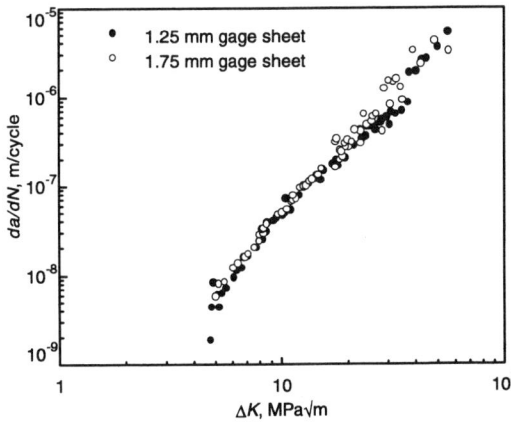

Specimens were tested in the T-L orientation. Sheet was aged at 540 °C (1000 °F), 8 h. $R = 0.1$; frequency, 30 Hz, at 22 °C (72 °F). Source: *Beta Titanium Alloys in the 1980's*, R.R. Boyer and H.W. Rosenberg, Ed., TMS/AIME, 1984, p 223

Fracture Properties

Ti-15-3: RT fracture toughness of sheet

Gage		Specimen	Fracture toughness (K_c)	
mm	in.	orientation	MPa√m	ksi√in.
1.27	0.050	L-T	100	91
		T-L	100	91
1.78	0.070	L-T	113	103
		T-L	107	97

Note: Yield strength of 1035 MPa (150 ksi) at RT. Directionality is low, 3 to 4 MPa√m (3 to 4 ksi√in.). Lot-to-lot variations can be up to 11 MPa√m (10 ksi√in.). Source: *Beta Titanium Alloys in the 1980's*, R.R. Boyer and H.W. Rosenberg, Ed., TMS/AIME, 1984, p 416

Ti-15-3: Fracture toughness of STA plate

Heat treatment	Orientation	Tensile yield strength		Ultimate tensile strength		Elongation, %	Fracture toughness (K_{Ic})	
		MPa	ksi	MPa	ksi		MPa√m	ksi√in.
800 °C (1470 °F), 20 min, AC, 480 °C (895 °F), 14 h, AC	L-T	1253	182	1376	199	6.2	44.3	40.3
	T-L	1304	189	1421	206	6.6	46.8	42.6
800 °C (1470 °F), 20 min, AC, 510 °C (950 °F), 14 h, AC	L-T	1213	176	1337	194	7.8	42.1	38.3
	T-L	1263	183	1382	200	6.9	43.4	39.5

Note: Hot rolled plate had a chemical composition (wt%) of 3.37 Al, 0.004 C, 3.36 Cr, 0.17 Fe, 0.0061 H, 0.0080 N, 0.14 O, 3.04 Sn, and 15.10 V. It was solution treated at 800 °C (1470 °F) for 20 min, air cooled, then aged at 510 °C (950 °F) for 8 or 14 h. Source: C. Ouchi, H. Suenaga, H. Sakuyama, and H. Takatori, Effects of Thermomechanical Processing Variables on Mechanical Properties of Ti-15V-3Cr-3Sn-3Al Alloy Plate, in *Designing With Titanium*, 1986, p 130

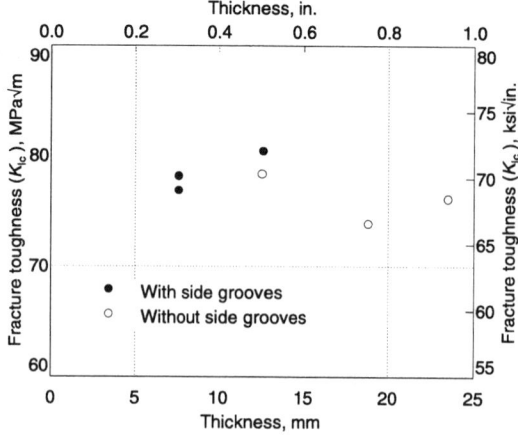

Ti-15-3: Fracture toughness vs sheet thickness

Alloy was heat treated to achieve strength levels of 1055 MPa (153 ksi) yield strength and 1117 MPa (162 ksi) tensile strength, with 7.5% ductility. Compact tension specimens with and without V-shaped side grooves (depth <10% original thickness) were tested on a servohydraulic machine in load control.
Source: P. Poulose, Determination of Fracture Toughness from Thin Side-Grooved Specimens, *Eng. Fract. Mech.*, Vol 26, 1987, p 203

Plastic Deformation

Strain Hardening. Annealed Ti-15-3 does not follow the usual strain hardening laws as the spread between the tensile yield and ultimate tensile strengths is quite small. Once stable flow is established (post yield strains), an approximately exponential stress-strain relationship is observed. The stress-strain "curve" at large strain may be estimated as:

$$\text{Flow stress} = \frac{\sigma_{max}}{A_o} \exp \varepsilon$$

where ε is the true plastic strain; σ_{max} is the maximum load; and A_o is the initial area.

Strain-Rate Sensitivity. Ti-15-3 is quite insensitive to strain rate in the annealed condition, as indicated by tensile strain rates ranging from 0.005 to 0.1 min^{-1} (see table). Strain rate may have an effect on ductility, particularly at high strain rates.

Ti-15-3: Effect of strain rate on tensile properties

Strain rate, min^{-1}	Ultimate tensile strength		Tensile yield strength		Elongation, %
	MPa	ksi	MPa	ksi	
0.005	745	108	738	107	29
0.005	758	110	752	109	30
0.050	752	109	724	105	25
0.050	752	109	731	106	24
0.100	765	111	738	107	24
0.100	758	110	731	106	23

Note: Specimens were annealed at 760 °C (1400 °F). Source: H.W. Rosenberg, Ti-15-3 Property Data, *Beta Titanium Alloys in the 1980's*, TMS/AIME, 1984

Tensile Stress-Strain

Ti-15-3: Typical tensile stress-strain curves for sheet

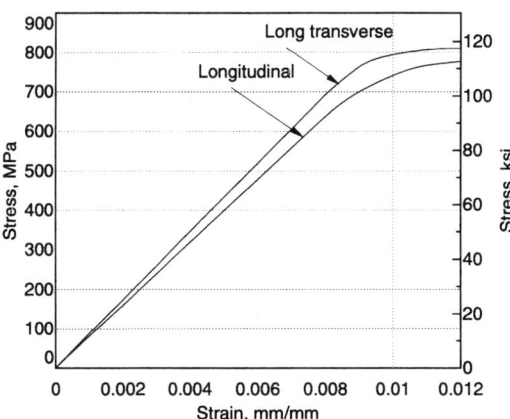

0.53 to 3.17 mm (0.021 to 0.125 in.) solution treated sheet.
Source: MIL-HDBK 5, 1 Dec 1991

Ti-15-3: Typical tensile stress-strain curves for sheet

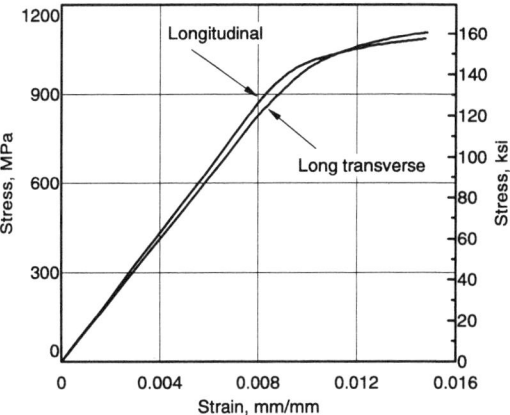

0.5 to 1.9 mm (0.020 to 0.076 in.) solution treated and aged sheet.
Source: MIL-HDBK 5, 1 Dec 1991

Compressive Stress-Strain

Ti-15-3: Typical compressive properties

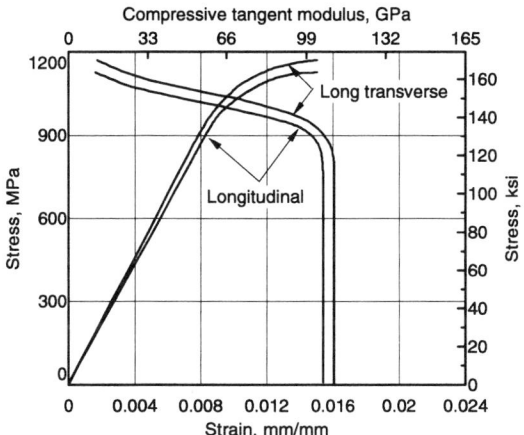

0.5 to 1.9 mm (0.020 to 0.076 in.) sheet aged at 540 °C (1000 °F).
Source: MIL-HDBK 5E, 1988

Ti-15-3: Tangent modulus

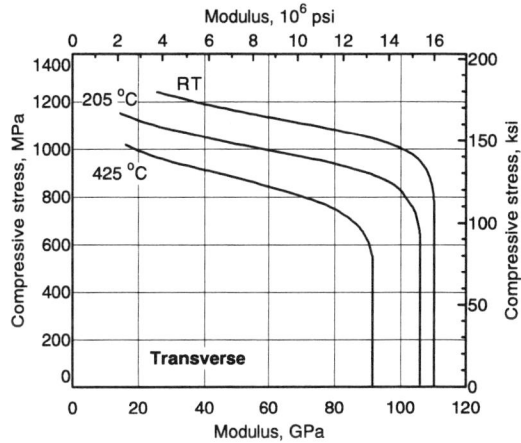

Typical compressive tangent modulus for solution treated and aged Ti-15-3.
Source: Collected Engineering Data Sheets, AFML-TR-78-179, 1978

Flow Stress*

Ti-15-3 has been studied at temperatures of 850 and 760 °C to determine flow characteristics and potential superplastic formability. Specimens were tested in the as-received condition and, as such, had a grain size of about 80 µm. It is found that this alloy exhibits deformation behavior quite different from the two-phase alloys in that it tends to flow soften with about 2 to 4% strain (see figure). The softening is reflected in the stress vs strain-rate characteristics, where the prestrain can be seen to reduce the flow stresses and increase the strain-rate sensitivity observed over a wide range of strain rates.

The microstructural changes that occur during softening include the development of a subgrain structure. This deformation behavior is characteristic of the conventional hot deformation behavior of other alloys. An interesting aspect of the deformation of the large grained β titanium alloy is that, once the subgrain structure has developed with the 2 to 4% strain, the flow stress is constant with strain, and the strain-rate sensitivity, m, also remains nearly constant (see figure).

---
*Adapted from C.H. Hamilton, *Superplastic Forming*, ASM International, 1985

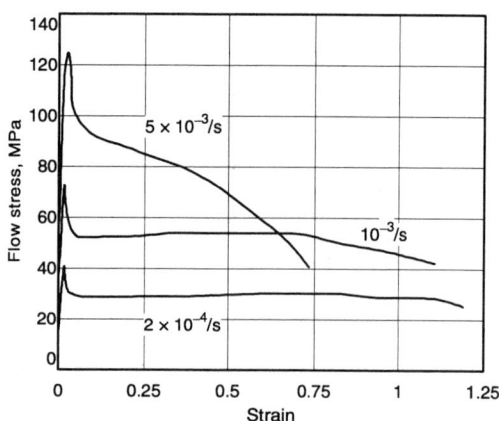

Flow stress vs strain at constant strain rate.
Source: C.H. Hamilton, Superplasticity in Titanium Alloys, *Superplastic Forming*, ASM International, 1985, p 13-22

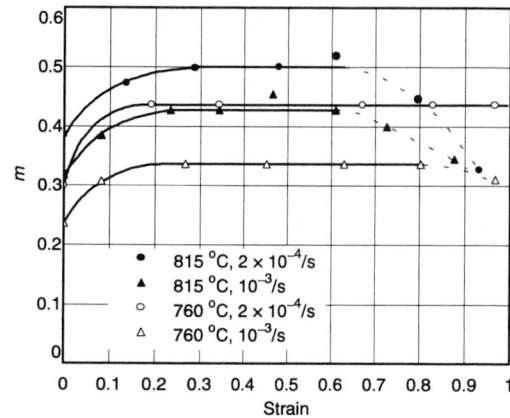

Strain-rate sensitivity, m, as a function of strain for specimens deformed under constant strain rate.
Source: C.H. Hamilton, Superplasticity in Titanium Alloys, *Superplastic Forming*, ASM International, 1985, p 13-22

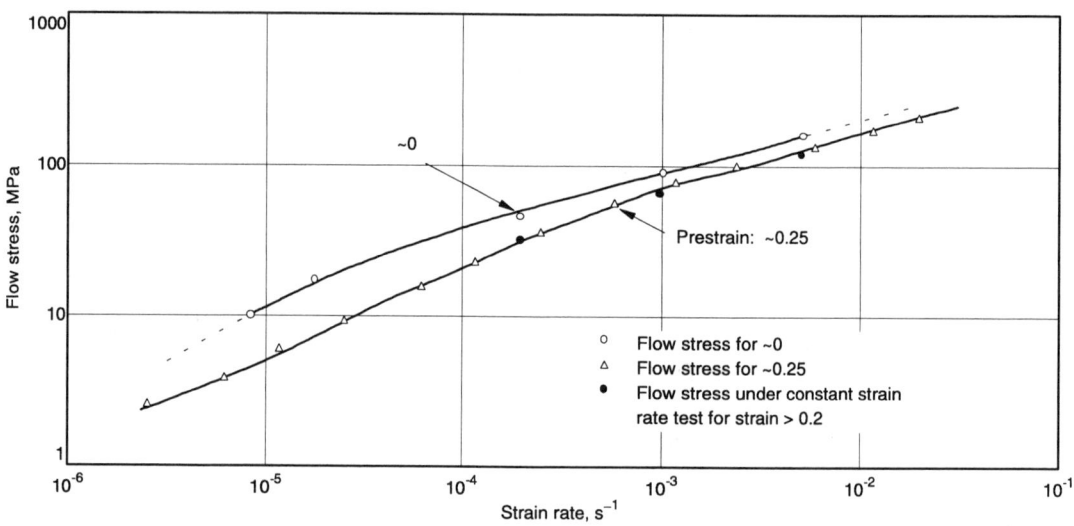

Flow stress vs strain rate at 850 °C showing the effect of a small prestrain.
Source: C.H. Hamilton, Superplasticity in Titanium Alloys, *Superplastic Forming*, ASM International, 1985, p 13-22

Forging

G.W. Kuhlman, ALCOA, Forging Division

Ti-15-3 is a very high strength, metastable beta alloy whose primary commercial applications in forgings are aerospace and missile components, pressure vessels, and corrosion-resistant applications. The alloy can be fabricated into all forging product types, although closed die forgings predominate.

Ti-15-3 is a moderately forgeable alloy (when forged above the β transus), with higher unit pressures (flow stresses), improved forgeability, and less crack sensitivity in forging than the α-β alloy Ti-6Al-4V. Due to the high alloying content of Ti-15-3, flow stresses are almost double that of the near-β alloy Ti-10V-2Fe-3Al. The desired final microstructure from thermomechanical processing of Ti-15-3 during forging manufacture is fine transformed β, with limited grain boundary films and with a fine recrystallized prior β grain size in preparation for final thermal treatments. The very fine microstructural features of Ti-15-3 achieved in forgings are responsible for its excellent strength, corrosion, and fatigue properties. Thus, Ti-15-3 is typically forged above the β transus through one or more forgings operations. Reheating for subsequent forging operations recrystallizes the alloy, thus refining prior β grain size.

Ti-15-3 is generally not subtransus (α + β) forged, because there is no microstructural advantage and there is a significant increase in unit pressure requirements.

Final thermal treatments for Ti-15-3 forging include solution annealing and aging. Forgings may be supplied in the solution annealed (ST) condition and/or fully aged (STA). In the ST condition, Ti-15-3 has lower strengths, but much higher ductility and toughness than in the STA condition. Solution treatment is conducted at 785 °C (1450 °F), followed by air cooling or equivalent cooling rates. Aging is conducted at 480 to 620 °C (900 to 1150 °F).

Beta forging working histories for Ti-15-3 require imparting enough hot work to reach final macrostructure and microstructure objectives. Generally, reductions in any given forging process are 30 to 50% to achieve desired dynamic and static recrystallization. Very low levels of β reduction are not recommended. Although Ti-15-3 is cold worked in other product forms (sheet), cold working is not used for forgings. Ti-15-3 may be successfully isothermally or hot die forged. However, it has lower strain-rate sensitivity than α-β and near-β alloys, and thus unit pressure reductions and forging shape sophistication improvements through these technologies are modest in comparison to other titanium alloys.

Ti-15-3, as with all β alloys has a higher affinity for hydrogen than other alloy classes. Although Ti-15-3 forms less α case from heating operations than other alloy classes, therefore requiring less metal removal in chemical pickling (milling processes), control of chemical removal processes is essential to preclude excessive hydrogen pickup.

Ti-15-3: Forging process temperatures

Process	Metal temperature °C	°F
Beta forge	790-925	1450-1700
Temperature relative to transus(a)	30 to 65 above transus	55 to 120 above transus

Note: See "Technical Note 4: Forging" for recommended die temperatures. (a) Subtransus forging of Ti-15-3 has not been found advantageous due to high unit pressures.

Ti-15-3: Effect of post-forge heat treatment on properties

Post-forge treatment	Tensile yield strength MPa	ksi	Ultimate tensile strength MPa	ksi	Elongation, %	Reduction of area, %	Fracture toughness (K_{Ic}) MPa√m	ksi√in.	Critical crack length(a) mm	in.	ΔK_{th}(b) MPa√m	ksi√in.	Smooth fatigue(c) MPa	ksi
STA 785 °C, AC + 510 °C, 8 h	1192	173	1275	185	9	22	57	52	2.5	0.09	...	...	840	122
STA 785 °C, AC + 535 °C, 8 h	1055	153	1151	167	11	30	61	55	3.6	0.14	4.4	4.0	810	117
DA 510 °C, 8 h	1165	169	1234	179	10	24	59	53	2.8	0.11	4.4	4.0	830	120
DA 535 °C, 8 h	1096	159	1158	168	12	32	67	61	4.1	0.16	...	...	810	117

Note: Like most commercial β alloys, Ti-15-3 does not respond to thermomechanical processing to improve fracture properties because α morphology is not significantly modified. (a) Critical crack length ≈ 1.1 $[K_{Ic}/YS]^2$. (b) ΔK_{th} is the threshold stress-intensity in fatigue crack growth rate tests. (c) Smooth fatigue stress at 10^7 cycles, tests conducted at R = 0.1 to 0.3, F = 30 to 125 Hz. Source: G.W. Kuhlman, "Critical Appraisal of Thermomechanical Processing (TMP) on Structural Titanium Alloys"

Forming

Ti-15-3 approaches the formability of the higher strength commercial-purity grades of titanium. Extensive work has been done to evaluate shop forming of Ti-15-3. Studies have included flange, bead, draw, joggle and bending forming of sheet materials at room temperature followed by aging. Ti-15-3 has been fabricated using a cold shear spinning operation to produce low-cost tank replacements of Ti-6Al-4V. It is not very good in stretch forming operations (due to the low degree of strain hardening), but the alloy can be drawn quite deep. It can also be hot formed, but this degrades the economics. The forming properties of Ti-15-3 are rather insensitive to minor variations in reduction, annealing time, and temperature.

Ti-15-3: Room-temperature formability comparison of annealed alloys

Alloy	Tensile yield strength MPa	ksi	Gage mm	in.	Minimum bend radius	Stretch form limit, %	Hydroform limit, % Stretch	Shrink
Ti-15-3	758	110	1.0	0.040	2.4t	...	12	...
			2.0	0.080	2.4t	28	20	...
CP Ti	275	40	1.0	0.040	3.0t	20	12	1.0
			2.0	0.080	3.0t	20	12	1.5
Ti-6Al-4V	827	120	1.0	0.040	4.7t	7	8	1.0
			2.0	0.080	5.5t	7	8	...

Source: R. Kaneko and C.A. Woods, "Low-Temperature Forming of Beta Titanium Alloys," NASA Report 3706, 1983

Ti-15-3: Shear spinnability comparison of aerospace materials

Material	Maximum room-temperature reduction, %
17-7 PH	38
Haynes 25	40
Ti-6Al-4V	50
6061	63
C103	66
Ti-8V-7Cr-3Al-4Sn-1Zr	69
1100	83
Ti-8V-4Cr-2Mo-2Fe-3Al	83
Type 301 stainless steel	83
Ti-15V-3Cr-3Al-3Sn	84

Source: A.E. Leach, Formed Ti-15V-3Cr-3Al-3Sn Tankage, *Beta Titanium Alloys in the 1980's*, TMS/AIME, 1984

Ti-15-3: Typical formability of annealed alloy

Bend radius	2.0t
Cup height	7.62 mm (0.30 in.)
Draw	46%(a)
Flange	
Stretch	20%
Shrink	1%(b)
Joggle	
L/d	2
d/t	5
Springback	15° at 90°

(a) Limit of dies, not of material. (b) Special procedures can improve this significantly. Source: H.W. Rosenberg, Ti-15-3 Property Data, *Beta Titanium Alloys in the 1980's*, TMS/AIME, 1984

Ti-15-3: Hydrogen content vs room-temperature formability

Hydrogen content, ppm	0.89-mm gage material Annealed bend results(a)		1.78-mm gage material Annealed cup tests(b)
	Longitudinal	Transverse	
90	...	...	9.40
129	0.44t	0.74t	...
970	...	...	9.14
1575	0.65t	0.94t	...
2000	...	...	9.40
3300	...	...	8.89
4000	>13t	>13t	...

(a) Minimum bend radius expressed as times thickness (t) for 105° bend. (b) Cup height of Olsen-type cup test in millimeters. Source: *Beta Titanium Alloys in the 1980's*, R.R. Boyer and H.W. Rosenberg, Ed., TMS/AIME, 1984

Ti-15-3: Tensile properties of foil

Gage		Direction	Ultimate tensile strength		Tensile yield strength		Elongation, %
mm	in.		MPa	ksi	MPa	ksi	
0.25	0.010	L	827	121	807	117	16
		T	841	122	814	118	16
0.075	0.003	L	827	121	800	116	4
		T	827	121	821	119	7

Source: H.W. Rosenberg, Ti-15-3 Property Data, *Beta Titanium Alloys in the 1980's*, TMS/AIME, 1984

Bending

Ti-15-3: Effect of aging on bend angle brake formed parts

Bend angle, °	Punch radius	Specimen configuration	Bend axis	Angle change after aging (a), °			
				End 1(b)		End 2(b)	
				a_1	a_2	a_1	a_2
45	2t	Z	L	+0.8	+1.2	+1.2	+1.2
90	2t	Z	L	−0.8	−1.3	−0.9	−1.2
90	2t	Z	T	−1.4	+1.1	−0.8	0
90	2.3t	Z	L	−1.2	−1.4	−1.2	−2.1
90	2.3t	L	L	+0.1	...	−0.2	...
90	2.7t	Z	L	+0.7	−1.2	+1.7	−0.5
135	2t	Z	L	+2.0	+0.5	−0.4	−0.9

Note: 2.0 mm (0.080 in.) with a_1 of 25 mm (1 in.) for L specimen and a_1 and a_2 of 25 mm (1 in.) for Z specimen. L specimen 457 mm (18 in.) long; Z 150 mm (6 in.) long. (a) Aging treatment: 505 °C (940 °F) for 12 h. (b) Angle measured 25 mm (1 in.) from each end. Source: R. Kaneko and C.A. Woods, "Low-Temperature Forming of Beta Titanium Alloys," NASA Report 3706, 1983

Ti-15-3: Springback

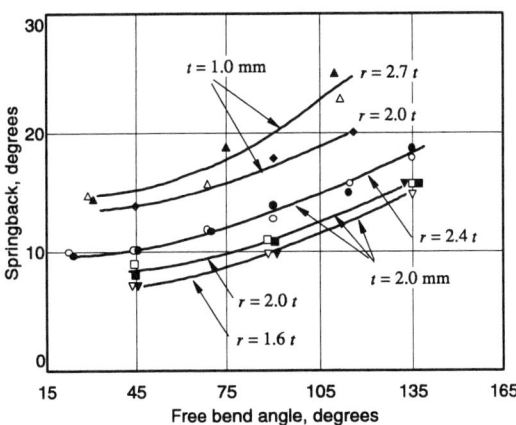

Room-temperature springback of annealed Ti-15-3.
Source: R. Kaneko and C.A. Woods, "Low-Temperature Forming of Beta Titanium Alloys," NASA Report 3706, 1983

Ti-15-3: Peak strain vs bend angle

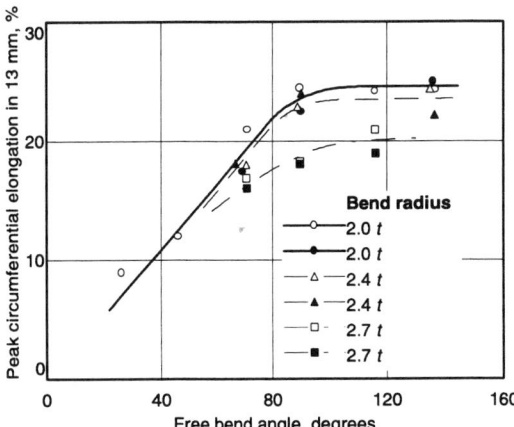

Influence of bend angle and radius on circumferential strain of annealed 2.0 mm (0.080 in.) sheet. Specimen width, 150 mm (6 in.). Open symbols represent longitudinal data. Closed symbols represent transverse data.
Source: R. Kaneko and C.A. Woods, "Low-Temperature Forming of Beta Titanium Alloys," NASA Report 3706, 1983

Ti-15-3: Minimum bend radii in cold press brake bending of annealed or solution treated alloys

	Minimum bend radius as a function of sheet thickness, t	
Alloy	$t <$ 1.75 mm (0.069 in.)	1.75 mm (0.069 in.) $< t <$ 4.76 mm (0.1875 in.)
CP titanium		
ASTM grade 1	2.5	3.0
ASTM grade 2	2.0	2.5
ASTM grade 3	2.0	2.5
ASTM grade 4	1.5	2.0
β alloys		
Ti-13V-11Cr-3Al	3.0	3.5
Ti-11.5Mo-6Zr-4.5Sn	3.0	3.0
Ti-3Al-8V-6Cr-4Mo-4Zr	3.5	4.0
Ti-8Mo-8V-2Fe-3Al	3.5	3.5
Ti-15V-3Cr-3Sn-3Al	2.0	2.0

Source: G.A. Lenning, J.A. Hall, M.E. Rosenblum, and W.B. Trepel, "Cold Formable Titanium Sheet Material Ti-15-3-3-3," Report AFWAL-TR-82-4174, Air Force Wright Aeronautical Laboratories, Dec 1982; and Military Standard MIL T 9046J, US Government Printing Office

Ti-15-3: Forming limit diagram

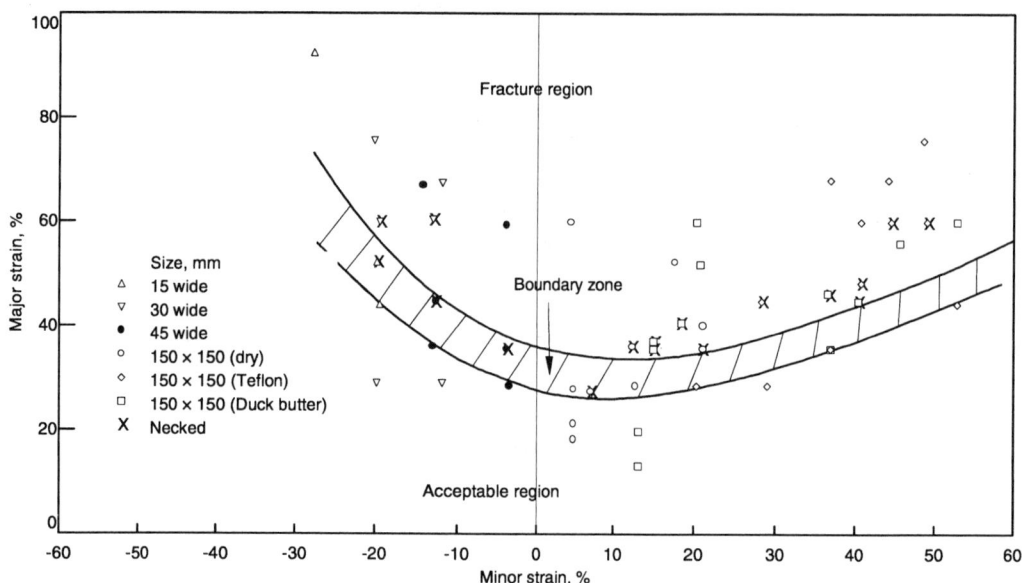

Forming limit diagram for 2.0 mm (0.080 in.) Ti-15-3 sheet. Unlubricated (dry) punches except as noted with teflon (Withrodraw 525) and duck butter of 1.0 mm (0.040 in.) thick film on punch side. Punch speeds were 25 mm/min (1.0 in./min) except for some 150 × 150 mm (6 × 6 in.) specimens run at 510 mm/min (20 in./min).
Source: R. Kaneko and C.A. Woods, "Low-Temperature Forming of Beta Titanium Alloys," NASA Report 3706, 1983

Heat Treatment

Ti-15-3: Stress relief and solution annealing treatments

Heat treatment	Temperature °C	Temperature °F	Time, min	Cooling method
Typical stress relief	650	1200	12	AC
Solution treating range	790-815	1450-1500	5-30	AC
Typical solution treatment	790	1450	15	AC

Ti-15-3: Aging treatments

Aging temperature °C	Aging temperature °F	Aging time, h
565	1050	8-16
540	1000	8-24
510	950	8-32
480	900	16-48

Note: Use of the longest aging times is recommended when it is desirable to minimize property scatter. In many cases, use of the shortest times will be quite satisfactory. Source: *Beta Titanium Alloys in the 1980's*, R.R. Boyer and H.W. Rosenberg, Ed., TMS/AIME, 1984

Ti-15-3: Effects of cold work on strength and ductility

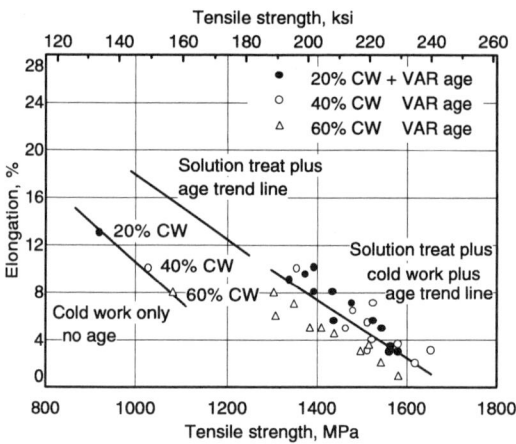

Source: H.W. Rosenberg, Property Scatter in Beta Titanium: Some Problems and Solutions, *Beta Titanium Alloys in the 1980s*, TMS/AIME, 1984

Ti-15-3: Effect of aging on tensile properties

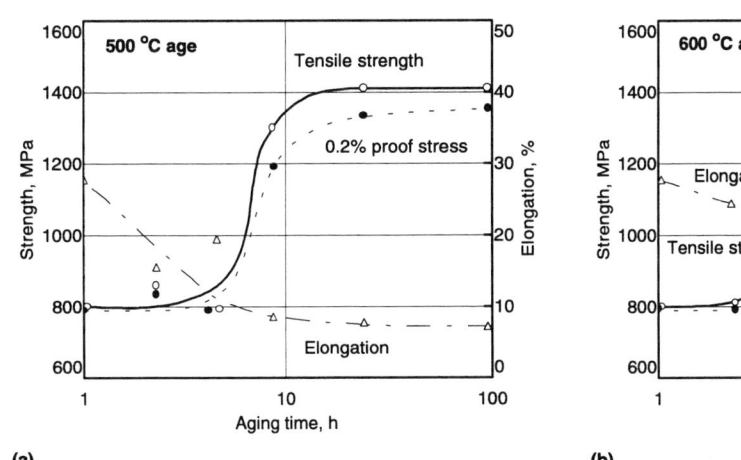
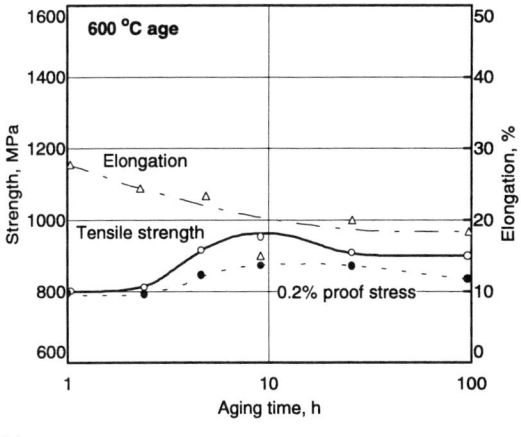

(a) (b)

2 mm (0.08 in.) sheet solution treated at 800 °C.
Source: M. Okada, D. Banerjee, and J.C. Williams, *Titanium Science and Technology*, Vol 3, Deutsche Gesellschaft fur Metallkunde, Germany, 1985, p 1836

Ti-15-3: Effect of aging on property scatter

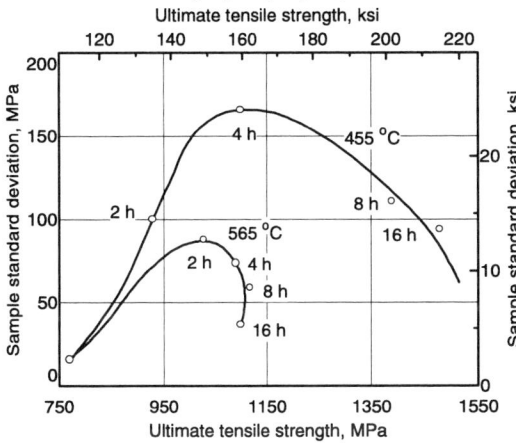

Effects of aging temperature and time on scatter of transverse tensile strength.
Source: H.W. Rosenberg, Property Scatter in Beta Titanium: Some Problems and Solutions, *Beta Titanium Alloys in the 1980s*, TMS/AIME, 1984

TIMETAL® 21S

Ti-15Mo-3Al-2.7Nb-0.25Si
Common Name: Beta-21S
UNS Number: R58210

Tom O'Connell, TIMET

Beta-21S is a very recently developed metastable β alloy that offers the high specific strength and good cold formability of a metastable β alloy, but has been specifically designed for improved oxidation resistance, elevated temperature strength, creep resistance, and thermal stability. Developing commercial applications in forgings include aerospace components and prosthetic devices. For the latter application, with appropriate thermomechanical processing, Beta-21S modulus is comparable to bone. For the former, Beta-21S may be processed to very high strengths with excellent oxidation and corrosion resistance.

Strip is the main product form. Beta-21S is also well suited for metal matrix composites because it can be economically rolled to foil and is compatible with most fibers. Strip is available in gages from 0.3 to 2.5 mm (0.012 to 0.100 in.).

Chemistry. The composition of Beta-21S is based on the objective of obtaining a cold rollable, strip-producible alloy for economical processing into foil form. The key to processing an alloy to foil form is cold rolling of strip product. If an alloy cannot be cold rolled as strip, a hot process on a hand-mill using cover sheets to form packs for heat retention is the only other viable option. Although the pack process offers the opportunity to cross-roll to minimize texture, it is nonetheless labor intensive and inherently a lower yield process.

In light of the fact that a cold rollable, strip-producible alloy was of primary importance, it was decided that a metastable β alloy was the best approach. This meant that the ordinary obstacles to overcome were the poor oxidation resistance and elevated-temperature mechanical properties of this class of alloy. The initial approach was to concentrate on the Ti-Mo and Ti-Cr systems. Although the Ti-V system is most commonly used for metastable β alloys (e.g., Ti-15V-3Cr-3Sn-3Al and Ti-3Al-8V-6Cr-4Zr-4Mo), vanadium is well known for its detrimental effects on oxidation resistance. Conclusions of chemistry screening on oxidation resistance were as follows:

- Silicon, niobium, hafnium, and tantalum were beneficial additions to the Ti-Mo system, as well as palladium, aluminum, and iron.
- Tin, zirconium, cobalt, yttrium, and iron were not beneficial additions to a Ti-Mo base.
- 20% Mo provides no advantage in oxidation resistance over 15% Mo.
- No additions were found that improve the corrosion resistance of the Ti-Cr series.

Effect of Oxygen. In a study on the effect of oxygen, oxygen levels up to 0.25% were found to have no significant effect on the strength/ductility relationship of aged Beta-21S. Higher oxygen levels degrade ductility. Increasing oxygen decreases the work-hardening capability of annealed sheet material, which could adversely affect some aspects of formability.

Oxygen absorption at the surface during exposure in air at elevated temperature degrades tensile ductility. The magnitude of the effect in sheet is dependent on the exposure time and temperature and on sheet thickness. After a suitable heat treatment, Beta-21S is metallurgically stable for at least 1000 h up to 615 °C (1140 °F).

Product Forms and Conditions. Beta-21S is available as cut sheet, strip, plate, bar, billet, and bloom. It is typically provided in the beta solution treated condition, which precipitates α to provide strengthening on aging. The morphology and distribution of the α depend on the heat-treatment temperature and the oxygen content. Lower heat-treatment temperatures and higher oxygen contents result in homogeneous spheroidal α; higher aging temperatures and lower oxygen result in lath-type α.

Applications. Beta-21S is most useful for applications above 290 °C (550 °F), with thermal stability up to 625 °C (1160 °F) and creep resistance comparable to Ti-6Al-4V. Developing commercial applications include forged prosthetic devices and cold rolled foil for metal matrix composites. Special properties include a modulus that is comparable to bone, improved oxidation resistance up to 650 °C (1200 °F), and resistance to aerospace hydraulic fluids (e.g., Skydrol). The latter properties have led to a number of aircraft engine applications. Excellent corrosion and hydrogen embrittlement resistance have led to chemical and offshore oil use.

Selected References

1. W.M. Parris and P.J. Bania, "Beta-21S: A High-Temperature Metastable Beta Titanium Alloy," Proc. 1990 TDA Int. Conf., Orlando, 1990
2. W.M. Parris and P.J. Bania, "Oxygen Effects on the Mechanical Properties of TIMETAL 21S," Proc. 7th Int. Titanium Conf., San Diego, 1992
3. J.S. Grauman, "A High-Strength Corrosion-Resistant Titanium Alloy," Proc. 1990 TDA Int. Conf., Orlando, 1990
4. J.S. Grauman, "Corrosion Behavior of TIMETAL-21S for Nonaerospace Applications," Proc. 7th Int. Titanium Conf., San Diego, 1992

Beta-21S: Typical composition range

	Composition, wt%									
	Al	Nb	Mo	Si	Fe	C	O_2	N_2	H_2	Ti
Minimum	2.5	2.4	14.0	0.15	0.2	...	0.11	...	...	...
Maximum	3.5	3.0	16.0	0.25	0.4	0.05	0.15	0.05	0.015	...
Aim	3.0	2.8	15.0	0.20	0.3	...	0.13	...	...	bal

Physical Properties

Phases and Structures. The microstructure of Beta-21S consists of recrystallized β grains with occasional unrecrystallized β grains. In addition, titanium silicides are present. The principal aging product is α (close-packed hexagonal α). Omega (ω) also has been observed, though it would not be a problem with proper heat treatment.

Beta-21S: Summary of typical physical properties

Beta transus	~793 to 810 °C (1460-1490 °F)
Melting range(a)	1672 to 1747 °C (3041 to 3177 °F)
Density(b)	4.94 g/cm^3 (0.178 lb/in.^{3})
Elastic modulus,	
Beta annealed	74 to 85 GPA (10.7 to 12.3 × 10^6 psi)
Beta annealed + aged	96.5 to 103.5 GPa (14 to 15 × 10^6 psi)
Electrical resistivity(b)	1.35 μΩ · m
Magnetic permeability	Nonmagnetic
Specific heat capacity(b)	710 J/kg · K (0.17 Btu/lb · °F)
Thermal conductivity(b)	7.5 W/m · K (4.3 Btu/ft · h · °F)
Thermal coefficient of linear expansion(c)	8.5 × 10^{-6}/°C (4.7 × 10^{-6}/°F)

(a) Calculated solidus and liquidus temperatures, respectively. (b) Typical values at room temperature of about 20 to 25 °C (68 to 78 °F). (c) Mean coefficient from room temperature to 200 °C (390 °F)

Corrosion Properties

Molybdenum improves corrosion resistance in reducing media, and this well-known effect is apparent when the corrosion rate of Beta 21S and grade 2 Ti are compared in HCl solution (see figure). However, the increased resistance from molybdenum in reducing media generally comes at the expense of resistance in oxidizing media. In this regard, the possible additive or synergistic effect of alloying on oxidation resistance was considered during the development of Beta 21S. The best overall oxidation resistance occurred with aluminum-silicon additions (see table). This alloying results in a slightly higher repassivation potential compared to other molybdenum-containing titanium alloys (see table on next page).

Crevice corrosion resistance improves with molybdenum additions, and a chloride crevice corrosion test (5% NaCl at 90 °C, pH adjusted to 0.5 and 1.0) indicated a chloride crevice corrosion threshold between pH 0.5 and 1.0.

Hydrogen Damage. Beta-21S retains ductility up to hydrogen levels of 2000 ppm. The percent of retained ductility versus hydrogen content is shown (see figure on next page). High hydrogen levels (2000 ppm) will slow down aging kinetics.

Beta-21S: Electrical resistivity vs temperature

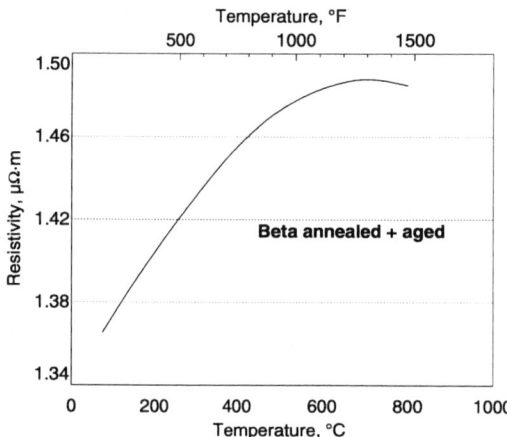

Electrical resistivity (R) between 25 and 750 °C (77 and 1380 °F) fits the expression:
$R(10^{-8}\Omega \cdot m) = 134 + 0.035\ T - 1.11 \times 10^{-5}\ T^2 - 1.27 \times 10^{-8}\ T^3$
Beta annealed plus 600 °C (1110 °F) for 8 h

Beta-21S: Corrosion rate as a function of HCl concentration

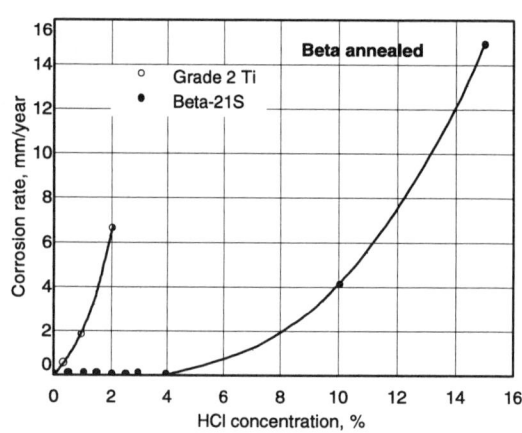

Boiling HCl, 72-h test.
Source: J.S. Grauman, "A New High Strength, Corrosion Resistant Titanium Alloy," TDA Int. Conf., Orlando, 1990

Beta-21S: Oxidation results from alloy development

	Alloy	Weight gain, %
250-g heat		
	Ti-15Mo-5Fe-2Hf	2.40
	Ti-15Mo-5Fe-0.2Si	1.52
	Ti-15Mo-5Fe-2Nb	1.17
	Ti-15Mo-5Fe-2Nb-0.2Si	0.94
	Ti-15Mo-3Nb-1.5Ta-3Al	0.83
	Ti-15Mo-5Nb-0.5Si	0.71
	Ti-15Mo-5Nb-3Al-0.5Si	0.60
8.2-kg heat		
	Ti-15Mo-5Nb-3Al-0.50Si	0.90
	Ti-15Mo-5Nb-0.5Si	0.73
	Ti-15Mo-3Nb-1.5Ta-3Al-0.2Si	0.67
	Ti-15Mo-2Nb-3Al-0.2Si	0.62
250-g buttons		
	Ti-15V-3Cr-3Sn-3Al	>65
	Commercially Pure Ti	7.70
	Ti-15Mo-5Zr	7.70
	Ti-15Mo-3Sn	5.37
	Ti-15Mo-5Co	2.89
	Ti-15Mo-0.1Y	2.73
	Ti-15Mo-5Re	2.68
	Ti-15Mo	2.63
	Ti-15Mo-5Fe	2.10
	Ti-15Mo-3Al	2.00
	Ti-15Mo-0.2Pd	1.79
	Ti-15Mo-0.1Si	1.45
	Ti-15Mo-5Hf	1.41
	Ti-15Mo-0.2Si	1.27
	Ti-15Mo-0.5Si	1.17
	Ti-15Mo-3Ta	1.04
	Ti-20Mo-2Nb	0.99
	Ti-15Mo-2Nb	0.98
	Ti-15Mo-5Nb	0.95
	Ti-15Cr-2Pd	9.76
	Ti-15Cr-3Ta	9.44
	Ti-15Cr-5Nb	7.62
	Ti-15Cr-0.5Si	7.00
	Ti-15Cr-3Sn	4.11
	Ti-15Cr-3Al	3.68
	Ti-15Cr-5Mo	2.90
	Ti-15Cr	2.27

Note: Initial oxidation results from a 48-h exposure at 815 °C (1500 °F) on 1.5 mm (0.02 in.) strip cold rolled from 250-g heat, 8.2-kg heat, and 250-g buttons. Source: W.M. Parris and P.J. Bania, "Beta-21S: A High Temperature Metastable Beta Titanium Alloy," TDA Int. Conf., Orlando, 1990

Beta-21S: Repassivation potential comparison

Alloy	Repassivation potential vs Ag/AgCl
Beta-21S	2.8
Ti grade 2	6.2
Ti-6Al-4V	1.8
Ti-6Al-2Sn-4Zr-6Mo	2.5
Ti-3Al-8V-6Cr-4Zr-4Mo	2.7

Note: The galvanostatic method, boiling 5% NaCl solution pH adjusted to 3.5, was used to measure repassivation. After approximately 1 h of exposure, the test specimen was subjected to a constant current density of 200 mA/cm^2. Source: J.S. Grauman, "A New High Strength, Corrosion Resistant Titanium Alloy," TDA Int. Conf., Orlando, 1990

Beta-21S: General corrosion behavior

Medium	Corrosion rate, mm/year
3% boiling H_2SO_4	0.16
10% $FeCl_3$, boiling	0.01
0.5% HCl, boiling	0.00254
1% HCl, boiling	0.00508
1.5% HCl, boiling	0.01016
2% HCl, boiling	0.01778
2.5% HCl, boiling	0.02794
3% HCl, boiling	0.04064
4% HCl, boiling	0.127
10% HCl, boiling	4.0
15% HCl, boiling	15.0
28% HCl, boiling, deaerated	55.0
10% formic acid, 10% acetic acid, boiling, deaerated	0.0

Note: Beta annealed material. Source: J.S. Grauman, "A New High Strength, Corrosion Resistant Titanium Alloy," TDA Int. Conf., Orlando, 1990

Thermal Properties

Heat Capacity

The specific heat (C_p) for beta-annealed plus aged Beta-21S between 25 and 750 °C (77 and 1380 °F) (see figure) fits the expression:

$$C_p (\text{cal/g} \cdot °C) = 0.116 + 4.83 \times 10^{-5} (T)$$

Thermal Expansion

The thermal coefficient of linear expansion (α) for beta-annealed plus aged Beta-21S between 25 and 750 °C (77 and 1380 °F) (see figure) follows the expression:

$$\alpha (\text{ppm/°C}) = 6.75 + 1.28 \times 10^{-2} T - 2.27 \times 10^{-5} T^2 + 1.52 \times 10^{-8} T^3$$

Thermal Conductivity

The thermal conductivity between 25 and 750 °C (77 and 1380 °F) for beta-annealed plus aged Beta-21S (see figure) fits the equation:

$$Q (\text{W/m} \cdot °C) = 7.33 + 1.66 \times 10^{-2} T$$

Beta-21S: Effect of hydrogen on residual ductility

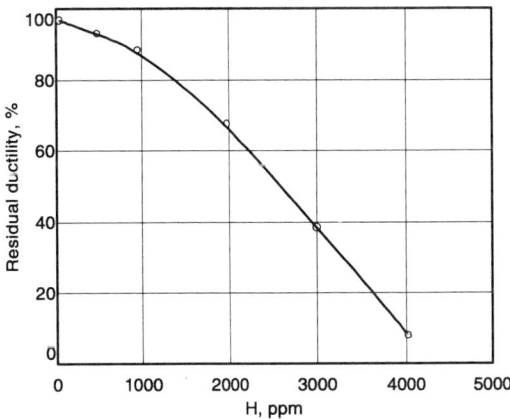

Cathodically charged 1.5 mm (0.06 in.) sheet in the beta-annealed condition. Based on bend radius tests on sheet material cathodically charged with H. Residual ductility determined from initial ductility without H charge, which was equivalent to 15% ± 3%.

Beta-21S: Specific heat vs temperature

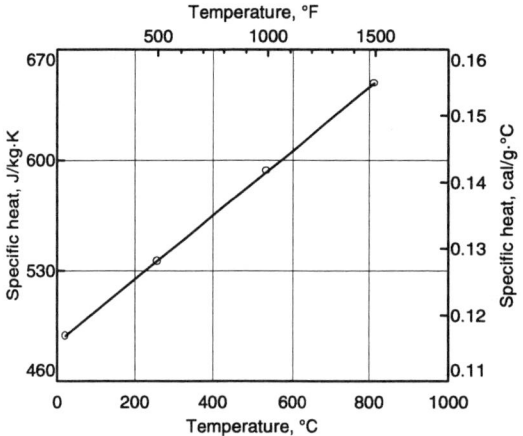

Beta annealed plus 600 °C (1110 °F) for 8 h

Beta-21S: Thermal coefficient of linear expansion

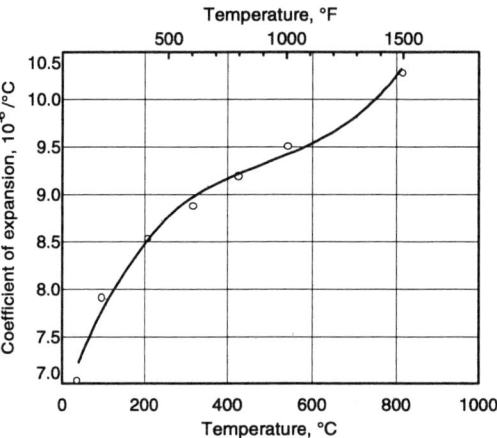

Beta annealed plus 600 °C (1110 °F) for 8 h

Tensile Properties

See also "Processing" for tensile data.

Although oxygen levels below 0.33 wt% do not appear to significantly affect the strength/ductility relationship, results of tests (see table) on sheet from two heats containing 0.14 and 0.25 wt% oxygen showed a deleterious effect on ductility for the higher oxygen content in the series aged at 595 °C (1100 °F). In the annealed condition, there is another effect of oxygen, which could be important in certain types of forming operations. In the annealed condition, the difference between yield, and ultimate tensile strengths decreased as the oxygen level increased from 42 MPa (6.1 ksi) at 0.09% oxygen to 12 MPa (1.7 ksi) at 0.33% oxygen. This behavior implies a decrease in work-hardening capability with increasing oxygen and, concomitantly, an increase in the tendency to neck locally and fail during stretching or drawing operations.

Beta-21S: Thermal conductivity

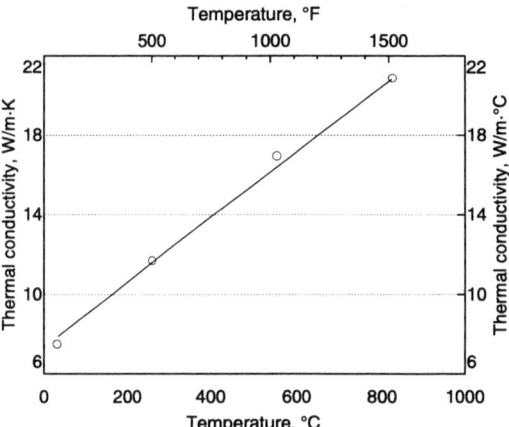

Beta annealed plus 600 °C (1110 °F) for 8 h

Beta-21S: RT tensile properties of sheet vs oxygen content

Aging temperature °C	°F	Aging time, h	Oxygen content, %	Ultimate tensile strength MPa	ksi	Tensile yield strength MPa	ksi	Elongation, %
None		...	0.14	880.5	127.7	860.5	124.8	12.0
			0.25	931.5	135.1	914.3	132.6	15.0
480	895	4	0.14	1093.8	158.6	983.9	142.7	11.5
			0.25	1011.5	146.7	975.0	141.4	14.0
		8	0.14	1257.0	182.3	1145.3	166.1	4.0
			0.25	1143.9	165.9	1114.9	161.7	5.0
		16	0.14	1383.2	200.6	1276.9	185.2	5.0
			0.25	1473.5	213.7	1373.5	199.2	4.5
		24	0.14	1428.6	207.2	1319.7	191.4	3.5
			0.25	1529.3	221.8	1454.8	210.9	3.0
540	1000	4	0.14	1297.0	188.1	1199.7	174.0	8.0
			0.25	1381.8	200.4	1297.6	188.2	5.5
		8	0.14	1269.4	184.1	1185.3	171.9	5.0
			0.25	1388.7	201.4	1303.2	189.0	3.5
		16	0.14	1289.4	187.0	1205.3	174.8	6.0

Note: Cold rolled 50% prior to annealing. 0.14% oxygen annealed at 815 °C (1500 °F), 5 min, AC; 0.25% oxygen annealed at 857 °C (1610 °F), 5 min, AC. Prior to tensile testing, all sheet specimens were descaled and pickled to remove 0.05 mm (0.002 in.) from each surface to remove any material contaminated by oxygen and/or nitrogen during heat treatment. Tensile testing was carried out according to ASTM E8. Gage section for the sheet specimens was 6 mm (0.25 in.) × 25 mm (1 in.) and for the bar specimens 6 mm (0.25 in.) diameter × 25 mm. Source: W.M. Parris and P.J. Bania, "Oxygen Effects on the Mechanical Properties of TIMETAL® 21S," 7th Int. Titanium Conf., July 1992

(continued)

Beta-21S: RT tensile properties of sheet vs oxygen content (continued)

Aging temperature °C	°F	Aging time, h	Oxygen content, %	Ultimate tensile strength MPa	ksi	Tensile yield strength MPa	ksi	Elongation, %
			0.25	1409.3	204.4	1341.1	194.5	3.5
		24	0.14	1268.0	183.9	1192.2	172.9	6.0
			0.25	1388.7	201.4	1336.9	193.9	3.5
595	1100	4	0.14	1103.8	160.1	1024.6	148.6	11.0
			0.25	1180.4	171.2	1108.7	160.8	6.0
		8	0.14	1063.9	154.3	996.3	144.5	11.0
			0.25	1172.2	170.0	1103.9	160.1	7.5
595	1100	16	0.14	1074.2	155.8	999.8	145.0	10.0
			0.25	1194.2	173.2	1128.7	163.7	5.0
		24	0.14	1116.3	161.9	1059.1	153.6	8.0
			0.25	1199.7	174.0	1128.7	163.7	6.5

Note: Cold rolled 50% prior to annealing. 0.14% oxygen annealed at 815 °C (1500 °F), 5 min, AC; 0.25% oxygen annealed at 857 °C (1610 °F), 5 min, AC. Prior to tensile testing, all sheet specimens were descaled and pickled to remove 0.05 mm (0.002 in.) from each surface to remove any material contaminated by oxygen and/or nitrogen during heat treatment. Tensile testing was carried out according to ASTM E8. Gage section for the sheet specimens was 6 mm (0.25 in.) × 25 mm (1 in.) and for the bar specimens 6 mm (0.25 in.) diameter × 25 mm. Source: W.M. Parris and P.J. Bania, "Oxygen Effects on the Mechanical Properties of TIMETAL® 21S," 7th Int. Titanium Conf., July 1992

Beta-21S: RT tensile properties of sheet and bar vs oxygen content

Heat treatment(a)	Oxygen, %	Simulated strip(b) Ultimate tensile strength MPa	ksi	Tensile yield strength MPa	ksi	Elongation, %	Hot rolled bar Ultimate tensile strength MPa	ksi	Tensile yield strength MPa	ksi	Reduction of area, %	Elongation, %
845 °C (1550 °F), 10 min, AC	0.090	813.6	118.0	771.6	111.9	19.8	837.1	121.4	795.7	115.4	66.8	22.5
	0.120	859.8	124.7	819.8	118.9	20.1	847.4	122.9	815.7	118.3	66.2	24.0
	0.130	874.3	126.8	847.4	122.9	17.3	874.3	126.8	843.9	122.4	61.8	23.0
	0.183(c)	900.5	130.6	888.8	128.9	18.4	882.6	128.1	853.6	123.8	63.6	23.3
	0.229(d)	930.8	135.0	912.9	132.4	17.4	899.1	130.4	890.1	129.1	66.1	27.0
	0.334(e)	970.8	140.8	958.4	139.0	21.5	917.0	132.9	913.6	132.5	61.4	26.5
845 °C (1550 °F), 10 min, AC + 480 °C (900 °F), 14 h, AC	0.090	1336.9	193.9	1258.3	182.5	6.8	...	...	...	...	...	...
	0.120	1443.1	209.3	1341.8	194.6	4.1	1431.4	207.6	1352.1	196.1	15.8	7.0
	0.130	1391.4	201.8	1306.6	189.5	3.0	1434.2	208.0	1346.6	195.3	15.9	6.5
	0.183(c)	1447.3	209.9	1375.6	199.5	2.8	1494.8	216.8	1415.5	205.3	12.5	4.5
	0.229(d)	1541.7	223.6	1470.7	213.3	2.3	1583.1	229.6	1501.7	217.8	10.4	4.5
	0.334(e)	1579.6	229.1	1462.4	212.1	3.0	1540.3	223.4	1443.1	209.3	5.0	2.0
845 °C (1550 °F), 10 min, AC + 540 °C (1000 °F), 8 h, AC	0.090	1157.0	167.8	1024.6	148.6	9.6	1202.5	174.4	1037.0	150.4	47.6	10.9
	0.120	1314.2	190.6	1232.8	178.8	5.8	1325.2	192.2	1253.5	181.8	24.3	8.3
	0.130	1320.4	191.5	1243.2	180.3	5.8	1326.6	192.4	1254.9	182.0	24.4	8.5
	0.183(c)	1421.7	206.2	1319.7	191.4	1.4	1395.5	202.4	1329.5	192.8	19.0	7.0
	0.229(d)	1434.8	208.1	1377.6	199.8	4.3	1467.3	212.8	1388.0	201.3	19.2	7.8
	0.334(e)	1461.1	211.9	1359.7	197.2	3.4	1425.2	206.7	1332.8	193.3	8.0	4.0
845 °C (1550 °F), 10 min, AC + 595 (1100 °F), 8 h, AC	0.090	937.0	135.9	822.6	119.3	16.8	1045.3	151.6	947.4	137.4	44.2	11.5
	0.120	1068.0	154.9	986.0	143.0	12.5	1103.2	160.0	1010.1	146.5	35.5	14.0
	0.130	1060.5	153.8	987.4	143.2	9.0	1099.8	159.5	1011.5	146.7	35.2	13.5
	0.183(c)	1152.8	167.2	1081.8	156.9	7.9	1166.6	169.2	1084.6	169.2	26.9	12.0
	0.229(d)	1223.2	177.4	1148.0	166.5	8.0	1232.1	178.7	1146.6	166.3	22.5	10.3
	0.334(d)	1289.4	187.0	1194.2	173.2	6.5	1259.0	171.2	1180.4	171.2	16.0	8.3

(a) Annealing time for sheet was 10 min, for bar 1 h. (b) Cold rolled 50% prior to annealing. (c) Annealed 857 °C. (d) Annealed 870 °C. (e) Annealed 885 °C. Source: W.M. Parris and P.J. Bania, "Oxygen Effects on the Mechanical Properties of TIMETAL® 21S," 7th Int. Titanium Conf., July 1992

Beta-21S: Typical room-temperature aged tensile properties

Aging temperature(a) °C	°F	Test direction	Tensile yield strength MPa	ksi	Ultimate tensile strength MPa	ksi	Elongation, %
540	1000	L	1288	189	1353	196	9.0
		L	1326	192	1394	202	7.5
		T	1346	195	1422	206	6.5
		T	1379	200	1438	208	7.0
		L	1100	159	1179	171	11.0
		T	1185	172	1243	180	11.0
		T	1165	169	1240	179	10.0
Duplex(b)		...	856	124	920	133	18.0
		...	840	122	914	132	20.0

(a) Aged 8 h after beta anneal. (b) 8 h at 690 °C (1275 °F), AC, + 650 °C (1200 °F) for 8 h, AC

Beta-21S: Typical RT beta-annealed tensile properties

Tensile yield strength		Ultimate tensile strength			Tensile yield strength		Ultimate tensile strength		
MPa	ksi	MPa	ksi	Elongation, %	MPa	ksi	MPa	ksi	Elongation, %
Longitudinal					Transverse				
869	126	924	134	11.0	910	132	952	138	10.0
869	126	896	130	9.0	903	131	931	135	9.0
834	121	862	125	12.0	876	127	910	132	10.0
855	124	876	127	12.0	889	129	910	132	10.0
869	126	896	130	12.0	896	130	931	135	10.0
869	126	896	130	14.0	903	131	931	135	11.0
869	126	876	127	15.0	896	130	903	131	12.0
876	127	883	128	12.0	903	131	910	132	12.0
903	131	952	138	11.0	945	137	1007	146	11.0
862	125	896	130	13.0	903	131	938	136	11.0
855	124	896	130	11.0	896	130	938	136	11.0
862	125	896	130	12.0	896	130	924	134	11.0
924	134	952	138	10.0	965	140	993	144	10.0
938	136	986	143	14.0	972	141	1014	147	7.0
938	136	986	143	11.0	979	142	1027	149	8.0
896	130	924	134	10.0	952	138	979	142	9.0

High-Temperature Strength

Beta-21S: High-temperature tensile properties (aged at 540 °C)

Test temperature		Test direction	Tensile yield strength		Ultimate tensile strength		Elongation, %
°C	°F		MPa	ksi	MPa	ksi	
24	75	L	1288	187	1353	196	9.0
		L	1326	192	1394	202	7.5
		T	1346	195	1422	206	6.5
		T	1379	200	1438	208	7.0
205	400	L	1105	160	1200	174	8.5
		L	1096	159	1204	175	9.5
		T	1127	163	1233	179	8.0
		T	1154	167	1249	181	6.0
315	600	L	1041	151	1149	166	8.0
		L	1019	147	1156	167	8.0
		T	1089	158	1197	173	6.0
		T	1050	152	1158	168	7.0
425	800	L	976	141	1090	158	8.0
		L	969	140	1077	156	9.0
		T	1016	147	1132	164	7.0
		T	1005	145	1122	162	6.0
540	1000	L	576	83	838	121	22.0
		L	674	97	849	123	22.0
		T	616	89	867	125	25.0
		T	648	94	886	128	24.5

Beta-21S: High-temperature tensile properties

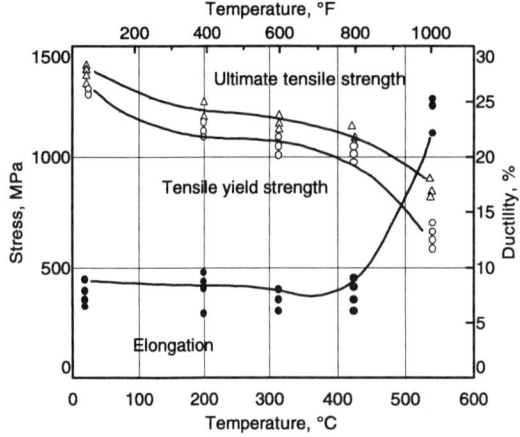

Beta annealed at 845 °C (1550 °F) for 10 min + 540 °C (1000 °F), 8 h

Beta-21S: High-temperature tensile properties (aged at 540 °C)

Test temperature		Test direction	Tensile yield strength		Ultimate tensile strength		Elongation, %
°C	°F		MPa	ksi	MPa	ksi	
24	75	L	1100	159	1179	171	11.0
		T	1185	172	1243	180	11.0
		T	1165	169	1240	179	10.0
205	400	L	893	129	1011	146	12.0
		L	903	130	1020	148	10.0
		T	907	131	1036	150	10.0
		T	944	137	1069	155	10.0
315	600	L	832	121	955	138	10.0
		L	830	120	969	140	10.0
		T	861	125	1001	145	9.0
		T	875	127	994	144	9.0
425	800	L	776	112	909	132	10.0
		L	807	117	925	134	10.0
		T	818	118	946	137	9.0
		T	856	124	967	140	7.0
540	1000	L	598	86	741	107	26.9
		L	587	85	751	109	24.0
		T	613	89	773	112	28.5
		T	633	92	822	119	12.0
		L	598	86	741	107	26.9
		L	587	85	751	109	24.0
		T	613	89	773	112	28.5
		T	633	92	822	119	12.0

Beta-21S: High-temperature tensile properties

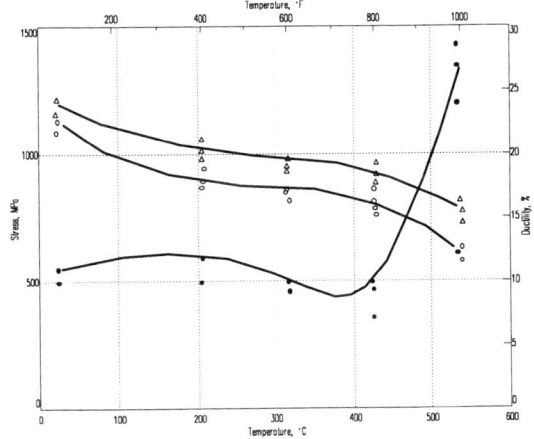

Beta annealed at 845 °C (1550 °F) for 10 min + 595 °C (1100 °F), 8 h

Beta-21S: Creep results in STA material

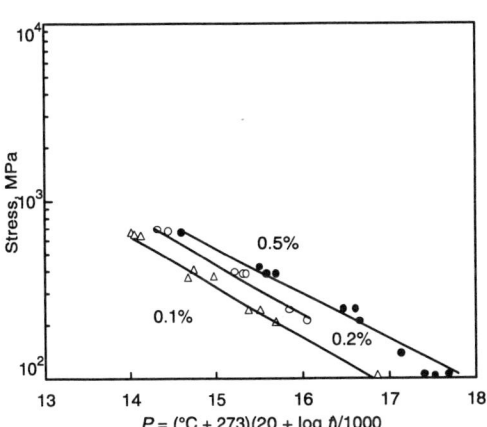

1.5 mm (0.06 in.) sheet; beta annealed plus aged 8 h at 540 °C (1000 °F)

Crack Resistance

Beta-21S: Fracture toughness

Heat treatment	Tensile yield strength		Ultimate tensile strength		Elongation, %	Test medium	Fracture toughness (K_c)	
	MPa	ksi	MPa	ksi			MPa√m	ksi√in.
Beta annealed	865	124	879	127	15	Air	107.5	97.8
							107.5	97.8
						Salt water	107.5	97.8
							106.7	97.1
Beta annealed + 540 °C (1000 °F), 8 h	1220	177	1320	191	6	Air	75.4	68.6
							72.6	66.0
						Salt water	68.5	62.3
							67.8	61.7
Beta annealed + 595 °C (1100 °F), 8 h	1040	151	1130	164	7.5	Air	100.8	91.7
							100.8	91.7

Note: Center notch sheet specimens 1.4 mm (0.05 in.) thick

Beta-21S: Fatigue crack growth

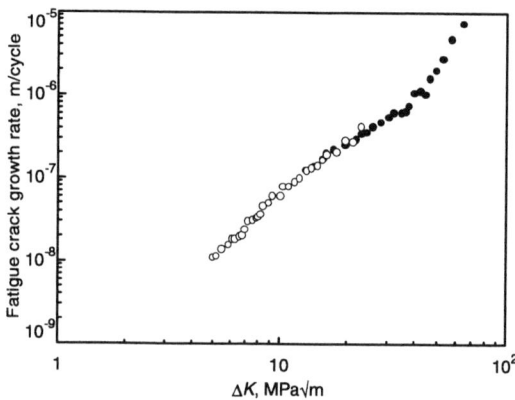

As in other beta alloys, microstructure and heat treatment seem to have virtually no effect on crack growth rates of Beta-21S. Solution treated and aged for 8 h at 540 °C (1000 °F). Room-temperature tests at lab air; 29 Hz; $R = 0.1$

Beta-21S: Fatigue crack growth

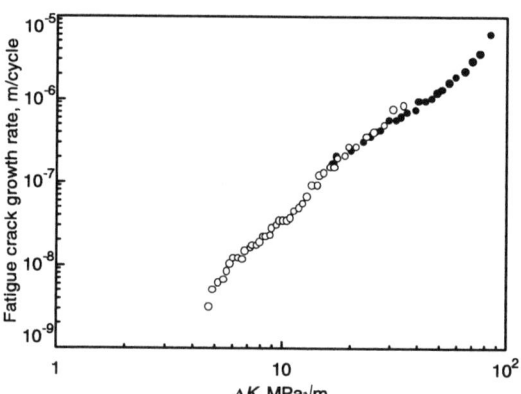

Beta annealed (790 °C, 10 min) sheet, 1.4 mm (0.05 in.) center notch sample, tested at RT; $R = 0.1$; 29 Hz in lab air

Processing

Formability. Limited forming data indicate a similarity to Ti-15V-3Cr-3Al-3Sn (Ti-15-3). In addition to tensile tests, some indication of sheet formability in annealed material was obtained by bend testing 25 mm (1 in.) wide strip. These specimens were bent 105° around successively smaller radii either until cracking visible at 20× magnification occurred or until the minimum radius of 0.75 mm (0.030 in.) was reached. Sheet in the annealed condition from all heats sustained a 105° bend around a 1.27 mm (0.050 in.) radius without cracking. This translates to a bend ductility of 1 T or less for sheet at all oxygen levels. Thus, oxygen contents up to 0.33% had no significant effect on this criterion for formability. However, as shown in the previous section on tensile properties, tensile data indicated a possible oxygen effect on other aspects of sheet formability.

Machining, welding, and brazing of Beta-21S is typical of beta alloys and is considered similar to that of Ti-15-3.

Heat Treatment. In cases where high-temperature exposure is anticipated, a duplex overage is used to retain ductility. The high temperature age "weakens" the grains relative to the grain boundaries and the second age stabilizes the grains against embrittlement. (See table and figures.)

Beta-21S: Selected heat treatments

Treatment	Temperature °C	Temperature °F	Duration	Cooling method
Solution treatment (beta anneal)	800-815	1470-1500	4 min, minimum	...
Aging(a)	480	900	24 h	AC
	540	1000	8 h	AC
	595	1100	8 h	AC
Duplex age				
First stage	690	1275	8 h	AC
Second stage	650	1200	8 h	AC

(a) Three selected aging treatments that cover strength levels likely to be used in commercial applications

Beta-21S: Tensile yield strength vs aging time

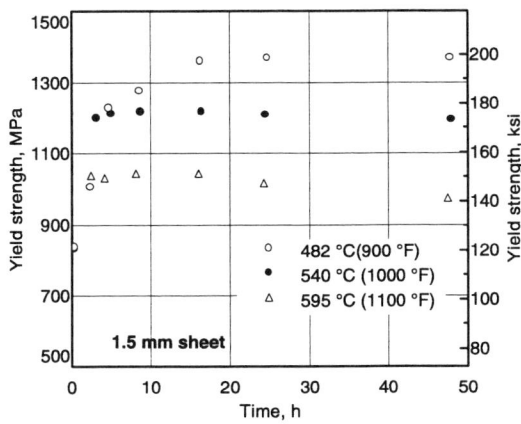

Beta annealed 1.5 mm (0.06 in.) sheet aged for indicated times and temperatures

Beta-21S: Ultimate tensile strength vs aging time

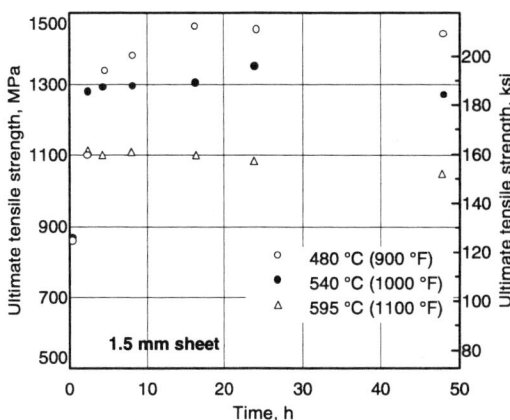

Beta annealed 1.5 mm (0.06 in.) sheet aged for indicated times and temperatures

Forging

Beta-21S can be fabricated into all forging product types, although current closed die forging predominates. Beta-21S is a reasonably forgeable alloy when forged above its beta transus, with higher unit pressures (flow stresses), improved forgeability, and less crack sensitivity in forging than the α–β alloy Ti-6Al-4V. Due to the high alloying content of Beta-21S, flow stresses are higher than those of the near-beta alloy Ti-10V-2Fe-3Al. The desired final microstructure from thermomechanical processing of Beta-21S during forging manufacture is a fine transformed β, with limited grain boundary films and a fine, recrystallized prior β grain size in preparation for final thermal treatments.

Thermomechanical Processing. The very fine microstructural features of Beta-21S achieved in forgings are responsible for its excellent mechanical properties and fatigue resistance. Reheating for subsequent forging operations recrystallizes the alloy from prior hot work, refining the grain size. Beta-21S is generally not subtransus forged because there is no microstructural advantage, and there is a significant increase in unit pressures.

Final thermal treatments for Beta-21S include a simple anneal (or solution anneal) for low-modulus applications or solution anneal and aging for higher strength levels. Forgings may be supplied annealed, solution annealed, and/or fully aged (STA). Annealing or solution annealing generally is conducted at 815 to 870 °C (1500 to 1600 °F). Aging is conducted at 535 to 595 °C (1000 to 1100 °F).

Beta forging working histories for Beta-21S require imparting enough hot work to reach final macrostructure and microstructure objectives. Generally, reductions in any given forging process are 30 to 50% to achieve desired dynamic and static recrystallization. Very low levels of beta reduction are not recommended.

Hydrogen. Beta-21S, as with all beta alloys, has a high affinity for hydrogen. Although Beta-21S forms less α case from heating operations than other alloy classes, therefore requiring less metal removal in chemical pickling (milling) processes, control of chemical removal processes is essential to preclude excessive hydrogen pickup.

Recommended forging metal temperatures range from 790 to 850 °C (1450 to 1560 °F). Recommended die temperatures are summarized in "Technical Note 4: Forging."

Ti-5Al-2Sn-4Zr-4Mo-2Cr-1Fe Beta-CEZ®

Compiled by Y. Combres, CEZUS Centre de Recherches, Ugine, France

Beta-CEZ® is a multifunctional near-β titanium alloy exhibiting high strength, high toughness, and intermediate-temperature creep resistance. Its processing flexibility makes it suitable for a wide range of applications.

Product Forms and Conditions. Typical product forms consist of forged billets in diameters ranging from 150 to 300 mm (6 to 12 in.) and forged or rolled bar in diameters ranging from 10 up to 110 mm (0.4 to 4.3 in.). Rolled plate and sheet are also available in thicknesses ranging from 25 to 3 mm (1 to 0.1 in.) and 500 mm (20 in.) wide. Products are supplied in the forged or solution treated conditions. The microstructure is fine and equiaxed.

Applications. Typical applications include heavy section forgings used for medium-temperature compressor disks in which an optimum combination of strength, ductility, toughness, and creep resistance is required. Beta-CEZ® is a structural alloy with very high strength and a good combination of strength, ductility, and toughness. Near-net shape forgings are possible due to the excellent formability of the alloy. Component applications are as forged parts, springs, and fasteners.

Physical Properties

Crystal Structure. In the solution treated and aged condition, the microstructure consists of α + β phases. The lattice parameters of the close-packed hexagonal α phase are $a = 2.9287$ Å and $c = 4.6606$ Å, whereas the lattice parameter of the body-centered cubic β phase is $a = 3.2040$ Å.

Grain Structure. The microstructure is typical of β metastable alloys and may be β or α + β, either equiaxed or lamellar. Highest strength and ductility are achieved with an equiaxed primary α phase and a finely precipitated secondary α phase microstructure. Optimum toughness is obtained with lamellar primary α microstructures.

Transformation Products. The continuous cooling (CCT) diagram is similar to that of Ti-17. Alpha precipitation occurs first at β grain boundaries and secondly inside the grains. For instance, the time difference between grain boundary and intragranular precipitation is about 1 h when cooled at 1 °C/min (1.8 °F/min) from the β field. The transformation of samples cooled from the β field exhibits a coarse α precipitation above 700 °C (1290 °F) and fine acicular precipitation between 700 and 400 °C (1290 and 750 °F). A temperature of > 750 °C (1380 °F) is recommended for solution treatments below the transus, whereas aging treatments are performed below 700 °C (1290 °F)

Chemical Corrosion Resistance. Corrosion resistance in acid or seawater, as well as hydrogen uptake and embrittlement are currently being studied. Data are not available yet.

Beta-CEZ®: Chemical composition

Element	Composition, wt%
Aluminum	4.5-5.5
Tin	1.5-2.5
Zirconium	3.5-4.5
Molybdenum	3.5-4.5
Chromium	1.5-2.5
Iron	0.5-1.5
Oxygen	800-1300 ppm
Hydrogen	<150 ppm

Beta-CEZ®: Summary of typical physical properties

Beta transus	890 °C (1634 °F)
Solidus	~1550 °C (2820 °F)
Melting (liquidus point)	1602 °C (2916 °F)
Density(a)	4.69 g/cm³
Specific heat capacity(a)	580 J/kg · K (0.14 Btu/lb · °F)
Thermal conductivity(a)	6.7 W/m · K (3.8 Btu/ft · h · °F)
Thermal coefficient of linear expansion(b)	10×10^{-6}/°C (5.5×10^{-6}/°F)

(a) Typical values at room temperature of about 20 to 25 °C (68 to 78 °F). (b) Mean coefficient from room temperature to 600 °C (1110 °F). See figure.

Beta-CEZ®: Thermal coefficient of linear expansion vs temperature

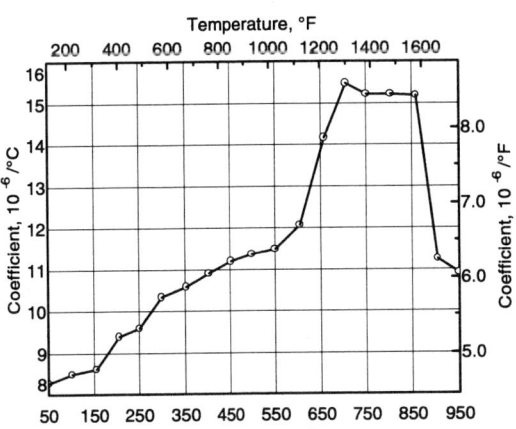

Mechanical Properties

Tensile Properties

Tensile properties depend strongly on microstructure (see table). Forged or rolled bars exhibit an equiaxed microstructure, whereas β processed and "through the β transus" processed pancakes exhibit lamellar and necklaced microstructures, respectively. Beta-CEZ® can maintain a high strength level at high temperatures for both the equiaxed or lamellar microstructures (see figure).

Beta-CEZ®: Young's modulus vs temperature

Temperature		Young's modulus	
°C	°F	GPa	10^6 psi
20	68	122	17
300	570	106	15
400	750	100	14

Beta-CEZ®: Typical tensile properties

Product form	Heat treatment	Ultimate tensile strength MPa	ksi	0.2% yield strength MPa	ksi	Elongation, %
150 mm (6 in.) diam forged bar	As forged	1040	150	960	139	18
	830 °C (1525 °C), 1 h, WQ + 550 °C (1020 °F), 8 h, AC	1601	232	1518	220	2
	830 °C (1525 °F), 1 h, WQ + 600 °C (1110 °F), 8 h, AC	1283	186	1208	175	11
	860 °C (1580 °F), 1 h, WQ + 550 °C (1020 °F), 8 h, AC	1557	226	1478	214	2
	860 °C (1580 °F), 1 h, WQ + 600 °C (1110 °F), 8 h, AC	1370	198	1304	189	5
12.7 mm (½ in.) diam rolled bar	As rolled	1490	216	1345	195	11
	830 °C (1525 °F), 1 h, WQ + 550 °C (1020 °F), 8 h, AC	1506	218	1460	211	13
	830 °C (1525 °F), 1 h, WQ + 600 °C (1110 °F), 8 h, AC	1373	199	1349	195	15
	860 °C (1580 °F), 1 h, WQ + 550 °C (1020 °F), 8 h, AC	1723	250	1683	244	7
	860 °C (1580 °F), 1 h, WQ + 600 °C (1110 °F), 8 h, AC	1540	223	1485	215	9
25 mm (1 in.) thick rolled plate	As rolled, L	1222	177	1124	163	15
	As rolled, T	1260	182	1163	168	11
	830 °C (1525 °F), 1 h, WQ + 600 °C (1110 °F), 8 h, AC, L	1334	193	1287	186	13
	830 °C (1525 °F), 1 h, WQ + 600 °C (1110 °F), 8 h, AC, T	1351	196	1300	188	12
	860 °C (1580 °F), 1 h, WQ + 600 °C (1110 °F), 8 h, AC, L	1405	203	1338	194	10
	860 °C (1580 °F), 1 h, WQ + 600 °C (1110 °F), 8 h, AC, T	1418	205	1340	194	6
300 mm (12 in.) diam β-processed pancake	600 °C (1110 °F), 8 h, AC	1608	233	1472	213	2
	830 °C (1525 °F), 1 h, WQ + 570 °C (1060 °F), 8 h, AC	1357	197	1171	170	5
	830 °C (1525 °F), 1 h, WQ + 600 °C (1110 °F), 8 h, AC	1326	192	1188	172	6
300 mm (12 in.) diam "through the transus" processed	600 °C (1110 °F), 8 h, AC	1227	178	1138	165	10
	830 °C (1525 °F), 1 h, WQ + 570 °C (1060 °F), 8 h, AC	1314	190	1200	174	10
	830 °C (1525 °F), 1 h, WQ + 600 °C (1110 °F), 8 h, AC	1263	183	1170	169	11

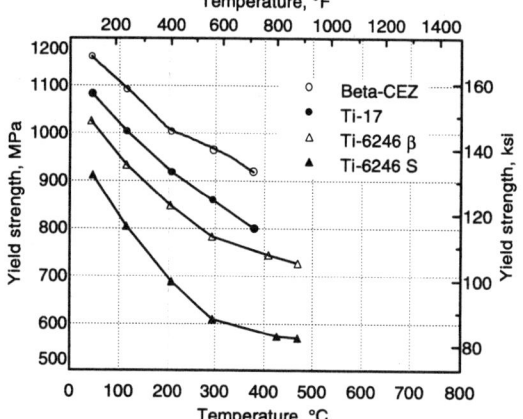

Beta-CEZ®: Yield strength comparison
Specimens were 300 mm (12 in.) diam β processed pancakes with a lamellar microstructure.
Source: Ti-17 and Ti-6246 data courtesy of SNECMA.

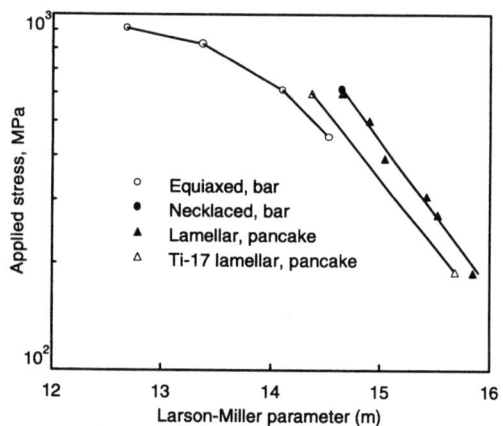

Beta-CEZ®: Creep property comparison
Beta-CEZ® had lamellar, necklaced, and equiaxed microstructures; Ti-17 specimens had lamellar β processed structures. All specimens were 300 mm (12 in.) diameter. $m = T(20 + \log t)10^{-3}$ (T in K; t in h).

Beta-CEZ®: Low-cycle fatigue for equiaxed microstructures aged at 600 °C

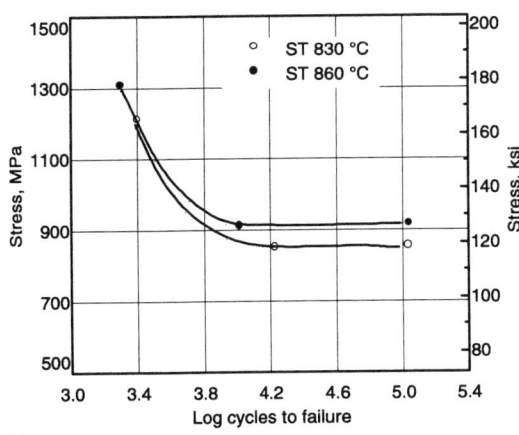
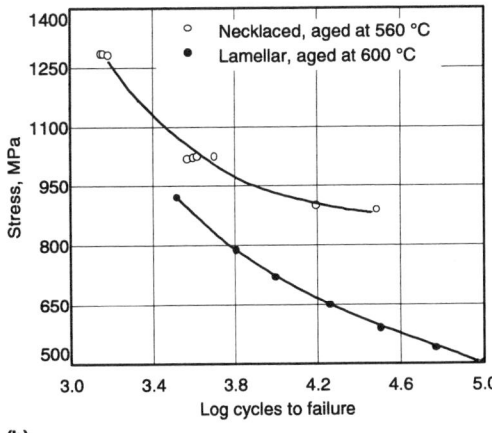

(a) Specimens were from 150 mm (6 in.) diam forged bar. b) Beta-CEZ®: Low-cycle fatigue for lamellar and necklaced microstructures after solution treating at 830 °C. b) Necklaced microstructures were 80 mm (3.1 in.) diam "through the β transus" forged bar; lamellar microstructures were 300 mm (12 in.) diam β-processed pancake.

Fatigue

The alloy behaves very well in low-cycle fatigue conditions between 20 °C (68 °F) (700 MPa, or 101 ksi) for 10^4 cycles) and 400 °C (750 °F) (600 MPa, or 87 ksi for 10^4 cycles). For high-cycle fatigue, equiaxed microstructures have a fatigue limit of 900 MPa (130 ksi) for 10^5 cycles at 20 °C (68 °F).

Crack Propagation Resistance

Products with the best toughness exhibit good fatigue crack propagation resistance at 20 °C (68 °F). Typical da/dN characteristics are shown here.

Fracture Toughness

Equiaxed microstructures are characterized by toughness ranging from 45 to 55 MPa√m (40 to 50 ksi√in.). Toughness of the lamellar structure ranges from 60 to 90 MPa√m (54 to 82 ksi√in.), whereas the necklaced microstructure has a toughness ranging from 65 to 95 MPa√m (59 to 86 ksi√in.)(see figure).

Low-temperature toughness usually ranges from 30 to 45 MPa√m (27 to 41 ksi√in.) at −253 °C (−423 °F).

Beta-CEZ®: Fatigue crack propagation for lamellar or necklaced microstructures

ΔK		da/dN,
MPa√m	ksi√in.	mm/cycle
6	5.4	10^{-8}
10	9.1	2×10^{-5}
60	54	10^{-3}

Note: Specimens were from 300 mm (12 in.) diameter pancakes in the β processed or "through the β transus" processed condition.

Beta-CEZ®: Fracture toughness vs yield strength comparison

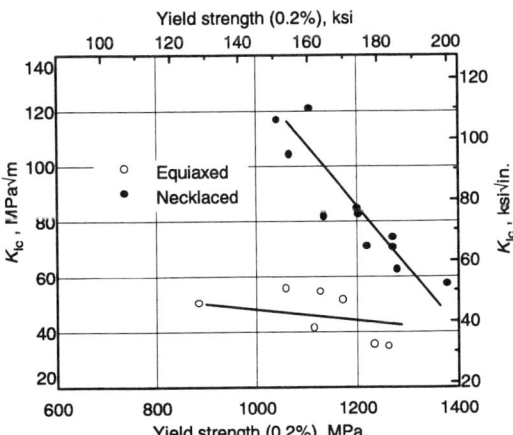

Specimens were 70 mm (2.7 in.) diam α + β rolled bar (equiaxed structure) and 80 mm (3.1 in.) diam "through the β transus" forged bar (necklaced structure).

Fabrication

Forming

Because the strain-rate sensitivity exponent of Beta-CEZ® is rather high compared to conventional alloys (0.3 for Beta-CEZ® versus 0.2 for Ti-6Al-4V), plastic flow is more stable and enhances formability. The metastable nature of the alloy lowers its sensitivity to temperature.

Hot working in the α + β range is recommended at 800 to 860 °C (1470 to 1580 °F) to maintain a fine equiaxed microstructure. In the β range, a temperature around 920 °C (1690 °F) is suggested to obtain a lamellar structure by β processing.

"Through the Transus" Processing is a patented technique that results in a "necklaced" microstructure. It is applied to a 100% β metastable structure below 890 °C (1635 °F). Lamellae in the core of the grains and fine equiaxed grains at the boundaries are thus obtained, which leads to an excellent combination of strength, ductility, and toughness.

Superplastic Forming. The alloy displays superplastic properties between 725 and 775 °C (1340 and 1430 °F); 1000% ductility can be reached at strain rates as high as 8×10^{-4} s^{-1}. Diffusion bonding is being studied.

Superplastic properties of Beta CEZ alloy are obtained at temperatures as low as 725 °C (*Scripta Met*, Vol 29, No. 4, 1993, p 503-508). The as-forged material exhibits a complex microstructure, including the usual β and the globular primary alpha phases, but also a significant amount of acicular alpha. The superplastic behaviour of this unusual microstructure is associated with the breaking up of the acicular alpha in the first steps of deformation, which leads to a very fine mean grain size. The origin of superplasticity at these low temperatures is not yet clearly understood. More detailed investigations are needed, particularly to determine the effective diffusion coefficients in the β phase, since slow and fast diffusing elements (in comparison to Ti atoms) are present in this alloy.

Beta CEZ: Strain-rate sensitivity at 760 °C

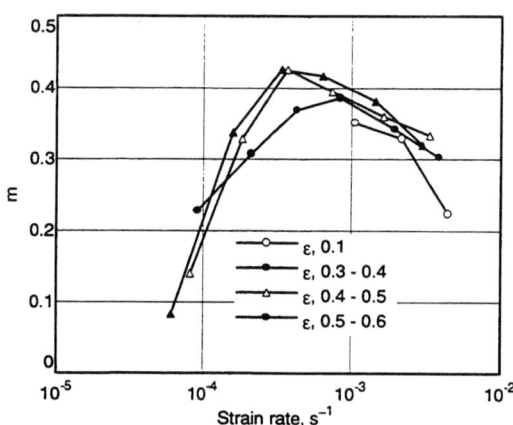

Deformation at 760 °C
Source: *Scripta Met*, Vol 29, p 503-508

Heat Treatment

Solution treatment is recommended between 750 and 860 °C (1380 to 1580 °F) from 1 to 4 h.

Aging is recommended between 525 and 650 °C (980 to 1200 °F) from 30 min to 8 h. As a function of aging time, the hardness evolves rapidly (see table). Maximum hardness is about 560 HV.

Beta-CEZ®: Hardness kinetics for equiaxed microstructures

Product form	Heat treatment	Aging time (t), min	Hardness (30 kg), HV
150 mm (6 in.) diam forged bar	As forged	0	345
	860 °C (1580 °F), 2 h, WQ + 550 °C (1020 °F), t, AC	1	380
		3	440
		10	470
		30	485
		100	480
		300	465
		1000	460
		3000	460

Ti-8Mo-8V-2Fe-3Al

Common Name: Ti-8823

Reviewed by R. Boyer, Boeing Commercial Airplane Co.

Ti-8823 was developed by Timet primarily as a high-strength formable sheet alloy, but it also possesses hardenability in sections up to 100 mm (4 in.) and possibly up to 150 mm (6 in.). As such, it has been considered as a possible fastener alloy, spring alloy, and for structural forgings.

Producers have found the metallurgical behavior of Ti-8823 to be more predictable than Ti-13V-11Cr-3Al, because Ti-8823 recrystallizes uniformly during hot working and ages (hardens) without forming an intermetallic compound phase. Also, it is superior to the Ti-13-11-3 beta alloy in aging kinetics because less aging time is required to reach high strength levels (see figure).

Additionally, Ti-8823 has good fracture toughness, notch fatigue strength, modulus of elasticity, stress-corrosion resistance, and thermal stability at least to 315 °C (600 °F). The good formability of Ti-8823 in the annealed condition, its deep hardenability, and capability for heat treatment to very high strength levels are consistent with the general advantages of beta titanium alloys.

The limitations of Ti-8823 are consistent with some of the disadvantages in general of beta titanium alloys, such as higher density (than for example alpha-beta alloys), relatively poor creep strength at moderately high temperatures, and marginal weld properties in the heat-treated condition. Also, smooth specimens of the Ti-8823 alloy do not appear to have the fatigue strength that might be expected from such a high tensile strength material. Ti-8823 also suffers from melting problems due to its high molybdenum content.

Selected References

- R.A. Wood, *Beta Titanium Alloys*, MCIC 72-11, Battelle Columbus Laboratories, 1972
- *Aerospace Structural Metals Handbook*, Vol 4, Code 3721, Battelle Columbus Laboratories, 1970

Ti-8823: Typical composition range

	Al	V	Fe	Mo	C	O_2	N_2	H_2
Minimum	2.6	7.5	1.6	7.5	...	0.1	...	...
Maximum	3.4	8.5	2.4	8.5	0.05	0.16	0.05	0.015
Nominal	3.0	8.0	2.0	8.0	...	...	...	...

Physical Properties

Modulus of elasticity of Ti-8823 alloy varies with the heat treatment condition. For example, values of 70 to 75 GPa (10 to 11 × 10⁻⁶ psi) have been measured for solution heat treated condition of Ti-8.5Mo-8.5V-2.25Fe-2.5Al 0.10O compositions. Solution heat treated plus aged bar of 25, 50, 75, and 100 mm (1, 2, 3, and 4 in.) section thickness had measured modulus values of 115, 114, 106, and 109 GPa (16.7, 16.5, 15.4, and 15.8 × 10⁻⁶ psi), respectively. Longitudinal and transverse elastic modulus values were about the same for bar of 75 and 100 mm (3 and 4 in.) section thickness. Temperature effects are also shown (see table on next page).

Ti-8823: Summary of typical physical properties

Beta transus	775 °C (1425 °F)
Density(a)	4.85 g/cm³ (0.175 lb/in.³)
Magnetic permeability	Nonmagnetic

(a) Typical values at room temperature of about 20 to 25 °C (68 to 78 °F)

Ti-8823: Aging response vs Ti-13V-11Cr-3Al

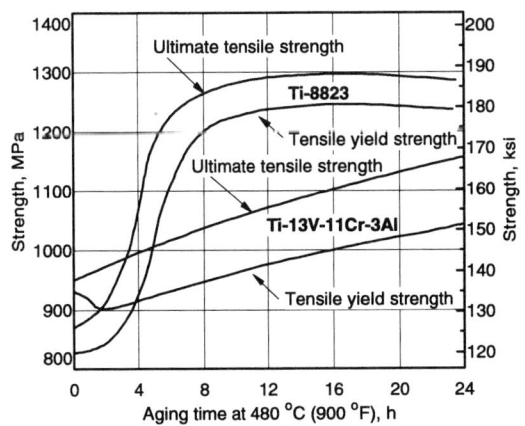

Aging response of alloys solution treated at 815 °C (1500 °F) for 10 min, AC, and cold rolled 25%
R.A. Wood, *Beta Titanium Alloys*, MCIC 72-11, Battelle Columbus Laboratories, 1972

Ti-8823: Modulus of elasticity

Heat treated condition	Modulus, GPa (10^6 psi) at:		
	–55 °C (–65 °F)	21 °C (70 °F)	315 °C (600 °F)
785 to 800 °C (1450 to 1475 °F), WQ or AC	...	86 (12.5)	...
785 °C (1450 °F), 10 min, AC + 480 °C (900 °F), 8 h, AC	107 (15.6)	118 (17.2)	98 (14.2)
785 °C (1450 °F), 10 min, AC + 595 °C (1100 °F), 24 h, AC	108 (15.7)	110 (15.9)	95 (13.8)

Source: R.A. Wood, *Beta Titanium Alloys*, MCIC 72-11, Battelle Columbus Laboratories, 1972

Ti-8823: Oxidation resistance

Temperature		Exposure time, h	Weight gain, g/cm²
°C	°F		
815	1500	2	0.003
		4	0.014
		8	0.040
980	1800	2	0.040
		4	0.082
		8	0.210
1040	1900	4	0.106
1200	2200	4	0.152

Source: R.A. Wood, *Beta Titanium Alloys*, MCIC 72-11, Battelle Columbus Laboratories, 1972

Mechanical Properties

See also "Processing" for mechanical property data.

Tensile property data for Ti-8823 are a function of thermomechanical history, product form, test temperature, etc. The aging response, as with all beta titanium alloys, can be significantly influenced by thermomechanical history (see figure). Thus, mechanical property data is included in the heat treatment section at the end of this datasheet.

Ti-8823 is cold workable, and this characteristic allows an excellent combination of strength and ductility to be achieved for certain applications. High strength levels can be obtained in this alloy, which translate to high structural efficiency. As with most beta titanium alloys, usage is limited to about 315 °C (600 °F).

Ti-8823: Effect of cold work and cold work plus aging

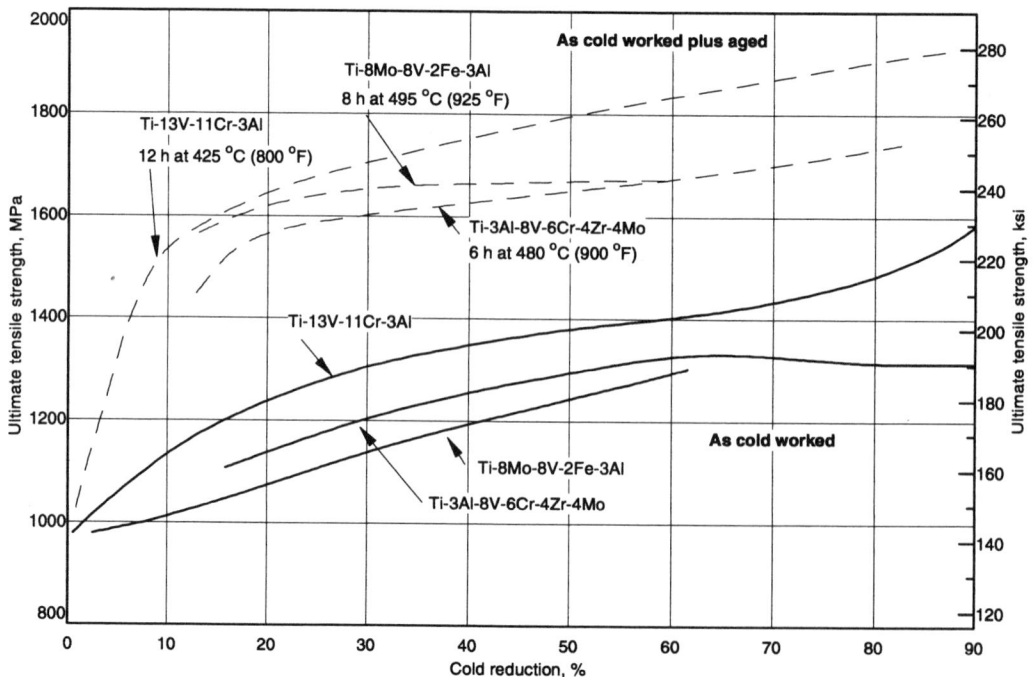

W.H. Heil, "Ti-8Mo-8V-2Fe-3Al Rod and Bar for Fastener Application," Titanium Metals Corporation of America, Technical Data Sheet, Mar 1969

Sheet

Ti-8823: Tensile properties

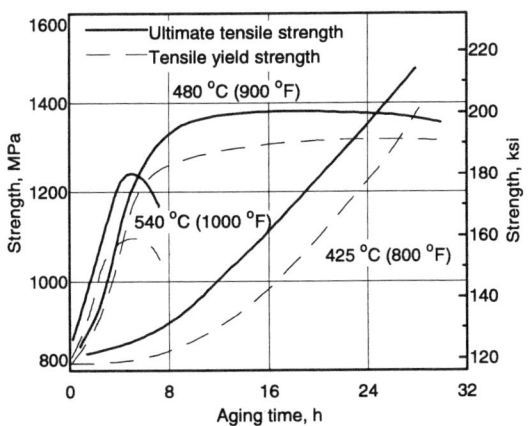

Effect of aging heat treatment variables on the tensile properties of 1.27-mm (0.050-in.) sheet solution heat treated 15 min at 815 °C (1500 °F), water quenched.
R.A. Wood, *Beta Titanium Alloys*, MCIC 72-11, Battelle Columbus Laboratories, 1972

Ti-8823: Vickers hardness

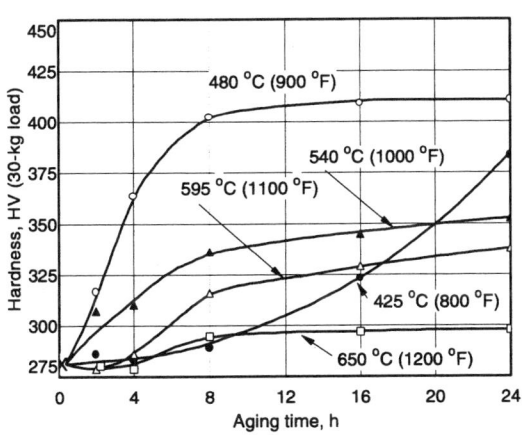

Effect of aging heat treatment hardness of 1.5 mm (0.060 in.) sheet solution heat treated 30 min at 800 °C (1475 °F) and air cooled.
R.A. Wood, *Beta Titanium Alloys*, MCIC, Battelle Columbus Laboratories, 1972

Ti-8823: Tensile properties vs aging temperature

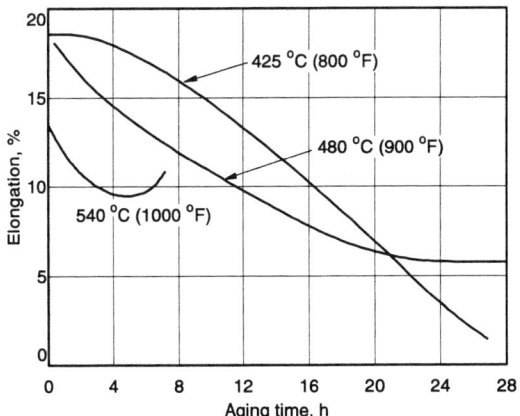

Effect of 8-h aging temperature on tensile properties of 6.5 mm (0.256 in.) diameter rod solution heat treated 15 min at 785 °C (1450 °F), air cooled

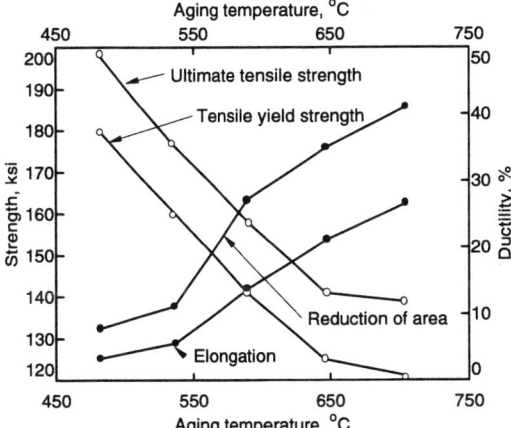

Fastener Stock

Ti-8823: Mechanical properties of fastener stock

Heat treatment	Tensile yield strength		Ultimate tensile strength		Elongation, %	Shear strength	
	MPa	ksi	MPa	ksi		MPa	ksi
15 min, 785 °C (1450 °F), AC + 8 h, 650 °C (1200 °F), AC	960	139	1027	149	22	703	102
20 min, 775 °C (1425 °F), AC + 8 h, 650 °C (1200 °F), AC	875	127	930	135	24	662	96
Unspecified, except 650 °C (1200 °F) aging	...	...	...	...	...	682-703	99-102

Source: R.A. Wood, *Beta Titanium Alloys*, MCIC 72-11, Battelle Columbus Laboratories, 1972

Ti-8823: Room-temperature shear and tensile properties vs cold work or heat treatment

Condition/ heat treatment	Diameter		Tensile yield strength		Ultimate tensile strength		Elongation, %	Reduction of area, %	Double shear strength	
	mm	in.	MPa	ksi	MPa	ksi			MPa	ksi
20 min, 775 °C (1425 °F), AC	7.9	0.312	862	125	889	129	29	58	655	95
20 min, 775 °C (1425 °F), AC + 8 h, 495 °C (925 °F), AC	7.9	0.312	1268	184	1344	195	8	21	834	121
15 min, 785 °C (1450 °F), AC	4.8	0.190	889	129	930	135	17	58	662	96
15 min, 800 °C (1475 °F), AC	7.9	0.312	862	125	868	126	28	64	620	90
Hot rolled	7.9	0.312	937	136	972	141	24	68	...	...
HR + 14% CR	7.2	0.284	993	144	1034	150	21	64	655	95
HR + 36% CR	6.2	0.244	1151	167	1179	171	12	57	696	101
HR + 53% CR	5.3	0.208	1220	177	1255	182	12	57	703	102
HR + 53% CR + 8 h, 495 °C (925 °F), AC	5.3	0.208	1599	232	1661	241	6	18	924	134
HR + 61% CR	4.8	0.191	1282	186	1303	189	10	52	724	105
HR + 61% CR + 8 h, 495 °C (925 °F), AC	4.8	0.191	1613	234	1668	242	9	35	951	138
15 min, 785 °C (1450 °F), AC + 8 h, 650 °C (1200 °F), AC	4.8	0.190	958	139	1027	149	22	...	703	102
20 min, 775 °C (1425 °F), AC + 8 h, 650 °C (1200 °F), AC	7.9	0.312	875	127	930	135	24	...	662	96

Source: R. Woods and R. Favor, *Titanium Alloys Handbook*, MCIC HB-02, Battelle Columbus Laboratories, 1972

Bar and Billet

Ti-8823: Room-temperature tensile properties of forged bar and billet (STA)

Bar stock section	Orientation and location	Tensile yield strength (0.2%)		Ultimate tensile strength		Elongation, %	Reduction of area, %	Modulus of elasticity	
		MPa	ksi	MPa	ksi			10^6 GPa	psi
25 mm (1 in.) square	L-C	1151	167	1255	182	8.0	13.9	116	16.8
		1151	167	1268	184	8.5	13.2	114	16.6
50 mm (2 in.) square	L-O	1172	170	1248	181	7.0	16.8	114	16.6
	L-C	1172	170	1186	172	9.5	19.0	98	14.3
	T-O	1241	180	1296	188	6.0	16.0	...	...
	T-C	1193	173	1220	177	6.0	11.5	...	...
75 mm (3 in.) square	L-O	1137	165	1199	174	10.0	16.1	107	15.6
	L-C	1144	166	1165	169	9.5	15.3	101	14.7
	T-O	1165	169	1241	180	5.0	10.2	110	16.0
	T-C	1158	168	1179	171	7.0	15.4	100	14.5
100 mm (4 in.) square	L-O	1096	159	1179	171	10.0	24.1	110	16.0
	L-MR	1151	167	1199	174	8.0	14.6	107	15.6
	L-C	1158	168	1186	172	9.0	16.6	100	14.6
	T-O	1165	169	1220	177	5.0	4.8	107	15.6
	T-MR	1179	171	1241	180	6.5	11.7	106	15.4
	T-C	1172	170	1234	179	7.0	10.9	107	15.6
100 mm (4 in.) round	L-O	1124	163	1213	176	10.5	17.0	109	15.9
	L-MR	1137	165	1213	176	8.5	14.7	109	15.9
	L-C	1117	162	1213	176	8.5	17.5	108	15.8
	L-MR	1144	166	1227	178	8.5	14.6	109	15.9
	L-O	1130	164	1213	176	9.0	15.5	111	16.1
	T-C	1199	174	1261	183	5.5	7.8	112	16.3
150 mm (6 in.) square	L-O	1172	170	1227	178	4.5	8.6	...	...
	L-MR	1165	169	1227	178	5.0	7.8	...	...
	L-C	1130	164	1193	173	5.0	6.3	...	...
	T-O	1172	170	1248	181	4.0	4.0	...	...
	T-MR	1151	167	1220	177	3.5	4.7	...	...
	T-C	1137	165	1206	175	4.0	7.8	...	...

Note: Specimens from 25 and 50 mm (1 and 2 in.) bar were 4.7 mm (0.187 in.) in diameter, 25 mm (1 in.) gage length. Remainder of specimens were 6.4 mm (0.252 in.) in diameter, 50 mm (2 in.) gage length. L, longitudinal; T, transverse; C, center; O, outside; MR, mid radius. Source: R.A. Wood, *Beta Titanium Alloys*, MCIC 72-11, Battelle Columbus Laboratories, 1972

High-Temperature Strength

Ti-8823: Tensile properties of welded sheet at 315 °C (600 °F)

Heat treatment	Ultimate tensile strength MPa	ksi	Tensile yield strength MPa	ksi	Elongation in: 50 mm (2 in.), % Local	Uniform	Total	13 mm (0.5 in.), %	Young's modulus GPa	10^6 psi
800 °C (1475 °F), 30 min, AC	593	86	579	84	35	0	1.5	14	101	14.7
+480 °C (900 °F), 8 h, AC	605	88	600	87	30	0	3.5	12	106	15.4
+weld	605	88	593	86	45	0	5	18	102	14.9
800 °C (1475 °F), 30 min, AC	1275	185	1193	173	15	0	2	6	105	15.2
+weld+480 °C (900 °F), 8 h, AC	1296	188	(a)	(a)	5	0	2	2	117	17.0
	1255	182	(a)	(a)	5	0	1	2	110	16.0
800 °C (1475 °F), 30 min, AC	613	89	600	87	25	0	5	10	69	10.0
+595 °C (1100 °F), 16 h, AC	593	86	579	84	40	0	5.5	16	104	15.1
+weld	613	89	586	85	35	0	4	14	82	11.9
800 °C (1475 °F), 30 min, AC	848	123	731	106	20	0	3.5	8	96	13.9
+595 °C (1100 °F), 16 h, AC	813	118	675	98	5	2.5	4.5	2.5	90	13.1
+weld	820	119	799	116	20	2.5	4.5	8.5	102	14.8
800 °C (1475 °F), 30 min, AC	1103	160	1013	147	5	0	2	2.5	101	14.7
+595 °C (900 °F), 8 h, AC	1096	159	1041	151	10	2.5	3	4	94	13.7
+weld+480 °C (900 °F), 3 h, AC	1110	161	1041	151	5	0	2	2.5	104	15.1
	1151	167	1082	157	10	2.5	3	4	109	15.9
	1110	161	1061	154	10	2.5	3	4	100	14.6
	1110	161	1041	151	10	2.5	3	4	95	13.8

(a) Broke before reaching yield stress

Ti-8823: Tensile strength at 315 °C (600 °F)

Condition	Tensile yield strength MPa	ksi	% of RT strength
480 °C (900 °F), 8 h, AC	944-979	137-142	74-79
480 °C (900 °F), 24 h, AC	1061	154	81

Ti-8823: Notched tensile strength

Condition	Notched tensile strength MPa	ksi	NTS/UTS ratio(a)
Room temperature			
480 °C (900 °F), 8 h, AC	1068-1186	155-172	0.77-0.87
480 °C (900 °F), 24 h, AC	1130	164	0.83
315 °C (600 °F) test			
480 °C (900 °F), 8 h, AC	1172-1268	170-184	1.10-0.92
480 °C (900 °F), 24 h, AC	1296	188	1.10

(a) $K_t = 8$

Ti-8823: Creep deformation at 315 °C (600 °F)

Exposure conditions/ Age treatment	Exposure time, h	Total elongation after exposure, %	Creep Deformation, %
880 MPa (128 ksi) load at 315 °C (600 °F)			
480 °C (900 °F), 8 h, age	150	5.5	0.29
	500	4	0.38
950 MPa (138 ksi) load at 315 °C (600 °F)			
480 °C (900 °F), 24 h, age	500	5	0.7
	150	6	0.27
	500	5	0.40

Fatigue and Fracture

Ti-8823: Notched fatigue strength

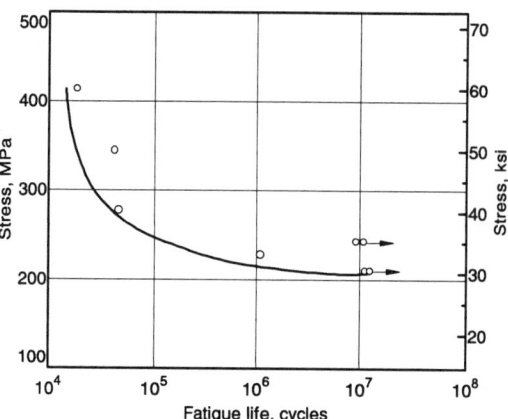

(K_t = 3.5); cold rolled 25%; heat treated 785 °C (1450 °F) 10 min, AC + 480 °C (900 °F), 8 h, AC; R = 0.25; A = 0.6

Ti-8823: Low-cycle torsional fatigue

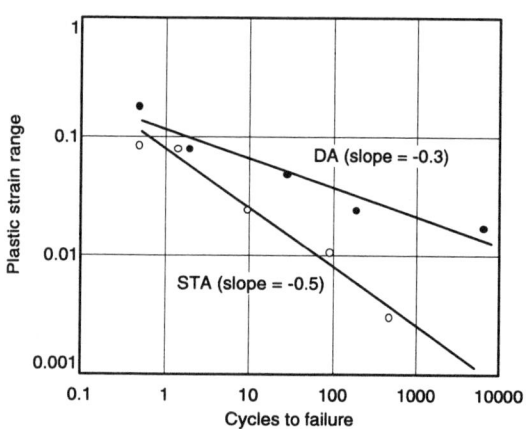

Ti-8823 was given either the generally recommended solution treatment and age (STA) heat treatment of 800 °C (1475 °F), 1.5 h, WQ, followed by aging at 540 °C (1000 °F) ~ 8 h; or direct age (DA) after hot working at 510 °C (950 °F), ~ 8 h. Solid cylindrical specimens approximately 0.200 in. in diameter were tested.
P.T. Lum and R. Chait, Proc. 2nd Int. Conf. Mechanical Behavior of Materials, 1976, p 796-800

Processing

Forging. Ti-8823 should be hot forged and hot rolled above its beta transus temperature of 775 °C (1425 °F). The recommended hot working temperature is 760 to 980 °C (1400 to 1800 °F) with breakdown at the higher temperatures and finishing at the lower temperatures. Cold working is used to finish such products as sheet, strip, plate, rod, and wire. Intermediate annealing may be necessary during these cold working operations.

Forming. The formability of the Ti-8823 alloy in the annealed or solution heat treated condition is excellent. It has relatively low yield strength, tolerably low work-hardening characteristics, and high ductility in tension and compression. Bending of flat-rolled product and upsetting of rod or wire at room temperature may be accomplished

Ti-8823: Hot-salt corrosion comparison of sheet

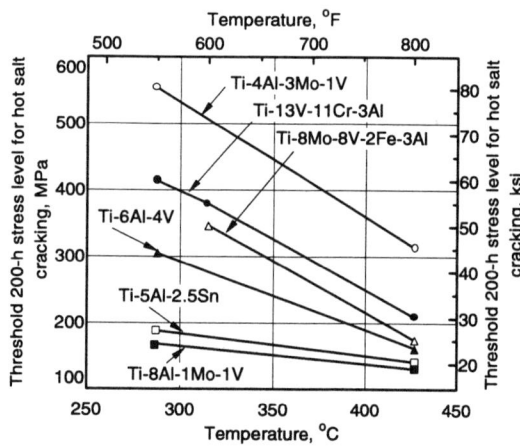

Ti-8823: Effect of alloy content on aging

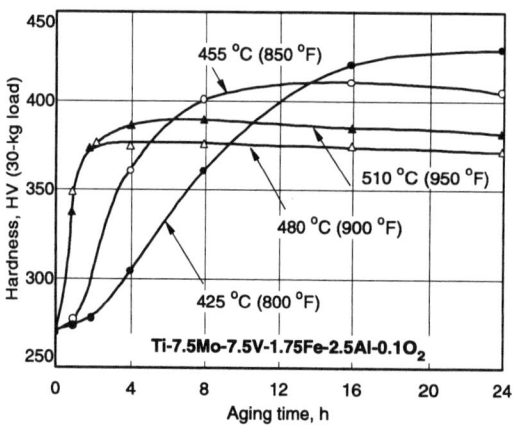

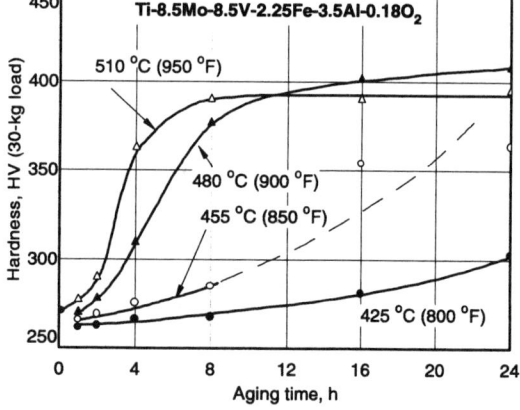

Aging kinetics of solution annealed sheet within normal minimum (a) and normal maximum (b) alloy content for indicated aging temperatures.
R.A. Wood, *Beta Titanium Alloys*, MCIC 72-11, Battelle Columbus Laboratories, 1972

with relative ease to produce formed products. Aged sheet may be hot formed with good results and overaged (stabilized). Rivet wire may be upset at room temperature to form highly satisfactory rivet heads.

Machining. Ti-8823 is machinable by most conventional techniques. It is more easily machined in the solution treated condition than in the aged condition. It requires rigid set up, heavy feed, slow speed, and adequate coolant.

Welding. Ti-8823 has fairly good weldability and weld stability. Through the use of conventional titanium welding practices, this alloy can be joined and useful mechanical properties maintained. The as-welded material exhibits properties similar to those of solution treated annealed material, except for lowered tensile ductility. Generally, properties of welded material using a postweld aging treatment are somewhat less desirable than those of unwelded solution treated plus aged material.

Heat Treatment. Stress relief anneal can be accomplished using a full annealing treatment or, in the case of aged material, a stress relief may be obtained by partial reaging at the same aging temperature used initially (see table).

Ti-8823: Recommended heat treatments

Treatment	Temperature °C	°F	Duration	Cooling method
Solution treatment (anneal)	785-800	1450-1475	(a)	AC or OQ(b)
Aging	480-510	900-950	8 h	AC
Overage (stabilization)	650	1200	...	AC

(a) Duration depends on thickness. (b) Sufficient rate to prevent α formation

Ti-8823: Bend radii of welded 1.5 mm (0.060 in.) gage sheet

Aging treatment after solution treatment (a)	Orientation of weld bead to bend axis	Passed	Failed
Room temperature			
Weld	None (control)	2.2T	1.8T
	Transverse	2.1T	1.8T
	Parallel	3.0T	2.4T
480 °C (900 °F), 8 h, AC + weld	None (control)	7.4T	6.6T
	Transverse	2.4T	2.1T
	Parallel	2.5T	2.4T
Weld + 480 °C (900 °F), 8 h, AC	None (control)	9.1T	8.0T
	Transverse	...	>9.5T
	Parallel	...	>8.3T
595 °C (1100 °F), 16 h, AC + weld	None (control)	4.2T	3.6T
	Transverse	2.3T	2.0T
	Parallel	1.6T	1.0T
Weld + 595 °C (1100 °F), 16 h, AC	None (control)	4.1T	3.6T
205 °C (400 °F)	Transverse	4.3T	3.7T
	Parallel	11.9T	11.0T
480 °C (900 °F), 8 h, AC, weld	Transverse	0.9T	...
	Parallel	2.0T	1.5T
Weld + 480 °C (900 °F), 8 h, AC	Transverse	...	>6.5T
	Parallel	(b)	(b)
595 °C (1100 °F), 16 h, AC + weld	Transverse	2.6T	1.6T
	Parallel	3.0T	2.5T
Weld + 595 °C (1100 °F), 16 h, AC	Transverse	3.0T	2.5T
595 °C (1100 °F)	Parallel	7.8T	7.3T
Weld + 595 °C (1100 °F), 16 hrs, AC	Transverse	...	>2.0T
	Parallel	...	>8.0T

(a) 800 °C (1475 °F) solution treatment for 30 min, AC. (b) Insufficient material

Ti-8823: Tensile properties of heavy sections vs section size and location

Section size/ heat treatment	Location	Ultimate tensile strength MPa	ksi	Tensile yield strength (0.2%) MPa	ksi	Elongation (4D), %	Reduction of area, %
100 mm (4 in.) square, 800 °C	L-O	1210	175.6	1133	164.4	10.0	24.1
(1475 °F) 1 h, WQ + 540 °C	L-MR	1197	173.7	1150	166.9	8.0	14.6
(1000 °F), 8 h, AC	L-C	1182	171.5	1160	168.3	9.0	16.6
150 mm (6 in.) square, 800 °C	L-O	1186	172.1	1120	162.5	8.0	14.8
(1475 °F), 1h, WQ + 540 °C	L-MR	1232	178.7	1165	169.0	6.0	12.1
(1000 °F), 8 h, AC	L-C	1213	176.0	1146	166.3	7.0	14.0
200 mm (8 in.) square, 800 °C	L-O	1183	171.6	1103	160.0	6.0	10.1
(1475 °F), 1 h, WQ + 540 °C	L-MR	1213	176.0	1148	166.5	5.5	6.6
(1000 °F), 8 h, AC	L-C	1226	177.8	1155	167.6	4.0	5.5

Source: E. Bohanek, Deep Hardenable Titanium Alloys for Large Airframe Elements in *Titanium Science and Technology*, Vol 3, R. Jaffe and H. Burte, Ed., Plenum Press, 1973, p 1993. L, longitudinal; O, outside; C, center; MR, mid radius

Ti-8823: Tensile properties of foil vs cold rolling and heat treatment

Condition	Tensile yield strength		Ultimate tensile strength		Elongation, %	Modulus of elasticity,	
	MPa	ksi	MPa	ksi		GPa	10⁶ psi
93% cold rolled	1268	184	1365	198	1	82	12
Annealed 2 min, 785 °C (1450 °F)(a)	937	136	944	137	7	96	14
Stabilized 4 h, 675 °C (1250 °F)(b)	1110	161	1179	171	8	124	18
Annealed + stabilized(c)	930	135	986	143	13	110	16
Annealed + 6 h, 510 °C (950 °F)(d)	1441	209	1530	222	5	131	19

(a) Fast cool in vacuum. (b) Slow cool in vacuum. (c) (a) treatment plus (b) treatment. (d) (a) treatment plus the 510 °C (950 °F) treatment, vacuum cooled. Source: W.H. Heil and J.M. Partridge, "Ti-8Mo-8V-2Fe-3Al - Its Development and Sheet Properties," Pacific Northwest Metals and Minerals Conference, Portland, Oregon, Apr 5, 1971, reported in *Beta Titanium Alloys*, R. Wood, MCIC 72-11, Battelle Columbus Laboratories, 1972, p 73-101

Ti-8823: Tensile properties of rod and wire vs processing and heat treatment

Processing/ heat treatment	Diameter		Cold work, %	Tensile yield strength		Ultimate tensile strength		Elongation(a), %	Reduction of area %
	mm	in		MPa	ksi	MPa	ksi		
As hot rolled	7.9	0.312	None	937	136	972	141	24	68
HR + CR	7.2	0.284	14	993	144	1034	150	21	64
HR + CR + 8 h, 480 °C (900 °F), AC	7.2	0.284	14	1489	216	1586	230	10	33
HR + CR	6.2	0.244	36	1151	167	1186	172	13	57
HR + CR + 24 h, 425 °C (800 °F), AC	6.2	0.244	36	1751	254	1827	265	3	11
HR + CR + 4 h, 480 °C (900 °F), AC	6.2	0.244	36	1572	228	1675	243	10	30
HR + CR	5.3	0.208	53	1227	178	1255	182	12	57
HR + CR + 8 h, 480 °C (900 °F), AC	5.3	0.208	53	1599	232	1661	241	6	18
HR + CR + 8 h, 510 °C (950 °F), AC	5.3	0.208	53	1468	213	1523	221	10	38
HR + CR	4.8	0.191	61	1282	186	1303	189	9	52
HR + CR + 8 h, 480 °C (900 °F), AC	4.8	0.191	61	1613	234	1668	242	10	35
HR + CR + 8 h, 510 °C (950 °F), AC	4.8	0.191	61	1489	216	1544	224	12	46
HR + CR	2.3	0.090	91	...	...	1324	192	0.19(b)	58
	1.6	0.063	96	...	...	1330	193	0.15(b)	65
	0.9	0.036	99	...	...	1365	198	0.14(b)	59

(a) In 4D except as indicated. (b) In 250 mm (10 in.). Source: J.A. Reed, "High Strength Ti-8Mo-8V-2Fe-3Al," TMCA Internal Progress Report on Project 99-6, reported in *Beta Titanium Alloys*, R. Wood, MCIC 72-11, Battelle Columbus Laboratories, 1972, p 73-101

Ti-15Mo-5Zr

T. Nishimura, Special Metals Laboratory, Kobe Steel LTD

Ti-15Mo-5Zr is a metastable β type alloy that exhibits good cold formability and age hardenability. It is β stabilized by molybdenum to enhance corrosion resistance to reducing atmospheres. Zirconium is added to (1) enhance corrosion resistance above the level achieved by molybdenum, (2) suppress ω transformation to prevent embrittlement, and (3) to improve thermal stability of the β phase. Zirconium additions of 5% minimum are used to enhance thermal stability.

Product Form. Forging billet and bar, hot rolled plate and bar, cold rolled sheet, and cold drawn wire are available. Cold rolled sheet is available in thicknesses up to 0.1 mm (0.004 in.). The standard cold drawn wire diameter minimum is 1.0 mm (0.04 in.).

Product Condition and Microstructure. Ti-15Mo-5Zr typically is processed to plate, billet, or bar in the β temperature fields. Solution treatments in the β region are used to obtain low flow stress and high ductility for cold processing. Products usually are supplied in the annealed condition. Annealing is carried out just above the β transus temperature for a fine-grained recrystallized microstructure.

Applications. Ti-15Mo-5Zr is used in the chemical industry because of its high strength, good cold formability, and high corrosion resistance. In addition, it is used as an erosion-resistant overlay for steam turbine blades in which ω phase is intentionally used to obtain an extremely high hardness in spite of being brittle.

Ti-15Mo-5Zr: Chemical composition of bar, wire, and sheet

			Chemical composition, %			
H(a)	O(a)	N(a)	Fe(a)	Zr	Mo	Ti
0.020	0.20	0.05	0.35	4.5-5.5	14.0-16.0	rem

(a) Maximum. Source: Kobe Steel

Physical Properties

Crystal Structure. Body-centered cubic single-phase β is obtained in the solution treated condition. Close-packed hexagonal ω phase and cph phase α are precipitated during aging below 400 °C (750 °F) and above 450 °C (840 °F), respectively. Omega phase is usually avoided because it causes embrittlement.

Grain Structure. The grain structure and distribution of phases depend on thermomechanical history. The β grain size after annealing ranges from approximately 20 to 100 μm.

Aging Transformations. See "Heat Treatment" at the end of this datasheet.

Elastic Properties

Modulus of Elasticity. At 25 °C (77 °F), 78 GPa (11×10^6 psi) in as-quenched state.

Modulus of Rigidity. At 25 °C (77 °F), 34 GPa (5×10^6 psi) in as-quenched state.

Corrosion Properties

Ti-15Mo-5Zr has high corrosion resistance to reducing atmospheres. It has better corrosion resistance in boiling hydrochloric acid or sulfuric acid solutions than commercially pure titanium. Additionally, Ti-15Mo-5Zr has higher erosion resistance compared to Ti-6Al-4V or other β titanium alloys.

Ti-15Mo-5Zr: Physical property summary

Density	5.06 g/cm³
Specific heat at 70 °C	0.13 cal/g · °C
Thermal coefficient of liner expansion	8.5×10^{-6}/°C
Magnetic permeability	1.000
Beta transus	730 °C (1345 °F)

Source: Kobe Steel

Ti-15Mo-5Zr: Corrosion resistance at 90 °C

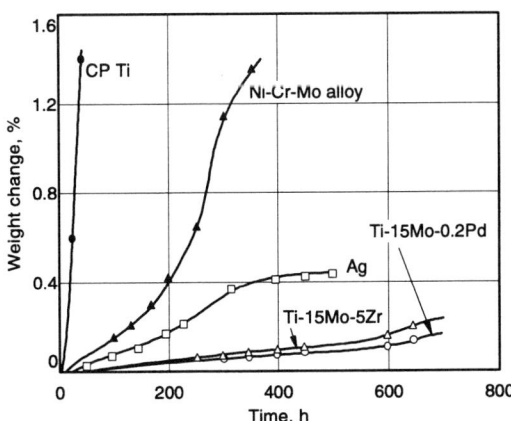

Typical corrosion in 12 wt% H_2SO_4, 20 wt% Na_2SO_4, 2 wt% $ZnSO_4$, and 66 wt% H_2O.
Source: Kobe Steel

Ti-15Mo-5Zr: Erosion resistance comparison

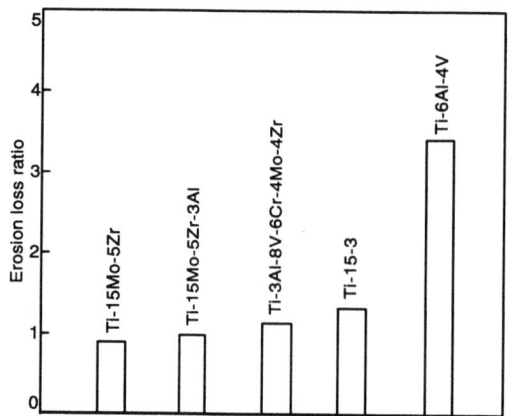

	UTS, MPa (ksi)	TYS (0.2%), MPa (ksi)	El, %	RA, %	Hardness, HV
Ti-15Mo-5Zr-3Al	1492 (216)	1419 (205.8)	8.3	18.6	418
Ti-3Al-8V-6Cr-4Mo-4Zr	1497 (217)	1403 (203.5)	4.4	3.5	421
Ti-15Mo-5Zr	1356 (196.7)	1333 (193.3)	2.5	4.9	434
Ti-15V-3Al-3Sn-3Cr	1410 (204.4)	1291 (187.2)	4.8	7.8	402
Stellite	1110 (161)	633 (91.8)	9.7	9.6	410

Source: J. Hoashi, et al., "Material Aspects on 40 Inch Long Titanium Alloy Blade for Steam Turbines," Titanium Steam Turbine Blading, Workshop Proceedings, R.I. Jaffee, Ed., 1988

Tensile Properties

A comparison of mechanical properties with other titanium materials was carried out with hot rolled and annealed or solution treated bar. The ductility of Ti-15Mo-5Zr is approximately twice that of Ti-6Al-4V at the same strength level. A maximum tensile strength is obtained by aging at around 400 °C (750 °F), but elongation is very low because of ω phase precipitation. The practical high-strength aging temperature ranges from 450 to 500 °C (840 to 930 °F).

Ti-15Mo-5Zr: Mechanical properties of hot rolled bar at room temperature

Alloy	Ultimate tensile strength		Tensile yield strength (0.2%)		Elongation, %	Reduction of area, %	Charpy impact toughness, J/cm²	Hardness, HV
	MPa	ksi	MPa	ksi				
Ti-15Mo-5Zr(a)	961	139	922	133	25	65	59	283
Ti-15Mo-0.2Pd(a)	1118	162	1069	155	20	55	39	278
CP Ti(b)	412	59	323	47	41	70	176	140
Ti-6Al-4V(b)	961	139	892	129	13	35	39	330
Ti-5Al-2Cr-1Fe(b)	1000	145	951	138	19	40	49	320

(a) Solution treated. (b) Annealed. Source: Kobe Steel

Ti-15Mo-5Zr: Isochronal curve of tensile properties vs aging temperature

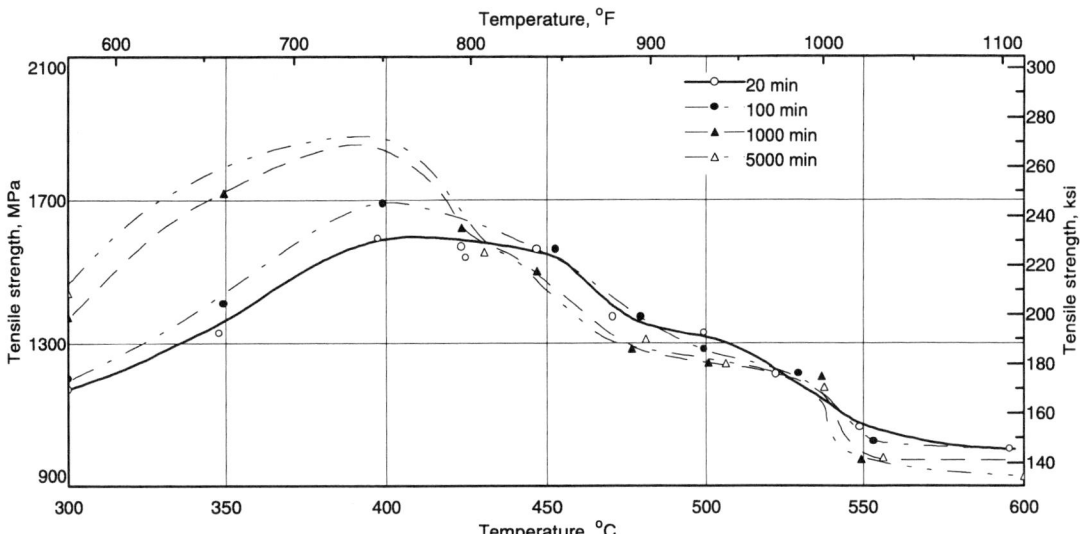

(a)

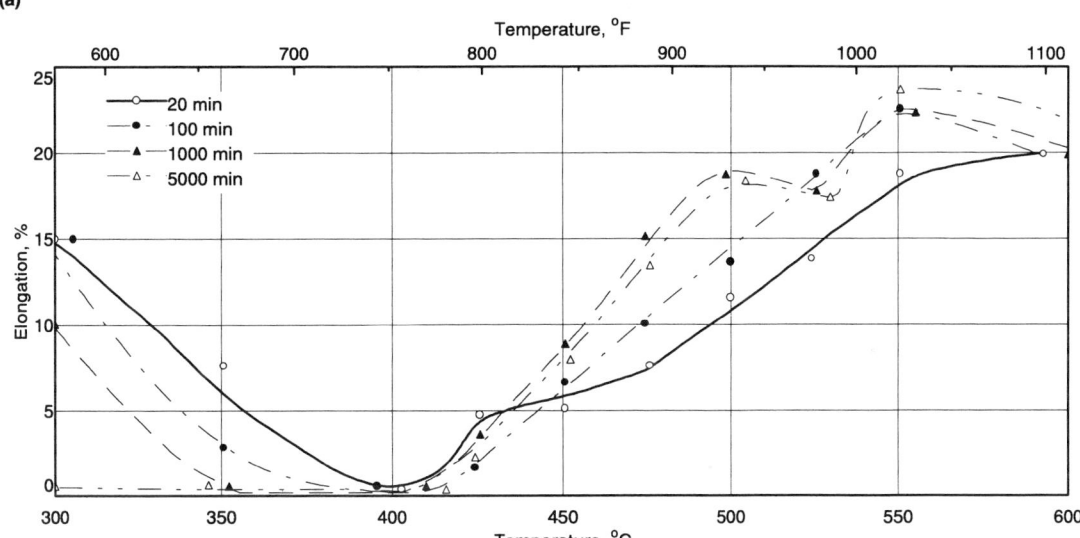

(b)

Specimens were 9.5 mm (0.35 in.) diameter hot rolled bar at 880 °C (1615 °F); 98% reduction. Solution treated at 730 °C (1345 °F), WQ. Aged as indicated.
Source: T. Nishimura, M. Nishigaki, and Y. Moriguchi, "Characteristics of Beta Titanium Alloy Ti-15Mo-5Zr-3Al," R & D Kobe Steel Engineering Reports, Vol 32, No. 1

Ti-15Mo-5Zr: Mechanical properties vs temperature

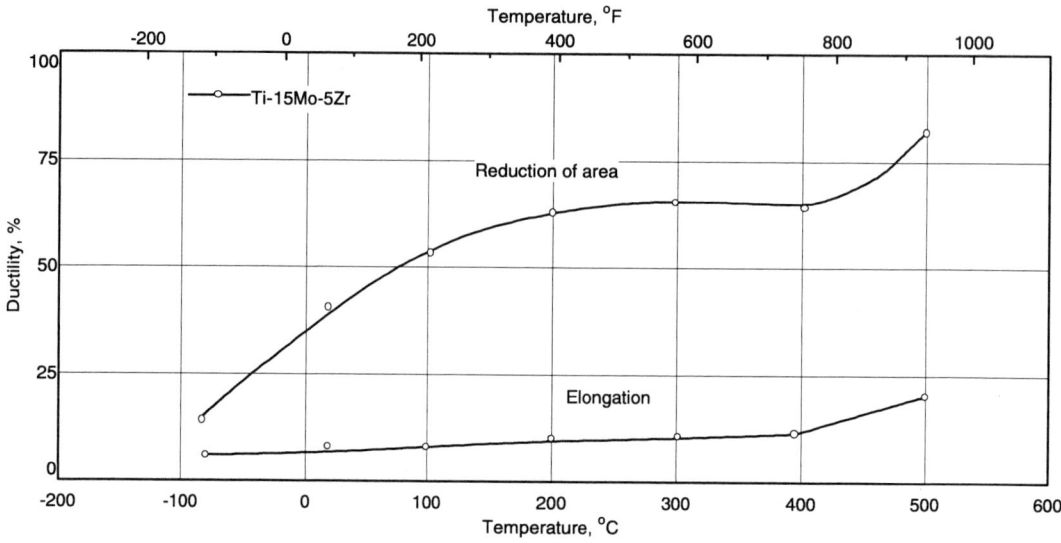

(a)

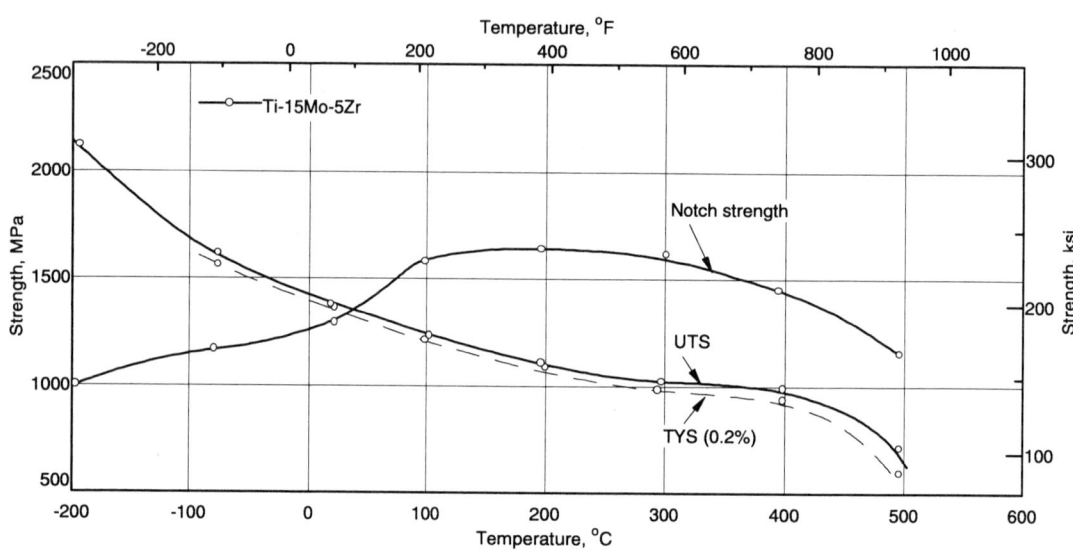

(b)

15 mm (0.6 in.) diameter hot rolled bar solution treated and aged at 730 °C (1345 °F), 1 h, WQ + 475 °C (890 °F), 100 min, AC. K_t = 5.3.
Source: Kobe Steel

Fatigue and Fracture

Ti-15Mo-5Zr: Repeated torsion fatigue (R = -1)

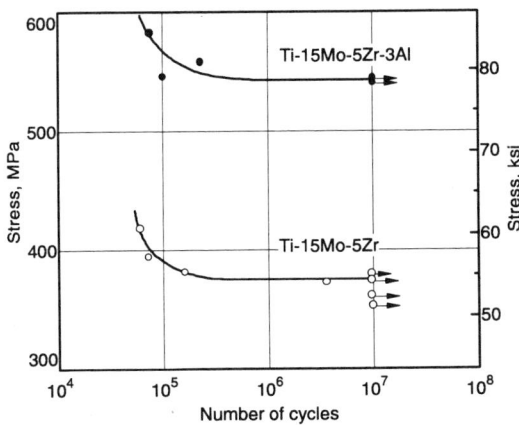

Smooth specimens of 6.5 mm (0.25 in.) hot rolled plate were tested at room temperature. Specimens were solution treated and aged as follows: Ti-15Mo-5Zr-3Al (UTS, 1421 MPa): 735 °C (1360 °F), 1 h, WQ + 500 °C (930 °F), 1000 min, AC. Ti-15Mo-5Zr (UTS, 1337 MPa): 730 °C (1345 °F), 1 h, WQ + 475 °C (890 °F), 100 min, AC. 25 Hz test frequency
Source: Kobe Steel

Ti-15Mo-5Zr: Smooth rotating bend fatigue (R = -1)

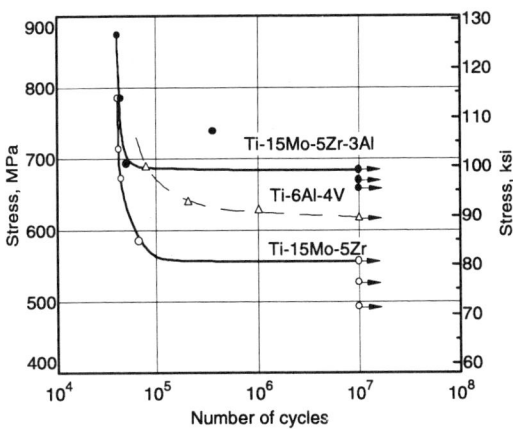

Smooth specimens of 15 mm (0.25 in.) diameter hot rolled bar were tested at 3400 rpm at room temperature. Specimens were solution treated and aged as follows. Ti-15-5-3 (UTS, 1455 MPa): 735 °C (1360 °F), 1 h, WQ + 500 °C (930 °F), 1000 min, AC. Ti-15-5 (UTS, 1377 MPa): 730 °C (1345 °F), 1 h, WQ + 475 °C (890 °F), 100 min, AC. Ti-6Al-4V (UTS, 1124 MPa): 925 °C (1700 °F), 1 h, WQ + 500 °C (930 °F), 4 h, AC. 3400 rpm test
Source: Kobe Steel

Fracture Properties

Ti-15Mo-5Zr: Delayed fracture properties

Fracture was not influenced by changes in environment (air, water, 3.5% NaCl)

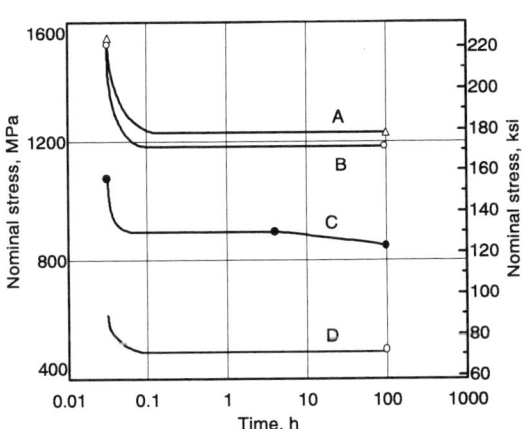

Notched specimen from 15 mm (0.6 in.) diameter hot rolled bar tensile tested at room temperature. $K_t = 7.6$. Specimens were solution treated and aged as follows. A: 730 °C (1345 °F), 1 h, WQ, 550 °C (1020 °F), 100 min, AC. B: Treatment A plus 730 °C (1345 °F), 1 h, WQ. C: 730 °C (1345 °F), 1 h, WQ, 475 °C (890 °F), 100 min, AC. D: 730 °C (1345 °F), 1 h, WQ, 425 °C (800 °F), 100 min, WQ.
Source: Kobe Steel

Ti-15Mo-5Zr: Charpy impact toughness vs temperature

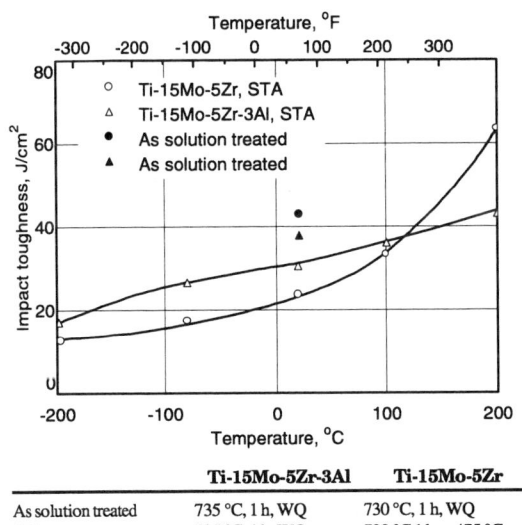

	Ti-15Mo-5Zr-3Al	Ti-15Mo-5Zr
As solution treated	735 °C, 1 h, WQ	730 °C, 1 h, WQ
STA	735 °C, 1 h, WQ + 500 °C, 1000 min, AC	730 °C 1 h, + 475 °C, 100 min, AC

Specimens were 15 mm (0.6 in.) diameter hot rolled bar heat treated as indicated.
Source: Kobe Steel

Fabrication

Forming. Ti-15Mo-5Zr is similar to Ti-15V-3Cr-3Sn-3Al or Ti-15Mo-5Zr-3Al.

Machining characteristics are similar to other β titanium alloys.

Heat Treatment. Solution treating at 730 °C (1345 °F), 1 h, followed by water quenching is recommended. With sheet thicknesses less than 3 mm (0.1 in.) or wire of diameters less than 3 mm (0.1

in.), air cooling is acceptable. Recommended aging temperatures are 450 to 500 °C (840 to 930 °F).

After aging at low temperatures of around 400 °C (840 °F), the alloy is highly strengthened but is extremely brittle because of ω phase precipitation. However, during aging over 450 °C (840 °F), α phase precipitates without embrittlement. The maximum amount of α phase is obtained at around 500 °C (930 °F) (see figure below). The optimum combination of strength and ductility can be obtained by solution treating and aging at 730 °C (1345 °F), 1 h, water quenching + 475 °C (890 °F) for 100 to 1000 min, AC.

Welding. With a filler rod of Ti-15Mo-5Zr, TIG welding can be carried out with minimal defects. Hardness increases minimally in the heat-affected and fusion zones. Ti-15Mo-5Zr can be welded with commercially pure titanium as well. Corrosion and wear resistance of parts made of commercially pure titanium can be enhanced by overlay welding and hardening.

Ti-15Mo-5Zr: Aging transformation diagram

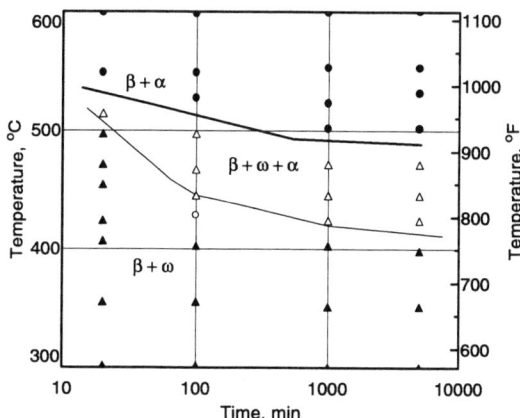

9.5 mm (0.35 in.) diameter bar hot rolled at 880 °C (1615 °F), 98% reduction. X-ray diffraction analysis.
Source: T. Nishimura, M. Nishigaki, and Y. Moriguchi, "Characteristics of Beta Titanium Alloy Ti-15Mo-5Zr-3Al," R & D Kobe Steel Engineering Reports, Vol 32, No. 1

Ti-15Mo-5Zr: Amount of ω and α phases during aging

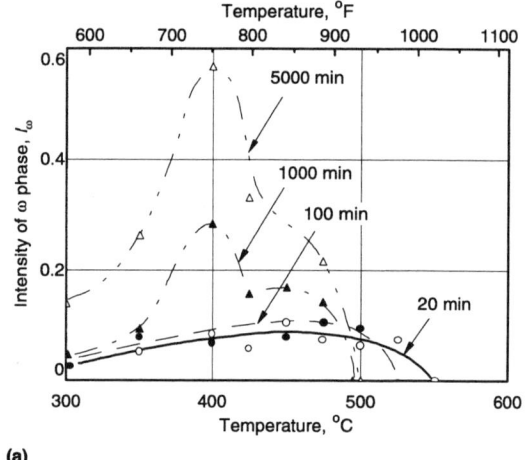

(a)

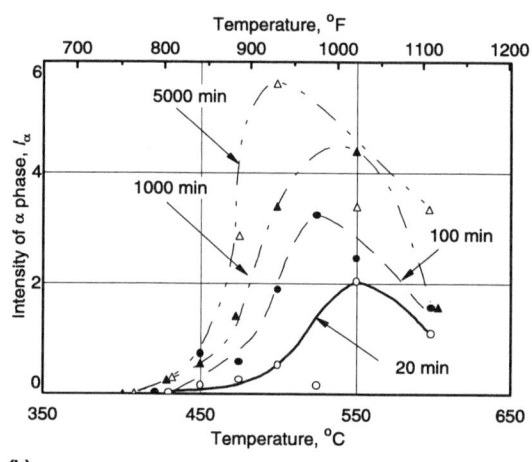

(b)

9.5 mm (0.35 in.) diameter bar hot rolled at 880 °C (1615 °F), 98% reduction. X-ray diffraction analysis. $I_\alpha = I\{10\bar{1}0\}_\alpha / I\{200\}_\beta$
$I_\omega = I\{11\bar{2}2\}_\omega / I\{200\}_\beta$

Ti-15Mo-5Zr-3Al

T. Nishimura, Special Metals Laboratory, Kobe Steel

Ti-15Mo-5Zr-3Al is a metastable β titanium alloy that is characterized by high strength, good cold formability, and in particular, high corrosion resistance to reducing atmospheres. Its corrosion resistance is superior to that of Ti-0.2Pd.

Chemistry. Molybdenum enhances corrosion resistance to reducing atmospheres. Zirconium is added to (1) further enhance corrosion resistance above that achieved by a single molybdenum addition, (2) suppress ω transformation to prevent ω embrittlement, and (3) to improve thermal stability of β phase. Zirconium additions of 5% minimum are required to enhance thermal stability. An aluminum addition of 3% is needed to suppress ω transformation effectively at lower temperatures and longer times. Moreover, aluminum enhances post-aging strength and resistance to oxidation as well.

Product Forms. Forging billet and bar, hot rolled plate and bar, cold rolled sheet, and cold drawn wire are available. Cold rolled sheet is available in thicknesses up to 0.1 mm. The standard cold drawn wire diameter minimum is 1.0 mm.

Conditions. Ti-15Mo-5Zr-3Al is hot worked or cold worked. Prior to cold working, the material is solution treated to obtain low flow stress and high ductility. Products usually are supplied in the solution treated condition. Solution treatment is carried out alternatively either in the β temperature field (at 800 to 850 °C, or 1470 to 1560 °F) for cold formability, or in the α-β field at 735 °C (1350 °F) for a good combination of strength and ductility after aging. In the former case, the microstructure consists of a small amount of α phase and recovered β phase.

Applications. In addition to its high corrosion resistance, Ti-15Mo-5Zr-3Al has high strength. It can be used in various applications where many other titanium alloys cannot be used. For example, it is a candidate material for sour gas well plants because of its high strength-to-density ratio and resistance to atmospheric stress-corrosion cracking. Moreover, it is currently used as an erosion shield material for 1015 mm (40 in.) titanium turbine blades in power plants.

Ti-15Mo-5Zr-3Al: Chemical composition

	Chemical composition, %						
H(a)	O(a)	N(a)	Fe(a)	Al	Zr	Mo	Ti
0.020	0.20	0.05	0.35	2.5-3.5	4.5-5.5	14.0-16.0	rem

(a) Maximum. All product forms bar, wire, sheet, plate. Source: Kobe Steel

Physical Properties

Crystal Structure. Body-centered cubic β phase is obtained after solution treating in the β temperature region and quenching. Close-packed hexagonal α phase and ω phase precipitate during aging above and below 425 (795 °F), respectively. Compared with Ti-15Mo-5Zr, embrittlement caused by ω phase does not occur as predominantly because the amount of ω phase is reduced by the 3% aluminum additions.

Grain Structure. The grain structure and distribution of phases depend on the thermomechanical history of the material. The grain size obtained by solution treating above the β transus generally ranges from approximately 20 to 100 μm.

Ti-15Mo-5Zr-3Al: Physical property comparison

	Ti-15Mo-5Zr	Ti-15Mo-5Zr-3Al
Density, g/cm^3	5.06	5.01
Specific heat, cal/g · °C	0.13 at 70 °C	...
Thermal coefficient of linear expansion, 10^{-6}/°C	8.5	...
Magnetic permeability	1.000	1.000
Beta transus, °C	730	785

Ti-15Mo-5Zr-3Al: Thermal conductivity vs temperature

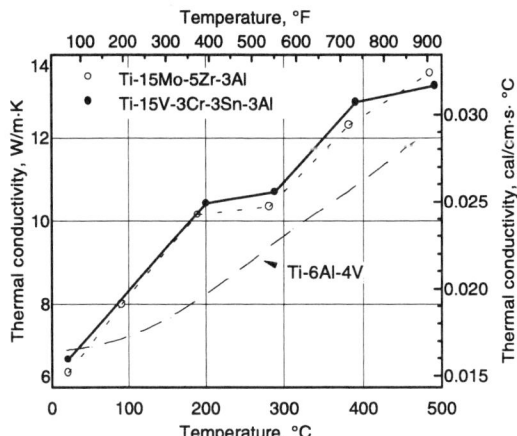

Specimens were solution treated 100 μm cold rolled foil, laser flash method.
Source: Advanced Aircraft Technology Development Center of Japanese Aerospace Companies, SJAC Report No. 6201, 1988

Modulus of Elasticity. At 25 °C (77 °F), 80 to 100 GPa (11 to 14.5×10^6 psi).

Modulus of Rigidity. At 25 °C (77 °F), 43 GPa (6.2×10^6 psi).

Corrosion Properties

Ti-15Mo-5Zr-3Al has high corrosion resistance to reducing atmospheres. Its erosion resistance is somewhat inferior to that of Ti-15Mo-5Zr. However, the strength, ductility, and toughness of Ti-15Mo-5Zr-3Al are superior to Ti-15Mo-5Zr, and it is used as an erosion shield as well as Ti-15Mo-5Zr. Stress-corrosion cracking properties in a H_2S-saturated solution with 5% NaCl and 0.5% CH_3COOH for solution treated and aged samples are shown (see figure).

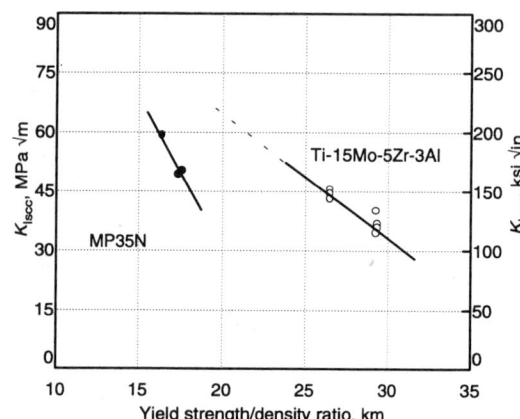

Ti-15Mo-5Zr-3Al: Relationship between yield strength/density ratio and K_{Iscc}

Double cantilever beam. Specimens were 10 mm thick hot rolled plate solution treated and aged 735 °C (1360 °F), 1 h, WQ + 500 °C (930 °F), 1000 min, FC in vacuum. Test atmosphere, H_2S saturated 5% NaCl + 0.5% CH_3COOH water.
Source: T. Nishimura, M. Nishigaki, and Y. Moriguchi, "Characteristics of Beta Titanium Alloy Ti-15Mo-5Zr-3Al," R & D Kobe Steel Engineering Reports, Vol 32, No. 1

Mechanical Properties

Tensile properties of Ti-15Mo-5Zr-3Al under optimum heat treatment can obtain a high tensile strength of 1470 MPa (213 ksi) with an elongation of 15% (see table). A higher tensile strength of over 1570 MPa (227 ksi) can be obtained by duplex aging (see table).

Ti-15Mo-5Zr-3Al: Mechanical properties of aged specimens

Tensile strength, MPa (ksi)	1475 (214)
Elongation, %	14
Reduction of area, %	59
Hardness, HV (10 kg)	412
NTS/TS ratio (with K_t=5.3)	1.06(a)
Charpy impact strength, kg · m/cm^2	3.06
Fatigue strength, MPa (ksi)	685 (100)

Note: Specimens aged at 735 °C (1350 °F), WQ + 500 °C (930 °F), 1000 min. 9.5 mm (0.35 in.) diameter bar hot rolled at 880 °C (1615 °F), 98% reduction. (a) Notched tensile strength/tensile strength ratio with notch factor (K_t) indicated.
Source: T. Nishimura, M. Nishigaki, and Y. Moriguchi, "Characteristics of Beta Titanium Alloy Ti-15Mo-5Zr-3Al," R & D Kobe Steel Engineering Reports, Vol 32, No. 1

Ti-15Mo-5Zr-3Al: Mechanical properties of duplex aged specimens

Aging treatment(a)	Tensile strength MPa	ksi	Elongation, %	Reduction of area, %	Hardness (10 kg), HV
425 °C (795 °F), 1000 min + 475 °C (890 °F), 1000 min	1585	230	10	52	413
425 °C (795 °F), 1000 min + 500 °C (930 °F), 1000 min	1558	225	10	58	417

Note: 9.5 mm (0.35 in.) diameter bar hot rolled at 880 °C (1615 °F), 98% reduction. (a) Solution treatment: 735 °C (1350 °F), 60 min, WQ.
Source: T. Nishimura, M. Nishigaki, and Y. Moriguchi, "Characteristics of Beta Titanium Alloy Ti-15Mo-5Zr-3Al," R & D Kobe Steel Engineering Reports, Vol 32, No. 1

ST Condition

Tensile strength is relatively higher after solution treatment (ST) at 735 °C (1350 °F), where α phase exists, than after solution treatment above the β transus (785 °C, or 1445 °F). As for material solution treated above the β transus, an elongation of over 20% and a reduction of area of 60% can be obtained, with cold rollability up to over 90%. Strength decreases with increasing solution temperature because of grain coarsening.

Ti-15Mo-5Zr-3Al: Effect of solution temperature on tensile properties

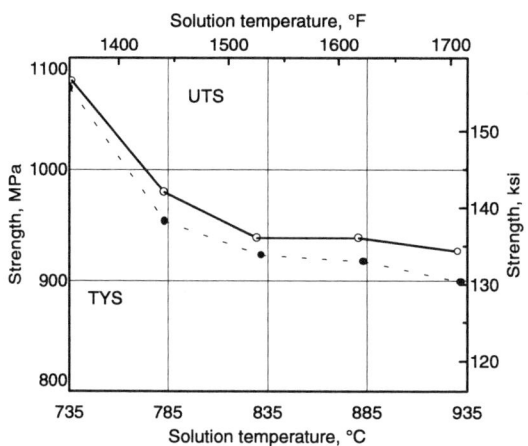
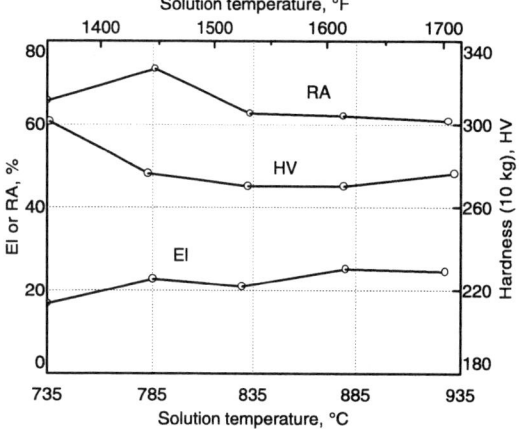

β transus is 785 °C. 9.5 mm (0.35 in.) diameter bar hot rolled at 880 °C (1615 °F); 98% reduction. Condition: Solution treated.
Source: T. Nishimura, M. Nishigaki, and Y. Moriguchi, "Characteristics of Beta Titanium Alloy Ti-15Mo-5Zr-3Al," R & D Kobe Steel Engineering Reports, Vol 32, No. 1

STA Condition

Tensile properties after aging are affected significantly whether solution treatments are carried out below or above the β transus. In material solution treated at 735 °C (1350 °F), age hardening begins in the early stages of aging and strength increases to 1470 MPa (213 ksi) accompanied by a slight decrease in ductility after aging for 1000 min. In material solution treated above the β transus, however, age hardening is more sluggish, and ductility tends to decrease after aging, up to 1370 MPa (199 ksi).

Ti-15Mo-5Zr-3Al: Effect of solution temperature on tensile properties

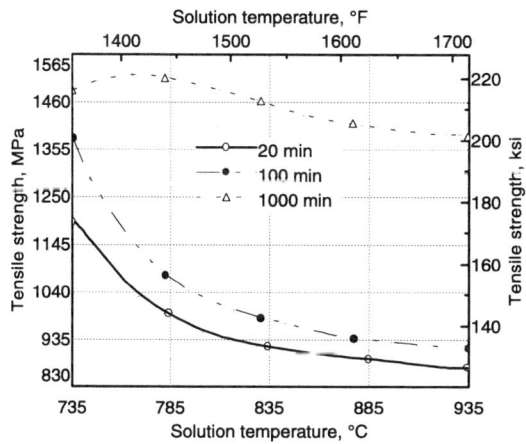
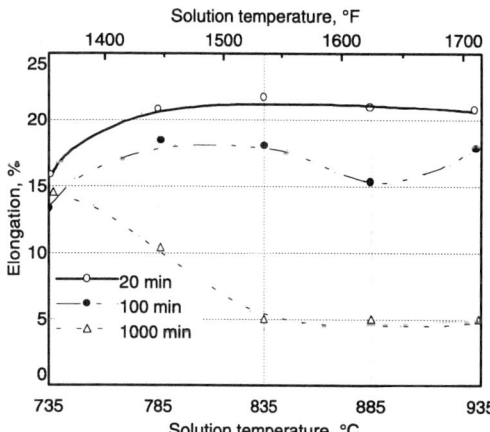

9.5 mm (0.35 in.) diameter bar hot rolled at 880 °C (1615 °F); 98% reduction. Specimens were solution treated for one hour and aged at 500 °C (930 °F) for indicated times.
Source: T. Nishimura, M. Nishigaki, and Y. Moriguchi, "Characteristics of Beta Titanium Alloy Ti-15Mo-5Zr-3Al," R & D Kobe Steel Engineering Reports, Vol 32, No. 1

Ti-15Mo-5Zr-3Al: Isochronal curve of tensile properties at various aging temperatures

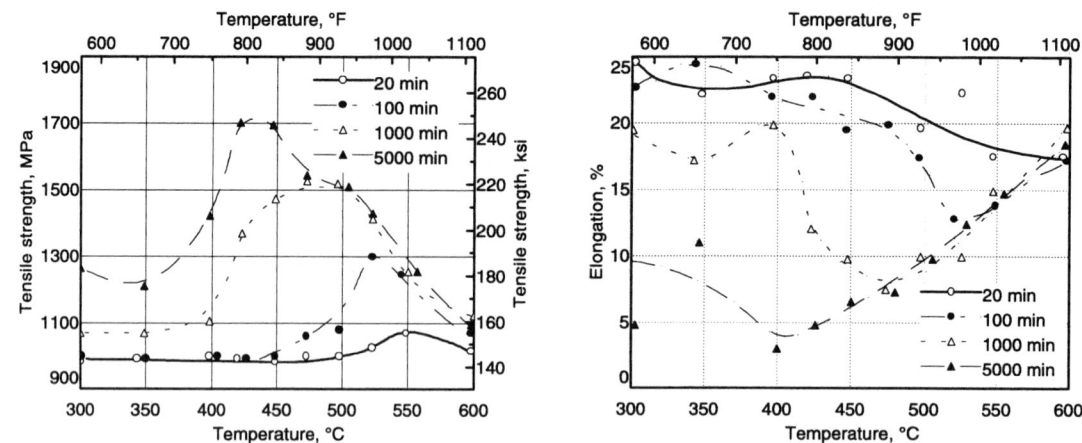

9.5 mm (0.35 in.) diameter bar hot rolled at 880 °C (1615 °F), 98% reduction. Specimens were solution treated at 785 °C (1450 °F) and water quenched and aged at indicated times.
Source: T. Nishimura, M. Nishigaki, and Y. Moriguchi, "Characteristics of Beta Titanium Alloy Ti-15Mo-5Zr-3Al," R & D Kobe Steel Engineering Reports, Vol 32, No. 1

Effect of Temperature

The accompanying figure in this section illustrates the dependence of tensile properties and notch strength on temperature for material that is solution treated at 735 °C (1350 °F) for 1 h, followed by water quenching and aging at 500 °C (930 °F) for 1000 min.

Ti-15Mo-5Zr-3Al: Temperature dependence of mechanical properties

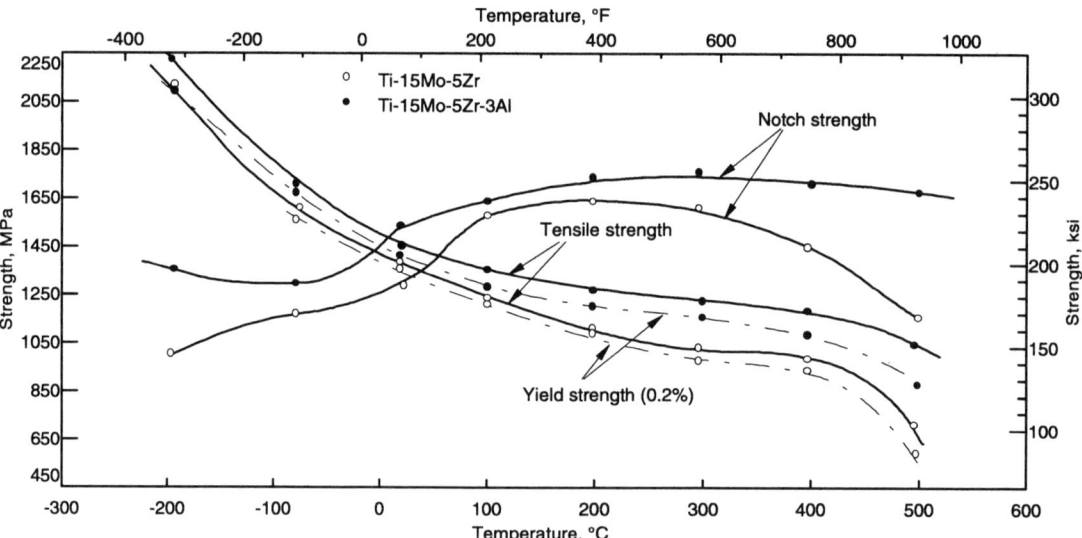

15 mm (0.6 in.) diameter hot rolled bar. Ti-15-5-3: 735 °C (1360 °F), 1 h, WQ + 500 °C (930 °F), 1000 min, AC. Ti-15-5: 730 °C (1345 °F), 1 h, WQ + 475 °C (890 °F), 100 min, AC. Exposure time, 20 min smooth tensile specimen.
Source: Kobe Steel

Fatigue Properties

Ti-15Mo-5Zr-3Al: Smooth specimen rotating bend fatigue

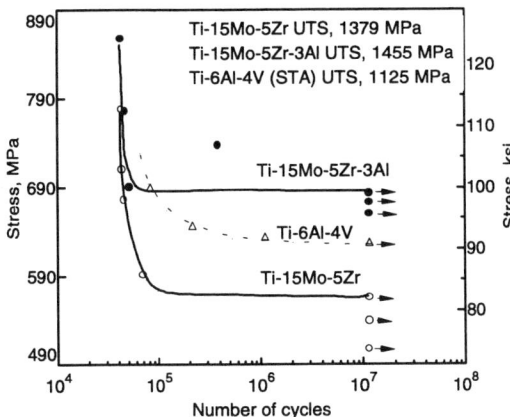

15 mm (0.6 in.) diameter hot rolled bar tested at 3400 rpm at room temperature. Ti-15-5-3: 735 °C (1360 °F), 1 h, WQ + 500 °C (930 °F), 1000 min, AC. Ti-15-5: 730 °C (1345 °F), 1 h, WQ + 475 °C (890 °F), 100 min, AC. Ti-6Al-4V: 925 °C (1700 °F), 1 h, WQ + 500 °C (930 °F), 4 h, AC.
Source: Kobe Steel

Ti-15Mo-5Zr-3Al: Smooth specimen repeated torsion fatigue

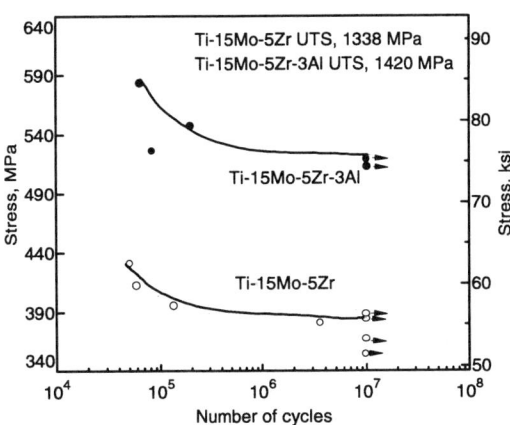

6.5 mm (0.25 in.) hot rolled plate at room temperature. Ti-15Mo-5Zr-3Al: 735 °C (1360 °F), 1 h, WQ + 500 °C (930 °F), 1000 min, AC. Ti-15Mo-5Zr: 730 °C (1345 °F), 1 h, WQ + 475 °C (890 °F), 100 min, AC.

Ti-15Mo-5Zr-3Al: Crack propagation rate data

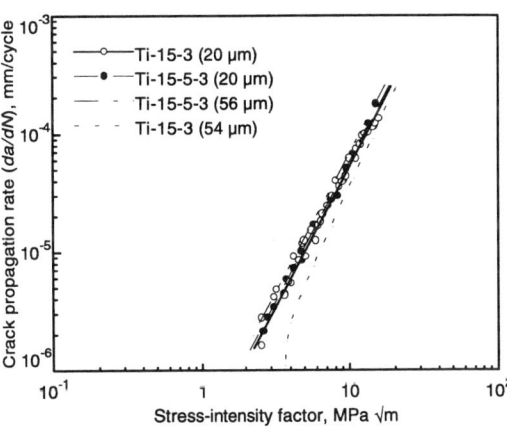

1 mm (0.04 in.) thick cold rolled sheet solution treated and aged. Ti-15Mo-5Zr-3Al: ST + 500 °C (930 °F), 8 h, AC. Ti-15V-3Cr-3Sn-3Al: ST + 540 °C (1000 °F), 8 h, AC. 3 to 5 Hz; $R = 0.1$.
Source: Advanced Aircraft Technology Development Center of Japanese Aerospace Companies, SJAC Report No. 6201, 1988

Fracture Properties

Ti-15Mo-5Zr-3Al: Delayed fracture properties

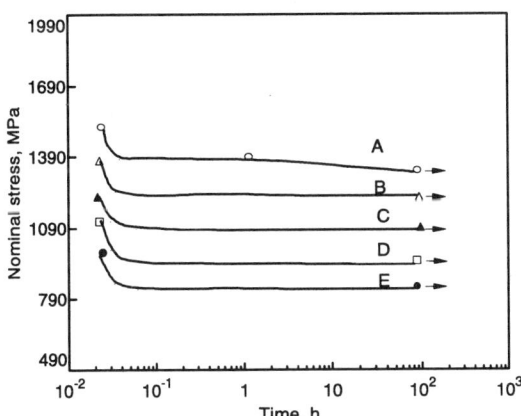

Notched specimen from 15 mm (0.6 in.) diameter hot rolled bar tensile tested at room temperature. $K_t = 7.6$; $R = 0.04$. A: 735 °C (1360 °F), 1 h, WQ. B: 735 °C (1360 °F), 1 h, WQ, 450 °C (840 °F), 6000 min, AC. C: 735 °C (1360 °F), 1 h, WQ, 500 °C (930 °F), 1000 min, AC. D: 735 °C (1360 °F), 1 h, WQ, 525 °C (975 °F), 300 min, AC. E: 735 °C (1360 °F), 1 h, WQ, 550 °C (1020 °F), 300 min, AC.
Source: Kobe Steel

Ti-15Mo-5Zr-3Al: Effect of β grain size on fracture toughness

Grain size, μm	Fracture toughness			
	K_{app}		K_c	
	MPa√m	ksi√in.	MPa√m	ksi√in.
20(a)	48	53	52	57
56(b)	40	44	43	47

Note: 1 mm (0.04 in.) cold rolled sheet solution treated and aged. (a) 850 °C (1560 °F), 2.5 min, AC + 500 °C (930 °F), 8 h, AC. (b) 850 °C (1560 °F), 30 min, AC + 500 (930 °F), 8 h, AC. Source: Advanced Aircraft Technolgy Development Center of Japanese Aerospace Companies, SJAC Report No. 6201, 1988

Ti-15Mo-5Zr-3Al: Effect of notch factor on notch tensile strength

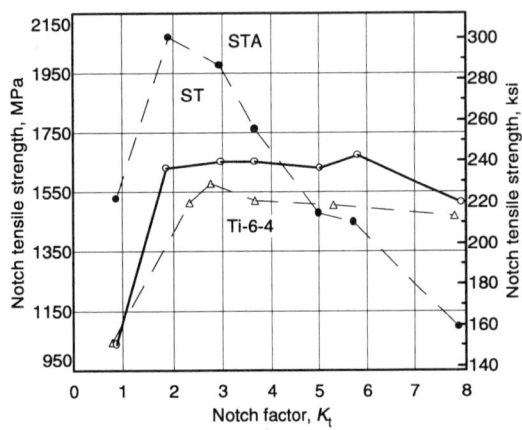

Ti-15-5-3: 15 mm (0.6 in.) diameter hot rolled bar: Solution treated at 735 °C (1360 °F), 1 h, WQ; and solution treated and aged at 735 °C (1360 °F), 1 h, WQ + 500 °C (930 °F), 1000 min, AC. Ti-6Al-4V annealed at 700 °C (1290 °F), 2 h, AC.
Source: Kobe Steel

Flow Stress

Ti-15Mo-5Zr-3Al: Flow stress vs temperature

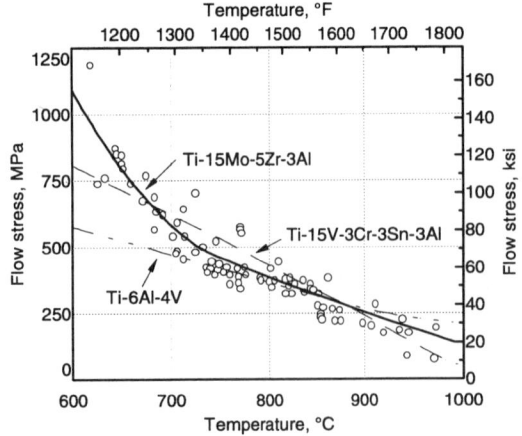

Specimens were 7 mm (0.275 in.) plate hot rolled at 1100 °C (2010 °F).
Source: Advanced Aircraft Technology Development Center of Japanese Aerospace Companies, SJAC Report No. 6010, 1986

Ti-15Mo-5Zr-3Al: Flow stress comparison

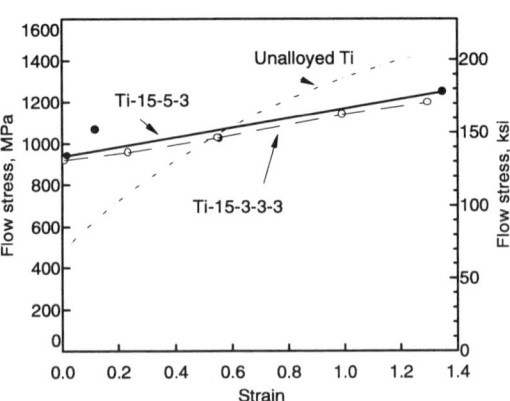

Cold rolled mill sheet. Ti-15-5-3 and Ti-15-3: solution treated. Unalloyed titanium: annealed.
Source: Advanced Aircraft Technology Development Center of Japanese Aerospace Companies, SJAC Report No. 6010, 1986

Forming Properties

Ti-15Mo-5Zr-3Al: Effect of solution treatment temperatures on *n*-value

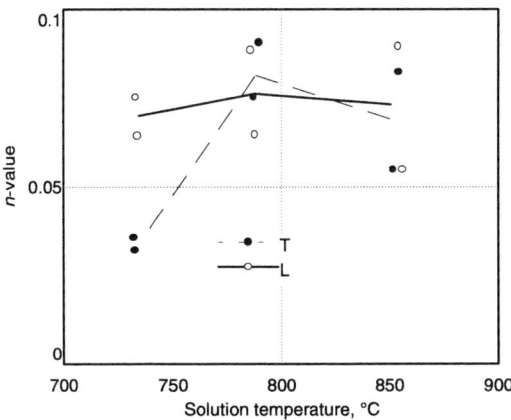

1 mm (0.04 in.) thick cold rolled sheet heat treated at 1100 °C (2010 °F), 90%, hot rolled + ST, 30 min, WQ + 86%, cold rolled + ST, 30 min, WQ.
Source: Advanced Aircraft Technology Develpment Center of Japanese Aerospace Companies, SJAC Report No. 6010, 1986

Ti-15Mo-5Zr-3Al: Effect of solution treatment temperatures on *r*-value

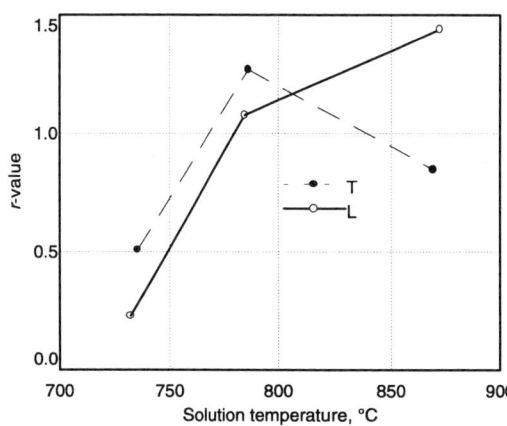

1 mm (0.04 in.) thick cold rolled sheet heat treated at 1100 °C (2010 °F), 90%, hot rolled + ST, 30 min, WQ + 86%, cold rolled + ST, 30 min, WQ.
Source: Advanced Aircraft Technology Development Center of Japanese Aerospace Companies, SJAC Report No. 6010, 1986

Sheet Forming

Ti-15Mo-5Zr-3Al: Relationship between bend angle and springback

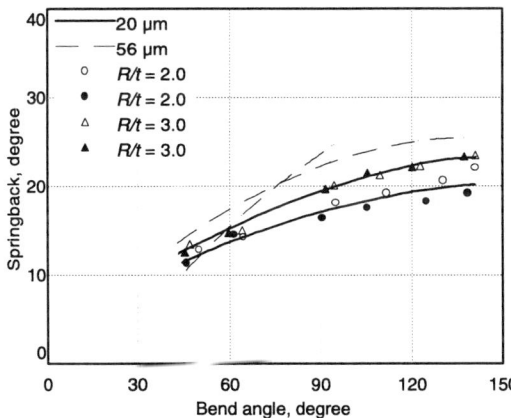

Specimens were 1.0 mm (0.04 in.) thick cold rolled sheet solution treated as indicated: 800 °C (1470 °F), 2.5 min, AC (20 μm grain size) and 800 °C (1470 °F), 30 min, AC (56 μm grain size). Open symbol, longitudinal; closed symbol, transverse.
Source: Advanced Aircraft Technology Development Center of Japanese Aerospace Companies, SJAC Report No. 6103, 1987

Ti-15Mo-5Zr-3Al: Erichsen value of cold rolled sheet

Thickness, mm	Erichsen value		
0.5	7.5	6.3	5.8
1.0	7.5	8.4	...

Note: Specimens were 70 × 70 mm (2.75 × 2.75 in.), solution treated at 850 °C (1560 °F), WQ. β grain size was 56 μm; punch diameter, 20 mm (0.78.); graphite grease was used as a lubricant. Source: Advanced Aircraft Technology Development Center of Japanese Aerospace Companies, SJAC Report No. 6010, 1986

Heat Treatment

Solution treatment conditions depend on subsequent product application. When cold formability is required, the material should be solution treated just above the β transus (785 °C, or 1450 °F). To obtain a better combination of strength and ductility after aging, the material should be solution treated at 735 °C (1350 °F) for 0.5 to 1 h. Water quenching is preferable as a cooling treatment after solution treating.

Aging should be carried out at temperatures of 425 to 500 °C (795 to 930 °F). Maximum strength can be obtained after aging at temperatures of 425 to 450 °C (795 to 840 °F), but long times are required. However, age hardening occurs relatively rapidly during aging at temperatures of 475 to 500 °C (885 to 930 °F).

To obtain a higher strength, duplex aging is sometimes used. The first aging is carried out at 425 °C (795 °F) for α phase to precipitate finely, and the second aging at 475 to 500 °C (885 to 930 °F) is used to accelerate the growth of the α precipitates.

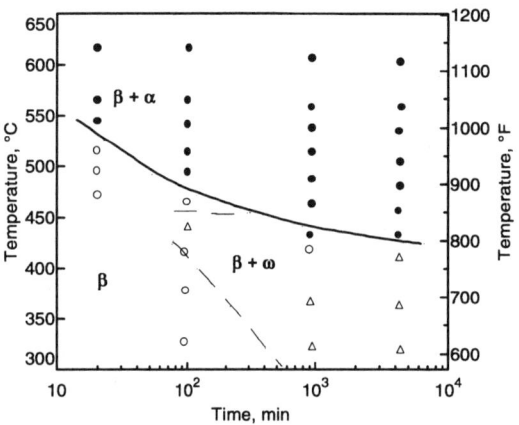

Ti-15Mo-5Zr-3Al: Aging transformation diagrams

9.5 mm (0.35 in.) diameter bar hot rolled at 880 °C (1615 °F); 98% reduction. X-ray diffraction.
Source: T. Nishimura, M. Nishigaki, and Y. Moriguchi, "Characteristics of Beta Titanium Alloy Ti-15Mo-5Zr-3Al," R & D Kobe Steel Engineering Reports, Vol 32, No. 1

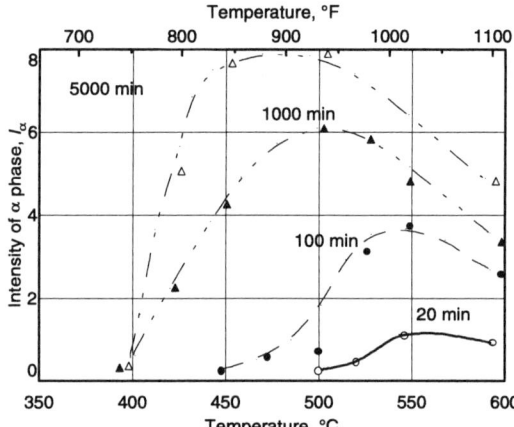

Ti-15Mo-5Zr-3Al: Amount of α phase during aging

$I_\alpha = I\{10\bar{1}0\}_\alpha / I\{200\}_\beta$. 9.5 mm (0.35 in.) diameter bar hot rolled at 880 °C (1615 °F), 98% reduction. X-ray diffraction.
Source: T. Nishimura, M. Nishigaki, and Y. Moriguchi, "Characteristics of Beta Titanium Alloy Ti-15Mo-5Zr-3Al," R & D Kobe Steel Engineering Reports, Vol 32, No. 1

Ti-11.5V-2Al-2Sn-11Zr

Common Name: Transage 129
Trade Names: Transage 129, T129
UNS Number: Unassigned

Compiled by Frank A. Crossley, Retired

Transage 129 is a martensitic, high-hardenability, age-hardenable, high-strength titanium-base alloy. It is a noncommercial, experimental alloy intended to improve structural efficiency in chemical and air frame applications. Transage 129 is less sensitive to the usual impurities than other (non-Transage) titanium alloys and has excellent fatigue resistance.

Product Forms and Fabrication. Transage 129 can be produced in all mill product forms. It has good formability and can be drawn at room temperature into deep cups with reasonable die radii. Transage 129 is especially recommended for cold formable sheet applications and it can be welded by all methods. Weld efficiency of 100% has been demonstrated to strength levels to 1446 MPa (210 ksi) for two-pass electron beam transverse weldments on 1.4 mm (0.056 in.) sheet. The alloy has excellent net-shape capability by isothermal forging, which can be done at temperatures as low as 650 °C (1200 °F), although beta forging at 760 to 815 °C (1400 to 1500 °F) is considered optimum.

Product Condition. The typical condition for the application of Transage 129 is aged to strength levels of 1240 MPa (180 ksi) or higher. In common with other Transage alloys, it has exceptionally high hardenability. Uniform age hardening is obtainable in heavy sections that are air cooled from beta solution heat treatment to achieve strengths of 1240 MPa (180 ksi) or higher. In the age-hardened condition, Transage 129 under a triaxial stress state, e.g., at the tip of a crack loaded in tension, undergoes strain-initiated, stress-induced transformation. This is an energy-absorbing phenomenon that increases resistance to crack propagation.

Transage 129: Composition limits of wrought alloy

Element	Composition, wt%
Aluminum	1.7-2.7
Carbon	0.08 max
Iron	0.20 max
Nitrogen	0.05 max
Oxygen	0.15 max
Tin	1.5-2.5
Vanadium(a)	10.5-12.5
Zirconium	10.0-12.0
Boron	0.03 max
Hydrogen	0.015 max
Yttrium	0.005 max
Residual elements	
Each	0.10 max
Total	0.40 max
Titanium	bal

(a) The vanadium-aluminum master alloy (nominally 15 to 17 wt% aluminum) addition is to be calculated to obtain the nominal vanadium content of 11.5 wt%.

Selected References

- F.A. Crossley and R.W. Lindberg, Microstructural Analysis of a High-Strength Martensite-Beta Titanium Alloy, *Proc. 2nd Int. Conf. Strength of Metals and Alloys*, Vol 3, Asilomar, American Society for Metals, 1970, p 841-845

- F.A. Crossley, R.L. Boorn, R.W. Lindberg, and R.E. Lewis, Fracture Toughness of Transage 129 Alloy, *Titanium Science and Technology*, Vol 3, R.I. Jaffee and H.M. Burte, Ed., TMS/AIME, Plenum Press, 1973, p 2025-2039

- F.A. Crossley and J.M. Van Orden, A New Titanium Alloy for Forms and Weldments, *Met. Eng. Quart.*, Vol 13, 1973, p 55-61; also *Source Book on Materials Selection*, Vol 2, American Society for Metals, p 170-176

- F.A. Crossley, A New Cost and Weight Saving Titanium Alloy, *Met. Prog.*, Vol 114 (No. 3), Aug 1978, p 60-64

- F.A. Crossley and R.H. Jeal, Fatigue and Fracture Behavior of the High Hardenability Martensitic Transage Titanium Alloys, *21st Structures, Structural Dynamics, and Materials Conf.*, Paper No. 81-0535-CP reprinted from CP811, A Bound Collection of Technical Papers, AIAA, Apr 1981, p 134-140; also *J. Aircraft*, Vol 18 (No. 8), Aug 1981, p 683-686

- F.A. Crossley and N.E. Paton, Superplastic Behavior of the Martensitic Transage Titanium Alloys: Hot Working Verifications of Theoretical Predictions, *Experimental Verification of Process Models*, C.C. Chen, Ed., ASM International, 1983, p 53-69

- F.A. Crossley, The Martensitic Transage Titanium Alloys for Improved Structural Efficiency and Reduced Cost, *Materials and Processes—Continuing Innovations*, SAMPE, 1983, p 1352-1367

- F.A. Crossley, The Martensitic Transage Titanium Alloys: Their Metallurgy, Processing Characteristics and Potential Applications, *Beta Titanium Alloys in the 1980's*, R.R. Boyer and H.W. Rosenberg, Ed., TMS/AIME, 1984, p 349-386, 485-496

- F.A. Crossley, E. Walden, and J.M. Van Orden, "Evaluation of Transage 129 (Ti-2Al-11V-2Sn-11Zr) Alloy for Heavy Section Forgings," unpublished paper

- F.A. Crossley, "Transage 129 (Ti-2Al-11V-2Sn-11Zr): Properties of Forgings and Extrusions," unpublished paper, 1973

- F.A. Crossley, private communication, July 1992

- H.M. Flower, A.I.P. Nwobu, and D.R.F. West, Age Hardening Reactions in Transage 129 and 134, *Titanium Science and Technology*, Vol 3, G. Lütjering, U. Zwicker, and W. Bunk, Ed., Deutsche Gesellschaft für Metallkunde, Germany, 1985, p 1567-1574

Phases and Structures

Beta Transus. 720 °C (1325 °F)

Hardening. Because of the dominant influence of the martensitic transformation on the age hardening response, Transage 129 (in common with other Transage alloys) has exceptionally good reproducibility of mechanical properties within a given heat and from heat to heat. In wrought products in both the solution heat treated and the aged conditions, microstructures are so fine that the use of transmission electron microscopy is necessary to resolve them. Upon cooling from beta solution heat treatment at rates slower than a water quench, the alloy transforms partially to a submicroscopic martensite (α'). Complete transformation is apparently blocked by the generation of forest dislocations during transformation. Upon heating into the age hardening temperature range of 425 to 565 °C (800 to 1050 °F), dislocations rapidly annihilate, and the martensitic α grows at the expense of retained β. This results in a very fine and uniform Widmanstätten distribution of α in a β matrix. Water quenching from beta solution heat treatment produces some stressed-induced orthorhombic martensite (α''), which makes age hardening mechanics more complex.

Segregation during ingot solidification is eliminated by subjecting the product to an anneal of 925 °C (1700 °F), or higher, for 1 h somewhere in the processing schedule following initial ingot breakdown.

Transage 129: Occurrence of ω phase

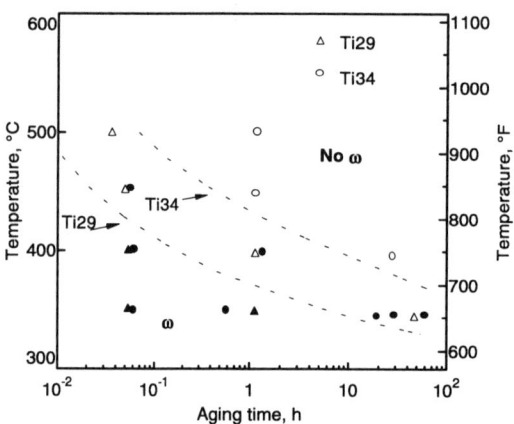

Temperature vs aging time for the occurrence of ω phase in 25 mm (1.0 in.) Transage 129 and 134 alloy plate, beta solution heat treated and water quenched.
Source: A.I.P. Nwobu, "Decomposition of Beta-Phase in Transage 134 and 129 Titanium Alloys," Ph.D. dissertation, Department of Metallurgy and Materials Science, Imperial College, London, 1979

Transage 129: Estimated martensite start temperature

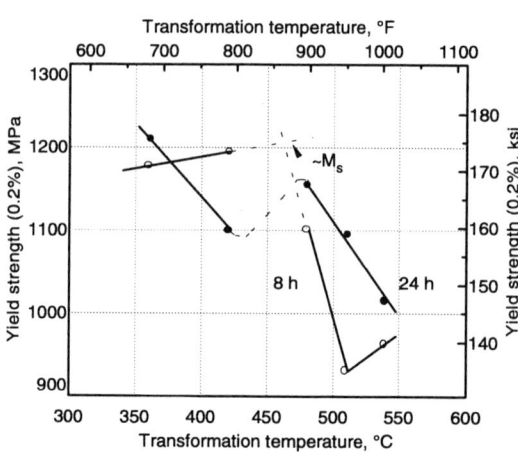

Yield strength vs transformation temperature for 8- and 24-h isothermal transformations of Transage 129, indicating a change in age hardening mechanism occurring at approximately 470 °C (875 °F). Transition from nucleation and growth transformation to athermal martensitic transformation was responsible. 25 mm (1.0 in.) plate from a 45-kg (100-lb ingot) forged to a 76 mm (3.0 in.) plate and continuously rolled to 25 mm plate from 955 °C (1750 °F). Specimen blanks isothermally transformed from 800 °C (1450 °F) for 20 min.
Source: F.A. Crossley, private communication, July 1992

Physical Properties

Transage 129: Summary of typical physical properties

Beta transus	~720 °C (1325 °F)
Density(a)	4.816 g/cm^3 (0.174 lb/in.^{3})
Magnetic permeability	Nonmagnetic

(a) Typical values at room temperature of about 20 to 25 °C (68 to 78 °F)

Transage 129: Modulus of elasticity

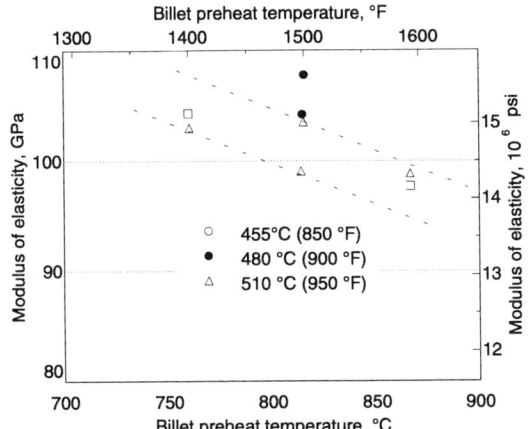

Modulus of elasticity vs billet preheat temperature for age hardened specimens from 13 mm (1/2 in.) diameter bar extrusions. Extrusion billets were prepared from a 45 kg (100 lb) ingot. The 34 mm (1.35 in.) diameter by 50 mm (2.0 in.) length billets were coated with protective glass. The billet chamber was heated to 315 °C (600 °F). The reduction ratio was 10:1.
Source: F.A. Crossley, "Transage 129 (Ti-2Al-11V-2Sn-11Zr): Properties of Forgings and Extrusions," unpublished paper, 1973

Transage 129: Resistance to crevice corrosion

Anodic breakdown residual potential (ABRP) of selected titanium alloys. The higher the ABRP test value, the less susceptible the material to pitting corrosion and crevice corrosion.

Material	ABRP, V
Transage 129(a)	8.2
CP Ti	8.1(b)
Ti-4Al-4V	6.5
Ti-4Al	6.2
Ti-8Al-1Mo	4.0
Ti-8Al	3.9

(a) 25 mm (1 in.) plate; ultimate tensile strength = 1296 MPa (188 ksi); elongation, 7%; reduction of area, 23%. (b) Post-publication correction: 8.1 V is the value for very high-purity Ti; CP Ti has values from 7.3 to 7.9 depending upon purity. Source: F.A. Crossley and J.M. Van Orden, A New Titanium Alloy for Forms and Weldments, *Met. Eng. Quart.*, Vol 13, 1973, p 55-61; also *Source Book on Materials Selection*, Vol 2, American Society for Metals, p 170-176

Tensile Properties

Transage 129: Tensile properties for various product forms

Material/ heat treatment	Orientation	Tensile yield strength MPa	Tensile yield strength ksi	Ultimate tensile strength MPa	Ultimate tensile strength ksi	Elongation, %	Reduction of area, %
1—25 mm plate from 820-kg ingot; 815 °C, 1 h, AC, 620 °C, 1 h, AC, 455 °C, 24 h, AC	...	1124	163	1255	182	9.8	24
2—1.5 mm sheet; 760 °C, 20 min, fan air cooled, aged at:							
650 °C, 1 h, AC, 425 °C, 24 h, AC	L	1280	186	1390	202	6	...
650 °C 1 h, AC, 480 °C 24 h, AC	L	1100	160	1190	173	8	...
480 °C, 24 h, AC	L	1200	174	1310	190	8	...
3—13 mm diameter extrusion, 10:1 reduction from 815 °C, aged at:							
510 °C, 24 h, AC	L	1240	180	1350	196	8	20
650 °C, 1 h, AC, 425 °C, 24 h, AC	L	1250	181	1390	202	7.5	22.5
4—25 mm diameter extrusion, 10.2:1 reduction from 815 °C, AC, aged at:							
510 °C, 24 h, AC	L	1170	169	1230	178	9.4	30
650 °C, 1 h, AC, 425 °C, 24 h, AC	L	1160	168	1280	185	8.7	22
5—115 × 180 × 610 mm forged billet; 790 °C, 24 h, AC, 675 °C, 1 h, AC, 455 °C, 24 h, AC	ST	1293	188	1344	195	2.0	2.5
	L	1251	182	1338	194	4.3	9.9
	T	1269	184	1358	197	2.3	3.5
6—175 mm diameter by 90 mm thick disc, upset 50% from 815 °C by press forging; 815 °C, 1 h, fan air cooled, 510 °C, 24 h, AC	R1(a)	1265	184	1320	192	3.8	11.9
	R2	1245	180	1284	186	5.5	14.7
	Axial	1276	185	1327	192	3.0	5.5

(a) R1, R2, and axial are radial and axial directions in the disc corresponding to short transverse, longitudinal, and transverse directions, respectively, in the 115 × 175 × 610 mm forged billet from which the 114 × 180 × 100 mm workpiece was cut. Source: Material 1—F.A. Crossley and R.H. Jeal, "Fatigue and Fracture Behavior of the High Hardenability Martensitic Transage Titanium Alloys," *21st Structures, Structural Dynamics, and Materials Conf.*, Paper No. 81-0535-CP reprinted from CP811, A Bound Collection of Technical Papers, AIAA, Apr 1981, p 134-140; also *J. Aircraft*, Vol 18 (No. 8), Aug 1981, p 683-686. Material 2—F.A. Crossley and J.M. Van Orden, A New Titanium Alloy for Forms and Weldments, *Met. Eng. Quart.*, Vol 13, 1973, p 55-61; also *Source Book on Materials Selection*, Vol 2, ASM International, p 170-176. Material 4, 5, and 6—F.A. Crossley, "Transage 129 (Ti-2Al-11V-2Sn-11Zr): Properties of Forgings and Extrusions," unpublished paper, 1973

Transage 129: Elevated temperature tensile property comparison

Alloy	Final aging treatment(a)	Test temperature °C	Test temperature °F	Ultimate tensile strength MPa	Ultimate tensile strength ksi	Tensile yield strength MPa	Tensile yield strength ksi	Elongation, %	Reduction of area, %
Transage 129	565 °C (1050 °F), 24 h, AC	24	75	1274	185	1196	173	5	17.1
		315	600	1078	156	946	137	8.5	26.5
		480	900	905	131	725	105	13	41.0
		24	75	1087	158	995	144	8	16.3
		480	900	811	118	651	94	22	76.5
Transage 162	593 °C (1100 °F), 24 h, AC	24	75	1212	175	1108	161	7	10.0
		205	400	1120	162	995	144	9	26.0
		315	600	1040	151	881	128	11.5	33.5
		425	800	1029	149	872	126	6	37.0
		480	900	976	142	994	122	6	45.0

Note: 16 mm (5/8 in.) diameter bar produced by hammer forging pieces of 25 mm (1.0 in.) plate from a 45 kg (100 lb) ingot. Transage 162 (Ti-2Al-11.5V-9Sn-2.5Zr) was the early prototype for Transage 175. (a) All specimens given the preliminary heat treatment of 815 °C (1500 °F), 30 min, AC, 510 °C (950 °F), 24 h, AC. Source: F.A. Crossley, "Transage 129 (Ti-2Al-11V-2Sn-11Zr): Properties of Forgings and Extrusions," unpublished paper, 1973

Transage 129: Yield strength vs hardness

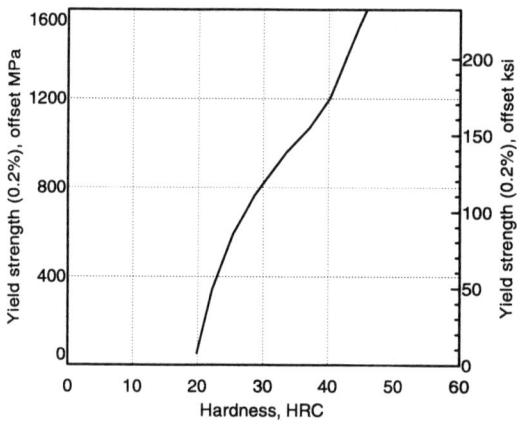

Yield strength vs Rockwell C hardness curve fit from 13 and 25 mm (1/2 and 1.0 in.) plate of Transage 129 alloy after STA (solution heat treat and age) and STIT (solution heat treat and isothermally transform) treatments.
Source: F.A. Crossley, private communication, July 1992

Sheet   Transage 129: Solution anneal temperature vs tensile strength of sheet

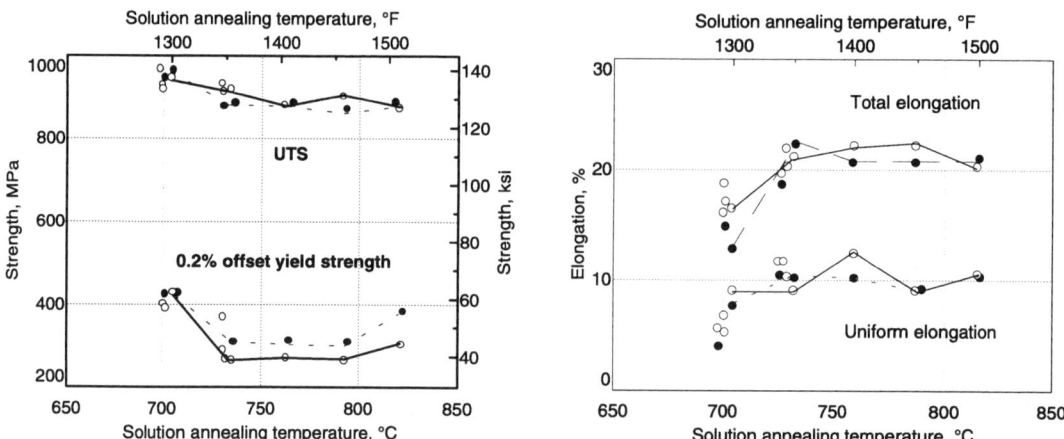

Tensile properties of cold rolled (50% reduction) Transage 129 alloy 1.5 mm (0.060 in.) sheet as solution annealed from 705 to 815 °C (1300 to 1500 °F) and fan air cooled. Final reduction of 50% unidirectional at room temperature. Note that the beta transus temperature is 720 °C (1325 °F).
Source: F.A. Crossley, private communication, July 1992

Transage 129: Solution annealing temperature vs aged tensile strength

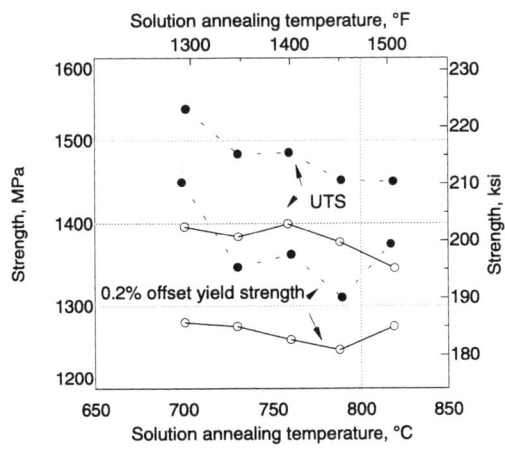
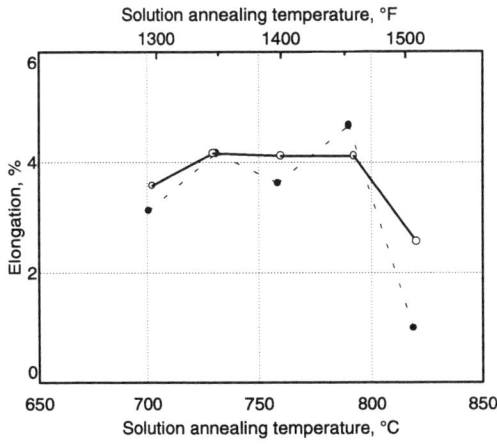

(a) (b)

Tensile properties of cold rolled (50% reduction) Transage 129 alloy 1.5 mm (0.06 in.) sheet as solution annealed at various temperatures, fan air cooled, and aged at 480 °C (900 °F), 1 h, AC. Final reduction of 50% unidirectional at room temperature. Open symbols, longitudinal; closed symbols, transverse.
Source: F.A. Crossley, private communication, July 1992

Transage 129: Effect of strain on age hardening response
Effect of uniaxial strain on age hardening response of hot rolled Transage 129 alloy sheet

Aging treatment(a)	Tensile yield strength at strain of:						Ultimate tensile strength at strain of:						Elongation, at strain of: (%),		
	0%		5%		10%		0%		5%		10%		0%	5%	10%
	MPa	ksi	MPa	ksi	MPa	ksi	MPa	ksi	MPa	ksi	MPa	ksi			
None	736	106.8	...	...	805	116.7	777	112.7	...	...	866	125.6	24	...	16
480 °C (900 °F), 24 h	1197	173.6	1014	147.0	954	138.3	1313	190.5	1202	174.4	1180	171.2	8	8	8
480 °C (900 °F), 24 h(b)	...	...	1011	146.7	1049	153.2	...	...	1220	177.0	1200	174.1	...	6	8
455 °C (850 °F), 24 h	1463	212.2	1123	163.0	1122	162.7	1571	227.9	1348	195.5	1334	193.5	5	3	4
455 °C (850 °F), 24 h(b)	1451	210.5	1144	165.9	1043	151.3	1578	228.8	1326	192.3	1314	190.6	2	4	3
650 °C (1200 °F), 1 h, 480 °C (900 °F), 4 h	1096	159.0	882	128.0	917	133.0	1718	176.6	1040	150.9	1065	154.5	10	14	11
650 °C (1200 °F), 1 h, 455 °C (850 °F), 24 h	1233	178.8	976	141.5	969	140.5	1377	199.7	1160	168.3	1176	170.5	5	10	9
650 °C (1200 °F), 1 h, 425 °C (800 °F), 24 h	1252	181.6	987	143.2	992	143.8	1390	201.6	1201	174.2	1182	171.4	6	10	10

Note: 1.5 mm (0.060 in.) sheet produced by hot rolling from a 136 kg (300 lb) ingot. (a) Solution heat treated at 760 °C (1400 °F), 20 min, fan air cooled. (b) Oxide and contamination layer due to aging treatment removed by sandblasting and pickling while they were left intact on all other specimens. Source: F.A. Crossley and J.M. Van Orden, A New Titanium Alloy for Forms and Weldments, Met. Eng. Quart., Vol 13, 1973, p 55-61; also Source Book on Materials Selection, Vol 2, American Society for Metals, p 170-176

Creep Properties

Transage 129: Stress-rupture

Test temperature		Stress		Rupture time, h	Deformation, %
°C	°F	MPa	ksi		
480	900	379	55	36.2	28
480	900	345	50	77.4	45
480	900	903	131	0.1	...
315	600	1076	156	0.1	...

Note: 815 °C (1500 °F), 1 h, AC, 510 °C (950 °F), 24 h, AC

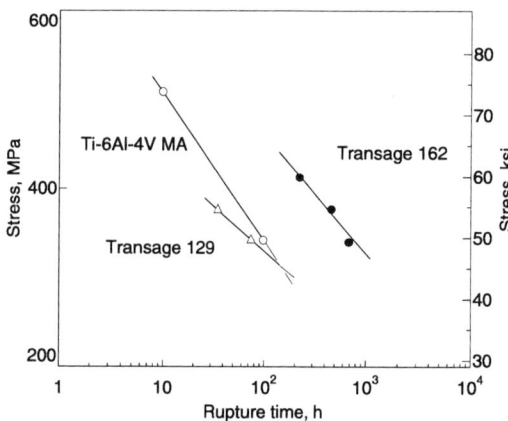

Transage 129: Stress-rupture comparison at 480 °C (900 °F)

16 mm (5/8 in.) diameter bar produced by hammer forging pieces of 25 mm (1.0 in.) plate from a 45 kg (100 lb) ingot. Transage 162 (Ti-2Al-11.5V-9Sn-2.5Zr) was the prototype for Transage 175 (Ti-2.7Al-13V-7Sn-2Zr). Source: F.A. Crossley, "Transage 129 (Ti-2Al-11V-2Sn-11Zr): Properties of Forgings and Extrusions," unpublished paper, 1973

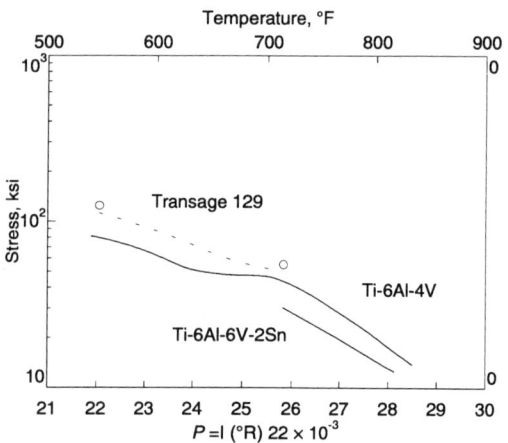

Transage 129: Larson-Miller plot for 0.1% creep deformation

Creep comparison of 16 mm (5/8 in.) diameter bar produced by hammer forging pieces of 25 mm (1.0 in.) plate from a 45 kg (100 lb) ingot after 100-h exposure at temperature. Source: F.A. Crossley, private communication, July 1992

Transage 129: Tensile properties following creep exposure

300-h creep conditions				Strain, %	Tensile properties				Elongation, %	Reduction of area, %
Temperature		Stress			Ultimate tensile strength		Tensile yield strength			
°C	°F	MPa	ksi		MPa	ksi	MPa	ksi		
No exposure					1276	185	1192	173	5	17
315	600	827	120	0.84	1469	213	1386	201	4	7

Solution treated at 815 °C (1500 °F), 1/2 h, AC, 510 °C (950 °F), 24 h, AC. Note: 16 mm (5/8 in.) diameter bar produced by hammer forging pieces of 25 mm (1.0 in.) plate from a 45 kg (100 lb) ingot. Source: F.A. Crossley, "Transage 129 (Ti-2Al-11V-2Sn-11Zr): Properties of Forgings and Extrusions," unpublished paper, 1973

Fatigue Properties

Transage 129: Room temperature load-controlled fatigue test properties

Material condition	Type of loading	K_t	R	Maximum stress at 10^7 cycle runout	
				MPa	ksi
1	Axial	2.7	0.1	400	58
2	Axial	2.7	0.1	345	50
3	Rotating bend	1.0	−1.0	600	87

Material 1—1.5 mm (0.060 in.) sheet produced by hot rolling from a 136 kg (300 lb) ingot heat treated at 760 °C (1400 °F), 20 min, fan air cooled, 650 °C (1200 °F), 1 h, AC, 425 °C (800 °F), 24 h, AC. Material 2—1.5 mm (0.060 in.) sheet produced by hot rolling from a 136 kg (300 lb) ingot heat treated at 760 °C (1400 °F), 20 min, fan air cooled, 480 °C (900 °F), 1 h, AC. Material 3—25 mm (1.0 in.) plate produced from an 820-kg (1800-lb) ingot. Source: F.A. Crossley and J.M. Van Orden, A New Titanium Alloy for Forms and Weldments, *Met. Eng. Quart.*, Vol 13, 1973, p 55-61; also *Source Book on Materials Selection*, Vol 2, American Society for Metals, p 170-176; and F.A. Crossley and R.H. Jeal, "Fatigue and Fracture Behavior of the High Hardenability Martensitic Transage Titanium Alloys," *21st Structures, Structural Dynamics, and Materials Conference*, Paper No. 81-0535-CP reprinted from CP811, A Bound Collection of Technical Papers, AIAA, Apr 1981, p 134-140; also *J. Aircraft*, Vol 18 (No. 8), Aug 1981, p 683-686

Transage 129: Rotating bend fatigue comparison

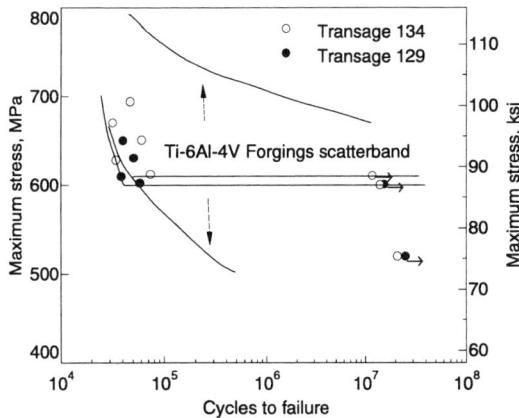

25 mm (1.0 in.) plate produced from an 820 kg (1800 lb) ingot. $R = -1.0$; $A = \infty$. Smooth specimens.
Source: F.A. Crossley and R.H. Jeal, "Fatigue and Fracture Behavior of the High Hardenability Martensitic Transage Titanium Alloys," *21st Structures, Structural Dynamics, and Materials Conference*, Paper No. 81-0535-CP reprinted from CP811, A Bound Collection of Technical Papers, AIAA, Apr 1981, p 134-140; also *J. Aircraft*, Vol 18 (No. 8), Aug 1981, p 683-686

Transage 129: Notch fatigue comparison

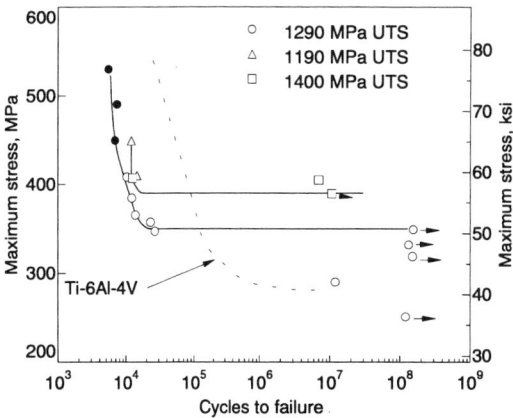

Axial, center-hole notch, high-cycle fatigue of 1.5 mm (0.060 in.) Transage 129 sheet produced by hot rolling from a 136 kg (300 lb) ingot. Heat treated at 760 °C (1400 °F) 20 min, fan air cooled, and aged to strengths indicated. Load controlled, $R = 0.1$; $K_t = 2.7$. Ti-6Al-4V STA sheet ($K_t = 2.8$). Closed symbols denote tests rerun at higher stress.
Source: F.A. Crossley and J.M. Van Orden, A New Titanium Alloy for Forms and Weldments, *Met. Eng. Quart.*, Vol 13, 1973, p 55-61; also *Source Book on Materials Selection*, American Society for Metals, Vol 2, p 170-176

Fracture Properties

Transage 129: Yield strength vs fracture toughness

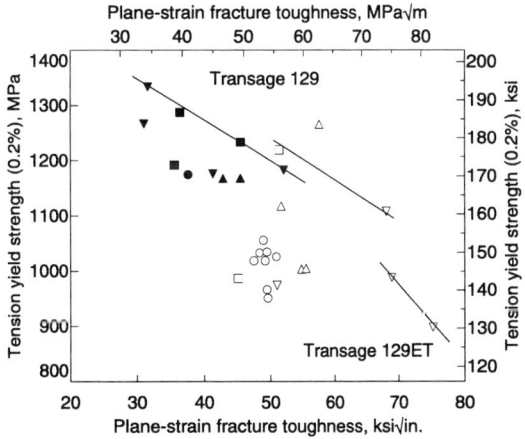

25 mm (1.0 in.) compact tension specimens tested according to ASTM E-399-70T from 125 mm (5 in.) diam bar (129ET) and 25 mm (1 in.) plate. Open symbols, isothermally transformed; closed symbols, solution treated and aged.
Source: F.A. Crossley, R.L. Boom, R.W. Lindberg, and R.E. Lewis, Fracture Toughness of Transage 129 Alloy, *Titanium Science and Technology*, Vol 3, R.I. Jaffee and H.M. Burte, Ed., TMS/AIME, Plenum Press, 1973, p 2025-2039; and F.A. Crossley, private communication, July 1992

Transage 129: Crack growth rate comparison at 20 °C (68 °F)

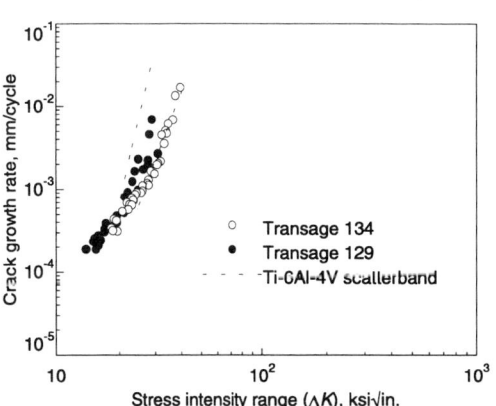

25 mm (1.0 in.) plate produced from an 820 kg (1800 lb) ingot. $R = 0.1$, 0.25 Hz.
Source: F.A. Crossley and R.H. Jeal, "Fatigue and Fracture Behavior of the High Hardenability Martensitic Transage Titanium Alloys," *21st Structures, Structural Dynamics, and Materials Conference*, Paper No. 81-0535-CP reprinted from CP 811, A Bound Collection of Technical Papers, AIAA, Apr 1981, p 134-140; also *J. Aircraft*, Vol 18 (No. 8), Aug 1981, p 683-686

Transage 129: Fracture and impact toughness

Processing temperature				Tensile yield strength		Ultimate tensile strength		Elongation, %	Reduction of area, %	Charpy impact toughness(a)		Fracture toughness(a) (K_{Ic})	
Forging		Rolling											
°C	°F	°C	°F	MPa	ksi	MPa	ksi			J	ft·lbf	MPa√m	ksi√in.
815 °C (1500 °F), 1 h, AC, 540 °C (1000 °F), 4 h, AC													
955	1750	955	...	1217	176	1284	186	6.5	15.1	...	...	...	...
1040	1905	870	...	1144	166	1223	178	8.5	26.6	6.5	4.8	45.2	41.1
1040	1905	1040	...	1158	168	1220	177	7.0	15.3	12.2	8.9	...	...
955	1750	870	...	1134	164	1207	175	8.5	21.7	8.9	6.5	42.6	38.7
955	1750	955	...	1164	169	1238	179	7.8	21.6	8.9	6.5	49.8	45.3
955	1750	870	...	1178	171	1294	187	8.5	18.9	8.7	6.4	57.1	51.9
815 °C (1500 °F), 1 h, isothermal transformation to 455 °C (850 °F), 24 h, AC													
955	1750	955	...	1138	165	1258	182	10	22.5	9.8	7.2	...	...
1040	1905	870	...	1109	161	1196	173	12	26.8	11.4	8.4	75.4	68.6
955	1750	870	...	1116	162	1179	171	10	31.2	8.1	5.9	...	...
955	1750	955	...	1002	145	1128	163	13	43.3	...	...	49.5	45.0
955	1750	955	...	1095	159	1220	177	9	30.4	...	...	...	...
955	1750	870	...	1080	156	1183	171	11.5	36.1	...	...	...	...
815 °C (1500 °F), 1 h, isothermal transformation to 455 °C (850 °F), 16 h, AC													
955	1750	870	...	1212	175	1335	193	6	16.0	9.3	6.8	56.0	50.9
955	1750	955	...	1257	182	1334	193	5	20.3	10	7.3	62.9	57.2
955	1750	870	...	1180	171	1306	189	6.5	20.0	7.2	5.3	...	...

Note: 25 mm (1.0 in.) compact tension specimens used for fracture toughness tests performed according to ASTM E-399-70T. Plate produced by forging billet to 75 mm (3 in.) and then rolling at indicated temperatures to 25 mm (1 in.). (a) At 0 °C (32 °F). Source: F.A. Crossley, private communication, July 1992; and F.A. Crossley, R.L. Boorn, R.W. Lindberg, and R.E. Lewis, Fracture Toughness of Transage 129 Alloy, *Titanium Science and Technology*, Vol 3, R.I. Jaffee and H.M. Burte, Ed., TMS/AIME, Plenum Press, 1973, p 2025-2039

Working

Transage 129: Flow stress vs reciprocal temperature

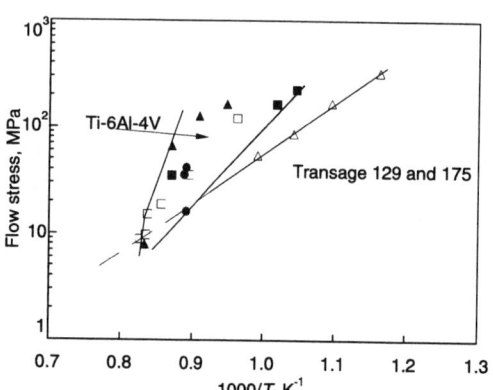

Dependence of flow stress on reciprocal temperature of 25 mm (1.0 in.) Transage 129 plate compared with Transage 175 and Ti-6Al-4V at a constant strain rate of 10^{-3}/s.
Source: F.A. Crossley and N.E. Paton, Superplastic Behavior of the Martensitic Transage Titanium Alloys: Hot Working Verifications of Theoretical Predictions, *Experimental Verification of Process Models*, C.C. Chen, Ed., ASM International, 1983, p 53-69

Transage 129: Flow stress vs strain rate

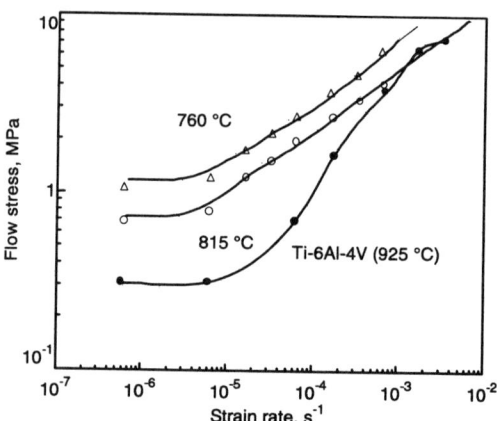

Flow stress dependence on strain rate of 25 mm (1.0 in.) Transage 129 plate at 760 and 815 °C (1400 and 1500 °F) compared with Ti-6Al-4V at 925 °C (1700 °F).
Source: F.A. Crossley and N.E. Paton, Superplastic Behavior of the Martensitic Transage Titanium Alloys: Hot Working Verifications of Theoretical Predictions, *Experimental Verification of Process Models*, C.C. Chen, Ed., ASM International, 1983, p 53-69

Transage 129: Strain rate sensitivity of 25 mm plate

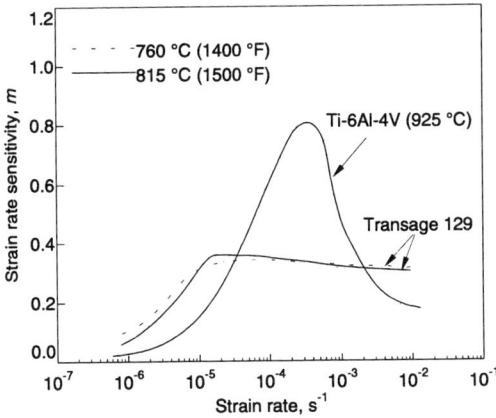

Source: F.A. Crossley and N.E. Paton, Superplastic Behavior of the Martensitic Transage Titanium Alloys: Hot Working Verifications of Theoretical Predictions, *Experimental Verification of Process Models*, C.C. Chen, Ed., ASM International, 1983, p 53-69

Forging

Transage 129 has net-shape and near-net-shape capabilities by isothermal forging at temperatures as low as 650 °C (1200 °F). However, considering relevant factors such as load requirements and die life, the optimum temperature is 760 to 815 °C (1400 to 1500 °F). Forging above the β transus temperature of 720 °C (1325 °F) minimizes directionality of mechanical properties.

Transage 129: Effect of upset reduction on tensile properties

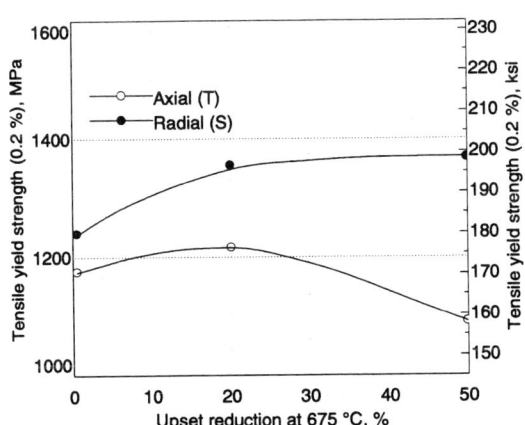 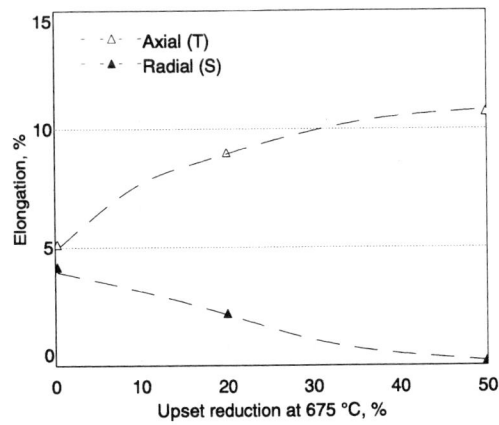

Effects of forging upset reduction from 675 °C (1250 °F) on directionality of tensile properties; forgings heat treated at 800 °C (1450 °F), 1 h, fan air cooled, 510 °C (950 °F), 3 h, AC.
Source: F.A. Crossley, E. Walden, and J.M. Van Orden, "Evaluation of Transage 129 (Ti-2Al-11V-2Sn-11Zr) Alloy for Heavy Section Forgings," unpublished paper

Extrusion

Transage 129: Extrusion constant vs temperature

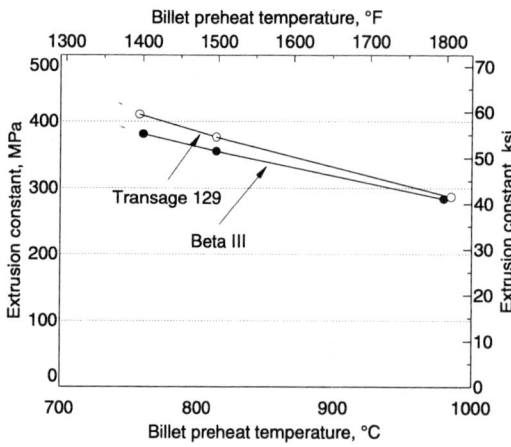

25 mm (1.0 in.) diameter extrusions were made from billets produced from an 820 kg (1800 lb) ingot. Ram speed of 76 to 127 mm/s (3 to 5 in./s) and 60° conical dies were used; 10.2:1 reduction. Source: F.A. Crossley, "Transage 129 (Ti-2Al-11V-2Sn-11Zr): Properties of Forgings and Extrusions," unpublished paper, 1973

Transage 129: Mechanical properties of extrusions

Extrusion temperature		Cooling after extrusion	Heat treatment(a)	Tensile yield strength		Ultimate tensile strength		Elongation, %	Reduction of area, %
°C	°F			MPa	ksi	MPa	ksi		
760	1400	WQ	A	1160	168	1270	184	9.5	28
815	1500	WQ	A	1140	166	1260	183	7.8	25
980	1800	WQ	A	1210	175	1330	193	4.8	8.7
760	1400	AC	A	1140	166	1290	184	8.4	26
815	1500	AC	A	1160	168	1280	185	8.7	22
980	1800	AC	A	1170	169	1290	187	4.6	7.6
760	1400	WQ	B	1230	178	1280	186	9.5	26
815	1500	WQ	B	1210	175	1270	184	10.4	29
980	1800	WQ	B	1200	174	1250	182	9.2	23
760	1400	AC	B	...	...	1230	179	10.0	34
815	1500	AC	B	1170	169	1230	178	9.4	30
980	1800	AC	B	1140	166	1210	175	7.7	18

Note: 25 mm (1.0 in.) diameter extrusions from billets produced from an 820 kg (1800 lb) ingot. Ram speed, 76 to 127 mm/s (3 to 5 in./s); 60° conical dies; 10.2:1 reduction. (a) Heat treatment A: 650 °C (1200 °F), 1 h, AC, 425 °C (800 °F), 24 h, AC. Heat treatment B: 510 °C (950 °F), 24 h, AC. Source: F.A. Crossley, "Transage 129 (Ti-2Al-11V-2Sn-11Zr): Properties of Forgings and Extrusions," unpublished paper, 1973

Forming

Transage 129: Forming limit comparison

Material	Olsen Cup height		Transverse bend radius (R/t)	Tensile yield strength		Hardness, HRC
	mm	in.		MPa	ksi	
Transage 129, solution treated, fan air cooled	7.14	0.281	2	734	105	27.3
Ti-70A, commercial purity	7.37	0.290	2	483(a)	70(a)	26
B120VCA, solution treated	6.35	0.25	3	827(a)	120(a)	32-36
Beta III						
Solution treated, water quenched	8.38	0.330	2	738	107	...
Solution treated, air cooled	6.35	0.25	3.5	883	128	...
Ti-6Al-4V, mill annealed	3.05	0.12	4.5	827	120(a)	36

Note: 1.5 mm (0.060 in.) sheet produced by hot rolling from a 136 kg (300 lb) ingot. Transage 129 sheet thickness is 1.5 mm (0.060 in.). (a) Guaranteed minimum. Source: F.A. Crossley and J.M. Van Orden, A New Titanium Alloy for Forms and Weldments, *Met. Eng. Quart.*, Vol 13, 1973, p 55-61; also *Source Book on Materials Selection*, Vol 2, American Society for Metals, p 170-176

Transage 129: Cold head ratio (D_f / D_i)

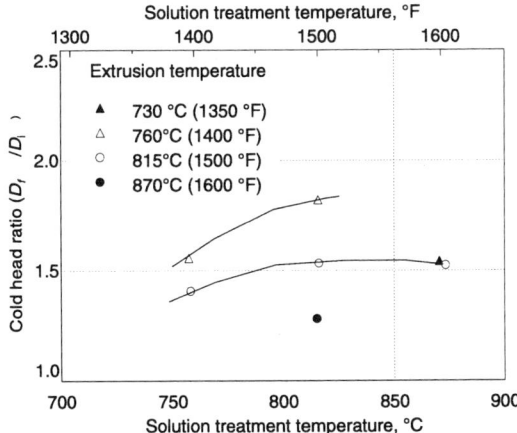

13 mm (1/2 in.) diameter extrusions were produced from a 45 kg (100 lb) ingot. The 34 mm (1.35 in.) diameter by 50 mm (2.0 in.) length billets were coated with protective glass. The billet chamber was heated to 315 °C (600 °F). The reduction ratio was 10:1. Crosshead speed, 0.5 cm/min (0.2 in./min).
Source: F.A. Crossley, "Transage 129 (Ti-2Al-11V-2Sn-11Zr): Properties of Forgings and Extrusions," unpublished paper, 1973

Heat Treatment

Transage 129: Recommended heat treatments

Treatment	Temperature °C	Temperature °F	Duration, h	Cooling method
Solution (β) anneal of sheet	760	1400	1/3	Fan air cool
Solution (β) anneal of heavy sections	815	1500	1	Water quench(a)
Aging (after solution anneal)	455-565	850-1050	24	Air cool
Isothermal transformation(b)	425-540	800-1000	24	Air cool
Triplex treatment(c)				
1st stage (β anneal)	See above		...	Fan air cool
2nd stage (α-β anneal)	650	1200	1	Air cool
3rd stage (age)	425-480	800-900	24	Air cool

(a) Water quenching for heavy section for maximum formability. If aging follows solution anneal, any convenient cooling rate may be used. (b) Solution treatment and isothermal transformation (STIT) produces higher toughness than STA. (c) For superior fatigue resistance

Solution Temperature Effect

Transage 129: Mechanical properties vs solution annealing

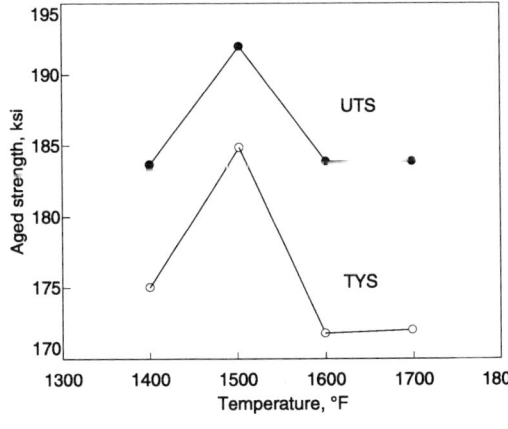

(a)

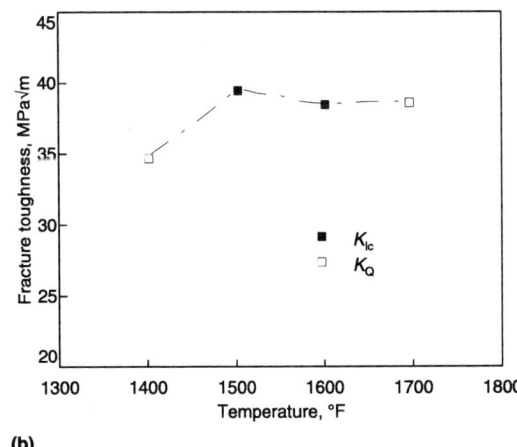

(b)

Effect of solution heat treatment temperature on tensile properties and fracture toughness of 25 mm (1.0 in.) Transage 129 plate. The solution heat treatments were followed by a water quench. Aged at 510 °C for 24 h, AC.
Source: F.A. Crossley, R.L. Boom, R.W. Lindberg, and R.E. Lewis, Fracture Toughness of Transage 129 Alloy, *Titanium Science and Technology*, Vol 3, R.I. Jaffee and H.M. Burte, Ed., TMS/AIME, Plenum Press, 1973, p 2025-2039

(continued)

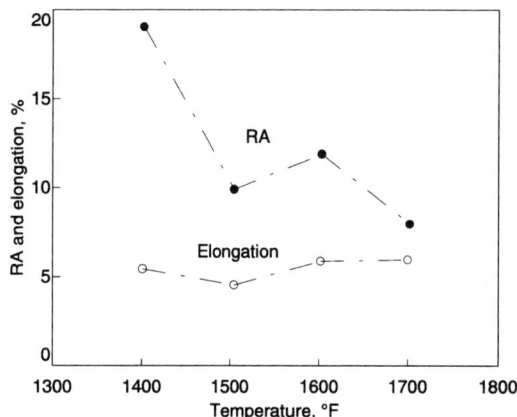

Transage 129: Ductility vs solution annealing

(c)
Effect of solution heat treatment temperature on tensile properties and fracture toughness of 25 mm (1.0 in.) Transage 129 plate. The solution heat treatments were followed by a water quench. Aged at 510 °C for 24 h, AC.
Source: F.A. Crossley, R.L. Boom, R.W. Lindberg, and R.E. Lewis, Fracture Toughness of Transage 129 Alloy, *Titanium Science and Technology*, Vol 3, R.I. Jaffee and H.M. Burte, Ed., TMS/AIME, Plenum Press, 1973, p 2025-2039

Oxidation

Transage 129: Oxidation and contamination
Oxide and α case depth vs Larson-Miller parameter for heating in air for 1-h exposure

Temperature		Oxide film thickness		Contamination depth		Total oxide and contamination	
°C	°F	μm	mils	μm	mils	μm	mils
677	1250	0.1	2	1.0	25	1.1	27
732	1350	0.3	8	2.1	53	2.4	61
760	1400(a)	0.36	9	2.2	56	2.6	66
788	1450	0.4	10	2.3	58	2.7	68
815	1500	0.6	14	2.4	62	3.0	76
871	1600	0.6	15	3.3	83	3.8	98

Note: From 730 to 815 °C (1350 to 1500 °F) intergranular diffusion of oxygen is faster than intragranular. At 870 °C (1600 °F), the two rates are about equal, and oxide dissolution at the interface is faster than oxide formation. (a) Interpolated values

Transage 129: Effect of α case on tensile properties

Quenching	Milled(a)	Ultimate tensile strength		Tensile yield strength		Elongation, %	Reduction of area, %
		MPa	ksi	MPa	ksi		
Fan air cool	No	781 ± 6	113.3 ± 0.9	593 ± 16	86.0 ± 2.3	9.0 ± 0.0	11.6 ± 0.9
Fan air cool	Yes	834 ± 0	121.0 ± 0.0	520 ± 8	75.4 ± 1.2	18.7 ± 1.2	33.7 ± 2.7
Water	No	788 ± 4	14.3 ± 0.6	365 ± 27	52.9 ± 3.8	13.3 ± 0.6	27.1 ± 3.0
Water	Yes	824 ± 2	119.5 ± 0.3	295 ± 14	42.8 ± 2.0	19.0 ± 0.0	39.7

Note: 6.6 mm (0.26 in.) plate produced from 820 kg (1800 lb) ingot. Beta solution annealed at 815 °C (1500 °F), 1 h. Sand blasted to remove oxide scale. Tensile test values given are average and standard deviation for three tests. (a) Milled 0.25 mm (0.010 in.) from surfaces. Source: F.A. Crossley, private communication, July 1992

Welding

Transage 129: Tensile properties of GTA weldments
Comparison of tensile properties of base metal and two-pass, GTAW transverse welded specimens

Aging treatment Temperature °C	°F	Time, h	Specimen(a)	Tensile yield strength MPa	ksi	Ultimate tensile strength MPa	ksi	Elongation in 13 mm (1/2 in.), %
	None		Base metal	365	53	869	126	20
			Weld metal(a)	283	41	738	107	20
				303	44	731	106	16
540	1000	4	Base metal	1048	152	1096	159	14
			Weld metal(a)	952	138	972	141	9
				1034	150	1048	152	6
510	950	24	Base metal	1089	158	1186	172	14
			Weld metal(a)	1103	160	1124	163	8
				1062	154	1076	156	6
480	900	24	Base metal	1200	174	1244	180	11
				1248	181	1262	183	7
				1196	172	1200	174	7

Note: 1.7 mm (0.066 in.) solution treated and aged sheet, laboratory produced from 25 mm (1.0 in.) plate produced from a 45 kg (100 lb) ingot; final reduction of 50% unidirectional at room temperature. (a) Transverse weld. All specimens longitudinal. Source: F.A. Crossley and J.M. Van Orden, A New Titanium Alloy for Forms and Weldments, *Met. Eng. Quart.*, Vol 13, 1973, p 55-61; also *Source Book on Materials Selection*, Vol 2, American Society for Metals, p 170-176

Transage 129: EB tensile properties of EBW plate

Condition	Ultimate tensile strength MPa	ksi	Tensile yield strength MPa	ksi	Elongation in 50 mm (2 in.), %	Reduction of area, %
As-welded	780	116	313	45.4	12.8	36
Aged: 540 °C (1000 °F), 4 h, AC	1172	170	1110	161	6.5	22
Aged: 675 °C (1250 °F), 2 h, AC, 430 °C (900 °F), 4 h, AC	1275	185	1207	175	5.2	20

Note: Bead-on-plate electron-beam, transverse-welded 6.4 mm (0.25 in) plate, 815 °C (1500 °F), 1 h, WQ. Sand blasted and pickled to remove 0.089 mm (0.0035 in.), welded and age hardened. Surfaces milled to remove 0.25 mm (0.010 in.) to achieve constant cross section and to remove oxygen-contaminated surface layer. Electron-beam settings: 125 kV, 13 mA, and travel speed of 12.7 mm/s (30 in./min). Source: F.A. Crossley, private communication, July 1992

Ti-12V-2.5Al-2Sn-6Zr

Common Name: Transage 134
Trade Names: Transage 134, T134
UNS Number: Unassigned

Compiled by Frank A. Crossley, retired

Transage 134 is a noncommercial, experimental high-strength titanium-base alloy recommended for applications where high strength and high fracture toughness are desired in heavy sections. Like Transage 129 and 175, Transage 134 is an age-hardenable β alloy that partially transforms to martensite during quenching.

Transage 134 can be produced in all mill product forms. Weldability and castability are good, and the alloy has net-shape capability by isothermal forging, which can be performed at temperatures as low as 650 °C (1200 °F), although 815 °C (1500 °F) is considered optimum.

For typical applications, Transage 134 can be aged to strength levels of 1140 MPa (165 ksi) or higher. At high strength levels it has the highest fracture toughness of the Transage alloys. In common with other Transage alloys, Transage 134 has exceptionally high hardenability. Uniform age hardening is obtainable in heavy sections air cooled from β solution heat treatment to achieve strengths of 1140 MPa (165 ksi) or higher.

Transage 134 (wrought, cast): Composition (wt%)

Element	wt%
Aluminum	2.0-3.0
Carbon	0.08 max
Iron	0.20 max
Nitrogen	0.05 max
Oxygen	0.15 max
Tin	1.5-2.5
Vanadium(a)	11.0-13.0
Zirconium	5.5-6.5
Boron	0.03 max
Hydrogen	0.015 max
Yttrium	0.005 max
Other	
Each	0.10 max
Total	0.40 max
Titanium	bal

(a) The vanadium-aluminum master alloy (nominally 15 to 17 wt% Al) addition is to be calculated to obtain the nominal vanadium content of 12.0 wt%.

Physical Properties

Transage 134: Summary of typical physical properties

Beta transus	750 °C (1385 °F)
Melting range	Not available
Density(a)	4.733 g/cm^3 (0.171/lb/in.^{3})
Electrical resistivity(a)	Not Available
Magnetic permeability	Nonmagnetic
Specific heat capacity(a)	Not available
Thermal conductivity(a)	Not Available
Thermal coefficient of linear expansion	Not Available

(a) Typical values at room temperature of about 20 to 25 °C (68 to 78 °F)

Phases and Structures. Like other Transage alloys, Transage 134 transforms partially to α′ martensite upon cooling from β solution heat treatment at any rate slower than a water quench. Complete transformation is apparently blocked by the generation of forest dislocations during transformation. Upon heating into the age-hardening range of 425 to 565 °C (800 to 1050 °F), dislocations rapidly become annihilated and the martensitic α grows at the expense of the retained β. This results in a very fine and uniform Widmanstätten distribution of α in a β matrix.

Transage 134: Occurrence of ω phase

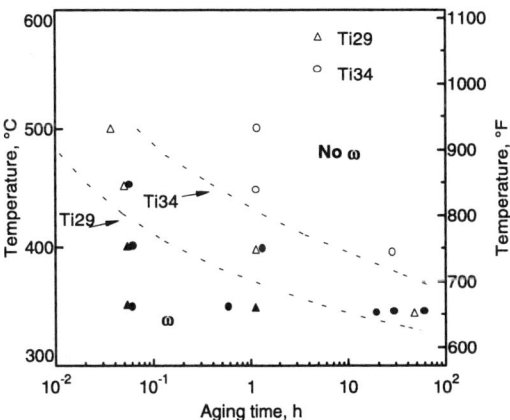

Temperature versus aging time diagram for the occurrence of ω phase in Transage 129 and 134 alloys β solution heated and water quenched.
Source: A.I.P. Nwobu, "Decomposition of Beta-Phase in Transage 134 and T129 Titanium Alloys," Ph.D. dissertation, Imperial College, London, Dec 1979

Transage 134 is less β stabilized than Transage 129 and 175 alloys. As a consequence, water

Transage 134: Transformations from β solution treatment(a)

Air cool, fan air cool, or oil quench
Quench transformation: $\beta \rightarrow \alpha'$ (<25 vol%) + β
Age transformation: $\beta + \alpha' \rightarrow \beta + \alpha$

Water quench
Quench transformation: $\beta \rightarrow \alpha''$ (100%)
Age transformation sequence:

(1) $\alpha''\beta_r \rightarrow (\omega, \alpha''_L, \alpha) + \beta$ (see note)

(2) $\alpha'' \xrightarrow{a} \alpha''_L + \alpha''_R \xrightarrow{b} \alpha + \beta$

(3) β or $\beta_r \xrightarrow{a} \omega + \beta \xrightarrow{b} \alpha''_L + \beta \xrightarrow{c} \alpha + \beta$

(4) β or $\beta_r \rightarrow \alpha + \beta \geq 450$ °C (840 °F)

Note: Cold working of α'' stabilizes it against reversion to β_r. α', hcp martensite; α'', orthorhombic martensite stress induced by water quenching or cold working (controversial as some investigators consider α'' to form athermally); α''_L, alloy-lean α''; α''_R, alloy-rich a"; β_r, beta formed by reversion from α''; ω, omega phase. Source: H.M. Flower, A.I.P. Nwobu, and D.R.F. West, Age Hardening Reactions in Transage 129 and 134, Titanium Science and Technology, Vol 3, G. Lütjering, U. Zwicker, and W. Bunk, Ed., Deutsche Gesellschaft für Metallkunde e.V., Germany, 1985, 1567-1574

---

quenching from β solution heat treatment produces 100% orthorhombic, deformation martensite (α''). Age hardening of this product is more sluggish and produces significantly lower ductility for a given strength level than age hardening of α' martensite.

Segregation during ingot solidification can be eliminated by subjecting the product to an annealing temperature of 925 °C (1700 °F) or higher for 1 h at some point in the processing schedule. Castings can be homogenized by annealing at 900 °C (1650 °F) for 2 h; this anneal may coincide with hot isostatic processing (HIP).

Corrosion

Transage 134: Cyclic polarization curve

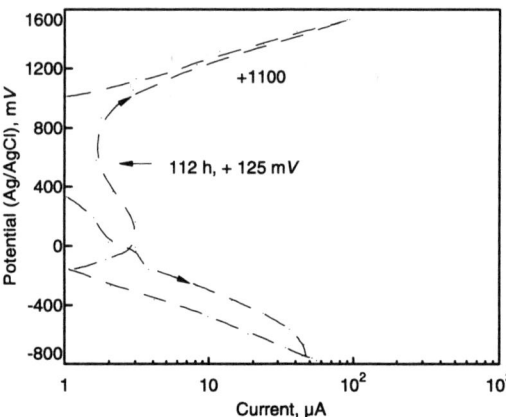

Localized corrosion resistance, measured by *in vitro* electrochemical tests in Tyrode's solution. Transage 134 exhibited a breakdown potential of +1100 mV, which was better than Hastelloy C-276 and Nitronic 50 (breakdown potentials of +460 and +40 mV, respectively).
Source: T.A. Bednarowicz, "Use of Transage 134 Titanium for Application of Human Body Implant Material," Report No. TM 373, Cameron Iron Works, 1 Apr 1981

Transage 134: Anhydrous methanol stress corrosion

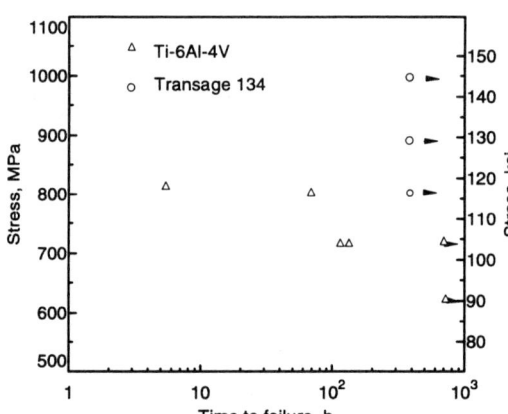

203 mm (8 in.) diam Transage 134 billet; heat treated at 815 °C (1500 °F) for 1 hr, followed by air cooling; then at 550 °C (1025 °F) for 2 hr, followed by air cooling; yield strength, 1014 MPa (147 ksi). 38 mm (1.5 in.) Ti-6Al-4V plate: heat treated at 1040 °C (1900 °F), followed by water quenching; then at 705 °C (1300 °F), followed by air cooling; yield strength, 917 MPa (133 ksi). Arrows indicate runout.
Source: F.A. Crossley, The Martensitic Transage Titanium Alloys: Their Metallurgy, Processing Characteristics and Potential Applications, *Beta Titanium Alloys in the 1980's*, R.R. Boyer and H.W. Rosenberg, Ed., TMS/AIME, 1984, p 349-386, 484-496

Transage 134: Potentials in Tyrode solution

Material	Breakdown potential		Potential, mV (Ag/AgCl)(a) Protection potential		Protection range		Corrosion potential (96 hr)
	40 V/hr	2.5 V/hr	40 V/hr	2.5 V/hr	40 V/hr	2.5 V/hr	
Transage 134	>+1250(b)	>>+1100(b)	No hysteresis		+1500	+1300	+125
C-276	+580	+460	+500	+440	+1160	+1060	−100
Nitronic 50	+300	+40	+20	−60	+680	+220	+30

(a) Potentials as a function of scan rate in Tyrode solution at 50 °C (120 °F) with a continuous purge of 95% air and 5% CO_2. (b) Water oxidation/oxygen evolution

Typical Tensile Properties

Transage 134: Ultimate tensile strength vs elongation

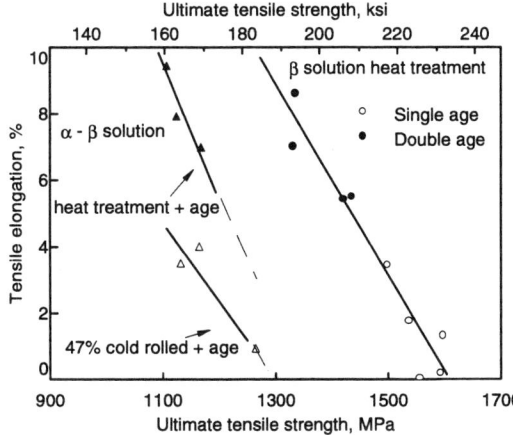

1.3 to 1.4 mm (0.050 to 0.56 in.) sheet prepared by hot rolling until final cold rolling reduction of 28 or 47%.
Source: F.A. Crossley, "220 ksi Yield Strength Transage Titanium," Final Technical Report No. LMSC/D-058737, 9/26/79-5/31/80, U.S. Naval Underwater Systems Center, Contract/PR 66604-9169-2291

Transage 134: Ultimate tensile strength vs reduction of area

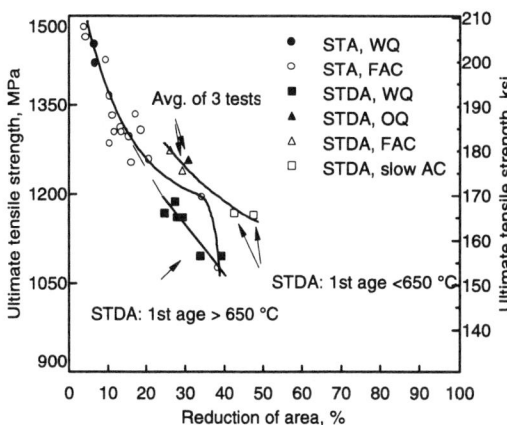

45 kg (100 lb) ingot hot rolled to 25 mm (1 in.) plate from 870 °C (1600 °F) and solution treated at 815 °C (1500 °F) for 1 hr, followed by air cooling. 273 kg (600 lb) ingot extruded to 137 mm (5.4 in.) bar from 815 °C (1500 °F), turned down to 127 mm (5 in.), then hammer forged from 870 °C (1600 °F) to 19 × 102 × 1220 mm (3/4 × 4 × 48 in.) bar. In the case of double aging, when the first age is higher than 625 °C (1160 °F), it is in fact an α-β solution heat treatment.
Source: F.A. Crossley, The Martensitic Transage Titanium Alloys: Their Metallurgy, Processing Characteristics and Potential Applications, *Beta Titanium Alloys in the 1980's*, R.R. Boyer and H.W. Rosenberg, Ed., TMS/AIME, 1984, p 349-386, 485-496

Transage 134: Typical tensile properties

Material and solution heat treatment	Aging treatment First °C (°F)	hr	Second °C (°F)	hr	No. of specimen	Ultimate tensile strength MPa	ksi	Tensile yield strength MPa	ksi	Elongation, %	Reduction of area, %	Ref(a)
1.4 mm (0.056 in.) sheet, cold rolled 28% reduction, 760 °C (1400 °F)/1/2 hr/FAC	440 (825)	24	...	...	3	1593	231	1551	225	1.3	...	1
	454 (850)	24	...	...	2	1531	222	1489	216	1.8	...	
25 mm (1 in.) plate rolled from 870 °C (1600 °F), 45 kg (100 lb) ingot, 815 °C (1500 °F)/1 hr/AC	595 (1100)	1	480 (900)	4	2	1138	165	1060	154	14	45	2
13 mm (1/2 in.) plate rolled from 815 °C (1500 °F), 273 kg (600 lb) ingot, 815 °C (1500 °F)/1 hr/AC, first aging followed by WQ	595 (1100)	1	480 (900)	4	2	1193	173	1135	165	8.5	25	3
19 × 102 × 1220 mm (3/4 × 4 × 48 in.) hammer forged bar 273 kg (600 lb) ingot, 815 °C (1500 °F)/1 hr/FAC, first aging followed by FAC	595 (1100)	1	480 (900)	4	3	1205	175	1135	165	10	29	4
51 × 51 × 76 mm (2 × 2 × 3 in.) block isothermally forged initially at 815 °C (1500 °F) and finished at 732 °C (1350 °F), 273 kg (600 lb) ingot, 815 °C (1500 °F)/1 hr/FAC	524 (975)	4	...	...	2	1197	174	1142	166	5	9	5
127 mm (5 in.) bar extruded from 815 °C (1500 °F) from 273 kg (600 lb) as-cast ingot, 815 °C (1500 °F)/24 hr/AC; 815 °C (1500 °F)/1 hr/AC (average cooling rate from 650 to 315 °C (1200 to 600 °F) 8 °C/min (14 °F/min)	705 (1300)	1	482 (900)	4	2	1289	187	1172	170	4	6.5	4
Cast-to-size and HIP bars, 6.4 mm (1/4 in.) reduced section, 815 °C (1500 °F)/	649 (1200)	1	538 (1000)	4	2	1151	167	1082	157	6.5	9.0	6
1 hr/AC, average cooling rate 35 °C/min (64 °F/min)	538 (1000)	24	...	...	2	1227	178	1165	169	5.0	6.0	
203 mm (8 in.) diam 45 kg (100 lb) forged billet, 815 °C (1500 °F)/1 hr/AC in order to estimate the age-hardening response of a 1364 kg (3000 lb) stress joint	552 (1025)	2	...	...	...	...	...	1014	147	...	...	4

(a) Ref 1: F.A. Crossley, "220 ksi Yield Strength Transage Titanium," Final Technical Report No. LMSC/D-058737, 9/26/79-5/31/80, U.S. Naval Underwater Systems Center, Contract/PR-66604-9169-2291. Ref 2: F.A. Crossley and R.H. Jeal, Fatigue and Fracture Behavior of the High Hardenability Martensitic "Transage" Titanium Alloys, *21st Structures, Structural Dynamics, and Materials Conference*, Part 2, AIAA, 1980, p 572-577; also *J. Aircraft*, Vol 18(No. 8), Aug 1981, p 683-686. Ref 3: F.A. Crossley, Effects of Process and Heat Treatment Variables on the Mechanical Properties of Transage 134 Alloy (Ti-2.5Al-12V-2Sn-6Zr), *Overcoming Material Boundaries*, Vol 17, SAMPE, 1985, p 190-199. Ref 4: F.A. Crossley, The Martensitic Transage Titanium Alloys: Their Metallurgy, Processing Characteristics and Potential Applications, *Beta Titanium Alloys in the 1980's*, R.R. Boyer and H.W. Rosenberg, Ed., TMS/AIME, 1984, p 349-386, 485-496. Ref 5: T. Gannon and S.W. McClaren, "Development of Advanced Navy Aircraft Landing Gear Structures," Final Report, Naval Air Systems Command, Contract N00019-82-0318, 28 Feb 1984. Ref 6: F.A. Crossley and W.J. Barice, Mechanical Properties of Two Cast and Isostatically Processed Martensitic Transage Titanium Alloys, *J. Met.*, Vol 33(No. 2), Feb 1981, p 26-32

Transage 134: Tensile yield strength versus reduction of area

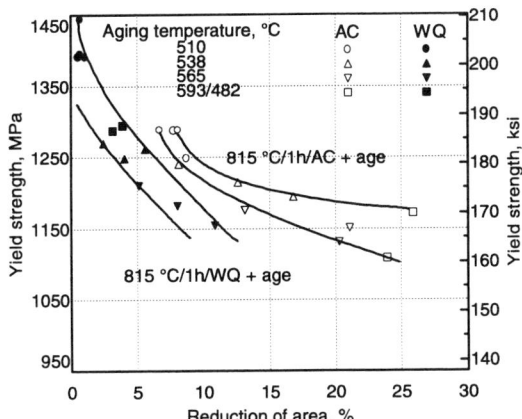

Beta solution anneal with water quench and age is compared with β solution anneal, air cool, and age. In the former case, the quench product is orthorhombic (stress-induced) martensite (α″); in the latter case, it is hcp martensite (α′). 273 kg (600 lb) ingot extruded to 137 mm (5.4 in.) bar from 815 °C (1500 °F), turned down to 127 mm (5 in.), then rolled to 13 mm (1/2 in.) plate from 815 °C (1500 °F).
Source: F.A. Crossley, "Vought Titanium Landing Gear Program Analysis of Results for Transage 134 (Ti-2.5Al-12V-2Sn-6Zr)," 1985

Transage 134: Comparison of age-hardening kinetics

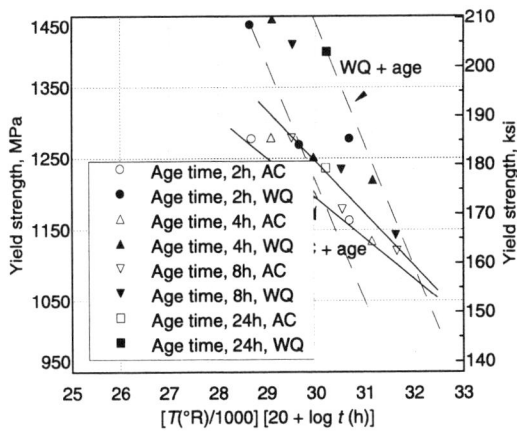

Yield strength versus the Larson-Miller parameter (for combining temperature and time in a single parameter to compare aging behaviors of Transage 134 as water quenched and as air cooled from the β phase field. 273 kg (600 lb) ingot extruded to 137 mm (5.4 in.) bar from 815 °C (1500 °F), turned down to 127 mm (5.0 in.), then rolled to 13 mm (1/2 in.) plate from 815 °C (1500 °F).
Source: F.A. Crossley, private communication, June 1992

Fatigue Properties

Transage 134: Smooth-specimen axial fatigue data ($R = 0$, $A = 1.0$)

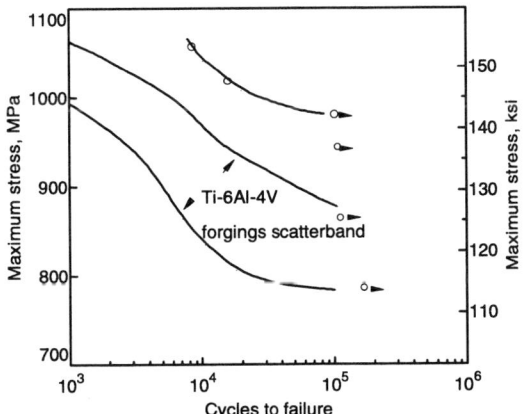

45 kg (100 lb) ingot hot rolled to 25 mm (1 in.) plate from 870 °C (1600 °F); solution treated at 815 °C (1500 °F) for 1 hr, followed by air cooling, then aged at 595 °C (1100 °F) for 1 hr, followed by air cooling, plus 480 °C (900 °F) for 4 hr, followed by air cooling. Data determined by Rolls-Royce, Aero Division.
Source: F.A. Crossley, The Martensitic Transage Titanium Alloys: Their Metallurgy, Processing Characteristics and Potential Applications, *Beta Titanium Alloys in the 1980's*, R.R. Boyer and H.W. Rosenberg, Ed., TMS/AIME, 1984, p 349-386, 485-496

Transage 134: V-notch, round bar fatigue data ($R = -1.0$, $A = \infty$)

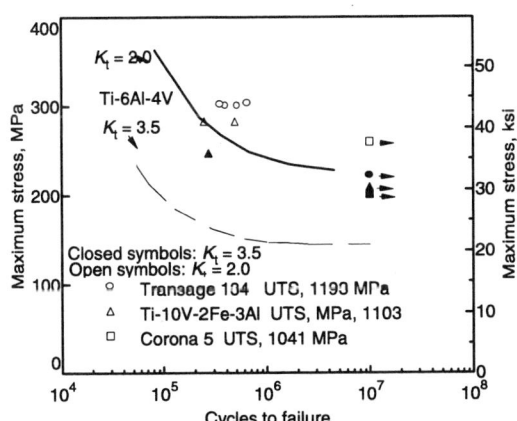

All materials were 51×51×76 mm (2×2×3 in.) isothermally forged blocks heat treated as follows: Corona 5: 857 °C (1575 °F)/4 hr/FAC, 538 °C (1000 °F))/4 hr/AC; Ti-10V-2Fe-3Al: 779 °C (1435 °F)/1 hr/FAC, 771 °C (1420 °F)/ 1 hr/WQ, 510 °C (950 °F)/ 8 hr/AC; Transage 134: 815 °C (1500 °F)/1 hr/FAC, 524 °C (975 °F)/, 4 hr/AC.
Source: T. Gannon and S.W. McClaren, "Development of Advanced Navy Aircraft Landing Gear Structures," Final Report, Naval Air Systems Command, Contract N00019-82-0318, 28 Feb 1984

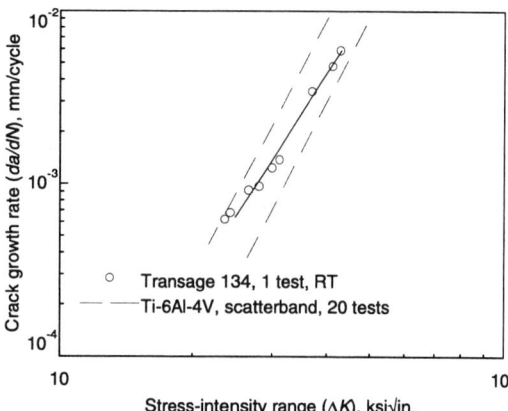

Transage 134: Crack growth rate under 300-s hold condition (R = 0.1)

45 kg (100 lb) ingot hot rolled to 25 mm (1 in.) plate from 870 °C (1600 °F); heat treated at 815 °C (1500 °F)/1 hr/slow AC, 595 °C (1100 °F)/1 hr/AC, 480 °C (900 °F)/5 hr/AC.
Source: F.A. Crossley and R.H. Jeal, Fatigue and Fracture Behavior of the High Hardenability Martensitic "Transage" Titanium Alloys, *21st Structures, Structural Dynamics, and Materials Conference*, Part 2, AIAA, 1980, p 572-577; also *J. Aircraft*, Vol 18(No. 8), Aug 1981, p 683-686

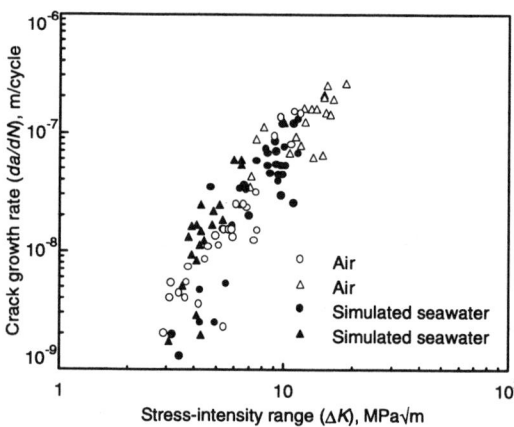

Transage 134: Crack growth rates in air and seawater (R = 0.1)

273 kg (600 lb) ingot extruded to 137 mm (5.4 in.) bar from 815 °C (1500 °F), turned down to 127 mm (5 in.), and homogenized at 815 °C (1500 °F) for 24 hr, followed by air cooling. Yield strength, 1170 MPa (170 ksi). Frequency, 3 to 6 Hz heat treated at 815 °C (1500 °F) / hr / AC, 700 °C (1300 °F) / hr / AC, 480 °C (900 °F) / 4 hr / AC.
Source: F.A. Crossley, The Martensitic Transage Titanium Alloys: Their Metallurgy, Processing Characteristics and Potential Applications, *Beta Titanium Alloys in the 1980's*, R.R. Boyer and H.W. Rosenberg, Ed., TMS/AIME, 1984, p 349-386, 485-496

Fracture Toughness

Transage 134: Fracture toughness for various heat treatments

Primary anneal		Cooling method	Final age(a)	Ultimate tensile strength		Yield strength		Elongation, %	Reduction of area, %	Fracture toughness	
Temperature °C (°F)	Time hr			MPa	ksi	MPa	ksi			MPa√m	ksi √in.
593 (1100)	2	WQ	1	1151	167	1082	157	9	25	66.9	60.8
690 (1275)	1	WQ	1	1276	185	1200	174	4	7	31.2	28.4
690 (1275)	1	AC	1	1310	190	1186	172	7	15	36.4	33.1
705 (1300)	1	AC	2	1303	189	1227	178	3	5	...	...
565 (1050)	1	AC	2	1269	184	1213	176	4	7	55.4	50.4

(a) Final age: (1) 454 °C (850 °F)/24 hr/AC; (2) 482 °C (900 °F)/4 hr/AC. 13 mm (1/2 in.) rolled plate, solution heat treated at 815 °C (1500 °F) for 1 hr, followed by air cooling; mechanical properties for secondary α-β solution anneal plus age versus double age. Source: F.A. Crossley, Effects of Process and Heat Treatment Variables on the Mechanical Properties of Transage 134 Alloy (Ti-2.5Al-12V-2Sn-6Zr), *Overcoming Material Boundaries*, Vol 17, SAMPE, 1985, p 190-199

Transage 134: Impact toughness and fracture toughness

Specimen No.	Direction(a)	Ultimate tensile strength		Yield strength		Elongation in 50 mm (2 in.), %	Reduction of area, %	Fracture toughness		Charpy impact energy	
		MPa	ksi	MPa	ksi			MPa√m	ksi √in.	J	ft · lbf
1	L	1197	173.6	1150	166.8	4.6	6.9	59.7	54.3	13.6	10.0
2	L	1182	171.4	1133	164.3	5.9	10.9	58.6(b)	53.3(b)	16.3	12.0
3	T	1207	175.1	1160	168.3	5.5	11.2	52.5	47.5	...	...
4								56.8	51.7	...	...

(a) L, longitudinal; T, transverse. (b) Value questioned since P_Q taken as 2338 kg (5144 lb), while load versus deflection curve indicates that P_Q should have been abut 3273 kg (7200 lb). Extruded bar turned down to 127 mm (5 in.); then isothermally forged initially at 815 °C (1500 °F) and finally at 730 °C (1350 °F) to 51 × 51 × 76 mm (2 × 2 × 3 in.) block. Heat treatment: 815 °C (1500 °F)/1 hr/FAC; 524 °C (975 °F)/4 hr/AC. Source: F.A. Crossley, "Vought Titanium Landing Gear Program Analysis of Results for Transage 134 (Ti-2.5Al-12V-2Sn-6Zr)," 1985

Transage 134: Tensile properties of extruded bar

	Ultimate tensile strength		Yield strength		Elongation,	Reduction of area,
	MPa	ksi	MPa	ksi	%	%
Aged: 635 °C (1175 °F)/2 hr/AC						
	896	130	834	121	11.5	33
Aged: 705 °C (1300 °F)/1 hr/C; 538 °C (1000 °F)/2 hr/AC						
	1089	158	1020	148	6	16
Aged: 705 °C (1300 °F)/1 hr/AC; 482 °C (900 °F)/4 hr/AC						
	1289	187	1172	170	4	6.5

Extruded bar turned down to 127 mm (5.0 in.); 815 °C (1500 °F)/24 hr/AC. Solution heat treated as 127 mm diam by 279 mm (5.0 × 11 in.) 6 bar: 815 °C (1500 °F) for 1 hr, followed by air cooling (average cooling rate from 650 to 315 °C, or 1200 to 600 °F, of 8 °C/min, or 14 °F/min). Aged as 67 mm (2 5/8 in.) thick sections. Source: F.A. Crossley, The Martensitic Transage Titanium Alloys: Their Metallurgy, Processing Characteristics and Potential Applications, *Beta Titanium Alloys in the 1980's*, R.R. Boyer and H.W. Rosenberg, Ed., TMS/AIME, 1984, p 349-386, 485-496

Transage 134: Notched tensile properties

Condition	Ultimate tensile strength		0.2% yield strength		Elongation,	Reduction of area,	Fracture toughness		Notched ultimate tensile strength ($K_t \approx 5$)		Notched UTS/UTS	Notched UTS/YS
	MPa	ksi	MPa	ksi	%	%	MPa√m	ksi√in.	MPa	ksi		
Oil quenched												
Longitudinal	1225	178	1155	168	10.0	30.5	59(a)	53(a)	1620	235	1.32	1.40
Transverse	1280	186	1225	178	7.7	20.1	...	...	1640	238	1.28	1.33
Fan air cooled												
Longitudinal	1205	175	1135	165	10.3	29.0	...	...	1670	242	1.38	1.47
Transverse	1260	183	1230	178	5.0(b)	22.7	...	...	1670	242	1.32	1.35

Note: Tension test properties are averages of three specimens. (a) LT orientation. (b) All three specimens broke near punched gage marks; individual values were 5, 5, and 5. Extruded bar turned down to 127 mm (5 in.), then hammer forged from 870 °C (1600 °F) to 19 × 102 mm (3/4 × 4 in.). Heat treatment: 815 °C (1500 °F)/1 hr/OQ or FAC; 595 °C (1100 °F)/1 hr/OQ or FAC; 480 °C (900 °F)/4 hr/AC. Source: F.A. Crossley, The Martensitic Transage Titanium Alloys: Their Metallurgy, Processing Characteristics and Potential Applications, *Beta Titanium Alloys in the 1980's*, R.R. Boyer and H.W. Rosenberg, Ed., TMS/AIME, 1984, p 349-386, 485-496

Mechanical Properties of Bar

Transage 134: Tensile properties of cast-to-size bars

Test temperature		Aging treatment					Time, hr	No. of tests	Ultimate tensile strength		0.2% yield strength		Elongation, %	Reduction of area, %
°C	°F	Temperature		Time, hr	Temperature				MPa	ksi	MPa	ksi		
		°C	°F		°C	°F								
25	77	...	...	...	538	1000	3	3	1310	190	1269	184	2.3	3.7
		...	...	...	566	1050	2	3	1234	179	1186	172	3.5	6.0
		...	...	...	538	1000	24	2	1227	178	1165	169	4.0	6.0
		593	1100	1	482	900	4	3	1220	177	1165	169	3.7	6.3
		649	1200	1	482	900	24	3	1179	171	1138	165	3.7	7.3
		649	1200	1	538	1000	4	2	1151	167	1082	157	6.5	9.0
					593	1100	2	2	1110	161	1041	151	6.5	9.0

Cast-to-size and hot isostatically processed at 815 °C (1500 °F) for 2 hr; at 103 MPa (15 ksi); 6.4 mm (1/4 in.) reduced section. Heat treatment: 815 °C (1500 °F) for 1 hr, followed by air cooling (average cooling rate from 815 to 315 °C (1500 to 600 °F) 35 °C/min (64 °F/min). Source: F.A. Crossley and W.J. Barice, Mechanical Properties of Two Cast and Isostatically Processed Martensitic Transage Titanium Alloys, *J. Met.*, Vol 33(No. 2), Feb 1981, p 26-32

Heat Treatment

Although less than that of Transage 129 and 175 alloys, Transage 134 has exceptional hardenability. For example, a 45 kg (100 lb), 200 mm (8 in.) diam billet was air cooled from β solution heat treatment and age hardened uniformly to a yield strength of 1014 MPa (147 ksi).

For maximum fracture toughness and fatigue resistance at strength levels higher than 1240 MPa (180 ksi), continuous grain-boundary must be controlled. Continuous α in prior-β grain boundaries results when a cooling rate from β solution annealing slower than a water quench is followed by an α-β solution anneal. This grain-boundary α is detrimental to fracture toughness. Continuous grain-boundary α does not result when an α-β solution anneal follows a water quench from β solution anneal because of the differences in the mechanisms of age hardening between orthorhombic (stress-induced) martensite (α″) and hexagonal close-packed (hcp) (α′).

Transage 134: Recommended heat treatments

Treatment	Temperature °C	Temperature °F	Duration	Cooling method
Beta solution anneal	815	1500	0.25-1 hr	AC or OQ(a)
Age for maximum strength (1170-1585 MPa, or 170-230 ksi)	440-525	825-975	24 hr below 480 °C (895 °F), or 4 hr for higher temperatures	AC
Double age for maximum toughness:				
First age	550-595	1025-1100	2 hr	WQ
Second age	455-485	850-900	24 hr	AC
Treatment for maximum toughness and fatigue resistance:				
Solution anneal	815	1500	0.25-1 hr	WQ
α/β anneal	650-700	1200-1300	1 hr	AC or OQ(a)
Age	455 or 480	850-900	24 hr	AC

(a) Oil quench (OQ) for heavier sections, but do not water quench (WQ).

Transage 134: Effects of heat treatment variables on double-aged mechanical properties

Variable	Direction(a)	Ultimate tensile strength MPa	Ultimate tensile strength ksi	Yield strength MPa	Yield strength ksi	Elongation, %	Reduction of area, %	Fracture toughness MPa√m	Fracture toughness ksi√in.
870 °C vs 815 °C/1 hr/AC; 565 °C/1 hr/AC; 480 °C/4 hr/AC									
870 °C (1600 °F)	L	1200	174	1138	165	7.0	17	53.1	48.3
815 °C (1500 °F)	L	1269	184	1213	176	4.0	7	55.4	50.4
Net change	...	−69	−10	−76	−11	+3.0	+10	−2.3	−2.1
870 °C/1 hr/AC; 565 °C/1 hr/WQ vs AC; 480 °C/4 hr/AC									
WQ	L	1248	181	1200	174	7.5	12	55.3	50.4
AC	L	1200	174	1138	165	7.0	17	52.7	48.0
Net change	...	+48	+7	+62	+9	+0.5	−5	+2.6	+2.4
815 °C/1 hr/FAC; 565 °C vs 595 °C/2 hr/WQ; 455 °C/24 hr/AC									
565 °C (1050 °F)	T	1207	175	1131	164	9.0	33	51.1	46.5
595 °C (1100 °F)	T	1131	164	1069	155	10.0	33	50.3	45.8
Net change	...	+76	+11	+62	+9	−1.0	0	+0.8	+0.7

Note: Conclusions from this limited study are that for higher strength and slightly higher fracture toughness, (1) solution anneal at 815 °C (1500 °F) rather than at 870 °C (1600 °F), (2) first temperature age at 565 °C (1050 °F) rather than at 595 °C (1100 °F), and (3) follow with water quenching rather than air cooling. (a) L, longitudinal; T, transverse. Source: F.A. Crossley, "Effects of Process and Heat Treatment Variables on the Mechanical Properties of Transage 134 Alloy (Ti-2.5Al-12V-2Sn-6Zr)," *Overcoming Material Boundaries*, Vol 17, SAMPE, 1985, p 190-199

Ti-13V-2.7Al-7Sn-2Zr

Common Name: Transage 175
Trade Names: Transage 175 and T175
UNS Number: Unassigned

Frank A. Crossley

Transage 175 is an experimental age-hardenable, high-strength titanium-base β alloy. It can be produced in all mill product forms and has good castability and excellent weld repair capability by titanium alloy standards. The alloy has net shape isothermal forging capability.

Typically, Transage 175 is aged to strength levels of 1240 MPa (180 ksi) or higher. In common with other Transage alloys, Transage 175 has exceptionally high hardenability. Uniform age hardening is obtainable in heavy sections that are air cooled from (beta) solution heat treatment to achieve strengths of 1240 MPa (180 ksi) or higher. Development of Transage 175 has been pursued to improve structural efficiency.

In common with other Transage alloys, Transage 175 has a unique age hardening mechanism. In wrought products in both the solution heat treated and the aged conditions, microstructures are so fine that transmission electron microscopy is needed to resolve them. On cooling from beta solution heat treatment, whether by water quenching or a relatively rapid furnace cool, the alloy partially transforms to a submicroscopic martensite (α′). Complete transformation is apparently blocked by the generation of forest dislocations during transformation. On heating into the age hardening temperature range of 425 to 565 °C (800 to 1050 °F), dislocations rapidly annihilate, and the transformation product α grows at the expense of retained β. This results in a very fine and uniform Widmanstätten distribution of α in a β matrix.

Segregation during ingot solidification is eliminated by subjecting the product to an anneal of 925 °C (1700 °F) or higher in the processing schedule. Castings can be homogenized by an anneal of 900 °C (1650 °F) for 2 h; this anneal may coincide with hot isostatic processing (HIP).

Transage 175 can be used to improve structural efficiency and reduce cost in all applications for which titanium alloys are used. Compared to the most commonly used titanium alloys, it can extend the temperature range of application. Transage 175 exhibits particularly good fatigue resistance in wrought and cast forms. It has demonstrated true endurance limits under various types of fatigue loading and at various temperatures. For example, wrought Transage 175 under load-controlled axial fatigue, $R = 0$, exhibited the same 4×10^5 cycle runout stress of 830 MPa (120 ksi) at 425 °C (800 °F) as at room temperature.

Transage 175: Chemical composition

Composition, wt%	Specification requirements	
	Transage 175, wrought	Transage 175C, castings
Aluminum	2.2-3.2	2.0-3.0
Carbon	0.08 max	0.08 max
Iron	0.20 max	0.20 max
Nitrogen	0.05 max	0.05 max
Oxygen	0.15 max	0.15 max
Tin	6.5-7.5	6.5-7.5
Vanadium(a)	12.0-14.0	11.0-13.0
Zirconium	1.5-2.5	1.5-2.5
Boron	0.03 max	0.03 max
Hydrogen	0.015 max	0.015 max
Yttrium	0.005 max	0.005 mx
Residual elements, each	0.10 max	0.10 max
Residual elements, total	0.4	0.40 max
Titanium	bal	bal

(a) The vanadium-aluminum (nominally 15 to 17 wt% aluminum) master alloy addition is to be calculated to obtain the nominal vanadium content of 13.0 wt% for wrought products and 12.0 wt% for electrode stock for castings.

Mechanical Properties

Transage 175 bar: Typical tensile properties

Typical tensile properties of 13 mm (0.5 in.) extruded bar, solution treated and aged at 815 °C (1500 °F), 1 h, air cooled, 480 °C (900 °F), 24 h, air cooled

Test temperature		Tensile yield strength (0.2%)		Ultimate tensile strength		Elongation, %	Reduction of area, %
°C	°F	MPa	ksi	MPa	ksi		
24	76	1248	181	1304	189	10	39
260	500	992	144	1136	165	10	50
427	800	926	134	1081	157	10	56

Cast Transage 175: Typical tensile properties

Test sample	Temperature		Tensile yield strength (0.2%)		Ultimate tensile strength		Elongation, %	Reduction of area, %
	°C	°F	MPa	ksi	MPa	ksi		
Base of 10 in. (255 mm) diameter impeller	24	76	1158	168	1200	174	4	5.5
Cast-to-size tensile test bars	24	76	1103	160	1150	167	6	11
	121	250	1031	150	1114	162	4	16
	260	500	999	145	1100	160	4	18
	427	800	841	122	1026	149	6	23

Transage 175: Modulus of elasticity

Material	Aging treatment	Measurement(a)	Test temperature		No. of specimens	Modulus of elasticity	
			°C	°F		GPa	10^6 psi
Cast-to-size bars	540 °C (1000 °F), 24 h, AC	Ext/strain gage	25	77	3	105	15.2
		Strain gage	120	250	2	92	13.4
			260	500	2	86	12.5
		Ext	425	800	2	77	11.2
12 mm (0.47 in.) extruded bar	510 °C (950 °F), 24 h, AC	Ext	25	77	2	99.3	14.4
	540 °C (1000 °F), 24 h, AC		540	1000	2	71.7	10.4
					2	97.8	14.0
	482 °C (900 °F)	Strain gage	25	77	7	100.0 ± 2.0	14.5 ± 0.3
			425	800	6	83.0 ± 1.0	12.03 ± 0.14

(a) Ext, extensometer. Source: See figure on next page.

Transage 175: Directionality of properties

Beta solution heat treatment followed by subsequent martensitic transformation tends to annihilate prior texture. For example, Transage 175 sheet that was unidirectionally cold rolled for the final 47% reduction and heat treated at 760 °C (1400 °F), 1/2 h, fan air cooled, 480 °C (900 °F), 48 h, AC had tensile properties as follows:

Orientation	Ultimate tensile strength		Tensile yield strength		Elongation, %
	MPa	ksi	MPa	ksi	
Longitudinal	1475	214	1344	195	6
Transverse	1393	202	1310	190	5

Source: F.A. Crossley, private communication, June 1992

Transage 175: Typical mechanical properties

Specimens	Ultimate tensile strength		Tensile yield strength		Elongation, %	Reduction of area, %	Hardness	
	MPa	ksi	MPa	ksi			HRA	HRC
Cast-to-size + HIP bar 815 °C (1500 °F), AC, 540 °C (1000 °F), 24 h, AC	1151	167	1103	160	5.7	11.0	...	39±1
Sheet 760 °C (1400 °F), 1/2 h, fan air cooled, 440 °C (825 °F), 4 h, AC	1407	204	1289	187	2	...	72.5	44(a)
760 °C (1400 °F), 1/2 h, fan air cooled, 440 °C (825 °F), 8 h, AC	1434	208	1379	200	1.7	...	73	45(a)
440 °C (825 °F), 24 h, AC	1455	211	1413	205	1.8	...	73.5	46(a)

(a) Conversion from Rockwell A scale

Transage 175: Typical tensile properties

Material	Aging temperature, °C	Time, h	No. of specimens	Ultimate tensile strength MPa	ksi	Tensile yield strength MPa	ksi	Elongation, %	Reduction of area, %
368 mm (14.5 in.) dome isothermally forged at 815 °C (1500 °F) and 0.21 mm/s (0.5 in./min)	510	24	41(a)	1418 ± 34	206 ± 5	1364 ± 32	198 ± 5	5.6 ± 0.9	11.5 ± 2.6
229 mm (9 in.) wheel isothermally forged at 815 °C (1500 °F) and 0.16 mm/s (0.38 in./min)	510	24	4	1338	194	1284	186	5.8	15.6
356 mm (14.0 in.) diameter by 1.8 mm (0.070 in.) cylinder spin forged from 815 °C (1500 °F)	510+ 524	2 +22	2	1341	194	1270	184	5	...
12 mm (0.5 in.) diameter extruded bar	510	24	2	1254	182	1185	172	15	44
Cast-to-size and HIP bar, 6.4 mm (0.25 in.) reduced section; cooled from solution heat treatment at 0.59 °C/s (64 °F/min)	538	24	3	1151	167	1103	160	5.7	11
Cast-to-size and HIP bar, 6.4 mm (0.25 in.) reduced section; cooled from solution heat treatment at 4.2 °C/s (450 °F/min)	538 579	24 24	3 2	1291 1213	187 176	1264 1124	183 163	2.0 3.0	4.3 10
267 mm (10.5 in.) diameter cast and HIP impeller	538	2	5	1216	176	1208	175	1.1	2.2
Same as impeller above except that solution heat treated to homogenize at 900 °C (1650 °F), 2 h	538	2	3	1201	174	1160	168	4.3	5.5
102 mm (4.0 in.) diameter bar produced from 818 kg (1800 lb) ingot, water quenched	510	24	2	1282	186	1227	178	6.5	...

Note: All materials were solution heat treated at 815 °C (1500 °F) followed by air cooling or slower rate, except as indicated. (a) The data represent two 818 kg (1800 lb) 483 mm (19 in.) diameter ingots extruded in the as-cast state from 815 °C (1500 °F) to 173 mm (6.8 in.) round (87% reduction) and processed as follows: Ingot No. 1: Upset 20 to 30% and redrawn from 1095 °C (2000 °F); upset 20 to 30% and redrawn from 925 °C (1700 °F); upset 20 to 30% and redrawn from 720 °C (1325 °F); upset 20 to 30% and redrawn from 925 °C (1700 °F); and finally upset 20 to 30% and redrawn from 720 °C (1325 °F). Annealed at 925 °C (1700 °F), 1 h, AC. Upset to 305 mm (12 in.) diameter pancake workpiece from 815 °C (1500 °F) (73% reduction). Ingot No. 2: Annealed at 925 °C (1700 °F), 1 h, AC. Upset to 305 mm (12 in.) diameter pancake workpiece from 815 °C (1500 °F) (68% reduction). The extensive processing of Ingot. No. 1 made only slight improvements in some tensile properties. For example, the property most improved was reduction of area, which was increased by 16%. On the other hand, yield strength was decreased by a negligible 0.13%. Therefore, the tensile data were treated collectively. Also, the specimens represent radial and tangential directions in the dome and the axial direction in the skirt of the dome. Extensive primary breakdown processing is unnecessary and unwarranted.

High-Temperature Strength

Transage 175: Tensile strength vs temperature for extruded bar

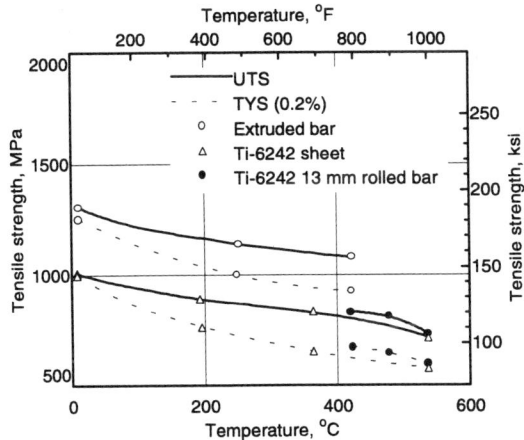

Specimens were cast-to-size test bars and 13 mm (0.5 in.) diameter bar extruded from 815 °C (1500 °F) at a reduction ratio of 20:1, heat treated at 815 °C (1500 °F), 1/4 h, AC, 480 °C (900 °F), 24 h, AC.
Source: F.A. Crossley, Elevated Temperature Mechanical Properties of Transage 175 Alloy [Ti-2.7Al-13V-7Sn-2Zr], SAMPE Quart., Vol 17 (No. 3), Apr 1986, p 5-12

Transage 175: Modulus of elasticity vs temperature

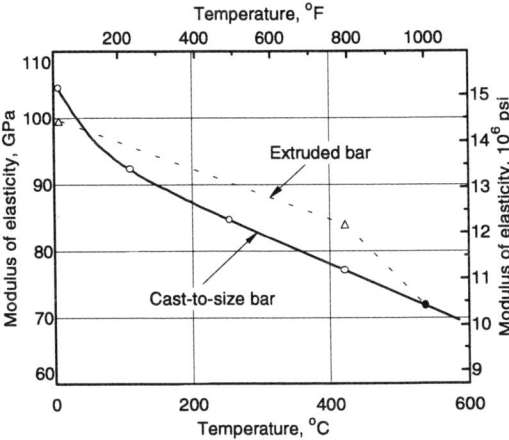

Specimens were cast-to-size test bar and 13 mm (0.5 in.) diameter extruded bar. Cast-to-size 6.4 mm (1/4 in.) reduced section, test bars hot isostatically processed at 815 °C (1500 °F), 2 h, and 103 MPa (15 ksi) and heat treated at 815 °C (1500 °F), 1 h, AC, 540 °C (1000 °F), 24 h, AC. 13 mm (0.5 in.) bar extruded from 815 °C at a reduction ratio of 20:1, heat treated at 815 °C (1500 °F), 1/4 h, AC, 480 °C (900 °F), 24 h, AC.
Source: F.A. Crossley and W.J. Barice, "Cast Transage 175 Alloy for Durability Critical Structural Components," 22nd Structures, Structural Dynamics, and Materials Conference, Paper No. 81-0535-CP reprinted from CP 811, A Bound Collection of Papers, AIAA, New York, 1981, p 134-140; also J. Aircraft, Vol 20 (No. 1), Jan 1983, p 66-69; and F.A. Crossley, "Elevated Temperature Mechanical Properties of Transage 175 Alloy [Ti-2.7Al-13V-7Sn-2Zr]," SAMPE Quart., Vol 17 (No. 3), Apr 1986, p 5-12; Data at 540 °C (1000 °F), private communication, June 1992

Transage 175: Elevated-temperature tension test properties

Test temperature		Ultimate tensile strength		Tensile yield strength (0.2%)		Elongation, %	Reduction of area, %
°C	°F	MPa	ksi	MPa	ksi		
24(a)	76	1300	189	1220	177	12	36
24	76	1292	187.4	1238	179.6	9.2	42.0
		1316	190.9	1259	182.6	8.8	39.3
Avg		1304	189	1248	181	9	41
260	500	1136	164.7	1017	147.5	10.1	54.7
		1136	164.7	967	140.2	9.1	46.0
Avg		1136	165	992	144	9.5	50
425	800	1089	158.0	923	133.8	10.2	56.0
		1073	155.6	929	134.7	10.7	56.9
Avg		1081	157	926	134	10.5	56

Note: Specimens were 12 mm (0.47 in.) extruded bar heat treated at 815 °C (1500 °F), 1/4 h, AC, 480 °C (900 °F), 24 h, AC. (a) Lockheed tests, average of two test results; all others by Metcut Research Associates, Inc., Cincinnati, Ohio. Source: *SAMPE Quart.*, Apr 1986

Transage 175: Tensile strength vs temperature

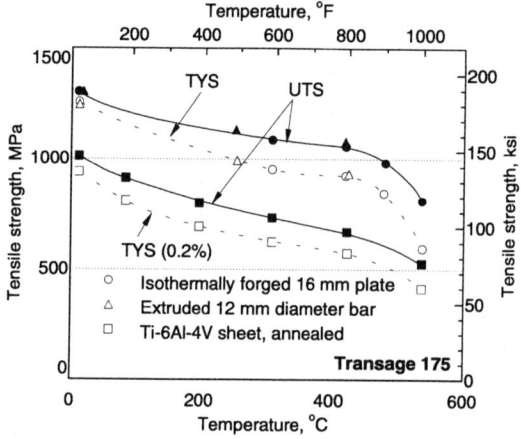

Comparison of tensile strength vs temperature of Transage 175 with Ti-6Al-4V annealed sheet. Specimens were 16 mm (5/8 in.) plate isothermally forged at 815 °C (1500 °F), heat treated at 815 °C (1500 °F) 1 h, AC, 510 °C (950 °F), 24 h, AC. Extruded bar heat treated at 815 °C (1500 °F), 1/4 h, AC, 480 °C (900 °F), 24 h, AC. Source: F.A. Crossley, private communication, June 1992

Transage 175: Tensile properties vs temperature

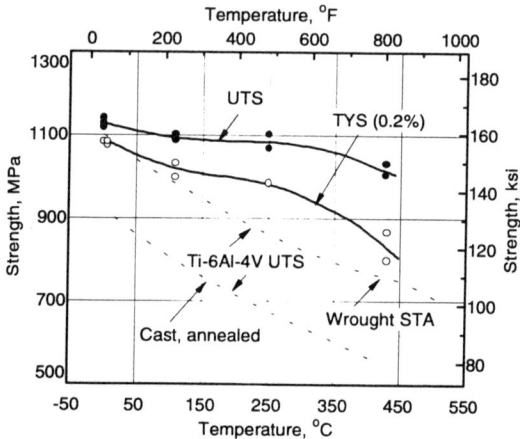

Tensile test properties of cast plus HIP Transage 175 vs temperature compared with the ultimate tensile strength of Ti-6Al-4V in cast annealed, and wrought, solution heat treated and aged forms. Specimens were cast-to-size bar, 6.4 mm (0.25 in.) reduced section, HIP at 815 °C (1500 °F), 2 h, 103 MPa (15 ksi), and heat treated at 815 °C (1500 °F), 1 h, cooling rate of 0.59 °C/s (64 °F/min), 540 °C (1000 °F), 24 h, AC.
Source: F.A. Crossley and W.J. Barice, "Cast Transage 175 for Durability Critical Structural Components," 22nd Structures, Structural Dynamics, and Materials Conference, Paper No. 81-0535-CP reprinted from CP 811, A Bound Collection of Technical Papers, AIAA, New York, Apr 1981; also *J. Aircraft*, Vol 20 (No. 1), Jan 1983, p 66-69

Transage 175: Creep-rupture comparison at 425 °C

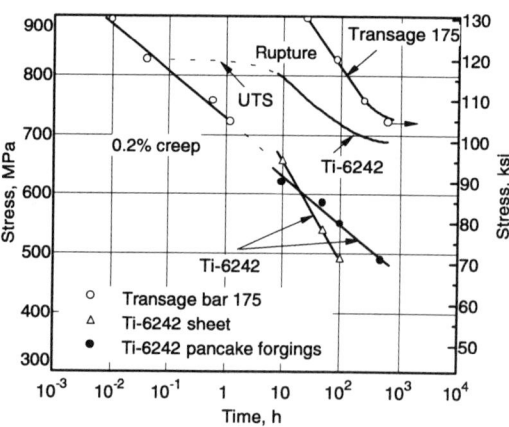

Creep-rupture and creep curves for Transage 175 compared with Ti-6Al-2Sn-4Zr-2Mo at 425 °C (800 °F). Specimens were 13 mm (0.5 in.) diameter bar extruded from 815 °C (1500 °F) at reduction ratio of 20:1, heat treated at 815 °C (1500 °F), 1/4 h, AC, 480 °C (900 °F), 24 h, AC.
Source: F.A. Crossley, Elevated Temperature Mechanical Properties of Transage 175 Alloy [Ti-2.7Al-13V-7Sn-2Zr], *SAMPE Quart.*, Vol 17 (No. 3), Apr 1986, p 5-12

Transage 175: High-temperature tensile properties

Test temperature		First age temperature		Age time, h	Second age temperature		Age time, h	No. of tests	Ultimate tensile strength		Tensile yield strength (0.2%)		Elongation, %	Reduction of area, %
°C	°F	°C	°F		°C	°F			MPa	ksi	MPa	ksi		
25	77	...	...	...	540	1000	24	3	1151	167	1103	160	5.7	11.0
		595	1100	1	480	900	8	3	1131	164	1041	151	7.0	14.7
					565	1050	24	3	1048	152	979	142	6.7	10.3
121	250	...	...	...	540	1000	24	2	1117	162	1034	150	4.0	16.0
260	500	...	...	...	540	1000	24	2	1103	160	1000	145	4.2	17.6
427	800	...	...	...	540	1000	24	2	1027	149	883	128	6.1	22.5

Note: Tension test properties of cast-to-size plus hot isostatically processed bar. Solution heat treated at 815 °C (1500 °F), 1 h, AC; average cooling rate from 815 to 315 °C (1500 to 600 °F) 0.59 °C/s (64 °F/min). Source: *JOM*, Feb 1981, p 27

Fatigue

Transage 175: Axial fatigue test properties

Material	Test temperature		K_t	Control	A	R	Maximum stress, MPa (ksi), or strain (%) for cycles:			Frequency, Hz
	°C	°F					10^5	10^6	10^7	
1—13 mm (0.5 in.) bar extruded from 815 °C (1500 °F) at a reduction ratio of 20:1	427	800	1.0	Load	1.0	0	...	840 (122)	...	60
				Strain	1.0	0	1.22	...	...	0.33
2—230 mm (9 in.) diameter wheel isothermally forged from α-β preform at 705 °C (1300 °F) and 0.42 mm/s (1.0 in./min) platen speed; heat treated at 720 °C (1325 °F), 2 h, AC, 480 °C (900 °F), 24 h, AC	24	76	1.0	Load	0.98	0.01	910 (132)	910 (132)	...	60
3—230 mm (9 in.) diameter wheel isothermally forged from β preform at 815 °C (1500 °F) and 0.21 mm/s (0.5 in./min) platen speed; heat treated at 815 °C (1500 °F), 1 h, AC, 510 °C (950 °F), 24 h, AC	24	76	1.0	Load	0.98	0.01	840 (122)	840 (122)	...	60
4—Cast-to-size test bar, HIP 815 °C (1500 °F), 2 h, 103 MPa (15 ksi); heat treated at 815 °C (1500 °F), 1 h, AC 4.2 °C/s (450 °F/min), 580 °C (1075 °F), 2 h, AC	120	250	1.0	Load	∞	−1.0	...	378 (55)	378 (55)	60
			3.0	Load	∞	−1.0	...	189 (27.5)	189 (27.5)	60
			1.0	Strain	1.0	0	0.67	...	...	0.33
	260	500	1.0	Strain	1.0	0	0.97	...	...	0.33

Source: 1. F.A. Crossley, Elevated Temperature Mechanical Properties of Transage 175 Alloy [Ti-2.7Al-13V-7Sn-2Zr], *SAMPE Quart.*, Vol 17 (No. 3), Apr 1986, p 5-12. 2. F.A. Crossley, G.L. Tingley, J.R. Becker, and V.A. Shende, "Process Development and Mechanical Properties of a Near-Net-Shape Forging of Transage 175 Alloy (Ti-2.7Al-13V-7Sn-2Zr)," presented in closed session at the 17th Nat. SAMPE Tech. Conf., Oct 22-24, 1985, Kiamesha Lake, NY; and G.L. Tingley, "Isothermal Forging and Shear Spinning of Transage 175 Titanium Alloy," Report No. LMSC/D941061, Lockheed Missiles and Space Company, Inc., Sunnyvale, CA, under contract N00123-81C-0349, Naval Sea Systems Command, Washington, DC; technical monitor, P. Jung, Naval Ocean Systems Center, San Diego, CA, 30 April 1985, not released. 3. Same as Source 2. 4. F.A. Crossley and W.J. Barice, Cast Transage 175 Titanium for Durability Critical Structural Components, *J. Aircraft*, Vol 20 (No. 1), Jan 1983, p 66-69.

Wrought Bar

Transage 175: Low-cycle strain-controlled axial fatigue at 400 and 425 °C for STA extruded bar

Temperature (°C) °F	Strain at termination, %			Stress at termination (MPa) ksi			Cycles to failure
	Total	Elastic	Plastic	Total range	Tensile	Compressive	
(400) 750	1.22	1.190	0.025	(992) 144	(717) 104	(276) 40	>112,000
(425) 800	1.05	1.030	0.020	(868) 126	(613) 89	(255) 37	>115,000
	1.11	1.095	0.015	(896) 130	(620) 90	(276) 40	108,000
	1.22	1.200	0.023	(1006) 146	(627) 91	(379) 55	142,000

Note: Specimens were extruded bar, 13 mm (0.5 in.) diameter, from 815 °C (1500 °F) at a reduction ratio of 20:1. Heat treated at 815 °C (1500 °F), 1/4 h, AC, 480 °C (900 °F), 24 h, AC. Triangular waveform at 0.33 Hz. Smooth bar testing (R = 0). Source: F.A. Crossley, Elevated Temperature Mechanical Properties of Transage 175 Alloy [Ti-2.7Al-13V-7Sn-2Zr], *SAMPE Quart.*, Vol 17 (No. 3), Apr 1986, p 5-12

Transage 175: Load-controlled fatigue at 425 °C

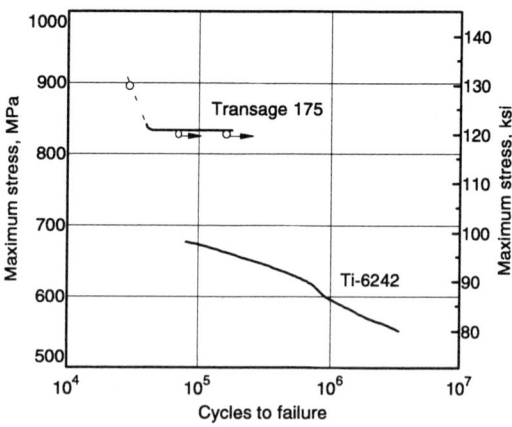

Load-controlled, smooth-bar axial fatigue for Transage 175 at 425 °C (800 °F) compared with Ti-6Al-2Sn-4Zr-2Mo. Specimens were extruded bar, 13 mm (0.5 in.) diameter, from 815 °C (1500 °F) at a reduction ratio of 20:1. Heat treated at 815 °C (1500 °F), 1/4 h, AC, 480 °C (900 °F), 24 h, AC. Triangular wave form. $K_t = 1$. Transage 175 tested at $R = 0$, $A = 1$, 0.33 Hz. Ti-6242 specimens were 1.0 mm (0.040 in.) sheet tested at $R = 0.01$, $A = 0.98$, 41.7 Hz.
Source: F.A. Crossley, Elevated Temperature Mechanical Properties of Transage 175 Alloy [Ti-2.7Al-13V-7Sn-2Zr], *SAMPE Quart.*, Vol 17 (No. 3), Apr 1986, p 5-12

Transage 175: Effect of processing on axial high-cycle fatigue

Specimen no.	Stress level		Life(a), cycles
	MPa	ksi	
$\alpha + \beta$ preform(b)			
Rerun 5	1172	170	14,800
1	1103	160	21,600
2	1034	150	45,000
Rerun 4	965	140	4.96×10^4
3	930	135	3.68×10^4
4	896	130	1.201×10^6 (runout)
5	827	120	5.375×10^5 (runout)
β preform(b)			
1	1100	160	7,800
2	1035	150	10,500
3	965	140	15,400
Rerun 6	895	130	27,900
Rerun 5	860	125	26,300
4	825	120	15,010(c)
5	825	120	4.28×10^5 (runout)
6	760	110	4.53×10^5 (runout)

(a) $R = 0.01$. To maximize specimen usage, runout specimens were rerun at higher stress. (b) See accompanying figure for processing details of $\alpha + \beta$ and β preforms. (c) Premature fracture due to defect

Forgings

Transage 175: Effect of processing on high-cycle fatigue

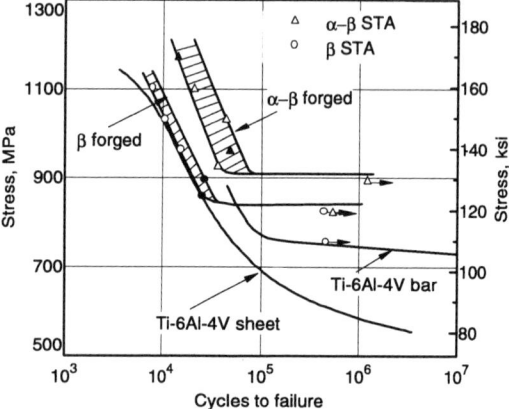

High-cycle fatigue for α-β forged and heat treated and β forged and heat treated conditions compared with Ti-6Al-4V. Specimens were isothermally, near-net-shape forged wheels, 230 mm (9 in.) in diameter. α-β processed material consisted of α-β preforms isothermally forged at 705 °C (1300 °F) at a platen speed of 0.42 mm/s (1.0 in./min) and heat treated at 720 °C (1325 °F), 2 h, AC, 480 °C (900 °F), 24 h, AC. β processed material consisted of β preforms isothermally forged at 815 °C (1500 °F) and 0.21 mm/s (0.5 in./min) and heat treated at 815 °C (1500 °F), 1 h, AC, 510 °C (950 °F), 24 h, AC. Ti-6Al-4V bar was 31.7 mm (1.25 in.) in diameter, annealed. Ti-6Al-4V sheet was 1.6 and 3.1 mm (0.063 and 0.125 in.) thick. Closed symbols are retests of runout specimens. $R = 0.01$.
Source: F.A. Crossley, G.L. Tingley, J.R. Becker, and V.A. Shende, "Process Development and Mechanical Properties of a Near-Net-Shape Forging of Transage 175 Alloy (Ti-2.7Al-13V-7Sn-2Zr)," presented in closed session at the 17th Nat. SAMPE Tech. Conf., Oct 22-24, 1985, Kiamesha Lake, NY

Castings

Transage 175: Stability of tensile properties under conditions of elevated-temperature fatigue

Exposure conditions	Tensile yield strength		Ultimate tensile strength		Elongation %	Reduction of area, %
	MPa	ksi	MPa	ksi		
None (avg 3)(a)	1103	160	1145	167	6	11
Strain-controlled fatigue at 120 °C (250 °F) to 100,000 cycles runout (avg 2)	1230	178	1272	184	4	7.5
None (single test)	1123	163	1213	176	3	10
Strain-controlled fatigue at 260 °C (500 °F) to 100,000 cycles runout (avg 3)	1134	164	1185	172	4	10

(a) Number of tests averaged

Transage 175: Low-cycle axial fatigue of cast-to-size bars at 120 and 260 °C

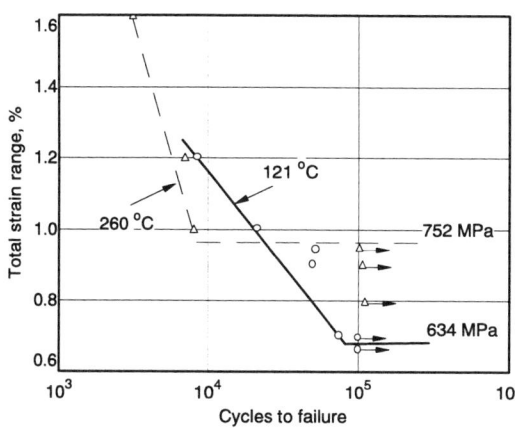

Strain-controlled, smooth bar, low-cycle fatigue curves for cast-to-size plus HIP specimens heat treated at 815 °C (1500 °F), 1 h, cooled at rate of 4.2 °C/s (450 °F/min), 580 °C (1075 °F), 2 h, AC. $R = 0$, $A = 1.0$, $K_t = 1.0$.
Source: F.A. Crossley and W.J. Barice, Cast Transage 175 Titanium Alloy for Durability Critical Structural Components, *J. Aircraft*, Vol 20 (No. 1), Jan 1983, p 66-69; and F.A. Crossley, The Martensitic Transage Titanium Alloys: Their Metallurgy, Processing Characteristics and Potential Applications, *Beta Titanium Alloys in the 1980's*, R.R. Boyer and H.W. Rosenberg, Ed., TMS/AIME, 1984, p 349-386, 485-496

Transage 175: High-cycle fatigue of cast-to-size bars at 120 °C

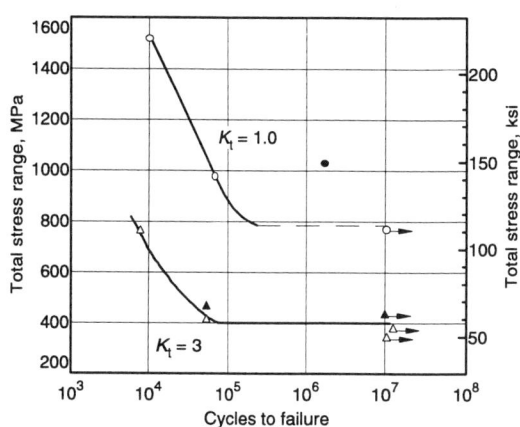

Load-controlled, smooth and notched bar, axial, $R = -1.0$ high-cycle fatigue for cast-to-size plus HIP at 815 °C (1500 °F), 2 h, 103 MPa (15 ksi) specimens heat treated at 815 °C (1500 °F), 1 h, cooled at rate of 4.2 °C/s (450 °F/min), 580 °C (1075 °F), 2 h, AC. Closed symbols are retests of 10^7 cycle run outs.
Source: F.A. Crossley and W.J. Barice, Cast Transage 175 Titanium Alloy for Durability Critical Structural Components, *J. Aircraft*, Vol 20 (No. 1), Jan 1983, p 66-69, and F.A. Crossley, The Martensitic Transage Titanium Alloys: Their Metallurgy, Processing Characteristics and Potential Applications, *Beta Titanium Alloys in the 1980's*, R.R. Boyer and H.W. Rosenberg, Ed., TMS/AIME, 1984, p 349-386

Forming

Extrusion. Transage 175 has excellent extrudability. Cameron Iron Works, Inc., Houston, extruded two 818 kg (1800 lb), 483 mm (19 in.) diameter, as-cast ingots (having no surface preparation) to 173 mm (6.3 in.) diameter bar for a reduction ratio of 6.8:1. The ingots were canned in copper and extruded from 815 °C (1500 °F). Machined 50 mm (2.0 in.) diameter billets, jacketed in copper, were extruded from 815 °C to 12 mm (0.5 in.) diameter bar using a graphite and oil lubricant. The reduction ratio was 20:1, and the extrusion constant was 510 MPa (37 tsi).

Sheet Forming. All cold and hot forming methods generally used for titanium alloys may be applied. It also has exceptional superplastic forming capability. Optimum temperature of superplastic forming is 815 °C (1500 °F) and optimum strain rates are 1.7×10^{-2} s^{-1} or higher.

Forging. Transage 175 has excellent forgeability. It has near-net-shape and net-shape capabilities by isothermal forging at temperatures as low as 650 °C (1200 °F). However, considering relevant factors such as load requirements, die life, and uniformity of mechanical properties, the optimum parameters are a temperature of 815 °C (1500 °F) and a platen speed of 0.21 mm/s (0.5 in./min). In hammer forging, the workpiece temperature should not fall below 700 °C (1300 °F).

Transage 175: Recommendations for the production of near-net-shape and net-shape forgings

Equipment	Isothermal press
Temperature of dies and work piece	815 °C ± 15 (1500 ± 25 °F)
Platen speed	0.21 mm/s (0.5 in./min)
Solution heat treatment	
Temperature	815 °C ± 15 °C (1500 ± 25 °F)
Time	1/2 to 1 h
Cooling rate	
Heavy sections	Fan air cool, or water quench
Light sections	Air cool
Aging treatment	
Temperature	450 to 540 ± 5 °C (850 to 1000 ± 10 °F), depending on strength desired. Strength range from 1520 to 1170 MPa (220 to 170 ksi)
Time	4 h for short-time strength; 24 h for long-time strength at elevated temperatures and aging temperatures below 480 °C (900 °F)
Cooling rate	Air cool

Note: At 815 °C (1500 °F), Transage 175 has a flow stress of 43 MPa (6200 psi). The alloy will flow at constant load as long as the load per unit plan area exceeds the flow stress. If shape is to be net, chemical mill to remove surface contamination and to meet drawing dimensions.

Transage 175: Effect of platen speed on mechanical properties

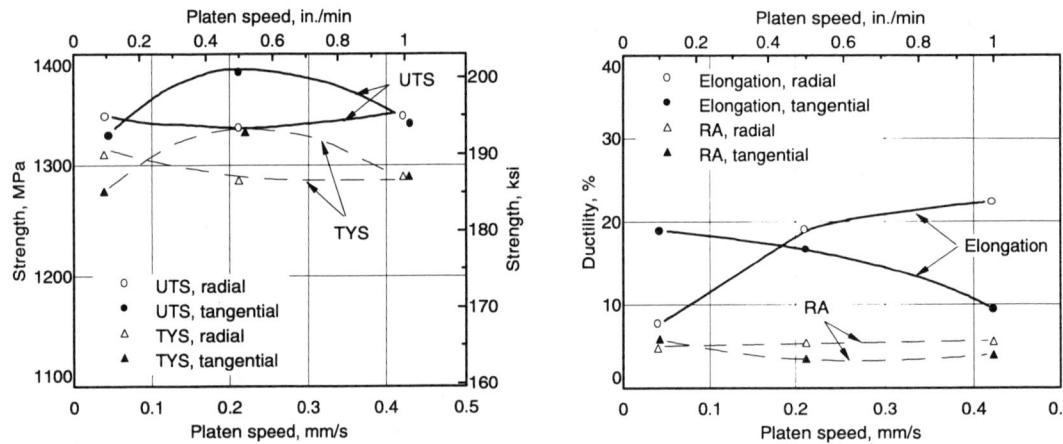

Tension test properties versus platen speed for Transage 175 β preforms isothermally forged at 815 °C (1500 °F) upset to a reduction of 62%. Specimens were 818 kg (1800 lb) ingots processed to 160 mm (6.3 in.) bar, then cogged to 100 mm (4.0 in.) bar at 815 °C (1500 °F) for 58% reduction and annealed 815 °C, 1 h, to produce β preform stock. Preforms isothermally forged at 815 °C at various platen speeds and heat treated at 815 °C, 1 h, WQ, 525 °C (975 °F), 24 h, AC.
Source: G.L. Tingley, "Isothermal Forging and Shear Spinning of Transage 175 Titanium Alloy," Report No. LMSC/D941061, Lockheed Missiles and Space Company, under contract N00123-81C-0349, Naval Sea Systems Command, Washington, DC, technical monitor, Peter Jung, Naval Ocean Systems Center, San Diego, CA, 30 Apr 1985, not released; and F.A. Crossley, The Martensitic Transage Titanium Alloys: Their Metallurgy, Processing Characteristics and Potential Applications, *Beta Titanium Alloys in the 1980's*, R.R. Boyer and H.W. Rosenberg, Ed., TMS/AIME, 1984, p 349-386, 485-496

Transage 175: Effect of forging temperature on mechanical properties

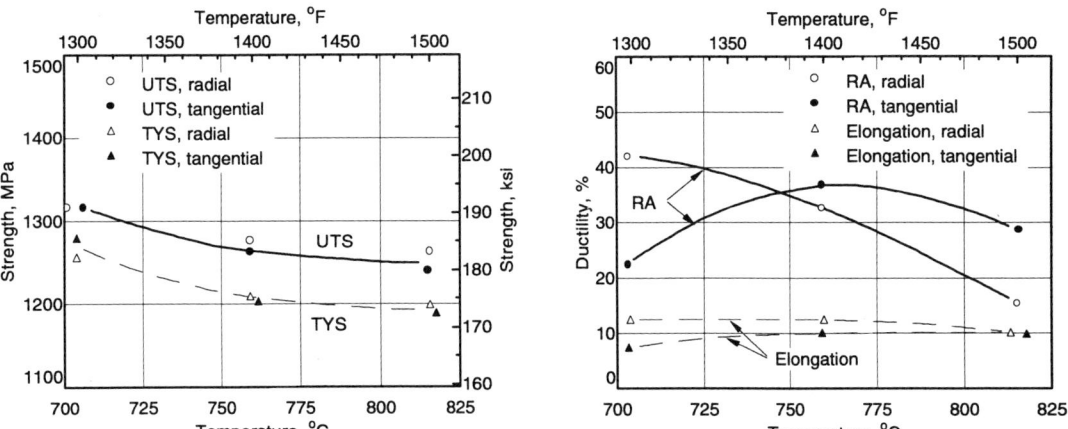

Tension test properties versus isothermal forging temperature of α + β preforms upset to 62% reduction. Specimens were 818 kg (1800 lb) ingots processed to 160 mm (6.3 in.) round, then cogged to 100 mm (4.0 in.) round at 730 °C (1350 °F) for 58% reduction to make α-β preform stock. Preforms were upset isothermally at various temperatures and 0.42 mm/s (1.0 in./min) platen speed. Heat treated at 720 °C (1325 °F), 2 h, WQ, 480 °C (900 °F), 24 h, AC.
Source: G.L. Tingley, "Isothermal Forging and Shear Spinning of Transage 175 Titanium Alloy," Report No. LMSC/D941061, Lockheed Missiles and Space Company, under contract N00123-81C-0349, Naval Sea Systems Command, Washington, DC, technical monitor, Peter Jung, Naval Ocean Systems Center, San Diego, CA, 30 Apr 1985, and F.A. Crossley, The Martensitic Transage Titanium Alloys: Their Metallurgy, Processing Characteristics and Potential Applications, *Beta Titanium Alloys in the 1980's*, R.R. Boyer and H.W. Rosenberg, Ed., TMS/AIME, 1984, p 349-386, 485-496

Transage 175: Tensile flow stress vs temperature

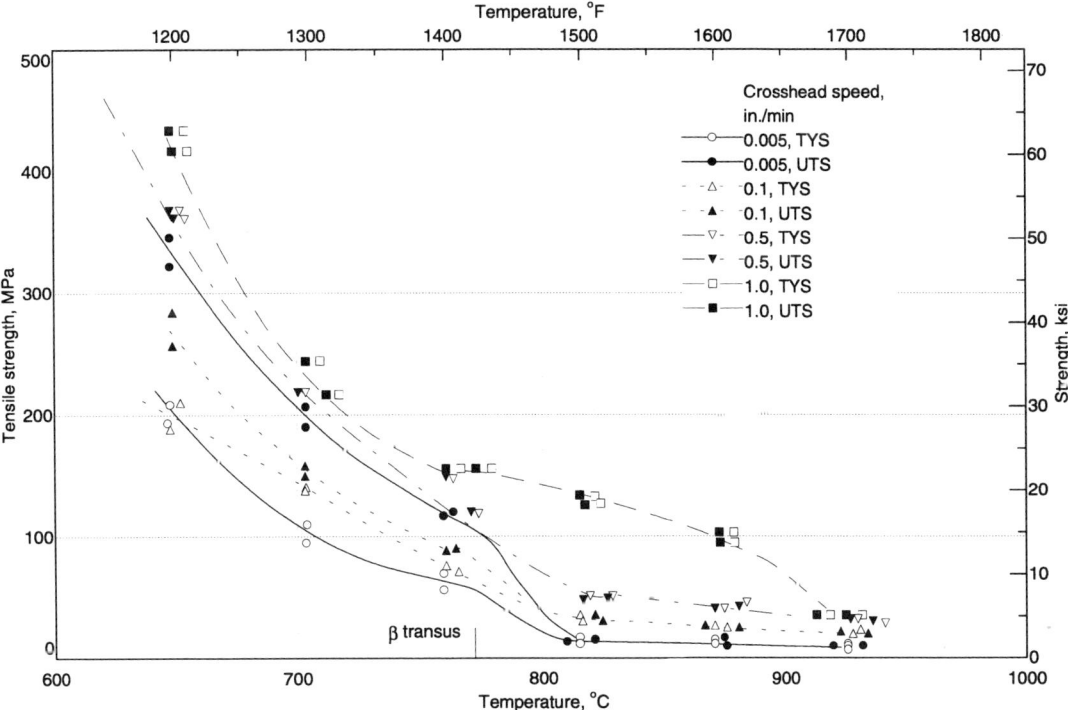

Ultimate tensile and tensile yield strength vs temperature with varying crosshead speed for β preform material. Specimens were 160 mm (6.3 in.) diameter bar processed from 818 kg (1800 lb) cast and extruded ingots, and annealed at 925 °C (1700 °F), 1 h, AC.
Source: F.A. Crossley, G.L. Tingley, J.R. Becker, and V.A. Shende, "Process Development and Mechanical Properties of a Near-Net-Shape Forging of Transage 175 Alloy (Ti-2.7Al-13V-7Sn-2Zr)," presented in closed session at the 17th Nat. SAMPE Tech. Conf., Oct 22-24, 1985, Kiamesha Lake, NY

Transage 175: Flow stress vs strain rate for various temperatures

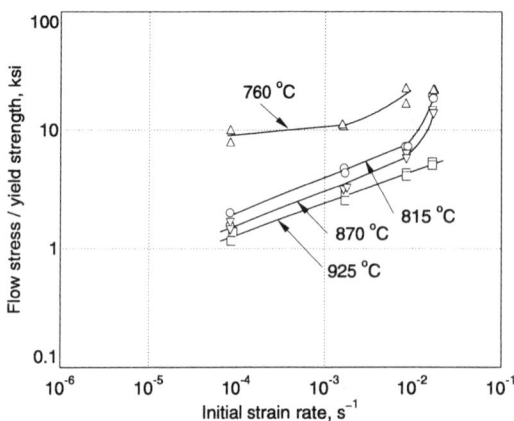

Flow stress (0.2% yield strength) vs initial strain rate for constant crosshead speed hot tensile tests. Specimens were 160 mm (6.3 in.) diameter bar processed from 818 kg (1800 lb) cast and extruded ingots and annealed at 925 °C (1700 °F), 1 h, AC. β transus is 765 °C (1410 °F).
Source: F.A. Crossley, G.L. Tingley, J.R. Becker, and V.A. Shende, "Process Development and Mechanical Properties of a Near-Net-Shape Forging of Transage 175 Alloy (Ti-2.7Al-13V-7Sn-2Zr)," presented in closed session at the 17th Nat. SAMPE Tech. Conf., Oct 22-24, 1985, Kiamesha Lake, NY

Forging Properties

Transage 175: Mechanical properties of α-β forged wheels

Room-temperature mechanical properties of wheel forgings produced using an α-β preform forged at 705 °C (1300 °F) and 0.42 mm/s (1.0 in./min) and heat treated at 720 °C (1325 °F), 2 h, AC, 480 °C (900 °F), 24 h, AC

Tensile specimen location	Forging number	Tensile yield strength (0.2%) MPa	ksi	Ultimate tensile strength MPa	ksi	Elongation, %	Reduction of area, %
Hub tang	1	1258	182.4	1318	191.2	2.0	5.5
Hub axial		1230	174.8	1296	188.0	6.5	23.8
Rim axial		1302	188.8	1382	200.5	2.6	7.0
Web tang		1247	180.8	1327	192.5	6.9	23.8
Web radial		1244	180.4	1327	192.5	7.5	21.0
Web radial		1258	182.4	1330	192.9	8.1	22.4
Web radial	2	1219	176.8	1282	186.0	8.5	29.2
Web radial(a)	3	1258	182.4	1313	190.5	7.5	20.3

(a) Fracture toughness (K_{Ic}) of 23.1 MPa√m (21.0 ksi√in.) in R-T direction and 23.5 MPa √m (21.4 ksi√in.) in T-R direction. Source: F.A. Crossley, G.L. Tingley, J.R. Becker, and V.A. Shende, "Process Development and Mechanical Properties of a Near-Net-Shape Forging of Transage 175 Alloy (Ti-2.7Al-13V-7Sn-2Zr)," presented in closed session at the 17th Nat. SAMPE Tech. Conf., Oct 22-24, 1985, Kiamesha Lake, NY

Transage 175: Mechanical properties of β forged wheels

Room-temperature mechanical properties of wheel forgings produced by using a β preform forged at 815 °C (1500 °F) and 0.21 mm/s (0.5 in./min) and heat treated at 815 °C (1500 °F), 1 h, AC, 510 °C (950 °F), 24 h, AC

Tensile specimen location	Forging number	Tensile yield strength (0.2%) MPa	ksi	Ultimate tensile strength MPa	ksi	Elongation, %	Reduction of area, %
Hub tang	1	1321	191.6	1393	202.1	2.0	3.9
Hub axial		1313	190.4	1380	200.1	4.2	12.3
Rim axial		1312	190.4	1382	200.5	4.3	18.1
Web tang		1258	182.4	1327	192.5	5.9	16.7
Web radial		1285	186.4	1341	194.5	6.1	13.8
Web radial		1274	184.8	1341	194.5	6.3	18.1
Web radial	2	1271	184.4	1327	192.5	5.5	16.7
Web radial(a)	3	1299	188.4	1344	194.9	5.5	13.8

(a) Fracture toughness (K_{Ic}) of 31.4 MPa√m (28.5 ksi√in.) in R-T direction and 30.9 MPa√m (28.1 ksi√in.) in T-R direction. Source: F.A. Crossley, G.L. Tingley, J.R. Becker, and V.A. Shende, "Process Development and Mechanical Properties of a Near-Net-Shape Forging of Transage 175 Alloy (Ti-2.7Al-13V-7Sn-2Zr)," presented in closed session at the 17th Nat. SAMPE Tech. Conf., Oct 22-24, 1985, Kiamesha Lake, NY

Heat Treatment

Two types of heat treatments may be applied to Transage 175: (1) solution anneal, preferably at 815 °C (1500 °F) for 1/4 to 1 h followed by fan air cooling for thin sections to be cold formed, or followed by any convenient cooling rate in preparation for age hardening; and (2) solution anneal followed by aging at 425 to 563 °C (800 to 1050 °F) depending on strength level desired. For a given aging temperature, the slower the cooling rate from the solution anneal, the lower the yield strength obtained. This type of heat treatment produces yield strength up to 1450 MPa (210 ksi).

The annealing temperature required to achieve the lowest strength state is 815 °C (1500 °F) for 1/4 to 1 h followed by rapid air cool or water quench depending on section size. However, from the standpoint of stress relief, age hardening, per se, eliminates residual stress due to the unique age-hardening mechanism of Transage titanium alloys. Consequently, partial or total age hardening facilitates machining because the workpiece is more stable geometrically as metal is removed.

Yield strength can vary from 895 to 1450 MPa (130 to 210 ksi) in inverse relation to age hardening temperature. For a given component, it may be necessary to determine the aging temperature by trial to achieve a desired combination of strength and ductility and/or toughness.

Transage 175 has exceptional hardenability. The alloy readily age hardens to strength levels of 1170 MPa (170 ksi) or higher, following the slow cooling rates imposed by hot isostatic processing facilities and by superplastic forming operations. In steel terms, the ideal round size exceeds 200 mm (8.0 in.).

Transage 175: Effect of forging parameters on mechanical properties

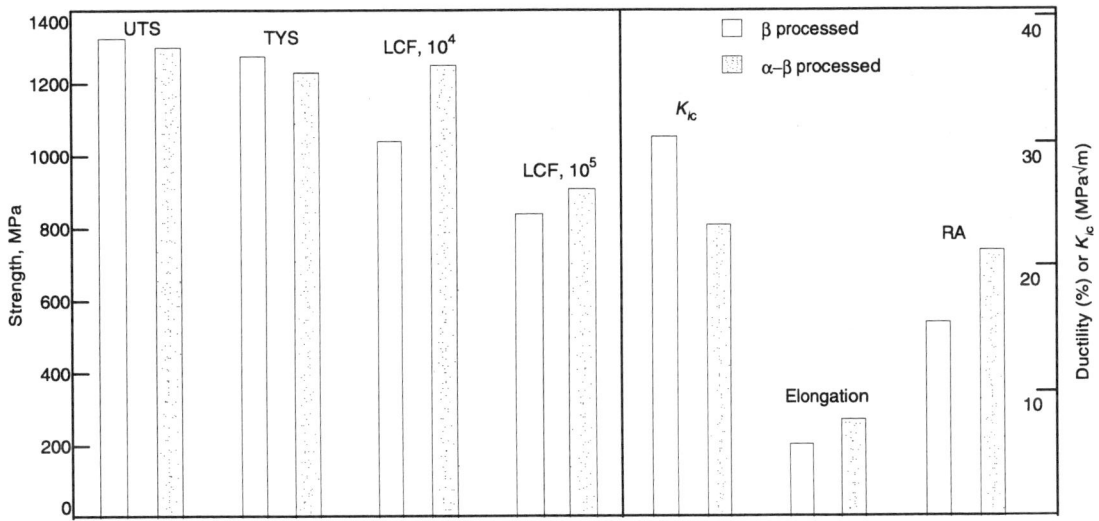

Comparison of mechanical properties between α-β processed and β processed wheel forgings.
α-β processed material consisted of α-β preforms isothermally forged at 705 °C (1300 °F) and 0.42 mm/s (1.0 in./min) and heat treated at 720 °C (1325 °F), 2 h, AC, 480 °C (900 °F), 24 h, AC. β processed material consisted of β preforms, isothermally forged at 815 °C (1500 °F) and 0.21 mm/s (0.5 in./min) and heat treated at 815 °C (1500 °F), 1 h, AC, 510 °C (950 °F), 24 h, AC.
Source: F.A. Crossley, G.L. Tingley, J.R. Becker, and V.A. Shende, "Process Development and Mechanical Properties of a Near-Net-Shape Forging of Transage 175 Alloy (Ti-2.7Al-13V-7Sn-2Zr)," presented in closed session at the 17th Nat. SAMPE Tech. Conf., Oct 22-24, 1985, Kiamesha Lake, NY

Transage 175: Typical heat treatments

STA wrought bar
815 °C (1500 °F) for 1 h, cooling rate optional from air cool to water quench depending on section size, age at 455 °C (850 °F) or higher temperature, depending on strength desired and application temperature, for 2 to 24 h

Castings
STA: 900 °C (1650 °F) for 2 h, air cool, fan air cool, or forced gas cool, age at 540 °C (1000 °F) for 2 h, air cool
HIP: 900 °C (1650 °F), 103 MPa (15 ksi) for 2 h, forced gas cool, age at 540 °C (1000 °F) for 2 h, air cool

Transage 175: Effect of solution temperature
RT tensile properties of cast impeller after aging at 540 °C (1000 °F), 2 h, AC

Ultimate tensile strength		Tensile yield strength (0.2%)		Elongation, %	Reduction of area, %
MPa	ksi	MPa	ksi		
No homogenization(a)					
1193	173	1186	172	0.3	0.8
1261	183	1261	183	0.2	0.8
1193	173	1172	170	3.9	9.4
1165	169	1165	169	0.4	0.0
1268	184	1255	182	0.9	0.0
845 °C (1550 °F), 2 h(b)					
1152	167.1	1137	164.9	3.9	5.2
1133	164.3	1096	159.0	3.5	4.8
1215	176.3	1194	173.2	3.0	3.2
870 °C (1600 °F), 2 h(b)					
1110	161.0	...	...	3.1	2.4
1211	175.7	1195	173.4	2.9	2.4
1262	183.1	1235	179.1	3.3	3.2
890 °C (1650 °F), 2 h(b)					
1202	174.4	1161	168.5	4.1	3.2
1196	173.5	1163	168.7	4.6	7.8
1204	174.7	1155	167.5	4.3	5.6
925 °C (1700 °F), 2 h(b)					
1135	164.7	1089	158.0	3.1	4.0
1147	166.4	1107	160.6	3.6	3.2
1226	177.9	1192	172.9	3.7	3.2

(a) Baseline; no homogenization; 815 °C (1500 °F), 2 h, gas fan cool. (b) Plus furnace cooled to 540 °C (1000 °F), then removed from furnace and air cooled. Solution heat treated at 815 °C (1500 °F), 2 h, gas fan cooled and aged 540 °C (1000 °F), 2 h, AC

Weldments

Transage 175: Tensile properties of electron-beam welded specimens
Tensile properties of specimens containing transverse EB weldments, post-weld aged at 525 °C (975 °F), 22 h, AC

No. of EB passes	Tensile yield strength (0.2%)		Ultimate tensile strength		Elongation, %	Fracture site
	MPa	ksi	MPa	ksi		
None	1256	182.1	1330	192.9	5	...
	1284	186.3	1352	196.1	5	...
Avg.	1270	184.2	1341	194.5	5	...
One	1293	187.5	1340	194.4	4	BM
	1340	194.4	1374	199.3	4	BM
Avg.	1317	191.0	1357	196.8	4	...
Two	1317	191.0	1350	195.8	5	BM
	1310	190.0	1384	200.7	5	BM
Avg.	1313	190.5	1367	198.2	5	...

Note: The material was a 356 mm (14 in.) diameter spin-forged cylinder machined to 1.8 mm (0.070 in.) thickness. The material was preaged 510 °C (950 °F), 2 h, AC for stability in machining, welded, and post-weld aged 525 °C (975 °F), 22 h, AC. Test specimens were axial, and the gage dimensions were 6.4 × 25 mm (1/4 × 1 in.). Source: F.A. Crossley, G.L. Tingley, J.R. Becker, and V.A. Shende, "Process Development and Mechanical Properties of a Near-Net-Shape Forging of Transage 175 Alloy (Ti-2.7Al-13V-7Sn-2Zr)," presented in closed session at the 17th Nat. SAMPE Tech. Conf., Oct 22-24, 1985, Kiamesha Lake, NY

Transage 175: Tensile properties vs solution treatment temperature of cast impeller

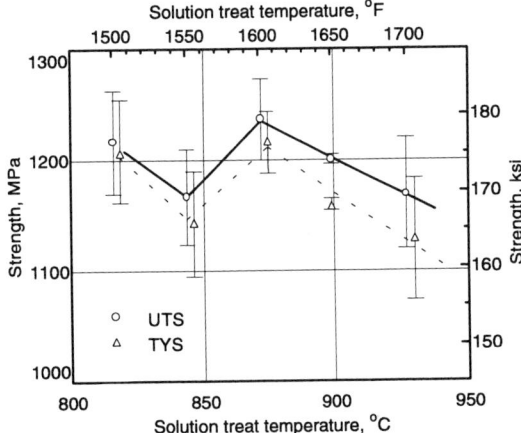

Tensile properties with standard deviation bars for specimens taken from the base of 267 mm (10.5 in.) diameter cast impeller vs temperature of 2-h solution heat treatments; 900 °C (1650 °F) appears to be optimum for solution heat treatment or HIP temperature. Processing: HIP 815 °C (1500 °F), 2 h, 103 MPa (15 ksi), heat treated at 900 °C (1650 °F), 2 h, furnace cooled to 540 °C (1000 °F), then AC, 540 °C (1000 °F), 2 h, AC.

Source: F.A. Crossley, The Martensitic Transage Titanium Alloys: Their Metallurgy, Processing Characteristics and Potential Applications, *Beta Titanium Alloys in the 80's*, R.R. Boyer and H.W. Rosenberg, Ed., TMS/AIME, 1984, p 349-386, 485-496; and F.A. Crossley, private communication, June 1992

Transage 175: Power requirements for EB butt weldments

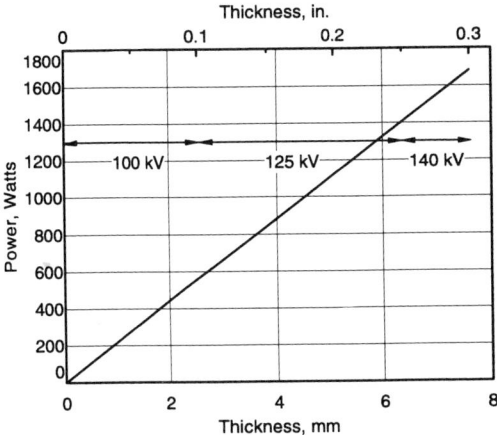

Power requirements for electron-beam butt weldments of Transage 175 sheet and plate for travel speed of 13 mm/s (30 in./min). Data are for single-pass, 100% penetration welds. As-welded weldments are ductile. The recommended sequence for welded structures of Transage 175 is β solution heat treat, e.g., 815 °C (1500 °F), 1 h, AC, weld, age harden to desired strength level. Transage 175 has demonstrated 100% weld efficiency at a strength of 1365 MPa (198 ksi).

Source: F.A. Crossley, private communication, June 1992

Ti-8V-5Fe-1Al

UNS: Unassigned

Compiled by P. Russo, RMI Titanium Company

Ti-8V-5Fe-1Al is a metastable β-titanium alloy that is capable of achieving an ultimate tensile strength of more than 1380 MPa (200 ksi) and a shear strength of more than 795 MPa (115 ksi). Special precautions must be taken when melting the alloy because of the segregation tendency of iron. It is available only by special order.

Density: 4.65 g/cm^3 (0.168 lb/in.^{3})

Product Forms and Conditions. Ti-8V-5Fe-1Al is generally supplied as bar and billet. The alloy can be hardened by solution treating and aging cycles. Welding is not recommended.

Applications. Ti-8V-5Fe-1Al has been used for aerospace fasteners and has potential for use in applications where high ultimate and shear strengths are critical concerns.

Typical composition of Ti-8V-5Fe-1Al

	Al	V	O	N	C	Fe	Ti
Min, wt%	0.8	7.5	0.25	...	...	4.0	...
Max, wt%	1.5	8.5	0.50	0.07	0.05	6.0	bal

Physical Properties

Beta Transus. 830 ± 14 °C (1525 ± 25 °F)

Ti-8V-5Fe-1Al: Isothermal TTT diagram

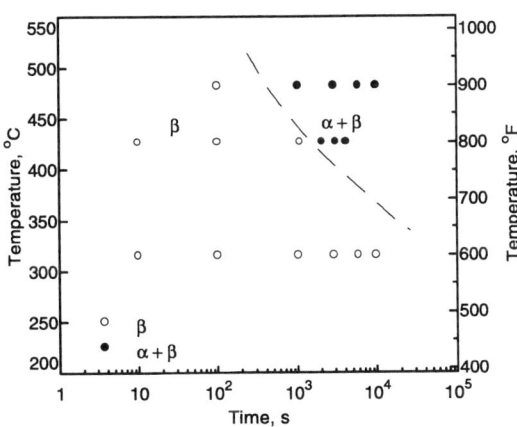

955 °C (1750 °F), ½ h, isothermal transformation, water quench.
Source: S.R. Seagle, "Initial Work on Isothermal Transformation Studies of MST 185," Research Report No. 1000R261, Project No. 25004, RMI Titanium Company

Ti-8V-5Fe-1Al: Elastic modulus

Temperature		Elastic modulus			
		Annealed		Solution treated + aged	
°C	°F	GPa	10^6 psi	GPa	10^6 psi
	RT	114	16.5	114	16.5
315	600	101	14.7	100	14.5

Source: R.A. Wood, "A Tabulation of Designations, Properties, and Treatments of Titanium and Titanium Alloys," DMIC Memorandum 171, Battelle Columbus Laboratories, 15 July 1963

Mechanical Properties

See also "Heat Treatment" for tensile data.

Ti-8V-5Fe-1Al: Guaranteed STA room-temperature properties

Property	Minimum value(a)
Ultimate tensile strength, MPa (ksi)	1448 (210)
Tensile yield strength, MPa (ksi)	1380 (200)
Elongation (in 4D), %	6.0
Reduction in area, %	12.0
Ultimate shear strength, MPa (ksi)	793 (115)

(a) Minimums for sizes up to 27 mm (1 1/16 in.) in diameter. Source: RMI Titanium Company

Ti-8V-5Fe-1Al: Typical room-temperature tensile properties of mill-annealed bar

Bar diam		Typical room-temperature properties(a)				Elongation, %	Reduction of area, %
		Ultimate strength		Yield strength			
mm	inch	MPa	ksi	MPa	ksi		
9.5	3/8	1209	175.2	1167	169.1	13.0	37.6
13	1/2	1212	175.7	1164	168.7	14.0	49.0
19	3/4	1161	168.3	1120	162.3	18.0	37.1
27	1 1/16	1228	178.0	1145	166.0	17.0	43.0

(a) 6.4 mm (1/4 in.) diam specimens. Source: *Aerospace Structural Metals Handbook*, Vol 4, Code 3719, Battelle Columbus Laboratories, 1968

Ti-8V-5Fe-1Al: Annealed room-temperature tensile properties

Annealing treatment	Ultimate strength		Yield strength		Elongation %	Reduction of area, %
	MPa	ksi	MPa	ksi		
675 °C (1250 °F), 1/2 h, AC	1263	183.0	1233	178.7	15.0	47.5
720 °C (1325 °F), 1/2 h, AC	1210	175.3	1170	169.5	16.0	47.7
760 °C (1400 °F), 1/2 h, AC	1179	170.9	1151	166.8	13.5	43.0
675 °C (1250 °F), 1/2 h, FC to 480 °C (900 °F), AC	1236	179.1	1214	175.9	20.5	52.9

Tensile properties of 6.4 mm (1/4 in.) specimens from 9.5 mm (3/8 in.) bar annealed as indicated. Source: *Aerospace Structural Metals Handbook*, Vol 4, Code 3719, Battelle Columbus Laboratories, 1968

Ti-8V-5Fe-1Al-0.14O: Room-temperature tensile properties of hot isostatically pressed gas-atomized powder

HIP temperature		Heat treatment(a)	Tensile strength		0.2% offset yield strength		Elongation, %	Reduction of area, %
°C	°F		MPa	ksi	MPa	ksi		
725	1340	As-HIP	1063	154	1037	150	6.3	29.1
		A	1007	146	960	140	17.2	39.8
		B	1407	204	1397	188	6.3	8.0
		C	1217	177	1160	170	12.5	27.8
780	1440	As-HIP	1135	165	1075	155	9.4	19.9
		A	1015	145	960	140	16.4	37.2
		B	1414	205	1315	190	4.7	9.9
		C	1175	170	1111	161	10.9	24.3

(a) Prepared using rapid solidification techniques. A, 675 °C (1250 °F), 1 h, air cooled; B, 675 °C (1250 °F), 1 h, water quenched + 440 °C (825 °F), 4 h; C, 675 °C (1250 °F), 1 h, water quenched + 495 °C (925 °F), 4 h. Source: C.F. Yolton and J.H. Moll, Evaluation of a High Strength Rapidly Solidified Beta Titanium Alloy, *Prog. Powder Metall.*, Vol 43, 1987, p 49-63

Ti-8V-5Fe-1Al: Effect of aging on RT tensile properties

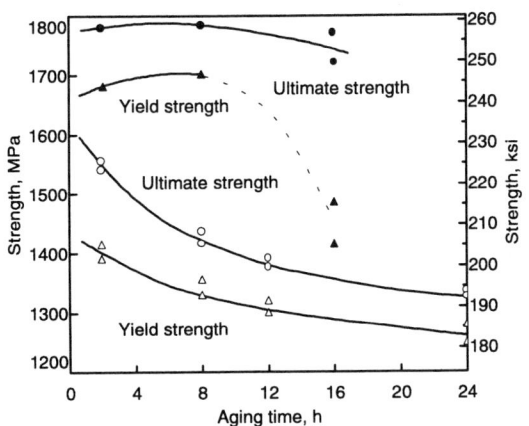
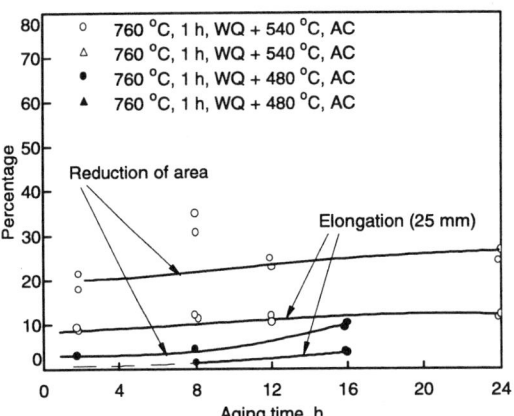

16 mm (5/8 in.) diam bar.
Source: J.R. Gross, "Creep and Stability of MST 185," Research Report No. 1000R298, Project No. 93003, RMI Titanium Company

---

Notched Tensile Strength

Ti-8V-5Fe-1Al: Effect of stress-concentration factor on room-temperature notch stress rupture

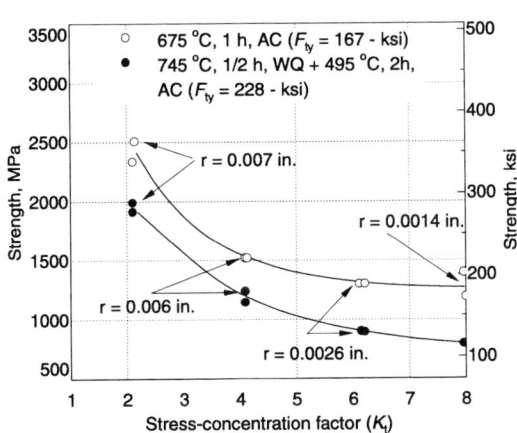

Specimens from 13 mm (1/2 in.) diam bar; room-temperature tests; 60 degree notch with a point radius (r) as indicated.
Source: L.J. Bartlo, "The Effect of Stress Concentration on Notch Tensile and Static Notch Properties of Ti-8V-5Fe-1Al," RMI Titanium Company, Research Report No. 1000R459, MO 75003, Reactive Metals, Inc., 30 Dec 1964

Ti-8V-5Fe-1Al: Effect of stress-concentration factor on notched strength

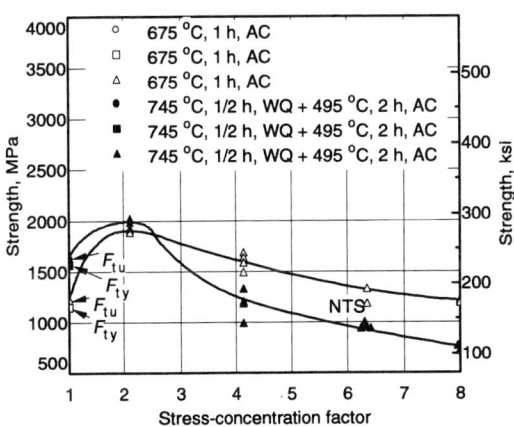

Specimens from 13 mm (1/2 in.) diam bar; room-temperature tests; 60 degree notch with a point radius (r) as indicated in other figure.
Source: L.J. Bartlo, "The Effect of Stress Concentration on Notched Tensile and Static Notch Properties of Ti-8V-5Fe-1Al," RMI Titanium Company, Research Report No. 1000R459, MO 75003, Reactive Metals, Inc., 30 Dec 1964

Shear Strength

The shear strength will vary with the tensile strength of the material. One study indicated that the shear strength is higher if the entire cross section is tested, as opposed to machining material from the surface to test the core. This is in contrast to what is observed for the tensile strength.

Ti-8V-5Fe-1Al: Room-temperature tensile tests and double shear test data

Test diam		0.2% Y.S.		U.T.S.		Elongation, %	R.A. %	Double shear	
mm	in.	MPa	ksi	MPa	ksi			MPa	ksi
4.75	0.187	1445	209.7	1500	217.7	13	41.2	771	111.9
4.75	0.187	1440	208.8	1504	218.2	12	39.8	809	117.4
6.37	0.251	1418	205.6	1470	213.1	12	34.7	808	117.2
6.37	0.251	1418	205.6	1468	212.9	12	37.3	851	123.5

Room-temperature tensile tests and double shear tests were run in the STA condition (1400 °F/1 h, WQ + 950 °F/2 h, AC). Source: RMI Titanium Company

Fatigue Strength

Ti-8V-5Fe-1Al: Room-temperature fatigue of ⅜-24 bolts

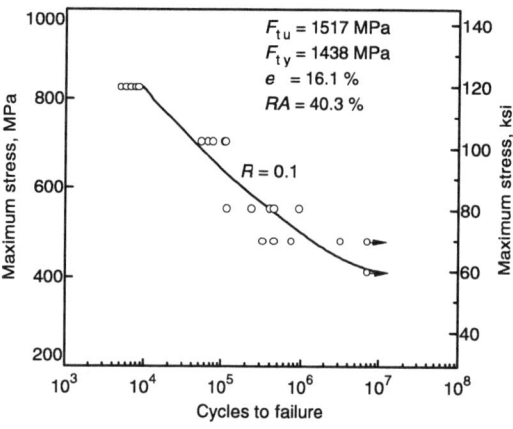

Axial loaded, room-temperature test. Notch factor of bolt threads were unspecified.
Source: *Aerospace Structural Metals Handbook*, Vol 4, Code 3719, Battelle Columbus Laboratories

High-Temperature Strength

Tensile Properties

Ti-8V-5Fe-1Al: Effect of temperature on tensile strength

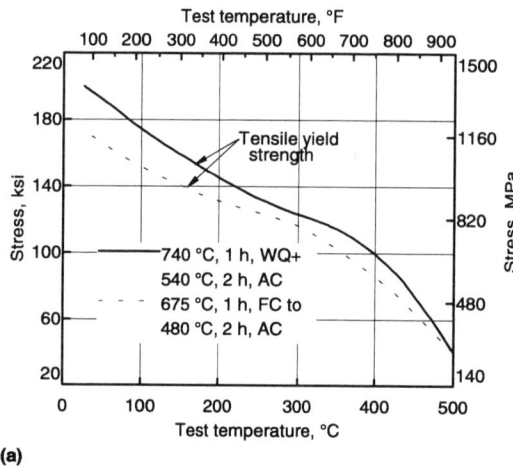

(a)

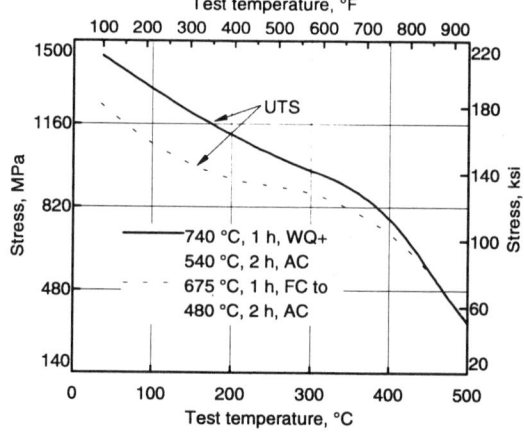

(b)

16 mm (⅝ in.) diam bar.
Source: Private communication, Reactive Metals, Inc.

Ti-8V-5Fe-1Al: Thermal stability of bar

Effect of exposure to elevated temperature with load on room-temperature tensile properties of annealed and heat-treated 16 mm (⅝ in.) diam bar

Creep exposure temperature		Stress		Time, h	Permanent deformation, %	Subsequent room-temperature tensile properties(a)				Elongation, %	Reduction of area, %
°C	°F	MPa	ksi			Ultimate strength MPa	ksi	Yield strength MPa	ksi		
Annealed bar											
				None		1228	178	1180	171	18.3	44.8
425	800	140	20	10	0.155	1220	177	1187	172	23.5	52.3
		345	50	10	0.83	1220	177	1200	174	23.5	52.5
		140	20	313.3	1.29	1220	177	1180	171	7.5(b)	10.2(b)
		210	30	310.5	2.79	1220	177	1170	170	23.0	54.0
315	600	760	110	310.5	0.78	1220	177	1152	167	22.5	51.1
Heat-treated specimen											
				None		1435	208	1366	198	12.0	37.5
425	800	310	45	212.5	...	1455	211	1394	202	15.0	38.9

(a) 6.4 mm (¼ in.) diam specimens. (b) Stress-corrosion crack. Source: *Aerospace Structural Metals Handbook*, Vol 4, Code 3719, Battelle Columbus Laboratories, 1968

Creep Properties

Ti-8V-5Fe-1Al: Creep deformation in 300 h for annealed bar

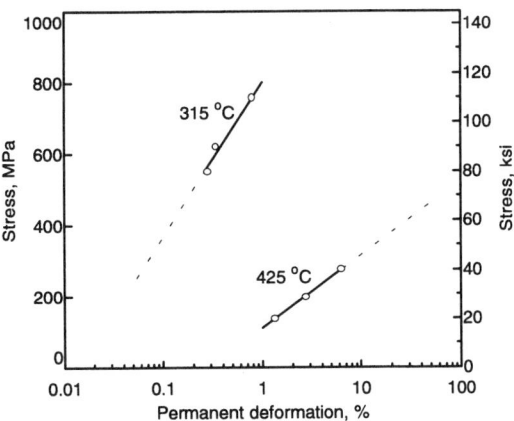

16 mm (⅝ in.) diam bar, 675 °C (1250 °F), 1 h, furnace cooled to 480 °C (900 °F), air cooled.
Source: J.R. Gross, "Creep and Stability of MST 185," Research Report No. 1000R298, Project No. 93003, Mallory-Sharon Titanium Corporation, 19 June 1958

Ti-8V-5Fe-1Al: Minimum creep rate for annealed bar

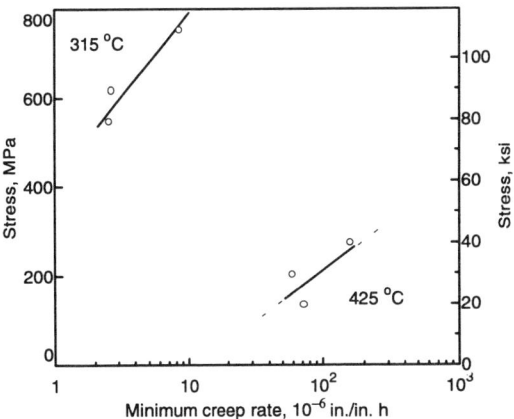

16 mm (⅝ in.) diam bar, 675 °C (1250 °F), 1 h, furnace cooled to 480 °C (900 °F), air cooled.
Source: J.R. Gross, "Creep and Stability of MST 185," Research Report No. 1000R298, Project No. 93003, Mallory-Sharon Titanium Corporation, 19 June 1958

Fabrication

Forging. Ti-8V-5Fe-1Al is readily forgeable. However, the material should receive final reductions of at least 50% below 815 °C (1500 °F) to optimize ductility.

Welding of Ti-8V-5Fe-1Al is not recommended.

Heat Treatment

Ti-8V-5Fe-1Al: Typical heat treatments

Treatment	Temperature °C	Temperature °F	Duration	Cooling
Stress relief	540-590	1000-1100	1 hour	Air cool
Anneal	675	1250	1 hour	Furnace cool
Alternate production anneal	675-730	1250-1350	1-2 hours	Air cool
Solution treat	760	1400	1 hour	Water quench
Alternate solution annealing range	730-790(a)	1350-1450(a)	...	...
Aging	480-540	900-1000	2 hours	Air cool

(a) Depending on product form and desired properties

Ti-8V-5Fe-1Al: Effect of cooling on STA properties

Solution treatment	Ultimate strength MPa	Ultimate strength ksi	Yield strength MPa	Yield strength ksi	Elongation, %	Reduction of area, %
745 °C (1375 °F), 1 h, WQ	1421	206	1339	194	15	39
745 °C (1375 °F), 1 h, AC	1413	205	1339	194	16	41
760 °C (1400 °F), 1 h, WQ	1497	217	1408	204	12	30
760 °C (1400 °F), 1 h, AC	1456	211	1366	198	13	33

Room-temperature properties of 6.4 mm (¼ in.) diam specimens taken from 14 mm (9/16 in.) diam bar, solution treated as indicated, plus a 540 °C (1000 °F) age for 2 h, air cooled. Source: *Aerospace Structural Metals Handbook*, Vol 4, Code 3719, Battelle Columbus Laboratories, 1968

Ti-8V-5Fe-1Al: Effect of solution treatment temperature on tensile properties

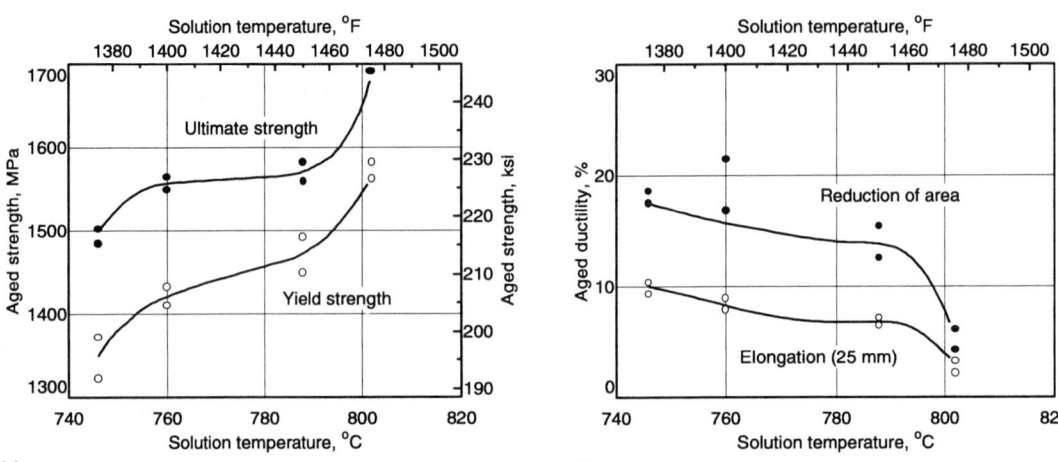

16 mm (⅝ in.) diam bar, solution treated 1 h, water quenched + 540 °C (1000 °F), 8 h, air cooled.
Source: J.R. Gross, "The Effects of Heat Treatment on the Mechanical Properties of MST 185 ⅝ Inch Diameter Bar," Research Report No. 1000R226, Project No. 34002, RMI Titanium Company

Ti-8V-5Fe-1Al: Effect of solution treatment and aging on tensile properties

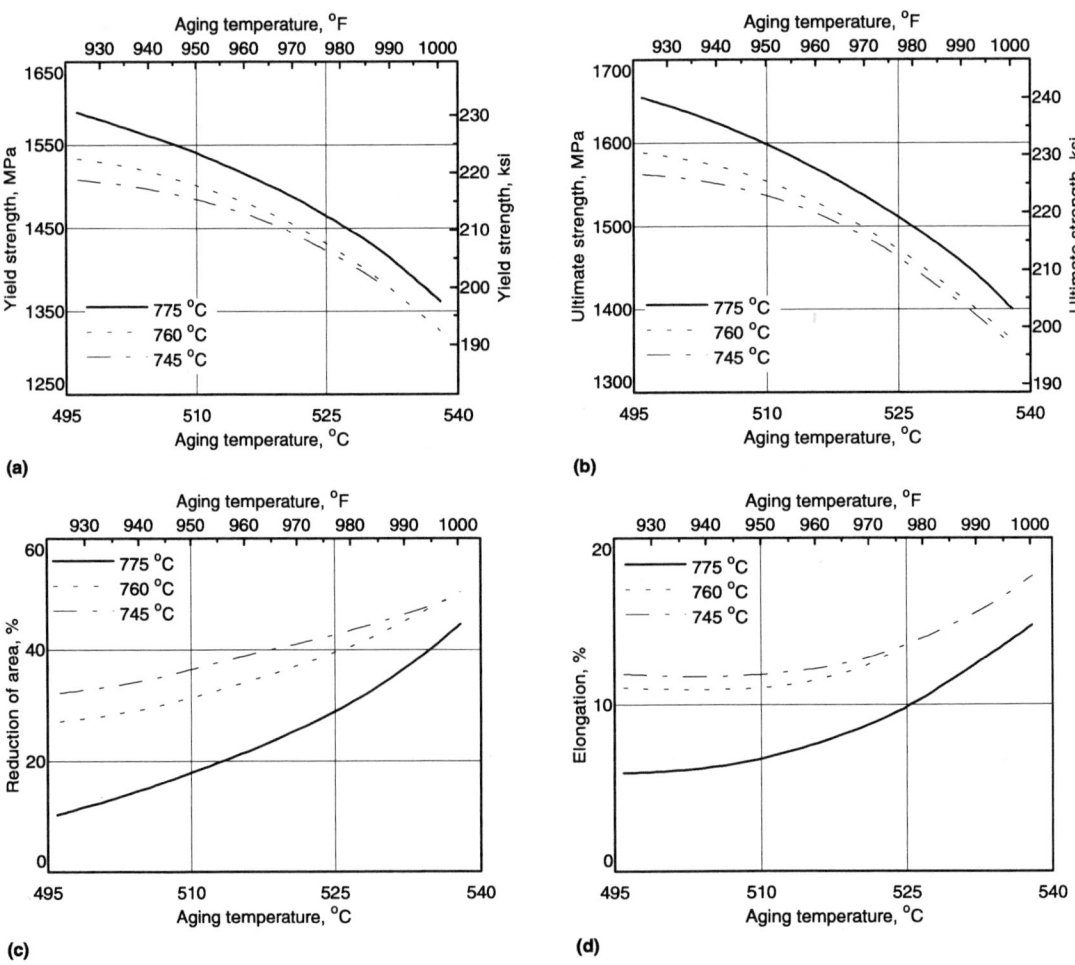

14 mm (9/16 in.) diam bar, solution treated ½ h, water quenched, aged 2 h, air cooled; room-temperature tests.
Source: RMI Titanium Company

Ti-16V-2.5Al

UNS Number: Unassigned

Compiled by P. Russo, RMI Titanium Company

Ti-16V-2.5Al is a metastable titanium alloy developed by RMI Titanium Company for use in high-strength sheet applications. The alloy can be strengthened by solution treating and aging, and it has been used as high-strength sheet for aircraft applications. The alloy is available only by special order.

Phases and Structures. In the solution-treated condition, Ti-16V-2.5Al consists of all β phase. The alloy is relatively lean in β-stabilizers and can form stress-induced martensites when deformed in the solution annealed condition. After aging, the microstructure consists of fine α in a β matrix.

Forging. Like most β titanium alloys, Ti-16V-2.5Al has good forgeability. Final forging temperatures should be maintained in the α-β phase field.

Machining. Ti-16V-2.5Al is expected to have machinability similar to other metastable β titanium alloys, such as Ti-15V-3Sn-3Zr-3Cr and Ti-3Al-8V-6Cr-4Mo-4Zr.

Selected Reference

- Determination of Design Data for Heat Treated Titanium Alloy Sheet, Report No. ASD-TDR-62-335, Vol 1, Lockheed-Georgia Company, Contract AF 33(616)-6346, Dec 1962

Ti-16V-2.5Al: Summary of typical physical properties

Beta transus	Not available
Melting (liquidus point)	Not available
Density(a)	4.65 g/cm^3 (0.168 lbf/in.^{3})
Electrical resistivity(a)	Not available
Magnetic permeability(a)	Nonmagnetic
Specific heat capacity(a)	527 J/kg · K (0.126 Btu/lb · °F)
Thermal conductivity	8.7 W/m · K (5 Btu/ft · h · °F)
Thermal coefficient of linear expansion(b)	9×10^{-6}/°C (5.0×10^{-6}/°F)

(a) Typical values at room temperature of about 20 to 25 °C (68 to 78 °F). (b) Mean coefficient from 40 to 95 °C (100 to 200 °F)

Elastic Properties

Young's Modulus

Ti-16V-2.5Al: Room-temperature moduli of STA sheet

Sheet thickness		Direction	Tensile modulus		Compressive modulus	
mm	in.		GPa	10^6 psi	GPa	10^6 psi
0.5	0.02	L	99.3	14.4	...	...
		T	96.5	14.0	...	...
1.6	0.063	L	95.8	13.9	97.2	14.1
		T	97.9	14.2	97.2	14.1
3.2	0.125	L	95.2	13.8	102	14.8
		T	96.5	14.0	103	15.0

Solution treated at 750 to 765 °C (1380 to 1410 °F) for 30 min, air cooled and aged at 525 to 530 °C (975 to 990 °F) for 4 to 6 h. Source: "Determination of Design Data for Heat Treated Alloy Sheet," Report ASD-TDR-62-335, Vol 1, Contract AF 33(616)-6346, Lockheed-Georgia Company, Dec 1962

1000 / Beta and Near-Beta Alloys

Ti-16V-2.5Al: Effect of temperature on tensile modulus

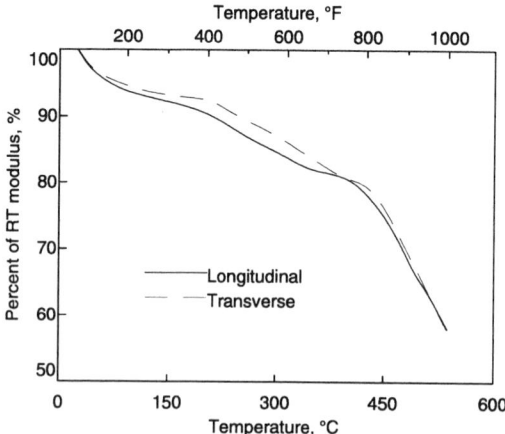

1.6 mm (0.063 in.) STA sheet.
"Determination of Design Data for Heat Treated Titanium Alloy Sheet," Report No. ASD-TDR-62-335, Vol 1, Contract AF 33(616)-6346, Lockheed-Georgia Company, Dec 1962

Ti-16V-2.5Al: Effect of temperature on compressive modulus

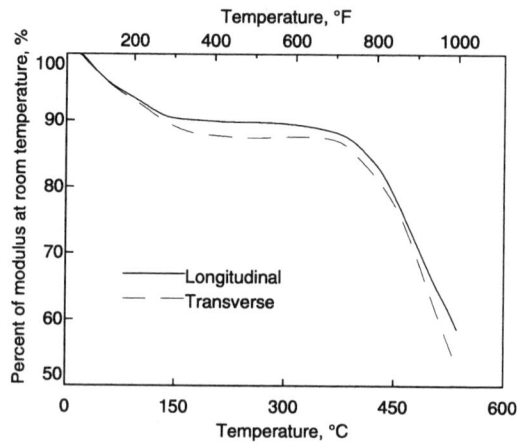

1.6 mm (0.063 in.) STA sheet.
"Determination of Design Data for Heat Treated Titanium Alloy Sheet," Report No. ASD-TDR-62-335, Vol 1, Contract AF 33(616)-

Tangent Moduli

Ti-16V-2.5Al: Typical compressive tangent modulus curve

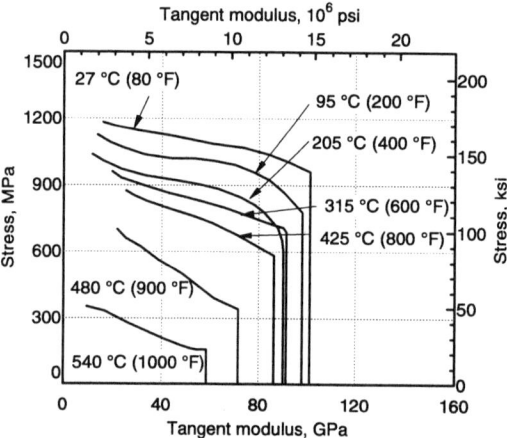

1.6 mm (0.063 in.) STA sheet; test direction, longitudinal.
Source: "Determination of Design Data for Heat Treated Titanium Alloy Sheet," Report No. ASD-TDR-62-335, Vol 1, Contract AF 33(616)-6346, Lockheed-Georgia Company, Dec 1962

Ti-16V-2.5Al: Typical compressive tangent modulus curves

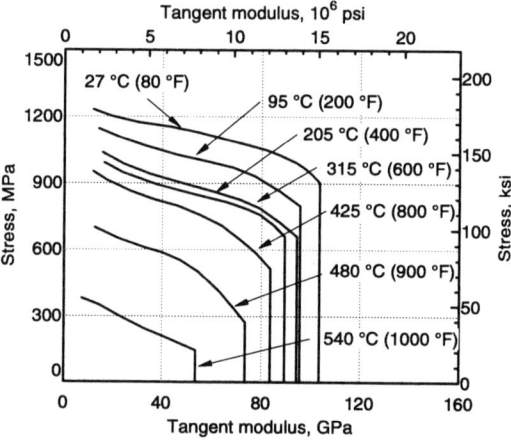

1.6 mm (0.063 in.) STA sheet; test direction, transverse.
Source: "Determination of Design Data for Heat Treated Titanium Alloy Sheet," Report No. ASD-TDR-62-335, Vol 1, Contract AF 33-(616)-6346, Lockheed-Georgia Company, Dec 1962

Secant Moduli

Ti-16V-2.5Al: Typical compressive secant modulus curves

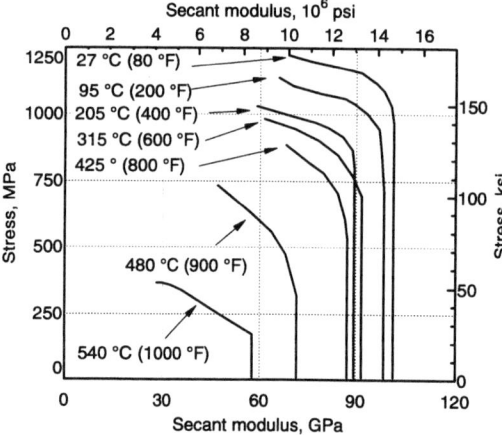

1.6 mm (0.063 in.) STA sheet; test direction, longitudinal.
Source: "Determination of Design Data for Heat Treated Titanium Alloy Sheet," Report No. ASD-TDR-62-335, Vol 1, Contract AF 33-(616)-6346, Lockheed-Georgia Company, Dec 1962

Ti-16V-2.5Al: Typical compressive secant modulus curves

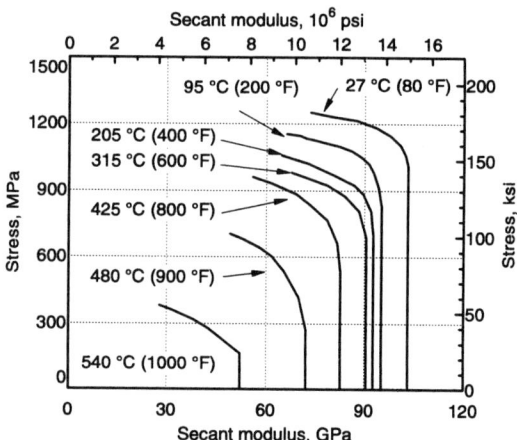

1.6 mm (0.063 in.) STA sheet; test direction, transverse.
Source: "Determination of Design Data for Heat Treated Titanium Alloy Sheet," Report No. ASD-TDR-62-335, Vol 1, Contract AF 33-(616)-6346, Lockheed-Georgia Company, Dec 1962

Room-Temperature Strength

Ti-16V-2.5Al is an age-hardenable alloy. A range of tensile properties is attainable, with yield strength values greater than 1240 MPa (180 ksi) possible. The formation of stress-induced martensite in solution-annealed material can result in a relatively low ratio of yield to tensile strengths.

Ti-16V-2.5Al: Mechanical design properties (B-basis)

Property(a)	Sheet thickness, mm (in.)		
	0.5 (0.02)	1.6 (0.063)	3.2 (0.125)
Ultimate tensile strength, MPa (ksi)			
L	1172 (170)	1110 (161)	1158 (168)
T	1158 (168)	1131 (164)	1158 (168)
Yield strength, MPa (ksi)			
L	1089 (158)	1000 (145)	1076 (156)
T	1082 (157)	1034 (150)	1076 (156)
Compressive yield strength, MPa (ksi)			
L	...	1044 (151)	1117 (162)
T	...	1076 (156)	1151 (167)
Shear strength, MPa (ksi)			
L	717 (104)	703 (102)	731 (106)
T	703 (102)	690 (100)	724 (105)
Ultimate bearing strength, MPa (ksi)			
$e/D = 1.5$			
L	1531 (222)	1613 (234)	1620 (235)
T	1482 (215)	1613 (234)	1606 (233)
$e/D = 2.0$			
L	1738 (252)	1944 (282)	1951 (283)
T	1724 (250)	1951 (283)	1889 (274)
Bearing yield strength, MPa (ksi)			
$e/D = 1.5$			
L	1400 (203)	1434 (208)	1455 (211)
T	1448 (210)	1434 (208)	1476 (214)
$e/D = 2.0$			
L	1558 (226)	1593 (231)	1710 (248)
T	1565 (227)	1593 (231)	1696 (246)
Elongation, %			
L	3.7	6.2	6.9
T	3.8	5.4	6.3
Young's modulus, GPa			
L	99.3 (14.4)	95.8 (13.9)	95.2 (13.8)
T	96.5 (14.0)	97.9 (14.2)	96.5 (14.0)
Compressive modulus, GPa			
L	...	97.2 (14.1)	102 (14.8)
T	...	97.2 (14.1)	103 (15.0)

(a) L, longitudinal; T, transverse. 750 to 765 °C (1380 to 1410 °F), 30 min, air cooled + 525 to 530 °C (975 to 990 °F), 4 to 6 h, air cooled. Source: "Determination of Design Data for Heat Treated Titanium Alloy Sheet," Report No. ASD-TDR-62-335, Vol 1, Contract AF 33(616) 6346, Lockheed-Georgia Company, Dec 1902

High-Temperature Strength

Tensile Strengths vs Temperature

Ti-16V-2.5Al: Effect of temperature on yield strength

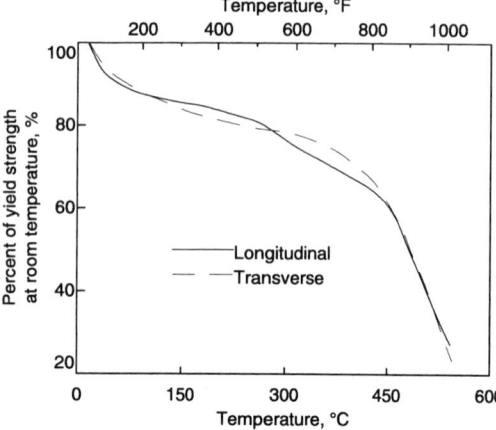

1.6 mm (0.063 in.) STA sheet.
Source: "Determination of Design Data for Heat Treated Titanium Alloy Sheet," Report No. ASD-TDR-62-335, Vol 1, Contract AF 33-(616)-6346, Lockheed-Georgia Company, Dec 1962

Ti-16V-2.5Al: Effect of temperature on ultimate tensile strength

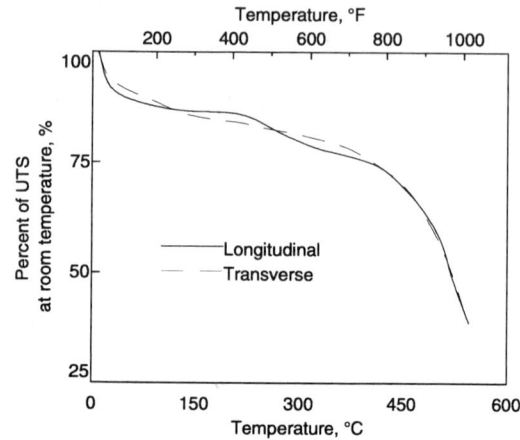

1.6 mm (0.063 in.) STA sheet.
Source: "Determination of Design Data for Heat Treated Titanium Alloy Sheet," Report No. ASD-TDR-62-335, Vol 1, Contract AF 33-(616)-6346, Lockheed-Georgia Company, Dec 1962

Creep Properties

Ti-16V-2.5Al: Creep and rupture data

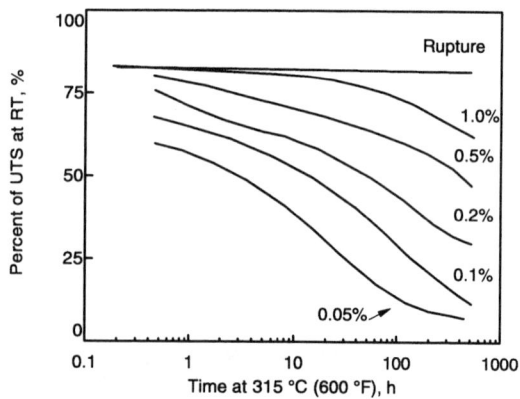

1.6 mm (0.063 in.) STA sheet; test temperature, 315 °C (600 °F).
Source: "Determination of Design Data for Heat Treated Titanium Alloy Sheet," Report No. ASD-TDR-62-335, Vol 1, Contract AF 33-(616)-6346, Lockheed-Georgia Company, Dec 1962

Ti-16V-2.5Al: Creep and rupture data

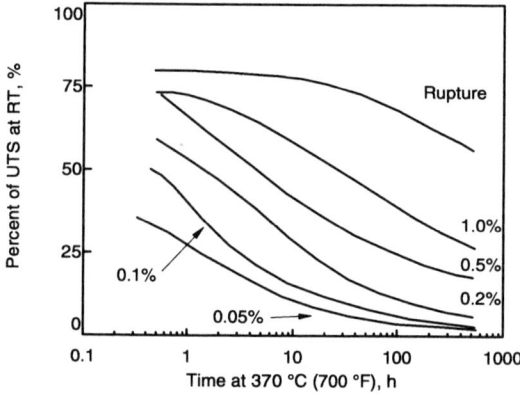

1.6 mm (0.063 in.) STA sheet; test temperature, 370 °C (700 °F).
Source: "Determination of Design Data for Heat Treated Titanium Alloy Sheet," Report No. ASD-TDR-62-335, Vol 1, Contract AF 33-(616)-6346, Lockheed-Georgia Company, Dec 1962

Ti-16V-2.5Al: Creep and rupture data

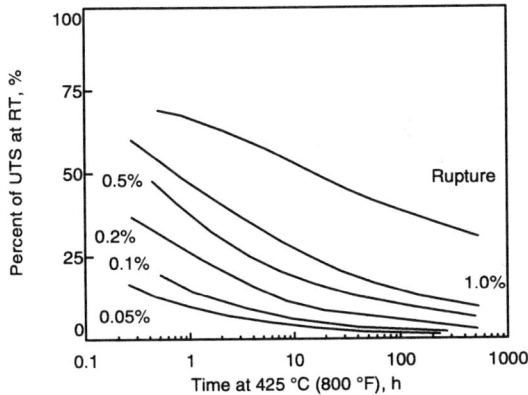

1.6 mm (0.063 in.) STA sheet; test temperature, 425 °C (800 °F).
Source: "Determination of Design Data for Heat Treated Titanium Alloy Sheet," Report No. ASD-TDR-62-335, Vol 1, Contract AF 33-(616)-6346, Lockheed-Georgia Company, Dec 1962

Compressive and Shear Strengths vs Temperature

Ti-16V-2.5Al: Compressive yield strength

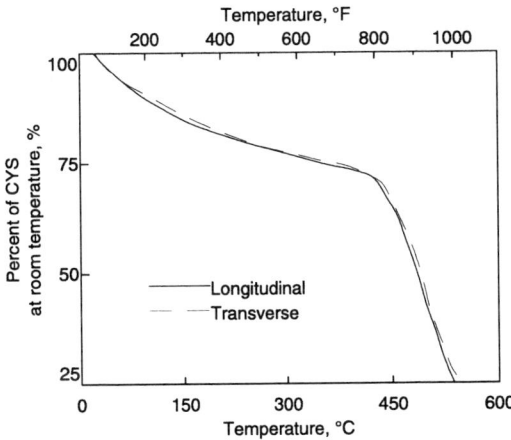

1.6 mm (0.063 in.) STA sheet.
Source: "Determination of Design Data for Heat Treated Titanium Alloy Sheet," Report No. ASD-TDR-62-335, Vol 1, Contract AF 33-(616)-6346, Lockheed-Georgia Company, Dec 1962

Ti-16V-2.5Al: Effect of temperature on ultimate shear strength

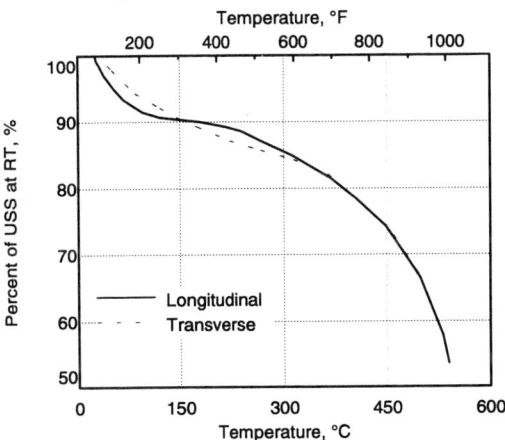

1.6 mm (0.063 in.) STA sheet.
Source: "Determination of Design Data for Heat Treated Titanium Alloy Sheet," Report No. ASD-TDR-62-335, Vol 1, Contract AF 33-(616)-6346, Lockheed-Georgia Company, Dec 1962

Fatigue Strength

Unnotched Fatigue

Ti-16V-2.5Al: Fatigue properties of sheet at RT

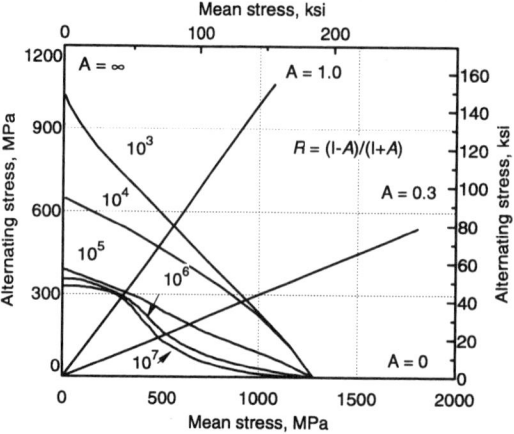

1.6 mm (0.063 in.) STA sheet; tested at room temperature; $K_t = 1.0$.
Source: "Determination of Design Data for Heat Treated Titanium Alloy Sheet," Report No. ASD-TDR-62-335, Vol 1, Contract AF 33-(616)-6346, Lockheed-Georgia Company, Dec 1962

Ti-16V-2.5Al: Fatigue properties of sheet at 200 °C

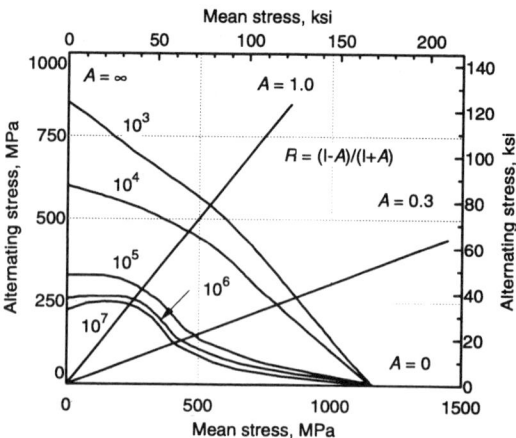

1.6 mm (0.063 in.) STA sheet; test temperature, 200 °C (400 °F); $K_t = 1.0$.
Source: "Determination of Design Data for Heat Treated Titanium Alloy Sheet," Report No. ASD-TDR-62-335, Vol 1, Contract AF 33-(616)-6346, Lockheed-Georgia Company, Dec 1962

Ti-16V-2.5Al: Fatigue properties of sheet at 315 °C

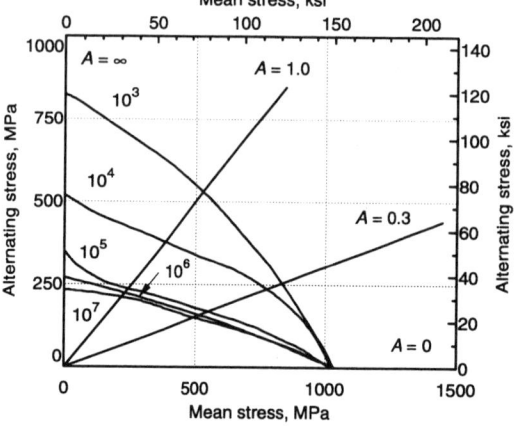

1.6 mm (0.063 in.) STA sheet; test temperature, 315 °C (600 °F); $K_t = 1.0$.
Source: "Determination of Design Data for Heat Treated Titanium Alloy Sheet," Report No. ASD-TDR-62-335, Vol 1, Contract AF 33-(616)-6346, Lockheed-Georgia Company, Dec 1962

Ti-16V-2.5Al: Fatigue properties of sheet at 425 °C

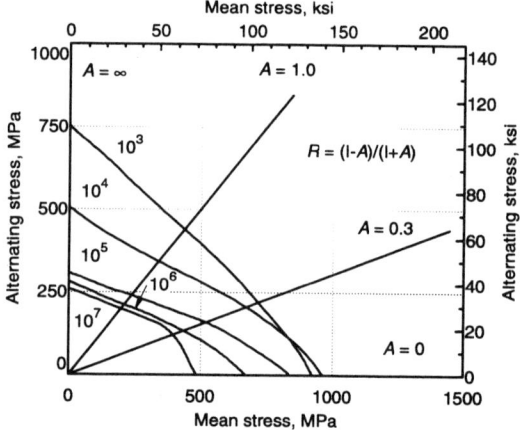

1.6 mm (0.063 in.) STA sheet; test temperature, 425 °C (800 °F); $K_t = 1.0$.
Source: "Determination of Design Data for Heat Treated Titanium Alloy Sheet," Report No. ASD-TDR-62-335, Vol 1, Contract AF 33-(616)-6346, Lockheed-Georgia Company, Dec 1962

Ti-16V-2.5Al: Fatigue properties of sheet at 480 °C

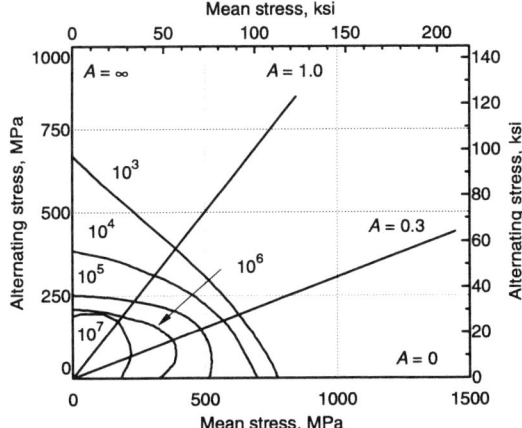

1.6 mm (0.063 in.) STA sheet; test temperature, 480 °C (900 °F); $K_t = 1.0$.
Source: "Determination of Design Data for Heat Treated Titanium Alloy Sheet," Report No. ASD-TDR-62-335, Vol 1, Contract AF 33-(616)-6346, Lockheed-Georgia Company, Dec 1962

Ti-16V-2.5Al: Notched fatigue of sheet at RT

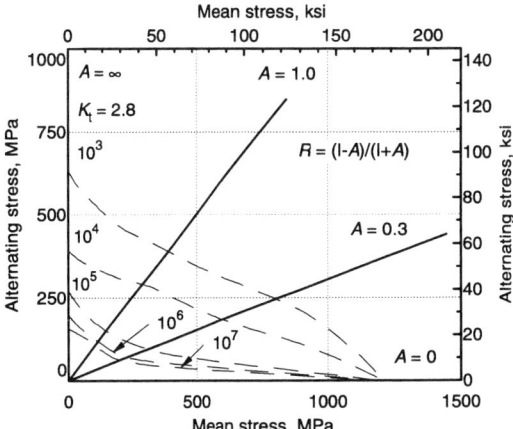

1.6 mm (0.063 in.) STA sheet; tested at room temperature, $K_t = 2.82$.
Source: "Determination of Design Data for Heat Treated Titanium Alloy Sheet," Report No. ASD-TDR-62-335, Vol 1, Contract AF 33-(616)-6346, Lockheed-Georgia Company, Dec 1962

Notched Fatigue

Ti-16V-2.5Al: Notched fatigue of sheet at 200 °C

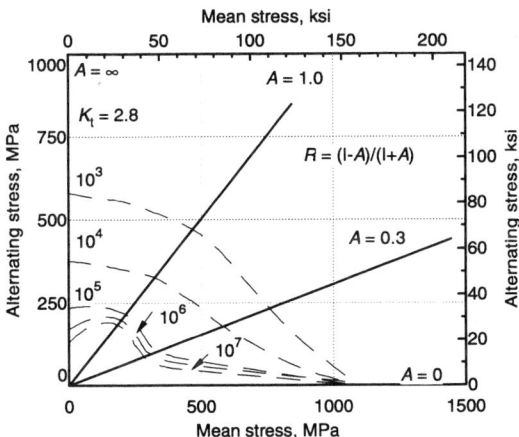

1.6 mm (0.063 in.) STA sheet; test temperature, 200 °C (400 °F); $K_t = 2.82$.
Source: "Determination of Design Data for Heat Treated Titanium Alloy Sheet," Report No. ASD-TDR-62-335, Vol 1, Contract AF 33-(616)-6346, Lockheed-Georgia Company, Dec 1962

Ti-16V-2.5Al: Notched fatigue properties of sheet at 315 °C

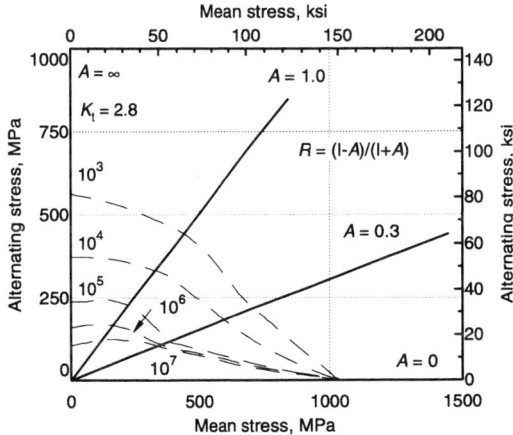

1.6 mm (0.063 in.) STA sheet; test temperature, 315 °C (600 °F); $K_t = 2.82$.
Source: "Determination of Design Data for Heat Treated Titanium Alloy Sheet," Report No. ASD-TDR-62-335, Vol 1, Contract AF 33-(616)-6346, Lockheed-Georgia Company, Dec 1962

Ti-16V-2.5Al: Notched fatigue properties of sheet at 425 °C

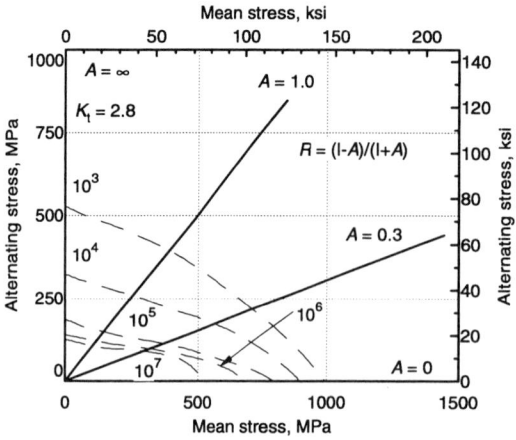

1.6 mm (0.063 in.) STA sheet; test temperature, 425 °C (800 °F); K_t = 2.82.
Source: "Determination of Design Data for Heat Treated Titanium Alloy Sheet," Report No. ASD-TDR-62-335, Vol 1, Contract AF 33-(616)-6346, Lockheed-Georgia Company, Dec 1962

Ti-16V-2.5Al: Notched fatigue properties of sheet at 480 °C

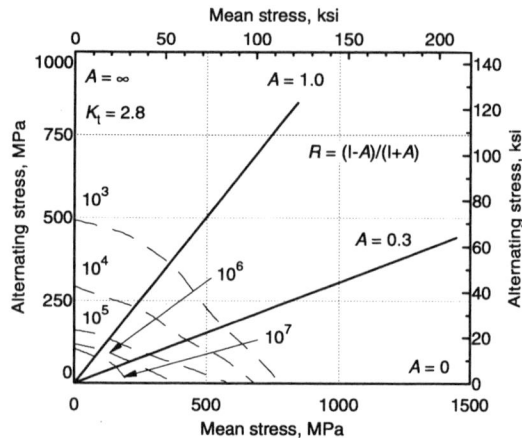

1.6 mm (0.063 in.) STA sheet; test temperature, 480 °C (900 °F); K_t = 2.82.
Source: "Determination of Design Data for Heat Treated Titanium Alloy Sheet," Report No. ASD-TDR-62-335, Vol 1, Contract AF 33-(616)-6346, Lockheed-Georgia Company, Dec 1962

Plastic Deformation

Ti-16V-2.5Al: Typical compressive stress-strain curves

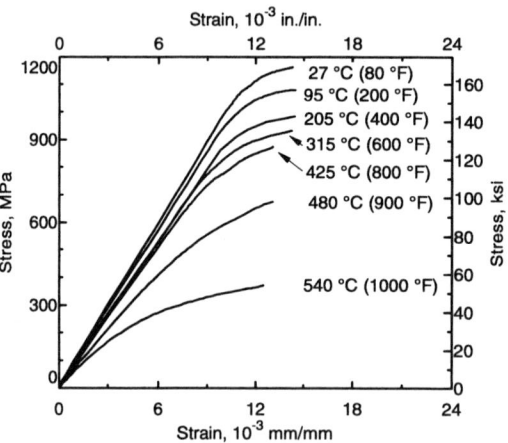

1.6 mm (0.063 in.) STA sheet; test direction, longitudinal.
Source: "Determination of Design Data for Heat Treated Titanium Alloy Sheet," Report No. ASD-TDR-62-335, Vol 1, Contract AF 33-(616)-6346, Lockheed-Georgia Company, Dec 1962

Ti-16V-2.5Al: Typical compressive stress-strain curves

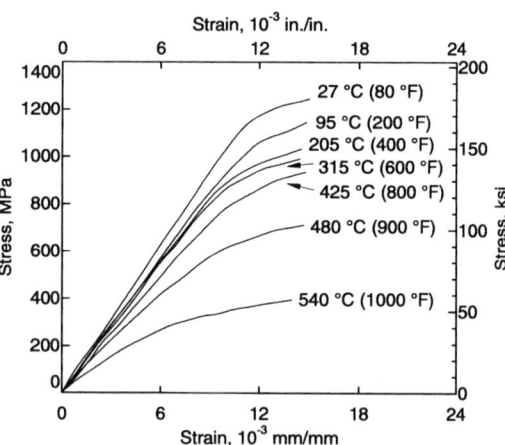

1.6 mm (0.063 in.) STA sheet; test direction, transverse.
Source: "Determination of Design Data for Heat Treated Titanium Alloy Sheet," Report No. ASD-TDR-62-335, Vol 1, Contract AF 33-(616)-6346, Lockheed-Georgia Company, Dec 1962

Ti-16V-2.5Al: Typical tensile stress-strain curves

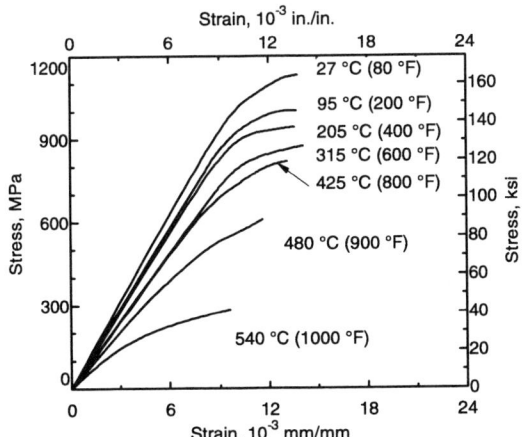

1.6 mm (0.063 in.) STA sheet; test direction, longitudinal.
Source: "Determination of Design Data for Heat Treated Titanium Alloy Sheet," Report No. ASD-TDR-62-335, Vol 1, Contract AF 33-(616)-6346, Lockheed-Georgia Company, Dec 1962

Ti-16V-2.5Al: Typical tensile stress-strain curves

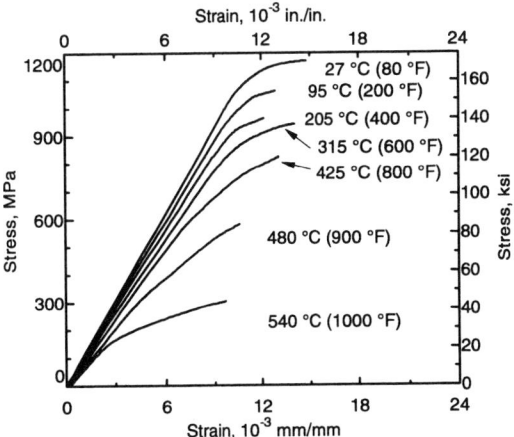

1.6 mm (0.063 in.) STA sheet; test direction, transverse.
Source: "Determination of Design Data for Heat Treated Titanium Alloy Sheet," Report No. ASD-TDR-62-335, Vol 1, Contract AF 33-(616)-6346, Lockheed-Georgia Company, Dec 1962

Advanced Materials

Titanium Aluminides*

In recent years, alloying and processing have been used to control the ordered crystal structure, microstructural features, and grain boundary structure and composition to overcome the brittleness inherent to ordered intermetallics (Ref 1-3). Success in this work has inspired parallel efforts aimed at improving strength properties. The results have led to the development of a number of attractive intermetallic alloys with useful ductility and strength.

Alloy design work has centered primarily on aluminides of nickel, iron, and titanium (Ref 1-3). These materials possess a number of attributes that make them attractive for high-temperature applications. They contain sufficient amounts of aluminum to form, in oxidizing environments, thin films of alumina (Al_2O_3) that often are compact and protective (Ref 4). These materials have low densities, relatively high melting points, and good high-temperature strength properties.

Many of the aluminides exist over a range of compositions, but the degree of order decreases as the deviation from stoichiometry increases. Additional elements can also be incorporated without losing the ordered structure. In many instances, the so-called intermetallic compounds can be used as bases for alloy development to improve or optimize properties for specific applications.

Because of their low density, titanium aluminides based on Ti_3Al and TiAl have been considered attractive candidates for applications in advanced aerospace engine components. Despite a lack of fracture resistance (low ductility, fracture toughness, and fatigue crack growth rate), the titanium aluminides Ti_3Al (α-2) and TiAl (γ) have potential for enhanced performance.

In addition to low ductility at ambient temperatures, the oxidation resistance of titanium aluminides is lower than desirable at elevated temperatures (Ref 5-7). The titanium aluminides are characterized by a strong tendency to form TiO_2 rather than the protective Al_2O_3 at high temperatures. Because of this tendency, a key factor in increasing the maximum use temperatures of these aluminides is enhancing their oxidation resistance while maintaining adequate levels of creep and strength retention at elevated temperatures.

---

*Information in the datasheets "Titanium Aluminides," "Ti_3Al Alloys," and "Gamma Alloys" was extracted from the article "Ordered Intermetallics" by C.T. Liu, J.O. Stiegler, and F.H. Froes, in *ASM Handbook*, Vol 2, 1990, with additional information from S.C. Huang, B. Marquardt, and R.G. Rowe of General Electric Company.

Properties of nickel, iron, and titanium aluminides

Alloy	Crystal structure(a)	Critical ordering temperature °C	Critical ordering temperature °F	Melting point °C	Melting point °F	Material density, g/cm³	Young's modulus GPa	Young's modulus 10⁶ psi
Ni_3Al	$L1_2$ (ordered fcc)	1390	2535	1390	2535	7.50	179	25.9
NiAl	B2 (ordered bcc)	1640	2985	1640	2985	5.86	294	42.7
Fe_3Al	DO_3 (ordered bcc)	540	1000	1540	2805	6.72	141	20.4
	B2 (ordered bcc)	760	1400	1540	2805	...	...	...
FeAl	B2 (ordered bcc)	1250	2280	1250	2280	5.56	261	37.8
Ti_3Al	DO_{19} (ordered hcp)	1100	2010	1600	2910	4.2	145	21.0
TiAl	$L1_0$ (ordered tetragonal)	1460	2660	1460	2660	3.91	176	25.5
$TiAl_3$	DO_{22} (ordered tetragonal)	1350	2460	1350	2460	3.4	...	...

(a) fcc, face-centered cubic; bcc, body-centered cubic; cph, close-packed hexagonal. Source: *ASM Handbook*, Vol 2, 10th ed., 1990

Properties of titanium aluminides, titanium-base conventional alloys, and nickel-base superalloys

Property	Conventional titanium alloys	Ti_3Al	TiAl	Nickel-base superalloys
Density, g/cm³	4.5	4.1-4.7	3.7-3.9	8.3
Modulus, GPa (10⁶ psi)	96-117 (14-17)	100-145 (14.5-21)	160-176 (23.2-25.5)	206 (30)
Yield strength, MPa (ksi)(a)	380-1150 (55-167)	700-990 (101-144)	400-650 (58-94)	...
Tensile strength, MPa (ksi)(a)	480-1200 (70-174)	800-1140 (116-165)	450-800 (65-116)	...
Creep limit, °C (°F)	600 (1110)	760 (1400)	1000 (1830)	1090 (1995)
Oxidation limit, °C (°F)	600 (1110)	650 (1200)	900 (1650)	1090 (1995)
Ductility at room temperature, %	20	2-10	1-4	3-5
Ductility at high temperature, %	High	10-20	10-60	10-20
Structure	cph/bcc	DO_{19}	$L1_0$	fcc/$L1_2$

(a) At room temperature. Source: *ASM Handbook*, Vol 2, 10th ed., 1990, p 926

Titanium aluminides: Property comparison of titanium aluminides with titanium alloys and superalloys

Property	Ti alloys	α-2	γ	Superalloys
Density, g/cm^3 (lb/in.^{3})	4.54 (0.16)	4.84 (0.17)	4.04 (0.14)	8.3 (0.30)
Stiffness, GPa (10^6 psi)	110 (16)	145 (21)	176 (25)	207 (30)
Creep temperature (max), °C (°F)	540 (1000)	730 (1345)	900 (1650)	1090 (1995)
Oxidation temperature (max), °C (°F)	590 (1095)	705 (1300)	815 (1500)	1090 (1995)
Ductility, RT, %	15	2-4	1-3	3-10
Ductility, operating temperature, %	15	5-12	5-12	10-20

Source: J.C. Chesnutt, Titanium Aluminides for Aerospace Applications, in *Superalloys 1992*, S.D. Antolovich, R.W. Stusrud, R.A. MacKay, D.L. Anton, T. Khan, R.D. Kissinger, and D.L. Klarstrom, Ed., The Minerals, Metals and Materials Society, 1992

Titanium aluminides: Processing capability

Alloy	Ingot metallurgy	Forging	Sheet rolling	Casting
Conventional Ti	Yes	Yes	Yes	Yes
Alpha 2	Yes	Yes	With difficulty	Limited
Gamma	With difficulty	With difficulty	With difficulty	Yes

Titanium aluminides: Mechanical property comparison

Alpha-2	Gamma
1100 MPa (159 ksi) UTS with 2-3% tensile ductility at room temperature	620 MPa (90 ksi) with 3% tensile ductility at room temperature
Up to 6% tensile ductility at room temperature	550 MPa (80 ksi) UTS at 760 °C (1400 °F); 380 MPa (55 ksi) UTS at 870 °C (1600 °F)
620 MPa (90 ksi) UTS at 760 °C (1400 °F)	
Good HCF for $K_t = 1$	Excellent oxidation resistance
Good oxidation resistance	More fire resistant than conventional titanium alloys

Source: *Superalloys 1992*, S.D. Antolovich, R.W. Stusrud, R.A. MacKay, D.L. Anton, T. Khan, R.D. Kissinger, and D.L. Klarstrom, Ed., The Minerals, Metals and Materials Society, 1992

References

1. *High-Temperature Ordered Intermetallic Alloys*, Materials Research Society Symposia Proceedings, Vol 39, C.C. Koch, C.T. Liu and N.S. Stoloff, Ed., Materials Research Society, 1985
2. *High-Temperature Ordered Intermetallic Alloys II*, Materials Research Society Symposia Proceedings, Vol 81, N.S. Stoloff, C.C. Koch, C.T. Liu, and O. Izumi, Ed., Materials Research Society, 1987
3. *High-Temperature Ordered Intermetallic Alloys III*, Materials Research Society Proceedings, Vol 133, C.T. Liu, A.I. Taub, N.S. Stoloff, and C.C. Koch, Ed., Materials Research Society, 1989
4. E.A. Aitken, *Intermetallic Compounds*, J.H. Westbrook, Ed., Wiley, 1967, p 491-516
5. N.S. Choudhury, H.C. Graham, and J.W. Hinze, in *Properties of High Temperature Alloys with Emphasis on Environmental Effects*, Electrochemical Society, 1976, p 668-680
6. M. Khobaib and F.W. Vahldiek, in *Space Age Metals Technology*, Vol 2, F.H. Froes and R.A. Cull, Ed., Society for the Advancement of Material and Processing Engineering, 1988, p 262-270
7. J. Subrahmanyam, Cyclic Oxidation of Aluminated Ti-14Al-24Nb Alloy, *J. Mater. Sci.*, Vol 23, 1988, p 1906-1910

Titanium aluminides: Creep behavior comparison

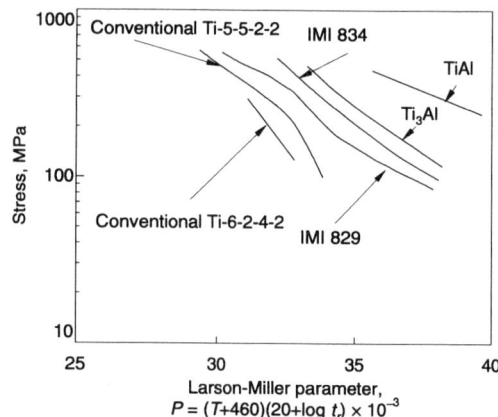

Comparison of the creep behavior of conventional titanium alloys and titanium aluminide intermetallics.
Source: F.H. Froes, *Mater. Edge*, No. 5, May 1988

Ti₃Al (α₂ or α-2)

Ti$_3$Al has an ordered DO_{19} structure that contains three independent slip systems that account for dislocation motion on the basal {0001}, prism {1010}, and pyramidal {0221} planes (Ref 1, 2). Prism slip requires only a single dislocation without creating a near-neighbor antiphase boundary, and additional slip requires movement of two dislocations (superdislocations) (Ref 3). In addition, two independent slip systems involving $\langle c + a \rangle$ slip occur to satisfy the Von Mises criterion for uniform deformation.

Compositional Stability. The α-2 (Ti$_3$Al) alloy has a wide range of compositional stability with aluminum contents of 22 to 39 at.%. The compound is congruently disordered at a temperature of 1180 °C (2155 °F) and an aluminum content of 32 at.%. The stoichiometric composition, Ti-25Al, is stable up to about 1090 °C (1995 °F) (Ref 4). Ternary phase diagrams centered around the α-2 phase have been a subject of research and debate (Ref 5-8); the Ti-Al-Nb ternary for α-2 alloys at 900 °C is well established (see "Ti$_2$Al-Nb (O-Phase)" in this introduction).

References

1. W.J.S. Yang, Observations of Superdislocation Networks in Ti3Al-Nb, *J. Mater. Sci. Lett.*, Vol 1, 1982, p 199-202
2. W.J.S. Yang, "C" Component Dislocations in Deformed Ti3Al, *Metall. Trans. A*, Vol 13, 1982, p 324
3. H.A. Lipsitt, D. Schechtman, and R.E. Schafrik, *Metall. Trans. A*, Vol 11, 1980, p 1369
4. T.B. Massalski, Ed., *Binary Alloy Phase Diagrams*, Vol 1 and 2, American Society for Metals, 1986
5. H. Bohm and K. Lohberg, Uber eine Uberstrukturphase vom CsCl-Typ im System Titan-Molybdan-Aluminum, *Z. Metallk.*, Vol 49, 1958, p 173-178
6. T.J. Jewett et al., in *High-Temperature Ordered Intermetallic Alloys III*, Materials Research Society Symposia Proceedings, Vol 133, C.T. Liu, A.I. Taub, N.S. Stoloff, and C.C. Koch, Ed., Materials Research Society, 1989, p 69-74
7. M.J. Kaufman et al., in *Sixth World Conference on Titanium*, Part II, P. Lacombe et al., Ed., Les Editions de Physique, 1989, p 985-990
8. R.G. Rowe, *High Temperature Aluminides and Intermetallics*, S.H. Whang, C.T. Liu, and D. Pope, Ed., TMS, 1990

Tensile data for P/M single-phase Ti₃Al

Temperature °C	°F	Powder type(a)	Elastic modulus GPa	10⁶	Tensile yield strength MPa	ksi	Ultimate tensile strength MPa	ksi	Fracture stress MPa	ksi	Elongation, %	Reduction of area, %	Specimen type(b)
25	77	R	...	...	...	...	...	...	533.5	77.3	0.5	0	A
		R	...	...	...	...	...	...	473.7	68.7	...	0	A
		R	...	...	...	...	...	...	337.0	48.8	...	0.4	A
		M	...	...	...	...	...	...	564.0	81.8	0.4	0	B
		M	...	...	...	...	...	...	679.9	98.6	0.3	0.2	B
		M	...	...	...	...	...	...	598.8	86.8	...	...	B
		M	...	...	...	...	...	...	576.4	83.6	...	...	B
		M	147.2	21	...	...	...	...	655.0	95.0	0	0	A
400	750	M	133.1	19	...	...	...	...	590.9	85.7	0.1	0.4	A
		R	...	...	...	...	...	...	608.9	88.3	1.6	2.2	A
		R	...	...	...	...	...	...	581.6	84.3	0.6	1.1	A
600	1110	R	...	...	570.4	82.7	...	...	571.8	82.9	...	0.8	A
		M	...	...	...	...	...	...	553.9	80.3	...	0.8	B
		M	111.7	16	...	...	...	...	514.4	74.6	0.3	0.5	A
		M	111.7	16	564.0	81.8	...	...	586.1	85.0	0.3	0.7	A
700	1290	R	...	...	397.3	57.6	...	...	642.4	93.1	3.0	3.5	A
		R	...	...	523.7	75.9	...	...	692.2	100.4	3.7	4.8	A
		R	...	...	311.7	45.2	507.5	73.6	...	...	1.8	4.7	A
		R	...	...	347.7	50.4	...	...	533.6	77.4	...	4.4	A
		M	...	...	443.8	64.3	498.9	72.3	...	...	2.7	3.6	B
		M	94.6	13.7	442.7	64.2	598.5	86.8	626.8	90.9	3.9	4.6	A
		M	94.1	13.6	412.7	59.8	476.4	69.1	...	...	0.9	2.1	A
		M	96.2	13.9	425.4	61.7	523.5	75.9	...	...	3.0	4.2	A
800	1470	R	...	...	386.2	56.0	519.9	75.4	537.2	77.9	6.7	9.1	A
		R	...	...	411.2	59.6	496.0	71.9	474.0	68.7	8.3	10.3	A
		R	...	...	247.5	35.9	410.9	59.5	...	...	7.0	9.5	A
		R	...	...	375.1	54.4	479.9	69.6	...	...	3.6	4.5	A
		R	...	...	384.7	55.8	495.6	71.9	...	...	...	4.7	A
		M	73.1	10.6	345.4	50.1	393.0	56.9	...	...	1.7	2.0	A
900	1650	R	...	...	271.2	39.3	320.6	46.5	...	...	9.0	13.6	A

(a) M indicates type A specimen manufactured from machine chip powder (Ti-15.2Al); R indicates Type A specimen manufactured from rotating electrode powder (Ti-17.1Al). (b) Specimen Type A is 57.15 mm (2.25 in.) long; strain rate, 0.036 cm/min. Specimen Type B is 22.9 mm (0.9 in.) long; strain rate, 0.064 cm/min. Source: *Metall. Trans. A*, Vol 11A, Aug 1980, p 1371

Ti₃Al: Crystal structure

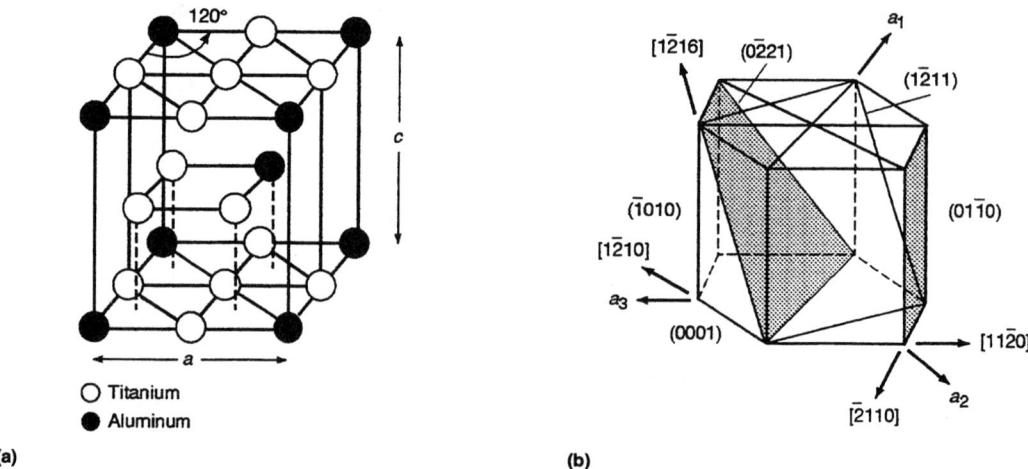

(a) Do_{19} hexagonal superlattice structure of Ti₃Al with lattice constants of c = 0.420 nm and a = 0.577 nm. (b) Possible slip planes and slip vectors in the structure.
Source: W.J.S. Yang, *J. Mater. Sci. Lett.*, Vol 1, 1982, p 199-202; W.J.S. Yang, *Metall. Trans. A*, Vol 13, 1982, p 324

Ti₃Al: Fracture mechanism map

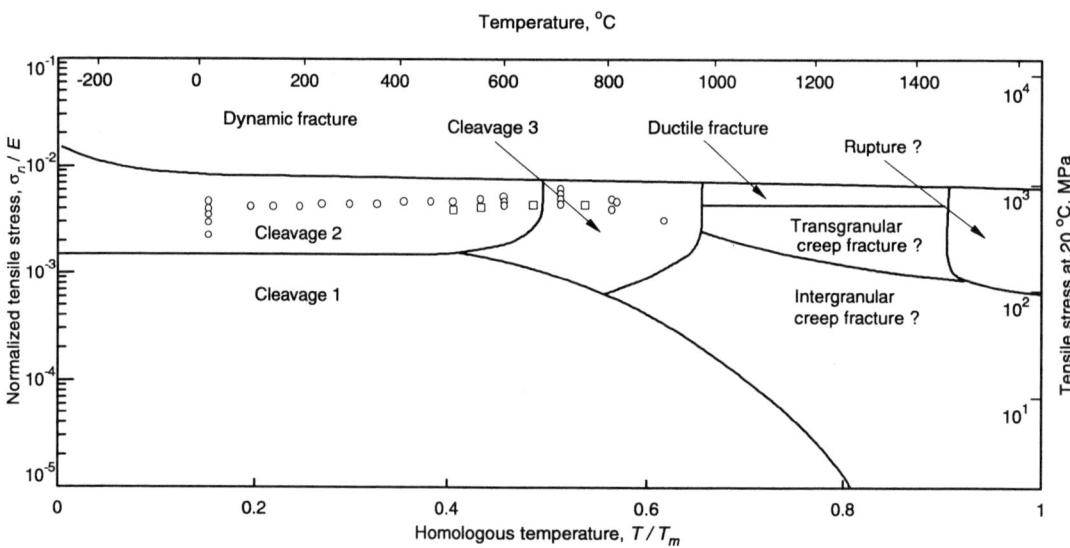

Normalizing parameters: T_m = 1928 K; E = 147.0–0.0537 T for Ti-17.1 Al and E = 148.96–0.057 T for Ti-15.2Al, where T is °C.
Source: Krishnamohanrao et al., Fracture Mechanism Maps for Ti, *Acta Metall.*, Vol 34, 1986, p 1783-1806

Ti-Al

The γ-TiAl phase has an $L1_0$ ordered face-centered tetragonal structure (Ref 1-3), which has a wide range (49 to 66 at.% Al) of temperature-dependent stability (Ref 1). At the equiatomic TiAl composition, the c/a ratio is 1.02; tetragonality increases up to c/a = 1.03 with increasing aluminum concentration (Ref 4-6). Within the compositional range specified at off-stoichiometric compositions, excess titanium or aluminum atoms occupy antisites without creating constitutional vacancies (Ref 7). The γ-TiAl phase apparently remains ordered up to its melting point of approximately 1450 °C (2640 °F).

The layered arrangement of titanium and aluminum atoms on successive (002) planes and the slight tetragonality of c/a = 1.02 gives rise to two types of dislocations with 1/2⟨110⟩-type Burgers vectors on {111} in γ-TiAl: ordinary dislocations 1/2⟨110⟩ and superdislocations ⟨011⟩ = 1/2⟨011⟩ + 1/2⟨011⟩ that will leave the superlattice undisturbed. Another superdislocation, with a Burgers vector of 1/2⟨112⟩, has also been suggested (Ref 8, 9). The superdislocation core can dissociate further into other complex partial dislocations, which are energetically more favorable, involving planar defects such as stacking faults and antiphase boundaries (Ref 8-10). The 1/6⟨112⟩ on {1̄11} partials form twin dislocations, but 1/6⟨2̄11⟩ partials

Tensile data for single-phase TiAl (60Ti-40Al at. %)

Temperature °C	Temperature °F	Remarks or pretreatment	Elastic modulus GPa	Elastic modulus 10^6 psi	Tensile yield strength (0.2%) MPa	Tensile yield strength (0.2%) ksi	Ultimate tensile strength MPa	Ultimate tensile strength ksi	Fracture stress MPa	Fracture stress ksi	Elongation, %	Reduction of area, %
25	77	As extruded	161.0	23.3	...	...	...	...	445.4	64.6	~0.1	0.75
			176.9	25.6	...	...	...	...	486.1	70.5	...	0.69
		Electropolished	174.1	25.2	...	...	...	...	487.5	70.7	...	...
600	1110	As extruded	162.0	23.5	417.8	60.6	...	...	477.8	69.3	0.47	1.65
			151.7	22.0	404.0	58.6	...	...	473.0	68.6	0.78	2.08
700	1290	As extruded	159.6	23.1	359.2	52.1	...	...	426.8	61.9	0.62	1.03
			147.2	21.3	368.2	53.4	...	...	450.9	65.4	0.78	1.95
800	1470	As extruded	128.0	18.5	301.0	43.6	352.3	51.1	359.9	52.2	10.45	13.8
			133.8	19.4	286.1	41.5	324.1	47.0	348.9	50.6	9.21	19.35
		High strain rate	132.8	19.2	335.1	48.6	...	...	375.8	54.5	0.31	0.67
900	1650	As extruded	115.3	16.7	148.2	21.5	219.3	31.8	216.5	31.4	27.6	39.0
			115.3	16.7	147.5	21.4	220.3	31.9	191.7	27.8	35.9	39.7
		High strain rate	116.0	16.8	286.1	41.5	301.3	43.7	322.7	46.8	8.42	9.45
		Exposed to air for: 168 h at 900 °C	115.3	16.7	164.4	23.8	174.1	25.2	194.1	28.1	55.0	67.5
		100 h at 900 °C	...	...	172.7	25.0	179.6	26.0	...	...	46.8	62.3
		100 h at 1000 °C	...	...	158.9	23.0	182.7	26.5	137.9	20.0	28.1	32.4
1000	1830	As extruded	91.0	13.2	111.7	16.2	129.3	18.7	(a)		(a)	
			95.8	13.9	131.7	19.1	167.5	24.3	...	...	...	37.2

(a) Retested at room temperature. Source: *Metall. Trans. A*, Vol 6, Nov 1975, p 1993

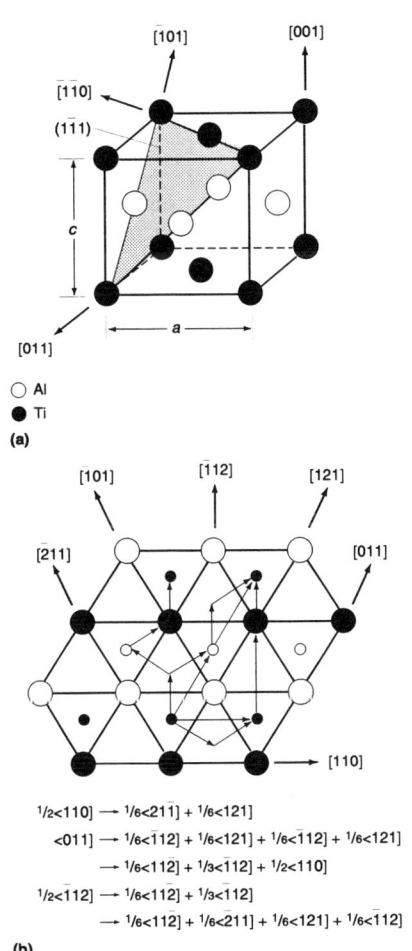

γ-TiAl: Crystal structure

(a) Ordered face-centered tetragonal ($L1_0$) TiAl structure. Shaded area represents the (111) plane. (b) Slip dislocations on (111) plane, ordinary dislocations 1/2⟨110⟩, superdislocations ⟨011⟩ and 1/2⟨112⟩, and twin dislocations 1/6⟨112⟩ with possible dissociations. Source: E.S. Bumps et al., *Trans. AIME*, Vol 194, 1952, p 609-614

are forbidden as twinning dislocations in the $L1_0$ structure (Ref 11, 12).

In single-phase γ alloys containing 52 to 54 at.% Al, deformation at room temperature occurs by motion of both ordinary and superdislocations; however, the superdislocations [011] and [101] are largely immobile because segments of the trailing 1/6[112]-type superpartials form faulted dipoles that must be extended as deformation progresses (Ref 8-10, 12, 13). Increasing temperature and decreasing aluminum content increase the 1/2⟨110⟩ slip activity as the faulted dipoles disappear and twinning dominates (Ref 13, 14). In two-phase Ti-48Al, the deformation modes of primary γ grains are twinning with ⟨112⟩ twin dislocations and slip by 1/2[110]-type dislocations.

The extremely low ductility values at ambient temperature and the increased ductility with increasing temperatures strongly influence the observed fracture mode. Tensile and fatigue specimens indicate that the predominant fracture modes are cleavage at low temperatures due to dislocation pile-up and intergranular fracture above the brittle-ductile transition temperature (Ref 13-17).

References

1. H.R. Ogden et al., Constitution of Titanium-Aluminum Alloys, *Trans. AIME*, Vol 191, 1951, p 1150-1155
2. D. Clark, K.S. Kepson, and G.I. Lewis, A Study of the Titanium-Aluminum System up to 40 at.% Aluminum, *J. Inst. Met.*, Vol 91, 1962-1963, p 197
3. Y.-W. Kim, Intermetallic Alloys Based on Gamma Titanium Aluminide, *J. Met.*, Vol 41 (No. 7), 1989, p 24-30
4. E.S. Bumps, H.D. Kessler, and M. Hansen, Titanium-Aluminum System, *Trans. AIME*, Vol 194, 1952, p 609-614
5. P. Duwez and J.L. Taylor, Crystal Structure of TiAl, *J. Met.*, 1952, p 70
6. S.C. Huang, E.L. Hall, and M.F.X. Gigliotti, in *High-Temperature Ordered Intermetallic Alloys II*, Materials Research Society Symposia

TiAl: Fracture mechanism map

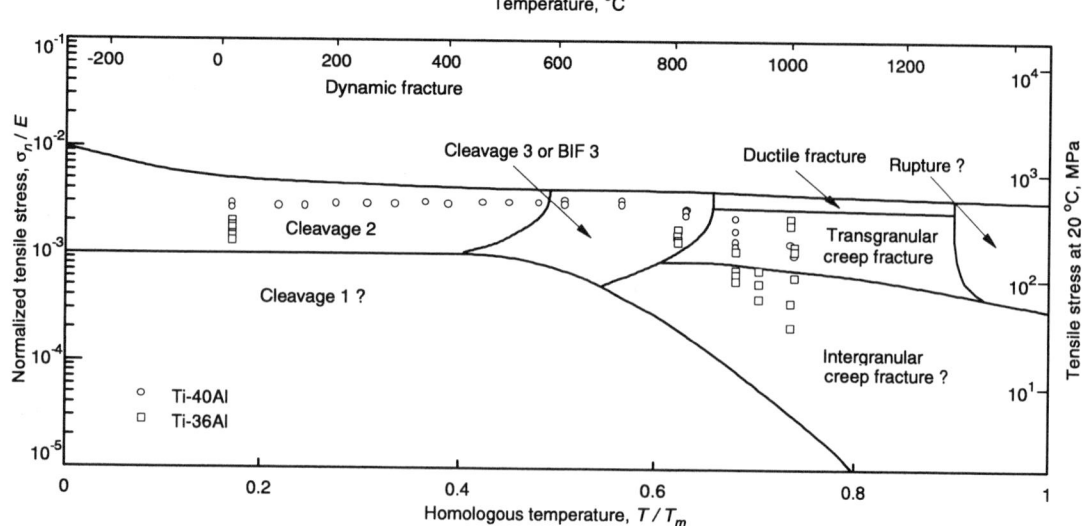

Normalizing parameters: T_m = 1743 K, E = 173.6–0.034 T (in °C).
Source: Krishnamohanrao et al., Fracture Mechanism Maps for Ti, Acta Metall., Vol 34, 1986, p 1783-1806

Proceedings, Vol 81, N.S. Stoloff, C.C. Koch, C.T. Liu, and O. Izumi, Ed., Materials Research Society, 1987, p 481-486
7. R.P. Elliott and W. Rostoker, The Influence of Aluminum on the Occupation of Lattice Sites in the TiAl Phase, Acta Metall., Vol 2, 1954, p 884-885
8. G. Hug, A. Loiseau, and A. Lasalmonie, Nature and Dissociation of the Dislocations in TiAl Deformed at Room Temperature, Philos. Mag. A, Vol 54 (No. 1), 1986, p 47-65
9. T. Kawabata and O. Izumi, Dislocation Structures in TiAl Single Crystals Deformed at 77K, Scr. Metall., Vol 21, 1987, p 433-434
10. G. Hug, A. Loiseau, and P. Veyssiere, Weak-Beam Observation of a Dissociation Transition in Ti-Al, Philos. Mag. A, Vol 57 (No. 3), 1988, p 499-523
11. D.W. Pashley, J.L. Robertson, and M.J. Stowell, The Deformation of CuAuI, Philos. Mag. A, 8th series, Vol 19, 1969, p 83
12. D. Schechtman, M.J. Blackburn, and H.A. Lipsitt, The Plastic Deformation of TiAl, Metall. Trans., Vol 5, 1974, p 1373
13. H.A. Lipsitt, D. Schechtman, and R.E. Schafrik, The Deformation and Fracture of TiAl at Elevated Temperatures, Metall. Trans. A, Vol 6, 1975, p 1991
14. E.L. Hall and S.-C. Huang, in High-Temperature Ordered Intermetallic Alloys III, Materials Research Society Symposia Proceedings, Vol 133, C.T. Liu, A.I. Taub, N.S. Stoloff, and C.C. Koch, Ed., Materials Research Society, 1989, p 693-698
15. T. Kawabata and O. Izumi, Dislocation Reactions and Fracture Mechanism in TiAl $L1_0$ Type Intermetallic Compound, Scr. Metall., Vol 21, 1987, p 435-440
16. T. Kawabata et al., Bend Tests and Fracture Mechanisms of TiAl Single Crystals at 293-1083 K, Acta Metall., Vol 36 (No. 4), 1988, p 963-975
17. S.M.L. Sastry and H.A. Lipsitt, Fatigue Deformation of TiAl Base Alloys, Metall. Trans. A, Vol 8, 1977, p 299

Ti_2Al-Nb (O Phase)

At higher niobium levels, the α_2 phase evolves to a new ordered orthorhombic structure that is based on the composition Ti_2AlNb (O phase), which has been observed in titanium aluminides with compositions from Ti-25Al-12.5Nb to Ti-25Al-30Nb (Ref 1-4). The crystal structures of the $\alpha2$ and ordered orthorhombic phases are compared in the accompanying figure, which shows the basal planes and atomic positions in the lanes above and below the plane of sheet.

In the ordered orthorhombic Ti_2AlNb phase, Banerjee and Mozer et al., have independently obtained results that are consistent with a preferential occupation of one of the DO_{19} titanium subsites by niobium atoms (Ref 3, 5). This results in rearrangement to a structure with orthorhombic rather than hexagonal symmetry. The triangular regions superimposed on the diagram reveal this distortion in the atomic positions of the next layer of atoms in each structure.

Single-phase alloys of ordered orthorhombic have excellent creep resistance, particularly after beta heat treatment. However, a fine two-phase structure of O and ordered beta (β_o) leads to significant strengthening and better room-temperature ductility and fracture resistance than single-phase O alloys. The O phase exhibits better strength and toughness than Ti_3Al. Additional information on alloys with O phase is provided in the following article "Ti_3Al Alloys."

References

1. R.G. Rowe, The Mechanical Properties of titanium Aluminides Near Ti-25Al-25Nb, in *Microstructure/Property Relationships in Titanium Alloys and Titanium Aluminides*, Y. Kim and R.R. Boyer, Ed., TMS/AIME, 1990, p 387-398
2. R.G. Rowe, "Recent Developments in Ti-al-Nb Alloys," in *High Temperature Aluminides and Intermetallics*, S.H. Whang, D.P. Pope, and J.O. Stigler, Ed., TMS/AIME, 1990, p 375-401
3. D. Banerjee, A New Ordered Orthorhombic Phase in a Ti_3Al-Nb Alloy, *Acta Metall.*, Vol 36, 1989, p 871-882
4. M.J. Kaufman, T.J. Broderick, C.H. Ward, and R.G. Rowe, Phase Relationships in the Ti_3Al + Nb System, in *Sixth World Conference on Titanium*, P. Lacombe, R. Tricot, and G. Beranger, Ed., Les Editions de Physique, 1989, p 985-990
5. B. Moser, L.A. Bendersky, W.J. Boettinger, and R.G. Rowe, Neutron Powder Diffraction Study of the Orthorhombic Ti_2AlNb Phase, *Scr. Metall. Mater.*, Vol 24, 1990, p 2363-2368

Basal plane schematics for the α_2 (DO_{19} and O) (ordered orthorhombic) phases

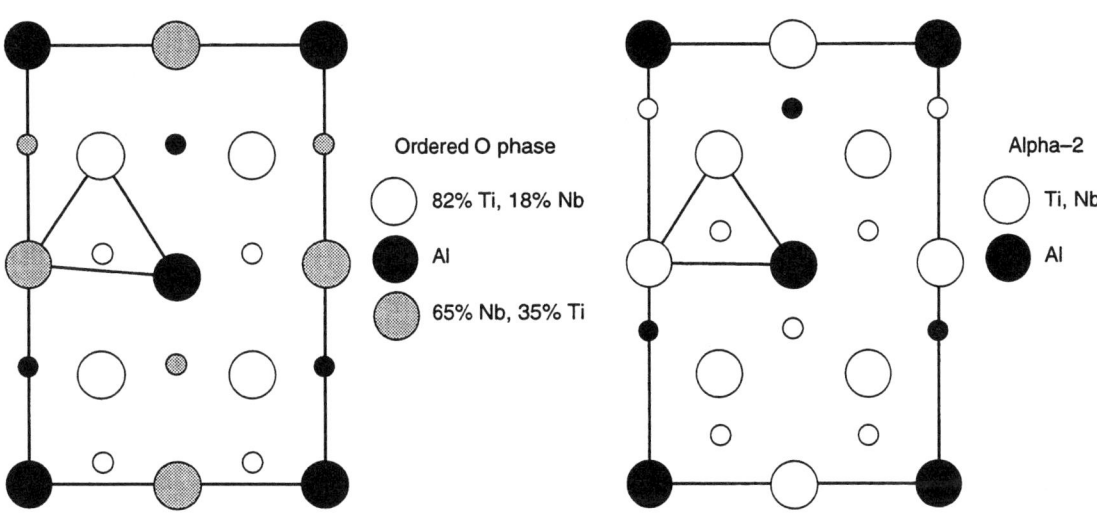

The large circles indicate atomic positions in the plane of the paper. Smaller circles represent atoms in planes above and below the plane of the paper.
Source: B. Mozer et al., *Scr. Metall. Mater.*, Vol 24, 1990, p 2363-2368

Ti-Al-Nb ternary section at 900 °C

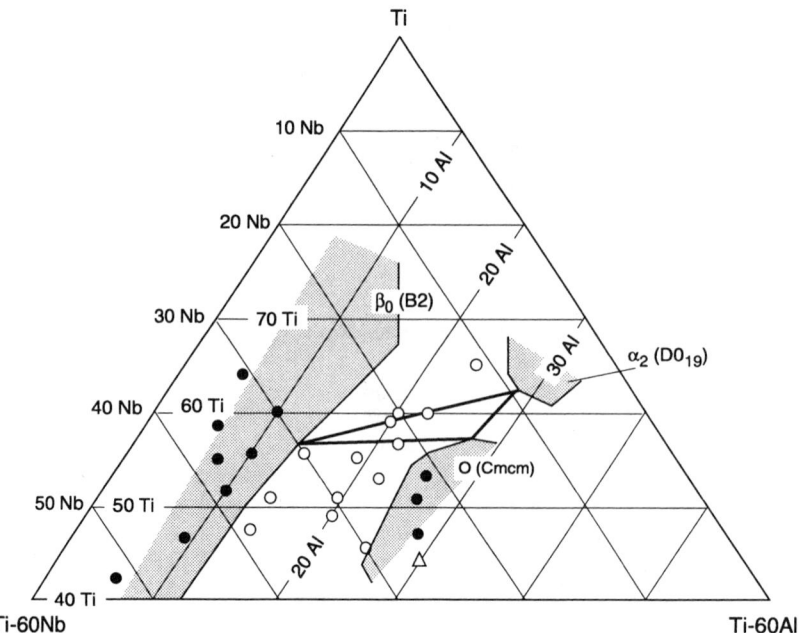

The solid points were single phase at 900 °C, the open points had two phases, and open triangles divided into three segments contained three phases at 900 °C, an ordered orthorhombic phase with Cmcm symmetry lies between 25 and 28 at.% Al for Nb contents of 15 to 30 at.%. The β_o phase had an orderd B2 (csCl structure) crystal structure for compositions near the aluminum solubility limit for this phase (approximately 13-15 at.% Al). Disagreement between the β_o phase boundary compositions and β_o tie-line compositions suggested that equilibration of the β_o phase was not achieved in 600 hours at 900 °C over the 10 to 12 at.% Al composition range between the β_o and O phases.
Source: R.G. Rowe, et al., Phase Equilibria in Ti-Al-Nb Alloys near Ti$_2$NbAl, Seventh World Conference on Titanium, 1993

Ti₃Al Alloys

Common Name: Alpha-2 Aluminide Alloys

The semicommercial and experimental Ti₃Al alloys developed to date are two phase (α-2 + β/B2), with contents of 23 to 25 at.% Al and 11 to 18 at.% Nb. Alloy compositions with current engineering significance are Ti-24Al-11Nb (Ref 1, 2), Ti-25Al-10Nb-3V-1Mo (Ref 3), Ti-25Al-17Nb-1Mo (Ref 4), and modified alloy compositions such as Ti-24.5Al-6Nb-6(Ta, Mo, Cr, V).

The current understanding of Ti₃Al, based from a recent review article (D. Banerjee et al., Physical Metallurgy of Ti₃Al Based Alloys, International Symposium on Structural Intermetallics, TMS/AIME) suggests the following:

- The O phase should be the major constituent of tough alloys and therefore these should contain typically greater than 15% Nb
- Nb is also desirable for high-temperature strength and stress rupture capability
- Al contents greater than 25% are unacceptably detrimental to toughness
- Mo additions of about 1% are desirable for high-temperature strength
- Only lath structures provide adequate creep resistance but must be optimised for strength, ductility, and creep
- No real problems associated with processing exist

Much of the early work on the α-2 titanium aluminides was conducted on Ti-24Al-11Nb (at.%). More recently, work has focused on characterizing Ti-25Al-10Nb-3V-1Mo, which has higher tensile strength and creep resistance than Ti-24Al-11Nb. Of late, some work has focused on Ti-24.5Al-12.5Nb-1.5Mo, which has improved oxidation resistance over Ti-25Al-10Nb-3V-1Mo with similar mechanical properties. Alloys with a two-phase microstructure of ordered orthorhombic aluminide (Ti₂Al-Nb) and ordered beta (β₀) are also being studied at higher niobium contents (Ti-25Al-25Nb, at.%). Ordered orthorhombic (O) appears to improve fracture toughness and creep resistance. In contrast niobium additions to Ti₃Al generally enhance most material properties, although excessive niobium can degrade creep performance of two-phase Ti₃Al alloys (α-2 + β/B2). Niobium can be replaced by specific elements for improved strength (molybdenum, tantalum, or chromium), creep resistance (molybdenum), and oxidation resistance (tantalum, molybdenum).

Material Processing. Microstructural features that can be varied by thermomechanical processing include primary α-2 grain size and volume fraction, secondary α-2 plate morphology and thickness, and the presence of secondary β grains (Ref 5). Beta processing generally results in elongated Widmanstätten α-2 in large primary β grains in a manner similar to that in conventional titanium alloys (Ref 6).

Up to 4 wt% H can be dissolved in titanium alloys at elevated temperatures. This hydrogen can then be used to improve processability, and final mechanical properties are enhanced after its removal. Removal of the hydrogen can be easily achieved by vacuum annealing (Ref 7, 8). This thermomechanical processing technique allows titanium aluminides to be processed at reduced temperatures (Ref 7-10) and results in a finer microstructure (Ref 7, 8, 10, 11).

Product Forms/Applications. Although a number of demonstration parts have been produced and tested, it is not currently used on a production basis. The material has been produced in most forms including castings, forgings, extrusions, and rolled sheet. Its mechanical properties are strongly effected by microstructure and processing. For example, the room-temperature tensile ductility of rolled sheet material may exceed 20% elongation, whereas that of cast or β heat treated material may be less than 1% elongation. Microstructural response to various processing paramcters have been identified (Met. Trans., 23A, 295-305; and P. Sagar Mater Sci & Engr, 1993).

Phases and Structures[*]

Ternary phase diagrams centered on the Ti₃Al phase and the pseudobinary Ti₃Al-Nb (O-phase) system have been a subject of considerable research (Ref 12-15). The ternary Ti-Al-Nb section at 900 °C has been well established (see the previous introductory article "Titanium Aluminides" in this Volume). The O phase is stable over a large range of Nb contents, and it can dissolve a greater fraction of beta stabilizers such as Mo and V into solid solution than can the α-2 phase. Solid-solution effects on O phase properties may be a topic of research.

The alloys that are being studied for potential engineering use generally contain high levels of β stabilizers such as niobium and, as a result, contain a small volume fraction of β phase, which also may be ordered. When cooled rapidly from heat treatment or processing temperatures, the microstructure often contains an ordered orthorhombic

---

[*] The section "Microstructural Stability" was extracted from an article by R.G. Rowe, "Advanced Ti3Al-Phase Alloy Property Comparison," *Synthesis, Processing and Modelling of Advanced Materials*, ASM International, 1991

(O) phase in addition to the ordered hexagonal and β phases. At niobium contents of less than 5 at.%, a martensitic transformation to the α phase occurs during rapid cooling from the high-temperature β phase (Ref 14, 16, 17).

Increased niobium contents suppress the martensitic transformation, and β or $B2$ can be retained at room temperature (Ref 14, 16, 18). This quenched-in $B2$ phase is metastable, and it contains transitional microstructures and phases such as antiphase boundaries (Ref 16, 17), "tweed" microstructures (Ref 14, 16, 19, 20), the O phase and fine ω phase particles (Ref 19). The ω transforms to α-2 on heat treatment in a similar manner to the ω → α transformation that occurs in conventional terminal titanium alloys (Ref 6).

Microstructural Stability.[*] Ti_3Al-based alloys all appear to have microstructures that are kinetically stable, but not necessarily in equilibrium at potential service temperatures. This occurs because of the slow redistribution of aluminum and niobium between $α_2$ and $β_o$ phases at low temperatures. It has been shown in Ti-25Al-12.5Nb, however, that there is a large difference in the $α_2$ and $β_o$ or β phase compositions between service temperatures (650 °C, or 1200 °F) and heat treatment temperature (900 to 1000 °C, or 1650 to 1830 °F) (Ref 21, 22). In Ti-25Al-12.5Nb, annealing for 1 h at 1040 °C (1905 °F) led to a compositional separation into $β_o$ phase of the composition Ti-22.8Al-18.1Nb and an $α_2$ phase of the composition Ti-28.1Al-11.3Nb. This heat treatment was chosen because it is characteristic of the two-phase annealed microstructures for many titanium aluminides and represents the microstructure in which many alloys are evaluated. After aging at 650 °C (1200 °F) for 68 h, disordered β phase precipitation at grain boundaries had the composition Ti-6.2Al-44.5Nb.

Titanium aluminides heat treated at temperatures much higher than their service temperature hence appear to have a relatively large chemical driving force to shift to new $α_2$ and β phase compositions. In the case of Ti-25Al-12.5Nb, the $β_o$ phase at elevated temperatures changed to disordered low aluminum bcc β phase after aging at 650 °C (1200 °F). The slow kinetics of bulk diffusion at 650 °C (1200 °F) appears to have lead to discontinuous precipitation.

It has been observed that aging Ti-25Al-10Nb-3V-1Mo at 870 °C (1600 °F) after oil quenching from 1085 °C (1990 °F) leads to cellular recrystallization or autorecrystallization (Ref 23). Given the large phase composition shifts observed in Ti-25Al-12.5Nb between 1040 and 650 °C (1905 and 1200 °F), discontinuous precipitation may also occur in Ti-25Al-10Nb-3V-1Mo. Discontinuous precipitation has been observed in Ti-25Al-30Nb after aging 100 h at 870 °C (1600 °F) (Ref 24). It was not observed in Ti-22Al-27Nb aged for the same time. This suggests that high-niobium O + $β_o$ alloys also have the potential for discontinuous precipitation, but that for Ti-22Al-27Nb, the kinetics may be slower than for Ti-25Al-30Nb or Ti-25Al-10Nb-3V-1Mo.

Ti-24.5Al-12.5Nb-1.5Mo: Beta approach curve

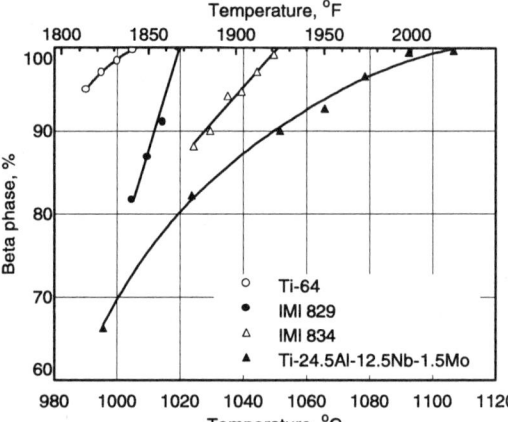

Source: B. Marquardt (General Electric), *7th World Conference on Titanium*

Ti-24.5Al-12.5Nb-1.5Mo: TTT diagram

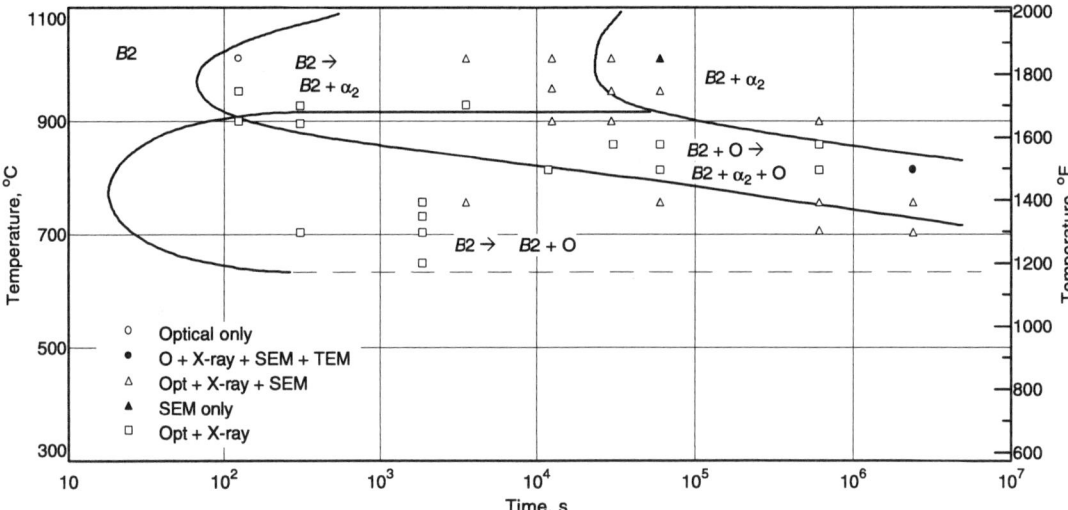

The symbols represent the actual time-temperature combinations that were evaluated.
Source: A. Bartz, "The Time-Temperature-Transformation Kinetics of Ti-24.5Al-12.5Nb-1.5Mo Alloys," Masters thesis, Wright State University, 1992

Oxidation

Isothermal oxidation behavior of Ti_3Al base alloys is not very different from IMI 834 at 650 to 700 °C.

Isothermal oxidation in air at 705 °C

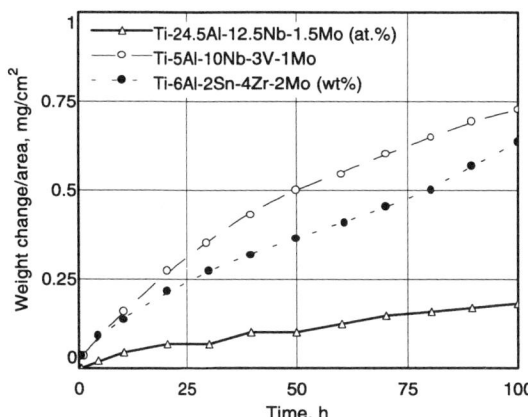

Source: J.C. Schaeffer, Isothermal Oxidation Behavior of Alpha-2 Titanium Aluminide Alloys, *Scr. Metall.*, submitted for publication

Isothermal oxidation in air at 650 °C

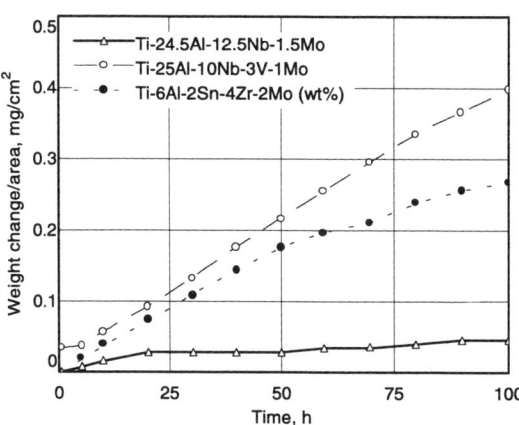

Source: J.C. Schaeffer, Isothermal Oxidation Behavior of Alpha-2 Titanium Aluminide Alloys, *Scr. Metall.*, submitted for publication

Mechanical Properties

Typical mechanical properties for a number of α-2 alloys are listed in the accompanying table. High-niobium (O phase) alloys with the orthorhombic Ti_2Al-Nb ordered phase are listed separately. Large gains in toughness can be achieved when the orthorhombic (O) phase forms as the major constituent rather than the α-2. The best combinations of strength and ductility are obtained when $α_2$/O phase particles are completely encased by B2 films in the microstructure (D. Banerjee et al., Physical Metallurgy of Ti_3Al Based Alloys, International Symposium on Structural Intermetallics, TMS/AIME, 1993).

Properties of α-2 Ti_3 Al alloys with various microstructures

Alloy	Microstructure(a)	Tensile yield strength MPa	ksi	Ultimate tensile strength MPa	ksi	Elongation, %	Plane-strain fracture toughness (K_{Ic}) MPa√m	ksi √in.	Creep rupture(b)
Ti-25Al	E	538	78	538	78	0.3	...	...	...
Ti-24Al-11Nb	W	787	114	824	119	0.7	...	...	44.7
	FW	761	110	967	140	4.8	...	...	...
Ti-24Al-14Nb	W	831	120	977	142	2.1	...	...	59.5
Ti-25Al-10Nb-3V-1Mo	W	825	119	1042	151	2.2	13.5	12.3	>360
	FW	823	119	950	138	0.8	...	...	...
	C+P	745	108	907	132	1.1	...	...	...
	W+P	759	110	963	140	2.6	...	...	...
	FW+P	942	137	1097	159	2.7	...	...	...
Ti-24.5Al-17Nb	W	952	138	1010	146	5.8	28.3	25.7	62
	W+P	705	102	940	136	10.0	...	...	...
Ti-25Al-17Nb-1Mo	FW	989	143	1133	164	3.4	20.9	19.0	476

(a) E, equiaxed α-2; W, Widmanstätten; FW, fine Widmanstätten; C, colony structure; P, primary α-2 grains. (b) Time to rupture, h, at 650 °C (1200 °F) and 380 MPa (55 ksi). Source: Y.-W. Kim and F.H. Froes, in *Proc. Symp. High-Temperature Aluminides and Intermetallics*, TMS, in press

Ti_3Al Alloys

Production of two-phase alloys by alloying Ti_3Al with β-stabilizing elements results in up to a doubling of strength. Interface strengthening of the two-phase mixture appears to be predominantly responsible for the increased strength, but other strengthening factors, such as long-range order, solid solution, and texture effects, also contribute (Ref 5, 25). A fine Widmanstätten microstructure with a small amount of primary α-grains exhibits better ductility than microstructures with a coarse Widmanstätten microstructure or an aligned acicular α-2 morphology (Ref 5).

The fatigue properties of titanium alloys are strongly influenced by microstructure, and work

Ti-24Al-11Nb: Tensile and fracture properties

Microstructure	Temperature °C	Temperature °F	Tensile yield strength MPa	Tensile yield strength ksi	Ultimate tensile strength MPa	Ultimate tensile strength ksi	Young's modulus GPa	Young's modulus 10^6 psi	Total elongation, %	Fracture strain, %	K_{Ic} MPa√m	K_{Ic} ksi √in.
Equiaxed $\alpha_2 + \beta$	25	77	648.8	94.1	692.9	100.5	193.8	13.6	2.11	4.2	20.4	18.5
	600	1110	378.8	54.9	636.5	92.3	53	7.7	38.8	64.1	12.5	11.3
Coarse basketweave	25	77	461.9	66.9	688.7	100	90	13.0	3.8	4.0	22.5	20.4
	600	1110	344.9	60.0	681.6	98.8	60	8.7	20.0	29.5	18.2	16.5
Fine basketweave	25	77	688.4	99.8	842.6	122.2	77	11.1	3.1	3.2	15.0	13.6
	600	1110	442.1	64.1	767.8	111.3	76	11.0	24.4	35.1	19.8	18.0

Source: K.S. Chan, Influence of Microstructure on Intrinsic and Extrinsic Toughening in an Alpha-Two Titanium Aluminide Alloy, *Metall. Trans. A*, Vol 23, Jan 1992, p 188

Ti-24.5Al-12.5Nb-1.5Mo: Room-temperature tensile properties

Young's modulus GPa	Young's modulus 10^6 psi	Tensile yield strength MPa	Tensile yield strength ksi	Ultimate tensile strength MPa	Ultimate tensile strength ksi	Elongation, %	Reduction of area, %
94.4	13.7	563	81.6	806	116.9	7.0	9.9
100.2	14.5	672	97.4	...	...	...	6.8
115.1	16.7	726	105.3	958	138.9	6.2	9.4
113.7	16.5	745	108.0	896	129.9	1.6	10.7
104.8	15.2	581	84.2	821	119.0	6.5	8.7
104.5	15.1	632	91.6	725	105.1	0.8	9.2
105.9	15.3	753	109.2	967	140.2	5.9	6.7
106.1	15.4	711	103.1	904	131.1	2.1	4.0
90.2	13.1	556	80.6	745	108.0	3.5	6.7
88.6	12.8	618	89.6	656	95.1	0.3	1.1
104.0	15.0	809	117.3	996	144.4	3.6	7.6
101.6	14.7	727	105.4	812	117.7	0.6	4.5

Note: Property variations for different microstructural conditions as reported in C.H. Ward et al., "Microstructure, Tensile Ductility, and Fracture Toughness of Ti-25Al-10Nb-3V-1Mo," *Seventh World Conference on Titanium*, to be published

on these alloys (Ref 6) suggests that high-ductility alloys perform best under low-cycle fatigue (LCF) conditions (Ref 26). The low ductility exhibited in material with Widmanstätten α plates in α + β alloys is responsible for reduced high-temperature LCF strength. Early data (Ref 27) suggests that fatigue crack growth rate is relatively insensitive to microstructure, although the coarse Widmanstätten microstructure exhibits the slowest fatigue crack growth rate at low stress intensities. Fracture toughness appears to depend on microstructures as well as alloy composition, but the precise relationship is yet to be defined (Ref 5).

Ti_2Nb-Al Alloys

Ti-22Al-27Nb: Air tensile tests

Test temperature °C	Test temperature °F	Aging treatment	Tensile yield strength MPa	Tensile yield strength ksi	Ultimate tensile strength MPa	Ultimate tensile strength ksi	Elongation, %
22	72	None	1056	153	1152	167	3.4
		None	1028	149	1083	157	2.2
		540 °C (1000 °F), 100 h	1083	157	1166	169	3.3
		540 °C (1000 °F), 100 h	1090	158	1159	168	2.8
		650 °C (1200 °F), 100 h	1090	158	1145	166	2.6
		650 °C (1200 °F), 100 h	1076	156	1145	166	2.5
		760 °C (1400 °F), 100 h	987	143	1076	156	5.2
		760 °C (1400 °F), 100 h	966	140	1083	157	5.0
540	1000	None	849	123	1007	146	14.3
		None	856	124	1049	152	14.3
		540 °C (1000 °F), 100 h	876	127	1049	152	17.9
		540 °C (1000 °F), 100 h	890	129	1070	155	16.1
650	1200	None	794	115	938	136	14.3
		None	807	117	945	137	12.5
		650 °C (1200 °F), 100 h	794	115	938	136	10.7
		650 °C (1200 °F), 100 h	807	117	952	138	10.7
760	1400	None	559	81	787	114	10.7
		None	593	86	766	114	14.3
		760 °C (1400 °F), 100 h	462	67	649	94	21.4
		760 °C (1400 °F), 100 h	552	80	731	106	16.1

Note: Two-phase O-β_o alloy (Ti-22Al-27Nb) heat treated at 1000 °C (1830 °F), Ar + 760 °C (1400 °F), 50 h. Source: R.G. Rowe et al., *Mater. Res. Soc. Symp. Proc.*, Vol 213, 1991, p 703-708

O-phase alloys: Vacuum tensile properties

Heat treatment, °C	Test temperature °C	Test temperature °F	Tensile yield strength MPa	Tensile yield strength ksi	Ultimate tensile strength MPa	Ultimate tensile strength ksi	Ultimate elongation, %	Elongation to failure, %
Ti-25.3Al-21Nb (O + α₂)								
1050 + 815	22	72	847	123	881	127	0.4	0.4
1050 + 815	650	1200	680	98	936	135	17.5	18.1
1175 + 760	22	72	...	...	924	134	0.1	0.1
1175 + 760	315	600	894	129	946	137	0.4	0.6
1175 + 760	480	895	765	111	960	139	2.1	2.1
1175 + 760	650	1200	730	106	945	137	2.5	2.5
1175 – 760	760	1400	615	89	778	113	2.5	2.9
Ti-21.7Al-25.3Nb (O + β₀)								
1000 + 815	22	72	1245	180	1415	205	4.6	4.6
1000 + 815	650	1200	1005	145	1110	161	3.3	9.9
1125 + 815	22	72	1134	164	1175	170	0.9	0.9
1125 + 815	315	600	977	141	1163	168	7	7.2
1125 + 815	480	895	929	134	1132	164	8.7	10.4
1125 + 815	650	1200	878	127	1014	147	2.4	3.1
1125 + 815	760	1400	734	106	822	119	1	1.2
Ti-21.8Al-27Nb (O + β₀)								
815	22	72	1294	187	1415	205	3.5	3.6
815	316	600	1181	171	1428	207	7.9	8.3
815	480	895	1127	163	1359	197	7.3	10.2
815	480	895	1210	175	1439	208	7.5	9.3
815	650	1200	1116	162	1257	182	4.5	8.3
815	650	1200	1125	163	1292	187	5.5	8.8
815	760	1400	867	125	954	138	1.2	11.3

Note: Compositions are given in at.%. Source: R.G. Rowe et al., Mater. Res. Soc. Symp. Proc., Vol 213, 1991, p 703-708

The titanium aluminide Ti_2AlNb with an ordered orthorhombic crystal structure rather than the ordered hexagonal DO_{19} structure of Ti_3Al was stronger and has higher fracture toughness than conventional $α_2$ alloys. This is also the case for two-phase O + $β_o$ alloys. The fracture toughness of the two-phase O + $β_o$ alloy Ti-22Al-27Nb has been shown to be 28 MPa√m (25 ksi √in.)

High-Temperature Strength

Specific yield strengths at high temperatures are a principal advantage of titanium aluminides relative to nickel-base superalloys (see figure). However, only the recent high-niobium (O + $β_o$) phase alloys exceed the specific strength of IN718 over the entire temperature range. Ti-22Al-27Nb had the highest specific strength and had 3.3% tensile ductility at room temperature.

In the direct aged condition, it had a two-phase fine Widmanstätten O + $β_o$ microstructure between prior single-phase O grains. A similar alloy, Ti-24.6Al-23.4Nb, also had a two phase O + $β_o$ microstructure, but with a very low volume fraction of $β_o$ phase. The direct aged microstructure of this alloy consisted of single-phase O grain surrounded by $β_o$ phase. Its strength was less than that of Ti-22Al-27Nb, but it exhibited comparable room-temperature ductility. Its specific yield strength was comparable to that of Ti-24Al-17Nb-1Mo. The properties of Ti-24.6Al-23.4Nb reflect the properties of a nearly single-phase O alloy.

Creep Resistance. The creep-rupture properties of Ti-25Al-10Nb-3V-1Mo and other advanced titanium aluminide alloys are comparable, on a density-corrected basis, to that of nickel-base superalloys (Ref 28). However, titanium aluminides, like most titanium alloys, exhibit relatively large primary creep strains, often above 0.2% strain. Thus, although they may exhibit low steady-state creep rates, titanium aluminide creep lifetimes are thus determined largely in the primary creep domain (Ref 21). A comparison of the creep resistance of recent titanium aluminides with that of the nickel-base superalloy IN718 is shown by the plot of creep stress vs Larson-Miller parameter for 0.2% creep strain (see figure). The Larson-Miller parameter for the plot was computed using a constant of 12.5, which has been found to produce the best fit over a wide range of creep stresses for titanium aluminide alloys (Ref 29). Correcting IN718 for its density by dividing its creep stress by the ratio of the density of IN718 to the average density of the titanium aluminide alloys, shows that IN718 has considerably greater creep resistance than these titanium aluminide alloys. Therefore, applications where specific creep resistance is controlling appear to favor superalloys when section thicknesses are adequate. For this reason, much of

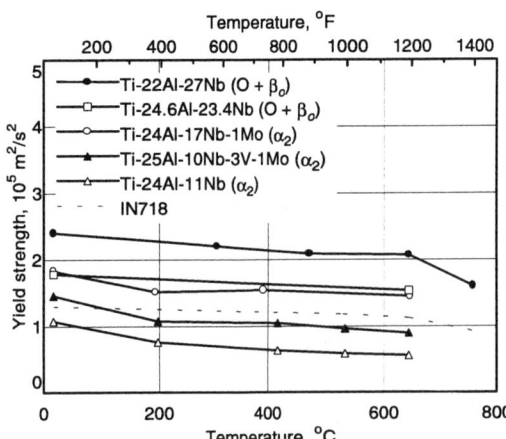

Specific yield strength

Strength-to-weight ratios at various test temperatures for ordered orthorhombic alloys (Ti-22Al-27Nb and Ti-24.5Al-23.5Nb), both conventionally cast; three current α_2 titanium aluminides; and Alloy 718 (UNS N07718), a nickel-base superalloy. Note that Ti-22Al-27Nb has a 45% advantage over Ti-24Al-17Nb-1Mo at 650 °C (1200 °F).
Source: R.G. Rowe, Advanced Ti$_3$Al-Base Alloy Property Comparison, *Processing and Modelling of Advanced Materials*, ASM International, 1991

the effort in recent alloy development studies has been to improve titanium aluminide creep resistance while maintaining or improving room-temperature fracture toughness (Ref 29, 30).

Titanium aluminides: Larson-Miller diagram 0.2% creep strain

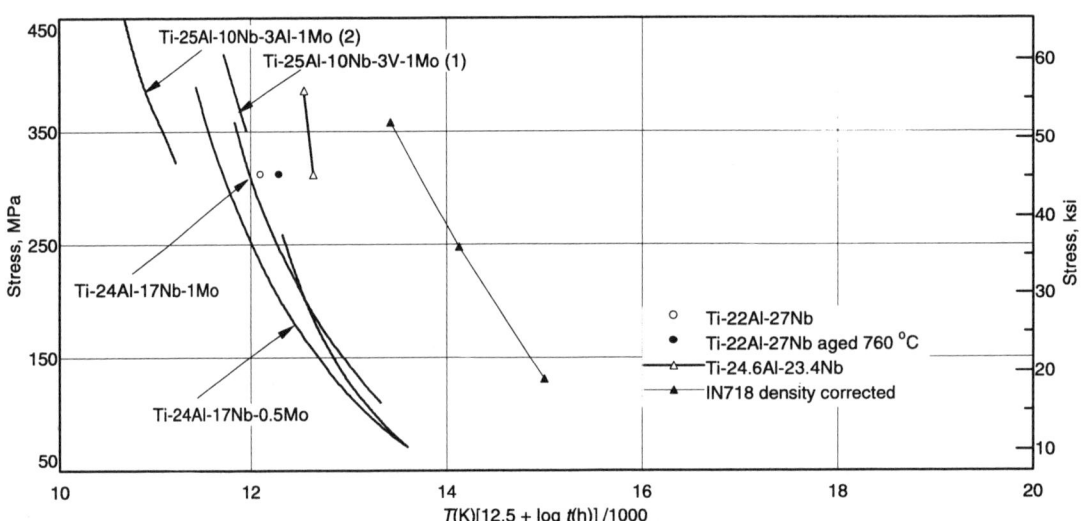

The two-phase O + β_o alloy Ti-22Al-27Nb had a 0.2% creep resistance (based on measurements at 380 MPa) that was comparable to the creep-resistant alloys Ti-25Al-8Nb-2Ta-2Mo and Ti-24Al-17Nb-1Mo. The high volume fraction O phase alloy Ti-24.6Al-23.4Nb had the highest 0.2% creep lifetime of all of the titanium aluminides.
The creep resistance of Ti-25Al-10Nb-3V-1Mo produced by two different processing routes has been plotted because of the large differences in creep resistance. Following the work on Ti-25Al-10Nb-3V-1Mo, Ti-24Al-17Nb-1Mo was developed, which had creep resistance that was comparable to Ti-25Al-10Nb-3V-1Mo alloys, but with higher fracture toughness. Independently developed Ti-25Al-8Nb-2Ta-2Mo had a creep resistance comparable to that of Ti-24Al-17Nb-1Mo, but the same fracture toughness as Ti-25Al-10Nb-3V-1Mo. Its alloying constituents were expected to lead to improved environmental resistance.
Source: R.G. Rose, Advanced Ti$_3$Al-Base Alloy Property Comparison, *Processing and Modelling of Advanced Materials*, ASM International, 1991

Ti$_3$Al Alloys

Recent detailed investigation into the effect of microstructure on creep behavior in Ti-25Al-10Nb-3V-1Mo has shown that the colony-type microstructure exhibits better creep resistance than other microstructures (Ref 31). Creep resistance of Ti-25-10-3-1 is raised by a factor of ten in the steady-state regime over that of conventional alloy Ti-1100 (Ti-6Al-3Sn-4Zr-0.4Mo-0.45Si) and two orders of magnitude over that of Ti-6Al-2Sn-4Zr-2Mo-0.1Si (Ref 31). However, 0.4% creep strain in Ti-25-10-3-1 is reached within 2 h.

Additions of silicon and zirconium appear to improve creep resistance (Ref 32), but the most significant improvement is attained by increasing the

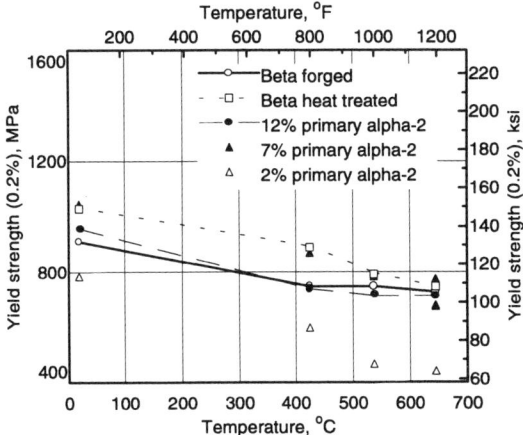

Ti-24.5Al-12.5Nb-1.5Mo: Yield strength vs temperature

Source: B. Marquardt (General Electric), *7th World Conference on Titanium*

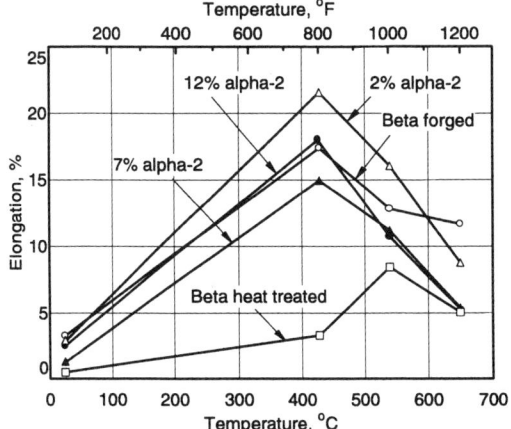

Ti-24.5Al-12.5Nb-1.5Mo: Elongation vs temperature

Source: B. Marquardt (General Electric), *7th World Conference on Titanium*

aluminum content to 25 at.% and limiting β-stabilizing elements to about 12 at.% (Ref 2, 33). However, the Ti-24.5Al-17Nb-1Mo alloy exhibits a rupture life superior to that of other α-2 alloys (Ref 5). Some recent work (see figures) has focused on the properties of Ti-24.5Al-12.5Nb-1.5Mo, which has improved oxidation resistance over Ti-25Al-10Nb-3V-1Mo.

Ti₂Nb-Al Alloys

Strength and ductility of $O + \beta_o$ alloys are strongly dependent on heat treatment, with the highest strength observed in the as-heat-treated condition, and the highest ductility after extended aging. The ordered orthorhombic alloys having the best combination of tensile, creep, and fracture toughness properties are two-phase $O + \beta_o$ alloys such as Ti-22Al-27Nb.

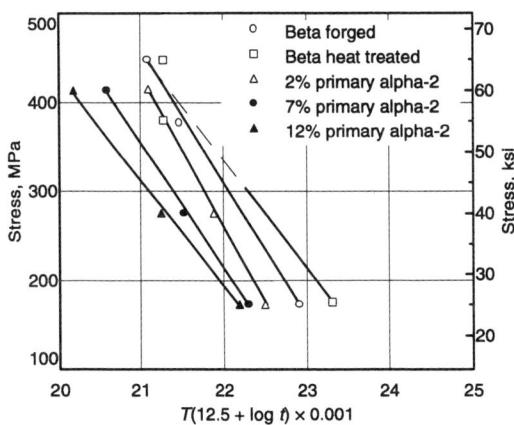

Ti-24.5Al-12.5Nb-1.5Mo: Larson-Miller diagram for time to 0.2% creep

T is in °R; t is in hours.
Source: B. Marquardt (General Electric), *7th World Conference on Titanium*

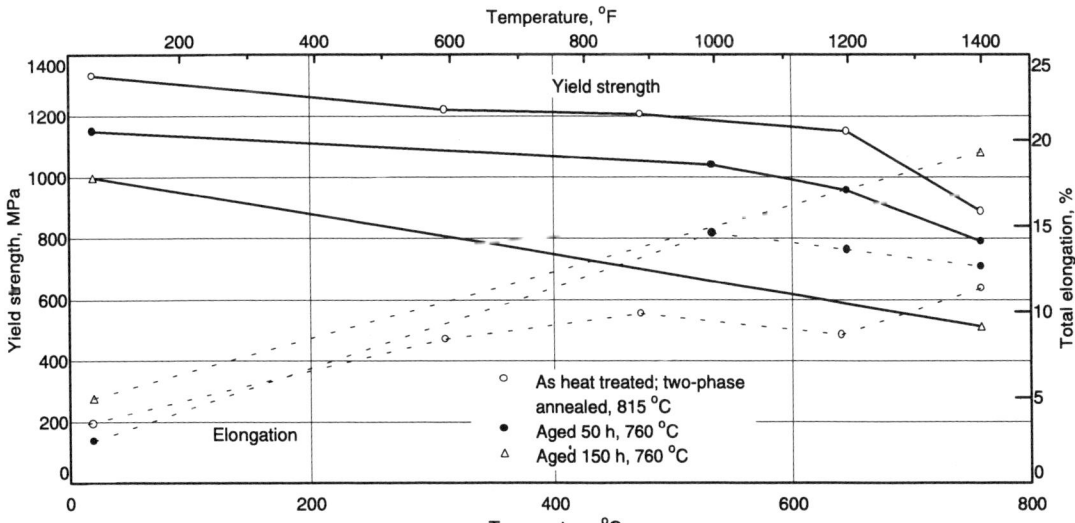

Ti-22Al-27Nb: Yield strength and ductility vs temperature

Conventionally cast Ti-22Al-27Nb with two-phase ($O + \beta_o$) structure.
Source: *Adv. Mater. Process.*, Vol 141, Mar 1992, p 33-35

Properties of single (O) and two-phase (O + β_o) alloys

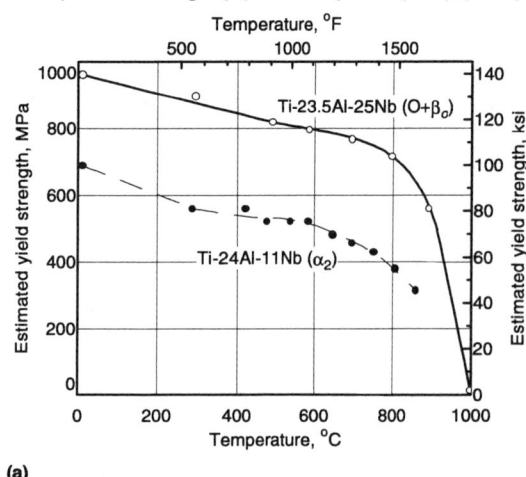
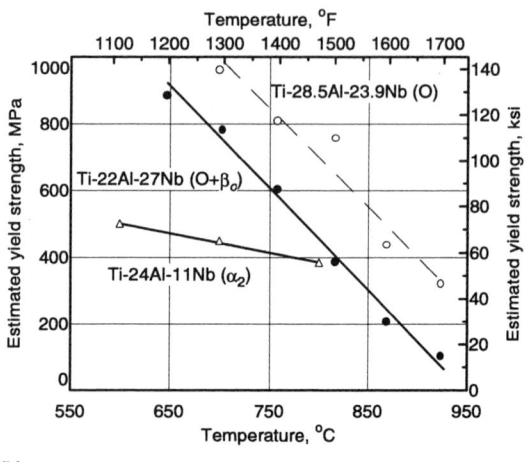

(a) (b)

Ti$_2$AlNb-base alloys having two-phase, ordered O + β_o microstructures are stronger than conventional α_2 titanium aluminides. Yield strength curves (a) were estimated from Vickers microindentation hardness data. Ti-23.5Al-25Nb is an O + β_o alloy prepared by rapid solidification. Ti-24Al-11Nb is an α_2 alloy prepared by conventional casting. Both alloys were beta heat treated. Yield strength curves (b) were estimated from three-point bend test data for beta heat treated materials. The high aluminum Ti-28.5Al-23.9Nb alloy is the strongest of the three, but its fracture toughness and ductility at room temperature are unacceptably low. The two ordered orthorhombic alloys were rapidly solidified; the α_2 alloy was conventionally cast.

Source: *Adv. Mater. Process.*, Vol 141, Mar 1992, p 33-35

Fracture Toughness

Adapted from "The Physical Metallurgy of Ti$_3$Al Based Alloys" by D. Banerjee *et al*, International Symposium on Structural Intermetallics, TMS/AIME, 1993

The presence of some B2 or β phase in the microstructure is necessary for ductility and toughness in Ti$_3$Al base alloys (see figure). However, microstructures containing greater than 10 to 15% B2 are not likely to provide adequate creep resistance. The most important microstructural features which control ductility and toughness are (1) the continuity of B2 phase around α_2 particles, in preference to α_2/α_2 contact, (2) the size of α_2 particles, since grain size effects on cleavage crack initiation are well established and (3) the volume fraction of the B2 phase, in that adequate thickness of B2 between α_2 particles is required to retard microcrack extension out of α_2.

The typical variations in α_2/O morphology and distribution obtained through thermomechanical processing are similar to the range of structures that can be obtained in conventional titanium alloys. An equiaxed morphology is obtained by processing in the two-phase α_2 + B2 or O + B2 region to recrystallize both B2 and α_2 or O. A lath morphology occurs on solution treatment in the single phase β region followed by cooling through the α_2/β transus. A range of properties are associated with

Effect of B2 on ductility/toughness

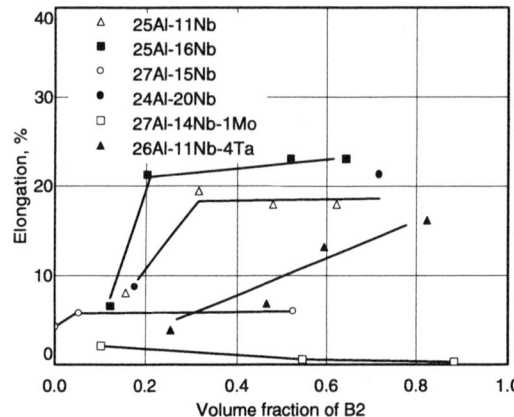
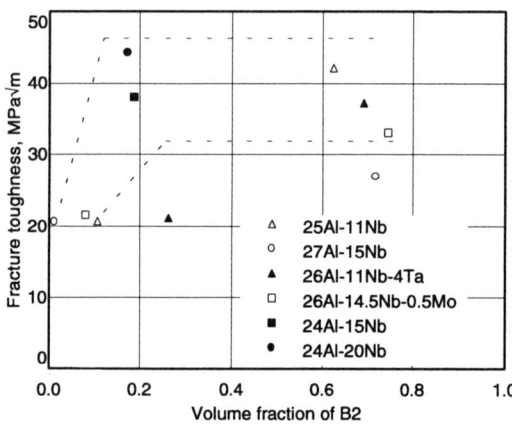

(a) Ductility as a function of B2 volume fraction; (b) fracture toughness as a function of B2 volume fraction

such microstructures (see figure). The strength of the lath structures increases with cooling rate from the single phase β region, that is, with decreasing lath size, while the ductility maximizes at some intermediate cooling rate. These trends are observed for all alloys, irrespective of composition. The maximum in ductility is always associated with a basketweave lath structure without any noticeable prior β grain boundary product.

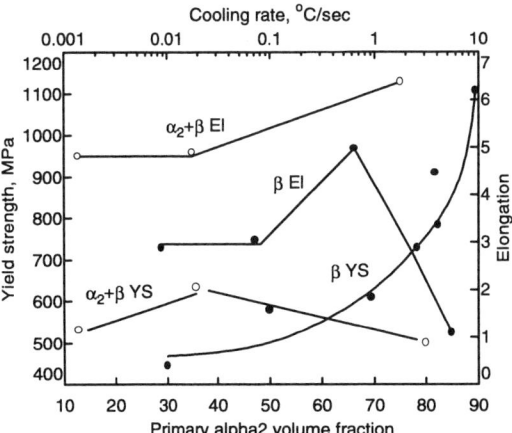

Ti-25Al-16Nb: Typical tensile property variations

Tensile properties at room temperature for a Ti-25Al-16Nb alloy; lath structures as function of cooling rate (β) and equiaxed structures (α2 + β) as a function of primary, equiaxed α2 volume fraction.

References

1. M.J. Blackburn, D.L. Ruckle, and C.E. Bevau, "Research to Conduct an Exploratory Experimental and Analytical Investigation of Alloys," Technical Report AFML-TR-78-18, U.S. Air Force Materials Laboratory, 1978
2. M.J. Blackburn and M.P. Smith, "Research to Conduct an Exploratory Experimental and Analytical Investigation of Alloys," Technical Report AFML-TR-81-4046, U.S. Air Force Wright Aeronautical Laboratories, 1981
3. M.J. Blackburn and M.P. Smith, "R&D on Composition and Processing of Titanium Aluminide Alloys for Turbine Engine," Technical Report AF-WAL-TR-82-4086, U.S. Air Force Wright Aeronautical Laboratories, 1982
4. M.J. Blackburn and M.P. Smith, "Development of Improved Toughness Alloys Based on Titanium Aluminides," Interim Technical Report FR-19139, United Technologies, 1988
5. Y.-W. Kim and F.H. Froes, in *Proceedings of the Symposium on High-Temperature Aluminides and Intermetallics*, TMS, in press
6. F.H. Froes, D. Eylon, and H.B. Bomberger, Ed., *Titanium Technology: Present Status and Future Trends*, Titanium Development Association, 1985
7. F.H. Froes and D. Eylon, *Hydrogen Effects on Materials Behavior*, A.W. Thompson and N.R. Moody, Ed., TMS, 1990
8. F.H. Froes, D. Eylon, and C. Suryanarayana, *J. Met.*, March 1990
9. W.H. Kao et al., in *Progress in Powder Metallurgy*, Vol 37, Metal Powder Industries Federation, 1982, p 289-301
10. C.H. Ward et al., in *Sixth World Conference on Titanium*, Part II, P. Lacombe, R. Tricot, and G. Beranger, Ed., Les Editions de Physique, 1989, p 1009-1014
11. W.J.S. Yang, Observations of Superdislocation Networks in Ti$_3$Al-Nb, *J. Mater. Sci. Lett.*, Vol 1, 1982, p 119-202
12. H. Bohm and K. Lohberg, Uber eine Uberstrukturphase vom CsCl-Typ im System Titan-Molybdan-Aluminum, *Z. Metallk.*, Vol 49, 1958, p 173-178
13. T.J. Jewett et al., in *High-Temperature Ordered Intermetallic Alloys III*, Materials Research Society, Symposia Proceedings, Vol 133, C.T. Liu, A.I. Taub, N.S. Stoloff, and C.C. Koch, Ed., Materials Research Society, 1989, p 69-74
14. M.J. Kaufman et al., in *Sixth World Conference on Titanium*, Part II, P. Lacombe et al., Ed., Les Editions de Physique, 1989, p 985-990
15. R.G. Rowe, *High Temperature Aluminides and Intermetallics*, S.H. Whang, C.T. Liu, and D. Pope, Ed., TMS, 1990
16. R. Strychor, J.C. Williams, and W.A. Soffa, Phase Transformations and Modulated Microstructures in Ti-Al-Nb Alloys, *Metall. Trans. A*, Vol 19 (No. 2), 1988, p 225-234
17. S.M.L. Sastry and H.A. Lipsitt, Ordering Transformations and Mechanical Properties of Ti$_3$Al and Ti$_3$Al-Nb Alloys, *Metall. Trans. A*, Vol 8, 1977, p 1543
18. W.A. Baeslack III, M.J. Cieslak, and T.J. Headley, Structure, Properties and Fracture of Pulsed Nd:YAG Laser Welded Ti-14.8 wt% Al-21.3 wt% Nb Titanium Aluminide, *Scr. Metall.*, Vol 22, 1988, p 1155-1160
19. J.C. Williams, in *Titanium Technology: Present Status and Future Trends*, F.H. Froes, D. Eylon, and H.B. Bomberger, Titanium Development Association, 1985, p 75-86
20. A.G. Jackson, K. Teal, and F.H. Froes, in *High-Temperature Ordered Intermetallic Alloys II*, Materials Research Society Symposia Proceedings, Vol 81, N.S. Stoloff, C.C. Koch, C.T. Liu, and O. Izumi, Ed., Materials Research Society, 1987, p 143-149
21. R.G. Rowe, Creep and Discontinuous Precipitation in a Ti$_3$Al-Nb Alloy at 923 K, *Scr. Metall. Mater.*, Vol 24, 1990, p 1209-1214
22. R.G. Rowe and E.L. Hall, Stress-Assisted Discontinuous Precipitation During Creep of Ti$_3$Al-Nb Alloys, *Mater. Res. Soc. Symp. Proc.*, Vol 213, 1991, p 449-454
23. C.H. Ward, J.C. Williams, and A.W. Thompson, Microstructural Instability in the Alloy Ti-25Al-10Nb-3V-1Mo, *Scr. Metall. Mater.*, Vol 24, 1990, p 617-622
24. R.G. Rowe and D. Banerjee, "Cellular Precipitation in Ti-25Al-30Nb at.%," to be published
25. C.H. Ward et al., in *Sixth World Conference on Titanium*, Part II, P. Lacombe, R. Tricot, and G. Beranger, Ed., Les Editions de Physique, 1989, p 1103-1108
26. R.W. Hertzberg, *Deformation and Fracture Mechanics of Engineering Materials*, 2nd ed., John Wiley & Sons, 1983
27. M.S. Stucke and H.A. Lipsitt, in *Titanium Rapid Solidification Technology*, F.H. Froes and D. Eylon, Ed., TMS, 1986, p 255-262
28. R.G. Rowe, Recent Developments in Ti-Al-Nb Alloys, in *High Temperature Aluminides and Intermetallics*, C.T. Liu, S.H. Whang, D.P. Pope, and J.O. Stigler, Ed., TMS/AIME, 1990, p 375-401

29. B.J. Marquardt, G.K. Scarr, J.C. Chesnutt, C.G. Rhodes, and H.L. Fraser, "Research and Development for Improved Toughness Aluminides," U.S. Air Force, WRDC Report No. WRDC-TR-89-4133, June 1990
30. M.J. Blackburn and M.P. Smith, "Improved Toughness Alloys Based Upon Titanium Aluminides," U.S. Air Force, WRDC Report No. WRDC-TR-89-4095, Oct 1989
31. W. Cho, "Effect of Microstructure on Deformation and Creep Behavior of Ti-25Al-10Nb-3V-1Mo," Technical Report, U.S. Air Force Office of Scientific Research, Oct 1988
32. C.G. Rhodes, in *Sixth World Conference on Titanium*, Part I, P. Lacombe, R. Tricot, and G. Beranger, Ed., Les Editions de Physique, 1989, p 119-204
33. M.G. Mendiratta and H.A. Lipsitt, Steady-State Creep Behavior of Ti_3Al-Base Intermetallics, *J. Mater. Sci.*, Vol 15, 1980, p 2985-2990

Gamma (Ti-Al) Alloys

Currently available γ alloys contain approximately 46 to 52 at.% Al and 1 to 10 at.% M, with M being at least one of the following: vanadium, chromium, manganese, niobium, tantalum, tungsten, and molybdenum (Ref 1 to 7). These alloys can be divided into two categories: single-phase (γ) alloys and two-phase (γ + α-2) materials (Ref 2). The (α-2 + γ)/γ phase boundary at 1000 °C (1830 °F) occurs at an aluminum content of approximately 49 at.%, depending on the type and level of solute M. Single-phase γ alloys contain third alloying elements such as niobium or tantalum that promote strengthening and further enhance oxidation resistance (Ref 8, 9). Third alloying elements in two-phase alloys can raise ductility (vanadium, chromium, and manganese) (Ref 1, 2, 5-7), increase oxidation resistance (niobium and tantalum) (Ref 8 and 9), or enhance combined properties (Ref 2).

Material Processing. The microstructure of the nominally γ alloys can be single-phase γ, or in slightly leaner compositions, two-phase γ + α-2. By appropriate thermomechanical processing (TMP), the morphology of the phases can be adjusted to produce either lamellar or equiaxed morphologies, or a mixture (duplex structure) of the two (Ref 2, 5, 10).

The lamellar structure can lead to refinement of the microstructure, improved ductility (Ref 5, 7) and a decreased microstructure scale by recrystallizing the fine γ grains (Ref 11). Optimum ductility occurs at about 10 vol% α-2. When the α-2 phase content exceeds 20 vol%, ductility can be degraded (Ref 6). This ductility behavior is consistent with the fact that α-2 becomes increasingly brittle with increasing aluminum content over 25 at.% (Ref 12). The α-2 plates contain approximately 35 at.% Al.

Control of the microstructure in single-phase γ alloys requires optimization of grain size and morphology. In two-phase alloys, the volume ratio of lamellar to equiaxed gamma (LG/γG) must also be controlled (Ref 2, 5, 6, 10, 13). A lamellar volume fraction of about 30% gives rise to the optimum combination of properties with desirable high-temperature creep resistance and acceptable levels of tensile strength and ductility (Ref 13). Heat treatment temperature and time strongly affect the LG/γG volume ratio. Thermomechanical processing refines the microstructure when processing is conducted in such a way that both the α and γ grains are recrystallized in the (α + γ) phase field.

Grain morphology varies considerably depending on composition, solution treatment temperature and time, cooling rate, and stabilization temperature and time (Ref 2). Grain size decreases with reduced aluminum content and with additions of vanadium, manganese, and chromium (Ref 6, 7). The number of annealing twins in the γ phase increases as aluminum content decreases, or when manganese or vanadium levels are increased (Ref 6). Chromium additions increase the volume fraction of the lamellar structure (Ref 7).

Typical processing of gamma alloys

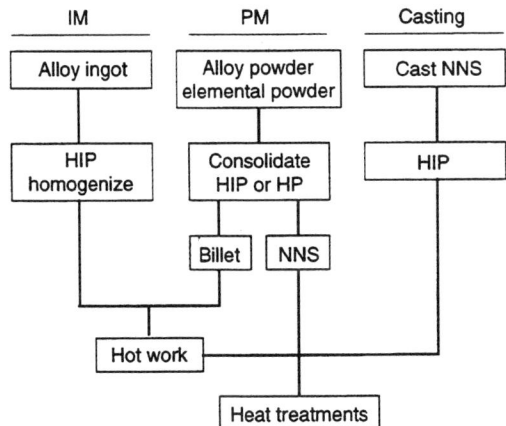

HIP, hot isostatic pressing; HP, hot pressing; NNS, near-net shape. Source: Y.-W. Kim and D.M. Dimiduk, Progress in the Understanding of Gamma Titanium Aluminides, *JOM*, Aug 1991, p 40-47

Mechanical Properties

The strength and ductility of γ alloys are strongly dependent on alloy composition and TMP conditions (Ref 2), although the Ti-52Al (at.%) alloy demonstrates the lowest hardness value at room temperature, regardless of the TMP treatment (Ref 14 to 18). At 1000 °C (1830 °F), strength tends to decrease gradually with increasing aluminum levels (Ref 2). Tensile strength and hardness vary in the same fashion with variations in aluminum content (Ref 2). Room-temperature tensile elongation is maximum at a composition of approximately Ti-48Al (at.%) (Ref 18).

Ternary alloys of composition Ti-48Al with approximately 1 to 3% of vanadium, manganese, or chromium exhibit enhanced ductility, but Ti-48Al alloys with approximately 1 to 3% of niobium, zirconium, hafnium, tantalum, or tungsten exhibit lower ductility than binary Ti-48Al (Ref 2). The brittle-ductile transition (BDT) occurs at 700 °C (1290 °F) in Ti-56Al, and it occurs at lower temperatures with decreasing aluminum levels. Increased room-temperature ductility generally results in a reduced BDT temperature. Above the BDT temperature, ductility increases rapidly with temperature, approaching 100% at 1000 °C (1830 °F) for the most ductile γ alloy compositions. The elastic moduli of γ alloys range from 160 to 176 GPa (23×10^6 to 25.5×10^6 psi) and decrease slowly with temperature (Ref 2, 10).

Low-cycle fatigue experiments (Ref 13) suggest that fine grain sizes increase fatigue life at temperatures below 800 °C (1470 °F). Fatigue crack growth rates for γ alloys are more rapid than those for superalloys, even when density is normalized (Ref 4). Both fracture toughness and impact resistance are low at ambient temperatures, but fracture toughness increases with temperature; for example, the plane-strain fracture toughness (K_{Ic}) for Ti-48Al-1V-0.1C is 24 MPa√m (21.8 ksi√in.) at room temperature (Ref 13). Fracture toughness is strongly dependent on the volume fraction of the lamellar phase. In a two-phase quaternary γ alloy, a fracture toughness of 12 MPa√m (10.9 ksi√in.) is observed for a fine structure that is almost entirely γ; K_{Ic} is greater than 20 MPa√m (18.2 ksi√in.) when a large volume fraction of lamellar grains is present (Ref 2, 10). Creep properties of γ alloys, when normalized by density, are better than those of superalloys, but they are strongly influenced by alloy chemistry and TMP. Increased aluminum content and additions of tungsten (Ref 19) or carbon (Ref 13) increase creep resistance. Increasing the volume fraction of the lamellar structure enhances creep properties (Ref 13), but lowers ductility. The level of creep strain from elongation on initial loading and primary creep is of concern because it can exceed projected design levels for maximum creep strain in the part.

Wrought Ti-48Al-2Cr-2Nb: Microstructural effects on mechanical properties

Mechanical property	Test temperature °C	Test temperature °F	Duplex structure	Transformed structure
Plastic ductility, %	RT		3.1	0.4
	760	1400	50.0	2.8
Yield strength, MPa (ksi)	RT		480 (69)	455 (66)
	760	1400	406 (59)	403 (58)
Toughness, MPa√m (ksi √in.)	RT		14.3 (13)	28.3 (25.7)
	760	1400	19.2 (17.4)	...
Alternate stress at 10^7 cycles, MPa (ksi)	650	1200	276 (40)	276 (40)
	760	1400	172 (25)	207 (30)
Creep rate(a), per h	760	1400	3.3×10^{-5}	4.0×10^{-7}
Creep time to 0.2% strain(a), h	760	1400	25	800
100-h rupture strength, MPa (ksi)	540	1000	580 (84)	525 (76)
	650	1200	480 (69)	510 (74)
	760	1400	230 (33)	370 (53)

(a) Tested at 105 MPa. Source: *Intermetallic Compounds—Structure and Mechanical Properties*, O. Ozumi, Ed., The Japan Institute of Metals, 1991, p 363-370

Ti-47Al-1Cr-1V-2.5Nb: Tensile properties

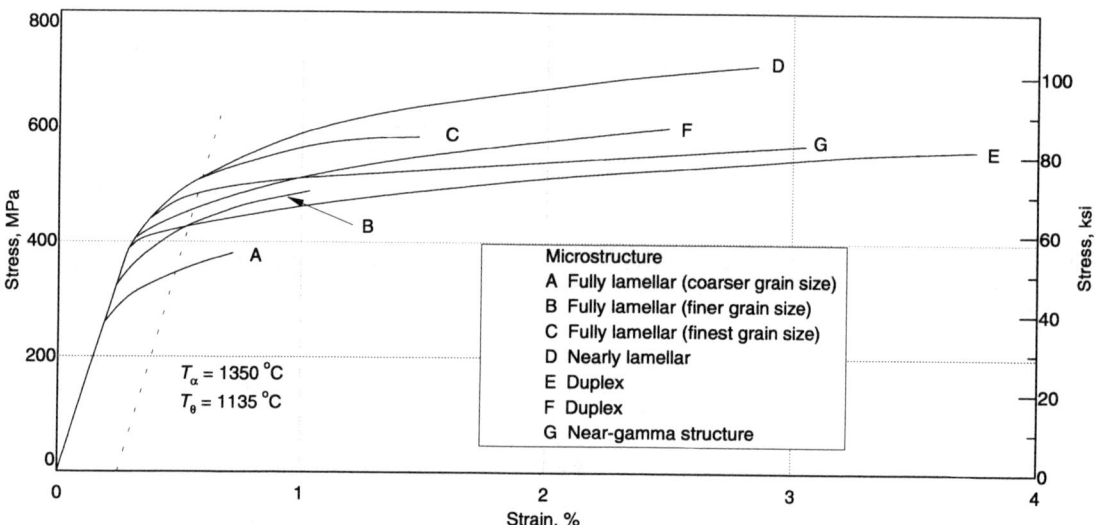

Room-temperature tensile properties of a two-phase γ alloy in various post-forging heat treatment conditions. A: 1370 °C, 2 h, CC to 900 °C, 6 h, AC. B: 1360 °C, 1 h, CC to 900 °C, 24 h, AC. C: 1350 °C, 1 h, CC to 1000 °C, 5 h, AC. D: 1330 °C, 5 h, CC + 900 °C, 6 h, AC. E: 1280 °C, 3 h, CC to 900 °C, 4 h, AC. F: 1280 °C, 2 h, CC + 900 °C, 6 h, AC. G: 1000 °C, 24, AC.
Source: Y.W. Kim and D.M. Dimiduk, Progress in Understanding of Gamma Titanium Aluminides, *JOM*, Aug 1991, p 40-47

Effect of aluminum content

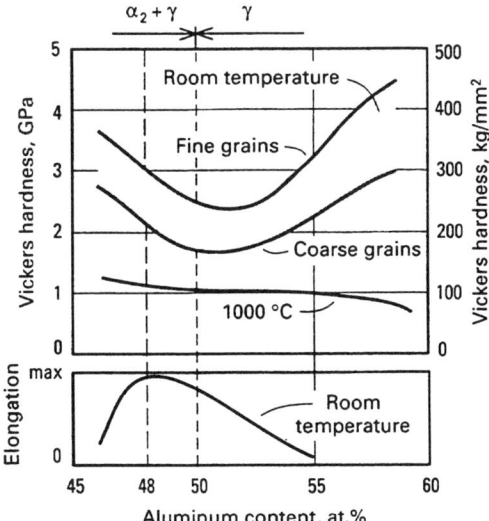

Effect of aluminum content on room-temperature tensile elongation and hardness of binary γ titanium aluminide alloys. Hardness values at 1000 °C (1830 °F) are also shown. Note the single-phase γ region and the two-phase (α_2 + γ) region.
Source: *ASM Handbook*, Vol 2, 10th ed., 1990, p 928

High-Temperature Strength

Tensile properties vs temperature

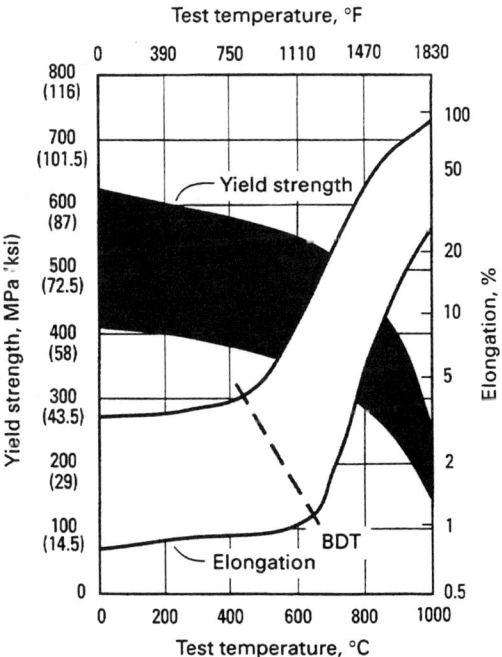

Ranges of yield strength and tensile elongation as functions of test temperature for γ-TiAl alloys. BDT, brittle-ductile transition.
Source: *ASM Handbook*, Vol 2, 10th ed., 1990, p 929

Ti-48Al-2Cr-2Nb: Yield strength (0.2%) vs temperature

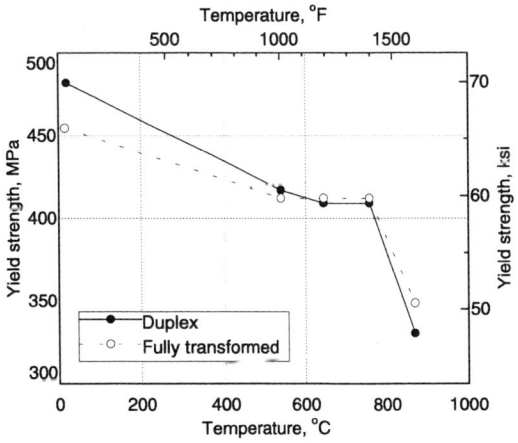

Source: *Microstructure/Property Relationships in Titanium Aluminides and Alloys*, Y.-W. Kim and R.R. Boyer, Ed., The Minerals, Metals and Materials Society, 1991, p 135-148

Ti-48Al-2Cr-2Nb: Stress-rupture properties

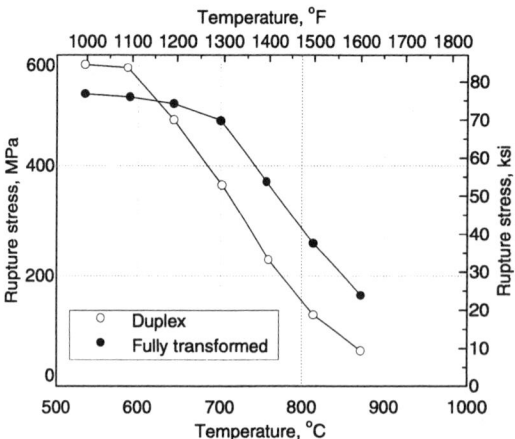

100-h rupture strengths as a function of temperature for duplex and fully transformed microstructures.
Source: *Microstructure/Property Relationships in Titanium and Alloys*, Y.-W. Kim and R.R. Boyer, Ed., The Minerals, Metals, and Materials Society, 1991, p 135-148

Fatigue Strength

Ti-48Al-2Cr-2Nb: Load-controlled high-cycle fatigue

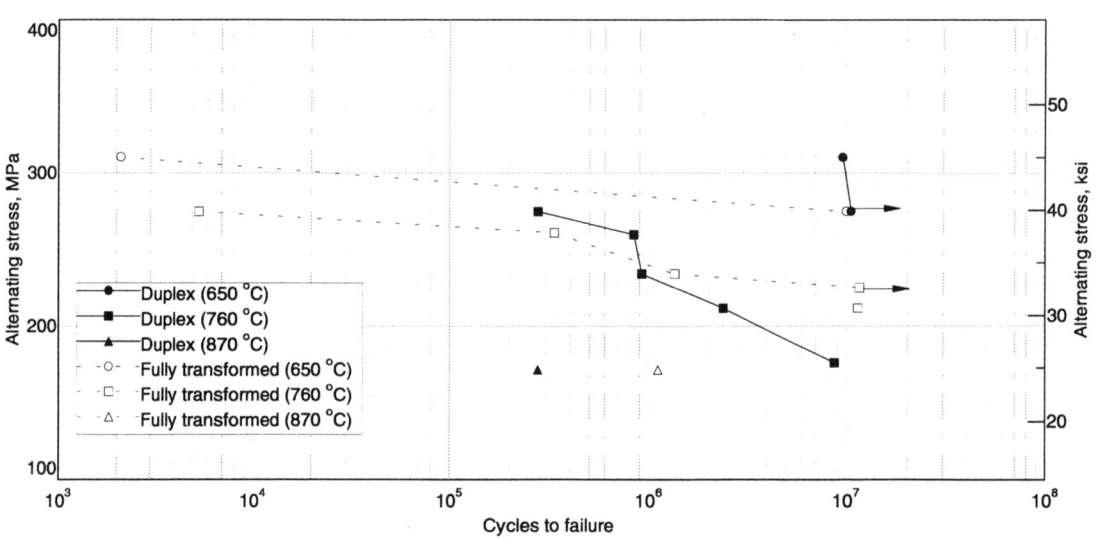

S-N curves for the duplex and fully transformed microstructures. Arrows denote runouts; runouts occurred at 275 MPa for both structures. Axial bend, $A = 1.0$, 30 Hz.
Source: *Microstructure/Property Relationships in Titanium Aluminides and Alloys*, Y.-W. Kim and R.R. Boyer, Ed., The Minerals, Metals and Materials Society, 1991, p 135-148

Environmental Resistance

Ti-50Al is known to be highly resistant to oxidation. However, its desirable tendency to form Al_2O_3 preferentially to TiO_2 exists only up to 850 °C (1560 °F). Oxidation resistance is reduced with decreasing aluminum content or increasing volume fraction of the α-2 phase. Alloying elements that improve ductility (vanadium, chromium, and manganese) typically decrease oxidation resistance; consequently, the most ductile γ alloys require additional elements, such as niobium, which improve the oxidation resistance. However, even in alloys with such a combination, the tendency to preferentially form Al_2O_3 is not expected to exist at 900 °C (1650 °F) or higher temperatures. Apparently, alloys with a high aluminum content and substantial additions of niobium or tantalum will be more desirable for oxidation resistance if approaches to provide these alloys with reasonable ductility can be identified. As a minimum, this requires knowledge of the γ/α$_2$ + γ phase boundary in the Ti-Al-Nb and Ti-Al-Ta ternary systems.

Although there is little solubility for hydrogen in γ-TiAl, the hydrogen solubility in α$_2$-(Ti$_3$Al) is substantial. Thus, when two-phase γ alloys dissolve hydrogen, the amount is likely to be highly dependent on the α-2 volume fracture. Hydrogen absorption can be reduced by reducing the amount of α-2 phase or lamellar structure, or by adding such elements as niobium and tantalum that improve oxidation resistance. In either case, the typical reduction of room-temperature ductility has been a barrier to these using alloys in hydrogen environments.

Oxidation of gamma alloys at 850 °C

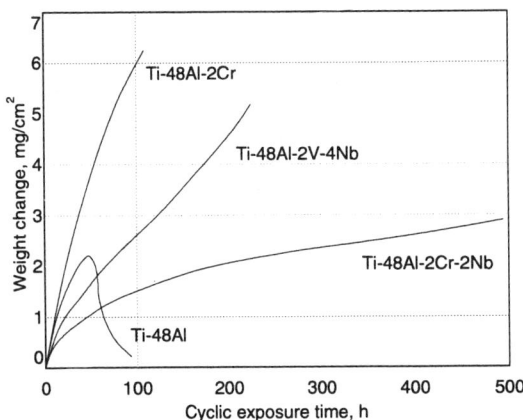

Weight changes versus time for titanium alloys during rapid thermal cycling in air at 850 °C (1560 °F). The curve for Ti-48Al-2V-4Nb is included for comparison 1-h cycles in air, 300 ml/min.
Source: *Intermetallic Compounds—Structure and Mechanical Properties*, O. Ozumi, Ed., The Japan Institute of Metals, Y.-W. Kim and R.R. Boyer, Ed., The Minerals, Metals and Materials Society, 1991, p 363-370

References

1. M.J. Blackburn and M.P. Smith, "Research to Conduct an Exploratory Experimental and Analytical Investigation of Alloys," Technical Report AFWAL-TR-80-4175, U.S. Air Force Wright Aeronautical Laboratories, 1980
2. Y.-W. Kim, Intermetallic Alloys Based on Gamma Titanium Aluminide, *J. Met.*, Vol 41 (No. 7), 1989, p 24-30
3. M.J. Blackburn and M.P. Smith, "Titanium Alloys of the TiAl Types," U.S. patent 4,294,615, 1981
4. M.J. Blackburn, J.T. Hill, and M.P. Smith, "R & D on Composition and Processing of Titanium Aluminide Alloys for Turbine Engines," Technical Report AFWAL-TR-84-4078, U.S. Air Force Wright Aeronautical Laboratories, 1984
5. S.-C. Huang and E.L. Hall, in *High-Temperature Ordered Intermetallic Alloys III*, Materials Research Society Symposia Proceedings, Vol 133, C.T. Liu, A.I. Taub, N.S. Stoloff, and C.C. Koch, Ed., Materials Research Society, 1989, p 373-383
6. T. Tsujimoto and K. Hashimoto, in *High-Temperature Ordered Intermetallic Alloys III*, Materials Research Society Symposia Proceedings, Vol 133, C.T. Liu, A.I. Taub, N.S. Stoloff, and C.C. Koch, Ed., Materials Research Society, 1989, p 391-396
7. T. Kawabata, T. Tamura, and O. Isumi, in *High-Temperature Ordered Intermetallic Alloys III*, Material Research Society Symposia Proceedings, Vol 133, C.T. Liu, A.I. Taub, N.S. Stoloff, and C.C. Koch, Ed., Materials Research Society, 1989, p 329-334
8. D.J. Maykuth, "Effects of Alloying Elements in Titanium," DMIC Report 136B, Battelle Memorial Institute, May 1961
9. I.A. Zelonkov and Y.N. Martynchik, Oxidation Resistance of Alloys of Compound TiAl with Niobium at 800 and 1000C, *Metallofiz., Nauk. Dumka*, Vol 42, 1972, p 63-66
10. Y.-M. Kim and F.H. Froes, in *Proceedings of the Symposium on High-Temperature Aluminides and Intermetallics*, TMS, in press
11. C.R. Feng, D.J. Michel, and C.R. Crowe, in *High-Temperature Ordered Intermetallic Alloys III*, Materials Research Society Symposia Proceedings, Vol 133, C.T. Liu, A.I. Taub, N.S. Stoloff, and C.C. Koch, Ed., Materials Research Society, 1989, p 669-674
12. M.J. Blackburn, D.L. Ruckle, and C.E. Bevau, "Research to Conduct an Exploratory Experimental and Analytical Investigation of Alloys," Technical Report AFML-TR-78-18, U.S. Air Force Materials Laboratory, 1978

13. M.J. Blackburn and M.P. Smith, "R & D Composition and Processing of Titanium Aluminide Alloys for Turbine Engine," Technical Report AFWAL-TR-82-4086, U.S. Air Force Wright Aeronautical Laboratories, 1982
14. E.S. Bumps, H.D. Kessler, and M. Hansen, Titanium-Aluminum System, *Trans. AIME*, Vol 194, 1952, p 609-614
15. H.R. Odgen *et al.*, Mechanical Properties of High Purity Ti-Al Alloys, *J. Met.*, Feb 1952
16. M.J. Blackburn and M.P. Smith, "The Understanding and Exploitation of Alloys Based on the Compound TiAl (Gamma Phase)," Technical Report AFML-TR-79-4056, U.S. Air Force Materials Laboratory, 1979
17. T. Tsujimoto *et al.*, Structures and Properties of an Intermetallic Compound TiAl Based Alloys Containing Silver, *Trans. Jpn. Inst. Met.*, Vol 27 (No. 5), 1986, p 341-350
18. S.-C. Huang, E.L. Hall, and M.F.X. Gigliotti, in *Sixth World Conference on Titanium*, Part II, P. Lacombe, R. Tricot, and G. Beranger, Ed., Les Editions de Physique, 1989, p 1109-1114
19. S.M. Barinov *et al.*, Temperature Dependence of Strength and Ductility of the Decomposition of Titanium Aluminide, *Izv. Akad. Nauk SSSR*, Vol 5, 1983, p 170-174

Ti-Ni Shape Memory Alloys

T.W. Duerig and A.R. Pelton, Nitinol Development Corporation

This datasheet describes some of the key properties of equiatomic and near-equiatomic titanium-nickel alloys with compositions yielding shape memory and superelastic properties. Shape memory and superelasticity *per se* will not be reviewed; readers are referred to Ref 1 to 3 for basic information on these subjects. These alloys are commonly referred to as nickel-titanium, titanium-nickel, Tee-nee, Memorite™, Nitinol, Tinel™, and Flexon™. These terms do not refer to single alloys or alloy compositions, but to a family of alloys with properties that greatly depend on exact compositional make-up, processing history, and small ternary additions. Each manufacturer has its own series of alloy designations and specifications within the "Ti-Ni" range.

A second complication that readers must acknowledge is that all properties change significantly at the transformation temperatures M_s, M_f, A_s, and A_f (see figure on the right and the section "Tensile Properties"). Moreover, these temperatures depend on applied stress. Thus, any given property depends on temperature, stress, and history.

Effect of phase transformation

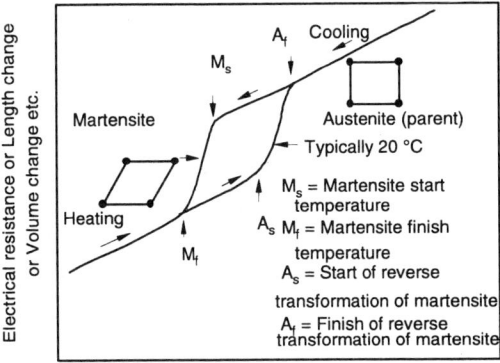

Schematic illustration of the effects on a phase transformation on the physical properties of Ti-Ni. All physical properties exhibit a discontinuity, characterized by the transformation temperatures shown.
Source: C.M. Wayman and T.W. Duerig, *Engineering Aspects of Shape Memory Alloys*, T.W. Duerig, et al., Ed., Butterworth-Heinemann, 1990, p 10

Product Forms and Applications

Titanium-nickel is most commonly used in the form of cold drawn wire (down to 0.02 mm) or as barstock. Other commercially available forms not yet sold as standard product would include tubing (down to 0.3 mm OD), strip (down to 0.04 mm in thickness), and sheet (widths to 500 mm and thicknesses down to 0.5 mm). Castings (Ref 4), forgings and powder metallurgy (Ref 5) products have not yet been brought from the research laboratory.

Typical Conditions. Titanium-nickel is most commonly used in a cold worked and partially annealed condition. This partial anneal does not recrystallize the material, but does bring about the onset of recovery processes. The extent of the post-cold worked recovery depends on many aspects of the application, such as the desired stiffness, fatigue life, ductility, recovery stress, etc. Fully annealed conditions are used almost exclusively when a maximum M_s is needed. Although the cold worked condition does not transform and does not exhibit shape memory, it is highly elastic and has been considered for many applications (Ref 6).

Response to Heat Treatment. Recovery processes begin at temperatures as low as 275 °C (525 °F). Recrystallization begins between 500 and 800 °C (930 and 1470 °F), depending on alloy composition and the degree of cold work.

Aging of unstable (nickel-rich) compositions begins at 250 °C (525 °F), causing the precipitation of a complex sequence of nickel-rich precipitates (Ref 7), as these products leach nickel from the matrix, their general effect is to increase the M_s temperature. The solvus temperature is about 550 °C (1020 °F).

Applications for titanium-nickel alloys can be conveniently divided into four categories (Ref 8):

- *Free recovery (motion)* applications are those in which a shape memory component is allowed to freely recover its original shape during heating, thus generating a recovery strain (Ref 9).

- *Constrained recovery (force)* applications are those in which the recovery is prevented, constraining the material in its martensitic, or cold, form while recovering (Ref 9). Although no strain is recovered, large recovery stresses are developed. These applications include fasteners and pipe couplings and are the oldest and most widespread type of practical use.

- *Actuators (work)* applications are those in which there is both a recovered strain and stress during heating, such as in the case of a titanium-nickel spring being warmed to lift a ball (Ref 10). In these cases, work is being done. Such applications are often further categorized according to their actuation mode, e.g., electrical or thermal.

- *Superelasticity (energy storage)* refers to the highly exaggerated elasticity, or springback, observed in many Ti-Ni alloys deformed above A_s and below M_d (Ref 11). The function of the material in such cases is to store mechanical energy. Although limited to a rather small temperature range, these alloys can deliver over 15 times the elastic motion of a spring steel.

Special Properties

Many shape memory-related properties are discussed in subsequent sections (transformation temperatures, superelasticity, etc.). Some properties, however, are strictly peculiar to shape memory alloys and cannot be conveniently categorized in standard outline forms. The more important of these properties are discussed below.

Free-recoverable strain in polycrystalline titanium-nickel can reach 8%, but is limited to a maximum of 6% if complete recovery is expected.

Applied stresses opposing recovery reduce recoverable strain. Clearly, stronger alloys will be affected less by opposing stresses. Work output is maximized at intermediate stresses and strains.

Recoverable stresses generally reach 80 to 90% of yield stress. In fact, alloy behavior depends on numerous factors, including the compliance of the resisting force and the constraining strain (Ref 9 and 12). Typical values are as follows:

Condition	Recovery stress, MPa
Annealed barstock	400
Cold worked barstock annealed at 500 °C (930 °F)	700
Cold worked wire annealed at 400 °C (750 °F)	1000

Free recovery behavior

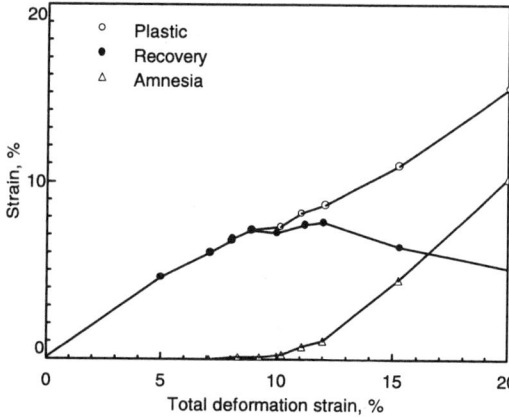

Ti-Ni-Fe barstock with 50 at.% Ni and 3% Fe fully annealed, tested in uniaxial tension. After deforming Ti-Ni to various total strains (x-axis), the material springs back to the plastic strain levels shown by the open circles. After heating above A_f, most of the strain is recovered, but some amnesia persists. The difference between the plastic strain and the amnesia is the recoverable strain (closed circles). Source: J.L. Proft and T.W. Duerig, *Engineering Aspects of Shape Memory Alloys*, T.W. Duerig et al., Ed., Butterworth-Heinemann, London, 1990, p 115

Effects of opposing stresses on recovery strain

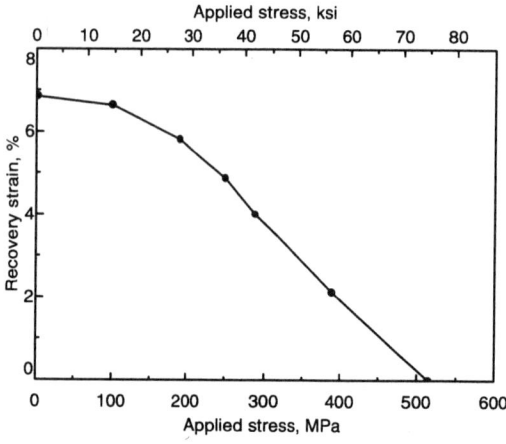

Ti-Ni-Fe barstock with 50 at.% Ni and 3% Fe fully annealed, tested in uniaxial tension.
Source: J.L. Proft and T.W. Duerig, *Engineering Aspects of Shape Memory Alloys*, T.W. Duerig et al., Ed., Butterworth-Heinemann, London, 1990, p 115

Work output of a Ti-Ni alloy

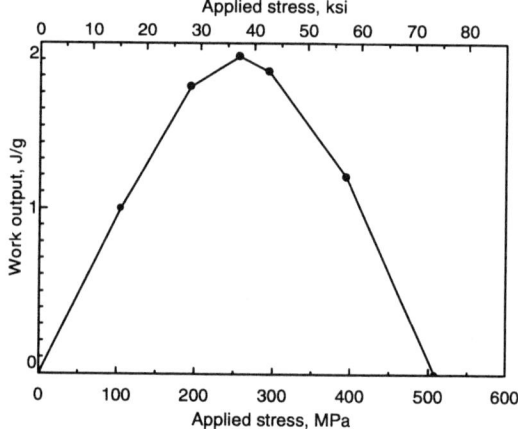

Ti-Ni-Fe barstock with 50 at.% Ni and 3% Fe in a work-hardened condition, tested in uniaxial tension.
Source: J.L. Proft and T.W. Duerig, *Engineering Aspects of Shape Memory Alloys*, T.W. Duerig et al., Ed., Butterworth-Heinemann, London, 1990, p 115

Chemistry and Density

Density, 6.45 to 6.5 g/cm^3

Titanium-nickel is extremely sensitive to the precise titanium/nickel ratio (see figure below). Generally, alloys with 49.0 to 50.7 at.% titanium are commercially common, with superelastic alloys in the range of 49.0 to 49.4 at.% and shape memory alloys in the range of 49.7 to 50.7 at.%. Binary alloys with less than 49.4 at.% titanium are generally unstable. Ductility drops rapidly as nickel is increased.

Binary alloys are commonly available with M_s

temperatures between −50° and +100 °C (−58 to 212 °F). Commercially available ternary alloys are available with M_s temperatures down to −200 °C (−330 °F). Titanium-nickel is also quite sensitive to alloying additions.

Oxygen forms a $Ti_4Ni_2O_x$ inclusion (Ref 13), tending to deplete the matrix in titanium, lower M_s, retard grain growth, and increase strength. Levels usually are controlled to <500 ppm. Nitrogen forms the same compound and has an additive effect to oxygen.

Fe, Al, Cr, Co, and V tend to substitute for nickel, but sharply depress M_s (Ref 14 to 16), with V and Co being the weakest suppressants and Cr the strongest. These elements are added to suppress M_s while maintaining stability and ductility. Their practical effect is to stiffen a superelastic alloy, to create a cryogenic shape memory alloy, or to increase the separation of the R-phase from martensite.

Pt and Pd tend to decrease M_s in small quantities (~5 to 10%), then tend to increase M_s, eventually achieving temperatures as high as 350 °C (660 °F) (Ref 17).

Zr and Hf occasionally have been reported to increase M_s, but are generally neutral when substituted for titanium on an atomic basis.

Nb and Cu are used to control hysteresis and martensitic strength. Nb is added to increase hysteresis (desirable for coupling and fastener applications), and copper (Ref 19) is added to reduce hysteresis (for actuator applications).

Effect of composition on M_s

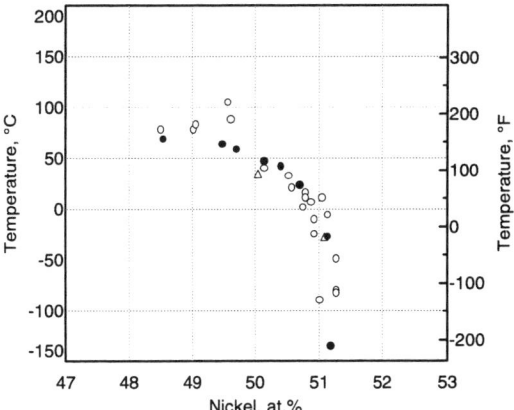

M_s temperatures in nickel-titanium alloys are extremely sensitive to compositional variation, particularly at higher nickel contents.
Source: K.N. Melton, *Engineering Aspects of Shape Memory Alloys*, T.W. Duerig et al., Ed., Butterworth-Heinemann, 1990, p 10

Central Portion of Ti-Ni Phase Diagram

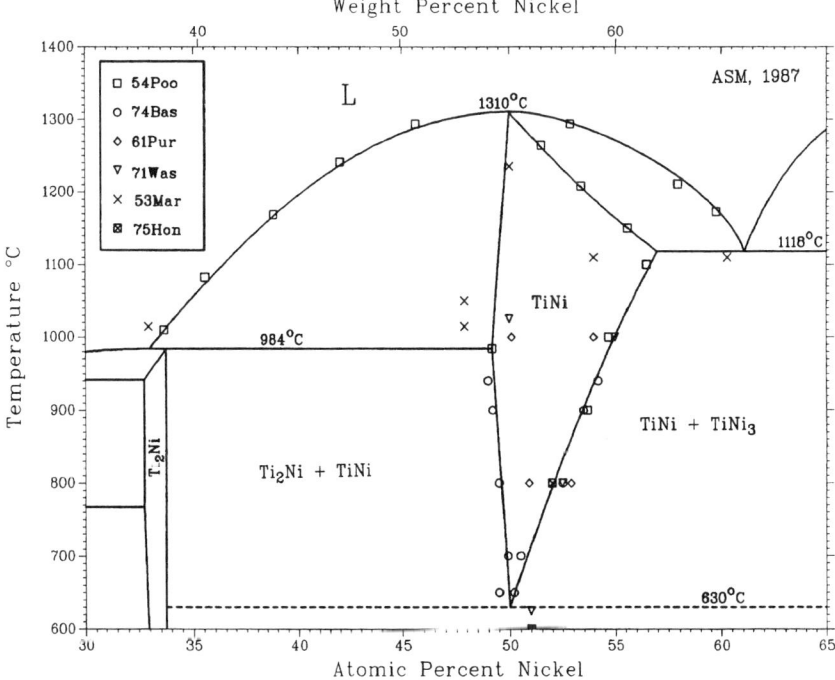

Phases and Structures

Crystal Structure

The high-temperature austenitic phase (β) has a B2, or CsCl ordered structure with a_o = 3.015 Å. The most common martensitic structure (B19′) has a complex monoclinic structure with a = 2.889 Å, b = 4.120 Å, c = 4.622 Å, and β = 96.8° (Ref 20). The M_s can range from <−200 to +100 °C (−328 to 212 °F). It is worth noting that there is also a "transition" structure that preceded the martensite, called the R-phase with a rhombohedral structure (Ref 21). Although this R phase exhibits a number of interesting properties, it will not be reviewed extensively here.

Time-temperature-transformation curve

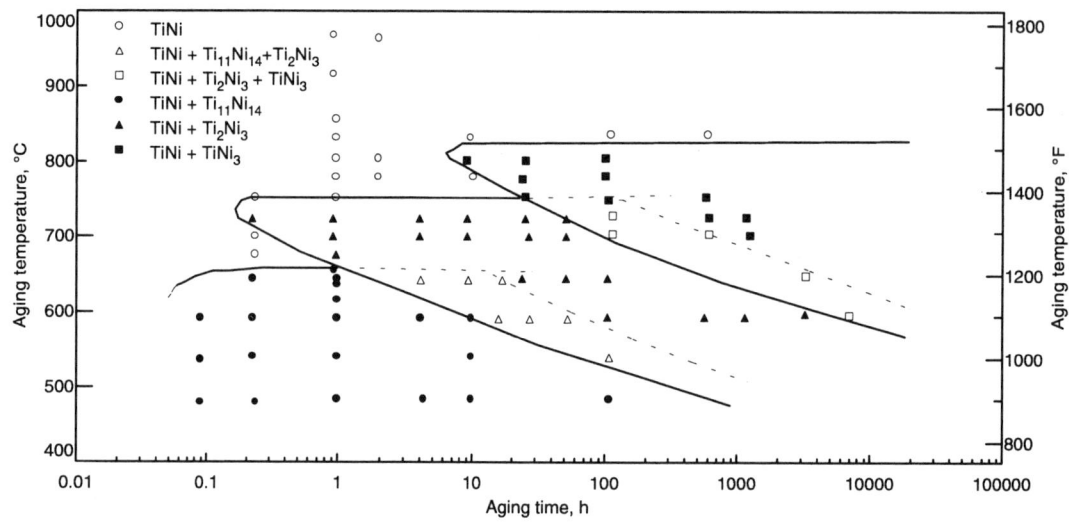

Time-temperature-transformation curve for Ti-51Ni, which shows precipitation reactions as a function of temperature and time.
Source: M. Nishida, C.M. Wayman, and T. Honma, *Metall. Trans. A*, Vol 17, 1986, p 1505

Transformation Products

The T-T-T diagram shows the aging reactions in unstable (>50.6% Ni) titanium-nickel alloys (Ref 7). In general, TiNi → $Ti_{11}Ni_{14}$ → Ti_2Ni_3 → $TiNi_3$ as the aging temperature increases or as time increases at a constant temperature. These precipitation reactions can be readily monitored via transformation temperature or mechanical-property measurements.

Physical Properties

Damping Characteristics

Internal friction and damping of titanium-nickel alloys are dramatically affected by temperature changes (see figure on left). Cooling (or heating) produces peaks, which correspond to the transformation temperatures. At higher temperatures, a very sharp increase is observed during cooling through the M_s. These usually high damping characteristics (Ref 22) have been studied for some time, but have not been used on a commercial basis due to their limited temperature range and rapid fatigue degradation.

Elastic Constants

Dynamically measured moduli (Ref 23 and 24) change markedly with the martensitic transformation and premartensitic effects (see figure on right). Typical values of elastic moduli are 40 GPa (5.8×10^6 psi) for martensite and 75 GPa (10.8×10^6 psi) for austenite. From a practical point of

Ti-Ni-Cu alloy damping characteristics

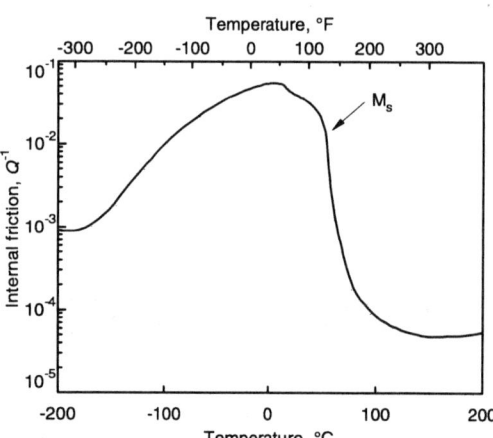

Internal friction of 44.7Ti-29.3Ni-26 Cu (wt%) during cooling with measurement frequency of ~1 Hz.
Source: O. Mercier and E. Török, *J. Phys.*, Vol C-4 (No. 43), 1982, p C-4270

Dynamic Young's modulus vs temperature

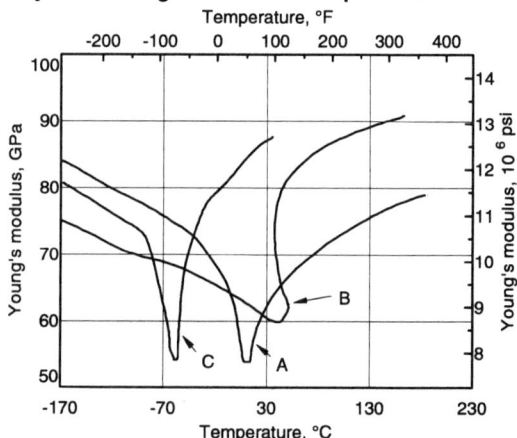

A: Ti-55Ni (wt%). B: 44.7Ti-29.3Ni-26Cu (wt%). C: 44.9Ti-51.7Ni-3.4Fe (wt%).
Source: O. Mercier, K.N. Melton, R. Gotthardt, and A. Kulik, *Proc. Int. Conf. Solid-Solid Phase Transformations*, H.I. Aaronson, D.E. Laughlin, R.F. Sekerka, and C.M. Wayman, Ed., AIME, 1982, p 1259

view, however, modulus is of little value; apparent elasticity is more controlled by the transformation and by mechanical twinning. Poisson's ratio, μ, is 0.33 (Ref 25).

Electrical Resistivity

General values for electrical resistivity of the two primary phases are as follows (Ref 26):
ρ (martensite) = $76 \times 10^{-6} \Omega \cdot$ cm
ρ (austenite) = $82 \times 10^{-6} \Omega \cdot$ cm

Variations in resistivity with temperature are complex functions of composition and thermomechanical processing (see figure). Note also the pronounced effect of the R-phase on resistivity.

Magnetic Characteristics

Magnetic susceptibility also undergoes a discontinuity during phase transition (Ref 26). Typical values are:
χ (martensite) = 2.4×10^{-6} emu/g
χ (austenite) = 3.7×10^{-6} emu/g

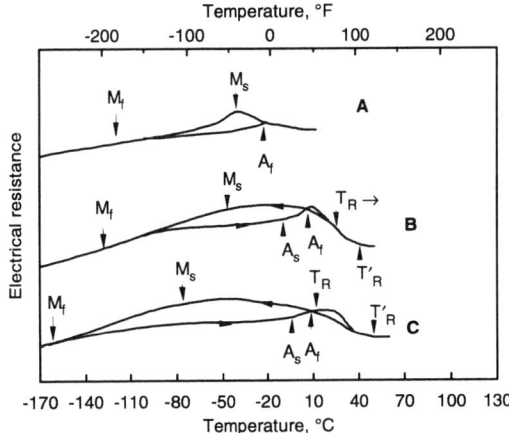

Electrical resistance vs temperature

Electrical resistance vs temperature curves for a Ti-50.6Ni (at.%) alloy that was thermomechanically treated as indicated. A: Quenched from 1000 °C (1830 °F). B: Quenched from 1000 °C (1830 °F), aged at 400 °C (750 °F). C: Directly aged at 400 °C (750 °F). T_R is the transition temperature from austenite to the rhombohedral R phase. T'_R is the shifted transition temperature from processing effects. Arbitrary units for electrical resistance.
Source: S. Miyazaki and K. Otsuka, *Metall. Trans. A*, Vol 17, 1986, p 53

Corrosion

Titanium-nickel generally forms a passive TiO_2 (rutile) surface layer (Ref 27). Like titanium alloys, there is a transition temperature of about 500 °C (930 °F), above which the oxide layer will be dissolved and absorbed into the material. Unlike titanium alloys, however, no α case is formed. Titanium-nickel will also react with nitrogen during heat treatments, forming a TiN layer.

General Corrosion

The rest potential of titanium-nickel in a dilute sodium chloride solution is around 0.23 V (SCE), which compares with 0.38 V for type 304 stainless steel. This puts titanium-nickel on the noble or protected side of stainless steel in the galvanic series. A passive oxide/nitride surface film is the basis of the corrosion resistance of titanium-nickel alloys, similar to stainless steels. Specific environments can cause the passive film to break down, thus subjecting the base material to attack. A summary of titanium-nickel reactions in various environments follows (Ref 28).

Seawater. Titanium-nickel is not affected when immersed in flowing seawater; however, in stagnant seawater, such as found in crevices, the protective film can break down, which results in pitting corrosion.

Acetic acid (CH_3COOH) attacks titanium-nickel at a modest rate of 2.5×10^{-2} to 7.6×10^{-2} mm/year (mpy) over the temperature range 30 °C (86 °F) to the boiling point and over the concentration range 50 to 99.5%.

Methanol (CH_3OH) appears to attack titanium nickel only when diluted with low concentrations of water and halides. This impure methanol solution leads to pitting and tunneling corrosion similar to that found in titanium alloys.

Cupric chloride ($CuCl_2$) at 70 °C (160 °F) attacks titanium-nickel at 5.5 mpy.

Ferric chloride ($FeCl_3$) at 70 °C (160 °F) and 8% concentration attacks titanium-nickel at 8.9 mpy. Titanium-nickel is attacked at a rate of 2.8 mpy in a solution of 1.5% $FeCl_3$ with 2.5% HCl.

Hydrochloric acid (HCl) has a variety of effects on the corrosion of titanium-nickel alloys depending on temperature, acid concentration, and specific alloy composition. With 3% HCl at 100 °C (212 °F) and a range of alloy compositions, the rate of attack was as low as 0.36 mpy and as high as 3.3 mpy. At 25 °C (77 °F) and 7M solution, titanium-nickel-iron alloys can lose up to 457 mpy.

Nitric acid (HNO_3) is more aggressive toward titanium-nickel than type 304 stainless steel. At 30 °C (86 °F), 10% HNO_3 attacks at a rate of 2.5×10^{-2} mpy; 60% solution attacks at 0.25 mpy; 5% HNO_3 at its boiling point attacks at 2 mpy.

Biocompatibility studies have been conducted in various media chosen to simulate the conditions of the mouth and the human body. In general, no corrosion of titanium-nickel alloys has been reported. For example, in tests where coupons of titanium-nickel were sealed at 37 °C (97 °F) for 72 h, the mass corrosion rate was on the order of 10^{-5} mpy for such media as synthetic saliva, synthetic sweat, 1% NaCl solution, 1% lactic acid, and 0.1% $HNaSO_4$ acid (Ref 29); see also Ref 30.

Hydrogen Damage

The interaction between hydrogen and titanium-nickel is sensitive to both concentration and temperature (Ref 31). In general, hydrogen levels in excess of 20 ppm by weight can be considered detrimental to ductility, with levels in excess of 200 ppm severely impairing. Under certain conditions, hydrogen can be absorbed during pickling, plating, and caustic cleaning. The exact conditions required for hydrogen absorption are not well defined, so it is advisable to exercise care when performing any of these operations.

Substantial amounts of hydrogen also can be absorbed in hydrogenated water at elevated temperatures and pressures, such as would be found in pressurized water reactor primary water systems. Relatively short exposure times have been shown to produce hydrogen levels well in excess of 1000 ppm (Ref 32).

Thermal Properties

Heat Capacity

A typical plot of specific heat (C_p) versus temperature for a 50.2% Ti alloy (see figure) shows a discontinuity at the M_s temperature of 90 °C (195 °F) (Ref 3). The peak and onset temperatures for the peaks are often used to characterize the transformation temperatures of an alloy. Care must be taken however, (Ref 33), because the presence of an R-phase prior thermal cycling, and residual stresses from sample cutting can tend to complicate the curves and introduce spurious peaks.

Latent Heats

The latent heat of the martensitic transformation strongly depends on the transformation temperature and stress rate ($d\sigma/dT$) through the formula

$d\sigma/dT = \Delta H/(\Delta \varepsilon T)$

Typical values for ΔH are 4 to 12 cal/g and values for $d\sigma/dT$ range from 3 to 10 MPa/°C.

The latent heat of fusion can be expressed as:
$\Delta H = -34{,}000$ J/mol (Ref 34).

Thermal Expansion

The thermal coefficient of linear expansion can be expressed as (Ref 23):

α (martensite) = 6.6×10^{-6}/°C
α (austenite) = 11×10^{-6}/°C

The volume change on phase transformation (ΔV) (from austenite to martensite) is –0.16% (Ref 35).

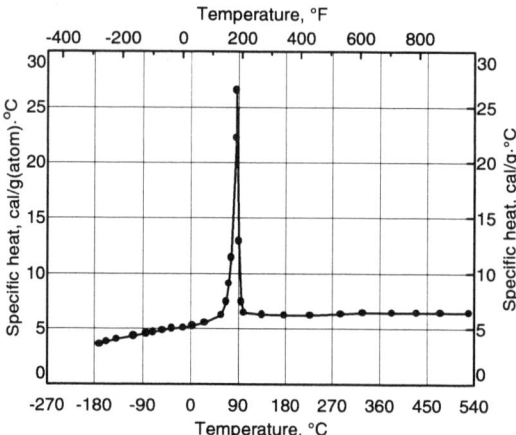

Specific heat (C_p)

Specific heat of Ti-49.8Ni (at.%), with a sharp peak in the specific heat at 90 °C (195 °F) corresponding to the M_s temperature.
Source: C.M. Jackson, H.J. Wagner, and R.J. Wasilewski, NASA Report, NASA-SP 5110, 1972

Transition Temperatures

Melting Point

$T_m = 1310$ °C (2390 °F)

Martensitic Transformation Temperatures

Characteristic transformation temperatures depend strongly on composition (see table on next page and the previous section on chemistry). Typical hysteresis widths range from 10 °C (18 °F) for certain titanium-nickel-copper alloys, to 40 to 60 °C (72 to 108 °F) for binary alloys, to 100 °C (180 °F) for titanium-nickel-niobium alloys.

Transformation temperatures are measured by a number of techniques, including electrical resistivity, latent heat of transformation by differential scanning calorimetry, elastic modulus, yield strength, and strain. However, the most useful measurement technique is to monitor the strain on cooling under a constant load and the recovery on heating.

Other important relationships of transformation temperatures are as follows. Applied stresses increase transformation temperatures according to the stress rate (see the next section on tensile properties). Martensitic deformations increase the stress-free A_s temperatures, particularly in alloys with low yield stresses. The increase is temporary, returning to the previous value after the first heating cycle. Increasing cold work tends to reduce transformation temperatures. The R-phase transformation temperature is much more constant than those for martensite, typically 20 to 40 °C (68 to 105 °F) in binary alloys.

M_d, which is defined as the temperature above which martensite cannot be stress-induced, may be about 25 to 50 °C (50 to 100 °F) higher than A_f.

Strain of a Ti-Ni-Nb specimen

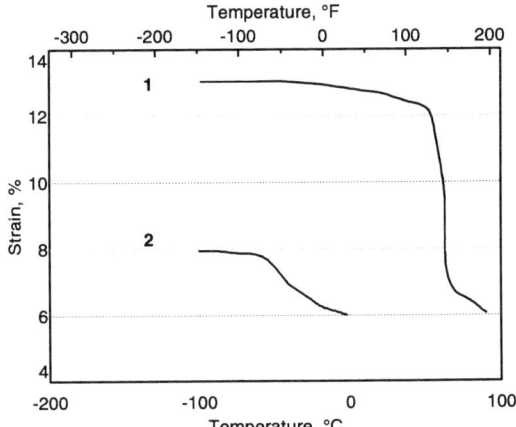

Strain after deforming and unloading, measured on the first and second heating cycles. Note the change in A_s and the recovery strain.
Source: K.N. Melton, J.L. Proft, and T.W. Duerig, MRS Int. Meeting on Advanced Materials, Vol 9, K. Otsuka and K. Shimizu, Ed., Materials Research Society, 1989, p 165

M_s temperature as a function of cold working

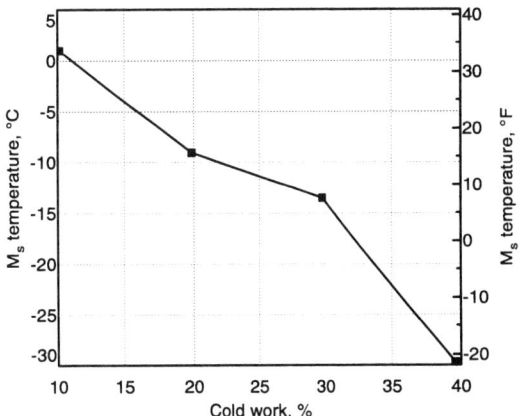

Change in the M_s temperature of a Ti-50.6Ni alloy cold worked 9.2 to 40% and subsequently annealed at 500 °C (930 °F) for 30 min.
Source: G.R. Zadno and T.W. Duerig, unpublished research

Ti-Ni shape memory transformation temperatures

M_d, which is defined as the temperature above which martensite cannot be stress induced, may be about from 25 to 50 °C (50 to 100 °F) higher than A_f.

Reference(a)	Composition at. % Ni	M_s	M_f	A_s	A_f	Technique
[71 Kor]	46.6	57	12	81	117	Dilatometry
	47.6	37	18	79	134	
	49.6	33	13	75	114	
	50.2	−51	30	33	32	
	51	−136	−178	0	−94	
	51.5	−4	−38	−12	46	
	52.8	28	−14	44	278	
[81 Mel]	49.4	57	5	63	106	Dilatometry
	49.7	20	−20	39	77	
	50.4	−30	−53	−12	0	
[80 Mil]	49.7	45	...	67	...	DTA (as-received material)
	50	44	...	120	...	
	50.1	10	...	52	...	
	50.5	−9, −29	−29	...	21	*,
[68 Wan]	51	20 to 25	...	60	...	Electrical, magnetic properties
[79 Che]	48.1	100	60	123	140	Electrical resistivity
	48.6	101	74	178	153	
	49.0	66	16	56	93	
	49.5	47	19	53	80	
	50.5	5	−31	8	44	
	51.0	−52	−85	−39	−34	

Compilation from *Phase Diagrams of Binary Titanium Alloys*, (J.L. Murray, Ed.), ASM International, 1987, p 203. (a) Cited references are as follows: 71 Kor: I.I. Kornilov, Ye. V. Kachur, and O.K. Belousov, "Dilatation Analysis of Transformation in the Compound TiNi," *Fiz. Met. Metalloved.*, 32(2), 420-422 (1971) in Russian; TR: *Phys. Met. Metallogr.*, 32(2), 190-193 (1971). 81 Mel: K.N. Melton and O. Mercier, "The Mechanical Properties of NiTi-Based Shape Memory Alloys," *Acta Metall.*, 29, 393-398 (1981). 80 Mil: R.V. Milligan, "Determination of Phase Transformation Temperatures of TiNi Using Differential Thermal Analysis," Titanium '80, Ti Sci. Tech., Proc. Int. Conf. Kyoto, Japan, May 18-22, T. Kimuzi, Ed., 1461-1467 (1980). 68 Wan: F.E. Wang, B.F. DeSavage, and W.J. Buehler, "The Irreversible Critical Range in the TiNi Transition," *J. Appl. Phys.*, 39(5), 2166-2175 (1968). 79 Che: D.B. Chernov, Yu.I. Paskal, V.E. Gyunter, L.A. Monasevich, and E.M. Savitskii, "The Multiplicity of Structural Transitions in Alloys Based on TiNi," *Dokl. Akad. Nauk SSSR*, 247, 854-857 (1979) in Russian; TR: *Sov. Phys. Dokl.*, 24(8), 664-666 (1979)

Tensile Properties

In general, a superelastic curve is characterized by regions of nearly constant stress upon loading (referred to the loading plateau stress) and unloading (unloading plateau stress). These plateau stress values are better indicators of mechanical strength than the traditional yield stress. Typical values are shown (see table).

Ti-Ni shape memory: Typical loading and unloading characteristics

Loading plateau	450 to 700 MPa
Unloading plateau	Up to 250 MPa
Maximum springback	11%
Maximum deformation with % permanent set	6%
Maximum stored energy	40-50 J/cm^3

Source: T.W. Duering and G.R. Zadno, *Engineering Aspects of Shape Memory Alloys*, T.W. Duerig et al., Ed., Butterworth-Heinemann, London, 1990, p 369

Above M_d

M_d is defined as the temperature above which martensite cannot be stress-induced. Consequently, titanium-nickel remains austenite throughout an entire tensile test above M_d. Tensile strengths depend strongly on alloy condition, and the ultimate tensile strength, yield strength, and ductility of cold worked titanium-nickel wire depend on final annealing temperatures (see figure).

Ductility drops sharply as compositions become nickel-rich. A review of other factors controlling ductility can be found in Ref 36.

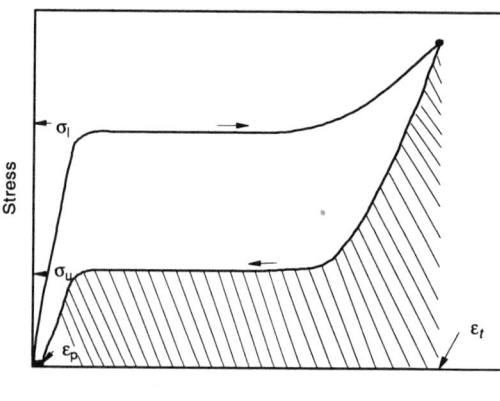

Schematic of superelasticity descriptors

Schematic diagram showing key descriptors of superelasticity: σ_u (unloading plateau measured as the inflection point), σ_l (loading plateau measured as the inflection point), ε_t (total deformation strain), ε_p (permanent set, or amnesia) and the stored energy (shaded area).
Source: T.W. Duerig and G.R. Zadno, *Engineering Aspects of Shape Memory Alloys*, T.W. Duerig et al., Ed., Butterworth-Heinemann, London, 1990, p 369

Aging of nickel-rich alloys increases austenitic strength to a typical peak strength of 800 MPa (116 ksi). Surprisingly, ductility is also increased during the aging process.

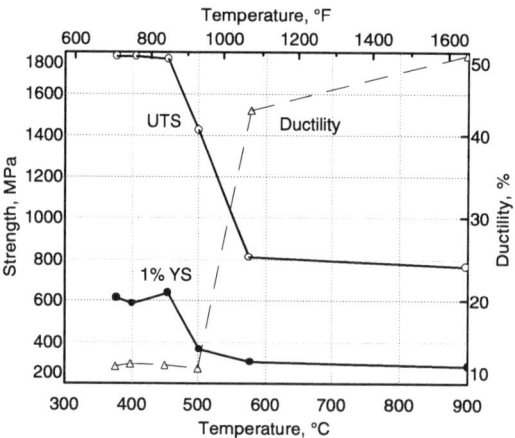

The influence of annealing temperature on mechanical properties of 0.5 mm (0.02 in.) Ti-50.6Ni wire with 40% cold work and annealed 30 min at temperature.
Source: G.R. Zadno and T.W. Duerig, unpublished research

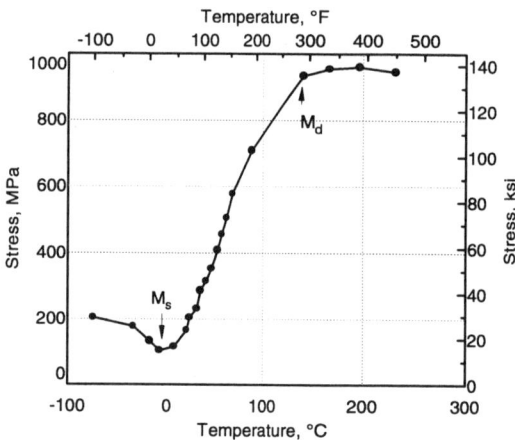

The influence of annealing temperature on mechanical properties of Ti-50.6Ni with 40% cold work annealed 30 min at temperature.
Source: G.R. Zadno and T.W. Duerig, unpublished research

Below M_s

Titanium-nickel yield stresses are controlled by the "friction" of the martensite twin interfaces. Typical yields stresses are 120 to 160 MPa (17 to 23 ksi) for binary alloys and as low as 60 to 90 MPa (9 to 13 ksi) for titanium-nickel-copper alloys. Ultimate tensile strengths and ductilities are similar to austenitic values.

Superelasticity (Between M_s and M_d)

Effect of Temperature

Between M_s and M_d, the material transforms from austenite to martensite during tensile testing. Yield strengths vary continuously from M_s to M_d (see figure). The rate of stress increase is called the stress rate, varying from 3 to 20 MPa/°C, with rates generally increasing with M_s.

Superelasticity, or pseudoelasticity, is an enhanced elasticity when unloading between A_s and M_d. The M_d transition is generally defined as the temperature above which stress-induced martensite can no longer be formed. On a stress-temperature graph, M_d is the temperature where the stress begins to level off.

Superelasticity is also highly temperature dependent (see figures). Changing alloy composition and heat treatment can shift the temperature range of superelastic behavior from –100 to +100 °C (–148 to 212 °F).

This datasheet describes some of the key properties of equiatomic and near-equiatomic titanium-nickel alloys with compositions yielding shape memory and superelastic properties. Shape memory and superelasticity *per se* will not be reviewed; readers are referred to Ref 1 to 3 for basic information on titanium-nickel, Tee-nee, Memorite™, Nitinol, Tine™, and Flexon™. These terms do not refer to single alloys or alloy compositions, but to a family of alloys with properties that greatly depend on exact compositional make-up, processing history, and small ternary additions. Each manufacturer has its own series of alloy designations and specifications within the "Ti-Ni" range.

A second complication that readers must acknowledge is that all properties change significantly at the transformation temperatures M_s, M_f, A_s, and A_f (see figure). Moreover, these temperatures depend on applied stress. Thus any given property depends on temperature, stress, and history. Superelasticity is an enhanced elasticity occurring when unloading between A_s and M_d (see the section "Tensile Properties" in this datasheet.)

Permanent set of superelastic wire

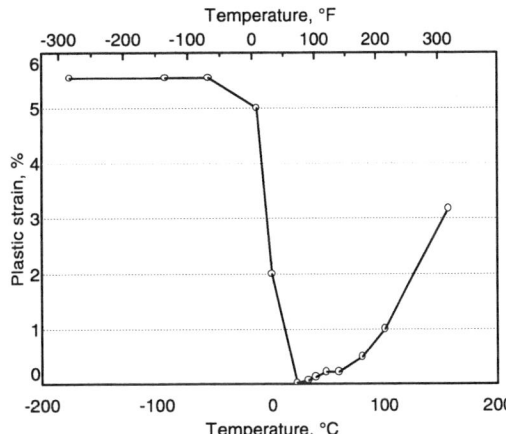

Permanent set of superelastic binary titanium-nickel wire deformed 8.3% and unloaded at various temperatures. Ti-Ni wire with 50.8 at.% Ni cold worked 40% and annealed at 500 °C (930 °F) for 2 min. The superelastic window is roughly 40 °C (70 °F) in width.
Source: T.W. Duerig and G.R. Zadno, *Engineering Aspects of Shape Memory Alloys*, T.W. Duerig et al., Ed., Butterworth-Heinemann, London, 1990, p 369

Loading and unloading plateau heights in Ti-Ni wire

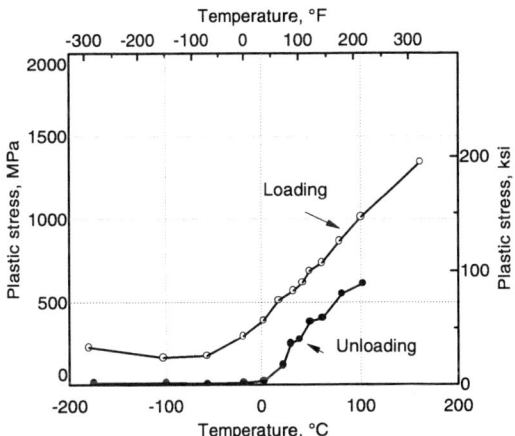

50.8 at.% Ni cold worked 40% and annealed at 500 °C (930 °F) for 2 min. Stress rate = 5.7 MPa/°C.
Source: T.W. Duerig and G.R. Zadno, *Engineering Aspects of Shape Memory Alloys*, T.W. Duerig et al., Ed., Butterworth-Heinemann, London, 1990, p 369

High-Temperature Behavior

Creep. Very few creep measurements have been made on titanium-nickel, although creep mechanisms have been proposed (Ref 37).

Stress relaxation is critical in many constrained recovery applications. Several measurements have been made (Ref 9, 12, 38). To summarize, relaxation of stable titanium-nickel alloys is extremely slow below 350 °C (660 °F) and becomes very rapid by 425 °C (795 °F).

Stress relaxation

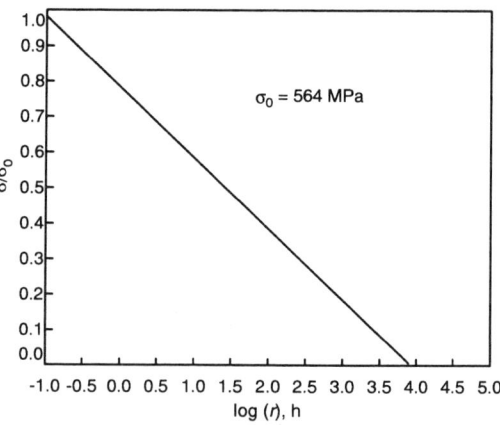

Stress relaxation of a $Ni_{47}Ti_{50}Fe_3$ alloy at 375 °C (705 °F) measured in a dynamically controlled creep machine. Round specimens with gage length of 6 mm, fully annealed.
Source: J. Proft and T.W. Duerig, *Engineering Aspects of Shape Memory Alloys*, T.W. Duerig et al., Ed., Butterworth-Heinemann, 1990, p 115

Fatigue Properties

The fatigue behavior of titanium-nickel alloys is extremely complex and encompasses many different topics. To summarize, titanium-nickel excels in low-cycle, strain-controlled environments, and does relatively poorly in high-cycle, stress-controlled environments. The superelastic mechanisms accommodate high strains without excessive stresses, but cannot accommodate high stresses without excessively high strains.

Stress Controlled

Isothermal stress-controlled testing is important in many constrained recovery applications in which the shape memory effect is only used as a mode of installation, e.g., fasteners and couplings, in which the alloys are used exclusively above the M_d temperature. Although fatigue has been extensively characterized by the coupling and fastener industries, much of the data remains largely unpublished or highly application specific. Typical S-N behavior for various alloy compositions tested well above their M_d temperatures is shown below (see figure on next page).

Strain Controlled

Isothermal strain-controlled behavior is highly dependent on the testing temperature relative to the alloy transformation temperature. Fracture follows the Coffin-Manson relationship:

$$N^\beta \Delta \varepsilon_p = C$$

where C and β are constants; N is the number of cycles to failure; and $\Delta \varepsilon_p$ is the applied strain amplitude (see figure). One specific example of particular interest is superelastic cycling (strain-controlled testing between A_s and M_d). There are several modes of degradation that occur under these circumstances (see figure). Again, these depend strongly on specific alloy conditions and the needs of the application. Further data are provided in Ref 39.

Thermal Fatigue

Thermal cycling both with and without applied loads such as found in actuators or heat engines, is certainly the most complex mode to analyze. Although it has been studied extensively, it remains difficult to predict failure, largely because failure can consist of fracture, ratcheting, migration of transformation temperatures, changes in reset force, etc. Moreover, damage accumulation depends on stress, strain, temperature change, heating methods, heating and cooling rates, and even orientation (horizontal or vertical).

Failure too is nebulous. Various degradation modes can occur, including a shift in transformation temperature, a reduction in the available strain, walking (or ratcheting), and fracture itself. Thermal cycling with no applied load can even result in some degradation. See Ref 40 to 43 for further information.

Typical S-N curves

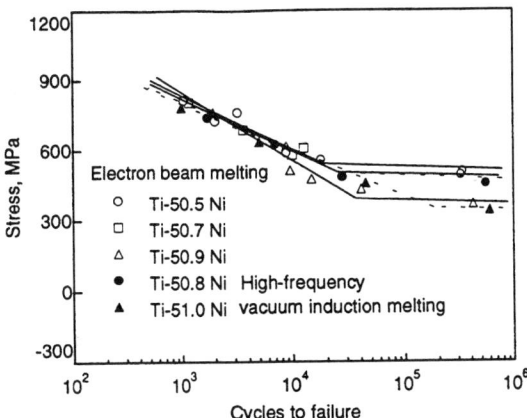

Titanium-nickel alloys of various compositions (at.%) prepared in various fashions is isothermally tested in sheet form. Testing is done well above M_d. Sheet specimens loaded and unloaded ($R = 0$). Cycled at 60 °C (140 °F).
Source: S. Miyazaki, Y. Sugaya, and K. Otsuka, MRS Int. Meeting on Advanced Materials, Vol 9, K. Otsuka and K. Shimizu, Ed., Materials Research Society, 1989, p 257

Low-cycle fatigue behavior

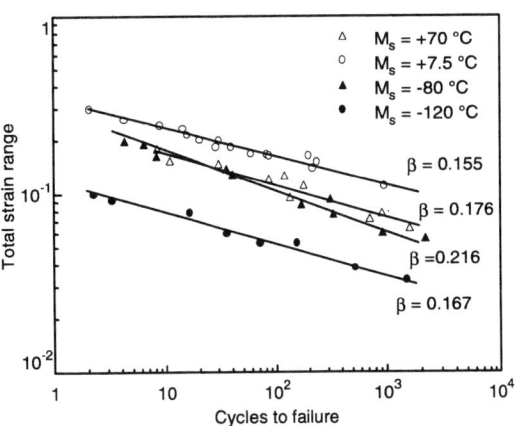

Low-cycle fatigue behavior of titanium-nickel alloys at room temperature tested in tension-compression ($R = -1$). The M_s temperatures of the alloys are indicated.
Source: K.N. Melton and O. Mercier, *Strength of Metals and Alloys*, P. Haasen et al., Ed., Pergamon Press, 1979, p 1246

Effects of isothermal superelastic cycling

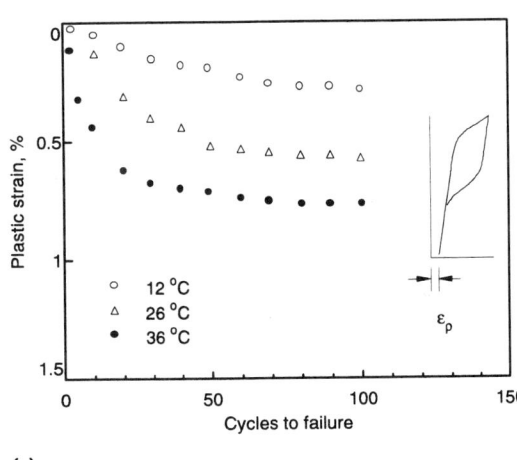

(a)

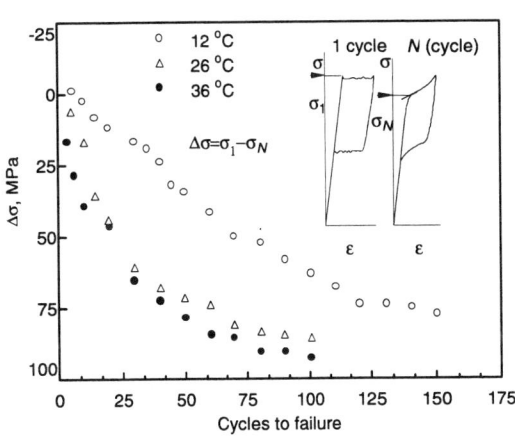

(b)

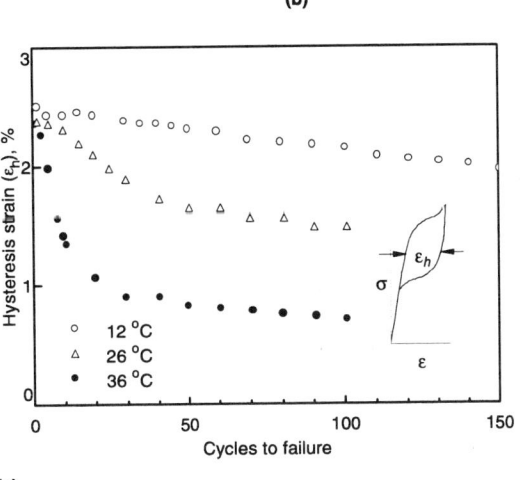

(c)

Tests of a superelastic titanium-nickel wire at three temperatures. Alloy contains 50.6 at.% nickel. Three modes of degradation occur simultaneously during the isothermal superelastic cycling of titanium-nickel alloys: walking, or an accumulation of permanent set (top), a change in yield stress (middle) and a reduction in the hysteresis width (bottom).
Source: S. Miyazaki, *Engineering Aspects of Shape Memory Alloys*, T.W. Duerig et al., Ed., Butterworths, 1990, p 403

Fracture

Fatigue Crack Propagation

Stress-inducing martensite at the tip of a propagating crack has been shown to significantly slow crack growth (Ref 44). Other sources of data include Ref 45.

Toughness

Although Charpy impact testing on titanium-nickel has been conducted (Ref 3, 46), very little K_{Ic} or J_{Ic} data exist. Indications are strong that a sharp toughness minimum exists at M_d (see figure).

Fatigue crack propagation

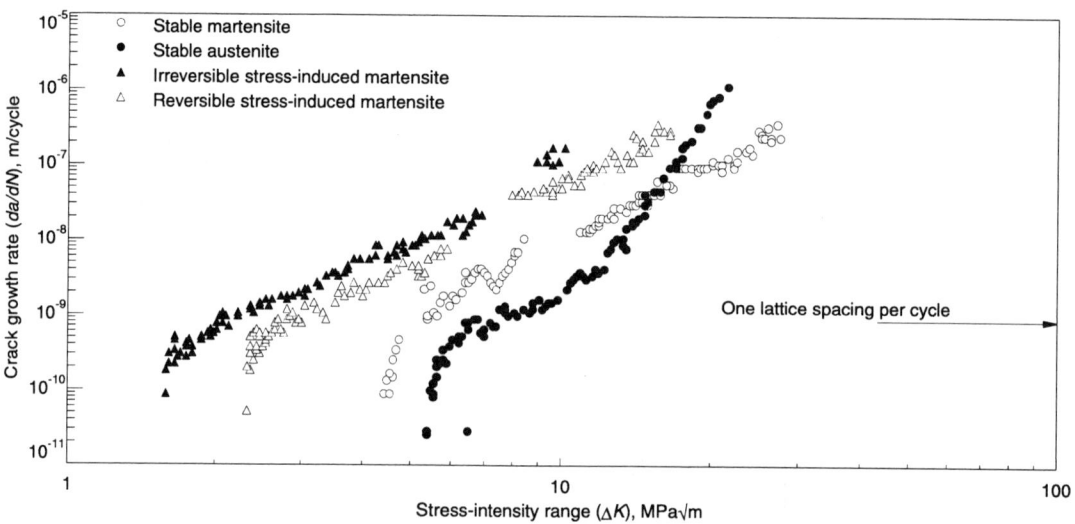

Fatigue crack propagation rates in four titanium-nickel alloys, representing *stable martensite* with M_s > room temperature, *stable austenite* with M_d << room temperature, *irreversible stress-induced martensite* with M_s < room temperature < A_s, and *reversible stress-induced martensite* with A_f < room temperature < M_d. Tested with $R = 0.1$ on 10 mm thick CT specimens at 50 Hz under conditions of decreasing ΔK.
Source: R.H. Dauskardt, T.W. Duerig, and R.O. Ritchie, MRS Int. Meeting on Advanced Materials, Vol 9, K. Otsuka and K. Shimizu, Ed., Materials Research Society, 1989, p 243

Processing

Bulk Working

Flow Stress. Upset forging tests conducted on seven different alloys at strain rates ranging from 10^{-5} to 10^2 indicate that the material has a very high hot ductility and is not highly strain rate dependent. The flow stresses can be characterized by the relation:

$$\sigma = k\dot{\varepsilon}^m \exp(-Q/RT)$$

where σ is the flow stress; $\dot{\varepsilon}$ is the strain rate; Q is an activation energy (50,000 cal/mole); and m and k are constants. Values for m were found to range from 0.1 to 0.17 with increasing temperature (see figure for typical data of an equiatomic binary alloy).

Extrusion. Both solid bar and large tube (50 mm diameter with 8 mm wall thickness) have been extruded from titanium-nickel. Parameters remain proprietary.

Forging. Titanium-nickel has been successfully forged into large cups. Parameters remain proprietary.

Rolling. Titanium-nickel can be hot rolled with relative ease, but is difficult to cold roll, especially in thin, wide sections.

Fracture energy vs temperature

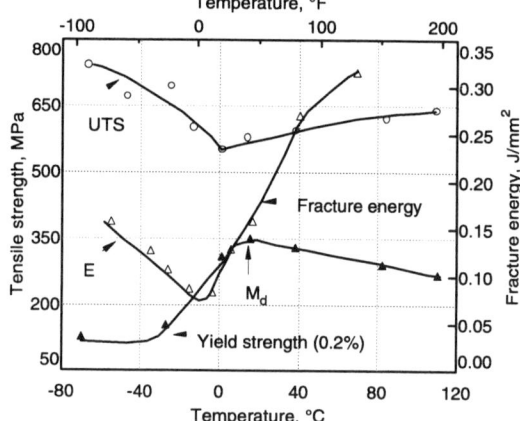

Vacuum induction melted alloy hot worked and annealed for 1 h at 950 °C and air cooled. Standard Charpy V-notch impact specimens were machined. The influence of temperature on fracture toughness of 44Ti-49Ni-5Cu-2Fe (wt%). Note that the minimum in the fracture energy occurs just below M_d, as determined from the corresponding 0.2% yield strength and ultimate tensile strength measurements.
Source: K.N. Melton and O. Mercier, Acta Metall., Vol 29, 1982, p 393

Fabrication

Casting. Although some unreported experiments have been conducted, casting has not been done on a commercial level.

Powder Metallurgy. Although a great deal of experimentation has taken place with both elemental and prealloyed powders, nothing has reached near-production levels. Reference 5 provides a review of methods.

Forming. Titanium-nickel sheet has been successfully formed into a range of complex shapes, both in the martensite and austenitic phases. Springback is high, as is die wear and friction. Parameters remain proprietary.

Machining. Titanium-nickel is very difficult to machine. Very low speeds and a great deal of coolant is required, and tool wear is very rapid. Milling and drilling are particularly difficult. Producers of couplings have demonstrated that large-scale machining production is possible.

Heat Treatment. Titanium-nickel can be heat treated in air up to ~500 °C (930 °F). No α case is formed, but a surface oxide of rutile develops quickly. Above 500 °C (930 °F), the oxide layer begins to flake (depending on time). Nitrogen and hydrogen atmospheres are not recommended. Argon, helium, and vacuum heat treatments are commonly used to preserve bright finishes.

Recrystallization is extremely rapid above 700 °C (1290 °F). Solution treatment requires temperatures of at least 550 °C (1020 °F). Stress relief is usually accomplished at temperatures as low as 300 °C (570 °F). A TTT diagram is shown in the previous section "Phases and Structures."

Fully annealed bar stock has a typical hardness of 60 HRA. Vicker numbers range from 190 HV in the annealed condition to 240 HV after 15% cold work (Ref 3).

Flow stress measurements

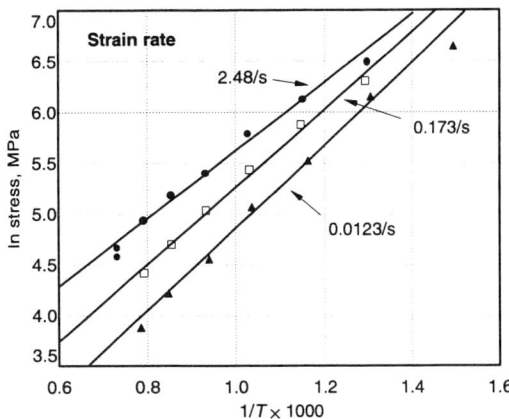

Natural log (ln) of flow stress is shown to vary linearly with the inverse of absolute temperature. 12 mm × 8 mm diameter specimens tested in compression under isothermal conditions (heated dies) at the temperatures and strain rates shown. Source: T.W. Duerig, unpublished data

Joining. Titanium-nickel is difficult to join because most mating materials cannot tolerate the large strains experienced by the alloy. Most applications rely on crimped bonds. It can be welded to itself with relative ease by resistance and TIG methods. Welding to other materials is extremely difficult, although proprietary methods do exist and are practiced in large volumes in the production of eyeglass frames.

Brazing can only be accomplished after re-enforcement and plating. Again, methods are proprietary; large-scale production is practiced by the eyeglass frame industry.

References

1. T.W. Duerig et al., Ed., *Engineering Aspects of Shape Memory Alloys*, Butterworth-Heinemann, London, 1990
2. J. Perkins, Ed., *Shape Memory Effects in Alloys*, Plenum Press, 1975
3. C.M. Jackson, H.J. Wagner, and R.J. Wasilewski, NASA Report, NASA-SP 5110, 1972
4. J. Takahashi, M. Okazaki, K. Hiroshi, and Y. Furuta, private communication, 1984
5. T.W. Duerig, in *Advanced Synthesis of Materials*, J. Moore, Ed., 1993, to be published
6. G.R. Zadno and T.W. Duerig, *Engineering Aspects of Shape Memory Alloys*, T.W. Duerig et al., Butterworth-Heinemann, London, 1990, p 414
7. M. Nishida, C.M. Wayman and T. Honma, *Metall. Trans. A*, Vol 17, 1986, p 1505
8. T.W. Duerig and K.N. Melton, Proc. of SMA'86, C. Youyi et al., Ed., China Academic Publishers, 1986, p 397
9. J.L. Proft and T.W. Duerig, *Engineering Aspects of Shape Memory Alloys*, T.W. Duerig et al., Butterworth-Heinemann, London, 1990, p 115
10. A. Keeley, D. Stöckel, and T.W. Duerig, *Engineering Aspects of Shape Memory Alloys*, T.W. Duerig et al., Ed., Butterworth-Heinemann, London, 1990, p 181
11. T.W. Duerig and G.R. Zadno, *Engineering Aspects of Shape Memory Alloys*, T.W. Duerig et al., Ed., Butterworth-Heinemann, London, 1990, p 369
12. C.R. Such, "The Characterization of the Reversion Stress in NiTi," M.S. Thesis, Naval Post Graduate School, Monterrey, CA, 1974
13. M.V. Nevitt, *Trans. Met. Soc. AIME*, Vol 218, 1960, p 327
14. D.M. Goldstein, W.J. Buehler, and R.C. Wiley, Effects of Alloying upon Certain Properties of 55.1 Nitinol, NOLTR 64-235, 1964
15. H.K. Eckelmeyer, Sandia Labs Report 74-0418, 1974
16. D.B. Chernov et al., *Dokl. Akad. Nauk. SSSR*, Vol 245 (No. 2), 1979, p 360
17. P.G. Lindquist and C.M. Wayman, *Engineering Aspects of Shape Memory Alloys*, T.W. Duerig et al., Butterworth-Heinemann, London, 1990, p 58
18. T.W. Duerig and K.N. Melton, *The Martensite Transformation in Science and Technology*, E. Hornbogen and H. Jost, Ed., Deutsche Gesellschaft für Metallkunde, Germany, 1988, p 191
19. W. Moberly and K. Melton, *Engineering Aspects of Shape Memory Alloys*, T.W. Duerig et

al., Ed., Butterworth-Heinemann, London, 1990, p 46
20. O. Matsumoto, S. Miyazaki, K. Otsuka, and H. Tamura, *Acta Metall.*, Vol 35, 1987, p 2137
21. H.C. Ling and R. Kaplow, *Metall. Trans. A*, Vol 12, 1981, p 2101
22. L. Kaufman, S.A. Kulin, P. Neshe, and R. Salzbrenner, *Shape Memory Effects in Alloys*, J. Perkins, Ed., Plenum Press, 1975, p 547
23. S. Spinner and A.G. Rozner, *J. Acoust. Soc. Am.*, Vol 40, 1966, p 1009
24. R.J. Wasilewski, *Trans. AIME*, Vol 233, 1965, p 1691
25. O.K. Belousov, *Russ. Metall.*, Vol 2, 1981, p 204
26. J.E. Hanlon, S.R. Butler, and R.J. Wasilewski, *Trans. AIME*, Vol 239, 1967, p 1323
27. C-M. Chan, S. Trigwell, and T.W. Duerig, *Surf. Interface Anal.*, Vol 15, 1990, p 349
28. J.D. Harrison, Raychem Report, 1991
29. S. Lu, *Engineering Aspects of Shape Memory Alloys*, T.W. Duerig et al., Ed., Butterworth-Heinemann, London, 1990, p 445
30. L.S. Castleman, S.M. Motzkin, F.P. Alicandri, V.L. Bonawit, and A.A. Johnson, *J. Biomed. Mater. Res.*, Vol 10, 1976, p 695
31. R. Burch and N.B. Mason, *J.C.S. Faraday I*, Vol 75, 1979, p 561; R.B. Burch and N.B. Mason, *J.C.S. Faraday I*, Vol 75, 1979, p 578
32. A.R. Pelton and T.W. Duerig, "An Analysis of Crofit Couplings After One-Year Service at Seabrook," Raychem Proprietary Report, 1991
33. G. Airoldi and B. Rivolta, *Phys. Scripta*, Vol 37, 1988, p 891
34. J.C. Gachon and J. Hertz, *CALPHAD*, Vol 7, 1983, p 1
35. K. Otsuka, T. Sawamura, K. Shimizu, and C.M. Wayman, *Metall. Trans.*, Vol 2, 1971, p 2583
36. S. Miyazaki et al., *Mater. Sci. Forum*, Vol 56-58, 1990, p 765
37. A.K. Mukherjee, *J. Appl. Phys.*, Vol 39 (No. 5), 1968, p 2201
38. A. Terui et al., *Nip. Kikai Gakkai Ronbunshu, A. Hen*, Vol 51 (No. 462), 1985, p 488
39. G.R. Zadno, W. Yu, and T.W. Duerig, *Materials Science Forum*, Vol 56-58, B.C. Muddle, Ed., 1990, p 771
40. H. Tamura, Y. Suzuki and T. Todoroki, Proc. Int. Conf. Martensitic Transformations, Japan Institute of Metals, 1986, p 736
41. Y. Furuya et al., MRS Int. Meeting on Advanced Materials, Vol 9, K. Otsuka and K. Shimizu, Ed., Materials Research Society, 1989, p 269
42. Y. Suzuki and H. Tamura, *Engineering Aspects of Shape Memory Alloys*, T.W. Duerig et al., Ed., Butterworth-Heinemann, London, 1990, p 256
43. J.L. Proft, K.N. Melton, and T.W. Duerig, MRS Int. Meeting on Advanced Materials, Vol 9, K. Otsuka and K. Shimizu, Ed., Materials Research Society, 1989, p 159
44. R.H. Dauskardt, T.W. Duerig, and R.O. Richie, MRS Int. Meeting on Advanced Materials, Vol 9, K. Otsuka and K. Shimizu, Ed., Materials Research Society, 1989, p 251
45. S. Miyazaki, Y. Sugaya, and K. Otsuka, MRS Int. Meeting on Advanced Materials, Vol 9, K. Otsuka and K. Shimizu, Ed., Materials Research Society, 1989, p 251
46. K.N. Melton and O. Mercier, *Acta Metall.*, Vol 29, 1982, p 393

Technical Notes

Technical Note 1: Metallography and Microstructure

Reviewed by Gerhard Welsch, Case Western Reserve University, and Rodney Boyer, Boeing Commercial Airplane Company

This Technical Note describes some of the basic microstructural features in titanium alloys and methods of specimen preparation for metallographic investigation. The microstructural features can be categorized as alpha microstructures of hexagonal α phase, beta microstructures of body-centered cubic β phase, and microstructures of various transition products from beta decomposition (which would be primarily α and β). These microstructural features affect mechanical properties as shown in Table 1.

Alpha Structures

Depending on the alloy composition and the thermal-mechanical processing history, the α-type (hexagonal) structures in α, α-β, and β titanium alloys may include:

- Primary alpha
- Secondary alpha (transformed beta from cooling or subtransus aging)
- Alpha structures hardened by α_2(Ti$_3$Al)
- Martensitic microstructures based on hexagonal alpha phase (labelled α')
- Martensitic microstructures based on orthorhombic phase (labeled α″)

Primary and secondary alpha generally are unaffected by aging, unless their aluminum concentration exceeds the solubility limit. In the latter case, α_2-precipitates may develop upon aging. The martensites readily decompose at aging temperatures into more stable α or α + β microstructures. Metastable beta alloys also form an unstable Type-1 α precipitate that decomposes into a Type-2 α during aging.

Primary alpha is the hcp phase that persists during heat treatment in the α-β phase field. The morphology of primary alpha is influenced by thermomechanical history. Microstructure and crystallographic texture are influenced most by the last working operation or heat treatment.

Morphology of primary alpha can be lamellar, equiaxed, or mixed. The lamellar morphology is obtained from β transformation during medium-to-slow cooling (of a prior heat treatment, in the case of primary α) from above the transus or high in the α - β phase field. Heavy working at temperatures below the transus are needed to transform lamellar alpha into an equiaxed alpha morphology. Annealing or solution treating below the beta transus does not significantly change the morphology of primary alpha. However, if the material has been heavily cold worked, then recrystallization to an equiaxed microstructure can occur when heat treated in the α + β phase field.

Precipitation of α_2. The primary α-phase, whether in lamellar or equiaxed form, is usually little affected by aging. However, this is true only if the oxygen concentration is well below 0.2 wt.% and the aluminum concentration is not higher than the solubility limit (approximately 6.0 wt%). Otherwise, fine α_2 (Ti$_3$Al) precipitates can occur. Also, there is indirect evidence from observation of superdislocation pairs, that ordering of (oxygen) interstitials can occur, which has a profound effect on dislocation glide systems, with lattice hardening and embrittlement as a consequence.

Secondary alpha (or transformed beta) refers to the local or continuous alpha structures that arise during annealing below the transus (or during cooling

Kroll's reagent (192)

Fig. 1 Example of acicular alpha with prior beta grains are outlined by the grain boundary alpha that was first to transform. Ti-5Al-2.5Sn, hot worked below the alpha transus, annealed 30 min at 2150 °F (1177 °C), which is above the beta transus and air cooled.

Kroll's reagent (192)

Fig. 2 Example of plate-like alpha. Alloy Ti-5Al-2.5Sn with same annealing temperature as previous figure but furnace cooled instead of air cooled. 85x

Kroll's reagent (192)

Fig. 3 Example of serrated alpha plates; the particles of TiH and retained beta show up black between the plates of alpha. Unalloyed Ti (99.0%) annealed 2 hours at 1000 °C (1830 °F), air cooled. 250x

Table 1 Relationships between critical microstructural features and mechanical properties of titanium alloys(a)

Feature	Enhances	Degrades
Equiaxed α	Strength Ductility Fatigue Initiation Resist. Low Cycle Fatigue Resist.	Fracture Toughness Fatigue Crack Growth Resist. Notched Fatigue Resist.
Elongated α	Fracture Toughness Notched Fatigue Resist. Fatigue Crack Growth Resist.	Ductility Fatigue Initiation Resist. Low Cycle Fatigue Resist.
Widmanstätten α	Fracture Toughness Notched Fatigue Resist. Fatigue Crack Growth Resist. Creep	Ductility Fatigue Initiation Resist. Low Cycle Fatigue Resist. Strength
Bi-Modal α	Strength Ductility Fatigue Initiation Resist. Low Cycle Fatigue Resist.	Fatigue Crack Growth Resist. Fracture Toughness
Colony α	Fatigue Crack Growth Resist. Fracture Toughness Notched Fatigue Resist.	Strength Ductility Fatigue Initiation Resist. Low Cycle Fatigue Resist.
Secondary α	Strength Ductility	Fracture Toughness
Grain shape (elongated)	Fracture Properties Fatigue Crack Growth Resist. Notched Fatigue Resist.	Fatigue Initiation Resist.
Coarse prior β grains	Fracture Toughness Creep	Strength Ductility Low Cycle Fatigue Resist. Fatigue Initiation Resist.
Fine prior β grains	Strength Fatigue Initiation Resist. Ductility	Fracture Toughness Notched Fatigue Resist.
Mixed-mode grain size	Strength Fatigue Initiation Resist.	Fracture Toughness
Alpha films	Fatigue Initiation Resist. Notched Fatigue Resist.	Fatigue Crack Growth Resist.
Grain boundary α	Fracture Toughness Fatigue Crack Growth Resist. Notched Fatigue Resist.	Ductility Fatigue Initiation Resist. Low Cycle Fatigue Resist.

Resist. = Resistance. (a) These general relationships do not necessarily address specific comparisons between some microstructural features. For example, the fatigue crack growth resistance of bi-modal alpha is generally better than equiaxed alpha, but less than that of recrystallized annealed equiaxed alpha.

through the subtransus region), by nucleation and growth in previously quenched-in martensite or metastable beta regions. Transformed beta typically consists of submicron alpha platelets which are usually separated by beta phase. Secondary alpha is produced by aging retained beta or martensitic structures, or by cooling from high in the α + β field at a rate slower than that required to form martensite.

The secondary alpha has various appearances and may be acicular or lamellar, platelike, serrated, or Widmanstätten. *Acicular or lamellar* α (Fig. 2) is the most common transformation product formed through the transus. It is a result of nucleation and growth of α with a specific crystallographic orientation (Burgers orientation) relative to the crystallographic planes of the prior β matrix.

Precipitation of lamellar alpha normally occurs on multiple variants or orientations of prior β habit planes. Upon cooling through the transus temperature with intermediate rates (air-cooling) or slow rates (furnace-cooling), the beta phase transforms by diffusion-controlled partitioning of alpha-stabilizing alloy elements into alpha lamellae and beta-stabilizing elements into the remaining volume. The beta-phase regions have the same crystal orientation as the original large beta grains, and the alpha-phase lamellae have orientations that are related to the beta phase by the Burgers relation. As a result, *colonies* or *lamella packets* of up to 12 different orientations can be generated within a prior beta grain. These multiple orientations of α often have the basketweave appearance characteristic of a *Widmanstätten* structure. Alpha-lamella packets forming from small beta grains may have a singular orientation.

A packet or cluster contains acicular or lamellar α grains aligned in the same orientation. They are also referred to as a "colony." Each colony of lamellae behaves like a single crystal or grain, because the orientation difference between neighbor lamellae is very small (less than a degree). Therefore, the effective grain size for mechanical properties is that of the colony size, and colony size is often regarded as an important microstructural feature when correlating this type of microstructure with properties such as fatigue strength or fracture toughness.

The cooling rate affects the number of variants of alpha products, the width of the individual lamellae, the degree of partitioning of the alloy elements, and if it has a Widmanstätten or colony microstructure. Under some conditions, the alpha lamellae take on a wide, platelike appearance (see Fig. 2 for furnace-cooled Ti-5Al-2.5Sn). Under other conditions, grains of irregular size with jagged boundaries, called "serrated α," are produced (see Fig. 3 of unalloyed titanium specimen).

Martensitic Structures. If the cooling rates from the β field are sufficiently rapid, an acicular or lath-like martensite is formed. In alloys with low concentrations of beta-stabilizing elements, the martensite has a distorted hexagonal crystal lattice, similar to that of α, and is referred to as α'. Because of the fine lath width, α' is stronger than lamellar α, but not necessarily brittle. The other principal type of martensite (orthorhombic α") occurs upon the quenching of a β phase that contains intermediate concentrations of beta-stabilizing alloy elements (e.g., between 10 and 15 wt% vanadium). The α" martensite has an orthorhombic structure with a very fine, internally twinned microstructure. The orthorhombic α" martensite is mechanically soft.

Hexagonal martensite (α') is an athermal transformation product, which can only be formed by quenching. The hexagonal martensites can be formed in either β isomorphous or β eutectoid systems, and the product of the decomposition reaction during tempering depends on the specific alloy system. In β-isomorphous alloys, aging of α' results in the formation of alpha laths (with thin layers of β). Optical microscopy cannot distinguish between aged and unaged martensite in most cases. In β eutectoid systems (such as Ti-Cu alloys), the martensite may decompose into an alpha phase and intermetallic compounds.

The crystal structure of α' martensite is hcp and exhibits the same Burgers orientation relationship with the β-phase as does the α phase:

- $(110)_\beta // (0001)_{\alpha'}$
- $<111>_\beta // <11\bar{2}0>_{\alpha'}$

The variation in the lattice parameters of α' martensite in Ti-V alloys de-

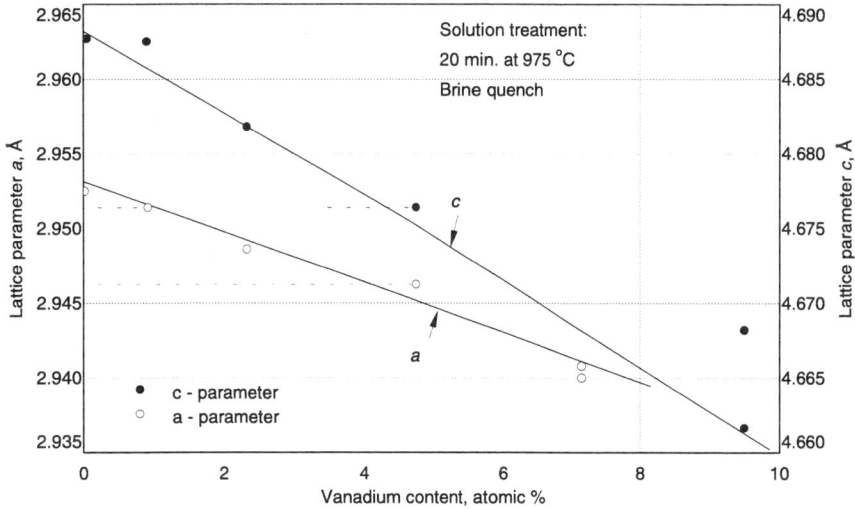

Fig. 4 Lattice parameters of α' versus vanadium content.

Table 2 Formation and decomposition of orthorhombic martensite (α")

Transformation	Process
Formation(a)	i) Decomposition of metastable β during quenching $\beta \rightarrow \alpha'' + (\beta)$ ii) Decomposition of retained β by intermediate (bainitic) transformation during isothermal aging(b) $\beta \rightarrow \beta_{lean} + \beta_{rich} \rightarrow \alpha'' + \beta_{rich}$ iii) Stress-induced transformation from retained β $\beta \rightarrow \alpha'' + $ Twinned β
Decomposition	i) Decomposition of α" by both spinodal decomposition and reverse martensitic transformation during cooling and aging $\alpha'' \rightarrow \alpha''_{lean} + \alpha''_{rich} \rightarrow \alpha''_{lean} + \beta \rightarrow \alpha + \beta$ ii) Decomposition of α" in the normal way $\alpha'' \rightarrow \beta$ upon heating

(a) Lean or rich refer to localized concentration of beta-stabilizing alloy element. (b) Formation of orthorhombic martensite during aging is considered doubtful unless aging temperatures are low enough to preclude diffusion-controlled transformation.
Source: Y. Murakami, Critical Review Phase Transformations and Heat Treatments, in *Titanium 80 Science and Technology*, Metallurgical Society of AIME, 1980, p 160

pends on alloying (see Fig. 4). Vanadium has a smaller atom diameter than titanium, thus the lattice parameters of α' martensite decrease with increasing vanadium content.

Orthorhombic martensite (α'') can be formed by (see Table 2):

- Transformation of metastable β during quenching
- Stress-induced transformation of retained metastable β

This martensite occurs in some binary alloys (such as Ti-Mo and Ti-Nb, but not Ti-V), ternary and higher order alloys, and in many commercial α + β alloys (such as Ti-6Al-4V).

The α'' martensite is important in two respects: (1) In the quenched state it is rather soft and increases room-temperature ductility. (2) As a precursor for subsequently aged or (α+β)-heat treated products, α'' enables the formation of a very fine and uniform distribution of α-precipitates, in β phase.

The formation of α'' is largely dependent on the concentration of the β stabilizing elements in the β phase at solution treatment temperature. In Ti-6Al-4V alloy, α'' martensite (~20 vol%) forms in localized regions containing a concentration range of about 9 to 13 wt% V, which occurs when Ti-6Al-4V is quenched from 800 to 850 °C (1470 to 1560 °F). In binary titanium alloys, the concentration ranges lie within 3-5 at.% Mo, 9-10 at.% V, and 7.6-14 at.% Nb. The α'' phase exhibits the following orientation relationship with the β phase:

- $(110)_\beta//(002)_{\alpha''}$
- $[111]_\beta//[110]_{\alpha''}$

The volume fraction of α'' martensite in α + β structures increases with increasing content of β stabilizer, cooling rate, and the β stabilizing potential of the alloying element. Furthermore, the quenching-temperature region for α'' is extended towards the β transus temperature as the β stabilizing capacity of the alloying element increases. In Ti-6Al-4V alloy, oxygen reduces the β-phase stability and promotes the formation of α'' martensite upon quenching. Lattice parameters for various compositions are shown (see Table 3).

Stress-Induced Martensite. Martensitic transformation can also be induced in a quenched metastable β-phase by an externally applied stress. The structure of the stress-induced (or assisted) martensitic products has been reported to be fcc (or fct), hcp, and orthorhombic (distorted hcp). However, previously reported hexagonal (α') stress-induced martensites were orthorhombic (α'') and that the misinterpretation arose from the overlap of α' and α'' reflections. The triggering stress to induce β-to-α'' transformation is as low as 150 MPa (22 ksi). Hydrogen, a β-phase stabilizer, tends to increase the triggering stress.

Face-Centered Cubic (fcc) and Face-Centered Orthorhombic (fco) Martensites. Face-centered cubic martensite has been found during TEM studies in binary Ti-Cr, -Fe, -Mo, -Mn alloys, as well as in other more complex systems. This martensite is not typical of the bulk material. It seems to be an artifact produced during thin-foil electropolishing as a result of hydrogen contamination, although this cause is not definitely established.

Face-centered cubic martensite contains fine twins of about 10 nm thick on $[111]_{fcc}$ type planes and has been confined to sections of TEM foils. The orientation relationship between the β-phase and these martensites are as follows:

- $(101)_{bcc}//(111)_{fcc}$
- $[11\bar{1}]_{bcc}//[01\bar{1}]_{fcc}$
- $[10\bar{1}]_{bcc}//[11\bar{2}]_{fcc}$

Slip Modes in the α-Phase. Various modes of slip can occur in α-Ti or in the α-phase of titanium alloys (see Table 4). In general, slip can occur on prismatic, pyramidal and basal planes by the movement of <a>, [c] and <c + a>-type dislocations. Since the <$11\bar{2}0$> slip directions are common to all three planes, the <a>-type dislocations can glide on prism, pyramid, and basal planes. The <c + a>-type slip can take place on prismatic and pyramidal planes. The [c]-type glide is restricted to only prismatic planes and generally does not occur.

At least five independent slip systems are required for extensive ductility in polycrystalline materials. The operation of <a>-type slip on prismatic, pyramidal, and basal planes provides only four independent slip systems. They do not allow shear straining along the c-direction. The displacement in the c-direction can be achieved by the movement of [c]- or <c + a>-type dislocations, or even by twins.

Twinning in α Phase. Twinning also contributes to the plastic deformation of α-Ti. It has been reported to occur on a variety of pyramidal planes: {$10\bar{1}1$}, {$10\bar{1}2$}, {$11\bar{2}1$}, {$11\bar{2}2$}, {$11\bar{2}3$}, and {$11\bar{2}4$} type planes (see Fig. 5 for examples). Twins on {$10\bar{1}2$}, {$11\bar{2}1$}, and {$11\bar{2}3$} planes allow an extension along the c-axis compression of α-Ti. {$11\bar{2}2$} twins initiate yielding and account for about 90% of plastic deformation between 25 and 300 °C (80 and 570 °F) for

Table 3 Lattice parameters of orthorhombic α'' martensite for various titanium alloys

Alloy, wt.%	a, Å	b, Å	c, Å
Ti-4Mo	3.001	4.998	4.657
Ti-8Mo	2.994	4.99	4.644
Ti-4W	3.	4.996	4.655
Ti-10V-2Fe-3Al	3.01	4.82	4.62
Ti-2Al-16V	3.027	4.898	4.624

Table 4 Modes of slip in the α-phase

Burgers vector type	Slip direction	Slip plane type	Number of slip systems Total	Independent
a	<$11\bar{2}0$>	basal (0001)	3	2
a	<$11\bar{2}0$>	prism {$10\bar{1}0$}	3	2
a	<$11\bar{2}0$>	pyramidal {$10\bar{1}0$}	6	4
c	[0001]	prism {$10\bar{1}0$}	3	2
c	[0001]	prism {$11\bar{2}0$}	3	2
c + a	<$11\bar{2}3$>	prism {$10\bar{1}0$}	6	5
c + a	<$11\bar{2}3$>	prism {$10\bar{1}0$}	6	5
c + a	<$11\bar{2}3$>	pyramidal {$11\bar{2}2$}	6	5

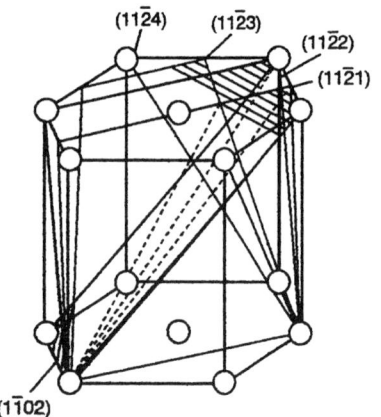

Fig. 5 Twinning planes in α Ti.

Table 5 Characteristics of Type-1 and Type-2 alpha

Characteristic	Type-1 α	Type-2 α
Orientation relation	Obeying Burgers relation $(110)_\beta // (0001)_\alpha$ $<111>_\beta // <11\bar{2}0>_\alpha$	Not obeying Burgers relation $\{10\bar{1}2\}<10\bar{1}1>$ twin orientation to Burgers orientation α
Formation	Initially formed during aging (metastable relative to Type 2)	Transformed from Type 1 after longer holding
Morphology	Monolithic plate (ternary alloy) Monolithic needle (binary alloy)	Colony of fine particles
Transformation mechanism	Nucleation and growth more likely than mechanical twinning	

Source: Y. Murakami, Critial Review Phase Transformations and Heat Treatment, in *Titanium 80 Science and Technology*, Metallurgical Society of AIME, 1980, p 158

α-Ti. Compression normal to the c-axis, however, is accommodated by $\{10\bar{1}2\}$ twins and prismatic slip of $<a>$-type dislocations at 25 °C (80 °F).

The volume fraction of twins (in α-Ti) increases significantly with an increase in grain size and strain. The effects of purity and temperature on the frequency of twinning are not yet well established. However, the trend is such that decreasing temperature and increasing purity generally increase the volume fraction of twins in α-Ti.

Type 1 and Type 2 alpha refer to the two types of alpha that form during the aging of a beta matrix. Type 1 is characterized as the Burgers α, while the other alpha (Type 2) does not obey the Burgers orientation relation (see Table 5). Details of kinetics of their formation and resulting effects on properties are the subject of ongoing research.

Beta Structures

Equilibrium Beta. In α + β and β alloys, some equilibrium β is present at room temperature. (Some isolated "islands" of β are also normally found in commercially pure and α- or near-α alloys due to the presence of tramp iron.) This retention of a stable β phase is due in large part to the partitioning of the alloying elements at temperatures that are high in the α-β phase field. Beta stabilizers concentrate in the β-phase regions and are rejected from α, and vice-versa. Consequently, if cooling rates are slow, a certain volume fraction of stable β phase can form after partitioning. For example, a heavily worked (equiaxed) material may have islands of intergranular β after annealing high in the α-β phase field.

Beta flecks are regions enriched in a β-stabilizing element due to segregation during ingot solidification. Their occurrence in α + β alloys is uncommon. Flecking becomes more of a problem with β alloys, which have much higher amounts of β-stabilizing additions. The problem is most prevalent in iron- and chromium-bearing alloys. The enrichment of a localized region with β stabilizers lowers the β transus locally, changing the microstructure, and thereby enabling their detection.

This microstructural modification can take two forms. In α + β alloys, such as Ti-6Al-6V-2Sn, vanadium enrichment lowers the β transus, but is not sufficient to stabilize the β to room temperature. When working or heat treating the material high in the α + β phase field, the microstructure observed (after cooling back to room temperature) will consist of primary α and transformed β. However, the β fleck region is actually above the transus due to β transus suppression from insufficient homogenization. Upon cooling, a β fleck results in the absence of primary α in this region. The fleck appears as an all-transformed β (Fig. 6) for α/β alloys.

A β fleck could go undetected if the final processing and heat treatment are conducted at a temperature low enough that the β-transus suppression is not sufficient to cause a microstructural perturbation. The effects of β flecks on properties in such alloys as Ti-6Al-4V and Ti-6Al-6V-2Sn are still in question, but the effect is not a major one.

Beta flecks are more of a problem with near-β alloys: they are observed macroscopically as shiny spots or flecks. Their appearance is similar in α + β alloys. The localized enrichment of β-stabilizing alloy concentration in the flecked regions of β alloys, however, is sufficient to stabilize the β down to room temperature. To guarantee material that will be fleck-free, producers must solution heat treat samples at a specific temperature below the β transus. The material will then contain α, but β-fleck regions will remain above or much nearer the transus. Therefore, they will be void of α or contain a significantly lower volume fraction of α upon cooling to room temperature. These regions in β alloys will be harder and have lower ductility.

Metastable beta can be retained at room temperature, if an alloy with sufficient β stabilizers is cooled rapidly enough from high in the α-β phase field or from above the β transus. The decomposition of this retained β (or of martensite, if it forms) is the basis for heat treating titanium alloys to high strengths. The α precipitates that form upon aging of retained

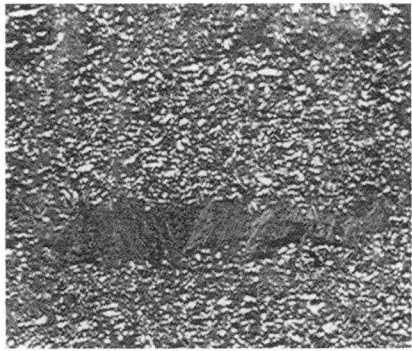

Fig. 6 Example of beta fleck from alloy segregation. Ti-6Al-6V-2Sn hand forging, forged at 1700 °F (927 °C), solution treated for 2 h at 1600 °F (871 °C), water quenched, aged for 4 h at 1100 °F (593 °C), and air cooled. Structure: "primary" alpha grains (light) in a matrix of transformed beta containing acicular alpha, except that alloy segregation has resulted in a dark "beta fleck" (center of micrograph) that shows no light "primary" alpha. 57×

β are often too fine to be resolved by optical microscopy, particularly with β and near-β alloys.

Alpha-beta alloys that contain enough β-stabilizing elements can retain metastable β phase at room temperature on rapid cooling from intermediate temperatures in the α + β phase field. However, the composition of the alloy must be such that the temperature for the start of martensite formation is depressed to below room temperature. In operational terms, metastable beta alloys are those alloys which can be quenched into ice water from its β transus without martensitic decomposition. As mentioned earlier, metastable β can actually form at slower cooling rates or can be retained by rapid quench from a lower solution temperature.

Transition Phases of Metastable Beta Alloys. The decomposition of metastable beta into a Type 1 or Type 2 alpha may be preceded by the formation of an isothermal omega (ω_{iso}) phase or the splitting of the beta phase into solute-rich and solute-lean regions. Occurrence of these transition products indicates the difficulty of the direct beta-to-alpha decomposition, possibly because the hexagonal-close-packed structure of the alpha phase

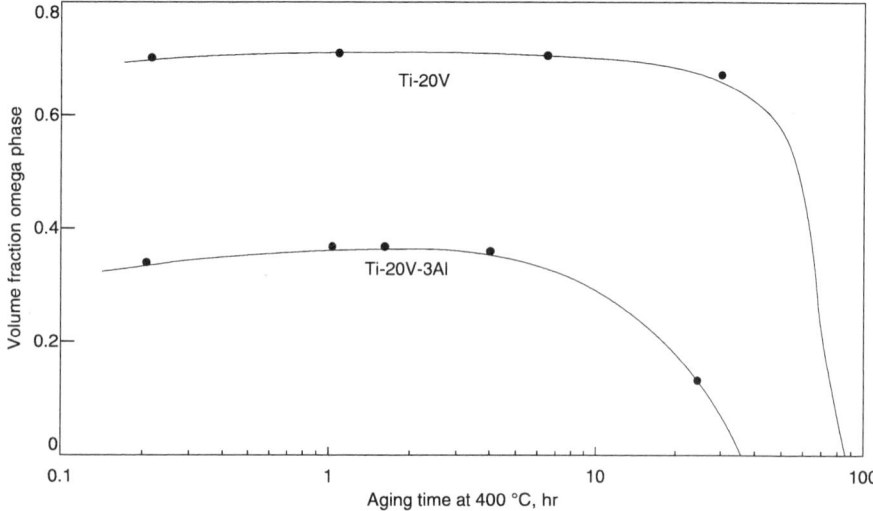

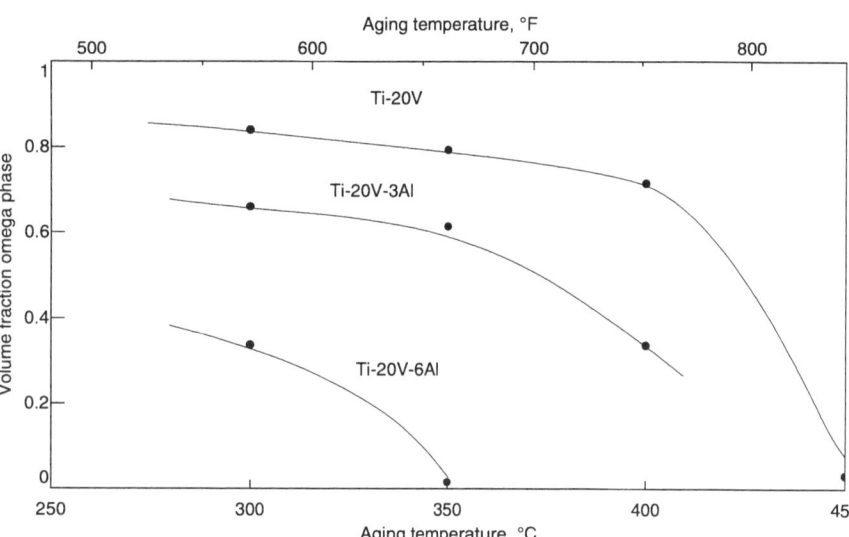

Fig. 7 Influence of alloying on precipitation of omega. (a) Volume fraction of ω_{iso} vs time at 400 °C (750 °F). (b) Volume fraction vs temperature.
Source: J.C. Williams, Critical Review of Kinetics and Phase Transformations, *Titanium Science and Technology*, Vol 3, Plenum Press, 1973

Table 6 Characteristics of omega in metastable beta

Orientation relation	$(11\bar{2}0)_\omega//(110)_\beta$, $<0001>_\omega//<111>_\beta$	
Morphology of isothermal ω	Very fine particle	Ellipsoidal (low misfit, long axes // $<111>_\beta$), or Cuboidal (high misfit, cube face // $(100)_\beta$),
Formation mechanism	(1) Displacement controlled reaction for athermal omega, or solute-lean $\omega \rightarrow$ Type 1 $\rightarrow$ Type 2	(2) Diffusional controlled nucleation followed by displacement fluctuation for isothermal omega
Relation between α and ω-phase		

Source: Y. Murakami, Critical Review of Phase Transformations and Heat Treatment, in *Titanium 80 Science and Technology*, Metallurgical Society of AIME, 1980, p 156

cannot be readily nucleated in the beta matrix.

Isothermal omega (ω_{iso}) forms during aging (at temperatures up to about 475 °C, or 890 °F) for alloys containing intermediate amounts of beta stabilizers. Ellipsoidal and cuboidal morphologies have been observed, depending on the misfit between the coherent ω_{iso} precipitates and the beta matrix. The stability of ω_{iso} depends on the relative concentration of alpha and beta stabilizers such as aluminum and vanadium (see Fig. 7).

Isothermal ω forms as a high density of small coherent precipitates and usually results in the severe degradation of ductility and fracture toughness. However, the relatively small size and high particle density of ω_{iso} precipitation provides desirable nucleation sites for the subsequent precipitation of Type 1 or Type 2 alpha. The decomposition sequence of ω_{iso} into alpha depends on the misfit between the coherent ω structure and the beta matrix. High-misfit systems (such as Ti-V) result in a much more uniform and highly dense distribution of alpha than low-misfit systems (such as Ti-Mo and Ti-Nb). (An appendix to this Technical Note contains a detailed, theoretical discussion on omega precipitation, its growth on different variants, and volume fractions for binary Ti-Mo alloys.)

Athermal omega (ω_{ath}) forms during quenching (like martensite) but occurs as extremely small particles (~2 to 4 nm) with a very high particle density (like ω_{iso}). It doesn't have a deleterious effect on mechanical properties. These unique features are attributed to a class of transformations known as displacement controlled transformations. Athermal ω can be reversibly formed on cooling beneath room temperature, which is consistent with a mechanism of displacement-controlled transformation (J.C. Williams, "Critical Review of Kinetics and Phase Transformations," *Titanium Science and Technology*, Vol 3, Plenum Press, 1973, p 1447). Athermal ω forms over a limited composition range and is affected by oxygen contents (see Fig. 8). The sequence of decomposition is like ω_{iso}, where solute-lean ω becomes Type 1 and Type 2 alpha (see Table 6).

Phase splitting or phase separation only occurs in the solute-rich β alloys; $\beta \rightarrow \beta_r + \beta_l$ where β_r is solute-rich β and β_l is solute-lean β. The solute-lean β is designated β', which occurs as uniformly distributed, coherent bcc zones in the beta-matrix (see Fig. 9). Phase separation occurs in a temperature range of about 200 to 500 °C (400 to 950 °F) in a wide variety of alloys that contain sufficient beta-stabilizers to preclude the formation of ω_{iso} during aging. However, β' is not an important decompo-

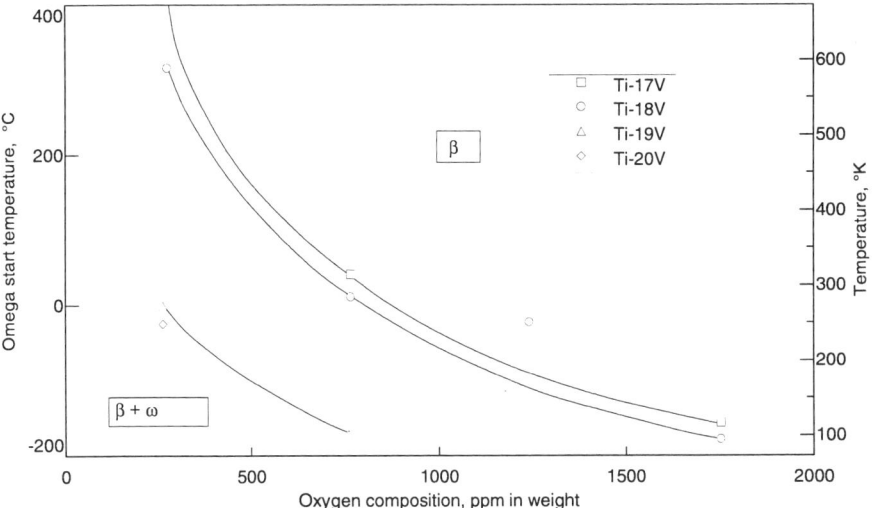

Fig. 8 Onset temperature of athermal omega versus oxygen content.
Source: J.C. Williams, Critical Review of Kinetics and Phase Transformations, *Titanium Science and Technology*, Vol 3, Plenum Press, 1973

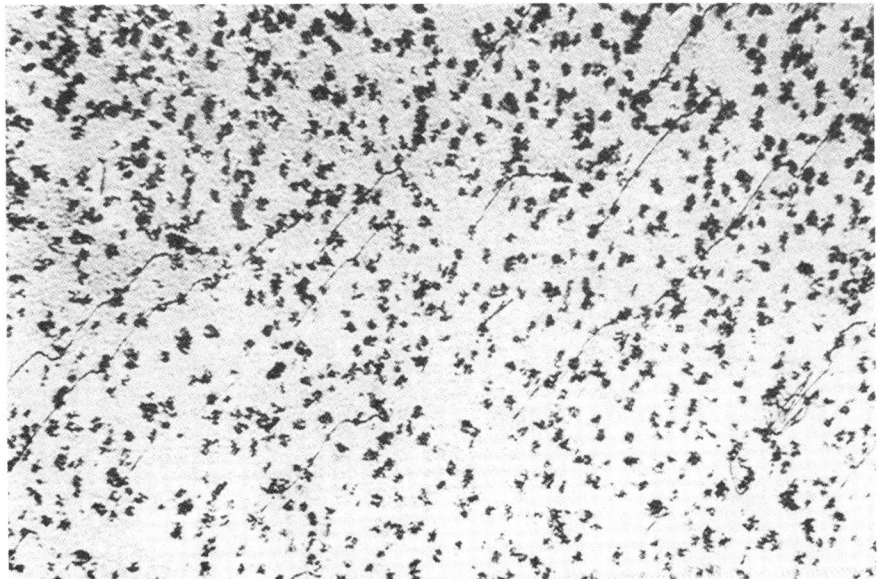

Fig. 9 β′ (solute-lean β) in a solute-rich matrix of Ti-40 at.% Nb solution treated at 900 °C (1650 °F), water quenched, then aged at 400 °C (750 °F) for 24 h. The dark precipitate is β′. Thin-foil transmission electron micrograph (TEM) 31,000 ×.
Source: J.C. Williams, Critical Review of Kinetics and Phase Transformations, *Titanium Science and Technology*, Vol 3, Plenum Press, 1973

The interface phase has shown both face-centered cubic and hexagonal forms. The phase seems to form athermally during slow cooling from above ~950 °C (~1740 °F).

Metallography

The scale of microstructural information ranges from about 2×10^{-10} m (atomic dimension) to more than about 1 mm. To capture the important information it is sometimes necessary to inspect microstructure at several magnifications ranging from inspection with the eye to high resolution electron microscopy.

The principal microstructural features may be grouped according to size from large to small, or from low to high magnification (see Table 7). The largest features are the beta grains. They are often observable with the naked eye after phase transformation during cooling. Next smaller in size are the microstructures dominated by the morphology of alpha and/or martensite phase. Next are phase transformation products and precipitates within beta phase and alpha phase, then come lattice defects such as dislocations, and finally there are the atomic structures of grain or phase boundaries of the phases themselves.

Sectioning of titanium and titanium alloys during specimen preparation follows conventional procedures, but deformation and overheating must be avoided as they can both cause changes in microstructures. Deformation can result in mechanical twinning and strain-induced transformation products, and overheating can change the phase distribution or the phases present. Abrasive cutting with silicon carbide wheels is satisfactory if adequate coolant is used. Hacksawing is recommended for small specimens to avoid overheating.

Mounting. Titanium and its alloys are mounted in common materials such as Bakelite, Lucite, epoxy resins, and diallyl phthalate. The selection of the mounting medium depends on equipment available, specimen size, and the metallographic features of interest. For example, when edge preservation is important, Bakelite or diallyl phthalate is recommended. Nickel plating specimens before mounting also assists in edge preservation, as does adding 50 vol% of silica (SiO_2) or a similar hard material powder, between 200 and 325 mesh, to the epoxy resin mounting material. In general, the temperatures encountered using Bakelite, Lucite, or diallyl phthalate do not cause problems. However, when the metallographic examination involves the hydride phase it may be best to leave the specimen unmounted or to mount it in a room-temperature-setting epoxy resin,

sition product because it does not occur in commercial beta alloys with heat treatments that are used. Like the ω phase, phase splitting can only be observed using electron microscopy.

Interface Phase

The interface phase is a reaction product that forms at the interface between the α and β phases of alloys such as Ti-6Al-4V (see Fig. 10). The formation of the interface phase appears to be an artifact, which occurs when specimens absorb hydrogen during electropolishing. The phase does not occur when specimen preparation techniques are designed to minimize or eliminate hydrogen absorption. The interface phase is not present in bulk at the hydrogen levels normally encountered in commercial alloys (D. Banerjee et al., A Resolution of the Interface Phase Problem in Titanium Alloys, *Acta Metall.*, Vol 36 (No. 1), 1988, p 125-141).

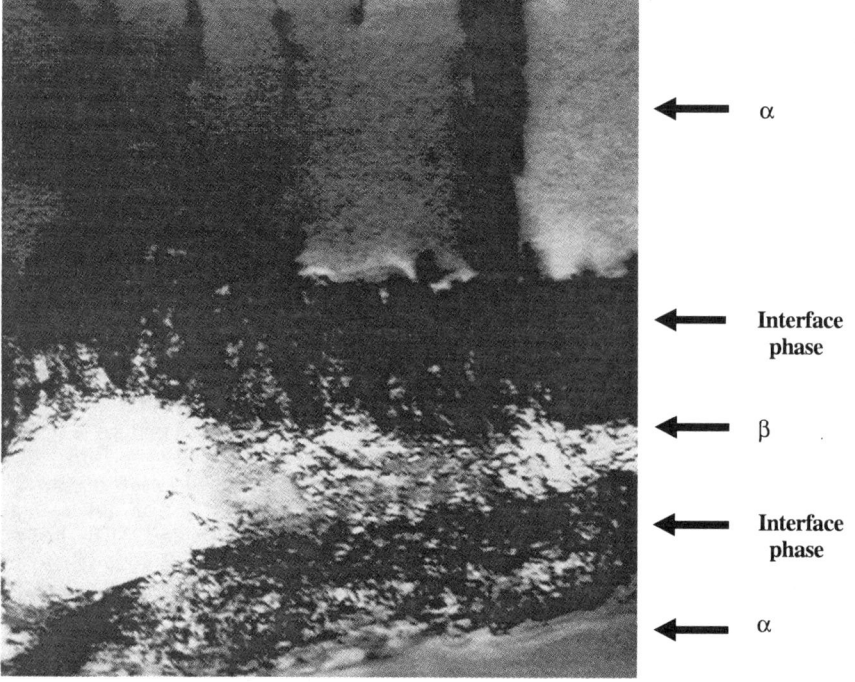

Fig. 10 TEM micrograph showing so-called interface phase in Ti-6Al-4V, ×60,000. Source: G. Welsch, Case Western Reserve University

Table 7 Hierarchy of microstructural features

Microstructure	Useful magnification for observation
Prior β-grain size	1 to 50× Naked Eye, Optical
Colonies of α-lamellae	50 to 1000× Optical, SEM, TEM
Basketweave α-lamellae	50 to 1000× Optical, SEM, TEM
Acicular or Widmanstätten α/α′	50 to 5000× Optical, SEM, TEM
Acicular α′-martensite	100 to 10 000× Optical, SEM, TEM
Precipitates and phase transformation products in β, α″, α′ and α	5000 to 100 000× SEM, TEM
Lattice defects in α, α′, α″, β	5000 to 100 000× TEM
Atomic structure of grain and phase boundaries	>500 000× TEM
Atomic structure of phases	Indirect information by X-ray, electron, and neutron diffraction

because of the increased solubility of hydrogen in titanium at only moderately increased temperatures. Mounting in a thermosetting material could cause the hydride to go back into solution and, upon cooling, re-precipitation of the hydride in an altered form, usually a fine dispersion. Care must also be exercised with epoxies. The exothermic reaction of some room-temperature-setting epoxies can generate enough heat to cause the hydrogen of the hydride to go into solution.

Grinding. The procedure for grinding titanium specimens is similar to that for grinding steel specimens. The specimens are ground on successive grades of silicon carbide paper, starting with 180-grit and proceeding to 240-, 400-, and 600-grit papers, using water or kerosene to keep the specimens cool and to flush away loose particles of metal and abrasive. It is possible to start grinding with a paper as coarse as 80 grit, provided pressure is light to minimize cold working. Hand, disk, and belt grinding are used. To avoid embedding abrasive particles in the ground surface, the papers may be dressed with solid stick wax. Wax is also used in dry final grinding.

Manual polishing consists of three stages: rough, intermediate, and final polishing. Rough polishing is performed on a high-speed polishing wheel covered with a lintless rayon or silk cloth using medium pressure and levigated 1- or 3-μm α-alumina (Al_2O_3). The levigated Al_2O_3 may be placed on the wheel and wetted with water or an acid solution if an etching action is desired. It may also first be mixed with the desired reagent, acid, or water and applied to the wheel as a slurry. A typical 500-mL slurry may contain 15 g levigated Al_2O_3, 10 mL nitric acid (HNO_3), 1.5 mL hydrofluoric acid (HF), with the balance consisting of water. An alternative rough polishing technique is the use of a nylon cloth and a 15- to 6-μm diamond paste.

Intermediate polishing is accomplished by proceeding to a finer abrasive, such as 0.3-μm levigated Al_2O_3 or 1-μm diamond paste, with a possible intermediate step of 3-μm diamond paste. Intermediate polishing begins with heavy pressure, which decreases as polishing progresses. Several cycles of polishing and etching may be needed to remove smeared metal. Etching should be light to avoid pitting. The specimen is unetched when final polishing begins.

Final polishing is performed on a low-speed wheel covered with Microcloth using 0.05-μm Al_2O_3, water, or an acid solution and light pressure on the specimen. Again, the Al_2O_3 may be placed directly on the wheel or in a slurry. A water solution would consist of 15 g 0.05-μm levigated Al_2O_3 and 35 mL H_2O. Colloidal silica is also an effective final polishing abrasive. Several cycles of polishing and etching may be necessary during final polishing.

Edge retention can be enhanced by directional polishing through each polishing stage. The specimen is held in position on the polishing wheel with the edge to be maintained. The mount should be rotated in place about 30 to 40° to minimize streaking the mount surface.

An alternate rough and intermediate polishing uses a wax wheel covered with 6 to 13 mm (1/4 to 1/2 in.) of paraffin. This procedure is somewhat faster than that described above, but can cause metal smearing. Consequently, polishing is normally performed using an acid slurry. Final polishing is accomplished as described above using the Microcloth. Rough polishing is performed using 3-μm Al_2O_3; intermediate, 0.3-μm Al_2O_3. A typical wax wheel slurry may consist of approximately 50 vol% Al_2O_3 and 50 vol% H_2O.

Automatic polishing provides control of the pressure exerted on the specimen. Specimens are ground through 180-, 240-, 400-, and 600-grit silicon carbide paper using water or kerosene as the coolant. Final polishing is carried out using Microcloth and 0.05-μm γ-Al_2O_3 for about 10 min. Several specimens can be prepared simultaneously.

Vibratory polishing, although a slower process than electrolytic or mechanical polishing, produces good results. It is a two-stage operation that follows grinding on abrasive papers. Preliminary polishing is performed on a canvas cloth for 2 to 4 h using a slurry of 0.05-μm Al_2O_3 or on a silk cloth for 30 to 45 min using a slurry of 0.3-μm Al_2O_3. Final polishing is performed on a short-nap cloth for 15 min to 4 h using a slurry of 0.05-μm Al_2O_3. Colloidal silica can also be used. Several short polishing and etching cycles may be used to remove disturbed metal.

Table 8 Etchants for examination of titanium and titanium alloys

Etchant	Comments
Macroetchants	
50 mL HCl, 50 mL H_2O	General-purpose etch for $\alpha + \beta$ alloys
30 mL HNO_3, 3 mL HF, 67 mL H_2O (slow) to 10 mL HNO_3, 8 mL HF, 82 mL H_2O (fast)	Used at room temperature to 55 °C (130 °F) for 3-5 min. Reveals grain size and surface defects
15 mL HNO_3, 10 mL HF, 75 mL H_2O	Etch about 2 min. Reveals flow lines and defects
Two-stage etch(a) consisting of: (1) 8 mL HF, 10 mL HNO_3, 82 mL H_2O and (2) 18 g/L (2.4 oz/gal) of NH_4HF_2 (ammonium bifluoride) in H_2O	Reveals α and β segregation (aluminum segregation)
Microetchants	
1-3 mL HF, 10 mL HNO_3, 30 mL lactic acid	Reveals hydrides in unalloyed titanium
1 mL HF, 30 mL HNO_3, 30 mL lactic acid	Reveals hydrides in unalloyed titanium
Kroll's reagent: 1-3 mL HF, 2-6 mL HNO_3, H_2O to 1000 mL	General-purpose etch for most alloys
10 mL HF, 5 mL HNO_3, 85 mL H_2O	General-purpose etch for most alloys
1 mL HF, 2 mL HNO_3, 50 mL H_2O_2, 47 mL H_2O	Removes etchant stains for most alloys
10 mL HF, 10 mL HNO_3, 30 mL lactic acid	Chemical polish and etch for most alloys
2 mL HF, 98 mL H_2O	Reveals α case for most alloys
98 mL saturated oxalic acid in H_2O, 2 mL HF	Reveals α case (interstitial contamination) for most alloys
6 g NaOH, 60 mL H_2O, heat to 80 °C (180 °F), add 10 mL H_2O_2	Good α-β contrast, general microstructures for most alloys
2 mL HF, 98 mL H_2O, then 1 mL HF, 2 mL HNO_3, 97 mL H_2O	General-purpose etch for near-α alloys(b)
10 mL KOH (40%), 5 mL H_2O_2, 20 mL H_2O	Stains α, transformed β
18.5 g benzalkonium chloride, 33 mL ethanol, 40 mL glycerol, 25 mL HF	General-purpose etch for titanium-aluminum-zirconium and titanium-silicon alloys
2 mL HF, 4 mL HNO_3, 94 mL H_2	Reveals microstructure in aged Ti-13V-11Cr-3Al
50 mL 10% oxalic acid, 50 mL 0.5% HF with H_2O	Etch 12-20 s. General-purpose etch for β alloys
10 s with Kroll's, then 10-15 s with 50 mL, 10% oxalic acid, 50 mL, 0.5% Hf with H_2O	Brings out aged structure in Ti-10V-2Fe-3Al

(a) Two-stage etch procedure: Degrease (if necessary) and clean, making sure the surface is water-break free. Immerse in solution (1) at 45-55 °C (110-135 °F) for 2-3 min and rinse thoroughly in clean cold water. Immerse in agitated bath solution (2) at room temperature for 1-2 min. Rinse thoroughly in clean cold water, rinse thoroughly in clean hot water at 90-100 °C (190-210 °F), blow dry with clean compressed air. Solutions must be used fresh. (b) First etchant stains α phase; second etchant removes stain.

Electropolishing is considerably faster than mechanical polishing. An electrolyte recommended for electropolishing contains 600 mL methanol, 360 mL ethylene glycol, and 60 mL perchloric acid ($HClO_4$). Polishing time is 15 to 25 s at a current density of 1 to 1.5 A/cm^2 (6.5 to 9.7 A/in.^{2}), depending on specimen size and polishing area. The electrolyte given above, having a low concentration of $HClO_4$, is nonexplosive and can be stored for several weeks. However, care should be exercised in handling concentrated $HClO_4$, because it can react explosively with organic materials. Electrolytes without perchloric acid are described in the section "Electropolishing" of Technical Note 9 in this Volume.

Etching. Several etchants are used for macroetching and microetching of titanium and titanium alloys. The choice is often arbitrary for a given type of alloy, but if one does not succeed the other etchants should be tried. Macroetching is usually performed by immersion. Microetching is accomplished by swabbing or immersion. Etching times are usually short, ranging from 3 to 10 s. The specimen should be examined after light etching to avoid overetching.

Nearly all etchants for titanium and titanium alloys contain HF and an oxidizing agent, such as HNO_3 (see Table 8). Kroll's reagent, a dilute aqueous solution containing HF and HNO_3, is the etchant most widely used for commercial titanium alloys.

Macroexamination. Macrostructural examination of titanium alloys provides useful information about material processing, both melting and metal-working. It is used for detection of melting defects or anomalies, qualitative assessment of grain refinement and uniformity, as well as determination of grain flow in forged products.

Three principal defects are to be found in macrosections of ingot, forged billet, or other semifinished product forms. These include high-aluminum defects (HADs or Type II defects), high-interstitial defects (HIDs, also referred to as Type 1 defects or low-density interstitial [LDI] defects), and β flecks. High-aluminum defects are areas containing an abnormally high amount of aluminum. These are soft areas of material and are also referred to as "α segregation." Defects referred to as "β segregation" are sometimes associated with α segregation. These are areas in which aluminum is depleted. The high interstitial defects are normally high in oxygen and/or nitrogen, which stabilize the α phase. These defects are hard and brittle; they are often found near porosity and in surface layer.

Tree rings are another macrostructural anomaly observed in titanium alloy macrosections. This phenomenon represents very minor composition variations that occur during melting. The appearance of tree rings is only a cosmetic nuance, not a cause for concern.

Grain flow of forgings is useful for evaluating the forging process. For high-quality forgings, the grain flow should conform to the general shape of the part. There should be no forging laps, seams, or areas of grain flow that appear as though they could produce forging laps in subsequent operations. In addition, the part should be uniformly recrystallized and sufficiently worked in all areas.

Special Metallographic Techniques. Several metallographic techniques have been developed for specific purposes, including recrystallization studies and microstructure/fracture topography correlations. Decoration aging was developed to study the extent of recrystallization of β alloys. After recrystallization annealing, the material is given a partial age at a time and temperature appropriate for the alloy of interest. The incompletely recrystallized grains retain some dislocation substructure (stored energy) that accelerates the aging process, resulting in a more rapidly aged grain. These grains then etch darker than the recrystallized ones, making it easy to identify the extent of recrystallization.

Another recently developed technique utilizes deep macroetching and thermal etching. The deformed specimen is polished, then subjected to overetching to produce deep grooves at the deformed grain boundaries. Next, the specimen is subjected to the recrystallization cycle of interest in a hard vacuum (10^{-6} torr), followed by oil quenching. The material recrystallizes and thermal etching occurs, which differentiates between different grains, because surface atoms evaporate or sublimate at different rates on different crystallographic planes. (Different grains will have different crystallographic planes at the exposed surface.)

The original beta grain boundaries are observable as ghost boundaries, due to the deep macroetching used previously. Therefore, the recrystallized α and original β microstructures can be observed simultaneously. This permits studying not only the recrystallized structure but also the recrystallization nucleation sites. The ghost boundaries can be removed by repolishing and chemical etching.

Subgrain boundaries can be revealed using a relatively simple technique. The specimen is electropolished and viewed in the scanning electron microscope in the backscattered electron mode. The contrast and delineation of subgrains are due to differences in crystallographic orientation. Electropolishing occurs at different rates on different crystallographic planes, as with the thermal etching phenomenon.

Several techniques have been developed to observe fracture topography and microstructure simultaneously in the scanning electron microscope using its large depth of field. A very simple method involves selective polishing and etching of the fracture face. The fracture face and machined surfaces are first masked with a suitable maskant, such as a stop-off lacquer, which can be applied with a small paint brush. Selected areas of the fracture face are left unmasked. The specimen is then electropolished, which will affect only the unmasked areas, and etched. Studying the interface between the polished and etched and the masked area permits a correlation of microstructural features and fractographic details. This technique is useful for correlating general microstructural details, but it may be difficult to pinpoint a specific area to study.

Precision sectioning techniques have also been developed. The area of interest on the fracture face, such as crack origin, is first located. The specimen is then cut on a plane perpendicular to the fracture face close to the area of interest. The distance from the cut face to the area of interest is measured. Next, the specimen is placed in a metallurgical mount, then ground and polished the measured distance for metallurgical analysis of the precise area of interest and correlation of microstructure to fractographic features. The microstructure and fracture face can be observed simultaneously using the scanning electron microscope by carefully dissolving the mount material.

Technical Note 1 Appendix: Example of ω_{ISO} Formation

A Study of the Growing Process of Thermal ω Phase in Ti-20wt%Mo Single Crystals Aged at 623 K using Dark-Field Electron Microscopy

E. Sukedai, Faculty of Engineering, Okayama University of Science, Okayama, Japan

It is well known that ω phase particles that occur as a precursor to the β to α transformation of β titanium alloys during aging (Ref 1) have a significant effect on mechanical properties (Ref 2, 3). Many investigations of the aged ω phase in other alloy systems have been conducted (Ref 3, 4). However, the process of ω phase particle formation has not been clarified. To understand the formation process, it is important to investigate the growth process and the distribution and volume fraction of the respective ω phase variants. These subjects have been studied using X-ray (Ref 5 to 7) and transmission electron microscopy (TEM) (Ref 8, 9). However, there is a limit to the effectiveness of X-ray methods in that ω phase variants cannot be visualized and although some types of ω phase variants can be observed using TEM, all four ω phase variants have not been classified (Ref 4, 10).

In the present work, a dark-field technique of transmission electron microscopy was used to observe respective ω phase variants directly. The growth process and the distribution and volume fraction of the respective ω phase variants caused by isothermal aging at 350 °C (660 °F) were investigated. The results were evaluated by comparing Vickers hardness values and yield strengths estimated from the volume fraction of ω particles.

Experimental Procedure

Ti-20wt%Mo alloy single crystals were grown in an argon atmosphere using the zone melting method. Plate specimens of about 1 mm thick and with a ($10\bar{1}$) surface were cut. The plate was solution treated at 950 °C (1740 °F) for 4.5 ks, quenched in ice water and polished mechanically to 0.3 mm thickness. Disks with 3-mm diameters were cut from the plate using a spark erosion machine. Some disks were aged isothermally at 350 °C (660 °F) for 3.6, 7.2, 12.6, 28.8, and 100.8 ks in air.

Specimens for electron microscopy were prepared using an electropolishing machine, Tenupol 2. The electrolyte was a 1:6:10 (by volume) solution of 70% perchloric acid, n-butyl alcohol, and methanol. Polishing was performed at about 223 K at 47 V. A JEM 2000-EX operated at 200 kV was used for taking dark-field images using an aperture 8 μm in diameter and tilted illumination; a JEM 4000-EX operated at 400 kV also was used for taking atomic structure images.

Results and Discussion

To observe four ω phase variants using dark-field imaging, the structure factors of each ω variant based on the atomic model of ω phase proposed by de Fontaine and Buck (Ref 11) was calculated for incident beams parallel to [$10\bar{1}$], [$31\bar{1}$], and [$31\bar{1}$] directions. Omega phase variants formed by the atom displacements along [111], [$1\bar{1}1$], [$\bar{1}1\bar{1}$], and [$11\bar{1}$] are referred to as ω_1, ω_2, ω_3, and ω_4, respectively. The results are shown in Fig. 1, in which the large black circles are fundamental diffraction spots from the bcc matrix structure, and 1, 2, 3, and 4 denote ω_1, ω_2, ω_3, and ω_4, respectively. Dark-field images of the respective ω phase variants at the same regions of each specimen aged for 12.6, 28.8, and 100.8 ks were taken using isolated spots from the ω phase variants. Dark-field images of ω particles and aged for 3.6 and 7.2 ks also were taken.

Examples of dark-field images taken from the same region of a specimen aged for 28.8 ks are shown in Fig. 2. The arrows indicate contamination on the specimen. Four ω phase variants are clearly visible. [It was found that each ω phase variant (ω_1, ω_2, ω_3, and ω_4) grew in the same direction as the atom movement for each of the ω variants according to the model (Ref 11)].

To calculate volume fractions of each ω phase variant in the same region outlined by white in Fig. 2, the volume fraction of the specimen in the region, V, was estimated by predicting the specimen thickness using a pair of stereomicrographs that were taken using the ω_1 spot in Fig. 2(a) and the ω_1 spot in Fig. 2(b). Second, volume fraction of individual particles of the ω_1 variant in this region was calculated by measuring the lengths of the major and minor axes, because the shape of the ω particle is an ellipsoid (Ref 12). The total volume of ω_1 variants (v_1) was then estimated. The volume fraction of ω_1 variant was obtained as the ratio of v_1/V. Using the same method, the volume fractions of ω particles of the other variants were obtained. For the specimen aged for 28.8 ks shown in Fig. 2, the volume fractions of ω_1, ω_2, ω_3 and ω_4 variants were 1.51, 1.92, 2.51, and 2.23%, respectively.

Using the same method, ω particles of each variant in five other specimens aged for 12.6, 28.8, and 100.8 ks were observed, and their volume fractions were estimated. For specimens aged for 3.6 and 7.2 ks, ω particles like those shown in Fig. 2 were not observed; only small dots were observed. After aging for 12.6 ks, ω particles were observed. Therefore, it seems apparent that ω particles appear after hardening occurs on the isothermal aging curve (Ref 3). It was also found that with increasing aging time, the size and density of ω particles increased.

The volume fractions in specimens aged for 12.6, 28.8, and 100.8 ks are summarized in Table 1. The volume fraction was found to increase with an increase in aging time. There was no systematic order to the volume fraction of each ω variant. In the present work, the surface normal to the specimen is [10$\bar{1}$]; consequently, ω_1 and ω_3 particles grew parallel to the specimen surface, and ω_2 and ω_4 particles grew at an angle to the surface. Therefore, if the free surface of the specimen has any effect on the growth of ω particles, there should be some systematic order to the volume fraction. It appears that the free surface has no effect on the growth of ω particles. The results shown in Table 1 suggest that the nucleation site of each ω variant is not homogeneous in each region. The formation mechanism of ω phase is not clear.

Volume fractions of similar Ti-Mo alloys with the same composition and aged at the same temperature have been investigated using X-ray diffraction (Ref 6) and small-angle scattering methods (Ref 7). Comparison of these results with the same aging times is summarized in Table 2. Volume fractions measured using dark-field electron microscopy are lower than those measured using X-ray methods.

The results obtained by dark-field technique were evaluated by comparing the dark-field images to a high-resolution structure image of the same region. Figure 3(a) and (b) are dark-field images of the regions designated ω_1 and ω_2 in Fig. 2 and Fig. 3(c) shows a high-resolution image. The seven ω particles observed in these dark-field images correspond to the seven particles in the high-resolution image. Therefore, it appears that the existence of ω particles can be observed more accurately in dark-field images.

Volume fractions measured by dark-field techniques were also compared using Vickers hardness values and yield strengths based on the precipitation-hardening theory estimated by volume fractions of specimens aged for each period. According to the precipitation-hardening theory (Ref 13), yield stress τ_y is given as $\tau_y = Gb/L$, where G is shear modulus, b is Burgers vector, and L is the mean distance among precipitations. In this case, a line tension of dislocation T was assumed as $T = Gb^2/2$. In the present work, the mean distance in each specimen was determined based on the assumption that ω particles were homogeneously distributed in the specimens. For example, the values of each specimen aged for 12.6, 28.8, and 100.8 ks were 32, 24, and 14 nm, respectively. Therefore, the ratio of the three yield stresses based on the precipitation-hardening theory was 31:42:71.

On the other hand, there is a well-known linear relationship between the Vickers hardness and yield strength of materials (Ref 13). Vickers hardness values of this single crystal aged at 350 °C (660 °F) have been reported as 350 ± 5, 373 ± 5, and 409 ± 5 kgf/mm^2, respectively, for 12.6, 28.8, and 100.8 ks. A linear relationship between Vickers hardness and inverse ω particle spacing was also found. The increase in yield stress due to an increase in volume fraction of ω particles is explained by the precipitation-hardening theory.

Experimental disagreement about volume fractions reported in Table 2 have been explained as follows. Gysler et al. (Ref 7) reported that the difference between their findings and those of Hickman (Ref 6) was due to the difference in oxygen contents of the specimens used in both experiments (Ref 7). The oxygen contents were 500 ppm in the Hickman work and 3000 ppm in the Gysler study. The oxygen content in as-rolled sheet in the current work was 2000 ppm. However, aging treatment in the present work was carried out in air, whereas Gysler et al. aged their specimens in an argon atmosphere. In the present work, the oxygen content may increase to more than 3000 ppm due to the aging treatment in air. Also, the contents of other elements in the specimens used in the other experimental work were unknown. One explanation of the disagreement of volume fractions in Table 2 is due to the difference in oxygen content as well as other unknown elements. However, it must also be remembered that, when ω phase particles are very fine, diffraction peaks from the particles become broad. Therefore, X-ray methods may easily include experimental errors, and the results obtained by dark-field electron microscopy are thought to be more accurate.

References

1. B.S. Hickman, *J. Inst. Met.*, Vol 96, 1968, p 330
2. H. Ikawa, S. Shin, M. Miyagai, and M. Morikawa, *J. Jpn. Inst. Met.*, Vol 34, 1970, p 673
3. M. Hida, E. Sukedai, and H. Terauchi, *Acta Metall.*, Vol 36, 1988, p 1429
4. F.H. Froes, C.F. Yolton, and J.M. Capenos, M.G.H. Wells, and J.C. Williams, *Metall. Trans. A*, Vol 11, 1980, p 21
5. B.A. Hatt, J.A. Roberts, and G.I. Williams, *Nature*, Vol 180, 1957, p 1406
6. B.S. Hickman, *Trans. Metall. Soc.*, Vol 245, 1969, p 1329
7. A. Gysler, W. Bunk, and V. Gerold, *Z. Metallkde.*, Vol 65, 1974, p 411

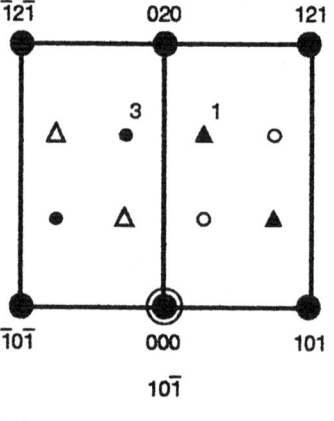
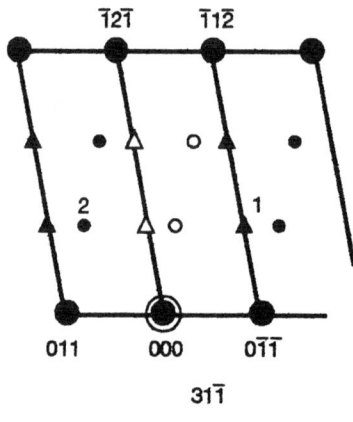
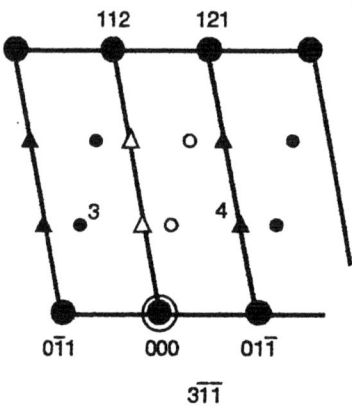

Fig. 1 Diffraction patterns taken with incident beam parallel to [10$\bar{1}$] in (a), [31$\bar{1}$] in (b), and [3$\bar{1}\bar{1}$] in (c). Spots 1 through 4 are formed by ω_1, ω_2, ω_3, and ω_4, respectively. The large black circles are the fundamental diffraction spots from the bcc matrix.

8. E.S.K. Menon, S. Banerjee, and R. Krishman, *Trans. India Inst. Met.*, Vol 31, 1978, p 305
9. J.C. Williams and M.J. Blackburn, *Trans. Metall. Soc. AIME*, Vol 245, 1969, p 2352
10. S.L. Sass, *J. Less-Common Met.*, Vol 28, 1972, p 157
11. D. de Fontaine and O. Buck, *Philos. Mag.*, Vol 27, 1973, p 967
12. E. Sukedai and H. Hashimoto, *J. Jpn. Inst. Met.*, Vol 56, 1992, p 1392
13. N.F. Mott and F.R.N. Nabarro, Report on the Strength of Solids, *Phys. Soc. London*, 1948, p 1
14. D. Tabor, *Proc. Roy. Soc.*, Vol A192, 1948, p 247

Table 1 Volume fractions of each ω variant in different specimens aged at 350 °C for different aging times

Aging time, ks	Volume fraction, %				Total
	ω_1	ω_2	ω_3	ω_4	
12.6	0.22	0.17	0.26	0.24	0.90
12.6	0.52	0.43	0.48	0.43	1.86
28.8	1.51	1.92	2.51	2.23	8.2
28.8	1.26	2.14	1.86	...	6.9(a)
100.8	2.96	2.22	2.95	2.55	10.68
100.8	2.51	2.16	3.32	2.22	10.21

(a) Predicted value

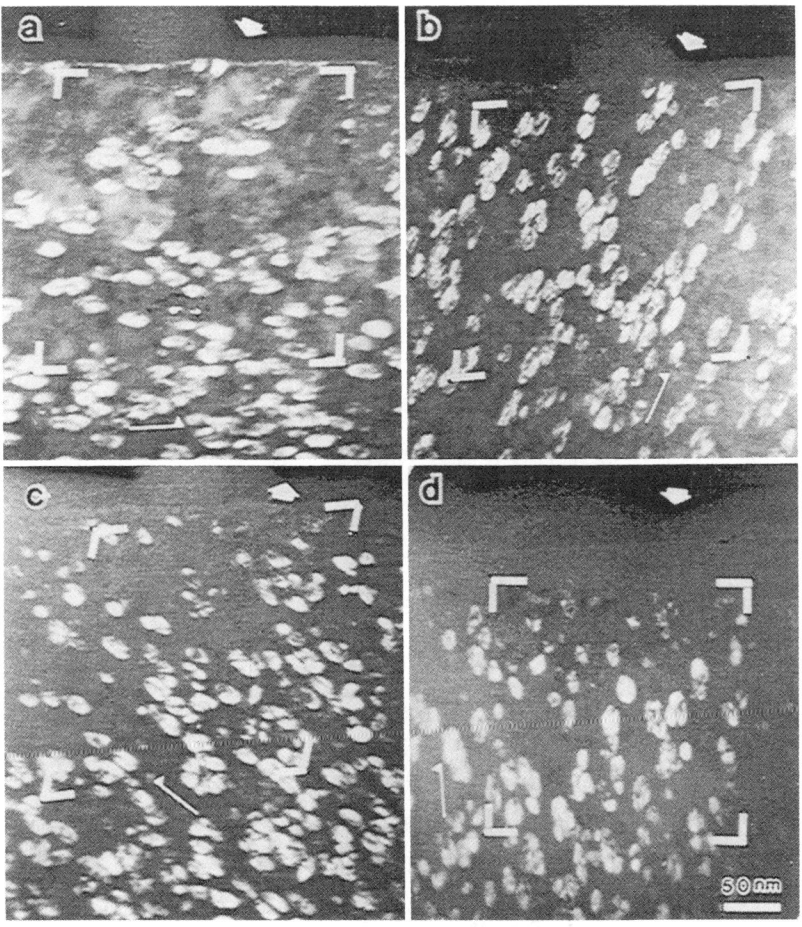

Fig. 2 Four ω phase variants in the same region of a specimen aged for 28.8 ks at 350 °C. (a), (b), (c), and (d) were taken using the isolated spots 1,3,2, and 4 in Fig. 1, respectively. Volume fractions of ω particles were estimated in the region framed by white lines. Thick arrows show the same point of the specimen. Thin arrows denote ⟨111⟩ directions.

Table 2 Comparison of volume fractions of ω particles in three specimens with the same composition aged at 350 °C estimated by different techniques

Aging time, ks	Volume fraction, %, determined by:		
	X-ray diffraction	X-ray small-angle scattering	TEM dark-field techniques
3.6	28	5	...
12.6	...	10	1.2
28.8	51	15	8.2
100.8	...	25	10.68

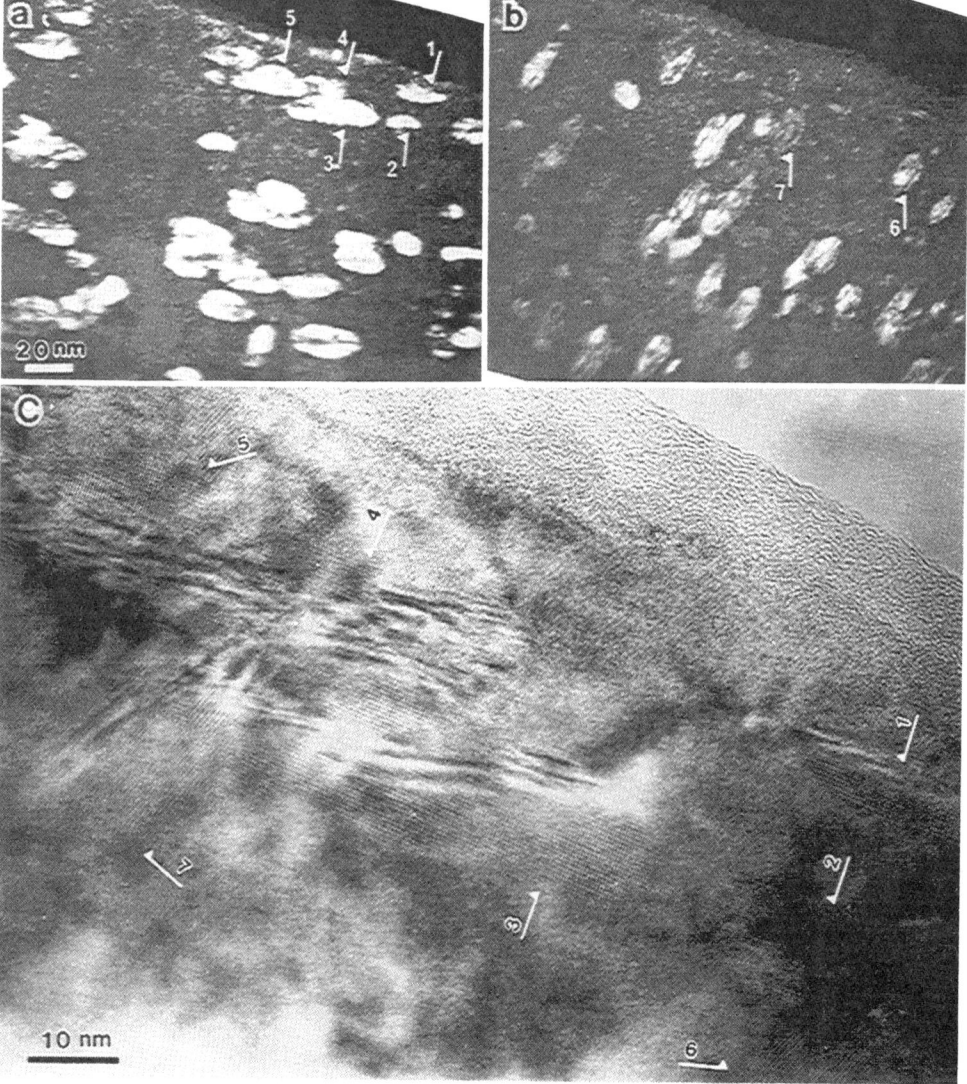

Fig. 3 Dark-field images (a) and (b) in the same region of a specimen aged at 350 °C for 28.8 ks taken using isolated spots 1 and 3 in Fig. 1(a), respectively. (c) HREM image in the same region as shown in (a) and (b). Omega particles 1 through 7 shown in (a) and (b) can be seen in (c).

Technical Note 2: Corrosion*

Reviewed by Bruce Craig, Metallurgical Consultants, Inc.

Successful application of titanium and its alloys can be expected in mildly reducing to highly oxidizing environments in which protective oxide films spontaneously form and remain stable. Titanium exhibits excellent resistance to atmospheric corrosion in both marine and industrial environments. Titanium and its alloys also resist H_2S and CO_2 gases at temperatures up to 260 °C (500 °F).

On the other hand, hot, concentrated, low-pH chloride salts corrode titanium; warm or concentrated solutions of hydrochloric, phosphoric, and oxalic acids also are damaging. In general, all acidic solutions that are reducing in nature corrode titanium, unless they contain inhibitors. Strong oxidizers, including anhydrous red fuming nitric acid and 90% hydrogen peroxide, also cause attack. Ionizable fluoride compounds, such as sodium fluoride and hydrogen fluoride, activate the surface and can cause rapid corrosion. Dry chlorine gas is especially harmful.

Most acidic solutions (except those containing soluble fluorides) can be inhibited by the presence of even small amounts of oxidizing agents and heavy metal ions. Thus, titanium can be used in certain industrial process solutions (including hydrochloric and sulfuric acids) that otherwise would be corrosive. Attack by red fuming nitric acid and chlorine gas can be inhibited by small amounts of water.

The major corrosion problems with titanium alloys appear to be crevice corrosion, which occurs in locations where the corroding media are virtually stagnant. Pits, if formed, may progress in a similar manner. A general comparison of corrosion resistance for titanium is provided in Fig. 1(a).

Protective Oxide Layer. The excellent corrosion resistance of titanium alloys results from the formation of a very stable, continuous, highly adherent, and protective oxide film on the surface. Because titanium metal itself is highly reactive and has an extremely high affinity for oxygen, these beneficial surface oxide films form spontaneously and instantly when fresh metal surfaces are exposed to air and/or moisture. In fact, a damaged oxide film can generally reheal itself instantaneously if at least traces (that is, parts per million) of oxygen or water (moisture) are present in the environment (Ref 1).

However, if titanium is exposed to strongly oxidizing or reducing environments, severe attack of the metal may ensue. In the complete absence of moisture under oxidizing conditions, any surface film that is formed is not protective and oxidation in depth may take place, often in the form of a violent exothermic reaction. Breakdown of the passive layer can also occur from dry oxidants (such as red fuming nitric acid), nonoxidizing aqueous environments (as defined by a Pourbaix diagram), and pitting or crevice attack in near-neutral aqueous solutions (particularly in the presence of halides). Finally, under continuous wear or sliding contact with other metals, the protective oxide may not reform, thereby allowing accelerated corrosion of the titanium (Ref 2).

The nature, composition, and thickness of the protective surface oxides that form on titanium alloys depend on environmental conditions. In most aqueous environments, the oxide is typically TiO_2, Ti_2O_3 or TiO (Ref 1). High-temperature oxidation tends to promote the formation of the chemically resistant, crystalline form of TiO_2 known as rutile, whereas lower temperatures often generate the more amorphous form of TiO_3, anatase, or a mixture of rutile and anatase (Ref 1). Although these naturally formed films are typically less than 10 nm thick (Ref 3) and are invisible to the eye, the TiO_2 oxide is generally chemically resistant and is attacked by very few substances, including hot, concentrated HCl, H_2SO_4, NaOH, H_3PO_4, and (most notably) HF. This thin surface oxide is also resistant to hydrogen permeation, as discussed in the section "Hydrogen Damage" in this Technical Note.

The TiO_2 film is an n-type semiconductor and thus can conduct electronic charge, depending on the potential drop across the semiconductor film (Ref 5). As a

---

* Adapted from "Corrosion of Titanium and Titanium Alloys," *Metals Handbook*, 9th ed., Vol. 13, *Corrosion*

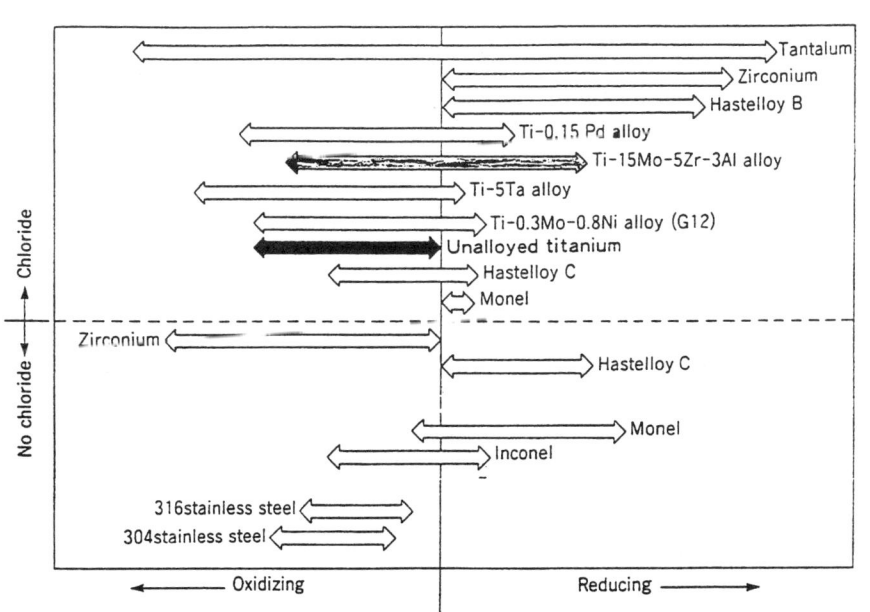

Fig. 1(a) Range of corrosion resistance of metals

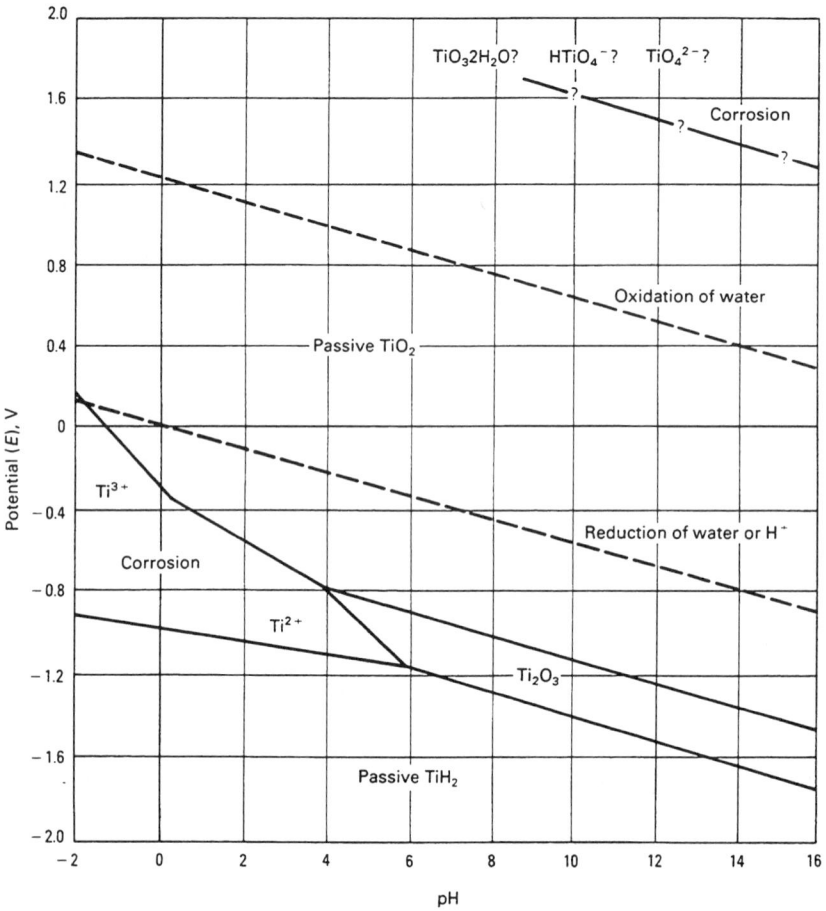

Fig. 1(b) Pourbaix diagram for water-titanium system at 25 °C. Source: Ref 4

cathode, titanium permits electrochemical reduction of ions in an aqueous electrolyte. On the other hand, very high resistance to anodic current flow through the passive oxide film can be expected in most aqueous solutions. Because the passivity of titanium stems from the formation of a stable oxide film, an understanding of the corrosion behavior of titanium is obtained by recognizing the conditions under which this oxide is thermodynamically stable. The Pourbaix (potential-pH) diagram for the titanium-water system (see Fig. 1b) depicts the wide regime over which the passive TiO_2 film is predicted to be stable, based on thermodynamic (free energy) considerations. Oxide stability over the full pH scale is indicated over a wide range of highly oxidizing to mildly reducing potentials, whereas oxide film breakdown and the resultant corrosion of titanium occur under reducing acidic conditions. Under strongly reducing (cathodic) conditions, titanium hydride formation is predicted.

Alloy Composition Effects. The nature of the oxide film on titanium alloys basically remains unaltered in the presence of minor alloying constituents; thus, small additions (<2 to 3%) of most commercially used alloying elements or trace alloy impurities generally have little effect on the basic corrosion resistance of titanium in normally passive environments. For example, despite small differences in interstitial elements (carbon, oxygen, and nitrogen) and iron content, all unalloyed grades of titanium possess the same useful range of resistance in environments in which corrosion rates are normally very low (Ref 6). However, under active conditions in which titanium exhibits significant general corrosion, certain alloying elements may accelerate corrosion. Increasing the iron and sulfur content, for example, increases corrosion rates when corrosion rates exceed 0.13 mm/yr (5 mils/yr) (Ref 6). Thus, minor variations in alloy chemistry may be of concern only under conditions in which the passivity of titanium is borderline or when the metal is fully active. On the other hand, minor nickel and palladium additions are highly effective in expanding the corrosion resistance of titanium alloys under reducing conditions. Moreover, small palladium additions can significantly increase crevice corrosion resistance in hot aqueous chlorides.

The influence of certain major alloying elements on the general and crevice corrosion behavior of various commercial titanium alloys has been determined in reducing aqueous acid media (Ref 7). Results indicate that vanadium and, especially, molybdenum additions (≥4% Mo) improve corrosion resistance but that increasing the aluminum content appears to be detrimental. The influence of alloying elements on the resistance of titanium alloys to pitting and stress-corrosion cracking (SCC) is addressed in subsequent sections of this Technical Note.

Effect of Product Form and Welding. Weldments (Ref 6) and castings (Ref 8, 9) of commercially pure grades and α-β alloys such as Ti-6Al-4V generally exhibit corrosion resistance similar to that of their unwelded, wrought counterparts. These titanium alloys contain so little alloy content and second phase that metallurgical instability and thermal response are not significant. Therefore, titanium weldments and associated heat-affected zones generally do not experience corrosion limitations in welded components when normal passive conditions prevail for the base metal. However, under marginal or active conditions (for corrosion rates ≥0.10 mm/yr, or 4 mils/yr), weldments may experience accelerated corrosion attack relative to the base metal, depending on alloy composition (Ref 6). The increasing impurity (iron, sulfur, oxygen) content associated with the coarse, transformed-β microstructure of weldments appears to be a factor. Few published data are available concerning the corrosion resistance of other α-β and β titanium alloy weldments and castings.

General Corrosion

General corrosion is characterized by a relatively uniform attack over the exposed surface of a metal. At times, general corrosion in aqueous media may take the form of mottled, severely roughened metal surfaces. This often results from variations in the corrosion rates of localized surface patches due to variations in process scales, corrosion products, or gas bubbles. When titanium is in the fully passive condition, corrosion rates are typically much lower than 0.04 mm/yr (1.5 mils/yr)—well below the 0.13-mm/yr (5-mils/yr) maximum corrosion rate commonly accepted by designers. This very small, acceptable corrosion is attributable to the thin steady state film on titanium alloy surfaces. As a result, titanium is often designed with a zero corrosion allowance in normal passive environments. In many environments in which titanium is fully resistant, slight surface oxide growth may occur; this oxide growth manifests itself as colored surfaces and very slight weight gain by test coupons.

General corrosion becomes a concern in reducing acid environments, particularly as acid concentration and temperature increase. In strong and/or hot reducing acids (in the absence of inhibitors), the oxide film of titanium can deteriorate and dissolve, and the unprotected metal is oxidized to the soluble trivalent ion (Ti → $Ti^{3+} + 3e^-$). This ion has a characteristic violet color in acid solutions. If dissolved oxygen or other oxidizing species are present in hot acid, the Ti^{3+} ion is readily oxidized to the less soluble (pale yellow) Ti^{4+} ion, which may subsequently hydrolyze to form insoluble TiO_2 precipitates (scales). Titanium ion hydrolysis often produces highly colored metal surfaces, involving thin titanium oxide films that may inhibit subsequent corrosion. Gray-matte or dull silver surface finishes can also be observed in reducing acid exposures involving severe corrosion attack. In reducing media, these are titanium hydride surface films, which are typically of the order of 0.05 mm (2 mils) thick.

Enhancing General Corrosion Resistance. Successful use of titanium alloys can be expected in mildly reducing to highly oxidizing environments in which protective TiO_2 and Ti_2O_3 films form spontaneously and remain stable. On the other hand, uninhibited, strongly reducing acidic environments may attack titanium, particularly as temperature increases. However, shifting the alloy potential in the noble (positive) direction by various means can induce stable oxide film formation, often overcoming the corrosion resistance limitations of titanium alloys in normally aggressive reducing media.

The methods of expanding the corrosion resistance of titanium into reducing environments include:

- Increasing the surface oxide film thickness by anodizing or thermal oxidation
- Anodically polarizing the alloy (anodic protection) by impressed anodic current or galvanic coupling with a more noble metal in order to maintain the surface oxide film
- Applying precious metal (or certain metal oxides) surface coatings
- Alloying titanium with certain elements
- Adding oxidizing species (inhibitors) to the reducing environment to permit oxide film stabilization

Of these five methods, the last two have been very practical, effective, and most widely used in actual service.

Alloying titanium with precious metals (such as palladium), nickel, and/or molybdenum or coating with certain precious metals (or their oxides) facilitates cathodic depolarization by providing sites of low hydrogen overvoltage on alloy surfaces and by shifting alloy potential in the noble (positive) direction where oxide film passivation is possible. Relatively small concentrations of certain precious metals (of the order of 0.1 wt%) are sufficient to expand significantly the corrosion resistance of titanium in reducing acid media.

Beneficial alloying elements include precious metals (≥0.05 wt% Pd) (Ref 10-13), nickel (≥0.5 wt%) (Ref 10 13, 14-16), and/or molybdenum (≥4 wt%) (Ref 7, 13, 17, 18). These beneficial alloying additions have been incorporated into several commercially available titanium alloys, including the titanium-palladium alloys (grades 7 and 11), Ti-0.3Mo-0.8Ni (grade 12), Ti-3Al-8V-6Cr-4Zr-4Mo, Ti-15Mo-5Zr, and Ti-6Al-2Sn-4Zr-6Mo. These alloys all offer expanded application into hotter and/or stronger HCl, H_2SO_4, H_3PO_4, and other reducing acids as compared to unalloyed titanium. The high-molybdenum alloys offer a unique combination of high strength, low density, and superior corrosion resistance, especially higher pitting resistance.

Inhibitor Additions. Various oxidizing species can effectively inhibit the corrosion of titanium in reducing acid environments when present in very small concentrations. The dissolved oxidizing species serve to depolarize cathodic reactions on titanium alloy surfaces; this passivates the alloy by shifting the alloy potential in the noble direction. Many of these species, which include a host of multivalent transition metal ions, are very potent inhibitors and may be effective at concentrations of 100 ppm or less (see Tables 1 and 2).

These inhibiting species often occur as natural process stream constituents or contaminants and need not be intentionally added to achieve complete titanium alloy passivation. However, because the performance of titanium alloys in reducing acids is highly influenced by the presence of many inhibiting species, the nature of background chemistry of a reducing acid environment should be thoroughly examined before determining alloy suitability. Titanium is often selected for normally aggressive reducing acid solutions, such as the hydrometallurgical acid-leaching process streams for metallic ores, because of the beneficial effect of these inhibitive ions (Ref 21).

Precious Metal Surface Treatments. Precious metals such as platinum and palladium have been ion plated, ion implanted, or thermal diffused into titanium alloy surfaces to achieve improved resistance to reducing acids (Ref 22). This approach has not been used commercially for industrial components because of high cost, coating application limitations, and the limitations (mechanical and corrosion damage) normally associated with very thin surface films. However, ion plated platinum or gold surface films also impart significant improvements in titanium alloy oxidation resistance at temperatures up to 650 °C (1200 °F) (Ref 23, 24).

Thermal Oxidation. Protective thermal oxide films can form when titanium is heated in air at temperatures of 600 to 800 °C (1110 to 1470 °F) for 2 to 10 min. The rutile TiO_2 film formed measurably improves resistance to dilute reducing acids as well as absorption of hydrogen under cathodic charging (Ref 25) or gaseous hydrogen conditions. Corrosion studies in hot, dilute HCl solutions have confirmed its superior protective benefits as compared to as-pickled, polished, or anodized surfaces on unalloyed titanium (Ref 25, 26). Corrosion and hydrogen uptake resistance was afforded by thermal oxidation in molten urea at 200 °C (390 °F) (Ref 26). Enhanced protection from dry chlorine attack can also be expected. Like anodizing, thermal oxidation offers no improvements in titanium resistance

Table 1 Species that inhibit the corrosion of titanium alloys in reducing acids

Inhibitor category	Species	Relative inhibitor potency
Oxidizing metal cations	$Ti^{4+}, Fe^{3+}, Cu^{2+}, Hg^{2+}, Ce^{4+}, Sn^{4+}, VP_2^+$,	High
	$Te^{4+}, Te^{6+}, Se^{4+}, Se^{6+}, Ni^{2+}$	High
		Low
Oxidizing anions	$ClO_4^{2-}, Cr_2O_7^{2-}, MoO_4^{2-}, MnO_4^{2-}, WO_4^-, IO_3^-,$	Very high
	$VO_4^{3-}, VO_3^-, NO_3^-, NO_2^-, S_2O_3^{2-}$	Very high
		Moderate
Precious metal ions	$Pt^{2+}, Pt^{4+}, Pd^{2+}, Ru^{3+}, Ir^{3+}, Rh^{3+}, Au^{3+}$	High
		High
Oxidizing organic compounds	Picric acid, o-dinitrobenzene, 8-nitroquinoline, m-nitroacetanilide, trinitrobenzoic acid, and certain other nitro, nitroso, and quinone organics	Moderate-high
Others	$O_2, H_2O_2, ClO_3^-, OCl^-$	Moderate

Source: Ref 19, 20

Table 2 Effect of certain multivalent metal ions on the corrosion of titanium in boiling reducing acids

Inhibiting ion	Concentration of inhibiting ion, ppm	Corrosion rate			
		Boiling 5% HCl		Boiling 10% H_2SO_4	
		mm/yr	mils/yr	mm/yr	mils/yr
Fe^{3+}	0	29	1142	>76.2	>3000
	100	0.025	1	0.208	8.2
	500	0.02	0.8	0.069	2.7
Cu^{2+}	0	29	1142	>76.2	>3000
	100	0.033	1.3	0.419	16.5
	500	nil		0.361	14.2
Mo^{6+}	0	29	1142	>76.2	>3000
	100	nil		0.001	0.04
	500	nil		nil	
Cr^{6+}	0	29	1142	>76.2	>3000
	100	nil		0.001	0.04
	500	nil		0.001	0.04
V^{5+}	0	29	1142	>76.2	>3000
	100	0.02	0.8	0.005	0.2
	500	0.008	0.3	0.005	0.2

Source: Ref 21

in highly alkaline or oxidizing aqueous media.

Although the thermal oxide has proved to be protective in relatively short-term tests in dilute reducing acids, long-term performance has not been fully demonstrated. Mechanical damage and plastic strain of thermally oxidized components must be avoided for effective protection. The oxide has been successfully applied on tubing and small components, but may be impractical for large components or where component distortion may occur during service.

Anodic protection is another effective means of passivating and protecting titanium alloys in reducing acids (Ref 27, 28, 29). Generally, an increase in anodic potential will decrease the corrosion rate as long as the anodic pitting potential is not exceeded for titanium in the electrolyte. Sustained impressed potentials in the range of +1 to +4 V versus the standard hydrogen electrode (SHE) are usually adequate to ensure full passivation of titanium in many acids (see Table 3). Limited use of anodic protection by impressed currents has been made in concentrated H_2SO_4 and H_3PO_4 solutions in which a very wide range of impressed potentials can be applied. The added cost of impressed current systems, challenges with protecting complex component geometries, and stray current problems have limited its applications. Also, titanium surfaces exposed to alternating wet/dry or vapor-phase conditions cannot be protected by this method.

Other Surface Treatments. Surface films of titanium nitrides and carbides are highly resistant to reducing acids. Studies have shown that the dense adherent nitride films produced by reactive plasma ion plating provided superior protection in deaerated H_2SO_4 solutions when compared to several other film-forming methods (Ref 30). Methods of applying nitride surface films to titanium include ion implantation (Ref 31), ion plating, sputter deposition, or thermal diffusion (nitrogen gas or molten cyanide bath). Because of the cost and limitations of film application and the inherent thin film performance limitations, these films are generally not used for corrosion resistance only. The improved wear resistance offered by these hard films is generally the primary incentive for application.

General Corrosion in Specific Media

Water and Seawater. Titanium and its alloys are fully resistant to potable water, natural waters, and steam to temperatures in excess of 315 °C (600 °F) (Ref 32). Slight weight gain is usually experienced in these benign environments, along with some surface discoloration at higher temperatures from passive film thickening. The immunity to attack of α alloys is observed regardless of oxygen level or in high-purity water, such as that normally used in nuclear reactor coolant systems (Ref 33-37). The typical contaminants encountered in natural water streams, such as iron and manganese, oxides, sulfides, sulfates, carbonates, and chlorides, do not compromise the passivity of titanium. In media containing chloride levels greater than 1000 ppm (for example, seawater) at temperatures about 75 °C (165 °F), consideration should be given to possible crevice corrosion when tight crevices exist in service.

Oxidation Media. Titanium alloys are generally highly resistant to oxidizing media and oxidizing acids over a wide range of concentrations and temperatures. Common chemicals in this category include chromic, nitric, perchloric, and hypochlorous acids and salts of these acids. Other oxidizing salts include thiosulfates, vanadates, permanganates, and molybdates. Corrosion rates at and below the boiling point of these aqueous salt solutions over the full range of concentration will typically be less than 0.03 mm/yr (1.2 mils/yr).

Titanium is also unique among the common engineering alloys in its immunity to general and pitting corrosion in oxidizing chloride environments. These comments also apply to bromine and iodine-containing media. Halide salts of oxidizing cationic species also enhance the passivity of titanium alloys such that negligible corrosion rates can be expected. Examples include $FeCl_3$, $CuCl_2$, and $NiCl_2$ solutions and their bromide counterparts.

Limited corrosion testing of α-β and β titanium alloys in boiling HNO_3 indicates that increasing aluminum and β alloying elements tend to decrease corrosion resistance. Alpha alloys are generally most resistant to hot HNO_3. Other studies have shown that high-purity (low iron, sulfur, and so on) unalloyed titanium does not experience the significant accelerated weldment attack in high-temperature HNO_3 exhibited at times by the less pure unalloyed grades and the near-β alloys.

However, dangerous and violent pyrophoric reactions may occur with titanium alloys exposed to dry oxidants such as red fuming nitric acid or to nitrogen tetroxide. The attack is intergranular and results in a surface residue of finely divided titanium particles that are highly reactive (high surface-to-volume ratio). The critical variables are the nitrogen dioxide (NO_2) and water contents of the acid (see Fig. 2). Fuming nitric acid containing less than 1.4 to 2.0% water or more than 6% NO_2 may cause this rapid impact-sensitive reaction to occur (Ref 38, 39, 43). Both water and NO are effective inhibitors for this attack, but increasing oxygen and NO_2 is detrimental in this situation. Corrosion rate data in red fuming nitric acid for various alloys as a function of NO_2 and water content also can be found in Ref 2.

Reducing Acids. The corrosion resistance of titanium alloys in reducing acid media is very sensitive to acid concentration, temperature, background chemistry, and purity of the acid solution, in addition to titanium alloy composition. When the temperature and/or concentration of pure (uncontaminated) reducing acid solutions exceed certain values, the protective oxide film of titanium may break down, which would result in severe general corrosion. Included in this category are hydrochloric, sulfuric, hydrobromic, hydriodic, hydrofluoric, phosphoric, sulfamic, oxalic, and trichloroacetic acids. Because the performance of titanium alloys in reducing

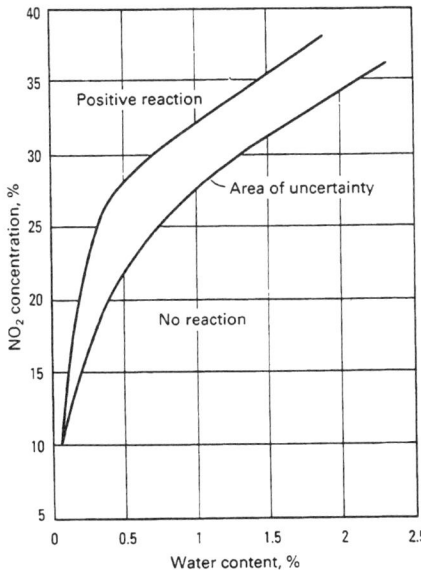

Fig. 2 Water content to avoid combustion of titanium in red fuming nitric acid. Source: Ref 32

Table 3 Effect of impressed anodic potentials on the corrosion of unalloyed titanium in hot reducing acids

Acid	Concentration, %	Temperature °C	Temperature °F	Applied potential, V versus SHE	Corrosion rate mm/yr	Corrosion rate mils/yr	Reduction in corrosion rate
Sulfuric	40	60	140	+2.1	0.005	0.2	11 000×
	40	90	195	+1.4	0.07	2.8	896×
	40	114	237	+2.6	1.8	71	189×
	60	60	140	+1.7	0.035	1.4	662×
	60	90	195	+3.0	0.10	4	163×
Hydrochloric	37	60	140	+1.7	0.068	2.7	2080×
Phosphoric	60	60	140	+2.7	0.018	0.7	307×
	60	90	195	+2.0	0.5	20	100×
Formic	50	Boiling		+1.4	0.083	3.3	70×
Oxalic	25	Boiling		+1.6	0.25	10	350×
Sulfamic	20	90	195	+0.7	0.005	0.2	2710×

Source: Ref 27, 28, 29

acids is influenced by inhibitor species, the nature of the acid environment must be considered carefully.

Hydrofluoric acid solutions can aggressively attack titanium alloys over the full range of concentrations and temperatures, because the fluoride ion (F^-) forms highly stable, soluble complexes with titanium. Although the addition of oxidizing species, such as HNO_3, will tend to reduce corrosion and retard hydrogen uptake in HF solutions, significant rates of attack still prevail. Inhibition of corrosion can be achieved in very dilute acid fluoride solutions when an excess of complexing metal ions (for example, Fe^{3+}, Al^{3+}, and Cr^{6+}) is present. In the absence of these complexing metal ions, solutions containing more than 20 ppm F^- may attack titanium when solution pH falls below 6 to 7.

Beneficial alloying additions have been incorporated into several commercially available titanium alloys, including the titanium-palladium alloy (grades 7 and 11), Ti-0.3Mo-0.8Ni (grade 12), Ti-3Al-8V-6Cr-4Zr-4Mo, Ti-15Mo-5Zr, Beta 21S, and Ti-6Al-2Sn-4Zr-6Mo. These alloys all offer expanded application into hotter and/or stronger HCl, H_2SO_4, H_3PO_4, and other reducing acids as compared to unalloyed titanium. The high-molybdenum alloys offer a unique combination of high strength, low density, and superior corrosion resistance.

Inhibitor Additions. Various oxidizing species can effectively inhibit the corrosion of titanium in reducing acid environments when present in very small concentrations. Typical potent inhibitors for titanium alloys in aggressive reducing acids are listed in Table 1. Many of these inhibitors are effective at levels as low as 20 to 100 ppm, depending on acid concentration and temperature. If not normally present in a given corrosive acid stream, minute additions of a process-compatible inhibitive species may be considered to protect titanium components. These can be especially practical when process streams are recycled.

Anodic Protection. Titanium alloys can be effectively protected in reducing acid media by impressed anodic (direct current) potentials. Sustained impressed potentials in the range of +1 to +4 V versus the standard hydrogen electrode (SHE) are usually adequate to measure full passivation of titanium in many acids. Limited use of anodic protection by impressed current has been made in concentrated H_2SO_4 and H_3PO_4 solutions in which a very wide range of impressed potentials can be applied. The added cost of impressed current systems, challenges with protecting complex component geometries, and stray current problems have inhibited its application. Also, titanium surfaces exposed to alternating wet/dry or vapor-phase conditions are not protected by this method.

Salt Solutions. Titanium alloys are highly resistant to practically all salt solutions over the pH range of 3 to 11 and to temperatures well in excess of boiling. Titanium withstands exposure to solutions of chlorides (Ref 44, 45), bromides, iodides, sulfites, sulfates, borates, phosphates, cyanides, carbonates, bicarbonates, and ammonium compounds. Corrosion rate values for titanium alloys in these various salt solutions are generally less than 0.03 mm/yr (1.2 mils/yr).

Titanium alloys are frequently selected because of their superior resistance to the chlorides typically found in many process streams, brines, and seawater. In hot chloride media, susceptibility to pitting is usually not an issue, but crevice corrosion may be possible, depending on pH and temperature. Special attention must be given to nonoxidizing acidic or hydrolyzable salt solutions as temperatures and concentrations increase. To avoid general or localized HCl attack resulting from salt hydrolysis, special concentration-temperature guidelines for titanium should be observed for concentrated $AlCl_3$ (Ref 44, 45), $ZnCl_2$ (Ref 46), $MgCl_2$ (Ref 47), and $CaCl_2$ solutions.

Alkaline Media. Titanium alloys are generally very resistant to alkaline media, including solutions of NaOH, KOH, $Ca(OH_2)$, $Mg(OH)_2$, and NH_4OH (Ref 44). Near-nil corrosion rates can be expected in boiling solutions of the latter three alkalies up to saturation. Titanium also exhibits low corrosion rates in NaOH and KOH solutions at subboiling temperatures (see Table 4). However, significant increases in corrosion are noted as the concentrations of these two strong alkalies increase at higher temperatures. Potassium hydroxide tends to be more aggressive than sodium hydroxide under these conditions.

Although corrosion rates are relatively low in alkaline media, titanium alloys may experience excessive hydrogen pickup and eventual embrittlement under certain conditions. For α and near-α alloys, hydrogen embrittlement is possible when temperatures exceed 80 °C (175 °F) and pH is 12 or more. The presence of dissolved oxidizing species in hot caustic solutions, such as chlorate, hypochlorite, or nitrate compounds, can extend resistance to hydrogen uptake to somewhat higher temperatures.

Organic Compounds. Titanium alloys are highly resistant to corrosion from most organic compounds, including alcohols, ketones, ethers, aldehydes, and hydrocarbons. The traces of moisture (ppm levels) normally present in industrial organic process streams are sufficient to maintain the protective oxide film of titanium. Totally anhydrous organic streams may prevent oxide film repair and should be avoided. In the special case of absolute methanol, at least 1.5% H_2O must be added to prevent depassivation and stress-corrosion cracking (Ref 48-51). Higher molecular weight alcohols are generally quite benign toward titanium alloys.

Chlorinated hydrocarbons generally do not pose any problems for most titanium alloys. A few high-strength alloys may be susceptible to stress-corrosion cracking under specific circumstances. If significant quantities of water are also present, many chlorinated hydrocarbons may undergo hydrolysis to form HCl at high temperatures. Titanium alloy performance will depend on the temperature and the extent of HCl formation and concentration in the aqueous phase.

General Corrosion Testing

General corrosion rates for titanium alloys can be determined from weight loss data, dimensional changes, and electrochemical methods. Electrochemical polarization testing is often used to supplement weight loss testing. Polarization testing can identify whether the alloy is truly fully passive or possibly metastable; this is often not discernible from weight loss tests alone. The immersion test procedures described in ASTM G 1 and G 31 apply, provided several modifications are observed (Ref 52). These modifications focus on test sample surface preparation and post test sample-cleaning procedures. Serious consideration must also be given to the presence of contaminants that may significantly affect corrosion.

Surface Preparation of Test Sample. The type of surface finish tested should resemble the one expected in service. For titanium alloys, this will often be the pickled finish, although sandblasted or ground surfaces are also common. The initial degreasing of test samples should avoid chlorinated organic solvents (with higher-strength titanium alloys), anhydrous methanol, or hot alkaline cleaners, if possible. Acceptable cleaning solvents include methyl ethyl ketone (MEK), acetone, most alcohols, benzene, and most detergent solutions. The pickled finish can be prepared by pickling the metal in a 35 vol% HNO_3-5 vol% HF (balance water) solution at 20 to 55 °C (70 to 130 °F) for several minutes or more. Typically, 0.02 to 0.05 mm (0.8 to 2 mils) of surface is removed in this process, depending on surface requirements. More dilute solutions, such as 12 vol% HNO_3-1 vol% HF, can also be used if slower pickling rates are desired. In any case, a minimum 7/1 HNO_3/HF vol% ratio should be maintained to avoid excessive uptake of hydrogen in titanium alloys during pickling. After pickling, a quick rinse in deionized water leaves a shiny specimen that is ready for weighing after air drying. Blasted and abraded surfaces are prepared by procedures similar to those used for other metals.

Post-test Sample Cleaning. After laboratory or *in situ* test exposure, titanium samples can be coated with tenacious, insoluble corrosion product (TiO_2) films or scales, which require removal before final weighing. Because titanium oxides are not soluble in common mineral acids, very light (<5-s exposure) sandblasting has been found to be most effective. If scaling consists of silicaceous, carbonaceous, sulfate, or other typical process stream deposits, then acids or alkaline solutions that are properly inhibited with oxidizing species must be used; common amine inhibitors are not effective on titanium. Recommended cleaning solutions for these scales are discussed in detail in Ref 53.

Equivalent Weight. Corrosion rates (mm/yr) can be calculated from electrochemical measurements (ASTM G5) by:

$$\text{Corrosion rate} = \frac{(0.0033)(i_{\text{corr}})(\text{EW})}{d}$$

where i_{corr} is the measured corrosion current (in milliamps per square centimeter), d is alloy density (in grams per cubic centimeter), and EW is the equivalent weight for titanium. The equivalent weight for titanium is approximately 16 under reducing acid conditions and 12 under oxidizing conditions. The value of i_{corr} is typically determined from Tafel slope extrapolation or linear polarization methods.

Crevice Corrosion

Titanium alloys may be subject to localized attack in tight crevices exposed to hot (>70 °C, or 160 °F) chloride, bromide, iodide, fluoride, or sulfate-containing solutions. Crevices can stem from adhering process stream deposits or scales, metal-to-metal joints (for example, poor weld joint design or tube-to-tube sheet joints), and gasket-to-metal flange and other seal joints.

The mechanism for crevice corrosion of titanium is similar to that for stainless steels, in which oxygen-depleted reducing acid conditions develop within tight crevices (see Fig. 3). Dissolved oxygen or other oxidizing species in the bulk solution are depleted in the restricted volume of solution in the crevice. Finite surface oxidation in crevices consumes these species faster than diffusion from the bulk solution can replenish them (Ref 54). As a result, metal potentials in crevices become active (negative) relative to metal surfaces exposed to the bulk solution. This creates an electrochemical cell in which the crevices become anodic and corrode, and the surrounding metal surface is cathodic.

Titanium chlorides formed within the crevice are unstable and tend to hydrolyze, forming hydrochloric acid (HCl) and titanium oxide/hydroxide corrosion products. Because of the small, restricted volumes of solution in these crevices, crevice pH levels as low as 1 or below can develop. These local reducing acidic conditions can result in severe and rapid localized active corrosion within crevices, depending on alloy resistance and temperature.

Although dissolved oxidizing species such as oxygen, chlorine, ferric ion (Fe^{3+}), and cupric ion (Cu^{2+}) tend to inhibit the general corrosion of exposed titanium surfaces, most of these species tend to accelerate the onset and propagation of titanium alloy crevice corrosion. These species are excellent cathodic depolarizers and thus accelerate cathodic reduction kinetics, which often are rate controlling. On the other hand, certain anionic oxidizing species, such as NO^-_3, ClO^-_3, OCl^-, CrO^{2-}_3, ClO^-_4, and MnO^-_4, inhibit crevice attack when present in halide solutions.

Crevice corrosion on titanium typically generates irregularly shaped pits. Microstructural examination of hand-polished and etched sections of crevices often reveals a surrounding layer of precipitated titanium hydride in α alloys. These are a by-product of hydrogen reduction at cathodic sites surrounding the crevice.

Although frequently interpreted as a pitting phenomenon, smeared surface iron pitting of unalloyed titanium in hot brines appears to be a special case of crevice corrosion (Ref 55). It results when iron, carbon steel, or low-alloy steel is gouged, scratched, smeared, and embedded into a titanium surface, breaching the titanium oxide film. During hot (80 °C, or 175 °F) brine exposure, the embedded iron can either corrode off the surface and permit repassivation or develop local acidic conditions if occluded by titanium metal smears or laps. Localized attack initiated by this mechanism creates a very characteristic circular pit morphology and can involve local hydrogen absorption. Pit initiation has not been observed with copper, nickel, or austenitic stainless steel alloys smeared into titanium surfaces. Titanium grades 7 and 12 appear to be much more resistant to this form of localized attack.

Enhancing Crevice Corrosion Resistance. Several effective strategies for preventing titanium alloy crevice corrosion and smeared iron pitting are: alloying titanium, precious metal surface treatments, metallic coatings, thermal oxidation, noble alloy contact, and surface pickling (for smeared surface iron). In all cases, the basic remedy aims at maintaining creviced metal surfaces at sufficiently noble potentials where titanium alloy passivity is maintained.

Crevice Corrosion in Specific Media. Titanium alloys generally exhibit superior resistance to crevice corrosion as compared to stainless steel and nickel-base alloys. Nevertheless, the susceptibility of titanium alloys to crevice corrosion should be considered when tight crevices exist in hot aqueous chloride, bromide, iodide, or sulfate solutions. Crevice test results indicate that the initiation of crevice corrosion often lacks reproducibility, consistency, and regularity. These test data must be judged relative to their statistical significance (that is, number of data points). Factors that affect crevice attack significantly include alloy composition, pH, temperature, halide concentration, presence of oxidizing species (cathodic depolarizers), sample surface condition, type of gasket, type of crevice (gasket-to-metal, metal-to-metal, deposit-to-metal), and the crevice geometry, particularly crevice gap (tightness). Crevice corrosion testing of titanium alloys generally aims at determining go/no-go performance information. The rate of crevice corrosion is of little practical interest, because crevice attack is generally very rapid. Many crevice test assemblies have been used, including the multiple-crevice washering (Ref 56).

Chlorides. The susceptibility of titanium alloys to crevice corrosion in hot, concentrated chloride solutions increases significantly as temperatures increase and pH decreases. Crevice attack of titanium alloys will generally not occur below a temperature of 70 °C (160 °F) regardless of solution pH or chloride concentration, or when solution pH exceeds 10 regardless of temperature. ASTM grade 12 provides crevice corrosion resistance when brine pH falls between 3 and 11 to temperatures as high as 300 °C (570 °F).

Bromides and Sulfates. Unpublished test results suggest that the pH temperature guidelines for crevice corrosion of titanium in saturated NaCl are applicable in saturated NaBr solutions. However, the rate of crevice attack is measurably lower than that in chlorides at corre-

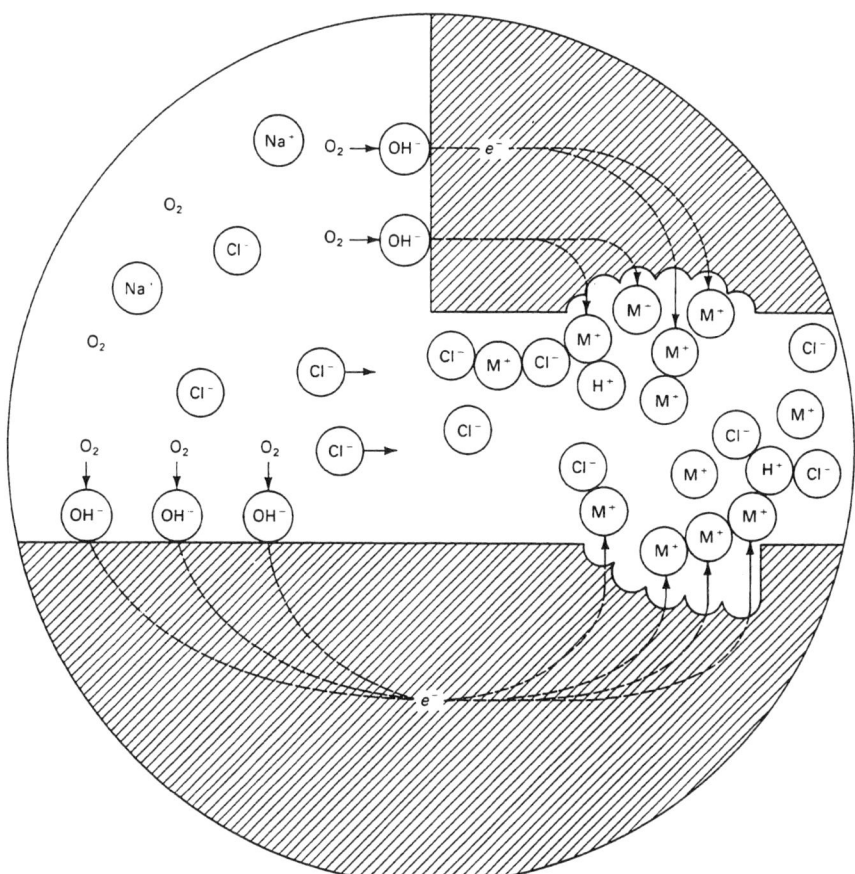

Fig. 3 Schematic of crevice corrosion mechanism

Table 4 Corrosion of unalloyed titanium in highly alkaline solutions

Medium	Concentration, %	Temperature °C	°F	Corrosion rate mm/yr	mils/yr
Ammonium hydroxide	28	26	79	0.002	0.08
Ammonium hydroxide	70	Boiling		nil	
Sodium carbonate	20	Boiling		nil	
Sodium hydroxide	28	25	75	0.003	0.12
Sodium hydroxide	10	Boiling		0.02	0.8
Sodium hydroxide	40	66	150	0.038	1.5
Sodium hydroxide	40	93	200	0.064	2.5
Sodium hydroxide	40	121	250	0.13	5
Sodium hydroxide	50	66	150	0.018	0.7
Sodium hydroxide	50-73	188	370	>1.1	>43.3
Sodium hydroxide	73	110	230	0.05	2
Sodium hydroxide	73	Boiling		0.13	5
Potassium hydroxide	10	Boiling		0.13	5
Potassium hydroxide	25	Boiling		0.3	12
Potassium hydroxide	50	25	75	0.010	0.4
Potassium hydroxide	50	Boiling		2.7	106

Source: Ref 1, 29, 37, 44

sponding pH and temperature values.

Pitting

Pitting is defined as localized corrosion attack occurring on exposed metal surfaces in the absence of any apparent crevices. This pitting occurs when the potential of the metal exceeds the anodic breakdown potential of the metal oxide film in a given environment. When the anodic breakdown (pitting) potential of the metal is equal to or less than the corrosion potential under a given set of conditions, spontaneous pitting can be expected.

Because of its protective oxide film, titanium exhibits anodic pitting potentials, E_b, that are very high (>1 V); thus, pitting corrosion is generally not of concern for titanium alloys. For example, pitting potentials exceed +80 V versus the saturated calomel electrode (SCE) in sulfate and phosphate solutions and are typically in the +5 to +10 V range for chlorides. Although pitting is normally not a limiting factor in titanium performance, pitting potential values provide useful guidelines for titanium for anode applications in which impressed anodic potentials may be high.

The pitting potential of titanium is dependent on alloy content, medium chemistry, temperature, potential scan rate, and, especially, surface condition.

In sulfate and phosphate media, anodic pitting potentials of titanium alloys are typically in the range of +80 to +100 V (versus Ag/AgCl reference electrode).

In halide salt solutions, titanium alloys exhibit somewhat lower but yet reasonably high pitting potentials. Values of +9 to +10.5 V (versus Ag/AgCl) can be expected in room-temperature chloride solutions, decreasing to approximately +1.2 V at 175 to 250 °C (345 to 480 °F). Pitting potentials of titanium can be raised in chloride solutions by addition of sulfate ions (Ref 57).

Anodic pitting potential values are significantly lower in bromide solutions, and they decrease with increasing temperature. Thus, pitting of titanium alloys may be possible in pure bromide solutions at higher temperature if highly oxidizing conditions prevail. However, additions of various oxidizing anions may inhibit pitting in NaBr solutions by significantly raising anodic pitting potentials (Ref 58). Critical concentrations of the inhibitive anions have been determined, and the relative efficiency of inhibition decreases in the order $SO_4^{2-} > NO_3^- > CrO_4^{2-} > PO_4^{3-} > CO_3^{2-}$.

Repassivation Potentials. Another important alloy property is the repassivation (protection) potential, which is defined as the minimum potential at which pitting can be maintained. Repassivation potentials (E_{p2}) represent conservative measures of anodic pitting tendency because they represent minimum potentials below which pitting cannot be sustained. This pitting parameter is not sensitive to surface condition or measuring technique artifacts, and it represents a more conservative design guideline than the anodic breakdown potential. The repassivation potentials of titanium alloys are also very high relative to other alloy corrosion potentials, and this explains why titanium alloys are generally resistant to pitting attack. Like pitting potentials, repassivation potentials are significantly lower in bromide and iodide media.

Repassivation potentials are readily determined by using the galvanostatic method (Ref 59) or the constant potential-surface scratch test (Ref 59, 60). The galvanostatic method involves impressing an anodic current density of approximately 200 mA/cm^2 (1290 mA/in.^{2}) on the specimen for at least several minutes before measuring the repassivation potential of the sample. Reproducible, unambiguous repassivation potentials are more difficult to derive by using reverse scan potentiodynamic techniques.

Hydrogen Damage

Titanium alloys are widely used in hydrogen-containing environments and under conditions in which galvanic couples or cathodic charging (impressed current) causes hydrogen to be reduced on metal surfaces. In most cases, these alloys display excellent resistance to damage.

Alpha and α-β titanium alloys suffer hydrogen damage primarily by hydride-phase formation. Pure α-titanium is relatively unaffected by small concentrations (<200 ppm) of hydrogen; however, the purity of the α-titanium is important to its behavior in hydrogen. Commercially pure titanium is much more sensitive to hydrogen than is pure titanium. The amount of hydrogen necessary to induce ductile-to-brittle transitions behavior in commercially pure titanium is much less than one-half the amount needed in pure titanium. Severe embrittlement can occur in the commercial grades at hydrogen levels as low as 30-40 ppm in the presence of a high residual stress or a stress riser, and elevated temperature. These conditions induce migration of the hydrogen to the stress riser, resulting in a much higher local concentration of hydrogen and the precipitation of hydrides.

Modes of Hydrogen-Assisted Failure. Hydrogen damage can occur at high or low strain rates. Hydrogen damage at high strain rates is the result of hydrides that precipitate after the high-temperature exposure of titanium to hydrogen. This results in a loss of impact toughness and is sometimes referred to as impact embrittlement.

The other mode of failure for titanium alloys in the presence of hydrogen predominates under slow strain rate loading. The low strain rate embrittlement is related to hydride formation caused by strain-enhanced precipitation, but embrittlement under impact is caused by hydride-phase formation after fabrication or heat treatment. Unlike many hydride-forming systems, titanium forms a stable hydride, but the kinetics of precipitation are slow compared to the Group Vb metals. Therefore, embrittlement is more prone to occur at low strain rates at which precipitation can proceed at a rate that is sufficient to provide a brittle crack path.

Both types of failure for titanium alloys in hydrogen are attributed to hydride-phase precipitation. Because hydrogen solubility increases with temperature for these alloys, hydride embrittlement typically decreases as the temperature increases. Additionally, at higher temperatures, the hydride may become more ductile, reducing brittle crack initiation. As expected, the threshold stress intensity for crack propagation is also a function of the hydrogen content decreasing with increasing hydrogen.

Sustained-Load Cracking in Inert Environments. High-strength titanium alloys for use in highly stressed components for military aircraft and other similar applications may be susceptible to sustained-load cracking in inert environments (including dry air). Sustained-load cracking is similar to SCC except that it is much slower and occurs in the total absence of a reactive environment. Sustained-load cracking is caused by, or is greatly aggravated by, hydrogen dissolved in the titanium during processing. Vacuum annealing can reduce the hydrogen level to less than 10 ppm, at which concentration the tendency toward sustained-load cracking is greatly reduced.

Effect of Microstructure and Oxide Layer. The role of microstructure in the hydrogen damage of titanium is quite complex and is not fully understood. However, it has been determined that under slow strain rates the α-β alloys fail by intergranular separation along boundaries but that completely α alloys fracture by transgranular cleavage. Embrittlement is not as severe in α-β alloys with a continuous equiaxed α matrix as for those alloys with a continuous acicular β matrix. However, this behavior is a function of hydrogen pressure and may be reversed at lower pressures.

The surface oxide of titanium is highly effective in reducing hydrogen penetration. Traces of moisture or oxygen in hydrogen gas containing environments very effectively maintain this protective film, thus avoiding or limiting hydrogen uptake (Ref 44, 61-62). On the other hand, anhydrous hydrogen gas atmospheres may lead to absorption, particularly as temperatures and pressures increase.

In α and α-β alloys, excessive hydrogen uptake can induce the precipitation of titanium hydride in the α phase. These acicular-appearing hydride platelets are brittle and have been well characterized in the literature (Ref 63-65). Small amounts of hydride precipitates are not detrimental from an engineering standpoint in most cases, but cause severe reduction in alloy ductility and toughness when present in greater amounts. For example, hydride precipitates can be observed in grade 2 titanium microstructures at hydrogen concentrations above bulk hydrogen approximately 100 ppm, depending on the amount of β phase present, but these precipitates do not necessarily result in gross embrittlement of grade 2 titanium until levels in excess of 500 to 600 ppm are achieved. Severe embrittlement has been observed in Grade 3 titanium with bulk hydrogen contents on the order of 30 ppm by weight.

Although uniaxial tensile properties may experience little effect from increasing hydrogen levels, biaxial or triaxial stress properties, such as bend ductility, cup (cold-drawing) formability, and impact toughness, in α and near-α alloys are very sensitive to hydrogen levels (Ref 64-69). In α and, especially, α-β alloys, hydrogen contents above critical levels can result in sustained-load cracking, which dramatically reduces useful maximum service loads in notched or cracked components under slow strain rate or constant tensile load situations (Ref 65-70).

Beta titanium alloys have a very high solubility for hydrogen such that embrittlement is generally not a result of hydriding (Ref 66, 69). Significant losses in ductility or formability may not occur below levels of several thousand parts per million of hydrogen (Ref 66). The tolerance to hydrogen decreases somewhat in the aged (high-strength) condition. This increased tolerance of the β alloys must be weighed against the significantly higher hydrogen uptake rates that result from the much larger hydrogen diffusion coefficient for β titanium (Ref 69, 71).

Prevention in Aqueous Media. Factors that can lead to hydrogen uptake and possible embrittlement of α and near-α titanium alloys in aqueous media have been identified from field and laboratory experience. The three general conditions that must exist simultaneously for the hydrogen damage of α alloys are (Ref 44, 62):

- A mechanism for generating nascent (atomic) hydrogen on a titanium surface. This may be from a galvanic couple, an impressed cathodic current, corrosion of titanium, or severe continuous abrasion of the titanium surface in an aqueous medium
- Metal temperature above approximately 80 °C (175 °F), where the diffusion rate of hydrogen into α titanium is significant (Ref 72-79)
- Solution pH less than 3 or greater than 12, or impressed potentials more negative than –0.70 V (SCE) (Ref 62, 74-77)

The key to preventing hydrogen damage is simply to avoid one or more of these conditions.

Galvanic couples between titanium and certain active metals and excessive cathodic charging from impressed-current cathodic protection systems are the usual causes of excessive hydrogen absorption. In near-neutral electrolytes such as seawater, active metals such as zinc, magnesium, and aluminum can lead to hydrogen uptake and eventual embrittlement when coupled to titanium above 80 °C (175 °F) (Ref 62, 78). A similar problem occurs when titanium is in galvanic contact with carbon steels or active stainless steels in aqueous media above 80 °C (175 °F) (Ref 63). Arsenic, antimony, and cyanide species and sulfides act as a hydrogen recombination poison (that is, they prevent the recombination of atomic hydrogen to form molecular hydrogen) and enhance hydrogen uptake in this situation.

No hydrogen uptake and embrittlement problems occur when titanium is coupled to fully passive materials in a given environment. These compatible materials may include other titanium alloys, passive stainless steels, copper alloys, and nickel-base alloys, depending on conditions.

Cathodic charging of hydrogen onto unalloyed titanium surfaces is not recommended when temperatures exceed 80 °C (175 °F). At metal temperatures below 80 °C (175 °F), thin surface hydride films may form on α titanium alloys; these are usually not detrimental from the standpoint of corrosion or mechanical properties. However, very high cathodic current densities may lead to enhanced hydride film growth and eventual wall penetration and embrittlement even at room temperature. Surface thermal oxides on titanium appear to inhibit hydrogen uptake effectively under moderate cathodic charging conditions, but can break down at high current densities.

High-temperature alkaline conditions may also result in excessive hydrogen uptake and embrittlement of titanium alloys. The nascent (atomic) hydrogen generated on titanium surfaces from small but finite general corrosion in hot (>80 °C, or 175 °F), strongly alkaline (pH ≥ 12) media appears to be responsible.

Hydrogen Testing. Galvanic coupling tests or cathodic charging tests can be conducted to evaluate susceptibility to hydrogen uptake. For a given environment, an active metal (iron, aluminum, etc.) sample is galvanically coupled to the titanium alloy sample such that a specific anode-to-cathode surface area is established. Impressed cathodic charging tests are performed in electrolytic cells containing a specific electrolyte. A power supply (potentiostat or galvanostat) impresses a constant potential or current on the cell such that the titanium is cathodic relative to an inert counterelectrode such as graphite or platinum. A reference electrode can also be used to control or to measure the polarization potential of the test cathode.

The surface condition of the coupon is a critical variable in all hydrogen uptake tests. Studies have shown that abraded or sandblasted surfaces absorb hydrogen more readily than as-pickled surfaces. Thickening of the surface oxide film by anodizing or thermal oxidation further retards absorption. The actual surface finish anticipated in service should be evaluated.

Titanium alloys tend to exhibit greater susceptibility under biaxial or triaxial stress states; therefore, bend tests, cup tests, or notched tensile tests are generally more sensitive to hydrogen effects. Impact toughness testing can be an especially sensitive indicator of hydrogen effects in α alloys, whereas slow strain rate methods are very suitable for α-β alloys (Ref 64, 67, 68). Because hydrogen content has relatively little effect on alloy hardness, hardness testing is not a good indicator of hydrogen absorption.

Hydrogen analysis of coupons is performed by the hot vacuum extraction method. In the hot vacuum extraction apparatus, a small sample is heated to 1100 to 1400 °C (2010 to 2550 °F) for several minutes to reversibly release the absorbed hydrogen, followed by evolved gas measurements.

Stress-Corrosion Cracking

Stress-corrosion cracking (SCC) is a fracture, or cracking, phenomenon caused by the combined action of tensile stress, a susceptible alloy, and a specific corrosive environment. The metal may show little evidence of general corrosion attack, although slight localized attack in the form of pitting or crevice corrosion may be visible. Usually, only specific combinations of metallurgical and environmental conditions cause SCC. This is important because it is often possible to eliminate or reduce SCC sensitivity by modifying either the metallurgical characteristics of the metal or the makeup of the environment. Another important characteristic of SCC is the requirement that tensile stress be present. These stresses may be provided by cold work, residual stresses from fabrication, or externally applied loads.

It is also important to distinguish between the two classes of titanium alloys. The first class, which includes ASTM grades 1, 2, 7, 11, and 12, is immune to SCC except in a few specific environments. These specific environments include anhydrous methanol/halide solutions, nitrogen tetroxide (N_2O_4), red fuming HNO_3, and liquid or solid cadmium. The second class of titanium alloys, including the aerospace titanium alloys, has been found to be susceptible to several additional environments, most notably aqueous chloride solutions. However, this susceptibility is often associated with high stress concentrations typical of laboratory testing with loaded, precracked specimens, and generally is not observed with smooth specimens.

Mechanisms of SCC. Over the years, a variety of mechanisms or models have been proposed to explain SCC phenomena in titanium alloys (Ref 80). In general, SCC is the anodic dissolution in highly localized areas that, aided by an applied tensile stress, propagates cracks into the metal. Crack advance occurs by discontinuous rupture of the oxide film at the crack tip.

The SCC generally begins from a corrosion pit or a crevice. In the presence of a tensile stress, the pit will produce a crack if corrosion is not so rapid so as to blunt the advancing crack tip. Once a crack initiates, the balance among the crack tip corrosion rate, the crack tip environment, and the repassivation kinetics are critical to either continued crack propagation or crack arrest.

Stress-Corrosion Testing. Because there is tremendous diversity in SCC behavior of metals and environments that cause SCC, one must rely upon experience for successful application of metals in corrosive environments, or from laboratory tests designed to reveal susceptibility to SCC. It is often the case that service experience is not available, and laboratory tests are the sole basis for material selection.

In testing metals for SCC resistance, two important considerations are: the environment to be employed and the specimen configuration to be selected. An environment must be selected that is representative of that expected in service. More often than not, this choice is fixed by the intended application. Artificial environments selected to accelerate the test must be recognized as compromises and will not produce wholly reliable information, although they can be used for assessing the relative susceptibilities of various alloys.

Selection of specimen configuration is somewhat different from choosing the test environment in that the investigator is free to choose from a multitude of previously designed configurations. Unfortunately, these configurations do not always produce the same result, nor do they even evaluate the same properties.

Most specimens configurations used for SCC testing fall into three categories:

- *Category 1*: smooth, statically loaded specimens: such as U-bends, C-rings, bent beams, and dead-load tensile bars (ASTM G 30, G 35, G 36, G 38, G 47)

- *Category 2*: notched and precracked specimens: such as cantilever beams, compact tension specimens, and double-cantilever beams (ASTM E 399)

- *Category 3*: smooth, dynamically loaded specimens: primarily the slow strain rate tensile specimens

Each of these categories is used to evaluate different characteristics of the SCC process, so comparisons between specimen types can often lead to incorrect conclusions.

Category 1 specimens are used to evaluate the susceptibility of a material to both initiation and propagation of SCC. Because the samples are smooth and subjected to a static load, they represent the most favorable conditions a material would experience in service. These samples are least effective when used in metal/environment combinations in which localized corrosion (for example, pitting) is unlikely, because SCC initiation may not occur for a long time. Thus, the test duration chosen is of paramount importance.

This configuration also provides limited quantitative information because the test gives simply all-or-nothing results. Once the specimen is stressed and exposed to an environment, it either cracks or it does not. If it does not crack, it is not ensured that SCC cannot occur; the test duration may have been too short, or the stress too low, or the orientation incorrect. Data typically reported in addition to pass or fail usually consist of time to failure at a given stress level or a threshold stress for cracking.

Category 2 specimens are different from Category 1 in that they primarily measure crack propagation or crack arrest. The specimen may be statically or dynamically loaded, depending on the information required, but is normally statically loaded. Because a crack is assumed to be present from the start, this configuration approaches worst-case conditions and is quite conservative. In addition, precracked (fracture mechanics) specimens provide quantitative information that can be used to rank alloys in a specific environment. However, these data cannot be used for design purposes (i.e. fracture mechanics).

With the category 2 specimen, it is often important that a fatigue precrack be initiated in the environment in which the SCC susceptibility test is to be performed. This is especially true for titanium, because the highly passive nature of the metal may repair the precracked area before the environment of interest can be introduced. Individual evaluations are required to determine whether precracking should be performed in the corrosive environment.

Category 3 specimens were developed from category 1 specimens in order to remove the uncertainty associated with SCC initiation and accelerate SCC testing for quicker answers. Because the smooth specimen is dynamically loaded to failure, the investigator is assured of two things. First, the specimen will always fail, although not always as the result of SCC. Second, at some point during the test, a crack will be mechanically produced that may serve as an initiation site for SCC. This specimen configuration also represents a worst-case situation because most materials are not intentionally subjected to such large plastic strains during real-life exposure. The major difficulty with this configuration stems from the data produced; such data are far from quantitative. The quantitative data derived from these tests, such as the ratio of time to failure in air versus time to failure in the test environment, provide some comparative information, but are of limited value to the materials selector. Therefore, testing with category 3 specimens resolves some of the deficiencies of testing with the category 1 specimen, although it creates several new problems. Unfortunately, like category 1 specimens, the category 3 specimens are also pass/fail and produce essentially no quantifiable data.

Given the advantage of quantifiable data of category 2 specimens and the difficulty of initiating SCC on titanium category 1 specimens, it is not surprising that almost all of the laboratory testing on titanium alloys has been performed with category 2 specimens. In the following sections, metallurgical, environmental, and

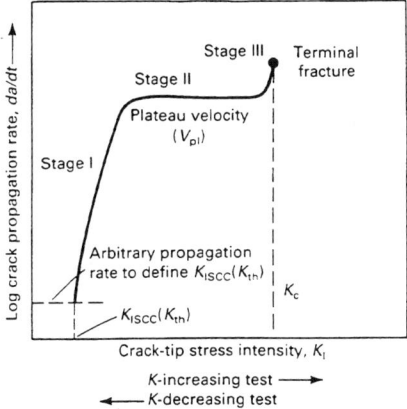

Fig. 4 Effect of stress on SCC kinetics. Stages I and II may not always be straight lines, and one or more may be absent in some systems.
Source: ASM Handbook, Vol 13, p 246

stress effect on the SCC susceptibility of titanium alloys will be discussed. The data presented are primarily derived from fracture mechanics test procedures in corrosive environments and, as such, are quantitative in nature. In many cases, results are compared with those attained in noncorrosive environments (often air) to give the reader some appreciation for the degree to which exposure to the environment reduces fracture toughness.

Most fracture toughness data are presented as crack velocity versus applied stress intensity (K_I). The stress intensity below which no measurable crack growth occurs is defined as the threshold stress intensity for SCC, K_{ISCC}.

Testing in a Hot Salt Environment. The hot salt test consists of exposing a stressed salt-coated test specimen to an elevated temperature for various predetermined lengths of time. The exposure periods are determined by the alloy, stress level, temperature, and selected damage criterion (that is, embrittlement, cracking, or rupture, or a combination of these phenomena). Exposures are typically carried out in laboratory ovens or furnaces equipped with loading equipment for stressing specimens. Environmental conditions, the degree of control required, and the means for obtaining control are described in ASTM G 41.

No actual service failures of titanium have been ascribed to hot salt SCC. Hot salt testing can be used for alloy screening to determine the relative susceptibility of Ti alloys to embrittlement and cracking and to determine the time-temperature-stress threshold levels for the onset of embrittlement and cracking. However, certain types of specimens are more suitable for each of these types of characterization. Precracked specimens are unsuitable for testing of titanium alloys, because cracking reinitiates at salt/metal/air interfaces and results in many small cracks that extend independently. Therefore, smooth specimens are recommended.

Testing in Water and Aqueous Solutions. Water, seawater, and almost any neutral aqueous solution (except atmospheric water vapor) can cause SCC in many titanium alloys in the presence of preexisting cracklike flaws, although susceptibility in these environments cannot be detected by smooth specimens. Therefore, fracture mechanics type characterization is necessary. For some titanium alloys, the extremely rapid growth of stress-corrosion cracks in salt water and the dependency on specimen geometry preclude the possibility of using crack growth rate data for design purposes.

Therefore, ranking of materials is based on K_{ISCC} values, although a true threshold stress intensity for SCC may not exist (Ref 81). Titanium alloys do not exhibit stage 1 type crack growth kinetics (see Fig. 4) in neutral aqueous solutions. Tests have been performed for sufficient periods of time to allow detection of crack growth rates of 10^{-9} m/s (1.4×10^{-4} in./h), but SCC has not been observed. The slowest crack velocity that has been detected is 10^{-8} m/s (1.4×10^{-3} in./h). Therefore, in neutral aqueous solutions, a threshold K_{ISCC} exists at which SCC will not propagate (Ref 81). The above rates, however, are not as slow as those observed in high-susceptibility aluminum alloys. Tests are commonly performed in water containing about 3.5% sodium chloride, artificial seawater, or natural seawater unless specific environments are being tested.

Testing in Organic Fluids. A wide variety of organic fluids can cause SCC in some titanium alloys under specific test conditions (see Table 5). Most of these fluids attack the passive surface film that is characteristic of titanium alloys. Consequently, precracked specimens do not have to be used to initiate SCC. A standard environment does not exist; for evaluation of SCC in organic fluids therefore, test conditions must be selected with appropriate consideration given to the type of environmental service required.

Other Forms of Corrosion

Galvanic Corrosion. The coupling of titanium with dissimilar metals usually does not accelerate the corrosion of titanium. The exception is in strongly reducing environments in which titanium is severely corroding and not readily passivated. In this uncommon situation, accelerated corrosion may occur when titanium is coupled to more noble metals. In its normal passive condition, titanium is beneficially influenced by materials that exhibit more noble (positive) corrosion potentials. In this regard, graphite and various precious metals (such as platinum, palladium, ruthenium, iridium, and gold) provide anodic protection when coupled to titanium by further stabilizing the oxide film of titanium at more noble potentials.

The corrosion potential of titanium under normally passive conditions is quite noble, but similar to stainless steel or nickel-base alloys in the passive condition. The small potential differences between these passive engineering alloys generally mean negligible galvanic interactions and good galvanic compatibility as long as passive conditions prevail for the alloys involved.

However, when titanium is coupled to a metal that is active (anodic) in an environment, accelerated anodic attack of the active metal may result. Moreover, hydrogen charging of the titanium may occur causing hydriding of titanium alloys. The rate of accelerated attack depends on many factors, including the cathode-to-anode surface area ratio, concentration of dissolved cathodic depolarizers (for example, oxygen, or atomic hydrogen), temperature, solution flow velocity, and medium chemistry. Depending on environmental conditions, active metals may include carbon or low-alloy steels, aluminum, zinc, copper alloys, or stainless

Table 5 Environments and temperatures conducive to SCC of titanium alloys

Environment	Temperature, °C (°F)
Hot dry chloride salts	260-480 °C (500-900 °F)
Seawater, distilled water, and aqueous solutions	Ambient
Nitric acid, red fuming	Ambient
Nitrogen tetroxide	Ambient to 75 °C (165 °F)
Methanol, ethanol	Ambient
Chlorine	Elevated
Hydrogen chloride	Elevated
Hydrochloric acid, 10%	Ambient to 40 °C (105 °F)
Trichloroethylene	Elevated
Trichlorofluoroethane	Elevated
Chlorinated diphenyl	Elevated

steels that are active (depassivated) or pitting. When galvanic corrosion is unacceptably high, consideration should be given to all-titanium component design, coupling to more compatible (passive) alloys, use of dielectric (insulating) joints, or cathodic protection of the active metal.

Erosion-corrosion is defined as the acceleration in metal degradation as a result of the combined effects of corrosion and mechanical damage of the surface from erosion. This form of attack is highly dependent on fluid velocity and angle of impingment and is favored in areas where high local turbulence, impingement, or cavitation of the fluid occur on metal surfaces. Suspended solids in fluid can also result in abrasion, which can drastically accelerate metal loss.

In normal passive environments, the hard, tenacious TiO_2 surface film of titanium provides a superb resistance barrier to erosion-corrosion. For this reason, titanium alloys can withstand flowing water or seawater velocities as high as 30 m/s (100 ft/s) with insignificant metal loss. The ability of the oxide film to repair itself when damaged and the intrinsic hardness of titanium alloys both contribute to their excellent resistance to erosion-corrosion. Therefore, inlet turbulence in shell and tube exchangers, entrained gas bubble impingement, and pump cavitation effects are generally not of concern in titanium tubing, piping, and pump components.

Titanium alloys also exhibit relatively high resistance to fluids containing suspended solids. Critical velocities for excessive metal removal depend on the concentration, shape, size, and hardness of the suspended particles, in addition to fluid impingement angle, local turbulence, and titanium alloy properties. The typically low concentrations of silt entrained in seawater are generally of little consequence, but continuous exposure to high-velocity slurries of hard particles can lead to finite metal removal. The harder high-strength titanium alloys may offer improved erosion resistance when marginal erosion of the softer unalloyed titanium grades is observed. When abrasive conditions are severe, application of hard surface coatings should be considered. In potential applications involving high-velocity slurries or suspended solids, it is advisable to conduct erosion tests whenever possible.

Corrosion fatigue refers to the reduction in fatigue resistance of a metal due to the presence of a corrosive medium. Because of the protective oxide film of titanium, the smooth or notched fatigue strength of the more common titanium alloys and their weldments is not significantly affected by water, seawater, and many other aqueous chloride media. These alloys typically exhibit smooth fatigue run-out stress to tensile strength ratios in the range of 0.5 to 0.6, which remain unchanged in 3.5% sodium chloride (NaCl) solutions and in seawater. The Ti-6Al-4V datasheet includes some corrosion-fatigue results.

References

1. N.D. Tomashov and P.M. Altovskii, *Corrosion and Protection of Titanium*, Government Scientific-Technical Publication of Machine-Building Literature (Russian translation), 1963
2. B.D. Craig, T.E. Ferg, and C.S. Aldrich, Corrosion/Wear of Titanium Against Steel in Water and Oil Base Drilling Mud, OMAE Conf, ASME, Calgary, Alberta, Canada, Paper No. 92-897, (1992)
3. V.V. Andreeva, *Corrosion*, Vol 20, 1964, p 35
4. M. Pourbaix, *Atlas of Electrochemical Equilibria in Aqueous Solutions*, NACE, 1974, p 217
5. B.D. Craig, *Fundamental Aspects of Corrosion Films in Corrosion Science*, Plenum Press, N.Y., 1991
6. L.C. Covington and R.W. Schutz, Effects of Iron on the Corrosion Resistance of Titanium, *Industrial Applications of Titanium and Zirconium*, STP 728, ASTM, 1981, p 163-180
7. R.W. Schutz and J.S. Grauman, Fundamental Corrosion Characterization of High-Strength Titanium Alloys, *Industrial Applications of Titanium and Zirconium: Fourth Volume*, STP 917, ASTM, 1986, p 130-143
8. J. Newman, Fighting Corrosion with Titanium Castings, *Chem. Eng.*, 4 June 1979
9. J.P. Dippel, Manufacturing Titanium Castings for the Pump Industry, *World Pumps*, Dec 1982
10. H.B. Bomberger and L.F. Plock, Methods Used to Improve Corrosion Resistance of Titanium, *Mater. Prot.*, June 1969, p 45-48
11. M. Stern and H. Wissenberg, The Influence of Noble Metal Alloy Additions on the Electrochemical and Corrosion Behavior of Titanium, *J. Electrochem. Soc.*, Vol 106 (No. 9), Sept 1959, p 759
12. M. Stern and C.R. Bishop, The Corrosion Behavior of Titanium-Palladium Alloy, *Trans. ASM*, Vol 52, 1960, p 239
13. N.D. Thomashov et al., Corrosion and Passivity of the Cathode-Modified Titanium-Based Alloys, *Titanium and Titanium Alloys—Scientific and Technological Aspects*, Vol 2, Plenum Press, 1982, p 915-925
14. A.J. Sedriks, Further Observations on the Electrochemical Behavior of Ti-Ni Alloys on Acidic Chloride Solutions, *Corrosion*, Vol 29 (No. 2), 1973, p 64
15. J.C. Griess, Jr., *Corrosion*, April 1968, p 99-103
16. L.C. Covington and H.R. Palmer, "A New Corrosion Resistant Titanium Alloy Ti-38A for High Temperature Brine Service," paper presented at the AIME Titanium Committee Session on Corrosion and Biomedical Applications of Titanium, Detroit, MI, AIME, Oct 1974
17. M. Stern and C. Bishop, The Corrosion Resistance and Mechanical Properties of Titanium-Molybdenum Alloys Containing Noble Metals, *Trans. ASM*, Vol 54, Sept 1961, p 286-298
18. N. Thomashov, R. Al'tovskii, and G. Chernova, Passivity and Corrosion Resistance of Titanium and Its Alloys, *J. Electrochem. Soc.*, Vol 108, Feb 1961, p 113-119
19. J.A. Petit et al., *Corros. Sci.*, Vol 21 (No. 4), 1981, p 279-299
20. V.P. Gupta, Process for Decreasing the Rate of Titanium Corrosion, U.S. Patent 4,321,231, 1982
21. R.W. Schutz and L.C. Covington, Hydrometallurgical Applications of Titanium, *Industrial Applications of Titanium and Zirconium: Third Conference*, STP 830, ASTM, 1984, p 29-47
22. T. Fukuzuka, K. Shimogori, H. Satoh, and F. Kamikubo, Protection of Titanium against Crevice Corrosion by Coating with Palladium Oxide, *Titanium '80—Science and Technology*, TMS, 1980, p 2631-2638
23. S. Fujishiro and D. Eylon, *Thin Solid Films*, Vol 54, 1978, p 309-315
24. S. Fujishiro and D. Eylon, *Metall. Trans. A.*, Vol 11A, Aug 1980, p 1261-1263
25. R.W. Schutz and L.C. Covington, *Corrosion*, Vol 37 (No. 10), Oct. 1981, p 585-591
26. T. Fukuzuka et al., *Industrial Applications of Titanium and Zirconium*, STP 728, ASTM, 1981, p 71-84
27. J.C. Cotton, *Chem. Eng. Prog.*, Vol 66 (No. 10), 1970, p 57
28. J.B. Cotton, *Chem. Ind.*, Vol 3, Jan. 1958, p 68-69
29. *Corrosion Resistance of Titanium*, Technical Handbook, Imperial Metals Industries (Kynoch) Ltd., Birmingham, UK
30. A. Erdemir et al., *Mater. Sci. Eng.*, Vol 69, 1985, p 89-93
31. P. Sioshansi, Surface Modification by Ion Implantation, *Mach. Des.*, 20 March 1966
32. P.C. Hughes and I.R. Lamborn, Contamination of Titanium by Water Vapour, *J. Inst. Met.*, Vol 89, 1960-1961, p 165
33. C.R. Breden, *Met. Prog.*, Vol 64, 1953, p 194

34. S.C. Datski, Report ANL 5354, U.S. Atomic Energy Commission, 1954
35. D. Schlain, "Corrosion Properties of Titanium," Bulletin 619, U.S. Bureau of Mines, 1964
36. J.D. Tkach and R.H. Meservey, "Corrosion of Thermocouple Sheath Materials in a Pressurized Water Reactor Environment," Aerojet Nuclear Company, 1971
37. R.L. Kane, The Corrosion of Titanium, *The Corrosion of Light Metals*, The Corrosion Monograph Series, John Wiley & Sons, 1967
38. L.L. Gilbert and C.W. Funk, Explosions of Titanium and Fuming Nitric Acid Mixtures, *Met. Prog.*, Nov 1956, p 93-96
39. B.R. Brown, "Stress Corrosion Cracking in High Strength Steels and in Titanium and Aluminum Alloys," Naval Research Laboratory, 1972
40. H.B. Bomberger, Titanium Corrosion and Inhibition in Fuming Nitric Acid, *Corrosion*, Vol 13 (No. 5), May 1957, p 287-291
41. R.L. Wallner et al., *Mater. Prot.*, Jan 1965, p 55-56
42. W.K. Boyd, "Stress Corrosion Cracking of Titanium Alloys—An Overview," paper presented at the International Symposium on Stress Corrosion Mechanisms in Titanium Alloys, Georgia Institute of Technology, Jan 1971
43. J.B. Rittenhouse and C.A. Popp, Inhibition of Corrosion in Fuming Nitric Acid, *Corrosion*, Vol 14, June 1958, p 283-284
44. L.C. Covington and R.W. Schutz, "Corrosion Resistance of Titanium," TIMET Corporation, 1982
45. R.L. LaQue and H.R. Copson, *Corrosion Resistance of Metals and Alloys*, 2nd ed., ACS Monograph, Reinhold, 1963, p 646-661
46. R.E. Smallwood, Corrosion of Titanium and Zirconium Alloys in Zinc Chloride Solutions, *Industrial Applications of Titanium and Zirconium*, STP 728, ASTM, 1981, p 147-162
47. R.W. Schutz and J.S. Grauman, Selection of Titanium Alloys for Concentrated Seawater, NaCl and MgCl₂ Brines, *Titanium 1986—Titanium Products and Applications*, proceedings of the Technical Program from the 1986 International Conference, San Francisco, CA, Titanium Development Association, 1986
48. E.G. Hanes, G. Goldberg, R.E. Emsberger, and W.T. Brehm, "Investigation of Stress Corrosion Cracking of Titanium Alloys," Second Progress Report, NASA, Grant N6R-39-008-014, Mellon Institute, May 1967
49. C.M. Chern, H.B. Kirkpatrick, and H.L. Gegel, "Cracking of Titanium Alloys in Methanolic and Other Media," paper presented at the International Symposium on Stress Corrosion Mechanisms in Titanium Alloys, Georgia Institute of Technology, Jan 1971
50. E.G. Haney and W.R. Wearmouth, Effect of Pure Methanol on the Cracking of Titanium, *Corrosion*, Vol 25 (No. 2), Feb 1969, p 87
51. A.J. Sedriks and J.S.A. Green, Stress Corrosion of Titanium in Organic Liquids, *J. Met.*, April 1971, p 48
52. R.W. Schutz and L.C. Covington, Guidelines for Corrosion Testing of Titanium, *Industrial Applications of Titanium and Zirconium*, STP 728, ASTM, 1981, p 59-70
53. R.W. Schutz, Titanium, *Process Industries Corrosion—The Theory and Practice*, NACE, 1986, p 503
54. J.C. Griess, Jr., Crevice Corrosion of Titanium in Aqueous Salt Solutions, *Corrosion*, Vol 24 (No. 4), April 1968, p 96-109
55. L.C. Covington, Pitting Corrosion of Titanium Tubes in Hot Concentrated Brine Solutions, *Galvanic and Pitting Corrosion—Field and Laboratory Studies*, STP 576, ASTM, 1976, p 147-154
56. B.J. Moniz, Field Coupon Corrosion Testing, *Process Industries Corrosion—The Theory and Practice*, NACE, 1986, p 67-83
57. I. Dugdale and J.B. Cotton, *Corros. Sci.*, Vol 4, 1964, p 397
58. T. Koizumi and S. Furuya, *Titanium—Science and Technology*, Vol 4, proceedings of the Second International Conference, Plenum Press, 1973, p 2383-2393
59. L. Szlarska-Smialowska and M. Janik-Czachor, *Corros. Sci.*, Vol 11, 1971, p 901-914
60. E.L. Liening, Electrochemical Corrosion Testing Techniques, *Process Industries Corrosion*, NACE, 1986, p 85-122
61. L.C. Covington, "Factors Affecting the Hydrogen Embrittlement of Titanium," paper presented at Corrosion/75, Toronto, Canada, NACE, April 1975
62. L.C. Covington, *Corrosion*, Vol 35 (No. 8), Aug 1979, p 378-382
63. V.A. Livanov et al., *Hydrogen in Titanium*, Israel Program for Scientific Translation Ltd., Catalog No. 2163, Daniel Davey & Company, Inc., 1965
64. N.E. Paton and J.C. Williams, Effect of Hydrogen on Titanium and Its Alloys, *Titanium and Titanium Alloys—Source Book*, ASM, 1982, p 185-207
65. R.R. Boyer and W.F. Spurr, Characteristics of Sustained-Load Cracking and Hydrogen Effects in Ti-6Al-4V, *Metall. Trans. A*, Vol 9A, Jan 1978, p 23-29
66. R. Bourcier and D. Koss, *Acta Metall.*, Vol 32 (No. 11), 1984, p 2091-2099
67. G.A. Lenning et al., Effect of Hydrogen on Alpha Titanium Alloys, *Trans. AIME*, Oct 1956, p 1235
68. C.M. Craighead et al., Hydrogen Embrittlement of Beta-Stabilized Titanium Alloys, *Trans. AIME*, Aug 1956, p 923
69. J.J. DeLuccia, "Electrolytic Hydrogen in Beta Titanium," Report NADC-76207-30, Air Vehicle Technical Department, Naval Air Development Center, June 1976
70. D.A. Meyn, Effect of Hydrogen on Fracture and Inert-Environment Sustained Load Cracking Resistance of Alpha-Beta Titanium Alloys, *Metall. Trans.*, Vol 5, Nov 1974, p 2405
71. W.R. Holman et al., Hydrogen Diffusion in a Beta Titanium Alloy, *Trans. AIME*, Vol 233, Oct 1965, p 1836
72. I.I. Phillips, P. Pool, and L.L. Shreir, Hydride Formation During Cathodic Polarization of Ti.-11. Effect of Temperature and pH of Solution on Hydride Growth, *Corros. Sci.*, Vol 14, 1974, p 533-542
73. R.L. Jacobs and J.A. McMaster, Titanium Tubing: Economical Solution to Heat Exchanger Corrosion, *Mater. Prot. Perform.*, Vol 11 (No. 7), July 1972, p 33-38
74. "Get More Advantages by Applying Titanium Tubing not only for Power Plants but also for Desalination Plants!!," Technical Brochure, Japan Titanium Society, May 1984
75. H. Satoh, T. Fukuzuka, K. Shimogori, and H. Tanabe, "Hydrogen Pickup by Titanium Held Cathodic in Seawater," paper presented at the Second International Congress on Hydrogen in Metals, Paris, June 1977
76. S. Sato, K. Nagata, and M. Magayama, "Experiences of Welded Titanium Condenser Tubes in Japan," Technical Research Laboratory, Sumitomo Light Metal Industries Ltd.
77. T. Fukuzuka, K. Shimogori, H. Satoh, and F. Kamikubo, "Corrosion Problems and Countermeasures in MSF Desalination Plant Using Titanium Tube," Kobe Steel Ltd., 1985
78. L.A. Charlot and R.E. Westerman, Low-Temperature Hydriding of Zircaloy-2 and Titanium in Aqueous Solutions, *Electrochem. Technol.*, Vol 6 (No. 3-4), March/April 1968
79. J. Lee and P. Chung, "A Study of Hydriding of Titanium in Seawater under Cathodic Polarization," paper 259, presented at Corrosion/86, Houston TX, NACE, March 1986
80. R.H. Jones, *Metals Handbook*, ASM International, Vol 13, 1987, p 145
81. M.J. Blackburn, W.H. Smyrl, and J.A. Sweeny, Titanium Alloys, *Stress Corrosion Cracking in High Strength Steels and in Titanium and in Aluminum Alloys*, B.F. Brown, Ed., Naval Research Laboratory, 1972, p 246-363

Technical Note 3: Casting

The titanium castings industry is relatively young by most industry standards, which is related to the youth of the titanium industry. The high reactivity of titanium, especially in the molten state, presents a special challenge to the foundry. Special, relatively expensive, methods of melting, moldmaking, and surface cleaning are required to maintain metal integrity. Thus, the number of titanium casting suppliers is limited.

Melting and Pouring Practice. The dominant, almost universal, method of melting titanium is with a consumable titanium electrode; the molten metal is contained in a water-cooled copper crucible while confined in a vacuum chamber. This "skull melting" technique prevents the highly reactive liquid titanium from dissolving the crucible because it is contained in a solid skull frozen against the water-cooled crucible wall. When an adequate melt quantity has been obtained, the residual electrode is quickly retracted, and the crucible is tilted for pouring into the molds. A skull of solid titanium remains in the crucible for reuse in a subsequent pour or for later removal.

Superheating. The consumable electrode practice affords little opportunity for superheating the molten pool because of the cooling effect of the water-cooled crucible. Limited superheating capabilities creates problems filling the molds; it is common either to pour castings centrifugally, forcing the metal into the mold cavity by centrifugal force, or to pour statically into preheated molds to obtain adequate fluidity. Post-cast cooling takes place in a vacuum or in an inert gas atmosphere until the molds can be safely removed to air without oxidation of the titanium.

Electrode Composition. Consumable titanium electrodes are either ingot metallurgy forged billet, consolidated revert wrought material, selected foundry returns, or a combination of all of these. Casting specifications or user requirements can dictate the composition of revert materials used in electrode construction.

Molding Methods

Shape casting of titanium was first demonstrated in the United States in 1954 at the U.S. Bureau of Mines using machined high-density graphite molds. Since this early work, semipermanent, reusable molds made from machined graphite have been used successfully on relatively simple-shaped parts that allow metal volumetric shrinkage to occur without restriction. However, the method is economical only when reasonably high volumes are required, that is, thousands of parts, because of the high cost of the solid mold material. The more common molding methods for titanium castings include rammed graphite and lost wax investment molding (see Table 1 for a sample of foundries).

Rammed Graphite Molding. The earliest commercial application of complex titanium shapes cast from rammed graphite molding was for corrosion-resistant components of valves and pumps. These applications continue to dominate the rammed graphite production method; however, in more recent years, some users have justified the expense of lost wax investment tooling for some commercial corrosion-resistant casting applications.

The traditional rammed graphite molding process uses powdered graphite mixed with organic binders. Patterns typically are made of wood. The mold material is pneumatically rammed around the pattern and cured at high temperature in a reducing atmosphere to convert the organic binders to pure carbon. The molding process and the tooling are essentially the same as for cope and drag sand molding in ferrous and nonferrous foundries. In the 1970's, derivations of rammed graphite mold materials were developed using components of more traditional sand foundries along with inorganic binders. This resulted in more dimen-

Table 1 Status and capacity of titanium foundries in the United States, Japan, and Western Europe in 1992

Foundry	Maximum pour weight		Approximate maximum envelope size			
			Rammed graphite		Investment casting	
	kg	lb	mm	in.	mm	in.
Howmet Corp. (MI and VA)	730	1600	...	...	1525 diam × 1525	60 diam × 60
Oremet Corp. (OR)	900	2000	1525 diam × 1830	60 diam × 72	...	...
PCC (OR)	770	1700	...	...	1525 diam × 1220	60 diam × 48
Rem Products (OR)	180	400	...	...	815 diam × 508	32 diam × 20
IMI Titanium Inc. (TiLine) (OR)	180	400	...	...	1370 diam × 610	54 diam × 24
IMI Titanium Inc. (TiTech) (CA)	400	875	915 diam × 610	36 diam × 24	915 diam × 610	36 diam × 24
PCC France (France)	270	600	990 diam × 990	39 diam × 39	1220 diam × 1220	48 diam × 48
Tital (West Germany)	180	400	1145 diam × 760	45 diam × 30	1015 diam × 635	40 diam × 25
Settas (Belgium)	820	1800	1525 diam × 1220	60 diam × 48	610 diam × 610	24 diam × 24
VMC (Japan)	180	400	1270 diam × 635	50 diam × 25	(a)	(a)
Mitsui (Japan)	(a)	(a)	(a)	(a)	(a)	(a)
ICC (Japan)	(a)	(a)	(a)	(a)	(a)	(a)
Kobe (Japan)	(a)	(a)	(a)	(a)	(a)	(a)
Wyman Gordon (CT)	90	200	...	...	910 diam × 455	36 diam × 18
Schlosser (OR)	200	450	...	...	1065 diam × 915	42 diam × 36
Ruger (AZ)	(a)	(a)	...	...	(b)	(b)

(a) Capacity size data unavailable. (b) Investment only, capacity unavailable

Table 2 Comparison of cast titanium alloys

Alloy	Estimated relative usage of castings	Nominal composition, wt %								Special properties(a)
		O	Al	Fe	V	Cr	Sn	Mo	Other	
Ti-6Al-4V	88%	0.18	6	0.13	4	...	...	...	...	General purpose
Ti-6Al-4V ELI	2%	0.11	6	0.10	4	...	...	...	...	Cryogenic toughness
Commercially pure titanium Gr 2 & 3	7%	0.25	...	0.15	...	...	...	...	...	Corrosion resistance
Ti-6Al-2Sn-4Zr-2Mo-0.1Si	2%	0.10	6	0.15	...	...	2	2	4 Zr, 0.1Si	Elevated-temperature creep
Ti-6Al-2Sn-4Zr-6Mo	<1%	0.10	6	0.15	...	...	2	6	4Zr	Elevated-temperature strength
Ti-5Al-2.5Sn	<1%	0.16	5	0.2	...	...	2.5	...	...	Cryogenic toughness
Ti-3Al-8V-6Cr-4Zr-4Mo	<1%	0.10	3.5	0.2	8.0	6	...	4	4Zr	Strength
Ti-15V-3Al-3Cr-3Sn	<1%	0.11	3	0.2	15	3	3	...	...	Strength
IMI-829	<1%	0.13	5.4	0.05	...	...	3.2	0.3	3.0 Zr, 1.0 Nb, 0.3 Si	Elevated-temperature creep

(a) Superior, relative to Ti-6Al-4V

Table 3 Standard industry specifications applicable to titanium castings

MIL-T-81915	Titanium and titanium alloy castings, investment
AMS-4985A	Titanium alloy castings, investment or rammed graphite
AMS-4991	Titanium alloy castings, investment
ASTM B 367	Titanium and titanium alloy castings
MIL-STD-2175	Castings, classification and inspection of
MIL-STD-271	Nondestructive testing requirements for metals
MIL-STD-453	Inspection, radiographic
MIL-Q-9859	Quality program requirement
MIL-I-6866B	Inspection, penetrant method of
MIL-H-81200	Heat treatment of titanium and titanium alloys
ASTM E-1320	Reference radiographs for titanium castings
ASTM E 120	Standard methods for chemical analysis of titanium and titanium alloys
ASTM E 8	Methods of tension testing of metallic materials
AMS-2249B	Chemical-check analysis limits for titanium and titanium alloys
AMS-4954	Titanium alloy welding wire Ti-6Al-4V
AMS-4956	Titanium alloy welding wire Ti-6Al-4V, extra low interstitial

sionally stable and less costly molds that were capable of containing molten titanium without undue metal/mold reaction.

Lost Wax Investment Molding. The principal technology that allowed the proliferation of titanium alloy castings in the aerospace industry was the investment casting method, introduced in the mid-1960's. The adaptation of this method to titanium casting technology required the development of ceramic slurry materials with minimum reaction with the extremely reactive molten titanium.

Proprietary lost wax ceramic shell systems have been developed by the several foundries engaged in titanium casting manufacture. Of necessity, these shell systems must be relatively inert to molten titanium and cannot be made with the conventional foundry ceramics used in the ferrous and nonferrous industries. Usually, the face coats are made with special refractory oxides and appropriate binders. After the initial face coat ceramic is applied to the wax pattern, more traditional refractory systems are used to add shell strength from repeated backup ceramic coatings. Regardless of face coat composition, some metal/mold reaction inevitably occurs from titanium reduction of the ceramic oxides. The oxygen-rich surface of the casting stabilizes the α phase, usually forming a metallographically distinct α case layer on the cast surface, which is removed later by means of chemical milling.

Foundry practice focuses on methods to control both the extent of the metal/mold reaction and the subsequent diffusion of reaction products inward from the cast surface. Diffusion of reaction products into the cast surface is time-at-temperature dependent. Depth of surface contamination can vary from nil on very thin sections to more than 1.5 mm (0.06 in.) on heavy sections. On critical aerospace structures, the brittle α case is removed by chemical milling. The depth of surface contamination must be taken into consideration in the initial wax pattern tool design. Hence, the wax pattern and castings are made slightly oversize, and final dimensions are achieved through careful chemical milling. Metal superheat, mold temperature and thermal conductivity, g force (if centrifugally cast), and rapid post-cast heat removal are other key factors in producing a satisfactory product. These parameters are interrelated, that is, a high g force centrifugal pour into cold molds may achieve the same relative fluidity as a static pour into heated molds.

Alloys

All production titanium castings to date are based on traditional wrought product compositions. As such, the Ti-6Al-4V alloy dominates structural casting applications (see Table 2). This alloy similarly has dominated wrought industry production since its introduction in the early 1950's, becoming the benchmark alloy against which others are compared. However, other wrought alloys have been developed, for special applications, with better room-temperature or elevated-temperature strength, creep, or fracture toughness characteristics than those of Ti-6Al-4V. These same alloys are also being cast when net shape casting technology is the most economical method of manufacture. As with Ti-6Al-4V, other cast titanium alloys have properties (see Table 6 and Fig. 1) generally comparable to their wrought counterparts.

Specifications. Industry-wide specifications (see Table 3) provide mechanical property guarantees and process control features. In addition, most major aerospace companies have comparable specifications. MIL Handbook 5, Aerospace Design Specifications, does not presently include titanium alloy castings, but it is expected that such information will be incorporated in the near future. As with wrought products, commercially pure titanium castings are used almost entirely in corrosion applications. Commercially pure titanium pumps and valves are the principal components made using titanium casting technology for the corrosion resistance field, and are used extensively in chemical and petrochemical plants, as well as standard fire fighting pumps for the U.S. Navy.

Newer Alloys. As aircraft engine manufacturers seek to use cast titanium at higher operating temperatures, Ti-6Al-2Sn-4Zr-2Mo and Ti-6Al-2Sn-4Zr-6Mo

Table 4 General linear and diametrical tolerance guideline for titanium castings

Size		Total tolerance bands(a)	
mm	in.	Investment cast	Rammed graphite process
25 to <102	1 to <4	0.76 mm (0.030 in.) or 1.0%, whichever is greater	1.52 mm (0.060 in.)
102 to <305	4 to <12	1.02 mm (0.040 in.) or 0.7%, whichever is greater	1.78 mm (0.070 in.) or 1.0%, whichever is greater
305 to <610	12 to <24	1.52 mm (0.060 in.) or 0.6%, whichever is greater	1.0%
≥610	≥24	0.5%	1.0%
Examples			
254 mm	10	1.78 mm (0.070 in.) total tolerance band or ±0.89 mm (±0.035 in.)	2.54 mm (0.100 in.) total tolerance band or ±1.27 mm (±0.050 in.)
508 mm	20	3.05 mm (0.120 in.) total tolerance band or ±1.52 mm (±0.060 in.)	5.08 mm (0.200 in.) total tolerance band or ±2.54 mm (±0.100 in.)

(a) Improved tolerances may be possible depending on the specific foundry capabilities and overall part-specific requirements.

Table 5 Surface finish of titanium castings

Process	NAS 823 surface comparator	RMS equivalent μm	RMS equivalent μin.
Investment			
As-cast	C-12	3.2	125
Occasional areas of	C-25	6.3	250
Rammed graphite			
As-cast	C-30-40	7.5-10	300-400
Occasional areas of	C-50	12.5	500
Hand finished	C12-25	3.2-6.3	125-250

RMS, root mean square

and IMI 829 are being specified more frequently. Extra low interstitial grade Ti-6Al-4V has been used for critical cryogenic space shuttle service where fracture toughness is an important design criteria. The most recent alloys to receive attention in the casting industry are the metastable β alloys Ti-3Al-8V-6Cr-4Zr-4Mo (Beta C) and Ti-15V-3Cr-3Al-3Sn (Ti 15-3). The latter was developed as a highly cold-formable and subsequently age-hardened sheet material. These alloys are highly castable and are readily heat treated to a 1170 MPa (170 ksi) strength level, making them serious candidates for the replacement of high-strength precipitation-hardened stainless steels such as 17-4PH. The full density advantage of titanium of about 40% is preserved because strength levels are comparable in both materials. Titanium-aluminide castings are being developed for application in the compressor sections of aircraft gas turbine engines and other high-temperature applications. Compositions based upon both the α_2 (Ti_3Al) and γ(TiAl) ordered phases have been cast experimentally, with the former being closer to limited-production status. The low ductility of these alloys at room temperature has been the major producibility challenge. It is anticipated that the service potential for titanium aluminides in the 595 to 925 °C (1100 to 1700 °F) temperature range will eventually be realized. The difficulty in forging and machining shapes in these brittle alloys may increase the advantage of net shape methods such as castings.

Casting Design

Titanium castings present the designer with few differences in design criteria, compared with other metals. Ideal designs do not contain isolated heavy sections or uniform heavy walls of large area so that centerline shrinkage cavities and regions with a coarse microstructure may be avoided. From a practical sense, however, ideal tapered walls to promote directional solidification are not usually a reality. The advent of hot isostatic pressing to heal internal as-cast shrinkage cavities has offered the designer much more freedom; however, there still is a practical limit to the size of internal cavity that can be healed through hot isostatic pressing without producing significant surface or structural deformation due to the collapse of internal pores.

The lost wax investment process provides more design freedom for the foundry to properly feed a casting than does the traditional sand or rammed graphite approach. It is normal practice to gate and riser hot isostatic pressed investment castings to achieve reasonably good internal x-ray quality so that hot isostatic pressing will not cause extensive surface or structural deformation.

The usual required minimum practical wall thickness for investment castings is 2.0 mm (0.080 in.); however, local sections as small as 1.1 mm (0.045 in.) are routinely made. Even thinner walls may be achieved by chemical milling beyond that required for α case removal; however, as-cast wall variation is only made worse by extensive chemical milling and wall thickness tolerances become wider. Sand or rammed graphite molded castings have a usual minimum wall thickness of 4.75 mm (0.190 in.), although 3.0 mm (0.12 in.) is not unreasonable for short sections.

Fillet radii should be as generous as possible to minimize the occurrence of hot tears. While 0.76 mm (0.030 in.) radii are produced, the preferred minimum is 3.0 mm (0.12 in.). A rule of thumb is that a fillet radius should be 0.5 times the sum of the thicknesses of the two adjoining walls.

With proper tool design, zero draft walls are possible. To promote directional solidification, a 3° included draft angle may be preferred. Hot isostatic pressing will close any centerline shrinkage cavities in zero draft walls, making it unnecessary to provide draft. Draft requirements are also dependent upon foundry practice, with rammed graphite tooling usually requiring draft, and investment casting typically not requiring draft.

Tolerances. Typically, the major area of concern is true position of a thin-section surface with respect to a datum. Surface areas of approximately 129 cm^2 (20 $in.^2$) or greater in sections of less than approximately 3.0 mm (0.120 in.) thickness are susceptible to distortion, depending on adjoining sections. The high strength of titanium compared with aluminum, and low elastic modulus compared with steel present challenges in straightening and in maintaining extremely tight, true positions (see Table 4 for general tolerance band capabilities for linear dimensions).

Hot sizing fixtures have been increasingly used to help control critical casting

Table 6 Typical room-temperature tensile properties of titanium alloy castings (bars machined from castings)
Specification minimums are less than these typical properties.

Alloy(a)(b)	Yield strength		Ultimate strength		Elongation, %	Reduction of area, %
	MPa	ksi	MPa	ksi		
Commercially pure (grade 2)	448	65	552	80	18	32
Ti-6Al-4V, annealed	855	124	930	135	12	20
Ti-6Al-4V ELI	758	110	827	120	13	22
Ti-1100, Beta-STA(c)	848	123	938	136	11	20
Ti-6Al-2Sn-4Zr-2Mo, annealed	910	132	1006	146	10	21
IMI-834, Beta-STA(c)	952	138	1069	155	5	8
Ti-6Al-2Sn-4Zr-6Mo, Beta-STA(c)	1170	170	1240	180	1	1
Ti-3Al-8V-6Cr-4Zr-4Mo, Beta-STA(c)	1241	180	1330	193	7	12
Ti-15V-3Al-3Cr-3Sn, Beta-STA(c)	1200	174	1275	185	6	12

(a) Solution-treated and aged (STA) heat treatments may be varied to produce alternate properties. (b) ELI, extra low interstitial. (c) Beta-STA, solution treatment in β-phase field followed by aging

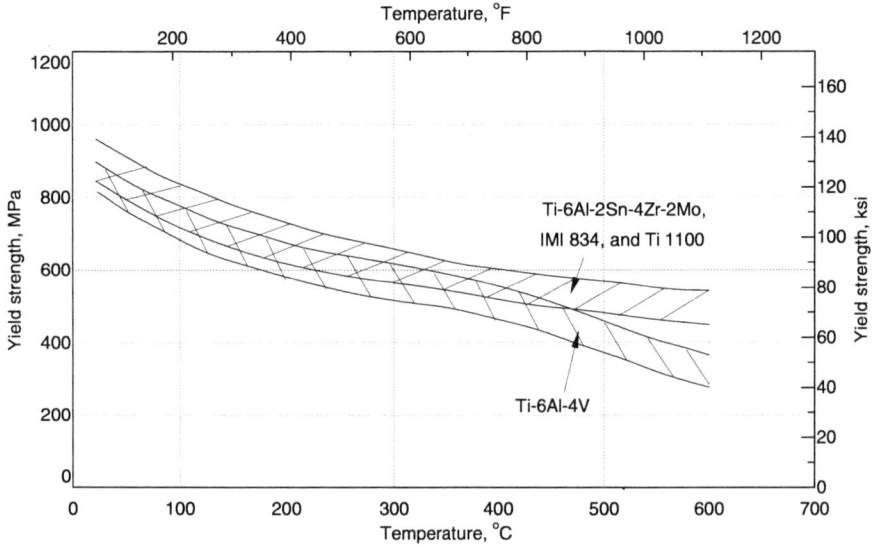

Fig. 1 Plot of yield strength versus temperature to compare elevated-temperature properties of cast Ti-6Al-2Sn-4Zr-2Mo, IMI 834, and Ti 1100 alloys with standard cast Ti-6Al-4V alloy. See separate alloy compilations in this Volume for additional cast property data.

dimensions. This technique typically involves the use of steel fixtures to "creep" the casting into final tolerances in an anneal or stress relief heat treatment by the weight of the steel or the use of differential thermal expansion of the steel relative to the titanium.

Standard casting industry thickness tolerances of ±0.75 mm (±0.030 in.) for rammed graphite and ±0.25 mm (±0.010 in.) for investment cast walls are more difficult to maintain with titanium primarily because of the influence of chemical milling (for critical applications it is necessary to mill all surfaces chemically to remove the residue α case). This operation is subject to variation because of part geometry and bath variables, and because it is usually manually controlled. Standard industry surface finishes are shown in Table 5.

Technical Note 4: Forging

Titanium alloy forgings are produced by all forging methods currently available, including open-die (or hand) forging, closed-die forging, upsetting, roll forging, orbital forging, spin forging, mandrel forging, ring rolling, and forward and backward extrusion. Selection of the optimal forging method for a given forging shape is based on the desired forging shape, the sophistication of the design of the forged shape, the cost and the desired mechanical properties and microstructure. In many cases, two or more forging methods are combined.

Titanium alloys generally are more difficult to forge than steel and aluminum, because their deformation resistance can increase dramatically with small changes in metal temperatures and strain rates (see comparisons in Fig. 1, 2, and 3). Therefore, the dies used in the conventional forging of titanium alloys are heated to facilitate the forging process and to reduce metal temperature losses during the forging process—particularly surface chilling, which may lead to inadequate die filling and/or excessive cracking. Heated dies also may allow reductions in strain rates, which further improve deformation characteristics.

However, in the conventional forging of titanium alloys (which involves nonisothermal die temperatures of 540 °C or less), the temperature losses encountered by such techniques far outweigh the benefits of forging at slow strain rates. Therefore, in the conventional forging of titanium alloys with relatively cool dies, intermediate strain rates are typically employed as a compromise between strain-rate sensitivity and metal temperature losses in order to obtain the optimal deformation possible with a given alloy.

Major reductions in resistance to deformation of titanium alloys can be achieved by slow-strain-rate forging techniques under conditions where losses of temperatures are minimized by heating dies to temperature at or close to the metal temperature. This requires the use of hot-die and isothermal forging technologies which are special categories of forging processes having die temperatures sig-

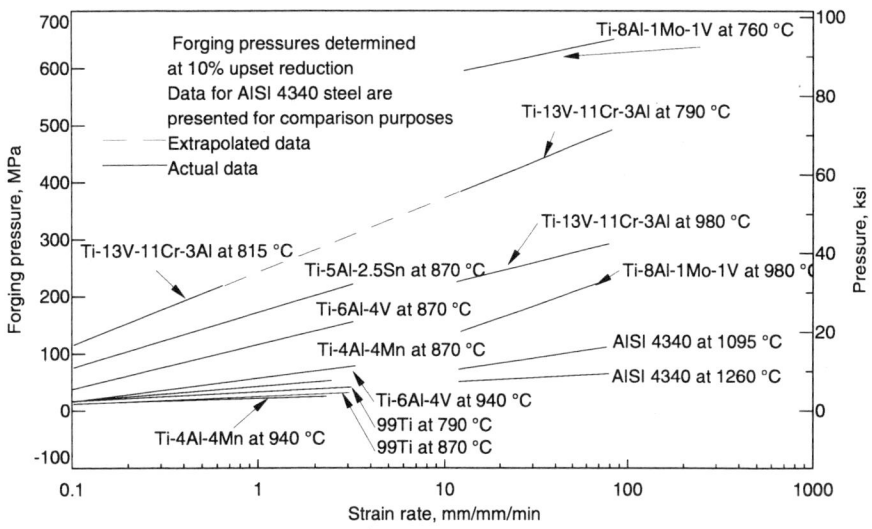

Fig. 1 Effect of strain rate on forging pressure.
Source: A.M. Sabroff, F.W. Boulger and H.J. Henning, *Forging Materials and Practices*, Reinhold, New York, 1968

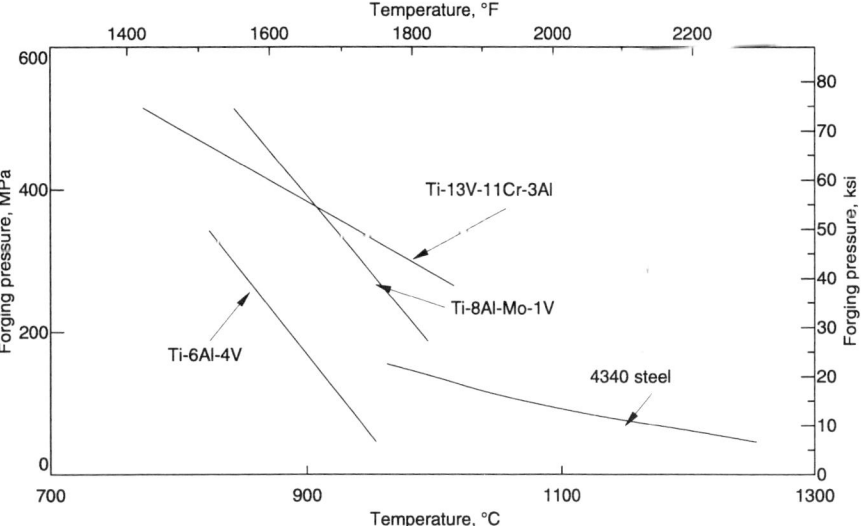

Fig. 2 Forging pressure vs forging temperature

nificantly higher than those used in conventional hot-forging processes. These processes are also referred to as near-net shape forging processes, and they have been used with titanium alloys such as Ti-6Al-4V, Ti-6Al-2Sn-4Zr-2Mo, Ti-10V-2Fe-3Al, and others.

With rapid-deformation-rate forging techniques, such as the use of hammers and/or mechanical presses, deformation heating of titanium alloys during the forging process becomes important. Because titanium alloys have relatively poor coefficients of thermal conductivity, temperature nonuniformity may then result, giving rise to nonuniform deformation behavior and/or excursions to temperatures that are undesirable for the alloy and/or final forging mechanical properties. As a result, in the rapid-strain-rate forging of titanium alloys, metal temperatures are often adjusted to account for in-process heat up, or the forging process (sequence of blows, and so on) is controlled to minimize undesirable temperature increases, or both.

Metal Temperatures

Metal temperatures for optimal titanium alloy forging conditions depend on the type of forging equipment to be used, the strain rate to be employed, and the design of the forging. In addition, the working history and forging parameters used in titanium alloy forging have a significant impact on the final microstructure (and therefore the resultant mechanical properties) of the forged alloy—perhaps to a greater extent than in any other commonly forged material. The design of the working process history from ingot to billet to final forging, and particularly the selection of metal temperatures and deformation conditions during the forging process(es), significantly affect the morphology of the allotropic phases which in turn dictate the final mechanical properties and characteristics of the alloys.

Conventional (α–β) forging of titanium alloys, in addition to implying the use of die temperatures of 540 °C (1000 °F) or less, is the term used to describe a forging process in which most or all of the forging deformation is conducted at temperatures below the β transus of the alloy. Alpha-beta forging is typically used to develop optimal strength/ductility combinations and optimal high/low-cycle fatigue properties of α + β alloys.

For conventional subtransus forged α, α + β, and near-β alloys, deformation imparted through one or more forging operations is cumulative. In these alloys, α + β forging does not modify the prior β grain size but modifies other microstructural features, particularly α grain size and morphology, through static and dynamic recrystallization and through residual strain that is exploited in the final thermal treatment. In conventional subtransus forging, billet or bar stock reductions in the course of forging operations greater than 50% are preferred. Subtransus forging processes are usually not employed for β and metastable β alloys.

Beta Forging. Heating the workpiece above the β transus temperature recrystallizes titanium. Thus, the working influences on microstructure are not fully cumulative; with each working-cooling-reheating sequence above the β transus, the effects of the prior working operations are at least partially if not totally lost because of recrystallization.

Beta forging techniques are, however, used to develop microstructures in α, α + β, and near β-alloys characterized by Widmanstätten or acicular primary α morphology in a transformed β matrix. This forging process is typically used to enhance fracture-related properties, such as fracture toughness and fatigue crack propagation resistance, and to enhance the creep resistance of α and α-β alloys. In fact, some α alloys (such as Ti-1100) are designed to be β forged to develop the desired final mechanical properties.

There is often a loss in strength and ductility with β forging as compared to α-β forging of some titanium alloys. Beta forging, particularly of α and α-β alloys, has the advantage of lower forging unit pressures and reduced cracking tendency. However, it must be done under carefully controlled forging process conditions to avoid nonuniform working, excessive grain growth, and/or poorly worked structures, all of which can result in final forgings with unacceptable or widely variant mechanical properties within a given forging or from lot to lot of the same forging.

In commercial practice, β forging techniques typically involve supratransus forging in the early and/or intermediate stages, with controlled amounts of final deformation below the β transus of the alloy. For α, α + β, and near-β alloys, in multiple-step β forging processes, early operations (e.g. preforming) are conducted subtransus. Supratransus working is generally limited to a single operation (e.g., blocking or first finishing), with dies designed to impart a high level of deformation (>50%). The supratransus working is followed by subtransus deformation (e.g. finish forging) in the range of 10 to 25% reduction. The final subtransus reduction is critical to the development of reasonable ductility in β forging processes of α, α + β, and near-β alloys.

For β and metastable β alloys, all forging operations are generally conducted

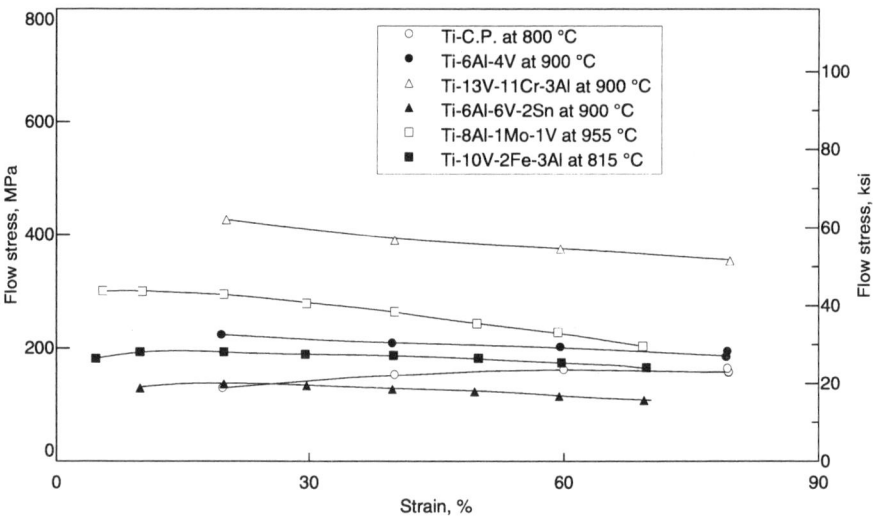

Fig. 3 Flow stress at a high strain rate (10/s) for commonly forged titanium alloys. Although actual forging pressures or unit pressure requirements may significantly exceed the pure flow stress of any given alloy under similar deformation conditions, flow stress information is useful in comparing alloys and process variables. At this rapid strain rate (10/s), the β alloy Ti-13V-11Cr-3Al has the highest flow stress even at a temperature well above the β transus of the alloy; at rapid strain rates, very highly alloyed titanium alloys retard dislocation glide and other mechanisms that influence deformation behavior. The α alloy Ti-8Al-1Mo-1V has the next highest flow stress and is typical of this class of titanium alloy. The α-β alloys Ti-6Al-4V and Ti-6Al-6V-2Sn have intermediate flow stresses at temperature below their β transus, with the more highly β stabilized Ti-6Al-6V-2Sn having lower flow stresses than Ti-6Al-4V. Finally, at a temperature slightly above its β transus, the metastable β alloy Ti-10V-2Fe-3Al has flow stresses lower than those of the α-β alloy Ti-6Al-4V. However, at this high strain rate, the flow stress reduction achieved by deforming β alloys above their β transus is less than the flow stress reduction achieved by deforming α-β alloys above their β transus.

Table 1 Recommended die temperatures for forging titanium alloys

Forging process/equipment	Die temperature °C	°F
Open Die Forging		
Flat Die Forging	150-260	300-500
Ring Rolling	95-260	200-500
Closed-Die Forging		
Hammers	95-260	200-500
Upsetters	150-260	300-500
Mechanical Presses	150-315	300-600
Screw Presses	150-315	300-600
Orbital Forging	150-315	300-600
Spin Forging	95-315	200-600
Roll Forging	95-260	200-500
Hydraulic Presses	315-480	600-900
Hot Die/Isothermal Forging		
Hot Die Forging	650-815	1220-1500
Isothermal Forging	815-985	1500-1800

supratransus, with the temperatures and forging reductions controlled to recrystallize prior β grains and to achieve a final product with a fine grain size, which is essential for satisfactory properties, particularly ductility. For β alloys, supratransus forging reductions of billet or bar stock greater than 50% are generally preferred.

Heating Methods. Titanium is heated for forging using a variety of commercially available furnace equipment, including induction, resistance, electric (radiant), oil, and natural gas. Titanium may be heated to required metal temperatures at any convenient rate. However, titanium has a low coefficient of thermal conductivity. Therefore, commercial forging preheat practices allow 20 to 30 min per 25 mm (per inch) of section thickness (bar, tubular or forging) to ensure that the workpieces reach the desired temperature throughout. Titanium is heated in ambient to oxidizing atmospheres so hydrogen pickup is retarded from furnace products of combustion. Heating titanium forgings in an oxidizing atmosphere creates a brittle surface layer (α case) that must be removed from finished forgings by pickling (or chemical milling) in appropriately constituted and controlled solutions of hydrofluoric and nitric or other acids.

Electric preheating furnaces are preferable to other types since hydrogen pickup is virtually eliminated and temperature control is good. If gas- or oil-fired units are used, the atmosphere should be maintained in as oxidizing a condition as possible; direct flame contact should be strictly avoided, and the time at high temperature limited to the minimum consistent with thorough heating. As a rough guide, a total heating time of one hour per 50 mm of section (0.5 h/in.) should suffice, assuming that the furnace capacity is adequate for the mass of metal involved. If the above precautions are taken, oxidation will not be unduly severe.

Forging Equipment

Titanium forgings are produced on all types of forging equipment, including hammers (standard and counterblow), upsetters, mechanical presses, screw presses, high-energy-rate forming (HERF) machines, hydraulic presses, ring mills, etc. Hammers and small to intermediate (3 to 9000 Mg, or 3 to 10,000 tons) hydraulic presses are preferred for open-die forging of titanium. Hammers, mechanical presses and screw presses are uniquely suited to the manufacture of small to intermediate-size closed-die forgings, frequently to very precise designs. Mechanical and screw presses are uniquely suited to certain types of precision titanium forgings, including turbine engine blades and prosthetic devices. Large closed-die forgings are predominantly manufactured on hydraulic presses. Hot-die and isothermal forging of titanium is restricted to small to intermediate (up to 9000 Mg, or 10,000 tons) hydraulic presses that possess required speed and pressure controls, die heating systems, and programmable press operation. Annular shapes in titanium are best produced on vertical or horizontal ring mills, although rings are manufactured by mandrel forging in open-die presses or hammers. Recommended die temperatures are shown in Table 1.

Forging Dies and Die Materials. Die design, die materials, and die sinking are important parts of the manufacturing and cost of titanium alloy forging products. Open-die and ring-rolling tooling components are constructed (in order of increasing cost) from commercially available ASM 6F or 6G type die steels, hot-work die steels (H12, H13, H19, or others), and high-temperature grades IN 718, etc.). Selection of die material for open-die forging or ring rolling of titanium alloys is predicated upon the alloys being forged, and the forging types and volumes. For closed-die and precision titanium forgings, dies are generally designed using computer-aided design (CAD) and frequently sunk using computer-aided manufacturing (CAM) techniques. Die-sinking techniques vary with the die material employed, but encompass hand and copy milling, computer and direct numerical control (CNC and DNC), electrical discharge machining (EDM), and others. Computer numerical control (CNC) and EDM die-sinking techniques are often integrated with CAM.

For closed-die and some precision forgings, ASM 6F and 6G die steels are widely used. Hot-work die steels (H12, H13, etc.) are employed for high-volume, small to intermediate size closed-die forgings and some precision forgings, and as inserts in larger dies. For hot-die and isothermal forging, special die materials are employed, including Ni- and Co-base superalloys (IN 100, Udimet 700/720, IN 713, and others) and molybdenum alloys (TZM and others) where press systems have necessary protective atmospheres.

Titanium alloys may be very abrasive to forging dies. Die lives are improved with surface treatments such as welding (MIG, TIG, etc.), gas and ion nitriding, carbon nitriding, plasma deposition, and surface alloying.

Workpiece Lubrication. Precoats, proprietary combinations of a variety of ceramic compounds and other additives, are applied prior to furnacing to retard reaction with the furnace atmosphere. They also act as parting agents, as insulators, and/or assist in lubrication. Ceramic precoats are available from commercial sources such as Acheson Colloids, Ferro Corporation, A.O. Smith, J.G. Smith Corporation and others. In closed-die forgings, die lubricants are typically proprietary graphite compounds in water or mineral oil/spirit or petroleum oil carriers with other additives. Die lubricants are available from Acheson Colloids, Wynns, Castrol, Graphite Products, Dylon Industries, and other sources.

Post-Forging Cooling. In commercial practice, post-forging cooling rates for most titanium forging processes are not critical, other than ensuring that the forgings cool uniformly to prevent undue distortion and rapidly enough to avoid undesirable microstructures such as coarse or blocky α. Most titanium forgings are air cooled. Rapid post-forging cooling using

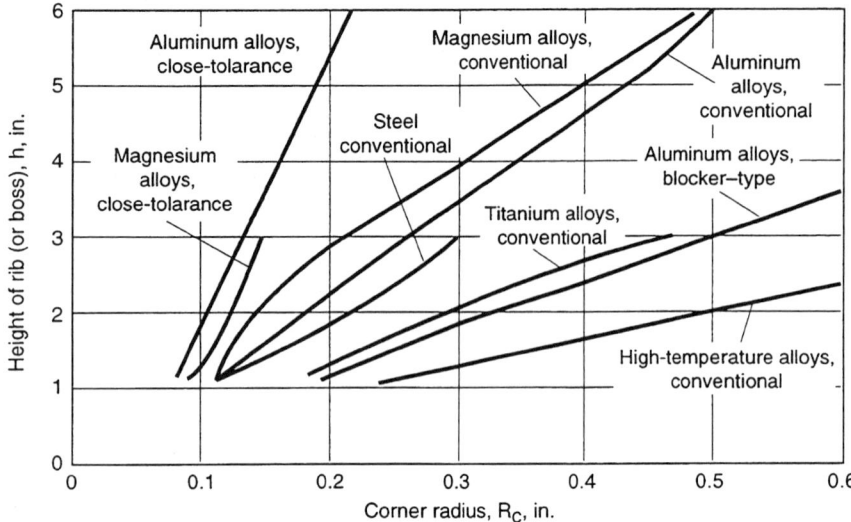

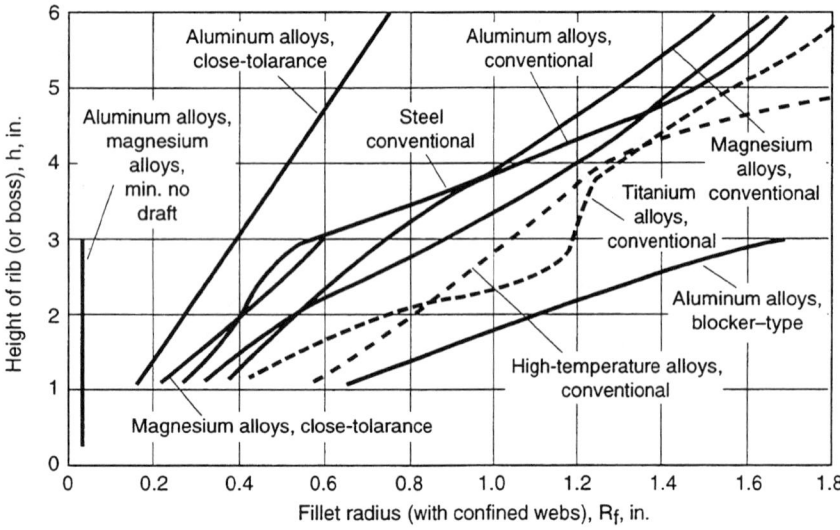

Fig. 4 Corner and fillet radii for various materials recommended by forging users and producers. Source: T. Altan et al., *Forging Equipment, Materials, and Practices*, MCIC HB-03, Battelle, 1973

Ancillary Procedures

Most titanium forgings are manufactured by multiple-step processes consisting of forging stock fabrication, forging stock preparation (cutting), one or more forging operations (frequently several operations in two or more sets of forging dies), intermediate cleaning and repairing processes, final thermal treatments and final cleaning, repairing, and inspection processes. Titanium open-die, ring, closed-die, and some precision forgings may then be machined, either by the forger or the user, to the final component configuration.

Forging Stock. Titanium alloy forging stock (billet, bar, or slab) is prepared by either the metal producer or the forger from ingots. Ingot production requires carefully controlled multiple melting processes. Initial melting may include, for purposes of increased scrap utilization, nonconsumable melting techniques, including the electron-beam or plasma-cold-hearth processes. The latter two melting processes have been found to enhance titanium ingot quality by reducing high- and low-density inclusions. The final melting process(es) is vacuum arc remelting. Forging stock is prepared from ingots by multiple primary hot-working processes (forging, rolling, etc.) that refine macrostructure (prior β grain size) and microstructure to provide material suitable for subsequent forging.

Stock Cutting. Titanium forging stock is cut by all commercially available techniques, including band sawing, abrasive sawing, shearing, and flame cutting.

Cleaning and Repairing. Titanium alloys are crack sensitive in most forging processes, thus intermediate cleaning and repairing processes are necessary to ensure satisfactory forging quality. Cleaning of titanium alloy forgings is a two-step process consisting of scale removal, using mechanical (abrasive blasting) or chemical (molten salt) techniques, followed by pickling in HF-HNO_3 or other acid solutions. Surface cracks are repaired or removed by abrasive grinding or other techniques. Similar processes are applied to finishing forgings prior to shipment.

Heat Treatment. Most titanium alloy forgings are thermally treated after forging, with heat treatment processes ranging from simple stress-relief annealing to multiple-step processes of solution treating, quenching, aging, and/or annealing designed to modify the microstructure of the alloy to meet specific mechanical property criteria. The furnaces used to heat treat titanium alloy forgings use the same general types of commercially available furnace equipment described in the previous section "Heating Methods." In addition, such furnaces may have water, oil, or other quenching media capabilities and are controlled to meet stringent temperature uniformity requirements of military and/or federal specifications.

Nondestructive Evaluation. Most titanium forgings are NDE inspected for internal quality using ultrasonic techniques. Because the forging process generally does not introduce internal discontinuities (such as those usually incurred in ingot casting processes), preferred practice is to perform ultrasonic inspection on forging stock, particularly for closed-die and precision forging products. The simple geometric shapes of bar, billet, and slab permit detailed and high-resolution ultrasonic evaluation of the internal quality that is then carried through the forging process into the final product. Ultrasonic inspection may also be performed on final forgings. The surface quality and highly configured shapes of closed-die and precision forgings of airframe components

quenching techniques (water, oil, etc.) may be employed for both conventional and β forging processes when fine transformation products and/or reduced grain boundary α are preferred microstructural features.

may preclude reliable inspection. For jet engine components, the finish forgings are machined to inspectable configuration. Air frame titanium forgings are frequently ultrasonic inspected after initial machining processes. In the case of precision forgings, surface quality is also evaluated using sensitive liquid-penetrant techniques.

Machining. Titanium alloy forgings can be machined by all commercially available traditional and nontraditional (chemical) techniques. Precision forgings (and, in some cases, other forging products) are frequently chemically milled, machined, and/or surface treated (by conversion processes, painting, etc.) prior to shipment as finished components.

The design of dies for the manufacture of forgings of aluminum alloys are a basis for basic die design for other materials. However, the exact design parameters are very dependent on specific material, determined largely by its deformation resistance and forgeability. Therefore, in this section, design guidelines for titanium alloys are summarized briefly. Additional information on these aspects is contained in *Forging Handbook* (T.G. Byrer, Ed., Forging Industry Association, 1985).

As for aluminum forgings, basic design considerations include specification of the following:

- Parting line
- Allowance for machining (forging envelope)
- Draft, corner, and fillet radii, and minimum section thicknesses and maximum rib heights
- Die closure/thickness tolerances
- Length and width/die wear tolerances
- Match/mismatch tolerances

Parting line should be selected to allow ease of part removal following forging. The parting line location and accompanying flash design affect overall metal flow and thus the ability to forge parts successfully. The choice of parting line may have important economic impacts as it affects the ease of machining, die material losses, and possible die breakage.

Finish Allowance. Because of oxidation, titanium alloys require large forging envelopes so that surface defects and other metal flow irregularities can be removed. Finish allowances also generally increase with increasing forging size because of longer heating times, added operations, and a greater chance for the introduction of defects during handling.

Draft. Forging projections are typically tapered to allow easy part removal from the die cavity. The most common draft angles are between 5 and 7° for conventional steel forgings. Titanium and nickel-base forgings generally require 7° or greater drafts. Ejectors often permit the use of lower draft angles (2 to 4° for steels).

Corner and Fillet Radii. Proper selection of corner and fillet radii is critical in avoiding metal flow problems such as laps, cold shuts, and flow-through defects in structural rib-web forgings and other parts with deep cavities.

As with draft angles, exact values of corner and fillet radii cannot be quoted, as

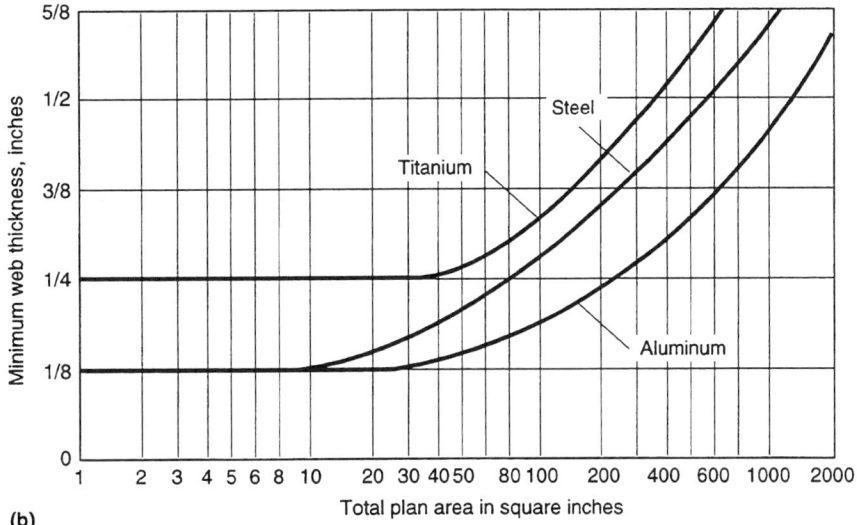

Fig. 5 Relation of minimum web thickness to total plan area for three materials, showing how the materials affect web thickness and total plan area.

Table 2 Minimum web thickness

Total plan area(a), in.²	Conventional forgings (7°, 5°, 3°)		Blocker forgings	
	Carbon and alloy steels	Titanium, Inconel(b)	Carbon and alloy steels	Titanium, Inconel(b)
To 10	0.13	0.20	...	...
30	0.16	0.25	0.25	0.38
60	0.20	0.32	0.28	0.43
100	0.25	0.40	0.31	0.47
200	0.32	0.48	0.38	0.56
300	0.37	0.58	0.45	0.64
500	0.44	0.70	0.50	0.75
800	0.50	0.80	0.56	0.85
1200	0.56	0.90	0.62	1.00
1600	0.62	1.00	0.69	1.13
2000	0.70	1.13	0.75	1.25
2500	0.80	1.25	0.81	1.38
3000	0.88	...	0.88	1.50
3500	1.13	...	1.19	...
4000	1.38	...	1.38	...
5000	2.25	...	2.50	...

(a) When required plan area falls between those listed, determine web thickness by interpolating in increments of 0.01 in. Specify thickness to two decimal places. (b) Also nickel- and cobalt-base superalloys. Source: *Forging Handbook*, T.G. Byrer, Ed., Forging Industry Association, 1985, p 76

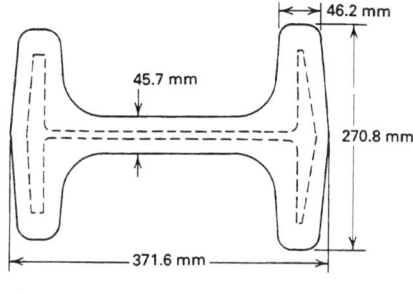

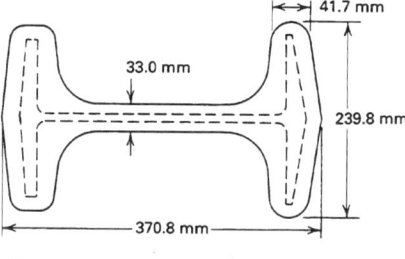

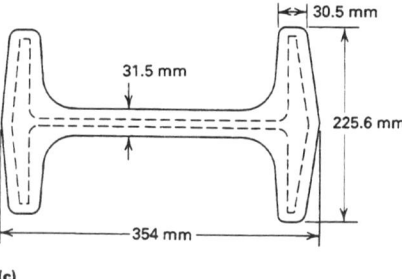

Fig. 6 Cross sections of Boeing 757 part illustrating design and tolerance criteria for the 272 kg (600 lb) machined weight forging obtained from three closed-die forging methods, along with their respective forging weights. (a) Blocker type, 1364 kg (3007 lb). (b) Conventional, 1087 kg (2397 lb). (c) High definition, 879 kg (1937 lb)

Table 3 Net titanium alloy precision forging design/tolerance criteria for selected parts and processes for metastable β and α–β alloys

Feature	Current	Goal
PVA, m² (in.²)	Up to 0.193 (300)	0.290 (450)
Length, mm (in.)	Up to 1015 (40)	1525 (60)
Length/thickness tolerance, mm (in.)	+0.5, −0.25 (+0.020, −0.010)	+0.75, −0.25 (+0.030, −0.010)
Contour tolerance, mm (in.)	±0.38 (±0.015)	±0.63 (±0.025)
Draft		
Outside	0°; +30 min, −0°	Same
Inside	1°; +30 min, −1°	Same
Corner radii, mm (in.)	1.5; +0.75, −1.5 (0.060; +0.030, −0.060)	Same
Fillet radii, mm (in.)	3.3; +0.75, −1.5 (0.130; +0.030, −0.060)	Same
Straight within, mm (in.)	0.25 each 254 mm (0.010 each 10 in.)	Same
Minimum web thickness, mm (in.)	2.3 (0.090)(a)	2.5 (0.100)
Minimum rib thickness, mm (in.)	2.3 (0.090)(a)	2.5 (0.100)

(a) In some designs and under some processing conditions, minimum web thickness can be as thin as 1.5 mm (0.060 in.) and minimum rib thickness can be as thin as 2.0 mm (0.080 in.).

they depend on part design and material. However, fillet radii are generally about twice as large as corner radii (see Fig. 4). Corner and fillet radii generally increase with the height of ribs or bosses in a forging geometry to enhance metal flow and to avoid defects.

Minimum Section Size and Maximum Rib Height. Minimum web and rib thicknesses that can be achieved by conventional hot forging depend on the minimum plan dimension (see Fig. 5 and Table 2). The maximum achievable rib height decreases as the rib thickness decreases.

Tolerances

For a given die design, tolerances determine when forgings must be rejected or when dies must be removed from service. By far the greatest tonnage of conventionally forged titanium alloys is produced in closed dies. Closed-die titanium alloy forgings can be classified similarly to other materials, such as aluminum, as blocker-type (achieved with single set of dies or block/finish dies), conventional (achieved with two or more sets of dies), high-definition (also requiring two or more sets of dies), and precision forgings (frequently employing hot-die isothermal forging techniques). Precision titanium alloy forgings are discussed below. Blocker-type titanium alloy forgings are typically produced in relatively less expensive dies, with design and tolerance criteria between those of open-die and conventional forgings. Conventional closed-die titanium forgings cost more than blocker-type, but the increase in cost is usually justified because of reduced machining costs. Finally, high-definition titanium alloy forgings are also more costly than conventional forging, but may also be justified by reduced machining. Preforming using open-die upsetting, and/or roll forging frequently precedes all types of titanium alloy closed-die forging processes.

In comparison with aluminum alloy closed-die forgings, all types of closed-die forgings in titanium alloys are typically produced to more generous design and/or tolerance criteria, reflecting the increased difficulty in forging these alloys. The example below describes a large main landing gear beam forging produced in the α-β alloy Ti-6Al-4V.

Example: A relatively high-volume main-landing gear beam has been fabricated with a progression of closed-die forging designs in an effort to reduce the overall cost of the final machined part. One cross section from this forging and the three types of closed-die forging approaches used to manufacture this part is shown (see Fig. 6). The original blocker-type configuration (designed prior to fi-

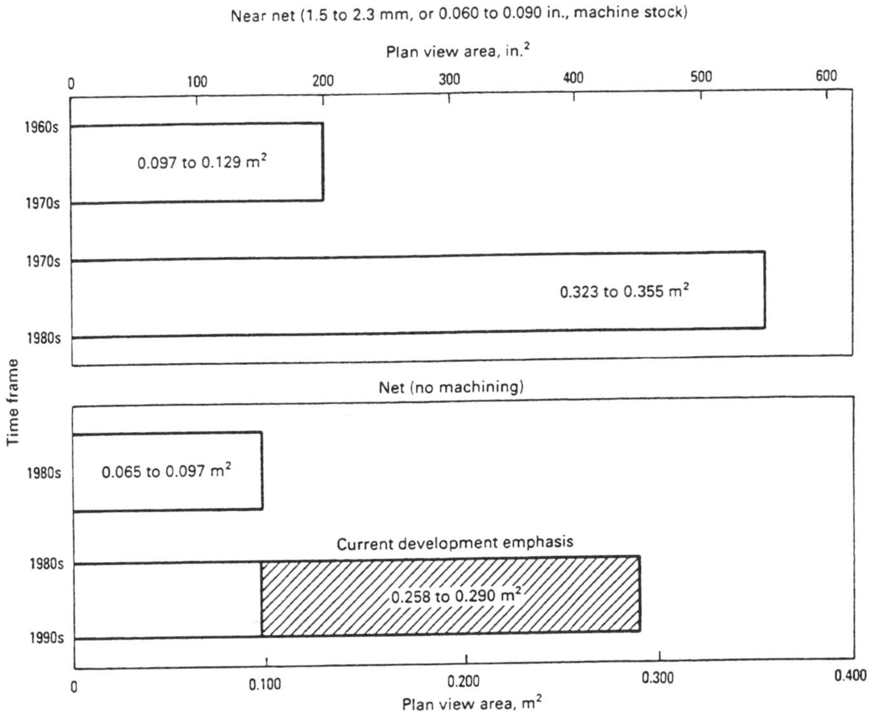

Fig. 7 Past and future near-net and net titanium alloy precision forging capabilities gaged in terms of plan view area. This figure differentiates between net and near-net precision titanium alloy forging because not all titanium alloys are equally producible under either conventional or hot-die methods.

nalization of the machined part) produced in two sets of dies is shown. As a blocker-type part, the forging weighed 1364 kg (3007 lb) versus a machined part weight of 272 kg (600 lb) for an overall recovery from the raw forgings of 20% (or a buy-to-fly ratio of 5 to 1). When the final machine part geometry had been better defined, the part was redesigned to a conventional forging (part b of figure) weighing 1087 kg (2397 lb), increasing the recovery from the raw forging of 25% (buy-to-fly of 4 to 1). Sufficient machining and metal cost savings were realized through this redesign to justify the costs of construction of a new set of dies. Finally, after some additional final machined part refinements, the part was redesigned to a high-definition shape (part c of figure), reducing the as-forged weight to 879 kg (1937 lb) and increasing the overall recovery to 31% (buy-to-fly of 3.3 to 1). Again, a cost savings was realized that justified the construction of new dies. Therefore, from blocker-type to close tolerance, the as-forged weight was reduced by nearly 500 kg (1100 lb), and the forged part/machined part recovery was increased by 11%—a significant cost savings.

Precision Forgings

As with aluminum alloys, titanium alloy precision forgings can be identified by a variety of terminologies; however, in each case, this product form requires significantly reduced machining on the part of the user. Precision forged titanium alloys are a significant commercial forging product that is undergoing major growth in usage and has been the subject of major forging process technology development and capital investment by the forging industry. In many instances, the term net precision titanium forging will be defined as a product that requires no subsequent machining by the user, and the term near-net precision titanium forging will be defined as a product requiring some metal removal (typically accomplished in a single machining operation) by the user. The viability of fabrication of net or near-net titanium alloy precision forgings is determined by the alloy being forged and by value analysis for fabrication of the most cost-effective precision forged product.

The first precision forged titanium alloy products commercially produced were turbine engine compressor and fan blades; conventional forging process techniques were used. With hot-die/isothermal forging techniques, very complex cross-section precision forged airframe components are being manufactured. Titanium alloy precision forgings are produced with very thin webs and ribs; sharp corner and fillet radii; undercuts, backdraft, and/or contours; and, frequently, multiple parting planes (which may optimize grain flow characteristics) in the same manner as aluminum alloy precision forgings.

Design Criteria. The design and tolerance criteria for precision titanium forgings are similar to those for aluminum alloy precision forgings and have been established to provide a finished product suitable for assembly or subsequent fabrication by the user. Precision titanium alloy forgings, with the exception of airfoils, do not necessarily conform to the same tolerances provided by machining of other product forms; however, design and tolerance criteria for titanium precision forgings are highly refined in comparison to other titanium alloy forging types and are suitable for the intended application of the product (see Table 3). If the standard precision forging design and tolerance criteria are not sufficient for the final component, then the forging producer frequently combines conventional and/or hot-die/isothermal forging with machining to achieve the most cost-effective method of fabrication to the required tolerances on the finished part.

The titanium precision forging design and tolerance criteria achievable may vary with the alloy type because all titanium alloys are not necessarily equivalent in workability using either conventional forging techniques or hot-die/isothermal forging technology. Generally, the net titanium precision forging design parameters given (see Table 3) apply to more readily workable β and metastable β alloys (such as Ti-10V-2Fe-3Al) and selected designs and forging processes for α-β alloys (such as Ti-6Al-4V and Ti-6Al-6V-2Sn). However, with more difficult-to-fabricate α titanium alloys and certain forging designs and/or forging processes for α-β alloys, the most cost effective forging technique may be near-net titanium precision forgings with modified design criteria (for example, typically 1.5 to 2.3 mm, or 0.060 to 0.090 in., machining allowance per surface), and modified rib/web thickness, fillet radii, corner radii, and so on but with the same dimensional tolerances outlined in the table. The table also indicates that as the size of the net titanium precision forging is increased to 0.290 m^2 (450 in.^{2}), some modification in design and tolerance criteria is appropriate.

Tooling and Design. Precision titanium forging uses several tooling concepts to achieve the desired design shape, with the specific tooling concept based on the design features of the precision forging and the forging process used. Similar tooling design concepts for aluminum alloys are also used with titanium alloys. For conventional forging processes for titanium precision forgings, of which tur-

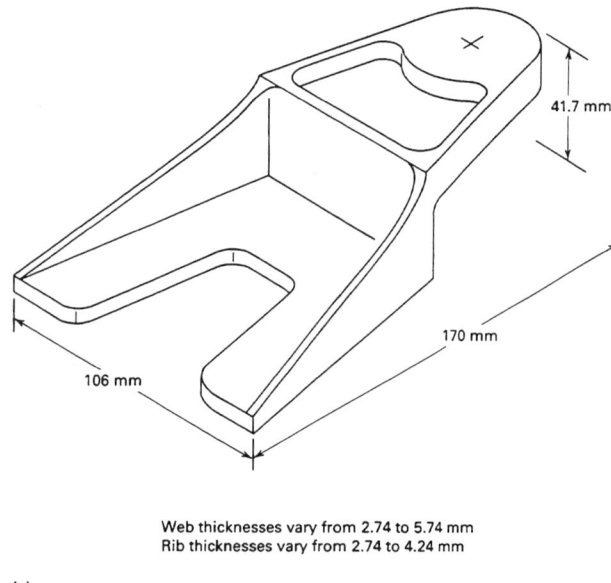
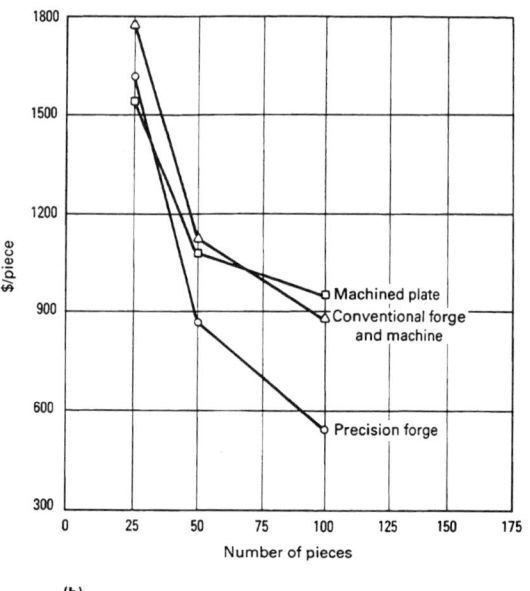

Fig. 8 Cost comparison for an engine mount part. (a) Net-shape precision forged Ti-10V-2Fe-3Al engine mount produced by hot-die/isothermal forging. (b) Cost compression of the engine mount shown to illustrate the cost-effectiveness of precision forging. This figure shows a cost comparison for an engine mount manufactured by 1) machining from Ti-6Al-4V plate, 2) machining from Ti-6Al-4V conventional forging, and 3) precision forged Ti-10V-2Fe-3Al.

bine airfoils are a primary example, a two-piece upper and lower die concept is the predominant approach.

For conventional titanium precision forgings, the die materials employed in tooling are either 6F2 or 6G types, or hotwork die materials such as H12 and H13. Tooling for conventional titanium precision forgings is designed and produced using the same techniques as those described above for other forging types; however, CNC direct die sinking and/or EDM electrode manufacture from CAD forging and tooling databases has been found to be particularly effective for the manufacture of the close-tolerance tooling demanded by precision titanium forgings.

The die materials used for the hot-die/isothermal forging of titanium alloys are based on the alloy to be forged, necessary forging process conditions (for example, metal/die temperatures, die stresses, strain rate, and total deformation), forging part design, and cost considerations. Cast, wrought, and/or consolidated powder techniques are used to fabricate die blocks/inserts from superalloy materials, including Alloy 718, Waspaloy, Udimet 700, Astroloy, Alloy 713LC (Ni-12Cr-6Al-4.5Mo-2Nb-0.6Ti-0.1Zr-0.05C-0.01B), and Alloy 100 (Ni-15.0Co-10.0Cr-5.5Al-4.7Ti-3.0Mo-1.0V-0.6Fe-0.15C-0.06Zr-0.015B), with these materials listed in order of increasing temperature capability from 650 to 980 °C (1200 to 1800 °F). Most of these die materials require more expensive nonconventional machining techniques for die sinking, with electrode discharge machining being the most prevalent technique. Computer-aided design part and tooling databases have also been effectively combined with CAM-driven CNC EDM electrode manufacturing techniques to reduce the cost of die manufacture. Typically, the manufacture of a set of dies for titanium precision forging with hot-die/isothermal forging costs up to seven times that required for the dies for the manufacture of the same part in aluminum. Heated holder and insert techniques can reduce the cost factor for titanium hot-die/isothermal precision forging dies to three times the cost of the same dies for an aluminum alloy.

Forging Processing. Conventional and hot-die/isothermal forging processes for precision titanium forgings use the same steps as those outlined above for other forging types. Precision titanium forgings can be produced from wrought stock, preformed shapes, or blocker shapes, depending on the complexity of the part, the tooling system being employed, and cost considerations. For example, for the conventional forging of airfoil shapes such as blades, multiple forging processes are used (because of the high cost of raw materials) to prepare the preshape necessary for the successful fabrication of the precision part in order to conserve input material and to facilitate the precision forging process. Precision titanium forging stock fabrication and inspection criteria are similar to those described above for other titanium alloy forging types.

Unlike aluminum alloy precision forging shapes, conventionally forged titanium alloy precision forgings are usually not produced in multiple operations in finish dies, but rather by a progression of processes in multiple die sets. However, with hot-die/isothermal forging processes for precision titanium parts, multiple operations in a given die set are used. Conventionally forged titanium precision forgings are usually produced on mechanical and/or screw presses, although hammers or hydraulic presses are occasionally used for certain designs. For hot-die/isothermally fabricated precision titanium forgings, hydraulic presses are used exclusively to obtain the desired slow strain rates and controlled deformation conditions. The mechanical and/or screw presses currently used for the fabrication of conventional titanium precision forgings range up to 150 MN (17 000 tonf) (maximum press capability of up to 280 MN, or 31 000 tonf, for the largest screw press), and hydraulic presses for the hot-die/isothermal precision forging processing of titanium alloys range up to 90 MN (10 000 tonf). Other large hydraulic presses, up to 310 MN (35 000 tonf), with necessary forging process capabilities are available for the hot-die isothermal forging of titanium (as well as aluminum alloy precision forging) as this titanium alloy precision forging technology is scaled-up in size.

Conventional and hot-die/isothermal forging process criteria for the precision forging of titanium alloys are similar to those described above for the titanium alloy forging types. With conventional forging, the metal and die temperatures used are usually controlled to be near the upper limits of recommended forging temperature ranges to enhance producibility and to minimize unit pressures. Die temperature selection in hot-die/isothermal forging is based on the alloy, die material/die heating system, specific forging process demands (for example, the viability of near-isothermal/hot die versus isothermal conditions), sophistication of the forging design, and thermomechanical processing criteria.

Because of the stringent dimensional tolerances associated with conventionally and hot-die/isothermally forged titanium precision forgings, dies are typically heated using state-of-the-art on press heating systems, such as resistance and/or induction heating. These heating systems maintain uniform die temperatures, typically ±15 °C (±25 °F) or better, in order to reduce dimensional variations. As with other forging types, precoating and die lubrication are critical elements in the forging of titanium precision forgings, and the precoats and die lubricants used are similar to those for other forging types, although lubricant materials are often specially formulated for an individual forging design and forging process. Insulative blankets are generally not used for the forging of precision titanium forgings, because such materials may adversely affect the dimensional integrity of the forged parts. The precoats used in the hot-die/isothermal forging of titanium alloys are selected or formulated for specific metal/die temperature conditions. Under some conditions, parting agents such as boron nitride are used on the dies to facilitate part removal with minimum distortion.

Straightening is often a critical process in the manufacture of conventionally or hot-die/isothermally forged titanium precision forgings. The straightening techniques used, with airfoils as a critical example, are predominantly die straightening procedures with the metal and dies at elevated temperatures. In this process, time-temperature-pressure parameters are controlled, usually with small-to-intermediate size hydraulic presses, to achieve the desired deformation conditions and therefore the dimensional conformance. Hot-die or isothermal forming techniques (with dies at temperatures from 700 to 925 °C, or 1300 or 1700 °F) are often used to straighten conventionally or hot-die/isothermally forged titanium alloy precision forgings, particularly large airfoil shapes.

Forging stock preparation, thermal treatments, in-process cleaning, trimming, and repair, and in-process and final inspection and thermal treatment verification processes, with the exception of nondestructive evaluation, are the same as those for other titanium alloy forging types. Because of the highly configured nature and thin sections typical of precision titanium parts, ultrasonic inspection cannot be used on finished parts; the exception is turbine engine disks, which are usually inspected using highly sophisticated, automated ultrasonic inspection equipment. Frequently, for airframe precision titanium forgings, airfoils, and other precision titanium shapes, the detailed ultrasonic inspection performed on the forging stock before fabrication is sufficient to ensure satisfactory internal quality in the final part. Unlike other titanium alloy forging types, precision titanium forgings, which are used in service with most (if not all) of the as-forged surfaces intact, are frequently inspected by sensitive liquid penetrant inspection techniques to ensure adequate surface quality.

Precision titanium forgings are frequently supplied as a completely finished product that is ready for assembly by the user. In such cases, the forging producer can use both conventional milling and unconventional machining techniques, such as chemical milling and electrode discharge machining, along with forging, to achieve the most cost-effective finished titanium shape. Further, the forging producer can apply a wide variety of surface finish and/or coating processes to this product as specified by the purchaser.

Technology Development Effectiveness. As a result of both conventional and hot-die/isothermal forging technology developmental efforts, the size of the net titanium precision forging that can be fabricated to the previously listed design and tolerance criteria (see Table 3) has tripled—from 0.081 m^2 (125 in.^{2}) to over 0.194 m^2 (300 in.^{2}) in terms of plan view area (see Fig. 7). The critical elements in projected changes in the state-of-the-art for titanium precision forgings, both in terms of size and cost effectiveness, are enhanced precision forging process control, CAD/CAM/CAE technologies, advanced and/or integrated manufacturing technologies, enhanced die heating systems, improved lubrication systems, and the availability of large superalloy die blocks necessary for the hot-die/isothermal forging of these alloys.

The selection of precision titanium forgings from the various methods available for achieving a final titanium shape is based on the value analyses conducted for each individual shape in question (see Fig. 8 showing a cost comparison of a precision forging in Ti-10V-2Fe-3Al using hot-die/isothermal forging). Analyses of other parts have also shown that titanium precision forged shapes are highly cost effective in comparison with other fabrication approaches, particularly when the other methods require multiple-axis machining techniques to achieve the final part geometry.

Forging industry and user evaluations of precision titanium alloy forgings have indicated that final part costs can be reduced by 80 to 90% or more in comparison to machined plate, and by 60 to 70% or more in comparison to machined conventional forgings. With potential cost reductions such as these, it is evident that further growth in precision titanium forging usage can be anticipated.

Selected References

- A.M. Sabroff, F.W. Boulger, and H.J. Henning, *Forging Materials and Practices*, Reinhold, 1968
- T.G. Byrer, Ed., *Forging Handbook*, Forging Industry Association and American Society for Metals, 1985, p 69-78
- "Approval and Control of Premium-Quality Titanium Alloys," AMS 2380, Aerospace Material Specification
- G.W. Kuhlman, Forging of Titanium Alloys, *Metals Handbook*, 9th ed, Vol 14, 1988

Technical Note 5: Forming

Adapted from "Forming of Titanium and Titanium Alloys," *Metals Handbook*, Ninth Edition, Volume 14, *Forming and Forging*

Titanium is more difficult to form than steel and aluminum alloys, and titanium alloys generally have less predictable forming characteristics than steel and aluminum alloys. In particular, the springback from room-temperature forming of titanium alloys is not easily predictable. The wide variations in yield strength among different heats magnified by a low modulus of elasticity can give a wide spread in springback angle, especially if the bend angle of the part is fixed by the forming tool. Therefore, to reduce the effect of springback variation a majority of formed titanium parts are made by hot forming or by cold preforming and then hot sizing.

Other characteristics adversely affecting the formability of titanium alloys include:

- Notch sensitivity, which may cause cracking and tearing, especially in cold forming
- Galling (more severe than with stainless steel)
- Relatively poor ability to shrink (a disadvantage in some flanging operations)
- Potential embrittlement from overheating and from absorption of gases, principally oxygen (scale and the surface layer adversely affected by the diffusion of oxygen can be removed readily)
- Limited workability
- Tendency toward nonuniformity in sheet

All titanium alloys also resist sudden movement i.e., are strain rate sensitive; therefore, it is recommended that equipment with the capability of controlling the rate of load application be used for stretching and pressing operations. The slower the forming speed, the better the formability at room temperature. At elevated temperatures some titanium alloys, like Ti-6Al-4V, have better formabilities at higher forming speeds while others such as 13V-11Cr-3Al exhibit less ductility at higher forming speeds. From an economic viewpoint, faster speeds may be necessary, and even tolerable, if larger radii can be accommodated in the design. The formability of titanium is poor in operations characterized by shrink flanges such as found in rubber press forming. Consequently areas that require gathering of material should be minimized when designing parts.

If these limitations are recognized and established guidelines for hot and cold forming are followed, titanium and titanium alloys can be successfully formed into complex parts. Titanium and its alloys can be formed to standard machine tolerances similar to those obtained in the forming of stainless steel. Very complex shapes can be formed to tight tolerances by superplastic forming, discussed in detail in the next Technical Note in this Volume.

Formability Ratings. The formability of materials can be ranked approximately based on yield strength/tensile strength ratios, and ductility. General rankings cannot predict whether or not a particular shape can be formed, but formability ratings for various materials at varying temperatures in different processes have been based on the material parameters and types of forming failures experienced (see Table 1). Relative for-

Table 1 Types of failures in sheet-forming processes and material parameters controlling deformation limits
The parameters can be determined in tensile and compressive tests. However, single values of conventional mechanical properties only test a small portion of sheet and do not assess uniformity.

Process	Cause of failure Splitting	Buckling	Ductility parameter(a)	Buckling parameters(b)
Brake forming	x		ε in 0.25 in.(c)	
Dimpling	x		ε in 2.0 in.(d)	
Beading				
Drop hammer	x		ε in 0.5 in.(c)	
Rubber press	x		$(\varepsilon$ in 2.0 in.$)(S_u)$	
Sheet stretching	x		ε in 2.0 in.	
Joggling	x	x	ε in 0.02 in.	E_c/S_{cy}
Liner stretching	x	x	ε in 2.0 in.(e)	E_t/S_{ty}
Trapped rubber, stretching	x	x	ε in 2.0 in.(f)	E_t/S_{ty}
Trapped rubber, shrinking		x		E_c/S_{cy} and $1/S_{cy}$
Roll forming		x		E_t/S_{ty}(g) and E_c/S_{cy}(h)
Spinning		x		E_c/S_{cy} and E_t/S_u
Deep drawing		x		E_c/S_{cy} and S_{ty}/S_{cy}

(a) ε indicates natural or logarithmic strain; the dimensions indicate the distance over which it should be measured. (b) E_c = modulus in compression; E_t = modulus in tension; S_{cy} = compressive yield strength; S_{ty} = tensile yield strength; S_u = ultimate tensile strength. (c) Corrected for lateral contraction. (d) For a standard 40-degree dimple. (e) The correlation varies with sheet thickness. (f) The correlation is independent of sheet thickness. (g) For roll forming heel-in sections. (h) For roll forming heel-out sections. Source: R.A. Wood and R.J. Favor, *Titanium Alloys Handbook*, MCIC-HB-02, Battelle, 1972

mability of some titanium alloys for four common aircraft forming operations is also shown (Table 2).

The Bauschinger Effect. In all forming operations, titanium and its alloys are susceptible to the Bauschinger effect—a drop in compressive yield strength subsequent to tensile straining in the same or another direction. The Bauschinger effect, unlike the strain-hardening behavior observed in other metals, involves stress-strain asymmetry that results in hysteresis loops in the metal's stress-strain behavior.

The Bauschinger effect is most pronounced at room temperature; plastic deformation (1 to 5% tensile elongation) at room temperature always introduces a significant loss in compressive yield strength (see Fig. 1), regardless of the initial heat treatment or strength of the alloys. At 2% tensile strain, for example, the compressive yield strength of Ti-6Al-4V drops to less than one-half the value for solution-treated material. Increasing the deformation temperature reduces the Bauschinger effect; subsequent full thermal stress relieving completely removes it.

Temperatures as low as the aging temperature will remove most of the Bauschinger effect in solution-treated titanium alloys. Heating or plastic deformation at temperatures above the normal aging temperature for solution-treated Ti-6Al-4V may eliminate the Bauschinger effect, but may, depending on temperature and time, cause overaging.

Preparation of Sheet for Forming

Before titanium sheet is formed, it should be inspected for flatness, uniformity, and thickness. Some plants test incoming material for hardness, strength, and bending behavior. Critical regions of titanium sheet should not be nicked,

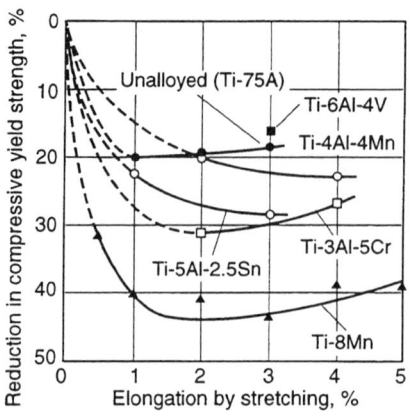

Fig. 1 Effect of cold stretching on compressive yield strengths.
Source: R.A. Wood and R.J. Favor, *Titanium Alloys Handbook*, MCIC-HB-02, Battelle, 1972

Table 2 Relative formability of four selected titanium alloys

Alloy	Condition	Brake press: Minimum bend radius at room temperature	Drop hammer: Maximum stretch at 450-510°C (850-950 °F)	Hydropress (trapped rubber) Stretch maximum at 315-370 °C (600-700 °F)	Hydropress (trapped rubber) Shrink maximum at 315-370 °C (600-700 °F)	Joggle (runout joggle-depth ratio) at: Room temperature	Joggle (runout joggle-depth ratio) at: 315-370 °C (600-700 °F)	Maximum stretch wrap at room temperature	Maximum skin stretch at 450-810 °C (850-950 °F)
Ti-13V-11Cr-3Al	Solution treated	1.5T	16%	10%	6%	1.25	1	5.5%	13.5%
Ti-8Mn	Annealed	3T	16%	7.5%	5%	4	3	8%	18%
Ti-5Al-2.5Sn	Annealed	3.5T	13%	5%	3%	4	4.5	8%	12.5%
Ti-6Al-4V	Annealed	4.5T	13%	5%	4%	4.5	3	3.5%	17%

Source: R.A. Wood and R.J. Favor, *Titanium Alloys Handbook*, MCIC-HB-02, Battelle, 1972

Table 3 Tool materials and lubricants used for forming titanium alloys

Operation(s)	Tool materials	Lubricants
Cold forming		
Press forming, drawing, drop hammer forming	Cast zinc die or punch with stainless steel caps	Graphite suspension in a suitable solvent
Press-brake forming	4340 steel (36–40 HRC)	Graphite suspension in a suitable solvent
Contour roll forming, three-roll forming	AISI O2 tool steel	SAE 60 oil
Stretch forming	Epoxy-faced cast aluminum, cast zinc, cast bronze	Grease-oil mixtures, wax; 10:1 wax-graphite mixture
Hot forming		
Press forming, drawing, drop hammer forming	High-silicon cast iron, stainless steels, heat-resistant alloys	Graphite suspension, boron nitride
Sizing	Low-carbon steel, high-silicon gray or ductile iron, AISI H13 tool steel, stainless steels, heat-resistant alloys	Graphite suspension, boron nitride
Press-brake forming	AISI H11 or H13 tool steel, heat-resistant alloys	Graphite suspension, boron nitride
Contour roll forming, three-roll forming	AISI H11 or H13 tool steel	Graphite suspension, boron nitride
Stretch forming	Cast ceramics, AISI H11 or H13 tool steel, high-silicon gray iron	Graphite suspension, 10:1 wax-graphite mixture, boron nitride
Superplastic forming	Ceramics, 22-4-9 stainless steel, 49M heat-resistant steel	Boron nitride

scratched, or marred by tool or grinding marks, because the metal is notch sensitive. All scratches deeper than the finish produced by 180-grit emery should be removed by sanding the surface. Edges of the workpieces should be smooth, and scratches, if any, should be parallel to the edge of the blank to prevent any concentration of stress that could cause the workpiece to break. To prevent difficulty in forming, as by increased notch sensitivity, surface oxide or scale should be removed before forming.

Cleaning. Grease, oil, stencils, fingerprints, dirt, and all chemicals or residues that contain halogen compounds must be removed from titanium before any heating operation. Salt residues on the surface of the workpiece can cause hot-salt cracking in service or in heat treating; even the salt from a fingerprint can cause problems. Therefore, titanium is often handled with clean cotton gloves after cleaning and before hot forming, hot sizing, or heat treatment.

Ordinary cleaners and solvents such as isopropyl alcohol and acetone are used on titanium. Halogen compounds, such as trichlorethylene, should not be used, unless the titanium is pickled in acid after cleaning. Titanium that has been straightened or formed with tools made of lead or low-melting alloy should be cleaned in nitric acid. (These tool materials are not recommended for forming titanium).

Removal of Tool Marks. Tool and grinding marks in titanium can be softened in an aqueous acid bath containing (by volume) 30% concentrated nitric acid and not more than 3% hydrofluoric acid. Failure to keep the ratio of nitric to hydrofluoric acid at 10 to 1 or greater (to suppress the formation of hydrogen gas during pickling), or the use of any pickling bath that produces hydrogen, can result in excessive hydrogen pickup. The acid bath should remove 0.025 to 0.075 mm (0.001 to 0.003 in.) of thickness from each surface to eliminate the marks made by abrasives. Titanium should be washed or cleaned before it is immersed in acid.

Removal of Scale. Heavy gray and black scale and similar hard oxides that form on titanium at temperatures of 540 °C (1000 °F) and higher can be removed chemically or by wet or dry mechanical methods that use fine abrasives. Wire brushing and coarse abrasives are generally not used, because they can leave stress-raising marks; if these techniques are used, the damaged surface layer can be removed by pickling in nitric-hydrofluoric acid, as described above.

Thin oxides that form at temperatures below 540 °C (1000 °F) can be removed by acid pickling. Very tenacious oxides may require gritblasting prior to pickling.

Tool Materials and Lubricants

Tool materials for forming titanium are chosen to suit the forming operation (see Table 3), forming temperature, and expected quantity of production. The cost of tool material is generally only a small fraction of the cost of tools, unless forming temperature is such that heat-resistant alloy tooling is required.

Cold forming can be done with epoxy-faced aluminum or zinc tools. Hot-forming tools are fabricated from ceramic, cast iron, tool steel, stainless steel, and nickel-base alloys.

Tool materials for the superplastic forming of titanium alloys are a special case (see the separate Technical Note "Superplastic Forming" in this Volume). They must be able to withstand the high temperatures (870 to 925 °C, or 1600 to 1700 °F) required for superplastic forming, but must not contain more than about 6% Ni, because of the possibility of nickel migration into the work metal at superplastic forming temperatures. Cast ceramics, 22-4-9 stainless steel (Fe-0.5C-22Cr-9Mn-4Ni), and 49C steel are used for this purpose.

Lubricants. Galling is the most severe problem to be overcome in hot forming. Lubricants may react unfavorably with titanium when it is heated, although molybdenum disulfide suspended in a volatile carrier, colloidal graphite, and graphite-molybdenum disulfide mixtures have been successfully used. Boron nitride slurries also are used. If the lubricant reacts with oxidation products to produce a tenacious surface coating, it must be removed by sandblasting with garnet grit or 120-mesh aluminum oxide, followed by acid pickling.

Boron nitride is the preferred temperature-resistant lubricant because of its higher lubricity, as well as ease of application and removal. Other lubricants used for hot forming have a graphite or molybdenum disulfide base. Zinc phosphate conversion coatings are sometimes first produced on the work metal surface to aid in the retention of lubricants during severe forming.

Lubricants for the cold forming of titanium are generally similar to those used for the severe forming of aluminum alloys. Frequently overlays of steel sheet or plastic sheet are used with an auxiliary lubricant.

Blank Preparation

Most of titanium alloy sheet 6.4 mm (¼ in.) thick or less is done in a punch press. As with other metals, maximum blank size depends on stock thickness, shear strength, and available press capacity.

Shearing. Titanium sheet up to 3.56 mm (0.140 in.) thick can generally be sheared without difficulty; with extra care, titanium sheet as thick as 4.75 mm (0.187 in.) can be sheared. Shears intended for low-carbon steel may not have enough hold-down force to prevent titanium sheets from slipping. A sharp shear blade in good condition with a capacity for cutting 4.8 mm (³⁄₁₆ in.) thick low-carbon steel can cut 3.2 mm (⅛ in.) thick titanium sheet. Cutters should be kept sharp to prevent edge cracking of the blank.

Sheared edges, especially on thicker work metal, can have straightness deviations of 0.25 to 5 mm (0.01 to 0.20 in.), usually because the shear blade is not stiff enough. Shearing can cause cracks at the edges of some titanium sheet thicker than 2.0 mm (0.080 in.). If cracks or other irregularities develop in a critical portion of the workpiece, an alternative method of cutting should be used, such as band sawing, abrasive waterjet cutting or laser cutting.

Slitting of titanium alloy sheet can be done with conventional slitting equipment and with draw-bench equipment. Slitting shears are capable of straight cuts only; rotary shears can cut gentle contours (minimum radius: ~250 mm, or 10 in.). The process can be used for sheet thicknesses to 2.54 mm (0.100 in.). However, an individual machine must be restricted to titanium thicknesses for which the machine is rated.

Band sawing prevents cracking at the edges of titanium sheet but causes large burrs. Band sawing is generally used to cut titanium sheet that is 3.18 mm (0.125 in.) or more in thickness. Abrasive blades are normally used.

Nibbling can be used to cut irregular blanks of titanium, but most blanks need filing or grinding after nibbling.

Edge Preparation. All visual evidence of a sheared or broken edge on a part should be removed by machining, sanding, or filing before final deburring or polishing. All rough projections, scratches, and nicks must be removed. Extra material must be allowed at the edges of titanium blanks so that shear cracks and other defects can be removed. On sheared parts, a minimum of 0.25 mm (0.010 in.) must be removed from the edge; on punched holes, 0.35 mm (0.014 in.). On parts cut by friction band sawing or abrasive sawing, 6.35 mm (0.25 in.) or one thickness of sheet should be removed from sheared edges, whichever is the smaller.

The lay of the finish on the edges of the sheet metal parts should be parallel to the edge surface of the blank, and sharp edges should be removed. Edges of shrink flanges and stretch flanges must be polished before forming. To prevent scratching the forming dies, edges of holes and cutouts should be deburred on both sides and should be polished where they are likely to stretch during forming.

Forming Temperatures

Close tolerances and acceptable mechanical properties can be achieved either by hot forming or by cold preforming followed by hot sizing. The relative advantage of each depends on equipment capabilities (see Table 4) and the severity of forming requirements. Hot-forming techniques are used on titanium and its alloys to increase formability, minimize springback, reduce variations in waviness between sheets, and produce maximum deformations with minimum interstage annealing.

Cold Forming. Titanium and titanium alloys are commonly stretch formed without being heated, although the die is sometimes warmed to 150 °C (300 °F). Simple brake forming of straight sections also can be done at room temperature if adequate bend radii are designed into the tool. When formed at room temperature, commercially pure titanium and titanium alloys behave like cold-rolled stainless steel. Shapes that can be successfully press formed in ¼-hard stainless steel usually can be press formed in commercially pure titanium, although titanium may require hot sizing to produce severe contours.

Springback tends to follow the yield strength/ultimate tensile strength ratio with a higher ratio indicating a higher degree of springback. However, the high springback of titanium (up to 20 or 30% of the bend angle at room temperature) is not the major concern because allowances can be made in tooling. The major concern is the variation of springback. Thus, hot sizing may be required on some percentage of parts, or the degree of springback can be partially overcome by warm forming (260 to 315 °C, or 500 to 600 °F).

Commercially pure titanium and the most ductile titanium alloys, such as Ti-15V-3Sn-3Cr-3Al and Ti-3Al-8V-6Cr-4Zr-4Mo, can be formed cold to a limited extent. Alloy Ti-8Al-1Mo-1V sheet can be cold formed to shallow shapes by standard methods. The cold forming of other alloys generally results in excessive springback, requires stress relieving between operations, and requires more power. The only true cold-formable titanium alloys are Ti-15V-3Sn-3Cr-3Al, Beta 21S, and Beta C. Hot sizing is usually not used for this alloy; however, properties must be developed with an aging treatment (8 h at 540 °C, or 1000 °F, is typical). Because of the high springback rates encountered with this alloy, more elaborate tooling must be used.

Hot Forming. Heating titanium increases formability, reduces springback, takes advantage of a lesser variation in yield strength, and allows for maximum deformation with minimum annealing between forming operations. Severe forming must be done in hot dies, generally with preheated stock. For applications in which the utmost ductility is required, temperatures below 315 to 425 °C (600 to 800 °F) are usually avoided. The formability of most titanium alloys at 650 °C (1200 °F) is comparable to that of annealed stainless steel at room temperature.

The greatest improvement in the ductility and uniformity of properties for most titanium alloys is at temperatures above 540 °C (1000 °F). Therefore, most hot-forming operations are done at temperatures above 540 °C (1000 °F). At still higher temperatures, some alloys exhibit superplasticity. However, contamination is also more severe at the higher temperatures. If forming is done above 540 °C (1000 °F) a protective coating should be used (which also serves as a lubricant) to minimize oxidation during forming. Forming done in a protective atmosphere, such as argon, or in a vacuum would eliminate the contamination problem. Temperatures generally must be kept below 815 °C (1500 °F) to avoid marked deterioration in mechanical properties. Superplastic forming, however, is performed at 870 to 925 °C (1600 to 1700 °F) for some alloys, such as Ti-6Al-4V.

Scaling and Embrittlement. Titanium is scaled and embrittled by oxygen-rich surface layers formed at temperatures higher than 540 °C (1000 °F). Generally, for heating in air, 1 h is the longest time at 700 °C (1300 °F) that should be permitted, and 20 min at 870 °C (1600 °F) should be the limit; these times are cumulative and include all time that the metal is at that temperature for all the operations on a given workpiece. The subsequent removal of scale and embrittled surface, or a protective atmosphere, should be considered for any heating above 540 °C (1000 °F). Argon gas is a predominately used atmosphere for superplastic forming.

Aging. Some hot-forming temperatures are high enough to age a titanium alloy. Heat-treatable β and α-β alloys may be formed at the aging temperatures. If the forming temperature is above the aging temperature then re-solution treatment will probably be required. Alpha-beta alloys should not be formed above the β transus temperature.

Because of aging, scaling, and embrittlement, as well as the greater cost of working at elevated temperatures, hot forming is ordinarily done at the lowest temperature that will permit the required deformation. When maximum formability is required, the forming should be done at the highest temperature practical that will retain the mechanical properties and serviceability required of the workpiece.

Tools. Titanium alloys are often formed hot in heated dies in presses that have a slow, controlled motion and that can dwell in the position needed during the press cycle. Hot forming is sometimes done in dies that include heating elements or in dies that are heated by the press platens. Press platens heated to 650 °C (1200 °F) can transmit enough heat to keep the working faces of the die at 425 to 480 °C (800 to 900 °F). Other methods of heating include electrical-resistance heating and the use of quartz lamps and portable furnaces.

Accuracy. Hot forming has the advantage of improved uniformity in yield strength, especially when the forming or sizing temperature is above 540 °C (1000 °F). However, care must be taken to limit the accumulation of dimensional errors resulting from:

Table 4 Advantages and disadvantages of hot forming and cold forming with hot sizing

Hot Forming	Cold Forming-Hot Sizing
Advantages	
(1) Single operation	(1) Forming can be accomplished on all available types of forming machines
(2) Lower forming pressures	(2) Parts are stress-relieved on sizing
(3) Material is at elevated temperature for shorter time	(3) Can use lower cost tooling materials in cold forming
Disadvantages	
(1) Requires temperature resistant tool materials	(1) Requires additional equipment (hot sizing presses)
(2) Tools must be adapted for heating	(2) Long dwell times in hot sizing press (30 min)
(3) Requires use of slow press with some dwell time (5 min)	(3) Long exposure times to elevated temperatures
(4) Limited to forming operations on equipment which can use heated tools	(4) Requires two sets of dies (one set heat resistant)

- Differences in thermal expansion
- Variations in temperature
- Dimensional changes from scale formation
- Changes in dimensions of tools
- Reduction in thickness from chemical pickling operations

Hot sizing is used to correct inaccuracies in shape, dimensions, and springback in cold preformed parts. Hot sizing uses the creep-forming principle to force irregularly shaped parts to assume the correct shape against a heated die by the controlled application of horizontal and vertical forces over a period of time. Buckles and wrinkles can be removed from preforms in this way. A combination of creep and compression forming is used when reducing bend radii by hot sizing. Hot sizing is often combined with stress relieving.

The correction of springback by hot sizing depends on time and temperature; the higher the temperature, the shorter the time for processing. However, the effect of temperature on the properties of the metal limits the maximum useful temperature. The pressure applied to the part during hot sizing should be high enough to keep the part firmly against the fixture or die. Any additional pressure above the clamping requirement has no effect on the part and can cause deformation of the tooling.

Hot platen presses are commonly used for the hot sizing of titanium. The tooling is designed for the hot sizing of preforms; that is, it must only hold the workpiece to the required shape for the necessary time at temperature. Hot sizing in hot platen presses is done in the following sequence of operations:

- The preformed parts are loaded on hot form blocks that are heated by the platen in the press
- The press is closed, and it heats the parts without applying the forming force
- Force is applied by the upper platen and auxiliary side rams, and is held for as long as necessary to complete the forming

Forming Methods

Press-Brake Forming. Titanium alloys cold formed in a press brake behave like work-hardened stainless steel, except that springback is considerably greater. If bend radii are large enough, forming can be done cold. However, if bend radii are small enough to cause cracking in cold forming, either hot forming or the process of cold forming followed by hot sizing must be used.

The setup and tooling for press-brake air bending are relatively simple because the ram stroke determines the bend angle. The only tooling adjustments are the span width of the die and the radius of the punch. The span width of the die affects the formability of bend specimens and is determined by the punch radius and the work metal thickness (see Fig. 2).

The minimum bend radius obtainable in press-brake forming depends on the alloy, work metal thickness, and forming temperature. Springback in press-brake forming depends on the ratio of punch radius (bend radius) to stock thickness and on forming temperature.

Deep Drawing. General guidelines for the deep drawing of titanium alloy dome shapes at room temperature are:

- Lubrication and tool size are key
- Dedicated, double-action draw press with blank holding pressure is preferred
- Draw radii depend on gauge, and should not be less than 5t
- The workpiece should be clean and blank edges should be smooth
- An overlay can be used to prevent wrinkles, but plastic overlays may rip or tear. Oils are preferred
- Severe forming and localized deformation should be avoided; forming pressure should be applied slowly
- The punch should be polished to prevent galling, regardless of lubrication
- Multiple reductions require intermediate anneals. Domes and cans are formable in one pass for some alloys (e.g. Ti-15-3).

More difficult drawing can be done at high temperature, and drawing temperatures to 675 °C (1250 °F) have been used. However, high-temperature deep drawing of titanium alloys has been largely replaced by the superplastic forming process.

Power (Shear) Spinning. Most titanium alloys are difficult to form by power spinning. Alloys Ti-6Al-4V and Ti-13V-11Cr-3Al and some grades of CP titanium are the most responsive to forming by this method. Most tools for the power spinning of titanium are made of high-speed steel and hardened to 60 HRC. Mandrels are heated for hot spinning. Tube preforms can be heated by radiation. The hot power spinning of titanium is done at 200 to 980 °C (400 to 1800 °F) depending on the alloy and the operation.

Lubricants for the power spinning of titanium depend on the forming temperature used. At temperatures up to 200 °C (400 °F), heavy drawing oils, graphite-containing greases, and colloidal graphite are used. Colloidal graphite and molybdenum disulfide are employed at temperatures to 425 °C (800 °F); above this temperature, colloidal graphite, powdered mica, and boron nitride are used.

Rubber-Pad Forming. The cold forming of titanium in a press with tooling that includes a rubber pad is used mostly for flanging thin stock (≤1 mm, or 0.04 in.) and for forming beads and shallow recesses. The capacity of the press controls the range in size, strength, and thickness of blanks that can be formed. Within this range, however, additional limits will be set by buckling and splitting.

Auxiliary devices, such as overlays, wiper rings, and sandwiches, are usually needed in rubber-pad forming to improve the forming and to reduce the amount of wrinkling and buckling. Rubber-pad forming is generally done at room temperature or with only moderate heat. Forming is very often followed by hot sizing to remove springback, to sharpen radii, to smooth out wrinkles and buckles, and to complete the forming. Hand work

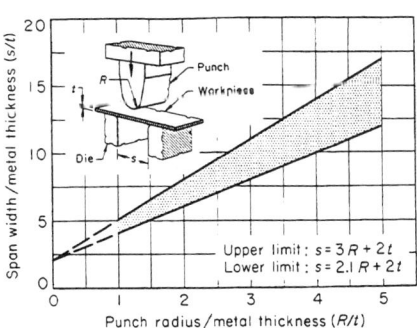

Fig. 2 Limits for press-brake forming of titanium alloys. Shaded area indicates acceptable forming limits

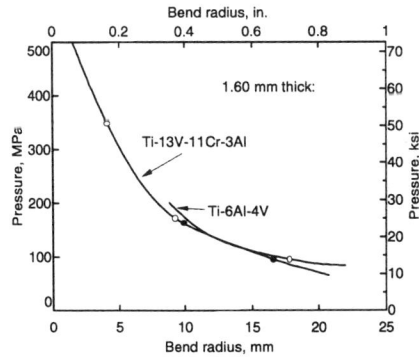

Fig. 3 Effect of pad pressure on bend radii

is sometimes needed to complete the forming. The cold-formed workpiece should be stress relieved or hot sized within 24 h after forming. Sharp bends can be made at higher forming pressures (see Fig. 3).

Springback behavior of titanium and its alloys in rubber-pad forming differs somewhat from that observed in other methods of forming. Springback can be difficult to predict. In general, springback in forming titanium varies directly with the ratio of bend radius to work metal thickness, and inversely with forming temperature. Springback is also inversely proportional to forming pressure.

Beads can be formed to a limited extent in titanium alloy sheet by rubber-pad forming. However, beads are readily formed by superplastic forming.

Stretch Forming. Tooling that is generally used for the stretch forming of stainless steel is also suitable for the cold stretch forming of titanium, when used with a high clamping force that will prevent slipping and tearing. Titanium may exhibit irregular incremental stretch under tension loads; therefore, optimal results are obtained when titanium is stretch formed at slow strain rates. The rate of wrapping around a die should be about 200 mm/min (8 in./min).

In the stretch forming of angles, channels, and hat-shaped sections, deformation occurs mainly by bending at the fulcrum point of the die surface; compression buckling is avoided by applying enough tensile load to produce about 1% elongation in the inner fibers. The outer fibers elongate more, depending on the curvature of the die and on the shape of the workpiece. It is sometimes preferable or required (especially if sufficient forming power is not available) to stretch wrap at elevated temperature. Again, the wrapping speed must be slow to prevent local overheating or necking.

Formability limits can be extended by permitting small compression buckles to occur at the inner fibers and removing them later by hot sizing. The buckled region represents a condition of overforming and should be limited to the amount that can be effectively removed by hot sizing.

Compression buckling is not a problem when sheet is stretch formed to produce single or compound curves. The ductility of sheet varies with orientation and is generally better in the direction of rolling. In the stretch forming of compound curves, the stretching force should be applied in the direction of the smaller radius. Stretch forming is being replaced in many applications by superplastic forming.

Three-roll forming is an economical method of forming titanium alloy sheet into aircraft skins, cylinders, or parts of cylinders. The sheet should be flat within 0.15 mm (0.006 in.) for each 50 mm (2 in.) of length. The corners of the sheet should be chamfered to prevent marking of the rolls.

The upper roll of the three-roll assembly can be adjusted vertically. The radius of the bend is controlled by the roll adjustment. Premature failure will occur if the contour radius is decreased too rapidly; however, too many passes through the rolls may cause excessive work hardening of the work metal. Several trial parts must sometimes be made in a new material or shape to establish suitable operating conditions.

Three-roll forming is also used to form curves in channels that have flanges of 38 mm (1.5 in.) or less. Transverse buckling and wrinkling are common failures in the forming of channels.

Contour Roll Forming. Titanium sheet can be contour roll formed like any other sheet metal, but with special considerations for allowable bend radius and for the greater springback that is characteristic of titanium. Springback is affected to some extent by roll pressure. Often, hot rolling must be done on heated work metal with heated rolls.

Creep Forming. In creep forming, heat and pressure are combined to cause the slow forming of titanium sheet into various shapes, such as double-curve panels, channel sections, Z-sections, large rings, and small joggles. The metal flows plastically at a stress below its yield strength. At low temperature, creep rates are ordinarily very low (for example, 0.1% elongation in 1000 h), but the creep rate of titanium accelerates sharply with increasing temperature.

Creep forming can be done by three different methods:

- A blank is clamped at the edges, as for stretch forming, and a heated male tool is loaded to press against the unsupported portion of the blank; the metal yields under the combination of heat and pressure and slowly creeps to fit the tool
- A set of dies containing heating elements or coils is used in a hydraulic press in a manner similar to hot sizing
- A heated female die is used with a vacuum diaphragm, as in vacuum forming

Temperatures for creep forming are the same as those used in hot forming. Generally, titanium must be held at the creep-forming temperature for 3 to 20 min per operation; creep forming sometimes takes as long as 2 h.

Vacuum Forming. Large panels (some as much as 18 m, or 60 ft, long) for aircraft are sometimes vacuum formed from titanium alloy sheet. Vacuum forming, however, has been largely replaced by superplastic forming. For vacuum forming, the blank is laid on a die of heated concrete, ceramic, or metal, and a somewhat larger flexible diaphragm is laid on top of the blank to provide a seal around its edges. After the blank has been heated to forming temperature, the air is pumped out from between the blank and the die so that atmospheric pressure is used to form the work. This method, a kind of creep forming, cannot be used to form to sharp radii.

Drop Hammer Forming. Titanium should not be permitted to rub against lead, zinc, or other low-melting metals that could contaminate it and cause embrittlement. Drop hammer tools can be capped with sheet steel, stainless steel, or nickel alloy, depending on the expected tool life. Nickel-base alloys, in thicknesses of 0.635 to 0.813 mm (0.025 to 0.032 in.), have the longest life.

Drop hammer forming is done at about 500 to 800 °C (900 to 1500 °F), depending on the alloy and the severity of forming. Thermal expansion of the dies must be considered in the design. The approximate rate of expansion for steel dies is 0.006 mm/mm (0.006 in./in.) as temperature is increased from 20 to 540 °C (70 to 1000 °F).

Multistage tools can be used if the part shape is complex and cannot be formed in one blow. Workpieces are then finished by hot sizing.

The minimum thickness of titanium sheet for drop hammer forming is 0.635 mm (0.025 in.); thicker sheet is used for complex shapes. Total tolerance on parts formed in drop hammers is usually 1.6 mm ($\frac{1}{16}$ in.).

Joggling is frequently done on titanium alloy sheet. A joggle is an offset in a flat plane, consisting of two parallel bends in opposite directions at the same angle. Generally, the joggle angle is less than 45°.

Depending on joggle depth, joggles can be either formed completely at room temperature or at elevated temperature in press brakes and mechanical or hydraulic presses. Common practice is to preform at room temperature and then hot size ("set" the joggle) in a heated die. The sizing operation is usually done under conditions that result in stress relieving or aging.

In press-brake formed or stretch-formed angles and channels, and in machined extrusions, joggles with radii smaller than the minimum bend radii for the metal at room temperature, or joggles with length-to-depth ratios of less than about 6 to 1, are more successfully formed at elevated temperature. Forming temperature varies between 315 and 650 °C (600 and 1200 °F), depending on the alloy and its heat-treated condition. Annealed alloys are joggled at 315 to 425 °C (600 to 800 °F) or hotter. Heat-treated or partly heat-treated alloys are joggled at, or near, their aging temperature.

Dimpling produces a small conical flange around a hole in sheet metal parts that are to be assembled with flush or flathead fasteners. Dimpling is most commonly applied to sheets that are too thin for countersinking. Sheets are always dimpled in the condition in which they are to be used because subsequent heat treatment may cause distortion of the holes or dimensional changes in the sheet.

The hot ram-coin dimpling process is generally used, although dimples have been produced at room temperature by swaging. In hot ram-coin dimpling, force in excess of that required for forming is applied to coin the dimpled area and to reduce the amount of springback.

The titanium is dimpled at up to 650 °C (1200 °F) with tool steel dies. If higher temperatures are required, heat-resistant alloy or ceramic tooling is needed in order to prevent deformation of the dies during dimpling. The work metal is usually heated by conduction from the dimpling tools, which are automated to complete the dimpling stroke at a predetermined temperature.

Pilot holes must be drilled, rather than punched, and must be smooth, round, cylindrical, and free of burrs. Because of the notch sensitivity of titanium, care must be taken in deburring the holes.

The amount of stretch required to form a dimple varies with the head and body diameters of the fastener and the bend angle. If the metal is not ductile enough to withstand forming to the required shape, cracks will occur radially in the edge of the stretch flange, or circumferentially at the bend radius. Circumferential cracks are more common in thin sheet; radial cracks are more common in thick stock.

Explosive Forming. Within the limits set by its mechanical properties, titanium can be explosive formed like other metals. Explosive forming is most commonly used for cladding titanium to other metals. Titanium is explosive formed using techniques similar to those used for other metals and alloys.

Technical Note 5A: Superplastic Forming of Titanium Alloys

Murray W. Mahoney, Rockwell International Science Center

Superplastic forming (SPF) of titanium alloys is a relatively new process that offers unique advantages over conventional forming operations. For example, characteristics of superplastic forming include low flow stresses, reduced machining, no springback, uniform metal flow, no resultant residual stresses, in general no cavitation, and formability of shapes not possible by any other approach. However, superplastic forming is not applicable to all titanium alloys, and there is probably no other forming process in which the relationship between materials properties and consistently successful forming operations is more critical than in superplastic forming. As such, a detailed presentation of the requirements for superplastic forming, including microstructural prerequisites and the significance of forming parameters, are presented in detail.

Superplasticity is the ability of certain materials, primarily metals, to undergo unusually large amounts of uniform plastic deformation before local necking occurs. The remarkable formability of superplastic materials is due to their high strain-rate sensitivity, defined by:

$$m = d \ln\sigma/d \ln\dot{\varepsilon} \quad \text{(Eq 1)}$$

where σ is the true flow stress, and $\dot{\varepsilon}$ is the true strain rate. It is generally accepted that materials, with m-values of 0.5 or higher can be superplastically deformed. However, m has been shown to be a complex function of a number of microstructural and forming parameters and may exhibit significant change during the forming operation. Titanium alloys are particularly complex in their forming characteristics due to their two-phase nature. Thus, a thorough understanding of the relationships between superplasticity and microstructure is necessary to form titanium alloys successfully.

Results of investigations on titanium alloys are presented, with representative illustrations showing how superplasticity is influenced by the microstructural con-

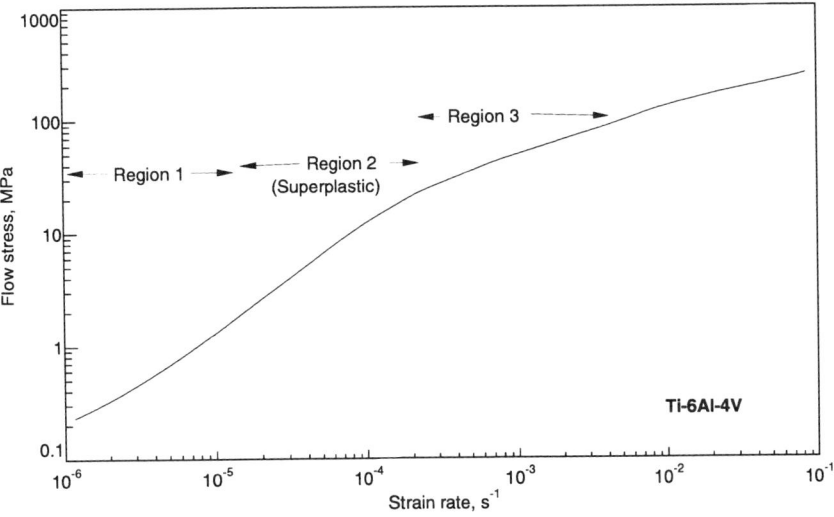

Fig. 1 Typical log stress vs log strain rate curve for superplastic Ti-6Al-4V at 870 °C (1600 °F) showing three regions of different deformation characteristics. Grain size, 5.3 μm.
Source: C.H. Hamilton and A.K. Ghosh, *Metall. Trans. A*, Vol 11A, 1980, p 1494

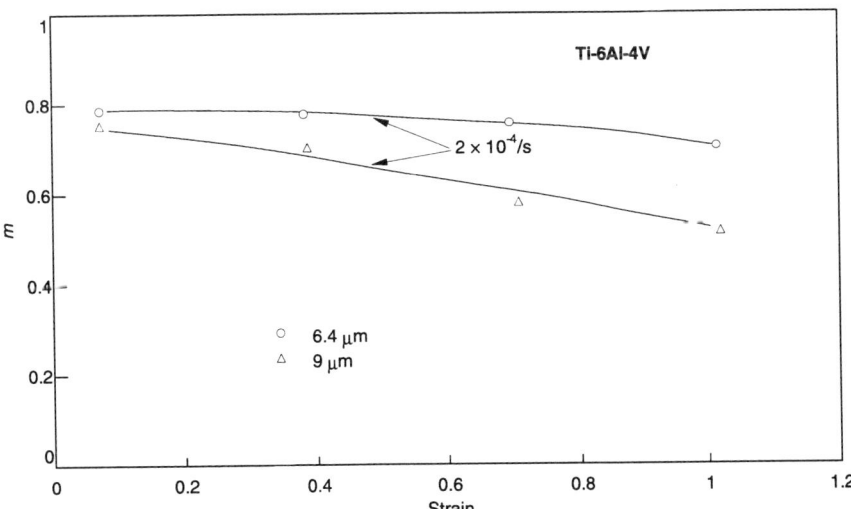

Fig. 2 Instantaneous strain-rate sensitivity of Ti-6Al-4V during tensile testing at 925 °C (1700 °F) showing decreasing m with increasing strain.
Source: C.H. Hamilton and A.K. Ghosh, *Proc. 4th Int. Conf. Titanium*, Kyoto, Japan, May 19-22, 1980, p 1001

ditions of grain size, grain stability, grain size distribution, grain shape, phase ratio and alloy composition, and forming parameters of temperature, strain rate, and accumulated strain. These microstructural conditions and forming parameters are discussed with regard to how they influence the superplastic response of titanium alloys via changes in strain-rate sensitivity and flow properties. However, results from one alloy to another cannot be either generalized or extrapolated. Although in some instances there are similarities within alloy classes, even generalizations within alloy classes would be a dangerous practice. Also, there is considerable synergism between forming and microstructural variables. Thus, the superplastic forming user must be acutely aware of both microstructural changes, e.g., lot-to-lot variations of incoming materials, and any changes that may occur in forming parameters. Briefly, to understand superplasticity, it is important first to understand how superplastic properties are determined.

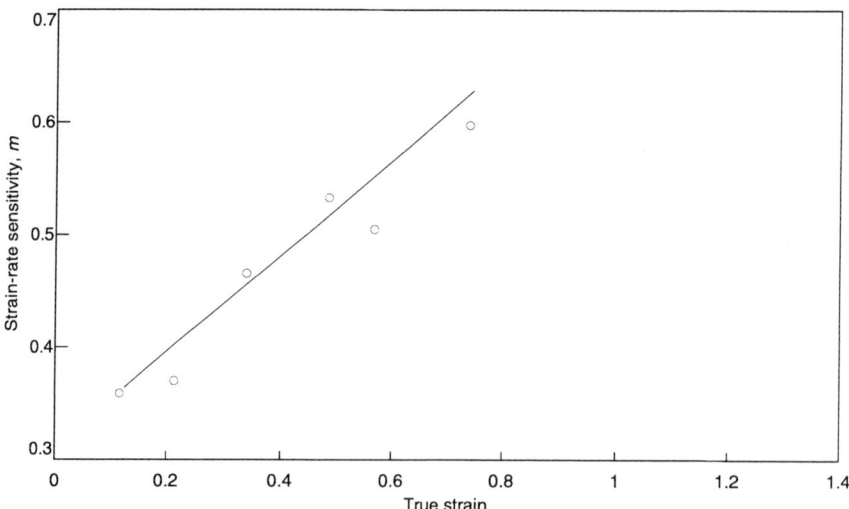

Fig. 3 Strain-rate sensitivity versus strain at a true strain rate of 5.33×10^{-4}/s for Ti-6Al-4V in which the microstructure improves during testing.
Source: R.R. Boyer and J.E. Magnuson, *Metall. Trans. A*, Vol 10, Aug 1979, p 1191

Strain-Rate Sensitivity

The value of the strain-rate sensitivity index m is the primary material parameter indicating the ability of a material to resist plastic instability or necking. When m is low, an increase in stress at a neck leads to a large increase in strain rate at that location and a low elongation to fracture. Conversely, when m is large, the strain rate increases slowly in response to increased stress in the neck region, and the neck thus forms gradually leading to a high elongation to failure. The strain-rate sensitivity index m, however, changes both with a change in forming parameters as well as during the forming operation itself. Thus, m must be determined as a function of strain, strain rate, temperature, and microstructure.

A number of approaches that have been used to determine m have produced inconsistent results. Based on common usage and thus acceptance within the superplastic forming community, the step strain-rate test method has proven to offer the truest measure of strain-rate sensitivity for establishing m and subsequent forming parameters. In this test, the strain rate is increased in successive steps, and an attempt is made to measure the corresponding flow stress. Herein lies potential error, and the investigator must be cautious as to how the flow stress is measured. One problem in assessing flow stresses for the lower strain rates ($<5 \times 10^{-4}$ s^{-1}) is that strain hardening does not permit the establishment of a load maximum. In most superplastic materials, strain hardening is due to concurrent grain growth. Thus, to separate flow stress from grain growth effects, stress must be selected soon after the elastic portion. From the slope of the stress versus strain-rate curve, the strain-rate sensitivity m can now be determined as a function of strain rate. An example of flow stress as a function of strain rate for Ti-6Al-4V at 870 °C (1600 °F) is shown in Fig. 1. Region II, the location of greatest slope, is where maximum strain-rate sensitivity occurs. The criticism against use of strain-rate step tests is that strain accumulates during the test and thus automatically becomes a variable and accordingly, to minimize error, must be kept low.

Additionally, m, can be measured as a function of increasing strain, and in fact, this test is more representative of an actual forming operation in which high levels of strain accumulate during forming. In this test, the sample experiences a constant strain rate $\dot{\varepsilon}_1$ until a small amount of strain accumulates and the flow stress is established. At this time, the strain rate is increased (e.g., 40%) to $\dot{\varepsilon}_2$ for a short period until a maximum in flow stress is established for $\dot{\varepsilon}_2$, at which time the strain rate is again decreased to $\dot{\varepsilon}_1$. This process is repeated with increasing strain, resulting in a determination of m as a function of strain. The value of m is then determined from:

$$m = d \ln(\sigma_2/\sigma_1)/d \ln(\dot{\varepsilon}_2/\dot{\varepsilon}_1) \qquad (Eq\ 2)$$

where σ_2 is the flow stress corresponding to a strain rate of $\dot{\varepsilon}_2$, and σ_1 is the flow stress at $\dot{\varepsilon}_1$. Generally, the value of m decreases with increasing strain. Results for this type test are illustrated in Fig. 2 for Ti-6Al-4V at 925 °C (1700 °F) for grain sizes of 6.4 and 9 μm. However, this trend

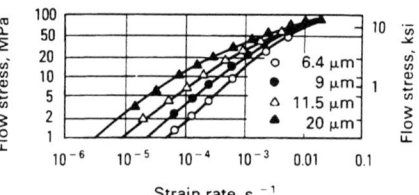

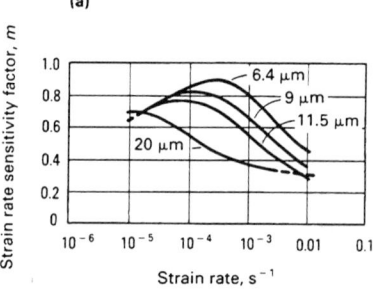

Fig. 4 Effect of grain size on flow stress and m as functions of strain rate for Ti-6Al-4V. Step strain-rate tests at 925 °C (1700 °F).
Source: N.E. Paton, *Titanium, Science and Technology*, G. Lütjering, U. Zwicker, and W. Bunk, Ed., Deutsche Gesellschaft für Metallkunde, Germany, 1985

is not always followed and is a function of the starting microstructure. For example, Fig. 3 illustrates an increase in m with an increase in strain for Ti-6Al-4V. In this case, the material was not recrystallized and had a residual elongated worked structure that would be expected to exhibit poor superplastic behavior (note the low initial m value). The combination of strain and temperature exposure experienced during the test caused recrystallization, producing an equiaxed microstructure with improved superplasticity and hence an increasing m value.

Each of these tests, stress versus strain rate and flow stress with increasing strain, should be performed as a function of temperature and at constant strain rates and not at constant crosshead speeds. Once the strain rate and temperature ranges at which m is a maximum are established, and the value of m as a function of strain has been determined, these parameters can be evaluated for acceptability for the forming operation of interest.

Grain Size

In all superplastic materials, grain size is one of the single most important parameters. However, because of the complexity of microstructures that can be obtained in almost all titanium alloys, grain size by itself is an inadequate description of microstructure. These additional microstructural variations are described below. The first consideration in establishing the effect of grain size on properties is to define a uniform method of measurement. Titanium alloys that have useful degrees of superplasticity are generally two-phase alloys, making the task of grain size measurement particularly important. In establishing grain size in titanium alloys, most investigators attempt to count all interphase boundaries as grain boundaries. However, difficulties arise when, as frequently occurs, a thin layer of one phase lies between two larger grains of a second phase, and this layer is counted as a single boundary rather than two, as it should be.

Grain size is known to strongly influence flow stress behavior and correspondingly m value and superplasticity in Ti-6Al-4V. Figure 4 illustrates flow stress and m as functions of strain rate for a range of grain sizes. As is typically found for most superplastic materials, increasing grain size increases the flow stress and tends to reduce the maximum m value as well as reduce the strain rate at which the maximum m is observed. A similar result is illustrated in Fig. 5 for a superplastic Ti-6.3Al-2.7Mo-1.7Zr alloy at 900 °C (1650 °F) for grain sizes ranging from 3.5 to 8.1 μm.

Grain Size Stability. Two-phase titanium alloys have some degree of inherent stability imparted by the equilibrium volume fraction of the phases present. Assuming a duplex alloy containing α and β phases, many α-α and β-β boundaries are present. Grain growth initially proceeds by elimination of these like boundaries through normal boundary migration processes. However, when substantial numbers of isolated α and β grains remain, further grain growth requires long-range diffusion. Examples of thermally activated grain growth in Ti-6Al-4V at temperatures from 870 to 955 °C (1600 to 1750 °F) are illustrated in Fig. 6. Additionally, stress- or strain-rate-assisted grain boundary mobility has been shown to accelerate grain growth during the superplastic forming process, leading to growth rates greater than those measured during static temperature exposure. Figure 7 shows this effect quantitatively for Ti-6Al-4V at 915 °C (1680 °F) as a function of true thickness strain measured from different locations of a formed component. Because temperature, time at temperature, and the strain rate were the same for all locations, the grain size effect is entirely due to the magnitude of strain. Similarly, Fig. 8 shows strain-rate-enhanced grain growth for Ti-6Al-4V at 925 °C (1700 °F) for a wide range of strain rates from 5×10^{-5} s^{-1} to 5×10^{-3} s^{-1}. However, in this case, the time required at temperature to accumulate the same level of strain is different for each strain rate.

Observed increases in flow stress during superplastic forming can be accounted for by the increase in grain size. However, both strain hardening and strain softening have been observed in Ti-6Al-4V, with hardening attributed to grain growth and softening a result of grain refinement. Due to the synergistic effect of temperature and strain rate on hardening and softening, this subject is better discussed in greater detail within the sections on forming parameters important to superplastic forming.

Grain Shape and Size Distribution. The nature of the grain size distribu-

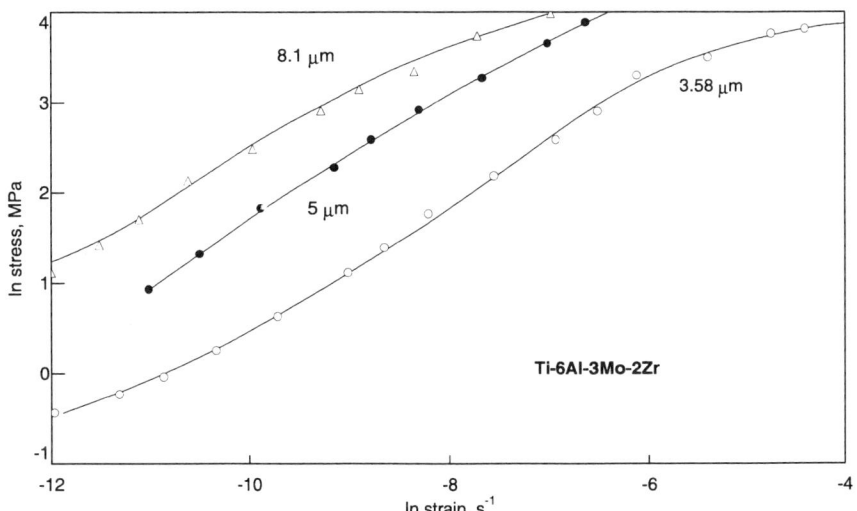

Fig. 5 ln stress versus ln strain rate plots for Ti-6Al-3Mo-2Zr at 900 °C (1650 °F) for three grain sizes.
Source: A. Dutta and A.K. Mukherjee, *Mater. Sci. Eng.*, Vol A138, 1991, p 221

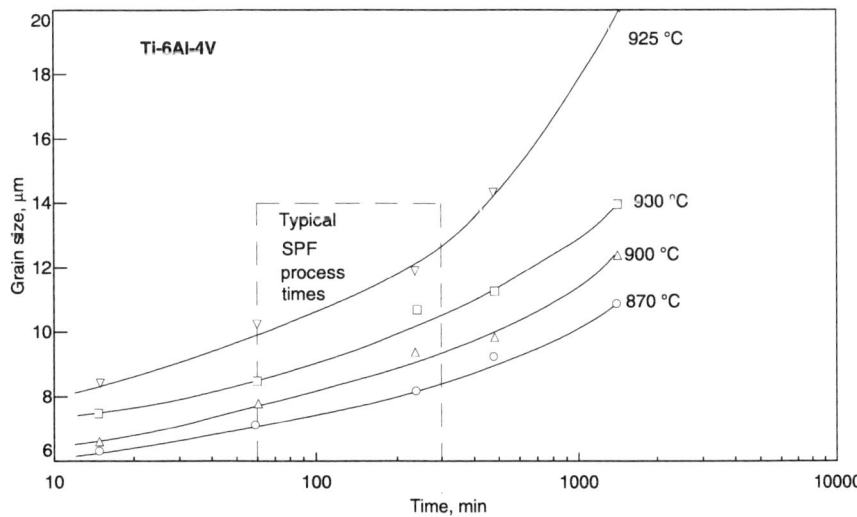

Fig. 6 Grain size versus time of exposure to elevated temperatures for Ti-6Al-4V.
Source: Unpublished research, Rockwell International

tion also has been shown to influence the stress-strain-rate curve and accordingly the m value. Due to the mechanism of superplasticity, i.e, grain rotation and migration, an equiaxed structure is preferable for superplastic forming. Thus, grain shape, characterized by a grain aspect ratio defined as the ratio of the interphase spacing along the major axis, is also a microstructural variable to be considered. Grain aspect ratios close to 1.0 are clearly favored for optimum superplasticity. In titanium alloys, the grain shape can become very complex. For example, although the α phase can exist as equiaxed grains developed by cold working followed by annealing above the recrystallization temperature, α can also exist in several other transformed and untransformed modifications. Of these, acicular α and blocky or primary α are the most common. Acicular α is a transformation product brought about through nucleation and growth. It has the appearance of fine needles and has been shown to exhibit no indication of superplasticity. Blocky α and lamellar α are detrimental to superplastic behavior, because they make grain boundary sliding difficult, resulting in higher flow stresses and lower m values. Control and understanding of each of these microstructural features, i.e., grain shape and size distribution, are important considerations for achieving optimum superplasticity in titanium.

Phase Ratio

Although grain size effects are a major factor in the superplastic behavior of a titanium alloy, the α/β phase ratio appears to be equally important. Phase ratio is a variable controlled by a combination of test temperature and alloy composition, with higher concentrations of β stabilizers such as vanadium or molybdenum leading to higher volume fractions of β phase at a given temperature. The importance of phase ratio is associated with the significant differences in properties between the α and β phases. For instance, at superplastic forming temperatures, the β phase is considered to be softer and to have a diffusivity two orders of magnitude higher than the α phase—properties of considerable importance to the mechanisms associated with superplastic flow.

As with many superplastic alloys, the presence of a second phase is key to the development of superplastic behavior. A finite concentration of the α phase significantly restricts grain growth because of the long-range diffusion necessary to transfer highly partitioned alloy elements. Without the presence of the second phase, the β phase will rapidly grow to diameters well in excess of 50 μm at temperatures above about 760 °C (1400 °F).

As shown in Fig. 9 and 10, the ratio of the volume fractions of α and β in many titanium alloys is strongly reflected in the superplastic properties of strain-rate sensitivity and elongation. For these alloys, the highest ductilities are observed with the β concentration in the range of 15 to 50 vol%. The maximum in m value also occurs at a significant concentration of the β phase, but there is no clear correspondence between this and the total elongation, because the maxima of each are indicated at different β volume fractions.

Results illustrated in Fig. 9 and 10 are for different temperatures, and accordingly, competing processes are occurring. As temperature increases, the volume fraction of β phase increases at the expense of the α phase. Consequently, rapid grain growth occurs in the β phase, thus reducing the superplastic capability. Also, results have shown the volume fraction of β phase can increase during superplastic deformation. Figure 11 shows the volume fraction of β in Ti-6Al-4V at temperatures from 750 to 850 °C (1380 to 1560 °F) as a function of superplastic strain. At the lower temperatures, it is hypothesized that the diffusivities are too low, and even with strain enhancement, no additional stabilization of β phase takes place. However, at 850 °C (1560 °F), results suggest that the transformation from α phase to β phase is induced by the strain of the su-

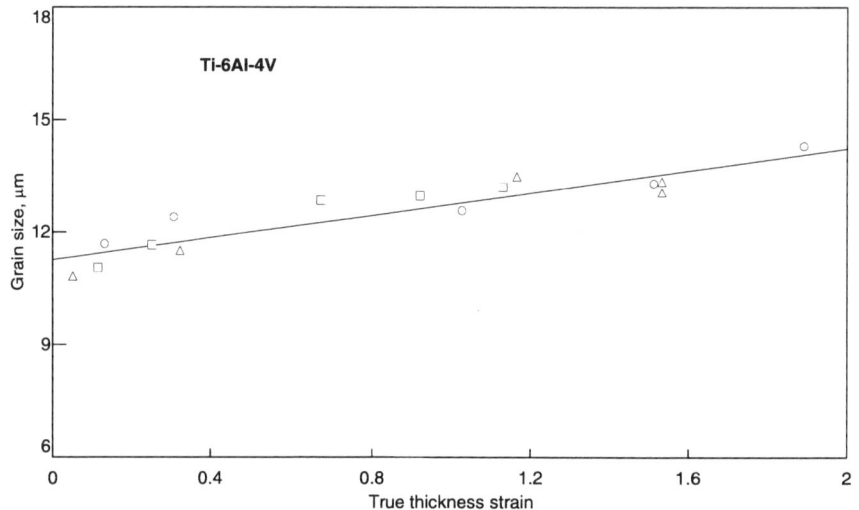

Fig. 7 Effect of SPF deformation on grain size at 915 °C (1680 °F).
Source: Unpublished research, Rockwell International

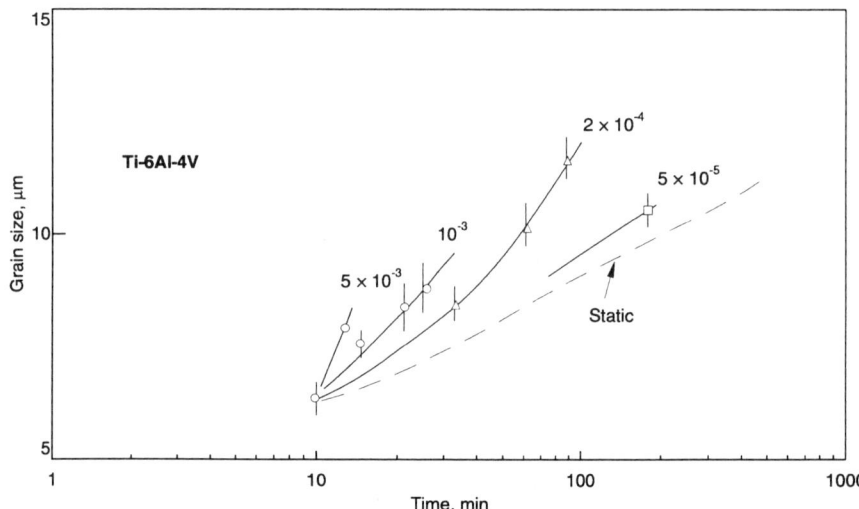

Fig. 8 Grain growth kinetics at 925 °C (1700 °F) and varying tensile strain rates (s^{-1}) compared with static kinetics for an initial grain size of 6.4 μm.
Source: C.H. Hamilton and A.K. Ghosh, *Titanium '80, Science and Technology*, H. Kimura and O. Izumi, Ed., TMS, 1980

perplastic deformation. Because of the increasing amount of the β phase during the deformation process, the softening that occurs balances the hardening caused by strain-enhanced grain growth.

In contrast, results on the metastable β alloys Ti-11.5Mo-6Zr-4.5Sn and Ti-10V-2Fe-3Al show that optimum conditions are obtained at temperatures just below the β transus where a small volume of α phase inhibits β grain growth. At lower temperatures with decreasing β diffusivities, superplasticity is lost. Such results emphasize that there is no optimum volume fraction of the phases for superplasticity in two-phase alloys; the often-cited equal volume fraction rule is not valid. Clearly, the relationship among phase ratio, superplastic strain, temperature, and alloy chemistry is complex. This complexity requires the superplastic practitioner to have an understanding of the synergism between microstructure and superplastic forming parameters to superplastically form titanium alloys successfully and consistently.

Alloy Composition

Most of the results presented here illustrate the superplastic properties for Ti-6Al-4V. This is due to the wide use of this alloy in the aerospace industry and because Ti-6Al-4V has been found to exhibit excellent superplastic properties in the conventionally produced form. Table 1 summarizes other titanium alloy compositions evaluated for superplasticity. A number of these alloys are not superplastic in spite of their initially high strain-rate sensitivity values. As discussed above, m changes during forming due to grain growth and a redistribution of the α/β phase ratio. Rapid grain growth is particularly true in all of the β alloys, where above the β transus, growth rates are very rapid.

In further studies on Ti-6Al-4V, Fig. 12 shows that lower flow stresses and somewhat higher m values were developed by the extra-low-interstitial (ELI) grade alloy than by the regular grade alloy, even though the ELI grade material had a somewhat larger grain size. Although the difference in volume fractions of α and β need to be considered, the greater slope and thus higher m value of the ELI material would indicate a more superplastic material.

Strain-rate sensitivity and flow stress results at 990 °C (1815 °F) for a near-α titanium alloy IMI 834 (Ti-6Al-4Sn-3.5Zr-0.7Nb-0.5Mo-0.33Si) are shown in Fig. 13 for longitudinal and transverse orientations. IMI 834 is a high-temperature alloy with good tensile strength and creep resistance up to 600 °C (1110 °F). Compared to similar results for Ti-6Al-4V (Fig. 4), IMI 834 has a lower m value, lower optimum strain rate, and higher flow stress, even at this higher temperature. However, even this high-temperature, high-strength alloy demonstrated uniform elongations of over 300%.

Superplasticity also has been demonstrated in the ordered titanium aluminide alloys including α$_2$-Ti$_3$Al (Ti-13.5Al-21.5Nb), super α$_2$ (Ti-14Al-19.5Nb-3.2V-2Mo), and even the very high temperature γ-TiAl alloy (Ti-36Al). Figure 14 shows the flow stress behavior and strain-rate sensitivity of super α$_2$ as a function of temperature. Superplasticity in the α$_2$ alloy is comparable to super α$_2$. An increase in temperature from 950 to 1010 °C (1740 to 1850 °F) causes a significant drop in the flow stress, but the m value remains in the range of 0.5 to 0.6, an indication of an expected high degree of superplasticity. Above this temperature, flow stresses at the low strain rates increase with temperature (Fig. 14a), indicating grain growth, which also reduces the m value (Fig. 14b) for the higher temperature.

A comparison of the superplastic elongation of Ti-6Al-4V with the different titanium aluminides is illustrated in Fig. 15. The titanium aluminides exhibit considerably less superplasticity and demonstrate higher forming stresses even at significantly higher forming temperatures. In spite of the increased degree of difficulty of forming the titanium aluminides, considerable success has been achieved even in scale-up to very large structures. An issue unique to titanium alloys is the observation of cavitation in γ-TiAl with increasing superplastic strain. Cavitation is occasionally observed in conventional titanium alloys, but only after considerable strain and then only near the final fracture site. However, as shown in Fig. 16, cavitation can occur in a γ-TiAl alloy at low levels of true strain (<1).

Temperature

Temperature is a fundamentally important forming parameter for superplasticity, which generally occurs above about $0.5 T_m$, where T_m is the melting point. The effect of temperature on the diffusion kinetics of titanium alloys is perhaps more complex than on other alloy systems because of variations in microstructural phase content. Depending on the alloy and temperature, α, β, or α + β phases may exist. This is important because the flow properties of the phases are different, and

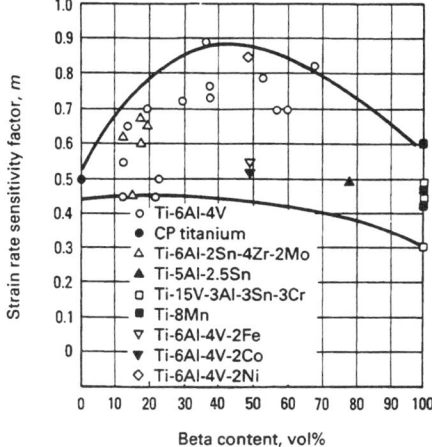

Fig. 9 Strain-rate sensitivity parameter (m) as a function of β phase content for several titanium alloys. Source: N.E. Paton, *Titanium, Science and Technology*, G. Lütjering, U. Zwicker, and W. Bunk, Ed., Deutsche Gesellschaft für Metallkunde, Germany,

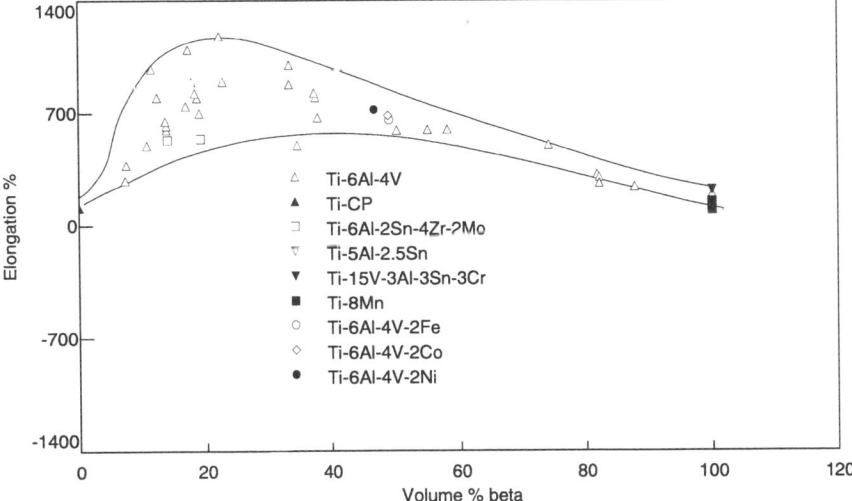

Fig. 10 Elevated temperature ductillity as a function of β phase content for several titanium alloys. Source: N.E. Paton, *Titanium, Science and Technology*, G. Lütjering, U. Zwicker, and W. Bunk, Ed., Deutsche Gesellschaft für Metallkunde, Germany, 1985

Table 1 Superplastic characteristics of titanium alloys

Alloy	Test temperature °C	°F	Strain rate, s^{-1}	m	Elongation, %
Ti-6Al-4V	840-870	1545-1600	1.3×10^{-4} to 10^{-3}	0.75	750-1170
Ti-6Al-5V	850	1560	8×10^{-4}	0.70	700-1100
Ti-6Al-2Sn-4Zr-2Mo	900	1650	2×10^{-4}	0.67	538
Ti-4.5Al-5Mo-1.5Cr	870	1600	2×10^{-4}	0.63-0.81	>510
Ti-6Al-4V-2Ni	815	1500	2×10^{-4}	0.85	720
Ti-6Al-4V-2Co	815	1500	2×10^{-4}	0.53	670
Ti-6Al-4V-2Fe	815	1500	2×10^{-4}	0.54	650
Ti-5Al-2.5Sn	1000	1830	2×10^{-4}	0.49	420
Ti-15V-3Cr-3Sn-3Al	815	1500	2×10^{-4}	0.5	229
Ti-13Cr-11V-3Al	800	1470	...	...	<150
Ti-8Mn	750	1380	...	0.43	150
Ti-15Mo	800	1470	...	0.60	100
Ti-6Al-4Sn-3.5Zr-0.7 Nb-0.5Mo-.3Si	990	1815	1×10^{-4}	0.60	300
Ti-14Al-20Nb- 3V-2Mo	955	1750	2×10^{-4}	0.60	900

the β phase has a self-diffusivity two orders of magnitude higher than the α phase. Changes in elongation with temperature are shown in Fig. 17 for fine-grain Ti-6Al-5V. There is a limited temperature range over which superplastic ductility is observed, a characteristic typical of superplastic titanium alloys, although the superplastic temperature range does vary with the alloy. At lower temperatures and at higher temperatures, the high strain-rate sensitivity of the flow stress is lost, and the related superplastic ductility is absent. The upper limit of superplasticity corresponds to the transformation to all β phase (temperatures above the β transus) and subsequent rapid grain growth. The lower temperature limit is associated with limited diffusivities. For many titanium alloys, a maximum in tensile ductility occurs at temperatures near, but below, the β transus temperature.

As an example of temperature effects, consider the general microstructural characteristics of Ti-6Al-4V. Between the superplastic temperatures of 775 and 925 °C (1430 and 1700 °F), the proportion of β phase varies from about 10 to 50%. The remainder is the α phase, which is generally equiaxed in the superplastically formable condition. At the lower temperatures, relatively little β phase is present. What is present can be found at triple points and long grain boundaries of the α phase and essentially fills the gaps in a contiguous network of α grains. At higher temperatures, where there is a higher volume fraction of β phase, the network of α grains begins to break up. The β phase is no longer solely a grain boundary phase. Above 875 °C (1610 °F), the fraction of the β phase increases rapidly with increasing temperature up to the β transus (990 to 1010 °C, or 1815 to 1850 °F). As the amount of α phase decreases, it retains its granular shape and does not become a grain boundary phase as the β phase was at lower temperatures. Throughout this range of tem-

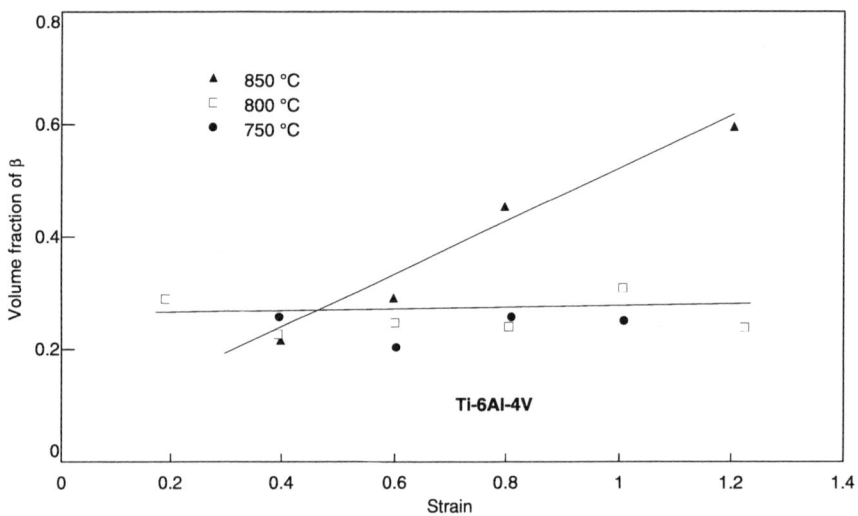

Fig. 11 Volume fraction of β phase versus strain at a strain rate of 10^{-4}/s.
Source: H.S. Yang, G. Gurewitz, and A.K. Mukherjee, *Mater. Trans. JIM*, Vol 32 (No. 5), 1991, p 465

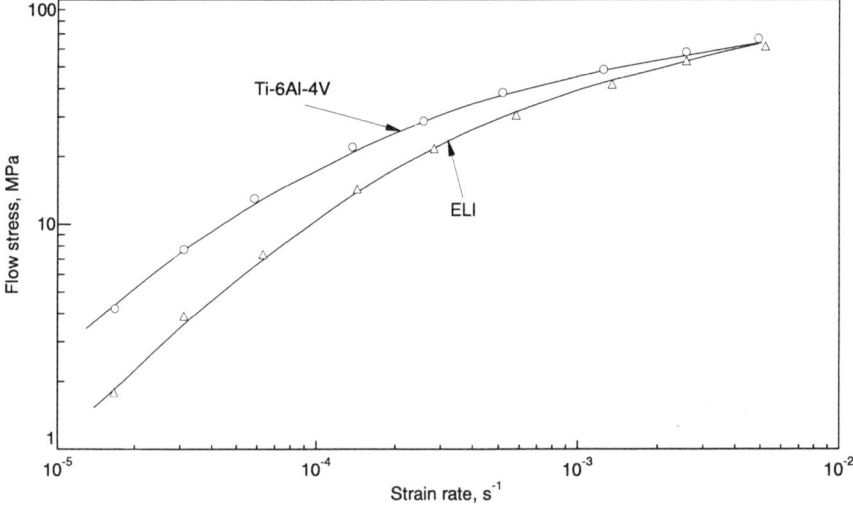

Fig. 12 Log flow stress versus log strain rate for regular grade (7.7 μm) and ELI (8.8 μm) at 850 °C (1560 °F).
Source: T.L. Mackay, S.M.L. Sastry, and C.F. Yolton, Air Force Technical Report AFWAL-TR-80-4038, Sept 1980

peratures, Ti-6Al-4V will exhibit varying degrees of superplasticity. Materials characterization tests, as described above, are necessary to establish the range of temperatures for satisfactory superplastic performance.

Forming Processes

Once the superplastic materials characterization is complete, practical forming issues must still be considered. These include surface preparation/descaling (see Technical Note 9), die material, lubricant, and the pressure/time relationship to maintain a constant strain rate for the geometry being formed. Because of thermal stresses developed during heating and cooling, and the exposure to press and gas pressure forces during forming for extended periods of time, die materials must be able to resist elevated temperature creep deformation. Also, die materials are in contact with the air, parting compounds, argon, and the titanium alloys. Die materials should be inert to each of these components for die life considerations and to prevent titanium contamination.

Die Materials. The industry standard for tooling is ESCO 49C (Fe-22Cr-4Ni-9Mn-5Co). All the U.S.A. SPF houses use 49C, with the exception of Murdock which uses HN (Fe-21Cr-25Ni-2Mn-2Si). 49M (Fe-22Cr-4Ni-9Mn-3Mo) is not recommended. The molybdenum (in 49M) causes catastrophic oxidation to occur resulting in severe pitting on the surface of the tool. Stainless steel 316 behaves the same, which also suggests a deleterious effect from molybdenum.

Another common die alloy for temperatures up to about 1000 °C (1830 °F) is a wrought stainless steel alloy of nominal composition Fe-22Cr-4Ni-9Mn. This alloy exhibits both creep and oxidation resistance for satisfactory operation up to 1000 °C (1830 °F) and is appropriate for forming of most titanium alloys. For higher temperatures, such as is necessary for superplastic forming of γ-TiAl, ceramic dies must be used.

Lubrication. Proper use of lubrication is necessary to reduce friction at rubbing surfaces such as over die radii and to facilitate part removal. This is of particular importance in superplastic forming where proper use of lubricants can prevent preferential thinning at die/formingsheet contact points. Typical lubricant materials include boron nitride and yttria. Each of these materials comes in fine powder form and can be spray coated onto dies and forming materials using isopropyl alcohol as a carrier. For production operations, buildup of lubricants in the die must be avoided.

Rockwell has a patent (Agarwal/Weisert) on "controlling friction" by using yttria (to increase friction) and boron nitride (to reduce friction). These are applied selectively to cause the metal to stick or slide, thus affecting thinning. This approach has been used with limited success. Boeing uses Everlube (a colloidal form of graphite) followed by a *light* application of boron nitride paint. Rohr uses Nicrobraze Orange brazing stopoff (boron nitride with some kind of orange pigment added). Jet Die and Flameco use T50 (colloidal graphite) with boron nitride powder rubbed onto the surface.

Gas pressure techniques are the most common method of superplastic forming of titanium alloys. Aside from materials characterization and the mechanics of forming, the most important component of the forming process is the establishment of a pressure/time profile to maintain strain rates within established limits, i.e., minimize thinning by maximizing the strain-rate sensitivity. The development of the pressure profile depends on part geometry, changes in flow stress as a function of strain, strain rate, and temperature, material thinning, and frictional effects. In general, pressure/time curves for critical areas of the

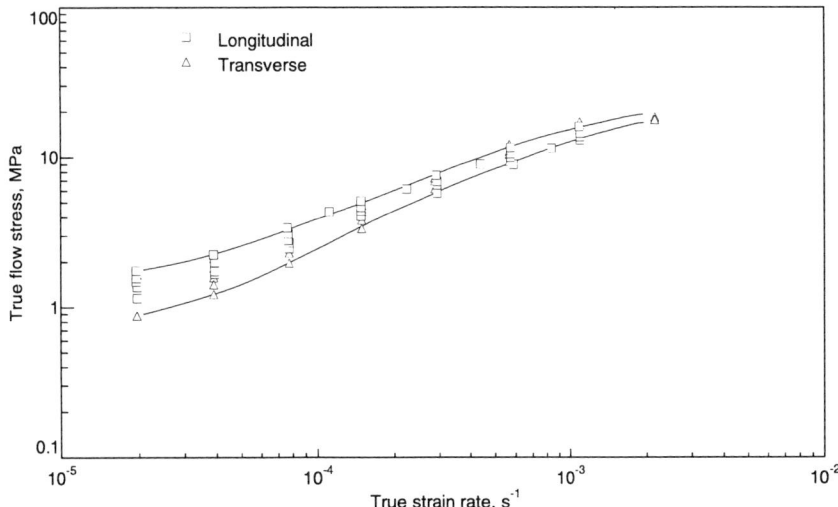

Fig. 13(a) Flow stress versus true strain rate for IMI 834 sheet at 990 °C (1815 °F).
Source: A. Wisbey and P.G. Partridge, "Superplasticity in Advanced Materials, Proceedings: International Conference on Superplasticity in Advanced Materials (ICSAM-91)," S.F. Hari, M. Tokizane, and N. Furushiro, Ed., Osaka, Japan, June 3-6, 1991, p 465

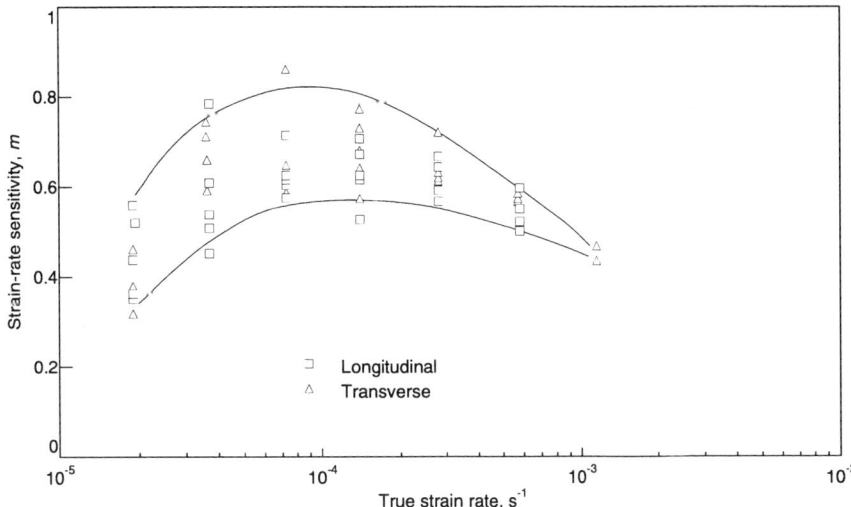

Fig. 13(b) Strain-rate sensitivity versus true strain rate for IMI 834 sheet at 990 °C (1815 °F).
Source: A. Wisbey and P.G. Partridge, "Superplasticity in Advanced Materials, Proceedings: International Conference on Superplasticity in Advanced Materials (ICSAM-91)," S.F. Hari, M. Tokizane, and N. Furushiro, Ed., Osaka, Japan, June 3-6, 1991, p 465

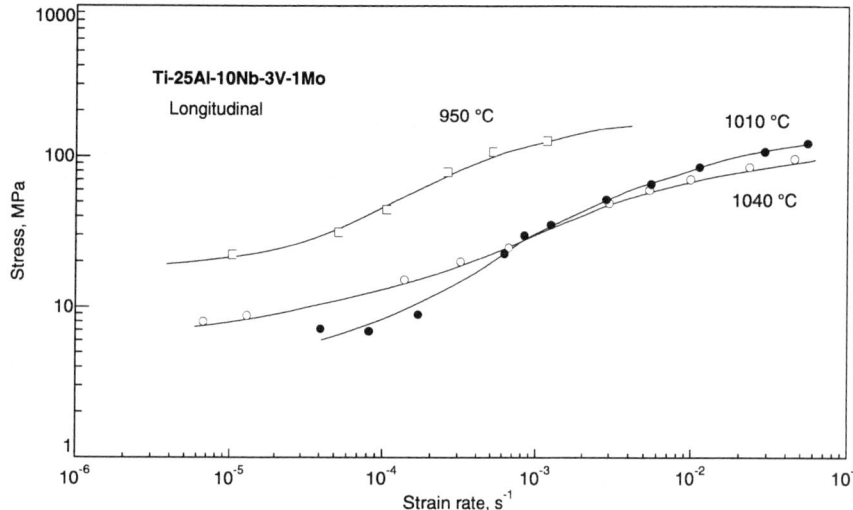

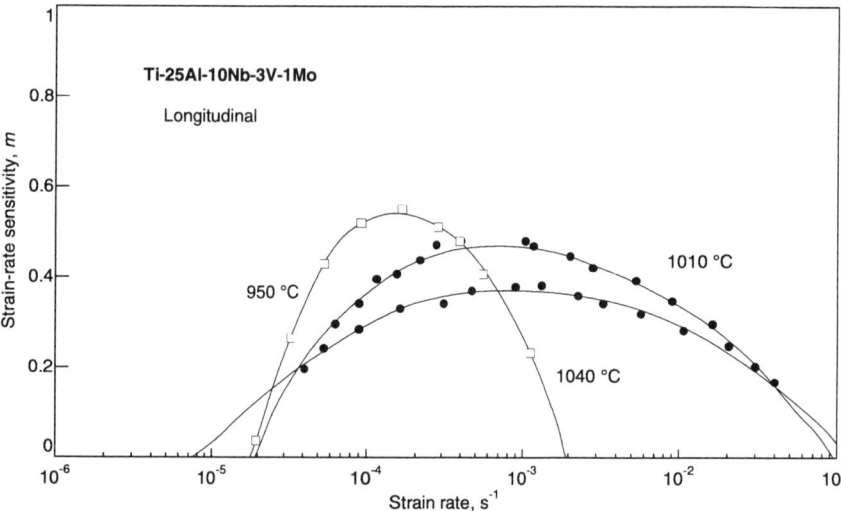

Fig. 14 Effect of test temperature on the stress-strain rate curves for super α_2.
Source: A.K. Ghosh and C.H. Cheng, "Superplasticity in Advanced Materials, Proceedings: International Conference on Superplasticity in Advanced Materials (ICSAM-91)," S.F. Hari, M. Tokizane, and N. Furushiro, Ed., Osaka, Japan, June 3-6, 1991, p 465

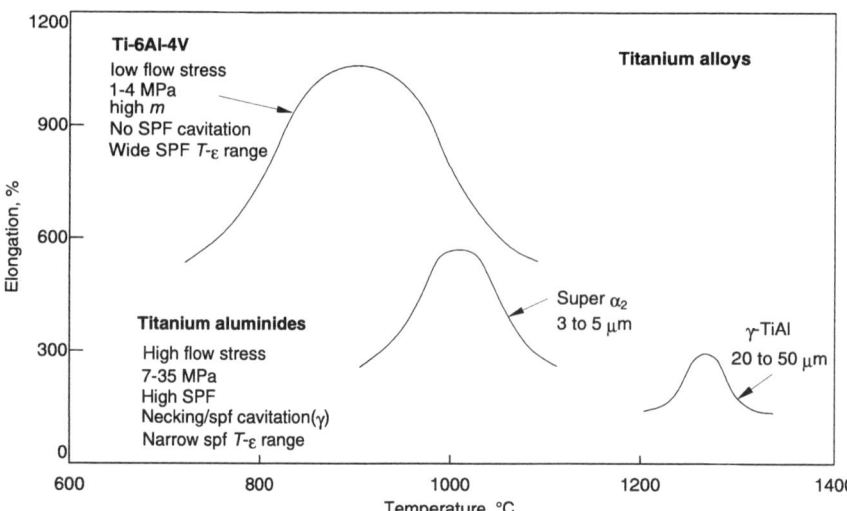

Fig. 15 Superplastic elongation of Ti-6Al-4V and aluminides.
Source: Unpublished research, Rockwell International

forming component are blended to yield a master profile that will reflect the pressure requirement for a critical area at given time. Critical area is defined as the location on the forming component that must remain within strain rate limits to prevent excessive thinning or premature fracture. Depending on component geometry, the critical area will change as forming progresses.

Superplastic Forming and Diffusion Bonding

The versatility of the superplastic forming process for titanium can be enhanced by combining it with diffusion bonding (solid-state joining). Both processes require similar conditions, that is, heat, pressure, clean surfaces, and an inert environment. The combined process is referred to as superplastic forming/diffusion bonding (SPF/DB). Diffusion bonding is carried out simultaneously with superplastic forming, thus eliminating the need for welding or brazing for complex parts.

The SPF/DB process has greatly extended the applicability of superplastic forming. Using SPF/DB, a sheet can be formed onto pre-placed details and diffusion bonded, or two or more sheets can be formed and bonded at selected locations. Figure 18 illustrates the SPF/DB process for three-sheet parts.

Diffusion bonding can be applied only to selected areas of a part by using a stop-off material (Fig. 18) that is placed between the sheets at locations where no bonding is desired. Suitable stop-off materials depend on the alloy being bonded and the temperatures used; yttria and boron nitride have been successfully used.

Applications. Superplastic forming and SPF/DB are rapidly gaining acceptance in the aircraft/aerospace industry. Applications range from simple clips and brackets to major airframe components and other load-bearing structures.

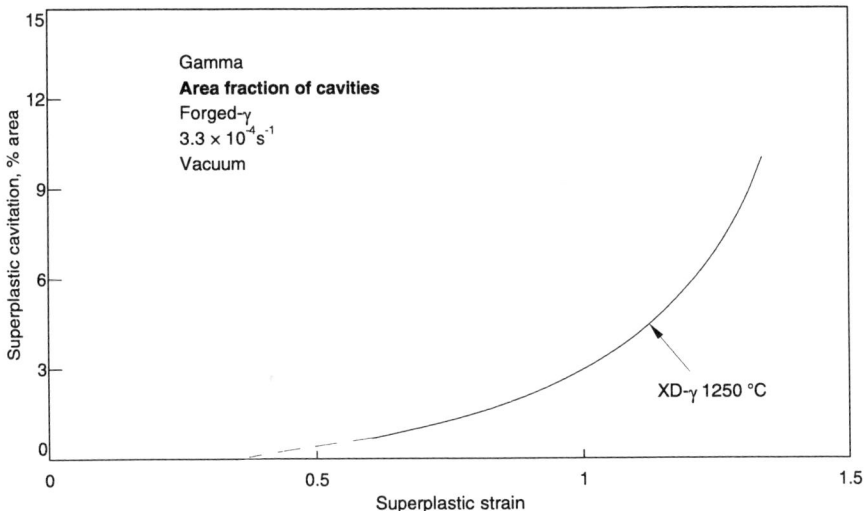

Fig. 16 Cavitation vs strain of gamma Ti-Al.
Source: Unpublished research, Rockwell International

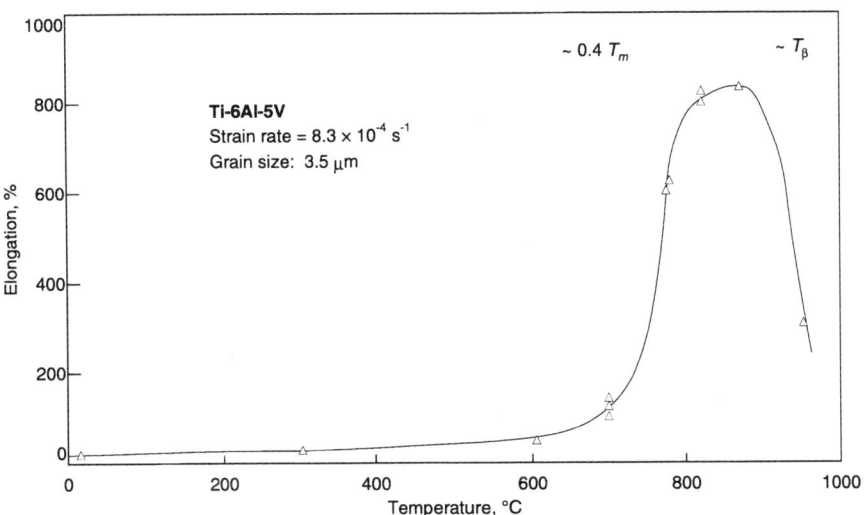

Fig. 17 Elongation vs temperature of Ti-6Al-5V.
Source: O.A. Kaibyshev, I.V. Kazachkev, and R.M. Galeev, *J. Mater. Sci.*, Vol 16, 1991, p 2501

Table 2 Activation energies for superplastic deformation and self-diffusion in titanium alloys

Alloy	Temperature range °C	°F	Activation energy (Q), kcal/mol
Ti-5Al-2.5Sn	800-950	1470-1740	50-65
Ti-6Al-4V	800-950	1470-1740	45
Ti-6Al-4V	850-910	1560-1670	45-99
Ti-6Al-4V	815-927	1500-1700	45-52
Ti-6Al-2Sn-4Zr-2Mo	843-900	1550-1650	38-58
Self-diffusion, α phase	...	...	40.4
Self-diffusion, β phase	...	...	36.5
Self-diffusion, β phase	...	...	31.3

Source: *Metals Handbook*, Vol 14, 9th ed., 1988, p 844

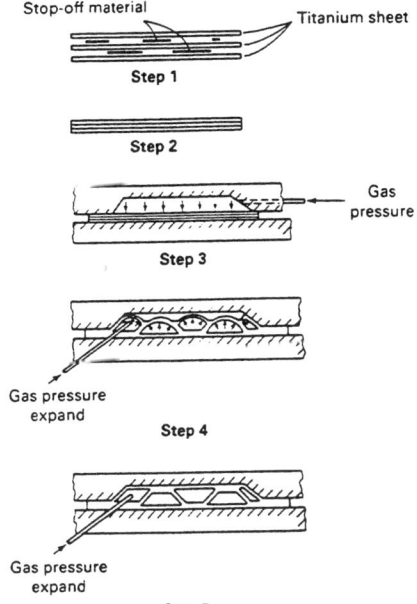

Fig. 18 Schematic showing the sequence of operations for SPF/DB of three-sheet titanium parts.
Source: *Metals Handbook*, Vol 14, 9th ed., p 844

Technical Note 6: Heat Treating

Adapted from *ASM Handbook*, Volume 4

Titanium and titanium alloys are heat treated in order to:

- Reduce residual stresses developed during fabrication (stress relieving)
- Produce an optimum combination of ductility, machinability, and dimensional and structural stability (annealing)
- Increase strength (solution treating and aging)
- Optimize special properties such as fracture toughness, fatigue strength, and high-temperature creep strength

Various types of annealing treatments (single, duplex, β, and recrystallization annealing, for example), and solution treating and aging (or overaging) treatments, are imposed to achieve selected mechanical properties. Stress relieving and annealing may be employed to prevent preferential chemical attack in some corrosive environments, to prevent distortion (a stabilization treatment), and to condition the metal for subsequent forming and fabricating operations.

Key considerations in heat treating of titanium and its alloys — practices that are to be followed and those that should be avoided — are summarized below.

- Provide sufficient stock for post-treatment metal-removal requirements (contaminated metal removal)
- Clean components, fixtures, and furnaces prior to heat treatment (*Caution*: Do not use ordinary tap water in cleaning of titanium components)
- Use temperature controls with an upper cutoff to prevent temperature from exceeding beta transus
- Charge cold components into furnaces operating at the required temperature
- Stack and support components to allow free access of heating and quenching media
- Observe quench-delay requirements to ensure hardening response during aging
- Review property requirements and select optimum heat-treating procedure
- Review strength requirements and select proper aging cycle
- Remove alpha case after all heat treating is complete
- Check for the presence of hydrogen after all processing is complete
- Do not nest components
- Do not allow temperature to exceed beta transus (unless it is specified as a beta anneal process)
- Do not rely on inert atmosphere or vacuum for prevention of oxygen contamination
- Do not rely on hardness tests for measurements of the effects of heat treatment
- Do not pickle assemblies with faying surfaces
- Control pickle/chemical milling bath to prevent excessive hydrogen pickup

Alloy Types and Response to Heat Treatment

The response of titanium and titanium alloys to heat treatment depends on the composition of the metal and the effects of alloying elements on the α-β crystal transformation of titanium, for example, alpha and near-alpha titanium alloys can be stress relieved and annealed, but high strength cannot be developed in these alloys by any type of heat treatment. In contrast, strengthening of beta alloys occurs when the retained beta phase decomposes during aging after solution treatment. For beta alloys, stress-relieving and aging treatments can be combined, and annealing and solution treat-

Table 1 Summary of heat treatments for α–β alloys

Heat-treatment designation	Heat-treatment cycle	Microstructure
Duplex anneal (or overage)	Solution treat at 50–75 °C (90–135 °F) below T$_β$(a), air cool and age for 2–8 h at 540–675 °C (1000–1250 °F)	Primary α, plus Widmanstätten α–β regions
Solution treat and age	Solution treat at ~40 °C (70 °F) below T$_β$, water quench(b) and age for 2–8 h at 535–675 °C (995–1250 °F)	Primary α, plus tempered α' or α β–α mixture
Beta anneal	Solution treat at ~15 °C (30 °F) above T$_β$, air cool and stabilize at 650–760 °C (1200–1400 °F) for 2 h	Widmanstätten α–β colonies
Beta quench (Beta STOA)	Solution treat at ~15 °C (30 °F) above T$_β$, water quench and temper at 650–760 °C (1200–1400 °F) for 2 h	Tempered α'
Recrystallization anneal	925 °C (1700 °F) for 4 h, cool at 50 °C/h (90 °F/h) to 760 °C (1400 °F), air cool	Equiaxed α with β at grain-boundary triple points
Mill anneal	α–β hot work plus anneal at 705 °C (1300 °F) for 30 min to several hours and air cool	Incompletely recrystallized α with a small volume fraction of small β particles

(a) T$_β$ is the beta transus temperature for the particular alloy in question. (b) In more heavily β-stabilized alloys such as Ti-6Al-2Sn-4Zr-6Mo or Ti-6Al-6V-2Sn, solution treatment may be followed by air cooling. Subsequent aging causes precipitation of α phase to form an α–β mixture.

ing may be identical operations.

Because alpha-beta alloys are two-phase alloys, they are the most common and the most versatile of the three types of titanium alloys. Phase compositions, sizes, and distributions can be manipulated by heat treatment (see Table 1) within certain limits to enhance a specific property or to attain a range of strength levels up to moderate thicknesses.

Not all heat-treating cycles are applicable to all titanium alloys, because the various alloys are designed for different purposes. Alloys Ti-5Al-2Sn-2Zr-4Mo-4Cr (commonly called "Ti-17") and Ti-6Al-2Sn-4Zr-6Mo are designed for strength in heavy sections; Ti-6Al-2Sn-4Zr-2Mo for creep resistance; Ti-6Al-2Nb-1Ta-1Mo and Ti-6Al-4V-ELI for resistance to stress corrosion in aqueous salt solutions, and for high fracture toughness; Ti-5Al-2.5Sn for weldability and cryogenic properties; and Ti-6Al-6V-2Sn, Ti-6Al-4V, and Ti-10V-2Fe-3Al for high strength at low-to-moderate temperatures.

Stress Relieving

Titanium and titanium alloys can be stress relieved without adversely affecting strength or ductility. Stress-relieving treatments decrease the undesirable residual stresses that result from (a) nonuniform hot forging deformation, from cold forming and straightening, (b) asymmetric machining of plate (hogouts) or forgings, (c) welding, and (d) thermal stresses from heat treatment. Removal of such stresses helps maintain shape stability and eliminates unfavorable conditions, such as the loss of compressive yield strength commonly known as the Bauschinger effect.

When symmetrical shapes are machined in the annealed condition, employing moderate cuts and uniform stock removal, stress relieving may not be required. Compressor disks made of Ti-6Al-4V have been satisfactorily machined in this manner, conforming with dimensional requirements. In contrast, thin rings made of the same alloy could be machined at a higher production rate to more stringent dimensions by stress relieving 2 h at 540 °C (1000 °F) after rough machining.

Separate stress relieving may be omitted when the manufacturing sequence can be adjusted to employ annealing or hardening as the stress-relieving process. For example, forging stresses may be relieved by annealing prior to machining. Large, thin rings have been effectively processed with minimum distortion by rough machining in the annealed state, followed by solution treating, quenching, partial aging, finish machining, and final aging. Partial aging relieves quenching stresses, and final aging relieves stresses developed during finish machining.

Combinations of time and temperature that are used for stress relieving titanium and titanium alloys are listed in the "Heat Treatment" section of each alloy datasheet. More than one time-temperature combination may yield satisfactory results (see, for example, the stress-relief nomograph in the Ti-6Al-4V datasheet). The higher temperatures usually are used with shorter times, and the lower temperatures with longer times, for effective stress relief. During stress relief of solution-treated and aged titanium alloys, care should be taken to prevent overaging to lower strength. This usually involves selection of a time-temperature combination that provides partial stress relief. The parts, in bulk or in fixtures, may be charged directly into a furnace operating at the stress-relief temperature. If a part is mounted in a massive fixture, a thermocouple should be attached to the largest part of the fixture.

The rate of cooling from the stress-relief temperature down to about 480 °C (900 °F) is important. Cooling rates of less than about 50 °C/h (100 °F/h) are recommended down to this temperature. Below this temperature the cooling rate is optional. Oil or water quenching should not be used to accelerate cooling, however, because this can induce residual stresses by unequal cooling. Furnace or air cooling is acceptable.

Stress-relieving treatments must be based on the metallurgical response of the alloy involved. Generally, this requires holding at a temperature sufficiently high to relieve stresses without causing an undesirable amount of precipitation or strain aging in alpha-beta and beta alloys, or without producing undesirable recrystallization in single-phase alloys that rely on cold work for strength.

Stress relieving of beta alloys and the more highly alloyed alpha-beta compositions should be done using a thermal exposure that is compatible with annealing, solution-treating, stabilization, or aging process.

There are no nondestructive testing methods that can measure the efficiency of a stress-relief cycle other than direct measurement of residual stresses by x-ray diffraction. No significant changes in microstructure due to stress-relieving heat treatments can be detected by optical microscopy.

Table 2 Minimum metal removal after thermal exposure of titanium alloys

Heat-treating temperature		Time at temperature, h	Minimum stock removal per surface(a)	
°C	°F		mm	in.
480 to 593	900 to 1100	Up to 12	0.005	0.0002
594 to 648	1101 to 1200	Up to 4	0.008	0.0003
		4 to 12	0.015	0.0006
649 to 704	1201 to 1300	Up to 1	0.013	0.0005
		1 to 8	0.020	0.0008
		8 to 12	0.025	0.0010
705 to 760	1301 to 1400	Up to 1	0.025	0.0010
		1 to 4	0.036	0.0014
		4 to 8	0.038	0.0015
		8 to 12	0.043	0.0017
761 to 787	1401 to 1450	Up to 1	0.030	0.0012
		1 to 2	0.038	0.0015
		2 to 4	0.046	0.0018
		4 to 8	0.051	0.0020
		8 to 12	0.056	0.0022
788 to 815	1451 to 1500	Up to ½	0.036	0.0014
		½ to 1	0.041	0.0016
		1 to 2	0.051	0.0020
816 to 871	1501 to 1600	Up to ½	0.058	0.0023
		½ to 1	0.066	0.0026
		1 to 2	0.076	0.0030
872 to 898	1601 to 1650	Up to ½	0.058	0.0023
		½ to 1	0.081	0.0032
		1 to 2	0.089	0.0035
899 to 926	1651 to 1700	Up to ½	0.086	0.0034
		½ to 1	0.091	0.0036
		1 to 2	0.107	0.0042
927 to 954	1701 to 1750	Up to ½	0.097	0.0038
		½ to 1	0.107	0.0042
		1 to 2	0.122	0.0048

(a) Values shown are typical, actual values may vary with alloy type. Values are only a guide.

Temperatures used for stress relieving complex weldments of alpha or alpha-beta alloys should be near the high ends of the ranges given in the data sheets. Complex weldments may be defined as those having multiple welds in complex configurations, possibly involving combinations of machine and manual welding. Simple weldments of commercially pure titanium (non-structural applications) often are used without stress relief. If a weldment is used at elevated temperatures in service, a stress relief is recommended to reduce chances of hydrogen embrittlement.

Annealing

Annealing of titanium and titanium alloys serves primarily to increase fracture toughness, ductility at room temperature, dimensional and thermal stability, and creep resistance. Many titanium alloys are placed in service in the annealed state. Because improvement in one or more properties generally is obtained at the expense of some other property, the annealing cycle should be selected according to the objective of the treatment. Common annealing treatments are:

- Mill annealing
- Duplex annealing
- Triplex annealing
- Recrystallization annealing
- Beta annealing

Mill annealing is a general-purpose treatment given to all mill products. It is not necessarily a full anneal and may leave traces of cold or warm working in the microstructures of heavily worked products (particularly sheet). Duplex and triplex annealing alter the shapes, sizes, and distributions of phases to those required for improved creep resistance and fracture toughness. Both recrystallization and beta annealing treatments are used to improve fracture toughness. Beta annealing is done at temperatures above the beta transus of the alloy being annealed.

Straightening, sizing, and flattening may be combined with annealing by use of appropriate fixtures. The parts, in bulk or in fixtures, may be charged directly into a furnace operating at the annealing temperature.

Either air or furnace cooling may be used, but the two methods may result in different levels of tensile properties. For example, air cooling of Ti-6Al-6V-2Sn from the mill-annealing temperature results in lower tensile strength than that obtained by furnace cooling. If distortion is a problem, the cooling rate should be uniform down to 315 °C (600 °F).

Annealing for Thermal Stability. In alpha-beta titanium alloys, thermal stability is a function of beta-phase transformations. During cooling from the annealing temperature, beta may transform and, under certain conditions and in certain alloys, may form the brittle intermediate omega phase. A stabilization annealing treatment is designed to produce a stable beta phase capable of resisting further transformation when exposed to elevated temperatures in service. Alpha-beta alloys that are lean in beta, such as Ti-6Al-4V, can be air cooled from the annealing temperature without impairing their stability. Furnace (slow) cooling may promote formation of Ti_3Al, an ordering reaction that can degrade resistance to stress corrosion. Slight increases in strength (up to 34 MPa, or 5 ksi) can be gained in Ti-6Al-4V and in Ti-6Al-6V-2Sn by cooling from the annealing temperature to 540 °C (1000 °F) at a rate of 56 °C/h (100 °F/h).

To obtain maximum creep resistance and stability in the near-alpha Ti-8Al-1Mo-1V and Ti-6Al-2Sn-4Zr-2Mo, a duplex annealing treatment is employed. This treatment begins with solution annealing at a temperature high in the alpha-beta range, usually 30 to 55 °C (50 to 100 °F) below the beta transus for Ti-8Al-1Mo-1V and 19 to 56 °C (35 to 50 °F) below the beta transus for Ti-6Al-2Sn-4Zr-2Mo. Forgings are held for 1 h (nominal) and then air or fan cooled depending on section size. This treatment is followed by stabilization annealing for 8 h at 595 °C (1100 °F). Final annealing temperature should be at least 55 °C (100 °F) above the maximum anticipated service temperature. Maximum creep resistance can be developed in Ti-6Al-2Sn-4Zr-2Mo by beta annealing or beta processing.

Straightening During Annealing. It may be difficult to prevent distortion of close-tolerance thin sections during annealing. Straightening of bar to close tolerances, and flattening of sheet, present major problems for titanium producers and fabricators. Because of springback and resistance to straightening at room temperature, it is necessary to employ elevated-temperature forming. At annealing temperatures, many titanium alloys have creep resistance low enough to permit straightening during annealing. With proper fixturing, and in some instances judicious weighting, sheet-metal fabrications and thin, complex forgings have been straightened with satisfactory results. Again, uniform cooling to below 315 °C (600 °F) can improve results.

Various jigs and processing techniques have been proposed for annealing titanium in a manner that will yield a flat product. "Creep flattening" and "vacuum creep flattening" are two such techniques. Creep flattening consists of heating titanium sheet between two clean, flat sheets of steel in a furnace containing an oxidizing or inert atmosphere. Vacuum creep flattening is used to produce stress-free flat plate for subsequent machining. The plate is placed on a large, flat ceramic bed that has integral electric-heating elements. Insulation is placed on top of the plate, and a plastic sheet is sealed to the frame. The bed is slowly heated to the annealing temperature while a vacuum is pulled under the plastic. Atmospheric pressure is used to creep flatten the plate.

Solution Treating and Aging

A wide range of strength levels can be obtained in alpha-beta or beta alloys by solution treating and aging. With the exception of the unique Ti-2.5Cu alloy (which relies on strengthening from the classic age-hardening reaction of Ti_2Cu precipitation similar to the formation of Guinier-Preston zones in aluminum alloys), the origin of heat-treating responses of titanium alloys lies in the instability of the high-temperature beta phase at lower temperatures. Heating an alpha-beta alloy to the solution-treating temperature produces a higher ratio of beta phase. This partitioning of phases is maintained by quenching (air cooling may be sufficient for beta or metastable beta alloys, at least in thin sections); on subsequent aging, decomposition of the unstable beta phase occurs, providing high strength. Commercial beta alloys, often supplied in the solution-treated condition, need only be aged.

After being cleaned, titanium components should be loaded into fixtures or racks that will permit free access to the heating and quenching media. Thick and thin components of the same alloy may be solution treated together, but the time at temperature (soaking time) is determined by the thickest section. For most alloys, the rule is 20 to 30 min per inch of thickness to get the required temperature, followed by the required soak time.

Solution Treating α and α-β Alloys). To obtain high strength with adequate ductility, it is necessary to solution treat at a temperature high in the alpha-beta field, normally 28 to 83 °C (50 to 150 °F) below the beta transus of the alloy. If high fracture toughness or improved resistance to stress corrosion is required, beta annealing or beta solution treating may be desirable for α or α-β alloys. A change in the solution-treating temperature of alpha-beta alloys alters the amount of beta phase and consequently changes the response to aging. Selection of solution-treating temperature usually is based upon practical considerations such as the desired level of tensile properties and the amount of ductility to be obtained after aging.

Because solution treating involves heating to temperatures only slightly below the beta transus, proper control of temperature is essential. If the beta transus is exceeded, tensile properties (especially ductility) are reduced and cannot be fully restored by subsequent thermal treatment. A load may be charged directly into a furnace operating at the solution-treating temperature. Although preheating is not essential, it may be used to minimize distortion of complex parts.

Solution Treatment (β Alloys). Beta alloys normally are obtained from producers in the solution-treated condition. If reheating is required, soak times should be only as long as necessary to obtain complete solutioning. Solution-treating temperatures for beta alloys are above the beta transus; because no second phase is present, grain growth can proceed rapidly.

Quenching. The rate of cooling from the solution-treating temperature has an important effect on strength. If the rate is too low, appreciable diffusion may occur during cooling, and decomposition of the altered beta phase during aging may not provide effective strengthening.

For alloys relatively high in beta-stabilizer content, and for products of small section size, air or fan cooling may be adequate; such slow cooling, where allowed by specified mechanical properties, is preferred because it minimizes distortion. Beta alloys generally are air quenched from the solution-treating temperature.

Water or a 5% brine or caustic soda solution is preferred for quenching alpha-beta alloys, because these quenchants provide cooling rates necessary to prevent decomposition of the beta phase obtained by solution treating, to provide maximum response to aging. The need for rapid quenching is further emphasized by short quench-delay-time requirements. Depending on the mass of the sections being heat treated, some alpha-beta alloys can only tolerate a maximum delay of 7 s, whereas more highly beta-stabilized alloys can tolerate quench delay times of up to 20 s. Less sensitive to delayed quenching are alloys such as Ti-6Al-2Sn-4Zr-6Mo and Ti-5Al-2Sn-2Zr-4Mo-4Cr, in which fan air cooling develops good strength through 100 mm (4 in.) sections.

Section size influences effectiveness of quenching and, in turn, response to aging. The amount and type of beta stabilizer in the alloy determine depth of hardening or strengthening. Thick sections exhibit lower tensile properties unless the alloy is highly alloyed with beta stabilizers.

Aging. The final step in heat treating titanium alloys to high strength consists of reheating to an aging temperature between 425 and 650 °C (800 and 1200 °F). Aging causes decomposition of the supersaturated beta phase retained on quenching, or transformation of the martensite to α + β. The time/temperature combination selected depends on the alloy and the required strength.

Aging at or near the annealing temperature will result in overaging. This condition, called solution treated and overaged, or STOA, is sometimes used to obtain modest increases in strength while maintaining satisfactory toughness and dimensional stability.

Although the aged condition is not necessarily one of equilibrium, proper aging produces high strength with adequate ductility and metallurgical stability. Heat treatment of alpha-beta alloys for high strength frequently involves a series of compromises and modifications, depending on the type of service and on special properties that are required, such as ductility and suitability for fabrication. This has become especially true where fracture toughness is important in design and strength is lowered to improve design life.

During aging of some highly beta-stabilized alpha-beta alloys, beta may transform first to a metastable transition phase referred to as omega phase. Retained omega phase, which produces brittleness unacceptable in alloys heat treated for service, can be avoided by severe quenching and rapid reheating to aging temperatures above 425 °C (800 °F). Because a coarse alpha phase forms, however, this treatment might not produce optimum strength properties. An aging practice that ensures sufficient time and temperature for complete transformation of omega into alpha usually is employed. Aging above 425 °C (800 °F) generally is adequate to complete the reaction.

The metastable beta alloys may not require solution treatment. Final hot working, followed by air cooling, leaves these alloys in a condition comparable to a solution-treated state. In some instances, however, solution treating at 790 °C (1450 °F) has produced better uniformity of properties after aging. Aging at 480 °C (900 °F) for 8 to 60 h produces tensile strengths of 1.10 to 1.38 GPa (160 to 200 ksi). Aging for times longer than 60 h may provide higher strengths, but will decrease ductility and fracture toughness if the alloy contains chromium and titanium-chromium compounds are formed. Short aging times can be used on cold-worked material to produce a significant increase in strength over that obtained by cold working. Slow heat-up rates for the age cycle can cause a significant increase in strength of the beta alloys relative to placing the material in a hot furnace. This is due to the formation of omega or a very fine alpha as the material heats up to the aging temperature. These serve as nuclei when the aging temperature is reached, and a very fine α dispersion, with increased strength, results. Use of beta alloys at service temperatures above 315 °C (600 °F) for prolonged periods is not recommended, because the loss of ductility caused by metallurgical instability is progressive.

Other Special Thermal Treatments. Certain physical properties, such as notch strength, fracture toughness, and fatigue resistance can be enhanced in some alloys by special thermal treatments. Four such treatments are:

- *Solution treating and overaging of Ti-6Al-4V*: Heat 1 h at 955 °C (1750 °F), water quench, hold 2 h at 705 °C (1300 °F), air cool. Advantages: improved notch strength, fracture toughness, and creep strength at strength levels similar to those obtained by regular annealing

- *Recrystallized annealing of Ti-6Al-4V or Ti-6Al-4V-ELI*: Heat 4 h or more at 925 to 955 °C (1700 to 1750 °F), furnace cool to 760 °C (1400 °F) at a rate no higher than 56 °C/h (100 °F/h), cool to 480 °C (900 °F) at a rate no lower than 370 °C/h (670 °F/h), air cool to room temperature. Advantages: improved fracture toughness and fatigue-crack-growth characteristics at somewhat reduced levels of strength

- *Beta annealing of Ti-6Al-4V, Ti-6Al-4V-ELI, Ti-6Al-2Sn-4Zr-2Mo*. Ti-6Al-4V or Ti-6Al-4V-ELI: Heat 5 min to 1 h at 1010 to 1040 °C (1850 to 1900 °F), air cool to 650 °C (1200 °F) at a rate of 85 °C/min (150 °F/min) or higher, then heat 2 h at 730 to 790 °C (1350 to 1450 °F), air cool. Advantages: improved fracture toughness, fatigue crack growth resistance, high cycle fatigue strength and resistance to aqueous stress corrosion. Ti-6Al-2Sn-4Zr-2Mo: Heat ½ h at 1020 °C (1870 °F), air cool, then hold 8 h at 595 °C (1100 °F), air cool. Advantages: improved creep strength at elevated temperatures, and improved fracture toughness. Disadvantages: Beta annealing of α-β alloys produces relatively low tensile ductility and fatigue strength.

- *High α-β solution treatment of Ti-5.8Al-4Sn-3.5Zr-0.7Nb-0.5Mo-0.3Si (IMI 834)*: Determine β transus approach curve on small samples by quenching from temperature and plotting the percent of β phase against temperature. Choose solution treatment temperature to give

85 to 88% β. Heat 2 h at temperature, oil quench, age at 700 °C (1290 °F) for 2 h, air cool. Advantages: Excellent combination of creep and fatigue properties with good room-temperature properties

Post Heat-Treating Requirements. Titanium reacts with the oxygen, water, and carbon dioxide normally found in oxidizing heat-treating atmospheres and with hydrogen formed by decomposition of water vapor. Unless the heat treatment is performed in a vacuum furnace or in an inert atmosphere, oxygen will react with the titanium at the metal surface and produce an oxygen-enriched layer commonly called "alpha case." This brittle layer must be removed before the component is put into service. It can be removed by machining, but certain machining operations may result in excessive tool wear. Standard practice is to remove alpha case by other mechanical methods or chemical methods, or by both.

Although temperature and total time at temperature can be used as a general guide to determine how much metal should be removed (see Table 2), oxidation rates of alloys do vary. One method to check for complete removal of alpha case is to etch the component with a solution composed of 18 g ammonium bifluoride per liter of water (2.4 oz/gal). The presence or absence of alpha case is detected by the difference in etching characteristics: light gray shows the presence of alpha case; dark gray indicates its absence. If the component has been machined, such as a forging, the ammonium bifluoride treatment must be preceded by etching in a solution consisting nominally of 5% HF, 30% min HNO_3, balance water. For other mill products, such as plate, microexamination of representative samples removed from the plate is commonly used.

Small amounts of hydrogen (100 to 200 ppm) can be tolerated in titanium alloys with the specific limiting amount determined by the type of alloy. High hydrogen content can lead to premature failure of a component. Hydrogen pickup occurs not only during heat treatment but also during pickling or chemical cleaning operations used to remove the alpha case. The amount of hydrogen pickup can only be determined by chemical analysis. If high hydrogen content is found, vacuum annealing is required. A typical vacuum annealing cycle consists of heating at or close to the annealing temperature for 2 to 4 h in a vacuum of not less than 10 μm.

Hardness testing is not recommended as a nondestructive method of checking the efficiency of heat treatment. The correlation between strength and hardness is poor. Whenever verification of a property is required, the appropriate mechanical test should be used.

Contamination During Heat Treatment

Before being subjected to any thermal treatment, titanium components should be cleaned and dried. *Caution*: Do not use ordinary tap water in cleaning titanium components. Oil, fingerprints, grease, paint, and other foreign matter should be removed from all surfaces. Cleaning is required because the chemical reactivity of titanium at elevated temperatures can lead to its contamination or embrittlement and can increase its susceptibility to stress corrosion. After cleaning, parts should be handled with clean gloves to prevent recontamination. If a component is to be sized, straightened, or heat treated in a fixture, the fixture also should be free of any foreign matter and loosely adhering scale.

Oxidation is not of primary concern in heat treating of titanium, although it may be a problem in sheet-forming operations. Oxygen pickup during heat treatment results in a surface structure composed predominantly of an oxygen enriched alpha phase, which can result in the formation of scale. This condition is detrimental because of the brittle nature of the oxygen-enriched alpha structure, which also is very abrasive to either carbide or high-speed steel machine tools. At 955 °C (1750 °F), the alpha structure can extend 0.2 to 0.3 mm (0.008 to 0.012 in.) below the surface and must be removed. An antioxidant spray coating may be applied to clean sheet-metal parts in order to minimize oxygen pickup. Such coatings work effectively at temperatures up to about 760 °C (1400 °F), but their use does not fully eliminate the need for removing the surface structure after heat treating.

Oxidation rates may vary considerably for different alloys. For commercially pure titanium the oxide film is barely perceptible after exposure at 315 °C (600 °F) in air, but it becomes darker and thicker with increasing temperature and time. Changes in surface color can be used as a rough guide of exposure temperature in air (see Table 3), while time at temperature becomes a more significant factor above about 500 °C (see Fig. 1).

Hydrogen pickup can occur during heat treatment (this is not normally a problem when using properly maintained furnaces), but pickling or chemical cleaning operations used to remove the alpha case is a more likely source of hydrogen contamination. Current specifications limit hydrogen content to a maximum of 125 to 200 ppm, depending on alloy and mill form. Above these limits, hydrogen embrittles some titanium alloys, thereby reducing impact strength and notch tensile strength and causing delayed cracking.

With the exceptions of high vacuum, salt baths, and chemically inert gases, such as argon, all heat-treating atmospheres contain some hydrogen at tem-

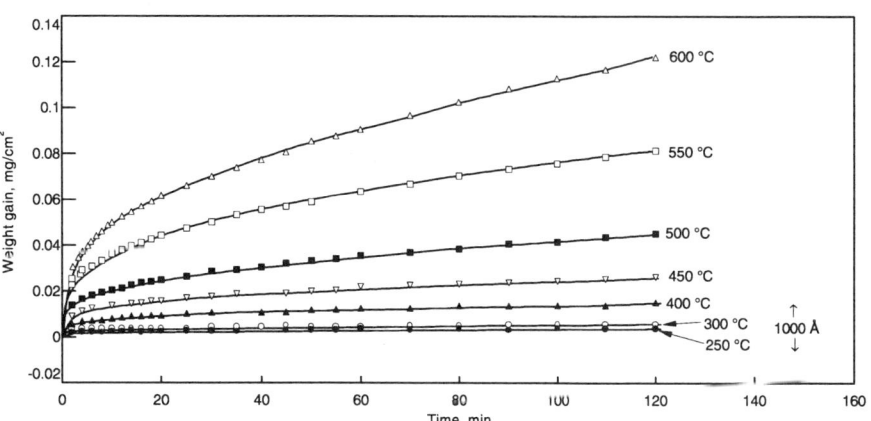

Fig. 1 Oxidation of unalloyed titanium in air

Table 3 Effect of air temperature on titanium's appearance

Air temperature		
°C	°F	Appearance of oxide film(a)
370	700	Straw yellow color
480	900	Blue color
650	1200	Dull gray

(a) Alloying elements and surface contaminants also influence color.

peratures used for annealing titanium. Hydrocarbon fuels produce hydrogen as a by-product of incomplete combustion, and electric furnaces with air atmospheres contain hydrogen from breakdown of water vapor. However, because small amounts of hydrogen can be tolerated in titanium and because inert media are expensive, most titanium heat-treating operations are performed in conventional furnaces employing oxidizing atmospheres with at least 5% excess oxygen in the flue gas, which normally eliminates hydrogen or at least keeps it to an acceptable level.

An oxidizing atmosphere serves in two ways to reduce hydrogen pickup: it reduces the partial pressure of hydrogen in the surrounding atmosphere, and it provides the titanium with a protective surface oxide that retards hydrogen pickup.

Nitrogen is absorbed by titanium during heat treatment at a much slower rate than oxygen and thus does not present a serious contamination problem. Dry nitrogen has been used successfully as a lower-cost protective atmosphere for heat treating of titanium forgings that are to be fully machined after treatment. If absorbed in sufficient quantities, however, nitrogen forms a hard, brittle compound.

Carbon monoxide and carbon dioxide decompose in the presence of hot titanium and produce surface oxidation.

Chlorides. Titanium alloys are subject to stress corrosion when parts with high residual stress are exposed to chlorides at temperatures above 290 °C (550 °F). Salt from fingerprints, and the chlorides contained in some degreasing solutions, may cause stress-corrosion cracking at temperatures above 315 °C (600 °F). Although this phenomenon is readily produced in laboratory testing, and is known to occur during heat treatment, hot-salt cracking has not been a problem in service. Care is required during thermal processing to ensure freedom from chloride contamination.

Growth During Heat Treatment

Solution treating of large parts requires allowances for growth during heat treatment. The growth due to heating may be retained after cooling, and this growth may be increased either by longer holding times at solution temperature or by lower heating rates.

Atmospheres. An oxidizing atmosphere should be maintained during any thermal treatment of titanium. Furnaces normally operated with exothermic atmospheres, endothermic cracked-ammonia atmospheres, or hydrogen atmospheres, because of the danger of hydrogen pickup, should be thoroughly "burned out" before being used for processing of titanium. If dimensions, shape, or size do not permit removal of scale by subsequent pickling or machining, a vacuum or an inert gas such as argon also can be used. Antioxidant coatings suitable for use to 760 °C (1400 °F) can be employed to minimize contamination, though the resulting contamination still should be removed.

Furnaces. Titanium usually is annealed or stress relieved in conventional furnaces constructed for annealing of steel. These furnaces are electric, gas fired, or oil fired, in order of decreasing popularity. The temperature-control equipment for these operations should have an accuracy of ±5.5 °C (±10 °F) and should be capable of controlling and recording the desired temperature within ±14 °C (±25 °F), except where tighter controls are required by the heat treatment specification in use. Some specifications require control of the aging temperature to ±5.5 °C (±10 °F).

Vacuum annealing furnaces are of either the cold-wall or the hot-wall type and may be heated by gas or electricity. Cold-wall electric vacuum furnaces are used most commonly with titanium. Maximum furnace operating temperature depends on the heating elements and radiation shields, but usually these furnaces are designed for a maximum temperature of 980 °C (1800 °F) and are adequate for all titanium alloys. Hot-wall electric furnaces and gas-fired vacuum furnaces have been used in production. When the furnace employs a metallic retort, operating temperatures are held below 980 °C (1800 °F); higher temperatures can be achieved with ceramic retort tubes.

Laboratory vacuum annealing furnaces usually are operated at pressures of 0.1 μm or less, whereas production furnaces are designed to operate at pressure of 0.5 to 3.0 μm.

Vacuum annealing is expensive, and generally it is used only when: (a) a reduction in hydrogen content is required, (b) further hydrogen contamination is prohibited, or (c) allowances that can be made for stock removal are insufficient to permit surface contamination resulting from annealing in air. Hydrogen outgassing at 700 °C (1300 °F) and below is so slow that its cost may be prohibitive. A temperature of 730 °C (1350 °F) is recommended as a minimum, and temperatures from 760 to 790 °C (1400 to 1450 °F) are preferred. At a temperature of 760 °C, removal of 100 ppm of hydrogen from 13 to 25 mm (½ to 1 in.) sections of Ti-6Al-4V alloy required approximately 2 h at a pressure of <10 μm. Actual time at temperature may vary widely depending on the capacity of the furnace to maintain a vacuum.

Solution-treating equipment can vary from a simple furnace with accurate temperature control and a water-quench tank to specialized installations for treating complex parts. Electrically heated furnaces are preferred because they minimize hydrogen pickup, although fuel-fired furnaces with slightly oxidizing conditions or with muffles that protect the metal from combustion products have been used successfully. Resistance and induction heating also have been used to reduce heating times and to minimize contamination during solution treatment. Accuracy of temperature-control equipment should be within ±2.8 °C (±5 °F), and the desired temperature should be controlled within ±14 °C (±25 °F).

To reduce distortion in long, thin products such as sheet or extrusions, in hollow cylinders and in long forgings during immersion quenching, parts often are suspended vertically in an electrically heated drop-bottom furnace. In addition, weights usually are attached to the bottom ends of sheet to improve flatness during heating and to facilitate lowering of the sheet into the quench tank.

Quenching Media. Because rapid cooling is required after solution treating of most titanium alloys, either water or a 5% brine or caustic soda solution is most widely used as the quenching medium. Low-viscosity oil with a high flash point has been used effectively in vertical immersion quenching of sheet to reduce distortion. Quenching oils used with steel provide rapid cooling to 370 to 425 °C (700 to 800 °F), and these oils are satisfactory. Their use, however, should be limited to thin sections to avoid degradation of strength compared to that obtained by water quenching from the same solution temperature. Various concentrations of glycol in water will produce quench rates between those of water and those of oil.

Aging Furnaces. Because they do not involve combustion by-products, furnaces of the electrical-resistance type are preferable for aging titanium and its alloys. Retorts, however, may be used with oil-fired or gas-fired furnaces to avoid contamination. Aging furnaces normally are equipped with internal fans to promote circulation of air or other atmosphere throughout the work zone. Temperature-control equipment should be accurate to ±1.1 °C (±2 °F) and should be capable of controlling temperature within ±8 °C (±15 °F).

At normal aging temperatures of 480 to 595 °C (900 to 1100 °F), a protective atmosphere is not required. Aging in air produces a superficial scale that can be removed easily by mechanical or chemical means.

Fixtures. In fixturing titanium components or assemblies to prevent distortion, the thermal-expansion characteristics of both the titanium alloy and the fixture itself must be considered. Ideally, both the alloy and the fixture will have equivalent thermal expansion characteristics within the intended aging-temperature range. Mild steel is commonly used because it is low in cost and can be made reasonably resistant to oxidation at aging temperatures through use of coatings such as electroless nickel. When mild steel fixtures are used, allowances must be made for the slight difference between the thermal expansion of the mild steel and that of titanium to avoid undesirable growth or distortion of the treated part.

In some applications, it is necessary to reduce or eliminate existing distortion in a part or assembly. This distortion may have resulted from water quenching, from relief of residual stresses during machining, from stresses induced by welding, or from uncontrollable springback after forming. Proper fixturing during aging can be used to minimize such distortion. Fixtures also must guard against sagging; for example, Ti-6Al-4V has a tendency to sag at 955 °C (1750 °F) during solution heat treating. Because titanium alloys exhibit creep behavior within the normal range of aging temperatures, it is possible to fixture and "creep form" components or assemblies to desired shape. Parts also may be sized by fixturing during aging.

Technical Note 7: Machining

Titanium is one of the more difficult metals to machine, but reasonable production rates and excellent surface finish can be obtained with conventional machining methods if the unique characteristics of this metal (such as its reactivity) are taken into account. Relative cutting resistance is shown in Fig. 1(a). Success in machining titanium depends largely on overcoming several of the inherent properties of the metal:

- Titanium is chemically reactive and therefore has a tendency to weld to the tool during the machining process, thus leading to chipping and premature tool failure. Additionally, its low heat conductivity increases the temperature at the tool/workpiece interface, thereby also affecting tool life adversely
- The low elastic modulus of titanium permits greater deflections of workpieces, and proper backup may be required. Greater clearances of cutting tools are also required due to these deflections
- Maintaining a sharp tool is very critical. Susceptibility to surface damage during machining operations, particularly during grinding, is a disadvantage. Even properly processed grinding operations can result in surfaces that lower fatigue life appreciably

Machining Methods. Conventional machining methods include turning; face milling, peripheral end milling, and climb cutting; drilling; tapping; reaming; wheel grinding, belt grinding, abrasive cutting, and hand abrasive grinding; hack sawing; and band sawing. Widely used nontraditional methods include electrochemical machining (ECM), chemical milling (CHM), and laser beam machining (LBM).

Guidelines

The following six guidelines contribute to the efficient machining of titanium:

- *Use of low cutting speeds*: Tool tip temperature is strongly affected by cutting speed. A low cutting speed helps to minimize tool edge temperature and maximize tool life. Lower speeds are required for alloys such as Ti-6Al-4V than are necessary for unalloyed titanium
- *Maintain high feed rates*: Tool temperature is affected less by feed rate than by speed. The highest rate of feed consistent with good practice should be used. The depth of cut should be greater than the work-hardened layer resulting from the previous cut
- *Use a generous quantity of cutting fluid*: A coolant provides heat transfer in addition to washing away chips and reducing cutting forces, thus improving tool life
- *Maintain sharp tools*: Tool wear results in a buildup of metal on the cutting edges and causes poor surface finish, tearing, and deflection of the workpiece
- *Never stop feeding while tool and work are in moving contact*: Permitting a tool to dwell in moving contact with titanium causes work hardening and promotes smearing, galling, and seizing, which may lead to a total tool breakdown
- *Use rigid setups*: Rigidity of the machine tool and workpiece ensures a controlled depth of cut

Tool Materials. Cutting tools used to machine titanium require abrasion resistance and adequate hot hardness. Carbide tools (such as grades C-2 and C-3), if feasible, optimize production rates. General-purpose high-speed tool steels (such as grades M1, M2, M7, and M10) are often suitable. However, best results are generally obtained with more highly alloyed grades, such as T5, T15, M33, and the M40 series.

In recent years, new tool materials such as ceramics, coated carbides, and cubic boron nitride have increased the rate of metal removal of steels, cast irons, and heat-resistant alloys. Also, polycrystalline diamonds have made it possible to machine high-silicon aluminum alloys much more economically. None of these newer developments in cutting tool materials has found application in increasing the productivity of titanium machined parts. Some improvement, however, has been noted by the use of cemented carbides having submicron grain size.

Cutting tool performance is influenced by many factors other than grade selection. Setup, processing methods, grinding techniques, material quality, and the condition of the machine tool and fixturing all influence cutter performance. For example, a setup using climb milling improves tool life as compared with conventional milling or having the work on center. With climb milling, there is less tendency for chipping because the chip leaving the tool is thinner than a chip produced in conventional milling.

Cutting Fluid. The correct use of coolants during machining operations greatly extends cutting tool life. Chemically active cutting fluids transfer heat efficiently and reduce cutting forces between tool and workpiece.

Large quantities of cutting fluid are needed to keep the workpiece and cutting tool cool during high-speed machining operations. Water-base fluids are more efficient than oils. A weak solution of rust inhibitor and/or water-soluble oil (5 to 10%) is the most practical fluid for high-speed cutting operations. Slow speed and complex operations may require chlorinated or sulfurized oils to minimize frictional forces and the galling and seizing tendency of titanium. The best tool life in intermediate-speed operations may be obtained by using a good coolant containing a chemically active additive.

Chlorine-Containing Fluids. If chlorinated cutting fluids are used on alloys that may be subject to stress-corrosion cracking, carefully controlled postmachining cleaning operations must be followed. The general prohibition against the use of chlorine-containing cutting fluids is not universally observed. A number

Table 1 Nominal speeds and feeds for turning titanium and titanium alloys with high-speed tool steel and carbide tools

Material	Hardness, HB	Condition	Depth of cut, mm (in.)(a)	High-speed tool steel Speed, m/min (sfm)	Feed, mm/rev (in./rev)	Tool material grade, AISI	Carbide tool, uncoated Speed, m/min (sfm) Brazed	Indexable	Feed, Tool mm/rev (in./rev)	Material grade
Commercially pure Ti (99.0)	110-170	Annealed	1.0 (0.040)	76 (250)	0.13 (0.005)	T15, M42(b)	160 (525)	172 (565)	0.13 (0.005)	C-3
			4.0 (0.150)	67 (220)	0.25 (0.010)	T15, M42(b)	137 (450)	148 (485)	0.25 (0.010)	C-2
			7.5 (0.300)	53 (175)	0.38 (0.015)	T15, M42(b)	104 (340)	110 (360)	0.38 (0.015)	C-2
			16 (0.625)	...	...	...	52 (170)	55 (180)	0.50 (0.020)	C-2
	140-200	Annealed	1.0 (0.040)	58 (190)	0.13 (0.005)	T15, M42(b)	137 (450)	152 (500)	0.13 (0.005)	C-3
			4.0 (0.150)	52 (170)	0.25 (0.010)	T15, M42(b)	119 (390)	130 (425)	0.25 (0.010)	C-2
			7.5 (0.300)	46 (150)	0.38 (0.015)	T15, M42(b)	88 (290)	98 (320)	0.38 (0.015)	C-2
			16 (0.625)	...	...	...	44 (145)	49 (160)	0.50 (0.020)	C-2
	200-275	Annealed	1.0 (0.040)	35 (115)	0.13 (0.005)	T15, M42(b)	88 (290)	113 (370)	0.13 (0.005)	C-3
			4.0 (0.150)	32 (105)	0.25 (0.010)	T15, M42(b)	76 (250)	98 (320)	0.20 (0.008)	C-2
			7.5 (0.300)	29 (95)	0.38 (0.015)	T15, M42(b)	58 (190)	73 (240)	0.38 (0.015)	C-2
			16 (0.625)	...	...	...	29 (95)	37 (120)	0.50 (0.020)	C-2
Alpha alloys Ti-5Al-2.5Sn, Ti-5Al-2.5Sn-ELI, Ti-6Al-2Nb-1Ta-0.80Mo	300-340	Annealed	1.0 (0.040)	24 (80)	0.13 (0.005)	T15, M42(b)	66 (215)	76 (250)	0.13 (0.005)	C-3
			4.0 (0.150)	21 (70)	0.25 (0.010)	T15, M42(b)	56 (185)	66 (215)	0.20 (0.008)	C-2
			7.5 (0.300)	18 (60)	0.38 (0.015)	T15, M42(b)	43 (140)	49 (160)	0.25 (0.010)	C-2
			16 (0.625)	...	...	...	21 (70)	24 (80)	0.38 (0.015)	C-2
Alpha-beta alloys Ti-6Al-4V, Ti-6Al-4V-ELI, Ti-6Al-2Sn-4Zr-2Mo, Ti-6Al-2Sn-4Zr-2Mo-0.25Si, Ti-6Al-2Sn-4Zr-6Mo	310-350	Annealed	1.0 (0.040)	21 (70)	0.13 (0.005)	T15, M42(b)	52 (170)	69 (225)	0.13 (0.005)	C-3
			4.0 (0.150)	18 (60)	0.25 (0.010)	T15, M42(b)	44 (145)	59 (195)	0.20 (0.008)	C-2
			7.5 (0.300)	15 (50)	0.38 (0.015)	T15, M42(b)	34 (110)	44 (145)	0.25 (0.010)	C-2
			16 (0.625)	...	...	...	17 (55)	21 (70)	0.38 (0.015)	C-2
	320-380	Solution treated and aged	1.0 (0.040)	20 (65)	0.13 (0.005)	T15, M42(b)	49 (160)	58 (190)	0.13 (0.005)	C-3
			4.0 (0.150)	17 (55)	0.25 (0.010)	T15, M42(b)	41 (135)	50 (165)	0.20 (0.008)	C-2
			7.5 (0.300)	14 (45)	0.38 (0.015)	T15, M42(b)	26 (85)	37 (120)	0.25 (0.010)	C-2
			16 (0.625)	...	...	...	15 (50)	18 (60)	0.38 (0.015)	C-2
Beta alloys Ti-3Al-8V-6Cr-4Mo-4Zr, Ti-8Mo-8V-2Fe-3Al, Ti-11.5Mo-6Zr-4.5Sn, Ti-10V-2Fe-3Al, Ti-13V-11Cr-3Al	275-350	Annealed or solution treated	1.0 (0.040)	12 (40)	0.13 (0.005)	T15, M42(b)	38 (125)	49 (160)	0.13 (0.005)	C-3
			4.0 (0.150)	9 (30)	0.25 (0.010)	T15, M42(b)	32 (105)	41 (135)	0.20 (0.008)	C-2
			7.5 (0.300)	7 (25)	0.38 (0.015)	T15, M42(b)	24 (80)	26 (85)	0.25 (0.010)	C-2
			16 (0.625)	...	...	...	12 (40)	15 (50)	0.38 (0.015)	C-2
	350-440	Solution treated and aged	1.0 (0.040)	11 (35)	0.13 (0.005)	T15, M42(b)	36 (110)	38 (125)	0.13 (0.005)	C-3
			4.0 (0.150)	7 (25)	0.25 (0.010)	T15, M42(b)	27 (90)	32 (105)	0.20 (0.008)	C-2
			7.5 (0.300)	...	...	...	21 (70)	24 (80)	0.25 (0.010)	C-2
			16 (0.625)	...	...	...	11 (35)	12 (40)	0.38 (0.015)	C-2

Note: ELI, extra-low interstitial. (a) *Caution:* Check power requirements on heavier depths of cut. (b) Any premium high-speed tool steel can be used. Source: Metcut Research Associates Inc.

of tests have been run in an attempt to develop specifics concerning the prohibition and when it should and should not be observed.

In one study, the U.S. Air Force Materials Laboratory (AFML Technical Report 69-144, 1969) arrived at certain conclusions such as:

- Sulfurized and chlorinated soluble-oil emulsions used in low-stress grinding and end milling/end cutting did not degrade the high-cycle fatigue strength of annealed Ti-6Al-4V (34 HRC) at 25 °C (75 °F) and 315 °C (600 °F) relative to results from a neutral soluble-oil emulsion
- Sulfurized and chlorinated soluble-oil emulsions used in abusive grinding did not degrade the 25 °C (75 °F) high-cycle fatigue strength of Ti-6Al-4V relative to results from a neutral soluble-oil emulsion
- Sulfurized and chlorinated oils and soluble-oil emulsions at crack tip environments did not accelerate 25 °C (75 °F) fatigue crack propagation rates in Ti-6Al-4V at 1 cpm and 1800 cpm relative to results in laboratory air environment
- A 100 h exposure under stress to sulfurized and chlorinated soluble oil emulsions did not affect 25 °C (75 °F) bend test results from low-stress ground and end milled end cut Ti-6Al-4V relative to results from a neutral soluble-oil emulsion

Another series of tests are reported (in German in *WTZ Industr.*, Vol 69, 1979, p 79-82). It was found through Auger analysis that when a cutting fluid containing chlorine was used, surface films were developed that had a thickness equal to or less than 150 nm (1500 Å) and a chlorine content of 3 at.% at the most. Similar films with 1.5 at.% and 100 to 150 nm (1000 to 1500 Å) thickness were obtained by machining titanium with demineralized water. The engineers concluded that the prohibition of machining titanium with lubricants containing chlorine additives can no longer be maintained.

Fire Prevention. Fine particles of titanium can ignite and burn. Use of water-base coolants or large volumes of oil-base coolants generally eliminates the danger of ignition during machining operations. However, an accumulation of titanium fines can pose a fire hazard. Chips, turnings, and other titanium fines should be

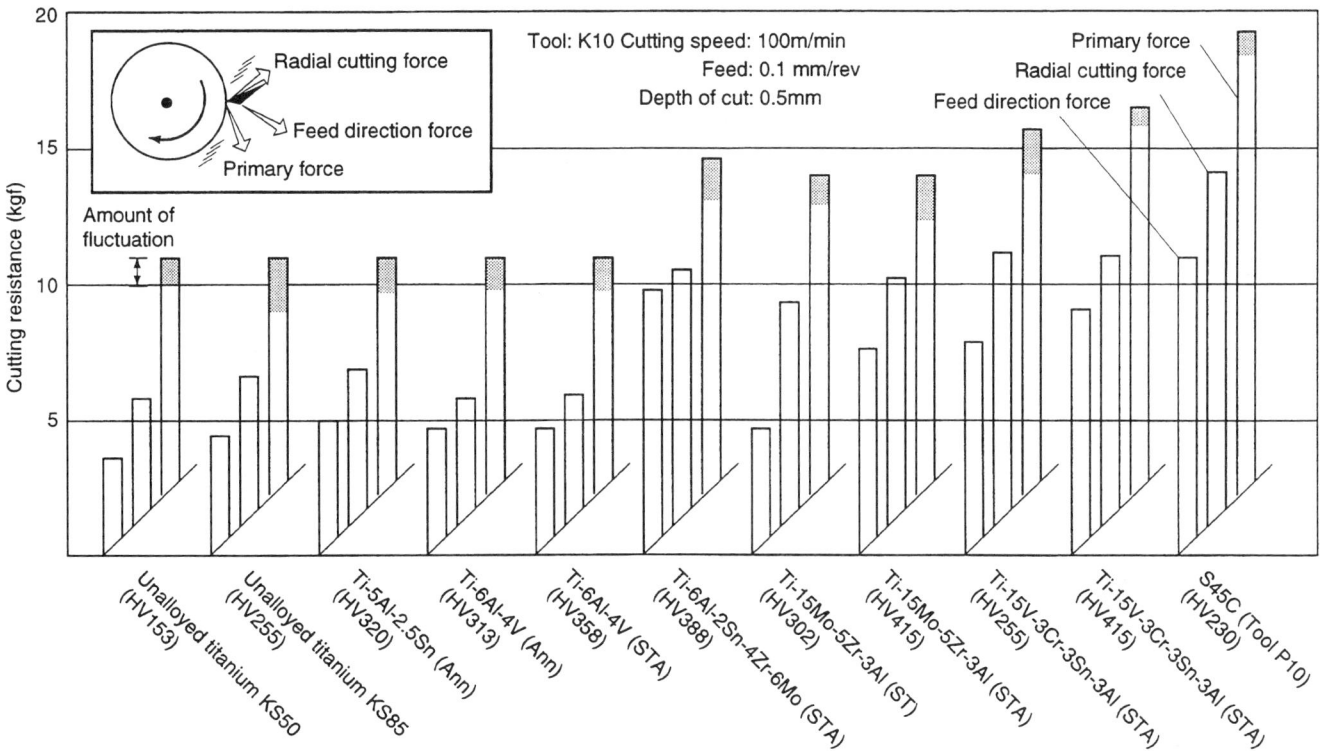

Fig. 1(a) Cutting resistance of various titanium alloys

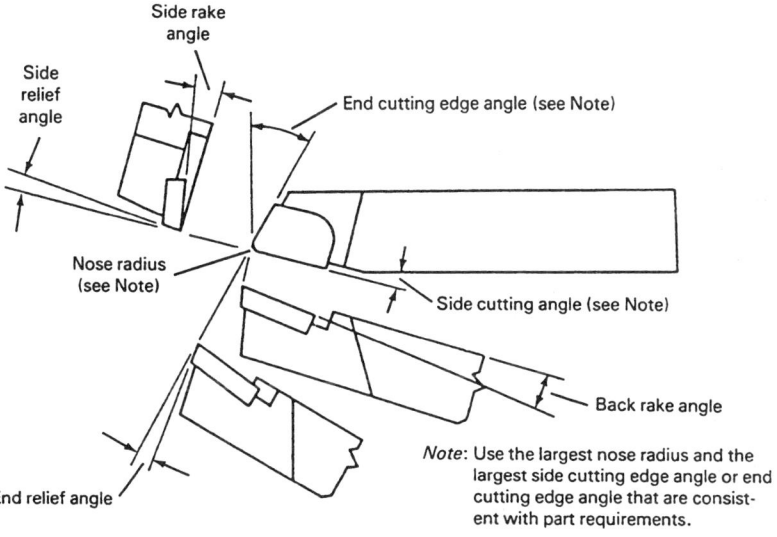

Tool material	Back rake angle, degrees	Side rake angle, degrees	Relief angle, degrees
High-speed steel	0	5	5
Brazed carbide	0	6	7
Indexable carbide	−5(a)	−5(a)	5

(a) For heavy cuts. All rake and relief angles are measured in the normal direction. Wrought and cast titanium alloys having hardness of 110 to 440 HB. Source: Adapted from *Machining Data Handbook*, Vol 2, 3rd ed., Metcut Research Associates Inc., 1980

Fig. 1(b) Typical tool geometries for single-point turning of titanium alloys

collected regularly to prevent undue accumulation and should always be removed from machines at the end of the day.

Salvageable material should be placed in covered, labeled, clean, dry, steel containers and stored, preferably in an outside yard area. Nonsalvageable fines should be properly disposed. Titanium sludge should not be permitted to dry out before being removed to an isolated, outside location.

Dry powders developed for extinguishing combustible metal fines are recommended for the control of titanium fires. For maximum safety, such extinguishers should be readily available to each machinist working with titanium. Dry sand retards but does not extinguish titanium fires. Carbon dioxide and chlorinated hydrocarbons are not recommended. Water should never be applied directly to a titanium fire.

Turning and Boring

Turning is the simplest of machining operations for titanium and its alloys. By means of proper machine parameters and the use of a coolant, surface finishes of 0.50 to 0.75 μm (20 to 30 μin.) rms are obtainable with ±0.025 mm (±0.001 in.) tolerances. Turning, facing, and boring operations on titanium are essentially the

Table 2 Tool geometries for negative rake throwaway inserts

Solid-base tool holders with negative rake tool geometry as follows: back-rake angle, 5 degrees; side-rake angle, 5 degrees; end-relief angle, 5 degrees; side-relief angle, 5 degrees.

Tool holder style	Type insert	End-cutting-edge angle, degrees	Side-cutting-edge angle, degrees
0-degree turning	Triangle	5	0
0-degree turning	Triangle	3	0
0-degree turning	Round	8	0
15-degree turning	Triangle	23	15
15-degree turning	Triangle	18	15
15-degree turning	Square	15	15
15-degree turning	Triangle	20	15
30-degree turning	Triangle	35	30
45-degree turning	Square	45	45
Facing	Triangle	0	0
Facing	Square	15	0
0-degree offset turning	Triangle	3	0

Source: Machining and Grinding of Titanium and Titanium Alloys, NASA Technical Memorandum No. X53312, 1965

Table 3 Tool geometries for positive rake throwaway inserts

Solid-base tool holders with positive rake tool geometry as follows: back-rake angle, 0 degrees; side-rake angle, 5 degrees; end-relief angle, 5 degrees; side-relief angle, 5 degrees.

Tool holder style	Type insert	End-cutting-edge-angle, degrees	Side-cutting-edge angle, degrees
0-degree turning	Triangle	3	0
0-degree turning	Triangle	5	0
15-degree turning	Triangle	23	15
15-degree turning	Square	15	15
15-degree turning	Triangle	20	15
30-degree turning	Triangle	35	30
Facing	Triangle	0	0
Facing	Square	15	0
0-degree offset turning	Triangle	3	0

Source: Machining and Grinding of Titanium and Titanium Alloys, NASA Technical Memorandum No. X53312, 1965

same, and no unusual difficulties are experienced with any of them. They give less trouble than milling, especially when cutting is continuous rather than intermittent. The same speeds used for turning (see Table 1) can be used for boring and facing cuts. However, in most cases, the depths of cut and feeds will have to be reduced during boring because of an inherent lack of rigidity of the operation. The problems to be minimized in turning-type operations include high tool-tip temperatures, and the galling and abrasive properties of titanium toward tool materials. During machining, chips should be expelled from the work area as promptly as possible, particularly during boring. Chips lying on the surface tend to produce chatter and poor surface finishes.

The tool should be examined frequently for nicks or worn flanks. These defects promote galling, increase cutting temperature, accelerate tool wear, and increase residual stresses in the machined surface. Sharp edges of turned titanium surfaces are potential sources of failure and should be removed with a wet file or wet emery. This operation should not be done dry or with oil because of a potential fire hazard. Large amounts of water-base soluble oils (5 to 10% solution) or chemically active coolants (5% sodium nitrate in water) are recommended. Sulfochlorinated oils may be used, if necessary, at low cutting speeds.

Tooling. Carbide tools provide the highest production rates for continuous turning operations. Interrupted cuts, plunge cuts, and grooving are best performed by the softer but tougher high-speed tool steels. Tools must be resharpened or replaced before final tool failure occurs. A 0.38 mm (0.015 in.) wear land for carbide tools and 0.75 mm (0.030 in.) wear land for high-speed tool steel or cast alloy tools can be used as a guide for halting turning operations. High-speed steel and cast alloy tools should be ground on a tool grinder to the tool geometry needed. The same is true for carbide tools; however, off-the-shelf brazed and throwaway carbide tools may fit the rake, lead, and relief angle requirements and are convenient to use. Cutting tools should be carefully ground and finished so that the tool surfaces over which chips pass possess a good finish, with the direction of finishing corresponding to the chip-flow direction. A rough surface can cause a properly designed tool to deteriorate rapidly. The life of a carbide tool can be extended if the sharp cutting edge is slightly relieved by honing.

Substantial reductions in costs are achieved by the use of throwaway tooling inserts, and the performance of mechanically clamped inserts is at least as good as that of brazed tools. Therefore, carbide inserts generally are used for turning operations because of their lower cost per cutting edge. Factors contributing to this saving include:

- Reduced tool-grinding costs
- Reduced tool-changing costs
- Reduced scrap
- Increased use of harder carbides for longer tool life or increased metal-removal rates
- Savings through tool standardization
- Maximum carbide utilization per top dollar

Disposal-type carbide (C-2) inserts in heavy-duty negative rake tool holders provide maximum metal removal at minimum cost in turning. Throwaway carbide inserts are designed to be held mechanically in either positive- or negative-rake tool holders (see Tables 2 and 3). The tool geometries for brazed tools are also shown (see Fig. 1b) although inserts are more common.

Tool angles are important for controlling chip flow, minimizing smearing or chipping, and maximizing heat dissipation. The rake angles and the side-cutting edge angle determine the angle of inclination and chip flow. Relief angles, together with the rake angles, control chipping and smearing.

Positive, zero, or negative rake angles can be used, depending on the alloy, heat-treated condition, the tool material, and machining operation. A negative side rake is recommended for rough turning with carbide tools, while a positive rake is best for finish and semifinish turning or when high-speed tool steels are used.

The side-cutting edge angle influences the cutting temperature near the cutting zone. Larger angles reduce cutting pressure and present longer tool edges. The reduced pressure minimizes heat generation; longer cutting edges allow a greater amount of heat dissipation. Higher values of the side-cutting edge angle generally permit greater feeds and speeds, unless chipping occurs as the cutting load is applied or removed.

Table 4 Nominal roughing speeds and feeds for reaming titanium and titanium alloys

Material	Hardness, HB	Condition	Speed, m/min (sfm)	Feed, mm/rev (in./rev)(a) Reamer diameter						Tool material grade, AISI or C grade
				3 mm (1/8 in.)	6 mm (1/4 in.)	13 mm (1/2 in.)	25 mm (1 in.)	38 mm (1 1/2 in.)	50 mm (2 in.)	
CPTi	110-170	Annealed	53 (175)	0.10 (0.004)	0.20 (0.008)	0.30 (0.012)	0.45 (0.018)	0.55 (0.022)	0.64 (0.025)	M1, M2, M7
			114 (375)	0.10 (0.004)	0.20 (0.008)	0.30 (0.012)	0.45 (0.018)	0.55 (0.022)	0.64 (0.025)	C-2
	140-200	Annealed	43 (140)	0.10 (0.004)	0.20 (0.008)	0.30 (0.012)	0.45 (0.018)	0.55 (0.022)	0.64 (0.025	M1, M2, M7
			114 (375)	0.10 (0.004)	0.20 (0.008)	0.30 (0.012)	0.45 (0.018)	0.55 (0.022)	0.64 (0.025)	C-2
	200-275	Annealed	37 (120)	0.10 (0.004)	0.20 (0.008)	0.30 (0.012)	0.45 (0.018)	0.55 (0.022)	0.64 (0.025)	M1, M2, M7
			91 (300)	0.10 (0.004)	0.20 (0.008)	0.30 (0.012)	0.45 (0.018)	0.55 (0.022)	0.64 (0.025)	C-2
Alpha alloys Ti-5Al-2.5Sn, Ti-5Al-2.5Sn ELI, Ti-6Al-2Nb-1Ta-0.80Mo	300-340	Annealed	21 (70)	0.08 (0.003)	0.18 (0.007)	0.23 (0.009)	0.30 (0.012)	0.38 (0.015)	0.43 (0.017)	T15, M42(b)
			76 (250)	0.08 (0.003)	0.20 (0.008)	0.30 (0.012)	0.40 (0.016)	0.50 (0.020)	0.58 (0.023)	C-2
Alpha-beta alloys Ti-6Al-4V, Ti-6Al-4V-ELI, Ti-6Al-2Sn-4Zr-2Mo, Ti-6Al-2Sn-4Zr-2Mo-0.25Si, Ti-6Al-2Sn-4Zr-6Mo	310-350	Annealed	20 (65)	0.08 (0.003)	0.15 (0.006)	0.25 (0.010)	0.30 (0.012)	0.35 (0.014)	0.40 (0.016)	T15, M42(b)
			61 (200)	0.08 (0.003)	0.15 (0.006)	0.25 (0.010)	0.30 (0.012)	0.35 (0.014)	0.40 (0.016)	C-2
	320-380	STA	15 (50)	0.08 (0.003)	0.18 (0.007)	0.25 (0.010)	0.30 (0.012)	0.35 (0.014)	0.40 (0.016)	T15, M42(b)
			49 (160)	0.08 (0.003)	0.18 (0.007)	0.25 (0.010)	0.30 (0.012)	0.35 (0.014)	0.40 (0.016)	C-2
Beta alloys Beta C Ti-8Mo-8V-2Fe-3Al, Ti-11.5Mo-6Zr-4.5Sn Ti-10V-2Fe-3Al, Ti-13V-11Cr-3Al	275-350	Annealed or solution treated	9 (30)	0.05 (0.002)	0.13 (0.005)	0.18 (0.007)	0.25 (0.010)	0.30 (0.012)	0.35 (0.014)	T15, M42(b)
			23 (75)	0.05 (0.002)	0.13 (0.005)	0.18 (0.007)	0.25 (0.010)	0.30 (0.012)	0.35 (0.014)	C-2
	350-440	STA	6 (20)	0.05 (0.002)	0.10 (0.004)	0.15 (0.006)	0.20 (0.008)	0.25 (0.010)	0.30 (0.012)	T15, M42(b)
			15 (50)	0.05 (0.002)	0.10 (0.004)	0.15 (0.006)	0.20 (0.008)	0.25 (0.010)	0.30 (0.012)	C-2

Note: ELI, extra-low interstitial. (a) Based on four flutes for 3 and 6 mm (1/8 and 1/4 in.) in reamers, six flutes for 13 mm (1/2 in.) reamers, and eight flutes for 25 mm (1 in.) and larger reamers. (b) Any premium high-speed tool steel can be used. Source: Metcut Research Associates Inc.

Relief angles between 5 and 12 degrees can be used on titanium. Angles less than 5 degrees encourage smearing of titanium on the flank of the tool. Relief angles around 10 degrees are better, although some chipping can occur. Chip breaking devices should be used for good chip control.

Reaming and boring are related processes with overlapping applications. Reaming to a tolerance of +0.05 to −0.000 mm (+0.002 to −0.000 in.) is possible. Reamer margins tend to gall and seize, but proper tool design and operating conditions effectively eliminate the problem. Sufficient stock must be available to provide continuous cutting and prevent galling and work hardening.

High-speed steel and carbide reamers are generally effective. However, tool steels deteriorate rapidly after tool wear starts. Spiral-flute reamers generally provide longer life than do straight-flute reamers. Sulfochlorinated oils appear to be the best cutting fluid. However, water-base oil emulsions are also effective, particularly with the softer, unalloyed titanium grades. Nominal cutting speeds are shown for roughing and finishing (see Tables 4 and 5).

Milling

The principal methods of milling are classified as peripheral, face, and end. These terms refer to the type of cutter used (see Fig. 2) and to the relationship of the spindle to the surface being milled. In some cases, the differences between the three methods are clearly defined, but more often a given milling operation is a combination of two methods. This is particularly true of end milling, which is almost invariably a combination of peripheral and face milling.

The milling method selected for a specific application depends largely on the amount of metal to be removed, workpiece size and shape, and the configuration to be milled. Total quantity to be produced, production rate (quantity per unit of time), work metal hardness, and cost are more likely to influence modifications of procedure within a given method than to determine selection of the method itself.

Milling machines are capable of machining holes and locating them with a fair degree of accuracy; tolerances range from 25 to 50 μm (0.001 to 0.002 in.). A milling machine is economical for doing such work in small quantities without additional equipment. If the holes do not need to be located accurately, a drill press will perform the task more quickly and easily. For large quantities, the milling machine is usually slower and unable to compete with jigs on drilling machines or with production boring machines. Jig boring machines are necessary where holes must be located more precisely than can be done with milling machines. Large pieces require the capacity and range of horizontal boring machines. Flat, straight, and many curved and irregular surfaces can be shaped, planed, or broached as well as milled.

Tool Wear. Welding and edge chipping are the basic tool wear problems when milling titanium. The amount of titanium welded on cutter edges is proportional to the chip thickness as each tooth leaves the cut. The weld metal and part of the underlying cutting edge then chips off as each tooth re-enters the cut, thus start-

Table 5 Nominal finishing speeds and feeds for reaming (see notes for Table 4)

Material	Hardness, HB	Condition	Speed, m/min (sfm)	3 mm (1/8 in.)	6 mm (1/4 in.)	13 mm (1/2 in.)	25 mm (1 in.)	38 mm (1 1/2 in.)	50 mm (2 in.)	Tool material grade, AISI or C grade
CPTi	110-170	Annealed	18 (60)	0.10 (0.004)	0.15 (0.006)	0.25 (0.010)	0.30 (0.012)	0.38 (0.015)	0.50 (0.020)	M1, M2, M7
			23 (75)	0.10 (0.004)	0.15 (0.006)	0.25 (0.010)	0.30 (0.0212)	0.38 (0.015)	0.50 (0.020)	C-2
	140-200	Annealed	15 (50)	0.10 (0.004)	0.15 (0.006)	0.25 (0.010)	0.30 (0.012)	0.35 (0.014)	0.40 (0.016)	M1, M2, M7
			20 (65)	0.10 (0.004)	0.15 (0.006)	0.25 (0.010)	0.30 (0.012)	0.35 (0.014)	0.40 (0.016)	C-2
	200-275	Annealed	14 (45)	0.10 (0.004)	0.15 (0.006)	0.25 (0.010)	0.30 (0.012)	0.35 (0.014)	0.40 (0.016)	M1, M2, M7
			18 (60)	0.10 (0.004)	0.15 (0.006)	0.25 (0.010)	0.30 (0.012)	0.35 (0.014)	0.40 (0.016)	C-2
Alpha alloys Ti-5Al-2.5Sn, Ti-5Al-2.5Sn-ELI, Ti-6Al-2Nb-1Ta-0.80Mo	300-340	Annealed	6 (20)	0.10 (0.004)	0.15 (0.006)	0.25 (0.010)	0.30 (0.012)	0.35 (0.014)	0.40 (0.016)	T15, M42(b)
			11 (35)	0.10 (0.004)	0.15 (0.006)	0.25 (0.010)	0.30 (0.012)	0.35 (0.014)	0.40 (0.016)	C-2
Alpha-beta alloys Ti-6Al-4V, Ti-6Al-4V-ELI,	310-350	Annealed	6 (20)	0.10 (0.004)	0.15 (0.006)	0.25 (0.010)	0.30 (0.012)	0.35 (0.014)	0.40 (0.016)	T15, M42(b)
			11 (35)	0.10 (0.004)	0.15 (0.006)	0.25 (0.010)	0.30 (0.012)	0.35 (0.014)	0.40 (0.016)	T15, M42(b) C-2
Ti-6Al-2Sn-4Zr-2Mo, 320-380	STA		6 (20)	0.08 (0.003)	0.13 (0.005)	0.20 (0.008)	0.25 (0.010)	0.30 (0.012)	0.35 (0.014)	T15, M42(b) C-2
Ti-6Al-2Sn4Zr-2Mo-0.25Si, Ti-6Al-2Sn-4Zr-6Mo			11 (35)	0.08 (0.003)	0.13 (0.005)	0.20 (0.008)	0.25 (0.010)	0.30 (0.012)	0.35 (0.014)	T15, M42(b) C-2
Beta alloys Ti-3Al-8V-6Cr-4Mo-4Zr, Ti-8Mo-8V-2Fe-3Al,	275-350	Annealed or solution treated	5 (15)	0.08 (0.003)	0.13 (0.005)	0.20 (0.008)	0.25 (0.010)	0.30 (0.012)	0.35 (0.014)	T15, M42(b)
			9 (30)	0.08 (0.003)	0.13 (0.005)	0.20 (0.008)	0.25 (0.010)	0.30 (0.012)	0.35 (0.014)	C-2
Ti-11.5Mo-6Zr-4.5Sn, Ti-10V-2Fe-3Al, Ti-13-113	350-440	STA	3 (10)	0.08 (0.003)	0.13 (0.005)	0.20 (0.008)	0.25 (0.010)	0.30 (0.012)	0.35 (0.014)	T15, M42(b)
			8 (25)	0.08 (0.003)	0.13 (0.005)	0.20 (0.008)	0.25 (0.010)	0.30 (0.012)	0.35 (0.014)	C-2

ing a wearland. The welding/chipping behavior described can be minimized by providing thin exit chips characteristic of down (climb) milling. Slower speeds and light feeds also reduce chipping and permit lower cutting temperatures.

Other tool wear problems include heat, deflection, and abrasion. High cutting temperatures soften chips, which tend to clog and load milling cutters. Deflection of thin parts and slender milling cutters promotes rubbing and adds heat. Abrasive oxide surfaces on titanium can notch the cutter at the depth-of-cut line. When forging scale is present, the nose of each tooth must be kept below the scale to avoid rapid tool wear.

The surface finish obtainable by milling depends on work metal composition and condition, speed, feed, tool material, tool design, and cutting fluid. Peripheral and face milling produce different types of surfaces because of the different relations of cutter rotation to the workpiece in the two methods. A finish of 3.20 µm (125 µin.) can usually be obtained with either carbide or high-speed steel cutters without stringent control of process variables. Finishes of 1.60 µm (63 µin.) or less are often obtained in milling with carbide tools. Values as low as 0.4 µm (17 µin.) rms are possible in finishing cuts.

Power Requirements. The power required for a milling operation is usually computed from the metal removal rate, as follows:

$$P_n = uvdw \qquad (\text{Eq 1})$$

where P_n is the power required at the cutter, u is the specific energy (in

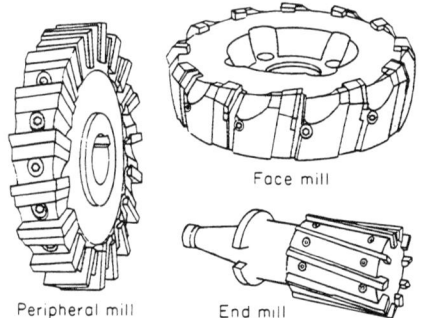

Fig. 2 Three types of milling cutters

Table 6 Typical tool geometries for face milling of titanium alloys

Tool material	Axial rake, degrees	Radial rake, degrees	Axial relief, degrees	Radial relief, degrees
High-speed steel	5	5	10 to 12	10 to 12
Indexable carbide	0 to –5	0 to –5	10 to 12	10 to 12
Brazed carbide	0 to –5	–10	10 to 12	10 to 12

Wrought and cast titanium alloys having hardness of 110 to 440 HB. Source: *Machining Data Handbook*, Vol 2, 3rd ed., Metcut Research Associates Inc., 1980

Table 11 Nominal speeds and feeds for end milling (peripheral) of titanium alloys with C-2 carbide

Material	Hardness, HB	Condition	Radial depth of cut, mm (in.)(a)	Speed, m/min (sfm)	Carbide tool Feed, mm/tooth (in./tooth) Cutter diameter			
					9 mm (3/8 in.)	13 mm (1/2 in.)	19 mm (3/4 in.)	25-50 mm (1-2 in.)
Commercially pure Ti (99.0)	110-170	Annealed	0.5 (0.020)	130 (425)	0.025 (0.01)	0.08 (0.003)	0.13 (0.005)	0.18 (0.007)
			1.5 (0.060)	119 (390)	0.05 (0.002)	0.10 (0.004)	0.15 (0.006)	0.20 (0.008)
			diam/4	73 (240)	0.038 (0.0015)	0.08 (0.003)	0.13 (0.005)	0.18 (0.007)
			diam/2	44 (180)	0.025 (0.001)	0.05 (0.002)	0.10 (0.004)	0.15 (0.006)
	140-200	Annealed	0.5 (0.020)	122 (400)	0.025 (0.001)	0.08 (0.003)	0.13 (0.005)	0.18 (0.007)
			1.5 (0.060)	113 (370)	0.05 (0.002)	0.10 (0.004)	0.15 (0.006)	0.20 (0.008)
			diam/4	70 (230)	0.038 (0.0015)	0.08 (0.003)	0.13 (0.005)	0.18 (0.007)
			diam/2	53 (175)	0.025 (0.001)	0.05 (0.002)	0.10 (0.004)	0.15 (0.006)
	200-275	Annealed	0.5 (0.020)	107 (350)	0.025 (0.001)	0.05 (0.002)	0.13 (0.005)	0.18 (0.007)
			1.5 (0.060)	99 (325)	0.05 (0.002)	0.08 (0.003)	0.15 (0.006)	0.20 (0.008)
			diam/4	61 (200)	0.038 (0.0015)	0.05 (0.002)	0.13 (0.005)	0.18 (0.007)
			diam/2	46 (150)	0.025 (0.001)	0.025 (0.001)	0.10 (0.004)	0.15 (0.006)
Alpha alloys Ti-5Al-2.5Sn, Ti-5Al-2.5Sn-ELI, Ti-6Al-2Nb-1Ta-0.80 Mo	300-340	Annealed	0.5 (0.020)	91 (300)	0.025 (0.001)	0.05 (0.002)	0.13 (0.005)	0.18 (0.007)
			1.5 (0.060)	84 (275)	0.05 (0.002)	0.08 (0.003)	0.15 (0.006)	0.20 (0.008)
			diam/4	52 (170)	0.038 (0.0015)	0.05 (0.002)	0.13 (0.005)	0.15 (0.006)
			diam/2	40 (130)	0.025 (0.001)	0.025 (0.001)	0.10 (0.004)	0.13 (0.005)
Alpha-beta alloys Ti-6Al-4V, Ti-6Al-4V-ELI, Ti-6Al-2Sn-4Zr-2Mo, Ti-6Al-2Sn-4Zr-2Mo-0.25Si, Ti-6Al-2Sn-4Zr-6Mo	310-350	Annealed	0.5 (0.020)	88 (290)	0.025 (0.001)	0.05 (0.002)	0.13 (0.005)	0.18 (0.007)
			1.5 (0.060)	79 (260)	0.05 (0.002)	0.08 (0.003)	0.15 (0.006)	0.20 (0.008)
			diam/4	49 (160)	0.038 (0.0015)	0.05 (0.002)	0.13 (0.005)	0.15 (0.006)
			diam/2	38 (125)	0.025 (0.001)	0.025 (0.001)	0.10 (0.004)	0.13 (0.005)
	320-380	Solution treated and aged	0.5 (0.020)	69 (225)	0.025 (0.001)	0.05 (0.002)	0.10 (0.004)	0.15 (0.006)
			1.5 (0.060)	61 (200)	0.05 (0.002)	0.08 (0.003)	0.13 (0.005)	0.18 (0.007)
			diam/4	38 (125)	0.038 (0.0015)	0.05 (0.002)	0.13 (0.005)	0.15 (0.006)
			diam/2	30 (100)	0.025 (0.001)	0.025 (0.001)	0.10 (0.004)	0.13 (0.005)
Beta alloys Ti-3Al-8V-6Cr-4Mo-4Zr, Ti-8Mo-8 3Al, V-2Fe-Ti-11.5Mo-6Zr-4.5Sn, Ti-10V-2Fe-3Al, Ti-13V-11Cr-3Al	275-350	Annealed or solution treated	0.5 (0.020)	46 (150)	0.025 (0.001)	0.05 (0.002)	0.13 (0.005)	0.18 (0.007)
			1.5 (0.060)	40 (130)	0.038 (0.0015)	0.08 (0.003)	0.15 (0.006)	0.20 (0.008)
			diam/4	23 (75)	0.025 (0.001)	0.05 (0.002)	0.13 (0.005)	0.15 (0.006)
			diam/2	15 (50)	0.018 (0.0007)	0.025 (0.001)	0.10 (0.004)	0.13 (0.005)
	350-440	Solution treated and aged	0.5 (0.020)	38 (125)	0.018 (0.0007)	0.038 (0.0015)	0.08 (0.003)	0.13 (0.005)
			1.5 (0.060)	34 (110)	0.025 (0.001)	0.05 (0.002)	0.10 (0.004)	0.15 (0.006)
			diam/4	18 (60)	0.018 (0.0007)	0.038 (0.0015)	0.08 (0.003)	0.13 (0.005)
			diam/2	14 (45)	0.013 (0.0005)	0.025 (0.001)	0.05 (0.002)	0.10 (0.004)

Note: ELI, extra-low interstitial. (a) For standard-length end mills, the maximum axial depth may be one and one-half times the cutter diameter. (b) Any premium high-speed tool steel can be used. Source: Metcut Research Associates Inc.

General Guidelines. Several specific guidelines must be considered in milling titanium and its alloys:

- Climb milling should be used as an alternative to conventional milling, when possible, to minimize tool chipping
- Slow speeds and uniform, positive feeds help to minimize tool temperature and wear
- Tools should not be allowed to dwell in the cut or rub across the workpiece

Another problem in milling titanium, particularly in the case of extrusions, is distortion originating from the release of stresses originally imposed by the basic mill processing operation. Distortion occurs when unequal amounts of metal are removed from opposite surfaces, or by the machining operation itself.

Climb milling techniques are usually used for carbide and cast alloy cutters to encourage formation of a thin chip. Conventional milling is usually more suitable for high-speed steel tools and for removing scale. Climb milling can be used for most milling applications. Its widespread use has been prevented by the lack of rigid machines with backlash eliminators, which are essential for climb milling. With such equipment, climb milling has several advantages over conventional milling:

- Fixtures and holding devices are simpler and less costly because climb milling exerts a downward force on the workpiece
- Cutters with higher rake angles can be used, decreasing power requirements
- Chip disposal is easier because chips pile up behind the cutter rather than in front of it
- Cutter wear is less because chip thickness is maximum at the start of the cut
- Finishes are generally improved because the rubbing action in starting a chip is eliminated

The main advantage of conventional milling is the lower impact encountered at initial tooth-workpiece engagement (zero chip thickness).

Feed rates for milling titanium gener-

Table 10 Nominal speeds and feeds for end milling (peripheral) of titanium alloys with high-speed tool steel

Material	Hardness, HB	Condition	Radial depth of cut, mm (in.)(a)	Speed, m/min (sfm)	High-speed tool steel Feed, mm/tooth (in./tooth) Cutter diameter			25-50 mm (1-2 in.)	Tool material grade, AISI
					9 mm (3/8 in.)	13 mm (1/2 in.)	19 mm (3/4 in.)		
Commercially pure Ti (99.0)	110-170	Annealed	0.5 (0.020)	53 (175)	0.05 (0.002)	0.08 (0.003)	0.13 (0.005)	0.15 (0.006)	M2, M3, M7
			1.5 (0.060)	49 (160)	0.08 (0.003)	0.10 (0.004)	0.15 (0.006)	0.18 (0.007)	M2, M3, M7
			diam/4	26 (85)	0.038 (0.0015)	0.05 (0.002)	0.08 (0.003)	0.10 (0.004)	M2, M3, M7
			diam/2	18 (60)	0.025 (0.001)	0.038 (0.0015)	0.05 (0.002)	0.08 (0.003)	M2, M3, M7
	140-200	Annealed	0.5 (0.020)	52 (170)	0.038 (0.0015)	0.08 (0.003)	0.13 (0.005)	0.15 (0.006)	M2, M3, M7
			1.5 (0.060)	46 (150)	0.05 (0.002)	0.10 (0.004)	0.15 (0.006)	0.18 (0.007)	M2, M3, M7
			diam/4	26 (85)	0.038 (0.0015)	0.05 (0.002)	0.08 (0.003)	0.10 (0.004)	M2, M3, M7
			diam/2	18 (60)	0.025 (0.001)	0.038 (0.0015)	0.05 (0.002)	0.08 (0.003)	M2, M3, M7
	200-275	Annealed	0.5 (0.020)	46 (150)	0.025 (0.001)	0.05 (0.002)	0.10 (0.004)	0.13 (0.005)	M2, M3, M7
			1.5 (0.060)	40 (130)	0.05 (0.002)	0.08 (0.003)	0.13 (0.005)	0.15 (0.006)	M2, M3, M7
			diam/4	23 (75)	0.038 (0.0015)	0.05 (0.002)	0.08 (0.003)	0.10 (0.004)	M2, M3, M7
			diam/2	15 (50)	0.025 (0.001)	0.038 (0.0015)	0.05 (0.002)	0.08 (0.003)	M2, M3, M7
Alpha alloys Ti-5Al-2.5Sn, Ti-5Al-2.5 Sn-ELI, Ti-6Al-2Nb-1Ta-0.80Mo	300-340	Annealed	0.5 (0.020)	34 (110)	0.025 (0.001)	0.05 (0.002)	0.10 (0.004)	0.13 (0.005)	T15(b)
			1.5 (0.060)	30 (100)	0.05 (0.002)	0.08 (0.003)	0.13 (0.005)	0.15 (0.006)	T15
			diam/4	17 (55)	0.025 (0.001)	0.038 (0.0015)	0.05 (0.002)	0.08 (0.003)	T15
			diam/2	12 (40)	0.025 (0.001)	0.025 (0.001)	0.038 (0.0015)	0.05 (0.002)	T15
Alpha-beta alloys Ti-6Al-4V, Ti-6Al-4V-ELI, Ti-6Al-2Sn-4Zr-2Mo, Ti-6Al-2Sn-4Zr-2Mo-0.25Si, Ti-6Al-2Sn-4Zr-6Mo	310-350	Annealed	0.5 (0.020)	30 (100)	0.025 (0.001)	0.05 (0.002)	0.10 (0.004)	0.13 (0.005)	T15(b)
			1.5 (0.060)	27 (90)	0.05 (0.002)	0.08 (0.003)	0.13 (0.005)	0.15 (0.006)	T15
			diam/4	15 (50)	0.025 (0.001)	0.038 (0.0015)	0.05 (0.002)	0.08 (0.003)	T15
			diam/2	11 (35)	0.025 (0.001)	0.025 (0.001)	0.038 (0.0015)	0.05 (0.002)	T15
	320-380	Solution treated and aged	0.5 (0.020)	26 (85)	0.025 (0.001)	0.05 (0.002)	0.08 (0.003)	0.13 (0.005)	T15(b)
			1.5 (0.060)	23 (75)	0.05 (0.002)	0.08 (0.003)	0.10 (0.004)	0.15 (0.006)	T15
			diam/4	12 (40)	0.025 (0.001)	0.038 (0.0015)	0.05 (0.002)	0.08 (0.003)	T15
			diam/2	9 (30)	0.025 (0.001)	0.025 (0.001)	0.038 (0.0015)	0.05 (0.002)	T15
Beta alloys Ti-3Al-8V-6Cr-4Mo-4Zr Ti-8Mo-8V-2Fe-3Al, Ti-11.5Mo-, 6Zr-4.5Sn	275-350	Annealed or solution treated	0.5 (0.020)	15 (50)	0.025 (0.001)	0.05 (0.002)	0.10 (0.004)	0.13 (0.005)	T15(b)
			1.5 (0.060)	14 (45)	0.038 (0.0015)	0.08 (0.003)	0.13 (0.005)	0.15 (0.006)	T15
			diam/4	8 (25)	0.025 (0.001)	0.05 (0.002)	0.08 (0.003)	0.10 (0.004)	T15
			diam/2	6 (20)	0.018 (0.0007)	0.038 (0.0015)	0.05 (0.002)	0.08 (0.003)	T15
Ti-10V-2Fe-3Al Ti-13V-11Cr-3Al	350-440	Solution treated and aged	0.5 (0.020)	12 (40)	0.018 (0.0007)	0.038 (0.0015)	0.05 (0.002)	0.10 (0.004)	T15(b)
			1.5 (0.060)	11 (35)	0.025 (0.001)	0.05 (0.002)	0.08 (0.003)	0.13 (0.005)	T15
			diam/4	6 (20)	0.018 (0.0007)	0.038 (0.0015)	0.05 (0.002)	0.08 (0.003)	T15
			diam/2	5 (15)	0.013 (0.0005)	0.025 (0.001)	0.038 (0.0015)	0.05 (0.002)	T15

Note: ELI, extra-low interstitial. (a) For standard-length end mills, the maximum axial depth may be one and one-half times the cutter diameter. (b) Any premium high-speed tool steel can be used. Source: Metcut Research Associates Inc.

- angle for the tool large enough to provide the required strength and heat conduction
- To select an inclination for the cutting edge that will provide the desired direction of chip flow

In face milling, positive inclination directs the chip outward, and negative inclination directs the chip toward the center of the cutter. A positive inclination is therefore generally desired.

Radial rake has a major effect on the power efficiency of metal removal and on cutter life. Generally, zero to positive radial rakes are used on high-speed steel cutters, and negative radial rakes on carbide cutters. Negative rakes are less efficient, but are usually necessary for a satisfactory service life for the carbide cutters because carbide edges are brittle.

Axial rake controls chip flow, thrust force of cut, and the strength of the cutting edges. The axial rake of high-speed steel cutters is usually positive, except for end mills that cut only on the periphery, which often have negative axial rake to transfer cutter thrust back against the spindle bearings while simultaneously applying downward thrust to the workpiece. Carbide cutters have either positive or negative axial rake, depending on the workpiece material and hardness and on the type of cutter.

The use of a corner angle plus a small nose radius also provides a longer cutting edge. This distributes cutting forces over a greater area, causing less pressure. It also aids in dissipating the heat of cutting. A 30 to 45-degree chamfer also can produce a longer cutting edge and a wider, thinner chip; however, a corner angle is usually more effective than a chamfer.

Relief angles are probably the most critical of all tool angles when milling titanium. Relief angles around 12 degrees give longer tool life than the standard relief angles of 6 or 7 degrees. If chipping occurs, the 12-degree relief angles should be reduced toward the standard values. Generally, relief angles less than 10 degrees may lead to excessive smearing along the flank, while angles greater than 15 degrees weaken the tool and encourage "digging-in," as well as chipping of the cutting edge.

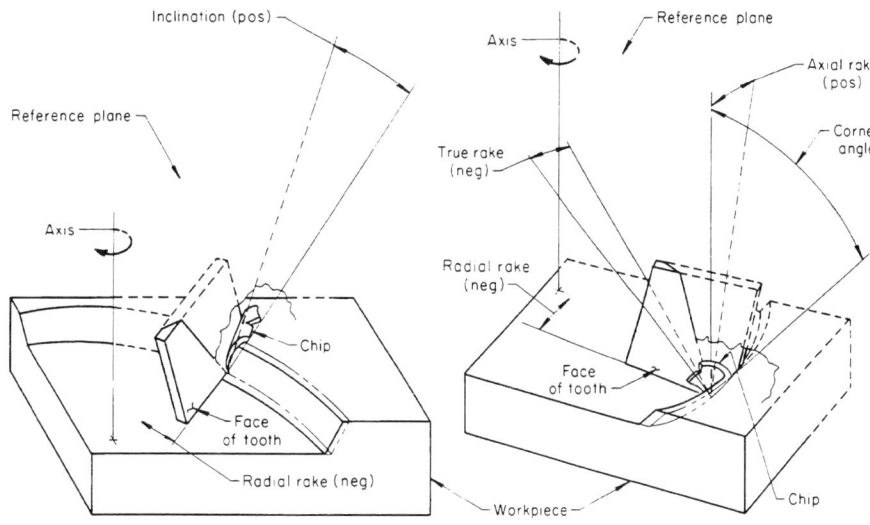

Fig. 4 Geometric relations of tool angles and reference plane.
Source: *Metals Handbook*, 9th ed., Volume 16, *Machining*, 1989, p 317

angle is measured with respect to a reference plane, which, in milling, passes through the axis of cutter rotation and the point of the tool. True rake (by definition) is measured in a plane perpendicular to the projection of cutting edge on reference plane.

The second angle of importance in machining is inclination. This is the angle that the cutting edge makes with the reference plane, which, by definition, is perpendicular to the direction of tool travel. Inclination determines the direction of chip curling. When the inclination is zero, chip flow is virtually in the plane of true rake.

Many combinations of axial rake, radial rake, and corner angle other than those listed in previous tables have been successfully used in practice. The requirements are:

- To select a true rake angle large enough for the particular cutting conditions to provide effective chip formation and yet leave an included

Table 9 Nominal speeds and feeds for face milling titanium and titanium alloys

Material	Hardness, HB	Condition	Depth of cut, mm (in.)(a)	High-speed tool steel Speed, m/min (sfm)	Feed, mm/tooth (in./tooth)	Tool material grade, AISI	Carbide tool, uncoated Speed, m/min (sfm) Brazed	Indexable	Feed, mm/tooth (in./tooth)	Tool material grade
CPTi	110-170	Annealed	1.0 (0.040)	53 (175)	0.15 (0.006)	T15, M42(b)	162 (530)	178 (585)	0.13 (0.005)	C-2
			4.0 (0.150)	41 (135)	0.23 (0.009)	T15, M42(b)	122 (400)	134 (440)	0.25 (0.010)	C-2
			7.5 (0.300)	32 (105)	0.30 (0.012)	T15, M42(b)	85 (280)	105 (345)	0.38 (0.015)	C-2
	140-200	Annealed	1.0 (0.040)	44 (145)	0.10 (0.004)	M2, M7	122 (400)	134 (440)	0.10 (0.004)	C-2
			4.0 (0.150)	34 (110)	0.15 (0.006)	M2, M7	91 (300)	101 (330)	0.15 (0.006)	C-2
			7.5 (0.300)	26 (85)	0.20 (0.008)	M2, M7	61 (200)	76 (250)	0.20 (0.008)	C-2
	200-275	Annealed	1.0 (0.040)	32 (105)	0.10 (0.004)	M2, M7	99 (325)	107 (350)	0.10 (0.004)	C-2
			4.0 (0.150)	24 (80)	0.15 (0.006)	M2, M7	84 (275)	91 (300)	0.15 (0.006)	C-2
			7.5 (0.300)	18 (60)	0.20 (0.008)	M2, M7	58 (190)	72 (235)	0.20 (0.008)	C-2
Alpha alloys Ti-5Al-2.5Sn, Ti-5Al-2.5Sn-ELI, Ti-6Al-2Nb-1Ta-0.80Mo	300-340	Annealed	1.0 (0.040)	21 (70)	0.10 (0.004)	T15, M42(b)	79 (260)	88 (290)	0.10 (0.004)	C-2
			4.0 (0.150)	17 (55)	0.15 (0.006)	T15, M42(b)	69 (225)	73 (240)	0.15 (0.006)	C-2
			7.5 (0.300)	12 (40)	0.20 (0.008)	T15, M42(b)	46 (150)	56 (185)	0.20 (0.008)	C-2
Alpha-beta alloys Ti-6Al-4V, Ti-6Al-4V-ELI, Ti-6Al-2Sn-4Zr-2Mo,	310-350	Annealed	1.0 (0.040)	17 (55)	0.10 (0.004)	T15, M42(b)	52 (170)	56 (185)	0.10 (0.004)	C-2
			4.0 (0.150)	14 (45)	0.15 (0.006)	T15, M42(b)	40 (130)	44 (145)	0.15 (0.006)	C-2
			7.5 (0.300)	11 (35)	0.20 (0.008)	T15, M42(b)	29 (95)	35 (115)	0.20 (0.008)	C-2
Ti-6Al-2Sn-4Zr-2Mo 0.25Si, Ti-6Al-2Sn-4Zr-6Mo	320-380	STA	1.0 (0.040)	17 (55)	0.08 (0.003)	T15, M42(b)	44 (145)	49 (160)	0.10 (0.004)	C-2
			4.0 (0.150)	15 (50)	0.13 (0.005)	T15, M42(b)	34 (110)	37 (120)	0.15 (0.006)	C-2
			7.5 (0.300)	12 (40)	0.18 (0.007)	T15, M42(b)	24 (80)	29 (95)	0.20 (0.008)	C-2
Beta alloys Ti-3Al-8V-6Cr-4Mo-4Zr, Ti-8Mo-8V-2Fe-3Al, Ti-11.5Mo-6Zr-4.5Sn, Ti-10V-2Fe-3Al, Ti-13V-11Cr-3Al	275-350	Annealed or solution treated	1.0 (0.040)	12 (40)	0.08 (0.003)	T15, M42(b)	40 (130)	44 (145)	0.10 (0.004)	C-2
			4.0 (0.150)	9 (30)	0.13 (0.005)	T15, M42(b)	30 (100)	34 (110)	0.15 (0.006)	C-2
			7.5 (0.300)	6 (20)	0.18 (0.007)	T15, M42(b)	21 (70)	26 (85)	0.20 (0.008)	C-2
	350-440	STA	1.0 (0.040)	9 (30)	0.05 (0.002)	T15, M42(b)	24 (80)	27 (90)	0.10 (0.004)	C-2
			4.0 (0.150)	8 (25)	0.10 (0.004)	T15, M42(b)	18 (60)	20 (65)	0.15 (0.006)	C-2
			7.5 (0.300)	6 (20)	0.15 (0.006)	T15, M42(b)	12 (40)	15 (50)	0.20 (0.008)	C-2

Note: ELI, extra-low interstitial. (a) Depth of cut is measured parallel to the axis of the cutter. (b) Any premium high-speed tool steel (T15, M33, M41-M47). Source: Metcut Research Associates Inc.

Table 7 Typical tool geometries for side milling of titanium alloys

Tool material	Axial rake, degrees	Radial rake, degrees	Axial relief, degrees	Radial relief, degrees
High-speed steel	10 to 15	5 to 10	5 to 7	5 to 11
Carbide	0 to –10	0 to –10	5 to 7	5 to 8

Wrought and cast titanium alloys having hardness of 110 to 440 HB. Source: *Machining Data Handbook*, Vol 2, 3rd ed., Metcut Research Associates Inc., 1980

Table 8 Tool geometry for high-speed steel end mills for peripheral and slot milling

General purpose—30-50° helix: steels, cast irons, copper alloys, titanium alloys, nickel alloys, high-temperature alloys, and zinc alloys

Nominal cutter diameter		Radial primary relief angle, degrees	Primary land width		Radial secondary clearance angle, degrees
mm	in.		mm	in.	
1.6	1/16	20-21	0.18-0.25	0.007-0.010	30-35
3	1/8	12-13	0.25-0.38	0.010-0.015	22-28
5	3/16	12-13	0.25-0.51	0.010-0.020	20-25
6	1/4	10-11	0.25-0.51	0.010-0.020	20-25
7	5/16	10-11	0.38-0.64	0.015-0.025	20-25
9.5	3/8	10-11	0.38-0.64	0.015-0.025	17-20
11	7/16	9-10	0.51-0.76	0.020-0.030	17-20
13	1/2	9-10	0.51-0.76	0.020-0.030	17-20
16	5/8	9-10	0.64-0.89	0.025-0.035	17-20
19	3/4	8-9	0.76-1.00	0.030-0.040	15-18
22	7/8	8-9	0.76-1.00	0.030-0.040	15-18
25	1	8-9	0.89-1.27	0.035-0.050	15-18
32	1 1/4	7-8	1.00-1.52	0.040-0.060	13-18
38	1 1/2	7-8	1.00-1.52	0.040-0.060	11-17
44	1 3/4	7-8	1.00-1.52	0.040-0.060	10-16
50	2	6-7	1.00-1.52	0.040-0.060	9-15

Source: Metcut Research Associates Inc.

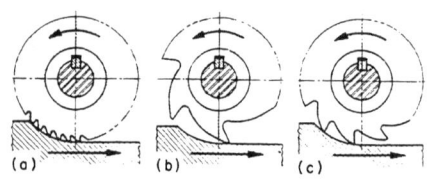

Fig. 3 Effect of number of teeth on a milling cutter. (a) Too many teeth, resulting in chip crowding and interference. (b) Too few teeth, resulting in intermittent contact. (c) Compromise for satisfactory milling

hp/in.^{3}/min), v is the table speed (in in./min), d is the depth of cut (in inches), and w is the width of cut (in inches).

If the metal removal factor is known, vdw can be divided by the factor to find the power required. Equation 1 gives the average power consumption. The instantaneous rate of doing work will vary with the chip thickness and may be considerably higher than the value calculated from Eq 1. However, the rotating parts of the machine and cutter act like a flywheel to smooth out power drawn from the motor. Therefore, metal removal rates for milling can be safely based on calculations of average power consumption. In general, 10 to 15 horsepower is usually sufficient for milling titanium.

Milling cutters vary widely in type and size and are broadly classified as peripheral mills, face mills, end mills, and special mills. Cutters can be of the solid, tipped, or inserted tooth types and have the same materials as single-point tools. Large cutters commonly have teeth of expensive material that are inserted and locked in place in a soft steel or cast iron body. All cutters should be ground and mounted to make sure all teeth are cutting the same amount of material. Typical tool geometries are shown in Tables 6, 7, and 8.

The choice of the milling cutter used depends on the type of machining to be done. Face mills, rotary face mills, plain milling cutters, and slab mills are used for milling plane surfaces. End mills are used for light operations such as profiling and slotting. Form cutters and gang-milling cutters are used to produce shaped cuts. All cutters need adequate body sections and tooth sections to withstand the cutting loads. Helical cutters are preferred for their smoother cutting action.

A milling cutter should have enough teeth to ensure uninterrupted contact with the work metal, and yet not so many as to provide too little space between teeth for chip disposal. On the other hand, cutter teeth should not leave the work before the next is engaged. This will cause vibration and chatter, resulting in poor finish, dimensional inaccuracy, and excessive tool wear. A nearly optimum compromise (see Fig. 3) is to use the smallest diameter cutter with the largest number of teeth without sacrificing chip removal and chatter.

Tool angles of a milling cutter should be chosen to promote unhampered chip flow and immediate ejection of the chip. The controlling angles in this regard include the axial rake, radial rake, and corner angles. These angles should be chosen to provide a positive angle of inclination to lift the chip from the machined surface.

The most significant angle in any machining operation is the true rake angle. True rake angle directly affects the shear angle in the chip-forming process and therefore greatly affects tool force, power requirement, and temperature. The larger the positive value of the true rake angle, the lower the force. It is limited in magnitude, however, by the strength required of the tool for a given machining operation. Negative rake or geometry is used when cutting harder materials such as Hy 80 and Hy 100 armor plate and titanium steels for aircraft applications.

In milling, the true rake angle is the resultant of the axial rake, radial rake, and corner angle (see Fig. 4). Each rake

Table 12 High-speed steel used for drilling titanium alloys

High-speed steel	Commercially pure	Titanium alloy			
		Ti-5Al-2.5Sn	Ti-8Al-1Mo-1V	Ti-6Al-4V	Ti-13V-11Cr-3Al
M1	S	S	S		G
M2					G
M3, Type 2	S	S	S		
M7	G, D	G, D	G, D		
M10	G, D, S	G, D, S	G, D, S	G, S	
M33	G, D	G, D	G, D		G
M34	G, D	G, D	G, D		
M36				S	
T4	G, D, S	G, D, S	G, D, S		
T5	G, D, S	G, D, S	G, D, S	G, S	

Note: G = general drilling; D = deep hole drilling; S = sheet drilling. Source: R. Wood and R. Favor, *Titanium Alloys Handbook*, MCIC-HB-02, Battelle, 1972

Table 13 Drill nomenclature and geometry for NAS 907 aircraft drills

Drill elements	NAS 907 drill type			
	C	D	B	E
Notch rake angle, degrees	4 to 7	20	4 to 7	10
Helix angle, degrees	23 to 30	28 to 32	23 to 30	12
Clearance angle, degrees	10 to 14	6 to 9	10 to 14	6 to 9
Point angle, degrees	118 ± 5	135 ± 5	135 ± 5	135 ± 5
Type point	P-5	P-1	P-3	P-2
Drilling application	Sheet	Hand-drilling sheet	Fixed feed	Fixed feed (dry)

ally lie in the range of 0.05 to 0.2 mm (0.002 to 0.008 in.) per tooth to avoid overloading the cutters, fixtures, and milling machine. Light feeds at slow speeds also help to reduce premature chipping. Delicate types of cutters and flimsy or nonrigid workpieces require smaller feeds. It is important to maintain a positive, uniform feed. Positive gear feeds without backlash are sometimes preferred over hydraulic feed mechanisms. Cutters should not dwell or stop in the cut.

Depth of Cut. The selection of cut depth depends on the part rigidity, the tolerances required, and the type of milling operation undertaken. For skin milling, light cuts of 0.25 to 0.5 mm (0.010 to 0.020 in.) seem to permit less warpage than deeper cuts of 1.0 to 1.5 mm (0.040 to 0.060 in.). When extrusions are being cleaned up, a 1.3 mm (0.050 in.) depth is usually allowed. However, depths of cut up to 3.8 mm (0.15 in.) can be used in other situations if sufficient power is available.

Cutting speed is a very critical factor in milling titanium. Excessive speeds will cause overheating of the cutter edges and subsequent rapid tool failure. When a new job is being started, it is advisable to try a cutting speed in the lower portion of the recommended range (see Tables 9, 10, and 11).

Drilling

Titanium is difficult to drill by techniques considered conventional for other materials. Thin chips flowing at high velocities are likely to fold and clog in the flutes of the drill. Also, the usual galling action of titanium, accentuated by high cutting temperatures and pressures, produces rapid tool wear. Out-of-round holes, tapered holes, or smeared holes are the apparent results, with subsequent tap breakage if the holes are to be threaded. These problems can be minimized by:

- Using short, sharp drills
- Supplying cutting fluids to the cutting zone
- Employing low speeds and positive feeds
- Supplying solid support to the workpiece, especially on the exit side of the drilled hole, where burrs otherwise would form

Successful drilling of titanium depends on the ability to reduce cutting temperatures and maintain rigidity and cutting speeds during drilling. Chlorinated or sulfochlorinated oils and soluble-oil emulsions are satisfactory cutting fluids. Oil-feeding drills may be required for deep holes.

When drilling holes over two diameters deep, the drill should be retracted frequently to clear the flutes and holes of chips. Retraction should be done to minimize dwell. Never allow the drill to ride in the hole without cutting metal.

Drill Materials. Most drilling is done with solid carbide tools or carbide-tipped drills with through-the-tool coolant for larger drills. Conventional high-speed steel drills are also used in various production applications (see Table 12).

The nature of the chips produced indicates the condition of the drill during drilling. A sharp drill produces tight-curling chips without difficulty. As the drill dulls, the cutting temperature rises and titanium begins to smear on the lips and margins. The appearance of feather-type chips in the flutes is a warning signal that the drill is dull and should be replaced. The appearance of irregular and discolored chips indicates that the drill has failed. Out-of-round holes, tapered holes, or smeared holes are results of poor drilling action.

Machine-ground points with fine finishes give the best tool life. A surface treatment such as chromium plating or a black oxide coating of the flutes may minimize welding of chips to the flutes. Large flutes reduce the tendency for chips to clog, and drill length should be as short as feasible.

Drill Design. The choice of drills depends on the drilling operation. Aircraft drills like NAS 907 Types C, D, and E (see Table 13) are usually used on sheet metal. For general drilling, conventional drill geometry and special point grinds are used. This means a normal helix of around 29 degrees, just enough relief to prevent rubbing and pickup, a thinned web to reduce drilling pressure, a correct point angle with its apex held accurately to the center line of the drill, and cutting lips of the

Table 14 Some specifications for drills used on titanium alloys

Drill diameter, in.	0.098	0.1285	0.1590	0.1850	0.1935	0.246	0.250
Overall length, in.	1 5/8	1 15/16	2 1/8	2 3/16	2 1/4	2 1/2	2 1/2
Flute length, in.	5/8	1 15/16	1 1/8	1 3/16	1 1/4	1 1/2	1 3/8
Helix angle, degrees	28 ±2	28 ±2	30 ±2	30 ±2	30 ±2	30 ±2	30 ±2
Lip relief angle, degrees	7 to 10	7 to 10	12 ±2	12 ±2	12 ±2	12 ±2	12 ±2
Point angle, degrees	135 ±3	135 ±3	135 ±3	135 ±3	135 ±3	135 ±3	135 ±3
Web thickness (below split), in.	0.029 to 0.032	0.038 to 0.042	0.047 to 0.052	0.055 to 0.061	0.057 to 0.063	0.060 to 0.065	0.060 to 0.065
Chisel edge angle, degrees	115 to 125	115 to 125	115 to 125	115 to 125	115 to 125	115 to 120	115 to 120
Chisel thickness (after splitting), in.	0.004 to 0.008	0.004 to 0.008	0.004 to 0.008	0.004 to 0.008	0.004 to 0.008	0.004 to 0.008	0.004 to 0.008

All drills have point angles of 135 ± 3 degrees and crankshaft (split) points, and are made from M33 high-speed steel. Source: Metal Removal Procedures, Lockheed Report SP479, Oct 1963

Table 15 Recommended starting feeds and speeds for drilling titanium alloys with indexable carbide inserts

Drill diam, in.	Cutting speed, sfm (m/min)	Feed rate, ipr (mm/rev)
13/16 to 1 1/8	100-135 (30-41)	0.003-0.004 (0.08-0.10)
1 to 1 3/8	100-150 (30-46)	0.004-0.007 (0.10-0.18)
1 1/4 to 1 5/8	115-165 (35-50)	0.005-0.008 (0.13-0.20)
1 1/2 to 2 1/2	130-175 (40-53)	0.006-0.009 (0.15-0.23)
2 3/8 to 3 1/2	135-190 (41-58)	0.006-0.010 (0.15-0.25)

Source: *Machining*, Vol 1, *Tool and Manufacturing Engineers Handbook*, Society of Manufacturing Engineers, 1983

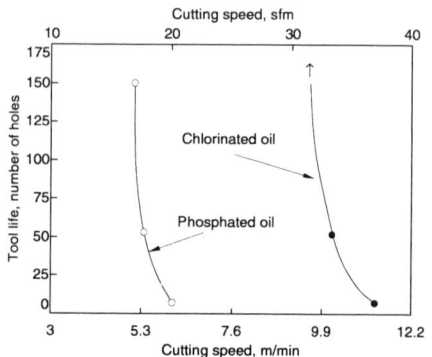

Fig. 5 Advantage of chlorinated oil in tapping. The effect of cutting speed and cutting fluid in tapping is shown for solution treated and aged Ti-6Al-4V with 375 HB hardness. Tap was a chromium-plated, spiral-point, three-flute, 8 mm (5/16 in.)-24 NF plug tap made of M1 high-speed tool steel. Percentage of thread was 75. Depth of through hole was 13 mm (1/2 in.). Tool life end point occurs with tap breakage or undersize thread.

same slope and of equal length.

Relief angles are of extreme importance to drill life. Small angles tend to cause excessive pickup of titanium, while excessively large angles will weaken the cutting edge. Relief angles between 7 and 12 degrees have been used, and recommended lip relief angles for high-speed steel twist drills are:

- 20° for No. 80 to 61 drills
- 18° for No. 60 to 41 drills
- 16° for No. 40 to 31 drills
- 14° for 3.2 to 6.4 mm (1/8 to 1/4 in.) drills
- 12° for 6.4 to 9.5 mm (1/4 to 3/8 in.) drills
- 10° for 9.5 to 13 mm (3/8 to 1/2 in.) drills
- 8° for 13 to 19 mm (1/2 to 3/4 in.) drills
- 7° for diameters 25 mm (1 in.) and up

Point angles also have a marked effect on drill life. The choice of 90, 118, or 135 degrees will depend on the feed, drill size, and the workpiece. Generally, blunt points (135 to 140 degrees) are superior on small-size drills (No. 40 to No. 31) and on sheet metal, while 118 degrees, 90 degrees, or the double angle (140 degrees or 118 degrees + 90-degree chamfer) seem best on larger sizes and bar stock.

Point Grinds. Grinding of spiral points produces a better drill for titanium than conventional chisel edge. Spiral points reduce the large negative rake angle of the chisel edge drill, provide a proper clearance angle along the entire surface of the cutting edge, and reduce thrust loading by 30%.

Another desirable point grind is the NAS 907 P-3 crankshaft (split-point) drill, which is considered better than the spiral point for sheet drilling. Examples of crankshaft (split-point) specifications have been summarized by Wood and Favor (see Table 14).

Feeds and Speeds. The selection of feeds depends largely on the size of the drill being used. Generally, a feed range of 0.025 to 0.13 mm/rev (0.001 to 0.005 ipr) is used for carbide and high-speed steel drills up to 6.3 mm (1/4 in.) in diameter. Drills 6.3 to 20 mm (1/4 to 3/4 in.) in diameter (see tables) will use a heavier feed range, 0.05 to 0.18 mm/rev (0.002 to 0.007 ipr). The choice of speed used will depend largely on the strength level of the titanium material and the nature of the workpiece (see Tables 15 and 16).

Tapping

Tapping titanium is a difficult operation. The limited chip flow inherent in taps, and the severe galling action of titanium can result in poor threads, improper fits, excessive tap seizures, and broken taps. Titanium also tends to shrink on the tap at the completion of the cut.

Holes to be tapped must be uniform and free of work hardening. As a first requirement, holes for tapping should have been produced by sharp drills operating under positive-feed drilling conditions. Dull drills produce surface-hardened holes, which will magnify tapping difficulties. A stiff nylon brush pressed against the tap on the return stroke will help to remove chips and has been reported to increase tap life by at least 50 percent.

The tapping operation itself requires

Table 16 Nominal speeds and feeds for drilling titanium and titanium alloys with high-speed steels

Material	Hardness, HB	Condition	Speed, m/min (sfm)(a)	Feed, mm/rev (in./rev)(a) Nominal hole diameter								Tool material grade, AISI
				1.5 mm (1/16 in.)	3 mm (1/8 in.)	6 mm (1/4 in.)	13 mm (1/2 in.)	19 mm (3/4 in.)	25 mm (1 in.)	38 mm (1 1/2 in.)	50 mm (2 in.)	
Commercially pure Ti (99.0)	110-170	Annealed	24 (80)	0.013 (0.0005)	...	...	...	...	...	...	...	M10, M7, M1
			34 (110)	...	0.05 (0.002)	0.13 (0.005)	0.20 (0.008)	0.25 (0.010)	0.30 (0.012)	0.38 (0.015)	0.43 (0.017)	M10, M7, M1
	140-200	Annealed	20 (65)	0.013 (0.0005)	...	...	...	...	...	...	...	M10, M7, M1
			27 (90)	...	0.05 (0.002)	0.13 (0.005)	0.20 (0.008)	0.25 (0.010)	0.30 (0.012)	0.38 (0.015)	0.43 (0.017)	M10, M7, M1
	200-275	Annealed	12 (40)	0.025 (0.001)	...	...	...	...	...	...	...	M10, M7, M1
			17 (55)	...	0.05 (0.002)	0.13 (0.005)	0.20 (0.008)	0.25 (0.010)	0.30 (0.012)	0.38 (0.015)	0.43 (0.017)	M10, M7, M1
Alpha alloys Ti-5Al-2.5Sn, Ti-5Al-2.5Sn-ELI, Ti-6Al-2Nb-1Ta-0.80Mo	300-340	Annealed	14 (45)	...	0.05 (0.002)	0.13 (0.005)	0.18 (0.007)	0.20 (0.008)	0.25 (0.010)	0.30 (0.012)	0.38 (0.015)	T15, M42(b)
Alpha-beta alloys Ti-6Al-4V, Ti-6Al-4V-ELI	310-350	Annealed	11 (35)	...	0.05 (0.002)	0.10 (0.004)	0.15 (0.006)	0.18 (0.007)	0.20 (0.008)	0.25 (0.010)	0.30 (0.012)	T15, M42(b)
Ti-6Al-2Sn-4Zr-2Mo, Ti-6Al-2Sn-4Zr-2Mo-0.25Si, Ti-6Al-2Sn-4Zr-6Mo	320-380	Solution treated and aged	9 (30)	...	0.05 (0.002)	0.08 (0.003)	0.13 (0.005)	0.15 (0.006)	0.18 (0.007)	0.23 (0.009)	0.25 (0.010)	T15, M42(b)
Beta alloys Ti-3Al-8V-6Cr-4Mo-4Zr, Ti-8Mo-8V-2Fe-3Al,	275-350	Annealed or solution treated	8 (25)	...	0.025 (0.001)	0.08 (0.003)	0.10 (0.004)	0.13 (0.005)	0.15 (0.006)	0.18 (0.007)	0.20 (0.008)	T15, M42(b)
Ti-11.5Mo-6Zr-4.5Sn, Ti-10V-2Fe-3Al, Ti-13V-11Cr-3Al	350-440	Solution treated and aged	6 (20)	...	0.025 (0.001)	0.05 (0.002)	0.08 (0.003)	0.10 (0.004)	0.10 (0.004)	0.13 (0.005)	0.15 (0.006)	T15, M42(b)

Note: ELI, extra-low interstitial. (a) For drilling deep holes with twist drills, reduce speed and feed as follows: If hole depth is three drill diameters, reduce speed by 10%, feed by 10%; four diameters, speed by 20%, feed by 10%; five diameters, speed by 30%, feed by 20%; six diameters, speed by 35%, feed by 20%; eight diameters, speed by 40%, feed by 20%. (b) Any premium high-speed steel can be used.
Source: Metcut Research Associates, Inc.

Table 17 Nominal speeds for tapping titanium and titanium alloys with high-speed tool steel taps

Material	Hardness, HB	Condition	Speed, m/min (sfm)(a) Pitch, mm (threads/in.)				Tool material grade, AISI
			>3 (≤7)	1.5-3 (8-15)	1-1.5 (16-24)	≤1 (>24)	
Commercially pure Ti (99.0)	110-170	Annealed	6.1 (20)	12.2 (40)	16.8 (55)	18.3 (60)	Nitrided M10, M7, M1
	140-200	Annealed	4.6 (15)	9.1 (30)	13.7 (45)	15.2 (50)	Nitrided M10, M7, M1
	200-275	Annealed	3.7 (12)	7.6 (25)	10.7 (35)	12.2 (40)	Nitrided M10, M7, M1
Alpha alloys Ti-5Al-2.5Sn, Ti-5Al-2.5Sn-ELI, Ti-6Al-2Nb-1Ta-0.80Mo	300-340	Annealed	3.0 (10)	4.6 (15)	6.1 (20)	7.6 (25)	Nitrided M10, M7, M1
Alpha-beta alloys Ti-6Al-4V, Ti-6Al-4V-ELI, Ti-6Al-2Sn-4Zr-2Mo, Ti-6Al-2Sn-4Zr-2Mo-0.25Si, Ti-6Al-2Sn-4Zr-6Mo	310-350	Annealed	2.1 (7)	4.6 (15)	5.5 (18)	6.1 (20)	Nitrided M10, M7, M1
	320-380	Solution treated and aged	0.9 (3)	2.1 (7)	2.7 (9)	3.0 (10)	Nitrided M10, M7, M1
Beta alloys Ti-3Al-8V-6Cr-4Mo-4Zr, Ti-8Mo-8V-2Fe-3Al,	275-350	Annealed or solution treated	1.5 (5)	3.0 (10)	4.3 (14)	4.6 (15)	Nitrided M10, M7, M1
Ti-11.5Mo-6Zr-4.5Sn, Ti-10V-2Fe-3Al, Ti-13V-11Cr-3Al	350-440	Solution treated and aged	0.6 (2)	0.9 (3)	1.2 (4)	1.5 (5)	Nitrided M10, M7, M1

Note: ELI, extra-low interstitial. (a) These speeds are for tapping 65 to 75% threads in shallow through holes. Reduce speed when tapping deep holes, blind holes, or higher percentages of thread. Source: Metcut Research Associates Inc.

sharp taps of modified conventional design, low tapping speeds (see Table 17) and an effective tapping lubricant. Paste-type cutting compounds (Lithopone paste, a mixture of zinc sulfide and barium sulfate) have given good results. Chlorinated or sulfochlorinated oils have also been used successfully in tapping titanium and its alloys (see Fig. 5).

Several suppliers design taps with greater clearance especially for titanium. Spiral-point interrupted-flute taps with alternate teeth omitted have given good results at slow speeds. Modification of the tap by grinding away the trailing edge of the thread is beneficial (see Fig. 6). Typical chamfer relief is 12°.

High-speed tool steel taps are generally used. Because 75% threads are difficult to obtain with normal cutting speeds, 65% types are recommended when possible to maximize tap life. Surface treatments such as black-oxide coatings or nitrating can assist in reducing galling tendencies and improving tap life.

Broaching

Broaching is similar to shaping and competes economically with milling and boring. Titanium can be broached under the general setup conditions required by the other machining operations. Because of the interrupted nature of the cut, welding of the chip to the cutting edge is quite troublesome as in milling. This tendency increases as the wearland develops. However, titanium can be broached successfully to a surface finish of 6 to 28 μin. (RMS) with proper tool designs and speeds.

Tool design is a very important factor affecting broaching. Titanium usually requires relief angles somewhat higher than the ½ to 2 degrees normally used in broaching other materials. If the relief angle is too small, metal pickup on the land relief surface can seriously affect the quality of the broached surface. Accordingly, relief angles between 3 and 5 degrees have been adopted and used successfully.

A rake or hook angle of 20 degrees is normally recommended for broaching conventional materials. For titanium, however, a reduction to +5 degrees will improve broaching performance to a marked degree. The smaller rake angle

Table 18 Broaching data for titanium and its alloys

Titanium alloy	Alloy condition	Type high-speed steel	Roughing (a) Cutting speed, fpm	Depth of cut, inch	Finishing (b) Cutting speed, fpm	Depth of cut, inch
Commercially pure	Annealed	T5	20-35	0.004-0.007	30-55	0.002-0.004
Ti-8Al-1Mo-1V	Annealed	T5	10	0.003-0.006	16	0.0015-0.003
Ti-5Al-5Sn-5Zr, Ti-5Al-2.5Sn, and Ti-7Al-2Cb-1Ta	Annealed	T5	15	0.003-0.006	22	0.0015-0.003
Ti-4Al-3Mo-1V	Annealed	T5	15	0.003-0.006	22	0.0015-0.003
	STA	T15	7	0.002-0.004	10	0.001-0.002
Ti-6Al-4V and Ti-8Mn	Annealed	T5	12	0.003-0.006	18	0.0015-0.003
	STA	T15	8	0.002-0.005	12	0.001-0.002
Ti-7Al-4Mo and Ti-6Al-6V-2Sn	Annealed	T5	10	0.003-0.006	16	0.0015-0.003
	STA	T15	7	0.002-0.004	10	0.001-0.002
Ti-13V-11Cr-3Al	Annealed	T5	11	0.003-0.006	17	0.0015-0.003
	STA	T15	6	0.002-0.004	9	0.001-0.002

(a) 3 to 4 degree relief angle. (b) 2 to 3 degree relief angle. Source: R.A. Wood and R.J. Favor, *Titanium Alloys Handbook*, MCIC-HB-02, Battelle, 1972

Table 19 Examples of low-stress grinding operating parameters

Grinding method	Material	Grinding motion	Classification	Grinding wheel speed, m/s (sfm)	Dressing technique	Lubricant	Workpiece speed, m/min (sfm)	Cross feed, mm/pass (in./pass)	Down feed or infeed(f), mm/pass (in./pass) Roughing	Finishing steps First	Second	Third	Spark-out passes
Surface grinding	Titanium alloys	Traverse or plunge	C60J8V	12-16 (2400-3100)	Coarse(a)	Sulfochlorinated oil............	18 (60)	1.3 (0.050)	0.05 (0.002)	0.23 at 0.013 (0.009 at 0.0005)	0.010 at 0.010 (0.0004 at 0.0004)	0.015 at 0.005 (0.0006 at 0.0002)	2
		Notch	C180R9B	25-30 (4800-6000)	Fine(b)	Sulfochlorinated oil............	18 (60)	0 (0)	0.025 (0.001)	0.0025 (0.0001)	...	0.0015 (0.00005)	0
Cylindrical grinding	Titanium alloys	Traverse	C60J8V	14-65 (2800-3250)	Coarse	Sulfochlorinated oil............	12-30 (40-100)(c)	180 (7)(d)	0.025 (0.001)	0.23 at 0.013 (0.009 at 0.0004)	0.010 at 0.010 (0.0004 at 0.0005)	0.015 at 0.005 (0.0006 at 0.00002)	2
		Plunge	C60J8V	14-65 (2800-3250)	Coarse	Sulfochlorinated oil............	12-30 (40-100)(c)	0 (0)	0.005 (0.0002)	0.0020 (0.00008) (d)	...	0.0005 (0.00002)(e)	5-20 rev
		Notch	C120R9B	23-38 (4700-5500)	Fine	Sulfochlorinated oil............	12-30 (40-100)(c)	0 (0)	0.005 (0.0002)	0.0020 (0.00008) (d)	...	0.0005 (0.00002)(e)	5-20 rev

(a) Single-point diamond at 25 mm (1 in.) in 7 s. (b) Single-point diamond at 25 mm (1 in.) in 21 s. (c) Use highest possible table speed (workpiece) up to 30 m/min (100 sfm). (d) Units are in mm/min (in./min). (e) Values are approximate; hand feed at a slow, steady rate; units are in mm/rev (in./rev). (f) Some finishing steps specify downfeed at a given infeed (e.g., 0.23 at 0.013 mm per pass). Source: Metcut Research Associates Inc.

provides greater support for the cutting edge, and improves heat transfer from the cutting zone. The maximum rake is about +10 degrees. An increase beyond this value invites tool failure.

The normal recommendation for the rise per tooth in broaching steel is 0.013 to 0.075 mm (0.0005 to 0.003 in.). Titanium materials, however, should be broached at 0.025 to 0.15 mm (0.001 to 0.006 in.) per tooth, depending on the alloy and its condition and the broaching operation. The lower values of this range should provide lower cutting forces and better surface finishes.

Broaches that have been wet ground may improve tool performance. Careful vapor blasting also may help tool life and finishes by reducing the tendency for smearing. Solid broaches are sometimes made slightly oversize (0.0005 in.) to compensate for the slight springback that will occur when the cut is completed.

The depth of cut is governed by the "rise per tooth" of the broach. A "rise per tooth" in the range of 0.002 to 0.005 inch per tooth has been used successfully when a +5 degree relief is employed. If a 3-degree relief is used, the rise should be reduced to 0.001 inch per tooth.

Cutting Speed. Some titanium alloys have shown a marked sensitivity to changes in cutting speed. Cutting speeds should be restricted to the range of 20 to 30 fpm for CP titanium (see Table 18) to 10 to 12 fpm.

Grinding

The surfaces of titanium alloys can easily be damaged during machining and grinding operations. Damage appears in the form of microcracks, built-up edges, plastic deformation, heat-affected zones, and residual tensile stresses. In service, failures can occur as a result of fatigue and stress corrosion.

Recommended gentle (low-stress) grinding parameters (see Table 19) in-

Table 20 Wheel specifications for precision grinding of titanium

Grinding operation	Conventional speed, 20-30 m/s (4000-6000 sfm)	Low speed, 7.5-10 m/s (1500-2000 sfm)
Centerless	37C54-M5B	...
Cylindrical	37C80-KVK	32A60-K5VBE
Internal	39C60-K8VK	32A60-L8VBE
Surface		
Horizontal spindle	39C80-K8VK	32A80-L5VBE
Vertical spindle		
Cylinder	37C60-HVK	...
Segments	32A24-H12VBEP	...
Thread	37C220-T9BH	...

Table 21 Operational information for abrasive cutting

Machine setting	Bar diameter(a) ≤760 mm (≤3.00 in.)	>760 mm (>3.00 in.)
Feed, mm²/min (in.²/min)	1290-2580	3225-3870
Speed, m/s (sfm)	35-60 (7000-12 000)	30-35 (6000-7000)
Cutting motion	Oscillating wheel	Oscillating wheel and work rotation

(a) A 10% water solution of rust inhibitor (nitrite-amine types) and/or 10% water solution of soluble oil may be used as coolants.

Table 22 Typical conditions for contour band sawing of titanium

Machining conditions	Cutting rates and band speeds for a workpiece thickness of:						
	6.4 mm (1/4 in.)	1.3 mm (1/2 in.)	25 mm (1 in.)	38 mm (1 1/2 in.)	75 mm (3 in.)	150 mm (6 in.)	300 mm (12 in.)
Linear cutting rates, mm/min (in./min)							
High-speed bands	58 (2.30)	38 (1.50)	15 (0.60)	7.5 (0.30)	5 (0.20)	3.8 (0.15)	...
Carbide tooth bands	...	...	...	...	9 (0.35)	6.4 (0.25)	3.8 (0.15)
Band speeds, m/min (ft/min)							
High-speed steel bands							
Pure titanium (99%)	37(a) (120)	37(a) (120)	27(b) (90)	27(b) (90)	27(b) (90)	...	...
Ti-6Al-4V, Ti-4Al-4Mn, ASTM Grade 2	24(a) (80)	24(a) (80)	20(b) (65)	20(b) (65)	20(b) (65)	...	...
Carbide tooth bands (2.5 pitch)							
Pure titanium (99%)	...	...	...	...	...	30 (100)	30 (100)
Ti-6Al-4V, Ti-4Al-4Mn, ASTM Grade 2	...	...	...	...	...	15 (50)	15 (50)

(a) Regular tooth form with a pitch of 10 (b) Regular tooth form with a pitch of 6

Table 23 Operational data for hacksawing titanium

Titanium grade	Speed, stroke/min	Machine setting Feed, mm/stroke (in./stroke) Workpiece size			
		100-150 mm (4-6 in.)	150-205 mm (6-8 in.)	205-255 mm (8-10 in.)	≥255 mm (≥10 in.)
Commercially pure	90-100	0.30 (0.012)	0.23 (0.009)	0.15 (0.006)	0.08 (0.003)
All alloy grades					
Annealed	60-90	0.23 (0.009)	0.15 (0.006)	0.08 (0.003)	0.08 (0.003)
Heat treated	30-60	0.23 (0.009)	0.15 (0.006)	0.08 (0.003)	0.08 (0.003)

Table 24 LBM cutting rates for titanium alloys

Type of cut	Work thickness		Cutting speed (a)	
	mm	in.	mm/min	in./min
Contour	0.5	0.020	5080	200
Contour	1.6	0.062	4060	160
Contour	3.1	0.125	3050	120
Straight	6	0.250	5080	200
Contour	6	0.250	1520	60
Contour	13	0.50	1020	40
Contour	25	1.0	380	15(b)
Contour	50	2.0	130	5(b)

(a) Data are based on use of a continuous wave CO_2 laser with an oxygen assist. (b) 16 kW (21 hp) laser was used. Source: Metcut Research Associates Inc.

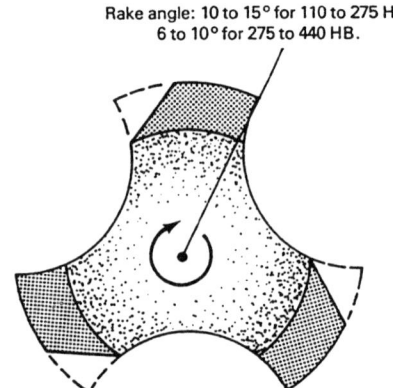

Fig. 6 Modified tap for titanium alloys. A stock tap modified by grinding away trailing edge of thread improves the tapping of titanium threads. It is essential to use taps with interrupted threads and with alternate teeth removed. The greatest chip clearance is obtained by grinding a large chamfer on the trailing edge of the tap. Typical chamfer relief angle is 12°.

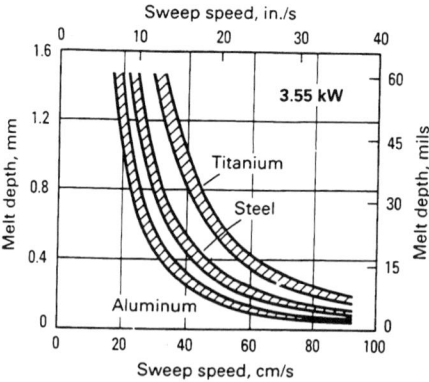

Fig. 7 Laser heating of titanium vs aluminum and steel

clude soft hardness wheels (except for grinding notches), the use of low wheel speeds, low down feeds or in feeds per pass, chemically active cutting fluids, and coarse wheel dressing with diamond. It is evident that low-stress practices decrease production rates compared to conventional methods. The most important consideration, however, is to use methods that make undamaged parts.

Both abrasive wheel and belt grinding are used. Metal removal is slow compared to that of carbon steel. However, under proper conditions, abrasive wear is reasonably low, and a surface finish of 0.4 μm (15 μin.) is possible.

Wheel Grinding. Selection of wheel, wheel speed, and fluid is important. For hard wheel grinding, vitrified bonded wheels are the most effective. Aluminum oxide (Al_2O_3) wheels give good results when limited to grinding speeds of 10 m/s (2000 sfm) or less. Silicon carbide (SiC) wheels can be used at 20 to 30 m/s (4000 to 6000 sfm) if higher speeds are desirable. A feed of about 0.025 mm/pass (0.001 in./pass) is generally suitable for all wheels. Abrasive grit size of 60 to 80 and wheel hardness medium grades J to L are commonly used.

No appreciable sparking accompanies Al_2O_3 wheel grinding. The workpiece can be flooded with standard grinding oils. Water-soluble nitrite amine solutions (rust inhibitors) also work well with aluminum oxide wheels. Silicon carbide wheels, however, operate best with sulfochlorinated grinding oils. Complete flooding of the workpiece minimizes the possibility of fire. A 10% solution of nitride rust inhibitor in water eliminates the risk of fire but is less effective than oil with SiC wheels. Water-soluble oils are also useful but are less effective. Some wheel specifications for various grinding operations and parameters are given in Table 20.

When it is necessary to grind by hand or if coolants cannot be used, care should be taken to provide protection (against hot sparks) for nearby personnel or equipment.

Belt Grinding. Coated-abrasive grinding requires proper selection of the belt, coolant, and operating parameters. Resin-bonded cloth belts with SiC abrasive generally provide the best performance. A 50-grit belt is typically used for coarse grinding, and a 120-grit or finer belt is used for finish grinding. A surface finish of 0.125 to 0.25 μm (5 to 10 μin.) is obtainable in commercial practice.

Fluids should always be used to protect the workpiece and eliminate sparks, which might cause fires. Spraying and flooding techniques are used. Water solutions of 15% tripotassium phosphate or 5% potassium or sodium nitrite have been effective.

Belt grinding performance generally improves with an increase in load and a decrease in speed. Speeds of the order of 5 to 10 m/s (1000 to 2000 sfm) and pressures in the vicinity of 0.7 MPa (100 psi) provide optimum productivity and belt life.

Abrasive Cutting. Rubber-bonded 60-grit SiC cutoff wheels flooded with a water solution of 10% nitrite amine (rust inhibitor) have been used. Machines with oscillating cutting heads give the best results. If the workpiece diameter is greater than 75 mm (3 in.), rotation is recommended to minimize wheel breakage and/or heat checking (see Table 21).

Hand Abrasive Grinding. A clean wheel used only on titanium is important. An open-type wheel containing large grains has been found to minimize clogging. Excessive buildup of heat should be avoided to minimize metal contamination. Ground surfaces should be filed or mechanically finished to remove abrasive particles and particularly any visible metal oxide (burns).

Sandpaper or steel wool should be avoided; wheel-type mechanical burrs (rotary files) should be operated at low speeds to avoid burning and to maximize tool life. Additionally, when titanium is ground, measures must be taken to protect adjacent titanium surfaces as well as the surroundings from the extremely hot sparks.

Sawing

Band Sawing. Coarse-pitched (6 teeth/25 mm, or 1 in.) high-speed tool steel blades, 25 mm (1 in.) wide, employed at speeds of 24 to 27 m/min (80 to 90 sfm) have given good results in cutoff operations. Band speeds for contour band sawing depend on part thickness (see Table 22). Cutting rates on the order of 650 mm^2/min (1 $in.^2$/min) are optimum for cutoff operations. Water-soluble or sulfochlorinated cutting fluids are required.

Hacksawing. Rigid setups and either water-soluble or sulfochlorinated cutting fluids are suggested. Low surface speeds and positive feed, combined with coarse-pitched (3, 4, and 6 teeth/25 mm, or 1 in.) high-speed tool steel blades have proved to be effective. Surface scale or contaminated surfaces cause rapid blade wear if not removed. Operating guidelines for hacksawing depend on workpiece size and heat treatment (see Table 23).

Nontraditional Machining Methods

Probably the most widely used of these methods for titanium and titanium alloys are electrochemical machining (ECM), chemical milling (CHM), and laser-beam machining (LBM).

Electrochemical Machining. The electrolyte for typical ECM operating conditions is a sodium chloride or potassium chloride solution of 0.12 kg/L (1 lb/gal) of water. Voltage must be greater than 11 V for KCl electrolytes. In one application, the maximum starting voltage was 3.2 V for annealed Ti-6Al-6V-2Sn in an NaCl electrolyte solution. A typical metal removal rate is approximately 1.64 cm^3/min/1000 A (0.10 in.^{3}/min/1000 A) at an electrolyte temperature of 40 °C (100 °F).

Chemical Milling. Typical operating conditions for titanium alloys are:

Principal etchant	Hydrofluoric acid
Etch rate, mm/min (in./min)	0.015-0.030 (0.0006-0.0012)
Optimum etch depth, mm (in.)	3.18 (0.125)
Etchant temperature, °C (°F)	46 ± 2.7 (115 ± 5)
Average surface roughness, R_a, μm (μin.)	0.40-2.50 (16-100)

Tolerances on depth of cut up to 12.7 mm (0.5 in.) for titanium alloys are:

Depth of cut		Tolerance	
mm	in.	mm	in.
0-1.25	0-0.050	0.05	0.002
1.25-2.55	0.050-0.100	0.075	0.003
2.55-6.35	0.100-0.250	0.10	0.004

Laser-Beam Machining. Contours can be cut rapidly with laser beams (see Table 24) as compared to conventional methods such as band sawing. For a given power level and sweep speed, the melt depth in titanium is slightly greater than that of steel and aluminum (see Fig. 7).

Technical Note 8: Powder Metallurgy

Adapted from *Metals Handbook*, Ninth Edition, Volume 7, and reviewed by J.P. Beckman, Crucible Compaction Metals

Titanium products fabricated by powder metallurgy (P/M) techniques can be divided into two general categories:

- Blended elemental P/M, in which a blend of elemental powders, along with master alloy or other desired additions, is cold pressed into shape and subsequently sintered to higher density and uniform chemistry
- Prealloyed P/M in which prealloyed powder produced by various techniques is consolidated by either cold pressing and sintering or HIP. The powder morphology determines the type of compaction method which can be used: Irregularly shaped powders can be cold pressed into green shapes and subsequently sintered. Spherical powders will not compact to green shapes and are typically containerized and consolidated by HIP

To date, most components produced by the various P/M methods have been made from the widely used Ti-6Al-4V structural alloy. However, P/M technology is also very well suited for other alloys, such as the high-strength β alloys and the high-temperature near-α alloys.

P/M technology also offers a promising solution to the poor workability of titanium aluminides by the technique of reactive powder processing (*Journal of Metals*, May 1993, p 52-56). In this process, elemental powders with the desired composition are compacted and worked into a Ti-Al composite form, which is easily machined or reformed into different shapes. Reactive sintering is then conducted to obtain the desired microstructure of TiAl and Ti₃Al intermetallics.

Other applications of P/M processing include mechanical alloying and the production of particulate-reinforced titanium composites such as CermeTi®. These and other developments are reviewed in a recent article by Froes and Suryanarayana (Powder Processing of Titanium Alloys, *Reviews in Particulate Materials*, Vol 1, MPIF, 1993, p 223-276). Full coverage of these techniques is beyond the scope of this Technical Note, which only provides a brief review of P/M technology and some property data for the P/M composite CermeTi®. P/M property data for more conventional Ti alloys such as Ti-6Al-4V are covered in the alloy datasheets. In general, the mechanical properties of titanium P/M products depend on alloy composition, density and final microstructure of the compact. The final density and microstructure depend on the nature of the powder, on the specific consolidation technique employed, and on postconsolidation thermal or mechanical treatments.

Elemental Powder

Elemental Powder Production. Historically, inexpensive titanium elemental powder has been produced from fines generated during the crushing of titanium sponge. Elemental titanium powder made from sponge fines is typically −100 mesh (<150 microns) and has an irregular angular shape (Fig. 1). Sodium reduced sponge yields larger amounts of fines and typically has lower chloride content than does magnesium reduced sponge. However, magnesium reduced sponge produced by the newer vacuum distillation process (VDP) has even lower levels of chloride. Residual chloride content causes porosity in the consolidated P/M parts. This porosity often results in a degradation of the mechanical properties as well as difficulties in welding. However, the quality of P/M parts made from blended elemental powders using sponge fines is acceptable for many applications except those which are fatigue-critical.

Due to plant closings, sponge fines are less readily available and alternative production methods for elemental titanium powders are being evaluated. These include various electrolytic and vapor phase reduction methods. Of course, elemental

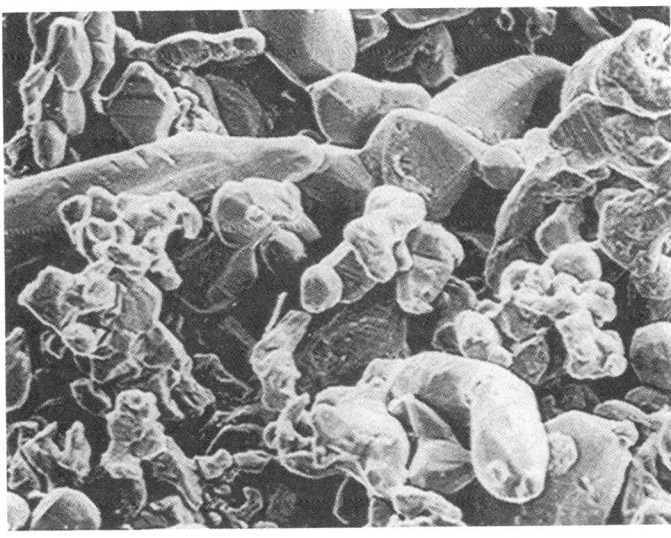

Fig. 1 SEM micrograph of sponge fines (-325 mesh RMI-020 commercially pure titanium powder) 500×

titanium powders can also be produced by the same methods used for producing prealloyed powders which are outlined below.

Prealloyed Powder

Each particle of the prealloyed (PA) titanium powder exhibits the composition of the alloy. Commercial production of PA powders involves either comminution of hydrogen embrittled alloy stock known as the Hydride Dehydride process (HDH), or various methods of melting and rapid solidification (RS). The HDH method produces irregularly shaped blocky particles (Fig. 2) which can be pressed and sintered, while the RS methods produce spherical powders (Fig. 3) which cannot.

Hydride Dehydride (HDH)

Comminution of brittle materials can be used to produce powder. Because titanium and its most common alloys are generally ductile, they must be converted to a brittle state prior to comminution. In the HDH process, this is accomplished by hydrogen embrittlement. Titanium (or a titanium alloy) is hydrided by heating in a hydrogen atmosphere. A variety of solids can be used as starting stock, although light-gage stock has greater surface area which speeds the solid state diffusion process. After hydriding, the product is extremely friable and either spontaneously reduces to powder or is easily pulverized. The powder is then dehydrided (degassed) by heating in a vacuum.

HDH powder is similar in morphology to sponge fines. Both have low packing density (20% - 40%) and low flow rates due to their irregular particle morphology but both are easily cold pressed into green shapes. However, depending on the starting stock, HDH powder may or may not be encumbered with residual chloride content and its resultant porosity on consolidation. The HDH process itself does tend to increase the oxygen content, and, depending on starting stock, other contaminants may be present which could make HDH powders unsuitable for critical applications.

Rapid Solidification (RS)

Rapid solidification produces spherical powder (Fig. 3), which flows readily with minimal bridging tendency, and packs to a very consistent density of approximately 65%. Rapid solidification methods can be categorized as either those that use centrifugal atomization, or those that don't. For those methods that do use centrifugal atomization, either the starting stock is rotated (e.g. PREP) or the collection device is rotated (e.g. RSR). On the other hand, gas atomization (GA) does not use centrifugal atomization. A molten metal stream flows through a nozzle into a ring of gas jets where the liquid metal is atomized by gas pressure.

A comprehensive overview of various RS methods is found in "Rapid Solidification Processing of Titanium Alloys," (C. Suryanarayana, F.H. Froes, R.G. Rowe, published in *International Materials Reviews*, 1991, Vol 36 (No. 3). Most of the RS processes for making Ti PA powder which are listed in this review article are not commercial. Commercial RS methods of powder production include PREP at Nuclear Metals Inc. and gas atomization (GA) at Crucible Materials Corporation.

PREP. A variety of centrifugal atomi-

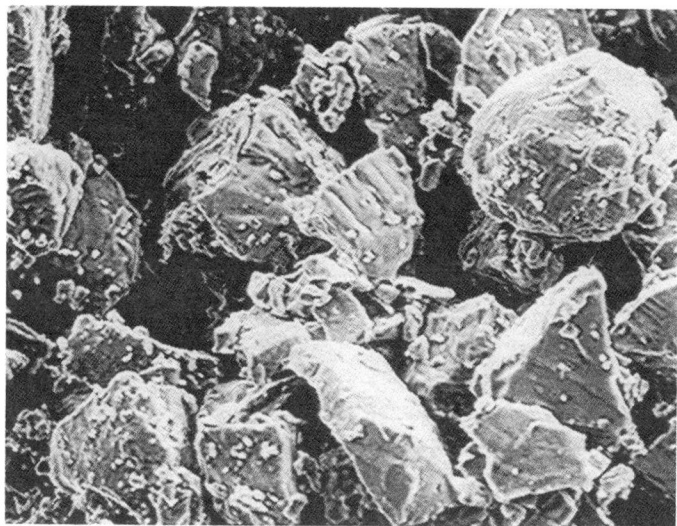

Fig. 2 SEM micrograph of -270 mesh WahChang HDH commercially pure titanium powder 500x

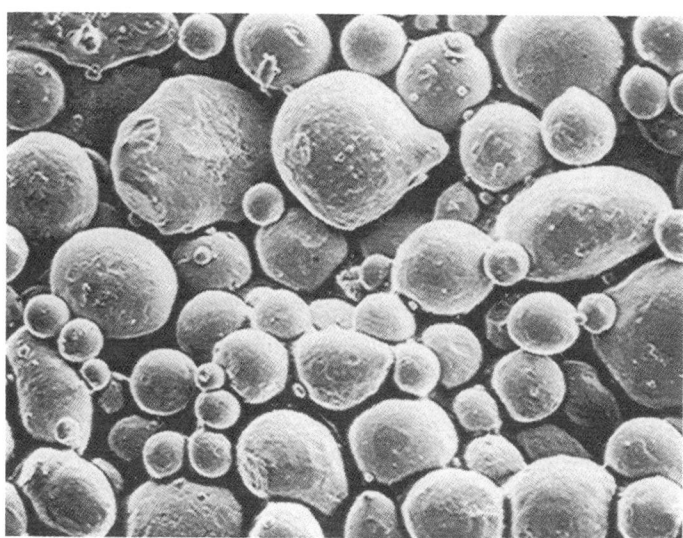

Fig. 3 SEM micrograph of -35 mesh gas atomized commercially pure titanium powder 100x

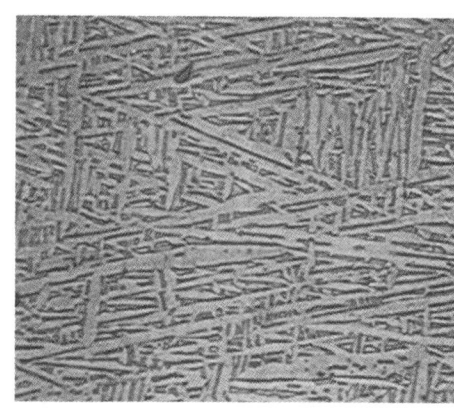

Fig. 4 Microstructure of a Ti-6Al-4V compact treated to produce a broken-up structure

zation techniques have been evaluated, but the plasma rotating electrode process (PREP) is the common commercial method. In the PREP process, a machined bar of the desired composition is used as an electrode which is rotated inside a controlled atmosphere chamber. A plasma torch placed opposite the rotating bar is focused on the bar end causing localized melting at the surface. Molten droplets are spun off the end of the rotating bar by the induced centrifugal force. The droplets solidify in flight producing spherical powder particles which are collected and withdrawn from the chamber. Because the liquid metal makes no contact with any refractory or other material, this process produces powder with no contamination.

Gas atomization (GA) is a non-centrifugal rapid solidification process which has been commercial since 1989. Raw material is melted in a vacuum/inert gas induction furnace with a "cold wall" water cooled copper crucible. As the charge is melted, a thin solid shell of the melt forms at the crucible wall. This shell is called a skull and the process is referred to as skull melting. It is refractory-free because the molten metal comes into contact with no material other than the skull of the alloy itself, thereby eliminating contamination. The molten metal flows through a metal nozzle which directs the liquid into a ring of high pressure jets of inert gas. The gas breaks up the metal stream into tiny spherical particles which solidify in flight, producing powder.

Gas atomization can produce a limitless range of compositions. Its versatility comes from the fact that no specially prepared prefabricated starting stock is required. Instead, raw materials are charged into the crucible to formulate the desired composition. Once the charge melts, it can be held in the molten state for a sufficient period of time to ensure complete homogenization. When induction heating is used for melting, the vigorous stirring action produced by the electromagnetic field also aids in homogenization. Direct atomization from the melt allows for production of highly alloyed compositions, even metastable solid solutions, and minimizes the possibility of segregation in the as-atomized powder. It should also mean lower raw material costs. The stock for PREP powder must be melted into an ingot, worked down to an appropriate size rod, and than machined to tight tolerances. Making powder directly from the melt saves all of these steps.

In contrast to GA, other RS atomization processes which involve localized melting of prealloyed bar stock are limited by the range of alloys available as conventionally produced starting stock. Some complex alloys, unavailable as barstock, may be cast into electrodes. Depending on the composition, there is a risk of thermal shock destroying the cast bar when the localized heat source impinges the electrode. Also, because localized melting methods liquify such a small portion of the metal at any time, compositional variations in the starting stock may directly translate into compositional variations in the resultant powder.

Other Methods of Powder Production. In addition to the HDH and RS methods, several other methods for PA powder production have been developed but none are currently in commercial use. These include:

- *The Hurd Shaker Process* in which the appropriate metal chlorides are heated together with sodium, and then vigorously shaken to exothermically react forming alloy powder and sodium chloride.
- *The ALTi-OXY Process* in which a low melting mixture of titanium dissolved in molten zinc is atomized by any of various methods, and the powder product is subsequently sublimed to remove the zinc.
- The *Goldschmidt Calciothermic Reduction* in which a blend of various metal oxides is essentially reduced with calcium, although several intermediate steps are required to produce the correct soichiometry in the final powder.

Consolidation

Titanium powders can be consolidated into preforms which will undergo further processing to mill products, or into near net shapes (NNS).

Blended elemental (BE) P/M technology involves the blending of titanium powder with a master alloy powder in proper proportions to yield the alloy composition. The blend is typically compacted by mechanical die pressing or cold isostatic pressing. Direct powder rolling and rapid omnidirectional compaction (ROC) may also be used. BE compacts typically have a green density of 85%. Vacuum sintering achieves densities >95%. Careful control of particle size and size distribution can produce compacts that are 99% dense. Porosity is a result of sodium chloride residues in the sponge from the reduction (Kroll) process. Essentially 100% density, and properties equivalent to that of wrought material, can be obtained by using low-Cl sponge.

Sintering is performed at temperatures in the range of 1150 to 1315 °C (2100 to 2400 °F) in a vacuum to prevent gas contamination that can severely degrade compact properties. The high sintering temperature is needed to provide particle bonding and to homogenize the chemistry. It is well above the β transus of all common titanium alloys. The resulting microstructure is much finer than ingot material treated at the same temperature.

Post-sinter hot isostatic pressing (HIP) may be used to further densify the P/M parts. After such treatments BE P/M conpacts are >99% dense and are used as near-net shape in many applications, or are used as preforms for further hotworking to forgings, bar, plate, sheet, or foil. Although not in commercial use, other hot pressing methods have been developed which can be used as an alternative to HIP achieving pseudo-hydrostatic pressure by forge pressing with sophisticated tooling configurations. These include VHP: vacuum hot pressing, ROC: rapid omnidirectional compaction, FDC: fluid die compaction, and the Ceracon process.

Prealloyed (PA) P/M Technology. PA powders produced by HDH have an irregular shape which allows them to be processed by the cold pressing and sintering techniques mentioned above. PA powders produced by RS are spherical and therfore will not cold compact. Hot isostatic pressing is the primary compaction method used for the consolidation of spherical powders. Powder compacts may also be consolidated by direct extrusion or other hot pressing methods developed as an alternative to HIP.

Compaction by hot pressing is typically carried out at a temperature below the beta transus to control microstructure and minimize reaction with the can. Powder cleanliness is one of the main factors governing the quality of PA compacts, because even a low level of contamination with foreign particles may lead to a substantial loss of inherent properties such as fatigue strength.

Postcompaction Treatments. P/M alloys that are subjected to postcompaction working (or to lower-temperature consolidation) may display improved tensile and fatigue strengths as a result of microstructure refinement. However, in most cases, process economics do not allow subsequent working because it nullifies the objectives of a true net-shape technology. Therefore, postcompaction methods leading to microstructure refinement without the use of working are more cost effective. Two approaches that meet this requirement are heat treatment and thermochemical processing.

Heat Treatment. Response to heat treatments after compaction depends on the particular alloy and the P/M processing method. In the case of BE Ti-6Al-4V, for example, the only successfully used heat treatment has been the broken-up structure (BUS) treatment in which a β quench is followed by 850 °C (1560 °F) long-term annealing. After such treatment, the microstructure of the alloy is showing broken-up α phase in a matrix of β (Fig. 4). This microstructure provides a significant improvement in both tensile and fatigue strengths.

The BUS method is an improvement over standard heat treatments, which typically provide higher tensile properties but no increase in fatigue strength properties. Strengthening by standard heat treatment (such as solution treating and aging) may also degrade the toughness of BE compacts to unacceptable levels. This limitation is possibly due to the porosity from chloride impurities. In contrast, PA compacts may have a better response to solution treatment and aging. Preallowed compacts of Ti-6Al-4V, for example, have responded well to solution treatment and aging (R.E. Peebles and C.A. Kelto, Investigation of Methods for the Production of High Quality, Low Cost Titanium Alloy Powders, *Powder Metallurgy of Titanium Alloys*, F.H. Froes and J.E. Smugeresky, Ed., The Metallurgical Society of AIME, 1980, p 47-58).

Hydrogenation. The use of hydrogen as a temporary alloying element is used for both postcompaction and precompaction treatment. Thermochemical treatment with hydrogen after compaction refines the microstructure. This refinement provides slightly higher strength levels. Also, it has been demonstrated that starting with powder in the hydrogenated condition provides advantages in processing (lower compaction pressure and/or temperature) and in control of the microstructure for compaction by both hot pressing and hot isostatic pressing. During vacuum hot pressing, the hydrogen is removed during the compaction cycle. In the case of hot isostatic pressing, the hydrogen is retained in the titanium during compaction. After both processes, a final vacuum anneal is required to bring the hydrogen content below specification levels (generally <125 ppm).

Applications

Porous BE Products. On the low end of the density scale, commercially pure (CP) titanium filters are produced for electrochemical and other corrosion-resistant applications. Titanium porous materials are available in sheet and disk form, with controlled porosity ranging from 3 to 40 μm and density ranging from 25 to 75% of theoretical. Sheet is available in widths ranging from 30 to 90 cm (1 to 3 ft). Porous titanium tube also is available in lengths up to 60 cm (2 ft) and with diameters up to 12.7 cm (5 in.).

Compacted BE Products. For more demanding applications, BE components with densities from 98% to close to 100% are produced by cold pressing or CIP (cold isostatic pressing), and sintering followed by HIP (hot isostatic pressing). Very complex shapes such as impellers, are produced by cold isostatic pressing with elastomeric molds. Dimensional tolerances appear to be about ±0.02 mm/mm in length; length is limited by the processing equipment. Part diameter is limited to 600 mm (24 in.) for the CIP method.

Larger parts are produced by press consolidation. Intricate shape capabilities of this method do not approach those of the CIP method, but press consolidation methods appear capable of 0.01 mm/mm in length. Press consolidation is not limited by size; presses up to 45,000 metric tons (50,000 tons) are available in the United States. Consequently, a part of almost 13,000 cm^2 (2,000 in.^{2}) can be produced.

PA powders are used to produce a variety of fully dense products ranging from simple mill products and forging preforms, to near-net shapes for medical aerospace applications. The properties of these products are comparable to wrought products and exceed those of cast products. PA powders are also used in plasma spray deposition to produce sheet, foil, and metal matrix composites.

Currently, P/M shapes from spherical PA powders are produced by the metal can method. The ceramic mold method and the fluid die process have also been developed but neither is used commercially. Both the metal can and ceramic mold processes use hot isostatic pressing consolidation, while the fluid die process is adaptable to either hot isostatic pressing or hot press compaction. The largest HIP chamber available to date is 1350 mm (54 in.) in diameter and 3000 mm (120 in.) in height.

Metal Can Method. The metal can is shaped to the desired configuration by state-of-the-art sheet metal methods, such as brake bending, press forming, spinning, or superplastic forming. Carbon steel is the best suited container material, because it reacts minimally with titanium, forming titanium carbide, which inhibits further reaction. Fairly complex shapes can be produced by this technique.

Ceramic Mold Method. The Crucible ceramic mold process utilizes a ceramic mold produced by the same process used by the investment casting industry. However, the mold materials have been modified to minimize reaction with titanium and to maximize shape capability. Hot isostatic pressing at 815 °C to 1240 °C (1500 °F to 2200 °F) and 103 MPa (15,000 psi) is used to consolidate the powder to full density and form the desired shape. Generally, titanium alloys are consolidated below the beta transus temperature. A prototype impeller made out of titanium aluminide powder has been produced by this method.

The fluid-die process (the ROC process) practiced by Kelsey-Hayes is an outgrowth of earlier work on glass containers. Dies are made from carbon steels, copper alloys, or ceramic materials and are of sufficient mass and dimension to behave as a viscous liquid under pressure at temperature. Dies are comprised of two halves, with inserts where necessary to simplify manufacture. The two halves are then welded together to form an hermetic seal. Powder loading and consolidation follow. The fluid die process combines the ruggedness and fabricability of metal with the flow characteristics of glass to generate a replicating container capable of producing extremely complex shapes. Like the ceramic mold process, a number of shapemaking steps are required to reach the final part configuration. Like the ceramic-mold method, the fluid-die process is capable of generating more complex and intricate shapes than the metal can method.

P/M Property Data CermeTi® Property Data

CermeTi is a trademark registered at the U.S. Office of Patents and Trademarks and is the subject of U.S. and foreign patents assigned to Dynamet Technology, Inc.

CermeTi metal matrix composites are particulate-reinforced titanium metal matrix composites developed by Dynamet

Technology, Inc. This family of materials includes a choice of several ceramic or intermetallic additions (e.g., TiC, TiB$_2$, or TiAl) at various loading levels in a select titanium alloy matrix such as Ti-6Al-4V. These materials are produced by blended elemental powder metallurgy processing to near-net shape by the proprietary CHIP process and can also be further consolidated/refined by standard forging or extrusion of the P/M preform.

Compared to titanium alloys, CermeTi materials offer improved tensile strength and elastic modulus, both at room temperature and elevated temperature, thereby increasing the use temperature at approximately the same low density. These composites also offer increased hardness and wear resistance over conventional titanium alloys.

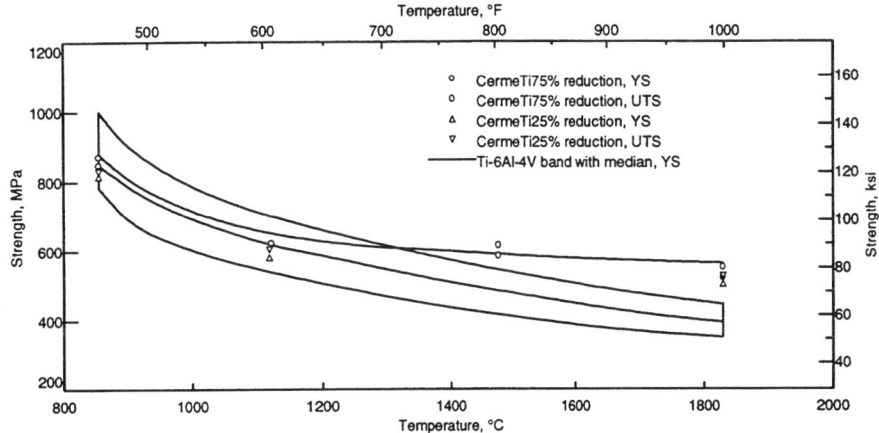

Fig. 1 CermeTi-15: Tensile property comparison vs test temperature

Table 1 CermeTi: Physical properties

Property	Ti-6Al-4V	CermeTi-C-10 10 wt% TiC/Ti-6Al-4V	CermeTi-C-20 20 wt% TiC/Ti-6Al-4V
Density, g/cm^3 (lb/in.^{3})	4.43 (0.160)	4.45 (0.16)	4.52 (0.162)
Tensile strength, MPa (ksi), at:			
RT	896 (130)	999 (145)	1055 (153)
540 °C (1000 °F)	448 (65)	551 (80)	620 (90)
Modulus, GPa (10^6 psi), at:			
RT	113 (16.5)	133 (19.3)	144 (21)
540 °C (1000 °F)	89 (13)	105 (15.3)	110 (16)
Fatigue limit (10^6 cycles), MPa (ksi)	517 (75)	275 (40)	...
Fracture toughness, MPa$\sqrt{m}$ (ksi $\sqrt{in.}$)	55 (50)	44 (40)	32 (29)
Coefficient of linear thermal expansion (RT to 540 °C, or 1000 °F), ppm/°C	8.5	8.1	8.0
Hardness, HRC	34	40	44

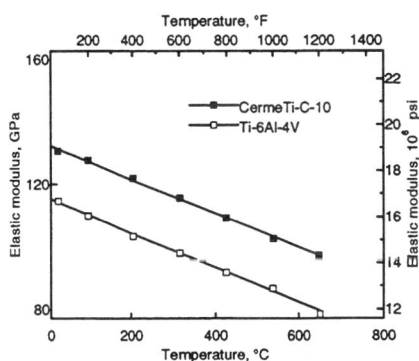

Fig 2 CermeTi with 15% wt% TiC compared to Ti-64. CermeTi-C-10: Dynamic (elastic) modulus vs temperature

Table 2 CermeTi-C: Mechanical property comparison

Material	Density g/cm^3	Density lb/in.^{3}	Yield strength MPa	Yield strength ksi	RT modulus GPa	RT modulus 10^6 psi	Fracture toughness MPa$\sqrt{m}$	Fracture toughness ksi$\sqrt{in.}$
6061-T6 aluminum	2.71	0.098	275	40	68.9	10.0	...	...
2014-T6 aluminum	2.79	0.101	413	60	72.4	10.5	30	28
4340 steel	7.75	0.280	827-1379	120-200	199	29.0	55-110	50-100
25 vol% SiC/6061-T6 Al	2.90	0.105	413	60	113.7	16.5	...	...
Ti-6Al-4V	4.43	0.160	896-1034	130-150	113.7	16.5	55-110	50-100
CermeTi-C-10	4.45	0.161	965	140	133.0	19.3	44	40
CermeTi-C-20	4.48	0.162	1034	150	144.8	21.0	32	29

Ti-1Al-8V-5Fe P/M Property Data

Tensile properties of various product forms of Ti-1Al-8V-5Fe
Room temperature tensile properties of various heat treatments of I/M and P/M Ti-1Al-8V-5Fe. Higher oxygen contents increase strength. Boron additions resulted in zero ductility.

Raw material	Processing	Heat treatment	UTS ksi	TYS ksi	Elongation, %	RA, %
I/M 4 in RCS	Forged	1450 °F/0.5 h/WQ + 950 °F/2 h	213.2	210.7	1.7	5.0
High O$_2$ (0.34 O wt%)		1450 °F/0.5h/WQ + 1000 °F/2 h	200.5	198.5	2.5	5.0
		1475 °F/0.5 h/WQ + 1050 °F/8 h	186.1	186.1	2.0	6.0
I/M 4 in RCS	Forged	1450 °F/0.5 h/WQ + 1000 °F/6 h	177.4	174.6	1.8	8.5
Low O$_2$ (0.15 O wt%)		1475 °F/0.5 h/WQ + 1050 °F/10 h	156.6	152.5	5.0	13.1
ALTi P/M	Extruded, 1200 °F, 10:1	1450 °F/0.5 h/WQ + 950 °F/2 h	204.7	194.5	3.0	7.1
		1450 °F/0.25 h/WQ + 925 °F/4 h	200.5	194.3	4.2	...
		1450 °F/8 h/WQ + 925 °F/4 h	198.3	191.0	8.0	15.1
GA PM, Lot 1	Extruded, 1500 °F, 6:1	1425 °F/1 h/WQ + 925 °F/4 h	220.0	212.4	7.8	19.5
(0.36 O wt%)	Extruded, 1200 °F, 10:1	1425 °F/0.5/WQ + 925 °F/4 h	225.0	221.0	2.0	4.0
	HIP, 1450 °F/15 ksi/6 h	1425 °F/1 h/WQ + 925 °F/4 h	217.9	207.7	7.0	15.5
GA PM, Lot 2	Extruded, 1500 °F, 6:1	1425 °F/1 h/WQ + 925 °F/4 h	266.5	251.4	...	2.0
(1% B)	Extruded, 1200 °F, 10:1	1425 °F/0.5 h/WQ + 925 °F/4 h	166.3	...	...	...
GA PM, Lot 3	HIP, 1500 °F/15 ksi/6 h	1425 °F/1 h/WQ + 925 °F/ 4 h	223.6	215.4	4.0	3.2
(0.362 O wt%)		1450 °F/1 h/WQ + 950 °F/4 h	223.8	216.0	4.0	5.7
GA PM, Lot 4	HIP, 1500 °F/15 ksi/6 h	1425 °F/1 h/WQ + 925 °F/4 h	165.3	...	...	...
(.25% B)		1450 °F/1 h/WQ + 950 °F/4 h	114.7	...	...	...

R.R. Boyer, E.R. Barta, Boeing Co., C.F. Yolton, Crucible Materials Corporation, Crucible Research Division, D. Eylon, University of Dayton, "P/M of High Strength Titanium Alloys," P/M in Aerospace and Defense Technologies, Vol 1, p 99-115

Tensile properties of gas atomized P/M Ti-1Al-8V-5Fe
Room temperature tensile properties of GA Ti-1Al-8V-5Fe powder with 0.36 oxygen content after various treatments. Transus: 850 °C (1565 °F)

Consolidations/ conditions	Heat treatment	UTS ksi	TYS (0.2%) ksi	Elongation, %	RA, %
HIP 1500 °F	1250 °F 1 hr AC	163	153	15.6	37.8
HIP 1500 °F	1425 °F 1 h WQ + 925 °F 4 h	219	208	4.7	14.4
HIP 1600 °F	1250 °F 1 h AC	162	152	10.9	19.9
HIP 1600 °F	1425 °F 1 h WQ + 925 °F 4 h	217	205	1.6	7.5
Extrude 1500 °F 6:1	1250 °F 1 h AC	154	142	21.9	48.3
Extrude 1500 °F 6:1	1425 °F 1 h WQ + 900 °F 4 h	222	212	9.4	15.0
Extrude 1500 °F 6:1	1425 °F 1 h WQ + 925 °F 4 h	220	212	7.8	19.5
Extrude 1500 °F 6:1	1500 °F 1 h WQ + 925 °F 4 h	227	219	6.3	7.0

C.F. Yolton, Crucible Materials Corporation, Crucible Research Division, unpublished data, 1992

Tensile properties of gas atomized P/M Ti-1Al-8V-5Fe
Room temperature tensile properties of GA Ti-1Al-5Fe powder with 0.14 oxygen content after various heat treatments. Transus: 750 °C (1390 °F)

HIP temperature °C	HIP temperature °F	Heat treatment(a)	Tensile strength MPa	Tensile strength (ksi)	Yield strength 0.2% offset MPa	Yield strength 0.2% offset (ksi)	Elongation, %	RA, %
725	1340	As-HIP	1063	(154.3)	1037	(150.5)	6.3	29.1
		A	1007	(146.2)	960	(139.4)	17.2	39.8
		B	1407	(204.2)	1397	(188.2)	6.3	8.0
		C	1217	(176.7)	1160	(168.4)	12.5	27.8
780	1440	As-HIP	1135	(164.7)	1075	(154.6)	9.4	19.9
		A	1015	(147.4)	960	(139.4)	16.4	37.2
		B	1414	(205.2)	1315	(190.8)	4.7	9.9
		C	1175	(170.5)	1111	(161.3)	10.9	24.3

(a) Heat treatment: A - 675 °C (1250 °F) 1 h, AC, B - 675 °C (1250 °F) 1 h, WQ/440 °C (825 °F) 4 h, C - 675 °C (1250 °F) 1 h, WQ/495 °C (925 °F) 4 h. Source: C.F. Yolton, J.H. Moll, Crucible Materials Corporation, Crucible Research Division, "Evaluation of a High Strength Rapidly Solidified Beta Titanium Alloy," Progress in Powder Metallurgy, 1987, Vol 43, MPIF 1987, p 49-63

truding, rolling, aging, hot forming, or a combination of several of these operations. With processing temperatures ranging from 425 to 1150 °C (800 to 2100 °F), the scale spectrum for titanium is far broader than for most other difficult-to-descale materials.

Coatings. Protective coatings are often used in titanium manufacturing operations. These coatings, which are an asset and a necessity in manufacturing operations, become a liability and a contaminant in cleaning operations. They are soluble and removable if the proper techniques are used. Protective coatings are applied to titanium surfaces during manufacturing operations to:

- Lubricate and aid in metal flow, good die contouring, and forming operations
- Act as barrier films, reducing gas contamination during high-temperature forming and heat-treating cycles
- Reduce surface flaws caused by nicking and scratching during manufacturing operations

The gas protective films are usually applied directly to the titanium surface. They are silicate-based materials that deposit uniform fusable films through solvent evaporation. These films form glassy barriers at treatment temperatures up to 815 °C (1500 °F) and are quite effective in reducing oxygen, hydrogen, and nitrogen contamination. Above 815 °C (1500 °F), most of these films are less effective, because spheroidizing creates voids in the film. Lubricant films or abrasion protective films are applied over a silica-based coating. This process has the advantage of providing double protection against scratching and scoring. During hot-forming operations and metal surface stretching, some voiding and penetration occurs, creating a titanium oxide on the surface.

Mechanical Removal of Scale

Scale is removed from titanium products by several mechanical methods. Abrasive methods, such as grinding and grit blasting, are preferred for removing heavy scale from large sections. Centerless grinding is used for finishing round bars, and wide-belt grinding is used for finishing sheet, strip, and plate. Grinding is usually most efficient when it is performed at low wheel and belt speeds.

Most alloy sheet materials with a high aluminum content such as Ti-5Al-2.5Sn are ground to eliminate pits and a rippled condition that develops in hot rolling as a result of discontinuous slip during plastic deformation. Grinding is frequently used to eliminate surface defects before cold rolling. Originally, strip was ground on standard strip grinders, using various oil lubricants; however, oils contributed to fire hazard and several grinding machines were partially or wholly destroyed when the oil ignited. When titanium was ground with aluminum oxide belts, a water lubricant was less effective than air. The water reacted with the aluminum oxide to form a weak hydroxide that proved ineffective as a grinding lubricant.

Belt Grinding. Dry belt grinding is dangerous because of the hazards of explosion and fire. It is uneconomical because of poor belt life. When stock is removed during dry grinding, small globules of molten metal and oxide roll along the sheet, causing a type of pitting by burning that is not removed by the grinding. Weld grit scratches and embedded grit result when titanium welds to the dry grit.

A 5% aqueous solution of potassium orthophosphate, K_3PO_4, is widely used as a grinding lubricant. It is applied as a flood at both the entrance and exit side of the contact line. Water-soluble oils, particularly highly chlorinated and sulfochlorinated oils, have also been successful as lubricants. These compounds should be used with care because of the possibility of chloride residues remaining as an integral part of the surface. Both types of lubricants improve grinding efficiency when the belts are coated with aluminum oxide or silicon carbide.

Flooding the work with lubricant is recommended; however, machines built for flooding are equipped with a recirculating and filtering system and waterproof cloth belts, and they are expensive. An alternative is to spray a water-soluble wax fog through atomizing nozzles on the line of contact at both the entrance and exit sides of the belt. The solution should not be sprayed through an ejector that mixes it with air and increases the fire hazard. Instead, it should be sprayed as an atomized liquid. Application of the spray can be controlled to volatilize the lubricant during grinding. This eliminates the need for waterproof belts. Care must be exercised to avoid a buildup of titanium chips that may cause a fire. Chips that would wash away in a flood are not removed by the spray.

Titanium should be ground at belt speeds not exceeding rates of 8 m/s (1500 ft/min). Using a 5% solution of potassium orthophosphate as a lubricant, maximum efficiency is achieved at about 6 m/s (1100 ft/min). Both billy roll and flat table grinding machines have been successful in grinding titanium. Sheet grinding machines, equipped with feed rolls, sometimes leave a ground line on the sheet where the feed rolls, on the exit side of the machine, interrupt uniform travel when they grip the sheet. A high degree of grinding uniformity is obtained on machines equipped with a flat table and vacuum chuck. On these machines, the table holding the sheet usually oscillates. Traveling-head machines are available also.

The belt grinding sequence (see Fig. 1) is usually begun with an 80-grit belt, when it is necessary to remove more than 0.07 mm (0.003 in.) of stock from the surface of the sheet. Descaling and pickling of the sheet before grinding prolongs belt life. Following the initial grit, each successive grit must remove enough stock to eliminate the scratches caused by the previous grit. The alpha alloys, such as Ti-5Al-2.5Sn, are less sensitive to surface condition than the alpha-beta alloys, such as Ti-6Al-4V. Surface pits on Ti-6Al-4V sheet, caused by weld grit scratches, seriously detract from bend ductility.

Abrasive blast cleaning techniques, either wet or dry, are convenient for removing scale from a variety of titanium products, ranging from massive ingots to small parts. Because it can be used at lower velocities and is less likely to be

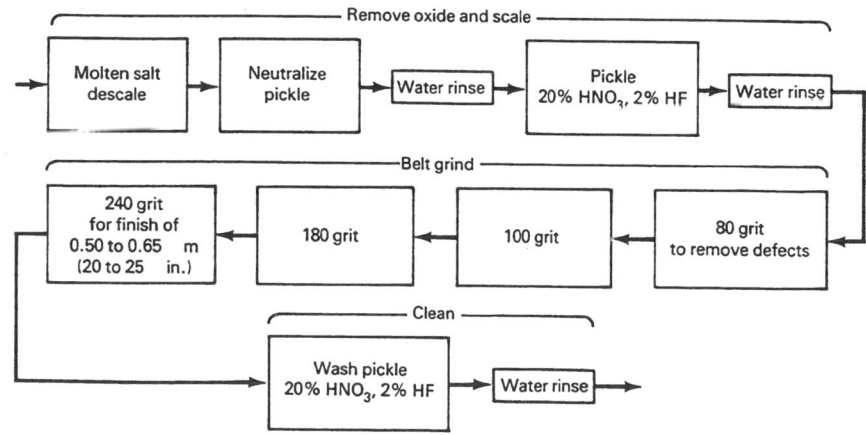

Fig. 1 Cleaning and belt grinding sequences for Ti-6Al-4V sheet

Technical Note 9: Descaling and Special Surface Treatments

Michael L. Bauccio, The Boeing Company

Surface treatments discussed in this article include scale removal, polishing, buffing, plating, conversion coatings, thermal spray, vapor deposition, and thermomechanical surface treatment. Titanium alloys generally do not require special surface treatments to improve corrosion resistance, because their resistance to many corrosive environments is excellent. However, titanium is coated or plated with corrosion-resistant metals (e.g., copper and platinum) as an alternative to oxide formation.

Before being subjected to any thermal treatment, titanium components should be cleaned and dried. *Caution: Do not use ordinary tap water in cleaning titanium components.* Oil, fingerprints, grease, paint, and other foreign matter should be removed from all surfaces. Cleaning is required because the chemical reactivity of titanium at elevated temperatures can lead to its contamination or embrittlement and can increase its susceptibility to stress corrosion. After cleaning, parts should be handled with clean gloves to prevent recontamination. If a component is to be sized, straightened, or heat treated in a fixture, the fixture also should be free of any foreign matter and loosely adhering scale.

Besides the descaling and cleaning of titanium alloys (which are often necessary preliminaries for other operations), wear resistance and/or lubricity are common concerns in the surface treatment of titanium alloys. Titanium alloys can have high wear rates when in contact with itself or other metal alloys, because its adherence causes high friction coefficients and galling. Methods of surface hardening and the application of coatings for lubrication are thus areas of interest.

Removal of Surface Oxides

Most surface treatments and operations require the removal of surface oxides, which can be in the form of α case scale or a thin oxide tarnish. Any oxide formed under 600 °C (1100 °F) can be removed with a nitric-hydrofluoric acid pickle bath (see "Removal of Tarnish Films" in this section). Oxides formed at temperatures over 600 °C (1100 °F) can be removed either mechanically or chemically by immersion in molten salt baths.

Cleaning and Descaling Problems

The metallurgical and chemical properties of titanium create a number of very special cleaning problems. These include:

- Affinity of titanium to common gases
- Galvanic effects caused by discontinuities in scaled surfaces
- Metallurgical restrictions on the temperature of the descaling media
- Variety of scales encountered in titanium descaling
- Protective coatings used in titanium manufacturing

Gas Absorption. The property that causes the most difficulty is the capacity of titanium to absorb common gases including oxygen, hydrogen, and nitrogen, all of which tend to embrittle the product. Because tightly packed hot rolling scale acts as partial protection against additional gas absorption, some mills perform two or three heat treatments over the scale. Each additional heat treatment toughens the scale and compounds descaling difficulties.

An additional problem is that heat-treating furnace atmospheres are maintained on the oxidizing side, because the diffusion rate of oxygen in titanium is so much lower than that of hydrogen. A layer of oxygen-rich metal develops beneath the resulting scale formation, varying in thickness from 0.05 to 0.07 mm (0.002 to 0.003 in.) in the heat-treated condition to 0.15 to 0.20 mm (0.006 to 0.008 in.) in the hot-rolled condition. This brittle surface is usually removed by acid or electrolytic chemical pickling.

Galvanic effects and discontinuities in the surface scale are encountered in all types of metal descaling, but appear to be more pronounced in titanium. Although the exact cause of small pits or cells formed in descaled material is debatable, possibilities include alloy or nonmetallic segregation, scale porosity, and surface contamination. A more severe galvanic attack problem is created by patch scale conditions on titanium surfaces, when areas of heavy scale flake away from an apparently uniform surface. The same problem has been observed with superimposed oxides, even though the surface layer may be quite thin and powderlike. Surface contamination with oil, grease, or fingerprints can also create a patch scale condition. All of these factors promote severe localized attack when areas of the basis metal are exposed selectively during descaling. Some producers have considered reoxidation of the product during processing as a possible solution.

Metallurgical Restrictions on Descaling. Solution treated, age-hardenable alloys of titanium are sensitive to time-temperature reactions, and the temperatures of descaling media. The metastable beta or beta alloys, which are solution treated and aged at temperatures ranging from 370 to 540 °C (700 to 1000 °F) for times of 8 h or more, may induce a subsequent aging effect and cause a change in mechanical properties if descaled at the higher temperatures. This is particularly true in thin-gage sheet materials and may cause property changes of as much as 70 MPa (10 ksi) in tensile strength. The alloys normally should not be allowed to exceed 260 °C (500 °F).

Variety of Scale. Another factor that contributes appreciably to titanium descaling problems is the wide variety of scale encountered, including scale formed by annealing, forging, solution treating, stress relieving, ex-

Typical P/M Property Data

Tensile properties of P/M beta Ti alloys
Each of the alloys was HIPed at 15 °F-25 °F above the transus or at 50 °F-60 °F below the transus. The HIP cycle was 12 h at 15 ksi.

Alloy	Heat treatment	Tensile strength (ksi)	Yield strength 0.2% offset (ksi)	Elongation (%)	Reduction of area (%)	Modulus 10^6 psi
Ti-10-2-3	As HIP 1400 °F	128.5	119.2	22.0	60.4	13.4
	1410 °F 1 h WQ + 900 °F 8 h	208.8	188.0	3.0	2.6	15.0
		207.3	186.7	4.0	6.2	13.5
	1375 °F 1 h WQ + 950 °F 8 h	173.5	158.1	10.0	29.6	15.0
	1410 °F 1 h WQ + 950 °F 8 h	186.7	171.7	8.0	15.9	14.9
		187.2	172.0	9.0	22.9	14.4
	1440 °F 1 h WQ + 950 °F 8 h	194.6	177.5	...	...	14.7
	As HIP 1475 °F	133.0	118.4	21.0	58.0	13.6
	1410 °F 1 h WQ + 900 °F 8 h	212.2	191.5	3.0	6.2	14.4
		211.0	192.6	2.0	6.2	14.4
	1375 °F 1 h WQ + 950 °F 8 h	172.5	155.1	11.0	31.8	14.7
	1410 °F 1 h WQ + 950 °F 8 h	194.4	180.1	6.0	16.4	15.0
		194.0	176.4	5.0	16.2	15.3
	1440 °F 1 h WQ + 950 °F 8 h	197.4	180.7	5.0	7.3	14.7
Beta III	As HIP 1325 °F	154.9	143.4	14.0	39.8	12.0
	1325 °F 5 h WQ + 900 °F 8 h	199.2	185.9	4.0	6.5	14.0
		200.3	186.8	6.0	9.2	14.0
	1275 °F 5 h WQ + 950 °F 8 h	176.7	165.0	9.0	22.7	14.1
	1325 °F 5 h WQ + 950 °F 8 h	185.6	174.3	6.0	9.0	14.1
	1375 °F 5 h WQ + 950 °F 8 h	191.7	179.8	7.0	12.9	14.0
	As HIP 1400 °F	152.3	140.1	12.0	32.5	10.8
	1325 °F 5 h WQ + 900 °F 8 h	201.3	186.4	3.0	4.4	13.7
		195.3	181.9	2.0	5.6	13.6
	1275 °F 5 h WQ + 950 °F 8 h	171.5	155.4	8.0	16.7	14.7
	1325 °F 5 h WQ + 950 °F 8 h	184.4	171.6	6.0	11.6	13.8
		184.3	170.9	6.0	12.1	13.9
	1375 °F 5 h WQ + 950 °F 8 h	181.7	166.4	8.0	18.3	14.0
Alloy 253	As HIP 1400 °F	127.0	90.5	20.0	45.0	12.0
	1425 °F 1 h WQ + 900 °F 8 h	209.6	196.7	2.0	4.4	14.7
		209.3	195.6	2.0	4.0	14.9
	1350 °F 1 h WQ + 950 °F 8 h	184.5	172.6	7.0	12.5	14.9
	1425 °F 1 h WQ + 950 °F 8 h	196.5	185.1	3.0	5.8	14.6
		197.8	186.1	3.0	6.4	14.8
	1450 °F 1 h WQ + 950 °F 8 h	198.4	185.7	4.0	6.0	15.8
	As HIP 1475 °F	127.2	121.1	15.0	41.2	11.3
	1425 °F 1 h WQ + 900 °F 8 h	195.0	181.2	2.0	4.6	14.3
		194.2	180.8	2.0	4.6	14.4
	1350 °F 1 h WQ + 950 °F 8 h	182.1	169.6	3.0	4.4	15.8
	1425 °F 1 h WQ + 950 °F 8 h	184.2	170.0	3.0	6.2	14.6
		185.6	172.2	4.0	6.8	14.6
	1450 °F 1 h WQ + 950 °F 8 h	185.6	171.9	4.0	7.0	15.4

Source: C.F. Yolton, Crucible Materials Corporation, Crucible Research Division, "P/M Beta Titanium Alloys for Landing Gear Applications," Progress in Powder Metallurgy, 1986, Vol 42, MPIF 1986, p 635-653

P/M Ti-6Al-2Sn-4Zr-2Mo-0.1Si
Gas atomized P/M Ti-6Al-2Sn-4Zr-2Mo-0.1Si meets tensile and creep properties for wrought material as specified by AMS 4975.

Test temperature	Room temperature	900 °F
UTS (ksi)	151	104
Yield (ksi)	135	82
% Elongation	19.0	12.0
% R.A.	42.5	20.2

Gas atomized powder, −35 and 80 mesh consolidated by HIP, HIP 1700 °F 14 ksi 4 h, solution treat of 1 h 1765 °F AC, plus age 8 h 1100 °F AC. Source: Unpublished data 1991, C.F. Yolton, Crucible Materials Corporation, Crucible Research Division

embedded in the surface, alumina sand is preferred to silica sand.

Sheet thickness to about 0.50 mm (0.020 in.) can be descaled without distortion if fine sand and low velocities are used. Mill scale on titanium semi-products can be removed with coarse high-carbon steel shot or grit, while finished compressor blades can be cleaned with zircon sand of 150 to 200 mesh. The type of product to be cleaned, the cleaning rate, and the cost of the abrasive must be balanced in the selection of a specific blast cleaning method.

Mineral abrasive particles, such as silica, zircon, or alumina sands, are used more commonly than metal abrasive for blasting finished or semifinished products. Although these abrasives are more expensive, they produce the finer finish that is required for the final product. Adequate safety precautions must be observed to avoid inhalation of fine sand particles. Air-circulating and dust-collecting systems must be cleaned frequently and equipped to cope with the fire hazard associated with titanium dust.

A fine dust remains on the titanium from the blasting operation, particularly when mineral abrasives have been used. This is not considered detrimental, although a washing or pickling cycle following the blast is desirable if the part is to be welded subsequently. The following describes a wet blasting procedure and a dry blasting procedure used for descaling titanium parts:

- *Wet blasting*: Parts are wet blast cleaned, using a slurry that consists of 400-mesh aluminum oxide, 40 vol% in water. Air pressure of 655 kPa (95 psi) is used to pump the slurry in a steady stream with a pressure of about 34 kPa (5 psi). The descaling rate, normally about 50 min/m^2 (5 min/ft^2), depends on the complexity of the part. Distortion and the need for planishing are held to a minimum by placing the blast nozzle at a distance of approximately 50 mm (2 in.) from the workpiece, and by using an angle of impingement of 60°.
- *Dry blasting*: Rocket motor case assemblies are dry blasted after final stress relieving at 480 to 540 °C (895 to 1000 °F). Blasting is accomplished with 100- to 150-mesh zircon sand at an air pressure of 275 kPa (40 psi). Each assembly is rotated at 2½ rev/min and is passed at a speed of 65 mm/min (2.5 in./min) between two diametrically opposed fixed position blasting nozzles. The nozzles blast the inside and outside surfaces simultaneously at the same wall location. To prevent distortion, each nozzle is placed at the same distance, 300 mm (12 in.), from the metal surface.

Molten Salt Descaling Baths

Molten salt descaling baths are primarily used for descaling bar, sheet products, and tubing. With the most effective barrier films available today, some gas penetration of titanium surfaces can be expected at the elevated temperatures required for working and heat treatment. The alpha case or oxygen-enriched layer resulting from this gas reaction is extremely hard and brittle and must be removed. Bar products used for machining finished parts must have this hard scale and oxide removed because these are very abrasive and cause rapid tool wear. Welding or forming stock must have these scales removed, or poor and small welds are made and forming (hot or cold) is virtually impossible without surface cracking or failure of parts. Removal presents no serious problems since chemical milling techniques have been perfected by the aircraft industry to effect weight savings.

One specific problem encountered in alpha case removal is that the titanium oxide formed is substantially more insoluble in the nitric hydrofluoric etchant than the base metal. Residues of oxide on the surface develop areas resembling craters on the finished product.

Where alpha case removal is a required part of a manufacturing operation, salt bath cleaning is specified because proper cycling practically guarantees a chemically clean surface. Conditioning salt baths fall into two basic categories of high-temperature salt baths and low-temperature salt baths. Alternatively, grit blasting may be used to break up the scale so the nitric-hydrofluoric etchant or chem-milling solution removes scale more evenly.

High-temperature salt baths may vary in chemical reaction and effectiveness depending on composition. All types operate at a range of 370 to 480 °C (700 to 895 °F). The temperature range is sufficiently high to produce the most rapid reaction possible for soiled and oxide films. High-temperature oxidizing salt baths are also capable of reacting chemically with organic films to destroy them. These baths are also excellent solvents for silicate barrier films. They do require special fixturing to reduce the strong galvanic effects present at these temperatures, and for this reason, they are used in cleaning primary forming operation products such as forgings, extrusions, rolled plate, and sheet. The major advantage of high-temperature oxidizing or reducing salt baths for titanium descaling is their great speed in removing extremely tenacious scale. Although reducing baths have the inherent disadvantage of promoting hydrogen absorption, this can be overcome or minimized by chemical additions. Vacuum degassing is another solution.

A primary producer of titanium sheet uses an oxidizing salt bath for removing the hot work scale in the following sequence of operations:

- Immerse in oxidizing salt for 5 to 20 min at 400 to 480 °C (750 to 895 °F)
- Quench with water, hold 1 min
- Immerse in sulfuric acid, 10 to 40 vol%, for 2 to 5 min at 50 to 60 °C (120 to 140 °F)
- Rinse with water, 1 min
- Repeat if necessary
- Pickle in nitric-hydrofluoric acid solution, time and concentration as required

The same producer also uses a sodium hydroxide reducing salt bath for descaling high beta or metastable beta alloys. A typical cycle using this type of salt is:

- Immerse in reducing salt for 1 to 3 min at 370 °C (700 °F)
- Quench in water 1 min
- Immerse in sulfuric acid 10 to 40 vol%, for 2 to 5 min at 50 to 60 °C (120 to 140 °F)
- Rinse in water
- Pickle in nitric-hydrofluoric acid solution, time and concentration as required
- Vacuum degas or decontaminate titanium beta alloys that absorb hydrogen in reducing baths

These baths are used by one of the major aerospace contractors for cleaning titanium blades for jet engines. Blade materials are Ti-6Al-4V and Ti-8Al-1Mo-1V. Descaling cycles for removing oxides and proprietary glass-like compounds from these blades are:

- Immerse in oxidizing salt for 15 min at 455 °C (850 °F)
- Rinse in cold water
- Pickle in solution of 35% nitric acid and 3.5% hydrofluoric acid for 1 min max at 20 °C (70 °F)
- Rinse in hot water

Low-Temperature Baths. The temperature range used for cleaning fabricated parts is 200 to 220 °C (390 to 430 °F). Descaling systems based on salts in this temperature range (see Table 1) eliminate

Table 1 Low-temperature salt bath and acid bath conditions

Sample composition	Scale formation temperature °C	°F	Salt bath immersion time(a), min	Acid cleaning bath time(b), min	Acid cleaning bath time(c), seconds
Ti-6Al-4V	650	1200	2	2	30
Ti-8Al-1Mo-1V	650	1200	2	2	30
Ti-8Al-1Mo-1V	820	1510	5	2	30
Ti-6Al-4V(d)	820	1510	5	5	30
Ti-6Al-4V(e)	950	1745	5	5	60
Ti-8Al-1Mo-1V(f)	950	1745	5	5	60

(a) Salt bath temperature 205 °C (400 °F). (b) Bath composition, 30% sulfuric acid. (c) Bath composition, 30% nitric acid, 3% hydrofluoric acid. (d) Sample recycled in salt bath for 5 min, in sulfuric acid bath for 5 min, in nitric acid-hydrofluoric acid bath for 30 s. (e) Sample recycled in salt bath for 5 min, in sulfuric acid bath for 5 min, in nitric acid-hydrofluoric acid bath for 60 s. (f) Sample recycled in salt bath for 5 min, in sulfuric acid bath for 5 min, in nitric acid-hydrofluoric acid bath for 60 s

some of the possible problems associated with higher temperature baths including:

- Age hardening
- Dissimilar metal reactions
- Chemical attack
- Metal distortion
- Hydrogen embrittlement

Salts in this range have a very limited composition because of the effect of various compounds on the melting point. Although they contain oxidizing agents, the effect of these materials is not as aggressive as it is in the high-temperature fused salts. Consequently, organic materials are not destroyed, but are saponified and absorbed. Silicate barrier films and molybdenum disulfide are soluble in these low-temperature salts. The temperature range permits cycling between salt and acid to reduce cleaning times and costs.

Aqueous caustic descaling baths have been developed to remove light scale and tarnish from titanium alloys, except for beta titanium alloys. This procedure is not recommended for beta alloys. They can absorb hydrogen so quickly in concentrated caustics that a 30 min exposure could exceed specification. Aqueous caustic solutions containing 40 to 50% sodium hydroxide have been used successfully to descale many titanium alloys. One bath containing 40 to 43% sodium hydroxide operates at a temperature near its boiling point, 125 °C (260 °F). Descaling normally requires from 5 to 30 min. Immersion time is not critical because little weight loss is encountered after the first 5 min. Caustic descaling conditions the scale so that it is removed readily during subsequent acid pickling.

A more effective aqueous solution contains either copper sulfate or sodium sulfate in addition to sodium hydroxide. This bath operates at a lower temperature, 105 °C (220 °F). A composition of this solution by weight is as follows:

50% sodium hydroxide, 10% copper sulfate pentahydrate ($CuSO_4 \cdot 5H_2O$), and 40% water. Using immersion times of 10 to 20 min, this bath has proved effective in descaling Ti-6Al-4V and Ti-2.5Al-16V alloys.

Removal of Tarnish Films

Tarnish films are thin oxide films that form on titanium at air temperatures between 315 and 650 °C (600 and 1200 °F), after exposure at 315 °C (600 °F). The film is barely perceptible, but with increasing temperature and time at temperature, it becomes thicker and darker. The film acquires a distinct straw yellow color at about 370 °C (700 °F), and a blue color at 480 °C (900 °F). At about 650 °C (1200 °F), it assumes the dull gray appearance of a light scale. Alloying elements and surface contaminants also influence the color and characteristics.

Tarnish films are readily removed by abrasive methods, and all but the heaviest films can be removed by acid pickling. Prolonged exposures at temperatures above about 600 °C (1100 °F), in combination with surface contaminants, result in heavier surface films that are not removed satisfactorily by acid pickling, but require descaling treatments for their removal.

Acid pickling removes a light amount of metal, usually a few tenths of a mil. It is used to remove smeared metal, which could affect penetrant inspection. Titanium and titanium alloys can be satisfactorily pickled by the following procedure:

- Clean thoroughly in alkaline solution to remove all shop soils, soap drawing compounds, and identification inks. If coated with heavy oil, grease, or other petroleum-based compounds, parts may be degreased in trichloroethylene before alkaline cleaning. Degreasing will not be harmful to the part in subsequent processing.
- Rinse thoroughly in clean running water after alkaline immersion cleaning.
- Pickle for 1 to 5 min in an aqueous nitric-hydrofluoric acid solution containing 15 to 40% nitric acid and 1.0 to 2.0% hydrofluoric acid by weight, and operated at a temperature of 24 to 60 °C (75 to 140 °F). The ratio of nitric acid to hydrofluoric acid should be at least 15 to 1. The preferred acid content of the pickling solution, particularly for alpha-beta and beta alloys, is usually near the middle of the above ranges.
- A good all-around pickle bath composition is 35 vol% HNO_3/5 vol% HF (48% acid). Compositions based on certain ratios can give drastically different hydrogen absorption rates. For instance, Ti-15-3-3-3 exhibits an absorption rate of about 5 ppm/mil in the 35/5 bath, while in a 7/1 bath (same ratio of acids), the rate is about 200 ppm/mil.
- A solution of 33.2% nitric acid and 1.6% hydrofluoric acid has been found effective. When the buildup of titanium in the solution reaches 12 g/L (2 oz/gal), discard the solution.
- Rinse the parts thoroughly in clean water.
- High-pressure spray wash thoroughly with clean water at 55 ± 6 °C (130 ± 10 °F).
- Rinse in hot water to aid in drying.
- Allow to dry.

To avoid excessive stock removal, the recommended immersion times for pickling solutions should not be exceeded. It is equally important to maintain the composition and operating temperature of the bath within the limits prescribed to prevent an excessive amount of hydrogen pickup. Gage loss from all acid pickling after descaling is estimated to be less than 0.025 mm/min (0.001 in./min), depending on the combination of variables used.

Hydrogen contamination is estimated to be 0 to 15 ppm/0.25 mm (0.001 in.) of metal removed, depending on alloy composition and gage material pickled. Hydrogen contamination can be held to a minimum by maintaining an acid ratio of 10 to 1 or greater of nitric acid to hydrofluoric acid. Hydrogen diffuses more rapidly into the beta phase. Alpha-beta alloys with a fully crystallized structure, with isolated β at grain boundary triple points, pick up less hydrogen than microstructures with transformed beta and/or simple mill annealed structures. The recrystallized annealed microstructure has only isolated β grains, rather than the more continuous β matrix of the transformed structure, which makes hydrogen ingress more difficult.

Mass Finishing (Barrel Finishing). Oxide films formed by heating to temperatures as high as 650 °C (1200 °F) for 30 min have been effectively removed from Ti-8Mn alloy parts by wet mass finishing. At barrel speeds of 43,000 to 51,000 mm/min (1700 to 2000 in./min), parts have been cleaned satisfactorily in about 1 h.

In mass finishing titanium parts, the ratio of finishing medium to parts should be between 10 and 15 to 1, depending on the size of the parts. Proportionately more medium is required as part size increases. Water is used to cover parts and medium. Surface finish is improved when more water is added, but cycle time required to obtain a given finish is increased. The rate of descaling increases directly with barrel speed but is limited by the fragility of the parts being processed. Parts are randomly loaded in the barrel, and rotated at relatively low barrel speeds to minimize distortion and nicking.

Aluminum oxide mediums are the most satisfactory. They do not contaminate the work and have a long useful life. For oxide removal, small well-worn mediums produce the highest finish. To avoid possible metallic contamination, the medium used for titanium should not be used in processing other metals. Strong acid-forming compounds are avoided, principally because they are corrosive and contribute to hydrogen embrittlement. Because of the fire hazard created by fine, dry titanium particles, dry mass finishing of titanium parts is not recommended.

Cleaning

Pickling Procedures Following Descaling. All advantages gained through proper conditioning and handling of titanium parts during cleaning can be lost if the composition of the final pickling acid is not controlled. Cold spent acid solutions have increased appreciably the time requirements for pickling and the possible quality problems experienced with hydrogen pickup. Highly concentrated hot acids can be overly aggressive, resulting in surface finish problems, such as a rough and pitted surface caused by preferential acid attack. Pickling solutions for cleaning can be weaker than descaling solutions.

A nitric-hydrofluoric acid solution, which is the final stage brightening in most alloy cleaning lines, should be maintained at a minimum ratio of 15 parts nitric acid to 1 part hydrofluoric acid to reduce hydrogen pickup effects; the concentration of hydrofluoric acid may vary from 1 to 5%, or even higher as long as the ratio is not exceeded; the activity of these pickle solutions is affected by titanium content, and the acids are frequently discarded at a level of 26 g/L (3 oz/gal); the solution used for final brightening can be used for the required alpha case removal also, with careful monitoring of titanium content.

Sulfuric acid, 35 vol% at 65 °C (150 °F), is recommended for pickling immediately following salt bath conditioning and rinsing to remove molten salt and residual softened scales. An acid of this formula has very little effect on titanium metal. Metal salts in the original and additional acid solutions further minimize these base metal attacks and hydrogen absorption.

Removal of grease, oil, and other shop soils from titanium parts is normally accomplished with the same type of equipment and the same cleaning procedures used for stainless steel and high-temperature alloy components. Nonchlorinated solvents or alkaline cleaners are recommended.

Vapor degreasing normally employs either trichloroethylene or perchloroethylene. At temperatures above 550 °F, these solvents are known to be a cause of stress-corrosion cracking in titanium alloys. Methylethyl ketone is used as a cleaner in situations where chlorinated solutions are not desired. All titanium parts should be acid pickled after vapor degreasing to remove residual chlorine.

Other cleaning methods use chemicals which, if they are left to dry on the part, may have a harmful effect on the properties of titanium. Among these are (a) soda ash, borates, silicates, and wetting agents commonly used in alkaline cleaners; (b) kerosene and other hydrocarbon solvents used in emulsion cleaners; and (c) mineral spirits employed in hand wiping operations. Residues of all these cleaning agents must be completely removed by thorough rinsing. To ensure a surface that is free of contaminants, rinsing is frequently followed by acid pickling.

Finishing

Wire brushing of titanium alloys is not recommended when other finishing methods, such as buffing, can accomplish the objective. Wire brushing used on titanium, in an attempt to remove surface scratches or oxide films, can result in serious defects. A stiff-bristled wire brush removed surface scratches and oxide films, but the surface was pitted by the wire tips. To avoid pitting, softer wire bristles were tried. The surface of the titanium acquired a burnished appearance; surface layers were cold worked; and grinding scratches, instead of being removed, were filled with smeared metal. Wire brushing with a silicon carbide abrasive grease has been used successfully to remove burrs, break sharp edges from edge radii, and blend chamfers.

The polishing and buffing of titanium is accomplished with the same equipment used for other metals. Polishing is frequently done wet, using mineral oil lubricants and coolants. Silicon carbide abrasive cloth belts have been effective. It is common to polish in two or more steps, using a coarser grit initially, such as 60 or 80, to remove gross surface roughness, followed by polishing with 120 or 150 grit to provide a smooth finish. Titanium tends to wear the sharp edges of the abrasive particles and to load the belts more rapidly than steel. Frequent belt changes are required for effective cutting. A good flow of coolant improves polishing and extends the life of the abrasive.

Dry polishing is more appropriate than wet for some applications. For these operations, belts or cloth wheels with silicon carbide abrasive may be used. Soaps and proprietary compounds may be applied to the belts to improve polishing and to extend belt life. Abrasive belt materials that incorporate solid stearate lubricants offer improved results for dry polishing operations.

Fine polishing of titanium articles for extremely smooth finishes requires several progressive polishing steps with finer abrasive until pumice or rouge types of abrasive are applied. With the softer grades of titanium, such as unalloyed material, fine polishing requires more time and care to prevent scratching. The harder alloy grades can be polished more readily to a surface of high reflectivity. If a matte finish is desired, wet blasting with a fine slurry may be used after initial polishing.

Titanium alloys can be buffed safely. The purpose of buffing is to improve the surface appearance of the metal and to produce a smooth tight surface. Buffing is used as a final finishing operation and is particularly adaptable to finishing a localized area of a part. Parts such as body prostheses, pacemakers, and heart valves require a highly buffed tight surface to prevent entrapment of particles. Close fitting parts for equipment, such as the modern guidance systems, and electronics applications require highly polished surfaces obtained by buffing. In addition, sheet sizes too large to be processed by other abrasive finishing methods, such as mass finishing or wet blasting, can be economically processed by buffing.

The principal limitations of buffing are (a) distortion, caused by the inducement of localized stress, (b) surface burning, resulting from prolonged dwell of the buff, (c) an inability to process inner or restricted surfaces, and (d) the feathering of holes and edges. Proper care of the buffing wheel is essential. Buffing with insufficient compound or a loaded wheel produces burning or distortion of the part. After buffing, no further cleaning of parts is required except degreasing to remove the buffing compound.

Electropolishing. Electrolytic polishing can completely remove all traces of worked metal remaining from mechanical grinding and polishing operations used in specimen preparation. When electropolishing is used in metallography, it is preceded by mechanical grinding (and sometimes polishing) and followed by etching.

The conditions and electrolytes required to obtain a satisfactory polished surface differ for different alloys. Even minor alloying additions to a metal may significantly affect the response of the metal to polishing in a given electrolyte. In developing a suitable procedure for electropolishing a metal or alloy, it is generally helpful to compare the position of the major component of the alloy with elements of the same general group in a periodic table and to study the phase diagram, if available, to predict the number of phases and their characteristics.

Single-phase alloys generally are easy to electropolish, whereas multiphase alloys are likely to be difficult or impossible to polish with electrolytic techniques. In multiphase alloys, the rates of polishing of different phases often are not the same. Polishing results depend significantly on whether the second or third phases are strongly cathodic or anodic with respect to the matrix. The matrix is dissolved preferentially if the other phases are relatively cathodic, thus causing the latter to stand in relief. Preferential attack may also occur at the interface between two phases.

For titanium and titanium alloys, electropolishing can be effectively done with mixtures of perchloric acid ($HClO_4$). However, mixtures of $HClO_4$ and acetic anhydride are extremely dangerous to prepare and are even more unpredictable to use. Many industrial firms and research laboratories, and some municipalities forbid the use of such potentially explosive mixtures, which have caused fatalities and property damage in some accidents. These mixtures also are highly corrosive to the skin, and the vapors of acetic anhydride can cause severe damage by inhalation. These hazards are considered sufficient reason for recommending that mixtures of $HClO_4$ and acetic anhydride not be used, despite their effectiveness as electropolishing electrolytes.

To avoid using mixtures of acetic and perchloric acid, electrolytes based on mixed acid or salts have been developed (see Table 2). For example, pure titanium (99.9% Ti) has been successfully electropolished with an electrolyte solution of 11.1% hydrofluoric acid, 59% lactic acid, 24.6% sulfuric acid, 3.6% dimethyl sulfoxide, and 1.7% glycerine. Polishing occurs with an applied voltage of 24 to 35 V at 97 mA/cm^2 (see Fig. 2).

Plating of Titanium

Titanium and titanium alloys can be difficult to plate because surface oxides can prevent good adhesion of the plating with the titanium substrate. Careful surface preparation is therefore essential. The type of property improvements (e.g., wear resistance, lubricity, corrosion resistance) varies with the different plating processes, which generally include electrodeposition methods (electroplating), electroless plating, and chemical conversion coatings.

Table 2 Electrolytes and voltages for electropolishing of titanium and titanium alloys

Electrolyte	Cell voltage	Time	Notes
Electrolytes composed of $HClO_4$ and alcohol with or without organic additions(a)			
700 mL ethanol (absolute), 120 mL distilled H_2O, 100 mL 2-butoxyethanol, 80 mL $HClO_4$ (60%)	30-65	15-60 s	(b)
600 mL methanol (absolute), 370 mL 2-butoxyethanol, 30 mL $HClO_4$ (60%)	60-150	5-30 s	...
590 mL methanol (absolute), 6 mL distilled H_2O, 350 mL 2-butyoxyethanol, 54 mL $HClO_4$ (70%)	58-66	45 s	(c)
11 mL $HClO_4$ (60%), 65 mL methanol (absolute), 24 ml butyl cellosolve	26-28	~3 min	(d)
Electrolytes composed of $HClO_4$ (60%) and glacial acetic acid			
940 mL acetic acid, 60 mL $HClO_4$	20-60	1-5 min	(e)
900 mL acetic acid, 100 ml $HClO_4$	12-70	1/2-2 min	...
800 mL acetic acid, 200 mL $HClO_4$	40-100	1-15 min	...
Electrolytes composed of mixed acids or salts			
995 mL ethanol (absolute), 100 mL n-butyl alcohol, 109 g $AlCl_3 \cdot 6H_2O$ (hydrated aluminum chloride), 250 g $ZnCl_2$ (zinc chloride) (anhydrous)	30-60	1-6 min	...
11.1% hydrofluoric acid, 59% lactic acid, 24.6% sulfuric acid, 3.6% dimethyl sulphoxide, 1.7% glycerine	24-35	...	(f)

(a) Chemical components of electrolytes are listed in the order of mixing. Except where otherwise noted the electrolytes are intended for use at ambient temperatures in the approximate range of 18 to 38 °C (65 to 100 °F), and with stainless steel cathodes. Absolute SD-3A or SD-30 ethanol can be substituted for absolute ethanol. (b) One of the best electrolytes for universal use. (c) Polish only. (d) Electrolyte and voltage for electropolishing as described in *Metals Handbook*, 9th ed., Vol 9, *Metallography and Microstructure*. (e) Good general-purpose electrolyte. (f) Source: J. Delleg, *Metallography*, Vol 7, 1974, p 357-360

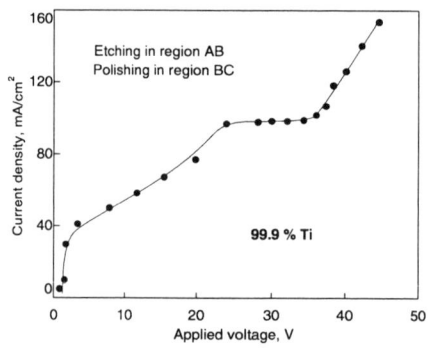

Fig. 2 Current-voltage curve for electropolishing in a mixed acid solution. Polishing is performed on the plateau. Electrolyte solution was 11.1% hydrofluoric acid, 59.0% lactic acid, 24.6% sulfuric acid, 3.6% dimethyl sulphoxide, and 1.7% glycerine.
Source: J. Pelleg, *Metallography*, Vol 7, 1974, p 357-360

Electroplating

Electroplating generally refers to the method of bath electrodeposition, where the part to be plated is made an electrode (the cathode) in an electrochemical cell. ASTM B 481 describes three separate processes for surface preparation and electroplating of titanium and titanium alloys (ASTM B 481, Standard Practice for Preparation of Titanium and Titanium Alloys, The American Society for Testing and Materials, 1990). Hard chromium is the most common plating for wear resistance, while copper and precious metal plating are useful for other reasons. The general concept of electrodeposition also includes the patented process of metalliding, which combines electrodeposition and high-temperature diffusion of plating into the substrate.

Hard chromium plating of titanium alloys requires surface pretreatment that removes surface oxides and allows good adhesion of plating. Pretreatment prior to the hard chromium plating of titanium alloys includes the application of an electroless nickel plate or a coating deposited from a high-chloride nickel strike bath. The following procedure also has given adherent chromium plating having good fatigue properties [C.G. John, Electroplated Coating of Titanium for Engineering Applications, *Titanium Science and Technology*, Vol 2, Lütjering, Zwicker, Bunk, Ed.), DGM, 1985, p 995-1001].

- Degrease areas to be coated in aqueous alkaline cleaners. (Trichloroethylene type solvents are prohibited by many specifications)
- Abrasive blast using a water slurry of aluminum oxide (190/220 mesh) at 415 kPa (60 psi) air pressure
- Etch and activation procedures
- Rinse with clean running deionized water
- Hard chromium plate to P.P.I./17 Iss.1
- Rinse, dry, and heat treat at 200 °C (400 °F) for 2 h

Copper Plating. The electrodeposition of copper on titanium and titanium alloys [see flow chart (Fig. 3)] provides a basis for subsequent plating. After cleaning and before plating, the surface of the titanium must be chemically activated by immersion in an acid dip and a dichromate dip to obtain adequate adhesion of the plated coating (see Fig. 3 for the compositions and operating temperatures of these activating solutions).

Water purity is critical in the composition of activating solutions, although chemicals of technical grade are as effective as, and may be substituted for chemicals of the chemically pure grade. In both the acid and dichromate baths, hydrofluoric acid content is most critical and must be carefully controlled.

After proper activation, titanium may be plated in a standard acid copper sulfate bath. The adhesion of the deposited copper is better than that of 60-40 solder to copper, and the deposit successfully withstands the heat of a soldering iron. The

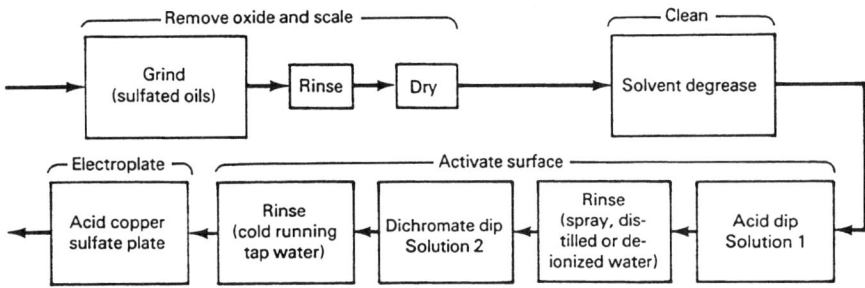

Fig. 3 Processing sequence for electroplating copper on titanium alloy parts

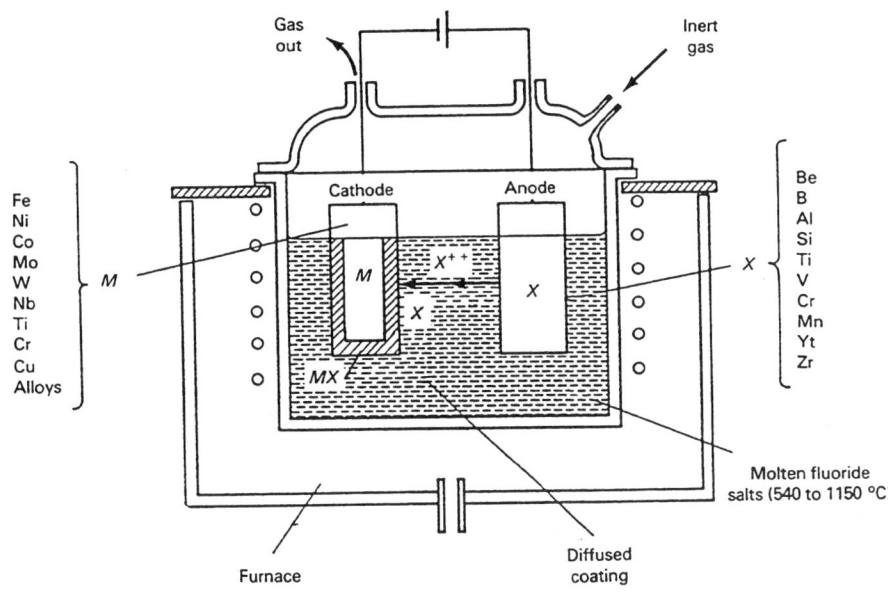

Fig. 4 Schematic of a metalliding bath

normal thickness of the plated deposit is about 25 μm (1 mil).

Copper-plated titanium wire is available commercially. The outstanding property of this material is the lubricity of its copper-plated surface. The wire can be drawn easily and can be threaded on rolls. Such wire has been used in applications that require electrical surface conductivity.

The titanium wire is plated continuously at a speed of about 60 m/min (200 ft/min) in a copper fluoroborate acid bath at a current density of 7.5 to 12.5 A/dm^2 (75 to 125 A/ft^2). The final copper deposit is a thin flash coating. Higher current densities up to 150 A/dm^2 (1500 A/ft^2) have been tried, but if the copper coating is too thick, adhesion is poor.

Platinum Plating. Although titanium is not satisfactory as an anode material because of an electrically resistant oxide film that forms on its surface, application of a thin film of platinum to titanium results in a material with excellent electrochemical properties. Theoretically, the thinnest possible film is sufficient to give the highly desirable low overvoltage characteristics of platinum; furthermore, the film need not be continuous or free of defects to be effective.

The greatest immediate use for platinum-coated titanium is for anodes in the chlorine-caustic industry. Some horizontal-type chlorine cells use expanded metal anodes. From 1.3 to 2.5 μm (0.05 to 0.1 mil) of platinum is applied to the anode surface. Replating of the anodes may be required after about 2 years, depending on the operating conditions. The attrition rate for platinum appears to be about 0.6 g/tonne (0.5 g/ton) of chlorine.

Several platinum and electrode suppliers have developed reliable methods for platinum plating titanium; most use proprietary solutions. A platinum diamino nitrite bath has been used successfully to apply platinum plating to titanium. In this and other procedures, certain precautionary steps are required to achieve adherent, uniform plating. The surface must be cleaned thoroughly, and etched in hydrochloric or hydrofluoric acid to produce a roughened surface. Some procedures also involve a surface activating treatment just before plating. Immersion for 4 min in a solution of glacial acetic acid 895 mL (30 fluid oz) containing hydrofluoric acid 125 mL (4 fluid oz) of 52% hydrofluoric acid, followed by a prompt rinse, appears to be an effective activating treatment if performed immediately before plating. A postplating treatment, consisting of heating to 400 to 540 °C (750 to 1000 °F) for a period of 10 to 60 min, stress relieves the plate, and improves adhesion. This treatment can be done in an air atmosphere, and a light oxide film forms on unplated areas.

Coatings for Emissivity. Electrodeposits and sprayed coatings of gold on titanium are being used to provide a heat-reflecting surface that reduces the temperature of the base metal. Low-emissivity coatings are also used to reduce heat radiated out to the surrounding aluminum structure in aircraft. Gold-coated titanium has been used for jet engine components. The gold coating is applied by spraying a gold-containing liquid on chemically clean titanium sheet. This is followed by a baking treatment. Normal coating thickness is about 25 μm (0.1 mil).

Metalliding, originally patented by General Electric, is a high-temperature electrodeposition process (see Fig. 4) that diffuses the plating elements into the substrate material. Controlling the purity of the salt electrolyte and plating atmosphere are difficult, but the process allows plating of elements on substrates that normally cannot be plated. Several elements have been successfully applied to various substrates (see Fig. 4).

Electroless Plating

Electroless plating baths have been developed for copper, silver, nickel, and other plating elements, but the nickel/phosphorus system is the most common for wear resistance. The deposit is achieved by a catalytic reduction of metal ions in the plating without the use of electricity.

Electroless nickel coating is used because of its wear and corrosion resistance. Deposition rates (10 μm/h, or 0.4 mil/h) are relatively slow compared to electrodeposition methods, but the electroless depositions are more uniform than those of electroplating. Normal thickness is about 50 μm (2 mil).

Adhesion of electroless nickel coatings to most metals is excellent. The initial replacement reaction, which occurs with catalytic metals, together with the associated ability of the baths to remove submicroscopic soils, allows the deposit to establish metallic as well as mechanical bonds with the substrate. With noncatalytic or passive metals, such as stainless steel, an initial replacement reaction does not occur, and adhesion is reduced. With proper pretreatment and activation, however, the bond strength of the coating usually exceeds 140 MPa (20 ksi) [G.G. Gawrilov, *Chemical (Electroless) Nickel Plating*, Redhill, England: Portcullis Press, 1979].

To increase the adhesion of the coatings, baking treatments are useful where pretreatment has been less than adequate and adhesion is marginal. With properly applied coatings, baking has only a minimal effect upon bond strength [R.N. Duncan, Properties and Applications of Electroless Nickel Deposits, *Finishers' Management*, Vol 26 (No. 3), 1981, p 5].

Chemical Conversion Coatings

Chemical conversion coatings are formed when the surface of a substrate material reacts with the surrounding environment. Simple examples of conversion coatings include anodized surfaces and the formation of surface oxides during exposure with air. Conversion coatings are also applied by immersing the material in a tank containing the coating solution. Spraying and brushing are alternate methods of application.

Generally, conversion coatings are thin (2.5 μm, or 0.1 mil) and not hard enough to be competitive for wear-resistant applications. However, chemical conversion coatings (particularly potassium titanate) are used on titanium to improve lubricity by acting as a base for the retention of lubricants. Phosphate-fluoride conversion coatings also improve paint adhe-

Table 3 Comparison of conversion coatings in wiredrawing of titanium

Coating	Drawing compound	Total reduction, %	No. of passes	No. of coats	Final condition
Bare	Molybdenum disulfide with grease	...	0(b)	...	Galled
Bare	Soapy wax	...	0(b)	...	Galled
Degreasing bath	Molybdenum disulfide with grease	85	8	2	Smooth
Pickling bath	Molybdenum disulfide with grease	94	17	7	Smooth
Pickling bath	Soapy wax	68	7	3	Galled
Pickling bath(a)	Molybdenum disulfide with grease	70	7	1	Smooth
Chemical immersion bath	Lacquer molybdenum disulfide	63	8	2	Smooth
Chemical immersion bath	Molybdenum disulfide	63	8	3	Smooth

(a) Coating heated for 1 h at 425 °C (795 °F). (b) First pass unsuccessful

Table 4 Conversion coating baths for titanium alloys

Bath No.	Bath solution	Composition	Amount g/L	Amount oz/gal	Temperature °C	Temperature °F	pH	Immersion time, min
1	Degreasing solution	$Na_3PO_4 \cdot 12H_2O$	50	6.5	85	185	5.1–5.2	10
		$KF \cdot 2H_2O$	20	2.6				
		HF solution(a)	11.5	1.5				
2	Pickling solution	$Na_3PO_4 \cdot 12H_2O$	50	6.5	27	81	<1.0	1–2
		$KF \cdot 2H_2O$	20	2.6				
		HF solution(a)	26	3.4				
3	Chemical immersion solution	$Na_2B_4O_7 \cdot 10H_2O$	40	5.2	85	185	6.3–6.6	20
		$KF \cdot 2H_2O$	18	2.3				
		HF solution(a)	16	2.1				

(a) Hydrofluoric acid, 50.3% by weight

sion to titanium (although grit blasting also provides good surface preparation for paint adhesion).

Results of extensive wiredrawing experiments (see Table 3) illustrate the effectiveness of conversion coatings when used with various lubricants. High-speed rotary tests also have indicated marked improvement in the wear characteristics of the metal after conversion coating and lubricating with one part of the molybdenum disulfide and two parts of thermosetting eponphenolic resin.

Reciprocating wear tests have shown that conversion coatings and oxidized surfaces provided some improvement in wear characteristics, but when conversion coated samples were also oxidized, a marked improvement was noted. The conversion coating increases the oxidation rate of titanium at about 425 °C (800 °F) and may increase oxidation rates at temperatures up to 595 °C (1100 °F). The original coating is retained above the titanium oxide layer.

Conversion coating baths (see Table 4) use various constituent amounts, immersion times, and bath temperatures. The resultant coatings are composed primarily of titanium and potassium fluorides and phosphates. One coating bath consists of an aqueous solution of sodium orthophosphate, potassium fluoride, and hydrofluoric acid, and can be used with various constituent amounts, immersion times, and bath temperatures. The resultant coatings are composed primarily of titanium and potassium fluorides and phosphates.

Cleanness of the part before immersion is critical, and all preliminary cleaning and handling operations [see flow chart (Fig. 5)] must be closely controlled for good results. Finger marks or residual grease on the surface of a part will interfere seriously with the coating process.

The appearance of the baths varies widely during the coating reaction, ranging from rapid bubbling to relative dormancy. Some coatings rub off when still wet; others are adherent. The solutions produce coatings of approximately the same dark gray or black appearance.

The control of pH and immersion time is important. Dissolved titanium and the active fluoride ion make it impossible to use glass electrodes for pH measurements. Indicator paper and colorimetry are the most satisfactory methods for measuring in the degreasing and chemical immersion baths, which are held in the pH range from 5 to 7. The pickling bath is quite acid, and titrametric analysis offers the most practical method of control. When the bath is in the proper coating range, a 20 mL (0.70 fluid oz) sample in 100 mL (3.4 fluid oz) of water will neutralize 11.8 to 12.0 mL (0.4 to 0.41 fluid oz) of normal sodium hydroxide, using a phenolphthalein indicator.

Coating thickness depends on immersion time. In all three baths, a specific time is reached after which the coating weight remains essentially constant. In the fluoride-phosphate baths, a maximum coating weight is reached at some time before this equilibrium point. The maximum coating weight is obtained in about 2 min in the low-temperature bath and in about 10 min in the two other baths. Coatings are easily removable without excessive loss of metal by pickling in an aqueous solution containing 20% nitric acid and 2% hydrofluoric acid by weight.

Anodizing refers to the application of conversion coatings by immersion of the part (as the anode) in an electrochemical cell. Titanium can be anodized, but the coatings are much thinner and softer than anodized aluminum. The oxide layers from anodizing titanium are limited to 0.1 to 0.2 μm (4 to 8 μin.) in a wide range of acid, neutral, and base electrolytes. Anodizing of commercially pure titanium in sulfuric and phosphoric acid at 110 volts produces up to 0.2 μm (8 μin.). In acids at elevated temperatures of 60 to 80 °C (140 to 175 °F), oxidation layers are on the order of 10 μm (0.4 mil).

Although anodizing generally produces TiO$_2$ layers less than 0.1 μm (4 μin.) thick, the process can provide a useful improvement for any titanium component subjected to a rubbing action. It is frequently used on titanium screws and fasteners. Anodizing is also used to prepare surfaces for adhesive bonding. Three anodizing processes for Ti-6Al-4V are outlined in Table 5.

Diffusion Treatments

Like steel, surface hardening is a common objective during the diffusion treatment of titanium alloys. In some cases, however, diffusion treatment is used to develop a chemical barrier at the surface. Oxidation is the most prevalent method of forming a protective surface barrier, although aluminization is an alternative treatment.

Surface hardening treatments can rely on the diffusion of hardening species such as nitrogen, carbon, and oxygen (see Fig. 6). Boron is another hardening species. The temperatures required for adequate diffusion depend on the alloy, the hardening species, and its concentration gradient (or delivery efficiency) at the sur-

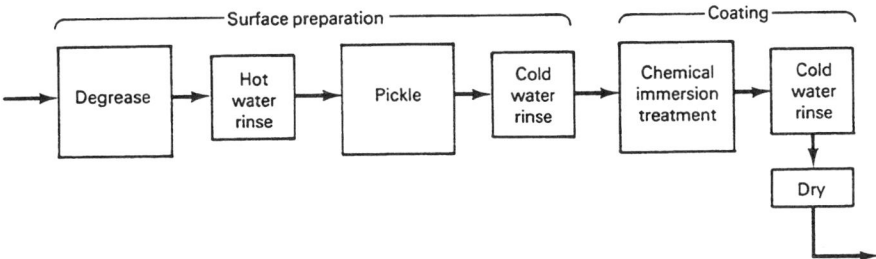

Fig. 5 Processing sequence used in the chemical coating of titanium alloys

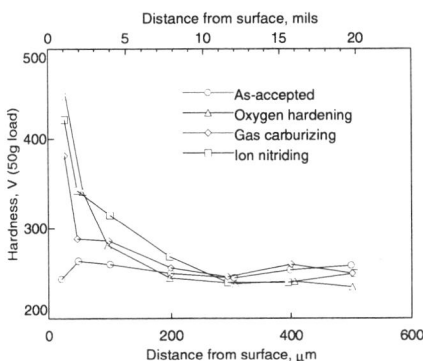

Fig. 6 Hardness profile of Ti-22V-4Al after various surface treatments. Because the cold-forged Ti-22V-4Al beta alloy was being evaluated as a candidate material for mass-produced valve-spring retainers, oxygen hardening in air was ultimately selected.
Source: M. Mushiake et al., Development of Titanium Valve Spring Retainers, SAE Technical Paper Series No. 910428, SAE, 1991

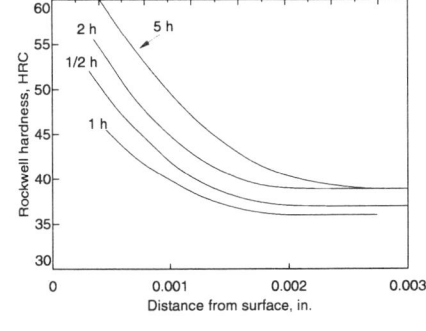

Fig. 7 Hardness profiles of Tiduran-treated Ti-6Al-4V.
Source: R.H. Shoemaker, New Surface Treatments for Titanium, Titanium Science and Technology, Vol 4, Jaffee and Burte, Ed., Plenum Press, 1973, p 2501-2516

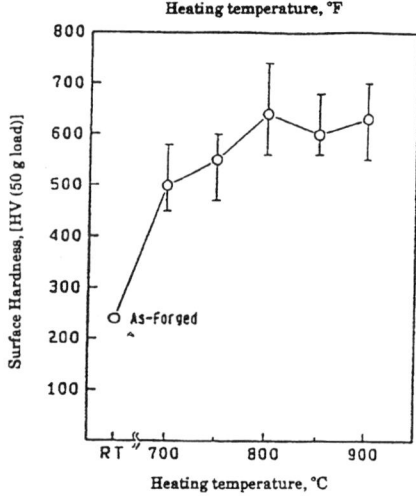

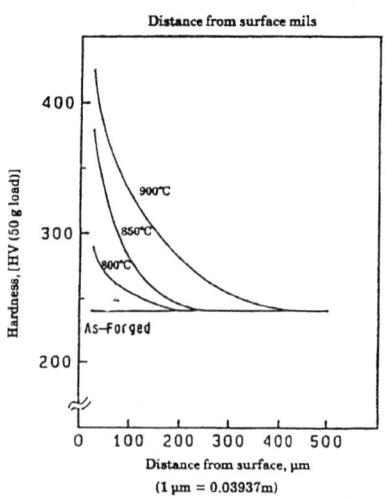

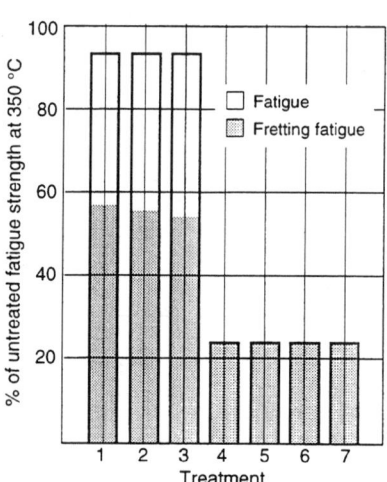

Fig. 8(a) Effect of temperature on oxygen hardening of Ti-22V-4Al. Cold-forged Ti-22V-4Al valve spring retainers heated in air for 30 min.
Source: M. Mushiake et al., Development of Titanium Valve Spring Retainers, *SAE Technical Paper Series No. 910428*, SAE, 1991

Fig. 8(b) Effect of temperature on oxygen hardening of Ti-22V-4Al. Cold-forged Ti-22V-4Al valve spring retainers heated in air for 30 min.
Source: M. Mushiake et al., Development of Titanium Valve Spring Retainers, *SAE Technical Paper Series No. 910428*, SAE, 1991

Fig. 9 A summary of effects of some surface treatments on the fatigue and fretting fatigue properties of Ti-6Al-4V at 350 °C. 1. Shot-peened Alman A16. High-velocity plasma flame-sprayed WC + 12% Co. 2. Electroplated Ni. Diffusion bonded + Shot-peened Al A7. 3. Electroplated Cr. Diffusion bonded + shot-peened Al A8-12 + electroplated Cr. 4. Hard Anodised. 5. Hard anodised + (epoxy resin + MoS_2. 6. Watervliet anodised + (polyimide resin + graphite). 7. Watervliet anodised + (eposy resin + MoS_2).

Table 5 Anodization procedures for Ti-6Al-4V

	CAA(a)	SHA(b)	NaTESi(c)
1. Degrease	Methyl ethyl ketone	Methyl ethyl ketone	Alkaline
2. Pickling			
15 vol% of 70% HNO_3, min	10	10	...
3 vol % of 49% HF at RT, s	30	30	
3. Rinse, deionized water, 25 °C (77 °F), min	1-5	1-5	1-5
4. Anodization	50 g/L (6.6 oz/gal) CrO_3, 1 g/L (0.13 oz/gal) NH_4HF_2, 10 V, 20 min, 20-25 °C (68-77 °F)	5M NaOH, 10 V, 20-30 min, 20-30 °C (68-86 °F)	7.50 mol/L NaOH, 0.33 mol/L Na-tartrate, 0.10 mol/L ethylene diamine tetraacetic acid, 10 V, 15 min, 30 °C (86 °F)
5. Rinse, deionized water, min	5-20	5-20	5-20
6. Air dry, °C (°F)	25-60 (77-140)	25-60 (77-140)	25-60 (77-140)

(a) Chromic Acid Anodization (CAA) process. (b) Sodium Hydroxide Anodization (SHA) process. (c) The NaTESi process is a variation of the SHA process, in which a tartrate titanium-complexing agent is used for chemical etching titanium prior to adhesive bonding [C. Matz, Optimization of the Durability of Structural Titanium Adhesive Joints, *Int. J. Adhesion and Adhesives*, Jan., 1988, p 17-24]. NaTESi stands for sodium hydroxide, sodium tartrate, ethylene diamine tetraacetic acid, and sodium silicate. Source: Clearfield, H.M., et al., Surface Preparation of Metals, in *Engineered Materials Handbook*, Vol 3, Adhesives and Sealants, ASM International, 1990, p 259-275

face. Plasma-assisted methods enable the use of lower diffusion temperatures, while salt bath treatments allow better control of oxygen as compared to gas or pack diffusion methods. Several salt-bath surface hardening treatments have been developed, but they are not used widely in industry. The most common salt-bath hardening treatment is the *Tiduran process* (Kolene Corp.), which involves immersion in a cyanide salt bath at about 800 °C (1475 °F). Hardness profiles depend on exposure time (see Fig. 7). The Tiduran process has been used on process control valves and high-performance parts in automotive engines and space vehicles.

Oxygen Surface Hardening. Although oxygen is usually regarded as an unwanted impurity, the excellent hardening by oxygen in solid solution with titanium also is used in its own right. The diffusion of oxygen at the surface produces three distinct regions in the hardened case:

- A brittle titanium-oxide (TiO_2) top layer, which is subsequently removed
- A TiO_2-TiO layer several microns thick
- A hardened region containing dissolved oxygen
- Usable hardening for wear resistance is provided by the oxygen-rich region and the layer of TiO-TiO_2

The surface hardness and the depth of the oxygen-rich case depend on the time and temperature of exposure. For example, heating of a Ti-22V-4Al alloy in air for 30 min increased surface hardness to more than 550 HV (see Fig. 8a). The extent of hardening reaches a plateau above about 800 °C (1470 °F), although higher temperatures produce a deeper case depth (see Fig. 8b). Higher treatment temperatures are therefore beneficial in terms of wear resistance.

However, when temperatures exceed about 800 °C (1470 °F), the ingress of gases can cause contamination and thus the degradation of mechanical properties. Below 800 °C (1470 °F), oxygen forms a passivation layer that prevents the ingress of gases such as hydrogen.

Nitriding produces a thin, gold-colored surface film of TiN (δ) above a thicker layer of Ti_2N (ε). Below these two layers, a hardened case containing dissolved nitrogen extends to a depth of 10 to 100 μm (0.4 to 4 mil) depending on the time and temperature exposure of the treatment. Surface hardness can be increased to more than 500 HV. Ion (plasma-assisted) nitriding is advantageous because the sputtering action

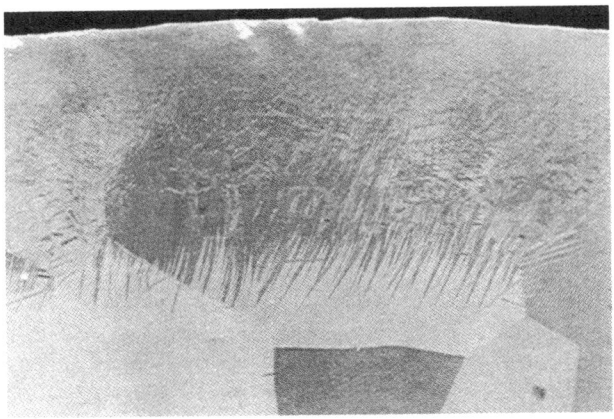
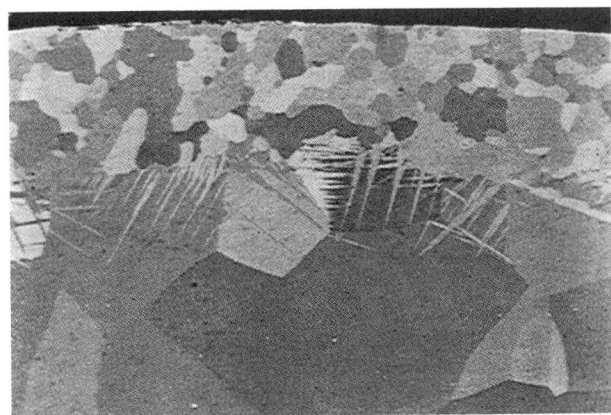

Fig. 10 Near-surface micrcostructures (250×) in Ti-8Al. (a) As-shot peened. (b) Shot-peened and locally recrystallized. Source: H. Gray et al, in Shot Peeing, (Ed., Wöhlfahrt, Kopp and Vöhringer), DGM, Oberursel, 1987, p 467. 250×

Table 6 Thickness of surface engineered layers on titanium

Thickness	Process	Examples
1 mm	Overlay coatings	Flame spray, plasma spray, D-gun
100 μm	Liquid state alloying	Laser, electron beam
	Thermochemical treatments	Nitriding, salt baths, CVD
10 μm	Plating	Hard chrome, electroless nickel
1 μm	PVD coatings	TiN
100 nm	Ion implantation anodizing	C^+, N^+

Table 7 Processing steps for the plasma-sprayed surface treatment

Step	Comment
1. Solvent degrease	Methyl ethyl ketone, wipe or vapor
2. Grit blast	Al_2O_3, 120 mesh
3. Substrate preheating	Plasma flame, one pass
4. Plasma spray, Ti-6Al-4V Power	500 A, 60 arc V
Gas pressure, kPa (psi)	
Primary (Ar)	690 (100)
Secondary (H_2)	345 (50)
Gas flow, m³/min (ft³/min)	
Primary	2.3 (80)
Secondary	0.42 (15)
Spray distance, mm (in.)	100-150 (4-6)
Coating thickness, mm (in.)	0.05 (0.002)
5. Air cool	...

Source: Cloarfield, H.M., et al., Surface Preparation of Metals, in *Engineered Materials Handbook*, Vol 3, *Adhesives and Sealants*, ASM International, 1990, p 259-275

generated by the plasma cleans and activates the surface and removes any oxide film which may be present or be formed during the process. This enables the nitriding to be carried out at lower temperatures and with greater reliability of bonding. Nitrogen migration down grain boundaries can occur, causing embrittlement.

Other Coatings

Thick Coatings

In order to achieve resistance to wear and galling under higher loads, it is necessary to increase the thickness of surface treatments beyond those obtained by nitriding and plating processes (see Table 6). Thicker coatings can be achieved by thermal spray methods and by liquid-state alloying methods such as laser or electron beam alloying.

Thermal Spray. A variety of refractory coatings have been deposited onto titanium by thermal spraying. The coatings are usually applied either by plasma spraying or by detonation-gun coating. Because there is little or no metallurgical bond between coating and substrate, the coating does have the tendency to spall off any sharp edges or under some types of loading. Fatigue strength is also somewhat reduced. However, as shown in Fig. 9, a high-velocity plasma flame-sprayed coating of tungsten carbide plus 12 percent cobalt demonstrated the best fretting fatigue resistance in an evaluation of several titanium surface treatments [Waterhouse, R.B. (Editor), *Fretting Fatigue*, Applied Science Publishers, London, p 186]. Table 7 summarizes the processing steps for producing plasma-sprayed coatings on titanium parts. See also the article "Friction and Wear of Titanium Alloys."

Molybdenum spraying has been used successfully on automotive components such as connector rods. Molybdenum spray coatings are not as hard as nitrided surfaces or other coatings, but the treated surface layer is thicker.

Liquid-state alloying involves surface melting (to a depth of about one-half millimeter or more) by rapid heating with an electron beam or laser. The liquid pool thus allows alloying with hardening species, which are either predeposited on the surface or deposited on the surface during heating. An example of liquid-state alloying is a laser-assisted gas nitriding process that produced case depths of 0.5 mm (0.02 in.) (T. Bell et al., Surface Engineering of Titanium with Nitrogen, *Surface Engineering*, Vol 2, 1986, p 133-143).

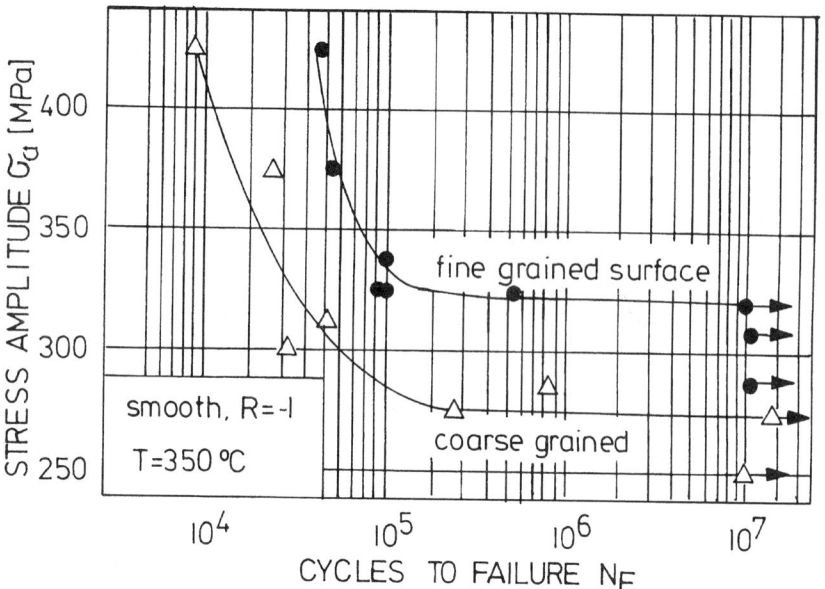

Fig. 11 S-N curves for coarse grained Ti-8Al at 350 °C with and without thermomechanical surface treatment for local grain refinement.
Source: Gray et al, ibid

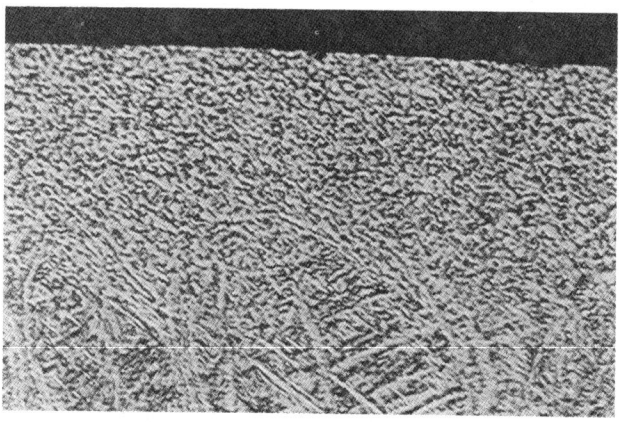

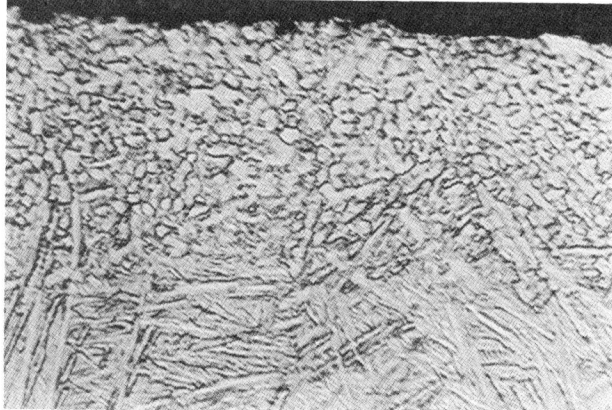

Fig. 12 Near-surface microstructures (500×) after thermomechanical surface treatment. (a) Ti-6Al-4V with lamellar core and equiaxed surface. (b) Ti-6Al-2Sn-4Zr-2Mo with lamellar core and duplex surface microstructure. 500×
Source: Gray et al, ibid

Vapor Deposition

Vapor deposition techniques include chemical vapor deposition (CVD) and physical vapor deposition (PVD). Ceramic coatings applied by CVD methods require high temperatures, and such processes are not used very often on titanium alloys.

Physical Vapor Deposition. The industrial use of PVD coatings is increasing for titanium components subjected to moderate loads. The most common PVD coating is TiN, although PVD coatings of CrN have also been considered for titanium alloys. As far as physical vapor deposition of TiN on titanium is concerned, the practice is almost always speeded up by use of glow-discharge techniques. The process is then termed "plasma-assisted PVD."

Thermomechanical Surface Treatment *

Mechanical surface treatments such as shot peening, polishing, or surface rolling can be utilized to improve the endurance limit in titanium-base materials by altering surface roughness, degree of cold work (dislocation density), and residual stresses. The surface roughness determines whether fatigue strength is primarily crack nucleation controlled (smooth) or crack propagation controlled (rough). For smooth surfaces, a work-hardened surface layer can delay crack nucleation owing to the increase in strength. In rough surfaces, the crack initiation phase can be absent, and a work-hardened surface layer is detrimental to crack propagation owing to the reduced ductility. Near-surface residual compressive stresses are clearly beneficial, as they can significantly retard crack growth once cracks are present (although subsequent stress relief or cyclic plastic deformation can reduce beneficial compressive residual stresses).

Another method of surface treatment is to modify the surface microstructure by thermomechanical processing. Thermomechanical treatments are widely used to optimize the bulk mechanical properties of high-strength titanium alloys. Likewise, it can make sense to "tailor" microstructural variations from the surface to the interior to meet the differing requirements (e.g., as in the carburizing of steels). As demonstrated in the following examples, cold-working induced by mechanical surface treatements can be utilized to develop a surface microstruc-

---

* Adapted from the invited lecture "Thermomechanical Surface Treatment of Titanium Alloys" by L. Wagner and J. Gregory presented at Second ASM Heat Treatment and Surface Engineering Conference in Europe, June 1993.

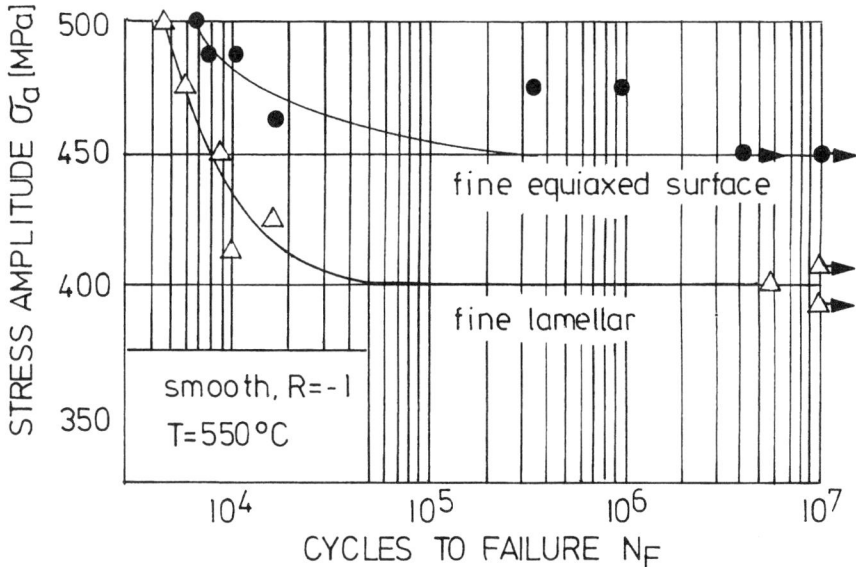

Fig. 13 S-N curves for Ti-6242 at 550 °C with a fine lamellar core microstructure with and without a thermomechanical surface treatment

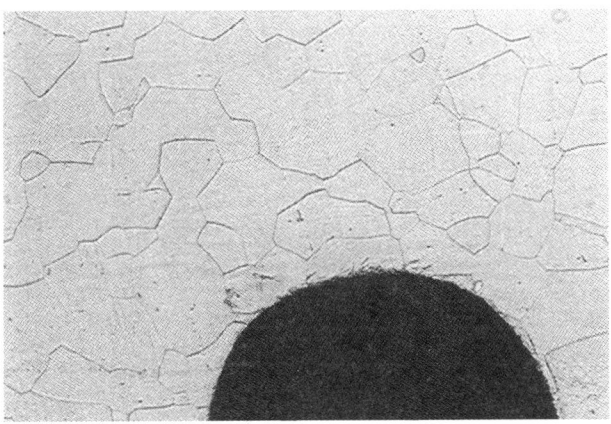

(a)

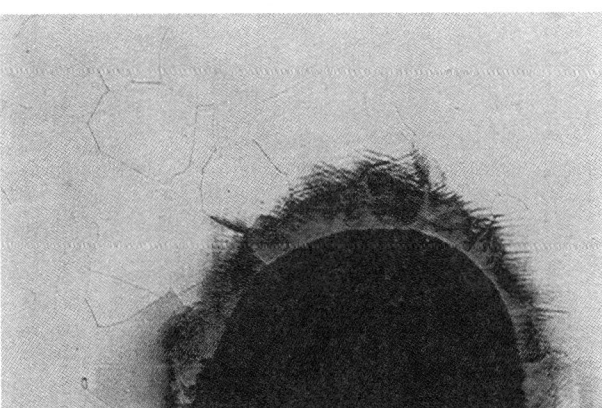

(b)

Fig. 14 Near-surface microstructure (50×) in Beta C (Ti-3Al-8V-6Cr-4Mo-4Zr). (a) After shot peening. (b) After selective surface aging following shot peening

ture which is different from that in the bulk, thus combining the optimum features of both, even in cases where conventional thermomechanical processing may not be practical, as in thick sections. A distinct advantage to be gained by altering the surface microstructure is that such alterations are more stable than those induced by mechanical surface treatments alone.

Alpha Alloys. A mechanical surface treatment in combination with subsequent recrystallization offers the possibility to combine the high strengths and endurance limits associated with fine grains with the superior long through-crack fatigue crack growth behavior and fracture toughness of the coarse grains. To maximize the total fatigue life in thicker sections, fine grains are needed ont he surface, where good resistance to crack initiation is critical, and coarse grains in the interior, where they can reduce the driving force for long crack growth. Shot peening followed by a heat treatment of 1h 820 °C was performed on coarse grained Ti-8Al to cold work and recrystallize the surface (Fig. 10). The improvement in fatigue limit owing to the fine (20 μm) is significant, roughly 50 MPa at 350 °C (Fig. 11).

Alpha-Beta and Near-Alpha Alloys. Because these alloys are often intended for high temperature service, for example in gas turbines, creep resistance is an important consideration. On this basis, lamellar microstructures would be preferable. However, these microstructures have poor fatigue resistance, particularly in the LCF regime, where surface crack growth determines fatigue life. In such cases, a variation in phase morphology between the surface and the core can be desirable. Figure 12 shows examples for an ($\alpha + \beta$) and a near-α alloy where fine surface microstructures were obtained by mechanically working the surface by shot peening and then heat treating. The improvement in S-N behavior (at high temperature) gained by this thermomechanical surface treatment is shown for Ti-6242 with a creep-resistant, fine lamellar core and a fatigue-resistant, fine equiaxed surface layer in Fig. 13.

Beta Alloys. Both shot peening and surface rolling in combination with specially developed aging treatments have been applied to Ti-3Al-8V-6Cr-4Mo-4Zr to selectively age harden only the surface. This new thermomechanical surface treatment shows promise for improving properties of high strength springs and fasteners. Figure 14 shows the near-surface region for shot-peened material both without and with a selective surface aging (SSA) treatment. The high strength of the surface increases the fatigue limit to values above those of a conventional bulk-aging treatment, while the high ductility of

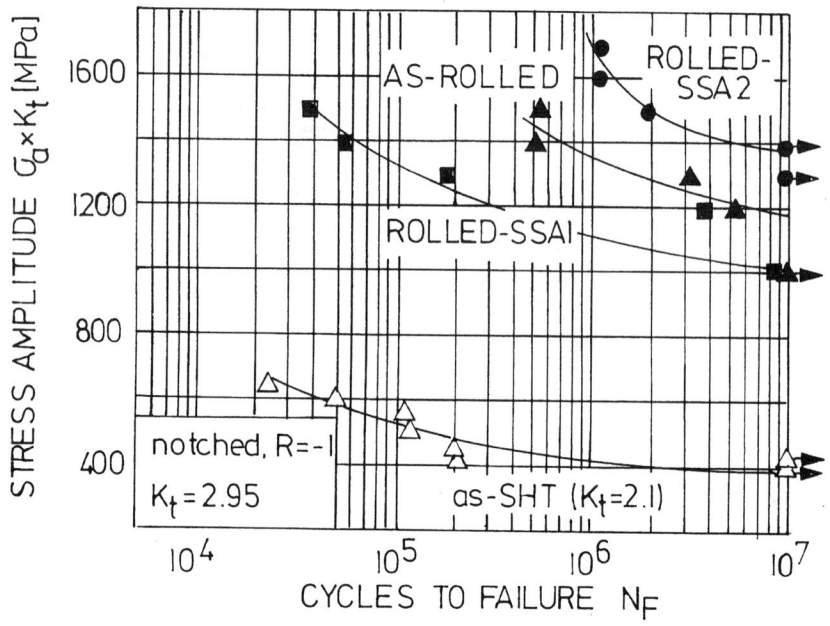

Fig. 15 S-N curves for Beta C in the as-SHT condition and after rolling with and without subsequent aging

the solution heat treated (SHT) condition in the interior provides good notched ductility and toughness. Fully reversed notched fatigue behavior for this alloy after surface rolling with and without SSA is shown in Fig. 15. Rolling alone increases the notched fatigue limit of the SHT condition (expressed as norminal stress times stress concentration factor) from the low value of 400 MPa to 1100 MPa. Depending on the subsequent aging treatment, the S-N behavior can deteriorate slightly (SSA1) or be even further improved (SSA2). These results suggest that the residual compressive stresses present in the as-rolled condition are significantly relieved by the SSA1, but not the SSA2 treatment.

Technical Note 10: Welding and Brazing

Titanium and most titanium alloys can be welded by arc, spot, seam, flash, pressure, friction and electron beam methods. Procedures and equipment are generally similar to those used for welding austenitic stainless steel or aluminum. However, because titanium and titanium alloys are extremely reactive above 540 °C (1000 °F), precautions must be taken to shield the joint from air. No flux is used when welding titanium.

Weldability

Unalloyed titanium and all alpha titanium alloys are weldable. All grades of pure titanium and alpha alloys are welded in the annealed condition. Welding of cold worked alloys anneals the heat-affected zone (HAZ) and negates the strength produced by cold working.

Alpha-Beta Alloys. Ti-6Al-4V and other weakly beta stabilized alloys can be welded in the annealed condition or in the solution-treated and partially aged condition, with aging completed during postweld stress relieving. Strongly beta-stabilized alloys are embrittled by welding. The low weld ductility of most alpha-beta alloys is caused by phase transformations in the weld zone or in the HAZ which promote a martensitic microstructure.

Metastable Beta Alloys. Most metastable beta alloys can be successfully welded, but because aged welds in beta alloys can be quite brittle, heat treatment to strengthen the weld by age hardening should be used with caution. Metastable beta alloys typically are weldable in the annealed or solution heat treated condition. In the as-welded condition, welds are low in strength but ductile. Beta alloy weldments are sometimes used in the as-welded condition. To obtain full strength, the metastable beta alloys are welded in the annealed condition; the weld is cold worked by peening or planishing, and the weldment is then solution treated and aged. This procedure also obtains adequate ductility in the weld.

Cleaning

To obtain a good weld, cleanliness of the welded surface is a major concern. The cleaning procedure depends on whether the oxide layer in the joint area is light or heavy (see Fig. 1).

Degreasing. Grease and oil accumulated during forming and machining must be removed before welding to avoid weld contamination. Scale-free metal requires only degreasing. Degreasing precedes descaling for metal with an oxide scale. Methods of degreasing include steam cleaning, alkaline cleaning, vapor degreasing, and solvent cleaning.

For vapor degreasing, toluene rather than a chlorinated solvent should be used, because residues from chlorinated solvents (and also from silicated solvents) may contribute to cracking of titanium weldments. Solvent cleaning is frequently used, especially for large components that cannot conveniently be placed in a vapor degreaser or washer for alkaline cleaning. Solvents applied include methylethyl ketone, toluene, acetone and other chlorine-free solvents. Because methyl alcohol has reportedly caused stress corrosion, it is prohibited for use on aerospace hardware. In solvent cleaning, the joint areas are hand-wiped with the solvent just before welding. All wiping should be done with clean, lint-free cloths or a cellulose sponge. Plastic or lint-free gloves should be worn; rubber gloves are likely to leave traces of plasticizer that can cause porosity in the weld metal. Handprints are also a source of contamination.

After a lightly oxidized joint area has been degreased, it should be pickled for a short time. A typical mixture is 4 wt% hydrofluoric acid and 40 wt% nitric acid. Because hydrogen is detrimental to the properties of titanium, causing embrittle-

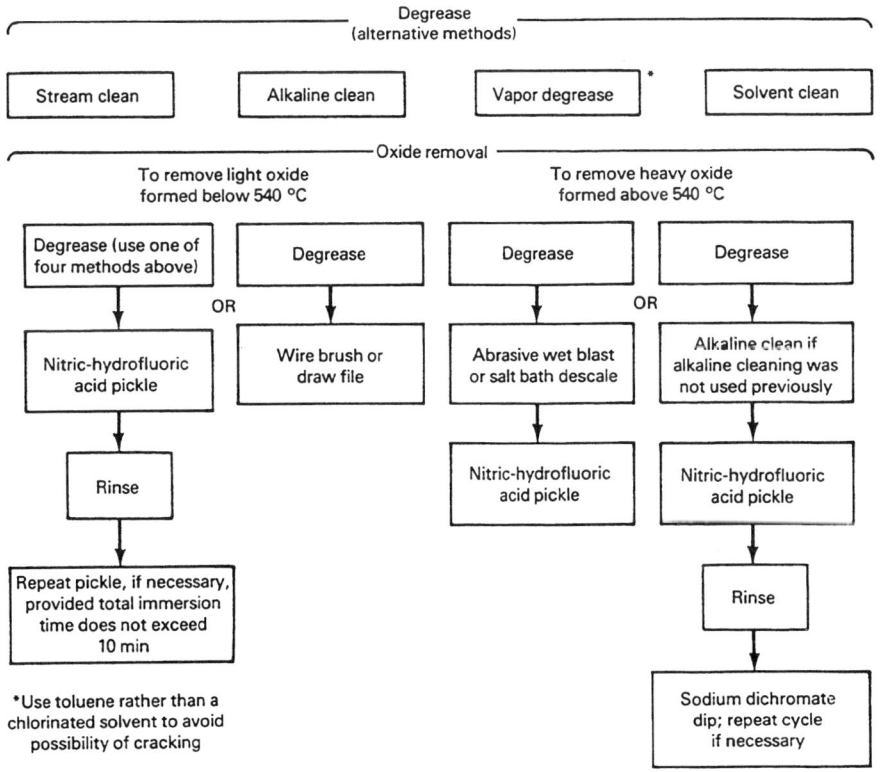

Fig. 1 Flow chart of procedures for cleaning titanium alloys

ment and sometimes contributing to weld porosity, pickling should be performed cautiously. Industrial practice usually can be to maintain the acid bath at a high oxidation potential of 30 wt% or more nitric acid, which simultaneously holds the ratio of nitric acid to hydrofluoric acid at 15 to 1 strictly as a factor of safety. For additional information on acid pickling of titanium, see "Technical Note 9: Surface Treatments" in this Volume. If the nitric acid content falls below 30 wt% and the ratio of nitric acid to hydrofluoric acid falls below 10 to 1, excessive hydrogen pickup is possible.

Oxide Removal. A lightly oxidized joint area may also be cleaned by brushing with a stainless steel wire brush or by draw filing. When weld corrosion resistance is important, however, these cleaning methods should be followed immediately by acid pickling. Steel wool or abrasives should never be used, because of the danger of contamination.

If grinding is required, the use of silicon carbide burrs is preferred, because wheels produce residues of rubber or resin on the surface that contaminate the weld. Excessive heat should be avoided while grinding and low rotating speeds should be used. Soft-backed grinding tools should be used.

If titanium has been exposed to temperatures above 540 °C (1000 °F) prior to welding, a more complex removal treatment such as chemical, salt bath, or mechanical, or combinations of these treatments, is required. The salt baths are basically sodium hydroxide to which oxidizing agents or hydrogen has been added to form sodium hydride (see "Technical Note 9: Surface Treatments" in this Volume).

Two alternative procedures for removing heavy scale are shown in the previous flowchart for cleaning titanium alloys. In one, the parts are subjected to liquid abrasive blasting or salt-bath descaling after degreasing. These treatments are usually followed by pickling in nitric-hydrofluoric acid, as the removal of light scale. When salt-bath descaling is used, oxide removal can be hastened by removing the workpieces from the bath, scrubbing them with brushes, and then re-immersing them. To prevent hydrogen pickup during salt-bath descaling, time cycles must be short (preferably no more than 2 min) and bath temperature must be carefully controlled.

In the second method for removing heavy scale (see Fig. 1) parts are alkaline cleaned after grease removal (unless alkaline cleaning was used for grease removal) and then pickled, rinsed, and dipped in a sodium dichromate solution. Selection of cleaning method depends largely on the size and shape of the parts and on the cleaning methods available in a particular plant.

Gas Shielding. For successful arc welding of titanium and titanium alloys, complete shielding of the weld is necessary, because of the high reactivity of titanium to oxygen and nitrogen at welding temperatures. Shielding is required for weldment areas that exceed about 540 °C (1000 °F) in air.

In welding titanium and titanium alloys, only argon and helium, and occasionally a mixture of these two gases, are used for shielding. Because it is more readily available and less costly, argon is more widely used. Argon shielding gas was used in the examples given in this article.

Because of high purity (99.985% min) and low moisture content, liquid argon is often preferred. The argon gas should have a dew point of –60 °C (–75 °F) or lower. The hose used for the shielding gas should be clean, nonporous, and flexible, made of Tygon or vinyl plastic. Because rubber hose absorbs air, it should not be used. Excessive gas flow rates that cause turbulence should be avoided; flowmeters are usually employed for all gas shields. Pressure (psi) gages may be employed for trailing and backup shields.

Welding in Chambers. Excellent welds can be obtained in titanium and its alloys in a welding chamber, where welding is done in a protective gas atmosphere, thus giving adequate shielding. Welding in a chamber, however, is not always practical. For example, in manual welding the location of the glove ports and the presence of a chamber wall impose limitations on visibility, movement, and accessibility. Welding of titanium was first done in metal chambers that can be evacuated and then backfilled with argon or helium. Such chambers are equipped with glove ports, so that the welder can handle the torch, separate filler metal (if used), and the weldment without admitting air to the chamber. Viewing ports enable the welder to see the welding operation. Although expensive to operate, especially for large weldments, metal chambers are frequently used in aerospace applications.

Generally, shielding gas is not supplied to the welding torch when welding titanium in a metal chamber, and excellent welds can be made if the chamber atmosphere is maintained properly. In some applications, however, where heavy or long welds are required, gas is supplied to the torch to improve shielding.

Rigid or collapsible chambers made of transparent plastic can be used where production runs are short, the assembly is large or complicated, and manual welding is required. Rigid plastic chambers are flow-purged with argon or helium, in volumes equal to five or ten times the volume of the chamber, before welding is started. Collapsible plastic chambers are first collapsed and then flow-purged with argon or helium; they require less gas for purging than do rigid chambers.

Advantages of plastic chambers (either rigid or collapsible) are low cost and good visibility of the work. Because there is generally a greater probability of leakage occurring in a plastic chamber than in a metal chamber, the atmosphere must be checked frequently to ensure that it is of proper purity. In addition, torch shielding is usually employed to make certain that the weld zone is adequately protected.

Out-of-Chamber Welding. With proper tooling, joints in titanium can be adequately shielded for welding without using a chamber. Both the weld and the HAZ must be shielded during welding and until the temperature of the metal in the area of the weld is below 540 °C (1000 °F). If shielding is inadequate, the welds are brittle due to oxygen or nitrogen embrittlement.

The welding torch (or electrode holder) is usually equipped to supply a trailing shield that provides a diffuse, nonturbulent flow of gas to the solidifying and cooling (see Fig. 2) weld. The length of the trailing shield must be adjusted to the speed of welding. Both straight and curvilinear welding can be shielded. In addition, the welding station must be shielded by curtains to prevent drafts.

Most shields are designed and/or handcrafted for the particular weld. Trailing shields must be capable of providing sufficient gas coverage to produce a sound, bright-silver weld deposit. In addition, heat-resistant tape is used around the trailing shield edges to contain the gas-shielding envelope.

Arc Welding

Gas tungsten arc welding (GTAW) is the most widely used process for joining titanium and titanium alloys, except in large thicknesses. Square-groove butt

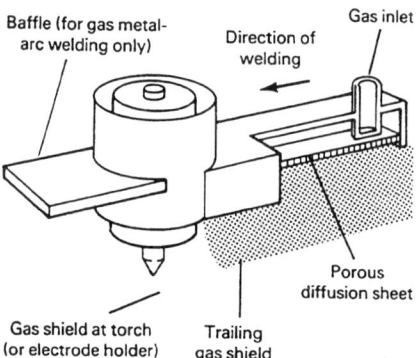

Fig. 2 Arrangement for shielding in automatic welding of titanium alloys in air. The baffle shown on the leading side of the torch (or electrode holder) is seldom used for GTAW, but is used for GMAW.

Table 1 Dimensions of typical joints for welding titanium and titanium alloys

Base-metal thickness, t, in.	Root opening	Groove angle, °	Weld-bead width
Square-groove butt joint			
0.010-0.090	0	...	...
0.031-0.125	0-0.10t	...	...
Single-V-groove butt joint			
0.062-0.125	0-0.10t	30-60	0.10-0.25t
0.090-0.125	(a)	90	...
0.125-0.250	0-0.10t	30-60	0.10-0.25t
Double-V-groove butt joint			
0.250-0.500	0-0.20t	30-120	0.10-0.25t
Single-U-groove butt joint			
0.250-0.750	0-0.10t	15-30	0.10-0.25t
Double-U-groove butt joint			
0.750-1.500	0-0.10t	15-30	0.10-0.25t
Fillet weld			
0.031-0.125	0-0.10t	0-45	0-0.25t
0.125-0.500	0-0.10t	30-45	0.10-0.25t

(a) Root face, 0.030 in. Source: J.J. Vagi, et al., "Welding Procedures for Titanium and Titanium Alloys," NASA TMX 53432, 1965

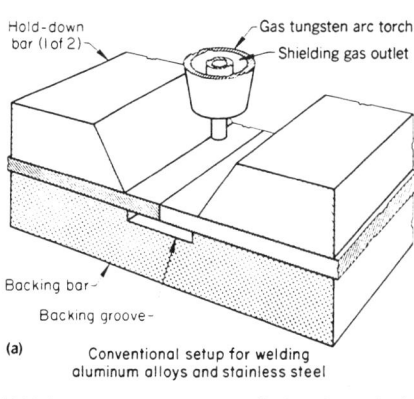

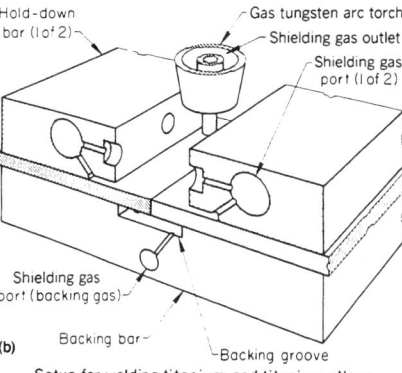

Fig. 3 Typical setups for inert gas shielding for GTAW. (a) Conventional setup for welding aluminum alloys and stainless steel. Gas shielding is from the torch. Use of shielding gas in the backing groove is optional. (b) Setup for welding titanium and titanium alloys outside a welding chamber. Gas shielding is from the torch and through parts in hold-down bars, backing bars, and from trailing and backup shields.

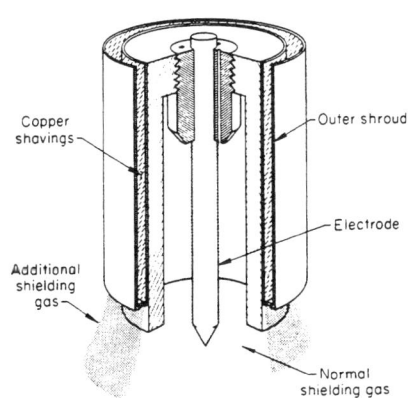

Fig. 4 Sectional view of torch nozzle equipped with an outer shroud. Copper shavings in the outer shroud provide additional gas shielding for manual inert gas welding.

joints can be welded without filler metal in base metal up to 2.5 mm (0.10 in.) thick. For thicker base metal, the joint should be grooved and filler metal is required. Where possible, welding should be done in the flat position.

Hot-wire GTAW can be used for welding titanium alloys more than 6 mm (¼ in.) thick. The hot-wire process combines conventional single-wire arc welding with a unique method of filler-metal additions and may be used for welding or surfacing. A 1.6 mm (1/16 in.) diam filler wire is preheated to the molten state as it enters the weld pool. The wire is mechanically fed into the weld pool through a holder from which argon gas flows to protect the liquid and solid metal from oxidation. No flux is used when welding titanium.

Joint Design

Joint designs for titanium welding are essentially the same as those found in the inert gas welding of other metals. If arc welding is done outside a controlled-atmosphere welding chamber, joints must be carefully designed so that both the top and the underside of the weld can be shielded (see Fig. 3). Dimensions of typical joints (see Table 1) require a tighter joint fit-up than for welding other metals, because of the possibility of entrapping air in the joint. The joint should be clamped to prevent separation during welding. To obtain a good weld, the joint and the surfaces of the workpieces at least 50 mm (2 in.) beyond the width of the gas trailing shield on each side of the weld groove must be meticulously cleaned.

GTAW Equipment

Power Supplies. Transformer-rectifiers and inverters are the preferred power supplies for welding titanium, because the current can be controlled more closely than with motor-generator sets; slight variations in welding current may cause variations in penetration. Direct current electrode negative is always used for GTAW of titanium because deeper weld penetration and a narrower bead can be obtained than with direct current electrode positive (DCEP). Also, in manual welding, DCEN is easier to control.

The power supply should include accessories for arc initiation because of the danger of tungsten contamination of the weld if the arc is struck by torch starting. If welding is to be done in air, controls for extinguishing the arc without pulling the torch away from the workpiece are needed, so that shielding-gas flow continues and the hot weld metal is not contaminated by air. For further details on power supplies and other equipment, see the article "Gas Tungsten Arc Welding" in Volume 6 of the 9th Edition *Metals Handbook*.

Electrodes. The conventional thoriated tungsten types of electrodes (EWTh-1 or EWTh-2) are used for GTAW of titanium. Electrode size is governed by the smallest diameter able to carry the welding current. To improve arc initiation and control the spread of the arc, the electrode should be ground to a point. The electrode may extend one and a half times the size of the diameter beyond the end of the nozzle.

Shielding. To ensure a diffuse, nonturbulent flow of shielding gas, nozzles of torches for welding titanium are larger than those used for welding other metals. With a 1/32 in.-diam electrode, a 9/16 in.-ID nozzle is ordinarily used, and with a 1/16 in.-diam electrode, a ¾ in.-ID nozzle is used. Phenolic or other plastic nozzles should not be used to avoid the danger of contaminating the weld with carbon.

Because titanium has low thermal conductivity, the area ahead of the arc does not get heated above 540 °C (1000 °F), therefore, leading shields are seldom required when welding is done by the

Table 2 Chemical compositions of titanium and titanium alloy filler metals AWS A5.16

AWS classification	Composition(a), %													
	C	O	H	N	Al	V	Sn	Cr	Fe	Mo	Nb	Ta	Pd	Ti
ERTi-1(b)	0.03	0.10	0.005	0.012	...	...	...	...	0.10	...	...	...	...	rem
ERTi-2	0.05	0.10	0.008	0.020	...	...	...	...	0.20	...	...	...	...	rem
ERTi-3	0.05	0.10-0.15	0.008	0.020	...	...	...	...	0.20	...	...	...	...	rem
ERTi-4	0.05	0.15-0.25	0.008	0.020	...	...	...	...	0.30	...	...	...	...	rem
ERTi-0.2Pd	0.05	0.15	0.008	0.020	...	...	...	...	0.25	...	...	...	0.15-0.25	rem
ERTi-3Al-2.5V	0.05	0.12	0.008	0.020	2.5-3.5	2.0-3.0	...	...	0.25	...	...	...	...	rem
ERTi-3Al-2.5V-1(b)	0.04	0.10	0.005	0.012	2.5-3.5	2.0-3.0	...	...	0.25	...	...	...	...	rem
ERTi-5Al-2.5Sn	0.05	0.12	0.008	0.030	4.7-5.6	...	2.0-3.0	...	0.40	...	...	...	...	rem
ERTi-5Al-2.5Sn-1(b)	0.04	0.10	0.005	0.012	4.7-5.6	...	2.0-3.0	...	0.25	...	...	...	...	rem
ERTi-6Al-2Nb-1Ta-1Mo	0.04	0.10	0.005	0.012	5.5-6.5	...	...	...	0.15	0.5-1.5	1.5-2.5	0.5-1.5	...	rem
ERTi-6Al-4V	0.05	0.15	0.008	0.020	5.5-6.75	3.5-4.5	...	...	0.25	...	...	...	...	rem
ERTi-6Al-4V-1(b)	0.04	0.10	0.005	0.012	5.5-6.75	3.5-4.5	...	...	0.15	...	...	...	...	rem
ERTi-8Al-1Mo-1V	0.05	0.12	0.008	0.03	7.35-8.35	0.75-1.25	...	...	0.25	0.75-1.25	...	...	...	rem
ERTi-13V-11Cr-3Al	0.05	0.12	0.008	0.03	2.5-3.5	12.5-14.5	...	10.0-12.0	0.25	...	...	...	...	rem

(a) Single values are maximum. (b) Extra-low interstitials for welding similar base metals

GTAW process. For welding operations where a trailing shield is not adaptable, the nozzle of the torch is fitted with a concentric outer shroud through which a supplementary supply of shielding gas is fed (see Fig. 4). A shielding gas is diffused through copper shavings contained in the outer shroud that cool the gas substantially, helping to protect the metal near the weld.

Shielding of the underside of a weld is provided by slotted backing bars, usually copper, through which a diffuse flow of argon or helium is maintained. Gas channels in the clamping fixtures also provide diffuse flow of inert gas to the weld area. These fixtures are placed close to the weld to avoid the danger of air contamination.

Fixtures. Copper fixtures are usually employed for GTAW. Other metals are used, but they should be nonmagnetic; arc blow tends to occur with magnetic metal fixtures. Metal fixtures are sometimes water cooled, but this method introduces the possibility of moisture from the air condensing on the fixtures.

Filler Metals. For welding titanium thicker than about 2.5 mm (0.10 in.) by the GTAW process, a filler metal must be used. Fourteen titanium and titanium alloy filler-metal (or electrode) classifications are given in AWS A5.16 (see Table 2). Five of these are essentially unalloyed titanium and the remainder are titanium alloy filler metals. Maximums are set on carbon, oxygen, hydrogen, and nitrogen contents.

Filler-metal composition is usually matched to the grade of titanium being welded. For improved joint ductility in welding the higher strength grades of unalloyed titanium, filler metal of yield strength lower than that of the base metal is occasionally used. Because of the dilution that occurs during welding the weld deposit acquires the required ductility. Unalloyed filler metal is sometimes used to weld Ti-5Al-2.5Sn and Ti-6Al-4V for improved joint ductility. The use of unalloyed filler metals lowers the beta content of the weldment, thereby reducing the extent of the transformation to fine alpha or martensite and improving ductility. Engineering approval, however, is recommended when employing pure filler metal with an alloy weldment to ensure that the weld meets strength requirements. Another option is filler metal containing lower interstitial content (oxygen, hydrogen, nitrogen, and carbon) or alloying contents that are lower than the base metal being used. The use of filler metals that improve ductility does not preclude embrittlement of the HAZ in susceptible alloys. In addition, low-alloy welds may enhance the possibility of hydrogen embrittlement.

Welding equipment for automatic hot-wire GTAW utilizes two separate power supplies. The power supply for the hot-wire addition is connected across the contact tube of the hot wire torch and workpiece, causing a current to flow through the wire. The wire feeder adds a continuous supply of filler wire to the weld deposit, constantly maintaining the electrical circuit that produces the molten metal.

Because the hot-wire addition was melted by its own alternating current power supply, the amount of hot wire addition can be controlled independently of the arc wire. The amount of hot wire addition depends on the quantity of metal that can be usefully deposited while maintaining good weld-bead geometry and uniform penetration.

GTAW Procedures

Generally, procedures for GTAW of titanium alloys are similar to those used for austenitic stainless steel.

Preheating is not required for titanium alloys, although it is sometimes used to prevent distortion. Cracking may occur in titanium alloy weldments, and it is most often related to contamination not prevented by preheating. Maintenance of a specific interpass temperature is not necessary. Preheating of 80 to 90 °C (180 to 200 °F) may be used to eliminate surface moisture. The usual method is a heat lamp, hot air gun, or infrared heater.

Preparation. The filler metal, as well as the base metal, should be clean at the time of welding. If filler metals have a large surface-to-volume ratio and are slightly contaminated, the weld may be severely contaminated. Some procedures require that the filler wire be cleaned immediately before use. The use of an acetone-soaked, lint-free cloth serves to remove surface contamination caused by the die lubricant used in the wire drawing operation, in addition to cleaning the filler wire. Pickling in nitric/hydrofluoric acid solution is also used for cleaning.

Tack welding is used to pre-position parts or subassemblies for final welding operations. Elaborate fixturing often can be eliminated when tack welds are used to their full advantage. Various tack welding procedures can be used, but in any procedure good cleaning practices and adequate shielding must be provided to prevent contamination of the welds. Contamination or cracks developed in tack welds can be transferred to the finish weld. One procedure is to tack weld in such a way that the finish weld never crosses over a previous tack weld. To accomplish this, sufficient filler metal is used in tack welding to completely fill the joint at a particular location. The finish weld beads are blended into the ends of the tack welds.

Arc length for welding without filler metal, as with stainless steel and the nickel-based alloys, should have a maximum size about equal to the electrode di-

Table 3 Welding procedure schedule for GTAW titanium

Material thickness(a), in.	Tungsten electrode diameter, in.	Filler rod diameter, in.	Nozzle ID, in.	Shielding gas flow, ft^3/h	Welding current(b), A	Number of passes	Travel speed(c), in./min
Square-groove and fillet welds							
0.024	1/16	...	3/8	18	20-35	1	6
0.063	1/16	...	5/8	18	85-140	1	6
0.093	3/32	1/16	5/8	25	170-215	1	8
0.125	3/32	1/16	5/8	25	190-235	1	8
0.188	3/32	1/8	5/8	25	220-280	2	8
V-groove and fillet welds							
0.25	1/8	1/8	5/8	30	275-320	2	8
0.375	1/8	1/8	3/4	35	300-350	2	6
0.50	1/8	5/32	3/4	40	325-425	3	6

Note: Tungsten used for the electrode: first choice 2% thoriated ETTh2, second choice 1% thoriated EWTh1. Adequate gas shielding is essential not only for the arc, but for heated material also. Backing gas is recommended at all times. A trailing gas shield is also recommended. Argon is preferred. For higher heat input, on thicker material use argon-helium mixture. Without backing or chill bar, decrease current 20%. (a) Or fillet size. (b) Direct current electrode negative. (c) Per pass

Table 4 Typical operating conditions for PAW of titanium alloys
Backing gas and trailing shield required; keyhole technique used with orifice-to-work distance of 3/16 in.

Thickness, in.	Travel speed, in./min	Current (DCEN), A	Arc voltage, V	Gas	Gas flow, ft^3/h Orifice gas	Gas flow, ft^3/h Shielding gas	Joint type
0.125	20	185	21	Argon	8	60	Square butt
0.187	13	175	25	Argon	18	60	Square butt
0.390	10	225	38	75He-25Ar	32	60	Square butt
0.500	10	270	36	50He-50Ar	27	60	Square butt
0.600	7	250	39	50He-50Ar	30	60	V-groove(a)

(a) 30° included angle; 3/8 in. root facer

ameter. With longer arc length, there is danger of turbulence, which may draw air into the weld pool. In addition, increasing the arc length produces wider weld beads. When filler metal is used, the maximum arc length should be about one and a half times the electrode diameter, depending on the thickness of the base metal.

Welding conditions or schedules for GTAW depend on workpiece thickness (see Table 3). In welding titanium alloys, the best heat input to use is that just above the minimum required to produce the weld. If heat input is greater, the possibility that the weld will become contaminated, distorted, or embrittled increases.

Avoiding porosity in welds is an important consideration in welding titanium alloys. If the joint and filler wire are properly cleaned and the tooling does not chill the weld too rapidly, porosity can be reduced or eliminated by using a lower welding speed, which will retard weld solidification and allow entrained gases to escape.

Effect of Shielding Gases on Arc Characteristics. The type of shielding gas used affects the characteristics of the arc. At a given welding current, the arc voltage is much greater with helium than with the argon. Because the heat energy liberated in helium is about twice that in argon, higher welding speeds can be obtained, weld penetration is deeper, and thicker sections can be welded more rapidly using helium shielding. However, when using pure helium for welding, arc stability and weld-metal control are sacrificed.

Argon is used in the welding of thin and thick sections where the arc length can be altered without appreciably changing the heat input. Argon-helium mixtures are also employed; particularly 75% argon, which improves arc stability, and 25% helium, which increases penetration. The 75%Ar-25%He mixture is also frequently utilized as the shielding gas at the torch in automatic operations. Furthermore, helium is used in shielding for out-of-position welds.

The greater voltage sensitivity of helium has a decided advantage over argon where automatic voltage controls are used. The greater voltage sensitivity permits feeding a signal to a motor which automatically raises and lowers the electrode to match changes in arc length. Therefore, it is possible to maintain a constant voltage during the continuous movement of the torch over the weld seam.

Other Welding Methods

Gas metal arc welding (GMAW) is employed to join titanium and titanium alloys more than 3 mm (1.8 in.) thick. It is applied using pulsed current or the spray mode and is less costly than GTAW, especially when base-metal thickness is greater than 13 mm (½ in.).

Metal transfer through the arc in GMAW can lead to difficulty in meeting stringent aerospace quality requirements. For example, weld spatter is often associated with inferior weld quality, and arc instability, which can occur in GMAW, is a potential cause of weld contamination and defect formation. Some users of titanium alloys prefer GTAW over GMAW (even for joining thick plate), because with the gas tungsten arc process more uniform and predictable transverse shrinkage is obtained.

Electrode wires for GMAW are available in several grades of unalloyed titanium and in titanium alloys that match the composition of the base metal. Shielding for out-of-chamber welding is provided by inert gas being fed through the nozzle of the electrode holder, through the backing bar or plate, and as a trailing shield, much as in GTAW. The electrode holder is basically the same as for GMAW of steel. To avoid contamination and porosity in GMAW, a leading shield is necessary, as well as a trailing shield and a suitable baffle added on the leading edge of the electrode holder. A leading shield prevents oxidation of spatter before it is melted in the weld metal.

Plasma arc welding (PAW) is also applicable to the welding of titanium and titanium alloys. It is faster than GTAW and can be used on thicker sections, such as one-pass welding of titanium alloy plate up to 13 mm (½ in.) thick, using square-groove butt joints and the key-hole technique.

The joining of titanium alloys is one of the major applications of PAW. Because titanium has a lower density, keyhole welds can be made through thicker titanium square butt joints than for steel. As with GTAW, PAW requires backing gas and a trailing gas shield to prevent atmospheric contamination of the weld and adjacent base metal (see Table 4). For PAW, a filler metal may or may not be used for welding material less than 13 mm (½ in.) thick.

Spot and Seam Welding. Because of their relatively low thermal and electrical conductivities, weldable titanium alloys are considered to spot and seam weld more readily than aluminum and some carbon steels. Joint designs (see Fig. 5) are similar to those used in steel.

The same equipment used to spot and seam weld stainless steel can be adapted to welding titanium. A good rule to follow in arranging welding conditions is to start with the same set-up used for like thicknesses of stainless steel and adjust the current time or force as needed. Spot and seam welding does not require inert gas shielding because the pressed surfaces exclude the air, and because there is a very short duration of the weld cycle. Electrode materials generally conform to the RWMA Class 2 and 3 alloys.

Flash and pressure welding both involve the application of heat and pressure. The major difference is the degree of heating. Flash welding requires actual melting of the metal at the joint, while the metal during pressure welding is heated to a plastic state.

Flash welding of titanium is much the same as other metals. Current is passed through a clamped joint, and pressure applied at the interface squeezes out molten metal during heating. Except for large or complex shapes, shielding generally is not required because the molten metal is squeezed out during the process.

Pressure welding requires careful preparation so that the joining faces are in tight alignment. The joint is heated to a plastic state by flame or induction methods, and the pieces are welded by an external pressure. Shielding normally is not required, except to prevent contamination of the inside of hollow parts.

Electron beam welding (EBW) is used to weld any metal that can be arc welded; weld quality in most metals is equal to or superior to that produced by gas tungsten arc welding (GTAW). Because this type of welding is performed in a high vacuum, atmospheric contamination of the weld is prevented. All of the commercial alloys of titanium that can be joined by arc welding can also be joined by EBW. Cost studies show that direct labor costs for EBW of titanium sections more than 25 mm (1 in.) thick are less than for arc welding, provided a suitably large vacuum chamber is available. Filler metal is not ordinarily used, and the work is not preheated. Tack welding, contrary to experience, in GTAW, presents no difficulties in EBW. For optimum results, welding is done in a high vacuum, but medium-vacuum welding is satisfactory for many applications.

Laser Beam Welding. Although electron beam welding is used more frequently than laser beam welding (LBW), laser beams can be transmitted for appreciable distances through the atmosphere without serious attenuation or optical degradation. Thus, the laser offers an easily maneuvered, chemically clean, high-intensity, atmospheric welding process that produces deep-penetration welds (aspect ratio greater than 1 to 1) with a narrow HAZ and subsequent low distortion.

Postweld Evaluation

The low weld ductility of most alpha-beta alloys is caused by phase transformation in the weld zone or in the HAZ. Alpha-beta alloys can be welded autogenously or with various filler metals. It is common to weld some of the lower-alloyed materials with matching filler metals. Where strength is not critical and more toughness is required, unalloyed alpha titanium filler metals may be used. The use of filler metals that improve ductility may not prevent embrittlement of the HAZ in alpha-beta alloys that are rich in beta stabilizers. In addition, low-alloy welds can be embrittled by hydride precipitation. It should be noted, however, that with proper joint preparation, filler-metal storage, and shielding, hydride precipitation can be avoided. The service conditions must also be considered, particularly at high temperatures, as hydrogen can migrate from the base alloy to the lighter alloyed filler metal.

Sheet thicknesses 2.5 mm (0.100 in.) and thinner can be GTAW without filler metal additions by the fused-root welding technique. Filler metal may be added to repair unfused and sunken weld-metal area. The lack of a joint line on the root face of the weld indicates 100% penetration.

Bend tests allow the evaluation of weldment ductility and the performance of gas shielding. For some alloys, ductility can be improved by special filler metals and postweld annealing operations.

Visual examination is the only non-destructive means of evaluating shield performance in maintaining ductility. However, surface appearance (see Table 5) is only an approximate indicator and does not always offer a reliable means of judging weld contamination.

Stress Relief. Most titanium weldments are stress relieved after welding to

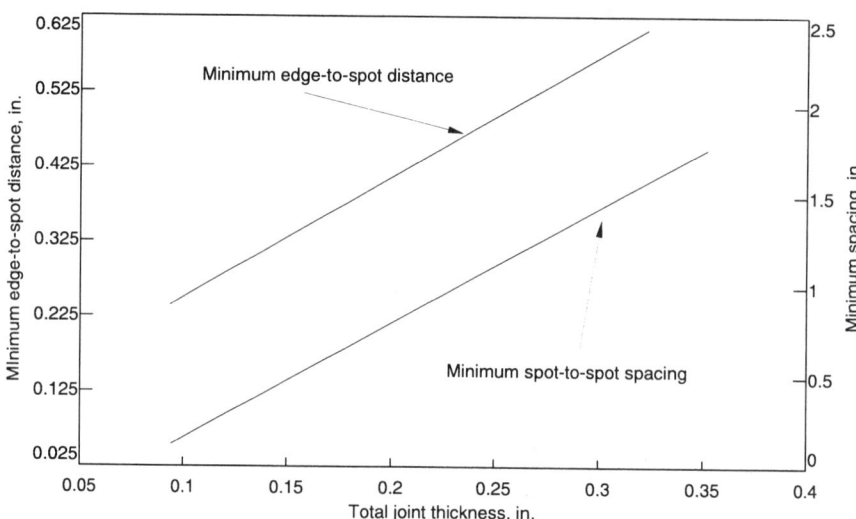

Fig. 5 Design recommendations for spot welds. If the spot welds are too close together, succeeding welds will dissolve the oxide films formed during previous welds and thus cause embrittlement.
Source: "Facts About Welding Titanium," RMI, Inc.

Table 5 Weldment color and conditions

Approximate condition color	
Silver	Bright like a new dime. Usually indicates correct shielding and sound, non-contaminated weld.
Straw	Pale yellow to golden. Acceptable welds of lessened ductility.
Blue/purple	Low ductility, unsuited to stressed application.
Gray/powdery	Not acceptable.

Table 6 Titanium braze alloys

Type	Form	Description
10Pd-Ag	Prealloyed Ring	Douglas developed—no brittle intermediate
95Ag-5Al	Powder, Wire, Foil	Considered ductile
9Pd-9Ga-Ag	Powder, Wire, Foil	Furnace braze at 1600 °F (870 °C) ductile
3003 Al	Foil	Considered ductile
15Cu-15Ni-Ti	Powder, Foil	Brittle when used only as a braze, useful diffusion bonding mechanism
20Cu-20Ni-Ti		
48Zr-48Ti-4Be	Powder, Ring	Brittle but strong
48Zr-48Ti-4Be with 2% Ni	Powder	Brittle but strong
76.5Al-16.5Cu-4.0Sn-3.0Si	Foil, Wire, Powder	Developed for bicycle & aerospace furnace braze at 1350 °F, ductile
66.5Al-2.5Cu-5Sn-3.5Si	Foil, Wire, Powder	Developed for bicycle & aerospace furnace braze at 1350 °F, ductile

Source: C.E. Forney, Jr., and J.H. Schemel, Ti-3Al-2.5V Seamless Tubins Engineering Guide, 2nd ed., Sandvik Special Metals Corporation, 1987, p 85

prevent weld cracking and susceptibility to stress-corrosion cracking in service. Stress relief also improves fatigue strength. An assembly subjected to a substantial amount of welding and severe fixturing restraint may require intermediate stress relieving of the partially welded structure, which should be done in an inert atmosphere; otherwise, the unwelded joints may have to be recleaned before being welded.

With unalloyed titanium and alpha titanium alloys, time and temperature should be controlled to prevent grain growth. If stress relief conditions are not known, tests should be conducted to ensure that stress relieving does not reduce fracture toughness, creep strength, or another property of importance. For example, stress relieving of Ti-13V-11Cr-3Al weldments causes aging and subsequent embrittlement of the weld and HAZ and, therefore, is not recommended.

Resolution heat treatment (re-annealing) may be used to relieve stress if the welded assembly is amenable to such treatment. All titanium surfaces should be free of dirt, fingerprints, grease, and residues before stress relieving. Contaminated surface metal must be removed from the entire weldment by machining or descaling and pickling to remove 0.025 and 0.05 mm (0.001 to 0.002 in.) per surface.

Repair welds should follow the established specification requirements for the original welds and be made prior to final heat treatment. Manual or automatic GTAW is generally used for repairing butt and fillet welds. Repairs can also employ a combination of welding processes such as GTAW and the initial welding process (GMAW, PAW, or hot wire welding).

Repair welds always must be carefully executed, and all traces of liquid-penetrant inspection material must be removed. Generally, inspection is performed on both faces of the repair weld and several inches beyond the repaired area.

Brazing

Because titanium can become embrittled by the interstitial absorption of hydrogen, nitrogen, and oxygen gases, brazing should be done in a vacuum at a pressure of 10^{-5} to 10^{-4} torr or in dry inert-gas atmosphere. Filler-metal selection in brazing titanium alloys is also critical, because they react with many of the constituents of brazing filler metals to form undesirable intermetallic compounds. Silver-base brazing alloys (see Table 6) are most often used, although some U.S. Air Force and NAVAIR specifications restrict their use.

Effects of Brazing on Properties. Pure titanium and alpha alloys are not heat treatable, and their material properties are not affected by brazing. In contrast, selection of filler metals and brazing cycles that are compatible with the heat treatment required for alpha-beta and beta-titanium base metals may present some difficulty. Ideally, brazing should be conducted 50 to 80 °C (100 to 150 °F) below the beta transus; the ductility of alpha-beta base metals may be impaired if this temperature is exceeded. The beta transus can be exceeded when beta-titanium metals are brazed; however, if the brazing temperature is too high, base-metal ductility after heat treatment may be impaired and considerable interaction between the filler metal and the base metal may occur. The tensile properties of heat treatable titanium alloys also may be adversely affected by brazing, unless the assembly can be heat treated after brazing. For example, alpha-beta titanium alloys must be solution treated, quenched, and aged to develop maximum strength. It is difficult to select a filler metal that is suitable for brazing and solution treating in a single operation. Similarly, it is not always possible to quench a brazed assembly at the desired cooling rate, and certain configurations, such as honeycomb sandwich structures, cannot be quenched rapidly without distortion. Brazing at the aging temperature is impractical, because few filler metals melt and flow at these temperatures.

The possibility of galvanic corrosion must be considered when filler metals are selected for brazing titanium-base metals. While titanium is an active metal, its activity tends to decrease in an oxidizing environment, because its surface undergoes anodic polarization similar to that of aluminum. Thus, in many environments, titanium becomes more chemically inactive than most structural alloys. The corrosion resistance of titanium is generally not affected by contact with structural steels, but other metals, such as copper, corrode rapidly in contact with titanium under oxidizing conditions. Thus, filler metals must be chosen carefully to avoid preferential corrosion of the brazed joint.

Equipment Precautions. When titanium is brazed, precautions must be taken to ensure that the brazing retort or chamber is free of contaminants from previous brazing operations. Mechanical properties of titanium may deteriorate because of gaseous contamination from the brazing furnace. Also, the choice of materials to be used in fixtures must be carefully considered. Nickel or materials containing high amounts of nickel generally should be avoided; nickel and titanium form a low-melting eutectic at about 942 °C, or 1728 °F (28.4 wt% Ni). If titanium workpieces come in contact with fixtures or a retort made of a nickel-base alloy, the parts may fuse together if the brazing temperature is in excess of 942 °C (1728 °F). If a fixture material, such as stainless steel, which may contain a high nickel content is used, it should be oxide coated. In most applications, coated graphite or carbon steel fixture materials are used.

Filler Metals. Braze filler metals initially used for brazing titanium and its alloys were silver with additions of lithium, copper, aluminum, or tin. Most of these brazed filler metals were used in low-temperature applications (540 to 600 °C, or 1000 to 1100 °F). Commercial braze filler metals (see Table 6), including silver-palladium, titanium-nickel, titanium-nickel-copper, and titanium-zirconium-beryllium, are now available that can be used in the 870 to 925 °C (1600 and 1700 °F) range. Higher strengths and improved resistance to crevice-type corrosion are desirable characteristics that current braze filler metals enjoy. For joining applications requiring a high degree of corrosion resistance, the 48Ti-48Zr-4Be and 43Ti-43Zr-12Ni-2Be braze filler metals were developed. A silver-palladium-gallium braze filler metal (Ag-9Pd-9Ga), which flows at 900 to 910 °C (1650 to 1675 °F), is another excellent filler metal with which to fill large gaps.

Methods to braze titanium honeycomb sandwich assemblies with aluminum braze filler metal have been developed. Aircraft structures up to 7 m (23 ft) in length are brazed successfully using 3003 brazing foils. Use of aluminum brazing filler metal (3003) provides satisfactory strength up to about 315 °C (600 °F). If temperatures of 540 to 600 °C (1000 to 1100 °F) are required, high-strength, corrosion-resistant titanium-zirconium-beryllium or titanium-zirconium-nickel-beryllium braze filler metals should be used (Hurwitz, D., Manufacturing Methods for Brazed Titanium Hybrid Structures, AFML-TR-76-119, Contr. F33615-74-C-5047, Final Report, June 1976).

Rolling

Alpha and Alpha-Beta Alloys

Rolling Temperature. The majority of specifications for near-α and α + β titanium alloys mandate final hot working in the α + β region. This means that the soaking temperature prior to rolling must be held below the β transus temperature for the alloy being rolled. Typical rolling temperatures for several common alloys are given in Table 1.

Rolling in the α + β region produces microstructures with ductilities and low-cycle fatigue properties substantially higher than microstructures produced by β rolling (rolling from temperatures above the β transus). For selected applications, β rolling may be used to enhance creep or fracture toughness properties. However, these properties are best optimized by rolling in the α + β range followed by heat treatments slightly above or below the β transus temperature.

For α + β processing, rolling temperatures (metal temperature at the start of the rolling operation) are typically selected at 30 to 55 °C (50 to 100 °F) below the β transus. This allows for rolling furnace temperature variations, usually ±15 °C (±25 °F), and inaccuracies in the β transus determination.

For β rolling, the temperature is generally 30 to 55 °C (50 to 100 °F) above the β transus. When β rolling, it is important that bars finish below the β transus temperature. Alpha-beta alloys should never experience slow cooling from a temperature above the β transus. This results in the formation of α phase at the prior β grain boundaries, which degrades both strength and ductility.

Commercially pure (CP) grades of titanium can be strengthened by warm working and are often rolled at 110 to 165 °C (200 to 300 °F) below the β transus.

Furnace Control. Gas-fired (controlled to maintain an oxidizing atmosphere), electric, or induction heating is recommended for processing titanium. Titanium and its alloys will rapidly absorb hydrogen when exposed to reducing atmospheres. There is no tolerance for even slight, short-duration overheating (heating above the β transus). Therefore, furnace surveys, control, and maintenance are essential.

Table 1 Typical rolling temperatures for titanium and titanium alloys

Alloy	Bar °C	Bar °F	Plate °C	Plate °F	Sheet °C	Sheet °F
Commercially pure titanium						
Grades 1-4	760-815	1400-1500	760-790	1400-1450	705-760	1300-1400
α and near-α alloys						
Ti-5Al-2.5Sn	1010-1065	1850-1950	980-1040	1800-1900	980-1010	1800-1850
Ti-6Al-2Sn-4Zr-2Mo	925-970	1700-1775	955-980	1750-1800	925-980	1700-1800
Ti-8Al-1Mo-1V	1010-1040	1850-1900	980-1040	1800-1900	980-1040	1800-1900
α-β alloys						
Ti-8Mn	...	...	705-760	1300-1400	705-760	1300-1400
Ti-4Al-3Mo-1V	925-955	1700-1750	900-925	1650-1700	900-925	1650-1700
Ti-6Al-4V	925-970	1700-1775	925-980	1700-1800	900-925	1650-1700
Ti-6Al-6V-2Sn	900-955	1650-1750	870-925	1600-1700	870-900	1600-1650
Ti-7Al-4Mo	955-1010	1750-1850	925-955	1700-1750	925-955	1700-1750
β alloy						

Soaking Time. When titanium alloys are heated for rolling, the time at temperature should be sufficient to heat the entire cross section. However, any additional soaking time should be minimized to prevent excessive oxidation (buildup of α case), hydrogen pick-up, and growth of primary α in the microstructure.

Starting Stock Requirements. It is important that the starting stock for rolling have a reasonably uniform α + β structure. Prior reduction of at least 75% in the α + β range is recommended. In particular, the structure should be as free as possible of α stringers. If the material has been slow cooled from above the β transus during the prior processing, α phase will have precipitated at the prior β grain boundaries. This α phase will persist as α stringers, which can be broken up only by substantial reduction in the α + β range. Reroll stock structure is usually enhanced by recrystallization from β annealing or other proprietary processes prior to commencing the α + β reduction. Intermediate recrystallization anneals are helpful in reducing the primary α particle size and breaking up elongated α particles.

Reduction Sequences. The major goal in α + β rolling of titanium alloys is to refine the microstructure by producing fine equiaxed primary α particles in a transformed β matrix. The total reduction required is directly related to the degree of structure refinement desired. During rolling, the primary α particles will elongate in the direction of rolling. These elongated particles will break up to some extent by dynamic recrystallization. Further refinement can be accomplished by intermediate recrystallization anneals or reheats (heating approximately 55 °C, 100 °F below the β transus temperature).

Reduction sequences and finishing temperatures are not as critical for titanium alloys as for austenitic-type alloys. There is no concern about phase precipitation, grain growth, or unrecrystallized structures, which are common problems when rolling austenitic-type alloys. Reduction between passes and reduction from the last reheat are not significant variables with regard to metallurgical properties. Edge cracking is one of the primary factors limiting the amount of reduction. Some aerospace specifications impose restrictions on the amount of banding (metal flow pattern) in the microstructure. This feature can be minimized by judicious use of reheats and minimized reduction after the last reheat.

Surface Tearing. Titanium alloys have a tendency to surface tear in areas where there is localized cooling and/or high strain rates, particularly along the corners of squares and rectangles. This problem is more pronounced in the near-α alloys such as Ti-8Al-1Mo-1V than in α-β alloys such as Ti-6Al-4V. Coatings can be beneficial; however, proper pass design and sequence are the best corrective actions. Contact with water during rolling should be eliminated or minimized. Maintaining rolling temperature as high as possible always improves workability and minimizes surface tearing.

Strain Rate. High strain rate hot working, such as hammer forging, can induce adiabatic shear bands and overheating in titanium alloys. The strain rates of most rolling operations are low enough to avoid this problem. However, precautions should be taken (control of the rolling speed) when rolling titanium on a high-speed, continuous mill. Excessive reduction rates can also suppress dynamic recrystallization in a structure of very fine, highly elongated primary α.

Ti-6Al-4V Rolling

Various sizes of sheet and plate of Ti-6Al-4V are made by rolling. Like Ti alloys, rolling temperatures are based on the beta transus of the material. The 75% α + β work previously stated is typical for starting stock and generally a minimum of 75% is given for final rolling. The temperatures for rolling are generally about 75 to 100 °F below the transus.

Plate Rolling. It is not easy to characterize Ti-6Al-4V rolling with a simple flow diagram. However, a basic flow for Ti-6Al-4V plate processing is:

- Ti-6Al-4V ingot (VAR)
- Roll at 1150 °C (2100 °F) to intermediate slab 200 to 300 mm (8 to 12 in.)
- Alpha + beta work, 925 - 950 °C (1700-1750 °F), to final slab 125 to 200 mm (5 to 8 in.)
- Roll at 925-950 °C (1700-1750 °F) to final plate size (with cross roll to balance properties)

Sheet processing is more complex, since both sheet bar and insert stock for the packs require intermediate rolling, conditioning and possibly vacuum degas. The insert stock is then packed for final rolling to sheet. Cross rolling is incorporated to balance longitudinal and transverse properties.

Beta Alloys

Beta III. The good hot and cold workability of Ti-11.5Mo-6Zr-4.5Sn makes it suitable for continuous strip processing. At 925 °C (1700 °F), the alloy has flow stresses equivalent to Ti-6Al-4V and about 60% that for Ti-13V-11Cr-3Al (see Fig. 1). It can be processed at reasonable roll loads at temperatures as low as 760 °C (1400 °F). The roll energy requirement for this alloy is comparable to type 304 stainless steel (see Fig. 2). Hot rolling of plate at 815 °C (1500 °F) appears to produce the best combination of strength and ductility for solution treated or solution treated and aged plate.

Cold rolling of this alloy is somewhat more difficult than commercially pure titanium, requiring about 10 to 20% additional rolling passes at normal roll loads to achieve gage. In terms of cold rollability without edge cracking, however, this alloy surpasses commercially pure titanium and many commercial alloys. Cold reductions exceeding 90% without edge cracking are possible.

Ti-15V-3Al-3Cr-3Sn can be extensively cold rolled as strip. From ingot form, the following principal operations are followed to generate strip product:

- Forge to nominally 10.2 to 15.2 cm (4 to 6 in.) thick slab and condition as required.
- Hot roll slab to nominally 3.0 to 4.6 mm (0.12 to 0.18 in.) thick hot band.
- Cold roll flatten, air anneal and blast, pickle, grind, and trim as required.
- Cold roll to desired gage. This may incorporate an intermediate continuous vacuum anneal, depending on final gage. Cold rolling reductions in the 60 to 70% range are possible.
- Continuous vacuum anneal for gages less than 1.8 mm (0.070 in.). For heavier gages, a batch vacuum anneal to reduce hydrogen followed by air solution annealing, grinding, and pickling is required.
- Cut to size, inspect, and ship.

The extensive use of cold rolling results in a closer gage tolerance product and has significant cost advantages, particularly at lighter gauges (similar processing and advantages would apply to Beta 21S).

Ti-13V-11Cr-3Al normally is fabricated to flat rolled products in the β-phase temperature field. However, the final fabrication of sheet by rolling to finish gages is often done cold to obtain improved flatness and gage uniformity. Similarly, in the production of rod and wire products aimed at spring manufacture, initial fabrication at elevated temperatures may be followed by cold working to improve the finished surface and the mechanical properties of the final product.

Beta C. Hot working operations such as forging or rolling are readily accomplished from furnace temperatures of 815 to 980 °C (1500 to 1800 °F). Finishing temperatures are often as low as 705 to 760 °C (1300 to 1400 °F). Beta C has very good hot workability and is highly resistant to cracking.

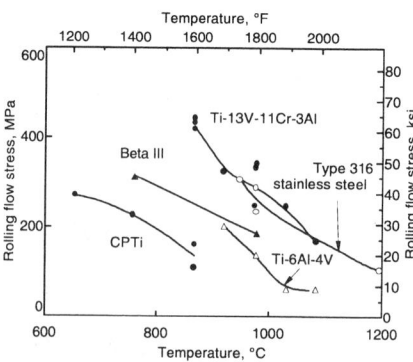

Fig. 1 Hot rollability of Ti-11.5Mo-6Zr-4.5Sn and several other alloys

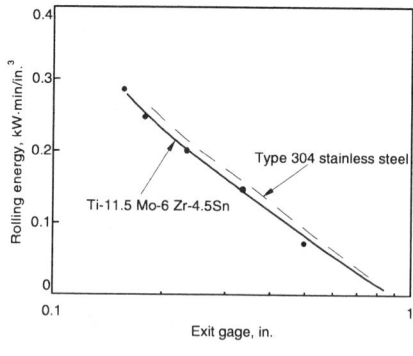

Fig. 2 Energy for hot strip rolling

Friction and Wear of Titanium Alloys*

F.M. Kustas and M.S. Misra, Martin Marietta Astronautics Group

TITANIUM ALLOYS offer an attractive combination of high specific mechanical properties, such as modulus/density and strength/density, because of their relatively low density. Titanium alloys also have toughness and corrosion resistance, making them useful materials for precision mechanism gears, turbine engine components, and biomedical prosthesis devices. However, these alloys have poor fretting fatigue resistance and poor tribological properties. From theoretical calculations, metals with low theoretical tensile and shear strengths exhibit higher coefficients of friction (μ) than higher-strength materials (Ref 1). Within the class of hexagonal close-packed (hcp) structures, titanium has relatively low values for these properties. Consequently, it is expected that titanium would exhibit high frictional values, which has been demonstrated for titanium sliding against itself in air (μ = 60, Ref 1) and vacuum (Fig. 1) (Ref 2).

Lower-tensile-strength materials, including titanium, also exhibit greater material transfer to nonmetallic counterfaces than higher-strength metals (Ref 1). The great affinity of titanium for oxygen results in the formation of an oxide surface layer, which is transferred to and adheres to nonmetallic materials, such as polymers, resulting in severe adhesive wear (Ref 2). Addition of a lubricant, such as polyperfluoroalkylether (PFPE) (Ref 3) or kerosene oil (Ref 4), reduces the coefficient of friction and wear damage somewhat, compared with unlubricated conditions (Fig. 2) (Ref 3-5), although reaction of titanium surfaces with these lubricants (generally under high-temperature conditions) can reduce lubricant performance (Ref 6).

Although titanium alloys offer attractive mechanical and physical properties, their surface properties are deficient, thus restricting the use of uncoated titanium alloys to nontribological applications. To realize the full benefit of titanium alloys in friction and wear applications, surface modification treatments are required to effectively increase near-surface strength, thereby reducing the coefficient of friction and lowering the tendency for material transfer and adhesive wear.

Surface Modification Treatments

Many surface treatments have been used to modify the tribological properties of titanium alloys. These treatments can be classified as physical vapor deposition (PVD), including ion implantation (Ref 5-13), plasma spray (Ref 15, 16), and evaporation (Ref 4); thermomechanical conversion treatments, including plasma nitriding (Ref 17, 18), gaseous nitriding (Ref 19), liquid nitriding (Ref 20), ionic nitriding (Ref 21), laser nitriding (Ref 22), ionic carburizing (Ref 21), and laser boriding (Ref 22); plating (Ref 17, 20, 23); and application of solid lubricants by resin bonding/burnishing (Ref 20, 23), direct ion beam deposition to produce diamond-

*Reprinted from *ASM Handbook*, Vol 18, 1992

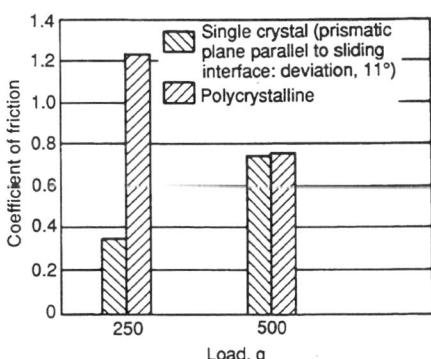

Fig. 1 Coefficient of friction for single-crystal and polycrystalline titanium sliding on polycrystalline titanium in vacuum. Pressure, 1.33 × 10⁻⁷ Pa (10⁻⁹ torr); sliding speed, 2.28 cm/s (0.90 in./s). Source: Ref 2

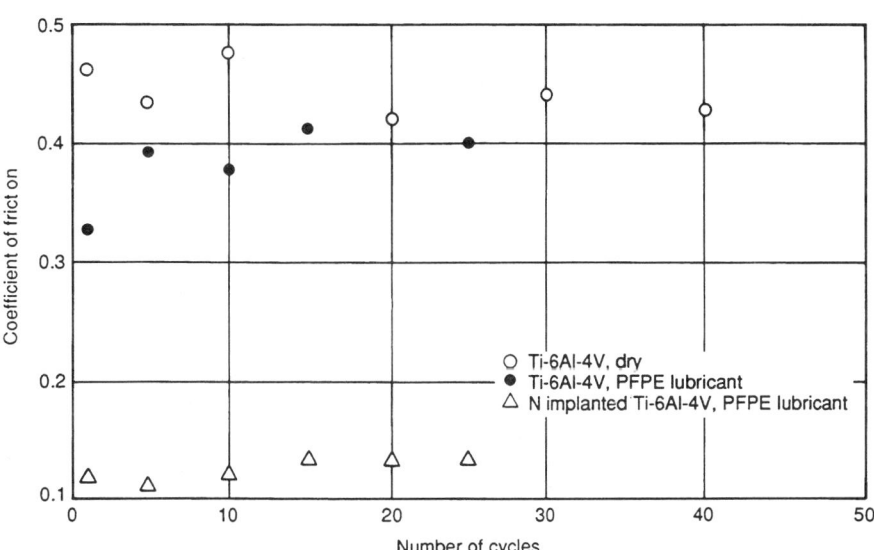

Fig. 2 Reduced coefficients of friction for PFPE-lubricated Ti-6Al-4V compared with unlubricated Ti-6Al-4V in air environment. Sliding material, WC-Co; sliding speed, 1 mm/s (0.04 in./s); load, 0.5 kgf. Source: Ref 3, 5

Table 1 Advantages and limitations of surface modification treatments for titanium alloys

Surface modification technique	Advantages	Limitations
Physical vapor deposition		
Ion implantation	Low-temperature process; effectively net shape; creation of unique near-surface alloys; gradual interface	Line of sight; shallow surface modification zone; high cost
Plasma spray	Rapid deposition rates; thick coatings; inherent porosity can trap lubricant; low cost	Coarse surface finish; need for bond layer, abrupt interfaces; hard-to-control coating composition; line of sight
Evaporation	Large range of available materials; tailorable compositions; high-purity coatings; low cost	Thin coatings; line of sight; abrupt interface (unless ion-beam assisted)
Sputtering	Low-temperature process; potential for layered coatings; tailored coating composition	Thin coatings; line of sight; abrupt interface (unless ion-beam assisted)
Thermochemical conversion		
Nitriding, carburizing, boriding	Gradual interface, conversion of substrate; thick surface modification zones	Potential for hydrogen embrittlement restricts selection of titanium alloys; high process temperatures can distort components and reduce properties; coarse surface finish
Plating	Not line of sight; low cost	Potential for hydrogen embrittlement restricts selection of titanium alloys; abrupt interface

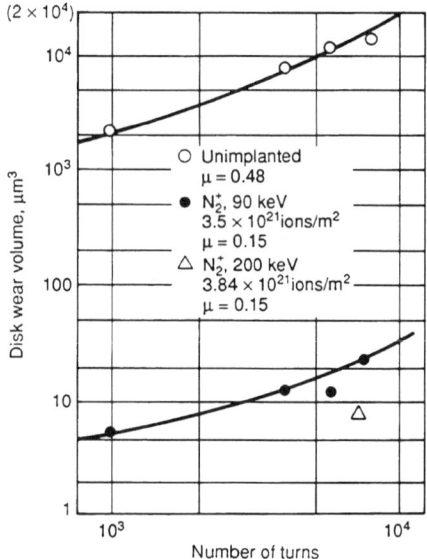

Fig. 3 Wear volume loss as a function of cycles for untreated and nitrogen-implanted Ti-6Al-4V, showing greater than two orders of magnitude reduction in volume loss for nitrogen-implanted material. Pin, 5 mm (0.2 in.) ruby ball; load, 2.61 N (0.266 kgf); velocity, 5.65 mm/s (2.2 in./s). Source: Ref 7

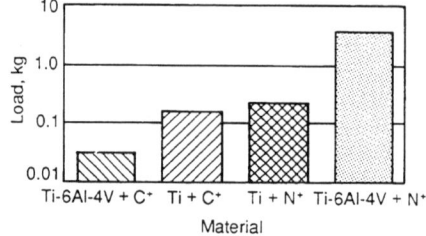

Fig. 4 Breakthrough load for four titanium substrate-ion combinations. Pin, 5 mm (0.2 in.) ruby ball; velocity, 56.6 mm/s (2.23 in./s). Source: Ref 7

like carbon (DLC) (Ref 24), and sputtering (of molybdenum disulfide and titanium nitride, for example) (Ref 18, 25, 26). These treatments differ dramatically with respect to induced surface temperatures, with nearly ambient conditions for low-dose-rate ion implantation and ion-beam deposition, while temperatures up to 900 °C (1650 °F) are common for plasma nitriding and carburizing treatments. Table 1 summarizes the major advantages and limitations of the various surface modification treatments for titanium alloys.

Physical Vapor Deposition

Physical vapor deposition includes ion implantation, sputtering, evaporation, and ion plating surface modification treatments.

Ion implantation is a low-temperature technique for modifying the near-surface region of a material. The modified zone is incorporated within the substrate, leaving no defined interface. The process is a line-of-sight technique and generally offers only shallow penetration depths.

Studies of ion-implanted titanium alloys have concentrated on carbon, nitrogen, and nitrogen-oxygen implantations to produce surfaces enriched in TiC, TiN, and Ti$_2$N, and Ti-O-N, respectively. From pin-on-disk friction wear tests, much greater (that is, two orders of magnitude) wear resistance has been shown (Fig. 3) for nitrogen-implanted Ti-6Al-4V compared with untreated material (Ref 7). Nitrogen or carbon implanted into pure titanium produces nearly equivalent results for breakthrough load to failure of the ion-implanted zone (defined as applied load for failure of the surface after fewer than 50 revolutions of the implanted disk (Fig. 4 (Ref 7). In contrast, nitrogen implantation of higher hardness and higher-strength Ti-6Al-4V is more effective than carbon implantation in increasing breakthrough load (Fig. 4).

Additional studies (Ref 8, 9) have shown that (1) an optimum heat treatment of about 400 to 470 °C (750 to 880 °F) after carbon implantation results in greater cycles to breakthrough (an increase by a factor of 30 compared with room-temperature implantation) because of formation of 60 nm diameter TiC particles and (2) a dual-energy implant (to produce deeper nitrogen penetration) will significantly increase the wear life (by at least one order of magnitude) (Ref 8) of the implanted zone. Carbon implantation increases fretting fatigue life (Fig. 5) by slowing the formation of debris that causes surface damage and subsequent crack initiation (Ref 10).

Modification of surface composition enhances the tribological properties of deficient materials. Quantitative analyses of worn surfaces on nitrogen-implanted Ti-6Al-4V have shown that oxygen and carbon are incorporated into the surface layer during either implantation or the wear event. These complex-composition surface layers help produce lower coefficients of friction (Ref 11, 12).

Implantation of Ti-6Al-4V has been performed to improve its corrosive wear resistance when in contact with ultra-high-molecular-weight polyethylene (UHMWPE) (Ref 13). This combination of materials is used for biomedical prostheses, such as artificial hips (Ref 28), where an integral stem and ball of Ti-6Al-4V is in contact with a UHMWPE acetabular cup. Flakes of TiO$_2$ formed during wear of titanium in contact with UHMWPE may result in adverse tissue response, including inflammation, infection, loosening of the prosthetic component, and possible carcinogenic reaction. Implantation of Ti-6Al-4V with nitrogen to produce a 20% surface concentration reduced corrosion current in an oxygen-saturated sodium chloride isotonic salt solution by 100 times compared with unimplanted Ti-6Al-4V. The wear surface of the nitrogen-implanted Ti-6Al-4V, in contact with UHMWPE pads, exhibited only a few black stripes after the test—in contrast to the wear surface of unimplanted Ti-6Al-4V, which was completely blackened early in the testing. The authors concluded that the black debris was TiO_2 embedded in the UHMWPE that was transferred to the Ti-6Al-4V surface (Ref 13). In a related study, nitrogen or carbon implantation of Ti-6Al-4V significantly reduced the wear volumes (by five times) of both the pin (Ti-6Al-4V) and the rotating disk (UHMWPE) under a screening environment of Ringer's solution (a lactated 0.9% NaCl solution with the same concentration of Cl^- ions as in body fluids) (Ref 14).

Recently, high-temperature nitrogen implantation of Ti-6Al-4V has been evaluated as a means of increasing the penetration depth of nitrogen into a titanium alloy substrate. It has been demonstrated that the most beneficial tribological properties occur for a 1000 °C (1830 °F) implantation of 10^{18} N_2^+-N^+/cm^2 at 60 keV. This treatment produced the deepest penetration of nitrogen (>750 nm) along with the formation of TiN and Ti_2N (Ref 29). Lowest unlubricated coefficient of friction values (down to about 0.12) were observed for this specimen, with μ decreasing with increasing applied load until a critical load of 98 N (22 lbf), at which implanted layer failure occurred (Ref 5). Under PFPE lubricated conditions, a threefold reduction in coefficient of friction was observed compared with nonimplanted Ti-6Al-4V for a 600 °C (1110 °F) nitrogen implantation treatment of 10^{17} N_2^+-N^+/cm^2 at 75 keV (Fig. 2) (Ref 5). The high ion-beam current densities employed during the processing (up to 500 μA/cm^2 versus <50 μA/cm^2 for conventional ion implantation) results in short-time beam exposures of only a few minutes at the elevated temperatures (Ref 29), which would not degrade bulk material properties.

Plasma Spray Coatings. Evaluation of the wear and fatigue performance of spherical bearing components with candidate coatings on Ti-10V-2Fe-3Al showed that plasma spray coatings of Cr_2O_3, Al_2O_3, and WC were the most wear resistant and exhibited the lowest torques (Table 2) (Ref 16). Test results using lubricants contaminated with road dust illustrated the superiority of plasma spray Cr_2O_3 on Ti-10V-2Fe-3Al compared to chromium-plated 15-5 PH steel where the chromium plating was completely removed during testing. Finally, grease-starved (after initial lubrication) tests showed superior performance for plasma-sprayed Al_2O_3 and WC coatings, which remained intact after 10,000 cycles without addition of grease. The porosity of these coatings served as lubricant reservoirs, providing grease replenishment during the test. As a result, Cr_2O_3-coated titanium components have been incorporated into forward trunnion spherical bearings for aircraft landing gear.

Evaporation. Hard coatings such as TiC, TiN, and Ti_2N deposited on titanium disks by activated reactive evaporation (ARE) have been evaluated against different mating rider materials such as 440C steel, TiC/440C steel, and TiN/440C steel under dry and lubricated (for example, kerosene oil) conditions (Ref 4). In general, the following conclusions can be drawn from the tribological data (Table 3):

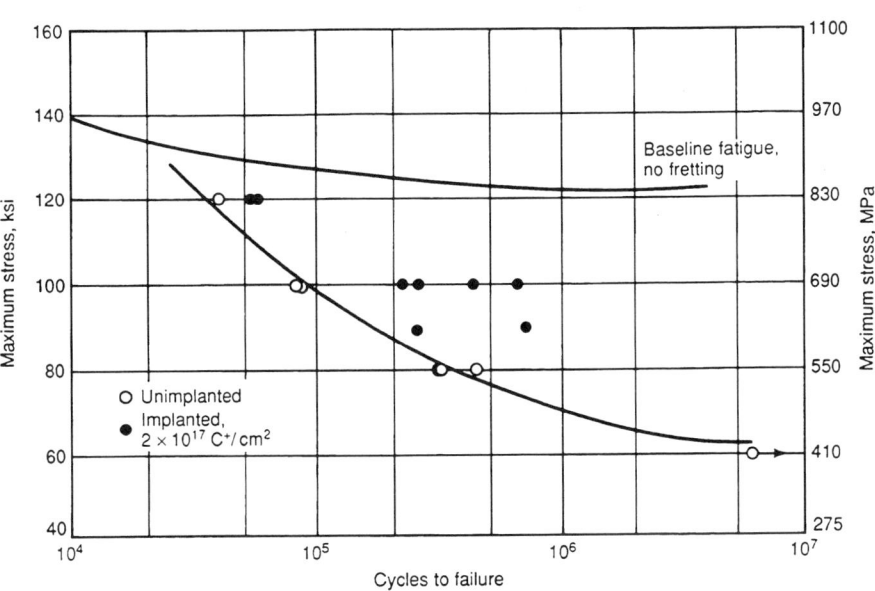

Fig. 5 Cycles to failure versus maximum fatigue stress for carbon-implanted and unimplanted Ti-6Al-4V. Normal stress, 690 MPa (100 ksi); baseline fatigue (no fretting) data, 20.7 MPa (3 ksi). Source: Ref 10, 27

Table 2 Normal lubrication bearing test results for plasma-spray-coated Ti-10V-2Fe-3Al
Bearings were lubricated every 900 cycles with clean grease per MIL-G-23827.

Bearing substrate coating	Bearing stress		Torque				Cycles	Comments
			At 100 cycles		At end of test			
	MPa	ksi	N·m	lbf·in.	N·m	lbf·in.		
Ti/Ni-B	41	6	220	1950	220	1950	103	Ball locked up in race; coating flaked off
	41	6	...	...	436	3855	71	Ball locked up in race; coating flaked off
Ti/ion nitride	41	6	220	1950	220	1950	4473	Coating worn off; ball diameter 0.3 mm (0.012 in.) smaller
Ti/Cr_2O_3(a)	41	6	259	2295	116.1	1027	10^5	No wear on ball; minimal wear on outer race
	62	9	330	2925	116.1	1027	10^5	No wear on ball; minimal wear on outer race
	90	13	275.5	2438	143.3	1268	10^5	No wear on ball; minimal wear on outer race; outer race shows evidence of heat discoloration
Ti/Cr_2O_3(b)	41	6	72.0	637	47.0	416	10^5	No wear on ball; minimal wear on outer race
Ti/Cr	41	6	165.3	1463	84.2	745	10^5	Copper pickup on ball
	62	9	248.3	2197	138.1	1222	10^5	Some surface scoring of chromium plate and copper pickup on ball
Ti/Al_2O_3	41	6	102	903	52.5	465	10^5	Copper pickup on ball
Ti/WC	41	6	144	1270	88.4	782	10^5	Copper pickup on ball
15-5 PH/Cr	62	9	225	2000	110	975	10^5	Copper pickup on ball

(a) Chromium oxide from Source 1. (b) Chromium oxide from Source 2. Source: Ref 16

Table 3 Coefficients of friction and rider and disk wear for various wear couple combinations for uncoated and hard-ceramic-coated titanium

All tests run for 500 m (1640 ft) under a load of 0.4 kg (0.9 lb), unless otherwise noted

Rider	Disk	Coefficients of friction(a)			Rider wear(b)		Disk wear(c)	
		μ_S	μ_D	μ_L	Dry	Lubricated(d)	Dry	Lubricated(d)
440C steel, uncoated	Ti, uncoated	0.765	0.45	0.425	High	High	High	High
	TiC/Ti	0.175	0.275	0.275	Medium	High	Low	Low
	TiN + Ti$_2$N/Ti	0.2	0.35	0.275	High	High	Low	Low
TiC/440C steel	Ti, uncoated	0.65	0.8	0.5	High	...	High	...
	TiC/Ti	0.275	0.175	0.175	Low	Low	Low	Low
	TiN + Ti$_2$N/Ti	0.275	0.2	0.1	Low	Low	Low	Low
TiN/440C steel	Ti, uncoated	0.5	0.765	0.4	High	Medium	High	High
	TiC/TiN	0.05	0.025	0.15	Low(e)	Very low	Low	Low
	TiN + Ti$_2$N/Ti	0.1	0.05	0.125	Low	Very low	Low	Low

(a) μ_S, static coefficient of friction measured when disk began to slide (velocity: 1.2-1.5 cm/s, or 0.5-0.6 in./s); μ_D, dynamic coefficient of friction measured after about 10 disk revolutions (velocity: 10-15 cm/s, or 4-6 in./s); μ_L, lubricated coefficient of friction measured after about 10 disk revolutions (velocity: 10-15 cm/s, or 4-6 in./s). (b) Wear test speed was 22-25 cm/s (8.5-10 in./s). Wear volumes are defined as: high >5 × 10^{-3} mm^3 (3 × 10^{-3} in.^{3}); medium, 10^{-4} to 5 × 10^{-3} mm^3 (0.6 × 10^{-4} to 3 × 10^{-3} in.^{3}); low, 10^{-5} to 10^{-4} mm^3 (0.6 × 10^{-5} to 0.6 × 10^{-4} in.^{3}); very low, <10^{-5} mm^3 (0.6 × 10^{-5} in.^{3}). (c) Wear test speed was 22-25 cm/s (8.5-10 in./s). Wear volumes are defined as high, >2.54 × 10^{-3} mm^3 (1.55 × 10^{-3} in.^{3}); medium, 2.54 × 10^{-4} to 2.54 × 10^{-3} mm^3 (1.55 × 10^{-4} to 1.55 × 10^{-3} in.^{3}); low, <2.54 × 10^{-4} mm^3 (1.55 × 10^{-4} in.^{3}). (d) Lubricated with kerosene oil. (e) Test run for 100 m (328 ft)

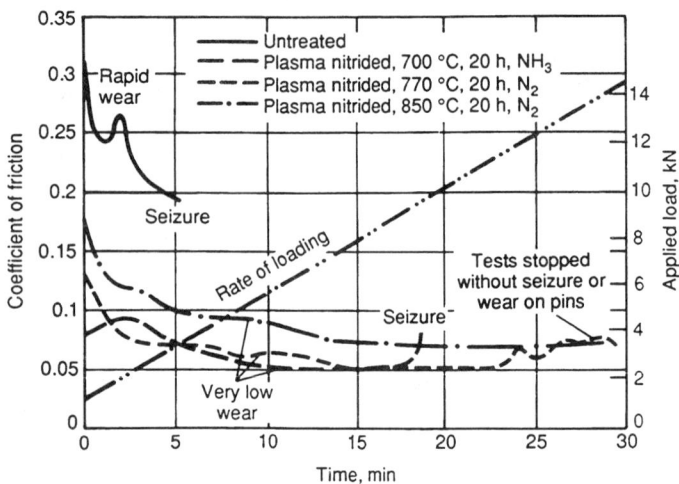

Fig. 6 Wear test results for untreated and plasma-nitrided Ti-6Al-4V. Source: Ref 18

- Coefficients of friction and wear volumes (for both rider and disk) are high for both bare and coated 440C steel versus bare titanium
- Coefficients of friction and titanium disk wear volumes are much lower for bare 440C steel versus coated titanium
- Coefficients of friction and wear volumes are lowest for hard coatings on both 440C steel riders and titanium disks

A material combination of TiN/440C rider versus TiC/Ti disk gave the best performance, although the reverse coating combination was not as effective (Table 3).

Thermochemical Conversion Surface Treatments

Nitriding, Carburizing, and Boriding. Ionic nitriding and carburizing offer the advantage of producing deeper nitride case depths (up to about 6 µm thick) due to thermal diffusion at elevated temperatures (750 to 900 °C, or 1380 to 1650 °F) (Ref 21). These treatments generally are applied to annealed Ti-6Al-4V and pure titanium. An adverse effect of irreversible grain growth was observed to be the most prevalent for a 900 °C (1650 °F) treatment (Ref 21). Large-grain microstructures generally have lower mechanical properties than fine-grain material.

Hydrogen-containing compounds, often used in the nitriding process, provide a source of hydrogen that can diffuse into the titanium alloy substrate during the high-temperature nitriding operation. For titanium alloys, the effects of hydrogen contamination on mechanical properties are highly dependent on the specific phase of the titanium alloy microstructure. For example, α-phase titanium alloys are prone to titanium hydride formation (Ref 21), which can embrittle the matrix and thereby reduce mechanical properties. In contrast, titanium alloys with a β-phase microstructure maintain hydrogen in a solid solution, which minimizes embrittlement but results in a lower hardening tendency (Ref 21).

Tribological properties of plasma-nitrided titanium alloys are very impressive. Low coefficients of friction (0.05 to 0.08) and low wear rates have been demonstrated for plasma-nitrided Ti-6Al-4V compared with untreated material (Fig. 6) (Ref 18). Fatigue performance of the nitrided material is highly dependent on prior substrate condition. Annealed Ti-6Al-4V that was subsequently nitrided showed a marked reduction (up to 21%) in endurance life compared with untreated Ti-6Al-4V; this can be attributed to coarsening of the α grains and production of a continuous α matrix (Ref 18). Use of a solution-treated and aged material, however, reduced the fatigue degradation to only 4% for the nitrided material (Ref 18).

Table 4 Wear tests using Cr-Mo-coated Ti-6Al-4V pins and uncoated steel V-blocks

Solid lubricant	Average time to failure at specific load, min				Standard deviation
	1330 N (300 lbf)	2220 N (500 lbf)	3330 N (750 lbf)	4450 N (1000 lbf)	
PWA 474	...	...	...	54	42
PWA 550	...	...	...	63	12
PWA 585	...	...	...	9	0.4
PWA 586	...	...	...	3	(a)
PWA 36035	...	...	...	227	29
PWA 36070	1	...	...	...	(a)
Lubeco K-350	...	...	...	48	25
Lubeco M-390	...	0.2	...	...	(a)
Esnalube 382	...	...	...	6	0.9
Ecoalube 642	...	...	...	74	76

(a) Standard deviations for these tests are misleading, because each set included at least one pin that failed to reach 4450 N (1000 lbf). Source: Ref 23

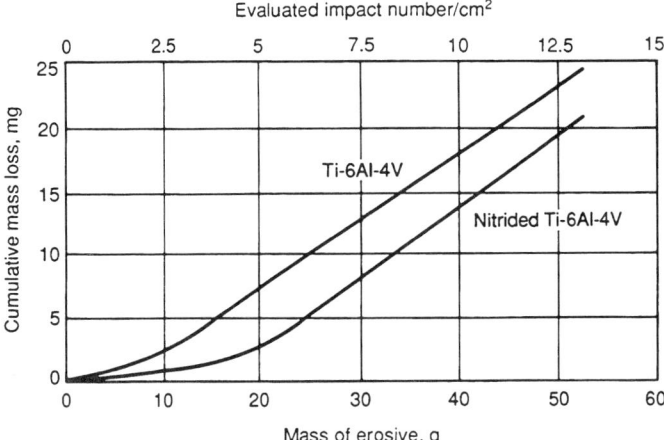

Fig. 7 Cumulative mass loss curves for Ti-6Al-4V and laser-nitrided Ti-6Al-4V. Source: Ref 22

Laser Surface Treatments. Laser surface heating and introduction of hard particles into the surface via reaction with gas (such as nitrogen or methane) or heating of surface deposits (graphite or boron nitride) have been used to produce thick nitride and boride surfaces (Ref 22). Erosion resistance (based on mass loss) of a laser-nitrided Ti-6Al-4V specimen against SiC particles was greater compared with untreated Ti-6Al-4V (Fig. 7). The laser-nitrided layer delays the onset of steady-state erosion that is characteristic for an uncoated Ti-6Al-4V surface. As shown in Fig. 7, the laser-nitrided layer is intact for a lower number of impacts, then suffers fracture and becomes embedded in the substrate as the number of impacts increases. Eventually, a sufficient number of impacts removes the laser-nitrided layer, exposing the uncoated Ti-6Al-4V substrate and resulting in steady-state erosion. In comparison, a TiN layer on a PVD-coated Ti-6Al-4V specimen was removed after only a few impacts.

Boride coatings on titanium alloys have shown good resistance to high-velocity particle impacts (Ref 22). Very thick (up to 80 μm), dense coatings have been produced on Ti-6Al-4V by laser boriding. Erosion rates of these thick coatings are reduced, because the material removal mechanism is modified to one of coating cracking rather than erosion of surface material.

Solid Lubrication

Solid lubricants are solid state materials (generally layered lattice compounds) that are applied to a surface by a burnishing operation (with an organic bonding agent) or deposited by plating or sputtering. Applications for solid-lubricant-coated titanium tribomaterials include components such as compressor blades, uniball bearings, and bearing seal rings for gas turbine engines (Ref 23) and gears for spacecraft mechanism actuators (Ref 17). For the former application, pulse plating deposition of Cr-Mo on Ti-6Al-4V and Ti-8Al-1Mo-1V was performed to produce a solid solution surface layer of Cr-Mo-Ti, which exhibited no degradation of substrate properties, increased creep life by up to 50%, and oxidation protection to 760 °C (1400 °F) (Ref 23). Commercial solid-lubricant coatings were applied to the Cr-Mo-coated/Ti-6Al-4V specimens. Unlubricated wear tests on the solid-lubricant-coated Cr-Mo-/Ti-6Al-4V specimens were performed by running them against AISI 1137 steel V-blocks. Test results showed excellent wear resistance for several of the solid-lubricant surface treatments (Table 4) (Ref 23). The best performer was PWA 36035 (Fel-Pro C-300: 50% solvent, 47% MoS_2, 3% graphite and other solids). Solvent compatibility tests, however, showed this coating to have poor resistance to jet fuel and oil. Cr-Mo-coated titanium alloy uniballs tested for 520 h in an engine simulator demonstrated no failures, in contrast to a 30% failure rate for standard steel uniballs.

Other potential aerospace applications of titanium alloys include drive mechanism components and launch locks (mechanisms used to distribute the load induced by an aerospace vehicle launch through an alternate load path). Replacement of steel gears with titanium alloy gears would reduce subsystem weight and result in reduced launch costs. The effects of various surface treatments on the wear life and efficiency of annealed Ti-6Al-4V gears have been studied by vacuum testing (Ref 17). As shown in Fig. 8, boronized (5 μm layer of metallic borides), Tifran (5 μm layer of complex oxides), and ion-plated aluminum coatings exhibited the best wear resistance and the lowest power losses.

The fabrication of solid lubricants such as diamondlike carbon (Ref 24) and molybdenum disulfide (Ref 25, 26) has also been studied. Diamondlike carbon coating exhibits very high hardness and solid lubricant behavior because of its graphite and amorphous carbon contents. To facilitate good DLC adhesion on Ti-6Al-4V, an intermediate bond layer of SiC is required (Ref 24). Unlubricated sliding wear tests in air of DLC/SiC/Ti-6Al-4V against a WC-Co pin showed a gradual reduction in coefficient of friction (down to 0.15) with increased load (Fig. 9) (Ref 24).

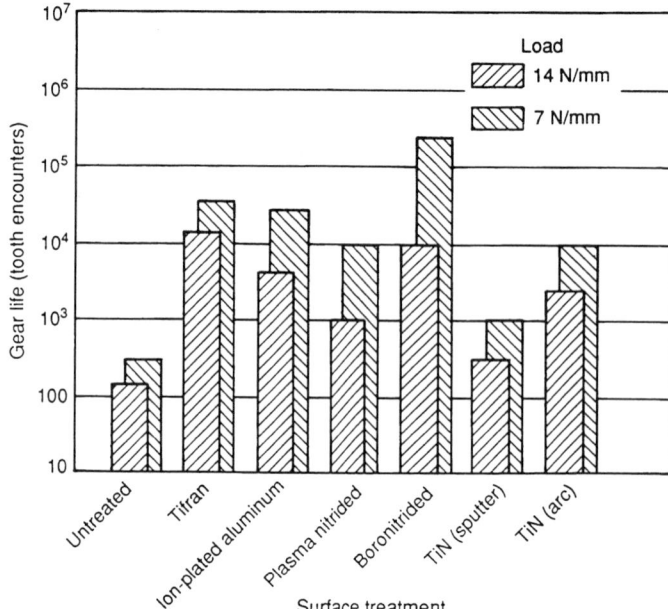

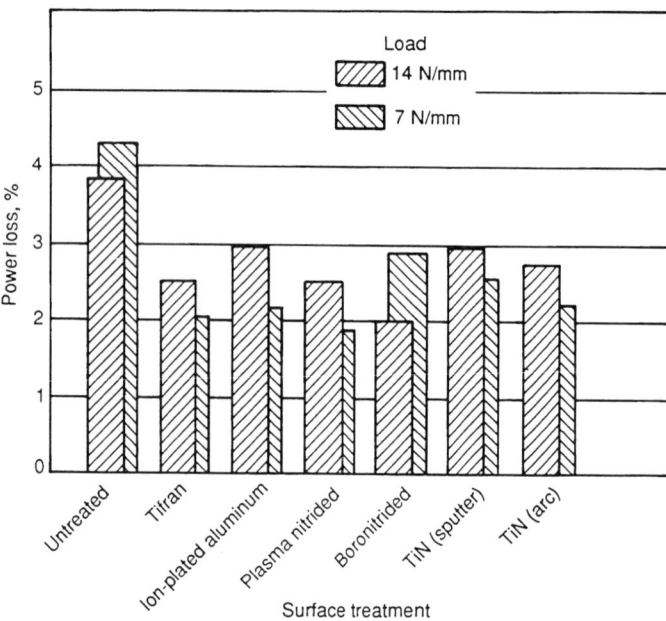

Fig. 8 Performance of surface-treated Ti-6Al-4V gears. (a) Effective life. (b) Effective gear efficiency. Source: Ref 17

Diamondlike carbon coating performed very well up to an initial stress level of 1.7 GPa (245 ksi) (52% above the yield strength of the Ti-6Al-4V substrate), at which DLC delamination was observed (Fig. 9).

Coating of titanium alloys with sputtered layered lattice solid lubricants, such as MoS_2, reduces their tendency to gall. Initial studies showed that MoS_2 adhesion to Ti-6Al-4V is very poor, as evidenced by the relatively low vacuum endurance life for MoS_2-coated Ti-6Al-4V loaded against itself (1.7 kgf load, 1.2 m/s sliding speed) of only about 60 revolutions (about 113 m sliding distance) (Ref 25). This is several orders of magnitude lower than for MoS_2-coated steel or ceramic (Si_3N_4) substrates (Fig. 10) (Ref 25).

Additional research has shown that pretreating the Ti-6Al-4V surface using a nitrogen implantation process produces a ceramic TiN/Ti_2N surface, which significantly increases the load-carrying capacity and high-load endurance lifetime of the sputtered MoS_2 coating on Ti-6Al-4V loaded against steel in an air environment (2 mm/s sliding speed) (Ref 26). Figure 11 shows the relative endurance lifetimes, as a function of applied load, for MoS_2-coated Ti-6Al-4V and MoS_2-coated nitrogen-implanted Ti-6Al-4V specimens. For MoS_2-coated Ti-6Al-4V, a gradual reduction in endurance life from more than 300 reciprocating cycles (5 and 6 kgf loads) to 33 cycles (6.5 kgf load) to 15 cycles (7 kgf loads) to nearly instantaneous failure (7.5 kgf load) was observed. In contrast, for the best nitrogen implantation pretreatment of $2 \times 10^{17} N_2^+$-N^+/cm^2 at 75 keV (400 °C or 750 °F, surface temperature), no MoS_2 coating failures were observed for test loads up to the 122.5 N (27.5 lbf) load limit of the test machine. This significant improvement in load-carrying capacity and high-load endurance life has been attributed to improved MoS_2 adhesion to the TiN/Ti_2N-rich Ti-6Al-4V surface through an irradiation-disrupted titanium oxide (Ref 26).

References

1. K. Miyoshi and D.H. Buckley, Correlation of Tensile and Shear Strengths of Metals with Their Friction Properties, *ASLE Trans.*, Vol 27 (No. 1), 1982, p 15-23
2. D.H. Buckley, *Surface Effects in Adhesion, Friction, Wear and Lubrication*, Elsevier Scientific, 1981, p 353
3. Independent Research and Development Brochure, Project D-81R, Martin Marietta Astronautics Group, 1990
4. T. Jamal, R. Nimmagadda, and R.F. Bunshah, Friction and Adhesive Wear of Titanium Carbide and Titanium Nitride Overlay Coatings, *Thin Solid Films*, Vol 73, 1980, p 245-254
5. F.M. Kustas, M.S. Misra, R. Wei, and P.J. Wilbur, High Temperature Nitrogen Implantation of Ti-6Al-4V; Part II: Tribological Properties, *Surf. Coat. Technol.*, 1992, accepted for publication
6. C.E. Snyder, Jr. and R.E. Dolle, Jr., Development of Polyperfluoroalkylethers as High Temperature Lubricants and Hydraulic Fluids, *ASME Trans.*, Vol 19 (No. 3), 1975, p 171-180
7. W.C. Oliver, R. Hutchings, J.B. Pethica, E.L. Paradis, and A.J. Shuskus, Ion Implanted Ti-6Al-4V, *Mater. Res. Soc. Symp. Proc.*, Vol 27, 1984, p 705-710
8. R.G. Vardiman, Wear Improvement in Ti-6Al-4V by Ion Implantation, *Mater. Res. Soc. Symp. Proc.*, Vol 27, 1984, p 699-704
9. R. Martinella, G. Chevallard, and C. Tostello, Wear Behavior and Structural Characterization of a Nitrogen

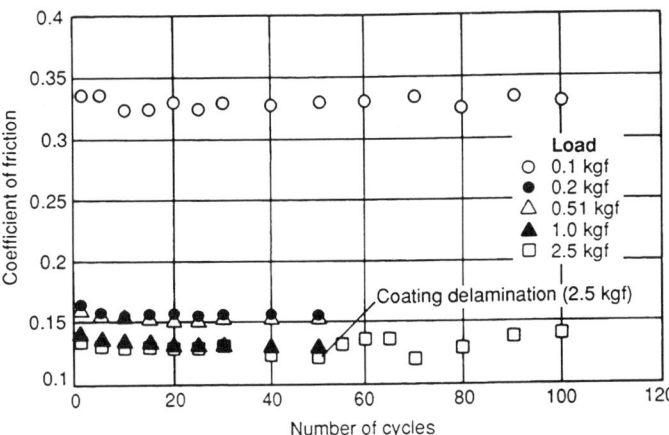

Fig. 9 Coefficients of friction versus wear cycles for 2.5 μm DLC/0.02 μm SiC interlayer coating on Ti-6Al-4V, showing reduction in μ with increased load. Sliding material, WC-Co; sliding speed, 2 mm/s (0.08 in./s). Source: Ref 24

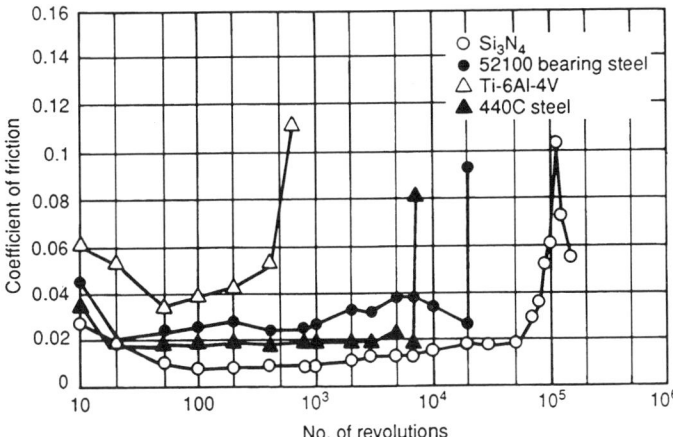

Fig. 10 Endurance lifetimes of MoS_2 on various substrates, showing longest lifetimes for Si_3N_4 and 52100 bearing steel. Sliding material, Ti-6Al-4V; sliding speed, 1.2 m/s (3.9 ft/s); applied load, 17 N/ball (1.7 kgf/ball). Source: Ref 25

Implanted Ti-6Al-4V Alloy at Different Temperatures, *Mater. Res. Soc. Symp. Proc.*, Vol 27, 1984, p 711-716

10. R.G. Vardiman, D. Creighton, G. Saliver, A. Effatian, and B.B. Rath, The Effect of Ion Implantation on Fretting Fatigue in Ti-6Al-4V, *Ion Implantation for Materials Processing*, F.A. Smidt, Ed., Noyes Data Corp., 1983, p 165-177

11. B.D. Barton and T.N. Wittberg, "Ion Implantation of Two Titanium Alloys," Report MLM-3603, U.S. Dept. of Energy, Office of Scientific & Technical Information, Aug 1989, p 3

12. F. Pons, J.C. Pivin, and G. Fargas, Inhibition of Tribo-Oxidation Preceding Wear, by Single-Phased TiN_x Films Formed by Ion Implantation into $TiAl_6V_4$, *J. Mater. Res.*, Vol 2 (No. 5), 1987, p 580-587

13. J.M. Williams, G.M. Beardsley, R.A. Buchanan, and R.K. Bacon, Effects of N-Implantation on the Corrosive-Wear Properties of Surgical Ti-6Al-4V Alloy, *Mater. Res. Soc. Symp. Proc.*, Vol 27, 1984, p 735-740

14. P. Sioshansi, R.W. Oliver, and F.D. Mathews, Wear Improvement of Surgical Titanium Alloys by Ion Implantation, *J. Vac. Sci. Technol.*, Vol A3 (No. 6), 1985, p 2670-2674

15. C. Chamont, Y. Honnorat, Y. Berthier, M. Goder, and L. Vicent, "Wear Problems in Small Displacements Encountered in Titanium Alloys Parts in Aircraft Turbomachines," presented at *Sixth World Conference on Titanium* (France), 1988, p 1883-1888

16. R.R. Boyer, Ti/Cr_2O_3 Grease Lubricated Spherical Bearings, *Proceedings of the Technical Program from the 1986 International Conference on Ti Alloys*, Vol 1, 1986, p 42-54

17. R.A. Rowntree, Surface-Treated Titanium Alloy Gears for Space Mechanisms, *Proceedings of the Second European Space Mechanisms Symposium* (West Germany), Oct 9-11, 1985, p 167-171

18. J. Lanagan, Properties of Plasma Nitrided Titanium Alloys, *Sixth World Conference on Titanium* (France), 1988, p 1957-1962

19. C. Jarbouli, V. Pellerin, D. Treheux, and L. Vicent, Study of the Fatigue Parameters of Titanium Alloys Treated for Wear Resistance, *Sixth World Conference on Titanium* (France), 1988, p 1859-1864

20. M. Thoma, Influence on Fretting Fatigue Behavior of Ti-6Al-4V by Coatings, *Sixth World Conference on Titanium* (France), 1988, p 1877-1881

21. B. Coll, P. Jacquot, M. Buvron, and J.P. Souchard, "Ionic Nitriding and Ionic Carburizing of Pure Titanium and Its Alloys," presented at ASM International Carburizing Conference, Sept 1989

22. J.P. Massoud and G. Coquerelle, High Power Laser Surface Treatments on Ti-6Al-4V in Order to Improve Its Erosion Resistance, *Sixth World Conference on Titanium* (France), 1988, p 1847-1852

23. P.L. McDaniel, R.E. Fisher, and V.G. Anderson, "Wear Resistant Coatings for Titanium," Report N00019-79-0544, Dept. of the Navy, Naval Air Systems Command, Oct 1980

24. F.M. Kustas, M.S. Misra, R. Wei, and P.J. Wilbur, Diamondlike Carbon Coatings on Ti-6Al-4V, *STLE Tribology Trans.*, 1992, accepted for publication

25. E.W. Roberts and W.B. Price, In Vacuo, Tribological Properties of "High-Rate" Sputtered MoS_2 Applied to Metal and Ceramic Substrates, *Mater. Res. Soc. Symp. Proc.*, Vol 140, 1989, p 251

26. F.M. Kustas and M.S. Misra, Improved Tribological Performance of MoS_2 on Ti-6Al-4V by Surface Pretreatment, *J. Vac. Sci. Technol.*, 1991, submitted for publication

27. G.L. Goss et al., *Wear*, Vol 24, 1973, p 77

28. J.E. Lemons, Biomaterial Surfaces and Biocompatibility, *Ion Implantation and Plasma Assisted Processes*, ASM International, 1988

29. F.M. Kustas, M.S. Misra, R. Wei, and P.J. Wilbur, High Temperature Nitrogen Implantation of Ti-6Al-4V; Part I: Microstructure Characterization, *Surf. Coat. Technol.*, 1992, accepted for publication

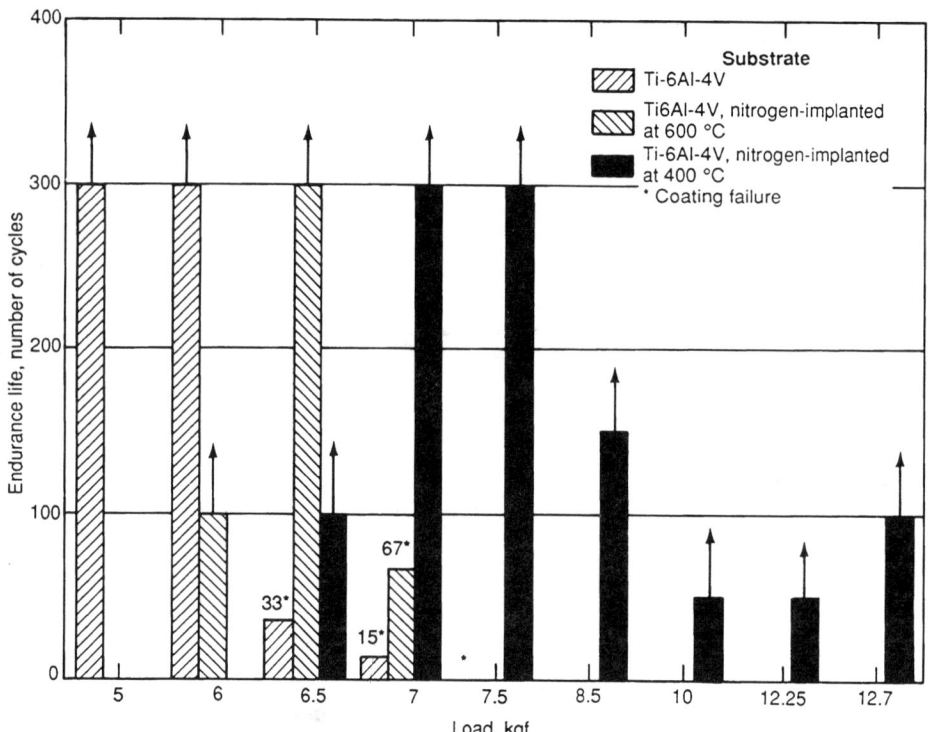

Fig. 11 Endurance lifetimes of MoS_2-coated Ti-6Al-4V specimens, showing significantly increased lifetimes (at high loads) for MoS_2 coating on nitrogen-implanted (processed at 400 °C, or 750 °F) Ti-6Al-4V substrate. Sliding material, bearing steel; sliding speed, 2 mm/s (0.08 in./s).
Source: Ref 26